www.prenhall.com/freeman/biology

BIOLOGICAL SCIENCE

BIOLOGICAL SCIENCE

SCOTT FREEMAN
UNIVERSITY OF WASHINGTON

PRENTICE HALL
Upper Saddle River, New Jersey 07458

Editor in Chief: Sheri L. Snavely
Project Development Editor and Editor in Chief of Development: Carol Trueheart
Production Editor: Donna Young
Project Manager: Karen Horton
Media Editor: Andrew T. Stull
Media Development Editors: Tracy Washburn, Karen Karlin, Deena Cloud
Executive Managing Editor: Kathleen Schiaparelli
Assistant Managing Editor, Science Media: Nicole Bush
Media Production Editor: Elizabeth Wright
Manager of Electronic Composition: Jim Sullivan
Electronic Production Specialist/Electronic Page Makeup: Joanne Del Ben
Managing Editor, Audio/Video Assets: Grace Hazeldine
Executive Marketing Manager: Jennifer Welchans
Director of Marketing Communications: Linda Taft
Vice President of Production and Manufacturing: David W. Riccardi
Director of Creative Services: Paul Belfanti
Manufacturing Manager: Trudy Pisciotti
Assistant Manufacturing Manager: Michael Bell
Director of Design: Carole Anson
Project Art Director: Kenny Beck
Electronic Production Support: Donna Marie Paukovits
Editorial Assistants: Nancy Bauer, Lisa Tarabokjia
National Sales Director for Key Markets: David Theisen
Marketing Assistant: Anke Braun
Assistant Managing Editor, Science Supplements: Dinah Thong
Supplements Production Editor: Veronica Malone
Copy Editor: Chris Thillen
Proofreader: Jennefer Vecchione
Cover Designers: Luke Daigle, Joseph Sengotta
Interior Designers: Lynne Stiles, Joseph Sengotta
Artist: Kim Quillin
Art Editor: Adam Velthaus
Art Project Support: Julie Nazario
Illustrators: Imagineering Scientific and Technical Artworks
Photo Editor: Beth Boyd-Brenzel
Photo Coordinator: Reynold Rieger
Photo Researcher: Yvonne Gerin
Cover photo: Hawksbill Sea Turtle © 2002 by Susan Middleton & David Liittschwager

The Hawksbill Sea turtle is an endangered species native to the tropical waters of the Atlantic Ocean. The individual shown on the cover was rescued from a fishing net and returned to the wild after being nursed back to health.

Other image credits appear in the rearmatter.

Upper Saddle River, NJ 07458

Printed in the United States of America

10 9 8 7 6 5 4 3 2

ISBN 0-13-081923-9 (Student Edition)
ISBN 0-13-009338-6 (Instructor's Edition)
ISBN 0-13-093205-1 (Volume 1)
ISBN 0-13-093206-X (Volume 2)
ISBN 0-13-093315-5 (Volume 3)

Pearson Education LTD.
Pearson Education Australia PTY, Limited
Pearson Education Singapore, Pte. Ltd
Pearson Education North Asia Ltd
Pearson Education Canada, Ltd.
Pearson Educación de Mexico, S.A. de C.V.
Pearson Education—Japan
Pearson Education Malaysia, Pte. Ltd

Brief Contents

For Students

It's difficult to imagine a more exciting time to launch a career related to biology. The advent of whole-genome sequencing and a rising interest in conservation biology are giving new momentum to a knowledge explosion that began several decades ago. From biochemistry, cell biology, and genetics to physiology, ecology, and evolution, the pace of discovery in the biological sciences is nothing short of astonishing. Your instructors are introducing you to what may currently be the most dynamic of all human endeavors.

Delving into biology through this introductory course should help you further two important goals. The first involves personal growth. The topics you'll be learning about pervade your life. Biology is about the food you eat and the air you breathe. It's about the history of life on Earth and the organisms that share the planet with us now. You'll be learning why we get sick, how we reproduce, how plants make food from sunlight. Biology is so basic that understanding it is a fundamental part of becoming an educated person. Taking this course can open your eyes, help you see and think about life in a new way, and fuel a lifelong curiosity about the natural world.

The second goal of a course like this involves a potential career path. By preparing you for more advanced classes and a major in the biological sciences, this introductory course will be a crucial first step in acquiring the background you'll need to enter a biology-related profession and help solve pressing problems in health, conservation, or agriculture. Many of the great challenges facing us today—from climate change and species extinctions to antibiotic resistance and emerging viruses—demand expertise in biology.

The purpose of this text is to help you make that important first step toward majoring in the biological sciences and pursuing a career related to biology. Its goal is not only to be a valuable reference for the fundamentals but also to introduce you to the excitement that drives this science. The presentation focuses on the questions that biologists ask about the natural world and how they go about answering them. Its objective is to introduce the core ideas that biologists use to make sense of the massive amount of information emerging from laboratories around the world.

The overall theme in this text is to help you learn how to think like a biologist. No matter what path your career takes, it is virtually certain that you will need to evaluate new hypotheses, analyze new types of data, and draw conclusions that change the direction of your work almost daily. Many of the facts you learn as an introductory student will change, but the analytical skills you learn in this course will serve you for life. Learning to think like a biologist will prepare you for upper-level courses and make you a better professional—whether you end up as a physician, pharmacist, educator, conservationist, or researcher.

If you approach this book with an open and inquiring mind, ready to challenge knowledge even as you absorb it, then you will have done your job. If this book communicates a sense of excitement about biological science and inspires you to keep learning more, it will have done its job. Thank you for joining a great adventure.

Scott Freeman

Preface

Scott Freeman received his Ph.D. in Zoology from the University of Washington and was nominated for an Excellence in Teaching Award in 1989. He was subsequently awarded an Albert Sloan Postdoctoral Fellowship in Molecular Evolution at Princeton University to investigate how generation time affects the rate of molecular evolution. Dr. Freeman's research publications explore a range of topics from the behavioral ecology of nest parasitism to the molecular systematics of the blackbird family. As an affiliate faculty member at the University of Washington, he has taught courses in evolution and has played an active role in the redesign of the general biology course. He is currently teaching the majors general biology course using an inquiry-based approach that emphasizes the logic of experimental design and the mastery of core concepts required for success in upper-level courses. Dr. Freeman is the co-author of *Evolutionary Analysis*, which presents evolutionary principles in the same spirit of inquiry that drives research.

Faculty who teach introductory biology may have the most exciting and difficult job on campus. The excitement springs from the breathtaking pace of advances in the biological sciences and the wide array of training and career options that are now open to prospective majors; the difficulty lies in introducing students to an already imposing and rapidly increasing number of facts and concepts.

When I took introductory biology as an undergraduate in 1975, faculty members were coping with the information explosion by extending the length of their introductory courses and using ever-larger textbooks. Today we don't have those options. Course length is capped at one year and most texts already run in excess of 1100 pages. Over the past decade in particular, presenting a fact-based, synoptic overview of what we know about biology has become increasingly untenable.

In short, the information explosion has changed our jobs. Instead of asking students to focus primarily on memorizing facts, more and more instructors are focusing their course on teaching students how to think like a biologist.

Why a New Introductory Textbook?

I wrote *Biological Science* to support professors who want their students to experience a more inquiry-driven approach in introductory biology. My goal was to write a book infused with the questions and the enthusiasm for learning that drive biological research. To help students understand how biologists think, each chapter is built around a series of questions that are fundamental to the topic being addressed. While exploring each question, the presentation incorporates data for students to interpret, offers evidence for competing hypotheses, introduces contemporary researchers, refers to work in progress, and highlights what researchers don't yet know. My aim was to help you teach biology the way you do biology—by asking questions and analyzing data to find answers.

At the same time, I made a strong commitment to covering the basics. We teach students who want to become doctors, researchers, science journalists, teachers, and conservationists. We have an obligation to prepare them for success in upper division courses, MCATs, and GREs, and to introduce the canon of facts and concepts that they must master to major in biology. Instead of listing these facts and concepts in an encyclopedic manner, however, *Biological Science* introduces them in the context of answering a question. In this textbook, facts become tools for understanding—not ends in themselves.

In addition to finding an appropriate balance between covering facts and exploring the scientific process, the level of the presentation is crafted to be appropriate for introductory students. Sections and sub-sections in the text begin with an overview of what question is being asked and end with commentary that helps students pull the material together. Instead of getting lost in the details of how an experiment was done, the text emphasizes why it was done and what the data mean.

Because beginning students are concerned about themselves and their world, most chapters explore how the topic relates to human welfare and all chapters end with an essay inspired by medical, commercial, or environmental concerns.

The Forest and the Trees: Helping Students to Synthesize and Unify

In addition to coping with an enormous amount of content in this course, instructors have to manage its diversity. In *Biological Science*, the emphasis on inquiry and experimentation provides a unifying theme from biochemistry through ecosystem ecology. In addition, the text highlights the fundamental how and why questions of biology. How does this event or process occur at the molecular level? In an evolutionary context, why does it exist?

The majority of chapters include at least one case history of an analysis done at the molecular level. Natural selection is introduced by exploring the evolution of antibiotic resistance via point mutations in the RNA polymerase gene of *Mycobacterium tuberculosis*. In the diversity unit, students learn about extracellular digestion in fungi by exploring experiments on the regulation of cellulase genes. A section of the behavior chapter features research on a gene involved in fruit fly foraging behavior. These are just three of many examples.

Similarly, evolutionary analyses do not begin and end with the evolution unit. Concepts like adaptation, homology, natural selection, and tree thinking are found in virtually every chapter. Unit 1, for example, presents traditional content in biochemistry—ranging from covalent bonding to the structure and function of macromolecules—in the context of chemical evolution and the origin of life. Meiosis is analyzed in terms of its consequences for generating genetic variation and making natural selection possible. Shared mechanisms of DNA repair and pattern formation are explained in the context of gene homologies. The overriding idea is that molecular and evolutionary analyses can help unify introductory biology courses, just as molecular tools and evolutionary questions are helping to unify many formerly disparate research fields within biology.

Supporting Visual Learners

Clear, attractive, and extensive graphics are critical to our success in the classroom. To emphasize the importance of analyzing figures in biology and to support students who learn particularly well visually, the book's art program is both extensive and closely interwoven with the manuscript. Each figure originated with rough sketches that I made while working on the first draft manuscript, which Dr. Kim Quillin then revised to increase clarity and improve appearance.

Throughout this process, our intent was to build an art program that is easy to read and that supports the book's focus on thinking like a biologist. A quick glance through the book should convince you that the art is as distinctive as the text. Color is used judiciously to highlight the main teaching points. Layouts flow from top to bottom and left to right, and extensive labeling lets students work through each figure in a step-by-step manner. Questions and exercises in the captions challenge students to actively interpret the graphics. The overall look and feel of the art is clean, clear, and inviting.

Serving a Community of Teachers

By de-emphasizing the encyclopedic approach to learning biology and focusing more on the questions and experimental tools that make biology come alive, our hope was to offer a book that is more readable and attractive to students and teachers alike. Embarking on an introductory course that launches a career in biology should be exciting, not anxiety-ridden. Learning concepts well enough to apply them to new examples and datasets may be more challenging for some students than memorizing facts, but it is also more compelling. By motivating the presentation with questions, and then using facts as tools to find answers, students of biology may come to think and feel more like the people who actually do biology.

I've always viewed working on this project as a gift, because it was a chance to serve the community of bright, enthusiastic, and dedicated people who teach this course. Thank you for your devotion to biology, for your commitment to your students, and for considering *Biological Science*. Teachers change lives.

Scott Freeman
University of Washington

Acknowledgments

This is the first new introductory biology textbook for majors to appear in over 15 years. The book and associated media were over five years in the making and reflect contributions from hundreds of teachers and researchers around the world. First and foremost among these individuals are the colleagues who contributed first-draft chapters in their areas of expertise. A commitment to scholarship and a passion for teaching resonated throughout these drafts and made an enormous impact on the published versions. The biologists who contributed draft chapters are:

Warren Burggren	University of North Texas
Kathleen Hunt	University of Washington
Kevin Kelley	University of California, Long Beach
Mary Rose Lamb	University of Puget Sound
Andrea Lloyd	Middlebury College
James Manser	Harvey Mudd College
Carol Reiss	Brown University
Thomas Sharkey	University of Wisconsin
DianeTaylor	Georgetown University
Carol Trent	Western Washington University
Susan Whittemore	Keene State College

Focus Group Participants and Reviewers

The contributors and I were guided and advised by introductory biology instructors who attended workshops held in Sundance, Utah while chapters for *Biological Science* were being drafted and revised. Many of the individuals who made up these focus groups were actively involved in efforts to reform introductory biology courses for majors; all helped to hone the look, feel, and content of the book. Workshop participants were:

Peter Berget	Carnegie Mellon University
Jack Burk	California State University, Fullerton
Mark Decker	University of Minnesota
Judith Heady	University of Michigan, Dearborn
Jean Heitz	University of Wisconsin
Carole Kelley	Cabrillo College
Judith Kjelstrom	University of California, Davis
Karen Koster	University of South Dakota
Dan Krane	Wright State University
Harry Nickla	Creighton University
Julie Palmer	University of Texas, Austin
Fred Singer	Radford University
Barbara Wakimoto	University of Washington
Charles Walcott	Cornell University
John Whitmarsh	University of Illinois, Urbana-Champaign
Dan Wivagg	Baylor University

The chapters themselves were thoroughly reviewed as they moved through the production process. Reviewers included experienced teachers who checked for scientific accuracy as well as issues such as level, pacing, and student comprehension. Other reviewers were experts in particular subfields who focused almost exclusively on making sure that chapters were accurate and current. In addition to multiple rounds of review in draft, all 52 chapters underwent a final review for accuracy just prior to publication.

To a person, our reviewers supplied exemplary attention to detail, expertise, and empathy for students. I am deeply indebted to all of the colleagues who reviewed chapters for *Biological Science*; it is simply impossible to overstate how crucial these individuals were to the success of this book. Their effort reflects a deep commitment to excellence in teaching and a profound belief in the importance of introductory courses for training the next generation of professionals. The individuals who reviewed chapters for *Biological Science* are listed at the end of this section.

Media and Supplements Authors

The media and supplements that accompany *Biological Science* were authored by a team of talented and dedicated introductory biology instructors led by Harry Nickla of Creighton University, Andrea Lloyd of Middlebury College, Julie Palmer of the University of Texas at Austin, and Warren Burggren of the University of North Texas. Our media and supplements authors brought an extraordinarily high level of creativity, experience, and ability to this project.

Jay Brewster	Pepperdine University
Brian Bagatto	University of Akron
Judith Heady	University of Michigan, Dearborn
Laurel Hester	South Carolina Governor's School for Science and Math
Carole Kelley	Cabrillo College
Heidi Picken-Bahrey	University of Washington
David Pindel	Corning Community College
Susan Rouse	Emory University

Our intent was to provide a media and supplements package that would be both original and tightly focused on solving key problems confronted by instructors and students in introductory biology. To meet this goal, our authors' efforts were guided by a group of innovative instructors who met as a focus group in New York City and again in Austin, Texas. The focus group participants followed up on these meetings by reviewing the media components and supplements as they were being produced.

John Bell	Brigham Young University
Peter Berget	Carnegie Mellon University
Ruth Buskirk	University of Texas, Austin
Judith Heady	University of Michigan, Dearborn
Kathleen Hunt	University of Washington
Mark Johnston	Dalhousie University
Carole Kelley	Cabrillo College
Julie Palmer	University of Texas, Austin
Robert Winning	Eastern Michigan University

The media and supplements authors were also assisted by biologists who contributed draft versions of the content. The media contributors for *Biological Science* are particularly experienced and creative instructors who use media extensively in their classrooms.

John Bell	Brigham Young University
Peter Berget	Carnegie Mellon University
Jack Burk	California State University, Fullerton
Shawn Gibbs	Wright State University
George Gilchrist	Clarkson University
Jim Hewlett	Fingerlakes Community College
Dave Hurley	University of Washington
Jennifer Katcher	Pima Community College
Andrea Lloyd	Middlebury College
Ric Matthews	San Diego Community College
Bill Russin	Northwestern University
Christina Trivett	Weber State University
Stuart Wagenius	Chicago Botanic Garden
Tracy Washburn	Sumanas, Inc.
Susan Whittemore	Keene State College

The Book Team

Textbook writing is a team effort. The look and feel of the art program is due to Dr. Kim Quillin, whose combination of creativity and content knowledge enabled her to invent a fresh and effective approach to the design of figures. Her fingerprints are on every single graphic in this book. The Prentice Hall team was equally resourceful, talented, and fun. Project Manager Karen Horton coordinated the contributor's efforts, the reviewer program, and the production of media and supplements with grace and aplomb. Photo researcher Yvonne Gerin was tireless in researching photographs to complement the text and Production Editor Donna Young expertly managed the thousands of details required to complete a project of this scope. Susan Middleton and David Liittschwager graciously provided the photographs for the cover, the title page, and the unit openers. Media Editor Andy Stull was responsible for assembling the media program accompanying this text and Jennifer Welchans, Executive Marketing Manager, is enthusiastically promoting *Biological Science* to ensure that professors have an opportunity to consider this textbook for their course. Development Editor Carol Trueheart acted as my conscience and taskmaster during the writing process; her insights, guidance, and ear have made me a much better author. More than any other individual, though, this book is a testament to the vision, courage, and talent of Sheri Snavely, Editor-in-Chief of Biology and Geosciences at Prentice Hall. *Biological Science* exists because of her devotion to excellence in biology publishing and her determination to offer teachers an innovative new option in introductory biology. She is a completely remarkable editor and person.

Finally, I thank Alex Davenport for help with library research and Susan, Ben, and Peter Freeman for their love and support. This book is dedicated to the memory of my mother and father, Elizabeth and William Freeman.

Reviewers

Lowell S. Adams	Weber State University
Venita F. Allison	Southern Methodist University
Jesper L. Andersen	Copenhagen Muscle Research Centre
David Asch	Youngstown State University
Mary V. Ashley	University of Illinois, Chicago
Karl Aufderheide	Texas A & M University
Tania Baker	Massachusetts Institute of Technology
Sandra L. Baldauf	Dalhousie University
Ronald A. Balsamo	Villanova University
Linda W. Barham	Meridian Community College
John D. Bell	Brigham Young University
Michel Bellini	University of Illinois, Urbana-Champaign
Malcolm J. Bennett	University of Nottingham
Spencer Benson	University of Maryland, College Park
Craig Berezowsky	University of British Columbia
Peter B. Berget	Carnegie Mellon University
J. Derek Bewley	University of Guelph
Ethan Bier	University of California, San Diego
Meredith Blackwell	Louisiana State University
Anthony H. Bledsoe	University of Pittsburgh
Robert Boyd	Auburn University
Michael R. Boyle	Seattle Central Community College
William S. Bradshaw	Brigham Young University
Eldon J. Braun	University of Arizona
Jay L. Brewster	Pepperdine University
Kent W. Bridges	University of Hawaii, Manoa
Judith L. Bronstein	University of Arizona
Carole Browne	Wake Forest University
Warren W. Burggren	University of North Texas
Jack Burk	California State University, Fullerton
Diane Caporale	University of Wisconsin, Stevens Point
John H. Carothers	Cabrillo College
Michael L. Christianson	University of Kansas
Deborah Clark	Quinnipiac College
E. Lee Coates	Allegheny College
Bruce J. Cochrane	University of South Florida
James M. Colacino	Clemson University
Linda T. Collins	University of Tennessee, Chattanooga
Scott Cooper	University of Wisconsin, LaCrosse
Victoria Corbin	University of Kansas
Catherine A. Coyle-Thompson	California State University, Northridge
Peter R. Crane	The Field Museum
Charles Creutz	University of Toledo
John L. Culliney	Hawaii Pacific University
Michael R. Cummings	University of Illinois, Chicago
Douglas A. Currie	University of Washington
Diana K. Darnell	Lake Forest College
Garry Davies	University of Alaska, Anchorage
Peter J. Davies	Cornell University
David W. Deamer	University of California, Santa Cruz
Mark Decker	University of Minnesota
Victor Defilippis	University of California, Irvine
Paula Dehn	Canisius College
Roger Del Moral	University of Washington
Fred Delcomyn	University of Illinois, Urbana
Esteban C. Dell'Angelica	University of California, Los Angeles
Edward DeLong	University of California, Santa Barbara
Charles F. Delwiche	University of Maryland
Alan Dickman	University of Oregon
G. Patrick Duffie	Loyola University, Chicago
Rick Duhrkopf	Baylor University
Thomas A. Ebert	San Diego State University
Kieran Elborough	ViaLactia Biosciences, Auckland New Zealand
David Eldridge	Baylor University
Jim Elser	Arizona State University
Franz Engelmann	University of California, Los Angeles
Joseph Erlichman	St. Lawrence University
Frederick B. Essig	University of South Florida
Susan Fahrbach	University of Illinois, Urbana
Daniel J. Fairbanks	Brigham Young University
Joseph O. Falkinham III	Virginia Polytechnic Institute & State University
Gerald Farr	Southwest Texas State University
Ross Feldberg	Tufts University
Richard Firenze	Broome Community College
Gary M. Fortier	Indiana University
Robert G. Fowler	San Jose State University
Carl S. Frankel	Pennsylvania State University, Hazleton
Arthur W. Galston	Yale University
Monica A. Geber	Cornell University
Jane Geisler	University of California, Riverside
Matt Geisler	University of California, Riverside
Erin Gerecke	Butler University
Rick Gillis	University of Wisconsin, Madison

John Godwin	North Carolina State University
Bruce Goldman	University of Connecticut
Michael A. Goldman	San Francisco State University
Elliott S. Goldstein	Arizona State University
Charles Good	Ohio State University, Lima
Nick Gotelli	University of Vermont
Linda Graham	University of Wisconsin, Madison
Robert T. Grammer	Belmont University
Alan Gray	Winfrith Technology Centre, UK
Erick Greene	University of Montana
Mark Groudine	Fred Hutchinson Cancer Research Center
Jessica Gurevitch	State University of New York, Stony Brook
Donald J. Hall	Michigan State University
Robert Hamilton	Mississippi College
Ray Hammerschmidt	Michigan State University
Georgia Ann Hammond	Radford University
Erin R. Hawkins	Louisiana State University
Judith E. Heady	University of Michigan, Dearborn
Harold Heatwole	North Carolina State University
Steven Heidemann	Michigan State University
Werner G. Heim	The Colorado College
William J. Heitler	University of St. Andrews
James W. Hicks	University of California, Irvine
John E. Hobbie	Marine Biological Laboratory, Ecosystems Center
Raymond B. Huey	University of Washington
Terry L. Hufford	George Washington University
William E. Hurford	Harvard Medical School
Steven W. Hutcheson	University of Maryland, College Park
Jeff Ihara	Mira Costa College
Mark O. Johnston	Dalhousie University
Walter S. Judd	University of Florida
Thomas C. Kane	University of Cincinnati
Tim Karr	University of Chicago
James Kasting	Pennsylvania State University
Judy Kaufman	Monroe Community College
Paul Keddy	Southwestern Louisiana University
Carole Kelley	Cabrillo College
Stephen T. Kilpatrick	University of Pittsburgh, Johnstown
Joel Kingsolver	University of North Carolina
Judith A. Kjelstrom	University of California, Davis
Loren W. Knapp	University of South Carolina
Andrew H. Knoll	Harvard University
Karen L. Koster	University of South Dakota
Dan E. Krane	Wright State University
Michael P. Labare	U. S. Military Academy
Mary Rose Lamb	University of Puget Sound
Marianne M. Laporte	Eastern Michigan University
Arlene Larson	University of Colorado at Denver
Allen Laughon	University of Wisconsin, Madison
Laura G. Leff	Kent State University
Paula P. Lemons	Duke University
Lynn O. Lewis	Mary Washington College
Anders Liljas	Lund University
Andrea Lloyd	Middlebury College
Robert G. Macbride	Delaware State University
Vivek Malhotra	University of California, San Diego
Richard Malkin	University of California, Berkeley
Peter Maloney	Johns Hopkins Medical School
James R. Manser	Harvey Mudd College
Joel A. Maruniak	University of Missouri
Ken Mason	Purdue University
David McClellan	Brigham Young University
Virginia McDonough	Hope College
Sarah Lea McGuire	Millsaps College
Pamela Metten	Oregon Health Sciences University
Leilani Miller	Santa Clara University
William J. Moody	University of Washington
Manuel A. Morales	University of Maryland
Lisa Nagy	University of Arizona
Robert K. Neely	Eastern Michigan University
Harry Nickla	Creighton University
Harry F. Noller	University of California, Santa Cruz
Stephen Nowicki	Duke University
Douglas Oba	Brigham Young University, Hawaii
Stephen O'Brien	National Cancer Institute
Karen Ocorr	University of Michigan
Laura J. Olsen	University of Michigan
Leslie E. Orgel	University of California, San Diego
William Osborne	University of Washington, Seattle
Marcy Osgood	University of Michigan
Stephen R. Overmann	Southeast Missouri State University
Thomas G. Owens	Cornell University
Norman R. Pace	University of Colorado
Julie M. Palmer	University of Texas, Austin
Matthew J. Parris	Arizona State University
D. J. Patterson	The University of Sydney
Craig L. Peebles	University of Pittsburgh

David Pfennig	University of North Carolina
Heidi Picken-Bahrey	University of Washington
Stuart Pimm	University of Tennessee, Knoxville
David Pindel	Corning Community College
Peggy E. Pollack	Northern Arizona University
Christopher Pomory	University of Southern California
F. Harvey Pough	Arizona State University
Mary Poulson	Idaho State University
Mitch Price	Pennsylvania State University
Gary P. Radice	University of Richmond
Deanna M. Rainer	University of Illinois, Urbana
Mani Ramaswami	University of Arizona
James B. Reid	University of Tasmania
Susan K. Reimer	Saint Francis University
Jonathan Reiskind	University of Florida
Carol Reiss	Brown University
William S. Reznikoff	University of Wisconsin, Madison
Eric J. Richards	Washington University
Laurel Roberts	University of Pittsburgh
Kenneth R. Robinson	Purdue University
Andrew Roger	Dalhousie University
Duke S. Rogers	Brigham Young University
Chris Romero	Front Range Community College
David S. Roos	University of Pennsylvania
Susan Rouse	Emory University
Stanley J. Roux	University of Texas, Austin
Jennifer Ruesink	University of Washington
Joanne Russell	Manchester Community College
Peter Russell	Reed College
Charles L. Rutherford	Virginia Technical University
Fred Sack	Ohio State University
Mark F. Sanders	University of California, Davis
K. Sathasivan	University of Texas, Austin
Daniel C. Scheirer	Northeastern University
Gary C. Schoenwolf	University of Utah School of Medicine
Janet L. Schottel	University of Minnesota
Michael Schultze	University of York, UK
Rodney J. Scott	Wheaton College
David L. Secord	University of Washington, Tacoma
Jan Serie	Macalester College
Marty Shankland	University of Texas, Austin
Thomas D. Sharkey	University of Wisconsin, Madison
Richard M. Showman	University of South Carolina
Barry Sinervo	University of California, Santa Cruz
Fred Singer	Radford University
J. M. W. Slack	University of Bath
John Smarrelli	Loyola University, Chicago
Paul Sniegowski	University of Pennsylvania
Sally Sommers-Smith	Boston University
Susan J.Stamler	College of DuPage
Gunther S. Stent	University of California, Berkeley
Lori Stevens	University of Vermont
Steven R. Strain	Slippery Rock University
Eric Swann	University of Minnesota
Naoko Tanese	New York University, School of Medicine
Ethan Temeles	Amherst College
Robert M. Thornton	University of California, Davis
Carol Trent	Western Washington University
Melvin T. Tyree	University of Vermont
Elizabeth Van Volkenburgh	University of Washington
Sara Via	University of Maryland
Stuart Wagenius	University of Minnesota
D. Alexander Wait	Southwest Missouri State University
Barbara T. Wakimoto	University of Washington
Charles Walcott	Cornell University
Lawrence R. Walker	University of Las Vegas, Las Vegas
Danny Wann	Carl Albert State College
Fred Wasserman	Boston University
Robert F. Weaver	University of Kansas
Elizabeth A. Weiss	University of Texas, Austin
Robin A. Weiss	Institute of Cancer Research, London
Thomas R. Wentworth	North Carolina State University
Olivia Masih White	University of North Texas
John Whitmarsh	University of Illinois
Susan Whittemore	Keene State University
Gabriele Wienhausen	University of California, San Diego
Robert S. Winning	Eastern Michigan University
Bill Wischusen	Louisiana State University
Rachel Witcher	Iowa Central Community College
Daniel E. Wivagg	Baylor University
Chris Wolverton	Ohio State University
Robin Wright	University of Washington
Lauren Yasuda	Seattle Central College
Paul G. Young	Queen's University
Nina C. Zanetti	Siena College
Miriam Zolan	Indiana University

Print and Media Resources for Instructors and Students

For the Instructor

Lecture Presentation Tools

INSTRUCTOR RESOURCE CD-ROM (IRCD)

The Instructor Resource CD-ROM for *Biological Science* simplifies your life by placing powerful, customizable tools at your fingertips. This comprehensive, easy-to-use electronic resource provides everything you need to both prepare for and present a lecture. It features all of the illustrations and and photographs from the book—both as exportable images and as prepared PowerPoint slides. The PowerPoint slides are fully editable, allowing you complete customization capabilities. The Instructor Resource CD-ROM also features hundreds of animations and activities that can be incorporated into your lecture presentation. These same animations and activities are part of a series of more comprehensive, chapter-specific animations and activities located on the Student CD-ROM. Each animation and activity can be presented either with or without text, audio narration, and self-grading quizzes. Imagine being able to put together a presentation using art from the textbook, video clips, and animations from the Student CD. Students are motivated to explore and use the media provided with their text when they see it presented to them in the classroom. This tool gives you the power to visually present and highlight a key concept from the text and then assign it as homework. All of the answers to the activities, end-of-chapter material, and website quizzes are included on the Instructor Resource CD-ROM.

TRANSPARENCY PACKAGE AND INSTRUCTOR RESOURCE KIT

Transparencies are an effective way to visually reinforce your lecture presentation. Every illustration from the text—including art, photographs, and tables—is available on four-color transparency acetates. We've put a lot of thought into how to deliver such a large number of acetates to you in a way that is easy for you to use and organize for lecture. The transparency set is three-hole-punched and organized by chapter in manila folders, which are stored in an Instructor Resource Kit file box along with the printed lecture tools from the Instructor Resource Guide. Some labels and all of the hand pointers in the test illustrations have been deleted from these transparencies to enhance projection. Labels and images have been enlarged and modifed to ensure optimal readability in a large lecture hall.

INSTRUCTOR GUIDE

Edited by Julie Palmer, University of Texas at Austin
Contributors: Carole Kelley, Cabrillo College; Judith Heady, University of Michigan at Dearborn; David Pindel, Corning Community College; Susan Rouse, Emory University

The Instructor Guide for *Biological Science* includes not only the traditional instructor support tools—lecture outlines and student objectives—but it also provides additional, more contemporary resources for today's teaching challenges—motivating students, reinforcing their understanding of the material, and helping them to develop critical thinking skills. These resources include chapter-by-chapter suggestions for inquiry-based classroom activities, simple demonstrations, and problems involving the data presented in a given chapter. Answers to all of the activities and problems—including answers to the figure caption questions and exercises and the end-of-chapter questions—are included in this Instructor Guide, making it easier to assign them to students. All content in the Instructor Guide is available in a printed volume or included electronically on the Instructor Resource CD-ROM.

Assessment Tools

TEST QUESTIONS (OVER 2600 QUESTIONS) AND TESTGEN EQ

The Test Questions for *Biological Science* have been written and edited by the author, Scott Freeman, and a team of talented instructors, to ensure the quality and accuracy of this important resource as well as its tight integration with the text. It contains a variety of questions compiled from our reviewers, top educators, and the author's own teaching experience. The Test Questions contain multiple choice questions in the following formats: factual recall, conceptual, and application/data interpretation questions that are in keeping with the most recent MCAT and GRE standards. The Test Questions are available as a printed volume and as part of the TestGen EQ Computerized Testing Software, a text-specific testing program that is networkable for administering tests. It allows instructors to view and edit electronic questions, export the questions as tests, and print them in a variety of formats.

Laboratory Support

SYMBIOSIS: THE PRENTICE HALL CUSTOM LABORATORY PROGRAM FOR BIOLOGY

With *Symbiosis*, you can custom-build a lab manual that exactly matches your teaching style, your content needs, and your course organization. You choose the labs you want from our extensive list of Prentice Hall publications or Pearson Custom

Publishing's own library of biology labs. You choose the sequence. Using the template tools provided in our unique **Lab Ordering and Authoring Kit**, you can create your own custom-written labs and then incorporate graphics from our biology graphics library. You can even add your course notes, syllabi, or other favorite materials. The result is a cleanly designed, well-integrated lab manual to share with your students.

Course Management Tools

Blackboard is a comprehensive and flexible eLearning software platform that delivers a course management system, customizable institution-wide portals, online communities, and an advanced architecture that allows for web-based integration of multiple administrative systems.

WebCT provides you with high-quality, class-tested material pre-programmed and fully functional in the WebCT environment. Whether used as an online supplement to either a campus-based or a distance-learning course, our pre-assembled course content gives you a tremendous headstart in developing your own online courses.

Course Compass is a dynamic, interactive eLearning program. Its flexible, easy-to-use course management tools allow you to combine Pearson Higher Education content with your own.

Instructors 1st For qualified adopters, Prentice Hall is proud to introduce Instructors 1st—the first integrated service committed to meeting your customization, support, and training needs for your course.

For the Student

Student CD-ROM

The Student CD-ROM for *Biological Science* provides resources to help students visualize difficult concepts, explore complex biological processes, and review their understanding of the most challenging material presented in this course. This comprehensive, easy-to-use electronic resource is integrated with the textbook, providing students either with rapid access to extended learning opportunities while reading the chapter or with detailed textbook references while working through an activity on the CD. These activities include animations to visualize elaborate biological concepts or processes and animated tutorials that allow students to explore more complex topics. The Student CD has an intuitive interface, a familiar chapter-based organization, and a powerful search engine, all designed to help students expertly navigate this resource. Each activity includes full audio narration, an integrated glossary, and an audio pronunciation guide. In addition, the CD also serves as a portal to the review and research tools provided on the Student Website, bringing together all of the resources to help students succeed in their course.

Student Website (www.prenhall.com/freeman/biology)

The Student Website for *Biological Science* provides students with the self-assessment, current research, and communication tools needed to help them succeed in their introductory biology course. Within each chapter on the Website, self-grading quizzes allow students to assess their understanding of the chapter material as well as providing an explanation should a student choose an incorrect answer to a question. Further, the Website includes a broad collection of science and research links for the subject areas described in each chapter. These links are outstanding tools for students wishing to explore a chapter's concepts or to extend their knowledge beyond the scope of the text. In combination with the Student CD, the Student Website provides a valuable set of resources to help students develop the skills that will help them in both their introductory biology course as well as in upper-division courses.

Student Study Guide

Edited by Warren Burggren, University of North Texas

Contributors: Jay Brewster, Pepperdine University; Laurel Hester, South Carolina Governor's School for Science and Mathematics; Brian Bagatto, University of Akron

The Student Study Guide helps students focus on the fundamentals chapter-by-chapter and contains additional resources to help students prepare for a career in the biological sciences. Each chapter presents a breakdown of chapter themes, key biological concepts, exercises, self-assessment activities, and quizzes. Additionally, the Study Guide features four introductory, stand-alone chapters: Introduction to Experimentation and Research in the Biological Sciences, Presenting Biological Data, Understanding Patterns in Biology and Improving Study Techniques, and Reading and Writing to Understand Biology.

Science on the Internet

Andrew Stull, Prentice Hall and Harry Nickla, Creighton University

This free, practical resource provides straightforward step-by-step directions for accessing regularly updated biology resource areas online as well as an overview of general online navigation strategies. This booklet is a helpful companion to the Student Website for *Biological Science*.

Contents

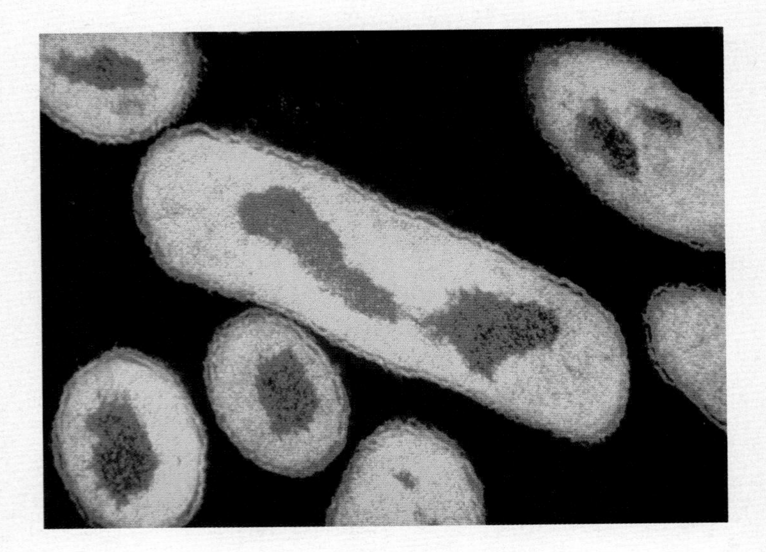

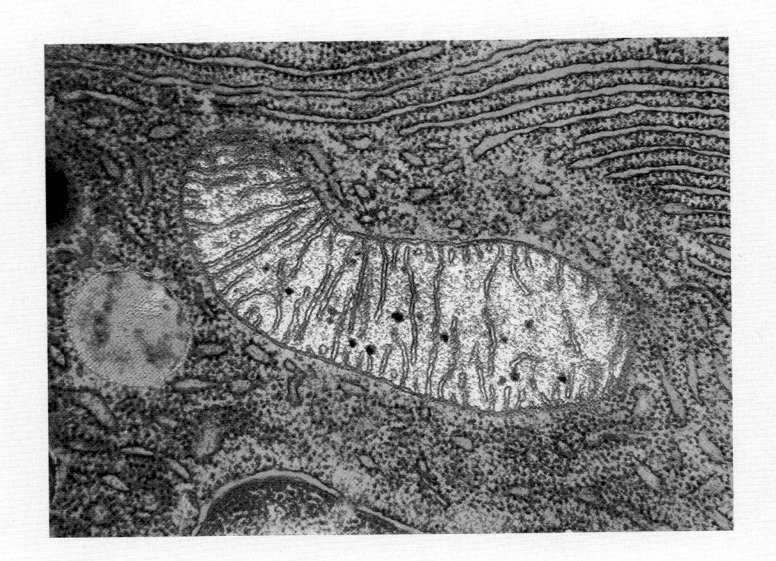

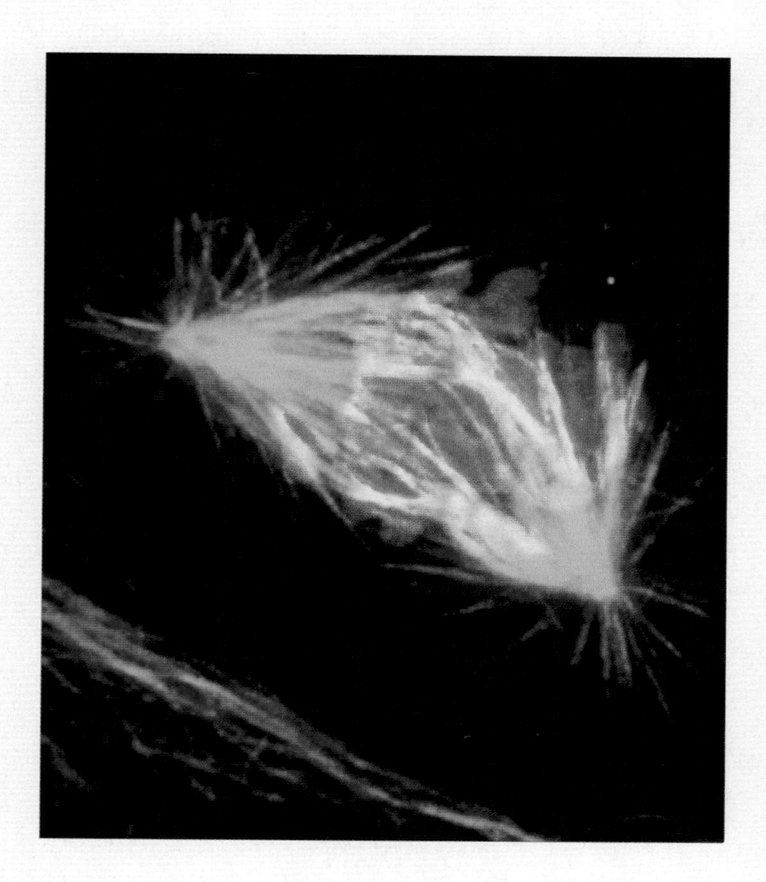

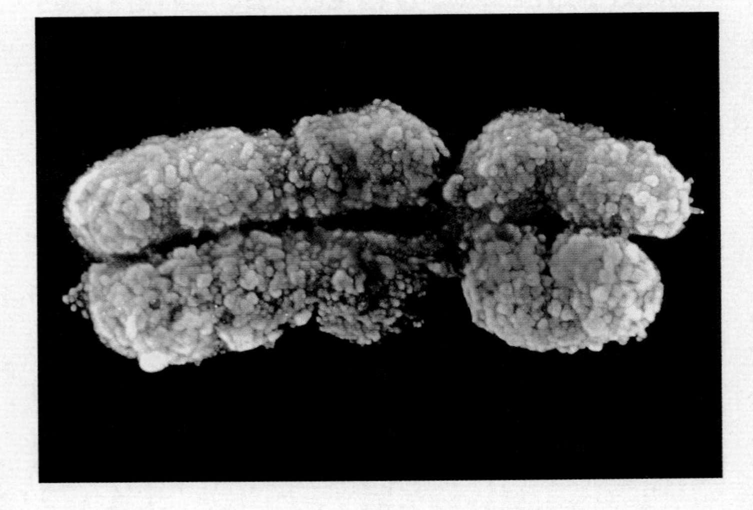

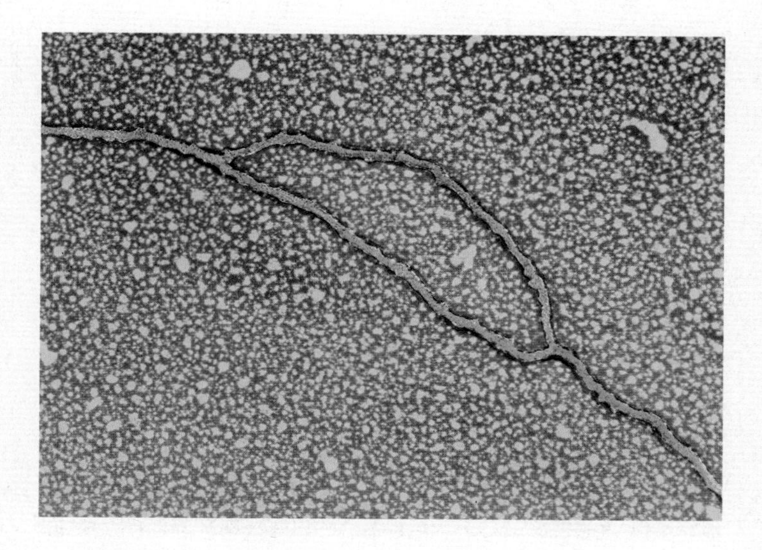

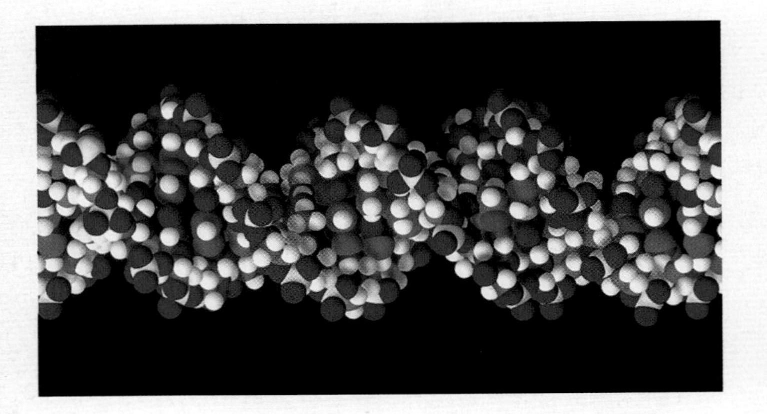

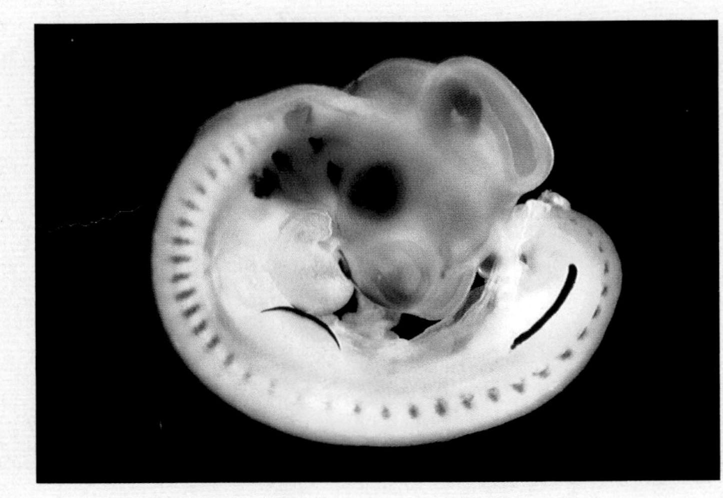

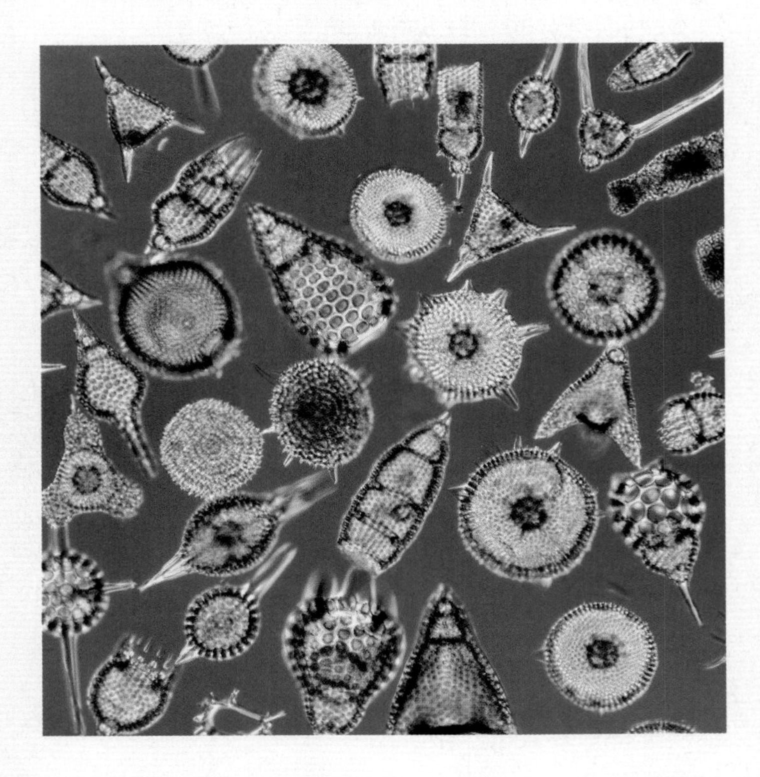

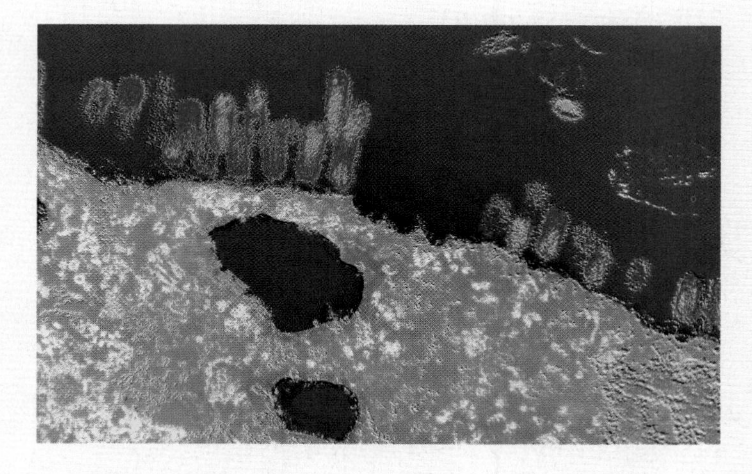

Biology and the Tree of Life

1

At its core, biological science is a search for the features and processes that unify the remarkable diversity of living organisms. Chapter 1 is an introduction to this search. Its goals are to convince you that the diversity of life is indeed remarkable, and that biologists have identified fundamental mechanisms and traits that unify all organisms. Appreciating the diversity of life—and understanding its underlying unity—is the first step in learning how to think like a biologist.

Chapter 1 begins with two of the greatest unifying ideas in all of science: the cell theory and the theory of evolution by natural selection. When these concepts emerged in the mid-1800s, they revolutionized how biologists understand the world. A major part of this revolution was the realization that all organisms are related to one another through common ancestry. Bacteria, mushrooms, roses, and robins are all part of a family tree, similar to the genealogies that connect individual people. How do biologists study this "tree of life"? What does it look like? The chapter concludes by exploring these questions.

In ancient and modern cultures, the tree of life, shown on this antique tapestry, has symbolized fertility, everlasting life, and peace among nations. This chapter explains that in modern biology, the tree of life is a literal representation of the relationships among species.

1.1 The Cell Theory

The initial conceptual breakthrough in biology—the cell theory—emerged only after some 200 years of work. In 1665 the Englishman Robert Hooke used a crude microscope to examine the structure of cork (a bark tissue) from an oak tree. The instrument magnified objects to just 30 times (30×) their normal size, but it allowed Hooke to see something extraordinary. In the cork he observed small, pore-like compartments that were invisible to the naked eye (**Figure 1.1a**, page 2). These structures came to be called cells.

Soon after Hooke published his results, a Dutch scientist named Anton van Leeuwenhoek succeeded in developing much more powerful microscopes—some capable of magnifications up to 300×. Leeuwenhoek inspected samples of pond water with these instruments and made the first observations of single-celled organisms (**Figure 1.1b**). He also observed and described the structure of blood cells and sperm cells from humans.

In the 1670s, Marcello Malphigi concluded that plant tissues like leaves and stems were composed of cells. Lorenzo Oken extended this observation in 1805 by claiming that all organisms consist of cells. In the late 1830s Oken's hypothesis was backed by Matthias Schleiden and Theodor Schwann, who independently reached the same conclusion after examining hundreds of plant and animal tissues under magnification. The conclusion was a classic example of inductive reasoning: Scientists made a broad generalization only after making hundreds of supporting observations.

Since the 1830s, however, microscopes have advanced dramatically. Biologists have discovered hundreds of thousands of new organisms. Has Schleiden and Schwann's conclusion held up? Is it true that all organisms are made of cells?

Are *All* Organisms Made of Cells?

The smallest organisms known today are bacteria that are barely 200 nanometers wide (**Table 1.1**). Lined up end to end, it would take 5000 of these organisms to span the distance between the smallest hash marks on a ruler (a millimeter). The tallest organisms known are sequoia trees from the Pacific Coast of the United States. Sequoias can be over 100 meters tall—the equivalent of a 20-story building. Bacteria and sequoias are composed of the same fundamental building block, however: the cell. Bacteria are unicellular (one-celled) organisms; sequoia trees are multicellular (many-celled) organisms.

Biologists are increasingly dazzled by the diversity and complexity of cells as advances in microscopy enable them to examine cells at higher magnifications. Oken, Schleiden, and Schwann's basic conclusion is intact, however. As far as is known, all organisms are made of cells. Today, a **cell** is defined as a water-based compartment filled with concentrated chemicals and bounded by a thin, flexible structure called a membrane.

However, as important as the claim made by Oken, Schleiden, and Schwann was, it formed only the first part of the cell theory. In addition to understanding what organisms are made of, scientists wanted to know the origin of that fundamental material—where does it come from?

Where Do Cells Come From?

Most scientific theories have two components. One component describes a pattern in the natural world, and the other component identifies a mechanism or process that is responsible for creating that pattern. Malpighi, Oken, Schleiden, and Schwann all articulated the pattern component of the cell theory. In 1858 Rudolph Virchow forcefully backed Oken's statement of the process component by stating that all cells arise from preexisting cells. The complete **cell theory**, then, can be stated as follows: All organisms are made of cells, and all cells come from preexisting cells.

The Oken-Virchow claim was a direct challenge to an alternative hypothesis called **spontaneous generation**. This is the proposition that organisms can arise spontaneously under certain conditions. For example, the bacteria and fungi that spoil foods like milk and wine were thought to simply appear in

(a) The first view of cells: Robert Hooke's drawing from 1665

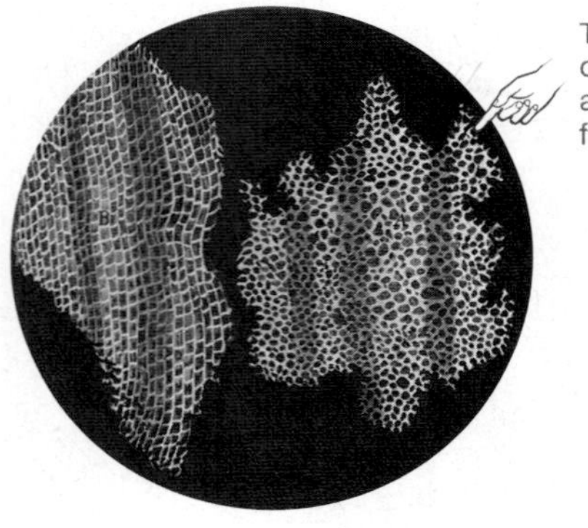

(b) Anton van Leeuwenhoek was the first to view single-celled "animalcules" in pond water.

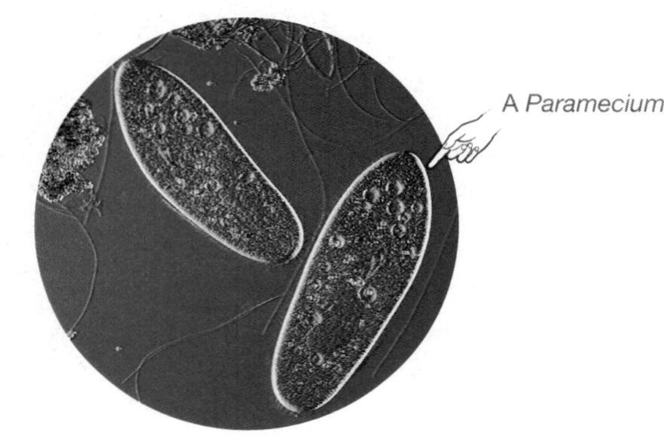

FIGURE 1.1 The Discovery of Cells

TABLE 1.1 Some Prefixes Used in the Metric System

Prefix	Abbreviation	Meaning	Example
giga-	G	10^9	1 gigameter (Gm) = 1×10^9 m
mega-	M	10^6	1 megameter (Mm) = 1×10^6 m
kilo-	k	10^3	1 kilometer (km) = 1×10^3 m
deci-	d	10^{-1}	1 decimeter (dm) = 1×10^{-1} m
centi-	c	10^{-2}	1 centimeter (cm) = 1×10^{-2} m
milli-	m	10^{-3}	1 millimeter (mm) = 1×10^{-3} m
micro-	μ*	10^{-6}	1 micrometer (μm) = 1×10^{-6} m
nano-	n	10^{-9}	1 nanometer (nm) = 1×10^{-9} m

*μ is the Greek letter mu (pronounced "mew").

these nutrient-rich media of their own accord—meaning they spring to life from nonliving materials. Virchow, in contrast, maintained that cells do not spring to life spontaneously but are produced when preexisting cells grow and divide.

Soon after Virchow's hypothesis appeared in print, the French bacteriologist Louis Pasteur set out to test its predictions experimentally. Specifically, Pasteur wanted to determine whether microorganisms can appear spontaneously in a nutrient broth, or whether they appear only when the broth is exposed to a source of preexisting cells. To accomplish this test, he created two treatment groups: a broth that was exposed to a source of preexisting cells and a broth that was not. The spontaneous generation hypothesis predicted that cells would appear in both treatments. Virchow's hypothesis predicted that cells would appear only in the treatment exposed to a source of preexisting cells.

Figure 1.2 shows Pasteur's experimental design. Note that the two treatments are identical in every respect but one. Both used glass flasks filled with the same amount of the same nutrient broth. Both were boiled for the same amount of time to kill any existing organisms such as bacteria or fungi. But because the flask pictured in Figure 1.2a had a straight neck, it was exposed to preexisting cells after sterilization by the heat treatment. (These cells are the bacteria and fungi that cling to dust particles in the air. They could drop into the nutrient broth because of the straight neck on the flask.) In contrast, the second flask (drawn in Figure 1.2b) had a long swan neck. Pasteur knew that water would condense in the crook of the swan neck after the boiling treatment, and that this pool of water would trap any bacteria or fungi that entered on dust particles. Thus, the swan-necked flask was isolated from any source of preexisting cells even though it was still open to the air. The design was effective because there was only one difference between the two treatments and because that difference was the factor being tested—in this case, exposure to preexisting cells. Reducing variables is a prized feature of an experimental design.

The result? As Figure 1.2a shows, the treatment exposed to preexisting cells quickly filled with bacteria and fungi. This treatment was important, however, because it showed that the heat sterilization step had not altered the nutrient broth's capacity to support growth. But the treatment in the swan-necked flask (see Figure 1.2b) remained sterile. Even when the broth was left standing for months, no organisms appeared in this treatment.

Because Pasteur's data were in direct opposition to the predictions made by the spontaneous generation hypothesis, the

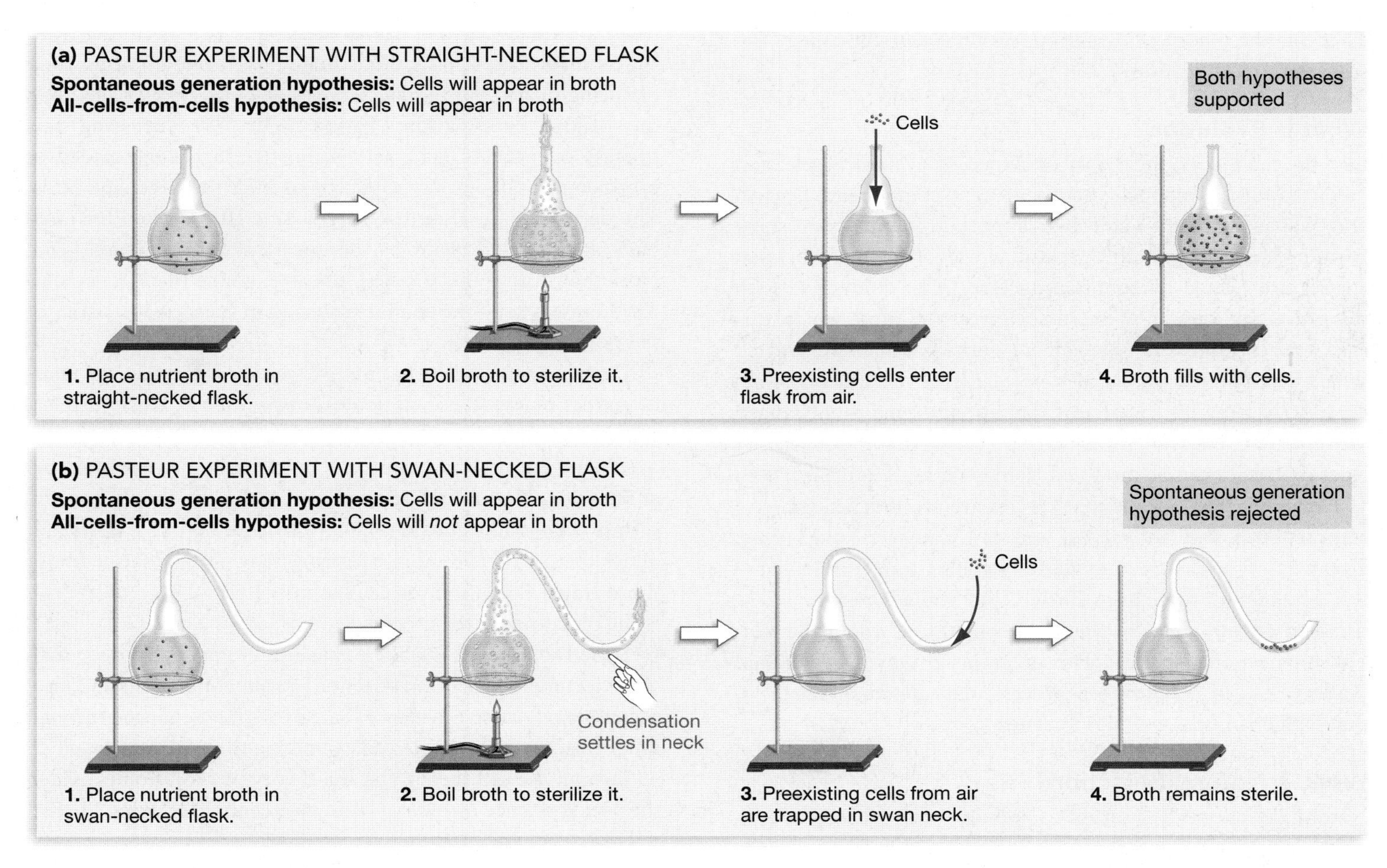

FIGURE 1.2 The Origin of Cells
The experiment diagrammed here was designed to test the predictions made by competing hypotheses for the origin of cells.

results persuaded most scientists that Virchow's hypothesis was correct.

The process component of the cell theory had an important implication. If all cells come from preexisting cells, then all individuals in a population of single-celled organisms are related by common ancestry, and all cells in a multicellular organism are descended from a fertilized egg.

The second great founding idea in biology is similar, in spirit, to the cell theory. It also happened to be published the same year as Virchow's hypothesis. This was the realization, made independently by Charles Darwin and Alfred Russel Wallace, that all *species*—meaning all distinct types of organisms—are connected by common ancestry.

1.2 The Theory of Evolution by Natural Selection

In 1858, short papers by Darwin and Wallace were read to a small group of scientists attending a meeting of the Linnaean Society of London. In their papers Darwin and Wallace argued that all species, past and present, are related by descent from a common ancestor. Their hypothesis was that species come from other, preexisting species and that species change through time. This was the theory of evolution, or what Darwin called "descent with modification."

What Is Evolution?

Like the cell theory, the theory of evolution by natural selection has a pattern component and a process component. Darwin and Wallace's theory made two important claims concerning patterns that exist in the natural world. The first was that species are related by common ancestry. This contrasted with the prevailing view in Western society at the time, which was that species represent independent entities that were created separately. The second claim was equally novel. Instead of accepting the popular hypothesis that species remain unchanged through time, Darwin and Wallace proposed that the characteristics of species can be modified from generation to generation.

Evolution, then, means that species are not independent and static entities, but are related and can change through time. This part of the theory of evolution—the pattern component—was actually not original to Darwin and Wallace. Several scientists had already come to the same conclusions about the relationships among species. The great insight by Darwin and Wallace was in proposing a process, called **natural selection**, that explained *how* evolution occurs.

What Is Natural Selection?

Natural selection, the process component of the theory of evolution, occurs whenever three conditions are met. The first condition is that individuals within a population vary in their characteristics. (A **population** is defined as a group of individuals of the same species living in the same area at the same time.) Darwin and Wallace had studied natural populations long enough to realize that variation among individuals is almost universal. In wheat, for example, some individuals are taller than others. The second condition is that the variable traits are **heritable**, meaning that they are passed on to offspring. As a result of work by wheat breeders, Darwin and Wallace knew that short parents tend to have short offspring. The third condition is that certain heritable traits help individuals survive better or reproduce more. For example, if tall wheat plants are easily blown down by wind, then shorter plants will tend to survive better and leave more offspring in windy environments.

If all three conditions are met, then a population's characteristics will change over time. In the example just given, populations of wheat that grow in windy environments would tend to become shorter from generation to generation. A change in the characteristics of a population, over time, is evolution.

To clarify how the process works, consider the origin of the vegetables called the "cabbage family plants." Broccoli, cauliflower, Brussels sprouts, cabbage, kale, savoy, and collard greens are all descended from the same species—the wild plant in the mustard family pictured in **Figure 1.3a**. To create the plant called Brussels sprouts, horticulturists selected individuals of the wild mustard species with particularly large side buds. In mustards, the size of side buds is a heritable trait. When the selected individuals mated with one another, their offspring turned out to have larger side buds, on average, than the original population (**Figure 1.3b**). By repeating this process over many generations, horticulturists succeeded in producing a population with extraordinarily large side buds. The derived population has been artificially selected for large buds and barely resembles the ancestral form (**Figure 1.3c**). Note that during this process, the size of side buds in each individual was set—the change occurred in the characteristics of the population.

Darwin pointed out that natural selection changes the characteristics of a wild population over time, just as artificial selection changes the characteristics of a domesticated population over time. But no horticulturist is involved in the case of natural selection. Natural selection occurs naturally, simply because certain individuals in wild populations have heritable traits that allow them to leave more offspring than individuals without those traits. Evolution, or change in the population over time, is the outcome of this process.

Since Darwin and Wallace published, biologists have succeeded in documenting natural selection in wild populations and have accumulated massive evidence that species have changed through time. Chapter 21 is devoted to examining these data in detail.

In sum, the cell theory and the theory of evolution provided the young science of biology with two central, unifying ideas:

- The cell is the fundamental structural unit in all organisms.
- All species are related by common ancestry and have changed over time in response to natural selection.

(a) Wild member of *Brassica oleracea*: small side buds

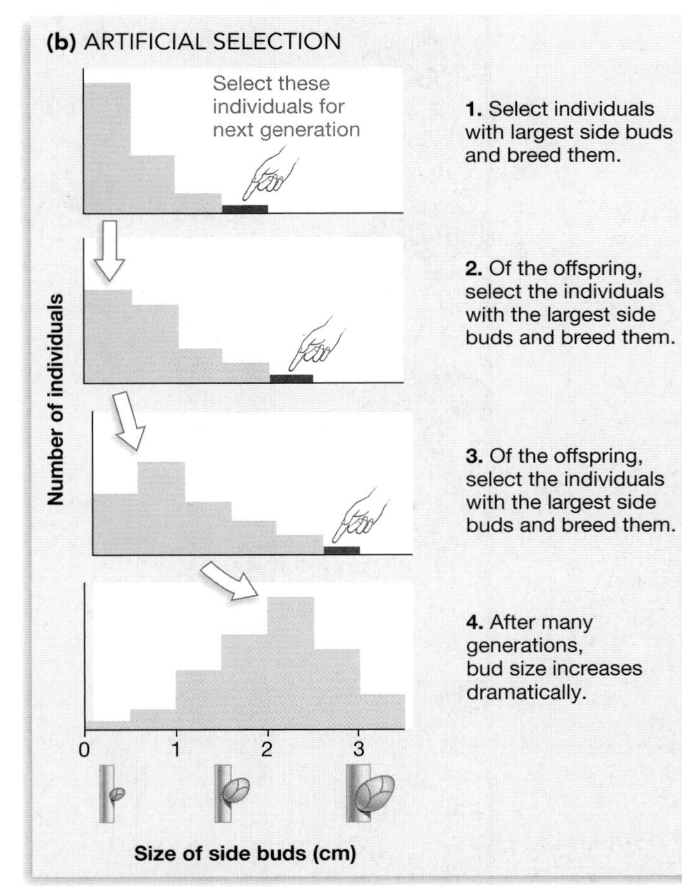

(c) Brussels sprouts: extremely large side buds

FIGURE 1.3 Artificial Selection Can Produce Dramatic Changes in Organisms
(a) This plant is a wild member of the mustard family. It is the progenitor of broccoli, Brussels sprouts, and other domesticated plants. **(b)** This diagram shows how artificial selection proceeded, generation after generation, and led to the development of Brussels sprouts. The charts are called histograms. For each generation, they show how many individuals in the population have side buds with a certain size. **(c)** Brussels sprouts are a domesticated strain developed by artificial selection, starting from the wild species shown in part (a).

CD ACTIVITY 1.1 Artificial Selection

1.3 The Tree of Life

The theory of evolution by natural selection makes an important prediction: All species, past and present, trace their ancestry back to a single common ancestor. If the theory is valid, biologists should be able to reconstruct a **tree of life.** This is a family tree of organisms, a diagram depicting their genealogical relationships with a single, ancestral species at its base.

Has this task been accomplished? If so, how? If the tree of life exists, what does it look like?

To begin answering these questions, we need to back up a step and review how biologists organized the diversity of organisms *before* the development of the cell theory and the theory of evolution.

Linnaean Taxonomy

In science, the effort to name and classify species is called taxonomy. This branch of biology began to flourish in 1735 when a Swedish botanist named Carolus Linnaeus set out to bring order to the bewildering diversity of organisms that were then being discovered.

The fundamental building block of Linnaeus' system is a two-part name that is unique to each type of organism. The first of the two names indicates the organism's **genus.** A genus is made up of a closely related group of species. For example, Linnaeus put humans in the genus *Homo*. Although humans are the only living species in this genus, several extinct organisms known from fossils were later also assigned to *Homo*. The second term in Linnaeus' two-part name identifies the species of organism. Earlier in the chapter a species was defined as a distinct, identifiable type of organism. More formally, a **species** is made up of individuals that regularly breed together. (Chapter 23 explores in more detail how species are defined and identified.) Linnaeus gave humans the specific name *sapiens*.

An organism's genus and species designation is called its scientific name or Latin name. Scientific names are always italicized, and genus names are always capitalized: *Homo sapiens*. Scientific names are based on Latin or Greek word roots or on "Latinized" words from other languages (see **Box 1.1**, page 6). Linnaeus gave a scientific name to every species then known to science.

Linnaeus also maintained that different organisms should not be given the same genus and species names. Other species may be assigned to the genus *Homo*, and members of other genera may be named *sapiens*, but only humans may be named

BOX 1.1 Scientific Names and Terms

Scientific names and terms are often based on Latin and Greek word roots that are descriptive. *Homo sapiens*, for example, is derived from the Latin *homo* for "man" and *sapiens* for "wise" or "knowing." The yeast that bakers use to produce bread and that brewers use to brew beer is called *Saccharomyces cerevisiae*. The Greek root *saccharo* means sugar, while *myces* refers to a fungus. *Saccharomyces* is aptly named "sugar fungus" because the domesticated strains used in commercial baking and brewing operations are commonly fed sugar. The specific name of this organism, *cerevisiae*, is Latin for *beer*. Loosely translated, then, the scientific name of brewer's yeast means "sugar fungus for beer."

Most biologists find it extremely helpful to memorize a few of the more common Greek and Latin roots. To aid you in this process, new terms in this text are often accompanied by a reference to their Latin or Greek word roots in parentheses.

Homo sapiens. As a result, each scientific name is unique. Linnaeus' system has stood the test of time. His two-part naming system is still the standard in biological science.

Taxonomic Levels To organize and classify the tremendous diversity of species that were being discovered in the 1700s, Linnaeus created a hierarchy of taxonomic groups. **Figure 1.4** shows how this nested, or hierarchical, classification scheme works, using humans as an example. Although our species is the sole living member of the genus *Homo*, humans are grouped with the orangutan, gorilla, common chimpanzee, and pygmy chimpanzee in a family called Hominidae. Linnaeus grouped members of this family with gibbons, monkeys, and lemurs in an order called Primates. The Primates are grouped in the class Mammalia with rodents, whales, and other organisms that have fur and produce milk. Mammals, in turn, join other animals with backbones in the phylum Chordata, and all other animals in the kingdom Animalia.

Each of these named groups—primates, mammals, or *Homo sapiens*—can be referred to as a taxon (plural: taxa). The essence of Linnaeus' system is that lower-level taxa are nested within higher-level taxa.

This hierarchical scheme is still in use. As biological science matured, however, several problems with Linnaeus' original proposal emerged.

How Many Kingdoms Are There? Linnaeus proposed that the diversity of species could be organized into just two kingdoms: plants and animals. Organisms that did not move and produced their own food were considered plants; organisms that moved and acquired food by eating other organisms were considered animals.

Not all organisms fell neatly into these categories, however. Molds, mushrooms, and other fungi make their living by absorbing nutrients from dead or living plants and animals. Even though they do not make their own food, they were placed in the kingdom Plantae because they do not move. The tiny, single-celled organisms called bacteria also presented problems. Some bacteria can move and many can make their own food. Initially they, too, were thought to be plants. Eventually, though, it became clear that the two-kingdom system was inadequate.

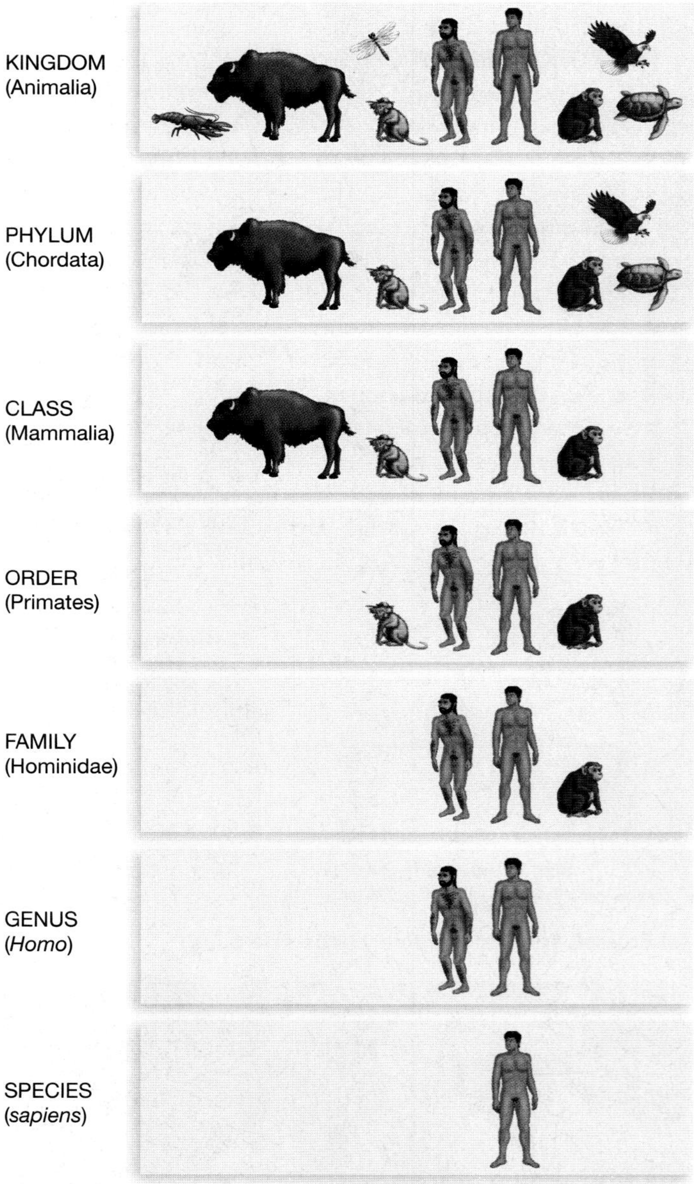

FIGURE 1.4 Linnaeus Defined Taxonomic Levels
In the Linnaean system, each animal species is placed into taxonomic hierarchy with seven levels. Note that lower levels are nested within higher levels, as illustrated here. Linnaeus proposed that these levels reflected the natural order of organisms.

Further, the development of the theory of evolution suggested a new goal for taxonomy. As evidence for evolution mounted, biologists concentrated on understanding how classification systems like the one invented by Linnaeus could be modified to reflect the genealogical relationships among organisms. The goal of taxonomy became an attempt to reflect **phylogeny** (meaning "tribe-source")—the actual, historical relationships among organisms.

In this context, it is important to recognize that Linnaeus proposed the two-kingdom system because he thought it reflected a fundamental pattern in the natural world. Linnaeus' two-kingdom system was a **hypothesis**—a proposed explanation for a phenomenon. In this case, the phenomenon that Linnaeus and other biologists were trying to explain was the relationship among the diverse forms of life found on Earth.

For example, Linnaeus had proposed that the fundamental division in organisms was between plants and animals. But when advances in microscopy allowed biologists to study the contents of individual cells in detail, a different fundamental division emerged. In plants, animals, and the complex unicellular and multicellular organisms that taxonomists call protists, cells contain a prominent membrane-bound component called a nucleus (**Figure 1.5a**). But in bacteria, cells lack this kernel-like structure (**Figure 1.5b**). Organisms with a nucleus were called **eukaryotes** (true-kernel); organisms without a nucleus were called **prokaryotes** (before-kernel).

When data began to conflict with Linnaeus' original proposal, biologists proposed alternative hypotheses. In 1969, for example, R. H. Whittaker suggested that a system of five kingdoms best reflected the actual patterns observed in nature. Whittaker's five-kingdom system is described in **Figure 1.6**.

(a) Eukaryotic cells have a membrane-bound nucleus.

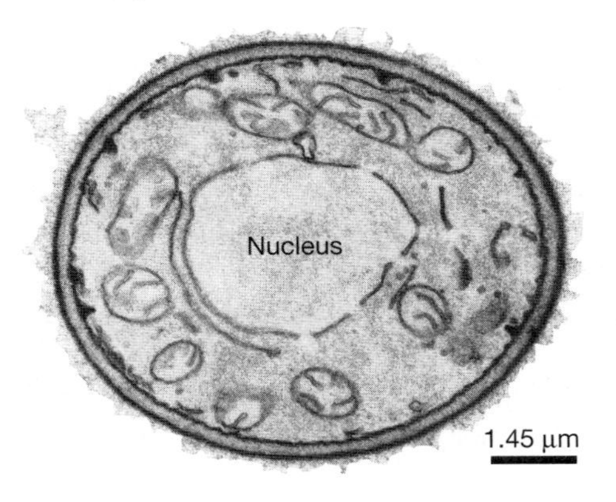

(b) Prokaryotic cells do *not* have a membrane-bound nucleus.

0.29 µm

FIGURE 1.5 Eukaryotes and Prokaryotes
These photos show cross sections of a eukaryotic and a prokaryotic cell.

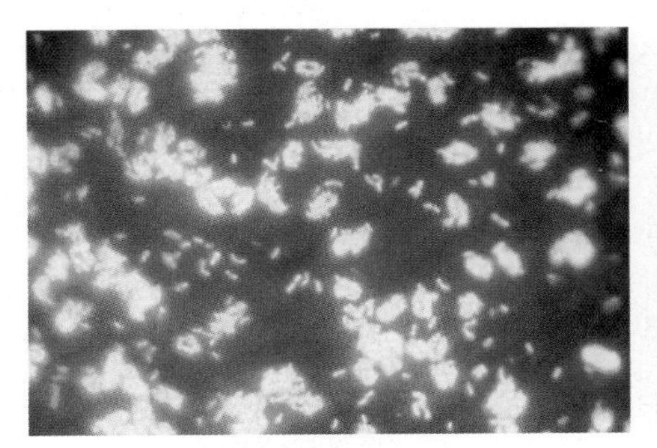

KINGDOM MONERA
(includes all prokaryotes)

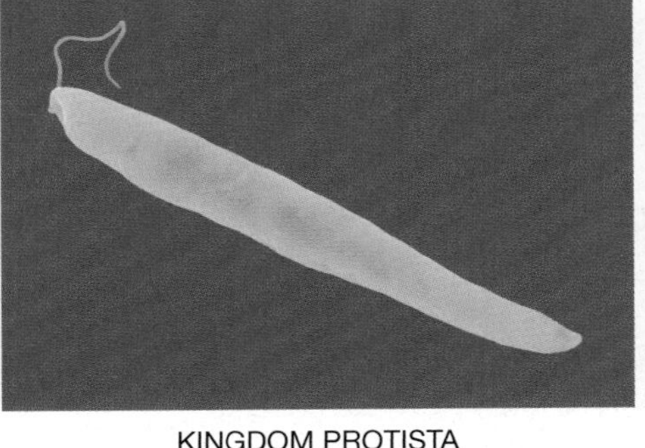

KINGDOM PROTISTA
(includes several groups of unicellular eukaryotes)

KINGDOM PLANTAE

KINGDOM FUNGI

KINGDOM ANIMALIA

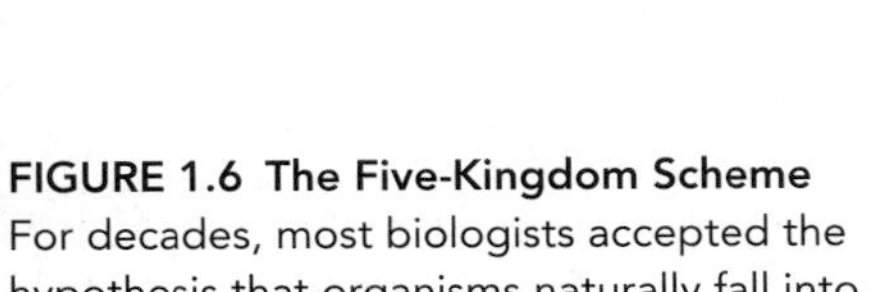

FIGURE 1.6 The Five-Kingdom Scheme
For decades, most biologists accepted the hypothesis that organisms naturally fall into the five kingdoms illustrated here.

Although the scheme was widely used, it represents just one proposal out of many. Other biologists proposed that organisms were organized into three, four, six, or eight kingdoms. But it was still not clear which of these schemes, if any, described the actual phylogeny of organisms accurately.

About the time that Whittaker published his proposal, however, biologist Carl Woese (pronounced "woes") and colleagues began working on the problem from a radically different angle. Instead of assigning organisms to kingdoms based on characteristics like the presence of a nucleus or the ability to move or to manufacture food, these researchers attempted to understand the relationships among organisms by analyzing their chemical components.

Using Molecules to Understand the Tree of Life

Woese and his co-workers had an explicit goal: To estimate the tree of life's shape. To accomplish this they studied a molecule that is found in all organisms. The molecule is called small subunit rRNA, or simply SSU RNA. It is an essential part of the machinery that cells use to grow and reproduce.

Although SSU RNA is large and complex, it is made up of four smaller chemical components called ribonucleotides. These ribonucleotides are symbolized by the letters A, U, G, and C. In SSU RNA, ribonucleotides are connected to one another in sequence, like beads on a string (**Figure 1.7**).

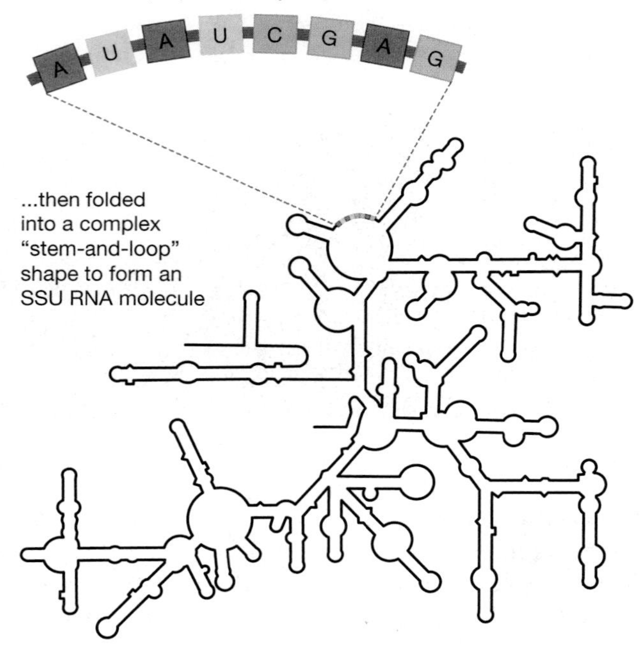

FIGURE 1.7 RNA Molecules Are Made Up of Sequences of Ribonucleotides
SSU RNA is a large, complex molecule that is made up of four smaller molecules called ribonucleotides. In most species, the sequence of ribonucleotides in SSU RNA molecules taken from the same individual is identical. But the sequence can vary between individuals and especially among species.

Why might SSU RNA be useful for understanding the relationships among organisms? Although it performs the same function in all organisms, the sequence of ribonucleotides in the molecule varies from species to species. In plants, for example, the molecule might start with the sequence A-A-G-U-A-C; in fungi the same section might contain A-A-G-U-A-G. The research program that Woese and co-workers pursued was based on a simple premise: If the theory of evolution is true, then SSU RNA sequences should be very similar in closely related organisms, but less similar in organisms that are less closely related. The logic here is that SSU RNA sequences change over time in response to mutation and other processes introduced in Chapter 22. As a result, the RNA sequences found in different species should become increasingly divergent over time.

After the researchers had determined the sequence of ribonucleotides in the SSU RNA of a wide variety of species, they analyzed what the similarities and differences in the sequences implied about relationships among the species. Their goal was to produce a diagram that described the phylogeny of the organisms in their study. A diagram that depicts evolutionary history in this way is called a **phylogenetic tree**. On a phylogenetic tree, branches that are close to one another represent species that are closely related; branches that are farther away represent species that are more distantly related.

The rRNA Tree To determine the tree implied by a data set, researchers run a computer program that is designed for the specific task of finding the arrangement of branches that is most consistent with the similarities and differences observed in the data. The tree produced by comparing SSU RNA sequences is shown in **Figure 1.8**. Because this tree includes species from many different kingdoms and phyla, it is often called the universal tree, or the tree of life.

The tree of life implied by SSU RNA data astonished biologists. According to this molecule:

- The fundamental division in organisms is not between plants and animals or even between prokaryotes and eukaryotes. Rather, *three* major groups occur. These are the bacteria; another group of prokaryotic, single-celled organisms called the Archaea; and the eukaryotes. To accommodate this new perspective on the diversity of organisms, Woese created a new taxonomic level called the domain. As the figure indicates, the three domains of life are now called the Bacteria, Archaea, and Eukarya.
- Some of the kingdoms that were defined earlier do not reflect the actual phylogeny of organisms. For example, the rRNA data contend that the multicellular eukaryotes known as brown algae, which live in marine environments, are not closely related to the plants that grow on land.
- Bacteria and Archaea are much more diverse than anyone imagined. If the differences among animals, fungi, and plants warrant placing them in separate kingdoms, then dozens of kingdoms exist among the prokaryotes.

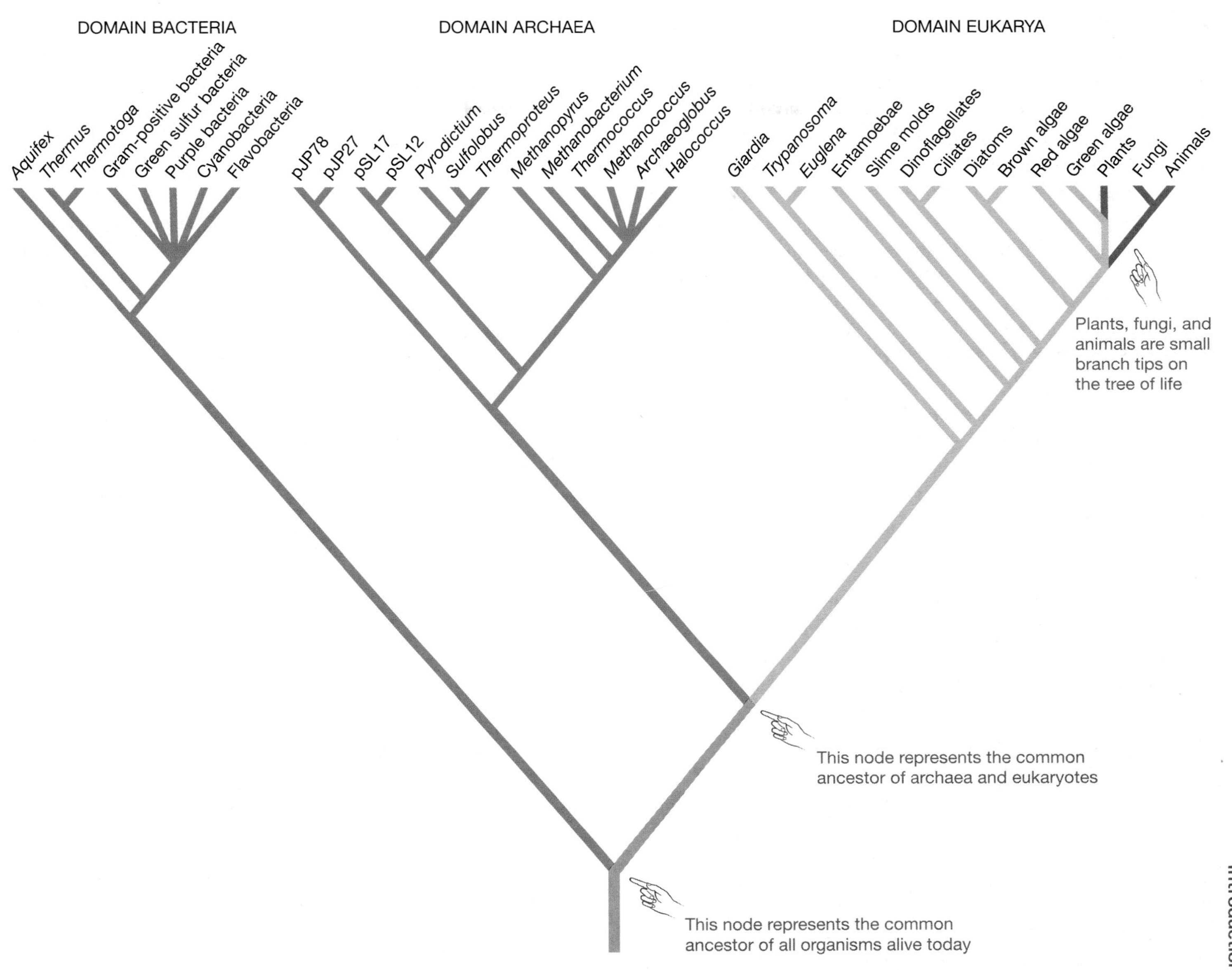

FIGURE 1.8 The Tree of Life
This "universal tree" was estimated from SSU RNA sequence data. The three domains of life revealed by this analysis are labeled. Common names are given for most of the lineages in the Bacteria and Eukarya included in this study. Genus names are given for members of the Archaea because these organisms have no common names. The archaeal species labeled with designations like "pSL17" were discovered so recently that they have not yet been given scientific names.

CD ACTIVITY 1.2 Introduction to Experimental Design

The Tree of Life Is a Work in Progress The SSU RNA tree inspired a flurry of research. Biologists in Woese's lab and in other research groups tested the conclusions by determining the sequence of other molecules found in cells and by reanalyzing older data sets in light of the new findings. In general, these studies have confirmed the major features of the SSU RNA tree. Most biologists now agree that the discovery of the Archaea and the placement of lineages like the brown algae qualify as authentic breakthroughs in our understanding of life's diversity (see **Box 1.2**, page 10).

Work on the tree of life continues at a furious pace, however, and the location of certain branches on the tree is hotly debated. As databases expand and as techniques for analyzing data improve, the estimate of the tree of life's shape presented in Figure 1.8 will undoubtedly change. Chapters 25–30 explore the major branches on the tree of life in much more detail.

1.4 Biological Science

This chapter has introduced some of the great ideas in biology. The development of the cell theory and the theory of evolution by natural selection provided cornerstones when the science was young; the tree of life is a relatively recent insight that has revolutionized how researchers understand the diversity of life on Earth.

These ideas are considered great because they explain fundamental aspects of nature, and because they are correct. They are considered correct because they have withstood extensive testing. How do biologists test ideas about how the natural world works? Two general strategies are essential. The first is to test the predictions made by alternative hypotheses; the second is to set up carefully designed experiments. To illustrate how these methods work, let's consider two questions currently being addressed by researchers.

Why Do Giraffes Have Long Necks? An Introduction to Hypothesis Testing

Why *do* giraffes have long necks? If you asked someone, they might tell you that long necks enable giraffes to reach food that is unavailable to other mammals. This hypothesis is expressed in African folk tales and has traditionally been accepted by many biologists.

The food competition hypothesis is so plausible that for decades no one thought to test it. Recently, however, Robert Simmons and Lue Scheepers set out to test it rigorously. Based on the data they assembled, they have concluded that the food competition hypothesis probably is not true. They also collected data that support an alternative hypothesis. Simmons and Scheepers propose that giraffes have long necks because their necks are effective weapons.

How did these biologists test the food competition hypothesis? What data support their alternative explanation? Before answering these questions, it's important to recognize that hypothesis testing is a two-step process. The first task is to state the hypothesis as precisely as possible and list the predictions that it makes. The second is to design an observational or experimental study that is capable of testing those predictions. If the predictions are accurate, then the hypothesis is supported. If the predictions are not met, researchers do further tests or begin to search for alternative explanations.

Stated precisely, the food competition hypothesis claims that giraffes compete for food with other species of mammals. In the giraffe population, neck length is a variable trait. It is also heritable, meaning that long-necked parents tend to have long-necked offspring. The food competition hypothesis claims that when food is scarce, during the dry season, giraffes with longer necks are able to reach food that is unavailable to other species and to giraffes with shorter necks. As a result, the longest-necked individuals in the population survive better than shorter-necked individuals. Because the long-necked giraffes leave more offspring, the average neck length of the population increases each generation. This type of natural selection has gone on so long that the population has become extremely long necked.

The Food Competition Hypothesis: Predictions and Tests

The food competition hypothesis makes several explicit predictions:

- Neck length is variable among giraffes.
- Neck length is heritable.
- Giraffes feed high in trees, especially during the dry season when food is scarce and the threat of starvation is high.

Simmons and Scheepers argue that the first prediction is correct. Studies in zoos and natural populations confirm that neck length is variable among individuals.

The researchers were unable to test the second prediction, however, because they studied giraffes in the field and were un-

BOX 1.2 Medicine, Economics, and the Tree of Life

Advances in biological science frequently have important practical applications. Even seemingly esoteric work, like efforts to understand the tree of life, can help direct research focused on improving human health and welfare.

For example, researchers are currently working to understand where organisms called microsporidians belong on the tree of life. The microsporidians represent a phylum in the Eukarya; all microsporidians are single celled and make a living by parasitizing animals. On the basis of SSU RNA data, Woese and colleagues concluded that microsporidians branched off early in the diversification of eukaryotes—near the base of the domain. But researchers who have examined other molecules and traits argue that microsporidians are actually extremely closely related to fungi, meaning that they branched off very late in the diversification of eukaryotes. These biologists contend that unusual changes in the SSU RNA of microsporidians have occurred, making that molecule an unreliable indicator of their phylogeny.

This debate has practical importance because microsporidians parasitize silkworms and honeybees. Untreated infections of these parasites can devastate commercial colonies of these insects and cause serious economic losses; microsporidians can also afflict AIDS patients. Because the drugs used to kill parasites like microsporidians work by disrupting specific chemicals inside their cells, a particular therapy is usually effective against a suite of closely related species (whose molecular composition is similar). The question is, should physicians try to treat microsporidian infections with drugs that are potent against eukaryotic cells from the base of the Eukarya, or with fungicides (fungus-killers)? Understanding where microsporidians occur on the tree of life is essential to designing effective treatments.

able to do breeding experiments. (If a researcher had access to breeding records from zoos, along with data on neck length in parents and offspring, this assumption could be tested. Chapter 21 will explain how this is done and provide an example.) As a result, Simmons and Scheepers simply had to accept this prediction as an assumption. In general, though, biologists prefer to test every assumption behind a hypothesis.

What about the prediction regarding feeding high in trees? According to Simmons and Scheepers, this is where the food competition hypothesis breaks down. Consider, for example, data collected by Truman Young and Lynn Isbell on the amount of time that giraffes spend feeding in vegetation of different heights. The graphs in **Figure 1.9a** show that in a population from Kenya, both male and female giraffes spend most of their feeding time eating vegetation that averages just 60 percent of their full height. The message in these data is that giraffes usually feed with their necks bent (**Figure 1.9b**). Only the largest bull giraffes break this rule. These males live with groups of females and offspring and normally feed with their necks fully extended (even while other individuals in the herd feed at low heights). However, Simmons and Scheepers propose that instead of reducing competition for food, the posture is primarily social in nature. It allows dominant males to search for predators and to detect the approach of bulls that might challenge their access to the breeding females in the herd.

Based on their own research and other studies in the published literature, Simmons and Scheepers claim that Young and Isbell's data are typical. For example, during the wet season in Tsavo National Park in East Africa, about 50 percent of giraffe browsing occurs below 2 meters—within reach of competing browsers such as gerenuk and lesser kudu. During the dry season at the same site, when competition for food should be most severe, giraffes still spent 37 percent of their time feeding at these low heights.

These data cast doubt on the food competition hypothesis. In response, Simmons and Scheepers offer an alternative explanation that is based on the mating system and social behavior of giraffes.

The Sexual Selection Hypothesis: Predictions and Tests

Giraffes have an unusual social system. Breeding occurs year round rather than seasonally. To determine when females are coming into estrus (or "heat") and are thus receptive to mating, the males perform an unusual behavior. They nuzzle the rumps of females. In response, the females urinate into the males' mouths. The males then tip their heads back and pull their lips to and fro, as if tasting the liquid. Anne Innis, who was one of the first to catalog this behavior, proposed that males taste urine to detect whether estrus has begun.

Once a female is in heat, males fight among themselves for the opportunity to mate. Combat is spectacular. The bulls stand next to one another, swing their necks, and strike thunderous blows with their heads. Innis saw a male knocked unconscious for 20 minutes after being struck; Simmons and Scheepers catalog numerous instances where the loser died. Giraffes are the only animals on Earth that fight in this way.

These observations inspired Simmons and Scheepers to propose an alternative explanation for why giraffes have long necks. The researchers hypothesize that the longest-necked giraffes are able to strike the hardest blows during combat. (In

(a) Most feeding is done below neck height.

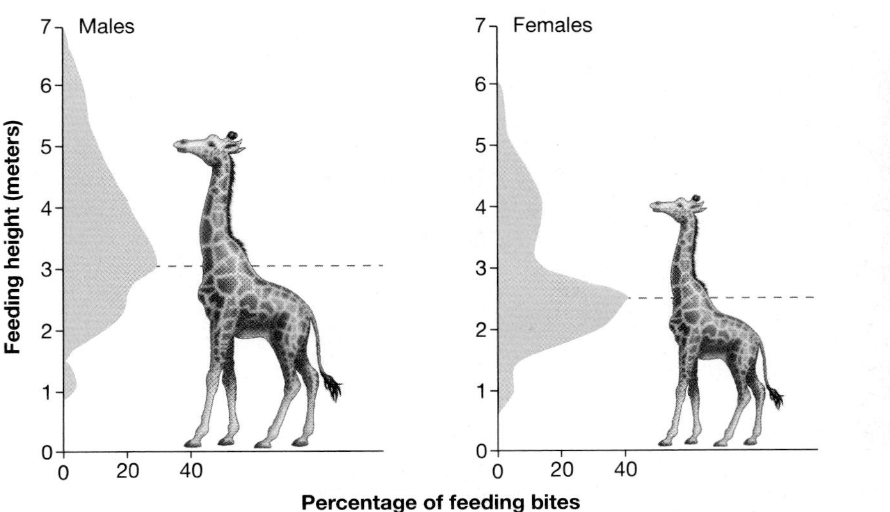

(b) Typical feeding posture in giraffes

FIGURE 1.9 Giraffes Do Not Usually Extend Their Necks to Feed
(a) These data plot the proportion of bites that male and female giraffes take at different heights. The drawings show the average size of a male and female. **(b)** Although it is common to see photos of giraffes straining to reach leaves high in trees, these almost always depict dominant bull giraffes, which feed high in vegetation much more frequently than do other individuals. Dominant bulls are a small proportion of the total population.

engineering terms, longer necks provide a longer moment arm. A long moment arm increases the force of the impact.) Simmons and Scheepers contend that males with longer necks win more fights and father more offspring than shorter-necked males. Based on the assumption that neck length is heritable, the average neck length in the population should increase over time. Because it occurs in the context of mating, this type of selection is called **sexual selection**. (Chapter 22 explores sexual selection in detail.)

What evidence do Simmons and Scheeper offer to support their hypothesis? They point to two types of data. First, several studies have shown that long-necked males are more successful in fighting, and that the winners of fights gain access to estrous females. Second, when bulls mature, their heads and necks continue to grow after the rest of their body has stopped growing. This secondary growth adds bone to the skull, additional horns, and musculature to the neck. By analyzing data on neck and head weight, Simmons and Scheeper were able to show that adult male giraffes tend to have much longer necks and heavier heads than adult females, even when scaled for differences in overall size. (Male giraffes are much larger than female giraffes, but Simmons and Scheepers showed that neck length and head weight in males are *proportionately* greater.)

Although these results are consistent with their hypothesis, Simmons and Scheepers are careful to point out that they are not definitive. In fact, the researchers end their paper by inviting other researchers to devise additional tests of the two hypotheses. Based on the data collected to date, most biologists would probably concede that the food competition hypothesis is in trouble and that the sexual selection hypothesis appears promising. More work needs to be done, however, before any solid conclusions can be drawn.

In many cases in biological science, "more work" involves experimentation. Experimenting on giraffes is difficult. But in the case study considered next, an experiment was able to definitively reject several hypotheses and confirm another.

A Sheep in Wolf's Clothing? An Introduction to Experimental Design

Experiments are a powerful tool in science for the simple reason that they allow researchers to isolate and test the effect that a single, well-defined factor has on the phenomenon in question. As an example of how experiments are designed, let's consider an unusual hypothesis about some unusual behavior in flies.

As **Figure 1.10** shows, the fly *Zonosemata vittigera* has distinctive dark bands on its wings. When one of these flies is disturbed, it responds by holding its wings perpendicular to its body and waving them up and down. This display appears to mimic the leg-waving, territorial threat display of *Zonosemata*'s major predator, jumping spiders (Figure 1.10).

Why do these flies have wing markings, and why do they wave them up and down when they are provoked? Erick Greene and colleagues designed an experiment to test three alternative hypotheses:

H_1—The wing-waving display is used in courtship. (This hypothesis is plausible because wing-waving courtship displays are common in flies.)

H_2—The flies are mimicking jumping spiders to scare off predators such as assassin bugs, preying mantises, lizards, and other kinds of spiders. (This hypothesis is also plausible because jumping spiders have a nasty bite.)

H_3—The flies mimic jumping spiders to deter predation by the jumping spiders themselves. (Some observers call this the "sheep in wolf's clothing" hypothesis.)

To test these alternatives, Greene and co-workers transplanted wings between houseflies (*Musca domestica*), which have

FIGURE 1.10 Science Starts with a Question
Why do *Zonosemata* flies have wing markings, and why do they wave their wings up and down when provoked?

clear and unmarked wings, and *Zonosemata*. They did this by cutting the wings off houseflies and using Elmer's glue to attach them to the wing stubs of amputated *Zonosemata*. The surgically altered *Zonosemata* flies were able to wave their new wings normally and even fly. Greene and colleagues also performed the reciprocal transplant, placing *Zonosemata* wings on houseflies.

As **Figure 1.11** shows, the surgical manipulations created five experimental groups. The important thing to note is that the three hypotheses make different predictions about what will happen when flies from each of the five groups encounter a jumping spider or another type of predator. Specifically, the courtship hypothesis predicts that all of the flies will be eaten by all of the predators. The other-predators hypothesis predicts that two of the groups—untreated *Zonosemata* and *Zonosemata* that have had their wings taken off and reglued—will be attacked by jumping spiders but avoided by other predators. The sheep-in-wolf's-clothing hypothesis predicts that these same two groups will be avoided by jumping spiders but attacked by other predators.

To test these predictions, the researchers needed to measure the response of jumping spiders and other predators to the five types of flies. When confronted with a test individual, did the spiders retreat or did they stalk and attack or actually kill? To answer this question, the investigators starved 20 jumping spiders from 11 different species for two days. Then they presented one of each of the five types of flies to each spider in random order. The researchers made these presentations in a test arena and recorded each jumping spider's most aggressive response during a 5-minute interval.

The result? There was a clear difference among groups. Jumping spiders tended to retreat from flies that gave the wing-waving display with marked wings, but usually attacked flies that lacked either wing markings, wing waving, or both (see Figure 1.11). Further, when the researchers tested flies from

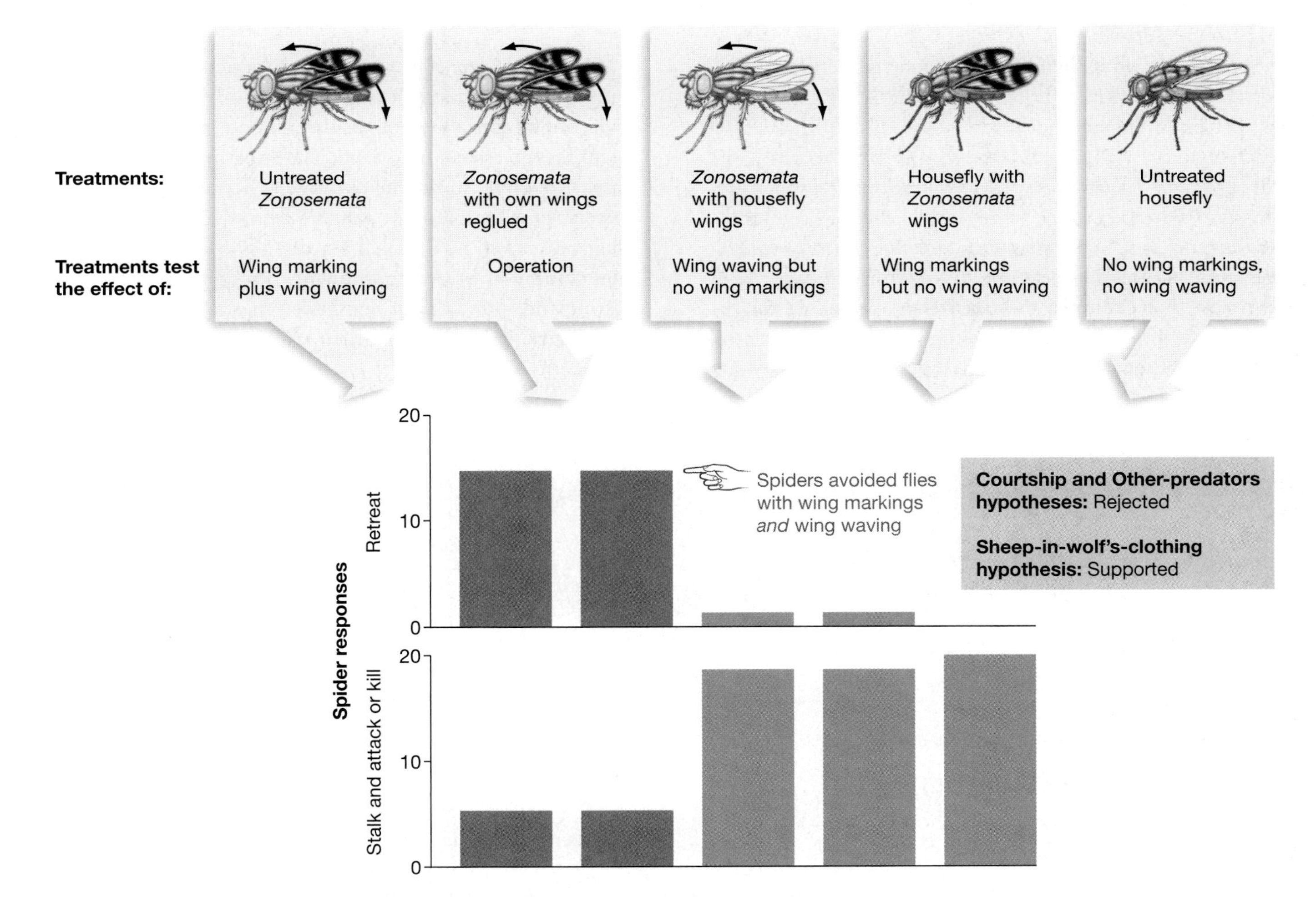

FIGURE 1.11 Experimental Evidence That Waving Marked Wings Deters Jumping Spiders
Researchers used *Zonosemata* flies and houseflies to create five treatment groups. Each group tested a different proposition about wing-waving display. Experimental results are presented as a type of graph, called a histogram. In this case, the histograms show how jumping spiders responded to the 20 flies in each treatment group. QUESTION The biologists who did this work claim that these data support the hypothesis that *both* wing markings and wing waving are essential to successful mimicry. Do you agree? Why or why not?

different treatments against other types of spiders, assassin bugs, mantises, and whiptail lizards, all of them were captured and eaten. In sum, the experiment provides strong support for the hypothesis that *Zonosemata* mimic their own predators to avoid being eaten by them.

As an experimental design, the study illustrates several important points:

- *It is critical to include control groups.* An **experimental control** is identical to a second treatment except for the one factor thought to be important. The name *control treatment* is used because it controls for factors—other than the one being tested—that might influence the experiment's outcome. For example, the individuals in the second group illustrated in Figure 1.11 act as a control for the effect of the surgery. Because jumping spiders responded to them just as they did to untreated *Zonosemata*, the researchers could claim that the wing surgery itself had no effect on the behavior of the flies or the spiders.
- *All of the treatments must be handled exactly alike.* It was critical that the investigators used the same test arena, the same time interval, and the same definitions of predator response in each test. What types of problems could arise if a different test arena were used for each of the five treatment groups?
- *Randomization is an important technique for equalizing other, miscellaneous effects among control and experimental groups.* For example, the researchers presented the different kinds of test flies to the spiders and other predators in random order. What problems could arise if they had presented the five types of flies in the same sequence to every spider?
- *Repeating the test on many individuals is essential.* It is almost universally true in research that larger sample sizes are better. This is because the goal of an experiment is to estimate a quantity—in this case, the likelihood that a jumping spider will attack a fly as a function of the fly's ability to wave marked wings. Replication reduces the amount of distortion in the estimate caused by unusual individuals or circumstances. For example, 10 *Zonosemata* with marked wings (out of the total of 40) were attacked. But four of these individuals were actually pounced on and killed before they even had a chance to display their wings. It is not acceptable to simply throw out these four data points, even though they might represent bad luck. If events like these really do represent bad luck, they should be rare. As long as the sample size is sufficiently large, luck will not confound the result.

To summarize, the experimental design was successful because it allowed the researchers to test the effect that predator type, wing type, and wing display had on the ability of *Zonosemata* flies to escape predation. The experiment is also a taste of things to come. In this text you will encounter hypotheses and experiments on questions ranging from why offspring resemble their parents to how global warming will affect plant and animal communities. A commitment to tough-minded hypothesis testing and sound experimental design is a hallmark of biological science. Understanding their value is an important first step in becoming a biologist.

Essay Where Do Humans Fit on the Tree of Life?

Given the vast diversity of organisms that make up the tree of life, where do humans fit in? Most of the major branches on the tree diagrammed in Figure 1.8 represent unicellular organisms. Familiar, multicellular species like grasses and mushrooms are part of the twigs labeled "Plants" and "Fungi." At the tree's tip, the label "Animals" identifies the kingdom that appeared most recently among the eukaryotes. Humans are found on this branch, along with millions of other species ranging from sponges and corals to insects and fish.

Our species is a tiny new twig on an enormous and ancient tree of life.

Just how recently did our species appear? Chapter 2 introduces a technique, called radiometric dating, that allows geologists to estimate the age of rocks containing bones, teeth, shells, and other traces of organisms that lived in the past. According to these data, the first traces left by our species appear about 100,000 years ago. To clarify just how recent this is in the sweep of Earth history, consider the calendar in **Figure 1**. The 12 months shown are scaled to represent the 4.6 billion years of Earth history. At this scale, 7 seconds make up a millennium, each day represents an interval of 12.6 million years, and each hour a span of 525,000 years. Note that the first unicellular organisms appear in late March and the first multicellular life in early October. Hominids walk upright for the first time on mid-afternoon of New Year's Eve. *Homo sapiens* appears an hour before the stroke of midnight.

The message of these analyses is clear: Our species is a tiny new twig on an enormous and ancient tree of life.

(Continued on next page)

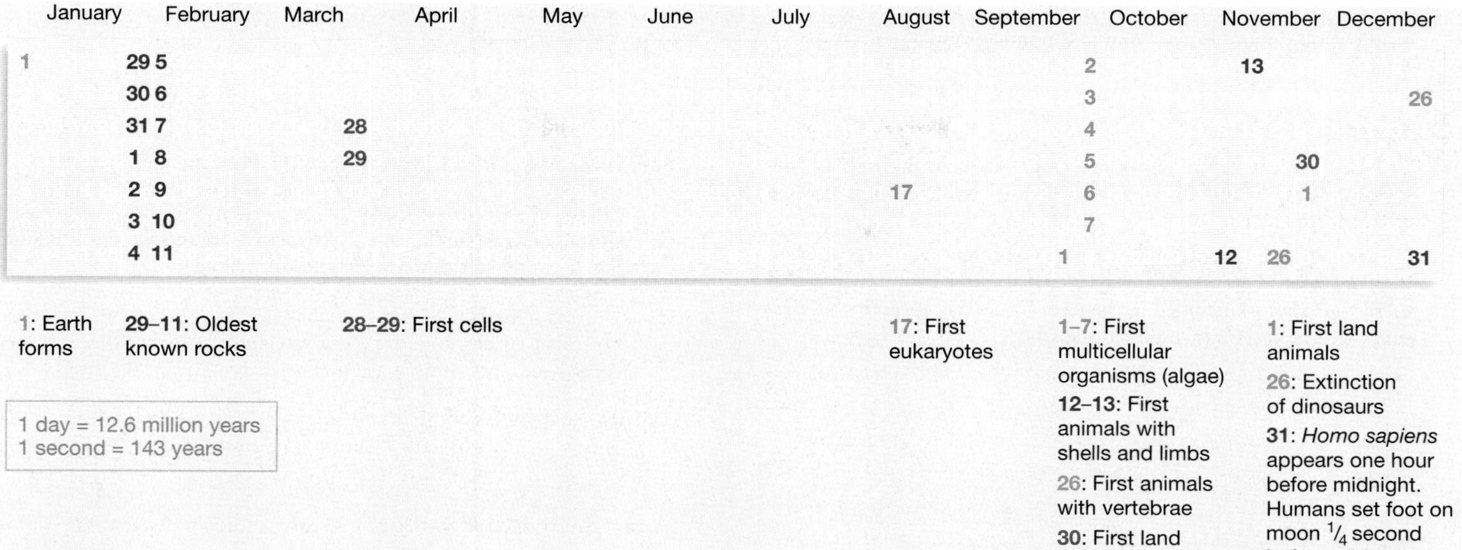

January February March April May June July August September October November December

1 | 29 5 | 30 6 | 31 7 | 1 8 | 2 9 | 3 10 | 4 11 | 28 | 29 | 17 | 2 | 3 | 4 | 5 | 6 | 7 | 1 | 12 | 13 | 30 | 1 | 26 | 26 | 31

1: Earth forms

29–11: Oldest known rocks

28–29: First cells

17: First eukaryotes

1–7: First multicellular organisms (algae)

12–13: First animals with shells and limbs

26: First animals with vertebrae

30: First land plants

1: First land animals

26: Extinction of dinosaurs

31: *Homo sapiens* appears one hour before midnight. Humans set foot on moon 1/4 second before midnight

1 day = 12.6 million years
1 second = 143 years

FIGURE 1

Chapter Review

Summary

A fundamental goal of biological science is to discover traits and processes that unify the diversity of living organisms. In the mid-1800s, the development of the cell theory and the theory of evolution identified some of these key unifying features and launched biology as a modern science. The cell theory states that all living organisms are made of cells and that all cells arise from preexisting cells. The theory of evolution states that all organisms are related by common ancestry and that the characteristics of populations change through time. Evolution is caused by natural selection, which occurs when certain heritable traits cause some individuals to leave more offspring than others.

The cell theory and theory of evolution predict that all organisms are part of a genealogy of species, and that all species trace their ancestry back to a single common ancestor. To reconstruct this phylogeny, biologists have analyzed the sequence of ribonucleotides in a molecule called SSU RNA that is found in all cells. A tree of life, based on similarities and differences in these sequences, has recently been constructed and is currently the focus of intense research. According to the information contained in SSU RNA, the tree of life's shape defines three major lineages among organisms: the Bacteria, Archaea, and Eukarya. Analyses of other molecules and traits have largely supported the conclusions made from SSU RNA data.

Another unifying theme in biology is a commitment on the part of biologists to hypothesis testing and to sound experimental design. Analyzing neck length in giraffes and the wing-waving displays of *Zonosemata* flies are case studies in the value of testing alternative hypotheses and conducting experiments.

Questions

Content Review

1. Anton van Leeuwenhoek made a huge contribution to the development of the cell theory. How?
   - **a.** He articulated the pattern component of the theory.
   - **b.** He articulated the process component of the theory.
   - **c.** He invented the first microscope, and saw the first cell.
   - **d.** He invented more powerful microscopes and was the first to describe the diversity of cells.

2. In the experiments with nutrient broth diagramed in Figure 1.2, what were the two treatments designed to test?
   - **a.** whether spontaneous generation occurs
   - **b.** whether heating could destroy cells effectively
   - **c.** how quickly cells would grow in flasks of different shapes
   - **d.** whether *all* living things are made of cells

3. What does the term *evolution* mean? (Circle all that apply.)
   - **a.** The characteristics of populations change through time.
   - **b.** The characteristics of an individual change through the course of its life, in response to natural selection.
   - **c.** "Modification with descent" occurs.
   - **d.** All species are related by common ancestry.

4. Which of the following results in evolution in natural populations?
   - **a.** artificial selection
   - **b.** radiometric dating
   - **c.** natural selection
   - **d.** spontaneous generation

5. What does it mean to say that a characteristic of an organism is heritable?
   - **a.** It evolves.
   - **b.** It can be passed on to the next generation.
   - **c.** It is not advantageous to the organism.
   - **d.** It does not vary in the population.

6. What is a phylogeny?
   a. the evolutionary history of a taxonomic group
   b. a tree
   c. a sequence of SSU RNA
   d. none of the above

Conceptual Review

1. The Greek roots in the term *taxonomy* can be translated as "arranging-rules." Explain why these roots were an appropriate choice for this term.
2. On the phylogenetic tree shown here, label the most recent common ancestor of species A and species B, and the common ancestor of all four species included in the tree. Circle the tips, which represent the four species included in the study.

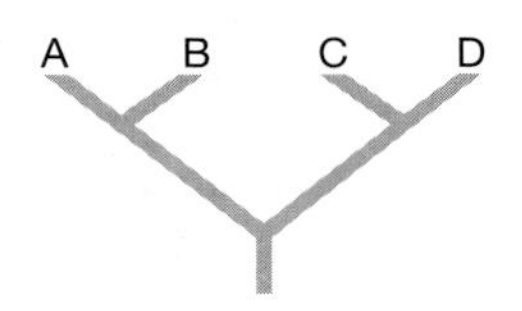

3. Why was it important for Linnaeus to establish the rule that only one type of organism can have a particular genus and species name?
4. Compare and contrast natural selection with the process that led to the divergence of a wild mustard plant into cabbage, broccoli, and Brussels sprouts.
5. Consider the following two quotations from the text:

   "If the theory of evolution is true, then SSU RNA sequences should be very similar in closely related organisms, but less similar in organisms that are less closely related."

   "On a phylogenetic tree, branches that are close to one another represent species that are closely related; branches that are farther away represent species that are more distantly related."

   These statements explain the logic behind using molecular sequence data to estimate evolutionary relationships. Is this logic sound? Why or why not?
6. In the introduction to this chapter, the author states that his goal is to "convince you that the diversity of life is indeed remarkable, and that biologists have identified fundamental mechanisms and traits that unify all organisms." Was this goal fulfilled? If so, what did you learn about the diversity of life? What did you learn about the traits and mechanisms that unify organisms?

Applying Ideas

1. A scientific theory is a set of propositions that defines and explains some aspect of the world. This definition contrasts sharply with the everyday English usage of the word *theory,* which often carries meanings like "speculation" or "guess." Explain the difference in the two definitions, using the cell theory and the theory of evolution by natural selection as examples.
2. Turn back to the tree of life shown in Figure 1.8. Note that Bacteria and Archaea are prokaryotes, while Eukarya are eukaryotes. Draw an arrow on the tree that points to the branch where the structure called the nucleus originated. Explain your reasoning.
3. The proponents of the cell theory could not "prove" that it was correct in the sense of providing incontrovertible evidence that all organisms are made up of cells. They could only state that all organisms examined to date were made of cells. Why was it reasonable for them to conclude that the theory was valid?
4. How do the tree of life and the taxonomic categories created by Linnaeus (kingdom, phylum, class, order, family, genus, and species) relate to one another?
5. Review Pasteur's experiment. Suppose a proponent of the spontaneous generation hypothesis criticized the results by claiming that life would appear in the swan-necked flask eventually. According to this view, Pasteur did not allow enough time to pass before reaching the conclusion that life does not originate spontaneously. Respond to this critic.
6. Suppose a friend takes issue with the section of the chapter on why giraffes have long necks. This friend claims that the food competition hypothesis and the sexual selection hypothesis are not mutually exclusive—that both could be true. Comment.
7. Erick Greene is still doing research on the behavior of *Zonosemata* flies. Write down three questions about their courtship or antipredator behavior that you would like to see answered.

CD-ROM and Web Connection

CD Activity 1.1: Artificial Selection *(Animation)*
(Estimated time for completion = 5 min)
Explore artificial selection as a tool for manipulating the characteristics of a population.

CD Activity 1.2: Introduction to Experimental Design *(Tutorial)*
(Estimated time for completion = 15 min)
Design an experiment to test your hypothesis about certain characteristics of the fly *Zonosemata vittigera.*

At your **Companion Website** (http://www.prenhall.com/freeman/biology), you will find self-grading exams and links to the following research tools, online resources, and activities:

History of Evolutionary Theory
This timeline lists scientists whose work has had a major impact on evolutionary theory.

The Tree of Life
Explore this latest construction of the phylogenetic tree of all living organisms.

Scientific Hypotheses
These short articles from *Discover* magazine present two hypotheses being tested now.

Additional Reading

Thomas, Lewis. 1974. *The Lives of a Cell* (New York: Viking Press). A wonderful collection of essays by a remarkable physician and researcher. Also see *The Medusa and the Snail* (New York: Viking Press) by Lewis Thomas, published in 1979.

Ford, B. J. 1998. The earliest views. *Scientific American* 278 (April): 50–53. What did van Leeuwenhoek see? A contemporary microscopist uses the original instruments to re-create his results.

The Origin and Early Evolution of Life

Unit 1

The chapters in this unit explore the chemical conditions that led to the origin of life. Their goal is to answer two fundamental questions. How were simple compounds present in the atmosphere and ocean of ancient Earth transformed into the first living entity? How did the first cells form, and what were they made of?

In addition to considering these specific questions, Unit 1 has a more general goal: to review the chemical principles that underlie much of biology. Life can be considered as a series of carefully orchestrated chemical reactions. The chemistry introduced in this unit lays the groundwork for understanding dozens of questions that are taken up later in the book, ranging from how human nerve cells relay messages to why global warming is occurring.

We begin in **Chapter 2: The Atoms and Molecules of Ancient Earth** by asking how and when Earth formed and by examining the chemical composition of the early atmosphere and ocean. This chapter provides an introduction to the structure and behavior of the atoms and simple molecules found in all living cells. **Chapter 3: Macromolecules and the RNA World** explores how simple carbon-containing compounds might have combined into a more complex molecule that led to the first living entity. **Chapter 4: Membranes and the First Cells** focuses on the chemical properties of the first cell membranes and provides an introduction to the cell—the fundamental structural unit of life.

In essence, this unit is about the tree of life's roots. The chapters trace the events that led to the origin of life and the evolution of the first cell. The effort to understand these events takes us back in time, into a realm where the boundaries of physics, chemistry, and biology intersect.

This Gulf sturgeon represents one of the most ancient of all lineages of fish. Fossilized sturgeon are found in rocks that are over 350 million years old. This unit examines the most ancient of all events in the history of life, which took place over 3.5 *billion* years ago. ©1994 Susan Middleton and David Liittschwager

The Atoms and Molecules of Ancient Earth

2

This chapter explores the earliest events in Earth history, from the formation of the planet through the advent of the first oceans and atmosphere.

As far as anyone knows, Earth is the only place where life exists. Given the size of the universe and its age, this is remarkable. How did life begin? The biologist Stefan Bengston calls this simple query "the mother of all questions."

This chapter introduces a hypothesis, called chemical evolution, that attempts to answer this question. **Chemical evolution** is the proposition that early in Earth's history, simple chemical compounds in the atmosphere and ocean combined to form larger, more complex substances. The hypothesis maintains that these chemicals reacted with one another to produce even more complicated compounds, and that continued chemical evolution eventually led to the origin of life and the start of biological evolution. Is the hypothesis plausible? What evidence do biologists have that chemical evolution actually occurred?

2.1 The Ancient Earth

Advocates of the chemical evolution hypothesis contend that when this process took place, Earth was a radically different place than it is today. To understand whether chemical evolution is plausible, then, it is important to understand what these claims about the nature of the early Earth are, and what evidence researchers use to support their argument. Let's take a closer look.

Studying the Formation of Planets

Astronomers and geologists pursue two strategies to study how Earth formed. They analyze how planets are being created elsewhere in the universe, and they perform computer simulations to model conditions that led to the development of Earth.

Direct studies of planet formation are in their infancy. Prior to 1992, astronomers had no firm evidence that planets existed anywhere outside of our solar system. Now numerous stars have been identified that appear to be surrounded by orbiting clusters of dust and rock. Although Earth-like planets have yet to be identified within these clusters, astronomers expect to find them. Their optimism is based on predictions made by computer simulations of how planets form.

In biology and the other sciences, computer simulations are an important tool for analyzing processes that are difficult to study directly. The general strategy is for researchers to write a computer program that specifies a set of starting conditions and the properties of the entities involved. The computer then records the outcome of interactions among these entities as the simulation proceeds.

For example, consider the outcome of simulations based on the nebula theory of planet formation. (A *nebula* is a cloud of dust and gas.) This theory maintains that the solar system originated when billions of dust-sized particles accumulated around the Sun. To test this idea, V. S. Safronov created a mathematic model that described how small particles behave in orbit around the Sun. The properties of the particles were based on the types of atoms found in the solar system and their abundance. Later, Safronov's equations were incorporated into computer simulations of planetary formation. In the initial stages of these simulations, dust and gas particles start colliding. Particles frequently stick together after the collisions due to electrical and gravitational attraction. As the simulation continues to run, the collision-and-sticking events begin producing small bodies called planetesimals. Collisions between these rocks eventually create larger bodies, called proto-planets, whose gravity attracts still more planetesimals. In this way, planet-sized bodies form.

Safronov's model is now widely accepted as the most plausible explanation for the formation of Earth. Because of the immense heat generated by the larger impacts during the "early accretion" phase of the simulation, most scientists conclude that Earth was molten initially. As Earth began to cool enough to form a crust of solid rock, two important events would have occurred. First, water would have rained out of the cooling atmosphere to form the first ocean. Second, volcanoes would have been common, as molten rock from below punctured the developing crust.

Several lines of evidence argue that heavy bombardment from space continued, however. For example, most astronomers believe that a particularly spectacular blow—a collision with a proto-planet the size of Mars—knocked the Earth into its present tilt and blasted debris into space that formed the moon. The extensive cratering on the moon's surface suggests that Earth was also bombarded long after this giant impact.

When Did Chemical Evolution Take Place?

Geologists and astronomers estimate when these events occurred using a procedure called **radiometric dating**. To understand this technique, it is essential to recognize that all atoms share the same basic structure. Extremely small particles, called electrons, orbit around a nucleus made up of larger particles called protons and neutrons (**Figure 2.1**). Protons have a positive electric charge, neutrons are electrically neutral, and electrons are negatively charged. Opposite charges attract, while like charges repel. Each of the elements found on Earth contains a characteristic number of protons. When the number of protons and the number of electrons in an atom or molecule are the same, the charges balance and the atom is electrically neutral.

Although each element has a characteristic number of protons, the number of neutrons can vary. Forms of an element

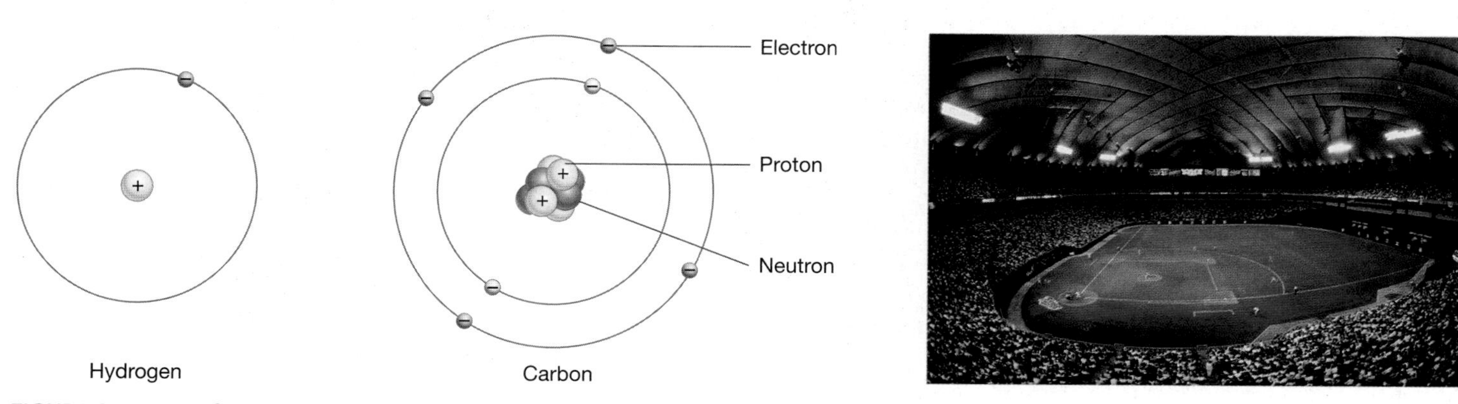

FIGURE 2.1 Parts of an Atom
These crude diagrams of the hydrogen and carbon atoms show how the nucleus, made up of protons and neutrons, is surrounded by orbiting electrons. Two aspects of these cartoons are misleading, however. First, electrons do not orbit the nucleus in circles; their actual orbits are complex. Second, the illustrations are not to scale. To get a better idea of the actual spatial relationships among the particles, consider that if an atom occupied the same volume as the stadium shown in the photo, the nucleus would be about the size of a pea. Most of an atom's volume is empty space.

with different numbers of neutrons are known as **isotopes**. For example, all atoms of the element uranium have 92 protons and 92 electrons. But naturally occurring isotopes of uranium can have either 143 or 146 neutrons, giving them a total of 235 or 238 protons and neutrons, respectively. The sum of the protons and neutrons in an atom is called its **mass number**. Thus, uranium occurs in two isotopes with different mass numbers: uranium 235 and uranium 238 (also called U-235 and U-238).

Atoms of U-235 and U-238 are unstable, however. Like other **radioactive isotopes**, their nuclei tend to change spontaneously by emitting either radiation or a particle. These changes are called radioactive decay. If the radioactive emission consists of a particle, the original, parent isotope is converted to a new isotope or element called a daughter isotope.

Radioactive decay has a distinctive feature: The proportion of atoms that decay in a sample of an isotope over a given period of time is constant. The rate at which decay occurs varies widely from isotope to isotope, however, and is usually reported in terms of a unit called a half-life. A **half-life** is the time it takes for half of the parent isotope to decay to a particular daughter isotope (**Figure 2.2**). The half-lives of many long-lived isotopes, such as U-235 and U-238, have been measured. The half-life of U-235 decaying to Lead-207, for example, is 713 million years.

The Age of Earth

To determine when a rock formed using radiometric dating, geologists must estimate two quantities:

- *The ratio of parent to daughter isotopes in a rock sample.* This can be measured directly, using instruments that identify the isotopes in a sample and record their abundance.
- *The ratio of parent to daughter isotopes that existed at the time the sample formed.* This ratio is estimated from information about how the rock formed. For example, when molten rock containing uranium atoms cools, the uranium forms crystals that do not contain lead. Based on this observation, geologists assume that rocks formed during Earth's early accretion phase started out with crystals containing 100 percent U-235 and 0 percent Lead-207. As time passes (and therefore half-lives pass), the crystals accumulate more and more Lead-207.

Once the ratio of parent and daughter isotopes has been determined in a sample, geologists can then calculate the number of half-lives that have passed and estimate the rock's age. To determine the age of Earth, geologists would simply need to find a uranium-bearing rock that formed during the early accretion phase.

Unfortunately, this is impossible—the planet was molten at the time. To estimate the age of Earth, researchers assume that all components of the solar system formed at about the same time. The uranium-lead ratios in meteorites, for example, indicate that they formed 4.6 billion years ago, or 4.6 Ga. (The phrase "billion years ago" is often abbreviated to Ga, for "giga-years ago.") The oldest moon rocks that have been analyzed date to 4.53 Ga; rocks from the youngest craters on the moon are 3.8 billion years ago.

The timeline presented in **Figure 2.3a** is based on these data. Note that scientists contend that Earth is 4.5–4.6 billion years old, and that the era of heavy bombardment ended about 3.9 billion years ago. The oldest rocks on Earth accord with these estimates. The most ancient rock crystals found thus far on Earth date to 4.4 Ga, indicating that some solid crust had begun to form by this time. Further, the oldest sedimentary rocks date to 3.8 Ga. Because sedimentary rocks form from deposits of sand, mud, or other particles carried by water, geologists are confident that an ocean existed by this time.

When Did Life Begin?

If chemical evolution took place, it must have occurred after Earth formed and before life began. To estimate when the origin of life took place, biologists examine traces of living organisms left in rocks—the **fossil record**. The most ancient fossil organisms that have been found to date are in rocks that formed a little less than 3.5 Ga

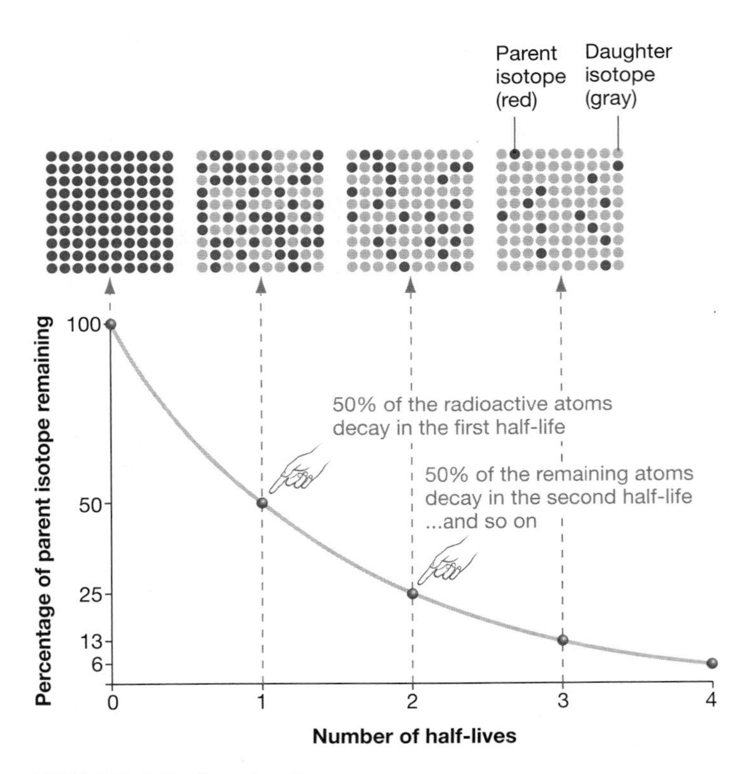

FIGURE 2.2 Radioactive Decay
This graph shows how a sample of radioactive atoms decays with time. At the time a sample forms, the ratio of parent to daughter isotopes is 100:0. After the amount of time represented by one half-life passes, the ratio is 50:50; after two half-lives it is 25:75, after three half-lives it is 12.5:87.5, and so on. EXERCISE Suppose you needed to determine the age of a rock. According to the data you've collected, the ratio of a certain parent isotope to daughter isotope is about 3:97. The half-life of the parent isotope is 100 million years. How old do you estimate this rock is?

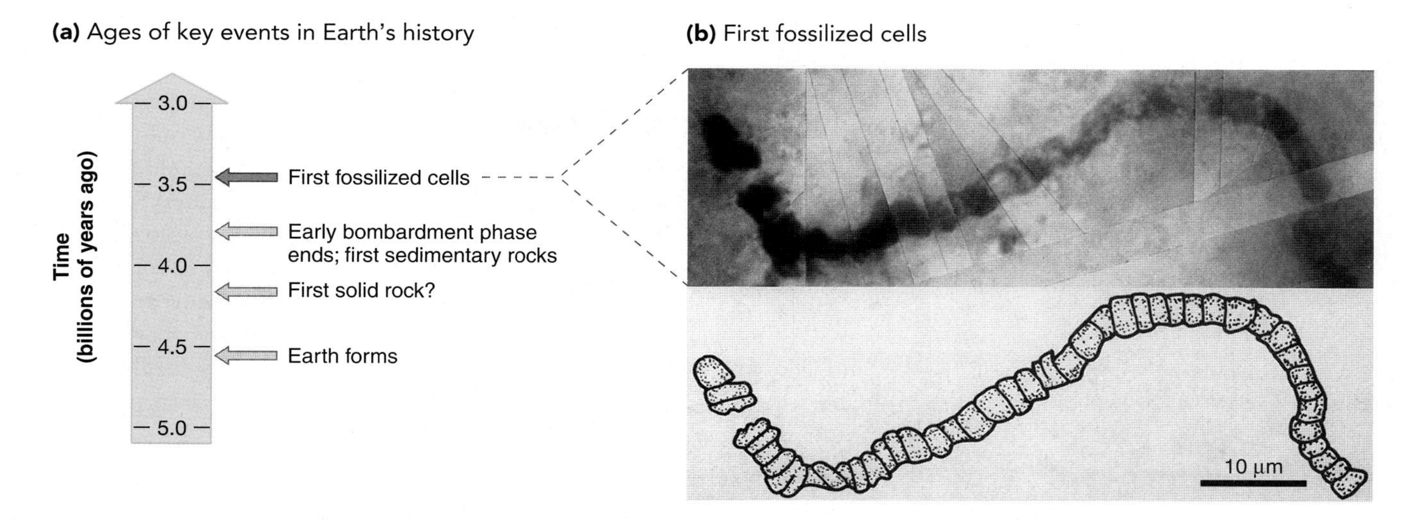

FIGURE 2.3 Early Events in the History of Earth
(a) Radiometric dating has assigned ages to several key events in early Earth history. **(b)** A researcher obtained this fossil by slicing samples of a rock called chert, quarried from a formation in Western Australia, into thin sections that could be examined under a microscope. Chert forms from ocean sediments. Note the scale bar next to the drawing of the specimen. The Greek letter μ ("mu") stands for *micro*, so μm indicates that the scale is in units of micrometers (1 μm is one thousandth of a millimeter and one millionth of a meter).
QUESTION The researcher who did this work claims that the organism pictured in part (b) lived in marine environments. What basis does he have for making this claim?

(**Figure 2.3b**). The biologist who found them, J. William Schopf, claims that the fossils were produced by small, filamentous life-forms remarkably similar to living organisms. Schopf also contends that the fossils have a well-developed cellular structure and represent at least 11 different species that can be distinguished on the basis of cell size and shape. Based on these observations, he suggests that life probably started after the end of massive bombardment and the formation of oceans 3.8 Ga, but well before 3.5 Ga. Although some researchers maintain that the structures in Figure 2.3b are not actual fossils but were instead formed by mineralization processes, enough other evidence has accumulated to support a consensus that chemical evolution occurred during a fairly narrow window of time, perhaps 300 million years or less.

2.2 The Building Blocks of Chemical Evolution

Just four types of atoms—hydrogen, carbon, nitrogen, and oxygen—make up 96 percent of the matter found in organisms today. The compounds found in living cells routinely contain thousands, or even millions, of these atoms. But early in Earth history, it is likely that these elements existed only in substances like water and carbon dioxide, which contain just three atoms.

The chemical evolution hypothesis maintains that simple compounds in the ancient atmosphere and ocean combined to form the larger, more complex substances found in living cells. To understand how the process could begin, we need to consider two questions:

- What is the physical structure of hydrogen, carbon, nitrogen, oxygen, and other atoms found in living cells?
- How do these atoms bond to one another, to form simple molecules like water and carbon dioxide—the building blocks of chemical evolution?

The Atoms Found in Organisms

Figure 2.4a (page 22) is an abstract of the periodic table of the elements that highlights the atoms found in living cells. Note that each atom has a symbol that represents it. Further, a superscript and a subscript accompany each symbol in the chart. The subscript is the element's **atomic number**, meaning the number of protons in its nucleus. The superscript is the mass number, which indicates the mass of the atom (see **Box 2.1**, page 23). The mass numbers given in Figure 2.4a represent the most common isotope of each element. For example, the most abundant isotope of carbon, ^{12}C, has six protons and six neutrons in its nucleus.

If an element is uncharged, its protons are matched by an equal number of electrons. Electrons carry a negative electric charge but are very light compared to protons. To understand how the atoms involved in chemical evolution behave, it is critical to understand how electrons are arranged around the nucleus.

Electrons move around atomic nuclei in specific areas called orbitals. Each electron orbital has a distinctive shape, and each orbital can hold two electrons. Orbitals, in turn, are grouped into levels called **electron shells**. These are numbered 1, 2, 3, and so on, to indicate their distance from the nucleus. The electrons of an atom fill the innermost shells first, before filling outer shells.

Figure 2.4b shows how electrons are distributed in the shells of H, C, N, O and other key elements. Two points are important to note:

- In each of the highlighted elements, the outermost electron shell contains at least one unpaired electron, meaning that one or more orbitals are not completely filled. Each of these unfilled orbitals has room for another electron.
- The number of unpaired electrons varies among elements. Carbon has four unpaired electrons; hydrogen has one.

These features are vitally important. Electrons in unfilled orbitals can participate in bonds that bind atoms together.

Covalent Bonding

Atoms are more stable when their orbitals are filled, meaning that each orbital in the outermost shell contains two electrons. Consider an atom of hydrogen, which has only one electron. This electron resides in an orbital that can hold two electrons. When an atom of hydrogen approaches another atom of hydrogen, these orbitals overlap and the two electrons are shared by the two nuclei (**Figure 2.5**). Each atom now has a partially filled orbital, and the bonded hydrogen atoms are more stable. The shared electrons "glue" the atoms together in a **covalent bond**. Substances that are held together by covalent bonds are called **molecules**. In this case, the bonded atoms form a molecule of hydrogen, written as H–H or H_2.

(a) The highlighted elements are the most important elements found in organisms.

Mass number (number of protons + neutrons)
Atomic number (number of protons)

$^{1}_{1}H$							$^{4}_{2}He$
$^{7}_{3}Li$	$^{9}_{4}Be$	$^{11}_{5}B$	$^{12}_{6}C$	$^{14}_{7}N$	$^{16}_{8}O$	$^{19}_{9}F$	$^{20}_{10}Ne$
$^{23}_{11}Na$	$^{24}_{12}Mg$	$^{27}_{13}Al$	$^{28}_{14}Si$	$^{31}_{15}P$	$^{32}_{16}S$	$^{35}_{17}Cl$	$^{40}_{18}Ar$

(b) Distribution of electrons in shells of these elements

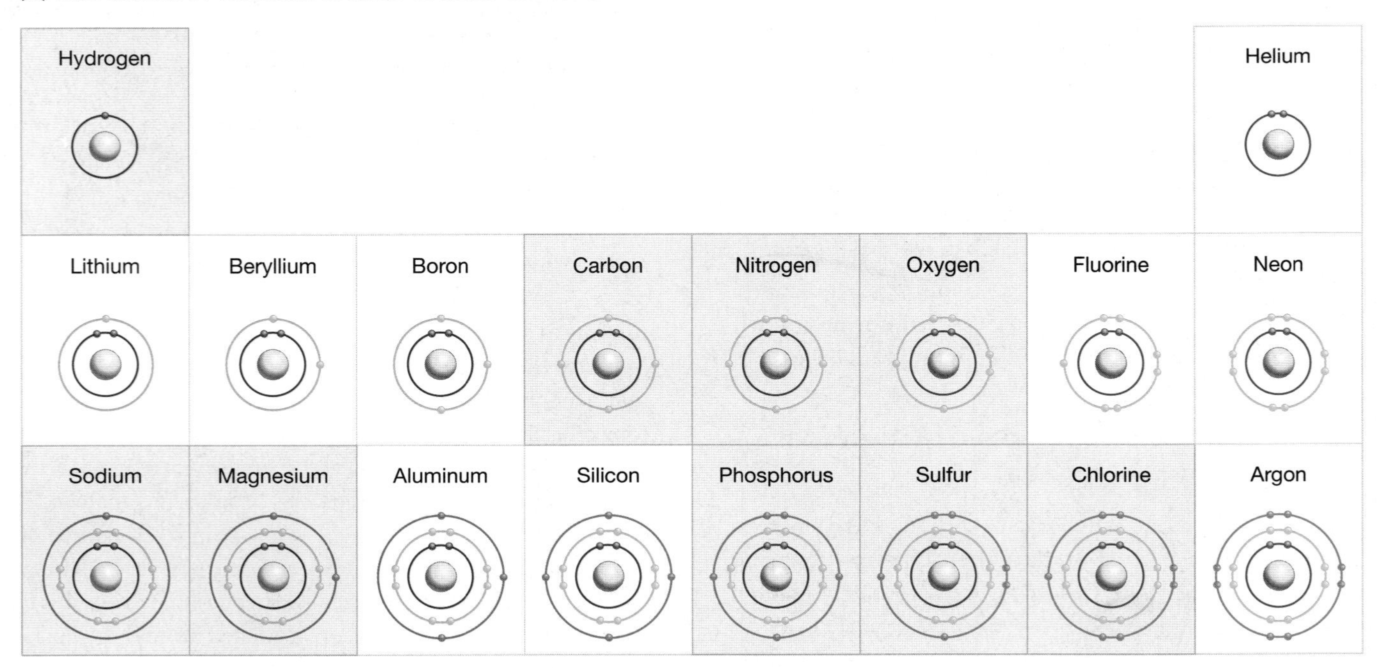

FIGURE 2.4 Characteristics of the Atoms Found in Organisms
(a) This segment from the periodic table highlights the elements commonly found in organisms. **(b)** In these diagrams the atomic nucleus is represented by a closed circle. The first energy shell is indicated by the red ring, and the second energy shell by the green ring; the dots on the rings represent electrons. Electrons are drawn in pairs if they occupy filled orbitals. Electrons are drawn by themselves if they occupy unfilled orbitals.
EXERCISE For each of the highlighted atoms, write down the number of bonds it is capable of forming based on the number of unpaired electrons in its outermost electron shell.

BOX 2.1 Atomic Mass

Mass is a quantity that represents the amount of matter in an object; *matter*, in turn, is simply a catch-all term for anything that has mass.

Although the mass of protons, neutrons, and electrons can be measured in grams, the numbers involved are so small—neutrons and protons each weigh about 1.7×10^{-24} grams (g)—that chemists and physicists prefer to use a special unit called the **atomic mass unit (amu)**, or dalton. By definition, 1 amu is exactly 1/12th the mass of a carbon atom containing 6 protons and 6 neutrons (this happens to be 1.6605×10^{-24} g). The masses of the two particles are virtually identical and are routinely rounded to 1 amu.

To determine the mass of an atom in amu, then, simply add up the number of protons and neutrons it contains. (Electrons have so little mass—about 1/2000th as much as a proton or neutron—that they can be ignored.) This sum is the atom's mass number. A carbon atom that contains 6 protons and 6 neutrons has a mass of 12 amu, and a mass number of 12.

Another way to think about covalent bonding is in terms of electrical attraction and repulsion. As two hydrogen atoms move closer together, their two positively charged nuclei repel each other, and their negatively charged electrons repel each other. But at the same time, each of the protons attracts both electrons, and each of the electrons attracts both protons. Covalent bonds form when the attractive forces are greater than the repulsive forces. This is the case when hydrogen atoms interact. The H_2 molecule results.

Some Simple Molecules Formed from H, C, N, and O

Look again at Figure 2.4b and count the number of unpaired electrons in the outermost, or **valence**, shells of carbon, nitrogen, and oxygen atoms. Carbon has four, nitrogen has three, and oxygen has two. Each unpaired electron can make up half of a covalent bond. As a result, a carbon atom can form a total of four covalent bonds. When each of these bonds is with a hydrogen atom, the molecule that results is written CH_4 and is called methane (**Figure 2.6a**). This is the most common molecule found in natural gas. When nitrogen's three unpaired electrons bond with hydrogen, the result is NH_3, or ammonia. Similarly, an atom of oxygen can form covalent bonds with two atoms of hydrogen, resulting in a water molecule (H_2O).

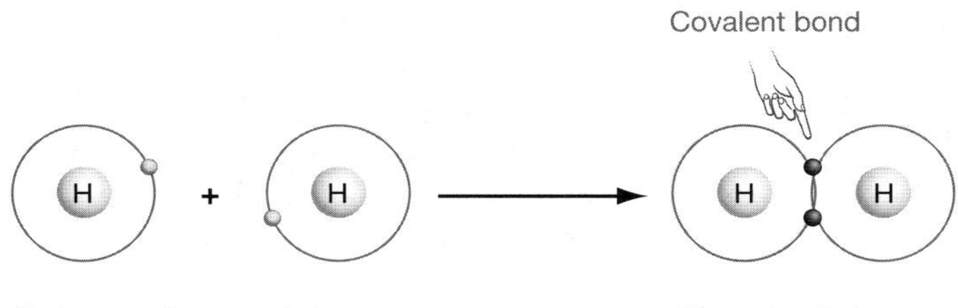

FIGURE 2.5 Covalent Bonds Result from Electron Sharing
When two hydrogen atoms come into contact, their electrons are attracted to the positive charge in each nucleus. As a result, their orbitals overlap, the electrons are shared by each nucleus, and a covalent bond forms.

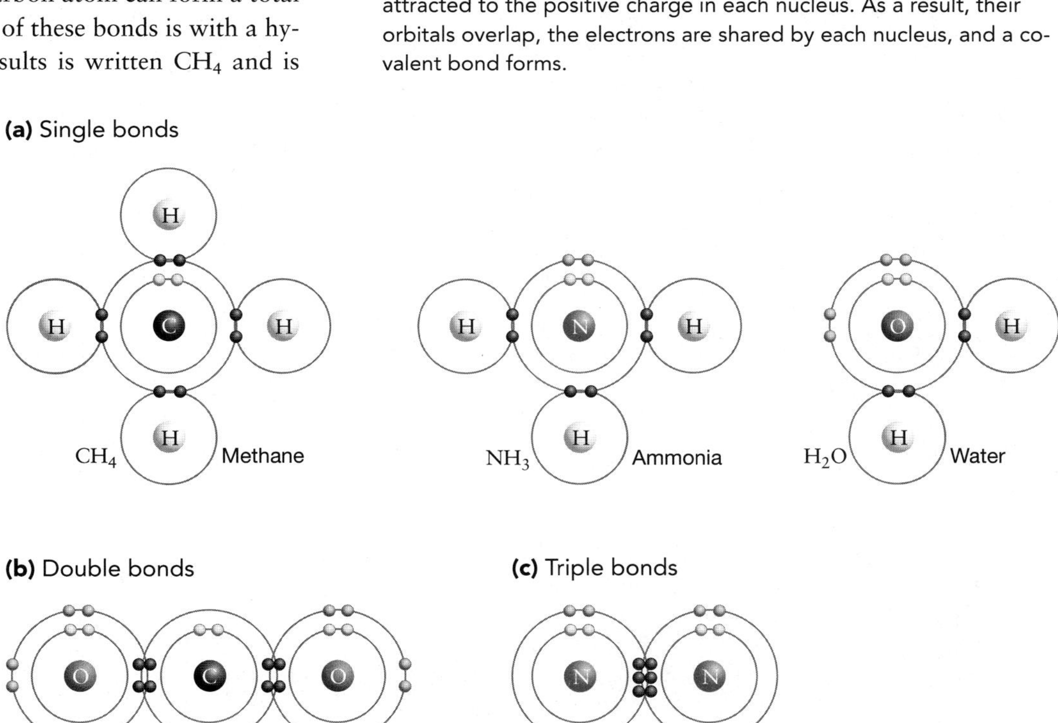

FIGURE 2.6 Unpaired Electrons in the Valence Shell Participate in Covalent Bonds
Covalent bonding is based on sharing of electrons in the outermost shell. Covalent bonds can be single (as in part a), double (as in b), or triple (as in c). QUESTION How do the electron configurations of H, C, O, and N in these molecules compare with their configurations shown in Figure 2.4b?

Double and Triple Bonds When more than one unpaired electron exists in a valence shell, double covalent bonds can form. **Figure 2.6b** shows how carbon forms double bonds with oxygen atoms to produce the molecule called carbon dioxide (CO_2).

Triple bonds occur when three pairs of electrons are shared. **Figure 2.6c** shows the structure of molecular nitrogen (N_2), which forms when two nitrogen atoms establish a triple bond.

Bond Angles and the Shape of Molecules Proponents of the chemical evolution hypothesis argue that the simple molecules introduced here were important components of Earth's ancient atmosphere and ocean. Their reasoning is based on the observation that these molecules are found in volcanic gases on Earth and in the atmospheres of nearby planets. If so, then molecules like carbon dioxide (CO_2) and hydrogen (H_2) provided the building blocks for chemical evolution. Before analyzing how they could have combined to form more complex compounds, however, we need to consider their shape, how scientists represent their composition and shape, and how their size and concentration are quantified.

The shape of a simple molecule is dictated by the geometry of its bonds. Bond angles, in turn, are determined by the orientation of the orbitals involved in the bond. **Figure 2.7a** illustrates the orbitals involved in bonds formed by carbon, nitrogen, and oxygen. These orbitals are oriented in a tetrahedron. In carbon, each of the four orbitals contains an unpaired electron and can form a covalent bond. As a result, methane (CH_4) is tetrahedral (**Figure 2.7b**). Ammonia (NH_3), in contrast, is shaped like a pyramid. **Figure 2.7c** shows that one of the four orbitals in nitrogen is filled with a pair of electrons; only the three orbitals with unpaired electrons bond with hydrogen. In oxygen, two of the four tetrahedral orbitals are filled with electron pairs. **Figure 2.7d** shows that the three atoms found in water (H_2O) are in the same plane, but the molecule is bent.

Representing Molecules Molecules can be represented in a variety of ways. The simplest representation is a **molecular formula**, which indicates the numbers and types of atoms in a molecule. Water has the molecular formula H_2O; methane has the molecular formula CH_4 (**Figure 2.8a**).

Molecular formulas are a compact way of stating a molecule's composition, but they contain no information about how the molecule is put together. **Structural formulas** help by indicating which atoms are bonded together. In a structural formula, single, double, and triple bonds are represented by single, double, or triple dashes. **Figure 2.8b** gives structural formulas for molecules that were important components of Earth's ancient atmosphere and ocean.

The limitation of structural formulas is that they are two dimensional, while molecules are three dimensional. More complex diagrams, such as ball-and-stick models and space-filling models, are based on the three-dimensional structure of molecules. **Figures 2.8c** and **2.8d** show how ball-and-stick and space-filling models indicate a molecule's geometry as well as the relative sizes of the atoms involved.

Quantifying the Concentration of Molecules Scientists quantify the number of molecules present in a sample using a unit called the **mole.** A mole refers to the number 6.022×10^{23} (just as the unit called the *dozen* refers to the number 12). The mole is a useful unit because the mass of one mole of any molecule is the same as its **molecular weight** expressed in grams. A molecular weight is the sum of the mass numbers of all of the atoms in a molecule. For example, summing the mass numbers of two atoms of hydrogen and one atom of oxygen gives $1 + 1 + 16$, or 18. This is water's molecular weight. It follows that if you weighed out a sample of 18 grams of water, it would contain 6.022×10^{23} water molecules, or one mole. When substances are dissolved in water, their concentration is expressed as **molarity** (symbolized by "M"). Molarity refers to the number of moles of the substance present per liter of solution.

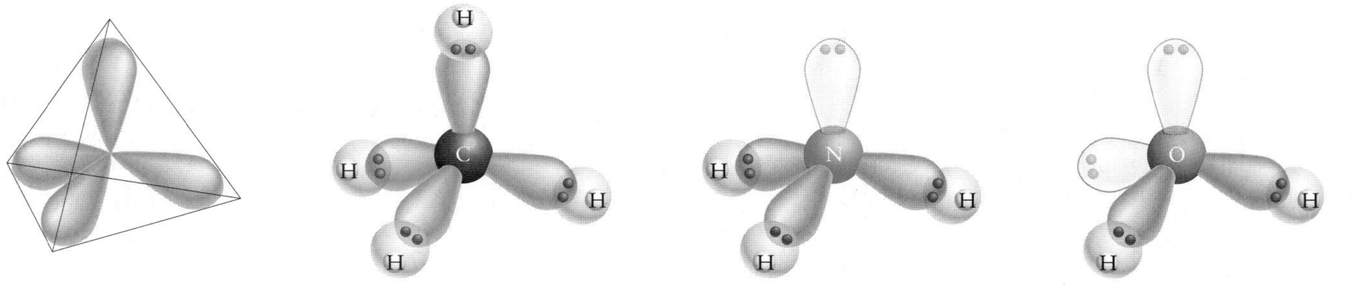

H H C H H H H N H H H O H

(a) Orbitals shaped like a tetrahedron

(b) Methane (CH_4)

(c) Ammonia (NH_3)

(d) Water (H_2O)

FIGURE 2.7 The Geometry of Methane, Ammonia, and Water
(a) The shaded areas represent the shapes of the orbitals where valence electrons are found in C, N, and O.
(b, c, d) The dots in the diagrams represent electrons. The shape of each molecule is radically different, even though some of the same orbitals are involved in bonding. EXERCISE In parts (b)–(d), label which molecule is bent and planar, which forms a tetrahedron, and which forms a pyramid.

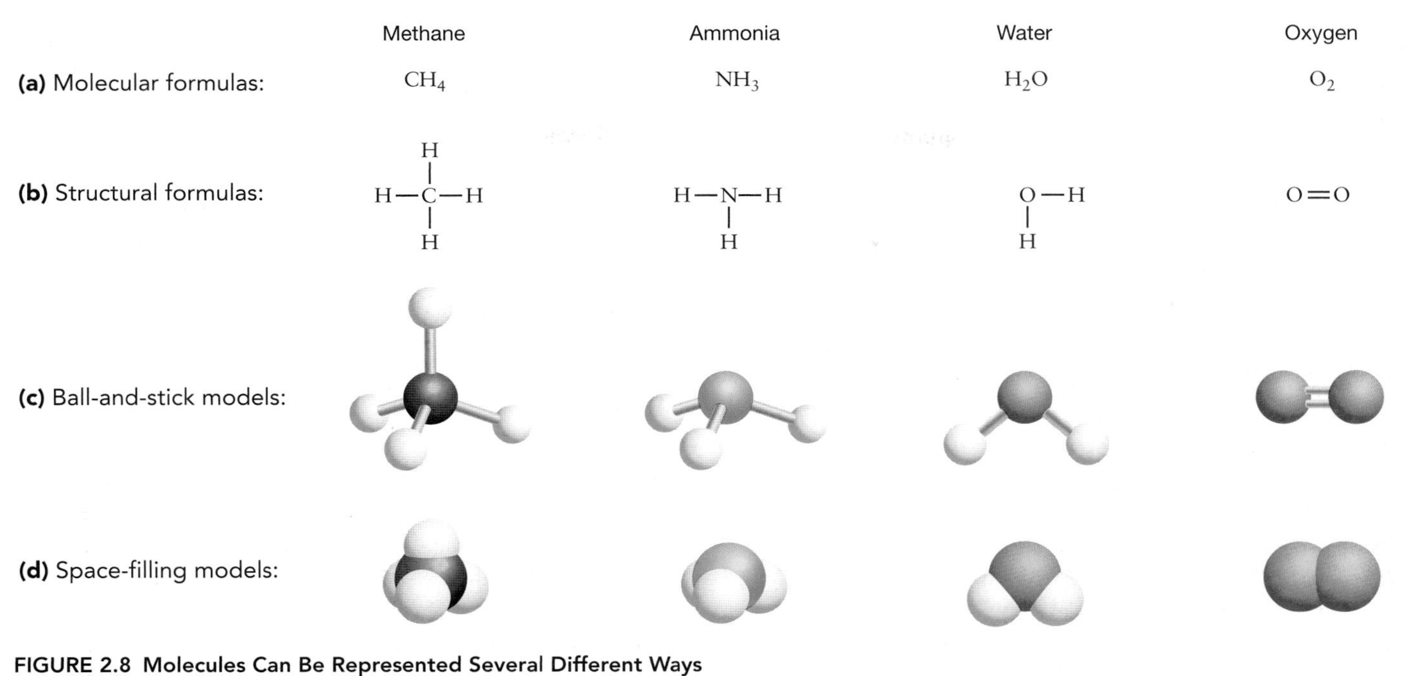

FIGURE 2.8 Molecules Can Be Represented Several Different Ways
Each method of representing a molecule has particular advantages. **(a)** Molecular formulas are compact and indicate the number and identity of atoms involved. **(b)** Structural formulas show which atoms are bonded to one another. **(c)** Ball-and-stick models take up more space than structural formulas, but include information about bond geometry. **(d)** Space-filling models are not as easy to read as ball-and-stick models, but offer a more accurate depiction of the spatial relationships between atoms.

2.3 Chemical Reactions, Chemical Energy, and Chemical Evolution

Proponents of the chemical evolution hypothesis contend that simple molecules present in the atmosphere and ocean of ancient Earth participated in chemical reactions that produced larger, more complex molecules. A **chemical reaction** is an event in which one substance is combined with others or broken down into another substance. According to one hypothesis, the early atmosphere was made up of gases ejected from volcanoes. Carbon dioxide, water, and nitrogen are the dominant gases ejected from volcanoes today; a small amount of molecular hydrogen (H_2) and methane (CH_4) may also be present. But if these molecules are placed in a glass tube together and allowed to interact, very little happens. Water vapor condenses to form liquid water as the mixture cools, but the simple molecules do not suddenly link together to create large, complex substances like those found in living cells. Instead, their bonds remain intact. How, then, did chemical evolution occur?

To answer this question we must explore two topics: how chemical reactions occur, and how conditions on ancient Earth made certain reactions possible.

Chemical Reactions

Chemical reactions are written in a format similar to mathematical equations, with reactant atoms or molecules on the left and the products of the reaction on the right. For example, the most common reaction in the mix of gases and water listed earlier would be:

$$CO_2(g) + H_2O(l) \rightleftharpoons H_2CO_3(aq)$$

The expression indicates that carbon dioxide reacts with water to form carbonic acid. Note that the state of each ion or molecule is indicated as gas (*g*), liquid (*l*), in aqueous solution (*aq*), or solid (*s*), and that the expression is balanced: One carbon, three oxygen, and two hydrogen atoms are present on each side of the expression. Also note that the expression contains a double arrow, meaning that the reaction is reversible. When the forward and reverse reactions proceed at the same rate, the concentration of reactants and products remains constant. A dynamic but stable state like this is termed a **chemical equilibrium.** A chemical equilibrium can be disturbed by changing the concentration of reactants or products. For example, adding CO_2 to the mixture would drive the reaction to the right, creating more product until the equilibrium proportions of reactants and products is reestablished. Adding carbonic acid to the solution would drive the reaction to the left.

A chemical equilibrium can also be altered by changes in temperature. For example, the water molecules in this system would be present as a combination of liquid water and water vapor:

$$H_2O(l) \rightleftharpoons H_2O(g)$$

If liquid water molecules absorb enough heat, they transform to the gas state. This is called an **endothermic** ("within heating") process because heat is absorbed during the process. In contrast, the transformation of water vapor to liquid water releases heat, and is called **exothermic** ("outside heating"). Raising the temperature of this system drives the equilibrium to the right; cooling the system drives it to the left.

In terms of chemical evolution, though, these reactions and state changes are not particularly interesting. Carbonic acid is not an important intermediate in the formation of more complex molecules. According to models developed by Joseph Pinto and colleagues, though, interesting things do begin to happen when energy is added to the system.

What Is Energy?

Energy can be defined as the capacity to do work or to supply heat. This capacity exists in one of two ways: as a stored potential or as an active motion.

Stored energy is called **potential energy.** An object gains or loses this capacity as a consequence of its position. An electron that resides in an outer electron shell will, if the opportunity arises, fall into a lower energy shell closer to the positive charges on the protons in the nucleus. As a result, an electron in an outer electron shell has more potential energy than an electron in an inner shell (**Figure 2.9a**).

Kinetic energy is the energy of motion. Molecules have kinetic energy because they are constantly in motion. This form of kinetic energy—the kinetic energy of molecular motion—is called **thermal energy.** The **temperature** of an object is a measure of how much thermal energy its molecules possess. If an object has a low temperature, its molecules are moving slowly (we perceive this as "cold"). If an object has a high temperature, its molecules are moving rapidly (we perceive this as "hot"). When two objects with different temperatures come into contact, thermal energy is transferred between them. We call this transferred energy **heat.**

There are many different forms of potential and kinetic energy, and energy can change from one form into another. To drive this point home, consider a water molecule sitting in a pool near the top of a waterfall, as in **Figure 2.9b**. This molecule has potential energy as a result of its position. If the molecule passes over the waterfall, its potential energy is converted to the kinetic energy of motion. When the molecule reaches the rocks below, it has experienced a change in potential energy because it has changed position. The bottom panel in Figure 2.9b shows that this change in potential energy is transformed into an equal amount of energy in other forms: mechanical energy that tends to break up the rocks, thermal energy that raises the temperature of the rocks and the water itself, and sound.

An electron in an outer electron shell is analogous to the water molecule at the top of a waterfall (**Figure 2.9c**). If the electron falls to a lower shell, its potential energy is converted to the kinetic energy of motion. After the electron occupies the lower electron shell, it experiences a change in potential energy. As step 3 in Figure 2.9c shows, the change in potential energy is transformed into an equal amount of energy in other forms—usually thermal energy, but sometimes light. These examples illustrate the first law of thermodynamics, which states that energy cannot be created or destroyed but only transferred and transformed.

According to the model of Earth's formation introduced in section 2.1, sources of energy were abundant and diverse at the time that chemical evolution occurred. As a result, the simple molecules present in the ancient Earth's atmosphere and ocean would have been exposed to massive amounts of thermal energy from volcanoes and asteroid impacts.

Chemical Evolution: A Model System

How would the simple molecules in Earth's ancient atmosphere be affected by inputs of energy? To answer this question, Joseph Pinto and colleagues created a computer model of the chemical reactions that occur when carbon dioxide, water, nitrogen, and hydrogen are exposed to energy.

The goal of this study was quite specific: The researchers wanted to determine whether a molecule called formaldehyde (H_2CO) could be produced. Along with hydrogen cyanide (HCN), formaldehyde is a key intermediate in the creation of the larger, more complex molecules found in cells. Forming formaldehyde and hydrogen cyanide is the critical first step in chemical evolution—a trigger that could set the process in motion.

Pinto's group proposed that the following reaction could take place:

$$CO_2(g) + 2\,H_2(g) \longrightarrow H_2CO(g) + H_2O(g)$$

Before exploring how they tested this hypothesis, however, it will be helpful to answer a basic question: Why doesn't this reaction occur spontaneously—that is, why doesn't it occur *without* an input of energy?

What Makes a Chemical Reaction Spontaneous? When chemists say that a reaction is spontaneous, they have a precise meaning in mind: Chemical reactions are spontaneous if they proceed on their own, without any continuous external influence like heat or pressure. What determines whether a reaction is spontaneous or nonspontaneous? There are two factors:

- *Reactions tend to be spontaneous if the products have lower potential energy than the reactants.* Reaction products have lower potential energy if their electrons are found in lower energy shells, closer to nuclei, than they are in the reactants. Because this difference in potential energy is given off as heat, the reaction is exothermic. For example, when natural gas burns, methane reacts with oxygen gas to produce carbon dioxide and water:

  $$CH_4(g) + 2\,O_2(g) \longrightarrow CO_2(g) + 2\,H_2O(g)$$

  The electrons involved in the C–O and H–O bonds of carbon dioxide and water are much closer to protons than they

(a) Potential energy of electrons

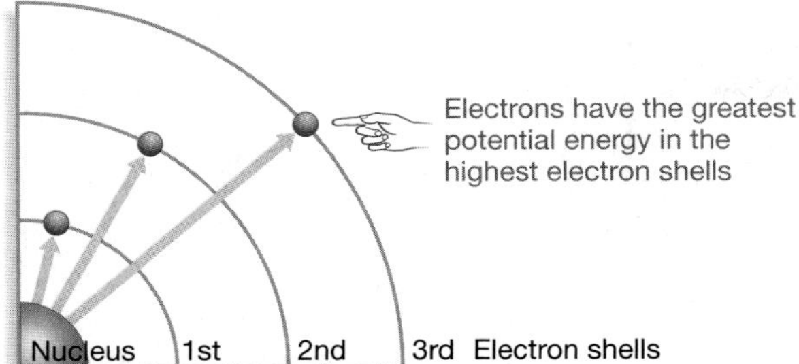

(b) ENERGY TRANSFORMATION IN A WATERFALL

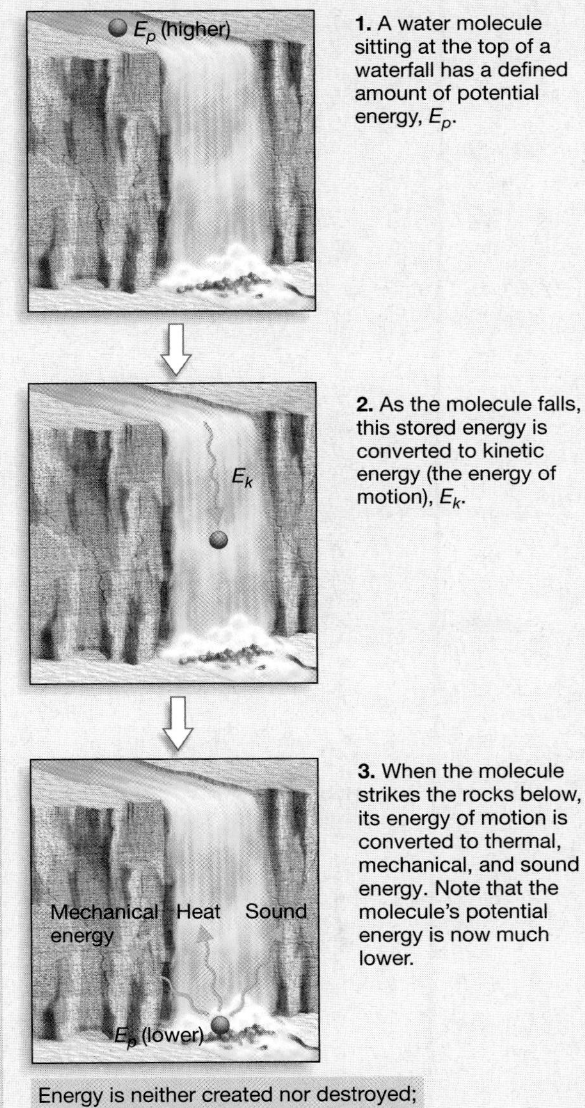

1. A water molecule sitting at the top of a waterfall has a defined amount of potential energy, E_p.

2. As the molecule falls, this stored energy is converted to kinetic energy (the energy of motion), E_k.

3. When the molecule strikes the rocks below, its energy of motion is converted to thermal, mechanical, and sound energy. Note that the molecule's potential energy is now much lower.

Energy is neither created nor destroyed; it simply changes form.

(c) ENERGY TRANSFORMATION IN AN ATOM

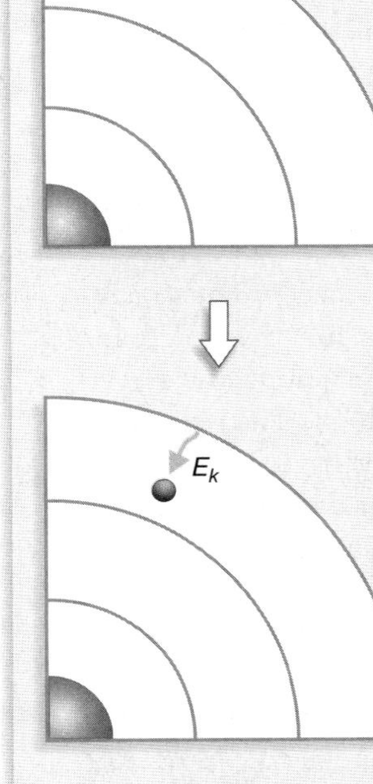

1. An electron in an outer shell has a defined amount of potential energy, E_p.

2. As the electron falls to a lower energy shell, its potential energy is converted to kinetic energy, E_k.

3. Once the electron arrives at a lower electron shell, the kinetic energy is converted to light or heat.

Energy is neither created nor destroyed; it simply changes form.

FIGURE 2.9 Energy Transformations
(a) Electrons in outer shells have more potential energy than electrons in inner shells because their negative charges are farther from the positive charges in the nucleus. Each shell represents a distinct level of potential energy. **(b and c)** During an energy transformation, whether it is a molecule of water falling down a waterfall or an electron dropping to a lower-energy shell, the total amount of energy in the system remains constant.

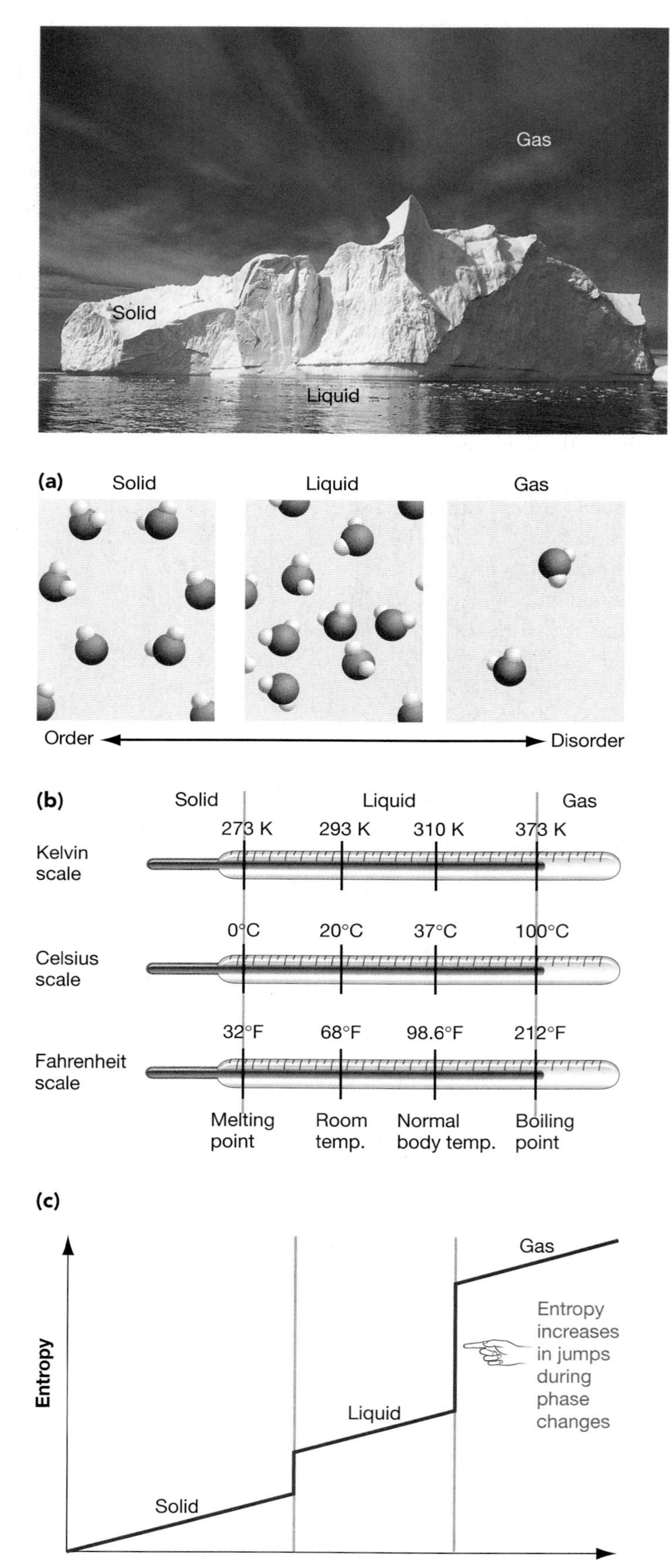

FIGURE 2.10 Changes in Entropy
(a) These space-filling models show water as a solid, liquid, and gas. Because ice forms a rigid, crystalline structure, the water molecules involved are highly ordered. In contrast, liquid water molecules exist in a loose network. Water vapor is less ordered still. **(b)** This diagram shows the relationship between the Kelvin, Celsius, and Fahrenheit scales. To calculate Gibbs free energy, temperature is measured on the Kelvin scale. **(c)** This graph shows a typical plot of entropy versus temperature. Note that entropy increases with temperature, and that it makes large jumps when molecules change state. To understand why this happens, recall that temperature measures the kinetic energy of molecular motion—or how much a molecule is moving around. Entropy increases with temperature because molecules that move around a lot are less ordered. Jumps occur at state changes because when molecules change from solid to liquid to gas form, they have many more ways to move.

were in the C–H and O–O bonds of methane and oxygen. In chemical reactions, the difference in potential energy between the products and the reactants is symbolized by ΔH. (The Greek letter Δ, named delta, is often used in chemical and mathematical notation to represent change.) When a reaction is exothermic, ΔH is negative.

- *Reactions tend to be spontaneous when the product molecules are less ordered than the reactant molecules.* Ice, for example, melts spontaneously at room temperature even though it must absorb heat energy to do so. The reaction is endothermic, yet occurs spontaneously because the molecules in liquid water are much less ordered than are the molecules in solid ice (**Figure 2.10a**). The amount of disorder in a group of molecules is called its **entropy** and is symbolized by S. When the products of a chemical reaction are less ordered than the reactant molecules, entropy increases and ΔS is positive.

In general, physical and chemical processes proceed in the direction that results in lower energy and increased disorder. The second law of thermodynamics, in fact, states that entropy always increases in a closed system—meaning one with no external source of energy. To determine whether a chemical reaction is spontaneous, then, the *combined* contributions of changes in heat and disorder must be assessed. To do this, chemists define a quantity called the **Gibbs free-energy change**, symbolized by ΔG:

$$\Delta G = \Delta H - T\Delta S$$

In this equation, T stands for temperature measured on the Kelvin scale (**Figure 2.10b**). The $T\Delta S$ term means that entropy becomes more important in determining free-energy change as the temperature of the molecules increases. To understand why this is so, think of water as a solid, a liquid, and a gas; that is, at low, medium, and high temperatures. Because gases can become much more highly disordered than can liquids or solids, entropy changes are more pronounced at high temperatures (**Figure 2.10c**).

To summarize, chemical reactions are spontaneous when ΔG is less than zero. Such reactions are said to be **exergonic.** In contrast, reactions are nonspontaneous when ΔG is greater than zero. Reactions like these are termed **endergonic.** When ΔG is zero, reactions are at equilibrium.

The reaction between carbon dioxide and hydrogen gas is nonspontaneous because it is endothermic and because it results in a decrease in entropy. For the reaction to occur, a large input of energy is required.

Energy Inputs and the Start of Chemical Evolution To explore how carbon dioxide and hydrogen gas could have reacted to form formaldehyde and trigger chemical evolution, Pinto and colleagues constructed a computer model of the ancient atmosphere. The model consisted of a list of all possible chemical reactions that can occur between CO_2, H_2O, N_2, and H_2 molecules. In addition to the spontaneous reactions, they included reactions that occur when these molecules are struck by sunlight. This was crucial because sunlight represents a source of energy.

The sunlight that strikes Earth is made up of packets of light energy called photons. The amount of light energy contained in a photon can vary widely. Today, most of the higher-energy photons in sunlight never reach Earth's lower atmosphere. Instead, they are absorbed by a molecule called ozone (O_3) in the upper atmosphere. But if Earth's early atmosphere was filled with volcanic gases, it is extremely unlikely that appreciable quantities of ozone existed. As a result, we can infer that when chemical evolution was occurring, large quantities of high-energy photons bombarded the planet.

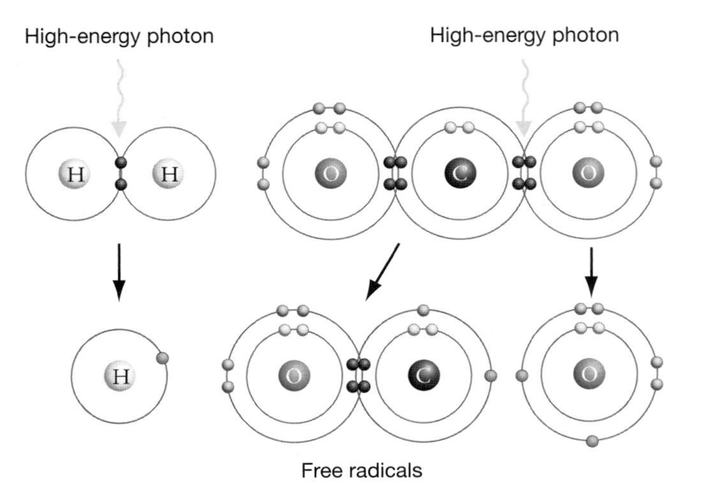

FIGURE 2.11 Free Radicals
When a high-energy photon strikes a hydrogen or carbon dioxide molecule, a variety of free radicals can be created. The possibilities include hydrogen atoms (left), the CO radical (center), and the oxygen atom (right). Free radicals are extremely reactive because they have one or more unpaired electrons.

Why was the energy in photons important? To understand the answer to this question, recall that the atoms in carbon dioxide and hydrogen have full valence shells. As a result, these molecules are largely unreactive. But energy from photons can break molecules apart by knocking electrons away from the valence shells of atoms (**Figure 2.11**). The atoms that result, called free radicals, have unpaired electrons and are extremely unstable. The model created by Pinto and associates included several reactions that produce highly reactive free radicals.

To understand which of the long list of possible reactions would actually occur, and to estimate how much formaldehyde could be produced in the ancient atmosphere, the researchers needed to consider the effects of two additional factors: temperature and concentration.

The Roles of Temperature and Concentration in Chemical Reactions Even if a chemical reaction occurs spontaneously, it does not necessarily happen quickly. For most reactions to proceed, one chemical bond has to break and another one has to form. For this to happen, the substances involved must collide in a specific orientation—in a way that brings the electrons involved into proximity.

The number of collisions occurring among the substances in a mixture depends on the temperature and the concentrations of the reactants. When the concentration of reactants is high, more collisions occur and reactions proceed more quickly. When their temperature is high, reactants move faster and collide more frequently. Higher temperatures tend to speed up chemical reactions.

To model the behavior of simple molecules in the ancient atmosphere, then, Pinto and co-workers needed to specify both the concentration of each molecule and the temperature. Then they were able to assign a rate to each of the reactions listed in their model based on the actual reaction rates observed in experiments conducted at controlled temperatures and concentrations.

Their result? Under temperature and concentration conditions accepted as reasonable approximations of early Earth conditions by most atmospheric scientists, they calculated that appreciable quantities of formaldehyde would be produced. Using a similar model, Kevin Zahnle showed that significant amounts of hydrogen cyanide (HCN) could also have been produced in the ancient atmosphere. According to these results, large quantities of the critical intermediates in chemical evolution would have been produced in the ancient atmosphere.

The Role of Energy in Chemical Evolution

The initial products of chemical evolution are important, for a simple reason: They have more potential energy than the reactant molecules. When formaldehyde is produced, an increase in potential energy occurs because the electrons that bond CO_2 and H_2 together are closer to protons than they are in H_2CO and H_2O. This form of potential energy—the potential energy located in chemical bonds—is called **chemical energy.**

This observation defines the basic outcome of the models explored by Zahnle and by Pinto and co-workers: The energy in sunlight was converted to chemical energy. This energy transformation is fundamentally important, because it explains *how* chemical evolution could occur. When small, simple molecules absorb energy, chemical reactions can occur that transform that energy into potential energy stored in chemical bonds. This increase in chemical energy makes the production of larger, more complex molecules possible.

To fully understand this point, the complete reaction that results in the formation of formaldehyde is written as

$$CO_2(g) + 2\ H_2(g) + \text{sunlight} \longrightarrow H_2CO(g) + H_2O(g)$$

The reaction is balanced in terms of the atoms *and* the energy involved. The energy in sunlight is converted to the potential energy in formaldehyde.

2.4 The Composition of the Early Atmosphere: Redox Reactions and the Importance of Carbon

In the models of the early atmosphere just reviewed, chemical evolution did not begin until energy was added to the system. But it's important to recognize that the start of chemical evolution also depended on the composition of the atmosphere. In the models analyzed by Zahnle and by Pinto and associates, formaldehyde and hydrogen cyanide are produced only if molecular hydrogen (H_2), ammonia (NH_3), or methane (CH_4) are present in the atmosphere. No matter what type of energy is added to the model systems, chemical evolution does not take place unless these molecules are present. What evidence do biologists have that these molecules actually existed in the ancient atmosphere?

The earliest hypotheses for the composition of Earth's first atmosphere, developed in the 1920s and 1940s by A. I. Oparin and Harold Urey, proposed that hydrogen, ammonia, and methane were abundant. Urey based this claim on the supposition that the intense gravity and lack of volcanic activity on Jupiter and Saturn had kept their atmospheres unchanged since the founding of the solar system. He thus inferred that Earth's ancient atmosphere was similar to the current atmosphere of Jupiter and Saturn. But in 1951 William Rubey argued that volcanic gases like CO_2, N_2, and H_2O dominated Earth's original atmosphere.

James Kasting and Lisa Brown point out that this controversy is difficult to resolve, because there is no direct evidence about the composition of the ancient atmosphere. The current consensus is based on recent models of the volcanic gases produced as Earth's crust formed. These models support Rubey's hypothesis with a key modification: Small but significant amounts of hydrogen (H_2), ammonia (NH_3), and methane (CH_4) were also present.

Why are these particular molecules essential to chemical evolution? Because they trigger the most important chemical reactions in biology: reduction-oxidation, or redox, reactions.

Redox Reactions

Reduction-oxidation reactions, or redox reactions, are a class of chemical reactions that involve the loss or gain of an electron. In a redox reaction, the substance that loses one or more electrons is said to be **oxidized**; the substance that gains electrons is said to be **reduced.** To help keep these terms straight, chemists use the simple mnemonic "LEO the lion goes GER"—Loss of Electrons is Oxidation; Gain of Electrons is Reduction. (An alternative is OIL RIG—Oxidation Is Loss; Reduction Is Gain.) A key point here is that oxidation events are always coupled with a reduction; if one substance loses an electron, another substance has to gain it.

The gain or loss of an electron can be relative, however. That is, during a redox reaction, electrons can be transferred completely from one molecule or atom to another, or the electrons can simply shift their positions in covalent bonds. For example, consider the burning of methane diagrammed in **Figure 2.12a.** The dots in the illustration represent the electrons involved in covalent bonds. As you study this figure, compare the position of the electrons in the reactant methane versus the product carbon dioxide. Note that the electrons have moved farther from the carbon nucleus in the carbon dioxide product. This means that carbon has been oxidized. It has "lost" electrons. Now compare the position of the electrons in the oxygen reactant and their position in the product water. These electrons have moved closer to the oxygen in the water molecule, meaning that the oxygen has been reduced. In this reaction, oxygen has "gained" electrons.

These shifts in electron position change the amount of chemical energy in the reactants and products. When methane burns, the electrons are closer to nuclei in the product molecules than they are in the reactant molecules, meaning that their potential energy has dropped. As a result, the reaction is exothermic.

Now consider the diagram in **Figure 2.12b,** which illustrates the reaction analyzed by Pinto and co-workers. Using the data in Figure 2.12a, you should be able to add the electron positions for each of the bonds involved in this reaction. Once you've accomplished this, you should be able to determine whether carbon is reduced or oxidized as a result of this reaction. Based on this analysis, the following claim should make sense: The key step in launching chemical evolution was the reduction of carbon.

Reducing Carbon

Life has been called a carbon-based phenomenon, and with good reason. With the exception of water, almost all of the molecules found in organisms contain carbon.

Understanding why carbon is so important in biology is straightforward: It is easily the most versatile atom on Earth. Because of its four valence electrons, it can form a large number of covalent bonds. With different combinations of single and double bonds, an almost limitless array of molecular shapes is possible. You have already examined the tetrahedral structure of methane; carbon dioxide is linear. When molecules contain more than one carbon atom, they can be bonded to one another in long chains,

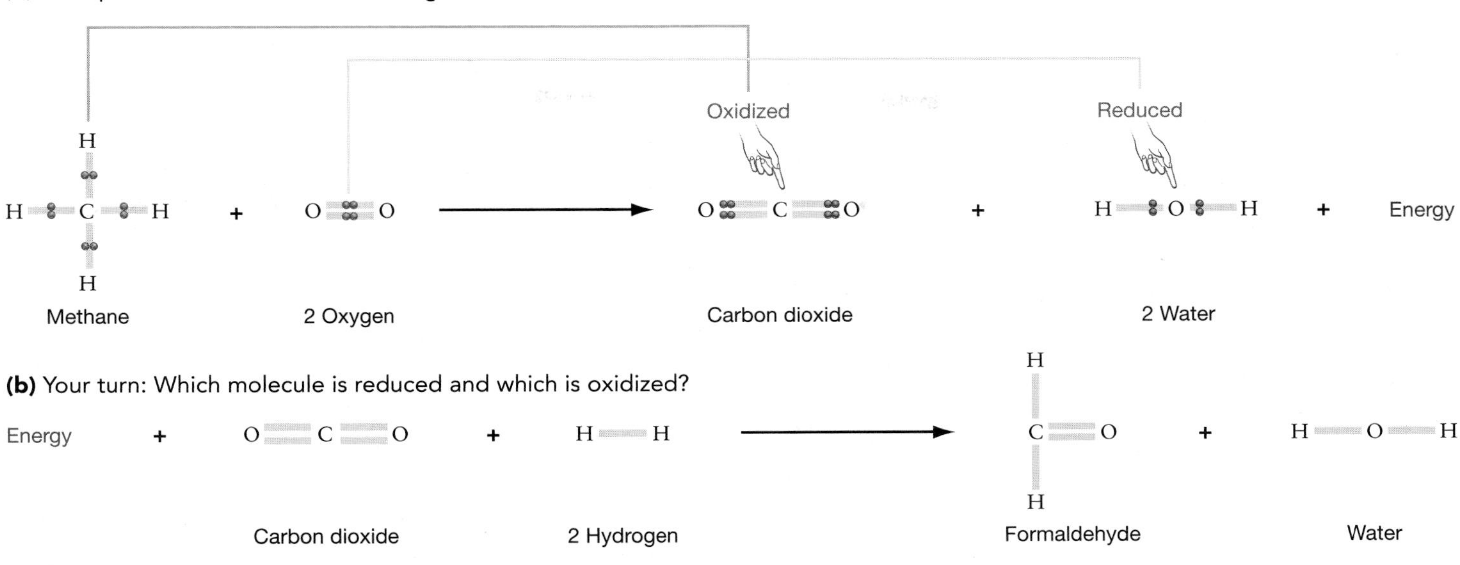

FIGURE 2.12 Changes in Potential Energy Occur During Chemical Reactions
(a) This equation illustrates what happens when methane burns. The lines between atoms indicate covalent bonds; the dots represent the relative positions of the electrons involved in those bonds. **(b)** This reaction results in the formation of formaldehyde—a building block of chemical evolution. EXERCISE Using the data in part (a) as a guide, add dots to each covalent bond in this equation to show the relative positions of the electrons involved.

as in the component of gasoline called octane (C_8H_{18}; **Figure 2.13a**), or in a ring, as in the sugar called ribose ($C_5H_{10}O_5$; **Figure 2.13b**). Molecules that contain a carbon-carbon bond are called **organic molecules.** Other types of molecules are referred to as inorganic compounds.

Linking Carbon Atoms Together The formation of carbon-carbon bonds was an important event in chemical evolution. It represented a crucial step toward the production of the types of molecules found in living organisms. Once reduced carbon compounds had formed, continued chemical evolution could occur by the addition of heat alone. For example, when molecules of formaldehyde are heated, they react with one another to form a molecule called acetaldehyde. Acetaldehyde contains a carbon-carbon bond. With continued heating, reactions between formaldehyde and acetaldehyde molecules can produce the larger carbon-containing compounds called sugars (see Figure 2.13b).

In sum, advocates of the chemical evolution hypothesis propose that two sources of energy—the potential energy in reduced inorganic compounds like H_2 and the energy in sunlight—made the production of reduced carbon-containing compounds possible. Subsequently, the potential energy in these carbon-containing molecules made the production of the first complex organic compounds possible (**Figure 2.14**, page 32).

Functional Groups Once chemical evolution was under way, small organic molecules would have accumulated on the ancient Earth. The characteristics of organic compounds are highly variable, yet predictable. Their behavior is dictated by groups of H, N, or O atoms that are bonded to carbon in a specific way. These groups of atoms are called **functional groups.**

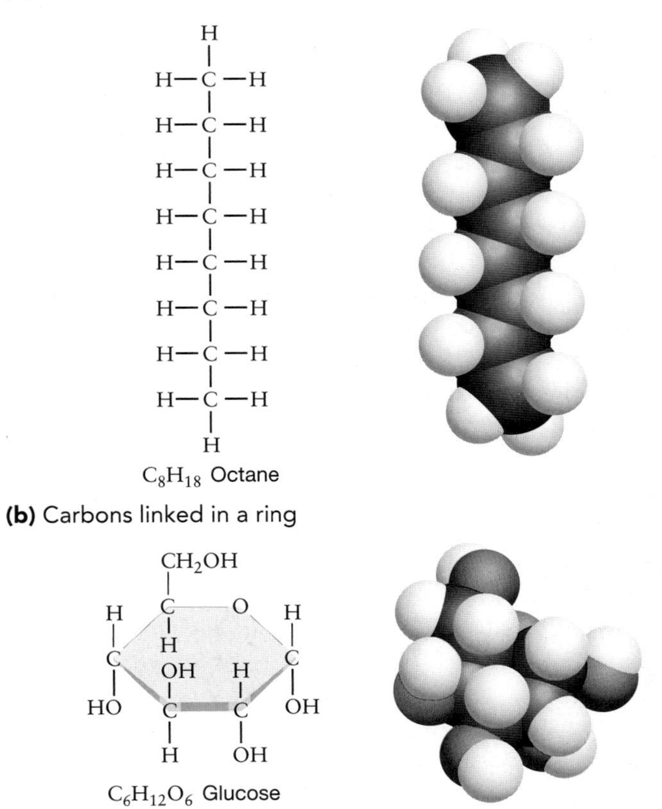

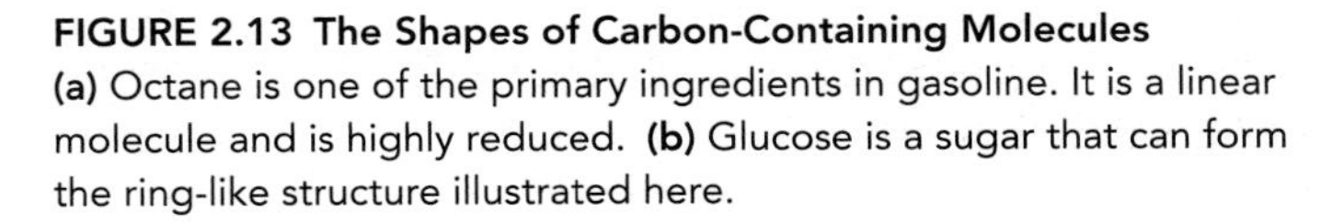
FIGURE 2.13 The Shapes of Carbon-Containing Molecules
(a) Octane is one of the primary ingredients in gasoline. It is a linear molecule and is highly reduced. **(b)** Glucose is a sugar that can form the ring-like structure illustrated here.

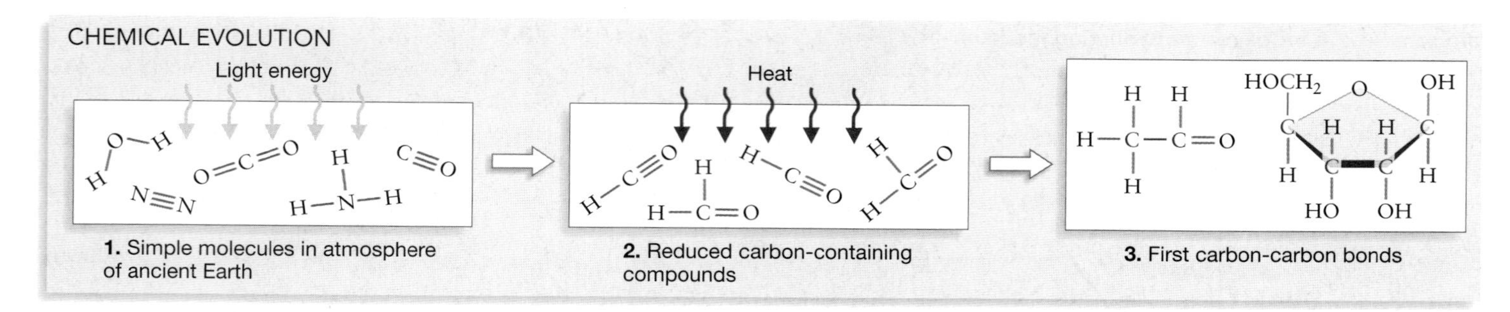

FIGURE 2.14 The Start of Chemical Evolution—An Overview
Chemical evolution is a process whereby simple molecules containing C, H, O, and N react to form reduced carbon-containing molecules, which then react to form organic compounds. The process is triggered by an energy source such as sunlight or the heat released in a volcanic eruption.

The carbonyl group, for example, is found on aldehyde molecules such as formaldehyde and acetaldehyde. This functional group is the site of the reaction that links these molecules together into larger, more complex compounds. The functional groups listed in **Table 2.1** confer equally distinctive properties. In effect, then, carbon atoms function as the skeleton of organic molecules. They provide shape and bonding capacity—a structural framework. The chemical properties of the molecule are determined by the functional groups attached to this skeleton.

Reactions among organic molecules launched the next phase of chemical evolution: The formation of the larger and more complex molecules found in living cells. Advocates of the chemical evolution hypothesis propose that these reactions occurred in water rather than in the atmosphere. Why?

2.5 The Early Oceans and the Properties of Water

The formation of Earth's first ocean was a turning point in chemical evolution, similar in impact to the formation of the first reduced carbon compounds.

Life is based on water. In a typical living cell, over 75 percent of the volume consists of this molecule. Virtually all researchers agree that most of the important steps in chemical evolution, including the origin of life itself, occurred in water. The claim is logical because as a dissolving agent, or **solvent**, water can contain more types of substances than any other molecule known. Chemical reactions depend on direct, physical interaction between the reactants; and substances are most likely to collide when they are dissolved. Life is based on water primarily because of its solvent properties.

CD ACTIVITY 2.2 Water and Ice

Water as a Solvent

Why is water such an efficient solvent? To understand the answer to this question, recall that water contains two hydrogen atoms bonded to an oxygen atom. Like most other atoms, oxygen and hydrogen differ in their ability to attract the electrons involved in covalent bonds. Chemists call this property an atom's **electronegativity.** Oxygen is among the most electronegative of all elements; it attracts covalently bonded electrons much more strongly than does the hydrogen nucleus. As **Figure 2.15a** shows, oxygen's electronegativity makes the distribution of charges on a water molecule asymmetrical. The side of the molecule containing the oxygen atom is slightly more negative, and the side with the hydrogen atoms is slightly more positive. Molecules that have partial charges like these are called **polar,** and these partial charges are symbolized by the lowercase Greek letter delta, δ.

Figure 2.15b illustrates how the polarity of water affects its interactions with other substances in solution. When two liquid water molecules approach one another, the partial positive charge on hydrogen attracts the partial negative charge on oxygen. This weak electrical attraction forms a **hydrogen bond** between the molecules.

In a water-based, or aqueous, solution, hydrogen bonds also form between water molecules and other polar molecules.

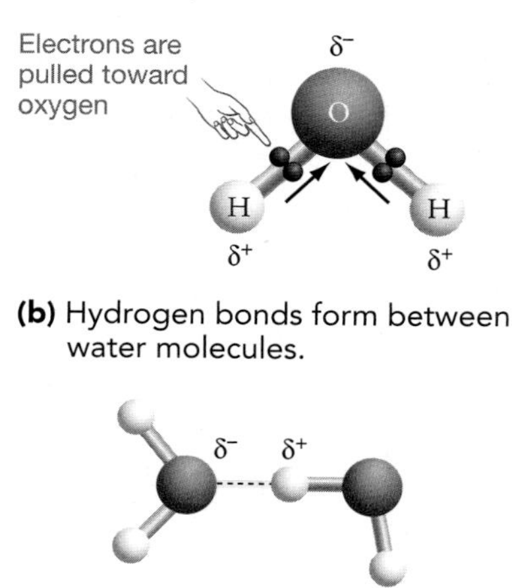

FIGURE 2.15 Water Is Polar and Participates in Hydrogen Bonds
(a) Because of oxygen's high electronegativity, the electrons that are shared when hydrogen and oxygen form a covalent bond are pulled toward the oxygen nucleus. The electrons spend more time close to the oxygen nucleus, so the oxygen atom has a slight negative charge and the hydrogen atom a partial positive charge. **(b)** The electrical attraction that occurs between the partial positive and negative charges on water molecules forms a hydrogen bond.
EXERCISE Label the hydrogen bond in part (b).

Similar interactions occur between water and atoms or molecules that carry an electric charge. Charged substances are called **ions.** Ions and polar molecules stay in solution because of their interactions with water's partial charges. In short, water is an efficient solvent because of its polarity.

Unusual Aspects of Water's Behavior

The hydrogen bonds that form between water molecules are responsible for two of its most unusual properties: It expands as it changes from a liquid to a solid, and it has an extraordinarily large capacity for absorbing heat. How do the hydrogen bonds

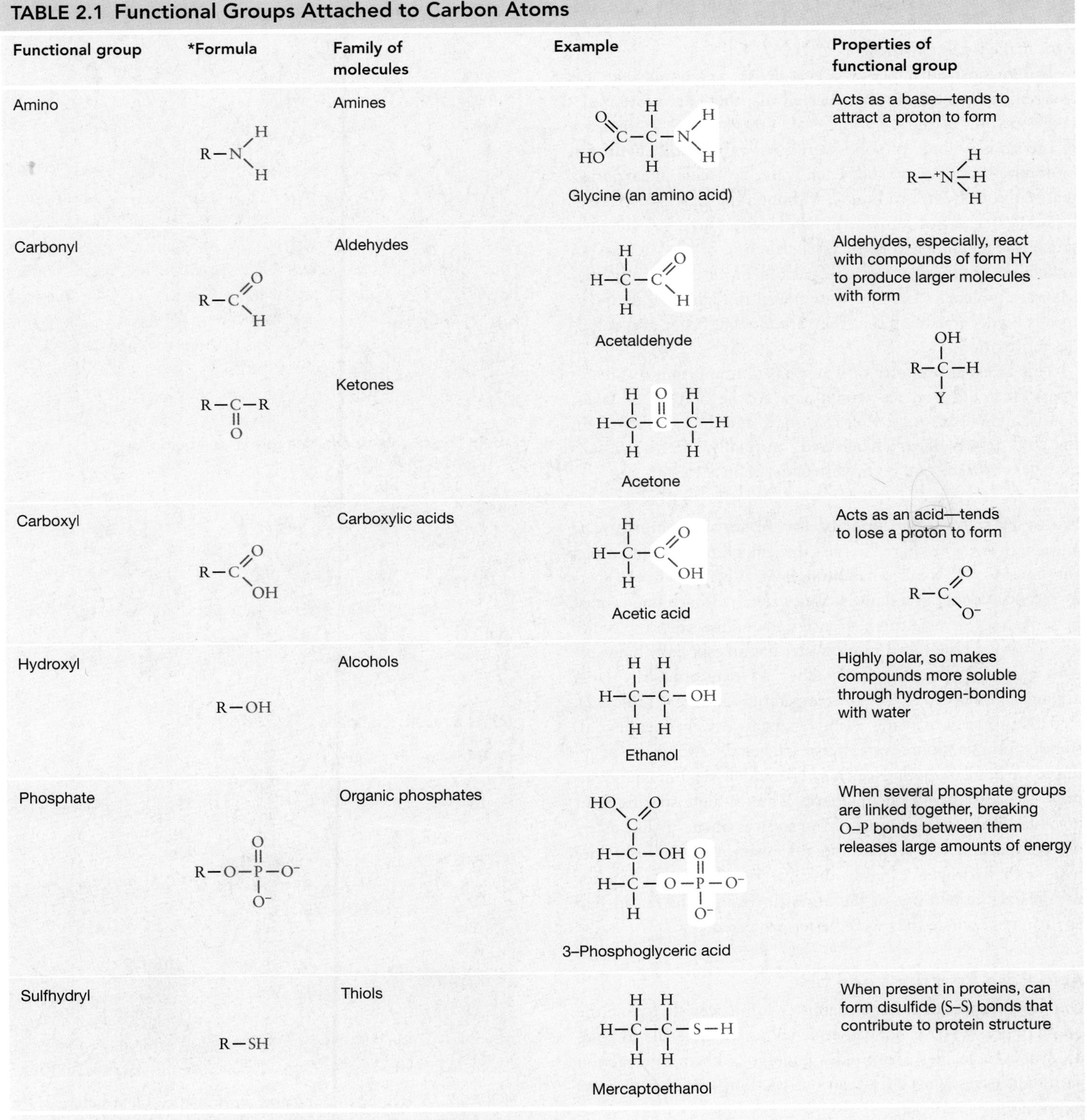

TABLE 2.1 Functional Groups Attached to Carbon Atoms

Functional group	*Formula	Family of molecules	Example	Properties of functional group
Amino	$R-NH_2$	Amines	$HOOC-CH_2-NH_2$ Glycine (an amino acid)	Acts as a base—tends to attract a proton to form $R-{}^{+}NH_3$
Carbonyl	$R-CHO$	Aldehydes	CH_3-CHO Acetaldehyde	Aldehydes, especially, react with compounds of form HY to produce larger molecules with form $R-CH(OH)-Y$
	$R-CO-R$	Ketones	$CH_3-CO-CH_3$ Acetone	
Carboxyl	$R-COOH$	Carboxylic acids	CH_3-COOH Acetic acid	Acts as an acid—tends to lose a proton to form $R-COO^-$
Hydroxyl	$R-OH$	Alcohols	CH_3-CH_2-OH Ethanol	Highly polar, so makes compounds more soluble through hydrogen-bonding with water
Phosphate	$R-O-PO_3^{2-}$	Organic phosphates	$HOOC-CH(OH)-CH_2-O-PO_3^{2-}$ 3-Phosphoglyceric acid	When several phosphate groups are linked together, breaking O-P bonds between them releases large amounts of energy
Sulfhydryl	$R-SH$	Thiols	CH_3-CH_2-S-H Mercaptoethanol	When present in proteins, can form disulfide (S-S) bonds that contribute to protein structure

*In these structural formulas, "R" stands for the rest of the molecule.

that form between water molecules explain this behavior, and how did these properties affect chemical evolution?

Water Is Denser as a Liquid than as a Solid When factory workers pour molten metal or plastic into a mold and allow it to cool to the solid state, the material shrinks. When molten lava pours out of a volcano and cools to solid rock, it shrinks. But when you fill an ice tray with water and put it in the freezer to make ice, it expands.

Unlike most substances, water is denser as a liquid than it is as a solid (**Figure 2.16**). In other words, there are more molecules of water in a given volume of liquid water than there are in the same volume of solid water. As water cools, hydrogen bonding produces a crystal. Each water molecule in ice participates in four hydrogen bonds. Although the exact structure of liquid water is not known, most evidence suggests that it is much less regular than the crystal structure of ice. As a result, molecules of liquid water are packed more closely together than are molecules of solid water, even though their temperature is higher (meaning that they are moving faster and colliding more often).

This unusual property of water has an important result: Ice floats. If it did not, ice would sink to the bottom of lakes, ponds, and oceans soon after it formed, and then stay frozen in the cold depths. If water behaved "normally," Earth's oceans may have frozen solid before life had a chance to start.

Water Has a Huge Capacity for Absorbing Energy The amount of energy required to raise the temperature of 1 gram of a substance by 1°C is called its **specific heat**. When a source of energy such as sunlight or a flame strikes water, hydrogen bonds must be broken before heat can be transferred and the water molecules begin moving faster. As a result, water has an especially high specific heat. In other words, it takes an extraordinarily large amount of energy to change the temperature of water (**Table 2.2**).

This property of water is important to the chemical evolution hypothesis. Sources of energy trigger the formation of reduced carbon compounds such as formaldehyde, but energy inputs can also break them apart. Water's high specific heat insulates dissolved substances from sources of energy like asteroid bombardment, sunlight, and volcanism. Because formaldehyde and hydrogen cyanide dissolve readily in water, they would have rained out of the atmosphere into the ocean—an environment where they were better protected.

Acid-Base Reactions and pH

One other aspect of water's chemistry influenced its role in the origin of life. Water is not a completely stable molecule. In reality, water molecules continually undergo a chemical reaction with themselves. This "dissociation reaction" can be written as follows:

$$H_2O \rightleftharpoons H^+ + OH^-$$

The double arrows in the expression indicate that the reaction proceeds in both directions.

The molecules on the right side of the expression are the hydrogen ion (H^+) and the hydroxide ion (OH^-). A hydrogen ion,

(a) In ice, water molecules form a crystal lattice.

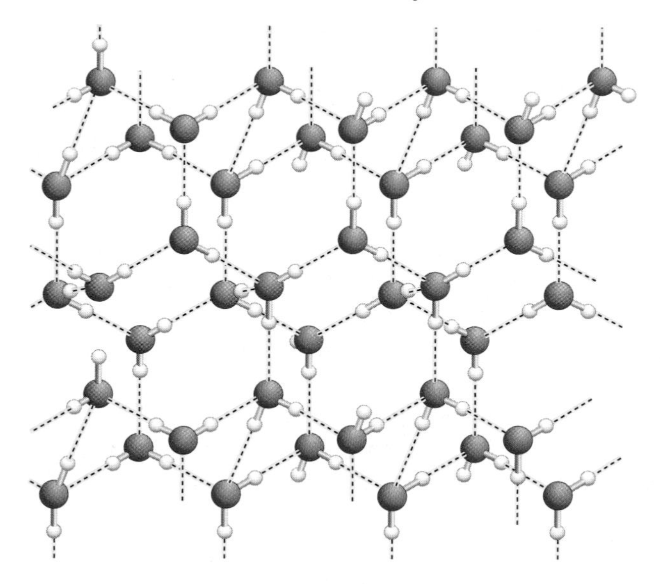

(b) In liquid water, no lattice forms, so liquid water is denser than ice.

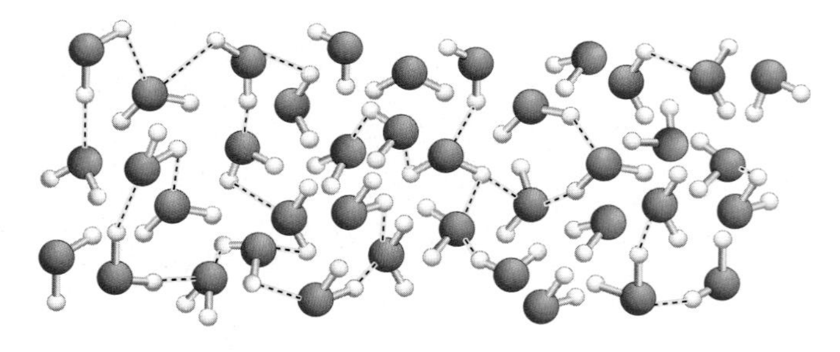

(c) As a result, ice floats.

FIGURE 2.16 Hydrogen Bonding Forms the Crystal Structure of Ice
Ice is less dense than water because the water molecules in ice form a crystal-like lattice. As a result, ice floats.

in turn, is simply a proton. Substances that give up protons during chemical reactions are called **acids**; molecules or ions that acquire protons during chemical reactions are called **bases.** A chemical reaction that involves a transfer of protons is called an **acid-base reaction.**

In water, protons associate with water molecules to form the hydronium ion, H_3O^+. Thus, the dissociation of water can also be written

$$H_2O + H_2O \rightleftharpoons H_3O^+ + OH^-$$

Note that one of the water molecules on the left side of the expression has given up a proton and acted as an acid, and that the other water molecule has accepted a proton and acted as a base. This illustrates another unusual property of water. Most acids act only as acids, and most bases act only as bases. But water can act as an acid and as a base.

In a solution, the tendency for acid-base reactions to occur is largely a function of the number of protons present. This raises a question about how many protons are present in water.

Chemists answer this question by measuring the concentration of protons directly. In a sample of pure water at 25°C, the concentration of H^+ is 1.0×10^{-7} M (M represents molarity, or moles per liter). Because this is such a small number (1 ten-millionth), the exponential notation is cumbersome. So chemists and biologists prefer to express the concentration of protons in a solution with a logarithmic notation called the **pH scale.** (The term *pH* is derived from the French *puissance d'hydrogène*, or "power of hydrogen.") By definition, the pH of a solution is the negative of the base-10 logarithm, or log, of the hydrogen ion concentration:

$$pH = -\log[H^+]$$

(The square brackets are a standard notation for indicating "concentration of.") Taking antilogs gives

$$[H^+] = \text{antilog}(-pH) = 10^{-pH}$$

Thus, the pH of pure water at 25°C is 7. **Figure 2.17** shows the pH scale and reports the pH of some common solutions. Note that pure water is used as a standard, or point of reference, on the pH scale. Solutions with a pH higher than 7 are considered basic, meaning that they are more likely to accept a proton than water. Solutions with a pH below 7 are acidic, meaning that they are more likely to give up a proton than water. Solutions with a pH of 7 are considered neutral solutions, neither acidic nor basic. Rainwater is almost pure water, meaning that its pH is close to 7.

Ions Present in the Early Ocean

The rains that formed Earth's first ocean would also have eroded the planet's newly formed rocks. On impact, the kinetic energy in falling raindrops is transformed into mechanical

TABLE 2.2 Some Specific Heats of Common Substances

The specific heats reported in this table were measured at 25°C (except for ice, which was measured at −11°C) and are given in units of joules per gram of substance per degree Celsius. (The joule is a unit of energy.) The most important thing to note in this table is that the specific heat of water, in both solid and liquid form, is extremely high. This means that water must absorb a great deal of thermal energy before it begins to change temperature.

Substance	Specific Heat
Air	1.01
Aluminum	0.902
Copper	0.385
Gold	0.129
Iron	0.450
Mercury	0.140
NaCl	0.864
Water (s)	2.03
Water (*l*)	4.179

Source: J. McMurry and R. C. Fay, *Chemistry*, 3rd ed. (Upper Saddle River, NJ: Prentice Hall, 2001), Table 8.1.

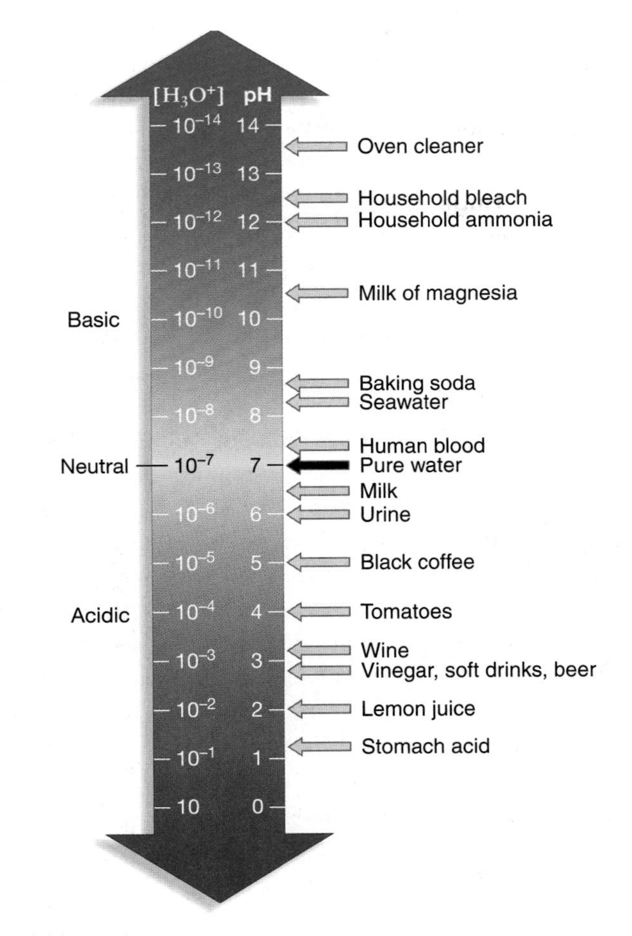

FIGURE 2.17 The pH Scale
Because the pH scale is logarithmic, a change in one unit of pH represents a change in the concentration of hydrogen ions equal to a factor of 10. Coffee is a hundred times more acidic than pure water.

energy that breaks molecules off the substrate or even creates small fragments. Streams carry these substances to the ocean. There, the ions and polar substances stay in solution while insoluble particles settle out. The particles drift to the ocean floor and form sediments.

Salts like the sodium chloride illustrated in **Figure 2.18a** are the most prominent dissolved substances that rainwater extracts from rocks. Salts are held together by **ionic bonds.** Ionic bonds are similar in principle to covalent bonds; but instead of sharing electrons between two atoms, the electrons in ionic bonds are completely transferred from one atom to the other. (**Box 2.2** explains the difference between ionic and covalent bonds in more detail.)

This electron transfer occurs because it gives the atoms involved a full valence shell. Sodium atoms (Na), for example, tend to lose an electron, leaving them with a full second shell. This is a much more stable arrangement, energetically, than having a lone electron in their third shell (**Figure 2.18b**). The sodium ion that results has a net charge of +1, because the atom has one more proton than it has electrons. This positively charged ion, or **cation**, is written Na^+. Chlorine atoms (Cl), in contrast, tend to gain an electron. When this occurs, all of the electrons in the atom's third shell are paired (**Figure 2.18c**). The atom has a net charge of −1, because the atom has one more electron than protons. This negatively charged ion, or **anion**, is written Cl^-. Sodium and chloride ions are the most common ions in today's oceans (**Table 2.3**).

Although the ionic composition of the ancient oceans is not known, advocates of the chemical evolution hypothesis assume that at least some components of today's seawater were present. Based on the arguments presented in this section, then, the environment from which chemical evolution proceeded was a salty solution of nearly neutral pH. The solution contained at least some reduced carbon compounds that resulted from redox reactions in the atmosphere, along with simple organic molecules produced when these reduced carbon-containing molecules were heated. In the ocean, however, these first products of chemical evolution were relatively well protected from further inputs of energy because of water's high specific heat.

These conclusions are based on models, however, and on logical arguments rather than direct data. Do biologists have experimental evidence that supports the chemical evolution hypothesis? If so, how did reduced carbon compounds like formaldehyde and hydrogen cyanide lead to the production of more complex compounds and the origin of life? Exploring these questions is the focus of Chapter 3.

TABLE 2.3 Major Ions Found in Present-Day Seawater

The concentrations of ions reported here are in moles per liter, or molarity (M).

Ion Name	Symbol	Concentration (M)
Chloride	Cl^-	0.550
Sodium	Na^+	0.470
Sulfate	SO_4^{2-}	0.028
Magnesium	Mg^{2+}	0.054
Calcium	Ca^{2+}	0.010
Potassium	K^+	0.010
Bicarbonate	HCO_3^-	0.002

BOX 2.2 Bond Types Form a Continuum

The degree to which electrons are shared in chemical bonds forms a continuum. As the right-hand side of **Figure 1** shows, covalent bonds between atoms with exactly the same electronegativity—for example, between atoms of hydrogen in H_2—represent one end of the continuum. The electrons in these nonpolar bonds are shared equally. In the middle of the continuum are bonds where one atom is much more electronegative than the other. In these asymmetric bonds, substantial partial charges exist on each of the atoms. These types of polar covalent bonds occur in water and ammonia because oxygen and nitrogen are much more electronegative than hydrogen. At the other extreme are molecules made up of atoms with extreme differences in their electronegativities. In this case electrons are transferred rather than shared, and the atoms have full charges. This situation is called ionic bonding. Common table salt, NaCl, is held together by ionic bonds.

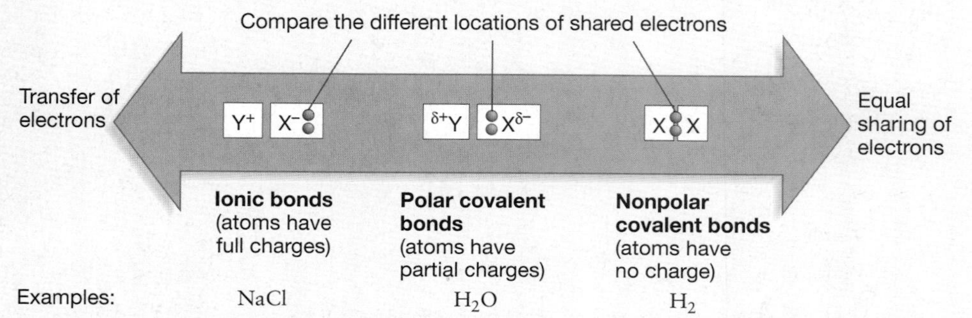

FIGURE 1 The "Electron Sharing Spectrum"

(a) Table salt is a crystal composed of two ions.

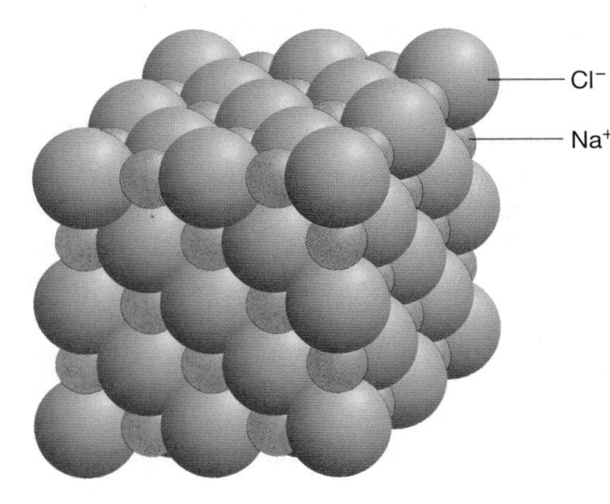

(c) A chloride ion being formed

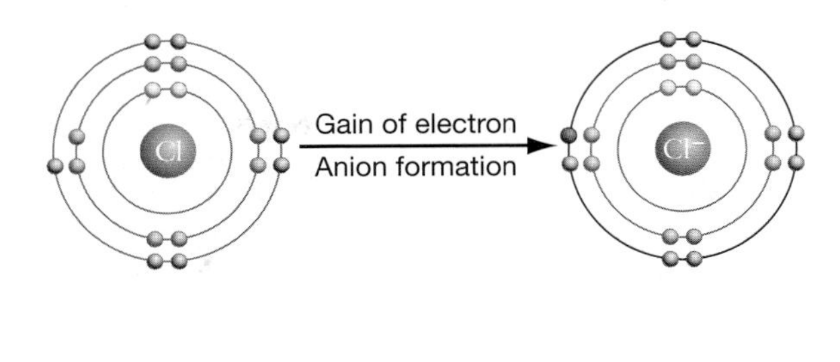

(d) Ionic solids dissolve readily in water.

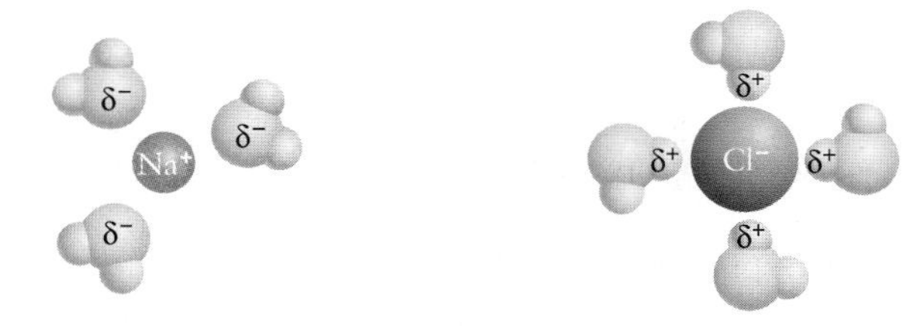

(b) A sodium ion being formed

Na
Loss of electron
Cation formation
Na⁺

FIGURE 2.18 Ionic Bonding and Ion Formation
(a) In table salt (NaCl), sodium and chloride atoms pack into a crystal structure. Salt crystals are difficult to break apart because the electrical attraction between the ions is so strong. **(b)** When an atom of sodium (Na) loses an electron to chlorine, it forms a sodium ion (Na^{+}). **(c)** When an atom of chlorine (Cl) gains an electron from sodium to form the chloride ion (Cl^{-}), it forms an ion with a full valence shell. **(d)** Ionic solids dissolve readily in water. Why? **EXERCISE** Ionic bonding can be thought of in terms of reduction and oxidation. In parts (b) and (c) of the figure, label which reaction results in reduction and which reaction results in oxidation.

Essay The Search for Extraterrestrial Life

Astronomers who search for life on other planets have recently been invigorated by two findings: (1) evidence that water may be present under the surface of Mars, Earth's moon, and one of Jupiter's moons; and (2) the discovery of cells living in rock formations hundreds of meters below Earth's surface.

Confirming that liquid water exists elsewhere in the solar system would be exciting, because life is probably impossible in its absence. If the location of water deposits can be validated and mapped, it would give scientists a promising place to look for extraterrestrial organisms—it would narrow the search.

Planets and moons that appear to be lifeless could actually be teeming with organisms below the surface.

Finding organisms in the deep subsurface of Earth suggested that life could also be found under the surface of Mars, the moon, or other bodies. Planets and moons that appear to be lifeless could actually be teeming with organisms below the surface. On Earth, cells have been found in rocks up to 860 meters (about half a mile) below the surface. These discoveries supported a general realization: Organisms are found in a wide variety of extreme environments on Earth. Species of the single-celled organisms called bacteria and archaea, and sometimes even multicellular (many-celled) animals, can thrive underneath glaciers, near superheated steam vents on ocean floors, in hot springs, and in ponds with extremely high levels of salt. The presence of life in extreme environments on Earth has encouraged astronomers to suggest that life may also exist in extreme environments in space.

As biologists gain a better understanding of chemical evolution and the diversity of life on ancient Earth, the search for extraterrestrial life has become more focused. Few astronomers expect to find the sophisticated types of life-forms favored by science fiction writers. Instead, they are looking for evidence of water, reduced carbon, and traces of single-celled life-forms similar to Earth's bacteria and archaea.

Finding life elsewhere in the solar system would instantly qualify as a tremendous scientific advance. It would refute the hypothesis that chemical evolution occurred just once and that life is unique to Earth. It would also open up the possibility of finding organisms elsewhere in the universe—perhaps in the recently discovered planetary systems that are forming around young stars.

Chapter Review

Summary

Chemical evolution is a hypothesis for the origin of the essential chemical components of life. It maintains that early in Earth history, large and complex molecules formed from simple precursor compounds in the ancient atmosphere and ocean, as chemical reactions converted the energy in sunlight and other sources into chemical energy. Based on data from radiometric dating, advocates of the hypothesis suggest that chemical evolution took place over a span of some 300 million years, beginning about 3.8 billion years ago.

Researchers have developed models for how the first step in chemical evolution occurred. This step was the formation of reduced carbon compounds, like formaldehyde and hydrogen cyanide, from oxidized molecules like CO_2. The models show that these reactions could occur only if a source of energy, like sunlight, *and* a source of reduced inorganic compounds, like H_2, were present in the ancient atmosphere. Although the composition of the ancient atmosphere is controversial, most investigators now agree that at least some reduced inorganic molecules existed and that sources of energy were abundant.

If reduced carbon compounds formed in the ancient atmosphere, they would have dissolved in water droplets and rained down into the early ocean. Advocates of the chemical evolution hypothesis propose that most subsequent chemical evolution took place in water, primarily because it is such an efficient solvent. Water is the most effective solvent known because it is polar, meaning that it has partial positive and negative charges. As a result, charged substances like ions and other polar molecules interact with it and stay in solution.

The general message of this chapter can be stated as follows: Chemical evolution was possible because the atmosphere and oceans of ancient Earth contained reduced inorganic compounds, abundant and diverse sources of energy, and a huge supply of water.

Questions

Content Review

1. Which of the following occurs when a covalent bond forms?
   a. The potential energy of electrons drops.
   b. Electrons in valence shells are shared between nuclei.
   c. Ions of opposite charge interact.
   d. Polar molecules interact.
2. If a reaction is exothermic, then which of the following statements is true?
   a. The products have lower potential energy than the reactants.
   b. Heat is released.
   c. The reverse reaction, if it exists, is endothermic.
   d. All of the above.
3. What is thermal energy? (Circle all that apply.)
   a. a form of potential energy
   b. a form of energy measured as heat
   c. mechanical energy
   d. the kinetic energy of molecular motion
4. What determines whether a chemical reaction is spontaneous?
   a. if it increases the disorder, or entropy, of the substances involved
   b. if it decreases the potential energy of the substances involved
   c. the combined effect of changes in potential energy and entropy
   d. the temperature only
5. Which of the following is an example of an energy transformation?
   a. A shoe drops, converting potential energy to kinetic energy.
   b. A chemical reaction converts the energy in sunlight into the chemical energy in formaldehyde.
   c. The electrical energy flowing through a light bulb's filament is converted into light and heat.
   d. All of the above.
6. What does the phrase "LEO the lion goes GER" stand for?
   a. Lower Electron Orbitals have Greater Electron Reactivity
   b. Lower Energy Originally; Greater Energy Recently
   c. Loss of Electrons is Oxidation; Gain of Electrons is Reduction
   d. All of the above.
7. When a compound is reduced, it
   a. gains entropy
   b. loses entropy
   c. gains potential energy
   d. loses potential energy
8. What property defines acids and bases?
   a. their pH
   b. their potential energy
   c. their tendency to transfer protons
   d. their tendency to transfer electrons
9. Which of the following best describes a polar substance?
   a. It is electrically neutral.
   b. It is electrically neutral overall, but contains partial positive and negative charges in different locations.
   c. It is negatively charged.
   d. It is positively charged.

Conceptual Review

1. The text claims that computer simulations are a productive way to explore processes that are difficult to study directly, such as planet formation or the early stages of chemical evolution. Do you agree? How willing are you to accept the results of the models reviewed in this chapter?
2. Section 2.2 describes the reaction between carbon dioxide and water, which forms carbonic acid:

$$CO_2(g) + H_2O(l) \rightleftharpoons H_2CO(aq)$$

In aqueous solution, carbonic acid immediately dissociates to form a proton and the bicarbonate ion, as follows:

$$H_2CO_3(aq) \rightleftharpoons H^+(aq) + HCO_3^-(aq)$$

Does this reaction raise or lower the pH of the solution? Does the bicarbonate ion act as an acid or a base? If an underwater volcano bubbled additional CO_2 into the ocean, would this sequence of reactions be driven to the left or the right? How would this affect the pH of the ocean?

3. When chemistry texts introduce the concept of electron shells, they emphasize that shells represent distinct potential energy levels. In introducing electron shells, this text emphasizes that they represent distinct distances from the positive charges in the nucleus. Are these two points of view in conflict? Why or why not?
4. To estimate when a rock formed using radiometric dating, geologists must know three quantities. What are they? How are they used to estimate the age of a rock sample? What sources of uncertainty are involved in measuring these three quantities?
5. Why does a reduced compound have more potential energy than an oxidized compound?
6. Why does ice float?

Applying Ideas

1. Suppose you wanted to estimate the age of an archaeological site using radiometric dating. Would you use the uranium-lead system described in section 2.1, or would you evaluate a radioisotope that had a shorter half-life? Explain your answer.
2. Hydrogen bonds form because the partial electric charges on polar molecules attract. Covalent bonds form because of the electrical attractions between electrons and protons. Covalent bonds are much stronger than hydrogen bonds. Explain why, in terms of the electrical attractions involved.
3. Oxygen is extremely electronegative, meaning that its nucleus pulls in electrons shared in covalent bonds. Because these electrons are close to the oxygen nucleus, they have lower potential energy. Explain the changes in electron position that are illustrated in Figure 2.12 in terms of oxygen's electronegativity.
4. When nuclear reactions take place, some of the mass in the atoms involved is converted to energy. The energy in sunlight is created during nuclear fusion reactions on the Sun. Explain what astronomers mean when they say that the Sun is burning down and that it will eventually burn out.

CD-ROM and Web Connection

CD Activity 2.1: Redox Reaction *(animation)*
(Estimated time for completion = 5 min)
Learn why redox reactions are central to metabolic processes of all living organisms.

CD Activity 2.2: Water and Ice *(animation)*
(Estimated time for completion = 5 min)
Ice floats—why are the unique properties of water essential to life on Earth?

At your **Companion Website** (http://www.prenhall.com/freeman/biology), you will find self-grading exams and links to the following research tools, online resources, and activities:

Biology and Chemistry
This site provides a wealth of resources for understanding the chemistry of life.

Energy
This study site reviews the dynamics and kinetics of chemical reactions and contains problem sets for review.

Atmospheric Chemistry of the Ancient Earth
This article describes how one scientist is investigating the early atmosphere using current geological techniques.

Additional Reading

Allègre, C. J., and S. H. Schneider. 1994. The evolution of the Earth. *Scientific American* 271 (October): 66–75. A review of how Earth's atmosphere, temperature, and continental landmasses have changed over 4.5 billion years.

Angel, J. R. P., and N. J. Woolf. 1996. Searching for life on other planets. *Scientific American* 274 (April): 60–66. An analysis of the techniques being used to search for, and characterize, planets outside our solar system.

Knoll, A. H. 1998. A Martian chronicle. *The Sciences*, July/August: 20–26. A look at efforts to find life on Mars.

Perkowitz, S. 1999. The rarest element. *The Sciences*, January/February: 34–38. A physicist examines the unusual properties of water.

Macromolecules and the RNA World

3

Chapter 2 introduced chemical evolution—the proposition that chemical reactions in the atmosphere and ocean of ancient Earth led to the formation of complex carbon-containing compounds. This chapter pursues chemical evolution to its conclusion. Is it possible that this process led to the origin of life?

The idea of chemical evolution was first proposed in 1923 by the Russian biochemist Alexander I. Oparin. The proposal was published again—independently and six years later—by the English evolutionary biologist J. B. S. Haldane. Today, the Oparin-Haldane proposal can best be understood as a formal scientific theory. Chapter 1 introduced the idea that scientific theories typically have two components: a statement about a pattern that exists in the natural world and a proposed mechanism or process that explains the pattern. In the case of chemical evolution, the pattern is that increasingly complex carbon-containing molecules formed in the atmosphere and ocean of ancient Earth. The process responsible for this pattern was the conversion of energy, from sunlight and other sources, into chemical energy in the bonds of large, complex molecules.

Scientific theories are continuously refined as new information comes to light, and many of Oparin and Haldane's original ideas about how chemical evolution occurred have been extensively revised. In its current form, the theory can be broken into four steps. Each of these steps requires an input of energy.

1. Chemical evolution began with the production of small, reduced, carbon-containing compounds like formaldehyde (H_2CO) and hydrogen cyanide (HCN). These molecules were introduced in Chapter 2.
2. These simple compounds reacted in the ocean to form the mid-sized molecules

This chapter reviews experiments on how lightning, volcanoes, sunlight, and other sources of energy can trigger the synthesis of the large molecules found in living cells.

called sugars, amino acids, and nitrogenous bases. Oparin and Haldane hypothesized that these building-block molecules accumulated in the shallow waters of the ancient ocean, forming a rich solution called the **prebiotic soup.**

3. The mid-sized building-block molecules linked together to form the types of large molecules found in cells. These large molecules are called proteins and nucleic acids.
4. Life became possible when a single molecule acquired the ability to make a copy of itself. This self-replicating molecule began to multiply through chemical reactions that it controlled. At this point, chemical evolution began to give way to biological evolution (**Box 3.1**).

Can these ideas be tested? One of the most important attributes of a scientific theory is that it makes predictions. When observations or experimental results conflict with the predictions made by a theory, scientists may reject the theory and look for a replacement. Alternatively, they may introduce modifications to make the theory more realistic and consistent with the data obtained.

Each of the four steps in the theory of chemical evolution is a prediction about an historical event. Unfortunately, these predictions cannot be tested by observing events directly in nature. According to the data reviewed in Chapter 2, chemical evolution took place some four billion years ago under physical conditions that were drastically different from today's Earth. As a result, the most productive strategy for testing Oparin and Haldane's theory has been to perform simulation experiments in the laboratory. The goal of these experiments is to mimic the chemical and physical conditions of ancient Earth and determine whether each of the four steps can be recreated. The goal of this chapter is to review these experiments and assess whether the theory of chemical evolution is plausible.

This chapter's investigation begins with research aimed at recreating the first step in chemical evolution—the production of reduced carbon-containing compounds. It ends with experiments aimed at recreating the endpoint of chemical evolution—the origin of life.

3.1 The Start of Chemical Evolution: Experimental Simulations

In 1953, a graduate student named Stanley Miller performed a breakthrough experiment in the study of chemical evolution. Miller wanted to answer a simple question: Can complex organic compounds be synthesized from the simple molecules present in Earth's early atmosphere and ocean? In other words, is it possible to recreate the first steps in chemical evolution by simulating early Earth conditions in the laboratory?

Miller based his experimental design on the assumption that Earth's atmosphere was dominated by reduced molecules when chemical evolution occurred. Reduced molecules are likely to give up electrons, so their presence means that redox reactions can occur.

BOX 3.1 How Do Biologists Define Life?

Like many simple questions, the one posed in the title of this box is difficult to answer. In reality, there is no precise definition of what constitutes life.

Instead of holding to a hard-and-fast definition, biologists usually point to two attributes to distinguish life from nonlife. The first attribute is the ability to reproduce. If something is alive, it can make a copy of itself. (In the case of sexually reproducing organisms, the copy is not exact. Instead, traits from male and female individuals are combined to produce an offspring.) In essence, this chapter is about the nature of the first molecule that could make a copy of itself. As you'll see, self-replication is sufficient to launch evolution by natural selection.

For something to be alive, however, most biologists insist that it have a second attribute: the ability to acquire particular molecules and use them in controlled chemical reactions. These chemical reactions, in turn, produce the molecules that make growth and reproduction possible. Organisms can exert precise control over chemical reactions because the reactants are bounded by a cell membrane. To many biologists, then, the presence of a cell membrane and controlled chemical reactions, or metabolism, is required for life.

This is an important point because the theory of chemical evolution maintains that the two attributes did not emerge simultaneously. Instead, the theory predicts that chemical evolution first led to the existence of a molecule that could make copies of itself. Later, a descendant of this molecule became enclosed in a membrane, creating the first cell. To distinguish this naked self-replicator from later, cellular forms of "true" life, then, the chapter refers to the self-replicator as the first living entity—but *not* as the first organism.

The distinction isn't trivial. Naked, self-replicating molecules undoubtedly existed early in Earth history, and researchers are almost certain to produce one in the laboratory within your lifetime. Section 3.4 explores how biologists are going about this work. Based on the results to date, however, it appears much less likely that anyone will create a cellular-based life-form in the near future.

Will humans be able to create life in a test tube? The answer depends, in part, on how you define life.

Miller's experimental setup, shown in **Figure 3.1**, was designed to produce a tiny microcosm of ancient Earth. The large glass flask in the figure represented the atmosphere and contained the highly reduced gases called methane (CH_4), ammonia (NH_3), and hydrogen (H_2). This large flask was connected to a smaller flask by glass tubing. The small flask held a tiny ocean—200 milliliters (mL) of liquid water. Miller boiled this water constantly so that water vapor was added to the mix of gases in the large flask. As this vapor cooled and condensed, it flowed back into the smaller flask, where it boiled again. In this way, water vapor circulated continuously through the system. This was important. If the reduced molecules in the atmosphere reacted with one another, the "rain" would carry them into the ocean, forming a simulated version of Oparin and Haldane's prebiotic soup.

Had Miller stopped there, however, little or nothing would have happened. Even at the boiling point of water (100°C), the molecules involved in the experiment are stable. They do not undergo spontaneous chemical reactions, even at high temperatures.

Something did start to happen in the apparatus, however, when Miller sent electrical discharges across the electrodes he'd inserted into the atmosphere. These miniature lightning bolts added a crucial element to the reaction mix: kinetic energy. After a day of continuous boiling and sparking, the solution in the boiling flask began to turn pink. After a week, it was deep red and cloudy. When Miller analyzed the molecules dissolved in the solution, he found that several complex carbon-containing compounds were present. The experiment, driven by the energy in the electrical discharges, had successfully recreated the start of chemical evolution.

To find out exactly which products resulted from the initial reactions in the simulated atmosphere, Miller drew samples from the apparatus at intervals (review Figure 3.1 to see how and where he did this). In these samples he found large quantities of a molecule called hydrogen cyanide (HCN) and a molecule called formaldehyde (H_2CO). This was exciting because hydrogen cyanide and formaldehyde are required for reactions that lead to the synthesis of more complex organic molecules such as sugars and amino acids.

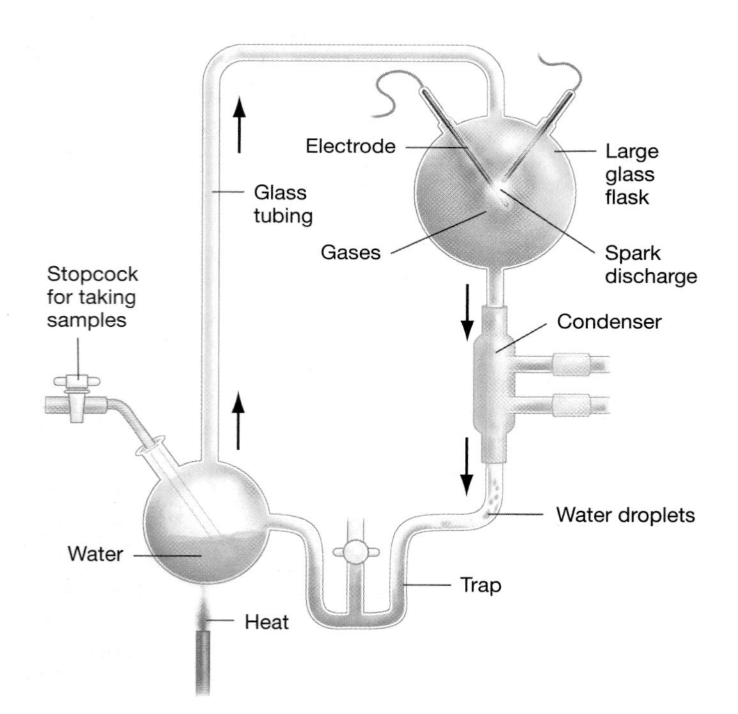

FIGURE 3.1 Miller's Spark-Discharge Experiment
This schematic diagram shows the important elements in Stanley Miller's apparatus for conducting spark-discharge experiments. The arrows indicate the flow of water vapor or liquid, starting with the 200 milliliters in the small, boiling flask. The large glass flask can contain any mixture of gases desired; when a voltage is applied across the electrodes in the flask, a spark jumps across the gap. The condenser consists of a jacket with cold water flowing through it.
EXERCISE Label the parts of the apparatus that mimic the ocean, the atmosphere, rain, and lightning.

Was Miller Correct About Conditions on Ancient Earth?

Miller's work is considered classic, primarily because he showed that hypotheses about chemical evolution can be tested experimentally. His research inspired attempts by others to recreate the key steps of chemical evolution in the lab. But the assumptions behind his experiments, and hence the results, have come under fire. Some researchers contend that Miller's simulated version of ancient Earth was unrealistic.

You may recall from Chapter 2 that most atmospheric scientists now propose that the early Earth had an atmosphere dominated by volcanic gases. If so, then most of the carbon in the atmosphere would have been in the form of oxidized molecules like carbon dioxide (CO_2) and carbon monoxide (CO), not reduced molecules like CH_4. Why is this important? The answer lies in the position of the electrons in these molecules. Look back at the diagram in Figure 2.12, and note that the electrons in the C–O bonds of CO_2 are held closer to the positive charges in nuclei compared to the electrons in the C–H bonds of CH_4. This makes the C–O bonds in CO_2 stronger than the C–H bonds of CH_4. As a result, oxidized molecules like CO_2 and CO should be much less reactive than reduced forms like CH_4, even in the presence of high-energy input from electrical discharges. This prediction has been confirmed experimentally. When CO_2, N_2, and H_2O are substituted for methane, ammonia, and hydrogen in Miller's spark-discharge experiment, virtually no chemical evolution takes place.

In response, biologists interested in the origin of life have begun to test new hypotheses for the start of chemical evolution. As you review these new hypotheses, ask yourself the same questions that researchers use to evaluate experiments on how chemical evolution began:

- Is the composition of the starting materials—the simple molecules present in the early atmosphere or ocean—plausible?

- Is the source of energy—required to trigger chemical reactions—credible?
- Are the products interesting? That is, are the molecules created in the experiment capable of participating in additional reactions so that chemical evolution would continue?

Reactions Triggered by Sunlight in an Oxidized Atmosphere Can chemical evolution occur in an oxidized atmosphere? To answer this question, atmospheric scientists Akiva Bar-Nun and Sherwood Chang performed a simplified version of Miller's experiment. They put water vapor and carbon monoxide into a glass flask, exposed the mixture to a lamp that emitted the types of high-energy radiation found in sunlight, and analyzed the reaction products. A wide variety of reduced carbon compounds formed, including methane (CH_4), formaldehyde, and acetaldehyde.

Bar-Nun and Chang also repeated the experiment and varied the reaction conditions. They altered the ratio of water vapor to carbon monoxide, changed the temperature, and extended the length of time that the mixture was exposed to light. In some samples they also added molecules like hydrogen (H_2), nitrogen (N_2), or carbon dioxide (CO_2) to the mix. In each case, a wide variety of reduced organic compounds formed.

These results reinforce the conclusions of the model developed by Joseph Pinto and colleagues that was reviewed in Chapter 2. Bar-Nun and Chang provided experimental evidence that the energy in sunlight can trigger chemical reactions that reduce the carbon found in volcanic gases. The reactions result in the production of molecules required for additional chemical evolution.

Redox Reactions at Hydrothermal Vents In sharp contrast to the hypotheses just reviewed, Gunter Wächtershäuser contends that the key events in chemical evolution did not begin in the atmosphere. Instead, he suggests that the process began far below the surface of the ancient ocean.

Specifically, Wächtershäuser hypothesizes that chemical evolution began near hydrothermal (hot-water) vents similar to the "black smokers" found at the bottom of today's oceans. **Figure 3.2a** diagrams how these unusual structures form. When cracks occur in Earth's crust, they fill with molten rock from below the surface. The thermal energy in the liquid rock heats water in the surrounding rocks to temperatures as high as 450°C. Because of the intense pressure at these depths, the "superheated" water does not boil. Instead, it rises up through the crust. As it passes through the crustal rocks it dissolves enough iron-, sulfur-, nickel- and carbon-containing compounds to form a blackened solution, which jets out into the surrounding water (**Figure 3.2b**). At the bottom of the ocean, this surrounding water is frigid—typically 4°C.

Proponents of the hydrothermal vent hypothesis point out that this environment provides the two ingredients required to trigger chemical evolution: kinetic energy and reduced carbon compounds. The hypothesis has also received some experimental support. One recent study was inspired by the observation that a carbon- and sulfur-containing molecule called methyl mercaptan (CH_3SH) has been detected near hydrothermal vents. Along with colleague Claudia Huber, Wächtershäuser investigated whether this molecule could react with carbon monoxide in solution to form carbon-carbon bonds and thus trigger chemical evolution. Their experiments confirm that, under the temperature and pressure conditions found near

(a) Formation of black smokers

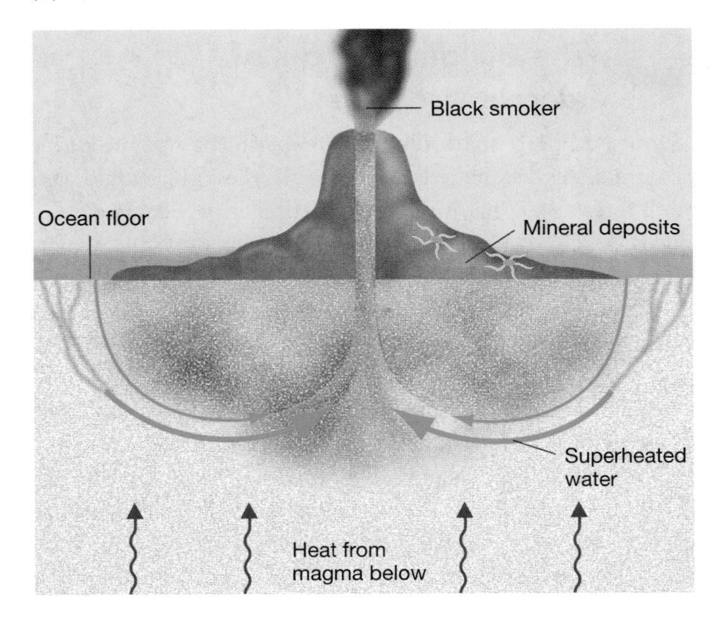

(b) Where hot and cold meet

FIGURE 3.2 Black Smokers
(a) Black smokers form where liquid rock, or magma, is located close to the ocean floor. **(b)** Black smokers are black because the superheated water that erupts from them contains large amounts of iron, carbon, and nickel. Chemical evolution may have begun at sites similar to this one.

black smokers, the two carbon-containing molecules react to form acetic acid (CH_3COOH). The result suggests that moderately complex organic molecules could have been synthesized at hydrothermal vents.

Based on these and other experiments simulating early Earth conditions, most researchers now agree that the first step in the Oparin-Haldane hypothesis is plausible. It is likely that early in Earth history, quantities of formaldehyde, hydrogen cyanide, acetic acid, and other reduced carbon-containing molecules were raining down from the skies and bubbling up from hydrothermal vents.

3.2 The Building Blocks of Macromolecules

If Stanley Miller's spark-discharge experiment had produced just formaldehyde and hydrogen cyanide, the result would have been interesting. What made his experiment particularly exciting was the discovery of larger, more complex molecules, called amino acids, in his miniature "ocean." Based on this result, Miller could claim that his experiment had begun to duplicate the second stage in chemical evolution—the formation of a prebiotic soup.

To understand the composition of this solution, biologists must work backward. The logic runs as follows: Living cells contain several types of large molecules, or **macromolecules.** The theory of chemical evolution contends that the entity that launched biological evolution consisted of one of these types of macromolecules, either a protein or a nucleic acid. Proteins and nucleic acids, in turn, are made up of smaller component molecules.

The prebiotic soup, then, is not an ill-defined jumble of organic compounds. Rather, Oparin and Haldane predicted that it would contain the building blocks of proteins and nucleic acids. Section 3.3 examines how these subunits could have bonded to one another during chemical evolution to form the first macromolecules. But here the focus is on the subunits themselves. What are they, and where did they come from?

Amino Acids

Miller's discovery of amino acids was noteworthy for a simple reason: Amino acids are the building blocks of proteins, and proteins are the workhorse molecules in cells. Proteins direct the chemical reactions that allow cells to grow and reproduce. They also form the structures that give cells their shape and make movement possible.

The amino acids that build proteins have a common structure. To understand how these molecules are put together, examine the generalized amino acid shown on the left in **Figure 3.3a.** Focus your attention on the carbon atom, shown in red. Note that it forms four bonds. One bond links the carbon to a hydrogen atom. A second bond connects the carbon to NH_2. An NH_2 is the amino functional group introduced in Chapter 2. A third bond links the carbon to COOH—the carboxyl functional group. The amino acid structure shown on the right in Figure 3.3a shows two important changes that occur in these functional groups when amino acids are in solution. In seawater, the concentration of hydrogen ions causes the amino group to act as a base. It attracts a proton to form NH_3^+. The carboxyl group, in contrast, acts as an acid. It loses a proton to form COO^-. The charges on these functional groups help amino acids stay in solution and add to their chemical reactivity.

Note that the final bond on the focal carbon atom is to something abbreviated as "R." This is the symbol that chemists use to indicate additional atoms, or a "side chain." In every amino acid, a carbon atom is linked to an R-group, a hydrogen atom, an amino group, and a carboxyl group. What distinguishes amino acids is the composition of the R side chain. The side chains of amino acids can be as simple as a hydrogen atom, or they can be more complex.

In **Figure 3.3b**, the 20 amino acids found in organisms are sorted according to the characteristics of their side chains. Note that some amino acids contain side chains consisting entirely of carbon and hydrogen atoms. These R-groups rarely participate in chemical reactions. The behavior of these amino acids depends primarily on their size and shape rather than their reactivity. In contrast, amino acids with acidic or basic side chains are fairly reactive.

In **Table 3.1** (page 46), these same amino acids are sorted according to how readily they interact with water. The important thing to note here is that amino acids with nonpolar side chains do not have charged or electronegative atoms capable of forming hydrogen bonds with water. These side chain groups are said to be **hydrophobic** (water-fearing) because, instead of interacting with water, they exhibit an aversion to water and tend to coalesce. In contrast, amino acids with polar side chain groups *do* contain charged or electronegative atoms. These side chains interact readily with water and are termed **hydrophilic** (water-loving).

The Problem of Chirality The amino acids produced in Miller's spark-discharge experiments were a breakthrough in research on chemical evolution. But they also raised a critical problem: the phenomenon known as chirality.

FIGURE 3.3 Amino Acids
(a) All amino acids have the same general structure. **(b)** The structural formulas of the 20 amino acids found in organisms are grouped here according to the characteristics of their side chains. The standard single-letter and three-letter abbreviations for each amino acid are given next to each name. Note that the carbon and hydrogen atoms in the ring structures of phenylalanine, tyrosine, and tryptophan are not drawn in. This makes the structures easier to read. There is a carbon at each bend in the rings. The lines inside the rings indicate double bonds. EXERCISE Study the composition of the side chains shown in part (b) and predict which of these side chains are most likely and least likely to interact with water. Then compare your answers to the data presented in Table 3.1.

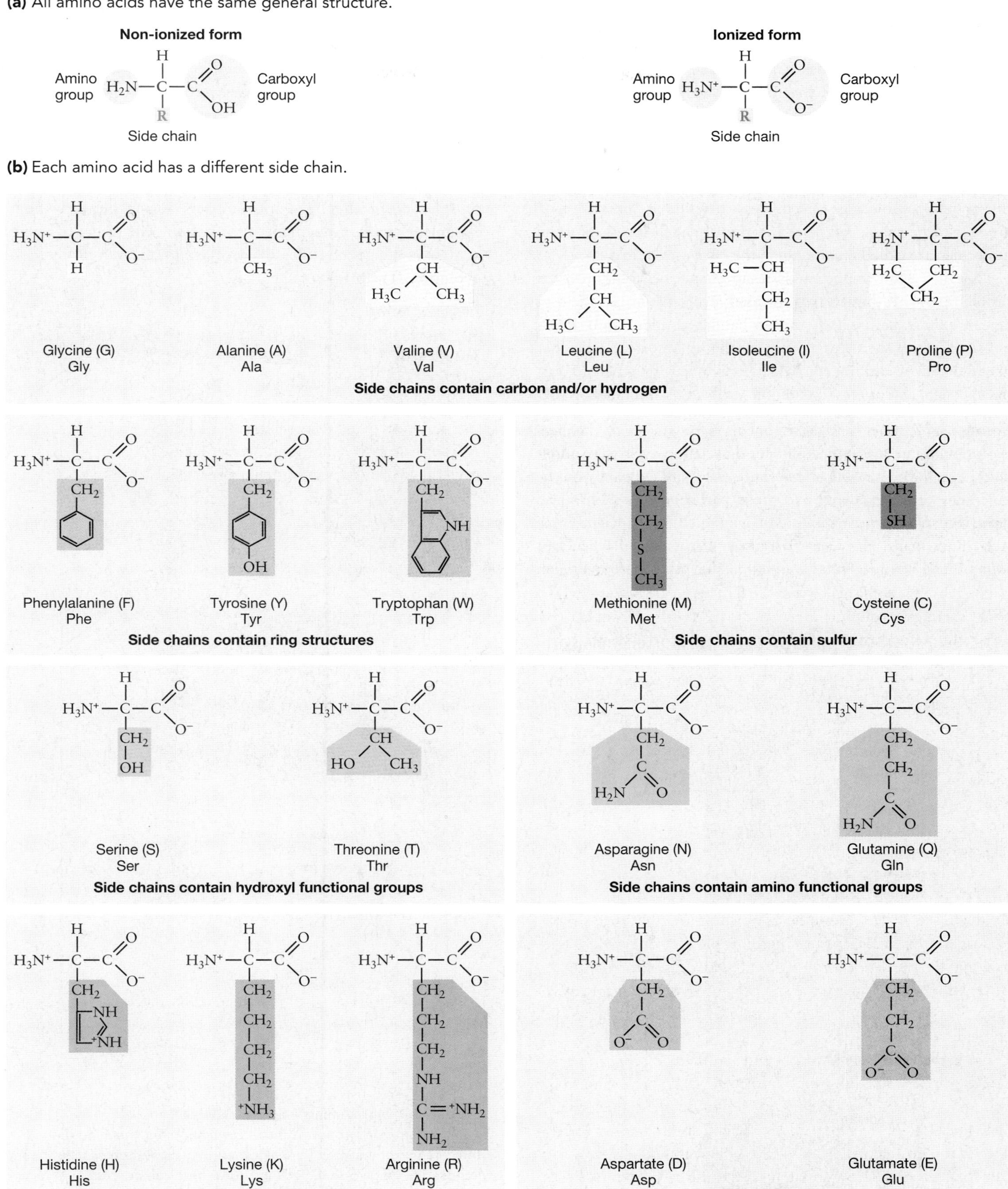

(a) All amino acids have the same general structure.
Non-ionized form
Amino group
Carboxyl group
Side chain
Ionized form
Amino group
Carboxyl group
Side chain
(b) Each amino acid has a different side chain.
Glycine (G) Gly
Alanine (A) Ala
Valine (V) Val
Leucine (L) Leu
Isoleucine (I) Ile
Proline (P) Pro
Side chains contain carbon and/or hydrogen
Phenylalanine (F) Phe
Tyrosine (Y) Tyr
Tryptophan (W) Trp
Side chains contain ring structures
Methionine (M) Met
Cysteine (C) Cys
Side chains contain sulfur
Serine (S) Ser
Threonine (T) Thr
Side chains contain hydroxyl functional groups
Asparagine (N) Asn
Glutamine (Q) Gln
Side chains contain amino functional groups
Histidine (H) His
Lysine (K) Lys
Arginine (R) Arg
Basic side chains
Aspartate (D) Asp
Glutamate (E) Glu
Acidic side chains

Every amino acid except glycine exists in two forms. These forms do not differ in their molecular formula or in the way that their atoms are bonded to one another. Rather, they differ in the spatial arrangement of their atoms. **Figure 3.4** provides an example of this. Note that the two forms of alanine in the figure are mirror images of one another, just as your left and right hands are mirror images of each other. Like your left and right hands, the left-handed and right-handed forms of alanine cannot be exactly superimposed. They also do not have a plane of symmetry. Molecules that possess this type of "handedness" are said to be **chiral**. The left- and right-handed forms of chiral molecules are called **enantiomers**; they are mirror images.

Chirality is an important issue in chemical evolution for a simple reason. Although two forms of every amino acid except glycine exist in nature, only the left-handed configuration occurs in organisms. If the Oparin-Haldane theory is correct, a mechanism must be found to explain why only one form emerged during chemical evolution. To date, the problem is still unresolved.

Sources of Amino Acids Proponents of the theory of chemical evolution are increasingly confident that amino acids were abundant in the prebiotic soup. This confidence is based partly on the outcomes of experiments like those performed by Miller and partly because amino acids could have rained onto ancient Earth from outer space. This latter hypothesis is not as far-fetched as it sounds. The key reactants in amino acid synthesis are hydrogen cyanide (HCN) and the class of carbon-containing compounds called aldehydes. These molecules are routinely found in interstellar dust and on celestial bodies such as the satellites of Jupiter. Further, a total of 18 different amino acids have been isolated from a meteorite that landed near Murchison, Australia, on September 28, 1969. The amino acids produced in Miller's apparatus are strikingly similar, in identity and in relative quantity, to the amino acids found in the Murchison meteorite.

In sum, amino acids could have formed in the ancient oceans or splashed in after meteorite impacts during the early bombardment phase of the planet's formation. The origin of amino acids is no longer considered a problem in research on chemical evolution. The origin of sugars is.

TABLE 3.1 How Amino Acids Interact with Water

In this table, the 20 amino acids are ranked according to how likely they are to interact with water. The ranking is from least likely to most likely.

Amino acid	Interaction with water
Highly hydrophobic	Least likely
Isoleucine	
Valine	
Leucine	
Phenylalanine	
Methionine	
Less hydrophobic	
Alanine	
Glycine	
Cysteine	
Tryptophan	
Tyrosine	
Proline	
Threonine	
Serine	
Highly hydrophilic	
Histidine	
Glutamate	
Asparagine	
Glutamine	
Aspartate	
Lysine	
Arginine	Most likely

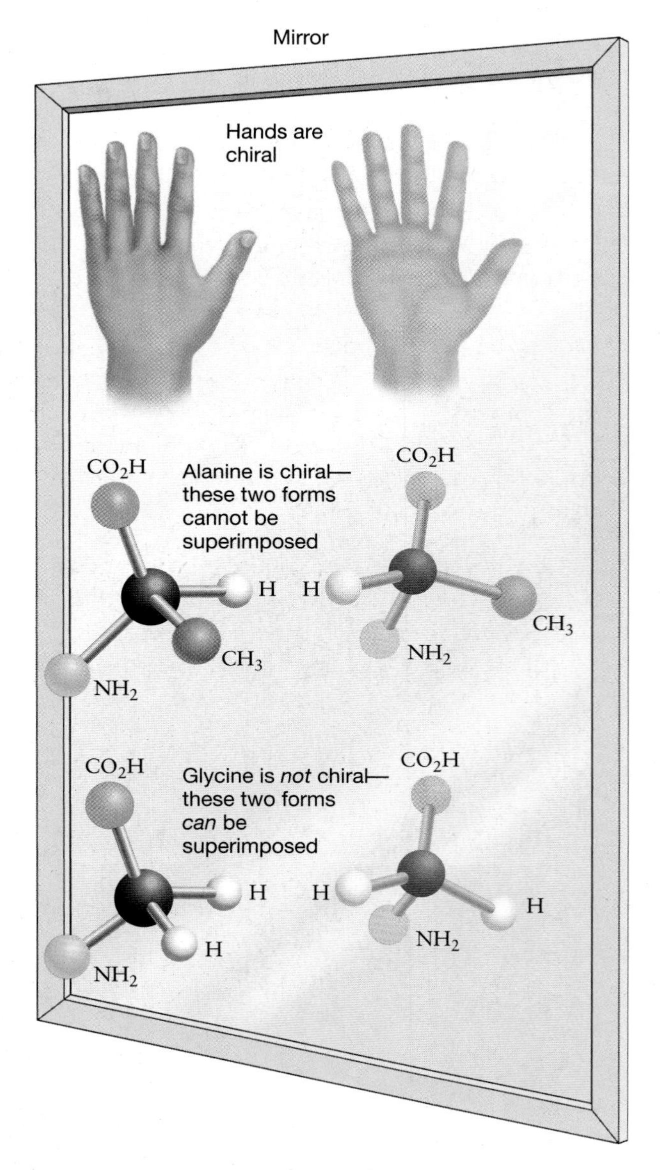

FIGURE 3.4 Most Amino Acids Are Chiral
Alanine comes in two forms that are enantiomers, meaning that they are mirror images of one another (they cannot be exactly superimposed). The glycine molecules shown here *can* be exactly superimposed, so they are not enantiomers. The 18 other amino acids also have mirror-image forms.

Sugars

In organisms today, the reduced organic compounds called sugars are an important source of energy. Later in this text, two chapters explore how organisms make sugars and then use the chemical energy stored in them. But sugars are also important building blocks for macromolecules. In this role, these molecules were key players in chemical evolution.

Sugars are organic molecules that have a distinctive set of functional groups: a carbonyl group and several hydroxyl (OH) groups. **Figure 3.5** shows the structure of a simple sugar. When sugars consist of a single subunit they are called a **monosaccharide** (one-sugar). A variety of monosaccharides exist; they can be linked together as subunits to form an enormous variety of **polysaccharides** (many-sugars).

Researchers studying the origins of life are particularly interested in the five-carbon monosaccharide called ribose. Ribose is a major component of the nucleic acid called RNA, and a related sugar called deoxyribose is found in the nucleic acid called DNA. In living cells, these nucleic acids are responsible for storing and transmitting the information required for making proteins. The theory of chemical evolution suggests that the first living entity may have consisted of RNA or DNA.

Laboratory simulations have shown that many sugars can be synthesized readily under conditions that mimic the prebiotic soup. Specifically, when formaldehyde (H_2CO) molecules are heated in solution, they react with one another to form almost all of the monosaccharides that have five or six carbons (these are called pentoses and hexoses). Ironically, the ease of forming these sugars creates two problems:

- *The various pentoses and hexoses are produced in approximately equal amounts.* But researchers suggest that ribose had to be the *predominant* sugar for RNA molecules to form in the prebiotic soup. (Why this is so will become clear later.)

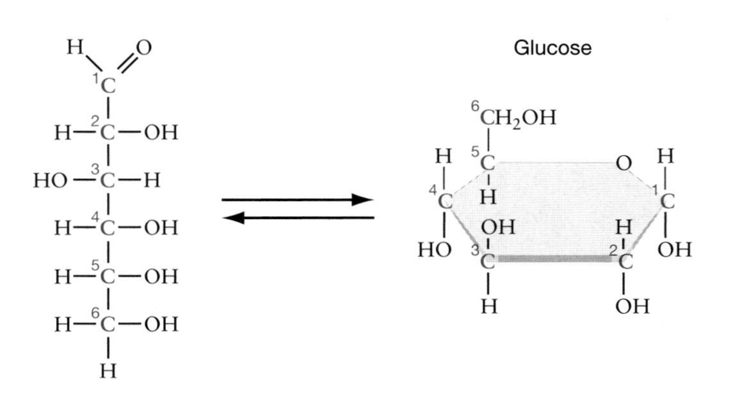

FIGURE 3.5 The Structure of Simple Sugars
This diagram shows the structure of glucose, the most common sugar found in humans and other vertebrates. Monosaccharides can exist in both linear and ring forms, but the ring form is much more common. Note that the carbons in sugars are numbered in a precise way.
QUESTION Compared to carbon dioxide, are sugars reduced or oxidized molecules?

- *Sugars, like amino acids, are chiral molecules—they come in left-handed and right-handed forms.* In the sugars found in organisms, right-handed forms predominate. Why? A successful theory for the origin of life should include a hypothesis to explain how one enantiomer came to be favored over the other.

Origin-of-life researchers refer to these two questions as "the ribose problem." As this book goes to press, these questions are still unanswered. The ribose problem is a serious challenge to the Oparin-Haldane theory. Research on how the right-handed enantiomer of ribose could have formed in the prebiotic soup continues.

Nucleotides

The nucleic acids called RNA and DNA are made up of subunits called nucleotides, just as proteins are made up of subunits called amino acids and polysaccharides are made up of subunits called monosaccharides.

Nucleotides are the most complex molecules considered thus far in the chapter. **Figure 3.6a** (page 48) diagrams the three distinct components that make up a **nucleotide**: a phosphate group, a 5-carbon sugar, and a nitrogen-containing (or "nitrogenous") base. The phosphate is bonded to the sugar molecule, which in turn is bonded to the nitrogenous base.

Although a wide variety of nucleotides are found in living cells, origin-of-life researchers concentrate on two types: the ribonucleotides and deoxyribonucleotides. In ribonucleotides, the pentose is ribose; in deoxyribonucleotides, it is deoxyribose (*deoxy-* refers to "lacking oxygen"). As **Figure 3.6b** shows, these two sugars differ by a single atom. Ribose has an –OH group bonded to the second carbon in the ring; deoxyribose has an H at the same location.

Four different ribonucleotides are found in organisms today. Each contains a different nitrogenous base. These bases, diagrammed in **Figure 3.6c**, belong to structural groups called purines and pyrimidines. The purines found in ribonucleotides are called adenine and guanine; the pyrimidines are cytosine and uracil. Chemists and biologists routinely abbreviate adenine and guanine to A and G, respectively; cytosine and uracil are abbreviated to C and U, respectively.

Similarly, four different deoxyribonucleotides are found in cells today and are distinguished by the structure of their nitrogenous base. Adenine, guanine, and cytosine are also found in deoxyribonucleotides. But instead of uracil, a closely related pyrimidine called thymine occurs in deoxyribonucleotides (**Figure 3.6c**). Thymine is frequently abbreviated as T.

To explain how these complex molecules came to populate the prebiotic soup, researchers need to understand the origin of all three components and how they bonded together to form the first nucleotides. This section reviewed why the formation of ribose is a daunting problem, but the origin of the pyrimidine bases is equally challenging. Simply put, origin-of-life

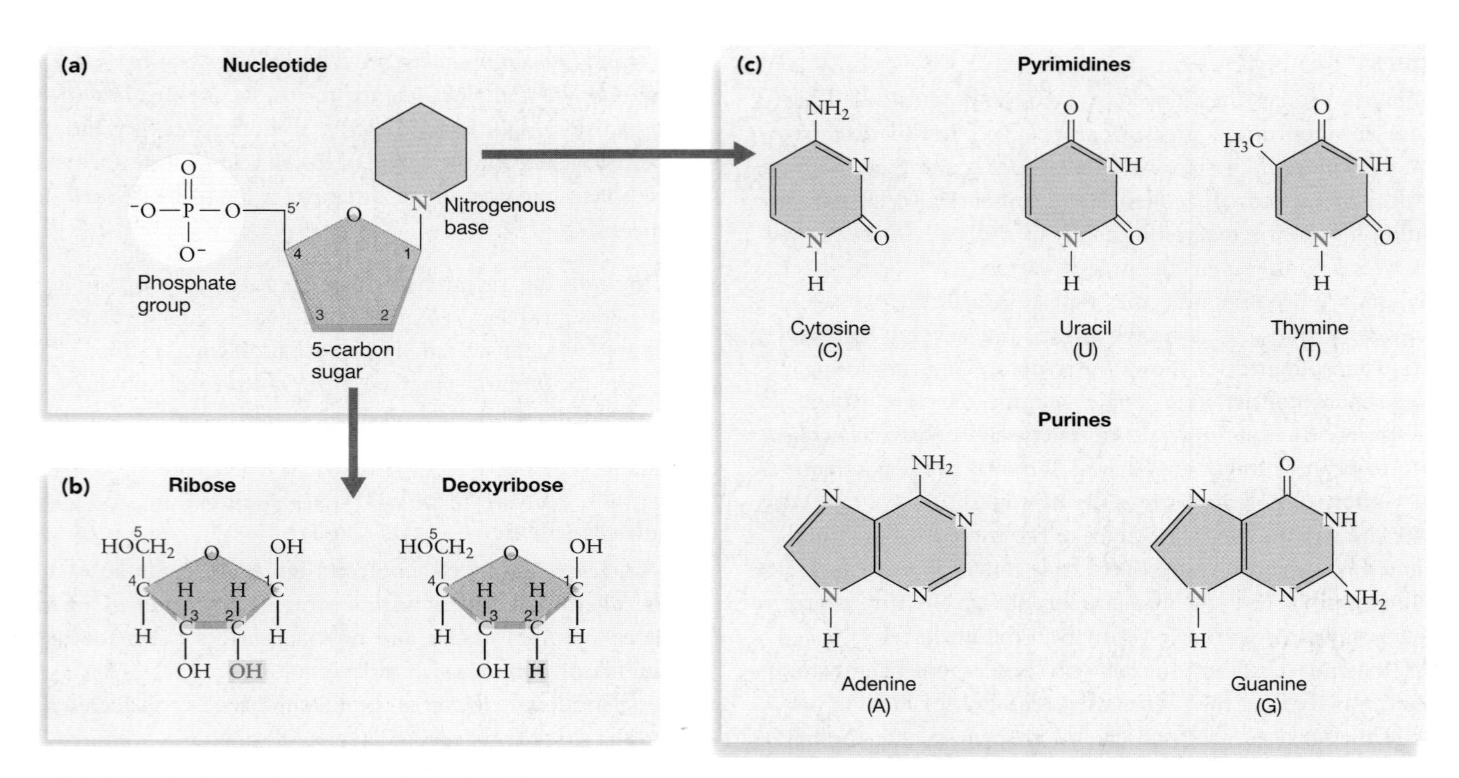

FIGURE 3.6 The General Structure of a Nucleotide
(a) This sketch shows the relationship between the phosphate group, the sugar, and the nitrogenous base found in a nucleotide. The numbers indicate the positions of the five carbons in the ring. Note that the nitrogenous base is bonded to carbon number 1 in the ring, while the phosphate is bonded to carbon number 5. The bond between the phosphate group and the sugar is called a 5′ linkage; the "prime" symbol indicates that the carbon being referred to is part of the sugar and not the attached nitrogenous base. Also notice that while hydrogen atoms are bonded to the carbon atoms in the ring (see part b), biologists routinely omit them to make the diagrams less cluttered. **(b)** Ribose and deoxyribose are similar sugars that are found in nucleotides. **(c)** Purines and pyrimidines are nitrogen-containing bases. A C–N bond links them to the sugar in a nucleotide. This bond forms at the nitrogen atom that is highlighted on each base. Note that purines are substantially larger molecules than pyrimidines.

researchers have yet to discover a plausible mechanism for the synthesis of cytosine, uracil, and thymine molecules prior to the origin of life. Purines, in contrast, are readily synthesized by reactions among HCN (hydrogen cyanide) molecules. Stanley Miller, for example, has found adenine and guanine in the solutions recovered after spark-discharge experiments.

With that introduction, you have now arrived at the frontier of research on early chemical evolution. To be satisfied that the second step in Oparin and Haldane's scenario occurred, biologists must still explain the origin of two key molecules: ribose and the pyrimidine bases.

3.3 The First Macromolecules

Oparin and Haldane's theory states that amino acids and nucleotides in the prebiotic soup became linked together to form proteins and nucleic acids, respectively. When a molecular subunit, or a **monomer** (one-part), bonds to other subunits to form a macromolecule, the process is called **polymerization** (Figure 3.7). A macromolecule that is made up of linked monomers is called a **polymer** (many-parts). Amino acids polymerize to form proteins; nucleotides polymerize to form nucleic acids.

What Oparin and Haldane did not address, however, is *how* polymerization reactions could have occurred prior to the origin of life. Solutions of amino acids do not spontaneously self-assemble into proteins. Nor do nucleotides spontaneously polymerize to form nucleic acids. This is not surprising, given the theory developed in Chapter 2. Complex and highly organized molecules are not expected to form spontaneously from simpler constituents. This is because polymerization organizes the molecules involved into a given structure and decreases their disorder, or entropy. In addition, polymers are energetically much less stable than their component monomers. As a result, polymerization reactions are nonspontaneous. Monomers must absorb energy in order to link together.

Could polymerization occur in the energy-rich environment of early Earth? When researchers have added heat or electrical discharges to solutions of either amino acids or nucleic acids, they

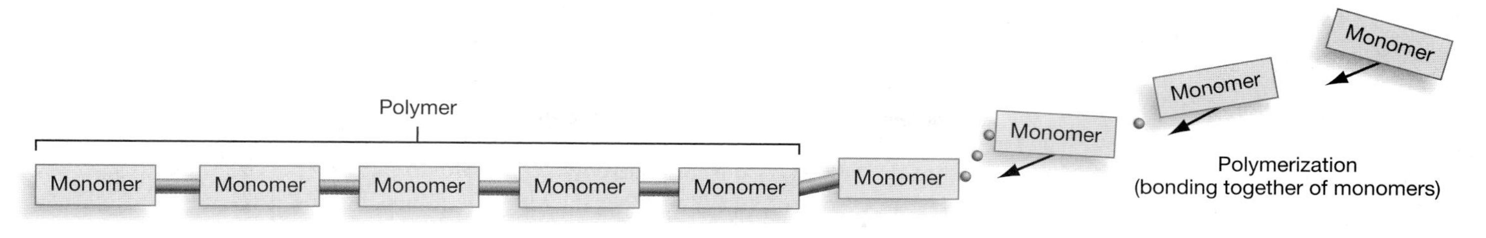

FIGURE 3.7 Monomers and Polymers
In this generalized diagram, the monomers could be amino acids, sugars, or nucleotides.

find that polymerization reactions proceed slowly, if at all. This is because monomers polymerize through **condensation reactions** or **dehydration reactions.** These reactions are aptly named, because the newly formed bond results in the loss of a water molecule (**Figure 3.8a**). The reverse reaction, called **hydrolysis,** breaks polymers apart by adding a water molecule (**Figure 3.8b**). The water molecule reacts with the bond linking the monomers together, separating a monomer from the polymer chain.

In a solution, such as the prebiotic soup, condensation and hydrolysis represent the forward and reverse reactions of a chemical equilibrium. Hydrolysis dominates because it increases the entropy of the molecules involved and because it is energetically favorable—it lowers the potential energy of the electrons involved.

This section has two objectives: (1) to explore how polymerization reactions might have occurred in the prebiotic soup, and (2) to review the structure and behavior of three key macromolecules—proteins, RNA, and DNA—that result from polymerization reactions. The long-term goal is to evaluate which of these macromolecules may have formed the first living entity, capable of self-replication.

Building Macromolecules: How Could Polymerization Reactions Occur on Ancient Earth?

Some recent experiments by James Ferris and colleagues suggest that the key to polymerizing macromolecules in the prebiotic soup was, quite literally, as common as mud. These researchers were able to create polymers by incubating monomers with clay-sized mineral particles. Just as the investigators predicted, the growing macromolecules were protected from hydrolysis because they clung, or adsorbed, to the mineral surfaces.

Ferris and co-workers designed their experiments to simulate events that could have occurred in the prebiotic soup. In one experiment, they put nucleotides in a solution with tiny mineral particles and allowed them to react for a day. (A **solution** is a homogeneous mixture of one or more substances dissolved in a liquid.) Then they separated the solid particles from the solution and added a fresh batch of nucleotides to the particles, along with a source of energy. They repeated this reaction-separation-reaction sequence for a total of 14 days, or fourteen additions of fresh nucleotides. Their idea was to simulate chemical evolution in an energy-rich, shallow-water environment, where rocks were periodically washed with seawater containing nucleotides. At the end of the two-week experiment, the group analyzed the mineral particles and found macromolecules up to 40 nucleotides long. (**Box 3.2** on page 50 introduces the technique they used to examine these polymers.) In similar experiments using amino acids, the researchers were able to produce polymers up to 55 amino acids long. Based on these results, Ferris and co-workers propose that at least some muddy tide pools and beaches became covered with macromolecules early in Earth history.

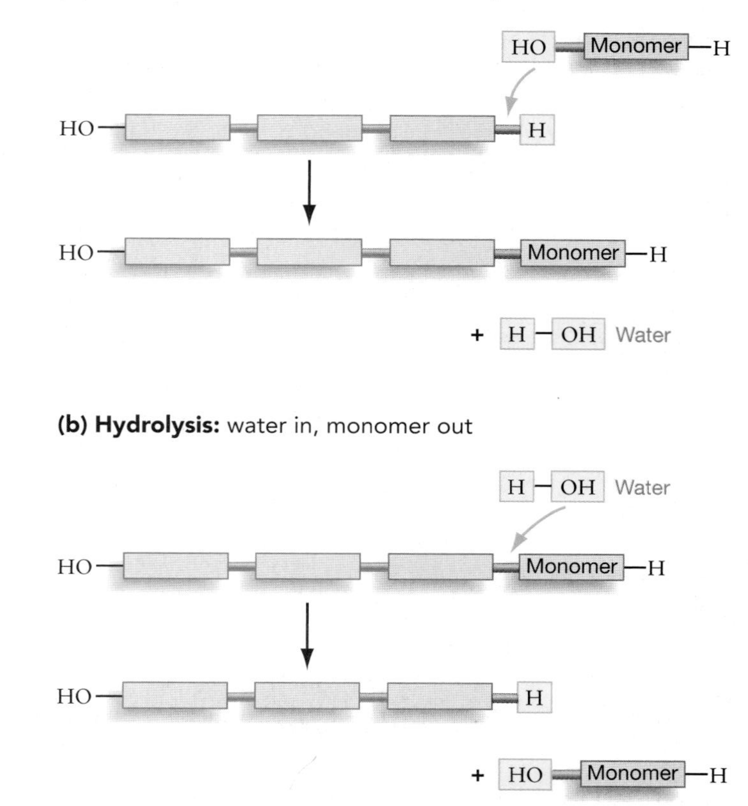

FIGURE 3.8 Polymerization Reactions
(a) In a condensation reaction, a monomer is added to a polymer to make a longer polymer. The new bond that forms results in the formation of a water molecule. **(b)** In hydrolysis, a water molecule reacts with the bond linking the monomers. A monomer is broken off the chain, resulting in a shorter polymer.

CD ACTIVITY 3.1 Condensation and Hydrolysis Reactions

BOX 3.2 Gel Electrophoresis and Autoradiography

In molecular biology, the standard technique for separating and analyzing macromolecules is called gel electrophoresis or, simply, electrophoresis. This is the technique that James Ferris and co-workers used to determine the size of the macromolecules in their experiment.

The principle behind the procedure is fairly simple. Both proteins and nucleic acids carry an electrical charge. As a result, the molecules move when placed in an electric field. Negative ions migrate toward the positive end of the field, and positive ions move toward the negative end. Researchers begin by adding a sample onto a gelatinous substance appropriately called a gel. Then they apply an electrical field across the gel. This causes the molecules to move through the gel toward an electrode. As they move, the molecules separate by size and by charge. Molecules that are smaller or more highly charged for their size move faster than larger or less highly charged molecules.

Figure 1 shows the type of electrophoresis setup used by Ferris and his colleagues. Step 1 in the figure shows how they loaded different samples of macromolecules, taken from different days during their polymerization experiment, into "wells" at the top of the gel slab. In step 2 they immersed the gel in a solution that conducts electricity and then applied a voltage across the gel. After the samples had run down the gel for a time (step 3 in the figure), they removed the electric field. Next the separated molecules had to be detected in some way. Often, proteins or nucleic acids can be stained or dyed. In this case, however, the researchers had attached a radioactive atom to the monomers used in the experiment, so the polymers that resulted could be visualized by laying x-ray film over the gel. The radioactive emissions expose the film, resulting in a black dot wherever a radioactive atom is located in the gel. This technique for visualizing macromolecules is called autoradiography.

The resulting picture of the gel shows the different bands of molecules (see step 4). The samples, taken on the 2nd, 4th, 6th, 8th, and 14th day of the experiment, are labeled along the top of this step. The far right lane contains macromolecules of known size called a size standard or "ladder," which is used to size the molecules in the experimental samples. The bands that appear in each sample lane represent the different polymers that had formed. Note that darker bands indicate the presence of many molecules and thus contain more radioactive marker; lighter bands contain fewer molecules.

Several conclusions can be drawn from these data. First, a variety of polymers formed at each stage. After the second day, for example, polymers from 12 to 18 monomers long had formed. Second, the overall length of polymers produced increased as the experiment continued. At the end of the 14th day, most of the polymers were between 20 and 40 monomers long. Ferris and colleagues concluded that the polymers were protected on the mineral surface and became longer as time passed. Based on these data, they claim that mineral surfaces were an important site of chemical evolution.

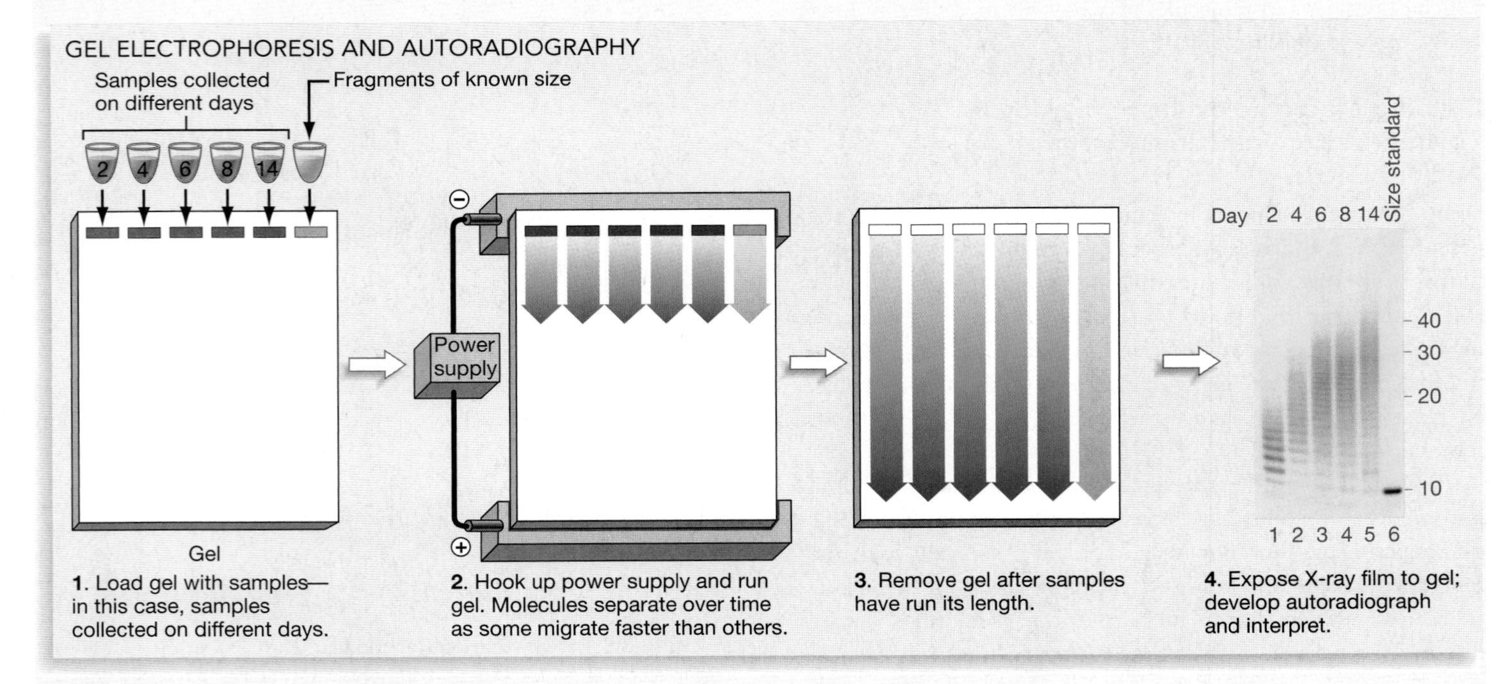

FIGURE 1 Running a Gel and Visualizing the Results
Gel electrophoresis is a technique for separating macromolecules; autoradiography is a method for visualizing them.

With this result in mind, let's take a closer look at the structure and behavior of macromolecules that could have populated these habitats: proteins, ribonucleic acid (RNA), and deoxyribonucleic acid (DNA). Which of these molecules was most likely to form the first self-replicating entity?

Proteins

Amino acids polymerize when a bond forms between the carboxyl group on one molecule and the amino group on a second molecule (**Figure 3.9a**). The C–N bond that results from this condensation reaction is called a **peptide bond**. When a series of amino acids are linked by peptide bonds into a chain, the amino acids are referred to as residues and the resulting molecule is called a **polypeptide.** Single polypeptides or several polypeptides that interact to form still larger molecules are known as **proteins.** Proteins are large polypeptides.

Figure 3.9b shows how the chain of peptide bonds in a polypeptide gives the molecule a structural framework, or a "backbone." Three points are important to note about this backbone. It is flexible, it has polarity, and the functional groups present in each residue extend out from it. The structure is flexible because the bonds on either side of the peptide linkage can rotate. The polarity exists because peptide bonds are asymmetrical: There is an amino group ($-NH_2$) on one side of every peptide bond and a carboxyl group ($-COOH$) on the other. By convention, biologists always write amino acid sequences in the same direction. The end of the sequence that has the free amino group is placed on the left and is called the N-terminus, and the end with the free carboxyl group appears on the right-hand side of the sequence and is called the C-terminus. The amino acids in the chain are always numbered starting with the N-terminal end (**Figure 3.9c**).

Biochemists call the unique sequence of amino acids in a protein the **primary structure** of that protein. With 20 amino acids available and the size of proteins varying from two amino acid residues to tens of thousands, the number of primary structures that are possible is practically limitless. (There are 20^n different polypeptides of length n.)

A protein's primary structure is important to its function. In some cases, even a single change in the sequence of amino acids can cause radical changes in how the molecule functions. As an example, consider the protein found in humans called hemoglobin. In some individuals, hemoglobin has a valine instead of a glutamic acid at the amino acid numbered 6 in the chain of 246 amino acids. People with this single change suffer from the debilitating disease called sickle-cell anemia.

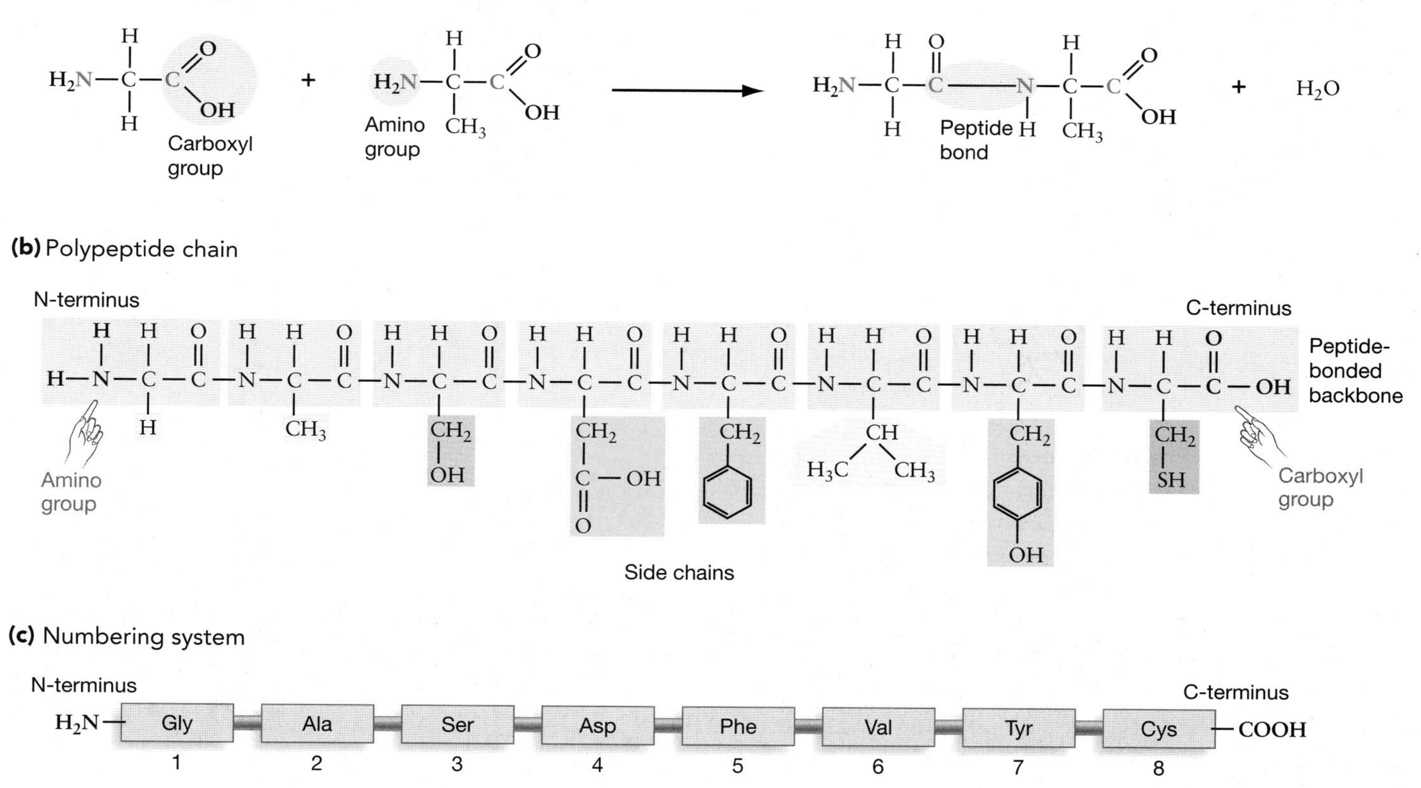

FIGURE 3.9 Amino Acids Polymerize to Form Polypeptides
(a) When the carboxyl group on one amino acid reacts with the amino group on a second amino acid, a peptide bond forms. **(b)** Amino acids can be linked into long chains by peptide bonds. **(c)** The sequence of amino acids in a polypeptide chain is numbered from the N-terminus (or amino-terminus) to the C-terminus (or carboxy-terminus).

In terms of generating diversity among proteins, however, variation in amino acid sequence is only the tip of the iceberg. When proteins are in aqueous solution, they fold in a way that places the hydrophobic side chains of the amino acids away from water and the hydrophilic side chains toward water. Because folding stabilizes the molecule, it releases free energy. As a result, folding occurs spontaneously. Folding gives a protein a complex, three-dimensional shape and makes additional levels of structure possible.

The **secondary structure** of proteins is created by hydrogen bonding between peptide groups—specifically between the carboxyl oxygen of one residue and the hydrogen on the amino group of another (**Figure 3.10a**). As **Figure 3.10b** shows, these hydrogen bonds form in distinct ways. Hydrogen bonds can cause the polypeptide's backbone to coil into a structure called an α-helix (alpha helix); or they can link two adjacent segments of a peptide chain together, causing the sequence to fold into a flattened, planar structure called a β-pleated sheet (beta-pleated sheet).

Alpha helices and β-pleated sheets give particular sections of proteins a distinct shape and structure. But interactions between R-groups—or between R-groups and the peptide backbone—also influence the overall three-dimensional form of a protein, or what biochemists call its **tertiary structure**. **Figure 3.11a** illustrates the variety of bonds and interactions that lock folds into place and give each protein a distinctive three-dimensional shape, or tertiary structure.

With so many interactions possible between side chains and peptide-bonded backbones, it's not surprising that polypeptides can vary in shape from rod-like filaments to globular masses (**Figure 3.11b**); later in the chapter you'll explore how the shapes of proteins affect their function. The correlation between the structure and function of proteins is a theme that will resonate throughout the text. In addition, some proteins contain several distinctive polypeptide subunits. This combination of subunits gives proteins a **quaternary structure**. Table 3.2 (page 54) summarizes the four levels of protein structure.

Polypeptides and the Prebiotic Soup The experiments by Ferris and colleagues suggest that reasonably large polypeptides could have been synthesized prior to the origin of life. As a result, it seems entirely plausible that the prebiotic soup contained a variety of proteins, which would have differed in size, shape, and composition. The question now becomes, could one of these polypeptides have become the self-replicating molecule? The answer to this question requires a closer look at the chemical properties required for a molecule to copy itself.

An Introduction to Catalysis The essence of chemical evolution is the production of larger and more complex molecules from simpler ones. This process was possible because solar, electrical, and heat energy are capable of driving nonspontaneous reactions. However, we have only begun to consider a critical factor in chemical evolution: reaction rates.

Even spontaneous chemical reactions are not necessarily fast. The oxidation of iron—the reaction known as rusting—is exothermic and spontaneous, but occurs slowly. For a macromolecule to make a copy of itself, the polymerization reactions involved must not only overcome a free energy barrier, but they must occur rapidly enough to dominate the reverse reaction—hydrolysis.

Reaction rates depend on how the chemical bonds involved are actually broken and made. Chapter 2 introduced the idea

(a) Hydrogen bonds form between peptide chains.

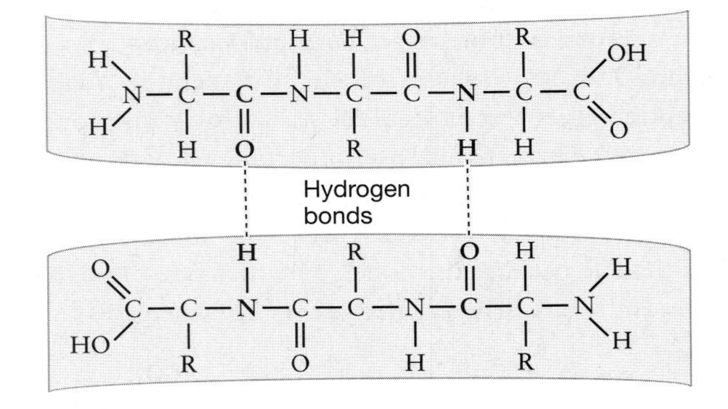

(b) Secondary structures of proteins result.

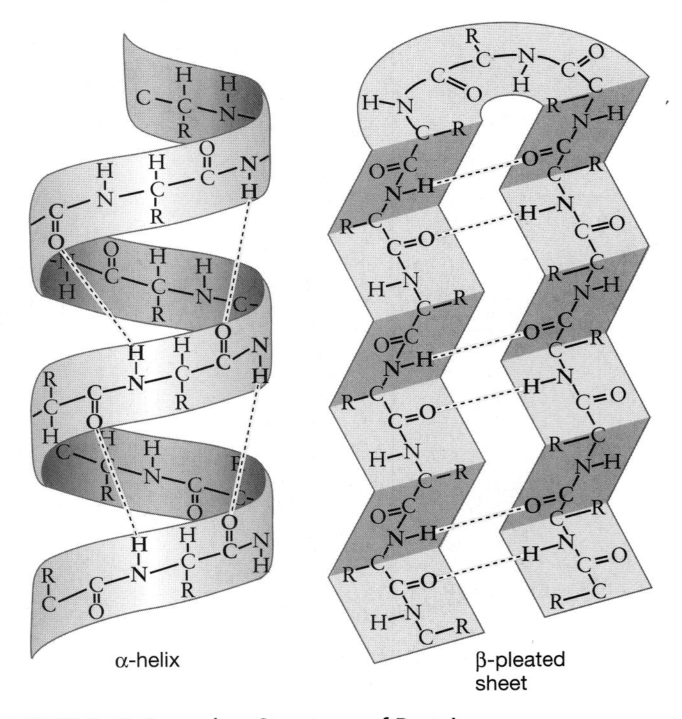

FIGURE 3.10 Secondary Structures of Proteins
(a) The peptide backbone of a protein can fold or coil on itself when hydrogen bonds form between peptide groups. **(b)** These sketches show the detailed structure of the coils called α-helixes and the folds called β-pleated sheets.

that molecules must collide with a specific orientation and a specific amount of energy for old bonds to break and new bonds to form. This is because the electrons involved in the reactions must interact. But these electrons also repel one another as they come into contact. In many cases, the kinetic energy of the collision must be large enough to overcome this repulsion for the reaction to proceed. If it is, then the collision creates a combination of old and new bonds called a **transition state** (see **Figure 3.12** on page 54). The amount of energy required to reach the transition state is called the **activation energy** of the reaction.

Figure 3.12 pulls these ideas together by showing the changes in potential energy that take place during the course of a chemical reaction. In this graph, ΔE indicates the overall change in energy. In this case the products have lower potential energy than the reactants, meaning that the reaction is exothermic. But because the activation energy for this reaction, symbolized by E_a, is high, the reaction may proceed slowly.

Reaction rates, then, depend on both the kinetic energy of the reactants and the activation energy. If the kinetic energy of the participating molecules is high, then collisions are likely to result in completed reactions. (The kinetic energy of molecules, in turn, is a function of their temperature. This is why chemical reactions tend to proceed faster at higher temperature.) But if the activation energy of a particular reaction is high, then collisions are less likely to result in completed reactions.

(a) Interactions that determine the tertiary structure of proteins

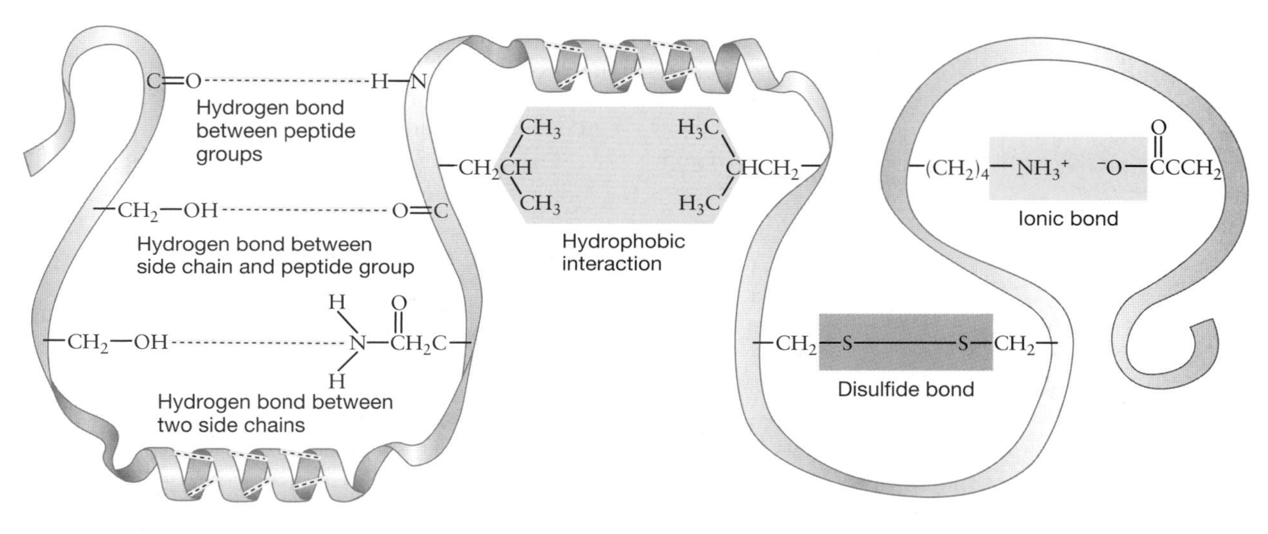

(b) Tertiary structures are diverse.

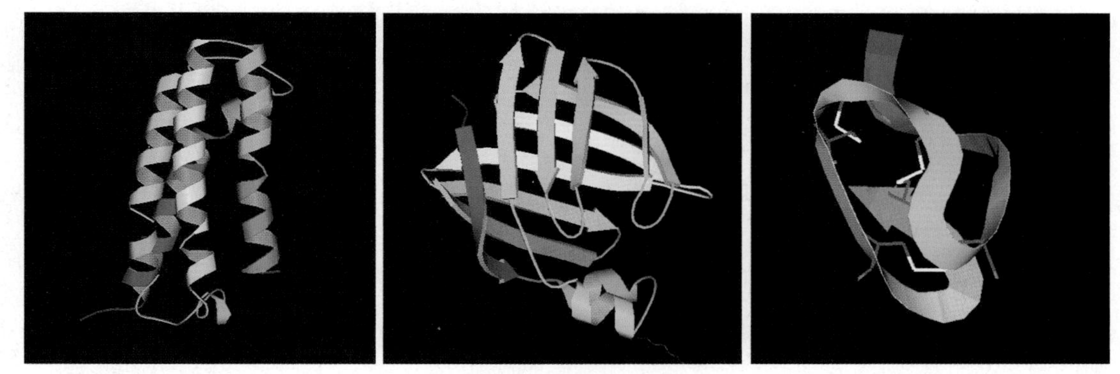

A tertiary structure composed mostly of α-helices

A tertiary structure composed mostly of β-pleated sheets

A tertiary structure rich in disulfide bonds

FIGURE 3.11 Tertiary Structure of Proteins
(a) An array of bonds and other interactions cause the peptide-bonded backbone of proteins to fold and bend in a precise way. The folds, in turn, give each protein the unique overall shape called its tertiary structure. Note that the length of the bonds illustrated here is exaggerated for clarity. **(b)** Proteins are diverse in shape. In these diagrams, the polypeptide chains are color-coded so that you can follow the chain from one end (red) to the other (dark blue).

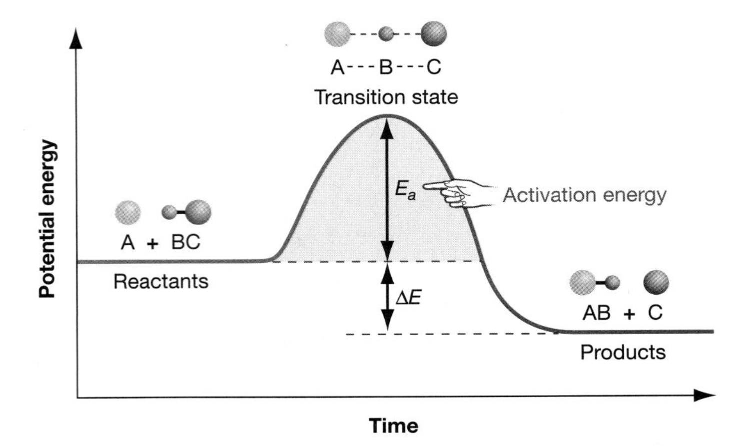

FIGURE 3.12 Changes in Energy During a Chemical Reaction This graph shows the changes in potential energy that occur over the course of a hypothetical reaction between an atom called A and a molecule containing atoms called B and C. The overall reaction would be written as A + BC $\rightarrow$ AB + C. E_a is the activation energy of the reaction, and ΔE is the overall change in potential energy. ΔE is calculated as the energy of the products minus the energy of the reactants. **EXERCISE** This graph illustrates an exothermic reaction. Draw the same type of graph for an endothermic reaction.

In many cases, however, the electrons in the transition state molecule can be stabilized when they interact with another ion, atom, or molecule. When this occurs, the activation energy required for the reaction drops and the reaction rate increases. A substance that lowers the activation energy of a reaction and increases the rate of the reaction is called a **catalyst.** A catalyst is not changed by a chemical reaction, even though it participates in the reaction. The composition of a catalyst is exactly the same after the reaction as it was before.

Figure 3.13 diagrams how catalysts lower the activation energy for a reaction by lowering the potential energy of the transition state. Note that the presence of a catalyst does not change the energy of the reactants or the products—only the transition state.

These details are important for two reasons: A self-replicating molecule would need to act as a catalyst during the assembly and polymerization of its copy, and proteins are the most efficient catalysts known. A protein that catalyzes a chemical reaction is called an **enzyme. Box 3.3** on pages 56–57 explains why proteins are so adept at catalysis.

These facts support the hypothesis that the self-replicating molecule was a polypeptide. In their efforts to create a self-replicating molecule, several labs have focused their attention on proteins. Most origin-of-life researchers are skeptical, how-

TABLE 3.2 A Summary of Protein Structure

Level	Description	Stabilized by:	Example: Hemoglobin
Primary	The sequence of amino acids	Peptide bonds	Gly Ser Asp Cys
Secondary	Formation of α-helices and β-pleated sheets	Hydrogen bonding between peptide groups along the peptide backbone	
Tertiary	Overall three-dimensional shape of a polypeptide	Bonds and other interactions between R-groups, or between R-groups and the peptide backbone	
Quaternary	Shape produced by combinations of polypeptides	Bonds and other interactions between R-groups, and between peptide backbones of different polypeptides	

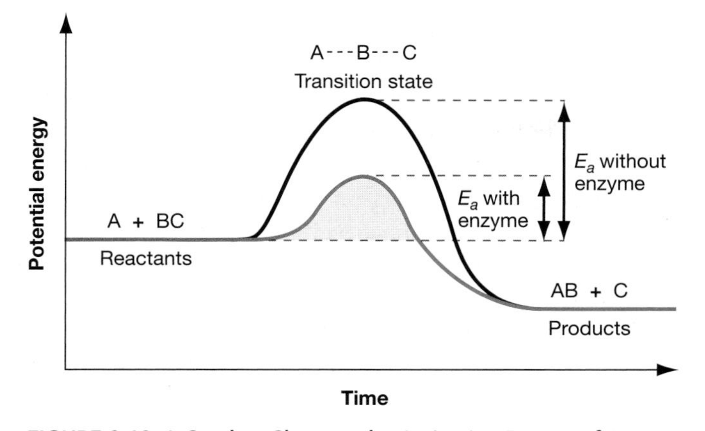

FIGURE 3.13 A Catalyst Changes the Activation Energy of a Reaction
This graph shows the energy profile for the same reaction diagrammed in Figure 3.12 when a catalyst is present. Even though the energy barrier to the reaction, E_a, is much lower, ΔE does not change.
QUESTION Can a catalyst make a nonspontaneous reaction occur spontaneously?

ever. To make a copy of something, a mold or template is required. Proteins cannot furnish this information. In contrast, RNA can.

RNA

Ribonucleic acid, or RNA, is the jackknife of macromolecules. Compared to proteins and DNA, it is exceptionally versatile. The primary function of proteins is to catalyze chemical reactions, and the primary function of DNA is to carry information. But RNA can do both.

RNA forms when ribonucleotides polymerize. As **Figure 3.14a** shows, this polymerization reaction occurs when a bond forms between the phosphate group of one nucleotide and the hydroxyl group on the sugar component of another. The result of this condensation reaction is called a **phosphodiester bond.** The polymer that is produced is called **ribonucleic acid,** or simply **RNA.**

Figure 3.14b shows how the chain of phosphodiester bonds in RNA gives the molecule a backbone, analogous to the peptide-bonded backbone found in proteins. The sugar-phosphate spine of RNA is polar, just like the peptide-bonded spine of polypeptides. The phosphodiester bond links the 5′ carbon on the ribose of one nucleotide to the 3′ carbon on the ribose of another nucleotide. In a strand of RNA, then, one end has an unlinked 5′ carbon while the other end has an unlinked 3′ carbon. By

(a) Formation of phosphodiester bond

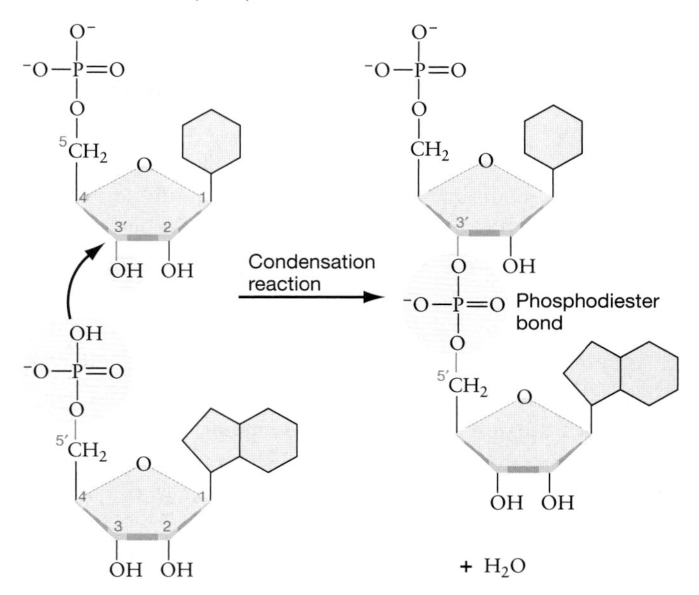

(b) The sugar-phosphate spine of RNA

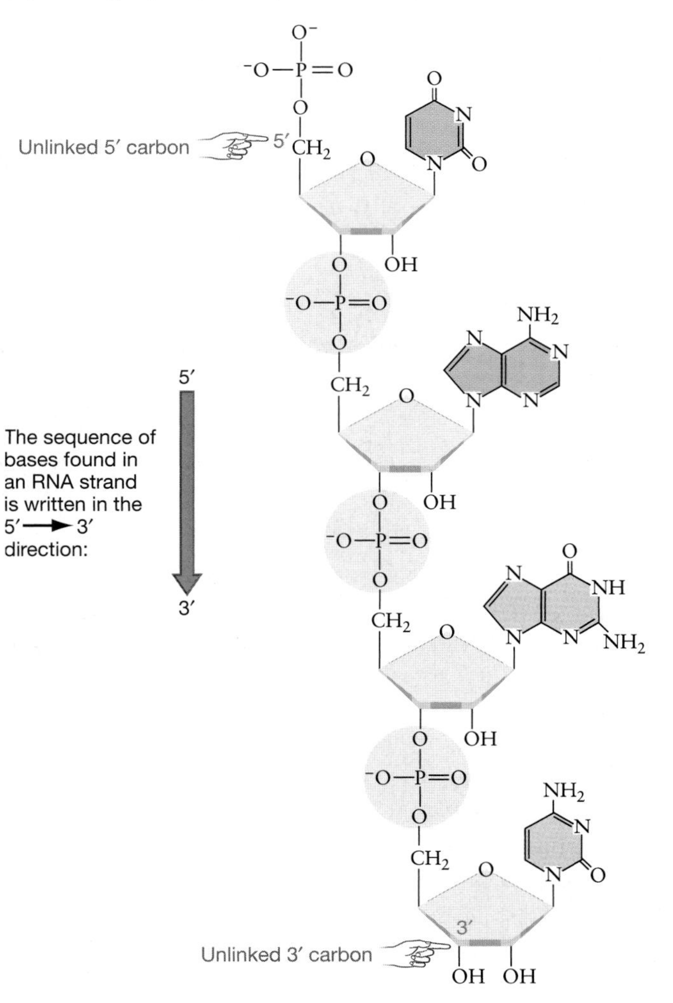

FIGURE 3.14 Ribonucleotides Polymerize to Form Ribonucleic Acid (RNA)
(a) Ribonucleotides can polymerize through condensation reactions. The linkage that results, between the 5′ carbon of one ribonucleotide and the 3′ carbon of another ribonucleotide, is called a phosphodiester linkage or phosphodiester bond. **(b)** This diagram shows the sugar-phosphate backbone of RNA. EXERCISE In part (b), identify the four bases using Figure 3.6c as a key. Then write down the sequence starting at the 5′ end.

convention, the sequence of bases found in an RNA strand is written in the 5′ → 3′ direction. This sequence forms the primary structure of the molecule.

Often, the sugar-phosphate backbone of RNA spontaneously folds into a specific shape determined by hydrogen bonding. This occurs because hydrogen bonds can form between the purine and pyrimidine bases located on the same strand. But as **Figure 3.15a** on page 58 shows, each of the purines can form a hydrogen bond with only one particular pyrimidine, and vice versa. Adenine forms hydrogen bonds only with uracil, and guanine forms hydrogen bonds with cytosine. (Guanine can also bond with uracil, but it does so much less effectively.) When A's and G's on one part of an RNA strand fold over and align with U's and C's on another segment, hydrogen bonding results in a stable stem-and-loop configuration. The result is a secondary structure called a **hairpin** (**Figure 3.15b**).

RNA as an Information-Containing Molecule The specificity of hydrogen bonding between purines and pyrimidines is called **complementary base pairing.** It is responsible for a completely new and revolutionary aspect of chemical evolution: the possibility that a molecule can be copied. This possibility exists because the RNA molecule itself can serve as the source of information for making a copy of itself. To see how this works, examine **Figure 3.16** (pages 58–59). In step 1 of this diagram, free ribonucleotides pair with complementary bases on a strand of RNA. In step 2, the ribonucleotides that have formed hydrogen bonds to the original strand—also called a template strand—polymerize as their sugar-phosphate groups form phosphodiester bonds. Note that the 5′ → 3′ directionality of this new strand is opposite to, or antiparallel to, the template strand. In step 3, the hydrogen bonds between the strands are broken by heating or by a catalyzed reaction. The new RNA molecule now exists independently. In steps 4–6, the pairing, polymerization, and bond-breaking steps are repeated using this new strand as a template. The molecule that results is a copy of the original molecule. The complementary sequence has served as a "mold."

The scenario diagrammed in Figure 3.16 is plausible, but could it actually occur in the prebiotic soup? Note that each of the steps in the diagram is actually a series of chemical reac-

BOX 3.3 How Proteins Catalyze Reactions

What do catalysts do? Catalysts lower the activation energy required for a reaction to take place. It's not difficult to appreciate why this is important. Most of the important reactions in biology do not occur at all, or else proceed at imperceptible rates, without a catalyst. In contrast, it is not unusual for a single enzyme molecule to catalyze over 500,000 reactions *per second.* Most enzymes are also quite specific in their activity—they catalyze a single reaction.

To understand how it is possible to speed up reactions this much, it's necessary to analyze the leading hypothesis for how enzymes work: the lock-and-key model. This hypothesis attempts to explain not only the efficiency of enzymatic reactions, but their specificity as well.

Contemporary versions of the lock-and-key model begin with observations about the tertiary structure of enzymes. Enzymes are huge, globular-shaped proteins. Many have a prominent cleft or cavity in their three-dimensional structure, as shown in **Figure 1a**. The lock-and-key model maintains that the reactant(s), or substrate, molecules fit into this cleft and bind to a specific location called the enzyme's active site. There the substrate molecule(s) are held in place through hydrogen bonding or other electrostatic interactions.

This binding step is one of the major reasons why enzyme catalysis works. Instead of occasionally colliding in a random fashion, enzymes bring reactants together in a precise orientation as illustrated in **Figure 1b**. After binding occurs, the enzyme often changes shape in response. This conformational change, called an induced fit, is noticeable in the protein illustrated in Figure 1a. The movement can strain bonds in the substrate and make it more reactive. In some cases the transition state is stabilized by the enzyme's conformational change, meaning that the transition state becomes a much more favorable configuration, energetically, than it was previously.

In other cases, the R-groups that line the active site interact directly with the bound substrate. Occasionally an R-group will form short-lived covalent bonds with the substrate and actually assist in transferring atoms or groups of atoms from one reactant to another. More commonly, the presence of acidic or basic R-groups allows the reactants to lose or gain a proton more readily.

In sum, then, enzymes speed the rate of reactions by bringing reactants together in the proper orientation and then decreasing the reaction's activation energy. Activation energies drop because enzymes destabilize bonds in the reactant, stabilize the transition state, make acid-base reactions more favorable, and/or change the reaction mechanism through a covalent bonding interaction.

The lock-and-key model has been modified extensively since it was first proposed by Emil Fischer in 1894. (Fischer, for example, thought that proteins were rigid structures, and that shape or "conformational" changes couldn't occur.) Even so, the model has been an enormously successful scientific idea. It not only inspired a large series of experimental tests but also has proved remarkably accurate, at least in broad outline.

(Continued on next page)

tions. Could an RNA molecule itself serve as a catalyst to facilitate these reactions?

RNA as a Catalytic Molecule RNA molecules do not begin to approach proteins in terms of the diversity of their primary structure and three-dimensional structure. The primary structure of RNA molecules is much more restricted because there are only 4 bases in RNA versus the 20 amino acids found in proteins, and RNA molecules cannot form the variety of bonds that give proteins their tertiary structure. Still, RNA molecules have distinct shapes due to their secondary structure, and their base sequences are variable enough to produce molecules with unique chemical behavior. Because of this structural and chemical complexity, RNA should be capable of stabilizing a few transition states and catalyzing at least a limited number of chemical reactions. Indeed, scientists have found that RNA *does* act as a catalyst. Sidney Altman and Thomas Cech shared the 1989 Nobel Prize in chemistry for showing that RNA enzymes, or **ribozymes**, exist in organisms. The ribozymes they isolated from a single-celled organism called *Tetrahymena* could catalyze both the hydrolysis and condensation of phosphodiester bonds.

This discovery was a watershed event in origin-of-life research. Prior to the results published by Altman and Cech, biologists thought that proteins were the only type of molecule capable of catalyzing chemical reactions in living organisms. But if a ribozyme can catalyze the polymerization reactions diagrammed in steps 2 and 5 of Figure 3.16, it raises the possibility that an RNA molecule or molecules could also catalyze the reactions in steps 3 and 6. A molecule capable of doing this could become the self-replicating molecule—the first living entity.

Section 3.4 examines recent experimental evidence that such a ribozyme could exist; but first, it's necessary to examine the structure and properties of DNA. Why haven't researchers proposed that the first self-replicating entity was a DNA molecule?

(Box 3.3 continued)

(a) Specificity between substrate and enzyme

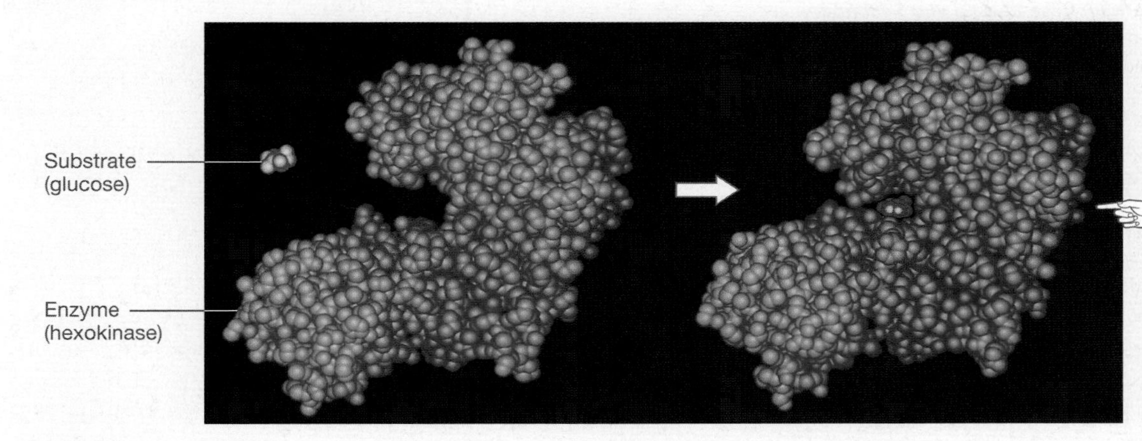

(b) The lock-and-key model of enzyme function

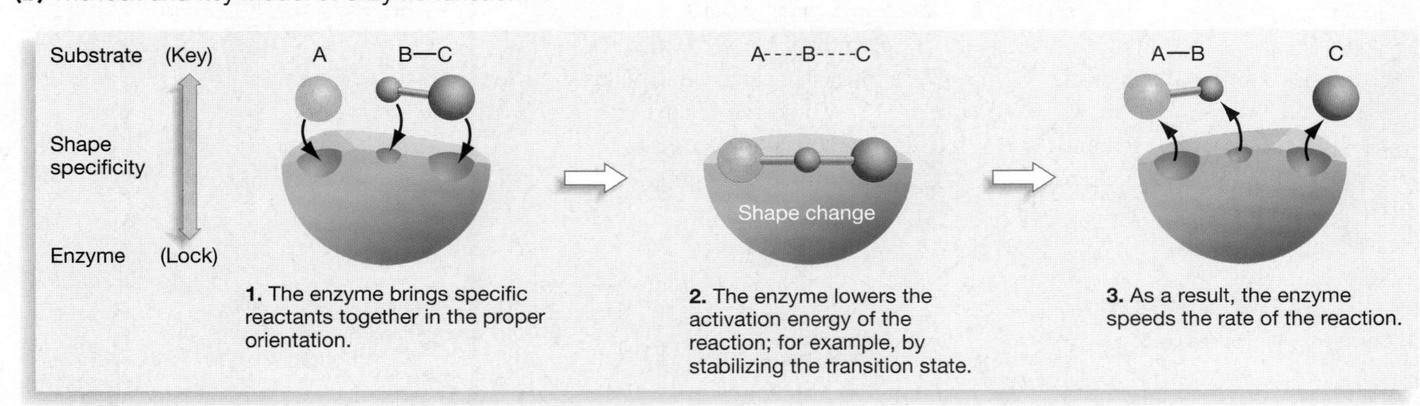

FIGURE 1 Structure and Function in Enzyme Action
The lock-and-key model is a hypothesis to explain how proteins act as efficient and specific catalysts.

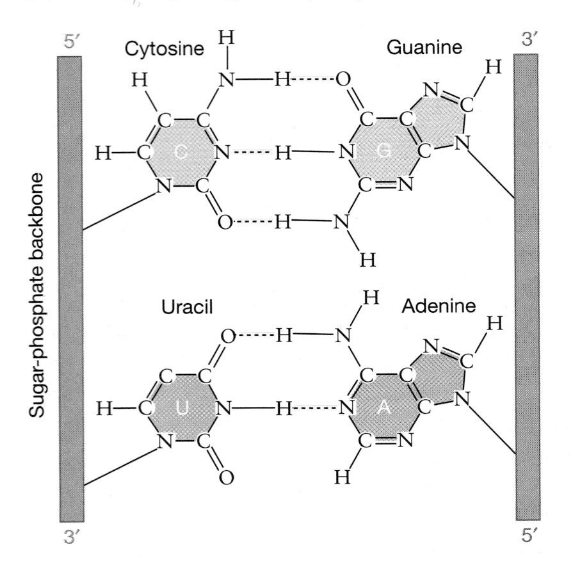

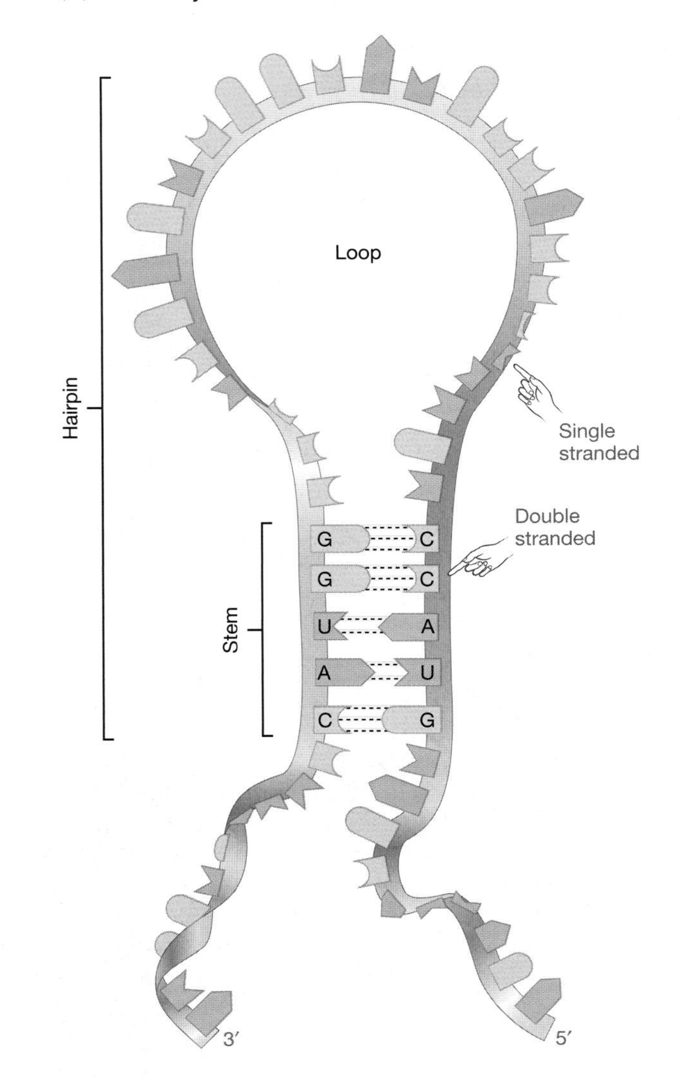

FIGURE 3.15 Complementary Base Pairing and Secondary Structure in RNA: Stem-and-Loop Structures
(a) When uracil and adenine are oriented correctly, two hydrogen bonds (symbolized by dotted lines) form between them. Three hydrogen bonds can form between cytosine and guanine. **(b)** This RNA molecule has secondary structure. The double-stranded "stem" and single-stranded "loop" form a hairpin. Note that the bonded bases in the stem are oriented in the opposite 5′ → 3′ direction. To capture this point, researchers say that they are antiparallel. EXERCISE Compare the hydrogen bonds in part (a). Which bond is stronger, the A–U bond or the C–G bond?

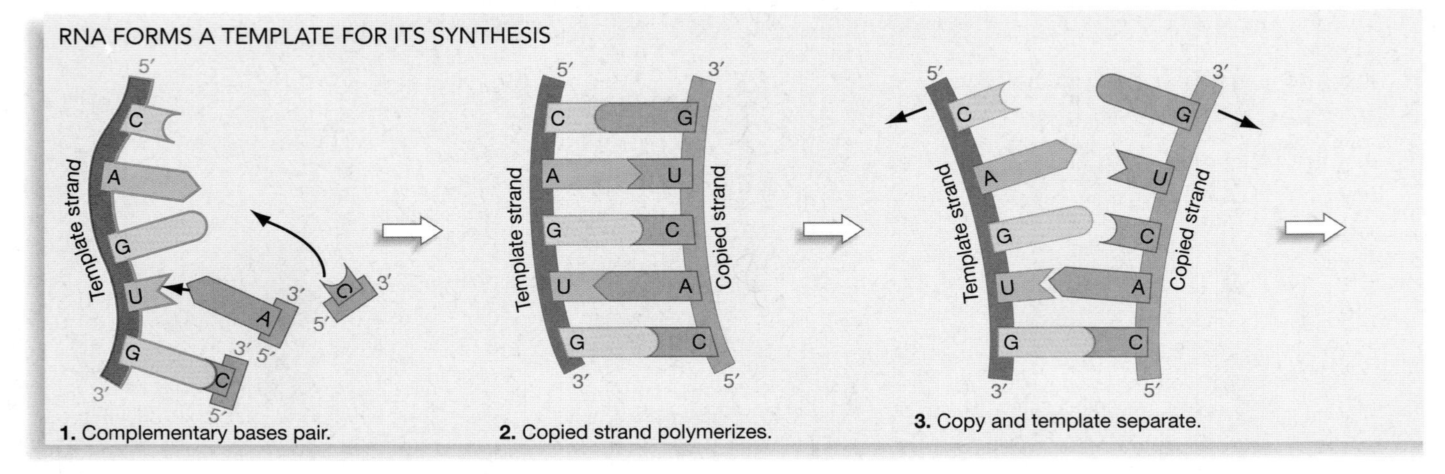

FIGURE 3.16 RNA Molecules Can Be Copied Because They Furnish a Template
This diagram is a hypothesis for how RNA molecules can be copied. The copying process is based on complementary base pairing between ribonucleotides.

DNA

DNA has been called the blueprint of life. In all cells, from bacteria to archaea to eukaryotes, DNA carries the information needed for growth and reproduction. How this information is encoded and translated into action is the primary focus of Chapters 11–15.

Deoxyribonucleotides polymerize in the same way that ribonucleotides do—through the formation of phosphodiester bonds. The polymer that results, called **deoxyribonucleic acid** or **DNA**, has a sugar-phosphate backbone with the same $5' \rightarrow 3'$ directionality as RNA. (As with RNA molecules, biologists always write DNA sequences in the $5' \rightarrow 3'$ direction.)

There are three important differences between RNA and DNA, however:

- The pyrimidine base uracil does not exist in DNA. Instead, DNA contains a closely related pyrimidine base called thymine. As uracil does in RNA, the thymine in DNA forms hydrogen bonds with adenine.
- The sugar in the sugar-phosphate backbone of DNA is deoxyribose, not ribose.
- DNA has a great deal more secondary structure than does RNA, and DNA normally exists as a double-stranded molecule. The purine and pyrimidine bases in DNA undergo hydrogen bonding with complementary bases on a *different* strand, rather than forming hydrogen bonds to complementary bases on the same strand, as occurs with RNA.

The last two differences are especially important. Because the –OH group on the 2′-carbon of ribose is reactive, it tends to participate in chemical reactions that tear RNA polymers apart. The absence of this –OH group makes DNA a much more stable molecule. The extensive secondary structure of DNA has the same effect—it increases the molecule's chemical stability relative to RNA.

The adenine-thymine (A-T) and guanine-cytosine (G-C) pairing between opposing strands of DNA, diagrammed in **Figure 3.17a** on page 60, is analogous to the base pairing found within single strands of RNA. More importantly, the hydrogen bonds between complementary base pairs twist each sugar-phosphate backbone into a spiral, or helix (**Figure 3.17b**). Because two strands are involved, this secondary structure is called a **double helix**. Biochemists like to say that DNA is put together like a ladder whose ends have been twisted in opposite directions. The sugar-phosphate backbone forms the supports, and the base pairs represent the rungs. Finally, the nitrogenous bases orient toward the interior of the DNA helix and stack tightly on top of each other. This packing forms a hydrophobic interior that is difficult to break apart. The sugar-phosphate backbones, which face the exterior of the molecule, are hydrophilic. **Box 3.4** (pages 61–62) introduces one of the great stories of twentieth-century biology: the discovery of the double helix.

In short, DNA is a highly structured, highly stable molecule. Compared to RNA and proteins, its shape is simple and its chemical reactivity extremely limited. For example, DNA has never been observed to catalyze reactions in an organism, even though researchers have been able to construct DNA molecules that catalyze reactions in the laboratory. Even counting the laboratory examples, though, the number of chemical reactions catalyzed by DNA is a small fraction of the reactions catalyzed by RNA and a minute fraction of the reactions catalyzed by proteins.

As a repository of the information needed to make copies of itself, however, DNA exceeds RNA and far exceeds proteins. The DNA double helix is more stable than RNA, and each strand in the DNA duplex can act as a template for the synthesis of its complementary strand. As **Figure 3.18** (page 60) shows, complementary base pairing allows each strand of a DNA double helix to be copied exactly, producing two daughter molecules. Still, it is unlikely that a DNA molecule could copy itself like the first living entity did.

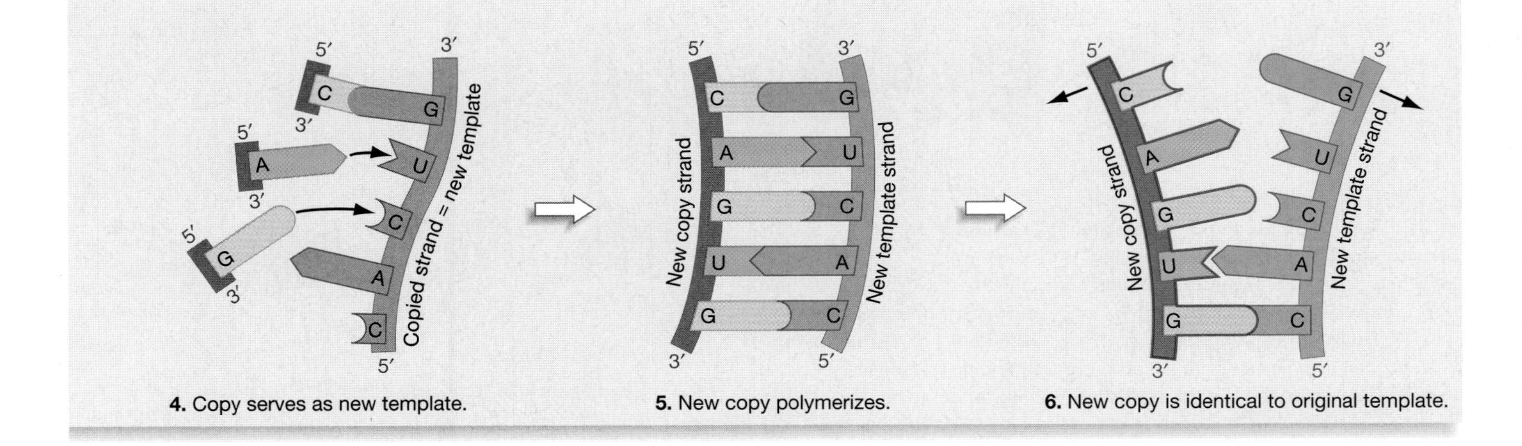

4. Copy serves as new template.

5. New copy polymerizes.

6. New copy is identical to original template.

(a) DNA base pairing

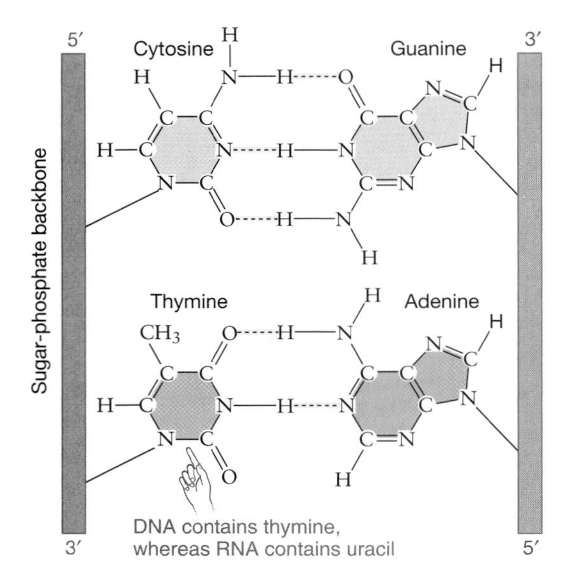

(b) DNA is a double helix.

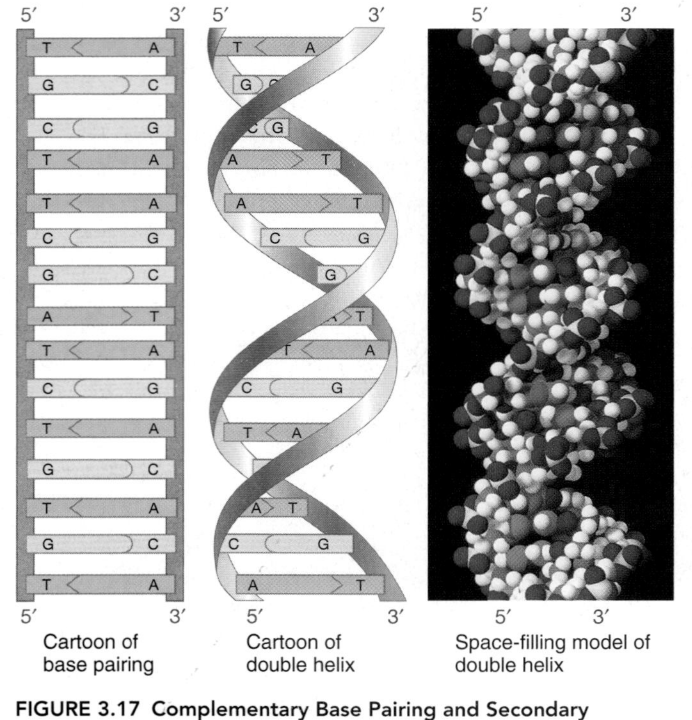

FIGURE 3.17 Complementary Base Pairing and Secondary Structure of DNA: The Double Helix
(a) When deoxyribonucleotides line up correctly, three hydrogen bonds form between each G–C pair. Only two bonds form between each A–T pair. (The analogous situation occurs in RNA.) As a result, G–C pairs in nucleic acids are linked more tightly than A–T pairs. **(b)** Complementary base pairing twists DNA into a double helix.

3.4 The First Living Entity

Oparin and Haldane proposed that chemical evolution led to a self-replicating molecule and the origin of life. To make a copy of itself, the first living molecule had to provide a template to copy. It also had to have the ability to catalyze a polymerization reaction. Proteins are extremely efficient catalysts, and DNA provides an extraordinarily stable template for making copies of

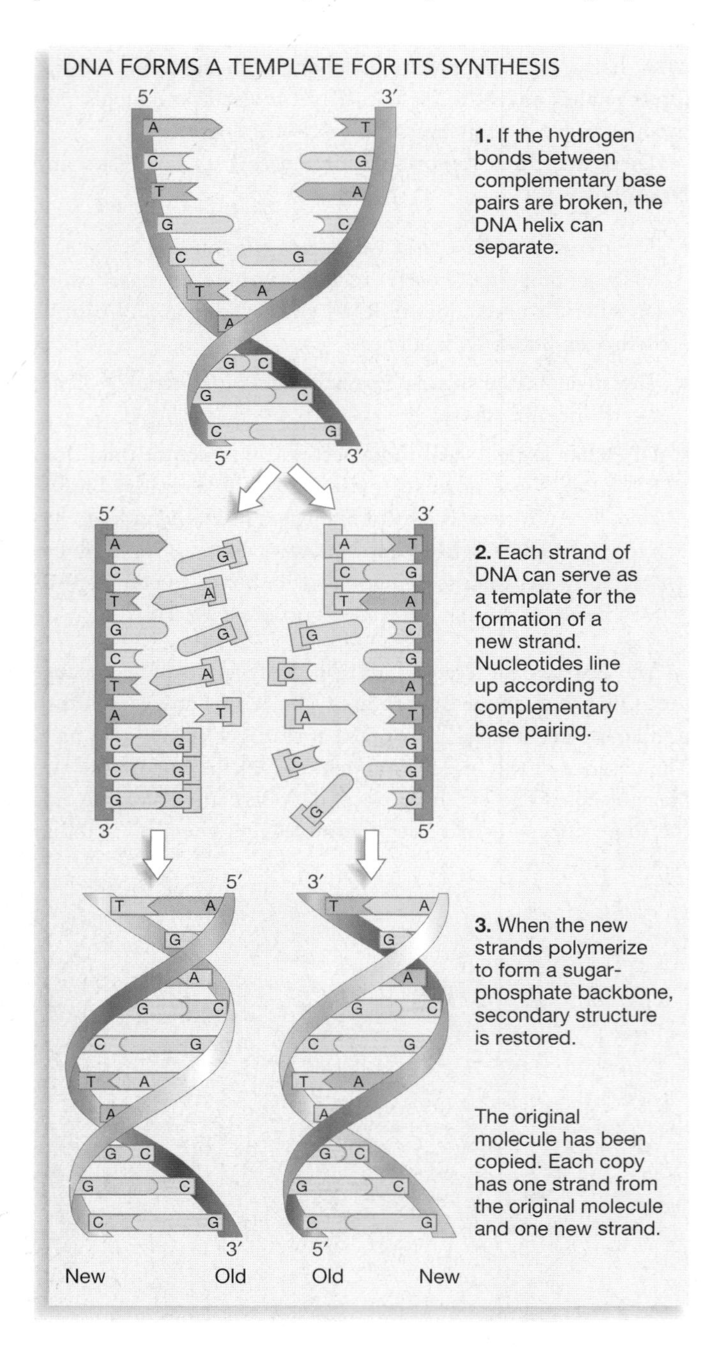

FIGURE 3.18 Making a Copy of DNA
If new bases are matched against each of the two strands of DNA via complementary base pairing, a copy of the DNA molecule can be produced.

itself through complementary base pairing. RNA, in contrast, is not particularly versatile as a catalyst, and it is not as dependable a template as DNA because it is less stable. However, RNA can furnish a template *and* catalyze polymerization reactions. This observation has led to a strong consensus among origin-of-life researchers that the first living entity was probably an RNA molecule. This conclusion is called the RNA world hypothesis.

Because no self-replicating RNA molecules exist in nature today, researchers test the RNA world hypothesis by trying to simulate key events in the laboratory. The eventual goal is to create an RNA molecule that can catalyze its own replication. A ribozyme capable of doing this would be called RNA replicase. If this goal is met, human beings will have created life in a test tube. As this book goes to press, several research teams are racing to do so.

Selection Experiments: The Search for RNA Replicase

To understand how researchers simulate early events in the RNA world, let's take a close look at recent work by David Bartel and Jack Szostak. Their goal was an ambitious one: They wanted to create a ribozyme capable of catalyzing a phosphodiester bond. This research has created considerable excitement among biologists interested in the origin of life, because catalyzing the formation of this bond is a key attribute of RNA replicase. Bartel and Szostak began their experiments by creating two sets of RNA molecules. (For now, don't worry about *how* Bartel and Szostak went about creating these molecules. Instead, focus on the logic and consequences of their experiment. Later in the text you'll have a chance to explore the techniques that biologists use to manipulate nucleic acids and proteins.) The first set of RNAs

BOX 3.4 Models in Biology: The Double Helix

The discovery of the secondary structure of DNA—two DNA strands joined in a double helix—was one of the great scientific breakthroughs of the twentieth century. James Watson and Francis Crick announced the discovery in 1953, in a one-page paper published in the journal *Nature*. At the time, Watson was a 25-year-old postdoctoral fellow and Crick was a 37-year-old graduate student.

Watson and Crick's finding was actually a model, not an experimental result. They had been looking for a secondary structure that could explain a series of recent observations about DNA:

- Chemists had worked out the structure of nucleotides and knew that DNA polymerized through the formation of phosphodiester bonds. Thus Watson and Crick knew that the molecule had a sugar-phosphate backbone.
- By analyzing the nitrogenous bases in DNA samples from different organisms, Erwin Chargaff had established two empirical rules: (1) the total number of purines and pyrimidines in DNA is the same; and (2) the numbers of T's and A's in DNA are equal, and the numbers of C's and G's in DNA are equal.
- By bombarding DNA with x-rays and analyzing how it scattered the radiation, Rosalind Franklin and Maurice Wilkins had calculated the distances between groups of atoms in the molecule. The scattering patterns showed that three distances were repeated many times in the secondary structure: 0.34 nanometers (nm), 2.0 nm, and 3.4 nm. From this, they inferred that DNA was helical in nature.

Watson and Crick also knew the sizes of the atoms and the lengths and angles of the bonds in deoxyribose, the phosphate group, and the nitrogenous bases. Based on these data, they were able to infer that the distance of 2.0 nm probably represented the width of the helix, and that 0.34 nm was likely to be the distance between bases stacked in the spiral. **Figure 1a** shows these proposed distances. Now they needed to make sense of Chargaff's rules and the 3.4 nm distance, which appeared to be exactly 10 times the distance between a single pair of bases.

Watson and Crick began building physical models, so they could tinker with different bond geometries and helical configurations. One of their models is shown **Figure 1b**. After many false starts, they had a fundamental insight: When two strands of DNA are arranged antiparallel to one another in a helical arrangement, only purine-pyrimidine pairs are able to fit in the space inside the coiled sugar-phosphate backbones (see **Figure 1c**). Further, the nitrogenous bases lined up in such a way that hydrogen bonds would form between complementary purines and pyrimidines.

Watson and Crick had discovered complementary base pairing, which revealed how macromolecules could hold the information required to make a copy of themselves. (The term *Watson-Crick pairing* is synonymous with the term *complementary base pairing*.) They ended their paper explaining the model with one of the classic understatements in scientific literature: "It has not escaped our notice that the specific pairing we have postulated immediately suggests a possible copying mechanism . . ." In Chapter 12 you'll review the experiments that were done to test the Watson-Crick model and explore how DNA is synthesized in living cells.

In recognition of their work, Watson, Crick, and Wilkins shared the 1962 Nobel Prize in medicine and physiology. Tragically, Rosalind Franklin died of cancer in 1958 at the age of 38. She never received the Nobel Prize, which cannot be awarded posthumously.

(Continued on page 62)

created by Bartel and Szostak consisted of many identical molecules. The composition of these RNAs is diagrammed in **Figure 3.19**. Each had a short "tag" sequence on its 5′ end and a short sequence of ribonucleotides ending with a cytosine on the 3′ end. The function of each portion of the sequence will become clear when you examine the experimental protocol. Following Bartel and Szostak, let's call these molecules the "oligo substrate." (Because *oligo* means "few," biologists call a short nucleic acid molecule an **oligonucleotide**, or simply an **oligo**.)

The other set of RNAs were all different; let's call this set the "pool RNAs." As the sketch in Figure 3.19 shows, these RNAs had two distinct sections—a hairpin loop and a randomly generated sequence of nucleotides. The hairpin contained the same sequence in each of the pool RNAs. The randomly generated sequence, however, was different in each pool RNA molecule. Each of these three regions had a special property:

- The sequences at the base of the hairpin loop were complementary to the bases on the oligo substrate's 3′ end. These sections of the two molecules "matched up" by complementary base pairing, as Figure 3.19 shows.
- The variable part consisted of 220 ribonucleotides that were generated randomly. None of these randomly generated se-

(Box 3.4 continued)

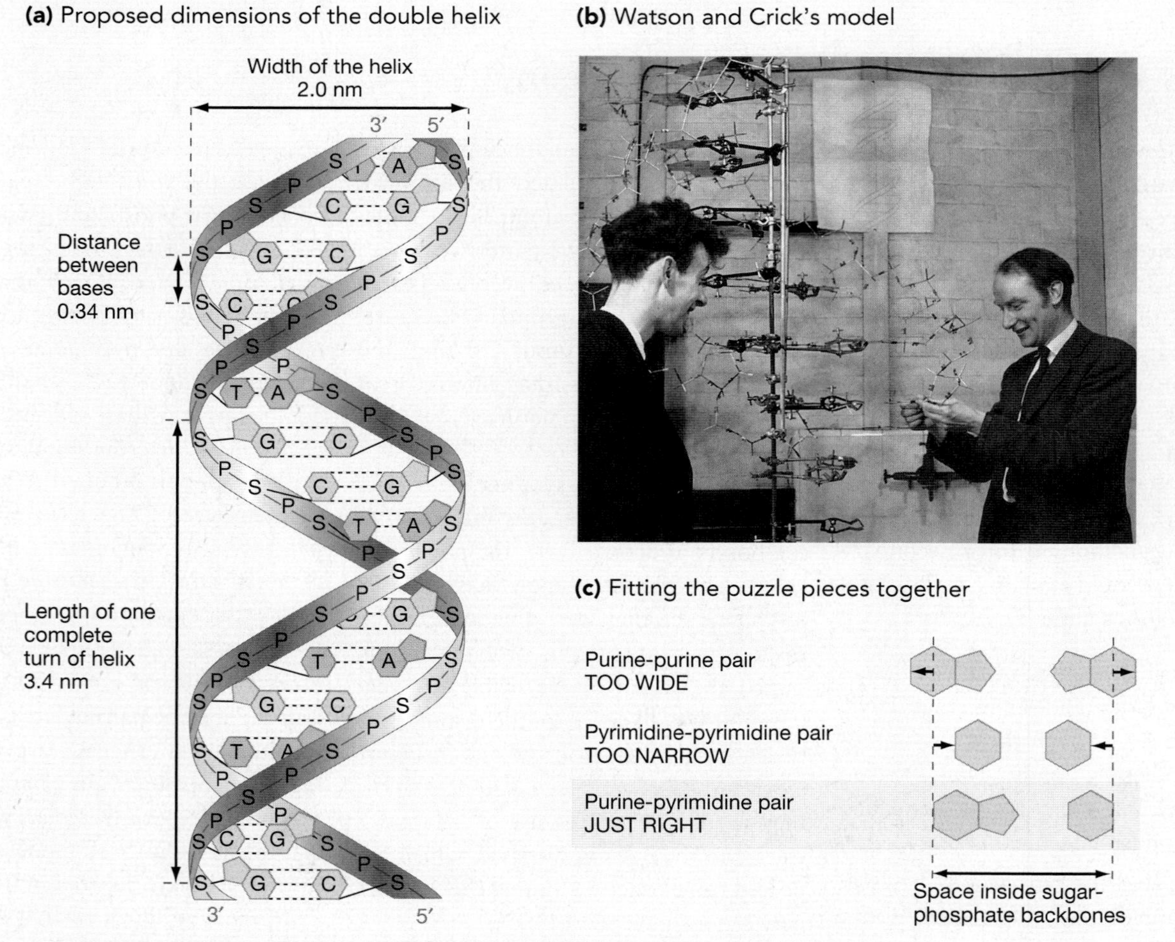

FIGURE 1 Data that Led to the Discovery of the Double Helix
(a) This is how the double helical model of DNA secondary structure explained the three distances observed in the molecule. **(b)** Watson and Crick did not know if the DNA helix was single-, double-, or triple-stranded, or exactly how the sugar-phosphate backbones in the helix were oriented. To help sort out the possibilities, Watson and Crick built physical models of the four deoxyribonucleotides using wires with precise lengths and geometries. Then they tested how the nucleotides fit together in different configurations. **(c)** The dashed lines represent the space available inside the double helix. Only purine-pyrimidine pairs fit inside this space.
QUESTION According to the model in part (a), how many base-pairs fit inside one complete turn of the double helix?

quences were alike, so each of the 1.6×10^{15} pool RNAs represented a unique sequence.

Bartel and Szostak posed the following question: Do any of these pool RNAs have the ability to catalyze the formation of a phosphodiester bond? If so, then the oligo substrate should end up linked to that particular pool RNA sequence (Figure 3.19).

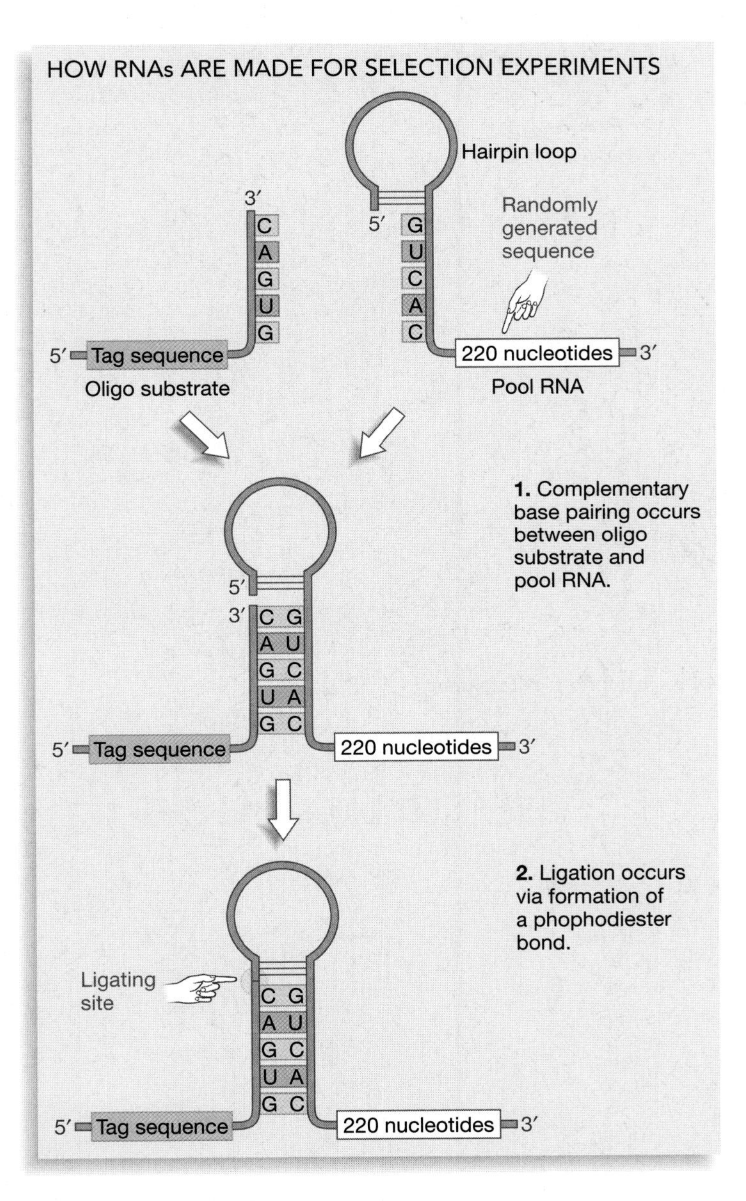

FIGURE 3.19 The RNAs Used in Selection Experiments
The oligo substrate used in the ribozyme selection experiment has two main features: a "tag sequence" and a short stretch of nucleotides with the sequence G-U-G-A-C. The pool RNAs in the experiment have a short stretch of nucleotides with the sequence C-A-C-U-G. They also have a sequence of 220 randomly generated nucleotides. The G-U-G-A-C sequence from the oligo substrate and C-A-C-U-G sequence from the pool RNAs are complementary. If the 220-nucleotide sequence in the pool RNA can catalyze the formation of a phosphodiester bond, then a ligation reaction will occur at the point indicated.

A reaction like this, in which two nucleic acid molecules are hooked together, is called a **ligation** reaction.

The design of the oligo substrates and pool RNAs gave Bartel and Szostak an elegant way of telling whether any of the randomly generated sequences demonstrated catalytic activity. **Figure 3.20** on page 64 shows how this was done. In step 1 of their experimental protocol, Bartel and Szostak let the pool RNAs react with the oligo substrates. If the pool RNA catalyzed a ligation reaction, the pool RNA and the oligo substrate would be joined together. In step 2, the researchers passed all the oligo substrates and pool RNAs through an "affinity column." This is a column containing beads that were coated with RNA sequences complementary to the tag sequences on the oligo substrates. The diagram shows how the affinity column "catches" the pool RNAs with catalytic activity. This protocol allowed the researchers to efficiently isolate the tiny handful of sequences with ribozyme-like properties from the 1.6×10^{15} pool RNAs.

Bartel and Szostak were just getting started, however. They repeated these same steps 10 times. But before going through each cycle again, they did something clever. They created a new set of pool RNAs equivalent in size to the original set, using the sequences selected from the previous round as a template. They did this by making many, many copies of the pool RNAs that had catalytic activity. These were not exact copies, however. Instead, Bartel and Szostak used a technique that ensured that most copies would have one or more random changes in its sequence. These inexact copies are a key feature of this experiment. Bartel and Szostak predicted that many of the ribozyme copies might be similar enough to the template to continue catalyzing the formation of phosphodiester bonds. But because each new molecule was slightly different from the original, they also predicted that some would have *improved* catalytic activity. This, in fact, is exactly what they found. As the histogram in **Figure 3.21** on page 64 shows, each round of selection and inexact copying resulted in improved catalytic activity. After 10 rounds, Bartel and Szostak had created a ribozyme that catalyzed the formation of phosphodiester bonds with reasonable efficiency. They had succeeded in creating a ribozyme with one of the attributes of an RNA replicase.

This experiment is considered a landmark in origin-of-life research, for three reasons. First, Bartel and Szostak's work showed that another key event in chemical evolution could be simulated in the laboratory. Second, the ribozymes produced after 10 rounds of selection are being used in further experiments designed to improve and expand the catalytic activity of RNA. Third, the experimental approach that Bartel and Szostak pioneered, now referred to as in vitro selection,* is being employed by other research teams with dramatic results.

*The term *in vitro* is Latin for "in glass." When biologists perform experiments outside of living cells, these experiments are done in vitro. The term *in vivo*, in contrast, is Latin for "in life." When biologists do experiments with living organisms, they are done in vivo.

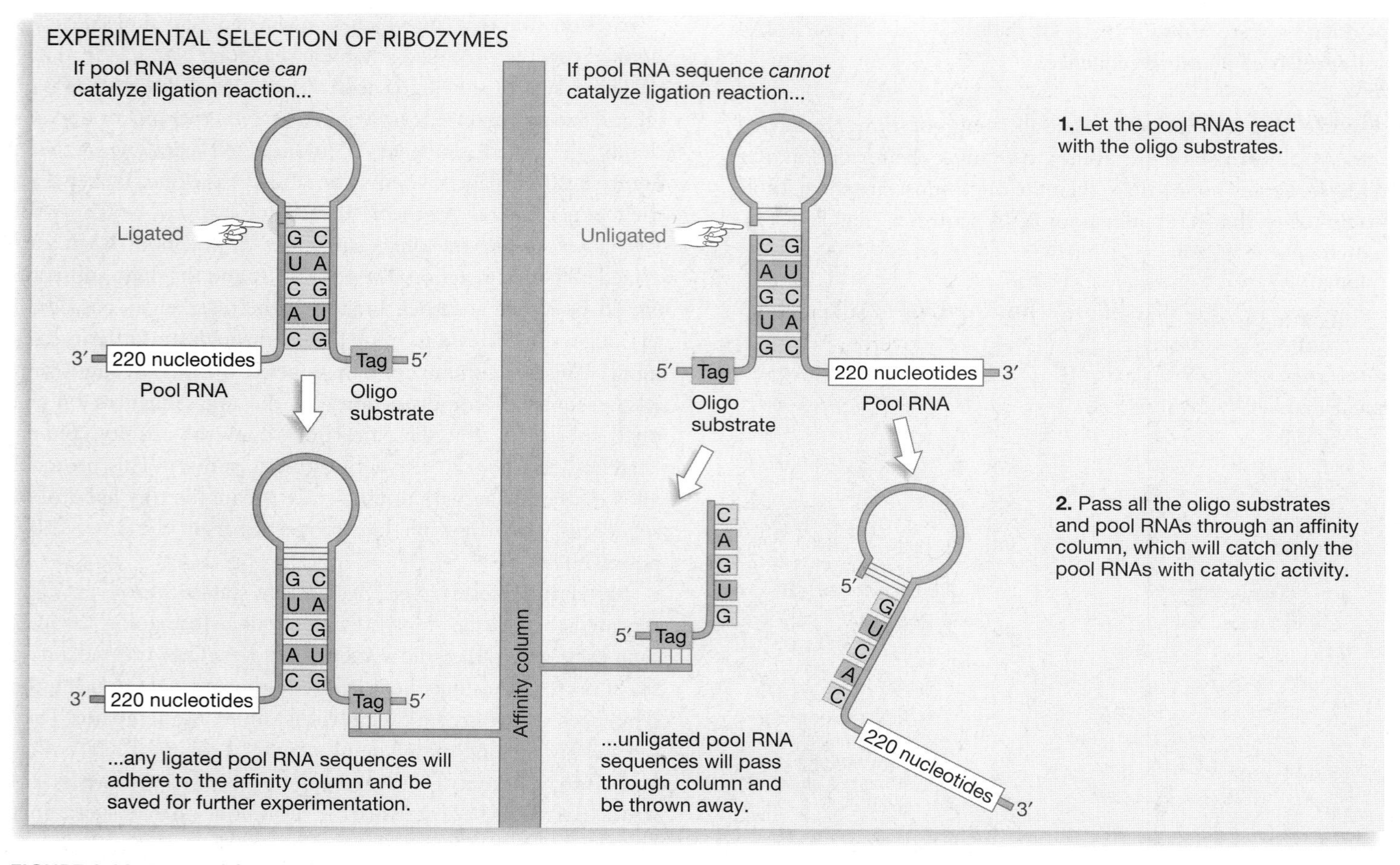

FIGURE 3.20 Protocol for the Experimental Selection of Ribozymes
Using the strategy illustrated here, researchers could select the pool RNAs that were capable of catalyzing the formation of a phosphodiester bond.

To date, in vitro selection experiments have produced ribozymes capable of catalyzing the amide (N–C) bonds found in amino acids, the peptide bonds found in proteins, and a phosphodiester bond leading to the polymerization of ribonucleotides to a growing strand.

In sum, origin-of-life researchers have a growing set of ribozymes capable of catalyzing key reactions in the formation of macromolecules. Each result strengthens the confidence that biologists have in the RNA world hypothesis. Each result also brings these research teams closer to the creation of an RNA replicase.

After Replication: Evolution by Natural Selection

Before closing this section, let's go back to the prebiotic soup, summarize the later steps in chemical evolution, and think carefully about what happened *after* the first replicator appeared. The research you've just reviewed supports two tentative conclusions:

- Proteins and nucleic acids could have polymerized on mineral clays, perhaps along the beaches or tide pools of the early ocean.
- An RNA molecule resulting from these polymerization reactions is a plausible candidate for the self-replicating molecule. RNA is a logical choice not only because of its abilities

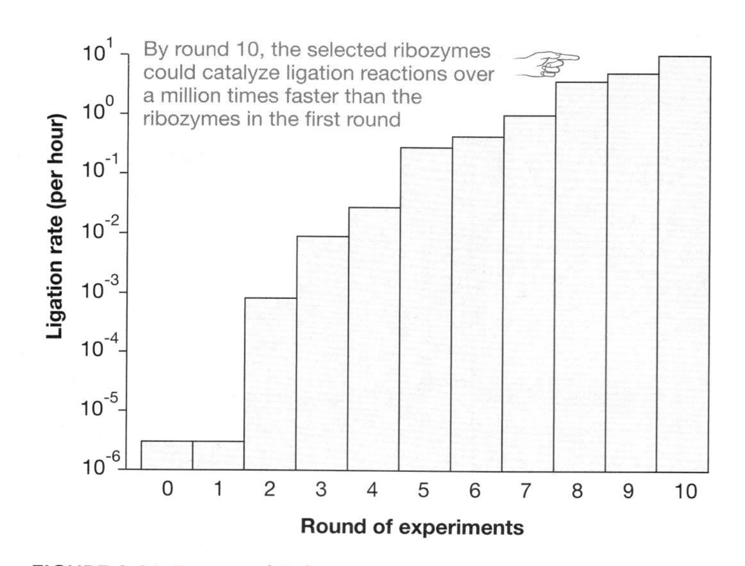

FIGURE 3.21 Repeated Selection Improves the Catalytic Ability of Ribozymes
After each round in their experiment, researchers took the ribozymes that adhered to the affinity column and made copies of them that included errors. Then they let these "next generation" ribozymes react with oligo substrates. In this bar chart, the number of ligation reactions that occurred per hour during each round in the experiment is plotted on the y-axis, using a logarithmic scale.

to act as a catalyst but also because it can act as a template. Ribozymes can catalyze the formation of phosphodiester bonds, and complementary base pairing provides a mechanism for making copies of the RNA sequence.

The work done to date, then, supports the Oparin-Haldane theory in general and the RNA world hypothesis in particular.

Once the self-replicating molecule appeared, though, what happened next? This question was answered by Charles Darwin, the nineteenth-century biologist introduced in Chapter 1. The evolutionary process that Darwin discovered, called **natural selection**, would have begun as soon as the self-replicating molecule appeared.

To understand how natural selection works, consider the fate of the self-replicating molecule. As **Figure 3.22** shows, an RNA molecule capable of self-replication would begin making copies of itself using nucleotides available in the prebiotic soup. Copies of the self-replicating molecule would begin to populate the prebiotic soup. These copies, however, would not be exact replicas of the original molecule. Instead, some copies would contain mistakes in the form of random changes in the sequence of nucleotides. These changes would occur because the ribozyme cannot match complementary bases perfectly every time a copy is made. A random change like this is called a **mutation**. Some or most copies of the self-replicating molecule would contain one or more mutations (see Figure 3.22).

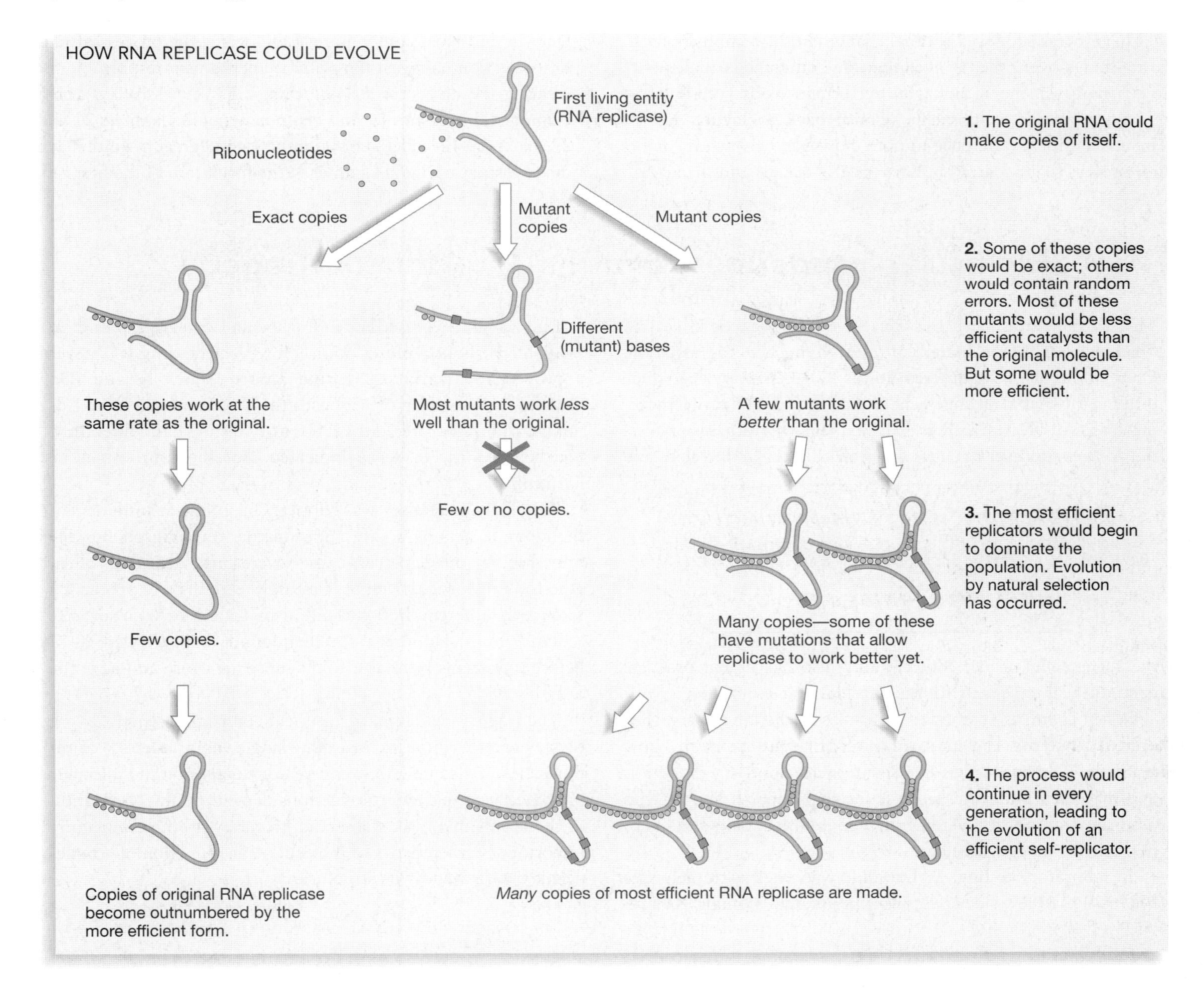

FIGURE 3.22 How an RNA Replicase Could Evolve
The process illustrated here is an example of evolution by natural selection. Natural selection was introduced in Chapter 1.

Like the ribozyme copies generated in Bartel and Szostak's experiment, these mutant molecules would vary in their ability to make copies of themselves. Some of the new sequences would make better copies faster than the original self-replicating molecule. Some would perform less well. As Figure 3.22 shows, the more efficient ribozymes would begin to dominate the population of self-replicating molecules. The less efficient molecules would die off as they were broken apart by ultraviolet radiation or hydrolysis before they had a chance to make a copy of themselves. In this way, certain ribozymes would be favored, or naturally selected, over others. Copies of these ribozymes represent their offspring. These offspring, in turn, would inherit any of the favorable variations that were copied correctly.

This process produces gradual changes in the composition of living things over time, or **evolution**. The outcome is analogous to the improvements in ribozyme performance observed in Bartel and Szostak's experiments (glance back at Figure 3.21). There is a critical difference to note, however, between the natural process that occurred in Earth's early oceans and the artificial process employed by Bartel and Szostak. Researchers perform **artificial selection** by setting up conditions designed to favor a certain trait (in Bartel and Szostak's case, this trait was the ability to catalyze the ligation of their oligo- and pool-RNAs). When natural selection occurs, no conscious entity is involved in setting up certain conditions. Instead, some mutants simply do better under the conditions existing in the natural environment. In the environment of the prebiotic soup, some replicators made more copies of themselves than others. As a result, these replicators increased in frequency in the population.

In sum, the current view of life's origins is that chemical evolution became less and less important after the self-replicating molecule appeared. Instead, biological evolution—or more formally, evolution by natural selection—took over. According to Darwin and most contemporary biologists, the process of natural selection has been the primary mechanism responsible for creating the diversity of living things that exists today. You'll examine the evidence for this claim in detail in Chapters 21 and 22. Next, however, let's consider the second great breakthrough in the history of life in Chapter 4: the formation of the first cell.

Essay Molecular Handedness and the Thalidomide Tragedy

This chapter introduced the concept of chirality, or molecular "handedness," in the context of research on chemical evolution. Chiral molecules have different forms called enantiomers, which differ in the spatial arrangement of their atoms. The problem facing origin-of-life researchers is to explain why organisms have only one enantiomer of chiral molecules like the amino acids and the sugar called ribose. But chirality also has urgent practical consequences, especially for the pharmaceutical industry.

People who used this tranquilizer were actually ingesting equal amounts of two mirror-image molecules.

When chemists synthesize molecules in the laboratory, they are usually produced as a mixture of enantiomers in equal amounts. This was true for a sedative developed in the 1950s and marketed under the trade name thalidomide. People who used this tranquilizer were actually ingesting equal amounts of two mirror-image molecules.

In laboratory testing, thalidomide was rated extremely safe because it had few side effects in experimental animals. As a result it was widely prescribed in Europe; in Germany it was even available without a prescription. However, the drug was never approved for marketing in the United States because Dr. Frances O. Kelsey (an official with the U.S. Food and Drug Administration) interpreted the laboratory testing data differently. She believed the evidence indicated that the drug might be harmful.

Tragically, Dr. Kelsey was correct. One of the enantiomers of thalidomide acts as a safe and effective tranquilizer. But the mirror-image molecule can cause severe birth defects. Women who took the drug during the first three months of a pregnancy frequently bore children without arms or legs or with severely reduced limbs. Hundreds of "thalidomide babies" were born before physicians were able to diagnose the cause and have the drug banned.

The tragedy made pharmaceutical companies acutely aware of the need to synthesize or isolate single enantiomers of chiral molecules. It also underscores a very general point in molecular biology: Because chemical reactions depend on the orientation of the atoms involved, shape and geometry are fundamentally important characteristics of molecules. The function of a molecule is determined by its structure.

Chapter Review

Summary

The theory of chemical evolution proposes that life began with the formation of a self-replicating molecule. Origin-of-life researchers are testing the steps outlined in the theory by designing experiments that simulate conditions found in the atmosphere and ocean of the early Earth.

Chemical evolution begins when simple molecules like methane, ammonia, water, carbon dioxide, and carbon monoxide react to form more complex organic molecules like formaldehyde and hydrogen cyanide. Experiments using electrical discharges or ultraviolet radiation as energy sources have shown that these nonspontaneous reactions could have occurred on ancient Earth.

Chemical evolution continues when formaldehyde and hydrogen cyanide react to form amino acids, sugars, and nucleotides. These molecules are the building blocks, or monomers, needed to synthesize the complex macromolecules found in living organisms. Experiments have shown that amino acids, sugars, and the nitrogenous bases called purines are easily produced under early Earth conditions. Many of these molecules are found on meteorites as well. Currently, research on this step in chemical evolution is focusing on how an important sugar called ribose could have been produced in large quantity, and how the nitrogenous bases called pyrimidines could have been synthesized under early Earth conditions.

The polymerization of monomers into macromolecules is the next key event in chemical evolution. Experiments have shown that these condensation reactions occur readily when growing polymers stick to clay minerals. Researchers have observed polypeptide formation and RNA formation from the polymerization of amino acids and ribonucleic acids, respectively, on clay particles.

The final step in chemical evolution is the formation of a self-replicating molecule. A self-replicating molecule must be able to catalyze polymerization reactions, and it must furnish the template necessary to make a copy of itself. Most origin-of-life researchers propose that the first self-replicating molecule was RNA, because RNA can catalyze a variety of chemical reactions and because complementary base pairing between nucleotides furnishes a mechanism for making a copy.

Once a self-replicating molecule was produced by chemical evolution, biological evolution—or evolution by natural selection—would begin. At least some copies of the self-replicating molecule would contain random errors, or mutations. If some of these mutations improved the catalytic ability of the self-replicator, the mutant molecule would copy itself more quickly. As a result, it would come to dominate the population of self-replicating molecules, or be "naturally selected." As this three-step process of copying, generating variation through errors, and selection continued, the catalytic ability of the first self-replicator would continuously improve.

Questions

Content Review

1. **What are the four nitrogenous bases found in RNA?**
   a. uracil, guanine, cytosine, thymine (U, G, C, T)
   b. adenine, guanine, cytosine, thymine (A, G, C, T)
   c. adenine, uracil, guanine, cytosine (A, U, G, C)
   d. alanine, threonine, glycine, cysteine (A, T, G, C)

2. What determines the primary structure of an RNA molecule?
   a. the sugar-phosphate backbone
   b. complementary base pairing and the formation of hairpin loops
   c. the sequence of deoxyribonucleotides
   d. the sequence of ribonucleotides

3. DNA attains a secondary structure when hydrogen bonds form between the nitrogenous bases called purines and pyrimidines. What are the complementary base pairs that form in DNA?
   a. A–T and G–C
   b. A–U and G–C
   c. A–G and T–C
   d. A–C and T–G

4. By convention, biologists write the sequence of bases in RNA and DNA in which direction?
   a. $3' \rightarrow 5'$
   b. $5' \rightarrow 3'$
   c. N-terminal to C-terminal
   d. C-terminal to N-terminal

5. In RNA, when does the secondary structure called a hairpin form?
   a. when hydrophobic residues coalesce
   b. when hydrophilic residues interact with water
   c. when complementary base pairing between ribonucleotides on the same strand creates a "stem-and-loop" structure
   d. when complementary base pairing forms a double helix

6. Twenty different amino acids are found in the proteins of cells. What distinguishes these molecules?
   a. the location of their carboxyl group
   b. the location of their amino group
   c. the composition of their side chains, or "R-groups"
   d. their ability to form peptide bonds

7. The secondary structure of DNA is called a double helix. Why?
   a. Two strands wind around one another in a helical, or spiral, arrangement.
   b. A single strand winds around itself in a helical, or spiral, arrangement.
   c. It is shaped like a ladder.
   d. It stabilizes the molecule.

8. What makes a molecule chiral?
   a. It has secondary structure.
   b. It has tertiary structure.
   c. It has a plane of symmetry.
   d. It has two mirror-image forms, which cannot be exactly superimposed on each other.

Conceptual Review

1. Explain the lock-and-key model for enzyme activity. Be sure to comment on what the active site of an enzyme does.
2. Isoleucine, valine, leucine, phenylalanine, and methionine are amino acids with highly hydrophobic side chains. Suppose a section of a protein contained a long series of these hydrophobic residues. How would you expect this portion of the protein to behave when the molecule is in aqueous solution?
3. Explain how complementary base pairing makes the copying of RNA and DNA molecules possible. Your answer should include diagrams.
4. Turn back to Figure 3.8, which shows a generalized cartoon of monomers undergoing condensation reactions to form a polymer. Label the type of bond formed when amino acids polymerize to a polypeptide. Label the type of bond formed when nucleic acids polymerize to form RNA or DNA. Does it take energy for polymerization reactions to proceed, or do they occur spontaneously? Why or why not?
5. A major theme in this chapter is that the structure of molecules correlates with their function. Explain why DNA's secondary structure limits its catalytic abilities compared to RNA. Why are proteins the most effective catalysts of all?

Applying Ideas

1. The essay at the end of Chapter 2 discussed efforts to find life on other components of our solar system. How does the research reviewed in this chapter inform the search for extraterrestrial life?
2. Suppose a selection scheme like the experiment reviewed in section 3.4 succeeded in producing a molecule that could make a copy of itself. According to the discussion in Box 3.1, would this molecule be alive? Write a one-page opinion piece for your local newspaper that explains the nature of the research and discusses the ethical and philosophical implications of the discovery.
3. Oparin and Haldane proposed that life began in a warm, shallow-water environment. Some researchers suggest that it is much more likely that life began on or near hydrothermal vents. Based on the research reviewed in the first three chapters of this text, which hypothesis do you favor? To organize your answer, make a chart with the four steps in chemical evolution (see the introduction to this chapter) as rows and the two hypotheses as columns. Fill the chart with notes on which steps were most likely to occur in each of the two environments.
4. Origin-of-life researcher Robert Crabtree maintains that experiments simulating early Earth conditions are a valid way to test the theory of chemical evolution. Crabtree maintains that if scientists working in the field agree that an experiment is a plausible reproduction of early Earth conditions, it is valid to infer that its results are probably correct—that the simulation effectively represents events that occurred some four billion years ago. Do you agree? Do you find the models presented in Chapter 2 or the experiments presented in Chapter 3 more convincing tests of the theory? Explain your answers.

CD-ROM and Web Connection

CD Activity 3.1: Condensation and Hydrolysis Reactions *(animation)*
(Estimated time for completion = 5 min)
Learn more about these two important reaction types, which affect the formation of RNA from nucleotides and the disassembly of proteins into their component amino acids.

CD Activity 3.2: Activation Energy and Enzymes *(animation)*
(Estimated time for completion = 5 min)
Learn how enzymes catalyze the thousands of chemical reactions that occur in our bodies each second.

At your **Companion Website** (http://www.prenhall.com/freeman/biology), you will find self-grading exams and links to the following research tools, online resources, and activities:

Stanley Miller on Exobiology
In this interview, Stanley Miller talks about how life may have originated on earth and elsewhere in the universe.

Altman, Cech, and Ribozymes
These pages at the Nobel Foundations website describe the prize-winning research of Sidney Altman and Thomas Cech into the enzymatic capabilities of Ribonucleic acids.

Self-Replicating RNA
This site describes David Bartel's current research and contains many links related to research on self-replicating RNA.

Additional Reading

Watson, J. D. 1980. *The Double Helix* (New York: Atheneum). A vivid personal account of the race to discover the secondary structure of DNA.

de Duve, C. 1995. The beginnings of life on Earth. *American Scientist* 83: 428–437. An overview of chemical evolution.

Gerstein, M., and M. Levitt. 1998. Simulating water and the molecules of life. *Scientific American* 279 (November): 100–105. An introduction to how computer modeling is helping researchers better understand how water interacts with proteins and nucleic acids inside cells.

Rebek, J., Jr. 1994. Synthetic self-replicating molecules. *Scientific American* 273 (July): 48–55. A review of work to produce self-replicating molecules not found in nature.

Membranes and the First Cells

4

Biological evolution began with the advent of the first self-replicating molecule. Another great milestone in the history of life occurred when a descendant of this replicator became enclosed within a membrane. This event created the first cell.

The cell membrane was a momentous development because it separated life from nonlife. Before cell membranes existed, self-replicating molecules probably floated free in the prebiotic soup, building copies of themselves as they randomly encountered the appropriate nucleotides. But the membrane made an internal environment possible—one that could have a chemical composition different from the external environment. This was important for two reasons. First, the chemical reactions necessary for life could occur much more efficiently in an enclosed volume, because reactants would collide more frequently. Second, the membrane could serve as a selective barrier. That is, it could keep compounds out of the cell that might damage the replicator but allow the entry of compounds required by the replicator. The membrane not only created the cell but also made it into an efficient and dynamic reaction vessel.

The goal in this chapter is to investigate how membranes behave, with an emphasis on how they define an internal environment distinct from the external environment. The chapter begins by reviewing the structure and properties of the most abundant molecules in cell membranes. These are the "oily" or "fatty" compounds called lipids. Section 4.2 expands on this introduction by analyzing how lipids behave when they form membranes. Which ions and molecules can pass through a membrane that consists of lipids? Which cannot, and why? The chapter ends by exploring how proteins that become inserted into a lipid membrane can control its permeability.

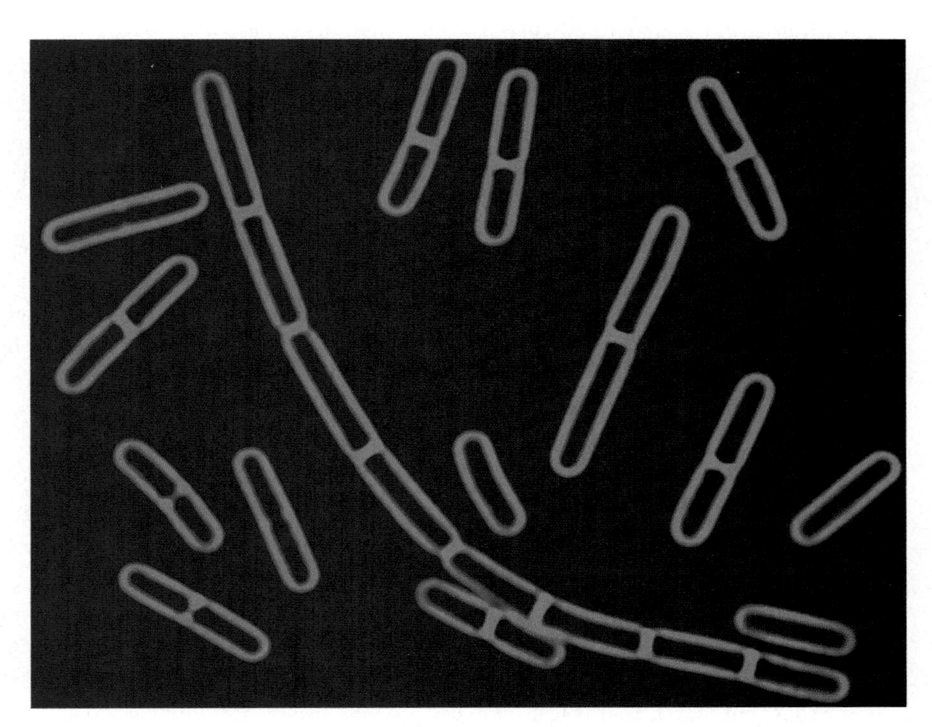

These bacterial cells have been stained with a red compound that inserts itself into the cell membrane. The cell membrane defines the basic unit of life. In single-celled organisms like those shown here, the membrane creates a physical separation between life on the inside and nonlife on the outside.

4.1 Lipid Chemistry

Most biochemists are convinced that the building blocks of membranes, called lipids, existed in the prebiotic soup. This conclusion is based on the observation that several different types of lipids have been produced in experiments that were designed to mimic the chemical and energetic conditions that prevailed early in Earth history. For example, Stanley Miller's spark-discharge experiments, reviewed in Chapter 3, succeeded in producing at least two types of lipids.

An observation made by biochemist A. D. Bangham illustrates why this result is interesting. In the late 1950s, Bangham performed a long series of experiments on how lipids behave when immersed in water. But until the electron microscope was invented, he had no idea what his lipid-water mixtures looked like. **Box 4.1** explains how electron microscopes magnify objects up to 250,000 times. When these instruments became available and Bangham was able to take high-magnification photographs of his experimental mixtures, he saw something astonishing: The lipids formed water-filled, cell-like, enclosed compartments, which he called vesicles (**Figure 4.1a**). These vesicles were reminiscent of actual cells (**Figure 4.1b**). Bangham had not done anything special to the lipid-water mixtures; he had merely shaken them by hand.

The experiment raises a series of questions. How could these structures have formed? Is it possible that vesicles like these ex-

BOX 4.1 The Transmission Electron Microscope

The transmission electron microscope, or TEM, is an extraordinarily effective tool for viewing cells at high magnification. A TEM forms an image from electrons that pass through a specimen, just as a light microscope forms an image from light rays that pass through a specimen.

Biologists who want to view a cell under an electron microscope begin by "fixing" the cell, meaning that they kill it with a chemical agent that disrupts the cell's structure and contents as little as possible. Then they permeate the cell with an epoxy plastic that stiffens the structure. Once this hardens, the cell can be cut into extremely thin sections with a glass or diamond knife. Finally, the sectioned specimens are impregnated with a metal—often lead (the reason for this last step is explained shortly).

Figure 1 shows what the microscope looks like and how it works. A beam of electrons is produced at the top of a column and directed downward. (All of the air is pumped out of the column so the electron beam isn't scattered by collisions with air molecules.) The electron beam passes through a series of lenses and the specimen. The lenses are actually electromagnets, which alter the path of the beam much like a glass lens bends light. The lenses magnify and focus the image on a screen at the bottom of the column. There the electrons strike a coating of fluorescent crystals, which emit visible light in response—just like a television screen. To photograph the image, the microscopist moves the screen out of the way and allows the electrons to expose a sheet of black-and-white film.

The image itself is created by electrons that pass through the specimen. If no specimen were in place, all the electrons would pass through and the screen (and photograph) would be uniformly bright. Unfortunately, cell materials by themselves would also appear fairly uniform and bright. This is because the ability of an atom to deflect an electron depends on its density. An atom's density, in turn, is a function of its atomic number. The hydrogen, carbon, oxygen, and nitrogen atoms that dominate biological molecules have low atomic numbers. This is why cell biologists must saturate cell sections with lead solutions. Lead has a high atomic number and scatters electrons effectively. Different macromolecules take up lead atoms in different amounts, so the metal acts as a "stain" that produces contrast. In the electron microscope, areas of dense metal scatter the electron beam most, producing the dark areas in the photographs.

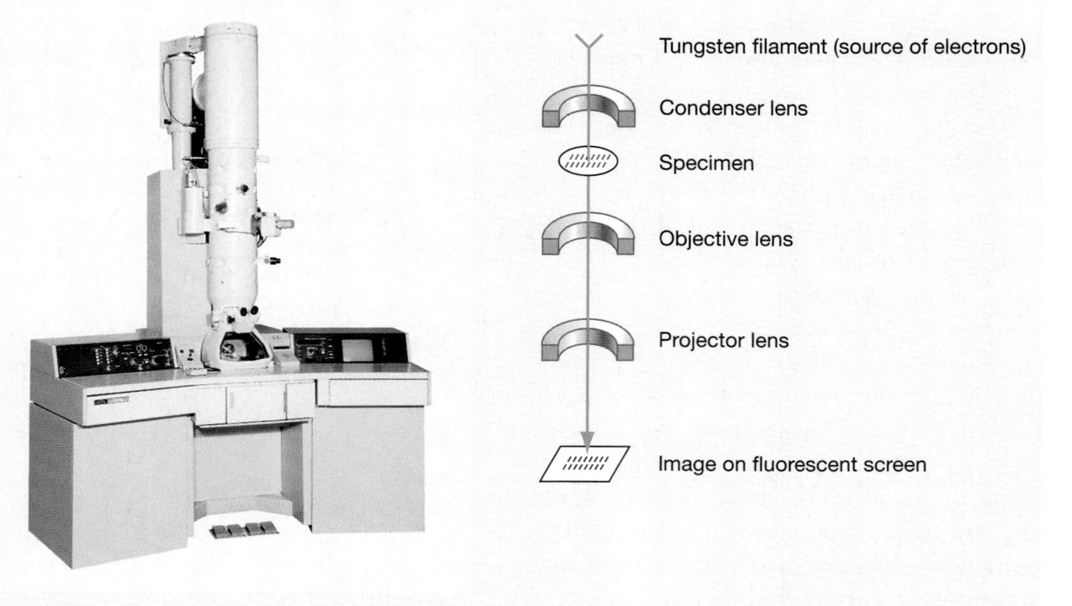

FIGURE 1 What Does a Transmission Electron Microscope Look Like?

isted in the prebiotic soup? If so, could they have surrounded a self-replicating molecule and become the first cell membrane? Before addressing these questions, though, it's necessary to investigate some basic issues.

What Is a Lipid?

Chapter 3 examined the structure of the organic molecules called sugars, amino acids, and nucleotides and explored how these monomers could have polymerized to form a variety of large molecules—such as proteins and nucleic acids—prior to the origin of life. Here we analyze another major type of mid-sized molecule found in living organisms: lipids.

Lipid is actually a catch-all term for carbon-containing compounds that are largely nonpolar and hydrophobic—meaning that they do not dissolve readily in water. Instead, lipids dissolve in liquids made up of nonpolar organic compounds. (Recall from Chapter 2 that water is a polar solvent.)

To understand why these molecules do not dissolve in water, examine the chemical structures diagrammed in **Figure 4.2**. The molecule on the left is a five-carbon compound called isoprene. The molecule on the right is a type of compound called a fatty acid. Isoprene and fatty acids are the building blocks of the lipids found in organisms. The critical thing to note is that they each contain a group of carbon atoms bonded to hydrogen atoms. Molecules that contain only carbon and hydrogen, like the octane molecule shown in the middle of the figure, are known as **hydrocarbons**. The entire isoprene subunit is a hydrocarbon, while **fatty acids** consist of a hydrocarbon chain bonded to a carboxyl (COOH) functional group. Because electrons are shared equally in carbon-hydrogen bonds, hydrocarbons are nonpolar. This property makes them hydrophobic. Lipids do not dissolve in water because they have a significant hydrocarbon component.

(a) Lipids and water form tiny compartments.

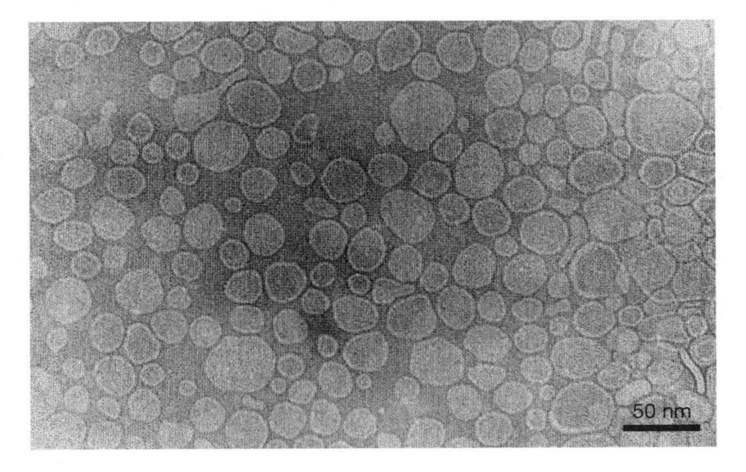

(b) Red blood cells

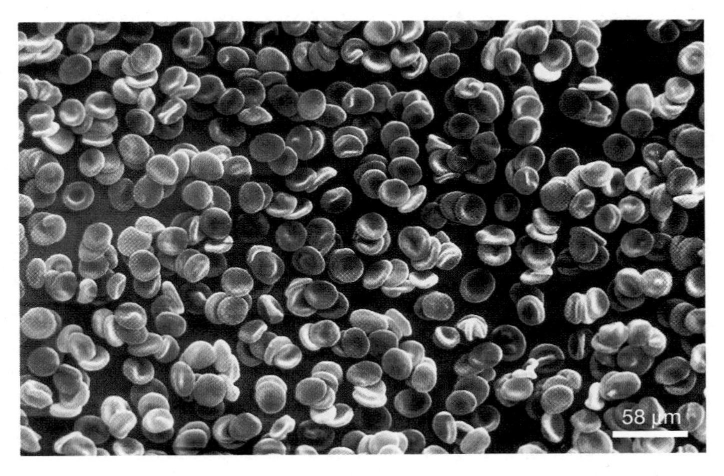

FIGURE 4.1 Lipids Can Form Tiny Compartments When Mixed with Water
(a) This electron micrograph shows the tiny, bag-like compartments that formed when a researcher shook a mixture of lipids and water. Note the scale bar. (b) These are red blood cells from humans. They are small, bag-like compartments reminiscent of the artificial vesicles in part (a).

A Look at Three Types of Lipids Found in Cells

Unlike sugars, amino acids, and nucleic acids, lipids are defined by a physical property—their solubility—instead of their chemical structure. As a result, the structure of lipids varies widely. To drive this point home, consider the three classes of lipids illustrated in **Figure 4.3** (page 72): steroids, phospholipids, and fats. These are the most important types of lipids found in cells. Note that each is constructed from either isoprene or fatty-acid subunits.

At the bottom left of the figure is the cholesterol molecule, which belongs to a family of compounds called steroids. Steroids are distinguished by the four-ring structure shown, which is constructed from chains of isoprene units. Note that cholesterol also has a hydrocarbon "tail" formed of isoprene subunits.

The middle section of the figure shows that phospholipids can form either from chains of isoprenes or from fatty acids. A **phospholipid** has a distinctive structure. A three-carbon molecule called glycerol is linked to a phosphate group (PO_4^{2-}) and to either two chains of isoprene or to two fatty acids. In some cases, the phosphate group is bonded to another small organic molecule (the amino acid serine is used as the example in the figure).

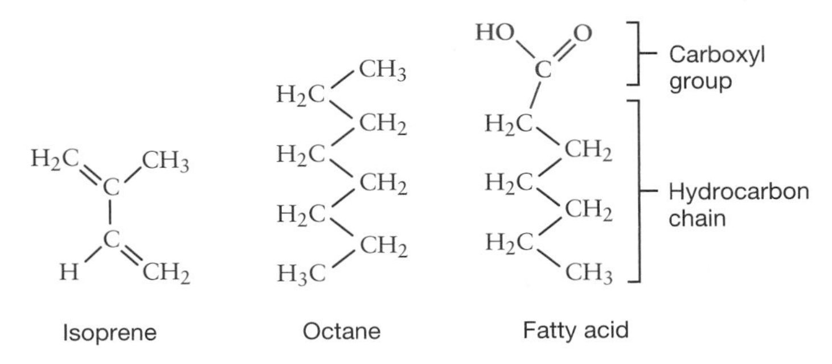

FIGURE 4.2 Hydrocarbon Groups Make Lipids Hydrophobic
Hydrocarbon is a general term for any molecule made up solely of carbon and hydrogen. Hydrocarbons are uncharged and nonpolar, so they do not interact with water. EXERCISE Circle the hydrophobic, hydrocarbon components of each molecule shown.

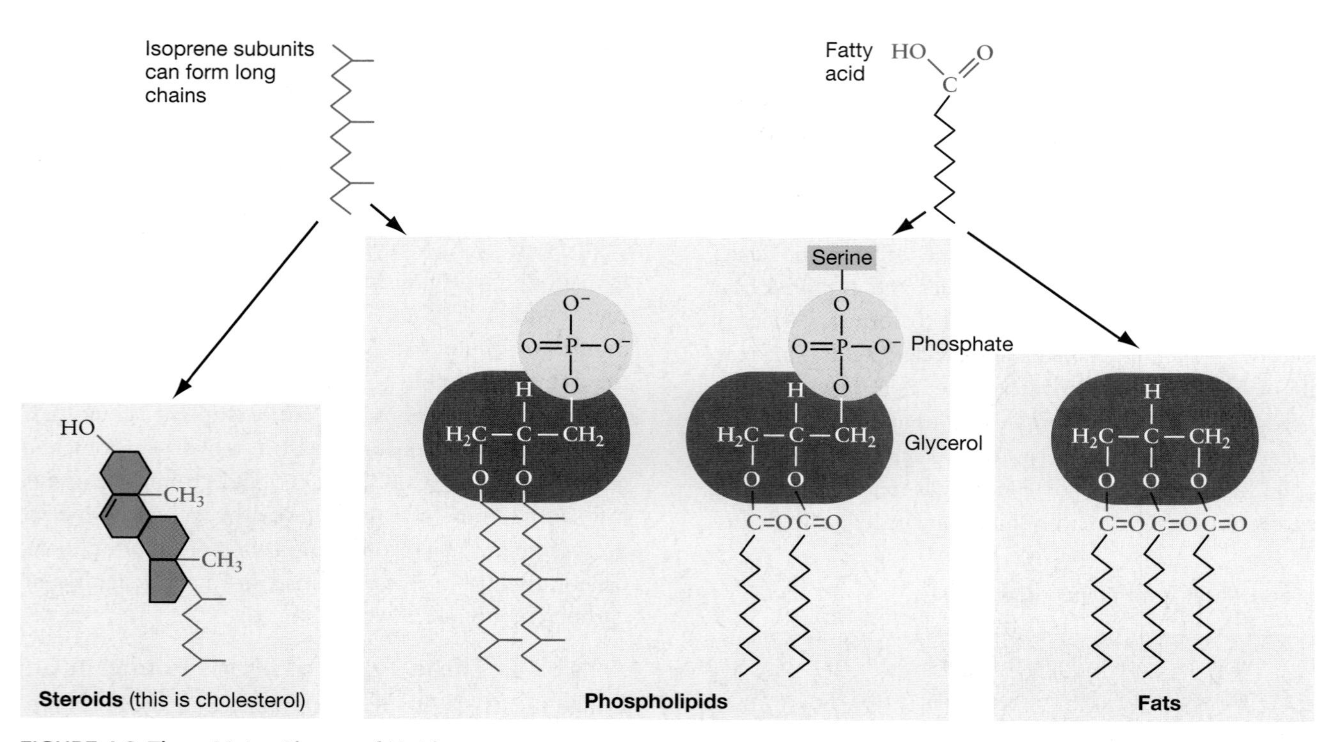

FIGURE 4.3 Three Major Classes of Lipids
When isoprene subunits are linked into the long chains shown at the upper left, they can become the building blocks of steroids and some types of phospholipids. Fatty acids are a building block for other types of phospholipids and for fats. In most sections of these diagrams the C and H symbols for carbon and hydrogen have been omitted, so the red or black lines indicate carbon-carbon bonds. This is a common practice among chemists and biologists because it helps make chemical structures easier to read and interpret.
EXERCISE On a separate sheet of paper, draw these molecules with the C and H symbols included. Remember that each carbon atom forms a total of four bonds.

Fats, such as the molecule shown at the right of Figure 4.3, are composed of three fatty acids linked to a glycerol molecule.

The classes of molecules illustrated in this figure will reappear throughout this text because lipids perform a wide variety of functions in cells—from storing energy to ferrying information. The question to be considered now is, which of these lipids are found in cell membranes?

The Structures of Membrane Lipids

Not all lipids can form the artificial membranes that Bangham and colleagues observed, and just two types of lipids dominate cell membranes. Membrane-forming lipids have a fundamental property in common. They have a polar, hydrophilic region in addition to the nonpolar, hydrophobic region common to all lipids. For example, examine the phospholipid illustrated in **Figure 4.4a.** Note that the molecule has a "head" region containing highly polar covalent bonds as well as positive and negative charges. (These charges balance, so the molecule has no overall, or net, charge.) These charges and polar bonds attract water molecules when a phospholipid is placed in solution. In contrast, the long hydrocarbon or isoprenoid tails of a phospholipid are nonpolar. Water molecules do not interact with this part of the molecule.

A compound that contains both hydrophilic and hydrophobic elements is called **amphipathic** ("dual sympathy"). As **Figure 4.4b** shows, cholesterol is also amphipathic. Phospholipids are the most common type of lipid in cell membranes, but cholesterol is also an important membrane component in some species.

The amphipathic nature of phospholipids is far and away their most important feature biologically. It is responsible for their ability to form cell membranes.

4.2 Phospholipid Bilayers

Phospholipids do not dissolve when they are placed in water. Like the lipids in the experiment by Bangham's group, they tend to form one of two types of structures instead. **Micelles** are tiny droplets created when the hydrophobic tails of phospholipids group together, away from the water, and the hydrophilic heads face the water (see **Figure 4.5a**, page 74). Phospholipid bilayers, or simply **lipid bilayers**, are created when two layers of lipid molecules align. The two sets of hydrophobic tails face one another inside the bilayer and are shielded from the solution by the two sets of hydrophilic heads, which face the solution (**Figure 4.5b**). Both types of structure share a common feature: The hydrophobic tails interact with each other, while the hydrophilic heads interact with water. Stated another way, these structures form because of the amphipathic nature of phospholipids.

Making Artificial Bilayers

When Bangham and colleagues agitated a lipid bilayer by shaking it, the bilayer sheets broke and re-formed as small, spherical structures. There was water on the inside as well as the outside of these vesicles because the hydrophilic heads of the lipids faced outward on each side.

Researchers have now produced these types of vesicles using dozens of different types of phospholipids. Artificial membrane-bound vesicles like these are called **liposomes** and support an important conclusion: If phospholipid molecules accumulated during chemical evolution early in Earth history, it is virtually certain that they formed water-filled vesicles.

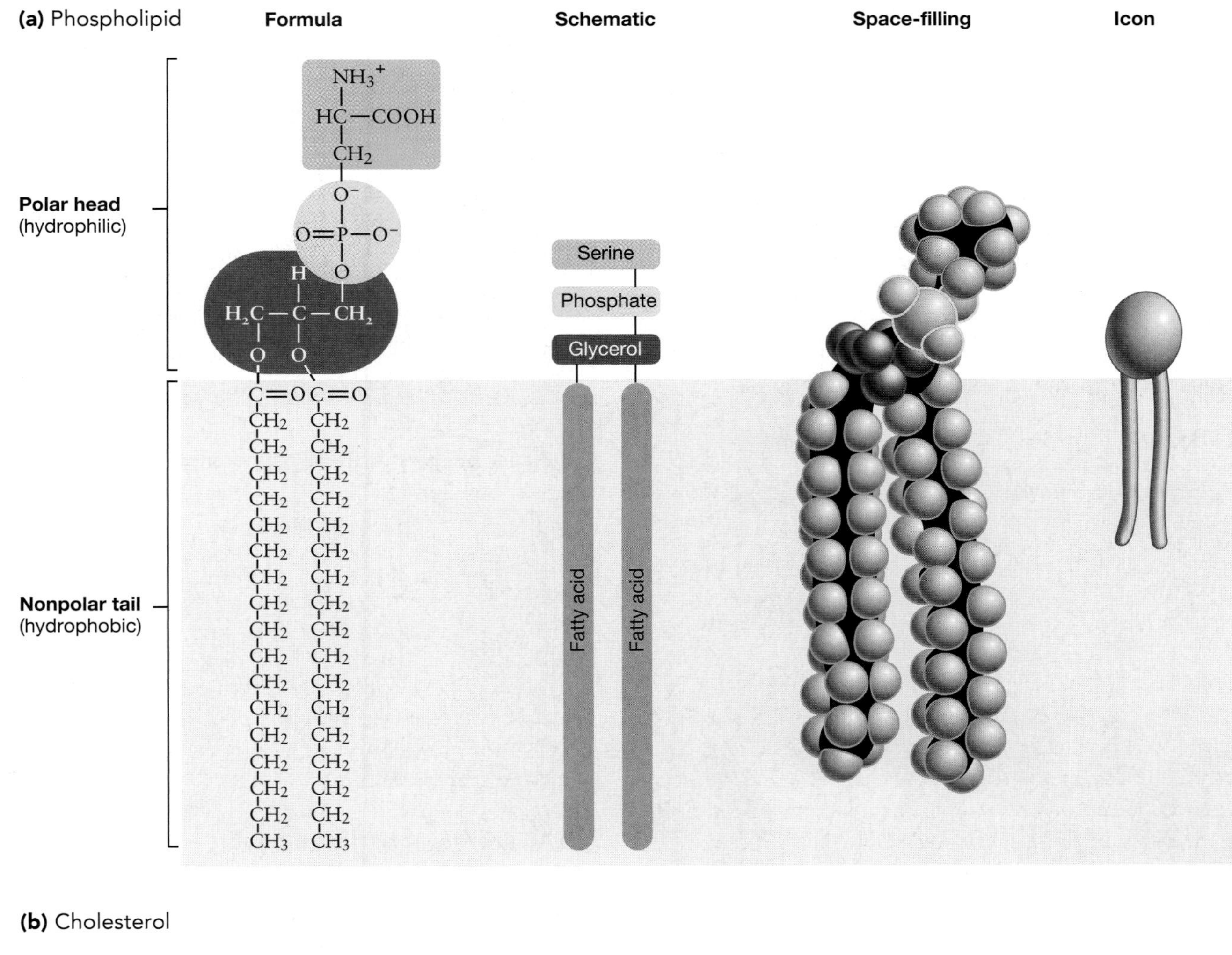

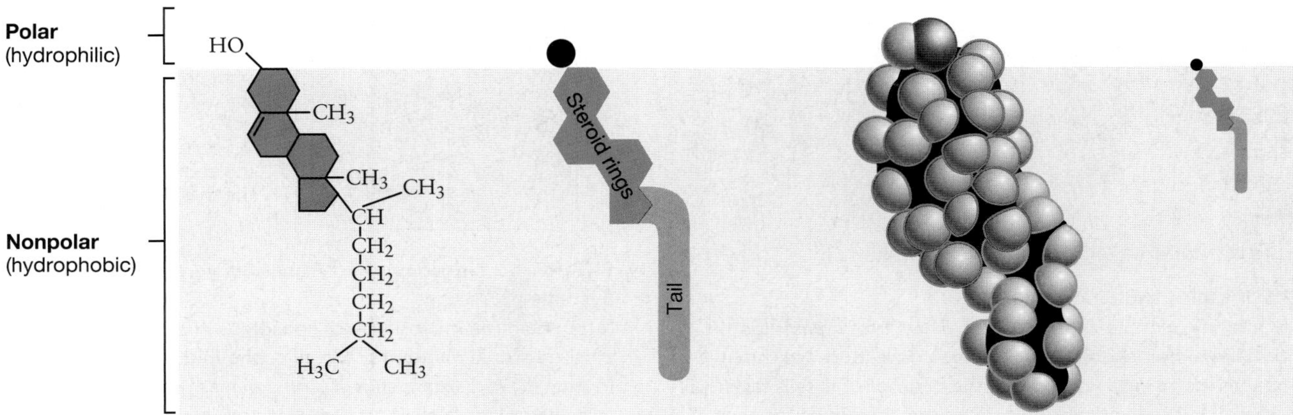

FIGURE 4.4 Amphipathic Lipids
Phospholipids (a) and cholesterol (b) are both amphipathic—they contain both hydrophilic and hydrophobic elements.

Another key aspect of micelle, bilayer, and liposome formation is that the structures form spontaneously. (*Spontaneously* is used here in the technical sense introduced in Chapter 2. It means that a chemical change occurs without an input of energy.)

The discovery of liposomes sparked intense interest among biologists because lipid bilayers also form the basic structure of cell membranes (**Figure 4.5c**). This interest led to an intensive effort to understand the properties of membranes by creating artificial bilayers and experimenting with them.

Some of the first questions posed by researchers concerned the permeability of lipid bilayers. Once a membrane forms a water-filled vesicle, can other molecules or ions pass in or out? If so, is this permeability selective in any way? The permeability of membranes is a critical issue because if certain molecules or ions pass through a lipid bilayer more readily than others, it means that the internal environment of a vesicle can become different from the outside. This difference between exterior and interior environment is a key characteristic of living cells. To explore the permeability of phospholipid bilayers, biologists began performing experiments on artificial membranes.

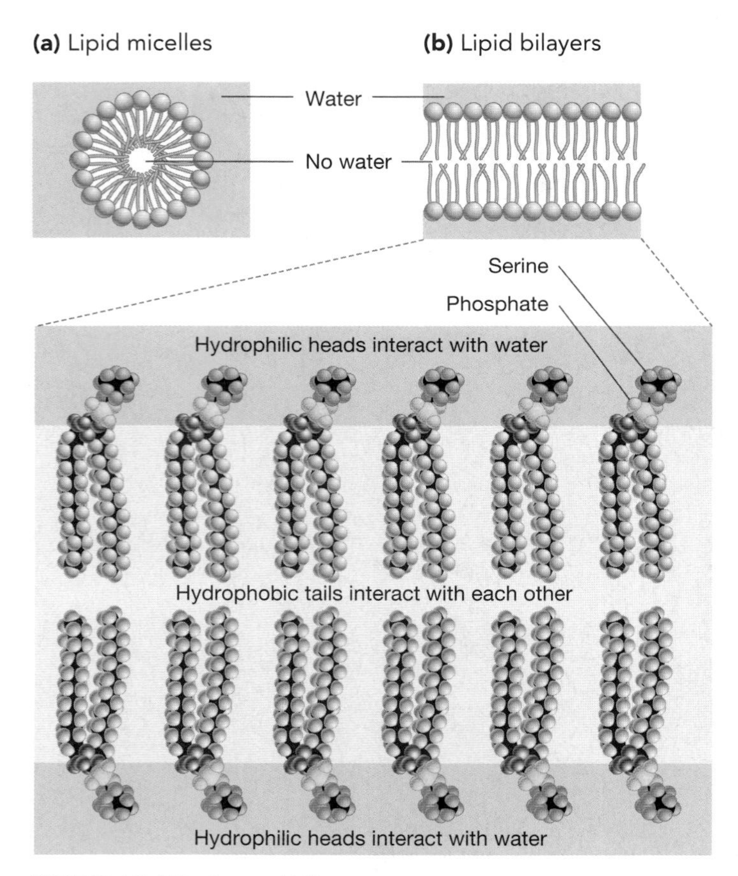

FIGURE 4.5 Micelles and Bilayers
(a) Micelles are tiny, globule-like structures. They form if amphipathic compounds have extremely short hydrophobic tails, because the hydrophilic heads face toward water while the hydrophobic tails face in. **(b)** In a lipid bilayer, the hydrophilic heads of phospholipids face out, toward water, while the hydrophobic tails face in, away from water. Cell membranes consist in part of phospholipid bilayers, as illustrated in the bottom drawing.

Artificial Membranes as an Experimental System

Researchers use two types of artificial membranes to study the permeability of lipid bilayers. **Figure 4.6a** shows the roughly spherical vesicles called liposomes. **Figure 4.6b** illustrates the lipid bilayers called planar bilayers, which are con-

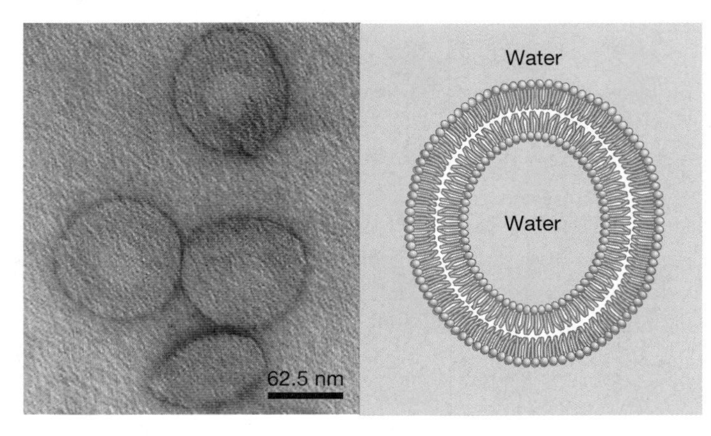

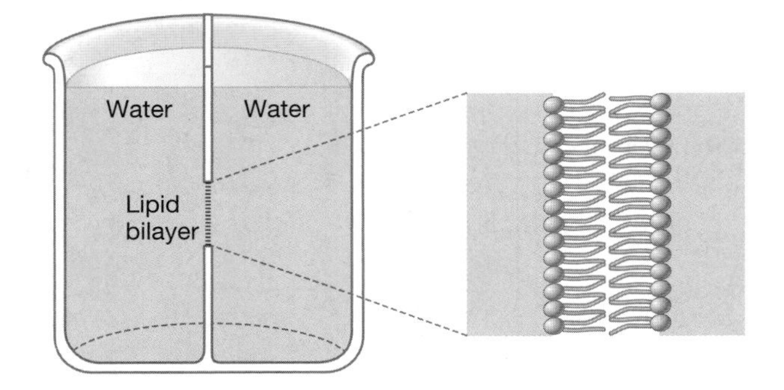

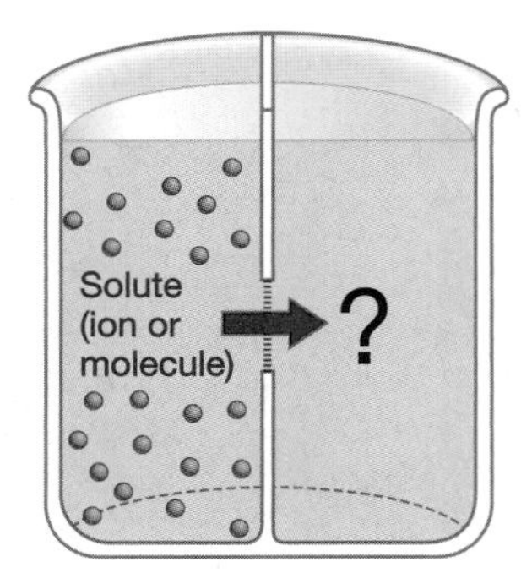

FIGURE 4.6 Liposomes and Planar Bilayers as Experimental Systems
(a) An electron micrograph of liposomes in cross section (left) and a cross-sectional diagram of the phospholipid bilayer in a liposome (right). **(b)** A diagram showing how planar bilayers are constructed across a hole in a glass wall separating two water-filled compartments (left), and a cartoon showing a close-up of the bilayer itself (right). **(c)** A wide variety of experiments are possible with planar bilayers and liposomes. A few are suggested here.

structed across a hole in a glass or plastic wall separating two aqueous solutions.

Using planar bilayers and liposomes, researchers can study what happens when a known ion or molecule is added to one side of a lipid bilayer (**Figure 4.6c**). Does the ion or molecule cross the membrane and show up on the other side? If so, how rapidly does the movement take place? What happens when a different type of phospholipid is used to make the artificial membrane? Does the membrane's permeability change when proteins or other types of molecules are added to it?

Biologists describe such an experimental system as elegant and powerful because it gives researchers precise control over which factor changes from one experimental treatment to the next. Control, in turn, is why experiments are such an effective way to explore scientific questions. A good experimental design allows researchers to alter one factor at a time and determine what effect, if any, each has on the process being studied.

Equally important, planar bilayers and liposomes provide a clear way to determine whether a given change in conditions has an effect. By sampling the solution on either side of the membrane before and after the treatment and then analyzing the concentration of ions and molecules in the samples, researchers have an **assay** to determine whether the treatment had any consequences.

What, then, have biologists learned about membrane permeability using such systems?

Selective Permeability of Lipid Bilayers

When researchers put molecules (or ions) on one side of a planar bilayer or a liposome and measure the rate at which the molecules arrive on the other side, a strong pattern emerges. Phospholipid bilayers are *highly* selective. Small, nonpolar molecules move across the bilayer quickly. In contrast, most charged species cross the membrane slowly, if at all. According to the data in **Figure 4.7**, small, nonpolar molecules like oxygen (O_2) move across membranes well over a billion times faster than do chloride ions (Cl^-). Larger molecules can also move rapidly if they are nonpolar. Indole, for example, is a large, nonpolar compound that moves across membranes 100 million times faster than do potassium ions (K^+). Very small and uncharged molecules like water (H_2O) can also cross membranes extremely rapidly, even though they are polar.

The leading hypothesis to explain this pattern is that large or charged compounds can't pass through the nonpolar, hydrophobic tails of a lipid bilayer. The reasoning here is that ions are more stable in solution than they are in the electrically neutral interior of membranes. To test this hypothesis, researchers have manipulated the size and structure of the tails in planar bilayers or liposomes.

Does the Type of Lipid in a Membrane Affect Its Permeability?

Theoretically, two aspects of a hydrocarbon chain could affect the way it behaves in a lipid bilayer—the number of double

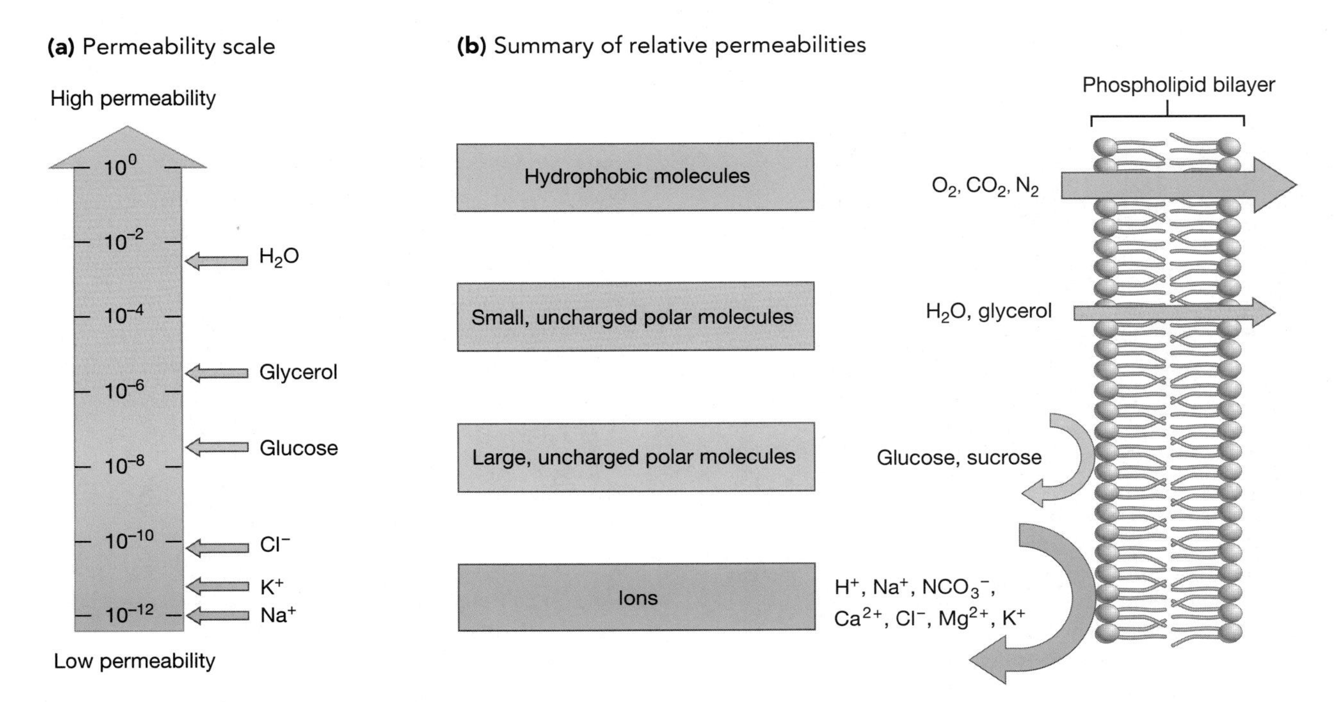

FIGURE 4.7 Permeability of Phospholipid Bilayers to Ions and Molecules
(a) The numbers inside the arrow represent "permeability coefficients," or the rate (cm/sec) at which an ion or molecule crosses a phospholipid bilayer. **(b)** This diagram summarizes the relative permeabilities of various molecules and ions, based on data like those presented in part (a). QUESTION About how fast does water cross the lipid bilayer?

bonds it contains and its length. Recall, from Chapter 2, that the shape of carbon-containing molecules changes when carbon atoms form a double bond. Instead of forming a three-dimensional tetrahedron, a double bond puts carbon and its attached atoms into a single plane (**Figure 4.8a**). The carbon atoms involved are also locked into place; they cannot rotate freely, as they do in carbon-carbon single bonds. As a result, a double bond between carbon atoms produces a "kink" in an otherwise straight hydrocarbon chain (**Figure 4.8b**).

(a) The angles of carbon bonds

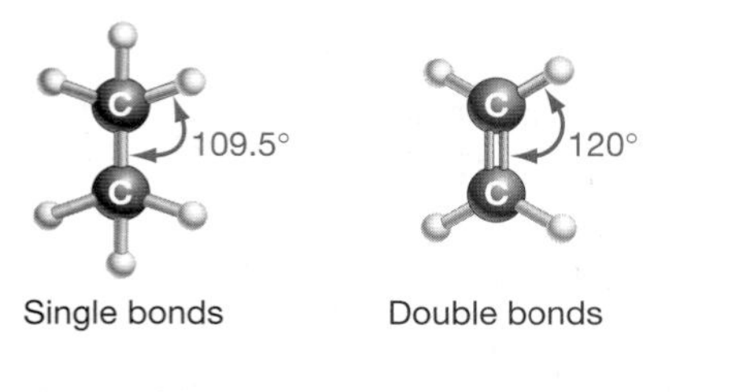

(b) Double bonds cause kinks in hydrocarbons.

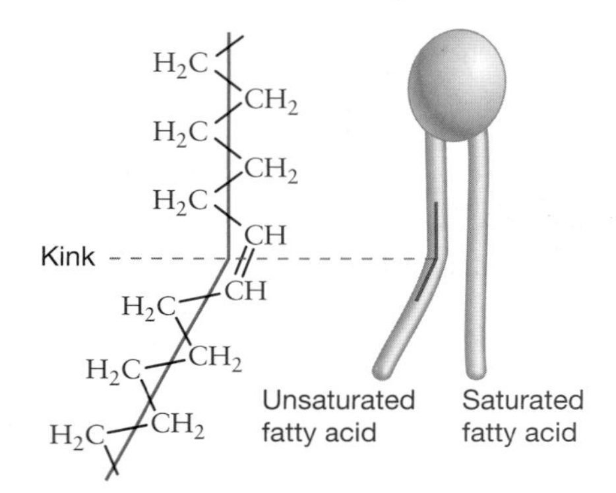

(c) Kinks change the permeability of membranes.

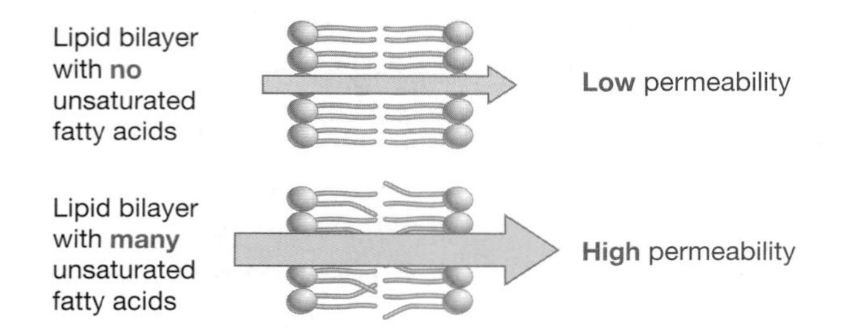

FIGURE 4.8 Unsaturated Hydrocarbon Chains
(a) When carbon forms bonds with four other atoms, the bonding orbitals are in a tetrahedral arrangement with each pair of bonds forming an angle of 109.5°. When carbon forms a double bond, as shown on the right, the bonding orbitals are in a planar arrangement with each pair of bonds forming an angle of 120°. **(b)** The double bond(s) found in an unsaturated hydrocarbon produce a "kink" (shown exaggerated) in an otherwise straight chain. The icon on the right indicates that one of the hydrocarbon tails in a phospholipid is unsaturated. **(c)** Phospholipid bilayers that contain many unsaturated fatty acids contain more gaps and should be more permeable.
EXERCISE Draw figures analogous to those in part (b) for a hydrocarbon chain containing two double bonds. (Hydrocarbon chains like these are called polyunsaturates because *poly* means "many.")

When a double bond exists between two carbon atoms in a hydrocarbon chain, the chain is said to be **unsaturated.** Conversely, hydrocarbon chains without double bonds are said to be **saturated.** This choice of terms is logical because if a hydrocarbon chain does not contain a double bond, it is saturated with the maximum number of hydrogen atoms that can attach to the carbon skeleton. If it is unsaturated, then fewer than the maximum number of hydrogen atoms are attached.

Why do double bonds affect the permeability of membranes? When hydrophobic tails are packed into a lipid bilayer, the kinks created by double bonds produce spaces between the tightly packed tails. These spaces reduce the strength of interactions among the hydrophobic tails. Because the interior of the membrane is "glued together" less tightly, the structure should become more fluid and more permeable (**Figure 4.8c**). In contrast, hydrophobic interactions should become stronger as saturated hydrocarbon tails increase in length. Membranes dominated by phospholipids with long, saturated hydrocarbon tails should be stiffer and less permeable because the interactions among the tails are stronger.

A biologist would predict, then, that bilayers made of lipids with long, straight, saturated tails should be much less permeable than membranes made of lipids with short, kinked, unsaturated tails. Experiments on liposomes by J. de Gier and colleagues have shown *exactly* this pattern. Phospholipids with long, saturated tails form membranes that are much less permeable than membranes consisting of phospholipids with shorter, unsaturated tails.

de Gier and co-workers also investigated how the addition of cholesterol molecules affects membrane permeability. **Figure 4.9a** shows a model of how cholesterol, as an amphipathic lipid found in many cell membranes, could fit between phospholipids. Because cholesterol is expected to fill spaces between phospholipids, adding cholesterol to a membrane should increase the density of the hydrophobic section. In accordance with this prediction, de Gier and colleagues found that adding cholesterol molecules to liposomes dramatically reduced their permeability. The data behind this claim are presented in **Figure 4.9b**. This graph makes another important point, however, which is that temperature has a strong influence on the behavior of lipid bilayers.

Does Temperature Affect the Fluidity and Permeability of Membranes?

At about 25°C, or "room temperature," the phospholipids found in cell membranes are liquid, and bilayers have the consistency of olive oil. This fluidity, as well as the membrane's permeability, decreases as the temperature decreases. As temperatures drop, the individual molecules in the bilayer move more slowly. As a result, the hydrophobic tails in the interior of membranes pack together more tightly. At very low temperatures phospholipid bilayers begin to solidify. As the data in **Figure 4.9b** indicate, low temperatures can make membranes impervious to molecules that would normally move across them readily.

The fluid nature of membranes also allows individual lipid molecules to move laterally within each layer, a little like a person moving about in a dense crowd. By tagging individual phospholipids and following their movement, researchers have clocked average speeds of 2 micrometers (μm)/second at room temperature. At these velocities, phospholipids travel the length of a small bacterial cell every second.

The overall message of these experiments on lipid and ion movement is that membranes are dynamic. Phospholipid molecules whiz around each layer while water and small, nonpolar molecules shoot in and out of the membrane. How quickly molecules move within and across membranes is a function of temperature and the structure of the phospholipid tails.

(a) Cholesterol fills spaces between phospholipids.

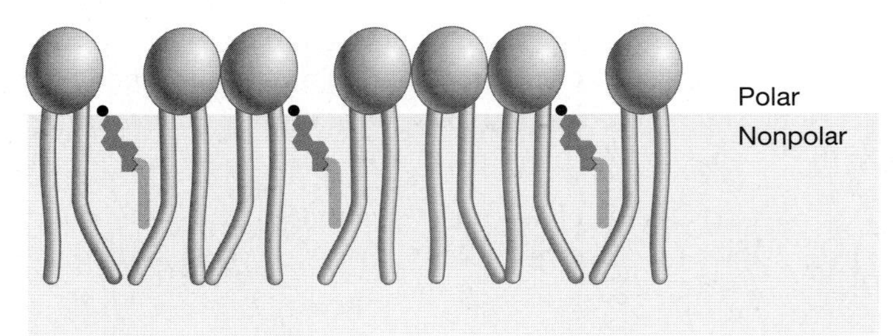

(b) As a result, membrane permeability decreases.

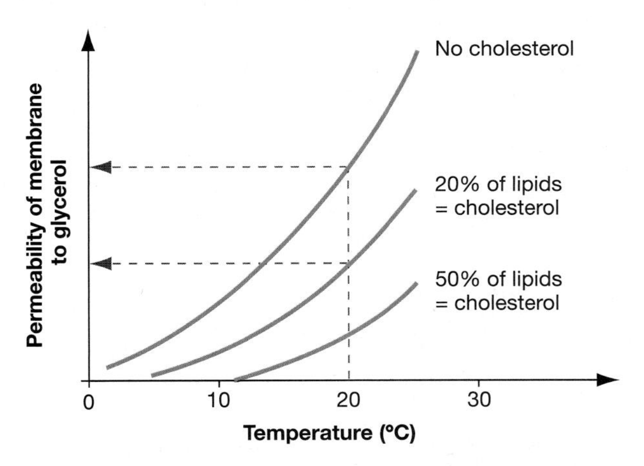

FIGURE 4.9 Adding Cholesterol to a Membrane Affects its Permeability
(a) This sketch is based on the relative sizes of phospholipids and cholesterol molecules. Cholesterol fills the gaps created by unsaturated (kinked) hydrocarbon tails. **(b)** On this graph the abscissa (x-axis) plots increasing temperature, and the ordinate (y-axis) plots increasing permeability of liposomes to glycerol molecules. Note that adding cholesterol reduces the permeability of membranes to glycerol, and that permeability increases with temperature.
EXERCISE Circle and label the hydrophilic and hydrophobic sections of cholesterol in part (a).

Given these insights into the permeability and fluidity of phospholipid bilayers, an important question remains. *Why* do certain molecules move across membranes spontaneously?

CD ACTIVITY 4.1 Diffusion and Osmosis

4.3 Why Molecules Move Across Lipid Bilayers: Diffusion and Osmosis

A thought experiment can help explain why molecules and ions are able to move across membranes spontaneously. Suppose you racked up a set of billiard balls. But instead of scattering them with the cue ball, you begin to vibrate the table. Because of the vibration, the balls will begin to move about randomly. They will also begin bumping into one another. After these collisions, some balls will move outward—away from their original position. In fact, the overall (or net) movement of balls will be outward. This occurs because the direction of the bumps between balls is non-random, even though the motion of individual balls is random. To be specific, the balls are much more likely to be bumped from the side where most of the other balls are. After still more moving and bumping, more balls will be found even further away. Eventually, the billiard balls will be distributed randomly across the table. The entropy of the billiard balls has increased.

This hypothetical example illustrates why molecules or ions placed on one side of a phospholipid bilayer move to the other side spontaneously. The molecules and ions have thermal energy and are in constant, random motion. Because they bump into one another more frequently in areas of high concentration than they do in areas of low concentration, they tend to move from a region of high concentration to a region of low concentration. This directed movement of molecules and ions is called **diffusion**. Diffusion occurs along a concentration gradient because molecules and ions move from regions of high concentration to regions of low concentration. Diffusion is a spontaneous process because it results in an increase in entropy.

Once the molecules or ions are randomly distributed throughout a solution, equilibrium is established. For example, consider the case of solutions separated by a phospholipid bilayer. **Figure 4.10** (page 78) shows how molecules that pass through the bilayer diffuse to the other side. At equilibrium, molecules continue to move back and forth across the membrane, but at equal rates—simply because each molecule or ion is equally likely to be bumped from any direction. This means that there is no longer a *net* movement of molecules across the membrane.

What about water itself? As the data in Figure 4.7 showed, water moves across phospholipid bilayers extremely quickly. Like other substances that diffuse, water moves along its concentration gradient—from higher to lower concentration. The movement of water is a special case of diffusion that is given its own name: **osmosis**. Osmosis occurs only when solutions are separated by a membrane that is permeable to some molecules but not others. Membranes like this are said to be selectively permeable.

The best way to think about water moving in response to a concentration gradient is to focus on the concentration of dissolved molecules and ions, or solutes, in the solution. Let's suppose the concentration of solutes is higher on one side of a selectively permeable membrane than it is on the other (**Figure 4.11a**). Further, suppose that the solutes in question cannot diffuse through the membrane to establish equilibrium. What happens then? Water will move from the side with a lower concentration of solutes to the side with a higher concentration of solutes (**Figure 4.11b**). It dilutes the higher concentration and equalizes the concentrations on each side. This movement is spontaneous.

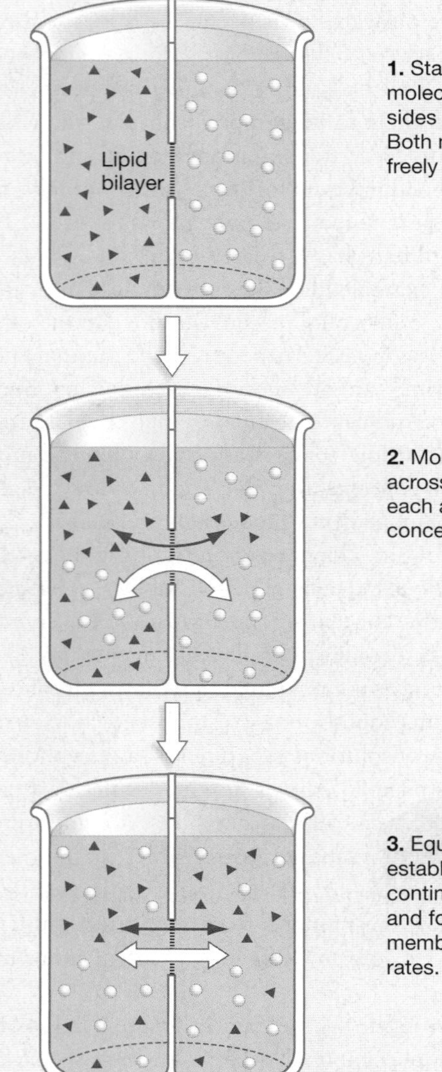

FIGURE 4.10 Diffusion Across a Semipermeable Membrane
QUESTION Suppose that in a follow-up to the experiment diagrammed here, you doubled the number of "purple triangle" molecules present on the left side of the semipermeable membrane (at the start). Would its rate of diffusion double?

It is driven by the increase in entropy achieved when solute concentrations are equal on either side of the membrane.

This movement of water is important because it can swell or shrink a membrane-bound vesicle. Consider the liposomes illustrated in **Figure 4.12**. If the solution outside the membrane has a higher concentration of solutes than the interior, and the solutes are not able to pass through the phospholipid bilayer, then water will move out of the vesicle and make the membrane shrivel. Conversely, if the solution outside the membrane has a lower concentration of solutes than the exterior, water will move into the vesicle and cause it to swell, or even burst. If solute concentrations are equal, the liposome will maintain its size.

Figure 4.12 also illustrates the vocabulary that biologists use to describe the relationships between solutions on either side of a membrane. A solution is **hypertonic** (excess-tone) if it causes a cell or liposome to shrink. A solution is **hypotonic** (insufficient-tone) if it causes the cell or liposome to swell. If the solution does not affect the membrane's shape, it is **isotonic** (equal-tone).

In sum, diffusion and osmosis are responsible for the movement of solutes and water across lipid bilayers. What does all this have to do with the first membranes floating around in the prebiotic soup? In effect, osmosis and diffusion *reduce* differ-

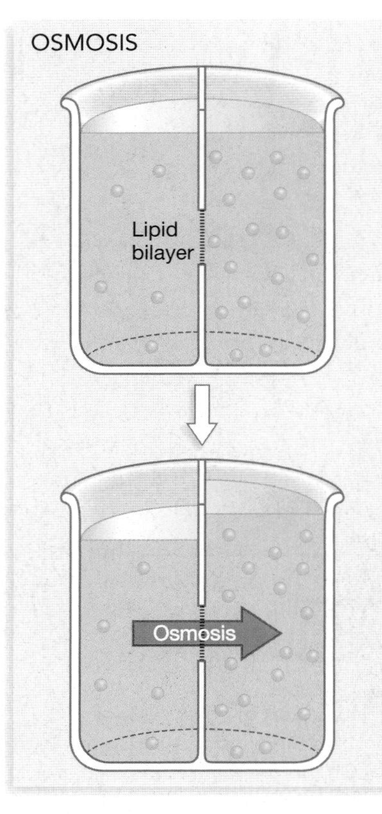

FIGURE 4.11 Osmosis
QUESTION Suppose you doubled the number of molecules on the right side of the membrane (at the start). At equilibrium, would the water level on the right side be higher or lower than what is shown here?

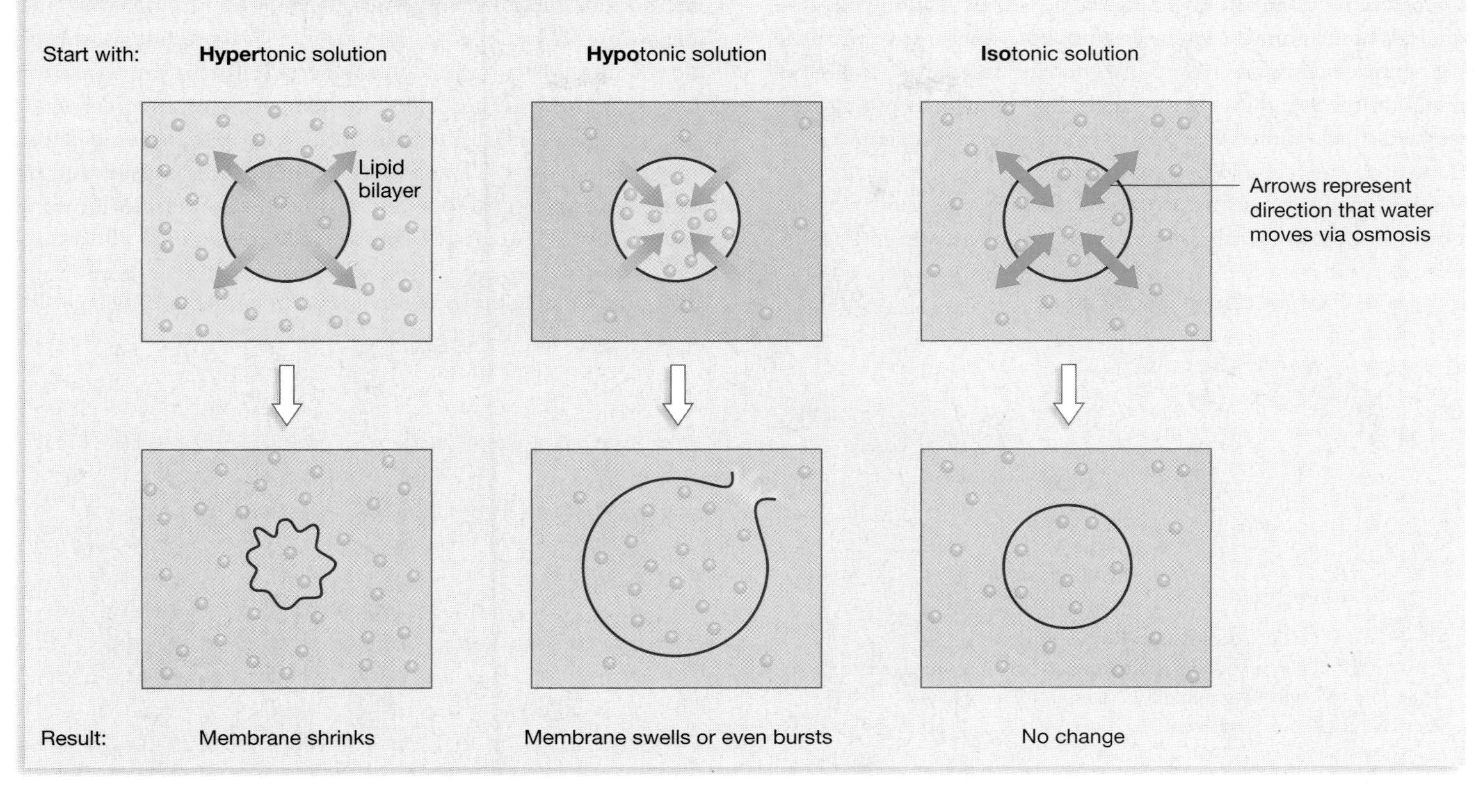

FIGURE 4.12 Osmosis Can Shrink or Burst Membrane-Bound Vesicles
QUESTION Some species of bacteria can live in extremely salty environments, like saltwater-evaporation ponds. Is this habitat likely to be hypertonic, hypotonic, or isotonic to the interior of the cells?

ences in chemical composition between the inside and outside of membrane-bound structures. If liposome-like structures were present in the prebiotic soup, it's unlikely that their interiors offered a radically different environment from the surrounding solution. In all likelihood, the primary importance of the first lipid bilayers was simply to provide a container for self-replicating molecules.

The task now is to investigate how a lipid bilayer can become a barrier capable of creating and maintaining a specialized internal environment—one that is conducive to life. How could a prebiotic bubble, like a liposome, become an effective cell membrane—one that admits ions and molecules needed by the replicator while excluding ions and molecules that might damage the replicator?

4.4 Membrane Proteins

What sort of molecule could become incorporated into a lipid bilayer that would affect its permeability? The title of this section gives the answer away. Proteins that are amphipathic can be inserted into lipid bilayers. Proteins can be amphipathic because they are made up of amino acids and because amino acids have side groups, or R-groups, that range from highly nonpolar to highly polar (or even charged; see Figure 3.3b and Table 3.1). It is conceivable, then, that a protein could have a series of nonpolar amino acids in the middle of the protein sequence and have polar or charged amino acids on either end of the protein sequence, as illustrated in **Figure 4.13a** on page 80. The nonpolar amino acids would be stable in the interior of a lipid bilayer, while the polar or charged amino acids would be stable alongside the polar heads and surrounding water (**Figure 4.13b**). Further, because the secondary and tertiary structures of proteins are almost limitless in their variety and complexity, it is possible to imagine that proteins could form tubes and thus function as some sort of channel or pore across a phospholipid bilayer.

Based on these theoretical considerations, it is not surprising that when researchers began analyzing the chemical composition of cell membranes in living organisms, they found that proteins were just as common, in terms of mass, than were phospholipids.

This section addresses two questions about these membrane proteins: How do biologists study them? How do these proteins affect the passage of ions and molecules?

Systems for Studying Membrane Proteins

When biologists begin studying a phenomenon, much of their early work is necessarily descriptive in nature. This is sensible because researchers usually need to understand *what* is happening before they can ask how and why it occurs. Accordingly, this analysis of membrane proteins begins with some straightforward, descriptive questions. Where are proteins located in biological membranes? Are they big or small? Can individual proteins be isolated so researchers can perform experiments on them?

Visualizing Membrane Proteins Many of biologists' basic questions about membrane proteins can be answered by simply looking at them. An innovative technique for visualizing the surface of cell membranes, developed in part by Daniel Branton and students in the early 1970s, provides a way to do just that. The process is called freeze-fracture electron microscopy, because the steps involve freezing and fracturing the membrane before examining it with a scanning electron microscope. (**Box 4.2** explains the difference between a scanning electron microscope and a transmission electron microscope.) As **Figure 4.14** shows, the technique allows re-

(a) Proteins can be amphipathic.

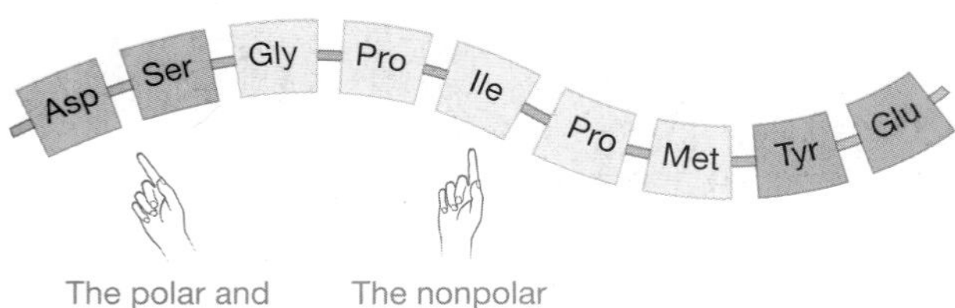

(b) Amphipathic proteins can integrate into lipid bilayers.

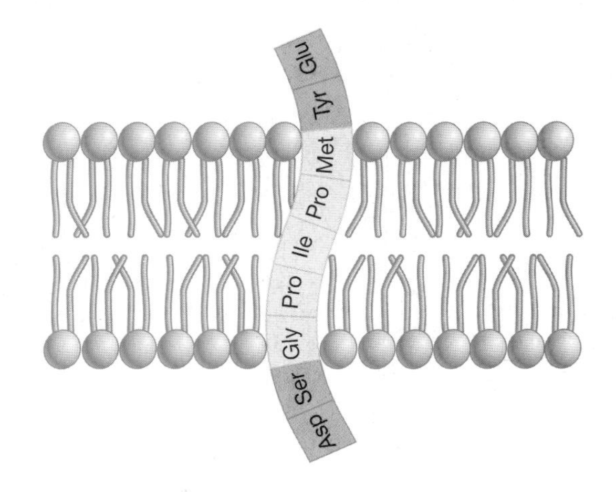

FIGURE 4.13 Proteins Can Be Amphipathic
Note that these drawings are conceptual and not to scale. It actually takes at least 20 amino acids to span a lipid bilayer. EXERCISE Label the types of molecular interactions that stabilize the membrane protein illustrated in part (b).

BOX 4.2 The Scanning Electron Microscope

The scanning electron microscope, or SEM, is the most useful tool biologists have for looking at the surfaces of cells. Materials are prepared for scanning electron microscopy by coating their surfaces with a layer of metal atoms. This is in contrast to TEM, which utilizes sectioned material impregnated with metal atoms. SEM allows researchers to inspect the surfaces of cells; TEM allows researchers to examine their interior.

"Scanning" is descriptive of how SEM works. To create an image of a surface, the instrument scans it with a narrow beam of electrons. Electrons that are reflected back from the surface or that are emitted by the metal atoms in response to the beam then strike a detector. The signal from the detector controls a second electron beam, which scans a TV-like screen and forms a magnified image of up to 20,000 times. Because SEM records shadows and highlights, it provides images with a three-dimensional appearance. It cannot magnify objects nearly as much as TEM, however.

BOX 4.3 The Fluid-Mosaic Model

Chemical analyses conducted in the 1930s confirmed that cell membranes consist of phospholipids and proteins. How were these two types of molecules arranged? In 1935, Hugh Davson and James Danielli proposed that cell membranes were structured like a sandwich, with hydrophilic proteins coating either side of a pure phospholipid bilayer. Early electron micrographs of cell membranes seemed to be consistent with the sandwich model, and by the 1960s it was widely accepted.

The observation that membrane proteins are amphipathic led S. Jonathon Singer and Garth Nicolson to propose an alternative hypothesis, however. In 1972, they proposed that at least some proteins span the membrane instead of being found only on either side of the lipid bilayer. They called their hypothesis the fluid-mosaic model of membrane structure. The name emphasized the dynamic, fluid nature of the structure and the hypothesis that it represented a mosaic of phospholipids and diverse proteins. As Figure 4.14 shows, data from freeze-fracture preparations conflicted with the Davson-Danielli hypothesis but were consistent with the fluid-mosaic model. Based on these and subsequent observations, the fluid-mosaic model is now widely accepted.

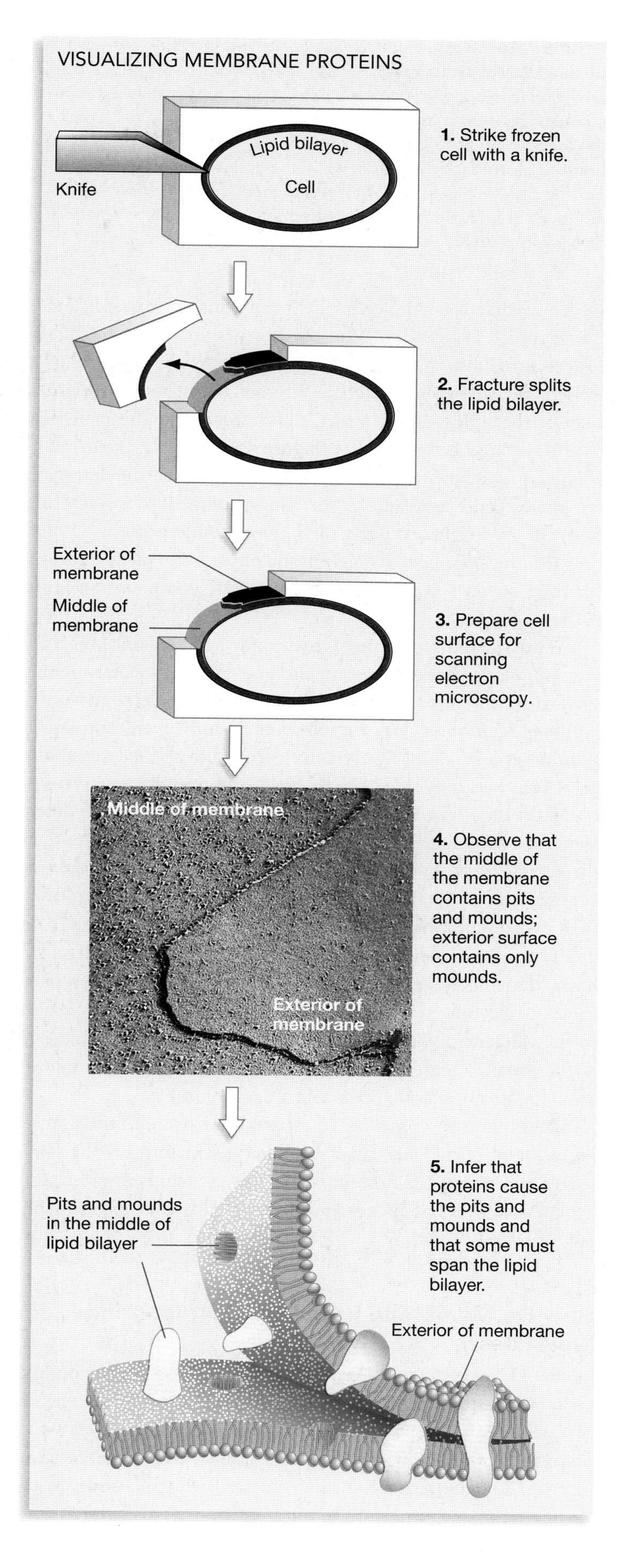

FIGURE 4.14 Freeze-Fracture Preparations Allow Biologists to View Membrane Proteins
Viewing freeze-fracture preparations with the scanning electron microscope confirmed that some proteins span cell membranes. Transmembrane proteins contact both the interior and exterior of the cell.

searchers to split a lipid bilayer and view the middle of the structure.

Photos of freeze-fractured membranes are revealing. The micrograph in Figure 4.14, for example, shows that pit- and mound-like structures stud the inner surfaces of the lipid bilayer. Researchers interpret these structures as the locations of membrane proteins. As step 5 in Figure 4.14 shows, the pits and mounds are hypothesized to represent proteins that span the lipid bilayer. **Box 4.3** explains how this finding helped confirm the current model of membrane structure.

Isolating Membrane Proteins from Living Cells Proteins might span cell membranes and yet have no effect at all on membrane permeability. They could be structural components that influence only membrane strength or flexibility. To test whether proteins can affect membrane permeability, researchers needed some way to isolate and purify membrane proteins for use in experiments.

An effective way to remove proteins from cell membranes and put the proteins into solution is to treat the cell membranes with a detergent. A **detergent** is a small, amphipathic

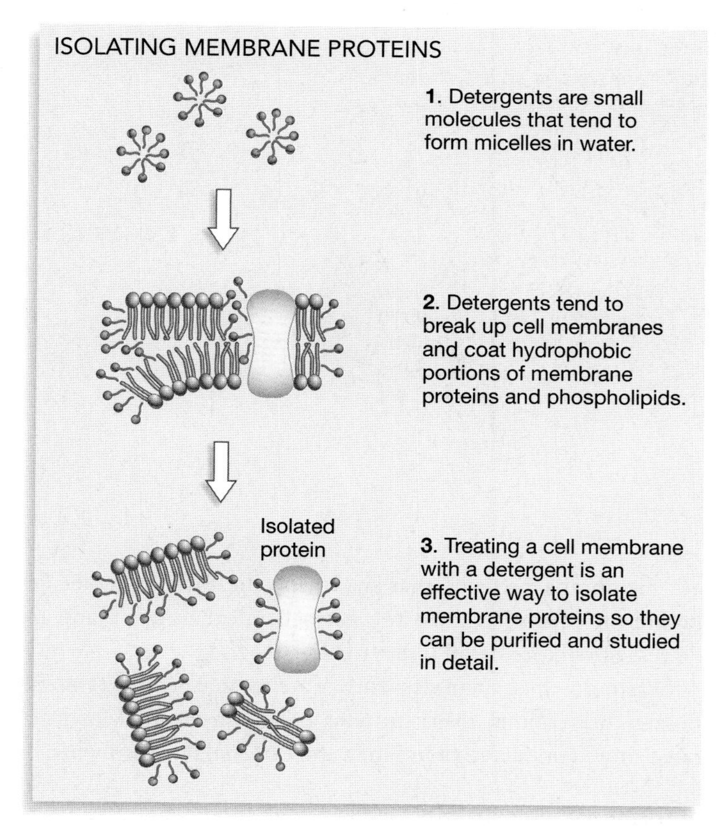

FIGURE 4.15 How Detergents Can Be Used to Get Membrane Proteins into Solution

FIGURE 4.16 Electrochemical Gradients and Ion Flow Through the Channel Called Gramicidin
(a) When ions build up on one side of a membrane, they establish a combined concentration and electrical gradient. **(b)** In this graph, the amount of electric current produced by ion flow across a membrane is plotted on the y-axis, while the concentration of ions is plotted on the x-axis. **(c)** When researchers built a ball-and-stick model of gramicidin, they discovered that it formed a hole, or pore. This is a top-down view of the peptide.

(a) Electrochemical gradients across membranes

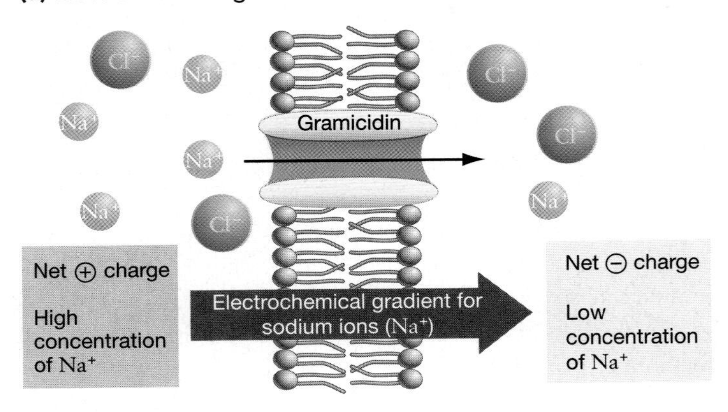

(b) Gramicidin increases ion flow across membrane.

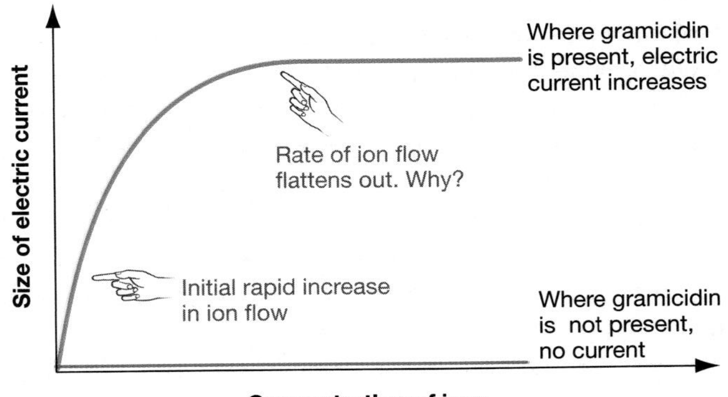

(c) Gramicidin is an ion channel.

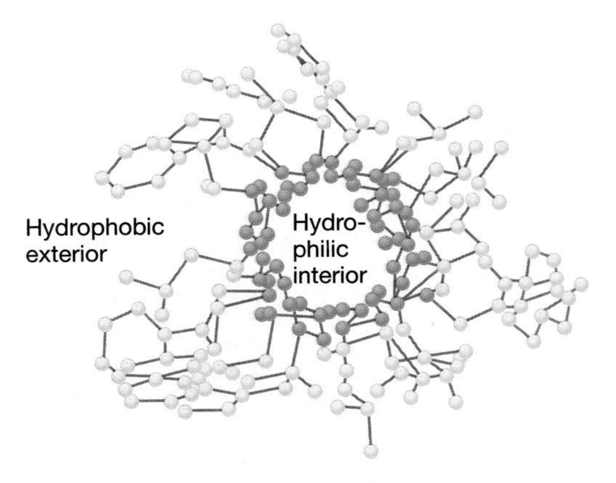

molecule that forms micelles in water—just as phospholipids do (**Figure 4.15, step 1**). When detergents are added to the solution surrounding a phospholipid bilayer, the hydrophobic tails of the detergent molecule interact with the hydrophobic tails of the lipids. In doing so, the detergent tends to disrupt the bilayer and break it apart (**Figure 4.15, step 2**). If the membrane contains proteins, the hydrophobic tails of the detergent molecules tend to displace the phospholipids by interacting with the hydrophobic parts of the membrane proteins. When they do, they displace the membrane phospholipids and end up forming water-soluble, detergent-protein complexes (**Figure 4.15, step 3**).

To isolate and purify these membrane proteins once they are in solution, researchers use a technique called gel electrophoresis (introduced in Box 3.2). **Gel electrophoresis** (electricity-moving) separates molecules based on their movement through a gelatinous substance (a gel) in an electric field applied across the gel. Molecules move through the gel—and thus are separated—based on their size and their charge. For example, if the detergent used to solubilize proteins carries a negative charge, then the detergent-protein complexes migrate toward the positively charged end of the gel. Larger molecules migrate more slowly than do smaller molecules, so the various proteins isolated from a cell membrane separate from each other with time. After a gel has had a voltage applied to it long enough for the protein molecules to become well separated, the voltage is turned off and the gel is stained with a dye that adheres to proteins. Each one of the proteins that was in the original solution has migrated to a unique position and appears as a band on the gel.

To obtain a pure sample of a particular protein, the appropriate band is cut out of the gel. That gel material is then dissolved to retrieve the protein. Once this protein is inserted into a planar bilayer or liposome, dozens of different experiments are possible.

How Do Membrane Proteins Affect Ions and Molecules?

In the 40 years since intensive experimentation on membrane proteins began, researchers have identified three broad classes of peptides or proteins—ionophores, transporters, and pumps—that affect membrane permeability. (A peptide consists of a small chain of amino acids.) The following subsections investigate the properties of ionophores and transporters. Pumps are membrane proteins that move specific ions or molecules into or out of a cell against an electrochemical gradient, and are discussed in more detail in later chapters.

What do these peptides and proteins do? Can cell membranes that contain these proteins create an internal environment more conducive to life than the external environment?

Ionophores: Facilitated Diffusion of Ions One of the first membrane peptides investigated in detail is called gramicidin. Gramicidin is produced by a bacterium called *Bacillus brevis* and is used medicinally as an **antibiotic** (against-life)—a molecule that kills bacteria. Gramicidin is extremely toxic to people if taken internally, but it can be used to fight skin infections. After observing that cells treated with gramicidin seemed to lose large numbers of ions, biomedical researchers became interested in understanding how the molecule works. At that point, membrane researchers became involved. Is it possible that gramicidin alters the flow of ions across cell membranes?

Paul Mueller and Donald Rubin answered this question by inserting purified gramicidin into planar bilayers. To assay whether gramicidin affected the permeability of phospholipid bilayers to ions, they measured the flow of electric current across the artificial membrane. Because ions carry a charge, the movement of ions produces an electric current. This gave Mueller and Rubin an elegant and clever test for assessing the bilayer's permeability to ions—one that was simpler and more sensitive than taking samples from either side of the membrane and determining the concentrations of solutes present.

Mueller and Rubin's technique points up an important fact about ion movement across membranes; that is, ions flow not only from regions of high concentration to low concentration but also from areas of like charge to areas of unlike charge (**Figure 4.16a**). Stated another way, ions move in response to a combined concentration and electrical gradient, or what biologists call an **electrochemical gradient.**

The result? The graph in **Figure 4.16b** shows Mueller and Rubin's data. When gramicidin was absent, no electric current passed through the membrane. But when gramicidin was inserted into the membrane, current began to flow. Based on this observation, Mueller and Rubin proposed that gramicidin is an **ionophore** (ion-mover). An ionophore is a peptide that makes phospholipid bilayers permeable to ions. Follow-up work by Valerie Myers and D. A. Haydon corroborated that gramicidin is selective. Only positively charged ions, or cations, pass through it. Negatively charged ions, or anions, cannot pass through gramicidin. Myers and Haydon also established that among cations, gramicidin is most permeable to hydrogen ions (or protons, H^+) and somewhat less permeable to other cations such as potassium (K^+) and sodium (Na^+).

Researchers gained additional insight into how gramicidin works when they determined its amino acid sequence and tertiary structure. **Figure 4.16c** gives a "top-down" view of the protein, showing that it forms a hole. The amino acids that line this channel are hydrophilic, while those on the exterior (in contact with the membrane phospholipids) are hydrophobic. Ionophores with this structure are called **ion channels.**

A variety of ionophores have now been characterized. Like gramicidin, they allow specific ions to pass through lipid bilayers and appear to be produced by bacteria as a weapon. Gramicidin, for example, is produced by *B. brevis* just before the bacterial cells form a resistant coating around their membranes. The ionophore wipes out competitors, giving *B. brevis* cells more room to grow when they emerge from the resistant phase. But unlike gramicidin, some ionophores do not form channels. As **Figure 4.17** shows, the antibiotic valinomycin acts as an ion carrier. The protein binds to a potassium (K^+) ion on the membrane's exterior, diffuses across the bilayer, and releases the cation on the interior.

It is important to recognize that molecules like gramicidin and valinomycin do not enable cell membranes to create an interior environment distinct from the exterior. Instead, they simply circumvent the lipid bilayer's impermeability to small, charged compounds. As a result, they *reduce* differences between the interior and exterior.

Can membrane proteins be more specific and admit only particular molecules needed by the replicator?

CD ACTIVITY 4.2 Facilitated Diffusion

Transporters: Facilitated Diffusion of Specific Molecules

Next to ribose, the six-carbon sugar called glucose is the most important sugar found in organisms. It is very likely that glucose was important to the earliest cells simply because virtually all cells alive today use it as a building block for important macromolecules or as a source of chemical energy. But as you saw in Figure 4.7, lipid bilayers are only moderately permeable

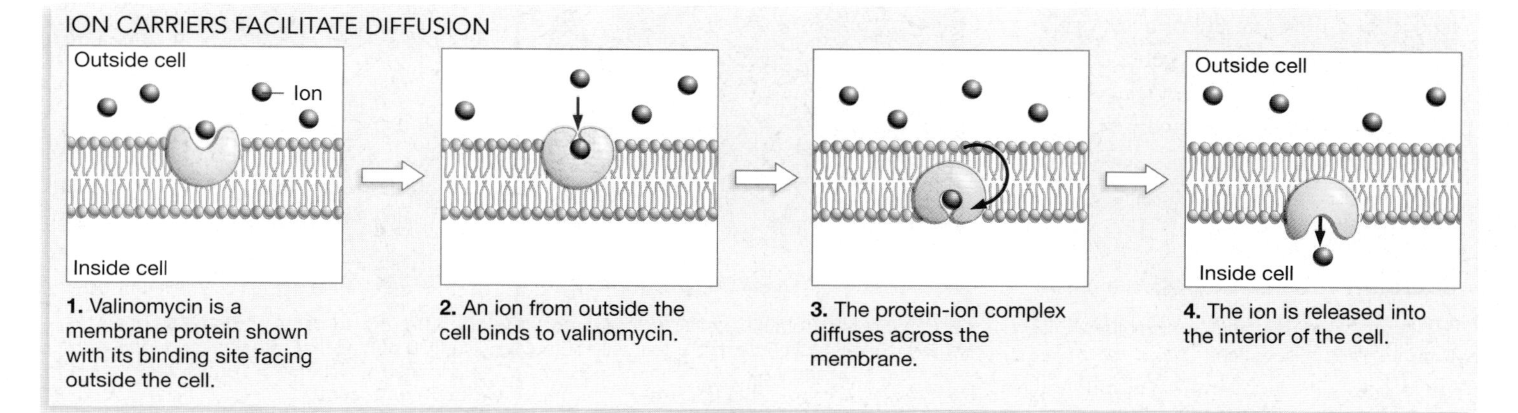

FIGURE 4.17 The Ionophore Valinomycin
Valinomycin is an ionophore that does not form a channel. It acts instead as an ion carrier.

to glucose. It is reasonable to expect, then, that cell membranes have some mechanism for increasing their permeability to glucose.

This expectation grew into a certainty when researchers compared the permeability of glucose across planar bilayers to its permeability across membranes from cells. The particular cell membrane studied in this case came from human red blood cells, which are among the simplest cells known. Mature red blood cells contain a membrane and hemoglobin molecules, and not much else (**Figure 4.18, step 1**). When these cells are placed in a hypotonic solution (**Figure 4.18, step 2**), water rushes into them by osmosis. As water flows inward, the cells swell. Eventually they burst, releasing the hemoglobin molecules and other cell contents. This leaves researchers with pure preparations of cell membranes called red blood cell "ghosts" (**Figure 4.18, step 3**). Experiments have shown that these membranes are much more permeable to glucose than are pure phospholipid bilayers. Why?

M. Kasahara and P. C. Hinkle were the first researchers to isolate the glucose transporter from red blood cells. The glucose transporter is the protein that increases the permeability of membranes to glucose. Kasahara and Hinkle added the purified protein to liposomes and demonstrated that it gave the artificial membrane the ability to transport glucose at the same rate as a membrane from a living cell. This experiment confirmed that a membrane protein was indeed responsible for transporting glucose across cell membranes. Follow-up work showed that the glucose transporter protein, called GLUT-1, transports the right-handed enantiomer of glucose, not the left-handed enantiomer. This is reasonable because cells use only the right-handed enantiomer of glucose.

Exactly how GLUT-1 works is a focus of ongoing research, however. Because glucose transport by GLUT-1 is so specific, researchers presume that the mechanism resembles the action of enzymes. One hypothesis, illustrated in **Figure 4.19**, proposes that glucose binds to GLUT-1 on the exterior of the membrane. This binding induces a conformational change in the protein that transports glucose to the interior of the cell. Recall from Chapter 3 that enzymes frequently change shape when they bind substrates, and that such conformational changes are often a critical step in catalysis of chemical reactions.

Importing molecules into cells via transporters is still powered by diffusion, however. When glucose enters a cell via GLUT-1, it does so because it is following its concentration gradient. If the concentration of glucose is the same on either side of the cell membrane, then no net movement of glucose occurs even if the membrane contains GLUT-1. Biologists refer to this type of transport as **facilitated diffusion** because the protein merely facilitates passage across the lipid bilayer. It is a form of passive transport because it does not require an expenditure of energy for transport.

It is also possible for cells to import molecules or ions *against* their concentration or electrical gradient. Accomplishing this task requires energy, however, because the cell must counteract the entropy loss that occurs when molecules are concentrated. It makes sense, then, that transport against a concentration gradient is called **active transport.** In addition to energy, active transport requires a machine; that is, a molecule capable of using energy to pump an ion or molecule against its concentration gradient.

What source of energy do cells use to accomplish this task? Chemical evolution may have been driven by heat energy from volcanoes, electric sparks, and ultraviolet radiation, but cells use chemical energy—the potential energy stored in chemical bonds—to do work. How do cells produce and store chemical energy? How do they produce the molecular machines required to perform tasks like active transport? Answering these and related questions is the focus of Unit 2.

HOW RESEARCHERS MAKE RED BLOOD CELL "GHOSTS"

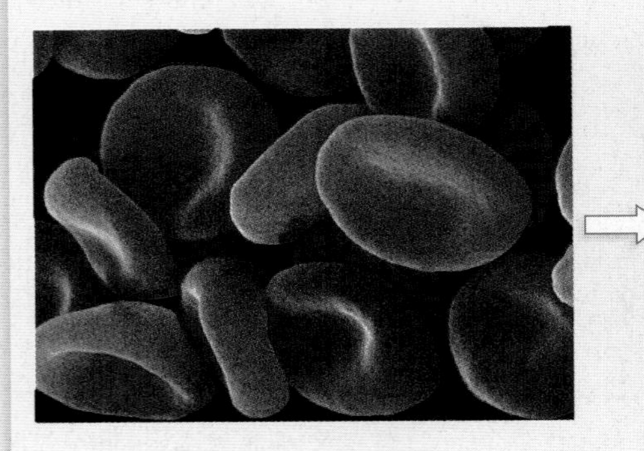

1. Normal blood cells in isotonic solution.

2. In hypotonic solution, cells swell as water enters via osmosis. Eventually the cells burst.

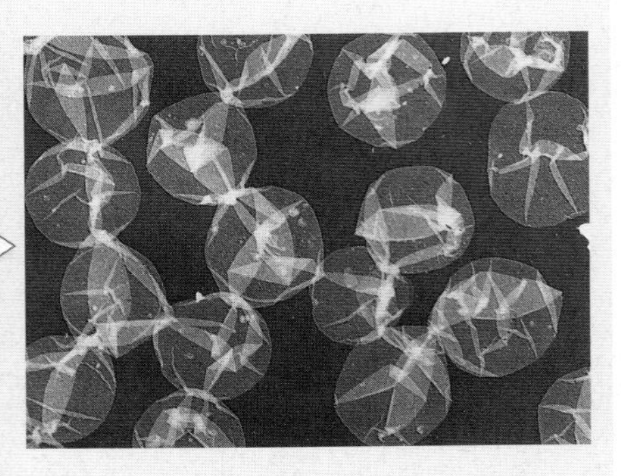

3. After the contents of the cells have spilled out, all that remains are cell "ghosts," which consist entirely of cell membranes.

FIGURE 4.18 Red Blood Cell "Ghosts"
Red blood cell ghosts are pure membranes that can be studied in detail.

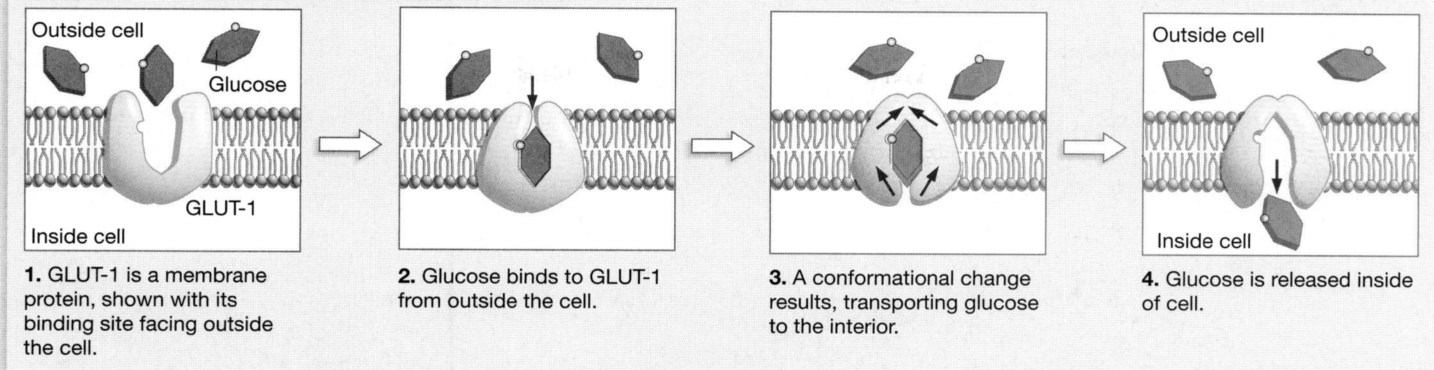

FIGURE 4.19 Hypothesis for Mechanism of Glucose Transport via GLUT-1
The model illustrated here suggests that the GLUT-1 transporter acts like an enzyme. It binds a substrate (in this case, a glucose molecule), undergoes a conformation change, and releases the substrate. QUESTION Hypotheses can be tested because they make specific predictions. State a prediction made by this model of GLUT-1 function. Describe how you would test the prediction.

Essay The Molecular Basis of Cystic Fibrosis

Cystic fibrosis (CF) is the most common genetic disease in humans of Northern European extraction. In these populations, one in every 2500 infants born has CF. Children affected by CF suffer from a progressive deterioration of their lungs, gastrointestinal tract, and reproductive tracts. Although the ability of physicians to manage the disease has improved markedly in recent years, the median life expectancy for patients with severe forms of CF is still only 29 years.*

Physicians now understand the molecular basis of cystic fibrosis.

CF produces a distinctive set of symptoms. Prominent among them are the production of salty sweat and the chronic development of thickened mucus in the linings of the lungs and associated air passages. The thickened mucus is extremely difficult to cough out and leads to two important complications: the direct obstruction of air passageways and rapid growth of disease-causing bacteria.

What causes these symptoms? Understanding the molecular basis of CF frustrated researchers for decades. Then an important clue came to light when P. M. Quinton showed that sweat-duct cells taken from CF patients are impermeable to chloride ions (Cl^-). As **Figure 1** on page 86 shows, sweat is produced as a salty solution. But as the solution moves through the duct toward the surface of the skin, chloride ions pass through the membrane and are reabsorbed by the body. This movement establishes an electrochemical gradient that causes sodium ions (Na^+) to follow. In CF patients, both ions stay in the interior of the sweat duct.

Quinton's result suggested that the disease might be caused by a defect in a membrane protein—one that normally allows chloride ions to move across lipid bilayers. Nine years later, Christine Bear and colleagues confirmed this result in spectacular fashion: Bear's group was able to isolate and purify quantities of the normal protein, which had come to be called cystic fibrosis transmembrane conductance regulator (CFTR). The researchers inserted the protein into planar bilayers and showed that it did indeed selectively permit the passage of chloride ions. Subsequent studies showed that people who suffer from CF produce abnormal forms of the protein. As a result, their chloride channels do not function.

Physicians now understand the molecular basis of cystic fibrosis. Cells that line the respiratory tract of humans normally release a steady stream of chloride ions, which when followed by sodium ions, makes the cells hypotonic relative to the exterior. This leads to an outflow of water. But when chloride ions are not released, water remains inside the cells. This leads to dramatically thickened mucus in the lungs and air passages, contributing to the fatal syndrome just described. Normal copies of CFTR are also involved in helping respiratory tract cells ingest and destroy certain disease-causing bacteria.

(Continued on page 86)

*The *median* is a value in a distribution of numbers, such as the life spans of people with CF. Specifically, half of the numbers in a distribution are found above the median value and half are found below. A median life expectancy of 29 years means that half of the people with CF live less than 29 years and half live more than 29 years.

(Essay continued)

The research that identified the protein's mode of action unleashed a torrent of new studies aimed at finding a cure. As this book goes to press, that work is still in progress. One strategy being tested is to synthesize large quantities of normal protein, package the protein channels into liposomes, and then deliver them to the respiratory lining of CF patients via an inhalator. The hope is that the liposomes will fuse with the cell membranes, introduce copies of CFTR that can function normally, and alleviate the symptoms of CF.

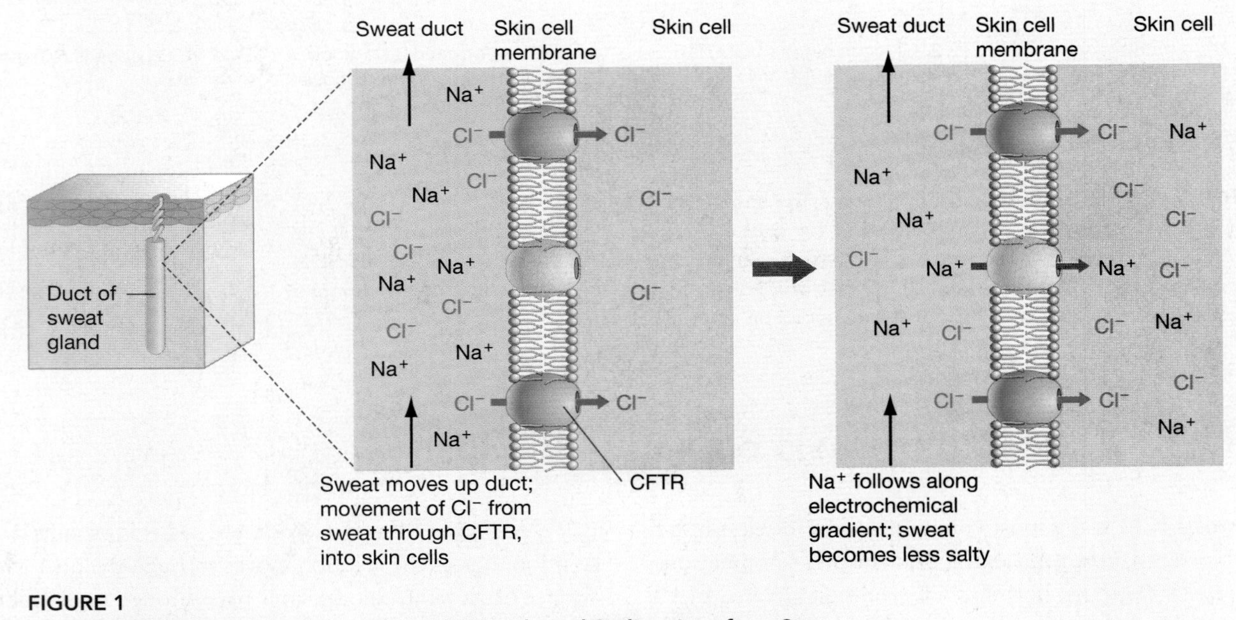

FIGURE 1
Normal CFTR Proteins Produce a Flow of Chloride and Sodium Ions from Sweat

Chapter Review

Summary

The cell membrane is a structure that forms a physical barrier between life and nonlife. Molecules called lipids, which do not dissolve in water, are a prominent component of cell membranes. Lipids called fatty acids and isoprenes, both of which have long hydrocarbon tails, are often found in larger molecules called phospholipids. Phospholipids have a polar head and nonpolar tail, and spontaneously form structures called micelles or bilayers when placed in water. The basic structure of cell membranes is created by a phospholipid bilayer.

Phospholipid bilayers are selectively permeable, meaning that only certain substances are able to pass through them. Small, nonpolar molecules tend to move across membranes readily; ions and other charged compounds cross rarely, if at all. The permeability and fluidity of phospholipid bilayers depends on temperature and on the types of phospholipids present. For example, because phospholipids that contain long, saturated fatty acids form a dense and highly hydrophobic membrane interior, they tend to be less permeable than phospholipids containing shorter, unsaturated fatty acids.

Molecules can move across membranes spontaneously because of diffusion. Diffusion is the directed movement of ions or molecules from a region of high concentration to low concentration, and is driven by an increase in entropy. Water also moves across membranes spontaneously if a molecule or ion that cannot cross the membrane is found in different concentrations on either side. In a process called osmosis, water moves from the region with a lower concentration of solutes to the region of higher solute concentration.

The permeability of phospholipid bilayers can be altered radically by membrane proteins. Ion channels, for example, are proteins that provide holes in the membrane, and facilitate the diffusion of certain species of ions into or out of the cell. Transporters are enzyme-like proteins that allow specific molecules to diffuse into the cell.

Questions

Content Review

1. What does the term *hydrophilic* mean, when it is translated literally?
   a. "oil loving"
   b. "water loving"
   c. "oil fearing"
   d. "water fearing"

2. If a solution is hypotonic relative to a cell, how will water move?
   a. It will diffuse into the cell.
   b. It will diffuse out of the cell.
   c. It will not move, because equilibrium exists.
   d. It will evaporate more rapidly.

3. If a solution is hypertonic relative to a cell, how will water move?
   a. It will diffuse into the cell.
   b. It will diffuse out of the cell.
   c. It will not move, because equilibrium exists.
   d. It will evaporate more rapidly.

4. When does a concentration gradient exist?
   a. when membranes rupture
   b. when solute concentrations are high
   c. when solute concentrations are low
   d. when solute concentrations differ on either side of a membrane

5. Which of the following must be true for osmosis to occur?
   a. Water must be at room temperature or above.
   b. Solutions with the same concentrations of solutes must be separated by a semipermeable membrane.
   c. Solutions with different concentrations of solutes must be separated by a semipermeable membrane.
   d. Water must be under pressure.

6. Why are the lipid bilayers in cells called "semipermeable?"
   a. They are not really all that permeable.
   b. Their permeability changes with their molecular composition.
   c. Their permeability is temperature dependent.
   d. They are permeable to some substances but not others.

Conceptual Review

1. Cooking oil is composed of lipids that consist of long hydrocarbon chains. These are not amphipathic molecules. Would you expect these lipids to spontaneously form membranes? Why or why not? Describe, on a molecular level, how you would expect these lipids to interact with water. Your answer should explain the saying, "oil and water don't mix."

2. Ethanol, the active ingredient in alcoholic beverages, is a small, polar, and uncharged molecule. Would you predict that this molecule crosses cell membranes quickly or slowly? Explain your reasoning.

3. Why can osmosis occur only if solutions are separated by a semipermeable membrane?

4. Examine the membrane in the accompanying figure. Label the molecules and ions that will pass through the membrane as a result of osmosis, diffusion, and facilitated diffusion. Draw arrows to indicate where each of the molecules and ions will travel.

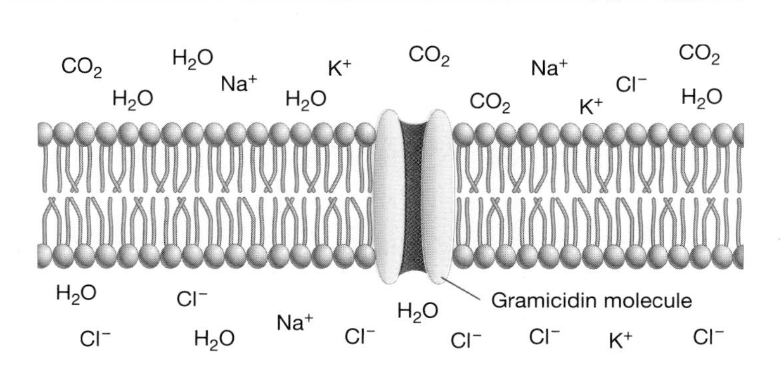

5. On page 79 the text claims that the part of membrane proteins that spans the hydrophobic tails of phospholipids is itself hydrophobic (see Figure 4.13b). Why does this make sense? Look back at Figure 3.3b and Table 3.1 and make a list of amino acids you would expect to find in these regions of transmembrane proteins.

Applying Ideas

1. When phospholipids are arranged in a bilayer, it is theoretically possible for individual molecules to flip-flop. That is, a phospholipid could turn 180° and become part of the membrane's other surface. Make a sketch of this process. Based on what you know about the behavior of polar heads and nonpolar tails, predict whether flip-flops are frequent or rare. Then design an experiment, using a planar bilayer made up partly of fatty acids that contain a dye molecule on their hydrophilic head, to test your prediction.

2. Wolfram Zillig has suggested that the membrane of the first cell that ever existed was actually made of protein, not phospholipid. He contends that membranes with lipid bilayers evolved later in the history of life. Comment on whether Zillig's hypothesis makes sense given what you've learned about diffusion, osmosis, the selective permeability of lipid bilayers, and the role of membrane proteins in passive and active transport. Is Zillig's hypothesis testable?

3. Unicellular organisms that live in extremely cold habitats have an unusually high proportion of unsaturated fatty acids in their cell membranes. Some of these membranes even contain polyunsaturated fatty acids, which have more than one double bond in the hydrocarbon chain. Researchers have proposed that the lack of saturation helps these membranes maintain a semifluid state at low temperatures instead of becoming semisolid or solid. Draw a picture of this type of membrane analogous to the sketch in Figure 4.8c. Using this picture as a resource, comment on the hypothesis that membranes with unsaturated fatty-acid tails function better at cold temperatures. How would you test the hypothesis? Make a prediction about the structure of fatty acids found in organisms that live in extremely hot environments.

4. When biomedical researchers design drugs that must enter cells to be effective, they sometimes add methyl ($-CH_3$) groups to make the drug molecules more likely to pass through cell membranes. Conversely, when researchers design drugs that act on the exterior of cell membranes, they sometimes add a charged group to decrease the likelihood that the drugs will pass through membranes and enter cells. Explain why these strategies make sense.

5. Advertisements frequently claim that laundry and dishwashing detergents "cut grease." What the ad writers actually mean is that the detergents surround oil droplets on clothing and dishes, making them water soluble. When this happens, the oil droplets can be washed away. Explain how this happens on a molecular level.

CD-ROM and Web Connection

CD Activity 4.1: Diffusion and Osmosis *(animation)*
(Estimated time for completion = 5 min)
Learn how diffusion and osmosis affect the integrity of cells.

CD Activity 4.2: Facilitated Diffusion *(animation)*
(Estimated time for completion = 5 min)
Membranes are not permeable to larger molecules and ions. How do ions and large, metabolically important molecules such as glucose enter your cells?

At your **Companion Website** (http://www.prenhall.com/freeman/biology), you will find self-grading exams and links to the following research tools, online resources, and activities:

Hydrocarbons and Lipids
This site contains two- and three-dimensional visualizations of hydrocarbons and lipids as well as many other molecules.

Membrane Structure
These pages explain the basics of membrane structure and the tools that scientists use to study them.

Membrane Protein Structure
Here, you can view the three-dimensional structures of a large set of membrane-bound cellular proteins.

Additional Reading

Bangham, A. D. 1995. Surrogate cells or Trojan horses. *BioEssays* 17: 1081–1088. The discoverer of liposomes reflects on their use in research and the pharmaceutical and cosmetic industries.

Bayley, H. 1997. Building doors into cells. *Scientific American* 277 (September): 62–67. A review of research on artificial pores, and their role in delivering drugs to targeted cells.

Singer, S. J. 1992. The structure and function of membranes: A personal memoir. *Journal of Membrane Biology* 129: 3–12. Singer is co-author of the fluid-mosaic model of membrane function, which predicted that cell membranes are composed of a fluid phospholipid bilayer sprinkled with a mosaic of membrane proteins. This paper reviews how the model was developed and how it inspired further research.

Cell Functions

The origin of the cell is still a mystery. Presumably, the original cell arose when a self-replicating molecule like those introduced in Chapter 3 became surrounded by a simple lipid bilayer or a case made of protein and lipid. But understanding how the first cells functioned remains one of the major tasks facing researchers interested in the origin of life.

In contrast, biologists know an enormous amount about how today's cells work. Early in the history of biology, researchers realized that all organisms are made of cells and that the cell is the fundamental structural and functional unit of life. Studying cells has always been a basic part of biological science.

The four chapters in this unit focus on three questions about the life of a cell. First, what does the cell look like and how does it function? To answer this question, **Chapter 5: Cell Structure and Function** reviews how cells are constructed and introduces recent work on how molecules are moved around inside the structure. Second, how do cells gain the chemical energy they need to stay alive? **Chapter 6: Respiration and Fermentation** and **Chapter 7: Photosynthesis** focus on this issue. Chapter 6 explores classical experiments on how cells gain chemical energy by oxidizing sugars or other reduced compounds; Chapter 7 investigates how photosynthetic organisms convert the energy in sunlight into chemical energy in the form of sugars. Finally, how do cells divide? **Chapter 8: Cell Division** answers this question and then explores how errors in the cell division process can lead to uncontrolled growth and the disease called cancer.

As with all but a handful of plant species, this running buffalo clover manufactures its own food by carrying out photosynthesis. ©1994 Susan Middleton and David Liittschwager

Cell Structure and Function

5

In Chapter 1 you were introduced to the cell theory, which states that all organisms consist of cells and that all cells are derived from preexisting cells. Since this theory was initially developed and tested in the 1850s and 1860s, an enormous body of research has confirmed that the cell is the fundamental structural and functional unit of life.

In a very real sense, then, understanding how an organism works is a matter of understanding how cells work. To drive this point home, recall from Chapter 1 that many members of the Eukarya and virtually all of the Bacteria and Archaea are unicellular. For researchers who study these species, understanding the cell is synonymous with understanding the organism as a whole. Even in plants, animals, and other multicellular eukaryotes, complex behavior originates at the level of the cell. For example, your ability to sense light and form a visual image of this page begins with changes in pigment molecules located in cells at the back of your eyes. The altered pigment molecules trigger changes in the membranes of nerve cells that connect your eyes to your brain. To understand complex processes like vision, then, researchers often begin by studying the structure and function of the individual cells involved—the parts that make the whole.

In Chapter 4 you were introduced to the cell by investigating the nature of the cell membrane. The fundamental message of Chapter 4 was that the cell membrane creates an environment inside the cell that is different from conditions outside. In this chapter, our goal is to describe the structures that are found inside cells and explore what they do. We'll focus on several particularly critical cell structures and processes and explore the experimental approaches that biologists use to understand them.

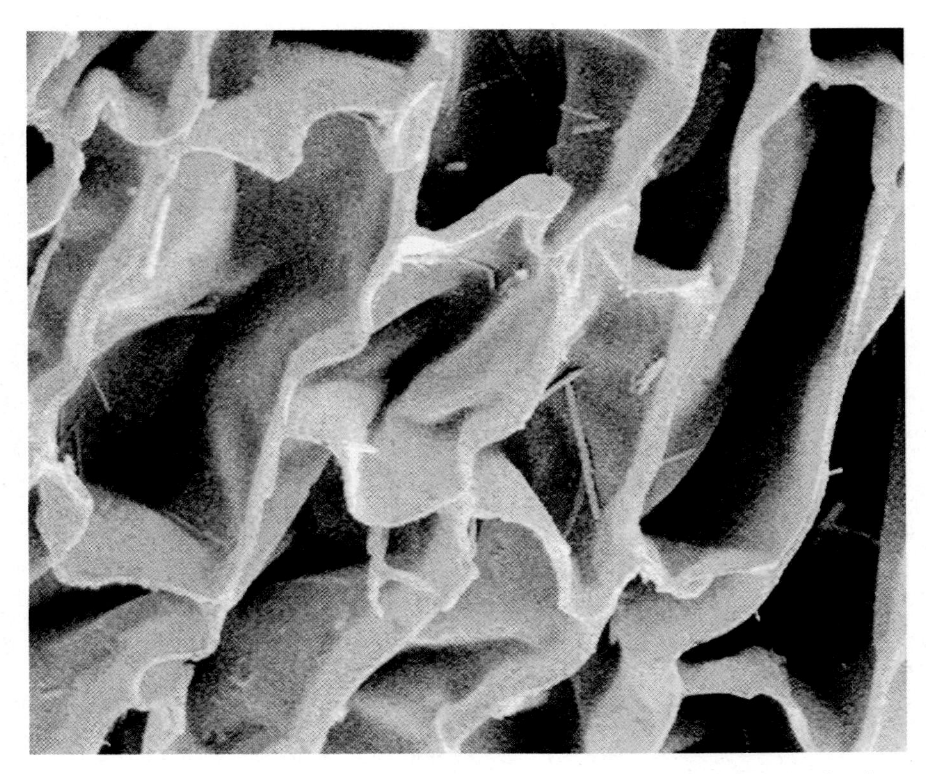

This is a scanning electron micrograph of cells in the cork from a wine bottle. Cork cells were the first cells ever observed.

To launch this investigation into how cells work, section 5.1 surveys the basic types of cells, cell structures, and cell processes that biologists have documented to date. The following three sections focus on the theme of how molecules move throughout cells to accomplish specific tasks. Section 5.2 considers traffic into and out of the membrane-bound compartment called the nucleus; section 5.3 examines how newly manufactured molecules move through an extensive system of internal membranes; and section 5.4 focuses on the system of fibers that transports materials inside cells.

5.1 A Tour of the Cell

In Chapter 1 you read about the two fundamental types of cells observed in nature. Recall that eukaryotic cells have a membrane-bound compartment called a nucleus, while prokaryotic cells do not. Species in the Bacteria and Archaea are prokaryotic; members of the Eukarya—including algae, fungi, plants, and animals—are eukaryotic.

In the late seventeenth century, biologists began studying the structure of cells with light microscopes. Over time, improvements in optics and cell preparation techniques allowed researchers to catalog the structures we'll review in this section. When electron microscopes became widely available in the 1950s, workers were able to describe the internal anatomy of these structures. As we'll see later in the chapter, recent advances in light microscopy have allowed investigators to videotape certain types of cell processes in living cells.

What have anatomical studies with light microscopes and electron microscopes revealed? Let's look first at the general anatomy of prokaryotic cells and eukaryotic cells, and then consider how the structures that have been identified help cells function and grow.

Prokaryotic Cells

Figure 5.1 shows the general structure of a prokaryotic cell. For bacterial and archaeal species, the cell membrane encloses a single compartment—meaning that the cell has few or no subdivisions delimited by internal membranes. The cell membrane is also called the **plasma membrane**, which consists of a phospholipid bilayer and membrane proteins. As you'll see in Chapter 25, virtually all bacteria and archaea also have a **cell wall** that surrounds the cell membrane.

The contents of a prokaryotic cell are collectively termed the **cytoplasm** (cell-substance) and include one or more chromosomes. **Chromosomes** are structures that consist of DNA complexed with specific proteins. Prokaryotic chromosomes are found in a localized area of the cell but are not separated from the rest of the cytoplasm by a membrane. By far the most extensive internal membranes observed in prokaryotes are found in photosynthetic species. As Chapter 7 will show, the internal membranes of photosynthetic bacteria contain the enzymes and pigment molecules required to convert the energy in sunlight into chemical energy in the form of sugar.

Two other other prominent cell structures found in prokaryotes are **ribosomes**, which manufacture proteins, and **flagella**, which power movement. Ribosomes are observed in all prokaryotic cells and are found throughout the cytoplasm. Bacterial ribosomes consist of three distinct RNA molecules and over 50 different proteins; it is not unusual for a single cell to contain 10,000 of these structures. Flagella may or may not be present on bacterial cells. They are usually few in number and are located on the surface of the cell. Over 40 proteins are involved in building and controlling bacterial flagella and, at top speed, flagellar movement can drive a bacterial cell through water at 60 cell lengths per second. In contrast, the cheetah qualifies as the fastest land animal but can sprint at a mere 25 body lengths per second.

Plant and Animal Cells

The generalized plant and animal cells shown in **Figure 5.2** (page 92) illustrate several fundamental points about the anatomy of eukaryotic cells. The first feature to note is size. Eukaryotic cells

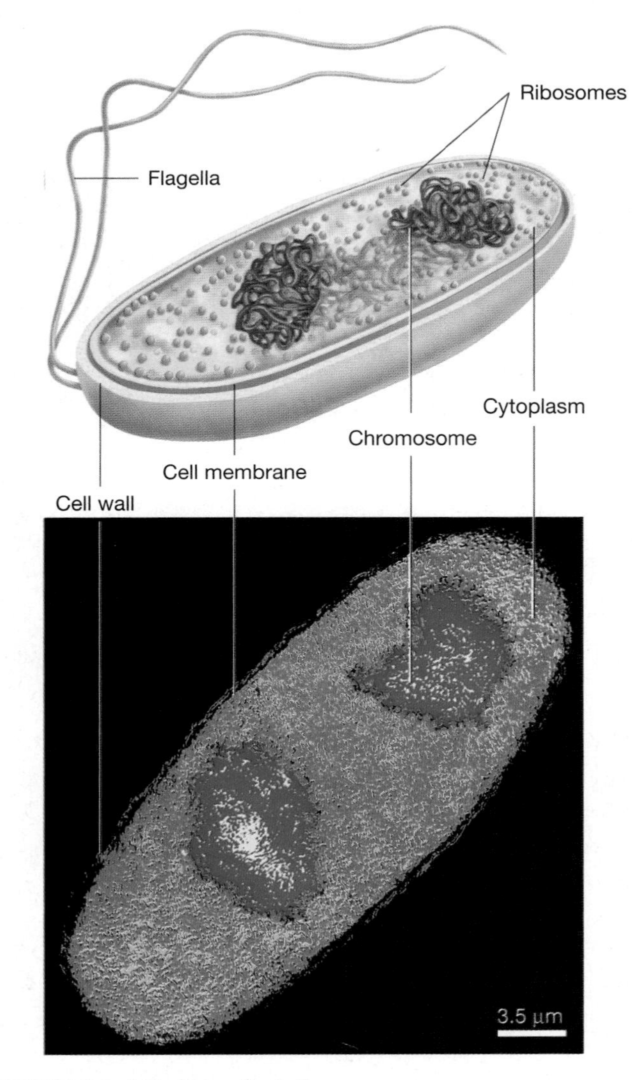

FIGURE 5.1 A Prokaryotic Cell
The sketch shows some of the major structural features of a typical prokaryotic cell. The transmission electron micrograph illustrates some of the same features.

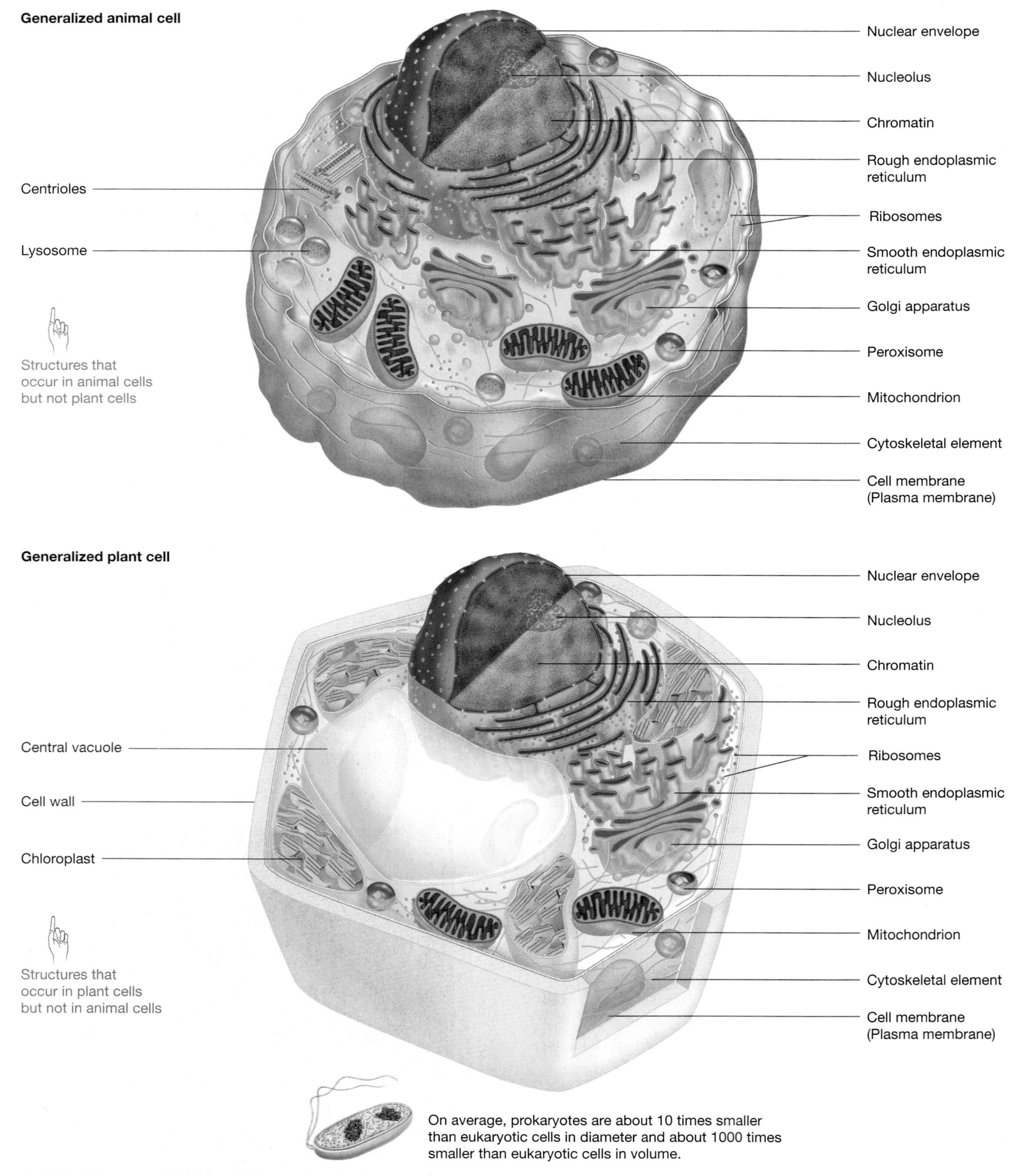

FIGURE 5.2 Animal and Plant Cells
These sketches depict generalized or "typical" animal and plant cells. **EXERCISE** Label the unlabeled organelles in each sketch.

range from about 10 to 100 μm in diameter while prokaryotic cells vary from about 1 to 10 μm in diameter. A second point is that eukaryotic cells contain numerous internal membranes. The membrane-bound compartments found in eukaryotic cells are called **organelles** (little organs). In effect, eukaryotic cells are compartmentalized. Why is this important?

- Compartmentalization allows incompatible chemical reactions to be separated. For example, fatty acid synthesis takes place in one organelle; fatty acids are degraded and recycled in a different organelle.
- Groups of enzymes that work together can be clustered on internal membranes instead of floating free in the cytoplasm. For example, many of the plant pigments, enzymes, and other molecules responsible for converting sunlight into chemical energy are arranged side by side in an internal membrane. The clustering of these molecules increases the efficiency of the reactions, because intermediary compounds have to diffuse shorter distances.
- Because organelles are membrane-bound containers, they are able to maintain high concentrations of molecules that are needed for specific chemical reactions. As a result, the reactions proceed efficiently.
- Compartmentalization makes large size possible. The size of prokaryotic cells is thought to be limited by the distance that molecules have to diffuse inside the cell. Because eukaryotic cells are subdivided, the molecules required for specific chemical reactions are often located within a given compartment and do not need to move long distances to interact.

Which Organelles Are Common to All Eukaryotic Cells? As the illustrations in Figure 5.2 show, the anatomy of plant and animal cells is similar in several important respects. Both types of cells are bounded by a cell membrane (also called the plasma membrane) and have chromosomes located inside a nucleus. The nucleus is the most prominent membrane-bound compartment in the cell and is enclosed by a double membrane called the **nuclear envelope.** In addition to chromosomes, the interior of the structure contains extensive protein scaffolding called the **nuclear matrix,** which stiffens the organelle and organizes the chromosomes. The nucleus also includes an area called the **nucleolus,** where ribosomes are assembled.

In eukaryotes, the cytoplasm consists of everything inside the cell membrane excluding the nucleus. Ribosomes are scattered throughout the cytoplasm and are the site of protein synthesis. In addition, an array of cytoplasmic organelles is observed in virtually all eukaryotic cells. These common organelles include a network of membrane-bound sacs and tubules called the **endoplasmic reticulum (ER).** Two types of ER exist: **rough ER** looks rough because it has ribosomes associated with it; **smooth ER** lacks ribosomes. The **Golgi apparatus** consists of a stack of flattened membranous sacs. The other organelles commonly found in eukaryotic cells include peroxisomes and mitchondria (singular: *mitochondrion*).

What do these structures do? The mitochondrion is explored in detail in Chapter 6, and the function of the rough ER and the Golgi apparatus is analyzed in section 5.3. Although smooth and rough ER are physically connected, their chemical composition and functions are distinct. Smooth ER is where hydrophobic toxins such as pesticides are converted to water-soluble compounds that can be secreted. Smooth ER is also the site of fatty acid and phospholipid synthesis. **Peroxisomes,** in contrast, contain enzymes that degrade fatty acids through the addition of molecular oxygen (O_2). These oxidation reactions produce small molecules that are used as substrates in reactions elsewhere in the cell, but also yield the corrosive molecule hydrogen peroxide (H_2O_2). Enzymes in the peroxisome degrade the H_2O_2 to water and O_2.

The other major structural feature that is common to plant and animal cells is an extensive system of protein fibers called the **cytoskeleton.** As we'll see in section 5.4, the cytoskeleton contains several distinct types of proteins and has an array of functions. In addition to giving the cell its shape, cytoskeletal proteins are involved in moving the cell itself and in moving materials within the cell.

How Do Plant and Animal Cells Differ? Although many organelles are found in both plant and animal cells, important anatomical differences occur in the two lineages. For example, plant cells possess an outer **cell wall** in addition to their cell membrane. They also possess an organelle called the **chloroplast,** where sunlight is converted to chemical energy in the process called photosynthesis. Chapter 7 explores the structure of the chloroplast and the chemistry of photosynthesis in detail. Plant cells also contain a large, membrane-bound structure called the **central vacuole.** The vacuole is a water-filled compartment that may take up as much as 90 percent of the volume inside the plant cell. Vacuoles help plant cells maintain stiffness and may store toxins like nicotine or caffeine that deter insects and other herbivores (plant-eaters). In addition, the central vacuole contains enzymes that degrade old organelles or macromolecules. In this respect, the vacuole is functionally related to an animal organelle called the lysosome. In animals, **lysosomes** are the site of hydrolytic reactions that degrade large molecules and worn-out organelles. Human diseases resulting from defects in lysosome function are featured in the end-of-chapter essay. Finally, animal cells contain structures called centrioles, which are involved in cell division. As Chapter 8 explains, **centrioles** are involved in organizing the cytoskeletal elements that are required for cell division.

Before delving into the details of how some particularly important organelles function, let's consider what all of these cell structures do in a very broad sense. What fundamental tasks must be accomplished for a cell to stay alive, grow, and divide?

What Are the Basic Cell Processes?

The cell is an oily sac. Inside, thousands of different chemical reactions are taking place. The products of these chemical reactions allow the cell to acquire resources from the environment, synthesize additional molecules, dispose of wastes, and reproduce. (How cells reproduce, by cell division, is the subject of Chapters 8 and 9.) Based on these observations, it is not surprising that biologists sometimes speak of cells as factories and routinely refer to organelles, ribosomes, and other structures as cellular machinery.

What does it take to maintain the living factory called the cell? The first requirement is information. In particular, cells need the instructions required to make enzymes. Enzymes, in turn, catalyze the chemical reactions that occur in cells. Recall from Chapter 3 that enzymes are proteins. As **Figure 5.3a** shows, the instructions for making enzymes and other cell components is archived in chromosomes in the form of DNA. These instructions are transcribed into short-lived messenger molecules made of RNA and then translated into proteins by ribosomes. Chapters 10 through 15 explore exactly how this flow of information occurs. For the purposes of understanding cell structure and function, the important point is that molecules need to move in and out of the nucleus in an ordered way. Messenger RNAs (mRNAs) need to move out of the nucleus to the ribosomes; enzymes and nucleotides must move into the nucleus in order for DNA to be transcribed and copied. In eukaryotes, this means that there is a great deal of molecular traffic into and out of the nucleus.

In addition to storing, retrieving, and using information, cells must take the proteins and other molecules that they produce and transport them to the correct location. Depending on the molecule involved, the destination could be an organelle like the nucleus or a mitochondrion, the cell membrane, the cell wall, or the exterior of the cell.

To move all of this cargo, cells utilize chemical energy stored in a molecule called adenosine triphosphate (ATP). As **Figure 5.3b** shows, potential energy is released when ATP is cleaved to produce an inorganic phosphate group (P_i) and adenosine diphosphate (ADP). Because a great deal of energy is required to add a phosphate group to ADP and form ATP, a great deal of energy is released when this bond is hydrolyzed. When a phosphate group from ATP is transferred to a protein, the potential energy change can power movement, make endothermic chemical reactions occur, and pump specific molecules across their membrane against electrochemical gradients. (Chapters 6 and 7 are devoted to exploring how cells manufacture and use ATP.)

Overall, the amount of chemical activity and the speed of molecular movement inside cells is nothing short of staggering. Every second, each cell in your body hydrolyzes an average of 10 million ATP molecules; each enzyme present catalyzes up to 25,000 reactions; each phospholipid molecule in the cell membrane whizzes across the breadth of the structure.

In this light, the static picture of cell structure presented earlier in this section is misleading. Cells are dynamic. They are the site of thousands of different and carefully controlled chemical reactions that take place at mind-boggling speed. The rest of this chapter focuses on this theme of dynamism. To begin, let's take a closer look at how molecules move in and out of the cell's control center: the nucleus.

(a) Molecules move in and out of the nucleus.

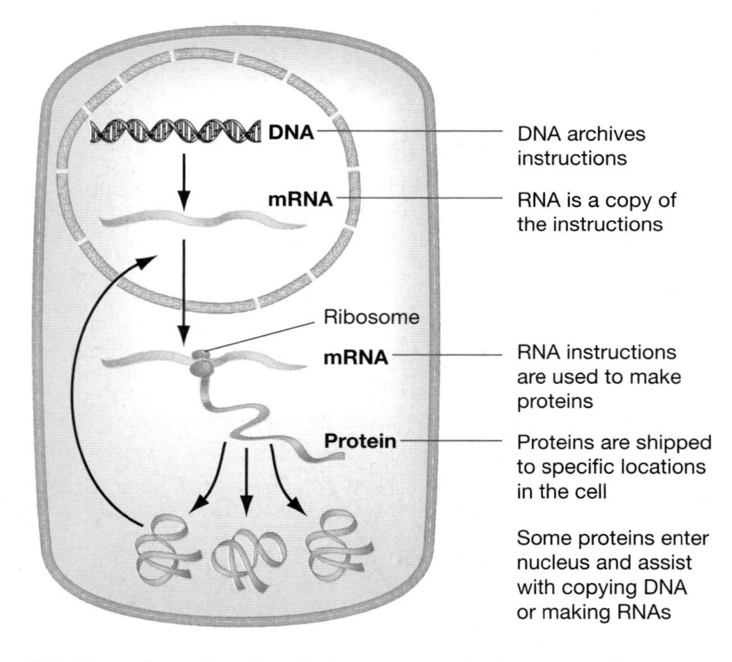

(b) Where does the chemical energy used to power cell activities come from?

$$\text{ATP} \xrightarrow{\text{Hydrolysis}} \text{ADP} + P_i + \text{Energy}$$

Adenosine triphosphate → Adenosine diphosphate + Inorganic phosphate group

FIGURE 5.3 Molecules Move Through Cells in a Controlled Manner
(a) The information that cells require to function is archived in chromosomes, in the form of DNA. This information is copied into RNA molecules that travel to the cytoplasm. The instructions in these messenger RNAs (mRNAs) are used to manufacture proteins. **(b)** In cells, movement and other types of work are made possible by the chemical energy stored in adenosine triphosphate (ATP).

5.2 The Nucleus and Nuclear Transport

The nucleus, which is the information archive in eukaryotic cells, is sequestered from the rest of the cell by the nuclear envelope. Biologists began to understand how the nuclear envelope is structured when electron microscopy became available in the 1950s. As the micrograph and sketch in **Figure 5.4a** show, the structure has two membranes, each consisting of a

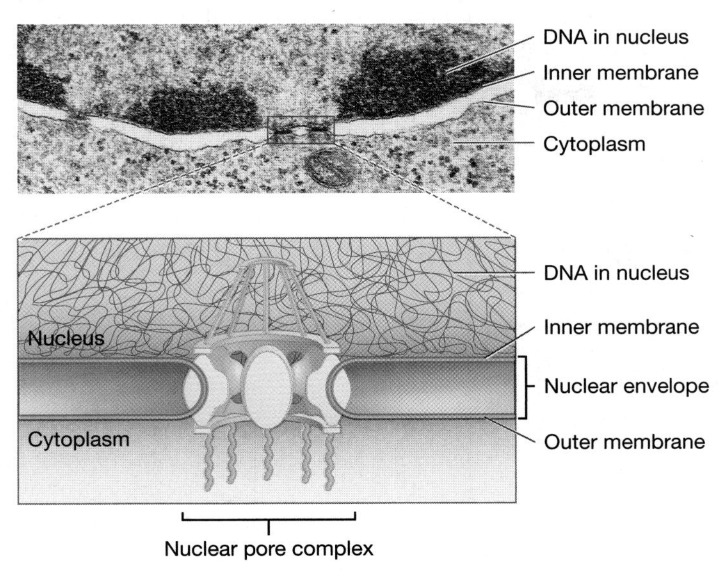

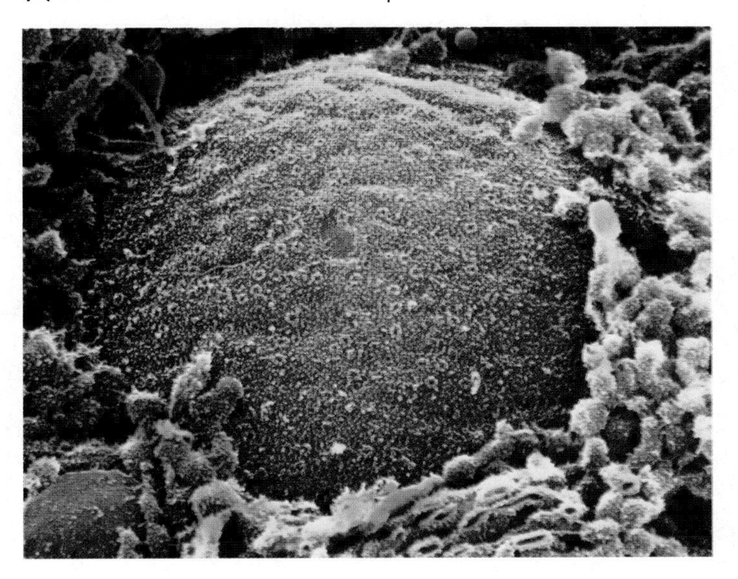

FIGURE 5.4 The Nuclear Envelope
(a) This transmission electron micrograph shows a cross section through the nuclear envelope. The interpretive drawing shows a detail from the area surrounding the nuclear pore. **(b)** This scanning electron micrograph shows that the surface of the nuclear envelope is studded with numerous nuclear pores.

lipid bilayer. The inner membrane and the outer membrane are separated by a space that is actually continuous with the interior of the endoplasmic reticulum. In addition, the scanning electron micrograph in **Figure 5.4b** shows that the nuclear envelope contains thousands of openings called **nuclear pores.** Because these pores extend through both inner and outer nuclear membranes, they connect the inside of the nucleus with the cytoplasm.

The messenger RNAs that carry instructions for making proteins travel out to the cytoplasm through the nuclear pores. Inbound traffic includes the enzymes and other molecules that are needed to make these RNAs and to copy DNA before the cell divides. In a typical nucleus, several thousand molecules pass through each of the 3000–4000 pores every minute. This traffic is regulated by a group of proteins called the **nuclear pore complex.** The complex has the elaborate structure illustrated in Figure 5.4a and contains dozens of different proteins. Functionally, the nuclear pore complex is a gatekeeper that controls the flow of molecules between the nucleus and the cytoplasm. How is this movement regulated?

Early Studies of Nuclear Transport

In the early 1960s, Carl Feldherr began to analyze how molecules are transported from the cytoplasm to the nucleus. He began by injecting gold particles into the cytoplasm of unicellular eukaryotes called amoebas. At intervals after making an injection, he prepared the cells for electron microscopy. In electron micrographs, gold particles show up as black dots. **Figure 5.5** shows some of Feldherr's data. One or two minutes after injection, some gold particles were associated with the nuclear pores. This observation was consistent with the hypothesis that transport from the cytoplasm to the nucleus occurs through these structures. Ten minutes after injection, particles appeared within the nucleus. Their concentration never exceeded the concentration observed in the cytoplasm, however. Based on this observation, Feldherr suggested that gold particles enter the nucleus by passive diffusion. Recall that passive diffusion does not require energy input and does not result in a higher concentration of molecules on one side of a membrane than the other.

Later, Feldherr extended these studies by injecting gold particles of varying sizes. He found that smaller particles passed into

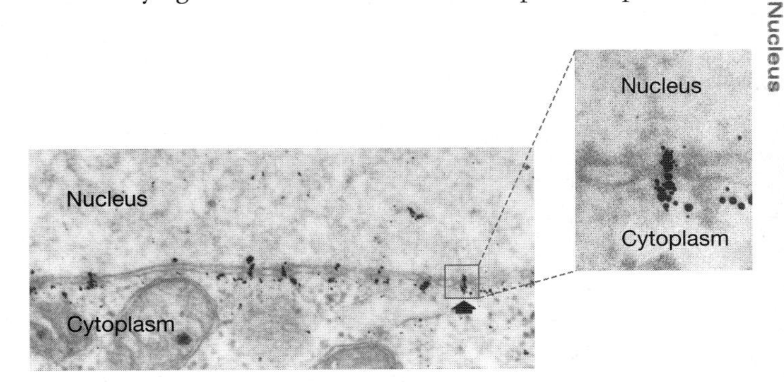

FIGURE 5.5 Molecules Enter the Nucleus Through the Nuclear Pores
These transmission electron micrographs show gold particles, which appear as black dots, congregating at nuclear pores soon after the particles were injected into the cell. **EXERCISE** Label the inner and outer membranes of the nuclear envelope. Circle the nuclear pores.

CD ACTIVITY 5.1 Transport into the Nucleus

the nucleus more efficiently than larger ones. Then to study transport in living cells, he injected proteins that contained a fluorescent label and followed their movement using a fluorescence microscope. These observations confirmed that size affects transport. Only proteins with molecular weights of less than approximately 60,000 daltons could diffuse efficiently into the nucleus. (The dalton or atomic mass unit was introduced in Box 2.1.)

Feldherr's research supported the hypothesis that small molecules enter the nucleus through passive transport. In contrast, other researchers showed that large proteins can be transported into the nucleus against a concentration gradient, and that messenger RNAs are transported out against a concentration gradient. Based on these observations, biologists realized that the movement of large molecules through the nuclear pore is an active or energy-demanding process. Then the question became, how is this movement directed? For example, proteins are manufactured in the cytoplasm. How are proteins that are needed inside the nucleus directed there?

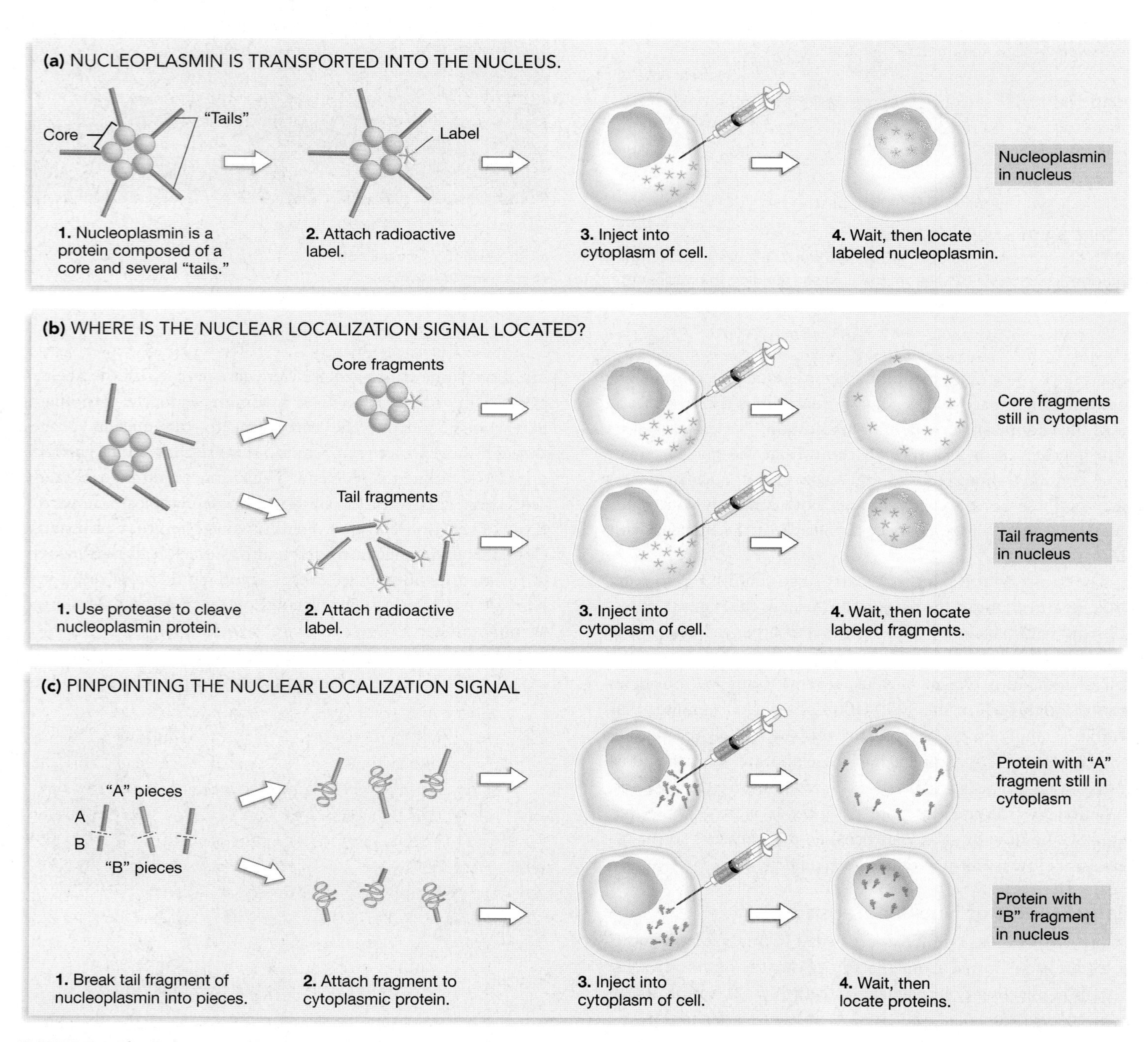

FIGURE 5.6 The Tail Region of Nucleoplasmin Contains a Nuclear Localization Signal
The experiments illustrated here succeeded in pinpointing a short stretch of amino acids that targets nucleoplasmin to the nucleus.

How Are Proteins Imported into the Nucleus?

In the 1980s, Ronald Laskey and colleagues began to study how a protein called nucleoplasmin is imported into the nucleus. Nucleoplasmin plays an important role in the assembly of chromosomes, and it has a distinctive structure. It consists of a core surrounded by a series of long "tails." As **Figure 5.6a** shows, the researchers labeled nucleoplasmin with a radioactive atom and injected it into the cytoplasm of living cells. After an interval, the researchers killed the cells, sliced them into sections, and processed them for autoradiology (see Box 3.2). The resulting data confirmed that labeled nucleoplasmin was found exclusively within the nucleus.

To explain their result, the biologists hypothesized that a specific section of the nucleoplasmin protein marks it for transport to the nucleus. In colloquial terms, they proposed that part of the polypeptide chain acts as a zip code with the message "send to nucleus."

If the signal actually exists, is it located in the core or the tails of nucleoplasmin? To answer this question, the researchers used enzymes called proteases to separate the core from the tails. As **Figure 5.6b** indicates, they separated the two components, labeled them with radioactive atoms, and then injected them into the cytoplasm of different living cells. When they used autoradiography to detect the location of the labeled protein fragments, the researchers found that tail fragments were transported to the nucleus while core fragments remained in the cytoplasm. This result suggested that a nuclear localization signal exists somewhere in the tail part of the protein.

To find out exactly what the signal is, the biologists did the experiments outlined in **Figure 5.6c**. They began by fusing small stretches of the nucleoplasmin tail region to proteins that are normally found in the cytoplasm. If one of the tail-region segments contained the signal, then the cytoplasmic protein it was attached to should be transported to the nucleus. To test this prediction, the researchers injected the "fusion proteins" into the cytoplasm of cells. They found that if a particular 17-amino-acid stretch of the nucleoplasmin tail is attached to a protein that is normally found in the cytoplasm, the molecule is transported to the nucleus. This specific stretch of amino acids was dubbed the nuclear localization signal.

Recently, workers in other laboratories have confirmed that other nuclear proteins contain similar localization signals. Although less is known about the signals that direct proteins and mRNAs out of the nucleus, recent studies suggest that specific export signals are involved.

To summarize, research on the transport of molecules through the nuclear pore has clarified how cells regulate traffic between the cytoplasm and nucleus. The key is that molecules have address labels that direct their transit into or out of the structure. Current research on traffic through the nuclear pore is focused on understanding how the proteins in the pore complex read these labels, and how ATP is used to power their transport against a concentration gradient.

Now let's turn to a related question. Do cytoplasmic proteins have addresses, just like nuclear proteins do? Suppose a protein is meant to be secreted from a cell or inserted into the plasma membrane. Once a protein is manufactured, how is it transported to its destination?

5.3 The Endomembrane System: Synthesis and Distribution of Cellular Products

Recent work on how proteins and other molecules are directed to their destinations in the cytoplasm has focused on two of the organelles introduced in section 5.1—the endoplasmic reticulum (ER) and the Golgi apparatus. In the 1950s, electron microscopy studies by Keith Porter and co-workers established that the ER is an elaborate network of membranes that extends from the nucleus throughout the entire cytoplasm, that ER is found in all eukaryotic cells, and that the organelle may or may not have ribosomes attached (**Figure 5.7a**, page 98). Although the Golgi apparatus was first described by Camillo Golgi in 1898, many researchers considered it an artifact of staining and other procedures used to prepare cells for microscopy. Its existence wasn't confirmed until the 1950s, when electron microscopic studies confirmed that it is present in all eukaryotic cells, and that it consists of a stack of flattened membrane-bound discs (**Figure 5.7b**).

The presence of chromosomes defined the nucleus as the cell's information archive. But the function of the rough ER and the Golgi apparatus was not immediately apparent from microscopic observations. Together, the organelles provide an extensive endomembrane (inner-membrane) system. What role does this system play in the life of a eukaryotic cell?

What Does the Endomembrane System Do?

In 1955, George Palade and colleagues began a series of studies to determine the function of the rough ER and the Golgi apparatus. The group concentrated on studying cells in the pancreas of guinea pigs. The pancreas is a gland that produces and secretes an array of enzymes and other molecules involved in the digestion and utilization of food. Palade was particularly interested in a type of pancreatic cell that contains a large amount of rough ER. Because ribosomes manufacture proteins, and because rough ER seemed to be most extensive in cells that actively secrete proteins or other types of molecules, Palade hypothesized that rough ER functions in protein synthesis and secretion. All proteins are synthesized in the cytoplasm and many remain there. But certain types of proteins are secreted from the cell or inserted into the cell membrane. Palade's hypothesis was that these proteins enter the ER from the cytoplasm, and that the rough ER and Golgi apparatus jointly form an endomembrane system that directs the production and secretion of specific proteins.

To test this idea, Palade and colleagues employed an experimental approach known as a pulse-chase experiment. The goal of a pulse-chase experiment is to study how molecules move from one location to another in the cell. The technique is based on providing the cell with a large concentration of a labeled molecule for a short time. For example, if a cell receives a large amount of labeled amino acid for a short time, virtually all of the proteins synthesized during that interval will be labeled. This "pulse" of labeled molecule is followed by chase—in the form of large amounts of an unlabeled version of the same molecule, provided for a long time. If the chase consists of unlabeled amino acid, then the proteins synthesized during the chase period will also be unlabeled. The general idea is to mark a population of molecules at a particular interval, and then follow their fate over time. The approach is analogous to adding a small amount of dye to a stream and then following the movement of the dye molecules.

To make pulse-chase experiments possible with pancreatic cells, the researchers developed a method for growing them in culture, or in vitro. They supplied the cells with a 3-minute pulse of the amino acid leucine, labeled with a radioactive atom, followed by a long chase with non-radioactive leucine. Because the radioactive leucine was incorporated into the proteins being produced at the time, it labeled them. Then the researchers killed the cells and examined them under the electron microscope. As **Figure 5.8a** shows, all of the newly synthesized proteins were associated with the rough ER. This result supported the hypothesis that proteins are synthesized on the rough ER.

Where do these newly synthesized proteins go? To answer this question, Palade's group waited varying periods after the radioactive leucine pulse before killing the cells and determining the location of the labeled proteins. After a 7-minute chase, labeled protein was found at the edge of the Golgi apparatus—on the side nearest the ER. (Recall from Figure 5.2 that one side of the Golgi apparatus faces the ER and the other side faces the cell membrane.) After 17 minutes, the labeled protein was located inside the Golgi apparatus. After 80 minutes, the proteins were either localized in structures called secretory granules on the far side of the Golgi apparatus or found outside the cell.

Because previous studies had shown that secretory granules are a carrying container for secreted proteins, the researchers were able to propose the model for protein secretion illustrated in **Figure 5.8b**. As a group, the pulse-chase experiments supported the hypothesis that the endomembrane system functions in protein secretion, and established that secreted proteins move from their site of manufacture in rough ER to the Golgi apparatus to secretory granules to the plasma membrane and cell exterior

(a) Rough ER (with ribosomes attached)

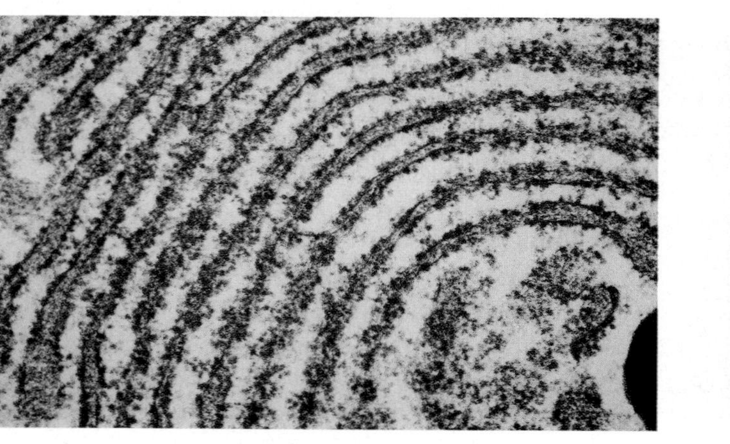

Smooth ER (no ribosomes attached)

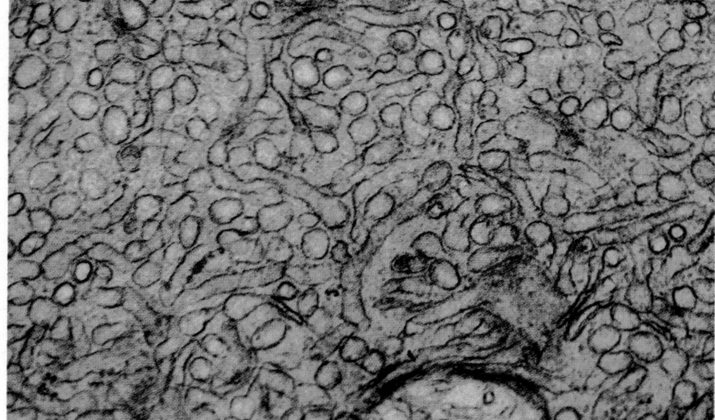

(b) Golgi apparatus

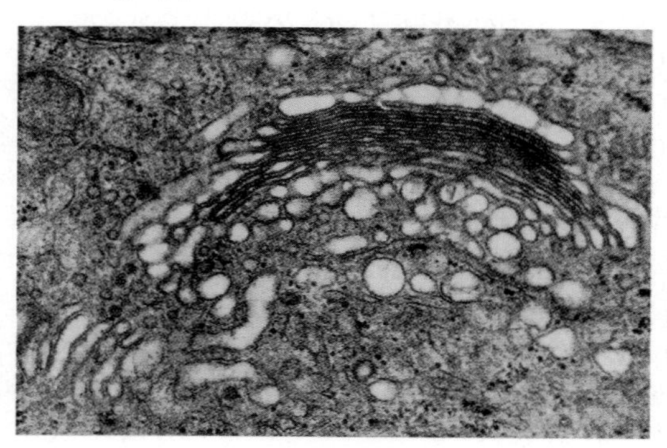

FIGURE 5.7 Rough ER, Smooth ER, and the Golgi Apparatus
(a) Rough ER often occurs in layers of flattened vesicles; smooth ER sometimes forms a mass of rounded tubules. **(b)** The Golgi apparatus consists of layers of flattened vesicles called cisternae. The side of the apparatus closest to the nucleus is called the *cis* (near) face; the side furthest from the nucleus is the *trans* (far) face.

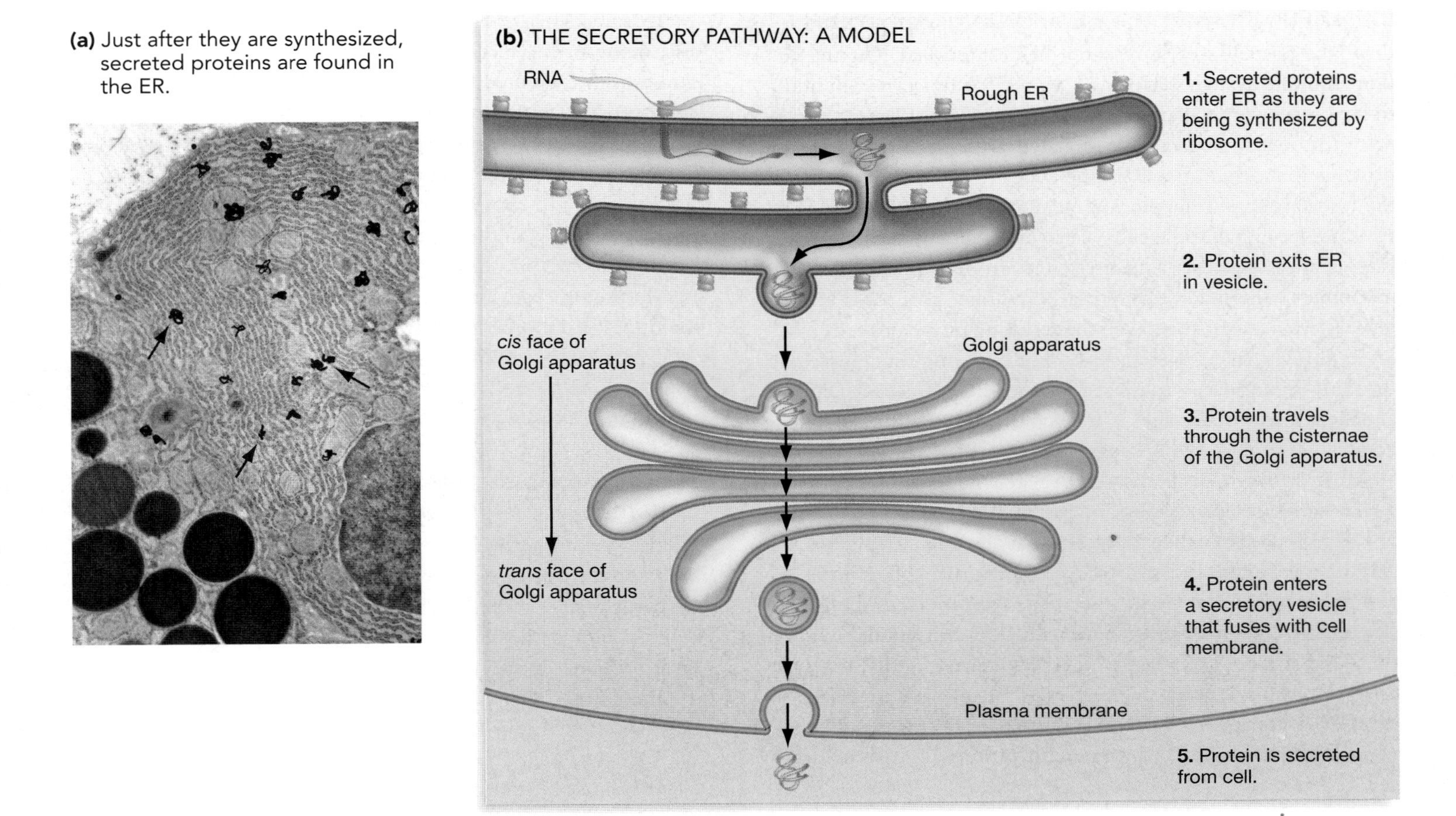

FIGURE 5.8 Secreted Proteins Follow a Pathway
(a) An electron micrograph of a pancreatic cell, 3 minutes after radioactive amino acids were added to it. The arrowheads point to black grains that represent radioactively labeled secreted proteins. Note that they are localized to the rough ER. The large, dark spheres in the lower left are secretory vesicles. **(b)** A model of the pathway by which secreted proteins are synthesized, processed, and transported out of the cell.

Directing Proteins to the Endomembrane System: The Signal Hypothesis

The work by Palade's group raised an important question: How are proteins that are meant to be secreted targeted for entry into the endomembrane system? In 1971, Gunter Blobel and colleagues proposed an answer to this question. Their **signal hypothesis** predicted that the first few amino acids of newly synthesized secreted proteins contain an address tag, analogous to the nuclear localization signal, which directs them into the ER.

The first indication that the signal hypothesis might be correct came from work by Cesar Milstein and colleagues. In the early 1970s, these researchers discovered that when certain secreted proteins were synthesized in a test-tube system that lacked ER, the proteins were 20 amino acids longer than the version of the same protein that is actually secreted by cells. When the researchers added ER to their test-tube protein-synthesis system, the protein that resulted was identical in length to the form that is actually secreted by cells.

To explain these results, Blobel proposed that the newly synthesized proteins contained an ER-specific signal sequence that is 20 amino acids long, and that the signal is removed in the ER prior to secretion. In 1975, his group published the results of experiments identifying the first of several ER signal sequences that have been documented. After establishing that the signal existed, Blobel's group focused on identifying how the signal was received. During the 1980s and 1990s, they documented that the ER membrane has protein receptors that interact with the signal sequence as it leaves the ribosome. Once the signal sequence binds to the receptor, the rest of the protein is translated and enters the organelle (**Figure 5.9**, page 100).

To summarize, results published over a span of three decades supported the signal hypothesis. Proteins that are destined for secretion from the cell have distinctive address labels, just like proteins that are destined for the nucleus do. For a signal or address to be effective, however, it must be received and interpreted. In the nucleus, nuclear localization signals interact with the proteins of the nuclear pore complex. In the ER, the address labels are read by receptors in the organelle's membrane. The overall message is that the transport of molecules throughout the cell is highly directed and controlled.

Other Functions of the ER and Golgi Apparatus

Over the past 40 years, research has established that the endomembrane system performs a variety of functions. In addition to serving as a manufacturing and transport center for secreted proteins, ribosomes associated with the ER synthesize proteins that are destined for the endomembrane system itself, the lysosomes, and the plasma membrane. Many of these proteins are modified in the ER by enzymes that add carbohydrate side chains. Because carbohydrates are polymers of sugar monomers, the addition of carbohydrate groups is called **glycosylation** (sugar-together); the resulting molecules are called **glycoproteins** (sugar-proteins; see **Figure 5.10**). Proteins are not the only molecules synthesized in the ER, however. The organelle is also a major site of lipid synthesis; many lipids undergo glycosylation in the ER as well.

Most ER products are shipped to the Golgi apparatus, which serves as a processing and dispatching center. As you saw in Figure 5.8b, ER products enter the Golgi apparatus from the side of the Golgi closest to the ER, pass through a series of compartments in the Golgi stack called **cisternae**, and exit the side of the Golgi farthest from the ER. Each cisternae within the Golgi stack contains a different set of enzymes that catalyze different glycosylation reactions. Thus the unidirectional movement of glycoproteins through the cisternae results in a precisely ordered series of chemical modifications to carbohydrate side chains.

To summarize, an array of molecules is synthesized in the ER; many of the molecules are subsequently modified by the addition of carbohydrate groups. From the ER, newly synthesized proteins pass through the components of the Golgi apparatus in sequence. There they undergo further glycosylation. Molecules are also sorted and stored in the Golgi apparatus until they can be shipped to their correct destination.

This overview of the endomembrane system raises two questions. How are proteins transported from one part of the system to the other in such an orderly way? If molecules that are manufactured and stored in the system are destined for such a wide array of locations, how are they sorted and given the correct "address label," so they are targeted to the correct destination? Let's consider each question in turn.

Transport of Materials Through the Endomembrane System

How do proteins travel from the ER to the Golgi apparatus? To answer this question, researchers have focused on understanding the movement of proteins that are eventually secreted by the cell.

Palade and colleagues offered the first hypothesis to explain how molecules move from the ER to the Golgi apparatus. During their experiments with radioactively labeled proteins, they observed that marked proteins appeared in a region between the ER and Golgi apparatus. Further, this region appeared to contain small membrane-bound vesicles. Based on these obser-

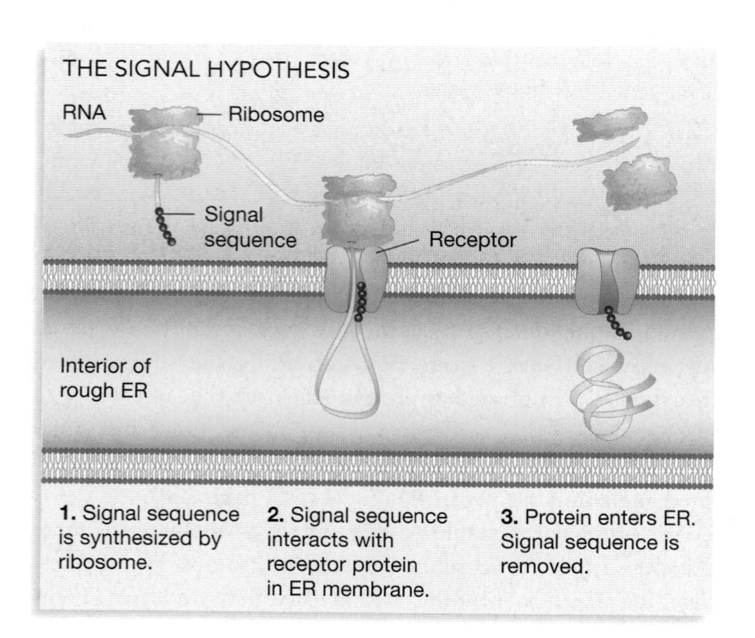

FIGURE 5.9 How Do Proteins Destined for Secretion Enter the Endomembrane System?
According to the signal hypothesis, proteins destined for secretion contain a short stretch of amino acids that interact with a receptor protein on the surface of the rough ER. Because of this interaction, the protein enters the ER. From there, the molecule follows the secretory pathway diagrammed in Figure 5.8b.

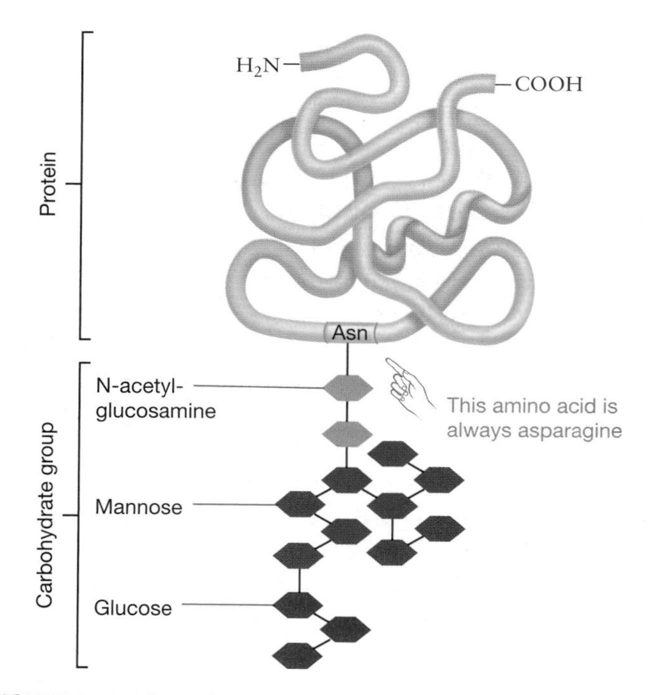

FIGURE 5.10 Glycosylation Adds Carbohydrate Groups to Proteins
When proteins enter the ER, most acquire the 14 sugar residues shown here. Some of these sugars may be removed or others added as proteins pass through the Golgi apparatus.

vations, the biologists suggested that proteins are transported between the two organelles in vesicles. As shown in Figure 5.8b, the idea was that vesicles bud off of the ER, move away, fuse with the membrane of the nearest cisternae in the Golgi apparatus, and dump their contents inside.

This hypothesis was controversial, however, because the researchers' electron micrographs were not detailed enough to confirm the claim that the labeled protein was actually inside the vesicles and not in the surrounding cytoplasm. In an attempt to clarify what was going on, Palade's group turned to a technique called differential centrifugation.

Subcellular Fractionation Studies Albert Claude and Christian de Duve developed differential centrifugation in the 1950s as a way to isolate particular cell components or fractions for further analysis. The technique, explained in detail in **Box 5.1** (page 104), is based on breaking cells apart to create a complex mixture and then separating various components through a series of spins in a centrifuge. The individual components can then be isolated and studied in detail.

Palade and co-workers were able to use this approach to study the transport of labeled proteins through the endomembrane system. When cells are homogenized, the ER and Golgi apparatus fragment and form small membrane-bound vesicles known as microsomes (small-bodies). Fragments from the rough ER form what researchers call rough microsomes; fragments from smooth ER, lysosomes or the vacuole, and the Golgi apparatus form smooth microsomes. Rough microsomes are denser than smooth microsomes because they contain ribosomes. As a result, the two are readily separated in a centrifuge.

When Palade and colleagues did a pulse-chase experiment to label newly synthesized proteins, they showed that the proteins initially localize to rough microsomes, but then appear in smooth microsomes after 7 minutes. This is the same time it took for labeled protein to appear in small transport vesicles in the electron micrographs the biologists had made earlier. Based on this correlation, Palade's group argued that proteins from the ER are indeed packaged into small vesicles for transport to the Golgi apparatus. Critics of the experiment were not convinced. They argued that smooth microsomes containing labeled protein might not be transport vesicles at all, but pieces of membrane derived from smooth ER or other cellular structures. Existing data could not rule out this explanation. Another approach was needed.

Identifying Three Classes of Transport Vesicles Barbara Pearse was the first biologist to positively identify a vesicle involved in transporting materials inside the cell. Her work was inspired by electron micrographs that revealed structures called coated vesicles in many different cell types. Pearse, who was able to isolate these vesicles using differential centrifugation, showed that they are coated with a distinctive protein called clathrin, which forms a cage-like structure around the vesicles.

Follow-up work by Pearse and others showed that clathrin-coated vesicles are involved in several different types of transport processes, including the movement of materials away from the Golgi apparatus and toward the cell membrane.

Subsequently, Vivek Malhotra and co-workers were able to study intact copies of the Golgi apparatus outside of cells (in vitro), and found that vesicles coated with a distinctive set of proteins called COPI shuttled material between the cisternae. Then Charles Barlowe and associates used a similar approach—studying intact rough ER in vitro—to isolate a third type of vesicle, which buds off intact rough ER and carries materials to the Golgi apparatus. These vesicles are coated with a protein called COPII.

To summarize, COPII vesicles bud from the rough ER and carry molecules to the Golgi apparatus. COPI vesicles then carry materials between the cisternae of the Golgi apparatus. Finally, clathrin-coated vesicles transport molecules to the plasma membrane. The overall message of these studies is that the mechanism proposed by Palade's group was essentially correct. Molecular cargo moves from the ER to the Golgi apparatus and beyond in membrane-bound vesicles.

Secretion Studies Using Live Cells In 1997, Jennifer Lippincott-Schwartz and colleagues published results that shed additional light on how proteins move from the ER to the Golgi apparatus. These biologists studied transport by labeling a secreted protein with a fluorescent molecule called **green fluorescent protein**, or **GFP**. As **Figure 5.11a** (page 102) shows, GFP is naturally synthesized in jellyfish that luminesce, or emit light.

Using a fluorescence light microscope, Lippincott-Schwartz and co-workers videotaped the movement of the labeled protein in living cells. As the photographs in **Figure 5.11b** show, the secreted protein was observed moving from the ER to the Golgi apparatus. High-magnification images indicated that the protein was inside tube-shaped transport elements. Follow-up studies showed that the secreted protein later moved to the cell surface in large, irregularly shaped vesicles that budded off from the Golgi apparatus.

Currently, researchers are working to characterize the tube-shaped transport elements and determine whether other types of vesicles are involved in shipping cargo inside the cell. Exciting progress has also occurred recently on several closely related questions. If materials are transported through the endomembrane system and the cytoplasm in several different types of carriers, how do the thousands of different products of rough and smooth ER find their way to the right container? Once vesicles assemble, how are the parcels directed to the correct destination?

Protein Sorting and Vesicle Targeting

The ER and Golgi apparatus are like a sprawling industrial complex—a series of gigantic production and shipping facilities where an array of products are manufactured and sent to dozens or even hundreds of different destinations in the cell. The scale and speed of the operation boggles the human imagination. How are all of the finished products put in the right shipping containers, and how are all of the different containers addressed?

Studies on enzymes that are manufactured in the ER and shipped to lysosomes have provided some answers to both questions. When researchers isolated lysosomal proteins from the Golgi apparatus and analyzed their structure, they found that carbohydrate side chains had been attached. In addition, one specific sugar group had a phosphate group attached, forming the compound mannose-6-phosphate. Follow up work showed that mannose-6-phosphate serves as the "zip code" portion of the address label for the protein. Like the nuclear localization and ER signals analyzed earlier, mannose-6-phosphate targets proteins to vesicles bound for a lysosome. These vesicles, in turn, have proteins on their surface that interact specifically with proteins in the lysosomal membranes. **Figure 5.12** pulls these observations together into a comprehensive model for how the products of the endomembrane system are loaded into specific vesicles and shipped off to their correct destination.

(a) Green fluorescent protein causes bioluminescence.

(b) Bioluminescence can be used to visualize the pathway of secreted proteins in living cells.

Proteins at the start

9 minutes later

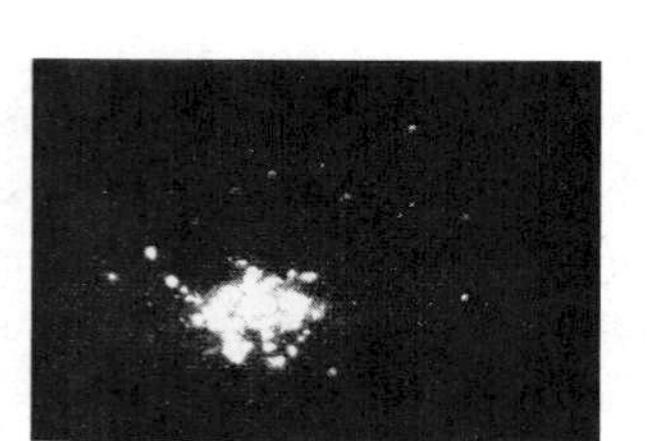

Interpretation: Individual proteins followed these paths from rough ER to Golgi apparatus.

Golgi apparatus

FIGURE 5.11 Using Green Fluorescent Protein to Track the Movement of Secreted Proteins
(a) Green fluorescent protein (GFP) is produced naturally by certain jellyfish. **(b)** Researchers have been able to visualize the movement of secreted proteins by attaching them to GFP. The photos show frames from a video sequence of protein movement; the drawing summarizes what happened over the entire time interval.

The data reviewed in this section present a convincing case for the existence of sophisticated cargo handling systems in cells. An element has been missing in this discussion of how molecules move from one location to another, however. If vesicles function like railroad cars or semi-trailer trucks, do they travel on tracks or roads? Or do they just float around? In general, how does a vesicle move to its destination?

5.4 The Cytoskeleton

Based on early observations with light microscopes, biologists viewed the cytoplasm of eukaryotic cells as a fluid-filled space devoid of structure. As microscopy improved, however, researchers realized that the cytoplasm actually contains a complex network of fibers. This cytoskeleton (cell-skeleton) helps to maintain cell shape by providing structural support. It's important to recognize, though, that the cytoskeleton is not a static structure like a bony skeleton. The cytoskeleton moves and changes in order to change the cell's shape, to move materials within the cell, or to move the cell itself. Like the rest of the cell, the cytoskeleton is dynamic.

As **Figure 5.13** shows, there are several distinct types of cytoskeletal elements. Microfilaments, intermediate filaments, and microtubules have a distinct size, structure, and function. Let's look at each of these types in turn.

Microfilaments

As Figure 5.13 indicates, microfilaments are long, fibrous polymers of a plate-shaped protein called actin. In many cells, actin is the most abundant of all proteins; each of your liver cells contains about half a billion of these molecules. Microfilaments grow and shrink as actin subunits are added or subtracted from either end.

Figure 5.14a (page 105) shows a fluorescence micrograph of the microfilaments in a mammalian kidney cell. Note that groups of microfilaments are organized into long bundles or dense networks. In both cases, the individual microfilaments are linked to one another by proteins. In combination, the bundles

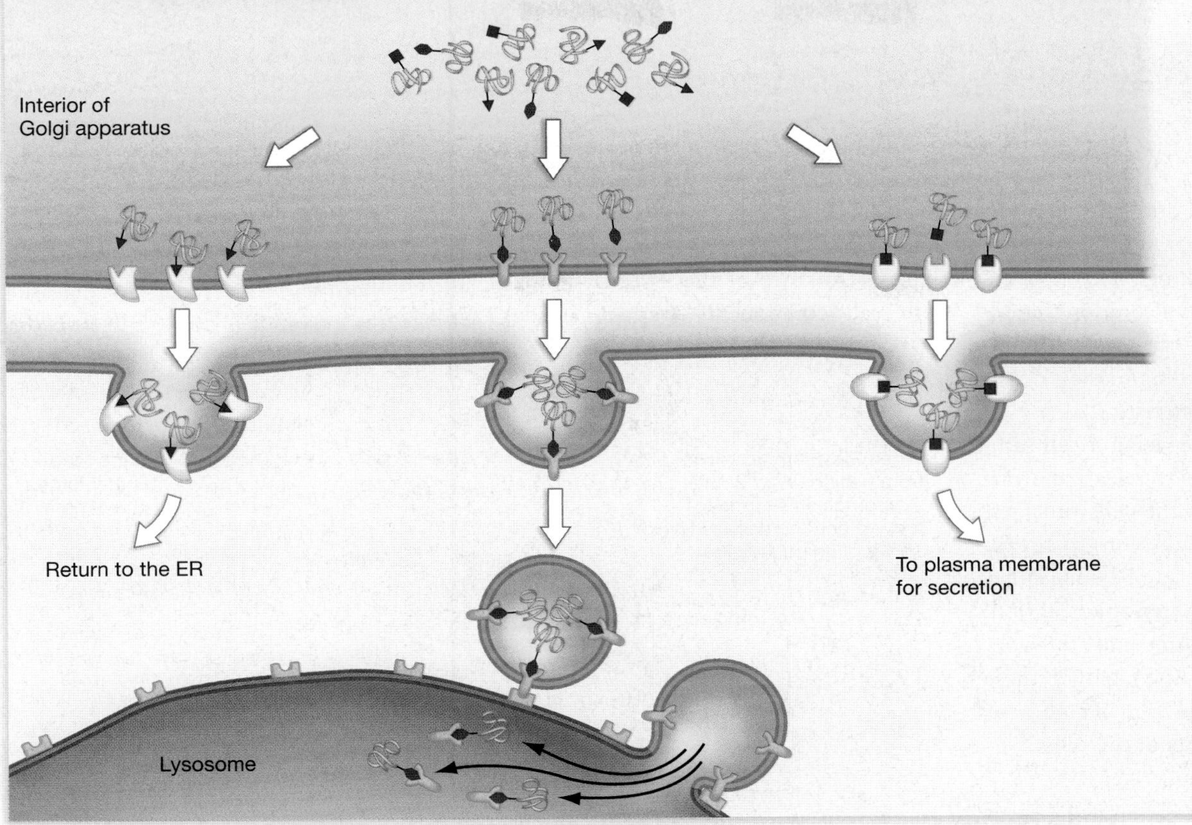

FIGURE 5.12 How Are Proteins Sorted into Vesicles? How Are These Vesicles Targeted to a Destination? This diagram summarizes the current model for how proteins are sorted into distinct vesicles in the Golgi apparatus, and how these vesicles are then targeted to their correct destination.

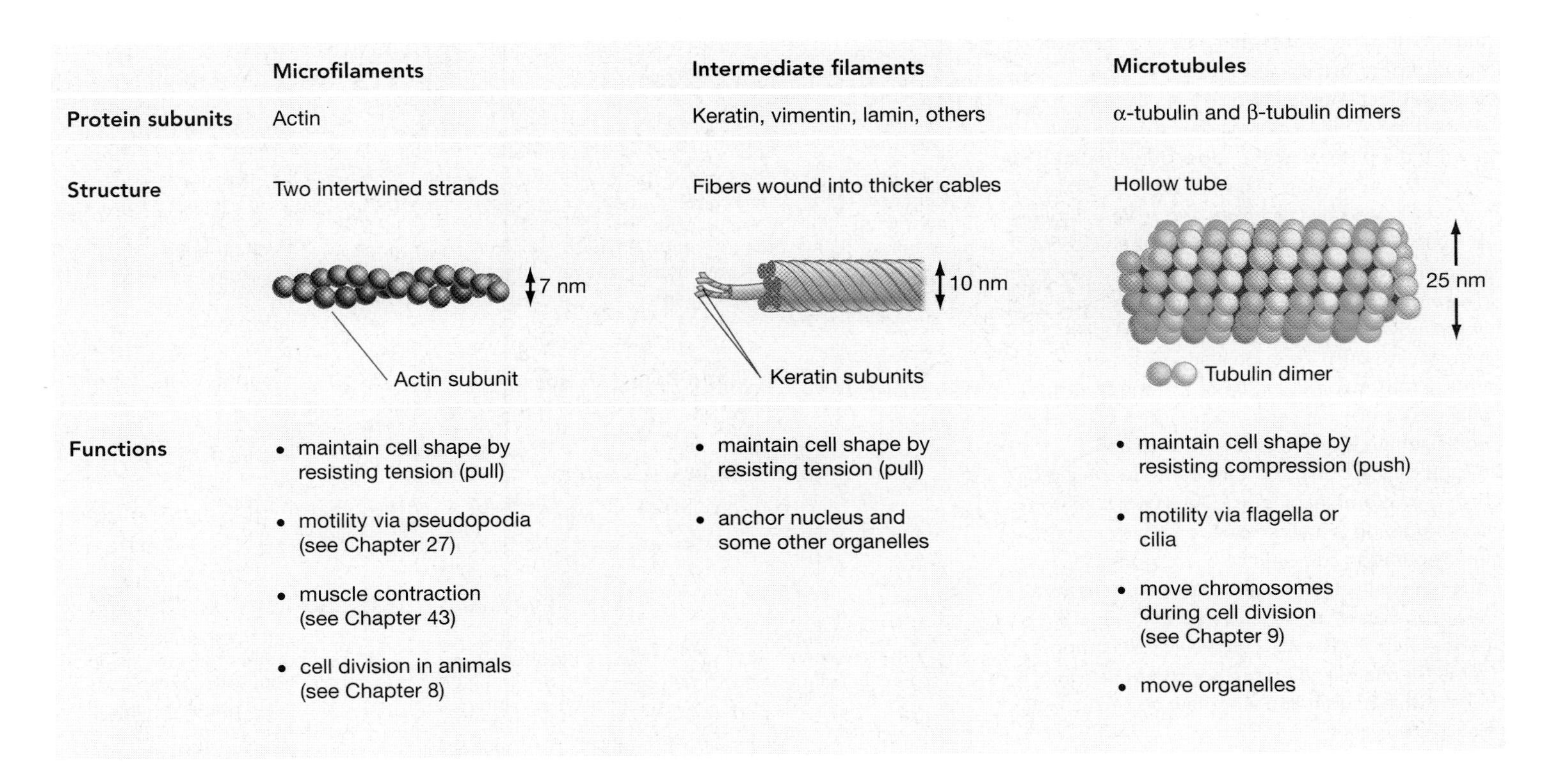

	Microfilaments	Intermediate filaments	Microtubules
Protein subunits	Actin	Keratin, vimentin, lamin, others	α-tubulin and β-tubulin dimers
Structure	Two intertwined strands	Fibers wound into thicker cables	Hollow tube
	7 nm Actin subunit	10 nm Keratin subunits	25 nm Tubulin dimer
Functions	• maintain cell shape by resisting tension (pull) • motility via pseudopodia (see Chapter 27) • muscle contraction (see Chapter 43) • cell division in animals (see Chapter 8)	• maintain cell shape by resisting tension (pull) • anchor nucleus and some other organelles	• maintain cell shape by resisting compression (push) • motility via flagella or cilia • move chromosomes during cell division (see Chapter 9) • move organelles

FIGURE 5.13 The Cytoskeleton Comprises Three Types of Filaments
The three types of filaments found in the cytoskeleton are distinguished by their size and structure, and by the proteins that serve as the subunit of the larger fiber.

BOX 5.1 An Introduction to Centrifugation

The centrifuge may be the most common tool used by biologists who study life at the level of molecules and cells. Its purpose is simple; it separates the components of cells. A centrifuge accomplishes this by spinning cells in a solution that allows molecules and other cell components to separate according to their density, size, or other characteristics.

The first step in preparing a cell sample for centrifugation is to release the various components from the cells (nucleus, ribosomes, proteins, and so on) by breaking the cells apart. This can be done either by putting them in a hypotonic solution, by exposing them to ultrasonic vibration, or by grinding them up. Each of these methods breaks apart cell membranes and releases the contents of the cells.

The pieces of cell membrane broken by these techniques quickly re-seal to form small vesicles. Thus the solution that results from this homogenization step is a mixture of these vesicles (often with cell components trapped inside), free-floating macromolecules released from the cells, and organelles. The combination of organelles, free macromolecules, and vesicles is called a cell extract or homogenate.

(Continued on next page)

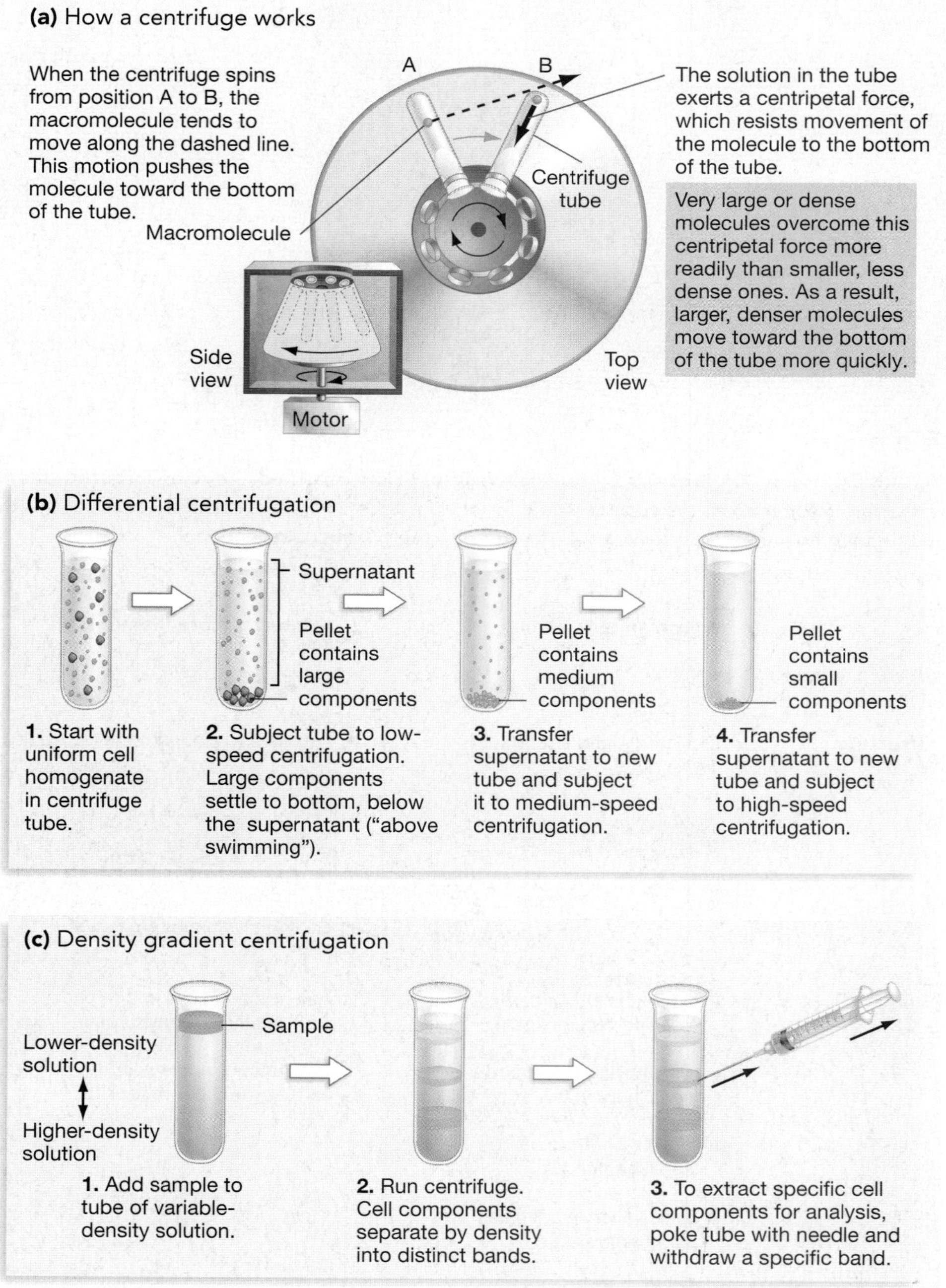

FIGURE 1 Using Centrifugation to Separate Cell Components
(a) This overhead view of a centrifuge illustrates why cell components separate by being spun. The green dot represents a macromolecule in solution inside a centrifuge tube. When the centrifuge spins and the tube moves from position A to position B, the macromolecule tends to move along the dashed line. This motion pushes the molecule toward the bottom of the tube, but is resisted by the solution, which exerts the centripetal force indicated by the broad arrow. Very large or dense cell components overcome this centripetal force more readily than smaller, less dense ones. Thus the larger, heavier components move toward the bottom of the tube more quickly. **(b)** Through a series of centrifuge runs made at increasingly high speed, an investigator can separate fractions of a cell homogenate by size. These types of centrifuge runs are an important early step in purifying macromolecules. **(c)** Density-gradient centrifugation is a technique for achieving fine separation between cell components. After the centrifuge has run, cell components have separated, by density, into distinct bands (step 2). To extract specific cell components for analysis, a researcher can poke the tube with a needle and withdraw a specific band (step 3).

(Box 5.1 continued)

The reason these structures become separated as a centrifuge spins is explained in part (a) of **Figure 1**. To separate the components of a cell extract thoroughly, researchers often perform a series of centrifuge runs. For example, part (b) of the figure illustrates how an initial spin, at low speed, causes the larger, heavier parts of the homogenate to move below the smaller, lighter parts. The subsequent runs illustrated in the diagram are often done at higher speeds and longer durations. These runs continue to separate cell components based on their size and density.

These types of centrifuge runs create fairly crude mixes of cell components, however. To accomplish finer separation, researchers frequently follow up with higher-speed centrifugation. In the version of high-speed centrifugation used by Palade and co-workers, the centrifuge tube is filled with a series of sucrose solutions of increasing density. The solutions are layered so that the highest-density material is at the bottom of the tube and the lowest-density material is at the top.

To separate cell components in this sucrose-density gradient, Palade and colleagues add the sample to the top of the centrifuge tube. As the centrifuge spins, components of the sample are driven toward the bottom of the tube. Separation is achieved because denser components move faster than less-dense components. When the centrifuge run is complete, each cell component has formed a band of material in the centrifuge tube. A researcher can collect the material in a particular band by puncturing the tube with a hypodermic needle and drawing out the contents of that band.

and networks of microfilaments in a cell form a matrix that helps stiffen the cell and define its shape. As future chapters will show, actin filaments are also involved in cell division and several types of cell movement in animal cells.

Intermediate Filaments

Because they are made from an array of different proteins, intermediate filaments are defined by size rather than composition. Unlike microfilaments and microtubules, which are made from the same protein subunits in all eukaryotic cells, intermediate filaments can be composed of an array of different proteins. In many cases, different types of cells in the same organism contain different types of intermediate filaments.

Intermediate filaments form a particularly dense network in the interior of the nucleus and near the inner membrane of the nuclear envelope. As **Figure 5.14b** shows, some intermediate filaments also project from the nucleus through the cytoplasm to the cell membrane, where they are linked to intermediate filaments that run parallel to the cell surface. In this way, intermediate filaments form a flexible skeleton that helps shape the cell surface and hold the nucleus in place.

The Role of Microtubules in Vesicle Transport

Microtubules are composed of protein subunits called α-tubulin and β-tubulin; each subunit consists of one molecule of α-tubulin and one molecule of β-tubulin. They are the largest

(a) Fluorescence micrograph of microfilaments in mammalian cells

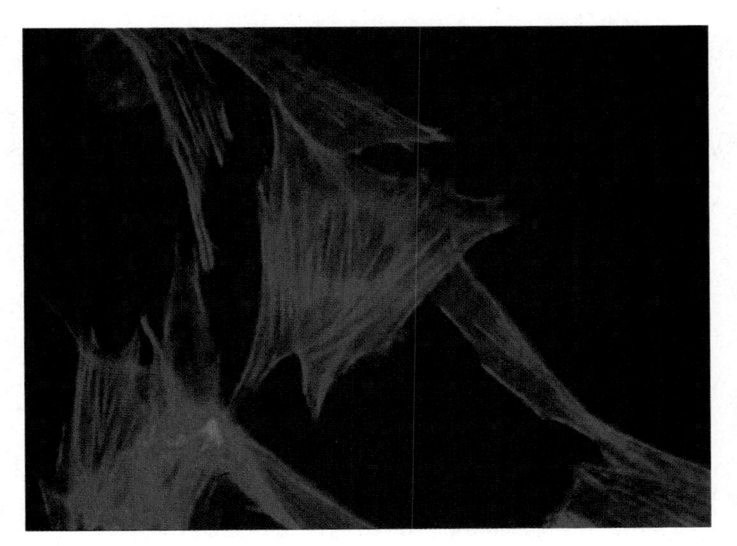

(b) Fluorescence micrograph of intermediate filaments in mammalian cells

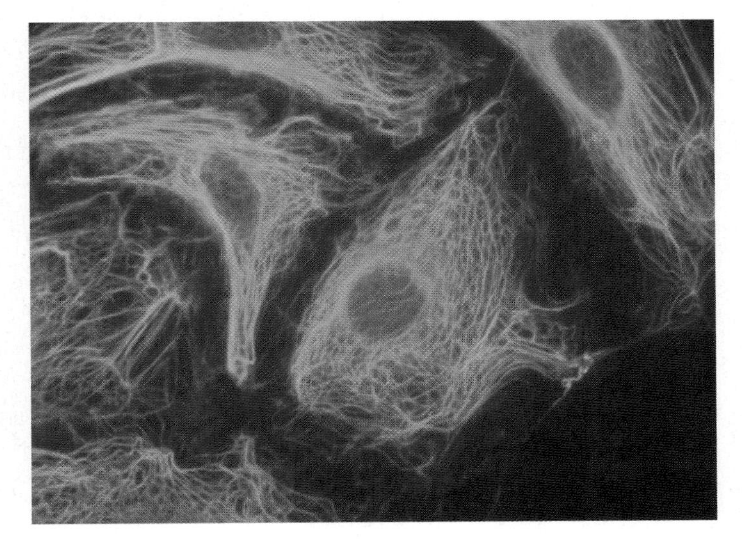

FIGURE 5.14 How Are Cytoskeletal Elements Distributed in the Cell?
To make these micrographs, researchers attached a fluorescent compound to actin (part a) and a protein found in intermediate fibers (part b).

cytoskeletal components in diameter, and they grow and shrink in length as tubulin subunits are added or subtracted from one end.

Like microfilaments and intermediate filaments, microtubules provide structural support for eukaryotic cells. But in addition, microtubules are involved in several different types of movement. For the rest of the chapter we'll focus on how microtubules function in moving materials inside cells as well as in moving the cell itself.

Using the Squid Giant Axon as a Model System Inside cells, materials are transported to a wide array of destinations. To study how this movement happens, Ronald Vale and colleagues focused on a cell known as the giant axon, which is found in squid. The giant axon is an extremely large nerve cell that runs the length of the animal's body. If the squid is disturbed, the giant axon carries an electrical signal to a band of muscle. At the junction where the nerve cell meets the muscle, the nerve cell releases molecules called neurotransmitters. The muscle tissue responds to the arrival of neurotransmitters by contracting. The contraction forces water out of a cavity in the body and propels the animal away from danger.

Vale and co-workers were interested in this system for three reasons. First, the giant axon is so large that it is relatively easy to see and manipulate. Second, neurotransmitters are synthesized in the cell's ER and then transported in vesicles down the length of the axon to the terminal, where they are stored until an electrical signal triggers their release. A large amount of cargo thus moves a long distance in this cell. Third, Vale and colleagues found that if they gently squeezed the cytoplasm out of the cell, vesicle transport still occurred in the cytoplasmic material. Once they'd made this observation, the researchers realized that they had a cell-free system with excellent prospects for exploring the mechanism of transport. What did they find out?

Microtubules Act as "Railroad Tracks" To study vesicle transport in the cytoplasm of squid giant axons, Vale and co-workers relied on video-enhanced microscopy techniques developed by Robert Allen. The set-up allows researchers to record light microscope images with a video camera, and to enhance contrast enough to make cytoskeletal filaments and other cell components visible. As **Figure 5.15** shows, this technique allowed them to document that vesicle transport occurred along a filamentous track present in the extruded cytoplasm. Vale and colleagues also found that if they depleted the amount of ATP in the preparation, vesicle transport stopped. Based on these observations, they concluded that vesicle transport is an energy-dependent process and that vesicles move along tracks.

To identify the type of filament involved, the biologists analyzed the chemical composition of these tracks and measured the diameter of the tracks using electron microscopy. Both types of data indicated that the tracks consisted of microtubules. Microtubules also appear to be required for movement of materials elsewhere in the cell. Recall from section 5.3 that Jennifer Lippincott-Schwartz and colleagues were able to videotape the movement of vesicles from the ER to the Golgi apparatus. When these investigators treated their experimental cells with a drug that disrupts microtubules, movement was abnormal.

The general message of these experiments is that transport vesicles move through the cell along microtubules. Now the question is, how? Do the tracks themselves move, like a conveyer belt? Or do the vesicles have some sort of motor?

A Motor Protein Generates Motile Forces To study how vesicles move along microtubules, Vale and colleagues set out to tear the axon's transport system apart and then put it back together. The idea was to strip the cell down to individual com-

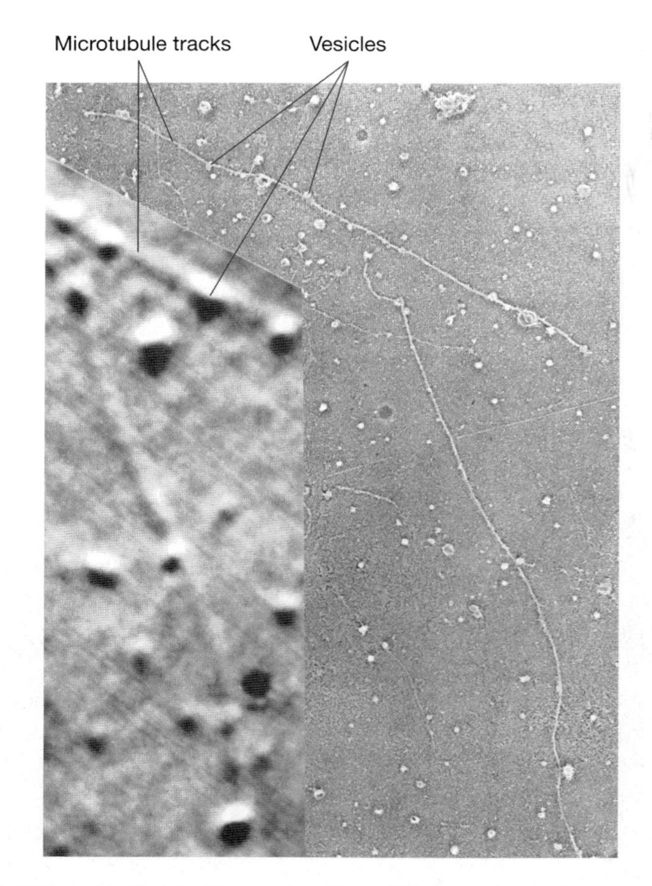

FIGURE 5.15 Vesicle Movement Along Microtubule Tracks
These photos show transport vesicles moving along two transport filaments in extruded cytoplasm from a squid giant axon. The photo on the left is from a videomicroscope, where researchers watched vesicles move. The higher-resolution electron micrograph on the right allowed researchers to measure the diameter of the filaments and confirm that they are microtubules.

ponents and then reconstitute the elements that are absolutely essential for transport. To begin, the researchers found that they could produce purified microtubules by assembling the fibers from purified α-tubulin and β-tubulin. They also found that they could isolate transport vesicles through the differential centrifugation technique described in section 5.3. But when they mixed purified microtubules with purified vesicles and ATP, no transport occurred. Something else was required. What was it?

To find the missing element or elements, Vale and co-workers began adding different subcellular components to their reconstituted transport system. Through trial and error, they found a soluble cellular fraction that triggered movement when added to the mixture of purified microtubules, transport vesicles, and ATP. This was a key discovery. The missing element had to be somewhere in the soluble fraction.

Through further purification steps, the researchers succeeded in isolating a protein in the soluble fraction that was responsible for generating vesicle movement. They named the molecule **kinesin**, from the Greek word *kinein* (to move). Kinesin can convert the chemical energy of ATP into mechanical work, resulting in vesicle movement. Kinesin is called a **motor protein** because it converts chemical energy into mechanical work, just like a car's motor does. So, how does the molecule work?

The Molecular Mechanism of Movement Biologists began to understand how kinesin works when x-ray crystallographic studies revealed its three-dimensional structure. As **Figure 5.16a** shows, the protein consists of two intertwined polypeptide chains and has three major regions: a head section with two globular pieces, a tail, and a stalk that connects the head and tail. Follow-up studies confirmed that the two globular components of the head bind to the microtubule while the tail region binds to the cargo. In this way, the kinesin molecule serves as a link between transport vesicles and microtubule "tracks."

How does kinesin move? More detailed studies of kinesin's structure indicated that each of the globular components of the molecule's head has a site for binding ATP as well a site that binds to the microtubule. To pull these observations together, biologists propose that kinesin transports vesicles by "walking" along a microtubule. The idea is that each part of the head region undergoes a conformational change when it binds ATP. As Chapter 3 showed, conformational or shape changes often alter the activity of a protein. As **Figure 5.16b** shows, the ATP-dependent conformational change in kinesin results in a "step." As each head alternately binds and degrades ATP, the protein and its cargo move down the microtubule track.

Follow-up work has shown that cells contain a number of different kinesin proteins. Each type of kinesin is specialized for carrying a different type of vesicle in the cell. In short, kinesins are the motors that move containers along microtubule tracks to destinations throughout the cell. Are kinesins also responsible for moving the entire cell?

Cilia and Flagella: Moving the Entire Cell

Flagella are long, hair-like projections that function in movement. Flagella are found in both bacteria and eukaryotes. The structure of flagella is different in the two groups, however. Bacterial flagella are made of a protein called flagellin; eukaryotic flagella are constructed from microtubules. Bacterial flagella move the cell by rotating like an airplane propeller; eukaryotic flagella move the cell by undulating.

To understand how cells move, we'll focus on exploring the structure and mechanism of action of eukaryotic flagella. As **Figure 5.17a** (page 108) shows, eukaryotic flagella are closely related to structures called cilia, also found in eukaryotes. Unicellular eukaryotes may have either flagella or cilia, while some multicellular organisms have both. In humans, for

(a) Structure of kinesin

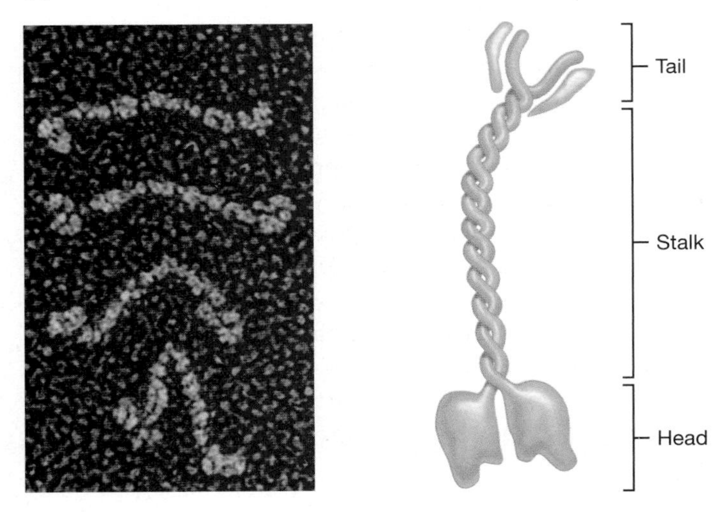

(b) Kinesin "walks" along a microtubule track

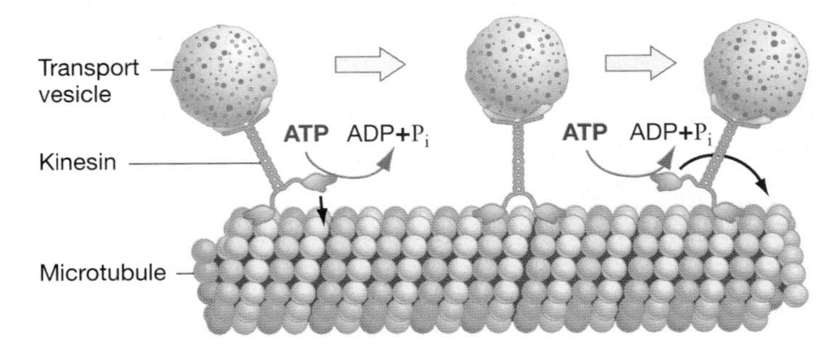

FIGURE 5.16 A Motor Protein Moves Vesicles Along Microtubules **(a)** As this electron micrograph and sketch show, kinesin has three major segments. **(b)** The current model depicting how kinesin "walks" along a microtubule track to transport vesicles. The two head segments act like feet that alternately attach and release in response to the gain or loss of a phosphate group.

example, the cells that line the respiratory tract have cilia; sperm cells have flagella.

There are several important differences between flagella and cilia. Flagella are generally longer than cilia, and cells generally have just one or two flagella but many cilia. When Keith Porter and Don Fawcett examined the two structures under the electron microscope, however, they found that the underlying organization of flagella and cilia is identical.

How Are Cilia and Flagella Constructed? In the 1950s, Porter and Fawcett's anatomical studies established that cilia and flagella have a characteristic "9 + 2" arrangement of microtubules. As **Figure 5.17b** shows, there are nine microtubule doublets surrounding two central microtubules.The doublets, consisting of one complete and one incomplete microtubule, are arranged around the periphery of the structure. The entire 9 + 2 structure is called the **axoneme** (axle-thread).

As electron microscopy improved, biologists gained a more detailed view of the structure. As the sketch in **Figure 5.17c** shows, spoke-like structures connect each doublet to the central microtubules. In addition, molecular bridges connect the nine doublets to one another. Finally, each of the doublets has a set of arms that project toward an adjacent doublet. How do these components generate motion?

A Motor Protein in the Axoneme In the 1960s, Ian Gibbons began studying the cilia of a common, unicellular eukaryote called *Tetrahymena* that lives in pond water. His goal was to uncover the cilia's mechanism of action. Gibbons found that by using a detergent to remove the plasma membrane that surrounds cilia and then subjecting the resulting solution to differential centrifugation, he could isolate axonemes. Further, the isolated structures would beat if Gibbons supplied them with ATP. These results confirmed that the beating of cilia is an energy-demanding process. They also provided Gibbons with a system for exploring the molecular mechanism of movement.

In an early experiment with isolated axonemes, Gibbons treated the structures with a molecule that affects the ability of proteins to bind to one another. His idea was to try to disrupt interactions among the proteins in the axoneme. As predicted, the axonemes that resulted from this treatment could not bend or hydrolyze ATP. When Gibbons examined the treated axonemes in the electron microscope, he discovered that the arms that projected from each doublet were missing. This observation suggested that the treatment had caused the arms to fall off the axoneme, and that the arms have a component that hydrolyzes ATP and is responsible for motion. Later, Gibbons identified the component as a large protein that he named **dynein** (from the Greek word *dyne*, meaning "force").

Like kinesin, dynein is a motor protein. Structural and chemical studies have shown that dynein undergoes conformational changes when ATP is hydrolyzed. As a result, the protein can walk along microtubules just like kinesin can. This walking motion allows the microtubule doublets to slide past one another. But when cilia and flagella beat, they don't just elongate—they bend. How does the walking action of dynein arms result in the bending of cilia and flagella?

The answer to this question lies partly in the structure of the axoneme, and partly in mechanisms that control which dynein arms along the axoneme are walking at any particular time. Recall that each peripheral doublet in an axoneme is connected to the central pair of microtubules by a spoke. In addition, adjacent doublets are connected by molecular bridges. These connections constrain the movement of microtubules as they slide

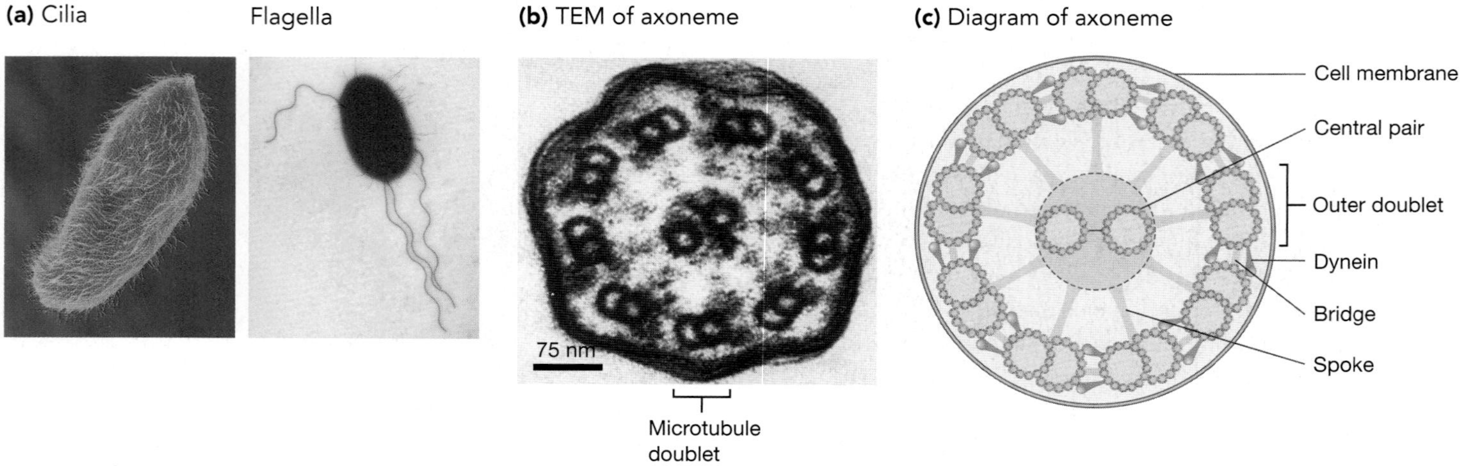

FIGURE 5.17 The Structure of Cilia and Flagella
(a) Cilia and flagella differ in length and number. **(b)** This transmission electron micrograph is a cross section through an axoneme. **(c)** A diagram of major structural elements in cilia and flagella.

past one another. When dynein arms on just one side of the axoneme walk, the result of the constrained, localized movement is bending (**Figure 5.18**).

The Dynamic Cell The role of microtubules in providing tracks for vesicle movement and in furnishing the structural backbone of cilia and flagella underscores the importance of movement in the life of the cell. Molecules inside the cell are constantly being moved from manufacturing and processing centers to specific destinations, and the cell itself may move in response to signals from the environment. Movement is just one aspect of the cell's dynamism, however. Information flows in and out of the nucleus in the form of RNAs and proteins; chemical reactions continually synthesize and degrade a mind-boggling array of compounds; ions and molecules traverse the plasma membrane with or without assistance from membrane proteins.

The effort to study cells is as dynamic as the cell itself. The earliest cell biologists depended on the static pictures furnished by the earliest light microscopes and focused on describing cell structure. Subsequently, electron microscopy furnished much more detailed views of how cells are put together. More recently, cell biologists have employed pulse-chase experiments, video-enhanced microscopy, differential centrifugation, and other techniques to study how cell components move and how organelles function. Cell biology began with an effort to describe the parts of cells; today, the field is focused on exploring how materials move between cell components and how cells respond to changing conditions.

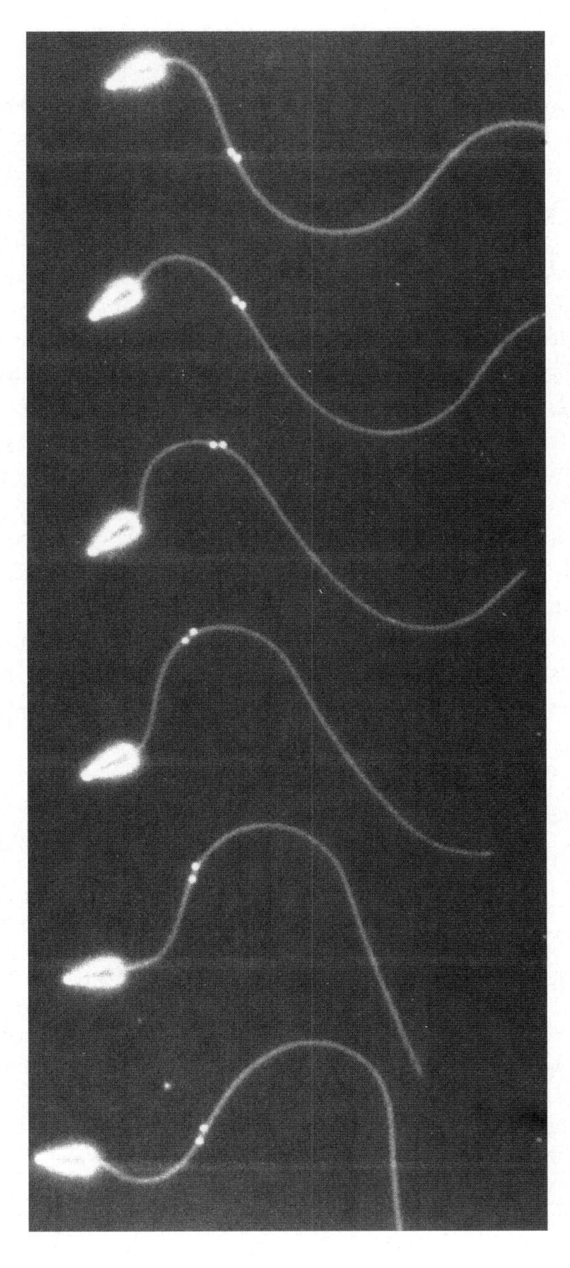

FIGURE 5.18 How Do Flagella Bend?
The photographs on the left show how the flagellum of a sperm bends as the cell moves from right to left. The researchers who took these photos attached a pair of gold beads to the flagellum; note that the light-colored beads move apart during bending. The drawing shows the position of dynein arms and microtubule doublets as a portion of a flagellum bends.

Essay Organelles and Human Disease

Given the importance of organelles in the function of eukaryotic cells, it is reasonable to predict that organelle malfunction would have serious consequences for the organism. Based on this logic, it's not surprising that a variety of human diseases are caused by abnormalities in organelles. Some of the best-understood organelle-based diseases involve lysosomes.

Lysosomes act as cellular recycling centers. Defunct organelles and old or damaged DNA, RNA, protein, carbohydrate, and lipid molecules are transported to lysosomes, where enzymes degrade them to simpler components that can be reused. It is essential that these degradative enzymes be confined to the lysosome because they are capable of destroying all of the cell's contents.

Can organelles malfunction?

Several human diseases result from leakage of lysosomal enzymes into the cytoplasm or the cell exterior. For example, asbestosis is a debilitating lung disease that occurs in asbestos miners and construction workers who handled large quantities of asbestos-containing insulation and fireproofing material. When these individuals breathed asbestos fibers into their lungs, the fibers were localized to the lysosomes of lung cells. Lysosomal enzymes cannot degrade asbestos fibers, however. Consequently, the fibers gradually accumulate and eventually result in the leakage of enzymes through the lysosomal membrane, disrupting the cells. The damage leads to the condition known as asbestosis, which is characterized by severe coughing and shortness of breath. Rheumatoid arthritis, an inflammatory joint disease, is partly caused by the leakage of lysosomal enzymes into the joint fluid.

About 40 different human disorders are caused by deficiencies in specific lysosomal enzymes. The most severe of these so-called lysosomal storage diseases is inclusion-cell disease, which causes facial and skeletal abnormalities as well as mental retardation. Inclusion-cell disease occurs because almost all of the degradative enzymes normally found in lysosomes are missing. Thus the lysosomes swell with undigested materials. The resulting structures are known as inclusions. The swollen lysosomes ultimately cause cell damage and disease.

Inclusion-cell disease is caused by a deficiency in a single enzyme required for attaching mannose-6-phosphate to proteins. Recall that this phosphorylated sugar serves as the address tag that targets proteins to the lysosome from the Golgi apparatus. In the absence of the mannose-6-phosphate tag, enzymes that are normally shipped to the lysosome are secreted from the cell instead. In fact, researchers originally discovered that mannose-6-phosphate serves as an address tag by analyzing the composition of lysosomal enzymes secreted by patients with inclusion-cell disease and comparing the structure of the proteins to lysosomal enzymes from normal individuals.

Tay-Sachs disease is another lysosomal storage disorder. Unlike inclusion-cell disease, Tay-Sachs results from the absence of a single enzyme from lysosomes. The missing enzyme normally degrades a specific type of glycolipid that is abundant in brain cells. In Tay-Sachs patients, the glycolipid accumulates in brain cells and disrupts their function. Symptoms include rapid mental deterioration after about 6 months of age, followed by paralysis and death within 3 years.

Although most organelle diseases cannot yet be cured, drug therapies are beginning to offer some hope. For example, the symptoms associated with some lysosomal storage diseases show improvement when the normal enzyme is provided in pill form. Some diseases caused by defects in lysosomal enzymes may also respond to the types of gene therapies introduced in Chapter 17.

Chapter Review

Summary

Because all organisms consist of cells, many questions in biology can be answered by understanding cell structure and function. There are two basic cellular designs—prokaryotic and eukaryotic. Eukaryotic cells are usually much larger and more functionally complex than are prokaryotic cells. Prokaryotic cells consist of a single membrane-bound compartment in which nearly all cellular functions occur. Eukaryotic cells contain numerous membrane-bound compartments called organelles; different organelles carry out different functions.

The defining organelle of eukaryotic cells is the nucleus, which contains the cell's chromosomes and serves as its control center. For a cell to function properly, the movement of molecules into and out of the nucleus must be carefully controlled. Traffic across the nuclear envelope occurs through nuclear pores, which contain a multiprotein nuclear pore complex that serves as gatekeeper. Both passive and active transport of materials occurs through these nuclear pore complexes. Active import and export of proteins and RNAs involves built-in signals that target cargo to the correct compartment.

The endomembrane system is an extensive system of membranes and membrane-bound compartments. The two principal organelles of the endomembrane system are the endoplasmic reticulum (ER) and the Golgi apparatus. Early studies revealed a role for the ER and Golgi apparatus in protein secretion.

Later work documented that the ER is the site of synthesis for a wide array of proteins and lipids. Most ER products are shipped to the Golgi apparatus, which serves as a processing and dispatching station. In many proteins and lipids the major processing step is glycosylation, or the addition of carbohydrate groups.

Materials move through the endomembrane system inside membrane-bound transport organelles called vesicles. When products leave the endomembrane system, they are sorted according to molecular tags and then enclosed in vesicles. These vesicles contain proteins that interact with receptor proteins on the surface of a target organelle or the cell membrane.

The cytoskeleton is an extensive system of fibers that serves as a structural support for eukaryotic cells. Elements of the cytoskeleton also provide the machinery for moving vesicles inside cells and for moving the cell as a whole through the beating of flagella or cilia. Cell motility and the movement of vesicles inside cells both depend on motor proteins, which can convert chemical energy stored in ATP into movement. Movement of intracellular transport vesicles occurs as the motor protein kinesin "walks" along microtubule tracks. Cilia and flagella are locomotor structures that bend as the motor protein dynein "walks" along microtubule tracks.

The data reviewed in this chapter provide a view of the cell as a dynamic reaction vessel, which synthesizes and ships an array of products in a highly regulated manner. Most of the reactions and all of the movement analyzed here require the expenditure of chemical energy stored by ATP. For a cell to stay alive and function normally, it has to hydrolyze massive amounts of this molecule. Understanding how cells synthesize ATP is the subject of Chapter 6.

Questions

Content Review

1. What is an organelle?
   a. any cellular structure whose function is essential for a cell to acquire resources, grow, and reproduce
   b. a membrane-bound structure inside a cell
   c. the site of ribosome synthesis
   d. the site of ATP synthesis
2. What organelles are the main components of the endomembrane system?
   a. the cell membrane and cell wall
   b. the nucleus, nuclear pores, and nucleolus
   c. mitochondria
   d. the ER and Golgi apparatus
3. Which of the following describe the nuclear envelope?
   a. It is continuous with the endomembrane system.
   b. It is continuous with the nucleolus.
   c. It is continuous with the cell membrane.
   d. It contains a single membrane.
4. What is a nuclear localization signal?
   a. a signal built into a protein that directs it to the nucleus
   b. a molecule that is attached to nuclear proteins, so that they are retained inside the structure
   c. a component of the nuclear pore complex
   d. a stretch of amino acids that directs proteins from the nucleus to the ER
5. Which of the following is not true of secreted proteins?
   a. They are synthesized in the rough ER.
   b. They are transported through the endomembrane system in membrane-bound transport organelles.
   c. They are transported from the Golgi apparatus to the ER.
   d. They contain a signal sequence that directs them into the ER.
6. What is a motor protein?
   a. one of the three types of fibers that comprise the cytoskeleton
   b. a protein that changes conformation when ATP is hydrolyzed and makes movement possible
   c. a molecule that "walks" along a microtubule when mannose-6-phosphate is attached
   d. an enzyme that catalyzes the formation of the ATP used in movement
7. What is the function of the cytoskeleton?
   a. It provides structural support for the cell, supplies tracks used for internal transport, and is used in cell motility.
   b. It is an important component in bacterial, archaeal, and plant cell walls.
   c. It surrounds the chromosomes of prokaryotic cells.
   d. It is the site of ribosome synthesis in eukaryotes.
8. Cilia and eukaryotic flagella differ in what respects?
   a. Only cilia are covered with membrane.
   b. Only flagella contain a 9 + 2 arrangement of microtubules.
   c. Cilia contain kinesin as a motor protein; flagella contain dynein.
   d. Cilia are shorter than flagella; cells have many cilia but only one or two flagella.

Conceptual Review

1. Compare and contrast the structure of a generalized plant, animal, and prokaryotic cell. What is the function of each structure? Which features are common to all cells? Which are specific to certain lineages?
2. Briefly explain the experimental strategy used by researchers to identify the signal that targets the nucleoplasmin protein to the nucleus.
3. Draw a diagram that traces the movement of a secreted protein from its site of synthesis to the outside of a eukaryotic cell. Identify all of the organelles that the protein passes through, and indicate the direction of movement.
4. Briefly describe the mechanisms that allow a eukaryotic cell to deliver different cellular products to different organelles or the cell membrane. Use diagrams where appropriate. Be sure to compare and contrast the molecular "addresses" used to direct proteins to the nucleus, ER, and lysosomes.
5. Describe how a motor protein like kinesin can move a transport vesicle down a microtubule track. Include all necessary steps and components.
6. Describe the logic of a pulse-chase experiment. How was this approach used to document the pattern of protein transport through the endomembrane system?
7. Briefly describe how researchers use differential centrifugation to isolate particular cell components for further study.

Applying Ideas

1. According to the experimental result summarized in Figure 5.6c, nucleoplasmin core fragments were not found in the nucleus because they were not transported there. It's equally plausible, though, that the fragments had entered the nucleus but were not retained there. To test this hypothesis, researchers injected core fragments directly into a series of cell nuclei. They found that the fragments were retained there. Do these data support a role for the nuclear localization signal in transport into the nucleus, retention within the nucleus, or both? Explain.
2. In addition to delivering cellular products to specific organelles, eukaryotic cells can take up material from the outside and transport it to specific organelles. For example, specialized cells of the human immune system ingest bacteria and viruses and then deliver them to lysosomes for degradation. Suggest a hypothesis for how this material is tagged and directed to lysosomes. How would you test this hypothesis?
3. The leading hypothesis to explain the origin of the nuclear envelope is that a deep infolding of the cell membrane occurred in an ancient prokaryote. Draw a diagram that illustrates this infolding hypothesis. Does your model explain the existence of the structure's inner and outer membrane? Explain.
4. When researchers compare different types of eukaryotic cells, they often find that the organelles in the cell correlate with that cell's function. For example, muscle cells in animals have large numbers of ATP-producing mitochondria; storage cells in plants have huge vacuoles. Similarly, structural cells in plants have thick secondary cell walls made of tough lignin and cellulose molecules. Based on this pattern, propose a function for cells that contain (a) a large number of lysosomes and (b) a large amount of rough ER.
5. Suggest a hypothesis or a series of hypotheses to explain why bacteria, archaea, algae, and plants have cell walls. Suppose that mutant individuals from each group were available that lacked a cell wall. How could you use these individuals to test your idea(s)?

CD-ROM and Web Connection

CD Activity 5.1: Transport into the Nucleus *(animation)*
(Estimated time for completion = 5 min)
How does a newly synthesized protein reach its proper destination in a cell?

CD Activity 5.2: Pulse-Chase Experiment *(animation)*
(Estimated time for completion = 5 min)
Follow an important experiment that traces the path of secreted proteins through the cell.

At your **Companion Website** (http://www.prenhall.com/freeman/biology), you will find self-grading exams and links to the following research tools, online resources, and activities:

The Plant Cell
Take an interactive tour of the organelles and other components of a plant cell.

Gaucher Disease
This website describes Gaucher Disease, in which lysosomes are unable to break down glucocerebroside, a fatty carbohydrate.

Muscle inside the Nucleus
This article discusses research that discovered myosin, a major protein component of muscle contraction.

Additional Reading

Claude, Albert. 1974. The coming of age of the cell. *Science* 189: 433–435. Claude's Nobel Prize address describing early studies on the structure and function of cells.

Gavin, R. H. 1999. Synergy of cytoskeletal components. *BioScience* 49: 641–655. Highlights dynamic interactions among components of the cell skeleton.

Hagmann, Michael. 1999. Protein ZIP codes make Nobel journey. *Science* 289: 666. A news report on the awarding of the Nobel Prize to Gunter Blobel for the signal hypothesis.

Ingber, D. E. 2000. The origin of cellular life. *BioEssays* 22: 1160–1170. Explores the nature of the first cells.

Palade, G. 1974. Intracellular aspects of the process of protein synthesis. *Science* 189: 347–358. Palade's 1974 Nobel Prize address describing early studies on the structure and function of the endomembrane system.

Rothman, J. E., and L. Orci. 1996. Budding vesicles in living cells. *Scientific American* 37 (March): 70–75. More on membrane-bound transport organelles.

Respiration and Fermentation

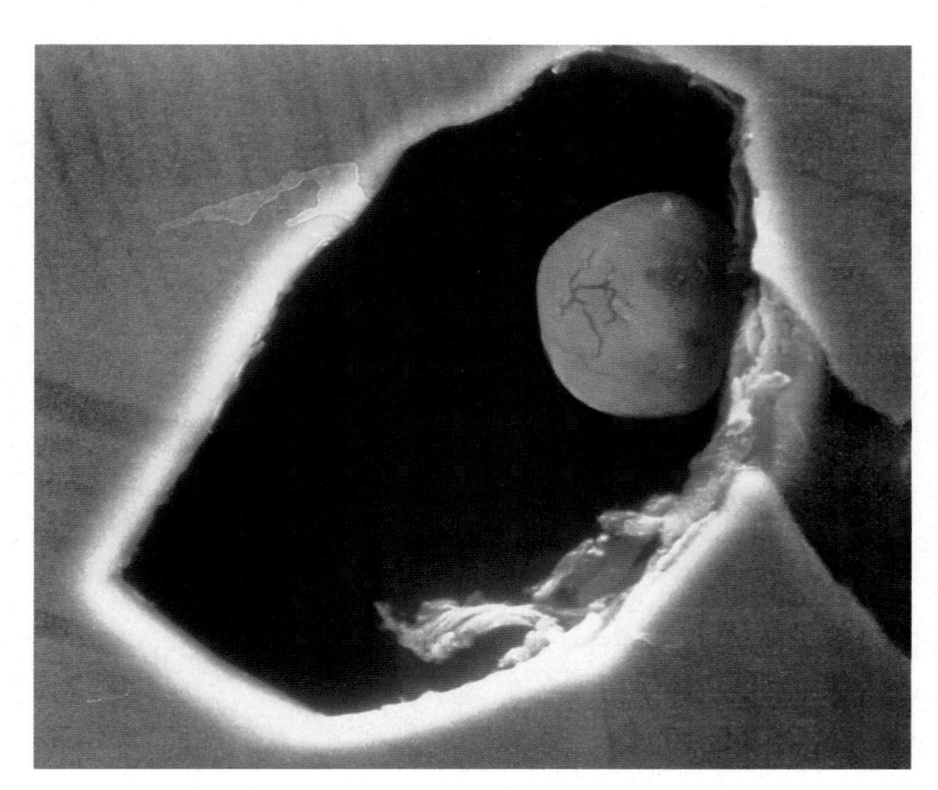

The black material in this photograph consists of carbon-containing compounds from organisms that lived over 3.8 billion years ago. The cells responsible for this carbon-rich deposit very likely used the biochemical pathways described in this chapter.

Cells are dynamic. Vesicles move cargo from the Golgi apparatus to the plasma membrane and other destinations, enzymes throughout the cell synthesize a complex array of macromolecules, and millions of membrane proteins pump ions and molecules to create an environment conducive to life. How does all this activity occur? The answer lies in a molecule called adenosine triphosphate (ATP).

In Chapter 5 we explored the role that ATP plays in moving materials through cells and in moving cells themselves through the environment. Recall that when ATP is hydrolyzed to yield adenosine diphosphate (ADP) and an inorganic phosphate ion (P_i), chemical energy is released. When a phosphate group is transferred to motor proteins like kinesin and dynein, they respond by moving. The transfer of a phosphate group can also drive the types of endergonic reactions introduced in Chapters 2 and 3, or to pump ions or molecules across the cell membrane against an electrochemical gradient (see Chapter 4). In short, ATP allows cells to do work. It takes work to stay alive, so there is no life without ATP.

The goal of this chapter is to investigate how cells make adenosine triphosphate by oxidizing reduced compounds such as sugars, molecular hydrogen (H_2), or hydrogen sulfide (H_2S). The energy released during the oxidation process is used to phosphorylate ADP and generate ATP. Section 6.1 introduces the events involved by reviewing the fundamental principles of reduction and oxidation and summarizing the reactions involved in oxidizing glucose, the most common fuel used by organisms. Then sections 6.2 through 6.4 delve into the glucose-oxidizing reactions in detail. Section 6.5 introduces an alternative route for ATP production—called fermentation—that occurs in many bacteria, archaea, and eukaryotes. In closing, the chapter examines how cells shunt certain carbon-containing compounds away from ATP production and into the synthesis of DNA, RNA, amino acids, and other molecules. This chapter is an introduction to **metabolism**—the chemical reactions that occur in cells.

6.1 An Overview of Cellular Respiration

Cells use chemical energy to move cargo, pump ions, complete endergonic reactions, and perform other types of work. Because reduced molecules like carbohydrates and fats contain a great deal of chemical energy, they make it possible for cells to do a great deal of work. But even if a cell has huge amounts of chemical energy stored in the form of carbohydrates and fats, no movement, endergonic reactions, or pumping by membrane proteins occur. Why? Just like the dollar, euro, and yen act as currencies in human economies, ATP acts as the currency in the economy of a cell. If you have a huge savings account at a bank, it won't do you any good when you walk up to a vending machine to get a soda. You need cash. Similarly, cells are not able to use the chemical energy stored in fats, carbohydrates, or H_2 to do work. To be useful, the chemical energy must first be converted to ATP.

It's important to recognize, though, that cells don't hoard large wads of cash in the form of ATP. In general, a cell contains only enough ATP to last from 30 seconds to a few minutes. Like many other cellular processes, the production and use of ATP is dynamic. Because ATP makes every other cell function possible, producing a steady supply of ATP is the most fundamental of all cell processes.

If fats, carbohydrates, or other types of reduced molecules act as a savings account for chemical energy, how do cells draw on this account to synthesize ATP? To get started on answering this question, let's review the nature of chemical energy and reduction-oxidation reactions.

The Nature of Chemical Energy and Redox Reactions

As you learned in Chapter 2, chemical energy is a form of potential energy. In cells, electrons are the most important source of chemical energy. The amount of potential energy that an electron has is based on its position relative to other electrons and to the positive charges in the nuclei of nearby atoms.

Why is so much potential energy released when ATP is hydrolyzed to ADP and inorganic phosphate (P_i)? To answer this question, look at the structure of ATP shown in **Figure 6.1a** and note that the phosphate groups in the molecule contain four negative charges. Because these negative charges repel each other, the potential energy of the electrons in the phosphate groups is high. When the bond between the outermost phosphate groups is broken by hydrolysis, ADP and P_i form. The amount of electrical repulsion occurring in each of these product molecules is much less than the amount occurring in ATP. In addition, the negative charges on ADP and P_i are stabilized much more efficiently by interactions with the partial positive charges on surrounding water molecules.

For these and other reasons, ADP and P_i have lower potential energy than does ATP. The difference in the potential energy of the reactant and the products is released as heat, motion,

(a) ATP consists of adenine, ribose, and three phosphate groups.

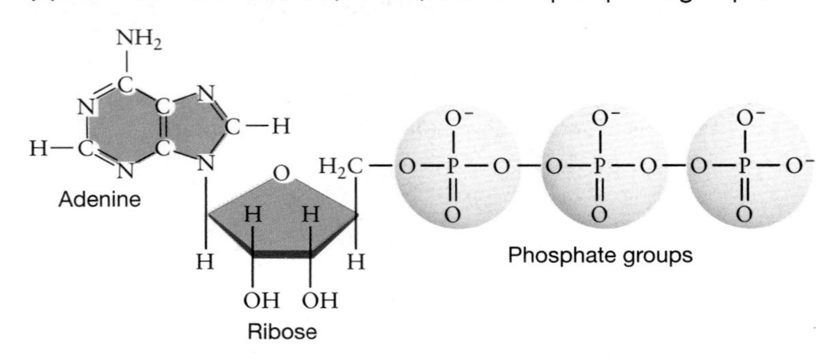

(b) Energy is released when ATP is hydrolyzed.

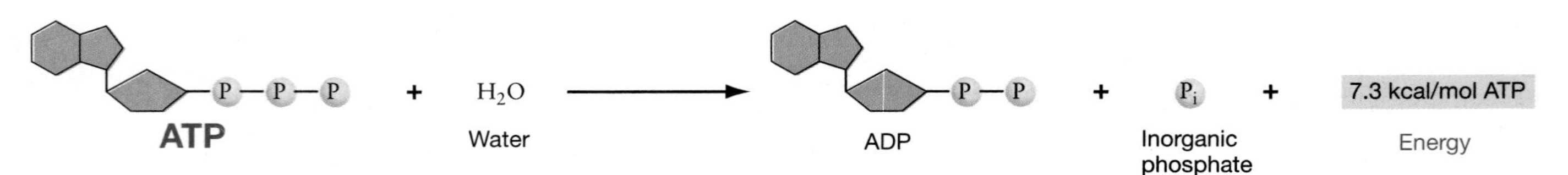

FIGURE 6.1 Adenosine Triphosphate (ATP)
(a) ATP has high potential energy in part because four negative charges are clustered in its three phosphate groups. The negative charges repel each other, raising the potential energy of the electrons. **(b)** When a mole of ATP is hydrolyzed to ADP and inorganic phosphate, a large amount of energy is released.
EXERCISE Draw the chemical structure of ADP and inorganic phosphate.

or some other form of energy. Under standard conditions of temperature and pressure, a total of 7.3 kilocalories (kcal) of energy is released per mole of ATP hydrolyzed (**Figure 6.1b**). In cells, this energy is used to transport materials, move flagella or cilia, pump ions, drive endergonic reactions, and complete other kinds of work.

If a great deal of energy is released when ATP loses a phosphate group, then a great deal of energy must be required to synthesize ATP. Where does this energy come from? The answer is reduction-oxidation (redox) reactions—the class of chemical reactions introduced in Chapter 2. Recall that redox reactions involve a transfer of electrons. When an atom or molecule is reduced, it gains an electron from an atom or molecule that becomes oxidized in the process. In addition to gaining an electron, the reduced compound gains potential energy; the oxidized compound loses potential energy. Stated another way, reduced compounds have high potential energy and act as electron donors; oxidized molecules have low potential energy and act as electron acceptors.

The most important set of redox reactions in biology involves the oxidation of the six-carbon sugar, glucose. To understand why these reactions are important, consider that when glucose is heated it undergoes the uncontrolled oxidation reaction called burning.

$$C_6H_{12}O_6 + 6\,O_2 \rightarrow 6\,CO_2 + 6\,H_2O + \text{energy}$$

In this reaction, the carbon atoms in glucose are oxidized to form carbon dioxide, and the oxygen atoms in the oxygen molecule (O_2) are reduced to form water. Glucose is the reduced molecule that acts as an electron donor and becomes oxidized; oxygen is the oxidized molecule that acts as an electron acceptor and thus becomes reduced.

A large drop in potential energy occurs during this redox reaction because the electrons in the product molecules are much more tightly bound than they are in the reactant molecules. If an electron is tightly bound in an atom or molecule, it has lower potential energy than it does in an atom or molecule where it is loosely bound. When glucose burns, the change in potential energy is converted to kinetic energy in the form of heat. When a mole of glucose burns, a total of 686 kcal of heat is released.

Glucose does not burn in cells, however. Instead, glucose is oxidized through a long series of carefully controlled redox reactions. These reactions are occurring, millions of times per minute, in your cells right now. Instead of being given off as heat, much of the energy released by the drop in potential energy is being used to make the ATP you need to think, move, and stay alive.

How does the oxidation of glucose take place in a way that supports the production of ATP? Let's get an overview of how the process occurs, and then pursue a more detailed analysis in sections 6.2 through 6.4.

Processing Glucose

An enormous diversity of organisms use glucose as a primary fuel. In virtually all species examined to date, the first step in the oxidation of glucose is a sequence of 10 chemical reactions that are collectively called **glycolysis** (sugar-loosen). These reactions occur in the cytoplasm of eukaryotic and prokaryotic cells. During glycolysis, one molecule of the six-carbon sugar glucose is broken into two molecules of a 3-carbon compound called pyruvate. Some of the potential energy released by this sequence of reactions is used to phosphorylate ADP molecules, forming ATP. In addition, one of the reactions in the sequence results in the reduction of a molecule called nicotinamide adenine dinucleotide, symbolized NAD^+. As **Figure 6.2** shows, the addition

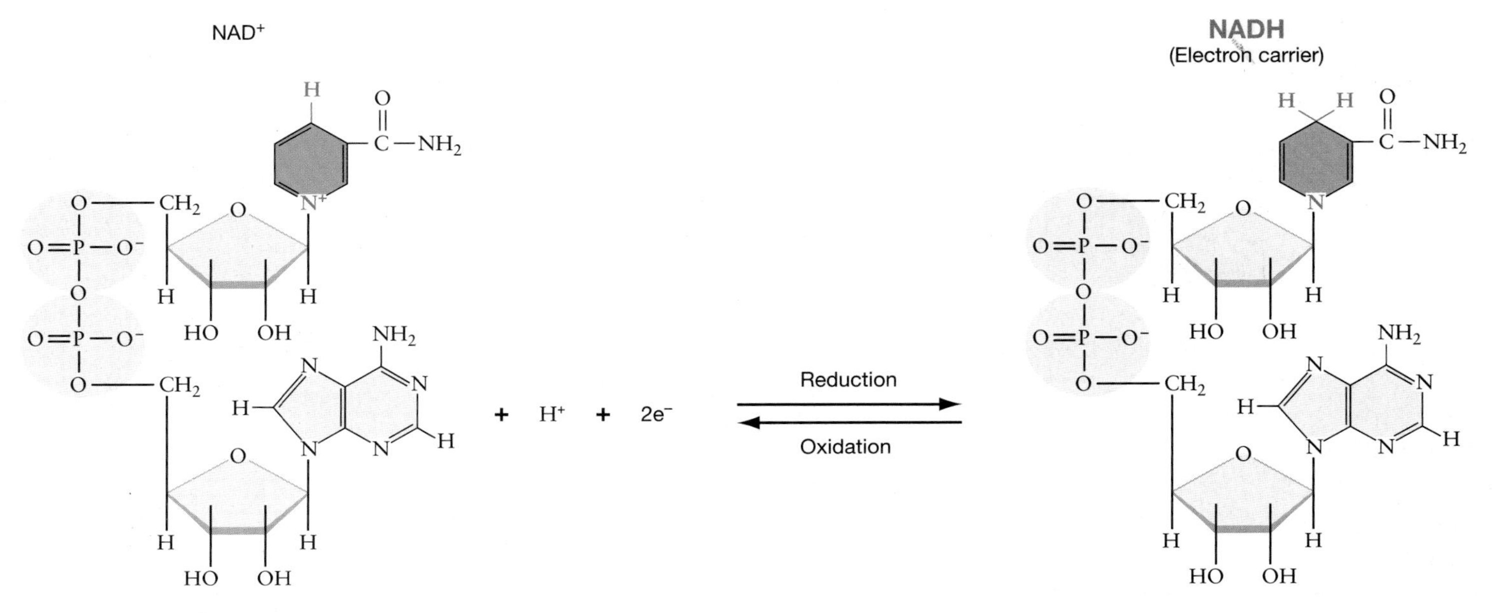

FIGURE 6.2 NAD$^+$ and NADH
NADH is the reduced form of NAD$^+$. QUESTION Compare the structure of NAD$^+$ with the structure of ATP in Figure 6.1. Which portions of the two molecules are identical?

of electrons to NAD^+ produces NADH. As we'll see, NADH is a reduced molecule that readily donates electrons to a more oxidized molecule. As a result, it is sometimes called an electron carrier and is said to have "reducing power."

The left panel in **Figure 6.3** summarizes the results of glycolysis. What happens to the pyruvate produced by this sequence of reactions? If oxygen is present in the cell, pyruvate enters a sequence of reactions called the Krebs cycle. In the reactions that lead up to and complete the Krebs cycle, each pyruvate is oxidized to three molecules of carbon dioxide. Some of the potential energy released by the reactions is used to reduce NAD^+ to NADH, to reduce another electron carrier called flavin adenine dinucleotide (FAD) to $FADH_2$, and to phosphorylate ADP, yielding ATP (see Figure 6.3, middle).

With the completion of the Krebs cycle, glucose has been completely oxidized to CO_2. This observation raises several questions. According to the overall reaction for the oxidation of glucose, molecular oxygen (O_2) is a reactant. Where does oxygen come into play? What happens to all of the reduced electrons carried by NADH and $FADH_2$?

Electron Transport and Oxidative Phosphorylation

In cells, the electrons carried by NADH and $FADH_2$ are gradually "stepped down" in potential energy by a series of molecules that acts as an electron transport chain. In eukaryotes, the components of the electron transport chain are located in the inner membrane of mitochondria; in prokaryotes they are found in the cell membrane. Electrons gradually fall from a higher to lower potential energy as they are passed from a more reducing to a more oxidizing molecule along the electron transport chain. The energy released is used by the transport chain molecules to pump protons to the outside of the membrane. The buildup of protons results in a strong electrochemical gradient that drives protons back across the membrane. The force generated by the incoming protons powers a membrane protein called ATP synthase, which phosphorylates ADP (see Figure 6.3, right). Once the electrons donated by NADH and $FADH_2$ have passed through the electron transport chain, they are accepted by oxygen. The addition of these electrons to oxygen, along with protons, results in the formation of water as a product.

Because the phosphorylation events catalyzed by ATP synthase are powered by an incoming stream of protons, which are pumped out by an electron transport chain that uses oxygen as an electron acceptor, this mode of ATP production is called **oxidative phosphorylation.** But when ATP is produced directly by the enzymes of glycolysis or the Krebs cycle the event is called **substrate-level phosphorylation.** As we'll see, substrate-level phosphorylation involves the enzyme-catalyzed transfer of a phosphate group from an intermediate substrate directly to ATP.

When oxygen accepts electrons from NADH and $FADH_2$ through the electron transport chain, the oxidation of glucose is complete. As a whole, this ATP production process is called **cellular respiration.** Cellular respiration refers to any ATP production process that involves a reduced compound that acts as an electron donor, an electron transport chain, and an electron acceptor.

If oxygen is lacking, however, no electron acceptor is available and cellular respiration cannot occur. In this case, alternative pathways called fermentation reactions take over. As we'll see in section 6.5, fermentation consists of reactions that allow glycolysis to continue even if pyruvate is not drawn off to the

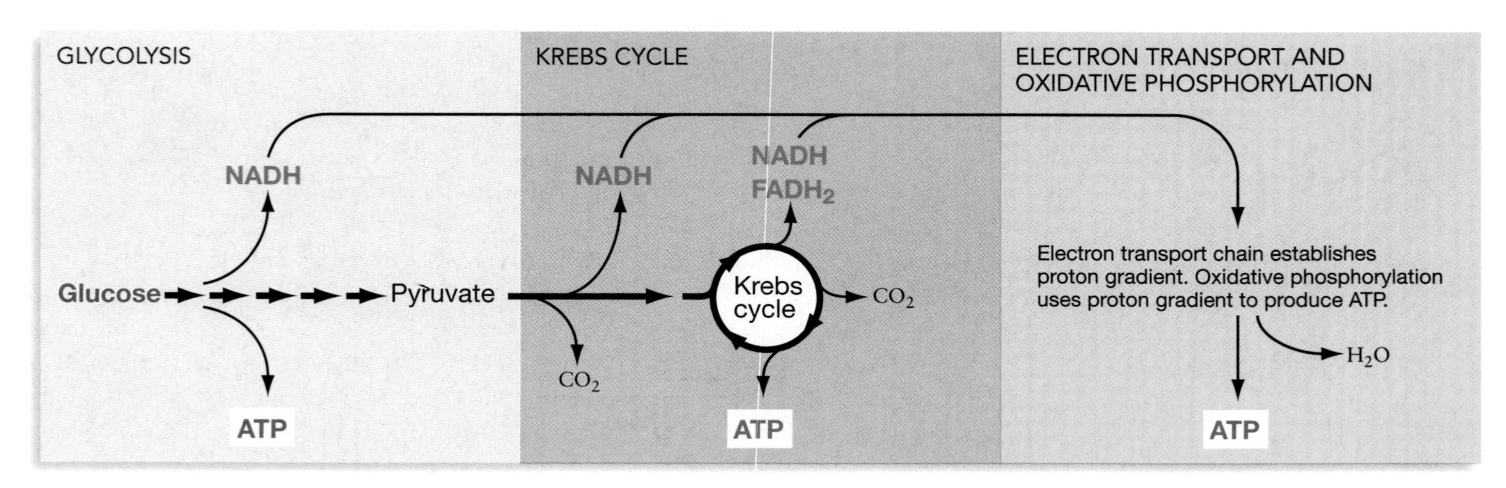

FIGURE 6.3 An Overview of Cellular Respiration
Cellular respiration consists of three component processes: glycolysis, the Krebs cycle, and electron transport combined with oxidative phosphorylation. Each component produces at least some ATP. Because the three components are connected, cellular respiration is an integrated metabolic pathway. Glycolysis and the Krebs cycle are connected by pyruvate; glycolysis and the Krebs cycle are connected to electron transport chains by NADH and $FADH_2$.

Krebs cycle. Ethanol (the active ingredient in alcoholic beverages), lactic acid, or other types of by-products result from different types of fermentation pathways.

The quick overview of cellular respiration summarized in Figure 6.3 is simply intended to introduce the three processes responsible for most ATP production in cells. To clarify the relationship between glycolysis, the Krebs cycle, and electron transport chains and oxidative phosphorylation, let's take a closer look at each process in turn.

6.2 Glycolysis

Glycolysis may be the most fundamental of all metabolic pathways, but it was discovered by accident. In the late 1890s Hans and Edward Buchner were working out techniques for manufacturing extracts of baker's yeast for therapeutic use. (Yeast extracts are still added to some foods today as a flavor enhancer or nutritional supplement.) In one set of experiments the Buchners added sucrose, or table sugar, to their extracts. Sucrose is a disaccharide consisting of glucose linked to a 6-carbon sugar called fructose. At the time the Buchners were working, sucrose was commonly used as a preservative. But the Buchners found that instead of preserving the yeast extracts, the sucrose was quickly broken down and fermented, with alcohol appearing as a by-product. This was a key finding because it showed that fermentation and other types of cellular metabolism could be studied in vitro—meaning, outside the cell. Until then, researchers had held that metabolism could take place only inside cells.

The Buchners and other researchers followed up on this observation by trying to determine how the sugar was being processed. One important early observation was that the reactions could be sustained much longer than normal if inorganic phosphate was added to the mixture. This finding implied that some of the compounds involved were being phosphorylated. Soon after, a molecule called fructose bisphosphate was isolated. The *bis-* prefix means that two phosphate groups are attached to the molecule at distinct locations. Follow-up work showed that all but two of the compounds involved in glycolysis are phosphorylated. The two exceptions are the starting and ending molecules, glucose and pyruvate.

A third major finding came to light in 1905, when researchers found that processing of sugar by yeast extracts stopped if the reaction mix was boiled. Because enzymes were known to be inactivated by heat, this discovery suggested that enzymes were involved in at least some of the processing steps. Later, investigators realized that each step in glycolysis is catalyzed by a different enzyme.

Over the next 35 years, each of the reactions and enzymes involved in glycolysis was gradually worked out by several different workers. Because Gustav Emden and Otto Meyerhof made particularly important contributions, glycolysis is sometimes called the Emden-Meyerhof pathway.

A Closer Look at the Glycolytic Reactions

Figure 6.4 (pages 118–119) details the 10 reactions involved in glycolysis. By breaking open cells, separating the components with the differential centrifugation techniques introduced in Chapter 5, and testing which components could sustain glycolysis, biologists discovered that all 10 reactions occur in the cytoplasm. Several points about the reaction sequence are important to note.

- Contrary to what most researchers expected, glycolysis starts by *using* ATP, not producing it. In the initial step, glucose is phosphorylated to form glucose 6-phosphate. After an enzyme rearranges this molecule to form fructose 6-phosphate, the third reaction in the sequence adds a second phosphate group, forming the fructose 1,6-bisphosphate observed by early researchers. Thus, two ATP are used up before any ATP are produced by glycolysis.
- Once the energy-investment phase of glycolysis is complete, the subsequent reactions represent an energy pay-off phase. The sixth reaction in the sequence results in the reduction of two NAD^+; the seventh produces two ATP and erases the energetic "debt" of two ATP invested early in glycolysis. The final reaction in the sequence produces another two ATP, giving a net yield of two NADH, two ATP, and two molecules of pyruvate for each molecule of glucose processed.
- Production of ATP during glycolysis is by substrate-level phosphorylation. In reactions 7 and 10 in Figure 6.4, an enzyme catalyzes the transfer of a phosphate group from a phosphorylated intermediate to ADP.

The discovery and elucidation of the glycolytic pathway ranks as one of the great achievements of biochemistry. Because the enzymes involved have been observed in nearly every bacterium, archaean, and eukaryote examined, biologists infer that the pathway evolved very early in the history of life. Stated another way, it is very likely that the ancestor of all organisms living today made ATP using the glycolytic pathway. The reactions outlined in Figure 6.4 are among the most ancient and fundamental of all life processes.

How Is Glycolysis Regulated?

Once the glycolytic pathway was worked out, researchers focused on understanding the structure and function of the enzymes involved and how the sequence of events is regulated. An important advance on both fronts occurred when H. A. Lardy and R. E. Parks Jr. observed that high levels of ATP inhibit a key glycolytic enzyme called phosphofructokinase. Phosphofructokinase catalyzes step 3 in Figure 6.4—the synthesis of fructose 1,6-bisphosphate from fructose 6-phosphate.

The discovery of phosphofructokinase inhibition by ATP was important because the enzyme catalyzes what biologists call the committed step of a reaction sequence. Step 3 is "committed" because its product, fructose 1,6-bisphosphate, cannot

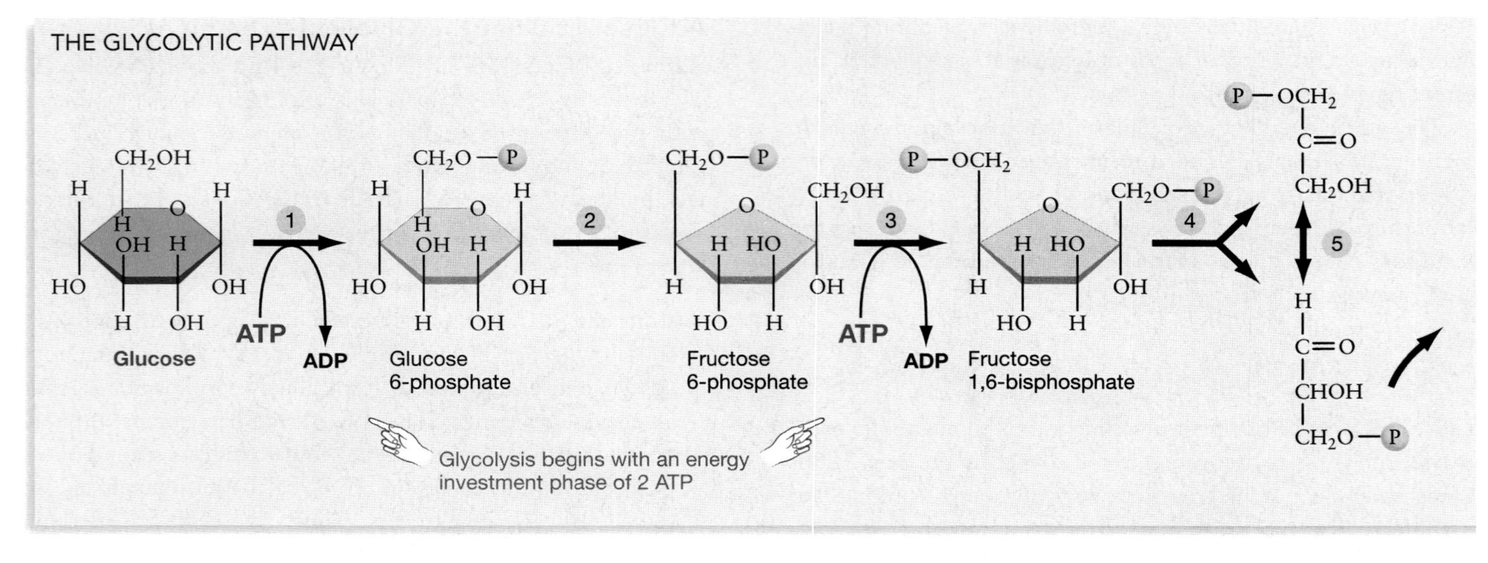

FIGURE 6.4 The Glycolytic Pathway
Glucose is oxidized to pyruvate through the sequence of 10 reactions shown here. Each reaction is catalyzed by a different enzyme, and the process begins with an investment of 2 ATP. Redox reactions later in the sequence result in the production of 2 NADH and 4 ATP, for a net gain of 2 ATP.

easily be converted back into fructose 6-phosphate. In contrast, the products of step 1 and step 2 in glycolysis are readily drawn off and used in other metabolic pathways. Once fructose 1,6-bisphosphate is synthesized, then, the rest of the glycolytic reactions have to follow. Before step 3, the sequence can be interrupted and the intermediates used elsewhere. In this light, it is logical that the pathway is turned on or off at the committed step.

Lardy and Parks's finding raised an important question, however. Why would a substrate that is required for the reaction in step 3 also inhibit the reaction? In the vast majority of cases, the addition of a substrate speeds the rate of a chemical reaction instead of slowing it.

To explain this unusual situation, the researchers hypothesized that high levels of ATP are a signal that the cell does not need to produce more ATP. According to this reasoning, cells that are able to stop glycolysis when ATP is abundant can conserve their stores of glucose for times when ATP is scarce. As a result, natural selection should favor the evolution of phosphofructokinase molecules that are inhibited by high concentrations of ATP.

This hypothesis provided a satisfying explanation for why ATP acts as an inhibitor of glycolysis's committed step. It did not, however, explain how high levels of the substrate are able to inhibit the enzyme. A resolution to this problem came later, when researchers were able to determine the three-dimensional structure of phosphofructokinase. As **Figure 6.5** shows, the structure confirmed that phosphofructokinase has two distinct binding sites for ATP. More specifically, ATP can bind at the enzyme's active site, where it is hydrolyzed to ADP and where the phosphate group is transferred to fructose 6-phosphate, resulting in the synthesis of fructose 1,6-bisphosphate. But when ATP concentrations are high, the molecule can also bind at a regulatory site. When ATP binds at this second location, the enzyme's conformation changes and the reaction rate at the active site drops dramatically.

Because of advances like these, glycolysis is among the best understood of all metabolic pathways. Now the question is, what happens to its product? How is pyruvate processed?

6.3 The Krebs Cycle

While the sequence of reactions in glycolysis was being worked out, several laboratories were focusing on a different set of oxidation reactions that take place in actively respiring cells. These reactions involve small organic (carboxylic) acids like citrate, malate, and succinate and result in the production of carbon dioxide. Carbon dioxide is a highly oxidized form of carbon, and it is the endpoint of glucose metabolism. Based on these results, researchers concluded that the oxidation of small carboxylic acids was an important component of cellular respiration.

Albert Szent-Györgyi and co-workers made several key observations about these reactions. First, a total of eight small carboxylic acids are oxidized rapidly enough to imply that they are involved in glucose metabolism. Second, when one of these molecules is added to cells, the rate of respiration increases. The added molecules are not used up, however. Instead, virtually all of the carboxylic acid added can be recovered later. Finally, Szent-Györgyi and colleagues were able to work out the order in which the eight acids were oxidized, resulting in the reaction pathway shown in **Figure 6.6a** (page 120).

These observations came together when Hans Krebs realized that the reaction sequence discovered by Szent-Györgyi's group

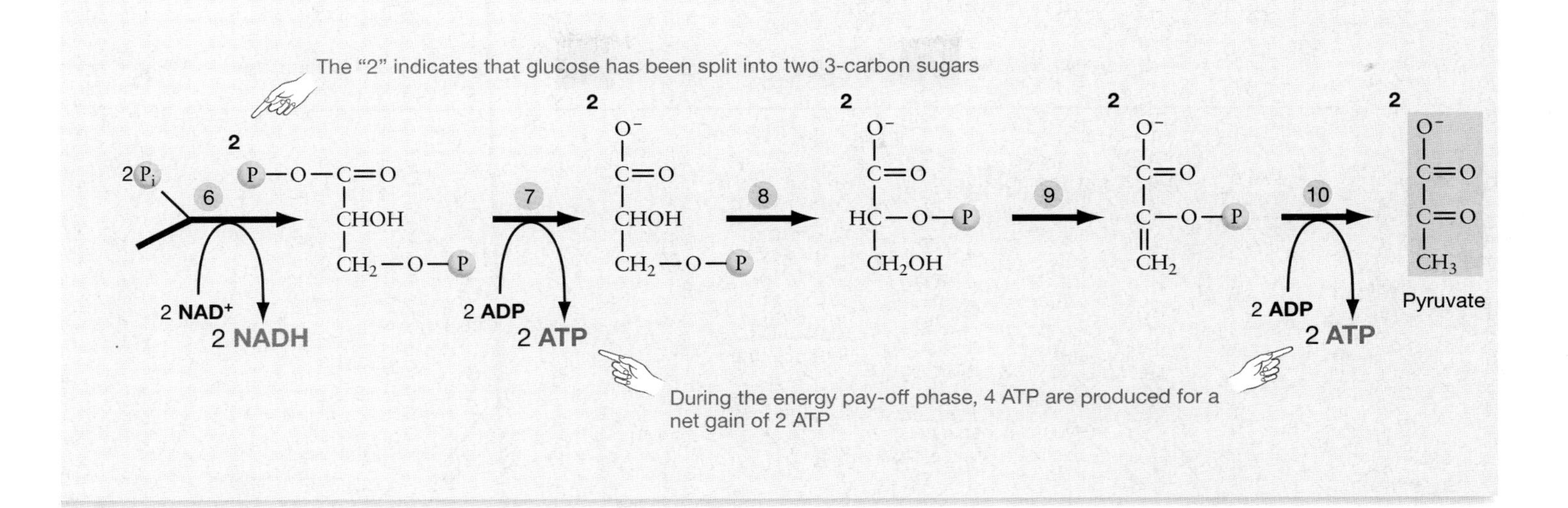

One subunit

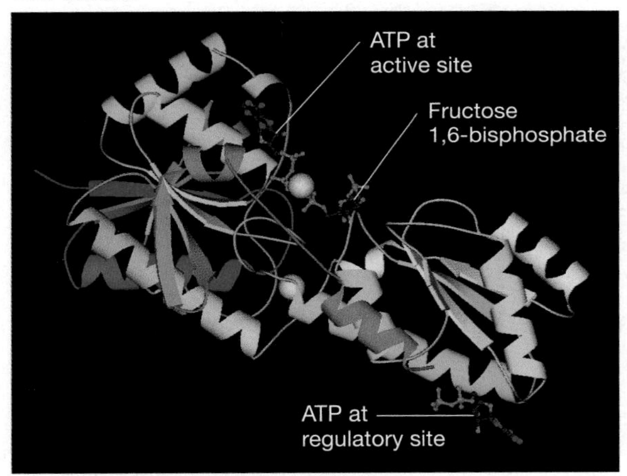

Complete enzyme

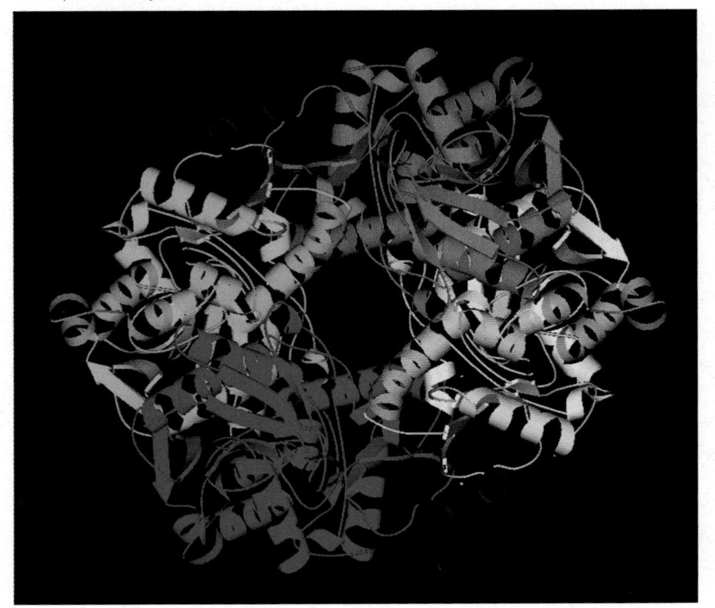

FIGURE 6.5 Phosphofructokinase Has Two Binding Sites for ATP The top panel shows one of the four identical subunits of phosphofructokinase; the bottom image shows how the four subunits interact to form the complete enzyme. Each subunit has an active site where a phosphate group is transferred from ATP to fructose 1,6-bisphosphate. Each subunit also has a regulatory site where ATP binds, however. When ATP binds at this site, the reaction is slowed.

might actually occur in a cyclical fashion, and be directly tied to the processing of pyruvate produced by glycolysis. To test this hypothesis, Krebs and W. A. Johnson set out to determine if pyruvate could react with the endpoint of the pathway in Figure 6.6a (oxaloacetate) to produce the starting point of the pathway—the six-carbon carboxylic acid called citrate. As predicted, citrate formed when Krebs and Johnson added oxaloacetate and pyruvate to cells. Based on this result, Krebs proposed that pyruvate is oxidized to carbon dioxide through the cycle of reactions diagrammed in **Figure 6.6b**. In honor of this insight, the pathway became known as the Krebs cycle.*

Converting Pyruvate to Acetyl CoA

When radioactive isotopes of carbon became available in the early 1940s, researchers in several different labs used them to confirm the cyclical nature of the Krebs cycle. For example, by adding radioactively labeled citrate or pyruvate to cells and analyzing the radioactive compounds that resulted, it was possible to show that carbon atoms cycled through the sequence of reactions just as Krebs had proposed.

Once Krebs's hypothesis was confirmed, the major question that remained was how pyruvate reacted with oxaloacetate to

*Because it begins with citrate, the Krebs cycle is also known as the citric acid cycle or the tricarboxylic acid (TCA) cycle. Citric acid is the protonated form of citrate; the term *tricarboxylic* is appropriate because citric acid has three carboxyl groups.

(a) A series of carboxylic acids is oxidized during cellular respiration.

Citrate → Isocitrate → α-ketoglutarate → Succinyl CoA → Succinate → Fumarate → Malate → Oxaloacetate

More reduced → More oxidized

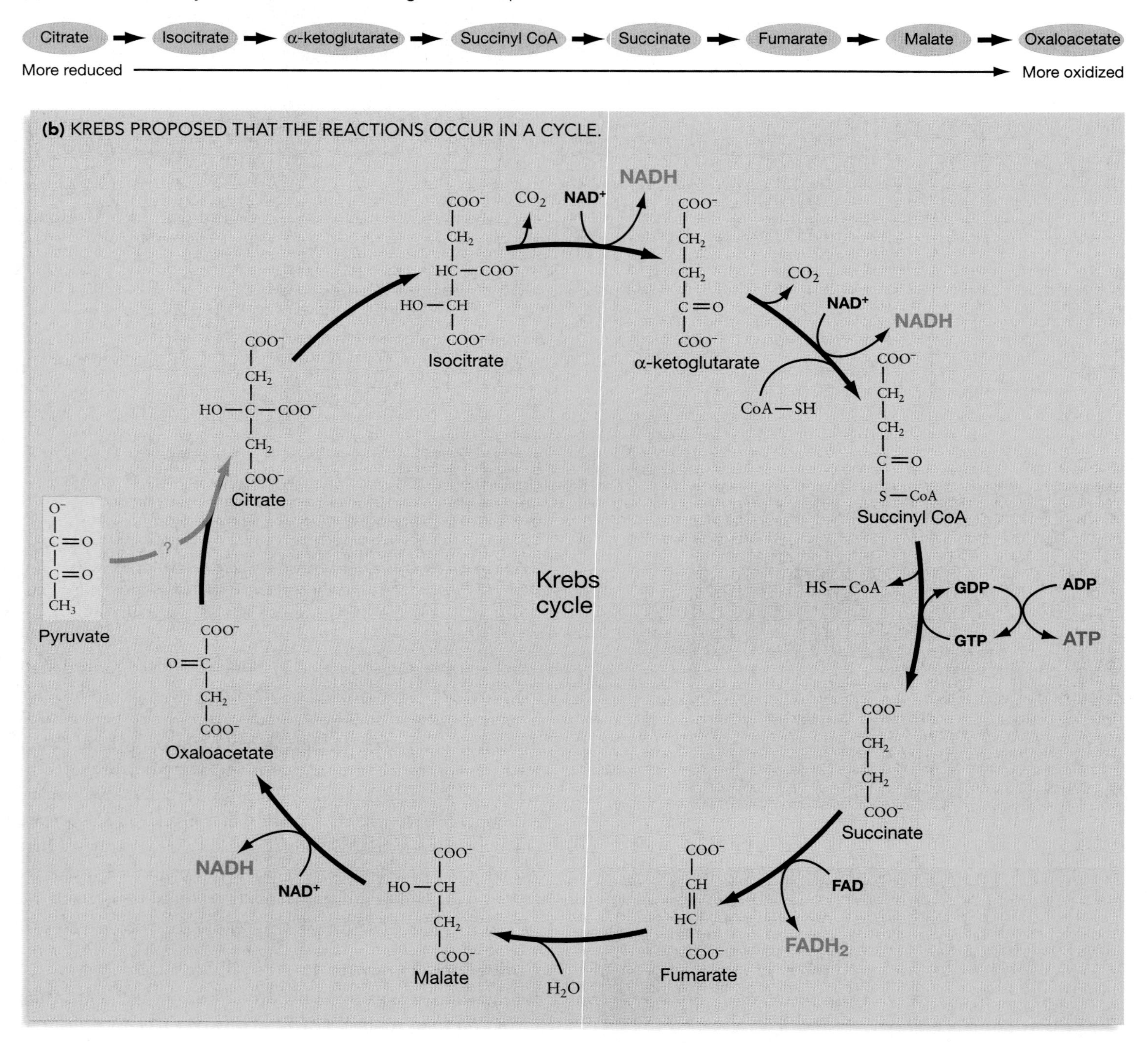

(c) Pyruvate is oxidized to acetyl CoA, which reacts with oxaloacetate to begin the Krebs cycle.

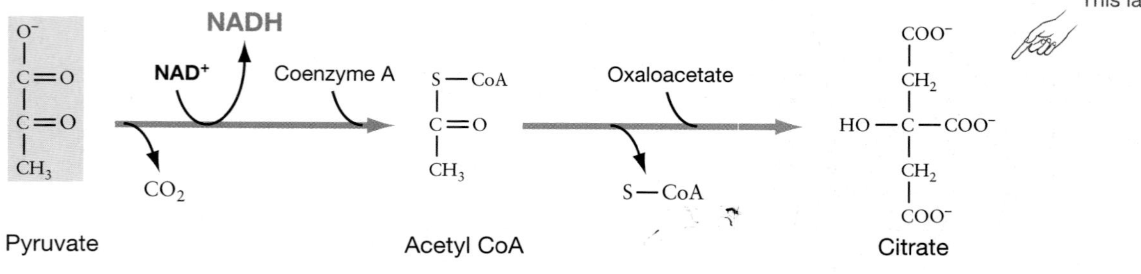

FIGURE 6.6 The Krebs Cycle
The three parts of this figure illustrate the historical development of the Krebs cycle. EXERCISE In part (b), mark one of the carbon atoms in citrate with a highlighter. Follow this atom from molecule to molecule around the cycle. What happens to it? Circle a different carbon atom found in citrate with a pencil. Follow this atom around the cycle. What happens to it?

form citrate and start the cycle. The issue remained unresolved until Fritz Lipmann discovered that a compound called coenzyme A (CoA) serves as a cofactor in a wide variety of reactions catalyzed by cellular enzymes. A **coenzyme** is a non-protein molecule that serves as a required cofactor in an enzyme-catalyzed reaction. Some coenzymes donate or accept electrons during the reaction; others transfer an acetyl ($-COCH_3$) group or carbon chains. Coenzyme A, for example, transfers an acetyl group to a substrate during many enzyme-catalyzed reactions. CoA is a large organic molecule with a sulfhydryl ($-SH$) group on one end. The "A" in CoA stands for *acetylation* because an acetyl group is often bonded to its sulfur atom.

Soon after Lipmann's finding, Severo Ochoa and associates found that pyruvate reacts with CoA to produce acetyl CoA, which then reacts with oxaloacetate to form citrate. Subsequent work showed that a series of reactions occurs as pyruvate and CoA are converted to acetyl CoA. The reactions are catalyzed by a large enzyme complex called pyruvate dehydrogenase and require a series of enzyme cofactors. These cofactors include thiamin, niacin, and pantothenic acid—molecules that are collectively called B-complex vitamins.

As **Figure 6.6c** shows, the reactions that convert pyruvate to acetyl CoA result in the production of one molecule of carbon dioxide, the reduction of NAD^+ to NADH, and the synthesis of acetyl CoA, which then reacts with oxaloacetate to form citrate and start the Krebs cycle. In this way, the reactions catalyzed by pyruvate dehydrogenase act as a preparatory step for the Krebs cycle. Once the Krebs cycle has been completed, all three carbons in pyruvate are oxidized to carbon dioxide. As Figure 6.6b and 6.6c indicate, the energy released by the oxidation of one molecule of pyruvate is used to produce four NADH, one $FADH_2$, and one ATP through substrate-level phosphorylation.

How Is the Krebs Cycle Regulated?

Like glycolysis, the Krebs cycle is carefully regulated. The reactions speed up when ATP supplies are low and slow down when ATP is abundant. Two of the major control points are highlighted in **Figure 6.7a**. Note that the pyruvate dehydrogenase complex is regulated. The enzyme that catalyzes the reaction between oxaloacetate and acetyl CoA to form citrate and start the cycle is also carefully controlled.

How does regulation occur? When supplies of ATP are abundant, the pyruvate dehydrogenase complex becomes phosphorylated. In this case, phosphorylation changes protein conformations in a way that inhibits catalytic activity. High concentrations of acetyl CoA and NADH also slow the reaction rate of the enzyme complex. This type of regulation is called **product inhibition** or **feedback inhibition.** Feedback inhibition occurs when the end product of a reaction pathway inhibits an enzyme early in the pathway (**Figure 6.7b**). The enzyme that

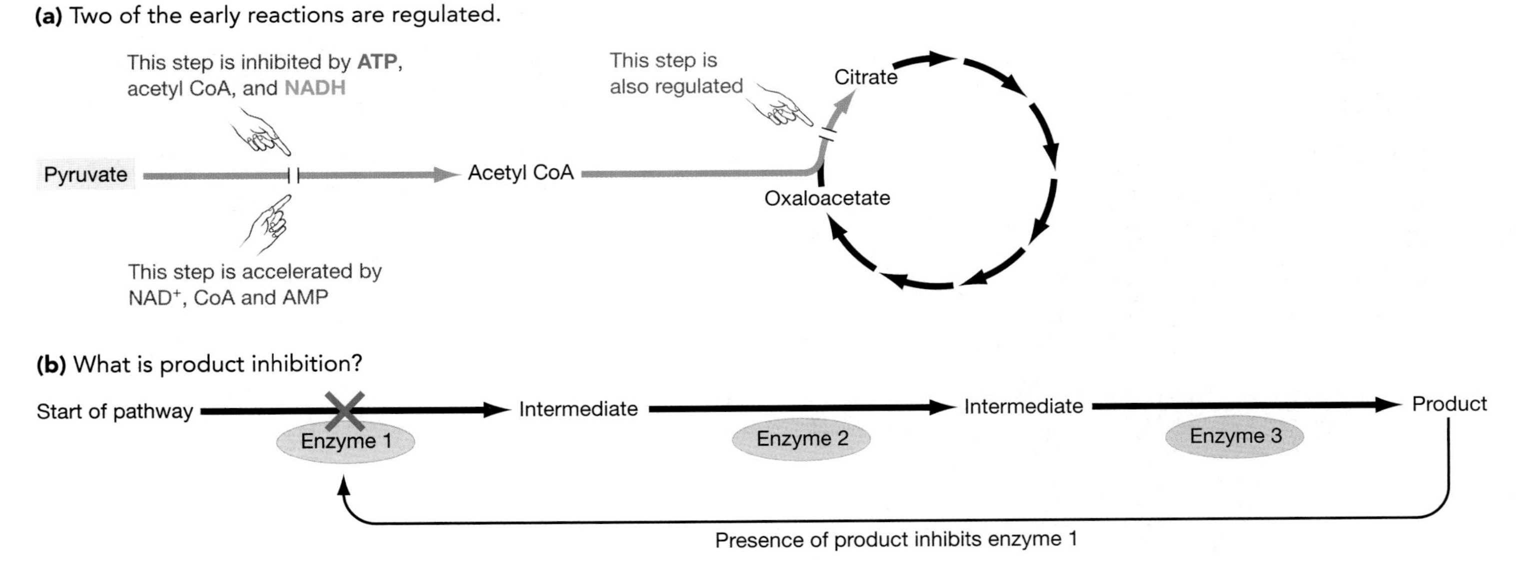

FIGURE 6.7 The Krebs Cycle Is Carefully Regulated
(a) The oxidation of pyruvate and formation of acetyl CoA is under both negative and positive control.
(b) Product inhibition occurs when the product of a metabolic pathway inhibits an enzyme active early in the pathway. Product inhibition occurs when acetyl CoA inhibits the oxidation of pyruvate. QUESTION Why does regulation occur early in the pathway? Why not later?

catalyzes the synthesis of citrate is not controlled by feedback inhibition, but by high levels of ATP.

In contrast to these examples of inhibitory control, the presence of NAD^+, CoA, or adenosine monophosphate (AMP)—which indicates low supplies of ATP—speeds the conversion of pyruvate to acetyl CoA. Stated another way, the key steps in the processing of pyruvate are under both positive and negative control. By speeding up the oxidation of pyruvate when ATP supplies are low and by slowing down the reactions when ATP is plentiful, cells carefully match the rate of cellular respiration to their energy requirements. Natural selection has favored the evolution of enzymes that allow cells to conserve pyruvate or acetyl CoA, just as it has favored glycolytic enzymes whose regulation allows cells to conserve glucose.

Where Does the Krebs Cycle Take Place?

In bacteria and archaea, the enzymes responsible for glycolysis and the Krebs cycle are located in the cytoplasm. In eukaryotes, though, the pyruvate produced by glycolysis is transported from the cytoplasm to mitochondria. Recall from Chapter 5 that mitochondria are organelles found in virtually all eukaryotes. **Figure 6.8** shows an electron micrograph of this organelle. As you can see, it has two membranes, and the inner membrane is folded to form invaginations called **cristae**. The solution inside the cristae is called the **mitochondrial matrix**. In eukaryotes, the enzymes responsible for the Krebs cycle are located in this matrix.

To pull these observations together, **Figure 6.9** reviews the relationship between glycolysis and the Krebs cycle and identifies where each process takes place in eukaryotic cells. As illustrated, for each molecule of glucose that is fully oxidized to carbon dioxide, the cell produces 10 NADH, 2 $FADH_2$, and 4 ATPs. These ATPs produced by substrate-level phosphorylation can be used immediately to drive endergonic reactions, power movement, or run membrane pumps.

Now the question is, what is the fate of the electrons carried by NADH and $FADH_2$? In effect, glycolysis and the Krebs cycle transfer electrons from glucose to NAD^+ and FAD. According to the overall reaction for the oxidation of glucose (glucose + $O_2 \rightarrow CO_2$ + H_2O + energy), oxygen serves as the final acceptor for the electrons in NADH and $FADH_2$. When oxygen accepts electrons, water is produced.

Based on these observations, questions about the fate of electrons carried by NADH and $FADH_2$ become more focused. If NADH and $FADH_2$ act as electron carriers, where do they carry them? How are the electrons transferred to oxygen? And finally, how is ATP generated as electrons are transferred from the reduced molecules NADH or $FADH_2$ to the oxidized molecule O_2?

In the 1960s—decades after the details of glycolysis and the Krebs cycle had been worked out—a startling answer to these questions emerged.

6.4 Electron Transport and Chemiosmosis

One fundamental question about the oxidation of NADH and $FADH_2$ turned out to be relatively straightforward to answer. To determine where the oxidation reactions take place, researchers isolated mitochondria using the differential centrifugation techniques introduced in Chapter 5. Then they broke the organelles open and separated the inner and outer membrane

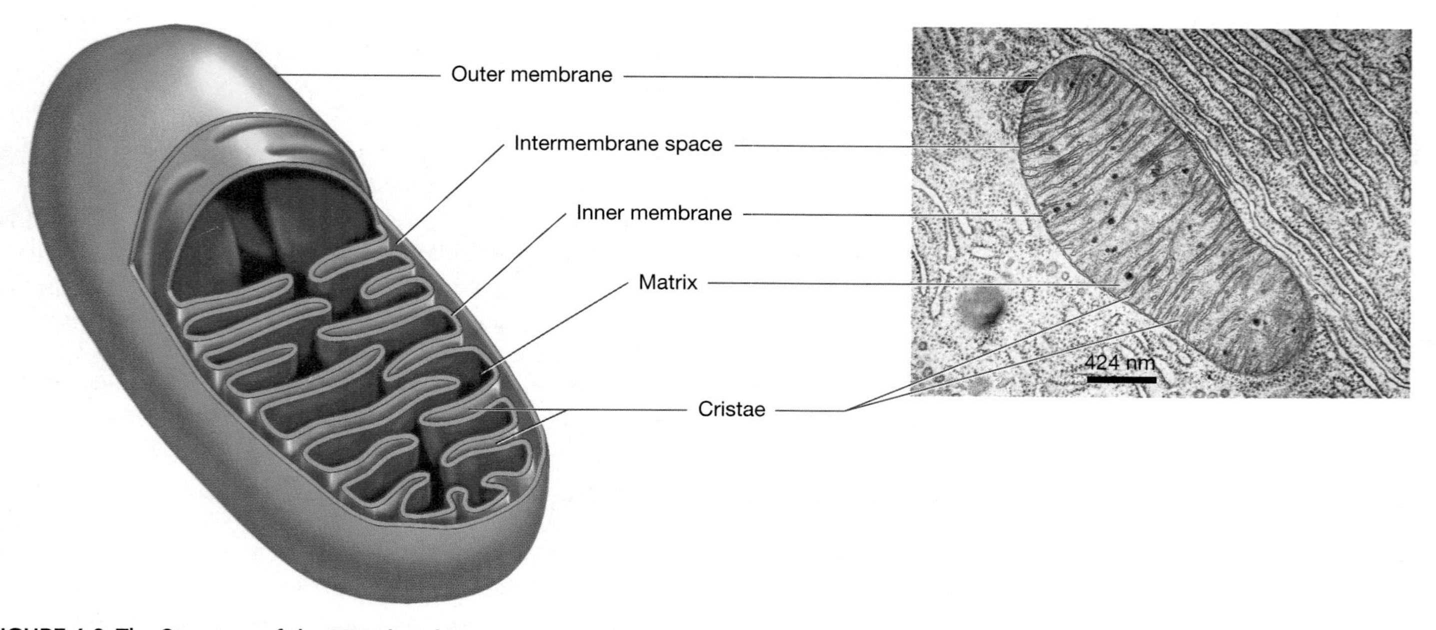

FIGURE 6.8 The Structure of the Mitochondrion
As this drawing and micrograph show, mitochondria have an inner and outer membrane. The folded portions of the inner membrane are called cristae.

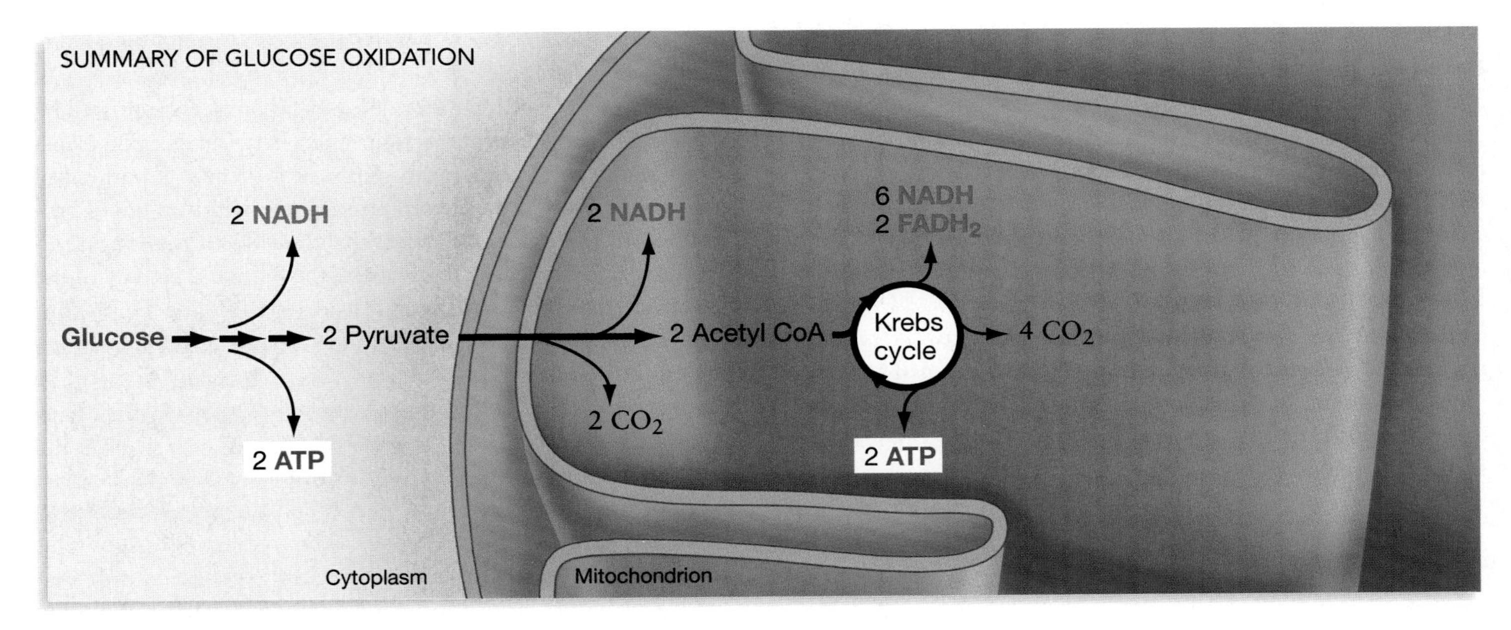

FIGURE 6.9 A Summary of Glucose Oxidation
In glycolysis and the Krebs cycle, glucose is completely oxidized to carbon dioxide. In eukaryotes, glycolysis occurs in the cytoplasm; oxidation of pyruvate and the Krebs cycle occur in the mitochondrial matrix.

from the matrix and the solution in the intermembrane space. The isolated membranes were capable of oxidizing NADH, but the matrix and other fluid were not. Follow-up work showed that the oxidation process takes place on the inner membrane of the mitochondria. In prokaryotes, the oxidation of NADH occurs in the cell membrane.

When biologists analyzed the components of the mitochondrial inner membrane, they were able to isolate a series of molecules that switch between a reduced and an oxidized state during respiration. What are these molecules, and how are they involved in the processing of NADH and $FADH_2$?

Components of the Electron Transport Chain

Researchers made two fundamental observations about the molecules involved in the oxidation of NADH and $FADH_2$ in mitochondria. First, there are five general types of compounds that participate in the redox reactions. Four of these types are proteins that contain distinctive chemical groups where the reduction-oxidation events actually take place. Depending on their structure, the active groups are called a flavin, an iron-sulfur complex, heme, or copper. Flavin is a ring-containing structure that also acts as the oxidized or reduced group in FAD and $FADH_2$. The iron-sulfur complexes found in mitochondrial membrane proteins participate in redox reactions because iron and sulfur groups they contain are readily reduced or oxidized. Similarly, the heme groups attached to mitochondrial proteins contain an iron atom that is readily reduced or oxidized. Several different heme-containing proteins occur in mitochondria; each is called a cytochrome. Finally, copper-containing proteins are important because the copper atom can be reduced or oxidized.

The fifth type of electron-carrying molecule in the inner membrane of mitochondria is not a protein. It is named ubiquinone because it belongs to a family of compounds called quinones and because it is nearly ubiquitous in organisms. It is also called coenzyme Q, or simply Q. Ubiquinone consists of a carbon-containing ring attached to long tail made up of isoprene subunits. This observation is important because the long, isoprene-rich tail is hydrophobic. As a result, Q is lipid soluble and can move through the mitochondrial membrane efficiently. In contrast, all but one of the flavin-, iron-sulfur, and heme-containing proteins are anchored in the membrane.

The second fundamental observation about the molecules involved in processing NADH and $FADH_2$ is that each differs in its tendency to become reduced or oxidized. By measuring the tendency of each molecule involved to gain electrons, researchers could place them in the order diagrammed in **Figure 6.10** (page 124). This drawing illustrates the concept of an electron transport chain. In an electron transport chain, the potential energy in electrons from NADH and $FADH_2$ is gradually stepped down through a series of reduction-oxidation reactions. Each successive molecule in the chain is slightly more oxidized. A highly oxidized molecule—such as oxygen—acts as the final electron acceptor.

Subsequent work showed that NADH donates an electron to a flavin-containing protein at the top of the chain, while $FADH_2$ donates electrons to an iron-sulfur containing protein that then passes them directly to Q. After passing through a series of cytochromes and another protein containing an iron-sulfur complex, the electrons are finally accepted by oxygen. From NADH to oxygen, the total potential energy difference is 53 kcal/mol.

CD ACTIVITY 6.2 Chemiosmosis

Once the nature of the electron transport chain became clear, biologists understood how the electron carriers NADH and $FADH_2$ are oxidized to NAD^+ and FAD and how oxygen acts as the final electron acceptor. Then the question became, how do these redox reactions generate ATP? Does substrate-level phosphorylation take place in the electron transport chain, like it does in glycolysis and the Krebs cycle?

The Chemiosmotic Hypothesis

Throughout the 1950s, most biologists interested in cellular respiration assumed that the electron transport chains in the inner membrane of mitochondria interact with enzymes that catalyze substrate-level phosphorylation. Despite intense efforts, however, no one was able to find an enzyme that used the energy released by the redox reactions to phosphorylate ADP and produce ATP.

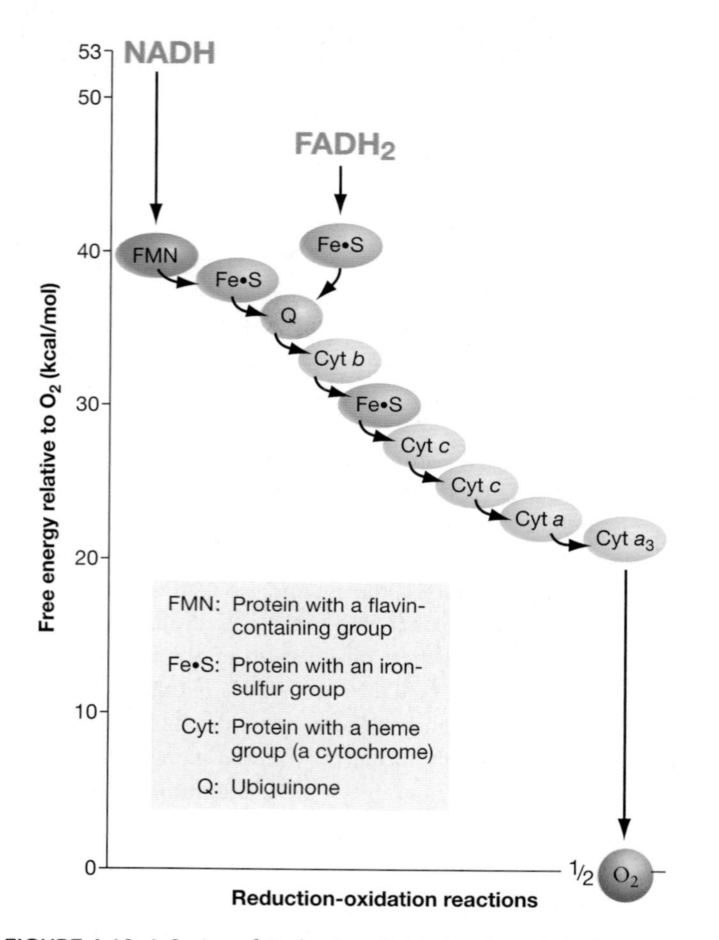

FIGURE 6.10 A Series of Reduction-Oxidation Reactions Occurs in an Electron Transport Chain
This diagram illustrates how electrons are stepped down in potential energy from the electron carriers NADH and $FADH_2$ through an electron transport chain to oxygen, which acts as an electron acceptor. The overall energy drop of 53 kcal/mol (from NADH to oxygen) is broken into a series of small steps. In each step, a member of the electron transport chain accepts an electron—meaning that it becomes reduced—and then passes the electron on to a slightly more oxidized molecule in the chain. In this way, molecules in the electron transport chain oscillate between a reduced state and an oxidized state. When oxygen accepts electrons, water is formed.

In 1961, Peter Mitchell made a radical break with prevailing ideas by proposing that the connection between electron transport and ATP production is indirect. Mitchell hypothesized that the real job of the electron transport chain is to pump protons from the matrix of the mitochondrion to the intermembrane space. According to Mitchell, the pumping activity of the electron transport chain would lead to a buildup of protons and a strong electrochemical gradient favoring the movement of protons back into the matrix. This **proton-motive force**, he hypothesized, is used by a protein in the inner membrane to synthesize ATP.

Mitchell's proposal of an indirect linkage between electron transport and ATP production is called the **chemiosmotic hypothesis.** Although proponents of substrate-level phosphorylation objected to Mitchell's idea, two key experiments supported the proposal. André Jagendorf and Ernest Uribe published the results of the first experiment in 1966. Jagendorf and Uribe's work focused on the inner membranes of plant chloroplasts, where the same type of proton-motive force was hypothesized to occur. To test whether a proton gradient across these membranes could result in the generation of ATP, the researchers ground up leaves, isolated the chloroplasts, and froze them briefly to disrupt them. The broken chloroplast membranes then formed small bag-like vesicles (introduced in Chapter 4). As **Figure 6.11a** shows, Jagendorf and Uribe put the vesicles into a solution of pH 4 and let them equilibrate until the insides of the vesicles were also at pH 4. When they transferred the vesicles into a solution at pH 8.0 that contained ADP and inorganic phosphate, the researchers observed ATP formation in the solution. This experiment suggested that a membrane protein was capable of synthesizing ATP using a proton-motive force.

Efraim Racker and Walther Stoeckenius published the results of a second test of the chemiosmotic hypothesis in 1974. Racker and Stoeckenius began their work by making vesicles from artificial membranes that contained an ATP-synthesizing enzyme found in mitochondria. Then they inserted a membrane protein (called bacteriorhodopsin) found in some bacteria into the vesicle membranes. Earlier work had shown that bacteriorhodopsin can capture the energy in sunlight and use it to pump protons across a membrane. As **Figure 6.11b** shows, Racker and Stoeckenius found that ATP was produced when their experimental vesicles were illuminated.

The experimental results of Racker and Stoeckenius supported the hypothesis that ATP synthesis is driven by an electrochemical gradient favoring a flow of protons across a membrane. The result was also important because it showed that NADH and electron transport are not required for ATP synthesis in mitochondria, provided that some other mechanism exists to pump protons. Based on these experimental tests, most biologists accepted the chemiosmotic hypothesis as valid.

How Is the Electron Transport Chain Organized?

Once the predictions of the chemiosmotic hypothesis were verified, researchers focused on understanding the three-dimensional structure of the electron transport chain components and detailing how electron transport is coupled to proton pumping. This work has shown that the components of the electron transport chain are organized into the four large complexes of proteins and cofactors shown in **Figure 6.12** (page 126). As that figure shows, protons are pumped by three of the complexes, and Q and cytochrome *c* act as shuttles that transfer electrons between complexes. Each complex contains proteins with flavin, iron-sulfur, or heme groups; cytochrome *c* contains a copper atom.

The structural studies completed to date confirm that in complex I and complex IV, protons actually pass directly through the assemblage of electron carriers. The exact route taken by the protons is still being worked out, however. It is also not clear how the redox reactions taking place inside each complex—as electrons are stepped down in potential energy—make the movement of the protons possible.

To date, the best-understood interaction between electron transport and proton pumping takes place in complex III. Research has shown that when Q accepts electrons from complex

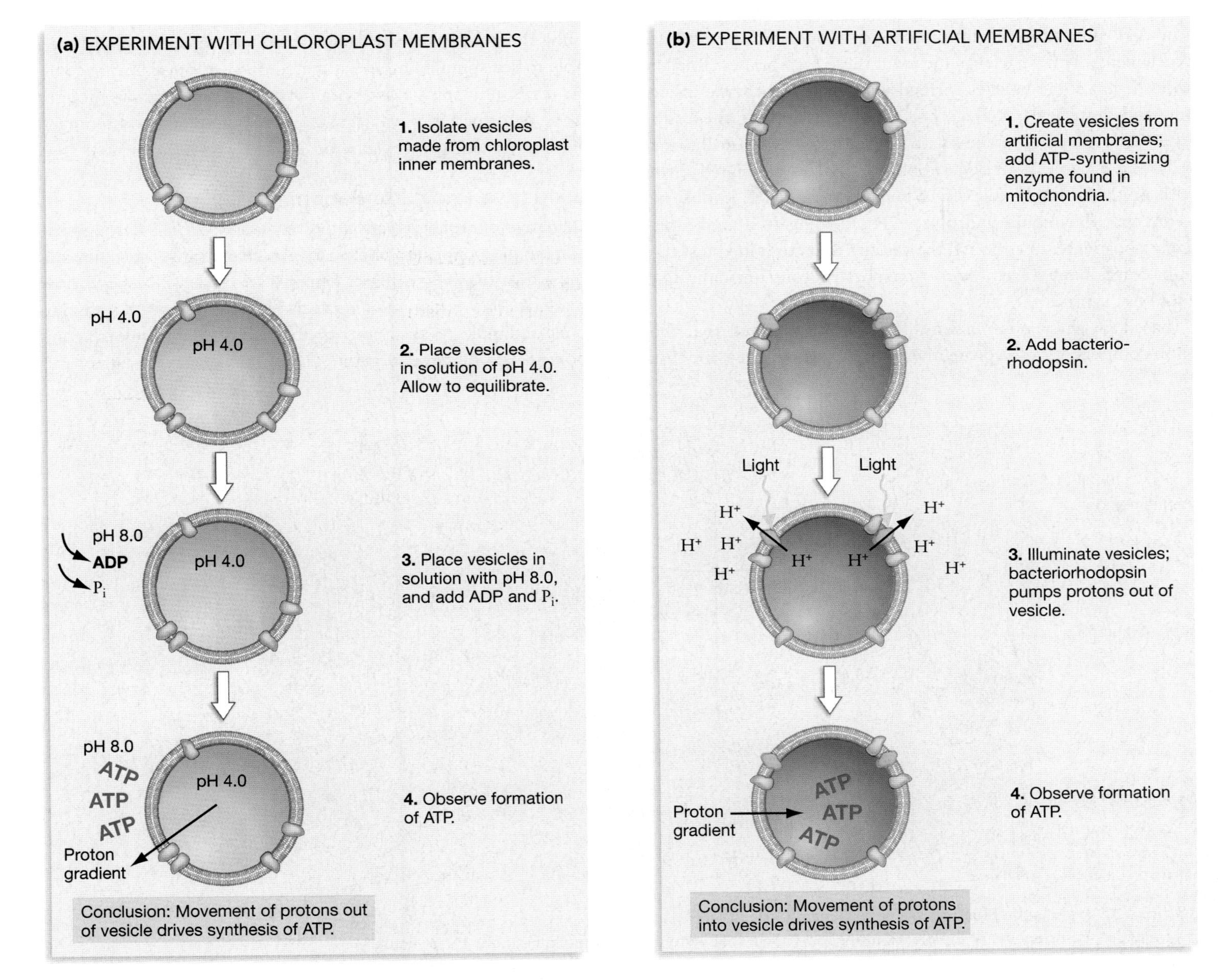

FIGURE 6.11 Experimental Evidence for the Chemiosmotic Hypothesis
The chemiosmotic hypothesis predicts that ATP is produced by a proton-motive force across a membrane. The two experiments illustrated here tested this prediction. QUESTION Do you regard these as convincing tests of the chemiosmotic hypothesis? Why or why not?

I or complex II, it also gains two protons. The reduced form of Q then diffuses to the other side of the inner membrane, where its electrons are used to reduce an iron-sulfur containing component of complex III near the intermembrane space. During this reaction the protons held by Q are released to the intermembrane space, where they contribute to the proton-motive force. In this way, the lipid solubility of Q allows it to diffuse across the membrane of the cristae and physically shuttle electrons and protons from one side of the membrane to the other. Now the question becomes, how does the proton-motive force make the production of ATP possible?

The Discovery of ATP Synthase

In 1960, Efraim Racker made several key observations about how ATP is synthesized in mitochondrial membranes. During experiments with vesicles made from these membranes, Racker noticed that some vesicles formed with their membrane inside out, and that they had large proteins studded along their surfaces. As **Figure 6.13a** shows, the proteins appeared to have a stalk and a knob. If Racker shook the vesicles or treated them with urea, the knobs fell off. When he isolated the knobs, he found that they could hydrolyze ATP, forming ADP and inorganic phosphate. In contrast, the vesicles that contained just the stalk component were able to transport protons normally but could not synthesize ATP.

Based on these observations, Racker hypothesized that the knob component of the protein was an ATPase—an enzyme that hydolyzes and synthesizes ATP. To test this hypothesis, he added the knob components back to vesicles that had been stripped of them and confirmed that the vesicles were then capable of synthesizing ATP (**Figure 6.13b**). Additional work showed that the membrane-bound stalk component of the complex is a protein that acts as a proton channel.

As **Figure 6.13c** shows, the three-dimensional structure of this protein complex is now reasonably well understood. The ATPase component is called the F_1 unit; the membrane-bound, proton-transporting component is designated the F_o unit. The entire complex is named **ATP synthase.** According to the current model for how this enzyme functions, a flow of protons through the F_o unit causes the rod connecting the two subunits to spin. By attaching long actin filaments to this rod and examining them under a microscope, researchers have actually been able to see spinning at an estimated 50 revolutions per second. As the F_1 unit spins along with the rod, its subunits are thought to be deformed in a way that catalyzes the phosphorylation of ADP and subsequent release of ATP.

Oxidative Phosphorylation

Oxidative phosphorylation is the formation of ATP through the combination of proton pumping by electron transport chains and the action of ATP synthase. **Figure 6.14** (page 128) shows how oxidative phosphorylation interacts with glycolysis and the Krebs cycle and indicates the approximate yield of ATP from each component of cellular respiration. The fundamental message of this

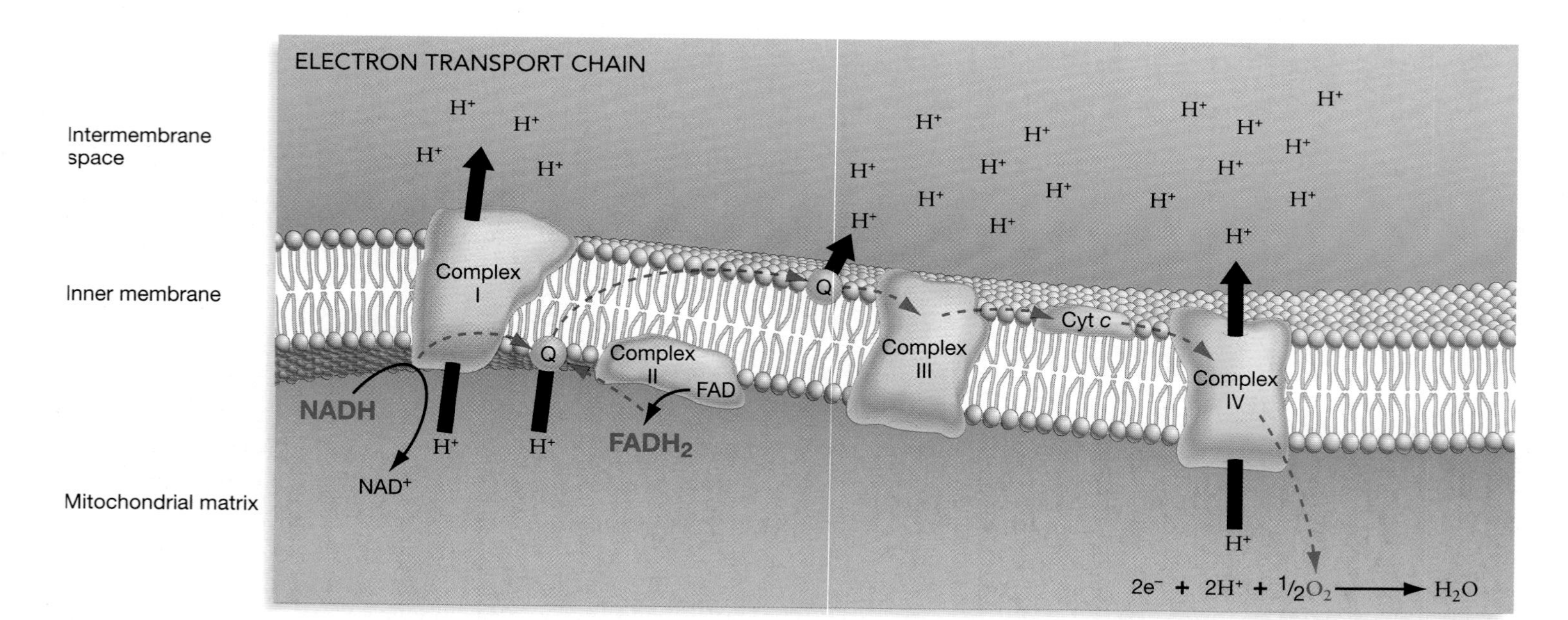

FIGURE 6.12 How Is the Electron Transport Chain Organized?
The individual components of the electron transport chain that were diagrammed in Figure 6.11 are actually grouped into the large multiprotein complexes shown here. Electrons are carried from one complex to another by Q and by cytochrome *c*. Complexes I, III, and IV use the potential energy released by the redox reactions to pump protons from the mitochondrial matrix to the intermembrane space. QUESTION What do the blue dashed lines and arrows indicate? Is the pH of the intermembrane space higher or lower than that of the mitochondrial matrix?

(a) A vesicle formed from an "inside-out" mitochondrial membrane

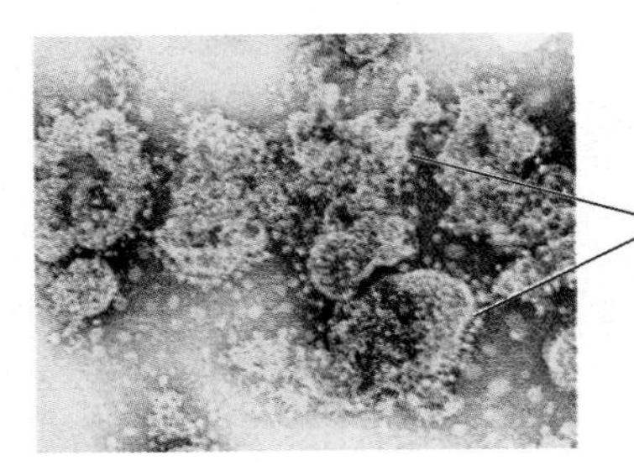

(b) What do the "stalk-and-knob" proteins do?

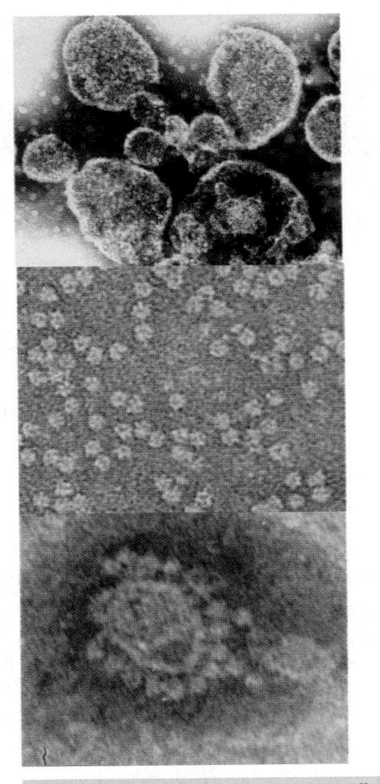

Conclusion: The "stalk-and-knob" protein is the ATP synthase.

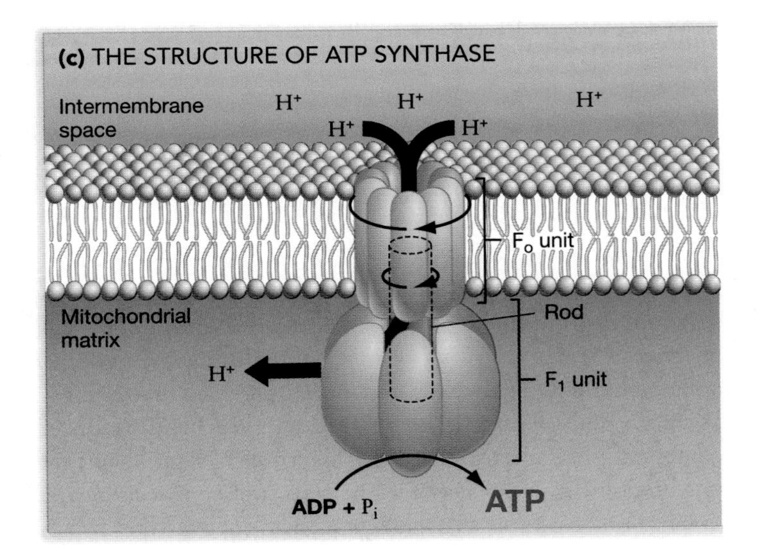

FIGURE 6.13 How Does the Proton-Motive Force Lead to Synthesis of ATP?
(a) When patches of mitochondrial membrane turn inside out and form vesicles, proteins that have a stalk-and-knob structure can be observed facing out. Normally, the knob on these proteins faces in, toward the mitochondrial matrix. **(b)** Experiments showed that the stalk-and-knob protein is an ATP synthase. **(c)** Structural studies have confirmed that ATP synthase has two major components, designated F_o and F_1. In response to passage of protons, the F_o unit and the rod that connects the F_o and F_1 units spins. The spinning motion alters the conformation of the F_1 unit in a way that drives phosphorylation of ADP.

diagram is that the vast majority of the "payoff" from the oxidation of glucose occurs through oxidative phosphorylation.

It is important to recognize, however, that cellular respiration can occur without oxygen. Oxygen is the electron acceptor used by all eukaryotes and a wide diversity of bacteria and archaea. Species that use oxygen as an electron acceptor are said to use **aerobic respiration.** Many thousands of bacterial and archaeal species rely on electron acceptors other than oxygen and electron donors other than glucose, however. As Chapter 25 will show, some bacteria and archaea use H_2, H_2S, CH_4, or other reduced inorganic compounds as electron donors; nitrate (NO_3^-) and sulphate (SO_4^{2-}) are particularly common electron acceptors for species that live in oxygen-poor environments. Cells that use electron acceptors other than oxygen are said to use **anaerobic respiration.** Even though the starting and ending points of cellular respiration are different, these cells are still able to use electron transport chains to create a proton-motive force that drives the synthesis of ATP.

Oxygen is the most effective of all electron acceptors, however, because the potential energy of its electrons is very low. As a result, the difference between the potential energy of electrons in NADH and oxygen is large (see Figure 6.10). The large differential in potential energy means that the electron transport chain can generate a large proton-motive force. Cells that do not use oxygen as an electron acceptor cannot generate such a large potential energy difference. They cannot make as much ATP as cells that use aerobic respiration and thus tend to grow much more slowly. If cells that use anaerobic respiration compete with cells using aerobic respiration, the cells that use oxygen as an electron acceptor almost always survive better and reproduce more.

What happens if oxygen or other electron acceptors are temporarily used up and are unavailable? Without oxygen or another electron acceptor in place, the electrons donated by NADH have no place to go. The electron transport chain stops, and all NAD^+ in the cell quickly becomes NADH. This situation is life threatening, because there is no longer any NAD^+ to supply the reactions of glycolysis. If NAD^+ cannot be regenerated somehow, all ATP production will stop and the cell will die. How do cells cope?

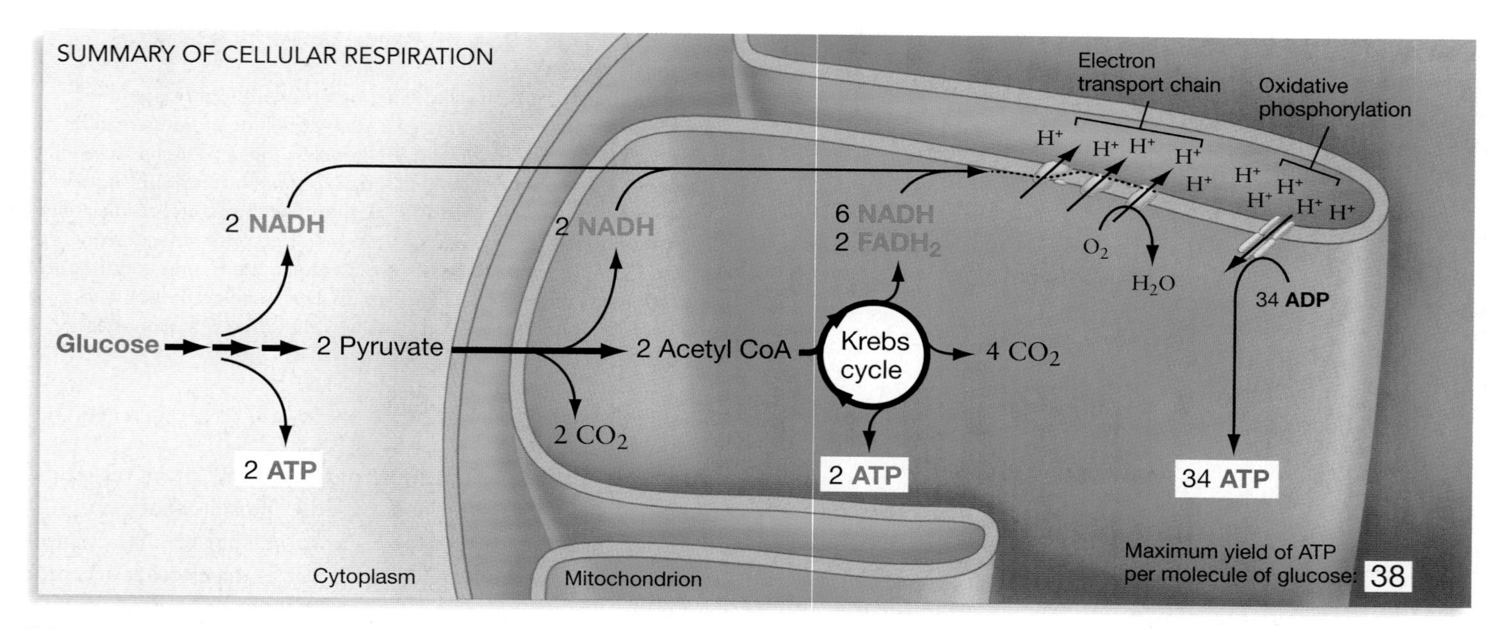

FIGURE 6.14 A Summary of Cellular Respiration
The ATP yields indicated in this figure are approximate, partly because the NADH produced by glycolysis does not enter mitochondria readily. The actual yield of ATP from these NADH depends on which of several mechanisms is used to transport the electron carriers in the mitochondrion. **EXERCISE** Label the electron transport chain and ATP synthase. Then across the top of the figure write the overall reaction for the oxidation of glucose. Explain what happens to each of the reactants. Explain where each of the products comes from.

6.5 Fermentation

Fermentation is defined as any set of redox reactions in which a reduced organic compound like glucose acts as an electron donor and another organic molecule acts as an electron acceptor, with ATP being produced by substrate-level phosphorylation. Fermentation occurs when the NADH produced by glycolysis is unable to donate high-energy electrons to the electron transport chain and produce NAD^+. In essence, fermentation converts NADH back to NAD^+ and allows glycolysis to continue (**Figure 6.15a**).

In organisms that usually use oxygen as an electron acceptor, fermentation offers an alternative mode of energy production if oxygen supplies temporarily run out. For example, if you sprint for an extended distance, your muscles begin metabolizing glucose so quickly that your lungs and circulatory system are unable to supply oxygen rapidly enough to keep electron transport chains active. In response, your muscle cells begin producing ATP as shown in **Figure 6.15b.** In this type of fermentation, the pyruvate produced by glycolysis accepts electrons from NADH, resulting in the formation of lactate and regeneration of NAD^+.

Figure 6.15c illustrates the type of fermentation pathway used by baker's and brewer's yeast—the fungus *Saccharomyces cerevisiae*. When these cells are placed in an environment like bread dough or a bottle of champagne and begin growing, they quickly use up all the available oxygen. They continue to use glycolysis to metabolize sugar, however, by enzymatically converting pyruvate to a 2-carbon compound called acetaldehyde. This reaction gives off carbon dioxide, which causes bread to rise and creates the bubbles in champagne and beer. Acetaldehyde then accepts electrons from NADH, forming the NAD^+ required to keep glycolysis going and the by-product ethanol. Ethanol is the active ingredient in alcoholic beverages.

A huge variety of other fermentative pathways exist among species of bacteria and archaea. For example, bacteria and archaea that exist exclusively through fermentation are present in phenomenal numbers in the oxygen-free environment of your small intestine, as well as in the rumen of cows. The rumen is a specialized digestive organ that contains over 10^{10} (ten billion) bacterial and archaeal cells per milliliter of fluid. The fermentations that occur in these cells result in the production of an array of fatty acids. Cows use these fermentation by-products as a source of energy. Other types of fermentations are used commercially in the production of sour cream, yogurt, cheese, vinegar, and other products.

Even though fermentation is a widespread and commercially important type of metabolism, it is important to recognize that it is extremely ineffective compared to cellular respiration. Recall that for each molecule of glucose that is metabolized, fermentation produces just two molecules of ATP, while cellular respiration produces as many as 38. Cellular respiration can produce as much as 19 times more energy per glucose molecule as fermentation. The reason for this disparity is that electron acceptors like pyruvate or acetaldehyde are highly reduced compared to an electron acceptor like oxygen. As a result, the potential energy drop between the start and end of fermentation is a tiny fraction of the potential energy change that occurs during cellular respiration. Based on these observations, it should not be surprising that if an organism is capable of both processes, it never uses fermentation when an appropriate electron acceptor is available for cellular respiration.

(a) Fermentation pathways allow cells to regenerate NAD^+ for glycolysis.

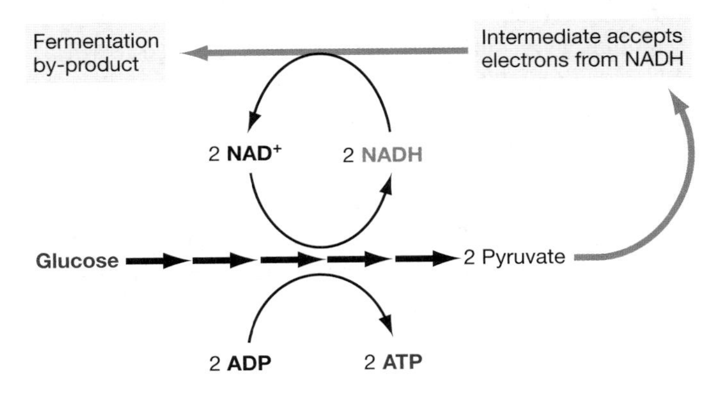

(b) Lactic acid fermentation occurs in humans.

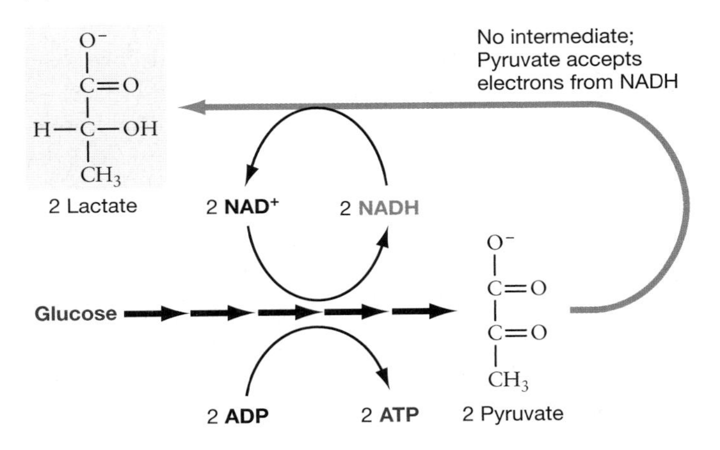

(c) Alcohol fermentation occurs in yeast.

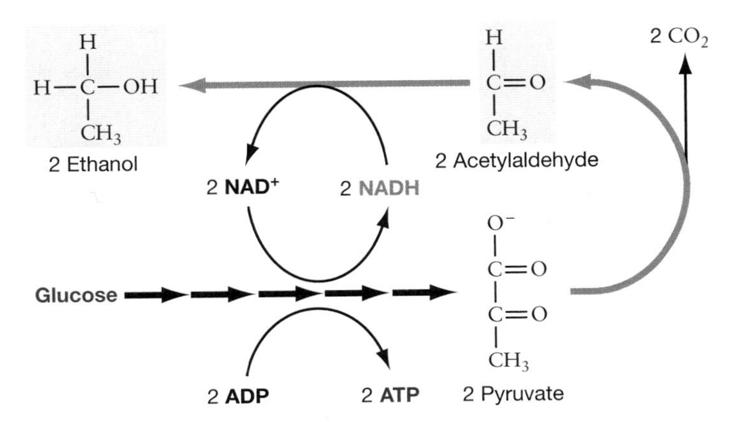

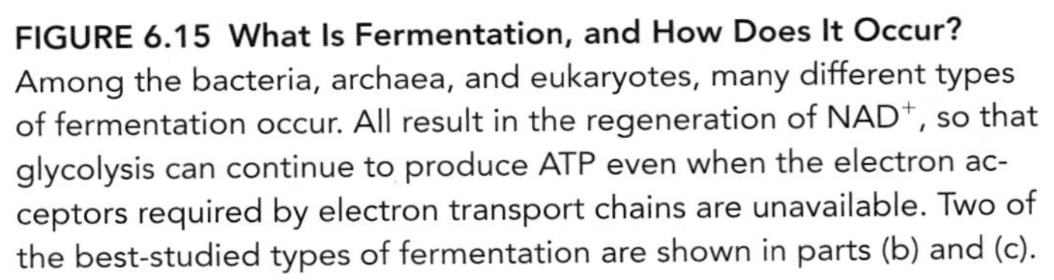

FIGURE 6.15 What Is Fermentation, and How Does It Occur? Among the bacteria, archaea, and eukaryotes, many different types of fermentation occur. All result in the regeneration of NAD^+, so that glycolysis can continue to produce ATP even when the electron acceptors required by electron transport chains are unavailable. Two of the best-studied types of fermentation are shown in parts (b) and (c).

6.6 How Does Cellular Respiration Interact with Other Metabolic Pathways?

The enzymes, products, and intermediates involved in cellular respiration and fermentation do not exist in a vacuum. Instead, they are part of a huge and dynamic inventory of chemicals inside the cell. Because metabolism involves thousands of different chemical reactions, the amounts and identities of molecules inside cells are constantly in flux (**Figure 6.16**). Fermentation pathways, electron transport, and other aspects of carbohydrate metabolism may be crucial to the life of a cell, but they also have to be seen as pieces of a larger puzzle.

To make sense of the chemical inventory inside cells and the full scope of metabolism, it is critical to recognize that cells have two fundamental requirements to stay alive, grow, and reproduce: a source of reduced electrons for generating chemical energy in the form of ATP, and a source of carbon-containing molecules that can be used to synthesize DNA, RNA, proteins, fatty acids, and other molecules. Reactions that result in the breakdown of molecules and the production of ATP are called **catabolic pathways**; reactions that result in the synthesis of larger molecules from smaller components are called **anabolic pathways.**

The goal of this section is to introduce how glycolysis and the Krebs cycle interact with other catabolic pathways and with

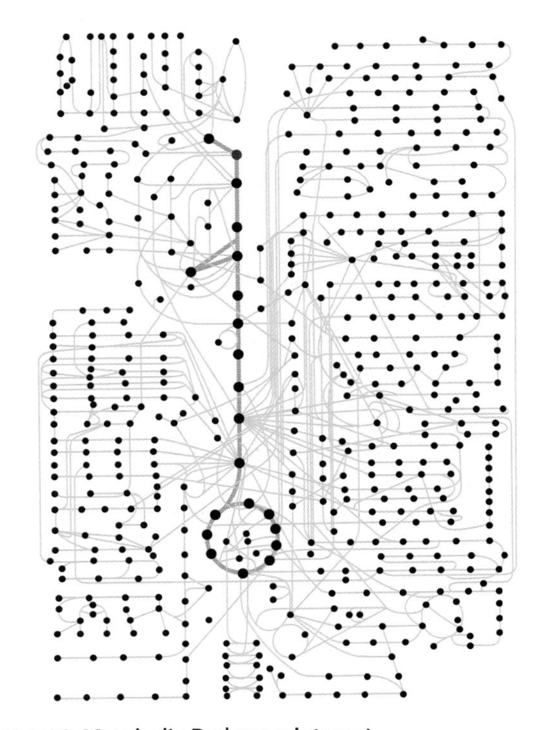

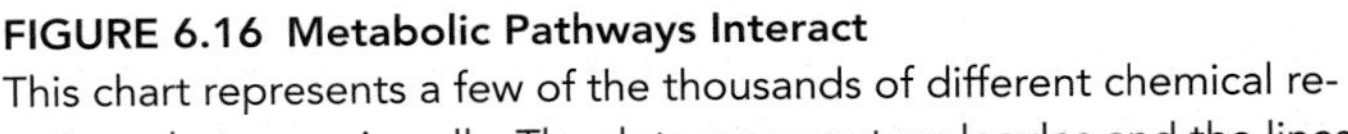

FIGURE 6.16 Metabolic Pathways Interact This chart represents a few of the thousands of different chemical reactions that occur in cells. The dots represent molecules and the lines represent enzyme-catalyzed reactions. **EXERCISE** Circle the 10 reactions of glycolysis. Draw a box around the Krebs cycle.

anabolic pathways. Let's start by considering how reduced molecules other than carbohydrates are used as fuel in eukaryotes, and move on to examine how molecules involved in glycolysis and the Krebs cycle are sometimes drawn off and used as building blocks in the synthesis of cell components.

Processing Proteins and Fats as Fuel

Most organisms ingest, synthesize, or absorb a wide variety of carbohydrates. These reduced molecules range from sucrose, maltose, and other simple sugars to large polymers such as glycogen and starch. **Glycogen** is the major form of stored carbohydrate in animals, while **starch** is the major form of stored carbohydrate in plants. Glycogen and starch are both polymers of glucose, but differ in the way long chains of glucose branch. Using enzyme-catalyzed reactions, cells can produce glucose from glycogen, starch, and most simple sugars. Glucose and fructose can then be processed by the enzymes of the glycolytic pathway.

Carbohydrates are not the only important source of reduced carbon compounds used in the catabolic pathways, however. Fats, which are highly reduced macromolecules consisting of glycerol bonded to chains of fatty acids, are routinely broken down by enzymes to form glycerol and acetyl CoA. Glycerol enters the glycolytic pathway once it has been phosphorylated to form glyceraldehyde-3-phosphate—one of the intermediates in the 10-reaction sequence. Acetyl CoA, in contrast, enters the Krebs cycle.

Proteins can also be catabolized. Once proteins are broken down to their constituent amino acids, the amino (–NH3) groups are removed in enzyme-catalyzed reactions and excreted. The carbon compounds that remain after this deamination step are converted to pyruvate, acetyl CoA, and other intermediaries in glycolysis and the Krebs cycle.

To pull all of these observations together, **Figure 6.17** summarizes how carbohydrates, fats, and proteins are catabolized and how their breakdown products feed an array of steps in cellular respiration. As different types of food become available, different sequences of catabolic reactions occur at different times during the life of a cell.

Anabolic Pathways Synthesize Key Molecules

Where do cells get the precursor molecules required to synthesize amino acids, RNA, DNA, phospholipids, and other cell components? Not surprisingly, the answer often involves intermediates in carbohydrate metabolism.

- If ATP is abundant, pyruvate and lactate (from fermentation) can be used as a substrate in the synthesis of glucose. Glucose, in turn, can be converted to glycogen and stored.
- In humans, about half of the 20 amino acids required by cells can be synthesized from molecules siphoned off from the Krebs cycle.
- Acetyl CoA is the starting point for anabolic pathways that result in the synthesis of fatty acids.
- The molecule produced by the first reaction in glycolysis can be oxidized to start the synthesis of ribose-5-phosphate—a key intermediate in the production of the ribonucleotides and deoxyribonucleotides required for manufacturing RNA and DNA.

Figure 6.18 pulls these observations together by showing how intermediates in carbohydrate metabolism are drawn off to make the synthesis of macromolecules possible. The take-home message of this figure is that the same molecules can serve an array of functions in the cell. As a result, catabolic and anabolic pathways are closely intertwined.

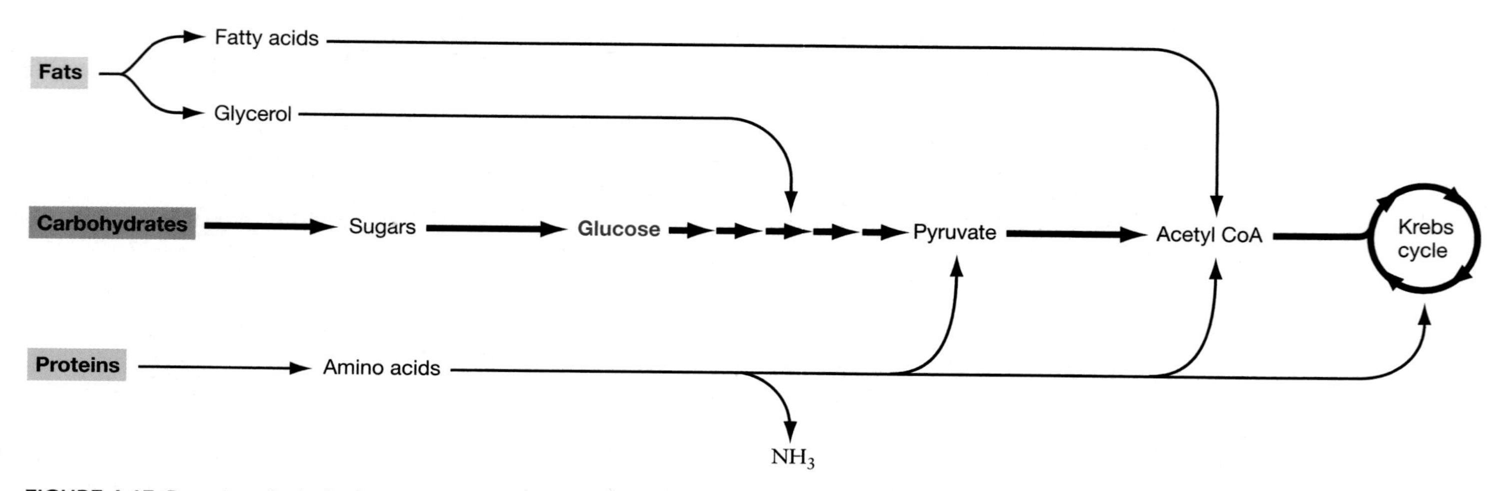

FIGURE 6.17 Proteins, Carbohydrates, and Fats Can All Furnish Substrates for Cellular Respiration
A variety of carbohydrates can be converted to glucose and processed by glycolysis. If carbohydrates are in short supply, cells can also use fats or proteins as a source of reduced compounds for ATP production.

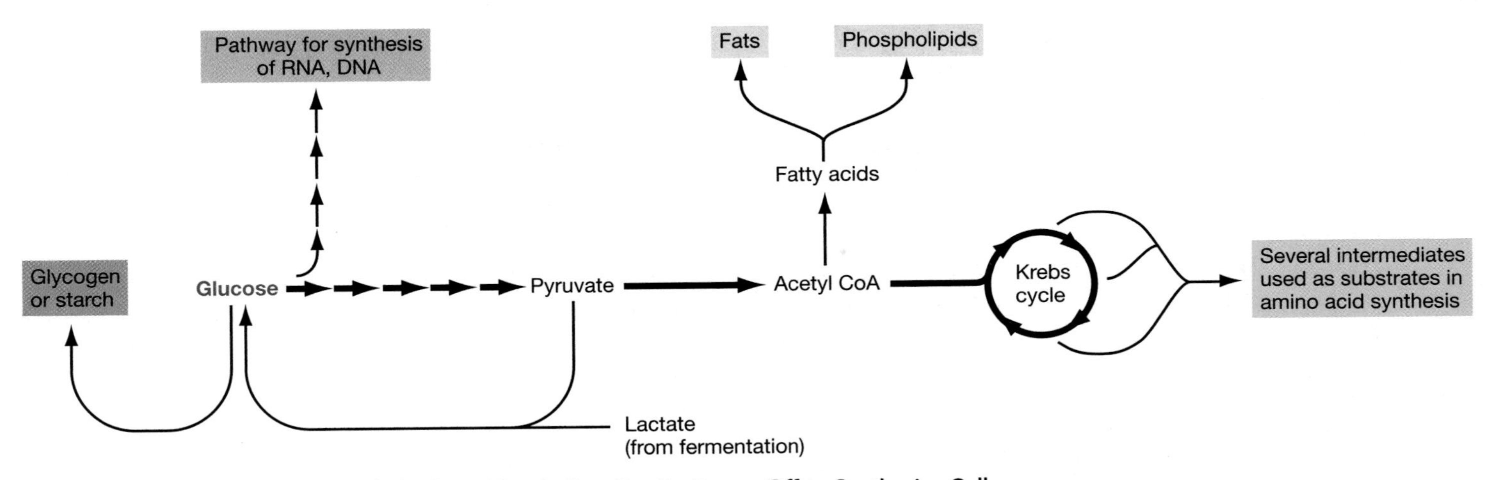

FIGURE 6.18 Intermediates in Carbohydrate Metabolism Can Be Drawn Off to Synthesize Cell Components
Several of the intermediates in carbohydrate metabolism act as precursor molecules in the synthesis of RNA, DNA, glycogen or starch, amino acids, and fatty acids.

Essay ATP Production During Exercise

In an adult animal, the highest demands for ATP are made by exercising muscles. As Chapter 43 will detail, muscles contain a motor protein called myosin. Like the motor proteins kinesin and dynein introduced in Chapter 5, myosin undergoes a conformational change in response to ATP hydrolysis. This movement by myosin is the physical basis of muscle contraction.

On average, a muscle contains only enough ATP to sustain about 15 seconds of intense exercise. For muscle contractions to continue, massive amounts of ATP are required. Depending on the level and duration of activity, the muscles being exercised may produce the ATP they need by cellular respiration or by fermentation. Sustained periods of submaximal activity like jogging are powered by aerobic respiration. In contrast, short periods of intense activity like sprinting are powered by a combination of aerobic and anaerobic respiration.

What happens when a marathon runner "hits the wall?"

To get a better understanding of the relationship between cellular respiration and fermentation in muscle cells, we'll focus on the constraints experienced by two types of runners. What aspects of cellular respiration affect the performance of a marathoner? What aspects of fermentation limit the duration of sprints? Let's consider each question in turn.

A marathon course is 26 miles, 385 yards long (42 km 195 m). The race was inaugurated with the first of the modern Olympic games, which took place in Athens, Greece, in 1896. The event was designed to commemorate Pheidippides, a Greek soldier who reportedly ran from the plains of Marathon in southeast Greece to Athens immediately after the Greek army had triumphed over the invading Persians in 490 B.C. According to legend, Pheidippides shouted "Victory!" when he arrived, then collapsed and died of exhaustion.

Contemporary marathon runners rarely collapse and die, but virtually all experience a phenomenon known as "hitting the wall." Marathon running is an aerobic activity fueled by cellular respiration. The glucose and glycogen stores in the body are only adequate to fuel about 20 miles of running, however. During the last five miles of the race, marathoners rely exclusively on fatty-acid metabolism to keep cellular respiration going. Hitting the wall is a feeling of exhaustion that occurs as the last of the glycogen stores are used up. Because fatty-acid metabolism is markedly slower than glycogen metabolism, ATP delivery is reduced and running becomes much harder.

Sprinters and weight lifters do not hit the wall. They do "feel the burn," however. As metabolism switches from aerobic to anaerobic metabolism during intense activity, muscles begin to feel a burning sensation. With continued production of ATP by fermentation, lactic acid levels can build up to the point where the pH of muscle cells drops. The acidification of the cells can damage enzymes and proteins and contribute to fatigue—the inability of a muscle to continue contracting with maximum force. In addition, anaerobic metabolism produces a great deal of heat, which can damage muscle proteins if it is not controlled.

(Continued on page 132)

(Essay continued)

For both sprinters and marathoners, an extended cooldown period after exercise is important for muscle health. Several important events occur during recovery from exercise. Lactic acid is transported from muscles to the liver, where it is converted to pyruvate and then to glucose. Oxygen-rich blood is delivered to muscles, where cellular respiration restores normal levels of ATP. In addition, heat is transferred from muscles to blood. The blood brings the heat to the body surface, where it is radiated to the environment.

Recently, research has shown that aerobic training can increase the number of mitochondria in muscle cells and thus increase ATP supplies during exercise. Similarly, sprint training can increase the number of glycolytic enzymes in muscle cells and thus the rate of ATP delivery during bursts of activity. Increasingly, an understanding of cellular respiration and fermentation is influencing the design of training programs for athletes.

Chapter Review

Summary

Cells pump ions, drive endergonic reactions, move cargo, and perform other types of work by hydrolyzing ATP. Cells produce ATP from sugars or other reduced compounds using one of two general pathways: cellular respiration or fermentation. Cellular respiration involves the transfer of electrons from a reduced compound like glucose to an oxidized molecule such as oxygen, through an electron transport chain. Fermentation involves the transfer of electrons from a reduced organic compound to an oxidized organic compound, without participation by an electron transport chain.

Cellular respiration has three components: glycolysis, the Krebs cycle, and electron transport coupled with oxidative phosphorylation. In eukaryotes, glycolysis takes place in the cytoplasm, the Krebs cycle occurs in the mitochondrial matrix, and electron transport and oxidative phosphorylation happen in the inner membranes of mitochondria.

Glycolysis is a 10-step reaction sequence in which glucose is broken down into two molecules of pyruvate. For each molecule of glucose processed during glycolysis, two ATP are produced by substrate-level phosphorylation and two NAD^+ are reduced to two NADH.

Prior to the Krebs cycle, each pyruvate is oxidized to acetyl CoA by a series of reactions that result in the production of a molecule of carbon dioxide and the synthesis of one NADH. The Krebs cycle begins when acetyl CoA reacts with oxaloacetate to form citrate. The series of reactions that ensues results in the regeneration of oxaloacetate and the production of two molecules of carbon dioxide, one ATP by substrate-level phosphorylation, and the reduction of one FAD to $FADH_2$ and reduction of three NAD^+ to three NADH.

In essence, glycolysis and the Krebs cycle use the energy released by the oxidation of glucose to reduce NAD^+ and FAD. The resulting electron carriers—NADH and $FADH_2$—then donate electrons to an electron transport chain, which gradually steps the electrons down in potential energy until they are finally accepted by oxygen, resulting in the formation of water. The components of the electron transport chain use the energy released by the oxidation of NADH and $FADH_2$ to pump protons across the inner mitochondrial membrane. The pumping activity creates an electrochemical gradient, or proton-motive force, that ATP synthase uses to produce ATP. The production of ATP by electron transport and generation of a proton-motive force is called oxidative phosphorylation.

If no electron acceptor is available, cellular respiration stops, because electron transport chains cannot work. As a result, all NAD^+ is converted to NADH and glycolysis cannot continue. Fermentation pathways regenerate NAD^+ when an organic molecule like pyruvate accepts electrons from NADH. Depending on the molecule that acts as an electron acceptor, fermentation pathways produce lactate, ethanol, or other reduced organic compounds as a by-product.

Fermentation and cellular respiration are referred to as catabolic pathways because they result in the breakdown of a reduced molecule and the generation of ATP. Anabolic pathways, in contrast, result in the synthesis of larger molecules from smaller precursors. In many cases, anabolic pathways begin with substrates that act as intermediates in carbohydrate metabolism.

Questions

Content Review

1. What is the "committed step" in a metabolic pathway?
   a. the only step that is regulated
   b. a step where an intermediate can be drawn off to other metabolic pathways
   c. a step that is catalyzed by an enzyme
   d. a step that is irreversible because the product can only be used in that pathway

2. What is feedback inhibition?
   a. when lack of an appropriate electron acceptor makes an electron transport chain stop
   b. when an enzyme that is active early in a metabolic pathway is inhibited by a product of the pathway
   c. when ATP synthase reverses and begins pumping protons out of the mitochondrial matrix
   d. when cellular respiration stops and fermentation begins

3. Where does the Krebs cycle occur in eukaryotes?
   a. in the cytoplasm
   b. in the matrix of mitochondria
   c. in the inner membrane of mitochondria
   d. in the intermembrane space of mitochondria

4. What does the chemiosmotic hypothesis claim?
   a. that substrate-level phosphorylation occurs in the electron transport chain
   b. that substrate-level phosophorylation occurs in glycolysis and the Krebs cycle
   c. that the electron transport chain is located in the inner membrane of mitochondria
   d. that electron transport chains generate ATP indirectly, by proton pumping and creation of a proton-motive force

5. What is an anabolic pathway?
   a. a series of chemical reactions that results in the breakdown of a large molecule and the release of chemical energy
   b. a series of chemical reactions that result in the production of NADH and $FADH_2$
   c. the synthesis of ATP, by a flow of protons across the inner mitochondrial membrane
   d. a series of chemical reactions that results in the synthesis of a large molecule from smaller precursor molecules

6. What is the basic task of the reactions in a fermentation pathway?
   a. to generate NADH from NAD^+, so electrons can be donated to the electron transport chain
   b. to synthesize pyruvate from lactate
   c. to generate NAD^+ from NADH, so glycolysis can continue
   d. to synthesize electron acceptors, so cellular respiration can continue

7. When do cells switch from cellular respiration to fermentation?
   a. when electron acceptors are not available
   b. when the proton-motive force runs down
   c. when NADH and $FADH_2$ supplies are low
   d. when pyruvate is not available

Conceptual Review

1. Explain why NADH and $FADH_2$ are called electron carriers, and why they are said to have "reducing power." Where do these molecules get electrons, and where do they deliver them? In eukaryotes, what molecule do these electrons reduce?

2. Compare and contrast substrate-level phosphorylation and oxidative phosphorylation.

3. What is the relationship between cellular respiration and fermentation? Why does cellular respiration produce so much more ATP than fermentation?

4. Make a diagram that shows the relationship between the three components of cellular respiration: glycolysis, the Krebs cycle, and electron transport. What molecules connect the three processes? Where does each process occur in a eukaryotic cell?

5. Explain the relationship between electron transport and oxidative phosphorylation. What does ATP synthase look like, and how does it work?

6. Hummingbirds and nectar-feeding insects eat a diet that is rich in sugars and extremely low in protein. How are these individuals able to synthesize the amino acids and proteins they require for growth?

7. Describe the relationship between carbohydrate metabolism, the catabolism of proteins and fats, and anabolic pathways.

Applying Ideas

1. Cyanide ($C{\equiv}N^-$) blocks complex IV of the electron transport chain. Suggest a hypothesis for what happens to the electron transport chain when complex IV stops working. Your hypothesis should explain why cyanide poisoning is fatal.

2. When membranes are folded, as in the cristae of mitochondria, the total surface area increases dramatically. Suppose that some mitochondria had unfolded cristae. How would their output of ATP compare to mitochondria with folded cristae? Explain your answer.

3. When yeast cells are placed into low-oxygen environments, the mitochondria in the cells become reduced in size and number. Suggest an explanation for this observation.

4. Most agricultural societies have come up with ways to ferment the sugars in barley, wheat, rice, corn, or grapes and produce alcoholic beverages. Historians argue that this was an effective way for farmers to preserve the chemical energy in grains and fruits in a form that would not be eaten by rats or spoiled by bacteria or fungi. Why does a great deal of chemical energy remain in the products of fermentation pathways?

5. To diagnose a bird's nutritional state, biologists feel the breast muscle. If the muscle mass is reduced enough that the sternum (breastbone) sticks out, biologists conclude the individual is near starvation. The logic here is that the individual is using muscle proteins as an energy source because all of its fat and carbohydrate stores have been used up. Why would proteins be the last source of stored chemical energy to be used by an organism?

CD-ROM and Web Connection

CD Activity 6.1: Glucose Metabolism *(tutorial)*
(Estimated time for completion = 10 min)
How is glucose converted to ATP such that the energy is available to the cell?

CD Activity 6.2: Chemiosmosis *(animation)*
(Estimated time for completion = 5 min)
What is the experimental support for the chemiosmotic hypothesis?

At your **Companion Website** (http://www.prenhall.com/freeman/biology), you will find self-grading exams and links to the following research tools, online resources, and activities:

ATP and Biological Energy
Discover how a group of researchers put together an artificial light-gathering ATP production cell.

Glycolysis and the Krebs Cycle
This site links to references on glycolytic metabolism, the intermediates of glycolysis, and a diagram of the electron transport chain.

Additional Reading

Harold, F. M. 2001. Gleanings of a chemiosmotic eye. *BioEssays* 23: 848–855.
Reviews the history of the chemiosmotic hypothesis of ATP synthesis.

Photosynthesis

7

About three billion years ago, a chance combination of light-absorbing pigment molecules and enzymes gave a bacterial cell the capacity to perform photosynthesis. **Photosynthesis** is the ability to convert light energy into chemical energy in the form of sugar. The origin of photosynthesis qualifies as one of the great events in the history of life. In terms of total numbers of individuals, photosynthetic organisms have long dominated the planet.

Photosynthesis is important because it creates the sugars that make life possible. It produces chemical energy as well as the building blocks needed to make cells. Photosynthetic organisms are termed **autotrophs** (self-feeders) because they make all of their own food. Organisms that eat photosynthetic organisms are called **heterotrophs** (different-feeders) because they obtain the sugars they need from other organisms. Without autotrophs, there are no heterotrophs. Photosynthesis is the process that underlies virtually all life on Earth.

This chapter's goal is to introduce how photosynthesis occurs. Section 7.1 begins with an overview of the process in plants. Section 7.2 details how chlorophyll and other pigment molecules found in photosynthetic organisms respond to sunlight; section 7.3 delves into the chemical reactions that convert the energy captured by these pigments into chemical energy in the form of ATP and an electron carrier called nicotinamide adenine dinucleotide phosphate (NADPH). Section 7.4 explores how ATP and NADPH are then used to reduce carbon dioxide to sugar. After studying these topics and the end-of-chapter essay, you should understand where the oxygen you breathe and the food you eat comes from. You will also appreciate why tree-planting programs may offer an effective way to slow global warming.

This is a close-up of a leaf—the major photosynthetic organ in plants. Plants and other photosynthetic organisms convert the energy in sunlight to chemical energy stored in the form of sugar. The sugar produced by photosynthetic organisms fuels cellular respiration and growth. Photosynthetic organisms, in turn, are consumed by animals, fungi, and a host of other species. Directly or indirectly, then, most organisms on Earth get their energy from photosynthesis.

7.1 What Is Photosynthesis?

Several important observations about photosynthesis occurred between the 1770s and the mid-1800s. Experiments by a series of different researchers showed that photosynthesis takes place only in the green parts of plants, that sunlight, carbon dioxide (CO_2), and water (H_2O) are required, and that oxygen (O_2) is produced as a by-product. By the early 1840s, enough was known about the process for Julius Robert von Mayer to propose that photosynthesis allows plants to convert sunlight into chemical energy in the form of carbohydrates. Soon after, the overall reaction for photosynthesis was understood to be

$$CO_2 + 2H_2O + \text{light energy} \rightarrow (CH_2O) + H_2O + O_2$$

where (CH_2O) stands for carbohydrate. To describe this reaction in words, carbon dioxide and water combine in the presence of light to form a carbohydrate, oxygen, and water. When glucose is the carbohydrate produced, the reaction can be written as

$$6\,CO_2 + 12H_2O + \text{light energy} \rightarrow C_6H_{12}O_6 + 6\,O_2 + 6\,H_2O$$

Exactly how does this reaction proceed? Based on the overall reaction, early investigators assumed that CO_2 and H_2O react directly to form CH_2O, and that the oxygen atoms in oxygen gas (O_2) come from carbon dioxide. Both assumptions proved to be incorrect, however. So, then, how do carbon dioxide and water interact, and where does the oxygen come from?

Photosynthesis: Two Distinct Sets of Reactions

During the 1930s, two independent lines of research converged to create a major advance in understanding photosynthesis. The first research program, led by Cornelius van Niel, focused on how photosynthesis occurs in cells from the purple sulfur bacteria. This research established that these organisms can grow on a culture medium that lacks sugars, meaning they must be autotrophs that manufacture their own carbohydrates. But to grow, the cells had to be exposed to sunlight and had to have access to hydrogen sulfide (H_2S). Further, van Niel showed that purple sulfur bacteria do not produce oxygen as a photosynthetic by-product. Instead, elemental sulfur (S) accumulates in the medium in which they grow. The overall reaction for photosynthesis in these organisms was

$$CO_2 + 2H_2S + \text{light energy} \rightarrow (CH_2O) + S$$

This observation was crucial for two reasons. First, it showed that CO_2 and H_2O do not combine directly during photosynthesis. Second, it showed that the oxygen atoms in CO_2 are not released as oxygen gas (O_2). In purple sulfur bacteria, carbon dioxide participates in the reaction but no oxygen is produced.

Based on these findings, researchers hypothesized that the oxygen atoms released during plant photosynthesis come from water. In support of this, Robert Hill showed in 1937 that isolated chloroplasts can produce oxygen in the presence of sunlight even if no CO_2 is present. The hypothesis was confirmed when heavy isotopes of oxygen—^{18}O compared to the normal ^{16}O—became available to researchers. Biologists exposed algae or plants to H_2O that contained ^{18}O, collected the oxygen gas given off as a by-product of photosynthesis, and confirmed that the oxygen gas contained the heavy isotope. The reaction that produced oxygen occurred only in the presence of sunlight:

$$2H_2O + \text{sunlight} \rightarrow 4H^+ + 4e^- + O_2$$

A second major line of research augmented these discoveries. When the radioactive isotope ^{14}C became available in the mid-1940s, Melvin Calvin and others began feeding labeled carbon dioxide ($^{14}CO_2$) to plants and identifying the sugars and other photosynthetic products that subsequently became labeled with ^{14}C. These researchers noticed that labeled carbon

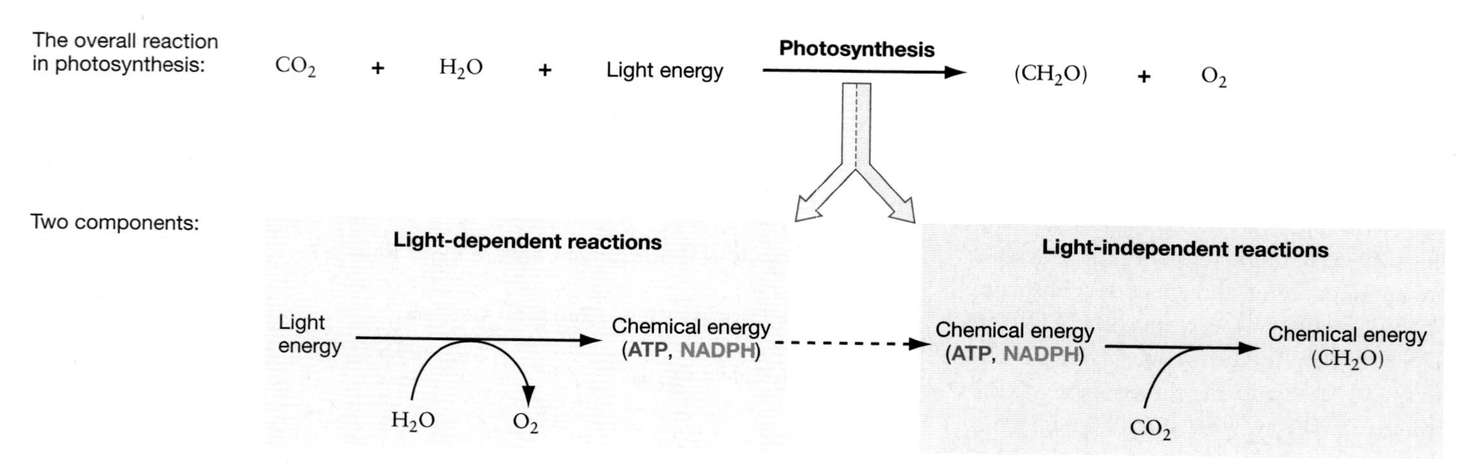

FIGURE 7.1 Photosynthesis Has Two Distinct Components
In the light-dependent reactions of photosynthesis, light energy is transduced to chemical energy. ATP is produced during these reactions. In addition, electrons are removed from water and transferred to NADPH, with oxygen (O_2) being released as a by-product. In the light-independent reactions, the ATP and NADPH produced in the light-dependent reactions are used to reduce carbon dioxide to carbohydrate.

dioxide could become incorporated into carbohydrates in the dark. In other words, the reduction of CO_2 to carbohydrate (CH_2O) did not depend directly on light. Because Calvin played an important role in detailing the exact sequence of reactions, the light-independent component of photosynthesis came to be known as the **Calvin cycle.**

To summarize, experiments with purple sulfur bacteria showed that the oxygen given off during photosynthesis comes from water, not carbon dioxide. In addition, experiments with labeled isotopes of oxygen and carbon indicated that the reaction in which oxygen atoms are split from water takes place only in the presence of sunlight. In contrast, the reactions in which carbon dioxide is reduced to sugar can take place in the dark. In other words, photosynthesis consists of two distinct sets of reactions: one set is dependent on light; the other set is independent of light. The light-dependent reactions result in the production of oxygen and electrons from water; the light-independent reactions result in the production of sugar from carbon dioxide.

How are these two sets of reactions connected? Researchers realized that the electrons released when water is split must be involved in the reduction of carbon dioxide. The leading hypothesis was that the energy in sunlight made it possible for electrons to be released from water and transferred to a phosphorylated version of NADH, called NADPH, that is abundant in photosynthesizing cells. Like NADH, NADPH is an electron carrier. As **Figure 7.1** shows, the light-independent reactions of the Calvin cycle were thought to use the chemical energy in NADPH and ATP to reduce CO_2 to carbohydrate (CH_2O). Where does all this activity take place?

The Structure of the Chloroplast

Once biologists established that photosynthesis takes place only in green portions of plants, microscopy suggested that the reactions occur inside organelles called **chloroplasts** (green-formed). As **Figure 7.2a** shows, leaf cells typically contain from 40 to 50 chloroplasts; on average, a square millimeter of leaf contains about 500,000 chloroplasts. When Hill showed that membranes derived from chloroplasts release oxygen when exposed to sunlight, the hypothesis that chloroplasts are the site of photosynthesis became widely accepted.

(a) Leaves contain millions of chloroplasts.

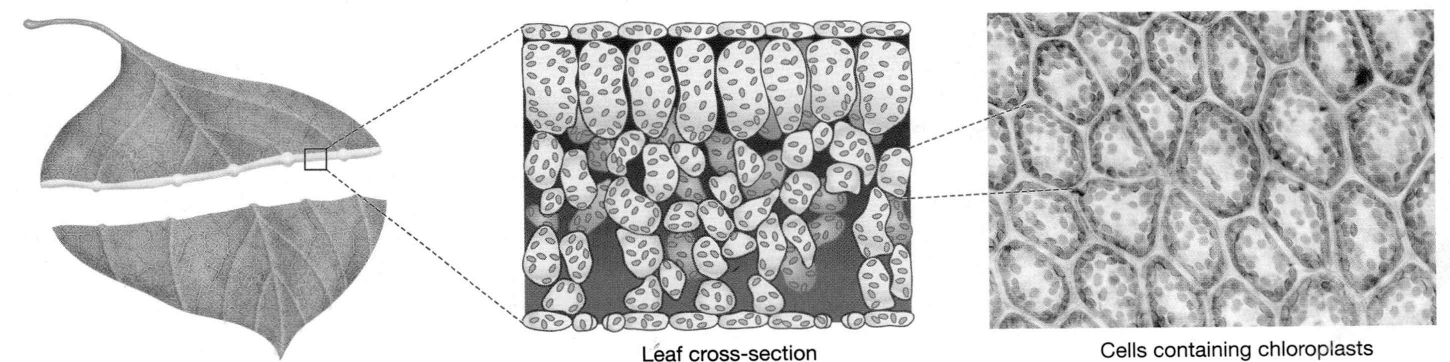

(b) Chloroplasts are highly structured, membrane-rich organelles.

Outer membrane
Inner membrane
Thylakoids
Granum
Stroma

(c) Photosynthetic bacteria have extensive internal membranes too.

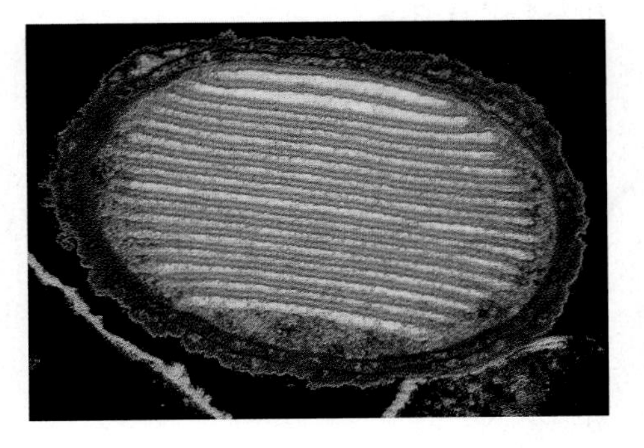

FIGURE 7.2 Where Does Photosynthesis Occur?
(a) In plants, photosynthesis takes place in organelles called chloroplasts. **(b)** The internal membranes of chloroplasts form flattened, vesicle-like structures called thylakoids, some of which form stacks called grana. **(c)** Cyanobacteria contain thylakoids. **EXERCISE** Using the drawing in part (b) as a guide, label the structures in the micrograph in part (b).

When electron microscopy became available in the 1950s, researchers observed that chloroplasts are extremely membrane-rich (**Figure 7.2b**). They have two organelle membranes and numerous stacks of vesicle-like structures called **thylakoids.** A stack of thylakoids is called a **granum** (plural *grana*). Grana are connected by membranes; the fluid-filled space outside this internal membrane system is referred to as the **stroma.** As **Figure 7.2c** shows, photosynthetic bacteria are also membrane-rich.

When researchers analyzed the chemical composition of thylakoid membranes, they found huge quantities of a pigment called **chlorophyll** (green-leaf). **Pigments** are molecules that absorb certain wavelengths of light and transmit others. As we'll see, chlorophyll is one of several pigments found in leaves. Because chlorophyll absorbs blue and red light and transmits green light, it is responsible for the green color of plants. Some species of photosynthetic bacteria also have chlorophyll in their internal membranes; other types of photosynthetic bacteria contain different pigments.

How are these structures and pigments involved in photosynthesis? As subsequent sections explore how chloroplasts capture light energy and transform it into chemical energy, it will become clear that there is a close connection between the structure of the chloroplast and its function as the photosynthetic center.

Developmental studies showed that chloroplasts are derived from colorless organelles called **proplastids.** Proplastids are found in the cells of embryonic plants and in the rapidly dividing tissues of mature plants. As plant cells grow and differentiate, proplastids mature into chloroplasts or the other types of plastids introduced in **Box 7.1** (page 142).

Before plunging into the details of how photosynthesis occurs, though, it's helpful to step back and appreciate just how astonishing the process is. Chemists have synthesized an amazing diversity of compounds from relatively simple starting materials, but their achievements pale in comparison to cells that can synthesize sugar from just sunlight, carbon dioxide, and water. Equally impressive is the observation that photosynthetic cells accomplish this feat in environments ranging from mountaintop snowfields to polar ice, the open ocean, and tropical rain forests.

In short, photosynthesis may rank as the most sophisticated chemistry on Earth. Without it, the only cells alive would be scratching out a living near hot springs and other specialized locations where abundant supplies of H_2, H_2S, and other highly reduced inorganic compounds make a limited amount of cellular respiration possible. Without photosynthesis, the tree of life would be a stunted, tiny version of the biodiversity we see today.

7.2 How Does Chlorophyll Capture Light Energy?

Photosynthesis begins with the light-dependent reactions, and the light-dependent reactions begin with the simple act of sunlight striking chlorophyll. To understand the consequences of this event, it's helpful to review the nature of light. Light is a type of electromagnetic radiation. Electromagnetic radiation, in turn, is a form of energy. Physicists describe light's behavior as both wave-like and particle-like.

Figure 7.3 illustrates the wave-like nature of electromagnetic radiation. As you can see, different types of electromagnetic radiation have a characteristic range of wavelengths—visible light, for example, ranges in wavelength from about 400 to about 750 nanometers (nm, or 10^{-9} m). In addition, shorter wavelengths of electromagnetic radiation contain more energy than longer wavelengths. To emphasize the particle-like nature of light, physicists point out that it exists in discrete packets called **photons.** In understanding photosynthesis, the important point is that each photon and wavelength of light has a characteristic amount of energy. Pigment molecules like chlorophyll absorb this energy. How?

Photosynthetic Pigments Absorb Light

Three things can happen when a photon strikes something. The photon can be absorbed, transmitted, or reflected. Most pigments absorb selected wavelengths of light. Because white light is a mixture of all wavelengths in the visible spectrum, pigments are colored. If a pigment absorbs all of the visible wavelengths, it appears black. If a pigment absorbs many or most of the wavelengths in the blue and green part of the spectrum but transmits or reflects red wavelengths, it appears red.

What wavelengths do plant pigments absorb? One approach to answering this question involves grinding up leaves and extracting the pigment molecules in a solvent, as shown in **Figure 7.4a**. The pigments in the raw extract can then be separated

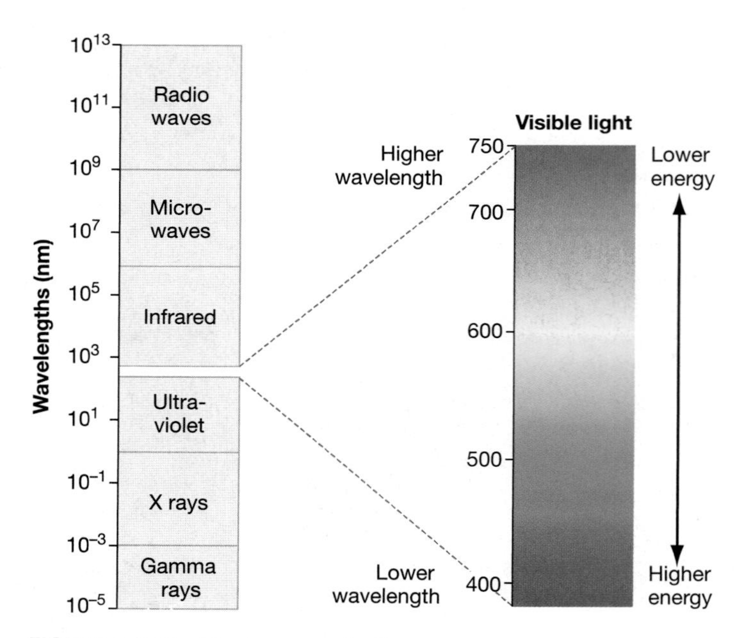

FIGURE 7.3 The Electromagnetic Spectrum
Electromagnetic energy radiates through space in the form of waves. Humans can see radiation at wavelengths between about 400 nm and about 750 nm. Note that the lower the wavelength of electromagnetic radiation, the higher its energy. QUESTION Does blue light contain more or less energy than red light?

from each other using a technique called **paper chromatography**. As step 2 in Figure 7.4a shows, paper chromatography begins when a raw extract is spotted near the bottom of a piece of filter paper. The filter paper is then placed in a solvent solution. As the solvent wicks up the paper, the molecules in the mixture are carried along. Because the size or solubility of the molecules in the extract varies, they are carried along with the solvent at different rates. Once the solvent has run to the top of the paper, the paper is dried, rotated 90°, and placed in a tank with a different solvent. The second solvent runs up the paper and separates the molecules in a second dimension.

Figure 7.4b shows a chromatograph from a grass-leaf extract that was run in a single solvent. Note that the leaf contains an array of pigments. By cutting out a specific region of the paper, extracting the pigment, and using a spectrophotometer to measure the light wavelengths it absorbs, researchers have produced data like those shown in **Figure 7.4c.** In studying this figure, note that there are two major classes of pigment in plant leaves. The chlorophylls, designated chlorophyll *a* and chlorophyll *b*, and the related pigment pheophytin absorb strongly in the blue and red regions of the spectrum and transmit green. Carotene and other carotenoids comprise a different family of pigments, which absorb blue and green wavelengths and thus appear yellow, orange, or red. Let's look quickly at the structure and function of these two classes of pigments, then analyze what happens during the absorption event itself.

The Role of Carotenoids and Other Accessory Pigments

The carotenoids found in plants belong to two classes, called **carotenes** and **xanthophylls. Figure 7.5a** (page 140) shows the structure of β-carotene, which gives carrots their orange color. A xanthophyll called zeaxanthin, which gives corn kernels their bright yellow color, is nearly identical to β-carotene except that the ring structures on either end of the molecule contain a hydroxyl (OH) group.

Both xanthophylls and carotenes are found in chloroplasts; in autumn, when the leaves of deciduous trees begin to die and their chlorophyll degenerates, the wavelengths scattered by carotenoids turn northern forests into spectacular displays of yellow and orange.

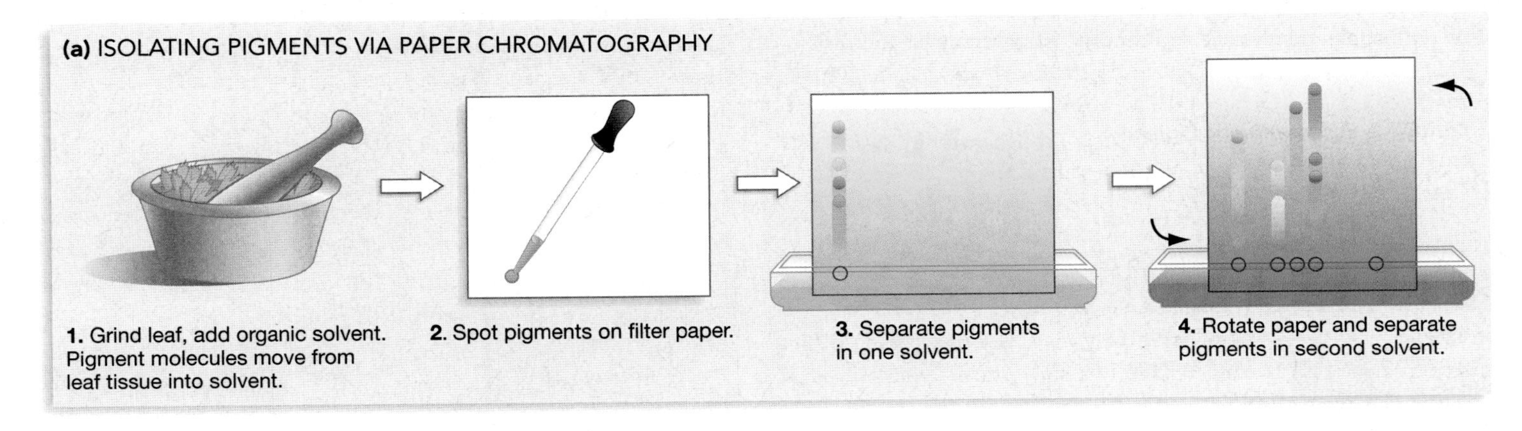

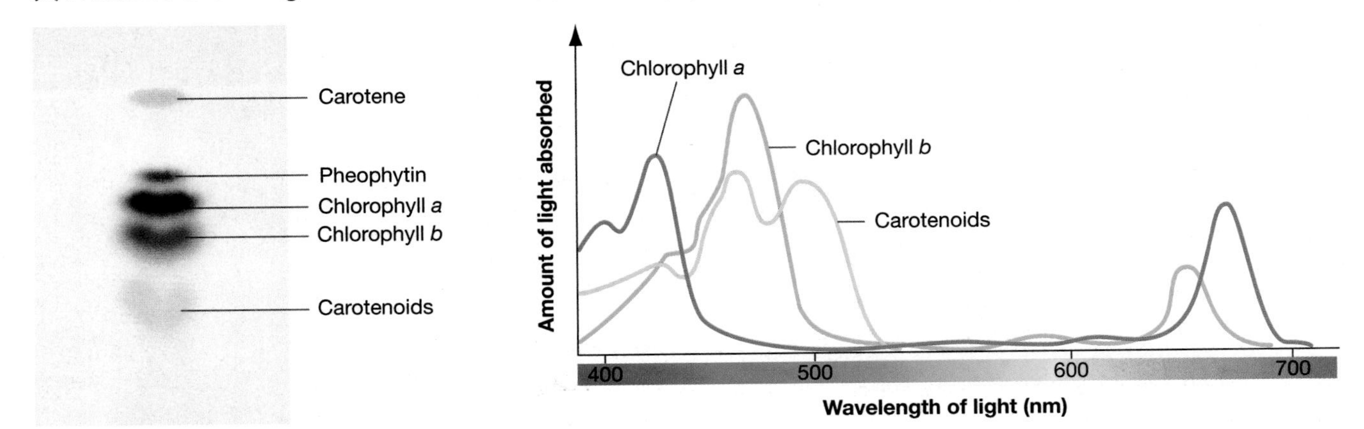

FIGURE 7.4 How Can the Absorption Spectrum of a Photosynthetic Pigment Be Determined?
(a) Paper chromatography is an effective way to isolate the various pigments in photosynthetic tissue.
(b) Photosynthetic tissues typically contain several different pigments. Different species of photosynthetic organisms may contain different types and combinations of pigments. **(c)** Each photosynthetic pigment has a distinct absorption spectrum.

What do carotenoids do? Because these pigments absorb wavelengths of light that are not absorbed by chlorophyll, they extend the range of wavelengths that can drive photosynthesis. But researchers discovered an even more important function by analyzing the fate of leaves in which carotenoids have been destroyed. Many herbicides, for example, work by inhibiting enzymes involved in carotenoid synthesis. Because individuals that lack carotenoids rapidly lose their chlorophyll and turn white, researchers have concluded that carotenoids serve a protective function. When carotenoids are absent, chlorophyll molecules are destroyed. As a result, photosynthesis stops and starvation occurs.

To understand the molecular basis of carotenoid function, recall from Chapter 2 that electromagnetic radiation—especially the high-energy, short-wavelength photons in the ultraviolet part of the spectrum—contain enough energy to knock electrons out of atoms and create free radicals. Free radicals, in turn, trigger reactions that quickly destroy nearby molecules. But because carotenoids can destroy free radicals, they protect chlorophyll molecules from destruction.

Carotenoids are not the only molecules that protect plants from the damaging effects of sunlight, however. Robert Last and colleagues have analyzed mutant individuals of the mustard plant *Arabidopsis thaliana* that are unable to synthesize compounds called flavonoids. Because flavonoids absorb ultraviolet radiation, individuals that lack flavonoids are highly susceptible to damage from UV light. In effect, flavonoids function as a sunscreen for leaves and stems.

The general message of research on carotenoids and flavonoids is that the energy in sunlight is a double-edged sword. It makes photosynthesis possible, but it can also lead to the formation of free radicals that damage the cell.

The Structure of Chlorophyll Figure 7.5b shows the structure of chlorophyll. As you can see, chlorophyll *a* and chlorophyll *b* are very similar in structure as well as in absorption spectra. In land plants, chloroplasts typically have about three molecules of chlorophyll *a* in their internal membranes for every molecule of chlorophyll *b*.

As the figure shows, a chlorophyll molecule has two fundamental parts: a long tail made up of isoprene subunits and a "head" that consists of a large ring structure with a magnesium atom in the middle. The head is where light is absorbed. But just what is "absorption?" What happens when a photon of a particular wavelength, say red light with a wavelength of 680 nm, strikes a chlorophyll molecule?

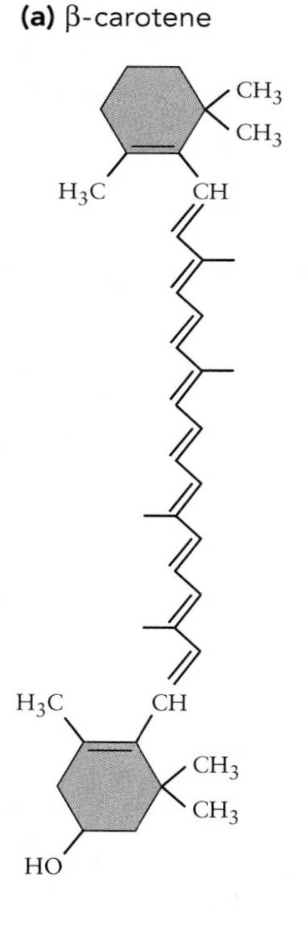

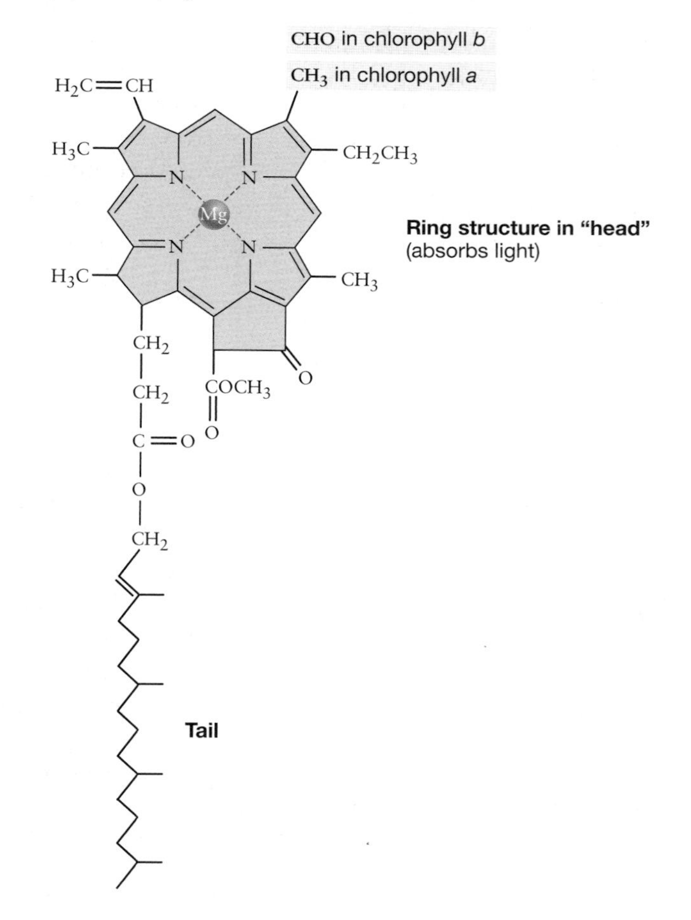

FIGURE 7.5 Photosynthetic Pigments Contain Ring Structures
(a) β-carotene is an orange pigment found in carrots and other plant tissues. Humans and other mammals have an enzyme that splits β-carotene in half, forming two molecules of vitamin A. Vitamin A, in turn, is oxidized to form a pigment used in vision. **(b)** Although chlorophylls *a* and *b* are very similar structurally, they have the distinctive absorption spectra shown in Figure 7.4c.

When Light Is Absorbed, Electrons Enter an Excited State

When a photon strikes chlorophyll, its energy can be transferred to an electron in the molecule's head region. In response, the electron is raised to a higher electron shell—one with greater potential energy. As **Figure 7.6a** shows, the excited electron states that are possible in a particular pigment are discrete and can be represented as lines on an energy scale. If the difference between the possible energy states is the same as the energy in the photon, then the photon can be absorbed. In chlorophyll, for example, the energy difference between state 0 and state 1 is equal to the energy in a red photon; the energy difference between state 0 and state 2 is equal to the higher energy in a blue photon. Thus chlorophyll can readily absorb red photons and blue photons. Chlorophyll does not absorb green light well because there is no step—no difference in possible energy states—that corresponds to the amount of energy in a green photon.

Once an electron is raised to a higher energy state, it can return to its normal or ground state in several ways. **Fluorescence** occurs when an electron falls back to its ground state, releasing a photon and heat in the process. As **Figure 7.6b** shows, a pure solution of chlorophyll fluoresces red when it is exposed to ultraviolet light; the extra energy in the high-energy UV photons is released as heat. But when chlorophyll is in a chloroplast, only about 2 percent of the red and blue photons that are absorbed produce fluorescence. What happens to the other 98 percent of excited electrons? The answer is that the energy in the excited electrons is passed among nearby chlorophyll molecules until it reaches a chlorophyll molecule in a protein complex called the reaction center (**Figure 7.6c**). At the reaction center, an excited electron undergoes a fundamentally important change. This change is the foundation of the energy transduction step of photosynthesis. What happens to excited electrons that enter a reaction center?

7.3 The Photosynthetic Reaction Centers

When a photon promotes an electron to a high-energy state, energy transduction occurs. In other words, the energy in light is converted to chemical energy in the form of an excited electron. But if fluorescence occurs and the electron falls back to the ground state, the energy transduction event is fleeting and no chemical energy is produced. For the energy transduction event to last, the electron must reach an electron acceptor that becomes reduced. This is exactly what happens in a photosynthetic reaction center.

What is the nature of a photosynthetic reaction center, and how does it work? During the 1950s, this was the central question facing biologists interested in photosynthesis. Robert Emerson and co-workers conducted an experiment that at first seemed puzzling, but eventually pointed the way to an important advance in understanding reaction centers. Emerson and colleagues were using green algae to study the photosynthetic

(a) Electrons can be promoted to discrete high-energy states.

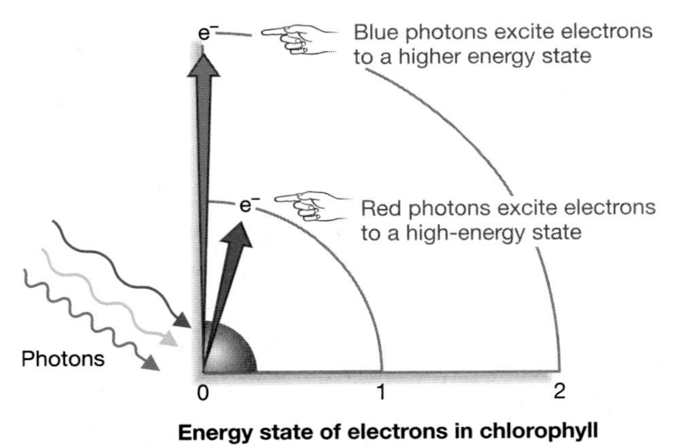

(b) If the excited electron is not captured, fluorescence occurs.

(c) Chlorophyll molecules transmit energy from excited electrons to a reaction center.

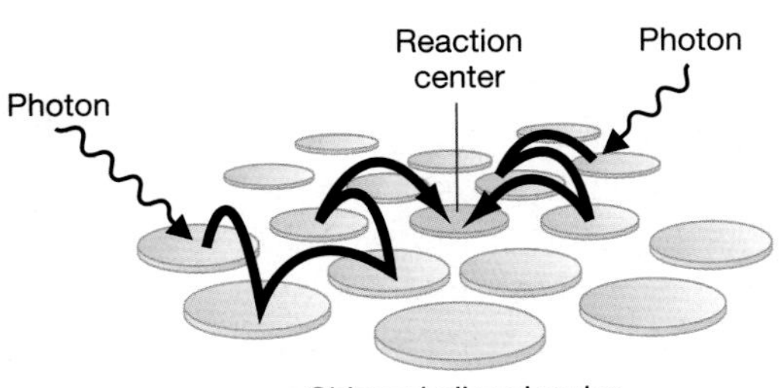

FIGURE 7.6 What Happens When a Photon Strikes Chlorophyll? **(a)** When a photon strikes chlorophyll, an electron can be promoted to a higher energy state, depending on the energy in the photon. **(b)** This photograph shows a pure solution of chlorophyll exposed to ultraviolet light. When the pigment absorbs blue photons, electrons are excited to a high-energy state. But the electrons immediately fall back to a low-energy state. As they do, the pigment emits red photons and heat. **(c)** In chloroplasts, excited electrons transmit energy to a reaction center. There the electron is captured by an electron acceptor that becomes reduced. As a result, fluorescence is rare.

response to various wavelengths of light. As the data in **Figure 7.7** show, they found that if they illuminated a sample of algae with either far-red light or red light, the photosynthetic response was about the same. Stated another way, wavelengths of 700 nm and 680 nm were equally effective at stimulating photosynthesis. But if the researchers used a combination of far-red and red light, the rate of photosynthesis increased dramatically. If wavelengths of both 700 nm and 680 nm were present, the photosynthetic rate was much more than double the rate produced by each wavelength independently.

What could explain this enhancement effect? After performing a series of follow-up experiments, Louis Duysens and colleagues suggested that green algae and plants actually have two distinct types of reaction centers, rather than just one. Duysen's group proposed that one reaction center, which came to be called **photosystem II**, is stimulated by wavelengths of 680 nm, while a different reaction center, now referred to as **photosystem I**, is activated by wavelengths of 700 nm. Further, the biologists hypothesized that the two photosystems have different functions.

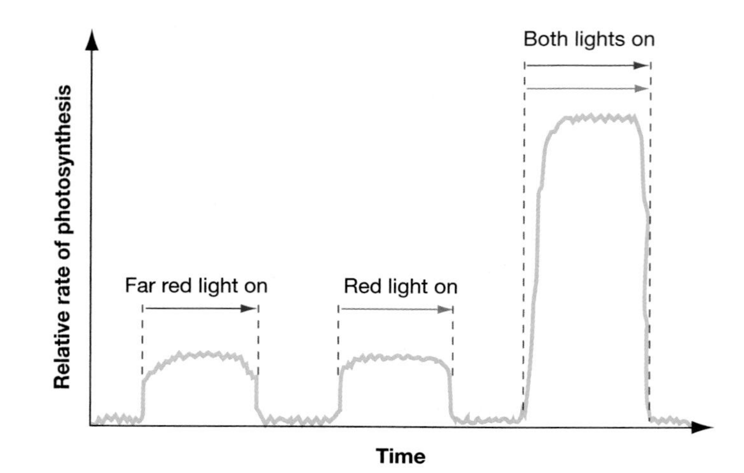

FIGURE 7.7 The Enhancement Effect
These data show the relative rate of photosynthesis when green algal cells are illuminated with far-red light (700 nm), red light (680 nm), or a combination of far-red and red light.

BOX 7.1 Types of Plastids

Plastids are a family of double membrane-bound organelles found in plants. As **Figure 1** shows, they arise from small, undifferentiated organelles called proplastids. As a cell matures and takes on a specialized function in the plant, its proplastids also differentiate. For example, if a developing cell is exposed to light, its proplastids are stimulated to develop into chloroplasts.

There are two major types of plastids in addition to chloroplasts. Leucoplasts (white-formed) often function as energy storehouses. More specifically, leucoplasts may store chemical energy in the form of highly reduced molecules. Some leucoplasts store oils; others synthesize and sequester the carbohydrate called starch. Chromoplasts (color-formed) are brightly colored because they synthesize and sequester large amounts of orange, yellow, or red pigments. High concentrations of chromoplasts are responsible for the bright colors of many flower petals and fruits.

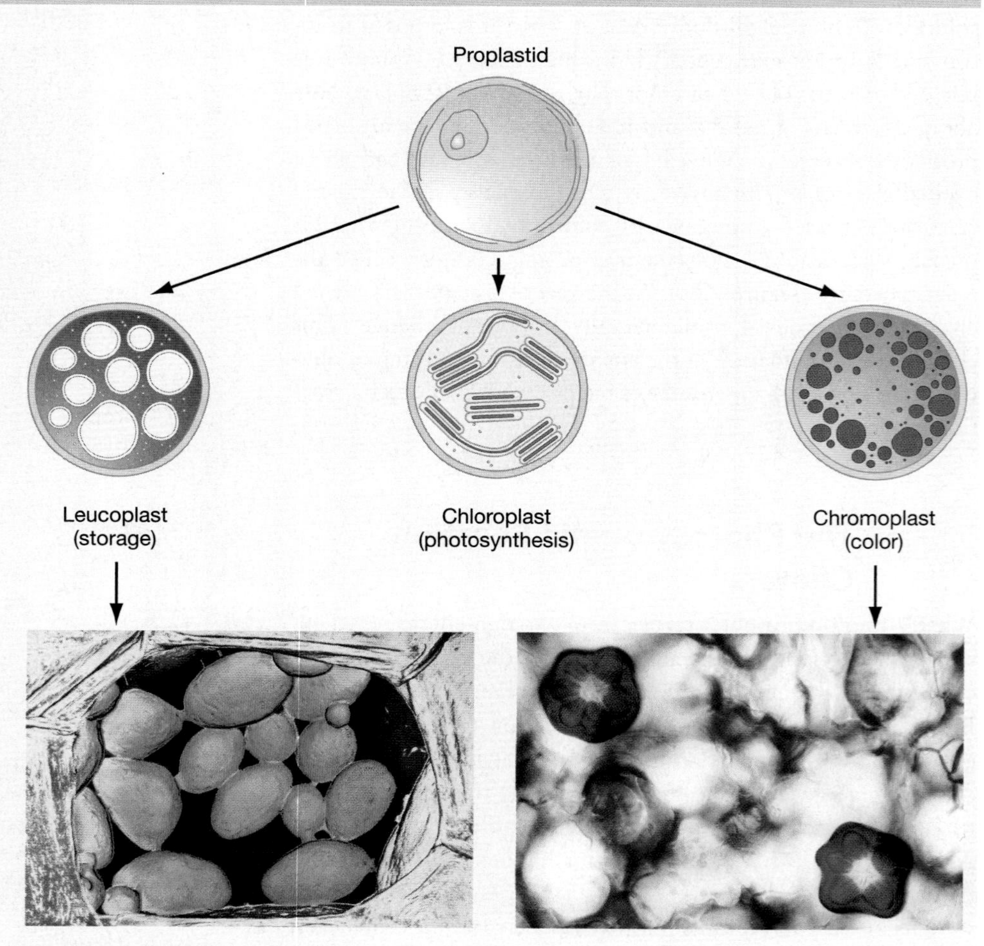

FIGURE 1 Types of Plastids
Although each type of plastid has a distinct structure and function, they all develop from small organelles called proplastids.

They proposed that excited electrons from one photosystem are added to $NADP^+$ to form NADPH, while excited electrons from the other photosystem are passed down an electron transport chain that is involved in ATP production. According to the two-photosystem hypothesis, the enhancement effect occurred because photosynthesis is most efficient when both photosystems are operating.

Research during the ensuing decades has confirmed the two-photosystem hypothesis. Let's take a closer look at each photosystem in isolation, and then consider how they are thought to work together.

How Does Photosystem II Work?

To work out the details of photosystem II, researchers focused on studying species from the purple nonsulfur bacteria and the purple sulfur bacteria. The single photosystem found in these cells has many of the same components observed in photosystem II of cyanobacteria (blue-green bacteria), algae, and plants. What are these components, and what do they do?

One particularly important component of the complex is a molecule called pheophytin. Structurally, pheophytin is similar to chlorophyll except that it lacks a magnesium atom in its head region. Functionally, though, pheophytin differs from chlorophyll. Instead of acting as a pigment that promotes an electron when it absorbs a photon, pheophytin acts as an electron acceptor. When an excited electron reaches a reaction center containing pheophytin, the electron binds to pheophytin. The chlorophyll molecule in the reaction center loses an electron and becomes oxidized. When pheophytin is reduced in this way, the energy transduction step that started with the absorption of light is completed. Electrons that reach pheophytin are never passed back to chlorophyll molecules. Instead they are passed down an electron transport chain, where they are gradually stepped down to a more oxidized state.

An Electron Transport Chain Pumps Protons The electron transport chain leading from photosystem II is similar to the electron transport chain found in mitochondria. Specifically, it contains several quinones and several cytochromes. Recall from Chapter 6 that quinones are small, hydrophobic molecules. One of the quinones found in photosystem II, called plastoquinone, is not anchored to a protein and is free to move from one side of the thylakoid membrane to the other. In doing so, it shuttles electrons that it receives from pheophytin to more oxidized molecules in the chain. These electron acceptors are found in a complex that contains a cytochrome similar to the cytochromes found in mitochondria. As **Figure 7.8a** shows, the quinones and the cytochrome complex in photosystem II create an electron transport chain. The excited electron captured by the reaction center is gradually stepped down in potential energy through a series of reduction-oxidation (redox) reactions.

As plastoquinone shuttles electrons from protein-bound quinones to the cytochrome complex, it also carries protons from one side of the thylakoid membrane to the other. As **Figure 7.8b** shows, the protons transported by plastoquinone

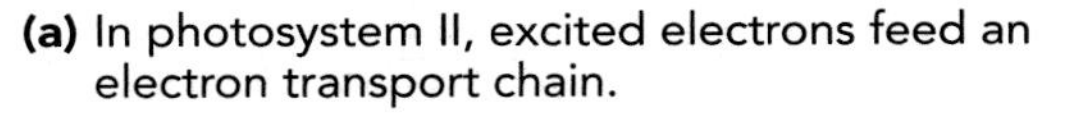
(a) In photosystem II, excited electrons feed an electron transport chain.

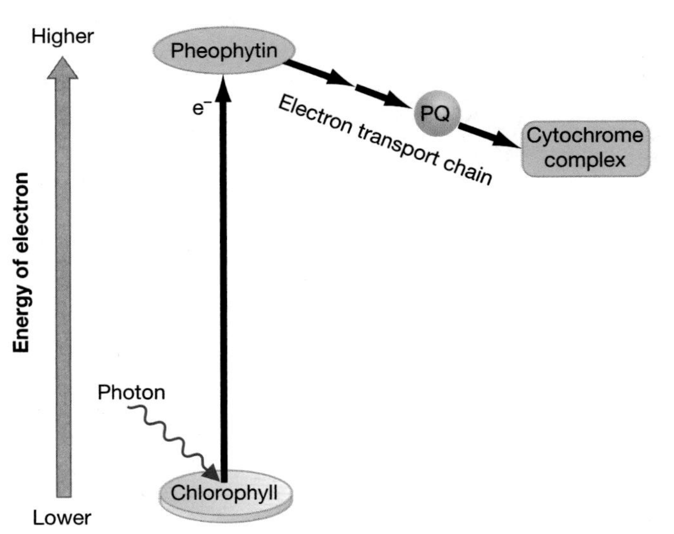

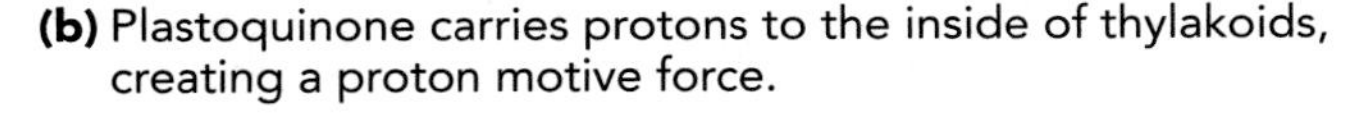
(b) Plastoquinone carries protons to the inside of thylakoids, creating a proton motive force.

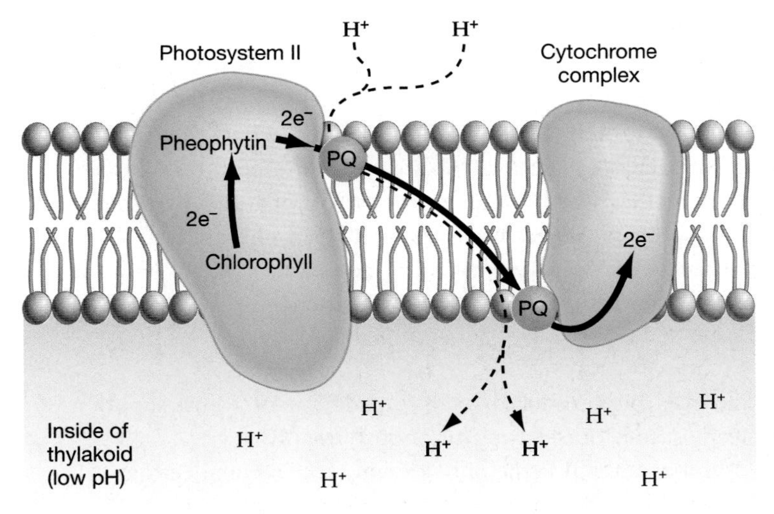

FIGURE 7.8 Photosystem II
(a) This diagram shows the energy state of electrons as they flow through photosystem II. When an excited electron leaves the chlorophyll molecule in the reaction center of photosystem II, it is stepped down in energy along an electron transport chain. **(b)** When plastoquinone receives two electrons, it also accepts two protons from the stroma. The reduced form of plastoquinone then diffuses across the thylakoid membrane to the cytochrome complex. When a cytochrome accepts the electrons from plastoquinone, the protons are released to the interior of the thylakoid. In this way, a large concentration of protons builds up inside the thylakoid.

result in a large concentration of protons inside the thylakoid. When photosystem II is active, the pH of the thylakoid interior reaches 5 while the pH of the stroma hovers around 8. Because the pH scale is logarithmic, the difference of 3 units means that there is $10 \times 10 \times 10 = 1000$ times higher concentration of H^+ in the interior of the thylakoid compared to the stroma. The net effect of electron transport, then, is to set up a large proton gradient that drives H^+ out of the thylakoid and into the stroma. Based on your reading of Chapter 6, it should come as no surprise that this proton motive force drives the production of ATP.

ATP Synthase Uses the Proton Motive Force to Phosphorylate ADP In mitochondria, recall that NADH and $FADH_2$ donate electrons to an electron transport chain that pumps protons out of the matrix and into the intermembrane space. In photosystem II, pheophytin donates reduced electrons to an electron transport chain that pumps protons out of the stroma and into the internal membrane space. In both instances, a stream of protons diffuses down the resulting electrochemical gradient through ATP synthase. The flowing protons make the enzyme spin, resulting in conformational changes that drive the phosphorylation of ADP. In this way, the light energy captured by chlorophyll is used to stockpile chemical energy in the form of ATP. This process is called **photophosphorylation.**

What happens to an electron at the end of the electron transport chain? Here ends the parallel between the photosystem found in the purple sulfur bacteria and the photosystem II of plants. In purple sulfur bacteria, cytochrome donates an electron back to the reaction center, where it is again promoted to a high-energy state when a photon is absorbed. In this way, electrons are cycled through the system. But as we'll see, photosystem II in plants donates electrons to photosystem I. This observation raises a critically important question. Where do electrons come from to replace those that leave photosystem II?

Photosystem II Obtains Electrons by Oxidizing Water Cyanobacteria, algae, and plants are said to perform oxygenic photosynthesis because they produce oxygen as a by-product of the process. Stated another way, part of photosystem II has the ability to "split" water and release oxygen. Recall that the oxygen-generating reaction can be written as

$$2H_2O \rightarrow 4H^+ + 4e^- + O_2$$

The electrons produced by the oxidation of water supply a steady source of electrons for photosystem II.

Photosystem II is the only known protein complex that can oxidize water. When it absorbs a series of photons and excited electrons activate the electron transport chain, the complex becomes so highly oxidized that it strips electrons away from water, leaving protons and oxygen. It is difficult to overstate the importance of this unique reaction. For example, all of the oxygen you breathe originates in this way. As **Box 7.2** explains, the rise of oxygenic (oxygen-producing) photosynthesis had a huge impact on the history of all life. In addition, chemists have long hoped to synthesize catalysts that would simulate the reaction and support the commercial production of O_2 and hydrogen gas (H_2) from water. If this could be accomplished, the H_2 produced could be used as a fuel for vehicles. Despite its fundamental importance, however, the mechanism responsible for the reaction is not completely understood.

What progress has been made thus far? Based on studies of the chemical structure of photosystem II, researchers have suspected that four manganese (Mn)-containing groups and a nearby tyrosine residue found in the protein complex might be involved in the reaction. The hypothesis was that the loss of an excited electron from the photosystem would create a free radical version of the tyrosine molecule, and that this free radical could then react with water molecules in conjunction with the Mn atoms.

Support for this hypothesis has recently come from successful efforts to synthesize molecules that contain Mn and tyrosine and that can catalyze the production of O_2 from water. In addition, the first x-ray crystallographic images of the oxygen-evolving complex have been published and confirm that the Mn-containing groups and tyrosine molecule are indeed in close physical proximity. Still, understanding exactly how photosystem II splits water remains one of the great challenges facing researchers interested in photosynthesis.

How Does Photosystem I Work?

In the same way that researchers figured out how photosystem II works by studying purple nonsulfur and purple sulfur bacteria, they turned to bacteria called heliobacteria (sun-bacteria) to study the structure and function of photosystem I. Using the energy in sunlight, heliobacteria produce NADPH from $NADP^+$. **Figure 7.9** shows how the system works. When the

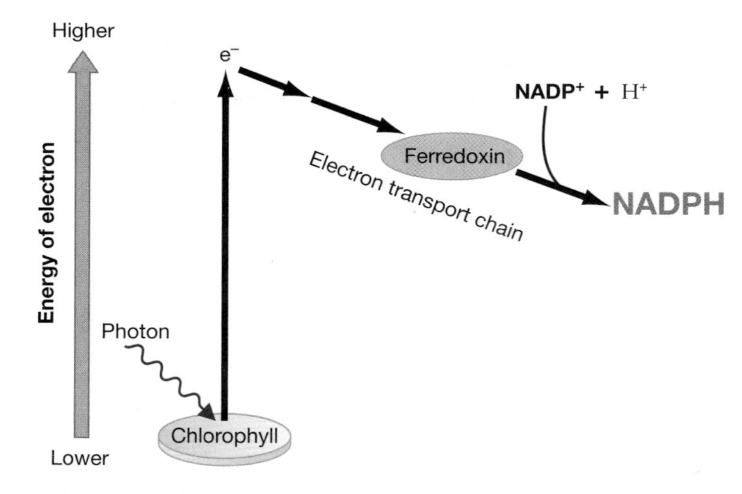

FIGURE 7.9 Photosystem I
This diagram shows the energy state of electrons as they flow through photosystem I. When an excited electron leaves the chlorophyll molecule in the reaction center of photosystem I, it passes through a series of iron- and sulfur-containing proteins until is it accepted by ferredoxin. In an enzyme-catalyzed reaction, the reduced form of ferredoxin reacts with NADP to produce NADPH.

reaction center absorbs a photon, the excited electron is passed through a series of iron- and sulfur-containing molecules to a related compound called ferredoxin. An enzyme called ferredoxin/$NADP^+$ oxidoreductase then transfers the electron and a proton to $NADP^+$, forming NADPH. NADPH, in turn, is analogous to the NADH and $FADH_2$ produced by the Krebs cycle. It is an electron carrier that can donate an electron to other compounds and thus reduce them.

To summarize, then, photosystem II results in the production of a proton motive force that drives the synthesis of ATP while photosystem I results in the production of NADPH. Several groups of bacteria have just one of the two photosystems. In contrast, the cyanobacteria, algae, and plants have both. In these organisms, how do the two photosystems interact?

The Z Scheme: Photosystems I and II Work Together

When Robert Hill and Fay Bendall realized that photosystem I and II have distinct but complementary functions, they proposed that they interact as shown in **Figure 7.10a**. This diagram illustrates a model known as the **Z scheme**. Note that when plotted on an axis showing the degree of oxidation or reduction, the path of electrons through photosystems I and II resembles the letter *Z* laid on its side.

Following the path of electrons through the Z scheme will help drive home how photosynthesis works. The process starts when a photon excites an electron in the chlorophyll molecules associated with photosystem II. When the energy is transmitted to the reaction center, a special pair of chlorophyll molecules called P680—so-called because they lose electrons most readily when they absorb wavelengths of 680 nm—passes an excited electron to pheophytin. From there the electron is gradually stepped down in potential energy through redox reactions among a series of quinones and cytochromes. Using the energy released by the redox reactions, plastoquinone carries protons across the thylakoid membrane. ATP synthase uses the resulting proton motive force to phosphorylate ADP.

Photosystem II is connected to the cytochrome complex by plastoquinone. The cytochrome complex, in turn, is connected to photosystem I by a small diffusible protein called **plastocyanin**. Plastocyanin is symbolized P_c in Figure 7.10. Note that the protein picks up an electron from the cytochrome complex,

(a) In the Z scheme, electrons flow from water to NADPH.

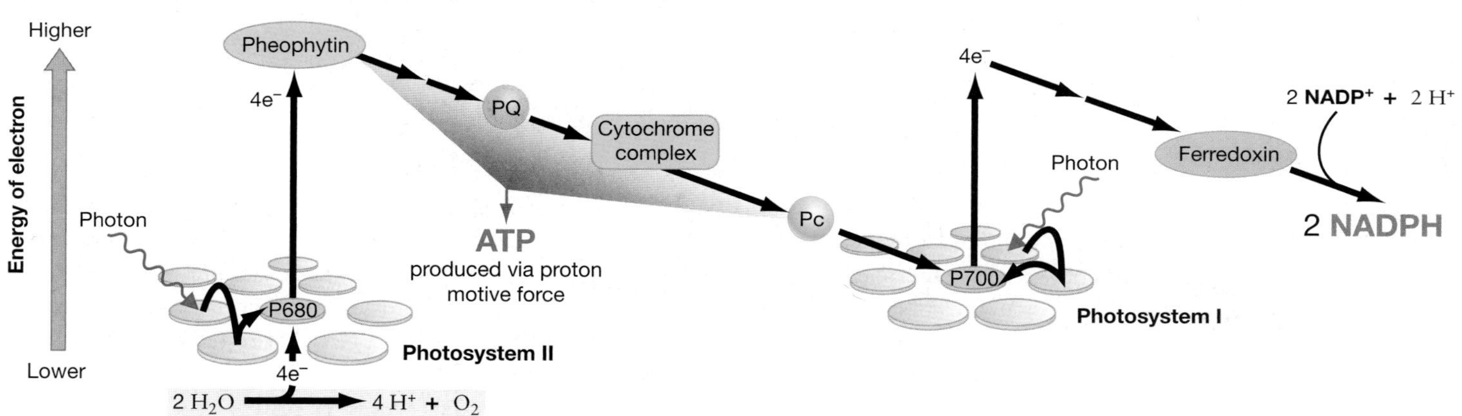

(b) In cyclic electron transport, which drives cyclic photophosphorylation, photosystem I transfers electrons to plastoquinone (PQ).

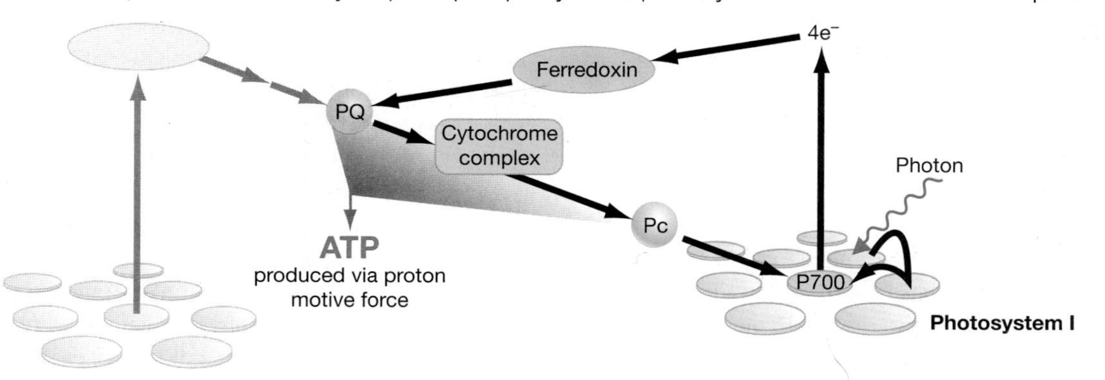

FIGURE 7.10 The Z Scheme
(a) The Z scheme is a model for how photosystem II and I interact. The model proposes that electrons from photosystem II enter photosystem I, where they are promoted to a high enough energy state to make the reduction of NADP possible. (b) Cyclic electron transport is an alternative to the Z scheme.

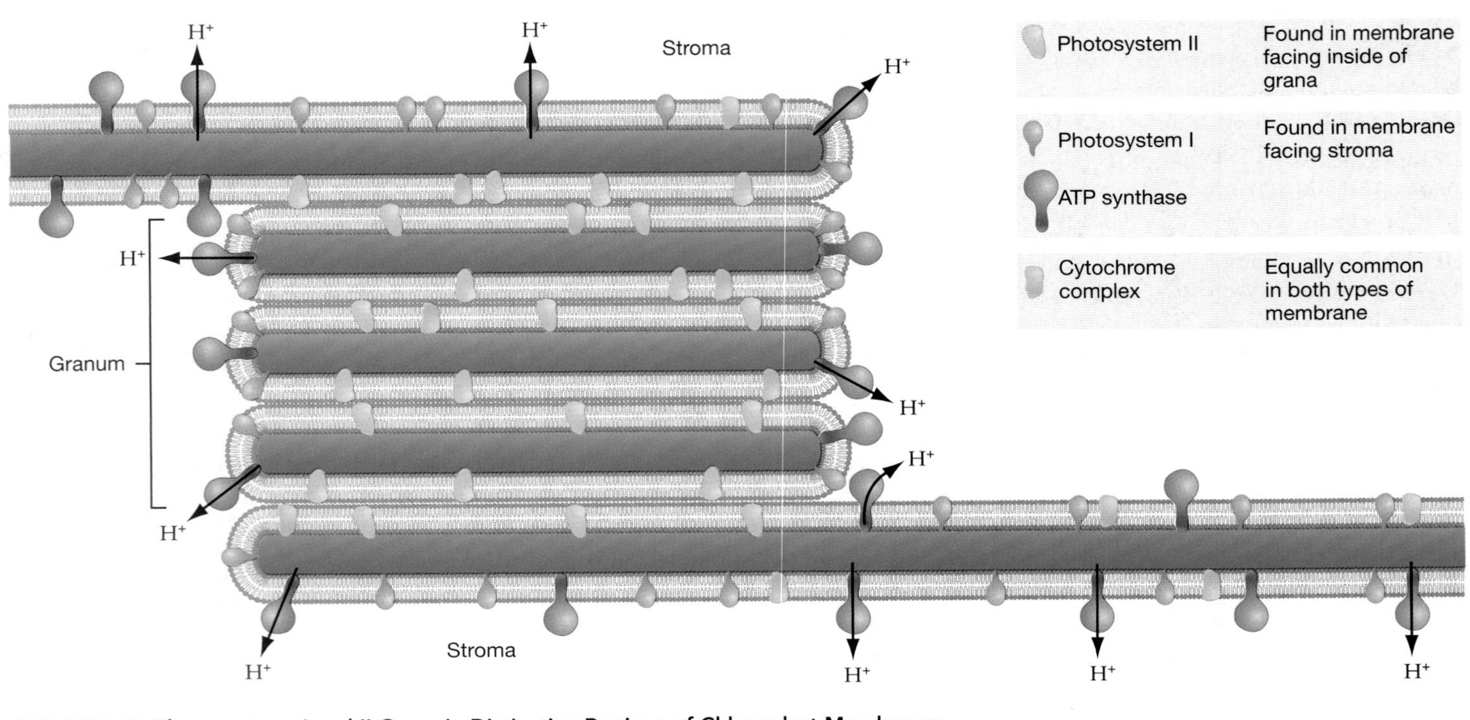

FIGURE 7.11 Photosystems I and II Occur in Distinctive Regions of Chloroplast Membranes
In the inner membranes of a chloroplast, virtually all of the active photosytem II is found in membranes facing the inside of grana. In contrast, virtually all of the photosystem I and ATP synthase is found in membranes that face the stroma. The cytochrome complex, plastoquinone, and plastocyanin are equally common in both types of membranes. EXERCISE On the figure, draw the path of an electron that follows the Z scheme from photosystem II to photosystem I. Then draw the path of an electron that participates in cyclic photophosphorylation.

diffuses along the thylakoid membranes, and donates the electron to photosystem I. A single plastocyanin molecule can shuttle over 1000 electrons per second between photosystems.

The flow of electrons from photosystem II to photosystem I, by means of plastocyanin, replaces electrons that are promoted away from a chlorophyll in the photosystem I reaction center. This chlorophyll molecule is called P700 because it loses electrons most readily when it absorbs wavelengths of 700 nm. The electrons that emerge from P700 are eventually transferred to the protein ferredoxin, which triggers the reduction of $NADP^+$ to NADPH in an enzyme-catalyzed reaction.

There is some circumstantial evidence that a different electron path occurs, at least intermittently. This scheme, called cyclic photophosphorylation, is illustrated in **Figure 7.10b**. The essence of this model is that photosystem I occasionally transfers electrons to photosystem II's electron transport chain, to augment ATP generation through photophosphorylation.

Although the Z-scheme model has held up well under experimental tests, several outstanding questions remain. Biologists are working to solve the three-dimensional structure of each photosystem. They are also trying to get a better understanding of how the two complexes are situated with respect to one another in the internal membranes. As the diagram in **Figure 7.11** shows, photosystems I and II occur in different parts of the chloroplast's internal membranes. Why this physical separation between the photosystems exists is not well understood and is still the focus of debate. In contrast, the fate of the ATP and NADPH produced by photosystems I and II is well documented. Chloroplasts use ATP and NADPH to reduce carbon dioxide to sugar. How?

7.4 The Calvin Cycle

A wide array of photosynthetic organisms use the chemical energy produced by the light-dependent reactions to manufacture sugar from carbon dioxide. The exact sequence of reactions involved was a mystery until just after War II, however, when radioactive isotopes of carbon became available for research purposes. Then between 1945 and 1955, a team led by Melvin Calvin carried out a groundbreaking series of experiments based on feeding green algae radioactively labeled carbon dioxide ($^{14}CO_2$). By isolating and identifying product molecules that contained ^{14}C, the researchers gradually documented which intermediate compounds are produced as carbon dioxide is reduced to sugar.

Figure 7.12a explains their general approach. First, the researchers fed green algae a pulse of labeled CO_2. After waiting a specified amount of time, they killed the cells, ground them up to form a crude extract, separated individual molecules in the extract using paper chromatography, and laid x-ray film over the filter paper. If radioactively labeled molecules were present on the filter paper, the energy they emitted would expose the film and create a black spot. The labeled compounds could then be isolated and identified.

By varying the amount of time between starting the pulse of labeled CO_2 and killing the cells, Calvin and co-workers began to piece together the sequence in which various intermediates were produced. For example, when they killed cells almost immediately after starting the $^{14}CO_2$ pulse, they found that a three-carbon compound called 3-phosphoglycerate (3PG) predominated (**Figure 7.12b**). This result suggested that 3-phosphoglycerate was the initial product of carbon reduction. Stated another way, it appeared that carbon dioxide reacted with some unknown molecule to produce 3PG.

This was an interesting result, in part because 3PG is one of the 10 intermediates in glycolysis. The finding that glycolysis and carbon reduction share intermediates was intriguing because of the relationship between the two pathways. The light-independent reactions manufacture glucose; glycolysis breaks it down. Because the two processes reverse one another, it was logical that at least some intermediates in glycolysis and CO_2 reduction would be the same.

As Calvin's group continued to piece together the sequence of events in carbon dioxide reduction, an important question remained. What compound reacted with CO_2 to produce 3PG? This was the key, initial step in the process. The group searched in vain for a two-carbon compound that might serve as the initial carbon dioxide acceptor and yield 3PG. Then, while Calvin was out running errands in his car one day, it occurred to him that the molecule that reacted with carbon dioxide might actually contain five carbons, not two. The idea was that adding CO_2 to a 5-carbon molecule would produce a 6-carbon compound, which could then split in half to form two molecules of 3PG.

Experiments to test this hypothesis confirmed that the 5-carbon compound ribulose bisphosphate (RuBP) acts as the initial reactant. Eventually the entire sequence of events in CO_2 reduction was worked out and became known as the Calvin cycle. As **Figure 7.13** (page 148) shows, the cycle begins when CO_2 reacts with RuBP. This step "fixes" carbon dioxide by attaching it to a more complex molecule. The resulting 3-phosphoglycerate molecules are phosphorylated by ATP and then reduced by electrons from NADPH. The sugar that results, glyceraldehyde 3-phosphate (G3P), is the product of the Calvin cycle. Some G3P is drawn off to manufacture glucose. The rest of the G3P keeps the cycle going by serving as the substrate for reactions

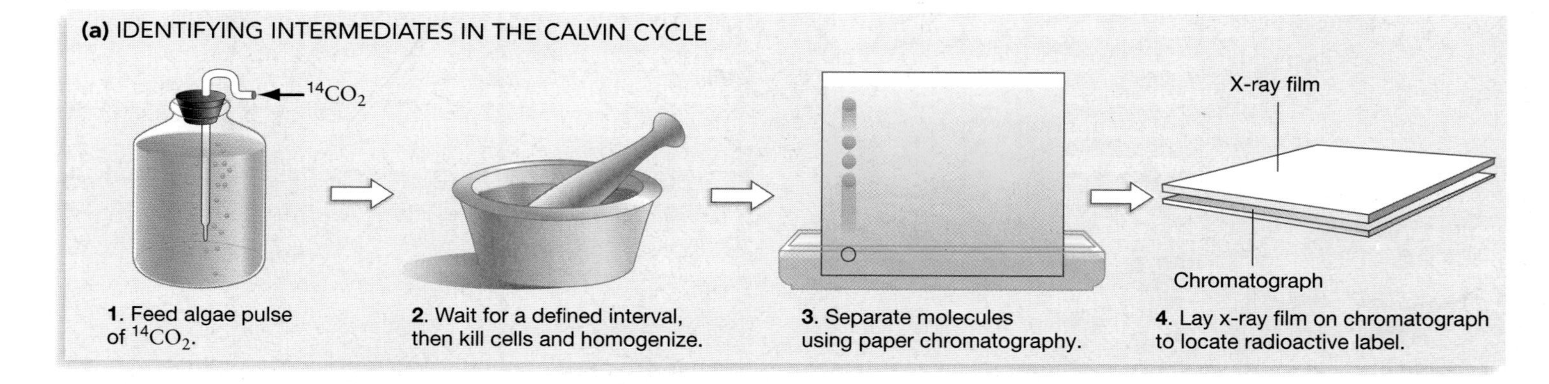

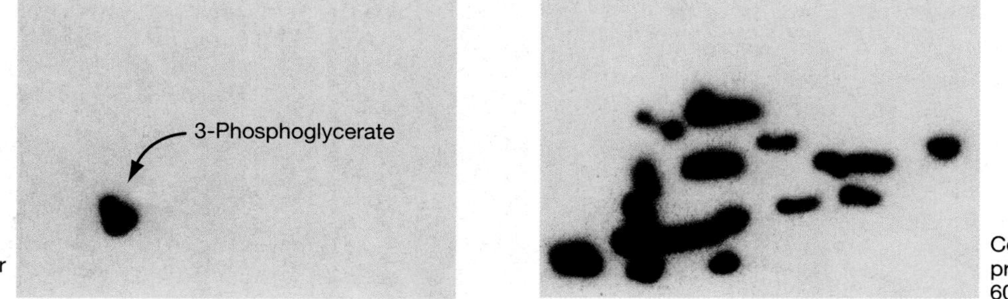

FIGURE 7.12 What Steps Are Involved in the Reduction of CO_2?
(a) By giving radioactive CO_2 to photosynthesizing cells, researchers were able to identify the compounds that are produced as carbon dioxide is reduced to carbohydrate. **(b)** Because 3-phosphoglycerate is the first molecule labeled with $^{14}CO_2$, researchers inferred that it was the initial product in the CO_2-reduction pathway.

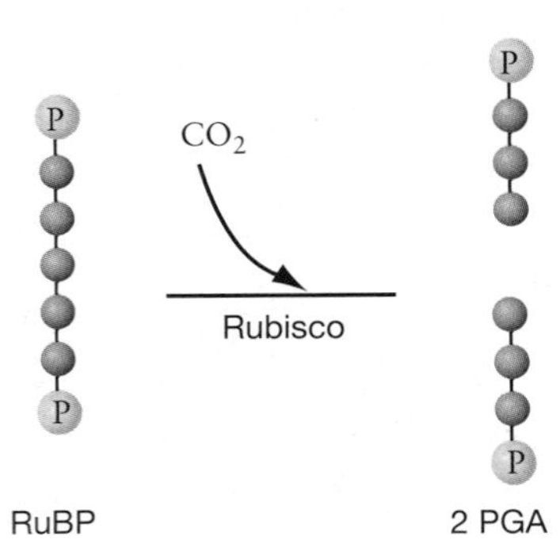

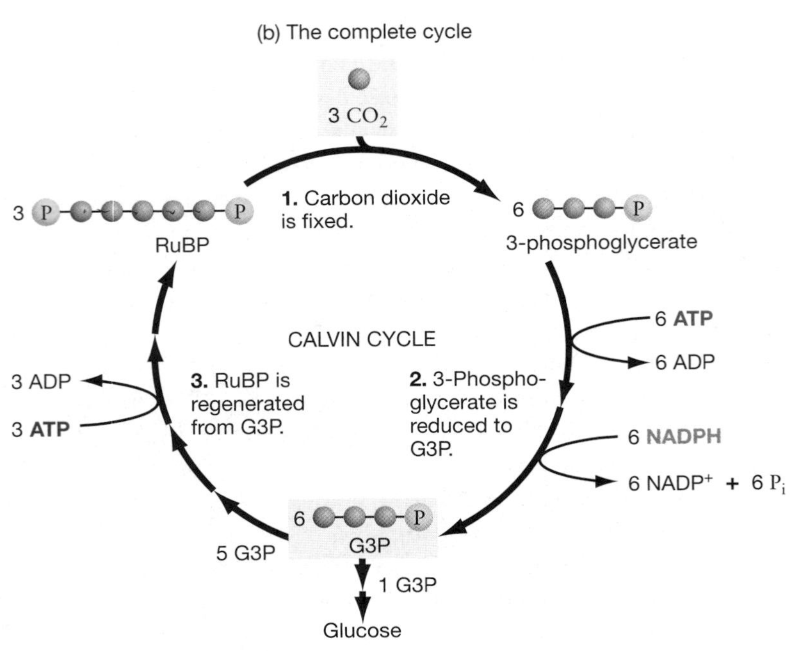

FIGURE 7.13 The Calvin Cycle
The Calvin cycle has three distinct phases. The first is a carbon-fixation step, in which CO_2 combines with RuBP to form two molecules of 3-phosphoglycerate. In the second phase, 3-phosphoglycerate is phosphorylated and reduced, in a series of steps, to form the sugar G3P. Some G3P is then drawn off to form sucrose and other products. In the final phase, the remaining G3P enters reaction pathways that regenerate RuBP, which continues the cycle.

that result in the regeneration of RuBP. The entire sequence of reactions takes place in the stroma of chloroplasts.

The discovery of the Calvin cycle clarified how the ATP and NADPH produced by light-dependent reactions allow cells to reduce CO_2 to carbohydrate (CH_2O). Once the reaction sequence of the Calvin cycle was confirmed, attention focused on the initial step—the reaction between RuBP and CO_2. This step is one of only two reactions that are unique to the Calvin cycle. Most reactions involved in reducing carbon dioxide also occur during glycolysis or other metabolic pathways.

The reaction between CO_2 and RuBP starts the transformation of carbon dioxide gas from the atmosphere to sugars that are used to build leaves, roots, and tree trunks. Millions of non-photosynthetic organisms—including fish, insects, fungi, and mammals—indirectly depend on this reaction to provide the sugars they need for cellular respiration. Ecologically, fixing carbon dioxide to RuBP may be the most important chemical reaction on Earth. How does it occur?

The Discovery of Rubisco

Arthur Weissbach and colleagues set out to find the enzyme that fixes CO_2 by grinding up spinach leaves and analyzing the resulting cell extracts. Their strategy was to purify a protein from the crude extracts and then determine whether it could catalyze the incorporation of $^{14}CO_2$ into 3PG. Eventually they were able to isolate an enzyme that catalyzes the reaction and that appeared to be extremely common in leaf tissue. Their data suggested that the enzyme constituted at least 10 percent of the total protein found in spinach leaves.

The CO_2-fixing enzyme was eventually purified and analyzed in detail. Its full name is ribulose-1,5-bisphosphate carboxylase/oxygenase, but it is commonly referred to as **rubisco.** Rubisco is found in all photosynthetic organisms that fix carbon by using the Calvin cycle, and it is widely acknowledged to be the most common enzyme on Earth. The three-dimensional structure of the protein has now been solved and is shown in **Figure 7.14.** The enzyme is square-shaped and has a total of eight active sites where CO_2 is fixed.

Despite the ecological importance and abundance of rubisco, however, biologists quickly realized that it is an inefficient enzyme. First, it is very slow. Each active site of rubisco can catalyze only about three reactions per second. Plants appear to make up for this lack of speed by synthesizing huge amounts of the enzyme. More troublesome is the observation that in addition to catalyzing the reaction between CO_2 and RuBP that fixes carbon dioxide, rubisco also catalyzes a reaction between oxygen and RuBP. The molecule that results from the reaction with oxygen is toxic. As it is broken down, ATP is consumed and CO_2 is produced. In effect, then, the reaction between oxygen and RuBP consumes energy and releases car-

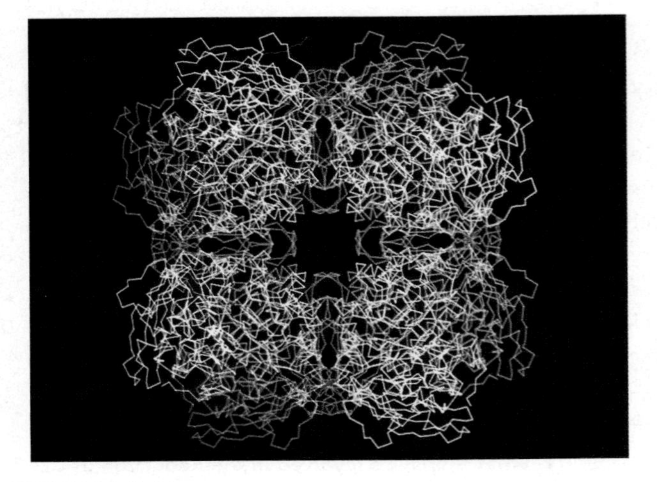

FIGURE 7.14 Rubisco
As this three-dimensional model shows, rubisco is a square-shaped enzyme with eight active sites.

BOX 7.2 Why Was the Oxygen Revolution Important?

A wide variety of bacteria and archaea perform what is known as anoxygenic (without oxygen formation) photosynthesis. These cells have photosystems that can use the energy in light to promote an electron to a high enough energy state that the electron can be used to reduce $NADP^+$ to NADPH. This observation raises an important question: Where do these cells get electrons to replace the ones used to manufacture NADPH?

As section 7.1 pointed out, most cells that perform anoxygenic photosynthesis get the electrons they need from highly reduced inorganic molecules like H_2 or H_2S. Because the electrons in H_2 or H_2S are in a fairly high-energy state, it is fairly easy for enzymes to remove them and pass them along to a photosystem. Unfortunately, cells that depend on these compounds as electron donors are restricted geographically. They can live only in specialized habitats like hot springs, where these compounds are abundant.

Based on these observations, it should be clear why the evolution of oxygenic photosynthesis was a revolutionary event. Cells with the ability to obtain electrons from water could live anywhere that water and light were available. Instead of being restricted to specialized habitats, photosynthetic organisms could spread throughout the world.

As Chapter 25 details, this explosion of life had a remarkable side effect. Because oxygen gas is produced as a by-product of oxygenic photosynthesis, oxygen levels began to build up in the oceans and atmosphere. (Prior to this time, oxygen gas had been virtually nonexistent on Earth.) Because oxygen is toxic to anaerobic organisms, the evolution of oxygenic photosynthesis probably resulted in the extinction of enormous numbers of species.

Once oxygen gas had built up in concentration, however, some bacterial cells evolved the ability to use it as an electron acceptor during cellular respiration. You might recall from Chapter 6 that because O_2 is highly oxidized, it creates a huge potential energy drop for the electron transport chains used in cellular respiration. As a result, organisms that use O_2 as an electron acceptor can produce much more ATP than can organisms that use other electron acceptors. Because aerobic organisms make ATP and grow so efficiently, they began to dominate the planet. The rest, as they say, is history.

bon dioxide. Because the pathway resembles respiration in the sense that it consumes oxygen and makes carbon dioxide, it is called **photorespiration.** Photorespiration effectively reverses photosynthesis.

As long as carbon dioxide concentrations in leaves are high, the CO_2-fixation reaction is favored and photorespiration is relatively rare. How do plants deliver carbon dioxide to rubisco?

Carbon Dioxide Enters Leaves via Stomata

Figure 7.15a shows a close-up view of a leaf surface. You can see that distinctive structures consisting of two bean-shaped cells and an opening are scattered among the outer cells of the leaf. The paired cells are called **guard cells,** the opening is called a **pore,** and the entire structure is called a **stoma** (plural **stomata**). If carbon dioxide concentrations inside the leaf are low as photosynthesis gets under way, chemical signals from cells in the leaf interior cause guard cells to change shape and create a pore. As **Figure 7.15b** shows, an open stoma allows carbon dioxide from the atmosphere to diffuse into the extracellular fluid surrounding photosynthesizing cells. From there the CO_2 diffuses along a concentration gradient into the

(a) Leaf surfaces contain stomata.

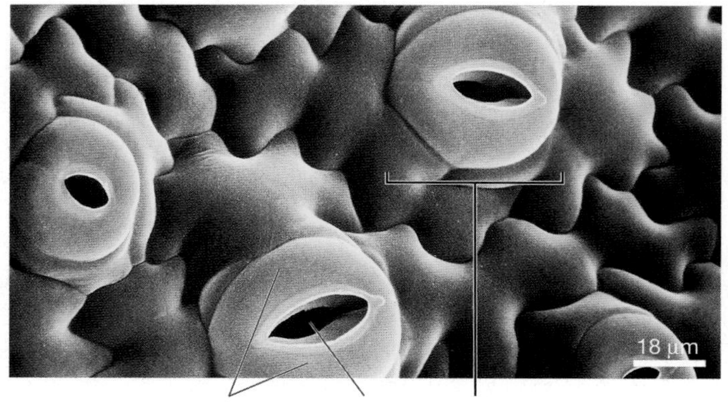

(b) Carbon dioxide diffuses into leaves through stomata.

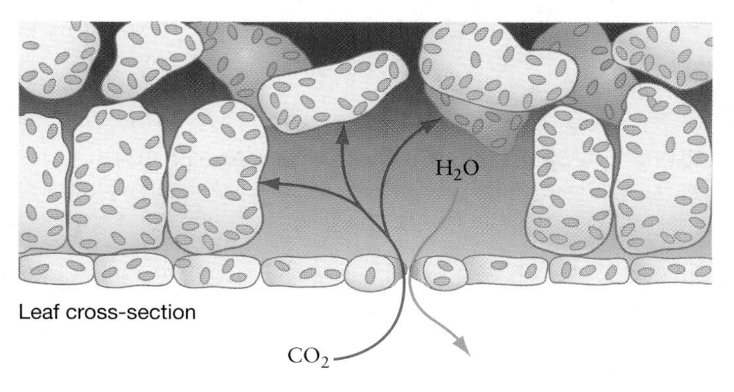

FIGURE 7.15 How Do Leaf Cells Obtain Carbon Dioxide?
(a) Stomata consist of two guard cells and a pore that opens to the leaf interior. **(b)** When a stoma is open, carbon dioxide diffuses into leaves along a concentration gradient.

chloroplasts of the cells. A strong concentration gradient favoring entry of CO_2 is maintained by the light-independent reactions, which constantly use up the CO_2 in chloroplasts.

Opening stomata to acquire CO_2 has a downside for plants, however. Because the atmosphere is usually much drier than the leaf interior, water tends to leave the leaf as CO_2 enters. Chapter 32 explains how the water lost in photosynthesizing leaves is replaced by water absorbed by the roots. But if the atmosphere is extremely hot and dry, and if soil water is scarce, then leaf cells lose so much water that they must either close their stomata and halt photosynthesis or die of dehydration. When conditions are hot and dry and water loss is severe, photosynthesis and growth stop. To grow successfully, then, plants constantly monitor CO_2 concentrations and water conditions inside the leaf, and they open and close stomata accordingly.

All land plants can regulate the opening and closing of their stomata. In addition, Chapter 28 explains in detail how novel reactions that occur before the Calvin cycle begins have allowed certain lineages of plants to adapt to particularly hot and dry habitats. Two types of specialized variations on the standard carbon fixation pathway have been discovered to date. These pathways are called C_4 photosynthesis and crassulacean acid metabolism (CAM). Both of them begin with enzymes that catalyze reactions between CO_2 and small organic acids. Because the molecules produced in these reactions bind carbon dioxide, C_4 and CAM plants stockpile large supplies of CO_2 that are used to fuel the Calvin cycle.

How do the C_4 and CAM pathways differ? As **Figure 7.16a** shows, the reactions in C_4 plants between CO_2 and small organic acids occur in cells adjacent to the cells where rubisco is active. If CO_2 levels become low in the cells containing rubisco, CO_2 can diffuse in from the adjacent cells where it was fixed to an organic acid. In this way, C_4 plants limit photorespiration by maintaining high levels of CO_2. As **Figure 7.16b** shows, the reactions between CO_2 and small organic acids in CAM plants occur in the same cells where rubisco is active, but these reactions take place at night. By fixing and sequestering CO_2 in this way, CAM plants can open their stomata during cool, moist, nighttime conditions and then close their stomata during the day to minimize water loss. The stockpiled CO_2 fuels the Calvin cycle when the light reactions are producing ATP and NADPH.

Obtaining and reducing CO_2 is fundamental to photosynthesis. In a larger sense, photosynthesis is fundamental to the millions of species that depend on plants, algae, and cyanobacteria for food. Now the question is, what do photosynthetic organisms do with the sugar they synthesize? What happens to the G3P that is drawn off from the Calvin cycle to fuel growth and reproduction?

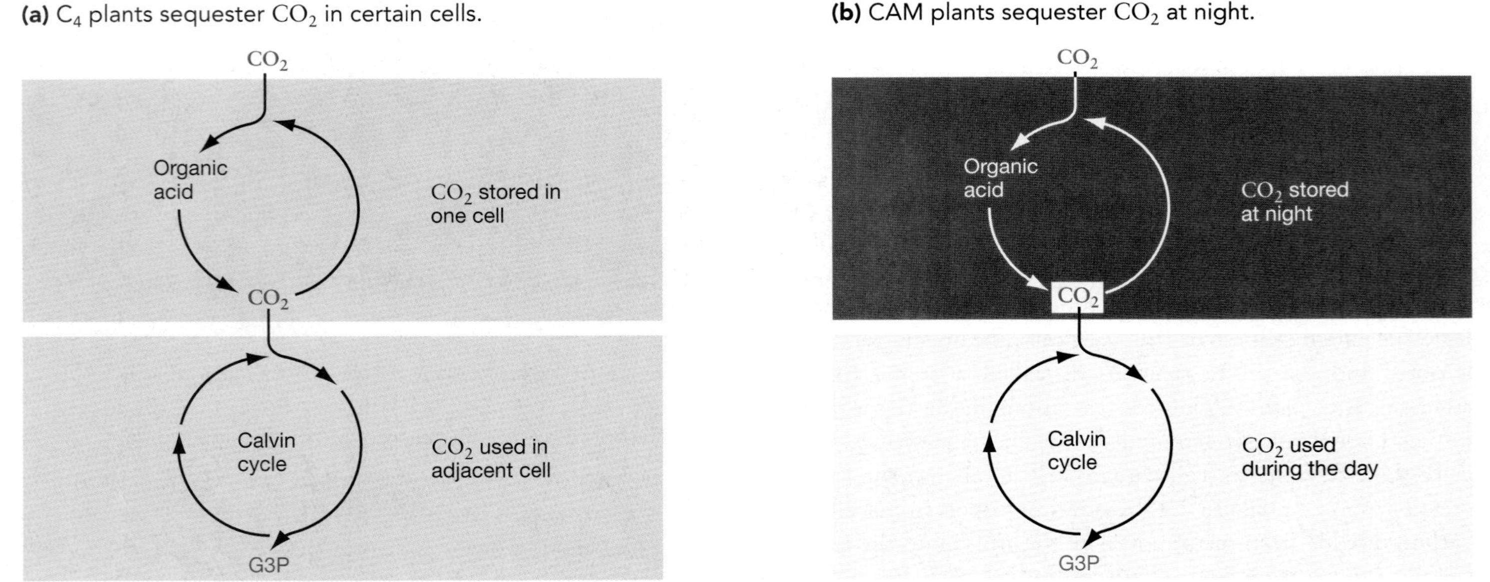

FIGURE 7.16 Specialized Pathways for Sequestering CO_2
(a) C_4 plants limit photorespiration by keeping CO_2 levels high. The CO_2 is attached to organic acids in some cells and then released to other cells where the Calvin cycle enzymes are located. In this way, CO_2 is concentrated in photosynthesizing cells and photorespiration is limited. **(b)** CAM plants open their stomata at night and fix CO_2 to organic acids. During hot, dry days, their stomata close to limit water loss. Photosynthesis can continue, however, because CO_2 is released from its storage place in organic acids. Because CO_2 concentrations in photosynthesizing cells are maintained at high levels, photorespiration is limited.

What Happens to the Sugar That Is Produced by Photosynthesis?

The G3P molecules that exit the Calvin cycle may enter one of several different reaction pathways. The most important of these pathways end in the production of sucrose and starch. Sucrose is a disaccharide (two-sugar) that consists of a glucose molecule bonded to a fructose molecule (**Figure 7.17a**), and starch is a polymer of glucose (**Figure 7.17b**).

When photosynthesis takes place slowly, almost all of the G3P produced is used to make sucrose that fuels respiration and growth in the photosynthetic cells and elsewhere. But if photosynthesis is proceeding rapidly and sucrose is abundant, then G3P is used to synthesize starch in the chloroplasts. This starch acts as a temporary sugar storage product. At night, the starch is broken down and used to manufacture sucrose molecules. In this way, chloroplasts provide sugars for the rest of the plant both by day and by night.

Sucrose that leaves photosynthetic tissues is either transported to young cells to fuel their growth or to storage areas in roots. (Chapter 32 explores how this transport takes place.) If the starch is eaten by an herbivore (plant-eater), however, it is broken into constituent sugars and used to fuel their growth and reproduction. In this way, virtually all cell growth and reproduction traces back to the chemical energy originally captured by photosynthesis. How does cell division and growth occur? That is the question taken up in Chapter 8.

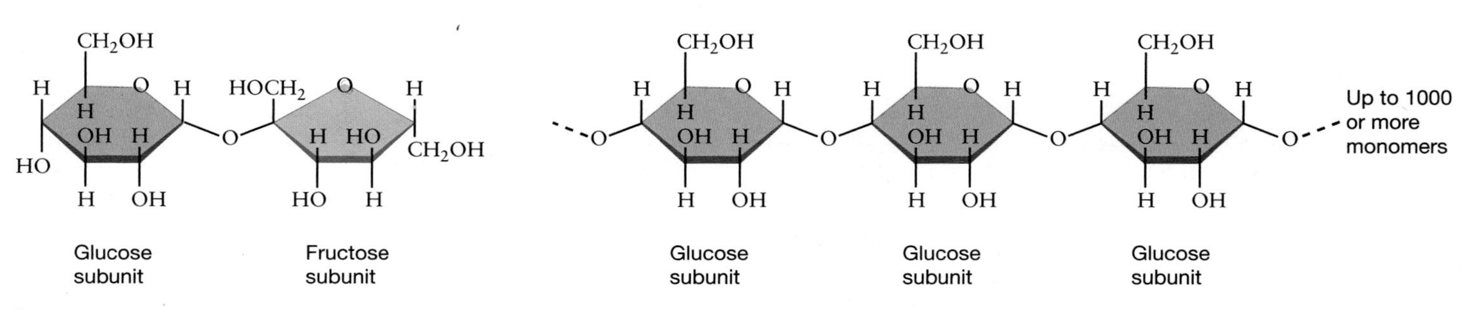

FIGURE 7.17 The Products of Photosynthesis
(a) In plants, sugars are usually transported in the form of sucrose. **(b)** Plants store sugars in the form of starch, which consists of long chains of glucose molecules.

Essay Are Rising CO_2 Levels in the Atmosphere Affecting Plant Productivity?

The concentration of carbon dioxide in the atmosphere has increased dramatically over the past 100 years. In the late 1800s the atmospheric carbon dioxide concentration is thought to have been about 280 μL/L. Today, CO_2 is present in the atmosphere at 360 μL/L. Most of the increase is due to CO_2 that was released when people burned natural gas, gasoline, coal, and other fossil fuels; or when forests were burned to convert them to agricultural use. If present trends in fossil fuel use and deforestation continue, atmospheric CO_2 levels are expected to increase to 560 μL/L by the year 2050. If this prediction is correct, then CO_2 levels would have doubled in just 150 years.

Can we counteract global warming?

Chapter 51 explains in detail why rising carbon dioxide concentrations have caused the dramatic global warming that is now under way. Here, though, let's consider a different question. How are increases in CO_2 affecting plants?

According to the overall reaction for photosynthesis, increases in a reactant such as CO_2 should lead to increased rates of photosynthesis. Stated another way, plant productivity should rise. Biologists have confirmed this prediction experimentally by increasing CO_2 levels in controlled environmental chambers and in natural habitats. For example, Evan DeLucia and colleagues set up a series of experimental plots in a 13-year-old pine forest in North Carolina, in the southeast United States. Each plot was ringed with towers that emitted either enough CO_2 to bring average levels inside the plot to 560 μL/L, or normal air as a control treatment. As predicted, the growth rate of pine trees in the CO_2-augmented plots increased 25 percent relative to controls.

Increased growth of trees and shrubs removes carbon dioxide from the atmosphere and sequesters the carbon in wood. As a result, increased plant growth should act as a feedback mechanism that helps counteract rising CO_2 levels in the atmosphere. Based on this logic, biologists have suggested that tree

(Continued on page 152)

(Essay continued)

planting and forest restoration programs could be an important part of a coordinated, worldwide effort to counteract global warming.

Biologists caution that plant growth rates will eventually stop responding to increased CO_2 availability, however. Because many plants respond strongly to augmented CO_2, it is clear that carbon dioxide can be an important limiting nutrient in plant growth. But at some point, plants should stop responding to increased carbon dioxide levels because some other nutrient—perhaps nitrogen or phosphorus—will become limiting.

The other major prediction regarding plant responses to increased CO_2 concentrations focuses on desert-dwelling species. To understand this prediction, recall that plants lose water when stomata are open to admit carbon dioxide for photosynthesis. If carbon dioxide levels rise, then plants should have to open their stomata less to obtain the CO_2 they need, thus losing less water to the atmosphere. As a result, they should be able to grow faster. Consistent with this prediction, Stanley Smith and co-workers found that when they artificially increased CO_2 levels in experimental plots in the Mojave Desert of southwestern North America, plant growth increased. The effect occurred only during a wet year, however—no increased growth was observed in a drought year. What will be the long-term effect of rising CO_2 on the desert ecosystem? The answer is not known. Research by Smith's team and other groups continues.

Chapter Review

Summary

Photosynthesis is the production of chemical energy, stored in the form of sugar, from light energy. Early work on photosynthesis showed that the process has two distinct components. In the light-dependent reactions, light energy is transduced to chemical energy in the form of ATP and NADPH. In the light-independent reactions, called the Calvin cycle, carbon dioxide (CO_2) is reduced to carbohydrate (CH_2O). In plants and algae, both process take place in chloroplasts. The light-dependent reactions occur in internal membranes of the chloroplast that are organized into structures called thylakoids and grana. The light-independent reactions take place in a fluid portion of the organelle called the stroma.

The energy transduction step of photosynthesis begins when chlorophyll absorbs a photon in the blue or red part of the visible spectrum. When absorption occurs, the energy in the photon is transferred to an electron in the chlorophyll molecule. The electron is raised to an excited state equivalent to the energy in the photon. If the electron is allowed to fall back to the normal or ground state, energy is given off as light (fluorescence) and heat. But in photosynthetic organisms, the energy in the excited electron is transferred to a chlorophyll molecule that acts as a reaction center. There the high-energy electron is transferred to an electron acceptor, which becomes reduced. In this way, light energy is transduced to chemical energy.

Researchers interested in the light-dependent reactions of photosynthesis realized that plants and algae have two distinct types of reaction centers, which are part of larger complexes called photosystem I and photosystem II. Each photosystem consists of 200–300 chlorophyll and carotenoid molecules, a reaction center, and an electron acceptor that completes energy transduction.

In photosystem II, high-energy electrons are accepted by pheophytin and then passed along an electron transport chain that includes a series of quinones. As electrons move through this chain, they are gradually stepped down in potential energy. The energy released by these reduction-oxidation (redox) reactions is used to pump protons across the thylakoid membrane. The resulting proton gradient drives the synthesis of ATP by ATP synthase. This method of producing ATP is called photophosphorylation.

In photosystem I, high-energy electrons are accepted by iron- and sulfur-containing proteins and passed to ferredoxin. In an enzyme-catalyzed reaction, the reduced form of ferredoxin passes the electron to $NADP^+$ to form NADPH. NADPH is a highly reduced compound required for redox reactions that result in the synthesis of sugars and other cell materials.

The Z scheme describes how photosystems I and II are thought to interact by tracing the route of electrons through the two photosystems. At the end of the electron transport chain in photosystem II, electrons are carried to photosystem I by the protein plastocyanin. There the electrons are promoted to a very high-energy state in response to the absorption of a photon, and they are subsequently used to reduce $NADP^+$. Some evidence suggests that electrons from photosystem I may occasionally be passed to the electron transport chain in photosystem II instead of being used to reduce $NADP^+$, resulting in a cyclic flow of electrons between the two photosystems.

The light-independent reactions of photosynthesis are called the Calvin cycle. The process of reducing carbon dioxide to sugar begins when CO_2 is attached to a 5-carbon compound called RuBP. This reaction is catalyzed by the enzyme rubisco.

The six-carbon compound that results immediately splits in half to form two molecules of 3-phosphoglycerate (3PG). Subsequently, 3PG is phosphorylated by ATP and reduced by NADPH to form a sugar called glyceraldehyde 3-phosphate (G3P). Some G3P is used to synthesize sucrose; the rest participates in reactions that regenerate RuBP so the cycle can continue.

The glucose generated by photosynthesis fuels cellular respiration and supplies a substrate for the synthesis of complex carbohydrates, amino acids, fatty acids, and other cell components. Because photosynthetic organisms are the primary food source for a diverse array of heterotrophs, they are the original source for the energy that sustains most life on Earth.

Questions

Content Review

1. What is the stroma of a chloroplast?
 a. the inner membrane
 b. the pieces of membrane that connect grana
 c. the interior of a thylakoid
 d. the fluid inside the chloroplast but outside of thylakoids

2. Why is chlorophyll green?
 a. because it absorbs all wavelengths in the visible spectrum
 b. because it absorbs wavelengths in the green part of the visible spectrum
 c. because it absorbs wavelengths in the blue and red parts of the visible spectrum
 d. because it absorbs wavelengths only in the blue part of the visible spectrum

3. What does it mean to say that CO_2 becomes "fixed?"
 a. It becomes bonded to an organic compound.
 b. It is released during cellular respiration.
 c. It acts as an electron acceptor.
 d. It acts as an electron donor.

4. What do the light reactions of photosynthesis produce?
 a. G3P
 b. RuBP
 c. ATP and NADPH
 d. plastoquinone

5. What does it mean to say that photosystem II "splits" water?
 a. Water is broken into monomers.
 b. A condensation reaction occurs.
 c. Water is oxidized to yield protons, electrons, and oxygen.
 d. Water is reduced to yield hydrogen gas.

6. Why is pheophytin an important component of photosystem II?
 a. It carries electrons to photosystem I.
 b. It absorbs photons with a wavelength of 680 nm.
 c. It protects chlorophyll from destruction by free radicals.
 d. It transduces light energy by acting as the initial electron acceptor.

Conceptual Review

1. Compare and contrast mitochondria and chloroplasts. In what way are their structures similar and different? What molecules or systems function in both types of organelles? Which enzymes or processes are unique to each organelle?

2. Explain how the energy transduction step of photosynthesis occurs. How is light energy converted to chemical energy in the form of ATP and NADPH?

3. Explain how the carbon reduction step of photosynthesis occurs. How is carbon dioxide "fixed?" Why are ATP and NADPH required to produce sugar?

4. Why does a pure solution of chlorophyll fluoresce when it is exposed to blue light? Why doesn't the chlorophyll in chloroplasts fluoresce very much when it is exposed to blue light?

5. Make a sketch of the Z scheme. Explain how photosystem I and photosystem II interact by tracing the path of an electron through the Z scheme. What molecule connects the two photosystems?

6. How does plastoquinone shuttle electrons and protons across the thylakoid membrane? Why is this shuttling action important?

7. In what sense does photorespiration "undo" photosynthesis?

8. What evidence suggested that 3-phosphoglycerate is the first product of the Calvin cycle?

9. Why do plants need both chloroplasts and mitochondria?

Applying Ideas

1. The Calvin cycle and rubisco are found in lineages of bacteria and archaea that evolved long before the origin of oxygenic photosynthesis. Based on this observation, biologists infer that rubisco evolved in an environment that contained little, if any, oxygen. Some biologists propose that this fact explains why photorespiration occurs today. Do you agree with the hypothesis that photorespiration is an evolutionary "holdover?" Why or why not?

2. In addition to their protective function, carotenoids also absorb certain wavelengths of light and pass the energy on to the reaction centers of photosystem I and II. Based on their function, predict exactly where carotenoids are located in the chloroplast. Explain your rationale. How would you test your hypothesis?

3. Consider plants that occupy the top, middle, or ground layer of a forest, and algae that live near the surface of the ocean or in deeper water. Would you expect the same photosynthetic pigments to be found in species that live in these different habitats? Why or why not? How would you test your hypothesis?

CD-ROM and Web Connection

CD Activity 7.1: Photosynthesis *(tutorial)*
(Estimated time for completion = 15 min)
Where does the body of a plant come from—soil, air, or water?

At your **Companion Website** (http://www.prenhall.com/freeman/biology), you will find self-grading exams and links to the following research tools, online resources, and activities:

Photosynthesis Center
Explore a variety of links to educational and research sites about photosynthesis.

Dual Nature of Light
Read an essay describing the wave-particle nature of light, which is essential for energy harvesting by photosynthetic pigments.

Capturing Light Energy
This site describes the physical and chemical reactions that occur in the different energy-harvesting pigments of photosynthesis.

Additional Reading

Szalai, V. A., and G. W. Brudvig. 1998. How plants produce dioxygen. *American Scientist* 86: 542–551. Explores recent results on electron transfer during photosynthesis.

Cell Division

8

Chapter 1 introduced the cell theory, which maintains that all organisms are made of cells and that all cells come from preexisting cells. These hypotheses grew out of work by the plant and animal anatomists Matthew Schleiden and Theodor Schwann, who concluded in the late 1830s that all living organisms are composed of cells, and by Rudolph Virchow, who proposed that all cells arise from preexisting cells. Schleiden and Schwann's hypothesis was confirmed by studies on the structure of organisms from throughout the tree of life; Virchow's hypothesis was confirmed by experiments that discredited the alternative hypothesis of spontaneous generation.

Although the cell theory was widely accepted among biologists by the 1860s, a great deal of confusion remained about how cells reproduced. Most proponents of the cell theory—including Schwann—believed that new cells arose within preexisting cells by a process that resembled crystallization. But Virchow proposed that all new cells arise through division of preexisting cells.

In the late 1800s, careful microscopic observations confirmed Virchow's proposal. For example, embryologists discovered that multicellular organisms are derived from single-cell embryos through a series of cell divisions. As better microscopes became available, biologists were able to describe the details of the process. In particular, researchers focused their attention on the nuclei of dividing cells and the fate of the chromosomes, which are the carriers of hereditary information—the instructions for building and operating the cell. These studies revealed that there are two fundamentally different kinds of cell division—meiosis and mitosis.

During the cell divisions that lead to the production of sperm and eggs, the amount of hereditary information found in the nucleus is reduced by one–half. As a result, the daughter

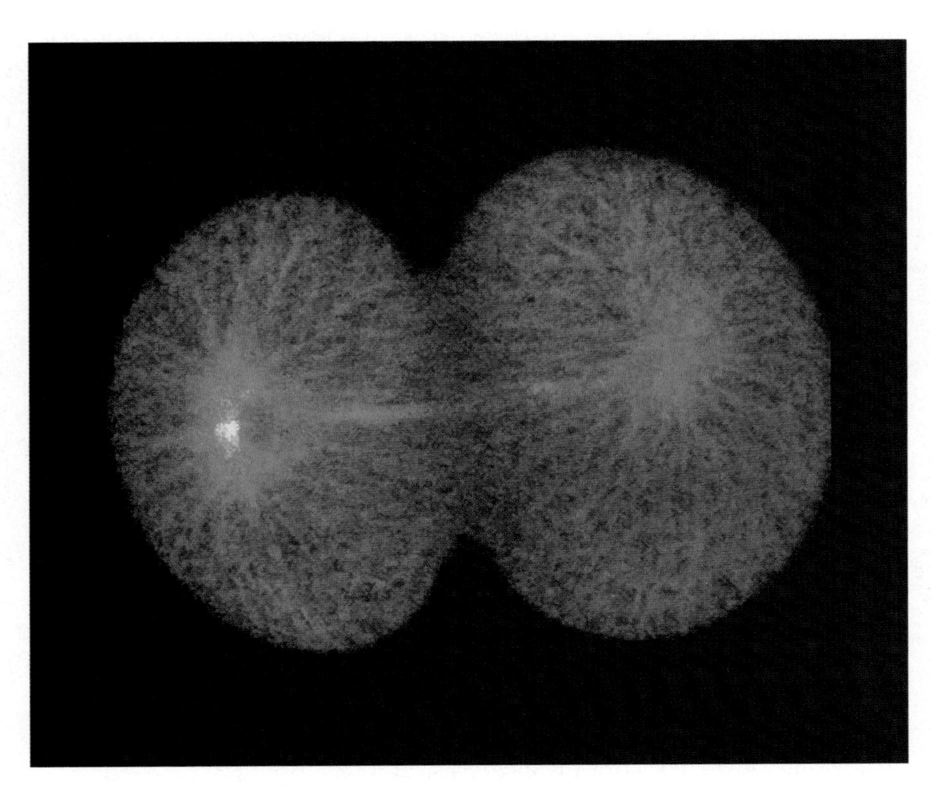

This sea urchin egg is undergoing its first cell division after being fertilized. Many thousands more cell divisions will occur as the individual develops into an adult.

cells that become sperm or eggs are genetically unlike the parent cell. This type of cell division is called meiosis and is involved only in the production of reproductive cells. Meiosis is the basis of sexual reproduction and is the subject of Chapter 9. During the divisions that produce all other types of cells, however, the amount of hereditary information remains constant. This type of division is the basis of growth and asexual reproduction and is called mitosis. **Mitosis** produces daughter cells that are genetically identical to the parent cell.

In multicellular organisms, mitosis is responsible for growth and the replacement of lost or damaged cells. In humans, for example, the billions of genetically identical cells that make up the adult body trace their ancestry back through a series of mitotic divisions to the fertilized egg. When you suffer a scrape or other insult that damages your skin, mitosis is responsible for generating the cells that repair the skin and heal the wound. In some species, mitosis is also responsible for reproduction. When yeast cells greatly increase in number in a pile of bread dough or in a vat of beer, they are reproducing by mitosis.

This chapter's goals are to explore how mitosis occurs and analyze how the process is regulated. The first section introduces the relationship between mitosis and other major events in a cell's life cycle. The next two sections provide an in-depth look at each event in mitosis and explore how mitosis and other events in the cell cycle are regulated. The chapter concludes by examining why uncontrolled cell division and cancer can result when regulatory systems break down.

8.1 Mitosis and the Cell Cycle

In the course of studying cell division, nineteenth-century biologists found that certain chemical dyes made thread-like structures visible in the nuclei of dividing cells. In 1882, Walther Flemming followed up on this discovery by publishing observations on how the thread-like structures change as the cells in a salamander embryo divide. As **Figure 8.1** shows, one of Flemming's most important observations was that the threads are paired when they first appear, just before the act of cell division. As the nucleus divides, however, each pair of threads splits longitudinally to produce single, unpaired threads. Flemming introduced the term *mitosis*, from the Greek *mitos* (thread), to describe this division process.

In 1883, Edouard van Beneden reported similar observations with the roundworm *Ascaris*. In addition to noting that each doublet split longitudinally, he reported that the total number of threads in a cell remained constant during subsequent divisions. Thus, all of the cells in a roundworm's body have the same number and types of threads.

In 1888, W. Waldeyer introduced the term **chromosome** (colored-body) to refer to the thread-like structures observed in the nuclei of dividing cells (**Figure 8.2**). As Chapters 10 and 11 will show, chromosomes are made up in part of deoxyribonucleic acid (DNA). DNA encodes the cell's hereditary information, or genetic material. By observing how chromosomes moved during mitosis, biologists realized that the purpose of mitosis was to distribute the parent cell's genetic material to daughter cells during cell division. Before delving into a detailed look at how mitosis occurs, though, let's first examine how it fits into the other events in the life of a cell.

The Cell Cycle

As early workers studied the fate of chromosomes during cell division, they realized that even rapidly growing plant and animal cells do not divide continuously. Instead, growing cells cycle between a dividing phase called the mitotic or **M phase**

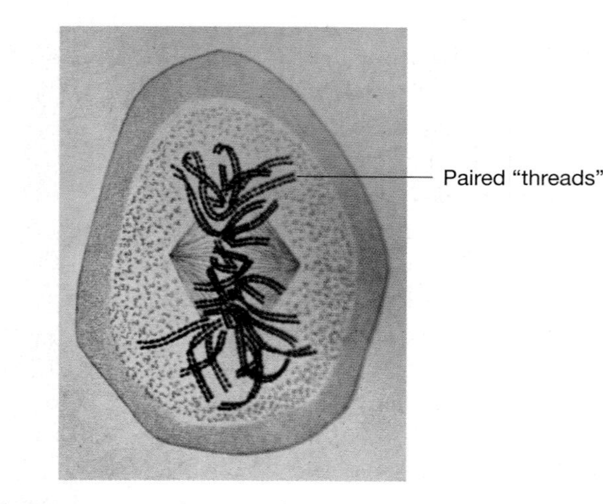

FIGURE 8.1 An Early View of Mitosis
This drawing of mitosis in the salamander larva was made in 1882. The black threads are chromosomes.

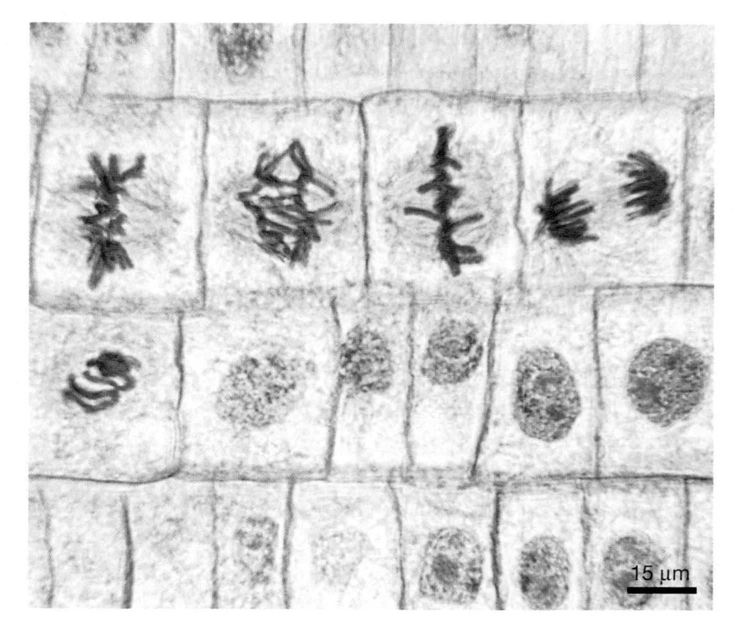

FIGURE 8.2 Mitosis in Progress
Chromosomes can be stained with dyes and observed in the light microscope. These cells are located in the tip of an onion root, where cell division is particularly rapid. Many of the cells shown here are in various stages of mitosis.

and a nondividing phase called **interphase**. Cells actually spend most of their time in interphase. No dramatic changes are observed in the nucleus during interphase, and individual chromosomes are not visible in the light microscope.

To describe the regular alternation between M phase and interphase, biologists began referring to the cell cycle. The **cell cycle** is the sequence of events that a dividing cell goes through from the time it is created, by division of a parent cell, to the time it undergoes division itself. In the cycle, the two most important events are the duplication of the hereditary material in chromosomes and the partitioning of chromosomes to the two daughter cells, during mitosis. Let's take a closer look.

When Does Chromosome Replication Occur?

Flemming, van Beneden, and other early workers could see chromosomes move to daughter cells in the light microscope during mitosis. Because each daughter cell ended up with the same number of chromosomes as the parent cell, it was logical to infer that the chromosomes were duplicated in the parent cell at some point in the cell cycle. Did the doublet threads that Flemming observed represent the replicated chromosomes? If so, when did this replication step occur? Because no dramatic changes to chromosomes are visible during interphase, some biologists hypothesized that replication must occur early in M phase. But the observation that chromosomes appear as doublets early in M phase suggested that replication occurred before the start of M phase, during interphase.

Which hypothesis is correct? This question was not answered until the 1950s, when two technical innovations provided the experimental tools needed to answer the question. The first advance was the availability of radioactively labeled nucleotides—the building blocks of DNA. If radioactive nucleotides are present as DNA is being synthesized, they will be incorporated into the new DNA and mark it. The second advance was the ability to grow eukaryotic cells outside of the source organism, in culture. Cultured cells are powerful experimental tools because they can be manipulated much more easily than cells in an intact organism (see **Box 8.1**).

In 1953, Alma Howard and Stephen Pelc used radioactive nucleotides and cultured cells to determine when chromosome replication takes place during the cell cycle. As **Figure 8.3a** shows, these researchers first provided a growing culture of mammalian cells with radioactive thymidine and nonradioactive adenosine, guanosine, and cytidine. These are the building blocks of DNA. After waiting 30 minutes, Howard and Pelc washed the radioactive thymidine out of the culture but continued to provide the cells with nonradioactive versions of the four DNA building blocks. (You might recall, from Chapter 5, that this strategy is called a pulse-chase experiment.) Then they removed a sample of cells from the culture, spread them out, and laid a sheet of x-ray film over them. As the photograph in Figure 8.3a shows, small black dots appeared where radioactive thymidine molecules had exposed the film.

The key observation in this experiment was that radioactive signals were not found in M-phase nuclei—only in interphase nuclei. Because only interphase cells had incorporated radioactive thymidine, and because only replicating DNA would incorporate radioactive thymidine, the researchers concluded that chromosome replication occurs during interphase.

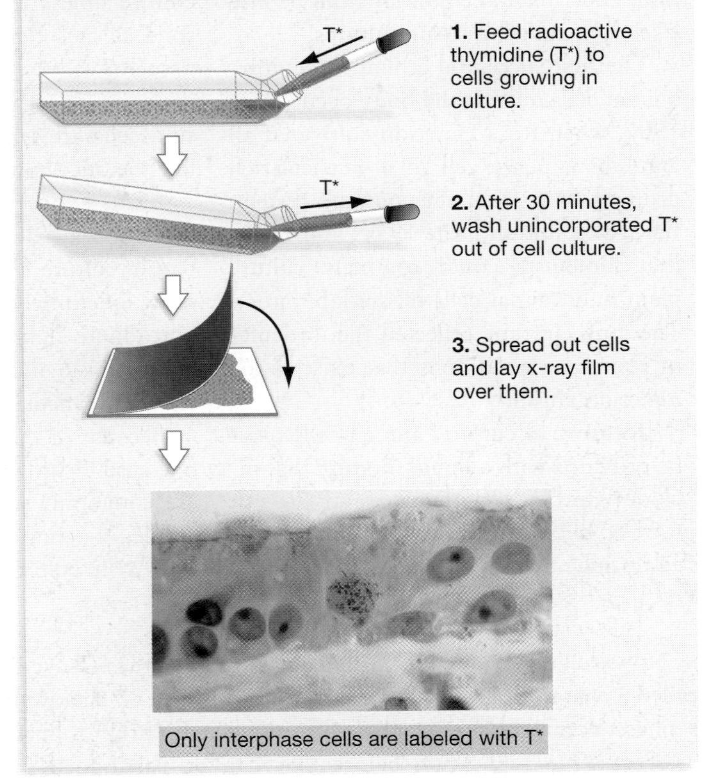

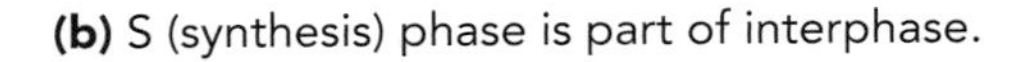
(b) S (synthesis) phase is part of interphase.

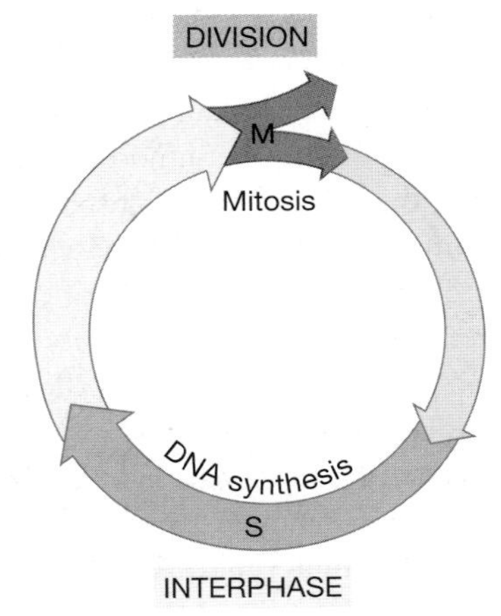

FIGURE 8.3 When Are Chromosomes Replicated?
(a) Because thymidine is incorporated into newly synthesized DNA, the experimental protocol shown here established when the synthesis (S) phase of the cell cycle occurs. **(b)** In the life of a cell, division alternates with interphase, when chromosomes are replicated.

BOX 8.1 Cell-Culture Methods

For researchers, there are several advantages to growing plant and animal cells outside of the host organism in laboratory culture. Cell cultures, which are called tissue cultures, provide homogenous populations of a single type of cell. In addition, experimental conditions can be precisely controlled in tissue cultures.

The first successful attempt to culture animal cells outside the body occurred in 1907, when Ross Harrison cultivated an amphibian nerve cell in a nutrient-rich drop of fluid from the immune system. But it was not until the 1950s and 1960s that biologists could routinely culture plant and animal cells in the laboratory. The long lag time reflected the difficulty of recreating conditions that exist in the intact organism.

To grow in culture, animal cells must be provided with a liquid mixture that includes nutrients, vitamins, and growth factors. Initially, this mixture was provided through the use of serum, or the liquid portion of blood; but now serum-free media are available for certain cell types. Serum-free media are advantageous because they are much more precisely defined chemically than serum. In addition, animal cells will not normally grow in culture unless they are provided with a solid surface that mimics the connective tissue or other types of surfaces that cells in the intact organisms adhere to. As a result, cells are typically cultured in flasks like the one shown in **Figure 1**.

Even under optimal conditions, though, normal cells display a finite lifespan in culture. But many cancerous cells grow indefinitely in culture. They also do not adhere tightly to the surface of the culture flask, and do not need growth factors in the media.

Because of their immortality and relative ease of growth, cultured cancer cells are commonly used in research. For example, the first human cell type to be grown in culture was isolated in 1952 from a malignant tumor of the uterine cervix. These cells are called HeLa cells in honor of their source, Henrietta Lacks, who is now deceased. HeLa cells continue to grow in laboratories around the world. They have been used in numerous studies on human cell function, including the experiments described in Figure 8.3, which revealed that chromosome replication occurs during interphase.

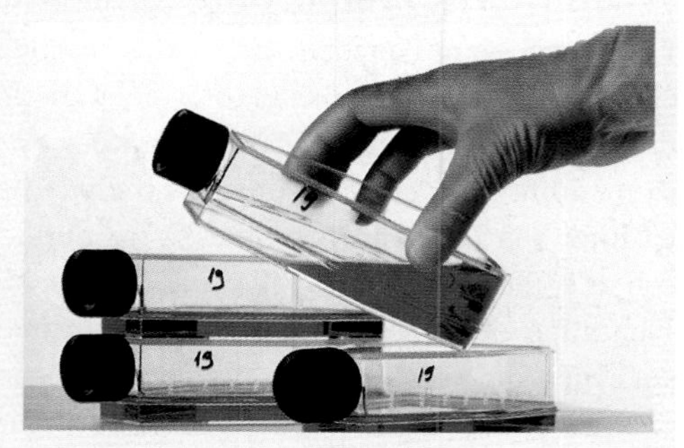

FIGURE 1 A Tissue Culture Flask
Because they need to adhere to a surface, cultured cells are grown in flat flasks with a liquid nutrient medium.

Howard and Pelc had identified a new stage in the cell cycle called **S phase**, for DNA synthesis phase. As **Figure 8.3b** shows, S phase occurs before M phase. As predicted, chromosomes must undergo replication before they can be partitioned to daughter cells during mitosis.

The Gap Phases

To determine how long it takes a cell to complete S phase, other investigators repeated Howard and Pelc's experiment but waited various lengths of time to examine the cells exposed to a pulse of radioactive thymidine. As the graph in **Figure 8.4** indicates, these experiments showed that no labeled, M-phase cells appeared until 4 to 5 hours after the pulse of radioactive thymidine ended. Sometime between 5 to 12 hours after the labeling period, though, all of the labeled cells had showed the chromosomal changes associated with mitosis.

The key to interpreting these data is to realize that the first labeled cells to enter mitosis must represent the cells that were just completing chromosome replication at the time of the labeling. Thus, the 4- to 5-hour time lag between the pulse and the appearance of the first labeled mitotic nuclei corresponds to the time lag between the end of S phase and the beginning of M phase. This gap in the cell cycle is called **G_2 phase**, for second gap. In the cultured mammalian cells used in these experiments, the G_2 phase lasts about 4 hours.

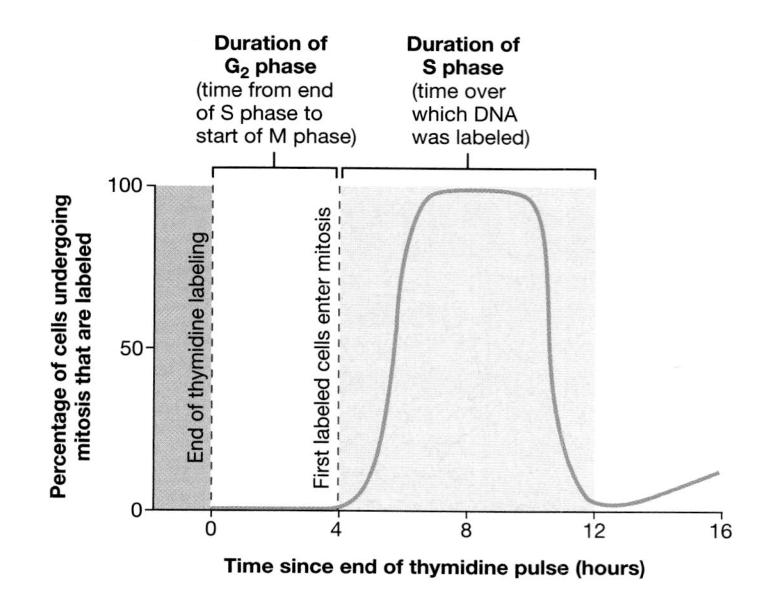

FIGURE 8.4 How Long are the G_2 and S Phases?
This graph plots the percentage of cells undergoing mitosis that are labeled versus the amount of time that has elapsed since the cells were fed a pulse of radioactive thymidine.

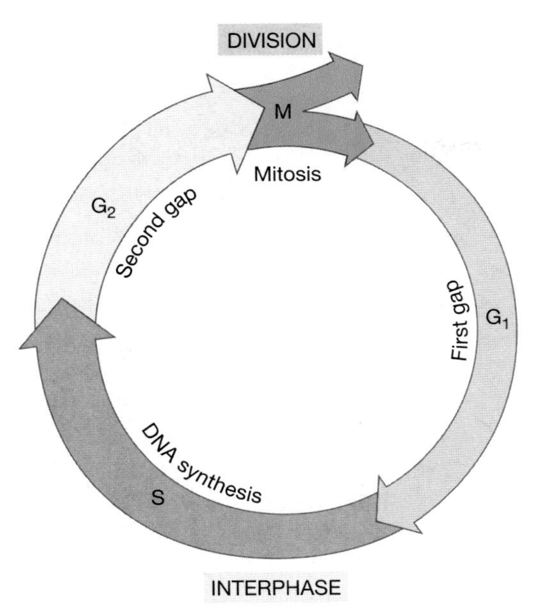

FIGURE 8.5 The Cell Cycle
This is a representative, or generic, cell cycle. The time devoted to the G_1 and G_2 phases varies dramatically among cells and organisms.
EXERCISE Label the length of each phase for the cultured mammalian cells discussed in the text.

How long are the other phases of the cell cycle? According to the data in Figure 8.4, labeled nuclei that are undergoing mitosis continue to be observed over a period of about 6 to 8 hours. Because all of these cells had to be somewhere in S phase when radioactive thymidine was available, so that the labeled molecule could be incorporated into their DNA, it is logical to conclude that their S phase lasts a total of 6 to 8 hours. The 30- to 45-minute duration of M phase could be estimated directly by watching mitosis in the light microscope. Similarly, biologists could estimate the time required to complete the total cell cycle by observing how long it takes cultured cells to double in number. For most mammalian cells growing in culture, the generation time is 18 to 24 hours.

When the times to complete the S, G_2, and M phases are added up and compared to the total cycle time, there is a discrepancy of 7 to 9 hours. This discrepancy represents another gap phase called **G_1 phase**, for first gap. As **Figure 8.5** shows, G_1 occurs after M phase and before S phase.

Why do the gap phases exist? In addition to copying their chromosomes during S phase, dividing cells must replicate organelles and manufacture additional cytoplasm. Before mitosis can take place, the parent cell must grow large enough and synthesize enough organelles that its daughter cells are normal in size and function. The two gap phases provide the time required to accomplish these tasks.

Given this overview of the major events in the life of a cell, let's turn now to M phase and the process of mitosis. How do cells divide? More specifically, how do they ensure that each daughter cell receives an identical complement of the genetic material in chromosomes?

CD ACTIVITY 8.1 The Cell Cycle

8.2 How Does Mitosis Take Place?

Flemming and van Beneden's early observations of cell division focused attention on the fate of the parent cell's chromosomes. Because chromosomes that are undergoing mitosis are visible under the light microscope when they are stained, investigators could watch mitosis occur. As a result, the major events in mitosis were well understood long before the cell cycle was fully described. Let's take a closer look at the events in mitosis, beginning with observations about the nature of a eukaryotic cell's chromosomes.

Events in Mitosis

Figure 8.6a shows the chromosomes found in a hypothetical plant or animal cell. The number of chromosomes in each cell varies widely among species; here, there are a total of four

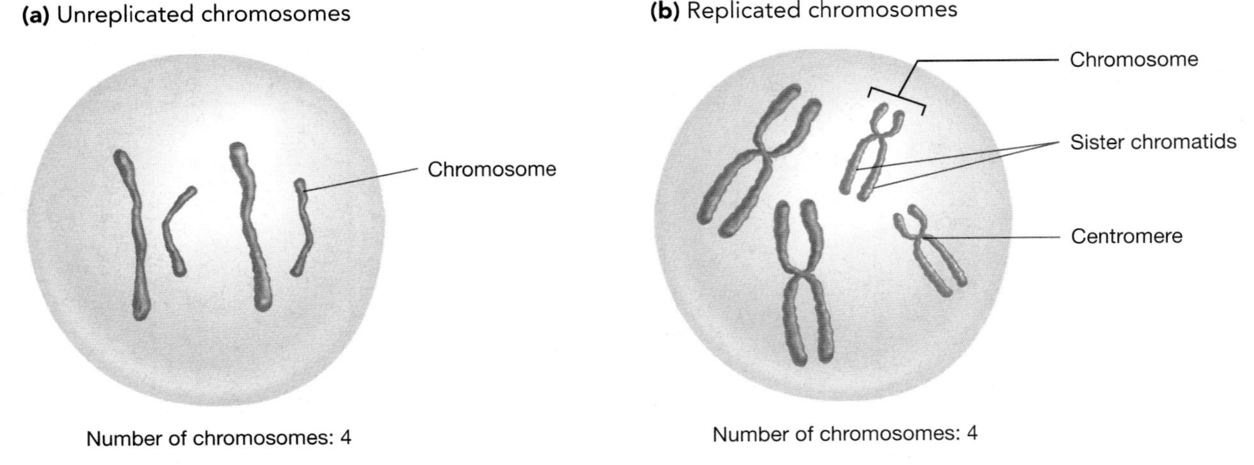

FIGURE 8.6 A Closer Look at Chromosomes
(a) Chromosomes consist of long strands of DNA that are associated with proteins. **(b)** When chromosomes have replicated, the two strands are called sister chromatids and are joined at a structure called the centromere.

CD ACTIVITY 8.2
Phases of Mitosis

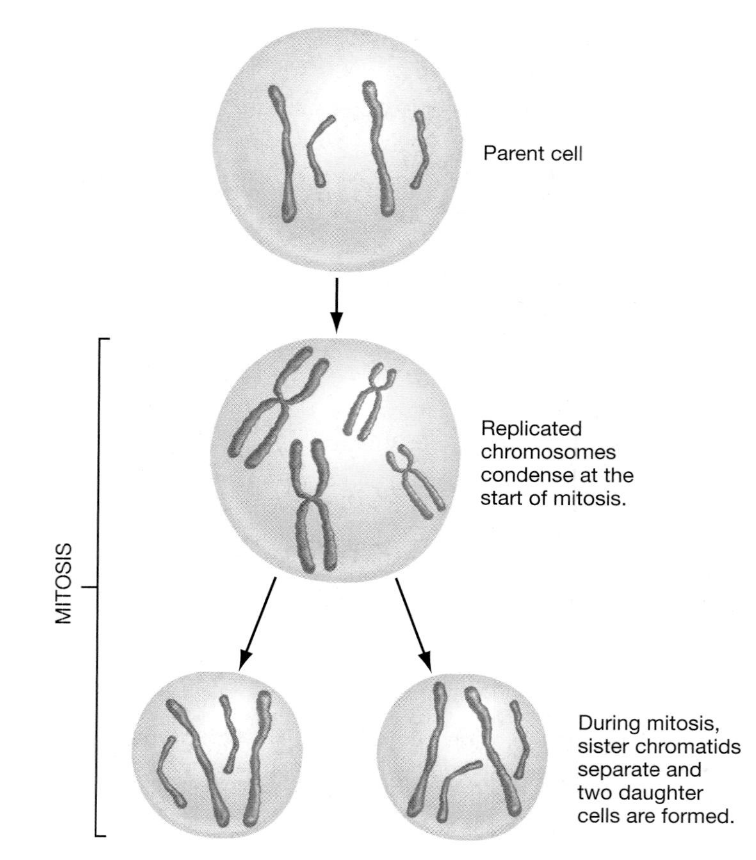

FIGURE 8.7 An Overview of Mitosis
Chromosomes are replicated prior to mitosis. During mitosis, the replicated chromosomes are partitioned to the two daughter cells.

chromosomes per cell. Under most conditions, chromosomes exist as extremely long, thread-like strands consisting of DNA associated with proteins. (The chromosomes in Figure 8.6a are shown condensed simply to make them visible.) The DNA-protein complex is called **chromatin.**

Prior to mitosis, the DNA in each chromosome is copied and chromatin condenses to form a much more compact structure. **Figure 8.6b** shows what chromosomes look like when they have been replicated. Replicated chromosomes correspond to the doublet threads that Flemming observed in 1882. Note that each of the DNA copies in a replicated chromosome is called a **chromatid,** that the two strands are joined together by a structure called a **centromere,** and that chromatids from the same chromosome are referred to as **sister chromatids.**

Chromosome replication occurs before mitosis begins. At the start of the cell division process, then, each chromosome consists of two sister chromatids joined at a centromere (**Figure 8.7**). During mitosis, the two sister chromatids separate to form independent chromosomes, and one copy goes to each of the two daughter cells. As a result, the daughter cells have ex-

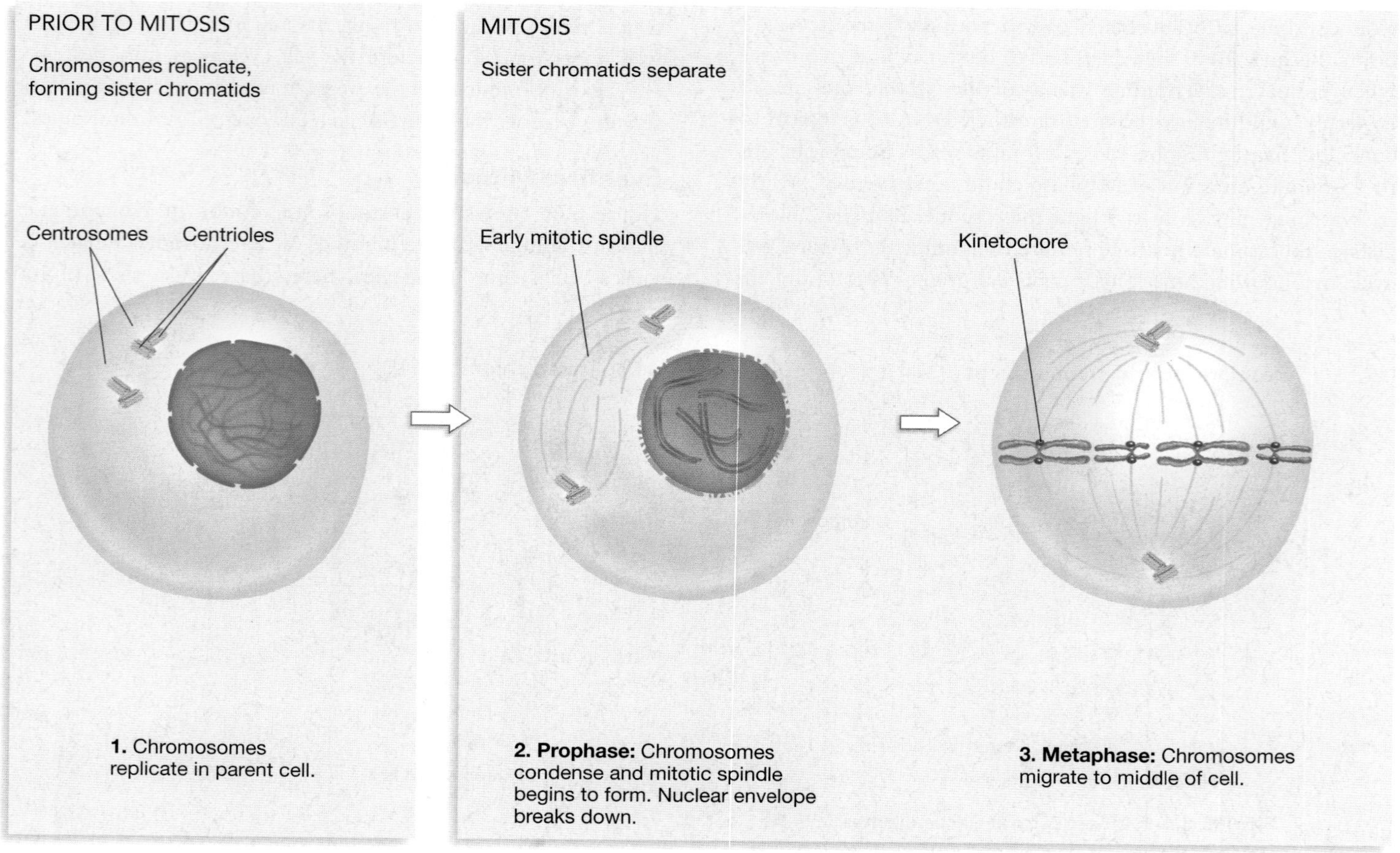

PRIOR TO MITOSIS

Chromosomes replicate, forming sister chromatids

Centrosomes

Centrioles

1. Chromosomes replicate in parent cell.

MITOSIS

Sister chromatids separate

Early mitotic spindle

2. Prophase: Chromosomes condense and mitotic spindle begins to form. Nuclear envelope breaks down.

Kinetochore

3. Metaphase: Chromosomes migrate to middle of cell.

FIGURE 8.8 The Subphases of Mitosis

actly the same complement of chromosomes as the parent cell did prior to replication.

Although mitosis is a continuous process, biologists routinely identify several subphases of M phase based on distinctive events that occur. To understand how mitosis proceeds, let's look at each subphase in turn. The discussion that follows describes mitosis in plant and animal cells; **Box 8.2** (page 164) describes cell division in bacteria.

Prophase Mitosis begins with the events of prophase (**Figure 8.8**). Recall that chromosomes replicate prior to this step; during prophase, they begin to condense into extremely compact structures. Chromosomes first become visible in the light microscope during prophase.

In the cytoplasm, prophase is marked by the formation of a structure called the mitotic spindle. The **mitotic spindle** produces mechanical forces that pull chromosomes into the daughter cells during mitosis. The mitotic spindle consists of an array of microtubules—one of the cytoskeletal elements introduced in Chapter 5. These microtubules are called spindle fibers. In animals, spindle fibers radiate between two structures called **centrosomes,** which contain microtubule-organizing bodies called centrioles.

As prophase continues, the nuclear envelope breaks down and a spindle fiber from each centrosome attaches to one of the two sister chromatids. The attachment between spindle fiber and chromatid is made at a structure called the **kinetochore.** The centrosomes then begin moving to opposite poles of the cell. At the same time, the kinetochore microtubules begin moving the chromosomes to the middle of the cell.

Metaphase During metaphase the centrosomes complete their migration to the opposite poles of the cell, and the kinetochore microtubules finish moving the chromosomes to the middle of the cell. When metaphase is complete, the chromosomes are lined up along an imaginary plane called the **metaphase plate.** At this point, the formation of the mitotic spindle is complete. Each chromatid is attached to a spindle fiber that runs from the kinetochore to one of the two poles.

Anaphase During anaphase, the kinetochore spindles begin to shorten. As they do, the sister chromatids are pulled apart to create independent chromosomes. As the kinetochore spindles continue to shorten, the sister chromatids are pulled to opposite ends of the cell. This is a critical step in mitosis because it assures that each daughter cell receives the same complement of chromosomes. When anaphase is complete, each end of the cell has an equivalent and complete collection of chromosomes that are identical to the one present in the parent cell prior to chromosome replication.

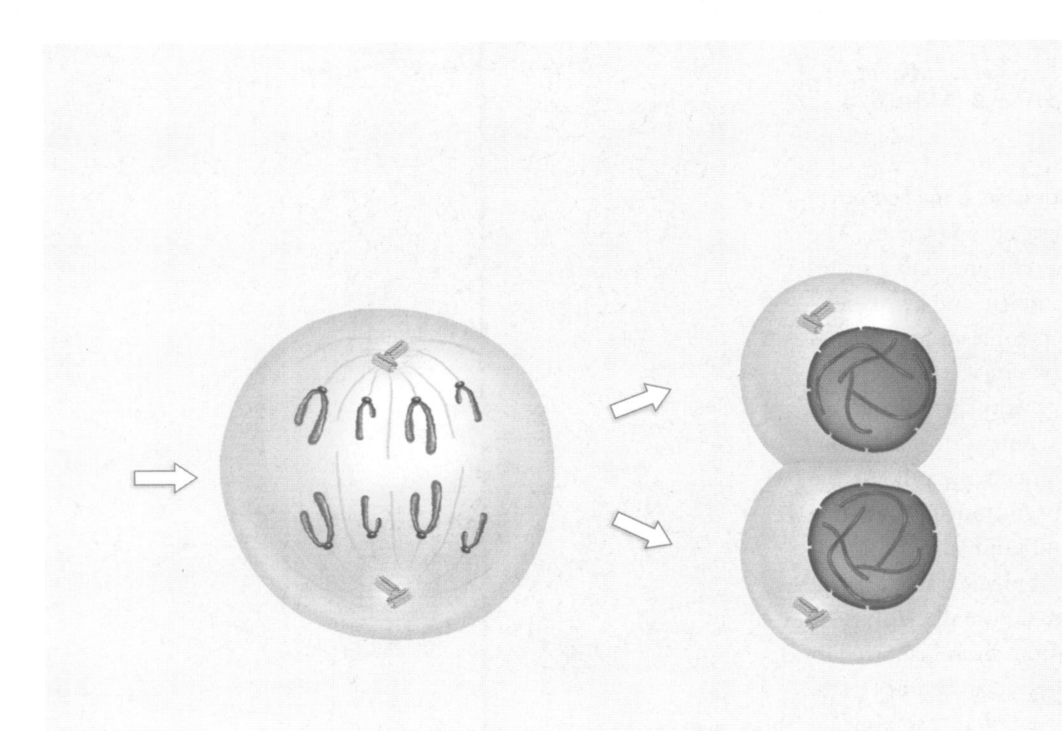

4. Anaphase: Sister chromatids separate. Chromosomes are pulled to opposite poles of the cell.

5. Telophase: The nuclear envelope reforms. **Cytokinesis:** The cell divides.

(a) Cytokinesis in animals

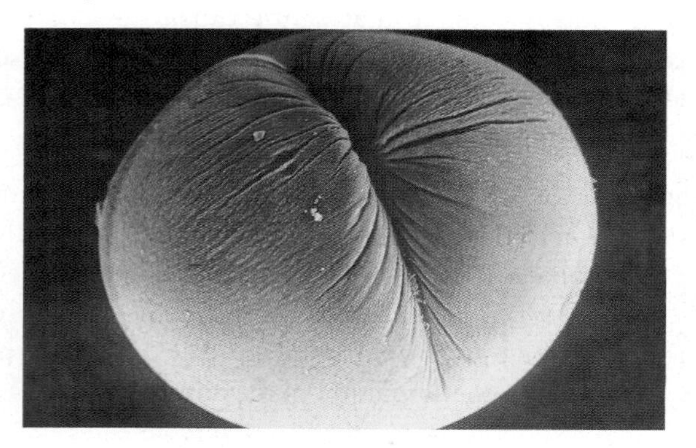

(b) Cytokinesis in plants

FIGURE 8.9 Cytokinesis
(a) In animals, the cytoplasm is divided by a cleavage furrow that pinches the cell in two. **(b)** In plants, the cytoplasm is divided by a cell plate that forms in the middle of the cell.

Telophase and Cytokinesis During telophase, a nuclear envelope begins to re-form around each set of chromosomes. The spindle apparatus disintegrates, and the chromosomes themselves begin to de-condense. Mitosis is complete and cytokinesis begins. **Cytokinesis** is the division of the cytoplasm and formation of two daughter cells. In animals, cytokinesis begins with the formation of a **cleavage furrow** (**Figure 8.9a**). The furrow appears because a ring of actin and myosin filaments forms just inside the cell membrane, in a plane that bisects the cell. As the actin and myosin filaments in the ring begin moving past one another, the cell membrane pinches in and gradually draws together. Eventually the original membrane is pinched in two, and cell division is complete. In plants, the mechanism of cytokinesis is different. Vesicles from the Golgi apparatus are transported to the middle of the dividing cell, where they form the cell plate shown in **Figure 8.9b**. The vesicles carry cell-wall components that gradually build up and divide the two daughter cells.

Interphase

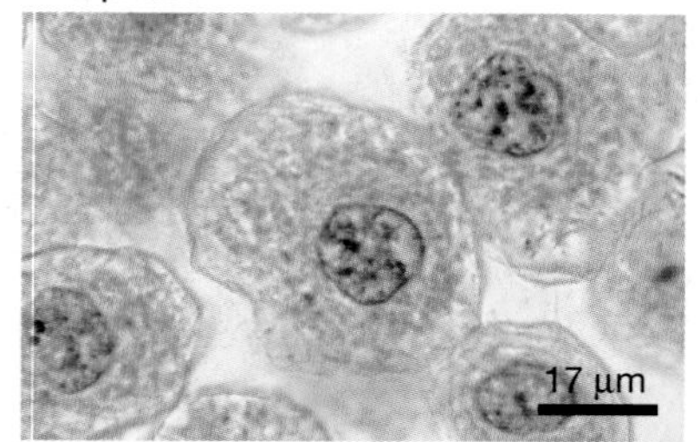

Prophase

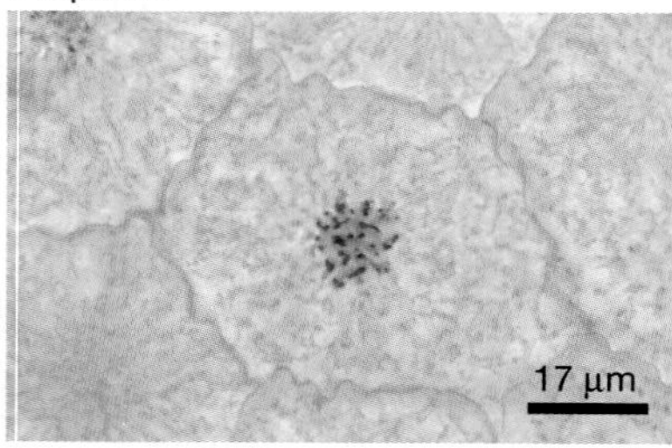

Metaphase

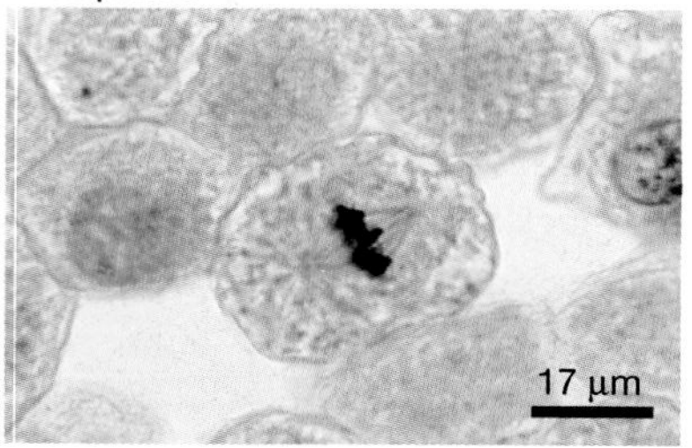

Anaphase

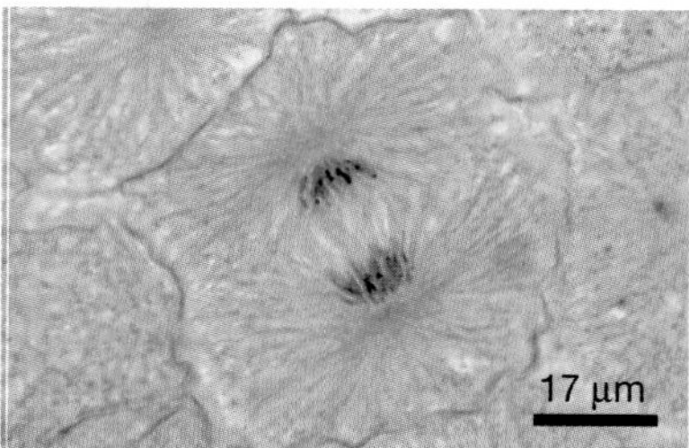

Telophase

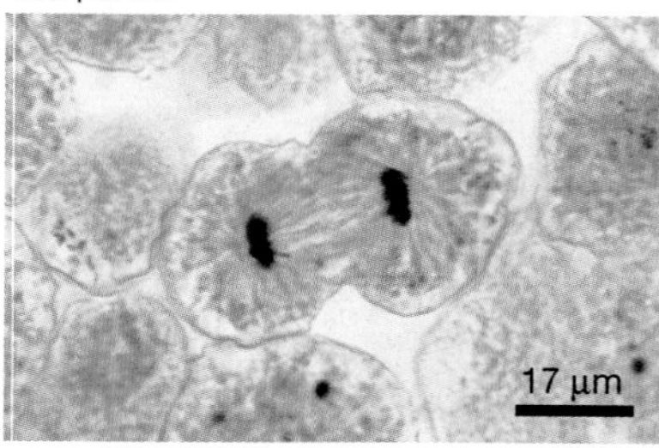

Cytokinesis

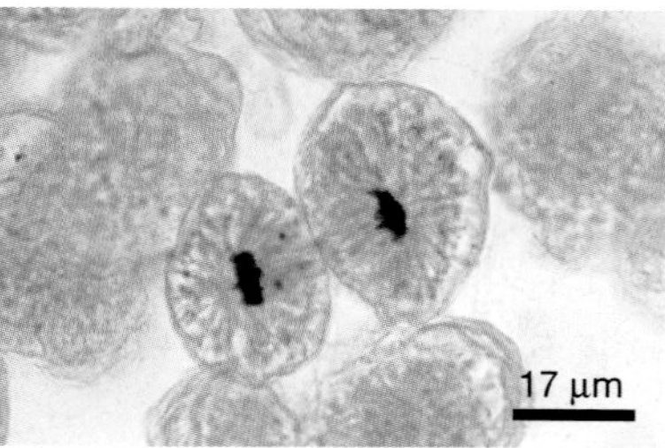

FIGURE 8.10 Mitosis and Cytokinesis in Action
These micrographs show a cell from a whitefish undergoing mitosis.
EXERCISE To the right of each image, write a 1-2 sentence description of what is happening.

Figure 8.10 shows a series of photographs of cells undergoing mitosis and cytokinesis. Once these processes had been described in detail, biologists turned their attention to understanding the molecular mechanisms involved. In particular, they wanted to know two things: How do sister chromatids separate to become independent chromosomes, and how do those chromosomes move to daughter cells? The exact and equal partitioning of genetic material to the two daughter cells is the most fundamental aspect of cell division. How does it occur?

Mechanics of Chromosome Partitioning

To understand how sister chromatids separate and move to daughter cells, biologists have focused on understanding how the mitotic spindle functions. Do spindle microtubules act as railroad tracks, the way they do in vesicle transport? Is some sort of motor protein involved? And what is the nature of the kinetochore, where the chromosome and microtubules are joined?

Mitotic Spindle Forces Recall that spindle fibers are composed of microtubules. You know from reading Chapter 5 that microtubules are polymers of the protein tubulin, and that the length of a microtubule is determined by the number of tubulin subunits it contains. This observation suggests a mechanism for the movement of chromosomes during anaphase. Is the spindle microtubule shortening due to a loss of tubulin subunits from one end?

To test this hypothesis, Gary Borisy and co-workers introduced fluorescently labeled tubulin subunits into prophase or metaphase cells. This treatment made the entire mitotic spindle visible when observed with a fluorescence microscope. Then, once anaphase had begun, the biologists marked a region of the spindle by irradiating it with a bar-shaped beam of laser light. Because the laser quenched the fluorescence in the exposed region, it became dark, or "photobleached." **Figure 8.11a** shows what happened. As anaphase progressed, the photobleached region remained stationary while the chromosomes moved toward this region.

To explain this result, Borisy and colleagues concluded that the microtubules themselves remain stationary during anaphase, and that chromosome movement occurs because tubulin subunits are lost from the kinetochore ends (see **Figure 8.11b**). As a result, the microtubule shortens and pulls the chromosome along.

A Kinetochore Motor If microtubules are shortening at the kinetochore end, how does the chromosome maintain its attachment to the microtubule? Although the answer to this question is not yet clear, the structure and function of the kinetochore is gradually becoming better understood. In particular, studies by a number of biologists have shown that the kinetochore contains motor proteins that appear to "walk" chromosomes down microtubules (**Figure 8.11c**). If so, then the mechanism of chromosome movement is reminiscent of the way that kinesin walks down microtubules during vesicle transport. As explained in Chapter 5, motor proteins convert the chemical energy of ATP into mechanical work in the form of movement. In some way, the kinetochore motor proteins detach and re-attach to the kinetochore microtubule, causing the microtubule to shorten and the chromosome to move. The nature of the kinetochore motors is still under investigation, however.

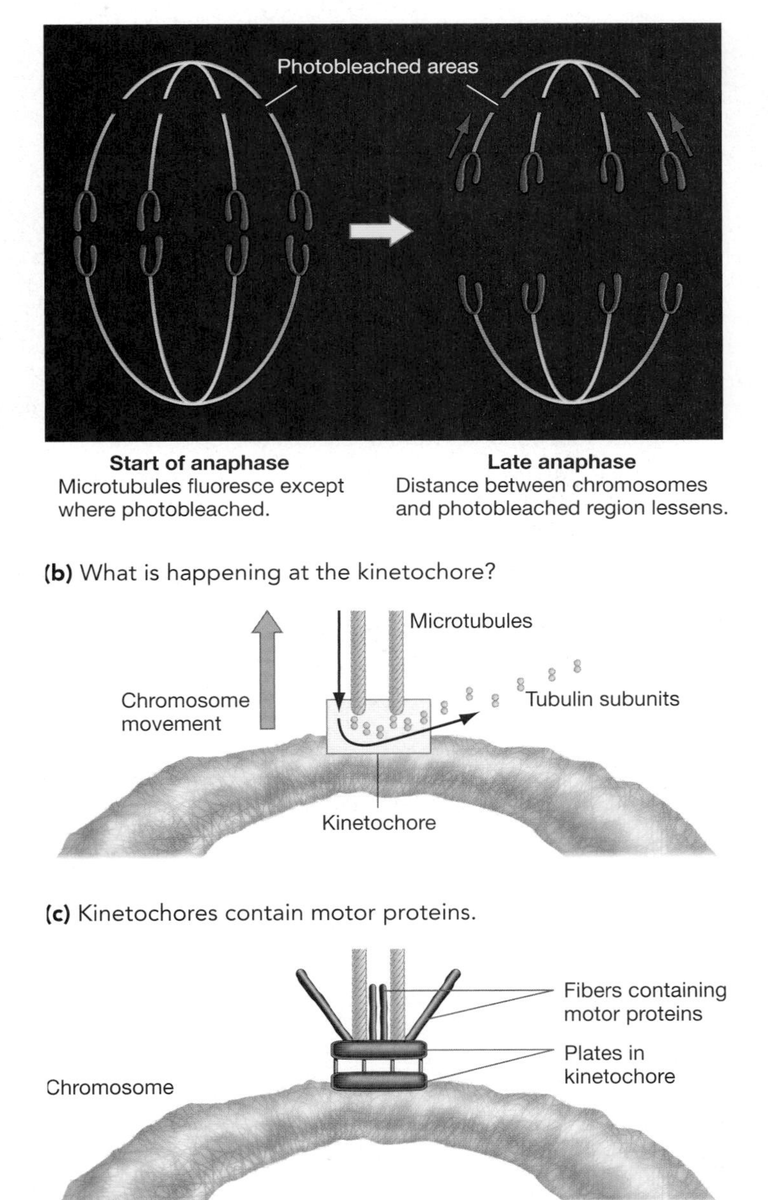

FIGURE 8.11 How Do Microtubules Move Chromosomes During Mitosis?
(a) The experiment shown here suggests that during anaphase, kinetochore microtubules shorten near the chromosome. **(b)** According to one hypothesis, microtubules shorten as tubulin subunits split off at the kinetochore. **(c)** The kinetochore consist of an inner plate, an outer plate, and associated fibers that contain motor proteins. The fibers are thought to act as "legs" that walk the chromosomes down the length of the kinetochore microtubules. **EXERCISE** In part (a), add drawings to show the outcome expected if microtubules shortened at the end opposite the chromosome.

Efforts to understand kinetochore motor proteins bring us to the frontier of research on mitosis. Having explored how the process occurs, let's focus on how it is controlled. When does a cell divide, and when does it stop dividing? How is cell division regulated?

8.3 Control of the Cell Cycle

Although the events of mitosis are virtually identical in all eukaryotes, other aspects of the cell cycle can be extremely variable. For example, the length of the cell cycle can vary enormously among different cell types in the same individual. In humans, intestinal cells routinely divide more than twice a day to renew tissue that is lost during digestion. In contrast, mature human nerve and muscle cells do not divide at all. Most of this variation is due to variation in the length of the G_1 phase. In rapidly dividing cells, G_1 is essentially eliminated. Nondividing cells, in contrast, are permanently stuck in G_1. Researchers refer to this arrested stage as the G_0 state, or simply "G zero."

A cell's division rate can also vary in response to changes in conditions. For example, human liver cells normally divide about once per year. But if part of the liver is damaged or lost,

BOX 8.2 Cell Division in Bacteria

Bacteria do not reproduce sexually. Instead, they reproduce asexually by division of a parent cell into two genetically identical daughters. This process is called binary fission. Although the structure of bacterial cells is very different from eukaryotic cells, bacteria face similar challenges in replicating and partitioning their hereditary information during cell division.

Figure 1 shows the major events in bacterial cell division. Note that most bacteria contain a single, circular chromosome composed of DNA. Because they lack a nucleus, the chromosome is located in the cytoplasm. Bacteria also lack a cytoskeleton, so they cannot form a mitotic spindle-like system to partition daughter chromosomes during division.

How are bacterial chromosomes partitioned to daughter cells after they are replicated? As the figure shows, the answer is that the chromosomes become attached to different sites on the plasma membrane. An infolding of the cell membrane between the attachment sites results in fission.

FIGURE 1 Steps in Bacterial Cell Division
QUESTION Are the daughter cells of bacterial cell division identical in chromosomal makeup, or are they different?

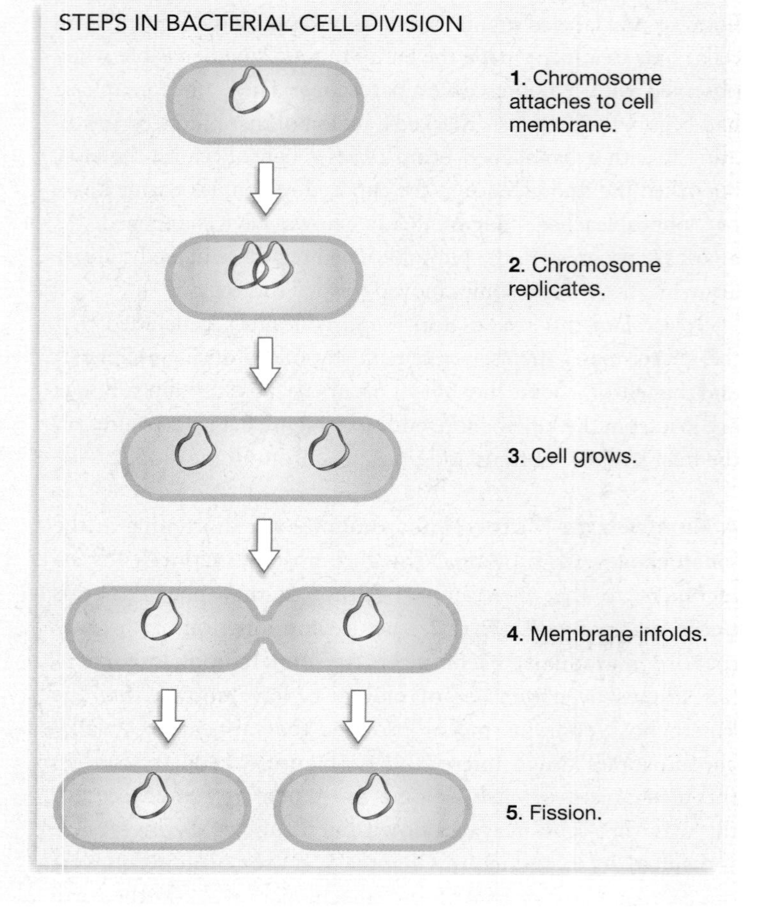

the remaining cells divide every day or two until repair is accomplished. Cells of unicellular organisms like yeasts, bacteria, or archaea divide rapidly only if the environment is rich in nutrients; otherwise, they enter a quiescent state.

To explain the existence of so much variability, biologists hypothesized that the cell cycle must be regulated in some way, and that regulation varies among cells and organisms. Understanding how the cell cycle is controlled is now the most prominent issue in research on cell division—partly because defects in control can lead to uncontrolled, cancerous growth. What evidence first suggested that regulatory molecules control the cell cycle?

The Discovery of Cell-Cycle Regulatory Molecules

The first solid evidence for cell-cycle control molecules came to light in 1970. In that year, Potu Rao and Robert Johnson published the results of experiments on fusing pairs of cultured mammalian cells (**Figure 8.12a**). In the presence of certain chemicals, viruses, or an electric shock, the membranes of two cells can be made to fuse. The hybrid cell that results has two nuclei and is called a **heterokaryon** (different-nuclei).

How did cell fusion experiments point to the existence of cell-cycle control molecules? When Rao and Johnson fused

(a) M phase cells (mitotic cells) induce interphase cells to begin M phase.

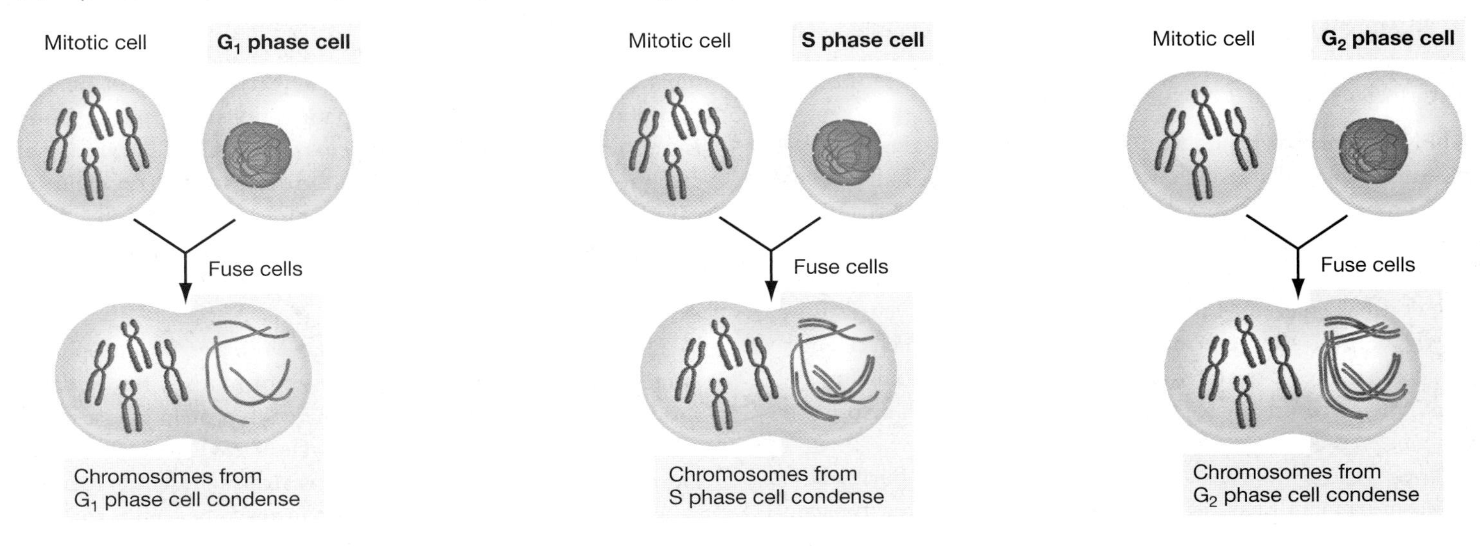

(b) S phase cells induce only G_1 cells to begin S phase.

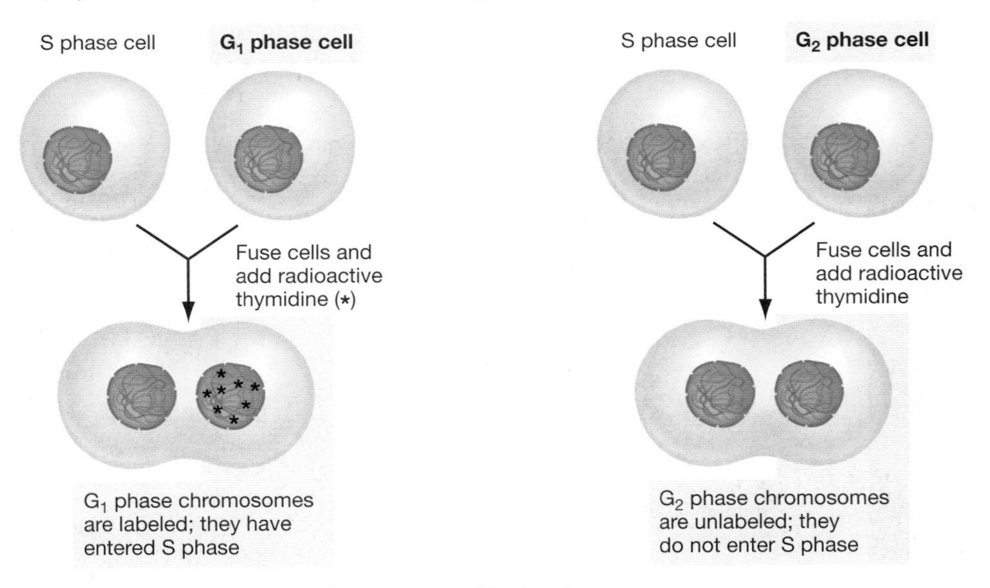

FIGURE 8.12 Evidence for Cell-Cycle Control Molecules
(a) When M-phase cells are fused with cells in interphase, the interphase chromosomes condense and begin M phase. **(b)** When S-phase cells are fused with G_1 cells, the G_1 chromosomes begin to replicate. But when S-phase cells are fused with G_2 cells, the G_2 chromosomes do not begin to replicate. QUESTION In part (a), the condensing chromosomes are single, partially replicated, or completely replicated. Why?

cells that were in different stages of the cell cycle, they found that one of the two nuclei changed phases. For example, when they fused a cell in M phase with an interphase cell, the nucleus of the interphase cell initiated M phase. To interpret this result, the biologists hypothesized that the cytoplasm of M-phase cells contains a regulatory molecule that induces interphase cells to enter M phase.

Similarly, when they fused an S-phase cell with a G_1 cell in the presence of radioactive thymidine, the nucleus of the G_1 cell immediately began incorporating the label—suggesting that it was starting S phase. In contrast, fusing an S-phase cell with a cell in G_2 had no effect on the G_2 nucleus (**Figure 8.12b**). Thus, S-phase cells contain a molecule that can induce G_1 cells, but not G_2 cells, to begin DNA synthesis.

To summarize, the cell fusion experiments suggested that regulatory molecules control entry into the S and M and phases of the cell cycle. What are these molecules?

M-Phase Promoting Factor The molecule that triggers mitosis was identified through studies on the eggs of the South African claw-toed frog, *Xenopus laevis* (**Figure 8.13a**). As the eggs of these frogs mature, they change from a cell called an oocyte, which is arrested in a phase similar to G_2, to a mature egg that has entered M phase. The eggs are attractive to study, partly because they are more than 1 mm in diameter. Their large size makes it relatively easy to purify large amounts of cytoplasm and to use microsyringes to inject eggs with cytoplasm from eggs in different stages of development.

In 1971, Yoshio Masui and Clement L. Markert purified cytoplasm from M-phase eggs and injected it into the cytoplasm of oocytes in the G_2-like phase. As **Figure 8.13b** shows, these injections always caused the immature oocyte to enter M phase. But when cytoplasm from interphase cells was injected into G_2 oocytes, the cells remained in the G_2-like phase. Masui and Markert concluded that the cytoplasm of M-phase cells—but not the cytoplasm of interphase cells—contains a factor that drives immature oocytes into M phase to complete their maturation.

This factor came to be called **M-phase promoting factor**, or MPF. Subsequent experiments showed that MPF induces mitosis in all eukaryotes. For example, injection of M-phase cytoplasm from mammalian cells into immature frog eggs results in egg maturation. Similarly, human MPF can trigger mitosis in yeast cells. How does it work?

MPF Contains a Protein Kinase and a Cyclin MPF was purified in 1988, and researchers subsequently found it is made up of two distinct polypeptide subunits. One of the polypeptide subunits is an enzyme that catalyzes a reaction called a protein phosphorylation. In protein phosphorylation, a phosphate group is transferred from ATP to a protein (**Figure 8.14**). Enzymes that catalyze the addition of a phosphate group to a target protein are called **protein kinases.** Protein kinases frequently act as regulatory molecules because the addition of a phosphate group tends to change the target protein's shape and activity. Proteins can be activated or inactivated by phosphorylation.

These observations suggested that MPF acts by phosphorylating a protein that triggers the onset of mitosis. But research showed that the concentration of the MPF protein kinase is relatively constant through the cell cycle. How can MPF trigger mitosis if the protein kinase subunit is always present in the cell?

The answer to this question lies in the other MPF subunit, which belongs to a family of proteins called the cyclins. Cyclins got their name because their concentrations fluctuate through-

(a) African clawed frog

(b) M phase cytoplasm induces oocyte to begin M phase.

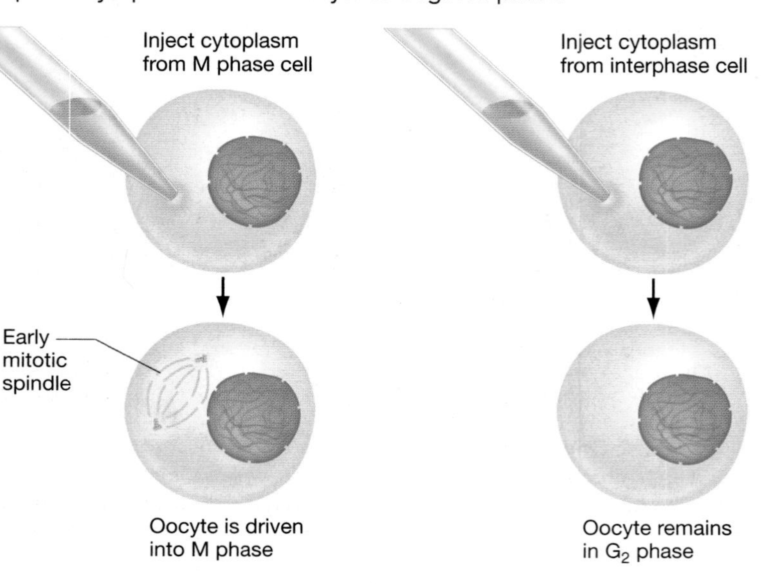

FIGURE 8.13 The Discovery of M Phase Promoting Factor (MPF) **(a)** The African clawed frog, *Xenopus laevis*, is a popular study organism because its eggs are large. **(b)** Microinjection experiments supported the hypothesis that a regulatory molecule induces the M phase.

out the cell cycle. As **Figure 8.15a** shows, the cyclin associated with MPF builds up during interphase and peaks in M phase. This increase in concentration is important because the protein kinase subunit in MPF is active only when it is bound to the cyclin subunit. As a result, the protein kinase subunit of MPF is called a **cyclin-dependent kinase,** or **Cdk.**

To summarize, MPF consists of a cyclin-dependent kinase subunit and a cyclin subunit. When cyclin concentrations are high, enough MPF is active to phosphorylate the array of proteins diagrammed in **Figure 8.15b**. Note that chromosomal proteins activated by MPF cause chromosomes to condense into the visible threads observed during M phase. In addition, microtubule-associated proteins that are phosphorylated by MPF may be involved in assembling the mitotic spindle apparatus. But MPF also activates an enzyme complex that promotes the degradation of MPF's cyclin subunit. Because the concentration of cyclin declines rapidly in response and then slowly builds up again during interphase, an oscillation in cyclin concentration is set up. This oscillation acts as a clock that drives the ordered events of the cell cycle.

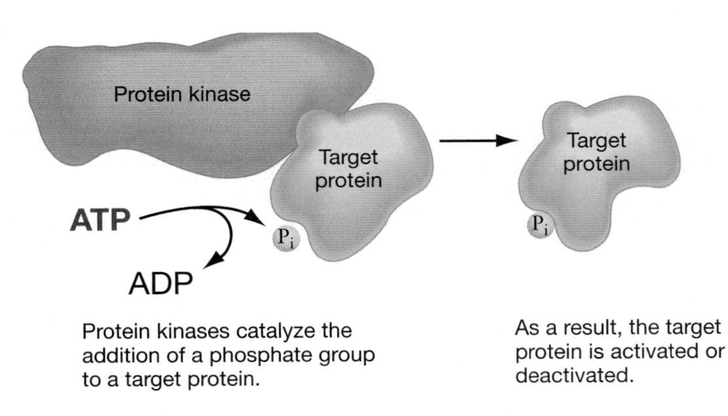

FIGURE 8.14 Protein Kinases Are Regulatory Molecules
Because the addition of a phosphate group tends to cause a change in conformation, phosphorylation changes the activity of many proteins.

Cell-Cycle Checkpoints

MPF is actually only one of many protein complexes involved in regulating the cell cycle. For example, a different cyclin and protein kinase are involved in triggering the passage from G_1 into S phase, and several regulatory proteins are involved in maintaining the G-zero state of quiescent cells. To pull these observations about cell-cycle control together, Leland Hartwell and Ted Weinert introduced the concept of cell-cycle checkpoints. A **cell-cycle checkpoint** is a critical point in the cell cycle that is regulated. Hartwell and Weinert identified checkpoints by analyzing mutant yeast cells with defects in the cell cycle. The mutants lacked a specific checkpoint and grew abnormally as a result. As **Figure 8.16** (page 168) indicates, biologists have now obtained evidence for three distinct checkpoints. In effect, a cell "decides" whether to proceed with division at each checkpoint.

The first checkpoint occurs late in G_1. For most cells, this checkpoint is the most important in establishing whether the cell will continue through the cycle and divide. What determines whether a cell passes through the G_1 checkpoint? Because a cell must reach a certain size before its daughter cells will be large enough to function normally, biologists hypothesize that some mechanism exists to arrest the cell cycle if the cell is too small. In addition, unicellular organisms arrest at the G_1 checkpoint if nutrient conditions are poor. As you'll see in section 8.4, cells in multicellular organisms pass through the G_1 checkpoint in response to signaling molecules from other cells.

The second checkpoint occurs at the boundary between the G_2 and M phases. Specifically, cells appear to arrest at the G_2 checkpoint if chromosome replication has not been completed properly. There are some data indicating that cells may also respond to external signals and their size at this checkpoint.

The final checkpoint occurs during metaphase in mitosis. If not all chromosomes are properly attached to the mitotic spindle, M phase arrests before sister chromatids begin to be pulled to opposite poles of the cell. Specifically, anaphase is delayed

(a) Cyclin concentrations regulate MPF activity.

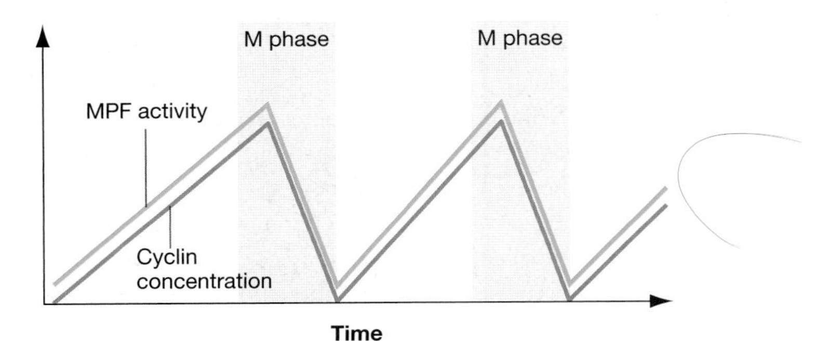

(b) Activated cyclin-dependent kinase has an array of effects.

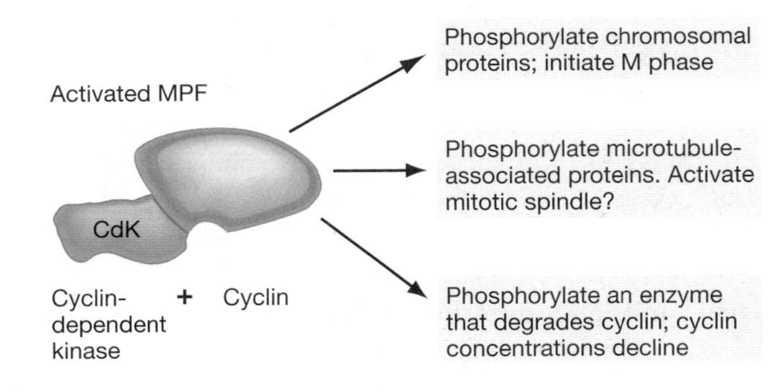

FIGURE 8.15 How Does the M Phase Promoting Factor Act?
(a) Cyclin concentrations cycle in dividing cells, reaching a peak in M phase. MPF activity peaks at that time because the cyclin-dependent kinase (Cdk) in MPF must be bound to cyclin to be active. **(b)** Cdk phosphorylates proteins that initiate the M phase and degrade cyclin.

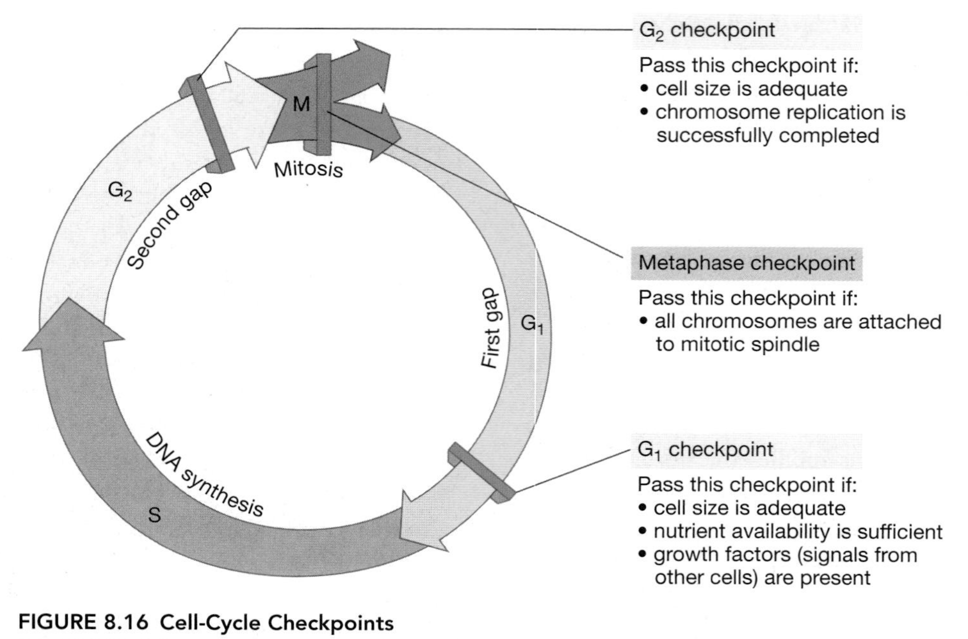

FIGURE 8.16 Cell-Cycle Checkpoints

QUESTION Why is it advantageous for a cell to arrest at the M-phase checkpoint if all chromosomes are not attached to the mitotic spindle?

until all kinetochores are properly attached to mitotic spindle fibers. If this checkpoint did not exist, some chromosomes might not separate properly and daughter cells would receive an incorrect number of chromosomes. The consequences for the daughter cells would be disastrous. Similarly, if cells do not respond normally to signals at the G_1 and G_2 checkpoints, they may begin growing in an uncontrolled fashion with dire consequences for the organism. Let's take a closer look.

8.4 Cancer: Out-of-Control Cell Division

Few diseases inspire more fear than cancer. The fear springs from the difficulty of treating many forms of cancer, the potentially fatal nature of many cancers, and their commonness. Most of us know someone who has had some form of cancer, and most of us know someone who has died from the disease. According to the American Cancer Society, 50 percent of American men and 33 percent of American women will develop cancer during their lifetime. In the United States, one out of four of all deaths are from cancer; it is the second leading cause of death, exceeded only by heart disease.

Humans suffer from at least 200 different types of cancer. Stated another way, cancer is not a single illness but a complex family of diseases that affect an array of organs, including the breast, colon, brain, lung, and skin. In addition, several different types of cancer can affect the same organ. Skin cancers, for example, come in multiple forms. Some are relatively easy to treat, while others are often fatal. Although cancers vary in their time of onset, growth rate, seriousness, and cause, all cancers have a unifying feature. They arise from cells in which cell-cycle checkpoints have failed.

To get a better understanding of why cancer occurs, let's first review some general characteristics of the condition and then delve into the details of why regulatory mechanisms become defective.

Properties of Cancer Cells

When even a single cell in a multicellular organism begins to divide in an uncontrolled fashion, a mass of cells called a **tumor** results. For example, normal cells in the adult human brain do not divide. But if a single abnormal brain cell begins unrestrained division, the growing tumor that results may seriously disrupt the brain's function. What can be done? If the tumor can be removed without damaging the affected organ, a cure might be achieved. Accordingly, surgical removal of the tumor is usually the first step in the treatment of a cancer. Often, though, surgery does not cure a cancer patient. Why?

In addition to growing quickly, cancer cells are able to spread throughout the body through the bloodstream or the lymphatic vessels introduced in Chapter 46. Invasiveness is a defining feature of a **malignant tumor**—one that is cancerous. Masses of noninvasive cells are noncancerous and form **benign tumors**, such as a wart. By spreading from the primary tumor site where uncontrolled growth originated, cancer cells can establish secondary tumors elsewhere in the body. This process is called **metastasis**. If metastasis has occurred by the time the original tumor is detected, surgical removal of the primary tumor will not result in a cure, and the disease may be very difficult to treat (**Figure 8.17**). As a result, healthcare workers stress early detection as the key to treating cancer effectively.

Cancer Involves Loss of Cell-Cycle Control

If cancer is caused by uncontrolled cell growth, what is the molecular nature of the disease? Recall that when many cells mature, they enter the G_0 state—meaning that their cell cycle is arrested at the G_1 checkpoint. In contrast, cells that do pass through the G_1 checkpoint are irreversibly committed to dividing. Based on this observation, biologists hypothesized that cancer involves a defect in the G_1 checkpoint. To understand the molecular nature of the disease, then, researchers focused on understanding the normal mechanisms that operate at this checkpoint. In this way, cancer research and research on the normal cell cycle have become two sides of the same coin.

Social Control In unicellular organisms, passage through the G_1 checkpoint is thought to depend primarily on cell size and the availability of nutrients. If nutrients are plentiful, cells pass through the checkpoint and grow rapidly. In multicellular organisms, however, cells receive a constant supply of adequate nutrients through the bloodstream. Most cells in multicellular organisms do not divide in response to the arrival of nutrients, though, so some other type of signal must be involved. Because these signals arrive from other cells, biologists refer to social control over cell division. The general idea is that individual cells should be allowed to divide only when their growth is in the best interests of the organism as a whole.

The most important signals involved in social control of the cell cycle are called **growth factors.** Growth factors were discovered in the course of working out techniques for growing cells in culture. When researchers isolated mammalian cells in culture and provided them with adequate nutrients, the cells arrested in G_1 phase. The cells began to grow only when biologists added **serum**—the liquid that remains after blood has been allowed to clot and the blood cells have been removed. Some component of serum allowed cells to pass through the G_1 checkpoint. What was it?

FIGURE 8.17 Cancers Spread
This photo shows the liver of a human who died of cancer. Note that a large number of tumors have formed due to migration of cancerous cells from the original tumor, which occurred in the patient's colon. The numbers provide a scale in centimeters.

In 1974, Russell Ross and colleagues identified one of the serum components responsible for stimulating cell division. The component was a protein called **platelet-derived growth factor** (or **PDGF**). As its name implies, PDGF is released by blood components called platelets, which promote blood clotting at wound sites. In response to the PDGF secreted by platelets, cells in the area are stimulated to divide. The increased cell numbers facilitate wound healing.

Researchers subsequently found that PDGF is produced by an array of cell types. Investigators also succeeded in isolating and identifying a large and diverse array of growth factors. For different types of cells to grow in culture, different combinations of growth factors must be supplied. Based on this result, biologists infer that different types of cells in an intact multicellular organism are controlled by different combinations of growth factors. Cancer cells, however, are another story. Cancerous cells can often be cultured successfully without externally supplied growth factors. This observation suggests that the normal social controls on the G_1 checkpoint have broken down in cancer cells.

Social Controls and Cell-Cycle Checkpoints How do growth factors stimulate division in normal cells? Recall that for a cell to begin mitosis or to pass through the G_1 checkpoint, a specific cyclin-Cdk complex must be activated. Because activation depends on the presence of a high concentration of the cyclin involved, investigators hypothesized that growth factors might trigger cyclin synthesis. As **Figure 8.18a** (page 170) shows, this hypothesis turned out to be correct. The link between growth factors and cyclin synthesis also turned out to be important in cancer biology. In some human cancers, the cyclin that is active at the G_1 checkpoint is overproduced. As a result, the Cdk that it binds to is constantly activated.

Why does overstimulation of the G_1 cyclin-Cdk complex lead to uncontrolled growth? To answer this question, recall that an activated Cdk phosphorylates target proteins. In many cancers, the key target protein is the retinoblastoma (Rb) protein. This protein was discovered by researchers interested in a childhood cancer called retinoblastoma, which occurs in about 1 in 20,000 children. The disease is characterized by the appearance of malignant tumors in the light-sensing tissue, or retina, of the eye.

The role of Rb protein in controlling normal cell division is diagrammed in Figure 8.18b. In quiescent (normal) cells, Rb binds to a protein called E2F. But if a growth factor stimulates production of cyclin and activates the G_1 cyclin-Cdk complex, the Rb protein becomes phosphorylated and releases from E2F. In response, E2F stimulates the production of molecules needed in S phase. As a result, the cell passes through the G_1 checkpoint. In effect, then, a cell's commitment to dividing depends on whether the Rb protein stays bound to the E2F protein. Because Rb acts as a brake on the cell cycle, it is known as a **tumor suppressor.**

Some cancers are caused by the presence of excessive amounts of growth factors, by cyclin production in the absence of growth signals, or by defects in the Rb protein itself. For example, defective Rb proteins are associated with some breast, bladder, and lung cancers as well as retinoblastoma. In many cases, the Rb protein is simply not produced. If Rb is not present, the result is constant activation of the E2F protein, a failure of the G_1 checkpoint, and uncontrolled cell division.

Cancer Results from Multiple Defects

One of the most general messages of research on the Rb protein is that a wide variety of defects can be responsible for the failure of the G_1 checkpoint and the onset of cancer. In addition, researchers have come to realize that cancer is seldom due to a single defect. In other words, several problems usually have to occur for control over the cell cycle to fail.

To drive this point home, consider research on children with the hereditary form of retinoblastoma. Hereditary retinoblastoma, which runs in families, is due to a defect in the retinoblastoma protein. The protein is defective because of an error in the DNA sequence that provides the instructions for making the protein. Recall that DNA is found in chromosomes. Each person has two of each type of chromosome. The two versions of the same type of chromosome are called homologous chromosomes or simply homologs. As a result, individuals with a defective version of the *Rb* DNA sequence usually have a normal copy of the sequence on the homologous chromosome (**Figure 8.19a**). This is a key point, because one copy of the normal DNA sequence produces enough normal Rb protein to allow the G_1 checkpoint to function normally. If so, then why do children with just one defective copy of *Rb* DNA tend to develop cancer?

Webster K. Cavenee and colleagues answered this question by examining the chromosomes of tumor cells and normal cells in hereditary retinoblastoma patients. They found that in some tumors, the cancer cells contain a solitary defective chromosome. But in other tumors, cells contained two copies of the defective chromosome.

To explain these observations, Cavenee and co-workers hypothesized that the abnormal chromosome complements in retinoblastoma tumor cells result from mistakes in mitosis. As **Figure 8.19b** shows, the sister chromatids of a normal chromosome may fail to separate normally during anaphase. More

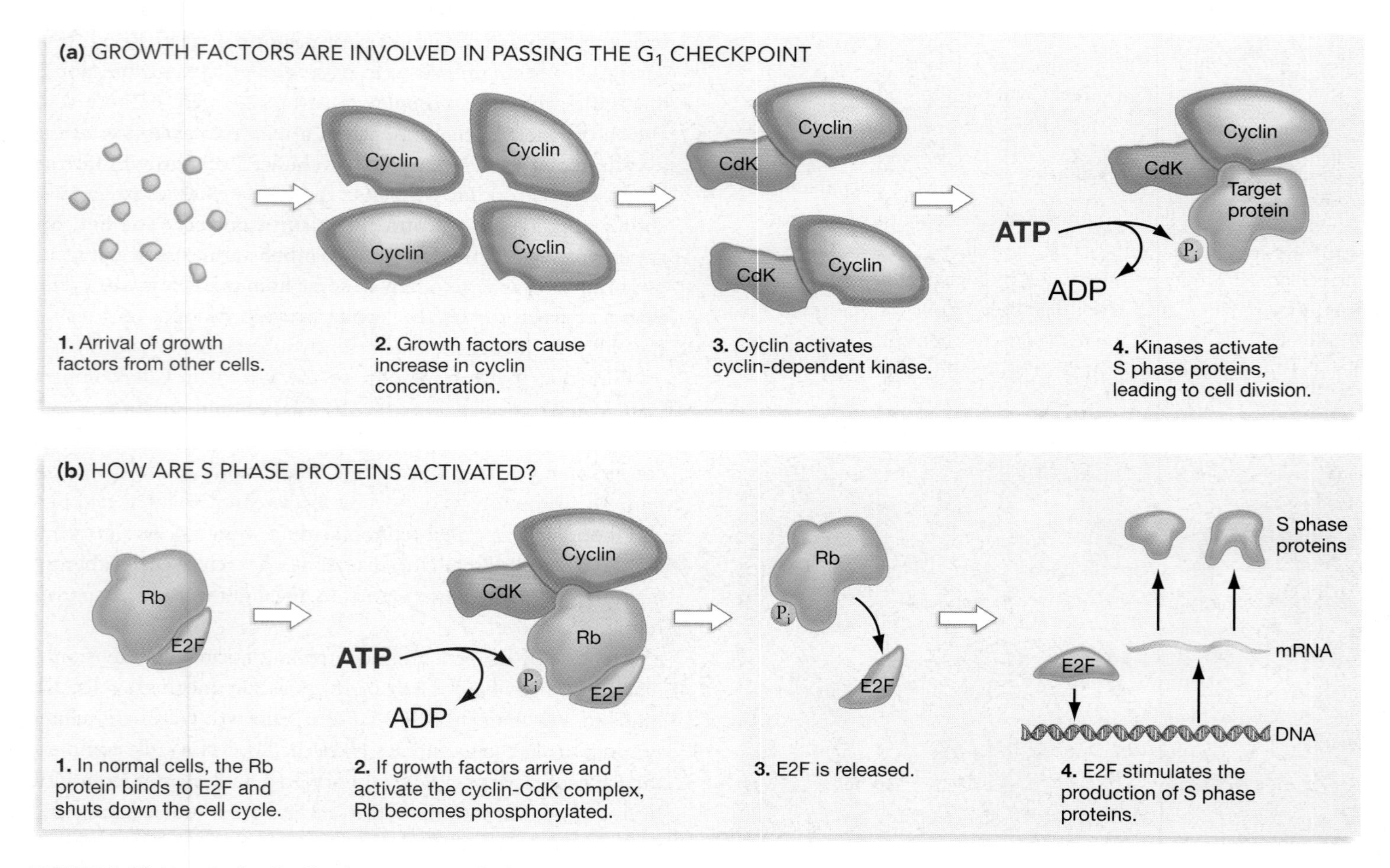

FIGURE 8.18 How Is the G_1 Checkpoint Controlled?
QUESTION What would happen if the Rb protein were defective, or not produced at all? What would happen if cyclin concentrations were high all the time, irrespective of the arrival of growth factors?

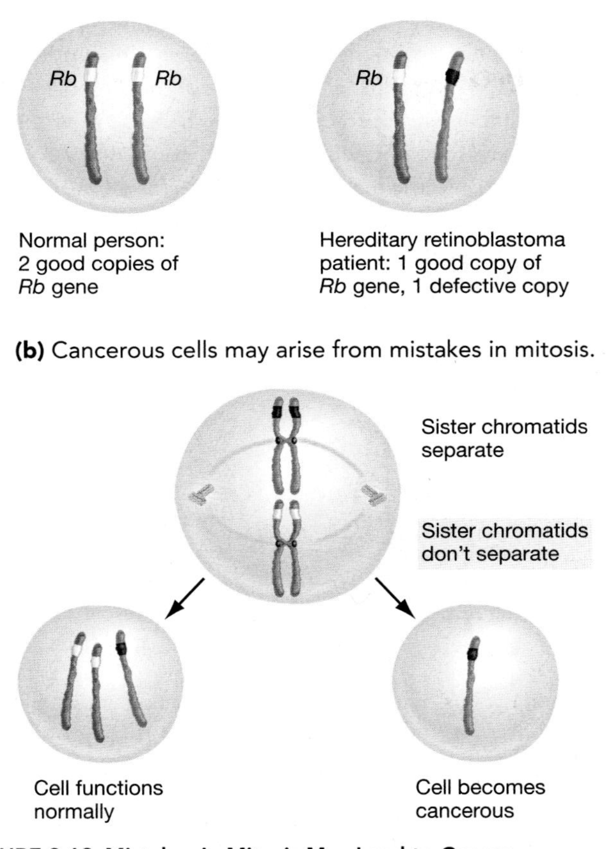

FIGURE 8.19 Mistakes in Mitosis May Lead to Cancer
(a) Individuals with hereditary retinoblastoma have a defective DNA sequence, or gene, for the Rb protein. **(b)** If chromosomes are not partitioned correctly during mitosis, as in hereditary retinoblastoma patients, daughter cells may end up with no functioning copies of the *Rb* gene—and thus no functioning Rb protein.

specifically, if one kinetochore of the replicated normal chromosome fails to attach to the spindle when anaphase begins, both sister chromatids will be pulled to the same daughter cell. But if the sister chromatids of the defective chromosome separate normally during the same anaphase, then one daughter cell will contain a solitary defective chromosome.

How can the tumors with two defective chromosomes be explained? Cavenee and co-workers propose that these tumors result from the same mistake in mitosis, followed by duplication of the defective chromosome. This hypothesis remains to be confirmed, however, and the mechanism of chromosome duplication is currently unknown.

As you may recall from section 8.3, the metaphase checkpoint normally ensures that the error diagrammed in Figure 8.19b does not happen. But the observations made by Cavenee and colleagues suggest that this checkpoint is not foolproof, and that mistakes in mitosis can contribute to the onset of cancer. Chromosomal abnormalities are actually extremely common in cancerous cells. Further, most cancers involve more than one defective gene. Most cancers develop only after a series of genetic errors combine to break cell-cycle control and induce uncontrolled growth and metastasis. Each type of cancer is due to a unique combination of errors. This observation explains why cancer is so difficult to treat, and why research on cancer appears to progress so slowly.

Because cancer is a family of diseases with a complex molecular basis, there will be no "magic bullet" that cures all forms of the illness. Still, progress in understanding the cell cycle and the molecular basis of cancer has been dramatic recently, and prevention and early detection programs are increasingly effective. The prognosis for many cancer patients is remarkably better now than it was even a single generation ago.

Essay Cancer Chemotherapy

If a malignant tumor is detected before metastasis occurs, doctors may be able to cure the patient by surgically removing the growth. But if the cancer has begun to spread through the body by the time it is detected, stray cancer cells could be hiding anywhere, regardless of the location of the original tumor. For the patient to survive, all of these cancerous cells must be eradicated.

Why does chemotherapy make people so sick?

In these cases, doctors often propose surgery followed by chemotherapy—treatment with anticancer chemicals or drugs. You may know someone who has undergone chemotherapy for cancer. If so, you've witnessed the side effects: weakness, weight loss, and hair loss. Why does chemotherapy make people so sick?

Most of the drugs currently approved for chemotherapy have the same mode of action—they kill rapidly dividing cells. To heighten the probability of killing all stray cancer cells, physicians usually prescribe high doses of these drugs. They are applied to the entire body, often through injection into the bloodstream. Besides killing cancer cells, though, anticancer drugs also kill normal cells that divide rapidly. These include blood cells, intestinal cells, and hair follicles. As a result,

(Continued on page 172)

(Essay continued)

chemotherapy leads to low blood-cell counts, loss of appetite, loss of hair, and other symptoms.

Physicians who prescribe chemotherapy have to achieve a balance between the probability of eradicating all cancer cells and the need to have the patient tolerate the treatment. One approach to achieving this balance is to mitigate the most dangerous side effects, such as low white-cell counts in the blood. Patients with low blood-cell counts may be given growth factors that stimulate the production of white cells. Because these cells fight infections, patients with depressed white-cell counts may also be given antibiotics.

Another approach in chemotherapy is to prevent side effects altogether by targeting the drugs more precisely. If the cancer has not spread far from the original tumor site, anticancer drugs may be applied to just a limited region of the body. This is rarely possible, however, so many researchers are focusing on an alternative strategy—attaching chemotherapy drugs to proteins that bind to specific membrane proteins on the surfaces of cancer cells. One idea is to package drugs inside liposomes that are coated with a protein that binds to cancer cells. The hope is that the liposomes will fuse with the cancer cell membrane and deliver the drugs directly to the malignant cell. If this research is successful, chemotherapy may eventually become largely free of side effects and patient suffering.

Chapter Review

Summary

In the late nineteenth century, careful microscopic observations revealed that cells reproduce by division of preexisting cells. During the type of cell division called mitosis, the chromosome content of the cell remains constant. As a result, the daughter cells of mitosis are genetically identical to each other and to the parent cell. Most unicellular and some multicellular organisms reproduce by mitosis, and mitotic divisions are responsible for building the bodies of multicellular organisms from a single-cell embryo.

Dividing cells alternate between the dividing phase, called M phase, and a resting phase known as interphase. This alternation describes the cell cycle—the ordered sequence of events that occurs during the life of a cell. Chromosome replication, or S phase, occurs during interphase. S phase and M phase are separated in time by gap phases called G_1 and G_2. Cell growth and replication of non-nuclear cell components occurs during the gap phases. The order of the cell-cycle phases is $G_1 \rightarrow S \rightarrow G_2 \rightarrow M \rightarrow G_1$ and so forth.

Progression through the cell cycle is regulated at three checkpoints. Passage through the G_1 checkpoint represents an irreversible commitment to divide, and is contingent upon cell size, nutrient availability, and/or growth signals from other cells. At the G_2 checkpoint, progression through the cycle is delayed until chromosome replication has been successfully completed. At the metaphase checkpoint during M phase, anaphase is delayed until all chromosomes are correctly attached to the mitotic spindle.

An array of different molecules regulates progression through the cell-cycle checkpoints. The most important of these regulatory molecules are cyclin-dependent kinases (Cdk's) and cyclins, which are found in all eukaryotic cells. Active Cdk's are enzymes that trigger progress through a checkpoint by phosphorylating important target proteins. Cyclins are proteins whose concentrations oscillate during the cell cycle, regulating the activity of Cdk's. In multicellular organisms, cyclin concentrations are partially controlled by growth factors from other cells. As a result, the G_1 checkpoint is said to be under social control.

Cancer is a common disease that is characterized by loss of social control over the G_1 checkpoint, resulting in cells that divide in an uncontrolled fashion. Cancer cells also have the ability to spread throughout the body, which often makes treatment difficult. In normal cells, growth-factor signals are required to activate G_1 cyclin-Cdk complexes and trigger division. An important target of activated G_1 cyclin-Cdk complexes is the retinoblastoma (Rb) protein, which regulates production of enzymes and other proteins required for S phase. Defects in G_1 cyclin and Rb protein are common in human cancers. Cancerous cells usually have multiple defects, however. For example, development of malignant tumors in some hereditary retinoblastoma patients involves not only a defective form of the Rb protein but also a mistake in chromosome partitioning during the mitotic divisions that produce the retina. In this and other cancers, disease is triggered by defects in several genes involved in cell-cycle control.

Mitosis represents only one of two major types of cell division. The other type of cell division is meiosis, which produces daughter cells that are genetically different from each other and from the parent cell. The mechanism and function of meiosis are the subject of Chapter 9.

Questions

Content Review

1. Which statement about the daughter cells of mitosis is correct?
   a. They differ genetically from one another and the parent cell.
   b. They are genetically identical to one another and to the parent cell.
   c. They are genetically identical to one another but different from the parent cell.
   d. One of the two daughter cells is genetically identical to the parent cell.
2. What are the two most important structures involved in moving chromosomes during mitosis?
   a. centrosomes and chromosomes
   b. kinetochores and chromosomes
   c. kinetochores and mitotic spindle fibers
   d. centrosomes and mitotic spindle fibers
3. During the cell cycle, chromosome replication occurs during which phase?
   a. G_1 phase
   b. S phase
   c. G_2 phase
   d. M phase
4. During the cell cycle, when are chromosomes partitioned to daughter cells?
   a. G_1 phase
   b. S phase
   c. G_2 phase
   d. M phase
5. Progression through the cell cycle is regulated by oscillations in the concentration of which type of molecule?
   a. centrosomes
   b. microtubules
   c. cyclin-dependent kinases
   d. cyclins
6. How do cancer cells differ from normal cells?
   a. uncontrolled division and invasiveness
   b. uncontrolled division only—many normal cell types are invasive
   c. invasiveness only—many normal cell types grow in an uncontrolled fashion
   d. interphase and M phase

Conceptual Review

1. According to the data in Figure 8.4, the first labeled mitotic cells appear about 4 hours after a labeling period ends. Labeled cells are then observed in mitosis over a span of 6–8 hours. From these data, researchers concluded that S phase lasts 6–8 hours and G_2 lasts about 4 hours. Explain their logic.
2. Sketch the phases of mitosis, listing the major events that occur in each phase. Identify at least two events that must be completed successfully for daughter cells to share an identical complement of chromosomes.
3. What are the consequences for the cell if the G_1 checkpoint fails? Answer the same question for the G_2 checkpoint and the metaphase checkpoint.
4. Explain how cell fusion and microinjection experiments supported the hypothesis that specific molecules are involved in the transition from G_2 to M phase and from G_1 to S phase.
5. Why are most protein kinases considered regulatory proteins?
6. Why are cyclins called cyclins? Explain their relationship to cyclin-dependent kinases and to growth factors.
7. Early detection is the key to surviving most cancers. Why?

Applying Ideas

1. In multicellular organisms, nondividing cells stay in G_1 phase. For the cell, why is it advantageous to be held in G_1 phase rather than S, G_2, or M phase?
2. The Rb protein helps regulate the cell cycle in many types of cells. Children with hereditary retinoblastoma have a defective version of the Rb protein, but get tumors only in their eyes—not elsewhere. Suggest a hypothesis to explain why tumors start only in the retinas of children. How could you test your hypothesis?
3. Predict the outcome of an experiment involving fusion of a cell in G_1 phase with a cell in G_2 phase. What would happen to the G_1-phase nucleus? To the G_2-phase nucleus? Why?
4. Cancer is primarily a disease of older people. Further, a group of individuals may share a genetic predisposition to developing certain types of cancer, yet vary a great deal in time of onset—or not get the disease at all. Discuss these observations in light of the claim made in this chapter that several defects usually have to occur for cancer to develop.

CD-ROM and Web Connection

CD Activity 8.1: The Cell Cycle *(tutorial)*
(Estimated time for completion = 15 min)
What are the events of the cell cycle, and how are those events regulated to ensure proper cell division?

CD Activity 8.2: Phases of Mitosis *(animation)*
(Estimated time for completion = 5 min)
How does each daughter cell receive a full set of chromosomes identical to those of the parent cell?

At your **Companion Website** (http://www.prenhall.com/freeman/biology), you will find self-grading exams and links to the following research tools, online resources, and activities:

The Cell Cycle and Mitosis
Examine a number of tutorials and study materials about the cell cycle and mitosis.

Spindle Formation
This research site from the Howard Hughes Medical Institute describes research on the cytoskeletal changes in cells undergoing mitosis.

Additional Reading

"Cancer." *Scientific American*, September 1996. An entire issue devoted to "What you need to know about cancer."

McIntosh, J. R., and K. L. McDonald. 1989. The mitotic spindle. *Scientific American* 261 (October): 48–56. A look at the changes that the spindle undergoes during each stage of mitosis, and how the stages are controlled.

Murray, A. W., and M. W. Kirschner. 1991. What controls the cell cycle. *Scientific American* 264 (March): 56–63. A review of early research revealing molecules involved in cell-cycle regulation.

Murray, A., and T. Hunt. 1993. *The Cell Cycle: An Introduction*. (New York: W. H. Freeman). Provides a concise yet thorough review of cell growth and division, emphasizing important experiments.

Gene Structure and Expression

Genetics is the study of how traits are inherited and expressed. Our exploration of this field begins with **Chapter 9: Meiosis** and **Chapter 10: Mendel and the Gene**. These two chapters introduce the observations and experiments that launched the study of genetics by establishing how and why offspring resemble their parents.

Chapter 11: How Do Genes Work?, **Chapter 12: DNA Synthesis, Mutation, and Repair**, and **Chapter 13: Transcription and Translation** introduce molecular genetics. Chapter 11 reviews classic experiments that confirmed DNA as the hereditary material and revealed how it encodes information. Chapter 12 explores how DNA is replicated, while Chapter 13 delves into research on the molecular machines responsible for translating the information encoded by DNA into action in the form of RNA and proteins.

Starting in the 1960s, experiments revealed the mechanisms that control which genes are expressed at different times and in what quantities. This body of work is surveyed in **Chapter 14: Control of Gene Expression in Bacteria** and **Chapter 15: Control of Gene Expression in Eukaryotes**.

The unit concludes with **Chapter 16: Genomes** and **Chapter 17: Genetic Engineering and its Applications**. These chapters explore cutting-edge research on whole-genome sequencing and efforts to introduce novel genes into organisms. The ethical and ecological implications of this research is a prominent theme in both chapters.

Most Mead's milkweed populations live on North American prairies that are mowed for hay before the plants set seed. Because the plants can only reproduce asexually, genetic diversity is extremely low. © 1994 Susan Middleton and David Liittschwager

Meiosis

9

Every organism alive today can claim a long, unbroken line of descent that begins with its parents, grandparents, and great-grandparents and stretches back billions of years—to the first cells that inhabited Earth. As an example of these relationships, consider the history of our own species, *Homo sapiens*. The earliest fossils of modern humans were unearthed in Africa and are about 100,000 years old. From then until now, perhaps 5000 generations of people have existed. Each of these generations is connected, like links in a chain, through the transmission of hereditary information. In this way you are directly related to the population of ancestral humans that lived in Africa about 100,000 years ago.

In species that reproduce sexually, the connection between generations is made when a sperm and egg unite to form a new individual. Oscar Hertwig and Herman Fol were the first biologists to observe this link being formed. Hertwig, who worked independently of Fol, studied the large, translucent eggs of a Mediterranean sea urchin. Due to the semitransparency of this cell, Hertwig was able to actually see the nuclei of a sperm and an egg fuse. Fol made similar observations in sea urchins.

When these results were published in 1876, they raised an important question. Cell biologists had already established that the number of chromosomes is constant from cell to cell within a multicellular organism. It was also accepted that chromosome number is the same in the parent and daughter cells of mitosis. Hertwig was able to confirm that all of the cells and nuclei in sea urchin embryos were the products of mitotic divisions, and that they were the direct descendants of the nucleus that forms at fertilization. The question is how can the chromosomes from a sperm cell and an egg

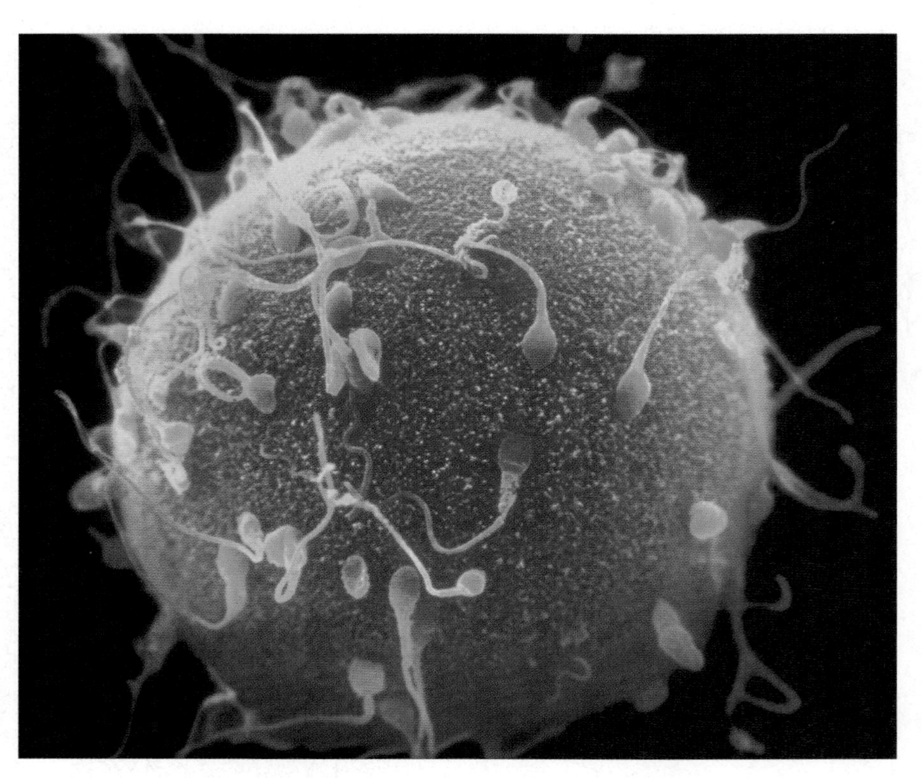

This scanning electron micrograph shows human sperm attempting to enter an egg. This chapter introduces the type of cell division called meiosis, which occurs prior to gamete formation.

cell combine, but form an offspring that has the same chromosome number as its mother and its father?

Edouard van Beneden hinted at the answer in 1883 when he noted that cells in the body of roundworms called *Ascaris* have four chromosomes, while their sperm and egg nuclei have only two chromosomes. Four years later, August Weismann formally proposed a hypothesis to explain the riddle: During the formation of **gametes**—reproductive cells like sperm and eggs—there must be a distinctive type of cell division that leads to a reduction in chromosome number. Specifically, if the egg and sperm contribute equally to the fertilized egg, Weismann reasoned that they must each contain half of the usual number of chromosomes. When the sperm and egg combine, the resulting embryo has the same chromosome number as its mother or its father.

In the decades that followed, cell biologists confirmed this hypothesis by observing gamete formation in a wide variety of plant and animal species. Eventually this type of cell division came to be called meiosis (lessening-act). **Meiosis** is a type of cell division that precedes the formation of eggs and sperm and leads to a halving of chromosome number. Meiosis creates the cells responsible for sexual reproduction.

9.1 How Does Meiosis Occur?

When cell biologists began to study the cell divisions that lead to gamete formation, they made an important observation: Each organism has a characteristic number of chromosomes. Consider the drawing in **Figure 9.1a**, taken from a paper published by Walter Sutton in 1902. It shows the chromosomes of the lubber grasshopper during the cell divisions leading up to the formation of sperm. There are a total of 24 chromosomes in the cell. Sutton realized, however, that they consist of just 12 distinct types of chromosomes; there are two chromosomes of each type. These different chromosomes can be identified on the basis of size and shape. Accordingly, Sutton designated 11 of the chromosomes with the letters *a* through *k* and the twelfth by the letter *X*. He referred to the pairs as **homologous chromosomes.** (They can also be called homologs.) The two chromosomes labeled *c*, for example, have the same size and shape and are homologous.

Figure 9.1b uses the colors red and blue to highlight the relationships among the 12 pairs of homologous chromosomes found in the cells of lubber grasshoppers. As Chapter 10 will show, later work demonstrated that homologous chromosomes are not only similar in size and shape, but also in content. Homologous chromosomes carry versions of the same genes. For example, each copy of chromosome *c* found in lubber grasshoppers might carry genes that influence eye formation, body size, call behavior, or other traits.

At this point in his study Sutton had succeeded in determining the number and types of chromosomes found in this species, or its **karyotype** (see **Box 9.1**, page 178). As karyotyping studies expanded, cell biologists realized that like lubber grasshoppers, the vast majority of plants and animals have more than one version of each type of chromosome. The biologists also invented terms to identify the number of copies observed. Organisms like lubber grasshoppers are called **diploid** (literally, "double-form") because they have two versions of each type of chromosome. Organisms like bacteria, archaea, and many algae are called **haploid** (single-form) because their cells contain a single chromosome or just one set of chromosomes. They do not contain homologous chromosomes. The total number of chromosomes found in some familiar plants and animals is listed in **Table 9.1**.

TABLE 9.1 The Number of Chromosomes Found in Some Familiar Organisms

Organism	Number of Unique Chromosomes (haploid set)	Diploid Chromosome Number
Humans	23	46
Domestic dog	36	72
Fruit fly	4	8
Chimpanzee	24	48
Ant	1	2
Garden pea	7	14
Corn (maize)	10	20

(a) 12 types of chromosomes in the lubber grasshopper

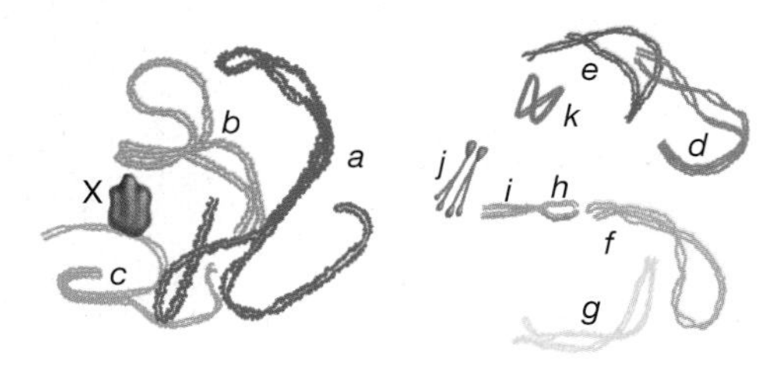

(b) Each type of chromosome has two homologs.

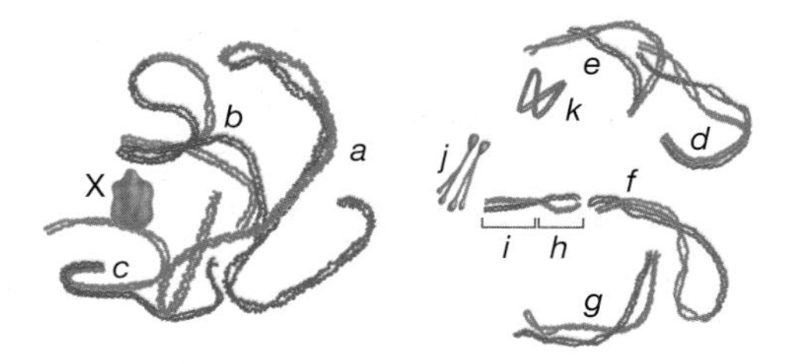

FIGURE 9.1 The Chromosomes of the Lubber Grasshopper
(a) Letters are placed next to each of the 12 distinct types of chromosomes found in lubber grasshopper cells. Note there are two of each type of chromosome. **(b)** The two members of a chromosome pair are called homologs. In this drawing, homologous chromosomes are indicated in blue and red.

BOX 9.1 Karyotyping Techniques

Although chromosomes maintain their individuality and physical integrity throughout the cell cycle, they are readily visible only during mitosis or meiosis. Early in cell division, chromosomes condense into highly compact structures that can be observed with the light microscope. To describe an individual's karyotype, then, biologists must study cells undergoing cell division.

Karyotyping is an important diagnostic procedure in humans. For example, early in a pregnancy a physician may obtain a sample of cells from a developing embryo. If the karyotype reveals an abnormal chromosome set, it means that the embryo may not develop properly.

To generate a karyotype, technicians begin by growing cells in culture. Once the cells are dividing rapidly they are treated with a compound called colchicine. Colchicine acts to stop mitosis. The chromosomes of colchicine-treated cells are then stained and examined with the light microscope.

Using traditional staining procedures, clinicians can distinguish chromosomes by size, centromere position, and striping or banding patterns. These data allow health-care workers to diagnose problems in embryos that involve missing, extra, or altered chromosomes.

Recently, however, researchers have developed a higher-resolution technique for karyotyping called chromosome painting. The "painting" is done with fluorescent dyes that are attached to short DNA molecules. The pieces of DNA bind to particular chromosomes. By using a combination of dyes, technicians can give each pair of homologous chromosomes a distinctive suite of colors (see **Figure 1**). The high-resolution image produced by this technique is important because it allows clinicians to diagnose subtle chromosomal abnormalities, such as small missing segments, in addition to the large-scale problems revealed by traditional stains.

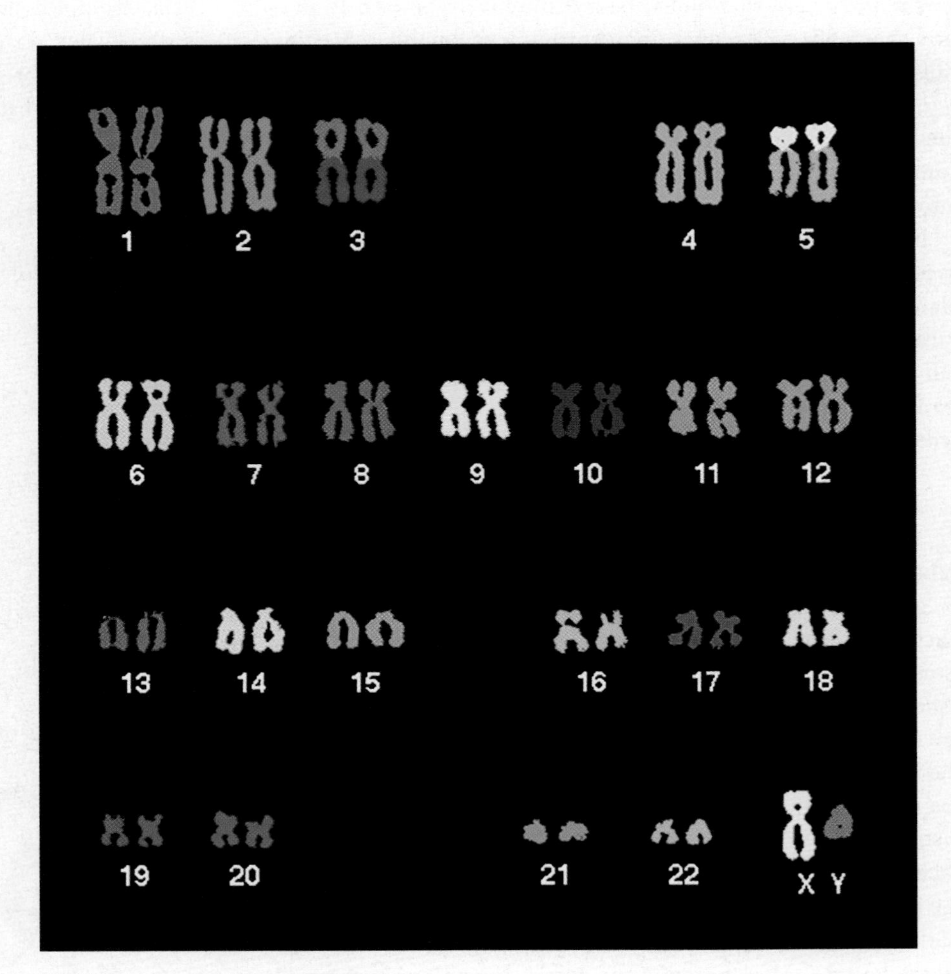

FIGURE 1 Human Chromosomes
Chromosomes undergoing mitosis are arranged randomly when first observed with the microscope. To determine a karyotype, a technician groups chromosomes by pairs and arranges them by number, as shown here. QUESTION Is this human karyotype normal or does it reveal the presence of an extra chromosome or the absence of a chromosome?

Researchers invented a compact notation, as well, to indicate the number of chromosome sets in a particular organism or type of cell. By convention, the letter n stands for the number of distinct types of chromosomes in a given cell. A number placed before the n represents the number of complete chromosome sets observed. So, for example, diploid cells or species are designated $2n$ (there are n distinct types of chromosomes in a diploid cell and, further, there are 2 complete sets of n chromosomes); haploid cells or species are labeled n (there is one set of n types of chromosomes in a haploid cell—no homologs are observed).

Later work revealed that it is common for species in some lineages—particularly land plants such as ferns—to contain more than two sets of chromosomes. Instead of having just two homologous chromosomes, **polyploid** (many-form) species may have three, four, or eight homologs in each cell. Such species are called triploid ($3n$), tetraploid ($4n$), or octoploid ($8n$), respectively.

Sutton and the other early cell biologists did more than describe the types and numbers of chromosomes observed in their study organisms, however. Through careful examination, they were able to track how these chromosome sets change during meiosis. These studies confirmed Weismann's hypothesis that a special type of cell division occurs during gamete formation; by doing so, the research solved the riddle of fertilization.

An Overview of Meiosis

Cells replicate each of their chromosomes before undergoing meiosis. At the start of the process, then, chromosomes are in the same state they are prior to mitosis. Through DNA replication, each chromosome consists of two identical sister chromatids joined at a structure called the centromere (**Figure 9.2a**).

Meiosis begins after the chromosomes have been replicated and continues through two cell divisions called meiosis I and meiosis II. As **Figure 9.2b** shows, the two divisions differ sharply. During meiosis I, the homologs in each chromosome pair separate. One homolog goes to each of the two daughter cells. As a result, each of the daughter cells from meiosis I has one of each type of chromosome instead of two. The products of meiosis I have half as many chromosomes as the original cell. During meiosis II, sister chromatids from each chromosome separate to the two daughter cells. One sister chromatid goes to each cell. The cell that starts meiosis II has one of each type of chromosome; the cells produced by meiosis II also have one of each type of chromosome.

The early cell biologists realized that the outcome of meiosis is a reduction of chromosome number. In most plants and animals, the original cell is diploid and the four daughter cells are haploid. These four haploid daughter cells, each containing one of each homologous chromosome, eventually go on to form gametes. When two gametes fuse during fertilization, a full complement of chromosomes is restored (**Figure 9.2c**). The cell that results from fertilization is diploid—just like its parents—and is called a **zygote**.

To summarize, during meiosis I homologous chromosomes separate and chromosome number is halved. During meiosis II, sister chromatids separate. A parent cell with replicated chromosomes produces daughter cells with unreplicated chromosomes. Meiosis II is virtually identical to a mitotic division in a haploid cell.

Sutton, Theodor Boveri, T. H. Montgomery, and a host of other early cell biologists worked out this sequence of events through careful observation of cells with the light microscope.

(a) Each chromosome replicates prior to undergoing meiosis.

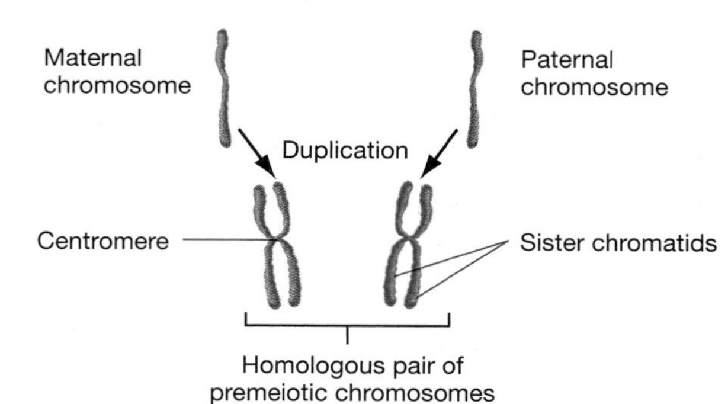

(b) During meiosis, chromosome number in each cell is reduced.

(c) A full complement of chromosomes is restored during fertilization.

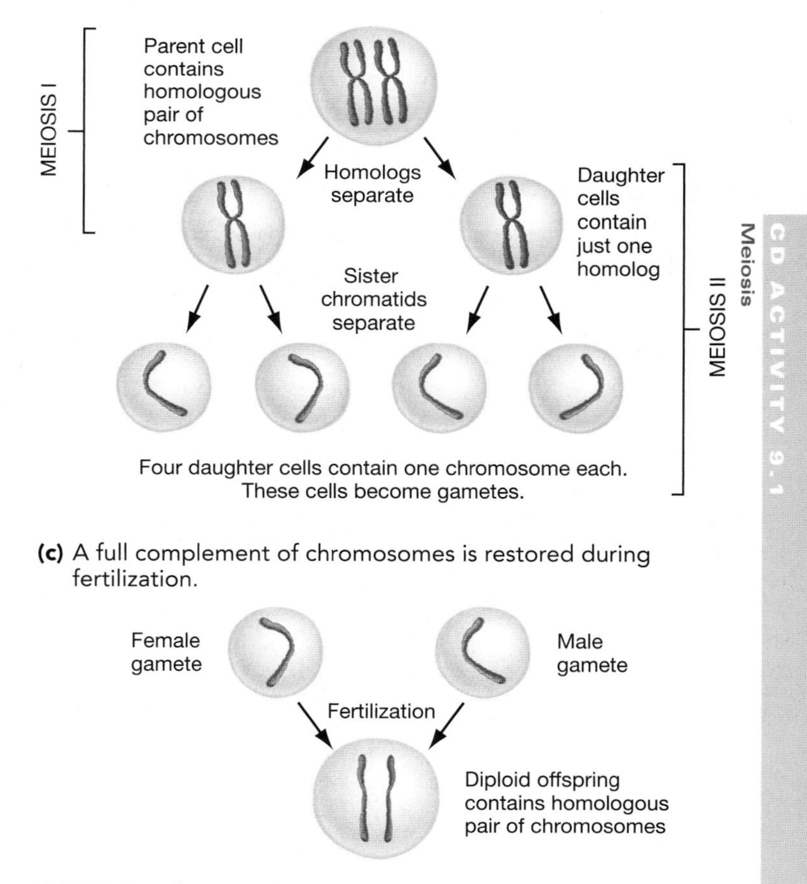

FIGURE 9.2 The Major Events in Meiosis
Meiosis reduces chromosome number by one-half. In diploid organisms, the products of meiosis are haploid.

CD ACTIVITY 9.1 Meiosis

Based on these studies, they came to a key realization: Diploid individuals receive a haploid chromosome set from their father and a haploid set from their mother. As a result, homologous chromosomes can be referred to as paternal or maternal in origin. This observation will become crucial as we examine meiosis in more detail.

A Closer Look at Meiosis I

Meiosis begins once chromosomes have been replicated. **Figure 9.3** diagrams the major steps that occur next, using a cell with a diploid chromosome number of 4 ($2n = 4$) as an example. In the figure, chromosomes originating from the individual's mother are shown in red while chromosomes derived from the individual's father are shown in blue.

The first event illustrated during meiosis I is crucial. In this step, homologous chromosome pairs come together physically in a process called **synapsis**. This forms a structure called a tetrad or bivalent. A tetrad consists of the four sister chromatids from two homologous chromosomes.

While they are synapsed, chromatids from homologous chromosomes can cross over one another. Each crossover forms an X-shaped structure called a **chiasma** (the plural is chiasmata). Normally, at least one chiasma forms in every pair of homologous chromosomes; often there are several. As Figure 9.3 shows, the chromatids involved in chiasma formation are homologous but they are not sisters. Consistent with this observation, Thomas Hunt Morgan proposed in 1911 that a physical exchange of paternal and maternal chromosomes occurs at chiasmata. According to this hypothesis, paternal and maternal chromatids break and rejoin at each chiasma, producing chromatids that have a mixture of paternal and maternal segments. Morgan called this process **crossing over**. In Figure 9.3, the hypothesized result is illustrated by chromosomes with both red and blue segments. **Box 9.2** (page 182) reviews one of the experiments that confirmed Morgan's hypothesis.

The next major stage in meiosis I occurs when pairs of homologous chromosomes migrate to the middle of the cell. Meiosis I concludes when each homologous pair of chromosomes separates and is distributed to two different daughter cells. A key point here is that the movement of each chromosome is independent of the movement of other chromosomes. As a result, each daughter cell gets a random assortment of maternal and paternal chromosomes.

After examining the diagram in Figure 9.3, it should make sense to you that at the end of meiosis I each daughter cell contains one member of each homologous pair of chromosomes. In each chromosome, however, the sister chromatids remain attached.

Although meiosis I is a continuous process, biologists summarize the events by identifying four distinct phases.

- Prophase I—Homologs synapse, forming pairs of homologous chromosomes (tetrads). Crossing over occurs.
- Metaphase I—Pairs of homologous chromosomes migrate to the middle of the cell.
- Anaphase I—Homologs separate and begin moving to one of two ends of the cell.

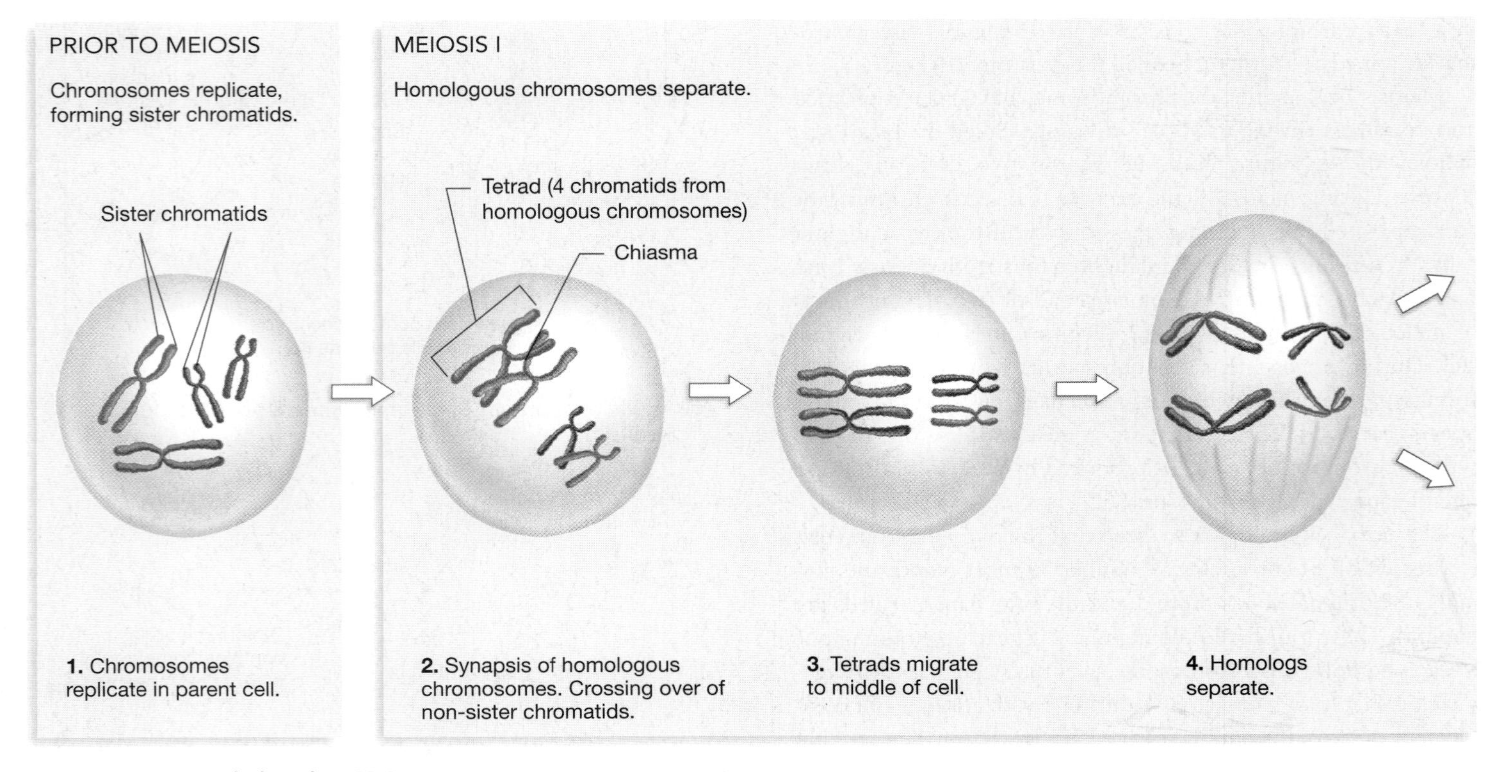

FIGURE 9.3 A Detailed Look at Meiosis

EXERCISE Label prophase, metaphase, anaphase, and telophase cells of meiosis I and meiosis II.

- Telophase I—Homologs finish moving to each end of the cell and the cytoplasm divides to form two daughter cells.

A Closer Look at Meiosis II

At the start of meiosis II, each chromosome consists of two sister chromatids attached to each other at the centromere. During the second division in meiosis, chromosomes migrate to the middle of the cell. From there, the sister chromatids of each chromosome separate and migrate to different daughter cells. Once they are separated, each chromatid becomes an independent chromosome. Four haploid cells, each with one chromosome of each type, are produced from meiosis II. It should make sense to you, after examining Figure 9.3 in detail, that meiosis II is virtually identical to a mitotic division in a haploid cell. (**Table 9.2** provides a detailed comparison of meiosis and mitosis.)

As in meiosis I, biologists routinely designate four distinct phases in meiosis II.

- Prophase II—Duplicated chromosomes, consisting of two sister chromatids, begin moving to the middle of the cell.
- Metaphase II—Chromosomes are arranged in the middle of the cell.

TABLE 9.2 A Comparison of Mitosis and Meiosis

Feature	Mitosis	Meiosis
Number of cell divisions	one	two
Number of chromosomes in daughter cells, compared to parent cell	same	half
Synapsis of homologs	does not occur	yes
Number of crossing-over events	none	at least one per pair of homologous chromosomes
Makeup of chromosomes in daughter cells compared to parent cell	identical	either paternal or maternal homolog is present, but not both; paternal and maternal segments mixed within chromosomes
Role in life cycle	reproduction in unicellular organisms; builds body of multicellular organisms	precedes production of gametes in sexually reproducing organisms

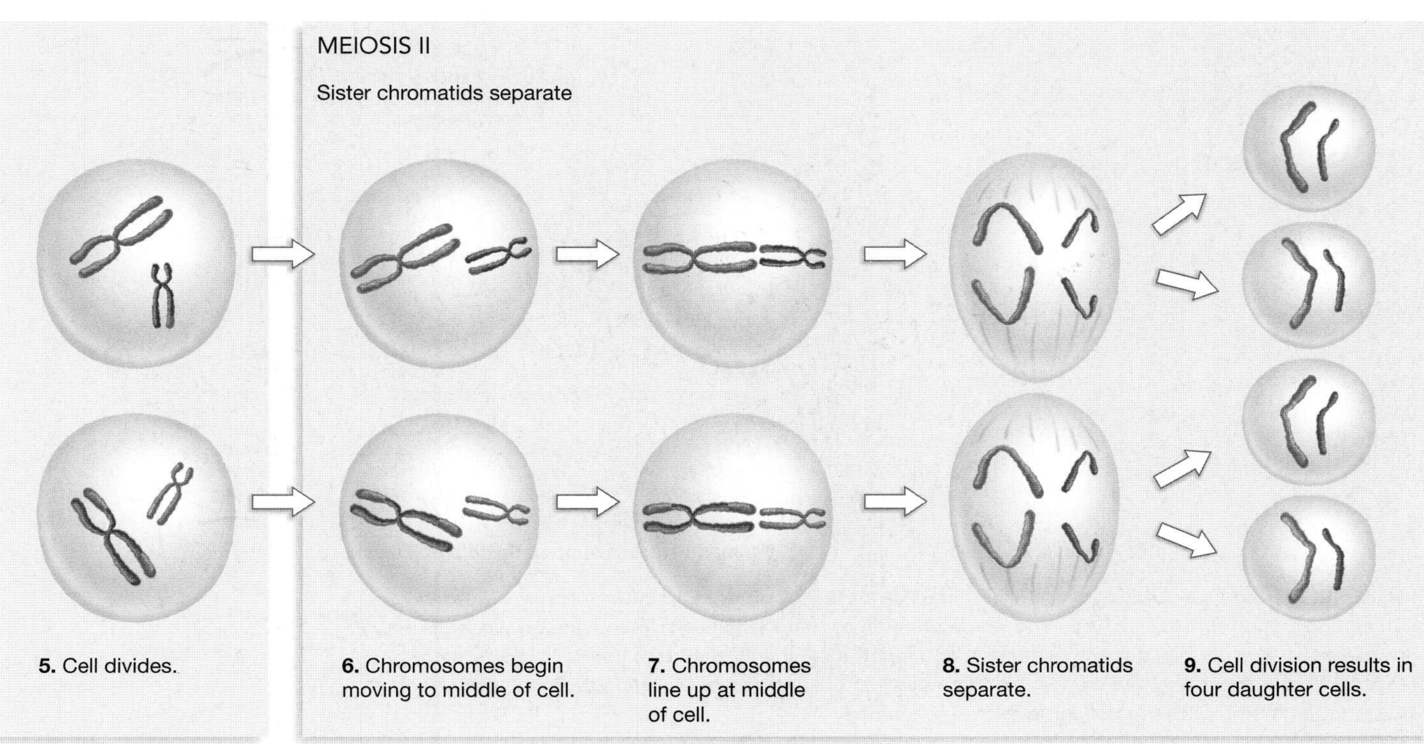

BOX 9.2 Experimental Evidence for Physical Exchange of Chromosome Segments During Crossing Over

Thomas Hunt Morgan proposed that segments of paternal and maternal chromatids cross over—or break and reanneal—during meiosis I. If Morgan was correct, it meant that chromosomes exchange genetic information when they cross over.

Was Morgan's hypothesis valid? Did crossing over imply a breaking and reattachment of chromosome segments?

An experiment that Harriet Creighton and Barbara McClintock performed in 1931 on *Zea mays* (maize or corn) helped to answer these questions. These researchers worked with strains of maize that differed in two ways. In one strain, chromosome 9 was longer than usual and had a knob at one end. In the other strain, chromosome 9 was of normal length and lacked this knob. **Figure 1a** illustrates these morphological differences in the chromosomes; **Figure 1b** shows how the two strains differed genetically. The knobby, long chromosome 9 consisted of genes that led to the production of colored, waxy kernels. The shorter, knobless chromosome 9 had genes that encoded instructions for colorless, starchy kernels.

Figure 1c shows the gametes that result when cells containing both of these chromosomes undergo meiosis. Note that Morgan's hypothesis predicts that if crossing over between the two genes occurs, the resulting gametes should have a new type of chromosome 9: short with a knob, or long with no knob. The offspring that have these chromosomes should also have a unique combination of traits: kernels that are colored and starchy, or kernels that are colorless and waxy.

Creighton and McClintock tested these predictions by arranging matings between individuals from each strain. Their offspring had one copy of the long, knobbed chromosome 9 and one copy of the short, knobless chromosome 9. What happened when the two distinctive homologs crossed over and produced gametes? When Creighton and McClintock compared the chromosomes observed in offspring with the morphology of their kernels, the results conformed *exactly* to the predictions in Figure 1c. Every individual with two copies of the short, knobbed chromosome 9 had colored, starchy kernels. Every individual with two copies of the long, knobless chromosome 9 had colorless, waxy kernels.

Based on these results and the outcome of similar experiments that Curt Stern performed on fruit flies, Morgan's hypothesis became widely accepted. A physical exchange of maternal and paternal chromatids occurs during crossing over, resulting in chromosomes with a combination of genetic information from the mother and father.

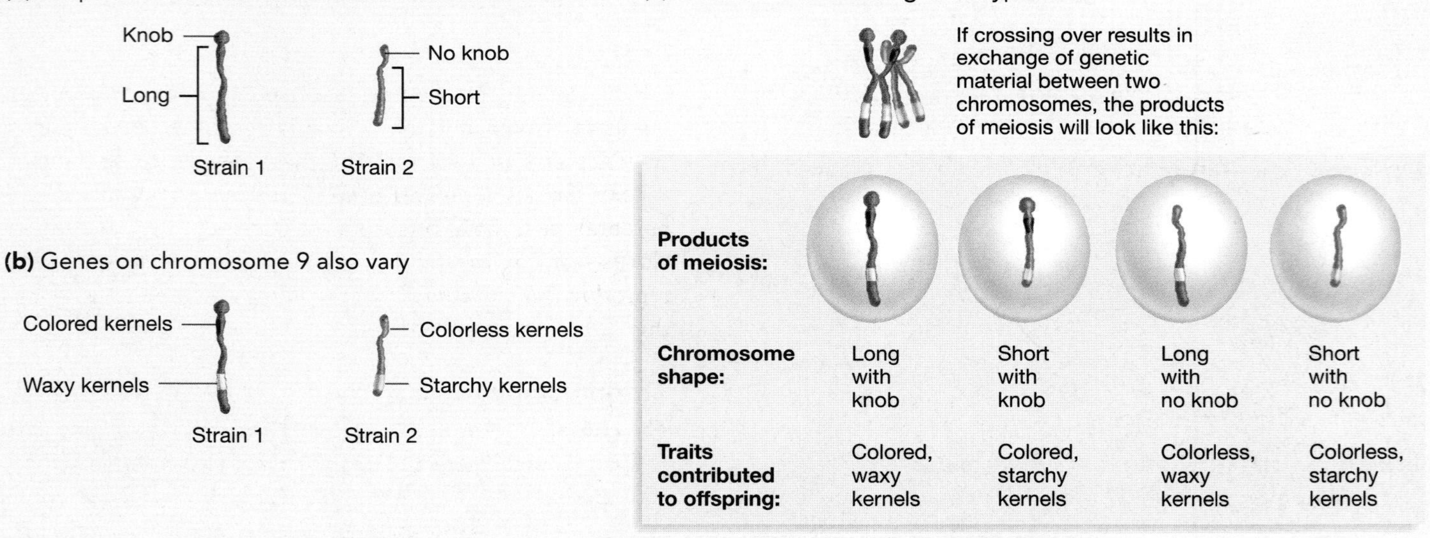

FIGURE 1 Crossing Over Leads to New Combinations of Genes on the Same Chromosome
(a) In certain strains of maize, the size and shape of chromosome 9 differs. **(b)** The distinctive versions of chromosome 9 also contain distinctive genes that affect the color and texture of kernels in adult maize plants. **(c)** These diagrams show the types of gametes and offspring traits that should occur if crossing over results in the physical exchange of chromosome segments.

(a)

(b)

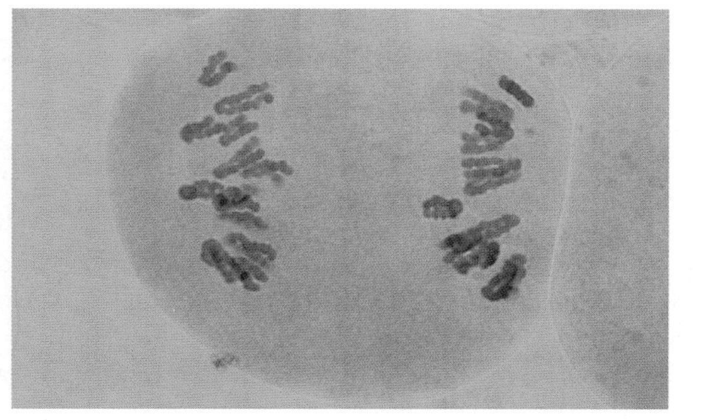

(c)

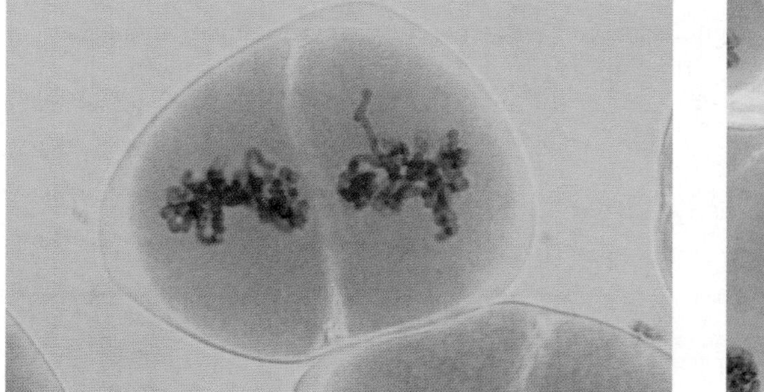

(d)

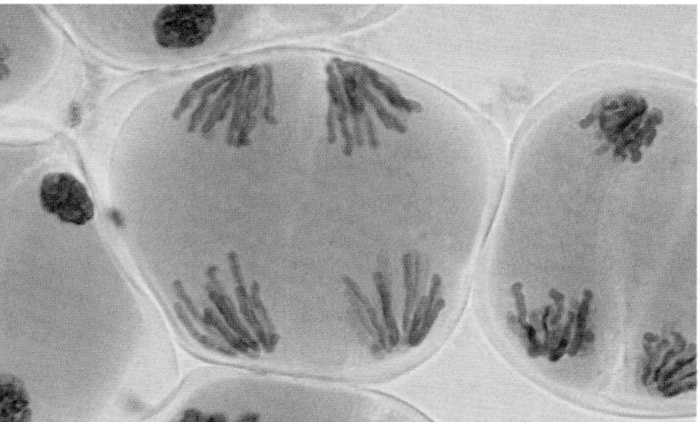

FIGURE 9.4 Meiotic Chromosomes in Easter Lilies
QUESTION AND EXERCISE **(a)** This cell is undergoing meiosis I. What is happening? **(b)** What stage of meiosis I is illustrated here? Is this cell haploid or diploid? **(c)** Is this meiosis I or II? What phase? **(d)** What is happening in this photograph?

- Anaphase II—Sister chromatids separate; each independent chromosome begins moving to either end of the cell.
- Telophase II—Chromosomes finish moving to each end of the cell; as this occurs, the cytoplasm divides to form two daughter cells.

Figure 9.4 offers you a chance to practice with the vocabulary and concepts reviewed in this section.

9.2 The Consequences of Meiosis

The cell biologists who worked out the details of meiosis in the late 1800s and early 1900s realized that the process solved the riddle of fertilization. Weissman's hypothesis—that a reduction division precedes gamete formation—was confirmed. But researchers also understood that meiosis has another important outcome. Thanks to crossing over during meiosis I, the chromosomes found in gametes are different than the chromosomes found in parental cells. Subsequently, fertilization brings haploid sets of chromosomes from a mother and father together to form a diploid offspring whose chromosomes are unlike those found in either parent.

This change in chromosomal makeup does not occur during asexual reproduction, which is usually based on mitosis. As Chapter 8 pointed out, the chromosomes in the daughter cells of mitosis are identical to the chromosomes found in the parental cell. The offspring of asexual reproduction are clones—or copies—of their parent. The offspring of sexual reproduction, in contrast, are unlike their parents. Why is this difference important?

Chromosomes and Heredity

The changes in chromosomes produced by meiosis and fertilization are significant because chromosomes contain the cell's heredity material. Stated another way, chromosomes contain the instructions for specifying what a particular trait might look like in an individual. These inherited traits range from eye color and height in humans to the number or shape of the bristles on a fruit fly's leg to the color or shape of the seeds found in pea plants.

In the early 1900s biologists began using the term **gene** to refer to the inherited instructions for a particular trait. Chapter 10 explores the experiments that confirmed that each chromosome found in an organism is composed of a series of genes encoding information for different traits. In humans, for example, a single chromosome might comprise genes that influence height, hair color, the spacing of teeth, and the tendency to develop colon cancer.

Because chromosomes are composed of genes, the offspring produced during asexual reproduction are genetically identical to one another as well as to their parent. But the offspring produced by sexual reproduction are genetically different from one another and unlike either their mother or their father. The goal of this section is to analyze three aspects of meiosis that create variation in the chromosomes—and hence the genetic makeup—of sexually produced offspring.

How Does the Segregation of Homologous Chromosomes Produce Genetic Variation?

The cells in your body each contain 23 homologous pairs of chromosomes and 46 chromosomes in total. Half of these chromosomes originated from your mother and half came from your father. Each chromosome is composed of genes that specify particular traits. For example, one gene that influences your eye color might be located on chromosome 2, while one of the genes that influences your hair color might be located on chromosome 6 (**Figure 9.5a**).

Suppose that the set of chromosomes you inherited from your mother contains genes that tend to produce brown eyes and black hair, but your paternal chromosomes possess genes that tend to specify blue eyes and red hair. (It's important to note, however, that in reality many genes interact in complex ways to produce human eye color and hair color.) Will the gametes you produce contain the genetic instructions you inherited from your mother or the genes you inherited from your father?

To answer this question, study the diagram of meiosis in **Figure 9.5b**. It shows that when pairs of homologous chromosomes line up during meiosis I and homologous chromosomes separate, a variety of combinations of maternal and paternal chromosomes result. In the example given here, meiosis results in gametes with genes for brown eyes and black hair, like your mother, and blue eyes and red hair, like your father. But two additional combinations also occur: brown eyes and red hair, or blue eyes and brown hair. A total of four different combinations of paternal and maternal chromosomes are possible when two chromosomes are distributed to daughter cells during meiosis I.

How many different combinations of maternal and paternal homologs are possible when larger numbers of chromosomes are involved? In an organism with three chromosomes per haploid set ($n = 3$), eight different types of gametes can be generated by randomly grouping maternal and paternal chromosomes. In general, a diploid organism can produce 2^n different combinations of maternal and paternal chromosomes, where n is the haploid chromosome number. This means that a human ($n = 23$) can produce 2^{23} or about 8.4 million gametes that differ in their combination of maternal and paternal chromosome sets. The message here is that the reassortment of whole chromosomes generates an impressive amount of genetic variation among gametes.

The Role of Crossing Over

Recall that segments of paternal and maternal chromatids exchange at each chiasma that forms during meiosis I. As a result, crossing over produces new combinations of genes on the same chromosome. This phenomenon is known as **recombination.** Recombination is a general term for any change in the combination of alleles found on a given chromosome. Recombination can occur in haploid organisms like bacteria as well as in diploid organisms that undergo meiosis. (**Box 9.3**, page 189, introduces the process responsible for recombination in bacteria.)

Crossing over and recombination are important because they dramatically increase the genetic variability of gametes

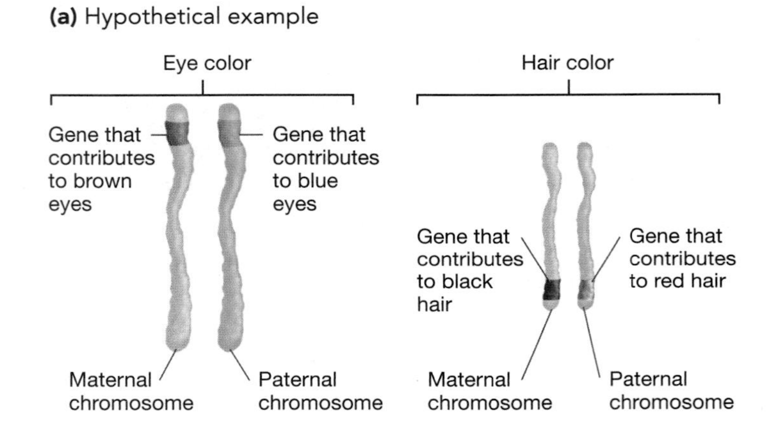

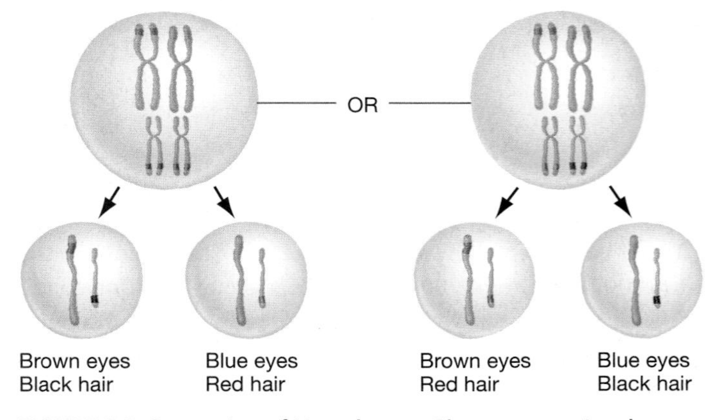

FIGURE 9.5 Separation of Homologous Chromosomes Leads to New Combinations of Genes
(a) In this hypothetical example, genes that influence hair color and eye color in humans are located on different chromosomes. **(b)** This diagram shows how gametes with different combinations of genes result from separation of homologous chromosomes during meiosis I.

produced by meiosis. In humans, for example, crossing over occurs at least once in all 23 pairs of chromosomes. As a result, the number of genetically different gametes that an individual can produce is much more than 8 million. It is incalculable. As section 9.3 will show, the genetic variation produced by meiosis has enormous consequences for the ability of offspring to survive and reproduce.

How Does Fertilization Affect Genetic Variation?

Crossing over and the random mixing of maternal and paternal chromosomes ensures that the chromosome contents of gametes are unlike one another and different from their parent. This is an important point. Even if two gametes produced by the same individual fuse to form a diploid offspring—meaning that **self-fertilization** or "selfing" takes place—the offspring are very likely to be genetically different from the parent (**Figure 9.6**).

In many sexually reproducing species, however, self-fertilization is rare or nonexistent. Instead, gametes from different individuals combine to form offspring. This is called **outcrossing.** Outcrossing increases the genetic diversity of offspring because it combines chromosomes from different individuals. How many genetically distinct offspring can be produced when outcrossing occurs? Let's answer this question using humans as an example. Recall that a human can produce about 8.4 million different kinds of gametes—even in the absence of crossing over. When a person mates with a member of the opposite sex, the number of different genetic combinations that can result is equal to the product of the number of different gametes produced by each parent. In humans this means that 8.4 million $\times$ 8.4 million $= 70.6 \times 10^{12}$ genetically distinct offspring can result from any particular mating. This calculation does not even take crossing over into account. In this light, it is not surprising that siblings are often very different from one another and from their parents.

When Does Meiosis Occur During the Life of an Organism?

Once biologists had described meiosis, they finally understood the mechanism behind sexual reproduction. The breakthrough

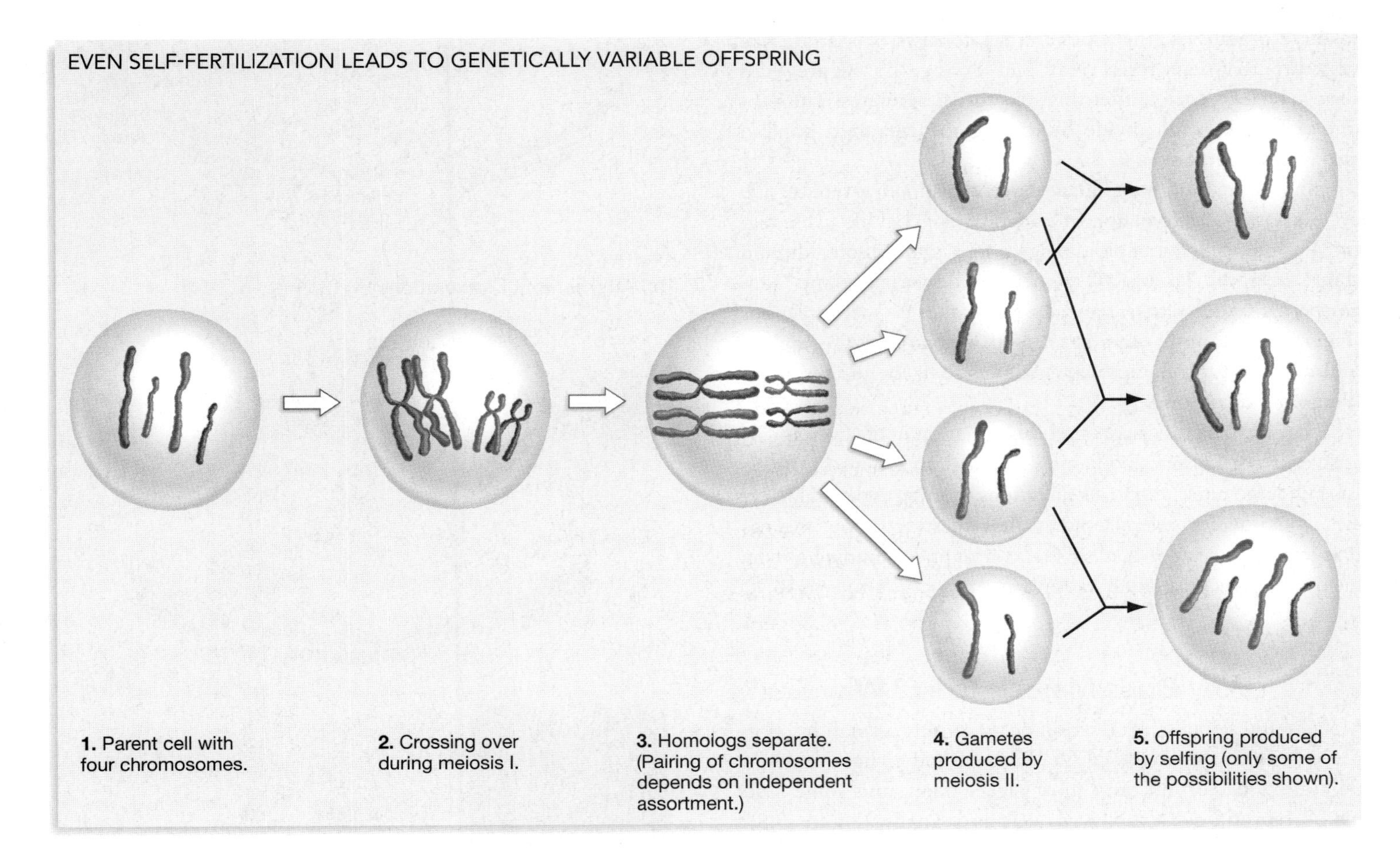

FIGURE 9.6 Even If Self-Fertilization Takes Place, Offspring Are Genetically Variable
This example shows some of the possible results of self-fertilization in an organism with four chromosomes ($2n = 4$). **EXERCISE** In step 2, label where a double crossover event is occurring. To the last line in this figure, add sketches showing the chromosome complements in additional offspring produced by selfing.

FIGURE 9.7 An Introduction to Life Cycles
(a) In most animals, the gametes are the only haploid cells. Meiosis occurs in special reproductive tissues. **(b)** In many algae, the fertilized egg is the only diploid cell. When this cell undergoes meiosis, the haploid cells that are produced go on to form a multicellular adult. **(c)** In land plants and some algae, there is a multicellular diploid stage and a multicellular haploid stage. Typically, one of these two stages is larger in size and longer in life span than the other. In ferns, the diploid stage is more prominent; in mosses the haploid stage is more prominent.

(a) Diploid dominant

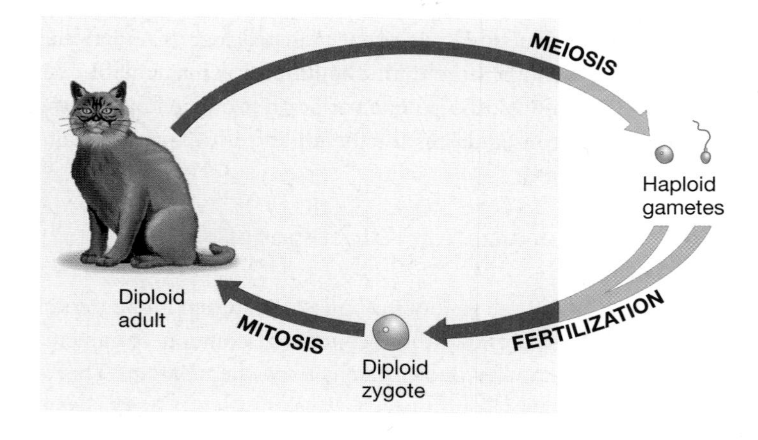

(b) Haploid dominant

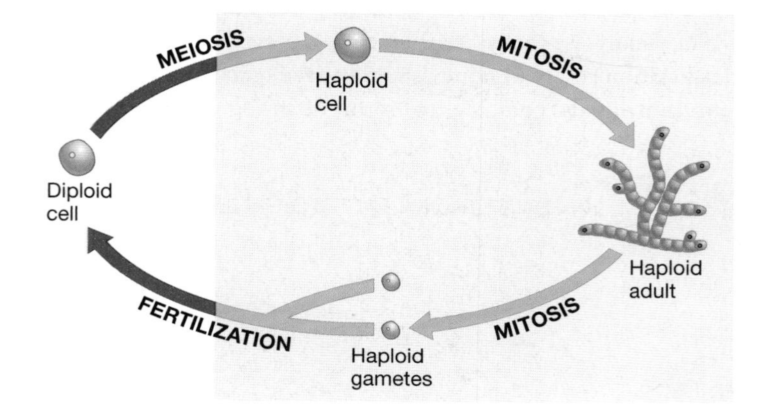

(c) Alternation of generations

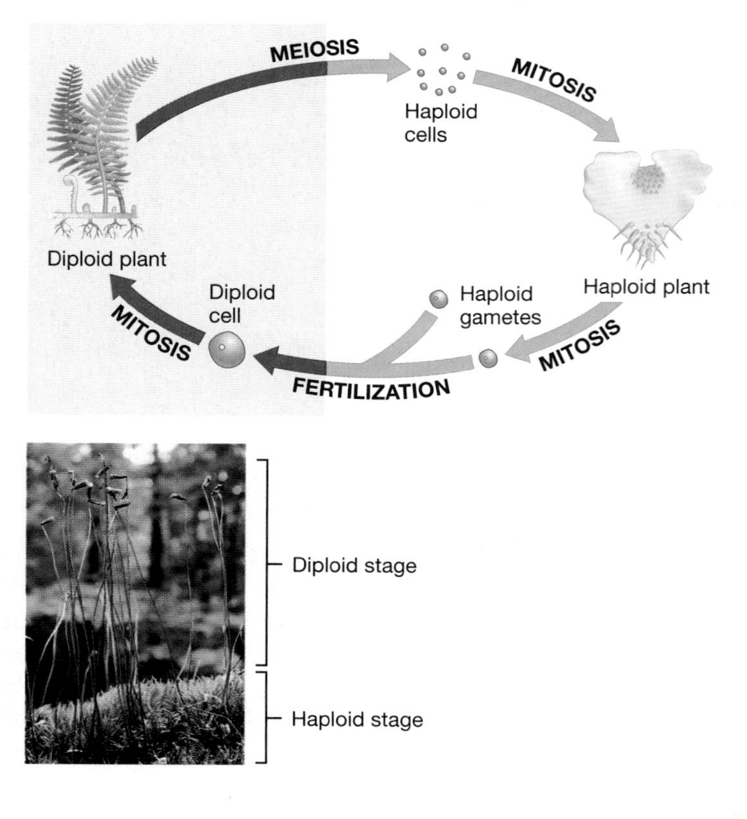

inspired biologists to take a closer look at the life cycle of sexually reproducing organisms. A **life cycle** is the sequence of events that occurs during the life span of an individual, from conception to the production of offspring.

The basic message of this research was that the life cycles of sexually reproducing organisms are highly variable. In humans, other vertebrates, and most invertebrate animals, individuals are diploid for most of their lives (**Figure 9.7a**). Meiosis occurs in the reproductive organs of an adult. Gametes represent the only haploid phase of the individual's life.

In contrast, many algae are haploid for most of their lives. In these organisms, the diploid stage consists of a spore-like structure that can resist heat and drying. As **Figure 9.7b** shows, meiosis occurs after this structure germinates. The haploid cells that result divide by mitosis to generate a haploid multicellular organism.

Land plants and some algae have a distinctive type of life cycle known as **alternation of generations.** This life cycle features a multicellular haploid stage and a multicellular diploid stage. In some land plants, such as flowering plants (angiosperms) and ferns, the haploid stage is inconspicuous (**Figure 9.7c, top**). In other species, such as mosses, the haploid multicellular form is more conspicuous than the diploid plant (**Figure 9.7c, bottom**).

What is the significance of all this variation in life cycles? The answer is not known. Explaining why certain groups of organisms undergo alternation of generations or other types of life cycles is just one of several fundamental questions about meiosis that remain to be addressed. In fact, a firm understanding of why meiosis and sex exist at all has only recently begun to emerge.

9.3 Why Does Meiosis Exist? Why Sex?

Meiosis and sexual reproduction occur in only a small fraction of the lineages on the tree of life. Bacteria and archaea undergo only asexual reproduction; most algae, fungi, and some land plants reproduce asexually as well as sexually (**Figure 9.8**). Asexual reproduction is found even among the vertebrates. Several fish species in the genus *Poeciliopsis*, for example, reproduce exclusively via mitosis.

Sexual reproduction is widespread among selected lineages, however, and predominates among multicellular organisms. In particular, it is the major mode of reproduction in species-rich lineages like the insects and vertebrates.

In 1978, John Maynard Smith pointed out that the existence of any sexual reproduction presents a paradox. Maynard Smith developed a mathematical model showing that because asexually reproducing individuals do not have to produce male offspring, their progeny can produce twice as many grand-offspring as individuals that reproduce sexually. **Figure 9.9** diagrams this result.

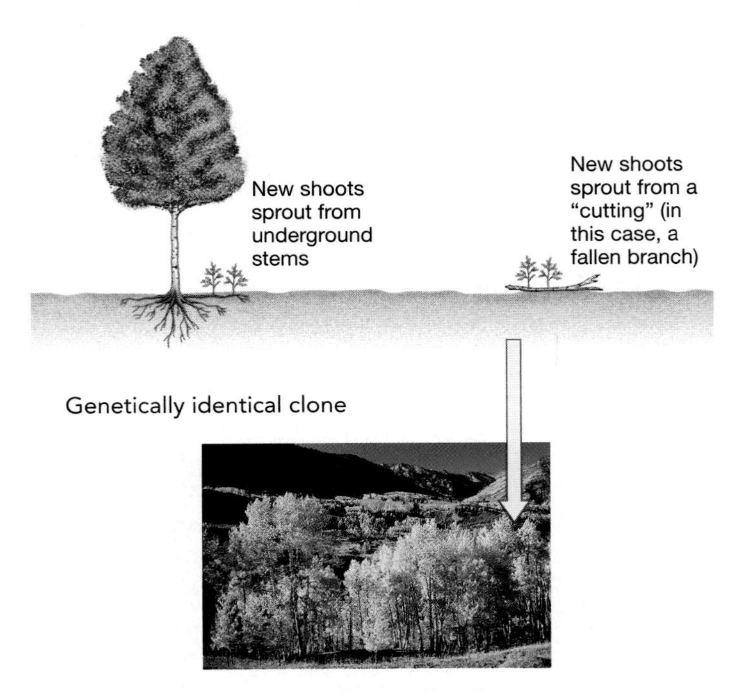

FIGURE 9.8 Asexual Reproduction in Quaking Aspen
Quaking aspen can reproduce asexually, leading to genetically identical clones. The photograph was taken in autumn, when quaking aspen leaves turn bright yellow.

Based on Maynard Smith's analysis, what will happen when asexual and sexual individuals exist in the same population and compete with one another? All other things being equal, individuals that reproduce asexually should increase in frequency while individuals that reproduce sexually should decline in frequency. In fact, Maynard Smith's model predicts that sexual reproduction should be eliminated by natural selection—the evolutionary process introduced in Chapter 1.

But because sex exists, the assumption that "all other things are equal" must be wrong. To resolve the paradox of sex, biologists began looking for ways that meiosis and outcrossing could lead to the production of offspring that survived better or reproduced more than offspring produced by mitosis.

The leading hypothesis to explain sexual reproduction focuses on the benefits of producing genetically diverse offspring. Specifically, in the late 1980s William Hamilton and co-workers proposed that genetically diverse offspring are better able to fight off disease-causing agents like bacteria, viruses, and parasites. (Chapters 37 and 46 explore why genetic diversity can increase disease resistance.) Do any data support this hypothesis?

Curtis Lively recently tested the disease-resistance hypothesis by studying a species of snail that is native to New Zealand (**Figure 9.10a**, page 188). The snails he chose to study live in freshwater habitats and are parasitized by over a dozen species of trematode worms. Snails that become infected are effectively killed because they cannot reproduce—the worms eat their reproductive organs. As a result, individuals that can

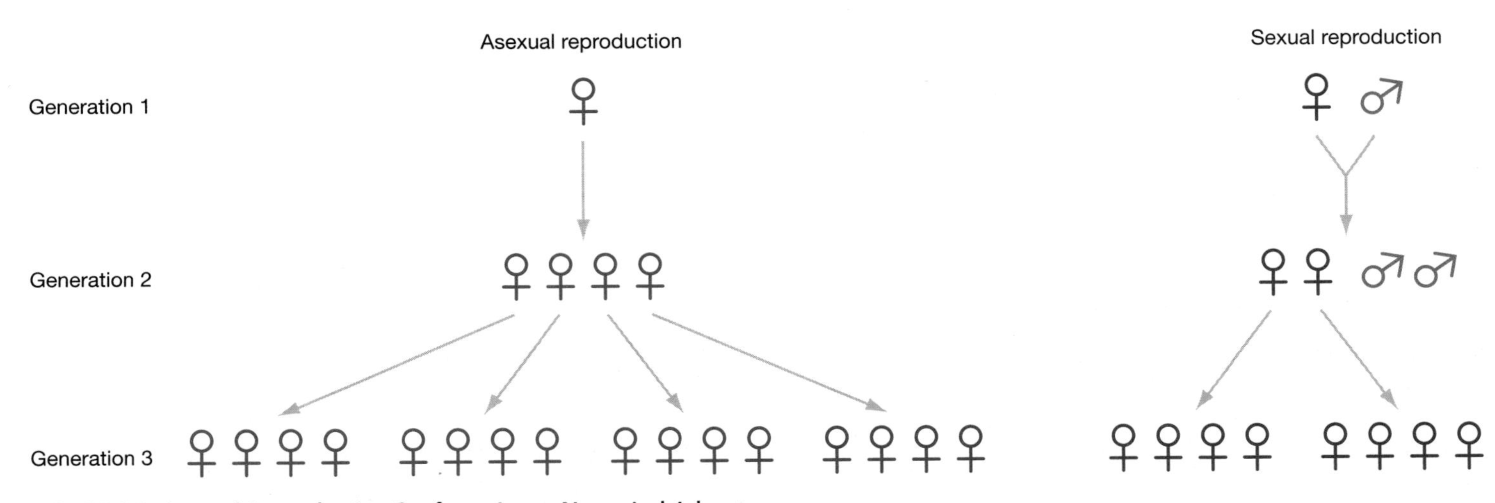

FIGURE 9.9 Asexual Reproduction Confers a Large Numerical Advantage
In this diagram, each female symbol (♀) and male symbol (♂) represents an individual. In the hypothetical example given here, every individual produces four offspring over the course of a lifetime, sexually reproducing individuals produce half males and half females, and all offspring survive to breed. QUESTION How many asexually produced offspring would be present in generation 4? How many sexually produced offspring?

resist infection should produce many more offspring than individuals that are susceptible to infection.

Lively was interested in working on this snail species because some individuals reproduce only sexually while others reproduce only asexually. Hamilton's disease-resistance hypothesis allowed him to make some testable predictions. Specifically, if the hypothesis is true, then sexually reproducing individuals should be common in habitats where trematode infection rates are high. Conversely, asexually reproducing individuals should be common in habitats where parasitic worms are rare. (Note, however, that these predictions are based on an important and untested assumption—that all other aspects of the two environments are identical.)

(a) Snails subject to parasitism by trematode worms

(b) Are genetically diverse populations more resistant to parasites?

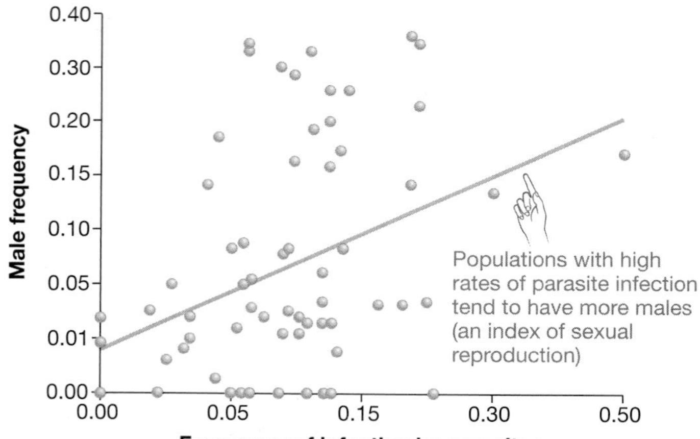

FIGURE 9.10 Are Sexually Produced Offspring Favored When Disease or Parasitism Rates Are High?
(a) *Potamopyrgus antipodarum* is a species of freshwater snail native to New Zealand. Some individuals in this species reproduce only sexually, while others reproduce only asexually. **(b)** The x-axis (abscissa) on this graph plots the proportion of snails infected with trematode worms in a population while the y-axis (ordinate) plots the proportion of individuals in the same population that are male. (The proportion of males is an index of how many individuals reproduce sexually.)

CD ACTIVITY 9.2 Mistakes in Meiosis

To test these predictions, Lively studied snails from a wide variety of habitats. By collecting a large number of individuals from different locations and examining them, he was able to calculate the frequency of individuals that were infected with trematodes at each location and the frequency of males at each location. (The frequency of males served as an index of the frequency of sexually reproducing individuals.) His results are plotted in **Figure 9.10b**. The data show that the frequency of sexually reproducing individuals tends to be higher in habitats where trematodes are more abundant.

Lively's result and a variety of other studies support the disease-resistance hypothesis. Although the paradox of sex remains an active area of research, an increasing number of biologists are becoming convinced that sexual reproduction exists because it allows the production of genetically diverse offspring with improved resistance to disease.

9.4 Mistakes in Meiosis

When homologous chromosomes separate during meiosis I, a complete set of chromosomes is transmitted to each daughter cell. But what happens if there is a mistake and the chromosomes are not properly distributed? What are the consequences for offspring if gametes contain an abnormal set of chromosomes?

In 1866 Langdon Down described a distinctive suite of co-occurring conditions observed in some humans. The syndrome was characterized by mental retardation, a high risk for heart problems and leukemia, and a degenerative brain disorder similar to Alzheimer's disease. Down syndrome, as the disorder came to be called, is observed in about 0.15 percent of live births (3 infants in every 2000). For over 80 years the cause of the syndrome was unknown. Then in the late 1950s Michel Lejeune published observations on the chromosome sets of nine Down syndrome children. His data suggested that the condition is caused by the presence of an extra copy of chromosome 21. This circumstance is called a trisomy (three-bodies) because there are three copies of the chromosome in each cell. To explain the anomaly, Lejeune proposed that the extra chromosome resulted from a mistake during gamete formation in one of the parents.

How Do Mistakes Occur?

In order to get one complete set of chromosomes into a gamete, two steps in meiosis must be perfectly executed. During the first meiotic division in humans, 23 pairs of homologous chromosomes must separate, or disjoin, from each other so that only one homolog ends up in each daughter cell. If both members of a chromosome pair happen to move to the same pole of the cell, the products of meiosis will be abnormal. This sort of meiotic error, illustrated in **Figure 9.11** (page 190), is referred to as **nondisjunction**. Note that one daughter cell in the figure has two copies of the same chromosome, while the other lacks that chromosome entirely. Gametes that contain an extra chromosome are symbolized as $n + 1$; gametes that lack one chromosome are symbolized as $n - 1$. If an $n + 1$ gamete is fertilized by a normal n gamete, the resulting offspring will be $2n + 1$. This is a trisomy. If the $n - 1$ gamete is fertilized by a normal n gamete, the resulting zygote will be $2n - 1$.

Occasionally, abnormal $n + 1$ and $n - 1$ gametes are produced during nondisjunction in the second meiotic division. If sister chromatids fail to separate and move to opposite poles of the dividing cell, then the resulting daughter cells will be $n + 1$ and $n - 1$.

Meiotic mistakes occur at a relatively high frequency. In humans, for example, researchers estimate that nondisjunction events may occur in as many as 10 percent of meiotic divisions. The types of mistakes vary, but their consequences are almost always severe. In a recent study of human pregnancies that resulted in early embryonic or fetal death, 38 percent of the 119 cases involved atypical chromosome complements that could be detected by karyotype analysis. Trisomy accounted for 36 percent of the abnormal karyotypes found, triploidy (from fusion of three gametes) for 30 percent, aberrantly sized or

BOX 9.3 Recombination in Bacteria

Bacteria and archaea have a single, circular chromosome. Because they are haploid, they cannot undergo meiosis. Bacteria do undergo recombination, however. In these single-celled organisms, recombination occurs when a chromosome from one individual is duplicated and then transferred to another individual. The transfer process is known as conjugation.

The steps in bacterial conjugation and recombination are outlined in the accompanying figure. The key to understanding the diagram is that the chromosomes of some bacterial cells contain genes that allow them to transfer some, or all, of their chromosome to another cell. Individuals with these genes are called Hfr cells, for High frequency of recombination. When an Hfr cell makes contact with a cell lacking these genes, the two cells may become joined (**Figure 1**). A copy of the Hfr chromosome moves into the other cell, and the two cells break contact. Later, the portions of the Hfr chromosome may become integrated into the main chromosome (see rightmost drawing in step 4). This results in recombination—a change in the combination of genes found on the same chromosome.

Although Chapter 14 explores the outcome of conjugation in more detail, three points are worth noting here. First, recombination in bacteria depends on a completely different mechanism for bringing homologous chromosomes together than recombination in eukaryotes. Second, in bacteria recombination occurs independently of reproduction. In contrast to eukaryotes, the event is not associated with cell division. Finally, recombination in bacteria is the one-way transfer of genetic material. During meiosis, in contrast, a reciprocal exchange occurs when chromosomes cross over.

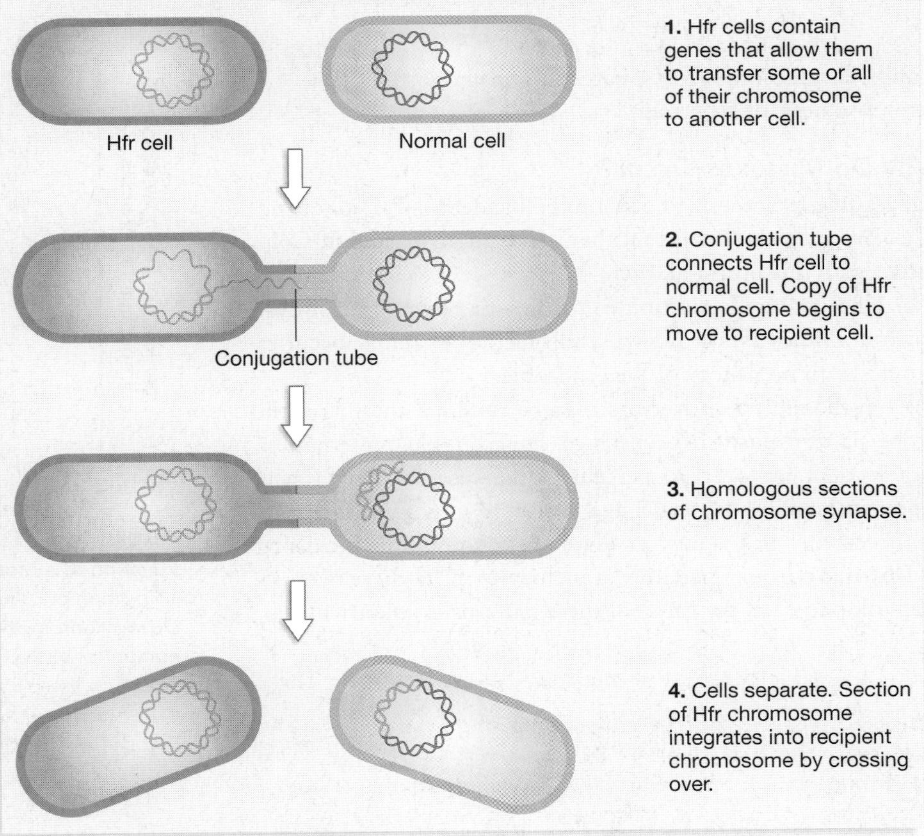

FIGURE 1 Recombination in Bacteria Is Based on Chromosome Transfer Between Cells Recombination occurs in bacteria when a section of an Hfr chromosome enters a non-Hfr cell and integrates into its chromosome via crossing over.

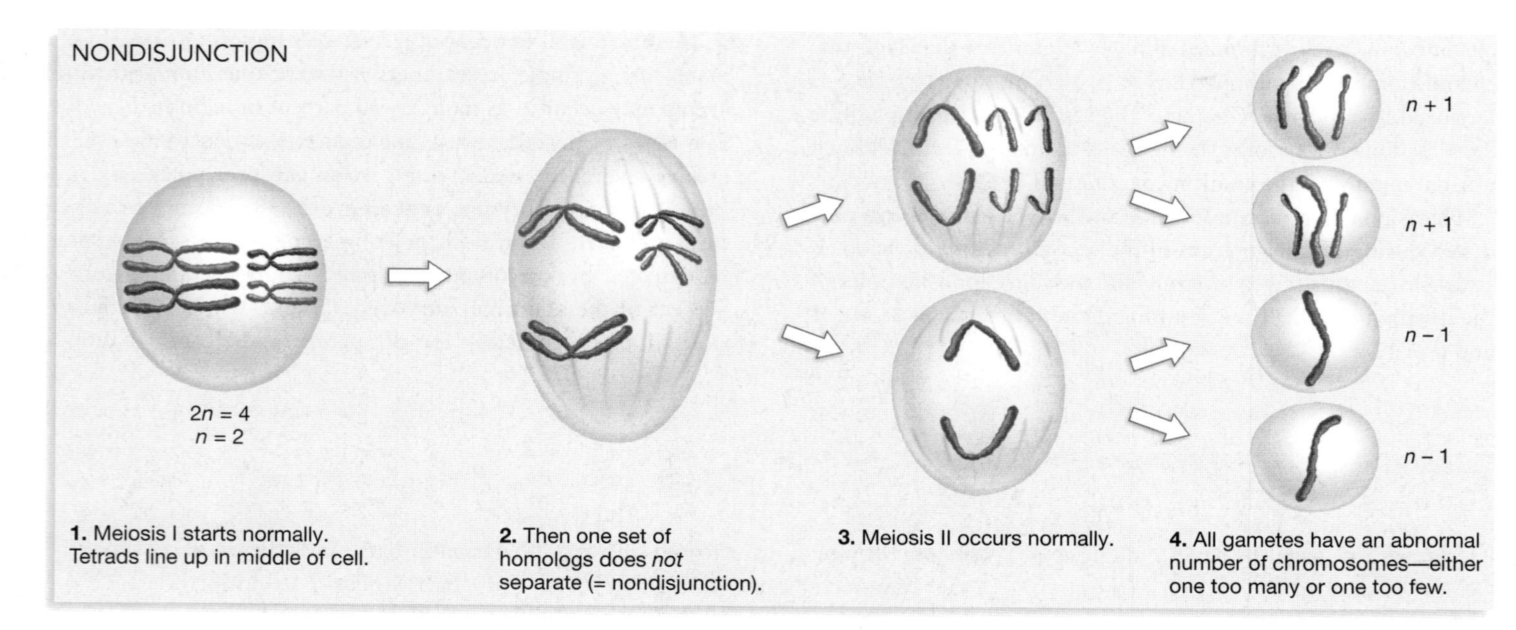

FIGURE 9.11 Nondisjunction Leads to Gametes with Abnormal Chromosome Numbers
If homologous chromosomes fail to separate during meiosis I, the gametes that result will have an extra chromosome or lack a chromosome. EXERCISE Nondisjunction can also occur during meiosis II, after meiosis I has proceeded normally. Starting with a parent cell like the one shown here, make a diagram showing each step in meiosis when sister chromatids fail to disjoin at meiosis II. How many of the resulting gametes are normal? How many have an extra chromosome or lack a chromosome?

shaped chromosomes for 4 percent, and monosomy ($2n - 1$ chromosomes) for 2 percent.

Why Do Mistakes Occur?

The leading hypothesis to explain the incidence of trisomy and other meiotic mistakes is that they are random accidents. Consistent with this proposal, there does not seem to be any genetic or inherited predisposition to trisomy or other types of dysfunction. Most cases of Down syndrome, for example, occur in families with no history of the syndrome.

Even though meiotic errors may be random, there are still strong patterns in their occurrence. For example, maternal errors account for over 90 percent of the cases of Down syndrome in humans. Maternal age is also an important factor in the occurrence of trisomy. As **Figure 9.12** shows, the incidence of Down syndrome increases dramatically in mothers over 35 years old. To date, the cause of these patterns is unknown.

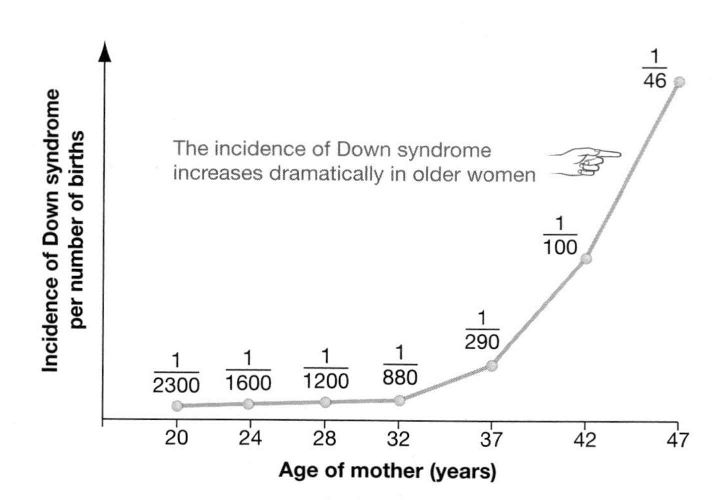

FIGURE 9.12 The Frequency of Down Syndrome Increases as a Function of a Mother's Age
This graph plots mother's age (on the x-axis) versus the incidence of Down syndrome, expressed as the number of affected infants per number of births. QUESTION Suppose that you are an obstetrician. Based on these data, at what age would you recommend that pregnant mothers undergo procedures to check the karyotype of the embryos they carry?

Essay Seedless Fruits

Have you ever sat outside on a hot summer night eating watermelon and having a seed-spitting contest? It is likely that your children will never have the experience because seedless varieties of watermelon have recently begun to dominate the market. New varieties of watermelon aren't the only example of a seedless fruit, either. Imagine biting into a banana and crunching down on hard, inedible seeds.

How did these seedless fruits come to be? One set of chromosomes makes the difference between the presence and absence of seeds in banana and watermelon. In both of these species, diploid ($2n$) varieties make seeds while triploid ($3n$) strains are sterile (and seedless).

What happens when an individual has an odd number of chromosome sets?

Many cultivated plant species are polyploid, meaning that they have more than two sets of chromosomes. As long as there is only one pairing partner for each chromosome during synapsis, meiosis I proceeds as it would in a diploid species and results in a halving of the chromosome number. For normal seed production, the key is that there must be an even number of chromosome sets. When the ploidy number is even, homologous chromosomes may be able to synapse and later separate correctly. The wheat used in bread making, for example is hexaploid ($6n$). It generates $3n$ gametes by meiosis. These fuse to form normal $6n$ offspring.

What happens when an individual has an odd number of chromosome sets? Consider the case of triploid bananas and watermelons. Early in meiosis I, each group of three homologous chromosomes tries to pair. The three homologs migrate to the middle of the cell together and separate. When the homologs separate, however, one daughter cell typically receives two copies of a given chromosome while the other daughter cell receives one copy.

This asymmetrical distribution of chromosomes occurs randomly for each chromosome in the cell. As a result, virtually all of the gametes that are produced have an unbalanced complement of chromosomes. Some will have two copies of some chromosomes and one copy of others.

In plants, as in animals, additions or deletions of specific chromosomes are almost always deleterious. In bananas and watermelons, gametes with unbalanced chromosome sets cannot produce viable embryos. Because they lack growing embryos, their seeds die and a seedless fruit results. Triploid bananas and watermelons are seedless because they can't complete meiosis correctly.

Where do these triploid, seedless strains come from in the first place? Triploid individuals originate in several different ways. For example, massive errors can occur during meiosis I in a diploid parent, and lead to all of the chromosomes migrating to a single daughter cell. If the $2n$ gamete that results fuses with a normal gamete, a triploid offspring results. Because the triploid is sterile, it has to be propagated through cuttings or some other mode of asexual reproduction.

Chapter Review

Summary

When biologists confirmed that sperm and egg nuclei fuse during fertilization, it led to the hypothesis that a special type of cell division must precede gamete formation. Specifically, the proposal was that a sperm and egg must each have half of the normal number of chromosomes found in cells. This hypothesis was confirmed when researchers observed meiosis and established that it results in gametes with half the normal chromosome number.

As the details of meiosis were being worked out, biologists realized that chromosomes exist in sets. In diploid organisms, individuals have two versions of each type of chromosome. One of the versions is inherited from the mother and one from the father; the similar, paired chromosomes are called homologs. Haploid organisms, in contrast, have just one of each type of chromosome.

Each chromosome is replicated before meiosis begins. At the start of meiosis I, then, each chromosome consists of a pair of sister chromatids joined at a centromere. Homologous pairs of chromosomes synapse early in meiosis I, forming a group of two homologous chromosomes, or a tetrad. After chromatids from the homologous chromosomes undergo crossing over, the pair of homologous chromosomes migrates to the middle of the cell. At the end of meiosis I the homologous chromosomes separate and are distributed to two daughter cells. During meiosis II, sister chromatids separate and are distributed to two daughter cells.

When meiosis occurs, the chromosome complement found in gametes and offspring differ from one another and from their parents for three reasons: (1) maternal and paternal homologs exchange segments during crossing over; (2) maternal and paternal homologs are mixed when chromosomes separate at the end of meiosis I; and (3) outcrossing results in a combination of chromosome sets from different individuals. The consequences of these changes became clear when biologists realized that chromosomes contain the hereditary material. Meiosis leads to genetic differences among offspring and between parents and offspring.

According to the disease-resistance hypothesis, meiosis exists primarily because genetically diverse offspring are better able to resist parasites. This hypothesis has been tested by studying populations of snails that contain both sexually reproducing and asexually reproducing individuals. The data from this research support the prediction that sex is more common in habitats where parasite infection is frequent.

Mistakes during meiosis lead to gametes and offspring with an unbalanced set of chromosomes. Children with Down syndrome, for example, have an extra copy of chromosome 21. The leading hypothesis to explain these mistakes is that they are random accidents that result in a failure of homologous chromosomes or sister chromatids to separate properly during meiosis.

Questions

Content Review

1. In the roundworm *Ascaris*, eggs and sperm have two chromosomes but all other cells have four. Observations like this inspired an important hypothesis. What is it?
   a. Before gamete formation, a special type of cell division leads to a quartering of chromosome number.
   b. Before gamete formation, a special type of cell division leads to a halving of chromosome number.
   c. After gamete formation, half of the chromosomes are destroyed.
   d. After gamete formation, either the maternal or the paternal set of chromosomes disintegrates.
2. What are homologous chromosomes?
   a. pairs of chromosomes that are similar in size, shape, and content
   b. similar chromosomes that are found in different species
   c. chromosomes that have knobs or no knobs on their ends
   d. chromosomes that are distinctly different in size and shape
3. What is meant by a paternal chromosome?
   a. the largest chromosome in a set
   b. a chromosome that does not separate correctly during meiosis I
   c. the member of a homologous pair that was inherited from the mother
   d. the member of a homologous pair that was inherited from the father
4. What is a tetrad?
   a. the "X" that forms when chromatids from homologous chromosomes cross over
   b. a group of four sister chromatids
   c. the point where homologous chromosomes synapse
   d. the group of four daughter cells produced by meiosis
5. Meiosis II is similar to what process?
   a. mitosis in haploid cells
   b. nondisjunction
   c. outcrossing
   d. meiosis I
6. What is genetic recombination?
   a. the new combination of maternal and paternal homologs that result when chromosomes separate at meiosis I
   b. the combination of maternal and paternal chromatids that results when homologs cross over
   c. the new combinations of chromosome segments that result when outcrossing occurs
   d. the combination of a prominent haploid phase *and* a prominent diploid phase in a life cycle

Conceptual Review

1. Triploid ($3n$) watermelons are produced by crossing a tetraploid ($4n$) strain with a diploid ($2n$) plant. Briefly explain why this mating produces a triploid individual. Why can mitosis proceed normally in triploid cells, but meiosis cannot?
2. Meiosis I is sometimes called a reduction division; meiosis II is sometimes called an equational division. Why?
3. Some plant breeders are concerned about the resistance of asexually cultivated plants, like seedless bananas and watermelons, to new disease-causing bacteria, viruses, or fungi. Briefly explain their concern by discussing the differences in the genetic "outcome" of asexual and sexual reproduction.
4. Explain why nondisjunction leads to trisomy and other types of abnormal chromosome sets. In what sense are these chromosome sets "unbalanced?"
5. Examine Figure 9.1, which shows the karyotype of the lubber grasshopper. The researcher drew these chromosomes as they were undergoing meiosis. What event is occurring in the drawing?

Applying Ideas

1. The gibbon has 44 chromosomes per diploid set and the siamang has 50 chromosomes per diploid set. In the 1970s a chance mating between a male gibbon and a siamang female produced a viable offspring. Predict how many chromosomes were observed in the somatic cells of the offspring. Do you predict that this individual would be able to form viable gametes? Why or why not?
2. Meiosis results in a reassortment of maternal and paternal chromosomes. If $n = 3$ for a given organism, there are 8 different combinations of paternal and maternal chromosomes. If no crossing over occurs, what is the probability that a gamete will receive *only* paternal chromosomes?
3. The data in the following table concern human fetuses and infants with various types of trisomy. The leftmost column indicates the extra chromosome in each case. (For example, the number 21 in the next-to-last row of this column indicates trisomy-21, which causes Down syndrome.) Note that the researchers were able to ascertain whether the gamete responsible for the trisomy came from the mother or the father. In many cases they were also able to determine whether the error occurred during meiosis I or meiosis II (if this point was unclear, the trisomy is recorded as "I or II").
   - What general conclusions can you draw from these data?
   - Note that far more data were collected on trisomy-21 (Down syndrome) than on any other trisomy. Does this mean that nondisjunction of chromosome 21 occurs more often than nondisjunction of other chromosomes? Or could other factors influence the number of trisomy-21 fetuses or infants that were available for study?

Trisomy	# Cases	Paternal			Maternal			% Paternal
		I	II	I or II	I	II	I or II	
2–12	16			3			13	19
13	7			2	1		4	29
14	8			2			6	25
15	11			3			8	27
16	62				51	1	10	0
18	73			3			70	4
21	436	5	24		306	101		7
22	11						11	0

4. Try to imagine what a human life cycle would be like if alternation of generations occurred. What might a multicellular haploid stage in our species look like? Make a diagram analogous to Figure 9.7c that illustrates your idea.
5. Lively's test of the disease-resistance hypothesis has been criticized because it was observational and not experimental in nature. As a result, he could not control for factors other than parasites that might affect the frequency of sexually reproducing individuals. To put this point another way, he observed a correlation between two factors, but a correlation does not imply causation.
   a. Design an experimental study that would provide stronger evidence that the frequency of parasite infection causes differences in the frequency of sexual versus asexual individuals in this species of snail.
   b. In Lively's defense, comment on the value of observing patterns like this in nature, versus in controlled conditions in the laboratory.

CD-ROM and Web Connection

CD Activity 9.1: Meiosis *(animation)*
(Estimated time for completion = 5 min)
How does a cell halve its chromosome number to prepare for sexual reproduction?

CD Activity 9.2: Mistakes in Meiosis *(tutorial)*
(Estimated time for completion = 10 min)
Learn how errors in meiosis produce gametes with too many or too few chromosomes.

At your **Companion Website** (http://www.prenhall.com/freeman/biology), you will find self-grading exams and links to the following research tools, online resources, and activities:

Meiosis
This tutorial steps through the process of meiosis and includes links to animations and information on the technique of karyotyping.

John Maynard Smith
In this interview, Smith discusses his use of mathematical models to understand the evolution of recombination.

Birth Defects
This site provides links to information on genetic errors such as trisomies and monosomies and their effects on human health.

Additional Reading

Haber, J. E. 1998. Searching for a partner. *Science* 279: 823–824. A look at recent research on how synapsis and recombination occur in yeast and fruit flies.

Wuethrich, B. 1998. Why sex? Putting theory to the test. *Science* 281: 1980–1982. A commentary on recent research designed to test different hypotheses for the advantage of sexual reproduction.

Yamamoto, A. and Y. Hiraoka. 2001. How do meiotic chromosomes meet their homologous partners?: lessons from fission yeast. *BioEssays* 23: 526-533. Focuses on the role played by the regions called telomeres at the ends of each chromosome.

Mendel and the Gene

10

The science of biology is built on a series of great ideas. Two of these—the cell theory and the theory of evolution—were introduced in Chapter 1. The cell theory describes the basic structure of organisms; the theory of evolution by natural selection clarifies why species change through time. These theories explain fundamental features of the natural world and answer some of our most profound questions about the nature of life. What are organisms made of? Where did species come from?

A third great idea in biology addresses an equally important question: Why do offspring resemble their parents? An Austrian monk named Gregor Mendel provided part of the answer in 1865 when he announced that he had worked out the rules of inheritance through a series of experiments on garden peas. The other part of the answer was provided by the biologists who described the details of meiosis during the final decades of the nineteenth century. In 1903, W. S. Sutton and Theodor Boveri linked these two results by formulating the chromosome theory of inheritance. This theory contends that the process of meiosis, introduced in Chapter 9, causes the patterns of inheritance that Mendel observed. It also asserts that the hereditary factors called genes are located on chromosomes.

This chapter focuses on the evidence for the chromosome theory of inheritance. Let's begin with a basic question. What are the rules of inheritance that Mendel discovered?

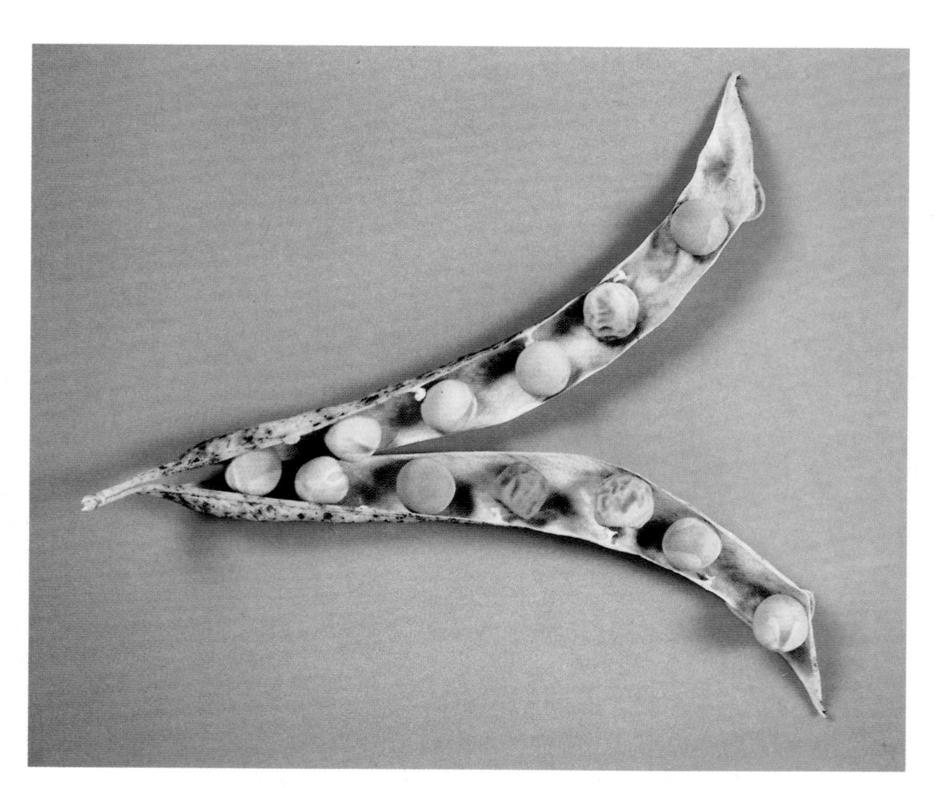

This garden pea pod shows how seeds can vary in shape and color. Gregor Mendel discovered the basic rules of inheritance by studying how variations in traits like these are transmitted from parents to offspring.

10.1 Mendel's Experiments with a Single Trait

Gregor Mendel was a monk who lived and worked in the city of Brün, located 70 miles north of Vienna. (Brün was then part of Austria. Today the city is called Brno and is part of the Czech Republic.) Mendel was educated at the University of Vienna in the natural sciences and also studied physics and mathematics under Christian Doppler, who discovered the Doppler effect in sound transmission.

In Mendel's day questions about **heredity**—or the transmission of traits from parents to offspring—were primarily the concern of animal breeders and horticulturists. In Brün, for example, there was a particular interest in how selective breeding could result in hardier and more productive varieties of sheep, fruit trees, and vines. To that end, an Agricultural Society had been formed. Its members emphasized the importance of research that would help breeding programs become more efficient. Mendel was an active member of this society; the monastery he belonged to was also devoted to scientific teaching and research.

What Questions Was Mendel Trying to Answer?

Mendel set out to address the most fundamental of all issues concerning heredity: What are the basic patterns in the transmission of traits from parents to offspring? At the time, two hypotheses had been formulated to answer this question. The first, called blending inheritance, claimed that the traits observed in a mother and father blend together to form the traits observed in their offspring. For example, blending inheritance contended that black sheep have hereditary determinants for blackness, and white sheep have hereditary determinants for whiteness. When these individuals mate, their hereditary determinants blend to form a new hereditary determinant for grayness. Hence, according to the theory of blending inheritance, their offspring should be gray. Mendel's scientific mentor, the widely respected botanist Carl Nägeli, was a proponent of this hypothesis.

The second hypothesis was called the inheritance of acquired characters, which claimed that traits present in parents are modified, through use, and passed on to their offspring in the modified form. The classical prediction of this hypothesis is that adult giraffes acquire longer necks by straining to reach leaves high in the tops of trees, and that they subsequently produce longer-necked offspring. The idea here is that the genetic determinants present in an individual are modified through use. Jean-Baptiste Lamarck originally formulated this hypothesis in the 18th century; in Mendel's day it was championed by Charles Darwin.

These hypotheses were being promoted by the greatest scientists of Mendel's time. But are they correct? What *are* the basic patterns of inheritance?

Garden Peas Serve as the First Model Organism in Genetics

Mendel was certainly not the first scientist interested in studying the basic mechanisms of heredity. Why was he successful where others failed? Two factors came into play. Mendel chose an appropriate model organism to study, and he mathematically analyzed the data that he collected. As we'll see, quantifying his results gave him the ability to recognize important patterns in the data.

Which model organism did he choose? After investigating and discarding several candidates, Mendel chose the pea plant *Pisum sativum*. His reasons were practical. Peas are inexpensive, easy to propagate, and have a relatively short reproductive cycle. These features made it possible for Mendel to continue experiments over several generations and to collect data from a large number of individuals. Because of his choice, garden peas became the first model organism in genetics. A **model organism** is easy to study and serves as a model for processes and patterns that occur in other species as well.

Two additional features of the pea made it possible for Mendel to design his experiments. He could control which parents were involved in a mating, and he could arrange matings between individuals that differed in easily recognizable traits like flower color or seed shape. Why was this important?

Arranging a Mating **Figure 10.1a** (page 196) shows the flower from a garden pea with the male and female reproductive organs. The male gametes are contained in pollen grains that are produced by structures called anthers. The female gametes are produced in structures called ovules. The fertilization process begins when pollen grains are deposited on the stigma, which is connected to the ovules by a column of tissue. A pollen tube then grows down the length of this structure until it reaches the ovules. Sperm cells from the pollen grain travel down the tube to the ovule, where fertilization takes place.

Under normal conditions, garden peas self-pollinate. Self-fertilization (or selfing) takes place when pollen from the anthers in a flower falls on the stigma of that same flower. Pollen from other plants rarely reaches the stigma because the petals form a compartment that encloses the male and female reproductive organs.

As **Figure 10.1b** shows, however, Mendel could circumvent this arrangement by removing the anthers from a flower before any pollen formed. Later he could dust the stigma with pollen from another plant. This type of mating is referred to as a cross-pollination, or simply a cross. Using this technique, Mendel could control the matings of his model organism.

Finding Variable Traits to Study Mendel conducted his experiments on strains of peas that varied in seven different traits. As **Figure 10.2** (page 196) shows, each trait exhibited one of two forms, or phenotypes. Biologists refer to the observable features of an individual, such as the eye color or hair color of a human, as its **phenotype** (literally, "show-type").

Mendel began his work by obtaining individuals from what breeders called pure lines or true-breeding lines. A pure line consists of individuals that produce progeny identical to themselves when they are self-pollinated or crossed to another member of the same line. For example, other breeders had developed pure lines for seed shape. During two years of trial experiments, Mendel confirmed that individuals with wrinkled seeds produced only wrinkled-seeded offspring when they were mated to themselves or to another pure-line individual with wrinkled seeds; individuals with round seeds produced only round-seeded offspring when they were mated to themselves or to another pure-line individual with round seeds.

Why is this important? Remember that Mendel wanted to find out how traits are transmitted from parents to offspring. Once he had confirmed that he was working with pure lines, he could predict how matings within each line would turn out; in other words, he knew what the offspring would look like. He could then compare these results to the outcomes of crosses between individuals from different pure lines. For example, suppose he arranged matings between an individual with round seeds and an individual with wrinkled seeds. He knew that one parent carried a hereditary determinant for round seeds, while the other carried a hereditary determinant for wrinkled seeds. But the offspring that resulted from this mating have both hereditary determinants. They would be **hybrids**—a mix of the two types. Would they have wrinkled seeds, round seeds, or a blended combination of wrinkled and round? What would be the seed shape in subsequent generations when hybrid individuals self-pollinated or were crossed with members of the pure lines?

Inheritance of a Single Trait

Mendel's first set of experiments involved crossing pure lines that differed in one trait, such as seed shape. Working with single traits was important because it made the results of the matings more interpretable. Once he understood how a single trait was transmitted from parents to offspring, Mendel could then

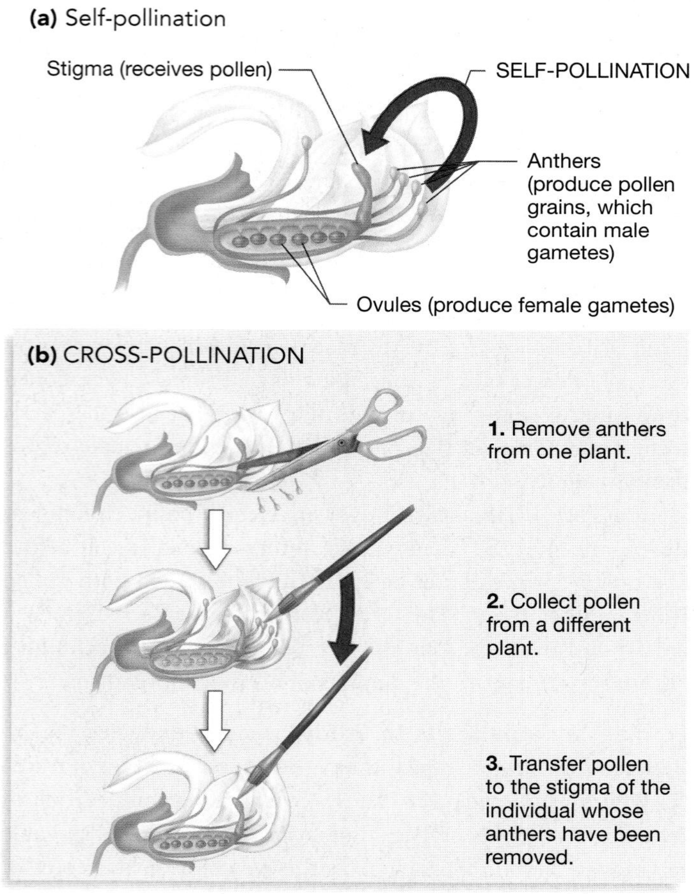

(a) Self-pollination

Stigma (receives pollen)

SELF-POLLINATION

Anthers (produce pollen grains, which contain male gametes)

Ovules (produce female gametes)

(b) CROSS-POLLINATION

1. Remove anthers from one plant.

2. Collect pollen from a different plant.

3. Transfer pollen to the stigma of the individual whose anthers have been removed.

FIGURE 10.1 Self-Pollination and Cross-Pollination in Peas
(a) The petals of a pea form an enclosed compartment. As a result, most fertilization takes place when pollen grains from the anther of a flower fall on the stigma of the same flower. **(b)** Mendel could arrange a mating between individuals by removing the anthers from one flower and then dusting its stigma with pollen collected from a different flower.

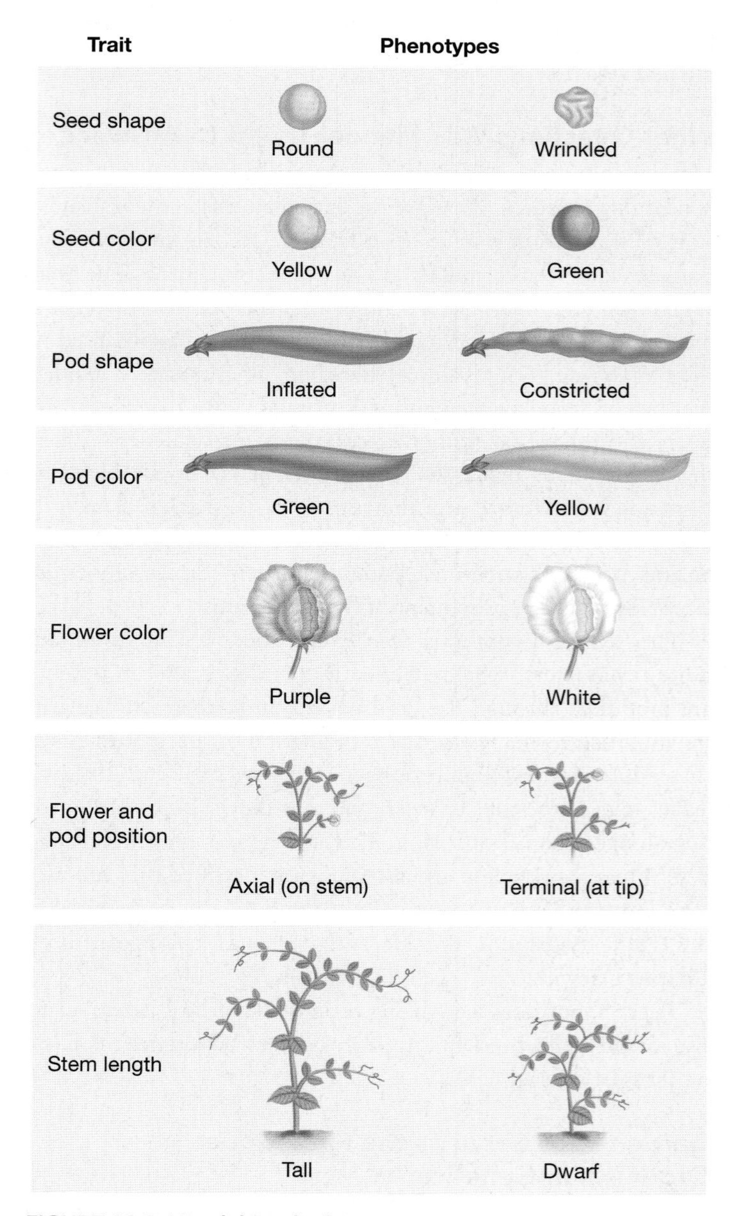

FIGURE 10.2 Mendel Studied Seven Traits That Were Variable in Garden Peas
Two distinct phenotypes existed for each of the seven traits that Mendel studied in garden peas.

explore what happened when crosses were performed between individuals that differed in two traits.

Mendel began by crossing individuals from round-seeded and wrinkled-seeded pure lines. The adults used in an initial experimental cross like this represent the **parental generation.** Their progeny are called the **F_1 generation.** (F_1 stands for "first filial"; *filial* refers to the relationship between parent and offspring.) Subsequent generations are symbolized as the F_2 generation, F_3 generation, and so on.

Reciprocal Crosses In his first set of crosses, Mendel took pollen from round-seeded plants and placed it on the stigmas of plants from the wrinkled-seeded line. As **Figure 10.3a** shows, all of the progeny seeds resulting from this cross were round. This was remarkable. The genetic determinant for wrinkled seeds seemed to have disappeared. Was it because this determinant was located in the egg instead of in the pollen? Did it matter which parent had the genetic determinants for round seeds?

To answer this question Mendel performed a second set of crosses—this time with pollen taken from a wrinkled-seeded individual (**Figure 10.3b**). This completed a **reciprocal cross**—a set of matings where the mother's phenotype in the first cross is the father's phenotype in the second cross; similarly, the father's phenotype in the first cross is the mother's phenotype in the second cross.

In this case the result of the reciprocal cross was identical. All of the F_1 progeny had round seeds. This second cross established that it does not matter whether the genetic determinants for seed shape are located in the male or female parent. But what had happened to the genetic determinant for wrinkled seeds?

Dominant and Recessive Traits Mendel planted the F_1 seeds and allowed the individuals to self-pollinate when they matured. He collected the seeds that were produced by many plants in the F_2 generation and observed that 5474 were round and 1850 were wrinkled. This observation was striking. The wrinkled seed shape reappeared in the F_2 generation after disappearing completely in the F_1 generation.

Mendel invented some important vocabulary to describe this result. He designated the genetic determinant for the wrinkled shape as **recessive.** This was an appropriate term because none of the F_1 individuals had wrinkled seeds—meaning that the determinant for wrinkled seeds appeared to temporarily become latent, or recede. In contrast, Mendel referred to the genetic determinant for round seeds as **dominant.** This term was apt because this determinant appeared to dominate over the wrinkled seed determinant when both were present.

What is the relationship between the two types of determinants? That is, how do they interact? Mendel made an important start in answering these questions when he noticed that the round and wrinkled seeds were present in a ratio of 2.96:1, or essentially 3:1. In other words, about 3/4 of the F_2 seeds were round and 1/4 were wrinkled.

Before trying to interpret these patterns, however, it was important for Mendel to establish that the results were not restricted to seed shape inheritance. So he repeated the experiments with each of the six other traits. In each case, he obtained similar results. The products of reciprocal crosses were the same; one form of the trait was always dominant regardless of the parent it came from; the F_1 progeny showed only the dominant trait and did not exhibit any intermediate phenotype; and in the F_2 generation, the ratio of individuals with dominant and recessive phenotypes was 3 to 1.

How could these patterns be explained? Mendel answered this question with a series of propositions about the nature and

CD ACTIVITY 10.1 Mendel's Experiments

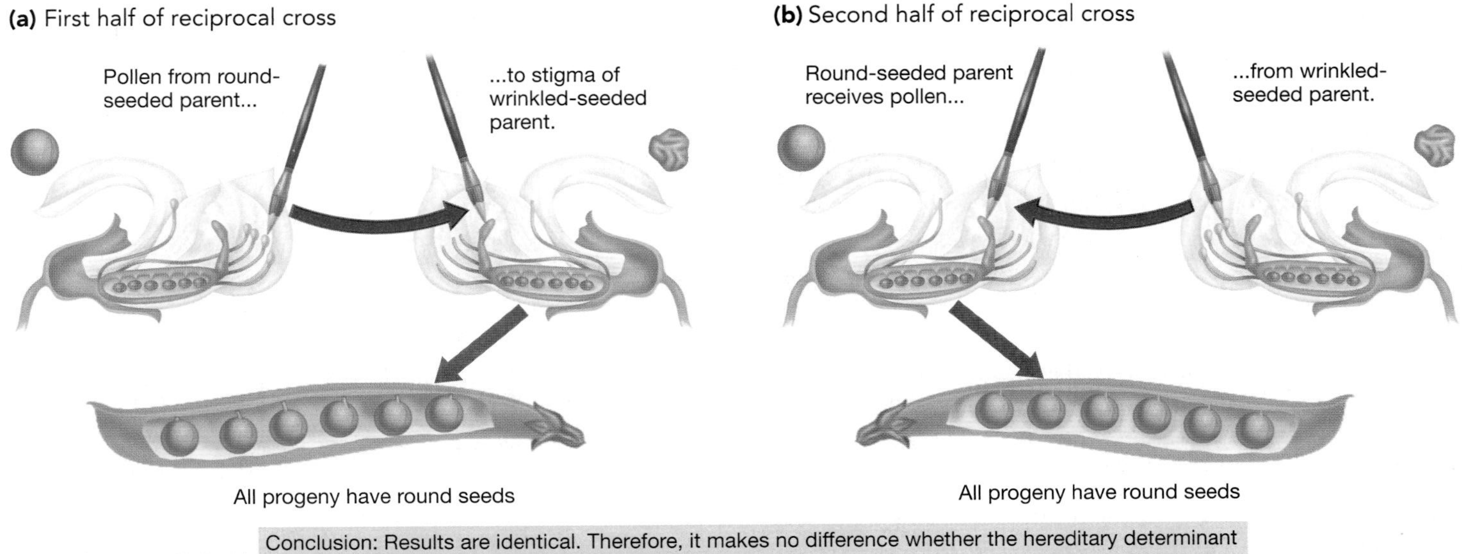

FIGURE 10.3 A Reciprocal Cross
QUESTION What is the purpose of a reciprocal cross?

behavior of the hereditary determinants. These hypotheses rank as some of the most brilliant insights in the history of biology.

The Nature and Behavior of the Hereditary Determinants

Mendel's results were clearly inconsistent with the hypothesis of blending inheritance. To explain the patterns that he observed, Mendel proposed a competing hypothesis called particulate inheritance. He maintained that the hereditary determinants for traits do not blend together or acquire new or modified characteristics through use. In fact, the hereditary determinants maintain their integrity from generation to generation. Instead of blending together, they act like discrete entities or particles.

Mendel's hypothesis was the only way to explain the observation that phenotypes disappeared in one generation and reappeared intact in the next. It also represented a fundamental break with ideas that had prevailed for hundreds of years.

Genes, Alleles, and Genotypes Today geneticists use the word **gene** to indicate the hereditary determinant for a trait. For example, the hereditary factor that determines the difference between round and wrinkled seeds in garden peas is referred to as the gene for seed shape.

Mendel's insights were even more penetrating, however. He also proposed that each individual has two versions, or **alleles**, of each gene. These alleles are responsible for the variation in the traits that he studied. In the case of the gene for seed shape, one allele of this gene is responsible for the round form of the seed while another is responsible for the wrinkled form. The alleles that are found in a particular individual are called its **genotype**. An individual's genotype has a profound influence on its phenotype—its physical traits.

The hypothesis that alleles exist in pairs was important because it gave Mendel a framework for explaining dominance and recessiveness. He proposed that some alleles are dominant and others are recessive. Dominance and recessiveness, then, simply describe a relationship between two particular alleles. Specifically, the terms identify which phenotype actually appears in an individual when both alleles are present. The allele for round seeds is dominant; the allele for wrinkled seeds is recessive. Therefore, when an allele for round seeds is present, seeds are round. When both alleles present are for wrinkled seeds, seeds are wrinkled.

These hypotheses explain why the phenotype for wrinkled seeds disappeared in the F_1 generation and reappeared in the F_2 generation. But why did round- and wrinkled-seeded plants exist in a 3:1 ratio?

The Principle of Segregation To explain the 3:1 ratio of phenotypes in F_2 individuals, Mendel reasoned that the two alleles of each gene must segregate—that is, separate—into different gamete cells during the formation of eggs and sperm in the parents. As a result, each gamete contains one allele of each gene. This is called the **principle of segregation.**

To show how this principle works, Mendel used a letter to indicate the gene for a particular trait. For example, *R* represents the gene for seed shape. He used uppercase (*R*) to symbolize a dominant allele and lowercase (*r*) to symbolize a recessive allele. (Note that the symbols for genes are always italicized.)

Using this notation, Mendel could describe the genotype of the individuals in the round-seeded pure line as *RR*. The genotype of the pure line with wrinkled seeds is *rr*. Because *RR* and *rr* individuals have two copies of the same allele, they are said to be **homozygous** for the seed shape gene (*homo* is the Greek root for "same," while *zygo* means "yoked together").

Figure 10.4 diagrams what happened to these alleles when Mendel crossed the pure lines. According to his model, *RR* parents produce eggs and sperm that carry the *R* allele, while *rr* parents produce gametes with the *r* allele. When two gametes—one from each parent—are fused together, they create offspring with the *Rr* genotype. Individuals that have two different alleles for the same gene are said to be **heterozygous** (*hetero* is the Greek root for "different"). Because the *R* allele is dominant, all of these F_1 offspring produced round seeds.

Why do the two phenotypes appear in a 3:1 ratio in the F_2 generation? Mendel proposed that during gamete formation in the F_1, the paired *Rr* alleles separate into different gamete cells. As a result, about half of the gametes carry the *R* allele and half carry the *r* allele (**Figure 10.5a**). During self-fertilization, a given sperm has an equal chance of fertilizing either an *R*-bearing egg or an *r*-bearing egg. What is the chance that an F_2 progeny would end up with the *rr* genotype and produce wrinkled seeds?

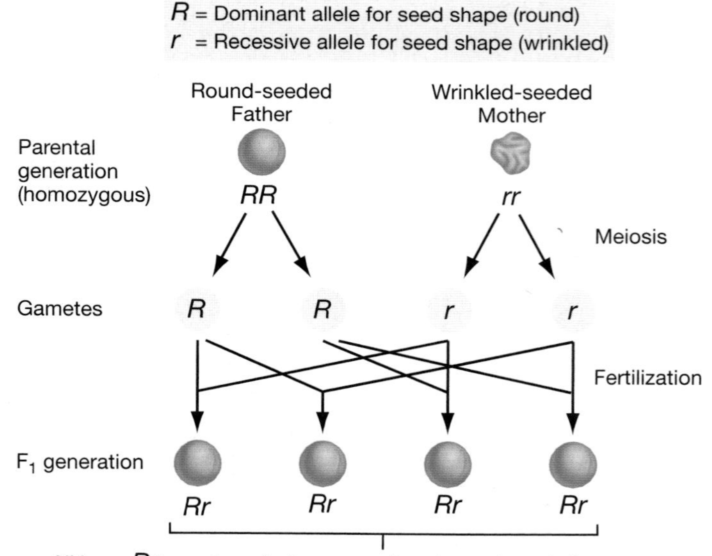

FIGURE 10.4 The F_1 Offspring of a Cross Between Pure Lines
QUESTION What types of gametes will be produced by the F_1 individuals in this experiment?

To predict the likelihood of each genotype (*RR*, *Rr*, and *rr*) in the F_2 generation, Mendel turned to the basic rules of probability that he had learned as an undergraduate in Vienna (**Figure 10.5b**). For example, for an offspring of *Rr* parents to be *rr*, it must inherit an *r* allele from each of its parents. How often does this happen? Mendel reasoned that because half of the sperm produced by an *Rr* individual has the *r* allele, the probability that any given offspring will inherit an *r* allele from this parent is 1/2, or 50 percent. Likewise, the probability that a given offspring will inherit an *r* allele from its mother is 1/2, or 50 percent. The probability that an offspring will inherit an *r* allele from both its mother and its father is the product of these two independent probabilities, or $1/2 \times 1/2 = 1/4$ (see **Box 10.1**, page 200). Therefore, Mendel reasoned that 1/4 of the F_2 offspring will be *rr* and should produce wrinkled seeds.

What is the chance that an F_2 progeny shows the dominant phenotype? A pea plant will produce round seeds if its genotype is *RR* or *Rr*. Using the same logic, then, offspring will inherit an *R* allele from both parents and have the *RR* genotype, $1/2 \times 1/2 = 1/4$ of the time (Figure 10.5b). So, what proportion of the offspring will be heterozygous? Progeny will be *Rr* if they inherit an *R* allele from their father and an *r* allele from their mother. This will happen $1/2 \times 1/2 = 1/4$ of the time. But they will also be *Rr* if they inherit an *r* allele from their father and an *R* allele from their mother. This will also happen $1/2 \times 1/2 = 1/4$ of the time. The probability of an offspring being *Rr* through either one or the other of these two routes is the sum of the two independent probabilities, or $1/4 + 1/4 = 1/2$ (see Box 10.1). Thus, half of the offspring produced will be *Rr*.

In sum, 1/2 of the F_2 offspring will be *Rr* and 1/4 will be *RR* (Figure 10.5b). This means that $1/2 + 1/4 = 3/4$ will have the dominant phenotype. Recall that 1/4 have the recessive phenotype. This is *exactly* what Mendel found in his experiments with peas. In the simplest and most elegant fashion possible, his model explains the 3:1 ratio of round to wrinkled seeds observed in the F_2 offspring and the mysterious reappearance of the wrinkled seeds.

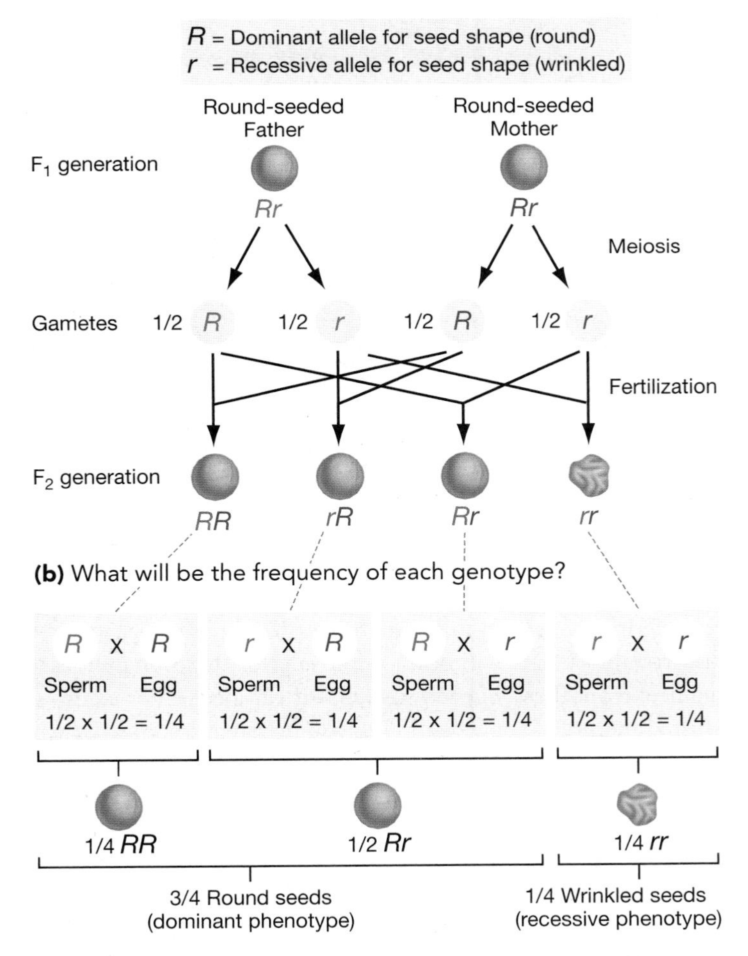

FIGURE 10.5 The F_2 Offspring of a Cross Between Pure Lines The cross shown here is a continuation of the experiment illustrated in Figure 10.4.

An Overview of Mendel's Model

Figure 10.6 summarizes the propositions that Mendel put forth to explain the basic patterns in the transmission of traits from parents to offspring. His model was clearly a radical break from the hypotheses of blending inheritance and inheritance of acquired characters that previously dominated scientific thinking about heredity.

MENDEL'S MODEL

1. **Peas have two versions, or alleles, of each gene.** This also turns out to be true for many other organisms.
2. **Alleles do not blend together.** The hereditary determinants maintain their integrity from generation to generation. They do not blend together, and they do not acquire characteristics in response to actions by an individual (like a giraffe stretching its neck).
3. **Each gamete contains one allele of each gene.** Pairs of alleles segregate during the formation of gametes.
4. **Males and females contribute equally to the genotype of their offspring.** When gametes fuse, offspring acquire a total of two alleles for each gene—one from each parent.
5. **Some alleles are dominant to others.** When a dominant and recessive allele for the same gene are found in the same individual, that individual exhibits the dominant phenotype.

FIGURE 10.6 Mendel's Model to Explain the Results of a Cross Between Pure Lines
QUESTION What is the difference between genes and alleles; a genotype and a phenotype; a homozygous versus heterozygous individual; and dominant and recessive alleles?

One of the most important features of Mendel's hypotheses was that they allowed biologists to predict the genotypes and phenotypes that should result from particular matings. To make these predictions efficiently, one of the leading geneticists of the early twentieth century, R. C. Punnett, invented the technique diagrammed in **Figure 10.7**. This system is called a Punnett square. The basic idea is to list the gametes produced by each parent along the tops and sides of the square. Once this is done, it is straightforward to fill in the boxes of the square with the offspring genotypes that result from fusion of these gametes. Finally, the proportions of each offspring genotype can be calculated from the proportions of each type of gamete produced by the parents.

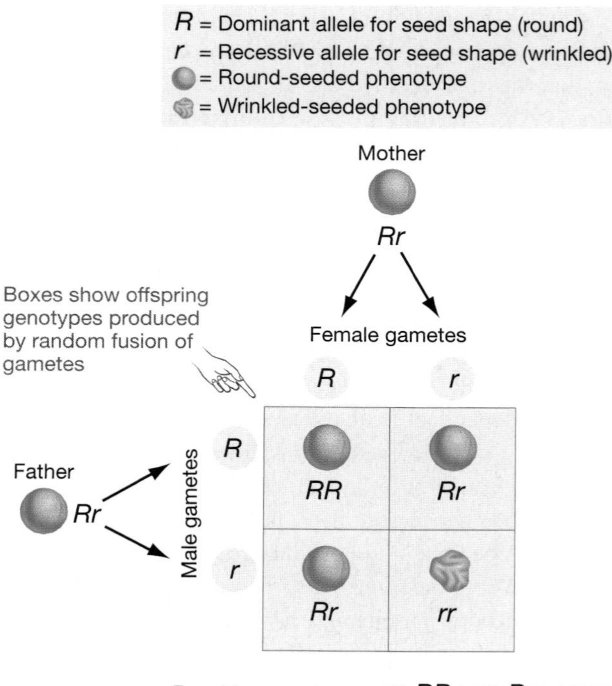

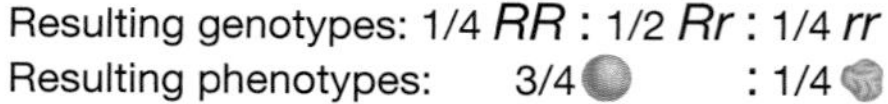

FIGURE 10.7 A Punnett Square
A Punnett square is an efficient way to predict the offspring produced by a particular mating. In this case, the mating shown is the same as the one analyzed in detail in Figure 10.5.

Testing the Model

Mendel's model explained his results in a logical way. But is it correct? To answer this question Mendel conducted a series of experiments with the F_2 progeny described previously. These experiments tested an important series of predictions about the offspring of these F_2 individuals. For example, consider the F_2 progeny of the cross between pure lines with round and wrinkled seeds.

- Plants with wrinkled seeds are *rr*. Thus, they should produce only *rr* offspring when they are self-pollinated or crossed with another individual with wrinkled seeds.
- Plants with the dominant phenotype may be either *RR* or *Rr*. These two genotypes should be present in the ratio 1:2. (That is, there should be twice as many heterozygotes as homozygotes among individuals with round seeds.)

BOX 10.1 Combining Probabilities

How often will two gametes, each carrying a particular allele, combine to form an offspring with a particular genotype? This is similar to asking how frequently *any* combination of independent events will occur. In this case the independent events are the probabilities of getting certain alleles from each parent. If the probability that each event will occur independently is known, then the combined probability—meaning the probability of getting a certain genotype—can be calculated by employing one of two rules. They are called the "both-and rule" and the "either-or" rule. Each rule pertains to a distinct situation.

The both-and rule applies when you want to know the probability that two or more independent events occur together. Let's use a deck of cards as an example. What is the probability of drawing the ace of hearts and then the ace of spades from a deck of 52 cards? These two events are independent, because the probability of drawing one of the cards has no effect on the probability of drawing the other. (In the same way, the probability of getting a certain type of gamete from one parent has no effect on the probability of getting a certain type of gamete from the other parent. Gametes fuse randomly.) The probability of drawing the ace of hearts is 1/52. Once the first card is gone, the probability of drawing the ace of spades is 1/51. The probability of drawing both of the aces, then, is $1/52 \times 1/51 = 1/2652$. In other words, if you drew two cards from a deck 2652 times, you would expect to get the ace of hearts and then the ace of spades once. It should make sense that the both-and rule is also called the multiplication rule.

The either-or rule, in contrast, applies when you want to know the probability of an event happening when there are several different ways for the event to occur. In this case, the probability that the event will occur is the sum of the probabilities of each way that it can occur. For example, suppose you wanted to know the probability of drawing either the ace of hearts or the ace of spades. The probability of drawing each is 1/52, so the probability of getting one or the other is $1/52 + 1/52 = 1/26$ (the either-or rule is also called the addition rule). If you drew one card from a deck 26 times, you'd expect to get one of the two aces once.

Are these predictions correct? To test them, Mendel planted the F_2 seeds and allowed the plants to self-pollinate when they matured. He then examined the phenotypes of the F_3 seeds. He quickly confirmed the first prediction: F_2 plants with wrinkled seeds always produced offspring with wrinkled seeds. This result was consistent with the claim that these F_2 plants have an *rr* genotype.

What about the offspring of parents with the dominant phenotype? Mendel let a total of 565 round-seeded plants self-pollinate. (**Box 10.2**, page 203, explains why Mendel used such large sample sizes in his experiments.) Of these, 193 produced only round-seeded offspring. Mendel inferred that these parents had the *RR* genotype. In contrast, 372 of the round-seeded parents produced offspring that made seeds that were either round or wrinkled. Mendel inferred that the round-seeded parental genotype was *Rr*. In this experiment, the ratio of *Rr* to *RR* parents—based on the number of round and wrinkled seeds produced—was 1.93:1. This is extremely close to the prediction of 2:1. Mendel observed the same patterns when he let F_2 individuals from the other six crosses (involving seed color and other traits) self-fertilize.

These results were a ringing confirmation of his model. They also verified that the genotype of a parent can be inferred by studying the phenotypes of its offspring. The research strategy that Mendel invented—of identifying genotypes from the phenotypes of offspring—was the basis of most research in genetics until the 1980s.

10.2 Mendel's Experiments with Two Traits

Working with one trait at a time allowed Mendel to recognize the principle of segregation. His next step was to extend this result. Does the principle of segregation hold true if parental lines differ with respect to two or more traits?

To explore this issue, Mendel crossed a parental line that had round yellow seeds with a parental line that had wrinkled green seeds (**Figure 10.8a**). Note that his earlier experiments had established that the allele for yellow seeds was dominant to the allele for green seeds.

As expected, all of the F_1 seeds showed the dominant yellow and round phenotypes (Figure 10.8a, bottom). Mendel then allowed the F_1 plants to self-pollinate. In the F_2 individuals that resulted he observed the four phenotypes shown in **Figure 10.8b**. Note that the four were present in a ratio of 9:3:3:1.

How do these data relate to the 3:1 ratio of phenotypes observed for single traits? To answer this question, Mendel summarized the data for seed color only. As **Figure 10.8c** notes, the ratio of round seeds to wrinkled seeds is 423 to 133, or 3.18 to 1. Likewise, the ratio of yellow to green plants is 416 to 140—almost exactly 3 to 1. When seed color and seed shape are considered separately, then, the F_2 seeds exhibited the same 3:1 (dominant:recessive) ratio observed in the first set of experiments. In each case, the dominant phenotype showed up in 3/4 of the F_2 offspring, while the recessive phenotype was observed in 1/4 .

How could Mendel use these data to calculate the probability of observing each of the four combined phenotypes—for color *and* shape? For example, if the probability that an F_2 individual will have round seeds is 3/4 and the probability that an F_2 individual will have yellow seeds is 3/4 , what is the probability that an F_2 individual will have round yellow seeds? Mendel knew that the probability of two independent events

(a) Cross peas that differ in two traits

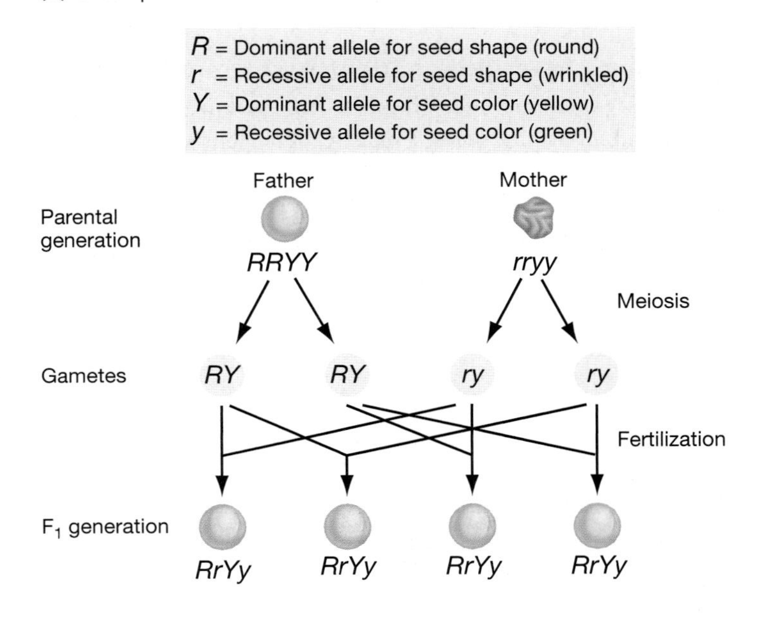

(b) Self-fertilize F_1 peas and count F_2 offspring

F_2 generation phenotype									
Number	315	+	101	+	108	+	32	=	556
Fraction of progeny	9/16	+	3/16	+	3/16	+	1/16	=	1

(c) How does the 9 : 3 : 3 : 1 ratio observed for two traits relate to the 3 : 1 ratio observed for one trait?

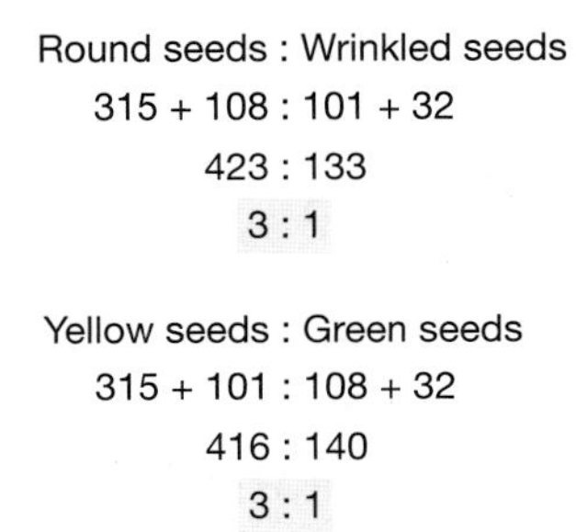

FIGURE 10.8 Analyzing the Inheritance of Two Traits

happening is the product of the probability of each occurring (see Box 10.1). Thus, the probability of observing round yellow seeds is $3/4 \times 3/4 = 9/16$. This is exactly what Mendel observed. The frequency of the other three phenotypes can be explained in the same way.

Mendel examined combinations of traits other than seed shape and color and obtained similar results. Each paired set of traits produced a 9:3:3:1 ratio of progeny phenotypes. He even did a limited set of crosses examining three traits at a time. These data also supported the claim that the frequency of combined phenotypes can be predicted from the probability of observing each phenotype separately. What does this result imply about how genes behave?

The Principle of Independent Assortment

Mendel's explanation for the data on combined phenotypes was inspired by the success of using the "both-and" rule of probability introduced in Box 10.1. This rule is based on the assumption that the events being analyzed are independent of one another. Because this assumption appeared to be valid in the case of the genetic traits he analyzed, Mendel proposed that during gamete formation, the segregation of each pair of alleles must occur independently.

This hypothesis is called the **principle of independent assortment.** The predictions that this hypothesis makes are explained in **Figure 10.9**, which diagrams the fate of alleles involved in Mendel's experiment with seed color and shape. Starting at top left in Figure 10.9, you can see the original mating between parents that are homozygous for the dominant traits (*RRYY*) and homozygous for the recessive traits (*rryy*). As the figure indicates, the F_1 offspring of this mating all have the same genotype: *RrYy*. They also have the same phenotype: round, yellow seeds.

The Punnett square in Figure 10.9 illustrates what happens when these F_1 individuals form gametes. If the alleles for each gene assort independently—meaning that a particular gamete has an equal chance of getting a *Y* or a *y* allele, irrespective of whether it also gets an *R* or *r* allele—then each *RrYy* parent will form four types of gametes in equal proportions. For example, the principle of segregation predicts that a gamete will receive an *R* allele from an *Rr* parent 1/2 of the time. The principle of independent assortment predicts that this gamete will also receive

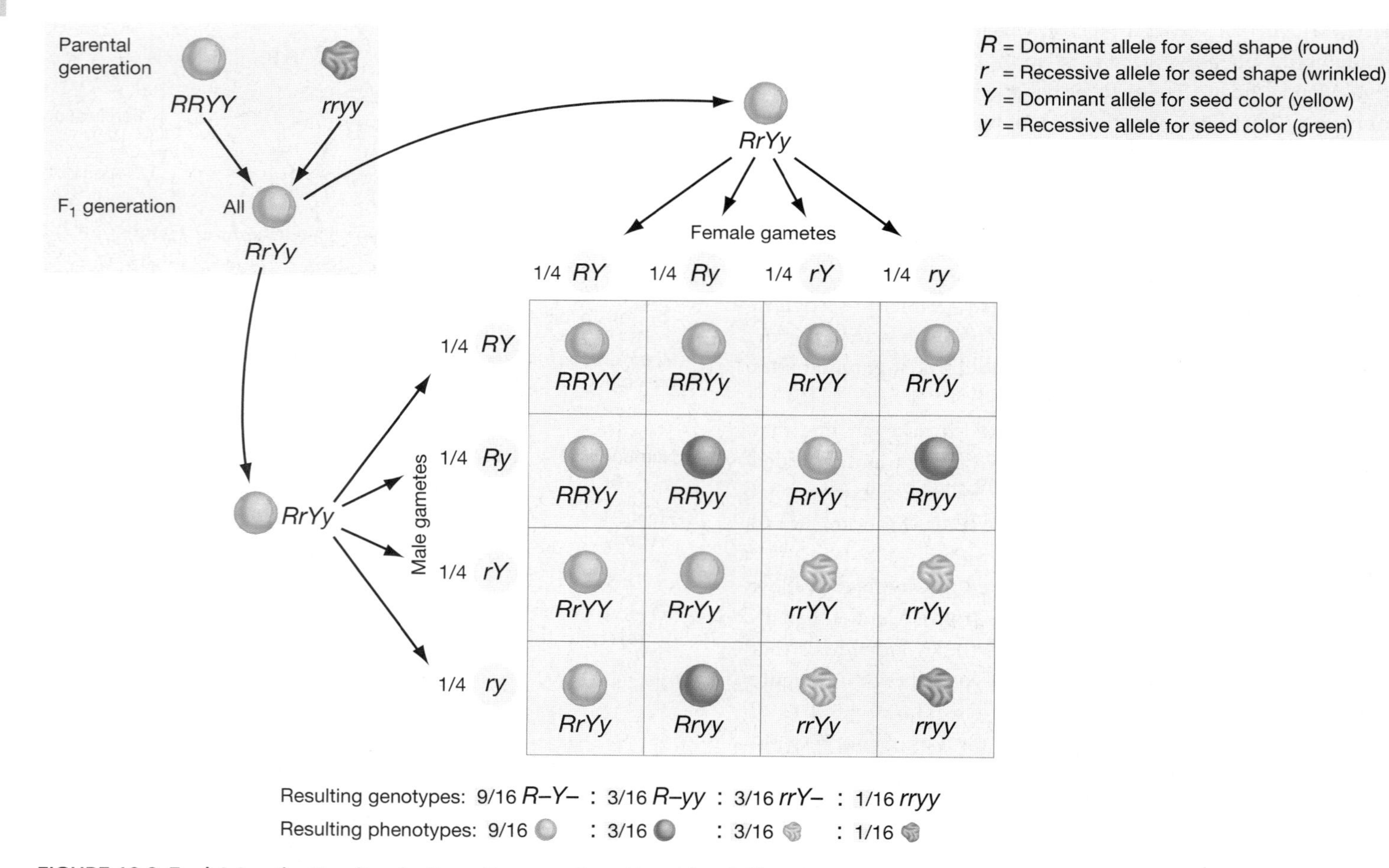

FIGURE 10.9 Explaining the Results of a Cross Between Pure Lines That Differ in Two Traits
To explain the results summarized in Figure 10.8, Mendel proposed that alleles for individual traits segregate independently and that genes for different traits assort independently. These diagrams show the types and proportions of gametes that result if these propositions are true.

BOX 10.2 Sample Size and Chance Fluctuations

Each time Mendel designed a cross to answer a question about the inheritance of a trait, he made sure to analyze the results in a large number of offspring. Collecting data from a large sample is a goal of almost all scientific studies. Large sample sizes reduce chance fluctuations in the outcome of an experiment and make it easier to recognize patterns in the data.

Scientists refer to chance fluctuations in data as "noise." Some noise is inevitable in the outcome of any experiment. For example, consider the data that Mendel collected on the seed-shape phenotypes observed in heterozygous parents that self-fertilized.

From parent to parent, the ratios of round and wrinkled seeds fluctuated from almost 1:1 to over 4:1. This is because in a small sample, there is a large element of chance in which gametes happen to combine. For example, even though only half of the gametes available when parent plant 4 self-fertilized contained an *r* allele, there happened to be a run of fertilizations between *r*- and *r*-containing gametes. Small samples frequently have skewed results like this, just due to chance. But when Mendel pooled the data into a large sample, it became clear that the overall ratio of phenotypes was about 3:1. If Mendel had been able to obtain a larger number of offspring from parent plant 4, it is very likely that the observed ratio of phenotypes would have been closer to 3:1.

Plant #	Round	Wrinkled	Ratio
1	45	12	3.75 : 1
2	27	8	3.37 : 1
3	24	7	3.42 : 1
4	19	16	1.19 : 1
5	32	11	2.91 : 1
6	26	6	4.33 : 1
7	88	24	3.66 : 1
8	22	10	2.20 : 1
9	28	6	4.66 : 1
10	25	7	3.57 : 1
Total	**336**	**107**	**3.14 : 1**

a *Y* allele from a *Yy* parent 1/2 of the time. An *RrYy* parent, then, will produce *RY* gametes $1/2 \times 1/2 = 1/4$ of the time.

What happens when the four types of gametes shown in Figure 10.9 fuse during self-fertilization? The boxes inside the Punnett square answer this question. They show all of the possible combinations of gametes when *RrYy* parents self-fertilize. The lines below the Punnett square summarize the results of this mating, by calculating the types and proportions of genotypes and phenotypes observed in the F_2 progeny.

Using a Testcross to Confirm Predictions

How can the predictions made by the principle of independent assortment be tested? For example, is it possible to confirm the prediction in Figure 10.9—that an *RrYy* plant produces four different types of gametes in a 1:1:1:1 ratio?

To answer these questions, Mendel invented a technique called a testcross. A **testcross** uses a parent that contributes only recessive alleles to the progeny. Testcrosses are powerful and useful because the genetic contribution of the homozygous recessive parent is easy to predict and analyze. As a result, a testcross allows experimenters to test the genetic contribution of the other parent.

In this case, Mendel performed a testcross between parents that were *RrYy* and *rryy*. The types and proportions of offspring that should result can be predicted with the Punnett square shown in **Figure 10.10**. If the principle of independent assortment is valid, there should be four types of offspring in equal proportions.

What were the actual proportions observed? Mendel did this experiment and examined the seeds produced by the progeny. He found that 31 were round and yellow, 26 were round and green, 27 were wrinkled and yellow, and 26 were wrinkled and green. As predicted, these numbers are nearly identical to the

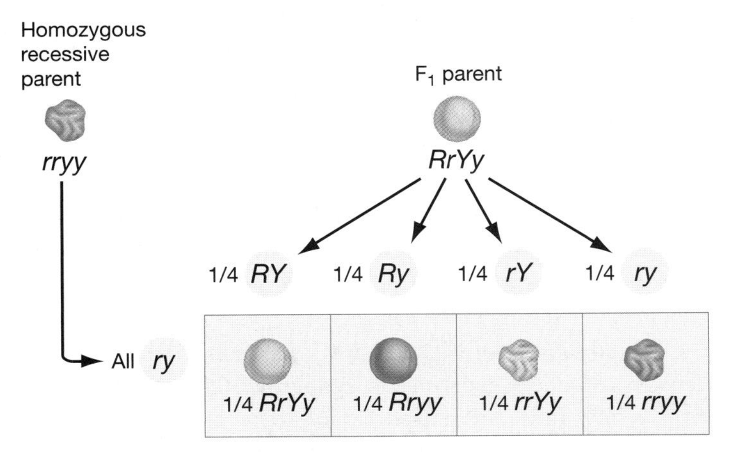

FIGURE 10.10 The Predictions Made by the Principle of Independent Assortment Can Be Evaluated in a Testcross
If the principle of independent assortment is true and *RrYy* parents produce four types of gametes in equal proportions, then a mating between *RrYy* and *rryy* parents should produce four types of offspring in equal proportions, as this Punnett square shows.

27.5 individuals expected with each genotype given the total of 110 individuals. The testcross had confirmed the principle of independent assortment.

Mendel's Contributions to the Study of Heredity

Mendel introduced an approach to studying heredity that is still in use today. This approach includes the following considerations:

- The thoughtful choice of an appropriate model organism
- Use of pure lines with discrete differences in traits to explore how traits are transmitted to offspring
- Examination of large numbers of progeny for each type of cross
- Application of the rules of probability to predict the numbers and types of progeny produced from crosses
- Use of a simple symbolism to represent genes and alleles

In short, Mendel's work provided a powerful conceptual framework for thinking about heredity. He was the first individual to correctly describe the basic rules of **transmission genetics**—the patterns that occur as genes pass from one generation to the next. All of the types and proportions of progeny that Mendel observed in his F_1 and F_2 generations could be explained as the consequence of two processes: The segregation of discrete, paired alleles into separate gametes and the independent assortment of alleles that control different traits.

Mendel's experiments were brilliant in design and execution. Unfortunately, they were ignored for 35 years.

10.3 The Chromosome Theory of Inheritance

Historians of science frequently debate why Mendel's work was overlooked for so long. It is almost undoubtedly true that his use of probability theory and his quantitative treatment of data were difficult for the biologists of the time to understand and absorb. It may also be true that the theory of blending inheritance was so well entrenched that there was a tendency to dismiss his results as peculiar or unbelievable. Whatever the reason, Mendel's work was not appreciated until other biologists, working with a variety of plants and animals, independently reproduced his results in the early 1900s.

The rediscovery of Mendel's work ignited the young field of genetics. Mendel's experiments established the basic rules that govern how traits are passed from parents to offspring. They described the pattern of inheritance. But what process is responsible for these patterns? Two biologists, working independently, came up with the answer. Walter Sutton and Theodor Boveri each realized that meiosis could be responsible for Mendel's rules. When this hypothesis was published in 1903, research in genetics exploded.

Meiosis is the type of cell division that precedes gamete formation. The details of the process, introduced in Chapter 9, were worked out in the final decades of the nineteenth century. What Sutton and Boveri grasped is that meiosis not only reduces chromosome number by half, it also explains the principle of segregation and the principle of independent assortment. **Figure 10.11** shows why. Independent assortment occurs because maternal and paternal chromosomes align themselves independently of one another during metaphase of meiosis I. As a result, chromosomes from the mother and father assort independently.

The cell nucleus drawn at the top of Figure 10.11a illustrates Sutton and Boveri's central insight—the hypothesis that chromosomes are composed of Mendel's hereditary determinants, or genes. In this example the gene for seed shape is shown at a particular position along a chromosome. This location is known as a **locus** (place). A genetic locus is the physical location of a gene. The paternal and maternal chromosomes shown in the figure each happen to possess a different allele of the gene for seed shape. One specifies round seeds (*R*) while the other specifies wrinkled seeds (*r*).

The subsequent drawings in Figure 10.11 show how these alleles segregate into different daughter cells during meiosis I. This physical separation of alleles produces Mendel's principle of segregation.

Figure 10.11b follows the fate of the alleles for two different genes—in this case, for seed shape and seed color—as meiosis proceeds. Because these genes are located on different chromosomes, they assort independently of one another at meiosis I. Four types of gametes, produced in equal proportions, result. This is the physical basis of Mendel's principle of independent assortment. Most of the genes that Mendel analyzed assort independently from one another *because they are each located on different chromosomes.*

Sutton and Boveri formalized these observations in the **chromosome theory of inheritance.** Like other theories in biology, the chromosome theory consists of a pattern—a set of observations about the natural world—and a process that explains the pattern. The chromosome theory states that Mendel's rules can be explained by the independent segregation of homologous chromosomes at meiosis I.

At the time that Sutton and Boveri published their findings, however, the hypothesis that chromosomes actually consist of genes was untested. What experiments confirmed that chromosomes contain genes?

10.4 Testing and Extending the Chromosome Theory

During the first decade of the twentieth century an unassuming insect rose to prominence as a model organism for testing the chromosome theory. This organism—the fruit fly *Drosophila melanogaster*—has been at the center of genetic studies ever since. *D. melanogaster* has all the attributes of a useful model organism to study in genetics: small size, ease of culture, short reproductive cycle (about 10 days), and abundant progeny (up to a few hundred per mating). The elaborate external anatomy of this insect makes it possible to identify interesting phenotypic variation among individuals.

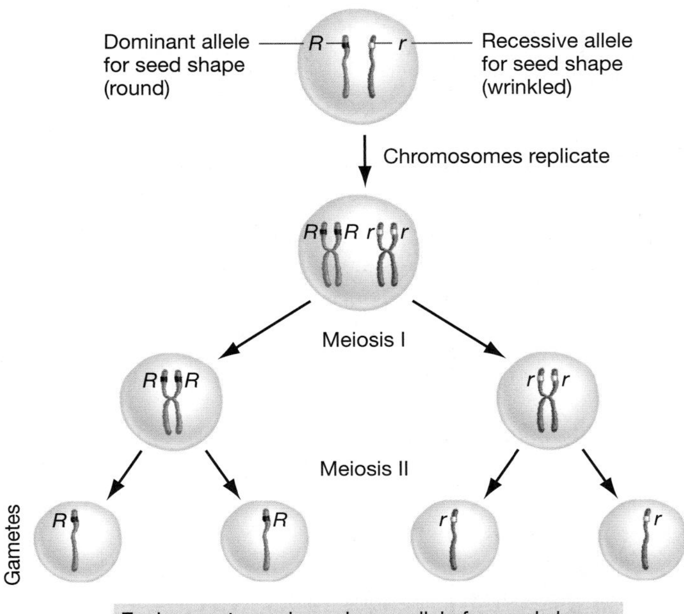

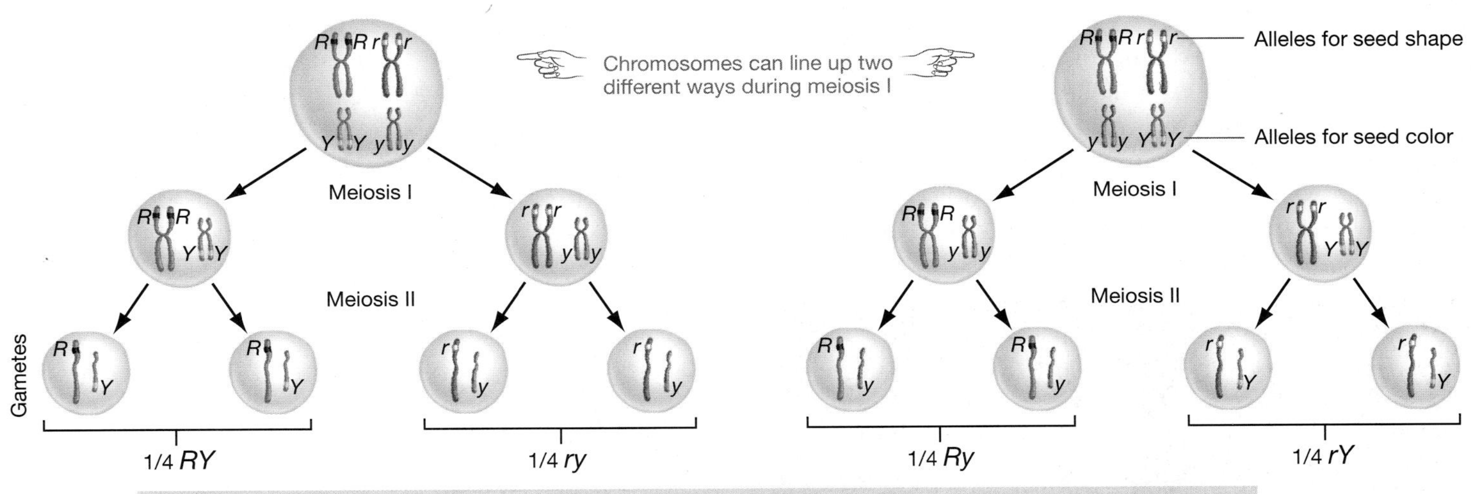

FIGURE 10.11 Meiosis Is Responsible for Both the Principle of Segregation and the Principle of Independent Assortment

In these diagrams, maternal and paternal chromosomes are shown in different colors, as in Chapter 9. The alleles for round (*R*) and wrinkled (*r*) seeds and the alleles for yellow (*Y*) and green (*y*) seeds are shown where they occur along the length of chromosomes of garden peas. **(a)** The two alleles found in a parent segregate into different gametes, just as Mendel hypothesized, because homologous chromosomes separate during meiosis I. **(b)** The alleles for different traits assort independently, again as Mendel hypothesized, because nonhomologous chromosomes assort independently during meiosis I.

The White-Eyed Fly

Drosophila research was pioneered by Thomas Hunt Morgan and his students. But because *Drosophila* is not a domesticated species like the garden pea, Morgan had no readily available phenotypic variants like Mendel's round and wrinkled seeds. Consequently, an early goal of his research was simply to find and characterize individuals with different phenotypes.

Morgan's group referred to individuals with the most common phenotype as **wild type**. But while examining his cultures one day, Morgan discovered a male fly that had white eyes rather than the wild-type red eyes (**Figure 10.12**, page 206). This individual had a discrete and easy-to-recognize phenotype that appeared spontaneously. Morgan inferred that the white-eyed phenotype resulted from a **mutation**—a change in a gene that specified eye color. Individuals with white eyes (or other traits attributable to mutation) are referred to as **mutants**.

To explore how eye color is inherited in flies, Morgan mated a red-eyed female fly with the mutant white-eyed male fly. Because all of the F_1 progeny had red eyes, Morgan tentatively concluded that the white-eyed phenotype was recessive and the wild-type red-eyed phenotype was dominant. To follow up on this result, Morgan allowed the F_1 males and females to breed with one another. As expected, he observed a 3 to 1 ratio of red-eyed to white-eyed phenotypes in the F_2 offspring. Morgan noted something peculiar, however. All of the white-eyed progeny were male. Half of the males in the F_2 generation had white eyes and half had red. But all of the F_2 females had red eyes.

Was there some sort of association between eye color and sex? To test this idea, Morgan crossed F_1 red-eyed females with white-eyed males. Some of the female offspring had white eyes. This told Morgan that it was possible for females to have white eyes. Obtaining white-eyed females also allowed Morgan to perform the second part of a reciprocal cross. (Recall that Mendel used reciprocal crosses to show that a parent's sex did *not* affect the inheritance of traits in peas.) When Morgan mated white-eyed females to red-eyed males from a pure line, the results were striking. All of the F_1 females had red eyes. But all of the males had white eyes.

Mendel's reciprocal crosses had always given similar results. But Morgan's reciprocal crosses did not. The experiment suggested a definite relationship between the sex of the progeny and inheritance of eye color. Even though the white phenotype appeared to be recessive in the first set of crosses, it appeared in the F_1 males when the reciprocal cross was performed. How could these observations be reconciled with Mendel's rules of inheritance? Thanks to a key finding by Nettie Stevens, Morgan was able to answer this question and confirm that genes do make up chromosomes.

The Discovery of Sex Chromosomes

Stevens began studying the karyotypes of insects about the time that Morgan began his work with *Drosophila*. One of her outstanding observations was a striking difference in the chromosome complements of males and females in the beetle *Tenebrio molitor*. In females of this species, diploid cells contain 20 large chromosomes. But diploid cells in males contained 19 large chromosomes and 1 small one. Stevens called the small chromosome the Y chromosome. As **Figure 10.13a** shows, this Y chromosome had a pairing partner at meiosis I, which Stevens called the X chromosome.

In addition to discovering the X and Y chromosome, Stevens observed that all eggs in this species had 10 large chromosomes. Sperm, however, could be divided into two categories. Because the X chromosome and Y chromosome paired up during meiosis and separated into different gamete cells, about 50 percent of the sperm contained 10 large chromosomes; the other 50 percent had 9 large chromosomes plus one small Y chromosome. Stevens observed similar patterns in other species of insects.

Based on these descriptive studies, Stevens developed a hypothesis to explain how sex determination occurs in these species. She proposed that a male is produced when an egg is fertilized by a sperm carrying a Y chromosome, but a female is produced when an egg is fertilized by a sperm carrying an X chromosome. As **Figure 10.13b** shows, the equal ratio of X- and Y-bearing sperm explains why the sexes are produced in nearly equal proportions.

X-Linked Inheritance and the Chromosome Theory

To explain the results of his crosses with white-eyed flies, Morgan put his genetic data together with Stevens's observations on sex chromosomes. *Drosophila* females, like *Tenebrio* females, have two X chromosomes; male fruit flies carry an X and a Y.

FIGURE 10.12 Eye Color Is a Variable Trait in the Fruit Fly *Drosophila melanogaster*
EXERCISE Label the phenotype that is considered wild type. Label the phentoype that is a rare mutant.

(a) Sex chromosomes from the beetle *Tenebrio molitor*

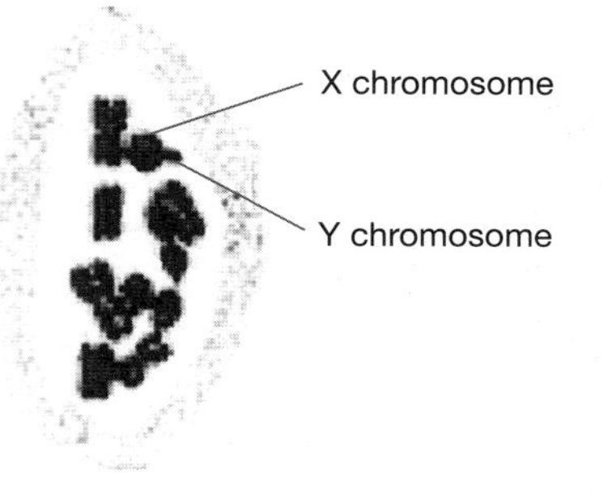

(b) Sex chromosomes pair at meiosis I

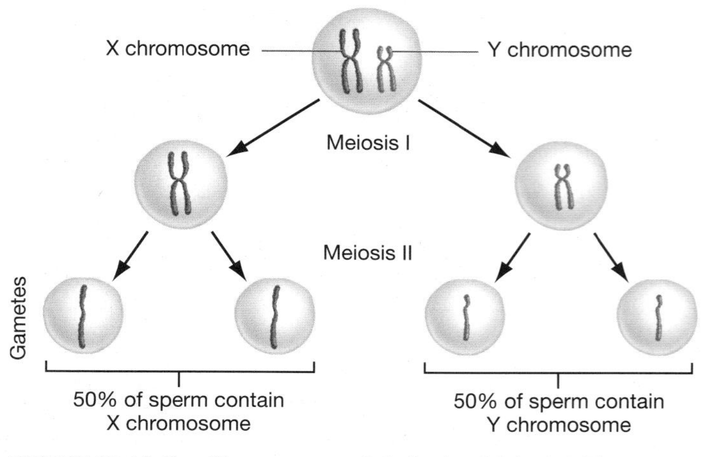

FIGURE 10.13 Sex Chromosomes Pair During Meiosis I, Then Segregate to Form X- and Y-Bearing Gametes
(a) As this drawing shows, the Y chromosome of the beetle *Tenebrio molitor* is much smaller than the X chromosome. **(b)** Sex chromosomes pair at meiosis I in male fruit flies, even though they are different morphologically. As a result, half of the sperm cells that result from meiosis bear an X chromosome, while half have a Y chromosome.

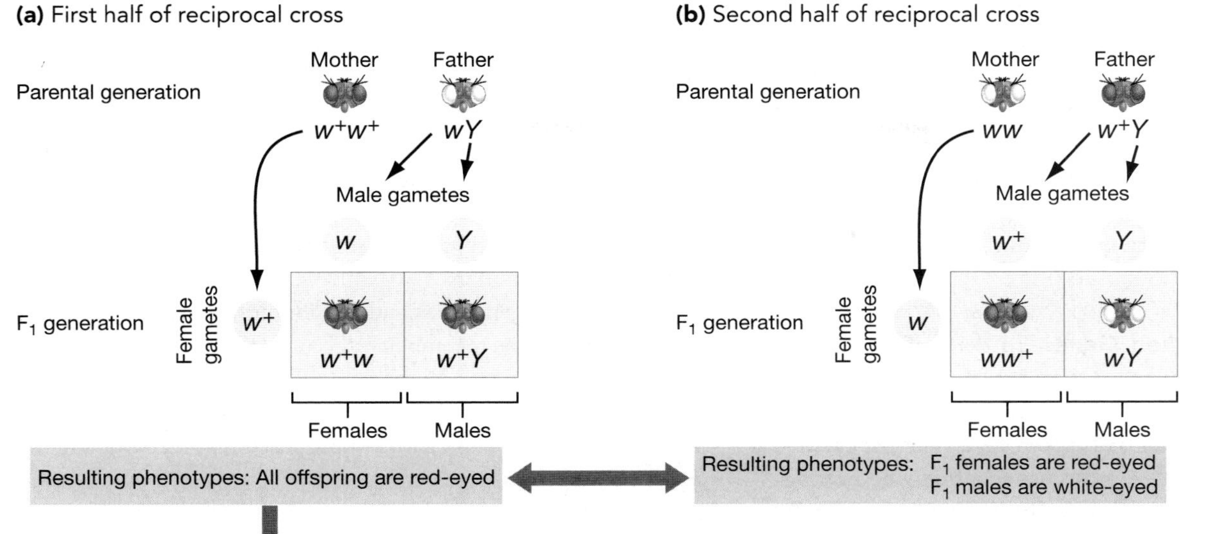

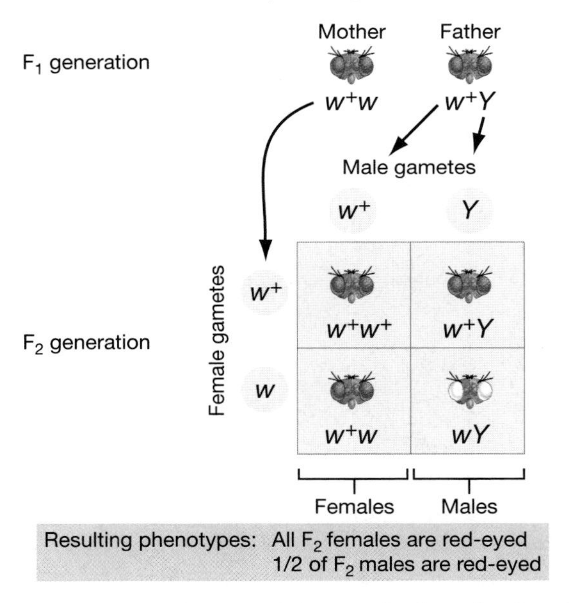

FIGURE 10.14 Reciprocal Crosses Confirm That Eye Color in *Drosophila* Is a Sex-Linked Trait
When Morgan crossed red-eyed females with white-eye males **(a)**, and then crossed white-eyed females with red-eyed males **(b)**, he observed strikingly different results. This was consistent with his hypothesis that eye color is a sex-linked trait in fruit flies. **EXERCISE** Make a Punnett square that summarizes a mating between the F_1 females and males in part (b).

As a result, half of the gametes produced by a male fruit fly should carry an X chromosome and the other half should carry a Y chromosome.

Morgan realized that the transmission pattern of the X chromosome in males and females could account for the results of his reciprocal crosses. Specifically, he proposed that the gene controlling eye color in flies is located on the X chromosome, and that the Y chromosome does not carry an allele of this gene.

According to this hypothesis, called **X-linked** or **sex-linked inheritance**, a female has two copies of the gene that specifies eye color—one contributed by each parent. A male, in contrast, has only one copy—contributed by his mother. The diagrams of crosses and the Punnett squares in **Figure 10.14** show how Morgan's hypothesis of X-linkage accords with the results of his experiments. As you work through this figure, note that the allele for red eyes is denoted w^+, while the allele for white eyes is denoted w. (In fruit-fly genetics, the + symbol always indicates the wild-type trait.) When reciprocal crosses give different results, such as those illustrated in Figure 10.14, it is likely that the gene in question is located on a sex chromosome. (Non-sex chromosomes are called **autosomes**, so genes located on non-sex chromosomes are said to show **autosomal inheritance**.)

Morgan's discovery of sex-linked inheritance carried an even more general message, however. In *Drosophila*, the gene for eye color is clearly correlated with inheritance of the X chromosome. This correlation was important evidence in support of the hypothesis that chromosomes contain genes.

What Happens When Genes Are Located on the Same Chromosome?

When later experiments confirmed that genes are indeed the physical components of chromosomes, the result prompted Morgan and other geneticists to reevaluate Mendel's principle of independent assortment. They realized that genes would not undergo independent assortment if they were on the same chromosome. The physical association of genes that are found on the same chromosome but influence different traits is called **linkage**.

The first examples of linked genes were loci found on the X chromosome of fruit flies. After Morgan established that the white eye gene was located on *Drosophila*'s X chromosome, he and colleagues established that body color was also an X-linked trait. Red eyes and gray body are the wild-type phenotypes in this species; white eyes and a yellow body occur as rare mutant phenotypes. The alleles for red eyes (w^+) and gray body (y^+) also are dominant to the alleles for white eyes (w) and yellow body (y).

Based on these observations, it seemed logical to predict that the linked genes would always be transmitted together during gamete formation. As **Figure 10.15a** shows, the two loci would not show independent assortment. Because the alleles for eye color and body color are inherited together, just two classes of gametes should be produced by the F_1—and produced in equal numbers. Is this what actually occurs?

The First Studies of Linked Genes To determine whether linked traits behave as predicted, Morgan performed the crosses described in **Figure 10.15b.** This figure introduces some new notation. When biologists analyze linked loci, they list the alleles found on homologous chromosomes separately by putting a slash between them. For example, the parental females in Morgan's experiment came from a pure line that had wy^+ alleles on one X chromosome and wy^+ alleles on the other; this is also written as the parental females having wy^+/wy^+ alleles on their two X chromosomes.

The interesting result in Figure 10.15b is contained in the table, which summarizes the phenotypes and genotypes observed in the experiment's F_2 generation of males. Most of these males carried an X chromosome with one of the two combinations of alleles found in their mothers. But a small percentage of males had novel genotypes and phenotypes. Morgan referred to these individuals as **recombinant** because the combination of alleles on their X chromosome was different from the combinations of alleles present in the parental generation.

To explain this result, Morgan proposed that recombinant wy and w^+y^+ gametes were generated when crossing over occurred during prophase of meiosis I in the wy^+/ w^+y females. (Crossing

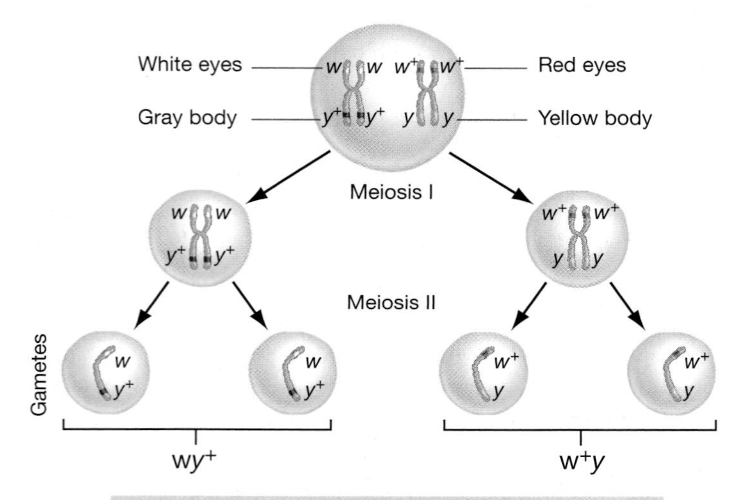

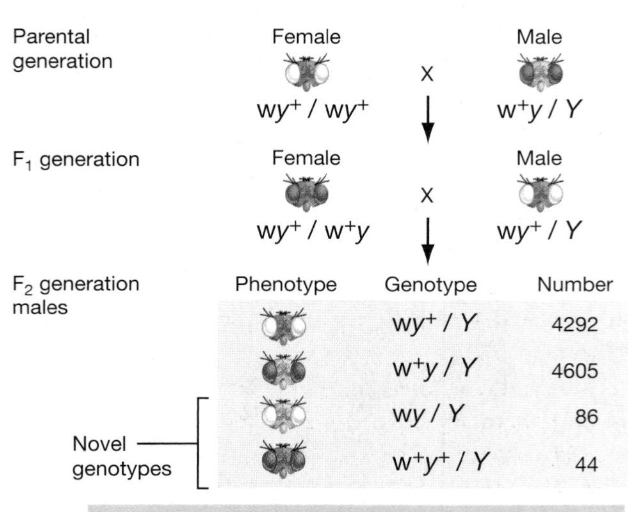

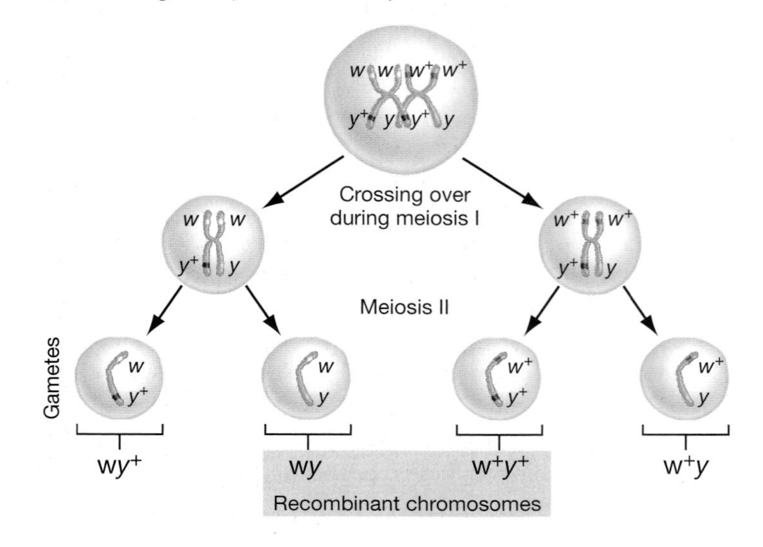

FIGURE 10.15 Linkage and Recombination
(a) Independent assortment does not occur when genes are linked. **(b)** Morgan crossed pure-line *Drosophila* females with white eyes and gray bodies with red-eyed, yellow-bodied pure-line males. When he let the F_1 offspring breed, he observed four phenotypes among the male F_2 progeny. Two of the phenotypes were not predicted by the model of complete linkage diagrammed in part (a). **(c)** To explain his results, Morgan proposed that crossing over occurred in a small percentage of the F_1 females during meiosis I. The recombinant chromosomes that resulted would produce the unusual phenotypes observed in F_2 males.

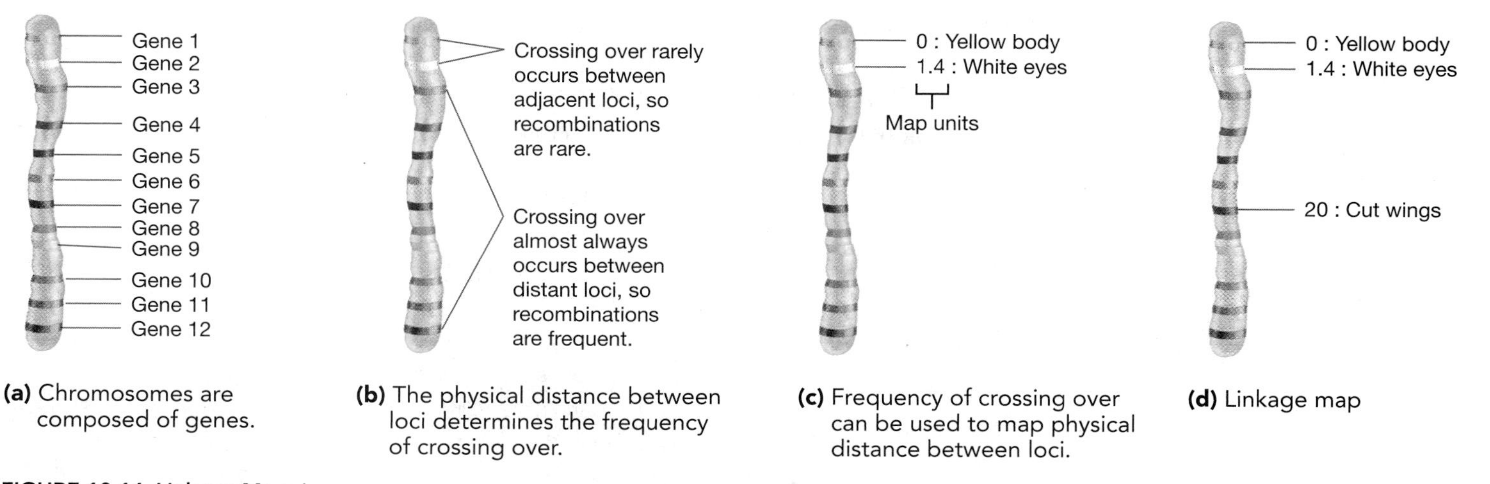

(a) Chromosomes are composed of genes.

(b) The physical distance between loci determines the frequency of crossing over.

(c) Frequency of crossing over can be used to map physical distance between loci.

(d) Linkage map

FIGURE 10.16 Linkage Mapping
(a) Morgan hypothesized that genes are found in a linear sequence along chromosomes. **(b)** If Morgan's proposal is true, then the frequency of recombination between two loci should be a function of the physical distance between them. **(c)** Recombinants between the yellow-body locus and the white-eye locus occur 1.4% of the time. Therefore, these loci are 1.4 map units apart on the chromosome. **(d)** Recombinants between the yellow-body locus and another X-linked gene called the cut-wing locus occur 20% of the time. But recombinants between the white-eye locus and the cut-wing locus occur just 18.6% of the time. Therefore, the loci must be arranged as shown here.

over, a physical exchange of segments from homologous chromosomes, was introduced in Chapter 9.) **Figure 10.15c** illustrates Morgan's hypothesis by showing how the alleles could be exchanged. This result cements the connection between the events of meiosis I and Mendel's laws. That is, linked genes segregate together unless there is a physical crossover between the chromosome they share and a homologous chromosome.

In this particular experiment, about 1.4 percent of the gametes were recombinant. But when Morgan and co-workers did the same types of crosses with different pairs of X-linked traits, they found that the fraction of recombinant gametes varied. For example, when crosses involved X-linked genes for a yellow body and an abnormality called cut wings (*ct*), recombinant gametes were produced about 20 percent of the time.

Morgan explained these observations by making an authentic conceptual breakthrough. He proposed that genetic loci are arranged in a linear array along a chromosome, as illustrated in **Figure 10.16a**. According to Morgan's hypothesis, the physical distance between loci dictates the frequency of crossing over between them. The smaller the distance between any two loci on a chromosome, the smaller the chance that crossing over will take place somewhere in between. The fundamental assumption is that greater physical distance between loci increases the chance that crossing over will take place between them (**Figure 10.16b**).

Linkage Mapping In 1911 A. H. Sturtevant, an undergraduate who was studying with Morgan, realized that variation in recombination frequency had an important implication. If genes are lined up along the chromosome and if the frequency of crossing over is a function of the physical distance between loci, then it should be possible to figure out where loci are in relation to each other based on the frequency of recombinants between various pairs. That is, it should be possible to create a **genetic map.**

Sturtevant proposed that in constructing a genetic map, the unit of distance along a chromosome should simply be the percentage of recombinants between loci. For example, he proposed that the eye-color and body-color loci are 1.4 map units apart on the X chromosome, because recombination between these loci occurs in 1.4 percent of meiotic events (**Figure 10.16c**). The *y* locus for body color and the gene for cut wings, in contrast, are 20 map units apart. Because recombinants occur in 18.6 percent of the gametes produced by females that are $w^{+}ct/wct^{+}$, Sturtevant inferred that the loci for white eyes, yellow body, and cut wings are arranged as illustrated in **Figure 10.16d**. Using data like these for many loci, Sturtevant constructed the first chromosome map.

Before moving on to consider some exceptions and extensions to Mendel's rules, let's step back and consider the events we've just surveyed—from the rediscovery of Mendel's work, around 1900, to Sturtevant's chromosome map. In 1899 geneticists did not understand the basic rules of heredity. But by 1911, they could map the locations of genes on chromosomes. A remarkable knowledge explosion had occurred.

10.5 Extending Mendel's Rules

Some of the most important early advances in genetics occurred when researchers found traits that did not seem to follow Mendel's rules. If experimental crosses produced F_2 progeny that did not conform to the expected 3:1 or 9:3:3:1 ratios in phenotypes, researchers had a strong hint that something interesting was going on. In many cases it turned out that unraveling the cause of the anomaly led to new insights into how genes and heredity worked.

This section is devoted to exploring some of these incongruities. How can traits that don't appear to follow Mendel's rules contribute to a more complete understanding of heredity?

Incomplete Dominance

Consider the flowers called four-o'clocks, pictured in **Figure 10.17a**. In this species horticulturists have developed a pure line that has red flowers and a pure line that has white flowers. When individuals from these strains are mated, however, all of their offspring are pink (**Figure 10.17b**). In Mendel's peas, crosses between purple- and white-flowered parents produced all purple-flowered offspring. Why the difference?

The answer can be inferred by examining the phenotypes of F_2 offspring in four-o'clocks. These are the progeny of self-fertilization in pink-flowered F_1 individuals. Of the F_2 plants, 1/4 have red flowers, 1/2 have pink flowers, and 1/4 have white flowers. These phenotypic ratios are unlike any we have seen to date. What causes them? The answer became clear when geneticists realized that the 1:2:1 ratio of red:pink:white is identical to the proportion of homozygotes and heterozygotes that would be produced if flower color is controlled by one gene with two alleles.

To convince yourself that this explanation is sound, study the genetic model shown in Figure 10.17b. According to the hypothesis shown in the diagram, the inheritance of flower color in four-o'clocks and peas is identical, except that the four-o'clock alleles show **incomplete dominance.** Neither red nor white alleles dominate; instead, the F_1 progeny—all heterozygous—show a phenotype intermediate between the two parental strains.

Incomplete dominance illustrates an important general point: Dominance is not necessarily an all-or-none phenomenon. Why this is so will become clear in Chapter 11, when we investigate the molecular basis of gene action.

Multiple Alleles

Mendel worked with traits for which phenotypic variation depended on the actions of a single gene with two alleles, complete dominance, and autosomal inheritance. This is the simplest possible genetic system. But as the examples of X-linked inheritance and incomplete dominance show, most loci are not this straightforward.

In most populations dozens or even hundreds of alleles can be identified at each genetic locus. The existence of more than two alleles of the same gene is known as **multiple allelism.** When different combinations of these alleles produce more than two distinct phenotypes, the trait is said to be **polymorphic** (many-forms).

The first polymorphic trait ever described is the ABO blood group in humans. Your own blood has almost certainly been analyzed and assigned an O, A, B, or AB phenotype. Karl Landsteiner discovered these phenotypes in 1900; much later it was established that they are caused by carbohydrates located on the membranes of red blood cells.

As **Table 10.1** shows, the ABO phenotypes result from three alleles called i, I^A, and I^B. The i allele is recessive and produces the O phenotype when it is homozygous. The A and B alleles, in contrast, are dominant with respect to the i allele but **codominant** with respect to each other. When codominance occurs, heterozygotes have the phenotype of both homozygotes. For example, AB heterozygotes have the AB phenotype instead of simply A or

(a) Four-o'clocks

(b) Incomplete dominance in flower color

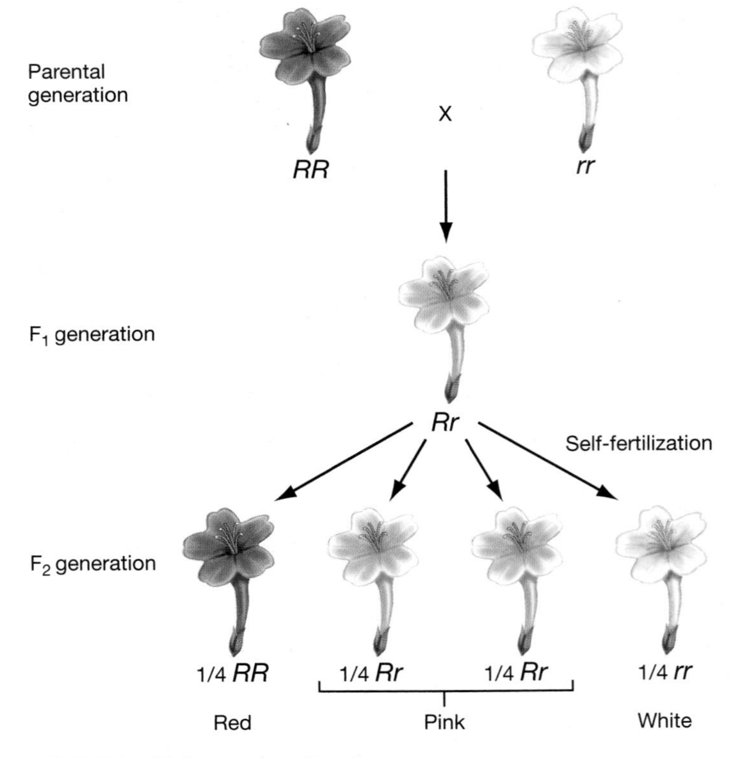

FIGURE 10.17 Incomplete Dominance
(a) Four-o'clocks got their name because their flowers open in the late afternoon. Flower color is highly variable. **(b)** The hypothesis illustrated here is that flower color is controlled by a single gene with two alleles, symbolized R and r. Heterozygotes have an intermediate phenotype because of a phenomenon known as incomplete dominance.

B. If you have the *AB* blood type, your red blood cells contain both the *A* and *B* carbohydrate types on their surfaces.

The ABO blood group involves three alleles, six genotypes, and four phenotypes. Would Mendel have discovered the principles of segregation and independent assortment had he worked with a trait like this? The patterns of inheritance that result from multiple allelism are complicated enough to suggest that the answer is no. Mendel was fortunate to work with the simplest genetic situation possible—traits whose inheritance were not complicated by linkage, multiple allelism, incomplete dominance, or codominance.

Genes Are Affected by Their Environment

When Mendel analyzed height as a trait in his experiments, he was careful to ensure that each plant received a similar amount of sunlight and grew in similar soil. This was important because individuals with alleles for tallness are stunted if they are deprived of nutrients, sunlight, or water—so much so that they look similar to individuals with alleles for dwarfing. In order for Mendel to analyze the hereditary determinants of height, he had to control the environmental determinants of height.

The Physical Environment Has a Profound Effect on Phenotypes The phenotypes produced by most genes and alleles are strongly affected by the physical environment. It thus can be argued that an individual's phenotype is as much a product of its environment as it is a product of its genotype. To drive this point home, consider the human genetic disease called phenylketonuria (PKU). People with PKU lack an enzyme that helps break down the amino acid phenylalanine. As a result, phenylalanine and a related molecule called phenylpyruvic acid accumulate in their bodies. The molecules interfere with the development of the nervous system and produce profound mental retardation. But if PKU individuals are identified at birth and placed on a low-phenylalanine diet, they develop normally.

Because of a simple change in their environment, individuals with a PKU genotype can have a normal phenotype. Genetic traits are influenced by more than the physical environment that an individual experiences, however. Alleles are also influenced by the action of other genes.

Interactions with Other Genes In Mendel's pea plants, a single locus influenced seed shape. Further, Mendel's data showed that this locus does not appear to be affected by the action of genes for seed color, seed-pod color, seed-pod shape, or other traits. The pea seeds he analyzed were round or wrinkled regardless of the types of alleles present at other loci.

In many cases, however, loci are not as independent as the gene for seed shape in peas. As an example, consider the inheritance of fruit color in bell peppers. As **Figure 10.18a** (page 212) shows, bell peppers come in a wild assortment of colors. Are the color variations due to multiple alleles of a single gene, or are several genes involved? To answer this question, consider the data presented in **Figure 10.18b** on the F_1 and F_2 offspring of pure-line parental strains with yellow and brown fruit.

The results are surprising. Like the experimental cross with four-o'clocks, the F_1 phenotype is different from either parental strain. But a fourth phenotype (green), not seen in either parent or in the F_1, appears in the F_2 generation. Further, the progeny ratios in the F_2 are in sixteenths rather than quarters.

Neither incomplete dominance nor multiple allelism appears to explain the F_2 data. Instead, recall that Mendel observed a *9:3:3:1* pattern in F_2 phenotypes when he studied the inheritance of two different traits. This suggests that two genes interact to produce pepper color. Further, we can infer that the red phenotype, present 9/16 of the time in the F_2, is due to a combination of the dominant alleles from each gene. Green might result from a combination of the recessive alleles for each gene, because this phenotype is present in 1/16 of the progeny.

The observation that pepper fruits start out with a green color due to the presence of chlorophyll suggests that one locus affects whether chlorophyll production stops or continues as a pepper develops. If it continues, a green or greenish fruit would result. The second gene might determine the nature of a second pigment in the fruit, such as red or yellow. The table in **Figure 10.18c** shows how several phenotypes could result from the alleles of these genes. The model nicely explains the data, and suggests that the yellow and brown parental lines differ with respect to two different genes.

The role of genes in the color of pepper fruit is more complicated than this, however. For example, an orange bell pepper results when red pigment is made in reduced amounts. Plants that produce orange fruits and plants that produce red fruits are both *R–Y–* in genotype (the dash in place of an allele means "either the dominant or recessive allele"), but differ with respect to a third, different gene. This is an example of an interaction among genes that affects the phenotype. When gene interactions occur, the phenotype produced by an allele depends on the action of genes at other loci. This phenomenon is known as **epistasis**.

TABLE 10.1 The ABO Blood Types in Humans

In humans, the four different ABO blood types are produced by the alleles present at a single locus. Three alleles are common in most populations: *i*, I^A, and I^B.

Phenotype (blood type)	Genotype
O	*ii*
A	I^AI^A or I^Ai
B	I^BI^B or I^Bi
AB	I^AI^B

i = recessive
I^A and I^B = codominant

(a) Fruit color is highly variable in bell peppers.

(b) Crosses between pure lines produce novel colors.

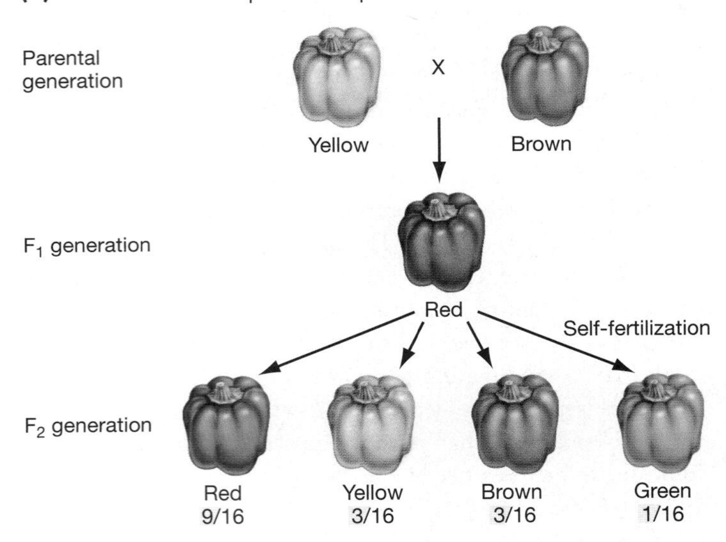

(c) Model to explain 9 : 3 : 3 : 1 pattern observed above: Two genes interact to produce pepper color.

Genotype	Color	Explanation of color
R–Y–	Red	Red pigment + no chlorophyll
rrY–	Yellow	Yellow pigment + no chlorophyll
R–yy	Brown	Red pigment + chlorophyll
rryy	Green	Yellow pigment + chlorophyll

Gene 1	Gene 2
R = Red	*Y* = Absence of green (no chlorophyll)
r = Yellow	*y* = Presence of green (+ chlorophyll)
(–) = *R* or *r*	(–) = *Y* or *y*

FIGURE 10.18 Inheritance of Fruit Color in Bell Peppers

Quantitative Traits

Mendel worked with **discrete traits.** In garden peas, seed color is either yellow or green—no intermediate phenotypes exist. But many traits in peas and other organisms don't fall into discrete categories. In humans, for example, height, weight, and skin color fall anywhere on a continuous scale of measurement. Characteristics like these are called **quantitative traits.**

Quantitative traits share a common characteristic. When the frequency of different values observed in a population are plotted on a line, they usually form a bell-shaped curve. **Figure 10.19** shows a classical example. In this case, the trait is human height and the population is a college class.

In the early 1900s Herman Nilsson-Ehle showed that if many genes each contribute a small amount to the value of a quantitative trait, then a continuous, bell-shaped distribution results. Nilsson-Ehle established this finding using strains of wheat that differed in kernel color. **Figure 10.20a** shows the results of a cross he performed between pure lines of dark red and white wheat. Note that the frequency of colors in F_2 progeny form a bell-shaped curve. To explain these results, Nilsson-Ehle proposed the model illustrated in **Figure 10.20b:**

- The parental strains differ with respect to three genes that control kernel color: *AABBCC* produces dark red kernels and *aabbcc* produces white kernels.
- The three genes assort independently. When the *AaBbCc* F_1 individuals self, white F_2s would occur at a frequency of $1/4 \times aa) \times 1/4\ (bb) \times 1/4\ (cc) = 1/64\ aabbcc$.
- The *a*, *b*, and *c* alleles do not contribute to pigment production, but the *A*, *B*, and *C* alleles contribute to pigment production in an equal and additive way. As a result, the degree of red pigmentation is determined by the number of uppercase alleles present. Each uppercase allele that is present makes a wheat kernel slightly darker red.

Later work showed that Nilsson-Ehle's model was correct in virtually every detail. Quantitative traits are produced by the independent actions of many genes. Each locus adds a small amount to the value of the phenotype.

In the decades immediately after the rediscovery of Mendel's work, then, analyses of phenomena such as sex linkage, incomplete dominance, multiple allelism, environmental effects, gene interactions, and quantitative inheritance provided a fairly comprehensive answer to the question of why offspring resemble their parents. By the 1940s, the burning question in genetics was no longer the nature of inheritance, but the nature of the gene itself. What are genes made of, and how are the instructions they contain translated into a functioning individual? These are the questions we turn to in Chapter 11.

(a) A "living histogram"—distribution of height in a college class

(b) Normal distribution = bell curve

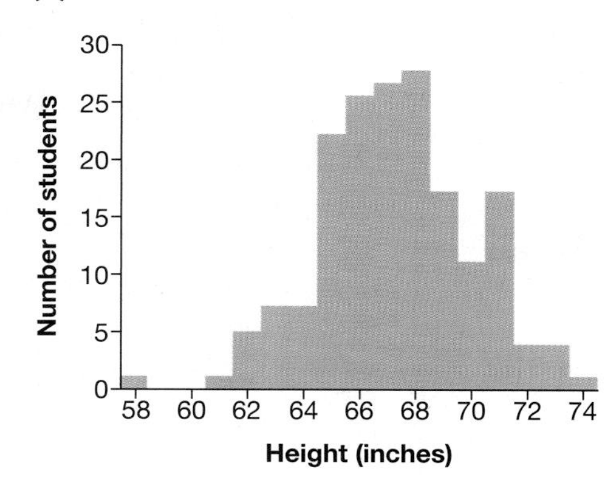

FIGURE 10.19 The Normal Distribution
(a) This photo was published in 1914 and shows male undergraduates at Connecticut Agricultural College sorted by height. **(b)** This histogram plots the heights of the students in part (a). The distribution of height in human populations forms a bell-shaped curve. This distribution is observed so frequently that it is often called a normal distribution. QUESTION The shortest student in the photo is 4′10″ (147 cm) and the tallest is 6′2″ (188 cm). How would a histogram of men in your class compare to the distribution shown here?

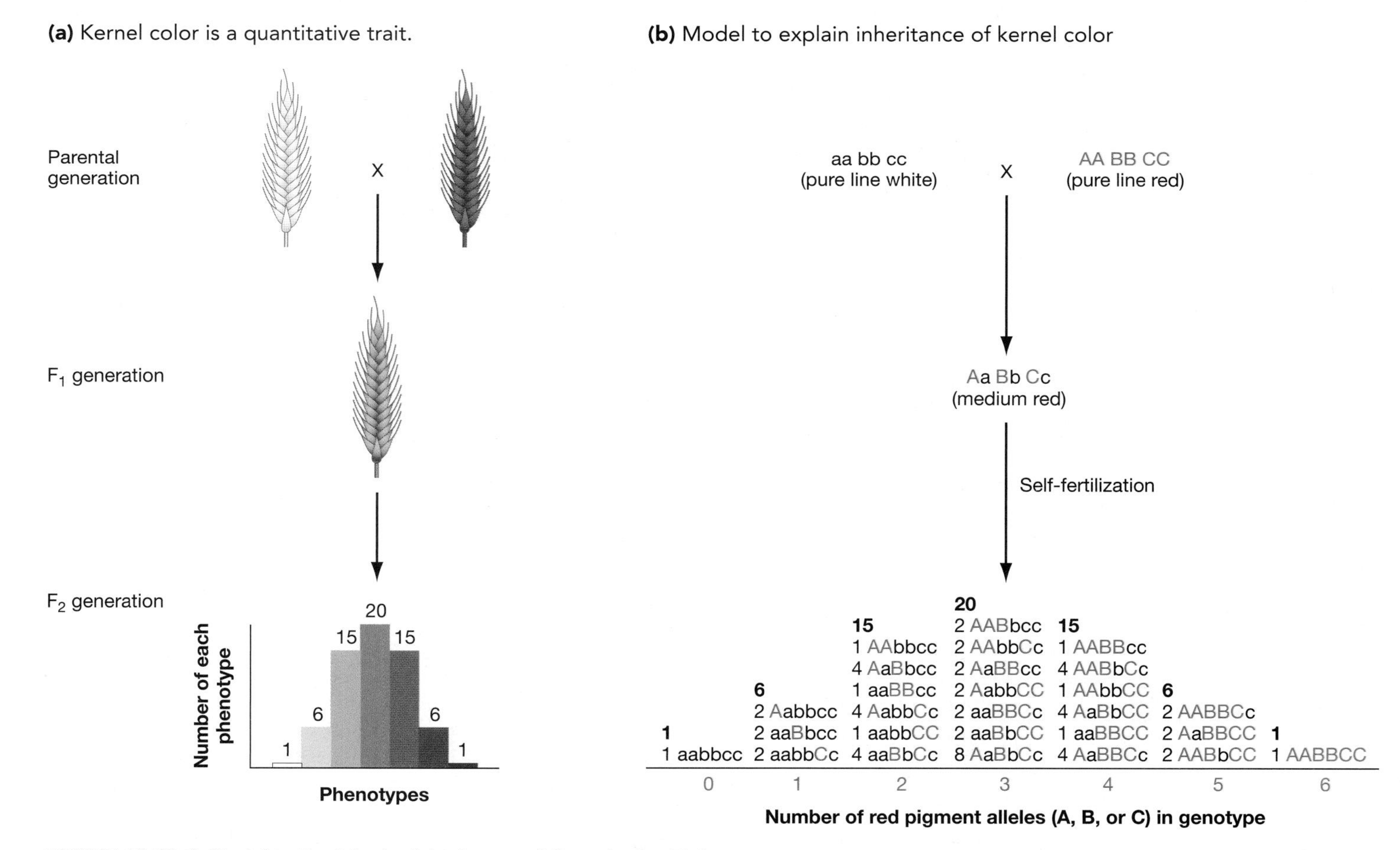

FIGURE 10.20 A Model to Explain the Inheritance of Quantitative Traits
(a) When wheat plants with white and red kernels were crossed, the F_2 offspring showed a range of phenotypes. The frequency of these phenotypes formed a normal distribution. **(b)** The model shown here attempts to explain these results. EXERCISE Confirm that the distribution of genotypes shown in part (b) is correct by drawing a Punnett square based on the gametes produced by the F_1 parents, and then filling in the F_2 offspring genotypes that result. (The square will be 8 × 8, and have 64 boxes!)

Essay Does "Genetic Determinism" Exist?

Is seed shape in garden peas genetically determined? Naïvely, the answer would appear to be yes. In Mendel's experiments, individuals with *RR* or *Rr* genotypes had round seeds while individuals with *rr* genotypes had wrinkled seeds. At the seed shape locus, then, there is a strong association between genotype and phenotype. There is also a strong resemblance between the phenotypes of parents and offspring. But what about traits in humans, where data confirm a strong resemblance between parents and offspring? Are traits like your height, weight, intelligence, and personality "genetically determined"?

Are traits like your height genetically determined?

The answer is no. In fact, no trait in any organism is "determined" in the sense of being fixed or invariant based on a certain genotype. Among pea seeds, for example, the degree of wrinkling varies widely. There are two reasons for this. First, traits like seed shape, seed color, flower color, and plant height are affected by many aspects of the physical environment. For example, the availability of water and nutrients and the presence of disease-causing fungi or viruses has an enormous effect on the phenotypes of these traits. Second, no gene acts alone. Mendel's seven traits also vary depending on the actions of alleles at other loci. Interaction among alleles, known as epistasis, is widespread.

Mendel was able to analyze the effect of a single locus on the traits he studied only because he controlled for these effects carefully. He grew individuals in similar environments and analyzed strains with similar genetic backgrounds. (In the process of creating pure lines through breeding, alleles at many loci become similar because they share common ancestors.)

What does all this have to do with human traits? Your characteristics, like seed shape in pea plants, are a combination of your genotype at loci that have a major effect on the trait, environmental influences, and the effect of alleles at dozens or hundreds of other loci. Environmental effects happen to be particularly strong in the case of human height and intelligence. Over the past 75 years, both human height and the aspect of intelligence measured by IQ tests have increased dramatically in the industrialized nations. Yet no biologists suggest that these changes are a result of changes in the genetic makeup of these populations.

Where does all this leave us? Most biologists would agree that it is sloppy at best—and misleading or even deceitful at worst—to refer to a trait as being genetically determined. Similarly, it would be equally careless to refer to a trait as being environmentally determined. All traits have a genetic basis, and all have an environmental component. The relative importance of genetics and environmental effects varies from trait to trait and from one population and environment to another.

In general, then, it is accurate to think of genotypes as contributing to phenotypic tendencies and predispositions or to the potential of an individual. But a gene can "determine" a trait only in the context of a carefully controlled environment and a precisely defined set of alleles at other loci. In the sense that the phrase is used in casual conversation, there is no such thing as genetic determinism.

Chapter Review

Summary

Gregor Mendel was the first individual to apply a modern scientific approach to the study of heredity. After extensively examining the inheritance of specific traits in the common garden pea, Mendel proposed two basic principles of transmission genetics. His principle of segregation proposed that each trait was specified by paired hereditary determinants (alleles of genes) that separate from each other during gamete formation. His principle of independent assortment stated that the segregation of one pair of genes—controlling a given trait—was not influenced by the segregation of other gene pairs.

Walter Sutton and Theodor Boveri realized that the movements of chromosomes during meiosis provide a physical basis for Mendel's principles of transmission genetics. Their chromosome theory of inheritance included the hypothesis that chromosomes are composed of genes.

Thomas Hunt Morgan and colleagues extended Mendel's work by describing X-linked inheritance and by showing that genes located on the same chromosome do not show independent assortment. Studies of X-linked traits helped confirm that genes are found on chromosomes, while studies of linked traits led to the first maps showing the locations of genetic loci on chromosomes. Later studies confirmed that many traits are influenced by the interaction of several genes, and that phenotypes are influenced by the environment an individual has experienced as well as by its genotype.

Genetics Problems

1. Name some attributes of a model organism. In what sense are garden peas or fruit flies a "model"?
2. Why is the allele for wrinkled seed shape in garden peas considered recessive?
3. The alleles found in haploid organisms cannot be dominant or recessive. Why?
4. Biologists no longer use the term *pure line* except in a historical context. Instead, they simply refer to "pure-line" individuals as homozygotes. Explain.
5. The genes for the traits that Mendel worked with are either located on different chromosomes or are so far apart on the same chromosome that crossing over almost always occurs between them. How did this circumstance help Mendel recognize the principle of independent assortment?
6. The text claims that Mendel worked with the simplest possible genetic system. Do you agree with this statement? Why or why not?
7. The artificial sweetener NutraSweet(tm) consists of a phenylalanine molecule linked to aspartic acid. The labels of diet sodas that contain NutraSweet include a warning to people with PKU. Why? Make a general comment on how aspects of an individual's environment affect the individual's phenotype using an example other than PKU.
8. When Sutton and Boveri published the chromosome theory of inheritance, research on meiosis had not yet established that paternal and maternal homologs actually do assort independently. Then in 1913, Elinor Carothers published a paper about a grasshopper species with an unusual karyotype. One chromosome had no homolog (meaning no pairing partner at meiosis I). Another chromosome had homologs that could be distinguished under the light microscope. If chromosomes assort independently, how often should Carothers have observed each of the four products of meiosis shown in the figure? Explain.

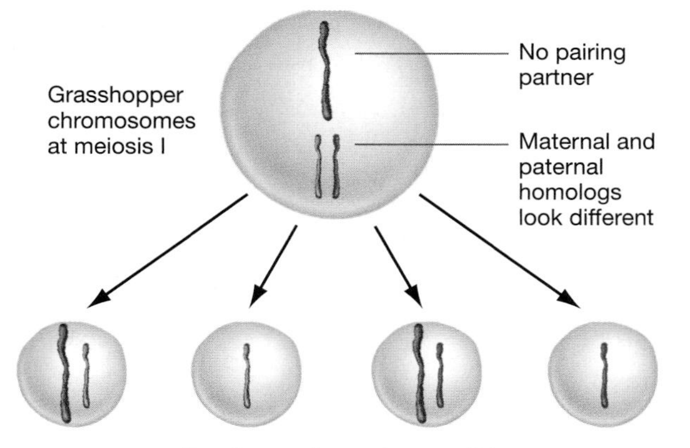

9. In humans, the ABO blood type is a polymorphic trait. Describe the alleles and genotypes responsible for the polymorphism. Suppose a woman with blood type O married a man with blood type AB. What phenotypes and genotypes would you expect to observe in their offspring, and in what proportions?
10. A plant with orange-spotted flowers was grown in the greenhouse from a seed collected in the wild. The plant was self-pollinated and gave rise to the following progeny: 88 orange with spots, 34 yellow with spots, 32 orange with no spots, and 8 yellow with no spots. What can you conclude about the dominance relationships of the alleles responsible for the spotted and unspotted phenotypes? Of orange and yellow? What can you conclude about the genotype of the original plant that had orange, spotted flowers?
11. As a genetic counselor, you routinely advise couples about the possibility of genetic disease in their offspring based on their family histories. This morning you met with an engaged couple, both of whom are phenotypically normal. The man, however, has a brother who died of Duchenne-type muscular dystrophy. This is an X-linked condition that results in death before the age of 20. The allele responsible for the disease is recessive. His prospective bride, whose family has no history of the disease, is worried that the couple's sons or daughters might be afflicted.
    - How would you advise this couple?
    - The sister of this man is planning to marry his fiancé's brother. How would you advise this second couple?
12. Study the experiment illustrated in Figure 10.8 and 10.9. To further test his hypothesis of independent assortment, Mendel let the F_2 progeny from this experiment self-fertilize. Then he examined the phenotypes of the F_3 generation.
    - Consider the F_2 individuals that have one dominant and one recessive trait. When they self-fertilize, what fraction of these individuals with will be "pure line" for these traits?
    - Consider the F_2 individuals that have both dominant traits. When they self-fertilize, what fraction of these individuals with will be "pure line" for these traits?
13. Suppose that you are heterozygous for two genes that are located on different chromosomes. You carry alleles *A* and *a* for one gene and alleles *B* and *b* for the other. Draw a diagram that illustrates what happens to these genes and alleles when meiosis occurs in your reproductive tissues. Label the stages of meiosis, the homologous chromosomes, sister chromatids, non-homologous chromosomes, genes, and alleles. Be sure to list all of the genetically different gametes that could form and how frequently each type should be observed. On the diagram, identify the events responsible for the principle of segregation and the principle of independent assortment.
14. Mr. Spock was the first officer of the starship *Enterprise* in the 1970s-era TV series called Star Trek. Mr. Spock's father came from the planet Vulcan and his mother from planet Earth. A Vulcan has pointed ears and a heart located on the right side of the chest. Mr. Spock has both of these traits, which are known to be determined by two different genes, with two alleles each. Suppose that Mr. Spock married an Earth woman and that they had many children. About half of their children look like Spock (pointed ears and right-sided heart) and about half look like their mother (rounded ears and left-sided heart).
    - What would Mendel predict the progeny phenotypes and ratios to be? Explain your answer using formal genetic terminology.
    - How do you explain the actual results?
15. In the television program Star Trek, a group of human-like organisms called Klingons are the arch-enemies of the *Enterprise* crew. In Klingons, one gene determines hair texture while another determines whether the individual will have a saggital crest (a protrusion on the forehead). The two genes are not linked.

    K = curly klingon hair (dominant)

    k = silky earthling-like hair (recessive)

    S = large saggital crest (dominant)

    s = smooth, flat earthling-like forehead (recessive)

The character Kayless is a half-human, half-Klingon with the genotype *KkSs*. He mates with Lieutenant Worf's sister, who is also heterozygous for both genes.

- Set up a Punnett square for this dihybrid cross.
- What are the four possible phenotypes that may result from this mating? In answering this question, include a description of both hair and forehead for each phenotype.
- What is the expected phenotypic ratio from the dihybrid cross?
- What fraction of the progeny are expected to be heterozygous for both genes?
- What fraction are expected to be homozygous for both genes?
- Are Kayless and his mate more likely to see an actual ratio close to the predicted values if they have 16 children or 160? Why? Explain why having a larger sample size does or does not affect observed phenotypic ratios.

16. The theory of blending inheritance proposed that the genetic material from the parents was irreversibly mixed in the offspring. As a result, offspring should always appear intermediate in phenotype to the parents. Mendel, in contrast, proposed that genes are discrete and that their integrity is maintained in the offspring and in subsequent generations. Suppose the year is 1890. You are a horse breeder and have just read Mendel's paper. You don't believe his results, however, because you often work with cremello (very light-colored) and chestnut (reddish-brown) horses. You know that if you cross a cremello individual from a pure-breeding line with a chestnut individual from a pure-breeding line, the offspring will be palomino—meaning that they have an intermediate golden-yellow body color. What additional crosses would you do to test whether Mendel's model is valid in the case of genes for horse color? List the crosses and the offspring genotypes and phenotypes you'd expect to obtain. Explain why these experimental crosses would provide a test of Mendel's model.

17. Two mothers give birth to sons at the same time in a busy hospital. The son of couple #1 is afflicted with hemophilia A, which is a recessive X-linked disease. Neither parent has the disease. Couple #2 has a normal son despite the fact that the father has hemophilia A. The two couples sue the hospital in court claiming that a careless staff member swapped their babies at birth. You appear in court as an expert witness. What do you tell the jury? Be sure to make a diagram that you can submit to the jury.

18. You have crossed two *Drosophila melanogaster* individuals that have long wings and red eyes—the wild-type phenotype. In the progeny, the mutant phenotypes called curved wings and lozenge eyes appear as follows:

Females	**Males**
600 long-wing, red eyes	300 long wing, red eyes
200 curved wing, red eyes	300 long wing, lozenge eyes
	100 curved wing, red eyes
	100 curved wing, lozenge eyes

- According to these data, is the curved wing allele autosomal recessive, autosomal dominant, sex-linked recessive, or sex-linked dominant?
- Is the lozenge eye allele autosomal recessive, autosomal dominant, sex-linked recessive, or sex-linked dominant?
- What is the genotype of the female parent?
- What is the genotype of the male parent?

19. In the parakeet, two autosomal genes that are located on different chromosomes control the production of feather pigment. Gene B codes for an enzyme that is required for the synthesis of a blue pigment and gene Y codes for an enzyme required for the synthesis of a yellow pigment. Recessive, loss-of-function mutations are known for each gene. Suppose that a bird breeder has two green parakeets and mates them. The offspring are green, blue, yellow, and albino.
    - Based on this observation, what are the genotypes of the green parents? What are the genotypes of each type of offspring? What fraction of the total progeny should exhibit each type of color?
    - Suppose that the parents were the progeny of a cross between two true-breeding strains. What two types of crosses between true-breeding strains could have produced the green parents? Indicate the genotypes and phenotypes for each cross.

CD-ROM and Web Connection

CD Activity 10.1: Mendel's Experiments *(tutorial)*
(Estimated time for completion = 10 min)
Cross pea plants that have different traits and predict the phenotypes of the resulting offspring.

CD Activity 10.2: The Principle of Independent Assortment *(animation)*
(Estimated time for completion = 5 min)
Explore Gregor Mendel's principle of independent assortment during the process of meiosis.

At your **Companion Website** (http://www.prenhall.com/freeman/biology), you will find self-grading exams and links to the following research tools, online resources, and activities:

Mendel Web
This site introduces the origins of classical genetics and provides information on current news, research, and ideas in Mendelian genetics.

Dihybrid Cross Problems
This problem set in Mendelian genetics provides a review of multiple allele crosses.

Beyond Genetic Determinism
This article focuses on the surprising discovery that there are far fewer genes in the human genome than originally predicted.

Additional Reading

Gould, L. 1996. *Cats Are Not Peas: A Calico History of Genetics* (Berlin: Springer-Verlag Press). An introduction to some classic studies in transmission genetics.

Mendel, G. 1866. Experiments in plant hybridization. Mendel's original paper, in English translation, can be found on pp. 7–17 in C. I. Davern, ed. 1981. *Genetics* (W. H. Freeman: San Francisco).

How Do Genes Work?

11

What are genes made of, and how do they work? This question dominated biology during the first half of the twentieth century. After the chromosome theory of inheritance was published and confirmed in the early 1900s, understanding the molecular nature of the gene became *the* burning question in biological science. But for decades the answer remained a mystery.

Biologists were not able to study genes directly during this period. Instead, they had to infer how genes work by studying the effects that particular alleles have on phenotypes. Researchers could study corn, for example, and identify individuals that transmitted alleles associated with increased height, growth rate, or oil content in seeds. By studying patterns in the inheritance of traits like these, biologists could deduce the genotypes of both parents and offspring.

From the start of the twentieth century through the early 1950s, then, the predominant research strategy was to conduct a series of experimental crosses, create a genetic model to explain the types and proportions of phenotypes that resulted, and then test the model's predictions through reciprocal crosses, testcrosses, or other techniques. This strategy was extremely productive. It led to virtually all of the discoveries analyzed in Chapter 10, including Mendel's rules, sex linkage, chromosome mapping, incomplete dominance, multiple allelism, and quantitative inheritance.

Entirely new research strategies came to the fore, however, when the molecular basis of inheritance was finally understood. How was the chemical nature of the gene eventually discovered? What observations and experiments revealed how genes act at the molecular level?

The goal of this chapter is to answer these questions. We begin with studies that identified

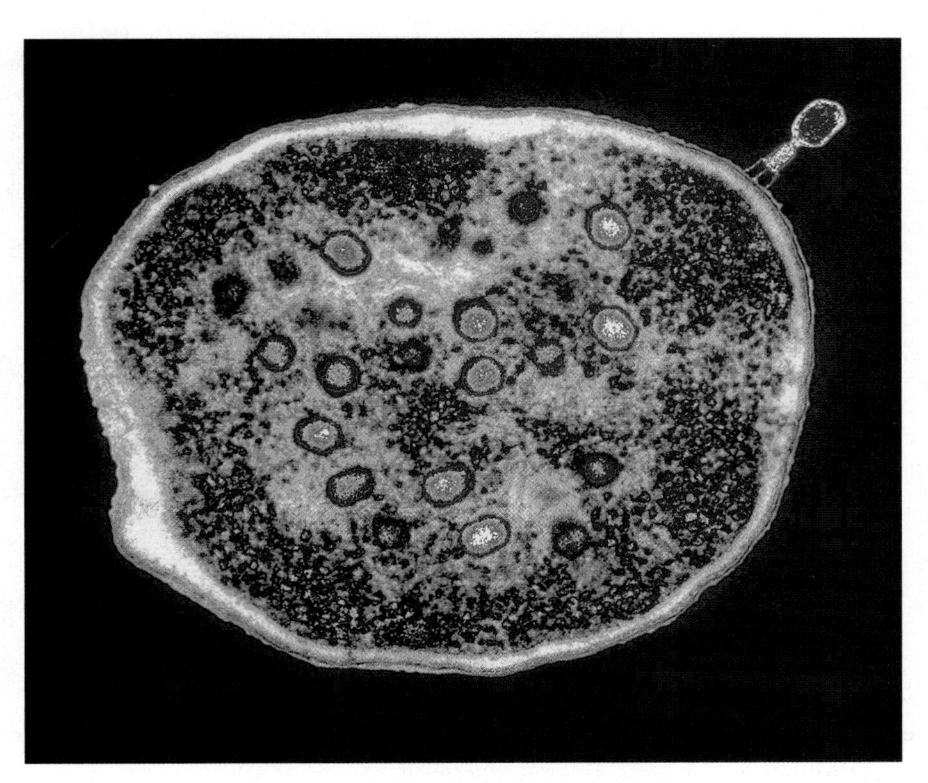

The bacteriophage in the upper right corner has just injected its DNA into a host cell—in this case, *Escherichia coli*. The experiments reviewed in this chapter established that DNA is the hereditary material and revealed how the information encoded in DNA is used to produce proteins and other cell components.

deoxyribonucleic acid (DNA) as the genetic material, and then we explore how genes act on a molecular level. The final two sections of the chapter introduce a series of classical experiments that led to a comprehensive model for how the information stored in DNA is translated into the RNA and protein molecules that actually build and run the cell.

11.1 DNA as the Hereditary Material

The chromosome theory of inheritance proposed that chromosomes are composed of genes. It had been known since the late 1800s that chromosomes are a complex of DNA and proteins, so after the theory was confirmed around 1920 researchers had just two candidates for the molecular basis of the gene. Genes were either made of DNA or proteins.

For decades most biologists backed the hypothesis that genes are made of proteins. The arguments in favor of this hypothesis were compelling. Hundreds, if not thousands, of complex and highly regulated chemical reactions occur in even the simplest living cells. The amount of information required to specify and coordinate these reactions is almost mind-boggling. With their almost limitless variation in structure and chemical behavior, proteins are complex enough to contain this much information.

DNA, in contrast, was known to be composed of just four types of deoxyribonucleotides. It was also thought to be a simple molecule with some sort of repetitive and uninteresting structure. So when Oswald Avery, Colin M. MacLeod, and Maclyn McCarty published experimental evidence that DNA was the hereditary material, most biologists had the same reaction: They didn't believe it.

Transformation Experiments

In the early 1940s, Avery, MacLeod, and McCarty set out to understand an experimental result that had been published years earlier by a bacteriologist named Frederick Griffith. In 1928, Griffith discovered a mysterious phenomenon involving hereditary traits that he referred to as "transformation." Griffith's transformation experiments appeared to isolate the hereditary material. Avery and his colleagues wanted to determine whether this material was protein or DNA.

Back in the 1920s, Griffith had been doing experiments designed to understand the *Streptococcus pneumoniae* bacterium, which is a leading cause of pneumonia, earaches, sinusitis, and meningitis in humans. For his experiments, however, he worked with strains of *S. pneumoniae* that infect mice. As is the case with the strains that affect humans, the strains that affect mice vary in their **virulence**—their ability to cause disease and death.

Virulent strains cause disease; benign strains do not. As **Figure 11.1a** shows, the strains that Griffith happened to work with can be identified by eye when grown on a nutrient medium in a petri dish in the lab. Cells from the virulent strain form colonies that look smooth, while cells from the benign strain form colonies that look rough. Logically enough, Griffith called this virulent strain S and the benign strain R.

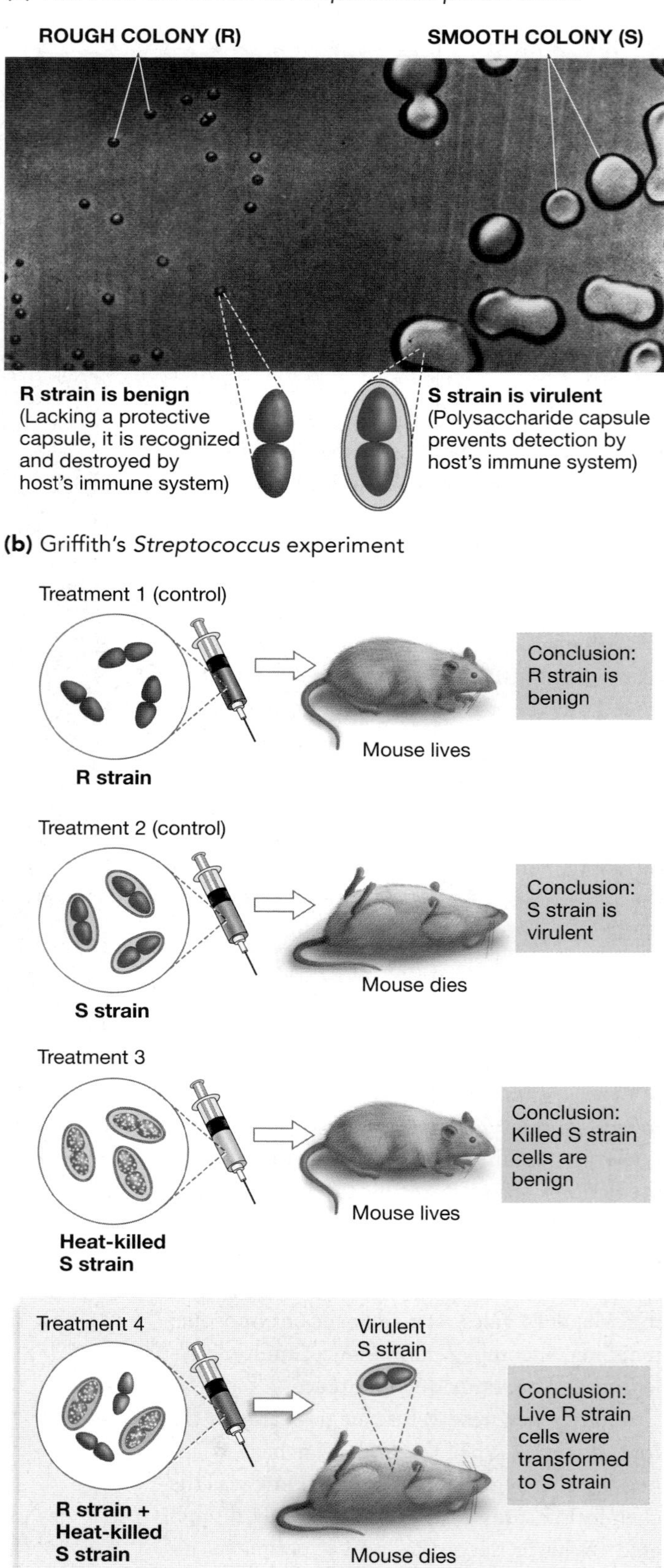

FIGURE 11.1 The Discovery of "Transformation"
(a) Virulent and benign strains of *Streptococcus* have differences in colony appearance and at the cellular level. **(b)** Mice died after being injected with a combination of live R cells and heat-killed S cells.

To better understand how the strains interact, Griffith carried out the experiment illustrated in **Figure 11.1b**. The figure shows four sets of experimental conditions that he designed. In the first treatment he injected mice with cells of the R strain. As he expected, these mice lived. In the second treatment he injected mice with cells of the S strain. Not surprisingly, these mice died of pneumonia. So far, so good: The first two treatments are controls showing the effect that each strain of *S. pneumoniae* has on mice. In the third treatment, Griffith killed cells of the S strain with heat and then injected them into mice. These mice also lived. This experimental treatment was interesting because it showed that dead S cells do not cause disease. In the final treatment, Griffith injected mice with heat-killed S cells *and* live R cells. Unexpectedly, these mice died. Autopsies confirmed pneumonia as the cause of death. When Griffith cultured *Streptococcus* from the dead mice, he found S cells, *not* R cells.

What was going on? Griffith proposed that something from the heat-killed S cells had "transformed" the benign R cells; something had changed the appearance and behavior of the R cells from R-like to S-like. Because this "something" was passed on to the offspring of the R cells (it had appeared in the cells Griffith cultured from the dead mice), it was clearly some sort of hereditary, or genetic, factor. Other researchers performed follow-up experiments showing that transformation could occur in vitro—in culture—as well. It wasn't necessary for the infection to occur in vivo—in an organism—for the heat-killed S cells to transfer this factor to the live R cells.

DNA Is the Transforming Factor

This brings us to the question that Avery and co-workers wanted to answer: What is the transforming factor? Avery and colleagues used an elegant experimental strategy to find the answer (**Figure 11.2**). They assumed that the transforming factor

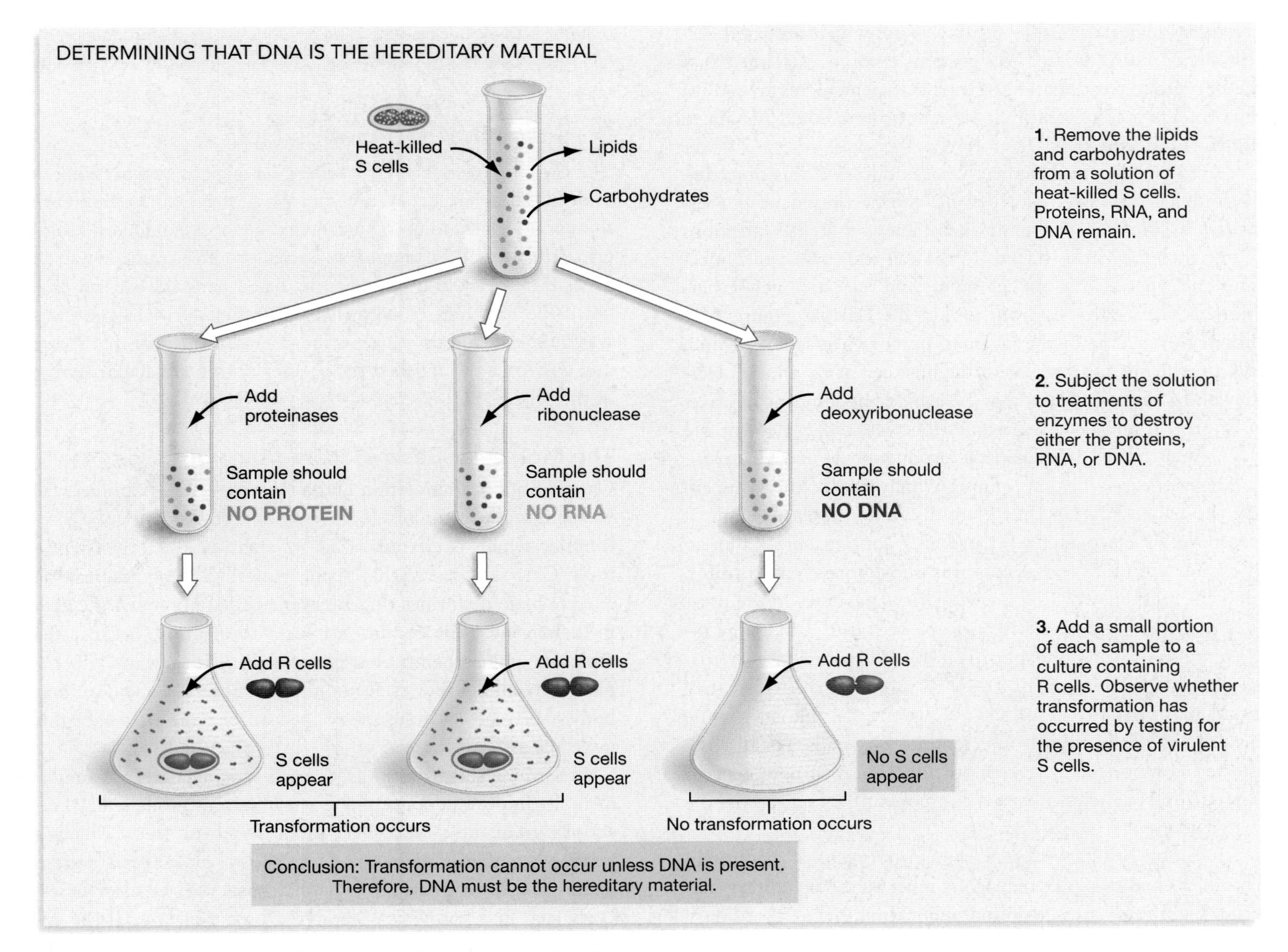

FIGURE 11.2 Experimental Evidence That DNA Is the Transforming Factor

EXERCISE Add labels to the three test tubes illustrated in the middle of this diagram, indicating which tubes contain protein + RNA, RNA + DNA, and DNA + protein.

consisted of protein, RNA, or DNA. To determine which type of molecule was responsible for transformation, they grew quantities of S cells in culture and heat-killed them. Then they used chemical solvents to extract the lipids and an enzyme treatment to remove the carbohydrates. As Figure 11.2 shows, they subjected the solution that remained to several different treatments. Some samples were treated with a series of enzymes that destroy proteins. (Enzymes that degrade proteins are called proteinases.) A second set of samples was treated with a ribonuclease (RNase)—an enzyme that destroys RNA. The final set of samples was treated with deoxyribonuclease (DNase), which hydrolyzes DNA. When they added small quantities of the three resulting solutions to cultures containing R cells, S cells appeared in all of the cultures except those treated with DNase. Avery and associates concluded that DNA must be present for transformation to occur.

This experiment demonstrated that DNA is the transforming principle, and it provided strong evidence that DNA is the hereditary material. But the result was not widely accepted. Researchers who advocated the protein hypothesis claimed that the enzymatic treatments were not sufficient to remove all of the proteins present, and that enough protein could remain to transform R cells.

Alfred Hershey and Martha Chase addressed this objection by studying how a virus called T2 infects the bacterium *Escherichia coli*. Hershey and Chase knew that T2 infections begin when the virus attaches to the cell wall of *E. coli*. Infections end when the *E. coli* cell bursts and releases a new generation of virus particles from within the cell. As **Figure 11.3** shows, the capsule of the original, parent virus is left behind, still attached to the exterior of the host cell as a "ghost." Hershey and Chase also knew that T2 is made up almost exclusively of protein and DNA. The question was, did protein or did DNA enter the cell and direct the production of new viruses?

Their strategy for determining which part of the virus enters the cell and acts as the hereditary material was elegant. It was based on two facts: (1) proteins contain sulfur but no phosphorus, and (2) DNA contains phosphorus but no sulfur. They began their work by growing virus particles in the presence of either the radioactive isotope of sulfur (^{35}S) or the radioactive isotope of phosphorus (^{32}P). When they used viruses with labeled protein or DNA to infect *E. coli* cells, they found that virtually all of the ^{35}S was in the "ghosts" while virtually all of the ^{32}P was injected into host cells. Because the injected component directs the production of a new generation of virus particles, it is this component that represents the virus's genes.

After these results were published, proponents of the protein hypothesis had to admit that DNA must be the hereditary material. Eight years after its initial publication, the extraordinary claim made by Avery and colleagues—that a seemingly simple molecule contained all the information for life's complexity—was finally accepted.

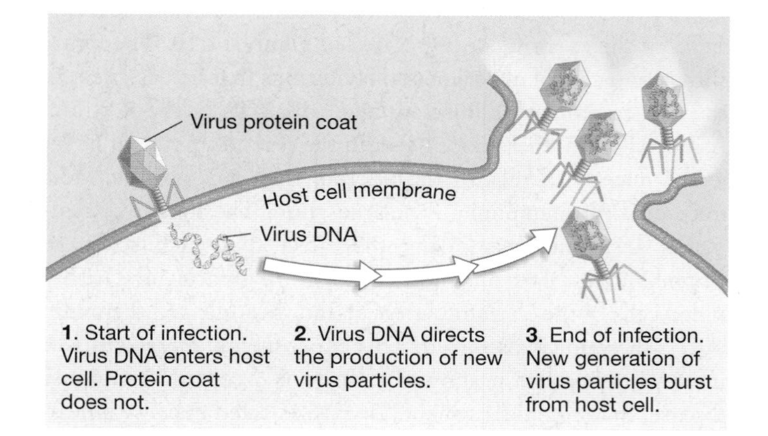

FIGURE 11.3 Later Experimental Evidence That DNA Is the Hereditary Material
When the virus called T2 attaches itself to the exterior of a host cell, it injects a strand of material and leaves a capsule outside. Experiments showed that the capsule consists of protein while the injected material is DNA. Because the injected DNA directs the production of a new generation of virus particles, it must be the hereditary material.

11.2 What Do Genes Do?

The work by Avery and co-workers was a giant step in trying to understand what genes are and how they work. But even though it was exciting to know what genes are made of, biologists still wanted to know *how* genes produce traits. What do genes do? In 1902, a physician named Archibald Garrod provided the first hint of an answer. The first formal hypothesis was published almost 40 years later—about the time that Avery and co-workers published their paper on the molecular nature of the gene.

The Molecular Basis of Hereditary Diseases

Garrod noticed that certain human diseases run in families. He was particularly interested in a condition known as alkaptonuria, which is characterized by arthritis and other symptoms. Garrod observed that people suffering from alkaptonuria excrete huge quantities of a molecule called homogentisic acid in their urine. (The condition is easy to diagnose because the molecule makes the affected person's urine turn black.) To explain why alkaptonurics accumulate homogentisic acid in their bodies, Garrod suggested they lack an enzyme that uses this molecule as a substrate.

In forming this hypothesis, Garrod was inspired by a fundamentally important concept. He knew that in cells, chemical reactions frequently take place in a series of steps called a **metabolic pathway.** As **Figure 11.4a** shows, molecules are often made or broken apart in a distinct sequence. In most cases, every step in a pathway is catalyzed by an enzyme. If the enzyme is not available, the reaction does not occur, and the substrate for that particular reaction accumulates. For example, **Figure 11.4b** shows the metabolic pathway that involves homogentisic acid. If the enzyme that catalyzes the reaction marked with a red "X" in the figure does not function correctly, homogentisic acid cannot be transformed into the next compound in the metabolic pathway. As a result, it accumulates in the body and causes the symptoms known as alkaptonuria. Using the same logic, Garrod identified the biochemical basis of other genetic diseases.

Garrod had correctly identified the cause of an inherited defect, or what he called an inborn error of metabolism. In doing so, he became the first researcher to draw a connection between genes and the chemical reactions that occur inside cells, or what biologists call **metabolism.**

Genes and Enzymes

Despite Garrod's brilliant work, an explicit hypothesis explaining what genes do did not appear until 1941. In that year, George Beadle and Edward Tatum published a series of breakthrough experiments on a bread mold called *Neurospora crassa.*

Beadle and Tatum's research program was organized around a simple idea, expressed concisely by Beadle: "one ought to be able to discover what genes do by making them defective." The researchers' goal was to find out how a gene acts by observing what happens when it *doesn't* act. This is actually an extremely common research strategy in genetics. The idea is to find or create mutants that lack a specific gene, analyze how the characteristics of the mutants differ from normal, or wild-type, individuals, and use the data to infer what the gene does. Archibald Garrod had used exactly this approach to study genetic diseases in humans.

Beadle and Tatum focused on studying *N. crassa* mutants that cannot grow on normal culture media. In the lab, wild-type individuals of *N. crassa* grow quite well on "minimal medium." In this case, the minimal medium consisted of a gelatinous solid containing only a few ions like sulfate (SO_4^{2-}) and phosphate (PO_4^{3-}), a vitamin called biotin, a source of nitrogen atoms, and a sugar that supplies carbon atoms and chemical energy. (For the nitrogen and energy requirements, Beadle and Tatum used ammonium chloride, NH_4Cl, and glucose.) Wild-type *Neurospora* can synthesize every other molecule they need on their own, using only the molecules and ions in the minimal medium as nutrients.

In one set of experiments, Beadle and Tatum set out to analyze individuals that cannot synthesize the amino acid arginine (arg). They needed two things to proceed: a way to create mutants and a way to select specific mutants—those that could not produce arginine. When genes are changed in a way that causes them to lose function, they are called **knock-out mutants, null mutants,** or **loss-of-function mutants.** To create these mutants, the researchers exposed *N. crassa* cultures to high-energy radiation like X rays or ultraviolet light. (Later in the text we'll explore why radiation causes mutations.) To find loss-of-function mutants for arginine, the biologists implemented the steps in **Figure 11.5a** (pages 222–223). They called the defective individuals that resulted from this screening procedure *arg* mutants. (In *Neurospora* and many other organisms, the names of genes are italicized and printed in lowercase.) Because the defects were heritable, Beadle and Tatum were confident that the mutations were located in the gene or genes responsible for arginine synthesis.

But did all of these mutants have exactly the same defect? Beadle and Tatum reasoned that the answer was probably no. They hypothesized that *N. crassa* probably synthesizes arginine through a series of steps, as illustrated in **Figure 11.5b.** Further, they suggested that molecules called ornithine and citrulline might be part of the metabolic pathway leading to arginine.

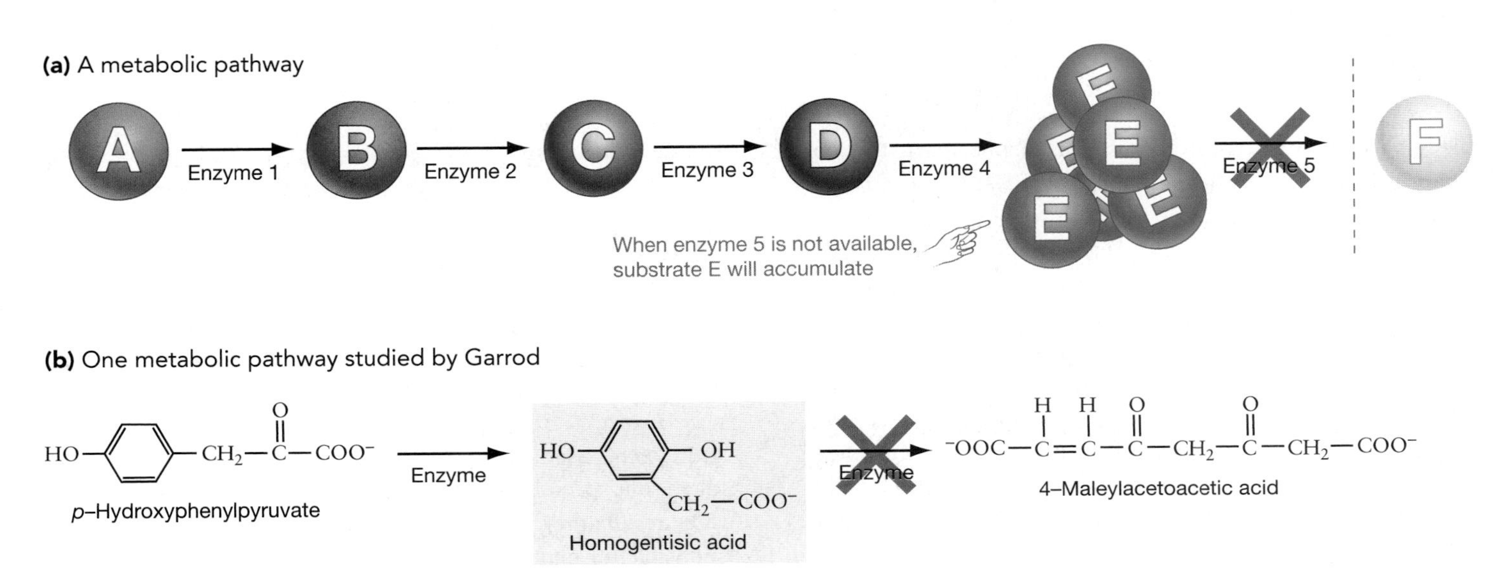

FIGURE 11.4 Enzymes Catalyze the Reactions in Metabolic Pathways
(a) A generalized metabolic pathway. The reactions at each step are catalyzed by a different enzyme.
(b) Garrod hypothesized that people suffering from alkaptonuria lack the enzyme that uses homogentisic acid as a substrate, thus causing the acid to accumulate.

CD ACTIVITY 11.1
The One-Gene, One-Enzyme Hypothesis

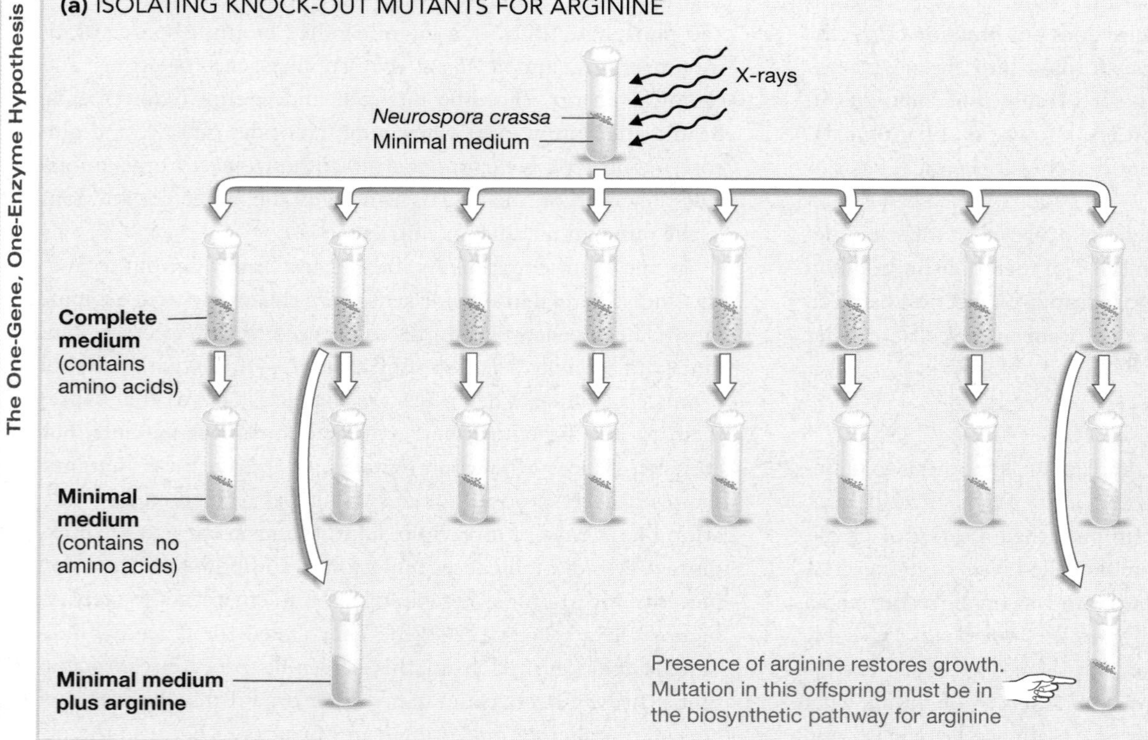

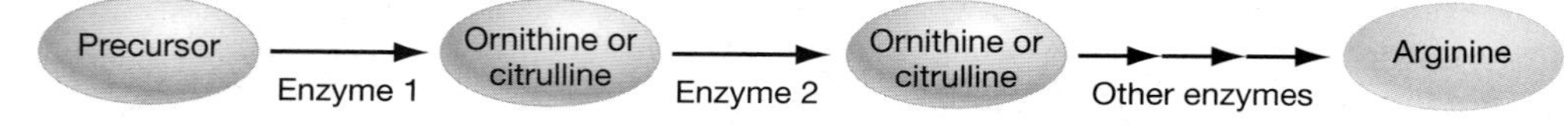

FIGURE 11.5 Experimental Evidence for the One-Gene, One-Enzyme Hypothesis
(a) Researchers first generated and isolated a large number of individuals with defects in arginine synthesis. **(b)** If a series of enzymes were involved in the synthesis of arginine, a defect in any one of them would shut down arginine synthesis. **(c)** To test whether different *arg* mutants had defects in different parts of the arginine pathway, biologists took spores from each *arg* mutant and added them to minimal medium supplemented with ornithine, citrulline, or arginine. **(d)** The results of the test allowed the researchers to deduce this pathway for the synthesis of arginine. **QUESTION** In part (a), why did they grow the irradiated spores on complete medium first?

This hypothesis was logical because the chemical structures of both ornithine and citrulline are extremely similar to arginine. Further, both molecules can be found in wild-type *N. crassa*.

To test the idea that *arg* mutants were not all alike, Beadle and Tatum took spores from *arg* mutants and put them into minimal media supplemented with either ornithine, citrulline, or arginine (**Figure 11.5c**). The results were dramatic. Different mutants were able to grow in the presence of different molecules. This observation confirmed that the mutations occurred in different genes, and supported the hypothesis that ornithine and citrulline are part of the synthetic pathway leading to arginine. But their data also identified the sequence of the reactions in the pathway. Beadle and Tatum's logic was as follows: *arg* mutants that could grow without ornithine or citrulline present must be able to synthesize these molecules themselves—meaning that their particular mutation must disable an enzyme needed late in the pathway of arginine biosynthesis. Conversely, *arg* mutants that could grow only with ornithine or citrulline present were not able to synthesize these molecules themselves—meaning that the mutation they carried disabled an enzyme needed early in the sequence of reactions. Because all mutants that grew when supplemented with ornithine also grew when they were supplemented with citrulline, but not vice-versa, they concluded that citrulline must come after ornithine in the pathway (**Figure 11.5d**).

In sum, Beadle and Tatum determined that *arg* mutants knock out specific steps in a biochemical pathway. To interpret these results they proposed that each gene in an organism is responsible for making a different protein, and that most of these proteins function as enzymes. Their **one-gene, one-enzyme hypothesis** explained what genes do in succinct terms: Genes contain the instructions for making proteins.

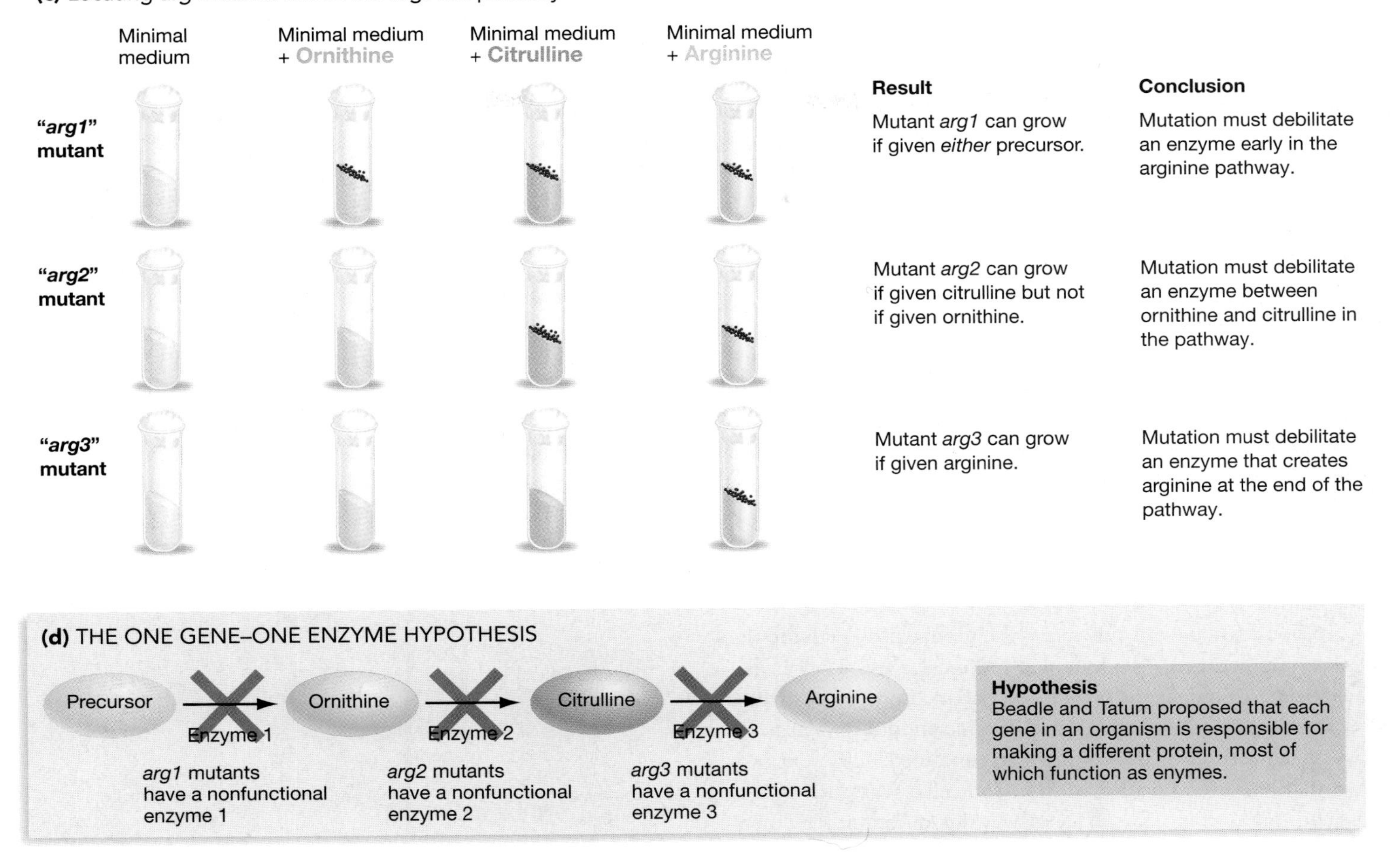

11.3 The Genetic Code

How does a gene specify the production of an enzyme? After Beadle and Tatum's hypothesis had been verified in a variety of organisms, this question became a central one. Part of the answer lay in understanding the molecular nature of the gene. Biochemists knew that the primary components of DNA were four nitrogen-containing bases: the pyrimidines thymine (abbreviated T) and cytosine (C), and the purines adenine (A) and guanine (G). They also knew that these bases were connected in a linear sequence by a sugar-phosphate backbone. Watson and Crick's model for the secondary structure of the DNA molecule, introduced in Chapter 3, revealed that two strands of DNA are wound into a double helix, held together by hydrogen bonds between the complementary base pairs A-T and G-C.

Given DNA's structure, it appeared extremely unlikely that it directly catalyzed the reactions that produce proteins. Its shape was too regular to suggest that it could bind a wide variety of substrate molecules and stabilize the transition states in chemical reactions. Instead, Crick proposed that the sequence of bases in DNA might act as some sort of genetic code. The idea was that DNA was *only* an information-storage molecule. The instructions it contained would have to be read and then translated into proteins.

Crick offered an analogy with Morse code. This is the system of transmitting messages using various combinations of dots and dashes to specify each letter of the alphabet. Crick's idea was that different combinations of bases could specify the 20 amino acids just as different combinations of dots and dashes specify the 26 letters of the alphabet. A particular stretch of DNA, then, could contain the information needed to specify the amino acid sequence of a particular enzyme. In code form, the tremendous quantity of information needed to build and run a cell could be stored compactly. This information could also be copied through complementary base-pairing and transmitted efficiently from one generation to the next.

It soon became apparent, however, that the link between DNA and proteins was an indirect one.

RNA as the Intermediary Between Genes and Proteins

The first clue that biological information does not flow directly from DNA to proteins came from data on the structure of cells. In eukaryotic cells, DNA is enclosed within a membrane-bound structure called the nucleus (see Chapter 5). But the structures called ribosomes, where protein synthesis takes place, are found outside the nucleus. This observation began to make sense after François Jacob and Jacques Monod suggested that RNA molecules act as a link between the genes inside the nucleus and the protein-manufacturing centers outside the nucleus. Jacob and Monod's hypothesis is illustrated in **Figure 11.6**. They predicted that short-lived molecules called "messenger RNA" carry information from DNA to the site of protein synthesis.

Research by Jerard Hurwitz and J. J. Furth confirmed the existence and function of messenger RNA, or **mRNA**. These investigators reasoned that if mRNA exists, then there must be an enzyme that catalyzes its synthesis. They called this hypothesized molecule **RNA polymerase** because its function would be to polymerize ribonucleotides to RNA. Then they set out to find it.

Hurwitz and Furth's first task was to create an assay for the enzyme's presence. They solved this problem when they realized that ribonucleotides are soluble in an acidic solution, but RNA is insoluble. As a result, RNA forms a precipitate—a solid that falls out of solution—while ribonucleotides stay suspended in solution.

To track down RNA polymerase, Hurwitz and Furth did the series of experiments diagrammed in **Figure 11.7**. They began by labeling ribonucleotides with radioactive isotopes of carbon or phosphorus. They added these ribonucleotides to a large series of acidic solutions, each of which contained a different protein that had been purified from *E. coli* cells. Was one of these proteins RNA polymerase? To answer this question, they incubated the mixtures and waited to see if a precipitate containing radioactivity formed. The approach was successful. Along with an enormous number of failures, they found a protein that produced very small yields of radioactively labeled precipitate. RNA had formed, and RNA polymerase had been discovered.

In follow-up experiments, Hurwitz and Furth found that much more RNA formed if they added DNA to the reaction mix. This result made sense because DNA was thought to act as a template for the synthesis of RNA. The presence of a template would make the polymerization reaction much more efficient. The researchers confirmed this conclusion when they added synthetic DNA containing only the base thymine to the reaction mix. The RNA that formed contained only adenine. The enzyme had synthesized RNA according to the rules of complementary base-pairing introduced in Chapter 3: Thymine pairs with adenine. Thus a sequence of DNA containing only thymine would direct the polymerization of a complementary sequence of adenine during RNA synthesis.

The link between DNA and proteins had been nailed down. It was clear that the genetic code specifies the sequence of RNA

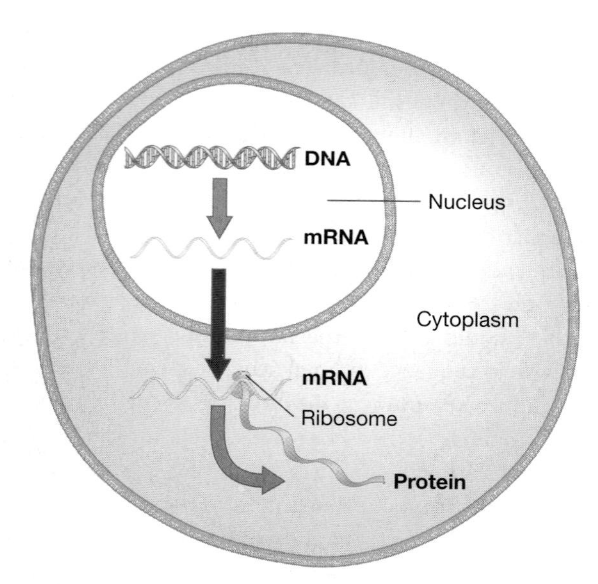

FIGURE 11.6 The Messenger RNA Hypothesis
In the cells of some organisms, DNA is found only in a membrane-bound structure called the nucleus. In these cells, proteins are manufactured outside the nucleus. Biologists proposed that the information coded in DNA is carried from inside the nucleus to ribosomes outside the nucleus by a molecule called messenger RNA (mRNA). QUESTION Why was the choice of the term *messenger* appropriate?

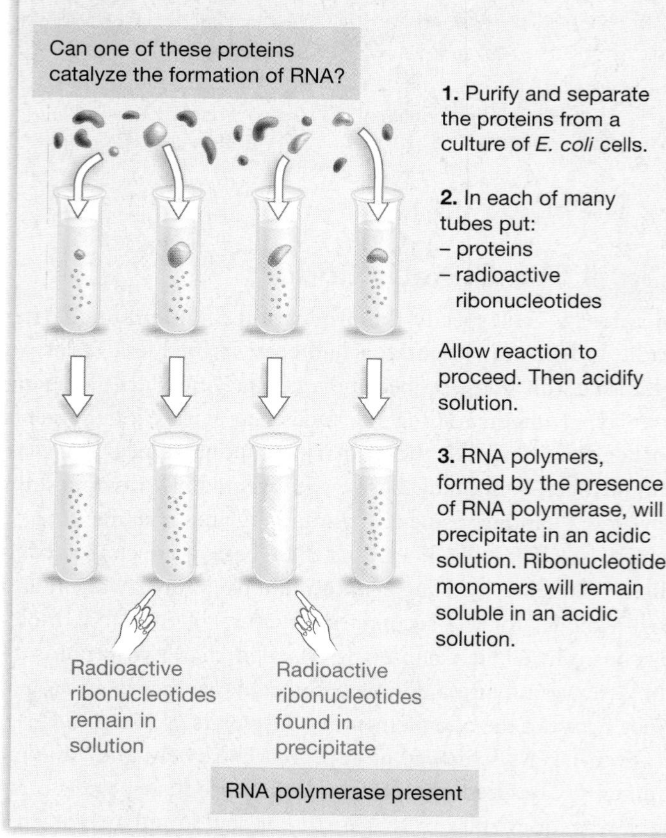

FIGURE 11.7 The Experiment That Led to Isolation of RNA Polymerase
QUESTION Experimental strategies that rely on trying all conceivable possibilities are sometimes called brute-force approaches. In what sense was the search for RNA polymerase a brute-force approach?

intermediaries, which, in turn, specifies the amino acid sequence of proteins.

The next challenge? Cracking the genetic code. Exactly how does the sequence of bases in a strand of DNA or mRNA code for the sequence of amino acids in a protein?

How Long Is a Word in the Genetic Code?

The first step in cracking the genetic code was to resolve how many bases make up a "word." In a sequence of mRNA, how long is a message that specifies an amino acid? Based on some simple math, George Gamow suggested that each code word contains three bases. His reasoning was based on the observation that 20 amino acids must be specified by a sequence of mRNA. Because there are only four different nucleotide bases, a one-base code could specify only four amino acids. Similarly, a two-base code could identify just 4 × 4, or 16 amino acids. A triplet code, though, could specify 4 × 4 × 4, or 64 different amino acids. As a result, there would be more than enough messages to code for the 20 amino acids (**Figure 11.8**). In fact, Gamow's hypothesis predicted that the genetic code is redundant—meaning that more than one triplet of bases, or a **codon**, could specify the same amino acid.

Work by Francis Crick and colleagues confirmed Gamow's hypothesis. These researchers used chemicals called acridines to induce a specific type of mutation in the DNA of a virus called T4. The mutation was the addition or deletion of a single base. As predicted by the triplet code, both addition and deletion mutants led to loss of function in the gene being studied. This is because the single mutation threw the sequence of codons, or the **reading frame**, out of register. To understand how additions or deletions affect a reading frame, consider the English sentence *The fat cat ate the rat*. In this case, the reading frame is a three-letter word and a space. If the fourth letter in this sentence—the "f" in fat—were deleted but the reading frame stayed intact, the sentence would be transformed into *The atc ata tet her at*. This is gibberish. **Figure 11.9a** (page 226) illustrates the same phenomenon for a sequence of codons. The top line in the figure shows the reading frame for a series of AAT codons. The next two lines in the figure illustrate how the deletion or addition of a base causes a different group of triplets to be read. In terms of the amino acid sequence of a protein, these mutations lead to gibberish.

Crick and co-workers found that when an addition of a base and a deletion of a base were both present in a single gene, the sequence of triplets was brought back into its appropriate reading register and normal function was restored. This result is diagramed in **Figure 11.9b**. The key experiment, though, is illustrated in **Figure 11.9c**. When the researchers created mutants that had either three additions or three deletions, normal function was also restored. Because the addition or deletion of either one or two bases did not restore normal function, these results confirmed that the code is read in triplets of bases. The results also launched a long, laborious, and ultimately successful effort to determine which amino acid is specified by each of the 64 codons.

How Did Biologists Decipher the Genetic Code?

The initial breakthrough in deciphering the genetic code came when Marshall Nirenberg and Heinrich Matthaei created a method for synthesizing RNAs of known sequence. The system they created was based on an enzyme called polynucleotide phosphorylase, which had been discovered by Mary Ann Grunberg-Manago. Polynucleotide phosphorylase randomly catalyzes phosphodiester bonds between ribonucleotides. By providing ribonucleotides containing only the base uracil to a reaction mix containing this enzyme, Nirenberg and Matthaei were able to create a long polymer of uracil-containing ribonucleotides. These

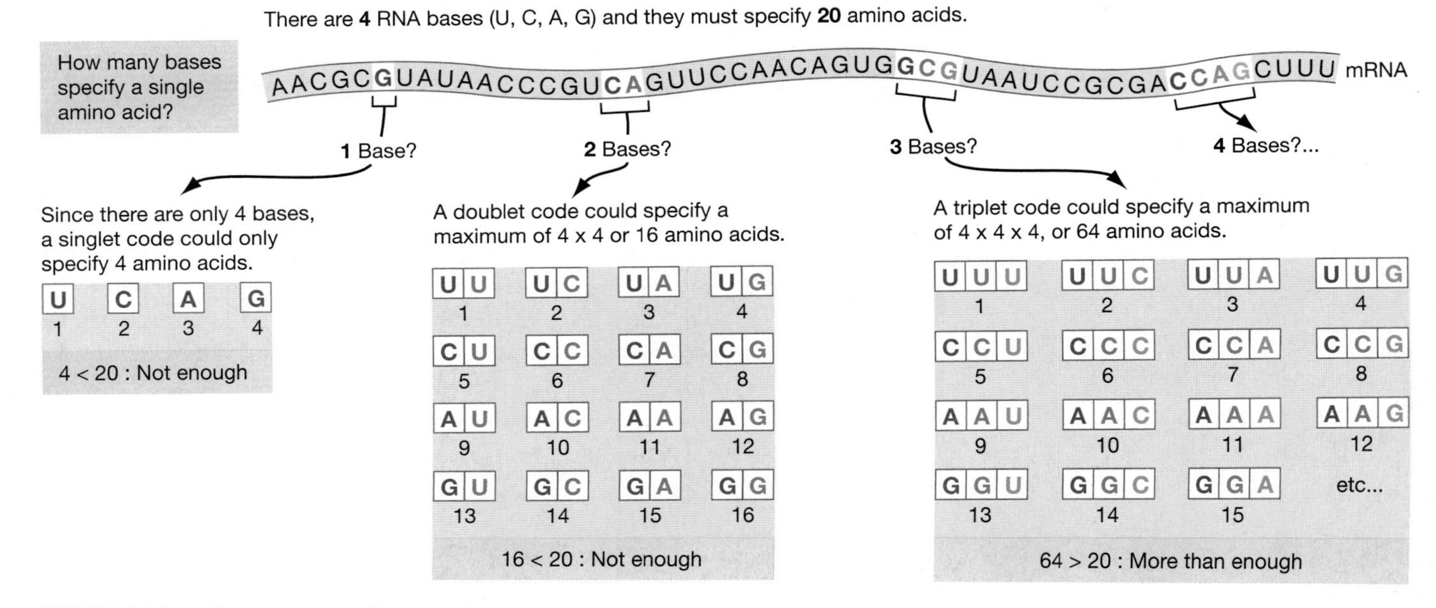

FIGURE 11.8 In the Genetic Code, How Many Bases Form a "Word"?

synthetic RNAs were added to an in vitro system for synthesizing proteins. The researchers analyzed the resulting amino acid chain and determined that it was poly-phenylalanine.

This result could only mean one thing: The RNA triplet UUU codes for the amino acid phenylalanine. By complementary base pairing, it was clear that the corresponding DNA sequence would be AAA. This initial observation was followed by experiments with RNAs consisting of only G, A, or C.

Researchers subsequently began to generate RNAs from pools containing two (or even three) ribonucleotides and analyze the amino acid chains that resulted. The key to these experiments is that the bases in the pool were present in unequal quantities. To understand why these experiments were informative, consider an experiment with a pool of ribonucleotides containing $\frac{3}{4}$ U and $\frac{1}{4}$ G. **Table 11.1a** shows the expected number of codons that would result when these ribonucleotides are polymerized to RNA based on probability. The logic behind the predicted values is straightforward. Every time polynucleotide phosphorylase adds a ribonucleotide from the pool to a growing strand of RNA, the probability of a U being added is $\frac{3}{4}$ (or 75 percent) and the probability of a G being added is $\frac{1}{4}$ (or 25 percent). The probability of producing the codon UUG, then, is the product of the three independent probabilities, or $(\frac{3}{4})(\frac{3}{4})(\frac{1}{4}) = \frac{9}{64}$.

What were the actual quantities of amino acids in the resulting proteins? **Table 11.1b** shows the data. By comparing the expected and observed values, investigators could infer that the codons UUG, UGU, and GUU code for leucine, valine, or cysteine. Why? The codons UUG, UGU, and GUU were each predicted to occur 33 percent as often as UUU; the observed proportions of the amino acids leucine, valine, and cysteine were each approximate-

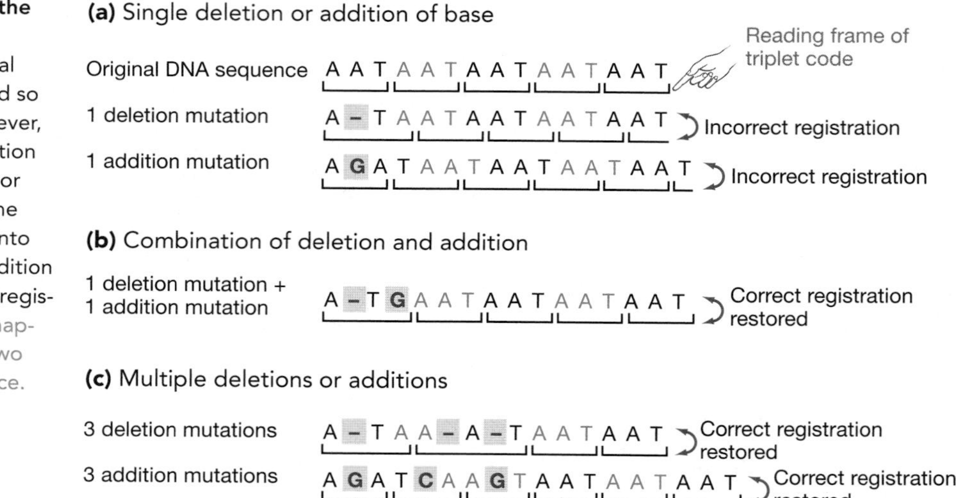

FIGURE 11.9 The Experiments That Confirmed the Triplet Code
(a) In the sequence of DNA shown here, the normal reading frame gives a message AAT, AAT, AAT, and so on. If a deletion or addition mutation occurs, however, the reading frame goes out of register. Here, deletion or addition mutations are indicated by a red dash or letter. (b) The combination of one deletion and one addition mutation brings the reading frame back into register. (c) Three deletion mutations, or three addition mutations, also bring the reading frame back into register. EXERCISE Add a diagram illustrating what happens to the reading frame when two deletion or two addition mutations take place in the same sequence.

FIGURE 11.10 The Genetic Code
To read the code, match the first base in an mRNA codon, on the left side, with the middle base in a codon, along the top, and the third base in a codon, along the right side. The 64 codons, along with the amino acid or the termination (stop) signal that they specify, are given in the boxes. EXERCISE Pick four codons at random. Which three bases in DNA specify each of these codons?

Second base

First base	U	C	A	G	Third base
U	UUU Phenylalanine	UCU Serine	UAU Tyrosine	UGU Cysteine	U
U	UUC Phenylalanine	UCC Serine	UAC Tyrosine	UGC Cysteine	C
U	UUA Leucine	UCA Serine	UAA Stop codon	UGA Stop codon	A
U	UUG Leucine	UCG Serine	UAG Stop codon	UGG Tryptophan	G
C	CUU Leucine	CCU Proline	CAU Histidine	CGU Arginine	U
C	CUC Leucine	CCC Proline	CAC Histidine	CGC Arginine	C
C	CUA Leucine	CCA Proline	CAA Glutamine	CGA Arginine	A
C	CUG Leucine	CCG Proline	CAG Glutamine	CGG Arginine	G
A	AUU Isoleucine	ACU Threonine	AAU Asparagine	AGU Serine	U
A	AUC Isoleucine	ACC Threonine	AAC Asparagine	AGC Serine	C
A	AUA Isoleucine	ACA Threonine	AAA Lysine	AGA Arginine	A
A	AUG Methionine start codon	ACG Threonine	AAG Lysine	AGG Arginine	G
G	GUU Valine	GCU Alanine	GAU Aspartic acid	GGU Glycine	U
G	GUC Valine	GCC Alanine	GAC Aspartic acid	GGC Glycine	C
G	GUA Valine	GCA Alanine	GAA Glutamic acid	GGA Glycine	A
G	GUG Valine	GCG Alanine	GAG Glutamic acid	GGG Glycine	G

CD ACTIVITY 11.2 The Triplet Nature of the Genetic Code

ly 33 percent as frequent as phenylalanine. Using similar results and logic, the investigators also inferred that the codons UGG, GGU, and GUG code for tryptophan or glycine.

Later experiments used other techniques to distinguish among these provisional assignments and determine exactly which amino acid is dictated by each codon. Researchers also discovered that certain codons were punctuation marks that signal "start of message" or "end of message." These codons relay information that the protein chain is complete or that protein synthesis should start at a given codon. There is one **start codon,** which codes for the amino acid methionine, to signal that protein synthesis should start at that point on the mRNA. There are also three different **stop codons** (also known as termination codons), which signal that the protein is complete.

TABLE 11.1 The Amino Acids Produced by Synthetic RNAs

(a) If a pool of ribonucleotides consists of $\frac{3}{4}$ uracil and $\frac{1}{4}$ guanine, the expected frequencies of codons produced are as follows:

Codon	Probability	Proportion*
UUU	$(\frac{3}{4})(\frac{3}{4})(\frac{3}{4}) = \frac{27}{64}$	1.00
UUG	$(\frac{3}{4})(\frac{3}{4})(\frac{1}{4}) = \frac{9}{64}$	0.33
UGU	$(\frac{3}{4})(\frac{1}{4})(\frac{3}{4}) = \frac{9}{64}$	0.33
GUU	$(\frac{1}{4})(\frac{3}{4})(\frac{3}{4}) = \frac{9}{64}$	0.33
UGG	$(\frac{3}{4})(\frac{1}{4})(\frac{1}{4}) = \frac{3}{64}$	0.11
GGU	$(\frac{1}{4})(\frac{1}{4})(\frac{3}{4}) = \frac{3}{64}$	0.11
GUG	$(\frac{1}{4})(\frac{3}{4})(\frac{1}{4}) = \frac{3}{64}$	0.11
GGG	$(\frac{1}{4})(\frac{1}{4})(\frac{1}{4}) = \frac{1}{64}$	0.03

*To compute these proportions, UUU is used as a baseline. Its frequency is assigned the value of 1, and then the frequency of the other codons is expressed as a fraction of its frequency. For example, in a sample of 64 codons, UUG is expected to appear 9 times while UUU is expected to appear 27 times. This means that UUG should be one-third as frequent as UUU; its frequency is 33 percent that of UUU. Similarly, UGG is expected to appear 3 times in a sample of 64 codons, meaning that it should be one-ninth as frequent as UUU.

(b) The following experimental results give the observed proportion of amino acids produced from RNAs when the ribonucleotide pool consists of $\frac{3}{4}$ uracil and $\frac{1}{4}$ guanine.

Amino acid	Proportion*
Phenylalanine	1.00
Leucine	0.37
Valine	0.36
Cysteine	0.35
Tryptophan	0.14
Glycine	0.12

*To compute these proportions, phenylalanine is used as a baseline. The frequency of phenyalanine in the product proteins is assigned the value of 1, and the frequency of the other amino acids is expressed as a fraction of its frequency.

The full genetic code, presented in **Figure 11.10**, is a tremendous achievement. It represents more than five years of work by several teams of researchers. As predicted, the genetic code is redundant. All amino acids except methionine and tryptophan are coded by more than one codon. Later work showed that the genetic code is also nearly universal. With a few minor exceptions, all codons specify the same amino acids in all organisms.

11.4 The Central Dogma of Molecular Biology

Even before the genetic code was fully revealed, Francis Crick articulated what became known as the central dogma of molecular biology. The central dogma summarizes the flow of information in cells. It simply states that DNA codes for RNA, which codes for proteins.

$$\text{DNA} \rightarrow \text{RNA} \rightarrow \text{proteins}$$

Besides clarifying information flow in the cell, the central dogma points out that information flow occurs in only one direction—from DNA to RNA to proteins. RNA does not code for the production of RNA or DNA, and proteins do not code for the production of RNA or DNA or proteins. (**Box 11.1**, page 228, points out important exceptions to these rules, however.)

This simple statement encapsulates much of the research reviewed in this chapter. DNA is the hereditary material. Genes consist of specific stretches of DNA. The sequence of bases in DNA specifies the sequence of bases in an RNA molecule. Groups of three bases within a sequence of mRNA specify the sequence of amino acids in a protein. In this way, genes ultimately code for proteins. Most proteins function as enzymes that catalyze chemical reactions in the cell. (As we'll see in Chapter 13, some genes code for RNA molecules that do not function as mRNAs. Instead, they have other functions.)

Biologists use specialized vocabulary to summarize this sequence of events. For example, biologists say that DNA is transcribed to RNA. In everyday English, the word *transcription* simply means making a copy of information. The scientific use of the term is appropriate because DNA acts as a permanent record—an archive or blueprint containing the information needed to build and run the cell. This permanent record is copied, via **transcription,** to a short-lived form called mRNA. This information is then transferred to a new form—a sequence of amino acids. In biology, the synthesis of protein from mRNA is called **translation.** In everyday English, the word *translation* refers to transferring information from one language to another; in biology, it means transferring information from one type of molecule to another (that is, from nucleic acid to protein). In diagrammatic form:

	transcription		translation	
DNA	$\longrightarrow$	mRNA	$\longrightarrow$	proteins
(information storage)		(information carrier)		(active cell machinery)

BOX 11.1 RNA Genomes: Exceptions to the Central Dogma

The old adage "Rules are made to be broken" applies to biology. In 1955, just three years after DNA had been confirmed as the hereditary material, Heinz Fraenkel-Conrat discovered that the virus he was studying (the tobacco mosaic virus, or TMV) does not contain DNA. The genes of this virus are made of RNA. Fraenkel-Conrat showed that when TMV infects a cell of a tobacco plant, its genes are translated directly into viral proteins. Thus, in RNA viruses such as TMV the flow of information is simply RNA → proteins. Over the years since Fraenkel-Conrat's work was published, researchers have identified hundreds of viruses with RNA genomes. These include the viruses that cause polio, hepatitis, flu, and the common cold in humans.

In terms of breaking rules, though, an even bigger bombshell exploded in 1970. In that year David Baltimore, Howard Temin, and Satoshi Mizutani announced the discovery of RNA viruses that violate the central dogma's tenet of one-way information flow. The genomes of these viruses include sequences that code for an enzyme called reverse transcriptase. This protein catalyzes the formation of DNA from an RNA template. When one of these viruses infects a host cell, reverse transcriptase "reverse transcribes" the viral RNA genome to DNA. The viral DNA usually incorporates itself into the host's chromosome. Messenger RNAs are then transcribed from the viral DNA and translated into proteins. In these viruses, information flow is RNA → DNA → RNA → proteins. As a result, they came to be called retroviruses (backward-viruses). The virus that causes AIDS is a retrovirus.

The experiments leading to this discovery were elegant. To confirm the existence of reverse transcriptase, Temin and Mizutani purified large quantities of the Rous sarcoma virus (a virus associated with cancer formation in chickens). Then they disrupted the lipid-containing envelope that surrounds the virus by treating the particles with a detergent. This step released the contents of the virus particles into solution. The researchers added deoxyribonucleotides that had been labeled with a radioactive isotope and incubated the mixture. Later they observed the formation of radioactive DNA molecules in the mixture. This was a smoking gun—strong evidence that the virus contains a molecule capable of synthesizing DNA from an RNA template.

To confirm this result, Temin and Mizutani did a control experiment. It was the same experiment just described, with one change. Along with the labeled deoxyribonucleotides, the researchers added ribonuclease, which breaks RNA molecules apart. In this treatment, no DNA formed. This control confirmed that the enzyme requires an RNA template.

With the discovery of reverse transcriptase and the retroviruses, biologists could no longer be quite so dogmatic about the central dogma. Although all cells conform to the central dogma, some viruses contradict it.

The central dogma elegantly describes at the molecular level what Mendel discovered at the whole-organism level. An organism's genotype is determined by the sequence of bases in its DNA; its phenotype is determined by the proteins it produces. Later work revealed that alleles of the same gene differ in their DNA sequence. As a result, the proteins produced by different alleles frequently differ in their amino acid sequence and in their function.

The central dogma also provided an important conceptual framework for the burgeoning field called molecular genetics, and inspired a series of fundamental questions about how genes and cells work. For example, what molecules are responsible for copying DNA—the permanent archive of hereditary information? How do transcription and translation proceed on a molecular level, and how are they regulated so that genes are expressed at an appropriate time and in an appropriate amount? The next four chapters are devoted to exploring these questions.

Essay How Do Viruses Work?

Viruses cause a great deal of human misery. They are responsible for colds, flu, AIDS, polio, and smallpox. The information in this chapter makes it possible to answer two questions about viruses: What are they, and why are the diseases they cause so difficult to treat?

One way to explain viruses is to consider what they are not. They are not cells, because they have no cell membrane. They are not alive, because they cannot make a copy of themselves without exploiting a host cell. Hence, viruses are not classified as living organisms. Biologists refer to them as "particles" instead.

Viruses are parasites.

Now let's get back to identifying viruses. Viruses are parasites. To be precise, they are obligate, intracellular (within-cell) parasites. To make copies of itself, a virus takes over the flow of information in the cell. A virus uses ribosomes and other machinery provided by the host cell to transcribe its genome and translate its proteins. Parasitism occurs only inside cells. Outside of cells, viruses exist only in an inactive, dormant form called the virion.

The general life cycle of a virus is analyzed in detail in Chapter 26. That chapter pays special attention to the human immunodeficiency virus—the agent that causes AIDS—and how it accomplishes the four steps common to all viral replication cycles. These common steps are illustrated in **Figure 1** using a bacterial virus as an example.

1. The virus attaches to the exterior of a cell and injects its genes.
2. The viral genes are transcribed and copied, using enzymes and ATP provided by the cell. At this point, the virus has taken over the flow of information in the cell.
3. Inside the cell, viral proteins and genes are assembled into a new generation of particles.
4. The particles burst out of the cell, or bud off by enveloping themselves in a piece of the cell membrane. (During this process, the host cell often dies.) If a virion encounters another host cell, another round of infection ensues.

After studying this generalized replication cycle, you can appreciate why viral diseases are so difficult to treat. Most drugs work by inhibiting or destroying enzymes that are made by the disease-causing agent and that are essential to its growth and reproduction. But because viruses use so many of the host cell's own enzymes, it is difficult to find drugs that block their replication without also destroying host cells in the process. (In cases like this, the cure can be worse than the disease.) This is why the most effective defense against viral diseases is vaccination—a topic treated in detail in Chapter 46.

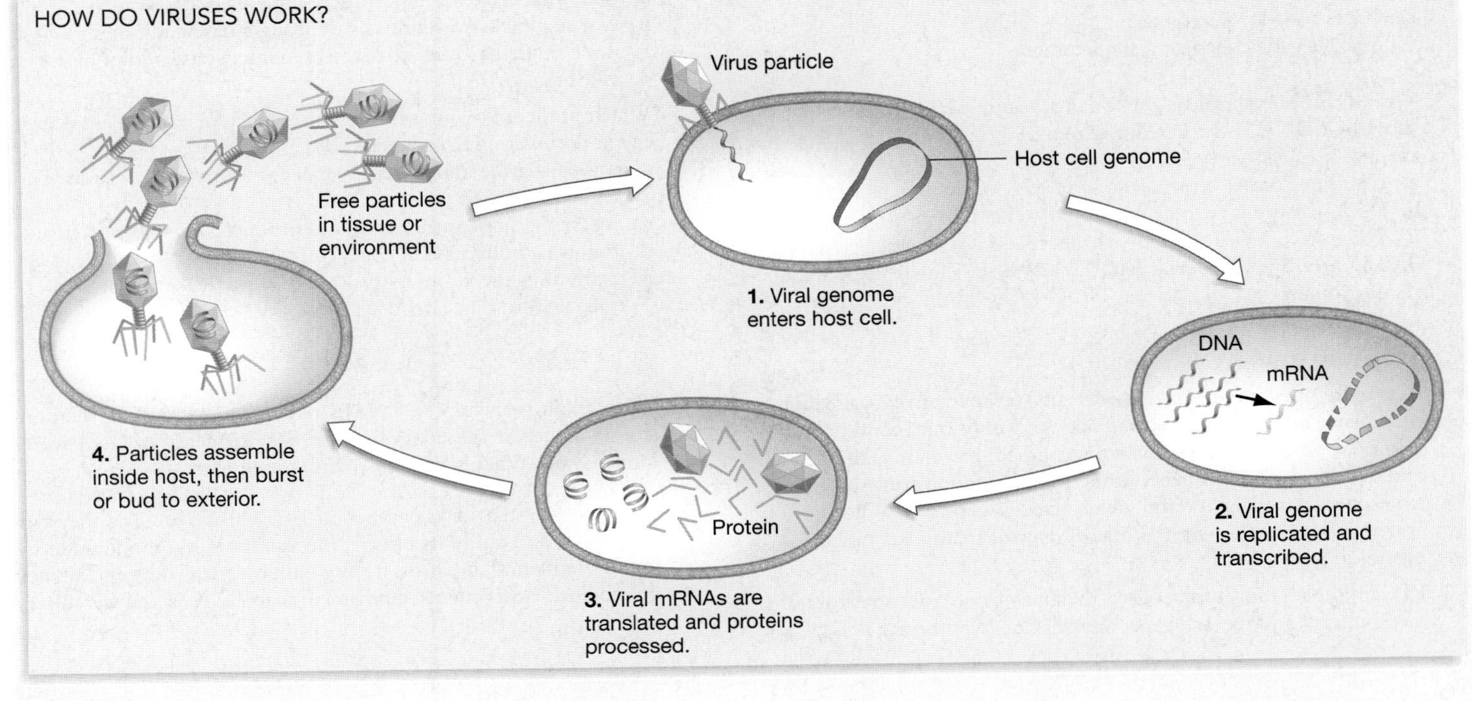

FIGURE 1
Some basic steps in the replication cycle of a bacterial virus that illustrate the concept that viruses work by controlling the flow of information in a host cell.

Chapter Review

Summary

The experiments that established DNA as the hereditary material were inspired by research showing that a benign strain of *Streptococcus* could be "transformed" to a virulent strain by contact with heat-killed cells of the virulent strain. Follow-up work demonstrated that the "transforming principle"—that is, the molecule that was transferred from the virulent to the benign strain and caused a heritable change in its characteristics—was DNA. This result was confirmed when researchers showed that DNA, and not proteins, enters cells during a viral infection.

The question of what genes actually do was solved in a series of experiments on a fungus called *Neurospora crassa*. Researchers isolated mutant individuals that cannot grow unless they are supplied with a specific amino acid, and then showed that different mutants were unable to make different chemical precursors of the amino acid. The one-gene, one-enzyme hypothesis explains these results by proposing that genes code for proteins, and that most proteins act as enzymes that catalyze specific reactions in metabolic pathways.

Experiments performed in several labs confirmed that DNA does not code for proteins directly. Instead, mRNA molecules are transcribed from DNA and then translated into proteins. One-way flow of information, from DNA to RNA to proteins, is called the central dogma of molecular biology. By synthesizing RNAs of known base composition and then observing the amino acid sequences that result when they are translated, researchers were able to unravel the genetic code. It is now established that the code is read in triplets, and that the code is "redundant"—meaning that most of the 20 amino acids are specified by more than one codon.

Questions

Content Review

1. What is metabolism?
   a. The chemical reactions that occur in organisms.
   b. The synthesis of macromolecules through spark discharges and other reactions during chemical evolution.
   c. The synthesis of macromolecules in a test tube, or in vitro.
   d. The synthetic compounds that mimic biological enzymes.

2. What does the one-gene, one-enzyme hypothesis state?
   a. Genes are composed of stretches of DNA.
   b. Genes are made of protein.
   c. Genes code for ribozymes.
   d. A single gene codes for a single protein.

3. Why did the experiment on *Streptococcus*, which purported to show that DNA is the hereditary material, fail to convince many skeptics?
   a. It had no control treatments, so it was poorly designed.
   b. DNA was thought to be too simple to contain information.
   c. The trait that was studied—virulence—is not genetic.
   d. The hereditary material had already been shown to be made of protein.

4. Why did researchers suspect that DNA does not code for proteins directly?
   a. In cells that have a nucleus, DNA is found inside the nucleus but proteins are produced outside the nucleus.
   b. In cells that do not have a nucleus, DNA and proteins are never found together.
   c. Messenger RNA was known to serve as an intermediate.
   d. The double helix could not unwind far enough.

5. To confirm that the genetic code is read in triplets, researchers created mutants with what attributes?
   a. one single-base deletion or one single-base addition
   b. two single-base deletions or two single-base additions
   c. three single-base deletions or three single-base additions
   d. all of the above

6. Which of the following describes an important experimental strategy in deciphering the genetic code?
   a. comparing the amino acid sequences of proteins with the base sequence of their genes
   b. analyzing the sequence of RNAs produced from known genes
   c. analyzing mutants that changed the code
   d. examining the proteins produced when RNAs of known sequence were translated

Conceptual Review

1. Researchers try to design experiments so that an experimental treatment shows the effect that one, and only one, condition or agent has on the phenomenon being studied. Examine Figure 11.1 and decide which treatments in the transformation study acted as experimental treatments, and which acted as controls. Write a general statement explaining the role of control treatments in experimental design.

2. Explain the relationship between the analysis of "inborn errors of metabolism" and the one-gene, one-enzyme hypothesis.

3. Throughout this chapter, DNA is referred to as an "information-storage molecule." Explain how the base sequence of DNA stores information. What is this information used for?

4. Scientists say that a phenomenon is a "black box" if they can describe it and study its effects, but don't yet know the underlying mechanism that causes it. In what sense was the gene a black box before experiments confirmed that DNA is the hereditary material?

Applying Ideas

1. It turns out that recessive alleles often contain a loss-of-function mutation. When a normal allele and a loss-of-function allele are paired in a heterozygous individual, the normal copy of the gene is frequently able to produce enough functional protein to give a normal phenotype. How do these facts relate to the phenomenon of dominance and recessiveness discovered by Mendel?
2. Examine the genetic code as illustrated in Figure 11.10. Note that when several codons specify the same amino acid, the first two bases in those codons are almost always identical. Does this observation support the hypothesis that the genetic code is arbitrary? That is, does the code represent a random assemblage of bases, like letters drawn out of a hat, or does it have distinct patterns? (If patterns exist, you could hypothesize that the code is structured in a way that helps transcription or translation occur more efficiently.) How would you test the hypothesis that the code is indeed random?
3. In Table 11.1, should the observed results correspond *exactly* to the predicted results? Why don't they?

CD-ROM and Web Connection

CD Activity 11.1: The One-Gene, One-Enzyme Hypothesis *(Tutorial)*
(Estimated time for completion = 15 min)
Examine the experiments on *Neurospora* that Beadle and Tatum used to develop their hypothesis.

CD Activity 11.2: The Triplet Nature of the Genetic Code *(Animation)*
(Estimated time for completion = 5 min)
Explore how scientists deduced the length of a genetic word called a codon.

At your **Companion Website** (http://www.prenhall.com/freeman/biology), you will find self-grading exams and links to the following research tools, online resources, and activities:

Oswald Avery
This profile describes the research of a key contributor to the discovery of DNA as the hereditary material.

Featured Genes
The National Center for Biotechnology Information provides a selected list of human diseases, their related genes, and their location in the human genome.

National Human Genome Research Institute
This site provides access to current research and issues related to the human genome.

Additional Reading

Davern, C. I. 1981. *Genetics: Readings from Scientific American* (San Francisco: W. H. Freeman). This volume contains popular accounts of many of the experiments reviewed in this chapter. Most of the articles are written by the scientists who performed the work.

Judson, H. F. 1996. *The Eighth Day of Creation: Makers of the Revolution in Biology* (Plainview, NJ: CSHL Press). Profiles of the researchers who ushered in biology's molecular revolution.

DNA Synthesis, Mutation, and Repair

12

The DNA of an organism is like an ancient text that has been painstakingly copied and handed down, generation after generation. But while the most ancient of all human texts contain messages that are thousands of years old, the DNA in living cells has been copied and passed down for *billions* of years. And instead of being copied by monks or clerks, DNA is replicated by molecular scribes.

What molecules are responsible for copying DNA, and how do they work? Do these molecular machines ever make mistakes? If so, what are the consequences? Finally, what happens if the genetic "document" is damaged by environmental insults like high-energy radiation or toxic chemicals?

The goal of this chapter is to analyze the experiments that answered these questions. We begin with classical work on the bacterium *Escherichia coli* and finish with an introduction to cutting-edge research on the genetics of cancer. In between we examine two of the most important technological advances in molecular biology.

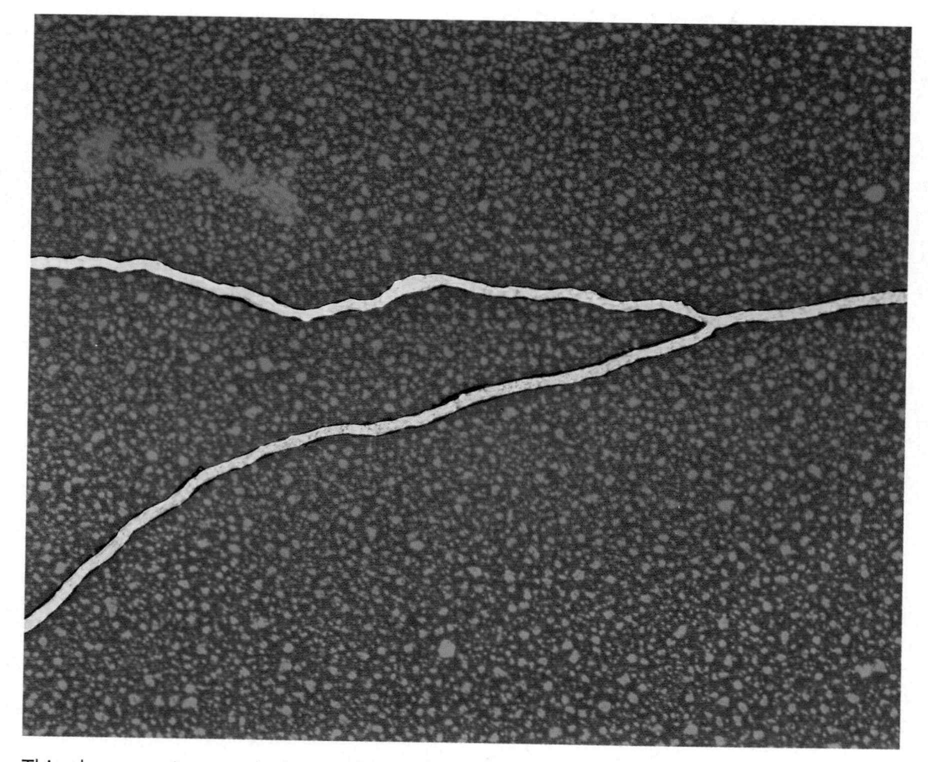

This electron micrograph shows DNA in the process of being replicated. The "Y" shape results from a structure called the replication fork, which is introduced in section 12.2.

12.1 Testing Early Hypotheses About DNA Replication

Chapter 3 introduced Watson and Crick's model for the secondary structure of DNA, which was proposed in 1953. DNA is a long, linear polymer that has two major components: a "backbone" made up of sugar and phosphate groups and a series of nitrogen-containing bases that project from the backbone (see Figure 3.19). Watson and Crick realized that if two of these long strands twist around each other, then certain of the nitrogen-containing bases fit together in pairs inside the spiral. The double-stranded molecule that results is called a double helix. The structure is stabilized by hydrogen bonds that form between the bases called adenine (A) and thymine (T) and the bases called guanine (G) and cytosine (C).

Watson and Crick realized that the A-T and G-C pairing rules suggested a way for DNA to be copied prior to mitosis and meiosis. They suggested that the existing strands of DNA served as a template for the production of new strands, with bases being added to the new strands according to the complementary base-pairing rules. For example, if the template strand contained a T then the new strand would add an A to pair with it; a G on the template strand would dictate the addition of a C on the new strand.

As **Figure 12.1** shows, however, biologists had three ideas about how the old and new strands might interact.

1. If the old strands of DNA separated, each could then be used as a template for the synthesis of a new, second strand (Figure 12.1a). This hypothesis is called semi-conservative replication because each new daughter DNA molecule would consist of one old strand and one new strand.
2. If the bases temporarily turned outward, they could serve as a template for the synthesis of an entirely new double helix all at once. Alternatively, a non-DNA intermediary might be produced, which would then serve as a template for synthesis of two new strands. This model is called conservative replication because the original molecule would be unchanged (Figure 12.1b). The newly formed molecule, in contrast, would be composed of two newly synthesized DNA strands.
3. If the original helix was cut and unwound in short sections before being copied and put back together, then new and old strands would intermingle as shown in Figure 12.1c. This possibility is called dispersive replication.

Which of the three hypotheses is correct?

The Meselson and Stahl Experiment

Matthew Meselson and Frank Stahl realized that if they could tag parental and daughter strands of DNA in a way that would make them distinguishable from each other, they could determine whether replication was conservative, semi-conservative, or

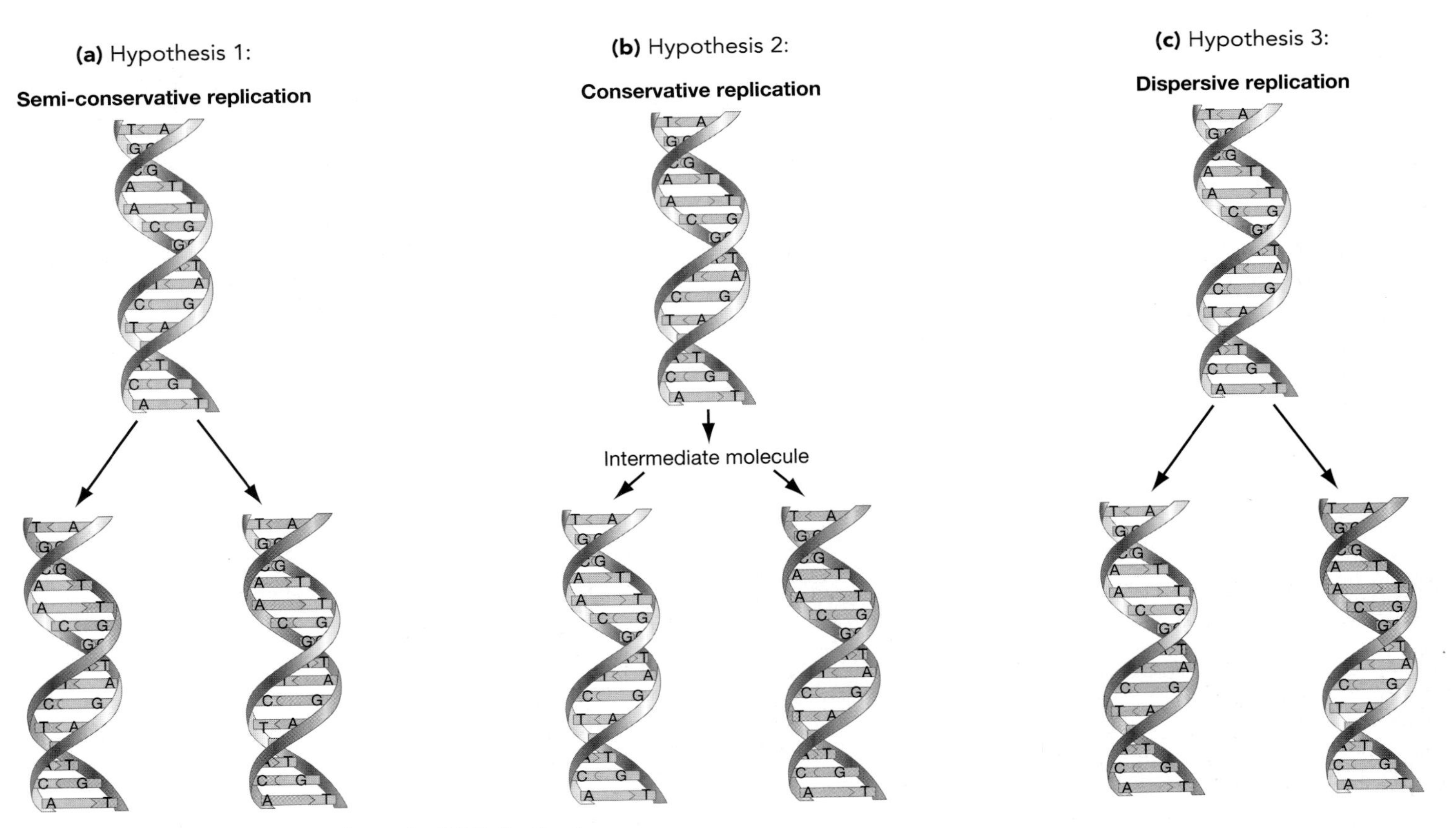

FIGURE 12.1 Alternative Hypotheses for DNA Synthesis
The phosphate-sugar backbone of DNA is shown as a ribbon in these diagrams (gray in an old strand and red in a newly synthesized strand). Note that the four nucleotide bases (A, T, C, and G) project from this backbone. **(a)** During replication, each strand in a DNA double helix could serve as the template for the synthesis of a daughter strand. **(b)** The original molecule could remain intact after synthesis, so that the product molecule would consist of two newly synthesized strands. **(c)** If dispersive replication occurred, each strand in the daughter DNA would consist of a mixture of old and new DNA.

dispersive. Before they could do any tagging, however, they needed to choose an organism to study. They decided to work with a common inhabitant of the human gastrointestinal tract, the bacterium *Escherichia coli*, because it is small and grows quickly and readily in the laboratory. These cells copy their entire complement of DNA, or their **genome**, before every cell division.

To distinguish parental strands of DNA from daughter strands when *E. coli* replicates, Meselson and Stahl grew the cells for many generations in the presence of one of two isotopes of nitrogen: either ^{15}N or ^{14}N. Because ^{15}N contains an extra neutron, it is heavier than the normal isotope ^{14}N. As a result, when DNA molecules are subjected to density-gradient centrifugation—a technique related to those introduced in Box 5.1—strands that contain ^{15}N form a band in the centrifuge tube that is separate from DNA containing ^{14}N.

This was the key to their experiment. Meselson and Stahl reasoned that if different nitrogen isotopes were available in the medium when parental and daughter strands of DNA were produced, then the two types of strands should behave differently during centrifugation. How could this tagging system be manipulated to test whether replication is conservative, semi-conservative, or dispersive?

Figure 12.2a summarizes their experimental strategy. They began by growing *E. coli* cells in the presence of ^{15}N as the sole source of nutrient nitrogen. They purified DNA from a sample of these cells and transferred the rest of the culture to a medium containing only ^{14}N. After enough time had elapsed for these experimental cells to divide once, they removed a sample and isolated the DNA. After the remainder of the culture had divided again, they removed a sample and purified the DNA.

As **Figure 12.2b** shows, the conservative and semi-conservative models make distinct predictions about the makeup of the molecules after the first and second replication. If replication is conservative, then the daughter cells should have DNA with either ^{14}N or ^{15}N—but not both. As a result, two distinct bands should form in the centrifuge tube. But if replication is semi-conservative or dispersive, then all of the experimental cells should contain a mix of ^{14}N and ^{15}N after one gen-

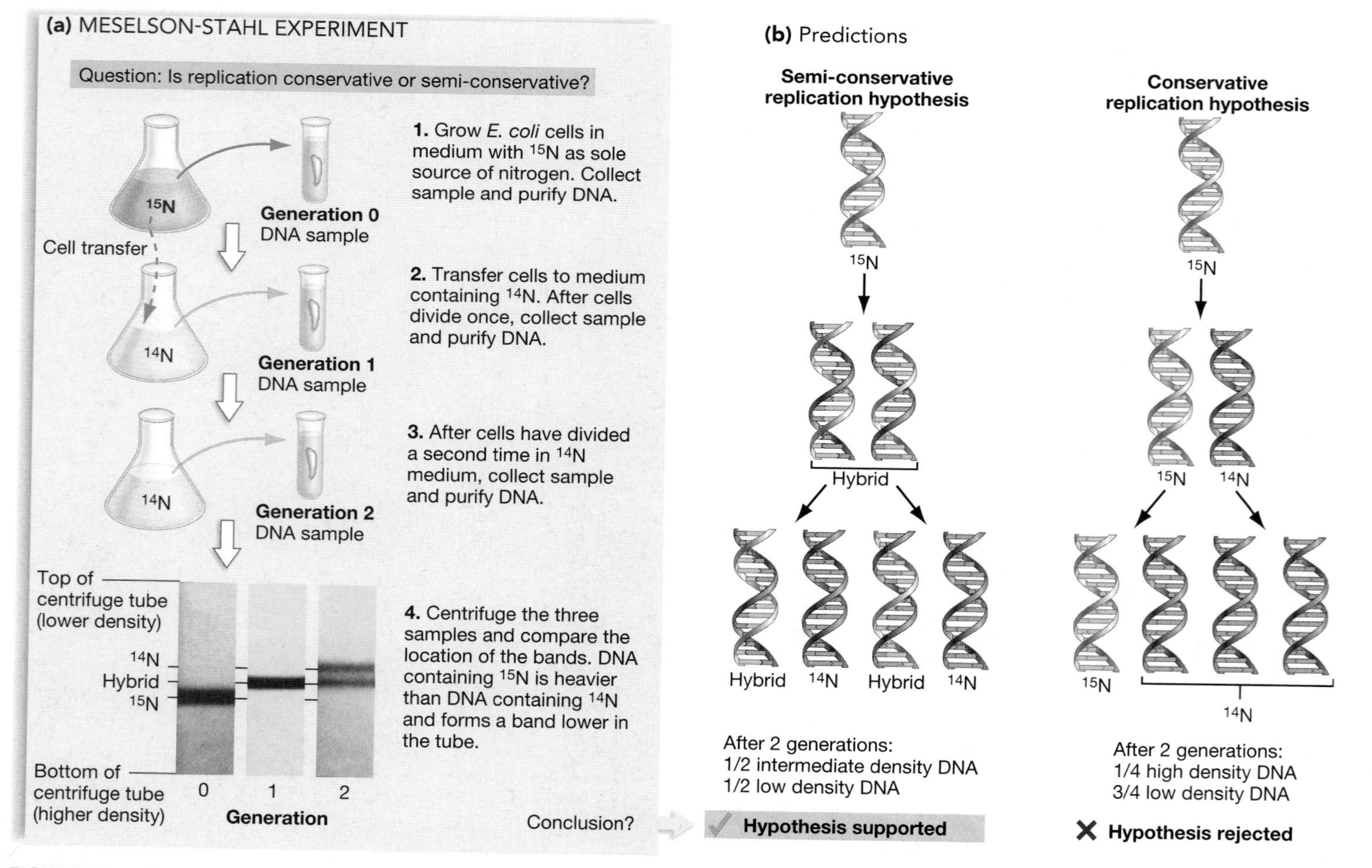

FIGURE 12.2 The Meselson-Stahl Experiment

EXERCISE Meselson and Stahl actually let their experiment run for four generations. Next to the photo at the bottom of the experiment, sketch what the data from fourth-generation DNA should look like; that is, where the band(s) should be.

eration. In this case, one intermediate band should form in the centrifuge tube. After two generations, however, half of the daughter cells should contain only ^{14}N if semi-conservative replication occurs—meaning that a second, lower-density band should appear in the centrifuge tube. This is in contrast to the prediction made by the conservative and dispersive model.

The photograph at the bottom of Figure 12.2a shows their results. After one generation, the density of the DNA molecules was intermediate; after two generations, a lower-density band appeared. Thus, after two generations there were two kinds of molecules—those with intermediate density and those with low density. These data offered strong support for the hypothesis that DNA replication is semi-conservative.

Then the question became: How does the DNA synthesis reaction proceed? Does it require an enzyme analogous to the RNA polymerase introduced in Chapter 11 or is it spontaneous?

DNA Polymerase

In the mid-1950s Arthur Kornberg and co-workers set out to study how DNA replicates by attempting to generate a cell-free, or in vitro, DNA replication system. Their goal was to resolve a long-standing paradox. If genes contain the information for making an organism, then what is responsible for making genes?

Watson and Crick had suggested that the nucleotides paired to a DNA template might be "zippered" together without assistance from an enzyme. Kornberg, in contrast, assumed that some sort of DNA-polymerizing enzyme existed. He hypothesized that a cell-free system capable of replicating DNA would require an enzyme to catalyze the formation of phosphodiester bonds between nucleotides in the newly formed strands.

To begin his search for this molecule, Kornberg and colleagues prepared proteins from *E. coli* cultures and added them to reaction mixtures containing DNA and monomers called deoxynucleoside triphosphates, or dNTPs. The general structure of a dNTP is diagrammed in **Figure 12.3a**. Note that the researchers used four different dNTPs; each carried an A, T, G, or C in addition to three phosphate groups. (The "N" in dNTP stands for any of the four bases. They used dATP, dTTP, dGTP, and dCTP.) The phosphate groups were important because the potential energy stored in the bonds between the phosphates could be used to drive the energy-demanding synthesis reaction (**Figure 12.3b**). The biologists also used radioactive versions of these monomers. This was vital because the radioactivity gave them an assay to detect whether dNTPs had been incorporated into newly synthesized DNA. If Kornberg and co-workers could detect radioactivity in a high-molecular-weight molecule instead of in low-molecular-weight monomers, it meant that DNA synthesis had occurred.

Kornberg and colleagues tested a variety of proteins with this in vitro assay. After many failed attempts, they finally succeeded in identifying an enzyme that allowed DNA synthesis to proceed. They called this molecule **DNA polymerase**. Its name was later changed to DNA polymerase I, or simply Pol I, when it became clear that there was more than one type of DNA polymerase in *E. coli* cells.

Template-Directed Synthesis Kornberg and co-workers confirmed that for DNA polymerase I to work, the reaction mix had to contain a template DNA and all four dNTPs. But was the template DNA actually directing synthesis, as predicted? Or were nucleotides merely being added randomly? To answer this question Kornberg and co-workers added the entire genome of a bacteriophage (bacterial virus, or phage) called ΦX174 (phi-X 174) to their in vitro system. After allowing the polymerization reaction to proceed, they isolated strands of ΦX174 DNA that had been synthesized by Pol I in vitro (**Figure 12.4**, page 236).

(a) Structure of dNTPs

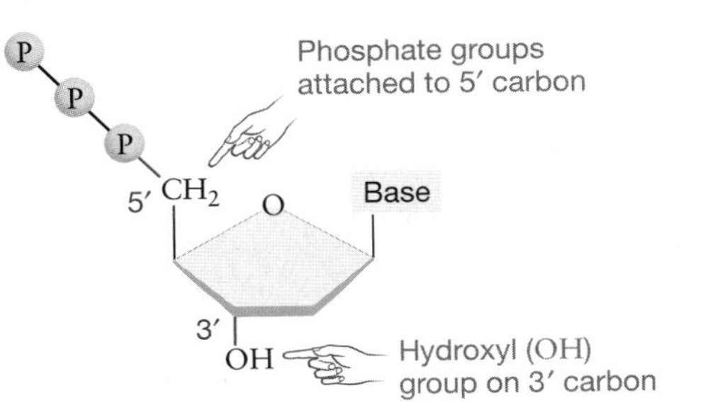

(b) DNA synthesis reaction

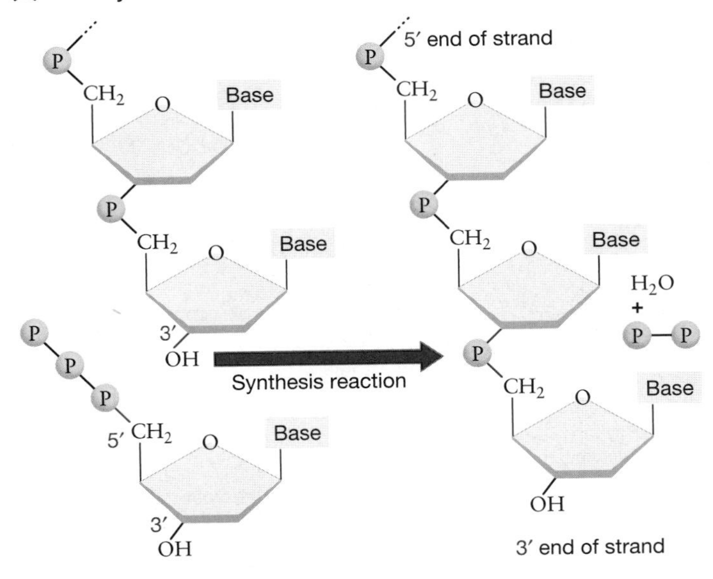

FIGURE 12.3 The DNA Synthesis Reaction
(a) In dNTPs (deoxynucleoside triphosphates), the base can be adenine (A), thymine (T), guanine (G), or cytosine (C). **(b)** A dNTP monomer is added to a DNA polymer when a phosphodiester bond forms between the 3′ carbon on the end of a DNA strand and the 5′ carbon on the dNTP. This is a condensation reaction. QUESTION According to the reaction shown here, does DNA synthesis proceed from the 5′ end of the new molecule toward the 3′ end, or in the 3′ to 5′ direction?

Then they added the product DNA to a culture of bacteria. When they manipulated conditions so that the product DNA could enter the host cells, the cells became infected and a new generation of phages was produced. This result confirmed that in vitro synthesis had indeed been directed by the original ΦX174 template, as predicted.

Despite this accomplishment, the Kornberg group had not yet demonstrated that DNA Pol I was responsible for the replication of the *E. coli* genome in vivo. More important, it appeared that something was wrong with their system. The in vitro reaction was almost 500 times slower than the replication rate observed in actual cells. What was going on?

The Cairns Mutant Paula de Lucia and John Cairns took a genetic approach to answering questions about how DNA Pol I functions in living cells. Their goal was to find a strain of *E. coli* with a mutation that knocked out this enzyme. By studying these mutant cells, they hoped to infer how *pol I* functions in vivo. This change in strategy was important. The researchers turned to genetic approaches—studying the effects of mutations in vivo—to complement the biochemical work that Kornberg and colleagues had done in vitro.

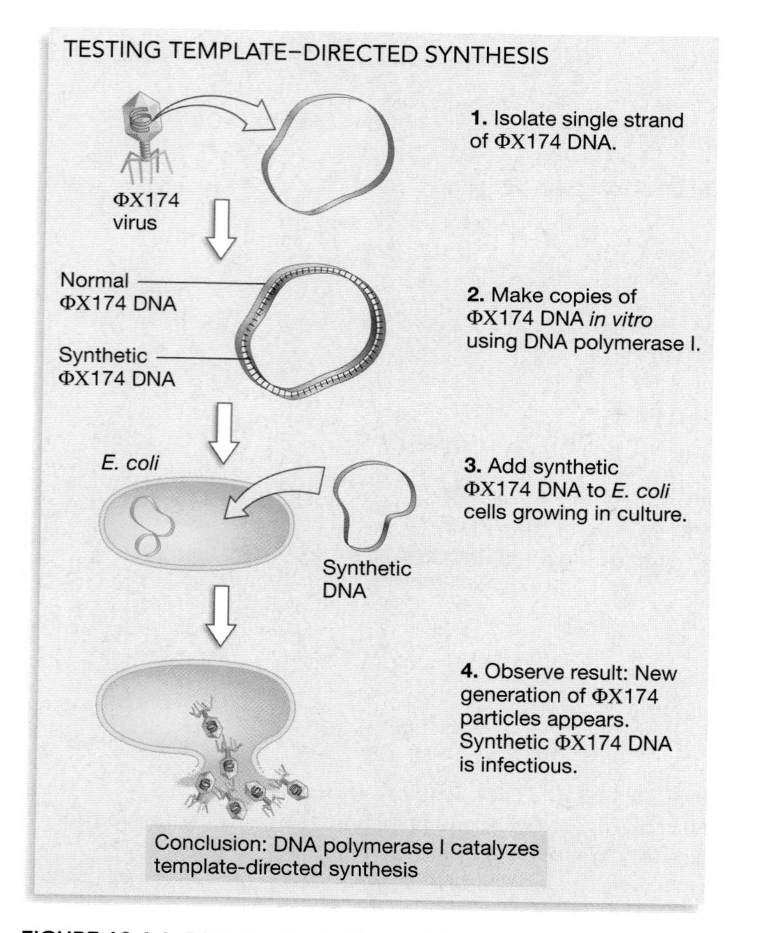

FIGURE 12.4 Is DNA Synthesis Directed by a Template?
This experiment tested the hypothesis that synthesis of DNA by DNA polymerase I is directed by the sequence of bases in a template strand of DNA. The alternative hypothesis is that the enzyme joins deoxyribonucleotides randomly.

Finding a *pol I* mutant presented a unique challenge, however. If *pol I* is required for a cell to replicate its DNA, and if *pol I* is defective, then the cell won't be able to divide and grow. How can geneticists isolate mutant strains to study if the mutations are almost guaranteed to be fatal?

Cairns and de Lucia were able to isolate and study a pol I mutant by using a trick. They found mutant cells that grow normally at low temperatures but cease to grow at high temperatures. These are called **temperature-sensitive mutants.** The proteins produced by temperature-sensitive mutations work well at low temperatures, but lose their structural integrity at high temperatures. Temperature-sensitive mutations are useful because they allow researchers to identify and study genes that are essential to cell function.

Analyzing mutants is actually one of the most important of all research strategies in genetics. T. H. Morgan used this approach to study the inheritance of eye color and other features in *Drosophila* (see Chapter 10); Beadle and Tatum used it to identify the enzymes required for arginine synthesis in *Neurospora* (see Chapter 11). The strategy is especially promising in *E. coli* because these cells are haploid. Because there is only one allele for each gene, dominance and recessiveness do not complicate the analysis—every replication mutant should have a defect in a gene required for DNA synthesis.

de Lucia and Cairns began their study by treating *E. coli* cells with a chemical that causes mutations. They grew up colonies of each type of treated cell at either the normal growing temperature of 37°C or a low temperature of 25°C or 30°C. (Recall that *E. coli*'s normal habitat is the human gastrointestinal tract; normal human body temperature is 37°C.) Then they added extracts from each colony to Kornberg's in vitro DNA synthesis system incubated at 45°C. After screening several thousand colonies in this way, they finally found a mutant that lacked pol I activity. Curiously, though, the *pol I* mutant cells were able to grow and divide normally at 45°C. This led Cairns and de Lucia to hypothesize that Pol I is *not* the main replication enzyme.

A follow-up study by Julian Gross and Marilyn Gross provided a hint to Pol I's actual function. These researchers showed that *pol I* mutants die off much faster than normal cells when they are exposed to ultraviolet (UV) light. Because UV light is known to cause mutations, a consensus began to build that Pol I's primary function was to repair damaged DNA.

These results explained why Pol I was so slow at synthesizing DNA in vitro. But if Pol I isn't the main replication enzyme, what is?

The Replication Apparatus

To find the main replication enzyme, researchers again turned to analyses of mutant *E. coli* cells that cannot replicate their DNA. The goal was to find the genes involved in DNA synthesis by studying mutant individuals that lack the trait.

CD ACTIVITY 12.1 DNA Synthesis

Several searches for temperature-sensitive mutations turned up cells that failed to grow at high temperature (39–41°C) but grew normally at low temperature (25–30°C). Did any of these cells have a defect in a molecule required for DNA synthesis? If these mutants could be identified, they would lead directly to the discovery of the main enzyme responsible for DNA replication.

To answer this question Masamichi Kohiyama and Alan Kolber fed radioactive thymidine molecules to replication mutants and monitored how much radioactivity was later incorporated into DNA polymers. **Figure 12.5** shows the results of experiments with one of these strains. The red data points on the graph indicate that at 30°C, the mutant cells produce more and more DNA over time. But when a sample of these cells was shifted to 41°C at the point marked by the blue arrow, DNA synthesis stopped. To explain this pattern, Kohiyama and Kolber proposed that the cells have a temperature-sensitive defect in a gene required for DNA replication.

DNA Pol III When researchers isolated the normal version of the enzyme encoded by the gene that Kohiyama and Kolber identified and added it to an in vitro DNA synthesis system, they found that synthesis proceeded just as rapidly as it did in growing cells. This enzyme, which has now been isolated and analyzed in detail, is called DNA polymerase III, or simply Pol III. (To date, a total of five DNA polymerases, named Pol I-IV, have been found in *E. coli*.) Pol III is the enzyme responsible for the vast majority of DNA synthesis in bacterial cells. When the double helix of bacterial DNA separates, Pol III adds the complementary bases and catalyzes the formation of phosphodiester bonds in the newly synthesized strand.

The discovery of Pol III ended one chapter in research on DNA synthesis and started another. The pioneering phase of work on DNA replication was over. Researchers had established that DNA replication is semi-conservative in nature, that it is enzyme-dependent, and that several different polymerases exist. The next challenge was to understand the mechanics of the process in detail.

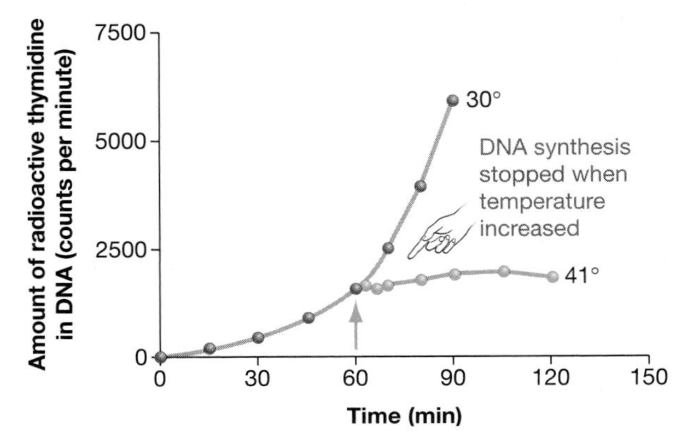

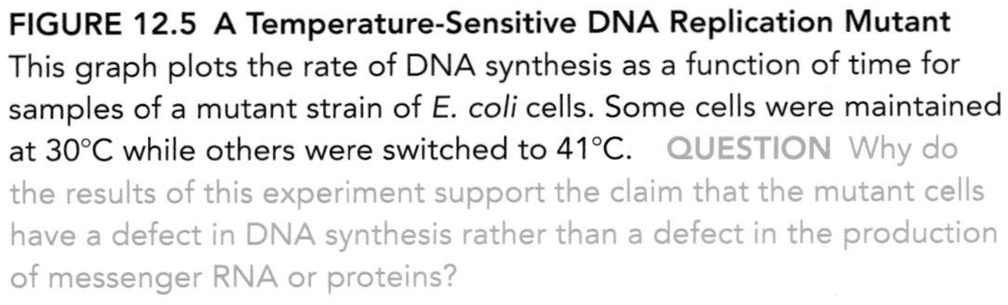

FIGURE 12.5 A Temperature-Sensitive DNA Replication Mutant
This graph plots the rate of DNA synthesis as a function of time for samples of a mutant strain of *E. coli* cells. Some cells were maintained at 30°C while others were switched to 41°C. QUESTION Why do the results of this experiment support the claim that the mutant cells have a defect in DNA synthesis rather than a defect in the production of messenger RNA or proteins?

12.2 A Comprehensive Model for DNA Synthesis

How does DNA polymerase III interact with other proteins to accomplish DNA synthesis? Two results from the pioneering phase of work on DNA replication helped focus this question and point the way to an answer.

The first result was an observation about the DNA polymerases. These enzymes catalyze the addition of a dNTP monomer only to the 3′ end of the DNA chain—never to the 5′ end (see Figure 12.3). As a result, DNA synthesis always proceeds in the $5' \rightarrow 3'$ direction.

The second result emerged from the first photograph taken of DNA replication in progress (**Figure 12.6**). John Cairns obtained this picture by feeding thymine that was labeled with radioactive hydrogen to *E. coli* cultures long enough for about two

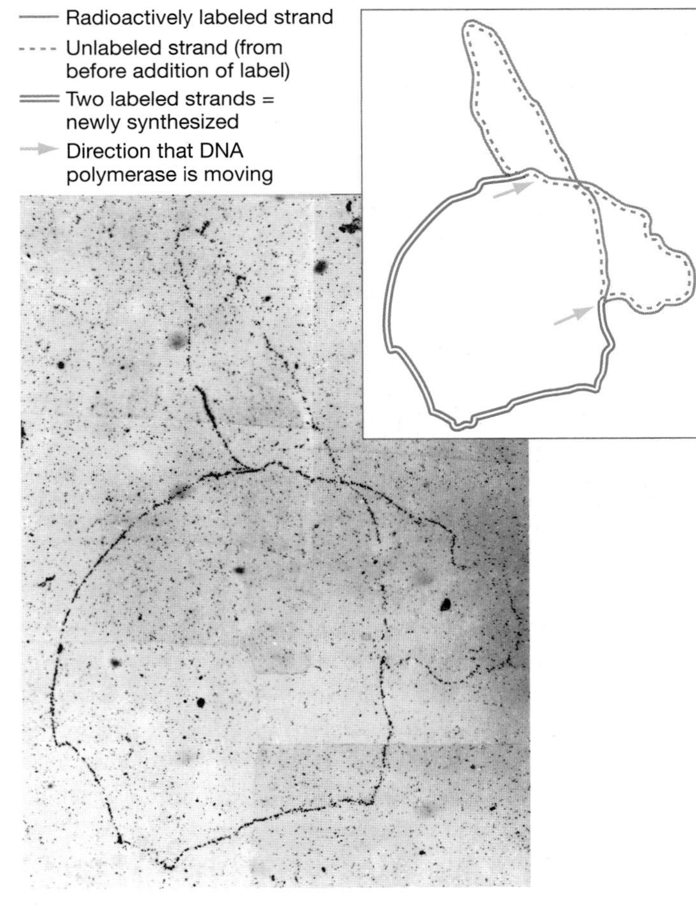

FIGURE 12.6 DNA Replication in Action
The photograph shows the original x-ray film referred to in the text; the diagram in the insert shows the researcher's interpretation. QUESTION The researcher who performed this experiment claimed that the replication event shown here was about two-thirds complete. Why?

replication cycles to take place. (Note that *E. coli* has one circular chromosome.) Then he purified DNA from the cells and laid x-ray film over the chromosomes for two months. A black dot was produced wherever a radioactive hydrogen atom decayed and sent its emission into the x-ray film. (This technique, called autoradiography, was introduced in Chapter 3; see Box 3.1.)

To interpret the photograph, Cairns simply counted the number of dots. These data allowed him to infer which sections of the chromosome had radioactive thymine on both DNA strands and which had radioactive thymine on just one strand. His result is shown in the insert of Figure 12.6. The dashed gray line represents an unlabeled, or "cold" strand. The solid red and gray lines represent radioactively labeled, or "hot" strands. The cold strand was synthesized before the addition of labeled thymine to the culture; the hot strands were synthesized afterwards. The photograph shows DNA replication in action. Loops of DNA meet at the sites, marked with green arrows, where DNA Pol III molecules are currently working. (In bacteria, replication starts at a specific point and then proceeds in both directions.)

Cairns's work helped to focus interest on the point where the parent DNA double helix was being split into two template strands. This structure, called the **replication fork**, is diagrammed in **Figure 12.7a**. As the illustration shows, DNA polymerase was thought to work into the fork, synthesizing DNA in the $5' \rightarrow 3'$ direction. The product is called the **leading strand** or **continuous strand** because it is synthesized continuously. But if DNA monomers can be added only to the $3'$ ends of the polymer, how could Pol III catalyze formation of a strand that is complementary to the other template strand? Recall from Chapter 3 that the two DNA strands in a double helix are said to be "antiparallel" because they are parallel to one another but oriented in opposite directions (one $5' \rightarrow 3'$ and the other $3' \rightarrow 5'$). The strand that is complementary to the leading strand is oriented in the $3' \rightarrow 5'$ direction away from the replication fork. How can DNA polymerase synthesize a new strand along this template?

Lagging Strand Synthesis

The paradox created by the antiparallel orientation of DNA templates and the unidirectional synthesis by DNA polymerase was resolved when Tuneko Okazaki and colleagues tested a hypothesis proposed by Walter Gilbert. Gilbert had suggested that Pol III synthesizes short fragments of DNA away from the replication fork in the $5' \rightarrow 3'$ direction, and that these fragments are later linked together to form a continuous whole called the **lagging strand or discontinuous strand** (**Figure 12.7b**).

To test this hypothesis, Okazaki's group set out to document the existence of short DNA fragments produced during replication. Their strategy was to add a short "pulse" of radioactive thymidine to *E. coli* cells followed by a large "chase" of nonradioactive thymidine. According to the discontinuous replication model, some of this radioactive thymidine should end up in short, single-stranded fragments of DNA.

As predicted, the researchers succeeded in finding these fragments when they purified DNA from the experimental cells and separated the molecules by centrifugation. A small number of labeled pieces of DNA, about 1000 base pairs long, were present. These short sections, which came to be known as **Okazaki fragments**, confirmed the validity of the discontinuous replication hypothesis.

RNA Primers

The discovery of Okazaki fragments raised another urgent question, however. In an in vitro system, polymerases are inactive if the DNA templates that are provided are completely double-stranded or completely single-stranded (**Figure 12.8**). The enzyme begins to work only when a $3'$ end *and* a single-stranded template are available. The template dictates which nucleotide should be added next, and the existing strand provides a free $3'$ hydroxyl (–OH) group that combines with the incoming dNTP. As a result, the existing strand is called a **primer**. But at the replication fork, only single-stranded DNA is available. How does Pol III begin to work?

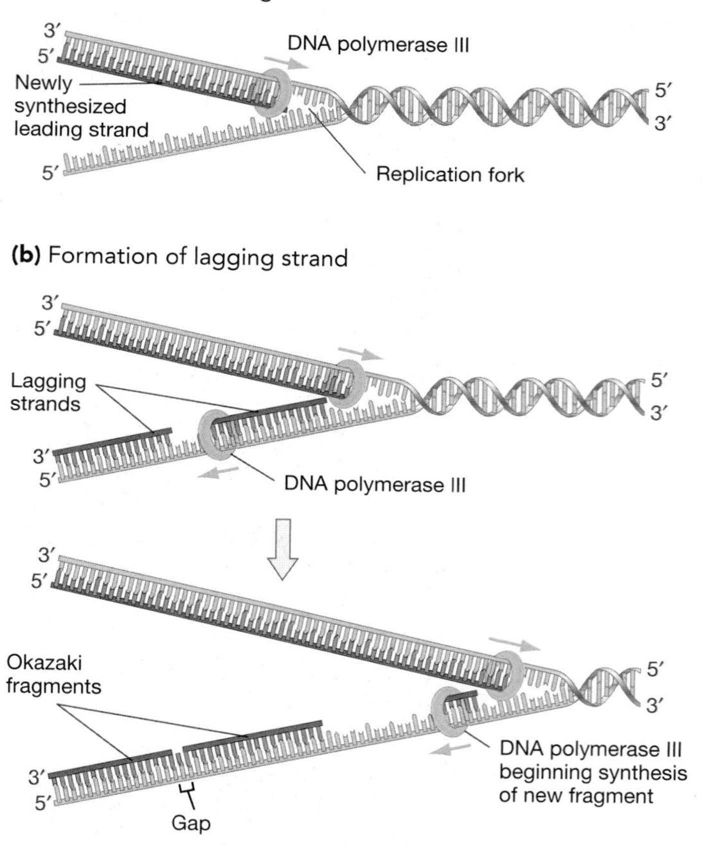

FIGURE 12.7 The Replication Fork
(a) DNA polymerase III synthesizes a leading strand of DNA in the $5' \rightarrow 3'$ direction at the structure called the replication fork. **(b)** It was hypothesized that DNA Pol III synthesized fragments of the lagging strand of DNA and that these fragments would later be linked into a continuous whole. **QUESTION** Why couldn't synthesis of the lagging strand by DNA Pol III be continuous, as it is in the leading strand?

Hurwitz and Furth's work with RNA polymerase, reviewed in Chapter 11, provided a hint. This enzyme is capable of synthesizing RNA from a completely single-stranded template. Unlike DNA polymerase, it does not need a primer with a 3′ end available. Based on this observation, Kornberg, Okazaki, and others proposed that an RNA polymerase might synthesize short RNA sequences to act as a primer, which are complementary to the single-stranded DNA on the lagging strand.

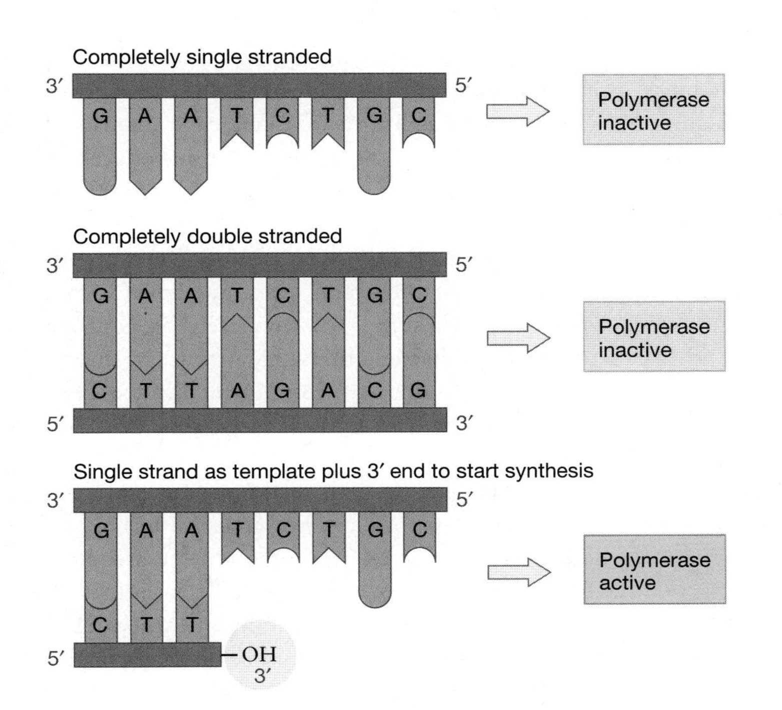

FIGURE 12.8 DNA Polymerase Requires a Primer and a Template When DNA templates are completely single-stranded or double-stranded, no synthesis occurs. The DNA polymerases work only when there is a single-stranded template *and* a 3′ hydroxyl group available, as shown here.

Laboratories from Japan, Europe, and the United States began working to test the RNA primer hypothesis. The clinching evidence came when Lee Rowen and Arthur Kornberg isolated an enzyme called primase from *E. coli*. Rowen and Kornberg offered several lines of evidence that primase synthesized the proposed RNA primers on the lagging strand. For example, the enzyme restored normal activity in assays from a strain of *E. coli* that had a temperature-sensitive defect in DNA synthesis. This experiment showed that the enzyme was required for DNA synthesis. They also added primase to an in vitro DNA synthesis reaction along with Pol III and radioactive ribonucleotides, and observed the production of short strands of radioactive RNA along with DNA. Later work showed that the leading strand has an RNA primer synthesized by RNA polymerase. With these discoveries, a coherent model for DNA synthesis began to emerge.

The Synthesis Machinery

The diagram of the replication fork in **Figure 12.9** introduces several new aspects of DNA synthesis and summarizes many of the results reviewed thus far. In reviewing this illustration, it is important to recognize that a specific enzyme catalyzes each of the numbered steps. Most of these enzymes were discovered when researchers were able to trace the defect in various temperature-sensitive replication mutants of *E. coli*. In this species at least 10 different proteins are involved in the replication machinery. Because several of the enzymes are constructed from a series of distinct polypeptides, at least 20 different genes are required to code for the complete apparatus in bacteria.

Let's do a quick walk-through of the process. The replication fork opens at the point indicated by the number 1 in Figure

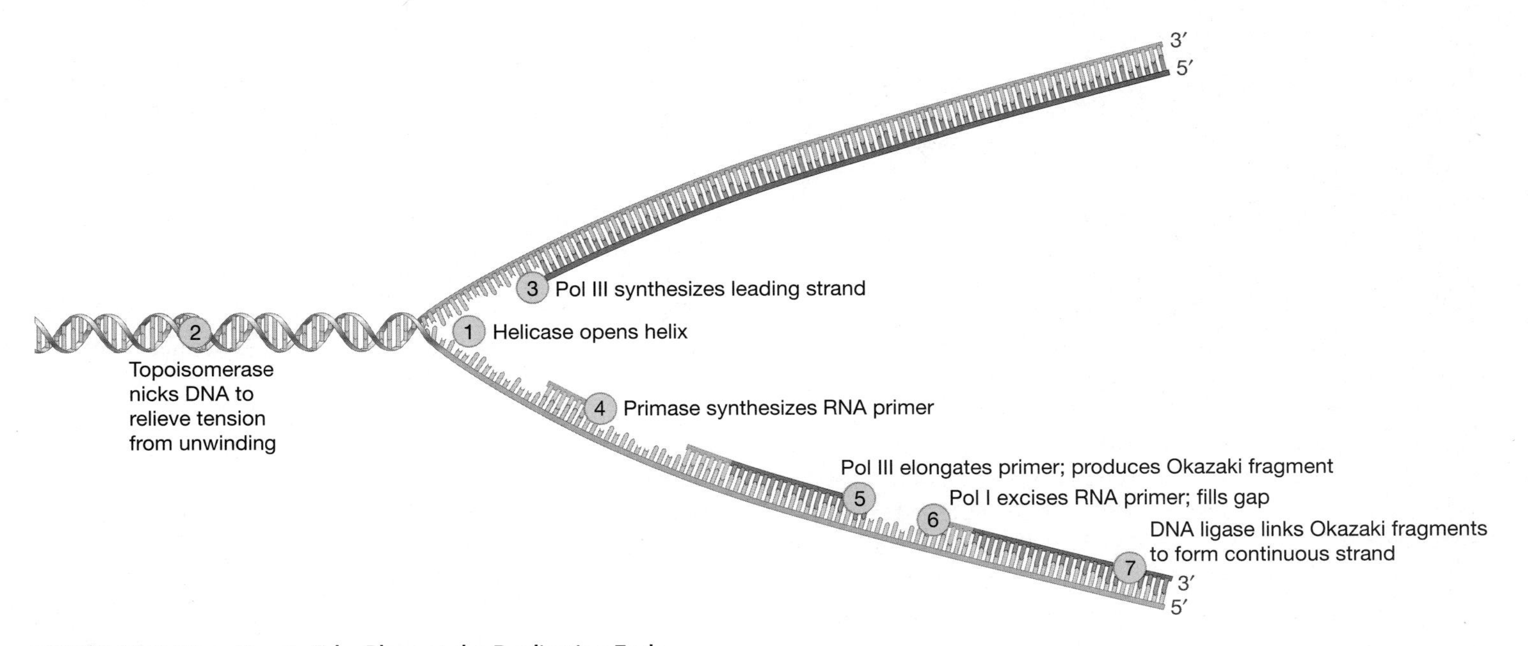

FIGURE 12.9 Many Events Take Place at the Replication Fork
Each of the steps diagrammed here requires the involvement of at least one enzyme.

12.9. Enzymes called helicases are responsible for breaking the hydrogen bonds between nucleotides; proteins called single-strand binding proteins stabilize the single strands that result by attaching to them. This "unzipping" step creates tension farther down the helix, however, just as pulling apart the strands of a rope will cause it to coil on itself and kink. This tension is relieved by enzymes called topoisomerases, which nick the DNA downstream to undo twists and knots (see step 2).

The point marked 3 in the figure is where DNA polymerase III adds nucleotides to the leading strand. Primase adds a short run of ribonucleotides to the other template strand at the point marked 4; Pol III extends this primer as shown by point 5. DNA polymerase I—the enzyme discovered by Kornberg and co-workers—then removes the ribonucleotides and fills in deoxyribonucleotides (see point 6). Finally, gaps between Okazaki fragments like those shown at point 7 are closed by an enzyme called ligase.

Given the number of enzymes and proteins involved in the synthesis machinery, it is not surprising that it took 25 years for biologists to assemble the results represented in Figure 12.9. The situation in eukaryotes is even more complex; most eukaryotic cells contain at least seven discrete DNA polymerases. Eukaryotic chromosomes also have multiple origins of synthesis instead of the single point of origin observed in bacteria. Synthesis is bidirectional in both types of cells, however (**Figure 12.10**).

The model of replication summarized in Figure 12.9 finally emerged in the early 1980s. Since then, biologists have concentrated on other questions about DNA synthesis. Can the reaction be manipulated to reveal the sequence of nucleotides in a particular section of DNA? What happens when errors occur during DNA replication or when DNA is damaged? These are the questions addressed in the remainder of the chapter.

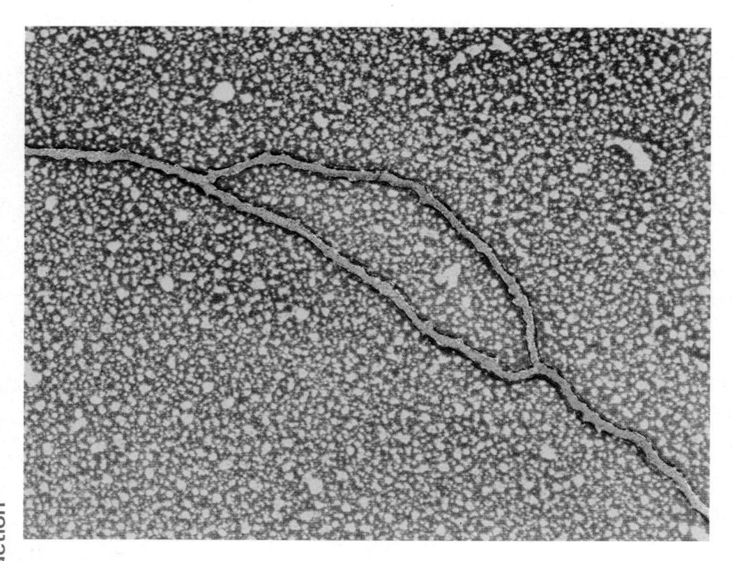

FIGURE 12.10 Bidirectional Replication
This photograph of DNA from a human cell illustrates an important similarity between bacteria and eukaryotes. In both types of cells, DNA synthesis proceeds in both directions from a point of origin. In most bacteria, chromosomes are circular and there is a single point of origin during DNA replication. Most eukaryotes, in contrast, have linear chromosomes; each contains several to many points of origin for DNA synthesis. EXERCISE Circle the two replication forks in the photograph. Add arrows showing the direction of DNA synthesis.

ACTIVITY 12.2 olymerase Chain Reaction

12.3 Analyzing DNA Sequences in the Laboratory

Knowing the DNA sequence of a gene is often critical to understanding its history and its role in the life of a cell. Once a gene's sequence is known, it is straightforward to infer the amino acid sequence of its product using the genetic code introduced in Chapter 11. The sequence can also be compared and contrasted to other alleles and other genes, both within and between species. These comparisons are often interesting. For example, the DNA polymerase genes of bacteria, yeast, and humans contain sections that are identical in DNA sequence. This is strong evidence that you and the bacteria that inhabit your gut share a common ancestor in the distant past. Experimental studies have shown that the DNA polymerization reaction is similar in bacteria and humans.

Given that sequence data are valuable, how can biologists obtain them? This section introduces two common techniques used to analyze DNA in the laboratory: the polymerase chain reaction (PCR) and dideoxy sequencing. Both are variations on the in vitro DNA synthesis reactions introduced earlier in the chapter. PCR is a way to create many copies of a particular stretch of DNA; dideoxy sequencing allows researchers to determine the exact sequence of bases. Along with the development of the light microscope and the electron microscope, PCR and dideoxy sequencing rank among the greatest of all technological advances in the history of biological science.

The Polymerase Chain Reaction

Researchers have to obtain many copies of a gene in order to analyze its molecular structure. The **polymerase chain reaction**, or **PCR**, is a technique for generating these copies. It is an in vitro DNA synthesis reaction in which a specific section of DNA is copied over and over to amplify the number of copies of that sequence.

Figure 12.11 illustrates the reaction protocol, which was originally designed by Kary Mullis. As Figure 12.11a shows, the reaction mix includes two short sequences of single-stranded DNA that act as primers. The primers used in PCR bracket the region that will be copied. One primer is complementary to a sequence on one strand of the target DNA; the other primer is complementary to a sequence on the other strand. This is a key feature of PCR. To perform the reaction, a researcher must know the sequences on either side of the region of interest so that the correct primers can be synthesized.

To begin the reaction, Mullis added the primers to a solution containing an abundant supply of the four dNTPs and copies of the template DNA (Figure 12.11b, step 1). Then he heated the

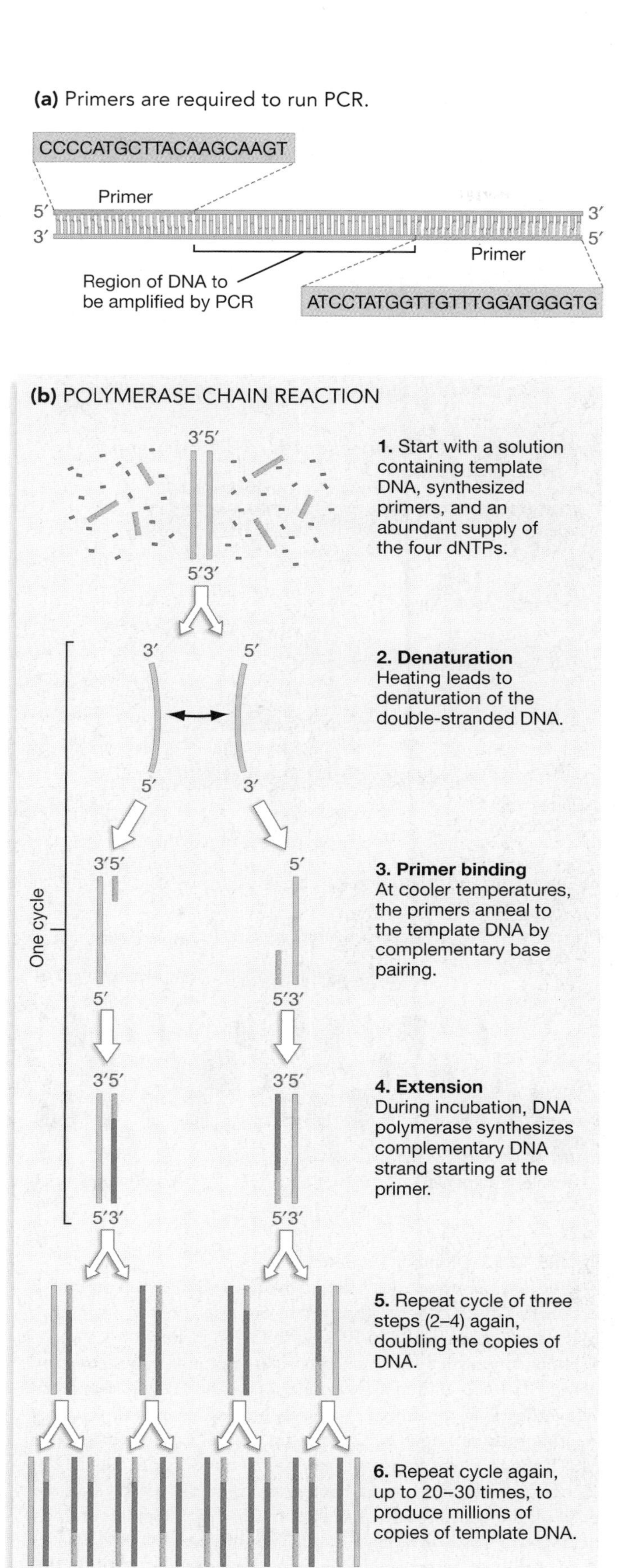

FIGURE 12.11 The Polymerase Chain Reaction (PCR)
(a) The orange sequences indicate a set of single-stranded primers, which bracket the region of DNA to be amplified. (These are the actual sequences used in a polymerase chain reaction experiment explained later in the chapter, where the sample DNA was from a Neanderthal fossil.) **(b)** Each PCR cycle (heating, primer binding, and extension) results in a doubling of the number of copies of the sequence between the primers. **EXERCISE** In part (a), label the 5′ and 3′ end of each primer. In part (b), draw in the sequence of events during the third cycle.

mixture to 95°C so that the double-stranded template DNA would denature (step 2). As the mixture then cooled to 60°C, the primers bonded to the complementary portions of the single-stranded template DNA (step 3). The next step (step 4) was to add DNA polymerase I and incubate the mixture at 37°C, so that Pol I could synthesize the complementary DNA strand starting at the primer.

These three steps—denaturation, primer binding, and extension—constitute a single PCR cycle. If one copy of the template sequence existed in the sample originally, then there would be two copies after the first cycle. Mullis repeated the cycle again and again. The amount of target sequence doubled each time because each newly synthesized segment of DNA served as a template. By performing 20–30 cycles, Mullis obtained huge numbers of copies of the target sequence.

The PCR protocol that is currently used in laboratories was inspired by the discovery of a bacterium called *Thermus aquaticus* in a hot spring inside Yellowstone National Park, Wyoming. This organism contains a heat-stable form of DNA polymerase known as *Taq* polymerase. This enzyme continues to function normally even when it is heated to 95°C. As a result, researchers no longer have to add fresh DNA polymerase at each cycle.

To understand why PCR is so valuable, consider a recent study by Svante Pääbo and colleagues. These biologists wanted to analyze DNA recovered from the 30,000-year-old bones of a fossilized human from the species *Homo neandertalensis*. Pääbo's goal was to determine the sequence of bases in the ancient DNA, compare it to DNA from modern humans (*Homo sapiens*), and determine the closeness of the relationship between the two species. If the sequences found in Neanderthals were similar to those found in modern humans, it would support the hypothesis that the two species interbred during the time that they coexisted in Europe 30,000 years ago. This would mean that some of our ancestors were Neanderthals.

The Neanderthal bone was so old, however, that most of the DNA in it had degraded. The biologists could recover only a minute amount of intact DNA. To have any hope of analyzing the sample, they needed to generate many copies. To accomplish this they turned to PCR.

Which Neanderthal gene should they try to amplify? Pääbo's group reasoned that because each human cell contains many copies of a locus called the control region of mitochondrial DNA, there might be an intact sequence from this locus somewhere in the Neanderthal DNA sample. To find out they added two short, single-stranded DNA primers to the sample. These primers had a sequence identical to sequences on either side of the control region in *Homo sapiens*. After adding the primers to the reaction mix, the researchers added the four types of dNTPs and *Taq* polymerase, and used a PCR machine to automatically change the temperature of the mix through a large number of cycles. The experiment was successful. Pääbo's group was able to synthesize many copies of the region—enough to perform a dideoxy sequencing reaction. Once researchers have many copies of a specific stretch of DNA, how can they determine the exact sequence of bases?

Dideoxy Sequencing

Frederick Sanger developed dideoxy sequencing as a clever variation on the basic in vitro DNA synthesis reaction. But saying clever may be an understatement. Sanger had to link three important insights to make his sequencing strategy work.

Sanger chose the name "dideoxy" because he was using monomers called dideoxynucleoside triphosphates, or ddNTPs. These molecules are identical to the dNTPs found in DNA, except that they lack a hydroxyl group at their 3′ carbon (**Figure 12.12a**). Sanger realized that if a ddNTP were added to a growing DNA strand, it would terminate synthesis. Why? A ddNTP has no hydroxyl group available on its 3′ carbon to link to the 5′ carbon on an incoming dNTP monomer. As a result, DNA polymerization stops once a ddNTP is added.

Sanger linked this property of ddNTPs to a second fundamental insight. Suppose, he reasoned, that a researcher attached a radioactive primer to a template DNA, like the control region in Neanderthals. If this labeled template were incubated with DNA polymerase, the four dNTPs, and ddGTP, the resulting daughter strands would comprise a limited set of lengths. To understand why, consider that the synthesis of each daughter molecule would start at the same point—the primer—but end whenever a ddGTP happened to be incorporated in the growing strand opposite a C in the template. The addition of a ddGTP stops further elongation. The collection of newly synthesized strands would vary in length as a result, with each specifying the distance from the primer to the successive C's in the template. Analogous reactions done using ddTTP, ddATP, and ddCTP give the distances between successive A's, T's, and G's, respectively. Finally, Sanger realized that when the fragments produced by the four reactions are lined up by size, they reveal the sequence of bases in the template DNA. For example, when Pääbo and co-workers had run all four reactions, they had a collection of fragments identifying where every A, T, C, and G occurred in the control region sequence. To line these fragments up in order of size, they separated them by gel electrophoresis (see Box 3.1). As **Figure 12.12c** shows, they could then read the sequence directly from the resulting autoradiograph.

Their result? The control region sequences found in Neanderthal DNA are highly distinct from the control region se-

(a) ddNTPs terminate DNA synthesis.

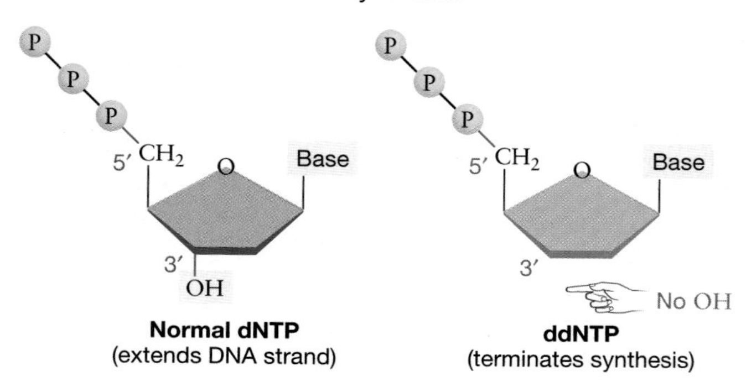

(b) Using ddNTPs, daughter strands of different length can be produced.

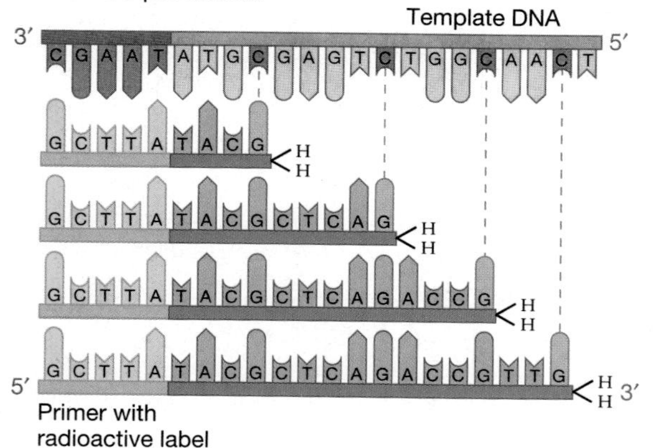

(c) Different-length strands can be lined up by size to determine DNA sequence.

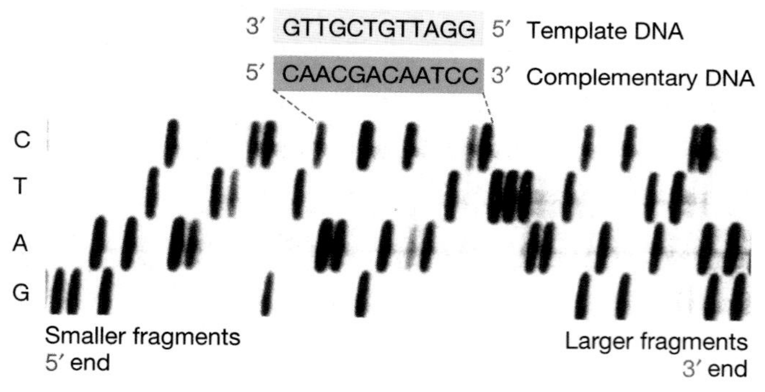

FIGURE 12.12 Dideoxy Sequencing
(a) Dideoxyribonucleotides have no hydroxyl (OH) group on their 3′ carbon, so DNA synthesis stops if they are inserted into a strand. **(b)** If a small amount of ddGTP is added to an in vitro DNA synthesis reaction along with a short, single-stranded primer that is complementary to the 3′ end of a target sequence, then DNA polymerase will synthesize fragments of varying length as shown here. **(c)** After separate reactions have occurred using ddGTP, ddTTP, ddATP, and ddCTP, the resulting fragments can be run out on a gel. The DNA sequence of the strand complementary to the target can be read directly off the gel as shown. (Recall from Chapter 3 that larger fragments move slowly in a gel and remain at the top while the smaller fragments move faster and move toward the bottom.) **EXERCISE** Start at the far left end of the gel pictured in part (c) and write down the entire sequence shown. Hint: the sequence starts with GG.

quences observed in living humans. Thus, there is no evidence that we inherited some of our DNA from a Neanderthal ancestor. Based on these data, Pääbo and co-workers doubt the hypothesis that Neanderthals and members of our own species interbred. A more recent study, of a Neanderthal fossil from the Caucasus region of Russia, came to the same conclusion.

12.4 The Molecular Basis of Mutation

DNA polymerases are fast, but they are accurate as well. In organisms ranging from *E. coli* to animals, the error rate during DNA replication averages less than one mistake per *billion* nucleotides. How can these enzymes be so precise? The answer is that DNA polymerase is selective, and it can proofread. But if mistakes remain after synthesis is complete, repair enzymes correct mismatched base pairs. Let's take a closer look.

Proofreading by DNA Polymerase

DNA polymerases are selective because the correct base pairing (A-T and G-C) is energetically the most favorable of all the pairing possibilities. As the enzyme marches along a DNA template, hydrogen bonding occurs between incoming nucleotides and the nucleotides on the template strand. Due to the precise nature of these interactions, the enzyme inserts the incorrect nucleotide just once in every thousand bases (**Figure 12.13a**).

What happens when a mismatch does occur? An answer to this question emerged when researchers found *pol III* mutants in

(a) Mismatched bases

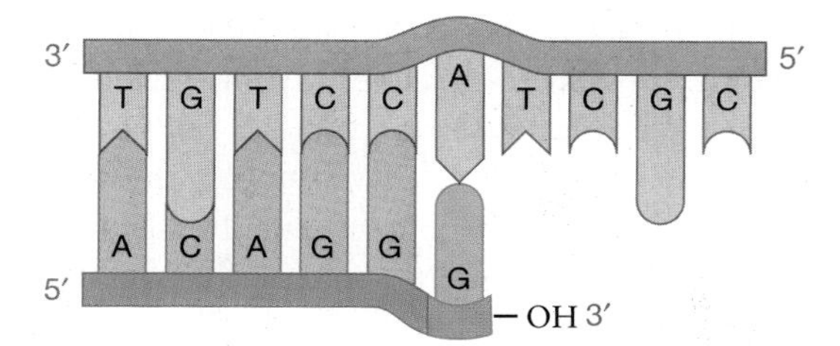

(b) Polymerase III can repair mismatches.

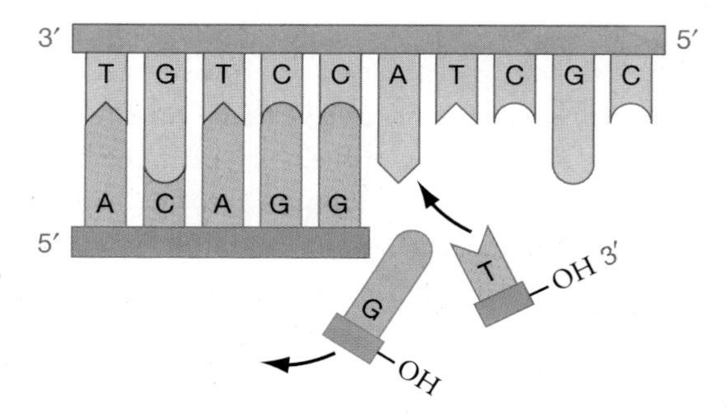

FIGURE 12.13 DNA Polymerase III Can Proofread
(a) Pol III adds an incorrect base to a growing strand of DNA about once in every thousand bases added. The result is a mismatch like the one shown here. **(b)** DNA Pol III can act as a 3′ → 5′ exonuclease, meaning that it can remove bases in that direction.

E. coli with error rates that were 100 times greater than normal. The defect was localized to a particular portion of the enzyme, called the ϵ (epsilon) subunit. Further analyses showed that this subunit acts as an **exonuclease**—meaning an enzyme that peels nucleotides off of DNA (**Figure 12.13b**). The Pol III exonuclease activity turned out to be directional. The enzyme removes nucleotides only in the 3′ → 5′ direction. These results led to the conclusion that Pol III can proofread. In other words, if the wrong base is added during DNA synthesis, the enzyme pauses, removes the mismatched base, and then proceeds with synthesis.

This proofreading activity reduces Pol III's error rate to about 1×10^{-7} (one mistake per 10 million bases). Is this accurate enough? The answer is no. If DNA replication were this sloppy, then at least 600 mistakes would occur every time a human cell replicated its DNA. Given that about a million-billion cell divisions take place in the course of a human's lifetime, an error rate this high would be problematic. Once replication is complete, other repair mechanisms come into play.

Mismatch Repair

When DNA polymerase leaves a mismatched pair behind in the DNA sequence, a battery of enzymes springs into action. The proteins responsible for mismatch repair were discovered in the same way that the proofreading capability of DNA Pol III was uncovered—by analyzing *E. coli* mutants. In this case, the mutants had normal Pol III but abnormally high mutation rates. The first mutant locus that caused a deficiency in mismatch repair was identified in the late 1960s and was called *mutS*. (The *mut* is short for "mutator.") By the late 1980s researchers had identified a total of 10 proteins involved in the identification and repair of base-pair mismatches in *E. coli*. Once these mutants were identified, researchers began exploring how their protein products functioned by attempting to reproduce the mismatch repair system in vitro.

As an example of this approach, consider experiments that Paul Modrich and co-workers conducted with small, circular, double-stranded DNA molecules they constructed. As **Figure 12.14a** (page 244) indicates, the experimental DNAs had two interesting features: a single base-pair mismatch and an adenosine base that had a methyl group ($-CH_3$) attached. The researchers added the methyl group in an attempt to simulate an event that occurs in *E. coli*. Several minutes after Pol III completes the synthesis of the bacterial genome, another enzyme adds methyl groups to several of the adenosines on the new strand. But until this reaction takes place, there is a marked asymmetry in the parent and daughter strand. The original template DNA is methylated, but the newly synthesized DNA is unmethylated.

This transient, half-methylation of DNA inspired a hypothesis to explain a fundamental mystery about mismatch repair: How do the enzymes know which of the mismatched bases is right and which is wrong? Clearly, the wrong base is found on the newly synthesized strand. Can the repair enzymes recognize which strand is methylated and use it as the template to fix the mistake?

To test this hypothesis, Modrich and co-workers compared how mismatch repair proceeded on experimental loops of DNA that either contained or lacked a methyl group. As predicted, repair proceeded only when a methyl group was present. Further, the repair proteins always removed the mismatched base on the unmethylated strand.

Based on results like these, researchers have been able to piece together the model of mismatch repair shown in **Figure 12.14b**. Recently, research on this repair pathway has acquired some urgency. As the essay at the end of the chapter explains, cancer can develop when the genes involved in this repair pathway are disabled by mutation.

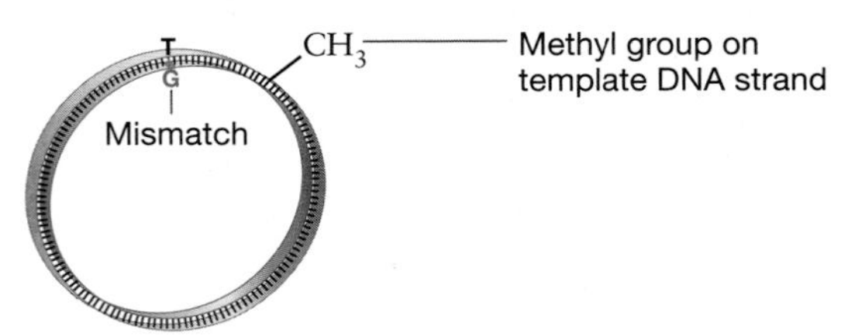

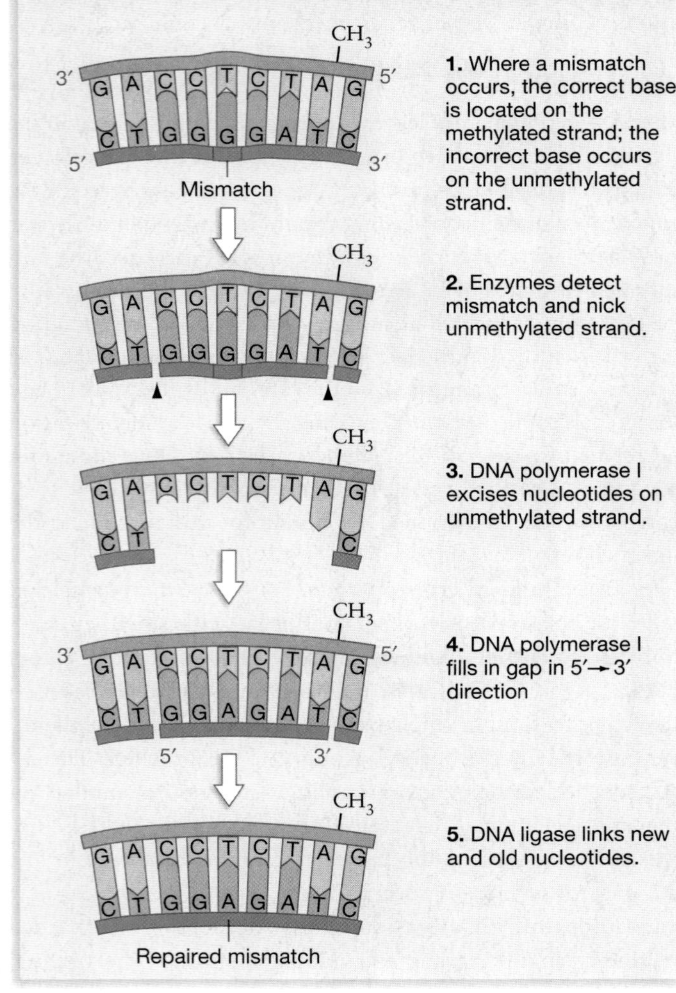

FIGURE 12.14 The Role of Methylation in Mismatch Repair
(a) The $-CH_3$ on this loop of double-stranded DNA is a methyl group. In bacteria, an enzyme attaches a methyl group to adenosine bases a few minutes after replication is complete. **(b)** This model is called methylation-directed mismatched base repair. **QUESTION** Why is it valid to say that the methylated parental strand serves as a template for repair?

Types of Mutation

What happens if proofreading by DNA polymerase and the mismatch repair system fail to correct a mismatched base? As **Figure 12.15** shows, uncorrected mismatch errors lead to a change in the sequence of bases in DNA. A single base change like this is called a **point mutation.**

What are the consequences of a point mutation? To understand the answer to this question, consider the first point mutation ever described. This mutation occurs in the human gene for a protein called hemoglobin. Hemoglobin is abundant in red blood cells and carries oxygen to our tissues.

Figure 12.16a shows the normal DNA sequence of the gene for hemoglobin along with the mutant form found in individuals who suffer from sickle-cell disease. Note that the mutant form has an adenine in place of a thymine at position 2 in the sixth codon specified by this gene. A glance back at the genetic code in Chapter 11 confirms why the change is significant: The mutant codon specifies valine instead of glutamic acid in the amino acid chain of hemoglobin (Figure 12.16a). The change causes the protein to crystallize when oxygen levels in the blood are low. When hemoglobin crystallizes, it causes red blood cells to become sickle-shaped (**Figure 12.16b**). When the misshapen cells get caught in blood vessels, intense pain and anemia result.

Clearly, point mutations can be deleterious. But point mutations may also be beneficial. For example, the thymine-to-adenine mutation in hemoglobin is actually advantageous in certain environments. Because the malaria parasite infects red blood cells, and because sickled cells are fragile and tend to have short life spans, individuals who have one copy of the mutant gene have lower parasite loads and tend to be much healthier than people with none.

In addition to having a negative or positive effect on the organism, point mutations may have virtually no effect at all. To understand why, suppose that at the next site in the hemoglobin locus (position 3 in the same codon), a cytosine had been substituted for the thymine. This point mutation would have no consequence. Why? The CTT and CTC codons both specify glutamic acid. In this case, the amino acid sequence of the gene product does not change even though the DNA sequence is altered due to mutation. This type of alteration in the base sequence is called a **silent mutation.**

The take-home message from this analysis of hemoglobin, sickle-cell disease, and malaria is simple. It is possible for point mutations to be deleterious, beneficial, or neutral. It is

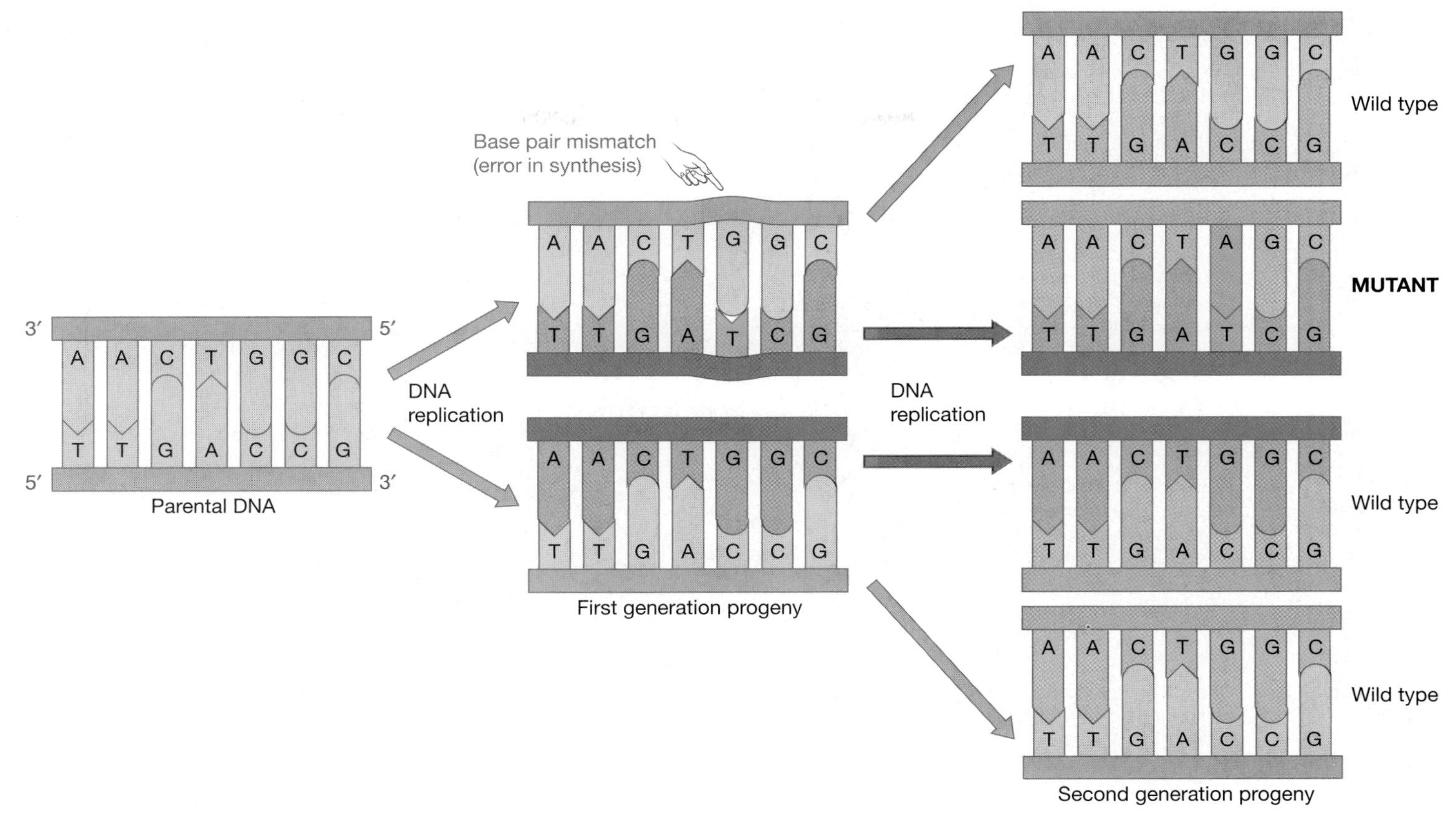

FIGURE 12.15 Unrepaired Mismatches Can Lead to Mutations
QUESTION Why is it logical that the type of mutation illustrated here is termed a point mutation?

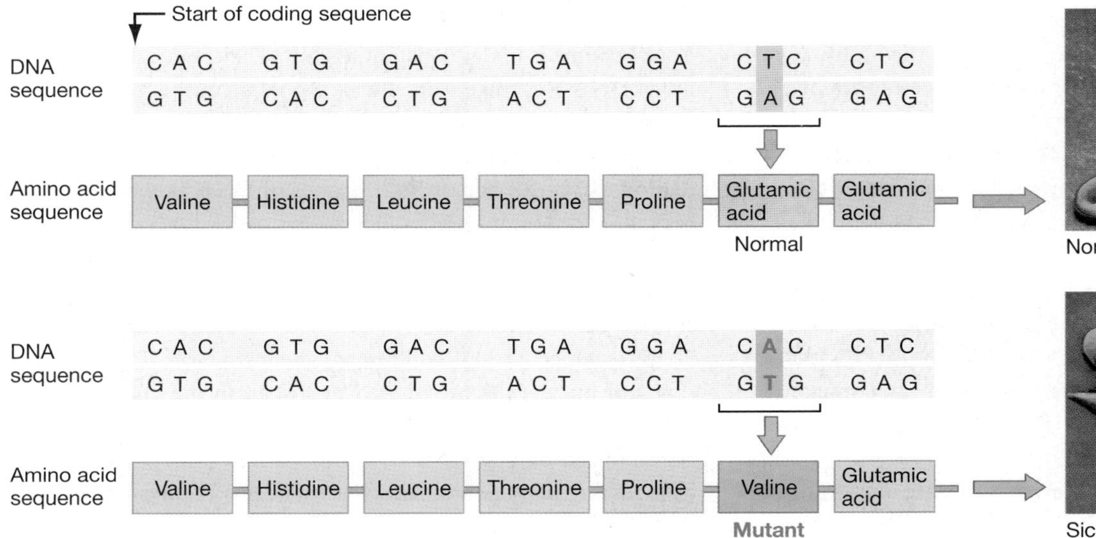

(b) Phenotype

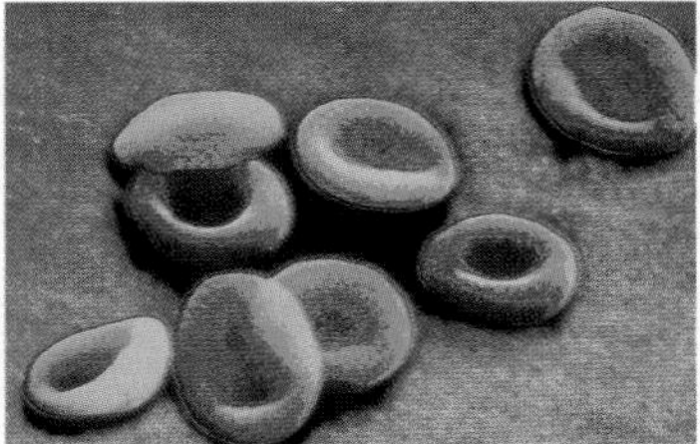

Normal red blood cells

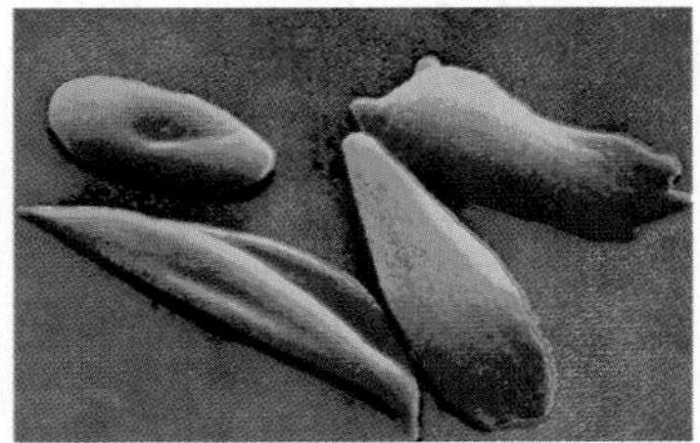

Sickled red blood cells

FIGURE 12.16 Sickle-Cell Disease Results from a Point Mutation in the β-Globin Gene
(a) These normal and mutant types of β-globin genes differ by a single nucleotide. The sequences shown here are just a small portion of the gene. **(b)** The mutation in part (a) changes the primary sequence of the globin gene product. The change in amino acid sequence causes hemoglobin molecules to crystallize when oxygen levels in the blood are low. As a result, red blood cells sickle and get stuck in small blood vessels.

Mutation type	Definition	Example	Consequence
Insertion	Addition of any number of nucleotides due to an error in DNA synthesis	Original sequence: ATA ACC GAT CAT GTA (G inserted) Mutant sequence: ATA ACC GAT CGA TGT A	Addition of 1 or 2 bases disrupts reading frame. Usually results in a dysfunctional gene product.
Deletion	Removal of any number of nucleotides due to an error in DNA synthesis	Original sequence: ATA ACC GAT CAT GTA (A removed) Mutant sequence: ATA ACC GTC ATG TA	Deletion of 1 or 2 bases disrupts reading frame. Usually results in a dysfunctional gene product.
Gene duplication	Addition of a small chromosome segment due to an error during crossing over at meiosis I	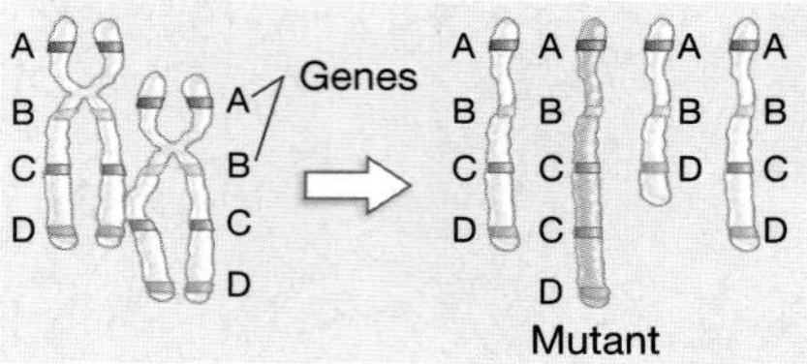 A B C D; Genes A B C D; Mutant: A B C C D	Produces an extra copy of one or more genes. If point mutations occur in extra DNA, it can produce a new product.
Chromosome inversion	Change in a chromosome segment when DNA breaks in two places, flips, and rejoins	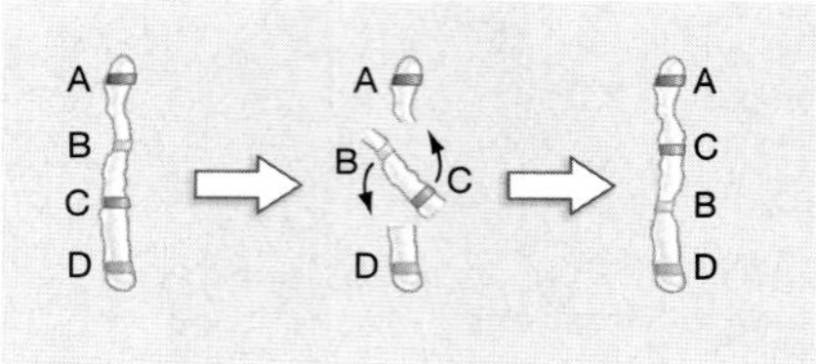 A B C D → A B C D → A C B D	Changes gene order along chromosome. Other types of chromosome breaks can lead to deletion or addition of chromosome segments.

FIGURE 12.17 A Mutation Is Any Change in an Organism's DNA
Many types of mutations occur. In addition to point mutations and the types listed here, Chapter 9 introduced polyploidy, trisomy, and other chromosomal changes that result in mutation.

important to recognize, however, that all mutations are random. As a result, many are deleterious. This should make sense. We would not expect a random change in the highly organized and efficient molecular machinery of a cell to be beneficial—any more than we would expect a random change in circuitry to improve a computer's processing performance.

Mutation is a much broader phenomenon than single base changes in a DNA sequence, however. A **mutation** is defined as *any* change in an organism's genome. As **Figure 12.17** shows, mutations result from the insertion or deletion of DNA sequences, changes in chromosome number or content due to errors in meiosis (see Chapter 9), and chromosome breaks that lead to segments of the genome being flipped—a phenomenon known as a chromosome inversion. Several of these types of mutation occur independently of the replication process.

The mutational mechanisms reviewed here are important. At the level of populations, they furnish the heritable variation that Mendel and Morgan analyzed and that makes evolution possible. At the level of individuals, they can cause disease and death. Currently, the mutations attracting the most attention from researchers are those that affect the genes responsible for repairing damaged DNA.

12.5 Repairing Damaged DNA

Your genome is under constant assault. The chemical bonds that hold DNA together can break spontaneously, sending nucleotides flying or snapping DNA's sugar-phosphate backbone in two. Even small amounts of x-ray, gamma ray, and ultraviolet radiation can break one or both strands of DNA as they bombard cells. In addition, molecules like the hydroxyl (OH) radicals produced during normal aerobic metabolism, the aflatoxin B1 found in moldy peanuts and corn, and the benzo[α]pyrene present in cigarette smoke actively attack and degrade nucleotides.

Even under normal conditions, chemists estimate that thousands of nucleotides are altered or lost from the DNA in every human cell every day. If this spontaneous, chemical, and radiation-induced damage is not repaired it may lead to permanent changes in the genome. The vast majority of these mutations would be harmful.

To cope with these insults, cells have a sophisticated battery of repair enzymes. These maintenance crews patrol DNA, identify damaged bases or sections of the sequence, and repair them. Collectively, these coordinated groups of molecules are called the excision repair systems.

DNA Nucleotide Excision Repair: An Overview

Excision repair systems acquired their name because they share a common mechanism: They excise a stretch of single-stranded DNA around a damaged site, then resynthesize a new strand based on the information in the intact, complementary strand. The mismatch repair system introduced in section 12.4, which corrects mistakes made during DNA replication, is an example of an excision repair system.

Many of these repair systems first came to light when researchers isolated strains of *E. coli* or yeast with abnormally high mutation rates and were later able to trace the cause of the defect. In particular, early work on how *E. coli* repairs damage induced by exposure to ultraviolet (UV) radiation inspired studies on DNA repair in an unusual model organism. In contrast to *E. coli* and the other experimental subjects introduced in this unit, this model organism reproduces slowly and is extremely expensive to maintain. The organism? Human beings.

Xeroderma pigmentosum: A Case Study

Xeroderma pigmentosum (XP) is a rare autosomal recessive disease in humans. Individuals with this condition are extremely sensitive to UV light. Their skin develops lesions after even slight exposure to sunlight. In normal individuals these kinds of lesions develop only after extensive exposure to ultraviolet light, x-rays, or other forms of high-energy radiation.

In 1968 James Cleaver proposed a connection between XP and DNA repair systems. He knew that in *E. coli*, mutations in the genes for DNA repair proteins result in increased sensitivity to radiation. Cleaver's hypothesis was that people with XP are extremely sensitive to sunlight because they are unable to repair the damage that occurs when DNA bases absorb UV light.

What is the nature of this damage? The most common defect caused by UV radiation is the formation of a covalent bond between adjacent pyrimidine bases. The thymine dimer illustrated in **Figure 12.18a** is an example. This defect creates a kink in the secondary structure of DNA, which blocks transcription and

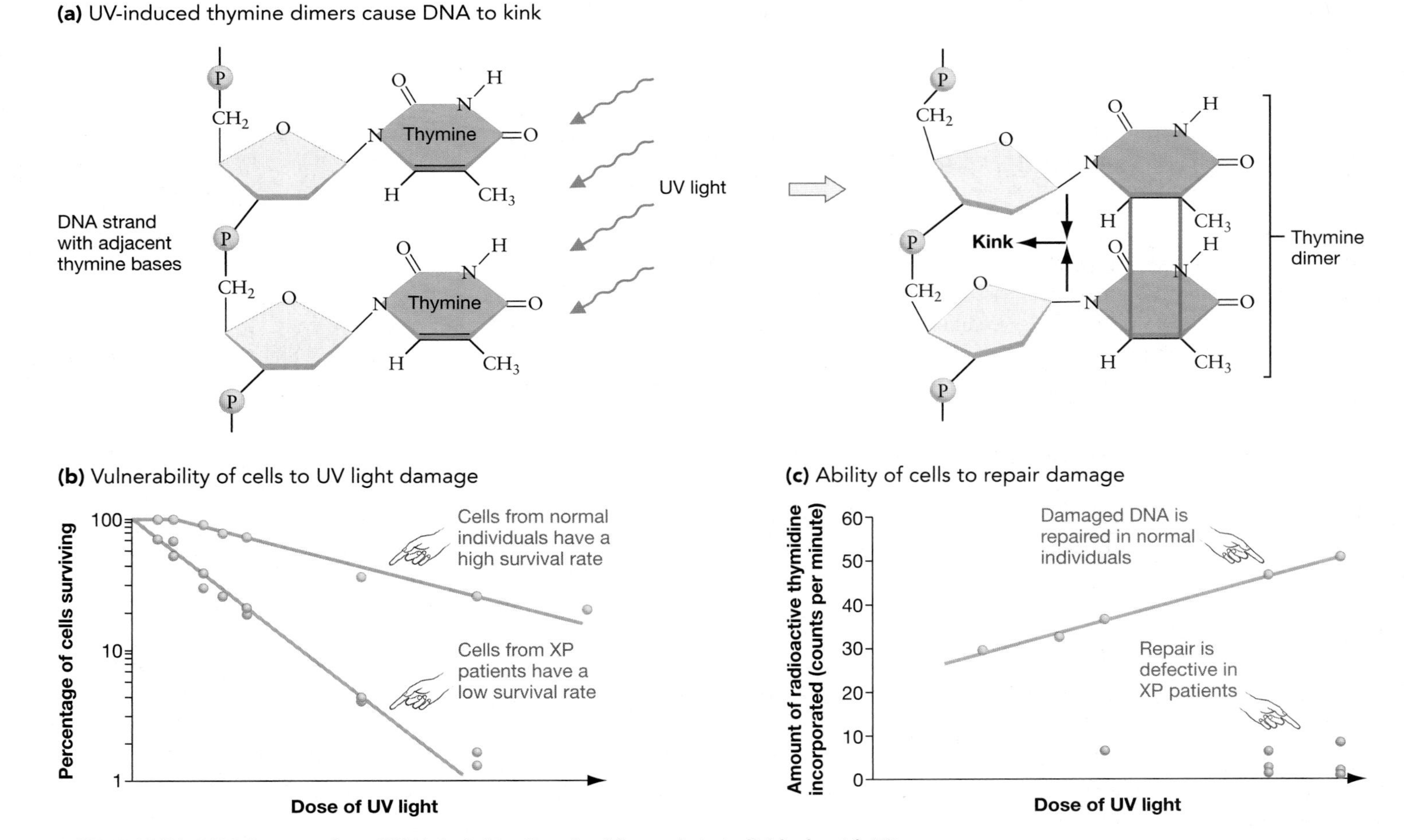

FIGURE 12.18 DNA Damage from UV Light Is Not Repaired Properly in Individuals with XP
(a) When UV light strikes a section of DNA with adjacent thymines, the energy can break bonds within each base and result in the formation of bonds *between* them. The thymine dimer that is produced causes a kink in DNA. **(b)** When cell cultures from normal individuals and from XP patients are irradiated with various doses of UV light, the percentage of cells that survive is strikingly different. **(c)** When cell cultures from normal individuals and from XP patients are irradiated with various doses of UV light and then fed radioactive thymidine, only normal individuals incorporate the labeled base.

stalls the movement of the replication fork during DNA replication. If the error is not repaired, the cell may die. Even if the cell lives, the defect may lead to a mutation because DNA polymerase typically substitutes just one base opposite the two disturbed thymines during the next cell division.

Cleaver's hypothesized connection between DNA damage, faulty error repair, and XP turned out to be correct. Much of the work that he and other investigators did relied on the use of **cell cultures.** A cell culture is a collection of cells derived from a specific type of plant or animal tissue that grows in a liquid suspension or on the surface of a dish on solid medium. Human skin cells, for example, can be sampled from individuals who suffer from XP and from people who have a normal phenotype. When these populations of cells are grown in culture and then exposed to increasing amounts of ultraviolet radiation, a striking difference emerges: Cells from XP individuals die off much more rapidly (**Figure 12.18b**).

The connection to repair systems was confirmed when Cleaver exposed samples of cells from normal and XP individuals to various amounts of UV light, then fed the cells radioactive thymidine. The purpose of the radioactive nucleotide was to label DNA synthesized during the repair period. If repair is defective in the XP individuals, then their cells should take up little of the radioactive base. Cells from normal individuals, in contrast, should incorporate large amounts of labeled thymidine. As the graph in **Figure 12.18c** shows, this is exactly what happens. These data are consistent with the hypothesis that repair synthesis is virtually nonexistent in XP individuals.

More recently, genetic analyses of XP patients have shown that the condition can result from mutations in any one of seven different genes. This result is not surprising in light of the large number of proteins involved in excision repair in bacteria.

Finally, the essay at the end of this chapter points out that defects in the genes responsible for excision repair are frequently associated with cancer. Individuals with xeroderma pigmentosum, for example, are 1000 to 2000 times more likely to get skin cancer than individuals with normal excision repair systems. To explain this pattern, biologists suggest that if mutations in the genes controlling the cell cycle (see Chapter 8) go unrepaired, the damaged proteins that result may allow uncontrolled growth and tumor formation. Stated another way, if the overall mutation rate in a cell is elevated because of defects in DNA repair genes, then the mutations that trigger cancer become more likely.

Essay The Genetic Basis of Cancer

Cancer geneticists have known for decades that families can have predispositions to certain types of cancers, including breast cancer and the skin cancers associated with XP. A common type of colon cancer called hereditary non-polyposis colorectal cancer (HNPCC) also runs in families. Affected individuals frequently develop tumors of the colon, ovary, and other organs before they are 50 years old. A major breakthrough in understanding HNPCC occurred in the early 1990s, when techniques introduced in Chapter 16 allowed researchers to determine that a gene associated with susceptibility to HNPCC mapped to a specific region of chromosome 2.

. . . cells from these patients had mutation rates 100 times that of normal cells.

Meanwhile, a completely different group of investigators was trying to determine if humans have mismatch repair genes similar to those found in *E. coli*. The pace of this work accelerated when the DNA sequence of the *mutS* gene was determined, and when a mismatch repair gene identified in the yeast *Saccharomyces cerevisiae* turned out to be extremely similar to *mutS*. The bacterial and yeast genes were so similar that they were considered homologous, meaning that they trace their ancestry to a gene in a common ancestor.

Using DNA sequence information from these two homologous genes, researchers were able to locate a similar sequence in the human genome. This locus came to be called *hMSH* (for human *mutS* homolog). Then the sparks really began to fly when biologists discovered that the *hMSH* gene mapped to the same region of chromosome 2 as the HNPCC susceptibility gene.

The link between a defect in mismatch repair and HNPCC was cemented when it was confirmed that individuals suffering from this cancer have mutated forms of *hMSH*. As predicted, cells from these patients turned out to have mutation rates 100 times that of normal cells when they are grown in culture. Follow-up studies confirmed that this increased mutation rate is indeed due to defective mismatch repair in cells derived from HNPCC patients.

(Continued on next page)

(Essay continued)

This breakthrough validates one of the themes in this unit: the use of model organisms to study basic questions in biology. When experiments on mutator strains of *E. coli* began, it appeared to be pure, or basic, research—without practical application. But because humans, yeast, and bacteria share a common evolutionary history, and because DNA synthesis and repair are so fundamental to the functioning of all cells, the results of those early experiments made a major advance in cancer biology possible. If individuals with mutant forms of *hMSH* can be identified early in life, dietary changes and therapy could significantly reduce their risk of developing cancer.

Chapter Review

Summary

Meselson and Stahl were able to validate the hypothesis that each strand of a parent DNA molecule serves as a template for the synthesis of a daughter strand by labeling DNA with ^{15}N or ^{14}N. Kornberg showed that DNA synthesis is an enzyme-catalyzed reaction by purifying DNA polymerase I from *E. coli* and using the protein to synthesize DNA in vitro. When researchers analyzed replication-deficient mutants in this species, however, it became clear that DNA polymerase III was the main replication enzyme and that a large series of proteins were essential for DNA replication.

DNA synthesis takes place only in the $5' \rightarrow 3'$ direction, requires both a template and a primer sequence, and takes place at the replication fork where the double helix is being opened. Synthesis of the leading strand, which is oriented in the $5' \rightarrow 3'$ direction, is straightforward; but synthesis of the lagging strand, which is oriented in the $3' \rightarrow 5'$ direction, is more complex. By feeding *E. coli* cells a short pulse of radioactive thymidine, Okazaki and co-workers were able to confirm that short DNA fragments form on the lagging strand. These fragments are primed by a short strand of RNA and are linked together after synthesis.

The polymerase chain reaction, or PCR, is an in vitro synthesis reaction that employs a heat-stable form of DNA polymerase called *Taq* polymerase. By selecting primers that bracket a certain stretch of DNA, PCR can amplify certain genes to extremely high copy number. These copies can then be analyzed by the dideoxy sequencing method. This is an in vitro synthesis reaction that employs dideoxyribonucleotides to stop DNA replication at each base in the sequence. By running the resulting DNA fragments out on a gel, the sequence of nucleotides in a gene can be determined.

In cells, DNA replication is remarkably accurate—because DNA polymerase proofreads, and because mismatch repair enzymes excise incorrect bases and replace them with the correct sequence. Occasionally, however, errors occur and the wrong base is substituted. This results in a point mutation. As analyses of mutations in human hemoglobin show, mutations can be beneficial, deleterious, or neutral. All mutations are random and most are deleterious.

DNA repair occurs after bases have been damaged by spontaneous breakage of bonds or by chemicals or radiation. Various excision repair systems exist to cut out damaged portions of genes and replace them with correct sequences. Several types of human cancers are associated with defects in the genes responsible for DNA repair.

Questions

Content Review

1. What comprises an individual's genome?
- **a.** all of its proteins
- **b.** all of its mRNAS
- **c.** all of its DNA
- **d.** all of its organelles

2. Why was Kornberg's discovery and analysis of DNA Pol I interesting?
- **a.** It showed that DNA synthesis is an enzyme-catalyzed reaction.
- **b.** It showed that DNA synthesis takes place only in the $5' \rightarrow 3'$ direction.
- **c.** It showed that DNA polymerases require a primer and a template.
- **d.** All of the above.

3. What is the most general definition of a mutation?
- **a.** any change in an individual's genome
- **b.** a single change in the base sequence of a gene
- **c.** a defect in a locus responsible for DNA synthesis
- **d.** a defect in a locus responsible for DNA repair

4. Where and how are Okazaki fragments synthesized?
- **a.** at the leading strand, oriented in a $5' \rightarrow 3'$ direction
- **b.** at the leading strand, oriented in a $3' \rightarrow 5'$ direction
- **c.** at the lagging strand, oriented in a $5' \rightarrow 3'$ direction
- **d.** at the lagging strand, oriented in a $3' \rightarrow 5'$ direction

5. In *E. coli*, an enzyme is involved in which aspect of DNA replication?
- **a.** synthesizing RNA primers on the lagging strand
- **b.** ligating, or binding together, Okazaki fragments
- **c.** opening up the DNA helix at the replication fork
- **d.** all of the above

6. How do dideoxyribonucleotides (ddNTPs) and deoxyribonucleotides (dNTPs) compare?
- **a.** dNTPs lack a hydroxyl (OH) group on their 2′ and 3′ carbons.
- **b.** ddNTPs lack a hydroxyl (OH) group on their 2′ and 3′ carbons.
- **c.** dNTPs contain a hydroxyl (OH) group on their 2′ and 3′ carbons.
- **d.** One of the four bases in ddNTPs is uracil instead of thymine.

Conceptual Review

1. Why are temperature-sensitive mutants useful?
2. The following diagram shows replication of a lagging strand. Label the 5′ and 3′ ends of each strand.

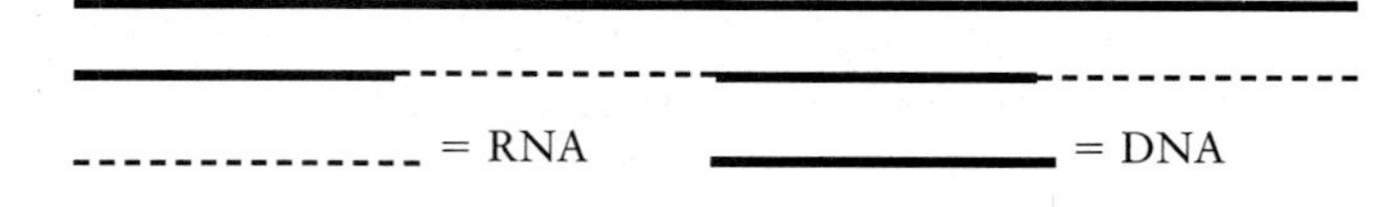

3. List the molecules that are required for a typical PCR reaction. Then list the function of each molecule required for replication of DNA in vivo (see Figure 12.8). For each molecule on the in vivo list, explain why it is or is not required for the in vitro reaction.
4. When researchers perform a sequencing reaction using the dideoxy method, why don't they get sequence information from both strands of the template DNA?

Applying Ideas

1. A friend of yours is doing a series of PCR reactions and comes to you for advice. She purchased three sets of primers, hoping that one set would amplify the template sequence shown below. (The dashed lines in the template sequence stand for a long sequence of bases.) None of the three primer pairs produced any product DNA, however.

	Primer a		Primer b
Primer Pair 1:	5′ GTCCAGC 3′	&	5′ CCTGAAC 3′
Primer Pair 2:	5′ GGACTTG 3′	&	5′ GCTGGAC 3′
Primer Pair 3:	5′ GTCCAGG 3′	&	5′ CAAGTCC 3′

Template

5′ ATTCGGACTTG - - - - - - - - - GTCCAGCTAGAGG 3′
3′ TAAGCCTGAAC - - - - - - - - - CAGGTCGATCTCC 5′

   a. Examine each primer pair and explain why it didn't work. Be sure to indicate whether both of the primers are at fault or if just one primer is the problem.
   b. Your friend doesn't want to buy new primers. She asks you whether she can salvage this experiment. What do you tell her to do?

2. In the late 1950s Herbert Taylor grew bean root-tip cells in a solution of radioactive thymidine and allowed them to undergo one round of DNA replication. He then transferred the cells to a solution without the radioactive nucleotide, allowed them to replicate again, and examined their chromosomes for the presence of radioactivity. His results are shown in the figure here.

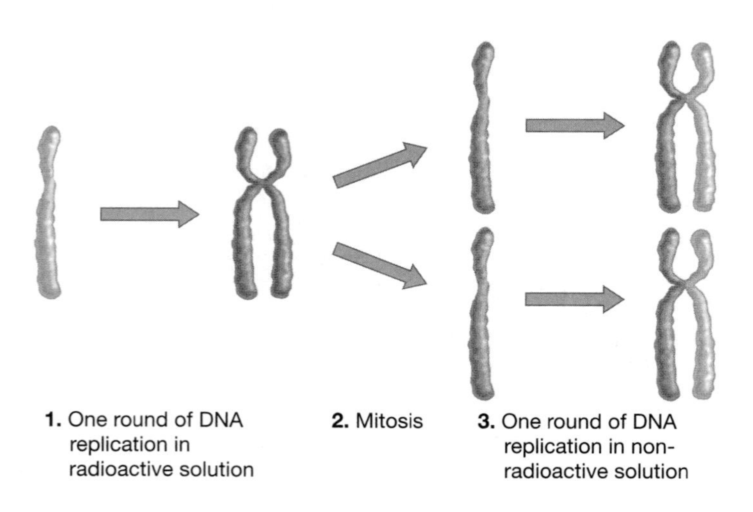

   a. Make diagrams explaining the pattern of radioactivity observed in the sister chromatids after the first and second rounds of replication.
   b. What would the results of Taylor's experiment be if eukaryotes used a conservative mode of DNA replication?

3. Examine the drawing of a ddNTP in Figure 12.12a. When a ddNTP molecule is present in a DNA synthesis reaction, it is added to the growing chain. Additional nucleotides cannot be added after this. Why does this result show that synthesis of the DNA chain occurs at the 3′ end of the growing chain? What would you expect to happen with a ddNTP if synthesis occurred at the 5′ end of the growing chain?

4. This graph shows the survival of four different *E. coli* strains after exposure to increasing doses of ultraviolet light. The wild-type strain is normal, but the other strains have a mutation in genes called *uvrA*, *recA*, or both *uvrA* and *recA*.

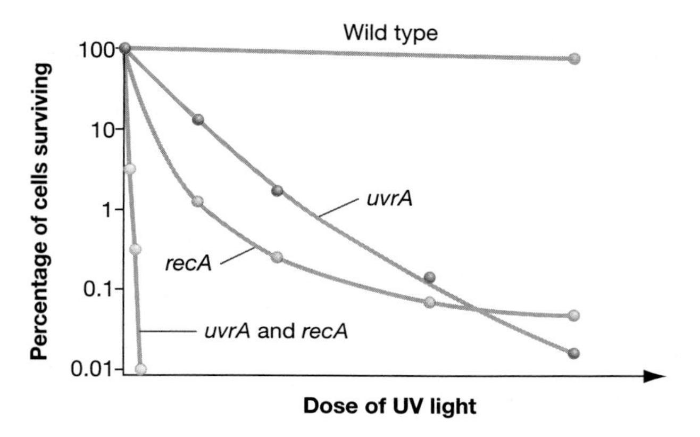

   a. Which strains are most sensitive to UV light? Which strains are least sensitive?
   b. What are the relative contributions of these genes to repair of UV damage?

5. One widely used test to identify whether certain chemicals, like pesticides or herbicides, might be carcinogenic involves exposing bacterial cells to the chemical and recording whether the exposure leads to an increased mutation rate. In effect, this test equates cancer-causing chemicals with mutation-causing chemicals. Why is this an informative test?

6. The essay at the end of this chapter emphasizes the importance of mutation in the development of certain types of cancer. Cancer is not simply a question of mutation rate, however, because people whose cells have an elevated mutation rate do not show an increased susceptibility to every kind of cancer. What processes other than mutation might be involved in the onset of cancer?

CD-ROM and Web Connection

CD Activity 12.1: DNA Synthesis *(animation)*
(Estimated time for completion = 5 min)
Learn how the enzymes of DNA synthesis work together to copy a molecule of DNA.

CD Activity 12.2: Polymerase Chain Reaction *(animation)*
(Estimated time for completion = 5 min)
Examine how the polymerase chain reaction produces millions of copies of DNA from a single DNA fragment.

At your **Companion Website** (http://www.prenhall.com/freeman/biology), you will find self-grading exams and links to the following research tools, online resources, and activities:

Classic Experiments in DNA Replication
Explore many of the early experiments in which biologists elucidated the molecular mechanisms of DNA and RNA replication.

Current Research on DNA Replication
This site describes one researcher's work on the regulation of DNA replication.

Cancer
Search for "DNA repair" to discover how mistakes in DNA repair are related to the formation of cancerous cells.

Additional Reading

Leffell, D. J., and D. E. Brash. 1996. Sunlight and skin cancer. *Scientific American* 275 (July): 52–59. Although most skin cancers appear in older people, the damage often begins decades earlier, when the Sun's rays mutate a key gene in a single cell.

Mullis, K. B. 1990. The unusual origin of the polymerase chain reaction. *Scientific American* 262 (April): 56–65. A personal account of how PCR came about.

Transcription and Translation

13

Proteins are the stuff of life. They control most of the chemical reactions, physiological responses, and developmental processes in the cell. In combination with lipids and carbohydrates, they also comprise the cell's structure. If DNA is the blueprint for an organism, then proteins are the carpenters, hammers, lumber, trucks, and plumbing.

A cell builds the proteins it needs from instructions encoded in its genome. Chapter 11 explored the basic strategy that cells use to retrieve these instructions and convert them into action in the form of proteins. The first step in the process is the transcription of a gene and the production of a messenger RNA, or mRNA. This is a short-lived version of the archived instructions in DNA. Then the sequence of nucleotides in the mRNA is translated into a sequence of amino acids. This series of events defines the central dogma of molecular biology:

$$\underset{\text{(information storage)}}{\text{DNA}} \xrightarrow{\text{transcription}} \underset{\text{(information carrier)}}{\text{mRNA}} \xrightarrow{\text{translation}} \underset{\text{(product)}}{\text{proteins}}$$

Elucidating this relationship between DNA and proteins was one of the great scientific advances of the twentieth century. But once the central dogma—summarized in **Figure 13.1**—was firmly established, biochemists puzzled over how cells actually accomplish the feat of converting genetic information into protein. How do the enzymes involved in transcribing mRNA from the DNA template know where one gene ends and another begins? Once an RNA message is produced, what molecular machines are responsible for translating its linear sequence of nucleotides into a linear sequence of amino acids?

This chapter explores the experiments that answered these questions. Because many antibiotics work by disrupting the translation

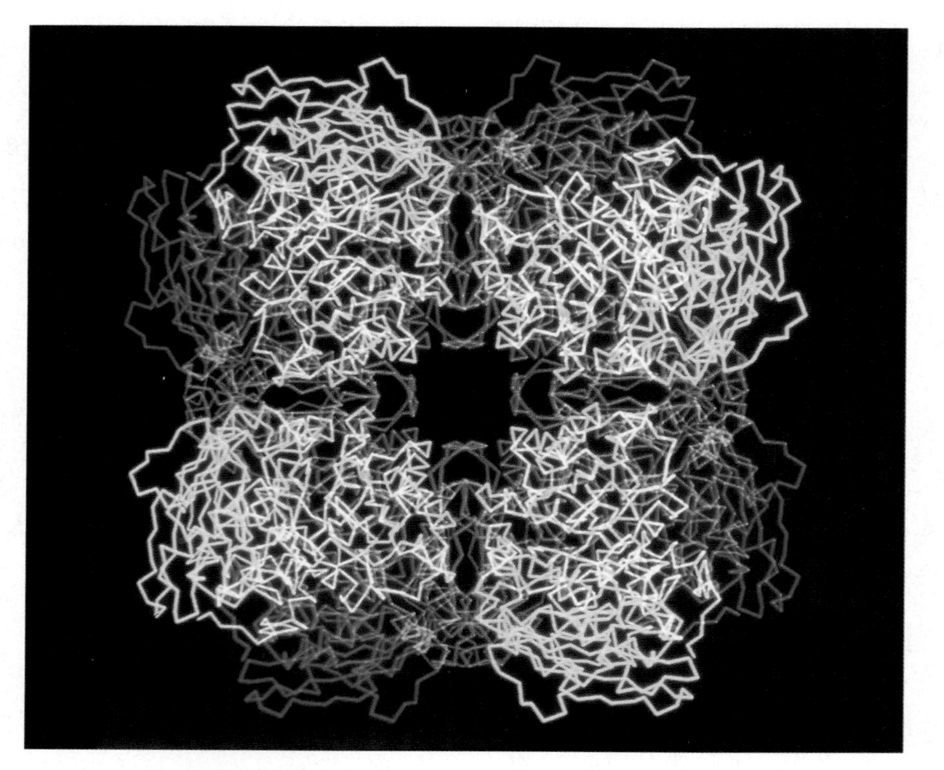

This is a structural model of the enzyme called rubisco, which catalyzes the carbon fixation reaction in photosynthetic organisms. Rubisco is the most abundant protein in the world. This chapter explores how mRNA is transcribed from DNA and how proteins are translated from mRNA.

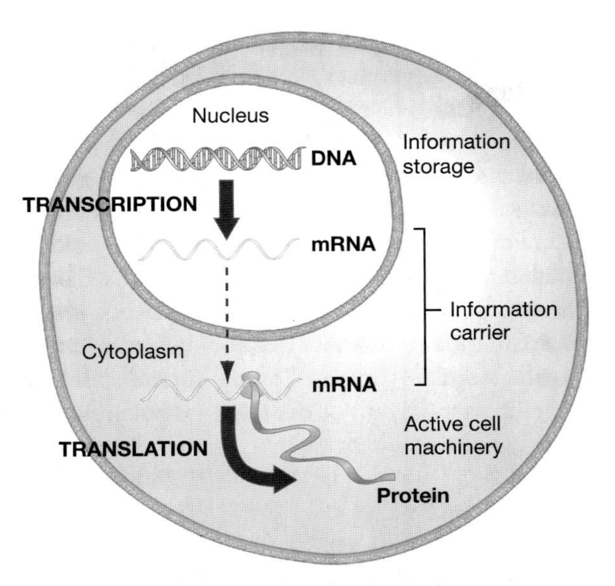

FIGURE 13.1 Central Dogma of Molecular Biology
In eukaryotes, transcription takes place in the nucleus and translation takes place in the cytoplasm. In all organisms, DNA, RNA, and proteins have distinct functions.

machinery, the research reviewed here has had a significant impact on biomedicine and drug development. Understanding transcription and translation at the molecular level also set the stage for an explosion of work, introduced in Chapters 14 and 15, on how protein production is regulated.

We begin by examining research on the mechanisms of transcription in bacteria and eukaryotes. After getting an overview of how translation occurs, we take a detailed look at the ribosome and other pieces of apparatus that make the process possible. The chapter concludes by examining how proteins are modified after translation is complete.

13.1 Transcription in Bacteria

The first step in converting genetic information into protein is to generate a messenger RNA version of the instructions. The enzyme RNA polymerase, introduced in Chapter 11, carries out this phase of protein production. In addition to synthesizing mRNA, RNA polymerase produces other types of RNA molecules; these are introduced later in the chapter.

As **Figure 13.2** shows, RNA polymerase catalyzes the formation of a phosphodiester bond between the growing 3′ end of an RNA chain and an incoming ribonucleotide triphosphate. Like the DNA polymerases introduced in Chapter 12, RNA polymerase performs a template-directed synthesis in the 5′ → 3′ direction.

Biochemists reached these conclusions by studying RNA synthesis in cell-free, or in vitro, systems. Once the basic chemical reaction was understood, an entirely new set of questions came to the fore.

- Where does RNA polymerase start and stop RNA synthesis on the DNA template?
- Which of the two DNA strands acts as a template?
- Are any other proteins or factors involved in transcription, or does RNA polymerase act alone?

To answer these questions, investigators focused on a model system—the bacteriophages (or just phages) that infect the bacterium *Escherichia coli*. The genomes of these bacterial viruses enter bacterial cells at the start of an infection, and at least some of the viral genes are transcribed by the bacterium's RNA polymerase. Studying how phage DNA is transcribed—instead of bacterial DNA—was attractive for purely practical reasons. The *E. coli* genome contains 4.6 million base pairs of DNA and over four thousand genes. In contrast, phages like λ (lambda), T4, and T7 contain less than 170,000 base pairs of DNA and from 20 to 100 genes. Their compact size made transcription and translation relatively easy to study. The hope was that studying the transcription of phage genes by bacterial

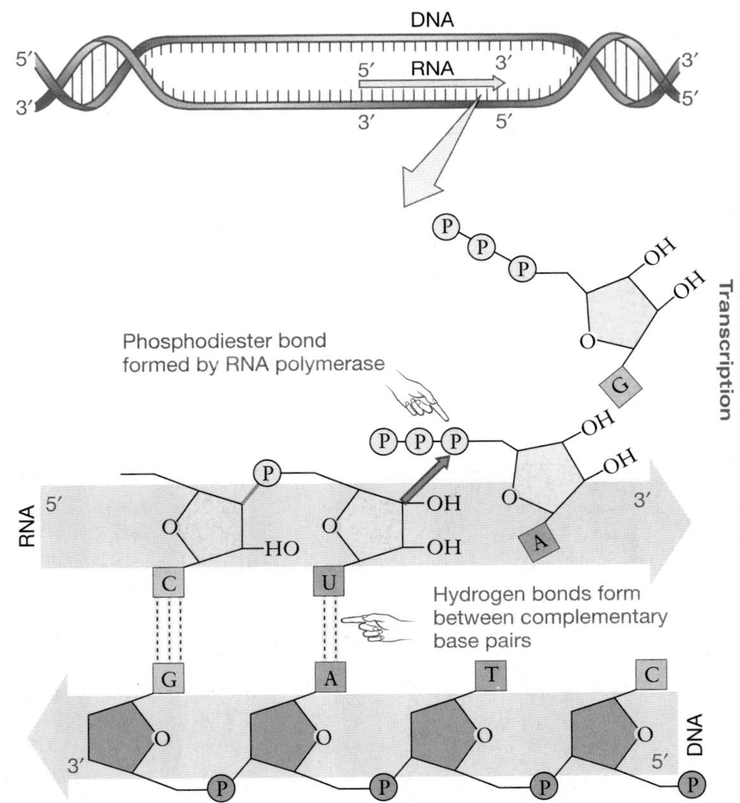

FIGURE 13.2 RNA Synthesis During Transcription
The reaction catalyzed by RNA polymerase results in the formation of a phosphodiester bond between ribonucleotides. RNA polymerase produces an RNA strand whose sequence is complementary to the bases in the DNA template. QUESTION In which direction is RNA synthesized, 5′ → 3′ or 3′ → 5′? In which direction is the DNA template "read"?

CD ACTIVITY 13.1 Transcription

machinery would reveal how the general process works in a wide variety of organisms.

How Does Transcription Begin?

In order for transcription to begin, it would seem logical that RNA polymerase must make physical contact with the DNA template. If so, where does this contact occur?

Biologists gained an important new tool to use in answering this question when filter papers made from a compound called nitrocellulose were developed. Nitrocellulose is a valuable research tool because it binds proteins but does not bind double-stranded DNA. As a result, researchers could label phage DNA with radioactive atoms, allow it to react with RNA polymerase, and then pass the mixture through a nitrocellulose filter (**Figure 13.3a**). If radioactive particles were found to bind to the filter, it meant that the RNA polymerase and DNA were bound together. If the radioactive DNA passed through the filter, it meant that the template was not in physical contact with RNA polymerase.

The Discovery of Promoters Researchers exploited this experimental setup in a variety of ways. In one set of experiments, they gradually increased the ratio of RNA polymerase to T7 DNA in the reaction mix and found that the amount of RNA polymerase-DNA retained on the filter also increased—up to a point. Virtually all of the RNA polymerase molecules in the reaction mix bound to DNA up until the ratio of the two

(a) Experimental evidence that RNA polymerase binds with DNA

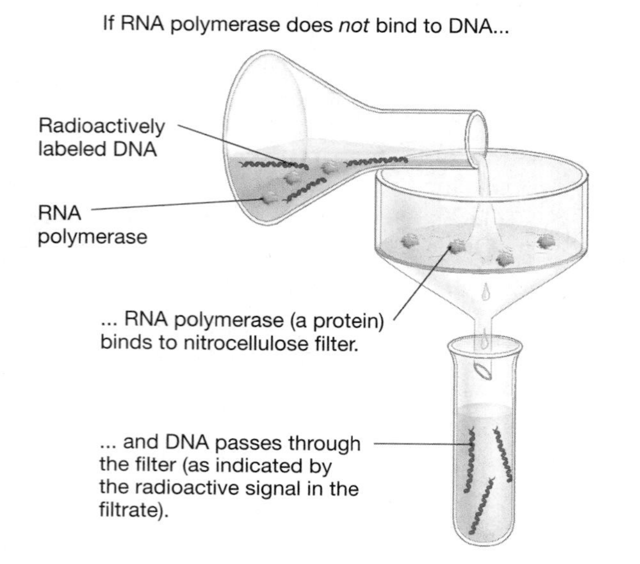

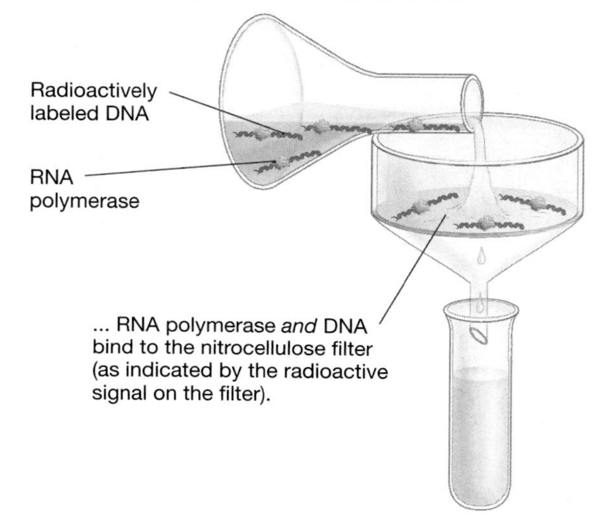

(b) Visual evidence

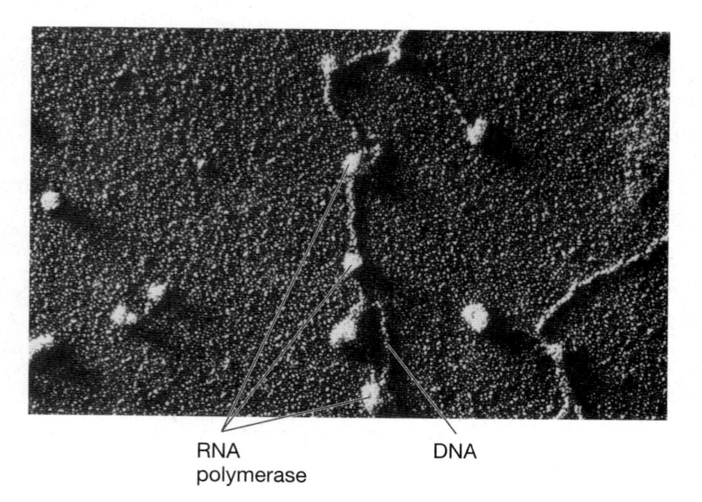

FIGURE 13.3 RNA Polymerase Binds to Promoter Regions in DNA
(a) Nitrocellulose filters retain proteins, but not double-stranded DNA (left). The figure on the right shows that if RNA polymerase binds to double-stranded DNA, the DNA will be retained on the nitrocellulose filter. **(b)** This electron micrograph shows a series of RNA polymerase molecules transcribing a stretch of DNA in a cell.

types of molecules was 8:1. But if the amount of RNA polymerase was increased beyond this point, no more binding occurred. Based on this observation, researchers estimated that the T7 genome contained about eight sites where RNA polymerase binds tightly.

The binding sites for RNA polymerase were named **promoters** and were hypothesized to be the sites where transcription starts. Not long after, the interaction between RNA polymerase and promoter sequences was observed directly with the electron microscope (**Figure 13.3b**).

At about the same time, evidence began to accumulate that a protein found in *E. coli* called sigma factor attaches itself to RNA polymerase before binding to DNA occurs. What does this protein do? David Hinkle and Michael Chamberlin suggested that sigma factor binds to the promoter sequence on the template DNA along with RNA polymerase and helps to initiate transcription. The idea here was that RNA polymerase does not initiate transcription by itself. Instead, other proteins tell the enzyme when and where to start synthesizing mRNA.

In support of this hypothesis, Hinkle and Chamberlin showed that when sigma factor was removed from RNA polymerase, the enzyme alone was no longer able to bind tightly to DNA in filter-binding experiments. Subsequent work confirmed that sigma is required for RNA polymerase to recognize the promoter region and initiate transcription. Once transcription is under way, however, sigma is released. Then the questions became, what sequences are recognized by sigma factors? What does a promoter actually look like?

The –10 Box and –35 Box An important advance in understanding the interaction between RNA polymerase, promoter regions, and sigma factors occurred when David Pribnow set out to determine the base sequence of the promoter sequences in T7 DNA. His first task was to isolate the promoter sequences from the rest of the genome. To do this, Pribnow allowed purified bacteriophage DNA and purified *E. coli* RNA polymerase-sigma complexes to interact and bind to one another. Then he treated the products with an enzyme that degrades DNA, called DNase (**Figure 13.4**).

The key to Pribnow's strategy was his assumption that the RNA polymerase would protect the promoter region from digestion by the DNase enzyme. Once the DNase was washed away, Pribnow treated the mixture with a chemical that degrades proteins; Pribnow was hoping that this treatment would degrade the RNA polymerase, leaving intact promoter DNA sequences. As the last step in Figure 13.4 shows, this step indeed removed the RNA polymerase and left purified promoter DNA.

Pribnow analyzed the promoter DNA by gel electrophoresis and found that the protected fragments of DNA were 40–50 base pairs long. When he sequenced the fragments and then compared them to promoters that had been analyzed in other phage and bacterial genomes, he found that a particular section in each promoter looked similar. This "consensus sequence" was located about 10 bases away from the start of the mRNA and had a series of bases identical or similar to TATAAT. The six-base pair sequence is now known as the **–10 box**, or **Pribnow box**, because it occurs about 10 bases from the point where RNA polymerase starts transcription. (DNA that is located away from the direction that RNA polymerase moves is said to be upstream; DNA located in the direction that RNA polymerase moves is said to be downstream. Thus, the –10 box is located 10 bases upstream from the transcription start site.)

Later, researchers recognized that a consensus sequence of TTGACA occurred in these same promoters about 35 bases upstream from the start of the mRNA. This finding led to a comprehensive model for how transcription begins in bacteria. As the diagram in **Figure 13.5a** (page 256) shows, the combined RNA polymerase-sigma complex binds to the –35 box and to the –10 box. This step is important because it identifies which DNA strand RNA polymerase transcribes and in which direc-

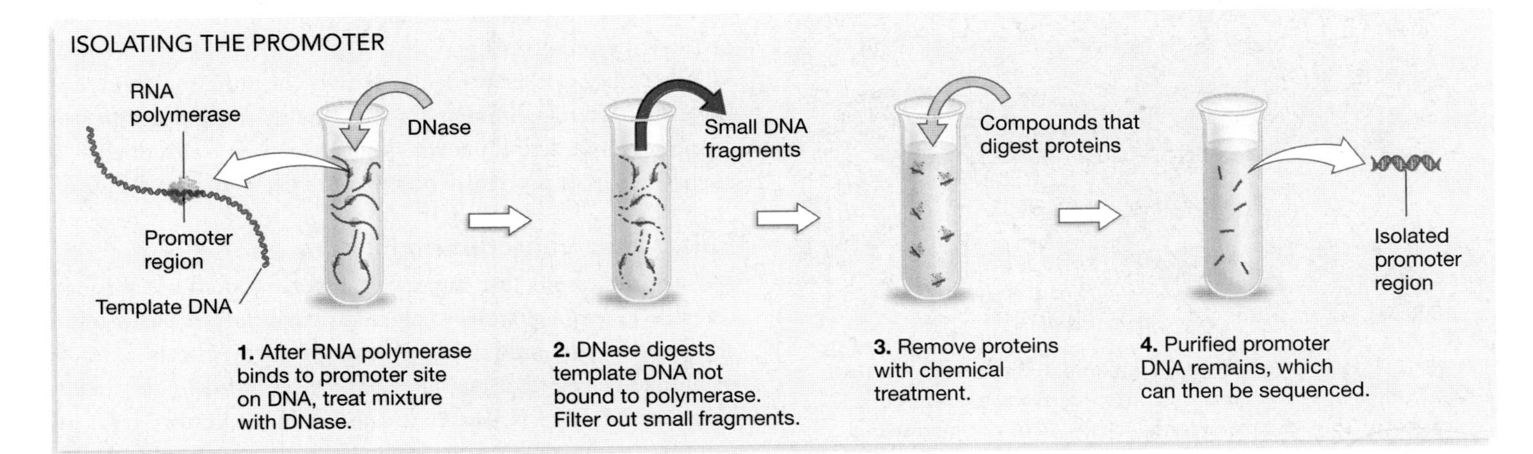

FIGURE 13.4 Isolating the Promoter
This protocol allowed a researcher to isolate pure preparations of promoter sequences.

tion the enzyme moves. RNA polymerase synthesizes mRNA in the $5' \rightarrow 3'$ direction, so DNA bases are read in the $3' \rightarrow 5'$ direction. In this way, the location of the promoter defines which strand functions as the template for RNA synthesis.

Recent research has also shown that bacteria typically have several different sigma factors. Each of these proteins recognizes promoters with distinctive base sequences. As a result, the identity of the sigma protein in the RNA polymerase complex determines which types of genes will be transcribed.

Once the recognition step is complete, RNA polymerase begins to synthesize RNA. The sigma factor is released, as shown in **Figure 13.5b**. Transcription initiation is complete.

Elongation and Termination

During the elongation phase of transcription, RNA polymerase moves along the DNA template, unwinding the double helix and catalyzing the addition of nucleotides to the 3′ end of the growing RNA molecule. As in DNA replication, synthesis is driven by the potential energy stored in ribonucleoside triphosphate monomers, or NTPs. (Recall from Chapter 12 that the deoxyribonucleoside triphosphate monomers used in DNA synthesis are abbreviated dNTPs). As you saw in Figure 13.2, the nucleotide sequence of the RNA molecule produced by transcription is determined by the nucleotide sequence of the DNA template via complementary pairing between bases. Because RNA has uracil rather than thymine, an A in the DNA template sequence specifies a U in the complementary RNA strand.

In most cases, termination of transcription occurs when the RNA polymerase reaches a specific sequence in the DNA that corresponds to a transcription termination site. At this site the polymerase stops synthesis, and both it and the RNA molecule dissociate from the DNA template. In other cases, termination of transcription is facilitated by interactions between RNA polymerase and specific termination proteins.

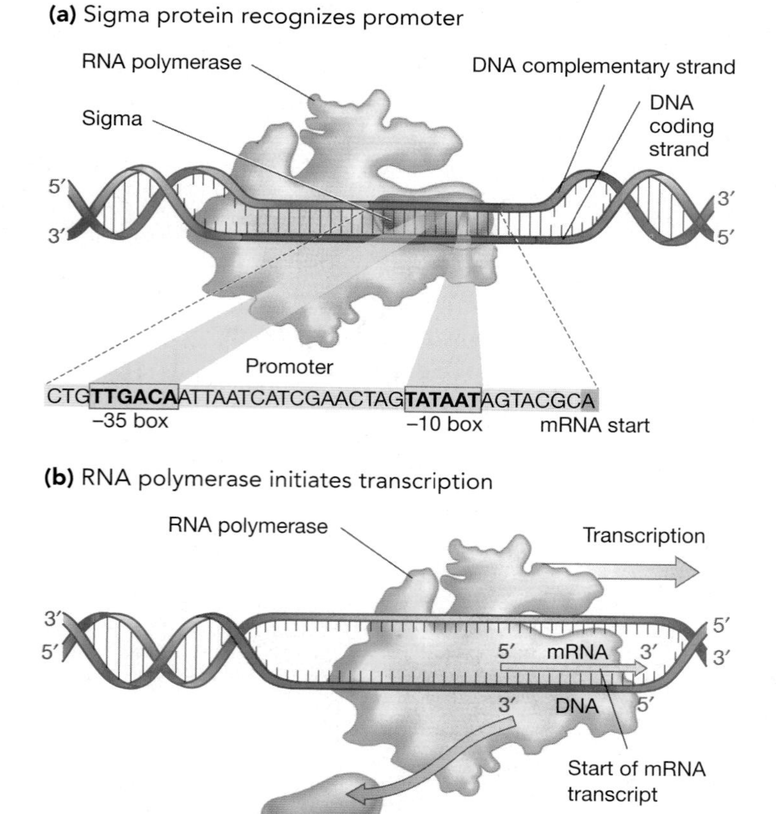

FIGURE 13.5 How Is Transcription Initiated in Bacteria?
(a) Many of the promoters found in bacteria have two consensus sequences called the –35 box and the –10 box. RNA polymerase initiates transcription when the sigma protein recognizes these sequences and interacts with them. **(b)** Once transcription is under way, the sigma protein disengages from RNA polymerase.

13.2 Transcription in Eukaryotes

The experiments reviewed in section 13.1 establish the basic outline of how transcription occurs in bacteria. How do these results compare with research on the mechanism of transcription in eukaryotes? What similarities and differences have come to light? Let's begin with the fundamentals. How does RNA polymerase work in eukaryotes, and how does it recognize the start of a gene?

In eukaryotes, RNA polymerase catalyzes the template-directed synthesis of RNA in the $5' \rightarrow 3'$ direction, just as in bacteria. Eukaryotes have three distinct RNA polymerases, however, instead of just one. As **Table 13.1** shows, RNA pol I, pol II, and pol III each transcribe a discrete type of RNA. The RNAs transcribed by two of the RNA polymerases do not function as messenger RNA, however. Only RNA pol II transcribes the genes that code for proteins. RNA pol II produces mRNA.

Like bacteria, eukaryotic genomes contain promoters that signal where transcription should begin. The eukaryotic promoters recognized by RNA polymerase II include a consensus sequence called the TATA box, located 30 base pairs upstream of the transcription start site, but in general eukaryotic promoters are more variable in sequence and more complex than are prokaryotic promoters.

What about other proteins involved in initiating transcription? In bacteria, different sigma proteins recognize different promoters and determine which genes will be transcribed. Do eukaryotes have proteins whose function is similar?

Eukaryotic Transcription Factors

Most of the results just summarized were worked out through cell-free experiments. To explore transcription in eukaryotes, researchers frequently used RNA pol II purified from extracts of human cells and a template genome provided by an adenovirus—a family of viruses that cause mild respiratory tract infections in humans. (You undoubtedly have adenovirus on your tonsils right now.) Because the adenovirus genome is small and is transcribed by the host cell's RNA polymerase, it furnished a convenient model system analogous to the bacteriophages that researchers used to study transcription in *E. coli*.

When RNA pol II was first purified and studied in these cell-free systems, however, it transcribed both strands of the template DNA instead of one. It also began copying at random locations in the adenovirus genome instead of at promoter regions (**Figure 13.6**). To understand why transcription was so inaccurate in this simplified system, Robert Roeder's group set out to determine if some component of the transcription apparatus was missing. Their working hypothesis was that eukaryotes might contain a sigma-like factor that recognizes the promoter.

To test this idea they added a crude extract from human cell nuclei—one that contained *all* of the soluble components found in the nucleus—to a simple transcription system consisting of adenovirus DNA and NTPs. As Figure 13.6 shows, the RNA polymerase in the extract was able to transcribe the virus genome starting at a promoter site. It was also able to use the correct strand of DNA as template. Roeder and co-workers concluded that this soluble extract must contain components that are required for accurate initiation of transcription. To identify these molecules they separated the components of the crude extract and added them either one at a time or in various combinations to the in vitro system. Using this approach, Roeder's group discovered several transcription factors that help RNA polymerase II recognize promoter sequences in DNA. As you'll see in Chapter 15, more recent studies have revealed that eukaryotic cells contain dozens of different transcription factors and that these proteins influence when genes are turned off and on.

The Startling Discovery of Eukaryotic Genes in Pieces

In the mid-1970s, two different groups of investigators discovered an unforeseen wrinkle in the production of eukaryotic mRNA. These teams turned up evidence suggesting that the protein-coding regions of eukaryotic genes are interrupted by stretches of noncoding DNA. The result implied that information processing and gene organization are strikingly different in bacteria and eukaryotes. It also meant that eukaryotic cells must somehow dispose of the noncoding sequences in order to make a functional protein from an mRNA.

TABLE 13.1 Eukaryotic RNA Polymerases

Name of Enzyme	Type of Loci Transcribed
RNA polymerase I (RNA pol I)	Genes that code for most of the large RNA molecules, or rRNAs, found in ribosomes (see section 13.5)
RNA polymerase II (RNA pol II)	Protein-coding genes (produce mRNAs)
RNA polymerase III (RNA pol III)	Genes that code for transfer RNAs (see section 13.4), and genes that code for one of the small RNA molecules, or rRNAs, found in ribosomes
RNA pol II and RNA pol III	RNA molecules found in snRNPs (see section 13.2)

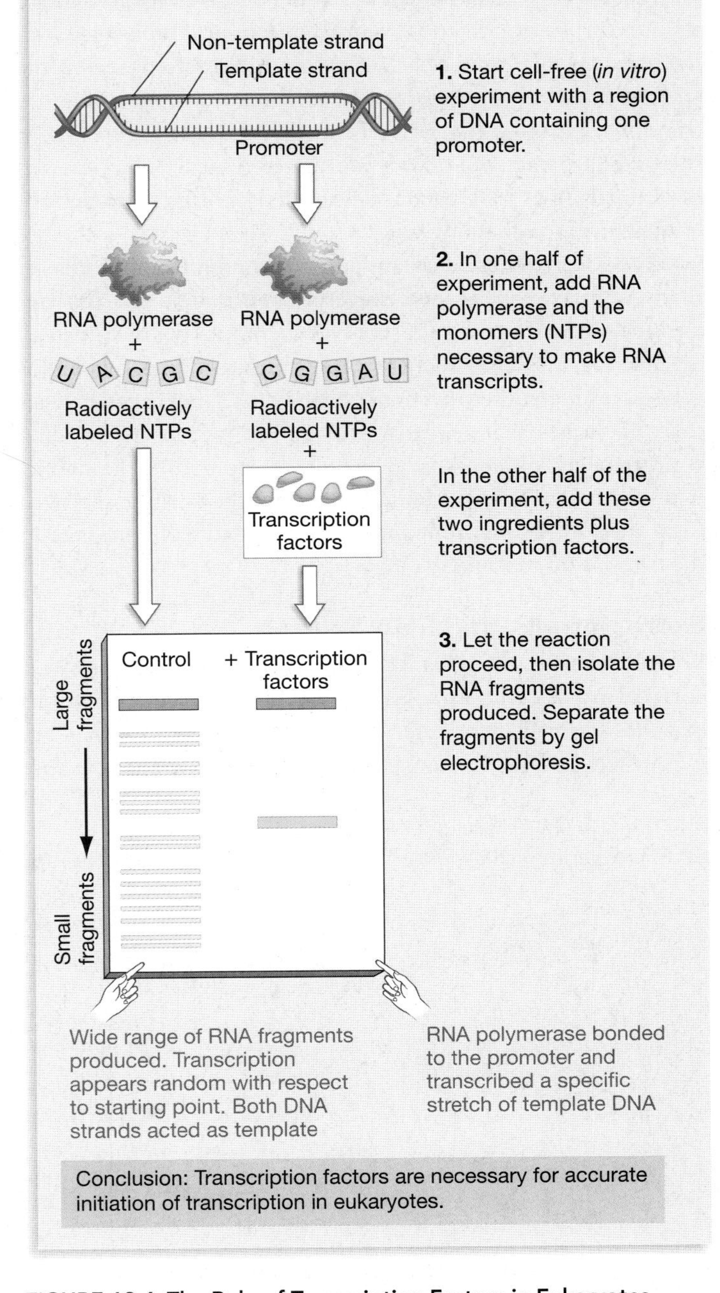

FIGURE 13.6 The Role of Transcription Factors in Eukaryotes
Researchers found that when eukaryotic RNA polymerase transcribes DNA in vitro, a wide range of DNA fragments are produced. Transcription appeared to be random with respect to starting point, and both DNA strands acted as the template. When transcription factors were added to the in vitro system, RNA polymerase bound to a promoter and transcribed a specific stretch of DNA.

What sort of data would provoke such a startling claim? As an example, consider work that Phillip Sharp and colleagues carried out on transcription of the adenovirus genome. One of their experiments consisted of purifying adenovirus mRNA, mixing it with adenovirus DNA that had been heated to separate the two strands, and incubating the molecules to promote base pairing between the mRNA and the single-stranded DNA. Under these conditions, the mRNA ought to form base pairs with the DNA sequences that act as the template for its synthesis. When they examined the RNA-DNA hybrid molecules with the electron microscope, however, the biologists observed the structure shown in **Figure 13.7a.** Instead of matching up smoothly, parts of the DNA formed loops. As **Figure 13.7b** shows, Sharp and co-workers interpreted these loops as stretches of nucleotides that are present in the DNA template but not in the corresponding mRNA.

According to these data and abundant confirming evidence, there is not a one-to-one correspondence between the nucleotide sequence of a eukaryotic gene and its mRNA. Instead of carrying messages such as "the protein-coding regions of genes are interrupted by noncoding DNA," eukaryotic genes carry messages that read something like "the protein-coding regiιντρονons of genes are inιντρονterrupted by noncoding DNA." These regions of noncoding sequence must be removed from the mRNA before it can carry an intelligible message to the translation machinery.

Exons, Introns, and RNA Splicing

When it became clear that the genes-in-pieces hypothesis was correct, Walter Gilbert suggested that the translated regions of eukaryotic genes be referred to as **exons** (because they are *ex*pressed) and the untranslated stretches as **introns** (because they are *int*ervening). As **Figure 13.8a** shows, transcription of eukaryotic genes by RNA polymerase generates a primary transcript that contains both the exon and intron regions. The introns are then removed from the transcript by a process known as **splicing.** In this phase of information processing, pieces of the primary transcript are removed and the remaining segments are joined together. Splicing results in an mRNA that contains an uninterrupted genetic message. (Splicing does not occur in bacteria because their genes do not contain introns.)

Figure 13.8b provides more detail about how introns are removed from genes. In the case illustrated here, the splicing event occurs inside the nucleus and is catalyzed by a complex of proteins and small RNAs known as **snRNPs.** This is an acronym for small nuclear ribonucleoproteins and is pronounced "snurps." After the snRNPs assemble on the initial or primary RNA transcript, a specific adenine nucleotide in the intron RNA sequence attacks the 5′ end of the intron and breaks the RNA at this point. As Figure 13.8b shows, this interaction is mediated by a group of snRNPs called a **spliceosome.** This free 5′ end of the intron becomes attached to the adenine nucleotide and forms a "lariat" of RNA. Then the free 3′ end of the first exon reacts with the 5′ end of the second exon. This reaction breaks the 3′ end of the intron and covalently joins the two exons into a contiguous coding sequence. In most cases, the released lariat is degraded to nucleotide monophosphates.

(a) Micrograph of DNA-RNA hybrid

(b) Interpretation of micrograph

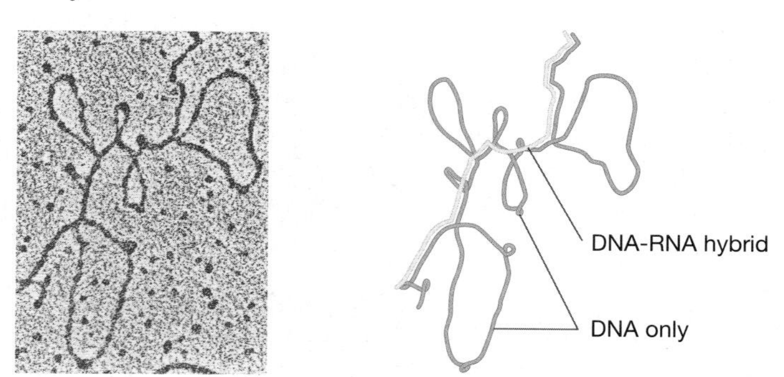

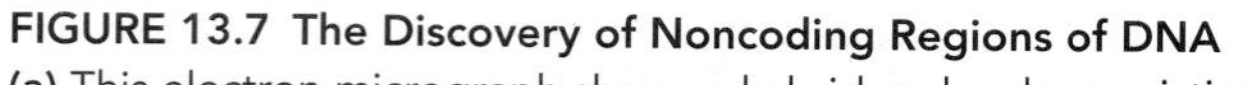

FIGURE 13.7 The Discovery of Noncoding Regions of DNA
(a) This electron micrograph shows a hybrid molecule consisting of a stretch of single-stranded DNA that is bonded via complementary base pairing to the mRNA it encodes. **(b)** This drawing shows how researchers interpreted the photograph in part (a). The loops represent regions of DNA that do not have an equivalent sequence in the mRNA. In other words there is intervening, "extra" DNA compared to the sequences in the mRNA.

Other Aspects of Transcript Processing: Caps and Tails

Intron splicing is not the only type of transcript processing that occurs in eukaryotes. As soon as they are transcribed, the 5′ ends of many eukaryotic mRNAs are chemically modified by the addition of a structure called "the cap" (**Figure 13.9**). In addition, an enzyme cleaves the 3′ end of the mRNA and other enzymes add a long tract of adenine nucleotides, 100–250 long. This sequence is known as the poly (A) tail.

Not long after the caps and tails on eukaryotic mRNAs were described, evidence began to accumulate that they function in protecting the message from degradation by ribonucleases and enhancing the efficiency of translation. It was not clear, however, whether the cap and tail were redundant or synergistic. Are both cap and tail needed for efficient translation?

Daniel Gallie pursued this question by comparing the fates of mRNAs with and without these structures. The mRNA he worked with codes for an enzyme found in fireflies called luciferase. Luciferase catalyzes a light-producing reaction when combined with ATP and a molecule called luciferin. Gallie obtained transcripts that did or did not contain a cap and/or tail, resulting in the four treatments displayed at the top of **Figure 13.10** (page 260). Then he introduced these experimental mRNAs into tobacco cells growing in culture. Using these cells as a "host" for luciferase mRNA was important because it allowed Gallie to infer that the only luciferase produced by the cells came from the mRNA he introduced.

The results of the experiment are shown at the bottom of the figure. Compared to transcripts without a cap and/or tail, luciferase mRNAs with both features last much longer in the cell before they are degraded by ribonucleases. They are also translated much more efficiently. In sum, Gallie's experiment confirmed that cap and tail act synergistically. Recent work has shown why. In at least some mRNAs, the cap and tail interact directly with the protein and RNA machinery responsible for translation.

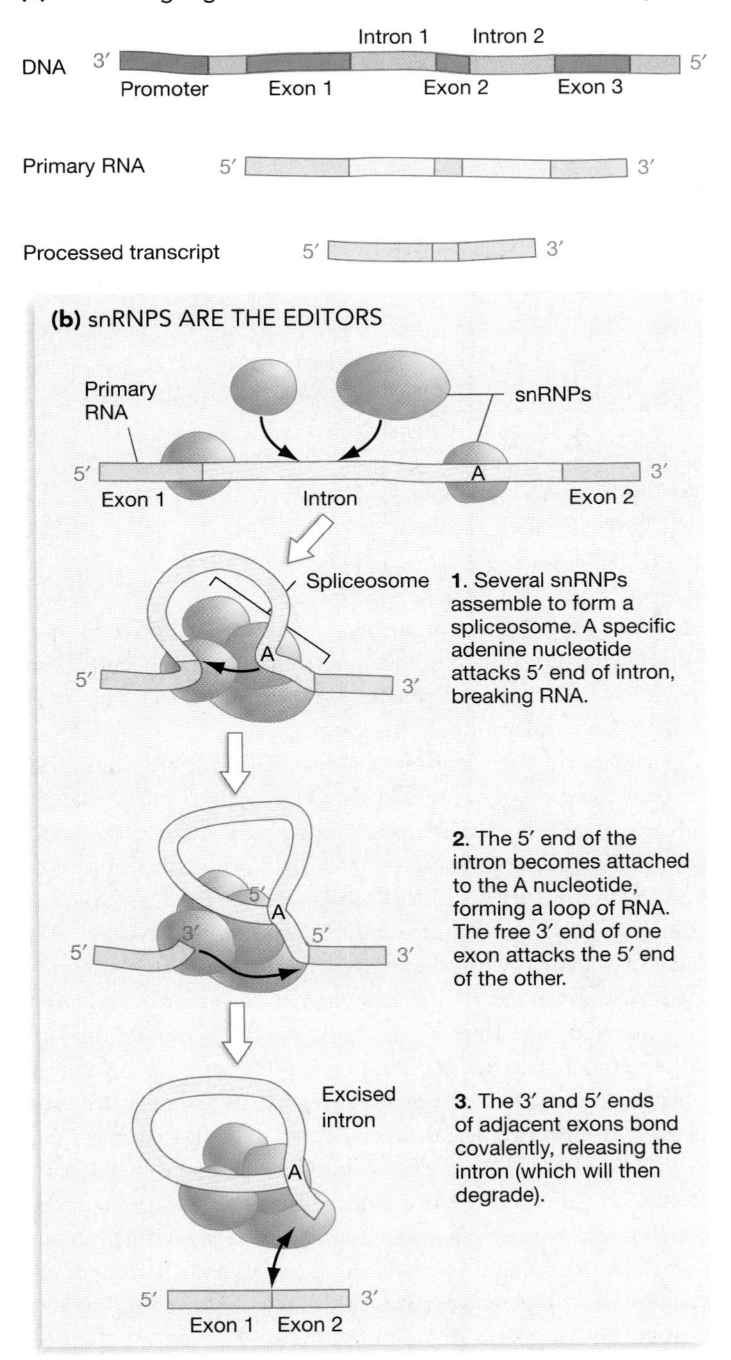

FIGURE 13.8 Introns Are Spliced Out of the Original mRNA
(a) RNA polymerase produces a primary transcript that contains exons and introns. Subsequent processing generates a mature transcript. **(b)** snRNPs are responsible for the splicing reactions that take place in the nucleus of eukaryotes.

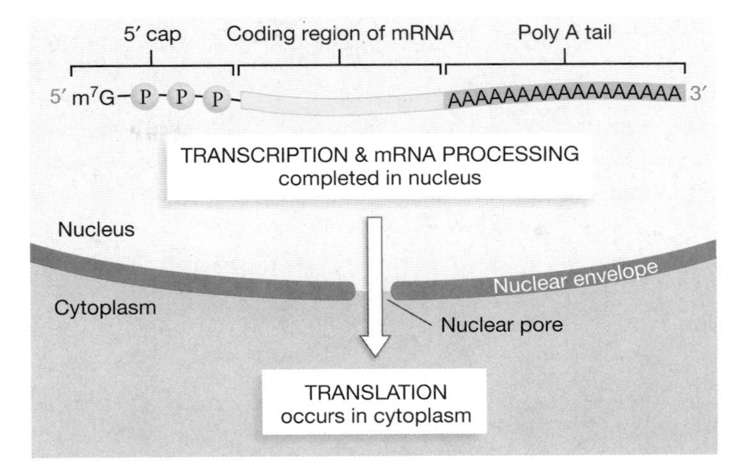

FIGURE 13.9 In Eukaryotes, mRNAs Are Given a Cap and a Tail
Eukaryotic mRNAs have a cap consisting of a molecule called 7-methylguanylate (symbolized m^7G in the diagram) bonded to three phosphate groups; the tail is made up of a long series of adenine residues. Once the cap and tail have been added, introns are spliced out and the processed message is exported from the nucleus to the cytoplasm, where translation takes place.

13.3 An Introduction to Translation

In translation, the sequence of bases in an mRNA is converted into the sequence of amino acids in a polypeptide. The genetic code specifies the relationship between the bases in a triplet codon and the amino acid it codes for (see Chapter 11). But how are the amino acids actually assembled into a polypeptide based on the information in messenger RNA?

Studying translation in cell-free systems proved to be an extremely productive approach to answering this question. Once in vitro translation systems had been derived from a variety of organisms, however, it became clear that the basic mechanisms of translation are fundamentally the same throughout the tree of life. Translation proceeds in much the same way from bacteria to eukaryotes.

Ribosomes Are the Site of Protein Synthesis

Where does translation take place? An answer to this question began to emerge in the 1930s, thanks to pioneering work done by T. Caspersson and by Jean Brachet. These researchers were interested in the small structures called ribosomes. By purifying and analyzing these structures, Brachet and Caspersson established that ribosomes contain a great deal of RNA along with a considerable amount of protein. Further, both researchers noticed that when cells from different animal tissues were analyzed, there was a strong correlation between the quantity of ribosomes

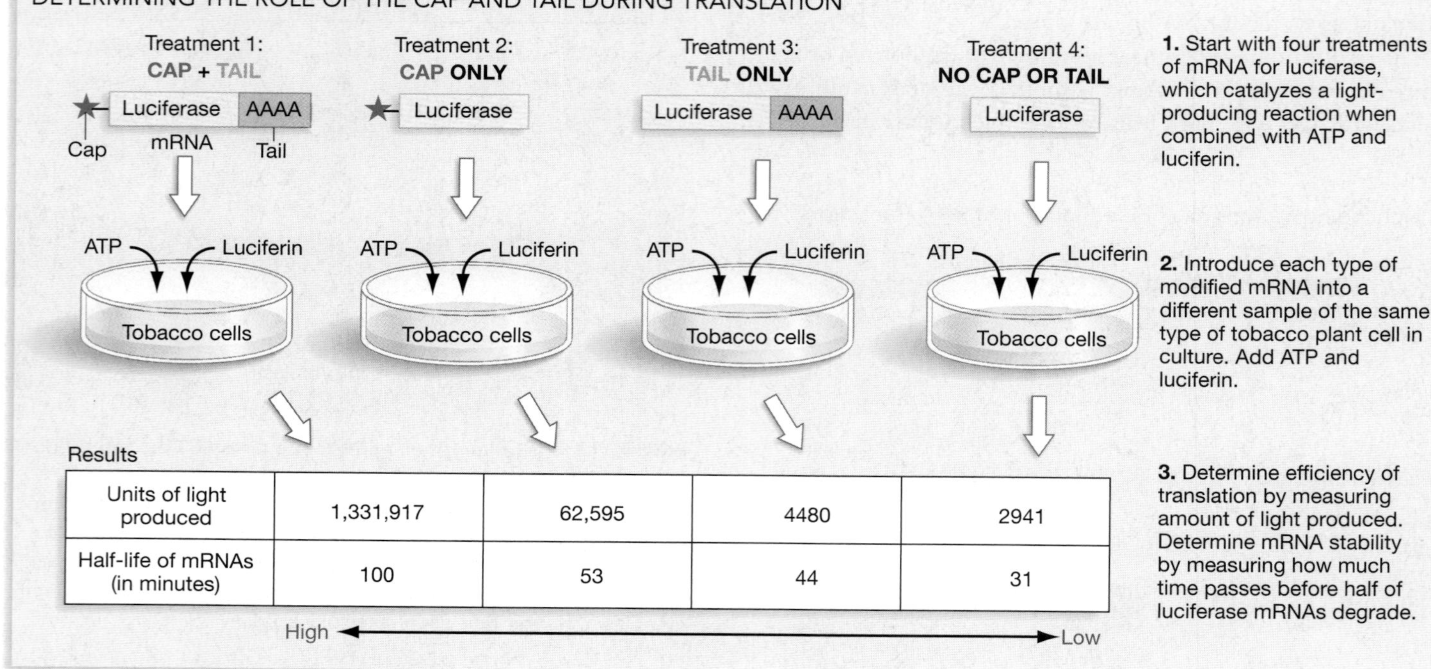

Results				
Units of light produced	1,331,917	62,595	4480	2941
Half-life of mRNAs (in minutes)	100	53	44	31

FIGURE 13.10 The Cap and Tail Make Translation More Efficient and Protect mRNAs from Degradation
The data at the bottom of the figure show that the cap and tail act synergistically—meaning that their combined effect is more pronounced than the effect that either has independently.

and the rate at which cells synthesize proteins. For example, mature blood cells have low rates of protein synthesis and very few ribosomes. But immature red blood cells divide rapidly, actively synthesize the protein hemoglobin, and contain huge numbers of ribosomes. Based on this correlation, Brachet proposed that ribosomes are the site of protein synthesis in the cell.

Roy J. Britten and collaborators set out to test Brachet's hypothesis directly. Their strategy was to tag proteins with radioactive atoms during translation, then assess whether the radioactivity was associated with the cell's ribosomes.

The Britten group began by feeding a "pulse" of radioactive sulfate ($^{35}SO_4^{-2}$) to growing cultures of *E. coli*. They expected the cells to incorporate the radioactive sulfur into the amino acids methionine and cysteine—which contain sulfur (see Chapter 3)—and then into newly synthesized proteins. Fifteen seconds after adding the radioactive sulfate, the researchers "chased" the label by adding a large excess of nonradioactive sulfate to the culture medium. The idea was to vastly reduce incorporation of radioactive sulfur into proteins by swamping the system with nonradioactive sulfate. This pulse-chase strategy defined a narrow window of time during which newly synthesized proteins were tagged with radioactivity. If Brachet's hypothesis was correct, then the radioactive signal should be associated with ribosomes for a short period. Later, all of the radioactivity should be found in proteins independent of the ribosomes.

To test this hypothesis, Britten and co-workers removed a sample of *E. coli* cells immediately after the addition of the nonradioactive sulfate, and again two minutes later. Then they broke the cells open and separated the contents using density-gradient centrifugation (see Chapter 12).

Was the radioactive protein associated with ribosomes? If so, then the radioactive signal should be found further down the centrifuge tube, where dense ribosomes sediment, than if it were present in the cytoplasm. The actual centrifugation data are shown in **Figure 13.11**. Immediately after the 15-second labeling period, a significant fraction of the radioactive sulfur is found associated with the ribosome fraction. Two minutes later, virtually none of the radioactive atoms are associated with the ribosome fraction. Instead, the radioactivity appears with the rest of the free cellular protein fraction.

These data confirmed Brachet's hypothesis. Proteins are synthesized at a ribosome and then released. About a decade later, O. L. Miller and colleagues published an electron micrograph like the one in **Figure 13.12a,** which shows bacterial ribosomes in action. The images confirm that in bacteria, ribosomes attach to mRNAs and begin synthesizing proteins even before transcription is complete (see **Figure 13.12b**). Transcription and translation can occur concurrently in bacteria because there is no nuclear envelope to separate the two functions. In eukaryotes, however, mRNAs are processed in the nucleus and exported to the cytoplasm before ribosomes attach to them and begin translation. Compartmentalization of transcription and translation is a major distinction between prokaryotes and eukaryotes.

How Does an mRNA Triplet Specify an Amino Acid?

During translation, genetic information is converted from nucleic acids to proteins. In effect, the hereditary instructions are translated from one distinctive chemical language into another. How does this conversion happen?

One early hypothesis was that the nucleic acids and amino acids might interact directly. The proposal was that the bases in a particular codon were complementary in shape or charge to the side group of a particular amino acid (**Figure 13.13a**). But Francis Crick pointed out that the chemistry involved didn't make sense. For example, how could the nucleic acid bases interact with a hydrophobic amino acid side group, which does not form hydrogen bonds?

Crick proposed an alternative hypothesis. As **Figure 13.13b** shows, he suggested that some sort of adapter molecule holds

(a) Location of radioactive amino acids immediately after radioactive pulse

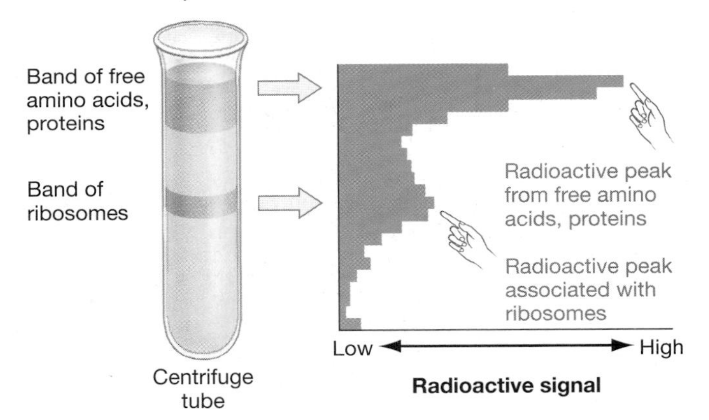

(b) Location of radioactive amino acids 2 minutes later

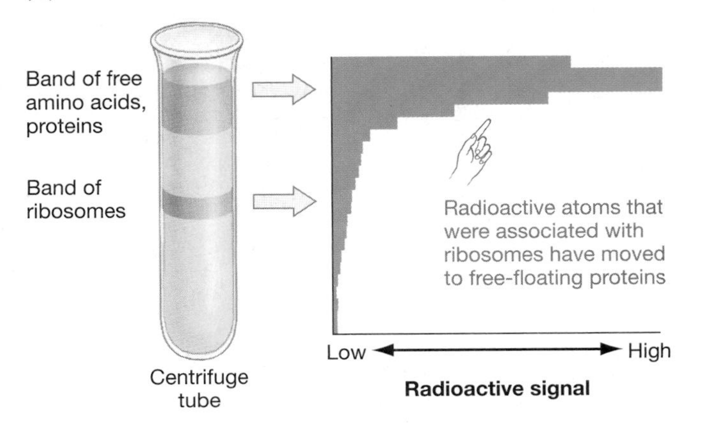

FIGURE 13.11 A Pulse-Chase Experiment Confirms That Ribosomes Are the Site of Protein Synthesis
These graphs indicate where radioactive amino acids are located when the contents of cells are subjected to density-gradient centrifugation. **(a)** Data from cells that were broken open after a 15-second pulse of radioactive sulfate. **(b)** Data from cells that were broken open two minutes after the radioactive pulse was chased with nonradioactive sulfur atoms.

(a) Micrograph of translation in action

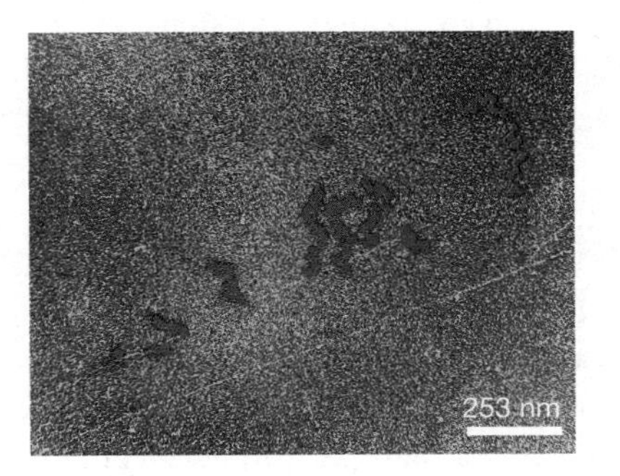

(b) Interpretation of micrograph

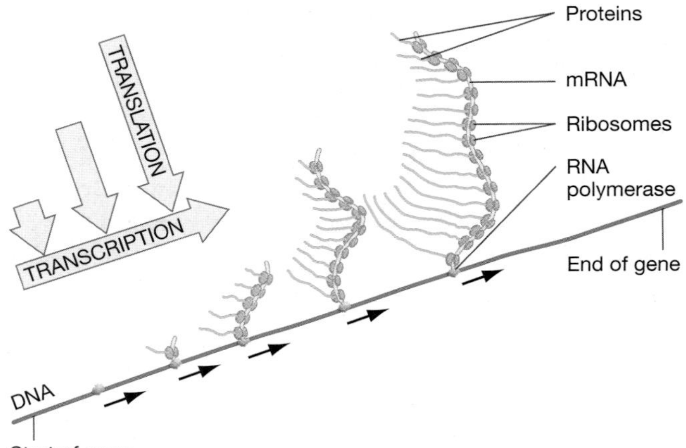

FIGURE 13.12 Translation in Action
Researchers interpreted the electron micrograph in part **(a)** as shown in the drawing in part **(b)**. Note that a series of RNA polymerase molecules are transcribing the DNA strand and producing a series of mRNAs. Then a series of ribosomes attaches to each of these mRNAs.

(a) HYPOTHESIS 1: Amino acids interact directly with mRNA codons.

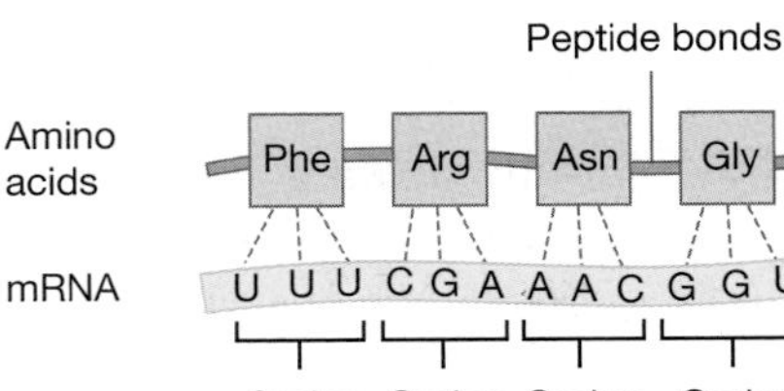

(b) HYPOTHESIS 2: Adapter molecules hold amino acids and interact with mRNA codons.

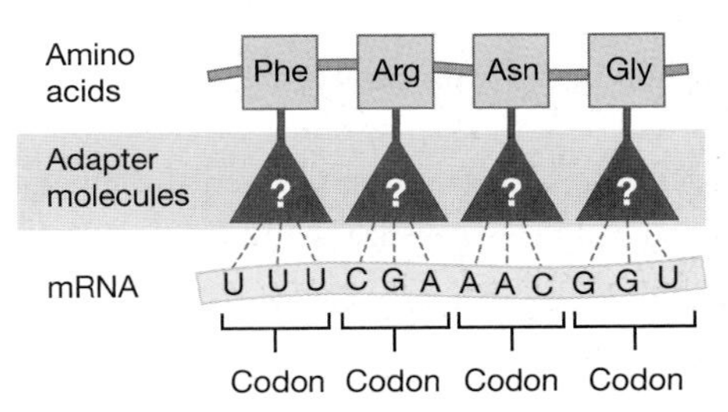

FIGURE 13.13 Two Hypotheses for How mRNA Codons Interact with Amino Acids

amino acids in place while interacting directly and specifically with a codon in mRNA via hydrogen bonding. In essence, Crick predicted the existence of a chemical go-between that produced a physical connection between the two types of molecules.

13.4 The Role of Transfer RNA

As Crick and others were developing the adapter-molecule hypothesis, biochemists were trying to identify the specific molecules required for protein synthesis. For example, Paul Zamecnik and co-workers worked out a cell-free protein synthesis system derived from mammalian liver cells. By process of elimination they showed that nuclei and mitochondria are not necessary for protein synthesis. In contrast, ribosomes, mRNA, amino acids, ATP, and a molecule called GTP are essential. (GTP is similar to ATP but contains guanosine instead of adenosine.) In addition, one indispensable component of the system was a crude cellular fraction that contained mostly proteins. On closer examination, however, Zamecnik discovered that it also contained a previously unknown type of RNA. Eventually, this class of molecules became known as **transfer RNA (tRNA)**.

What does tRNA do? A completely fortuitous discovery provided a clue. Zamecnik set up an experiment in which he added the amino acid leucine, labeled with the radioisotope ^{14}C, to his cell-free protein synthesizing system. The treatment was actually designed to serve as a control for an unrelated experiment. Zamecnik was amazed, however, to find that some of the RNA recovered from the reaction was radioactive. Intrigued by this result, he went on to show that the radioactive amino acids were attached to tRNA molecules. By manipulating the reaction conditions, he showed that the attachment of the amino acids to tRNAs requires an input of energy in the form of ATP.

A tRNA that becomes covalently linked to an amino acid, like the molecules that Zamecnik observed, is called an **aminoacyl tRNA**. More recent research has shown that enzymes called aminoacyl tRNA synthetases are responsible for catalyzing the addition of amino acids to tRNAs. For each of the 20 amino acids, there is a different aminoacyl synthetase and one or more tRNAs (**Figure 13.14**).

What is the function of aminoacyl tRNAs? By following the fate of radioactively labeled amino acids in Zamecnik's cell-free system, Mahlon Hoagland confirmed that aminoacyl tRNAs transfer their amino acids to proteins. Hoagland's evidence for this claim is shown in the graph in **Figure 13.15**. These data indicate the location of radioactive signal after Hoagland linked radioactive leucine to a tRNA, then added the labeled aminoacyl tRNA to a mixture of ribosomes and other components required for protein synthesis. The graph shows that over time, radioactivity was lost from the tRNA and incorporated into polypeptides synthesized on the ribosomes. This finding inspired the use of *transfer* in tRNA's name. (Before Hoagland's work, the molecules were known as soluble RNAs.)

Connecting Structure with Function

The experiments done in Zamecnik's lab confirmed that aminoacyl tRNAs act as the interpreter in the translation process. tRNAs are Crick's adapter molecules. They serve as the chemical go-betweens that allow amino acids to interact with an mRNA template. But precisely how does this connection occur?

The question was answered by research on tRNA's molecular structure. The initial studies established the sequence of nucleotides in various tRNAs, or what is termed their primary structure. Transfer RNAs are relatively short, ranging from 75 to 85 nucleotides in length. When Robert Holley and associates studied the primary sequence closely, however, they noticed that certain parts of the molecule could form distinctive secondary structures. (Recall from Chapter 3 that secondary structures form in proteins and DNA due to local interactions stabilized by hydrogen bonding.) Specifically, some of the nitrogenous bases in the tRNA could form hydrogen bonds with complementary bases in different regions of the molecule. If so, then short stretches of double-stranded RNA would form, similar to the DNA double helix. As a result, the entire tRNA

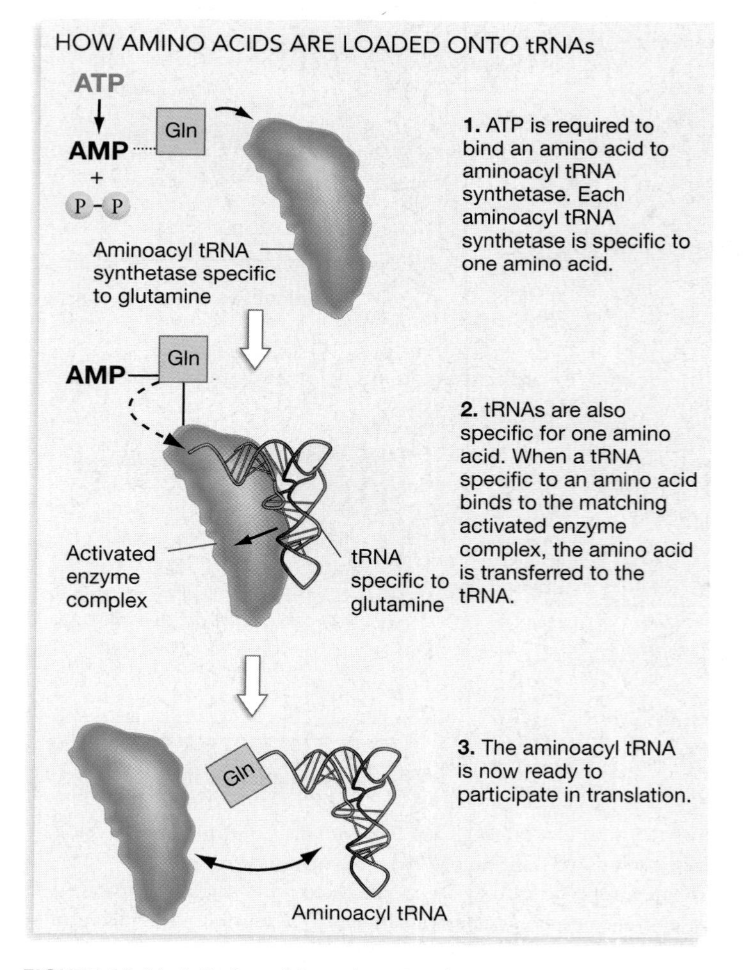

FIGURE 13.14 A Series of Reactions Produces Aminoacyl tRNAs
This diagram shows how an aminoacyl tRNA synthetase—in this case, for glutamine—becomes activated in a reaction that requires ATP.

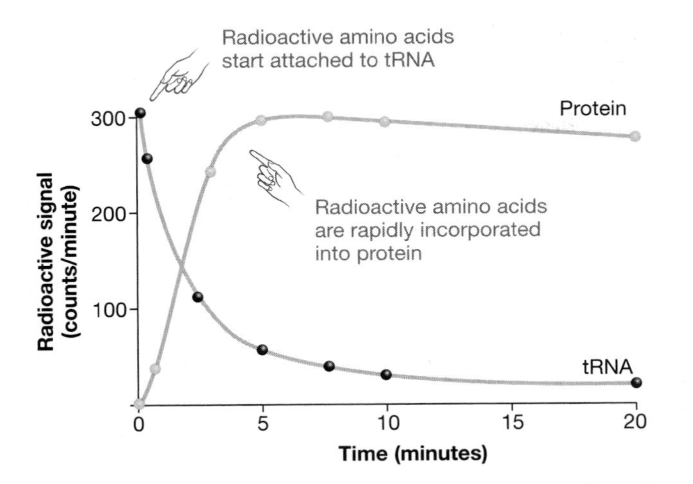

FIGURE 13.15 During Translation, Amino Acids Are Transferred from tRNAs to Proteins
Inside the ribosome, peptide bonds form between the polypeptide chain and the amino acid complexed to a tRNA.

molecule could assume the cloverleaf shape illustrated in **Figure 13.16a.** The stems in the cloverleaf are produced by complementary base pairing between different portions of the molecule, while the loops consist of single-stranded RNA.

Two aspects of this secondary structure proved especially interesting. The CCA sequence at the 3′ end of the molecule offered a binding site for amino acids, while the triplet on a loop of the cloverleaf on the opposite end could serve as an **anticodon.** The anticodon is a set of three nucleotides that forms base pairs with the mRNA codon specifying the amino acid carried by that tRNA. **Figure 13.16b** depicts an early model for how tRNAs connect mRNA codons with the appropriate amino acids.

This model had to be modified, however, when studies using a technique called x-ray crystallography (**Box 13.1**, page 264) revealed the tertiary structure of tRNAs. (Recall from Chapter 3 that the tertiary structure of a molecule is defined by the three-dimensional arrangement of its atoms.) According to these data, the cloverleaf structure folds over to produce a three-dimensional molecule shaped like an L (**Figure 13.16c**). The important thing to note about this structure is that it results in a precise separation between the anticodon and the attached amino acid. As we'll see in section 13.5, this separation is important for the positioning of the amino acid and the tRNA in the ribosome.

How Many tRNAs Are There?

When research succeeded in characterizing all the different types of tRNAs available in cells, a paradox arose. Sixty-one

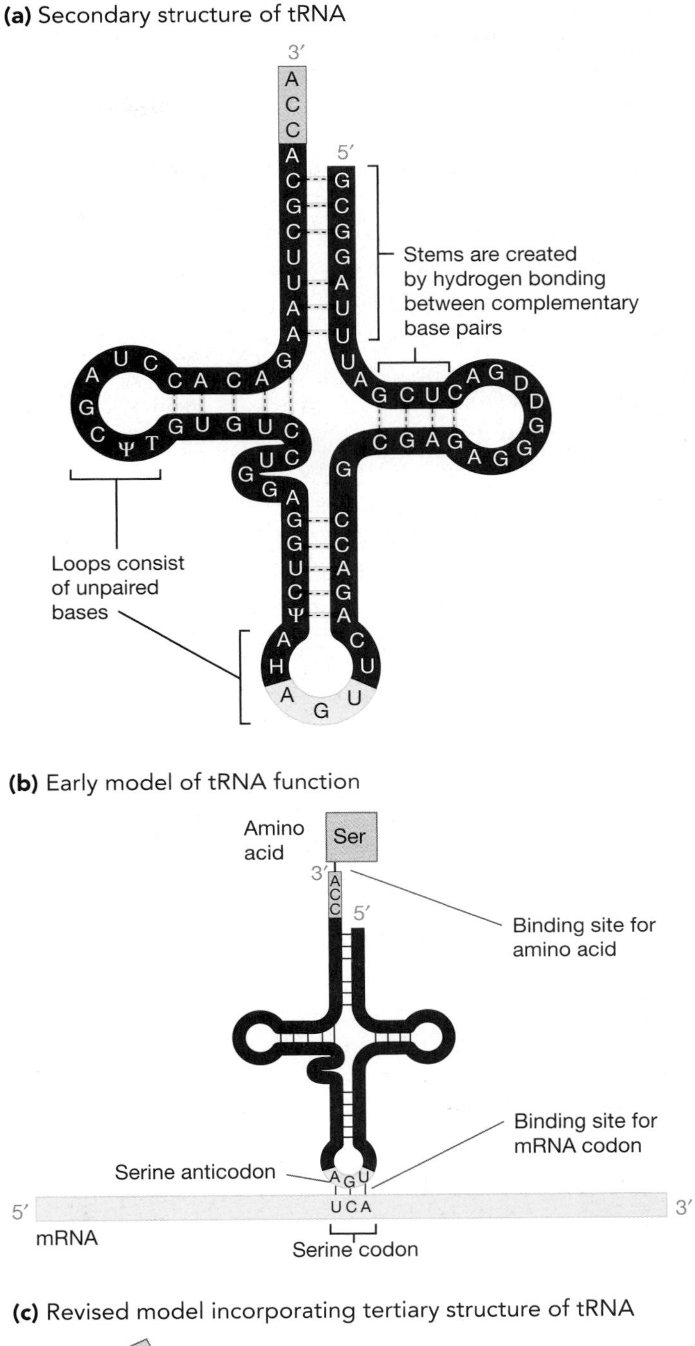

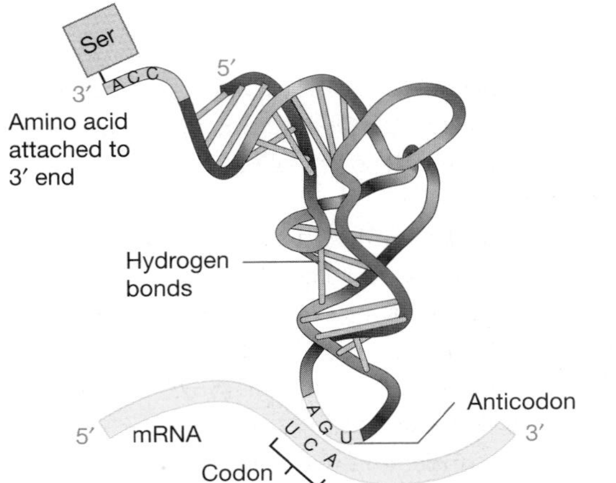

FIGURE 13.16 The Structure of Transfer RNA
(a) The secondary structure of tRNA resembles a cloverleaf. **(b)** If amino acids are attached to the 3′ end of a tRNA with an anticodon appropriate for that amino acid, then the tRNA will match the mRNA codon and insert the correct amino acid into the growing polypeptide chain during translation. **(c)** Recent data indicate that tRNAs have an L-shaped tertiary structure.

BOX 13.1 An Introduction to X-Ray Crystallography

X-ray crystallography is the most widely used technique for reconstructing the three-dimensional structure of molecules. As its name implies, the procedure is based on bombarding crystals of a molecule with x-rays. The crystals scatter the x-rays, producing a diffraction pattern that can be recorded on x-ray film or other types of detectors. (The approach is also called x-ray diffraction analysis.)

Although the technical details of the technique are far beyond the scope of this book, the basic principle is that x-rays are scattered in precise ways when they interact with the electrons surrounding the atoms in a crystal (**Figure 1**). By varying the orientation of the x-ray beam that strikes a crystal and documenting the diffraction patterns that result, researchers can record a complex set of patterns that reflect the three-dimensional structure of the crystal.

Using mathematical principles, the diffraction pattern can be used to construct a map representing the electron density in the crystal. By relating these electron-density maps to information about the primary structure of the protein, a three-dimensional model of the molecule can be built.

It is important to appreciate the amount of work involved in producing an accurate three-dimensional representation of a protein or RNA. It frequently takes hundreds or thousands of failed attempts before researchers produce the crystals of sufficient quality for x-ray diffraction analysis. But the information contained in an accurate three-dimensional model makes the effort worthwhile. Efforts to visualize polymerases, tRNAs, ribosomes, and ribosome subunits have produced important insights into the mechanisms of transcription and translation.

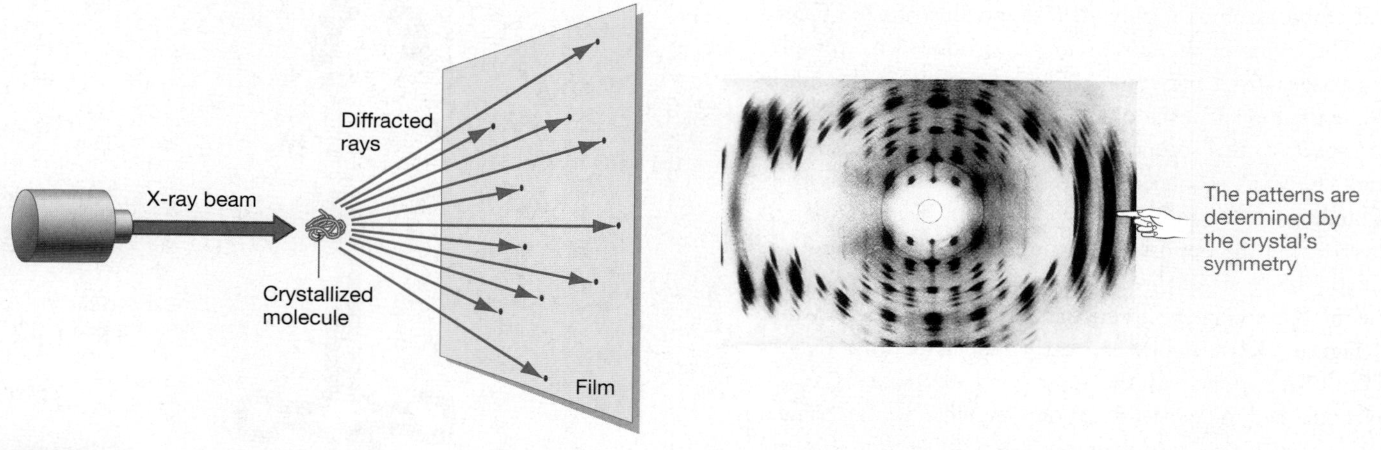

FIGURE 1
When crystallized molecules are bombarded with x-rays, the radiation is scattered in distinctive patterns. The photograph at the right shows an x-ray film that recorded the pattern of scattered radiation.

different mRNA codons specify an amino acid, but cells typically contain only about 40 different tRNAs. How can all of the messages in the genome be translated with only two-thirds of the tRNAs required?

In answer to this paradox, Francis Crick proposed what is known as the wobble hypothesis. To understand his logic, recall from Chapter 11 that many amino acids are specified by more than one codon. Further, recollect that codons for the same amino acid tend to have the same nucleotides at the first and second positions but a different nucleotide at the third position. For example, both CAA and CAG act as codons for the amino acid glutamine. (Codons are written in the $5' \rightarrow 3'$ direction.) Surprisingly, experimental data have shown that a tRNA with an anticodon of GUU can form base pairs with both codons in mRNA. (Anticodons are written in the $3' \rightarrow 5'$ direction.) The GUU anticodon matches the first two bases (C and A) in both cases, but the U in the third position forms a nonstandard base pair with a G in the CAG codon.

Crick proposed that the requirement for base pairing in the third position of the codon is not as strict as for the first two positions. If so, it would allow a limited flexibility or "wobbling" in the base pairing. According to the **wobble hypothesis,** a nonstandard G–U base pair in the third position is acceptable as long as it does not change the amino acid meaning of the codon.

13.5 The Ribosome

When it became clear that ribosomes are the site of protein synthesis, an intensive effort was begun to understand the details of the process. Some of the first questions were about the machinery itself. What are ribosomes made of and how are they put together?

Masayasu Nomura and co-workers approached this problem like kids who decide to take a bicycle apart. Their goal was to understand the system by separating the pieces, sorting through them, and putting them back together so that the machine worked. Nomura's group was able to separate the various components of ribosomes, identify them by size and chemical composition, and put them back together into a functional, reconstituted system. (As we'll see, other workers followed up on this research by exploring the effects of adding, deleting, or modifying various components.)

The earliest studies by Nomura and others were based on separating pieces of the ribosome by centrifugation. These studies established that ribosomes have two major units. In bacteria, the smaller subunit is designated 30S and the larger subunit 50S. (The "S" stands for a unit called the Svedberg, which describes how rapidly a particle sediments down a centrifuge tube.)

When researchers broke each ribosomal subunit apart and analyzed the components chemically, they found that each consists of **ribosomal RNA (rRNA)** and proteins. The large subunit in *E. coli*, for example, contains two different rRNA molecules and about 35 proteins. When Nomura and co-workers purified all of these components and added them one by one to an in vitro translation system, they made two fundamental observations. First, ribosomes self-assemble—no energy inputs or enzymes are required. Second, for translation to work at a normal rate, all of the components needed to be present.

Understanding the overall structure of ribosomes is helpful, but what about the protein synthesis process itself? How do tRNAs, mRNAs, and ribosomes come together for translation to take place?

How Does Translation Occur?

For a ribosome to translate an mRNA, it must begin at the start of the message, go to the end, and then stop. Biologists call these three phases of translation initiation, elongation, and termination. The details of each phase were unraveled over the course of several decades, by teams of biologists working at many different laboratories around the world. In many cases the key experimental approaches involved using antibiotics or other drugs to poison specific steps in the translation process. Let's take a closer look.

Initiation The key to understanding initiation is to recall, from Chapter 11, that the codon AUG is found at the start of all mRNAs and codes for the amino acid methionine. As **Figure 13.17** shows, translation begins when the small subunit of a ribosome binds to a sequence on the mRNA just upstream from the AUG that indicates the start site for translation. In bacteria, this region of the mRNA is called the Shine-Dalgarno sequence. The interaction between the small subunit and the message is mediated by proteins called initiation factors.

Once this initial binding event has occurred, an aminoacyl tRNA bearing methionine associates with other initiation factors and binds to the AUG codon. (In bacteria, the initial methionine has a formyl group attached at its amino end, so the first amino acid in an mRNA is *N*-formylmethionine.) Initiation is complete when the large ribosome joins the complex.

Before moving on, however, it is important to note that the AUG codon and the tRNA bearing methionine or *N*-formylmethionine interact in a specific location inside the ribosome. This location is called the **P-site** because it is the location where peptide bond formation takes place. You saw in Chapter 3 that amino acids become linked through peptide bonds during protein synthesis. The "P" in P-site stands for peptidyl. When peptide bonds form, the protein that is being synthesized elongates. How does peptide bond formation occur?

Elongation The peptide bonds that link amino acids into polypeptides form between the amino group on one amino acid and the carboxyl group on another (for more detail about this condensation reaction, see Chapter 3). The amino acids carried

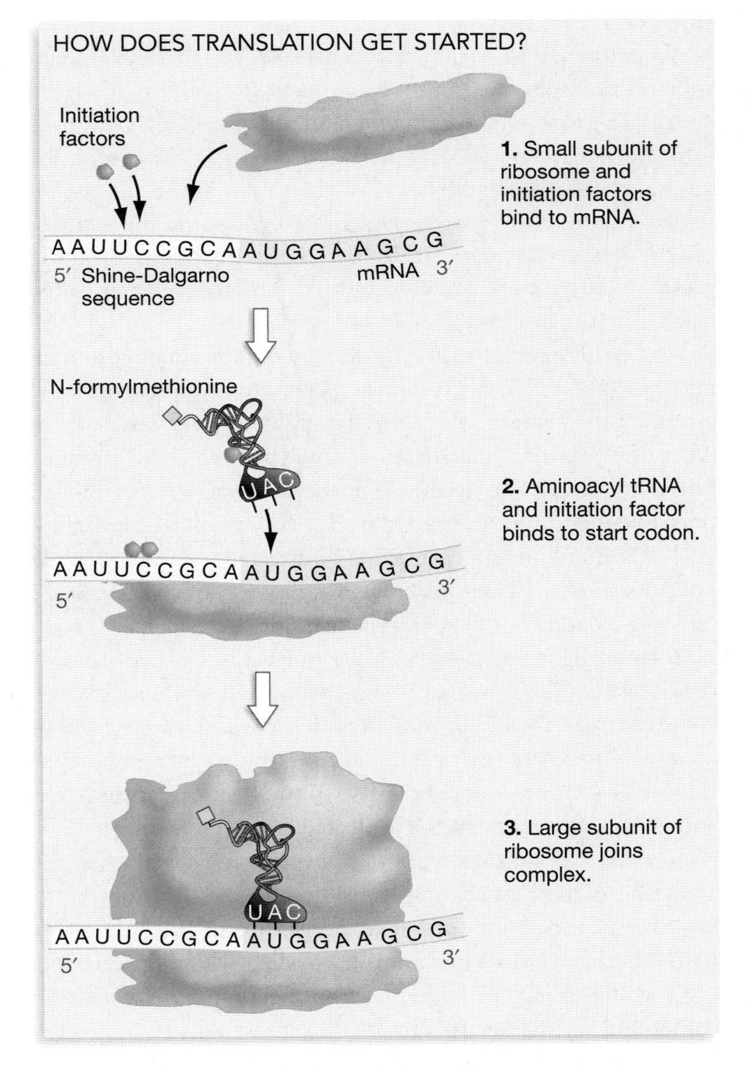

FIGURE 13.17 A Model of Translation Initiation in Bacteria
This diagram summarizes the current model for the sequence of events that occurs as a bacterial ribosome assembles and begins to translate an mRNA.

by tRNAs have an exposed amino group. It makes sense, then, to infer that they must join to the carboxy end of a growing polypeptide chain somewhere inside the ribosome.

Experiments with an antibiotic called puromycin provided the first insights into how and where peptide bonds form. Puromycin bears a close resemblance to the 3′ end of an aminoacyl tRNA; it also resembles an amino acid because it can form a peptide bond. When added to cells, puromycin poisons them by shutting down translation. To understand how this happens, researchers labeled amino acids with a radioactive isotope and added them to in vitro translation systems with an RNA template, tRNAs, and aminoacyl tRNA synthetases. Some experiments were carried out in the presence of puromycin and some without. As **Figure 13.18** shows, the results were striking. When puromycin was not present, labeled polypeptides were found attached to tRNAs inside ribosomes. When puromycin was present, however, the antibiotic formed peptide bonds with the polypeptide chains and split them from tRNAs.

To explain this result, researchers suggested that elongation proceeds in the four-step process illustrated in **Figure 13.19**. The first step occurs when an aminoacyl tRNA moves into the ribosome and occupies an area called the **A-site**, for aminoacyl. For this aminoacyl tRNA to remain in the A-site, its anticodon has to match the exposed codon on the mRNA. As the figure shows, a tRNA that holds methionine or the growing polypeptide chain occupies the adjacent region called the P-site (peptidyl site).

As a result of this first step in elongation, the amino group on an aminoacyl tRNA is positioned next to the exposed carboxyl group on the peptidyl tRNA. Step 2 in Figure 13.19 shows that when the peptide bond forms, the polypeptide chain is transferred to the tRNA in the A-site. According to this model, puromycin occupies the A-site because it resembles an aminoacyl tRNA, and the polypeptide is transferred to it after peptide bond formation. When puromycin is present, translation stops after this peptide bonding and the transfer step occurs.

What happens next? Step 3 in elongation is called translocation. The mRNA moves toward its 3′ end so that the ribosome can read the next codon. When it does, the empty tRNA moves to the **E-site** (exit site), and the freshly synthesized peptidyl tRNA moves from the A-site to the P-site. Translocation ends when the empty tRNA is ejected from the E-site.

Just as experiments with puromycin established that the growing polypeptide chain is transferred to the tRNA in the A-site after peptide bond formation occurs in step 2, experiments with other drugs were critical to unraveling what happens during translocation. For example, translocation stops when the compound fusidic acid is present because it poisons a protein called an elongation factor, which moves the mRNA along the ribosome. The antibiotic erythromycin stops translocation by plugging the exit channel where the growing polypeptide leaves the ribosome. When erythromycin is present, the freshly synthesized amino acid chain gets jammed inside the ribosome and translocation cannot occur.

These three steps—incoming aminoacyl tRNA, peptide bond formation, and translocation—repeat down the length of the mRNA. Studies with in vitro translation systems have confirmed that each elongation cycle depends on an input of energy from several GTP molecules as well as assistance from the proteins called elongation factors.

Termination The genetic code presented in Chapter 11 included three stop codons, which function as signals for terminating translation. When a ribosome reaches UAA, UAG, or UGA, proteins called release factors enter the A-site of the ribosome and terminate translation. The completed polypeptide

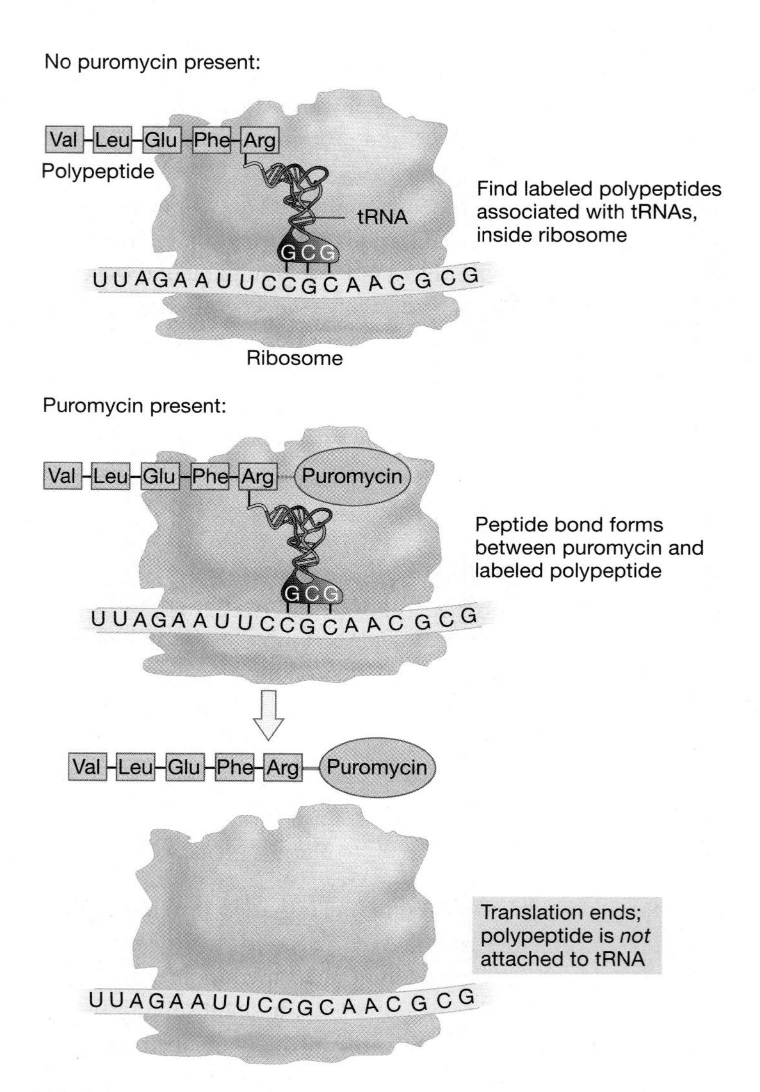

FIGURE 13.18 New Polypeptide Bonds Are Formed Within the Ribosome
Puromycin terminates translation prematurely because it resembles the 3′ end of an aminoacyl tRNA, and because it is capable of forming a peptide bond. When it is added to an in vitro translation system, it forms a peptide bond with the polypeptide held by a tRNA inside the ribosome.

is released from the last tRNA in the P-site, the ribosome separates from the mRNA, and the two ribosomal subunits dissociate. The subunits are ready to attach to the start codon of another message and start translation anew. Before going on to explore the fate of freshly minted proteins, let's examine some key aspects of ribosome structure and function in more detail.

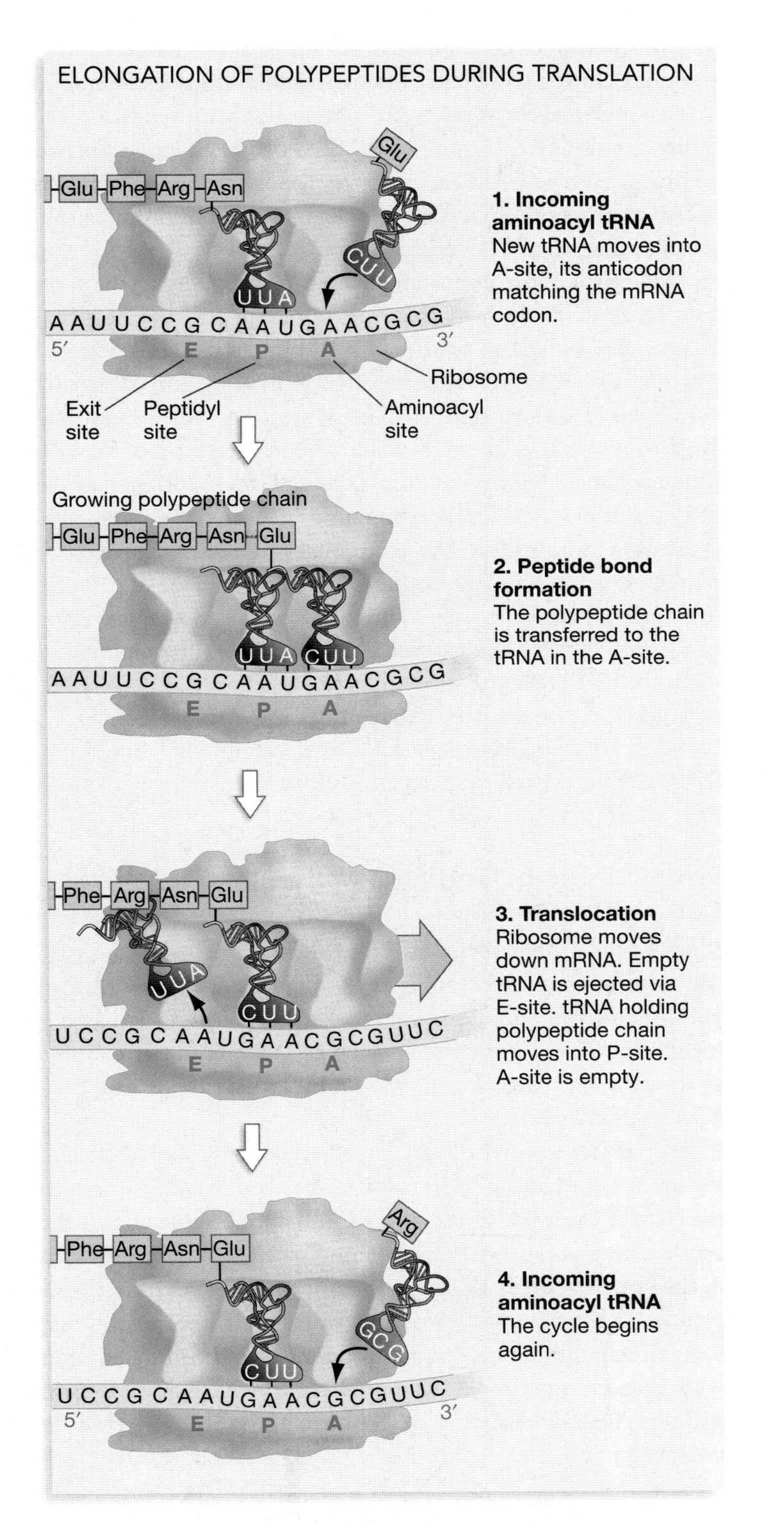

FIGURE 13.19 A Model for How Polypeptides Elongate
This model was inspired by data from experiments with antibiotics that poison each of the steps illustrated here. Puromycin occupies the A-site and poisons step 1.

How Is Peptide Bond Formation Catalyzed?

The RNA molecules in a ribosome account for 60 percent of its mass. Until the discovery of ribozymes, these rRNAs were viewed as an inert structural matrix. (Ribozymes are the RNA enzymes introduced in Chapter 3.) Once it was established that RNA molecules could catalyze reactions, an important question emerged. Is the formation of the peptide bond catalyzed by a ribozyme or by a protein enzyme?

Even before ribozymes were discovered, several types of data indicated that the active site might lie in an rRNA and not in a protein. When the translation reaction was reconstituted in vitro, catalytic activity did not appear or disappear with the addition or subtraction of any single ribosomal protein. Other observations suggested that the largest rRNA molecule (called 23S rRNA in bacteria and 28S rRNA in eukaryotes) contains the catalytic center. For example, an antibiotic called chloramphenicol inhibits catalysis. Some strains of *E. coli* can grow in the presence of this poison, however. The key observation is that the resistant cells have a mutation in their 23S rRNA, which suggests that rRNA is the target of the antibiotic. In other strains of *E. coli,* mutations in the nucleotide sequence of 23S rRNA result in loss of catalytic activity of the ribosome. These cells show a direct link between a defect in catalysis and a defect in rRNA.

After ribozymes were discovered, Harry Noller and colleagues began a series of experiments to test whether 23S rRNA can catalyze peptide bond formation in vitro in the absence of ribosomal proteins. In one recent study, the researchers treated *E. coli* ribosomes with three different agents that degrade proteins. Although chemical analyses confirmed that only a few tiny fragments of ribosomal proteins remained, the treated ribosomes still catalyzed peptide bond formation. Noller cautioned that this evidence was still not definitive, however, because it was possible that enough protein remained to catalyze the reaction.

Most researchers agreed that the issue of peptide bond catalysis would not be settled until the three-dimensional structure of the ribosome was understood in enough detail to know whether proteins or RNA occupied the space at the junction of the A-site and P-site, where peptide bond formation actually takes place. Late in the year 2000, papers published by two teams of biologists documented the ribosome's structure at the atomic scale and settled the question for good.

What Does the Ribosome Actually Look Like?

Ribosomes have been studied intensively for decades. But until very recently researchers had only a general idea of what ribosomes looked like. It has long been known that they are made up of a small subunit and a large subunit, and that the two pieces assemble on an mRNA to form a complete ribosome at the start of translation and disassemble when translation of

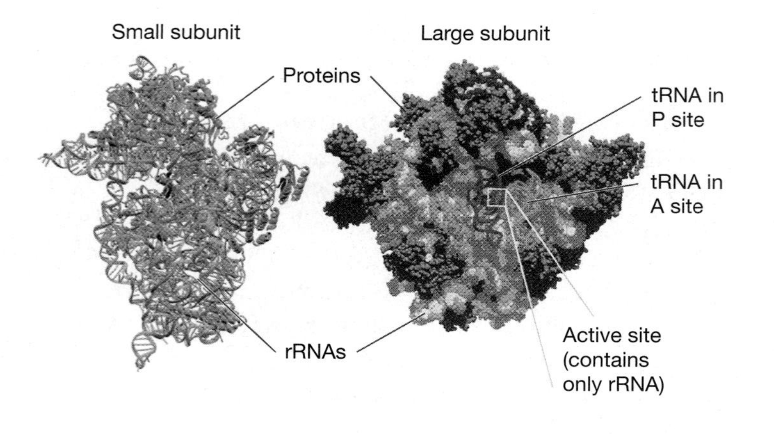

FIGURE 13.20 The Three-Dimensional Structure of the Ribosome X-ray crystallographic data confirmed that tRNAs sit in the A-site and P-site of the ribosome exactly as predicted by the model in Figure 13.19.

that mRNA is complete. It has also been established that mRNA attaches to the small subunit, while the catalytic center resides in the large subunit. But the details of ribosome structure were a mystery.

X-ray crystallographic data, painstakingly assembled since 1980 by biologists in laboratories around the world, gradually began to reveal more and more detail about the structure of the ribosome. **Figure 13.20** shows some of the most recent images available for the subunits of the bacterial ribosome. In the ribbon model of the small subunit (at left in the figure), the surface facing you is the one that contacts mRNA during initiation and elongation. Most of the structure that contacts the message consists of rRNA (shown in gray), and not protein.

At the right in Figure 13.20 is the large subunit, oriented so that you are looking up into the structure from the small subunit. The bright red molecule is a tRNA in the P-site, and the bright green molecule is a tRNA in the A-site. The region where peptide bond formation occurs is enclosed in a yellow box, and it consists entirely of rRNA. No proteins, which are shown in purple, are closer than 18 angstroms from the site where catalysis takes place. (One angstrom is one-tenth of a nanometer, or 10^{-10} m.) Based on these results, biologists are convinced that the ribosome indeed contains a ribozyme. The production of proteins is catalyzed by RNA.

13.6 Post-Translational Events

The function of proteins depends crucially on their shape. To help drive this point home, **Box 13.2** explores how a change in the shape of the prion ("PREE-on") protein found in mammals converts normal molecules into abnormally folded proteins that cause the brains of humans and other mammals to disintegrate. Because shape is so important to the function of proteins, researchers want to understand how the linear sequence of amino acids in a newly synthesized polypeptide is converted to the unique three-dimensional structure of an active enzyme.

Christian Anfinsen and colleagues provided a preliminary answer to this question through studies of the ribonuclease (RNase) enzyme. These researchers treated pure preparations of this protein with chemicals that eliminated its tertiary structure and converted it back into an unfolded string of amino acids. After the chemicals were removed, they found that the molecules slowly and spontaneously folded back into a normal conformation. This led Anfinsen to propose the thermodynamic hypothesis of protein folding, which holds that proteins fold into their most stable energetic state.

One of the most important observations made by these researchers, however, was that the refolding process occurred much more slowly in vitro than it did in cells. This led to the hypothesis that enzymes are involved in folding, and eventually led to the identification of a class of proteins called **molecular chaperones** that facilitate the folding process. Many of the molecular chaperones belong to a family of molecules called the heat-shock proteins. These compounds acquired their name because they are produced in large quantities after cells experience high temperatures or other treatments that make proteins lose their tertiary structure. The idea is that the newly synthesized heat-shock proteins speed the refolding of proteins into their normal shape after unfolding has occurred.

Folding occurs as new polypeptides are synthesized and released from ribosomes. Once folding is complete, some proteins are transported to specific locations while others are modified in some way.

Protein Sorting: Getting a Molecule to Its Correct Destination

Many of the proteins synthesized by the cell are destined for locations other than the cytoplasm. As **Figure 13.21** shows, there are three such destinations: the nucleus, organelles such as mitochondria or chloroplasts, and the cell membrane or exterior of the cell. How does a protein "know" where to go, and how does it get there?

As Chapter 5 showed, proteins destined for a site other than the cytoplasm have a sorting signal that directs the protein to the correct location. In the case of mitochondrial proteins, the signal consists of a short series of amino acids on the amino end of the polypeptide. By studying mutant forms of yeast that have defects in mitochondria formation, researchers have established that mitochondrial proteins are imported into the mitochondria with help from molecular chaperones in the heat-shock family, and that they fold into their active form after they are inside the organelle.

Post-Translational Modification

In some cases, proteins are synthesized in an inactive form. The vertebrate hormone insulin, for example, is not active when it

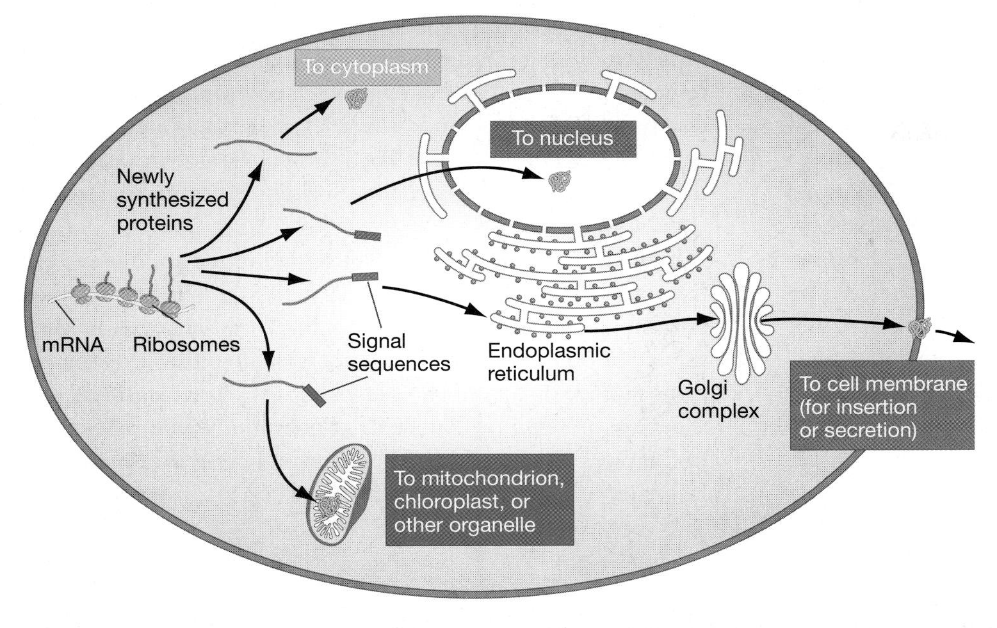

FIGURE 13.21 Newly Synthesized Proteins Have Several Destinations
Signal peptides on the ends of newly synthesized proteins mark them for transport to specific organelles or to the nucleus. Proteins bound for the cytoplasm do not have signal peptides.

BOX 13.2 Prions

Over the past several decades, evidence has accumulated that certain proteins can act as infectious, disease-causing agents. The proteins involved are called prions, for proteinaceous infectious particles, and their existence has been intensely controversial. The reason for the debate? Prions violate one of the central tenets of molecular biology, which is that only nucleic acids (RNA or DNA) can contain hereditary information.

How could a protein transmit information to another individual and cause a change in phenotype? According to the prion hypothesis developed by Stanley Prusiner, infectious proteins are improperly folded forms of normal proteins that are encoded in DNA and present in healthy individuals. The infectious and normal forms do not necessarily differ in amino acid sequence, however. Instead, their shapes are radically different. Further, the infectious form of a protein can induce the normal molecule to change its shape to the mutant form. In this way, the prion transmits information.

Figure 1 illustrates the shape differences observed in the normal and infectious forms of the first prion described. The molecule on the left is called the prion protein (PrP) and is a normal component of mammalian cells. Mutant versions of the protein, like the one on the right, are found in a wide variety of species and cause a family of diseases known as the spongiform encephalopathies—literally, "sponge-brain illnesses." Hamsters, cows, goats, and humans afflicted with these diseases suffer from massive degeneration of the brain. Cattle suffer from "Mad Cow disease"; sheep and goats acquire scrapie (so-called because the animals itch so badly that they scratch off their wool or hair); humans develop kuru or Creutzfeldt-Jakob disease. Some mutant forms of the proteins are more easily induced to adopt the infectious form. In these cases, the diseases are hereditary. In other cases the disease is transmitted when individuals eat tissues containing the infectious form of PrP. All are fatal.

Although the association between shape change in prion proteins and disease has been confirmed repeatedly, several important questions about the prion hypothesis remain to be answered. Perhaps the most important of these concerns the claim that protein itself, and not a virus associated with the protein, contains the information that produces the shape change. The critical experiment to test this hypothesis would involve synthesizing prion proteins in vitro, to assure that they are completely free of viruses, and then showing that the naked proteins fold improperly and cause infections and disease. To date, however, investigators have not been able to accomplish the synthesis and perform the experiment.

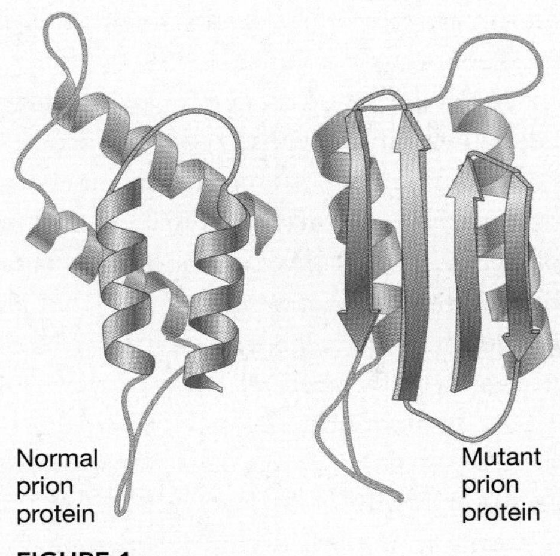

FIGURE 1
In this ribbon model of a normal prion protein (left) and the mutant form (right) that causes "Mad Cow disease" in cattle, the coils represent the secondary structures called α-helices and the arrows β-pleated sheets.

initially comes off the ribosome. In order for insulin to begin regulating sugar levels in the blood, the protein has to be activated by the removal of certain amino acids before it is secreted from the source cell. This step converts an inactive peptide called proinsulin to the active insulin molecule. Many proteins undergo some sort of chemical modification after translation. In many or most cases, these modifications change the molecule's activity.

One particularly important type of modification involves the addition of a phosphate group to an amino acid side chain. This phenomenon is known as **phosphorylation**. Because a phosphate group has two negative charges, phosphorylating a protein can cause major changes in shape and chemical reactivity. In eukaryotes, phosphorylation often occurs in response to signals from outside the cell, such as arrival of a hormone, and results in shape changes that turn the phosphorylated enzyme on or off.

Why are these types of chemical modification important? Post-translational modifications allow cells to quickly respond to new conditions. For example, when physiological conditions demand a decrease in blood-sugar levels, insulin can be activated extremely rapidly. Activating a protein via post-translational modification is much faster than waiting for transcription and translation.

In many cases, however, cells respond to intracellular or extracellular signals by altering which genes are being transcribed and in what amounts. Understanding how changes in gene transcription occur is vital for understanding how growing organisms develop and how diseases like cancer get started. How is gene expression regulated? That question, which we explore in Chapters 14 and 15, may be the most important of all issues facing molecular biologists today.

Essay Transcription, Translation, and Toxins

Although wild mushrooms can be delicious, dining on certain species can be lethal. For example, in late summer and fall the fruiting bodies of the death cap mushroom (*Amanita phalloides*) appear in forests, fields, and backyards in North America and Europe. The death cap is among the most poisonous of all mushrooms and is thought to account for 90 percent of mushroom poisoning deaths worldwide. Ingestion of a single mushroom cap can kill a healthy adult.

Toxins have been valuable tools for studying translation.

The active agent in death cap is a toxin called α-amanitin (alpha-amanitin), whose mode of action turned out to be a boon for researchers interested in understanding how transcription occurs in eukaryotes. Biologists observed that production of tRNAs and rRNAs continued in cells treated with low concentrations of α-amanitin even though protein synthesis stopped. At very high concentrations of the toxin, tRNA production also stopped, but transcription of rRNAs continued. These observations led to the discovery that eukaryotes have three RNA polymerases, each with a distinct role in the transcription of mRNA, tRNA, and rRNA. Low concentrations of α-amanitin inhibit RNA polymerase II, which synthesizes mRNAs, while high concentrations poison RNA polymerase III, which synthesizes tRNAs. RNA polymerase I is not affected by the poisons.

Toxins have also been valuable tools for biologists interested in studying translation. Several widely prescribed antibiotics work because they inhibit a particular stage of protein synthesis only in bacteria. Streptomycin binds to the small subunit of the ribosome and prevents accurate reading of the mRNA; tetracycline stops the elongation of growing polypeptides by preventing aminoacyl tRNAs from entering the A-site in the bacterial ribosome. Because these drugs do not harm eukaryotic ribosomes, they are effective agents for fighting bacterial infections in humans and domestic animals.

Unfortunately, bacterial ribosomes are not the only targets of translation inhibitors. A toxin produced by certain strains of the bacterium *Corynebacterium diphtheriae* binds to an elongation factor and inhibits the translocation step in eukaryotic translation. As a result, the toxin contributes to the disease called diptheria in humans. (You may have been vaccinated against this bacterium as an infant when you received a series of DPT shots for diptheria, pertussis, and tetanus.)

Diptheria toxin, streptomycin, tetracycline, and α-amanitin have all served as tools in the effort to understand transcription and translation. Toxins are useful research tools in biochemistry for the same reason that knock-out mutations, which completely disable genes, are useful in genetics. These types of compounds allow researchers to understand what happens when a molecule or a particular stage in a process does *not* work. This is similar, in concept, to learning how a car works by removing one part at a time and carefully recording the machine's response, or to learning how the brain works by studying the behavior of people whose brains have been damaged in specific, quantifiable ways by accidents. Toxins are a molecular dissecting kit. They let biologists explore the intricacies of cell function by inactivating one component at a time.

Chapter Review

Summary

The instructions for making and operating a cell are archived in DNA, transcribed into messenger RNA, and then translated into protein. Early experiments used filters that separated proteins from naked DNA to establish that RNA polymerase begins transcription by binding to specific sites, called promoter sequences, in DNA. In bacteria this binding occurs in conjunction with a protein called sigma factor. Sigma proteins recognize conserved sequences within promoters that are located 10 bases upstream from the start of the actual genetic message; RNA polymerase itself binds to a conserved sequence –35 bases from the start site. These binding sites ensure that the correct strand is transcribed in the correct direction. Eukaryotic promoters are complex and varied compared with bacterial promoters, and many more transcription factors are present.

After transcription of a eukaryotic mRNA is complete, several important events occur. Stretches of noncoding RNA called introns are spliced out by small complexes called snRNPs, a "cap" signal is added to the 5′ end, and a poly (A) tail is added to the 3′ end. Experiments with modified mRNAs established that both the 5′ cap and the poly (A) tail serve as recognition signals for the translation machinery, and they protect the message from degradation by RNases.

Experiments with radioactively labeled amino acids confirmed that ribosomes are the site of protein synthesis, and that transfer RNAs (tRNAs) serve as the chemical bridge between the RNA message and the polypeptide product. tRNAs have a predicted secondary structure that resembles a cloverleaf and an L-shaped tertiary structure. One leg of the L contains the anticodon that forms a base pair with the mRNA codon, while the other leg holds the amino acid appropriate for that codon. Because imprecise pairing—or "wobbling"—is allowed in the third positions of anticodons, only about 40 different tRNAs are required to translate the 61 codons that code for amino acids.

By using antibiotics that cause translation to stop at different stages, researchers were able to establish that the process of elongating a polypeptide involves three steps: (1) an incoming aminoacyl tRNA occupies a position in the ribosome called the A-site; (2) the growing polypeptide chain is transferred from a peptidyl tRNA in the ribosome's P-site to the tRNA in the A-site and a peptide bond is formed; and (3) the ribosome is translocated to the next codon on the mRNA, accompanied by ejection of the empty tRNA from the E-site. Recent data on the three-dimensional structure of the ribosome has confirmed that peptide bond formation is catalyzed by a ribozyme, not an enyzme.

While translation is in progress, proteins fold into their three-dimensional conformation (tertiary structure), sometimes with the aid of chaperone proteins. Some proteins are targeted to specific locations in the cell by the presence of signal sequences, while others remain in an inactive form until modified by phosphorylation or the removal of certain amino acids.

Questions

Content Review

1. How did the A-site of the ribosome get its name?
   a. It is where amino acids are affixed to tRNAs, producing aminoacyl tRNAs.
   b. It is where the amino group on the growing polypeptide chain is available for peptide bond formation.
   c. It is the site occupied by incoming aminoacyl tRNAs.
   d. It is surrounded by α-helices of ribosomal proteins.

2. How did the P-site of the ribosome get its name?
   a. It is where the promoter resides.
   b. It is where peptidyl tRNAs reside.
   c. It is the site where peptide bond formation takes place.
   d. It is the site where the growing polypeptide chain is phosphorylated.

3. What is a molecular chaperone?
   a. a protein that recognizes the promoter and guides the binding of RNA polymerase
   b. a protein that guides newly translated proteins to their correct destination in the cell
   c. a protein that helps newly translated proteins fold into their proper 3-D configuration
   d. a protein that is a component of the large ribosomal subunit, and that assists with peptide bond formation

4. The three types of RNA polymerase found in eukaryotic cells transcribe different types of genes. What does RNA polymerase II produce?
   a. rRNAs
   b. tRNAs
   c. mRNAs
   d. the small RNAs found in snRNPs

5. What is an anticodon?
   a. the part of an RNA message that signals the termination of translation
   b. the part of an RNA message that signals the start of translation
   c. the part of a tRNA that binds to a complementary codon in mRNA
   d. the part of a tRNA that accepts an amino acid, via a reaction catalyzed by tRNA synthetase

6. Which of the following questions about transcription were answered by the discovery of the –35 and –10 consensus sequences in bacterial promoters?
   a. Does RNA polymerase act alone to initiate transcription, or are other proteins involved?
   b. How do sigma proteins bind to RNA polymerase?
   c. How does RNA polymerase "know" when to terminate transcription?
   d. How does RNA polymerase "know" which strand of DNA acts as the template?

Conceptual Review

1. Explain the relationship between promoter sequences, transcription factors, and RNA polymerase.
2. According to the wobble rules, the correct amino acid can be added to a growing polypeptide chain even if the third base in an mRNA codon does not correctly match the corresponding base in a tRNA anticodon. How do the wobble rules and the redundancy of the genetic code relate?
3. Why does splicing occur in eukaryotic mRNAs? Where does it occur, and how are snRNPs involved?
4. An in vitro RNA synthesis reaction was set up and allowed to proceed in the presence of nonradioactive ribonucleotides (NTPs). After several minutes had passed, radioactive NTPs were added and RNA synthesis was allowed to continue. Then the RNA molecules were isolated from the reaction mixture and analyzed for the presence of radioactive nucleotides at the 5′ and the 3′ ends. Based on what you know about RNA synthesis, which end of the RNA molecules were found to be radioactive? (Assume that conditions were set so that each ribosome transcribed a single message.)
5. Sketch the structure of a tRNA. Label the anticodon and the CCA sequence where amino acids bind. Explain how a tRNA interacts with tRNA synthetase. Sketch the structure of an aminoacyl tRNA. Explain how an aminoacyl tRNA interacts with an mRNA in the ribosome.
6. How did the A-site, P-site, and E-site of the ribosome get their names? Explain the sequence of events that occurs during translation as a protein elongates by one amino acid.
7. What evidence supports the hypothesis that peptide bond formation is catalyzed by a ribozyme?

Applying Ideas

1. The 5′ cap and poly (A) tail that are attached to eukaryotic mRNAs appear to help the message last longer by protecting it from degradation by RNases. But why is an enzyme like RNase in the cell in the first place? What function would an enzyme that destroys messages serve?
2. Pribnow let RNA polymerase and DNA react before treatment with the DNase enzyme, and then recovered fragments of DNA about 50 nucleotides long. He also did an experiment in which polymerase was omitted at the first step. In this treatment no DNA fragments were obtained. In what sense does this second experiment serve as a control for the first experiment?
3. Look back at Figure 13.1, which shows the formation of a phosphodiester bond during RNA polymerization. Then study the structure of the nucleotide shown here, called cordycepin. If cordycepin is added to cell-free transcription reaction, it is added onto the growing RNA chain. This observation confirms that synthesis occurs by addition of monomers in the form of triphosphates to the 3′ end of the growing chain. Briefly explain why. Be sure to describe the expected result if synthesis occurred at the 5′ end of the polymer.

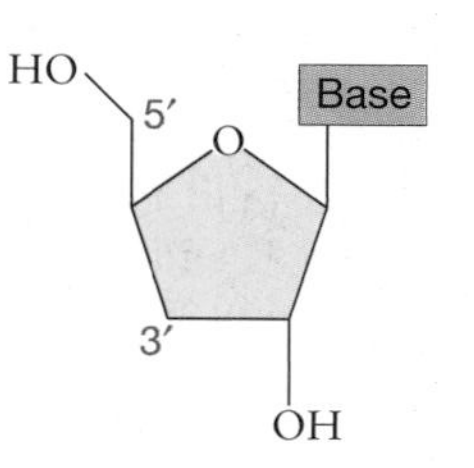

4. Carl Woese has determined the nucleotide sequence of rRNA molecules from a diverse array of organisms and compared them with each other. According to his data, certain portions of the rRNAs in the large subunit are very similar in all organisms. To make sense of this result, Woese suggests that the conserved sequences have an important functional role. His logic is that the conserved sequences are so important to cell function that any changes in the sequence cause death. Woese also claims that the existence of the conserved sequences supports other data indicating that peptide bond formation is catalyzed by the large rRNA. Explain the thinking behind his claim.

CD-ROM and Web Connection

CD Activity 13.1: Transcription *(tutorial)*
(Estimated time for completion = 5 min)
Explore how the cell converts the code in DNA into an RNA copy that can be used to direct protein synthesis.

CD Activity 13.2: The Events of Protein Synthesis *(animation)*
(Estimated time for completion = 5 min)
Learn how eukaryotic cells uses the code in messenger RNA molecules to make proteins.

At your **Companion Website** (http://www.prenhall.com/freeman/biology), you will find self-grading exams and links to the following research tools, online resources, and activities:

RNA World
This site discusses the structure, function, and possible evolution of RNA.

Ribosomes
Review the process of translation, focusing on the ribosome and transfer RNA.

The 30S Subunit
This site describes research on the 30S ribosomal subunit, in which RNA is now known to take an active part in amino acid polymerization.

Additional Reading

Frank, J. 1998. How the ribosome works. *American Scientist* 86: 428–439. A look at recent research on visualizing the three-dimensional structure of the ribosome.

Hoagland, M. 1990. *Towards the Habit of Truth: A Life in Science* (New York: W.W. Norton) An autobiography by a pioneer in the study of transfer RNAs.

Prusiner, S. B. 1997. Prion diseases and the BSE crisis. *Science* 278: 245–251. The primary proponent of the prion hypothesis reviews evidence that conformational changes can be inherited and cause disease.

Control of Gene Expression in Bacteria

14

Bacteria are found in virtually every habitat on Earth, from boiling hot springs to alpine snowfields and from the open ocean to rock crevices deep underground. As Chapter 25 will show, the millions of different species that exist have evolved a bewildering variety of ways to solve the fundamental problem of living—obtaining the carbon and energy required for growth and reproduction. Although some species specialize by using just one type of food, the vast majority of bacteria are able to switch among several distinct sources of carbon and energy depending on which food items are available in the environment.

The fundamental question addressed in this chapter is *how* this switching occurs. Each type of nutrient used by these cells requires a different suite of membrane-transport proteins and enzymes. How does a bacterial cell turn some genes off and others on so it can take advantage of alternative food sources? In general, how do bacteria regulate gene expression so that a cell makes only the products that it needs?

As a case study, we explore research on a common inhabitant of your intestine—the bacterium *Escherichia coli*. These cells are capable of using a wide array of sugars to supply the carbon and energy they need. As your diet changes from day to day, the availability of different sugars in your intestine varies. Precise control of gene expression gives *E. coli* cells the ability to respond to these changes in its environment and use the different sugars.

To appreciate why regulating gene expression is so important to *E. coli*, it is critical to realize that bacterial cells can be packed an inch thick along your intestinal walls. The organisms represent many different species, but they are all competing for space and for nutrients. For a cell to survive and reproduce in this environment, it must harvest and use resources

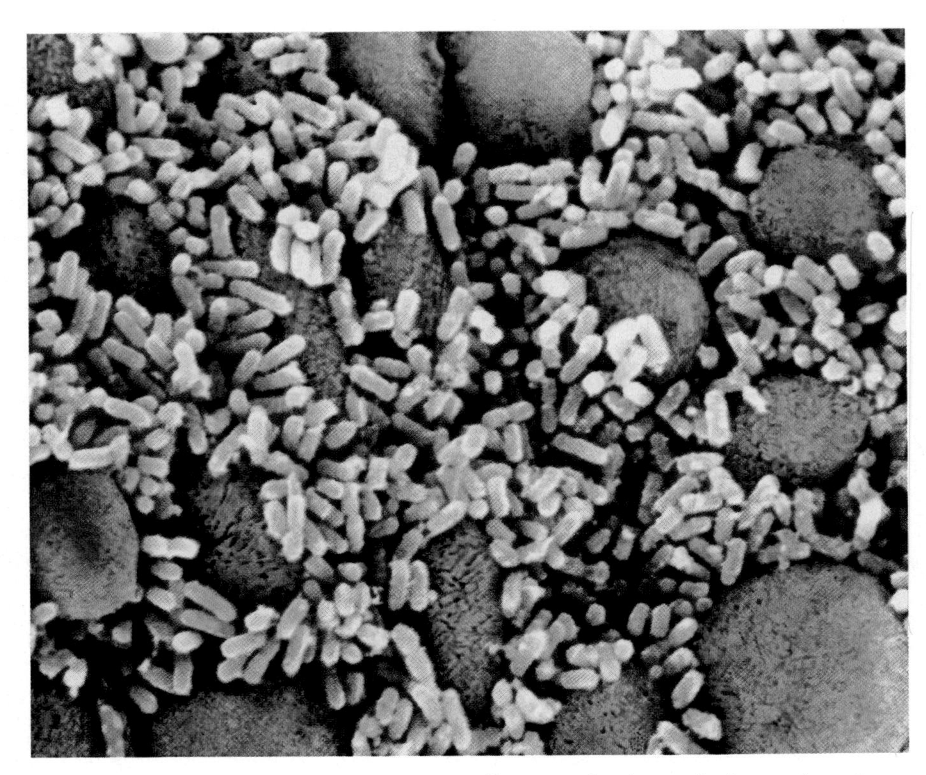

This photograph shows a few of the millions of bacteria that live in the human intestine—an environment where the types and amounts of available nutrients are constantly changing. This chapter explores how changes in gene expression help bacteria respond to environmental change.

efficiently. So even though *E. coli* is capable of metabolizing a wide variety of sugars, it would be energetically wasteful for a cell to produce all of the enzymes required to process all of these sugars all of the time. Instead, it is logical to expect that these cells produce only the enzymes needed to process the food molecules available at a given time. Based on this reasoning, transcription and translation in bacteria should be triggered by specific signals from the environment or the availability of specific sugars. In this way, controlling gene expression is a fundamental aspect of their ecology.

Understanding how genes are turned on and off has enormous practical consequences as well. In commercial and industrial applications, the ability to use *E. coli* and other bacteria as factories for producing vitamins, proteins, and antibiotics depends on a thorough grasp of how they regulate gene expression. As you'll see in the essay at the end of the chapter, biomedical researchers are exploring ways to block the expression of "virulence genes"—the alleles that differ between harmless bacteria and deadly ones.

To begin our exploration of how transcription and translation are controlled, let's review the flow of information in cells. Then we'll be ready to analyze the experiments that revealed the secrets of bacterial gene regulation.

14.1 Gene Regulation and Information Flow

Chapters 11 and 13 introduced the central dogma of molecular biology, which states that information flows from DNA to RNA to protein. The introduction to this chapter states that cells should be able to regulate which proteins they produce at any given time. For a cell to exert such control, the flow of information must be regulated in some way. Look at the central dogma again, this time as Francis Crick wrote it, and think of the ways that a cell might be able to produce a protein in response to some signal from the environment.

DNA → RNA → Protein

The first arrow, from DNA to RNA, represents transcription—the process of making a messenger RNA (mRNA). The second arrow, from RNA to protein, represents translation—the process whereby ribosomes read the information in an mRNA and synthesize the encoded protein.

The question before us is, how can a bacterial cell conserve its supply of ATP and amino acids and avoid producing proteins that it does not need at a particular time? Looking at the flow of information from DNA to protein, you might suggest three possible mechanisms. First, the cell could avoid making the mRNAs for particular enzymes. If there is no mRNA, then ribosomes cannot manufacture the gene product. Loci that are controlled in this way are said to undergo **transcriptional regulation.**

DNA —/→ RNA → Protein

Second, the cell might transcribe the mRNA for the enzyme, but it might have a way to prevent the mRNA from being translated into protein on the ribosome. Different messenger RNAs also have different lifetimes. Processes that alter mRNA life span or the efficiency of translation are referred to as **translational regulation,** or **post-transcriptional regulation.**

DNA ⟶ RNA —/→ Protein

Finally, Chapter 13 pointed out that some proteins are manufactured in an inactive form and have to be activated by chemical modification, such as the addition of a phosphate group. This type of regulation over gene activity is termed **post-translational control.**

DNA ⟶ RNA ⟶ Protein/

Which forms of control occur in bacteria?

Mechanisms of Regulation—An Overview

The short answer to the question just posed is "all of the above." As **Figure 14.1** shows, many different factors affect how much active protein is produced from a particular gene. Transcriptional control is particularly important because of its efficiency—it saves the most energy for the cell. As we'll

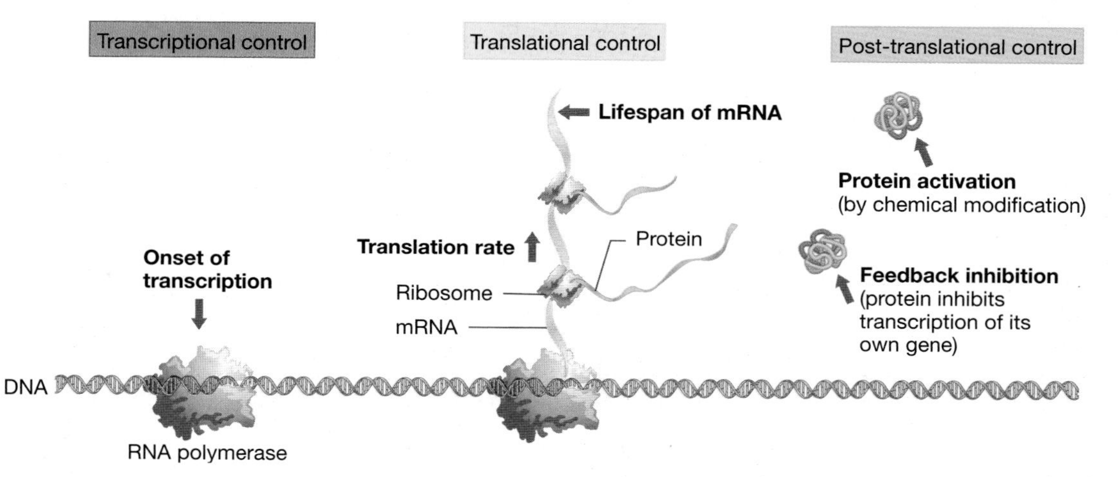

FIGURE 14.1 Gene Expression Is Regulated in Many Different Ways
Although this chapter focuses on how regulatory molecules affect the ability of RNA polymerase to initiate transcription, there are actually several important mechanisms of gene regulation in bacteria.

see in this chapter, a variety of regulatory proteins and molecules affect the ability of RNA polymerase to bind to a promoter and initiate transcription. Translational control is also significant because it allows a cell to change which proteins are being produced quickly. Translational control is based on how long mRNAs survive before they are degraded by ribonucleases, how efficiently translation of an mRNA is initiated, and how effectively elongation factors and other proteins interact with tRNAs and ribosomes during the translation process. Post-translational control is significant as well, and is the fastest of all three types of mechanisms. For example, bacterial proteins may be manufactured in an inactive state, and be activated by phosphorylation or some other type of chemical modification.

Although this chapter focuses almost exclusively on mechanisms of transcriptional control, it is important to keep in mind that a variety of processes affect gene expression in bacteria. In addition, it is critical to realize that these regulatory mechanisms affect both the amount and rate of gene expression—not just whether a certain gene is "on" or "off." Although most of the experiments reviewed in this chapter focus on the question of whether transcription does or does not occur at a particular locus, the amount and rate of expression is actually highly variable. Often, gene expression is not an all-or-none proposition. Finally, some genes—like those that code for the enzymes required for glycolysis—are transcribed all the time, or constitutively.

With these caveats in mind, we're ready to delve into research that furnished key insights into how bacteria regulate gene expression. *E. coli* has over 4,300 genes. What factors determine which loci are expressed at any particular time?

Metabolizing Lactose—A Model System

As Chapters 10 through 13 have shown, many of the fundamental advances in genetics have been achieved through the analysis of a model system. Studying the inheritance of seed shape in garden peas revealed the fundamental patterns of gene transmission; exploring transcription of phage and *E. coli* genomes led to the discovery of transcription factors and promoters. In the effort to understand gene regulation, one of the most productive model systems has been lactose metabolism in *E. coli*.

Escherichia coli use a wide variety of sugars for ATP production via glycolysis and the electron transport chain. These sugars also serve as a raw material in the synthesis of amino acids, vitamins, and other complex compounds. Lactose, the sugar found in milk, is among these sugars. Lactose is a disaccharide (two-sugar) made up of one molecule of glucose and one molecule of galactose.

For *E. coli* to use lactose, it must first transport the sugar into the cell from the environment. Once it is in the cytoplasm, *E. coli* can break lactose into its constituent monosaccharides (one-sugars) using an enzyme called β-galactosidase. (β is pronounced "beta.") The glucose monosaccharide released by this reaction is used directly via glycolysis; the galactose monosaccharide is further metabolized by other enzymes.

In the early 1900s biologists discovered that *E. coli* produces β-galactosidase only when lactose is present in the environment. No β-galactosidase is synthesized if lactose is absent. Because lactose appears to induce the production of the enzyme that breaks this sugar down, the hypothesis was that lactose regulates the gene for β-galactosidase by acting as an inducer. An **inducer** is a molecule that stimulates the expression of a specific gene.

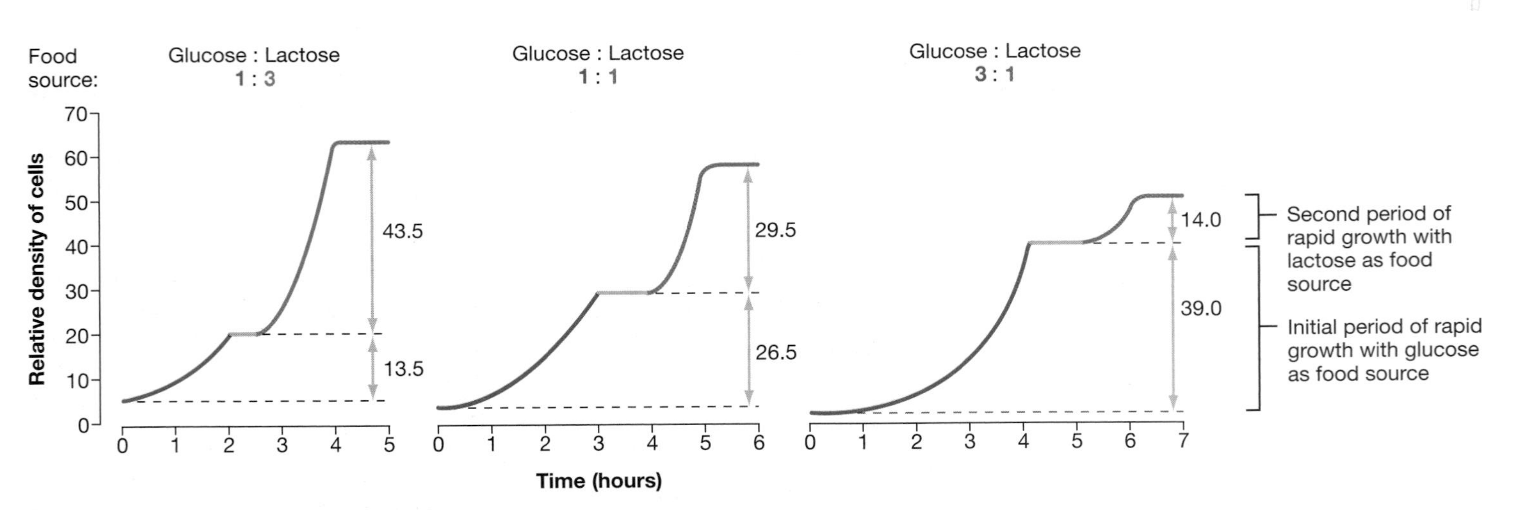

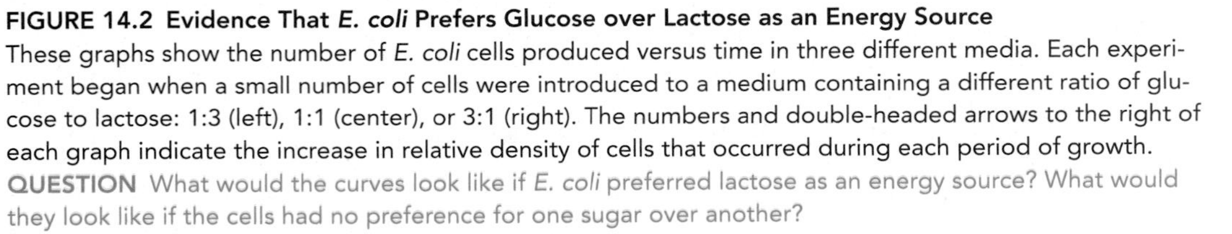

FIGURE 14.2 Evidence That *E. coli* Prefers Glucose over Lactose as an Energy Source
These graphs show the number of *E. coli* cells produced versus time in three different media. Each experiment began when a small number of cells were introduced to a medium containing a different ratio of glucose to lactose: 1:3 (left), 1:1 (center), or 3:1 (right). The numbers and double-headed arrows to the right of each graph indicate the increase in relative density of cells that occurred during each period of growth.
QUESTION What would the curves look like if *E. coli* preferred lactose as an energy source? What would they look like if the cells had no preference for one sugar over another?

In the late 1950s, Jacques Monod expanded on this observation by investigating whether lactose utilization also depends on other sugars that might be present. Specifically, Monod hypothesized that *E. coli* might actually prefer glucose to lactose as a food source because glucose plays a central role in glycolysis and because it produces more ATP per molecule than does any other sugar. In other words, would *E. coli* still produce β-galactosidase when glucose and lactose are both present in the surrounding environment?

To answer this question, Monod grew *E. coli* cells in media that contained glucose and lactose in three different ratios. **Figure 14.2** presents the data he collected on the number of cells produced in these experimental treatments over time. As you inspect these graphs, note that the general shape of the growth curves is similar in all three treatments. An initial period of rapid growth is followed by a period of no growth (about an hour long), which is followed by another, second period of rapid growth. The length and size of each growth period varies markedly, however. When glucose and lactose are present in a ratio of 1:3, the initial rapid growth period increases cell density by a factor of 13.5; the second period of rapid growth increases cell density by a factor of 43.5. The ratio of cells produced in each period of growth is roughly 1:3. When glucose and lactose are present in a ratio of 1:1, however, the ratio of cells produced in each period is also 1:1, and when the sugars are present in a 3:1 ratio the ratio of cells produced in the two growth periods is also 3:1.

Monod viewed these results as support for his hypothesis that glucose is the preferred energy source for *E. coli*. The fundamental observation is that the three growth curves track the food sources. Monod's interpretation was that bacteria grow on glucose until it is used up. This initial rapid growth period using glucose is followed by a period of no growth, because the cells need time to synthesize the molecules required to switch to lactose as a food source. Then growth resumes once lactose utilization begins, and the second period of rapid growth results.

Based on these preliminary findings, Monod teamed up with François Jacob to investigate how the genes responsible for lactose metabolism are controlled. As we'll see, research on this system is still continuing some 50 years later.

14.2 Identifying the Genes Involved in Lactose Metabolism

To understand how *E. coli* controls the production of the proteins required for lactose metabolism, Monod and Jacob first had to identify the genes involved. To do this they employed the same tactic that was used in the pioneering studies of DNA replication, transcription, and translation reviewed in earlier chapters. They isolated and analyzed mutant individuals. In this case, their goal was to find *E. coli* cells that were *not* capable of metabolizing lactose.

Finding mutants for a particular trait is a two-step process. Researchers begin by generating a large number of individuals with mutations at random locations in their genomes. Then they screen the mutants for those that have defects in the process or biochemical pathway under study. To produce mutant cells, Monod and other researchers exposed *E. coli* populations to x-rays, UV light, or chemicals that damage DNA and increase mutation rates. Then they set out to identify cells with defects in lactose metabolism.

Sometimes researchers can select for a particular type of mutant directly. For example, a biologist studying resistance to penicillin simply spreads mutagenized bacterial cells on a medium that contains penicillin. Only resistant mutant cells survive, grow, and form colonies; the others die. Thus, any colonies on the plate represent penicillin-resistant cells and can be used for further study.

In the case of lactose metabolism, however, the selection process is more difficult. Monod and colleagues were looking for cells that *cannot* grow in an environment that contains only lactose as an energy source. Normal cells grow well in this environment. How could they select cells on the basis of *lack* of growth?

Mutant Screens—Replica Plating and Indicator Plates

Two general techniques are used to identify mutants with defects in lactose metabolism. Replica plating is illustrated in **Figure 14.3** (page 278). In this procedure, mutagenized bacteria are spread onto a plate containing glucose. On this medium, each cell produces a single colony. Then a block covered with a piece of sterile velvet is pressed onto this master plate. Because of the contact, cells from each colony on the master plate are transferred to the velvet. Then the velvet is pressed onto a plate containing a medium with only lactose as a carbon and energy source. Cells from the velvet stick to the surface and produce a copy of the master plate. This copy is called the replica plate. After these cells grow, an investigator can compare the colonies that thrive on the replica plate's medium with those on the master plate. In this case, colonies that grow on the master plate but are missing on the replica plate represent mutants deficient in lactose metabolism. By picking these particular colonies from the master plate, researchers build a collection of lactose mutants.

Monod also used an alternative strategy based on **indicator plates**, where mutants with metabolic deficiencies are observed directly. Consider what happens when mutagenized bacteria are grown on agar plates containing lactose and subsequently sprayed with a solution containing a colorless compound called o-nitrophenyl-β-D-galactoside (ONPG). ONPG consists of two subunits—a galactose molecule and a molecule called o-nitrophenol—that are joined by the same bond that links the galactose and glucose subunits of lactose. Consequently, ONPG acts as an alternative substrate for β-galactosidase. When the enzyme cleaves ONPG, galactose and o-nitrophenol are released. This is important because o-nitrophenol is intensely yellow. When colonies exposed to ONPG turn yellow, it means that they have a normal copy of β-galactosidase.

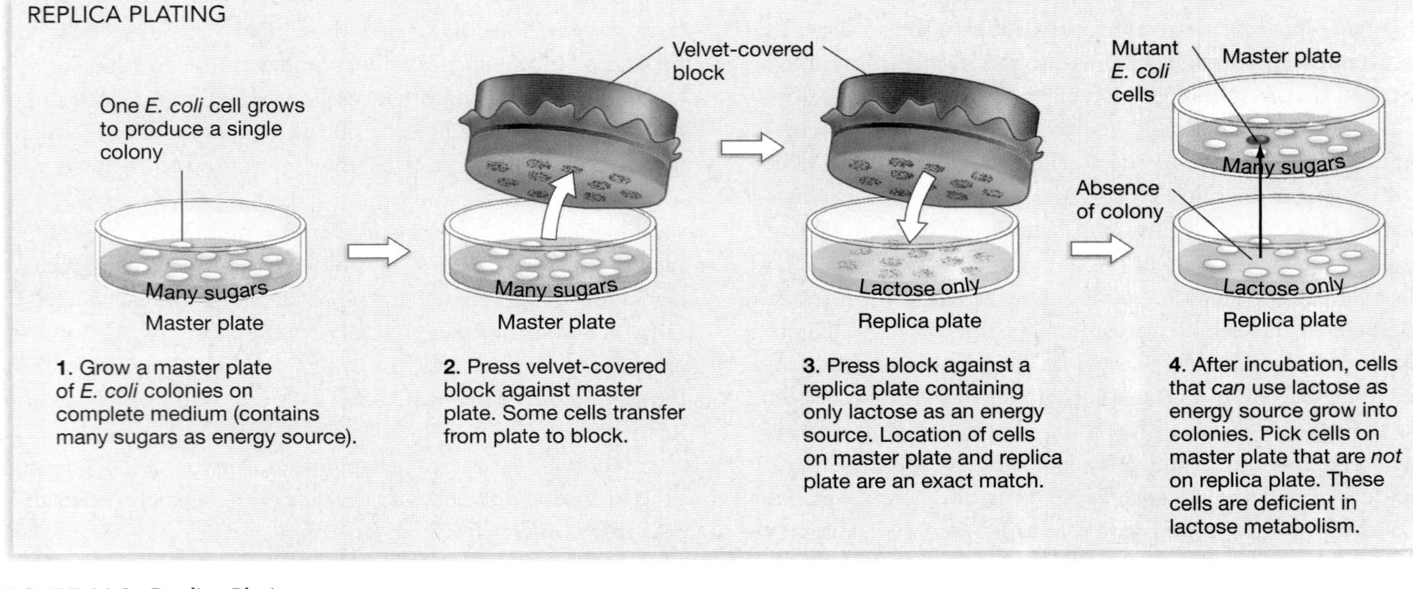

FIGURE 14.3 Replica Plating

In this example, replica plating is used to isolate mutant *E. coli* cells that have a deficiency in lactose metabolism. There are two keys to this technique: (1) the location of colonies on the master plate and replica plate must match exactly, and (2) the media in the two plates must differ by a single component.

QUESTION How would you alter this protocol to isolate mutant cells with a deficiency in the enzymes required to synthesize the amino acid tryptophan?

Colonies that stay white are unable to cleave ONPG, meaning that they have a mutant form of this enzyme.

Identifying Different Classes of Mutants

The work with ONPG and other lactose-like molecules allowed Monod to identify a distinct class of mutants—those with a defect in β-galactosidase. In these cells, synthesis of the enzyme is regulated normally, but the enzyme itself is inactive. The locus that codes for the enzyme was designated *lacZ* and the mutant protein was symbolized $LacZ^-$. Cells with normal forms of β-galactosidase are denoted $LacZ^+$. (Recall that abbreviations for proteins are written in normal script—$LacZ^+$—while gene names in *E. coli* are written in lowercase and italicized— *lacZ*.)

Monod used another lactose-like molecule, called methyl-β-D-1-thio-galactoside (TMG), to identify mutants at a second locus involved in lactose metabolism. To test the hypothesis that *E. coli* has some sort of protein involved in importing lactose into the cell, he measured how much radioactively labeled TMG is found in bacterial cells when they are grown in media containing a low concentration of this molecule. The data showed that normal cells accumulate about 100 times the concentration of TMG found in the surrounding environment. The result suggested that normal *E. coli* have a membrane protein involved in the transport of lactose into the cell.

When Monod did the same experiment with cells that are not able to grow on lactose, however, he found certain mutants that were unable to accumulate TMG. It was logical to infer that these cells have a mutant form of the membrane transport protein responsible for importing lactose. This protein is called **galactoside permease**; the locus that encodes it is designated ***lacY***.

What about other cells that have a functioning copy of β-galactosidase but are not able to metabolize lactose? What sorts of defects might these mutants have? When cells are grown on plates containing neither lactose nor glucose and then sprayed with ONPG, some colonies turn yellow. These individuals appear to have β-galactosidase and galactoside permease molecules available at all times. Cells like these are called **constitutive mutants** because the genes involved in lactose metabolism never shut down. Instead, the genes in question are always expressed. (The term *constitutive* refers to enzymes or genes that are always part of the cell's constitution. In normal cells certain genes, like those that code for enzymes in the glycolytic pathway mentioned earlier, are constitutively expressed.)

Jacob and Monod named the gene responsible for the constitutive β-galactosidase mutants *lacI*; the mutant phenotype was designated $LacI^-$. The use of "I" in the name was inspired by the observation that these mutants do not need an inducer to express the enzyme. Recall that in normal cells, gene expression is induced by the presence of lactose. But in cells with a damaged copy of *lacI*, gene expression occurs with or without lactose being present. This means that $LacI^-$ cells have a deficit in gene regulation. Based on these observations, it is logical to infer that the *lacI* gene product acts to prevent the transcription of *lacZ*. Because lactose acts as an inducer for the production of β-galactosidase, it is reasonable to expect that the *lacI* product or locus interacts with lactose itself in some way. (Later work

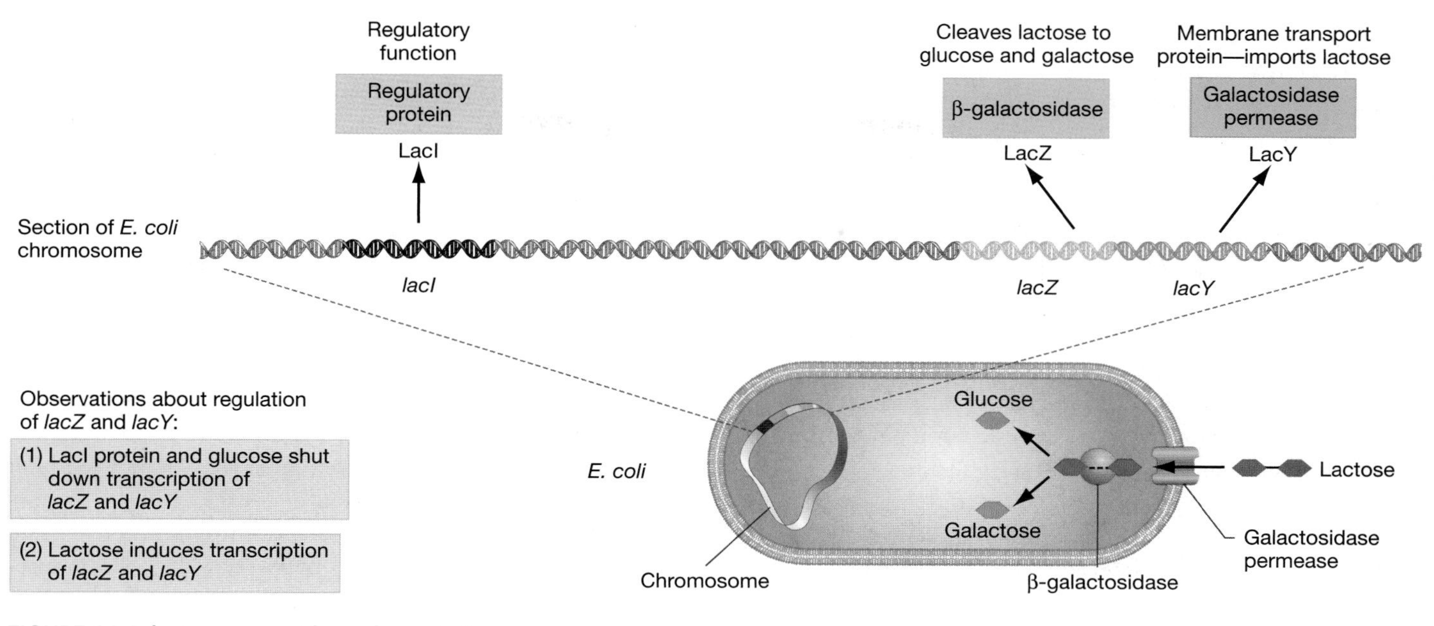

FIGURE 14.4 ***lac*** **Loci Are in Physical Proximity**
Early experiments on lactose utilization in *E. coli* identified these three genes and documented the function of their protein products. When researchers mapped the physical location of *lacI*, *lacZ*, and *lacY* on the *E. coli* chromosome, they found that they are positioned close together.

showed that the actual inducer is a derivative of lactose called allolactose. For the sake of historical accuracy and simplicity, however, this discussion refers to lactose itself as the inducer.)

The Elements of the Lactose Utilization System

Based on these early studies, Jacob and Monod had succeeded in identifying three genes involved in lactose metabolism: *lacZ*, *lacY*, and *lacI*. The first two genes code for proteins involved in the metabolism and import of lactose; *lacI* is responsible for some sort of regulatory function (**Figure 14.4**). When Jacob and Monod mapped the physical location of the three loci on *E. coli*'s circular chromosome, they found that the genes are close together. This was a hint that the regulation of *lacZ* and *lacY* might be coordinated in some way.

Jacob and Monod also knew that glucose and the *lacI* product repress the expression of *lacZ* and *lacY*. Lactose, in contrast, does the opposite—it induces their transcription. The question then became, how do these regulatory factors interact with one another? What are the actual mechanisms that control gene expression?

14.3 The Discovery of the Repressor

In the late 1950s the physicist Leo Szilard suggested to Monod that β-galactosidase synthesis is under **negative control**. His hypothesis was that the *lacZ* locus is normally shut off. According to this proposal, the enzyme is synthesized constantly in constitutive mutants because a factor that normally keeps enzyme synthesis shut off is inactive. This "shut-off factor" came to be called the **repressor** and was presumed to be the product of the *lacI* locus (**Figure 14.5a**, page 280). When a functional repressor is absent, transcription was thought to proceed normally (**Figure 14.5b**). The hypothesis of negative control also makes the important prediction illustrated in **Figure 14.5c**; that is, lactose acts an inducer because it interacts with the repressor in a way that triggers synthesis. How could researchers assess these ideas?

The test that confirmed the hypothesis of negative control is called the PaJaMo experiment in honor of the three collaborators who participated: Arthur *Pa*rdee, François *Ja*cob, and Jacques *Mo*nod. The experiment had two fundamental goals: (1) to separate the effects that the inducer (lactose) and repressor (predicted to be the *lacI* product) have on *lacZ* activity, and (2) to assess whether these effects conform to the model of negative control. The basic ideas were that adding repressor to a cell should shut down production of β-galactosidase, while adding inducer should trigger production.

The PaJaMo Experiment

The PaJaMo protocol exploited the discovery that genes can be transferred from one bacterial cell to another through a process called conjugation. As Chapter 9 pointed out, gene transfer in bacteria is unidirectional and involves just a portion of the donor cell's genome. Recall that certain cells are able to act as donor cells while others act as recipients. In effect, recipient cells become "partial diploids"—meaning that they have two copies of the alleles that were transferred. (**Box 14.1**, pages 284–285, provides more detail on how conjugation occurs. Box 9.2 explained how certain types of conjugation result in recombination.)

To test the hypothesis of negative control, Pardee and colleagues isolated donor cells that had normal copies of *lacI* and *lacZ* and that were sensitive to the antibiotic streptomycin (**Figure 14.6a**, step 1). As step 2 in Figure 14.6a shows, they also isolated recipient cells that had mutant copies of *lacI* and *lacZ* and that were resistant to streptomycin. (Note that the locus for streptomycin resistance is symbolized *str*; resistant alleles are denoted str^r and sensitive alleles str^s. Cells that are str^s die in the presence of streptomycin; str^r cells do not.)

With these cell types in hand, the researchers were able to perform the "mating" diagrammed in step 3 in Figure 14.6. When conjugation occurred, recipient cells that constitutively expressed nonfunctional copies of β-galactosidase received functional copies of the *lacI* and *lacZ* genes (see step 4).

As soon as the cell populations were mixed, Pardee and co-workers withdrew samples at regular intervals. Then they measured the level of β-galactosidase activity in each sample.

The data from this experiment are shown in Figure 14.6b. The graph shows that enzyme activity increased quickly about an hour after normal copies of the genes were introduced, then leveled off and began to slowly decline. Two observations are critical to interpreting this result. First, the researchers added streptomycin to the medium after an hour had passed. This is the amount of time required for conjugation to occur. The antibiotic treatment was important because it killed all the donor cells and ensured that β-galactosidase activity occurred only in recipient cells. Second, the experiment occurred in a medium that did not contain lactose. As a result, no induction occurred.

Why did β-galactosidase activity start and then stop? Pardee and co-workers interpreted these data as strong support for the model of negative control. They claimed that *lacZ* was transcribed after it entered recipient cells even in the absence of the inducer, because no repressor was present. As time passed and *lacI* transcription occurred, however, levels of repressor built up and shut down the production of β-galactosidase. Because *lacZ* was expressed without lactose present when there was no repressor, the data support the hypothesis that the inducer works by removing the repressor.

To test this interpretation, Pardee and colleagues repeated the experiment but added a molecule called isopropyl-1-thio-β-

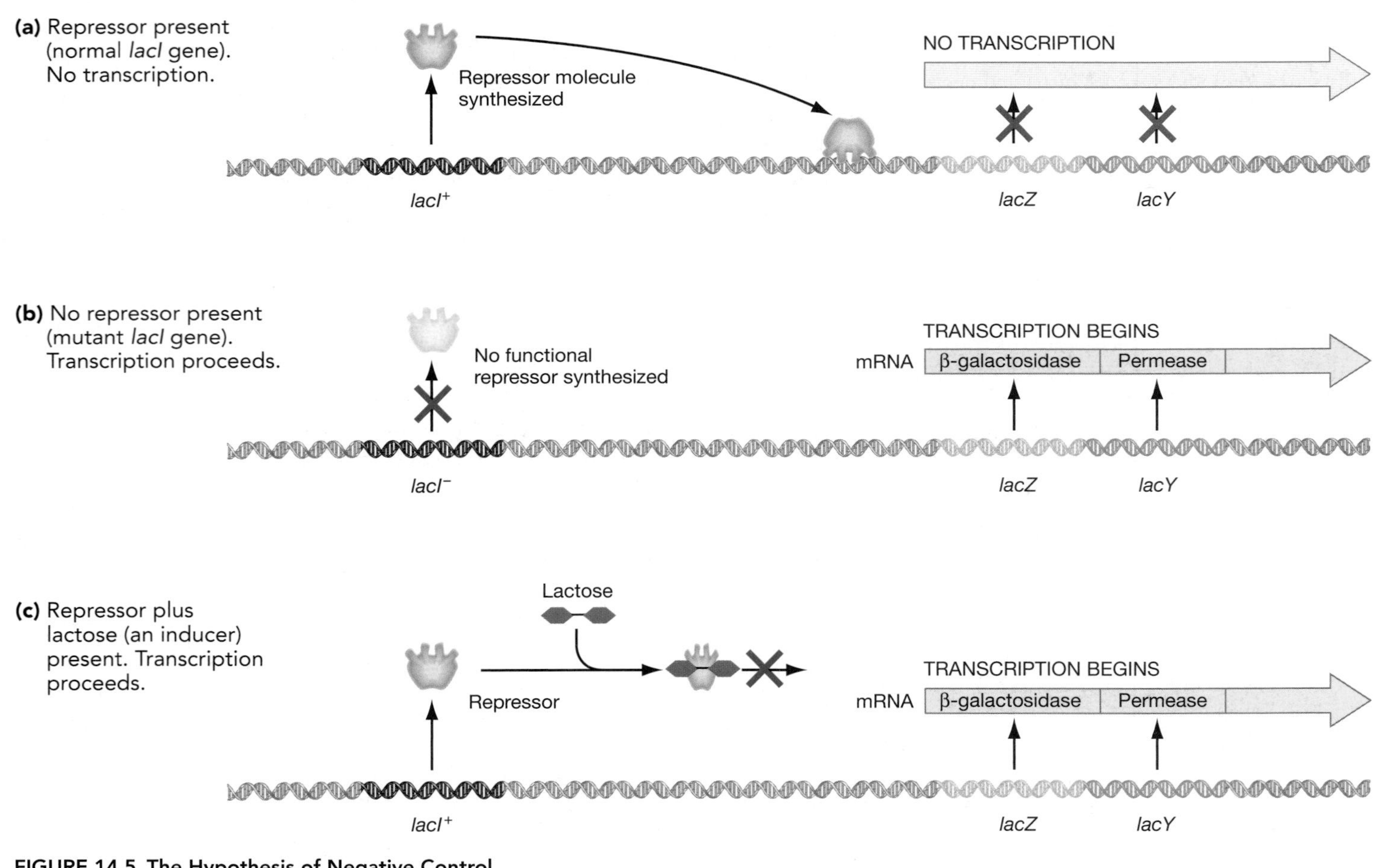

FIGURE 14.5 The Hypothesis of Negative Control
(a) The negative control hypothesis maintains that transcription of loci involved in lactose utilization is normally blocked by a repressor molecule. **(b)** When a functional repressor is absent, transcription proceeds normally (constitutive expression). **(c)** The negative control hypothesis contends that lactose acts as an inducer by interacting with the repressor and that this interaction prevents the repressor from functioning.

D-galactoside (IPTG) after β-galactosidase production was well under way. IPTG acts as an inducer. As the data in Figure 14.6c show, the addition of the inducer removed the effect of repressor buildup. The inducer worked to block repressor activity, and β-galactosidase activity increased steadily instead of being shut down.

These results are consistent with the model of negative control summarized in **Figure 14.7** (page 282). Jacob and Monod published this model in 1961. There are two key features to point out.

- They predicted that the repressor is a protein encoded by *lacI* that binds to DNA and prevents transcription of the *lacZ*

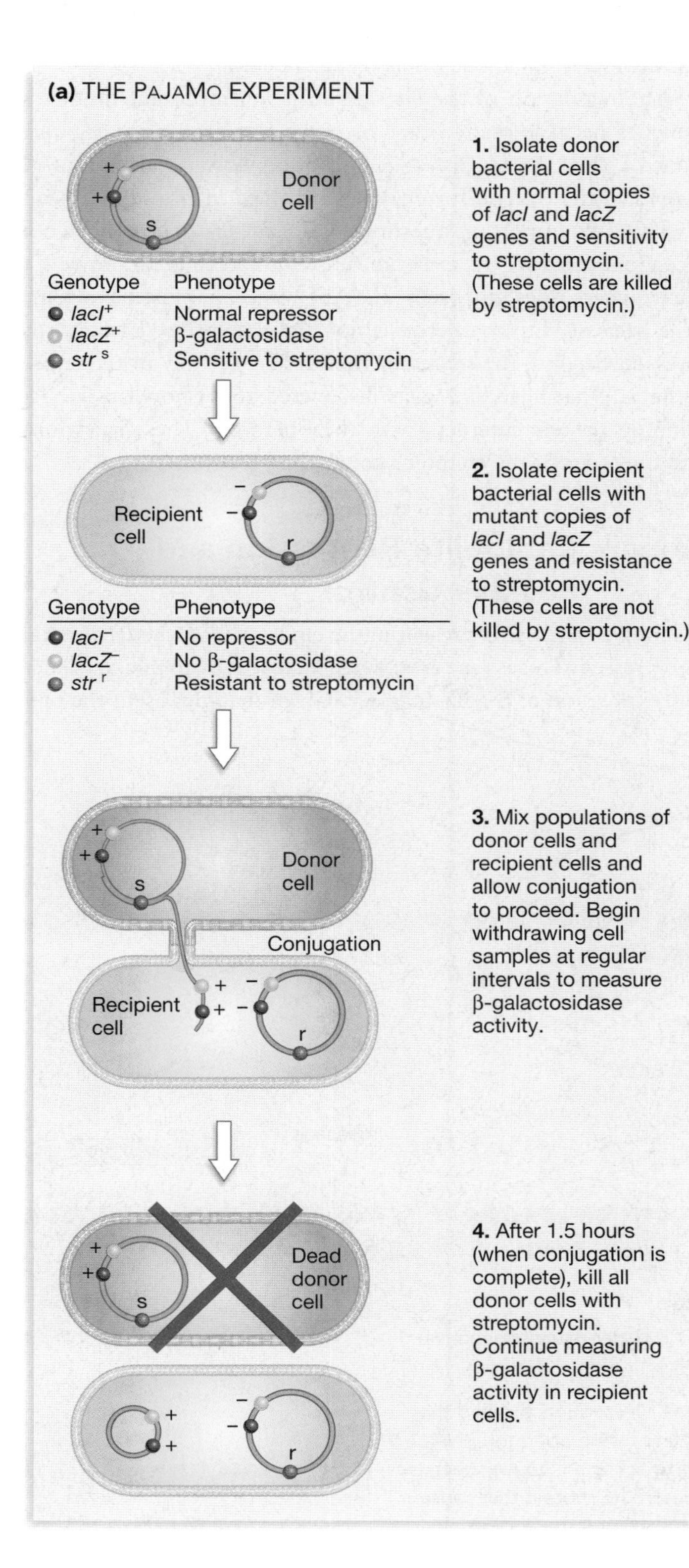

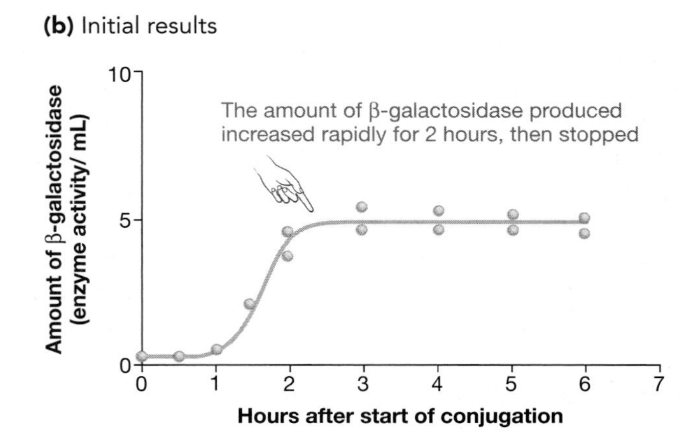

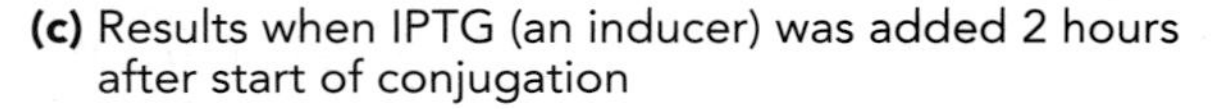

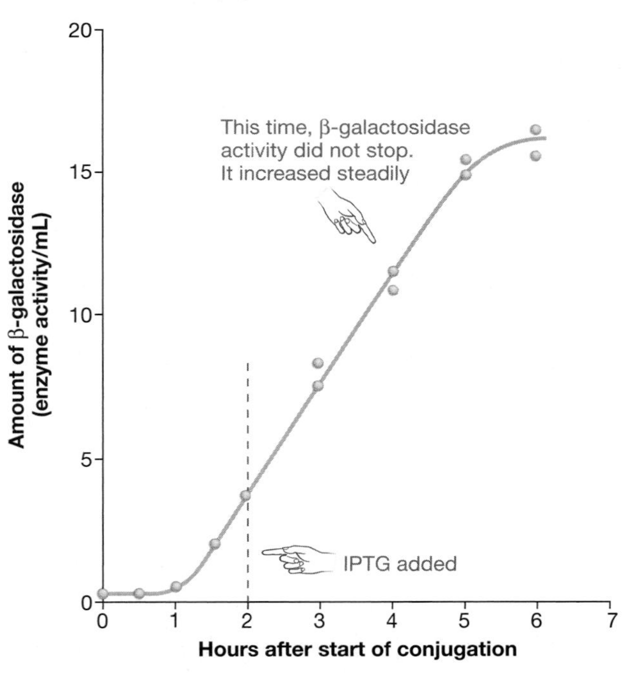

FIGURE 14.6 The PaJaMo Experiment
Pardee, Jacob, and Monod conducted this experiment to test the hypothesis of negative control for the *lacZ* locus.

locus. The binding site for the repressor is called the operator. Thus, they proposed that *lacI* is expressed constitutively, and that its product acts by blocking RNA polymerase from making contact with DNA and initiating transcription.

- The inducer (lactose) interacts directly with the repressor by binding to it. When it does, the repressor changes shape in a way that causes it to drop off the DNA. Transcription is then able to proceed. This is an example of allosteric (different-structure) regulation. **Allosteric regulation** occurs when a small molecule binds to a large protein and causes it to change its shape and activity.

The *lac* Operon

Jacob and Monod published their model on negative control of *lacZ* in 1961. Subsequently they were able to confirm the existence of the operator by finding mutant *E. coli* that expressed *lacZ* constitutively but that had normal forms of the repressor locus *lacI*. In these mutants the repressor protein is unable to function because the operator is abnormal. In 1967 Walter Gilbert and Benno Müller-Hill were able to tag copies of the repressor protein with a radioactive atom and show that they bound to DNA. This result confirmed that the operator is not a protein or an RNA product, but is part of the *E. coli* chromosome.

Biologists interested in the lac region also found a third gene, called *lacA*, that is tightly linked to *lacY* and *lacZ* and that codes for an acetylase enzyme. But after decades of study on this region, scientists still are not sure what role, if any, the *lacA* product plays in lactose metabolism. A more important discovery was that there is just one promoter upstream of *lacZ*, *lacY*, and *lacA* (**Figure 14.8a**. Because there is only one site for RNA polymerase to bind to the DNA and start transcription, the mRNA that is produced carries the coding information for all three proteins (**Figure 14.8b**). A region of bacterial DNA that codes for a series of functionally related genes under the control of the same promoter is called an **operon**. As a result, the *lac* loci became known as the ***lac* operon.**

The regulation of the *lac* operon is an important model in genetics because many other bacterial loci are under negative control. (**Box 14.2**, page 287, introduces how negative control works in the operon for enzymes involved in tryptophan synthesis.) Two important questions about the *lac* operon have not been answered by the experiments reviewed thus far, however. First, where does glucose fit in? The molecule must have some role because the *lac* operon is not highly expressed when it is present. Second, the repressor molecule is the key to the entire system. What have biologists discovered about how this DNA-binding protein interacts with the operator? These questions are the focus of the chapter's concluding sections.

14.4 Catabolite Repression and Positive Control

As long as glucose is present in the environment, the *lac* operon is repressed. This is true even when lactose is available to induce the expression of β-galactosidase. Given that glucose is the pre-

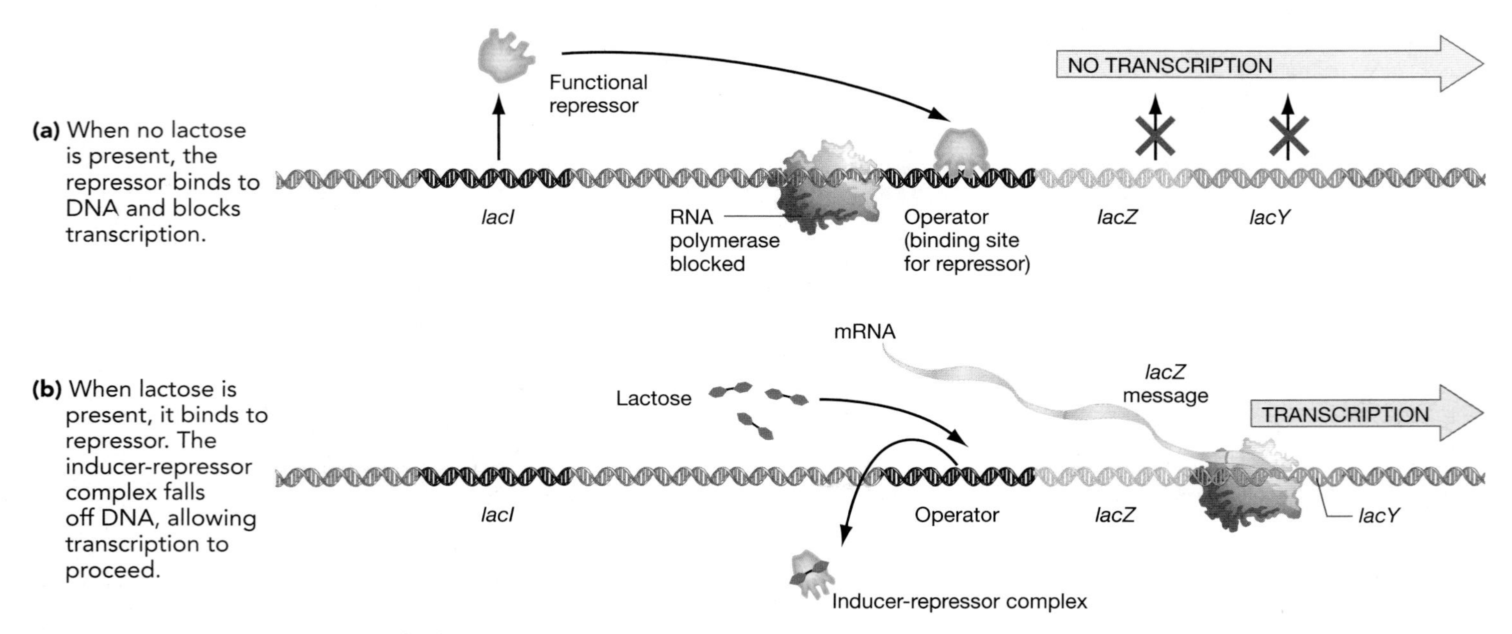

FIGURE 14.7 The Elements of Negative Control
(a) Jacob and Monod hypothesized that the repressor is a protein that binds to a DNA sequence called the operator, which is just "upstream" from the *lacZ* locus. When the repressor is bound to the operator, RNA polymerase is physically blocked from transcribing *lacZ*. **(b)** Lactose was thought to act as an inducer because it binds to the repressor, and because this binding causes conformational changes in the protein that cause it to fall off the operator.

ferred food source, this aspect of control over the *lac* operon is sensible ecologically and energetically. But the phenomenon also makes sense in terms of the chemical reactions involved. Recall that glucose is produced when β-galactosidase cleaves lactose. When glucose is present, it would not make sense for the cell to cleave lactose and produce still more glucose.

Chemists use the term *catabolism* to refer to reactions that result in the breakdown of a large molecule into simpler subunits. The hydrolysis of lactose into its glucose and galactose subunits is an example of catabolism (**Figure 14.9a**). In many cases, the operon that encodes the enzymes responsible for catabolism is inactivated when the end product of the reaction, or the catabolite, is abundant. This phenomenon is known as **catabolite repression.** In the case of the *lac* operon, glucose is the catabolite. When glucose is abundant in the cell, the *lac* operon is repressed (**Figure 14.9b**).

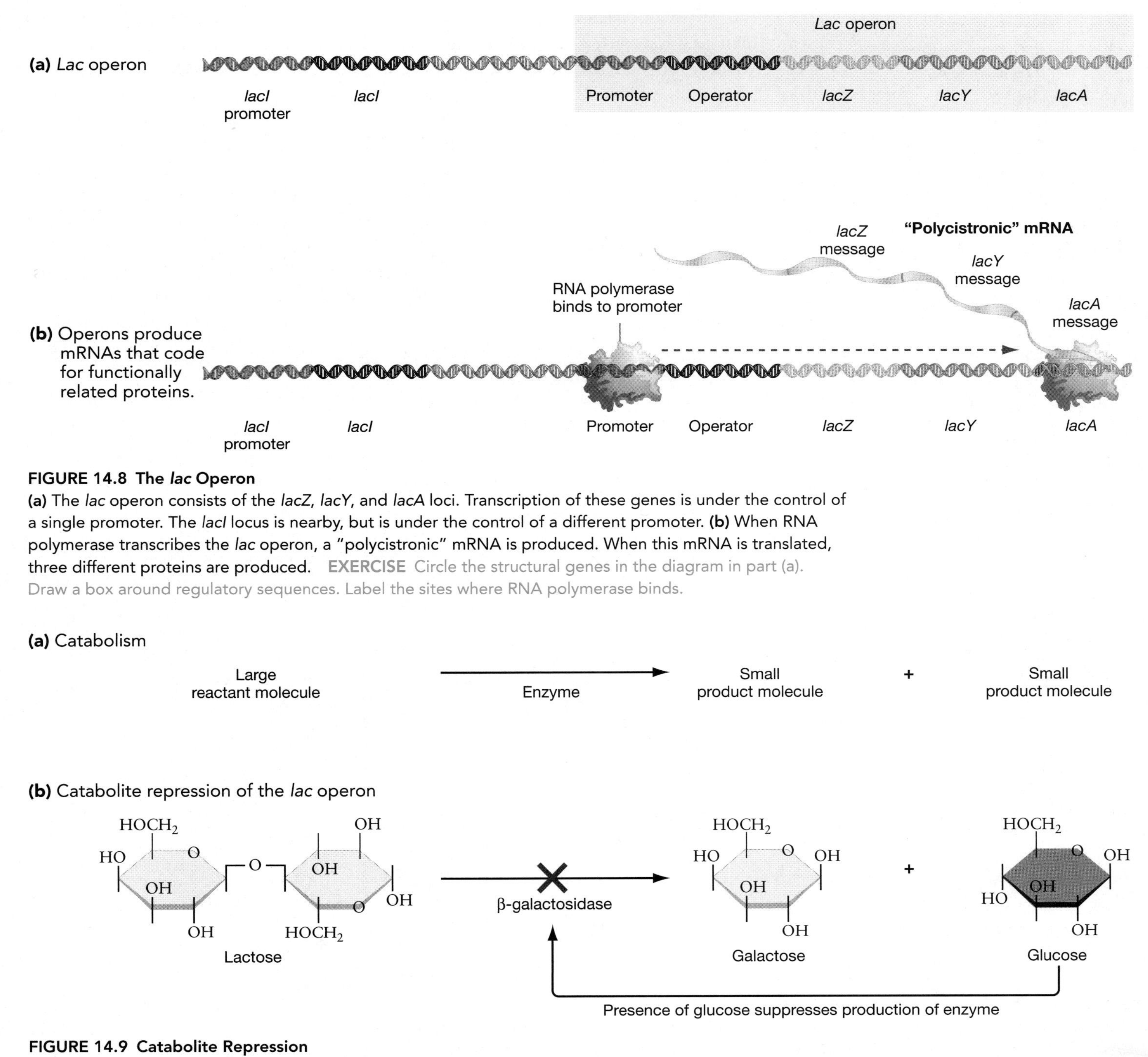

FIGURE 14.8 The *lac* Operon
(a) The *lac* operon consists of the *lacZ*, *lacY*, and *lacA* loci. Transcription of these genes is under the control of a single promoter. The *lacI* locus is nearby, but is under the control of a different promoter. (b) When RNA polymerase transcribes the *lac* operon, a "polycistronic" mRNA is produced. When this mRNA is translated, three different proteins are produced. EXERCISE Circle the structural genes in the diagram in part (a). Draw a box around regulatory sequences. Label the sites where RNA polymerase binds.

FIGURE 14.9 Catabolite Repression
(a) This reaction is a generalized example of catabolism. (b) Catabolite repression occurs when one of the small product molecules represses the production of the enzyme responsible for the reaction. In the case of lactose metabolism, the production of β-galactosidase is suppressed when glucose is present.

BOX 14.1 Gene Transfer in Bacteria

Bacteria are asexual. They reproduce by replicating their circular chromosome and dividing their cytoplasm in two. As a result, they produce daughter cells that are genetically identical. In fact, *all* of the bacteria descended from a progenitor cell are identical unless a new mutation occurs.

In 1946, however, Joshua Lederberg and E. L. Tatum demonstrated that genes can be transferred from one bacterial cell to another through a process that resembles sex. **Figure 1** diagrams their experiment, which used two different strains of *E. coli*. One strain was unable to grow unless the medium contained the amino acids leucine and threonine and the vitamin thiamine. The second strain contained mutations in genes required for synthesis of the amino acids cysteine, phenylalanine, and the vitamin biotin. The researchers mixed the two strains in a liquid medium containing all six of the nutrients that the cells couldn't synthesize for themselves. After allowing the cells to grow, they moved a sample to a plate containing a medium that lacked the six compounds (Figure 1b). As a control, cells from the original strains were also plated on this minimal medium. The result? No colonies appeared on the plates inoculated with the original strains. In contrast, colonies did appear on the plate inoculated with the bacteria that had been mixed. Because the growing cells had to have normal alleles at all six loci, the result suggested that genetic material had been transferred between strains.

In 1950 Bernard Davis showed that the result was not due to transformation. (Transformation, introduced in Chapter 11, is the process in which bacterial cells take up naked DNA present in the environment.) Davis's experiment employed the U-tube shown in **Figure 2**. The same bacterial strains used in Lederberg and Tatum's experiment grew on either side of the U, but were separated by a filter in the middle. The holes in the filter were big enough to let naked DNA to pass, but too small to allow cells to cross. Davis found that when samples of the bacteria were removed and plated on minimal medium, no colonies grew. This result showed that bacteria must be in physical contact for gene transfer to occur.

In bacteria, gene transfer that depends on physical contact is called conjugation. The event occurs because some *E. coli* cells possess an independent loop of DNA, or plasmid, called the F-plasmid (the "F" stands for fertility). Cells that have an F-plasmid are referred to as F^+; cells that lack it are F^-.

Conjugation begins when tubules on the surface of F^+ cells make contact with an F^- cell (**Figure 3**, step 1). After the tubule attaches, it retracts. The cell walls touch as a result, and a structure called a conjugation tube forms (step 2). Then a single strand of the plasmid DNA passes through the conjugation tube into the F^- cell (steps 2 and 3). The single strand of plasmid DNA in each cell serves as a template for the formation of double-stranded DNA, resulting in the formation of two F^+ cells (step 4).

If the F-plasmid involved happens to include a stretch of DNA called an inser-

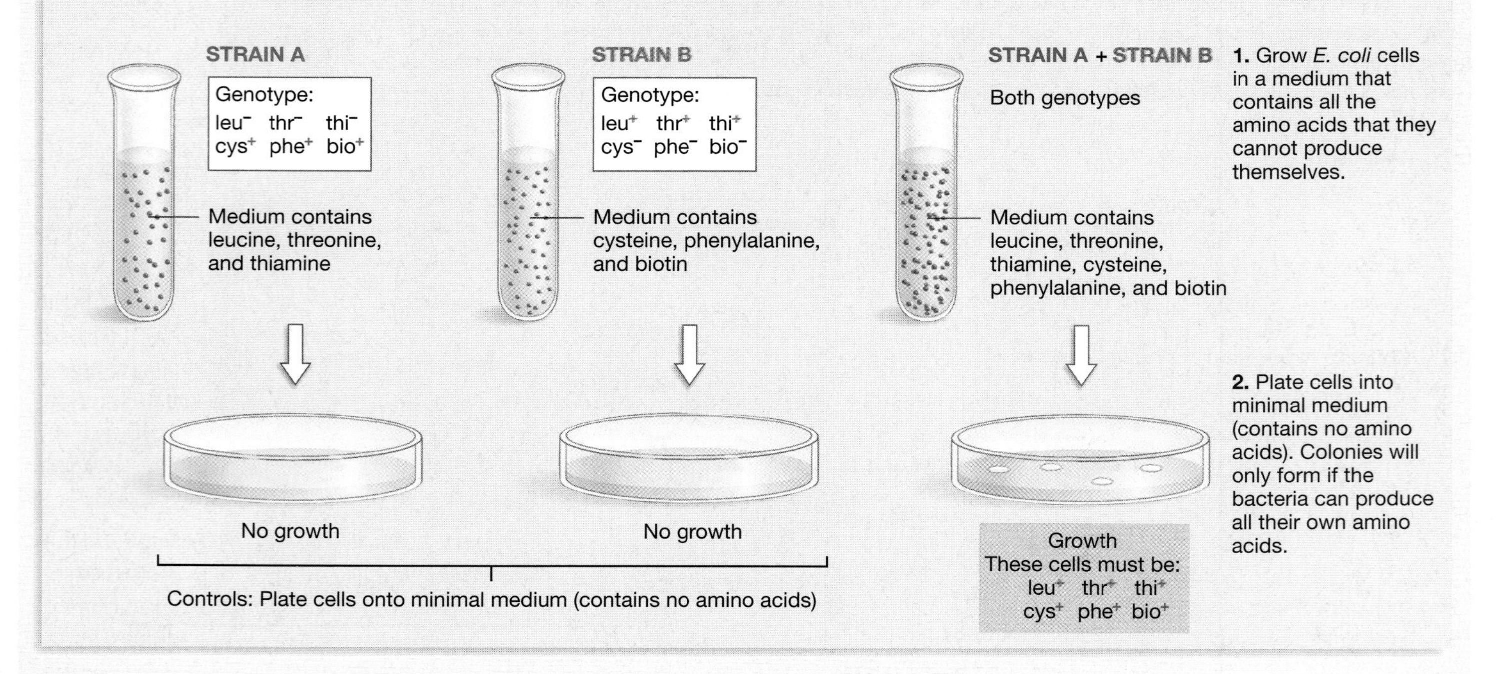

FIGURE 1 Experimental Evidence That Bacteria Can Transfer Genes Between Cells
Since no growth occurred in the two control plates, the researchers could defend their claim that the cells in the rightmost plate did not gain the ability to grow on minimal medium because of new mutations.

tion sequence (IS), it can integrate into the bacterial chromosome. Cells that have an F factor incorporated in their genome are called Hfr strains (for high frequency of recombination) because the plasmid DNA retains the ability to mobilize and travel through the conjugation tube. (Hfr strains were introduced in Box 9.2.) When conjugation occurs between an Hfr cell and an F^- cell, segments of chromosomal DNA frequently accompany the plasmid DNA during transfer. Sometimes the new alleles replace the old ones as a result of recombination, thus becoming part of the recipient cell's main chromosome. That's what happened in the Lederberg and Tatum experiment.

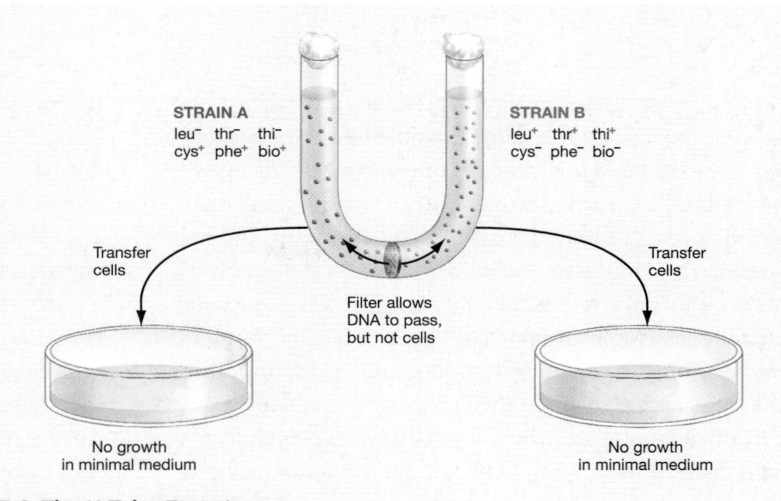

FIGURE 2 The U-Tube Experiment
This experiment supported the hypothesis that bacterial cells must be in physical contact to transfer genes. The results are inconsistent with the hypothesis that the results summarized in Figure 1 occurred because of transformation by naked DNA.

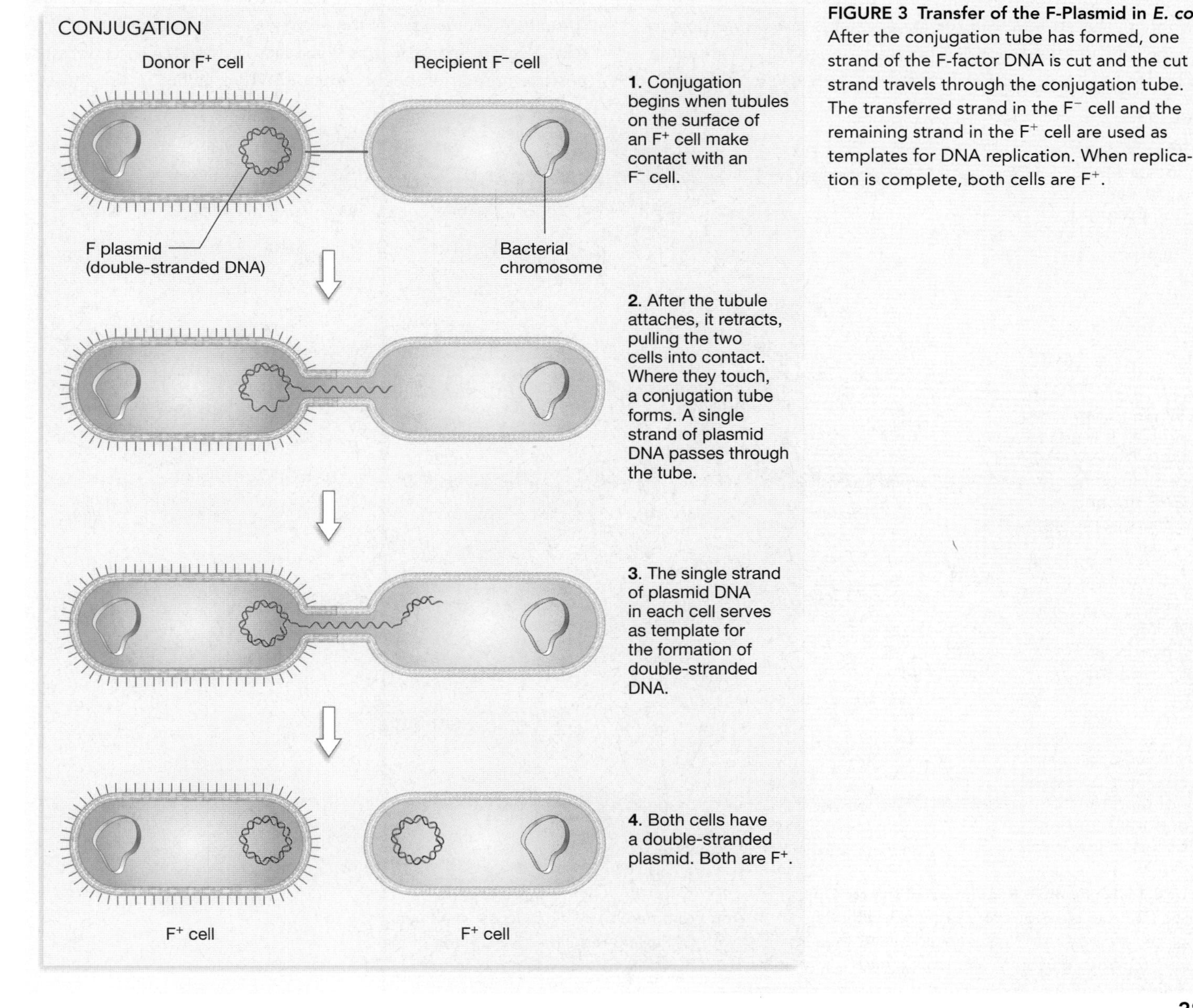

FIGURE 3 Transfer of the F-Plasmid in *E. coli*
After the conjugation tube has formed, one strand of the F-factor DNA is cut and the cut strand travels through the conjugation tube. The transferred strand in the F^- cell and the remaining strand in the F^+ cell are used as templates for DNA replication. When replication is complete, both cells are F^+.

How does glucose act to prevent expression of the *lac* operon? Answering this question became possible when researchers discovered a second major element of control over the *lac* operon. To understand how this second feature works, recall that the repressor normally shuts the operon off. But when the inducer is present, the repressor no longer binds to the operator. Transcription still does not occur, however, unless a protein called **catabolite activator protein (CAP)** binds near the promoter. The association between CAP binding and transcription is the second regulatory element in the *lac* operon. It is also an example of **positive regulation**, which occurs when a protein is required to activate transcription. The *lac* operon is under both positive regulation and negative control.

What Does CAP Do?

When investigators attempted to synthesize mRNAs from the *lac* operon in vitro, it became clear that, compared to other promoters, the promoter in the *lac* operon is not very efficient at attracting and binding RNA polymerase. For full expression of the *lac* loci to occur, CAP must bind to a sequence within the promoter called the CAP site (**Figure 14.10a**). When it does, the rates of *lacZ*, *lacY*, and *lacA* transcription increase dramatically. A long series of follow-up experiments explained why: CAP interacts with RNA polymerase in a way that makes the enzyme more likely to bind to the promoter and initiate transcription.

Like the repressor protein, CAP is allosterically regulated. It has alternate conformations that depend on whether another molecule is bound to it. In this case, the molecule that controls the regulatory protein is not lactose or IPTG but cyclic adenosine monophosphate (cAMP). cAMP is produced when glucose levels are low. When cAMP binds to CAP, CAP undergoes a conformational change. The CAP-cAMP complex is then able to recognize and bind to the CAP site. As a result, the *lac* operon is transcribed (**Figure 14.10b**). If there is no cAMP, then CAP has a conformation that prevents binding (**Figure 14.10c**). In both positive and negative regulation of the *lac* operon, then, a small molecule affects the activity of a large regulatory protein. Negative control ends when lactose binds to the repressor; positive regulation begins when cAMP attaches to the catabo-

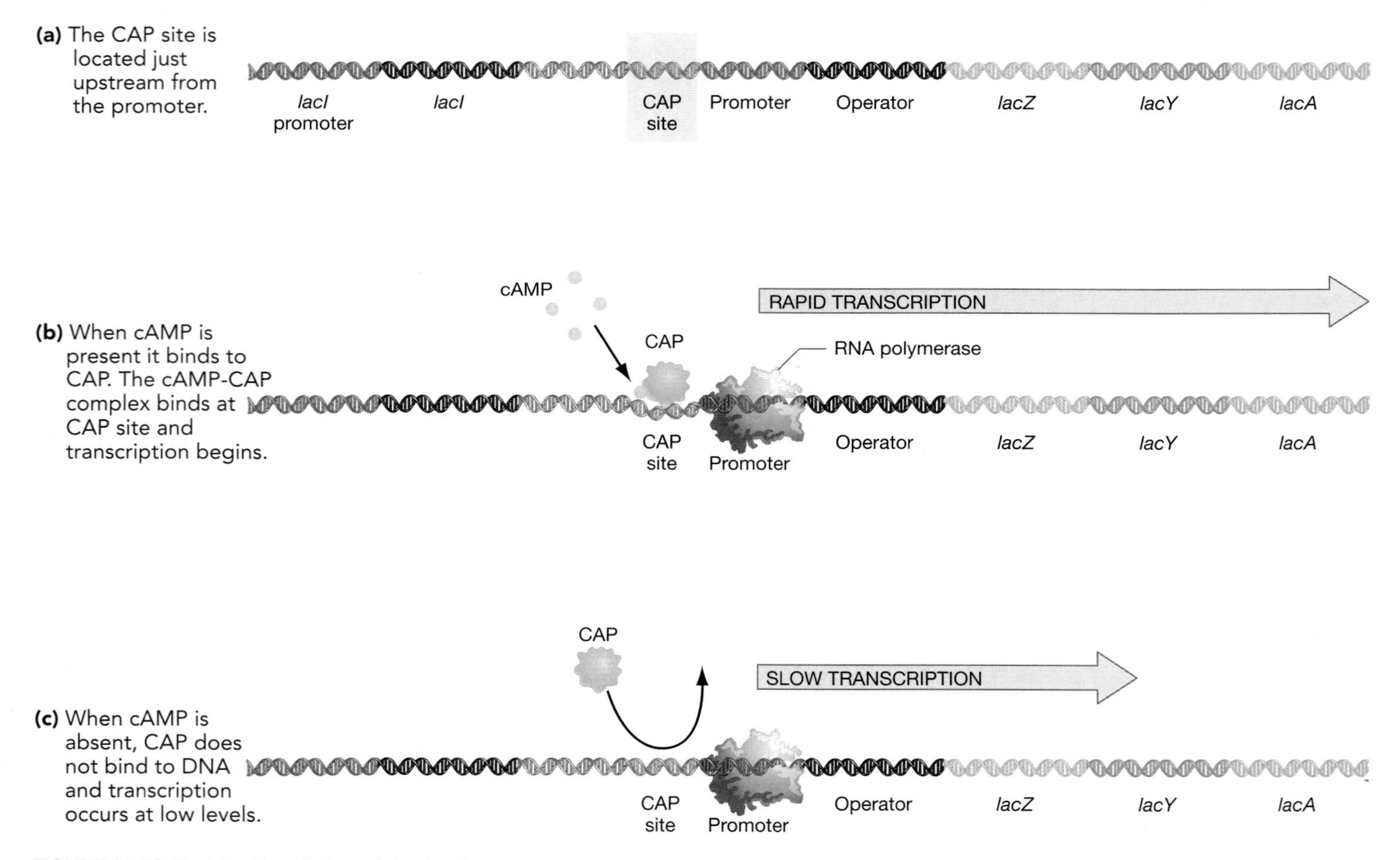

FIGURE 14.10 Positive Regulation of the *lac* Operon
(a) The CAP site is a regulatory sequence within the lac operon promoter. **(b)** When glucose levels in an *E. coli* cell are low, cAMP is produced. cAMP then interacts with CAP to up-regulate transcription of the *lac* operon. **(c)** When glucose is abundant, cAMP is rare in the cell. The CAP protein does not bind to the CAP site efficiently and transcription of the *lac* operon decreases.

lite activator protein and causes it to bind to the CAP site. The intricacy of these control systems should not be surprising. When an *E. coli* cell switches from one food source to another, it is a life-and-death matter.

How Does Glucose Affect the CAP-cAMP Complex?

Given this introduction to the mechanism of positive control, let's return to the question of glucose's role. Specifically, how

BOX 14.2 Negative Control and Attenuation in the *trp* Operon

In *E. coli*, the enzymes that are required to synthesize the amino acid tryptophan are located in an operon (**Figure 1a**). If RNA polymerase contacts the operon's promoter, the five *trp* genes are transcribed and translated together and synthesis of tryptophan begins. Like the *lac* operon, though, the *trp* operon has an operator sequence in the promoter. Tryptophan binds to a repressor protein when this amino acid is abundant in the cell. The tryptophan-repressor complex then binds to the operator and shuts down transcription so that no more tryptophan is synthesized (**Figure 1b**).

This type of negative control turns out to be common in operons that are responsible for biosynthesis. If the product molecule from a synthetic pathway is present in large quantities, it activates a repressor protein and transcription stops. The *lac* operon, in contrast, is responsible for breaking down a specific substrate and works the other way around. If a substrate molecule like lactose is present, it inactivates the repressor and leads to the initiation of transcription. The two styles of negative control correlate with the distinctive functions of the *trp* operon and *lac* operon.

Experiments by Charles Yanofsky revealed that negative control in the *trp* operon is supplemented by a type of control called attenuation. Attenuation means a reduction or lessening. It occurs because a leader sequence located between the promoter and the coding regions of the *trp* operon has an unusual property: When this leader is transcribed, the resulting mRNA loops back on itself and forms secondary structures through complementary base pairing. The exact type of secondary structure that forms depends on how rapidly a ribosome moves through the mRNA as translation begins. The ribosome's speed varies because the leader mRNA contains several codons for the amino acid tryptophan. If tryptophan is abundant in the cell, the ribosome adds it to the growing polypeptide and continues to move down the mRNA rapidly. As it does, however, the secondary structure shown in **Figure 2** forms in the mRNA downstream and leads to the termination of transcription. (Exactly how termination occurs is not known.) But if tryptophan is lacking in the cell, the ribosome stalls as it moves down the mRNA and waits for tryptophan to arrive. In this case, the termination signal does not form and transcription continues.

(a) When tryptophan is absent, transcription occurs.

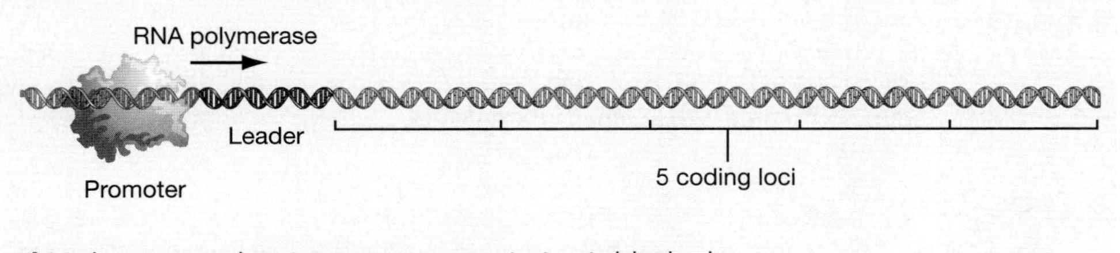

(b) When tryptophan is present, transcription is blocked.

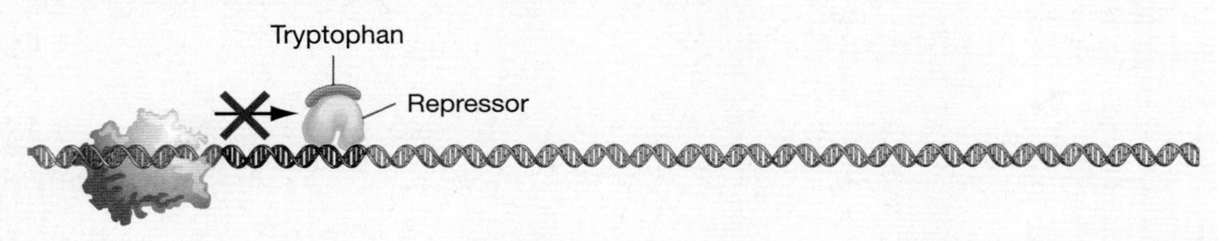

FIGURE 1 The *trp* Operon Is Under Negative Control
(a) The *trp* operon contains five coding loci. **(b)** When tryptophan is present it binds to a repressor protein. The tryptophan-repressor complex binds to the operator and shuts down transcription.

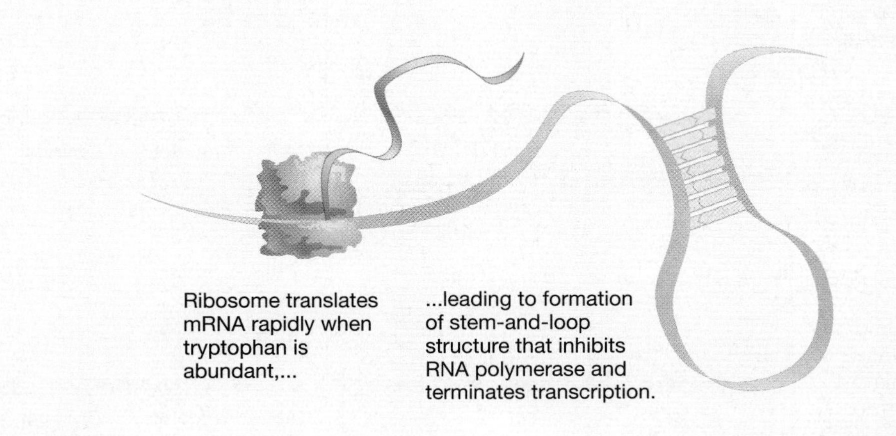

FIGURE 2 Attenuation Occurs when a Particular Secondary Structure Forms in mRNA
The mRNA shown here is from the leader sequence of the *trp* operon. If tryptophan is abundant in the cell and the leader sequence is translated rapidly, the mRNA forms a stem-and-loop structure that acts as a transcription termination signal.

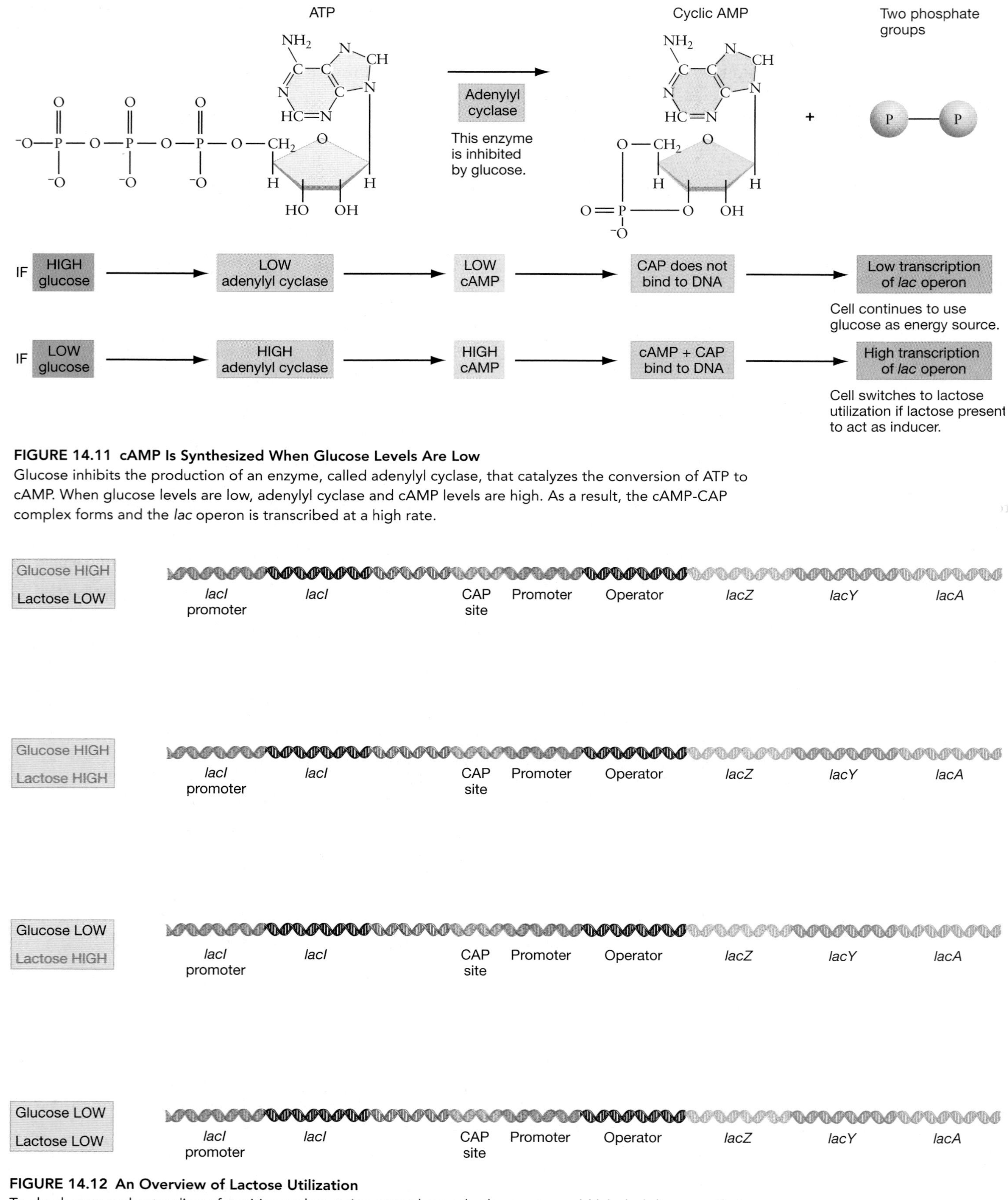

FIGURE 14.11 cAMP Is Synthesized When Glucose Levels Are Low
Glucose inhibits the production of an enzyme, called adenylyl cyclase, that catalyzes the conversion of ATP to cAMP. When glucose levels are low, adenylyl cyclase and cAMP levels are high. As a result, the cAMP-CAP complex forms and the *lac* operon is transcribed at a high rate.

FIGURE 14.12 An Overview of Lactose Utilization
To check your understanding of positive and negative control over the *lac* operon, add labeled diagrams that indicate the state of the operon under the four sets of conditions given here. Be sure to indicate the amounts and activity of the four regulatory molecules involved—lactose (as inducer), the repressor protein (*lacI* product), CAP, and cAMP.

does glucose affect the interaction between CAP and cAMP? The answer hinges on the two observations diagrammed in **Figure 14.11**. cAMP is produced from ATP by an enzyme called adenylyl cyclase, but adenylyl cyclase's activity is inhibited by the presence of glucose. As a result, the concentration of cAMP in *E. coli* is inversely proportional to the amount of glucose present. When the level of glucose is high, cAMP concentration is low. In this state, the CAP-cAMP complex does not bind to the *lac* operon and stimulate transcription. Conversely, when the concentration of glucose is low there is an increase in the level of cAMP. In this case, the CAP-cAMP complex forms, binds to the CAP site, and attracts RNA polymerase. Once transcription of *lacY*, *lacZ*, and *lacA* gets under way, lactose can be employed as an alternative energy source.

Figure 14.12 gives you a chance to review how the components of the lactose utilization system interact under various environmental conditions. To appreciate how these details affect the ability of a cell to grow and reproduce, it is critical to recognize that the CAP-cAMP system is not specialized for lactose utilization. Many loci and operons required for the metabolism of sugars other than glucose have CAP sites in or adjacent to their promoters. When glucose levels fall and cAMP concentrations rise, the effect on gene regulation is similar to ringing an alarm bell. The genes encoding the enzymes required for the cell to use lactose, maltose, glycerol, and other food sources are turned on. Conversely, when glucose supplies are adequate, cAMP levels are lower and genes for these enzymes are expressed only at low levels.

14.5 The Operator and the Repressor—An Introduction to DNA-Binding Proteins

The mechanisms that bacterial cells use to turn genes on and off have been a focus of research since the early 1950s. The first major result was an understanding of how negative control affects the *lac* operon in *E. coli*. Then in the late 1960s and early 1970s investigators worked out the details of positive control, using the effect of CAP on the *lac* operon as a model system. As other genes and operons were investigated in *E. coli* and other bacteria, these early results turned out to be typical of other genes and species. In bacteria, the transcription of genes is either constitutive or controlled. When regulation occurs, it is either negative (via a repressor protein such as the *lacI* product), positive (via a transcription activator such as CAP), or both. Catabolite repression is also extremely common.

Currently, research on the regulation of bacterial gene expression is focused on understanding how DNA-binding proteins work. In the *lac* operon of *E. coli*, for example, the key regulatory molecules are CAP and the repressor. How do these proteins interact with DNA at the molecular level? And how do small regulatory molecules such as cAMP and lactose induce changes in the conformation of these large DNA-binding proteins?

Finding the Operator

To understand how DNA-binding proteins work, a logical first step is to find and characterize the DNA sequences that they target. In the *lac* operon, the sequence of interest is the operator. By mapping mutants with defects in the operator, researchers were able to confirm that the target site is just downstream, or in the 3′ direction, from the promoter. But what is the actual DNA sequence?

To identify the sequences targeted by DNA-binding proteins, investigators use a technique called DNA footprinting. Footprinting is a strategy for identifying and purifying DNA sequences that serve as binding sites for proteins. It was the basis of the experiments introduced in Chapter 13 that resulted in the characterization of the first promoter ever discovered. **Box 14.3** (page 292) explains DNA footprinting in more detail.

Footprinting experiments with the *lac* repressor have yielded two outstanding results. First, the operon actually contains three sites where the repressor protein binds (**Figure 14.13a**). Second, the three operators have a similar structure—the sequences have an axis of symmetry. As **Figure 14.13b** shows, the bases form an

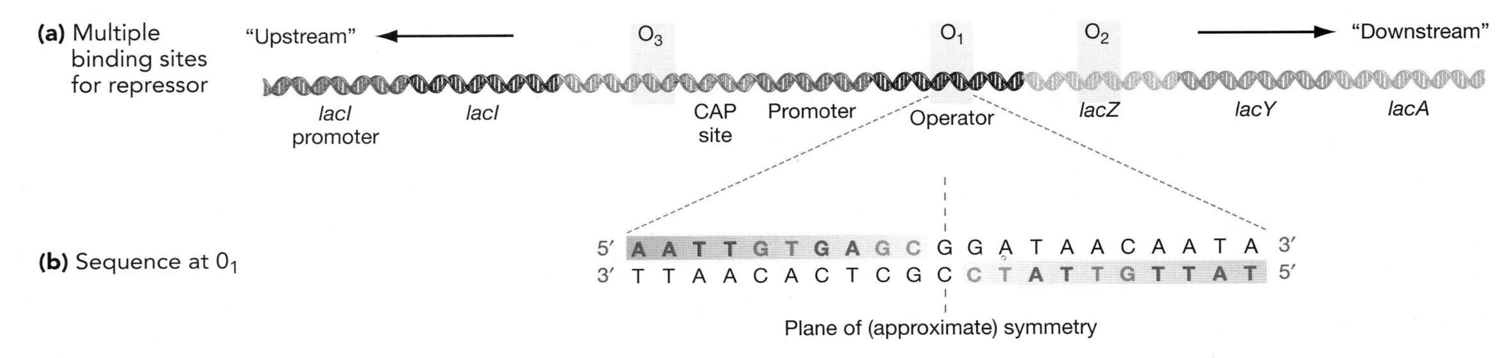

FIGURE 14.13 Operator Sequences
(a) The *lac* operon actually contains three distinct operator sequences where the repressor can bind. The sites are called O_1, O_2, and O_3. **(b)** Many sequences targeted by DNA-binding proteins have the structure shown here, which researchers refer to as dyad symmetry. Note that the sequences on either side of the axis of symmetry are similar. In this example, all but three bases are identical.

inverted-repeat sequence on either side of this axis. When the sequences are rotated around the axis, they match or nearly match. It turns out that many other DNA-binding sites, in a variety of species, share the same symmetrical configuration. To understand the significance of this result, we need to look more carefully at the repressor. How does this DNA-binding protein recognize specific sequences along the double helix?

DNA Binding via the Helix-Turn-Helix Motif

Researchers use the term **domain** to refer to sections of a protein that have a distinctive tertiary structure; the term **motif** is used to designate a domain that is found in many different proteins. When investigators began to purify repressor proteins from different bacteria and viruses and analyze their secondary and tertiary structures, they discovered that many have a "helix-turn-helix" structure like the one illustrated in **Figure 14.14a**. The motif consists of two α-helices connected by a series of amino acids that form a turn.

Is the helix-turn-helix domain the part of the *lac* repressor that makes physical contact with DNA? Physical models constructed during the 1980s suggest that the answer is yes. As **Figure 14.14b** shows, the exterior of the DNA double helix has two prominent, repeating features: a minor groove and a major groove. Careful measurements confirmed that the helix-turn-helix motif is the correct size to fit into the major groove (**Figure 14.14c**). This observation was consistent with the hypothesis that the amino acids in the helix-turn-helix motif interact directly with the bases in the major groove.

This hypothesis was recently corroborated in spectacular fashion by Mitchell Lewis and colleagues. These researchers used the technique called x-ray crystallography—introduced in Chapter 13—to determine the three-dimensional structure of the *lac* repressor and the repressor-operator complex. **Figure 14.15a** shows their structure for the repressor alone. The protein is called a tetramer (four-parts) because it consists of four copies of the polypeptide encoded by *lacI*. The most important point, however, is that helix-turn-helix motifs stick out from the main body of the repressor. When Lewis and associates analyzed the repressor-operator complex via x-ray crystallography, they confirmed that each of these helix-turn-helix motifs binds to operator DNA inside the major groove (**Figure 14.15b**).

How Small Molecules Affect the Conformation of Proteins

The *lac* operon is under negative control because the repressor binds to the operator and prevents transcription; when the helix-turn-helix motifs of the repressor are attached to DNA, RNA polymerase is physically blocked from contacting the promoter. When lactose is present, however, the repressor falls off the operator. How does the inducer affect the repressor in a way that disrupts DNA binding?

To answer this question, Lewis and co-workers analyzed the three-dimensional structure of the repressor when it was bound to the lactose-like inducer called IPTG. Their data showed that IPTG binds at locations in the middle of the tetramer, and that it produces an important change in the

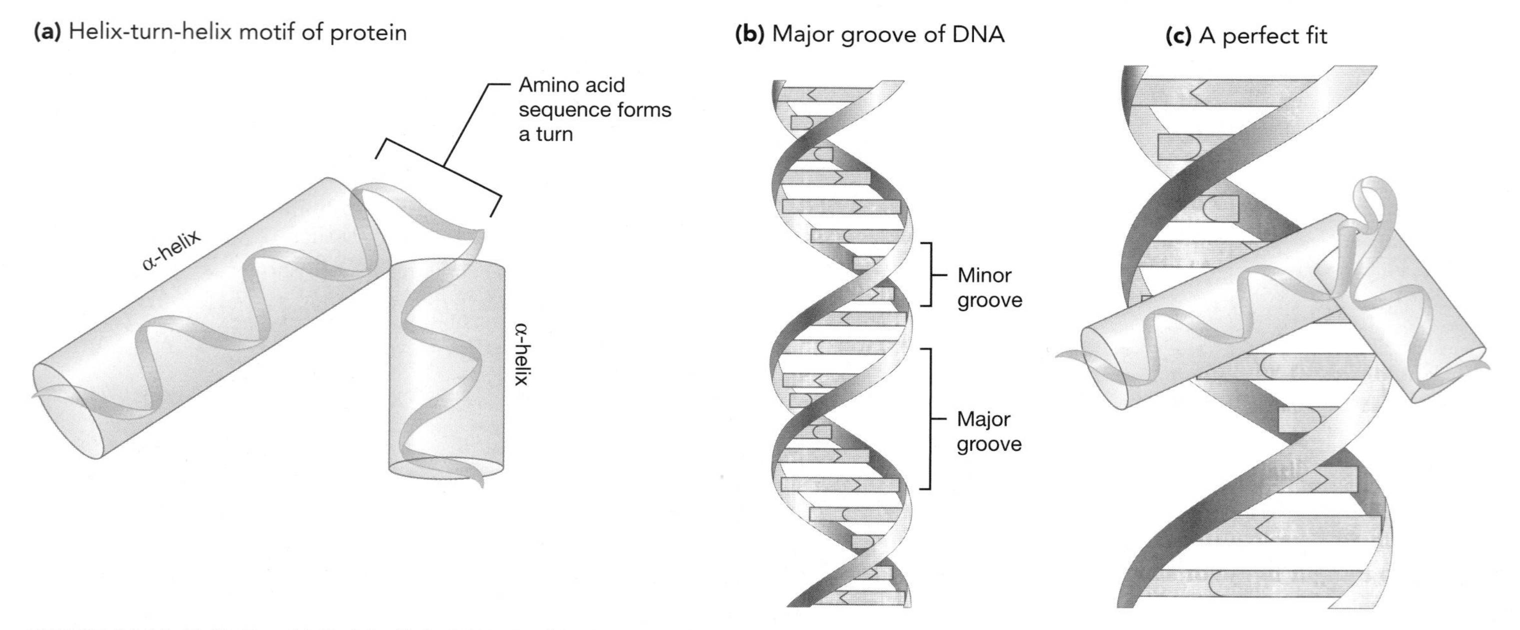

FIGURE 14.14 Helix-Turn-Helix Motifs in DNA-Binding Proteins
(a) This model shows a helix-turn-helix motif in a DNA-binding protein. **(b)** The outside of a DNA double helix has two distinct structures—a major groove and a minor groove. **(c)** One of the helices in helix-turn-helix motifs fits inside the major groove of DNA and interacts with the bases inside. **QUESTION** The part of a helix-turn-helix motif that interacts directly with DNA is called the recognition helix. Why?

shape of the protein near the helix-turn-helix domains. **Figure 14.15c** illustrates this change. The diagram on the left summarizes the repressor's shape under normal conditions. The important point to note is that structures called the hinge helices exist near the helix-turn-helix motifs, and that they lock the DNA-binding motifs into place. As the diagram on the right shows, however, the hinge helices uncoil when IPTG or another inducer binds to the repressor. As a result, the helix-turn-helix motifs release from the operator.

Future Directions

Now that the physical basis of negative control and induction is understood, what does the future hold? One prominent question is how the molecular mechanisms reviewed in this chapter affect the ability of bacteria to survive and reproduce in their natural habitats. For the past 50 years, biologists have studied gene regulation in cells growing in carefully controlled laboratory culture. But how does controlling gene expression help *E. coli* thrive in the human gut? How rapidly and dramatically do the concentrations of glucose, lactose, and other sugars change in this environment? What other species compete with *E. coli* for nutrients? Even as research in molecular biology reaches the level of understanding physical interactions between DNA and regulatory proteins at the atomic scale, can it also be focused at the level of ecological interactions, where millions of cells compete for resources in a complex environment?

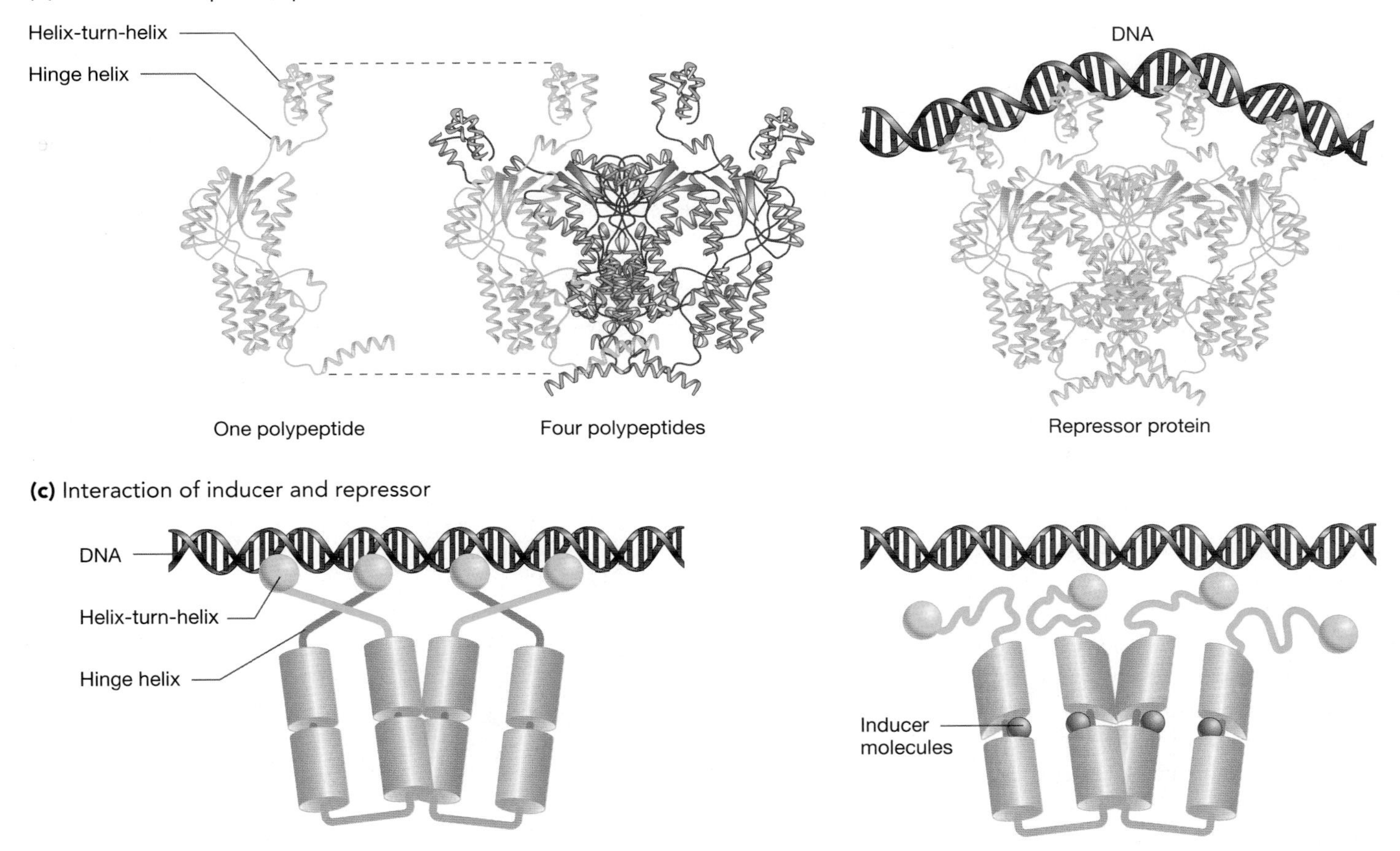

FIGURE 14.15 How Does the Repressor Interact with the Operator and the Inducer?
(a) These ribbon diagrams illustrate that the repressor is a complex of four polypeptides, each a product of the *lacI* locus. **(b)** The completed repressor protein has a total of four helix-turn-helix motifs to bind DNA at the operator. **(c)** X-ray crystallographic studies show that "hinge helices" in the repressor exist only when the inducer is not present (left). When the inducer binds to the repressor, the hinge helices relax and cause the protein to fall off the operator DNA (right).

BOX 14.3 DNA Footprinting

The goal of a DNA footprinting study is to determine the sequence of a region of DNA that is bound by a particular regulatory protein. To begin, researchers obtain a fragment of DNA from the region of interest—the *lac* operon of *E. coli*, for example—and generate a large number of copies. The fragments are then labeled using a radioactive tag at one end of one strand (**Figure 1**, step 1). After labeling, some of the labeled fragments are mixed with the DNA-binding protein being studied (step 2). Both sets of fragments are then treated with a nuclease—an enzyme that cuts DNA randomly. The concentration of nuclease used is so low, however, that each fragment is cut at most just once. You can see in step 3 that the nuclease treatment produces molecules of different lengths. In the sample containing the DNA-binding protein, however, the binding site is protected from the nuclease.

The researchers separate the resulting DNA fragments by gel electrophoresis and visualize them using autoradiography (see Chapter 3). The results of a footprinting experiment with the *lac* operon and repressor are shown in step 4 of the figure. Note that the sample with the repressor has missing bands, or a "footprint," in the segment protected by the protein. To determine the DNA sequence in this region, researchers frequently save a sample of the labeled fragments from step 1, use them in a dideoxy sequencing reaction (see Chapter 12), and run the resulting fragments out in the lanes adjacent to the footprint. They can then read the sequence directly at the binding site.

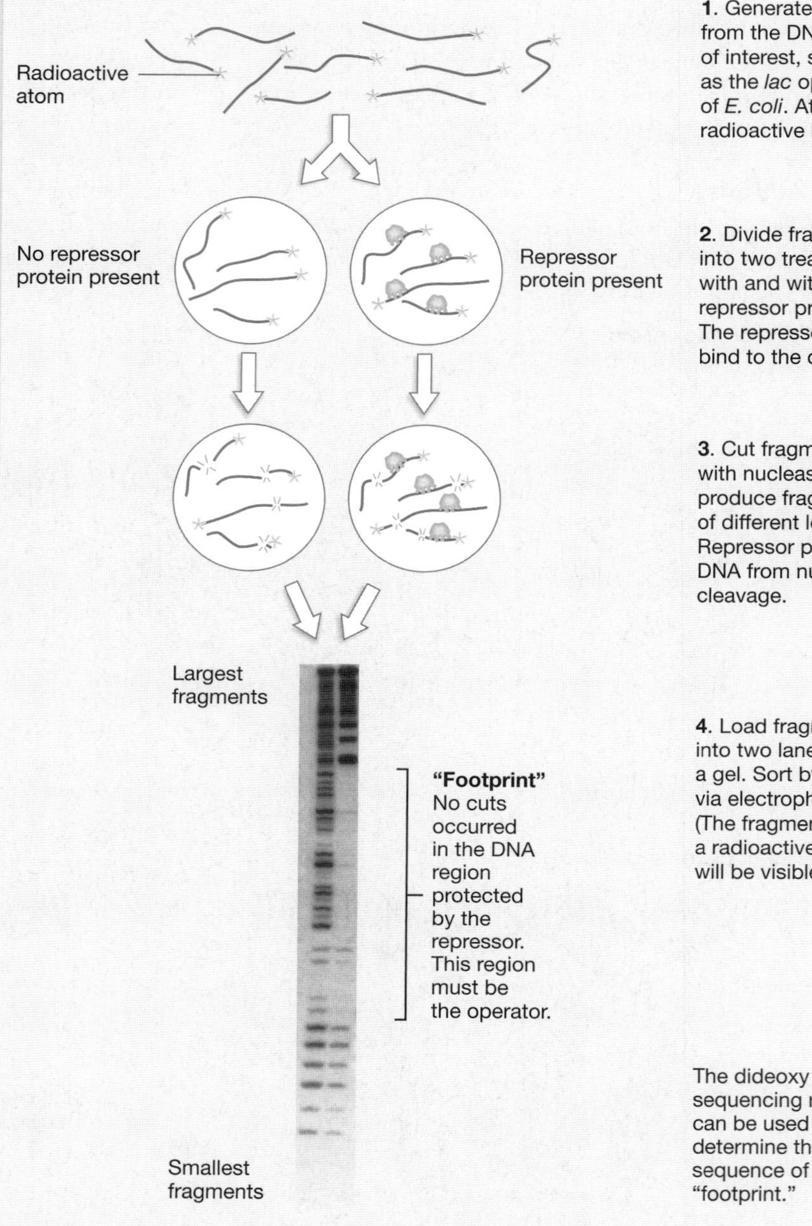

FIGURE 1
This diagram outlines the steps in a footprinting study.

Essay Controlling the Expression of Disease-Causing Genes

When researchers document the diversity of subtypes or strains that exist in species of disease-causing bacteria, they frequently find harmless varieties. Consider the bacterium *Haemophilus influenzae*. Even though this organism is a common cause of fevers and sore throats in humans, most strains do not cause illness. Of the 104 distinctive *H. influenzae* subtypes that have been identified, just six are responsible for causing 80 percent of the disease attributable to the organism. Similarly, only a handful of *E. coli* strains cause diarrhea; the vast majority of subtypes live in the human intestinal tract as harmless commensals. Further, the severity of illness caused by pathogenic forms of *E. coli* varies dramatically from strain to strain; the symptoms they cause range from mild to life threatening.

Virulence is a trait with a genetic basis.

Researchers today are often able to identify the genes and alleles that differ between pathogenic and benign strains of the same species. Not infrequently, the genes that are associated with virulence are clustered together physically on the main chromosome or on a plasmid in what investigators call a "pathogenicity island."

These observations have two messages. First, virulence is a trait with a genetic basis just as the ability to metabolize lactose is a trait with a genetic basis. Virulence usually results from the ability to grow rapidly and to infect certain types of cells. Alleles required for lactose metabolism are favored in environments where lactose is abundant; alleles associated with virulence are favored in environments where rapid growth is advantageous. Based on this reasoning, it is logical to expect that pathogenic strains will increase or decrease in the host population over time because of changes in the environment.

Second, understanding the regulation of pathogenicity loci might suggest novel approaches in drug or vaccine development. To drive this point home, consider recent research by Douglas Heithoff and colleagues. These investigators are interested in the organism responsible for causing typhoid fever in humans, *Salmonella typhi*, but perform their experiments with a closely related species that infects mice called *S. typhimurium*. They found that pathogenicity genes in *S. typhimurium* are regulated in part by an enzyme called DNA adenine methylase (Dam). Dam is the enzyme that adds a methyl group ($-CH_3$) to certain adenine residues after DNA replication is complete (see Chapter 12). Gene expression is altered dramatically in mutants of *S. typhimurium* that lack functioning copies of the Dam protein. In these Dam^- mutants, the cell's operator and promoter sequences are no longer methylated. As a result, binding by regulatory proteins and RNA polymerase is abnormal.

The researchers' most exciting result, however, was that Dam^- mutants do not cause illness when injected into mice. These data suggest that drugs capable of blocking Dam might be an effective treatment against typhoid fever. Further, Heithoff and colleagues found that when they injected mice with Dam^- *S. typhimurium* cells and later challenged the same individuals with virulent Dam^+ cells, they did not become sick. In effect, the mutant cells acted as a live vaccine. The message of these experiments is that finding ways to inhibit the expression of virulence genes may augment traditional approaches to drug and vaccine development.

Chapter Review

Summary

Transcription in bacteria is either constitutive or controlled. When control occurs, it results from one of two events—the binding of a repressor protein to a control sequence in DNA and the prevention of transcription, or the binding of an activator protein and the promotion of transcription. (In the case of the *lac* operon, both events occur.)

The regulatory proteins involved in negative control and positive regulation are often allosterically regulated, meaning that they have the ability to switch between two different conformations in response to binding by a small molecule—often a nutrient or product compound affected by the locus being regulated. The change in conformation is important because it affects the protein's ability to bind to DNA.

Precise regulation over gene expression is important to bacterial cells, because they frequently compete with other individuals for scarce resources and because they often switch between different sources of carbon and energy to take advantage of the nutrients available in the environment. The intestinal bacterium *E. coli*, for example, is capable of using lactose as a food source when glucose is not available. The loci involved in lactose metabolism are normally transcribed at a very low level because a repressor protein binds to a DNA sequence—called the operator—near their promoter. But when lactose is present, it binds to the repressor and induces a conformational change that causes the repressor to fall off the operator. Rapid transcription of the *lac* loci does not begin, however, unless glucose is also lacking. When glucose is scarce, levels of cAMP rise in the cell. This is important because the regulatory protein called CAP undergoes a conformation change when it is bound to cAMP. The CAP-cAMP complex binds to a control sequence near the *lac* promoter and induces transcription by facilitating binding by RNA polymerase.

Investigations into the regulation of gene expression in bacteria began almost 50 years ago, through an analysis of mutants with deficiencies in lactose metabolism. Research has progressed to the point where biologists are asking questions about interactions between regulatory proteins and DNA sequences at the atomic level. As we'll see in Chapter 15, studies on gene expression in eukaryotes have arrived at a similar point.

Questions

Content Review

1. Genes for enzymes in the glycolytic pathway are expressed constitutively. What does this mean?
   a. Their expression is controlled through modification of the mRNA products.
   b. Transcription occurs constantly.
   c. Transcription is normally off.
   d. Transcription occurs in response to an "activator" protein.
2. Why do researchers frequently begin a search for mutants by exposing organisms to UV light or x-rays?
   a. The treatment causes mutants to turn yellow.
   b. The treatment triggers gene expression.
   c. The treatment exposes constitutive mutants.
   d. The treatment increases the mutation rate by damaging DNA.
3. Monod found mutant *E. coli* that were unable to concentrate the lactose-like molecule TMG. What is the defect in these cells?
   a. They lack a functional copy of β-galactosidase.
   b. They lack a functional copy of acetylase.
   c. They lack a functional copy of galactoside permease.
   d. They lack a functional copy of the repressor.
4. Why are the loci involved in lactose metabolism considered an operon?
   a. They occupy adjacent locations on the *E. coli* chromosome.
   b. They have a similar function.
   c. They are all required for normal cell function.
   d. They are under the control of the same promoter.
5. What is catabolite repression?
   a. a mechanism that turns off the enzymes responsible for catabolic reactions when the product is present
   b. a mechanism that turns off the enzymes responsible for catabolic reactions when the product is absent
   c. repression that occurs because of allosteric changes in a regulatory protein
   d. repression that occurs because of allosteric changes in a DNA sequence
6. What is a helix-turn-helix motif?
   a. a protein domain involved in folding into the active conformation
   b. a protein domain involved in induction
   c. a protein domain involved in DNA binding
   d. a protein domain involved in catalysis

Conceptual Review

1. From the data in Figure 14.2, how could Monod conclude that *E. coli* cells use up glucose before beginning to utilize the lactose? Why does this result support the hypothesis that glucose is a preferred food source for this species?
2. In *E. coli*, rising levels of cAMP can be a considered a starvation signal. Explain.
3. Explain the difference between positive regulation and negative control. Why is it advantageous for the *lac* operon in *E. coli* to be under both positive regulation and negative control? What would happen if only negative control occurred? What would happen if only positive regulation occurred?
4. CAP is also known as the cAMP-receptor protein. Why?

Applying Ideas

1. You are interested in using bacteria to metabolize wastes at an old chemical plant and convert them into harmless compounds. You find bacteria that are able to tolerate high levels of the toxic compounds toluene and benzene, and you suspect that it is due to the ability of the bacteria to break these compounds into less-toxic products. If that were true, these toluene- and benzene-resistant strains would be valuable for cleaning up toxic sites. How could you find out whether these bacteria have enzymes that allow them to metabolize toluene?
2. Assuming that the bacteria you examined in exercise 1 do have an enzymatic pathway to break down toluene, would you predict that the loci involved are constitutively expressed, under positive regulation, or under negative control? Why? What experiments could you do to test your hypothesis?
3. The *lacI* mutants discussed in the text were constitutive because the repressor failed to recognize and bind to the operator region. Other repressor mutants have been isolated that are called $LacI^s$ mutants. These repressor proteins continue to bind to the operator, even in the presence of the inducer. How would this mutation affect the function of the *lac* operon? Specifically, how well would $LacI^s$ mutants do in an environment that has lactose as its sole sugar?
4. X-gal is a colorless lactose-like molecule that can be split into two fragments by β-galactosidase. One of these product molecules is blue. The photograph shows a close-up of *E. coli* colonies growing in a medium that contains lactose as the sole energy source. Draw a line to three colonies whose cells have functioning copies of β-galactosidase. Draw a line to three colonies whose cells have mutations in the *lacZ* locus or in one of the loci involved in regulation of the *lacZ*. Suppose you could analyze the sequence of the *β-galactosidase* gene from each of the mutant colonies. How would these data help you distinguish which cells are structural mutants and which are regulatory mutants?

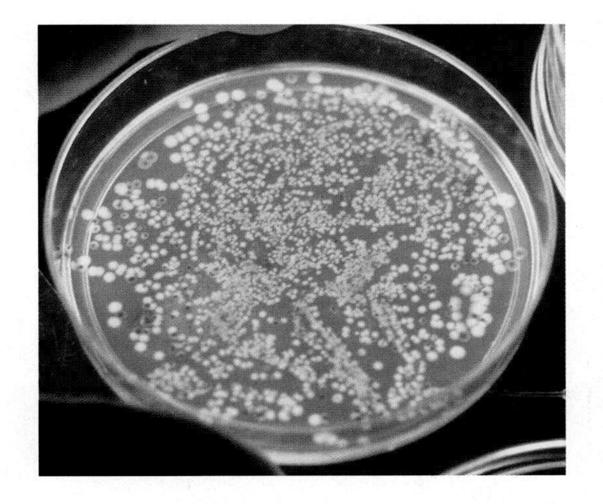

5. *Lactococcus lactis* is commonly used to process milk into yogurt, buttermilk, and cheese because it contains the enzymes required to convert lactose to lactate. Milk is the natural habitat of this organism. State two predictions about how control of the *lac* operon in *L. lactis* differs from the mechanisms that exist in *E. coli*. How would you test your predictions?

6. CAP has helix-turn-helix motifs. Based on your knowledge of the repressor's structure and the operator's sequence, predict (a) how CAP responds to binding by cAMP, and (b) what the DNA sequence at the CAP site looks like.

7. Why is glucose the preferred sugar in *E. coli*? After all, the galactose that is released when β-galactosidase cleaves lactose can also enter the glycolytic pathway, once a series of enzyme-catalyzed reactions have converted the galactose to glucose-6-phosphate.

CD-ROM and Web Connection

CD Activity 14.1: The *lac* Operon *(tutorial)*
(Estimated time for completion = 15 min)
Learn how the bacterium *Escherichia coli* regulates its switch from using glucose as a source of carbon and energy to using lactose.

At your **Companion Website** (http://www.prenhall.com/freeman/biology), you will find self-grading exams and links to the following research tools, online resources, and activities:

Catabolite Repression
This site describes the research on catabolite repression pathways.

The Lac Repressor
This page includes detailed information on the lac repressor, its molecular structure and binding characteristics, as well as CHIME images of its three-dimensional structure.

Jacob, Lwoff, and Monod
The Nobel site describes the lives and scientific work of Francois Jacob, Andre Lwoff, and Jacques Monod, the scientists credited with elucidating the regulation of gene expression.

Additional Reading

Miller, Robert V. 1998. Bacterial gene swapping in nature. *Scientific American* 278 (January): 66–71. Explores the mechanism of "sex" in bacteria.

Control of Gene Expression in Eukaryotes

15

Chapter 14 introduced the central problem that bacteria face in controlling gene expression: turning specific genes on or off in response to changes in food availability or other aspects of the environment. Multicellular eukaryotes face a different challenge in controlling gene expression, however. How can the structure and function of cells from the same individual vary so dramatically? For example, the muscle cells in your upper arm are packed with filaments that slide back and forth in response to electrical signals. The nerve cells that transmit these signals lack sliding filaments. In contrast, they feature membranes that are packed with proteins specialized for transmitting electrical currents. Instead of moving, nerve cells have dozens or hundreds of threadlike projections that transmit electrical signals to muscle cells and other nerve cells.

How can these two types of cells be so different? The differences in morphology and behavior are not the result of variation in gene content, because the DNA inside these cells is identical. Instead, the cells that make up multicellular eukaryotes are different because they express different genes and make different proteins. The striking differences observed among cells result from variation in gene expression. How do nerve cells produce nerve-specific proteins while adjacent muscle cells synthesize muscle-specific proteins?

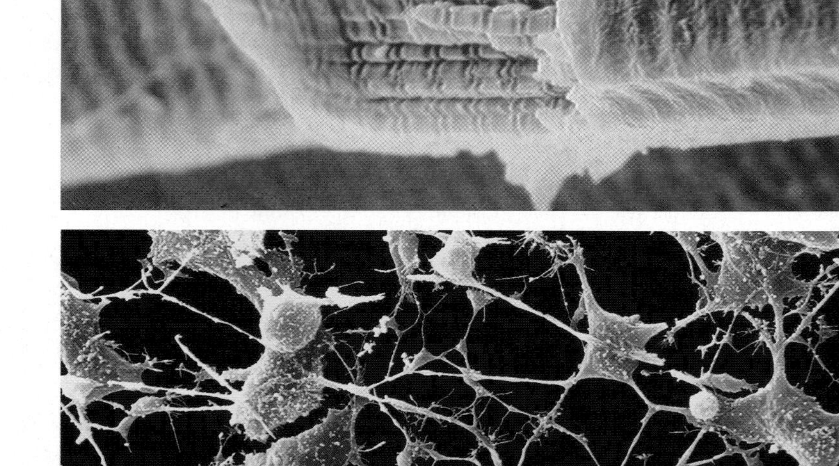

Mammalian muscle cells (above) and nerve cells (below) illustrate the diversity of cell types found within multicellular eukaryotes. This chapter focuses on how different types of cells are able to express different types of genes.

15.1 Mechanisms of Gene Regulation—An Overview

In multicellular eukaryotes, cell-specific gene expression frequently begins with signals from other cells. To drive this point home, consider the cells illustrated in **Figure 15.1a**. They are located in the developing limb of a chicken embryo. What determines whether a particular cell in this limb bud will become nerve, muscle, bone, skin, or some other type of cell? The answer hinges on the concentrations of a large and complex series of signaling molecules. One particularly important group of signals is responsible for a phenomenon known as pattern formation. Pattern formation is the process that causes embryonic cells to form muscles, bones, and other tissues in their proper spatial arrangement. Pattern formation is based on signaling molecules—often produced far from the cells that receive the signal—that identify where each cell is located in three-dimensional space. In the limbs of vertebrates such as chickens and humans, different signaling molecules identify where cells are along the three axes illustrated in **Figure 15.1b**.

Unit 4 explores the nature of the signaling molecules involved in embryonic development. The task before us now, however, is to understand how eukaryotic cells respond to these signals. What happens after a cell has received information about where it is located in the body and about how nearby cells are developing? For example, suppose that an embryonic cell receives a series of signals whose message could be interpreted as, "Become a muscle cell in the biceps." How are these signals translated into action—meaning, into changes in gene expression?

To understand the answer, it is important to return to the central dogma of molecular biology and analyze the steps that occur as information flows from eukaryotic DNA to mRNA to proteins. These steps are diagrammed in **Figure 15.2** (page 298). Each pattern-forming signal influences at least one of these steps. Further, because each of the events illustrated in the figure can be stopped, started, speeded up, slowed down, or altered in other ways, eukaryotic cells have a large and complex array of mechanisms available to control which proteins they produce and in what amounts.

As Figure 15.2 shows, regulation of eukaryotic gene expression begins with the physical state of DNA itself. Eukaryotic DNA is wrapped around proteins and coiled upon itself. As we'll see in section 15.2, this complex physical structure must be altered before transcription can begin. Once a section of eukaryotic DNA molecule is remodeled and available for transcription, a variety of regulatory sequences in DNA come into play. Promoters like those found in prokaryotes are located close to the gene and serve as binding sites for RNA polymerase. But eukaryotes also have other types of regulatory sequences, which may be located far from the promoter. In many cases, the rate of transcription depends on how these sequences interact with regulatory proteins called transcription factors.

Once transcription is complete, a variety of post-transcriptional events occur. These, too, may be influenced by signaling molecules. For example, most eukaryotic genes have introns that must be spliced out of the primary transcript before the mature mRNA is exported from the nucleus to the cytoplasm. In some cases, carefully regulated alterations in the splicing pattern occur. When the splicing sequence changes, a different message emerges. The altered message, in turn, leads to the production of an altered protein.

Finally, once an RNA message reaches the cytoplasm and is available for translation, its impact on cell function hinges on a series of important variables—whether the translational machinery is operational, how long the mRNA remains intact before it is degraded by RNases, how efficiently the protein is

(a) Positional information in the limb bud

(b) Differentiated cells in the limb

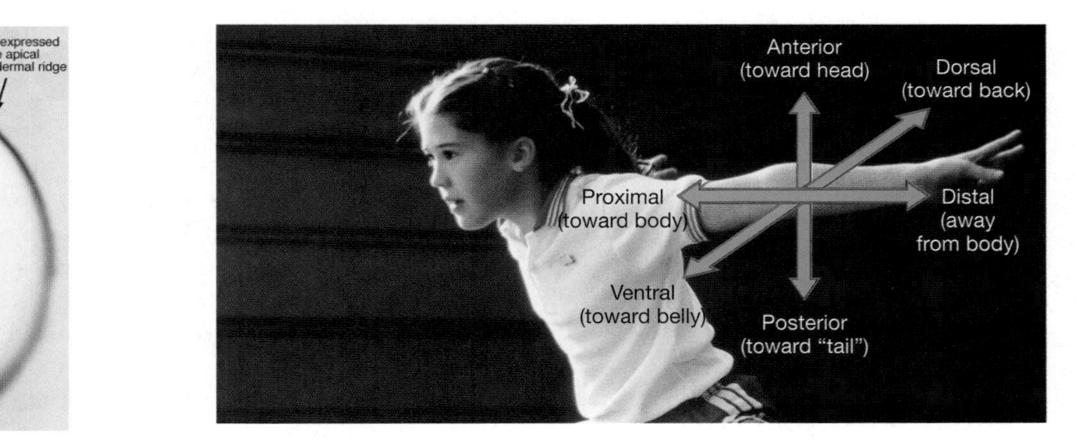

FIGURE 15.1 Embryonic Cells Differentiate Based on Signals from Other Cells
(a) This photograph shows a limb bud from a chicken embryo that has been stained for two important signaling molecules. Note that a protein called Fgf8 is present at the tip of the developing limb, while a protein called Shh occurs at its base. Because these molecules diffuse away from these locations, their concentration provides cells throughout the limb bud with information about their position in the limb. **(b)** As limbs develop, these and other signaling molecules provide information about where cells are located along the three axes indicated here.

folded into a functional conformation, whether the protein is activated by phosphorylation or some other chemical modification, and how long the finished product remains intact before it is degraded by enzymes called proteases.

To appreciate the breadth and complexity of gene regulation in eukaryotes, let's follow the series of events that occurs as a cell responds to a developmental signal. Suppose a molecule arrives that specifies the production of muscle-specific proteins. What happens next?

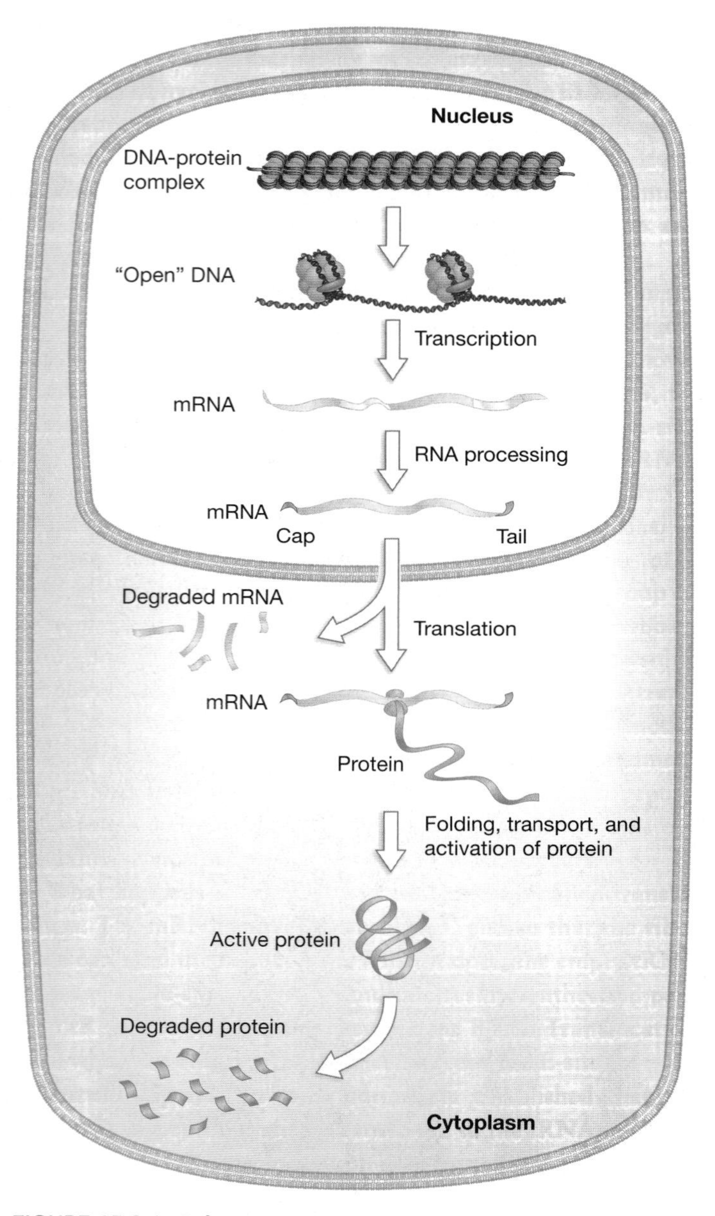

FIGURE 15.2 In Eukaryotes, Gene Expression Can Be Controlled at Many Different Steps
Numerous steps occur during information processing in eukaryotes. Each of these steps provides an opportunity for regulation of gene expression. **EXERCISE** Place a horizontal line through the diagram so that mechanisms of transcriptional control fall above the line and mechanisms of post-transcriptional control fall below.

15.2 Eukaryotic DNA and the Regulation of Gene Expression

If the arrival of a signaling molecule is going to result in the transcription of a particular gene, a drastic remodeling of the DNA around the target locus must occur. To appreciate why, consider that a typical cell in your body contains about six billion base pairs of DNA. Lined up end to end, these nucleotide pairs would form a double helix about 2.04 m (6.5 feet) long. But the nucleus that holds this DNA is thinner than a sheet of paper. How is the genome packed inside?

An answer to this question has emerged incrementally. Just after the turn of the nineteenth century, chemical analyses established that eukaryotic DNA is intimately associated with proteins. Follow-up work determined that the most abundant of these proteins belong to a group called histones. In the 1970s, electron micrographs like the one in **Figure 15.3a** revealed that the protein-DNA complex, or **chromatin**, has a regular, bead-like structure—it looks like beads on a string. The "beads" came to be called **nucleosomes.** More details emerged in 1984 when T. J. Richmond and colleagues estimated the three-dimensional structure of eukaryotic DNA using the technique called x-ray crystallography (introduced in Chapter 13). The x-ray crystallographic data indicated that each nucleosome consists of DNA wrapped twice around a core of eight histone proteins; between each pair of nucleosomes a "linker" stretch of DNA is associated with a histone called H1 (**Figure 15.3b**).

The intimate association observed between DNA and histones occurs because DNA is negatively charged and histones are positively charged. DNA has a negative charge because the phosphate group of nucleotides tends to release a proton at the pH found in cells; histones are positively charged because they contain many lysine and/or arginine residues (see Chapter 3).

Physical models of chromatin suggest that another layer of complexity exists in eukaryotic DNA. According to these models, H1 proteins interact with each other and with nucleosomes to produce the higher-order structure illustrated in **Figure 15.3c.** Based on its width, this structure is called the 30-nanometer fiber (a nanometer is one-billionth of a meter, and is abbreviated nm). In some or many cases, 30-nm fibers may also be packed into still higher order structures.

The general message of these studies is that chromatin is highly compacted to fit into the nucleus. The implication for the control of gene expression is profound. How can RNA polymerase transcribe a particular locus if the DNA sequence is wound into tight coils and complexed with proteins?

Evidence That Chromatin Structure is Altered in Active Genes

Once the nucleosome-based structure of chromatin was established, it seemed clear that the DNA-histone interaction must be altered somehow in order for RNA polymerase to make contact with DNA and for transcription to proceed. Harold Weintraub and Mark Groudine tested this hypothesis by comparing the structure of the β-globin (beta-globin) and ovalbumin genes in the same cell type. In chickens, the β-globin gene is transcribed at high levels in reticulocytes (the precursors of red blood cells). The ovalbumin gene, in contrast, is actively expressed in cells that line the female reproductive tract but not in reticulocytes. The contrast is logical because β-globin is a component of the oxygen-carrying molecule called hemoglobin, while ovalbumin is the major protein found in egg whites.

Weintraub and Groudine predicted that the interaction between histones and DNA is loosened in actively transcribed regions of the genome. To determine whether these predicted differences in chromatin structure actually occur, they treated chromatin from reticulocytes with an enzyme that degrades DNA. The enzyme they employed, called DNase I, rapidly breaks down DNA that is not extensively complexed with proteins. In contrast, the enzyme works very slowly on DNA that is tightly bound to histones—probably because the proteins physically protect the DNA from the enzyme—similar to the way that regulatory proteins protect DNA sequences in the "footprinting" studies introduced in Box 14.3. As a result, DNase I gave the researchers an effective assay for chromatin structure (**Figure 15.4a**, page 300). For example, if the β-globin gene is not tightly bound to histones in reticulocytes—because it is being actively transcribed—then it should be degraded by the enzyme treatment. In contrast, ovalbumin sequences from the same cells should remain intact because they are not being transcribed and remain tightly associated with histones.

After treating chromatin samples from reticulocytes with seven different concentrations of DNase I, the researchers used a technique called Southern blotting to visualize the β-globin and ovalbumin sequences. **Southern blotting** is a protocol for identifying specific regions of DNA. The procedure involves separating fragments of DNA by gel electrophoresis and visualizing specific gene segments via autoradiography. (See **Box 15.2**, pages 308–309 for a more detailed explanation of how Southern blotting and related techniques work.)

The results of the experiment are shown in **Figure 15.4b**. The upper autoradiograph shows how DNase I treatment affected the sequences of the β-globin gene. To understand what these data mean, start with the very dark band on the far left. This band represents fragments of DNA from the β-globin gene in reticulocytes that were *not* treated with DNase I. The untreated β-globin fragments created the black spot because they hybridized to a radioactively labeled piece of the β-globin gene. This result indicates that the fragment containing the β-globin gene is intact.

The lanes to the right of this dark spot contain DNA from reticulocytes treated with increasing concentrations of DNase I. Note that the black bands fade out as DNase I concentrations increase, meaning that less and less of the radioactively labeled β-globin sequence was able to bind to the experimental DNA. These data indicate that DNase I cut the gene into tiny fragments. At medium to high concentrations of enzyme, in fact, virtually *all* of the β-globin sequences are degraded. The data in the lower autoradiograph, in contrast, are from the fragment containing the ovalbumin gene. Here the pattern is different. At

(a) Nucleosomes in chromatin

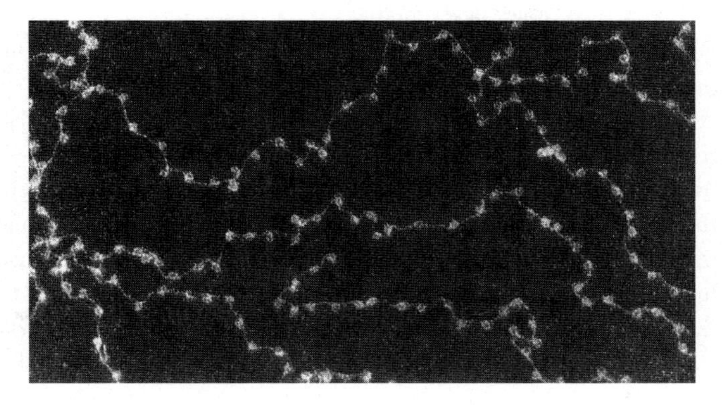

(b) Nucleosome structure

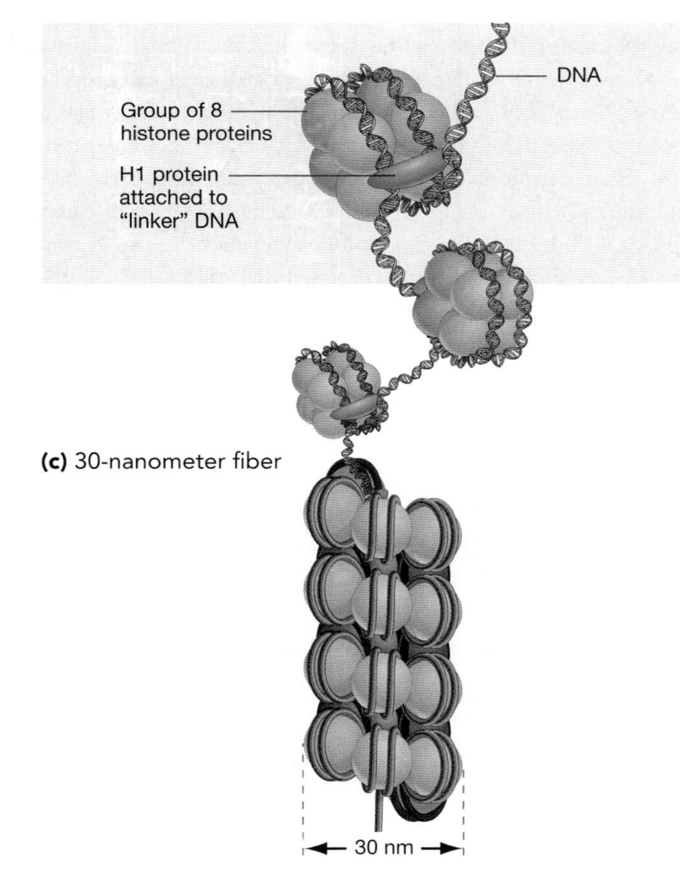

FIGURE 15.3 Chromatin Structure
(a) An electron micrograph of chromatin shows that bead-like structures, called nucleosomes, are linked by short stretches of DNA. **(b)** X-ray crystallographic data revealed that nucleosomes consist of DNA wrapped around a core of eight histone proteins. The linker DNA is associated with a particular histone called H1. **(c)** A model for the structure of the 30-nanometer fiber. It proposes that H1 proteins form a core in the center of the fiber. Note that the 8 histone proteins in each nucleosome are shown as one ball in the bottom drawing, to make the overall structure of the 30-nanometer fiber clearer.

medium concentrations of DNase I, the ovalbumin gene remains undamaged.

These patterns confirm that the interaction between histones and DNA is altered in actively transcribed genes. This conclusion was solidified when researchers discovered mutant forms of brewer's yeast, *Saccharomyces cerevisiae*, that do not produce the usual complement of histones. In these cells, many genes that are normally inactive are instead transcribed at high levels. Together, these observations suggest that histone-DNA interactions inhibit transcription. Further, the data imply that eukaryotic genes are normally turned off. If so, then the process of becoming a muscle cell must begin when 30-nanometer fibers and nucleosomes are "opened up," so that muscle-specific genes can be recognized by regulatory proteins and RNA polymerase and transcribed. How do these changes in chromatin structure happen?

(a) DNase assay for chromatin structure

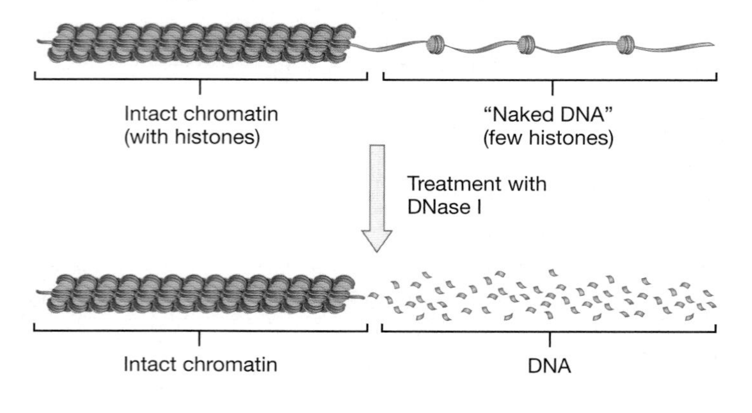

(b) Examples of assay results

DNase concentration (μg/mL)
0 0.01 0.05 0.1 0.5 1.0 1.5
Location of β-globin fragment

DNA fragment must be "naked"—even medium concentrations of DNase degrade it.

DNase concentration (μg/mL)
0 0.01 0.05 0.1 0.5 1.0 1.5
Location of ovalbumin fragment

DNA fragment must be associated with histones—only high concentrations of DNase degrade it.

FIGURE 15.4 Evidence That Chromatin Structure Is Altered in Actively Transcribed Loci
(a) The enzyme DNase I has little effect on tightly packed chromatin but cuts DNA that is only loosely associated with histones. **(b)** The first autoradiograph shows fragments from the β-globin gene in chicken reticulocytes; the second autoradiograph shows fragments from the ovalbumin gene in the same cells. The samples that were run in the 12 lanes were each exposed to a different concentration of DNase I. **QUESTION** Why did the researchers include a treatment with no DNase I added? (This is the far left lane in the autoradiographs in part b.)

How Is Chromatin Altered?

Determining how chromatin is modified from a condensed, inactive form into an open, actively transcribed configuration is currently the subject of intense interest. One important area of research is focusing on a large group of proteins called the Swi/Snf chromatin-remodeling complex. Researchers discovered these proteins by analyzing mutant yeast cells that were unable to transcribe their genes properly. Similar protein complexes have now been found in humans and fruit flies, suggesting that they might be found in many or all eukaryotes. The presence of Swi/Snf complexes increases the sensitivity of chromatin to digestion by DNase I. The complex requires ATP to work, however. Based on these observations, Swi/Snf appears to alter nucleosomes and expose DNA so that transcription can occur. Exactly how the chromatin remodeling occurs has yet to be determined.

Several lines of evidence also suggest a role for enzymes that chemically modify one or more of the histones within the nucleosome. Specifically, several teams of researchers have found that histone proteins associated with transcriptionally active loci tend to be acetylated—meaning that acetyl groups (CH_3COO^-) have been added to lysine residues in the histones. This observation is intriguing because the addition of a negatively charged acetyl group to the positively charged lysine would tend to reduce the electrostatic attraction between histones and DNA. As a result, the association between nucleosomes or between nucleosomes and DNA might loosen enough for RNA polymerase to initiate transcription.

Support for the acetylation hypothesis has increased recently. A key observation involves the proteins that act as positive regulators in eukaryotes. As we'll see in section 15.4, biologists have known for decades that certain proteins act as transcription activators in eukaryotes. What the recent findings suggest is that at least some of these transcription activators can catalyze the addition of an acetyl group to histones near the binding site for RNA polymerase. Does this reaction loosen the electrostatic attraction between nucleosomes or between histones and DNA, and thus open up chromatin and allow RNA polymerase to begin transcription? If so, then acetylation might be part of the mechanism behind transcription initiation. The acetylation hypothesis was first proposed in 1964, and is still the subject of ongoing research.

15.3 Regulatory Sequences in DNA

Even though researchers do not completely understand the mechanisms responsible for opening and closing chromatin, the data reviewed in section 15.2 show a clear link between chromatin structure and gene regulation. Here we consider the DNA sequences themselves. Do eukaryotes have regulatory sequences analogous to the CAP site and operator introduced in Chapter 14? If so, what role do they play in making cell-specific gene expression possible?

Regulatory Sites in 5′ Flanking Sequences

Chapter 12 introduced an important regulatory sequence in eukaryotes—the promoter. Promoters are located upstream, or in the 5′ direction, of the coding sequence; and they are the sites where RNA polymerase initially interacts with DNA to begin transcription. Most eukaryotic promoters include the bases TATA; in 1989 researchers in several laboratories independently discovered a protein in *Saccharomyces cerevisiae* (brewer's yeast) that binds to the TATA box. This TATA-binding protein, or **TBP**, is analogous to the *E. coli* sigma factor introduced in Chapter 12 in several respects. It binds to the promoter, it associates with RNA polymerase, and it helps initiate transcription. But, as explained in section 15.4, the analogy between TBP and sigma factor should not be taken too far. The interaction between TBP and eukaryotic RNA polymerase II is much more complex than the interaction between sigma factor and bacterial RNA polymerase.

What about binding sites for regulatory proteins that might make the production of cell-specific proteins possible? Do they exist in eukaryotic DNA? If so, where are they in relation to the promoter?

The first answers to these questions emerged in the late 1970s when a group of researchers led by Yasuji Oshima began to study how *S. cerevisiae* cells metabolize the sugar galactose. When galactose is absent from the environment, yeast cells produce tiny quantities of the enzymes required to metabolize it. But when galactose is present, production of these enzymes increases by a thousandfold. Oshima and co-workers focused their early work on mutants that were unable to respond to increased galactose availability. These mutants were unable to grow if galactose was the only sugar present in the culture medium.

By analyzing these mutant cells, Oshima and colleagues determined that they had a defect in a gene they called *GAL4*. (In *S. cerevisiae* and humans, the names of genes are written in full caps and italicized.) If the *GAL4* gene is mutated so that no active GAL4 protein is produced, there is no transcription of any of the five enzymes required to metabolize galactose—even if the cells are grown in an environment that contains the sugar. To make sense of this result, Oshima and colleagues hypothesized that the *GAL4* locus encodes a positive regulator. They predicted that the protein product of this gene binds to DNA and stimulates transcription of the genes involved in galactose metabolism.

When other researchers sequenced the GAL4 protein, they confirmed that it has a DNA-binding domain analogous to the helix-turn-helix motif introduced in Chapter 14. The DNA-binding domain in GAL4 is called a zinc-finger region. Later, biologists from other laboratories determined that the binding site for GAL4 is a 20-base-pair stretch of DNA called the upstream activator sequence. This sequence is analogous to the CAP (catabolite activator protein) site in the *lac* operon of *E. coli*, and is distinct from the "basal" promoter where TBP binds.

These results identified the first two components involved in transcription regulation in eukaryotes (**Figure 15.5**). The basal promoter contains the TATA box or other sequences that serve as the site where a complex containing TBP and RNA polymerase initially interacts with DNA. Most if not all loci also have control elements just upstream from this basal promoter. Proteins called regulatory transcription factors bind to these promoter-proximal elements and either activate or repress transcription. Researchers soon discovered, however, that an entirely different type of regulatory sequence also existed in eukaryotes.

The Discovery of Enhancers

At about the time that promoter and promoter-proximal elements were being characterized in yeast cells, research on viruses that infect eukaryotic cells suggested that some regulatory sequences were located far from the promoter. Experiments performed in Susumu Tonegawa's laboratory helped to confirm this result. These biologists were interested in how human cells regulate the expression of a gene involved in fighting viral and bacterial infections. The gene they were interested in is called the immunoglobulin heavy chain gene. As **Figure 15.6** (page 303, step 1) shows, this locus has several coding regions separated by a large noncoding region called an intron. Under certain circumstances, coding regions near this intron are transcribed more actively than coding regions farther away. Based on this observation, Tonegawa and co-workers hypothesized that sequences in the intron might be involved in enhancing gene expression.

The experiment that they designed to test this hypothesis was simple in concept. The idea was to remove portions of the intron and then observe whether transcription can still take place. To accomplish this, the researchers needed two things—a way to remove portions of the intron, and a system for assaying the levels of transcription from altered and unaltered copies of the gene.

Tonegawa and co-workers used restriction enzymes to remove different portions of the intron. **Restriction enzymes** are proteins produced by bacteria that cut double-stranded DNA at specific locations. In step 2 of Figure 15.6, you can see the

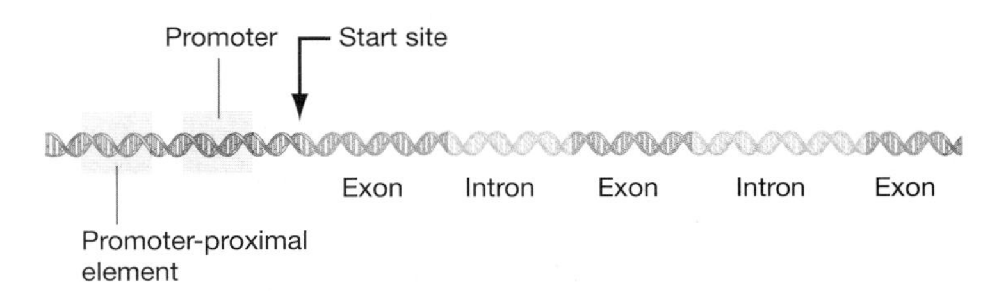

FIGURE 15.5 Promoter-Proximal Elements Regulate Expression of Some Eukaryotic Genes
All eukaryotic genes have promoters, which are the sites where RNA polymerase initially interacts with the gene. Note that exons and introns are not drawn to scale here or in other figures in the chapter. Exons and introns are typically very large compared to promoters and promoter-proximal (regulatory) elements.

locations of the cuts that different restriction enzymes made within the immunoglobulin-heavy chain intron. Step 3 illustrates how the DNA ligase enzyme, introduced in Chapter 12, linked the resulting fragments back together. In essence, then, Tonegawa and his colleagues cut up the sequence of DNA for this gene and joined certain fragments together to create recombinant genes. As **Box 15.1** explains, these two steps—cutting and joining fragments of DNA—are fundamental components of **recombinant DNA technology.**

To test whether the missing portions of the intron are required for the gene to be expressed, the biologists inserted copies of altered and unaltered genes into mouse cells that were growing in culture. As step 4 in Figure 15.6 shows, some cells received genes missing different portions of the intron,

BOX 15.1 Recombinant DNA Technology

The essence of recombinant DNA technology is to cut DNA into fragments with restriction enzymes, paste specific sequences together, and insert the resulting recombinant genes into a cell so that expression can take place. Let's take a closer look at each of these three steps.

Over 100 different restriction enzymes have now been identified. Many share an important feature—they cut DNA only at sites that form palindromes. In English, a word or sentence is a palindrome if it reads the same way backwards as it does forwards. (*Madam* and *gag* are examples.) In biochemistry, a stretch of double-stranded DNA forms a palindrome if the $5' \rightarrow 3'$ sequence of one strand is identical to the $5' \rightarrow 3'$ sequence on the complementary strand. **Figure 1, step 1** provides an example. In most cases, each type of restriction enzymes recognizes a unique type of palindromic DNA sequence.

When restriction enzymes make staggered cuts at palindromes like the ones illustrated in **step 2 in Figure 1**, the resulting fragments have "sticky ends." They are sticky because the single-stranded bases on one fragment are complementary to the single-stranded bases on the other fragment and will bind to them. The presence of these sticky ends allows fragments from different parts of the genome to be spliced together. For example, **step 3 in Figure 1** shows how a researcher might isolate a restriction fragment containing the coding region of one gene, and combine it with fragments (cut by the same restriction enzyme) containing the regulatory sequences from a different gene. After the sticky ends have linked the fragments via complementary base pairing (**Figure 1, step 4**), the gaps that remain can be sealed by adding DNA ligase (**Figure 1, step 5**). The result is a recombinant gene.

Researchers can choose from several different techniques to insert a recombinant gene into a cell so expression can take place. In some cases the DNA can be injected into the cell directly. Alternatively, some types of host cells can be treated with chemicals that make them permeable to DNA and allow the direct uptake of foreign genes. In other instances the recombinant gene can be inserted into a virus particle, which then carries the recombinant gene into the target cell.

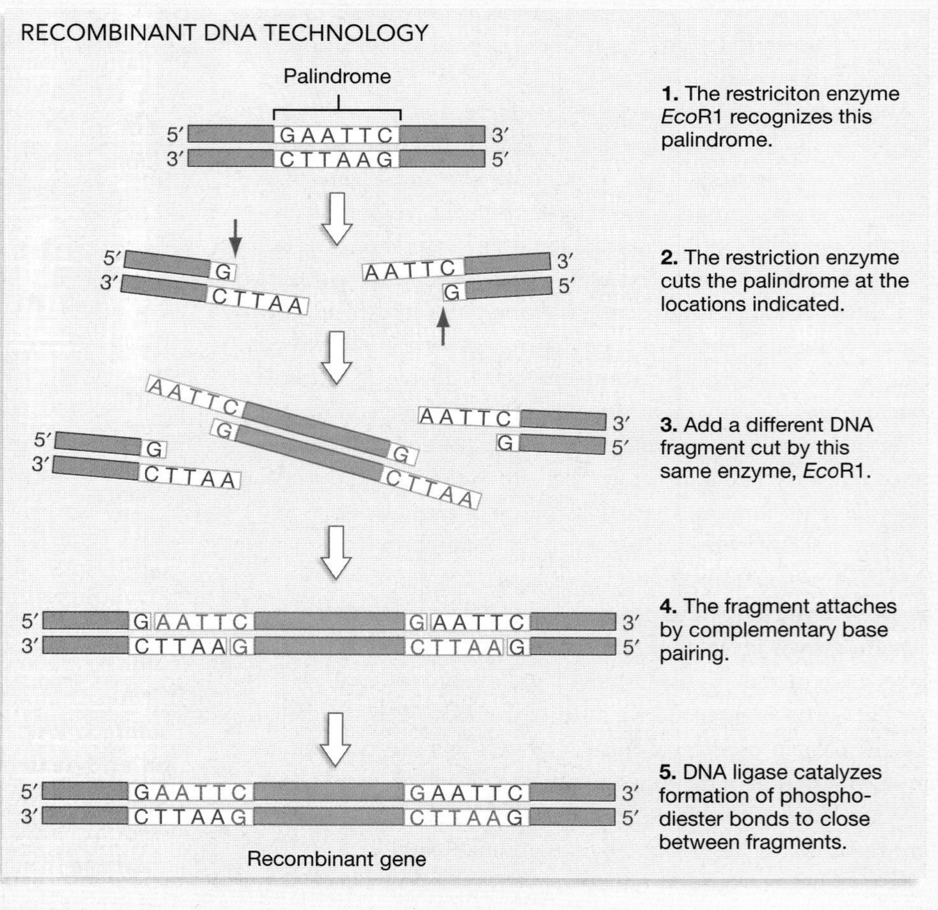

FIGURE 1
This diagram shows the DNA sequence that is recognized by an enzyme called *Eco*R1 (for *Escherichia coli* restriction 1). When *Eco*R1 is added to a DNA sample, it cuts wherever it finds the palindrome shown in step 1. Other restriction enzymes recognize different sequences, which can be four, five, six, or more bases long. **EXERCISE** In step 2, circle the parts of the sequences that researchers refer to as "sticky ends."

some cells received the intact gene, and some cells received no gene. After incubating the cells long enough for transcription to take place, the investigators extracted the mRNAs present in each population of cells, separated the molecules using gel electrophoresis, and tested for the presence of immunoglobulin heavy chain mRNA using the Northern blotting technique introduced in Box 15.2.

The data shown in step 5 of **Figure 15.6** have an important message—the gene is not transcribed if certain parts of the intron are missing. To make sense of this result, Tonegawa and

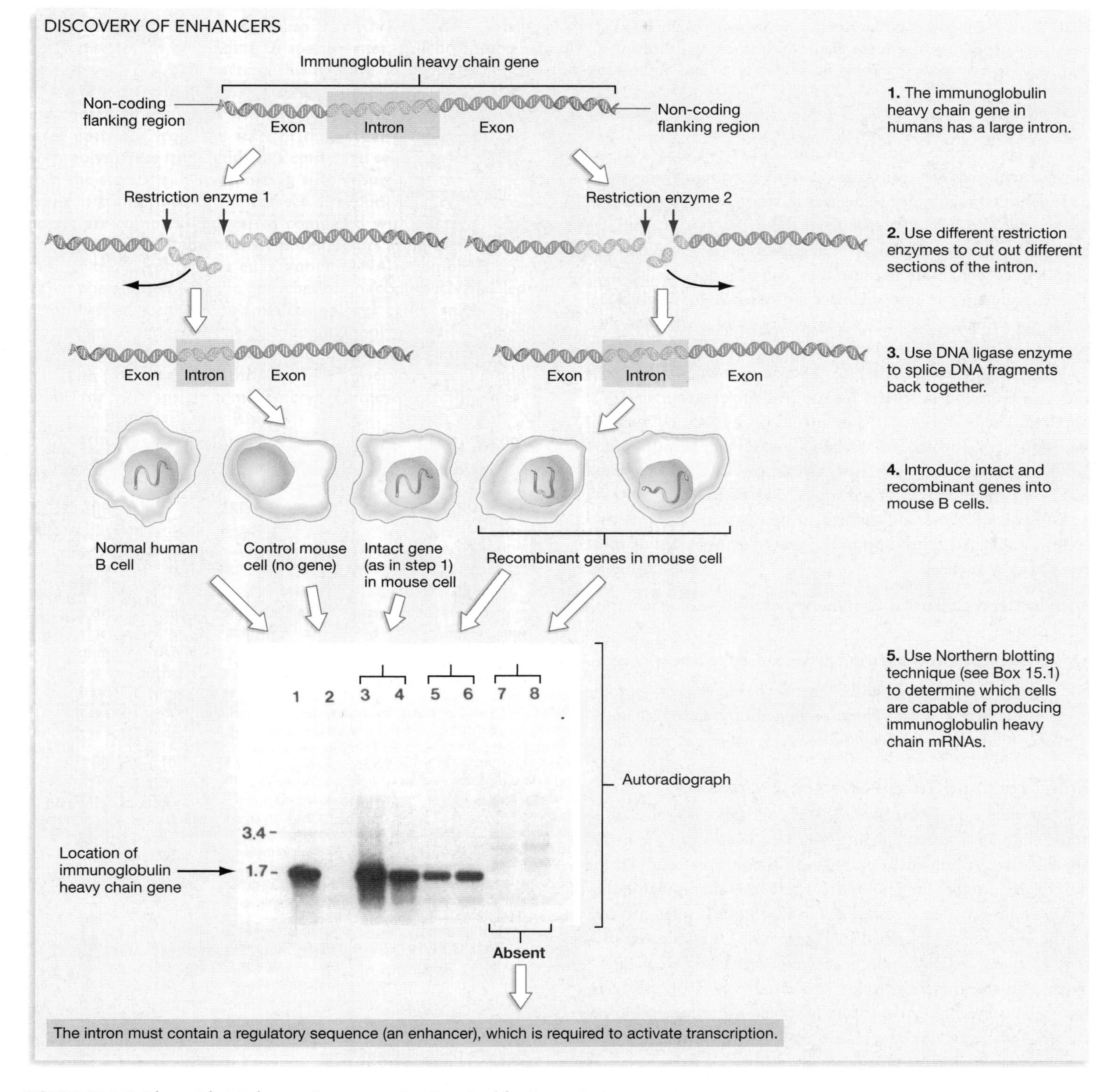

FIGURE 15.6 Evidence That Enhancer Sequences Are Required for Transcription

QUESTION Why did the researchers insert an intact gene? Why did they assess mRNA production in cells that did *not* receive a gene?

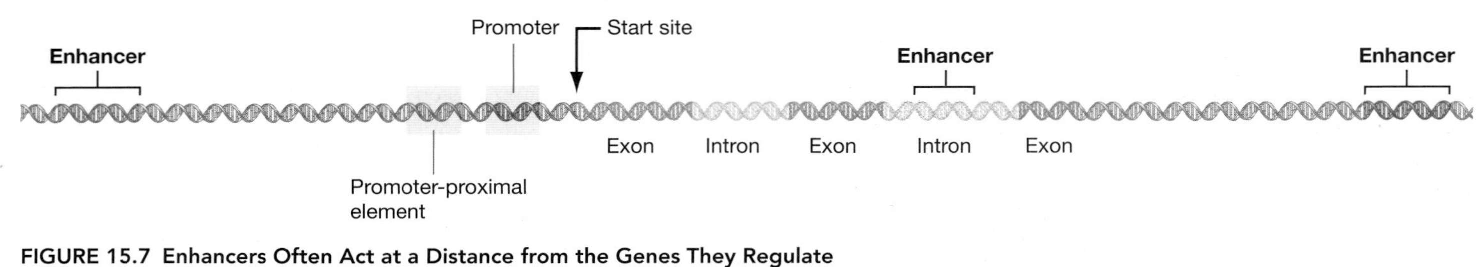

FIGURE 15.7 Enhancers Often Act at a Distance from the Genes They Regulate
Eukaryotic genes may have more than one enhancer. Enhancers can be located either 5′ or 3′ to the gene they regulate or in introns, and can be located up to tens of thousands of base pairs away from the promoter.

co-workers proposed that the intron contains a regulatory sequence required for transcription activation. Previous work had identified regulatory sequences in the genes of viruses that are not located just upstream from the promoter, in what are called 5′ flanking sequences. These regulatory elements had been named **enhancers** because they enhance gene expression. The experiment we just reviewed established that enhancers occur in human cells; subsequent work confirmed that they occur in all multicellular eukaryotes.

The salient characteristic of eukaryotic enhancers is that they can be found in several locations relative to the gene they regulate. They can be located in introns or in the 5′ or 3′ flanking sequences—up to 50,000 bases away from the promoter. Unlike upstream activator sites in yeast or the CAP site found in *E. coli*, enhancers act at a distance (**Figure 15.7**).

Subsequent work on human, roundworm, and fruit-fly genes established other important generalizations about these regulatory sequences:

- Enhancers can work even if their normal 5′ → 3′ orientation is flipped.
- Enhancers can work even if they are moved to a new location.
- Regulatory sequences with similar characteristics, but the opposite effect, exist. These sequences are called **silencers**. When silencers are active, they shut down transcription.

Enhancers and Tissue-Specific Expression

An experiment by Julian Banerji and colleagues established an additional point about enhancers—that the effect of a particular enhancer is limited to certain cell types. For example, the only place in the human body where the immunoglobulin-heavy chain gene is expressed is in a type of cell called a B cell. In the experiment described in Figure 15.6, the cultured cells that received copies of the gene were derived from B cells.* When the researchers inserted unaltered copies of that gene into cells from connective tissue, however, no transcription occurred. Further, when Banerji and co-workers cut the enhancer

*Tonegawa and co-workers used a population of B cells containing a mutant form of the immunoglobulin heavy chain gene. They were thus confident that any mRNAs detected in Northern blots were transcribed from inserted copies, and not from the cell's own genes.

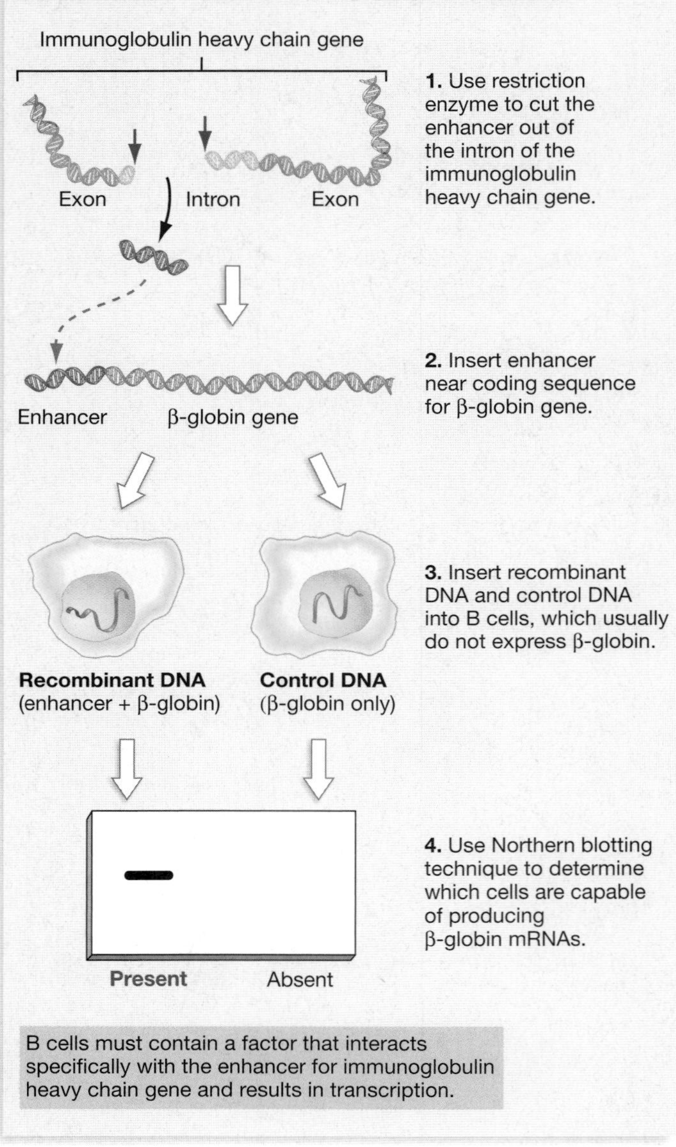

FIGURE 15.8 Evidence That Enhancers Are Involved in Tissue-Specific Expression of Genes
Researchers attached the enhancer sequence from the immunoglobulin-heavy chain gene to the coding sequence for the β-globin gene. The recombinant gene was expressed in B cells, which normally never produce β-globin.

sequence from the immunoglobulin-heavy chain gene, spliced it onto copies of the coding sequence for β-globin, and inserted the recombinant stretch of DNA into B cells, the B cells produced β-globin mRNA (**Figure 15.8**). Normally, the gene for β-globin is never expressed in B cells. The experiment showed that B cells contain some factor that interacts specifically with the enhancer for the immunoglobulin-heavy chain gene and results in transcription of whatever gene is nearby.

These results suggest an answer to the question of how cells produce cell-specific proteins. If promoter-proximal elements, enhancers, and silencers are the binding sites for regulatory proteins, and if certain regulatory proteins are present in certain cell types, then interactions between these regulatory sites and proteins would lead to cell-specific gene expression. For example, a signal molecule with the message "become a muscle cell" could act by triggering the production of regulatory proteins that bind to an enhancer or promoter-proximal element that in turn triggers the transcription of muscle cell proteins (**Figure 15.9a**).

Although this model is a compelling explanation for cell-specific gene expression, important questions remain. Do the regulatory proteins illustrated in Figure 15.9a actually exist? If so, how does the interaction between regulatory proteins and DNA sequences trigger transcription?

15.4 Regulatory Proteins

Researchers relied on a classic genetic strategy to identify the proteins involved in controlling the transcription of specific genes—they analyzed mutants. The first success came with the identification of the GAL4 protein in yeast, which binds to a promoter-proximal element (see section 15.3). Thanks to subsequent analyses of mutant roundworms, fruit flies, and yeast cells with defects in the transcription of specific genes, investigators have been able to assemble a huge catalog of regulatory proteins. The DNA footprinting technique introduced in Chapter 14 was important in confirming that these molecules bind to enhancers and other control elements; x-ray crystallography and other

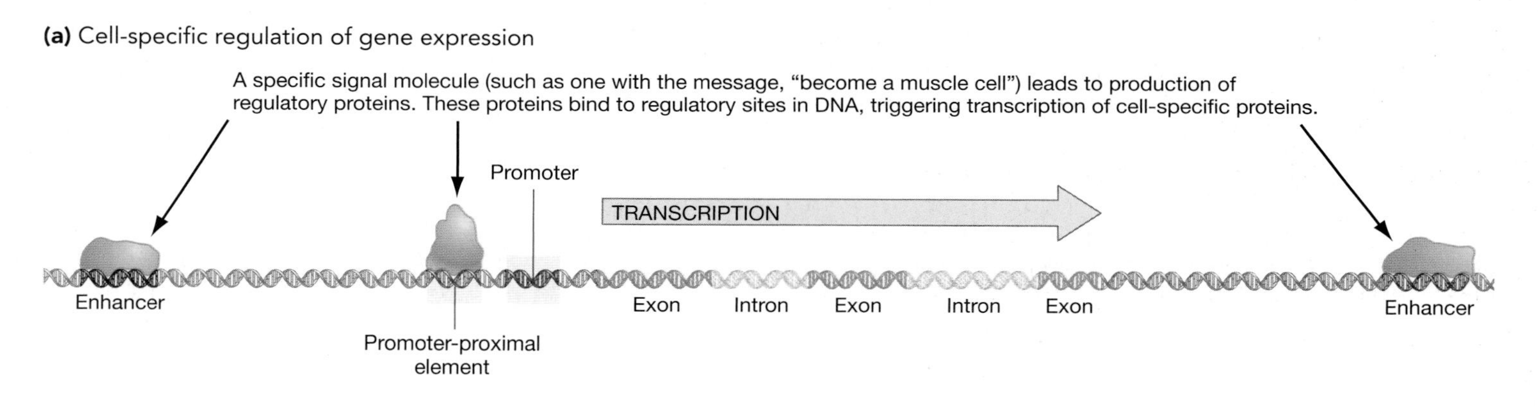

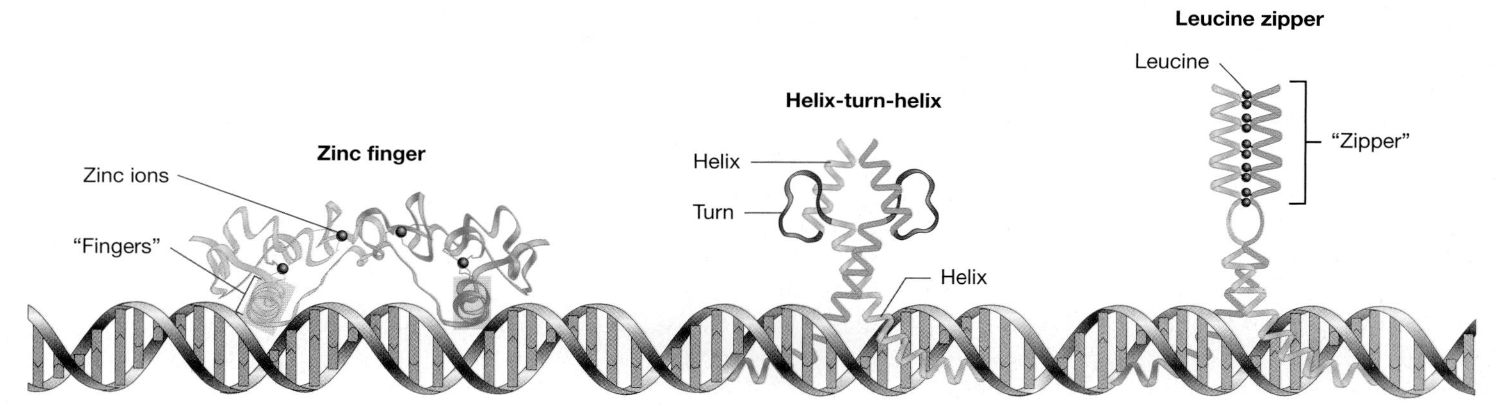

FIGURE 15.9 Proteins Regulate Transcription by Binding to Regulatory Sites in DNA
(a) Researchers hypothesized that signaling molecules lead to the production or activation of regulatory proteins, which then influence transcription by binding to enhancers, silencers, and/or promoter-proximal elements in DNA. **(b)** Structural models of the GAL4 protein, which activates transcription of yeast genes involved in galactose metabolism, showed that it has a DNA-binding domain called a zinc-finger motif (left). Other regulatory proteins found in eukaryotes have different DNA-binding domains, ranging from helix-turn-helix motifs (center) to leucine zippers (right).

structural analyses revealed the array of DNA-binding domains illustrated in **Figure 15.9b**, including the helix-turn-helix motif found in *E. coli*'s CAP and *lac* repressor proteins (see Chapter 14).

In addition, when researchers purified the proteins associated with the start site in active genes, they discovered that TBP and RNA polymerase II do not act alone. Instead, TBP and RNA polymerase are components of a huge apparatus (**Figure 15.10, top**). Before the start of transcription, eight to ten proteins become complexed with TBP, forming an integrated unit called TFIID (for transcription factor IID; there is also a TFIIA, TFIIB, etc.). TFIID then contacts the promoter. A different, even larger group of proteins associates directly with RNA polymerase II. TFIID, other transcription factors, and the proteins associated with RNA polymerase II together form the **core transcription complex**, which is responsible for remodeling chromatin and initiating transcription (**Figure 15.10, bottom**).

Regulatory and Basal Transcription Factors

Based on the results discussed thus far, biologists recognize that two major classes of proteins affect transcription in eukaryotes. **Regulatory transcription factors** bind to specific enhancers, silencers, or other sites in DNA and help activate or repress the transcription of particular genes. Some function as **transcription activators**; others function as **repressors.** In contrast, proteins called **basal transcription factors** are found in the core transcription complex at the promoter. Basal transcription factors are part of the huge molecular machine called the core transcription complex.

Currently, many researchers are focused on understanding how regulatory transcription factors interact with each other and with basal transcription factors to turn the transcription of specific genes on or off. The answer is far from obvious. For example, some regulatory transcription factors bind to enhancers that are tens of thousands of bases away from the promoter. How do these proteins affect other regulatory factors or the core transcription apparatus from such a distance? The outline of an answer is beginning to take shape as a result of research on the proteins that assemble at the promoter.

A Model for Transcription Initiation

How does the core transcription complex assemble at the promoter and initiate transcription? By adding intact or modified versions of molecules from the core transcription complex to in vitro systems, biologists have begun to answer this question. For example, when researchers in Robert Tjian's laboratory added various combinations of regulatory transcription factors and proteins from TFIID to an in vitro transcription system with fruit-fly DNA, they found that maximum production of mRNA occurred when particular combinations of TFIID components and enhancer-binding proteins were present. To make sense of this result, the researchers contend that TFIID proteins and regulatory transcription factors interact directly via physical contact.

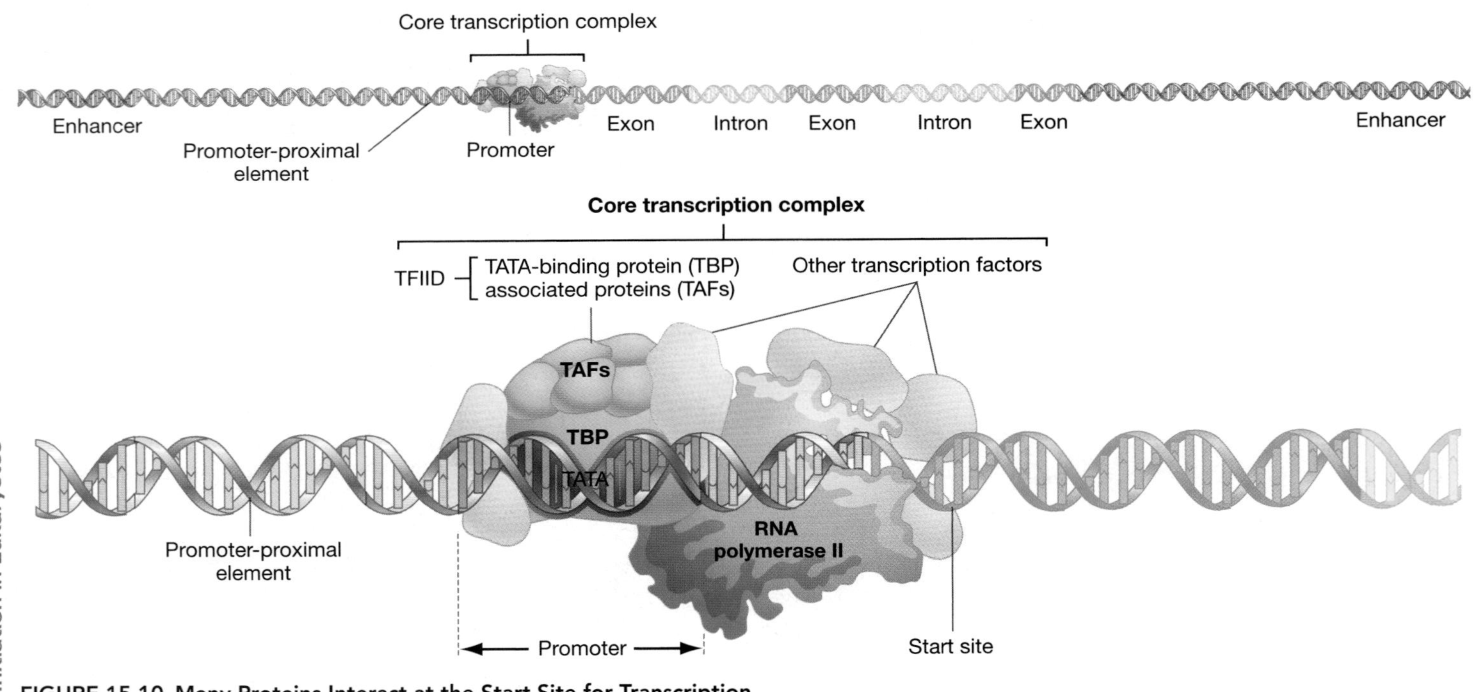

FIGURE 15.10 Many Proteins Interact at the Start Site for Transcription
(a) TBP and RNA polymerase II are components of molecular machines that interact at the promoter and transcription initiation site. TBP binds to the promoter; RNA polymerase II catalyzes the synthesis of mRNA from the DNA template. **(b)** TBP combines with 8–10 TBP-associated factors, or TAFs, to form the complex called TFIID. The combination of TFIID, other transcription factors, and the RNA polymerase II grouping is called the core transcription complex.

CD ACTIVITY 15.1 Transcription Initiation in Eukaryotes

Similar types of experiments have established that RNA polymerase does not interact with the promoter directly. Instead, the enzyme is guided to the transcription initiation site by TFIID and other basal transcription factors. These events appear to occur after chromatin remodeling complexes have modified nucleosomes and opened the DNA.

Figure 15.11 pulls these results together into a model of how transcription begins in eukaryotes. The fundamental idea here is that transcription initiation is a multistep process. The first event is the binding of regulatory transcription factors and the activation of chromatin remodeling complexes. This step modifies chromatin structure enough to make the promoter accessible. Subsequently, regulatory transcription factors contact TFIID and guide it to the transcription start site. As the diagram shows, this step forms loops of DNA around the active locus. Finally, regulatory transcription factors bind to other components of the core transcription complex, making it possible for the core transcription complex to assemble. Subsequently, RNA polymerase begins transcription.

15.5 Post-Transcriptional Processing

Currently, research on transcriptional control in eukaryotes is focused on understanding each of the steps outlined in Figure 15.11 in more detail. Transcriptional control of gene expression, via interactions between transcription activators and specific enhancer sequences, is clearly a key to the formation of muscle, nerve, bone, and other types of cells in multicellular eukaryotes.

In addition to controlling which genes are transcribed, eukaryotic cells also regulate gene expression by altering the rate of translation initiation or by modifying the nature or life span of mRNAs and proteins. For example, once muscle-specific genes have been turned on in a particular cell, the fate of the messages depends on a series of events. What are these post-transcriptional regulatory mechanisms? How do they affect the cell?

Alternative Splicing of mRNAs

Chapter 13 introduced how the noncoding sequences called introns are spliced out of eukaryotic mRNAs. This mRNA processing step provides an opportunity to regulate gene expression because introns can be removed and exons spliced together in multiple ways. For example, consider a protein found in the muscle cells of mammals called tropomyosin. As **Figure 15.12a** (page 308) shows, the initial mRNA transcript for this protein contains a large series of exons. In two of the major types of muscle found in mammals, called smooth and striated muscle, a distinctive subset of exons is spliced together to produce a message that is ready for translation (**Figure 15.12b**). As a result, the tropomyosin protein found in the two cell types is different.

Step 1: Regulatory transcription factors recruit chromatin remodeling complex.

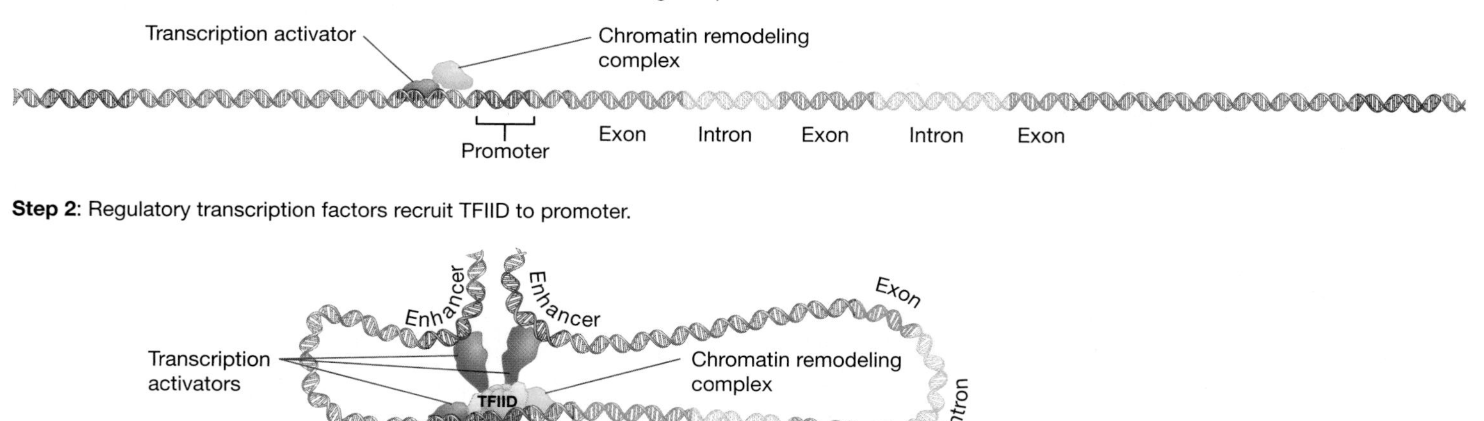

Step 3: RNA polymerase and associated proteins join TFIID to form core transcription complex; transcription begins.

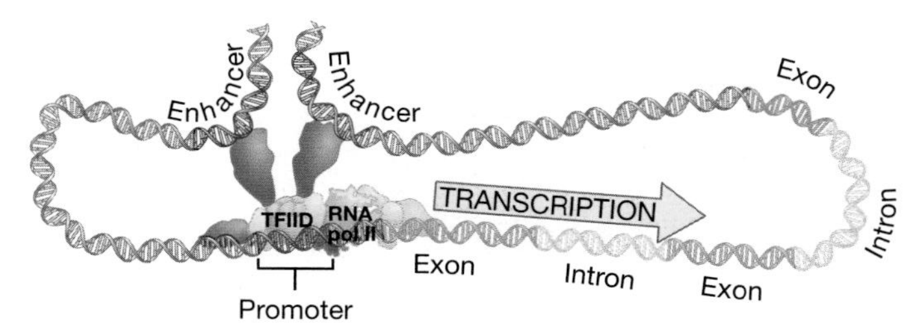

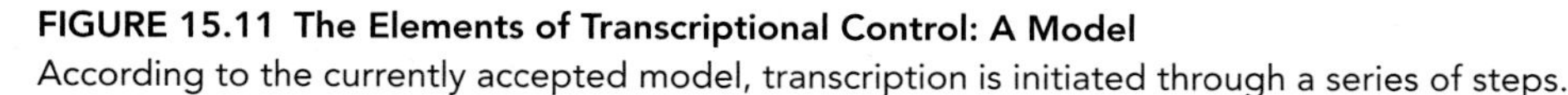

FIGURE 15.11 The Elements of Transcriptional Control: A Model
According to the currently accepted model, transcription is initiated through a series of steps.

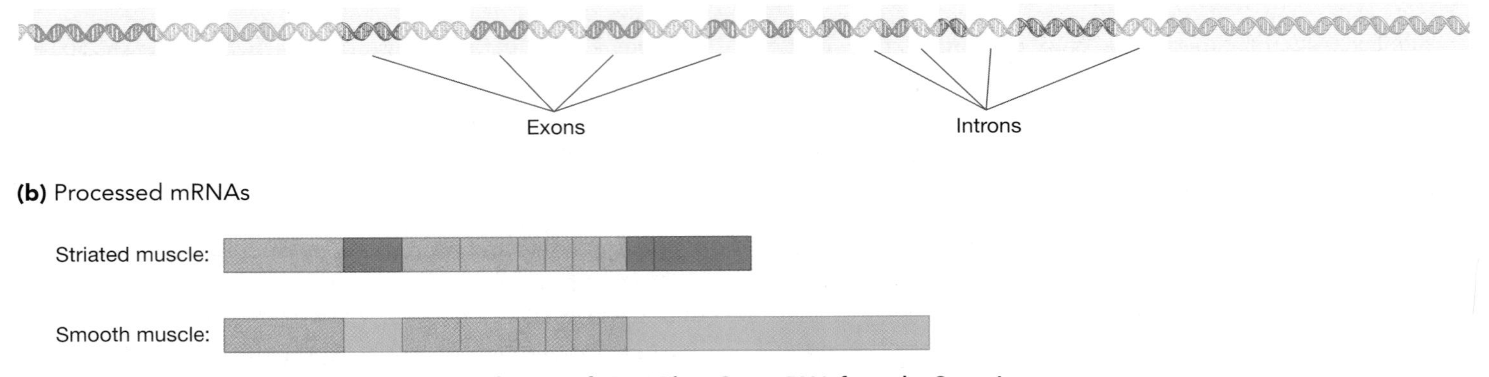

FIGURE 15.12 Alternative Splicing Allows Production of More Than One mRNA from the Same Locus
(a) The tropomyosin gene has a large series of exons and introns. **(b)** In striated and smooth muscle cells, alternative splicing leads to the production of distinct mRNAs for the tropomyosin protein.

BOX 15.2 Southern Blotting

Southern blotting, named after its inventor Edwin Southern, is among the most basic techniques in molecular biology. It is a multistep procedure that allows researchers to identify and characterize specific genes within an organism's genome.

The first step in Southern blotting is to obtain or amplify DNA from the cells of the organisms being studied and digest it with restriction enzymes. The bacteria that manufacture these enzymes use them to restrict, or cut up, DNA from invading viruses. Further, different bacterial species produce restriction enzymes that cut DNA at different sequences. As a result, researchers can use pure, commercially available preparations of these proteins to cut DNA at precisely defined locations (**Figure 1, step 1**).

The DNA fragments generated by a restriction-enzyme digest can then be separated by gel electrophoresis (**Figure 1, steps 2 and 3**). Once the fragments are sorted by size in this way, they are treated with chemicals that break the hydrogen bonds between base pairs, resulting in the formation of single-stranded DNA (**Figure 1, step 4**). Next the fragments of single-stranded DNA are transferred from the gel to a piece of nitrocellulose or nylon membrane, using the blotting technique illustrated in step 5.

The product of these steps is a series of single-stranded DNA fragments, sorted by size and permanently bound to the nylon membrane. In some cases, these fragments represent the entire genome of

(Continued on next page)

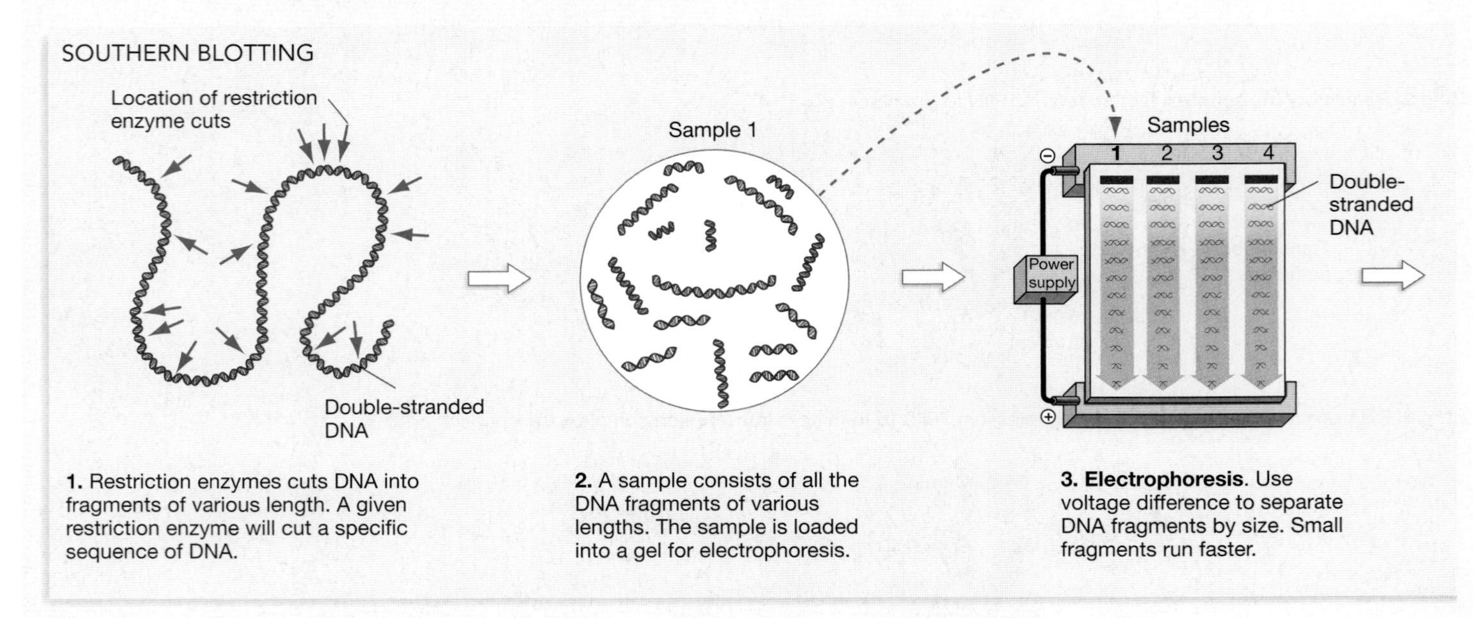

FIGURE 1
QUESTION The intensity of a band in an autoradiograph indicates how much target DNA was in the sample. The darker the band, the more DNA. Why?

Other examples of alternative splicing have been discovered in a variety of genes and organisms. In each case, the consequence is the same. Alternative splicing makes it possible to produce more than one protein product from a single locus. In female fruit flies, for example, a series of splicing events results in the production of a protein that represses the transcription of male-specific genes. In males, an alternative series of splicing events results in a version of the same protein that represses the transcription of female-specific genes.

Alternative splicing appears to be regulated by proteins that bind to the ribonucleoprotein molecules introduced in Chapter 13, which mediate splicing. Current research on alternative splicing is focused on understanding how these regulatory proteins act, and on documenting how widespread the phenomenon is.

Modifying mRNAs and Proteins in the Cytoplasm

Once splicing is complete and processed mRNAs are exported to the cytoplasm, several new regulatory mechanisms come into play. For example, when receptor molecules in the cytoplasm or on the surface of a cell detect a change of some sort—from a sudden increase in temperature, for example, or from infection by a virus—they add a phosphate group to a protein required by the translation apparatus. As Chapter 13 pointed out, phosphorylation frequently leads to changes in the shape and activity of a protein.

(Box 15.2 continued)

the organism being studied. To find the gene of interest within this collection of fragments, a researcher must have a DNA sequence that is complementary to some region in the gene in question. This sequence is called a probe. The probe DNA is labeled with a radioactive atom and made single-stranded by heating it. The probe is then added to a solution bathing the nylon sheet (**Figure 1, step 6**). During this incubation step, the labeled probe binds to the fragment or fragments on the nylon that have complementary base pairs. In this way, the probe identifies the gene of interest.

To visualize which fragments hybridized with the probe, a researcher lays x-ray film over the nylon membrane. As **Figure 1, step 7** shows, radioactive emissions from the probe DNA expose the film. The black band that results identifies the target gene.

A variation on Southern blotting is based on separating RNAs via gel electrophoresis, transferring them to a filter paper, and probing them with a single-stranded and radioactively labeled DNA probe. This technique is used to identify the RNA fragments produced by a particular gene. It is called Northern blotting, in a humorous tribute to the protocol it was patterned after. The variation called Western blotting involves separating proteins via electrophoresis, and then probing the resulting filter with an antibody that binds to the protein of interest. The use of antibodies in research is explored in detail in later chapters.

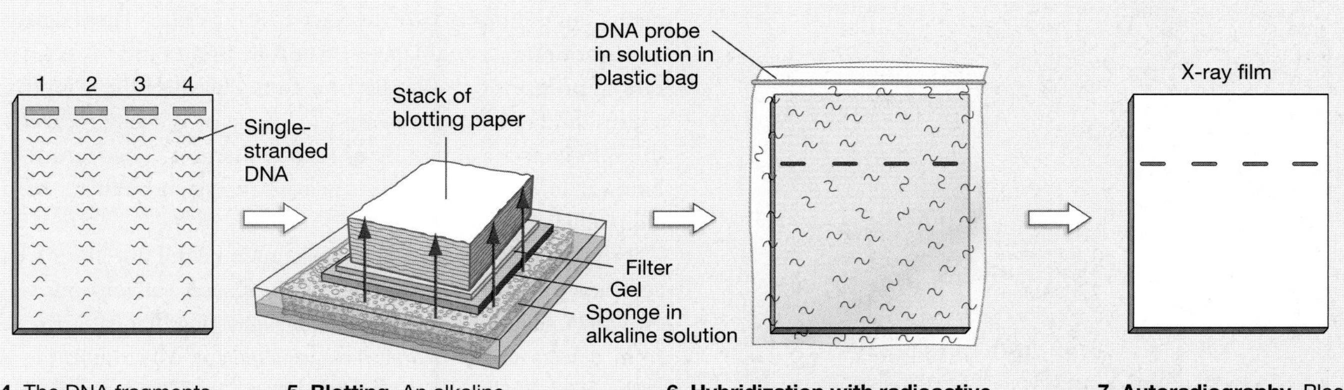

4. The DNA fragments are treated to make them single stranded.

5. Blotting. An alkaline solution wicks up into blotting paper, carrying DNA from gel onto nylon filter, where it becomes permanently bound.

6. Hybridization with radioactive probe. Incubate the nylon membrane with a solution containing labeled probe DNA. The radioactive probe base pairs to the fragments containing complementary sequences.

7. Autoradiography. Place membrane against X-ray film. Radioactive DNA fragments expose film, forming black bands that indicate location of target DNA.

In this case, the resulting shape change makes the protein nonfunctional. In response, translation slows or ceases. For the cell, this dramatic change in gene expression can mean the difference between life and death. If the danger is due to temperature, shutting down translation prevents the production of polypeptides that would be damaged by heat; if the insult is a virus, the cell avoids manufacturing viral proteins and succumbing to the infection.

Changes in regulatory proteins may increase or decrease the overall translation rate of mRNAs. But in addition, mRNAs can be altered in a way that leads to their destruction or increases their life span. These events change the amount of protein produced per message.

Control of gene expression continues even after translation occurs and a protein product is complete. Mechanisms of post-translational regulation are important because they allow the cell to respond to new conditions rapidly. Instead of waiting for transcription and translation to occur, the cell can respond to altered conditions by quickly activating or inactivating existing proteins. Regulatory mechanisms that occur late in the flow from information from DNA to RNA to protein involve a trade-off, however, because transcription and translation require energy.

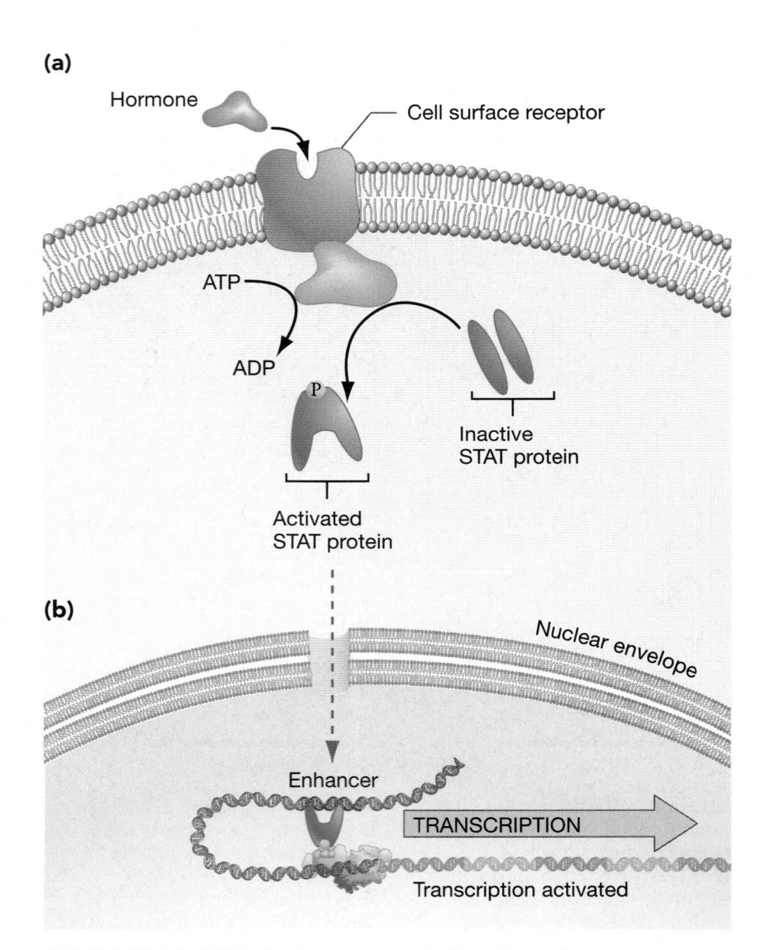

FIGURE 15.13 STATs Activate Transcription After They Are Phosphorylated
(a) A series of events, beginning with binding of a hormone to a cell-surface receptor, leads to the phosphorylation and activation of a STAT protein. **(b)** Phosphorylated STAT proteins move to the nucleus, bind to DNA, and activate transcription of selected target genes.

To explore post-translational control in more depth, consider a group of regulatory transcription factors found in mammals called STATs (signal transducers and activators of transcription). Under normal conditions, STATs sit in the cytoplasm in an inactive form. But as **Figure 15.13a** shows, a series of events is triggered when a hormone-like signaling molecule called a cytokine binds to receptors on the cell surface. Either the receptor itself or a molecule associated with the receptor responds to cytokine binding by adding a phosphate group to a STAT. In this case, phosphorylation changes the STAT to an active form. As **Figure 15.13b** shows, the functional protein moves to the nucleus, binds to an enhancer, and activates transcription. In this way the cell is able to respond in a precise way to a chemical signal from elsewhere in the body. Phosphorylation is a common mechanism of post-translational control over gene expression.

To summarize, mechanisms of post-transcriptional and post-translational control help cells mount a quick response to changing conditions. In many cases they work in close association with transcriptional control, which determines the long-term fate of the cell. Given the importance of both processes to normal cell function, it is no surprise that defects in gene regulation can be disastrous.

15.6 Linking Cancer with Defects in Gene Regulation

Normal regulation of gene expression results in the orderly development of an embryo and efficient responses to changes in environmental conditions. Abnormal regulation of gene expression can lead to birth defects and diseases like cancer. In humans alone, hundreds of distinct cancers exist. These diseases are enormously variable in terms of the tissues they affect, their rate of progression, and their outcome. Because the symptoms, causes, and consequences are so diverse, it is most accurate to think of cancer as a family of related diseases. The feature that binds them together is the disease-causing mechanism of uncontrolled cell growth.

From this perspective, the most fundamental question in cancer biology is a simple one. What causes uncontrolled cell growth? Chapter 9 introduced data suggesting that cancer results from defects in the proteins responsible for controlling the cell cycle. Further, Chapter 12 linked cancers like xeroderma pigmentosum and hereditary non-polyposis colorectal cancer with defects in DNA repair. This association suggests that increased mutation rates are involved in triggering cancer—a conclusion supported by research on people who have been exposed to high-energy radiation, the toxic chemicals called mutagens, and other agents that damage DNA and cause mutations. For example, cancer rates have been extremely high in people who were exposed to massive doses of radiation from the atomic bombs dropped on Hiroshima and Nagasaki, Japan, during World War II. Mutagenic chemicals are so closely linked with increased cancer rates that the cancer-causing compounds called **carcinogens** can be identified by their tendency to cause mutations.

The correlation between increased mutation rates and increased cancer rates makes cancer biology's fundamental question more focused. Which genes, when mutated, disrupt the cell cycle and trigger uncontrolled cell growth? Recent research suggests that most cancers are associated with mutations in regulatory proteins or DNA sequences that regulate transcription. Here we consider three examples. In the first, a mutation prevents a regulatory transcription factor from functioning; in the second, a mutation leads to constant activity by a regulatory protein and overexpression of the genes it controls; in the third, a mutation in an enhancer sequence leads to increased expression of a regulatory transcription factor.

The Guardian of the Genome—*p53*

The gene that currently qualifies as the most important locus associated with human cancers was discovered in a roundabout way. During investigations on how a virus called SV40 infects mammalian cells, researchers found that viral proteins inactivate certain cellular proteins. One of the targeted cellular proteins came to be called p53 because it has a molecular weight of 53,000 daltons. This discovery inspired a great deal of interest because cells infected with SV40 occasionally become cancerous. Because an SV40 protein inactivates the p53 protein, its activity mimics a loss-of-function mutation in the cell's genome. Based on these observations, researchers hypothesized that knock-out mutations in *p53* make the development of cancer more likely.

An association between *p53* and cancer was confirmed in spectacular fashion when sequencing studies revealed that over half of all human cancers are associated with mutant, nonfunctional forms of *p53*. Why are cancer and mutations at this locus linked? In mice that have two mutant, inactive copies of the *p53* gene, development takes place normally. This observation suggests that *p53* is not required for regular cell function. Yet Warren Maltzman and Linda Czyzyk found that when they exposed human cells to the damaging effects of UV radiation, levels of p53 protein increased markedly. Follow-up studies confirmed that there is a close correlation between DNA damage and the amount of p53 in a cell.

The observation that DNA damage is associated with p53 concentrations inspired an important hypothesis for how the protein functions. The fundamental idea, illustrated in **Figure 15.14a**, is that p53 serves as a brake on the cell cycle. According to this hypothesis, p53 is activated after DNA damage occurs and acts to arrest the cell cycle. As a result, the cell has time to repair its DNA or be targeted for destruction. But if mutations in *p53* make the protein product inactive, then damaged cells are not shut down or killed. They continue to grow, except that now they are likely to contain many mutations because of the damage they have incurred. In at least a few instances, one or more of these mutations encourages the start of cancerous growth.

According to this view, *p53* functions as a **tumor suppressor**—a gene that prevents cancer formation by stopping

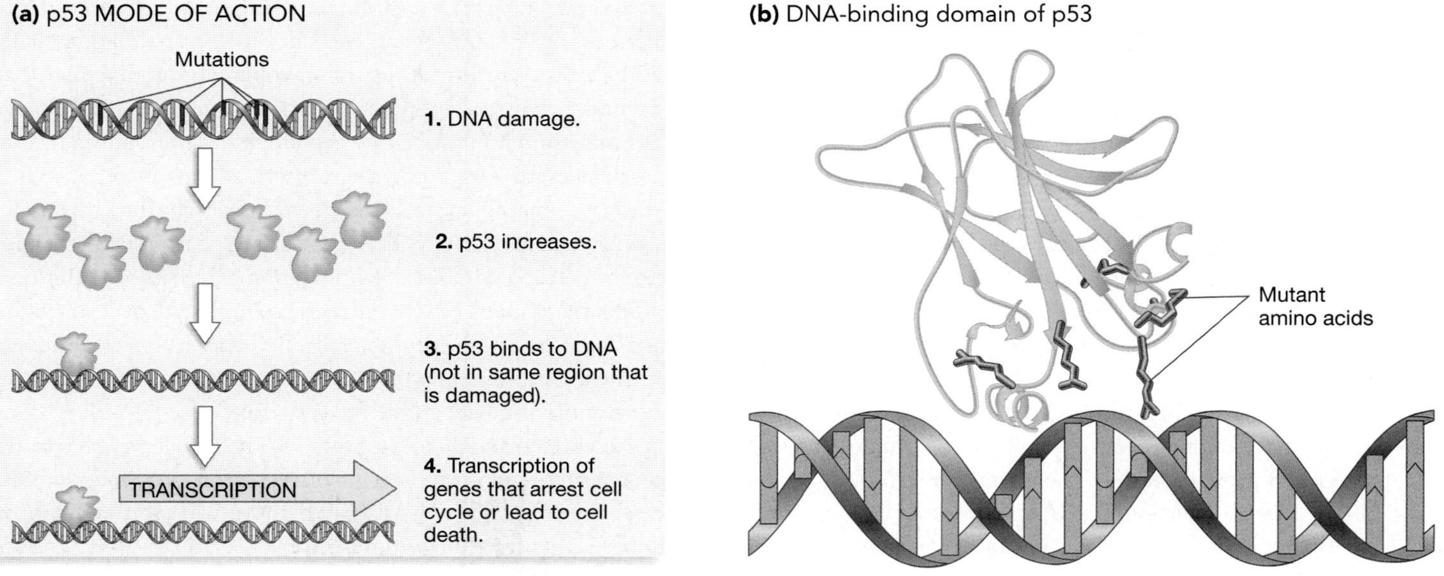

FIGURE 15.14 p53 Is a DNA-Binding Protein That Acts as a Tumor Suppressor
(a) This sequence of events summarizes the current model for p53's mode of action. **(b)** This ribbon model shows the DNA-binding domain of p53. The darkened regions indicate the six most common locations where *p53* genes isolated from cancer patients encode mutant amino acids. Note that the mutations change amino acids located at or near the DNA binding site. QUESTION Why would substituting one amino acid for another at one of the locations highlighted in part (b) affect a protein's ability to bind to DNA?

the cell cycle. The model inspired researchers to call *p53* "the guardian of the genome."

How does p53 actually act? The protein's amino acid sequence suggests that it has a DNA-binding domain. This provided a hint that p53 might be a transcription activator or repressor. This hypothesis was supported when a team led by Yunje Cho was able to estimate the three-dimensional structure of p53 bound to DNA. These studies also correlated beautifully with earlier sequencing studies of the *p53* alleles found in cancerous cells. By combining the sequencing and structural data, Cho and co-workers showed that cancerous cells tend to produce p53 proteins with mutations at or near the DNA binding site (**Figure 15.14b**). The inference here is that the proteins are inactive because they are not able to bind to DNA and regulate transcription as they normally would.

Currently, research on p53 is forging ahead on two fronts. Biologists are striving to identify the genes that are regulated by the protein, and they are attempting to find molecules that could act as anticancer drugs by mimicking p53's shape and activity.

Experimental Evidence That STAT Mutants Develop Cancer

The family of transcription activators called STATs, introduced earlier in the chapter, provides another example of a protein implicated in cancer. Recall that STATs activate genes in response to signals from outside of the cell. The genes targeted by STATs are involved in cell growth. Under normal conditions, STATs bind to regulatory sequences in DNA and activate transcription only if they are phosphorylated. But in some human cancers and in many cell cultures that show cancer-like uncontrolled growth, STATs bind to DNA all the time, or constitutively—not just when they are phosphorylated. These observations hint that certain types of STAT mutants might trigger cancer. The general point here is that constant activation of a regulatory transcription factor, because of a change in the normal post-translational control mechanism, can lead to cancer.

What sort of STAT mutants might bind to DNA constitutively and lead to uncontrolled growth? When researchers in James Darnell's laboratory succeeded in estimating the three-dimensional structure of a protein called STAT3, they made two important observations. First, the active form of the protein is a **dimer**—a combination of two proteins. Second, the phosphorylated amino acid residues occur at the point of contact between the two STAT3 proteins in the dimer (**Figure 15.15a**).

STAT3's three-dimensional structure inspired a hypothesis and experimental test about its role in cancer formation. Jacqueline Bromberg and co-workers reasoned that if mutations caused cysteine residues to be substituted in the portion of the protein where dimer formation takes place, then disulfide bonds like those described in Chapter 3 might form between STAT3 monomers (**Figure 15.15b**). If so, then mutant forms of STAT3 monomers might form dimers without being phosphorylated.

To test this hypothesis, Bromberg and co-workers created mutant STAT3 proteins with cysteines in the appropriate location, introduced the protein into normal cells, and analyzed the amount of transcription at a gene targeted by STAT3 (**Figure 15.15c**). Compared to cells with normal STAT3, the cysteine mutants produced ten times as much of the target gene product. This observation confirmed their prediction that cysteine mutants dimerize spontaneously and bind to DNA constitutively.

Next the biologists injected mice with cells that expressed cysteine-mutant STAT3 or with cells that expressed normal STAT3. After just four weeks, all of the mice injected with mutant cells had developed tumors at the site of injection. In contrast, none of the mice injected with normal cells showed evidence of tumors, even after eight weeks. These experiments show that if mutations cause cysteines to be substituted for the normal amino acids in the dimerization domain of STAT3, cancerous growth is likely. The general message? Any mutation that causes constitutive dimerization of a STAT protein can lead to constitutive activation of target genes that stimulate cell growth. The result is cancer.

Changes in Regulatory Sequences Can Lead to Cancer

One particularly general observation about cancerous cells is that they tend to have an abnormal number of chromosomes or contain chromosomes with translocations, meaning segments that have moved from one chromosome to another. Chromosome-level mutations like these occur because of mistakes during mitosis or because radiation has broken a piece of chromatin, allowing it to relocate.

To grasp why translocations can lead to disease, consider the cells responsible for a cancer called Burkitt's lymphoma. These tumor cells have a translocation involving the coding region for a regulatory transcription factor named *c-myc*. In most cases of Burkitt's lymphoma, the *c-myc* coding region becomes inserted near genes that are actively transcribed in the types of cells known as lymphocytes. Why does the insertion of this coding sequence next to an actively transcribed region lead to uncontrolled cell growth? The leading hypothesis to answer this question is that enhancers that activate lymphocyte genes begin activating *c-myc* in the mutant cells. Because the *c-myc* product is a transcription regulator that encourages cell growth, overexpression of the gene leads to cancer.

Research on *p53*, STAT proteins, and *c-myc* has shown that changes in regulatory proteins or regulatory sequences in DNA can trigger uncontrolled cell growth and the formation of tumors. In combination with research on the relationship between defective DNA repair systems and cancers like xeroderma pigmentosum and hereditary nonpolyposis colorectal cancer, these examples confirm that on the molecular level, cancer is not a single disease. In many cases, the fundamental defect can be traced to mechanisms of gene regulation at the transcriptional, post-transcriptional, or post-translational level.

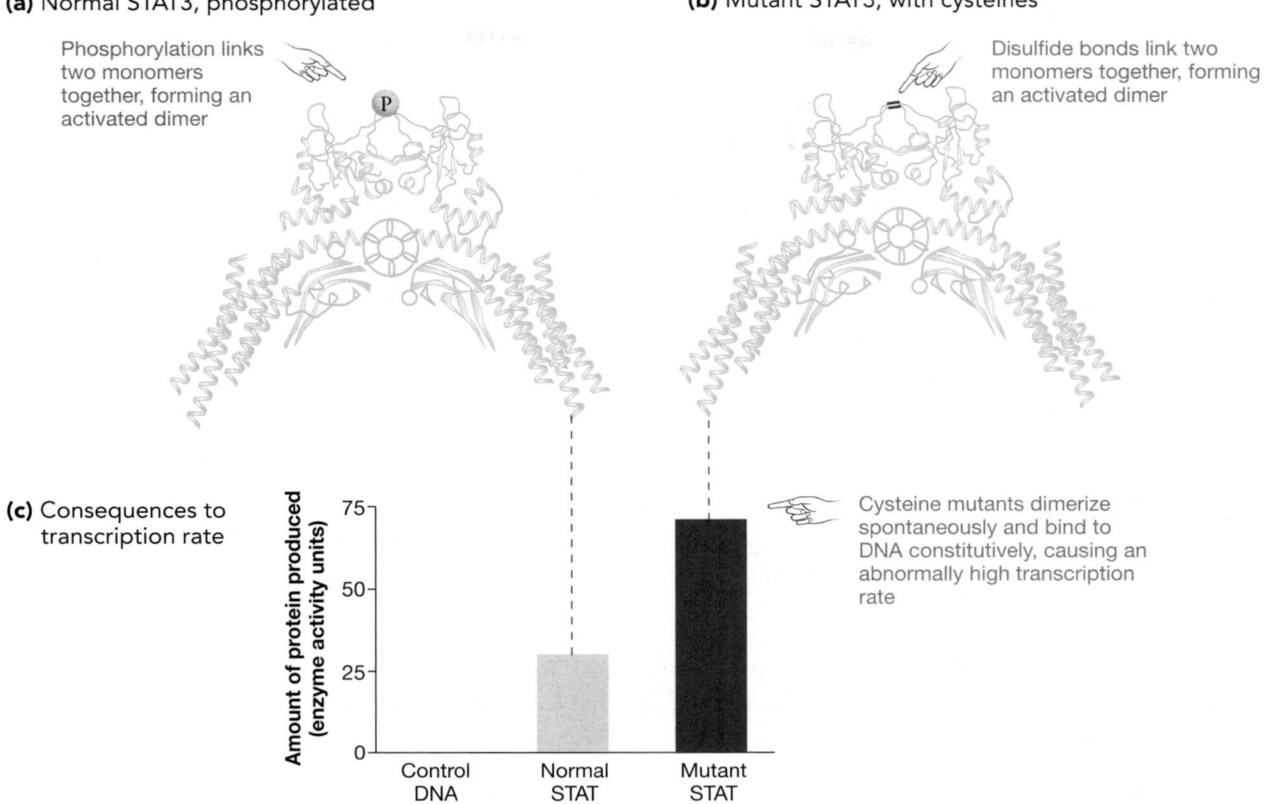

FIGURE 15.15 Generating a Cancer-Causing Mutation Experimentally
(a) Structural models show that the phosphorylated sections of STAT3 proteins bind to one another, forming the two-protein complex, or dimer, shown here. **(b)** Based on the model in part (a), researchers hypothesized that substitution of cysteines at the locations shown in red would result in the formation of disulfide bonds between STAT3 proteins. If so, STAT3 molecules would form dimers and activate transcription even if they were not phosphorylated. **(c)** These histograms show the transcriptional activity of a gene that is normally activated by STAT3. The data come from cells that received either a control sequence (one that does not encode a transcription activator), a normal STAT3 gene, or a copy of the cysteine-mutant version of the STAT3 gene.

Essay Gene Regulation and the Green Revolution

Although this chapter emphasized how basic research on gene regulation has led to exciting advances in the study of cancer, there is an equally important arena for applying this basic knowledge—the effort to increase crop yields and improve the nutritional status of the world's human population. Currently, much of this work focuses on using recombinant DNA technology to genetically modify domesticated plants, using techniques that are discussed in detail in Chapter 17. Here we consider an example of how changes in regulatory proteins affect the productivity of rice, wheat, and corn.

Can gene regulation affect the productivity of crops?

Beginning in the 1960s, worldwide distribution of new strains of rice and wheat, combined with more intensive use of fertilizers, produced a sharp increase in agricultural productivity known as the green revolution. Many of these more productive strains were short-stemmed, or dwarfed, varieties of traditional lines. Because dwarfed individuals invest fewer resources into developing long stems, the size and quality of their seeds tends to be higher; because they are shorter, they are less prone to wind damage.

Agricultural scientists produced these strains through traditional breeding methods that rely on controlled matings between parents with desirable traits. Which genes were affected by this breeding program? Specifically, which alleles contribute to dwarfing?

Jinrong Peng and colleagues began working toward an answer to this question by studying the genetic basis of dwarfing in a tiny member of the mustard family called *Arabidopsis thaliana*. Experiments showed that dwarfed individuals in this species are small because they do not respond to the growth hormones called gibberellins. By mapping and sequencing the loci responsible for the trait, Peng and co-workers found a gene they called *GAI*, for gibberellin-insensitive. Recent follow-up studies in wheat, rice, and corn have confirmed that dwarfed individuals in these species, including members of green-revolution strains, have mutant forms of the same gene.

What does *GAI* do? Although it is clear that the protein product of *GAI* interacts with gibberellins in some way to modulate stem length, the exact mechanism of action is still unknown. The sequencing data offer an important clue, however. The protein encoded by *GAI* has a phosphorylation domain, which suggests that its activity is under post-translational control. Research is now focused on understanding why mutants that lack a functional copy of *GAI* are dwarfed.

Could this discovery help keep the green revolution going? Using traditional breeding methods, agronomists have never been able to produce a dwarfing, highly productive strain of basmati rice—an important crop plant in northern India. But when Peng and associates used recombinant DNA techniques to introduce mutant copies of *GAI* into basmati rice embryos, the resulting adult plants were dwarfed. This result creates hope that recombinant DNA technology can be used to extend advances made by traditional breeding programs. As Chapter 17 explains, however, this research program is intensely controversial.

Chapter Review

Summary

Understanding how the cells in a multicellular organism can become radically different in size, shape, and function—even though they have the same DNA—is a central issue in research on eukaryotic gene regulation. How does cell-specific gene expression occur?

In embryos, cell-specific gene regulation is possible because signals from other cells identify where each cell is in the body and how nearby cells are differentiating. In mature cells, signals from other cells identify changes in the environment that require changes in gene expression. Signaling molecules can cause changes in gene expression in many different ways. They may initiate or repress the transcription of specific genes, change the life span of certain mRNAs or alter how they are spliced, or modify the activity or life span of particular proteins.

Transcription regulation occurs because signaling molecules transmit their signal to regulatory transcription factors that interact directly with DNA. Proteins that regulate transcription are called transcription activators or repressors. These molecules bind to regulatory sites called enhancers and silencers, which are located at a distance from the gene in question, or to promoter-proximal sequences near the start of the coding sequence. According to the currently accepted model, regulatory transcription factors have a physical interaction with the core transcription complex at the promoter. In addition, regulatory transcription factors recruit chromatin remodeling complexes. These events result in the assembly of transcription machinery at the promoter and the activation of transcription.

The core transcription complex may consist of as many as 50–60 protein subunits. Some of these subunits are associated with the TATA-binding protein (TBP), which binds to the promoter. Others are associated with RNA polymerase, which catalyzes the synthesis of RNA from the DNA template. How these proteins interact with each other, with DNA, and with transcription activators and repressors is currently the subject of intense research and debate.

In eukaryotes, DNA is wrapped around histone proteins to form a bead-like structure that is then coiled into 30-nm fibers. Transcription initiation is not possible until this structure is loosened and the interaction between DNA and histone proteins is relaxed. These changes depend on the acetylation of histone proteins and the action of chromatin remodeling complexes.

Once a message is transcribed, several other regulatory events come into play. Alternative splicing allows a single locus to produce more than one product. Mechanisms of post-translational control, such as the activation of transcription activators by phosphorylation, trigger rapid responses to signals from outside the cell or to environmental change.

Recent research has shown that mutations in transcription activators or repressors may trigger uncontrolled cell growth and tumor formation. For example, the cell-cycle regulator *p53* is associated with cancer formation when it fails to function due to mutations in its DNA-binding site. Mutant STAT proteins may lead to cancer if they bind to DNA constitutively and trigger overexpression of the genes they control. In addition, cancer may result from mutations in the relationship between enhancers and coding loci. The cells responsible for Burkitt's lymphoma have chromosome translocations that place the coding locus for *c-myc* next to enhancers for other genes.

Understanding how transcription activation and repression occur is fundamental to cancer biology, the gene therapy strategies introduced in Chapter 17, and the mechanisms of plant and animal development analyzed in Unit 4.

Questions

Content Review

1. What is chromatin?
   a. the protein core of the nucleosome, which consists of histone proteins
   b. the 30-nm fiber
   c. the DNA-protein complex found in eukaryotes
   d. the histone *and* non-histone proteins in eukaryotic nuclei

2. What is a tumor suppressor?
   a. a gene associated with tumor formation when its product does not function
   b. a gene associated with tumor formation when its product functions normally
   c. a gene that accelerates the cell cycle and leads to uncontrolled cell growth
   d. a gene that suppresses tumor formation by stopping the cell cycle

3. Which of the following statements about enhancers is correct?
   a. They contain a consensus sequence called a TATA box.
   b. They are located only in 5′ flanking regions.
   c. They are located only in introns.
   d. They are found in a variety of locations, and are active in any orientation.

4. In eukaryotes, why are certain genes expressed only in certain cell types?
   a. Regulatory transcription factors vary from cell to cell.
   b. The promoter sequence varies from cell to cell.
   c. The location of enhancers varies from cell to cell.
   d. RNA polymerase varies from cell to cell.

5. What is alternative splicing?
   a. changes in phosphorylation that lead to different types of post-translational regulation
   b. changes in mRNA processing that lead to different combinations of exons being spliced together
   c. changes in the snRNPs, or "snurps," found in the nucleus
   d. conformational changes that turn proteins on or off

6. What two major components comprise the core transcription complex?
   a. TFIID and the RNA polymerase complex
   b. TBP and its associated proteins
   c. enhancers and silencers
   d. the promoter and the start codon

Conceptual Review

1. Explain the relationship between enhancers, silencers, transcription activators, repressors, and the core transcription complex.
2. Explain why the "sticky ends" left by restriction enzymes allow biologists to paste specific DNA fragments together.
3. Why does chromatin need to be "remodelled" for transcription to occur?
4. Why is p53 considered a tumor supressor?
5. Make a diagram showing how restriction fragments containing regulatory sequences from one gene (the promoter and enhancers) can be combined with the coding sequence (introns and exons) from a different gene.

Applying Ideas

1. Histone proteins have been extremely highly conserved during evolution. The histones found in fruit flies and humans, for example, are nearly identical in sequence. Offer an explanation for this observation. (Hint: What are the consequences of a mutation in one of the histone proteins?)
2. Suppose you wanted to insert recombinant DNA into yeast and produce a line of cells that manufactured large quantities of the chicken ovalbumin protein. Based on your knowledge of how galactose metabolism is controlled, what regulatory sequences might you add to the coding region for the protein to create a recombinant DNA? How would you induce expression of the recombinant gene once it is inside yeast cells?
3. The text noted that levels of p53 protein in the cytoplasm increase after DNA damage. Design an experiment to determine whether this increase is due to increased transcription of the *p53* gene or to activation of preexisting p53 proteins, via a post-translational mechanism such as phosphorylation.

CD-ROM and Web Connection

CD Activity 15.1: Transcription Initiation in Eukaryotes *(animation)*
(Estimated time for completion = 5 min)
Explore how a eukaryotic cell unravels chromatin and uses cell-specific proteins to turn a gene on.

At your **Companion Website** (http://www.prenhall.com/freeman/biology), you will find self-grading exams and links to the following research tools, online resources, and activities:

Gene Regulation in Eukaryotes
This page describes the molecular mechanisms and genetics of eukaryotic gene expression.

Southern Blotting
This site describes the Southern blotting protocol often used in molecular research, including the research that led to the current model of eukaryotic gene expression.

Restriction Enzymes
This article describes the history of restriction enzyme research and their molecular interactions with DNA.

Additional Reading

Cavenee, W. K., and R. L. White. 1995. The genetic basis of cancer. *Scientific American* 272 (March): 72–79.

Tjian, R. 1995. Molecular machines that control genes. *Scientific American* 272 (February): 54–61.

Genomes

16

The first data sets describing the complete DNA sequence, or genome, of human beings were published in February 2001. The achievement was immediately hailed as a landmark in the history of science and the crown jewel of twentieth-century biology. It is important to recognize, though, that the Human Genome Project is part of a much larger and ongoing effort to sequence genomes from an array of bacteria, archaea, and other eukaryotes. The effort to sequence, interpret, and compare whole genomes is referred to as **genomics**. Currently, the pace of research in this field is nothing short of explosive.

Why is whole-genome sequencing attracting so much attention? How do researchers sequence the gene complement in an organism, and what have biologists learned from the first genomes that have been completed? Finally, what might the future hold in the way of research directions? These questions are the foundation of this chapter. As an introductory biology student, you are part of the first generation trained in the genome era. Genomics promises to revolutionize biological science and will almost certainly be an important part of your personal and professional life. Let's delve right in with a look at what genomics is and why it's being done.

At genome sequencing centers such as this one, automated sequencing machines churn out hundreds of thousands of base pairs of sequence data each day.

16.1 An Introduction to Whole-Genome Sequencing

Genomics has moved to the forefront of biology in part because of technological advances that have driven down the cost of sequencing DNA. Recently, the cost of sequencing has been decreasing by a factor of two every 18 months. As data have become less expensive and easier to obtain, the pace of whole-genome sequencing has accelerated. The result is that an almost mind-boggling amount of sequence is being generated. As this book goes to press, the primary international repository for DNA sequence data contains over 10^9 (10 billion) nucleotides. The size of this database—a publicly funded online service called GenBank—doubles every year.

Which genomes are being sequenced, and why are certain species chosen? Reviewing a little history can shed some light on these questions. The first organismal genome to be sequenced came from a bacterium that lives in the human upper respiratory tract. This bacterium, *Haemophilus influenzae*, has one circular chromosome and a total of 1,830,138 base pairs of DNA. It was an important genome to explore because the organism causes ear ache and respiratory tract infections in children, and because one particular strain is capable of infecting the membranes surrounding the brain and spinal cord and causing meningitis. The genome of *H. influenzae* was also small enough to sequence completely with a reasonable amount of time and money.

The publication of the *H. influenzae* genome in 1995 was quickly followed by the completion of genome sequences from an assortment of bacteria and archaea. The first eukaryotic genome, from the yeast *Saccharomyces cerevisiae*, was finished in 1996 and was followed by the publication of genomes from a variety of protists, plants, and animals. The array of genomes published to date and in progress is illustrated in **Figure 16.1**.

Several of the early sequencing projects focused on the need to work out techniques for the Human Genome Project, which was started in 1988. In other cases, organisms were selected for whole-genome sequencing because they cause disease or have unique or interesting biological properties. For example, genomes of bacteria and archaea that inhabit extremely hot environments have been sequenced in the hopes of discovering enzymes that might be useful for high-temperature industrial applications. Other bacteria and archaea were chosen for sequencing because they produce methane (CH_4; natural gas) or other interesting compounds as a by-product of respiration. Finally, species such as the fruit fly *Drosophila melanogaster* and the mustard plant *Arabidopsis thaliana* were analyzed because they serve as important model organisms in biology, and because data from well-studied organisms promised to help researchers interpret the human genome once it was completed.

Before exploring the results of these studies in more detail, it will be helpful to review some of the technological advances that have made whole-genome sequencing possible. How is this work done?

Recent Technological Advances

In Chapter 12 you read about how researchers determine the sequence of bases in a particular stretch of DNA. Recall that dideoxy sequencing is based on in vitro DNA synthesis reactions in which the gene being sequenced serves as the template. The sequencing reactions produce a series of DNA fragments that end with an A, C, T, or G. These fragments are then separated via electrophoresis and visualized by autoradiography or by other techniques. This method is still the basis of today's genome sequencing projects, though several important modifications have been adopted. For example, instead of separating the fragments generated by sequencing reactions through electrophoresis in hand-prepared gels that are poured between glass plates, researchers now perform the electrophoresis step using mass-produced, gel-filled capillary

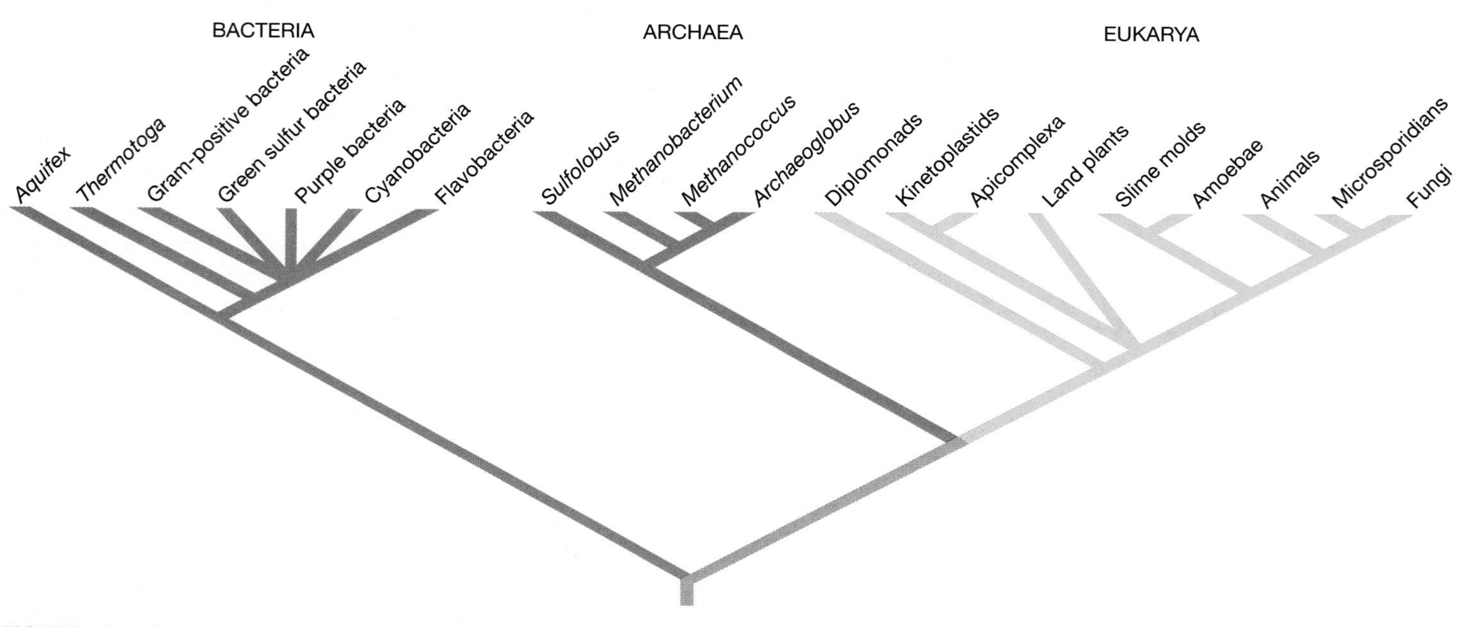

FIGURE 16.1 Genome Sequencing Projects
This version of the tree of life is made up of species whose genomes have been sequenced and lineages with representatives that have been or are being sequenced.

tubes. In addition, recall that researchers attached a radioactive isotope to the primer in the reaction mix to label the fragments of DNA produced by a sequencing reaction. Now, however, fluorescent markers are bonded to the dideoxyribonucleoside triphosphates (ddNTPs) used in a sequencing reaction. As **Figure 16.2** shows, the switch to fluorescing markers was important for two reasons: Each stretch of DNA could be sequenced with one dideoxy reaction instead of four separate ones, and machines were developed to detect the fluorescence produced by each fragment and read the output of the sequencing reaction.

Based on these and other innovations, researchers used automated sequencing machines to establish factory-style DNA sequencing centers such as the one shown at the start of this chapter. Over 20 such centers have now been established in the United States, the United Kingdom, Germany, France, Japan, and China. Some of these laboratories employ dozens of biologists and can conduct 100,000 sequencing reactions in a 12-hour period.

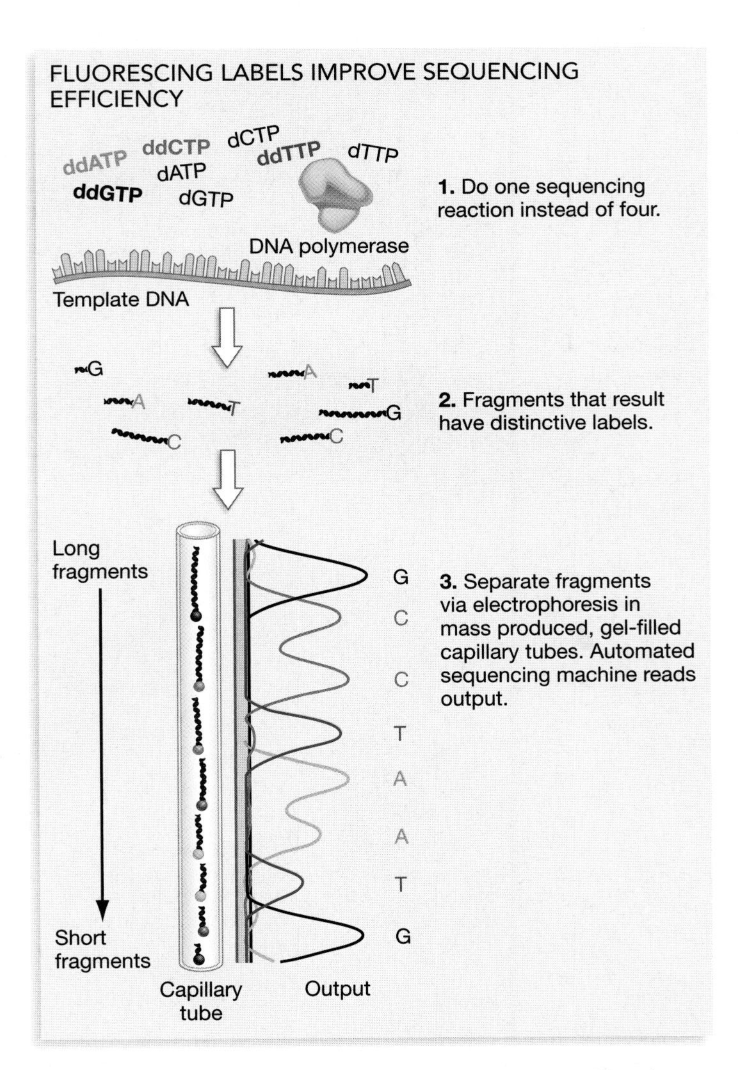

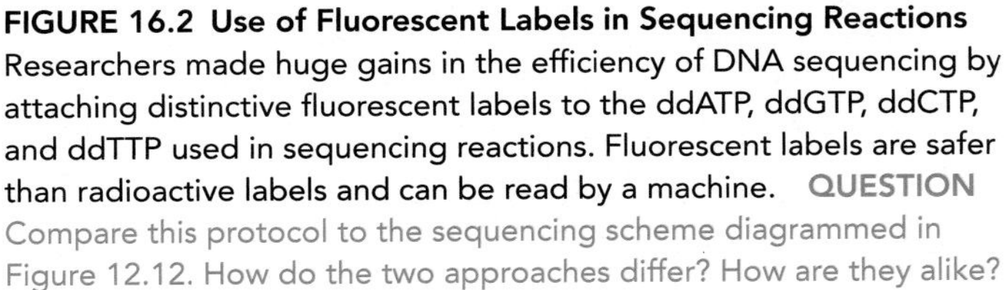
FIGURE 16.2 Use of Fluorescent Labels in Sequencing Reactions Researchers made huge gains in the efficiency of DNA sequencing by attaching distinctive fluorescent labels to the ddATP, ddGTP, ddCTP, and ddTTP used in sequencing reactions. Fluorescent labels are safer than radioactive labels and can be read by a machine. QUESTION Compare this protocol to the sequencing scheme diagrammed in Figure 12.12. How do the two approaches differ? How are they alike?

CD ACTIVITY 16.1 Human Genome Sequencing Strategies

Sequencing Strategies

How do investigators generate the approximately 1000-base-pair-long stretches of DNA that are used as templates in sequencing reactions? When researchers decide to sequence a large genome, this is a crucial question. The human genome, for example, contains about 3 billion base pairs. How can a genome this large be broken up into manageable pieces, the sequence of each piece determined, and the pieces then put back together in the correct order?

One prominent strategy is called map-based sequencing. As **Figure 16.3** (page 320) shows, map-based sequencing begins by fragmenting a genome into pieces about 160 kilobases (kb) long using restriction enzymes, then inserting each piece into a loop of DNA called a **bacterial artificial chromosome (BAC)**. Each BAC is then inserted into a different *Escherichia coli* cell, creating what researchers call a BAC library. By growing the bacterial cells in culture tubes, researchers can obtain many copies of each BAC. Copies of the 160-kb fragments are then analyzed, and distinctive regions of sequence documented. The presence of known or distinctive DNA sequences allows researchers to place each BAC on a physical map of the original chromosome. (Although the details of how this mapping is done are beyond the scope of this text, several of the relevant ideas and techniques are introduced in Chapter 17.)

Once a BAC library has been constructed and the fragments placed on a map of the genome, each 160-kb segment is cut into pieces about 1000 base pairs long. These small fragments are then inserted into the small loops of DNA called plasmids and placed inside bacterial cells. As Figure 16.3 shows, the plasmids are copied many times as the bacterial cells grow into a large population. As a result, large numbers of each 1000-base-pair fragment can be isolated and used in a sequencing reaction. Once the results of the sequencing reactions are analyzed, the small fragments can be placed in order to create the entire 160-kb sequence from the BAC by analyzing stretches where sequences from different 1000 base-pair fragments overlap.

An alternative strategy, called shotgun sequencing, is identical to this strategy except that the early step of mapping fragments from BACs is skipped. In shotgun sequencing, about 500 base pairs of DNA are sequenced from each end of every BAC in the library. By finding regions where these ends overlap with the ends of sequences in other BACs, the series of 160-kb segments can be placed in their correct order along the chromosome.

Although the logic of sequencing and assembling fragments sounds straightforward, the process is often fraught with difficulty. The primary problem is that many genomes—including ours—are riddled with or even dominated by long stretches of simple repeated sequences such as ATATATATAT . . . and so

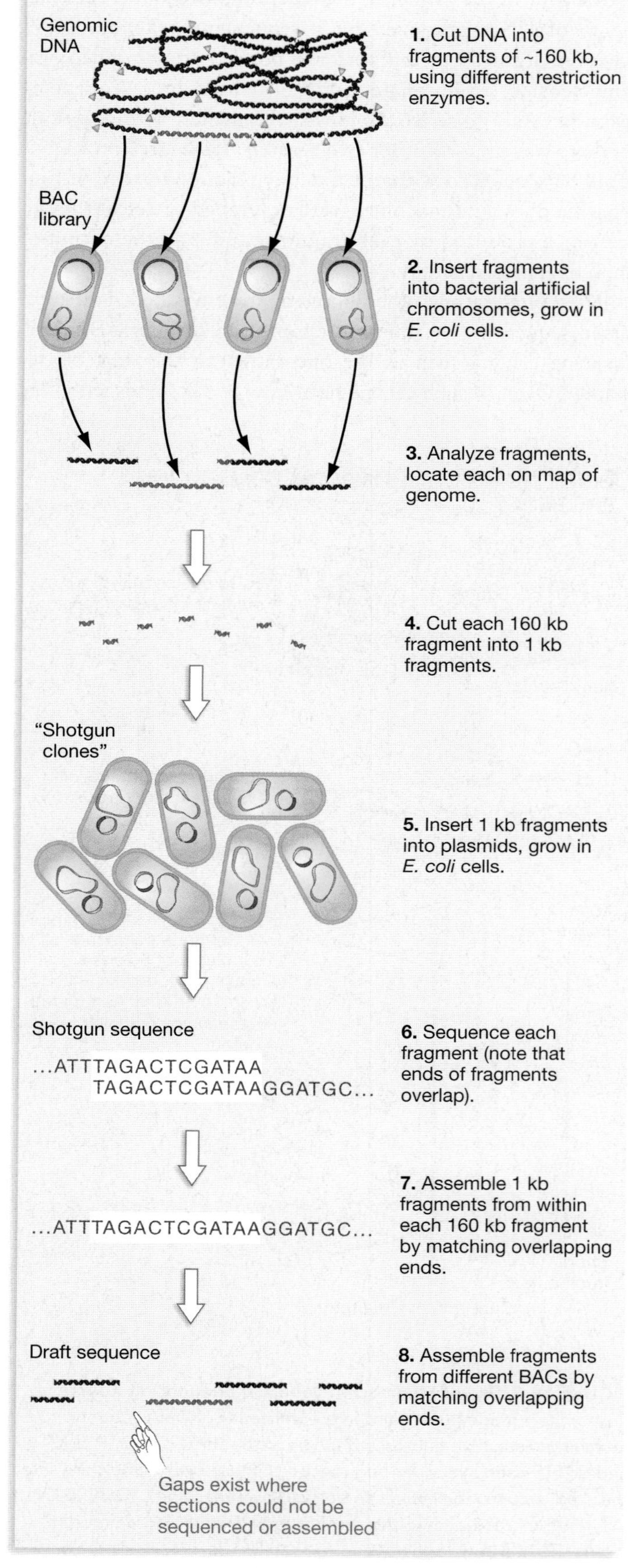

FIGURE 16.3 Sequencing Strategies
This diagram shows the sequence of steps used in determining a whole-genome sequence.

on. In addition, sequences that are identical and yet longer and more complex than these simple AT repeats are found at hundreds or thousands of different locations in the genomes of humans and many other eukaryotes. If sequences scattered throughout the genome are identical, how can researchers determine where each fits in relation to the others?

Aligning fragments that are dominated by repeated sequences is a thorny problem. To heighten the probability that the resulting fragments are placed in the proper order along a chromosome, researchers determine the DNA sequence on either side of repeated segments. Then sophisticated computer programs mix and match the resulting DNA segments until an alignment consistent with all available data is obtained (see **Box 16.1**, page 322). Particularly troublesome stretches are re-sequenced until they can be placed into the full data set with confidence.

Annotating Genomes

Obtaining raw sequence data is just the beginning of the effort to understand a genome. As Stanley Fields has pointed out, raw sequence data are analogous to the parts list for a house, except that the list has no punctuation: "windowwabeborogovestaircasedoorjubjub . . ." Where do the genes for "window," "staircase," and "door" start and end? Which sequences have a regulatory function—or no known function at all? Which actually code for rRNAs, tRNAs, or proteins?

When annotating a genome, a biologist basically identifies the locations of genes. In bacteria and archaea, identifying genes is relatively straightforward. Biologists use computer programs that scan the sequence in each direction. The program tests each of the three-codon reading frames that are possible until it finds an ATG start codon followed by a gene-sized stretch of sequence and a stop codon (**Figure 16.4**). In addition, the computer programs look for sequences typical of promoters, operators, or other regulatory sites. Loci identified in this way are called open reading frames, or ORFs.

Once a putative ORF is found, a computer is asked to check its sequence against existing catalogs of known genes. If an exact or near-exact match is found, the sequence can be assigned a name and perhaps a function. If a close match does not occur, researchers use the computer to find known genes that have stretches of similar or identical sequence to the ORF. A "hit"—meaning a known gene that has a degree of sequence similarity to the ORF—usually inspires the hypothesis that the ORF and known gene have the same or a similar function. Finally, if the ORF appears to be unique, it is a promising candidate for the types of follow-up studies you'll read about in section 16.4.

Unfortunately, this method of finding and analyzing genes is not as straightforward in eukaryotes. Most coding regions in

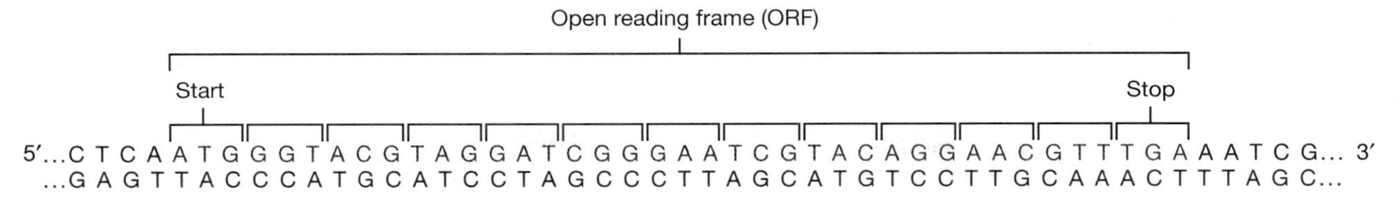

FIGURE 16.4 Finding Open Reading Frames
By searching a genome for start and stop codons that are separated by a stretch of codons in the correct reading frame, a computer program can identify putative protein-coding sequences called open reading frames (ORFs).

eukaryotes are broken up by introns and, as section 16.3 explains, many or even most regions in eukaryotic genomes do not actually code for a product. As a result, scanning eukaryotic genomes for open reading frames is not as useful as it is in bacteria and archaea.

One of the most effective gene-finding techniques in eukaryotes begins with the mRNAs produced by coding regions. As **Figure 16.5** shows, an investigator can isolate all of the mRNAs in a certain type of cell—perhaps a human skin cell or fruit-fly egg—and then use the enzyme reverse transcriptase to produce a set of DNA fragments that are complementary to each mRNA in the sample. As detailed in Chapter 26, reverse transcriptase catalyzes the synthesis of DNA from an RNA template. A DNA fragment that is synthesized from mRNA by reverse transcriptase is called a **complementary DNA (cDNA).** DNA polymerase is then used to make the cDNAs double-stranded. A collection of cDNAs from a particular organism or cell type is called a **cDNA library** (Figure 16.5).

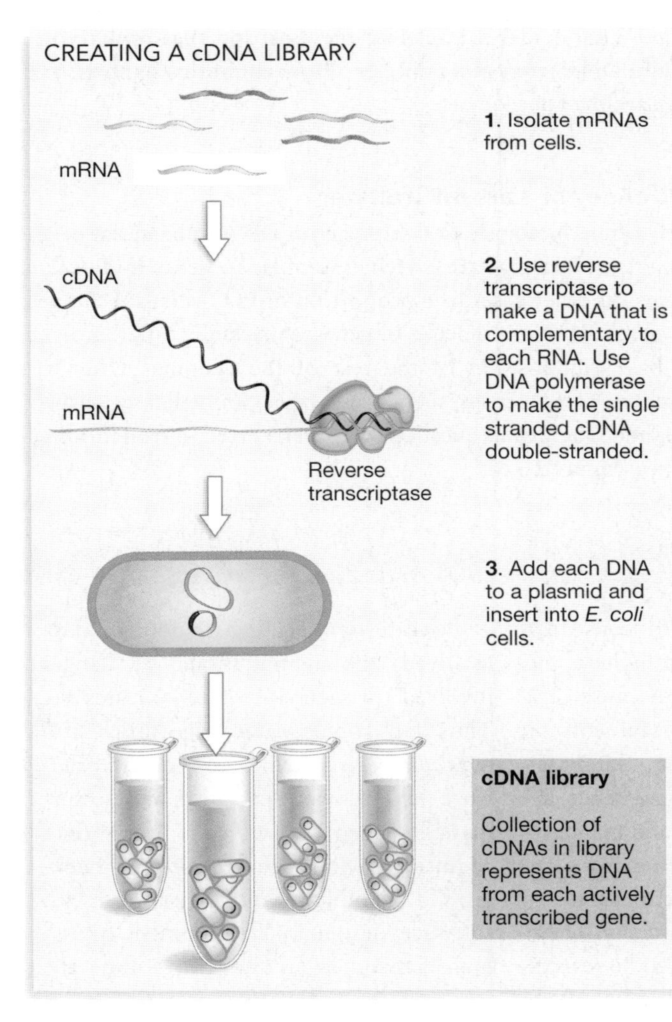

FIGURE 16.5 Creating a cDNA Library
By creating a complementary DNA (cDNA) for each mRNA in a cell, researchers build up an inventory of coding sequences.

Using cDNAs to identify genes is a two-step process. To begin, researchers build up an inventory of the coding sequences found in a particular organism by sequencing cDNAs. Then when the entire genome is sequenced, a computer can scan it and pinpoint where each cDNA is located. Using cDNA sequences and other approaches, investigators gradually build up a catalog of confirmed or suspected coding sequences in eukaryotic genomes.

Given this introduction to how researchers sequence and annotate genomes, we're now ready to delve into the data. Let's first consider what genome sequencing has revealed about the nature of bacterial and archaeal genomes and then move on to eukaryotes. Is the effort to sequence whole genomes paying off?

16.2 Bacterial and Archaeal Genomes

By the time you read this paragraph, the genomes of over 75 different bacterial and archaeal species will have been completely sequenced, and projects focused on sequencing an equal number of additional prokaryotic genomes will be under way. In addition to this impressive array of different species, complete genome sequences are now available for several different strains of the same bacterial species. For example, researchers have sequenced the genome of a harmless laboratory strain of *Escherichia coli* as well as the genome of a form that causes severe disease in humans. As a result, researchers can now compare the genomes of cells that are adapted to distinct habitats.

This section of the chapter focuses on a simple question. Based on the initial wave of data published between 1995 and 2001, what general observations have biologists been able to make about the nature of bacterial and archaeal genomes?

The Natural History of Prokaryotic Genomes

In a sense, genome sequencers can be compared with the great naturalists of the eighteenth and nineteenth centuries. These biologists explored the globe, collecting the plants and animals they encountered. Their goal was to describe what existed. Similarly, the first task of a genome sequencer is to catalog what is in a genome—specifically, the number, type, and organization of genes. Several interesting conclusions can be drawn from relatively straightforward observations about the data obtained thus far:

- In bacteria, there is a general correlation between the size of a genome and the metabolic capabilities of the organism. In general, parasites have tiny while non-parasitic organisms have relatively large genomes. The smallest genomes are found in bacteria from the genus *Mycoplasma*, which are obligate parasites. These organisms acquire almost all of their nutrients from their hosts and lack the enzymes required to manufacture many essential compounds. In contrast, the genomes of *E. coli* and *Pseudomonas aeruginosa* are 8 to 10 times as large. Their genes code for enzymes that synthesize virtually every molecule needed by the cell. Based on this observation, it is not surprising that *E. coli* is able to grow under a wide variety of environmental conditions. Using similar logic, researchers hypothesize that the large genome of *Pseudomonas* explains why it is able to occupy a wide array of soil types, including marine and marshy habitats.
- Many of the genes that have been identified have no known function. *E. coli* probably qualifies as the most intensively studied of all organisms, but the function of 38 percent of its genes is unknown.
- There is tremendous genetic diversity among prokaryotes. About 15 percent of the genes in each prokaryotic genome appear to be unique to that species.
- Redundancy among genes is common. *E. coli* has 86 pairs of genes whose sequences are nearly identical—meaning that the proteins they produce are nearly alike in structure and presumably in function. The significance of this redundancy is unknown. Presumably, slightly different forms of the same protein are produced in response to slight changes in environmental conditions.
- Multiple chromosomes are more common than anticipated. Of the prokaryotes sequenced to date, there is a significant percentage with two circular chromosomes instead of one.
- Many species contain plasmids. A plasmid is a small, extrachromosomal loop of DNA that contains a small number of genes, though not genes that are absolutely essential for growth. In many cases, plasmids can be exchanged between cells of the same or different species (see Chapter 14).

Perhaps the most surprising observation of all, though, is that in many bacterial and archaeal species, a significant proportion of the genome appears to have been acquired from other, often distantly related, species. This is a remarkable claim. What evidence backs up the assertion that prokaryotes acquire DNA from other species? How could this happen, and what are the consequences?

Evidence for Lateral Transfer

Fairly often, biologists find stretches of DNA in bacterial or archaeal genomes that are extremely similar to genes in different species. In other cases, the proportion of G-C versus A-T pairs in a particular gene or series of genes is markedly different from the base composition of the rest of the genome. When researchers come across such genes, they immediately suspect that the sequences in question were transferred in from another species (**Figure 16.6**).

BOX 16.1 Bioinformatics

Bioinformatics is a general term for any effort to manage, analyze, and interpret information in biology. More specifically, bioinformatics refers to a new field that is being created in response to the information processing demands of genome sequencing projects.

What are these demands? Early in the development of genomics, one of the most pressing needs was for computer programs that could arrange short stretches of completed sequences in the correct order along a chromosome. The resulting programs were designed to fit overlapping fragments into a consistent pattern and identify regions where the correct order was not clear because of a high frequency of repeated segments or other problems. In addition, databases that could hold completed sequence information had to be created and managed in a way that made the raw data and a variety of annotations available to the international community of researchers. These sequence databases also had to be searchable, so investigators could evaluate how similar newly discovered genes were to genes that have been studied previously.

Because of the large amount of data involved, the computational challenges involved in genomics are formidable. Thus far, sophisticated algorithms and continually improving computer hardware have allowed specialists in bioinformatics to keep pace with the rate of data acquisition. Individuals with a background in biology and experience with software design or development are in high demand in laboratories around the world.

Lateral transfer refers to the movement of DNA from one species to another. How does it occur? In at least some cases, plasmids appear to be involved. For example, most of the genes responsible for conferring resistance to antibiotics are found on plasmids. Although the actual physical mechanism involved is not known, researchers have recently documented the transfer of plasmid-borne, antibiotic-resistance genes between very distantly related species of disease-causing bacteria. In addition, it is apparent that genes from plasmids have occasionally become integrated into the main chromosome of a new species.

Some biologists hypothesize that lateral transfer also occurs when prokaryotes take up raw pieces of DNA from the environment—perhaps in the course of acquiring other molecules. This may have occurred in the bacterium *Thermotoga maritima,* which occupies the high-temperature environments near undersea hot springs. Almost 25 percent of the genes in this species are extremely closely related to genes found in archaea that live in the same habitats. The archaea-like genes occur in distinctive clusters within the *T. maritima* genome, which supports the hypothesis that the sequences were transferred from an archaean to the bacterium in large pieces.

Similar types of direct gene transfer are hypothesized to have occurred in the bacterium *Chlamydia trachomatis*. This organism is a major cause of blindness in humans from Africa and Asia, as well as the most common source of sexually transmitted bacterial disease in the United States. The *C. trachomatis* genome contains numerous eukaryote-like genes. Because *C. trachomatis* acts as an intracellular parasite, the most logical explanation for this observation is that the bacterium occasionally takes up DNA directly from its host cell, resulting in a eukaryote-to-bacterium transfer.

In addition to being transferred between species by means of plasmids or DNA fragments, genes can be transported by viruses. For example, when Frederick Blattner and co-workers compared the sequences of the laboratory and disease-causing strains of *E. coli*, they found that the pathogenic cells have almost 1400 "extra" genes. Compared to the rest of the genome, most of these genes have a distinctive G-C to A-T ratio. In addition, many are extremely similar to sequences isolated from viruses that infect *E. coli*. Based on these observations, most researchers support the hypothesis that at least some of the "pathogenicity loci" in *E. coli* were brought in by viruses.

To summarize, mutation and recombination are not the only source of genetic variation in bacteria and archaea. Lateral transfer is an important source of new genes and allelic diversity in these domains, and it can occur in a variety of ways. This insight would not have been possible without data from whole-genome sequencing. As it turns out, the result is also having an impact on biomedical research.

Comparative Genomics: Understanding Virulence

Currently, dozens of whole-genome sequencing projects are focused on disease-causing bacteria. When these projects are complete, biologists hope to compare harmless and pathogenic strains of bacteria and achieve a much more detailed understanding of the genetic basis of virulence.

As an example of this work, let's look more closely at what Blattner and colleagues found when they compared the

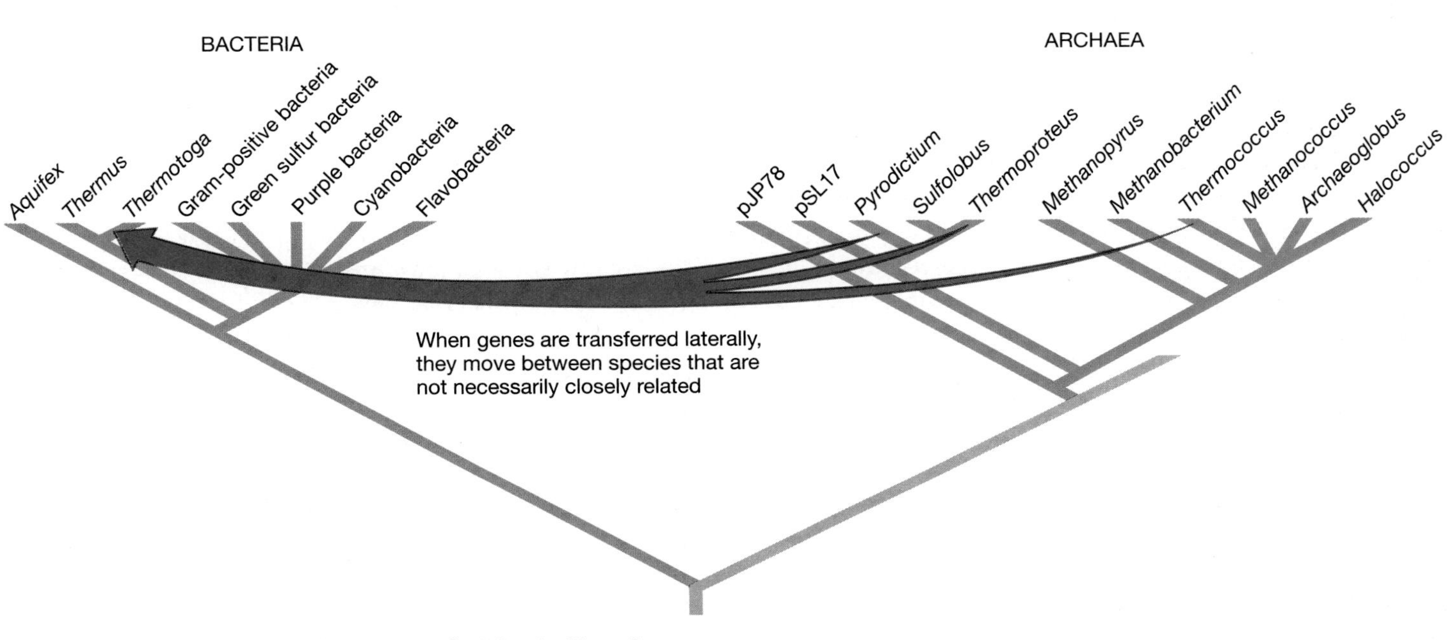

FIGURE 16.6 How Did "Lateral Gene Transfer" Get its Name?
Lateral gene transfer refers to the movement of DNA between species; as this figure shows, it can occur between very distantly related organisms. Vertical gene transfer refers to movement of DNA between generations.

genomes of benign and pathogenic strains of *E. coli*. Recall that the disease-causing strain, which is designated O157:H7, has a large number of genes not found in benign strains, and that the genes associated with disease may have been introduced by viruses. Blattner and associates were interested in exploring the genome of O157:H7 because the strain is a common cause of food poisoning. People who eat undercooked meat may develop serious, and occasionally fatal, infections by O157:H7.

When Blattner's group analyzed the novel genes found in O157:H7, they found that several are extremely similar to sequences from a closely related pathogenic bacterium called *Shigella*. In countries where public water supplies are untreated, *Shigella* is an important cause of dysentery. Some of the *Shigella*-like genes found in O157:H7 code for proteins that poison cells in the human intestine and cause severe diarrhea. This observation suggests that the same genes may be responsible for causing food poisoning by *E. coli* and dysentery by *Shigella*. If compounds can be developed that neutralize *Shigella* toxins, they should also be effective against infections by *E. coli* O157:H7.

The data available to date indicate that the *Shigella–E. coli* story may be typical. In the disease-causing organisms whose genomes have been sequenced thus far, biologists have found that the genes responsible for causing illness or antibiotic resistance tend to occur in discrete clusters and probably originate in lateral transfer events. In this way, genome sequencing is allowing biologists to reconstruct the events that led to the evolution of dangerous strains of bacteria.

Results like these are important to physicians and biomedical researchers. Have efforts to sequence eukaryotic genomes led to similar types of advances?

16.3 Eukaryotic Genomes

Sequencing eukaryotic genomes presents two huge challenges. The first is sheer size. Compared to the genomes of bacteria and archaea, which range from 580,070 base pairs in *Mycoplasma genitalium* to over 6.3 million base pairs in *Pseudomonas aeruginosa* (a bacterium that causes urinary and respiratory tract infections in humans), eukaryotic genomes are large. The genome of baker's yeast, which is a unicellular eukaryote, contains over 12 million base pairs. The roundworm *Caenorhabditis elegans* has a genome of 97 million base pairs, the fruit-fly genome contains 180 million base pairs, the mustard plant *Arabidopsis thaliana* genome has 130 million base pairs, and humans contain over 2.91 billion base pairs.

The second great challenge in sequencing eukaryotic genes is coping with repeated sequences. Many eukaryotic genomes are dominated by repeated DNA sequences that do not code for products used by the organism. These repeated sequences pose serious problems in aligning and interpreting raw sequence data. What are they? If they don't code for a product, why do they exist?

Natural History: Types of Sequences

In many eukaryotic genomes, the exons, introns, and regulatory sites associated with genes comprise a relatively small percentage of the genome. In humans, for example, exons actually constitute less than 5 percent of the total genome; repeated sequences account for well over 50 percent. In contrast, over 90 percent of the sequences in a bacterial or archaeal genome code for a product used by the cell.

Most of the repeated sequences observed in eukaryotes are derived from sequences known as transposable elements. **Transposable elements** are parasitic sequences that are capable of moving from one location to another in a genome. They are similar to viruses, except that they are not transmitted from one host individual to another by infection. Instead, transposable elements transmit copies of themselves to additional locations within the host genome and are passed along to offspring along with the rest of the genome.

How Do Transposable Elements Work? Transposable elements come in a variety of types and spread through genomes in a variety of ways. As an example of how these parasites work, let's consider a type found in humans and other eukaryotes, called a long interspersed nuclear element (LINE). A LINE consists of a stretch of DNA that contains a promoter for transcription by RNA polymerase II and two genes. One of the genes codes for the enzyme reverse transcriptase; the other codes for a protein called integrase.

Figure 16.7 shows how a copy of a LINE moves to a new location in the genome. If a LINE is transcribed and the mRNA is subsequently translated, the LINE mRNA moves to the nucleus along with the reverse transcriptase and integrase proteins. Once inside the nucleus, reverse transcriptase makes a cDNA version of the mRNA. DNA polymerase in the cell then makes the cDNA double-stranded. After integrase nicks the chromosomal DNA, the LINE cDNA becomes inserted into the chromosome. In this way, the parasitic sequence reproduces.

Because most of the LINEs observed in the human genome don't contain a promoter and both genes, however, researchers infer that the insertion process illustrated in Figure 16.7 is usually disrupted in some way. Analyses of the human genome sequence have revealed that only a handful of the LINEs in our genomes appear to be complete and potentially active; the vast majority are fossils. **Table 16.1** summarizes some other types of transposable elements found in humans.

Virtually every prokaryotic and eukaryotic genome examined to date contains a least some transposable elements. The number varies widely, however, and bacterial and archaeal genomes have relatively tiny numbers of transposable elements. This observation has inspired the hypothesis that bacteria and archaea either have efficient means of removing parasitic sequences or can somehow thwart insertion events.

Research on transposable elements and lateral transfer has had a huge impact on how biologists view the genome. Many

TABLE 16.1 Major Types of Transposable Elements in the Human Genome

Name	Description	Number of Copies	Estimated Percentage of Total Genome
SINE (short interspersed nuclear element)	Short RNAs (normally complexed with proteins involved in transport of polypeptides into rough ER) that are reverse transcribed and integrated into genome	1,558,000	13.14
LINE (long interspersed nuclear element)	Contain reverse transcriptase and move to new location in genome through RNA intermediate	868,000	20.42
LTR (long terminal repeat)	Similar to certain viruses; contain coding sequence for reverse transcriptase and move to new location in genome through RNA intermediate	450,000	8.29
DNA transposon	Many different subtypes; all code for an enzyme called transposase which moves element to new location in the form of DNA	294,000	2.84

Source: International Human Genome Consortium. 2001. *Nature* 409: 860–921.

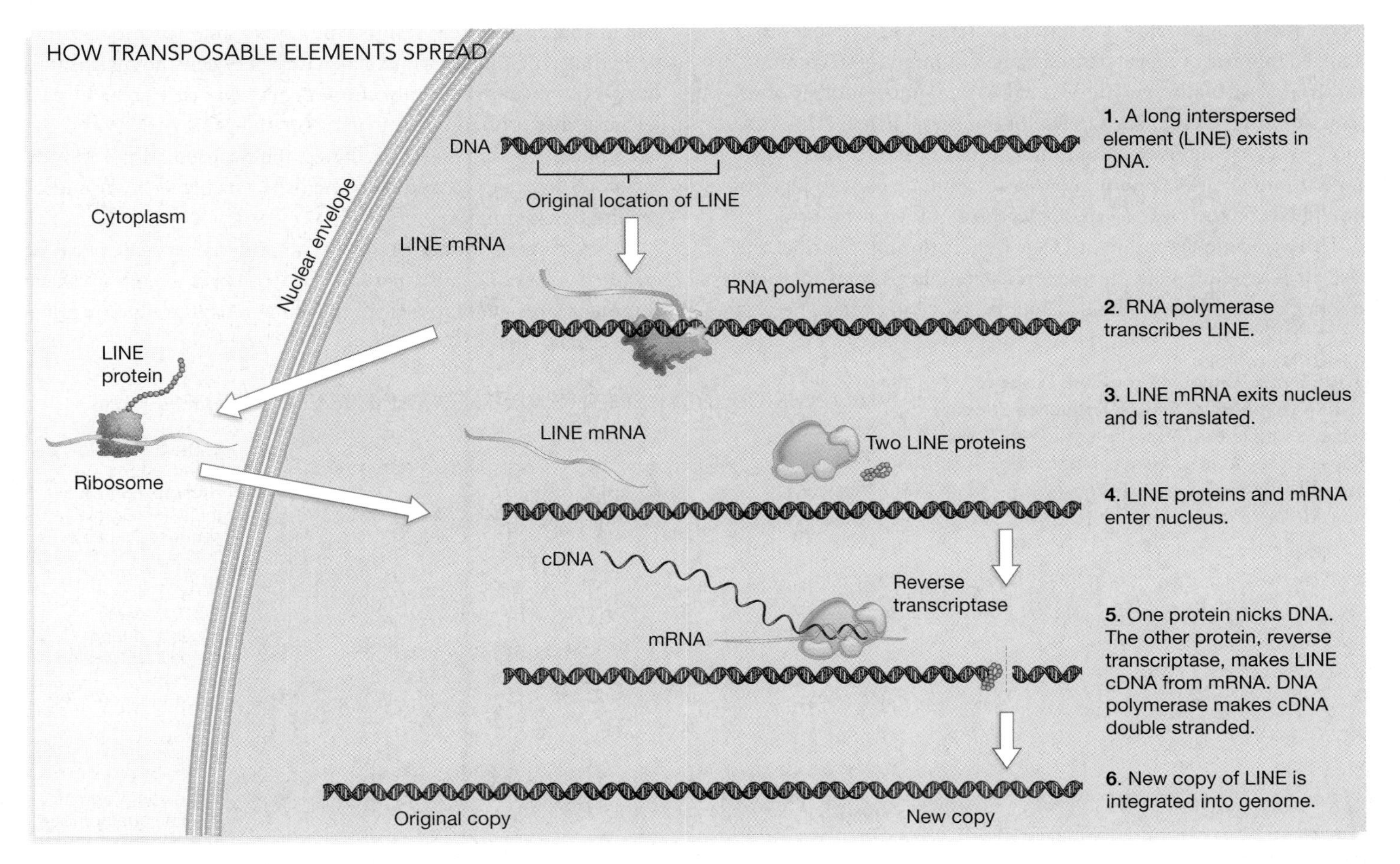

FIGURE 16.7 Transposable Elements
This sequence of events is the leading hypothesis for how LINEs spread.

genomes are riddled with parasitic sequences; others have undergone radical change in response to lateral transfer events. The general picture is that genomes are dynamic. Their size and composition can change dramatically over time.

Simple Sequence Repeats and DNA Fingerprinting In addition to containing repeated sequences from transposable elements, eukaryotic genomes have several hundred loci called simple sequence repeats. Repeating units that are just 1 to 13 bases long are referred to as **microsatellites;** repeating units that are 14 to 500 bases long are called **minisatellites.** Simple sequence repeats make up 3 percent of the human genome. The most common type is a repeated stretch of the dinucleotide AC, giving the sequence ACACACAC . . . and so on. Both types of satellite sequences are thought to arise when DNA polymerase skips or mistakenly adds extra bases during replication.

The discovery of satellite sequences turns out to have important practical applications. Soon after these sequences were first characterized, Alec Jeffreys and co-workers established that microsatellite and minisatellite loci are "hypervariable," meaning that they vary among individuals much more than any other type of sequence. As **Figure 16.8** shows, highly repetitive stretches often misalign when homologous chromosomes synapse and cross over in meiosis. DNA polymerase also tends to slip as it copies these sequences. As a result, mutations in the form of a changed number of repeats are extremely common—so common, in fact, that virtually every individual has a unique number of repeats at one or more microsatellite or minisatellite loci. This variation in repeat number among individuals is the basis of DNA fingerprinting. **DNA fingerprinting** is a technique used to identify individuals on the basis of unique features of their genomes.

To appreciate the value of DNA fingerprinting, consider the case of a woman who disappeared from her home in Richmond, Prince Edward Island, Canada. Not long after she was reported missing, police investigators found her abandoned car and documented that dried blood in the vehicle matched the woman's blood. Later, a bloodstained jacket with several white cat hairs on the lining was found in woods not far from her home. After several months had passed, the woman's body was recovered from a shallow grave. Subsequently, the woman's estranged boyfriend was arrested and charged with murder.

In an effort to build their case against the accused, police asked biologists whether DNA in the cat hairs from the bloody jacket matched DNA from the boyfriend's white cat, Snowball. To test this hypothesis, Marilyn Menotti-Raymond and co-workers isolated a tiny amount of DNA from the cat hairs on the jacket. Then they used the polymerase chain reaction (PCR) to amplify 10 different microsatellite loci from the DNA in the sample. Because they performed the PCR with fluorescently labeled dNTPs, the biologists could separate the resulting DNA fragments by using capillary gel electrophoresis and read the fragment sizes with an automated sequencing machine.

Menotti-Raymond and co-workers found that the sizes of the microsatellite loci from Snowball and from the hairs on the jacket matched exactly at all 10 loci. How likely is it that the hairs actually came from Snowball? To answer this question, the biologists documented the size of the same microsatellite loci in a randomly chosen sample of cats from Prince Edward Island. Based on their results, they were able to calculate the probability of finding an exact match between the microsatellite DNA of two randomly chosen cats from this area. In part because this probability was very small—2.2×10^{-8}—the police concluded that the data linked the accused to the bloody jacket. A jury agreed and convicted the estranged boyfriend of second-degree murder.

DNA fingerprinting has been helpful in a wide array of criminal cases. Because parents and offspring tend to share fragment sizes generated from the same microsatellite or mini-

FIGURE 16.8 Unequal Crossover Leads to Unique Numbers of Simple Sequence Repeats Because simple sequence repeats are so similar, they are likely to misalign when homologous chromosomes synapse at meiosis I.

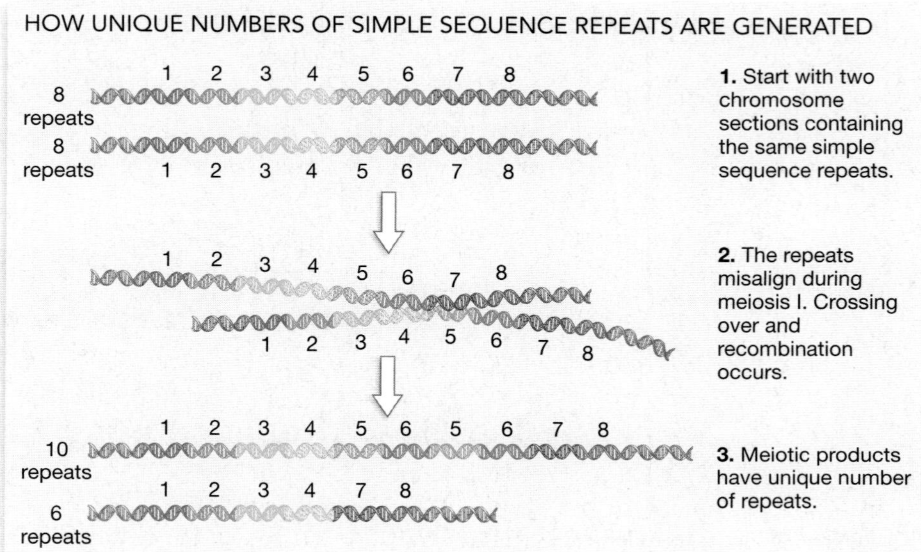

satellite loci, DNA fingerprinting also offers an accurate way to assign paternity in birds, primates, and other species as well as in humans.

Now that we've reviewed the characteristics of some particularly prominent types of noncoding sequences in eukaryotes, it's time to consider the nature of the coding sequences in these genomes. Let's start with the most basic question of all. Where do eukaryotic genes come from?

Gene Families

In both prokaryotic and eukaryotic genomes, biologists routinely find groups of similar genes clustered along the same chromosome. The genes may be similar not only in structural aspects, such as the arrangement of exons and introns, but also in sequence. Loci that are extremely similar to each other are considered to be part of the same **gene family**, and they are hypothesized to have arisen from a common ancestral sequence through gene duplication. In eukaryotes, where lateral transfer is rare, gene duplication may be the most important of all mechanisms that generate new genes.

The most common type of gene duplication results from the mistake in meiosis called unequal crossover. In addition to causing the duplication or deletion of a small section of repeated sequence as shown in Figure 16.8, unequal crossover can also produce duplications of gene-sized segments of DNA. As **Figure 16.9** shows, the duplicated segments that result are arranged in tandem, meaning one after the other. Because the original gene is still functional and produces a normal product, the duplicated stretches of sequence are redundant. If mutations in the duplicated sequence alter the protein product, and if the altered protein product performs a valuable function in the cell, then an important new gene has been created.

Alternatively, mutations in the duplicated region may make expression of the new locus impossible. For example, a mutation could produce a stop codon in the middle of an exon. When a gene is disabled in this way, the resulting sequence is referred to as a pseudogene. A **pseudogene** is a sequence that closely resembles a working gene but is not transcribed. Pseudogenes have no function.

The globin loci diagrammed in Figure 16.9 illustrate several important points about gene families. In humans, the globin gene family contains several pseudogenes and an array of loci that code for oxygen-transporting proteins. The various coding loci in the family serve slightly different functions, however. Some genes are active only in the fetus or the adult; others code for proteins that form different parts of the finished oxygen-carrying molecule, hemoglobin.

In addition to the gene duplication events that result from unequal crossover, an entire complement of chromosomes may be duplicated as a result of a mistake in either mitosis or meiosis. In this case, the resulting organism contains double the normal set of chromosomes and is considered **polyploid**. When polyploidy occurs, every gene in the duplicated genome is free to mutate and possibly to acquire a new function. As detailed in Chapter 23, polyploidy can lead to the formation of new species. Polyploidy has been a particularly important source of new genes in plants.

Comparative Genomics: Gene Number and the Importance of Alternative Splicing

Of all observations about the nature of eukaryotic genomes, perhaps the most striking has been that particularly complex organisms do not appear to have particularly large numbers of genes. **Table 16.2** (page 328) indicates the number of genes that are estimated to exist in selected genomes that have been sequenced to date. The basic observation is that the total number of genes in *Homo sapiens*, which is considered an extremely complex organism, is not that much higher than the total number of genes in fruit flies and roundworms and *Arabidopsis*. Before the human genome was sequenced, many biologists expected that humans would have at least 100,000 genes. Instead, we have only about 40,000.

How can this be? In prokaryotes, there appears to be a rough correlation between genome size, gene number, a cell's metabolic capabilities, and its ability to live in a variety of habitats. Why isn't there a stronger correlation between gene number and the morphological and behavioral complexity of eukaryotes?

The leading hypothesis to answer this question focuses on alternative splicing. Recall from Chapter 15 that the exons of a particular gene can be spliced in ways that produce distinctive mature mRNAs. As a result, a single eukaryotic gene can code for multiple transcripts and thus multiple proteins. The

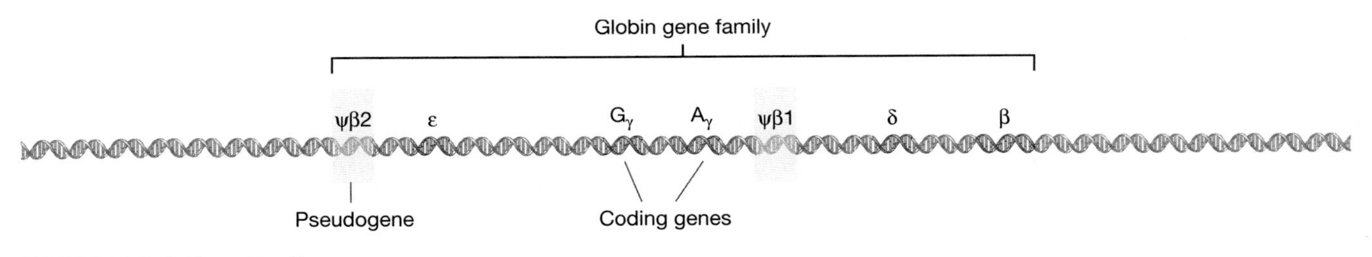

FIGURE 16.9 A Gene Family
Each member of the globin gene family has exons and introns that are extremely similar in length and number. Note that the members of the family are arranged one after the other, or in tandem. All of these loci are thought to have descended from a single ancestral sequence through a series of gene duplication events.

TABLE 16.2 Number of Genes in Selected Genomes

Species	Notes	Genome Size (Millions of Base Pairs)	Estimated Number of Genes
Mycoplasma genitalium	Parasitic bacterium; causes urogenital-tract infections in humans	0.58	517
Haemophilus influenzae	Bacterium; common resident of human upper respiratory tract; can cause earache and meningitis	1.83	1743
Thermotoga maritima	Bacterium that lives in extremely hot environments	1.89	1877
Vibrio cholerae	Bacterium that lives in saltwater marshes and coastal environments; also causes cholera	4.0	3885
Escherichia coli (laboratory strain)	Bacterium that is an important model organism in biochemistry and genetics	4.6	4288
Saccharomyces cerevisiae	Baker's yeast; a unicellular fungus; important model organism in biochemistry and genetics	12	6340
Plasmodium falciparum	Single-celled eukaryote; a parasite that causes malaria in humans	30	6500
Drosophila melanogaster	Fruit fly; important model organism in genetics and developmental biology	180	13,600
Caenorhabditis elegans	A roundworm; important model organism in developmental biology	97	19,000
Mus musculus	House mouse; important model organism in genetics and developmental biology	3100	$>$ 21,000
Arabidopsis thaliana	A mustard plant; important model organism in genetics and developmental biology	119	26,000
Homo sapiens	Humans	3000	30,000–40,000

alternative splicing hypothesis claims that complex eukaryotes do not need enormous numbers of distinct genes. Instead, alternative splicing creates different proteins from the same gene. The alternative forms might be produced at different developmental stages, or act as a response to changed environmental conditions.

In support of the alternative-splicing hypothesis, researchers have analyzed the mRNAs produced by specific regions of the human genome and have estimated that each coding locus produces an average of three distinct transcripts. If this result is valid for the rest of the genome, the actual number of different proteins that can be produced is more than triple the gene number.

Based on results like these, gaining a better understanding of how alternative splicing is regulated has become an urgent research priority. What other questions have been raised because of the explosion in whole-genome sequencing?

16.4 Future Prospects

To explain the impact of genomics on the future of biological science, Eric Lander has compared the sequencing of the human genome to the establishment of the periodic table of the elements in chemistry. Once the periodic table was established and validated, chemists focused on understanding how the elements combine to form molecules. Similarly, biologists now want to understand how the elements of the human genome combine to produce an individual.

In essence, a genome is just a parts list. Once that list is compiled, annotated, and analyzed in a preliminary way, researchers then delve deeper. The goal of this section is to explore some of the ways in which researchers are using whole-genome data to answer fundamental questions about how organisms work.

Functional Genomics

Biologists who are interested in the research field called functional genomics look at the lists of protein-coding regions in a genome and ask, "How do all these gene products interact?" After all, the products of genes do not exist in a vacuum. Instead, groups of proteins act together to respond to environmental challenges such as extreme heat or drought. Similarly, distinct groups of genes are transcribed at different stages as a multicellular eukaryote grows and develops.

FIGURE 16.10 How Do DNA Microarrays Work?
(a) To monitor changes in gene expression, investigators spot thousands of short, single-stranded DNA sequences from coding sequences onto a glass plate. By probing this microarray with labeled cDNAs synthesized from mRNAs, researchers can identify which coding sequences are being transcribed. **(b)** By probing a microarray with cDNAs from different times or conditions, researchers can identify when and how alternative splicing occurs.

(a) Determining which genes are being transcribed

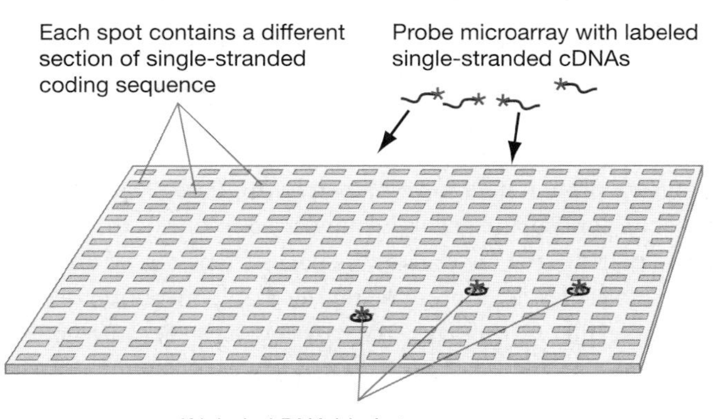

Output looks like this:

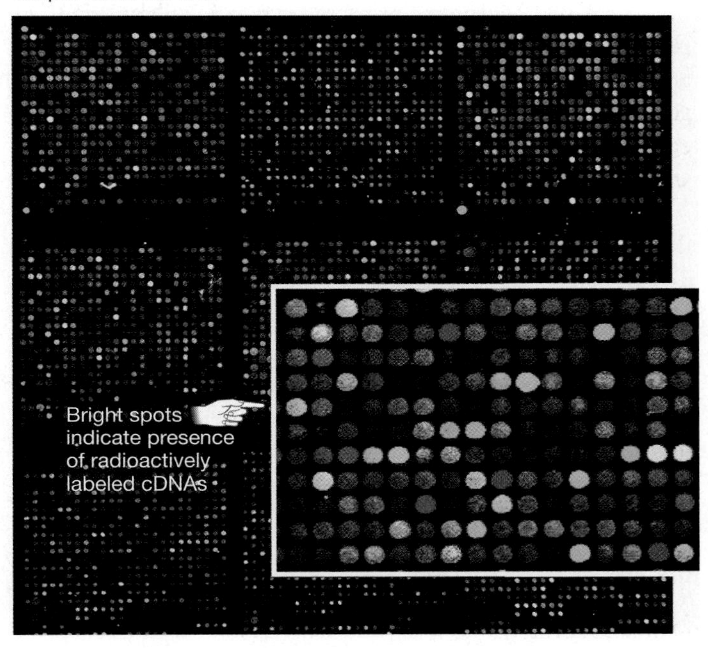

(b) Evidence for alternative splicing

Suppose 5 spots on microarray contain sequences from 5 exons of same locus. If hybridization with cDNAs gives...

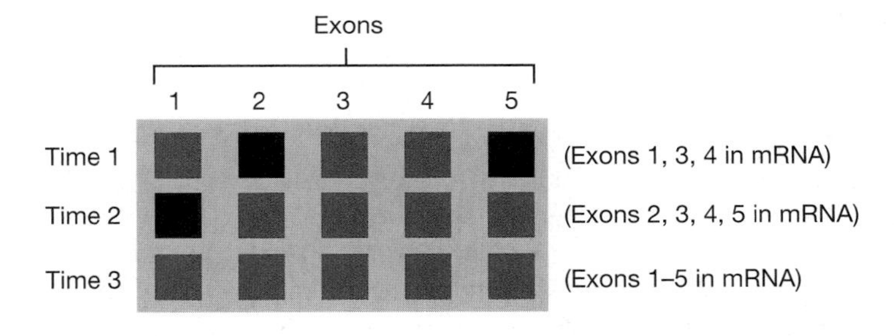

...then alternative splicing is occurring.

Whole-genome sequencing has inspired two important new approaches to study how genes interact. One method focuses on analyzing changes in gene expression; the other seeks to better understand protein-protein interactions. Let's take a closer look.

DNA Microarrays DNA-based microarrays are beginning to revolutionize how researchers analyze changes in gene activity. As **Figure 16.10a** shows, the approach is based on affixing a large series of different single-stranded DNA sequences to a glass slide in tiny but distinct spots. In many instances, the DNAs on the plate represent short stretches from each exon in a genome.

Once such a slide has been manufactured, a variety of experiments are possible. For example, mRNAs can be isolated at different stages in development and converted to cDNAs with reverse transcriptase. The resulting cDNAs are then labeled with a fluorescent marker and hybridized with the microarray (Figure 16.10a). Because the labeled cDNAs bind to the complementary sequences on the slide, the results give an accurate picture of which genes are being expressed at the time. The "probe" cDNAs can then be washed off and the slide re-probed with cDNAs from a later stage in development. In this way, researchers collect data on which genes are expressed at different intervals as the organism matures.

Analogous experiments can establish which genes are transcribed in response to heat, to the presence of an antibiotic, or to a viral infection. Using microarrays is exciting because researchers no longer have to study changes in the expression of genes one at a time. Instead, they can study the expression of thousands at a time, and identify which sets of genes are expressed in concert. As **Figure 16.10b** shows, microarrays can also document how alternative splicing affects the population of mRNAs present.

Proteomics The term **Proteomics** refers to the large-scale study of protein function. Now that biologists have a catalog of all the protein-coding genes in an organism, researchers are beginning to work out new techniques for examining protein-protein interactions on a massive scale. One approach is similar to the DNA microarrays just described, except that a large series of proteins is spotted on a glass slide. This microarray of proteins can then be treated with an assortment of proteins produced by the same organism. These "probe" proteins are marked with a fluorescent or radioactive tag. If any of the labeled proteins bind to the proteins in the microarray, it suggests that the two molecules may also interact in the cell.

Investigators have also begun to explore the consequences of adding ATP molecules that contain a radioactive phosphate group to a protein microarray along with probe proteins. If the radioactive phosphate group is transferred to one of the protein-protein pairs on the microarray, it suggests that one of the interacting molecules is a protein kinase that phosphorylates the other molecule. In this way, microarrays have become an efficient means for biologists to screen large numbers of proteins and determine how they interact.

Medical Implications

With the advent of microarray technology, the "periodic tables" provided by genome projects are having an important impact on research into gene expression and protein-protein interactions. But the governments and corporations that fund genome projects have undertaken the expense primarily because of the potential benefits to biomedical research. In this respect, is genomics living up to its promise?

Even though genome data became available only recently, the promise for improving human health and welfare is clear. To support this statement, consider how whole-genome analyses are informing efforts to find new drugs and new vaccines.

Rational Drug Design Many drugs act by disabling a particular enzyme. Because genome sequencing provides a comprehensive catalog of the proteins produced by an organism, it offers a comprehensive list of targets for drug therapy.

To illustrate how biologists are mining genome sequences for new drug targets, consider Alzheimer's disease. This illness affects over 4 million elderly people in the United States alone, and is characterized by a degeneration of brain tissue accompanied by progressive loss of memory and other mental faculties. One of the hallmarks of Alzheimer's disease is the formation of deposits, or plaques, of a protein called β-amyloid in the brain. The accumulation of β-amyloid plaques appears to be responsible for the onset of the illness.

Recent research has shown that a protein called β-site amyloid precursor protein-cleaving enzyme (BACE) is involved in the production of the β-amyloid protein. Once this result was confirmed, biologists began searching for a compound that might inactivate BACE and slow the formation of β-amyloid plaques. This effort was recently augmented by data from the Human Genome Project. As members of the International Human Genome Sequencing Consortium were working to annotate the sequence, they found a gene that is closely related to the locus that encodes BACE. The protein product of this gene has been named BACE2 and was found to be 52 percent identical in amino acid sequence to BACE. Follow-up experiments are attempting to test the hypothesis that BACE2 is involved in the formation of β-amyloid plaques. If so, researchers will begin searching for drugs that treat Alzheimer's disease by inactivating BACE *or* BACE2. Having the genome sequence available has doubled the number of potential drug targets.

Research like this is an example of a general approach that pharmaceutical companies call rational drug design. Instead of randomly screening thousands of compounds in an effort to turn up a drug that reduces β-amyloid plaques, investigators begin with data indicating how and why a particular therapy might be effective and then pursue a more focused avenue of research. In this way, rational design promises to make drug development faster and more efficient.

Rational Vaccine Design Although efforts to exploit genome data in rational drug design are still in their infancy, genomics has already inspired important advances in vaccine development. To illustrate how this work is proceeding, consider recent research on the bacterium *Neisseria meningitidis*. This species is a major cause of meningitis and blood infections in children, and it was one of the first prokaryotes to have its genome sequenced. Even though antibiotics can treat *N. meningitidis* infections effectively, the organism grows so quickly that it often injures or even kills the victim before a diagnosis can be made and drugs administered. As a result, biomedical researchers have been interested in developing a vaccine that would prime the immune system and allow children to ward off infections.

Vaccine development has been difficult in this case, however. The immune system usually responds to molecules on the outer surface of bacteria or viruses, and *N. meningitidis* is covered with a polysaccharide that is identical to a compound found on the surface of brain cells. A vaccine composed of this molecule would prime the immune system to attack brain cells as well as cells of *N. meningitidis*.

To circumvent this problem, Mariagrazia Pizza and coworkers analyzed the genome sequence of *N. meningitidis* and tested 600 protein-coding regions for the ability to encode vaccine components. As **Figure 16.11** shows, the biologists inserted the 600 DNA sequences into *E. coli* cells. Later they succeeded in isolating 350 different *N. meningitidis* proteins from the transformed cells. Then they injected these proteins into mice and analyzed whether an immune response occurred. Their results show that seven of the proteins tested evoked a strong immune response and represent promising vaccine components. Follow-up work is now under way to determine whether one or more of these molecules could act as a safe and effective vaccine in humans.

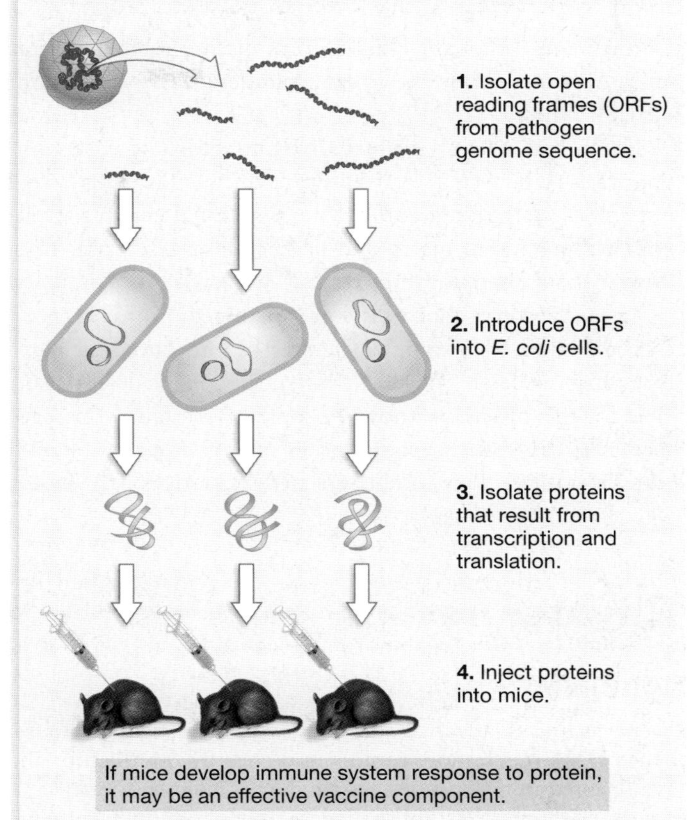

FIGURE 16.11 A New Approach to Vaccine Development
Because all potential genes are identified after whole-genome sequencing, virtually all proteins produced by a pathogen can be tested for their ability to provoke an immune response and act as a vaccine.

The general message of these case histories is that genome sequencing data are providing biomedical researchers with new drug targets and candidate vaccine components. Although early results are promising, only time will tell if genomics ushers in what some have claimed will be an entirely new era of medical research and practice.

Essay Genomics and Issues of Privacy

As efforts to annotate the human genome continue, it is very likely that biologists will be able to track down the genes responsible for many or even most inherited diseases. Suppose that this prediction is correct, and that the follow-up research now under way succeeds in identifying alleles that make people susceptible to disorders such as heart attack, high blood pressure, diabetes, certain cancers, and mental illness. Now suppose that in the near future, technology advances to the point where you will be able to walk into a clinic, request that your DNA be sequenced, and then receive a report detailing the alleles you carry that make you susceptible to certain diseases. Would you want to know? Suppose that health insurance companies, life insurance companies, your friends, and even prospective employers had access to these data. Would you want them to know?

Would you want to know which of your alleles make you susceptible to certain diseases?

Early in the debate about whether government agencies should support the Human Genome Project, questions like these were discussed extensively. As part of the funding proposals that allowed the project to move forward, money was set aside to study the ethical and legal implications of the project. During the rush to complete the human genome sequence, however, these questions received somewhat less attention than they had earlier. Now that the genome is complete, the discussion over the ethical and legal ramifications of the data may start anew.

To bring these issues into sharper focus, consider the situation faced by Nancy Wexler. After her mother died of an inherited disorder called Huntington's disease, Wexler spearheaded an intensive search for the gene that causes the illness. Chapter 17 details how this gene hunt was conducted and how follow-up work revealed the molecular nature of the disease. Thanks to the work of Wexler and her colleagues, it is now possible to test people for the presence of the defective allele.

(Continued on page 332)

(Essay continued)

Unfortunately, though, Huntington's disease remains untreatable. Affected people die in late middle age after a long and difficult illness.

If you had a family history of Huntington's disease, would you want to be tested? A negative result would remove worry, but a positive test might affect your emotional health as well as your willingness to marry and have children. To date, Nancy Wexler has chosen not to be tested.

The situation is much different for hereditary diseases that can be treated. For example, Chapter 12 introduced the hereditary cancer called HNPCC. Now that the genetic defect responsible for this disease is known, people with a family history of the illness can be tested to see if they might be susceptible. If the result is positive, they can be screened regularly for the onset of cancer and make dietary changes that make tumor development less likely.

With all genetic disorders, though, an important series of privacy questions comes to the fore. Should genetic testing be mandatory? If so, should people who carry recessive alleles for an illness be prohibited from marrying, so they do not produce affected offspring? If testing is voluntary, should people who get tested be obliged to provide the results to their spouse or siblings? Should insurance companies have access to the data? What if one genetically identical twin wants to be tested, but the other does not?

These questions, and many others like them, remain unanswered. Sorting out the legal and ethical issues associated with the Human Genome Project is a task that still lies before us.

Chapter Review

Summary

Recent technical advances have allowed investigators to sequence DNA much more rapidly and cheaply than before and have resulted in a flood of genome data. Once the sequence of an entire genome is established, researchers annotate the data by finding genes and determining their function. To identify putative genes in bacteria and archaea, researchers program computers to scan the genome for start and stop codons that are in the same reading frame and that are separated by gene-sized stretches of sequence. Finding such open reading frames (ORFs) is difficult in eukaryotes, because exons are interrupted by introns and because most eukaryotic DNA does not code for a product. One approach to finding eukaryotic genes is to analyze the sequences of complementary DNAs (cDNAs) synthesized from mRNAs, and then match these sequences to DNA found in the genome itself.

Species of bacteria and archaea are usually targeted for whole-genome sequencing because they cause disease or have interesting metabolic abilities. In these groups, a general correlation seems to exist between the size of an organism's genome and its complexity. Parasites tend to have small genomes; organisms that live in a broad array of habitats or that use a wide variety of nutrients tend to have larger genomes. Many of the genes identified in bacteria and archaea still have no known function, however, and a significant percentage of them are extremely similar to other genes in the same genome. Another generalization about prokaryotic genomes is that genes are frequently transferred laterally, or between species. Lateral transfer appears to be common in genes responsible for causing disease.

Compared to prokaryotes, eukaryotic genomes are large and contain a high percentage of transposable elements, simple sequence repeats, and other noncoding sequences. There is no obvious correlation between morphological complexity and gene number in eukaryotes, although the number of distinct transcripts produced may be much larger than the actual gene number in certain species because of alternative splicing. By analyzing loci that belong to closely related gene families, researchers have confirmed that gene duplication and polyploidy have been important sources of new genes in eukaryotes.

The availability of whole-genome sequences is inspiring a variety of new research programs. Biologists are affixing small amounts of proteins or cDNAs to microarrays in order to study protein-protein interactions and changes in mRNA populations over time. In addition, the availability of whole-genome data has allowed investigators to find novel genes that are closely related to existing drug targets, and to find new proteins that may serve as vaccine candidates.

Questions

Content Review

1. What is an open reading frame?
   a. a gene whose function is already known
   b. a putative gene identified by its similarity to a complementary DNA (cDNA)
   c. a putative gene identified by finding a start codon and stop codon in register—meaning in the correct three-codon reading frame
   d. any member of a gene family

2. What does the lateral transfer of genes refer to?
   a. the movement of genes from one species to another
   b. the movement of genes carried by viruses from one individual to another individual of the same species
   c. the movement of transposable elements to new locations in the genome
   d. the movement of an exon to a new location by gene duplication

3. What are minisatellites and microsatellites?
   a. small, extrachromosomal loops of DNA similar to plasmids
   b. parts of viruses that have become integrated into the genome of an organism
   c. incomplete or "dead" remains of transposable elements
   d. short and simple repeated sequences in DNA

4. What is the leading hypothesis to explain the paradox that large, complex eukaryotes such as humans have relatively small numbers of genes?
   a. lateral transfer of genes from other species
   b. alternative splicing of mRNAs
   c. polyploidy, or doubling of the entire chromosome complement
   d. expansion of gene families through gene duplication

5. What evidence do biologists use to infer that a locus is part of a gene family?
   a. Its sequence is exactly identical to another gene.
   b. Its structure—meaning its pattern of exons and introns—is identical to a gene found in another species.
   c. Its composition, in terms of percentage of A-T and G-C pairs, is unique.
   d. Its sequence, structure, and composition are similar to another gene in the same genome.

6. What is a pseudogene?
   a. a coding sequence that originated in a lateral transfer event
   b. a gene whose function has not yet been established
   c. a polymorphic gene—meaning that more than one allele is present in a population
   d. a gene whose sequence is similar to functioning genes, but is not transcribed

Conceptual Review

1. Explain how open reading frames are identified in the genomes of bacteria and archaea. Why is it more difficult to find open reading frames in eukaryotes?

2. Why is the observation that parasitic organisms tend to have relatively small genomes logical?

3. How does a LINE sequence transmit a copy of itself to a new location in the genome? Why are LINEs and other repeated sequences referred to as "genomic parasites?"

4. How does DNA fingerprinting work? Stated another way, how does variation in the size of microsatellite and minisatellite loci allow investigators to identify individuals?

5. Researchers can create microarrays of short, single-stranded DNAs that represent many or all of the exons in a genome. Explain how these microarrays are used to (a) test the hypothesis that alternative splicing is common, and (b) document changes in the transcription of genes over time or in response to environmental challenges.

Applying Ideas

1. Parasites lack genes for many of the enzymes found in their hosts. Most parasites, however, have evolved from free-living ancestors that had large genomes. Based on these observations, W. Ford Doolittle claims that the loss of genes in parasites represents an evolutionary trend. He summarizes his hypothesis with the quip "use it or lose it." What does he mean?

2. When data first revealed that many eukaryotic genomes were dominated by repeated sequences that did not code for products used by the organisms, many researchers began referring to the sequences as "junk DNA." Is this an appropriate term? Defend your answer.

3. According to eyewitness accounts, communist revolutionaries executed the last czar of Russia along with his wife and three children, the family physician, and several servants. Many decades after this event, a grave purporting to hold the remains of the royal family was identified. Biologists were asked to analyze DNA from each skeleton and determine if the bodies were indeed those of several young siblings, two parents, and several unrelated individuals. If the grave were authentic, describe what the DNA fingerprints of each person would look like relative to the fingerprints of other individuals in the grave.

4. Suppose that you are writing a proposal to a funding agency to request funding for a whole-genome sequencing project. Describe why you chose the organism you did, and how the data that result from your project will justify the expense involved.

CD-ROM and Web Connection

CD Activity 16.1: Human Genome Sequencing Strategies *(animation)* *(Estimated time for completion = 10 min)*
Delve into the strategies and steps involved in sequencing the entire human genome.

At your **Companion Website** (http://www.prenhall.com/freeman/biology), you will find self-grading exams and links to the following research tools, online resources, and activities:

National Center for Biotechnology Information
Here, you can search for a particular gene using the GenBank database or select the *human genome resources* link to learn more about human DNA.

Bacterial Artificial Chromosomes
Review several articles that discuss BACs as tools use to sequence large stretches of DNA.

Lateral Transfer and Antibiotic Resistance
This article explains the concept of lateral transfer and illustrates how and why this might allow a bacterial species to become resistant to an antibiotic.

DNA Microarrays
Explore this fantastic new technology, its applicability to science and medicine, and its potential shortcomings.

Additional Reading

Fraser, C. M., E. A. Eisen, and S. L. Salzberg. 2000. Microbial genome sequencing. *Nature* 406: 799–803. An overview of whole-genome data from bacteria and archaea.

International Human Genome Sequencing Consortium. 2001. Initial sequencing and analysis of the human genome. *Nature* 409: 860–921. Introduces one of the first drafts of the complete genome sequence of humans.

Meinke, D. W., J. M. Cherry, C. Dean, S. D. Rounsley, and M. Koornneef. 1998. *Arabidopsis thaliana*: A model plant for genome analysis. *Science* 282: 678–683. An introduction to the most important model organism in plant biology.

Pennisi, E. 2001. The human genome. *Science* 291: 1177–1180. A commentary on the completion of the first draft of the human genome.

Roos, D. S. 2001. Bioinformatics—trying to swim in a sea of data. *Science* 291: 1260–1261. An introduction to efforts by computational biologists to make the enormous data sets from genome sequencing projects readily available and interpretable.

Genetic Engineering and Its Applications

17

Biologists engineer DNA sequences by removing them from a particular organism, manipulating them, and inserting the sequences back into the same species or into a different species. These efforts are referred to as biotechnology, genetic engineering, or recombinant DNA technology. As Chapter 15 pointed out, genetic engineering became possible after the discovery of restriction enzymes and the enzyme DNA ligase in the 1960s and 1970s. Restriction enzymes allowed investigators to cut DNA at precise points; DNA ligase allowed them to put fragments of DNA back together in a recombined sequence. By mixing and matching sequences from different sources, biologists could create "designer genes."

During the 1980s and 1990s, researchers followed up on these discoveries by developing systems for transferring recombinant genes into various types of organisms and by working out techniques for controlling the expression of the introduced DNA. Since then, research has concentrated on applying these techniques to solve problems in medicine, industry, and agriculture.

The goal of this chapter is to introduce some of the key techniques and conceptual issues involved in genetic engineering, using a case history approach. The first example we'll review represents one of the initial efforts to use recombinant DNA technology to cure an inherited disease in humans. The disease is called pituitary dwarfism. The research involved relied on techniques for identifying genes, copying them, moving them into a new host organism, and then controlling their expression. It also raised ethical issues that are an important theme in genetic engineering.

Section 17.2 reviews how researchers found the gene responsible for Huntington's disease.

This tobacco plant has been genetically engineered to express a gene from fireflies. The firefly gene encodes an enzyme called luciferase, which acts on a substrate to produce a flash of light.

Chapter 16 introduced the basic logic that is used to find the physical location of genes responsible for certain traits; here, we delve into this topic in much more depth. The goal is to understand how pedigrees and genetic markers are used to pinpoint the DNA sequences responsible for inherited diseases.

The concluding two sections of the chapter focus on efforts to place novel genes into humans and plants. Gene therapy may make it possible to cure at least some genetic diseases in humans; the development of genetically modified foods may make it possible to boost standards of living in impoverished areas of the world. In addition to understanding the techniques used in this research, however, it is essential to consider the ethical, economic, ecological, and political issues involved. Gene therapy and the release of genetically modified plants are currently under intense scrutiny in the popular press, and legislation to regulate both efforts is being debated by lawmaking bodies around the world. What are the potential perils and benefits of introducing recombinant genes into human beings, food plants, and other organisms? This question represents one of the great challenges of the twenty-first century.

17.1 Using Recombinant DNA Techniques to Manufacture Proteins: The Effort to Cure Pituitary Dwarfism

To better understand the basic techniques and tools that are used by genetic engineers, let's consider the effort to treat pituitary dwarfism in humans. As Chapter 44 will detail, researchers discovered in the 1940s that a structure located near the base of the mammalian brain, called the pituitary gland, produces a protein that stimulates growth. The molecule, which was found to be just 191 amino acids long, was named growth hormone. The gene that codes for the protein is called *GH1*.

Once growth hormone was discovered, researchers immediately suspected that at least some forms of inherited dwarfism might be due to a defect in the GH1 protein. This hypothesis was confirmed when it was established that people with certain types of dwarfism produce little or no growth hormone. These people have defective copies of *GH1* and exhibit pituitary dwarfism, type I. By studying families in which dwarfism is common, several teams of researchers established that pituitary dwarfism, type I is a recessive trait. Stated another way, affected individuals have two copies of the defective allele; in contrast, people who have only one defective allele are carriers of the trait—meaning that they can transmit the defective allele to their offspring—but are not affected. Affected individuals grow more slowly than average people, reach puberty anywhere from two to ten years later than average, and are short in stature as adults—typically no more than 1.2 m (4 feet) tall.

Why Did Early Efforts to Treat the Disease Fail?

Once the molecular basis of pituitary dwarfism was understood, physicians began treating the disease with injections of growth hormone. This approach was inspired by the spectacular success that physicians had achieved in treating type I diabetes mellitus by injecting patients with insulin that had been isolated from pigs. Diabetes mellitus is due to a deficiency of the peptide hormone insulin. Early trials showed that people with pituitary dwarfism could be treated successfully with growth hormone therapy, but only if the protein came from humans. Growth hormones isolated from pigs, cows, or other animals were ineffective. Until the 1980s, however, the only source of human growth hormone was pituitary glands dissected from human cadavers. Up to 20,000 pituitaries needed to be collected from cadavers to supply enough growth hormone to treat the population of affected individuals. As a result, the drug was extremely scarce and expensive.

Meeting demand turned out to be the least of the problems with growth hormone therapy, however. To understand why, recall from Chapter 13 that certain degenerative brain disorders in mammals are caused by infectious proteins called prions. Some prion diseases are hereditary; however, all of the prion diseases can develop if an individual ingests prion proteins. Kuru and Creutzfeldt-Jakob disease are prion diseases that affect humans, and some forms of Creutzfeldt-Jakob disease are hereditary. When physicians found that some of the children treated with human growth hormone were developing Creutzfeldt-Jakob disease in their teens and twenties, they realized that the supply of growth hormone was contaminated with a disease-causing protein from the brains of the cadavers supplying the hormone. As a result, the use of growth hormone isolated from cadavers was banned in 1984. What happened next?

Using Recombinant DNA Technology to Produce a Safe Supply of Growth Hormone

Even as the problems with traditional hormone therapy for pituitary dwarfism became apparent, researchers began to work out a genetic engineering approach for producing uncontaminated supplies of growth hormone. The idea was to find the gene that encodes the human growth hormone protein and insert it into bacterial cells or yeast cells. The hope was that huge quantities of recombinant *Escherichia coli* or *Saccharomyces cerevisiae* cells could be grown, and that the recombinant cells would produce the hormone in sufficient quantities to meet demand at an affordable price.

To meet this goal, investigators looked for the human growth hormone gene using the steps outlined in **Figure 17.1**. The first task was to isolate the mRNAs produced in the pituitary gland. Presumably, one of these transcripts would encode growth hormone. To find the target transcript, the biologists used the enzymes reverse transcriptase to make a complementary DNA (cDNA) of each of the pituitary-gland mRNAs. Recall that re-

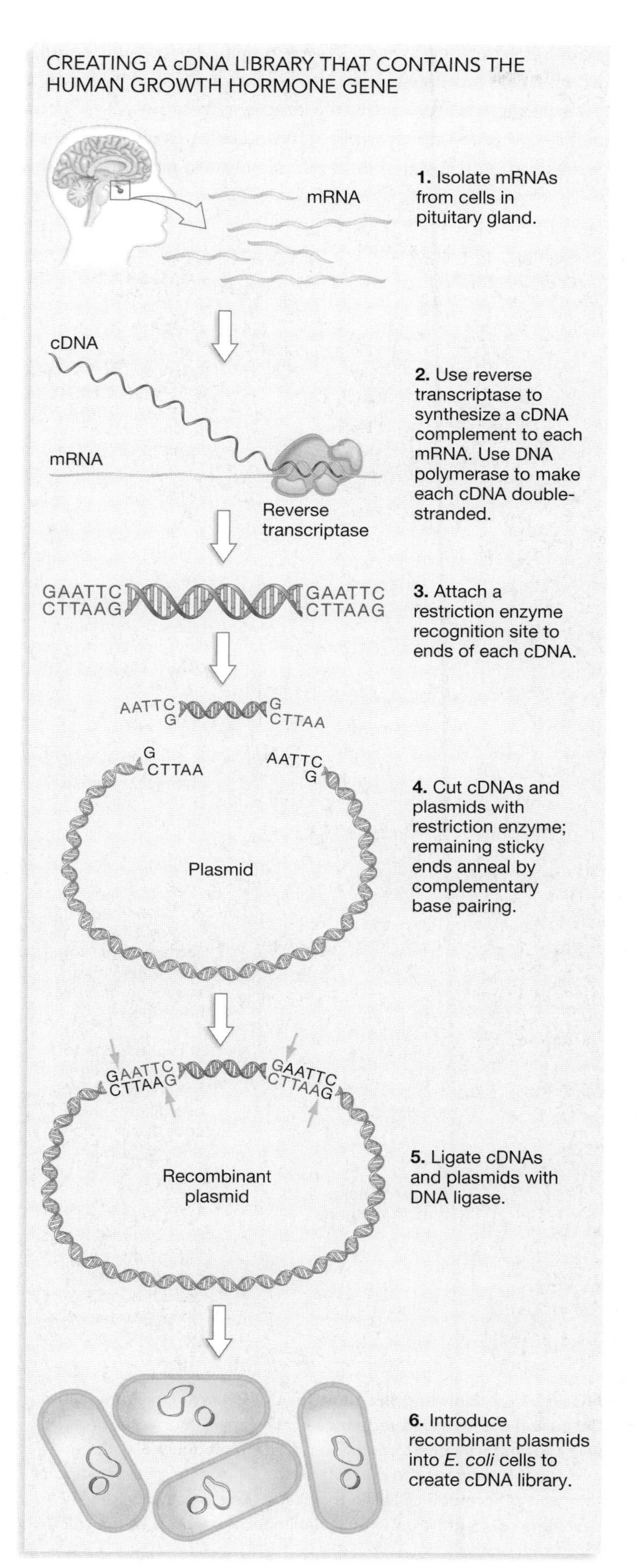

FIGURE 17.1 Starting the Hunt for the Growth Hormone Gene The hunt for the growth hormone gene began with the creation of a cDNA library from pituitary cells. Because growth hormone is produced in this tissue, researchers knew that the gene for growth hormone existed somewhere in the library.

verse transcriptase synthesizes a DNA strand from an RNA template; DNA polymerase can then be used to synthesize the complementary synthesize the other strand and yield a double-stranded DNA. The product of this step—step 2 in Figure 17.1—was thousands of different cDNAs. Then the question became, which of these cDNAs encodes the growth hormone protein?

The key to answering this question was to isolate each of the many cDNAs and then obtain enough copies that each of the isolated sequences could be analyzed. Steps 3 through 6 in Figure 17.1 show how this was done. The approach illustrates some of the most basic and widely used techniques in genetic engineering. Let's take a closer look.

Using Plasmids Plasmids are circular DNA molecules that are found in the cytoplasm of bacteria and in the nucleus of certain unicellular eukaryotes, including baker's yeast. Plasmids are physically distinct from the bacterial or yeast chromosome(s) and are not required by the cell for normal growth and reproduction. Because plasmids replicate independently of the chromosome(s), a bacterial cell can contain hundreds of copies of a plasmid.

Over the past several decades, naturally occurring plasmids have been modified by researchers to make them useful as carriers of recombinant genes. For example, the plasmids used in genetic engineering experiments carry at least one gene that allows researchers to easily identify and isolate the bacterial cells that carry these plasmids. In many cases, the gene involved codes for a molecule that confers resistance to a specific antibiotic. When bacterial cells are treated with that antibiotic, only cells carrying the plasmid are able to survive.

Using Restriction Endonucleases In addition to a gene for antibiotic resistance, the plasmids used in genetic engineering also contain sequences that can be cut by restriction endonucleases. As explained in Chapter 15, Box 15.2, restriction endonucleases are bacterial enzymes that cut DNA molecules at specific sequences.* To insert pituitary-gland cDNAs into plasmids, researchers attached the same restriction enzyme recognition site to the end of each cDNA (see step 3, Figure 17.1). Then they allowed the restriction enzyme to cut the plasmids and the cDNAs. As step 4 in the figure shows, the enzyme activity left single-stranded, "sticky ends" in the plasmids and cDNAs. As

*Restriction enzymes defend bacterial cells from viral attack. When a virus infects a bacterial cell, restriction enzymes cut the viral DNA, inactivating it. The cell's DNA is protected from being cut because its recognition sites have methyl ($-CH_3$) groups attached.

step 5 shows, the sticky ends from the plasmid and cDNAs attached to each other via complementary base pairing and formed a recombinant molecule. DNA ligase—the enzyme that connects Okazaki fragments during DNA replication—was then used to seal the recombinant pieces of DNA together.

Transformation: Introducing Recombinant Plasmids into Bacterial Cells Plasmids serve as a **vector** or vehicle for transferring recombinant genes to a new host. If a recombinant plasmid can be inserted into a bacterial or yeast cell, the foreign DNA will be copied and transmitted to new cells as the host cell grows and divides. In this way, researchers can obtain thousands or millions of copies of specific genes.

For bacterial cells to take up foreign DNA—a process known as transformation—they must be subjected to a specific chemical treatment or to an electrical shock. (No one knows why these methods disrupt the bacterial cell wall and membrane and allow foreign DNA to enter.) The recombinant plasmid can then enter the cell. The cells are subsequently spread out on a gelatinous medium (agar) that contains antibiotic and allowed to grow. Only bacterial cells that have successfully taken up a plasmid carrying the antibiotic resistance gene are able to grow in the presence of the antibiotic. The resulting collection of transformed bacterial cells represents a cDNA library (step 6, Figure 17.1).

Using Nucleic Acid Hybridization to Find a Target Gene Which of the cDNAs in the pituitary-gland cDNA library encodes growth hormone? To answer this question, investigators used the genetic code to infer the approximate DNA sequence of the growth hormone gene from the amino acid sequence of the growth hormone protein. They synthesized many copies of a short, single-stranded stretch of DNA that was complementary to this inferred sequence, attached a radioactive atom to the fragments, and used the resulting molecules to probe a copy of the cDNA library (**Figure 17.2**). The radioactively labeled probe DNA bound to its complementary sequence in the cDNA library, identifying the recombinant cell that contained the human growth hormone.

To accomplish their goal of producing large quantities of the human growth hormone, the investigators then transferred the growth hormone cDNA to a plasmid that contains the promoter recognized by the bacterial RNA polymerase. As a result, the transformed *E. coli* cells began to transcribe and translate human growth hormone gene. Bacterial cells containing the human growth hormone gene were grown in huge quantities. These cells have proved to be a safe and reliable source of the human growth hormone protein.

Ethical Concerns

As supplies of growth hormone increased, physicians used it in treating not only people with pituitary dwarfism but also children who happened to be short although they had no actual growth hormone deficiency. The treatment was popular because the molecule would increase the height of these children by a few centimeters. In essence, the drug was being used as a cosmetic—a way to improve appearance in cultures where height is deemed attractive. But should parents request growth hormone

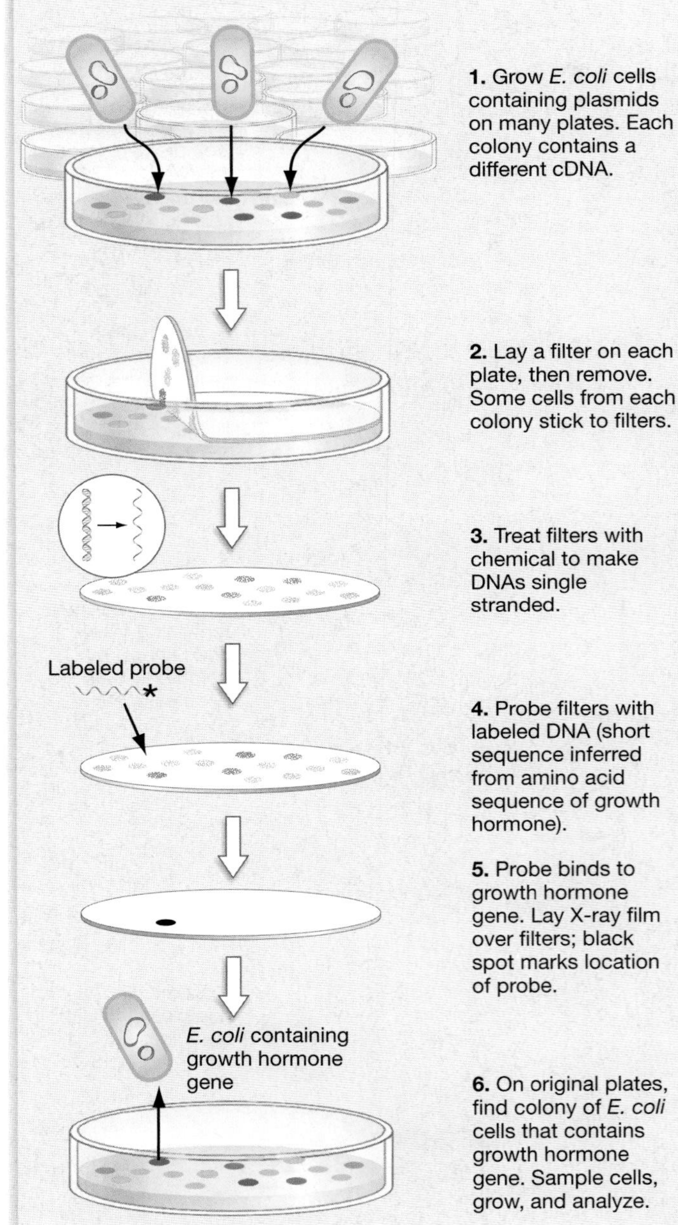

FIGURE 17.2 Finishing the Hunt for the Growth Hormone Gene Because the amino acid sequence of the growth hormone protein was known, researchers could infer the DNA sequence of the gene and use this sequence to create a probe DNA. Once the location of the growth hormone cDNA in the cDNA library was known, researchers could grow up virtually unlimited quantities of the cDNA for further analysis.

treatment for genetically normal children to change their appearance? What if parents want a tall child to be even taller to enhance her potential success as, say, a basketball player?

Growth hormone has also been found to enhance the maintenance of bone density and muscle mass. As a result, it has become a popular performance-enhancing drug for athletes. Part of its popularity stems from the fact that it is currently undetectable in the drug tests administered by governing bodies. Should athletes be able to enhance their physical skills by taking hormones or other types of drugs? This question remains unanswered. In the meantime, it is clear that while solving one important problem, recombinant DNA technology created others. Throughout this chapter, an important theme is that genetic engineering has costs that must be carefully weighed against its benefits.

17.2 Gene Hunting Based on Pedigree Analysis

Huntington's disease is a rare but devastating illness. Typically, affected individuals become symptomatic between the ages of 35 and 45. At onset, an individual appears to be clumsier than normal and tends to develop small tics and abnormal movements. As the disease progresses, however, uncontrollable movements become more pronounced. Eventually the affected individual twists and writhes involuntarily. Personality and intelligence are also affected—to the extent that the early stage of this disease is sometimes misdiagnosed as schizophrenia. The illness may continue to progress for 10 to 20 years and is eventually fatal.

Because Huntington's disease appeared to run in families, physicians suspected that it was a genetic disease. To confirm this hypothesis, researchers set out to identify the gene or genes involved and to document that one or more genes is altered in affected individuals. As has been the case with many genetic diseases, this simple goal turned out to be extraordinarily difficult to achieve. Let's take a closer look.

Analyzing Pedigrees

The first step in studying a genetic disease is to determine its pattern of inheritance. When studying a mutation in a fruit-fly or a pea-plant gene, a researcher can determine the pattern of inheritance by making a series of controlled crosses like those introduced in Chapter 10. Because this approach is not possible with humans, investigators must study the human crosses that currently exist. They do so by constructing the **pedigree** or family tree of affected individuals.

A pedigree, such as the one shown in **Figure 17.3**, records the genetic relationships among the individuals in a family along with the sex of each person and their phenotype with respect to the trait in question. If the trait is due to a single gene, analyzing the pedigree may reveal whether the trait is due to a dominant or recessive allele and whether the gene responsible is located on a sex chromosome or an autosome. What does this mean?

Sex-Linked Traits As Chapter 10 showed, sex-linked genes are located on one of the sex chromosomes. Human males have one X and one Y chromosome; females have two X chromosomes (**Figure 17.4a**, page 340). Males have just one copy of each X-linked gene; females have two. Because the complement of sex chromosomes differs in males and females, the pattern of inheritance in sex-linked traits is different.

Is Huntington's disease a sex-linked trait? To answer this question, let's consider the pedigree of a classic sex-linked trait—the occurrence of hemophilia in the descendants of Queen Victoria of England (**Figure 17.4b**). Hemophilia is caused by a defect in an important blood-clotting factor, which results in prolonged bleeding from even a minor injury; hemophiliacs are at high risk of bleeding to death. From the pedigree in Figure 17.4b, the key observation is that only males in this family developed the condition. In addition, note the affected male in generation II. His two sons were unaffected, but the trait reappeared in a grandson. Stated another way, hemophilia skipped a generation.

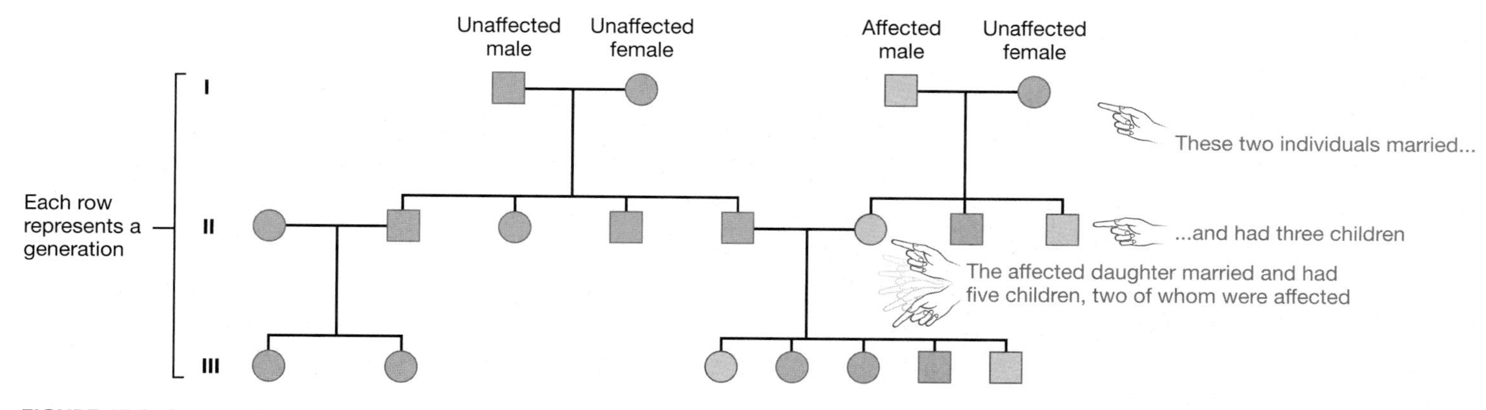

FIGURE 17.3 Constructing a Pedigree
Pedigrees are drawn using the notation shown here. They are a compact way of communicating a great deal of information about which individuals in a family exhibit a certain phenotype.

To explain this pattern, biologists hypothesize that hemophilia is due to an X-linked recessive allele. This hypothesis is logical because males have only one X chromosome. As a result, the phenotype associated with an X-linked recessive allele appears in every male that carries it. Females express the recessive phenotype only if they are homozygous for that allele. Because the defective allele is rare, it is unlikely to appear in the homozygous state.

Further, the appearance of an X-linked recessive trait skips a generation in a pedigree because the affected male passes his only X chromosome on to his daughters (his sons receive his Y chromosome). But because his daughters have two X chromosomes and are most likely to be heterozygous for this allele, they don't show the trait. They, however, will pass the defective allele on to about half of their sons—the probability is that about half will inherit the X chromosome with the defective allele and half will inherit the other X chromosome—so about half of their sons should develop the trait.

In contrast, if an X-linked trait is dominant it will appear in every individual who has the defective allele. A good indicator of an X-linked dominant trait is a pedigree in which an affected male has all affected daughters but no affected sons. Besides the inherited form of a bone disease called rickets, however, very few diseases are known to be due to X-linked dominant alleles. In addition, few inherited diseases are due to loci on the Y chromosome, simply because so few functional genes are located on this chromosome.

Patterns of Inheritance: Autosomal Recessive Traits **Figure 17.5a** shows the pedigree of a family affected by Huntington's disease. Because the malady appears in both males and females at about equal rates, the gene involved in Huntington's disease does not appear to be sex-linked. If a single gene is responsible for the trait, and if the gene is located on an autosome, how can we determine whether the disease-causing allele is recessive or dominant?

The key to answering this question is the recognition that, by definition, dominant alleles produce the phenotype in question when an individual is homozygous *or* heterozygous for the defective allele. In contrast, recessive alleles produce the phenotype in question only when the individual is homozygous for that allele. Heterozygous individuals who carry a recessive allele for an inherited disease are referred to as **carriers** of the condition. The term is appropriate because these individuals can carry the allele and transmit it even though they do not exhibit the disease.

Based on these observations, it should make sense that the pedigree in **Figure 17.5b** is from a family where an autosomal recessive disease occurs. Note that some children in this pedigree exhibit the trait even though their parents do not. Their parents were carriers. In addition, when an affected individual has children, the children did not necessarily have the trait. The phenotype shows up in offspring only when both parents carry the recessive allele that is responsible and both pass this allele on to their offspring.

Patterns of Inheritance: Autosomal Dominant Traits The pedigree of the family affected by Huntington's disease in

(a) Human sex chromosomes

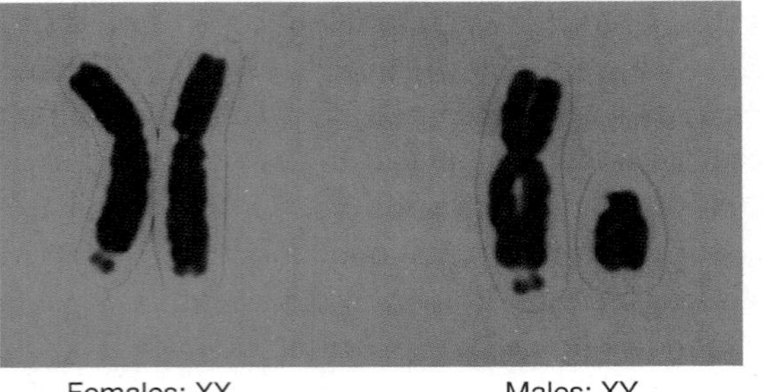

(b) Occurrence of hemophilia in royal families of Europe

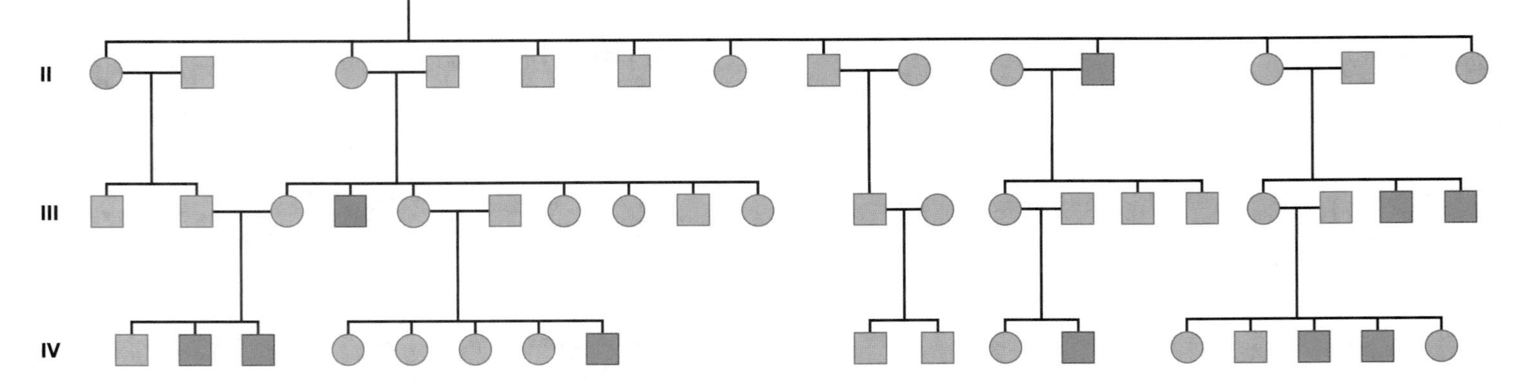

FIGURE 17.4 A Pedigree of an X-Linked Disease
(a) Humans have two types of sex chromosomes, called X and Y. Females are XX; males are XY. **(b)** This pedigree shows the occurrence of hemophilia in the descendants of Queen Victoria and Prince Albert of England, who lived in the late 1800s. Note that in this family, the condition shows up only in males.

Figure 17.5a has two features that indicate this disease is passed to the next generation as an autosomal dominant allele. First, if a child shows the trait, then one of its parents shows the trait as well. Second, if families have a large number of children, the trait usually shows up in every generation.

The pattern is explained by the hypothesis that most people who have the disease are heterozygous for the trait; their genotype for this allele would be symbolized as *Hh*. About half their children receive the mutant allele that causes the disease (*H*) from their affected parent, while the other half receive the normal allele (*h*). Children who possess the disease allele develop the disease. Other examples of autosomal dominant traits in humans include widow's peak, dimples, the presence of extra fingers or toes (polydactyly), and achondroplastic dwarfism, which is due to a defect in cartilage structure.

Finding the Gene for Huntington's Disease

Once biologists understood that Huntington's disease was caused by a dominant allele located on an autosome, their next goal was to find the gene and determine its DNA sequence. Identifying the gene would make it possible to reach several important objectives:

- *Improved Genetic Testing* Because the symptoms of Huntington's disease do not appear until middle age, affected people have often married and had children by the time they develop symptoms. If members of families with Huntington's patients could be tested for the defective allele early in life, they might make different decisions about starting their own families and passing on the illness.
- *Improved Therapy* Once the sequence of a gene is known, researchers are better able to understand the structure and function of the resulting protein. Information about a protein's mode of action can be extremely helpful in developing drugs or other treatments. As we'll see in section 17.3, finding a gene also makes it possible to consider replacing defective alleles with normal copies.
- *Improved Understanding of the Phenotype* Several fundamental questions about Huntington's disease might be resolved if the gene were found. How can one defective protein

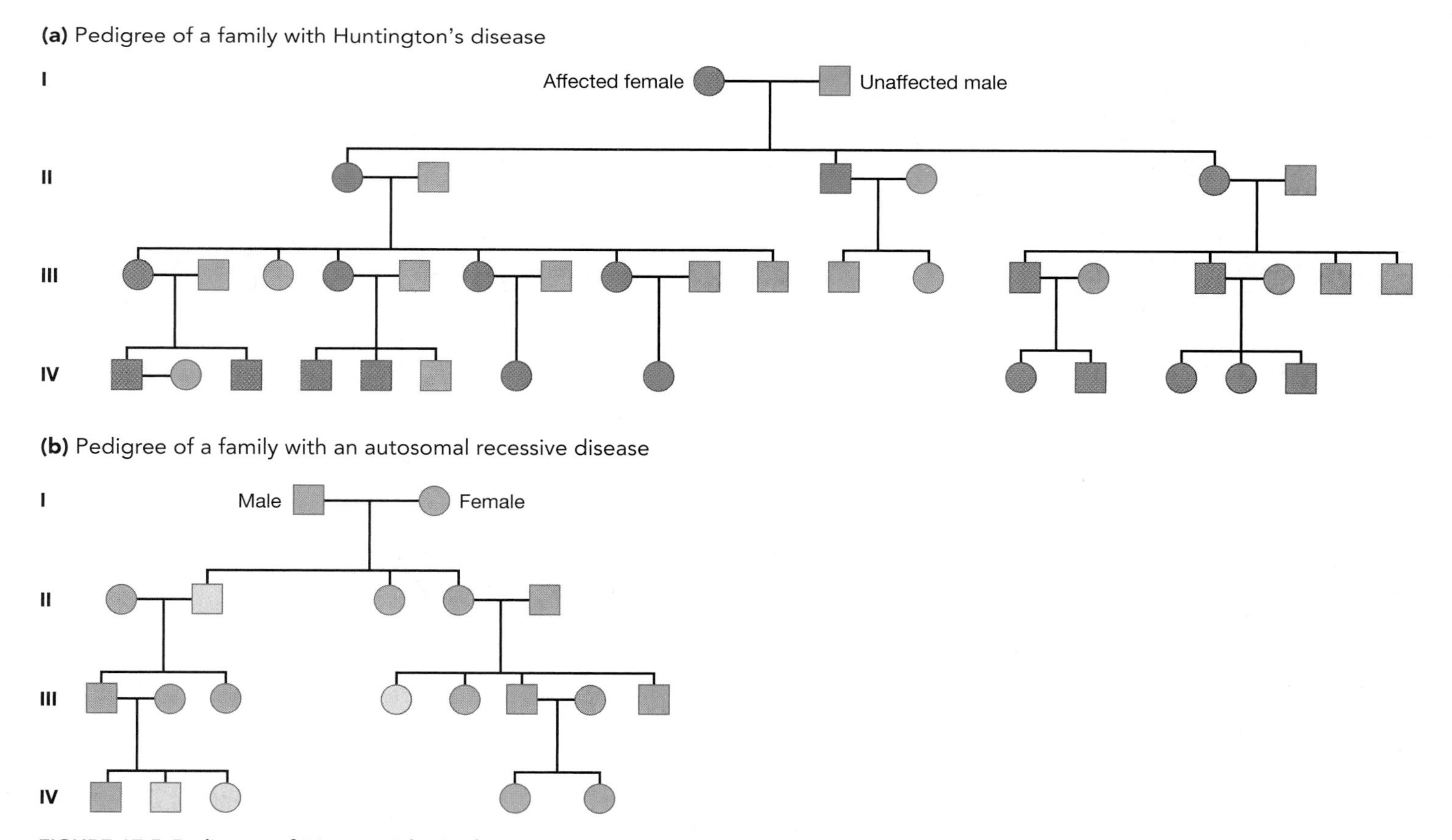

FIGURE 17.5 Pedigrees of Diseases Inherited on Autosomes
(a) Diseases that are inherited as autosomal dominants appear in both males and females and tend to appear in every generation. **(b)** Diseases that are inherited as autosomal recessives appear in both males and females. For an individual to be affected, both parents must carry the allele responsible. QUESTION In a pedigree where an autosomal recessive disease appears, an affected parent frequently has unaffected children. But unaffected parents can have affected offspring. Why?

lead to abnormal movements *and* personality changes? Why does it take so long for symptoms to appear, and why is the illness progressive?

Using Genetic Markers The basic logic that underlies efforts to locate unknown disease-causing genes is to combine analyses of established genetic markers and the pedigrees of affected people. The fundamental idea is to find a known genetic marker that is closely linked to the unknown gene. If the marker and the disease gene are located close to one another on the chromosome, then they are almost always inherited together. Coinheritance of the marker and the disease gene occur because they remain together as chromosomes segregate and assort independently during meiosis. The closer the marker and disease gene are to one another on the chromosome, the less likely they are to be separated by crossing over during meiosis.

The types of genetic markers used in gene hunts have changed over time. In the late 1970s and early 1980s, when biologists were searching for the Huntington's disease gene, the best genetic markers available were short stretches of DNA where specific restriction enzymes cut the double helix. These sequences are known as restriction enzyme recognition sites.

Figure 17.6 explains how restriction enzyme recognition sites work as genetic markers. The key idea is that the number of recognition sites varies among individuals because the base sequence of their DNA varies. For example, if a mutation alters the DNA sequence of a recognition site, the restriction enzyme will no longer cut at that location. When an investigator cuts DNA from many individuals with the same restriction enzyme, then, the size of the resulting fragments will vary. These size variants are called restriction fragment length polymorphisms (RFLPs). By separating the fragments from each individual via gel electrophoresis, probing the fragments with a piece of labeled DNA from a known gene, and visualizing them with autoradiography, a researcher can identify the recognition site "phenotype" of each person. The sizes and numbers of bands on the gel distinguish that individual. In addition, banding patterns are inherited. Individuals pass their particular banding pattern on to their offspring.

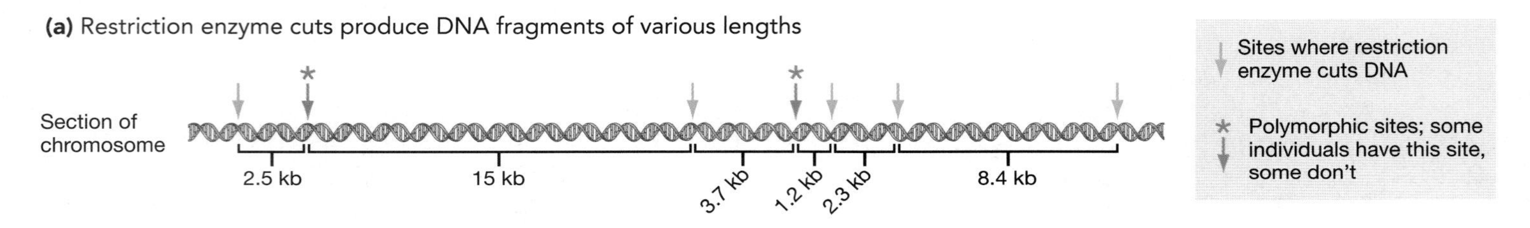

(b) There are two polymorphic recognition sites in this region of DNA. As a result, four different banding patterns are possible when the fragments are run out on a gel:

A ** B *– C –– D –*

17.5 kb
15.0 kb
8.4 kb
4.9 kb
3.7 kb
2.5 kb
2.3 kb
1.2 kb

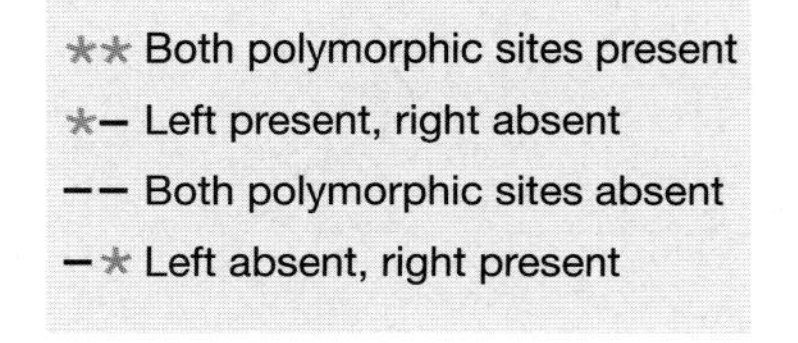

If relatives who share a particular banding pattern also share an inherited illness, then the gene responsible for the disease is located near this region of their genome.

FIGURE 17.6 Using Restriction Enzyme Fragments as Genetic Markers
Genetic markers identify certain locations in the genome and vary among individuals. This diagram shows how the fragments of DNA generated by restriction enzyme cuts can be used as geneticmarkers. The banding patterns in part (b) are those predicted for individuals who are homozygous for each pattern.

By cutting DNA with different restriction enzymes and probing the resulting gels with sequences from different locations in the genome, researchers gradually build a map showing the physical locations where recognition sites are variable among individuals.

Using a Pedigree Once a chromosome map containing many restriction enzyme recognition sites or other genetic markers has been assembled, biologists need help from families affected by an inherited disease to find the gene in question. Gene hunts are more likely to be successful if large families are involved. The Huntington's disease team was fortunate to find a large, extended family affected with the disease living along the shores of Lake Maracaibo, Venezuela.

From historical records, the researchers deduced that the Huntington's disease allele was introduced to the family by a European sailor or trader who visited the area in the early 1800s. When family members agreed to participate in the study, there were over 3000 of his descendants living in the area. One hundred of these people had been diagnosed with Huntington's disease. To help in the search for the gene, family members agreed to donate skin or blood samples for DNA analysis and to furnish information on who was related to whom.

This research resulted in the pedigree shown in **Figure 17.7**. Note that the diagram includes information about the disease phenotype and the particular restriction enzyme fragment pattern observed in each family member. To find the Huntington's disease gene, researchers looked for fragment patterns that were present in affected individuals but absent in unaffected individuals. Several restriction enzyme recognition sites appeared to be inherited along with Huntington's disease. Were these associations due to chance or to a close physical association between the recognition site and the defective allele? Stated another way, were the recognition site and Huntington's disease gene close to one another on the DNA double helix?

To answer this question, the researchers used statistical analysis to determine the likelihood that an apparent association between a genetic marker and the disease phenotype was due to chance. After testing 11 different restriction enzyme fragment patterns in this way, they finally found one—the 'C' pattern shown in Figure 17.6—that was inherited along with the defective allele often enough to indicate that the association was not due to chance. Stated another way, the statistical analysis supported the hypothesis that the marker—which happened to be on chromosome 4—was actually located close to the Huntington's disease gene. By examining associations between other markers on chromosome 4 with the Huntington's disease phenotype, the team succeeded in narrowing down the location of the Huntington's disease gene to a region about 500,000 base pairs long (see **Box 17.1**, page 344).

Pinpointing the Defect Once the general location of the Huntington's disease gene was known, biologists looked in that region for exons that encode a functional mRNA. They planned to sequence exons from diseased and normal individuals, compare the data, and pinpoint specific bases that were different between the two groups of individuals.

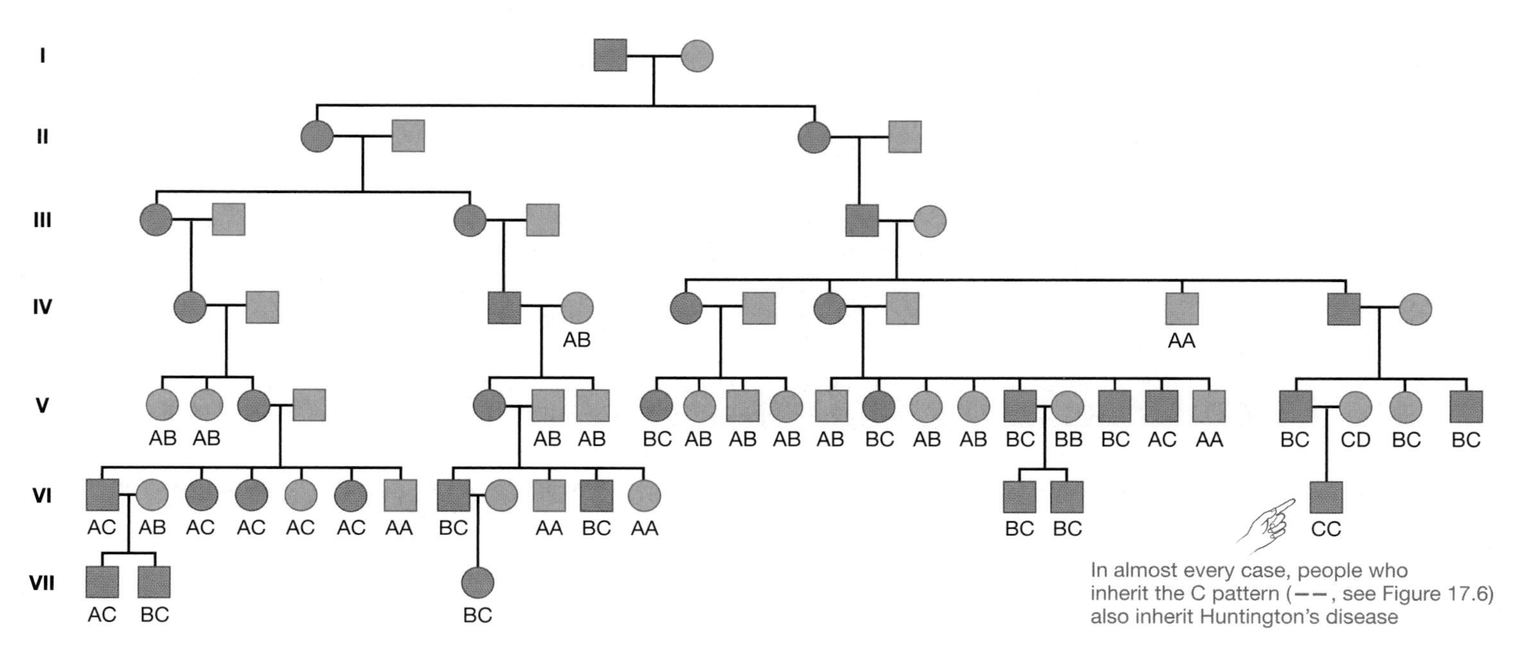

FIGURE 17.7 Are Certain Genetic Markers Inherited Along with the Huntington's Disease Gene? This is part of a pedigree from a large extended family living near Lake Maracaibo, Venezuela. Individuals affected by Huntington's disease inherited the defective allele from a common ancestor. QUESTION Why might genotype data be missing for so many individuals in this pedigree?

When this analysis was complete, the research team found that individuals with Huntington's disease have an unusual number of CAG codons near the 5′ end of a particular gene. CAG codes for glutamine. Healthy individuals have 11–25 copies of the CAG codon, while affected individuals have 42 or more copies. Although many genetic diseases are caused by single base changes that alter the amino acid sequence of a protein, diseases caused by the expansion of a particular codon repeat have also been observed. For example, the most common form of inherited mental retardation, fragile-X syndrome, is caused by an increase in the number of copies of a CGG codon at a specific site in the genome.

When the research team confirmed that the increase in the CAG codon was always observed in affected individuals, they concluded that the long search for the Huntington's disease gene was over. They named the newly discovered locus *IT15* and its protein product huntingtin.

New Approaches to Therapy How has the effort to locate the Huntington's disease gene helped researchers and physicians understand and treat the illness? First, people with a family history of Huntington's disease can now determine whether they carry the mutant allele through DNA testing. If tests confirm that they are susceptible to the disease, they can consider their alternatives (see **Box 17.2**).

The discovery also allowed investigators to understand the molecular nature of the disease for the first time. Autopsies of Huntington's patients showed that their brains actually decrease in size due to the death of neurons, and that the tissue contains insoluble aggregates of the huntingtin protein. These aggregates are thought to be a direct consequence of the changes in the number of CAG repeats in the gene, and thus the number of glutamine residues in the protein. Long stretches of polyglutamine are known to result in the formation of protein aggregates. The leading hypothesis to explain Huntington's disease proposes that a buildup of huntingtin protein triggers neurons to undergo apoptosis, or programmed cell death.

The discovery of the gene has also allowed scientists to introduce the defective allele into mice. These transgenic mice develop tremors and abnormal movements, exhibit higher levels of aggression toward litter and cage mates, and experience a loss of neurons in the brain. When researchers succeed in finding laboratory animals with disease symptoms that parallel a human disease, they say that they have an animal model of the disease.

Animal models are valuable in disease research because they can be used to test potential treatments before trying them on human patients. For example, some research groups are now looking for drugs that prevent the aggregation of the huntingtin protein. Other biologists are testing drugs that prevent the death of cells containing aggregated huntingtin proteins. Yet another research group has established that the progression of the disease in mice can be slowed if affected individuals live in cages with toys and other sources of stimulation once symptoms appear. Mice residing in bare cages

BOX 17.1 Genome Sequencing and Gene Hunting

As the Huntington's disease story shows, researchers need a large catalog of genes or sequences that are polymorphic in a population in order to hunt down disease genes. If two or more alleles of a gene are present in a population, then that locus can be used as a marker for a particular segment in the genome. Using pedigrees of families with many individuals affected by a particular disease, researchers can then look for correlations between particular polymorphic alleles or sequences and the occurrence of an illness caused by a defective gene. When analyses show that family members who share an inherited illness also tend to inherit a certain genetic marker, then researchers can infer that the gene responsible for the disease is located close to the marker.

Although analyzing restriction fragment patterns allowed investigators to find the gene responsible for Huntington's disease, the Human Genome Project has revolutionized the types of polymorphic markers used by gene hunters. Because DNA from many different individuals was used as source material during the human genome project and because overlapping segments of genes were routinely sequenced, researchers were able to identify 1.42 million sites where single bases vary among individuals. Where you might have a "C" at a particular site, others may have a "T." These variable sites are called **single nucleotide polymorphisms**, or **SNPs** (pronounced "snips"). The group that cataloged these polymorphic sites estimates that 60,000 lie within exons and that at least one polymorphic site lies within 5000 bases of every exon in the genome.

For disease gene hunters, this new catalog of SNPs is an enormously powerful resource. The possibility of analyzing the inheritance of millions of polymorphic sites all over the genome—instead of just a few hundred widely scattered ones such as restriction enzyme sites—makes it much more likely that researchers can locate genes associated with illnesses like Alzheimer's, certain cancers, and inherited forms of mental illness. The database promises to be particularly important in understanding the genetic basis of diseases where many different loci are involved.

progress to disease much more quickly than "stimulated" mice. If this result can be extended and confirmed, it could lead to techniques for stimulating the brains of people known to carry the Huntington's allele and delay illness until better therapies are available.

17.3 Can Gene Therapy Cure Inherited Diseases in Humans?

For biomedical researchers interested in curing inherited diseases like Huntington's disease, sickle-cell anemia, and cystic fibrosis, the ultimate goal is to replace or augment defective copies of the gene with normal alleles. This approach to treatment is called **gene therapy.**

For gene therapy to succeed, several steps must be completed. First, the wild-type allele for the gene in question must be sequenced and its regulatory sequences understood. Second, there must be a method for introducing a copy of the normal allele into affected individuals. Third, the normal allele must replace a dominant allele entirely or be expressed at the correct time in the correct cells to fix a recessive defect. Chapter 15 discussed aspects of gene regulation in humans and other eukaryotes; here, we focus on techniques for introducing DNA into humans. As a case study, let's consider the first successful application of gene therapy in humans.

How Can Novel Alleles Be Introduced to Human Cells?

Section 17.1 reviewed how recombinant DNA sequences are packaged into plasmids and taken up by *E. coli* cells. Humans and other mammals lack plasmids, however, and their cells do not take up foreign DNA in response to chemical or electric treatments as efficiently as bacterial cells.

How can foreign genes be introduced efficiently to human cells? To date, researchers have focused on packaging foreign DNA into viruses for transport into human cells. Recall from Chapter 11 that viruses insert their DNA directly into host cells. As Chapter 26 will detail, in some cases viral DNA becomes integrated into the host-cell chromosome where it is transcribed. As a result, viruses that infect human cells can be used as vectors or carriers of human DNA.

Two major classes of viruses have proven to be effective vectors for recombinant DNA. The first class consists of viruses that have an RNA genome. As described in Chapter 26, the genomes of these retroviruses encode the enzyme reverse transcriptase. When an RNA virus infects a human cell, this enzyme catalyzes the production of a DNA copy of the viral genome

BOX 17.2 Genetic Testing

When the Huntington's gene was found, biologists used the information to develop a test for the presence of the defective allele. The test involves obtaining a DNA sample from the test individual and using the polymerase chain reaction to amplify the region that contains the CAG repeats responsible for the disease. If the number of CAG repeats is 35 or less, the individual is normal. Forty or more repeats results in a positive diagnosis for Huntington's.

As more is known about the human genome, more and more tests like this will be developed for diseases that have some genetic component. What types of genetic testing are done currently?

- *Carrier Testing* People from families affected by a genetic disease frequently want to know whether they carry the allele responsible before starting a family. That is especially true for diseases like cystic fibrosis (CF) that are due to recessive alleles. If it turns out that only one of the prospective parents has the allele, then none of the children they have together should develop CF. But if both of them carry the allele, then each child they produce has a 25 percent probability of having CF.
- *Prenatal Testing* Suppose that two parents who each carry the CF allele decide to have children but want healthy children. Once the mother is pregnant, a physician can obtain fetal cells after 10 to 16 weeks of gestation. The fetal cells are cultured and DNA is isolated. The allele in question is amplified by PCR (polymerase chain reaction) and sequenced. Based on the results of this test, a couple may choose to continue or terminate a pregnancy.
- *Adult Testing* Huntington's disease isn't the only trait that appears in adulthood. For example, about 5 percent of the women who develop breast cancer do so because they inherited a faulty gene. To date, three different genes have been identified as causes of breast cancer. Women from families with a history of breast cancer may want to be tested for the presence of one of these mutant genes.

It is important to note, though, that the results of genetic tests may be difficult to interpret. If a person has an expanded CAG repeat in the *IT15* gene, it is virtually certain that he or she will develop the disease. But if a woman has one of the alleles associated with breast cancer, there is just an 80 percent chance that she will develop the illness. In cases like this, the benefit of testing is to make the person involved more vigilant about getting regular checkups. Then if breast cancer does develop, it can be caught at an early and potentially curable stage.

and inserts it into a host-cell chromosome. Using recombinant techniques, human DNA can be packaged into retroviruses and transferred into somatic cells.

The adenoviruses are a second class of virus that has been used to carry recombinant alleles into human cells. Adenoviruses have a DNA genome. Certain adenoviruses infect human lung cells and other tissues in the respiratory tract and cause colds or flu-like symptoms. Based on this observation, adenoviruses have been tested as vectors to treat individuals suffering from cystic fibrosis. (As the essay in Chapter 4 pointed out, the lungs are the organs most affected by the defective cystic fibrosis gene.) Unfortunately, adenoviruses do not integrate into host cells. As a result, they express foreign genes for a short time only.

There are serious problems associated with using viruses to transform human cells, however. Viruses usually cause disease. The RNA viruses include the human immunodeficiency virus (HIV), which causes AIDS, as well as an array of viruses that cause cancer in animals. Using these agents for gene therapy requires that sequences responsible for causing disease be inactivated or removed from their genomes. Even if they have been inactivated, the altered particles may be able to recombine with viral DNA that already exists in the individual being treated and lead to the formation of a new infectious strain. Finally, viral proteins trigger a response by the immune system. Despite these risks, viruses are still the best vectors currently available for gene therapy.

Using Gene Therapy to Treat X-Linked Immune Deficiency

Recently, Marina Cavazzana-Calvo and co-workers reported the successful treatment of a disease using gene therapy. The disease is called severe combined immunodeficiency (SCID). Children who are born with the disease lack a normal immune system and are unable to fight off infections.

The type of SCID treated by Cavazzana-Calvo and colleagues is designated SCID-X1 because it is caused by mutations in a gene on the X chromosome. The gene is required for the development of immune-system cells called T cells. T cells develop in the bone marrow, from undifferentiated cells called stem cells. The gene encodes a receptor (called γc) for a growth factor that stimulates the development of the T cells.

Traditionally, physicians have treated SCID-X1 by keeping the patient in a sterile environment, isolated from any direct human contact until the patient could receive bone-marrow tissue transplanted from a close relative. In most cases, the T cells that are needed by the patient differentiate from the transplanted bone-marrow cells and allow the individual to live normally. In some cases, though, no suitable donor is available. Could gene therapy cure this disease by furnishing functioning copies of the defective gene?

Before any protocol for gene therapy can be implemented in humans, researchers must show that the proposed strategy works in animals. To begin their work, then, Cavazzana-Calvo and colleagues introduced the γc gene into a retroviral vector and used the recombinant virus to infect bone-marrow cells from dogs. Follow-up analyses showed that the dog cells that took up the vector were long-lived and produced the receptor protein, and that the receptor recognized the proper growth factor. Another set of experiments was carried out on mice that lacked functioning copies of the γc gene. When the team transferred a normal copy of the gene into the marrow cells of these mice, immunodeficiency was prevented.

Based on these results, the team gained approval to treat two boys, each less than one year old, who had SCID-X1. No suitable bone-marrow donor was available to treat either patient. Even though they'd been isolated in germ-free rooms, both had developed illnesses such as pneumocystic pneumonia and thrush (oral yeast infections), which appear in people with malfunctioning immune systems.

To implement gene therapy, the researchers removed bone marrow from each patient and collected the stem cells that produce mature T cells. Each day for three days, researchers infected the stem cells with an altered mouse retrovirus that carried the normal γc gene. About 30 percent of the cells took up the γc gene and manufactured the receptor protein. The stem cells were then transferred back into the patients. One of the boys had detectable levels of functioning T cells just 30 days after reinsertion of the transformed marrow cells; the second had working T cells after 60 days.

T-cell concentrations continued to increase in both patients for 8 months. At that time, the research team had enough confidence in the two boys' immune systems to inoculate them with polio and tetanus vaccines. To their great relief, they found that the boys' immune systems succeeded in producing the normal response proteins, called antibodies, to the polio and tetanus vaccines. The children now had functioning immune systems. The infections that existed prior to gene therapy also began to clear up.

About three months after the experimental treatment began, the patients were removed from germ-free isolation rooms. At the time of the team's initial publication on the protocol, in April 2000, the boys had been residing at home for almost a year and were growing and developing normally. As long as the γc gene continues to be expressed in T-cell precursors, the boys will be able to produce a pool of T cells and mount an immune response to fight infection.

Controversy over Gene Therapy

Although the treatment of SCID-X1 qualifies as an important success story, gene therapy is controversial. Just a few months before the report on the promising treatment of SCID-X1 appeared, a major gene therapy center was closed when a patient being treated for a different enzyme deficiency disease died after adenovirus-mediated gene therapy. In this case, the death was directly attributable to the treatment. This incident raised serious questions about the use of adenoviruses as vectors for recombinant DNA.

Testing new drugs, new approaches to surgery, and new gene therapy protocols always carries a risk for the patients involved. The researchers who run the trials must explain the risks clearly and make every effort to minimize them. Although gene therapy holds great promise for the treatment of a wide variety of devastating inherited diseases, fulfilling that promise will require years of additional research and testing.

17.4 Biotechnology in Agriculture

Although progress in human gene therapy has been slow, progress in transforming crop plants with recombinant genes has been rapid. In the United States alone, more than 30 genetically modified or "transgenic" crops are now in commercial use. Currently over 30 percent of the total U.S. soybean crops and 20 percent of the total U.S. corn crops are planted with recombinant strains; globally, over 44 million hectares of transgenic crops were grown in 2000.

Recent efforts to develop transgenic plants have focused on three general objectives:

- Reducing losses to herbivore damage by introducing the gene for a naturally occurring insecticide. For example, researchers have moved a gene from the bacterium *Bacillis thuringiensis* into corn; the "Bt toxin" encoded by this gene protects the plant from corn borers and other caterpillar pests.
- Reducing losses to competition with weeds by introducing a gene whose product makes the crop plant resistant to an herbicide. Soybeans that have been genetically engineered for resistance to the herbicide glyphosate are a good example.
- Improving the quality of the product consumed by people. Soybeans and canola plants, for example, have been engineered to produce a higher percentage of unsaturated fatty acids.

To gain a better understanding of genetic engineering in plants, let's take a detailed look at efforts to produce transgenic rice with improved nutritional qualities. Although almost half the world's population depends on rice as its staple food, the grain is an extremely poor source of certain vitamins and essential nutrients. For example, rice contains no vitamin A at all. This is a serious issue because vitamin A deficiency causes blindness in 250,000 Southeast Asian children each year. Vitamin A deficiency also renders children more susceptible to diarrhea, respiratory infections, and childhood diseases such as measles.

Humans and other mammals synthesize vitamin A from a precursor molecule named β-carotene (beta-carotene). β-carotene, in turn belongs to a family of plant pigments called the carotenoids. Carotenoids are orange, yellow, and red in color and are especially abundant in carrots. Rice plants synthesize β-carotene in their chloroplasts but not in the part of the seed that is eaten by humans. Could genetic engineering produce a strain of rice that synthesized β-carotene in the carbohydrate-rich seed tissue called endosperm? Ingo Potrykus and colleagues set out to answer this question. If successful, their research could help to solve an important global health problem.

Synthesizing β-Carotene in Rice

Potrykus and co-workers began their work by searching for compounds in rice endosperm that could serve as precursors for the synthesis of β-carotene. They found that maturing rice endosperm contains a molecule called geranyl geranyl diphosphate, which is an intermediate in the synthetic pathway that leads to the production of carotenoids. As **Figure 17.8** shows, three enzymes are required to produce β-carotene from geranyl geranyl diphosphate. If genes that encode these enzymes could be introduced into rice plants along with regulatory sequences that would trigger their synthesis in endosperm, the researchers could produce a transgenic strain of rice that would contain β-carotene.

Fortunately, genes that encode two of the required enzymes had already been isolated from daffodil, and the gene for the third enzyme had been purified from a bacterium. Because the sequences had been inserted into plasmids and grown in bacteria, many copies were available for manipulation. To each of the coding sequences in the plasmids, Potrykus and colleagues added the promoter region from an endosperm-specific protein.

To develop transgenic rice strains that are capable of producing β-carotene, the three sets of sequences had to be inserted into rice plants. As you know from reading this chapter, introducing recombinant DNA into *E. coli* is fairly straightforward. But how are foreign genes introduced into plants?

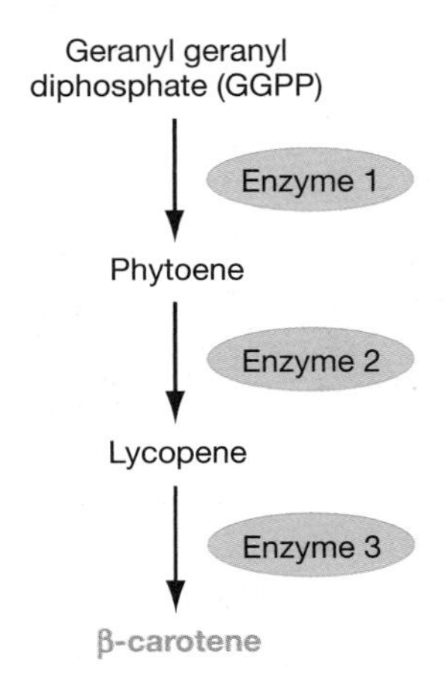

FIGURE 17.8 Synthetic Pathway for β-Carotene
GGPP is a molecule found in rice seeds. Three enzymes are required to produce β-carotene from GGPP.

The *Agrobacterium* Transformation System

Agrobacterium tumefaciens is a bacterium that infects plants. As **Figure 17.9a** shows, tissues that become infected with this parasite form a tumor-like growth called a gall. When researchers looked into how these infections occur, they found that a plasmid carried by *Agrobacterium* cells, called a Ti (tumor inducing) plasmid, plays a key role. Ti plasmids contain several functionally distinct sets of genes. One set encodes products that allow the bacterium to bind to the cell walls of a host. Another set, referred to as the virulence genes, encodes the proteins required to transfer part of the Ti DNA, called the T-DNA (transferred DNA), into the interior of the plant cell. The T-DNA then travels to the nucleus and integrates into the plant chromosomal DNA (**Figure 17.9b**). When transcribed, T-DNA induces the infected cell to begin growing and dividing. The result is the formation of a gall that protects a growing population of *Agrobacterium* cells.

(a) Plant with crown gall disease

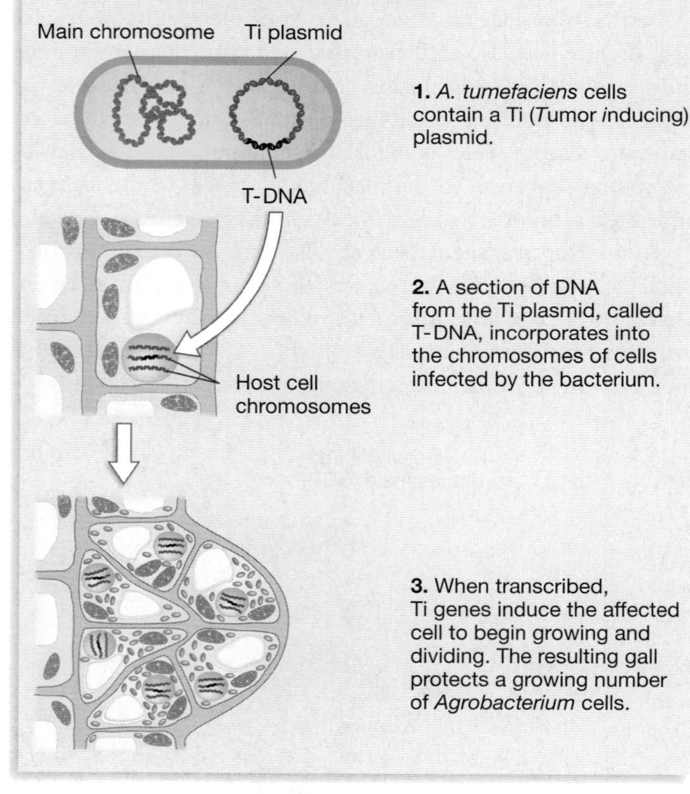

FIGURE 17.9 ***Agrobacterium* infections**
(a) The tumor-like growth at the bottom of this plant's stem is due to an infection by the bacterium *Agrobacterium tumefaciens.* **(b)** *A. tumefaciens* cells contain the Ti plasmid. A section of DNA from the Ti plasmid, called T-DNA, incorporates into the chromosomes of cells infected by the bacterium. Proteins produced from T-DNA induce the formation of a gall.

Once the Ti plasmid's mechanism of action was worked out, researchers realized that the plasmid offered an efficient way to introduce recombinant genes into plant cells. Follow-up experiments confirmed that recombinant genes could be added to the T-DNA that integrates into the host chromosome, that the gall-inducing genes could be removed, and that the resulting sequence is efficiently transferred and expressed in its new host plant. In addition to the *Agrobacterium* system, researchers have also been able to transfer foreign genes into crop plants directly. In this approach, tiny tungsten particles are coated with foreign genes and then shot directly into plant cells using an instrument called a "gene gun."

Golden Rice

To generate a strain of rice that produces all three enzymes needed to synthesize β-carotene in endosperm, Potrykus and co-workers exposed embryos to *Agrobacterium* cells containing genetically modified Ti plasmids like the one illustrated in **Figure 17.10a.** The resulting plants were then grown in the greenhouse. When the transgenic individuals had matured and set seed, the researchers found that some rice grains contained so much β-carotene that they appeared yellow. In **Figure 17.10b** you can see this "golden rice." It is still not clear, however, that the transgenic rice contains enough β-carotene to relieve vitamin A deficiency when normal amounts of rice are consumed.

While the production of golden rice in the laboratory is promising, there is still much to do before the advance has an impact on public health. The transgenic rice must be grown successfully in normal field conditions. In addition, most or all of the rice strains grown in different parts of the world will need to be crossed with transgenic stock to acquire the appropriate genes.

Although the future of golden rice is still to be determined, recent advances in agricultural and medical biotechnology promise to increase the quality and quantity of food and to alleviate at least some inherited diseases that plague humans. Each solution offered by genetic engineering tends to introduce new issues, however. As this chapter's essay points out, researchers and consumer advocates have expressed concerns about the increased numbers and types of genetically modified foods. During this debate, people who are well informed about the techniques and issues involved will be an important resource for the community.

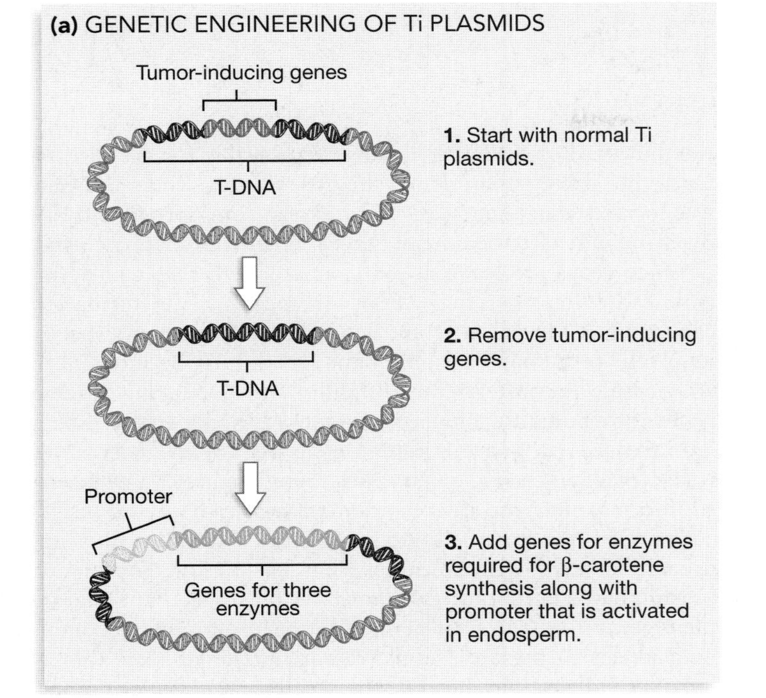

FIGURE 17.10 Producing "Golden Rice"
(a) To produce a strain of rice capable of producing β-carotene in the endosperm of their seeds, researchers constructed the Ti plasmid shown here. **(b)** These photos show seeds from individuals that were and were not transformed with the Ti plasmids illustrated in part (a). Chemical analyses confirmed that the golden color in the transformed seeds is due to large quantities of β-carotene.

Essay Controversies over Genetically Modified Foods

Enthusiasm about the potential benefits of golden rice and other genetically modified foods must be tempered by concern about the risks of releasing large numbers and types of recombinant crop plants. In Europe, public opinion has been extremely skeptical about the wisdom of developing genetically engineered crops. Protesters in Britain, the U.S., and elsewhere have even vandalized test plantings of genetically modified crops.

Are genetically modified foods safe to eat?

Proponents of genetic engineering in agriculture can point to several important success stories. U.S. farmers applied 8.2 million fewer pounds of active pesticide ingredient in 1998, in part because glyphosate-resistant and insecticide-producing crop strains were coming into widespread use. (Recall that glyphosate is an herbicide.) Use of glyphosate-resistant crops also makes conservation tillage practices—which radically reduce soil erosion—much more effective. If crop plants are glyphosate resistant, then the compound can be applied in small amounts to kill weeds after growth starts. If not, then other weed killers must be applied in larger amounts earlier in the growing season. Genetic engineering also made it possible to develop soybeans and corn strains that produce higher concentrations of the amino acid lysine, improving their nutritional value.

Opponents of genetically modified foods can point to several important concerns, however. Widespread adoption of insecticide-producing crop strains may lead to the evolution of pests that are resistant to the compounds. Research has shown that corn plants genetically engineered to produce the Bt toxin leak the molecule into the soil, where it might kill untargeted organisms. Research also shows that crop plants genetically engineered for glyphosate resistance produce pollen carrying the recombinant allele, and that pollen can subsequently be transferred to closely related weeds. Based on these data, there is legitimate concern that careless use of these crop strains could lead to the more rapid evolution of herbicide-resistant weeds. Finally, many people simply do not trust that genetically modified foods are safe to eat.

Are genetically modified crops good or bad? There is no easy or pat answer to this question. Like any technology, genetic engineering has costs and benefits. For citizens, the challenge is to become informed about the issues and insist that regulatory agencies strictly enforce measures to maximize the benefits of the technology and minimize the risks involved. For biologists, the challenge is to quantify those benefits and risks as carefully as possible and communicate their results clearly, so that good decisions can be made.

Chapter Review

Summary

Genetic engineering is based on inserting a modified version of an allele into an organism. Genetic engineering became possible after the discovery of restriction enzymes, DNA ligase, and plasmids. Restriction enzymes allowed researchers to cut DNA at specific locations and insert it into plasmids or other vectors with the help of DNA ligase.

To isolate and sequence a specific allele, investigators must determine its location in the genome. This chapter reviewed two techniques for finding a specific gene. Locating the locus for human growth hormone was relatively straightforward because the amino acid sequence of the protein and the tissue responsible for the producing the protein were known. As a result, investigators could create a cDNA library from cells that were transcribing growth hormone mRNA, probe the library with a DNA sequence that encoded a short stretch of the known amino acid sequence, and find the gene.

Locating the locus responsible for Huntington's disease was much more difficult because only the disease phenotype and the pattern of inheritance were known. To find the huntingtin protein, investigators analyzed a large number of genetic markers and a large pedigree of an affected family to find a marker that was inherited along with the allele responsible for the disease. Once this strategy pinpointed the general area where the gene was located, they sequenced many exons from normal and diseased individuals to determine exactly where the defect occurred.

Once genes are located and characterized, they can be introduced into other individuals or species. Transformation can occur in several different ways, depending on the species involved. In humans, recombinant DNA must be introduced by viruses. Because of the difficulty in introducing foreign genes into humans, and because such difficult ethical issues are involved, progress in human gene therapy has been slow. In bacteria and certain eukaryotes, foreign genes can be inserted into plasmids and the recombinant plasmids can be inserted back into a host cell. An important variant on this approach has made genetic engineering feasible for crop plants. Certain bacteria that infect plants have plasmids that integrate their genes into the host-plant genome. By adding recombinant alleles to these plasmids, researchers have been able to introduce alleles that improve the nutritional quality of crops, make them resistant to herbicides, or allow them to produce insecticides.

Questions

Content Review

1. What do restriction enzymes do?
   a. They cleave bacterial cell walls and allow viruses to enter.
   b. They ligate pieces of DNA by catalyzing the formation of phosphodiester bonds between them.
   c. They cut DNA at specific sites known as recognition sequences.
   d. They act as genetic markers and allow researchers to create the physical maps of chromosomes used in gene hunts.

2. What is a plasmid?
   a. an organelle found in many bacteria and certain eukaryotes
   b. a circular DNA molecule, found in many bacteria and certain eukaryotes, that replicates independently of the main chromosome(s)
   c. a type of virus that has a DNA genome and that infects certain types of human cells, including lung and respiratory tract tissue
   d. a type of virus that has a RNA genome, codes for reverse transcriptase, and inserts a cDNA copy of its genome into host cells

3. Once the gene that causes Huntington's disease was found, researchers introduced the defective allele into mice to create an animal model of Huntington's disease. Why was this valuable?
   a. It allowed them to test potential drug therapies without endangering human patients.
   b. It allowed them to study how the gene is regulated.
   c. It allowed them to produce large quantities of the huntingtin protein at will.
   d. It allowed them to study how the gene was transmitted from parents to offspring.

4. To begin the hunt for the human growth hormone gene, researchers created a cDNA library from cells in the pituitary gland. What did this library contain?
   a. only the sequence encoding growth hormone
   b. DNA versions of all the mRNAs in the pituitary-gland cells
   c. all of the coding sequences in the human genome, but no introns
   d. all of the coding sequences in the human genome, including introns

5. If a trait tends to skip a generation before it appears again in a pedigree, and if it occurs about equally often in males and females, what is its most likely pattern of inheritance?
   a. X-linked
   b. Y-linked
   c. autosomal dominant
   d. autosomal recessive

6. What does it mean to say that a genetic marker and a disease gene are closely linked?
   a. The marker lies within the coding region for the disease gene.
   b. The sequence of the marker and the sequence of the disease gene are extremely similar.
   c. The marker and the disease gene are found on different chromosomes.
   d. The marker and the disease gene are in close physical proximity and tend to be inherited together.

Conceptual Review

1. Explain how restriction enzymes and DNA ligase are used to insert foreign genes into plasmids and create recombinant DNA. Your answer should include a sketch illustrating some of the key events involved.
2. How do researchers get foreign DNA into cells? To answer this question, explain how recombinant plasmids are introduced into *E. coli* cells, how viruses are used to transport genes into human cells, and how the Ti plasmid in *Agrobacterium* cells is used to transport genes into plants.
3. What is a cDNA library? How is one created? Give an example of how a cDNA library can be used in research.
4. What information is contained in the human pedigrees that are used in hunting for disease genes? What are genetic markers, and how are they used to create a genetic map? Explain how researchers combine analyses of pedigrees and genetic markers to narrow down the location of disease genes.
5. Recall that researchers added the promoter sequence from an endosperm-specific gene to the Ti plasmids used in creating golden rice. Why was this important? Comment on the role that promoter and enhancer sequences have in genetic engineering in eukaryotes.
6. In reviewing work on human growth hormone and Huntington's disease, this chapter described two contrasting strategies for finding genes. Compare and contrast these strategies. Why was the hunt for the growth hormone gene much simpler than the hunt for the Huntington's disease gene?
7. In parasitology, a *vector* is an organism that carries a disease-causing organism to a new host. Why is it appropriate to apply this term to the plasmids and viruses used in genetic engineering?

Applying Ideas

1. The text posed the following questions about Huntington's disease: How can one defective protein lead to abnormal movements *and* personality changes? Why does it take so long for symptoms to appear, and why is the illness progressive? Based on your understanding of the molecular nature of the disease, offer hypotheses to answer these questions.
2. Discuss some of the ethical issues involved in human gene therapy. Specifically, should therapy be restricted to somatic cells, or should individuals be able to alter their germ-line cells (meaning that they would alter the alleles they pass on to their offspring)? Should gene therapy be approved for disease-causing alleles only, or should parents also be able to pay to transform their children with alleles associated with height, intelligence, hair color, eye color, athletic performance, musical ability, or similar traits?
3. Several organizations are actively trying to stop the development of transgenic crops and implement bans on the marketing of genetically modified foods. Suppose a representative from one of these organizations comes to your door seeking membership support. Explain why you will or will not support their efforts.

CD-ROM and Web Connection

CD Activity 17.1: Producing Human Growth Hormone *(animation)*
(Estimated time for completion = 10 min)
Retrace the steps for making human growth hormone using DNA technology.

At your **Companion Website** (http://www.prenhall.com/freeman/biology), you will find self-grading exams and links to the following research tools, online resources, and activities:

Harvest of Fear—PBS Presentation on Transgenic Crops
This Public Broadcasting Service (PBS) site discusses the major benefits and risks associated with consuming genetically modified products.

Transgenic Crops—A Balanced Overview
This site presents a balanced review of the arguments for and against using transgenic crops.

Genetic Engineering and its Dangers—Philosophy Essays
A collection of articles written on the religious, philosophical, and technical aspects of genetic engineering.

EPA and Biotechnology Crops
This site describes a pending review by the Environmental Protection Agency (EPA) on the registration process for all genetically modified crops.

Additional Reading

Anderson, W. F. 1998. Human gene therapy. *Nature* 392: 25–30. Discusses the promises and problems of gene therapy research, emphasizing the types of viruses being used to introduce genes to human cells.

Hanley, Z., T. Slabas, and K. M. Elborough. 2000. The use of biotechnology for the production of biodegradable plastics. *Trends in Plant Science* 5: 45–46. Introduces recent research on engineering plants that produce plastics.

Mann, C. C. 1999. Crop scientists seek a new revolution. *Science* 283: 310–314. Surveys efforts to improve the productivity of rice, corn, and wheat.

Wolfenbarger, L. L., and P. R. Phifer. 2000. The ecological risks and benefits of genetically engineered plants. *Science* 290: 2088–2093. Assesses the costs and benefits associated with using genetically modified crop varieties.

Developmental Biology

The study of animal and plant development centers on one fundamental question: How does a multicellular organism with a highly organized body develop from a fertilized egg? From a single, simple-looking cell, an individual with billions of highly specialized cells arises. Exploring how this process occurs is currently among the most vibrant of all areas in biological science.

Developmental biology is also one of the most synthetic fields in biology. As you explore the chapters in this unit, you'll draw on information about the structure and function of macromolecules introduced in Unit 1, data on cell function and cell division reviewed in Unit 2, and conclusions about information flow and gene expression presented in Unit 3. Because the chapters in this unit will require you to think about whole organisms as well as molecules, they are an effective bridge between the molecular and cellular focus of Units 1-3 and the whole-organism and ecological focus of Units 5-9.

To launch your look at this extraordinarily dynamic field, **Chapter 18: An Introduction to Development** surveys the series of events that take place as model organisms like mustard plants and fruit flies develop. **Chapter 19: Early Development** follows up with a more detailed look at how gametes form, how fertilization occurs, and how fertilized eggs begin dividing. **Chapter 20: Determining a Cell's Fate** concludes the unit by examining the events that are responsible for the formation of specific cell and tissue types and body structures.

These young chinook salmon developed from eggs laid in rivers and streams along the Pacific coast of North America. The parents of these individuals swam to these freshwater breeding grounds from the northernmost reaches of the Pacific Ocean, then died after mating.

An Introduction to Development

18

In 1859 Charles Darwin referred to the origin of species as "the mystery of mysteries." At that time the most urgent task confronting biologists was to explain how species come to be. Today, however, Darwin's theory of evolution by natural selection explains many of the most fundamental questions about how species are created and how they change through time. What question qualifies as the current mystery of mysteries in biological science? Although there are many candidates, one of the most compelling is the question of how a multicellular individual develops from a single cell—the fertilized egg.

Biologists began to study how plants and animals develop in the 1800s. Early investigators sought to catalog the events that occur as an individual develops from a fertilized egg to its juvenile and then adult forms. This early work was observational and descriptive in nature and carried out with microscopes.

A fundamental message of these studies was that development is highly variable among organisms. In humans, a fertilized egg develops into a newborn child who contains hundreds of different cell types and trillions of cells. This process takes nine months. In this species, rapid growth and development continue for another 15 to 18 years when individuals become adults. In contrast, a fertilized egg from a burr oak tree takes about four months to grow into an embryo that is encased in an acorn. After spending the winter in a dormant state, the embryo grows into a seedling. Over the next 250 to 300 years, the tree will continue to grow and develop new leaves, roots, and flowers. At the other extreme, a fertilized egg of the fruit fly *Drosophila melanogaster* transforms into a feeding larva in just a day. About five days later the larva stops feeding, secretes a surrounding case, and forms a pupa. Some four

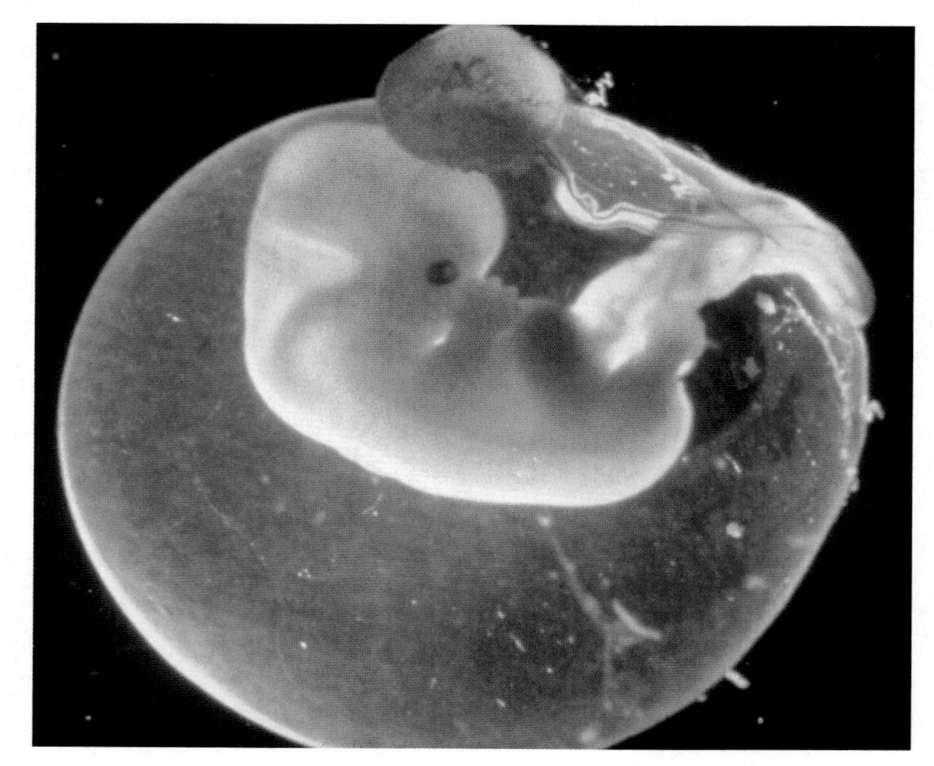

At five weeks after fertilization, a human embryo shows distinctive structures such as eyes, growing limbs, and the precursors of vertebrae. In addition to reviewing what happens during early development in a variety of species, this chapter explores experiments that gave biologists their first glimpse of how development occurs.

days later, an adult emerges. The adult fly flies, feeds, courts, mates, and starts the cycle anew.

How did biologists start to make sense of all this variability in the rate and pattern of development? By the late nineteenth century, researchers had started to use experimental approaches to ask how and why these changes occur. For study, they chose organisms that developed rapidly and produced large numbers of eggs that were easy to manipulate. Then they removed cells at various stages of development to see which were required for the formation of certain structures, injected dyes into cells to follow their fate through time, and analyzed mutant individuals that developed abnormally. After a century of poking and prodding embryos with these and other techniques, biologists have concluded that several unifying processes and themes underlie the bewildering variation in plant and animal development.

The goals of this chapter are to review how developmental patterns vary among different groups of plants and animals and to introduce the underlying unity of genetic and cellular processes responsible for development. Let's begin by doing what the earliest investigators did, which is to describe the stages in development of different organisms. Section 18.2 then explores one of the earliest and most fundamental questions about development. Do the cells in a developing organism differ in their genetic makeup or simply in which genes are expressed? The chapter closes with a look at how biologists began tracking the fate of individual cells within embryos. The effort to map the fates of particular cells during development led biologists to realize that in organisms as diverse as frogs, fruit flies, and cabbage, a cell's fate is determined by two common mechanisms. Let's delve in.

18.1 Developmental Stages and Patterns

What major events occur as a fertilized egg grows into an embryo? The answer to this question varies among species. To sample the diversity of developmental patterns observed in plants and animals, let's consider embryonic development in a plant, an invertebrate, and two vertebrates. More specifically, we'll look at the earliest events in the life of the wild mustard plant *Arabidopsis thaliana*, the fruit fly *Drosophila melanogaster*, frogs, and humans. These four species are particularly well studied. In addition, their developmental sequences are similar to those in many of their close relatives. Two vertebrates are featured to illustrate differences between developmental patterns in species that lay eggs versus those that give live birth.

As you study the developmental stages and patterns observed in these four species, it is important to keep in mind that questions about development focus not so much on growth as on the formation of distinctive cell types, tissues, and structures. In a sense, growth is relatively easy to understand. Cell number in the embryo increases as cells undergo mitosis. Although controlling the rate and pattern of cell division is important in development, the really difficult questions concern how one type of cell or tissue becomes different from another type, and how structures like brains, eyes, leaves, and flowers form. In looking at descriptions of embryonic development, then, a biologist is not so much concerned about growth as about changes that identify particular tissues or structures.

Embryonic Development in *Arabidopsis thaliana*

Arabidopsis thaliana is a flowering plant whose short life span and ease of culture has made it a favorite experimental subject for biologists interested in development and genetics (**Figure 18.1a**, page 356). **Figure 18.1b** shows the plant's entire life cycle, which can take just 6 weeks under laboratory conditions. The cycle begins with the formation of haploid sperm and egg cells, or **gametes**, via meiosis. Once meiosis is complete, the haploid cells that are produced undergo changes that lead to the production of mature sperm and eggs. The combination of meiosis and these maturation steps is called **gametogenesis**. In *Arabidopsis* and other flowering plants, sperm formation occurs in pollen grains while egg cell formation occurs inside a part of the flower called an **ovule**. The union of sperm and egg is called **fertilization** and takes place inside the ovule; the single-celled embryo that results is called a **zygote**.

When the zygote undergoes mitosis and its daughter cells continuing dividing mitotically, an embryo begins to develop inside the ovule. As embryonic development continues, the ovule's covering develops into a seed coat that encases both the embryo and a nutrient-rich tissue called **endosperm** (inside-seed). The mature seeds are usually dispersed away from the parent plant by wind, an animal, or water. At germination, the seed coat breaks and the embryo emerges to become a seedling. Mature embryos and seedlings have three prominent parts: a root, initial leaves called **cotyledons**, and a stem-like structure called the **hypocotyl** that joins the roots and cotyledons.

How do these embryonic structures form? As **Figure 18.1c** shows, the cells that result from the fertilized egg's first division are asymmetrical in size, orientation, and their eventual function or fate. The bottom, or basal cell, is large and gives rise to a column of cells called the suspensor, which anchors the embryo as it develops. The small apical cell, in contrast, is the progenitor of the mature embryo. Initially it gives rise to a simple ball of cells at the tip of the suspensor. As this group of cells grows and develops, the cotyledons begin to take shape. Later, groups of cells called the **shoot apical meristem** and **root apical meristem** form. A meristem consists of undifferentiated cells that divide repeatedly, with some of their daughter cells becoming specialized cells. Meristematic tissues are "forever young." They continue to produce cells that differentiate into adult tissues and structures throughout the individual's life. The shoot and root apical meristems, for example, produce all of the lengthwise growth in the main shoot and root. As they grow, the embryonic root and hypocotyl elongate to form the main axis of the body.

All of this growth takes place without the aid of cell movements. Because plant cells have stiff cell walls, they do not move. For the embryo to take shape, then, mitoses and cell divisions have to occur in precise orientations and the resulting cells have to exhibit differential growth. For example, cotyledons broaden out from the hypocotyl in part because of changes in the orientation of the mitotic spindles in the meristematic cells that contribute to the cotyledons, and partly because differential growth results in daughter cells with a distinctive shape. In contrast, animal embryos take shape largely as the result of directed cell movements.

The root, hypocotyl, and cotyledons are organs that form along the main axis of the embryo. The cross-sectional view in **Figure 18.2** shows that early development in *A. thaliana* also produces three embryonic tissues. As you can see, an embryo contains an outer covering of epidermal cells that protect the individual. Inside this layer of cells is a mass of ground tissue, which may later differentiate into cells that are specialized for photosynthesis, food storage, or other functions. The vascular tissue in the center of the plant will eventually differentiate into specialized cells that transport food and water between root and shoot. In the embryo, the three tissue systems are arranged in a radial pattern.

Although this pattern of development is typical of many plants, the sequence of events and the structures and tissues that form are very different from those observed during the development of animals. These differences are not surprising, given that both the fossil record and evolutionary trees suggest that plants and animals evolved from different single-celled ancestors. Stated another way, multicellular bodies evolved independently in plants and animals. As a result, their developmental pathways are distinct. Let's begin an analysis of animal development by looking at what may be the best-studied of all multicellular organisms.

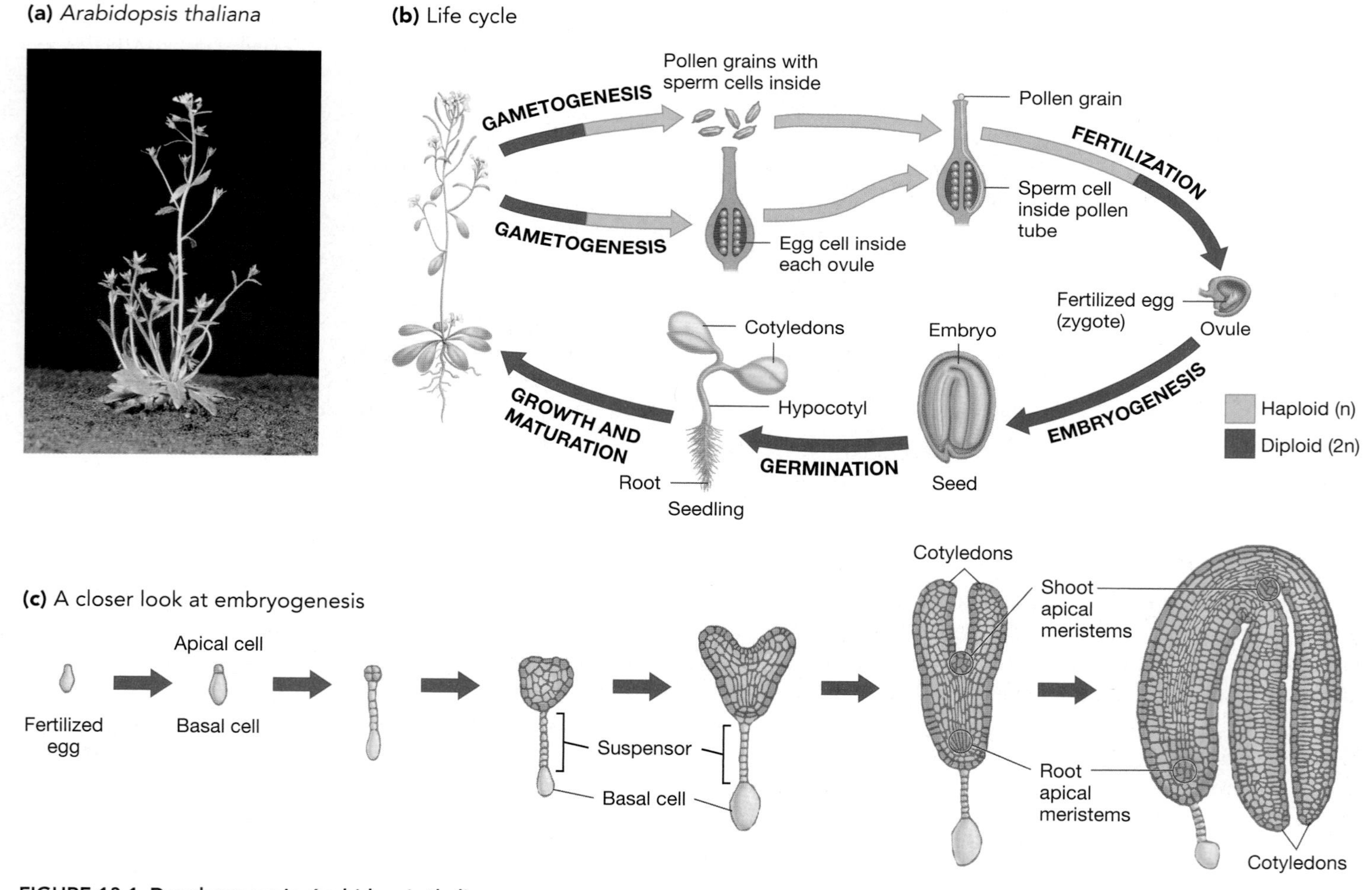

FIGURE 18.1 Development in *Arabidopsis thaliana*
(a) *A. thaliana* is an important model organism in developmental biology and genetics. **(b)** The life cycle of *A. thaliana* can take as little as 6 weeks under optimal conditions. As you can see, fertilization takes place inside the ovule. The embryo develops inside the ovule, which matures into a seed. **(c)** During embryogenesis, the root and shoot apical meristems form in addition to the first leaves (cotyledons) and root and a stem-like hypocotyl.

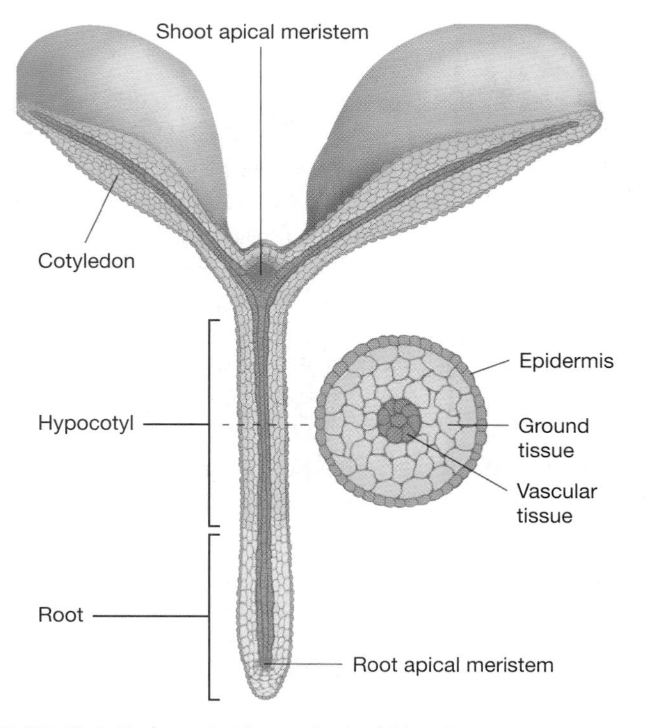

FIGURE 18.2 Embryonic Tissues in *Arabidopsis*
In addition to producing embryonic structures, embryogenesis in plants produces the three embryonic tissues illustrated here. The three tissue layers are organized in a radial fashion.

An Invertebrate Model: *Drosophila melanogaster*

Earlier chapters introduced the fruit fly *Drosophila melanogaster* and its role as a model organism in genetics. Here we consider how a fly is made.

In *Drosophila*, gametogenesis leads to the production of sperm and eggs in different individuals—not the same individual as in *Arabidopsis*. But fertilization occurs inside female reproductive structures, as it does in in *Arabidopsis*. A female fly then lays the fertilized eggs in rotting fruit or a similar medium so they develop outside of her body. In contrast, recall that in *Arabidopsis* the early stages of embryo formation occur while the seed is maturing and still attached to the parent plant.

As **Figure 18.3** shows, the nucleus of a fertilized fly egg repeatedly undergoes mitosis without cytokinesis occurring. The result is an embryo with many nuclei scattered throughout a cytoplasm filled with nutrient-rich **yolk**. Later, most nuclei migrate to the periphery of the embryo and become surrounded by a cell membrane.

Once the embryo is enclosed by a sheet of cells in this way, a dramatic change occurs. A long furrow forms on one side of the embryo, and cells on the periphery begin to move through it to the interior. These cell movements are called **gastrulation.** As gastrulation continues, another furrow forms and defines the head region; soon after, furrows form along the length of the body and demarcate the series of body regions called segments.

When gastrulation is complete and the larva matures, it has a functioning mouth and digestive tract and begins feeding on

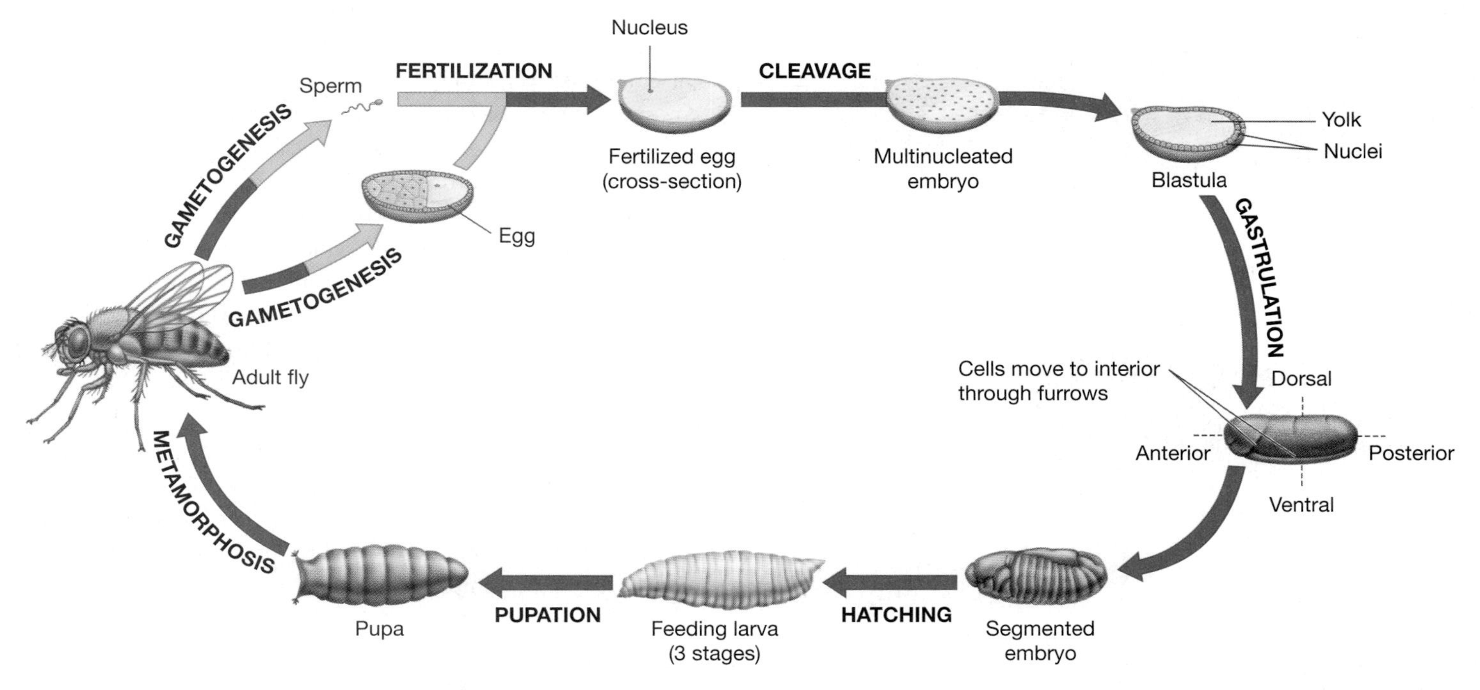

FIGURE 18.3 Development in *Drosophila melanogaster*
When fruit-fly embryogenesis is complete, the segmented embryo hatches into a feeding larva. After feeding for several days, the larva becomes quiescent, secretes a case, and becomes a pupa. During pupation, the entire body is remodeled into the adult form in a process called metamorphosis.

rotting fruit or other food sources. Then, during the pupa stage, its body is completely remodeled to form an adult in the process called **metamorphosis** (change-form).

Several general observations emerge from the details of these developmental steps. First, note that distinctive head and tail regions appear after gastrulation and define an anterior-to-posterior body axis that is reminiscent of the basal-to-apical or root-to-shoot axis of *Arabidopsis*. In addition, fly embryos have a strong back-to-belly or dorsal-to-ventral asymmetry that becomes apparent during gastrulation. The long furrow that initiates gastrulation identifies the ventral or belly side of the mature larva and the adult fly.

A second general observation about fly development is illustrated in **Figure 18.4**, which indicates a one-to-one correspondence between the segments that appear in the larva and the segments found in the adult. Despite the massive changes in shape and tissue organization that occur during metamorphosis, segmentation is fundamental to the organization of both the larval and adult body.

Finally, it is important to note that gastrulation is responsible for segregating cells into three embryonic tissues or germ layers. These tissues are called ectoderm, mesoderm, and endoderm. (Translated literally, the three terms mean "outside skin, middle skin, and inside skin.") Embryonic **ectoderm** forms the outer covering and nervous system of the larva and adult; **mesoderm** gives rise to muscle, internal organs, and connective tissues; and **endoderm** produces the digestive tract or gut and some of the associated organs.

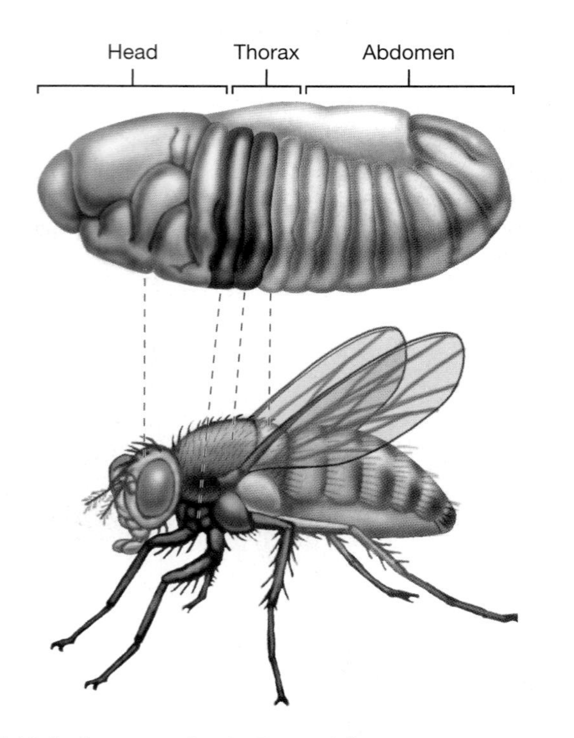

FIGURE 18.4 Segmentation in *Drosophila*
There is a one-to-one correspondence between the larval and adult segments of fruit flies. In addition to being segmented, the body of insects is organized into regions called the head, thorax, and abdomen.

How does fly development compare to the development of vertebrates such as the frog *Xenopus laevis*? Although the fossil record indicates that fruit flies and frogs last shared a common ancestor more than 600 million years ago, some features of fly and frog development are similar. Features like asymmetry, segmentation, and three embryonic tissues are also found in frog embryos. These are remarkable observations. Even though many features of development vary among animal species, certain aspects have been highly conserved during evolution. Let's take a closer look.

A Vertebrate Model: The Frog

Figure 18.5 shows some prominent stages in the development of a frog. At first glance, the sequence of events seems to bear little resemblance to development in an insect. Upon closer inspection, though, some important similarities emerge. For example, note that the first major event after fertilization of a frog egg is called cleavage. **Cleavage** consists of embryonic cell divisions that occur without cell growth. In effect, cleavage divisions partition the egg cytoplasm into several hundred cells. In many species, early cleavage divisions occur synchronously—meaning that all of the cells in the embryo divide at the same time.

As the drawings in Figure 18.5 show, cleavage in frogs results in a ball of cells which then hollows out at least partially prior to gastrulation. The hollow part of the spherical embryo is called a **blastocoel;** the embryo as a whole is called a **blastula;** the individual cells are called blastomeres.

In the frog, gastrulation begins when a round invagination called the **blastopore** forms. Sheets of cells from the exterior of the embryo move through the blastopore into the interior of the embryo. Cells that end up inside the embryo become mesoderm or endoderm; cells that remain on the exterior become ectoderm. The blastopore also makes the embryo asymmetrical. Because the blastopore eventually becomes the organism's anus, its formation is the first visible evidence of the body's anterior-to-posterior axis. As gastrulation ends, a structure called the neural tube begins to form along the embryo's back. The neural tube will become the spinal cord and brain of the larva and adult frog. In this way, the dorsal-to-ventral (back-to-belly) axis of the body takes shape. The formation of the neural tube is the key event preparatory to a developmental stage called **organogenesis**, which occurs after gastrulation.

The result of gastrulation in the frog, then, is identical to the result of gastrulation in the fly. Embryonic tissues separate into distinct layers or compartments and the major body axes take shape. Another striking similarity between fly and frog development emerges later, in the mesodermal tissues just below the neural tube of frogs. The band of mesoderm that forms in this region breaks into a series of segments called **somites**. These segments eventually give rise to the ribs and vertebrae of the larval and adult frog. Although the extent of segmentation in vertebrates is far less obvious than that observed in insects, it is noteworthy that embryonic tissues are organized in a similar way in the two groups.

The early stages of human development are also marked by cleavage, gastrulation, organogenesis, and the formation of body segments. In the case of humans and most other mammals, though, these events take place within the mother's body. Frogs, like flies, lay their eggs so that development takes place in the surrounding environment. But in some lizards, sharks, and fish—and in the vast majority of mammals—early development takes place inside the mother. In the case of humans, how does this happen?

Early Development in Humans

Like fruit flies and frogs, humans have separate sexes. Gametogenesis in males results in the production of the tiny, motile gametes called sperm; the analogous process in females results in larger, non-motile eggs. As **Figure 18.6** shows, fertilization and cleavage occur inside the reproductive tract of human mothers in a structure called the fallopian tube. The **fallopian tube**

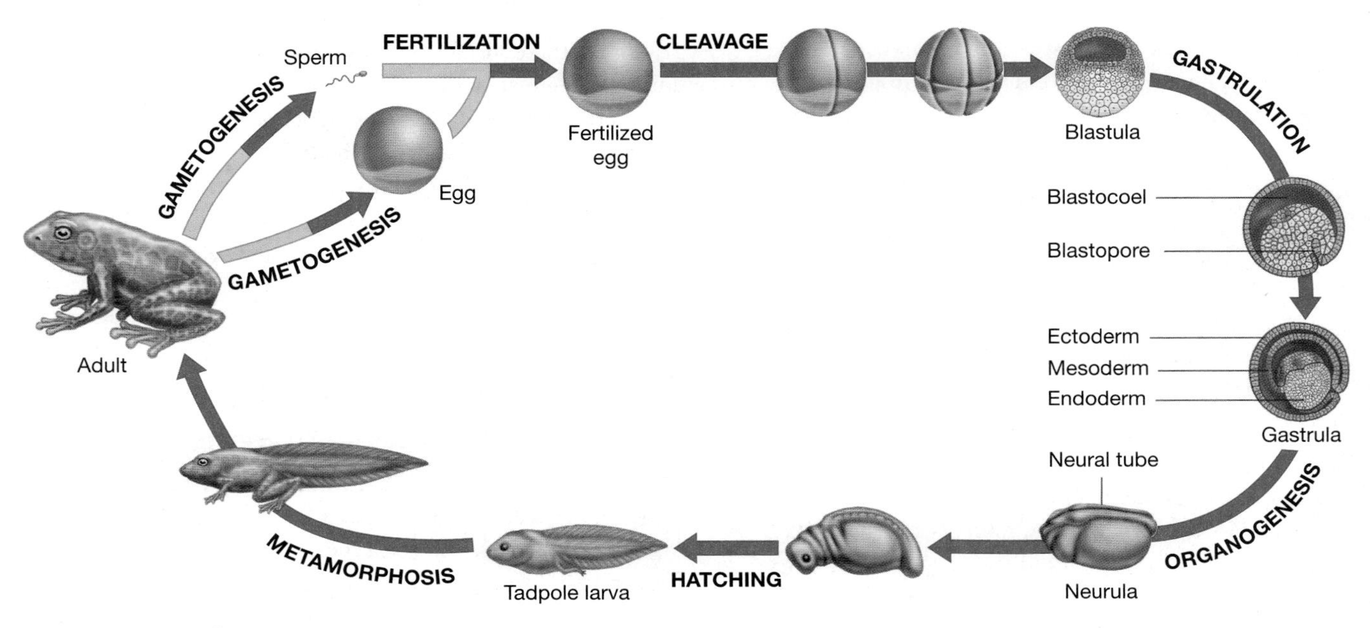

FIGURE 18.5 Development in a Frog
In vertebrates, early embryonic development consists of cleavage, gastrulation, and organogenesis. Frogs hatch into a larval stage called a tadpole. Although frogs undergo metamorphosis, they do not pupate.

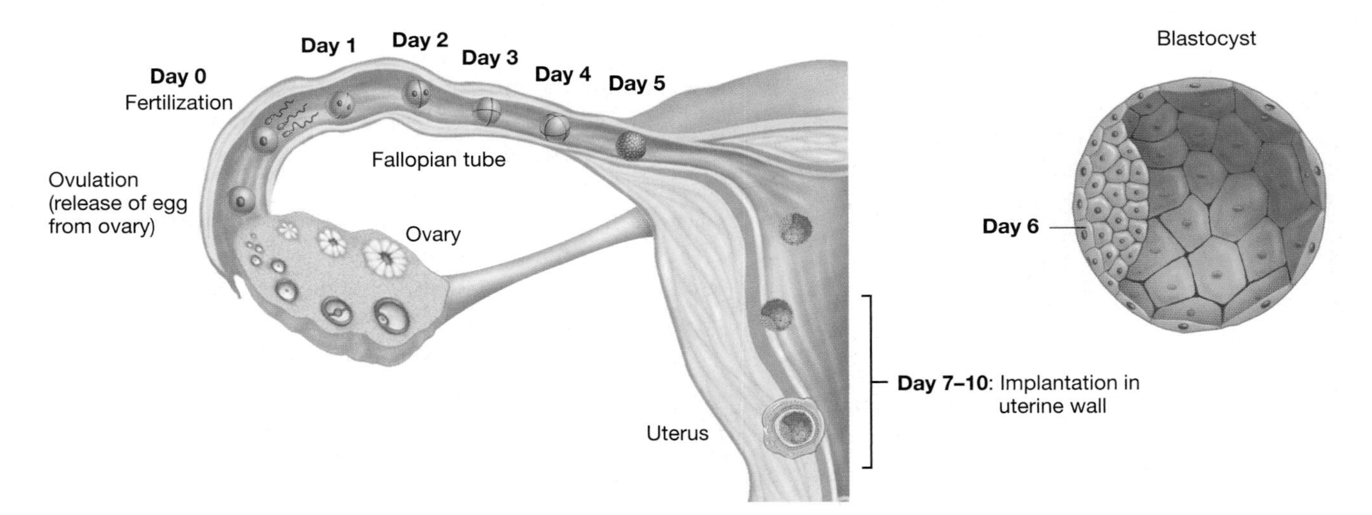

FIGURE 18.6 Cleavage in Mammals
In mammals, cleavage occurs prior to implantation of the blastocyst in the wall of the uterus. Once implantation occurs, gastrulation and neurulation occur and the placenta begins to form from a combination of embryonic and maternal cells.

connects the ovary and uterus. The **ovary** is where the egg matures; the **uterus** is where the embryo develops into a fetus. Cleavage occurs as the embryo travels down the length of the fallopian tube.

When cleavage is complete, the embryo becomes implanted into the wall of the uterus and the placenta begins to form. The **placenta**, derived from a mixture of maternal and embryonic cells, allows nutrients and wastes to be exchanged between the mother's blood and the embryo's blood. For this reason, human eggs are small compared to the eggs of frogs and largely lack a source of nutrients such as the endosperm of seed plants and the yolk of fly or frog eggs. Gastrulation and the formation of the neural tube occur after the embryo has become implanted into the uterine wall. Later events in pregnancy and birth are detailed in Chapter 45.

Making Sense of Variation and Similarity

Early descriptive studies of plant and animal development established that the sequence of events varies a great deal among species. At first glance, it seems puzzling that so much variation exists in a processes as fundamental as gametogenesis, fertilization, and early development. Why do so many differences exist? Part of the answer lies in the realization that the groups examined in this section have been evolving independently for hundreds of millions of years. Thus, it is not surprising that plants have a cell wall and do not undergo cleavage and gastrulation—even though all animals do. Even among animals, the patterns of cleavage and gastrulation vary markedly among groups that diverged long ago. As later chapters will show, insects and related groups have a very different way of undergoing cleavage and gastrulation than humans and their close relatives do.

Part of the variation observed in developmental patterns is undoubtedly due to natural selection. As you might recall from reading Chapter 1, natural selection occurs when individuals with certain genetically based traits produce more offspring than individuals without those traits. In the case of development, natural selection can favor certain patterns or events over others. For example, the evolution of larval forms in groups like insects and frogs was probably favored by natural selection because it allowed juveniles and adults of the same species to exploit different food sources and thus avoid competing with each other. The evolution of the placenta and pregnancy is thought to have been advantageous because of the protection it offers the developing embryo. Similarly, the seed is seen as a trait that protects the embryo during early development and allows it to be dispersed to a location away from the parent plant prior to germination.

Amid this variation, though, several common themes emerge about developmental patterns.

- In both plants and animals, the earliest events in development elaborate asymmetries in the embryo that establish fundamental aspects of the adult body plan. In plants, an asymmetrical initial division sets up the root-shoot body axis. In most animals, gastrulation makes the head-to-tail and back-to-belly axes of the embryonic and adult body visible.
- Although the details of how cleavage occurs vary among animal species, the outcome is the same. Cleavage divides the maternal cytoplasm and creates a sphere of cells that undergoes gastrulation.
- In both plants and animals, early development results in the formation of three embryonic tissues, which give rise to all adult tissues and structures. In animals, gastrulation is responsible for demarcating these three populations of cells.

Even as these principles were emerging from descriptive studies in the nineteenth century, biologists started to move beyond merely observing development. They wanted to know how and why certain events happened, and began using experimental approaches to find answers. One of the most fundamental questions about development inspired over 100 years of increasingly sophisticated experimentation. At a genetic level, are all of the cells in a multicellular organism alike?

18.2 Does the Genetic Makeup of Cells Change as Development Proceeds?

Cells become different from each other, or differentiate, as development proceeds. **Differentiation** is the process that results in the generation of diverse cell types. Cells can be identified as differentiated if they produce proteins that are specific to a cell type such as muscle, nerve, or skin. Among the most basic questions in developmental biology is how differentiation occurs.

Until the latter part of the nineteenth century, some investigators believed that differentiation was inconsequential in development; they thought that the organism is already completely preformed in miniature in the embryo. According to this **preformation** hypothesis, development of the mature organism involved only the growth of this smaller, completely formed version. In contrast, other biologists thought that the mature organism is produced gradually from an essentially formless embryo. This hypothesis is called **epigenesis.** Its earliest proponent was the Greek scholar Aristotle, who made primitive observations of embryonic development in the fourth century B.C. (Recall that microscopes were not available until the seventeenth century A.D.)

Two important events in the nineteenth century led to the acceptance of the epigenesis hypothesis and the rejection of preformation. First was a rapid improvement in microscopy, which allowed biologists to make detailed observations of embryonic development. Kasper Wolff, for example, observed early chick development and reported that recognizable anatomical structures are produced from relatively formless embryonic tissues. Specifically, he documented that the round intestinal tube forms from flat tissue. The second important event was the development of the cell theory. Once biologists accepted that all organisms consist of cells, they realized that preformation was not credible. Entire miniature organisms made of cells could not exist inside a single cell (the fertilized egg), and it was ludicrous to suppose that each of these miniature organisms contained even smaller versions of individuals within their own gametes. Based on these developments, epigenesis became widely accepted as the mechanism of embryonic development.

By the late nineteenth and early twentieth centuries, then, biologists had begun to view organisms as communities of cells and had started to focus on the behavior of cells within embryos. They verified that embryos are produced by a series of mitotic divisions, and they began to investigate how individual cells contribute to the maturing individual. The question of the day became, how does epigenesis occur? More specificially, how are the different cell types in the larval or adult body generated from a single-celled embryo?

The Germ Plasm Theory

One of the first attempts to explain the process of development was the **germ plasm theory**, proposed by August Weismann in 1883. According to Weismann, the cells of an embryo become specialized into different types because they inherit different sets of instructions called **determinants** during cleavage and subsequent cell divisions. Weismann referred to the complete set of instructions for an organism as the "germ plasm," and proposed that the determinants that make up the germ plasm are found in the chromosomes. Although Weismann published before the nature of the gene was understood, his proposal essentially claimed that different cells differ in their genetic makeup.

Figure 18.7 illustrates Weismann's ideas. He proposed that a single-celled embryo possesses a complete germ plasm. As development begins, however, the germ plasm breaks up with each cell division. The result is that each generation of daughter cells inherits progressively fewer determinants. According to Weismann, each cell in the body eventually contains just a single determinant. The determinant that remains specifies the production of a particular cell type.

Weismann also contended that a complete copy of the germ plasm is retained and sequestered in certain cells, which become the precursors to the sperm and eggs produced by the adult. These reproductive cells are called **germ-line cells**. According to Weismann, only the **somatic cells** that form the body of the organism undergo the progressive reduction in genetic information illustrated in Figure 18.7.

Weismann's theory predicted that because each daughter cell has less genetic information than the parent cell, cells are restricted in their developmental potential as early as the two-cell stage. After the first cleavage division, each cell is expected to contain only one-half of the determinants required to produce a complete embryo. Can this prediction be tested?

Early Tests of the Germ Plasm Theory Wilhelm Roux supported the germ plasm theory and set out to test its predictions experimentally. In 1888, Roux performed what may have been the first experiment exploring the mechanisms of embryonic development. He allowed frog embryos to undergo their first cleavage division and then attempted to kill one of the two resulting blastomeres by pricking it with a hot needle (**Figure 18.8**, page 362).

Roux found that the undamaged blastomere developed into a half-embryo. In contrast, the damaged blastomere did not develop and failed to produce the other half of the embryo. These results appeared to furnish strong support for the germ plasm theory. The production of a half-embryo from the undamaged cell suggested that it contained exactly one-half of the determinants necessary to produce a complete embryo.

Roux also made an observation that was inconsistent with the germ plasm theory, however. He found that cells derived from the undamaged blastomere could migrate to the damaged side of the embryo and develop into some of the missing structures. This observation suggested that rather than having

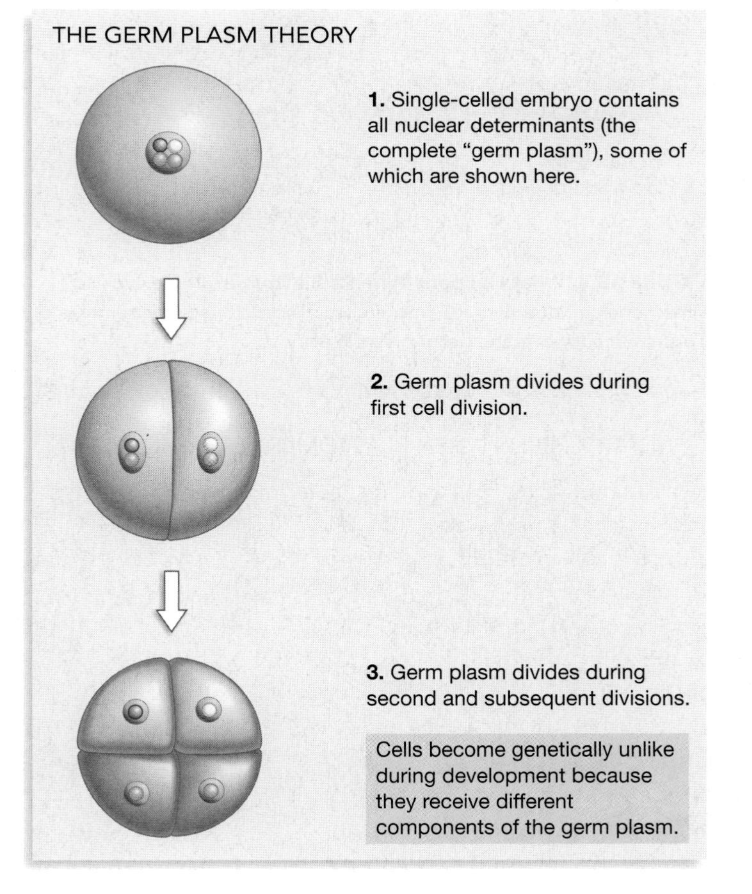

FIGURE 18.7 The Germ Plasm Theory
According to the germ plasm theory, nuclear determinants are progressively divided into the daughter cells of cleavage and subsequent cell divisions. The fate of a cell depends on the nuclear determinant that it receives.

just half of the germ plasm, as predicted, those blastomeres had all of the determinants (or genes) required for normal development. Despite this result, Roux continued to champion the germ plasm theory.

In 1892, however, Hans Driesch published additional evidence against the germ plasm theory. In experiments with sea urchin embryos, Driesch found that he could physically separate blastomeres after the first cleavage division to form two independent cells. He did this by shaking the embryo violently or by soaking it in seawater that lacked calcium. When he did so, each blastomere developed into a complete larva.

Driesch also found that blastomeres isolated from a four-cell embryo could form small but complete larvae (**Figure 18.9**). These experiments contradicted Weismann's hypothesis that nuclear determinants segregated into daughter cells. Instead, the blastomeres produced early in sea urchin development appeared to be equivalent in their potential to produce a complete organism. In other words, each blastomere appeared to have the same, complete complement of genes.

Genetic Equivalence During Cleavage The results published by Roux and Driesch were contradictory. Between 1901 and 1903, Hans Spemann attempted to clarify the situation by experimenting with newt embryos—an amphibian species like Roux's frogs. Spemann separated blastomeres at the two-cell stage by tying a loop of hair from a human baby around the egg and drawing it tight. When he did so, he found that each blastomere could form a complete tadpole. J. F. McClendon

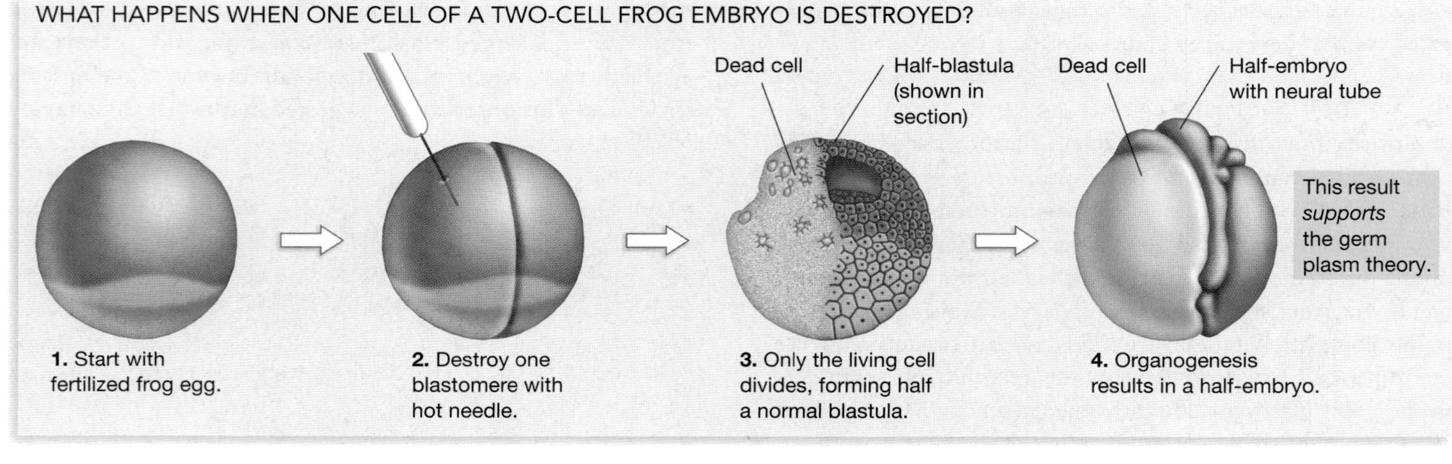

FIGURE 18.8 What Happens When a Blastomere Is Destroyed?
When one cell of a two-cell frog embryo is destroyed, the remaining cell develops into one-half of a larva. This result supports the germ plasm theory.

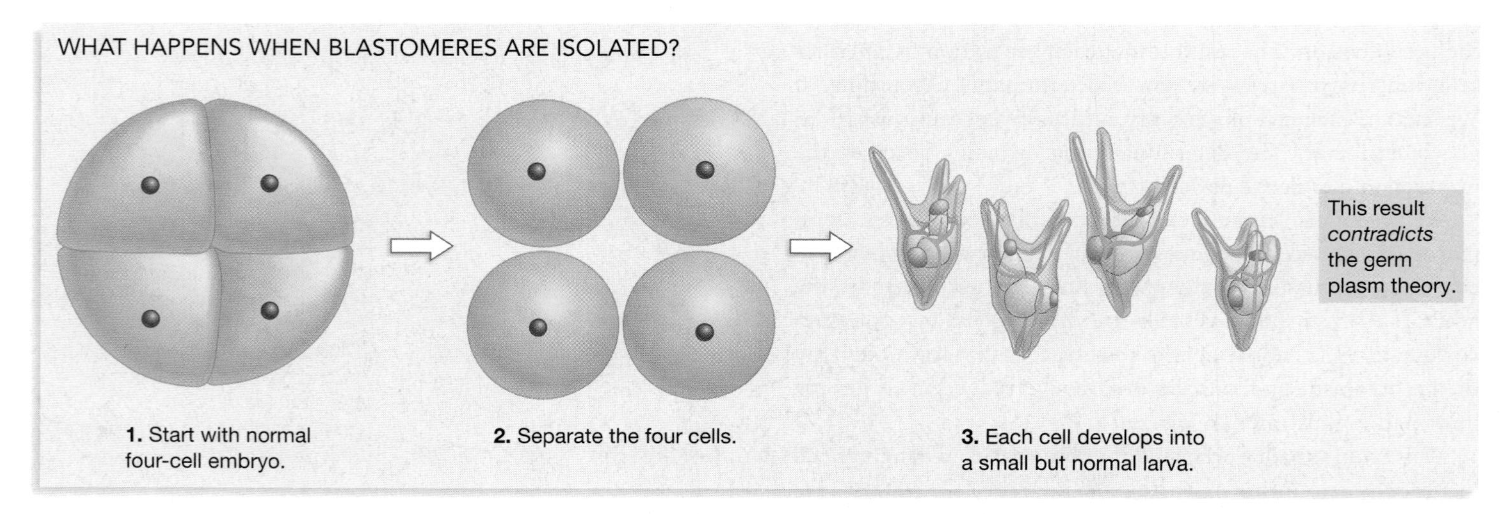

FIGURE 18.9 What Happens When Blastomeres Are Isolated?
When sea urchin blastomeres are separated at the 4-cell stage, each develops into a small but normal larva. This result contradicts the germ plasm theory.

reported similar results in 1910 using the same frog species that Roux studied. Unlike Roux's pricking experiment, the constriction experiments indicated that there is no difference in developmental potential between blastomeres at the two-cell stage. These results supported the conclusion that blastomeres have the same nuclear determinants.

If newt cells were equivalent at the two-cell stage, did they eventually become non-equivalent? To answer this question, Spemann began exploring the potential of nuclei produced at later stages of newt development. He tied a baby hair around a fertilized egg and tied a knot, but did not pull it tight. As **Figure 18.10** shows, this procedure caused a partial constriction and confined the nucleus to one side of the embryo. The side containing the nucleus underwent cleavage, while the side missing the nucleus did not. If the constriction was loose enough, however, a nucleus sometimes slipped through to the non-nucleated side. When this happened, Spemann pulled the knot tight, preventing further exchange of nuclei between the two halves. In one experiment, a nucleus from a cell at the 32-cell stage crossed over to the non-nucleated side. After Spemann pulled the knot tight and followed subsequent development, he found that both halves of the constricted embryo produced normal tadpoles (see Figure 18.10). He concluded that at the 32-cell stage, the nuclei of blastomeres still contain all the information required to produce an entire organism.

Reconciling Conflicting Results Based on his observations, Spemann rejected the germ plasm theory. Weismann's hypothesis predicted that differences in the information content of nuclei should exist as early as the two-cell stage, but Spemann observed no differences among nuclei at the 32-cell stage.

Why did Roux, Driesch, and Spemann produce such conflicting results? One explanation focuses on their experimental designs. Roux tried to isolate individual blastomeres by destroying adjacent cells. Driesch and Spemann attempted to achieve the same result by physically separating the two blastomeres. The essential difference is that the two blastomeres remained in contact in Roux's experiment; in Driesch's and Spemann's they did not. In light of Spemann's results, it became clear that some sort of interaction must be occurring between the live and dead blastomeres in Roux's experiment. In other words, somehow the pricked blastomere was still influencing the fate of the other blastomere.

Even though Roux's experimental design contained a flaw that produced an incorrect conclusion, his results revealed an important message. Contact between cells matters, because cells influence each other during development. How this cell-to-cell contact and influence occurs is a major theme of Chapters 19 and 20. In addition, one of Weismann's key proposals actually proved to be correct, even though his germ plasm theory was rejected. In animals, the cells destined to become the precursors of sperm and egg cells are sequestered very early in development. Stated another way, germ-line and somatic cells become distinct in most animal species soon after cleavage begins. In contrast, germ-line cells differentiate from somatic cells very late in plant development. In the case of

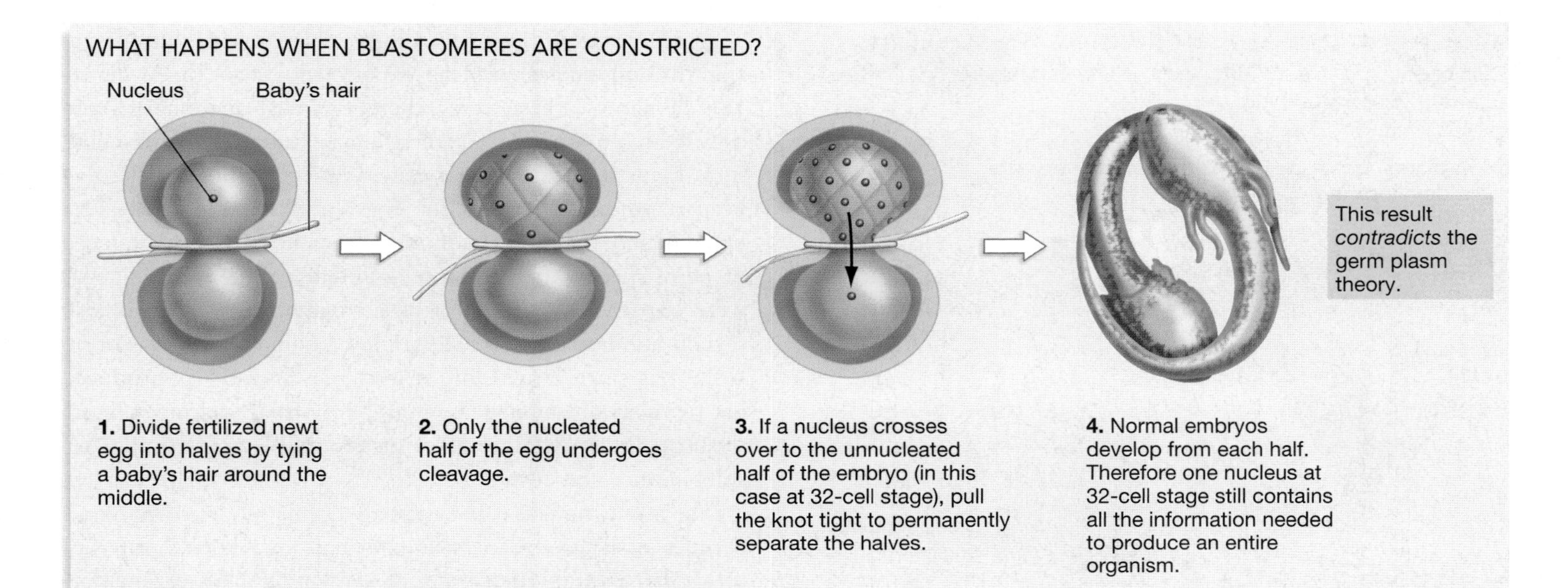

FIGURE 18.10 What Happens When Blastomeres Are Constricted?
By tying a baby's hair around a fertilized newt egg, a researcher was able to divide it into two halves. Only the nucleated half of the egg underwent cleavage. In this experiment, a nucleus crossed over to the non-nucleated half of the embryo at the 32-cell stage. The researcher then pulled the knot tight to permanently separate the two halves. Normal embryos developed from each half. This result contradicts the germ plasm theory.

Arabidopsis, germ cells arise from somatic cells during flower formation (**Figure 18.11**).

Despite the success of Spemann's work, it was still not clear that the genetic content of cells remained equivalent after cleavage is complete. Is the genetic makeup of differentiated cells such as muscle, nerve, and skin identical? In 1938, Spemann proposed an experiment to test this possibility. His idea was to remove the nucleus from a differentiated cell in an adult frog and transfer it to an egg whose own nucleus had been removed. Spemann predicted that the egg would develop normally. Unfortunately, years would pass before technological advances made this experiment possible.

Are Highly Differentiated Cells Genetically Equivalent?

In the 1950s, Robert Briggs and Thomas King developed techniques for isolating nuclei from leopard frog cells and transferring them into unfertilized eggs whose nuclei had been removed. In a series of experiments, they found that nuclei from an embryo undergoing cleavage or organogenesis could direct development of complete tadpoles. In contrast, donor nuclei taken from tadpoles were unable to do so. These results suggested that cells may lose at least some genetic information late in development.

(a) In animals, germline cells differentiate very early in development.

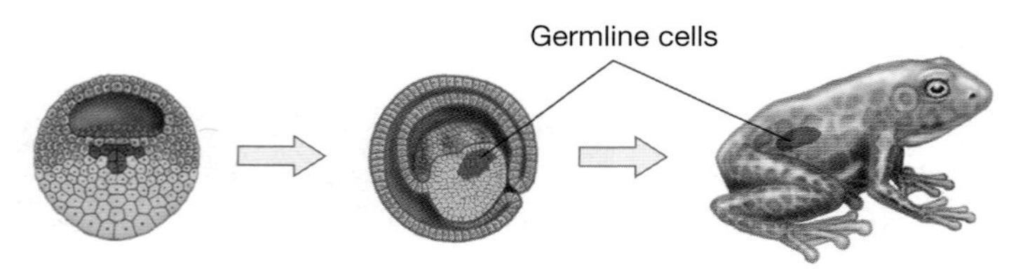

(b) In plants, germline cells differentiate very late in development.

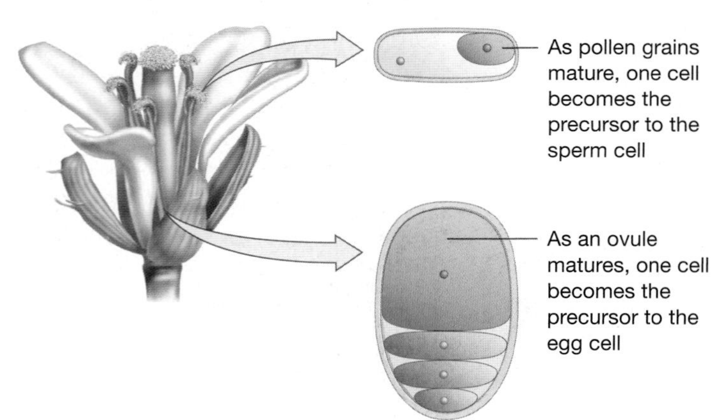

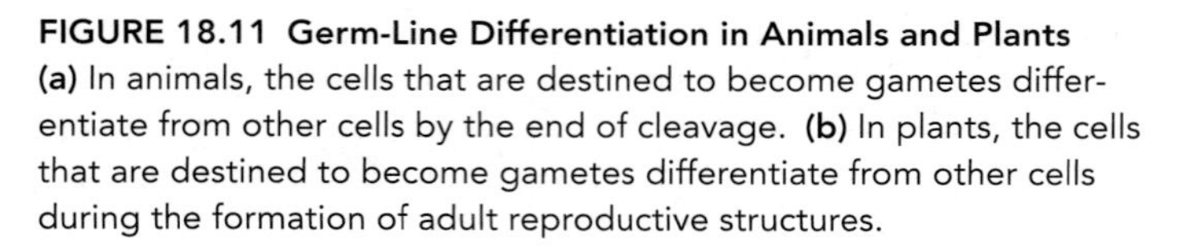

FIGURE 18.11 Germ-Line Differentiation in Animals and Plants
(a) In animals, the cells that are destined to become gametes differentiate from other cells by the end of cleavage. **(b)** In plants, the cells that are destined to become gametes differentiate from other cells during the formation of adult reproductive structures.

The Briggs and King experiment was criticized, however, because the lack of development directed by tadpole nuclei represented a negative result. A negative result can always be criticized on the grounds that something was wrong with the experimental technique. For example, some researchers contended that the older nuclei used by Briggs and King might have complete genetic information but fail to support early development because they were more easily damaged during transfer.

These criticisms were addressed in a series of experiments that John Gurdon and co-workers carried out during the 1960s using the African clawed frog *Xenopus laevis*. In one dramatic experiment, Gurdon's group showed that nuclei from the intestinal lining of young tadpoles could direct the development of complete tadpoles. In some cases, the tadpoles succeeded in metamorphosing into fertile, adult frogs. Although this experiment seemed to show conclusively that differentiation occurs without changes to the genetic makeup of cells, critics were not satisfied. Some researchers claimed that Gurdon and co-workers had not been careful enough in defining the source of the donor nuclei. This was a legitimate concern because germ-line cells, which do not undergo differentiation, can be closely associated with the intestine of tadpoles. Other investigators contended that intestinal cells from adults, rather than tadpoles, should have been used. To conclusively show that differentiation occurs without any loss of genetic information, Gurdon's group cultured skin cells that were actively producing a skin-specific protein called keratin—the same protein that makes up the bulk of human hair. When they succeeded in producing tadpoles from eggs that received skin-cell nuclei, most critics were convinced that mature cells are genetically equivalent.

In 1997, Ian Wilmut and colleagues reinforced this conclusion through nuclear transfer experiments in sheep. As **Figure 18.12a** shows, these researchers removed mammary-gland cells from a 6-year-old pregnant female and grew them in culture. Later they fused these cells with eggs whose nuclei had been removed. Note that the eggs came from a black-faced breed of sheep while the donor nuclei came from a white-faced breed of sheep. After developing in culture, the resulting eggs were implanted in the uteri of surrogate mothers. In one of several hundred such transfers, a lamb named Dolly was born. Dolly has since grown into a fertile adult and has produced her own lamb named Bonnie by normal mating (**Figure 18.12b**). In 1998, other research groups reported similar results in mice and cows.

The same outcome had actually been reported in plants decades earlier. Gardeners and farmers had known for centuries that in many plant species, new individuals can be produced from a section of root or shoot. Then during the 1950s and 1960s, a series of researchers succeeded in showing that in culture, single differentiated cells isolated from mature plants can develop into independent mature plants that function normally. Taken together, work on cloning plants and animals has shown that the process of cellular differentiation does not in-

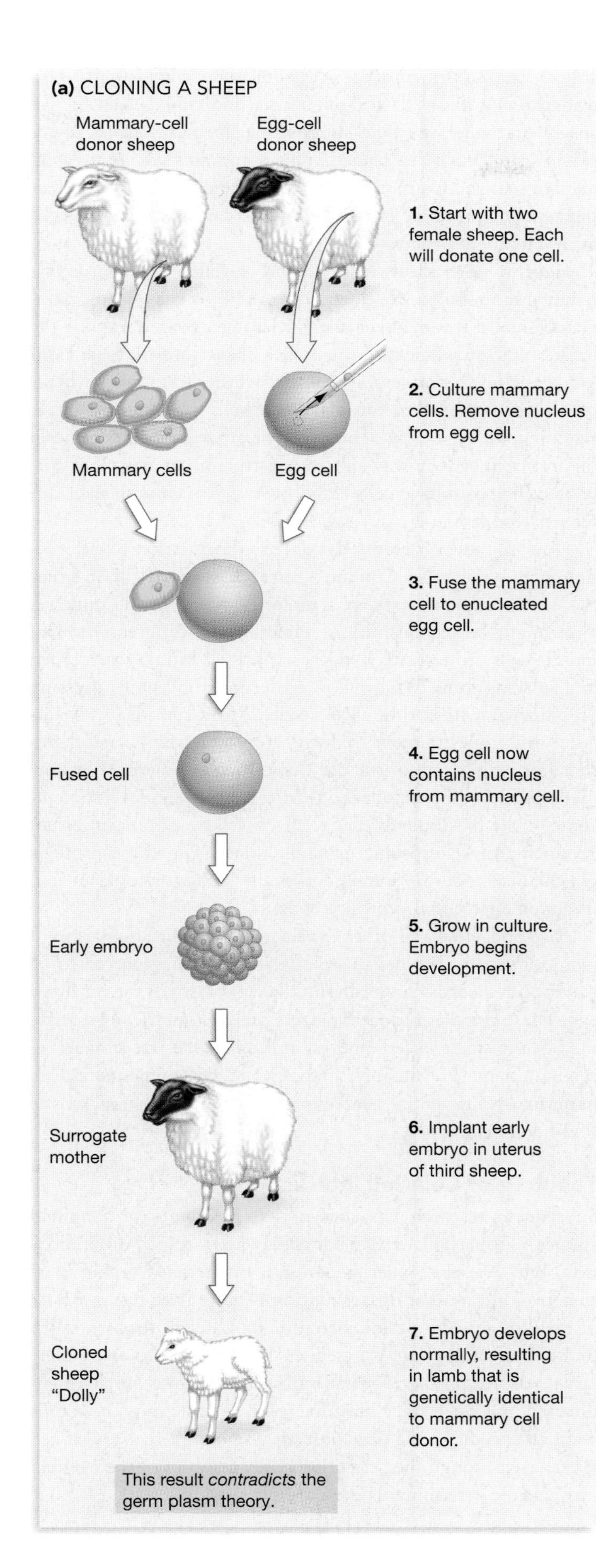

FIGURE 18.12 Cloning Mammals
(a) When the nucleus from a differentiated mammary-gland cell was transferred to an enucleated egg, normal development occurred. The resulting lamb resembled the breed that the donor nucleus came from, and not the breed of the egg donor or the surrogate mother. DNA testing also showed that the lamb was genetically identical to the individual that donated the nucleus. **(b)** Dolly, the sheep produced by cloning, was able to produce her own lamb through normal sexual reproduction.

volve changes in the genetic makeup of cells. But there is an important exception to this rule. In humans and other mammals, small stretches of DNA are rearranged in certain immune-system cells late in development. As a result, many immune cells are genetically unique.

To summarize, over a century of experimentation resulted in the rejection of the germ plasm theory and the realization that development does not normally involve changes in the genetic makeup of cells. Now the question is, if cells are genetically equivalent, how do different cell types arise? The answer is that cell types differ because they produce different sets of proteins. At its core, then, the process of differentiation involves the highly regulated expression of different subsets of genes from the common and complete set of genes found in all cells of the embryo. But what controls this **differential gene expression**? Important clues to an answer came from research in the early part of the twentieth century.

18.3 What Causes Differential Gene Expression?

The cells that make up an embryo must begin to express distinctive subsets of genes at some point during the progression through cleavage, gastrulation, and organogenesis. To understand when differential gene expression might begin and why, researchers needed to link specific cells with specific destinies or fates. For example, suppose an investigator could identify a particular cell in a blastula based on its position. Is this cell destined to become nerve, muscle, or skin? When the destiny of the cells in an embryo is described in this way, the result is called a **fate map**. Could compiling a fate map provide clues to the cause or causes of differential gene expression? The answer turned out to be yes.

The Discovery of Cytoplasmic Determinants

One of the first fate maps ever devised turned out to have important implications for understanding how differential gene expression is regulated in embryos. The investigator was Edwin Conklin, and his study organism was the species of sea squirt shown in **Figure 18.13a**. Sea squirts belong to a group of marine organisms that are closely related to the vertebrates, although they look quite different as adults. Besides being easy to collect, sea squirt embryos are somewhat transparent. This transparency allowed Conklin to see individual cells in live specimens under the light microscope. In addition, the cytoplasm inside each sea squirt embryo contains differently pigmented regions. Conklin realized that the presence of different pigments might allow him to follow the fate of individual cells as development progressed.

(a) The sea squirt *Stylea partita*

(b) Mapping the fate of pigment cytoplasm

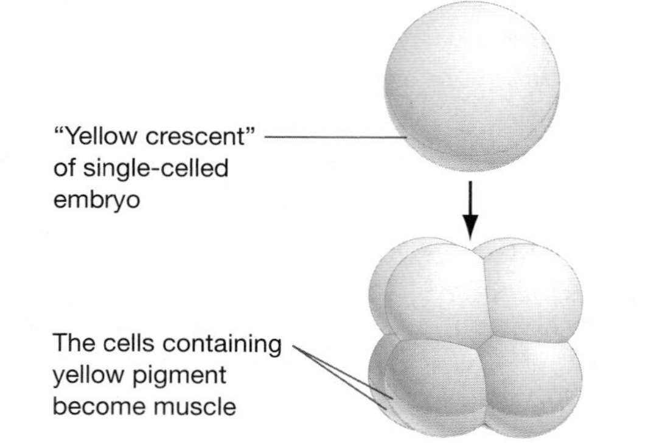

FIGURE 18.13 Do Molecules in the Egg Cytoplasm Influence Development?
(a) The sea squirt *Stylea partita* has eggs that contain distinctive pigments. **(b)** The yellow pigment found in sea squirt eggs is associated with the development of muscle cells.

Figure 18.13b shows how a band of yellow pigment is distributed in the single-celled embryo, and how certain embryonic cells inherit this pigment during cleavage. By determining the location of the pigment in the bodies of sea squirt larvae, Conklin concluded that embryonic cells inheriting yellow cytoplasm were destined to become muscle cells in the mature larva. By doing the same type of analysis for other pigments found in the fertilized egg, he was able to construct a fate map that correlated the position of cells early in development with their differentiated state in larvae.

How do these observations relate to the question of differential gene expression? Conklin's research suggested that either the yellow pigment itself or a molecule found in the same region might be responsible for directing the differentiation of muscle cells. To test this hypothesis, J. R. Whittaker succeeded in displacing some of the yellow crescent cytoplasm by pressing on embryos with a tiny glass needle. More specifically, Whittaker was able to move yellow cytoplasm from blastomeres that normally produce muscle tissue to blastomeres that normally produce only ectoderm. During subsequent development, some of the blastomeres that received yellow cytoplasm began producing muscle-specific proteins in addition to the proteins typical of ectoderm. Something in the yellow cytoplasm had triggered differential gene expression.

Based on these and other experiments, the differentiation of many or most cell types in sea squirt embryos appeared to be based on regulatory molecules in the cytoplasm of the fertilized egg. These **cytoplasmic determinants** are asymmetrically distributed in the single-celled embryo and direct the fate of cells receiving them. Presumably, cytoplasmic determinants induce patterns of differential gene expression that are characteristic of different cell types.

The Role of Cell-Cell Interactions

Subsequent research has shown that in an array of animal species, cytoplasmic determinants play a key role in differentiation. But Driesch's experiments with isolated sea urchin blastomeres indicate that differentiation is more than just a matter of dividing up cytoplasmic determinants. Recall that when sea urchin blastomeres are isolated, each develops into a complete larva. Also recall that Spemann observed normal development directed by 32-cell stage blastomeres in newt larvae. These cells were able to direct the development of a complete array of cell types, even though they contained only a subset of the original cytoplasmic determinants present in the egg.

What was going on? Instead of emphasizing the importance of cytoplasmic determinants, the experiments of Driesch, Roux, and Spemann revealed the importance of cell-cell interactions during development. Cell-to-cell interactions are the only reasonable explanation for why sea urchin blastomeres develop differently when they are isolated than they do in the intact embryo. A blastomere from an embryo at the two-cell stage produces only part of the larva, but when isolated it produces the entire structure. In frogs, newts, and sea urchins, the fate of a cell depends at least partly on cues from neighboring cells.

To summarize, work by an array of early biologists suggested that molecules inherited from the egg cytoplasm or received from neighboring cells cause profound changes in which genes are expressed in a particular cell, and thus determine its fate. Cues based on cytoplasmic determinants and cues based on interactions with neighboring cells represent the two basic mechanisms of cell differentiation. As the cytoplasm in the egg is divided during cleavage, cytoplasmic determinants are being sequestered in particular cells. As cells move to new locations during gastrulation or find themselves in a specific location in a plant embryo, the nature and extent of cell-to-cell interactions changes radically.

How are cytoplasmic determinants localized in the fertilized egg? What is the molecular nature of cell-to-cell signals? Are cytoplasmic determinants and cell-to-cell signals involved in fundamental events like the establishment of the primary body axes and the formation of embryonic tissues? Now that we've surveyed developmental patterns in an array of species and identified differential gene expression as the basic feature of development, we're ready to take up these more detailed questions.

Essay Human Cloning?

Biologists can harvest nuclei from the somatic cells of mice, cows, and sheep and use them to produce embryos. These nuclear transfer experiments produce offspring that are genetically identical to the individual providing the donor nucleus. A group of genetically identical organisms is called a clone, so the process of creating individuals by transferring nuclei into enucleated eggs is called cloning. Can humans be cloned?

Before addressing the technical feasibility of cloning humans, let's consider a more practical issue. Why would anyone want to clone a human? Proponents of the technology contend that cloning by nuclear transfer would allow otherwise infertile couples to have offspring that are genetically related—in fact, identical—to one of the parents. Another argument is that cloning would allow a family to replace a dying child. Opponents of the technology are concerned that, in an attempt at immortality, a dictator or wealthy eccentric could finance clones of themselves in perpetuity.

Can humans be cloned?

Even benign uses of human cloning create ethical dilemmas that are yet to be thoroughly debated or resolved. What is the effect on a family of having children that are genetically identical to a parent? Would the child be less likely to respect and obey that parent? Would the child suffer from the expectation that it should turn out exactly like that parent? Given the number and seriousness of the questions involved, several nations have proposed a moratorium on all research related to cloning humans.

In addition to the legal and ethical issues involved, there are significant technical hurdles to overcome before human cloning is feasible. Of the species that have been successfully cloned, monkeys and mice are most closely related to humans. Less than three percent of nuclear transfers result in viable mice, however. Most result in aborted pregnancies; some produce individuals that die soon after birth. Currently, the success rate is similar in cloning sheep and cows and production of deformed newborns has occurred. Most people regard a failure rate this high as unacceptable for humans, especially given the high likelihood of producing grossly deformed babies.

As cloning techniques improve, however, human cloning may become technologically feasible. As this book goes to press, a group of researchers are beginning to act on a proposal to attempt cloning in as many as 600 human couples. Meanwhile, the nations of the world are debating the merits of the technology and deciding whether it will be regulated or banned outright.

Chapter Review

Summary

The pattern and rate of embryonic development vary widely among species of plants and animals. Cleavage and gastrulation are nearly universal in animals, but gastrulation does not occur at all in plants. The fate of most animal cells is determined early in development, while specialized cells and tissues continue to differentiate from meristematic tissues throughout the life of a plant. Among animals, some species develop into distinct larval and adult stages while others do not. The time to adulthood may be extremely fast, as in *Arabidopsis thaliana* or *Drosophila melanogaster*, or slow as in humans.

Several common themes unify the diversity of ways that plants and animals develop, however. In both plants and animals, the earliest events in development make the major body axes visible and result in the formation of embryonic tissues, which later give rise to distinct larval or adult structures. In animals, cleavage divides the egg cytoplasm and produces a hollow or semi-hollow sphere of cells that undergoes gastrulation; gastrulation arranges embryonic tissues into layers and makes the major body axes visible. In some animal groups, all or part of the body is organized into distinctive segments.

Once the basic developmental events had been described in an array of organisms, biologists began using experiments to understand how different cell types arise from a single-celled embryo. One early challenge was to understand whether all of the cells in a mature larva or adult are genetically identical. The genetic equivalence of cells was established by isolating blastomeres from early or late-stage blastulas and showing that these nuclei could direct the development of a complete sea urchin, newt, or frog embryo. Later, a series of nuclear transplant experiments culminated in the cloning of sheep, mice, and cows from somatic cell nuclei taken from adults.

After establishing that virtually all cells in adult plants and animals are genetically identical, biologists focused on the question of how differential gene expression occurs in different cell types. Efforts to test the germ plasm theory highlighted the importance of cell-to-cell interactions in differentiation. The basic observation was that when cells are isolated from other cells, they develop differently from cells that are in contact with other cells. Based on this result, biologists inferred that cells exchange molecular signals among themselves, and that these signaling molecules affect the expression of genes and the fate of cells during development.

In contrast, efforts to map the fate of specific cells during development pointed to the importance of cytoplasmic determinants in the differentiation of some cell types in certain species. In combination, then, cytoplasmic determinants present in the fertilized egg and cell-to-cell signals direct development by causing changes in gene expression. Exploring how these two types of signals work is the subject of Chapters 19 and 20.

Questions

Content Review

1. What happens during cleavage?
   a. The neural tube—precursor to the spinal cord and brain—forms.
   b. Basal and apical cells form, which are the precursors to the suspensor and embryo, respectively.
   c. The fertilized egg divides without growth occurring, forming a ball of cells.
   d. Massive movements of cells make the primary body axes visible and organize the three embryonic tissues.

2. What happens during gastrulation?
   a. The neural tube—precursor to the spinal cord and brain—forms.
   b. Basal and apical cells form, which are the precursors to the suspensor and embryo, respectively.
   c. The fertilized egg divides without growth occurring, forming a ball of cells.
   d. Massive movements of cells make the primary body axes visible and organize the three embryonic tissues.

3. What happens during organogenesis?
   a. The neural tube—precursor to the spinal cord and brain—forms.
   b. Basal and apical cells form, which are the precursors to the suspensor and embryo, respectively.
   c. The fertilized egg divides without growth occurring, forming a ball of cells.
   d. Massive movements of cells make the primary body axes visible and organize the three embryonic tissues.

4. What is epigenesis?
   a. the belief that a mature organism is already present in miniature in the embryo or gametes
   b. the theory that early embryonic development is controlled by the zygote's genes
   c. the theory that early embryonic development is controlled by maternal genes
   d. the process by which the mature organism is produced gradually from an essentially formless embryo

5. What is a fate map?
   a. a description of an individual organism's fate
   b. a description of the fate of each embryonic region or cell
   c. a list of cell fates that are possible only in animal embryos
   d. a list of cell fates that are possible only in tunicate embryos

6. What is differential gene expression?
   a. the process by which different cell types express different combinations of genes from the common set found in nearly all cells
   b. the fact that different cell types contain different sets of genes in their DNA
   c. the asymmetric segregation of nuclear determinants during early embryonic cleavages
   d. the loss of chromosomes during development

Conceptual Review

1. How did efforts to map cell fates suggest that cytoplasmic determinants are important during development?
2. In humans, identical twins arise if a single embryo splits in two early in development. How does this observation compare with the results of the blastomere-isolation experiments performed on sea urchins and newts? Explain.
3. Why did acceptance of the cell theory in the nineteenth century led to rejection of the theory of preformation?
4. How does the cloning of Dolly the sheep show that differentiation occurs without loss of genetic information from the nucleus?
5. Explain why blastomere-pricking and blastomere-isolation experiments led to different results. How did efforts to reconcile the conflicting results lead researchers to realize that cell-to-cell interactions are important during development?
6. Compare and contrast the early development of *Arabidopsis* with *Drosophila melanogaster*. How are the rates and patterns of development similar? How are they different?

Applying Ideas

1. For many decades it was difficult for researchers to do cell-isolation, cell-destruction, or other types of manipulative experiments with plant embryos, because they are encased in an ovule or seed coat. Now, however, plant embryos can be grown in culture. Suggest a manipulative experiment to try with plant embryos. What question is your experiment trying to answer? How would your experiment help answer the question?
2. Suppose you could add a chemically inert dye to blastomeres. How could this dye help you to construct a fate map for a particular organism?
3. In a lizard species native to north-central Mexico, some populations lay eggs while others bear live young. How could you use these populations to test the hypothesis that bearing live young is favored by natural selection because developing offspring are better protected?
4. Suppose there is a drug that leads to direct development when administered to frog eggs—meaning that the larval (tadpole) stage is skipped. How could you use this drug as an experimental tool, to test the hypothesis that natural selection favors the development of larval forms because of reduced competition with adults for food?

CD-ROM and Web Connection

CD Activity 18.1: Early Stages of Animal Development *(tutorial)*
(Estimated time for completion = 5 min)
Compare the stages of development in amphibians, reptiles, birds, and mammals.

At your **Companion Website** (http://www.prenhall.com/freeman/biology), you will find self-grading exams and links to the following research tools, online resources, and activities:

Virtual Library of Developmental Biology
Explore the many aspects of developmental biology and embryology.

Virtual Embryo
This site provides a large set of resources on the development of embryology, and the key concepts explored in this chapter.

Society for Developmental Biology
The Society for Developmental Biology promotes the study, education, and dissemination of knowledge and research in developmental biology.

Additional Reading

McLaren, Anne. 2000. Cloning: Pathways to a pluripotent future. *Science* 288: 1775–1780. An essay on the prospects for cloning humans and other animals.

Silver, L. M. 1998. *Remaking Eden: How Genetic Engineering and Cloning Will Transform the American Family.* (New York: Avon Books)

Smith, B. R. 1999. Visualizing human embryos. *Scientific American* 280 (March): 76–81. An introduction to how magnetic resonance microscopy is being used to study the events of early human development.

Willier, B. J., and J. M. Oppenheimer. 1964. *Foundations of Experimental Embryology*. Englewood Cliffs, NJ: Prentice-Hall, Inc. A collection of classic research papers in experimental embryology, including those of Roux and Driesch, with editor's comments.

Early Development

19

For a biologist, few events inspire as much wonder as watching a fertilized egg go through early development. In animals, the zygote divides itself up during cleavage. The resulting cells then begin the massive, coordinated movements of gastrulation, and the embryo is dramatically reorganized. Just a few hours or days later, organogenesis proceeds and a recognizable creature with a head, eyes, back, belly, or other structures appears. In plants, the fertilized egg divides to form two simple-looking cells—one small and one large. The small cell divides repeatedly to form a seemingly disorganized clump of cells. But as this clump grows and lengthens, tiny leaves and a root appear. A mature seedling takes shape a short time later, ready to germinate and begin a life that in some species may span several thousand years.

Understanding how these events happen is among the greatest of all challenges facing biologists today. Chapter 18 introduced this challenge by providing an overview of key events in the life of *Arabidopsis*, *Drosophila*, frogs, and humans. As you learned in that discussion, the establishment of the major body axes and the formation of embryonic tissues are among the most important of all early developmental events. Chapter 18 also began exploring how these key events happen. In particular, the discussion focused on how genetically identical cells end up with very different fates. What determines the future of an embryonic cell? Early experimental and observational studies pointed to a role for molecules present in the egg cytoplasm and for cell-to-cell signals that occur later in development.

With this background in hand, we're ready to explore the events and processes of early development in detail. What are cytoplasmic determinants, and how are they localized to

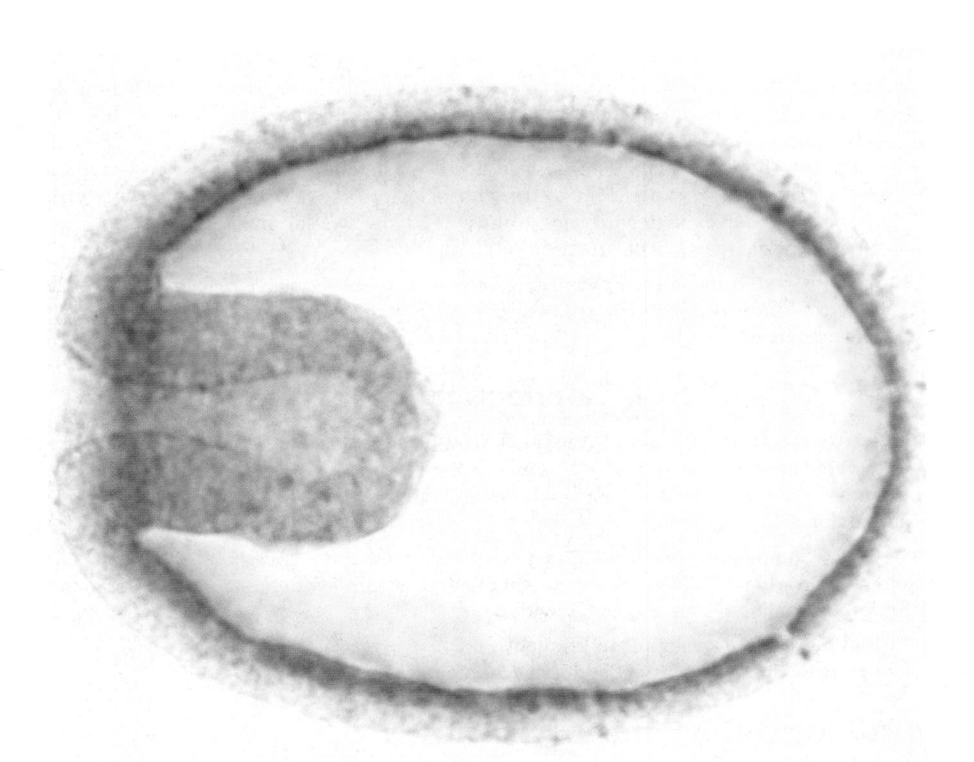

The photo shows a sea star embryo undergoing a process called gastrulation. The structure on the left consists of cells that are moving into the hollow center of the embryo. When the invagination is complete, the cells will begin to form the gut and other structures. This chapter introduces the critical early events of development, beginning with the structure of sperm and eggs and ending with gastrulation.

particular positions in the egg? At the molecular level, what happens during fertilization? How are cytoplasmic determinants segregated to certain cells during cleavage, and how do cells move in a coordinated way during gastrulation? Let's follow each step of early development in order, starting with gametogenesis and ending with gastrulation, and consider each question in turn.

19.1 Gametogenesis

The development of a new individual begins with the formation of an egg and sperm in adult organisms. The DNA and cytoplasm in these reproductive cells are the initial components of the new individual. Both sperm and egg contribute an equal amount of DNA to the offspring—usually a haploid genome containing one copy of each gene. But because egg cells are routinely hundreds or thousands of times larger than sperm cells, an egg contributes much more cytoplasm.

Later chapters will explore the sequence of mitotic and meiotic cell divisions that lead up to the production of sperm and eggs. Here, let's focus on the structure of the mature reproductive cells. Understanding how these specialized cells are put together will lay the groundwork for addressing questions about fertilization and early development.

Sperm Structure and Function

Figure 19.1a shows the structure of a mammalian sperm cell. As the labels on the diagram show, the cell has four main compartments. A head region contains the nucleus and a bag of enzymes called the **acrosome**, the neck encloses a centriole, the midpiece is packed with mitochondria, and the tail region consists of a flagellum. The cell is so highly specialized that it has been called "DNA with a propeller." The propeller is the flagellum, which is powered by ATP manufactured in the mitochondria. Once the sperm reaches an egg, the enzymes in the acrosome are responsible for digesting the outer coverings of the egg cell so that the two cell membranes can make contact. In most animal species, the entire sperm enters the egg after fusion occurs.

As **Figure 19.1b** shows, the sperm of flowering plants are even simpler in form. Once a pollen grain has germinated at the top of the female reproductive structure, two sperm nuclei move down a growing pollen tube toward the egg cell. The nuclei bud off of the pollen tube and move toward the egg. As we'll see in section 19.2, both sperm cells will participate in fertilization.

Egg Structure and Function

Eggs are large and nonmotile reproductive cells; sperm are small and motile. Eggs are large because they contain the nutrients and cytoplasmic determinants required for the embryo's early development. Even in mammals, where embryos start to obtain nutrition through the placenta within a week or two of fertilization, eggs are an important early source of nutrients. In animals where development takes place outside the mother's body, stores in the egg are the *only* source of nutrients until organogenesis is complete and a larva or juvenile hatches and begins to feed. In animals, the nutrients required for early development are provided by a fat- and protein-rich **yolk** that is loaded into egg cells as they mature. In contrast, the nutrients required for early plant development are not synthesized and stored until after fertilization has occurred. As we'll see in section 19.2, a flowering plant embryo's nutrient supply is created during seed maturation and not during gametogenesis.

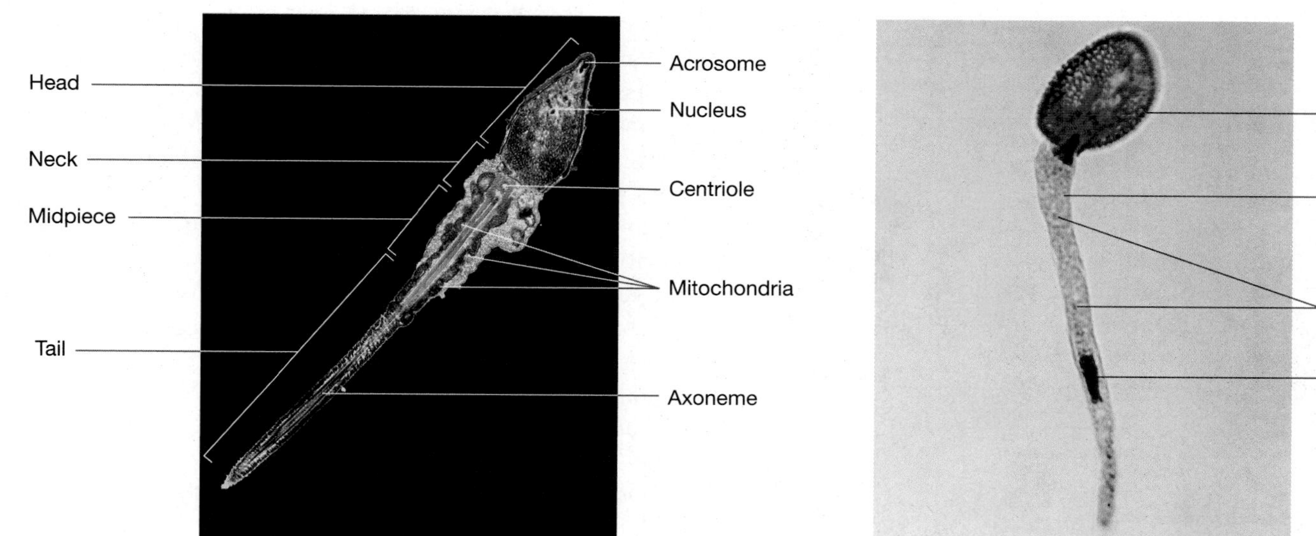

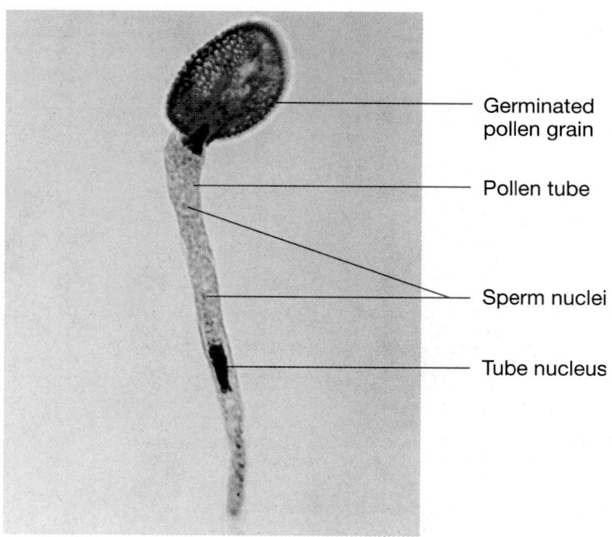

FIGURE 19.1 The Structure of Sperm
(a) The morphology of human sperm is typical of many animal species. **(b)** Two sperm cells are produced when pollen germinate. They travel down the pollen tube toward the ovule.

In addition to nutrients, many animal eggs contain organelles called cortical granules. **Cortical granules** are small vesicles filled with enzymes that are involved in fertilization. As the egg matures, the cortical granules are transported to the cell surface and localized just under the cell membrane. Just above the cell membrane a fibrous, mat-like sheet of glycoproteins called the **vitelline envelope** forms and surrounds the egg. As **Figure 19.2** shows, a large gelatinous matrix also encloses the egg in some species.

In terms of understanding how early development occurs, however, the most important of all molecules found in an egg may be the cytoplasmic determinants introduced in Chapter 18. In many species, eggs contain proteins and mRNAs that control early developmental events. Recall that molecules associated with the yellow cytoplasm of sea squirt eggs are associated with the differentiation of muscle cells; as we'll see, cytoplasmic determinants are also involved in the development of major body axes and embryonic tissues.

As the yellow pigment example showed, the key to cytoplasmic determinants is that they are localized to particular parts of the egg. Before looking at how this localization process occurs, though, it's important to recognize that the role of cytoplasmic determinants is highly variable among species. For example, cytoplasmic determinants do not appear to play a major role in the early development of mammal embryos. Instead of being produced by the maternal genome and sequestered in the egg, mouse and human proteins that have important roles in controlling early development are produced by the zygotic genome after fertilization is complete. It is also unclear at present whether cytoplasmic determinants play a significant role in early plant development.

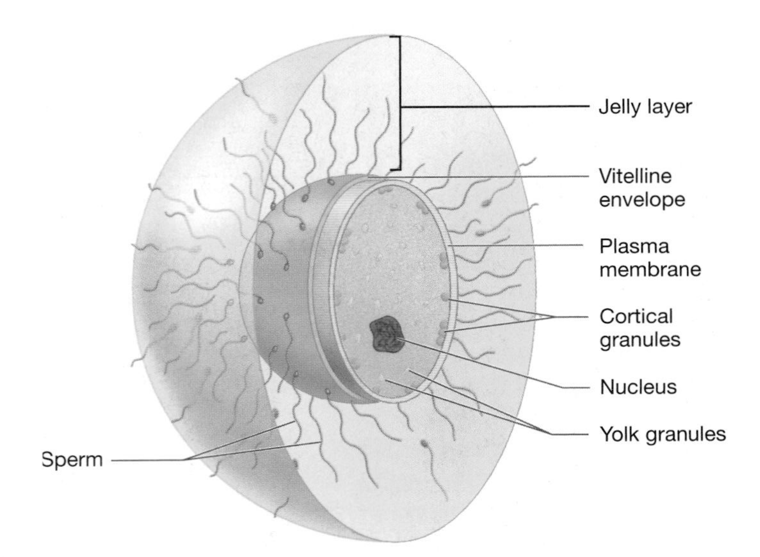

FIGURE 19.2 The Sea Urchin Egg
Animals that lay their eggs in water frequently surround their eggs with a gelatinous coat. Depending on the species being considered, yolk granules may be scattered throughout the egg or localized in one end.

In millions of animal species, however, molecules in the egg cytoplasm drive key events in early development. If their position in the egg is the key to their function, how do they become localized in a particular region? To answer this question, let's consider one of the best-studied of all cytoplasmic factors.

Localizing Cytoplasmic Factors

In the eggs of the African clawed frog *Xenopus laevis*, an array of key molecules are asymmetrically distributed. The most obvious of these substances are components of yolk. As **Figure 19.3a** shows, about 75 percent of the yolk is localized to one-half of the egg. The yolk-rich region is known as the **vegetal hemisphere**; the yolk-poor half is called the **animal hemisphere.**

In the 1980s, Douglas Melton discovered a less obvious asymmetrically distributed molecule in frog eggs—an mRNA called *Vg1*. Using in situ hybridization techniques introduced in **Box 19.1** (page 374), Melton was able to visualize the mRNA and confirm that it is localized to the vegetal pole of *Xenopus* eggs (**Figure 19.3b**). Subsequent work showed that cells inheriting this "vegetal 1" mRNA during cleavage eventually become endoderm and begin producing the Vg1 protein. Instead of remaining inside endodermal cells, however, the protein is secreted. In response to Vg1, nearby cells are induced to become mesoderm.

Based on these observations, the researchers concluded that the Vg1 protein functions as a cell-to-cell signal. It is produced by endodermal cells and acts on target cells that become mesoderm. More specifically, the cells that respond to Vg1 become mesoderm on the dorsal (back) side of the embryo. But *Vg1* mRNA can also be thought of as a cytoplasmic determinant-like molecule. It is sequestered in one region of the egg and influences the fate of cells later in development. Other cytoplasmic determinants have been found in frogs and other species as well. These molecules are localized in egg cells and direct the fate of cells that receive them. How are these types of molecules localized in the correct position in the egg? For example, how does *Vg1* mRNA end up in cells fated to become endoderm on the dorsal side of the embryo?

To make a start at answering this question, Melton and colleagues used in situ hybridization to visualize where *Vg1* mRNA is found as the development of a frog egg progresses. They found that *Vg1* mRNA is initially distributed in the egg-cell cytoplasm uniformly. Later, the mRNA becomes localized to the vegetal hemisphere; when the egg is mature, it is found in a tight band at the end of the vegetal pole. This is the position it occupies in the early embryo. These results let them define the question more precisely: How does *Vg1* mRNA move from throughout the cytoplasm to the vegetal pole?

How Do Cytoplasmic Determinants and Other Molecules Move to Their Correct Location? You might recall from Chapter 5 that molecules are frequently transported inside cells along the cytoskeleton. To determine whether the cytoskeleton is involved in *Vg1* localization, Melton and co-workers followed the experimental sequence outlined in **Figure 19.4a.** Their basic strategy was to treat frog eggs with chemicals that disrupt the function of specific cytoskeletal elements, and then to visualize the location of *Vg1* using in situ hybridization. More specifically, they treated eggs with two different drugs. Nocodazole disrupts microtubules by preventing tubulin subunits from polymerizing. Cytochalasin B disrupts microfilaments by preventing actin subunits from polymerizing.

When Melton and colleagues applied the drugs to egg cells midway through the maturation process, they observed that nocodazole prevented the concentration of *Vg1* mRNA in the vegetal hemisphere. This observation suggested that

(a) In frogs, yolk is concentrated in the vegetal hemisphere.

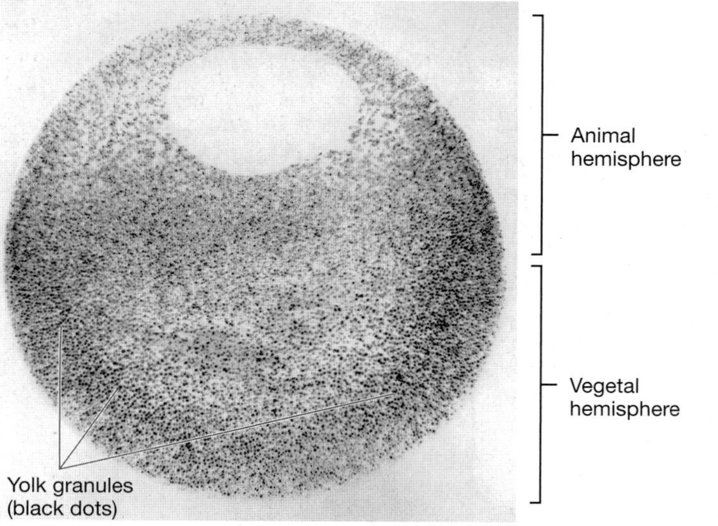

(b) *Vgl* mRNA is localized to the vegetal pole.

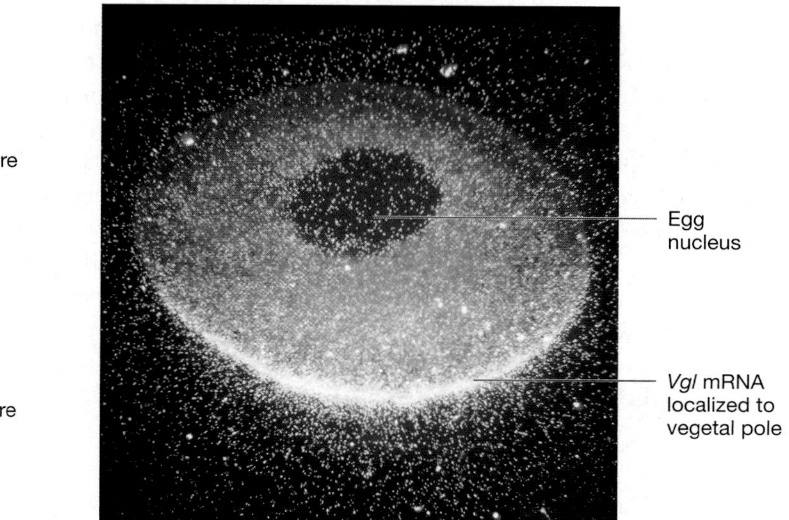

FIGURE 19.3 Eggs Are Highly Asymmetrical
(a) In frogs and in many other species, the yolk is concentrated in one half of the egg called the vegetal hemisphere. **(b)** The bright light spots on this autoradiograph mark in situ hybridizations between a DNA probe and *Vg*1 mRNA (for details on this technique, see Box 19.1). The *Vg*1 mRNA is localized to the cytoplasm immediately under the cell membrane in the vegetal hemisphere. The dark spot in the photo is the egg nucleus.

(a) Treating eggs with drugs can disrupt certain cytoskeletal elements.

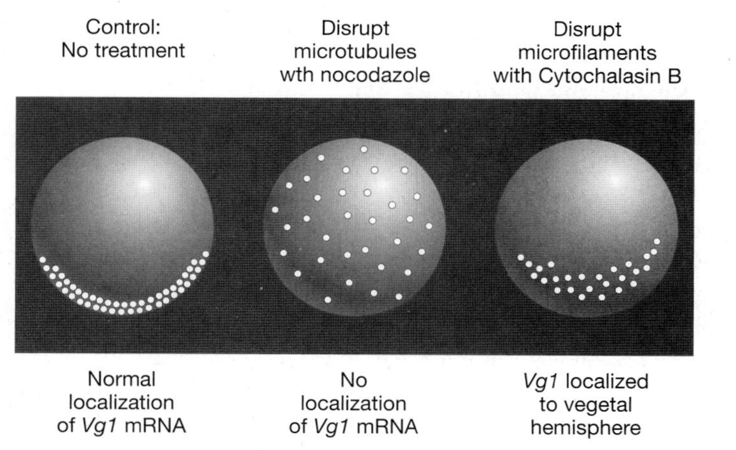

(b) A HYPOTHESIS TO EXPLAIN *Vg1* TRANSPORT

Microtubules

Microfilaments

1. *Vg1* mRNA is transported to vegetal hemisphere along microtubules.

2. *Vg1* mRNA binds to microfilaments just under cell membrane.

FIGURE 19.4 Localizing Cytoplasmic Factors
(a) This experiment tested the hypothesis that the cytoskeleton is involved in transporting *Vg*1 mRNA. **(b)** To explain the results in part (a), researchers hypothesized that both microtubules and microfilaments are involved in localization of *Vg*1 mRNA.

BOX 19.1 Visualizing mRNAs by in Situ Hybridization

Where are the mRNAs for a particular cytoplasmic determinant localized in an egg? Once the zygote's genome is activated, when and where are certain genes transcribed in the embryo? Answering questions like these became possible when Mary Lou Pardue and Joseph Gall developed a technique called in situ (in place) hybridization. In situ hybridization allows researchers to visualize the location of specific mRNAs in organisms. As a result, biologists are able to determine where particular genes are being transcribed at various times during development.

To perform in situ hybridization for a particular mRNA, a researcher must have many copies of the gene that encodes the message. For example, once Melton had isolated the mRNA called *Vg1*, he could use reverse transcriptase to make cDNA copies of the sequence (**Figure 1**). By inserting these cDNAs into plasmids inside growing *E. coli* cells, he obtained many copies of the DNA for the *Vg1* mRNA and protein.

The next step in the procedure is to attach some sort of label to the DNA sequences for the gene in question. The addition of a label gives researchers a way of detecting the location of the DNA sequence. Researchers used radioactive atoms as labels when the technique was first developed, but now most prefer to attach chemical groups that bind colored reagents.

Finally, the labeled DNA is made single stranded and added to a preparation containing the tissue. (In situ hybridization cannot be done in living organisms, however.) As it enters the cells, the single-stranded, labeled DNA binds to mRNA molecules whose sequence is complementary, forming a DNA-RNA hybrid molecule. In this way, the single-stranded, labeled DNA "lights up" a specific mRNA; excess labeled DNA that does not bind to complementary mRNAs is washed away. These final steps are analogous to the "probing" done by a labeled DNA during the Southern and Northern blotting procedures introduced in earlier chapters. When the in situ hybridization is complete, the location and intensity of the signal from the label indicate the location and amount of mRNA present in particular regions of the organism.

(Continued on next page)

microtubules are involved in the initial movement of *Vg1* toward the vegetal pole. In contrast, the mRNA moved to the vegetal half in egg cells treated with cytochalasin B, but never became concentrated in a tight band at the pole. This result suggested that microfilaments are involved in the final localization of the mRNA along the cell membrane. In support of this hypothesis, the researchers found that if they treated mature egg cells with cytochalasin B, *Vg1* mRNA diffused away from its normally tight peripheral distribution. Nothing happened in cells treated with nocodazole, however.

To pull these observations together, Melton's group proposed the two-step model for *Vg1* mRNA localization, as depicted in **Figure 19.4b**. In the first step, *Vg1* mRNA moves to the vegetal hemisphere along microtubules. In the second step, the mRNAs become anchored to the cell periphery on microfilaments.

Egg Development Involves a Series of Steps Subsequent work by Melton and others has shown that a specific region of the *Vg1* mRNA directs its localization to the vegetal pole. This "localization element" in the molecule functions as an address label, much like the signal sequences found on proteins that move through the endomembrane system introduced in Chapter 5. Recent research has shown that *Vg1*'s localization element binds to a protein attached to the endoplasmic reticulum (ER). According to the current view, then, *Vg1* mRNA attaches to a protein on the ER and is transported toward the egg's vegetal pole along microtubule tracks.

The general message of this work is that egg development involves a complex and highly organized series of events. The egg enlarges as it is packed with nutrients in the form of yolk. Meanwhile, cytoplasmic determinants in the form of maternal mRNAs and proteins are manufactured and transported to specific locations in the cell. The mature egg is an extremely specialized cell, ready to interact with sperm and direct early development. How do sperm enter the egg and get the process under way?

19.2 Fertilization

Fertilization seems like a simple process: A sperm cell fuses with an egg cell to form a zygote. Upon reflection, though, you should begin to appreciate that the process is actually extraordinarily complex. For fertilization to take place, sperm and egg cells must be in the same place at the same time and then must recognize and bind to each other. Next they must fuse—even though all of the other cells in the body are designed to prevent fusion. In most species fusion must also be limited to a single sperm so that the zygote does not receive extra chromosomes. Finally, the fusion of the two nuclei has to trigger the onset of development. Despite this complexity, the contact that takes place between sperm and egg at fertilization qualifies as the best-studied of all cell-to-cell interactions.

Work on fertilization began in earnest early in the twentieth century, when biologists began to study the sperm-egg interaction in sea urchins. Sea urchins are marine invertebrates that perform fertilization externally (**Figure 19.5a**, page 376). To maximize the probability that sperm and egg will meet, male and female sea urchins secrete huge quantities of gametes. As a result, sea urchins provided researchers with large numbers of

(Box 19.1 continued)

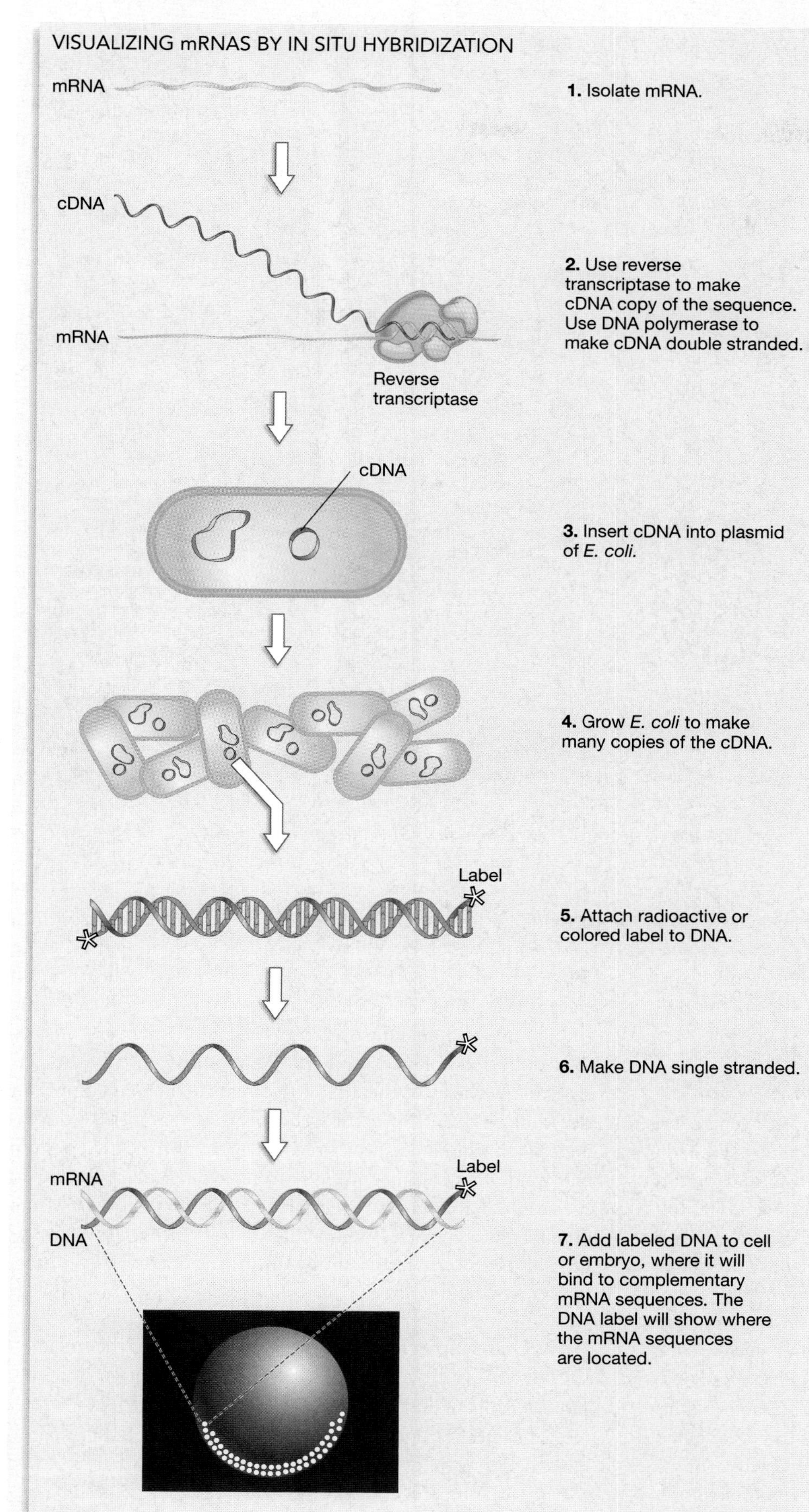

FIGURE 1 In Situ Hybridization
QUESTION According to Chapter 18, differential gene expression is fundamental to development. In situ hybridizations allow researchers to determine where specific mRNAs are located. Why is in situ hybridization currently one of the most important techniques used in developmental biology?

(a) Sea urchin releasing gametes

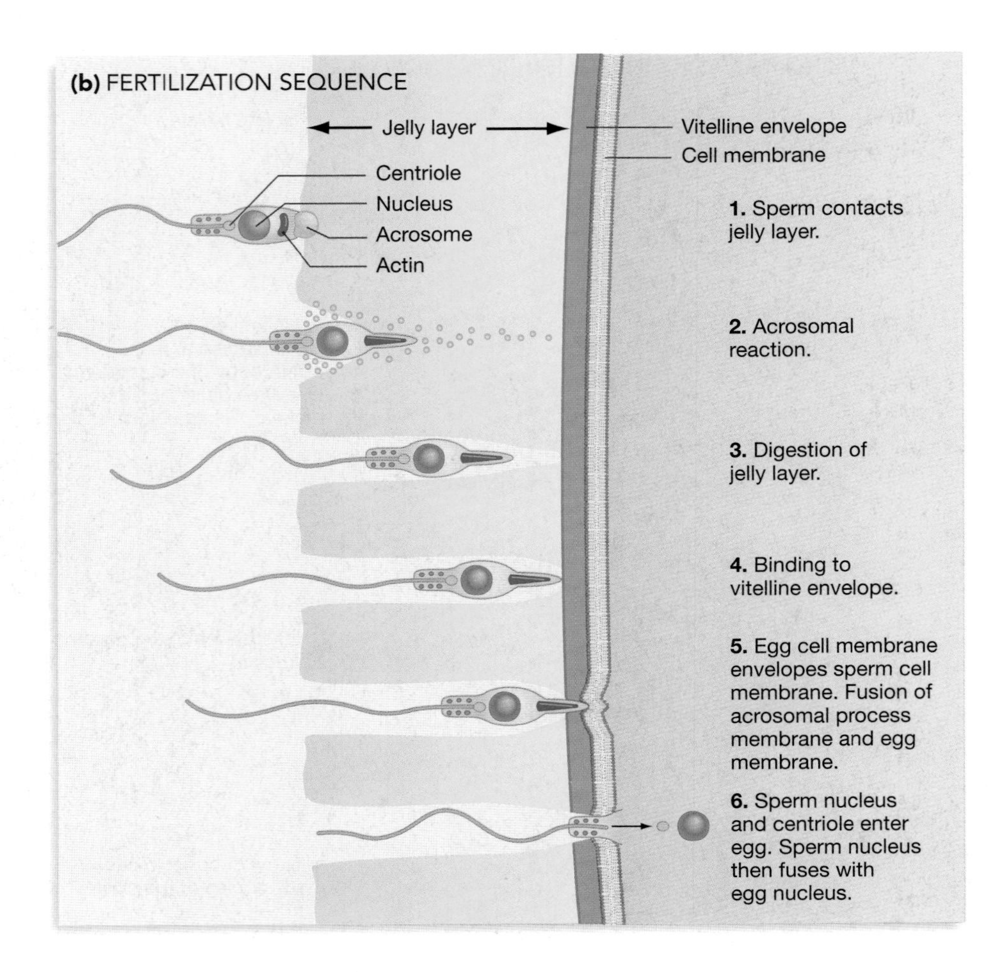

FIGURE 19.5 Fertilization in Sea Urchins
(a) Sea urchins release their gametes into the ocean. **(b)** Fertilization involves a complex sequence of events.

gametes that could be studied in a test tube or culture dish filled with seawater.

Sea urchins continue to be an intensively studied model system in research on fertilization. What have researchers discovered in a century of work?

An Overview of Sea Urchin Fertilization

In sea urchins, the gelatinous coat that surrounds the egg contains a molecule that attracts sperm. The attractant is a small peptide that diffuses away from the egg and into the surrounding seawater. Sperm respond to the attractant by swimming toward areas where its concentration is higher. The same type of interaction between sperm and eggs has been observed in a wide variety of animals that lay their eggs in seawater.

What happens when a sea urchin sperm and egg actually meet? As you can see in **Figure 19.5b**, the sperm head initially encounters the jelly layer of the egg cell. The contact triggers an **acrosomal reaction.** In the first part of this reaction, the contents of the acrosome are expelled from the cell. The enzymes that are released digest the gelatinous covering around the egg and allow the sperm to reach the vitelline envelope surrounding the egg. The second part of the reaction involves the polymerization of actin into microfilaments that form a protrusion, which extends until it makes contact with the egg cell membrane. Finally, the cell membranes of the egg and sperm fuse. The sperm nucleus and centriole enter the egg, the sperm and egg nuclei fuse to form the zygote nucleus, and fertilization is complete.

Frank Lillie was the first to recognize that an important question was hidden in this sequence of events. How do gametes from the same species recognize each other? After all, in many habitats sperm and eggs from a particular sea urchin species float about in seawater along with eggs and sperm from other sea urchin species and many other organisms. What prevents cross-species fertilization and the production of dysfunctional hybrid offspring?

How Does Species Recognition in Sea Urchins Occur?

Lillie was the first researcher to identify a substance on the surface of egg cells that appeared to be involved in binding sperm. He called the compound "fertilizin" and showed that it caused sperm to clump together. He also showed that clumping occurred only when fertilizin from the eggs of a particular sea urchin species was combined with sperm from the same species. Each species appeared to have its own version of fertilizin.

Based on these observations, Lillie proposed that fertilizin on the surface of the egg interacted with a substance on sperm in a lock-and-key fashion. Further, Lillie suggested that the interaction was favored by natural selection because it increased the probability of fertilization occurring between sperm and eggs of the same species. Does any evidence suggest that the lock-and-key hypothesis is valid?

The Discovery of Bindin Decades passed before Lillie's hypothesis was extended and tested. Then, in the 1970s, Victor Vacquier and co-workers succeeded in identifying a protein on sea urchin sperm heads that binds to the surface of eggs in a species-specific manner. They called this protein bindin. Follow-up work showed that the bindin proteins from even very closely related species are distinct. As a result, bindin should ensure that a sperm binds only to eggs from the same species. The next question was, what does bindin bind to? If bindin acts as a key, what acts as the lock?

An Egg Receptor for Sperm Kathleen Foltz and William Lennarz hypothesized that sea urchin eggs have a protein on their surface that binds to bindin. To find this bindin receptor protein, they attempted to isolate the part of the receptor that is exposed on the outside of the egg. They predicted that this region of the protein interacts with the bindin on sperm.

Figure 19.6a illustrates Foltz and Lennarz's experimental approach. They began by treating the surface of sea urchin eggs with a protease. Recall that proteases cleave peptide bonds. When they isolated the protein fragments that were released from the egg surface, they found one that bound to sperm and to isolated bindin molecules. Further, this binding occurred in a species-specific manner. A protein fragment from the eggs of one species bound to sperm of its own species, but did not bind to sperm of different species. Based on these observations, the biologists claimed that they had found the outward-facing portion of the egg-cell receptor for sperm.

To isolate the entire receptor, Foltz and Lennarz used techniques introduced in Chapter 46 to produce antibodies to their receptor fragment. An antibody is a kind of protein that is able to bind specifically to another molecule. By attaching a fluorescent dye to antibodies that reacted with the outer fragment of the egg-cell receptor, the researchers located the sperm receptors on intact sea urchin eggs. As **Figure 19.6b** shows, the fluorescent antibodies revealed that receptors were evenly distributed over the surface of the egg.

These experiments provided convincing evidence that an egg receptor for sperm exists on the surface of sea urchin eggs. The work also provided important support for Lillie's lock-and-key hypothesis. During sea urchin fertilization, species-specific bindin molecules on sperm interact with species-specific receptors on the surface of the egg. This interaction is required for the cell membranes of sperm and egg to fuse. As a result, cross-species fertilization is rare.

Blocking Polyspermy

Early in the history of studies on sea urchin fertilization, researchers noticed that only one sperm succeeded in fertilizing the egg, even when dozens or even hundreds of sperm were clustered around the vitelline envelope. From the standpoint of producing a viable offspring, this observation was logical. If multiple fertilization or

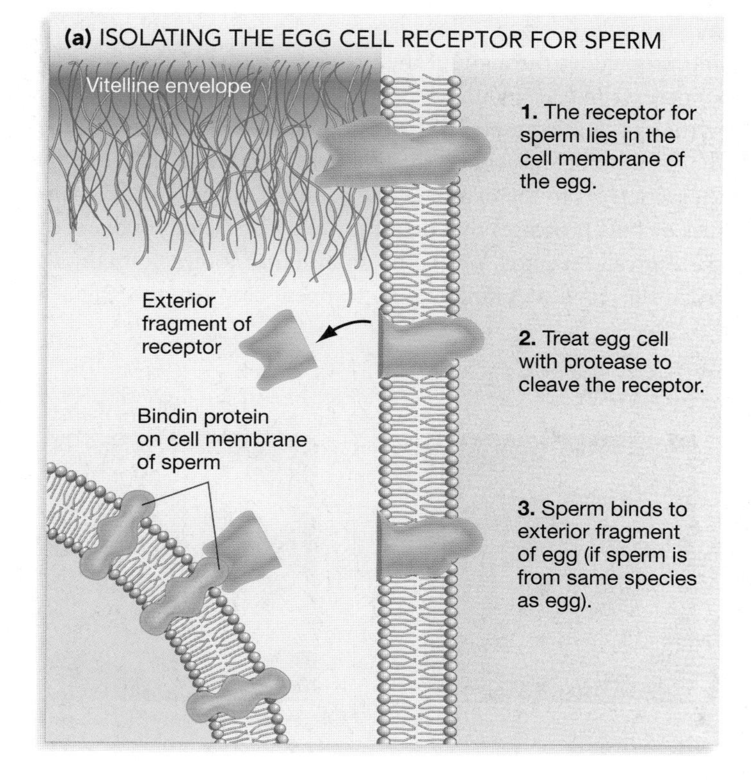

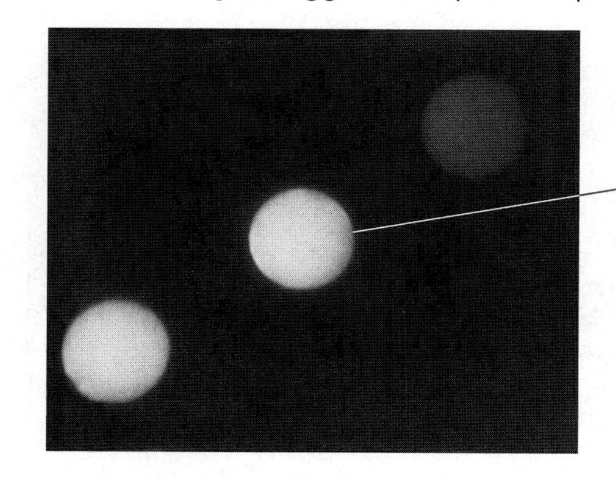

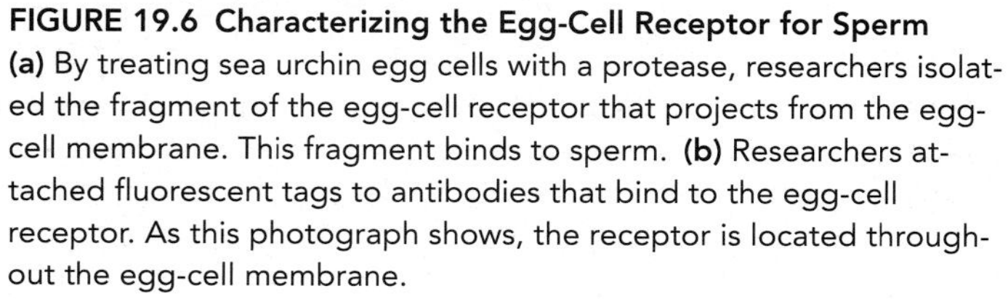
FIGURE 19.6 Characterizing the Egg-Cell Receptor for Sperm
(a) By treating sea urchin egg cells with a protease, researchers isolated the fragment of the egg-cell receptor that projects from the egg-cell membrane. This fragment binds to sperm. **(b)** Researchers attached fluorescent tags to antibodies that bind to the egg-cell receptor. As this photograph shows, the receptor is located throughout the egg-cell membrane.

polyspermy occurred, the resulting zygote would have more than two copies of each chromosome. As an adult, an individual with an abnormal number of chromosomes is largely or completely infertile because meiosis cannot be completed normally. If many sperm are present, how is polyspermy avoided?

Research over the past 70 years has revealed a wide array of mechanisms that block polyspermy in various animal species. In sea urchins, for example, fertilization results in the erection of a physical barrier to sperm entry via the mechanism illustrated in **Figure 19.7**. As Figure 19.7a shows, the entry of a sperm causes calcium ions (Ca^{2+}) to be released from storage areas inside the egg. A wave of ions starts at the point of sperm entry and propagates throughout the egg. In response to this dramatic increase in Ca^{2+} concentration, a wide array of events occur in the egg. For example, the cortical granules located just inside the membrane fuse with the egg-cell membrane and release their contents to the exterior. The contents of the cortical granules include proteases that digest the exterior-facing fragment of the egg-cell receptor for sperm. In addition, other compounds from the cortical granules are trapped between the egg-cell membrane and the envelope and cause water to flow into the space by osmosis. The influx of water causes the envelope matrix to lift away from the cell and form the **fertilization envelope** illustrated in Figure 19.7b. The fertilization envelope, in turn, keeps additional sperm from contacting the egg membrane.

To summarize, a century of research on sea urchin fertilization has produced a wealth of knowledge about one of the most important cell-to-cell interactions in nature. The interaction between sperm and egg triggers a series of remarkable events, ranging from the acrosome reaction and the fusion of haploid genomes to mechanisms for blocking polyspermy. Although sea urchins have acted as a productive model system, an important question remains. How relevant are these findings to other species? In particular, how similar is sea urchin fertilization to what goes on in mammals and plants?

Fertilization in Mammals

Unlike sea urchins, fertilization in humans and other mammals occurs internally. As section 19.1 showed, fertilization in humans takes place at the end of the fallopian tubes near the

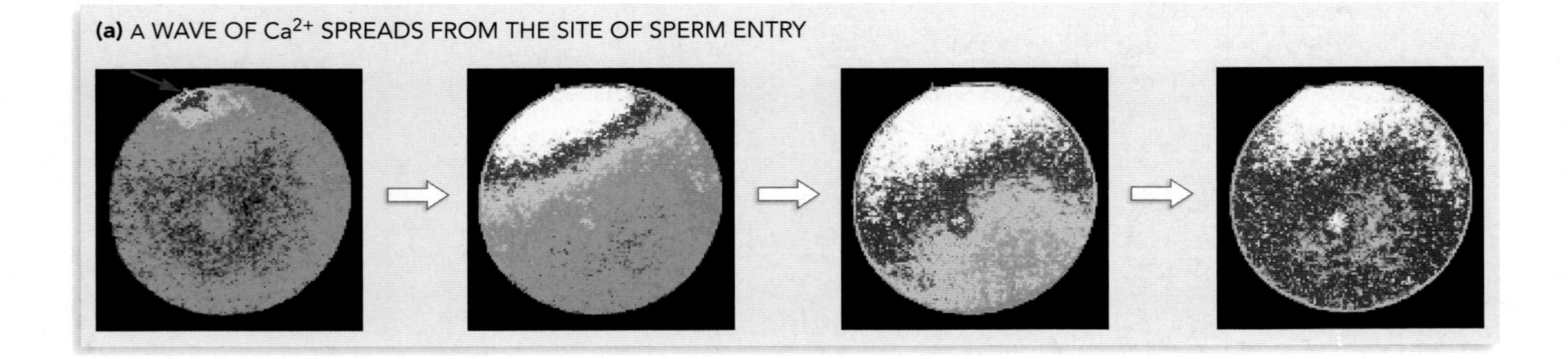

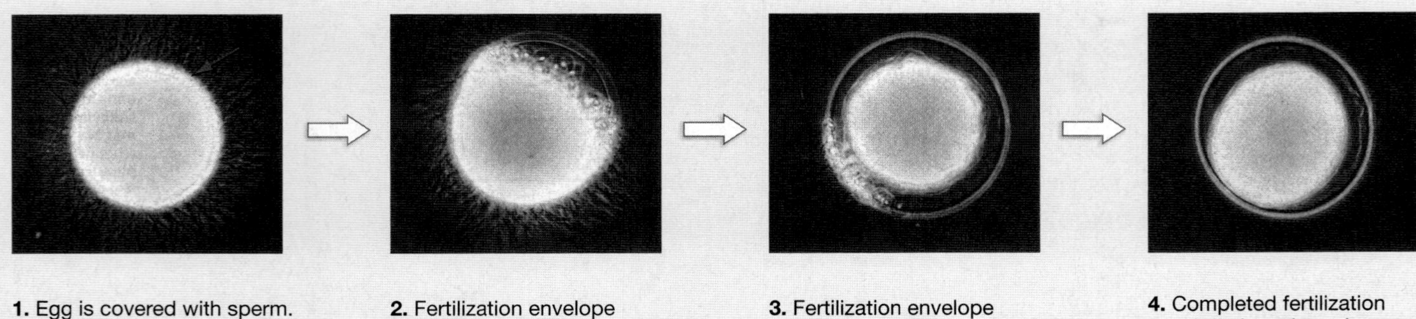

FIGURE 19.7 The Slow Block to Polyspermy Is a Physical Barrier
(a) During fertilization, a wave of Ca^{2+} begins at the point of sperm entry and spreads throughout the egg. The white dots in these photos are from a reagent that reacts with calcium ions. **(b)** In response to increased Ca^{2+} concentrations, cortical granules fuse with the egg-cell membrane and release their contents to the exterior, causing a fertilization envelope to rise from the egg-cell membrane and clear excess sperm.

ovary. This fact makes mammalian fertilization much more difficult to study than fertilization in sea urchins and similar species. With the advent of the in vitro fertilization (IVF) methods highlighted in this chapter's essay, however, biologists have finally acquired the ability to study mammalian fertilization under laboratory conditions. What have they found?

Because females have actively chosen a mate prior to the sperm-egg interaction, species recognition is not an issue in mammals and other species with internal fertilization. The acrosomal reaction still occurs, however, after the sperm head reaches the equivalent of the sea urchin vitelline envelope—an extracellular matrix known as the zona pellucida. The enzymes released from the acrosome digest the zona pellucida. As a result, the sperm head is able to reach the egg-cell membrane and fuse with it.

Even though species recognition was predicted to be unimportant, biologists have still attempted to identify specific proteins on the sperm- and egg-cell surfaces that mediate binding. The logic here is that some sort of specific interaction must take place so that sperm cells do not fuse with cells that line the reproductive tract. Although there is no convincing evidence for a bindin-like protein on the sperm head as yet, Paul Wassarman and colleagues have recently presented data suggesting that egg cells have a binding site for sperm. To search for this site in mouse eggs, the researchers analyzed the three glycoproteins found in the zona pellucida. They found that one of the three glycoproteins, called ZP3, binds to the heads of sperm. But as predicted, the binding of mammalian sperm to ZP3 is not species specific. Finally, researchers have found that enzymes released from cortical granules modify ZP3 in a way that prevents binding by additional sperm.

These results reinforce one of the themes highlighted in Chapter 18. Variation in developmental processes among species may result from natural selection, because certain traits confer an advantage in a particular species and environment. In this case, differences in the mechanisms of fertilization vary in a logical way between species with external versus internal fertilization.

Chapter 18 also pointed out that variation in developmental processes occurs among groups that have evolved independently for long periods. In this light, the marked differences in plant and animal fertilization are not surprising.

Fertilization in Flowering Plants

Fertilization in flowering plants has been exceptionally difficult to study, because the process takes place inside the ovule. Two important observations are worth noting, however. First, an event known as double fertilization takes place in flowering plants. As **Figure 19.8a** shows, two sperm nuclei enter the ovule. There they encounter the egg cell and a huge cell that contains two haploid nuclei. One of the sperm fuses with the egg to form the zygote, while the other sperm fuses with the two haploid nuclei to form a triploid ($3n$) cell. The triploid cell divides repeatedly to form a nutritive tissue called **endosperm**, which provides the proteins, carbohydrates, and fats or oils required for embryonic development, germination, and early seedling growth. In species with large seeds, the endosperm grows into a sizeable nutrient reservoir as the ovule matures (Figure 19.8b). When you eat wheat, rice, corn, almonds, or other grains or nuts, you are eating primarily endosperm.

A second major observation about plant fertilization involves the start of the process. Recall from Chapter 18 that the development of a new plant begins when a pollen grain germinates at

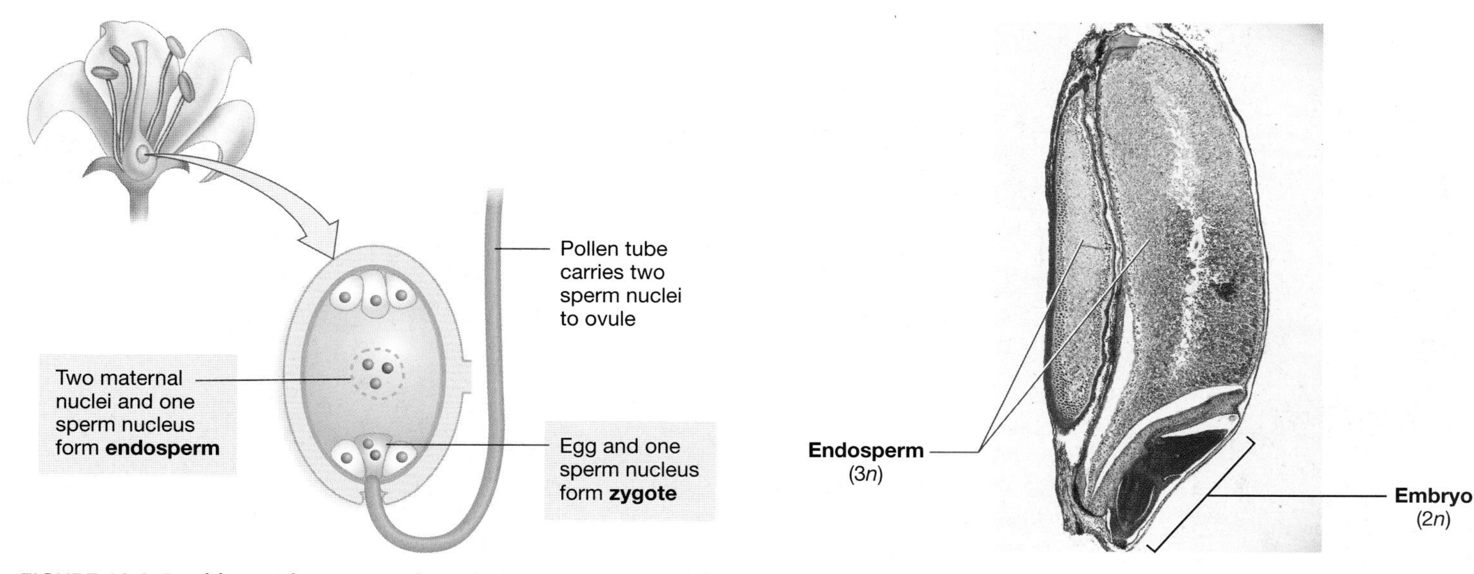

FIGURE 19.8 Double Fertilization Leads to the Formation of Endosperm
(a) In flowering plants, a sperm nucleus fuses with two haploid nuclei near the egg to form triploid ($3n$) endosperm tissue. **(b)** Endosperm is a nutritive tissue that is packed with fats or oils and carbohydrates.

the top of the female reproductive structure. Once a pollen grain germinates, sperm are produced and begin to move toward the ovule. Recent research has shown that the germination of a pollen grain involves complex interactions between proteins on the surface of the pollen grain and proteins on the walls and membranes of cells at the top of the female reproductive structure. Some of these interactions are involved in species recognition and ensure that only pollen from the same species is allowed to germinate. But in some species, the interactions also prevent pollen from the same individual from germinating on its own female parts. In this case, the result is that self-fertilization and inbreeding are avoided.

Later chapters will explore the molecular mechanisms involved in self-self recognition in plants as well as the damaging consequences of inbreeding. For now, though, let's turn our attention to the cell divisions that take place immediately after fertilization.

19.3 Cleavage

Cleavage refers to the cell divisions that take place in animals after fertilization. Cleavage divisions partition the egg cytoplasm without additional cell growth taking place. As **Figure 19.9a** shows, the zygote simply divides into two, then four, then eight cells, and so on, without concurrent growth. These divisions rapidly create a multicellular embryo. In *Drosophila*, cleavage produces about 5000 cells in 3 hours. When cleavage is complete, the embryo consists of a sphere of cells that is ready to undergo gastrulation.

The exact pattern of cleavage varies widely among species, however. Cells can divide at right angles to each other so that they form tiers, as in Figure 19.9a, or at oblique angles so that they pile up in the spiral arrangement shown in **Figure 19.9b.** In birds, fish, and other species whose eggs have large, membrane-bound structures filled with yolk, cleavage does not split the egg completely but produces a mound of cells around the yolk or on top of it, as shown in **Figure 19.9c.**

There are actually eight or more distinctive types of cleavage observed in animals. Each type is restricted to a particular group of closely related species. For example, the pattern shown in Figure 19.9a is observed in sea urchins and their relatives; the configuration in Figure 19.9b is typical of molluscs and most worms; the arrangement in Figure 19.9c is found in fishes, reptiles, and birds. How are these patterns controlled?

Snail-Shell Coiling and Cytoplasmic Determinants

In 1894, H. E. Crampton published a remarkable observation that provided an important clue about how cleavage is controlled. Crampton was studying coiling patterns in the shells of snails. Snail shells coil either to the left or the right of the opening

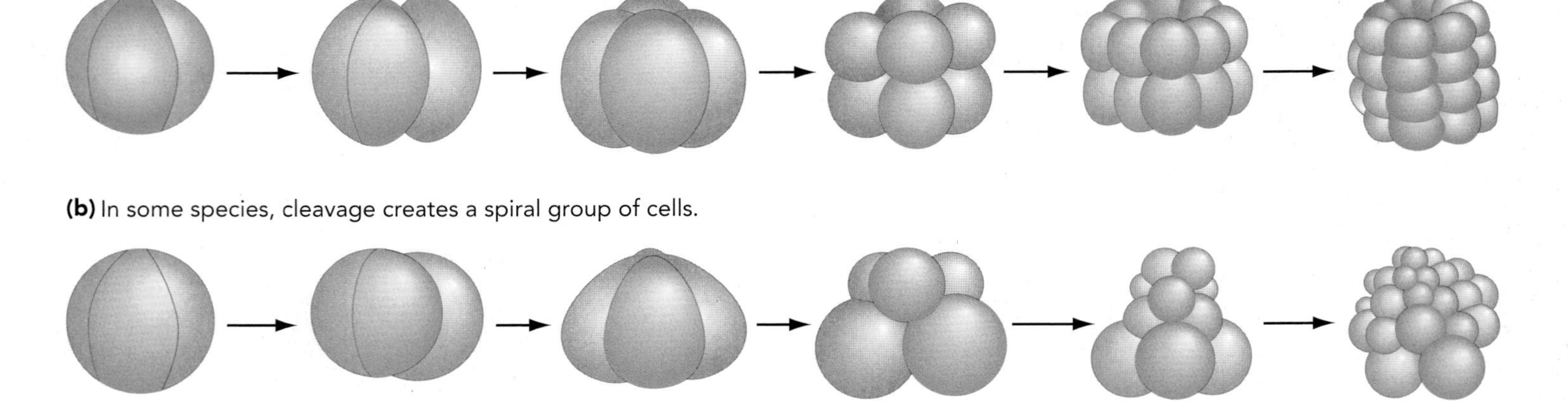

(c) In some species, a ball of cells is created on top of yolk.

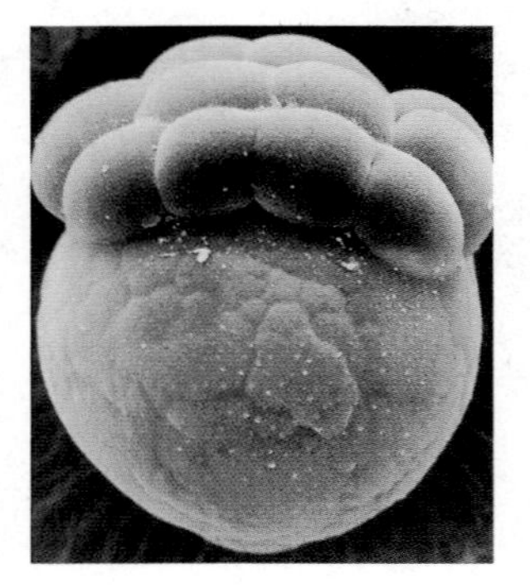

FIGURE 19.9 Cleavage Patterns Vary Among Species
(a) Growth does not occur during cleavage because the embryo cannot yet feed. Note that some of the cells in this figure are colored green simply to help you follow their fate. These cells are not actually green in the embryo. **(b)** These drawings illustrate how a pattern called spiral cleavage occurs. **(c)** Distinctive cleavage patterns occur in fishes, reptiles, birds, and other species with extremely large amounts of yolk.

where the individual's body emerges. Usually, all members of the same species have shells that coil in the same direction. Occasionally, though, Crampton found mutant individuals whose shells coil in the direction opposite the norm for its species. The remarkable observation he made is that the direction of cleavage in snail embryos mirrors the direction that their shells coil. As **Figure 19.10** shows, normal members of a species with left-handed coiling undergo cleavage in a way that leads to a left-handed coil of cells in the blastula. The mutant individuals, though, underwent cleavage in a way that leads to a right-handed coil of cells. The mutants also had shells with right-handed coiling.

Alfred Sturtevant followed up on this observation by arranging matings between right-handed and left-handed individuals and analyzing the phenotypes of their offspring. Sturtevant found that the pattern of inheritance could be explained by hypothesizing that a single gene with two alleles controlled coiling pattern. He called the alleles *D* and *d* and determined that *D* is dominant. Some crosses gave particularly interesting results, however. For example, when he crossed *DD* females and *dd* males, all of the offspring were *Dd* and had right-handed shells. But when he crossed *dd* females and *DD* males, all of the offspring were *Dd* but had left-handed shells. To explain these results, he proposed that the direction of cleavage and coiling depends on the mother's genotype and not on the offspring's genotype. That is, mothers with *dd* genotypes produce left-coiling shells in their offspring. If the mother's genotype is *Dd* or *DD*, however, they produce offspring with a right-coiled shell.

How is this "maternal effect" possible? The most likely answer is that the cleavage pattern in snails, and thus the coiling pattern of their shells, is established by a cytoplasmic determinant. The specific hypothesis here is that a molecule that is produced by the mother and localized in the egg cytoplasm determines the orientation of the mitotic spindle, as shown in Figure 19.10. The mother's genotype determines the coiling pattern because the gene responsible is expressed during egg maturation—not during cleavage. This result suggests that cleavage is controlled primarily by cytoplasmic determinants from the mother and not by proteins produced in the zygote. Is this conclusion valid in other species as well?

Activating the Zygotic Genome

In 1982, John Newport and Marc Kirschner published studies on the African clawed frog that helped answer the question of how cleavage is controlled. Newport and Kirschner injected *Xenopus* embryos with a compound derived from poisonous mushrooms called α-amanitin, which inhibits transcription. Even though their genome was not transcribed at all, the embryos progressed through early cleavage normally. The rate of cell division even slowed dramatically after the twelfth cleavage division in the treated embryos, just as it does in normal embryos.

A dramatic reduction in the rate of cell division during cleavage occurs in many other animal species as well, and is sometimes referred to as the midblastula transition. During this phase, Newport and Kirshner found that new RNAs begin being produced in *Xenopus* embryos. Taken together, their research supports the hypothesis that cleavage is directed

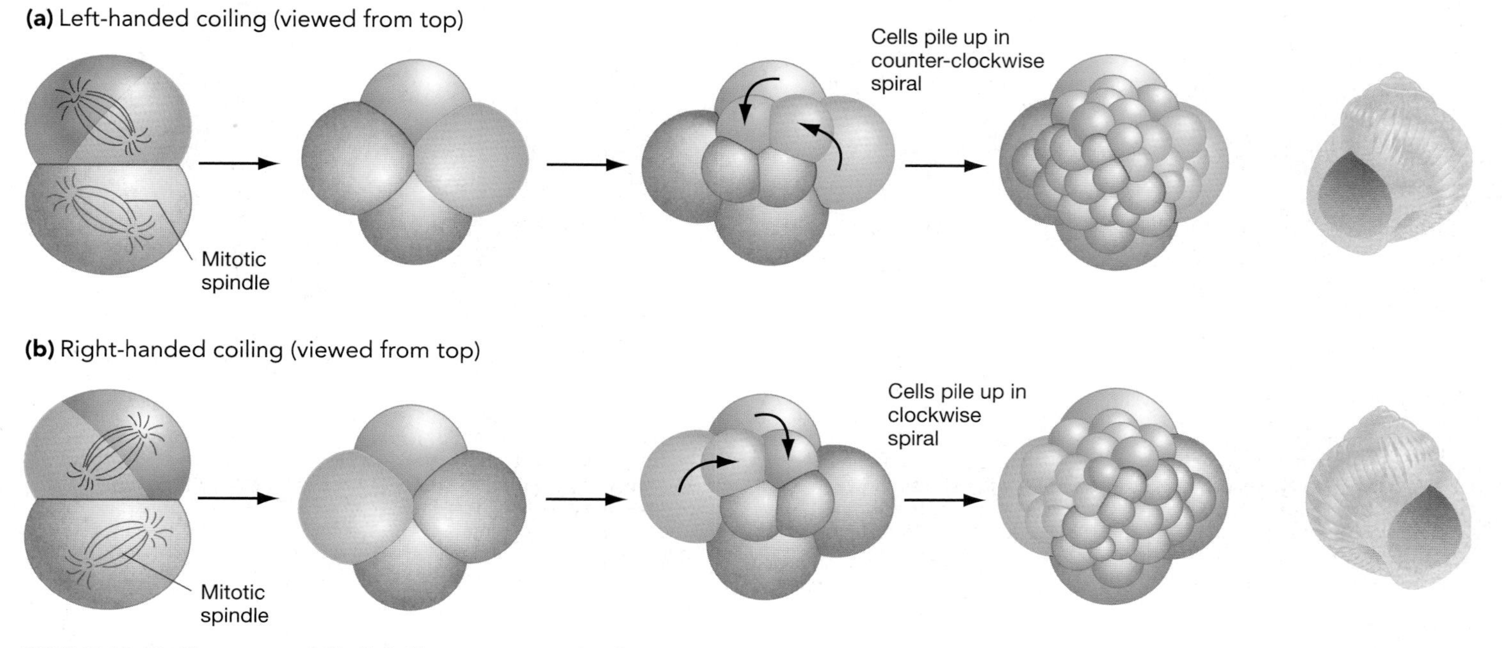

FIGURE 19.10 Cleavage and Shell-Coiling Patterns in Snails
In snails, a cytoplasmic determinant sets up the orientation of the mitotic spindles during cleavage. Mutant forms of this cytoplasmic determinant switch the orientation of the mitotic spindles, leading to blastomeres and shells that coil in a direction opposite of normal.

by cytoplasmic determinants in the egg, and that the zygotic genome is transcribed for the first time after cleavage is well underway. Follow-up work suggests that this conclusion is valid for most other animal species as well. Mammals, however, are an important exception to this rule. In mice and other mammals, mRNAs begin to be transcribed from the zygotic genome at the two-cell stage.

As cleavage continues, then, the embryo consists of a ball of cells that are actively making mRNAs and proteins. Cell division slows dramatically. At the onset of gastrulation, dramatic cell movements begin.

19.4 Gastrulation

Gastrulation refers to the massive cell movements that take place in animals after cleavage is complete. The process reorganizes the three embryonic tissue types—ectoderm, mesoderm, and endoderm. These conclusions grew in part out of efforts to map the fate of blastomeres in newt and frog embryos. This work was pioneered by W. Vogt in the 1920s and has continued, using various techniques, to the present. Vogt's approach was to soak blocks of agar with a nontoxic dye and then press the blocks against the surface of blastulas. The dyes marked

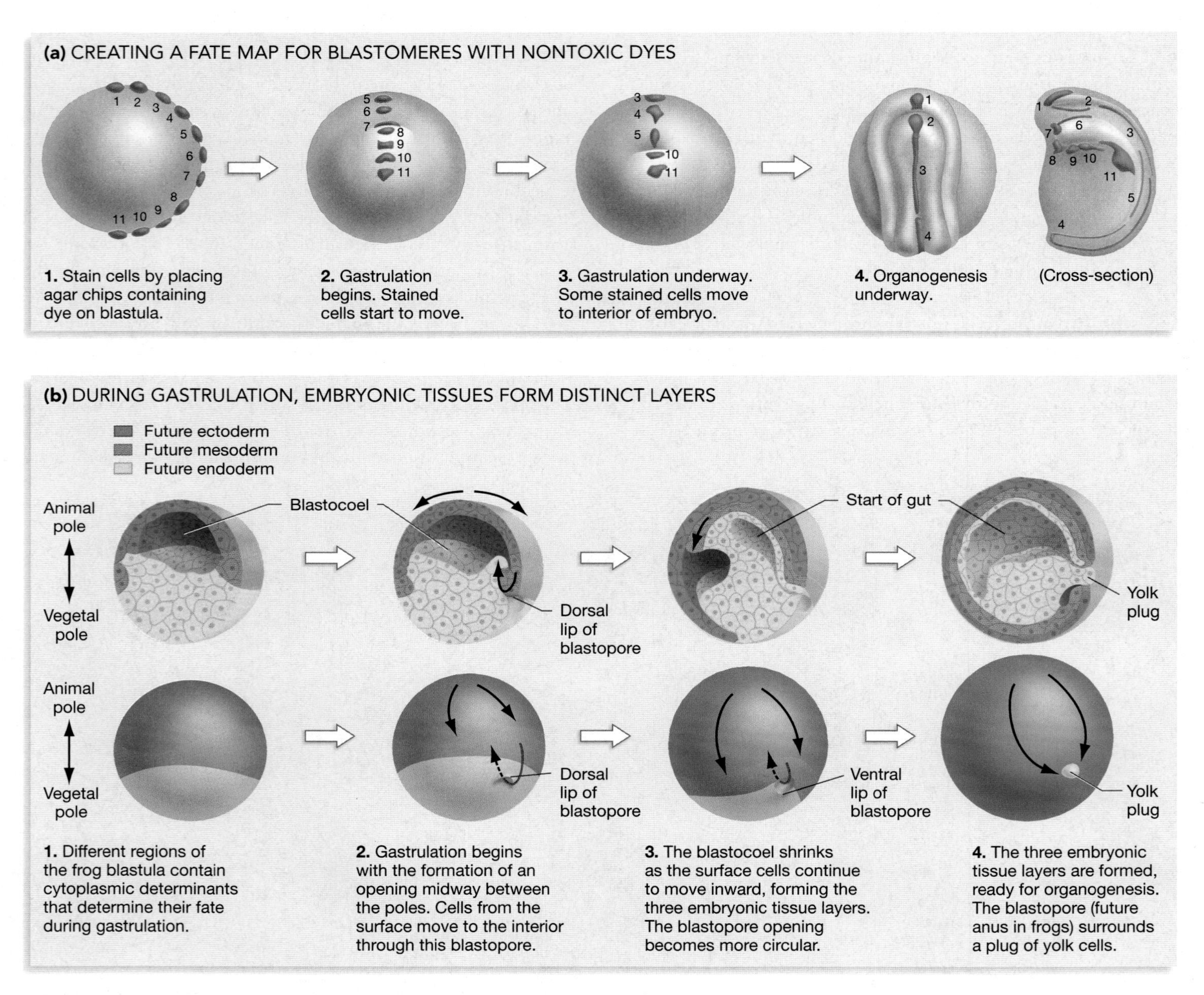

FIGURE 19.11 Documenting Cell Movements During Gastrulation

EXERCISE In frogs, the initial opening of the blastopore is on what will become the dorsal side of the larva and adult. The blastopore itself becomes the anus of the larva and adult. Using these cues, mark where the head, tail, back, and belly of the larval frog will develop on the last drawing of part (b).

blastomeres as shown in **Figure 19.11a**. By allowing marked embryos to develop and then examining them at intervals during gastrulation, Vogt was able to follow the movement of cells from the exterior of the embryo to the interior.

Subsequent work with dyed frog embryos allowed researchers to map the movements of blastomeres (**Figure 19.11b**). Note that some embryonic tissues are already determined in the blastula by cytoplasmic determinants present in the egg. The cells at the animal pole are destined to become ectoderm and produce skin and nerve tissues; the cells between the two poles will develop into mesoderm and generate muscle and internal organs; the cells at the vegetal pole will become endoderm and form the gut and associated organs. During gastrulation, these three populations of cells become completely rearranged.

Although the pattern of gastrulation varies among species almost as much as cleavage patterns, certain important features are shared. For example, note in Figure 19.11b that the process begins when an opening forms in the blastula. In frogs this opening becomes round and is called the blastopore; it appears about two-thirds of the way between the embryo's animal and vegetal poles. Cells from the periphery move to the interior of the embryo through this opening; these cells are colored in the figure to make their movements easier to follow.

At the end of gastrulation, the three embryonic tissues are arranged in layers and the gut has formed. In addition, the major body axes have become visible. In frogs the blastopore becomes the anus, and the region on the animal pole side of the original blastopore opening becomes the dorsal side of the embryo. In this way, the head-to-tail and back-to-belly axes of the body become apparent.

Descriptive studies like these are immensely informative, but they do not address how the body axes become defined. The formation of the anterior-posterior and dorsal-ventral axes is among the most important events in all of development. How are the head, tail, back, and belly of the embryo specified? To answer these questions, researchers turned to experimental approaches.

The Discovery of the "Organizer"

In 1924, Hans Spemann and Hilde Mangold published the results of a dramatic experiment on newt and salamander embryos. Spemann and Mangold wanted to understand how the body axes of the embryo are established. From fate mapping studies, they knew that the cells on the animal-pole side of the blastopore became part of the animal's back. Spemann and Mangold predicted that if these cells were determined to become back prior to gastrulation, then they should still form back tissue if they were transplanted to a different location in the embryo.

To test this prediction, they transplanted embryonic tissues between embryos of two differently pigmented species of newt. The differences in pigmentation allowed them to follow the fate of the transplanted cells. For example, **Figure 9.12** illustrates an experiment in which they took cells from the dorsal side of a blastopore in a nonpigmented embryo and added the cells to the ventral side of a pigmented embryo. Much to their amazement, the transplanted cells did not just become back tissue. Instead, a second, "twinned" embryo developed that was joined to the normal host embryo. The second embryo contained both pigmented and nonpigmented cells.

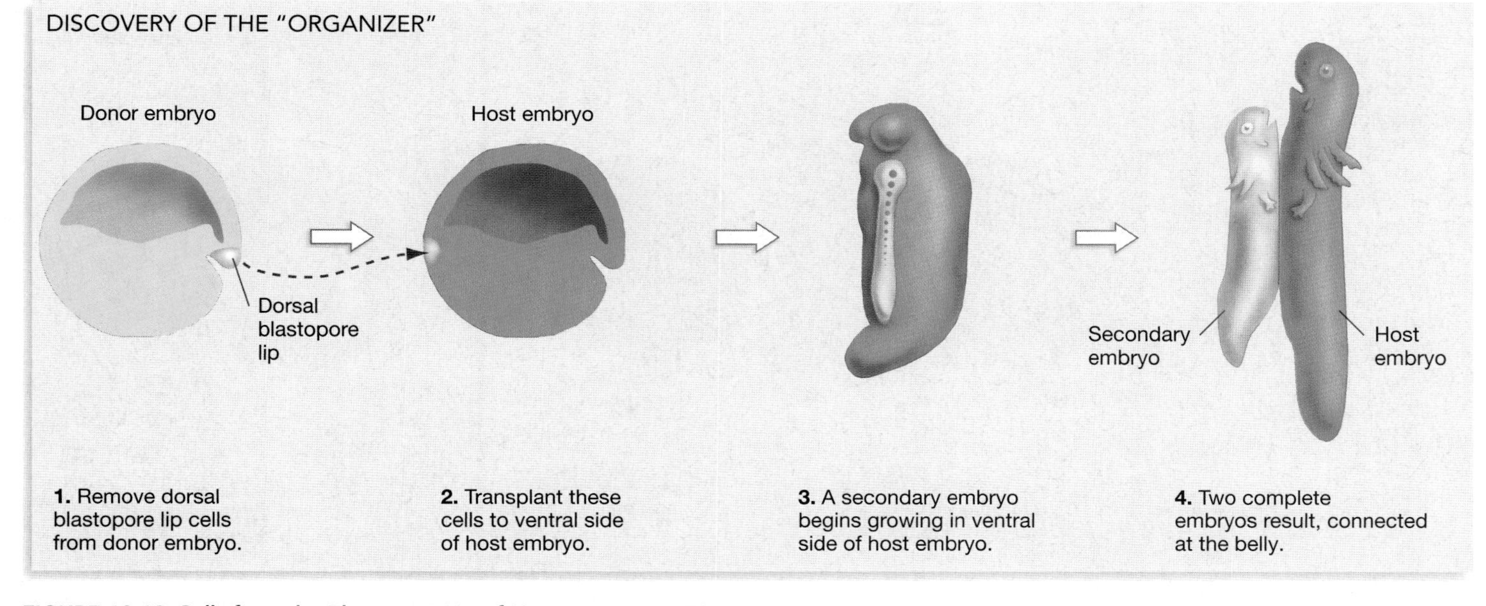

FIGURE 19.12 Cells from the Blastopore Lip of Newts Act as an "Organizer"
QUESTION Why do the twinned embryos end up belly to belly? Why do they have the same head-to-tail orientation?

CD ACTIVITY 19.1
The Gray Crescent in Frog Egg

To interpret this result, Spemann and Mangold reasoned that the transplanted cells must be able to alter the fates of host tissues. Their logic was that because the second embryo contained both donor and host cells, the transplanted cells must have recruited host cells to form part of the second embryo. Later, Spemann began calling the cells from the dorsal blastopore lip the **organizer** because of their ability to organize host and donor tissues into a second embryo. Spemann also coined the term **induction** to describe the ability of organizer cells to direct the development of other cells. His idea was that the organizer somehow induces the formation of a second embryo.

The organizer appeared to be the key to the movement of cells during gastrulation and the formation of the body axes. Further, embryos from other vertebrate species have an organizer-like region where gastrulation is initiated. How does this group of cells form, and how does it function?

Forming the Organizer An unusual observation about frog eggs provided an initial clue about the formation of the organizer. Just after fertilization, a region of gray cytoplasm appears in the zygotes of some frog species opposite the point of sperm entry. Because of its shape and color, the region is called the **gray crescent.** Several lines of evidence indicate that the gray crescent corresponds to the region where the organizer forms. For example, fate-mapping studies showed that cells derived from this region become the dorsal blastopore lip. In addition, Spemann found that when he artificially separated blastomeres at the two-cell stage, blastomeres that lacked the gray crescent developed as a mass of unorganized tissues. In contrast, blastomeres containing the gray crescent developed normally.

How does fertilization lead to the formation of the gray crescent and organizer? John Gerhart and colleagues have shown that in frogs, fertilization triggers a dramatic rearrangement of the egg cytoplasm that produces the gray crescent. As **Figure 19.13** indicates, the region of egg cytoplasm just beneath the plasma membrane is called the cortex. This cortical cytoplasm is pigmented in the animal hemisphere and unpigmented in the vegetal hemisphere. When fertilization occurs, the cortical cytoplasm rotates 30° toward the site of sperm entry. The rotation occurs because the sperm centriole orients an array of microtubules, which in turn orient the movement of cortical cytoplasm.

Why is this important? When the cortical cytoplasm rotates, a protein in the vegetal cortex is activated by being displaced toward the animal pole. Subsequently, this molecule triggers the formation of the organizer in the region where the vegetal cortex is juxtaposed with the animal-pole cytoplasm. In this way, the point of sperm entry determines the location of the blastopore on the far side of the embryo and thus the orientation of the major body axes. Work to date indicates that the key molecule is probably a protein called β-catenin, and that β-catenin may function in conjunction with Vg1 protein. Currently, research is focusing on how β-catenin is activated and how it and Vg1 trigger the formation of the organizer.

Molecules Involved in Organizer Function How does the organizer work? Because cells in the organizer can induce changes in other cells, biologists hypothesized that they must produce some sort of signaling molecule. The idea was that a cell-to-cell signal diffuses away from the organizer and affects

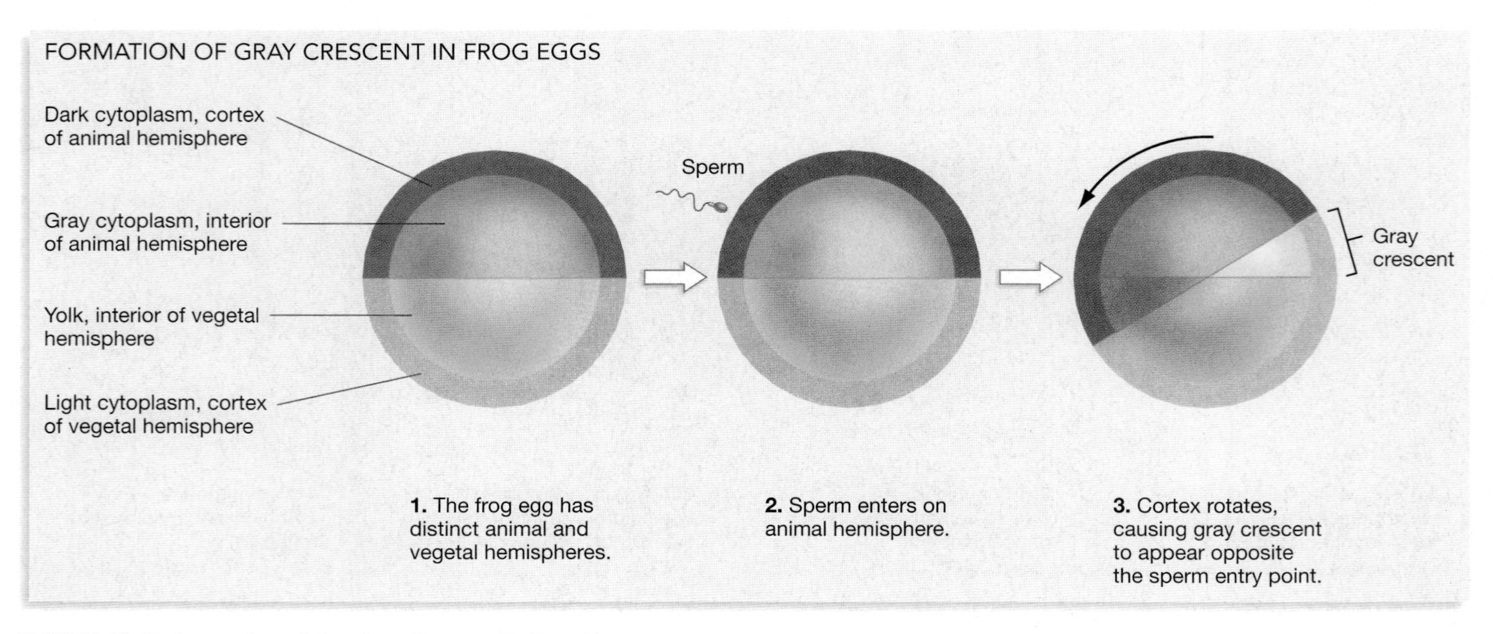

FIGURE 19.13 Formation of the Gray Crescent in Frog Eggs
As this diagram shows, the "gray crescent" in fertilized frog eggs forms opposite the point of sperm entry. The gray crescent will later form the organizer and be the site where gastrulation begins.

nearby cells. Recently, several teams of biologists have been actively searching for the molecules involved in organizer function. The research is driven by the hope that the molecules that are active in frogs will lead to the discovery of similar molecules active in the development of other species with organizer-like regions—including humans.

In frogs, the most common experimental strategy for identifying important organizer molecules is to inject blastulas with particular compounds and observe the response. For example, Bill Smith and Richard Harland began a search for organizer signals by isolating mRNAs found in frog gastrulas. They injected these candidate mRNAs one by one into embryos that had been irradiated with ultraviolet light. Because UV light inhibits the cortical cytoplasm from rotating at fertilization, irradiated embryos do not normally form organizer regions.

Among the many mRNAs that they tested, Smith and Harland found one that resulted in the formation of a complete embryo. This mRNA appeared to code for a protein that is an organizer product. To support this hypothesis, the researchers used in situ hybridization techniques to locate where the mRNA is produced in intact embryos. As predicted, the mRNA is found only in the organizer region. Smith and Harland also found that if they injected an excess of this mRNA, the treated embryos developed as an enlarged head region with no trunk or tail. As a result, Smith and Harland named the corresponding protein "noggin." They hypothesized that noggin functions in setting up the major body axes of the embryo.

Similar studies have identified other signaling molecules produced by the organizer region. Currently, research is focused on understanding exactly how noggin and these other organizer products act on target cells during and after gastrulation.

How Are Cell Movements Coordinated?

Coordinated movements of cells are a hallmark of gastrulation. In addition to understanding the molecules that determine the fate of cells during and after gastrulation, biologists are interested in understanding how the movements themselves occur. In particular, how do the cells destined to become ectoderm, mesoderm, and endoderm stay together?

An answer to this question began to emerge from research that had nothing to do with gastrulation. In 1907, H. V. Wilson published the results of a series of experiments on the dissociation and reassociation of adult sponge cells. Sponges are aquatic invertebrates with just two tissue types, ectoderm and endoderm. Wilson was able to dissociate cells from adult individuals. At first the individual cells settled to the bottom of the culture dish. But later, they began to move and stick to other cells. Eventually the cells reformed complete adult sponges. In his most dramatic experiment, Wilson dissociated the cells of adult sponges from two differently pigmented species. When he mixed them together in a culture dish, the individual cells sank to the bottom and formed mixed aggregates containing cells from both species. But as **Figure 19.14** shows, the cells sorted themselves out over time into aggregates containing cells from only one species or the other.

Wilson's work suggested that molecules on the cell surface are responsible for sponge cells adhering to one another. Because cells attached to each other in a species-specific fashion, the adhesion appeared to be highly selective. Does selective cell adhesion occur in embryos where gastrulation occurs?

Selective Adhesion Occurs in Embryonic Tissues In 1955, Philip L. Townes and Johannes Holtfreter published the results of cell-sorting experiments involving amphibian embryos. To begin their work, they dissected groups of cells from each of the three embryonic tissue layers—endoderm, mesoderm, and ectoderm. By incubating the cells in a culture medium with a high pH, they were able to dissociate the tissues into suspensions of individual cells. They then mixed cells from different germ layers in various combinations and normalized the pH to promote reaggregation.

What happened? At first, the individual cells were in random mixtures with all germ layers present. But over time, the cells began sorting themselves out. Eventually, cells from a given germ layer were associated preferentially with other cells

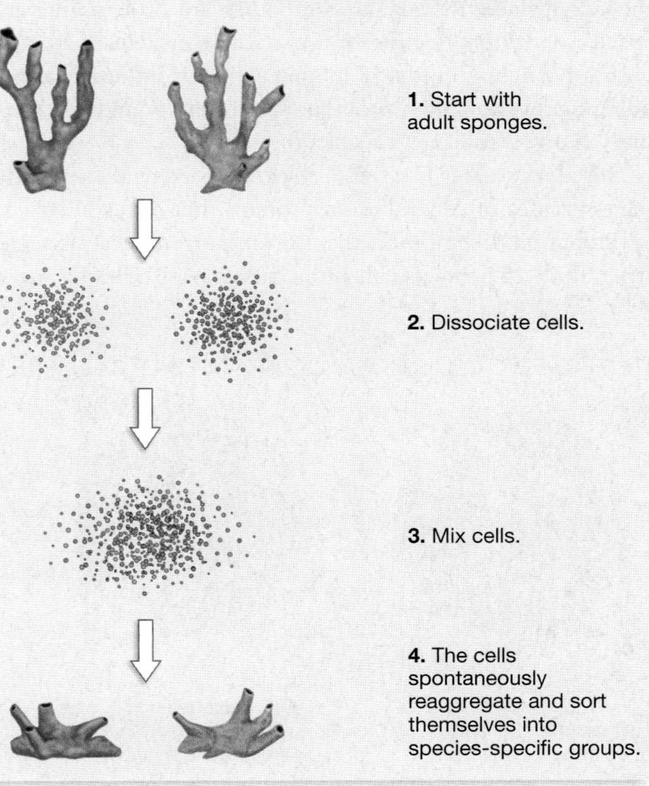

FIGURE 19.14 Sponge Cells Adhere in a Species-Specific Manner
The experiment depicted here suggested that cell-to-cell interactions are species-specific in sponges.

from that same germ layer. In addition, the spatial arrangement of cells in the experimental aggregates mimicked their arrangement in normal embryos. For example, when ectodermal and mesodermal cells were mixed, the ectodermal cells actively moved to the outside of the mass of cells while mesodermal cells moved to the inside. As **Figure 19.15** shows, ectodermal cells eventually covered the surface of an internal group of mesodermal cells, just like they do in an embryo at the end of gastrulation. When all three germ layers were mixed, ectodermal cells reassociated on the outside of the aggregate, with endodermal cells in the interior core and mesodermal cells in between.

Based on these observations, Townes and Holtfreter claimed that embryonic cells from different tissues display selective adhesion. The implication is that ectoderm, mesoderm, and endoderm stick together during gastrulation because they have molecules on their surfaces that adhere to each other. Although cells often move individually during development, molecules that promote adhesion might be important in coordinating the massive cell movements that occur during gastrulation. What are these molecules, and how do they work?

The Discovery of Cell Adhesion Proteins In the 1970s, biologists began searching for cell-surface proteins that are involved in cell-to-cell adhesion. The most common research strategy began with the production of antibodies to proteins on the cell surface. Recall that antibodies are proteins that bind specifically to one section of a particular protein. If treatment with a particular antibody inhibited cells from aggregating or adhering, biologists inferred that the protein bound by the antibody is involved in cell-to-cell adhesion.

Based on studies like these, researchers have identified three major classes of cell adhesion proteins that differ in structure and function. In embryos, the most important class appears to be a family of proteins called the **cadherins.** Cadherins are distinctive because they bind to one another in the presence of calcium (Ca^{2+}). Documenting a role for calcium was not surprising, because researchers had long known that cell adhesion in embryos depends on the presence of this ion. Chapter 18, for example, mentioned Driesch's discovery that early sea urchin blastomeres dissociated if they were placed in calcium-free seawater.

During the 1980s, Masatoshi Takeichi and colleagues obtained direct evidence for the role of cadherins in cell adhesion. These researchers experimented with a type of mouse cell that normally does not aggregate or express cadherins. When Takeichi and colleagues genetically engineered these cells to express a particular cadherin protein, they began to aggregate in the presence of Ca^{2+}.

Further, Takeichi's group and others have shown that several distinctive types of cadherins exist, and that each type is expressed on certain types of cells. For example, ectodermal cells in embryos and skin cells in adults express E-cadherin, while cells of the central nervous system express N-cadherin. Each type of cadherin binds only to its own type on the surface of other cells. Because cadherins bind in a like-to-like fashion, each type of cell adheres to its own type.

The Role of Cadherins in Early Development Takeichi and co-workers have also presented evidence that changes in cadherin expression play an important role in development. One of their most dramatic sets of studies involves the formation of the neural tube. Recall from Chapter 18 that the neural tube is the progenitor of the spinal cord and brain of vertebrates, and that it forms once gastrulation is complete and organogenesis is under way.

As **Figure 19.16** indicates, the biologists found that throughout gastrulation, the ectodermal layer in the embryo expresses E-cadherin. Later, however, cells that are destined to form the neural tube gradually turn off E-cadherin expression and turn on N-cadherin expression. Eventually they ex-

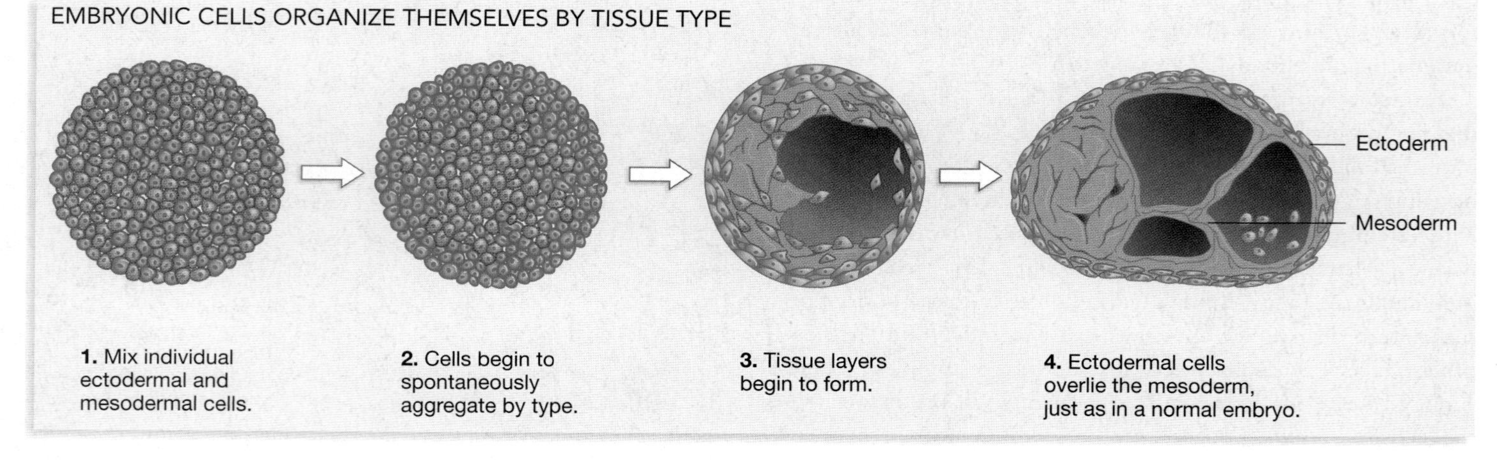

FIGURE 19.15 Embryonic Cells Organize Themselves by Tissue Type
EXERCISE Sketch drawings predicting the result of mixing mesodermal and endodermal cells. Do the same for a mix of ectodermal, mesodermal, and endodermal cells.

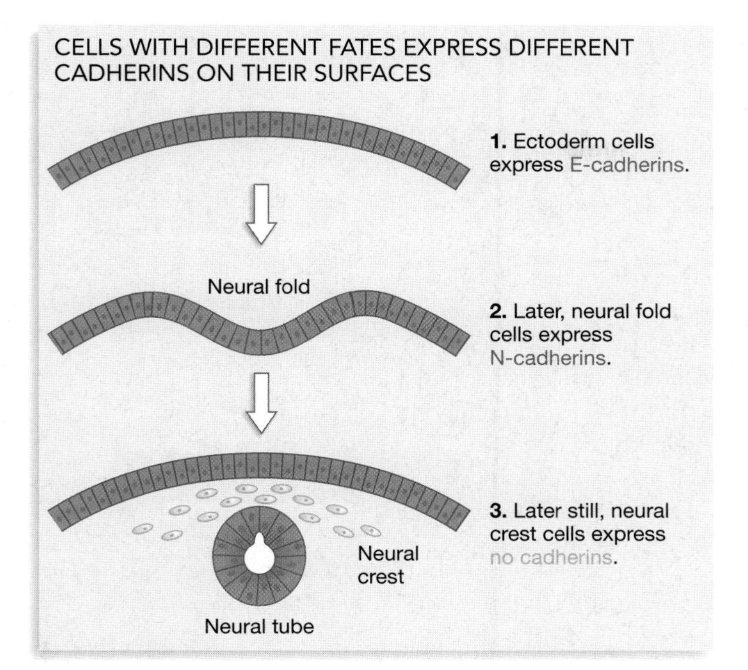

FIGURE 19.16 Cells with Different Fates Express Different Cadherins on Their Surfaces
Cells that switch from expressing E-cadherin to N-cadherin aggregate and form the neural tube. Cells that switch to expressing no cadherin at all become neural crest cells. Neural crest cells migrate throughout the embryo and eventually become nerve cells, bone and cartilage cells in the face, connective tissue, or pigment cells.
QUESTION Why is it logical that migrating cells stop expressing cadherins?

press only N-cadherin. Ectodermal cells that express different cadherin subclasses separate from each other. As a result, the neural tube is formed with an overlying layer of nonneural ectoderm.

This result could be criticized, however, because it is correlative in nature. Stated another way, a critic could argue that the change in cadherin expression is the result of neural tube formation, and not the cause. To address this issue, the biologists injected the mRNA for N-cadherin into one of the blastomeres present at the two-cell stage of development. By following this cell's descendants later in development, they were able to identify specific abnormalities associated with the expression of N-cadherin. For example, when N-cadherin was expressed in nonneural ectoderm as well as in the neural tube, the two cell populations did not separate normally. As a result, neural tube formation failed. Other researchers have shown that blocking of N-cadherin expression in the neural tube results in a similar defect.

These experiments underscore a prominent theme from Chapter 18. Plant and animal development is based on differential gene expression. In this case, the expression of a specific cadherin protein allows distinct cell types to aggregate and move together during gastrulation and organogenesis. What causes a particular cell to change its cadherin gene expression? Although work on cadherins and other cell adhesion molecules has advanced rapidly over the past several decades, this question remains unanswered. It is highly probable, though, that differential expression of cadherin genes depends on the type of signaling molecules and transcription factors introduced in Chapter 20.

Essay Treating Human Infertility

In 1978 the world welcomed the first "test-tube baby" when Louise Brown was born in Britain. Unlike babies born previously, Louise did not develop from a zygote produced by natural conception. Instead, she developed from a zygote created by fertilization in a laboratory dish. Since then, over 100,000 babies have been produced by the technique called in vitro fertilization (IVF).

What alternative strategies are available?

Many human couples are unable to conceive children naturally. For them, IVF can be an important alternative strategy for becoming pregnant. In IVF, egg cells isolated from the woman's ovaries are combined with sperm cells collected from the man's semen in small dishes, under conditions that attempt to duplicate the environment inside the female reproductive tract. Typically, 50,000 to 5 million sperm are used per egg. As in natural conception, fertilization occurs when a healthy sperm penetrates the zona pellucida and fuses with the egg. The resulting embryos are then surgically implanted in the mother's uterus.

IVF is often successful for couples whose infertility results from problems in the woman's reproductive system, such as damaged or absent fallopian tubes. But for up to 40 percent of infertile couples, the problem stems from low sperm counts, poor sperm motility, deformed sperm, or other male problems. Traditional IVF seldom works with such couples because it requires large numbers of high-quality sperm.

In the early 1990s, though, improvements to IVF were introduced that provided new hope for infertile men. In one technique, called subzonal injection or SUZI for short, a concentrated sample of sperm is examined under a microscope. Five to 10 of the healthiest sperm are drawn into a microneedle and

(Continued on page 388)

(Essay continued)

then injected directly into the zona pellucida. Depositing the sperm closer to the egg increases the chances of sperm-egg fusion. SUZI may not help men who produce very few sperm, however, and it leads to polyspermy and inviable embryos in about one-third of attempts. A more recent technique called intracytoplasmic sperm injection (ICSI, pronounced "ICK-see") overcomes these problems. In ICSI, a single sperm is drawn into a microneedle and injected into the cytoplasm of the egg. In a significant fraction of attempts, ICSI results in the fusion of sperm and egg nuclei and a viable embryo.

Even if IVF, SUZI, or ICSI are successful, only a minority of couples achieves pregnancy when an embryo is transferred to the uterus. In some cases, failure results from problems with implantation in the uterine wall. To develop in the mother's womb, the embryo must break out or hatch from the zona pellucida and burrow into the uterine wall. Eggs from women over age 40 often have problems hatching. To help overcome these problems, a physician may assist hatching by using a micropipet to deliver a small amount of acid or enzyme solution to the zona pellucida. This treatment facilitates hatching by dissolving the treated region of the zona.

Because of IVF and other forms of assisted reproductive technologies, the chances of naturally infertile couples conceiving their own children have greatly increased. An intensive research program undertaken over the past three decades has led to huge advances in our understanding of human fertilization and early development, and has allowed many couples to experience the joy of raising their own children.

Chapter Review

Summary

The development of an organism actually begins with the formation of gametes in its parents. In plants and animals, sperm cells contribute a haploid genome to the embryo; depending on the species involved the sperm cell may also contribute a few other cell components. Eggs, in contrast, contribute a haploid genome and a large amount of cytoplasm to the embryo. In flowering plants the egg cytoplasm is augmented by nutrient-rich endosperm tissue; in animals the egg cytoplasm often includes nutrient-rich yolk as well as cytoplasmic determinants that direct development. One factor found in the cytoplasm of frog eggs is called Vg1 and is localized to the yolk-rich or vegetal pole of the egg. After *Vg1* mRNA is transported to the vegetal pole along microtubules, it becomes associated with microfilaments just under the cell membrane. Later in development the Vg1 protein is involved in specifying mesoderm. The protein may also have a role in establishing the organizer region.

Fertilization is the best known of all cell-to-cell interactions, primarily because of observational and experimental studies on sea urchin eggs and sperm. When a sea urchin sperm contacts the jelly layer surrounding the egg, the acrosome in the sperm head releases digestive enzymes, and microfilaments polymerize to form a projection. When the sperm reaches the egg itself, a protein called bindin on the sperm-head membrane binds to a receptor on the envelope surrounding the egg. Because bindin and egg-cell receptors are species-specific, cross-species fertilizations are prevented. After the sperm- and egg-cell membranes make contact and fuse, a wave of calcium ions is released from stores inside the egg. The increase in Ca^{2+} concentration causes cortical granules to fuse with the egg-cell membrane. The contents of the cortical granules cause a fertilization envelope to rise off the egg cell membrane and provide long-term protection against multiple fertilization.

The fertilization sequence is similar in mammals, except that no bindin-like protein has yet been identified in sperm. Plants, in contrast, undergo a very different series of events. In flowering plants, two sperm nuclei leave the germinating pollen grain and migrate to the ovule. One of the sperm nuclei fertilizes the egg to form a zygote, while the other sperm fuses with two nuclei near the egg to form the nutritive tissue called endosperm.

In animals, development begins with a series of cell divisions that divide the egg cytoplasm into a large number of cells. The pattern of these cleavage divisions varies widely among species. In addition, evidence from work on snail cleavage suggests that in some species the pattern of cleavage is controlled by cytoplasmic determinants present in the egg.

Once cleavage is complete, the embryo consists of a sphere of cells. Fate-mapping studies with dyes and other tools have shown that cells undergo massive movements during gastrulation. These movements arrange the three embryonic tissues in layers and make the back-to-belly and head-to-tail axes of the body visible. In many species of vertebrates, a region of cells near the dorsal lip of the blastopore acts as an "organizer" during gastrulation. Cells in the organizer secrete proteins like noggin that induce changes in target cells.

The coordinated movements of cells during gastrulation are possible in part because cells of the same type have proteins on their surfaces that promote cell-to-cell adhesion. Changes in cadherins or other cell-to-cell adhesion proteins may be important in regulating cell movements during and after gastrulation. For example, when ectodermal cells differentiate to form the precursors of nerve cells, they switch from expressing E-cadherin to expressing N-cadherin. These types of changes in gene expression are a key to differentiation—an event explored in detail in Chapter 20.

Questions

Content Review

1. How are the vitelline envelope of sea urchins and the zona pellucida of mammals similar?
   a. They are a gelatinous coat that protects the egg.
   b. They hold the cortical granules, which activate the block to polyspermy.
   c. They hold stores of Ca^{2+}, which activate the block to polyspermy.
   d. They are an extracellular matrix that sperm bind to.
2. What happens during the acrosome reaction?
   a. Bindin binds to the egg-cell receptor for sperm.
   b. The sperm- and egg-cell membranes fuse.
   c. Enzymes that digest the egg jelly layer are released, and microfilaments in the tip of the sperm head polymerize to form a point.
   d. The centriole released from the sperm orients microtubules in the fertilized egg and causes the cortical cytoplasm to rotate 30°.
3. Many flowering plant species have elaborate mechanisms to prevent pollen from the same individual from germinating on its own female reproductive parts. Why?
   a. to prevent self-pollination and inbreeding
   b. to prevent polyspermy
   c. to prevent cross-species fertilization and the production of dysfunctional hybrid offspring
   d. to prevent double fertilization and the formation of endosperm
4. The text claims that cleavage simply divides the egg cytoplasm into a large number of cells. Which of the following facts supports this conclusion?
   a. Cleavage occurs extremely rapidly.
   b. Cleavage produces an enormous number of cells.
   c. The total volume of the egg and the cells that result from cleavage is the same.
   d. Cleavage results in a sphere of cells that is ready to undergo gastrulation.
5. What is a cadherin?
   a. a glycoprotein found in the zona pellucida of mammal eggs
   b. a molecule secreted by the organizer that induces changes in target cells
   c. a cell-adhesion protein found on the surface of animal cells
   d. a cytoplasmic determinant found in the cortical cytoplasm of frog eggs

Conceptual Review

1. How did experiments with drugs support the hypothesis that *Vg1* mRNA is localized to the vegetal hemisphere of frog eggs along microtubules and then bound to microfilaments at the vegetal pole?
2. Given their normal methods of reproduction, why is it logical that sperm-egg interactions are species-specific in sea urchins but not in mammals?
3. Compare and contrast yolk and endosperm. How are their structure and function similar and different? Be sure to compare when and how they are produced.
4. How did analyses of cleavage and shell-coiling patterns in snails support the hypothesis that cleavage is controlled by cytoplasmic determinants present in the egg? Does the observation that transcription does not start in some animals until cleavage is underway support this hypothesis or challenge it? Explain.
5. In frogs, how does gastrulation make the head-to-tail and back-to-belly axes of the body visible?
6. Why is it significant that cells from different tissue types express different cell adhesion molecules? What evidence suggests that these tissue-specific adhesion molecules are important in development?
7. Consider that transcription occurs at a low level until cleavage is well underway in some animal species, and that in vertebrates the location of the dorsal blastopore lip—and thus the major body axes—is determined at fertilization. Then explain Sydney Brenner's claim that "A description of the organism is already written into the egg."
8. Compare and contrast the structure and function of a sperm and an egg. How do sperm and egg have to interact in order for fertilization to occur?

Applying Ideas

1. Many questions about fertilization and early development remain unanswered:
   - Is there a protein on the sperm head of mammals that binds specifically to ZP3 in the zona pellucida?
   - How does the entry of sperm cause Ca^{2+} to be released from intracellular stores, so that polyspermy will be blocked?
   - How are cytoplasmic determinants and factors other than *Vg1* localized to specific regions of the egg? How is yolk localized to particular regions of the egg?
   - In frogs, why and how does the organizer form opposite the point of sperm entry?
   - If the zygotic genome is transcribed at the two-cell stage in mammals, do cytoplasmic determinants play a significant role in early development?
   - What forces cause cells to move during gastrulation?

   Choose one of these questions and design an experiment that would contribute to answering it.
2. Suppose you are a physician at a fertility clinic. A couple comes to you for assistance with becoming pregnant. The woman is 41 years old, and the man has a low sperm count. Based on the information in this chapter, which treatment would you recommend and why?

CD-ROM and Web Connection

CD Activity 19.1: The Gray Crescent in Frog Eggs *(animation)*
(Estimated time for completion = 5 min)
During fertilization, why is the site of sperm entry important to the embryo's development?

At your **Companion Website** (http://www.prenhall.com/freeman/biology), you will find self-grading exams and links to the following research tools, online resources, and activities:

Gametogenesis in Sea Urchins
Explore several aspects of early development, including the normal and abnormal production of sperm and eggs.

Genetic Control of Ovule Development
This article focuses on the interactions between cell layers in *Arabidopsis,* including the molecular interactions involved in oogenesis.

Hans Spemann
This Nobel site describes Hans Spemann's research leading to the elucidation of the cellular organizer responsible for the onset of gastrulation in amphibians.

Additional Reading

Kubiak, J. Z. and M. H. Johnson. 2001. Human infertility, reproductive cloning and nuclear transfer: a confusion of meanings. *BioEssays* 23: 359–364. Proposes new techniques for treating infertile couples.

Pedersen, R. A. 2001. Sperm and mammalian polarity. *Nature* 409: 473-474. Introduces recent data indicating that the point of sperm entry helps determine the location of the body axes in mammal embryos.

Villeneuve, A. M. 2001. How to stimulate your partner. *Science* 291: 2099-2101. Reviews recent data on sperm proteins that stimulate eggs to complete meiosis after fertilization has occurred.

What Determines a Cell's Fate?

20

When a physician and researcher named Lewis Thomas considered how a human being develops from a fertilized egg, he could only marvel. "You start out as a single cell derived from the coupling of a sperm and an egg, this divides into two, then four, then eight, and so on, and at a certain stage there emerges a single cell which will have as all its progeny the human brain. The mere existence of that cell should be one of the great astonishments of the earth. People ought to be walking around all day, all through their waking hours, calling to each other in endless wonderment, talking of nothing except that cell. It is an unbelievable thing, and yet there it is, popping neatly into its place amid the jumbled cells of every one of the several billion human embryos around the planet, just as if it were the easiest thing in the world to do.*"

This chapter focuses on how the cells in an embryo finish "popping into place." Earlier chapters explored the start of this process. When an animal completes gastrulation, the embryonic tissues are arranged in layers and the major body axes are visible. By the time a plant embryo has finished an early round of cell divisions, the root-to-shoot axis is established and embryonic tissue layers have begun to form. As this chapter shows, identifiable structures and organs then start to take shape. In a human embryo, a distinctive head and tail and the beginnings of nerves and muscle and other specialized tissues emerge. An orchid develops recognizable leaves, roots, and stem.

Once the earliest stages of development are complete, how does the overall body plan of an animal or plant take shape? Later, how do cells

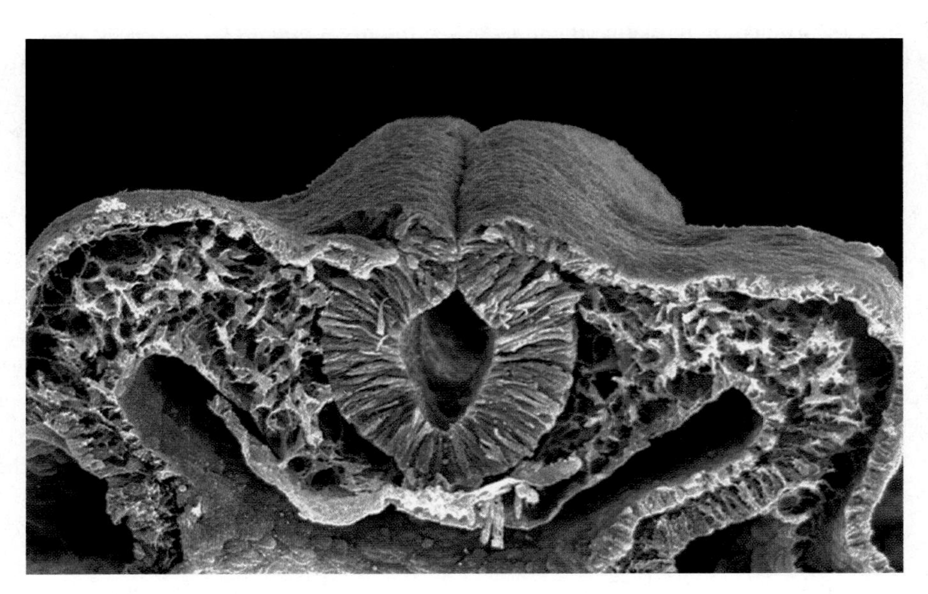

This is an artificially colored scanning electron micrograph of a chick embryo. Ectodermal cells along the back of the embryo, at the top of the photograph, have folded into an embryonic structure called the neural tube. The neural tube develops into the brain and spinal cord of the adult animal. This chapter explores how cells form specific structures based on their position in the embryo and then differentiate to form specialized cells like nerve or muscle.

*Thomas, L. 1979. *The Medusa and the Snail* (New York: Viking Press), p. 156. Note that Thomas was exercising some poetic license here. The brain actually arises from a group of cells in the embryo rather than a single cell.

become committed to becoming a specialized cell in a muscle or bone or root?

Questions like these are the heart of this chapter. To answer them, section 20.1 introduces work on embryonic development in *Drosophila melanogaster*—the best studied of all multicellular organisms. Research on fruit flies has emphasized the importance of a cell's position in determining its fate. Position is important because a cell's fate is specified by localized cytoplasmic determinants and signals from other cells. The same themes carry over to section 20.2, which examines recent research on the best studied of all plants, the mustard *Arabidopsis thaliana*.

Once the overall form of the body is established and cells are arranged in space, cells start becoming organized into tissues and organs. Eventually cells differentiate and begin to express tissue-specific genes. Section 20.3 introduces how this tissue-specific gene expression occurs by exploring research on the transformation of mesoderm into cells that express muscle-specific proteins.

The fundamental messages of this chapter are that a cell's fate during development depends on its position and is established in a progression of steps. Initially, a cytoplasmic determinant may mark a cell as mesoderm. After gastrulation, other molecules may signal that the cell is now part of the mesoderm on the embryo's dorsal (back) side. Subsequent signals might indicate that this particular cell will become part of the dorsal mesoderm that contributes to muscle or bone. Finally, the cell may receive signals that trigger the production of muscle-specific proteins. A generalized embryonic cell has become a specialized mature cell. How does this process begin? If a cell's fate depends on its position, how does it "know" where it is in the body?

20.1 Pattern Formation in *Drosophila*

Biologists refer to the events that determine the spatial organization of an embryo as **pattern formation**. If a molecule signals that a target cell is in the embryo's head, or tail, or dorsal side, or ventral side, it is involved in pattern formation. Pattern formation is the first step in determining a cell's fate. Before a cell becomes part of a muscle or nerve or gut lining, it receives precise information about where it is positioned in the body.

To understand how cells get information specifying their position inside the embryo, biologists turned to the fruit fly *Drosophila melanogaster*. Because fruit flies produce large numbers of offspring rapidly, researchers could survey laboratory populations for rare mutant embryos in which the normal spatial relationships among cells are disrupted. In other words, biologists took a genetic approach to studying development. The idea was that identifying individuals with abnormal pattern formation would make it possible to find the genes and gene products responsible for normal development. A *Drosophila* larva that lacks a head, for example, is likely to have a mutation in a gene that helps pattern the head.

In the 1970s, Christiane Nüsslein-Volhard and Eric Wieschaus undertook a massive effort to identify pattern-formation mutants in *Drosophila*. They began by exposing adult flies to treatments that cause mutations by damaging DNA. Later, they examined embryos or larvae descended from these individuals for body plan defects. After intensive effort, Nüsslein-Volhard and Wieschaus were able to identify over 100 genes that play fundamental roles in pattern formation. What are these genes, and what do their products do?

The Discovery of *bicoid*

One of the most dramatic mutations that Nüsslein-Volhard and Wieschaus analyzed is illustrated in **Figure 20.1**. The embryo on the right is missing all of the structures normally found in the anterior end. Instead of having a head, thorax, and several anterior abdominal segments, these mutants have duplicated posterior structures. The gene responsible for this phenotype is called *bicoid*, meaning "two-tailed." The mutation is lethal because larvae do not develop beyond the stage shown in the figure. Based on its phenotype, Nüsslein-Volhard and Wieschaus suspected that the *bicoid* product plays a role in pattern formation along the anterior-posterior body axis.

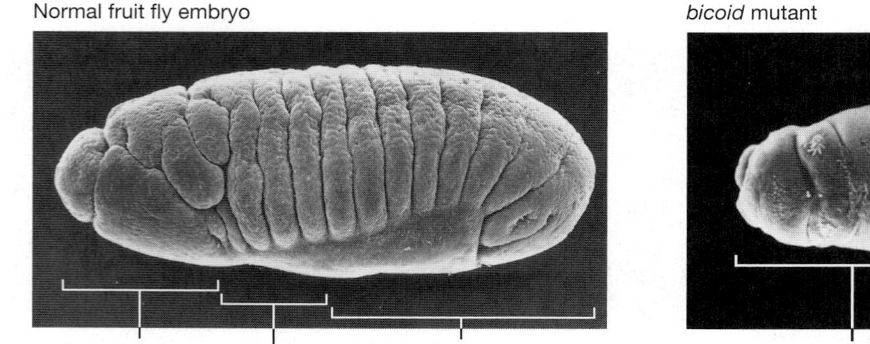

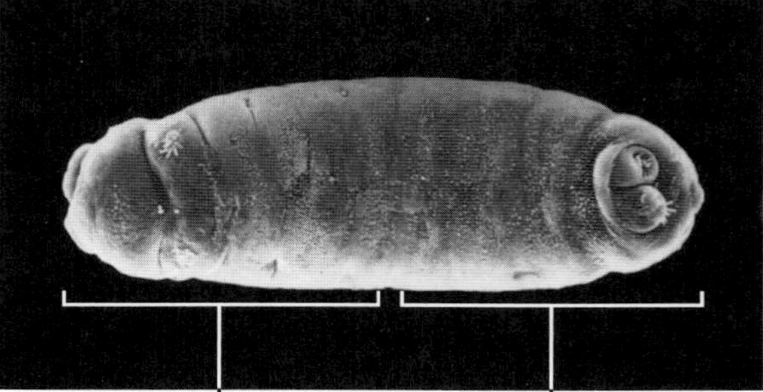

FIGURE 20.1 A Pattern-Formation Mutant
Normal embryos (left) have a distinctive head, thorax, and abdominal region. In *bicoid* mutants (right), embryos have no head or thoracic segments. Instead they have duplicated sets of the posteriormost regions, making them "two-tailed."

When Nüsslein-Volhard and Wieschaus did a series of mating experiments with flies that carried the mutant *bicoid* allele, they noticed that the gene displays a pattern of inheritance called **maternal effect inheritance**. Maternal effect inheritance means that a gene is transcribed in the mother and affects the phenotype observed in offspring. As **Box 20.1** (page 394) explains, it is the mother's genotype and not the offspring's genotype that determines the expression of the trait. Like the snail-shell-coiling gene introduced in Chapter 19, *bicoid* affects the development of an embryo because it is expressed in the mother's tissues during egg formation. The *bicoid* gene product is deposited in the egg along with other maternal determinants and yolk.

What Does *bicoid* Do? To determine how the *bicoid* product works, Nüsslein-Volhard and colleagues crossed individuals carrying the mutant form of *bicoid* with individuals carrying other known genetic markers. This technique, introduced in Chapter 17, allowed them to map where *bicoid* occurs on the *Drosophila* chromosomes. Subsequently, they were able to isolate the DNA sequence that encodes the *bicoid* product. This was a crucial step. Understanding the DNA sequence allowed the biologists to add fluorescent or radioactive markers to single-stranded copies of *bicoid* DNA. By treating *Drosophila* adults and embryos with these labeled copies, they could determine where *bicoid* mRNA is located. Chapter 19 introduced this in situ hybridization technique in more detail.

When Nüsslein-Volhard's group treated adult flies with labeled copies of *bicoid* DNA, they found that the corresponding mRNA is found only in the mother's ovary. This result was consistent with the results of the breeding experiments that identified *bicoid* as a maternal effect gene. As **Figure 20.2a** shows, the mRNA is deposited specifically in the anterior end of the developing egg.

What is the fate of this mRNA? To answer this question, the biologists set out to determine where and when the Bicoid protein is produced. As **Figure 20.2b** shows, they began by making the Bicoid protein from *bicoid* DNA. Using techniques that will be introduced later in the text, they produced antibodies that specifically bind to the Bicoid protein. Recall that antibodies bind to specific segments of a protein. When researchers attach a fluorescent or radioactive compound to an antibody, it can be used as a labeled probe to mark the location of a specific protein. As Figure 20.2b shows, this antibody-staining experiment allowed researchers to document the location and quantity of Bicoid protein in embryos.

By doing the experiment outlined in Figure 20.2b at different stages of development, Nüsslein-Volhard and co-workers found that Bicoid protein first appears at fertilization. Further, their data indicated that the protein diffused away from the site of translation at the anterior end of the embryo. As the photograph at the bottom of the figure shows, the result is a steep concentration gradient. The protein is abundant in the anterior end but declines to low concentrations in the posterior end. When the nuclei in the early embryo become surrounded by cell membranes, the Bicoid protein is trapped in the newly formed cells.

To pull these observations together, Nüsslein-Volhard and co-workers hypothesized that high concentrations of Bicoid protein inside cells lead to the formation of anterior structures such as the head, with progressively lower concentrations giving rise to thoracic segments and the first abdominal structures. Absence of Bicoid, in contrast, results in formation of posterior structures. Because the eggs of mothers homozygous for the mutant form of *bicoid* lack functional mRNAs, no protein is produced anywhere in the embryo. The result is a larva that lacks anterior structures.

(a) Where is *bicoid* mRNA located?

In the fly ovary, *bicoid* mRNA is deposited into the anterior end of the developing egg

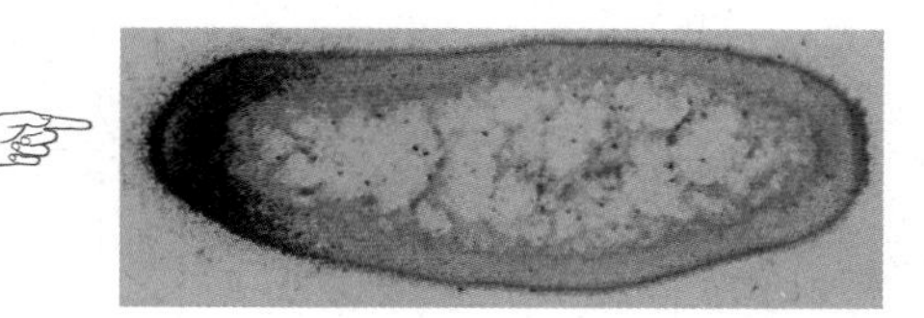

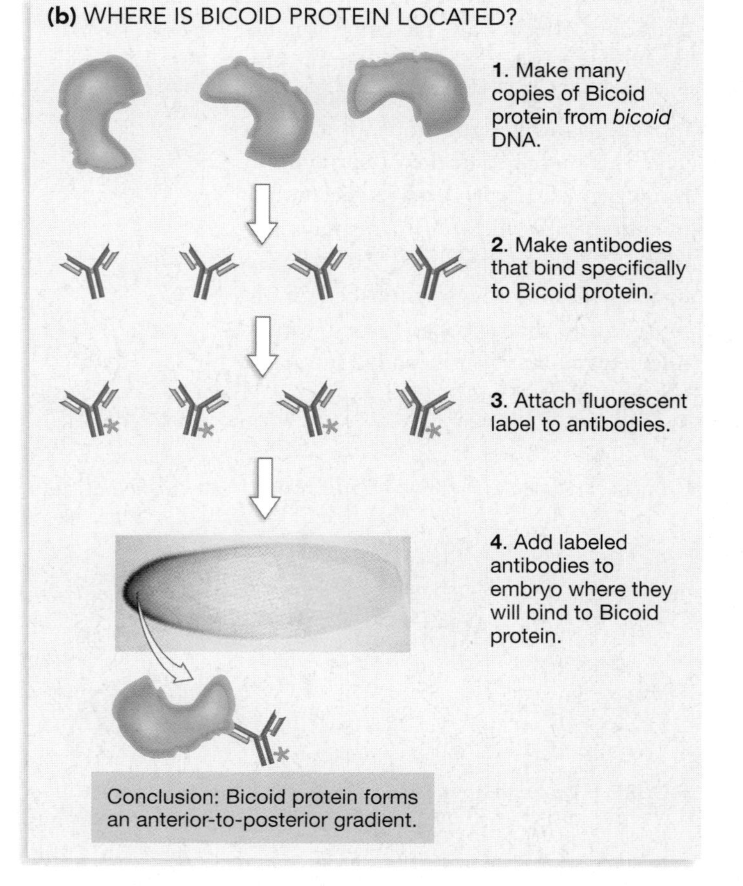

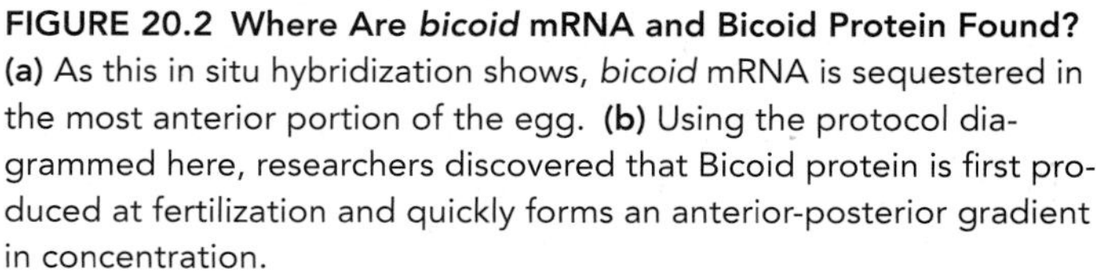

FIGURE 20.2 Where Are *bicoid* mRNA and Bicoid Protein Found? **(a)** As this in situ hybridization shows, *bicoid* mRNA is sequestered in the most anterior portion of the egg. **(b)** Using the protocol diagrammed here, researchers discovered that Bicoid protein is first produced at fertilization and quickly forms an anterior-posterior gradient in concentration.

BOX 20.1 Maternal Effect Inheritance

When researchers find a gene that appears to be important early in development, one of their first goals is to establish how the gene is inherited. Breeding experiments can answer a fundamental question about the gene. Is it expressed in the mother, meaning that it codes for a cytoplasmic determinant present in the egg, or is it expressed in offspring?

Figure 1 shows how crosses are set up to provide an answer. Note that the mutant gene being considered here is recessive, and that investigators arrange matings between parents that are heterozygous at this locus. The two sets of crosses show the contrasting patterns of inheritance in a maternal effect gene and a gene that is expressed in offspring. The key observation is that in traits with maternal effects, all offspring of homozygous mothers have the mutant phenotype—whether the offspring is homozygous or heterozygous. But if the gene is transcribed in the offspring, only homozygous offspring have the mutant phenotype. Because *bicoid* displays maternal effect inheritance, researchers concluded that it is transcribed in the mother and that it codes for a cytoplasmic determinant that is loaded into eggs.

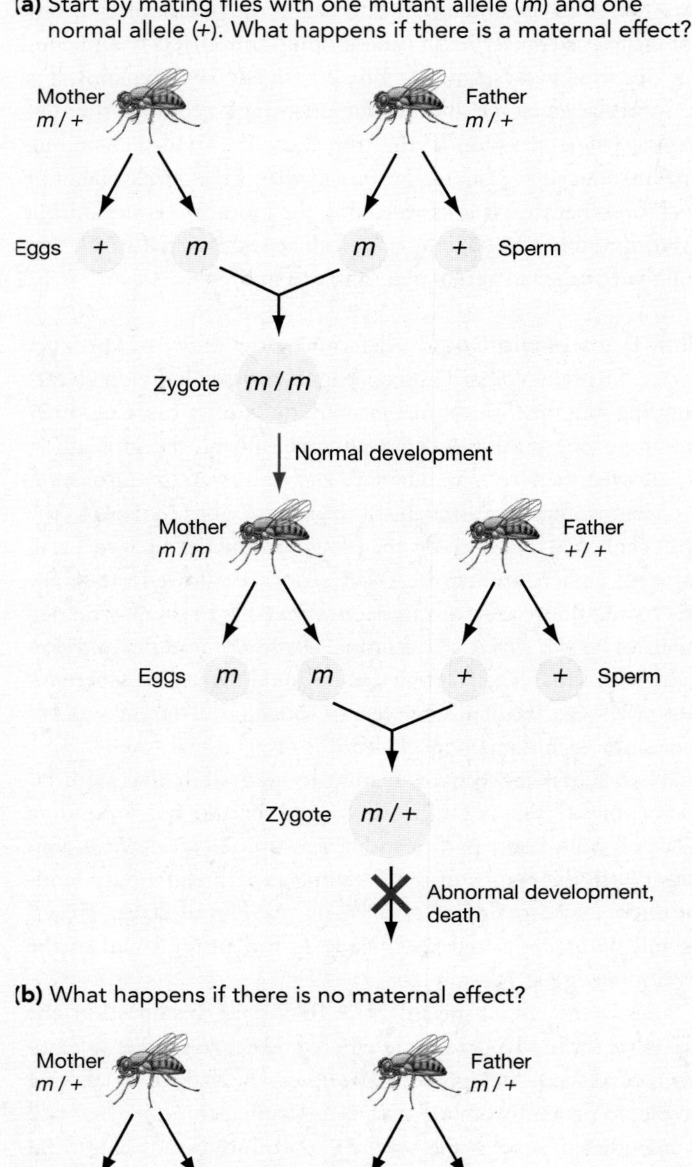

FIGURE 1 Documenting That a Gene Has a Maternal Effect

Testing the Gradient Hypothesis The hypothesis that a Bicoid gradient sets up the head-to-tail axis of the fly embryo is consistent with the data from antibody staining experiments. But can the proposal be tested more rigorously? To answer this question, Nüsslein-Volhard and colleagues isolated *bicoid* mRNA from wild-type individuals and injected it into eggs from females that lacked a functioning copy of the *bicoid* gene. Because the treated eggs developed normally, the researchers could infer that *bicoid* is indeed responsible for establishing a normal anterior-posterior gradient.

In addition to "rescuing" mutant larvae by injecting bicoid mRNA, the researchers tested the gradient hypothesis by artificially altering the distribution of Bicoid protein in developing embryos and determining the effects on the body plan. For example, the biologists produced purified *bicoid* mRNA and injected it into wild-type embryos. The treated individuals developed head structures at the site of injection—even at the posterior pole. Observations like these provided strong support for the gradient hypothesis.

Bicoid Is a Transcription Factor The Bicoid protein appears to provide cells with information about where they are along the anterior-posterior or head-to-tail axis of the embryo. How is this information translated into action? The in situ hybridization and antibody-staining experiments indicated that Bicoid was active when the fly embryo consists of many nuclei scattered throughout the egg-cell cytoplasm. How does Bicoid act on these target nuclei?

DNA sequencing studies helped answer this question. The *bicoid* gene contains sequences that are typical of regulatory transcription factors—the proteins that bind to enhancers or other regulatory sequences in DNA and control the transcription of specific genes (see Chapter 15). Based on this observation, biologists concluded that the Bicoid protein must enter nuclei, bind to DNA, and either increase or decrease the expression of specific genes. Bicoid either turns on genes responsible for forming anterior structures, shuts down the transcription of genes responsible for producing posterior structures, or both.

To summarize, a gradient in Bicoid protein concentration provides cells with information about their position along the anterior-posterior body axis. Because it is sequestered in a precise location in the egg and because it affects the fate of cells later in development, *bicoid* can be considered a cytoplasmic determinant. The absence of the *bicoid* product leads to a deformed body and death. As other research showed, however, *bicoid* is just one of dozens of pattern-formation genes found in fruit flies.

The Discovery of Segmentation Genes

Among the dozens of mutants isolated by Nüsslein-Volhard and Wieschaus, one called *hunchback* attracted a great deal of attention. Embryos with mutant forms of *hunchback* resemble *bicoid* mutants. Both types of embryos lack a series of consecutive segments from the anterior end. Breeding experiments like those described in Box 20.1 showed that the *hunchback* mRNA responsible for these mutations is produced in the embryo itself as well as the mother, however.

To understand what *hunchback* does, recall from Chapter 18 that the bodies of fly larva and adults are partitioned into a series of segments. The *hunchback* gene turned out to be one of many genes that affects the identity or arrangement of these segments. Nüsslein-Volhard and Wieschaus were eventually able to identify three general classes of genes that affect segmentation, all of which except for *hunchback* are transcribed for the first time in the embryo and not the mother. The three types of segmentation genes are called **gap genes**, **pair-rule genes**, and **segment polarity genes**. Gap gene mutants like *hunchback* lack several consecutive segments. Pair-rule mutants lack alternative segments and have just half of the normal total number of segments. Segment polarity mutants lack portions of each segment. In many pair-rule mutants, the missing portion is replaced by a mirror-image duplication of the intact part of the same segment.

Figure 20.3 shows the patterns of mRNA expression for segmentation genes in normal larvae. According to these in situ hybridizations, gap genes are expressed in broad regions

Where are the protein products of segmentation genes located?

Gap gene

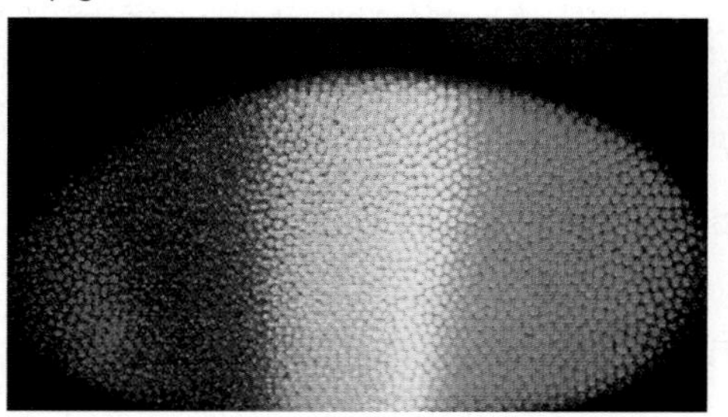

Pair-rule gene

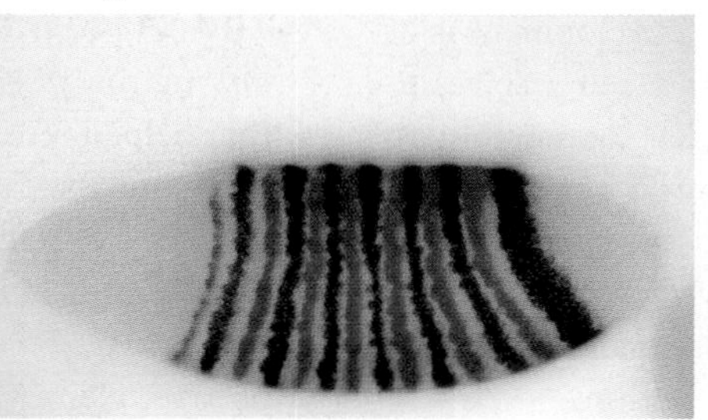

Segment polarity gene

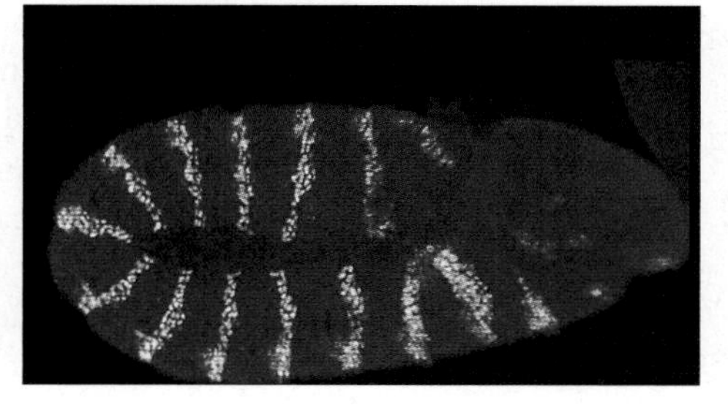

FIGURE 20.3 Segmentation Genes in Fruit Flies
Antibody staining shows the location of segmentation gene products in fly embryos. QUESTION From left to right, these photographs show embryos at progressively later stages of development. Why?

along the head-to-tail axis. Pair-rule genes, in contrast, are expressed in alternative segments, while segment polarity genes are expressed in restricted regions of each segment. If a segmentation gene does not function, the mutant individual lacks segments or parts of segments where the mRNA is found in normal individuals. Based on these data, Nüsslein-Volhard and Wieschaus and co-workers concluded that the segmentation genes are responsible for defining the segmented body plan of the fruit fly.

Do Segmentation Genes Interact? When researchers first identified the segmentation genes and described when and where they are expressed, they noticed that they are expressed in sequence. The gap genes are expressed first, followed by pair-rule genes, followed by segment polarity genes. This observation suggested that the segmentation genes might interact in some way. Specifically, investigators hypothesized that the gap genes might activate transcription of pair-rule genes, which in turn may trigger the expression of pair-rule and segment polarity genes.

Although research on interactions among segmentation gene products is continuing, an enormous amount of progress has already been made. One of the key early discoveries was that the Bicoid protein activates transcription of *hunchback*. As a result, the *hunchback* mRNA protein is found in high concentration in the anterior part of the embryo where the Bicoid protein is abundant. Subsequent work showed that *hunchback* and other gap genes encode transcription factors that regulate the expression of pair-rule genes. Each gap gene regulates the production of a defined set of pair-rule proteins. In addition, several of the gap genes' products regulate the expression of other gap genes or even their own expression. The pair-rule genes activated by the gap genes also encode transcription factors; some of the genes they regulate are segment polarity genes.

As **Figure 20.4** shows, this sequence of events can be described as a regulatory cascade or hierarchy. The *bicoid* gene, gap genes, pair-rule genes, and segment polarity genes each define a level in the cascade or hierarchy. As development proceeds, genes at levels further down in the cascade or hierarchy are progressively activated or repressed. The products of these regulatory genes may also interact with genes at the same level.

The relationships described in Figure 20.4 support the claim made in the introduction to this chapter that development is a step-by-step process. The key points are that a cell receives a different set of signals at each step in the developmental sequence, and that each signal causes a change in gene expression. Long before a cell begins expressing tissue-specific proteins that identify it as a muscle cell or a nerve cell, it expresses genes that identify it as a cell in a particular segment or region of the body. This is the essence of pattern formation.

Work on *Drosophila* segmentation genes did more than provide insights into the progressive nature of development, however. It also carried important messages about the nature of information delivery.

The Nature of Developmental Signals As the in situ hybridization and antibody staining results in Figures 20.2 and 20.3 show, the gene products that regulate development often form gradients in specific regions of the embryo. This observation suggests that two types of information are being delivered to target nuclei: Both the type and concentration of signals received by cells or nuclei are important. Further, both the identity and concentration of signaling molecules vary with respect to position in the embryo.

To grasp why these points are important, recall that differential gene expression is the essence of development and think back to the mechanisms of gene regulation introduced in Chapters 14 and 15. Recall that for a eukaryotic gene to be transcribed efficiently, the transcription activators that bind to enhancers must be present while transcription repressors that bind to silencers must be absent. Differential gene expression occurs because many of the developmental signals received by a cell alter the array of transcription factors that are present in the nucleus and active. These transcription factors bind to promoters, enhancers, silencers, or other types of regulatory sequences in DNA and affect the amount and rate of gene expression. In addition, other developmental signals may lead to alterations in the mRNAs found in target cells or induce changes in proteins that are already present. Which genes are expressed by developing cells at particular places and times? The answer depends on complex interactions among an array of developmental signals and thus transcription factors.

The differential gene expression triggered by regulatory proteins shapes embryos. For example, the major regions of a fly embryo become defined early in development as a result of interactions among gap gene products. Segments emerge slightly later as the products of pair-rule genes interact with gap gene products and with each other. Finally, segment polarity genes are expressed and demarcate cells in the anterior versus posterior end of each segment. In this way, the overall body plan of the fly is established. Now the question is, what makes a particular head segment different from, say, a particular thoracic segment?

The Discovery of Homeotic Genes

During the 1940s, Edward Lewis began a long series of studies that helped researchers pinpoint the genes that identify particular segments. Lewis studied mutations that alter the body pattern of adult flies, much as Nüsslein-Volhard and Wieschaus searched for mutations that affect the body pattern of embryos and larvae. Some of the mutants he and others studied had bizarre phenotypes. As **Figure 20.5** shows, some flies had legs where their antennae should be; others had two sets of wings instead of one.

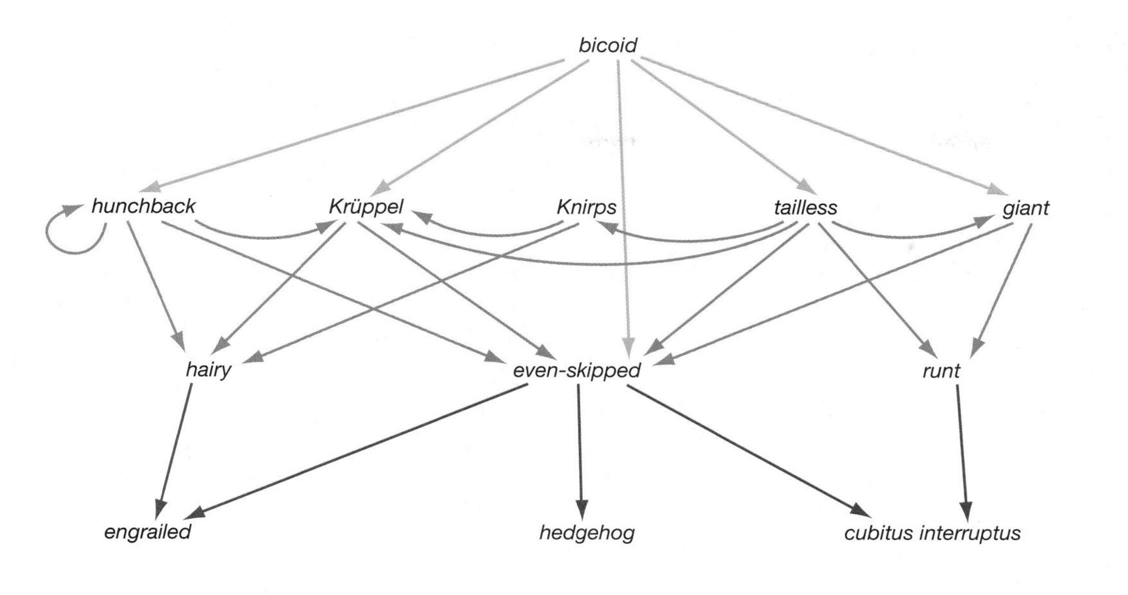

FIGURE 20.4 Segmentation Genes Regulate Each Other
The arrows in this diagram indicate a few of the known interactions among the segmentation genes and their products. To appreciate just how complex the regulation of these genes is, consider that the pair-rule gene *even-skipped* has at least five distinct promoters. One of these promoters has six binding sites for Krüppel protein, three binding sites for Hunchback protein, three binding sites for giant protein, and five binding sites for Bicoid protein.

FIGURE 20.5 Homeotic Mutants in *Drosophila*
An array of homeotic mutants has been observed in fruit flies. Among the most spectacular are individuals with legs growing where antennae should be (top) and individuals with wings growing where small, stabilizing structures called halteres should be (bottom).

Lewis realized that these mutants had an entire segment or structure that had been transformed into a related segment or structure. This phenomenon had actually been observed in plants by a series of workers starting in the late 1800s and is termed **homeosis** (like-condition). For example, a mutation or series of mutations could transform thoracic segment number 3 into thoracic segment 2. Instead of bearing a pair of small stabilizer structures called halteres, the transformed segment would bear a pair of wings. As a result, the mutant would have four wings instead of two.

The existence of homeotic mutants meant that there must be homeotic genes that specify the identity of each segment. To explain the existence of a four-winged fly, for example, Lewis suggested that the homeotic gene responsible for identifying thoracic segment number 3 was defective. Consequently, the cells in what would normally be thoracic segment 3 developed as if they were in thoracic segment 2. Stated another way, Lewis hypothesized that the products of homeotic genes specify the identity of segments along the anterior-posterior axis of the body.

The Homeotic Complex When techniques became available for identifying and sequencing genes, researchers confirmed the homeotic gene hypothesis in spectacular fashion. Studies conducted during the 1970s and 1980s initially identified eight genes in the *Drosophila* genome that lead to homeosis when they are defective. As **Figure 20.6a** shows, the eight genes are found in two clusters on the same chromosome. When investigators explored where these genes are expressed in *Drosophila* embryos, they found that five are activated in the anterior part of the embryo while three are expressed in the posterior sections. The five genes expressed in the anterior part of the embryo are known as the Antennapedia complex; the three expressed in the posterior regions are called the Bithorax complex. As a group, the genes became known as the **homeotic complex** (abbreviated **HOM-C**).

Studies on the timing of gene expression revealed the remarkable pattern illustrated in **Figure 20.6b**. In the embryo, genes in the homeotic complex are expressed in the same sequence as they are found along the chromosome. The gene called *lab*, for example, occurs at one end of the complex and is turned on in the most anterior segment of the embryo. Each subsequent gene in the complex is expressed in slightly more posterior segments, so that genes are expressed in the same "spatial" sequence in the embryo as their order on the chromosome. In addition, the timing of expression corresponds to the order of genes along the chromosome. The first gene expressed is *lab*, then *Dfd*, then *Scr*, and so on. Why this pattern occurs is still a mystery. The mechanism of gene action, in contrast, is increasingly well understood.

Genes in the Homeotic Complex Encode Regulatory Proteins Sequencing studies revealed that each gene in the homeotic complex has a 180-base-pair DNA sequence called the homeobox. The homeobox codes for a DNA binding domain in the proteins produced by these genes. Based on this observation, biologists hypothesized that HOM-C genes code for transcription factors that bind to DNA and regulate the activity of other genes. More specifically, products from the homeotic complex are thought to activate genes involved in the production of segment-specific structures such as legs, wings, antennae, and halteres. Subsequent work has identified the gene *Distal-less*, which is involved in the formation of legs and other appendages, as an example of a gene regulated by the HOM-C loci. Two areas are now the focus of intense research—finding and characterizing target genes other than *Distal-less*, and understanding how the genes within the complex are regulated.

(a) Two clusters of genes form the homeotic complex.

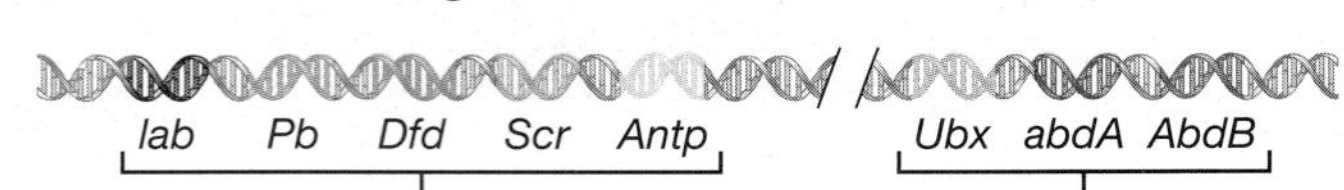

(b) The sequence of the homeotic genes on the chromosome correlates with where they are expressed in the embryo.

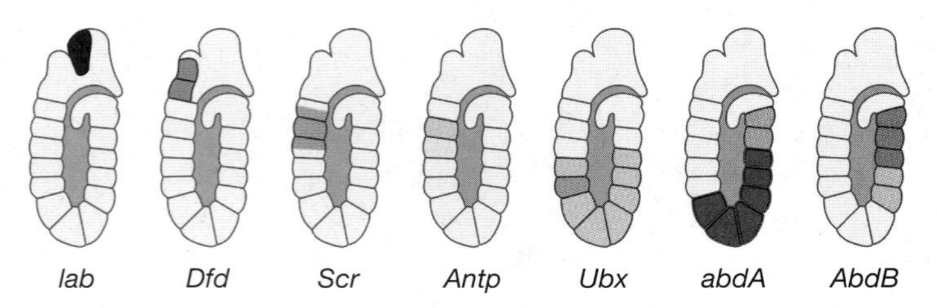

FIGURE 20.6 Organization and Expression of the Homeotic Complex
(a) The HOM-C genes are arranged on the chromosome as shown here. **(b)** This diagram shows where genes in the homeotic complex are expressed. The location of expression along the embryo's anterior-posterior axis matches the sequence of genes along the chromosome.

Where Do Homeotic Genes Fit in the Regulatory Cascade? The segmentation genes introduced earlier in the chapter demarcate segments along the anterior-posterior axis of the fly embryo. The products of HOM-C genes trigger the production of structures in particular segments of adults. Is there a direct relationship between the two groups of genes? The answer is yes. In flies that carry a mutation in a gap gene or a pair-rule gene, the expression of one or more genes in the homeotic complex is abnormal. Based on this observation, researchers have concluded that the gap and pair-rule genes regulate the transcription of the homeotic loci. The eight HOM-C genes could be added to the regulatory hierarchy featured in Figure 20.4.

Before leaving the topic of homeotic gene action and regulation, it's essential to note that not all homeotic loci are part of the Antennapedia or Bithorax complexes. Many other genes are involved in specifying the structures that develop in particular parts of the fly body; some of these cause homeosis when they are defective. The genes discussed in detail here are only part of a large and complex network of genes that regulate the later stages of pattern formation.

Drosophila as a Model Organism

Based on the experiments reviewed thus far, it is clear that biologists have invested an enormous amount of time and effort in understanding pattern formation in *Drosophila*. Why? In addition to being fascinating and important in its own right, research on *Drosophila* development was intended to be applicable to other species that are difficult to manipulate in the lab. *Drosophila* was used as a model organism—one that is studied intensively in the hope that knowledge learned about it can be applied to other organisms.

A variety of model organisms have been featured in earlier chapters. Recall that research on *Escherichia coli* allowed biologists to understand the mechanisms of DNA synthesis and repair in plants and animals. Similarly, work on the yeast *Saccharomyces cerevisiae* led to an understanding of how glycolysis works in virtually every organism on Earth and how the cell cycle is controlled in multicellular species. In the same way, research on pattern formation in *Drosophila* has led to a better understanding of how pattern formation works in many other animal groups. For example, researchers have found that a homeotic complex called the *Hox* gene complex occurs in frogs, crustaceans (crabs and their relatives), birds, various types of worms, mice, and humans. Although the number of *Hox* loci varies widely among species, their chromosomal organization is similar to the HOM-C genes of flies shown in Figure 20.6a.

Recent studies of *Hox* genes have shown that they are expressed along the head-to-tail axis of the mouse embryo in the same sequence as fruit flies. In addition, experiments have shown that when mouse *Hox* genes are altered by mutation, defects in pattern formation result. Based on these data, biologists conclude that in flies, mice, and probably humans and most other animals, the HOM-C and *Hox* genes appear to play a key role in defining the position of cells along the head-to-tail axis of the body. This conclusion was supported in spectacular fashion when researchers in William McGinnis's lab introduced the *Hoxb6* gene from mice into fruit fly eggs. The *Hoxb6* gene in mice is similar in structure and sequence to the *Antp* gene of flies. Because it was introduced without its normal regulatory sequences, the *Hoxb6* gene was expressed throughout the treated fly embryos. The larvae that resulted had identical defects to those observed in naturally occurring fly mutants where the *Antp* gene is mistakenly expressed throughout the embryo.

To interpret these observations, biologists hypothesize that the genes in the HOM-C/Hox complexes are related. Stated another way, genes ancestral to the HOM-C/*Hox* loci arose before the origin of animals or very early in their evolution. Since that time, the number of HOM-C/*Hox* genes has changed dramatically as animals diversified into the array of species we see today. But over this span of about a billion years, the same types of gene products have been involved in directing cells to develop according to their position inside the embryo. In other words, at least some of the molecular mechanisms of pattern formation have been highly conserved during animal evolution. The discovery of these shared mechanisms is among the most significant of all results that have emerged about animal development.

Even more remarkably, researchers have discovered genes that contain homeoboxes in fungi and plants. No genes similar to those found in the HOM-C or *Hox* complexes exist in these groups, however. In plants, the genes involved in pattern formation are different from the segmentation and homeotic genes of animals. Why? As you saw in Chapter 18, multicellular bodies evolved independently in plants and animals. Based on this observation, it is logical to predict that the mechanisms for specifying the positional identity of cells differ between the groups. How does pattern formation occur in plants?

20.2 Pattern Formation in *Arabidopsis*

Historically, the field of developmental biology has been dominated by the study of animals. The reason is simple. Compared to plant embryos, animal embryos are accessible and relatively easy to study. Recently, however, biologists have begun studying how pattern formation occurs in plants by analyzing mutant individuals. Although the specific genes involved in plant development have turned out to be different from those involved in animal development, the basic logic of pattern-formation mechanisms is often quite similar.

Most work on pattern formation in plants has been done on the weedy mustard plant *Arabidopsis thaliana*. Recall from Chapter 18 that *Arabidopsis* seedlings grow from a seed into a mature plant capable of producing gametes in just 6 to 8 weeks. Even though their life span is short, individuals are complex enough to be interesting to study.

An important advantage of studying plants is that pattern formation is not limited to the period of early development. Instead, complex structures such as leaves, roots, branches, and flowers are produced throughout a plant's life from the apical meristems located in roots and shoots. To investigate how pattern formation occurs in embryonic and adult plants, let's consider two questions. How is the root-shoot body axis established during embryogenesis, and what is the mechanism of pattern formation during flower development? Biologists are addressing both questions by analyzing mutant individuals with defective roots, shoots, or flowers.

The Root-Shoot Axis of Embryos

Gerd Jurgens and colleagues set out to identify genes that are transcribed in the zygote or embryo of *Arabidopsis* and that are involved in establishing the root-to-shoot axis of the body. It's no surprise that this effort was similar to the project that Nüsslein-Volhard and Wieschaus had undertaken with *Drosophila*—Jurgens had participated in the fly work.

The biologists' initial goal was to identify individuals with defects in pattern formation at the seedling stage. More specifically, they were looking for mutants that lacked specific regions along the root-to-shoot axis. As **Figure 20.7** (page 400) shows, they succeeded in finding several bizarre-looking mutants. Certain individuals lacked the first leaves or cotyledons, some lacked the embryonic stem or hypocotyl, and others lacked roots.

To interpret these results, the researchers suggested that each type of mutant had a defect in a different gene, and that each gene was involved in specifying the position of cells along the root-to-shoot axis of the body. The specific hypothesis here was that these genes are analogous to the gap genes, which specify the identity of cells within well-defined regions along the head-to-tail axis of fruit flies.

What are these *Arabidopsis* genes, and what do they do? To answer these questions, consider the gene responsible for the mutants lacking hypocotyls and roots. This gene has been mapped and sequenced and named *monopterous*. Because its DNA sequence indicates that the gene has a DNA-binding sequence, *monopterous* is thought to encode a transcription factor that regulates the activity of target genes. The *monopterous* protein, in turn, is manufactured in response to signals from a molecule called auxin that is produced in the apical meristem.

The genes and proteins involved in setting up the root-to-shoot axis of *Arabidopsis* are not yet understood in as much detail as the pattern-formation genes of *Drosophila*. Current evidence suggests that there are strong similarities, however. In both plants and animals, pattern formation is based on cell-to-cell signals. Concentration gradients of regulatory proteins are important, and regulatory cascades result in the step-by-step specification of a cell's position and fate.

Many questions remain about pattern formation in plant embryos. How is auxin production turned on as the apical meristem first begins to form in embryos? Once production of the *monopterous* gene product begins, what target genes are affected? What genes other than *monopterous* are found in the regulatory cascade responsible for development along the root-to-shoot axis? Are any of them active in adults as well? Research on pattern formation in plant embryos presents a host of interesting challenges.

Flower Development in Adults

When receptor proteins in *Arabidopsis* sense that days are getting longer and the temperature is favorable, the shoot apical meristem is stimulated to act as a floral meristem. The flowers produced by this modified meristem contain the individual's reproductive organs. During flower development, the floral meristem produces four distinctive organs, arranged as shown in **Figure 20.8**. As you can see, the sepals are located around the outside of the flower and provide protection. Inside the sepals is a ring of petals, which enclose the male and female organs. If insects or other animals pollinate the species in question, the petals may be colored to help advertise the reproductive structures. The male organs, or stamens, are located in a whorl inside the petals. In the center of the entire structure is the female reproductive organ, or pistil. The question is, how does the floral meristem produce these four organs in the characteristic pattern of whorls-within-whorls?

The first hint of an answer came in 1873, when A. Braun discovered an unusual mutant strain of *Arabidopsis*. In these individuals, one kind of floral organ was replaced by another. The mutant phenotype was homeotic, similar to the transformation of segments later observed in *Drosophila* homeotic mutants.

Over 100 years later, Elliot Meyerowitz and colleagues extended Braun's observation by assembling a large collection of *Arabidopsis* individuals with homeotic mutations in flower structure. Their goal was to identify and characterize the genes responsible for specifying the four floral organs.

Meyerowitz's group found that the mutants could be sorted into three general classes depending on the type of homeotic transformation that occurred. As **Figure 20.9a** shows, some mutants had only pistils and stamens. Others had only sepals and pistils. Individuals in the third group had only sepals and petals. The key observation was that each type of mutant lacked the elements found in two whorls.

What was going on? Presumably, each class of homeotic mutation was due to a defect in a single gene. Meyerowitz realized that if three genes are responsible for setting up the pattern of a flower, then the mutants suggested a hypothesis for how the three gene products interact. Because he referred to the three hypothetical genes as A, B, and C, his hypothesis is called the ABC model.

The ABC Model As **Figure 20.9b** shows, there are three basic ideas behind the ABC model of pattern formation in flowers. The first proposition is that each of the three genes involved is expressed in two adjacent whorls. The second postulate is that four different combinations of gene products result from this pattern of expression. The final idea is that each one of these

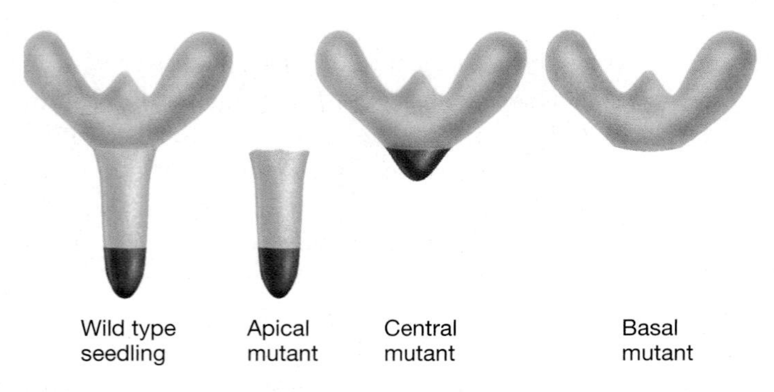

FIGURE 20.7 Pattern-Formation Mutants in *Arabidopsis* Embryos
Researchers have identified *Arabidopsis* mutants with the defects shown here. Note that each of the mutant individuals is missing a defined section of the body along the root-to-shoot axis.

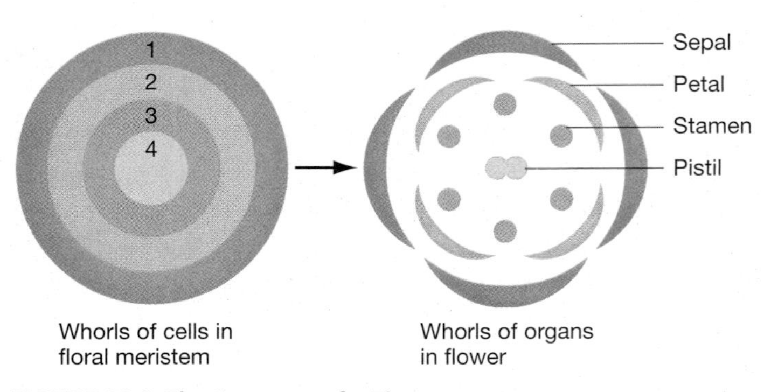

FIGURE 20.8 The Structure of a Flower
In a floral meristem, whorls of cells produce the four whorls of floral organs shown on the right.

four combinations of gene products triggers the development of a different floral organ. Specifically, Meyerowitz proposed that (1) the A protein alone causes cells to form sepals, (2) a combination of A and B products sets up the formation of petals, (3) B and C combined specify stamens, and (4) the C protein alone designates cells as the precursors of pistils.

Does this model explain how the three classes of homeotic mutants occur? The answer is yes, if we also assume that the presence of the A protein inhibits the production of the C protein, and that the presence of the C protein inhibits the production of the A protein. Then the patterns of gene expression and mutant phenotypes diagrammed in Figure 20.9b correspond.

(a) Distinct types of homeotic mutants occur in *Arabidopsis* flowers.

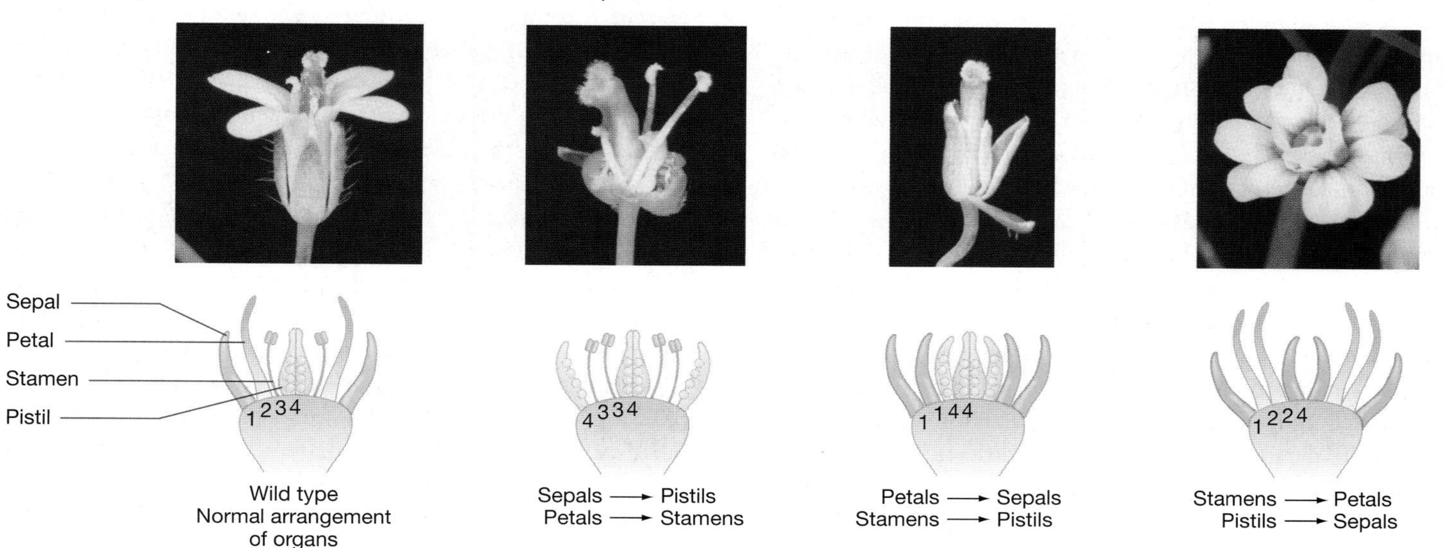

(b) A model to explain why the homeotic mutations occur

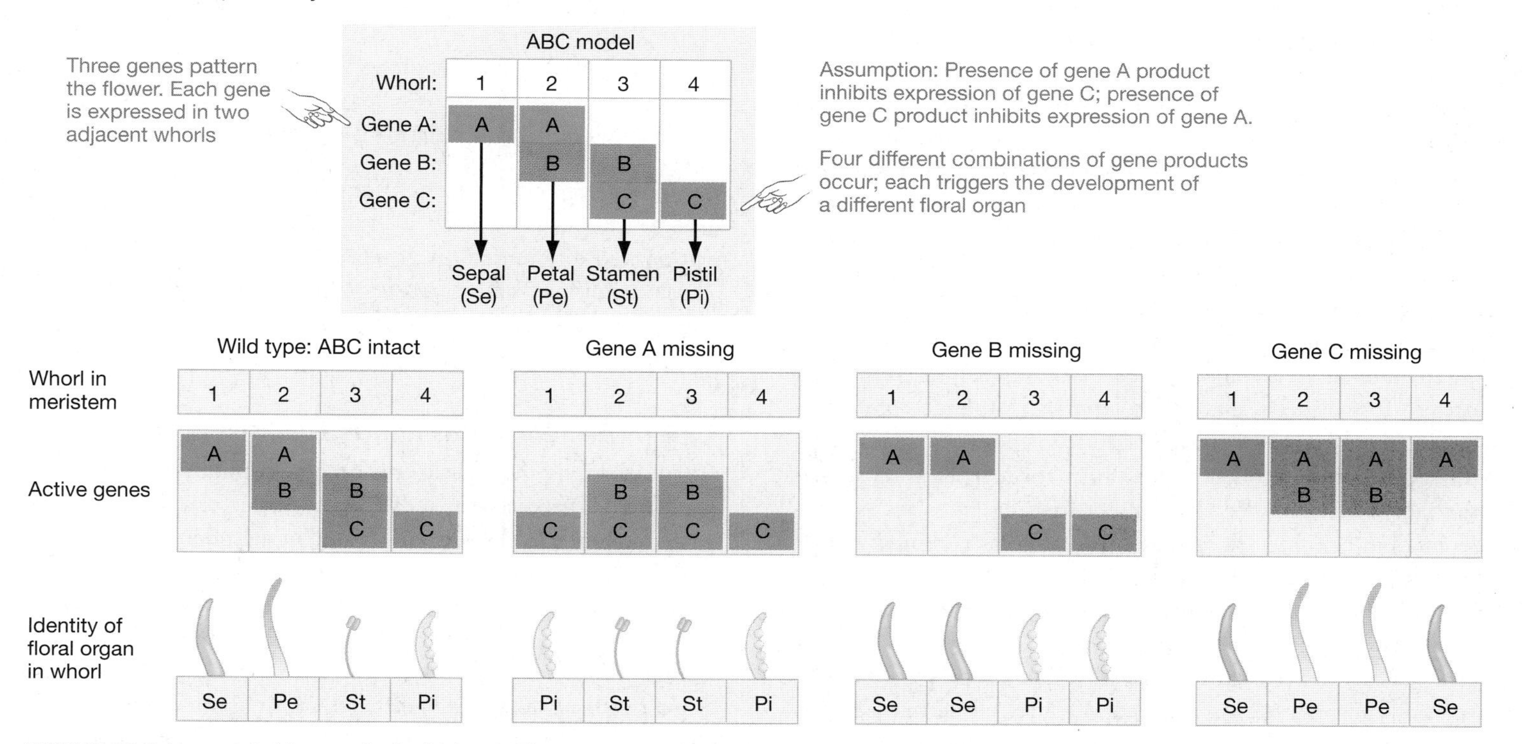

FIGURE 20.9 Homeotic Mutants in *Arabidopsis* Flowers
(a) In each type of homeotic floral mutant, two adjacent whorls of organs are transformed into different organs. Either sepals and petals, petals and stamens, or stamens and the pistil are transformed. **(b)** The ABC model is a hypothesis to explain why three types of homeotic mutants exist.

For example, if the A gene is disabled by mutation, then it no longer inhibits the expression of the C gene and all cells produce the C protein. As a result, cells in the outermost whorl express only C protein and develop into pistils, while cells in the whorl just to the inside produce B and C proteins and develop into stamens.

Even though the model is plausible and appeared to explain the data, it still needed to be tested directly. To accomplish this, Meyerowitz and co-workers mapped the genes responsible for the mutant phenotypes and identified the DNA sequences involved. Once they had isolated the genes, they were able to obtain single-stranded DNAs and use them to perform in situ hybridizations. Their goal was to document the pattern of expression of the A, B, and C genes and see if they corresponded to the predictions of the model. As anticipated, the mRNAs for each of the three genes showed up in the sets of whorls predicted by the model. The *A* gene is expressed in the outer two whorls, the *B* gene is expressed in the middle two whorls, and the *C* gene is expressed in the inner two whorls.

This result strongly supported the validity of the ABC model. Just as different combinations of HOM-C gene products specify the identity of fly segments, different combinations of floral identity genes specify the parts of a flower.

How Do the Floral Identity Genes Work? When Meyerowitz and others analyzed the sequence of the floral organ identity genes, they discovered that all three genes contained a segment that coded for the DNA-binding domain of a protein. Based on this observation, the researchers hypothesized that, like the homeotic genes found in *Drosophila* and other animals, the floral genes appeared to be transcription factors. But instead of containing a homeodomain, the genes that identify floral organs encode a long sequence of 58 amino acids, called the MADS box, that binds to DNA. Transcription factors with MADS boxes also occur in fungi and animals. As with research on *Drosophila*, biologists are currently working to identify the genes targeted by the ABC proteins.

To summarize, strong parallels exist between the process of pattern formation in plants and animals. Although completely different genes are involved, the logic of how the genes act is the same. In both groups, pattern formation depends on signals that act in a regulatory hierarchy or cascade. In each case, differences in cell fate are mediated by different combinations of transcription factors. These regulatory proteins are expressed in specific regions of the animal or plant and direct the development of cells located in that region.

Once the general pattern of an organism or structure has been established, the specific fate of each cell within a region or structure is determined. In a flower petal, the genes responsible for synthesizing pigments are activated. In a fruit fly's thorax, the cells destined to become flight muscles start producing muscle-specific proteins. How does this final step in development happen?

20.3 Differentiation: Becoming a Specialized Cell

According to data introduced in Chapter 18, virtually all of the cells inside multicellular organisms are genetically identical. Once this result was established, it became clear that cells are different from each other because they express different genes and manufacture different proteins. The generation of cellular diversity through differential gene expression is called **differentiation.** A cell is differentiated if it manufactures proteins that are specific to a particular cell type. Cells in a flower petal that make pigment-synthesizing enzymes are differentiated. So are cells in the thorax of an insect that produce the motor proteins found in flight muscles.

Pattern formation is an early step in differentiation and serves to organize cells in space. Later, cells become assembled into recognizable tissues and organs in a process called **morphogenesis.** During pattern formation and morphogenesis, changes in gene expression cause cells to become committed to a particular fate. **Determination** is the term used to indicate a cell's irreversible commitment to become a particular cell type. Finally, cells that are determined actually begin expressing tissue-specific proteins and begin functioning as muscle or nerve or bone cells. At that point a cell is differentiated. Although this series of events is not strictly linear and sequential, the following formulation expands on the idea that development is a progressive, step-by-step process:

Pattern formation → Morphogenesis →

Determination → Differentiation

To explore how the steps after pattern formation take place, let's consider a particularly well-studied example—the determination and differentiation of muscle cells in vertebrates. What causes undifferentiated mesodermal cells to begin producing the muscle-specific proteins that make breathing, walking, and opening textbooks possible?

Organizing Mesoderm into Somites

To understand how muscle cells become committed to their fate and then differentiated, recall from earlier chapters that the precursors of muscle cells are located in mesoderm. Further, recall that once gastrulation is complete, mesoderm is located in a layer sandwiched between endoderm and ectoderm. As **Figure 20.10** shows, the endoderm is located on the ventral (belly) side of the embryo. As morphogenesis proceeds, endodermal cells begin to form the gut or digestive tract. At about the same time, the ectoderm on the dorsal (back) side of the embryo folds and forms the neural tube, which is the precursor of the brain and spinal cord. Chapter 19 detailed how a change in the expression of adhesion proteins called cadherins is involved in the formation of the neural tube from dorsal ectoderm.

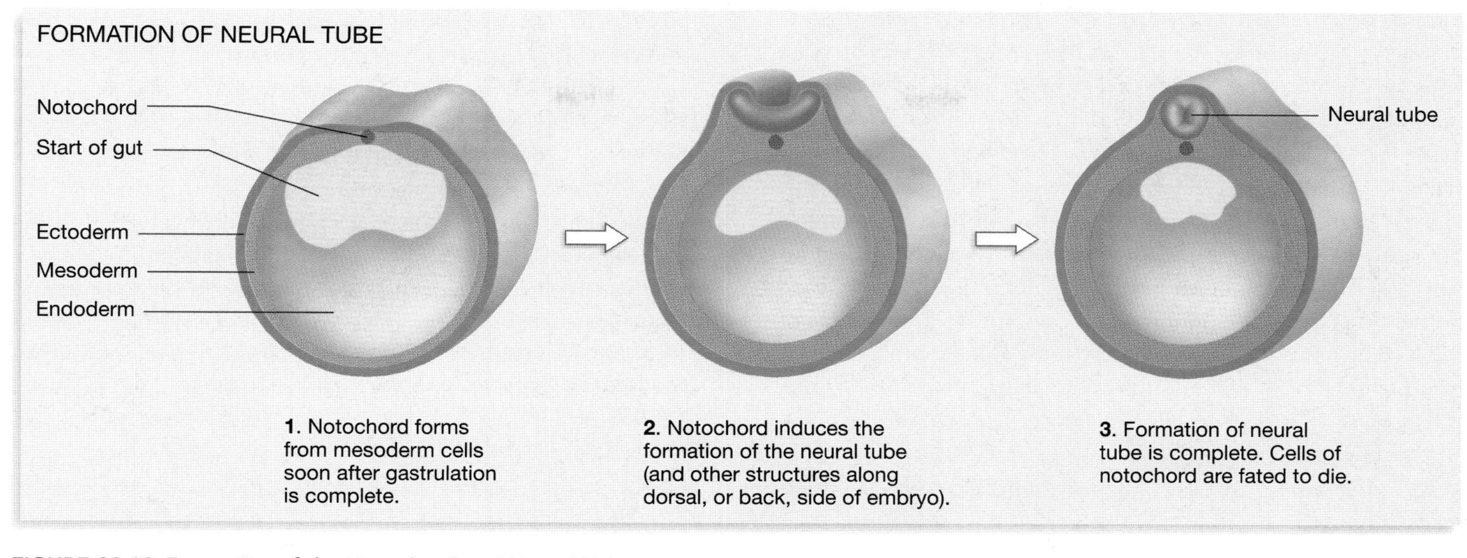

FIGURE 20.10 Formation of the Notochord and Neural Tube
In vertebrates, the notochord forms from mesoderm cells soon after gastrulation is complete. The notochord induces the formation of the neural tube and other structures along the dorsal (back) side of the embryo.

What is happening to mesoderm as these events take place? One of the first distinctive mesodermal structures to appear is a rod-like element called the **notochord**, shown in cross section in Figure 20.10. The notochord is unique to the group of animals called the chordates, which include humans and other vertebrates. It is also critical to the formation of the neural tube. The notochord not only produces signaling molecules that induce the neural tube to form but also acts as a base as the neural tube folds into its final configuration. The notochord is a transient structure in most species, however, appearing only in embryos. As **Box 20.2** (page 404) explains, many of the cells in the notochord are fated to die.

Once the neural tube forms, nearby mesodermal cells become organized into blocks of tissue called somites. As the scanning electron micrograph and sketch in **Figure 20.11** show, somites form on either side of the neural tube. The mechanism of somite formation is still not well understood, but it appears to involve a change in cadherins and other cell adhesion proteins.

Fate-mapping studies have shown that the mesodermal cells in a somite are destined for a variety of structures. Cells from somites build the vertebrae and ribs, the deeper layers of the skin that covers the back, and the muscles of the back, body wall, and limbs. By transplanting cells from one somite to another, though, researchers have found that initially any cell in a somite can become any of the somite-derived elements of the body. In other words, the cells that form the somite are not determined initially. As a somite matures, though, cells in certain sections of the structure do become committed to a specific fate. For example, when Charles Ordahl and Nicole Le Douarin transplanted cells from various locations in somites to the outer part of the structure early in somite formation, they

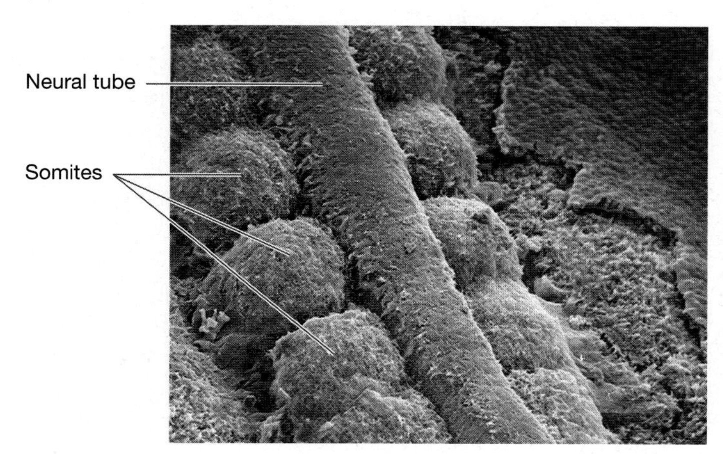

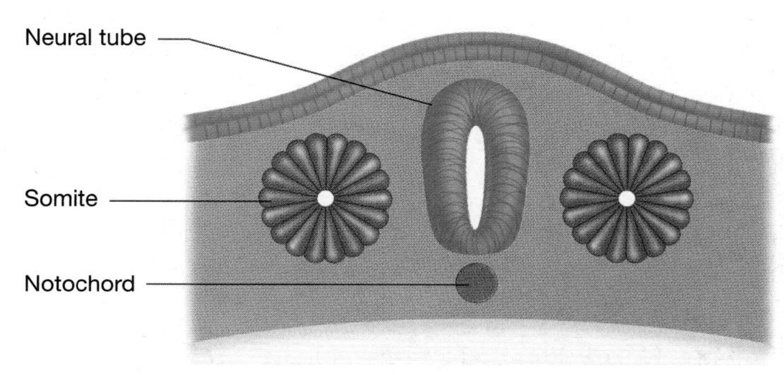

FIGURE 20.11 Formation of Somites
Somites, which are made of mesodermal cells, form on either side of the neural tube. EXERCISE Somites form from the anterior to posterior end of an embryo. Label the anterior and posterior end of the embryo in the micrograph.

found that the cells eventually became committed to form the muscles of the limbs (**Figure 20.12a**). This result supports the hypothesis that cells in a somite are initially not determined, but later become committed based on their position.

Like the notochord of many species, somites are transient structures that appear relatively briefly during development. Once determination occurs, the cells that make up somites break up into distinct populations and migrate to their final location in the developing embryo. As **Figure 20.12b** shows, each population of cells is committed to becoming a different type of cell. In the case of the muscle cells located at the outside of a somite, how does this determination event occur?

Determination of Muscle Cells

Recent studies have shown that an array of signals causes distinct populations of cells in a somite to become determined. **Figure 20.12c** illustrates how signaling molecules diffuse away from cells in the notochord, the neural tube, and nearby ectoderm and mesoderm and act on target cells in the somite. Each type of signal acts on a distinct population of cells within the structure.

Why do cells on the outside of somites become committed to producing muscle in response to the signals they receive? Harold Weintraub and colleagues answered this question by experimenting with **myoblasts**—cells that are committed to be-

BOX 20.2 Programmed Cell Death

As animal tissues and organs take shape, certain cells are fated to die. For example, selected cells in the notochord of vertebrates die and are replaced by cells that form the vertebrae. As the human hand and foot form, cells that are initially present between the fingers and toes die. As a result, independent digits form. In these and many other cases, cell death is a normal part of development. Programmed cell death is called **apoptosis**. Translated literally, the term means "falling away."

How does programmed cell death occur? An answer to this question emerged from studies of the roundworm *Caenorhabditis elegans*. This species is a popular research subject because it has a complex array of organs and tissues but less than a thousand cells. In addition, biologists are able to identify individual cells in both embryos and adults because the cells of *C. elegans* are transparent. Thanks to these features, the complete fate map of this species is known. Biologists know the destiny of each cell in the embryo at every stage of development.

As a *C. elegans* individual matures, 131 of its 1090 somatic cells undergo apoptosis. To explore how this process occurs, Hillary Ellis and Robert Horvitz set out to identify mutations that disrupt the normal pattern of cell deaths. Their initial work uncovered two genes that are essential for apoptosis. If a worm carries mutations that inactivate either gene, cells that would normally die survive. The result is abnormal development and death. Ellis and Horvitz named the genes *ced-3* and *ced-4* (for *ce*ll *d*eath abnormal) and proposed that they are part of a genetic program that executes apoptosis. Their hypothesis was that a cell dies if these genes are activated.

Subsequently, Horvitz and colleagues identified a gene that regulates the activity of *ced-3* and *ced-4*. Mutations that inactivate this sequence, called *ced-9*, result in the deaths of many cells that normally survive. Based on this observation, Horvitz's group suggests that the normal function of *ced-9* is to inhibit the suicidal activities of *ced-3* and *ced-4*.

By searching databases of known DNA sequences, biologists found genes in mice called *caspase-9* and *Apaf-1* that are similar to *ced-3* and *ced-4*, respectively. In addition, a mouse protein that inhibits apoptosis, called Bcl 2, is similar to the *ced-9* product. To test the hypothesis that these genes are important in mouse development, Keisuke Kuida and co-workers used genetic engineering techniques to produce mice in which both copies of the *caspase-9* gene are disrupted. As the photos in **Figure 1** show, the resulting embryos exhibited a severe malformation of the brain. The defect occurred because cells that would normally die early in development survived. The researchers noted that the engineered embryos had normal digit separation, however. Based on this observation, they suggest that genes other than *caspase-9* are involved in programmed cell death during the formation of the toes.

Are these genes found in humans? The answer to this question is not known. Normal apoptosis is important in the development of human embryos, however, and abnormal apoptosis has been implicated in certain diseases of adults. Inappropriate activation of the cell-death program is involved in some neurodegenerative diseases, including ALS (Lou Gehrig's disease). In addition, failure of negative regulators such as Bcl 2 has been implicated in the development of certain cancers.

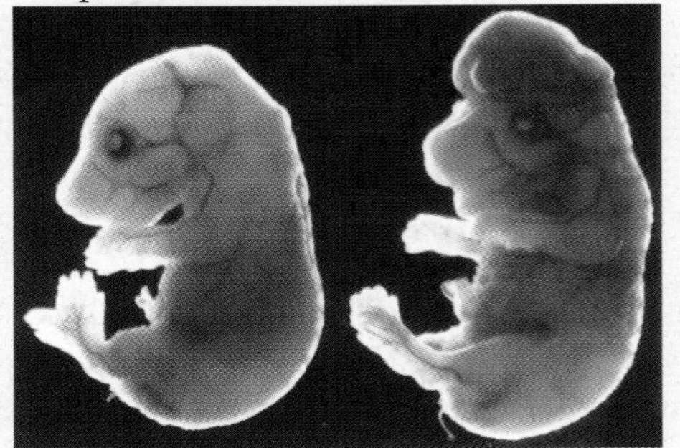

FIGURE 1 Defects Occur When Programmed Cell Death Fails
The mouse embryo on the left is normal; the individual on the right has two defective versions of the *caspase-9* gene.

coming muscle but that have not yet begun producing muscle-specific proteins. Specifically, Weintraub and co-workers hypothesized that myoblasts contain at least one regulatory protein that commits them to their fate.

Figure 20.13 (page 406) outlines how the biologists went about searching for this hypothetical protein. As the diagram shows, they began by isolating mRNAs from myoblasts and using reverse transcriptase to convert the mRNAs to cDNAs. Then they attached a type of promoter to the cDNAs that would ensure their expression in any cell. Finally, they introduced the recombinant genes into connective tissue cells called fibroblasts and monitored the development of the transformed cells. Just as they predicted, one of the myoblast-derived cDNAs converted fibroblasts into muscle cells. Follow-up experiments showed that the same gene could convert pigment cells, nerve cells, fat cells, and liver cells into cells that produced muscle-specific proteins.

Weintraub's group called the protein product of this gene **MyoD**, for *myo*blast *d*etermination. Follow-up work showed that *MyoD* encodes a transcription factor, and that the MyoD protein binds to enhancer elements located upstream of muscle-specific genes. In addition, the MyoD protein activates expression of the *MyoD* gene. This was a key observation because it meant that once *MyoD* is turned on, it remains on. Other researchers have found that genes closely related to *MyoD* are also required for the differentiation of muscle cells.

To summarize, cells are initially committed to become part of a vertebrate's back when determinants located in the egg are shifted by cortical rotation (see Chapter 19). In response to these signals, cells become committed to form mesoderm and then to become part of the notochord. Subsequently, signals from the notochord and nearby cells induce the production of MyoD and other muscle-determining proteins in certain populations of cells from somites. In response, these target cells are

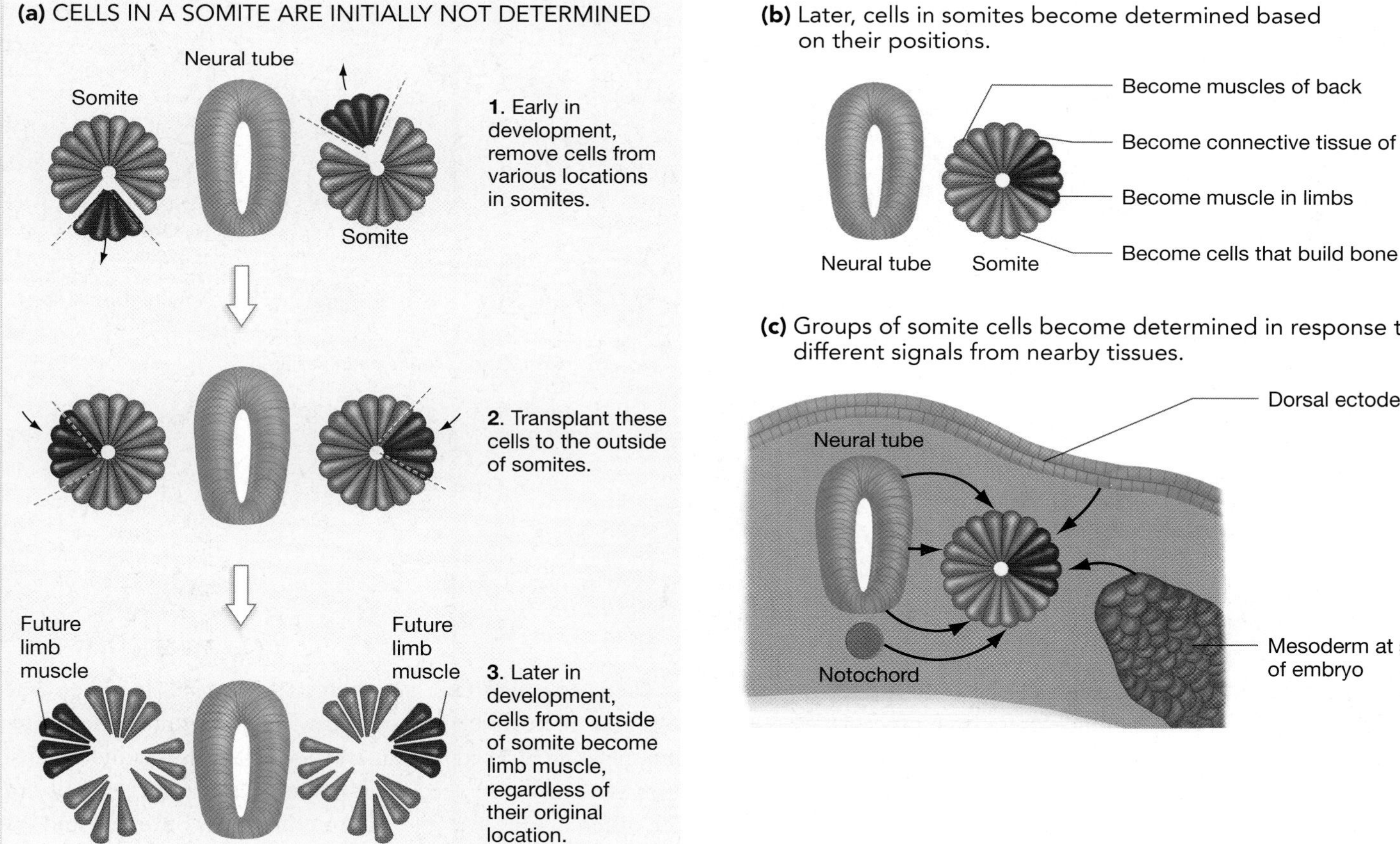

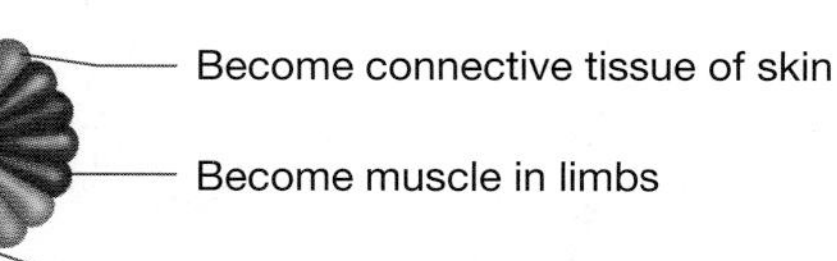
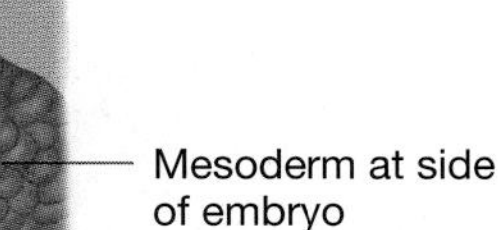

FIGURE 20.12 Cells in the Somite Become Determined
(a) By transplanting cells early in somite formation to new locations in the structure, researchers showed that they are not determined initially. **(b)** The somite eventually breaks up into four distinct populations of cells, each with a different fate. **(c)** Somite cells become determined in response to signals from the neural tube, notochord, and nearby ectoderm and mesoderm.

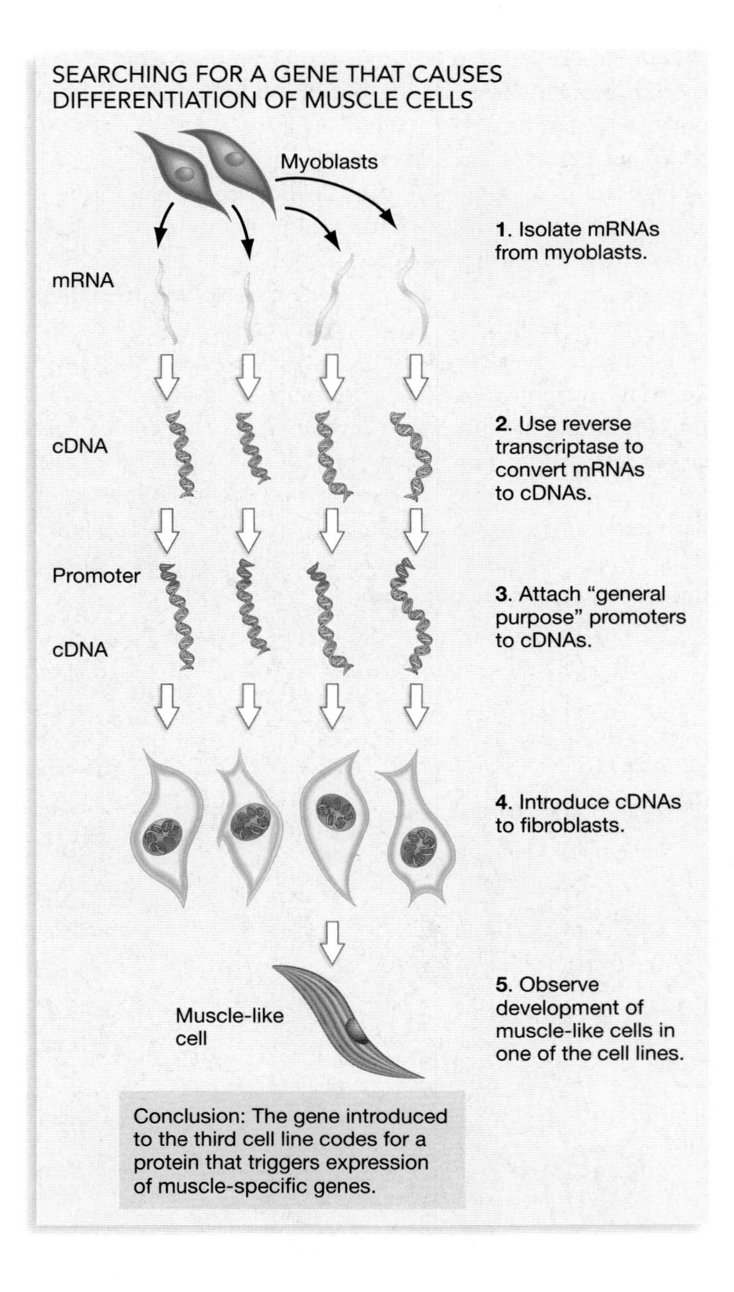

FIGURE 20.13 A Gene That Causes Differentiation
This protocol allowed researchers to find a gene involved in the differentiation of muscle cells. QUESTION Why did they have to attach a "general purpose promoter" to the cDNAs?

committed to becoming muscle. The commitment is reinforced by MyoD's ability to enhance its own production.

As a case history, the determination and differentiation of vertebrate muscle cells illustrates several principles common to all cell types. First, it confirms that differential gene expression is at the heart of development. Second, it supports the view that the development of an individual is a stepwise process. Determination occurs when the production of a specific regulatory protein like MyoD commits a cell to a particular fate. Differentiation occurs when the cell actually begins to produce cell-specific proteins. At that point, the process of development is largely complete. Growth takes over as the most important process in shaping the juvenile organism.

Essay Human Stem Cells

In most cases, the fate of a cell is sealed once it has "popped into place." At some point in the step-by-step process, development can become irreversible. When a muscle cell in a human

Why are stem cells used in research?

embryo differentiates, for example, it stops growing and begins producing muscle-specific proteins. This cell remains a muscle cell for as long as it lives. It cannot change fate and become a nerve cell.

There are important exceptions to this rule, however. The cells that make up plant meristems divide continuously and retain the ability to produce an array of cells, tissues, and organs. In many cases even specialized plant cells retain the ability to "de-differentiate" and give rise to other types of tissues and organs. Gardeners and farmers take advantage of this trait when they take cuttings from stems or roots and use them to propagate new individuals.

Even in adult mammals, certain populations of cells retain the ability to divide and produce an array of cell types. Cells with this

(Continued on next page)

(Essay continued)

ability are called stem cells. Stem cells can be found in your muscle, skin, liver, gut lining, and elsewhere. Given that you lose an average of 100 billion cells from the lining of your intestine and from your blood each day, it is not surprising that stem cells exist.

The stem cells in your body continue to divide throughout your life. Some of their daughter cells remain as stem cells; others differentiate to replace cells in nearby tissues that have been lost or damaged. One of the key attributes of your stem cells is that they behave as though they are determined but not differentiated. The stem cells in your skin, for example, can give rise only to skin cells. Although the stem cells in your bone marrow give rise to an impressive array of different cell types, most of their descendant cells become components of blood.

In contrast, stem cells that are isolated from embryos are not determined. Cells from the inside of a blastocyst, for example, can give rise to virtually any cell type in the body. Recently, researchers have started to explore whether large numbers of these embryonic stem cells could be grown in culture and then implanted into adults as a treatment for diseases and injuries caused by the loss of differentiated cells. Parkinson's disease and Alzheimer's disease are caused by a loss of brain cells; a prominent form of diabetes is caused by loss of certain cells in the pancreas; muscular dystrophy results from the death of muscle cells. Biomedical researchers hope that implanted stem cells would differentiate and at least partially make up for the loss of the original cells.

Research on human stem cells is intensely controversial, however. Because in vitro fertilization techniques frequently produce an excess of early embryos, the primary sources for embryonic stem cells has been fertility clinics. If these surplus embryos are not discarded, they can be dissected and used as a source of stem cells. Opponents of this practice argue that, because the embryos are sacrificed in this process, the procedure represents the taking of a human life.

Currently, many nations are debating rules to govern the use of embryonic stem cells in research. Although the practice is banned in Italy and Norway, it is legal or even actively supported in most countries of the European Union. In the United States, public funding for the establishment of new cultures of embryonic stem cells has been banned, but research on cultures of embryonic stem cells that had been established prior to the year 2001 is still being supported.

Chapter Review

Summary

A cell's fate is determined in a series of steps. During pattern formation, cells receive information about their location in the embryo and differentiate in a way that forms ordered arrangement of cells in space. Determination occurs when a cell becomes committed to a particular fate. A cell is differentiated when it begins producing tissue-specific proteins and functions as a mature, specialized component of the organism.

Researchers have used a genetic approach to study the mechanisms of pattern formation in fruit flies and *Arabidopsis*. By analyzing mutant embryos or adults with defects in pattern formation, biologists have been able to identify a series of genes and gene products that organize cells in space. The pattern-formation genes that have been discovered to date fall into two broad categories, based on the phenotypes of individuals with mutant forms of the genes: those that cause the loss of specific regions in the embryo—for example, the segmentation genes in *Drosophila* or *monopterous* and related loci in *Arabidopsis*—and those that cause homeotic transformations in adults.

When a homeotic transformation takes place, a structure that is usually found at a different place in the body is substituted for the normal structure at a particular location. Flies with legs instead of antennae and flowers with sepals instead of petals are examples of homeotic transformations. To explain why these phenotypes occur, biologists suggest that the products of homeotic genes identify particular locations in the developing organism. For example, a HOM-C gene product in *Drosophila* might signal, "this is thoracic segment 2," whereas a MADS-box protein in *Arabidopsis* might signal, "this is the second whorl of floral organs." In contrast, the segmentation genes found in fruit flies and the *monopterous*-like genes in mustard plants are thought to define particular segments or regions in the embryo. The *Drosophila* segmentation genes generate a series of segments along the head-to-tail axis of the body; the *monopterous*-like genes of *Arabidopsis* define the cotyledons, hypocotyl, and root along the root-to-shoot axis of the body. The protein products of these genes are thought to carry signals such as "this area is a segmental boundary in the thorax," or "this area is part of the root."

Research on pattern-formation genes has brought several general principles about development to light. Development is a step-by-step process based on differential gene expression. As pattern formation and other aspects of development proceed, changes in gene expression occur in response to cytoplasmic determinants or signals from other cells. A cell's response depends on the types of signals it receives and their concentration. Often this response occurs through the production of regulatory transcription factors. The type and quantity of signals received by a cell depend on its position in the embryo.

These same principles govern the processes of determination and differentiation. For example, the mesodermal cells that make up the somites found in vertebrate embryos take on one of several different fates, depending on their position in the somite and thus the signals that they receive from nearby structures. Cells on the outside of somites become determined to be muscle because they receive signals that trigger the production of MyoD and other regulatory proteins. In turn, these regulatory proteins cause the cell to differentiate by triggering the transcription of muscle-specific genes.

Questions

Content Review

1. What does an in situ hybridization allow researchers to do?
   a. Identify the genes responsible for pattern formation.
   b. Determine whether a gene codes for a transcription factor.
   c. Determine whether a gene has a maternal effect pattern of inheritance.
   d. Locate where a specific mRNA is found in an embryo.
2. In combination, what do the products of gap genes, pair-rule genes, and segmentation polarity genes of fruit flies do?
   a. They trigger the production of segment-specific structures like antennae.
   b. They define the segmented body plan of the embryo.
   c. They set up the back-to-belly axis of the larval body.
   d. They trigger determination in cells once segments have formed.
3. What does the ABC model of flower development attempt to explain?
   a. How different combinations of gene products trigger the formation of different floral organs.
   b. Why petals are found on the inside of the whorl of sepals instead of the outside.
   c. Why the four types of floral organs occur in whorls.
   d. Why apical meristems are converted to floral meristems.
4. What is a homeotic mutant?
   a. an individual with a structure located in the wrong place
   b. an individual with an abnormal head-to-tail axis
   c. in flies, an individual that is missing segments; in *Arabidopsis*, an individual that is missing a hypocotyl or other embryonic structure
   d. an individual with double the normal number of structures or segments
5. What evidence suggested the MyoD protein was involved in the differentiation of muscle cells?
   a. The MyoD gene is only expressed late in the development of somites—not early.
   b. Expression of *MyoD* DNA can convert non-muscle cells to muscle-like cells.
   c. *MyoD* mRNA was isolated from muscle cells.
   d. MyoD is part of a closely related family of transcription factors.
6. What is programmed cell death (apoptosis)?
   a. an experimental technique that biologists use to kill specific cells
   b. a cell suicide program required for normal development
   c. a pathological condition seen only in damaged or diseased organisms
   d. a developmental mechanism unique to the roundworm *C. elegans*

Conceptual Review

1. What evidence suggests that at least some of the molecular mechanisms responsible for pattern formation have been highly conserved over the course of animal evolution?
2. How did researchers succeed in identifying molecules that have important roles in *Drosophila* and *Arabidopsis* pattern formation?
3. Explain why mRNA injection experiments support the hypothesis that *bicoid* sets up the normal head-to-tail axis of a *Drosophila* embryo. Why did researchers bother to do this experiment? To accept the gradient hypothesis as valid, why wasn't it enough to observe the *bicoid* mutant phenotype, the distribution of *bicoid* mRNA, and the distribution of Bicoid protein?
4. How does pattern formation in the *Arabidopsis* flower resemble pattern formation in the *Drosophila* embryo? How is it different?
5. What is the difference between determination and differentiation? What is the relationship of these processes to pattern formation?
6. The development of in situ hybridization and antibody-staining technology had a huge impact on biologists' ability to study pattern formation and determination. Why?
7. Why is it significant that many of the genes involved in pattern formation and determination encode transcription factors?

Applying Ideas

1. Recent research has shown that the products of two different *Drosophila* genes are required to keep *bicoid* mRNA concentrated at the anterior end of the egg. In individuals with mutant forms of these proteins, *bicoid* mRNA diffuses farther toward the posterior pole than it normally does. First, predict what effect these mutations will have on segmentation of the larva. Second, suggest a hypothesis for how these proteins function at a molecular level. How would you go about testing your hypothesis? Finally, predict whether the mutations show a maternal effect pattern of inheritance. Explain your rationale.
2. In 1992, David Vaux and colleagues used genetic engineering technology to introduce an active human gene for the Bcl 2 protein into embryos of the roundworm *C. elegans*. When they examined the embryos, they found that cells that normally undergo programmed cell death survive. What is the significance of this observation?
3. Animal cells make massive movements during gastrulation. Similarly, the cells that make up somites break up into distinct populations and migrate to their final position in the embryo after determination has occurred. Plant cells, in contrast, do not move because they are surrounded by a stiff cell wall. Discuss how this lack of movement might affect the processes of pattern formation, differentiation, or determination in plants.
4. According to the data presented in this chapter, segmentation genes set up the segmented body plan of a fruit fly while HOM-C genes trigger the production of segment-specific structures like antennae, legs, and wings. Now consider the differences in the body plans of a spider, a fruit fly, a centipede, and an earthworm. Many biologists hypothesize that differences in the expression of segmentation or HOM-C genes are responsible for some of the morphological differences observed in these groups. Do you find this hypothesis plausible? Why or why not?

CD-ROM and Web Connection

CD Activity 20.1: Early Pattern Formation in *Drosophila* *(animation)*
(Estimated time for completion = 5 min)
How does a fertilized fly egg develop a head end and a tail end?

At your **Companion Website** (http://www.prenhall.com/freeman/biology), you will find self-grading exams and links to the following research tools, online resources, and activities:

***Drosophila* Pattern Formation**
This Nobel article describes the work for which Edward B. Lewis, Christiane Nüsslein-Volhard, and Eric F. Wieschaus were awarded the 1995 Nobel Prize for elucidating the molecular control of pattern formation.

Homeotic Genes
This site focuses on the study and molecular biology of homeotic genes.

Journal of Biochemistry
This site is a collection of research news and announcements concerning the study of cellular differentiation in development and embryogenesis.

Additional Reading

Izpisùa Belmonte, J. C. 1999. How the body tells left from right. *Scientific American* 280 (June): 46–51. A look at how normal left-right asymmetry is established.

Nüsslein-Volhard, C. 1996. Gradients that organize embyro development. *Scientific American* 275 (August): 38–43. Reviews how gradients in key molecules pattern the fruit fly embryo.

Pederson, R. A. 1999. Embryonic stem cells for medicine. *Scientific American* 280 (April): 69–73. Explores how stem cells from embryos might be used in the treatment of disease and injury.

Riddle, R. D. and C. J. Tabin. 1999. How limbs develop. *Scientific American* 280 (February): 74–79. Introduces some of the gene products involved in patterning the limbs of vertebrates.

Evolutionary Processes and Patterns

Evolution is the tie that binds all of biology together—from molecular genetics to ecosystem ecology. The chapters in this unit provide an in-depth look at how evolution happens. They have two goals: to deepen your understanding of the processes and events that were introduced in earlier chapters, and to lay the groundwork for an exploration of the diversity of life in Unit 6—or what evolution produces.

Chapter 21: Darwinism and the Evidence for Evolution examines the evidence for evolution and analyzes natural selection—Darwin's explanation for why organisms change through time. **Chapter 22: Evolutionary Processes** goes on to introduce processes other than natural selection that are responsible for evolution, including mutation, gene flow, a random process called genetic drift, and a type of natural selection called sexual selection. **Chapter 23: Speciation** asks how these processes create new species. **Chapter 24: The History of Life** concludes the unit with a look at some broad patterns in the history of life, such as the rapid diversification of species on islands and the phenomenon known as mass extinction.

As we analyze these fundamental questions about the tree of life we'll also have a chance to look at how evolutionary theory is applied to contemporary problems. Evolutionary thinking is helping biologists understand the emergence of drug-resistant diseases and determine whether a mass extinction event is now underway. As you move through this unit, you'll come to appreciate that evolutionary processes have not only shaped the tree of life for some 3.5 billion years, they continue to act—right before our eyes.

The 'Ae'o is a wetland bird native to the Hawaiian islands. These islands were created by volcanoes and were colonized by species that originally lived in North America or Asia. © 1994 Susan Middleton and David Liittschwager

Darwinism and the Evidence for Evolution

21

This chapter is about one of the great ideas in science. The theory of evolution by natural selection, formulated independently by Charles Darwin and Alfred Russel Wallace, explains how organisms have come to be adapted to environments ranging from arctic tundras to steaming rainforests. In an important sense, the ideas of Darwin and Wallace rank alongside Copernicus' theory of the Sun as the center of our solar system, Newton's laws of motion and gravitation, and Einstein's general theory of relativity. These are all examples of revolutionary breakthroughs in our understanding of the world.

Like most scientific breakthroughs, however, this one did not come easily. When Darwin published a full exposition of the theory in 1859 in a book called *On the Origin of Species by Means of Natural Selection*, it unleashed a firestorm of protest throughout Europe. At that time the leading explanation for the diversity of organisms was a theory called special creation. This theory held that all species were created independently, by God, as recently as 6000 years ago. The theory of special creation also maintained that these species were immutable, or unchanged, since the moment of their creation.

To grasp how Darwin's ideas contrasted with the theory of special creation, it's important to recall from Chapter 1 that scientific theories usually have two components. The first is either a claim about a pattern that exists in nature or a statement that summarizes a series of observations about the natural world. In short, the pattern component is about facts. The second component is a process that produces the pattern or set of observations. For example, the pattern component in the theory of special creation was that species were unchanged through time and that they were independent of one another. The process that explained this pattern

The Galápagos islands are located in the Pacific Ocean, off the coast of Ecuador. Different islands in the group are home to distinct species of marine tortoises, finches, mockingbirds, and other organisms. Darwin formulated the theory of evolution by natural selection in part to explain why closely related species are found on nearby islands.

was the instantaneous and independent creation of living organisms by a supernatural being.

This chapter begins by examining the pattern component of Darwin's and Wallace's theory. (**Box 21.1** discusses why the theory of evolution became associated primarily with Darwin's name.) Specifically, section 21.1 considers the evidence behind the claim that species are not independent but are instead related, and that they are not immutable but instead have changed through time. Darwin and Wallace proposed that natural selection explains this pattern; in section 21.2 we analyze the process. The concluding section reviews two recent studies of evolution by natural selection. These case studies illustrate how biologists test the Darwin-Wallace theory by studying evolution in action.

21.1 The Evidence for Evolution

In *On the Origin of Species*, Darwin repeatedly used the phrase "descent with modification" to describe evolution. By this he meant that the species existing today have descended from other, preexisting species, and that species change through time. This view was clearly a radical departure from the pattern of independently created and immutable species proposed by the theory of special creation. But what evidence do we have that Darwin was correct? How can biologists infer that species are indeed related—not independent—and that they have changed through time instead of being static entities? Let's consider each question in turn.

BOX 21.1 Why Darwin Gets Most of the Credit

Although Darwin and Wallace formulated the same explanation for how species change through time, Darwin's name is much more prominently associated with the theory because he thought of it first and provided massive evidence for it in *On the Origin of Species*. But historians of science speculate about whether Darwin would have published his theory at all had Wallace not threatened to scoop him (**Figure 1**).

Darwin actually wrote a paper explaining evolution by natural selection in 1842—a full 17 years before the first edition of *On the Origin of Species* came off the presses. He never submitted the work for publication, however. Why? Darwin claimed that he needed time to document all of the arguments for and against the theory and to examine its many implications. There is probably an element of truth in this—Darwin was a remarkably thorough thinker and writer. But many historians of science argue that Darwin held off largely out of fear. Because his theory removed any role for the divine in creation, Darwin knew that he would be exposed to scathing criticism from religious and scientific leaders. He was also an extremely private person, had had a strong religious upbringing, and was frequently in poor health. He responded to stressful situations or personal attacks by suffering long bouts of debilitating illness. The prospect of fighting for his ideas against the most powerful men in Europe was daunting. But Wallace forced his hand.

Alfred Russel Wallace was also a native of England, but had been making a living by collecting butterflies and other natural history specimens in Malaysia and selling them to private collectors. While recuperating from a bout of malaria there in 1858, he wrote a brief article outlining the logic of evolution by natural selection. He sent a copy to Darwin, who immediately recognized that they had formulated the same explanation for how populations change through time. The two had their papers read together before the Linnean Society of London, and Darwin then rushed *On the Origin of Species* into publication a year later. The first edition sold out in a day.

Fortunately for Darwin's health, a fellow biologist and friend named Thomas Huxley took charge of publicly defending the theory against criticism, which came from both scientific and religious quarters. Darwin continued to live quietly on his estate in Down, England, and actively continued a brilliant research and publishing career.

(a) Charles Darwin

(b) Alfred Russel Wallace

FIGURE 1
(a) This is a sketch of Charles Darwin in 1840, just four years after he returned from the voyage of the *Beagle* and two years before he drafted his first paper explaining evolution by natural selection. **(b)** This is Alfred Russel Wallace, who in 1858 formulated the theory of natural selection independently of Darwin.

Evidence That Species Are Related

Perhaps the seminal experience in Darwin's life was a five-year voyage he took aboard an English naval ship named the HMS *Beagle*. While fulfilling its mission to explore and map the coast of South America, the *Beagle* spent considerable time in the Galápagos Islands off the coast of present-day Ecuador. Darwin had taken over the role of ship's naturalist and gathered extensive collections of the plants and animals found in these islands. Among the birds he collected were what came to be known as the Galápagos mockingbirds, pictured in **Figure 21.1a**.

(a) The Galápagos mockingbirds differ only slightly in size, shape, and coloration.

Nesomimus macdonaldi

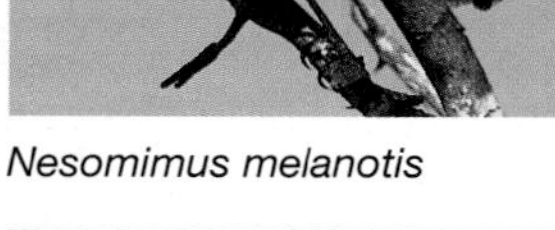

Nesomimus melanotis

Nesomimus parvulus

Nesomimus trifasciatus

(b) Darwin reasoned that they are similar because they share a common ancestor.

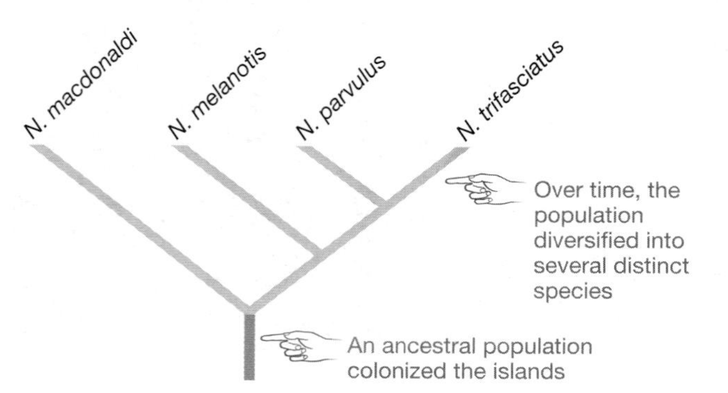

FIGURE 21.1 Close Relationships Among Island Forms Argues for Shared Ancestry
(a) Darwin collected mockingbirds from several different islands of the Galápagos. **(b)** This phylogeny illustrates Darwin's explanation for why mockingbirds from different islands are similar, yet distinct.

Several years after Darwin had returned to England, a naturalist friend in London pointed out that the mockingbirds he had collected on different islands were distinct species. This struck Darwin as remarkable. Why would species that inhabit neighboring islands be so similar, yet clearly distinct? This turns out to be a very general pattern. In island groups across the globe, it is routine to find closely related, but different, species on neighboring islands.

Darwin realized that this pattern—puzzling when examined as a product of special creation—made perfect sense when interpreted in the context of evolution, or modification with descent. He proposed that the mockingbirds were similar because they had descended from the same common ancestor. **Figure 21.1b** shows a phylogeny—a genealogy of populations—that illustrates this idea.

But island forms are not the only species that show strong similarities. Resemblances among species are widespread. These resemblances can be identified and analyzed at three different levels:

- **Structural homologies** are similarities in limbs, shells, flowers, or other structures of organisms, such as those shown in **Figure 21.2a**. A classic example is the common basic plan observed in the limbs of vertebrates. In Darwin's own words, "What could be more curious than that the hand of a man, formed for grasping, that of a mole for digging, the leg of the horse, the paddle of the porpoise, and the wing of the bat, should all be constructed on the same pattern, and should include the same bones, in the same relative positions?"
- **Developmental homologies** are similarities found at two levels in embryos: in their overall morphology and in the fate of embryonic tissues. For example, **Figure 21.2b** illustrates the strong general resemblance that exists among the early embryos of vertebrates. Further, in groups as different as fish and mammals, the same group of embryonic cells develops into the jaw tissue of the adult organism.
- **Genetic homologies** can be recognized in the amino acid sequences of proteins or the DNA sequences of genes. For example, a gene called *eyeless* in fruit flies and a gene called *Aniridia* in humans are both involved in the development of the eye. The two genes produce proteins that are nearly identical in amino acid sequence (**Figure 21.2c**). These genes also contain a section that is 90 percent identical in DNA sequence.

Although structural and developmental homologies were recognized and studied long before Darwin, no one was quite sure why such striking similarities existed among certain organisms but not others. Darwin, in contrast, saw them as a logical consequence of descent with modification. In fact, biologists now reserve the term **homology** for similarities that exist *because* of common descent. This distinction is important because traits found in different species can be similar even though they were *not* inherited from a common ancestor. These **analogous traits** result from a phenomenon known as **convergent evolution.** Convergent evolution is explored in detail in **Box 21.2**, pages 416–417.

Perhaps the ultimate similarity among organisms, however, is the genetic code. With a few minor exceptions, the same 64 mRNA codons specify the same amino acids in all organisms that have been studied. The variations that exist have been found primarily in the mitochondrial DNA of fungi and in the nuclear DNA of some protists. In these genomes, the changes are limited to one or two codons out of the 64.

Evidence That Species Have Changed Through Time

Strong similarities among certain organisms, which Darwin interpreted as evidence of evolution, extend to the fossil record. Early in the nineteenth century, for example, paleontologists reported striking resemblances between the fossils found in the rocks underlying certain regions and the living species found in the same area. The pattern was so widespread that it became known as the "law of succession." The general observation was that species in the fossil record were succeeded, in the same

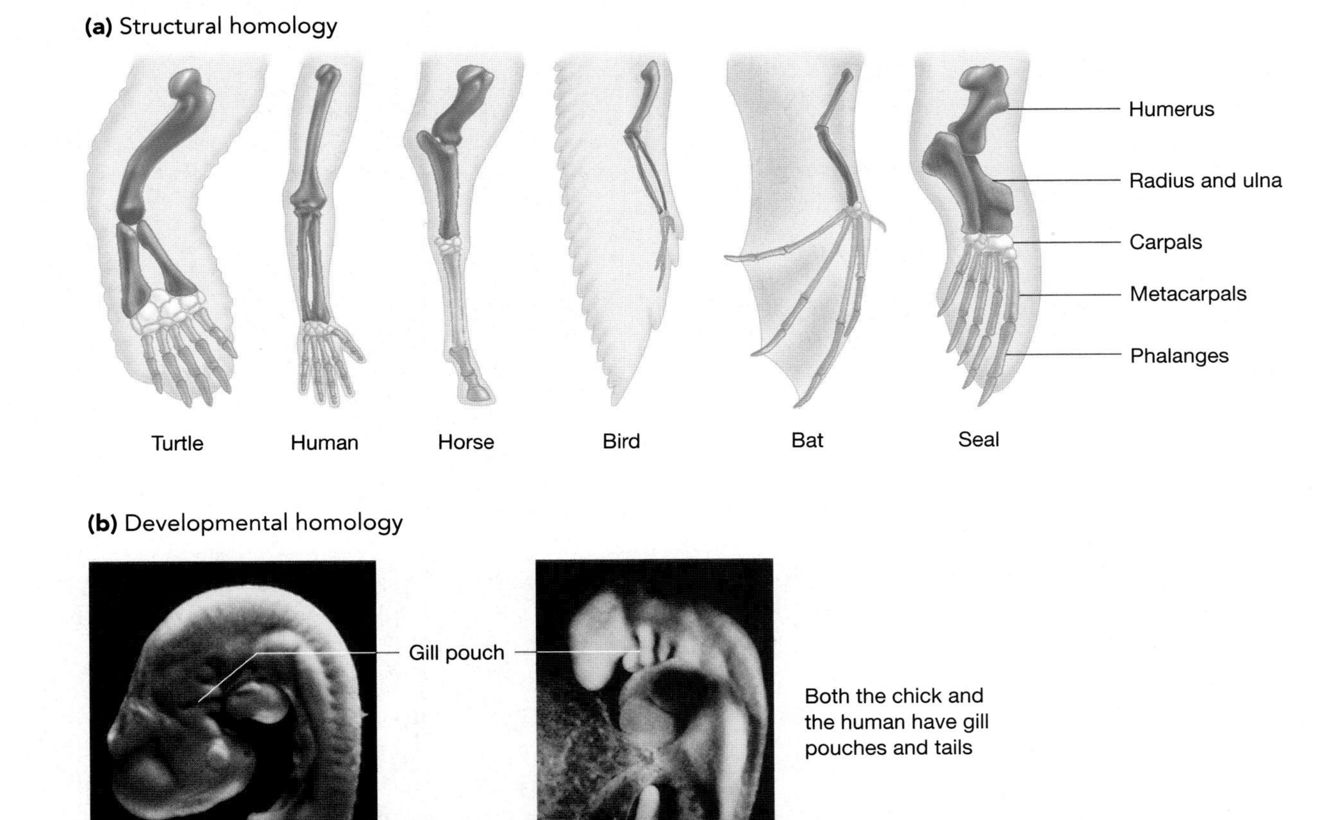

(c) Genetic homology

Gene:

LQRNRTSFTQE QIEA LEKEFERTHYPDVFARERLAA KID LPEARIQVWFSNRRAKWRREE *Aniridia* (Human)

LQRNRTSFTND QIDS LEKEFERTHYPDVFARERLAG KIG LPEARIQVWFSNRRAKWRREE *eyeless* (Fruit fly)

Only six of the 60 amino acids in these sequences are different. The two sequences are 90% identical.

FIGURE 21.2 Structural, Developmental, and Genetic Homologies
(a) Even though their function varies, all vertebrate limbs are modifications of the same basic configuration, or basic plan. Darwin interpreted structural homologies like these as a product of descent with modification. Note that the limbs are not drawn to scale. **(b)** These photos show the strong resemblance of the early embryonic stages of a chick and a human. **(c)** Each letter in the sequences given here represents an amino acid. The sequences are from a portion of the protein products of the *Aniridia* gene found in humans and the *eyeless* gene found in *Drosophila*.

region, by similar species (**Figure 21.3**). At first the pattern was simply reported and not interpreted. No one knew what sort of process might cause the pattern.

The fact that species had gone extinct, which was recognized in the early nineteenth century, was another insight relevant to the pattern called evolution. Paleontologists began discovering fossil bones, leaves, and shells that were unlike structures from any known living animal or plant. At first many scientists insisted that living examples of these species would be found in unexplored regions of the globe. But as exploration continued and the number and diversity of fossil collections grew, the argument became less and less plausible. After Baron Georges Cuvier published a detailed analysis of an extinct species called the Irish "elk" in 1812, most scientists accepted extinction as a reality. This gigantic deer was judged to be too large to escape discovery and too distinctive to qualify as a large-bodied population of an existing species.

Initially, extinct forms were interpreted in one of two ways. Advocates of the theory of special creation argued that they were

FIGURE 21.3 The Law of Succession
Living and fossil sloths are found only in Central America and South America. Darwin interpreted similarities among fossil and living species in the same geographic area as evidence for evolution.

BOX 21.2 How Do Biologists Distinguish Homology from Analogy?

Homology is a powerful concept in biology. Medical researchers, for example, study cancer in rats and vision in cats because homologous genes and structures are responsible for these phenomena in humans. Clearly, it is often vital to determine whether a homologous relationship exists among genes and other traits found in different species.

Determining whether homology exists is not always a simple matter, however, because not all similarities among organisms result from shared ancestry. The aquatic reptiles called icthyosaurs, for example, are strikingly similar to modern dolphins (**Figure 1a**). Both are large animals with streamlined bodies and large dorsal fins. Both chase down fish and capture them between elongated jaws filled with dagger-like teeth. But no one would argue that icthyosaurs and dolphins are similar because they recently shared a common ancestor. As the phylogeny in **Figure 1a** shows, analyses of other traits show that icthyosaurs are reptiles and dolphins are mammals.

Based on these data, it is sensible to argue that the similarities between icthyosaurs and dolphins result from convergent evolution. Convergent evolution occurs when natural selection favors similar solutions (like streamlined bodies and elongated jaws filled with sharp teeth) to the problems posed by a similar way of making a living (like chasing down fish in open water). The products of convergent evolution are known as analogous traits. Analogous traits are similar but they are not homologous.

In many cases, though, homology and convergence are much more difficult to distinguish than in the icthyosaur and dolphin example. How do biologists recognize homology? As an example, consider the *HOM* loci of insects and *Hox* loci of vertebrates introduced in Chapter 19. Even though insects and vertebrates last shared a common ancestor some 600–700 million years ago, biologists argue that *HOM* and *Hox* loci are derived from the same ancestral sequences because:

- The loci are organized in a similar way. The diagram in **Figure 1b** shows that in both insects and vertebrates, the loci are found in gene complexes, with related genes adjacent to one another on the chromosome.
- They share a highly conserved 180-bp homeobox sequence that binds to DNA and regulates the expression of other genes.
- Their products have similar functions—specifying locations in embryos—and are expressed in similar patterns in time and space.

In addition, many other animals, on lineages that branched off between insects and mammals, have similar genes. This is crucial: If genes found in distantly related lineages are indeed related by common ancestry, then similar genes should be found in many intervening lineages on the tree of life. (Examine the phylogeny in Figure 1a and ask yourself whether the analogous traits found in icthyosaurs and dolphins fulfill this criterion.)

In sum, loci in the *HOM* and *Hox* complexes are considered homologous because they are similar in organization, function, and composition. It is extremely improbable that convergent evolution could produce the degree of similarity observed in these genes.

(Continued on next page)

(Box 21.2 continued)

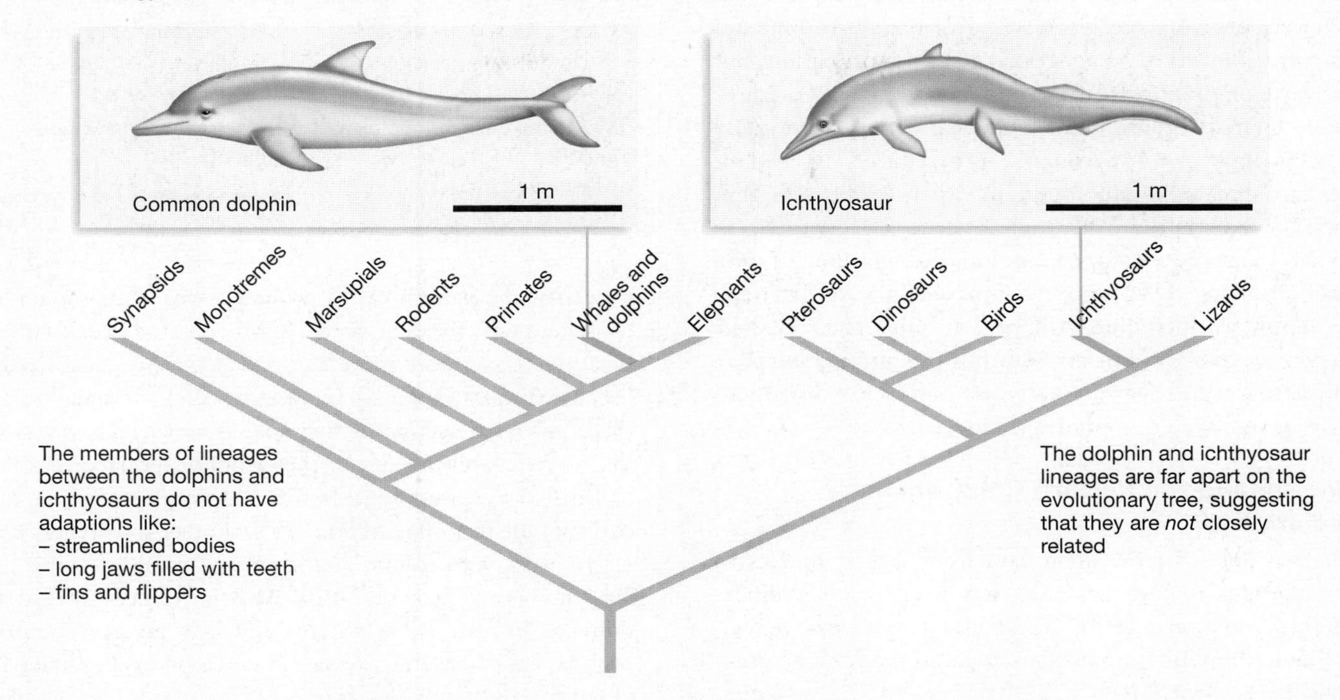

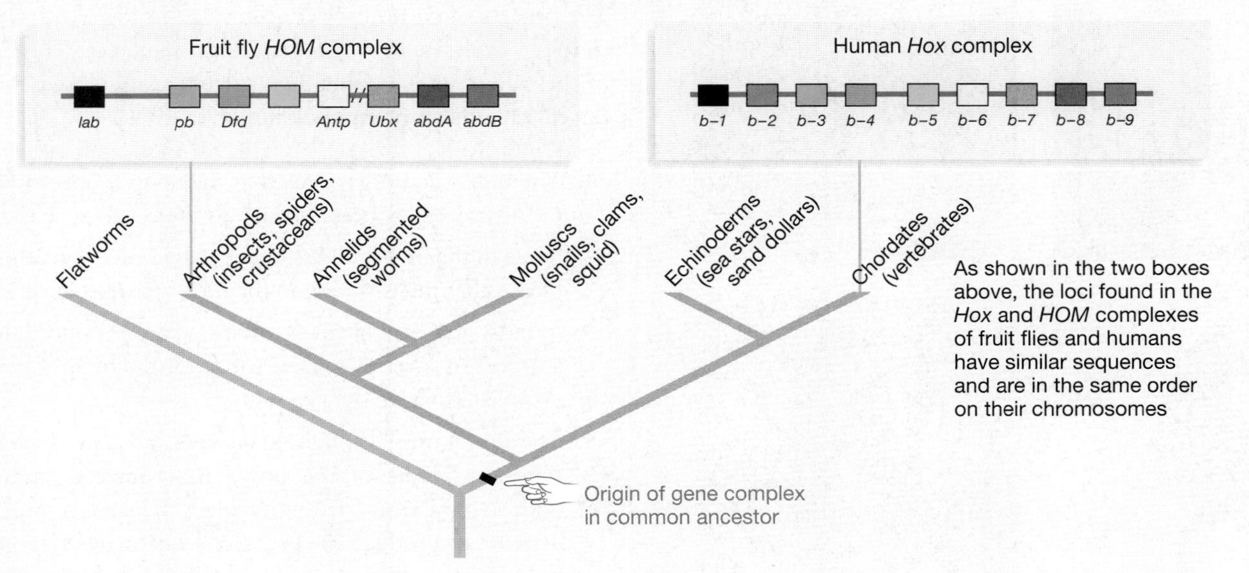

FIGURE 1
(a) Dolphins and icthyosaurs look similar, but they are not closely related. As a result, biologists infer that traits like streamlined bodies, sharp teeth, and flippers evolved independently in the two groups. **(b)** All of the animal groups illustrated on this phylogeny have *Hox/HOM* complexes that are similar to those illustrated for fruit flies and humans.

victims of the flood at the time of Noah. Darwin, in contrast, interpreted them as evidence that species are not static, immutable entities, unchanged since the moment of special creation.

Darwin contributed a third piece of evidence that species have changed through time. He described **vestigial traits** in a wide variety of living species. Vestigial traits are rudimentary structures that have no function or reduced function, but are clearly homologous to functioning organs or structures in closely related species. Humans have a variety of vestigial traits. Our appendix, for example, is a reduced descendant of the caecum—an organ found in other vertebrates that functions in digestion. The human coccyx illustrated in **Figure 21.4** is a rudimentary homolog of the tailbone found in other primates. Even goose bumps are a vestigial trait. Many mammals, including primates, are able to erect their hair when they are cold or excited. But our rudimentary fur does little to keep us warm, and goose bumps are largely ineffective in signaling our emotional state.

Darwinism and the Pattern Component of Evolution

Darwin was able to draw upon data from several sources to support the idea that species have descended, with modification, from a common ancestor; **Table 21.1** summarizes this evidence. But it's important to recognize that no single observation or grand experiment instantly "proved" the fact of evolution and swept aside belief in special creation. Rather, Darwin argued that the pattern called evolution was simply much more consistent with the data. Stated another way, descent with modification was a more successful and powerful scientific theory because it could explain observations such as vestigial traits and the close relationships among species on neighboring islands that were inexplicable under the theory of special creation.

TABLE 21.1 Evidence for Evolution

Prediction 1: Species Are Related, Not Independent	Prediction 2: Species Are Not Static, but Change Through Time
Closely related species often live in the same geographic area.	Fossils frequently resemble living species found in the same area today.
Structural, developmental, and genetic homologies are widespread.	Many species have gone extinct.
	Vestigial traits are common.

Darwin's greatest contribution did *not*, however, lie in recognizing the fact of evolution. Several other researchers had already proposed evolution as a pattern in nature long before Darwin began his work. **Box 21.3** introduces one of these thinkers. Instead, Darwin's crucial insight lay in recognizing a process, called natural selection, that could explain the pattern of modification with descent.

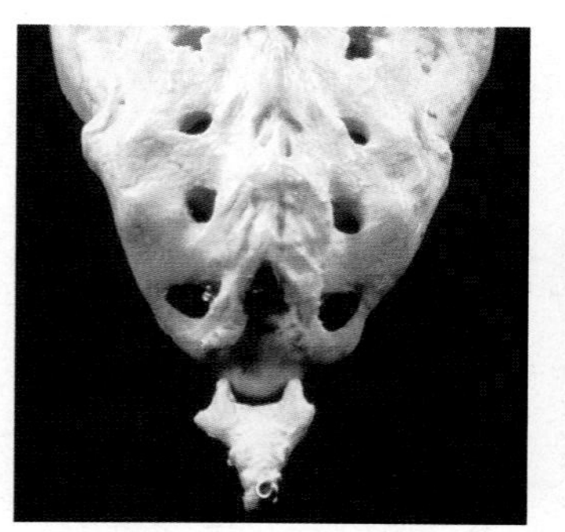

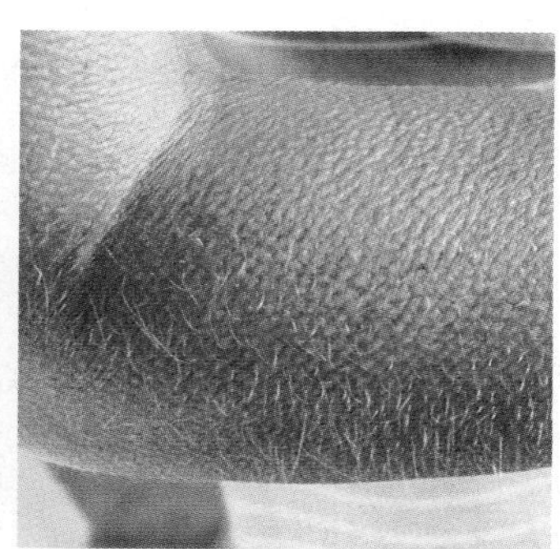

Capuchin monkey tail
(used for balance, locomotion)

Erect hair on chimp
(insulation, emotional display)

FIGURE 21.4 Vestigial Traits
The tailbone and goose bumps are human traits that are functionless. But they are homologous to functioning structures in related species.

21.2 How Natural Selection Works

The process of natural selection can be broken down into four simple postulates, or steps in a logical sequence. The outcome of these four steps is **evolution**—a genetically based change in the characteristics of a population over time.

1. The individual organisms that make up a population vary in the traits they possess, such as their size or shape.
2. Some of the trait differences are passed on to offspring. This means that the variation in the trait has some genetic basis.
3. In most generations, more offspring are produced than can survive. As a result, only a subset of the offspring that are produced survive long enough to reproduce.
4. The subset of offspring that survive and reproduce is not a random sample of the population. Instead, individuals with certain traits are more likely to survive and reproduce or to produce the greatest number of offspring. The individuals with these traits are, in Darwin's words, "naturally selected."

When variation in the favored traits is heritable (that is, genetically based), evolution results. When only certain individuals reproduce or reproduce the most, and when these favored traits are inherited by their offspring, the characteristics of the population change from one generation to the next. This is evolution by natural selection.

To summarize the process yet another way, biologists often combine postulates 1 and 2 by observing that heritable variation exists for most traits in most populations. Postulates 3 and 4 can also be combined by noting that individuals experience differential success in their ability to survive and reproduce. Evolution, or change in the characteristics of a population over time, results when differential success is based on heritable variation.

In explaining the process of natural selection, Darwin referred to the successful individuals as "more fit." In doing so, he gave the word fitness a definition quite different from its everyday English usage. **Darwinian fitness** is the ability of an individual to survive and reproduce. This is a quantity that can be measured. Researchers study populations in the lab or in the field and estimate the relative fitness of each individual by counting how many offspring it produces relative to other individuals.

The concept of fitness, in turn, provides a compact way of formally defining adaptation. The biological meaning of adaptation, like the biological meaning of fitness, is quite different from its normal English usage. In biology, an **adaptation** is a heritable trait that increases the fitness of an individual with that trait relative to individuals without that trait.

At least some of the ideas presented in this section should be familiar by now. Chapter 1 used artificial selection on populations of cabbage family plants to introduce how natural selection works. Chapter 3 explored how the process could act on a population of simple, self-replicating molecules and lead, over time, to increasingly complex molecules and the first self-replicating entity.

This section of the chapter has been heavy on theory and terminology and light on concrete examples, however. To help you understand each step in the process of natural selection more thoroughly, section 21.3 is devoted to data—specifically, to recent studies of how natural selection works in actual populations. Biologists accept the theory of evolution by natural selection not only because of its explanatory power but because it has been observed directly.

21.3 Evolution in Action: Recent Research on Natural Selection

Darwin's theory of evolution by natural selection is testable. If the theory is correct, biologists should be able to test the validity of each postulate and actually observe evolution in natural populations.

This section summarizes two examples in which evolution by natural selection has been, or is being, observed in nature. Literally hundreds of other case studies are available, involving a wide variety of traits and organisms. The examples here were chosen because they demonstrate *how* biologists go about studying evolution by natural selection. To begin, let's explore the evolution of drug resistance—one of the great challenges facing today's biomedical researchers and physicians.

BOX 21.3 Evolutionary Theory Before Darwin

In the introduction to *On the Origin of Species*, Darwin outlined some of the evidence for the fact of evolution, including many of the points summarized in section 21.1. Then he wrote that a biologist "might come to the conclusion that each species had not been independently created, but had descended . . . from other species. Nevertheless, such a conclusion even if well founded, would be unsatisfactory, until it could be shown *how* the innumerable species inhabiting this world have been modified" (emphasis added). Most scientists would agree with Darwin—recognizing a pattern in nature without understanding the process that caused it is inadequate.

In 1809, decades before Darwin began his work, the French biologist Jean-Baptiste de Lamarck proposed that species are not static and independently created entities, but have changed through time and are related by common ancestry. Lamarck was one of several writers who advocated the reality of evolution before Darwin published *On the Origin of Species*.

More important, Lamarck also offered a mechanism to explain how species evolved. He proposed that simple forms of life are continuously being created by spontaneous generation. Lamarck reasoned that these simple cells then evolved to progressively "higher" forms through a process known as the inheritance of acquired characters. Specifically, he proposed that progressive evolution occurs because individuals change as they develop, in response to the environment. Lamarck also proposed that these phenotypic changes are passed on to offspring. A classic scenario is that giraffes who stretched their necks to reach leaves high in treetops would produce offspring with elongated necks.

Lamarck's ideas were roundly rejected by the leading scientists of the day, partly because he was not as thorough as Darwin in amassing evidence for the fact of evolution. Today biologists reject Lamarck's ideas because they know that changes in phenotypes that develop as a result of an individual's actions do not produce changes in genotypes, and cannot be transmitted to offspring. (If you gained increased upper body strength through weightlifting, your offspring would not tend to be stronger as a result.) Still, Lamarck deserves credit for bringing evolution to the forefront of scientific debate. In this way, he helped to pave the way for Darwin.

How Did *Mycobacterium tuberculosis* Become Resistant to Antibiotics?

The bacterium that causes tuberculosis, or TB, has long been one of the great scourges of humankind (**Figure 21.5**). TB was responsible for almost 25 percent of all deaths in New York City in 1804; in nineteenth-century Paris, the figure was closer to 33 percent. To put these numbers in perspective, consider that all types of cancer, combined, currently account for about 30 percent of the deaths that occur in the United States. TB was once as great a public health issue as cancer is now.

Although tuberculosis still kills more adults than any other viral or bacterial disease in the world, TB attracted relatively little attention in the industrialized nations between about 1950 and 1990. As the map in **Figure 21.6** shows, during that time TB was primarily a disease of developing nations.

The decline of tuberculosis in Western Europe, North America, Japan, Korea, and Australia is one of the great triumphs of modern medicine. In these countries sanitation, nutrition, and general living conditions began to improve dramatically in the early twentieth century. When people are healthy and well nourished, their immune systems work well enough to stop most *M. tuberculosis* infections quickly—before the infection can harm the individual and before the bacteria can be transmitted to a new host. Equally important, antibiotics such as isoniazid and rifampin became available in the industrialized countries starting in the early 1950s. These drugs allowed physicians to stop even advanced infections, and they saved millions of lives.

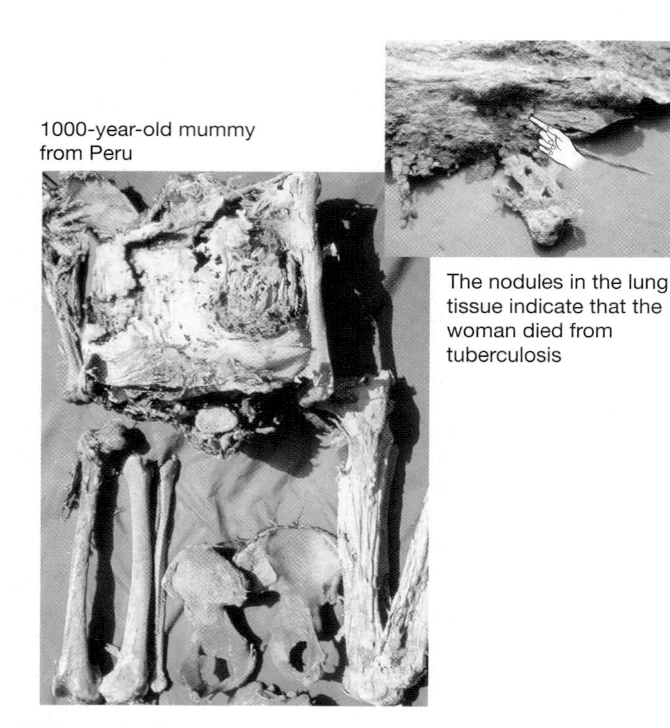

FIGURE 21.5 Tuberculosis Is an Ancient Disease of Humans
The photograph at left shows a 1000-year-old mummy of a middle-aged woman, from in a tomb found in the coastal deserts of Peru. The inset photo is a scanning electron micrograph of lung tissue from the mummy. The nodules in the tissue are strikingly similar to those found in contemporary TB victims. When biologists extracted DNA from nodules in the mummy, they found sequences diagnostic of *M. tuberculosis*. This is strong evidence that TB was present in the New World long before the arrival of Europeans.

In the late 1980s, however, rates of *M. tuberculosis* infection surged in many countries, and in 1993 the World Health Organization (WHO) declared TB a global health emergency. Physicians were particularly alarmed because the strains of *M. tuberculosis* responsible for the increase were largely or completely resistant to rifampin, isoniazid, and other antibiotics that were once extremely effective. How and why did the evolution of drug resistance occur? The case of a single patient—a young man with AIDS who lived in Baltimore—will illustrate what is happening all over the world.

The story begins when the HIV-positive individual was admitted to the hospital with fever and coughing. Chest x-rays followed by bacterial cultures of sputum ejected from the lungs showed that he had an active TB infection. He was given a battery of antibiotics for six weeks, followed by twice-weekly doses of rifampin and isoniazid for an additional 33 weeks. Ten months after therapy started, bacterial cultures from sputum indicated no *Mycobacterium tuberculosis* cells. His chest x-rays were also normal. The antibiotics seemed to have cleared the infection.

Just two months after the TB tests proved normal, however, the young man was readmitted to the hospital with a fever, severe cough, and labored breathing. Despite being treated with a variety of antibiotics, including rifampin, he died of respiratory failure 10 days later. Sputum cultures showed that *M. tuberculosis* was again growing actively in his lungs. But this time the bacterial cells were completely resistant to rifampin.

Drug-resistant bacteria had killed the patient. Where did they come from? Is it possible that a strain that was resistant to antibiotic treatment evolved *within* the patient? To answer this question, a research team led by William Bishai analyzed DNA from the drug-resistant strain and compared it to stored DNA from the drug-susceptible *M. tuberculosis*, cultured a year earlier from the same patient. After examining extensive stretches from each genome, the biologists were able to find only one difference: a point mutation in the *rpoB* gene. This locus codes for a component of RNA polymerase, the protein that transcribes DNA to mRNA. The mutation changed a cytosine to a thymine and altered the normal codon reading TCG to a mutant one reading TTG. As a result, the RNA polymerase produced by the drug-resistant strain had leucine instead of serine at the 153rd amino acid in the polypeptide chain.

This result is meaningful. Rifampin works by binding to the RNA polymerase of *M. tuberculosis*. When the drug enters an *M. tuberculosis* cell and binds to its RNA polymerase, it interferes with transcription. If sufficient quantities of rifampin are present for long enough and if the drug binds tightly, the cell will die. But apparently the substitution of a leucine for a serine prevents rifampin from binding efficiently. Consequently, cells with the C → T mutation discovered by Bishai's team continue to grow even in the presence of rifampin.

These results suggest that a chain of events led to the patient's death. Bishai and co-workers hypothesize that early in the course of the infection, while *M. tuberculosis* colonies were actively growing in the patient's lungs, a mutation occurred in the *rpoB* gene of one cell. This mutant cell continued to grow and divide. It is extremely unlikely that it could grow as rapidly as cells with normal RNA polymerase, however. Mutations are random changes, and random changes in proteins are unlikely to make them perform better in a normal environment. Thus, the mutant cells undoubtedly stayed at low frequency in the population occupying the patient's lungs—even while that population grew to the point of inducing symptoms that sent the patient to the hospital. At that point, therapy with rifampin began. In response, cells in the population with normal RNA polymerase began to grow much more slowly or to die outright. But cells with the C → T mutation had an advantage in the new environment. They began to grow more rapidly than the normal cells and continued to increase in numbers after therapy ended. Eventually the *M. tuberculosis* population regained its former numbers, and the patient's symptoms reappeared. However, drug-resistant cells now dominated the population. This is why the second round of rifampin therapy was futile. If a friend of the patient or a health-care worker had contracted TB from him, rifampin therapy would have been useless and the disease would continue to spread.

Does this sequence of events mean that evolution by natural selection occurred? One way of answering this question is to review the four postulates listed in section 21.2, and determine whether each was tested and verified.

1. *Did variation exist in the population?* The answer is yes. Due to mutation, both resistant and nonresistant strains of TB were present. Most *M. tuberculosis* populations, in fact, exhibit variation for the trait; studies on cultured *M. tuberculosis* show that a mutation conferring resistance to rifampin is present in one out of every 10^7 to 10^8 cells.
2. *Was this variation heritable?* Bishai's team showed that the variation in the phenotypes of the two strains—from drug susceptibility to drug resistance—was due to variation in their genotypes. Because the mutant *rpoB* gene is copied before a *Mycobacterium* replicates, the allele and the phenotype it produces—drug resistance—is passed on to offspring.
3. *Did natural selection occur?* That is, did some *Mycobacterium tuberculosis* individuals leave more offspring than other individuals? Clearly, only a tiny fraction of *M. tuberculosis* cells in the patient survived the first round of antibiotics long enough to reproduce—so few, in fact, that after the initial therapy the chest x-ray was normal and the sputum sample contained no *M. tuberculosis* cells.
4. *When selection occurred, did a non-random subset of the population survive better and reproduce the most?* Because the *M. tuberculosis* population present early in the infection was different from the *M. tuberculosis* population present at the end, it is clear that individuals with the drug-resistant allele had survived selection by rifampin better than cells with the normal allele. To use the terminology introduced earlier, *M. tuberculosis* individuals with the mutant *rpoB* gene had higher fitness in an environment where rifampin was present. The new allele is an adaptation to the antibiotic.

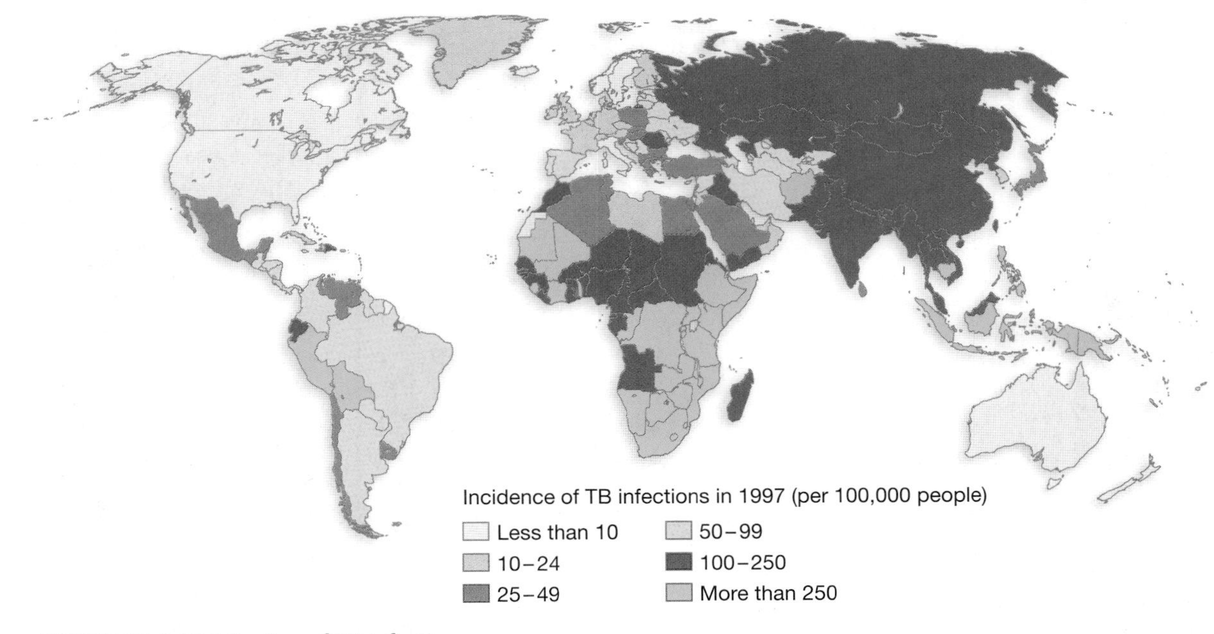

FIGURE 21.6 Distribution of TB Infections
The vast majority of TB infections occur in the developing nations. In these countries, TB is responsible for 7 percent of all deaths (children and adults).

The study by Bishai and co-workers verified all four postulates and confirmed that evolution by natural selection had occurred. The example also neatly illustrates the contemporary definition of **evolution**: a change in allele frequencies in a population over time. The characteristics of the *M. tuberculosis* population changed over time because the mutant *rpoB* gene increased in frequency. But note that the individual cells themselves did not evolve. When natural selection occurred, the individual cells did not change through time; they simply survived or died, or produced more or fewer offspring. This is a fundamentally important point: Natural selection acts on individuals, but only populations evolve.

The events just reviewed have occurred many times in other patients. Recent surveys indicate that drug-resistant strains now account for about 10 percent of the *M. tuberculosis*–causing infections throughout the world. And the emergence of drug resistance in TB is far from unusual. Resistance to a wide variety of insecticides, fungicides, antibiotics, antiviral drugs, and herbicides has evolved in hundreds of insects, fungi, bacteria, viruses, and plants. In many cases, the specific mutations that lead to a fitness advantage and the spread of the resistance alleles are known.

Why Do Some Populations of Alpine Skypilots Have Larger Flowers Than Others?

The TB example is particularly satisfying because the molecular basis of both heritable variation and differential success is understood. But can biologists still study natural selection when the alleles responsible for variation are not known, or the molecular mechanism of adaptation has not been identified? The answer is yes. As an example of how this work is done, and to illustrate the role of experiments in evolutionary research, let's review work by Candace Galen on variation in the flower size of a Rocky Mountain wildflower (**Figure 21.7**).

Galen's research on natural selection in the alpine skypilot (*Polemonium viscosum*) began with a simple observation: Individuals found in the treeless, tundra habitats above timberline have flowers that are substantially larger than individuals growing among the stunted trees at timberline. The tundra in-

(a) In tundra habitats above timberline, the alpine skypilot is pollinated primarily by bumblebees.

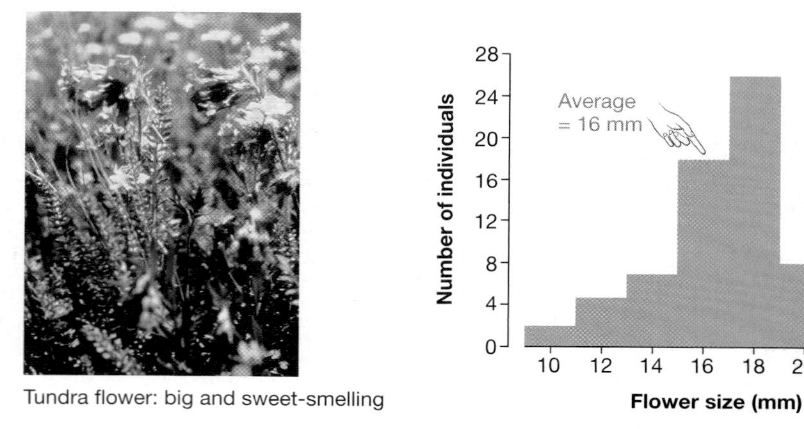

Tundra flower: big and sweet-smelling

Tundra pollinator: bumblebee

(b) In forested habitats below timberline, the alpine skypilot is pollinated primarily by flies.

Below-timberline flower: small and skunky-smelling

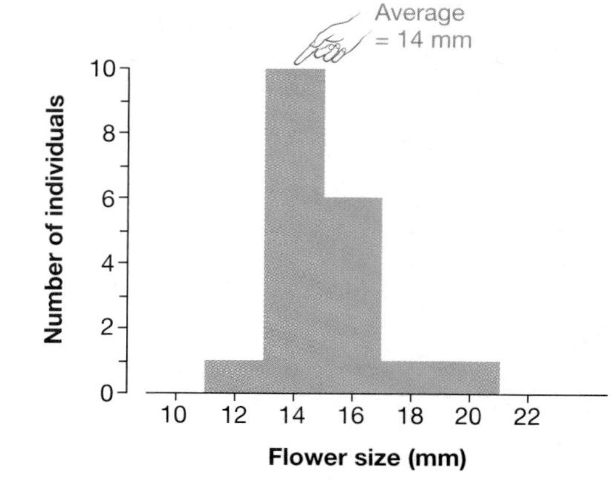

Below-timberline pollinator: fly

FIGURE 21.7 Alpine Skypilot Populations Differ in Important Traits
Bees and flies feed on pollen and nectar in alpine skypilot flowers. As they feed, they become dusted with pollen, which they carry to other flowers. QUESTION Why might bees be attracted to larger flowers and flies be attracted to smaller flowers?

CD ACTIVITY 21.2 Natural Selection in Alpine Skypilots

dividuals also have longer stalks, and most have sweet-smelling flowers. Flowers on plants growing at timberline have shorter stalks, and most have a smell that biologists describe as "skunky."

Galen also noticed that individuals growing below timberline are primarily pollinated by flies, while individuals above timberline are pollinated only by bumblebees. Because bumblebees are much larger than flies and are attracted to sweet smells, Galen thought that the differences in pollinators might be responsible for the difference in skypilot flowers. Specifically, she proposed that natural selection, caused by a preference by bumblebees to land on larger, sweet-smelling flowers and a preference by flies to land on smaller, skunky-smelling flowers, had acted on variation in flower size and produced the differences in the populations.

This hypothesis makes several predictions:

- *Individuals with larger flowers attract more bees.* To test this prediction, Galen randomly chose 100 sweet-smelling plants from the timberline population and transplanted them to an insect-proof enclosure. She measured the size of the flowers on each plant, and then released bees into the enclosure and observed which flowers they visited among the many available. The data in **Figure 21.8** show that plants with large flowers and taller stalks received many more bee visits.
- *Individuals that attract more pollinators have higher fitness.* Galen tested this prediction by counting the number of seeds produced by each plant used in the enclosure experiment. Seed number is a measure of fitness because it correlates with the number of offspring produced. Galen's data, plotted in **Figure 21.9**, show that the flowers that received more bee visits set more seed.
- *Differences among individuals in flower size occur in part because of differences in their genetic makeup.* This is a crucial point. The first two analyses showed that variation exists, and that bees can exert natural selection that favors larger flowers and taller stalks. But unless flower size is heritable, evolution will not occur as a result of this selection—no change in the population would occur over time. For average flower size to change in the population over time, alleles that produce large flowers and longer stalks have to increase in frequency in the population.

The genetic basis of flower size can't be determined directly, however. Unlike the situation with drug resistance in TB, no one knows exactly which loci code for flower size. As Chapter 10 pointed out, this situation is routine for quantitative traits like flower size. Traits like height, shape, and intelligence are not determined by a single gene, but result from interactions among many gene products.

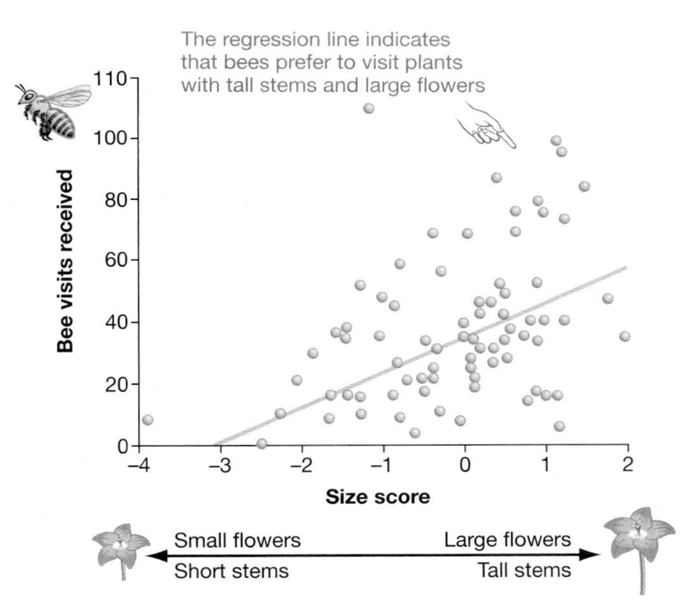

FIGURE 21.8 Bees Prefer to Visit Large Flowers
Each data point in this graph (called a scatterplot) represents an alpine skypilot plant. The quantity called size score, plotted on the horizontal axis, is a combined measure of the average stem height and the size of flowers on a plant. The line through the points results from a statistical procedure called regression analysis, and is called the best-fit line.

Note, however, that the data points are widely scattered. For example, look at the amount of variation in bee visits received by flowers with size scores close to 1. This scatter occurs because other factors besides size influence how likely bees are to land on a particular flower. These factors might include the flower's smell, wind strength and direction, where the bees began their foraging flights, and variation in the bees' ability to distinguish flower size.

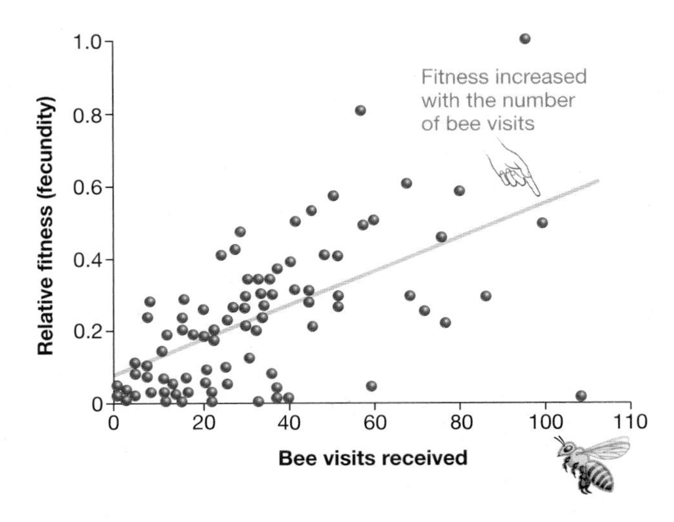

FIGURE 21.9 Bee Visits Lead to Increased Reproductive Success of Alpine Skypilots
Each data point on this scatterplot represents an alpine skypilot plant. The number of bee visits to each plant is plotted against fitness. Note that fitness is a relative quantity—the researcher assigned a fitness of 1.0 to the individual producing the most seeds, and then expressed the fitnesses of all other plants as a percentage of that number of seeds. QUESTION From these data, the researcher drew the conclusion stated in the title of this figure. Does this conclusion make sense to you? Why or why not?

It's also entirely possible that the variation observed in quantitative traits like weight or intelligence is *not* due to genetic differences. In plants it is common for quantitative traits like leaf size or stem height to vary primarily because of variation in environmental factors such as temperature or water availability or the richness of soil.

The most common way to determine whether a quantitative trait has a genetic basis, and thus to understand whether it can evolve in response to natural selection, is to estimate a quantity known as heritability (symbolized by h^2). The **heritability** of a trait is the proportion of the total variation observed in a population that is due to genetic variation. Because it is a proportion, h^2 varies from 0 to 1.0. A value of 0 means that none of the variation in the population is genetic; a value of 1.0 means that all of it is.

Heritabilities are estimated by comparing the values of traits among relatives, usually between parents and offspring. The logic of this approach is straightforward: The similarities between parents and offspring result, at least in part, from the alleles they share. To estimate the heritability of flower size in alpine skypilots, Galen collected seeds produced by the bumblebee-pollinated plants in the enclosure experiment and germinated them in the greenhouse. Then she planted 617 of the seedlings in randomly assigned locations in the timberline habitat. (**Box 21.4** explains why assigning offspring to randomly chosen environments was crucial to the success of Galen's experiment.) Seven years later, 58 of the seedlings had survived long enough to flower. Galen measured the size of their flowers and compared them to the flower size of their mothers. As **Figure 21.10** shows, a strong relationship existed: Small-flowered mothers tended to produce offspring with small flowers, while large-flowered mothers tended to produce offspring with large flowers. According to Galen's data, h^2 was close to 1.0. This is strong evidence that flower size is heritable in the population. Heritable traits can respond to natural selection and evolve.

To clinch her argument that tundra flowers have evolved large size in response to natural selection by bumblebees, Galen performed the experiment outlined in **Figure 21.11**. Examine the figure and note that she established two treatment groups, called controls and experimentals. The goal of the control treatment was to observe what happens when bees do *not* exert natural selection; the goal of the experimental treatment was to observe what happens when bees do exert natural selection.

To create the control, Galen pollinated a large group of randomly selected timberline skypilots by hand, using pollen from randomly selected individuals. She collected the resulting seeds, germinated them in the greenhouse, and planted the seedlings out into randomly assigned locations in the timberline habitat. When these individuals matured, she compared their flower sizes to the sizes in offspring of individuals that had been pollinated by bumblebees. The result? Examine the two histograms in Figure 21.11. They show that offspring of bumblebee-pollinated plants had flowers that were significant-

BOX 21.4 Problems in Estimating the Heritability of Traits

Demonstrating that a quantitative trait has a genetic basis, by showing that its heritability is greater than zero, is one of the most basic requirements in a study of evolution by natural selection. But it is also one of the most difficult.

To understand why, recall two points. First, heritability is the proportion of variation observed in a population that results from variation in the genetic makeup of the individuals. The rest of the observed variation results from variation in the environments experienced by the individuals. Second, heritability is usually estimated by comparing the values of traits in parents and offspring.

Here's the problem. Parents and offspring often share their environment as well as their genes. In alpine skypilots, for example, seed dispersal is limited and offspring tend to germinate near their parents. If a parent plant has small flowers because it is growing in a particularly dry location, its offspring are likely to have small flowers for the same reason. In general, any correlation that exists between the environments experienced by parents and offspring tends to *increase* the similarity of their traits. This correlation inflates the estimate of trait heritability.

To remove this source of bias, Candace Galen planted the offspring used in her heritability study in *random* locations in the environment. This procedure ensured that there would be no correlation between the environments experienced by parents and offspring. An alternative way to remove this bias is to rear offspring in an identical environment—usually in the lab or greenhouse. (This approach is called a common garden experiment.) In species where parents provide care, eggs or young can be placed with randomly chosen foster parents.

Estimating the heritability of human traits is difficult because these types of experiments are impossible. To control for the effects of shared environments, heritability studies with humans are often conducted by measuring similarities in monozygotic twins who were separated at birth and placed with different adoptive parents. (These studies attempt to simulate the foster parenting experiments mentioned earlier.) Monozygotic twins are genetically identical, so if they were randomly placed with respect to the environments in their adoptive homes, similarities between them would be largely a product of their shared alleles.

Based on studies that attempt to meet these criteria, heritabilities for a wide variety of human traits have now been estimated. For example, the aspect of intelligence measured by IQ tests is estimated to have a heritability between 0.5 and 0.7, while height is estimated to have a heritability as high as 0.86.

ly larger than flowers from offspring of hand-pollinated plants. Because flower size is heritable, Galen could conclude that the experimental population, pollinated by bees, was genetically different than the control population, which was randomly pollinated. The experimental population had evolved by natural selection.

What's the next step in this research program? One question is, if large flowers are so advantageous in the tundra environment, why aren't the flowers of tundra skypilots even larger? It's not difficult to generate plausible hypotheses to answer this question. Perhaps alleles capable of creating even larger flowers don't exist in the population. Perhaps some other selective force such as the energetic cost of producing even larger flowers reduces survival and counteracts selection by bumblebees. But the fact is, no one knows. Galen's research continues.

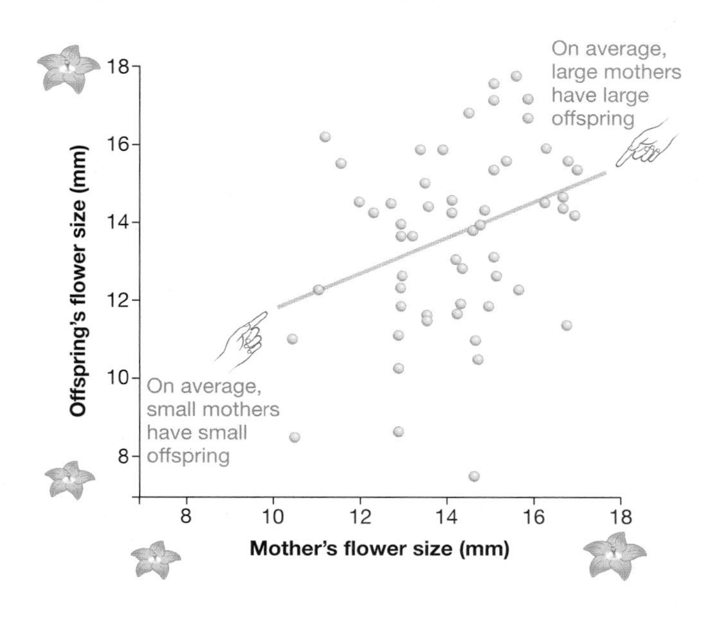

FIGURE 21.10 Flower Size Is a Heritable Trait in Skypilots
This scatterplot shows the relationship between a mother's flower size, plotted on the x-axis, and her offspring's flower size, plotted on the y-axis. The slope of the line fitted through the points is positive, indicating a relationship between the size of flowers in parents and offspring. Because the environments of the parents and offspring were random in this experiment, the similarity shown here must be the result of alleles shared by parents and offspring. **QUESTION** If there were no relationship between parent and offspring flower size, what would the scatterplot look like? Would the slope of the line fitted through the points be positive, negative, or flat?

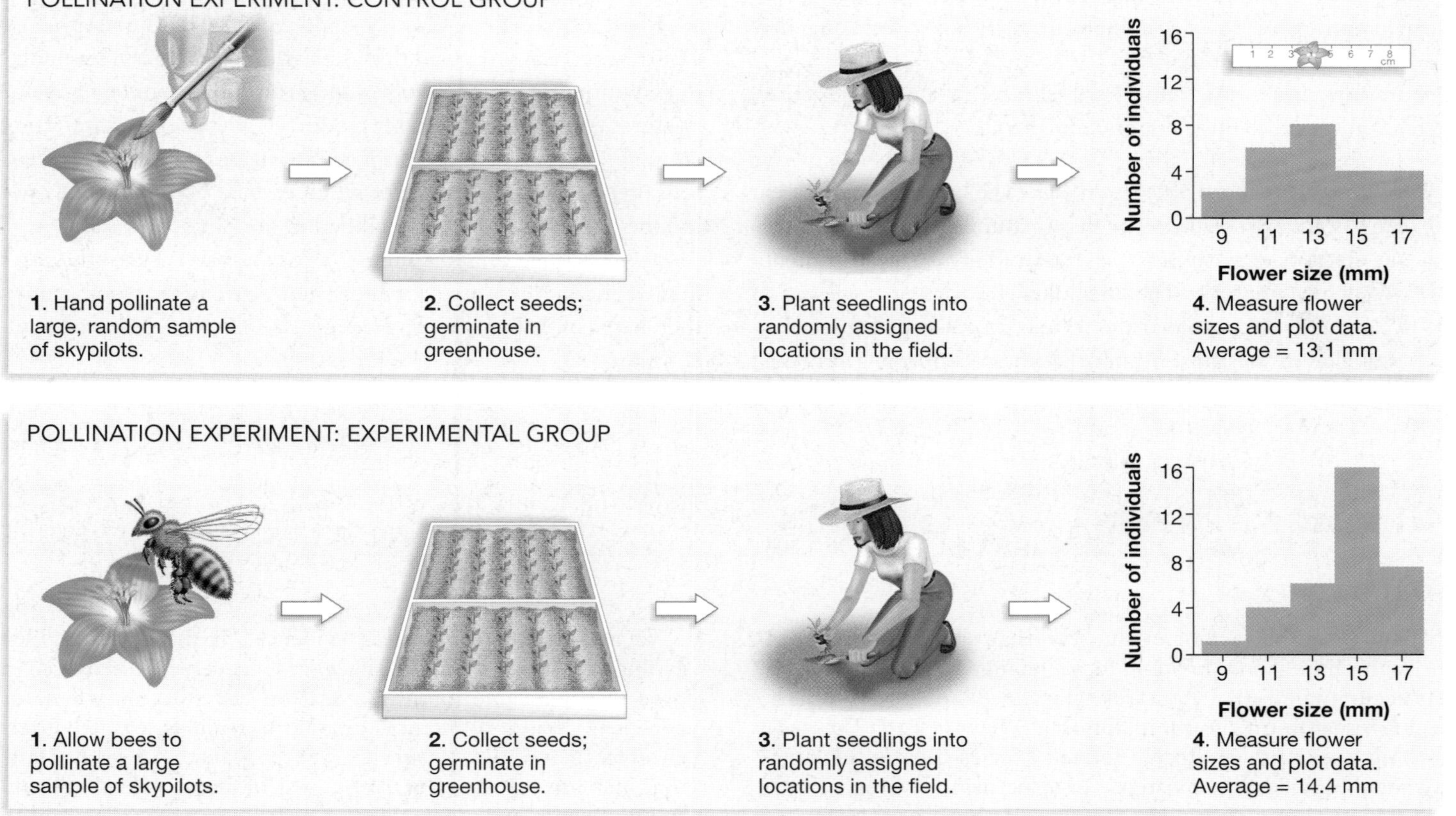

FIGURE 21.11 Offspring of Bee-Pollinated Plants Have Larger Flowers Than Offspring of Hand-Pollinated Plants
The histograms at right show the distribution of flower size in the offspring of alpine skypilots randomly pollinated by hand and in the offspring of those pollinated by bumblebees. The average size of flowers in the experimental treatment is over 1 mm larger than the average size of flowers in the control treatment.

Essay The Debate over "Scientific Creationism"

The theory of evolution by natural selection is the cornerstone of modern biology. Controversy over the pattern component of the theory (descent with modification) ended in the late nineteenth century, when mounting evidence for the fact of evolution silenced Darwin's critics. Controversy over the process component (natural selection) ended when biologists successfully integrated Mendelian genetics into Darwin's postulates. This extension of Darwin's theory, called "The Modern Synthesis," occurred in the 1930s. It was important because mutation and Mendel's laws explained *why* heritable variation exists.

Although scientific controversy over the theory of evolution by natural selection ended some 70 years ago, political and social controversy are alive and well. In the United States, advocates of "scientific creationism" and "intelligent design theory" lobby for a ban on teaching evolution in public schools, for equal time devoted to teaching the theory of special creation, or for disclaimers in textbooks declaring that evolution is "just a theory."

Where does the controversy stand today?

In objecting to the theory of evolution by natural selection, creationists make three major claims:

- *Earth is only about 6000 years old.* The assertion here is that life has not existed long enough for natural selection to produce the diversity of organisms seen today. Many creationists do not accept the principles behind radiometric dating, reviewed in Chapter 2, or the geological principle known as uniformitarianism. Uniformitarianism states that the rock-forming processes measured today, like the slow deposition of shells to form limestone, were responsible for rock formations produced in the past. Geologists working in the late 1700s realized that under uniformitarianism, Earth must be extremely old.
- *Organisms are "irreducibly complex."* According to this argument, adaptations like the vertebrate eye and metabolic pathways are so complex and well integrated that they could not have arisen from a gradual, incremental process like natural selection. Proponents of this view downplay the importance of the many fossils with characteristics that show transitions between simpler and more complex traits.
- *The theory is unproven.* Creationists argue that the evidence for evolution is inherently "soft" or unsatisfactory because no one was present when life began, and because no one actually witnessed the major branching events on the tree of life. In doing so, they deny that inferring historical events from contemporary evidence is a valid research program in science. (Consider for some 150 years after the atomic theory was proposed, no one had actually seen an atom. Yet the theory was widely accepted as correct during this time.)

Where does the controversy stand today? The U.S. Supreme Court has repeatedly ruled that legislation banning the teaching of evolution in public schools is unconstitutional on the basis of the First Amendment—the separation of church and state. The court's opinion is that scientific creationism and intelligent design theory promote a specific religious belief because they are founded on religious tenets codified in the Bible.

Despite the court's rulings, however, and even though many religious leaders as well as many scientists see no conflict between evolution and religious faith, the controversy continues.

Chapter Review

Summary

Evidence for the fact of evolution—that species are related through shared ancestry and have changed through time—has accumulated since Lamarck, Darwin, Wallace, and others began research on the topic almost 200 years ago. The geographic proximity of closely related species like the Galápagos mockingbirds; the existence of structural, developmental, and genetic homologies; the near-universality of the genetic code; resemblances of modern to fossil forms; the fact of extinction; and the presence of vestigial traits are all inconsistent with the alternative theory that species were formed instantaneously and independently, and have remained unchanged through time.

In addition to articulating evidence for the pattern called evolution, Darwin and Wallace independently discovered a process responsible for it. Natural selection occurs whenever genetically based differences among individuals lead to differences in their ability to survive and reproduce. Alleles or traits that increase the reproductive success of an individual are said to increase the individual's fitness. Alleles or traits that lead to higher fitness, relative to individuals without the allele or trait, are called adaptations.

Evolution is an outcome of natural selection, and is now defined as changes in allele frequencies in populations that occur from one generation to the next. Evolution by natural selection has been confirmed by a wide variety of studies, and is now considered to be *the* organizing principle of biology.

Natural selection is not, however, the only process that causes evolutionary change. Chapter 22 introduces three other processes that can change allele frequencies over time. When compared with natural selection, these processes have radically different consequences.

Questions

Content Review

1. How can Darwinian fitness be estimated?
   a. Document how long different individuals in a population survive.
   b. Count the number of offspring produced by different individuals in a population.
   c. Either a or b, or both.
   d. Fitness can't be measured; it's an abstract quantity.
2. What makes a trait "vestigial"?
   a. if it improves the fitness of its bearer, compared to individuals without the trait
   b. if it changes in response to environmental influences
   c. if it existed a long time in the past
   d. if it is rudimentary, and no longer functions
3. What is an adaptation?
   a. a trait that improves the fitness of its bearer, compared to individuals without the trait
   b. a trait that changes in response to environmental influences
   c. a trait that existed a long time in the past
   d. a trait that is rudimentary, and no longer functions
4. Heritability quantifies the proportion of variation in a population that is due to what?
   a. measurement error
   b. variation in the environments that individuals experience
   c. variation in the genetic makeup of individuals
   d. new mutations
5. What does descent with modification refer to?
   a. changes in populations through time
   b. changes in individuals through time
   c. embryonic development
   d. the number of new mutations that occur each generation
6. Why are homologous traits similar?
   a. They are derived from a common ancestor.
   b. They are derived from different ancestors.
   c. They result from convergent evolution.
   d. They are acquired characters that can be inherited.

Conceptual Review

1. The evidence supporting the pattern component of the theory of evolution can be criticized on the grounds that it is indirect. For example, no one has directly observed the formation of a vestigial trait over time. Because of the indirect nature of the evidence, it could be argued that structural and genetic homologies are coincidental, and do not result from common ancestry. Is indirect evidence for a scientific theory legitimate? Are you persuaded that modification with descent is the best explanation available for the data reviewed in section 21.1? Why or why not?
2. Some biologists encapsulate Darwinian evolution with the phrase "mutation proposes, selection disposes." Explain how this quip relates to the four postulates listed at the start of section 21.2.
3. Review the section on the evolution of drug resistance in *Mycobacterium tuberculosis*.

   In *M. tuberculosis*, how does heritable variation arise for the trait of drug resistance?

   What evidence do researchers have that a drug-resistant strain evolved in the patient analyzed in their study, instead of being transmitted from another infected individual?

   If the antibiotic rifampin were banned, would the mutant *rpoB* gene have lower or higher fitness in the new environment? Would strains carrying the mutation continue to increase in frequency in *M. tuberculosis* populations?
4. Review Box 21.4 and the section on the evolution of flower size in alpine skypilots.

   To estimate the heritability of flower size, a researcher planted offspring into randomly assigned locations in the environment. Why was this randomization step important?

   Consider the experiment that was designed to test whether any differences exist between the flower size of offspring from hand pollination between randomly chosen skypilots and the offspring from pollinations performed by bumblebees. In what sense did the randomly pollinated population serve as an experimental control?
5. Darwinism is sometimes criticized because it is "just a theory." In everyday English, the word *theory* refers to an idea or proposed explanation that is untested. (In the dictionary, some definitions of the word even offer *conjecture* and *speculation* as synonyms.) As a scientific term, however, the word theory has a far different meaning. Based on material in this chapter and Chapter 1, give a concise definition of a scientific theory. How has the conflict between the everyday and scientific usage of the word contributed to the controversy over the theory of evolution by natural selection?

Applying Ideas

1. The geneticist James Crow wrote that successful scientific theories have the following list of characteristics:

   They explain otherwise puzzling observations.

   They provide connections between otherwise disparate observations.

   They make predictions that can be tested.

   They are heuristic, meaning that they open up new avenues of theory and experimentation.

   Crow added two other elements, which he considered important on a personal, emotional level:

   They should be elegant, in the sense of being simple and powerful.

   They should have an element of surprise.

   How well does the theory of evolution by natural selection fulfill these six criteria? Think of a theory you've been introduced to in another science course, for example the atomic theory, and evaluate it using this list.

2. The average height of humans has increased steadily for the past 100 years in the industrialized nations. This trait has clearly changed over time. Most physicians and human geneticists believe that the change is due to better nutrition and less disease. Has human height evolved?
3. Genome sequencing projects may dramatically affect how biologists analyze evolutionary changes in quantitative traits. For example, suppose the genomes of many living humans are sequenced, and that genomes could be sequenced from many people who lived 100 years ago (this might be possible to do using preserved tissue). If 20 genes have been shown to influence height, how could you use the sequence data from these genes to test the hypothesis that human height has evolved in response to natural selection?
4. Examine Figure 21.6. How would you expect the data on this map to change over the next several decades as drug-resistant strains of *M. tuberculosis* increase in frequency in North America and Western Europe?

CD-ROM and Web Connection

CD Activity 21.1: Natural Selection for Antibiotic Resistance *(tutorial)*

(Estimated time for completion = 10 min)

Examine the results of natural selection in a population of *Mycobacterium tuberculosis*, the bacterium that causes tuberculosis.

CD Activity 21.2: Natural Selection in Alpine Skypilots *(tutorial)*

(Estimated time for completion = 15 min)

Investigate how researchers use experimental methods to determine whether natural selection is causing evolution in real-world populations.

At your **Companion Website** (http://www.prenhall.com/freeman/biology), you will find self-grading exams and links to the following research tools, online resources, and activities:

History of Evolution

This timeline introduces you to some evolutionary theorists and the roles they played in shaping evolutionary history.

Evolution Simulations

Evo-Tutor.org offers several simulations of evolutionary concepts in which you can change the parameters of a virtual population and simulate the evolutionary effects on that population.

Darwin's Finches and Speciation

Learn more about one of the most famous projects to document evolution in a wild population, including recent research in this area.

Additional Reading

Dawkins, Richard. 1989. *The Selfish Gene*. Oxford: Oxford University Press. A classic statement of how natural selection operates at the level of the gene.

Galen, C. 1999. Why do flowers vary? *BioScience* 49: 631–640. Reviews how different agents of natural selection affect the size and shape of flowers.

Nesse, R. M., and G. C. Williams. 1998. Evolution and the origins of disease. *Scientific American* 279 (November): 86–93. A look at how natural selection operates on organisms that cause disease in humans.

Wiener, Jonathan. 1995. *The Beak of the Finch*. New York: Vintage Books. This is a Pulitzer Prize–winning account of how biologist Peter Grant and colleagues have documented evolution in action on the Galápagos islands.

Evolutionary Processes

22

Chapter 21 defined evolution as a change in allele frequencies, and explored how natural selection is altering the frequencies of alleles responsible for drug resistance in the bacterium that causes tuberculosis and for increased flower size in alpine skypilots. An important concept in that chapter is that, even though natural selection acts on individuals, evolutionary change occurs in **populations.** A population is a group of individuals that live in the same area and regularly interbreed.

Natural selection is not the only process that causes evolutionary change, however. There are actually four mechanisms that shift allele frequencies in populations:

- Mutation modifies allele frequencies by continually introducing new alleles (recall from Chapter 12 that mutations are random variations on existing alleles).
- Gene flow produces allele frequency changes because individuals from elsewhere introduce alleles when they join a population, and because individuals in the population remove alleles when they leave the area.
- A process called genetic drift changes allele frequencies randomly.
- Natural selection increases the frequency of certain alleles—the ones that contribute to improved reproductive success.

This chapter has two fundamental messages: Natural selection is not the only agent responsible for evolution, and each of the four evolutionary processes has different consequences. Natural selection is the only mechanism that results in adaptation. Mutation, migration, and genetic drift do not favor certain alleles and do not lead to increased fitness. Mutation and drift introduce a non-adaptive component into the course of evolution.

Due to hunting and habitat destruction, cheetahs declined in number from perhaps 30,000 in the mid-1950s to about 15,000 in 1975. Although most populations have continued to shrink since the mid-1970s, the current census is unknown. Conservation biologists create management strategies for the cheetah that are based, in part, on the types of models introduced in this chapter.

The task before us now is to take a closer look at each of the four evolutionary processes. To illustrate their consequences we consider their impact on the cheetah, an endangered species native to Africa and the Middle East (**Figure 22.1**). Cheetahs are of special concern to conservation biologists because genetic surveys of the remaining populations indicate that only one or two alleles exist at many loci. Section 22.1 analyzes why low genetic diversity is a focus of concern. Section 22.2 introduces the mathematical tools that biologists use to study changes in allele frequencies. Subsequent sections explore the consequences of mutation, migration, drift, and natural selection. The chapter also investigates how inbreeding impacts the genetics of populations, and concludes by introducing a special type of natural selection called sexual selection.

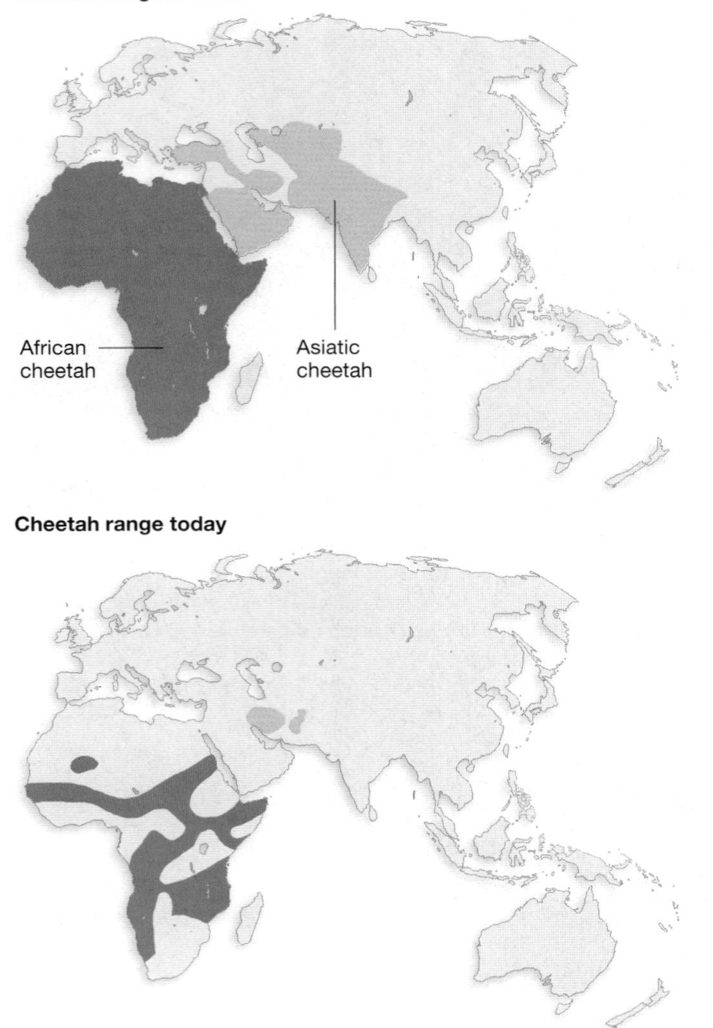

FIGURE 22.1 As Habitats Shrink, Cheetah Populations Become Isolated
The habitats occupied by cheetahs have been reduced from large, contiguous areas of Africa and the Middle East to small, isolated pockets in national parks and game reserves.

22.1 Why Is Genetic Diversity Important?

A major focus of this chapter is to understand how the four evolutionary forces affect the diversity of alleles within populations as well as their frequency. Why is genetic diversity important? This question can be answered in two ways.

First, evolution by natural selection presents a paradox. If selection favors certain alleles over others, then the favored alleles should increase in frequency until they reach fixation—a frequency of 1.0. When this occurs, there is no genetic variation left for selection to act on and evolution should stop. Without genetic diversity, there is no evolution. If cheetah populations lack genetic diversity, can they evolve in response to changes that occur in their environment? What forces are responsible for restoring genetic diversity?

Second, genetic diversity itself can be an important adaptation. In humans, for example, Mary Carrington and co-workers recently documented that individuals who are infected with HIV tend to get AIDS much less readily if they are heterozygous at the major histocompatibility complex, or MHC. These genes code for cell surface proteins involved in the immune response to invading bacteria and viruses. As the data graphed in **Figure 22.2** show, homozygous patients—who have little or no allelic variation at their MHC loci—progress to disease much more rapidly than do heterozygous individuals who are infected with

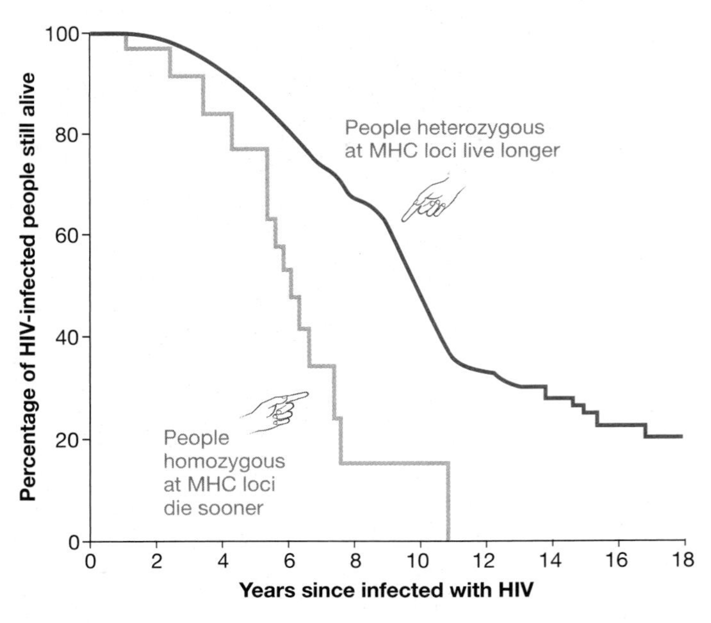

FIGURE 22.2 Allelic Diversity and Resistance to Disease
The y-axis (ordinate) on this graph plots the percentage of individuals in a study population of people infected with HIV who are alive. The x-axis (abscissa) plots the years since individuals in the study were initially infected with the virus. Note that almost 25 percent of heterozygous people infected with the virus were still alive even after 18 years, while all homozygous people died within 12 years of becoming infected.

the virus. The message of this study and many others like it is that genetic diversity at the MHC loci is strongly correlated with resistance to disease.

A biologist named Stephen O'Brien hints that a correlation between disease resistance and allelic diversity may characterize cheetahs as well. Unlike lions and other big cats, cheetahs have little to no allelic variation in the MHC. Most individuals are homozygous at all loci. O'Brien contends it is no coincidence that a disease epidemic recently wiped out many cheetahs at a captive breeding center, even though lions and other cats living at the same facility—which can be infected with the same disease—were not affected.

As we examine mutation, gene flow, genetic drift, and natural selection, then, we will ask: How might each of these processes affect allelic diversity in cheetah populations? Can conservation managers manipulate the four evolutionary forces in a way that will help cheetah populations survive?

22.2 Analyzing Allele Frequency Change: The Hardy-Weinberg Principle

To study how the four evolutionary processes affect populations, biologists take a two-pronged approach. First they create mathematical models that track the fate of alleles over time. Then they collect data to test the predictions made by the models. Throughout this chapter we adopt the same approach. We review what mathematical models say about each of the four evolutionary forces, examine data sets designed to test the predictions of the models, and apply the theory to problems in conservation biology and human genetics.

Perhaps the most powerful of these models was developed in 1908 by G. H. Hardy and Wilhelm Weinberg, working independently. Hardy's and Weinberg's model analyzes what happens to the frequencies of two alleles at a single genetic locus when the four evolutionary forces are *not* acting on a population. In this way, the Hardy-Weinberg model acts like the control treatment in an experiment on evolution. It lets biologists predict how allele frequencies change when the four forces are *not* operating. As a result, it provides a contrast to models and data sets that explore the effects of mutation, migration, drift, and selection. Hardy's and Weinberg's treatment is called a null model.

Hardy developed his version of the model in response to a claim made by G. Udny Yule that dominant alleles automatically increase in frequency as a result of meiosis. Weinberg developed his version to understand whether certain genetic diseases will increase in frequency in human populations.

Both researchers began by supposing that just two alleles of a particular gene exist in a population. Let's call these alleles A_1 and A_2. They could be any alleles at any locus, but let's suppose that they are alternative alleles at one of the MHC loci in cheetahs.

At this locus, then, three genotypes are possible: A_1A_1, A_1A_2, and A_2A_2. (Cheetahs are diploid, so each individual has two alleles at each locus.) Now we can ask, what will be the frequencies of the two alleles and the three genotypes in the next generation when the alleles follow Mendel's rules of independent segregation and assortment? Remember that evolution occurs if allele frequencies in a population change from one generation to the next.

The easiest way to analyze the fate of alleles in a population is to think about the gametes (eggs and sperm) that a parental generation produces when it breeds. Hardy and Weinberg used a novel approach to do this. Instead of thinking about the consequences of matings between parents with different genotypes, as we did with Punnett squares in Chapter 10, they wanted to know the consequences of matings when all of the parents in the population bred. To analyze this situation, they imagined that all of the gametes produced in each generation go into a single bin called the **gene pool**. Each gamete in this pool contains one allele for each locus. We'll call the frequency of A_1 alleles in the gene pool p_1 and the frequency of A_2 alleles in the gene pool p_2. Because p_1 and p_2 are frequencies, they can have any value between 0 and 1. Because there are only two alleles, the two frequencies must add up to 1; that is, $p_1 + p_2 = 1$.

To help you get a more concrete feel for the model, **Figure 22.3** (page 432) provides a numerical example. In this case, we suppose that A_1 has a frequency of 0.7 ($p_1 = 0.7$) and A_2 has a frequency of 0.3 ($p_2 = 0.3$).

To create offspring from the gametes in the gene pool, then, Hardy and Weinberg imagined picking two gametes (and thus, two alleles) from the pool at random. This corresponds to a mating in which two gametes join to form a zygote. To understand the genetic makeup of the entire offspring generation, they imagined repeating this procedure many times. By modeling the results of many random matings in which pairs of alleles are picked like this, we can calculate the frequencies of the three offspring genotypes.

For example, suppose you want to know the frequency of A_1A_1 genotypes in the offspring generation. This genotype results whenever you pick an A_1 allele out of the gene pool followed by another A_1 allele. The probability of drawing an A_1 from the gene pool is p_1. The probability of drawing another A_1 is also p_1. The probability of drawing both and obtaining an A_1A_1 genotype is the *product* of these probabilities. Thus, the frequency of the A_1A_1 genotype $= p_1 \times p_1 = p_1^2$. (See Box 10.1 to review the rules for combining probabilities.) In the example given in Figure 22.3, the frequency of the genotype is $0.7 \times 0.7 = 0.49$.

Calculating the frequency of A_1A_2 individuals in the next generation is slightly different, because there are two ways to obtain this genotype when you pick gametes from the gene pool. You can draw an A_1 and then an A_2, or you can draw an A_2 and then an A_1. The probability of drawing an A_1 and then an A_2 is $p_1 \times p_2 = p_1p_2$. The probability of drawing an A_2 and then an A_1 is $p_2 \times p_1 = p_2p_1$. The frequency of A_1A_2 genotypes

CD ACTIVITY 22.1 The Hardy-Weinberg Principle

in the next generation is the probability of drawing an A_1A_2 or A_2A_1. This is the sum of each independent probability, or $2p_1p_2$. In the example provided in Figure 22.3, the frequency of this genotype is $2 \times 0.7 \times 0.3 = 0.42$.

Figuring out the frequency of A_2A_2 individuals is analogous to the logic we used to calculate the frequency of A_1A_1 homozygotes. Only now the probability of drawing an A_2 is p_2, so the probability of drawing two in a row is $p_2 \times p_2 = p_2^2$. In the Figure 22.3 example, this frequency is $0.3 \times 0.3 = 0.09$.

Step back for a moment and take a look at what we just accomplished: We knew the allele frequencies in the parental generation; now we can predict the genotype frequencies in the offspring generation. The allele frequencies were p_1 (for the A_1 allele) and p_2 (for the A_2 allele); the genotype frequencies are p_1^2 (for the A_1A_1 genotype), $2p_1p_2$ (for the A_1A_2 genotype), and p_2^2 (for the A_2A_2 genotype). Note that the genotype frequencies in the offspring generation must add up to 1 (just as the allele frequencies in the parental generation must add up to 1). This means that $p_1^2 + 2p_1p_2 + p_2^2 = 1$. In the example outlined in Figure 22.3, $0.49 + 0.42 + 0.09 = 1$.

Our final task is to calculate the *allele* frequencies in the offspring generation from the genotype frequencies. Once we've done this, we can compare allele frequencies in the offspring and parental generations and determine if evolution occurred.

To calculate these allele frequencies, think about the gametes that the offspring produce when *they* breed. What is the fre-

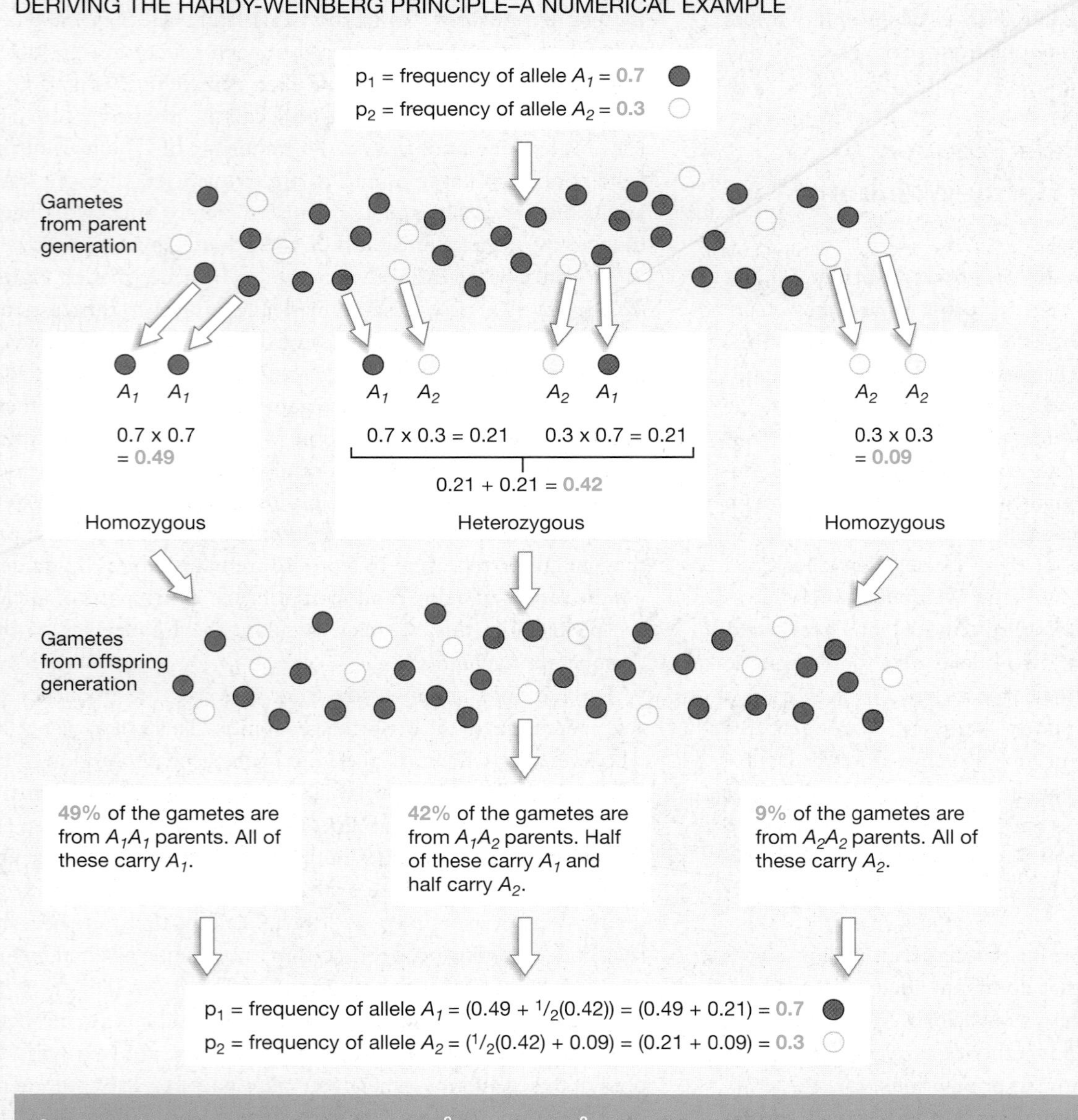

FIGURE 22.3 A Numerical Example of the Hardy-Weinberg Principle

quency of the A_1 and A_2 alleles in this generation's gene pool? To answer this question we need to consider the gametes produced by individuals with each of the three genotypes. A_1A_1 offspring constitute a proportion p_1^2 of the population, so they contribute a proportion p_1^2 of alleles to the gene pool. All of these alleles are A_1. But A_1A_2 individuals also produce gametes containing allele A_1. These individuals make up a proportion $2p_1p_2$ of the population, and one-half of the gametes they make contain allele A_1. This means that an additional proportion $1/2 \times 2p_1p_2 = p_1p_2$ of the alleles in the pool are A_1. If we add up these frequencies, a proportion $p_1^2 + p_1p_2 = p_1(p_1 + p_2) = p_1$ of all the gametes are A_1—because $(p_1 + p_2) = 1$. Figure 22.3 walks through these calculations for the numerical example.

To calculate the frequency of the other allele, A_2, note that the other half of the gametes made by A_1A_2 offspring are A_2. This gives p_1p_2 as the frequency of A_2 gametes thus far. In addition, all of the gametes produced by A_2A_2 individuals contain allele A_2. These individuals are p_2^2 of the population, so they add a proportion p_2^2 of A_2 alleles. The total frequency of A_2 alleles is $p_1p_2 + p_2^2 = p_2(p_1 + p_2) = p_2$—because $(p_1 + p_2) = 1$.

The allele frequencies in the offspring generation are the same as in the parental generation: p_1 and p_2. The analysis shows that allele frequencies have not changed from the parental generation to the offspring generation. No evolution has occurred.

This result is called the **Hardy-Weinberg principle**. It maintains that allele frequencies do not change over time when they are transmitted according to the rules of Mendelian inheritance. For evolution to occur, some other factor must come into play. More formally, the Hardy-Weinberg principle states that if allele frequencies in a population are given by p_1 and p_2, then genotype frequencies will be given by p_1^2, $2p_1p_2$, and p_2^2 for generation after generation as long as certain conditions apply. What are these conditions?

The Hardy-Weinberg Model Makes Important Assumptions

The model we just reviewed is based on important assumptions about how the population and alleles behave. Specifically, for a population to conform to the Hardy-Weinberg principle there must be:

1. *No mutation.* We didn't consider that new A_1's or A_2's or other, new, alleles might be introduced into the gene pool.
2. *No migration.* No new alleles were added by immigration or lost through emigration. As a result, all of the alleles in the offspring population came from the original population's gene pool.
3. *No "genetic drift," or random allele frequency changes.* We avoided this type of allele frequency change by assuming that we drew alleles in their exact frequencies p_1 and p_2, and not some different value caused by chance.
4. *Random mating.* We enforced this condition by picking gametes from the gene pool at random.
5. *No selection.* We assumed that all members of the parental generation survived and contributed equal numbers of gametes to the gene pool.

The Hardy-Weinberg principle tells us what to expect if none of the evolutionary forces are acting. But do any real-life data conform to the predictions made by the model?

Testing the Hardy-Weinberg Principle

One of the first loci that geneticists could analyze in natural populations was the MN blood group of humans. Most human populations have two alleles, designated M and N, at this locus. Because the gene codes for a protein found on the surface of red blood cells, researchers could determine whether individuals are MM, MN, or NN by treating blood samples with antibodies to each protein (this technique is introduced in Chapter 17). To estimate the frequency of each genotype in a population, geneticists obtain data from a large number of individuals. **Table 22.1** (page 434) shows the genotype frequencies for populations from throughout the world on the lines titled "observed." The allele frequencies in the right-hand columns are computed from these observed values (the caption explains how this is done). Using these allele frequencies, the expected genotype frequencies are computed according to the Hardy-Weinberg principle.

Note that in every case the observed and expected genotype frequencies are almost identical. (A statistical test shows that the differences observed are probably due to chance.) For every population surveyed, genotypes at the MN locus are in Hardy-Weinberg proportions. As a result, geneticists conclude that the assumptions of the Hardy-Weinberg model are valid for this locus. The data imply that the M and N alleles are not affected by the four evolutionary mechanisms, and that mating is random with respect to this locus—meaning that humans do not choose mates on the basis of their MN genotype.

Applying the Hardy-Weinberg Model

The Hardy-Weinberg principle acts as a null model by specifying the relationship between genotype and allele frequencies when none of the four evolutionary forces is operating and mating is random. If allele frequencies are given by p_1 and p_2, then genotype frequencies should be in the ratio $p_1^2 : 2p_1p_2 : p_2^2$. If biologists observe genotype frequencies that do *not* conform to Hardy-Weinberg proportions, it alerts them that something interesting is going on. If the null hypothesis is violated, it means that nonrandom mating is occurring or that one or more of the four forces is operating; further research would be needed to determine which of the four forces is acting.

To understand how investigators reach a conclusion like this, consider data collected by Therese Markow and colleagues on the genotypes of 122 individuals from the Havasupai tribe native to Arizona. Markow's group studied loci in the MHC called *HLA-A* and *HLA-B*. From the observed genotype frequencies, the researchers calculated the frequency of each

TABLE 22.1 Do Genotype and Allele Frequencies in the MN Blood Group of Humans Conform to the Hardy-Weinberg Model?

These data illustrate the four steps involved in testing whether a locus conforms to the Hardy-Weinberg principle: (1) Estimate genotype frequencies by observation—in this case, by testing many blood samples for the M and N alleles. (2) Estimate allele frequencies based on the observed genotype frequencies. In this case the frequency of the M allele is the frequency of MM homozygotes plus half the frequency of MN heterozygotes; the frequency of the N allele is the frequency of NN homozygotes plus half the frequency of MN heterozygotes. (3) Use the allele frequencies to calculate the expected genotypes under the Hardy-Weinberg proportions of $p_1^2 : 2p_1p_2 : p_2^2$. (4) Use statistical tests to determine whether the differences between the observed and expected genotype frequencies are due to chance or some other factor. **EXERCISE** Fill in the values for allele frequencies and expected genotype frequencies for the Ainu people of Japan.

			Genotype Frequencies		Allele Frequencies	
Population and Location		**MM**	**MN**	**NN**	**M**	**N**
Eskimos (Greenland)	Observed	0.835	0.156	0.009	0.913	0.087
	Expected	0.834	0.159	0.008		
Native Americans (U.S.)	Observed	0.600	0.351	0.049	0.776	0.224
	Expected	0.602	0.348	0.050		
Caucasians (U.S.)	Observed	0.292	0.494	0.213	0.540	0.460
	Expected	0.290	0.497	0.212		
Aborigines (Australia)	Observed	0.024	0.304	0.672	0.178	0.822
	Expected	0.031	0.290	0.679		
Ainu (Japan)	Observed	0.179	0.502	0.319		
	Expected					

allele. When they used these allele frequencies to calculate the expected number of each genotype according to the Hardy-Weinberg principle, they found that the observed and expected values did not match (**Table 22.2**). Specifically, there were too many heterozygotes and not enough homozygotes compared with the predicted values. The result indicates that one of the assumptions behind the Hardy-Weinberg principle is being violated. But which? The researchers argued that mutation, migration, and drift are negligible in this case and offered two competing explanations for their data:

- *Mating may not be random with respect to MHC genotype.* Specifically, people may prefer mates with MHC genotypes unlike their own and thus produce an excess of heterozygous offspring. This hypothesis is plausible. Claus Wedekind and colleagues have done experiments showing that college students can distinguish each others' MHC genotypes on the basis of body odor. Individuals in this study were also more attracted to the smell of MHC genotypes unlike their own. If this is true among the Havasupai, nonrandom mating would lead to an excess of heterozygotes.
- *Heterozygous individuals may have higher Darwinian fitness.* Carole Ober and co-workers collected data, from Hutterite people living in South Dakota, that support this hypothesis. In this population, married couples who share MHC alleles

TABLE 22.2 Do Genotype Frequencies at the MHC Loci of Humans Conform to the Hardy-Weinberg Model?

The data reported here are the observed number of homozygotes and heterozygotes at two loci in the MHC, called *HLA-A* and *HLA-B*, in a sample of 122 Havasupai people. The expected values of homozygous and heterozygous genotypes were calculated using the Hardy-Weinberg principle, based on observed allele frequencies. Statistical tests show that it is extremely unlikely that the difference between the observed and expected numbers could occur purely by chance. This means one of two things: either natural selection favors heterozygotes, or humans tend to mate with individuals who have a different MHC genotype.

	Observed Number	**Expected Number**
HLA-A		
Homozygotes	38	48
Heterozygotes	84	74
HLA-B		
Homozygotes	21	30
Heterozygotes	101	92

Source: T. Markow et al., HLA polymorphism in the Havasupai: Evidence for balancing selection. *American Journal of Human Genetics* 53 (1993): 943–952.

have more trouble getting pregnant and experience increased rates of spontaneous abortion compared to couples with unlike MHC alleles. These data suggest that homozygous fetuses have lower fitness than fetuses heterozygous at MHC loci. If this were true among the Havasupai, selection would lead to an excess of heterozygotes. This pattern of natural selection is called **heterozygote advantage.**

Which explanation is correct? It is possible that both are. But the fact is, no one knows. Research continues.

22.3 Mutation

The Hardy-Weinberg principle has an important message for conservationists concerned about preserving genetic diversity in cheetahs and other endangered species. If the assumptions behind the model hold, allele frequencies do not change and genetic variation is preserved in perpetuity. But the model's first assumption is almost certain to be violated: According to the data in Chapter 12, mutation occurs constantly.

When a DNA molecule is copied, errors by DNA polymerase result in random changes in the sequence of deoxyribonucleotides. If a mutation occurs in a stretch of DNA that codes for a protein, the changed codon may result in a polypeptide with a novel amino acid sequence. In this way, mutation constantly introduces new alleles into all populations at all loci. Mutation is an evolutionary mechanism that increases genetic diversity in populations.

In principle, then, mutation can change the frequencies of alleles through time and cause evolution. But does mutation occur often enough to make it an important factor for conservationists and human geneticists concerned about the direction of evolutionary change? The short answer is no.

Mutation as an Evolutionary Mechanism

To understand why mutation is not an important cause of evolutionary change, consider that the highest mutation rates that have been recorded at individual loci are on the order of 1 mutation in every 2000 gametes produced by an individual. This means that, at most, about 1 in every 1000 offspring carries a mutation at a particular locus. Will mutation affect allele frequencies in a species like the cheetah? If 18,000 cheetahs are alive, and if they produce 2000 surviving offspring each year, then mutation would introduce a maximum of two new alleles to the population at each locus. But in a population numbering 20,000, there are 40,000 allele copies at each locus. The change in allele frequency introduced by mutation—2 in 40,000, or 0.00005—is insignificant. As an evolutionary process, then, mutation does little to change allele frequencies on its own. Does this mean that mutation plays *no* role in evolution?

The Role of Mutation in Evolutionary Change

The theory we just developed focuses on the role of mutation at a specific locus. In contrast, Richard Lenski and Michael Travisano designed an experiment to evaluate the role that mutation plays at many loci over many generations. These researchers studied *Escherichia coli*, a bacterium introduced in Chapter 12 that is a common resident of the human intestine. **Figure 22.4** (page 436) diagrams how Lenski and Travisano set up a large series of populations, each founded with a single cell. Every day for over four years they transferred a small number of cells from the populations into a new batch of growth medium, so that each population grew continuously. Because *E. coli* is asexual and reproduces by cell division, mutation was the only source of genetic variation. The question is, was mutation an effective evolutionary force? Did the populations change through time?

Lenski and Travisano answered this question through the competition experiments diagrammed in the figure. They saved samples from the experimental populations at regular intervals, added a genetic marker to the cells (the gene's product allowed them to identify the older-generation cells), and preserved them by freezing. After many generations had passed, they put cells from the older and newer generations of each population together in the same flask, let them compete, and counted the cells that resulted. The more numerous population had grown faster, meaning that it was better adapted to the experimental environment. In this way the researchers could measure the fitness of descendant populations relative to ancestral populations. If relative fitness was greater than 1, it meant that recent-generation cells outnumbered older-generation cells when the competition was over.

The data from a series of competition experiments are graphed at the bottom of Figure 22.4. Note that Lenski and Travisano compared the fitness of the original and descendant populations every 100 generations, and that relative fitness increased dramatically—almost 30 percent—over time. But note also that fitness increased in fits and starts. This pattern is emphasized by the solid line on the graph, which represents a mathematical function fitted to the data points.

What caused this stair-step pattern? Lenski and Travisano hypothesize that each jump was caused by a novel mutation that conferred a fitness benefit. Their interpretation is that cells that happened to have the beneficial mutation grew rapidly and came to dominate the population. After a beneficial mutation occurred, the fitness of the population stabilized—sometimes for hundreds of generations—until another random but beneficial mutation occurred and produced another jump in fitness.

The experiment makes an important point, which is that mutation is the ultimate source of genetic variability. If mutation did not occur, evolution would have stopped. Without mutation, there is no variation for natural selection to act upon. On its own, however, mutation is ineffective at changing allele frequencies.

Further, consider that even though mutation rates are low when reported for individual loci, the genomes of many organisms contain tens of thousands of genes. The human genome,

FIGURE 22.4 An Experiment to Test the Role of Mutation in Evolutionary Change

EXERCISE The authors of this study claim that each jump in fitness recorded in the graph at the bottom of this figure was due to a beneficial mutation that spread rapidly in the population. Add large arrows to the graph that point to the times when these hypothesized mutations occurred.

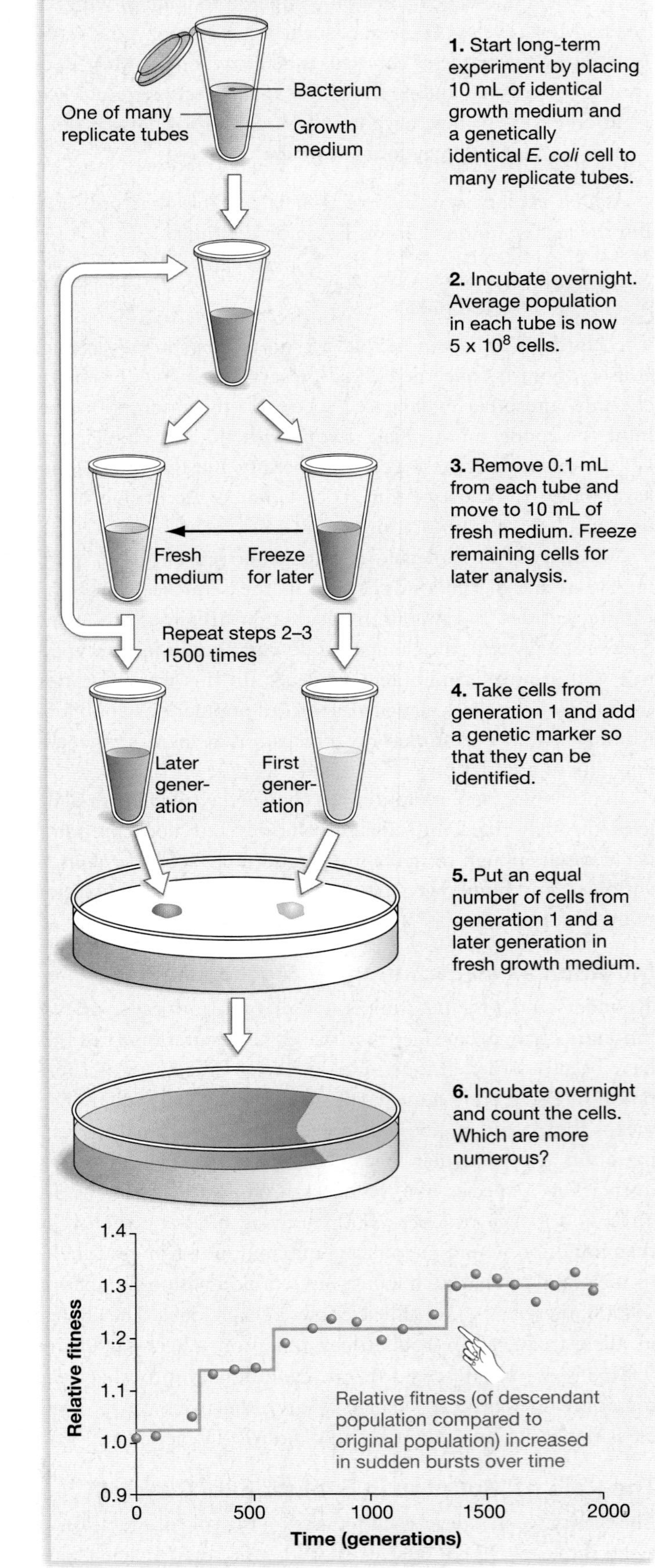

for example, is currently estimated to have about 40,000 genes. If mutations occur at rates of one per 10^5 loci per generation, then virtually every gamete produced contains at least one mutation.

The message here is that mutation may be inconsequential in changing allele frequencies at a particular locus; but when considered across the genome and when combined with natural selection, it becomes an important evolutionary force. Mutation is the ultimate source of the genetic variation that makes evolution possible.

22.4 Migration

The term **migration** has two definitions in biology. In the ecological sense, migration refers to the seasonal movement of individuals from one habitat to another. In the evolutionary sense, migration refers not to seasonal movement but to **gene flow**—the movement of alleles. Gene flow occurs when individuals disperse from one population to another, join the new population, and breed. Throughout this chapter, migration is used in the evolutionary sense.

In natural populations migration has an important consequence: It tends to eliminate genetic differences among populations. That is, gene flow tends to equalize allele frequencies among populations.

To see how this happens, suppose that young cheetahs from a captive population were going to be reintroduced to the wild at a national park in Iran. (This hypothetical example is interesting because a tiny, remnant population of cheetahs exists in Iran, completely isolated from the larger populations in sub-Saharan Africa.) To determine the evolutionary consequences of this gene flow, focus on an allele called A_1 and suppose that it is present in the captive population at a frequency p_c and in the wild population at a frequency p_w. If the captive population had been founded by individuals from southwestern Africa, p_c and p_w are likely to be very different. Let's consider an extreme case in which p_c is 1.0 and p_w is 0. If the first reintroduction resulted in the wild population being composed of 15 percent captive-bred individuals and 85 percent wild-bred individuals, the frequency of A_1 in the wild population would then be 0.15. A large change in allele frequencies occurred in the wild population simply due to migration. Specifically, allele frequencies are now much more similar to the captive population than they were before. If the reintroduction program continued, allele frequencies in the wild population would become increasingly similar to those in the captive population. In this way, migration tends to homogenize the genetic makeup of populations.

Conservation biologists are often cautious about reintroduction programs for exactly this reason. If a wild population contains alleles that perform well in the wild, then gene flow from captive stock might introduce alleles that are less well adapted to the wild habitat. Gene flow can result in a population that is much less well adapted to its environment if the proportion of migrants is large. In other cases, however, migration can be an important conservation tool—because migration can increase genetic diversity in isolated populations. Small, isolated populations tend to lose alleles due to an important evolutionary force called genetic drift.

22.5 Genetic Drift

Genetic drift is defined as any change in allele frequencies in a population that is due to chance. The process is aptly named because it causes allele frequencies to drift up and down randomly over time.

To understand why drift occurs, suppose that just two alleles are present at a particular locus in the MHC of cheetahs, and that each is present in the gene pool at a frequency of 0.5. Imagine that you draw a sample of gametes out of the gene pool to make 10 offspring, and that these 10 offspring make up the entire next generation. You can simulate this process of drawing a sample of gametes out of the gene pool by flipping a coin. Imagine that heads represents a gamete with one of the alleles, A_1, and tails a gamete with the other allele, A_2. The coin is fair, meaning that the probability of getting a heads or a tails (and thus an A_1 or A_2) is 0.5. Here are the results of 20 coin flips, made as this is written:

$$A_2, A_1, A_1, A_1, A_1, A_1, A_2, A_2, A_2, A_2, A_2, A_2,$$
$$A_1, A_2, A_2, A_1, A_1, A_2, A_1, A_2$$

According to the flips, we drew 9 A_1 alleles and 11 A_2 alleles from the gene pool. In the next generation, then, the frequency of allele A_1 is 0.45 and the frequency of allele A_2 is 0.55. Evolution has occurred, purely due to the vagaries of sampling. This is genetic drift.

This simple exercise makes two important points:

- *Genetic drift is random with respect to fitness.* The allele frequency changes it produces are not adaptive.
- *Genetic drift is more pronounced in small populations.* If population size were 3 instead of 10, the first six coin flips just listed would have produced an even more extreme change—an offspring generation with allele frequencies of 0.83 for A_1 and 0.17 for A_2. However, if the population size were 1000, the 2000 coin flips required would almost certainly have produced allele frequencies extremely close to 0.5.

Experimental Studies of Genetic Drift

What happens when drift continues in a small population, generation after generation? Biologists Warwick Kerr and Sewall Wright performed an experiment that dramatically illustrates the long-term consequences of drift. They started with a large laboratory population of fruit flies that contained a **genetic marker**—a specific allele that causes a distinctive phenotype. In this case, the marker was the morphology of leg bristles. Fruit flies have bristles on their legs that can be either straight or bent. The morphology of leg bristles depends on a single locus, and Kerr and Wright's lab population contained just two alleles—normal and "forked."

To begin the experiment the researchers set up 96 cages in their lab. Then they placed four adult females and four adult males of the fruit fly *Drosophila melanogaster* in each. They chose flies to begin these experimental populations so that the frequency of the normal and forked allele in each of the 96 starting populations was 0.5. The two alleles do not affect the fitness of flies in the lab environment, so Kerr and Wright could be confident that if changes in the frequency of normal and forked phenotypes occurred, they would not be due to natural selection.

After these first-generation adults bred, Kerr and Wright reared their offspring. Then they randomly chose four males and four females—meaning that they simply grabbed individuals without regard to whether their leg bristles were normal or forked—from each of the 96 offspring populations and allowed them to breed and produce the next generation. The researchers repeated this procedure until all 96 populations had undergone a total of 16 generations. During the entire course of the experiment no migration from one population to another occurred, and previous studies had shown that mutations from normal to forked are rare. The only evolutionary process operating during the experiment was genetic drift.

Their result? After 16 generations, the 96 populations fell into three groups. Forked leg bristles were found on all of the individuals in 29 of the experimental populations. (In a case like this, biologists say that the forked allele is "fixed," meaning that it has reached a frequency of 1.0.) The normal allele had been lost from these 29 populations due to drift. In 41 other populations, however, the opposite was true: All individuals had normal bristles. In these populations the other allele—forked—had been lost due to chance. Both alleles were still present in 26 of the populations. The punch line is startling: In 73 percent of the experimental populations (70 out of the 96), genetic drift had reduced allelic diversity to zero.

The experiment highlights a third important property of genetic drift: It can lead to the loss or fixation of alleles. Which alleles are fixed or lost is purely a matter of chance. Although drift operates in all populations at all times, its effects are most pronounced in small populations. Results like these make conservationists shudder.

FIGURE 22.5 A Method for Surveying Allelic Diversity in Populations
This diagram outlines the steps in protein electrophoresis. During the 1960s through 1980s, this technique was widely used for surveying allelic diversity in populations.

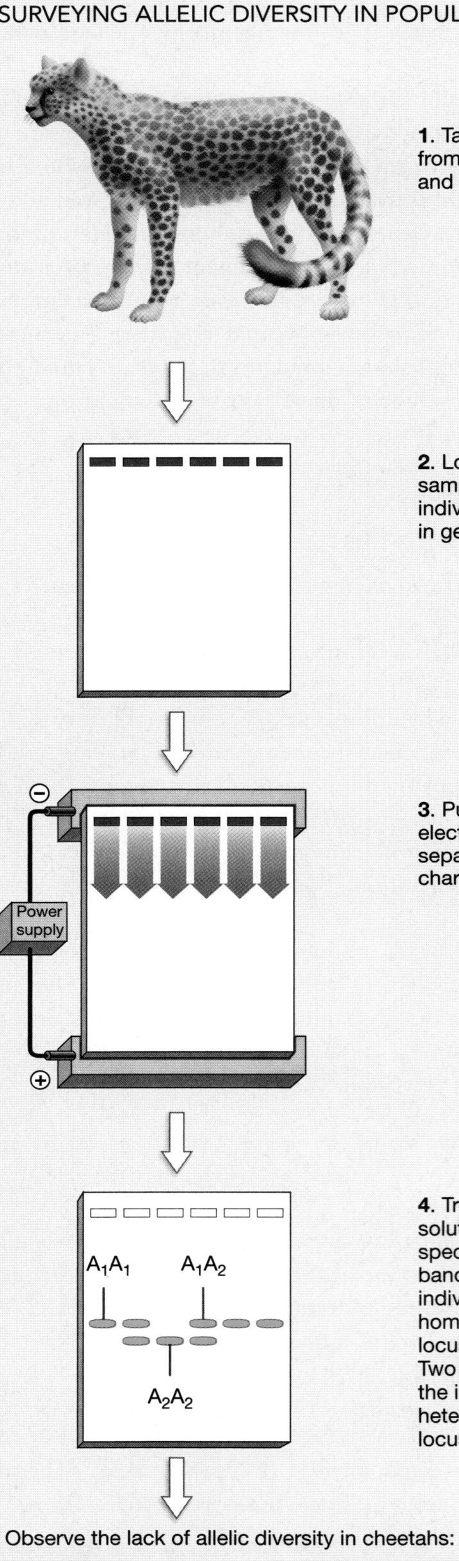

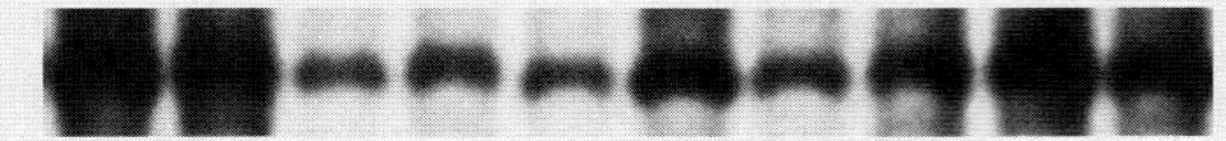

Genetic Drift in Natural Populations

In 1983 a team of researchers led by Stephen O'Brien published a study of genetic diversity in cheetah populations from eastern and southern Africa, based on the technique outlined in **Figure 22.5**. The researchers took blood samples from 55 individuals, isolated the proteins present, and separated the molecules by electrophoresis. (Recall, from Chapter 3, that electrophoresis separates molecules in an electric field on the basis of their size and charge as they move through a gelatinous substance, or gel.) Then they identified the location of 47 different proteins in the gels by using stains that bind to specific enzymes. How did this technique allow them to evaluate allelic diversity? When different alleles produce proteins with different charges, the molecules separate during electrophoresis and are distinguishable when stained, as the gel in Figure 22.5 shows. But to their astonishment, the biologists found zero allelic diversity in the loci they surveyed.

Why? O'Brien's group suggested that the extremely low genetic diversity of cheetahs is due to genetic drift. Specifically, they hypothesized that cheetahs went through an event called a **genetic bottleneck**—a sudden constriction in population size that reduces allelic diversity due to drift. The idea is that the world population of cheetahs was once large, say 100,000 breeding individuals. In this population there would be 200,000 allele copies at each locus. But O'Brien and co-workers propose that this population was suddenly reduced to a few thousand individuals at some time in the past. When this occurred, much of the original allelic diversity in the species was lost due to drift. These types of bottlenecks are more than a theoretical possibility; they have been documented recently in whooping cranes, Pere David's deer, the Tasmanian wolf, and in other endangered species.

O'Brien points out that genetic drift now threatens what little allelic diversity remains. As humans convert more and more land to agricultural use, cheetah populations are being fragmented into isolated "islands" of habitat. The numbers of breeding cheetahs within each population can be small—as few as 50. In most zoos and other captive breeding centers, cheetah populations are even smaller.

Fortunately, the negative impact of drift can be counteracted by manipulating another evolutionary force: migration. Biologists responsible for the design of wildlife reserves often try to link them using corridors of natural habitat. The hope is that individuals will disperse to neighboring populations, breed, and introduce their alleles. At captive breeding centers and zoos, managers routinely exchange semen samples or offspring of endangered species. This practice has two objectives: to introduce new alleles through gene flow, and to keep captive populations from becoming inbred.

22.6 Inbreeding

In the Hardy-Weinberg model, gametes were picked from the gene pool at random and paired to create offspring genotypes. In nature, however, matings between individuals are seldom, if ever, random. For example, in small populations that are isolated from gene flow, matings between relatives become common. Mating between relatives is called **inbreeding**.

To understand how inbreeding affects populations, let's follow the fate of alleles and genotypes when this type of nonrandom mating occurs. We'll again focus on a single locus with two alleles, A_1 and A_2. Suppose that these alleles initially have equal frequencies of 0.5. In **Figure 22.6**, the width of the boxes represents the frequency of the three genotypes, which start out at the Hardy-Weinberg ratio of $p_1^2 : 2p_1p_2 : p_2^2$. But now let's imagine that these individuals don't produce gametes that go into a gene pool. Instead, they self-fertilize. (Many flowering plants, for example, contain both male and female organs and can self-pollinate.) As the arrows in the figure show, homozygous parents produce all homozygous offspring, but heterozygous parents produce homozygous and heterozygous offspring in a 1:2:1 ratio. As a result, the homozygous proportion of the population increases each generation, while the heterozygous proportion is halved. At the end of the four generations illustrated in the figure, heterozygotes are rare. No evolution has occurred, however, because allele frequencies have not changed. Self-fertilization, or selfing, is the most extreme form of inbreeding; but the same outcome occurs, more slowly, with less extreme forms of inbreeding.

This outcome is important because of a phenomenon called inbreeding depression. **Inbreeding depression** is a loss of fitness that takes place when homozygosity increases. The key to understanding why these fitness losses occur is to recall from Chapter 12 that many recessive alleles represent loss-of-function mutations. These alleles have little or no effect when they occur in heterozygotes, because one normal allele usually produces enough functional protein to support a normal phenotype. But loss-of-function mutations are deleterious or even lethal when they are homozygous. As a result, the offspring of inbred matings are expected to have lower fitness than the progeny of outcrossed matings. This prediction has been verified in a wide variety of species, often through laboratory or greenhouse studies that compare the fitnesses of offspring from controlled matings. As **Table 22.3** shows, inbreeding

TABLE 22.3 Inbreeding Reduces Fitness in Humans

The percentages reported here give the mortality rate of children produced by first-cousin marriages versus marriages between nonrelatives. In every study, children of first-cousin marriages have a higher mortality rate.

Deaths	Period	Children of First Cousins (%)	Children of Nonrelatives (%)
Children under 20 (U.S.)	18th–19th century	17.0	12.0
Children under 10 (U.S.)	1920–1956	8.1	2.4
At/before birth (France)	1919–1950	9.3	3.9
Children (France)	1919–1950	14.0	10.0
Children under 1 (Japan)	1948–1954	5.8	3.5
Children 1-8 (Japan)	1948–1954	4.6	1.5

Source: C. Stern, *Principles of Human Genetics* (San Francisco: Freeman, 1973).

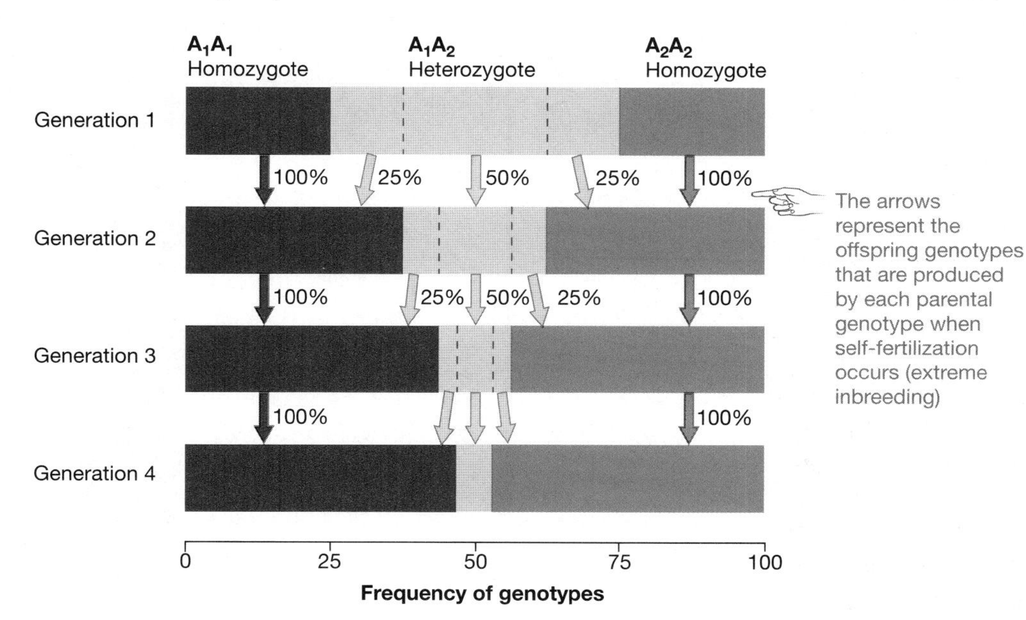

FIGURE 22.6 Why Inbreeding Increases Homozygosity
The width of the boxes corresponds to the frequency of each genotype. Note that homozygosity has almost doubled in just four generations.

FIGURE 22.7 Directional Selection
(a) When traits like size and shape are plotted they often form a bell-shaped curve, or normal distribution. When directional selection acts on these traits, individuals with one set of extreme values experience poor reproductive success. As a result, the distribution shifts in one direction and the average value of the trait changes. **(b)** The upper histogram shows the distribution of overall body size in cliff swallows that died of starvation during an extended cold snap. (The researchers used an index of overall body size that combined measurements of wing length, tail length, leg length, and beak size.) The lower histogram shows the distribution of size in individuals from the same population that survived the cold spell. In each case, N indicates the sample size.

(a) Directional selection changes the average value of a trait.

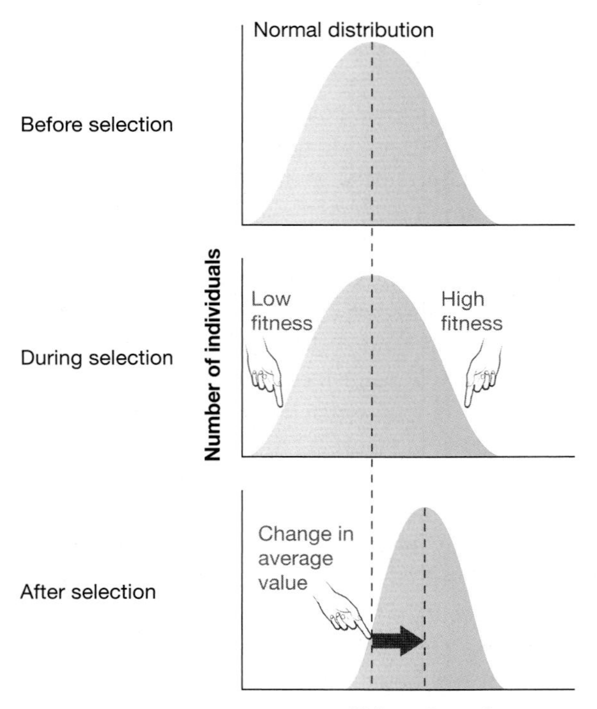

(b) For example, directional selection caused overall body size to increase in a cliff swallow population.

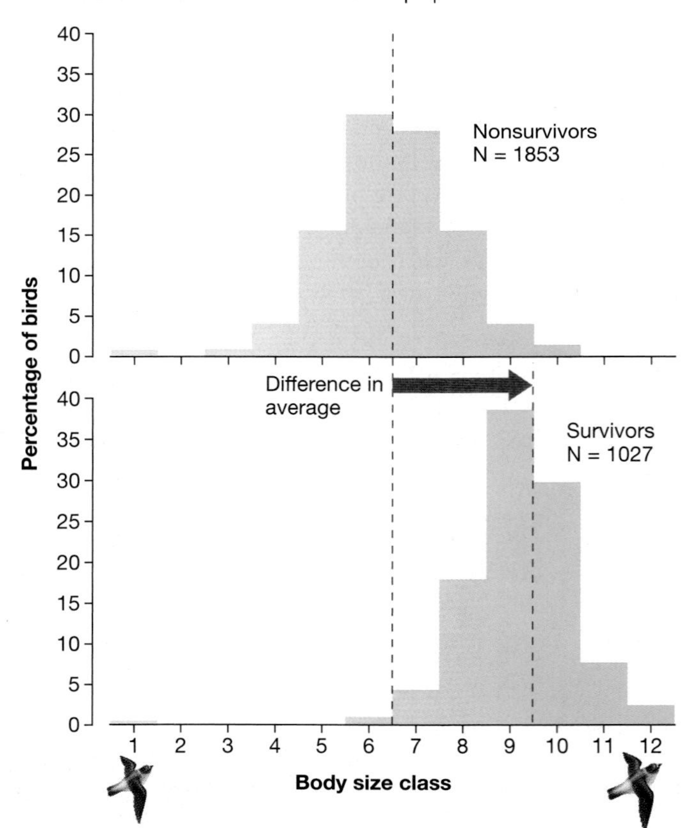

depression is also pronounced in humans. When natural selection against homozygous individuals is as strong as indicated in the table, allele frequencies in the population change rapidly. In this way, inbreeding is an indirect cause of evolutionary change. It increases the rate at which natural selection eliminates deleterious recessive alleles.

It should come as no surprise that many species have mechanisms to avoid inbreeding. Over 40 percent of flowering plants, for example, have "self-incompatibility" loci. As Chapter 36 details, if pollen from the same individual falls on the stigma, the proteins produced by these genes cause reactions that inhibit the pollen from germinating. In many contemporary human societies, it is against the law for individuals to marry who are more closely related than first cousins.

22.7 Natural Selection

Inbreeding and the three evolutionary mechanisms analyzed thus far have an important feature in common. The genetic changes that they produce are not adaptive. The average fitness of a population does not increase as a result of mutation, migration, drift, or inbreeding. In terms of Darwinian fitness, these evolutionary mechanisms produce random changes. Selection is the only evolutionary mechanism that leads to *non*-random changes in allele frequencies.

Natural selection occurs when individuals with certain phenotypes survive better or reproduce more than other individuals in a population. If certain alleles are associated with the favored phenotypes, they increase in frequency while other alleles decrease in frequency. If favored alleles reach fixation, then allelic diversity is eliminated and evolutionary change stops. But does natural selection *always* decrease the genetic diversity of populations?

Directional Selection

According to the studies introduced in Chapter 21, natural selection has increased the frequency of drug-resistant strains of tuberculosis and alpine skypilots with large, sweet-smelling flowers in certain habitats. This type of natural selection is called **directional selection** because allele frequencies change in one direction, as illustrated in **Figure 22.7a**. Directional selection tends to reduce the genetic diversity of populations.

CD ACTIVITY 22.2 Three Modes of Natural Selection

This simplistic view may be misleading, however. To appreciate why, consider recent data on the body size of cliff swallows native to the great plains of North America. In 1996, cliff swallow populations that were being studied by Charles Brown and Mary Bomberger Brown endured a six-day period of exceptionally cold, rainy weather. Cliff swallows are migratory and feed by catching mosquitoes and other insects in flight. Insects disappeared during this cold snap, however, and the biologists recovered the bodies of 1853 swallows that died of starvation. As soon as the weather improved, they caught and measured the body size of 1027 survivors from the same population. As the histograms in **Figure 22.7b** show, survivors were much larger on average than birds that died. Directional selection, favoring large body size, had occurred. To explain this observation, Brown and Brown suggest that larger birds survived because they had larger fat stores and did not get as cold.

If this type of directional selection continued, alleles that contribute to small body size would quickly be eliminated from the cliff swallow population. It is not clear that this will be the case, however. By examining weather records, Brown and Brown established that cold spells as severe as the one that occurred in 1996 are rare. Further, research on other swallow species suggests that smaller birds are more maneuverable in flight and thus more efficient feeders. If so, then selection for feeding efficiency could counteract selection by cold weather and help maintain genetic variation in body size. Because research is continuing, the issue is likely to be resolved soon.

Stabilizing Selection

In the case of the cliff swallows, selection greatly reduced one extreme in the range of phenotypes and resulted in a directional change in the average characteristics of the population. But selection can also reduce both extremes in a population, as illustrated in **Figure 22.8a**. This pattern of selection is called **stabilizing selection.** It has two important consequences: There is no change in the average value of a trait over time, and genetic variation in the population is reduced.

Mary Karn and L. S. Penrose collected a classical data set in humans illustrating stabilizing selection. When Karn and Penrose analyzed birth weights and mortality in 13,730 babies born in British hospitals, they found that babies of average size (slightly over 7 pounds) survived best. As **Figure 22.8b** shows, mortality was high for very small babies and very large babies. This is strong evidence that birth weight is under strong stabilizing selection in humans.

Disruptive Selection

A third pattern of selection observed in nature has the opposite effect of stabilizing selection. Instead of favoring phenotypes near the mean and eliminating the extreme phenotypes, **disruptive selection** eliminates phenotypes near the mean and favors the extreme phenotypes (see **Figure 22.9a**, page 442).

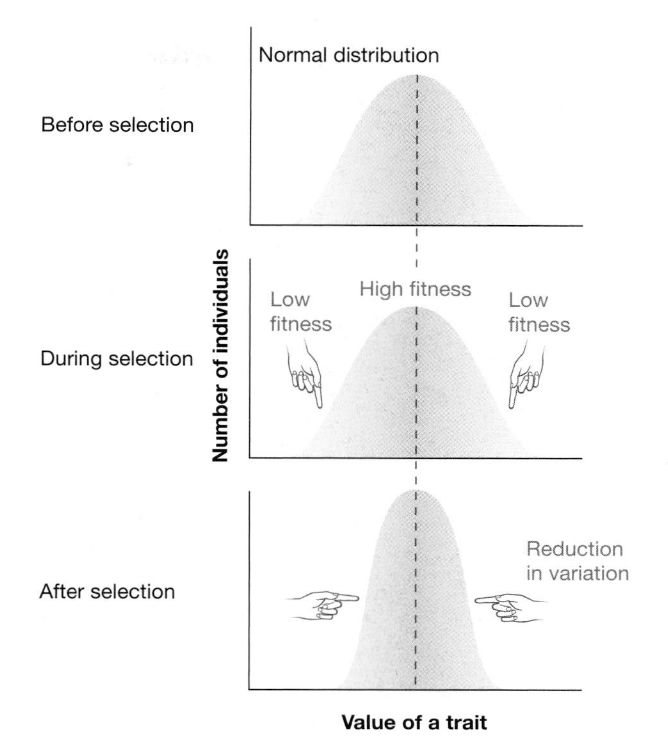

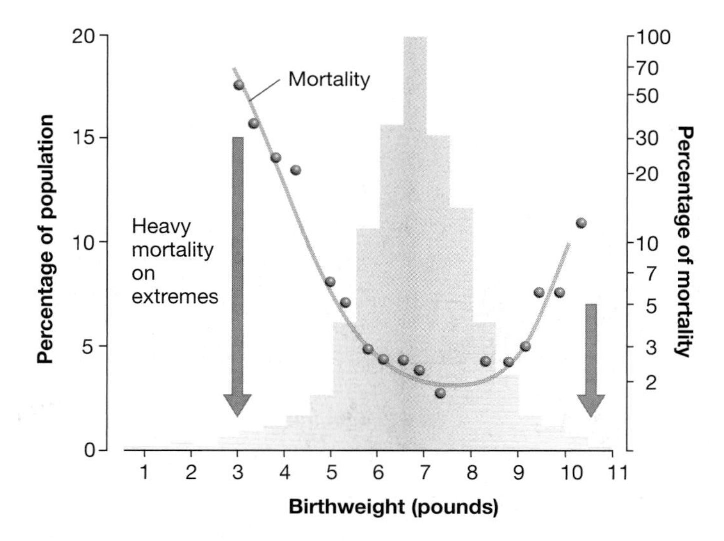

FIGURE 22.8 Stabilizing Selection
(a) When stabilizing selection acts on normally distributed traits, individuals with extreme phenotypes experience poor reproductive success. As a result, the overall variation in the trait decreases even though the average value of the trait stays the same. **(b)** The histogram shows the percentages of newborns with various birth weights. The gray dots are datapoints indicating the percentage of newborns in each weight class that died, plotted on the logarithmic scale shown on the right. The gray line is a function that fits the datapoints.

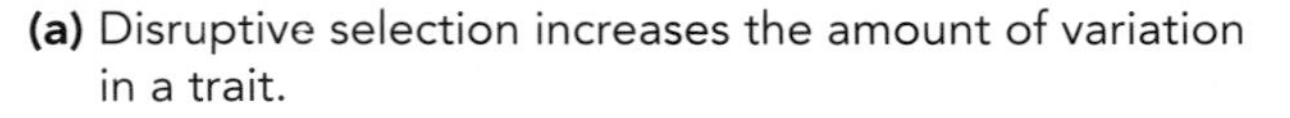

(a) Disruptive selection increases the amount of variation in a trait.

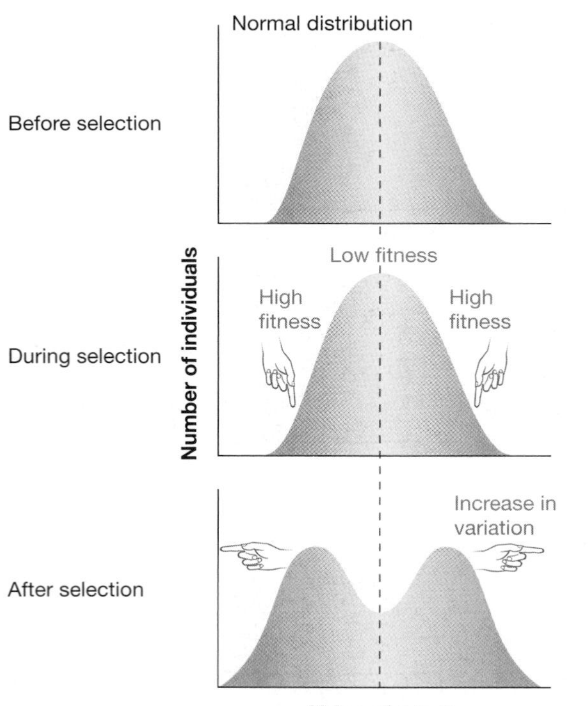

(b) For example, only juvenile blackbellied seedcrackers with very long or very short beaks survived long enough to breed.

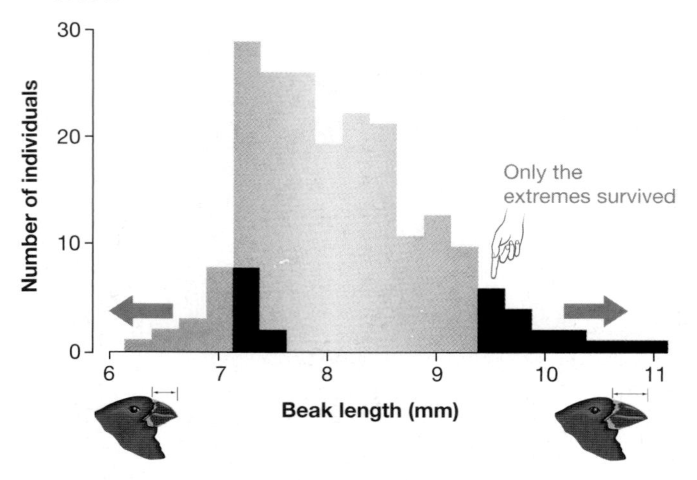

FIGURE 22.9 Disruptive Selection
(a) When disruptive selection occurs on traits with a normal distribution, individuals with extreme phenotypes experience high reproductive success. As a result, the overall variation in the trait increases even though the average value of the trait stays the same. **(b)** This histogram reports the distribution of beak length in a population of black-bellied seedcrackers from Cameroon in West Africa. The orange bars represent all juveniles, and the black bars juveniles that survived to adulthood. Only juveniles with the most extreme phenotypes survived long enough to breed.

Thomas Bates Smith has shown that disruptive selection maintains the striking bills of African seedcrackers pictured in **Figure 22.9b**. The data plotted in the figure show that individuals with very short or very long beaks survive best, and that birds with intermediate phenotypes are at a disadvantage. In this case, the selective agent is food. At his study site in south-central Cameroon, Smith found that only two sizes of seed are available to the seedcrackers: large and small. Birds with small beaks crack and eat small seeds efficiently. Birds with large beaks handle large seeds efficiently. But birds with intermediate beaks have trouble with both. Disruptive selection has a striking effect on genetic diversity in that it produces high overall variation in the population.

Many other patterns of natural selection have been documented in addition to the three reviewed here. The general messages from research on these patterns are that natural selection can increase adaptation, and that different patterns of selection produce different outcomes for allelic diversity and the direction of evolutionary change.

22.8 Sexual Selection

If female peacocks choose males with the longest and most iridescent tails as mates, then the frequency of alleles that contribute to long, iridescent tails will increase in the population. Charles Darwin was the first biologist to recognize that this form of nonrandom mating, called **sexual selection**, is a mechanism of evolutionary change. Sexual selection occurs when individuals within a population differ in their ability to attract mates. It is a special type of natural selection—selection for enhanced ability to obtain mates.

Like inbreeding, sexual selection violates the assumption of random mating that underlies the Hardy-Weinberg principle. But the similarities between inbreeding and sexual selection end there. Inbreeding affects all loci, while sexual selection affects genes involved in mate choice and other aspects of reproduction. Inbreeding results in changes in genotype frequencies, but sexual selection produces changes in allele frequencies.

What types of evolutionary change does sexual selection produce? To answer this question, consider the traits illustrated in **Figure 22.10**. They range from weapons that males use to fight over females, such as antlers and horns, to the elaborate ornamentation and behavior used in courtship displays. All the individuals pictured along the bottom row of this figure are male. The females of the corresponding species are shown in the top row. The examples highlight a key observation: Sexually selected traits often differ sharply between the sexes. The term **sexual dimorphism** (two-forms) refers to any trait that differs between males and females. Human beings, for example, are sexually dimorphic in size, distribution of body hair, and many other traits. Sexual selection often results in sexual dimorphism. Why?

The Theory of Sexual Selection

A. J. Bateman was the first biologist to understand why sexual selection is responsible for sexual dimorphism. Bateman's hypothesis, published in 1948, was elaborated by Robert Trivers in 1972. The theory contains two elements: a claim about a pattern in the natural world and a mechanism that causes the pattern.

The pattern component of the theory is that sexual selection usually acts on males much more strongly than females. As a result, traits that respond to sexual selection are much more highly elaborated in males. The mechanism that Bateman and Trivers proposed to explain this pattern can be summarized with a quip: "Eggs are expensive, but sperm are cheap." In most species, females invest much more in their offspring than do males. The energetic cost of creating a large egg is enormous. In contrast, a sperm contains few energetic resources.

This phenomenon is called the fundamental asymmetry of sex. The asymmetry occurs in the parental investment made by males and females to create offspring. Male-female differences in parental investment are especially pronounced in mammals because of pregnancy and nursing, but females invest more in offspring than males do in almost all sexual species. In species where males furnish a great deal of parental care, like humans and many birds, the fundamental asymmetry of sex is relatively small.

To understand the implications of this asymmetry, it's essential to recognize that both males and females have a limited amount of energy to invest in producing gametes and rearing offspring. Because eggs are large and energetically expensive, females produce relatively few young over the course of a lifetime. For females, fitness is limited primarily by an ability to gain the resources required to produce eggs and rear young—not by the ability to find a mate. In contrast, sperm are so simple to produce that a male can father an almost limitless number of offspring. In males, fitness is not limited by acquiring the resources needed to produce sperm but by the number of female mates.

The fundamental asymmetry of sex dictates that sexual selection will be more intense in males than in females. Intense sexual selection, in turn, leads to the exaggeration of sexually dimorphic traits.

Testing the Theory: Sexual Selection in Elephant Seals

Biologist Burney Le Boeuf and colleagues tested the theory of sexual selection during a long-term study on an elephant seal population breeding on Año Nuevo Island off the coast of California. Elephant seals feed on marine fish but haul themselves

(a) Beetle

Males

During the breeding season, males of the beetle *Dynastes granti* use their elongated mandibles to fight over females.

(b) Sage grouse

Each male sage grouse has a display territory. Males vie among themselves for the best territory. Females choose the male giving the best display as the father of their offspring.

(c) Cichlid

In this species and many other cichlids from the Rift Lakes of Africa, males are brightly colored and perform courtship displays.

FIGURE 22.10 Sexually Selected Traits
Males often have exaggerated traits that they use in fighting or courtship. In many species, females lack these traits.

out of the water during the breeding season. Most breeding habitats are on islands, where newborn pups are protected from terrestrial and marine predators. Within these islands females tend to congregate in small areas that are both safe from predators and suitable for hauling themselves out of the water.

As a result of these circumstances, a distinctive mating system is set up. Males fight over the ownership of patches of beach occupied by congregations of females (**Figure 22.11a**). Males that win battles monopolize matings with the females residing inside their territories. Males that lose battles are relegated to territories with few females or are excluded from the beach.

Le Boeuf and co-workers were confident that sexual selection occurs in this species, because sexual dimorphism is extreme. Males frequently weigh three tons (2700 kg) and are over four times larger, on average, than females. This difference in size is logical because fights between males are essentially slugging contests, with the larger male winning.

Because the biologists marked many of the individuals in the population on Año Nuevo, they were able to track the lifetime reproductive success of a large sample of males and females. These data allowed them to test both aspects of the theory of sexual selection.

Is Sexual Selection More Intense in Males than Females? To test the pattern component of the theory of sexual selection—that sexual selection is more intense in males than females—Le Boeuf's group plotted the lifetime reproductive success of males and females. If sexual selection is based on differential reproductive success, then variation in reproductive success should be much greater in males than in females. As the data in **Figure 22.11b** show, this is true in elephant seals. Among the females in the study, reproductive success varied from 0 to 10 offspring reared to weaning (meaning that pups were no longer nursing and left the beach). Among males in the study, reproductive success varied from 0 to over 90 offspring. Although variation in reproductive success was dramatic in females, it was even more pronounced in males. The largest males were extremely successful at transmitting their alleles to the next generation.

Is Differential Success Related to the Fundamental Asymmetry of Sex? The process component of the theory of sexual selection contends that sexual selection is most intense in the sex that makes the *least* investment in offspring. Le Boeuf and co-workers predicted that female elephant seals invest much more in offspring than do males, and that their lifetime reproductive success is limited not by access to mates but by access to the resources required to bear and rear young.

To test these predictions, the team weighed females and pups throughout the course of the breeding season. Adult females weigh about 650 kg and give birth to one pup, weighing 50 kg, each year. In their first five weeks of life, the pups routinely gain over 100 kg—entirely from mother's milk. The energy expenditure required to accomplish this growth takes a tremendous toll. The researchers found that mothers typically lost up to 200 kg, or almost a third of their body weight, during pregnancy and nursing. The father's parental energy investment, in contrast, was limited to a few minutes spent in copulation and the almost negligible energetic cost of producing sperm. To breed the following year, females had to regain their lost body weight. This observation supports the hypothesis that the life-

(a) Males compete for the opportunity to mate with females.

(b) Variation in reproductive success is greater for males than females.

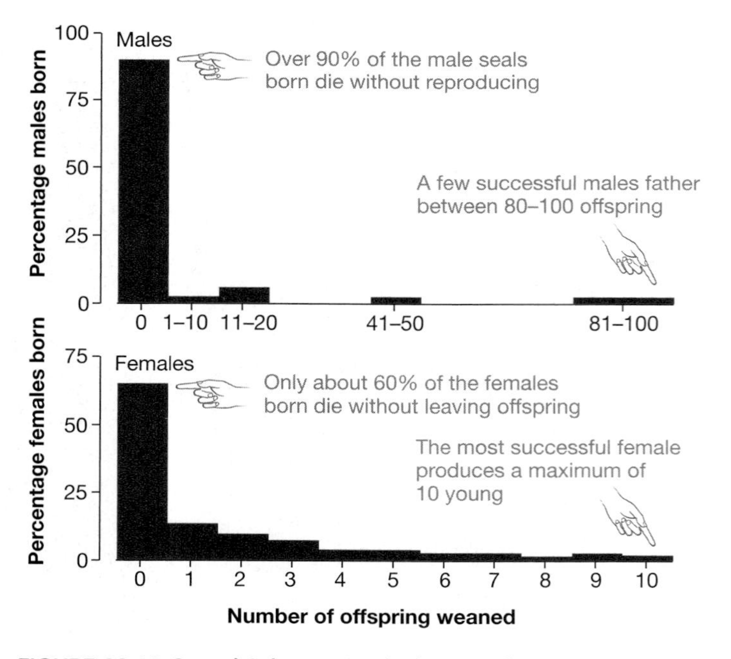

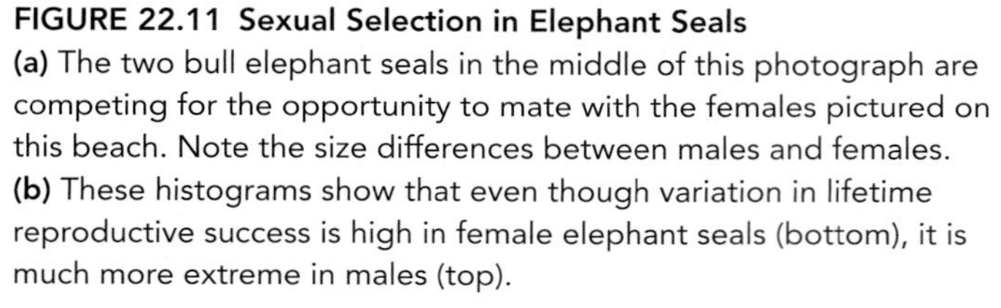

FIGURE 22.11 Sexual Selection in Elephant Seals
(a) The two bull elephant seals in the middle of this photograph are competing for the opportunity to mate with the females pictured on this beach. Note the size differences between males and females. **(b)** These histograms show that even though variation in lifetime reproductive success is high in female elephant seals (bottom), it is much more extreme in males (top).

time reproductive ability of elephant seal females is limited by their ability to acquire enough food to support a pregnancy, not by their ability to persuade a male to copulate.

Even though the theory of sexual selection has been verified in dozens of species besides elephant seals, sexual selection continues to be a hot topic of research. For example, male-male competition is rare in many species. Instead, sexual selection is dominated by female choice of mates. What aspects of male behavior and morphology do females use in making a choice of mates? In some bird species, male tail length has been shown to be an important criterion of mate choice. In some frog species, females prefer to mate with the deepest-voiced males. Do females benefit from these choices? If so, how? If you take a course in animal behavior or evolutionary biology later in your college career, you'll be able to explore these questions in depth.

Essay Evolutionary Theory and Human Health

This chapter is probably the most theory-rich in the text. The reason for this is partly historical. The field covered here, called population genetics, underwent its initial development in the early decades of the twentieth century. At that time it was extremely difficult, technically, for researchers to measure allele frequencies in natural or laboratory populations and perform observational or experimental studies of evolutionary change. Instead, biologists relied on the types of mathematical models introduced here to understand how mutation, migration, drift, selection, and nonrandom mating affect allele and genotype frequencies. Extensive surveys of allele frequencies were not possible until the 1960s, when the first techniques for assaying protein diversity—using the methods introduced in Figure 22.5—became widely available.

Does evolutionary theory have practical implications?

Evolutionary theory can be difficult, especially for people who are not accustomed to using mathematics. Even though the chapter is sprinkled with examples of how evolutionary theory can address public health issues and inform decisions about the management of endangered species, it is still legitimate to ask: Does evolutionary theory have practical implications? For example, can population genetic models help researchers, physicians, and conservationists grapple with real-life problems more successfully?

As an example of how evolutionary theory can directly assist researchers working on pressing medical and environmental problems, consider recent work on the genetic disease called cystic fibrosis (CF). The essay at the end of Chapter 4 points out that in human populations of northern European extraction, one in 2500 infants is born with CF. It is the most common genetic disease in these populations and is caused by a defect in a membrane protein. The incidence rate is very high, especially given that until recently people with CF rarely lived past their teens, and thus were unlikely to reproduce.

Models developed by J. Bertranpetit and F. Calafell in the mid-1990s suggested that the incidence of 1 in 2500 could not be caused by new mutations alone or by gene flow from other populations. Then why are mutant alleles at such relatively high frequency? After all, they are strongly selected against when homozygous. Based on their models of allele frequency change, the researchers hypothesized that the mutant allele(s) must have some selective advantage in northern European populations when found in a heterozygous state.

Shortly after Bertranpetit and Calafell's work was published, Gerald Pier and co-workers provided dramatic support for the prediction. Pier and associates showed that the bacterium that causes typhoid fever, called *Salmonella typhi*, is not able to enter cells expressing mutant forms of the membrane protein involved in CF. Their experiments suggest that defective alleles protect people against typhoid fever when heterozygous, but cause CF when homozygous. For CF researchers, physicians, and genetic counselors, this is an important insight.

It's been said that in science, theory and experiments interact like two people moving a long, heavy table forward by standing at either end. First one person lifts and moves an end ahead, then the other. Then the first moves again, then the other. A good theoretical model makes testable predictions that inspire highly focused experimental or observational studies. The results of these studies often reveal major or minor problems with the theory. The model must then be modified, prompting a new round of predictions and experiments. The more closely theoreticians and data collectors work together, the more rapid the advance.

Chapter Review

Summary

There are four forces of evolution. Each causes allele frequencies to change in populations, but each has different consequences. Only natural selection can result in adaptation. The other three mechanisms introduce a non-adaptive component into evolutionary change. In addition, inbreeding is a form of nonrandom mating that causes genotype frequencies, but not allele frequencies, to change. **Table 22.4** summarizes how each of the events and processes discussed in the chapter affects evolution.

Biologists study the consequences of the different evolutionary mechanisms through a combination of mathematical modeling and experimental or observational research. The Hardy-Weinberg principle is important in the study of evolution because it specifies what genotype and allele frequencies are expected to be if mating is random and none of the four evolutionary forces is operating.

Mutation is too infrequent to be a major cause of allele frequency change. But because mutation constantly introduces new alleles at all loci, it is essential to evolution. Without mutation, selection and drift would eventually eliminate all allelic variation, and evolution would cease.

In the evolutionary sense, migration represents gene flow, or movement of alleles, between populations. Gene flow tends to homogenize allele frequencies and decrease differentiation among populations, but may also serve as an important source of new variation in populations.

Genetic drift results from random sampling and is an important evolutionary force in small populations. Drift leads to the random fixation of alleles and tends to reduce overall allelic diversity.

Inbreeding, or mating among relatives, is a form of nonrandom mating. Inbreeding does not change allele frequencies, so it is not an evolutionary mechanism. It does, however, change genotype frequencies by leading to an increase in homozygosity. This can cause inbreeding depression.

Natural selection occurs in both a wide variety of patterns and a wide variety of intensities. Directional selection, for example, may lead to certain alleles becoming fixed—and thus reduces allelic diversity in populations. Disruptive selection, in contrast, eliminates phenotypes with intermediate characteristics and increases allelic diversity in populations. The rate of evolution under natural selection depends both on the intensity of selection and the amount of genetic variation available.

Sexual selection produces another form of nonrandom mating. Sexual selection is a type of natural selection. It is responsible for the evolution of phenotypic differences between males and females. Sexual selection occurs when certain traits help males succeed in contests over mates, or when certain traits are attractive to prospective mates.

TABLE 22.4 A Summary of Evolutionary Mechanisms

Event	Result
Mutation	Increases variation by introducing new alleles. This occurs too infrequently to be the sole cause of gene frequency change, however.
Migration (gene flow)	Reduces differences among populations. Can increase genetic variability of a population by introducing new alleles; can also reduce adaptedness of population by introducing poorly adapted alleles.
Genetic drift	Causes random changes in allele frequencies; tends to lower genetic variation by leading to loss or fixation of alleles
Inbreeding*	Changes genotype frequencies by increasing homozygosity; if inbreeding depression occurs, genetic variability is reduced
Selection	Increases adaptation; depending on pattern of selection, can lead to maintenance, increase, or reduction of genetic variation
None (Hardy-Weinberg conditions)	No changes in allele or genotype frequencies; genetic variability is maintained

* Because inbreeding does not change allele frequencies, it does not cause evolution and is not considered an evolutionary mechanism. (Inbreeding only changes genotype frequencies.)

Questions

Content Review

1. Why isn't inbreeding, or mating among relatives, considered an evolutionary mechanism?
   a. It does not change genotype frequencies.
   b. It does not change allele frequencies.
   c. It does not occur often enough to be important in evolution.
   d. It does not violate the assumptions behind the Hardy-Weinberg Principle.
2. What are four mechanisms that cause allele frequencies to change?
   a. migration, gene flow, nonrandom mating, and natural selection
   b. inbreeding, Hardy-Weinberg, genetic drift, and mutation
   c. sexual selection, natural selection, inbreeding, and genetic drift
   d. mutation, selection, migration, and genetic drift
3. Why is genetic drift aptly named?
   a. It causes allele frequencies to drift up or down, randomly.
   b. It is the ultimate source of genetic variability.
   c. It is an especially important mechanism in small populations.
   d. It occurs when populations drift into new habitats.
4. What does it mean if an allele reaches "fixation"?
   a. It is eliminated from the population.
   b. It has a frequency of 1.0.
   c. It has a frequency of 0.5.
   d. It is adaptively advantageous.
5. Sexual dimorphism refers to any trait besides genitalia, that
   a. is more exaggerated in females
   b. is more exaggerated in males
   c. differs between the sexes
   d. is used in courtship displays
6. When does sexual selection occur?
   a. when females mate with the first male that comes along
   b. when all individuals experience the same reproductive success
   c. when individuals experience differential success in survival
   d. when individuals experience differential success in obtaining mates
7. What is the only evolutionary mechanism that leads to adaptation?
   a. mutation
   b. migration
   c. genetic drift
   d. selection

Conceptual Review

1. In what sense is the Hardy-Weinberg principle a null model, similar to the control treatment in an experiment?
2. The text maintains that mutation has two roles as an evolutionary mechanism. It is too infrequent to produce significant changes in allele frequencies, but it is the ultimate source of genetic variability. Explain.
3. Make a table with three columns and seven rows, modeled after Table 22.4. Title the columns "Event," "Effect on Populations," and "Example." In the rows of the column titled "Event," write Mutation, Migration, Genetic drift, Inbreeding, Selection, and None—(Hardy-Weinberg conditions). Fill in the rest of the table.
4. Sexual selection, like inbreeding, is a form of nonrandom mating. How do these two processes differ? Why does sexual selection frequently lead to sexual dimorphism, with males showing exaggerated characters?
5. Directional selection is paradoxical in the sense that it can lead to the fixation of favored alleles. When this occurs, genetic variation is zero and evolution stops. Explain why this rarely occurs.

Applying Ideas

1. In humans, albinism is caused by loss-of-function mutations in genes involved in the synthesis of melanin, the dark pigment in skin. Only people homozygous for a loss-of-function allele have the phenotype. In Americans of northern European ancestry, albino individuals are present at a frequency of about 1 in 10,000 (or 0.0001). Knowing this genotype frequency, we can calculate the frequency of the loss-of-function alleles. If we let p_2 stand for this frequency, we know that $p_2^2 = 0.0001$, so $p_2 = \sqrt{0.0001} = 0.01$. By subtraction, the frequency of normal alleles is 0.99. If these loci conform to the conditions required by the Hardy-Weinberg principle, what is the frequency of "carriers"—or people who are heterozygous for the condition? Your answer indicates the percentage of Caucasians in the United States who carry an allele for albinism.
2. The text contends that conservation managers frequently use gene flow, in the form of transporting individuals or releasing captive-bred young, to counteract the effects of drift on small, endangered populations. Explain how gene flow can also mitigate the effects of inbreeding.
3. In 1789, a small band of mutineers led by Fletcher Christian took over the British warship HMS Bounty and fled to tiny Pitcairn Island in the South Pacific. There the sailors joined a small population of native Polynesians, married, and raised families. All contemporary Pitcairn islanders can trace at least part of their family tree back to the colonization event. What evolutionary forces were at work in this small, isolated population before, during, and after the arrival of the mutineers?
4. Imagine that you are a conservation biologist charged with creating a recovery plan for an endangered species of turtle. The turtle's habitat has been fragmented into small, isolated protected areas by suburbanization and highway construction. There is some evidence that certain populations are adapted to marshes that are normal, while others are adapted to acidic wetlands or salty habitats. Further, some populations number less than 25 breeding adults, making genetic drift and inbreeding a major concern. In creating a recovery plan, the tools at your disposal are captive breeding, capture and transfer of adults to create gene flow, or creation of habitat corridors between wetlands to make migration possible. Write a two-paragraph essay outlining the major features of your proposal.

CD-ROM and Web Connection

CD Activity 22.1: The Hardy-Weinberg Principle *(tutorial)*
(Estimated time for completion = 20 min)
How can you tell if a population is evolving?

CD Activity 22.2: Three Modes of Natural Selection *(tutorial)*
(Estimated time for completion = 10 min)
How does natural selection shape a population?

At your **Companion Website** (http://www.prenhall.com/freeman/biology), you will find self-grading exams and links to the following research tools, online resources, and activities:

Host Plant Resistance and Conservation of Genetic Diversity
A look at the importance of the diversity of agricultural crops worldwide.

Genetic Drift Simulation
A fictional animal is used to explain the concept of genetic drift.

How Females Choose Their Mates
This *Scientific American* article discusses the evolutionary applications of sexual selection.

Additional Reading

Eberhard, W. G. 1990. Animal genitalia and female choice. *American Scientist* 78: 134–141. An essay examining how sexual selection can produce changes in the size and shape of genitalia.

Emlen, D. J. 2000. Integrating development with evolution: a case study with beetle horns. *BioScience* 50: 403–418. Analyzes the remarkable diversity of horns in dung beetles in light of how developmental processes interact with sexual selection and other evolutionary forces.

Grant, P. R. 1991. Natural selection and Darwin's finches. *Scientific American*, October: 82–87. A review of research on the evolutionary forces at work in the finch populations of the Galápagos Islands.

Speciation

23

Although Darwin called his masterwork *On the Origin of Species by Means of Natural Selection*, he actually had little to say about how new species arise. Instead, his writing focused on the process of natural selection and the changes that occur within populations over time. He spent much less time considering changes that occur *between* populations.

Populations can change and diverge when they are isolated from one another in terms of gene flow. When gene flow ends, populations begin to evolve independently. If isolated populations diverge sufficiently over time to form distinct types, or **species**, the process of speciation has taken place. **Speciation** is a splitting event that creates two or more distinct species from a single ancestral group (**Figure 23.1**, page 450). Traditionally, biologists have thought that speciation occurs slowly—over tens of thousands of years. When the event is complete, a new branch has been added to the tree of life.

In essence, then, speciation is an outcome of isolation and divergence. Isolation and divergence, in turn, result from changes in the four evolutionary forces introduced in Chapter 22: gene flow, mutation, genetic drift, and natural selection. Isolation is created by reductions in gene flow; divergence is created when mutation, genetic drift, and selection act on populations separately. How do these four evolutionary processes interact to produce new species? If speciation events are slow and continuous, how do biologists study them?

This chapter is devoted to exploring these questions. Our first task is to examine how species are defined and identified. Subsequent sections focus on how speciation occurs in two contrasting situations: when populations occupy the same geographic area, and when they are separated into distinct regions. The chapter

These soapberry bugs belong to different populations. In this chapter we analyze data indicating that these populations are in the process of becoming distinct species.

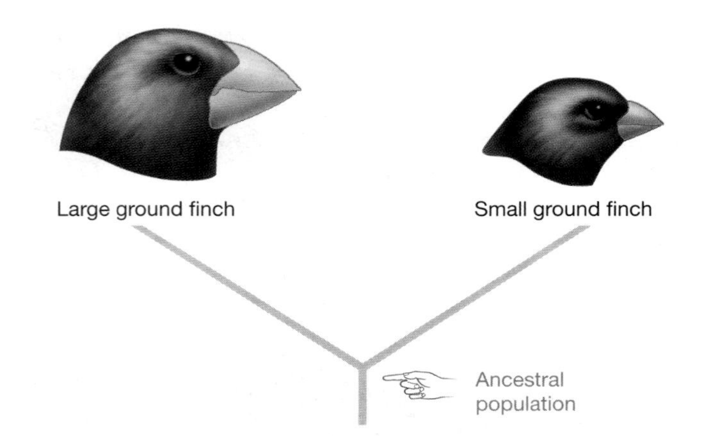

FIGURE 23.1 Speciation Creates Distinct Populations
When lack of gene flow isolates populations, they tend to diverge. This is because mutation, drift, and selection act on the populations independently. As a result, they acquire distinctive characteristics. For example, the large ground finch and medium ground finch shown here are derived from the same ancestral population. This ancestral population split into two populations that were isolated by lack of gene flow. Because the populations began evolving independently, they acquired the distinctive characteristics observed today.

concludes with a look at a classical question in speciation research: What happens when populations that have been isolated from one another come back into contact? Do they interbreed and merge back into the same species, or do they remain independent and form new species?

Speciation research was once considered an aspect of biological science with few practical applications. But with species losses mounting and with novel strains of bacteria and viruses causing punishing diseases in humans, the question of how new species arise has taken on new urgency.

23.1 Defining and Identifying Species

Species are distinct types of organisms because they represent evolutionarily independent groups. Like the Galápagos finches pictured in Figure 23.1, species are distinct from one another in appearance, behavior, habitat use, or genetic characteristics. These distinctions occur because mutation, drift, and selection act on each species independently of what is happening in other populations.

What makes a species independent? The answer begins with *lack* of gene flow. Chapter 22 showed that gene flow eliminates genetic differences among populations. Allele frequencies in populations, and thus their characteristics, become more alike when gene flow occurs between them. If gene flow between populations is extensive and continues over time, it eventually causes even highly distinct populations to coalesce into the unit known as a species. Conversely, if gene flow between populations stops, then mutation, selection, and drift begin to act on the populations independently. As a result, allele frequencies in the populations diverge. When allele frequencies change sufficiently over time, populations become distinct species.

Formally, then, species are defined as evolutionarily independent populations. Even though this definition sounds straightforward, it can be exceedingly difficult to put into practice. How can evolutionarily independent populations actually be identified in the field and in the fossil record? There is no single, universal answer. Even though biologists and paleontologists agree on the definition of a species, they frequently have to use different sets of criteria to identify them.

After reviewing the pros and cons of three criteria used for recognizing a distinct species, we consider how using different species guidelines can affect efforts to preserve biodiversity.

The Biological Species Concept

According to the biological species concept, the critical criterion for identifying species is reproductive isolation. This is a logical yardstick because no gene flow occurs between populations that are reproductively isolated from each other. Specifically, if two different populations do not interbreed in nature, or if they fail to produce viable and fertile offspring when matings take place, then they are considered distinct species. Biologists can be confident that reproductively isolated populations are evolutionarily independent.

The biological species concept has disadvantages, however. The criterion of reproductive isolation cannot be evaluated in fossils or in asexual species, and it is difficult to apply when closely related populations do not happen to overlap with each other. In the latter case, biologists are left to guess whether interbreeding and gene flow would occur if the populations happened to come into contact.

The Morphospecies Concept

How do biologists identify species when the criterion of reproductive isolation cannot be applied? Under the morphospecies (form-species) concept, researchers identify evolutionarily independent lineages by differences in size, shape, or other morphological features. The logic behind the morphospecies concept is that distinguishing features are most likely to arise if populations are independent and isolated from gene flow.

The morphospecies concept is compelling simply because it is so widely applicable. It is a useful criterion when biologists have no data on the extent of gene flow, and it is equally applicable to sexual, asexual, or fossil species. Its disadvantage is that the features used to distinguish species are subjective. In the worst case, different researchers working on the same populations disagree on the characters that distinguish species. Disagreements like these end in a stalemate because no independent criteria exist for resolving the conflict.

Phylogenetic Species

The phylogenetic species concept is a recent addition to the tools available for identifying evolutionarily independent lineages. It is based on reconstructing the evolutionary history of populations and is increasingly popular among biologists who study fossil, sexual, and asexual species. Proponents of the approach argue that it is widely applicable and precise.

To understand the reasoning behind the concept, recall Darwin's claim, explored in Chapter 21, that all species are related by common ancestry. That chapter also introduced the phylogenetic trees used to represent the genealogical relationships among populations. **Figure 23.2** shows such a tree; the logic that investigators use to reconstruct phylogenies is introduced in **Box 23.1**, pages 456–457. On any given evolutionary tree, there are many monophyletic (one-tribe) groups. A **monophyletic group** is an ancestral population, all of its descendants, and *only* those descendants. Several monophyletic groups are color-coded on the tree in Figure 23.2. Under the phylogenetic species concept, a species is defined as the smallest monophyletic group in a tree representing populations. In Figure 23.2, the groups marked A, B, C, and so on represent 10 phylogenetic species. When all known populations are included on the tree of life, each tip would be called a species.

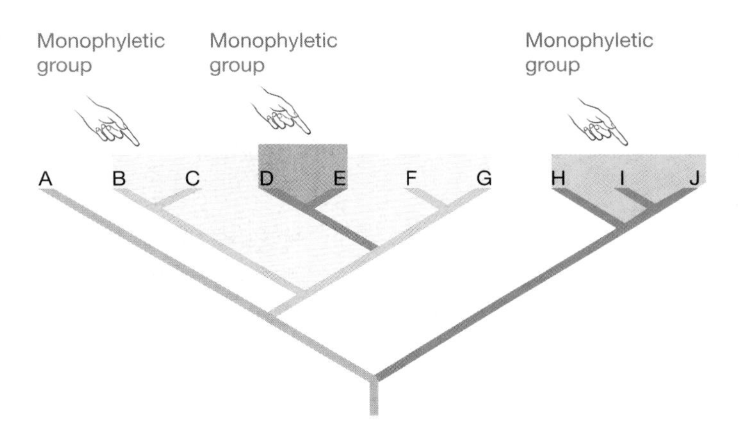

FIGURE 23.2 Monophyletic Groups
The color-coded lineages on this phylogenetic tree are all monophyletic because they contain all the descendants of a common ancestor. **EXERCISE** This tree has many monophyletic groups other than those indicated. Identify one by picking any branch point (node) on the tree. Then draw a circle around it and all of the branches descending from that point.

The phylogenetic species concept has two distinct advantages: (1) It can be applied to any populations (fossil, asexual, or sexual), and (2) it is logical because populations are only monophyletic if they are independent of one another and isolated from gene flow. The approach has a distinct disadvantage, however; carefully estimated phylogenies are available only for a tiny (but growing) subset of populations on the tree of life. Critics of the approach also point out that it would probably lead to recognition of many more species than either the morphospecies or biological species concepts. Proponents counter that far from being a disadvantage, recognizing increased numbers of species may better reflect the reality of life's diversity.

In actual practice, researchers use all three species concepts summarized here (**Table 23.1**). Conflicts have occurred, however, when different species concepts are applied to the real world of conservation action. To appreciate this point, let's consider the case of the dusky seaside sparrow.

Species Definitions in Action: The Case of the Dusky Seaside Sparrow

Seaside sparrows live in salt marshes along the Atlantic and Gulf coasts of the United States. The scientific name of this species is *Ammodramus maritimus*. (Recall, from Chapter 1, that scientific names consist of a genus name followed by a species name.) As the map, photos, and labels in **Figure 23.3a** (page 452) indicate, researchers had traditionally named a variety of seaside sparrow "subspecies" under the morphospecies concept. Subspecies populations have distinguishing features, like coloration or calls, but are not considered distinctive enough to be called separate species. Because salt marshes are often destroyed for agriculture or oceanfront housing, by the late 1960s biologists began to be concerned about the future of some sparrow populations. A subspecies called the dusky seaside sparrow (*Ammodramus maritimus nigrescens*) was in particular trouble; by 1980 only six individuals from this population remained. All were males.

At this point government and private conservation agencies sprang into action under the auspices of the Endangered Species Act, a law whose goal is to preserve biodiversity by

TABLE 23.1 A Summary of Species Concepts

Species Concept	Criterion for Recognizing Species	Advantages	Disadvantages
Biological	Reproductive isolation between populations (they don't breed and produce viable offspring)	Reproductive isolation = evolutionary independence	Not applicable to asexual or fossil species; difficult to assess if populations do not overlap geographically
Morphospecies	Populations are morphologically distinct	Widely applicable	Subjective (researchers often disagree about how much morphological distinction = speciation)
Phylogenetic	Smallest monophyletic group on evolutionary tree	Widely applicable; based on testable criteria	Few well-estimated phylogenies are currently available

preventing the extinction of species. The law uses the biological species concept to identify species and calls for the rescue of endangered species through active management. Because current populations of seaside sparrows are physically isolated from one another, and because young seaside sparrows tend to breed near where they hatched, researchers believed that little to no gene flow occurred among populations. The dusky seaside sparrow became a priority for conservation efforts because it was reproductively isolated.

To launch the rescue program, the remaining male dusky seaside sparrows were taken into captivity and bred with females from a nearby subspecies: *A.m. peninsulae*. Officials planned to use these hybrid offspring as breeding stock for a reintroduction program. The goal was to preserve as much genetic diversity as possible by reestablishing a healthy population of dusky-like birds. The plan was thrown into turmoil, however, when John Avise and William Nelson estimated the phylogeny of the seaside sparrows by comparing gene sequences. This tree, shown in **Figure 23.3b**, shows that seaside sparrows represent two distinct monophyletic groups: one native to the Atlantic Coast and the other native to the Gulf Coast. Far from being an important, reproductively isolated population, the phylogeny showed that the dusky sparrow is merely a typical Atlantic Coast sparrow. Further, officials had unwittingly crossed the dusky males with females from the Gulf Coast lineage. Because the goal of the conservation effort was to preserve existing genetic diversity, this was the wrong population to use.

Avise and Nelson maintained that the biological and morphospecies concepts had misled a well-intentioned conservation program. Under the phylogenetic species concept, officials should have allowed the dusky sparrow to go extinct and then concentrated their efforts on simply preserving one or more populations from each coast. In this way the two monophyletic groups of sparrows—and the most genetic diversity—would be preserved. Under the morphospecies concept, officials did the right thing by preserving distinct types.

When conservation funding is scarce, life-and-death decisions like these are crucial. In Chapter 52 we explore these issues in depth. Now our task is to consider an even more fundamental question: How do isolation and divergence produce the event called speciation?

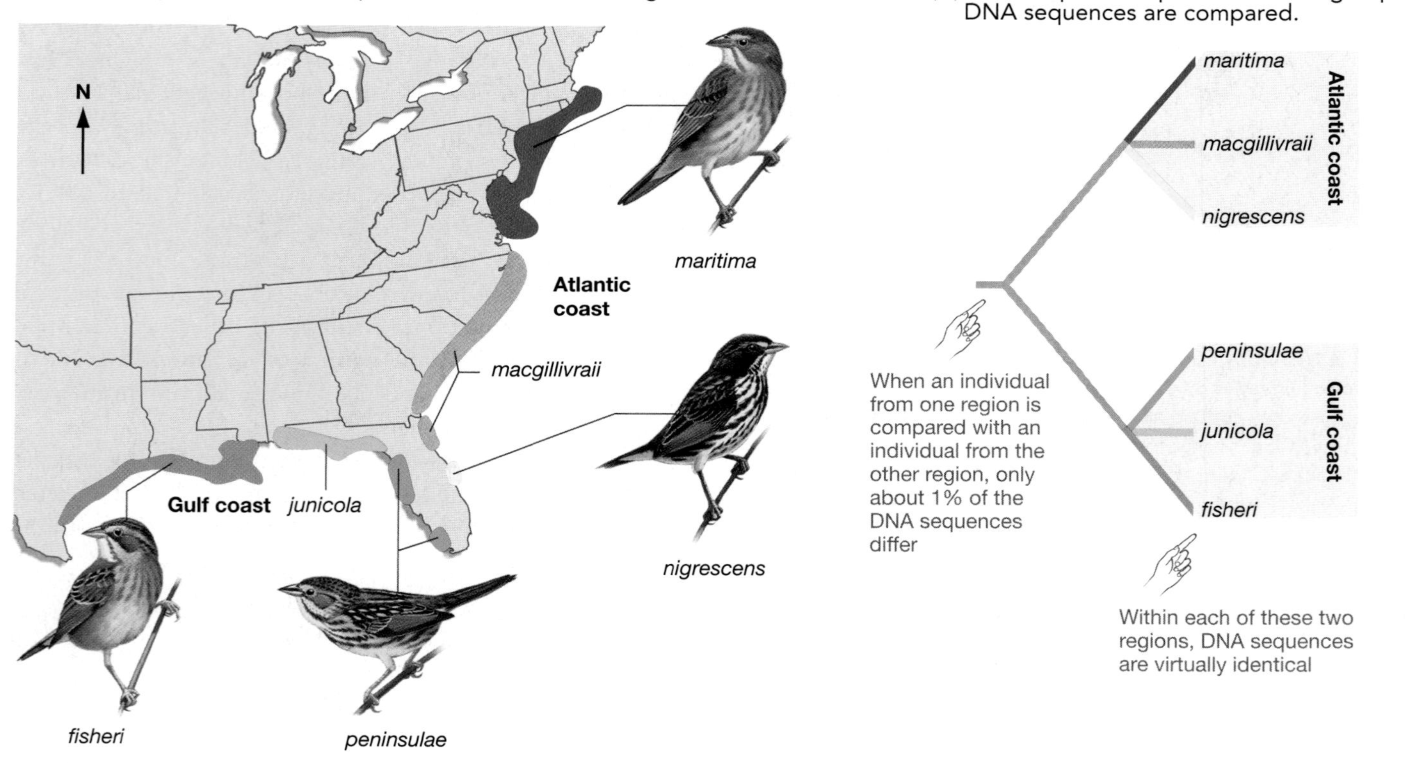

FIGURE 23.3 Seaside Sparrows
(a) The "subspecies" of seaside sparrows named on this map are distinguished by their distinctive coloration. **(b)** Each tip on this evolutionary tree represents an individual from a subspecies of seaside sparrow. The tree was estimated by comparing DNA sequences. The tree shows that seaside sparrows represent two distinct monophyletic groups, one native to the Atlantic coast and the other native to the Gulf coast. QUESTION If you were a conservation biologist and could save only two subspecies of seaside sparrows from extinction, why would it be ill-advised to choose *fisheri* and *peninsulae*?

23.2 Isolation and Divergence in Sympatry

When populations or species live in the same geographic area, or at least close enough to one another to make interbreeding possible, biologists say that they live in **sympatry** (together-homeland). Traditionally, researchers have predicted that speciation could *not* occur among sympatric populations, because gene flow is possible. The prediction was that gene flow would easily overwhelm any differences among populations created by genetic drift and natural selection.

To illustrate this point, consider research on the water snakes native to the Lake Erie region of North America (**Figure 23.4**). On mainland habitats near the lake, the vast majority of water snakes have the banded coloration shown at the bottom of the figure. But unbanded snakes predominate on islands. Why? In the late 1950s Joseph Camin and Paul Ehrlich noticed that island snakes bask on limestone rocks. They hypothesized that unbanded snakes are more difficult for predators to see while basking on this surface. Richard King recently confirmed this hypothesis by showing that, on islands, young unbanded snakes survive much better than do young banded snakes.

If natural selection favors different coloration patterns in the two populations, why haven't they diverged to become separate species? The answer is gene flow. Individuals from the mainland swim out and join the island populations on a regular basis. The alleles they introduce when they breed keep the banded coloration pattern at a reasonably high frequency on islands and maintain the two populations as a single species. In this way, gene flow prevents speciation because it overwhelms the diversifying force of natural selection. Is this always the case?

Can Natural Selection Cause Speciation Even When Gene Flow Is Possible?

Recently, several well-documented examples have upset the traditional view that sympatric speciation is rare or nonexistent. These studies are fueling a growing awareness that, under certain circumstances, natural selection that causes populations to diverge can overcome gene flow and cause speciation. As an example, let's consider research spearheaded by Scott Carroll on the speciation of soapberry bugs.

The soapberry bug is a species of insect, pictured in **Figure 23.5a** (page 454), native to the south-central and southeastern United States. The bugs make their living by feeding on plants in a family called Sapindaceae, including the soapberry tree, serjania vine, and balloon vine. As the figure shows, the bugs feed by piercing fruits with their beaks, reaching in to penetrate the coats of the seeds inside the fruit, and then sucking up the contents of the seeds.

The soapberry bug's story began to get interesting when horticulturists brought three new species of sapindaceous plants to North America from Asia in the mid-twentieth century. Two of these "exotic," or nonnative, species are now cultivated as ornamentals and one grows as a weed. Soon after these plants were introduced to the New World, soapberry bugs began using them as hosts. But as Figure 23.5a shows, the fruits of the nonnative species are much different from the fruits of native species.

In soapberry bug populations that feed on native host plants, beak length corresponds closely to the size of the host fruit. For example, bugs that feed on species with big fruit tend to have long beaks. The correlation between fruit size and beak length is logical because it should allow individuals to reach the

FIGURE 23.4 Mainland and Island Populations of Water Snakes Water snakes occupy most of the larger islands in Lake Erie, along with mainland habitats on either side of the lake. On island habitats, unbanded individuals like the one in the top photo are quite common. All or most of the water snakes found in mainland habitats are strongly banded, such as the individual in the bottom photo.
QUESTION Banded snakes are most common on the island closest to shore. Why?

FIGURE 23.5 Beak Length and Host Fruits of Soapberry Bugs
(a) The drawing at the left shows a soapberry bug feeding on an introduced plant, called the flat-podded golden rain tree, that is common in central Florida. The drawing on the right shows a soapberry bug feeding on a native plant, called the balloon vine, that is common in southern Florida. As the histograms show, soapberry bug populations that feed on the two host plants have very different beak lengths, which correspond to differences in the size of the host fruit. **(b)** These data show the beak lengths of soapberry bugs collected in Florida over a 90-year period. Beak length declined after the introduction of the small-fruited plant. EXERCISE Draw a line through the data that emphasize the decline in beak length over time.

(a) Beak length correlates with fruit size.

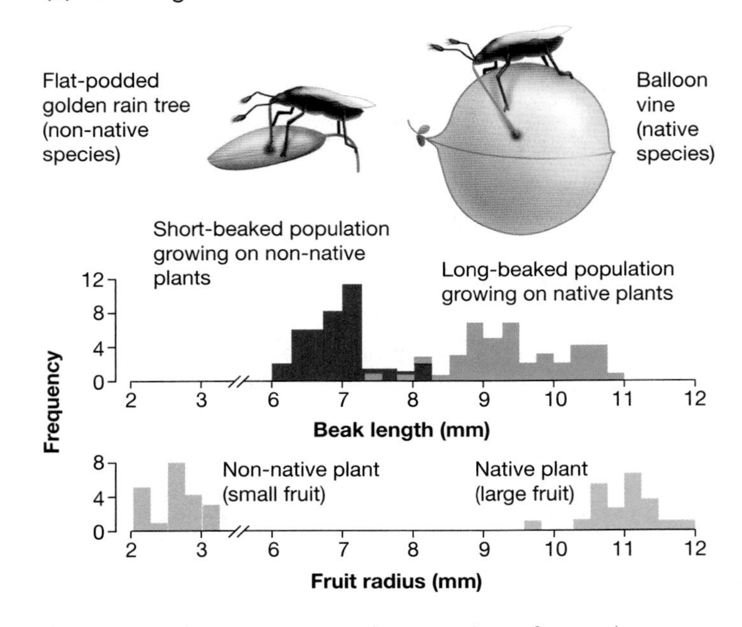

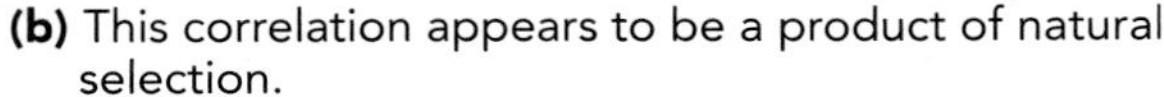

(b) This correlation appears to be a product of natural selection.

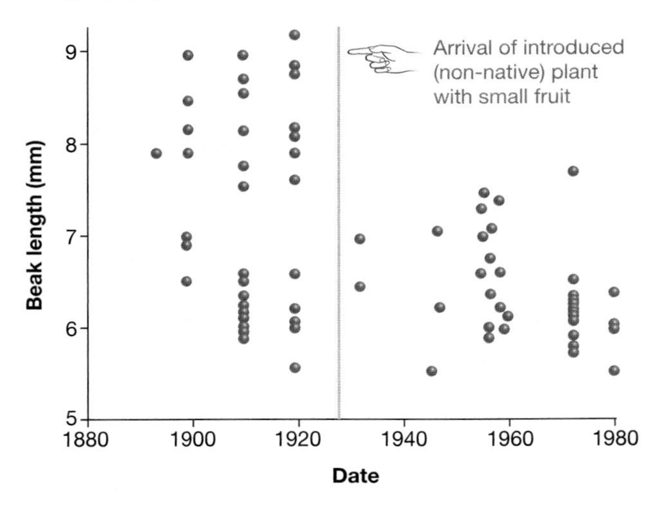

seeds inside the fruit efficiently. It also prompted Scott Carroll to ask a simple question: In populations of soapberry bugs that exploit the introduced plant species, have beak lengths evolved to match the sizes of the new fruits? If so, it would mean that natural selection is currently producing divergence in soapberry bug populations.

To answer this question, Scott Carroll and Christin Boyd measured large samples of bugs found on both native and non-native hosts. Some of the resulting data are shown in Figure 23.5a. The top histogram shows that bugs collected on native plants growing in south Florida have much longer beaks than those collected on nonnative plants growing in central Florida. The bottom histogram confirms that the fruits of the native species are much larger than the fruits of the introduced species. The data argue that soapberry bug populations exploiting exotic species have indeed changed—presumably in response to natural selection for efficient use of host fruits.

To test this conclusion, Carroll and Boyd measured the beaks of soapberry bugs preserved in museum collections that had been collected in Florida between 1890 and 1980. Upon plotting the relationship between beak length and time, they discovered that average beak length in Florida populations has declined sharply (**Figure 23.5b**). Because the older samples were collected long before the exotic species were introduced, the result supports the hypothesis that beak length has changed because some soapberry bug populations have switched to a new host.

Will the populations that exploit native and exotic host plants continue to diverge and eventually form new species? Only time, and further research, will tell. But Carroll and coworkers expect that the answer is yes. This prediction rests on an important observation. Because soapberry bugs mate on or near their host plants, switching to a new host species reduces gene flow among populations at the same time that it sets up natural selection for divergence. As a result, natural selection may be able to overwhelm gene flow and cause speciation, even when populations are sympatric.

Although the soapberry bug's story might seem localized and specific, the events may be common. Most insects are associated with specific host plants; biologists estimate that over three million insect species exist. It is reasonable to contend that switching host plants, as soapberry bugs have done, is a major mode of speciation.

Does Mutation Play a Direct Role in Speciation?

Based on the theory and data reviewed thus far, it is clear that gene flow and natural selection play important roles in speciation. Can a third evolutionary process—mutation—influence speciation as well? The answer might appear to be no. Chapter 22 emphasized that even though mutation is the ultimate source of genetic variation in populations, it is an inefficient mechanism of evolutionary change. If populations become isolated, it is unlikely that mutation, on its own, could cause them to diverge appreciably.

This view turns out to be naive, however. There is a particular type of mutation, relatively common in plants, that can trigger speciation among sympatric populations. The key is that

CD ACTIVITY 23.1
Speciation by Changes in Ploidy

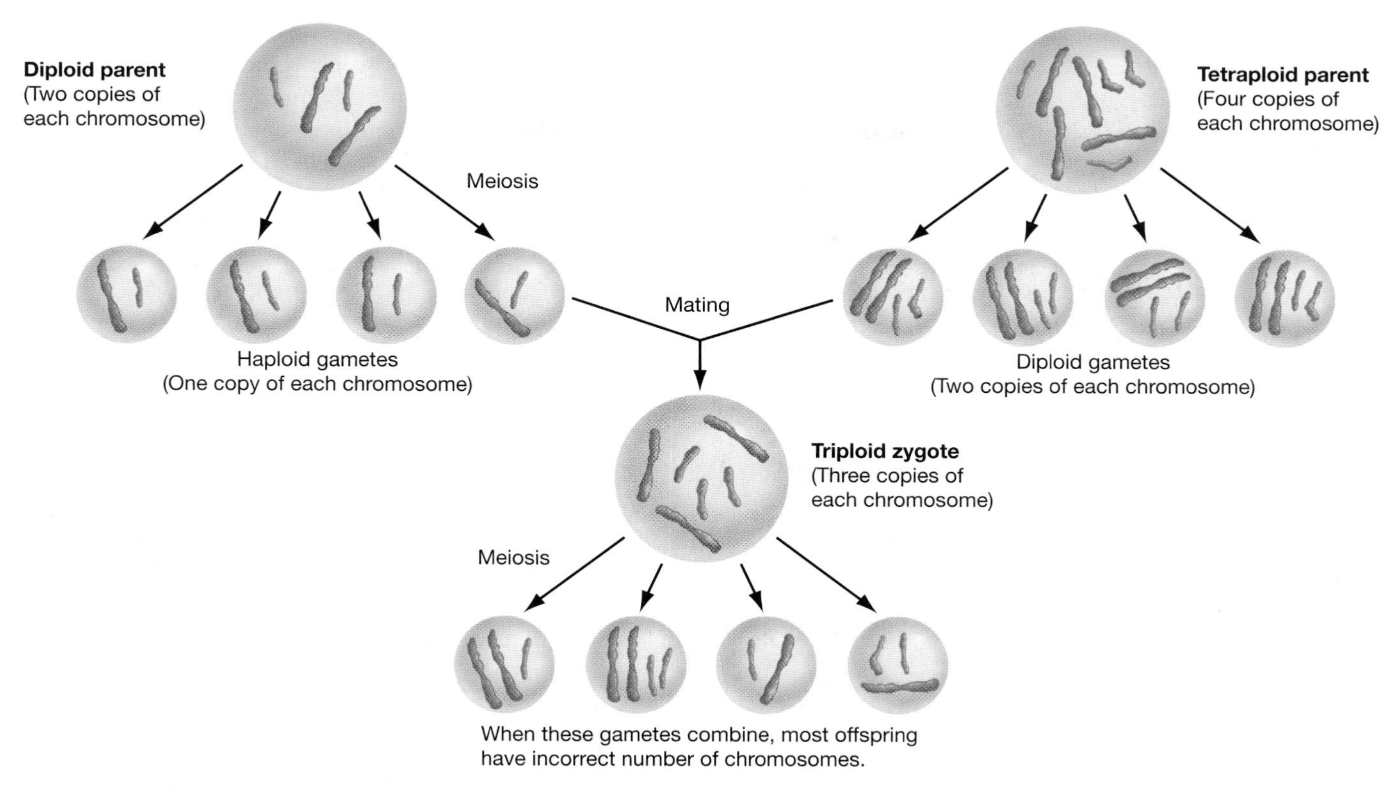

FIGURE 23.6 Why Polyploidy Can Lead to Reproductive Isolation
The mating diagrammed here shows why normal and polyploid individuals are reproductively isolated. The example illustrates just two chromosomes, although most organisms have at least four.

the mutation reduces gene flow between mutant and normal, or wild-type, individuals. It does so because mutant individuals have more than two sets of chromosomes. This condition is known as **polyploidy**.

To understand why polyploid individuals are genetically isolated from wild-type individuals, let's study the mating between a normal individual and a polyploid individual diagrammed in **Figure 23.6**. In the example given here, the normal individual is diploid, or $2n$, and the mutant is tetraploid, or $4n$. Normal individuals produce haploid gametes, while mutant individuals produce diploid gametes. These gametes unite to form a triploid ($3n$) zygote. Even if this offspring develops normally and reaches sexual maturity, it will not be able to form gametes. The sketch at the bottom of Figure 23.6 illustrates why: When meiosis occurs, homologous chromosomes cannot synapse correctly. Because virtually all of the gametes produced by the triploid zygote end up with a dysfunctional set of chromosomes, the individual is virtually sterile. As a result, the tetraploid and diploid individuals cannot mate and produce fertile offspring. The tetraploid and diploid populations are reproductively isolated.

What sort of mutation leads to polyploidy and triggers speciation? Eric Rabe and Christopher Haufler recently documented a "polyploidization mutant" in the maidenhair fern. This plant inhabits woodlands across North America and is illustrated in **Figure 23.7**. This diagram also shows that, during the normal life cycle of a fern, individuals alternate between a haploid (n) stage and a diploid ($2n$) stage. Rabe and Haufler initially set out

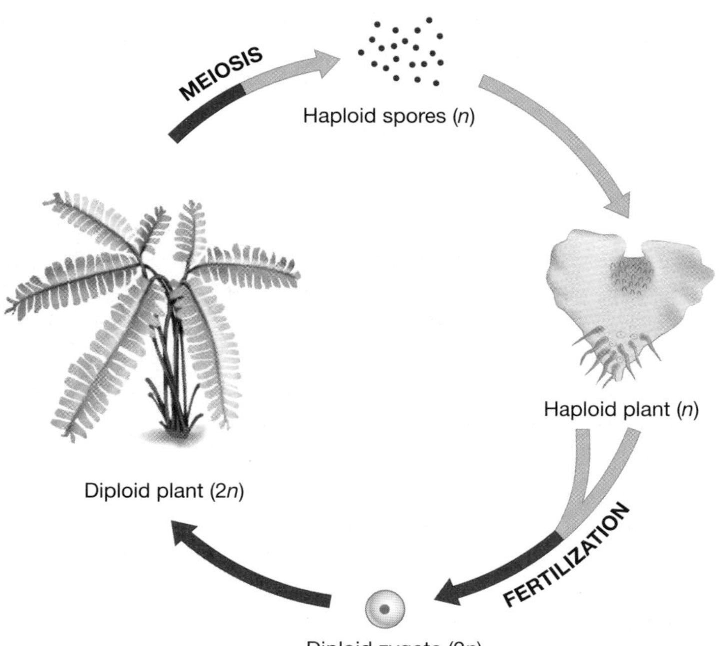

FIGURE 23.7 The Life Cycle of Maidenhair Ferns
Like all ferns, maidenhair ferns have a haploid stage and a diploid stage during their life cycle.

to do a routine survey of allelic diversity in a population of these ferns using the protein electrophoresis technique introduced in Chapter 22. They happened to be examining individuals in the haploid stage and found several individuals that had *two* sets of each allele instead of just one. These individuals were diploid even though they had the "haploid" growth form shown in Figure 23.7. Rabe and Haufler followed these individuals through their life cycle and confirmed that when they mated, they produced offspring that were tetraploid ($4n$). The researchers had stumbled upon a polyploid mutant within a normal population.

To follow up on the observation, they located the parent of the mutant individuals. This plant turned out to have a defect in meiosis. Instead of producing normal, haploid cells as a result of meiosis, the mutant individual produced diploid cells. These diploid cells eventually led to the production of diploid gametes and tetraploid offspring. If the process continued, a polyploid population of maidenhair ferns will be established. The polyploid individuals will be genetically isolated from the parental population and thus evolutionarily independent. If genetic drift and selection cause the two populations to diverge, speciation will be under way.

BOX 23.1 How Do Researchers Estimate Phylogenetic Trees?

Throughout this text evolutionary trees have been interpreted without comment on how they are put together. What are some of the central ideas used in estimating evolutionary relationships?

A phylogenetic tree summarizes the history of a group of species or populations. Like any other pattern or measurement in nature, the genealogical relationships among species cannot be known with absolute certainty. Instead, the relationships have to be estimated from data.

To infer the historical relationships among species, researchers analyze their morphological or genetic characteristics. For example, to reconstruct relationships among fossil species of humans, scientists routinely analyze aspects of tooth, jaw, and skull structure. To reconstruct relationships among contemporary human populations, investigators usually compare base sequences at a particular genetic locus.

What do these data say about who is most closely related to whom? Once the characteristics of different populations have been measured, researchers have several general strategies available for analyzing the data and inferring which species or populations branched off early in evolution and which diverged more recently.

One of these general approaches is based on computing a statistic that summarizes the overall similarity between each population, based on the data. This is called a phenetic approach to estimating trees. For example, researchers might use gene sequence data to compute an overall "genetic distance" between each population. (A genetic distance summarizes the average percentage of bases that differ between two populations.) A computer program then compares the similarities among populations and builds a tree that clusters the most similar populations and places more divergent populations on more distant branches.

A second general strategy for inferring trees is called the cladistic approach. Cladistic methods are based on the realization that relationships among species can be reconstructed by identifying shared derived characters, or synapomorphies (literally, "union-form"), in the species being studied. As an example of how this approach works, consider the evolutionary relationships of the whales and the group of mammals called the Artiodactyla. (Artiodactyls have hooves and an even number of toes. Cows, deer, and hippos are artiodactyls.)

Based on analyses of their overall morphology, whales and dolphins were not thought to be part of the Artiodactyla. But when Norihiro Okada and co-workers analyzed the distribution of the parasitic gene sequences called SINEs, which occasionally insert themselves into the genomes of mammals (see Chapter 16), they found that whales and hippos share several types of SINES that are not found in other groups.

To understand what this observation means, study the data for 20 SINE genes given in **Figure 1**. Note that whales and hippos share the SINEs numbered 4, 5, 6, and 7. Other SINE genes are present in some artiodactyls but not others; camels have no SINE loci at all. To explain these data, the biologists hypothesized that no SINEs were present in the population that is ancestral to all of the species in the study. Over the course of evolution, however, different SINEs became inserted into the genomes of descendant populations. As a result, the presence of a particular SINE represents a derived character. Because whales and hippos share four of these derived characters, they must be closely related.

The phylogeny in Figure 1 groups the species in the study according to the presence of these shared derived characters. According to this phylogeny, whales are artiodactyls and share a relatively recent common ancestor with hippos. This observation inspired the hypothesis that whales and dolphins are descended from a population of artiodactyls that fed in shallow water, much as hippos do today.

In actual practice, biologists routinely use both phenetic and cladistic approaches and compare the results. The technical details involved in reconstructing trees are extremely complex, and the merits of contrasting approaches are hotly debated. The important thing to remember is that every tree represents an estimate of the actual relationships, and that every estimate is only as good as the original data and the care used in data analysis.

(Continued on next page)

By creating polyploids, mutation can cause speciation. Botanists agree that this mode of speciation has probably been common because many diploid species of plants have close relatives that are polyploid. The event breaks two long-held "rules" about speciation: It occurs in sympatry, and it is not slow—speciation by polyploidy is virtually instantaneous.

23.3 Isolation and Divergence in Allopatry

Speciation begins when gene flow between populations is reduced or eliminated. Although this can happen when insects switch host plants or when plants become polyploid, it also happens routinely when populations become physically separated. Physical isolation, in turn, occurs in one of two ways. As **Figure 23.8** (page 458) illustrates, a population can colonize a new habitat, or a new physical barrier can split a widespread population into two or more isolated subgroups. Speciation that begins with physical isolation is known as **allopatric** (different-homeland) **speciation.**

The case studies that follow address two questions: How do colonization and range-splitting events occur? Once populations are physically isolated, how do genetic drift and selection produce divergence?

Dispersal and Colonization Isolate Populations

Peter Grant and Rosemary Grant recently witnessed a colonization event while working in the Galápagos Islands off the coast of South America. They had been studying several species of finch on an island called Daphne Major for years. In 1982 five members of a new species, called the large ground finch, arrived and began nesting. These colonists had apparently dispersed from a population that lived on a nearby island in the Galápagos. Because finches normally stay on the same island year round, the colonists represented a new population, allopatric with their source population.

(Box 23.1 continued)

Locus	1	2	3	4	5	6	7	8	9	10	11	12	13	14	15	16	17	18	19	20
Cow	0	0	0	0	0	0	0	1	1	1	1	1	1	1	1	1	1	1	0	0
Deer	0	0	0	0	0	0	0	1	?	1	1	1	1	1	1	?	1	1	0	0
Whale	1	1	1	1	1	1	1	0	?	1	0	1	1	0	0	0	?	1	0	0
Hippo	0	?	0	1	1	1	1	0	1	1	0	1	1	0	0	0	?	1	0	0
Pig	0	0	0	?	0	0	0	0	?	0	0	0	?	?	0	0	0	1	1	1
Peccary	?	?	?	?	?	?	?	?	?	?	?	?	?	?	?	?	?	?	1	1
Camel	0	0	0	0	0	0	0	0	0	0	0	0	0	0	0	0	0	0	0	0

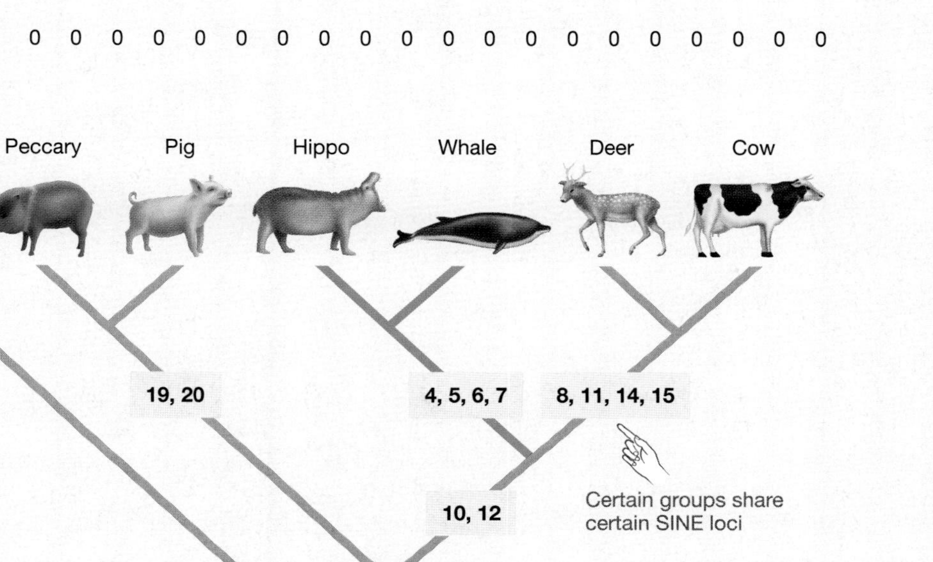

FIGURE 1
The phylogenetic tree shows the relationships implied by the data on the presence and absence of the SINE genes given in the table. Each branch on the tree is defined by a unique combination of shared, derived characters.

The finches' arrival gave the researchers a chance to test a long-standing hypothesis about dispersal and colonization. Decades ago Ernst Mayr suggested that colonization events are likely to trigger speciation, for two reasons: The physical separation between populations reduces or eliminates gene flow, and genetic drift will cause the old and new populations to diverge rapidly because the number of individuals involved in colonization events is usually small. (Recall from Chapter 22 that genetic drift can produce dramatic changes in allele frequencies in small populations.)

To evaluate whether genetic drift occurred when large ground finches colonized Daphne Major, Grant and Grant caught, weighed, and measured most of the parents and offspring produced on Daphne Major over the succeeding 12 years. When they compared these data with measurements from large ground finches in other populations, they discovered that the average bill size in the new population was much larger. As Mayr's hypothesis predicted, the colonists represented a nonrandom sample of the original population. Genetic drift produced a colonizing population with characteristics significantly different from those of the source population.

These results are important because the size and shape of finches' bills closely correlate with the types of seeds they eat. If large seeds are common on Daphne Major, then large ground finches with large beaks will survive and reproduce well, and a large-beaked population will evolve. The general point here is that the characteristics of a colonizing population are likely to be different than the characteristics of the source population. In addition, the novel environment experienced by the colonizers exposes them to novel selection pressures, which should extend the rapid divergence that begins with genetic drift. Colonization, followed by genetic drift and natural selection, is thought to be responsible for speciation in Galapagos finches and many other island groups.

Vicariance Isolates Populations

If a new physical barrier like a mountain range or river splits the geographic range of a species, biologists say that **vicariance** has taken place. Vicariance is a leading explanation for speciation in tropical rainforests, just as colonization is the leading hypothesis for speciation on islands.

In 1969 biologist Jürgen Haffer, inspired by his studies of birds in the Amazon basin of South America, published an influential paper on allopatric speciation. His central observation was that many Amazonian birds have geographic distributions, or ranges, that overlap. As **Figure 23.9a** shows, he proposed

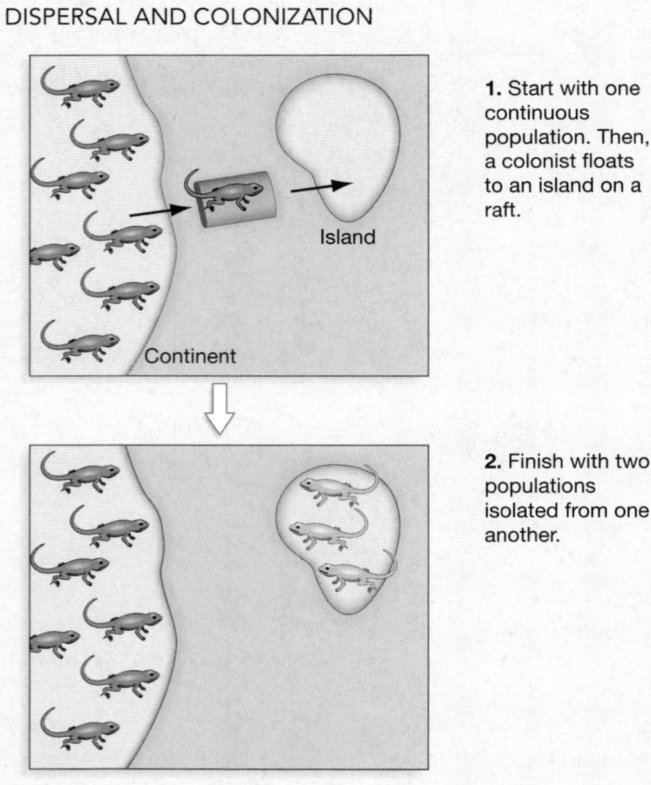

VICARIANCE

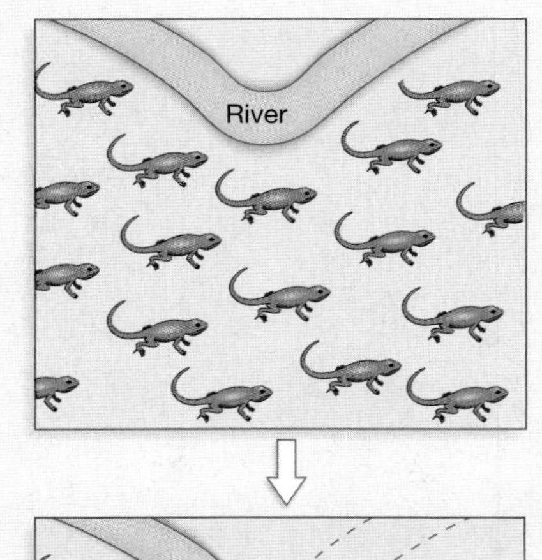

1. Start with one continuous population. Then a chance event occurs that changes the landscape (river changes course).

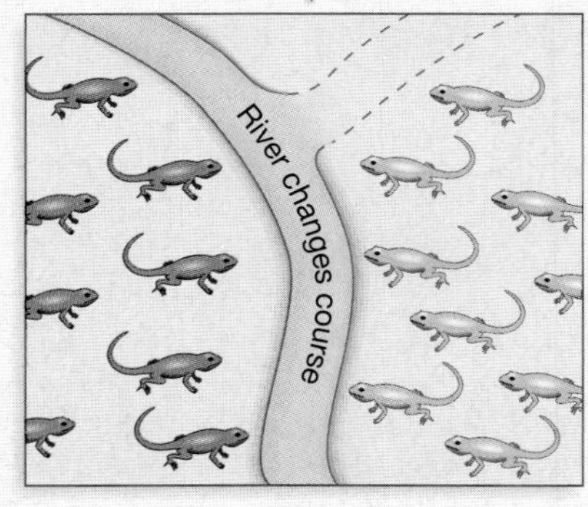

2. Finish with two populations isolated from one another.

FIGURE 23.8 Allopatric Speciation Begins in One of Two Ways
These diagrams illustrate the difference between dispersal and vicariance. When dispersal occurs, colonists establish a population in a novel location. In vicariance, a widespread population becomes fragmented into isolated subgroups.

that these overlapping ranges resulted from vicariance during the ice ages that occurred over the last two million years. Specifically, Haffer suggested that tropical bird species once had large, continuous ranges. But as the global climate cooled during the most recent ice age, these forests shrank into small pockets, or refuges, that were surrounded by large, dry expanses of grasslands and savannas. In this way, climate change split widespread forest-dwelling populations into a series of subgroups isolated in habitat islands. Haffer proposed that these events triggered allopatric speciation. He contends that overlapping ranges exist today because populations are gradually expanding their ranges out and away from the original Pleistocene refuges.

This proposal, called the Pleistocene refugium hypothesis, was recently supported by G. T. Prance's research on the distributions of flowering plants in the Amazon. Based on the data shown in **Figure 23.9b**, Prance claims that flowering plants have many of the same regions of range overlap as birds. To make sense of this result, he contends that forest-dwelling plants were isolated in the same refuges as forest-dwelling birds. Proponents of the refugium hypothesis argue that it is correct because the same pattern occurs in two different groups of organisms.

The Pleistocene refugium hypothesis is controversial, however, because a growing number of data sets indicate that many or most tropical species are older than the hypothesis predicts. Understanding why tropical rainforests are home to so many different species is still an important issue in speciation research.

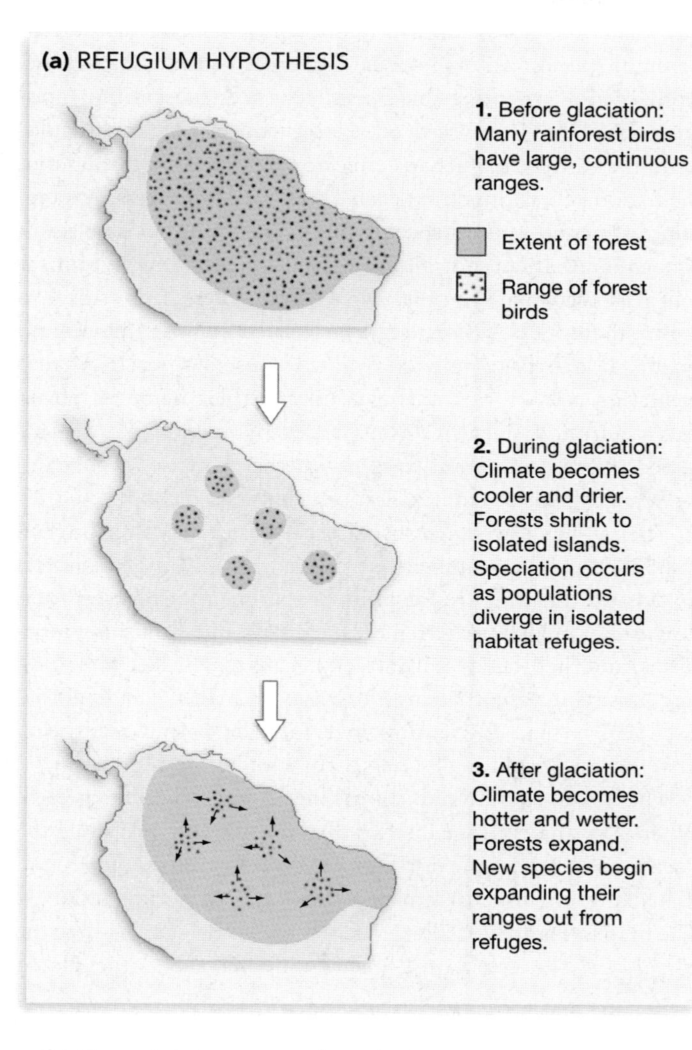

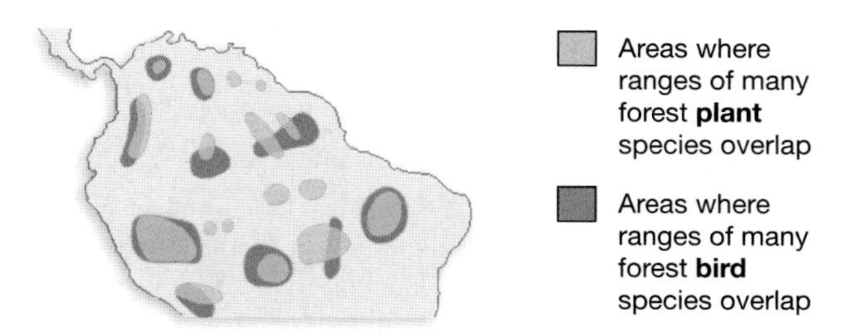

FIGURE 23.9 Vicariance in Tropical Rainforests: Theory and Data **(a)** This flowchart illustrates the Pleistocene refugium hypothesis. The key idea is that fragmentation of forest habitats led to speciation during the recent ice age. **(b)** Based upon how forest bird species are currently distributed, a researcher proposed that forest refuges existed in the red areas during the Pleistocene. Based upon how forest plant species are currently distributed, a different researcher proposed that forest refuges existed in the green areas during the Pleistocene. The extensive coincidence between the two data sets supports the hypothesis that forest refuges existed.

23.4 Secondary Contact

When populations that have been isolated come into contact again, what happens? If divergence has taken place and if divergence has affected when, where, or how individuals in the populations mate, then it is unlikely that interbreeding will take place. In cases like this, prezygotic (before-zygote) isolation exists. **Table 23.2** (page 460) lists some of the ways that prezygotic isolation can occur. When it does, intermating is rare, gene flow is minimal, and populations continue to diverge.

But what if prezygotic isolation does not exist, and the populations begin interbreeding? The simplest outcome is that the populations fuse over time, as gene flow erases any distinctions between them. Several other possibilities exist, however. Let's explore three of them.

Reinforcement

If two populations have diverged extensively and are distinct genetically, it is reasonable to expect that their hybrid offspring will struggle to survive and reproduce. A wide variety of examples support this prediction. In some cases hybrid offspring tend to die early in development; in other cases they survive to sexual maturity but are infertile. In cases like this, biologists say that postzygotic (after-zygote) isolation exists. When it occurs, there should be strong natural selection against interbreeding. The logic here is that hybrid offspring represent a wasted effort on the part of parents. Individuals who do not interbreed, due to a different courtship ritual or pollination system or other form of prezygotic isolation, should be favored because they produce more viable offspring.

Natural selection for traits that isolate populations in this way is called **reinforcement**. The name is descriptive because the selected traits reinforce differences that developed while the populations were isolated from one another.

Some of the best data on reinforcement come from laboratory studies of closely related fruit fly species in the genus *Drosophila*. Jerry Coyne and Allen Orr recently analyzed a large series of experiments that tested whether members of closely related fly species are willing to mate with one another. Coyne and Orr found an interesting pattern. If closely related species are sympatric, individuals from the two species are seldom willing to mate with one another. But if the species are allopatric, individuals are often willing to mate with one another. This is exactly the pattern to be expected if reinforcement were occurring, because natural selection can act to reduce mating between species only if their ranges overlap. There is a long-standing debate, however, over just how important reinforcement is in groups other than *Drosophila*.

Hybrid Zones

Hybrid offspring are not always dysfunctional. Frequently they are perfectly capable of mating and producing offspring, and have features that are intermediate between the two parental populations. When this is the case, hybrid zones can form. A **hybrid zone** is a geographic area where interbreeding occurs and hybrid offspring are common. Depending on the fitness of hybrid offspring and the extent of breeding between parental species, hybrid zones can be narrow or wide, and long- or short-lived. As an example of how researchers analyze the dynamics of hybrid zones, let's consider recent work on two bird species performed by Sievert Rohwer and colleagues.

Townsend's warblers and hermit warblers live in the coniferous forests of North America's Pacific Northwest. In southern Washington state, where their ranges overlap, the two species hybridize extensively. As **Figure 23.10a** shows, hybrid offspring have characteristics that are intermediate relative to the two parental species. To explore the dynamics of this hybrid zone, Rohwer and co-workers examined gene sequences in the mitochondrial DNA (mtDNA) of a large number of Townsend's, hermit, and hybrid warblers collected from forests throughout the region. They found that each of the parental species has certain distinctive, species-specific mtDNA sequences. This result was valuable because it allowed them to infer how hybridization was occurring. To grasp the reasoning here, it is critical to realize that mtDNA is maternally inherited in most animals and plants. If a hybrid individual has Townsend's mtDNA, that means its mother was a Townsend's warbler while its father was a hermit warbler. In this way, identifying mtDNA types allowed the research team to infer whether Townsend's females were mating with hermit males, or vice versa, or both.

Their data showed a clear pattern. Most hybrids form when Townsend's males mate with hermit females. Scott Pearson followed up on this result with experiments showing that Townsend's males are extremely aggressive in establishing territories, and that they readily attack hermit males. Hermit males, in contrast, do not challenge Townsend's males. The hypothesis, then, is that Townsend's males invade hermit territories, drive off the hermit males, and mate with hermit females.

The team also found something completely unexpected. When they analyzed the distribution of mtDNA types along the Pacific coast and in the northern Rocky Mountains, they found that many Townsend's warblers actually had hermit mtDNA. The map in **Figure 23.10b** shows that in some regions—like the islands off the coast of British Columbia called the Haida

TABLE 23.2 Prezygotic Isolation Can Occur Many Different Ways

Type of Prezygotic Isolation	Description	Example
Temporal	Populations are isolated because they breed at different times.	Bishop pines and Monterey pines release their pollen at different times of year.
Habitat	Populations are isolated because they breed in different habitats.	Parasites that begin to exploit new host species are isolated from the original population.
Behavioral	Populations do not interbreed because their courtship displays differ.	To attract males, female fireflies give a species-specific sequence of flashes.
Gametic barrier	Matings fail because eggs and sperm are incompatible.	In sea urchins, a protein called bindin allows sperm to penetrate eggs. Differences in the amino acid sequence of bindin cause matings to fail between closely related populations.
Mechanical	Matings fail because male and female genitalia are incompatible.	In many insects the male copulatory organ and female reproductive canal fit like a "lock and key." Changes in either organ initiate reproductive isolation.

Gwaii—*all* of the warblers had hermit mtDNA, even though they looked like full-blooded Townsend's. To explain this result, Rohwer's team hypothesized that hermit warblers were once found as far north as Alaska, and that Townsend's warblers have gradually taken over their range. Their logic is that repeated mating with Townsend's warblers over time made the hybrid offspring look more and more like Townsend's, even while maternally inherited mtDNA kept the genetic record of the original hybridization event intact.

If the hypothesis proposed by Rohwer's team is correct, the hybrid zone should continue moving south. If it does so, hermit warblers may eventually become extinct. In many cases, however, hybridization does not lead to extinction, but rather leads to the opposite—the creation of new species.

New Species through Hybridization

Recently Loren Rieseberg and colleagues examined the relationships of three sunflower species native to the American West, pictured in **Figure 23.11** (page 462): *Helianthus annuus*, *H. anomalus*, and *H. petiolaris*. The first two of these species are known to hybridize in regions where their ranges overlap. The third species, *H. anomalus*, resembles these hybrids. In fact,

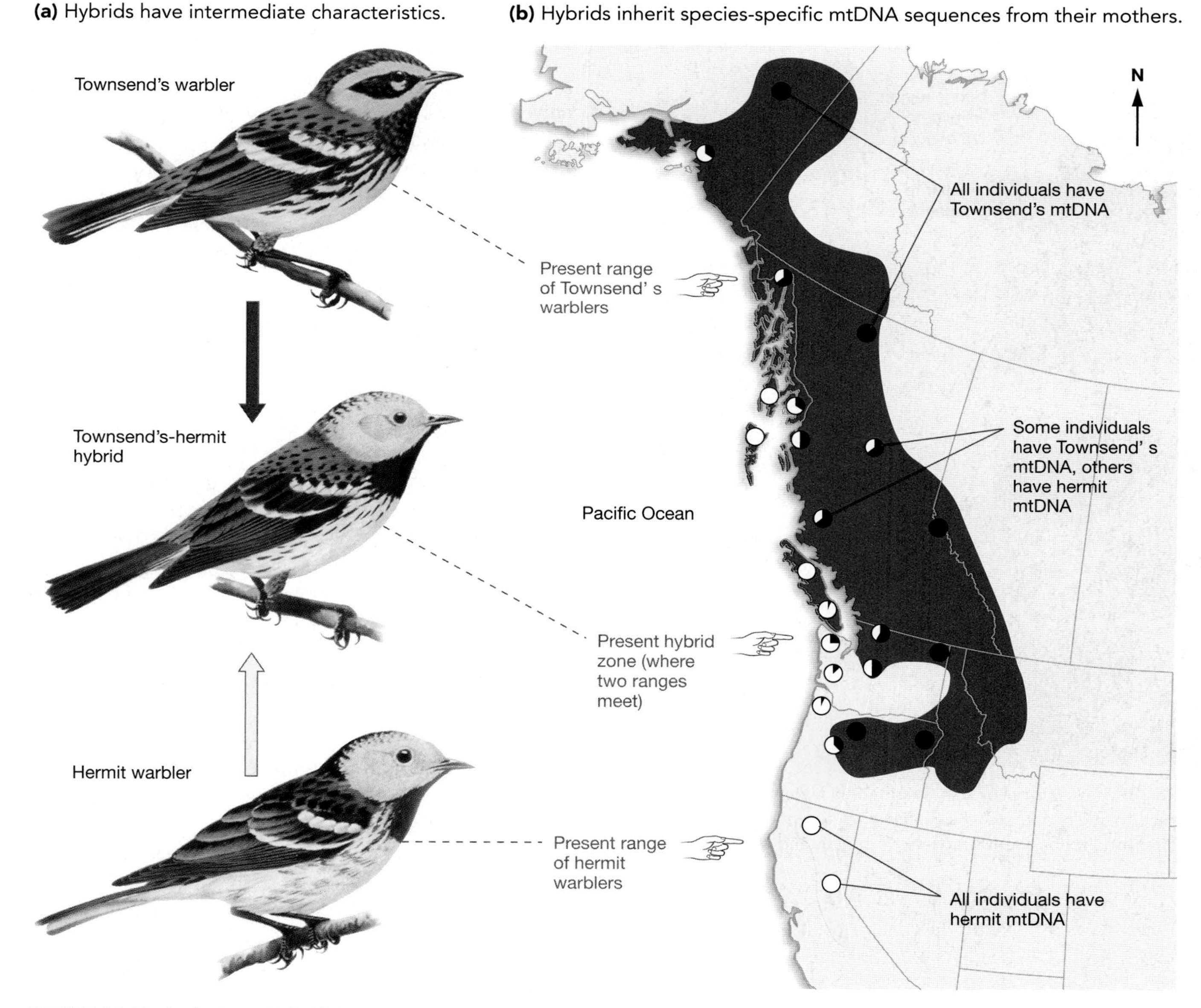

FIGURE 23.10 Analyzing a Hybrid Zone
(a) When Townsend's warblers and hermit warblers hybridize, the offspring have the intermediate characteristics illustrated here. **(b)** This map shows the current range of Townsend's and hermit warblers in the northern Rocky Mountains and northern Pacific coast of North America. Note that hermit warblers are currently found further south along the Pacific coast. The pie charts show the percentage of individuals with Townsend's warbler mtDNA (in black) and hermit warber mtDNA (in white).

because some gene sequences in *H. anomalus* are remarkably similar to those found in *H. annuus* while others are almost identical to those found in *H. petiolaris*, biologists have suggested that *H. anomalus* originated in hybridization between *H. annuus* and *H. petiolaris*.

If this interpretation is correct, then hybridization must be added to the list of ways that new species can form. The specific hypothesis here is that *H. annuus* and *H. petiolaris* were isolated and diverged as separate species, later began interbreeding, and as a result created a third, new species that had unique combinations of alleles from each parental species and therefore different characteristics. The hypothesis is supported by the observation that *H. anomalus* grows in much drier habitats than either of the parental species and is distinct in appearance.

Rieseberg and associates set out to test the hybridization hypothesis by trying to recreate the speciation event experimentally. Specifically, they mated individuals from the parental species and raised their offspring in the greenhouse. When these hybrid individuals were mature, the researchers either mated them to other hybrids or "backcrossed" them to *H. annuus* individuals. This breeding program continued for four more generations before the experiment ended. Ultimately, the experimental lines were backcrossed twice and mated to other hybrid offspring three times.

The experimental hybrids looked like the natural hybrid species, but did they resemble them genetically? To answer this question, the research team constructed genetic maps of each population using a large series of genetic markers, similar to the types of markers introduced in Chapters 16 and 17. Because each parental population had a large number of unique markers in their genomes, the research team hoped to identify which genes found in the experimental hybrids came from which parental species.

Their results are diagrammed in Figure 23.11. The vertical bar on the left represents a region called S in the genome of the naturally occurring species *Helianthus anomalus*. As the legend indicates, this region contains two sections of sequences that are also found in *H. petiolaris* and three that are also found in *H. annuus*. The vertical bar on the right shows the composition of this same region in the genome of the experimental hybrid lines. The genetic composition of the synthesized hybrids matches that of the natural species.

In effect, Rieseberg and co-workers succeeded in recreating a speciation event. Their results resoundingly confirm the hybridization hypothesis for the origin of *H. anomalus* and illustrate the dynamic range of outcomes from secondary contact: fusion of populations, reinforcement of divergence, founding of stable hybrid zones, extinction of one population, or the creation of new species.

Helianthus annuus

H. anomalus

DNA comparison of S region of genome:

H. anomalus (naturally occurring species)

Experimental hybrid (cross of *H. annuus* and *H. petiolarus*)

Gene region similar to *H. annuus*

Gene region similar to *H. petiolarus*

(Not all of this region was tagged)

H. petiolarus

There are two reasons to believe that *H. anomalus* arose from a hybridization event:

1. *H. anomalus* contains gene regions that are similar to both *H. annuus* and *H. petiolarus*.

2. When hybrids were created experimentally, their genetic composition matched that of *H. anomalus*.

FIGURE 23.11 New Species Can Originate in Hybridization Events
According to one hypothesis, the sunflower *Helianthus anomalus* originated in hybridization events between *H. annuus* and *H. petiolarus*.

Essay Human Races

This chapter is built around one fundamental theme: When populations are isolated by lack of gene flow, they can diverge as a result of mutation, genetic drift, and natural selection. Human populations are sometimes called races if they are differentiated by physical characteristics like facial features, the color and texture of hair, skin color, and in some cases height or body build. Presumably, these distinctions arose due to a lack of gene flow in the past. How does the divergence observed among human populations compare to the amount of divergence measured among populations of other organisms? In a biological sense, how profound is the divergence that has occurred among human populations?

. . . how profound is the divergence that has occurred among human populations?

To quantify the amount of genetic variation that exists among populations, researchers routinely collect DNA samples from individuals that represent populations from throughout the range of a species. In humans, for example, biologists usually survey genetic diversity by obtaining DNA from aboriginal Australian, African, Asian, European, Native American, and New Guinean people. Then they analyze the sequence at one or more loci in the nuclear or mitochondrial genomes of these individuals. To summarize the genetic differences that exist, they calculate the "average percentage of sequence divergence." This number represents the average percentage of bases that are different in two randomly chosen individuals of the species.

Because of the intense interest in genetic diversity in our own species, numerous studies of humans have been done utilizing different techniques or focusing on different loci. Because measuring the amount of genetic divergence among populations is fundamental to understanding a wide variety of questions about speciation and evolution, data sets from dozens, if not hundreds, of plant and animal species are available to compare with humans.

Two important conclusions have emerged from these studies:

- Genetic divergence among human populations is extremely low compared to other species of animals. For example, when Rebecca Cann and colleagues examined mitochondrial DNA in humans from all over the world, they observed an average sequence divergence of 0.32 percent. (This means that only one-third of 1 percent of bases are different in different populations, on average.) When John Avise and coworkers did a similar study of deer mice from North America, they observed an average sequence divergence of 3.1 percent. Deer mouse populations have experienced 10 times as much genetic divergence as human populations.
- Populations from sub-Saharan Africa have more genetic diversity among themselves than do non-African populations. For example, Sarah Tishkoff and associates examined the distribution of alleles at the CD4 locus in nuclear DNA, and found that all of the alleles present in humans can be found in African individuals. Populations in other parts of the world have subsets of the array of alleles found in Africa, but no unique alleles. Based on data like these, Svante Pääbo comments that "in a genetic sense, everyone on this planet looks like an African."

What is the overall message of these studies? Compared to other species, genetic differences among human populations are tiny.

Chapter Review

Summary

Speciation is a splitting event in which one lineage gives rise to two or more independent, descendant lineages. It involves at least two and sometimes three events:

- Populations of the same species become isolated when gene flow between them is eliminated or drastically reduced.
- Genetically isolated populations diverge as mutation, drift, and selection act on them independently.
- Recently diverged populations may come back into contact and again begin interbreeding. Depending on the fate of "hybrid" offspring, the populations either fuse into a single species or maintain their distinct identities.

Speciation forms new branches on the tree of life. It is an important research field because speciation events create biodiversity, and because speciation can lead to novel strains of bacteria and viruses that threaten human health and welfare.

Researchers use several different criteria to test whether populations represent distinct species. The biological species concept focuses on the degree of hybridization between species to determine whether gene flow is occurring. The morphospecies concept infers that speciation has occurred if populations have distinctive morphological traits. The phylogenetic species concept defines species as the smallest monophyletic groups on evolutionary trees.

Contrary to traditional expectations, evidence is mounting that speciation can occur even when populations are sympatric. Some soapberry bug populations, for example, have changed their primary food source. Instead of exploiting traditional host plant species native to North America, they now eat seeds from recently introduced exotic species. In populations making this host switch, natural selection has caused rapid changes in beak length. Research is now focused on confirming that the original and descendant populations are distinct species. Similarly, mutations that produce polyploidy can trigger speciation because they lead to reproductive isolation between mutant and normal populations.

Speciation can also begin when small groups of individuals colonize a new habitat or when a large, continuous population becomes fragmented into isolated habitats. Colonization is thought to be a major mode of speciation on islands, while range splitting is thought to be a major mode of speciation in tropical forests.

When populations that have diverged come back into contact, several outcomes are possible. If prezygotic isolation exists, the populations will probably continue to diverge. Alternatively, interbreeding can cause diverged populations to fuse into the same species. Researchers have also documented that secondary contact can lead to reinforcement and complete reproductive isolation, the formation of hybrid zones, or the creation of a new hybrid species.

Questions

Content Review

1. What distinguishes a morphospecies?
   a. It has distinctive characteristics such as size, shape, or coloration.
   b. It represents a distinct twig in a phylogeny of populations.
   c. It is reproductively isolated from other species.
2. When does speciation occur?
   a. only when populations are allopatric
   b. only when populations are sympatric
   c. only when populations hydridize
   d. when populations become isolated by lack of gene flow, then diverge as a result of mutation, drift, and selection
3. What does it mean to say that populations are sympatric?
   a. They are related as ancestors and descendants.
   b. They share a common ancestor.
   c. They occupy the same geographic area.
   d. They occupy different geographic areas.
4. What does it mean to say that populations are allopatric?
   a. They are related as ancestors and descendants.
   b. They share a common ancestor.
   c. They occupy the same geographic area.
   d. They occupy different geographic areas.
5. When does vicariance occur?
   a. when small populations coalesce into one large, continuous population
   b. when a large, continuous population is fragmented into isolated subpopulations
   c. when individuals colonize a novel habitat
   d. when individuals disperse
6. When do hybrid zones form?
   a. during prezygotic isolation
   b. during postzygotic isolation
   c. when reinforcement occurs
   d. when populations with different characteristics meet, interbreed, and produce intermediate offspring

Conceptual Review

1. Because speciation is an historical event, it can be difficult to study. Make an outline listing the sections and subsections in this chapter, and then fill it in by describing the experimental and analytical approaches used in the case studies provided for each topic. In your opinion, which approaches to studying speciation described in this chapter are most powerful? Why?
2. Why is "reinforcement" an appropriate name for the concept, introduced at the start of section 23.4, that natural selection should favor divergence if populations experience postzygotic isolation?
3. Explain how isolation and divergence are occurring in soapberry bugs, even though populations occupy the same geographic area. Of the four evolutionary processes (mutation, gene flow, drift, and selection), which two are most important in causing this event?
4. Why would genetic drift be an especially important evolutionary process causing divergence during colonization events? If colonists occupy a different habitat than the source population, why would natural selection also be an important process causing divergence?
5. Why is it significant that soapberry bugs mate on their host plant? How does this fact affect gene flow among populations that exploit different host plants?

Applying Ideas

1. A large amount of gene flow is now occurring among human populations due to intermarriage among people from different ethnic groups and regions of the world. Is this phenomenon increasing or decreasing racial differences in our species? Explain.
2. Speciation resulting from polyploidy appears to be much more common in plants (especially ferns) than animals. According to one hypothesis, mutations leading to polyploidy are more common in plants because the germ line and soma are not separated, as they are in animals. As a result, polyploidy can arise from an error during mitosis as well as an error in meiosis. Specifically, if nondisjunction occurs at anaphase during mitosis, it can result in a polyploid cell line. If these cells later differentiate into germ-line cells and undergo meiosis, and if the diploid gametes self-fertilize, viable polyploid offspring result. Does this scenario make sense? Comment.
3. Ellen Censky and co-workers recently documented a colonization event. After two hurricanes passed through the Caribbean Sea in the fall of 1995, a large raft made up of fallen trees and other debris drifted to the shore of an island called Anguilla. At least 15 individuals of the green iguana left the raft and went onto the island. Iguanas were not found on Anguilla prior to this event. Based on the wind patterns during the storms, the researchers propose that the 15 iguanas originally lived on the island of Guadalupe.

   Outline a short-term study designed to test the hypothesis that genetic drift produced allele frequency differences in the two populations.

   Outline a long-term study designed to test the hypothesis that natural selection will produce changes in the characteristics of the two populations over time.
4. All over the world, natural habitats are being fragmented into tiny islands as suburbs, ranches, and farms expand. In effect, vicariance is occurring at a scale never before seen in the history of life. Based on the data presented in this chapter, do you predict that speciation will take place in the newly isolated populations? Or do you predict that they will be wiped out before speciation can occur?

CD-ROM and Web Connection

CD Activity 23.1: Speciation by Changes in Ploidy *(animation)*
(Estimated time for completion = 10 min)
Take a look at examples of how a new plant species can form spontaneously through errors in meiosis or mitosis.

CD Activity 23.2: Allopatric Speciation *(animation)*
(Estimated time for completion = 5 min)
How can geographic isolation lead to speciation? Examine allopatric speciation and explore the consequences of reuniting divergent populations.

At your **Companion Website** (http://www.prenhall.com/freeman/biology), you will find self-grading exams and links to the following research tools, online resources, and activities:

Journey into Phylogenetic Systematics
This exhibit walks you through the basic concepts of cladistics.

Colonization
This article from the Smithsonian Institution describes research on the distribution of species on the Hawaiian Islands.

High-Speed Speciation
Examine recent research indicating that isolation and changes in a population's gene pool can occur rapidly under certain circumstances.

Additional Reading

Knowlton, N. 1994. A tale of two seas. *Natural History* (June): 66–68. A commentary on how the rise of the Isthmus of Panama, about 3 million years ago, led to speciation in Atlantic and Pacific populations.

Meffert, L. M. 1999. How speciation experiments relate to conservation biology. *BioScience* 49: 701–711. Discusses how recent work on the genetic diversity of small populations could effect captive breeding programs.

Meyer, A. 1993. Phylogenetic relationships and evolutionary processes in East African cichlid fishes. *Trends in Ecology and Evolution* 8: 279–284. An introduction to the cichlids of Africa's Rift Lakes—the most spectacular speciation story in fish.

Wayne, R. K., and J. L. Gittleman. 1995. The problematic red wolf. *Scientific American* 273: 36–39. An example, analogous to the dusky seaside sparrow, in which difficulties in identifying species have complicated conservation efforts.

The History of Life

24

The skull shown here is from a species called *Homo erectus* (the name means "upright man"). This individual is from a population that is closely related to the ancestors of modern humans.

Botanists study the geometry of trees to understand how they withstand the lateral forces exerted by windstorms and how their arrangement of branches, twigs, and leaves leads to the efficient capture of photons. This chapter focuses on the geometry of the tree of life—the wonderful diversity of organisms that has been produced since life began. How has the tree's shape changed over time? What patterns can be discerned in how the tree has grown, and why do these patterns exist?

Chapters 1 and 2 examined the geologic time scale, radiometric dating, and life's three domains: the Bacteria, Archaea, and Eukarya. These topics introduced the general size and shape of the tree of life. Chapter 23 considered speciation, or how individual twigs on the tree of life form. Now the focus falls between the very broad scope of the early chapters and the very fine level of Chapter 23. The task now is to find common themes and patterns in the history of life.

The goal of the chapter's first two sections is to introduce the data and analytical tools that researchers use to reconstruct *what* happened in the history of life. Then section 24.3 focuses on *how* key changes documented in the fossil record occurred—specifically, how gene duplication events, changes in gene expression, and natural selection cause evolutionary innovations. A prominent pattern in the tree of life, called adaptive radiation, is the subject of section 24.4. The chapter closes with a look at what may be the most influential of all events in the history of life—the ecological disasters known as mass extinctions.

24.1 Tools for Studying History

Biologists have two major tools to study the past—phylogenies and the fossil record. Because

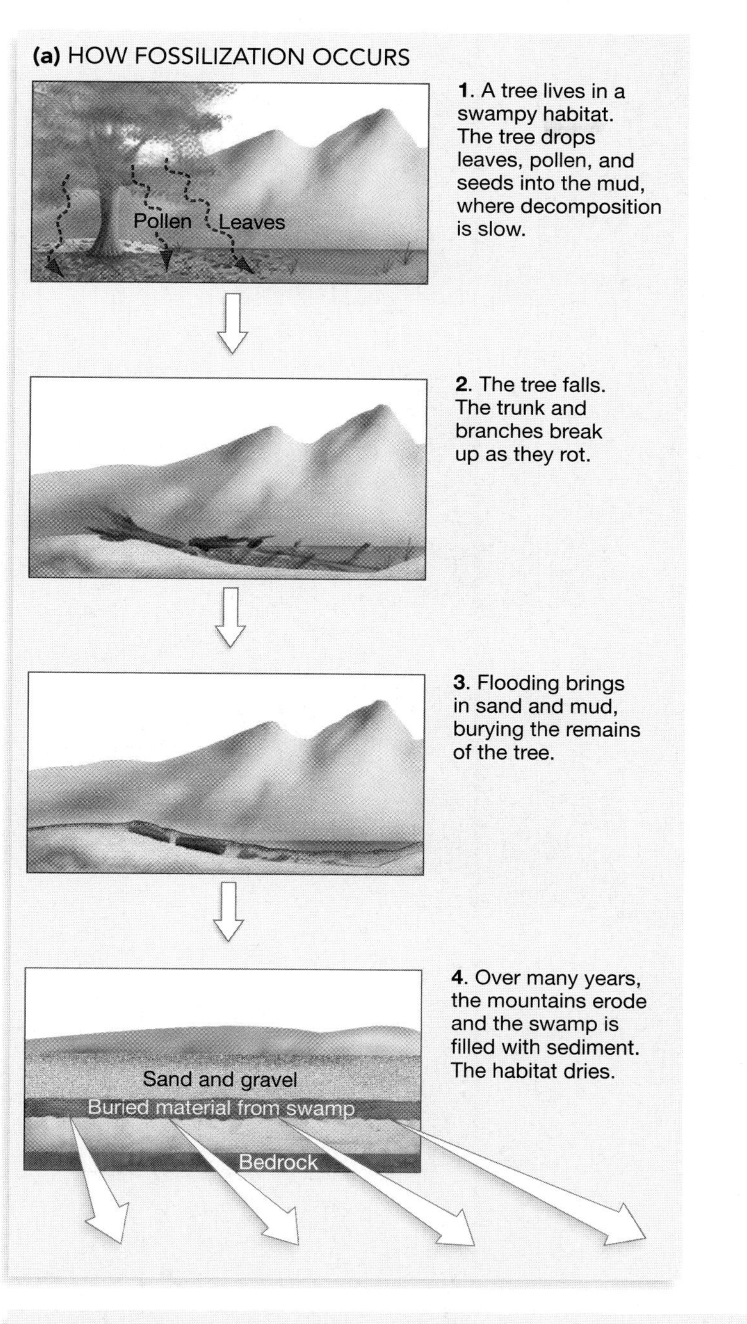

FIGURE 24.1 The Fossilization Process
Fossilization occurs most readily when the remains of an organism are buried in sediments, where decay is slow.

analyzing phylogenies has been a prominent theme throughout the text, let's concentrate here on how the fossil record is collected, interpreted, and used to answer questions.

The Nature of the Fossil Record

Most of the processes that form fossils begin when part or all of an organism is buried in ash, sand, mud, or some other type of sediment. Consider a series of events that begins when a tree falls into a swamp. **Figure 24.1a** illustrates the leaves of a tree falling onto a patch of mud, where they are buried by silt and other organic debris before they decay. Pollen and seeds settle into the muck at the bottom of the swamp, where decomposition is slow. The stagnant water is too acidic and too oxygen-poor to support large populations of bacteria and fungi, so much of this material is buried intact before it decomposes. The trunk and branches that sit above the water line rot fairly quickly, but as pieces break off they, too, sink to the bottom and are buried.

Once burial occurs, several things can happen. If decomposition does not occur, the organic remains can be preserved intact—like the fossil pollen shown in **Figure 24.1b**. Alternatively, if sediments accumulate on top of the material and become cemented into rocks like mudstone or shale, they can compress the organic material below into a thin, carbonaceous film. This happened to the leaf in Figure 24.1b. If the remains decompose *after* they are buried, the hole that remains can fill with dissolved minerals and faithfully create a cast of the remains—like the branch in Figure 24.1b. If the remains rot extremely slowly, dissolved minerals can gradually infiltrate the interior of the cells and then harden into stone, forming a fossil like petrified wood (Figure 24.1b).

(b) FOUR TYPES OF FOSSILS

Intact
The pollen was preserved intact because no decomposition occurred.

Compression
Sediments accumulated on top of the leaf and compressed it into a thin carbon-rich film.

Cast
The branch decomposed after it was buried. This left a hole that filled with dissolved minerals, faithfully creating a cast of the original.

Permineralized
The wood decayed very slowly, allowing dissolved minerals to gradually infiltrate the cells and then harden into stone.

FIGURE 24.2 Preparing a Fossil

EXERCISE Label the head, wings, neck, legs, toes, tail, and ribs on this *Archaeopteryx* specimen. Outline the extent of its feathers.

PREPARING A FOSSIL

1. Excavate fossil-bearing rock.

2. Split the rock to reveal the fossil.

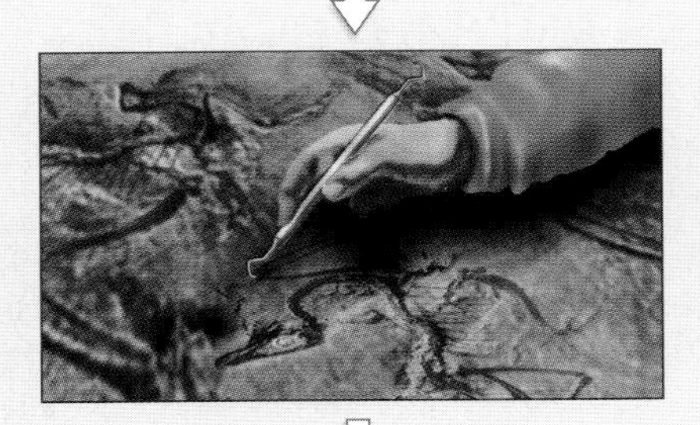

3. Clean away excess rock to expose as much of the fossil as possible.

4. The result.

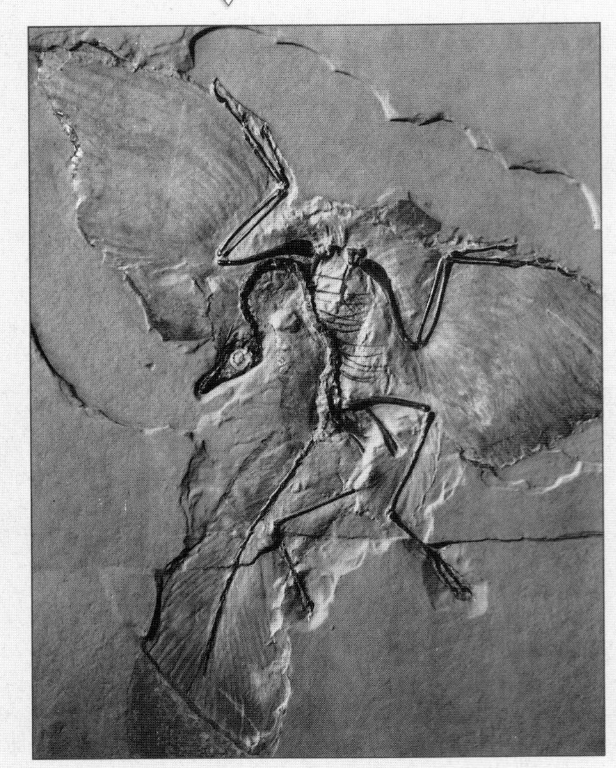

Archaeopteryx—the first bird in the fossil record

After many centuries have passed, fossils can be exposed at the surface by many mechanisms, including erosion, a road cut, or quarrying. If researchers find a fossil, they can prepare it for study by painstakingly clearing away the surrounding rock (**Figure 24.2**). If the species represented is new, researchers describe its morphology in a scientific publication, name the species, estimate the fossil's age based on dates assigned to nearby rock layers, and add the specimen to a collection so that it is available for study by other researchers. It is now part of the fossil record. This is the information database that supports the research reviewed in this chapter.

The scenario just presented is based on conditions that are ideal for fossilization: The tree fell into an environment where decomposition was slow and burial was rapid. In most habitats the opposite situation occurs; decomposition is rapid and burial is slow. In reality, then, fossilization is an extremely low probability event. To appreciate this point, consider that there are seven specimens of the first bird to appear in the fossil record, which is *Archaeopteryx*. All were found at the same site in Germany where limestone is quarried for printmaking (the bird's specific name is *lithographica*). If you accept an estimate that crow-sized birds native to wetland habitats in northern Europe would have a population size of around 10,000 and a life span of 10 years, and if you accept the current estimate that the species existed for about 2 million years, then you can calculate that about 2 billion *Archaeopteryx* lived. But as far as researchers currently know, only 1 out of every 286,000,000 individuals fossilized. For this species, the odds of becoming a fossil were almost 40 times worse than your odds are of winning the grand prize in a state lottery.

Limitations of the Fossil Record

Before looking at how the fossil record is used to answer questions about the history of life, it is essential to review the nature of the archive and recognize several features:

- Because burial in sediments is so crucial to fossilization, there is a strong habitat bias in the database. Organisms that live in areas where sediments are actively being deposited, like beaches, mudflats, and swamps, are much more likely to fossilize than others. Within these habitats, burrowing organisms like clams are already underground—pre-buried—at death and are therefore much more likely to fossilize. Organisms that live aboveground in dry forests, grasslands, and deserts are much less likely to fossilize.
- Slow decay is almost always essential to fossilization, so organisms with hard parts like bones or shells are much more likely to leave fossil evidence. This introduces a strong taxonomic bias in the record. Hard-shelled organisms such as

clams and snails have a higher tendency to be preserved than worms. A similar bias exists for tissues within organisms; because pollen grains are encased in a tough outer coat that resists decay, they fossilize much more readily than do flowers. Teeth are the most common mammalian fossil simply because they are so hard and decay resistant. Shark teeth are abundant in the fossil record but shark bones, which are made of cartilage, are extremely rare.

- The record has a strong temporal bias. Recent fossils are much more common than ancient fossils. **Figure 24.3** shows why. When two of Earth's tectonic plates converge, the edge of one plate may sink beneath the other plate. The rocks composing the plate edge that descends are either melted or radically altered by the increased heat and pressure they encounter as they move downward into the Earth. These alterations obliterate any fossils found in the rock. On continents, fossil-bearing rocks are constantly being broken apart and destroyed by wind and water erosion. The older a fossil is, the more likely it is to be demolished.
- Because fossilization is so improbable, the fossil record is weighted toward common species. Organisms that are abundant, widespread, and present on Earth for long periods of time leave evidence much more often than species that are rare, local, or ephemeral.

In sum, the fossil database represents a highly nonrandom sample of the past. Paleontologists recognize that they are limited to asking questions about tiny and scattered segments on the tree of life. Yet, as the data in this chapter show, the record is a scientific treasure trove. (**Box 24.1**, pages 470–471, introduces how molecular analyses can sometimes supplement the fossil record.) Analyzing fossils is the only way scientists have of examining the physical appearance of long-extinct forms and inferring how they lived. The fossil record is biology's book of the dead.

24.2 The Cambrian Explosion

To get a better feel for how biologists use the fossil record to understand the history of life, let's examine recent research on a momentous event—the origin and early diversification of animals. The first animals appear in the fossil record about 563 million years ago. Their early diversification occurred soon after that, at the start of the Cambrian period. This diversification happened so rapidly that it earned a nickname—the Cambrian explosion.

It is impossible to appreciate the drama of this event without understanding the incredible sweep of geologic time. As **Figure 24.4** (page 471) shows, almost all life-forms were unicellular for about 2.5 *billion* years after the origin of life. The exceptions were several lineages of multicellular algae, which show up in the fossil record about 1000 million years ago. At this time, and for tens of millions of years afterward, the organisms that would become animals were still unicellular, marine creatures with cilia or flagella—probably not too different from some protists living today. Then, about 565 million years ago, the first animals—sponges, jellyfish, and perhaps simple worms—appear in the fossil record. Just 40 million years later, virtually every major group of animals is represented. In a relatively short time, creatures with shells, exoskeletons, legs, heads, tails, and segmented bodies had evolved.

Section 24.3 explores how the Cambrian explosion occurred. Here the focus is on what happened. By combining evidence from the fossil record and phylogeny reconstruction, researchers have begun to clarify the timing and sequence of key events.

Pre-Cambrian and Cambrian Fossils: An Overview

The Cambrian explosion is documented by three major fossil assemblages that record the state of animal life at 570 million years ago (mya), at 565 to 544 mya, and at 525 to 515 mya. The species collected from each of these intervals are referred to respectively as the Doushantuo fossils (from the Doushantuo

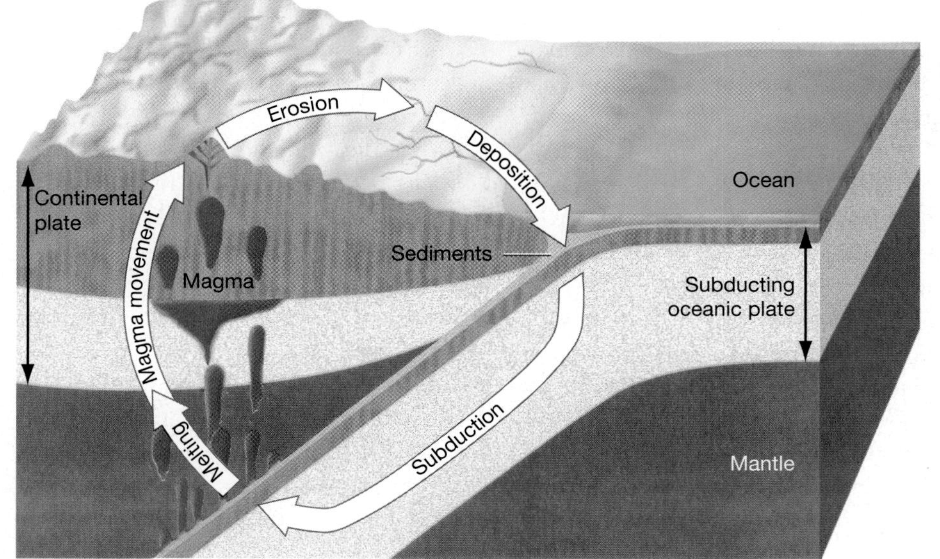

FIGURE 24.3 Why Are Very Ancient Fossils Particularly Rare?
When fossils on continents are exposed at the surface, they can be eroded into particles that are deposited as sediments. In the oceans, fossils are frequently subducted below continental crust, where they are melted or radically altered by heat and pressure. Because rocks and minerals constantly cycle in this way, fossils are continually being lost.

formation in China), Ediacaran fossils (Ediacara Hills, Australia), and Burgess Shale faunas (British Columbia, Canada).

Fortunately for biologists, the glimpses of life provided by the three faunas are extraordinarily clear. Each of the assemblages breaks a cardinal rule of fossil preservation. Soft-bodied animals, which usually do not fossilize efficiently, are well represented in all three. By sheer luck, a few habitats happened to exist at each of these times in which burial occurred so rapidly and decomposition occurred so slowly that organisms without shells were able to fossilize.

To appreciate why this extraordinarily efficient preservation is so important, consider the Burgess Shale. In several localities around the world dated to 525–515 million years ago, fossilization processes were typical—only shelled organisms are found. But soft-bodied organisms also fossilized in the atypical conditions of the Burgess Shale. At this locality, over *five times* as many species are represented. The database is 500 percent better than usual.

The presence of these exceptionally rich deposits before, during, and after the Cambrian explosion makes the fossil record for the event extraordinarily complete. Let's take a detailed look at the animals found in the Doushantuo, Ediacaran, and Burgess Shale faunas.

The Doushantuo Microfossils

Two papers, published within days of each other in 1998, introduced the world to the faunas preserved in the Doushantuo formation of southern China. These phosphate rocks, mined extensively for fertilizer, contain abundant submillimeter-sized fossils. Through careful preparation and microscopy Chia-Wei

BOX 24.1 The Molecular Clock

Allan Wilson, Motoo Kimura, and other researchers have proposed that the fossil record can be supplemented with information on the amino acid and DNA sequences found in living species. This hypothesis, called the molecular clock, holds that certain types of mutations increase to fixation in populations at a steady rate. As an example of this phenomenon, consider data that Kimura compiled on changes in the molecule called hemoglobin. He compared the amino acid sequence of this protein in vertebrates ranging from sharks to humans. When he plotted the number of amino acid differences between species against the date that the species diverged, according to the fossil record, he produced the graph shown in **Figure 1**. The evolution of hemoglobin seems to proceed at a constant rate over time.

To see how researchers use molecular clocks to supplement the fossil record, let's consider some recent analyses of human evolution. The seven fossil species of *Homo* that have been identified to date, and their relationships, are shown in **Figure 2**. Note that the lineage leading to *Homo erectus* and *H. ergaster* and the lineage leading to *H. sapiens* split almost 2 million years ago. (The relationship between *H. erectus* and *H. ergaster* is not known. Two other recent populations of hominids, called *H. neanderthalensis* and *H. heidelbergensis*, evolved after this split occurred. But it is not yet clear whether these two species represent branches off the *H. erectus* lineage or off the lineage leading to *H. sapiens*. Until more and better quality fossils can be found and interpreted, researchers are left with a question: When did the most recent common ancestor of our species live? The fossil record indicates that this population occurred somewhere between the last fossil *H. erectus*, 1.5 million years ago, and the first fossil *H. sapiens*, 100,000 years ago.

Satoshi Horai and colleagues recently used a molecular clock to answer this question. They compared the complete mitochondrial genomes sequenced from African, European, and Japanese individuals. These data allowed them to estimate the number of bases in mtDNA that had changed as human populations diverged.

To estimate how long it took these genetic differences to develop, the researchers also sequenced the mtDNA of our close relatives, the orangutan and African apes. According to the fossil record, orangutans and African apes diverged 13 million years ago. By dividing the number of bases that differ in the mtDNA of orangutans and African apes by 13 million years, the researchers arrived at an estimate that mtDNA changes at the rate of 7×10^{-8} per base per year.

When Horai and colleagues multiplied this rate by the number of bases in mtDNA that had changed as human populations diverged, they concluded that the last common ancestor of all living humans must have lived sometime between 125,000 and 161,000 years ago.

This date roughly agrees with data from a molecular clock identified in the nuclear genome of humans, which estimated the origin of humans at between 75,000 and 287,000 years ago. Because these two independent clocks are in rough agreement, given the uncertainties in the estimates, it appears that molecular clocks can help close the gap in the fossil record. Our species is almost certainly less than 250,000 years old. Large species of mammals typically last in the fossil record for about 1.5 million years prior to extinction. Humans, then, are a relatively young species.

What did hominid populations look like just before this ancestral population split off and the evolution of *Homo sapiens* began? The molecules cannot answer this question. Only the discovery of new fossils can. Even as you read this, it is virtually certain that researchers are at the fossil localities in Africa and Asia that have yielded evidence of our ancestors, looking for traces of the hominids who lived 250,000 years ago.

(Continued on next page)

Li and colleagues were able to identify several dozen individual sponges, ranging from 150 μm to 750 μm across, in samples dated at approximately 580 million years ago. Several different cell types are present in these fossils, and some cell bodies contain spicules made of silica-containing compounds. Both of these traits are typical of sponges living today.

In slightly younger deposits found in the formation, dated to about 570 million years ago, Shuhai Xiao and colleagues found clusters of cells that they interpreted as animal embryos. This conclusion was based on a simple observation: Their samples contained one-cell, two-cell, four-cell, and eight-cell fossils, along with individuals containing larger cell numbers whose

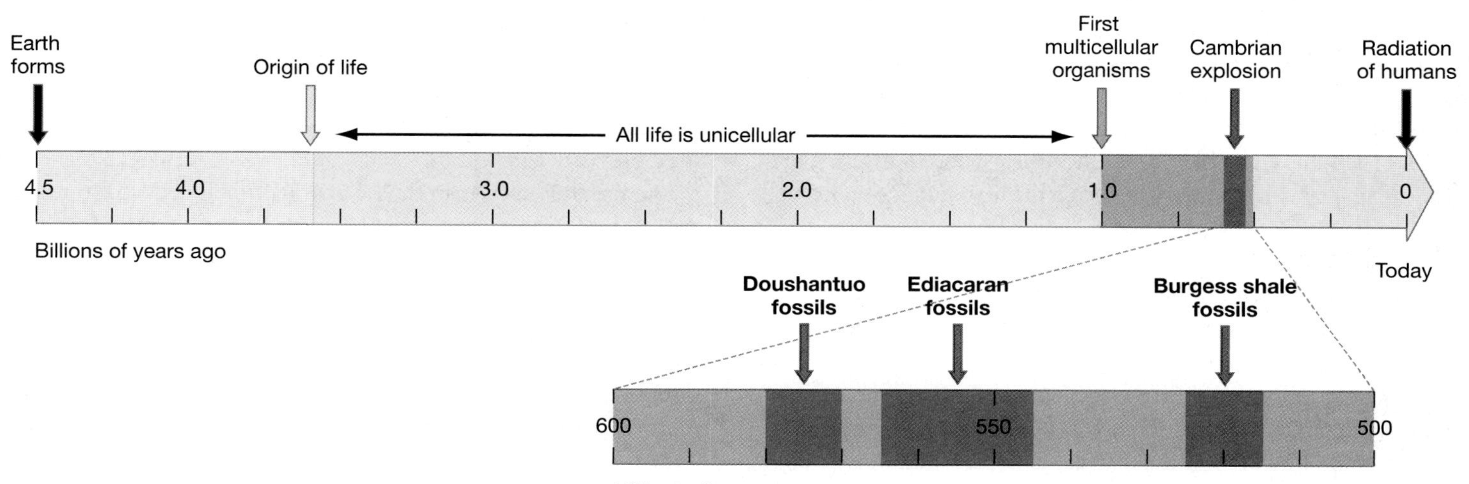

FIGURE 24.4 Life's Timeline
The dates given here were estimated from radiometric dating techniques introduced in Chapter 2.

(Box 24.1 continued)

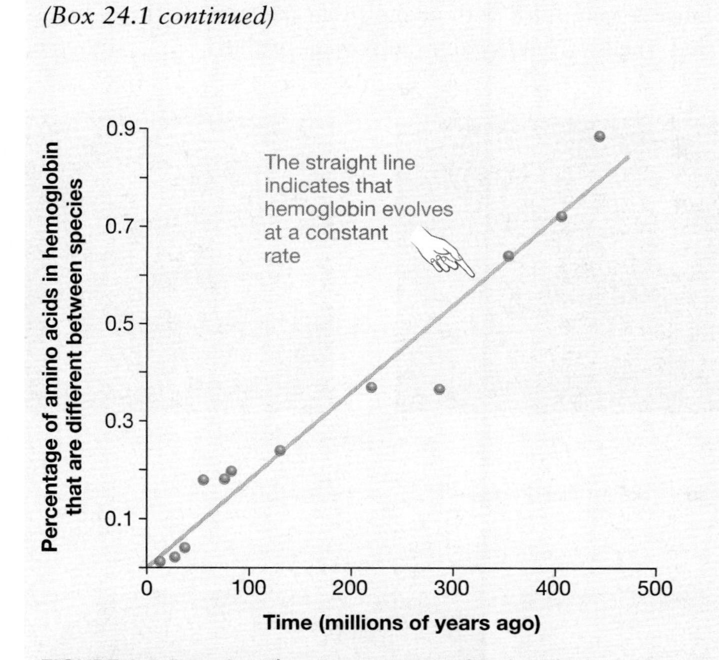

FIGURE 1 A Data Set That Supports a Molecular Clock
Each dot on this graph represents a comparison between a pair of species. The point at the upper right, for example, represents a comparison of hemoglobin proteins in sharks and mammals. According to the fossil record, sharks branched off from other animals about 450 million years ago. QUESTION How fast does this clock tick? That is, what percentage of amino acids in hemoglobin change per 100 million years, on average?

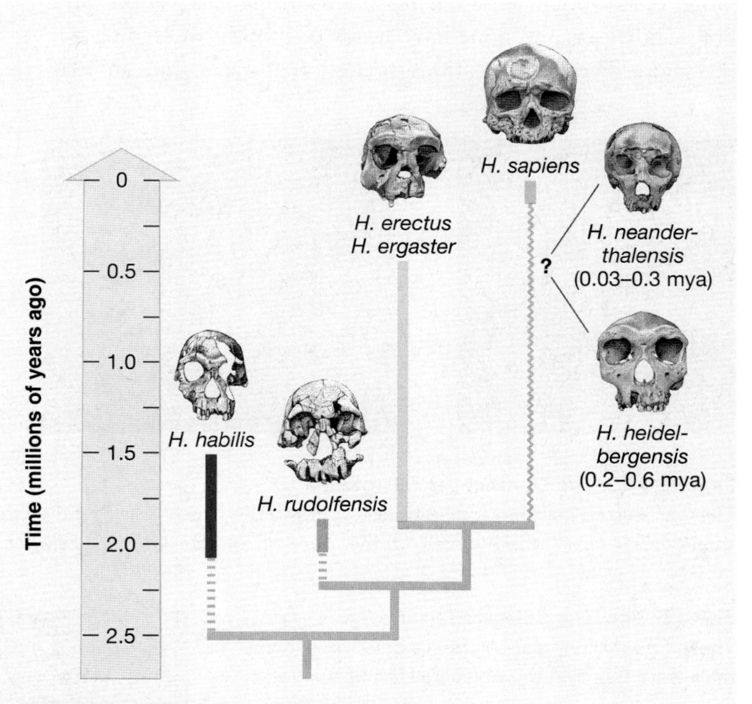

FIGURE 2 A Family Tree of Recent Humans
The colored branches on this phylogeny show the range of dates in which fossils of recent hominid species have been found. The dotted lines indicate dates for which researchers expect to find fossils of each species. The line leading to *Homo sapiens* is jagged because it is still not clear where *H. neanderthalensis* and *H. heidelbergensis* fit along this branch.

overall size was the same (**Figure 24.5**). Recall from Chapter 18 that this is exactly the pattern that occurs during cleavage in today's animals. In other words, cell number increases but total volume remains constant. What type of animal did these embryos develop into? The answer is still unknown.

The animals isolated from both samples were scattered among cyanobacteria as well as multicellular algae that may represent early members of the red, brown, and green algal lineages. These bacteria and "seaweeds" were undoubtedly photosynthetic. The composite picture, then, is of a shallow-water marine habitat dominated by photosynthetic organisms. Scattered among them were tiny animals who may have made their living by filtering organic debris from the water. They were the first animals on Earth.

The Ediacaran Faunas

In the 1940s paleontologists discovered a variety of animal fossils in the Ediacara Hills of southern Australia. The specimens included the compressed bodies of large sponges, jellyfish, and comb jellies, and many burrows, tracks, and other traces from unidentified animal species (**Figure 24.6**). No animals with shells were present, however. In the decades since the initial discovery, similar faunas dated between 565 and 544 million years ago have been found at sites around the world. Taken together, the fossils from this 20-million-year interval indicate that shallow-water marine habitats contained a diversity of animal species. None of the organisms that fossilized during this period have limbs, however, and none have heads or mouths or feeding appendages. These observations suggest that Ediacaran animals burrowed in sediments, sat immobile on the sea floor, or floated in the water, and that they did not actively hunt and capture food but simply filtered organic material from their surroundings.

The Burgess Shale Faunas

The discovery of fossils in the Burgess Shale of British Columbia, Canada, early in this century ranks as the most sensational addition ever made to the fossil record. Combined with the later unearthing of an extraordinary fossil assemblage in the Chengjiang deposits of China, the Burgess Shale gives researchers a compelling picture of life in the oceans 525–515 million years ago.

Few, if any, species in the Ediacaran faunas are also found in the Burgess Shale–type assemblages 20–40 million years later. New species of sponges, jellyfish, and comb jellies are abundant; but entirely new groups are present as well. Principal among these are the arthropods and molluscs. Today, the arthropods include the spiders, insects, and crustaceans (crabs, shrimp, and lobsters); molluscs include the clams and mussels. But echinoderms (sea stars and sea urchins), several types of worm and worm-like creatures, and even a chordate—the group ancestral to today's vertebrates—are found in these fossil faunas. In short, virtually every major animal lineage is documented.

This tremendous increase in the size and morphological complexity of animals, illustrated in **Figure 24.7**, was accompanied by diversification in how they made a living. The Cambrian seas were filled with animals that had eyes, mouths, limbs, and shells. They swam, burrowed, walked, ran, slithered,

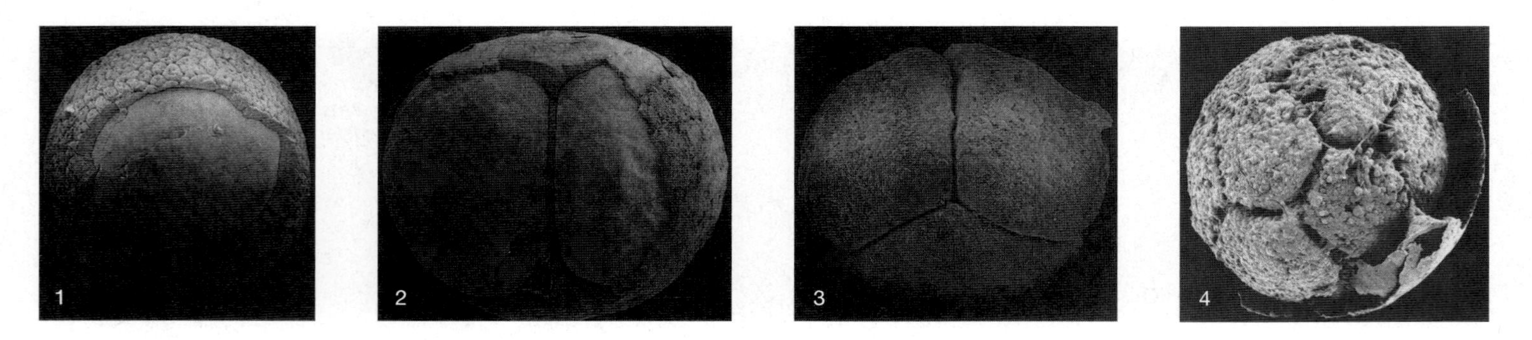

FIGURE 24.5 The Doushantuo Microfauna
These fossilized cells are arranged according to the hypothesis that they represent (1) a fertilized egg and then embryos at (2) the two-cell, (3) the four-cell, and (4) the 16-cell stages.

FIGURE 24.6 The Ediacaran Fauna
These 560 million-year-old fossils of worm-like animals were found in the Ediacara Hills of Australia.

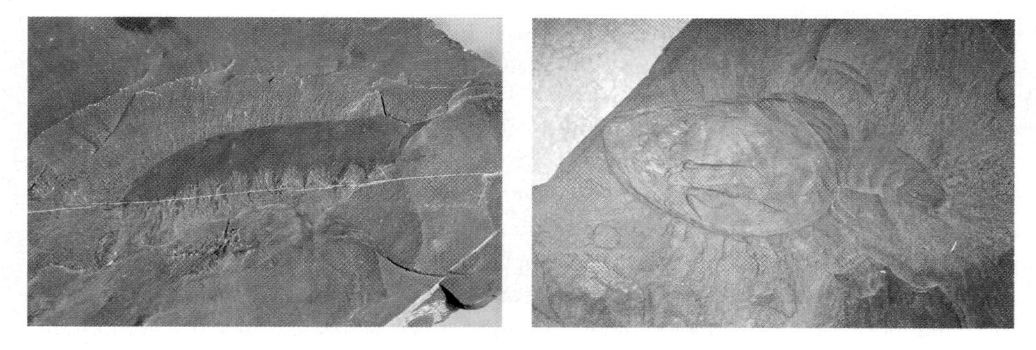

FIGURE 24.7 The Burgess Shale Fauna Similar to this shrimp-like arthropod (left) and crab (right), many of the animals fossilized in the Burgess Shale had heads, tails, shells or exoskeletons, and appendages.

clung, or floated; there were predators, scavengers, filter feeders, and grazers. The diversification of the animals created and filled many of the ecological niches found in shallow-water marine habitats today. The explosion still echoes. Animals that fill today's teeming tide pools, beaches, and mudflats trace their ancestry to species preserved in the Burgess Shale.

24.3 The Genetic Mechanisms of Change

The Doushantuo, Ediacaran, and Burgess Shale faunas document rapid and fundamental changes in the size and complexity of multicellular animals. Now the focus shifts to a new question: How did the major changes chronicled in the fossil record occur?

A remarkable coalition of scientists is assembling to answer this question. Paleontologists, comparative anatomists, developmental biologists, and molecular geneticists are all contributing data aimed at clarifying the genetic basis for novel structures like heads, tails, and limbs. This research program is often called "evo-devo" because it combines evolutionary and developmental studies. To explore this field let's consider two important types of mutation that can make major innovations possible.

Gene Duplications and the Cambrian Explosion

Chapter 19 emphasized that the proteins coded by homeotic loci are responsible for laying out the three-dimensional pattern of multicellular organisms as they develop. As soon as developmental biologists discovered how important these genes are in specifying the tissues and structures found in animals, they began asking about their role in evolution. Did the origin and elaboration of these genes trigger the origin and elaboration of animal body shapes and appendages that occurred during the Cambrian explosion?

To answer this question, biologists are working to determine the number and identity of homeotic loci found in different animal groups and compare them to the phylogeny of the same groups. The idea is to look for correlations between the evolutionary history of animal groups, their morphology, and their genetic make-up. Many researchers predicted there would be a strong association between the order in which animal groups appeared during evolution, the number of homeotic genes present in each group, and morphological complexity and body size. Some biologists even suggested that each major animal lineage would have unique homeotic loci associated with their unique body plans and appendages.

The logic behind this "new genes, new bodies" hypothesis was that gene duplication events could have occurred before and during the Cambrian explosion and produced new copies of existing homeotic genes. (Look back at Chapter 16 to review how gene duplication events occur.) These new genes would make possible the new body plans and appendages recorded in the Burgess Shale fauna. Again, the idea was that the number of homeotic loci present would correlate directly with morphological complexity.

To see if the predictions made by this hypothesis hold up, use **Figure 24.8** (page 474) to analyze the homeotic genes called the *Hox* loci. To help interpret the data, recall from Chapter 20 that *Hox* genes are always found in clusters, with loci lined up on the chromosome one after the other. Recall, too, that each gene in the *Hox* cluster has a distinct function as an embryo develops. Specifically, each *Hox* gene is involved in a different aspect of pattern formation—the events that organize cells in the space inside an embryo. Finally, note that each *Hox* locus in the figure is color-coded to indicate homology with other loci. If genes have the same color, it means their DNA sequences are so similar that researchers are confident the loci are related by common descent.

How do the genes found in different lineages map onto the animal tree? The figure makes the following points:

- As predicted by the new genes, new bodies hypothesis, the number and identity of *Hox* loci varies widely among animals.
- Groups that branched off early and have small, simple bodies—like the sponges—have fewer *Hox* genes than groups that branched off later—like the vertebrates. The *Hox* cluster appears to have expanded during the course of evolution. This observation also supports the new genes, new bodies hypothesis.
- It is sensible to argue that the new *Hox* loci were created by gene duplication events because genes within the cluster are similar in their structure and base sequence. For example, the locus colored orange in the figure, which appears first in flatworms, probably originated when a mutation resulted in a duplication of the locus colored in gold, which is present in sponges. The important point here is that biologists can use the phylogeny to understand the order in which different loci within the cluster appeared.

- Vertebrates, represented in the figure by mouse and zebra fish, have several copies of the entire *Hox* cluster. This observation implies that the complete set of 13 genes was duplicated several times. Because all vertebrates examined to date have the duplicated clusters, the mutations probably occurred close to the origin of the lineage. (Note from the figure that in some cases, copies of genes within these duplicated clusters have been lost. Mice, for example, have no pink-colored locus in their second set of *Hox* genes.) Because vertebrates include some of the largest and most complex animals, this observation also supports the new genes, new bodies hypothesis.

The data would certainly appear to support a simple-minded version of the hypothesis, except for one crucial observation: Within arthropods (represented by the fruit fly, centipede, and crustacean) and within vertebrates, there is no correspondence at all between the number of *Hox* loci and the complexity of the resulting organisms. For example, fish evolved much earlier than mammals, and mammal bodies are considered more complex than fish; but zebra fish have more *Hox* loci than mice do.

Clearly, the situation is more involved than initially predicted. Duplication of *Hox* loci has undoubtedly been important in making the elaboration of animal body plans possible. But new genes are not the whole story. Changes in the expres-

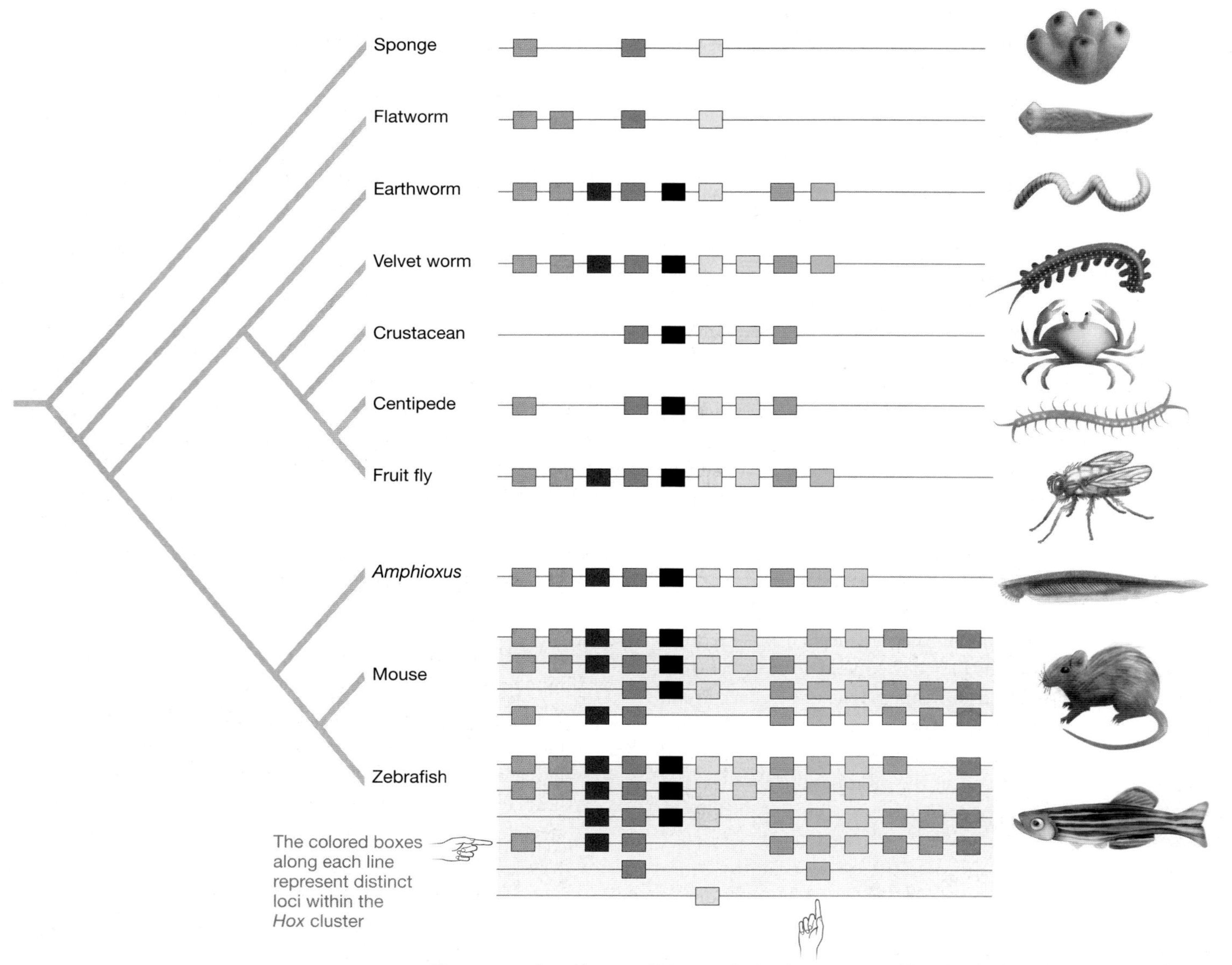

FIGURE 24.8 *Hox* Loci in Animals
In this diagram, *Hox* clusters are represented by a horizontal line (note that mice have four distinct *Hox* clusters and zebra fish have six). **EXERCISE** Next to the illustration of each animal on this evolutionary tree, write the number of *Hox* genes it has.

sion and function of existing genes have been equally or even more important.

Changes in Gene Expression: The Origin of the Foot

Recent research on the vertebrate foot provides a compelling example of how changes in gene expression can affect evolution. One of the major innovations during vertebrate evolution was the origin of a limb with feet. This led to the evolution of amphibians, reptiles, and mammals—the **tetrapods** (four-footed). The fossil record indicates that the tetrapod limb evolved from the fins of fish. But tetrapods actually have fewer *Hox* genes than fish. Might changes in the timing or location of homeotic gene expression be responsible for the fin-to-limb transition?

Paolo Sordino and associates explored this question by comparing the expression of two genes—the *hoxd-11* locus and a gene called *Sonic hedgehog* (*Shh*)—in the zebra fish and the mouse. In tetrapod embryos, *hoxd-11* is expressed as the limbs begin to bud and grow. The gene product marks locations along the long axis of the limb as it grows out. In contrast, the protein produced by *Shh* marks the front-to-back, or anterior-to-posterior, axis of the developing limb.

Are the timing and location of *hoxd-11* and *Shh* expression identical in fish and tetrapods? To answer this question Sordino and co-workers treated limb buds—the tissues that develop into fins and limbs—with a molecule that hybridizes to the gene transcript produced by *hoxd-11*. They also stained limb buds with a molecule that hybridizes to the gene transcript produced by *Shh*. These treatments allowed them to identify when and where these genes begin producing their protein product.

When Sordino and co-workers treated the fish and mammal limb buds early in development, they found no differences in the pattern of gene expression. As **Figure 24.9a** shows, *hoxd-11* gene transcripts appear in the hindmost part of the developing limb in both the fish and the mouse. Similarly, *Shh* transcripts are found in similar amounts and locations in fish and mammal limbs at this stage of development.

When the investigators performed the experiment late in development, however, a striking difference emerged. In mouse limbs, the location of *hoxd-11* transcripts shifts. Study the late limb bud from a mouse embryo, shown in **Figure 24.9b**, and note that the *hoxd-11* product localizes to the part of the structure that is away from the body and faces toward the head. There is also a late expression of *Shh*. But neither of these

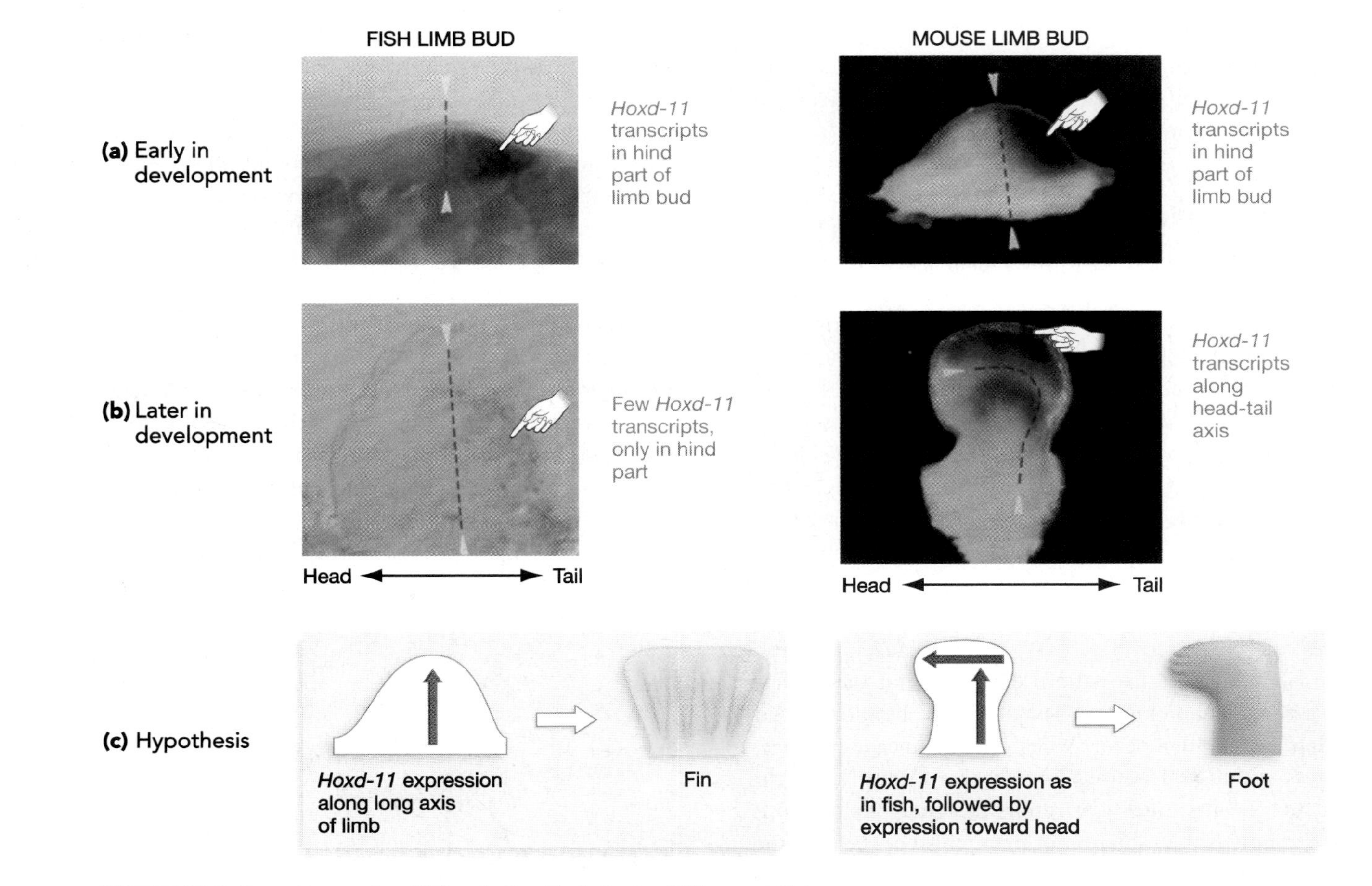

FIGURE 24.9 Gene Expression Differs in the Limb Buds of Mice and Fish
In early development, there is no difference in the pattern of gene expression in the fish and mouse limb buds. However, in later development there is a striking difference in gene expression. This difference may account for the origin of the hand and foot.

events occurs in fish. Figure 24.9b, for example, shows that *hoxd-11* is expressed in small amounts and only in the hindmost cells late in zebra-fish limb development. This is a dramatic difference in the timing and location of gene expression.

Sordino and colleagues propose that this difference was an innovation that produced the first hand and foot tissues in the history of life. The sketches in **Figure 24.9c** illustrate their hypothesis that the late and "reoriented" expression of *hoxd-11* and *Shh* added an entirely new element to the limb. More specifically, the researchers suggest that the mutations occurred in air-breathing, swamp-dwelling fish that lived about 400 million years ago. They proposed that the mutations were favored by natural selection because the new structure made movement on land more efficient.

The predictions made by the hypothesis are now being tested. For example, Sordino and colleagues predict that *hoxd-11* and *Shh* expression follows the zebra-fish pattern even in fish groups that branched off very early in their radiation. (Zebra fish belong to a group that emerged millions of years after the tetrapods had evolved from fish). Today the fish lineages most closely related to tetrapods are represented by several lungfish and an unusual species called the coelacanth. Some lungfish use their fins to pull themselves along a substrate; as coelacanths swim, they move their fins in a pattern similar to the walking gait of a tetrapod. How are *hoxd-11* and *Shh* expressed as the limbs of lungfish and coelacanths develop? The experiments needed to answer this question are under way.

24.4 Adaptive Radiations

This chapter is about the tree of life's shape, and how and why it has changed through time. Thus far we have examined the tools biologists use to study the tree of life's past, investigated a major suite of branching events in detail, and probed the genetic mechanisms that underlie major "growth spurts." Now it's appropriate to ask, if a biologist steps back and looks at the tree as a whole, what broad patterns can be discerned?

When the tree of life is examined from afar, one of the patterns that jumps out is that dense, bushy outgrowths are scattered among the branches. As **Figure 24.10** shows, this shape results when many large and distinctive groups of organisms branch off from a lineage in a short amount of time. Biologists sometimes call this pattern a star phylogeny, because of its starburst shape. Why does this pattern exist? One of the leading causes is a phenomenon known as an adaptive radiation.

An **adaptive radiation** occurs when a single lineage produces descendants with a wide variety of adaptive forms. When the diversification occurs quickly, it results in a star phylogeny. A classical example is the radiation of mammals between 65 and 60 million years ago. During this short interval the primates (monkeys and apes), bats, carnivores, deer, whales, horses, and rodents originated. The organisms resulting from this rapid divergence represent a remarkable array of adaptive forms. They swim, fly, glide, burrow, swing through trees, walk on four legs, or walk on two legs. They occupy habitats from mid-oceans to mountaintops and from rainforests to deserts, and they eat fruit, nuts, leaves, twigs, bark, insects, crustaceans, molluscs, fish, and other mammals.

CD ACTIVITY 24.1 Adaptive Radiation

The hallmark of an adaptive radiation is ecological diversification within a single lineage. What makes them occur?

Colonization Events as a Trigger

One of the most consistent themes in adaptive radiations is opportunity. The radiation of mammals, for example, occurred immediately after the extinction of the dinosaurs. As they diversified, mammals took over the ecological roles formerly filled by dinosaurs and the swimming reptiles (like the icthyosaurs and mosasaurs). Adaptive radiations frequently occur when habitats are unoccupied by competitors.

Recently Jonathon Losos and colleagues documented this process in detail. They did not study a radiation that followed an extinction event, however. Instead they analyzed radiations that were triggered by colonization events on islands that had distinct habitats and were free of competitors.

The study by Losos and co-workers focused on the *Anolis* lizards of the Caribbean. Biologists have interpreted this group of lizards as an adaptive radiation for two reasons: The lineage includes 150 species, and there is a strong correspondence between the size and shape of each species and the habitat it occupies. Species that spend most of their time clinging to broad tree trunks or running along the ground, for example, tend to have long legs and tails, while twig-dwellers have relatively short legs and tails (**Figure 24.11a**). These data suggest that lizard species have diversified in a way that allows them to occupy many different habitats.

Exactly how did the diversification occur? As the first step in answering this question, Losos and co-workers estimated the phylogeny of *Anolis* from DNA sequence data. Then they compared the habitats occupied by each species to their relationships on the tree. The results shown in **Figure 24.11b** for species found on two different islands are typical. Note that the

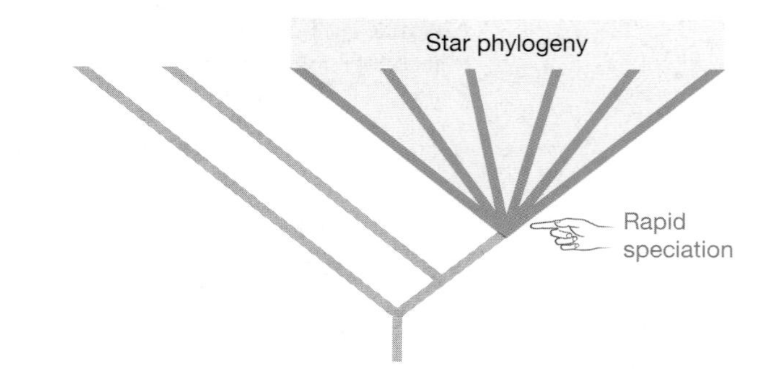

FIGURE 24.10 Adaptive Radiations Often Produce "Star Phylogenies"
A phylogeny has a starburst shape when rapid speciation has occurred. If speciation is followed by divergence into many different adaptive forms, then an adaptive radiation has taken place.

original colonist on each island belonged to a different ecological type. The initial species on Hispaniola, for example, lived in the trunks and crowns of trees while the original colonist on Jamaica occupied twigs. From different evolutionary starting points, then, an adaptive radiation occurred on each island. The key point is that on both islands, the same four ecological types eventually evolved. New species arose on each island independently, but because both islands had similar habitats, each ended up with a complement of species that was similar in lifestyle and appearance. Losos and co-workers found the same pattern on many other islands. In other words, a series of "miniature adaptive radiations" occurred, one on each island, within the overall *Anolis* radiation. Each was triggered by two conditions—opportunity in the form of available habitat and lack of competitors.

The Role of Morphological Innovation

Still other "triggers," besides extinction and colonization events, can ignite adaptive radiations. Foremost among them is morphological innovation, the phenomenon discussed in sections 24.2 and 24.3. Important new traits like multicellularity, shells, exoskeletons, and limbs were a driving force behind the adaptive radiation called the Cambrian explosion.

In sum, adaptive radiations are an important pattern in the history of life. They can be triggered by several different events, and they demonstrate that rapid speciation and morphological divergence are often tightly coupled during evolution.

24.5 Mass Extinctions

Mass extinction events are evolutionary hurricanes. They buffet the tree of life, snapping twigs and breaking branches. They are catastrophic episodes that wipe out huge numbers of species and lineages in a short time, giving the tree of life a drastic pruning. One mass extinction event, about 250 million years ago, nearly uprooted the tree entirely.

Mass extinction events need to be distinguished from background extinctions. A **mass extinction** refers to the rapid

(a) Morphological diversity

Short-legged lizard

Long-legged lizard

(b) Morphological diversity correlates with habitat diversity.

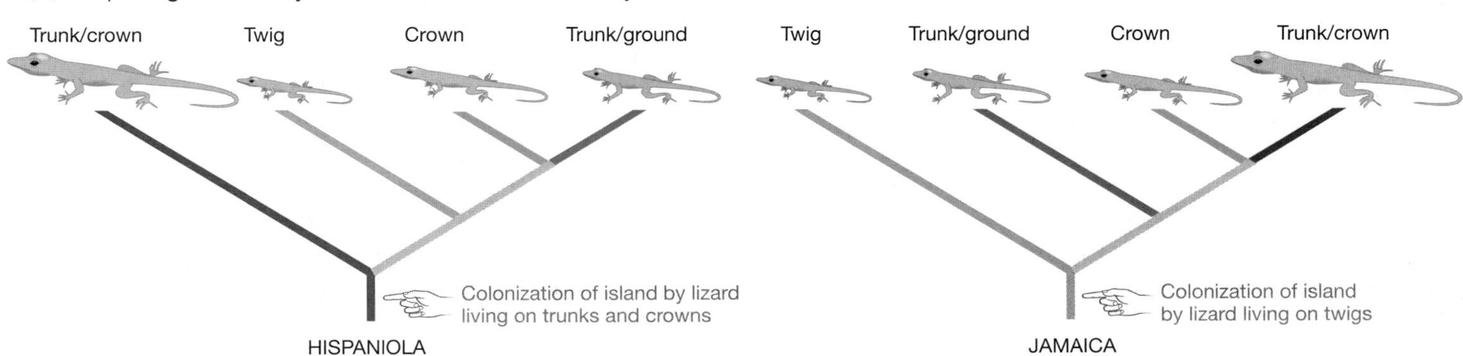

FIGURE 24.11 Adaptive Radiations of *Anolis* Lizards
(a) Short-legged lizard species (left) spend most of their time on the twigs of trees and bushes; long-legged species (right) live on tree trunks and the ground. **(b)** These diagrams show the evolutionary relationships among lizard species on the islands of Hispaniola and Jamaica. Note that the initial species to colonize each island was different. But in terms of how they look and where they live, a similar suite of four species evolved. According to these data, then, similar adaptive radiations took place independently on the two islands.

extinction of a large number of lineages scattered throughout the tree of life. **Background extinction** refers to the lower, average rate of extinction observed over the entire history of life. Although there is no hard and fast rule for distinguishing between the two extinction rates, paleontologists traditionally recognize and study five mass extinction events. In **Figure 24.12**, for example, M. J. Benton plotted the percentage of plant and animal groups that died out during each stage in the geologic time scale since the Cambrian explosion. Five prominent spikes in the graph—denoting an extraordinarily large number of extinctions within a short time—are drawn in red. These are referred to as "The Big Five."

How Do Background and Mass Extinctions Differ?

Paleontologists are interested in distinguishing between background and mass extinctions because they have contrasting causes and effects. Background extinctions are thought to occur when normal environmental change or competition with other species reduces certain populations to the point where they die out. Mass extinctions, in contrast, are thought to result from extraordinary, sudden, and temporary changes in the environment. During a mass extinction, species do not die out because individuals are poorly adapted to normal conditions. Rather, populations die out from exposure to exceptionally harsh, short-term conditions—like huge volcanic eruptions or massive, short-term sea-level changes.

To drive this point home, and to examine what happens after a mass extinction has occurred, let's examine one of The Big Five in detail. Although the event we'll analyze was not the largest in history—the mass extinction at the end of the Permian period may have wiped out 90 percent of the multicellular organisms alive at the time—it is among the most dramatic. It extinguished the dinosaurs and ushered in the diversification of mammals.

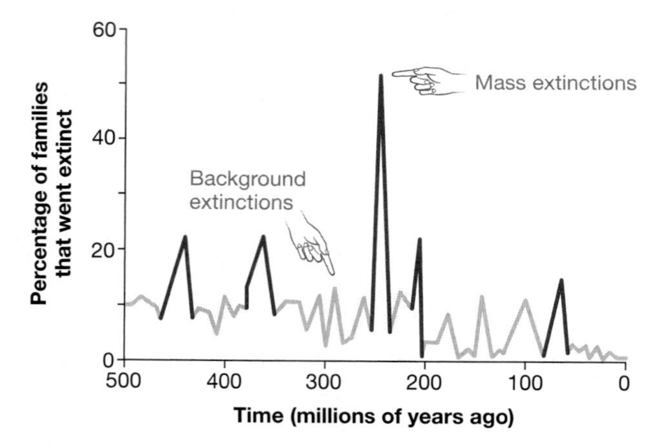

FIGURE 24.12 The Big Five
This graph shows the percentage of groups of species, called families, that went extinct over each interval in the fossil record since the Cambrian explosion. The five mass extinction events are drawn in red. (Note: When studying diversification and extinction over time, paleontologists often use the groups of species called families as a unit of analysis. No one has compiled the data required to show the percentage of species that went extinct over each interval in the fossil record.)

K-T Extinction and the Demise of the Dinosaurs

The end-Cretaceous extinction of 65 million years ago is as satisfying a murder mystery as you could hope for. An asteroid did it.

The impact hypothesis for the extinction of the dinosaurs claims that an asteroid struck Earth and caused widespread destruction. The hypothesis was once intensely controversial, but is now supported by overwhelming evidence.

- Worldwide, sedimentary rocks that formed at the Cretaceous-Tertiary (K-T)* boundary contain extraordinarily high quantities of the element iridium. Iridium is extremely rare in Earth rocks, but is an abundant component of asteroids and meteorites.
- Shocked quartz and microtektites are minerals that are found only at documented meteorite impact sites. Shocked quartz forms when shock waves from an asteroid impact alter the structure of sand grains. Microtektites form when minerals are melted at an impact site, and then cool and resolidify. In Haiti, both minerals are abundant in rock layers dated to 65 million years ago (**Figure 24.13**).
- A crater the size of Sicily exists just off the northwest coast of Mexico's Yucatán peninsula. Microtektites are abundant in sediments from its wall.

Astronomers and paleontologists estimate that the asteroid was about 10 km across—quite literally, the size of a mountain. The distribution of shocked quartz and microtektites indicates that the body hit Earth at an angle and splashed material over much of southeastern North America (**Figure 24.14**).

According to both computer models and geologic data, the impact unleashed a devastating chain of events. A fireball of hot gas would have spread from the impact site; large soot and ash deposits in sediments dated to 65 million years ago testify to extensive wildfires, worldwide. The impact site itself is underlain by a sulfate-containing rock called anhydrite. The SO_4^{2-} released by the impact would react with water in the atmosphere to form sulfuric acid (H_2SO_4), triggering extensive acid rain. Massive quantities of dust, ash, and soot would have blocked the Sun for extensive periods, leading to rapid global cooling and a crash in plant productivity. The upshot? The fossil record suggests that between 60 and 80 percent of all species went extinct.

*Geologists use *K* to abbreviate Cretaceous, because *C* refers to the Cambrian period.

Selectivity The asteroid impact did not kill indiscriminately. Certain lineages escaped virtually unscathed while others vanished. Among vertebrates, for example, the dinosaurs, pterosaurs (flying reptiles), and all of the large-bodied marine reptiles (mosasaurs, icthyosaurs, and plesiosaurs) expired; while mammals, crocodilians, amphibians, and turtles were largely unharmed.

Why? Answering this question has sparked intense controversy and debate. For years the leading hypothesis was that the K-T extinction event was size selective. The logic here was that the extended darkness and cold would affect large organisms disproportionately because they require more food than small organisms. But extensive data on the survival and extinction of marine clams and snails has shown no hint of size selectivity, and small-bodied and juvenile dinosaurs perished along with large-bodied forms. This aspect of the mystery is unsolved.

(a) Normal quartz and shocked quartz

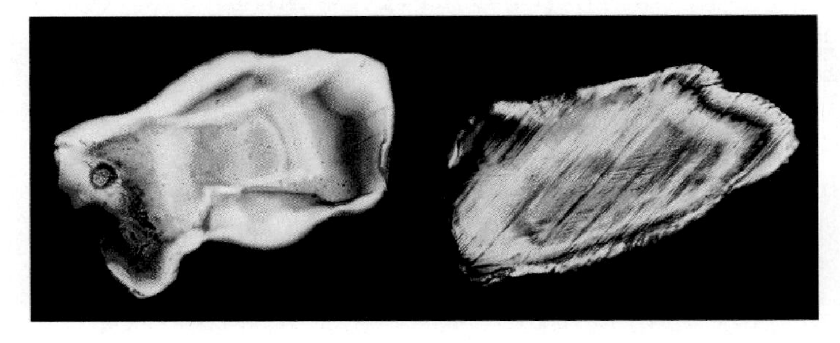

(b) Microtektites

FIGURE 24.13 Minerals That Form During Asteroid Impacts
(a) A normal quartz grain is markedly different from a shocked quartz grain. **(b)** Microtektites are small particles of glass that form when minerals melt at an impact site and then recrystallize. QUESTION Why is the presence of shocked quartz and microtektites in rock layers dated to 65 million years ago considered evidence that supports the impact hypothesis?

Recovery After the K-T extinction, fern fronds and spores dominate the plant fossil record from North America and Australia. These data suggest that extensive stands of ferns replaced diverse assemblages of woody and flowering plants after the impact, and that terrestrial ecosystems were radically simplified. In marine environments, some groups of invertebrates do not recover normal levels of species diversity in the fossil record for 4–8 million years past the K-T boundary. Recovery was slow.

The flora and fauna of the Tertiary was markedly different from the preceding period. One prominent change was that mammals took the place of the dinosaurs and marine reptiles. The lineage called Mammalia, which had consisted largely of rat-sized, nocturnal predators and scavengers in the heyday of the dinosaurs, exploded after the impact. Within 5–10 million years, all of the major mammalian orders observed today had appeared—from pigs to primates. Why? A major branch on the tree of life had been lopped off, and mammals flourished in its place. The Cretaceous is sometimes called the Age of Reptiles, and the Tertiary the Age of Mammals. The change was not due to a competitive superiority conferred by adaptations like fur and lactation. Rather, it was due to a chance event: a once-in-Earth's-lifetime impact event.

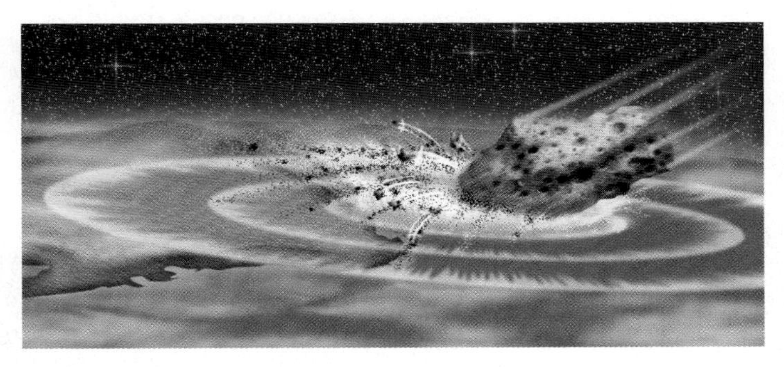

FIGURE 24.14 The Asteroid Impact
This is an artist's conception of what the impact event may have looked like. QUESTION Why did the artist show material splashing onto the southeastern portion of North America?

Essay Is a Mass Extinction Event Under Way Now?

As human populations expand, wildlife habitat shrinks. An expert on the K-T extinction, paleontologist Peter Ward, has reviewed data on accelerating rates of habitat loss around the globe and warned that a "human meteor" is about to strike the planet. The admonition raises a legitimate scientific question: Are humans currently causing a mass extinction event on the scale of The Big Five?

Data on extinction events over the past 400 years argue that the answer is no. Even though human populations have been expanding rapidly since 1600 (**Figure 1**), the percentage of species known to have gone extinct over this period does not begin to approach the 50–90 percent figures registered during The Big Five. As the following table shows for selected groups, the percentage of species that have gone extinct recently is quite low.

Lineage	Species Extinct since 1600	Total Named Species	Percentage of Extinct Species
Molluscs	191	100,000	0.20
Crustaceans	4	4,000	0.01
Insects	61	1,000,000	0.005
Vertebrates	229	47,000	0.50
Gymnosperms	2	758	0.30
Dicotyledons	120	190,000	0.20
Monocotyledons	462	52,000	0.20

The research group that compiled these data urges that they be interpreted carefully, however. There are two key points to note: (1) the number of extinct species listed is undoubtedly an underestimate, especially for poorly studied groups like plants and insects, and (2) virtually all of the extinct species recorded in the table were island forms. This second point is critical because island extinctions, which usually result from the introduction of nonnative predators and herbivores like rats and goats, have long since peaked in intensity. In other words, the data do not reflect current extinction rates. Instead of being triggered by the introduction of non-native species to islands, most current extinctions are thought to be occurring in response to widespread habitat destruction—particularly the conversion of tropical rainforests to pastures and cropland. The destruction of these species-rich habitats is a direct consequence of increasing human populations.

How rapidly is this new agent—humans—causing extinctions? This is a difficult question to answer because reliable data on current extinction rates are scarce. Biologists can get a

Are humans currently causing a mass extinction event . . . ?

hint, however, by extrapolating from very recent events. From 1986 to 1990, for example, 15 vertebrates were added to the list of extinct species. If this rate of extinction continued, it would take only 7000 years to eliminate half of all known vertebrates. Researchers have also attempted to answer the question by assuming that all species currently considered "threatened"—a formal status conferred by international conservation agencies—will go extinct in the next 100 years. If so, extinction rates in plant and animal lineages would vary from 10 to 100 times their background extinction rates.

The message of these admittedly crude analyses is that the human meteor is probably a real, not an imagined, threat. If present rates of human population growth and habitat destruction continue, a mass extinction event will almost undoubtedly occur.

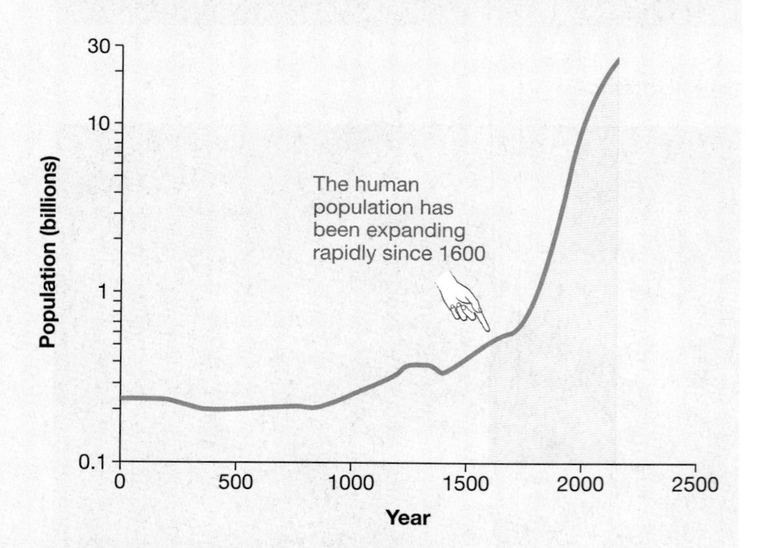

FIGURE 1 Human Population Growth
This graph shows the estimated world population size of humans from the year 1 to the year 2150. The line past 1990 is an extrapolation based on the estimate for continued high growth.

Chapter Review

Summary

Phylogenies and the fossil record are the two main tools that biologists use to study the history of life. Even though the fossil record has strong taxonomic, habitat, and temporal biases, it is an enormously valuable database.

Perhaps the best-studied event in the history of life is the Cambrian explosion: the diversification of animals over a 20–40 million-year period, starting about 540 million years ago. The animals that lived in marine environments just before and after the radiation are documented in the Doushantuo, Ediacaran, and Burgess Shale faunas. During this interval the first heads, tails, appendages, shells, exoskeletons, and segmented bodies evolved.

Data on the number and identity of homeotic genes in different animal lineages suggest that at least some of the animal radiation was possible because gene duplication events created new copies of *Hox* loci. But the correlation between morphological complexity and number of homeotic genes is far from perfect. Changes in the timing and location of existing genes were also responsible for many morphological innovations. For example, changes in gene expression may have been involved in the origin of the hand and foot in tetrapods.

Morphological innovations like large, complex bodies and limbs with feet and hands can initiate adaptive radiations. Adaptive radiations are a prominent pattern in the history of life and can also be triggered by colonization of a new habitat or the demise of competitors after a mass extinction. During adaptive radiations speciation events and morphological change occur rapidly, as a single lineage diversifies into a wide variety of ecological roles.

Mass extinctions have altered the course of evolutionary history at least five times. They prune the tree of life more or less randomly, and have marked the end of several prominent lineages and the rise of new branches. The Cretaceous-Tertiary extinction killed about 70 percent of existing species and was caused by an asteroid impact. After the devastation of a mass extinction, it can take 5–10 million years for ecosystems to recover their former levels of diversity.

Questions

Content Review

1. Choose the best definition of a fossil.
   a. any trace of an organism that has been converted into rock
   b. a bone, tooth, shell, or other hard part of an organism that has been preserved
   c. any trace of an organism that lived in the past
   d. the process that leads to preservation of any body part from an organism that lived in the past

2. Why is burial a key step in fossilization?
   a. It slows the process of decay by bacteria and fungi.
   b. It allows tissues to be preserved as casts or molds.
   c. It protects tissues from wind, rain, and other corrosive elements.
   d. All of the above.

3. Why do molecular clocks exist?
   a. Natural selection is not important at the molecular level.
   b. Homologous genes have the same structure.
   c. They can be calibrated.
   d. Some DNA sequences, in some lineages, change at a steady rate over time.

4. Why are the Doushantuo, Ediacaran, and Burgess Shale fossil deposits unusual?
   a. Soft-bodied animals are preserved.
   b. They are easily accessible to researchers.
   c. They are the only fossil-bearing rocks from that period.
   d. They include terrestrial, instead of just marine, species.

5. Which of the following best characterizes an adaptive radiation?
   a. Speciation occurs rapidly.
   b. As a single lineage diversifies, descendant populations occupy many habitats and ecological roles.
   c. Natural selection is particularly intense.
   d. Species recover after a mass extinction.

6. How are mass extinctions distinguished from background extinctions?
   a. They rapidly eliminate over half of all species.
   b. They are restricted to certain habitats or geographic areas.
   c. They affect just a few branches on the tree of life.
   d. They are always due to asteroid impacts.

Conceptual Review

1. The text claims that the fossil record is biased in several ways. What are these biases? If the database is biased, is it still an effective tool to use in studying the diversification of life? Explain.

2. The initial radiation of animals took place over some 40 million years, at the start of the Cambrian period. Why is the radiation called an "explosion?"

3. What is the "new genes, new bodies" hypothesis? Based on the data presented in this chapter, is the hypothesis correct?

4. Give an example of an adaptive radiation that occurred after a colonization event, a mass extinction, and a morphological innovation. In each case, provide a hypothesis for why the adaptive radiation occurred.

5. Summarize the evidence that supports the impact hypothesis for the K-T extinction.

Applying Ideas

1. Suppose that the dying wish of a famous eccentric was that his remains would fossilize. His family has come to you for expert advice. What steps would you recommend to maximize the chances that his wish will be fulfilled?
2. Summarize the nature of the fossils found in the Doushantuo, Ediacaran, and Burgess Shale deposits. (Note that the three fossil assemblages are listed in order, from most ancient to most recent.) According to the fossil record, what trends or general patterns occurred during the early evolution of animals? For example, did animals tend to get larger or smaller? More or less complex?
3. Experiments summarized in section 24.3 suggest that the tetrapod foot resulted from changes in the timing and location of expression of certain genes. Based on the information presented in Chapter 15 on the regulation of gene expression in eukaryotes, exactly what sort of mutations would lead to changes in when and where genes are transcribed?
4. The synthesis of results from highly disparate fields, like developmental genetics and fossil studies, is an exciting recent trend in research on the history of life. Comment on whether you find this synthesis valuable. For example, did reading about the "new genes, new bodies" hypothesis and the experiments on gene expression in tetrapod limbs help you understand how important changes have occurred during the diversification of life? Why or why not?

CD-ROM and Web Connection

CD Activity 24.1: Adaptive Radiation *(animation)*
(Estimated time for completion = 5 min)
How do adaptive radiations occur and what determines the characteristics of the resulting species?

At your **Companion Website** (http://www.prenhall.com/freeman/biology), you will find self-grading exams and links to the following research tools, online resources, and activities:

Follow a Fossil
Explore the basics of finding, collecting, preserving, and analyzing fossils.

Burgess Shale Fossils
This site describes the fossils found in the Burgess Shale deposits.

Searching for the First Animal
Discover how fossils from the Doushantuo deposits have sparked an entirely new way of looking at evolution.

Ediacaran Biota
A description of the Ediacaran fossil deposits found in Southern Australia.

***Hox* Genes: The Activators**
Explore how the *Hox* genes may have arisen and changed through evolutionary history.

The K-T Extinction
Examine the asteroid hypothesis for the Cretaceous-Tertiary extinction.

Additional Reading

Bengtson, S. (1998). Animal embryos in deep time. *Nature* 391: 529–530. An essay on the discovery of animal embryos in the Doushantuo deposits.

Meyer, A. (1998). *Hox* gene variation and evolution. *Nature* 391: 225–228. A commentary on the "new genes, new bodies" hypothesis and the diversification of animals.

Pennisi, E., and W. Roush (1997). Developing a new view of evolution. *Science* 277: 34–39. An overview of how biologists are getting a better picture of the history of life by integrating fossil evidence with results from developmental genetics.

The Diversification of Life

The early chapters in this book introduced the origin of life, how organisms work at the cellular and molecular level, and how multicellular organisms develop. Most of the research reviewed in those units was done on a relatively small number of model organisms, such as *Escherichia coli*, yeast, fruit flies, and *Arabidopsis thaliana*. Similarly, the chapters in Unit 5 analyzed how evolution occurs but only hinted at the diversity of organisms that evolution has produced.

This unit, in contrast, surveys the full sweep of life's diversity. The topic is important because the ability to think broadly about biology is grounded in a fundamental understanding of life's diversity. Even researchers who work at the sub-molecular level need what Barbara McClintock calls "a feeling for the organism"—an awareness of the problems that organisms encounter as they grow and reproduce and an appreciation for how they solve those problems.

What organisms make up the tree of life? What do they look like, how do they make a living, and how do they affect human health and welfare? Each chapter in the unit answers these questions for a major lineage on the tree of life. **Chapter 25** introduces the **Bacteria and Archaea**; **Chapter 26** reviews the parasitic agents called **Viruses. Chapter 27: Protists** introduces the eukaryotes by considering the lineages called protists. The unit concludes with chapters on the most visible and familiar organisms: **Chapter 28: Land Plants, Chapter 29: Fungi**, and **Chapter 30: Animals.**

As you study these organisms, it is helpful to maintain a slight sense of awe. They may very well be the only organisms in the universe.

Like most other amphibians, the Santa Cruz Long-toed Salamander lives on land but lays its eggs in water. Amphibians evolved from air-breathing fish that existed some 250 million years ago. © 1994 Susan Middleton and David Liittschwager

Bacteria and Archaea

25

Biologists who study the organisms at the base of the tree of life, in the domains Bacteria and Archaea, are exploring the most exciting frontier in biodiversity. So little is known about the extent of these domains that recent collecting expeditions have turned up entirely new **phyla**—a name given to major lineages within each domain. To a biologist, these findings are equivalent to suddenly discovering a new group of eukaryotes as distinctive as the ferns, sponges, or jellyfish.

Just how many bacterial and archaeal species are alive today? To date, a mere 5000 species have been formally named and described. In reality, it is virtually certain that tens of millions exist. Consider that in a habitat as well studied as the human mouth, just 300 of the 500 different bacteria that have been isolated have also been described and named. Norman Pace points out that there may be tens of millions of different insect species but notes that, "If we squeeze out any one of these insects and examine its contents under the microscope, we find hundreds or thousands of distinct microbial species." Most of these microscopic organisms (or **microbes**) are bacteria or archaea; virtually all are unnamed and undescribed.

The physical world may be mapped and explored, and many of the larger plants and animals named, but in **microbiology**—the study of organisms that can be seen only with the aid of a microscope—this is an age of exploration and discovery.

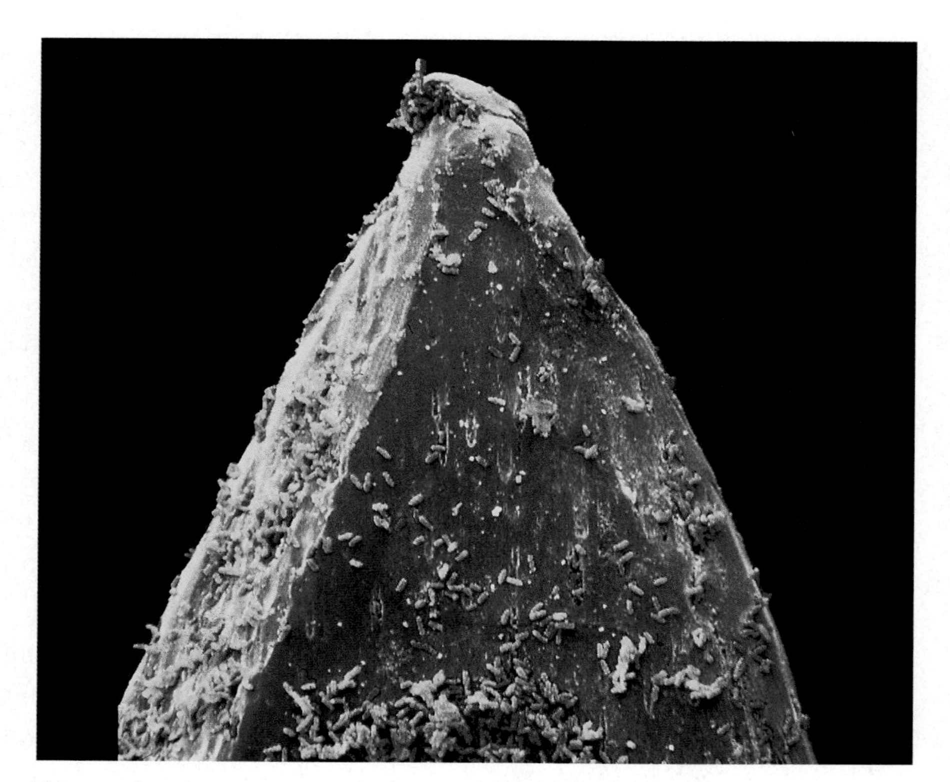

This scanning electron micrograph shows rod-shaped bacteria clustered on the tip of a syringe. Although bacteria and archaea are small and relatively simple morphologically, their ecological diversity, abundance, and metabolic capabilities are unrivaled.

25.1 What Are the Bacteria and Archaea?

The salient morphological and biochemical features of bacteria, archaea, and eukaryotes are summarized in **Table 25.1**. To distinguish the bacteria and archaea from the eukaryotes, biologists usually emphasize the nuclear envelope that encloses the genetic material of eukaryotes. To distinguish bacteria and archaea, biologists usually point first to the types of molecules that make up their cell membranes and cell walls. As the table indicates, these crucial parts of the cell have very different compositions in the two domains. The unique features of the cell wall and cell membrane are particularly important because virtually all bacteria and archaea are unicellular.

What other features characterize these organisms? Bacteria and archaea are ancient, abundant, ubiquitous, and diverse. The oldest fossil microbes found to date are 3.4-billion-year-old cells that are virtually indistinguishable from contemporary bacteria. Because eukaryotes don't appear in the fossil record until 1.75 billion years ago, and because archaea probably arose at about the same time as eukaryotes, biologists infer that bacteria may have been the only form of life on Earth for almost 1.7 billion years.

Although biologists can only guess at the total number of bacterial and archaeal species that exist today, their abundance is well documented. A teaspoon of good-quality soil contains *billions* of microbes. In sheer numbers, the bacterium *Prochlorococcus*—found in the plankton of the world's oceans—may be the dominant life-form on the planet. Oceanographers routinely find this organism at concentrations of 70,000 to 200,000 cells per milliliter of seawater. At these concentrations, a drop of seawater contains a population equivalent to a large city. Yet *Prochlorococcus* was first described and named only recently—in 1988.

Bacteria and archaea are also found in almost every conceivable habitat. On land, they live in environments as unusual as oxygen-free mud, hot springs, salt flats, the roots of plants, and the guts of animals. They have also been discovered living in bedrock to a depth of 1500 m below Earth's surface. In the ocean they are found from the surface to depths of 10,000 m, and at temperatures ranging from 0°C in Antarctic sea ice to over 110°C near submarine volcanoes.

How do biologists study this incredible diversity of organisms? The research strategies employed today range from classical approaches, such as microscopy, to the latest DNA sequencing technologies.

Morphological Diversity

Microbiologists use transmission electron microscopes, scanning electron microscopes, and an array of light microscopes to characterize the morphology of bacteria and archaea. Even

TABLE 25.1 Characteristics of Bacteria, Archaea, and Eukarya

	Bacteria	Archaea	Eukarya
Nucleus present?	No	No	Yes
Circular chromosome?	Yes	Yes	No
DNA associated with histone proteins?	No	Yes	Yes
Organelles (membrane-bound compartments) present?	No	No	Yes
Unicellular or multicellular?	Almost all unicellular	All unicellular	Many multicellular
Sexual reproduction?	No*	Not known	Common
Structure of lipids in cell membrane	Similar to eukaryotes—glycerol bonded to straight-chained fatty acids via ester linkage	Unique—glycerol bonded to branched fatty acids (synthesized from isoprene subunits) via ether linkage	Similar to bacteria—glycerol bonded to straight-chained fatty acids via ester linkage
Cell wall material	Almost all include the complex material peptidoglycan, which contains muramic acid	Varies widely among species, but no peptidoglycan and no muramic acid	When present, usually made of cellulose or chitin
Transcription and translation machinery	Unique—one relatively simply RNA polymerase; translation begins with formylmethionine; translation poisoned by several antibiotics that do not affect archaea or eukaryotes	Similar to eukaryotes—several relatively complex RNA polymerases, translation begins with methionine	Similar to archaea—several relatively complex RNA polymerases, translation begins with methionine

*Sexual reproduction begins with meiosis and often involves the exchange of haploid genomes between individuals of the same species. In bacteria, meiosis does not occur. Small numbers of genes can be transferred from one bacterial cell to another, however, and genetic recombination may occur. For more detail, see Chapters 9, 14, and 17.

though almost all are unicellular, morphological diversity does exist. For example, bacteria and archaea range in size from the smallest of all free-living cells—bacteria called mycoplasmas with volumes as small as 0.03 μm^3—to the largest bacterium known, *Thiomargarita namibiensis*, with volumes as large as 200×10^6 μm^3. Over a billion mycoplasma cells could fit inside an individual *Thiomargarita* (**Figure 25.1a**). Bacteria and archaea exhibit a variety of shapes as well, including filaments, spheres, rods, chains, and spirals (**Figure 25.1b**). Further, many cells are motile, with movement powered by flagella or by gliding over a surface (**Figure 25.1c**).

Compared to plants, animals, and other eukaryotes, however, the morphology of bacteria and archaea is simple. Consequently, it is difficult for biologists to understand which species are closely or distantly related just by comparing their morphological characteristics. To understand the diversity of these organisms in an evolutionary context, researchers have turned to comparing their genomes. Can similarities and differences in the DNA of bacteria and archaea help us understand which species are most closely related?

Molecular Phylogenies

The ribosome is a complex molecular machine that is found in all organisms. Ribosomes are the site of protein synthesis in cells and contain a molecule known as SSU (for small subunit) RNA. In the late 1960s Carl Woese and colleagues began a massive effort to determine and compare the base sequences of SSU RNAs from a wide diversity of species. (Chapter 1 introduced this research program in detail.) Their goal was to use similarities and differences in the sequence data to infer the evolutionary history of a wide diversity of organisms. The result of their analysis is shown in **Figure 25.2a**. This diagram came to be known as the universal tree, or the tree of life.

Woese's tree is now considered a classic result. Prior to its publication, biologists thought that the major division among organisms was between cells that lacked a membrane-bound nucleus (prokaryotes) and cells that possessed a membrane-bound nucleus (eukaryotes) (**Figure 25.2b**). But according to the SSU RNA tree, the major divisions among life-forms are between the three groups that Woese named the Bacteria, Archaea, and Eukarya. Follow-up work documented that Bacteria were the first lineage to diverge from the common ancestor of all living organisms, meaning that the Archaea and Eukarya are each other's closest relatives.

Since the first universal tree was published in 1977, however, technological innovations have produced data sets that have shaken the tree of life. What are these new methods, and how have they altered thinking about life's diversity?

Direct Sequencing Direct sequencing is a strategy for documenting the presence of bacteria and archaea that cannot be grown in culture and studied in the laboratory. The protocol

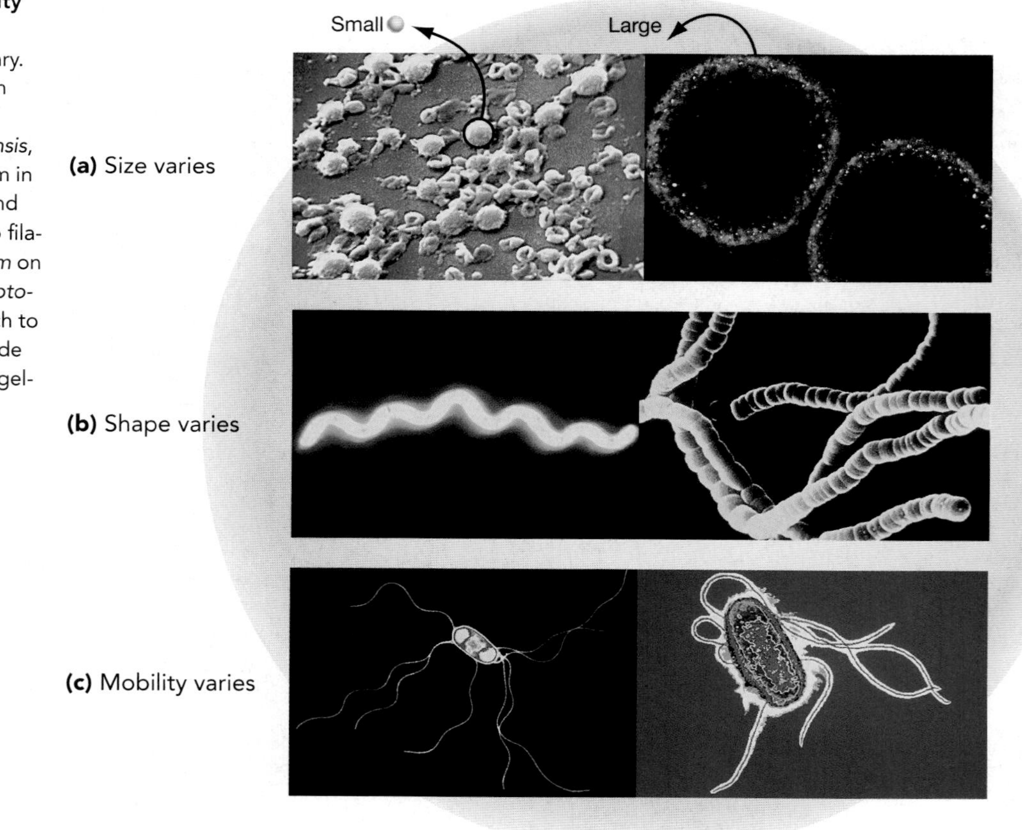

FIGURE 25.1 Morphological Diversity Among Bacteria and Archaea
(a) The size of bacteria and archaea vary. The cells of the *Mycoplasma* shown on the left are about 0.5 μm in diameter, while those of *Thiomargarita namibiensis*, shown on the right, are about 0.15 mm in diameter. **(b)** The shape of bacteria and archaea vary from rods and spheres to filaments or spirals like the *Rhodospirillum* on the left. In some species, like the *Streptococcus faecalis* on the right, cells attach to one another and form chains. **(c)** A wide variety of bacteria and archaea use flagella to power swimming movements.

actually allows biologists to name and characterize organisms that have never been seen.

The technique has had an enormous impact on our understanding of diversity, for a simple reason. Only a tiny fraction of the bacteria and archaea that exist can be grown in the lab using existing methods. The pioneer of direct sequencing, Norman Pace, estimates that over 99 percent of the bacteria and archaea in nature have never been cultured. The tree of life constructed by Woese and colleagues, however, was dominated by organisms that *can* be cultured. Because direct sequencing allows biologists to investigate previously unstudied bacteria and archaea, it has revealed huge new branches on the tree of life. Direct sequencing studies have also revolutionized thinking about the habitats where archaea are found.

Figure 25.3 on pages 488–489 outlines the steps performed in a direct sequencing study. To begin, researchers collect a sample from a habitat—a few microliters of water or a few micrograms of soil. Next, the cells in the sample are broken open (lysed) and the DNA is purified. The DNA sequences are amplified through the polymerase chain reaction (or PCR; see Chapter 12) using "universal" rRNA primers. These are sequences from the gene for SSU RNA that are extremely highly conserved throughout the tree of life. Because the universal primers hybridize with rDNA sequences from virtually any organism, researchers can be confident that virtually all of the rDNA genes in the water or soil sample are amplified in this step. The rDNA genes are purified via gel electrophoresis and then separated by inserting them into small, circular sequences of DNA called plasmids. Each plasmid now contains a different rDNA, each representing a different species. These plasmids are inserted into *Escherichia coli* cells, which are placed in optimal growing conditions. The plasmids replicate along with the expanding bacterial colonies, producing millions of copies of each rDNA. The plasmids can then be isolated and the rDNA genes purified and sequenced. By comparing these data with sequences in existing databases, biologists can determine whether the sample contains previously undiscovered archaea or bacteria.

Direct sequencing studies have produced new and sometimes startling results. For example, for two decades after the discovery of the Archaea, researchers thought that they could be conveniently grouped into four categories: extreme halophiles (salt-lovers), sulfate-reducers (which produce hydrogen sulfide, or H_2S, during metabolism), methanogens (methane-producers), and extreme thermophiles (heat-lovers). The existing data indicated that archaea were restricted to extreme environments, and that the four phenotypes corresponded to separate lineages within the domain. As a result of direct sequencing studies, this hypothesis has now been discarded. Beginning in the mid-1990s, direct sequencing revealed archaea in habitats as diverse as rice paddies and the Arctic Ocean. Some of these newly discovered organisms appear to belong to an entirely new lineage, tentatively called the Korarchaeota. The group's rDNA sequences are so distinctive that it may represent a "kingdom" analogous to plants or animals, yet it was identified and named before any of its members had been observed. Biologists know *nothing* about the newly discovered species except for their habitat and phylogenetic relationships.

Genome Sequencing: Evidence for Lateral Gene Transfer

Direct sequencing studies have unveiled major new branches on the tree of life, but another research program—whole-genome sequencing—has caused biologists to question the nature of the tree itself. To understand why, consider an analysis published by Michael Smith and Russell Doolittle in the early 1990s. When these researchers compared the DNA sequence that encodes an enzyme called glucose-6-phosphate isomerase from species throughout the tree, the result was confusing: The sequence from the bacterium *Esherichia coli* was much more similar to the sequence found in animals than it was to the sequence found in other bacteria. How could this gene be so similar to distant relatives like animals, yet so different among closely related bacteria?

Smith and Doolittle generated an hypothesis to solve the paradox. They proposed that the gene must have been transferred

CD ACTIVITY 25.1 The Tree of Life

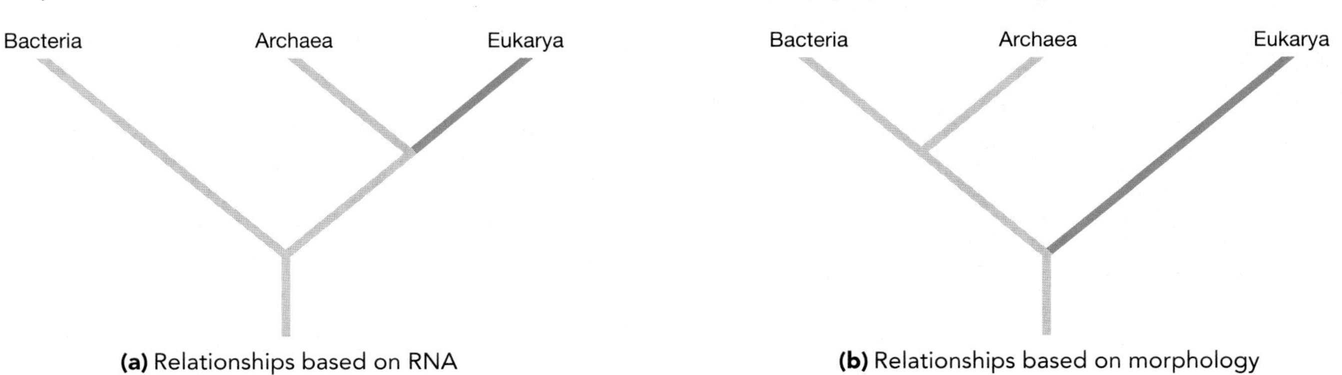

FIGURE 25.2 Relationships among Bacteria, Archaea, and the Eukaryotes
The tree in part **(a)** illustrates the relationships among species in the three domains of life according to similarities and differences in their SSU RNA sequences. The traditional tree of life, part **(b)**, is based on morphology. The traditional tree proposed that the major division among organisms was between those without a membrane-bound nucleus (the prokaryotes) and those with a membrane-bound nucleus (the eukaryotes).

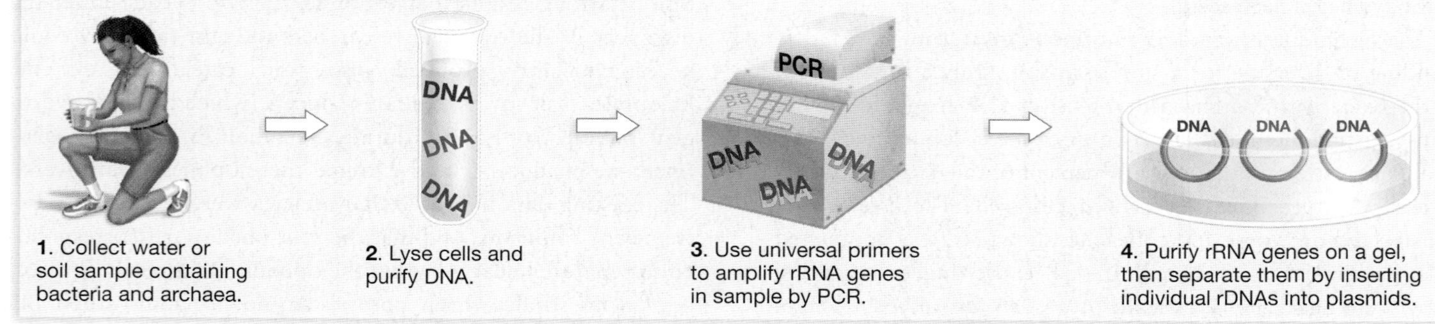

FIGURE 25.3 Direct Sequencing: An Experimental Protocol
Direct sequencing allows researchers to isolate specific genes from the organisms present in a sample and generate enough copies that the DNA can be sequenced.

directly from an animal to *E. coli* through one of the mechanisms of bacterial gene transfer analyzed earlier in the text. These mechanisms include the direct uptake of DNA from the environment (called transformation; see Chapter 11) and viral infection. The result is **lateral gene transfer**, the transmission of a gene from a species in one part of the tree of life to a distantly related organism. When genes are transferred laterally among bacteria and archaea, the primary mechanism is thought to involve the loops of mobile DNA called plasmids (see Chapters 14 and 17).

When biologists began sequencing entire genomes from bacteria and archaea during the 1990s, dozens of lateral transfer events emerged. The data indicated that extensive "gene swapping" has taken place among distantly related organisms throughout evolutionary history. Results like these have led Johann Peter Gogarten to propose that a "net of life" exists rather than a tree of life (**Figure 25.4**).

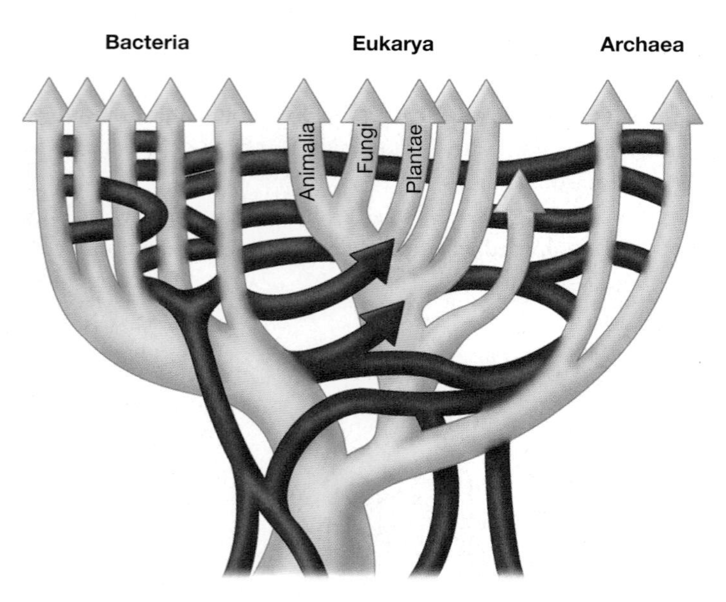

FIGURE 25.4 Is There a "Net of Life?"
This cartoon illustrates how lateral gene transfer affects the relationships among branches on the tree of life. The branches colored in gray show the diversification of species in the three domains of life, according to their SSU RNA. The branches shown in red indicate the movement of genes from species in one part of the tree to another via lateral gene transfer.

What do other biologists make of all this? W. Ford Doolittle and others have pointed out that when researchers construct a tree of life by comparing gene sequences involved in information processing (DNA replication, DNA transcription, and protein synthesis), the results usually agree with the tree implied by SSU RNA relationships. In contrast, comparison of genes involved in bacterial and archaeal metabolism often produce anomalous results. These loci seem to have been swapped extensively among bacterial and archaeal lineages. One interpretation is that the tree estimated from information processing genes accurately reflects evolutionary history. The enzymes used in energy and carbon metabolism, however, have been swapped back and forth among species over time. This interpretation is intriguing because metabolic diversity is a hallmark of the Bacteria and Archaea.

25.2 Metabolic Diversity in Bacteria and Archaea

Organisms in the domains Bacteria and Archaea are the masters of metabolism. As a group they can eat almost anything, from hydrogen molecules to crude oil. Taken together, they can transform carbon dioxide and other inorganic sources of carbon into sugar via six distinct pathways. Plants, in contrast, use just one biochemical system to synthesize sugar from carbon dioxide, and animals lack the ability altogether. Bacteria and archaea may be small and morphologically simple, but their biochemical capabilities are dazzling.

Just how varied are bacteria and archaea when it comes to finding and processing food? To appreciate the answer, recall that organisms have two fundamental nutritional needs: acquiring chemical energy in the form of adenosine triphosphate

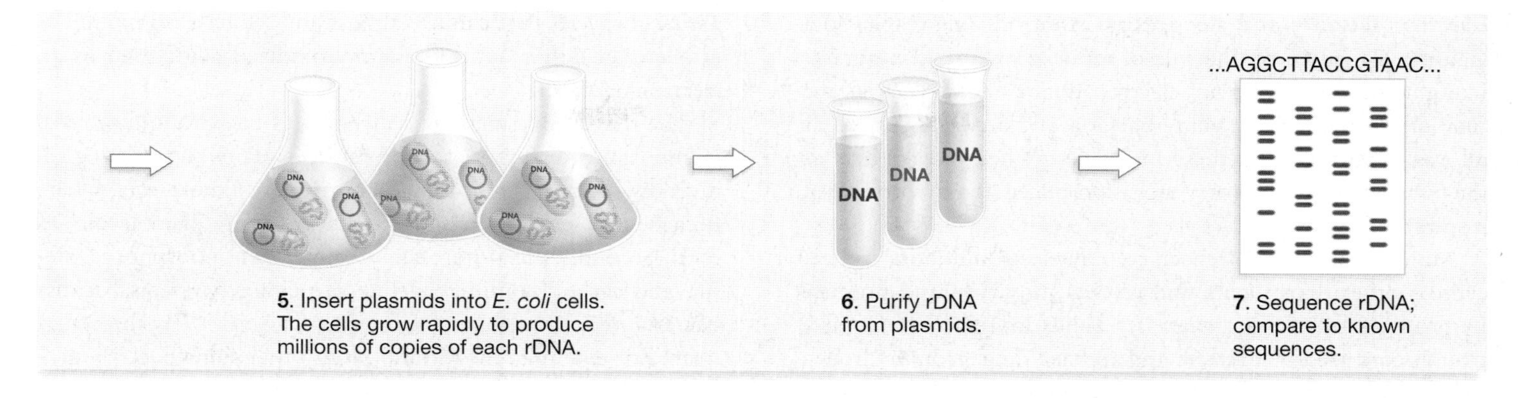

5. Insert plasmids into *E. coli* cells. The cells grow rapidly to produce millions of copies of each rDNA.

6. Purify rDNA from plasmids.

7. Sequence rDNA; compare to known sequences.

(ATP); and obtaining carbon in a form that can be used to synthesize fatty acids, proteins, DNA, RNA, and other molecules needed to build cells. Bacteria and archaea produce ATP in three ways. They can either use light energy to promote electrons to the top of electron transport chains, oxidize reduced organic molecules like sugars, or oxidize reduced inorganic molecules like ammonia (NH_3) or methane (CH_4). Bacteria and archaea obtain carbon either by processing inorganic sources like CO_2 and CH_4 or by absorbing ready-to-use organic compounds from their environment. **Table 25.2** summarizes the diversity of ways that these cells obtain energy and carbon.

What makes this remarkable diversity possible? Bacteria and archaea have evolved dozens of variations on the most basic themes of metabolism: extracting usable energy from reduced compounds, using light to produce high-energy electrons, and fixing carbon. This metabolic diversity is important because it allows bacteria and archaea to occupy diverse habitats. Let's take a closer look.

Obtaining Energy from Reduced Molecules

Millions of species obtain the energy required to make ATP by oxidizing reduced organic compounds like sugars, starch, or fatty acids. As Chapter 6 showed, cellular enzymes can strip electrons from reduced organic molecules and transfer them to electron carriers like NADH and $FADH_2$. These compounds, in turn, feed electrons to the molecules that make up electron transport chains. As electrons are stepped down from a high-energy state in a reduced compound to a low-energy state in an oxidized compound, a proton motive force is generated. As Chapter 6 pointed out, a flow of protons through ATP synthase results in the production of ATP (see **Figure 25.5**, page 490).

The essence of this process, called respiration, is that a highly reduced molecule serves as an original electron donor, while a highly oxidized molecule serves as a final electron acceptor. The potential energy difference between these reduced and oxidized compounds is transformed into chemical energy in the form of ATP.

TABLE 25.2 Strategies for Obtaining Energy and Carbon: An Overview

As the columns in this table show, organisms are called autotrophs if they obtain carbon-containing compounds by synthesizing them from inorganic sources like carbon dioxide (CO_2). If they acquire the carbon-containing molecules they need from organic molecules, they are called heterotrophs.

As the rows in the table indicate, organisms are called phototrophs if they use light as a source of energy to produce ATP, organotrophs if they produce ATP by oxidizing reduced organic molecules like glucose, and lithotrophs if they produce ATP by oxidizing reduced inorganic molecules like hydrogen (H_2).

Source of Energy	Source of Carbon: **Autotrophs** Synthesize their own reduced organic compounds from CO_2, CH_4, or other inorganic sources.	Source of Carbon: **Heterotrophs** Use reduced organic compounds produced by other organisms.
Light (phototrophs)	**Photoautotrophs** Cyanobacteria use photosynthesis to produce ATP. They fix CO_2 via the Calvin cycle.	**Photoheterotrophs** Heliobacteria use photosynthesis to produce ATP. They absorb reduced carbon molecules from the environment.
Reduced Organic Molecules (organotrophs)	**(no term)** *Clostridium aceticum* ferments glucose to produce ATP. It uses reactions called the acetyl-CoA pathway to fix CO_2.	**Chemoorganotrophs** *E. coli* uses fermentation or respiration to produce ATP. It absorbs reduced carbon molecules from the environment.
Reduced Inorganic Molecules (lithotrophs)	**Chemolithotrophs** Nitrifying bacteria like species of *Nitrosomonas* produce ATP via respiration, using ammonia (NH_3) as an electron donor. They fix CO_2 via the Calvin cycle.	**Chemolithotrophic Heterotrophs** *Beggiatoa* produces ATP via respiration, using hydrogen sulfide (H_2S) as an electron donor. It absorbs reduced carbon molecules from the environment.

Electron Donors and Acceptors In introducing respiration, Chapter 6 focused on the role of reduced organic compounds like glucose as the original electron donor and oxygen as the final electron acceptor. Many bacteria and archaea, as well as all eukaryotes, rely on these molecules. When oxygen acts as the final electron acceptor, water is produced as a by-product of respiration.

Many bacteria and archaea employ electron donors and acceptors other than sugars and oxygen, however, and produce by-products other than water. As **Table 25.3** shows, the electron donors used by bacteria and archaea range from hydrogen molecules (H_2) and hydrogen sulfide (H_2S) to ammonia (NH_3) and methane (CH_4). In addition to oxygen, some organisms use oxidized compounds like sulfate (SO_4^{2-}), nitrate (NO_3^-), carbon dioxide (CO_2), or ferric ions (Fe^{3+}) as electron acceptors.

Box 25.1 introduces a technique called enrichment culture that researchers use to determine which electron donors and acceptors are used by a particular species. Decades of studies using enrichment cultures have left researchers marveling at the metabolic diversity of bacteria and archaea. It is only a slight exaggeration to claim that researchers have found bacterial and archaeal species that can use almost any reduced compound as an electron donor and almost any oxidized compound as an electron acceptor.

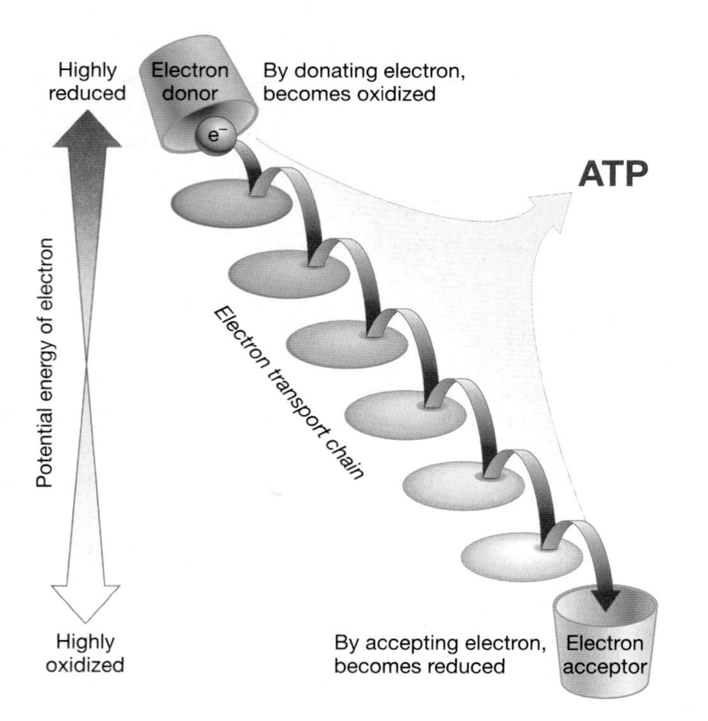

FIGURE 25.5 Respiration Requires an Electron Donor and an Electron Acceptor
Bacteria and archaea can exploit a wide variety of electron donors and acceptors. Some of these electron donors are more highly reduced than others, and some electron acceptors are more highly oxidized than others. **EXERCISE** After reading section 25.2 and studying Table 25.3, add the chemical formula for a specific electron donor, electron acceptor, and reduced by-product to the diagram. Do this for a particular species of bacteria or archaea. Then write in the electron donor, electron acceptor, and reduced by-product observed in humans.

The remarkable metabolic diversity of bacteria and archaea is important for two reasons. First, it explains their ecological diversity. Bacteria and archaea are found almost everywhere because they exploit an almost endless variety of molecules as food. Second, crucial nutrients like nitrogen, phosphorus, sulfur, and carbon continue cycling through ecosystems because bacteria and archaea can utilize them in almost any molecular form. For example, the nitrate (NO_3^-) that some bacteria produce as a by-product of respiration is used as an electron acceptor by other species and converted to molecular nitrogen (N_2). This molecule, in turn, is converted to ammonia (NH_3) by yet another suite of bacterial and archaeal species. In this way, bacteria and archaea are responsible for driving the global nitrogen cycle described in section 25.3.

Fermentation Chapter 6 introduced a strategy for making ATP from reduced organic compounds that does not involve electron transport chains. In fermentation, no outside electron acceptor is used; the redox reactions are internally balanced (see Chapter 6). Because fermentation is a less efficient way to make ATP compared to respiration, in many species it occurs as an alternative metabolic strategy when no electron acceptors are available to the cell. In other species, fermentation does not occur at all; in still other species, fermentation is the only way for cells to make ATP.

Although Chapter 6 focused on how glucose is fermented to ethanol or lactic acid, some bacteria and archaea are capable of using other reduced organic compounds as fermentable substrates. For example, the bacterium *Clostridium aceticum* can ferment nonsugars like ethanol, acetate, and fatty acids—as well as fermenting glucose. Other species of *Clostridium* ferment complex carbohydrates (including cellulose or starch), proteins, amino acids, or even purines. Species that ferment amino acids produce end products with names like cadaverine and putrescine. These molecules are responsible for the repugnant odor of rotting flesh. Other bacteria can ferment lactic acid, a prominent component of milk. This fermentation has two end products: propionic acid and CO_2. The propionic acid is responsible for the distinctive taste of Swiss cheese; the CO_2 is responsible for the holes.

The diversity of enzymatic pathways observed in bacterial and archaeal fermentations extends the metabolic repertoire of these organisms, and supports the claim that as a group, bacteria and archaea can use virtually any reduced molecule as a source of energy. Given this diversity, it is no surprise that bacteria and archaea are found in such widely varying habitats. Different environments offer different energy-rich molecules; various species of bacteria and archaea have evolved the biochemical machinery required to exploit most or all of these food sources.

TABLE 25.3 Some Electron Donors and Acceptors Used by Bacteria and Archaea

Electron Donor	Electron Acceptor	Product	Category*
H_2 or organic compounds	SO_4^{2-}	H_2S	sulfate-reducers
H_2	CO_2	CH_4	methanogens
CH_4	O_2	CO_2	methanotrophs
S or H_2S	O_2	SO_4^{2-}	sulfur bacteria
organic compounds	Fe^{3+}	Fe^{2+}	iron-reducers
NH_3	O_2	NO_2^-	nitrifiers
organic compounds	NO_3^-	N_2O, NO, or N_2	denitrifiers (or nitrate reducers)
NO_2^-	O_2	NO_3^-	nitrosifiers

*This column gives the name biologists use to identify species that use a particular metabolic strategy.

BOX 25.1 Culturing Techniques as a Research Tool

To study how bacteria and archaea obtain carbon and energy in a usable form, biologists rely heavily on their ability to culture the organisms in the lab. Researchers place specific electron donors and electron acceptors or certain fermentable substrates in the culture medium. Then they observe whether the organism they are studying can grow.

When this approach is used to isolate organisms from a natural environment, it is termed an enrichment culture. The logic is that by defining which energy sources are available to the organisms in a sample of water or soil, researchers can isolate strains or species that depend on that source, enrich their numbers, and obtain a population large enough to study in detail.

To appreciate how this strategy works in practice, consider recent research published by Shi Liu and colleagues. Inspired by reports that bacteria had been found in specimens of volcanic rock deep below Earth's surface, the team obtained samples of rock and fluid from drilling operations in the states of Virginia and Colorado. The samples came from sedimentary rocks at depths ranging from 860 to 2800 meters below the surface. At these depths, temperatures are between 42°C and 85°C. The questions posed by Liu and colleagues were simple: Is anything alive down there? If so, what do the organisms eat?

Liu and co-workers hypothesized that if organisms living in the deep subsurface were capable of respiration, they might use hydrogen molecules (H_2) as an electron donor and the ferric ion (Fe^{3+}) as an electron acceptor. Fe^{3+} is the oxidized form of iron and is abundant in the rocks they collected. It exists in the deep subsurface in the form of ferric oxyhydroxide. If an organism in the samples reduced the ferric ions during cellular respiration, the researchers predicted that a black, oxidized, and magnetic mineral called magnetite (Fe_3O_4) would start appearing in the cultures as a by-product of respiration.

What did their enrichment cultures produce? In some culture tubes, a black compound began to appear within a week. Using a variety of tests, Liu and associates confirmed that the black substance was indeed magnetite. Microscopy revealed the organisms themselves—previously undiscovered bacteria (**Figure 1**). Because they grow only when incubated at between 45°C and 75°C, these organisms are considered thermophilic (heat-loving).

This discovery was spectacular, because it hinted that Earth's crust may be teeming with organisms to depths of over a mile below the surface. These remarkable bacteria flourish at temperatures that would instantly kill a human being.

FIGURE 1 Bacteria from the Deep Subsurface
These bacteria grow in crevices of rocks hundreds of meters below Earth's surface.

Obtaining Energy from Light

Instead of using reduced molecules as a source of high-energy electrons, phototrophs (literally, "light-eaters") pursue a radically different strategy: They use the kinetic energy in light to raise electrons to high-energy states. As these electrons are stepped down to lower energy states by electron transport chains, the energy released is used to generate ATP.

Chapter 7 (Photosynthesis) introduced an important feature of phototrophy: the process requires a source of electrons. In the cyanobacteria and plants, the required electrons come from water. When these organisms "split" water molecules apart to obtain electrons, they generate oxygen as a by-product. In contrast, many phototrophic bacteria use a reduced molecule other than water as the source of electrons. Frequently the electron donor is hydrogen sulfide (H_2S); a few species can use the ion known as ferrous iron (Fe^{2+}). Instead of producing oxygen as a by-product of photosynthesis, these cells produce elemental sulfur (S) or the ferric ion (Fe^{3+}). They live in habitats, like the mud at the bottom of lakes, where oxygen is completely absent.

Chapter 7 also introduced the photosynthetic pigments found in plants and explored the light-absorbing properties of chlorophylls *a* and *b*. Cyanobacteria have these same two pigments. But as **Figure 25.6** shows, researchers have isolated seven additional chlorophylls, from bacterial species other than cyanobacteria. Further, research has confirmed that each of these pigments has a unique absorption spectrum, and that each major group of photosynthetic bacteria has one or more of these distinctive chlorophylls. Why are bacterial chlorophylls so diverse? The leading hypothesis is that photosynthetic species with different absorption spectra are able to live side by side without competing for light. If so, then the diversity of photosynthetic pigments observed in bacteria has been an important mechanism for generating biodiversity among phototrophs.

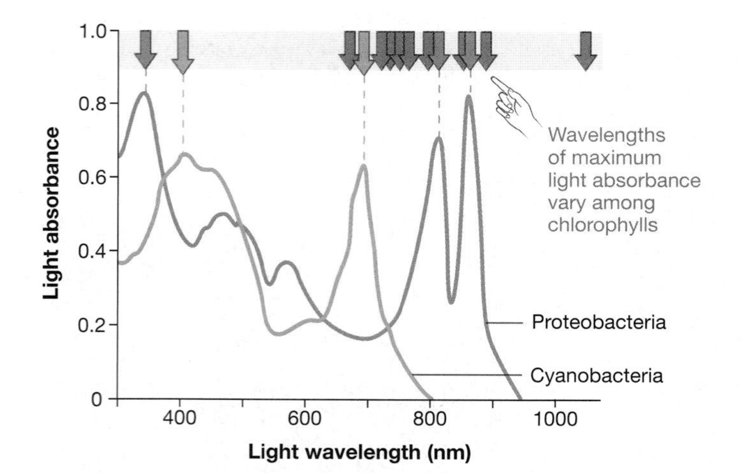

FIGURE 25.6 The Wavelengths of Light Absorbed by Chlorophylls Found in Photosynthetic Bacteria
This diagram shows the wavelengths of light that are absorbed by the various types of chlorophyll found in different species of photosynthetic bacteria. The complete absorbance spectrum is given for the chlorophyll found in cyanobacteria (green line) and in photosynthetic proteobacteria (blue line). The red arrows point to absorbance peaks observed in chlorophylls isolated from other groups of bacteria. QUESTION Why would it be advantageous for different species of photosynthetic bacteria to have chlorophylls that absorb different wavelengths of light?

Pathways for Fixing Carbon

In addition to acquiring energy, organisms must obtain building-block molecules that contain carbon-carbon bonds. Chapters 6 and 7 introduced the two mechanisms that organisms use to procure usable carbon—either making their own or getting it from other organisms. In cyanobacteria and plants, the enzymes of the Calvin cycle transform carbon dioxide (CO_2) to organic molecules that can be used in synthesizing cell material. Animals and fungi, in contrast, obtain carbon by eating plants or animals or by absorbing the organic compounds released as dead tissues decay.

Bacteria and archaea pursue these same two strategies, but with an interesting twist: Some species can fix carbon from sources other than CO_2. For example, certain proteobacteria are called methanotrophs (literally, "natural gas eaters") because they use methane (CH_4) as their primary electron donor and carbon source. Among the methanotrophs, CH_4 is assimilated via two different enzymatic pathways, depending on the species. Some bacteria can utilize other inorganic sources of carbon, ranging from carbon monoxide (CO) to methanol (CH_3OH).

Further, several groups of bacteria can fix CO_2 using pathways other than the Calvin cycle. To date, researchers have identified three additional biochemical processes that result in CO_2 fixation. What is the message of these observations? Although the enzymes of the Calvin cycle are widespread among the bacteria and archaea, several species have evolved different solutions to the problem of fixing carbon.

25.3 Bacteria, Archaea, and Global Change

The metabolic diversity of bacteria and archaea explains their tremendous taxonomic and ecological diversity. But the biochemical capabilities of bacteria and archaea, combined with their numerical abundance, also make these organisms a potent force for global change. Consider just two of the claims made by biologists: that metabolism by cyanobacteria gave Earth its first oxygen atmosphere, and that nitrogen-fixing bacteria have a profound effect on the productivity of both agricultural and natural ecosystems. What evidence supports these conclusions?

The Oxygen Revolution

Climatologists who study the composition of Earth's atmosphere are virtually certain that no free molecular oxygen (O_2) existed for the first 2.5 billion years after the planet formed. This hypothesis is based on two observations: (1) there was no plausible source of oxygen at the time the planet cooled to a solid state; and (2) the oldest Earth rocks indicate that for many years afterward, any oxygen that formed immediately reacted with iron atoms to produce iron oxides like hematite (Fe_2O_3) and magnetite (Fe_3O_4).

Scientists are confident that free oxygen was scarce when the planet formed because little, if any, is released when contemporary rocks are heated. In contrast, large amounts of nitrogen gas (N_2), carbon dioxide (CO_2), and water vapor are released by today's volcanoes. These molecules are thought to be the major components of the early atmosphere.

Oxygen was present in the early oceans, however, because the fossil record indicates that cyanobacteria were producing quantities of oxygen very early in Earth's history. If these organisms released oxygen into the ocean as a by-product of photosynthesis, what happened to all of this oxygen? The oldest sedimentary rocks on Earth, from the 3.75-billion-year-old Isua sequence of Greenland, suggest that whatever free oxygen existed immediately reacted with iron atoms. This is because the Isua rocks contain numerous banded iron formations, or BIFs (**Figure 25.7a**). These are layers of iron oxides like hematite (Fe_2O_3) and magnetite (Fe_3O_4). Geologists hypothesize that these sediments formed when iron atoms, liberated from Earth's crust by volcanic activity, dissolved in the ocean and reacted with oxygen released by cyanobacteria. The scenario is that insoluble iron oxides precipitated out of the water, collected on the ocean floor, and eventually become cemented into the sedimentary rocks called BIFs.

BIFs become increasingly common in Earth rocks through time, peaking in abundance about 2.5 billion years ago and then declining until their disappearance about 2.0 billion years ago. During that interval, a mineral called pyrite is also common in ocean sediments. This is important because pyrite (FeS_2, also called iron sulfide or fool's gold) is easily oxidized and destroyed. Because pyrite is common in rocks older than 2.0 billion years, geologists can infer that little molecular oxygen was available in the oceans during that interval.

According to the geologic record, then, quantities of free oxygen did not begin to build up in the oceans until 2.0 billion years ago. A mere 200 million years later, at 1.8 billion years ago, red beds begin to appear (**Figure 25.7b**). Like BIFs, red beds are rocks dominated by iron oxides, and therefore oxygen is clearly present by this point. Unlike BIFs, however, red beds form in terrestrial environments. The appearance of red beds suggests that oxygen began to be common in the atmosphere as well as the ocean about 1.8 billion years ago.

Once oxygen was common in the oceans, cells could begin to use it as an electron acceptor. Aerobic respiration was now a possibility. This was a crucial event in the history of life. Because oxygen is extremely electronegative, it is a powerful electron acceptor. Much more energy is released as electrons fall down electron transport chains with oxygen as the ultimate acceptor than with other substances as the electron acceptor. Once oxygen was available, much more ATP could be produced for each electron donated by NADH or $FADH_2$. As a result, the rate of energy production and metabolism could rise dramatically.

Coincidentally, 2.0 billion years ago is about when the first macroscopic algae appear in the fossil record. Biologists hypothesize that a causal link was involved with the availability of free oxygen. The claim is that multicellularity and large size were made possible by the high metabolic rates and rapid growth fueled by aerobic respiration.

To summarize, the data indicate that cyanobacteria were responsible for a fundamental change in Earth's atmosphere, from one dominated by nitrogen gas and carbon dioxide to one

FIGURE 25.7 Banded Iron Formations and Red Beds
(a) Banded iron formations (BIFs) form in marine environments. **(b)** Red beds form in terrestrial environments. Both types of rock consist of iron oxides, which form when iron reacts with oxygen. Biologists hypothesize that oxygen was not available to organisms for use as an electron acceptor until virtually all of the free iron in Earth's oceans and lakes was locked up in BIFs and red beds.

dominated by nitrogen gas and oxygen. Never before, or since, have organisms done so much to alter the nature of the planet.

Nitrogen Fixation

Chapter 51 introduces the phenomenon known as biogeochemical cycling and explores how molecules like carbon and oxygen move between abiotic and biotic components of ecosystems. Ecologists have been particularly interested in the nitrogen cycle, illustrated in **Figure 25.8**, because fertilizing agricultural and natural ecosystems with this nutrient frequently results in increased productivity. As a result, researchers infer that plant growth is often limited by the availability of nitrogen.

In Figure 25.8, the arrow labeled "Fixation by bacteria and archaea" marks the most critical step in the nitrogen cycle. Bacteria and archaea are responsible for this event. In this phase of the cycle, molecular nitrogen (N_2) is absorbed from the atmosphere and reduced to ammonia (NH_3). Ammonia, in turn, can be used to build nitrogen-containing compounds like amino acids and nucleic acids. Bacteria and archaea are the only organisms known that can fix nitrogen. Without them, nitrogen cycling would stop and ecosystems would collapse.

Which Bacteria and Archaea Fix Nitrogen? Do the nitrogen-fixing bacteria and archaea belong to a particular lineage or share other distinctive characteristics? The short answer is no. Nitrogen fixation has been observed in species from many different taxonomic groups. Nitrogen fixers can be phototrophs, chemoorganotrophs, or chemolithotrophs, meaning that they use the energy in sunlight, reduced organic compounds, or reduced inorganic compounds to synthesize ATP (see Table 25.2). Many live inside the tissues of eukaryotes like pea-family plants, but many others are free-living in water or soil.

The common thread among nitrogen-fixers is that their genomes contain the *nif* genes, which code for the enzyme complex called nitrogenase. Nitrogenase is responsible for catalyzing the reduction of N_2. Recent work by Thomas Hurek and colleagues suggests that at least some of the taxonomic diversity of nitrogen-fixers results from lateral gene transfer. These researchers studied the nitrogenase genes found in several species from the group called β-proteobacteria. When they generated a phylogeny of the *nif* sequences, they found that the sequences from most species clustered together in relationships identical to those implied by their SSU rRNA. But the *nif* genes of two β-proteobacteria species were anomalous. These genes turned out to be extremely closely related to sequences from distant relatives called the α-proteobacteria. To explain this pattern, the researchers suggest that *nif* genes were transferred laterally between the lineages early in their history. This observation suggests that many nitrogen-fixing bacteria and archaea may have obtained the *nif* complexes through lateral transfer as well.

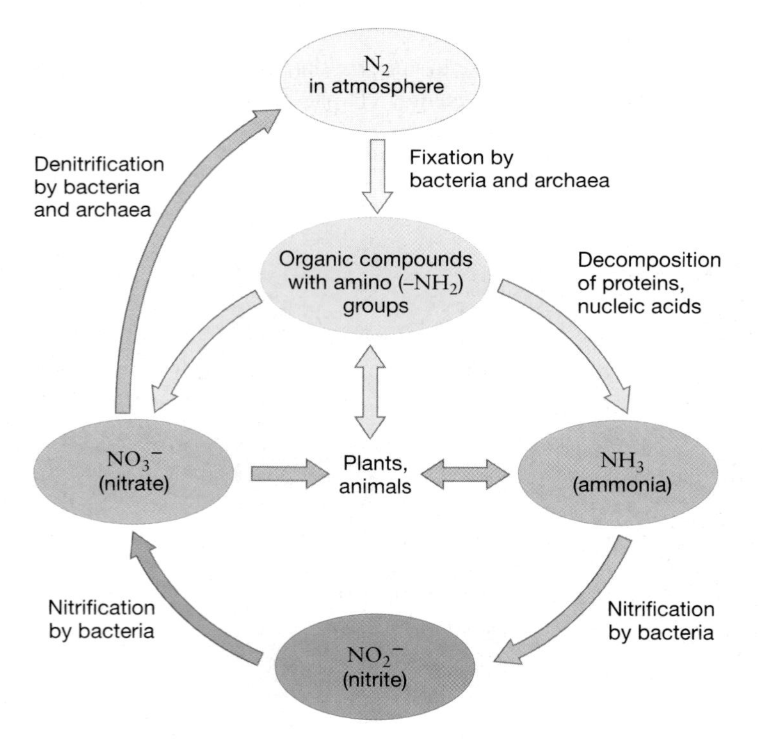

FIGURE 25.8 The Nitrogen Cycle

EXERCISE Suppose that bacteria and archaea were no longer capable of fixing nitrogen. Draw an X through the part of the cycle that would be affected.

Nitrates as a Pollutant Humans may not have *nif* genes, but people now affect the global nitrogen cycle on a scale that compares with the impact of bacteria and archaea. In the late 1940s, industrial chemists developed the capacity to produce ammonia from atmospheric nitrogen. In the decades that followed, farmers throughout the world began applying large quantities of ammonia-based fertilizers on their crops. This helped fuel an explosion of agricultural productivity and an increase in nutrition and quality of life in many countries.

Unfortunately, widespread use of ammonia-based fertilizers had an unforeseen side effect. As noted in section 25.2, ammonia is used as an electron donor by soil-dwelling bacteria and archaea, in addition to being used as a nutrient by corn and wheat. Because the nitrate (NO_3^-) produced as a by-product of bacterial ammonia metabolism dissolves readily in water, it runs off agricultural fields and enters aquatic ecosystems. There it sets off a series of events. As **Figure 25.9** shows, nitrate acts as a nitrogen fertilizer for cyanobacteria and algae, which respond by growing in massive numbers. When these organisms die, decomposers like heterotrophic bacteria and archaea grow in number. Metabolism by these organisms uses up all of the available oxygen in the water and produces anaerobic "dead zones," like the one now found in the Gulf of Mexico near the mouth of the Mississippi River.

Could other bacteria or archaea counteract the nitrate pollution that leads to dead zones? A recent finding by H. N. Schulz and colleagues suggests one potential approach. During a routine survey of species living in marine sediments off the coast of Namibia in southwest Africa, Schulz and co-workers discov-

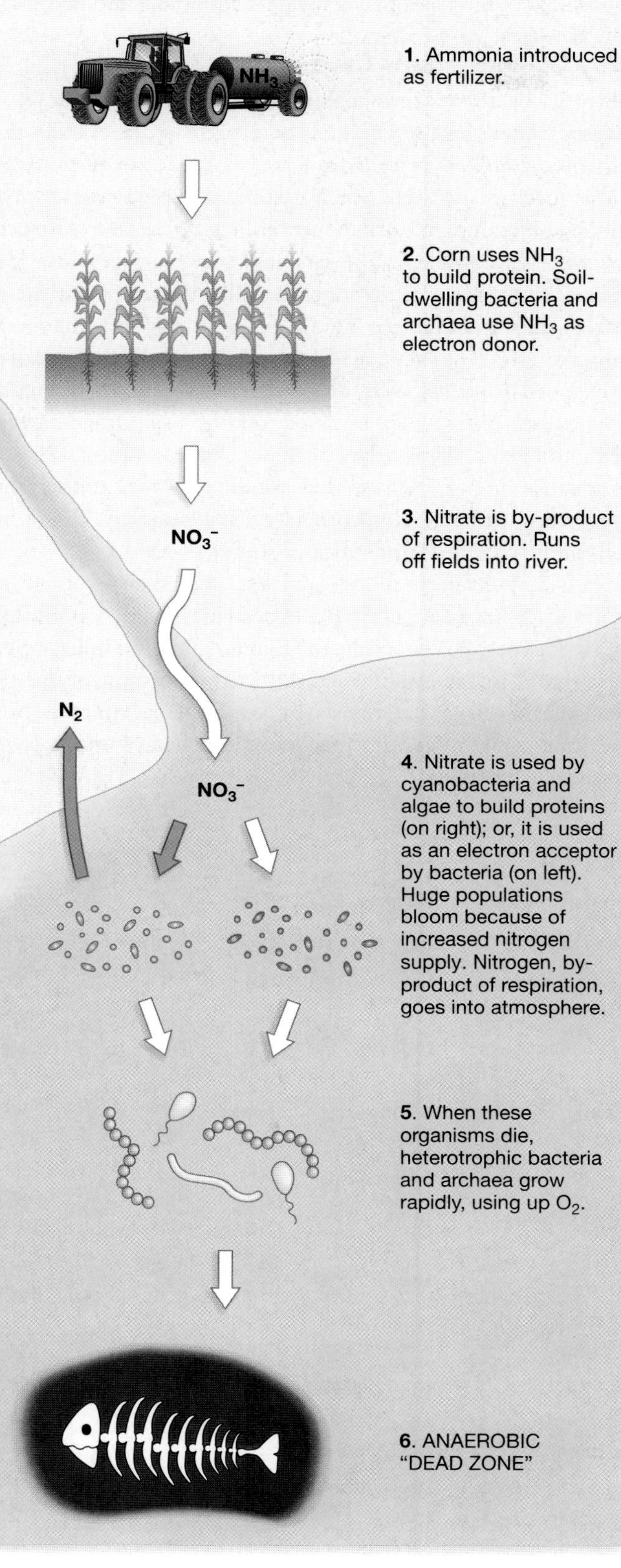

FIGURE 25.9 Why Nitrate Acts as a Pollutant in Aquatic Ecosystems

ered *Thiomargarita namibiensis*, the largest bacterium ever described. Although Schulz and co-workers have not yet been able to culture the organism and confirm its metabolic capacities in enrichment culture (see Box 25.1), several lines of evidence suggest that *T. namibiensis* uses hydrogen sulfide (H_2S) as an electron donor and nitrate as an electron acceptor. For example, large quantities of nitrate occur naturally in the sediments where the cells are common. The sediment also smells strongly of rotten eggs, which indicates the presence of H_2S. Growth of *T. namibiensis* increased rapidly when the researchers put additional nitrate and hydrogen sulfide in flasks containing sediment, cells, and seawater. Also, Schulz and associates were able to confirm by microscopy and direct chemical tests that the large size of the cells is due to an enormous central vacuole filled with nitrate.

Could *T. namibiensis* help mitigate this form of pollution by absorbing nitrate and reducing it to nitrogen gas? As Figure 25.9 shows, the possibility exists. Because *T. namibiensis* grows best in anoxic conditions, microbiologists are now exploring the possibility of introducing it to the dead zone in the Gulf of Mexico.

25.4 Bacterial Diseases

Bacteria and archaea affect human welfare indirectly through their role in global ecosystems. But bacteria also have a huge direct impact on human health and well-being. Although no archaea are known to cause disease in humans, a tiny fraction of bacterial species do.

In the late 1800s Robert Koch established that bacteria can cause disease. By the time Koch began his work, microscopists had confirmed the existence of the particle-like organisms we call bacteria, and Louis Pasteur had shown that microorganisms are responsible for spoiling foods like milk, wine, and broth. Koch hypothesized that microorganisms might also be responsible for causing infectious diseases, which spread by being passed from an infected individual to an uninfected individual.

Koch set out to prove his hypothesis by identifying the microorganism that was responsible for causing anthrax. Anthrax is a disease of cattle and other grazing mammals that can result in fatal blood poisoning. The disease also occurs, infrequently, in humans and mice.

To establish a causative link between a specific microbe and a specific disease, Koch proposed that four criteria had to be met:

- *The microbe must be present in individuals suffering from the disease and absent from healthy individuals.* By careful microscopy, Koch was able to show that a bacterium called *Bacillus anthracis* was always present in the blood of cattle suffering from anthrax, but absent from asymptomatic individuals.
- *The organism must be isolated and grown in a pure culture away from the host organism.* Koch was able to grow pure

colonies of *B. anthracis* in glass dishes on a nutrient medium, using gelatin as a substrate.

- *If organisms from the pure culture are injected into a healthy experimental animal, the disease symptoms should appear.* Koch demonstrated this in mice. The symptoms of anthrax infection appeared, and then the mice died.
- *The organism should be isolated from the diseased experimental animal, again grown in pure culture, and demonstrated by its size, shape, and color to be the same as the original organism.* Koch did this by purifying *B. anthracis* from the blood of diseased experimental mice.

These criteria, now called **Koch's postulates**, are still used to confirm a causative link between new diseases and a putative infectious agent. His experimental results also became the basis for the **germ theory of disease**. This theory, which laid the foundation for modern medicine, holds that infectious diseases are caused by bacteria and viruses. The list of bacteria that can cause illness in human beings is long and imposing; a small selection is given in **Table 25.4**. Pathogenic (disease-producing) forms come from many lineages in the domain.

If there is a trend in the table, it is that pathogenic bacteria tend to afflict tissues at the entry points to the body like wounds in the skin, pores in the skin, the respiratory and gastrointestinal tracts, and the urogenital canal. Perhaps the main message of the table is that pathogenic species are diverse. The immune system has to identify and eliminate hundreds if not thousands of disease-causing agents throughout the body.

Why Do Pathogens Cause Illness?

Why do bacteria cause disease? This question must be addressed at two levels. The first, or proximate, level focuses on explaining how host cells and tissues come to be damaged. These mechanisms vary. Some bacteria cause disease simply as a by-product of their normal metabolic activities. *Streptococcus mutans*, for example, is able to cling to the surface of teeth because it can synthesize an adhesive called dextran from sucrose (table sugar) found in the mouth. The cells also ferment sucrose to make ATP. The lactic acid produced by these fermentations is expelled from the cells, where it reacts with tooth enamel. This causes pitting that leads to tooth decay. Other bacteria, like pathogenic forms of the normal gut inhabitant *Escherichia coli*, cause disease because they produce specific toxins (these are usually proteins). Still others cause disease by killing host cells and using their contents as a source of food.

It is important to note, however, that many species and strains of bacteria are normal and even beneficial inhabitants of human tissues. For example, the human gut has a lining of bacteria that is several millimeters thick. At least some of these organisms may help their hosts by synthesizing vitamins or by breaking down molecules that are untouched by the host's digestive enzymes.

TABLE 25.4 Some Diseases Caused by Bacteria

Bacterium	Lineage	Tissues Affected	Disease
Chlamydia trachomatis	Planctomyces	urogenital canal	genital tract infection
Clostridium botulinum	Gram positives	gastrointestinal tract, nervous system	food poisoning (botulism)
Clostridium tetani	Gram positives	wounds, nervous system	tetanus
Haemophilus influenzae	Gram negatives	ear canal, nervous system	ear infections, meningitis
Mycobacterium tuberculosis	Gram positives	respiratory tract	tuberculosis
Neisseria gonorrhoeae	Proteobacteria (β group)	urogenital canal	gonorrhea
Propionibacterium acnes	Actinomycetes	skin	acne
Pseudomonas aeruginosa	Proteobacteria (β group)	urogenital canal, eyes, ear canal	urinary tract infections; eye and ear infections
Salmonella enteritidis	Proteobacteria (γ group)	gastrointestinal tract	food poisoning
Streptococcus pneumoniae	Gram positives	respiratory tract	pneumonia
Streptococcus pyogenes	Gram positives	respiratory tract	strep throat, scarlet fever
Treponema pallidum	Spirochetes	urogenital canal	syphilis
Vibrio parahaemolyticus	Proteobacteria (γ group)	gastrointestinal tract	food poisoning
Yersinia pestis	Gram negatives	lymph and blood	plague

The second, or ultimate, level of explanation for why bacteria cause disease focuses on why natural selection might favor pathogenic bacteria over strains that do not cause disease. Many parasitic bacteria display great variation in virulence among strains or species. Three species in the genus *Shigella*, which live in the human intestine, furnish a good example. *Shigella* infections can cause a severe form of diarrhea called dysentery. This disease killed hundreds of thousands of soldiers during both the American Civil War and World War I. Of the three species considered here, *S. dysenteriae* is by far the most virulent. It grows extremely quickly, produces paralysis and blood loss, and is frequently fatal. *S. flexneri*, in contrast, is much less virulent; and *S. sonnei* least virulent of all. These last two forms grow slowly and produce forms of dysentery that are relatively benign. What environmental conditions favor the virulent species at the expense of more benign forms?

Virulent Versus Benign Forms

Paul Ewald claims that when virulent and benign parasites victimize the same host and compete with each other, the long-term outcome of the competition depends on how rapidly the parasites are transmitted to new hosts. Ewald predicts that virulent strains are favored when transmission rates are high, whereas benign strains are favored when transmission rates are low. The logic here is simple: When transmission to a new host is rare, virulent strains are likely to kill their original host before they can be transmitted to a new host. They should die out, while benign forms should increase in frequency.

Ewald's transmission-rate hypothesis predicts that the deadly *S. dysenteriae* should be a serious medical concern only when public sanitation is poor. Why? *Shigella* cells spread when uninfected hosts drink water that has been contaminated with feces from infected hosts. If sanitation conditions are poor, meaning that many people are drinking water contaminated by feces, then patients infected with *S. dysenteriae* shed billions of bacterial cells into the water supply and infect many new hosts before their original host dies.

Do any data support this prediction? Ewald observes that in every region studied thus far, the most prevalent *Shigella* species has changed from the most virulent form (*S. dysenteriae*) to the intermediate form (*S. flexneri*) to the least virulent form (*S. sonnei*) as open sewers were gradually upgraded to modern, more hygienic sewer systems.

If Ewald's transmission-rate hypothesis is tested and confirmed in other species, it suggests a side benefit to improved standards of living: People are not only exposed to fewer pathogens, but the pathogens they are exposed to should become less virulent.

Essay Antibiotics and the Evolution of Resistance

Antibiotics are molecules that kill bacteria. A wide variety of naturally occuring antibiotics are produced by bacteria or fungi, presumably as a defense against bacterial species that compete with them for space and resources. These molecules work in several different ways. Some, such as penicillin, poison enzymes involved in the formation of bacterial cell walls. Others, like erythromycin, disrupt the enzymatic machinery involved in protein synthesis. Over 8000 antibiotics are known, and hundreds more are discovered each year. Only a tiny percentage, however, are useful in treating human diseases.

Drug resistance is now a fact of life . . .

The first antibiotic that brought important diseases under chemical control was discovered quite by accident. In 1928, Alexander Fleming was studying a pathogenic bacterium called *Staphylococcus* and noticed that cells began dying off in culture plates that had inadvertently been contaminated with a fungus called *Penicillium*. Fleming seized on the observation, confirmed that a molecule from the fungus was responsible for killing the cells, and called it penicillin. In the 1930s a group led by Howard Florey showed that the drug was effective against *Staphylococcus* and *Streptococcus* infections in humans. Florey then led a research consortium that worked out methods for producing the molecule in quantity by the start of World War II.

Penicillin was a "magic bullet"—a drug that killed many different infectious agents without harming the host. The drug saved many millions of lives that would have been cut short by *Staphylococcus* infections or pneumonia. Now, however, the magic is gone. In 1941, 10,000 units of penicillin administered four times a day for four days cured pneumonia completely. Now, pneumonia patients who receive 24 *million* units of penicillin a day have a good chance of dying. Similarly, in 1941 all strains of *Staphylococcus aureus* were treated effectively with penicillin. Today, 95 percent of the strains are unaffected by it.

To understand why these pathogens have become resistant to penicillin, recall from Chapter 21 that a random mutation in the gene for RNA polymerase can produce resistance to rifampin in *Mycobacterium tuberculosis*. In this case, resistance occurs because rifampin acts by binding to RNA polymerase and interfering with the transcription of *M. tuberculosis* genes;

(Continued on page 498)

(Essay continued)

the mutation in the gene for RNA polymerase makes rifampin bind less tightly. In the case of penicillin and many other antibiotics, resistance alleles are found not on the main bacterial chromosome but on plasmids. Penicillin, for example, acts by binding to the enzymes responsible for creating the cross-links in the main molecular building block of bacterial cell walls. This weakens the cell walls, which eventually rupture, causing cell death. But a plasmid-borne gene called β-*lactamase* codes for a protein that cleaves the penicillin molecule. This renders the drug useless, because it can no longer bind to the cross-linking enzymes. Through conjugation, or plasmid transfer, this resistance allele has spread rapidly among strains within pathogenic species and even *between* species.

Drug resistance is now a fact of life for drug companies, physicians, and patients. In response, medical agencies are trying to coordinate efforts to control how antibiotics are used. In Hungary, for example, a steep reduction in penicillin use in treating pneumonia, encouraged by the National Institute of Public Health, led to a sharp drop in the frequency of penicillin-resistant strains of *Pneumococcus*. If antibiotics are to continue being effective, they will have to be prescribed and taken much more carefully than they were in the past.

Chapter Review

Summary

Metabolic diversity is the hallmark of the bacteria and archaea, just as morphological complexity is the hallmark of the eukaryotes. Many bacteria and archaea can extract energy from reduced carbon compounds, such as sugars, through fermentation pathways or by transferring high-energy electrons to electron transport chains with oxygen as the final electron acceptor. But among the bacteria and archaea many different reduced inorganic or organic compounds serve as electron donors, and a wide variety of oxidized inorganic molecules serve as electron acceptors. Dozens of distinct organic compounds are fermented, including proteins, purines, alcohols, and an assortment of carbohydrates.

Photosynthesis is also widespread among bacteria. In cyanobacteria, water is used as a source of electrons during photosynthesis and oxygen gas is generated as a by-product. But in other species, the electron excited by photon capture comes from a reduced substance like ferrous iron (Fe^{2+}) or hydrogen sulfide (H_2S) instead of water (H_2O); the oxidized by-product is the ferric ion (Fe^{3+}) or elemental sulfur (S) instead of oxygen (O_2). These organisms also contain chlorophylls that are distinct from those found in plants and cyanobacteria.

To acquire molecules containing carbon-carbon bonds, some species use the enzymes of the Calvin cycle to reduce CO_2. But biologists have also discovered at least four additional biochemical pathways in bacteria and archaea that transform carbon dioxide (CO_2), methane (CH_4), or other sources of inorganic carbon into organic compounds like sugars or carbohydrates.

To study bacteria and archaea that cannot be cultured, biologists frequently take advantage of a research strategy called direct sequencing. In direct sequencing, DNA sequences are extracted directly from organisms in the environment—without culturing them in the lab. By analyzing where these sequences are placed on the tree of life, biologists can determine whether they represent new species. If they do, information on where the original sample was collected can expand our knowledge about the types of habitats used by bacteria and archaea.

Bacteria and archaea may be tiny, but they have a huge impact on global ecosystems and human health. Cyanobacteria were responsible for producing Earth's first oxygen-containing atmosphere, and nitrogen-fixing bacteria and archaea keep the global nitrogen cycle running. Bacteria also cause some of the most dangerous diseases in humans, including plague, syphilis, botulism, cholera, and tuberculosis. Disease results when bacteria kill host cells or produce toxins that disrupt normal cell functions. When bacterial strains vary in their ability to cause disease and compete with one another, the transmission-rate hypothesis makes the following predictions: Rapidly growing, highly virulent forms will be favored by natural selection when transmission to new hosts occurs frequently; while slow-growing, nonvirulent forms will be favored by natural selection when transmission to new hosts occurs rarely.

Questions

Content Review

1. How do molecules that function as electron donors and electron acceptors differ?
   a. Electron donors are organic molecules; electron acceptors are inorganic.
   b. Electron donors are inorganic molecules; electron acceptors are organic.
   c. Electron donors are more reduced; electron acceptors are more oxidized.
   d. Electron donors are more oxidized; electron acceptors are more reduced.

2. Why is molecular hydrogen (H_2) a common source of energy among bacteria and archaea?
   a. It is a highly reduced molecule.
   b. It is a highly oxidized molecule.
   c. It is readily available, because it is a gas.
   d. It is a common by-product of respiration.

3. What is a methanotroph?
   a. An organism that uses carbon monoxide (CO) as an energy source.
   b. An organism that uses methane (CH_4) as an electron donor.
   c. An organism that uses methane (CH_4) as an electron acceptor.
   d. An organism that produces methane as a by-product of respiration.

4. What do some photosynthetic bacteria use as a source of electrons instead of water?
   a. oxygen
   b. hydrogen sulfide (H_2S)
   c. organic compounds
   d. nitrate

5. What is distinctive about the chlorophylls found in different photosynthetic bacteria?
   a. their membranes
   b. their role in acquiring energy
   c. their role in carbon fixation
   d. their absorption spectra

6. What are organisms called that use light to promote electrons to high-energy states?
   a. phototrophs
   b. heterotrophs
   c. organotrophs
   d. autotrophs

7. How is direct sequencing used?
   a. to grow unknown organisms in laboratory culture
   b. to determine where organisms that cannot be cultured are placed on the tree of life
   c. to determine the metabolic capabilities of newly discovered prokaryotes
   d. to improve the accuracy of phylogenetic trees

8. What do Koch's postulates outline the requirements for?
   a. showing that an organism is lithotrophic
   b. showing that an organism is a bacterium, and not an archaea
   c. showing that an organism causes a particular disease
   d. showing that an organism has virulent as well as nonvirulent forms

Conceptual Review

1. Biologists often use the term *energy source* as a synonym for "electron donor." Why?
2. The text claims that the tremendous ecological diversity of bacteria and archaea is possible because of their impressive metabolic diversity. Do you agree with this statement? Why or why not?
3. Why is it logical to suppose that *Thiomargarita namibiensis* might help mitigate nitrate pollution in the Gulf of Mexico?
4. Suppose that universal PCR primers were available for genes involved in electron transport chains or for some of the bacteriochlorophylls. Why would it be interesting to use these genes in a direct sequencing study?
5. What is the evidence for lateral transfer of *nif* genes among bacteria and archaea?
6. The text claims that the evolution of an oxygen atmosphere paved the way for increasingly efficient respiration and higher-energy activities by organisms. Explain.
7. Look back at Table 25.3 and note that the by-products of respiration in some organisms are used as electron donors or acceptors by other organisms. In the table, draw lines between the dual-use molecules listed in the "Electron Donor," "Electron Acceptor," and "Product" columns.

Applying Ideas

1. The researchers who observed that magnetite was produced by bacterial cultures from the deep subsurface (see Box 25.2) carried out a follow-up experiment. They treated some of the cultures with a drug that poisons the enzymes involved in electron transport chains. In cultures where the drug was present, no more magnetite was produced. Does this result support or undermine their hypothesis that the bacteria in the cultures perform anaerobic respiration with Fe^{3+} serving as the electron acceptor? Explain your reasoning.
2. *Streptococcus mutans* obtains energy by oxidizing sucrose. This bacterium is abundant in the mouths of Western European and North American children and is a prominent cause of cavities. The organism, and cavities, are virtually absent in children from East Africa. Propose a hypothesis to explain this observation. Outline the design of a study that would test your hypothesis.
3. Use of condoms slows the spread of sexually transmitted diseases like syphilis, gonorrhea, and AIDS. According to the transmission-rate hypothesis, should the virulence of these diseases be increasing or decreasing in countries where condoms have recently increased in popularity?
4. Suppose that you've been hired by a firm interested in using bacteria to clean up organic solvents found in toxic waste dumps. Your new employer is particularly interested in finding cells that are capable of breaking a molecule called benzene into less toxic compounds. Where would you go to look for bacteria that might be able to metabolize benzene as an energy or carbon source? How would you design an enrichment culture capable of isolating benzene-metabolizing species?

5. Would you predict that disease-causing bacteria, like those listed in Table 25.4, obtain energy from light, reduced organic molecules, or reduced inorganic molecules? When they perform respiration, which substance would you predict that they use as an electron acceptor? Explain your answer.

6. Why would it be advantageous for certain strains of *Escherichia coli* to secrete a toxin that causes diarrhea? (This bacterium is a common and usually harmless inhabitant of the human gut.)

CD-ROM and Web Connection

CD Activity 25.1: The Tree of Life *(animation)*
(Estimated time for completion = 5 min)
Explore the concept of gene-swapping—or "lateral gene transfer"—and how it affects our understanding of the tree of life.

At your **Companion Website** (http://www.prenhall.com/freeman/biology), you will find self-grading exams and links to the following research tools, online resources, and activities:

Bizarre Life Forms Thrive Beneath Earth's Surface
This short article from the National Science Foundation traces the discovery of Archaea, likening it to Jules Verne's science fiction tale *A Journey to the Center of the Earth.*

Extremophiles
This *Scientific American* article focuses on the metabolic diversity of the Archaea, and the potential technological innovations that may arise from research on extremophile organisms.

Recapturing an Unused Resource
This paper from the Carnegie Mellon Research Institute explores the potential industrial and domestic uses of bacterial produced methane gas.

Bacterial Resistance
This site discusses how antibiotic resistance is an important factor in issues ranging from worldwide health to the use of anti-microbial cleaning products.

Additional Reading

Azam, F. 1998. Microbial control of oceanic carbon flux: The plot thickens. *Science* 280: 694–696. An introduction to the role of planktonic bacteria and archaea in the global carbon cycle.

Doolittle, R. F. 1998. Microbial genomes opened up. *Nature* 392: 339–342. A recent review of genome sequencing projects focused on Bacteria and Archaea.

Fredrickson, J. K., and T. C. Onstott. 1996. Microbes deep inside the Earth. *Scientific American* 275 (October): 68–73. A survey of biodiversity below the Earth surface.

Hoppert, M., and F. Mayer. 1999. Prokaryotes. *American Scientist* 87: 518–525. An introduction to how bacterial and archaeal cells are organized.

Jarrell, K. F., D. P. Bayley, J. D. Correia, and N. A. Thomas. 1999. Recent excitement about the Archaea. *BioScience* 49: 530–541. A summary of recent results on the distribution and ecology of archaea.

Madigan, M. T., and B. L. Marrs. 1997. Extremophiles. *Scientific American* (April): 82–87. A look at adaptations found in heat-, cold-, acid-, and salt-loving bacteria and archaea.

Pace, N. R. 1997. A molecular view of microbial diversity and the biosphere. *Science* 276: 734–740. A survey of recent data on the most ancient lineages in Bacteria and Archaea, by the scientist who developed direct sequencing.

Viruses

26

Viruses are parasites that afflict every twig on the tree of life. They are not cells. They cannot manufacture their own ATP or carbon-containing compounds, and they cannot make copies of themselves. Viruses enter cells, take over the cell's biosynthetic machinery, and use that machinery to manufacture a new generation of viruses.

Because they are not organisms, viruses are referred to as particles or agents and are not given scientific names. Most biologists would argue that they are not even alive. And yet viruses have a genome, they are superbly adapted to exploit the metabolic and synthetic capabilities of their host cells, and they evolve. **Table 26.1** (page 502) summarizes some of the characteristics of viruses.

The diversity and abundance of viruses almost defies description. Nearly all organisms examined thus far are parasitized by at least one kind of virus. The bacterium *Escherichia coli*, which resides in the human intestine, is afflicted by four different types of bacterial viruses or bacteriophage (literally, "bacteria eater"). In the human body virtually every system, tissue, and cell can be infected by one or more kinds of virus (**Figure 26.1**, page 502). The ocean's plankton teems with bacteria and archaea, yet viruses outnumber them in this habitat by a factor of 10 to 1.

Any study of life's diversity would be incomplete unless it included a look at the acellular parasites that exploit that diversity. But viruses are also important from a practical standpoint. To health-care workers, agronomists, and foresters, these parasites are a persistent—and sometimes catastrophic—source of misery and economic loss. After surveying the diversity of viruses in section 26.1, the rest of the chapter delves into a detailed account of the best-studied of all viruses—the human immunodeficiency virus. Section 26.2 introduces HIV; sections 26.3-26.6 analyze the four major phases of its life cycle.

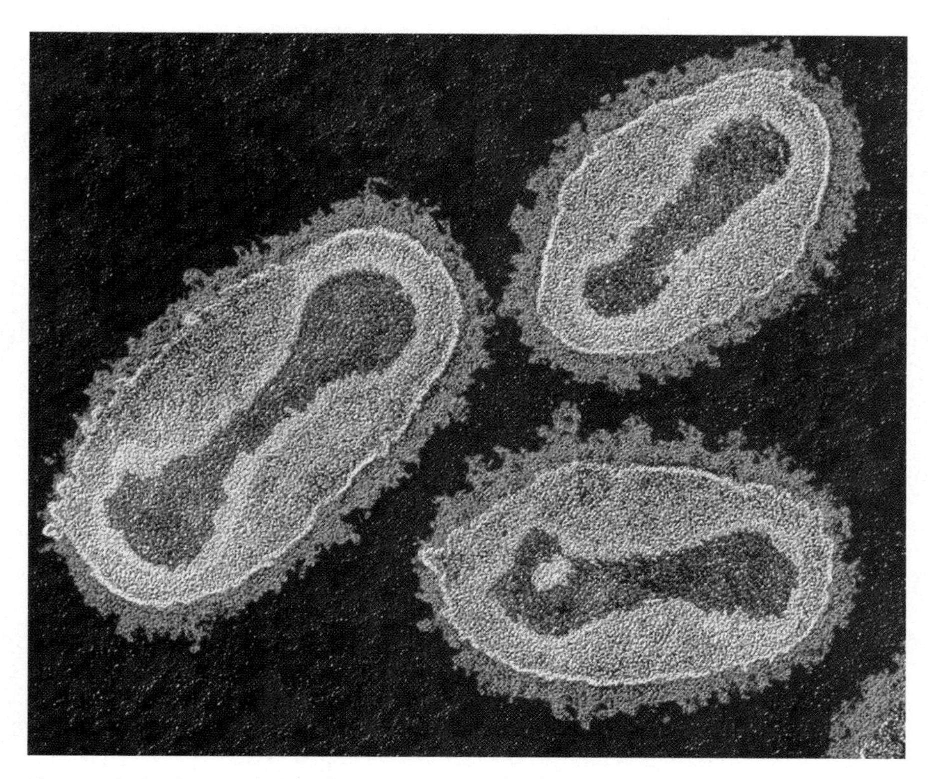

These colorized transmission electron micrographs show particles of smallpox virus. Although smallpox had been responsible for millions of deaths throughout human history, it is now eradicated.

TABLE 26.1 Characteristics of Viruses

	Viruses	Organisms
Hereditary material	DNA or RNA; can be single-stranded or double-stranded	DNA; always double-stranded
Cell membrane present?	No	Yes
Genetic recombination?	Yes	Yes
Can carry out transcription independently?	No—even if a polymerase is present, transcription of viral genome requires use of ATP and nucleotides provided by host cell	Yes
Can carry out translation independently?	No	Yes
Metabolic capabilities	Virtually none	Extensive—synthesis of ATP, reduced carbon compounds, vitamins, lipids, nucleic acids, etc.

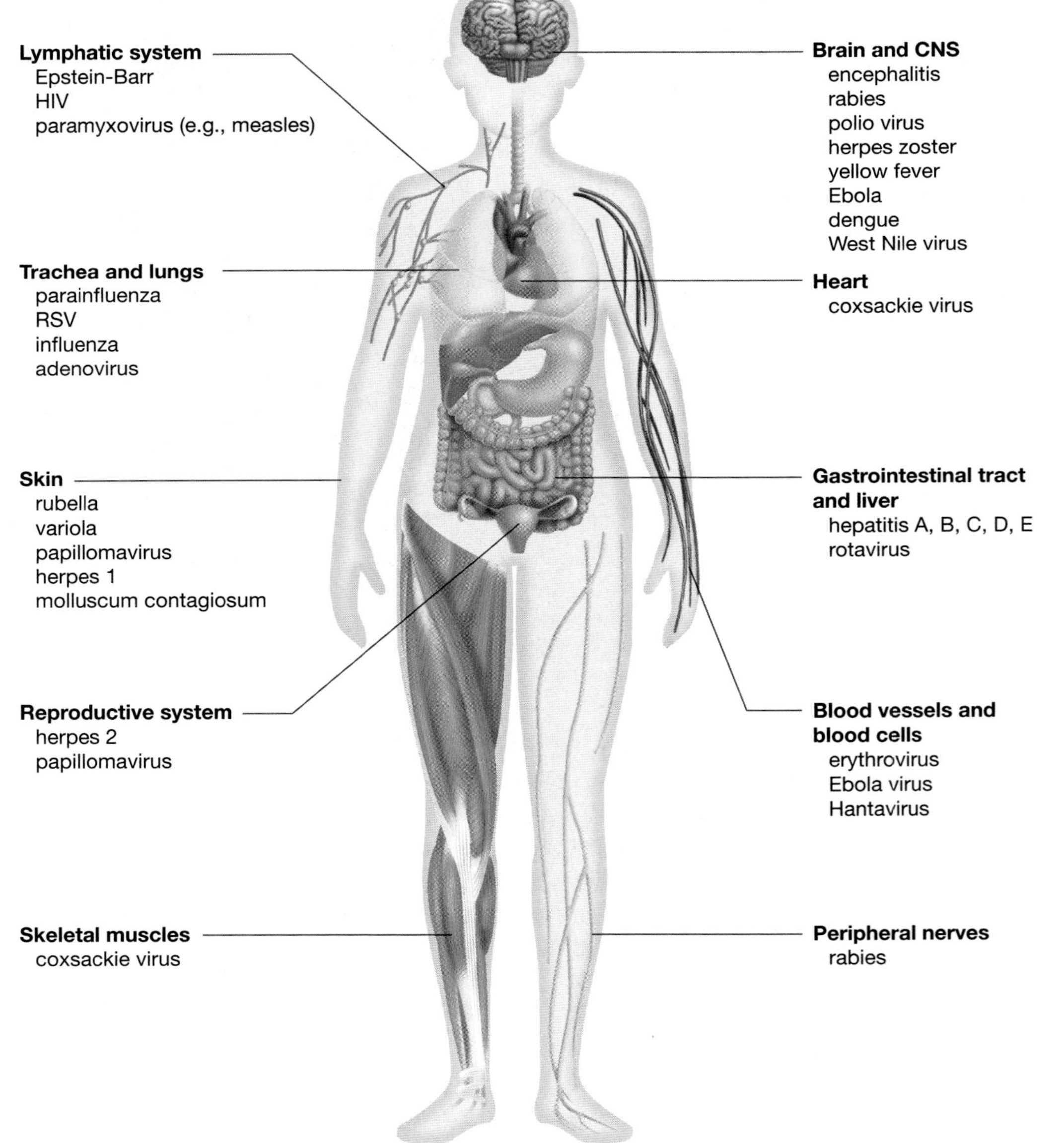

FIGURE 26.1 Human Cells, Tissues, and Systems That Are Parasitized by Viruses

EXERCISE Draw a box around the names of viruses for which you have been vaccinated. Circle viruses that have sustained infections in your body.

26.1 What Are the Viruses?

Studying viruses takes researchers into the realm of "nanobiology," where structures are measured in billionths of a meter. (One nanometer, abbreviated nm, is 10^{-9} meters.) Viruses are typically 50 to 100 nm in diameter; millions of viruses can fit on the head of a pin. A wine bottle filled with seawater taken from the ocean's surface contains about 10 *billion* virus particles—almost double the world population of humans (**Figure 26.2**). At these scales, transmission electron microscopy is the preferred tool for visualization. Only the very largest of the viruses, like the smallpox virus, are visible with a light microscope.

As the micrographs in **Figure 26.3a and b** (page 504) show, viruses are dwarfed by eukaryotic and bacterial cells, and viruses come in a wide variety of shapes. In regard to their morphological structure, however, they fall into just two general categories. Viruses can be enclosed by a shell of protein called a **capsid**, or by a capsid and a membrane-like **envelope** (see **Figure 26.3c**). In their morphology, then, the important distinction among viruses is whether they are enveloped or nonenveloped. Inside these structural elements, however, lies the true source of viral diversity—a bewildering variety of genomes. **Box 26.1** introduces hypotheses that attempt to explain this diversity.

What Is the Nature of the Viral Genome?

DNA is the hereditary material in all cells. As cells synthesize the molecules they need to function, information flows from DNA to mRNA to proteins. Although all cells follow this pattern, which is called the central dogma of molecular biology, some viruses break it. This conclusion traces back to work by H. Fraenkel-Conrat and Robley Williams, who were able to separate the protein and nucleic acid components of a particle known as the tobacco mosaic virus, or TMV. Surprisingly, the nucleic acid portion of this virus consisted of RNA, not DNA. Later experiments demonstrated that the RNA of TMV, by itself, could infect plant tissues and cause disease. This was a confusing result because it showed that in this virus, at least, RNA—*not* DNA—functions as the genetic material.

Subsequent research revealed an amazing diversity of viral genome types. In some groups of viruses, like the agents that cause measles and flu, the genome consists of RNA. In others, like the particles that cause herpes and smallpox, the genome is composed of DNA. Further, the RNA and DNA genomes of

FIGURE 26.2 Viruses Are Abundant in the Ocean
To create this photomicrograph, researchers treated seawater with a fluorescing compound that binds to nucleic acids. The smallest, most abundant dots are viruses. The larger, numerous spots are bacteria and archaea. The largest splotches are protists. QUESTION In this sample, viruses outnumber bacteria and archaea by about what factor?

BOX 26.1 Where Did Viruses Come From?

No one knows where viruses came from originally, but many biologists suggest that they are closely related to the plasmids and transposons introduced in Chapter 16. Viruses, plasmids, and transposons are all acellular, mobile genetic elements that replicate with the aid of a host cell. Simple viruses are actually indistinguishable from plasmids except for one feature: The viruses have a protein coat or membrane-like envelope.

Robin Weiss argues that simple viruses, plasmids, and transposons represent "escaped gene sets." Specifically, he proposes that these elements are descended from clusters of genes that escaped from bacterial or eukaryotic chromosomes long ago. According to this hypothesis, the escaped gene sets took on a mobile, parasitic existence because they happened to encode the information needed to replicate themselves at the expense of genomes that once held them.

Weiss contends that DNA viruses with large genomes originated in a very different manner, however. He proposes that the large DNA viruses trace their ancestry back to free-living bacteria that once took up residence inside eukaryotic cells. He hypothesizes that these organisms degenerated into viruses by gradually losing the genes required to synthesize ATP, nucleic acids, amino acids, and other compounds. Although the idea sounds speculative, it cannot be dismissed lightly. In Chapter 27 you'll review evidence that the organelles called mitochondria and chloroplasts, which reside inside eukaryotic cells, originated in much the same way. However, Weiss contends that instead of evolving into intracellular symbionts that aid their host cell, DNA viruses became parasites capable of destroying them.

To date, these hypotheses have not been tested rigorously. Until they are, there is no consensual view of where viruses came from.

(a) Virus particles are dwarfed by bacterial and eukaryotic cells.

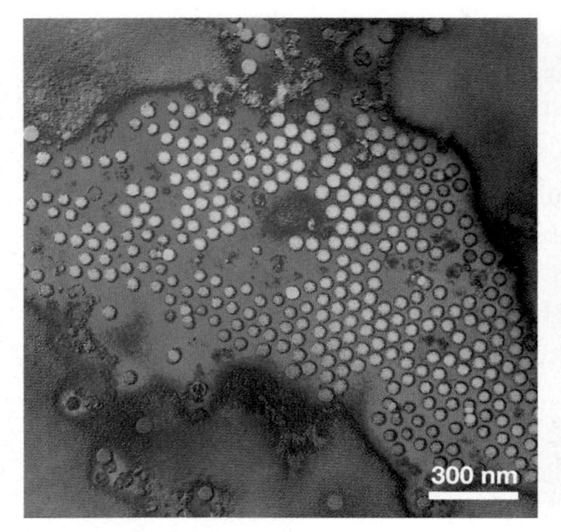

Virus particles (green dots)

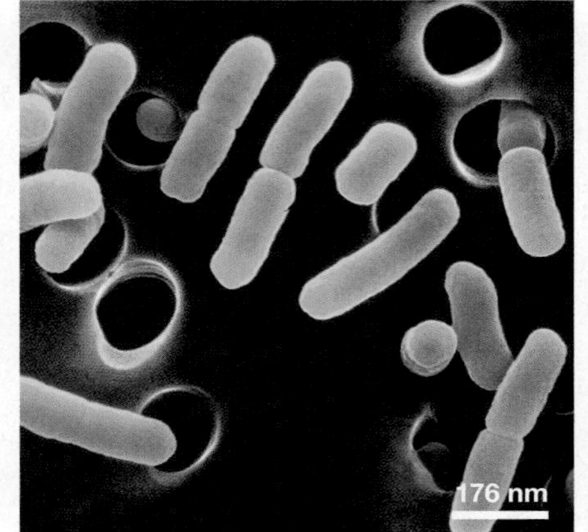

Bacterial cells

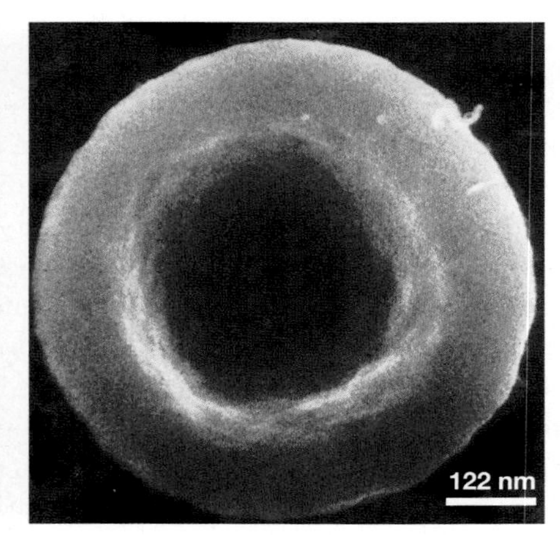

Eukaryotic cell

(b) Viruses come in a variety of shapes.

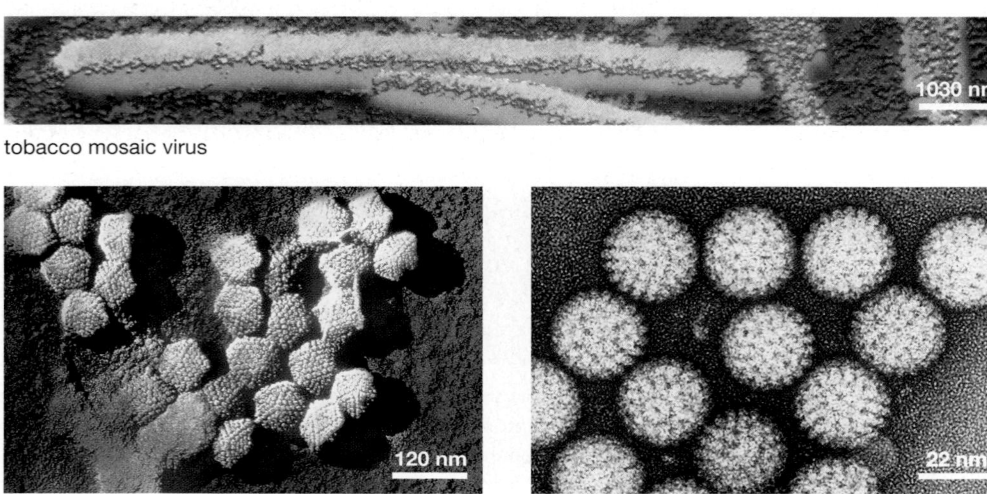

tobacco mosaic virus

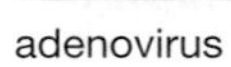

adenovirus

influenza virus

61 nm

bacteriophage T4

(c) An important distinction among viruses is whether or not they have envelopes.

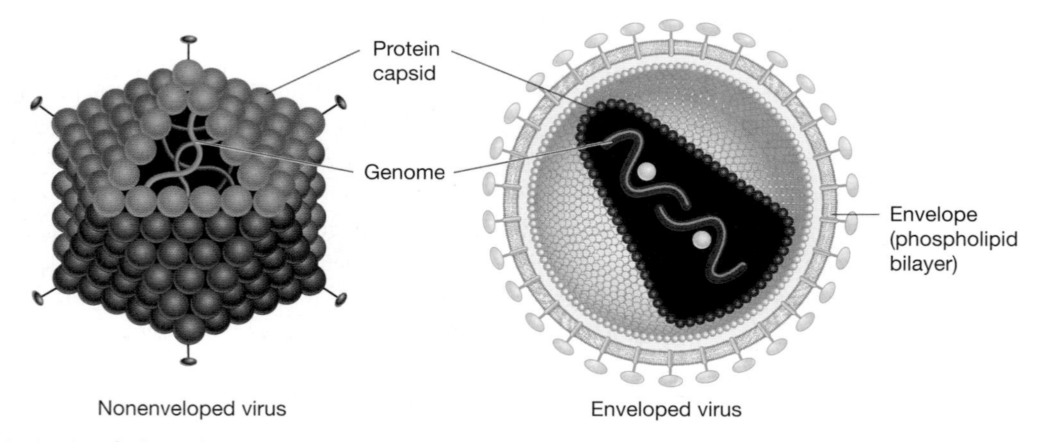

Nonenveloped virus

Enveloped virus

FIGURE 26.3 Morphological Diversity in Viruses
(a) Compared to bacterial cells and eukaryotic cells, viruses are tiny. Note the differences in the scale bars for each micrograph. **(b)** The shapes of viruses vary. Forms include rods, polyhedrons, spheres, and complex shapes with heads and tails. **(c)** Viruses can be nonenveloped or enveloped. A protein coat forms the exterior of nonenveloped viruses. In enveloped viruses, the exterior is composed of a membranous sphere. Inside this envelope, the hereditary material is enclosed by a protein coat.

viruses can be either single-stranded or double-stranded. The single-stranded genomes can also be classified as "positive sense" or "negative sense." In a positive-sense virus, the genome contains the same sequences as the mRNA required to produce viral proteins. In a negative-sense virus, the base sequences in the genome are complementary to those in viral mRNAs. (Some single-stranded genomes contain both positive-sense and negative-sense sequences!)

Finally, the number of genes found in viruses varies widely. The tymoviruses that infect plants contain as few as three loci, but the genome of smallpox can code for up to 263 proteins. **Table 26.2** summarizes the diversity of viral genome types.

How Do Viruses Copy Their Genomes?

Viruses must replicate their genomes to make a new generation of particles and continue an infection. Many DNA viruses accomplish this crucial step using their own DNA polymerase enzyme. This protein synthesizes copies of the viral genome using nucleotides provided by the host.

RNA viruses use completely different enzymes to replicate their genomes, however. In most of these agents, copies of the genome are synthesized by a viral enzyme known as RNA replicase. This is an RNA polymerase that synthesizes RNA from an RNA template, using ribonucleotides provided by the host cell. In other RNA viruses, however, the genome is transcribed from RNA to DNA by a viral enzyme called **reverse transcriptase.** This is a DNA polymerase that makes a single-stranded complementary DNA, or cDNA, from a single-stranded RNA template. Viruses that reverse-transcribe their genome in this way are called **retroviruses** (literally, "backward viruses"). The name is apt because the flow of information goes from RNA back to DNA. After DNA polymerase makes the **cDNA** double-stranded, another enzyme inserts it into a stretch of host-cell chromosome. From there the viral genes are transcribed to mRNA, which is translated into proteins by the host cell's ribosomes.

CD ACTIVITY 26.1 Herpes Virus Replication

Viral Replication Cycles

Although the hosts, morphology, and genomes of viruses vary widely, they all share the same basic replication cycle. As the top drawing in **Figure 26.4** (page 506) shows, a viral infection begins when a particle enters a host cell. Viral or host enzymes then make copies of the genome, using nucleotides and ATP provided by the host. The host cell also manufactures viral proteins. Once synthesis of the viral genome and proteins is complete, a new generation of particles assembles inside the host. Finally, the agents exit the cell. Once outside the host cell the particles die, infect a new host cell, or are transmitted to a new host individual.

This basic replication cycle is referred to as **lytic** growth, and usually results in the death of the host cell. In most cases, the killing mechanism is a massive disruption of the host-cell membrane as the new generation of virus particles leaves the cell.

Lysogeny Some DNA viruses perform an important variation on the lytic replication cycle. These particles, which include many of the "phages" that parasitize bacteria, are capable of inserting their DNA into the host's chromosome. Often this integration

TABLE 26.2 The Diversity of Viral Genomes

This table summarizes the major types of genomes found among viruses. Key: ss = single-stranded; ds = double-stranded; (+) = positive-sense (genome sequence is the same as viral mRNA); (−) = negative-sense (genome sequence is complementary to viral mRNA).

Genome	Example(s)	Host	Disease	Notes
(+)ssRNA	TMV	Tobacco plants	Tobacco mosaic disease (leaf wilting)	TMV was the first RNA virus to be discovered.
(−)ssRNA	Influenza	Many mammal and bird species	Influenza	The negative-sense ssRNA viruses transcribe their genomes to mRNA via a protein called RNA replicase.
dsRNA	Phytovirus	Rice, corn, and other crop species	Dwarfing	These viruses are transmitted from plant to plant by insects. Many can also replicate in their insect hosts.
ssRNA or (+)ssDNA that requires reverse transcription for replication	Rous sarcoma virus	Chickens	Sarcoma (cancer of connective tissue)	Rous sarcoma virus was identified as a cancer-causing agent in 1911, decades before any virus was seen.
ssDNA—can be (+), (−), or both	φX174 (phi X 174)	Bacteria	Death of host cell	The genome for φX174 is circular and was the first complete genome ever sequenced.
dsDNA	Baculovirus Smallpox Bacteriophage	Insects Humans Bacteria	Death Smallpox Death	These are the largest viruses in terms of genome size and overall size.

occurs without serious damage to the cell. Once the viral genome is in place, it is replicated by the host's DNA polymerase each time the cell divides. In this way, copies of the viral genome are passed on to descendant cells just like one of the host's own genes (see bottom drawing, **Figure 26.4**).

This integrated state is referred to as **lysogeny**. In the lysogenous state, a virus is often latent, or quiescent. This means that no new particles are being produced and no unrelated cells are being infected. The virus is simply transmitted from one generation to the next along with the host's genes.

A lysogenic phase in an infection cycle is possible if the virus's genome codes for an enzyme called integrase. This protein cuts the host DNA, inserts the viral DNA, and reanneals the broken ends. In some viruses integrase can also catalyze the reverse reac-

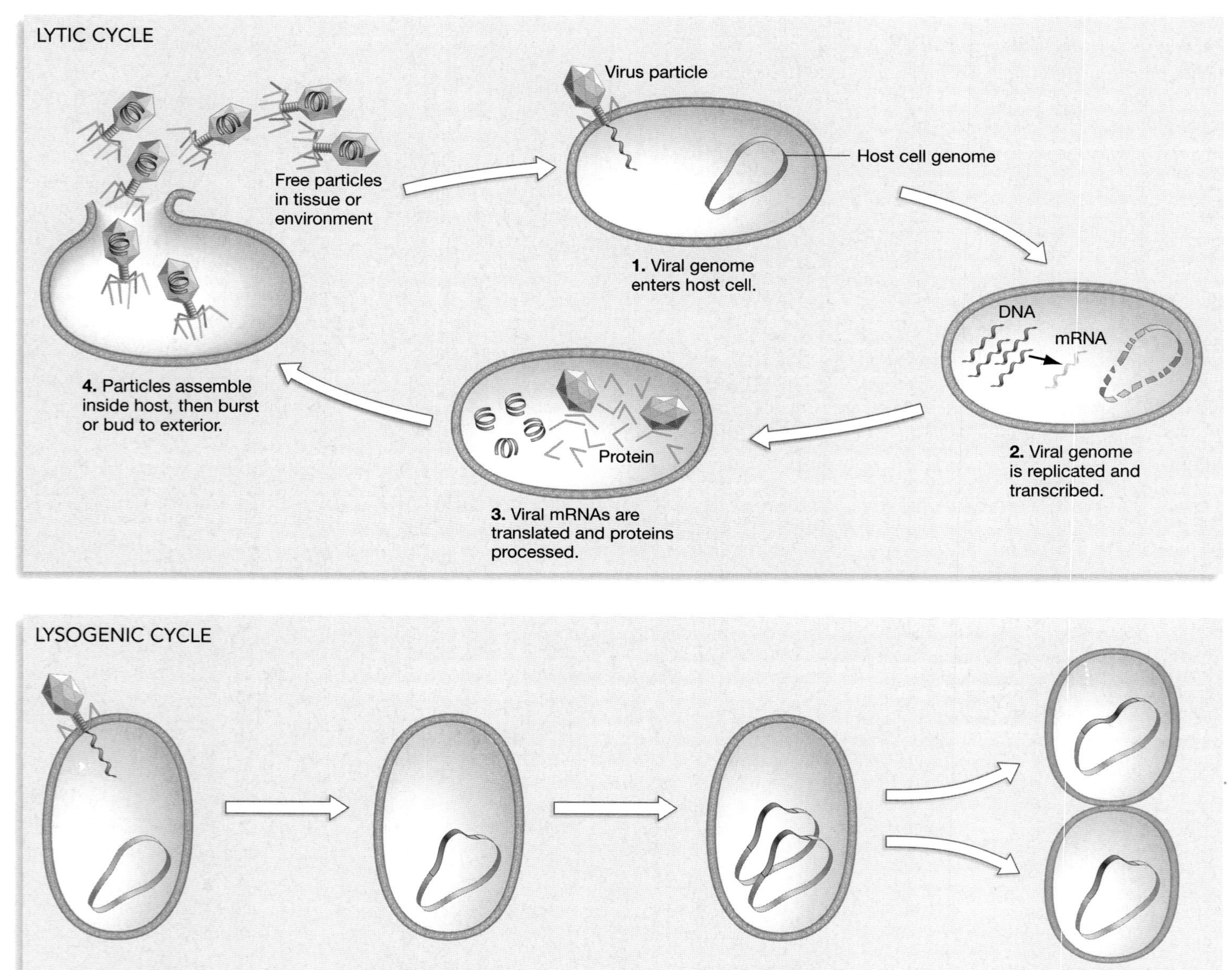

FIGURE 26.4 Viruses Replicate via Lytic and Lysogenous Cycles
All viruses follow the same general replication cycle, diagrammed at the top. As the bottom drawing shows, some viruses are also capable of lysogeny, meaning that their genome can become integrated into the host-cell chromosome. QUESTION Many of the viruses that infect bacteria are capable of lysogeny. If a particular host-cell population is growing and dividing rapidly, the viruses that infect the cells will tend to stay in the lysogenous state. If the cells begin to starve and stop dividing, however, the viruses tend to switch to lytic growth. From the virus's point of view, why is this switch advantageous?

tions. When such reactions occur, the viral genome exits the host chromosome and resumes an independent existence. In this way, lysogeny often acts as an alternative to the lytic replication cycle.

How Can Replication Cycles Be Disrupted? Viral infections are notoriously difficult to treat because the particles use so many of the host cell's enzymes during the replication cycle. Drugs that disrupt these enzymes usually harm the host much more than they harm the virus. But some enzymes *are* encoded in the genome of a virus. What are these proteins, and what do they do during the virus's infection cycle? Might they be vulnerable to treatment with drugs? Alternatively, is it possible to stop viral infections in the final phase of the replication cycle—during transmission to a new host cell or new host individual?

The rest of this chapter explores answers to these questions. But instead of surveying research on all viruses, the focus is on just one example—the human immunodeficiency virus, or HIV. This is the agent that causes the disease called acquired immunodeficiency syndrome, or AIDS.

There are three reasons why it makes sense to focus on HIV. First, HIV is the best studied of all viruses. Second, experiments on HIV illustrate how investigators go about analyzing the replication cycle of other viruses. Finally, understanding HIV is urgent. The virus is responsible for one of the deadliest epidemics in human history.

26.2 What Is HIV?

Most viruses are discovered when researchers isolate the agent responsible for a new disease. HIV is no exception. The search for HIV began soon after AIDS was first diagnosed and formally described by a physician who treated a young man from Los Angeles in 1981. When the same suite of symptoms appeared in people from New York, San Francisco, and other major cities, researchers became alarmed. Biologists launched an intensive search for the disease-causing agent, and in 1982 HIV was isolated in Françoise Barré-Sinoussi's lab.

Figure 26.5a shows a typical HIV particle. The virus is about 100 nm across and spherical in shape. The entire particle contains just eight different proteins, along with two copies of its single-stranded RNA genome (**Figure 26.5b**).

Once the pathogen was identified, researchers began testing frozen blood and tissue samples to estimate when the virus first began afflicting humans. The earliest documented infection occurred in a man from the Democratic Republic of Congo. A team led by David Ho found fragments of HIV genes in a blood sample taken from this patient in 1959. The result suggests that HIV first began parasitizing humans sometime in the 1950s. Because it began victimizing humans so recently, biomedical researchers call HIV an **emerging virus**. An important question is, where did HIV come from? Is it related to viruses that parasitize other mammals?

Where Did the Virus Come From?

The key to discovering where HIV comes from lies in understanding the evolutionary history, or phylogeny, of the virus. Paul Sharp, Feng Gao, and other researchers have reconstructed this history by comparing the composition of HIV genes to sequences from viruses that parasitize chimps, monkeys, and other mammals. The relationships among these gene sequences support several conclusions:

- HIV belongs to a group of viruses called the lentiviruses, which infect a wide range of mammals including house cats, horses, goats, and primates. (*Lenti* is a Latin root that means "slow"; here it refers to the long period observed between the start of an infection by these viruses and the onset of the diseases they cause.)
- Many of HIV's closest relatives also have *immunodeficiency* in their name. Like HIV, these agents parasitize cells that are part of the immune system. Several of them cause diseases with symptoms reminiscent of AIDS. Curiously, though, HIV's closest relatives don't appear to cause disease in their hosts. These viruses infect monkeys and chimpanzees and are called simian immunodeficiency viruses (SIVs).
- There are actually two distinct types of human immunodeficiency viruses, called HIV-1 and HIV-2. Although both can cause AIDS, HIV-1 is far more virulent and is the better studied of the two. It serves as the focus of this chapter.

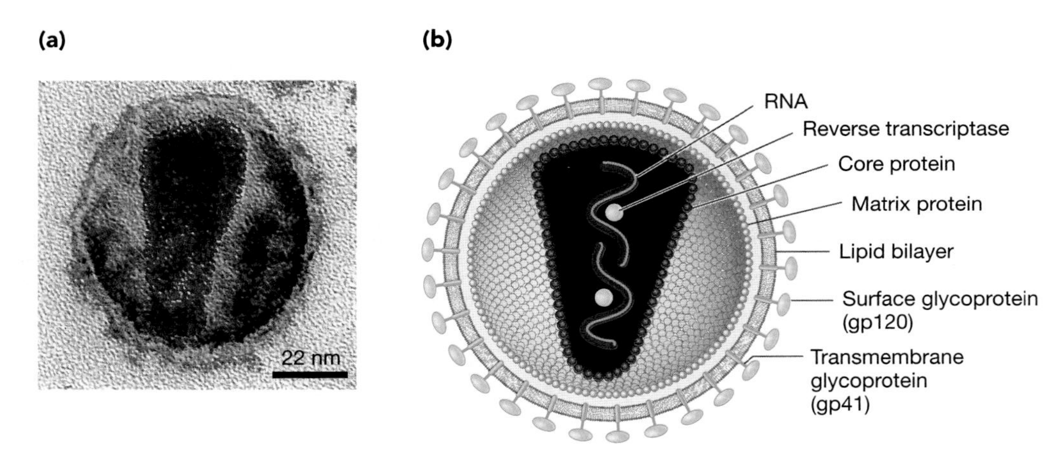

FIGURE 26.5 The Human Immunodeficiency Virus (HIV)
(a) In this photomicrograph of HIV, the envelope and protein core are clearly visible. **(b)** This diagram is a cutaway view of HIV. Note that the surface glycoproteins called gp41 and gp120 are somewhat visible in part (a). **EXERCISE** Draw lines connecting the structures visible in parts (a) and (b). According to this micrograph, did the artist represent the structure accurately?

- HIV-1's closest relatives are immunodeficiency viruses isolated from chimpanzees that live in central Africa. In contrast, HIV-2's closest relatives are immunodeficiency viruses that parasitize monkeys called sooty mangabeys. In central Africa, where HIV-1 infection rates first reached epidemic proportions, contact between chimpanzees and humans is extensive. Chimps are hunted for food and kept as pets. Similarly, sooty mangabeys are hunted and kept as pets in west Africa, where HIV-2 infection rates are highest.

To make sense of these observations, Sharp and Gao and colleagues suggest that HIV-1 is a descendant of viruses that infected chimps, and that HIV-2 is a descendant of viruses that infect sooty mangabeys. The "jumps" between host species probably occurred when a human cut up a monkey or chimpanzee for use as food.

Why Does HIV Cause Disease?

Like many viruses, HIV specializes in parasitizing a few particular types of cell. In HIV's case, the cells most affected are helper T-lymphocytes and macrophages. Chapter 46 explains just how crucial these cells are to the immune system's response to invading bacteria and viruses. When a protein on the surface of a helper T cell binds to a fragment of a bacterial or viral protein, the helper T cell begins to activate the immune system's response to the invading bacterium or virus. Part of this response involves macrophages. Macrophages are involved in destroying bacterial cells and viral particles.

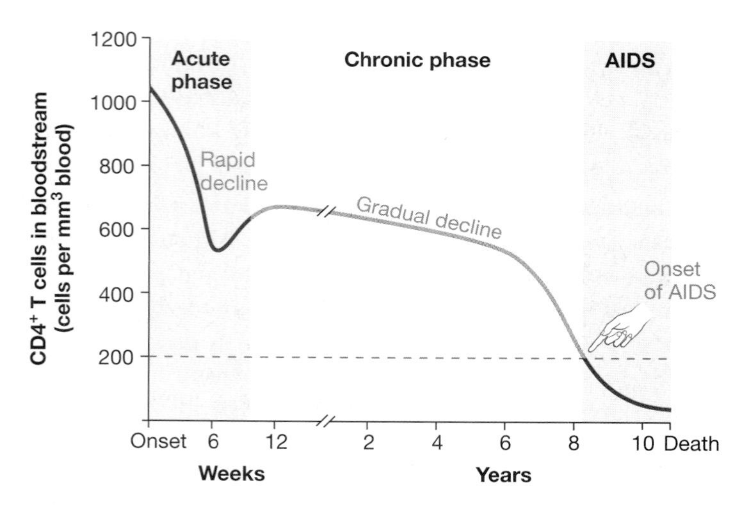

FIGURE 26.6 T-Cell Counts Decline over the Course of an HIV Infection
This graph is based on data from a typical patient infected with HIV. It shows changes in the number of T cells that are present in the bloodstream over time. The acute phase of infection occurs immediately after infection and is sometimes associated with symptoms like fever. Infected people usually show no disease symptoms in the chronic phase, even though their T-cell counts are in slow, steady decline. AIDS typically occurs when T-cell counts dip below $200/mm^3$ of blood.

As an HIV infection proceeds, however, the number of helper T cells in the bloodstream declines. The graph in **Figure 26.6** shows the decreasing number of T cells over the course of a typical infection. When the T-cell count drops, the immune system's responses to invading bacteria and viruses become less and less effective. Eventually, too few helper cells are left to trigger the immune response, and a variety of pathogenic bacteria and viruses multiply unchecked. In almost all cases, one or more of these infections will prove fatal.

To summarize, HIV does not kill its hosts directly. It kills them indirectly because it destroys the immune system cells called T cells. T cells are an essential component of the body's response to infections by bacteria, fungi, and other viruses. When HIV reduces the number of functioning T cells below a critical value, the immune system can no longer eliminate pathogens. As a result, hosts die of pneumonia, fungal infections, or unusual types of cancer.

The Scope of the AIDS Epidemic

Viruses have caused the most devastating epidemics in recent human history. During the eighteenth and nineteenth centuries, it was not unusual for Native American tribes to lose 90 percent of their members to viral diseases like measles and smallpox that had been contracted from European settlers. The "Spanish" influenza of 1918–1919 previously qualified as the most devastating epidemic recorded to date. This viral outbreak killed over 20 million people worldwide.

AIDS recently surpassed these events, however. Researchers with the United Nation's AIDS program estimate that AIDS has already killed 21.8 million people worldwide, and that an additional 36.1 million people are currently infected with HIV. If no vaccine is found, and if present rates of infection continue, researchers project that the epidemic may eventually kill as many as 100 million people. Because HIV is usually transmitted through sexual contact, the vast majority of AIDS victims are young adults.

HIV infection rates are highest in east and central Africa, where one of the greatest public health crises in history is now occurring (**Figure 26.7**). In Botswana and other localities, blood-testing programs have confirmed that up to a third of all adults carry HIV. Physicians, politicians, educators, and aid workers all use the same word to describe the epidemic's impact: staggering.

26.3 How Does an HIV Infection Begin?

The replication cycle of a virus begins when a free particle enters a target cell. This is no simple task. Cells are protected by surrounding membranes or by cell walls *and* membranes. How do viruses breach cell walls and membranes, insert themselves into the cytoplasm inside, and begin an infection?

Most plant viruses enter host cells after a sucking insect, like an aphid, has disrupted the cell wall with its mouthparts. Viruses that parasitize bacterial and animal cells, in contrast, gain entry by binding to a specific molecule on the cell wall or cell membrane. HIV also uses this mode of entry.

The T cells that HIV parasitizes have hundreds of membrane proteins on their surfaces. To which does HIV bind? A group of physicians led by Michael Gottlieb hinted at an answer in 1981. These doctors were among the first to treat young men suffering from AIDS—the suite of diseases that appear only in patients with severely impaired immune systems. Blood tests revealed why the young men experienced impaired immune function. They had few or no T cells possessing a particular membrane protein, called CD4. Gottlieb and co-workers hypothesized that their patients' immune systems were rendered ineffective by the loss of a critical component—cells that contained the CD4 protein. These cells are symbolized $CD4^+$.

The result also suggested that HIV might use the CD4 protein itself to gain entry to these cells. The logic here is simple. If $CD4^+$ cells are the specific cell type infected by HIV, it might be because HIV interacts with the CD4 protein when starting an infection. In 1984, research groups in France and England showed that CD4 does function as the "doorknob" that HIV uses to enter host cells. The two teams made this finding using the same general research strategy, which is diagrammed in **Figure 26.8** (page 510). They began by growing large populations of helper T cells in culture. Then they added HIV particles to a sample of the cultured cells, along with an antibody to a cell-surface protein found on helper T cells. (An antibody is a protein that binds with high specificity to another protein. Chapter 46 explains how they are produced.) They repeated this experiment 160 times, but added a different antibody each time. Why? Each of the 160 antibodies bound to and effectively blocked a different cell surface protein. If one of the antibodies used in the experiment happened to bind to the receptor used by HIV, the antibody would cover up the receptor. In this way the antibody would protect that cell from infection by HIV, because the antibody-bound receptor is inaccessible to the HIV particle. This approach provided a sensitive assay for identifying the receptor, and it led both research teams to reach exactly the same result. Only antibodies to CD4 protected the cells from viral entry.

This finding also seemed to explain why helper T cells are particularly vulnerable to infection by HIV: They have tens of thousands of copies of the CD4 protein in their membranes.

The Search for a Co-Receptor

Is binding to the CD4 doorknob on helper T cells sufficient for HIV to gain access to a cell? Work by Paul Maddon and colleagues suggested that the answer is no. This conclusion was based on experiments with mouse cells that carried CD4 proteins from humans on their surfaces. To create these cells, Maddon and co-workers introduced the human gene for CD4 into the genomes of mouse cells and grew the transfected cells in culture. Tests confirmed that the altered mouse cells produced

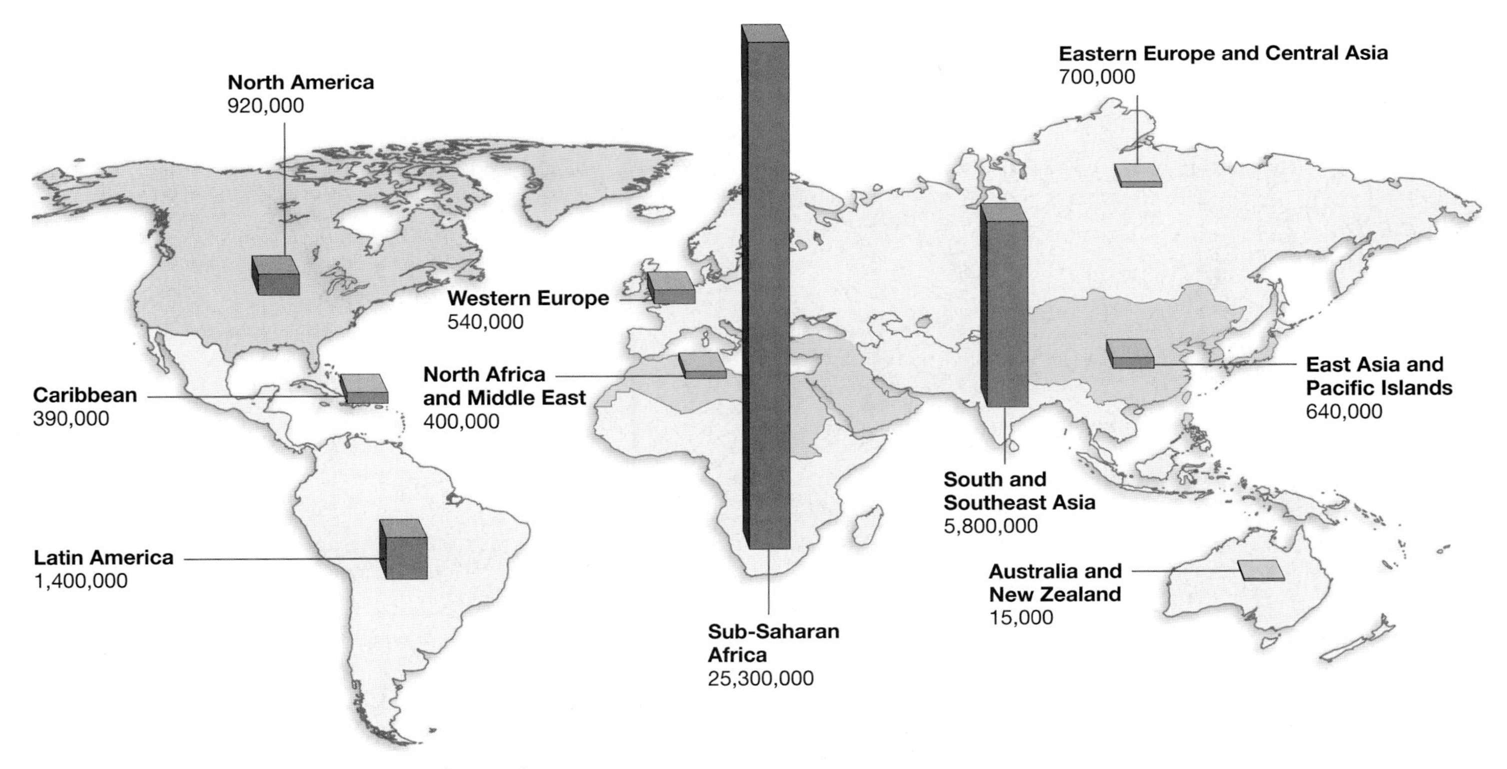

FIGURE 26.7 Geographic Distribution of HIV Infections
These data, compiled by the World Health Organization, show the numbers of people living with HIV at the end of 2000 by geographic area.

human CD4 proteins and inserted them into their membranes. When the researchers added HIV to the cultures, the virus was able to bind to the cell surface. Particles were not able to actually enter the CD4-bearing mouse cells, however. To make sense of this result, Maddon and co-workers proposed that HIV must interact with a second membrane protein, or co-receptor, in order to enter a cell. The idea was that this yet-to-be-identified receptor would be present on $CD4^+$ cells in people, but absent from the mouse cells the researchers had cultured.

The search for this co-receptor continued for a decade. Answering the question was crucial because researchers hoped to design drugs able to protect cells from HIV infection by interfering with the second receptor. (Unfortunately, blocking the CD4 protein with drugs produced terrible side effects.) Finally, a suite of experiments by Edward Berger and colleagues paid off. These experiments were based on a model system for studying interactions between the surface proteins found on HIV particles (called gp120 and gp41) and the surface proteins found on $CD4^+$ T cells (**Figure 26.9a**).

As **Figure 26.9b** shows, the model system was based on constructing two types of cells: human cells with HIV surface proteins on their membranes, and mouse cells with CD4 and one additional cell protein on their membranes. Why did they use human cells to mimic HIV's surface, and mouse cells to mimic the surface of human T cells? The answer has two parts. First, the system gave them a clear assay—a way of documenting an event. If the two cell types fused, it had to mean that HIV's gp120 and gp41 proteins were interacting with CD4 and the mysterious co-receptor. Second, using model cells allowed the researchers to test the effect of only one proposed co-receptor at a time. If they had used actual T cells in the experiment, competing researchers could argue that other proteins on the cell surface—besides the one being tested—were involved in successful fusion events. In short, the system allowed them to isolate the effect of a single protein on interactions between gp120, gp41, and CD4.

Which of the many membrane proteins found on $CD4^+$ T cells would work as this co-receptor? Berger's team tried dozens of candidate proteins before they hit on one that was recognized by HIV's gp120 and gp41 proteins. This molecule, which came to be called CXCR4, significantly increased the rate of membrane fusion in the experimental cells. Soon after the Berg-

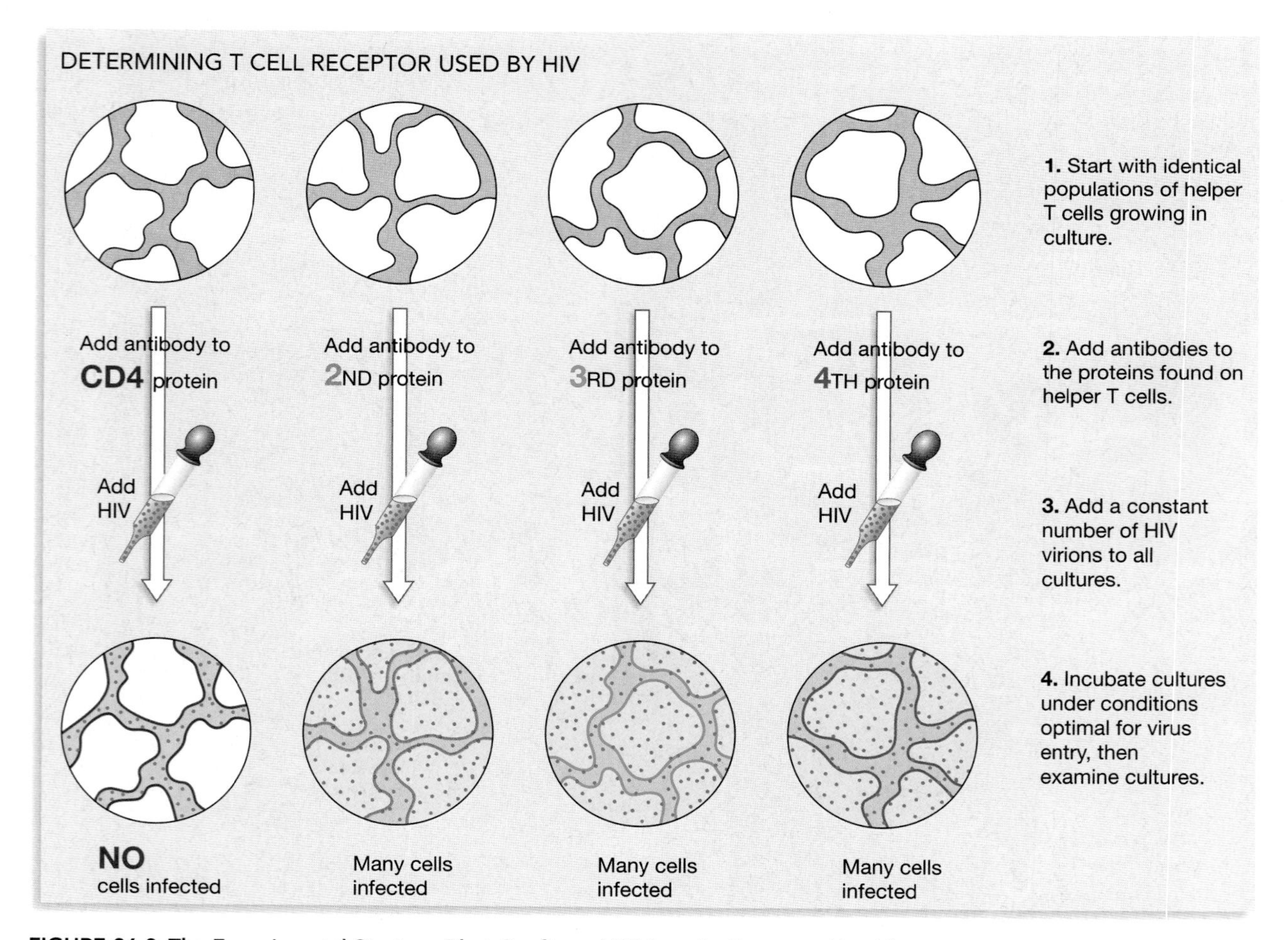

FIGURE 26.8 The Experimental Strategy That Confirmed CD4 as the Receptor Used by HIV
The antibodies added to each culture bound to a specific protein found on the surface of T cells. Antibody binding blocked the membrane protein, so the protein couldn't be used by HIV to gain entry to the cells.

er team's data were published, other workers announced the discovery of yet another co-receptor—a protein called CCR5. Later work confirmed that different strains of HIV rely on different co-receptors to enter CD4-bearing T cells.

In sum, the data indicate that the gp120 and gp41 proteins on HIV's surface bind to CD4 and either CXCR4 or CCR5 on the surface of a helper T cell. After binding occurs, the lipid bilayers of the particle membrane and the helper T cell fuse, and HIV has succeeded in opening the cell. The contents of the virus enter the cytoplasm, and infection proceeds.

Receptor Molecules and Resistance to Infection

As soon as CXCR4 and CCR5 were identified as co-receptors, researchers began to look at these proteins in people who are naturally resistant to HIV infection. Data sets poured in and confirmed a remarkable result: People who have been repeatedly exposed to HIV without being infected, or who carry HIV but are asymptomatic even after 15 years, often have mutant co-receptors. One of the most common mutations is a 32-base-pair deletion in the CCR5 gene. This mutation truncates the co-receptor and keeps it from being inserted into the cell surface. Strains of HIV that rely on this co-receptor cannot enter their target cell. As a result, people who are homozygous for the mutation are effectively protected against these strains of HIV. But do these people have impaired immune systems? To date, observations indicate that they do not. If this result is confirmed, a drug that blocks CCR5 or CXCR4 might confer immunity to HIV infection. Could such a drug be developed? Research continues. It is not likely that this question will remain unanswered for long.

(a) Molecules involved in HIV's entry into T cells

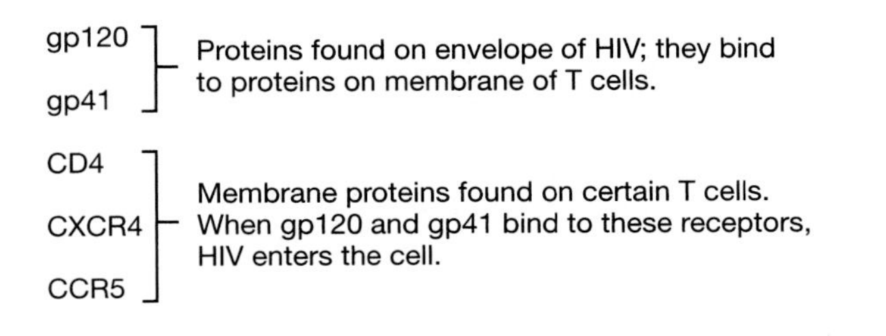

(b) Using test cells to identify surface receptors

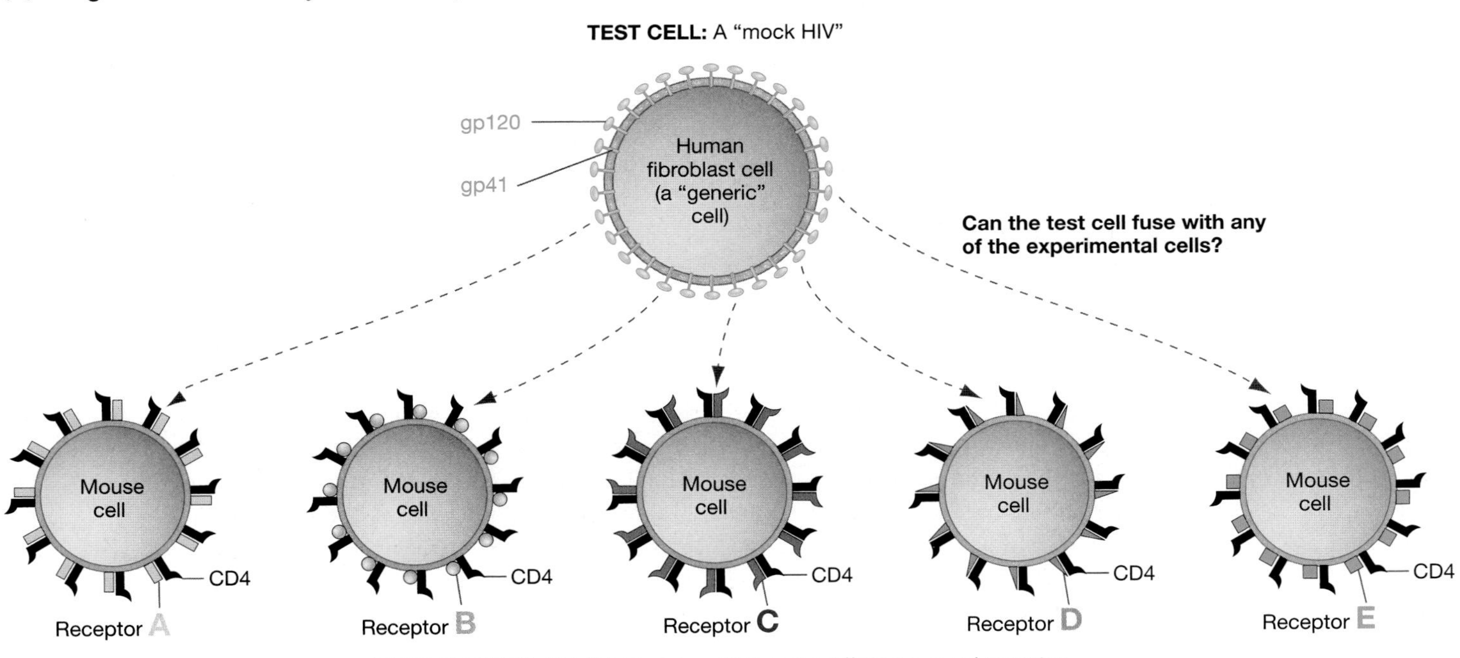

FIGURE 26.9 Experimental System Used to Identify the Co-Receptor for HIV Entry
Researchers introduced the genes for HIV's envelope proteins (gp41 and gp120) into human cells, which expressed these genes and inserted the proteins into their plasma membranes. Thus, their surfaces acted like the surface of HIV. The researchers also introduced the gene for the CD4 protein, along with the gene for another protein found on the surface of helper T cells, into mouse cells. Thus, the mouse cell surfaces acted like a helper T cell surface, except that each experimental mouse cell contained CD4 receptor proteins (shown in black on the experimental cells) along with just one other protein found on helper T cell surfaces.

26.4 How Does HIV Replicate Its Genome?

Once a virus has entered a cell, it cannot begin making copies of itself immediately. Only a subset of viruses contain all of the enzymes necessary for transcribing their own genome, and none can make the ribosomes and tRNAs necessary for translating mRNAs. For a particle to make more particles, it must exploit the host cell's biosynthetic machinery. How does this process begin? The answer depends on the nature of the virus's genome.

Like all other lentiviruses, HIV is a retrovirus. Two copies of the RNA genome and about 50 molecules of reverse transcriptase lie inside the core of each particle. Soon after these molecules enter a host cell, reverse transcriptase makes cDNA versions of the viral genome, using nucleotides and ATP from the cell. A viral protein called integrase then catalyzes the cutting of host DNA, the insertion of the viral genome into the host DNA sequence, and the reannealing of the DNA strand. A copy of the HIV genome is now integrated into the host chromosome.

When biomedical researchers first began the search for a drug capable of stopping HIV infections, they focused on disrupting this part of the infection cycle. Specifically, they set out to find molecules that would obstruct reverse transcriptase and prevent HIV's genome from being copied. The logic behind this research program was sound. Reverse transcriptase is not a normal part of a human cell's biosynthetic machinery, so a drug that interfered with the enzyme should kill the virus without harming the host.

The first widely prescribed anti-AIDS drug was a molecule called azidothymidine (AZT). This molecule is so similar in structure to the nucleotide thymidine that it is recognized by reverse transcriptase. When the virus's genome codes for a thymidine to be placed in a growing strand of cDNA, reverse transcriptase sometimes incorporates an AZT molecule instead. This terminates synthesis because AZT will not accept the addition of another deoxyribonucleotide. In this way, AZT blocks viral replication.

AZT was extremely effective against replication of HIV in the laboratory. Early tests with HIV-infected cell cultures, and later trials with AIDS patients, showed that the drug slowed the rate of infection dramatically. Researchers achieved similar results with two other reverse transcriptase inhibitors, called ddI and ddC. By the late 1980s, it seemed physicians had a battery of drugs that could stop HIV infections cold.

Not long after AZT, ddI, and ddC were put into widespread use, however, most patients stopped responding to treatment. Why? Is it possible that HIV evolved resistance to the drugs? Researchers tested this hypothesis by isolating HIV strains from patients who no longer responded to AZT. As feared, these strains were resistant to AZT in culture. When the biologists sequenced the reverse transcriptase gene from these HIV strains, they found a smoking gun. In many strains, an identical series of mutations had occurred in the part of the enzyme that binds thymidine. Because the mutations are correlated with resistance to AZT, the most logical conclusion was that these mutations made the enzyme less likely to be fooled by the drug. This is why the mutant HIV strains could complete their replication cycle, even when AZT was present.

This sequence of events should sound eerily familiar. As you may recall from Chapter 21, natural selection has recently favored strains of *Mycobacterium tuberculosis* that are resistant to the drug called rifampin, and these strains are currently responsible for the resurgence of tuberculosis in the industrialized world. A parallel sequence of events has occurred with HIV. Strains of the virus have now emerged that are resistant to more than one reverse transcriptase inhibitor. As a result, AZT, ddI, and ddC are often ineffective, even when they are used in combination. To help patients infected with HIV, researchers have turned to the next phase of the virus replication cycle: the production of viral proteins.

26.5 How Are Viral Proteins Translated and Processed?

In the third phase of the replication cycle, viral mRNAs and proteins are produced and processed. The molecules follow one of two routes through the cell, depending on whether the proteins end up in the outer envelope of a particle or in the capsid.

RNAs that code for a virus's envelope proteins follow a route through the cell identical to the RNAs of the cell's transmembrane proteins. **Figure 26.10a** diagrams how these viral mRNAs are translated by ribosomes attached to the endoplasmic reticulum (ER). Afterward they are transported to the Golgi complex, where carbohydrate groups are attached. The finished glycoproteins are then inserted into the plasma membrane, where they are ready to be assembled into new particles.

In contrast, note the route taken by RNAs for proteins that make up the inner core of a particle, illustrated in **Figure 26.10b**. These RNAs are translated by ribosomes in the cytoplasm, just like non-membrane-bound cellular mRNAs. The polyprotein sequences that result are later cut into functional proteins by a viral enzyme called protease. This enzyme cleaves viral polyproteins at specific locations—a critical step in the production of finished viral proteins. If a drug could function as a protease inhibitor, it could stop viral replication in time to prevent new cells from becoming infected. Could this viral enzyme become a target for drug therapy?

Protease—A Target for Rational Drug Design?

When researchers succeeded in visualizing the three-dimensional structure of HIV's protease, the picture confirmed the molecule's mode of action. The molecule has an opening in its interior that is adjacent to the active site (**Figure 26.11a**). Researchers immediately began searching for molecules that could fit into the opening and prevent the enzyme from functioning by binding to and blocking the active site (**Figure 26.11b**).

The search for such a molecule combined two strategies in drug development: rational drug design and trial and error. In a **rational drug design**, biochemists seek to synthesize a compound that can inhibit an enzyme with a known structure. In the case of HIV's protease, researchers wanted to design molecules that could fit into the enzyme's active site, bind to it, and

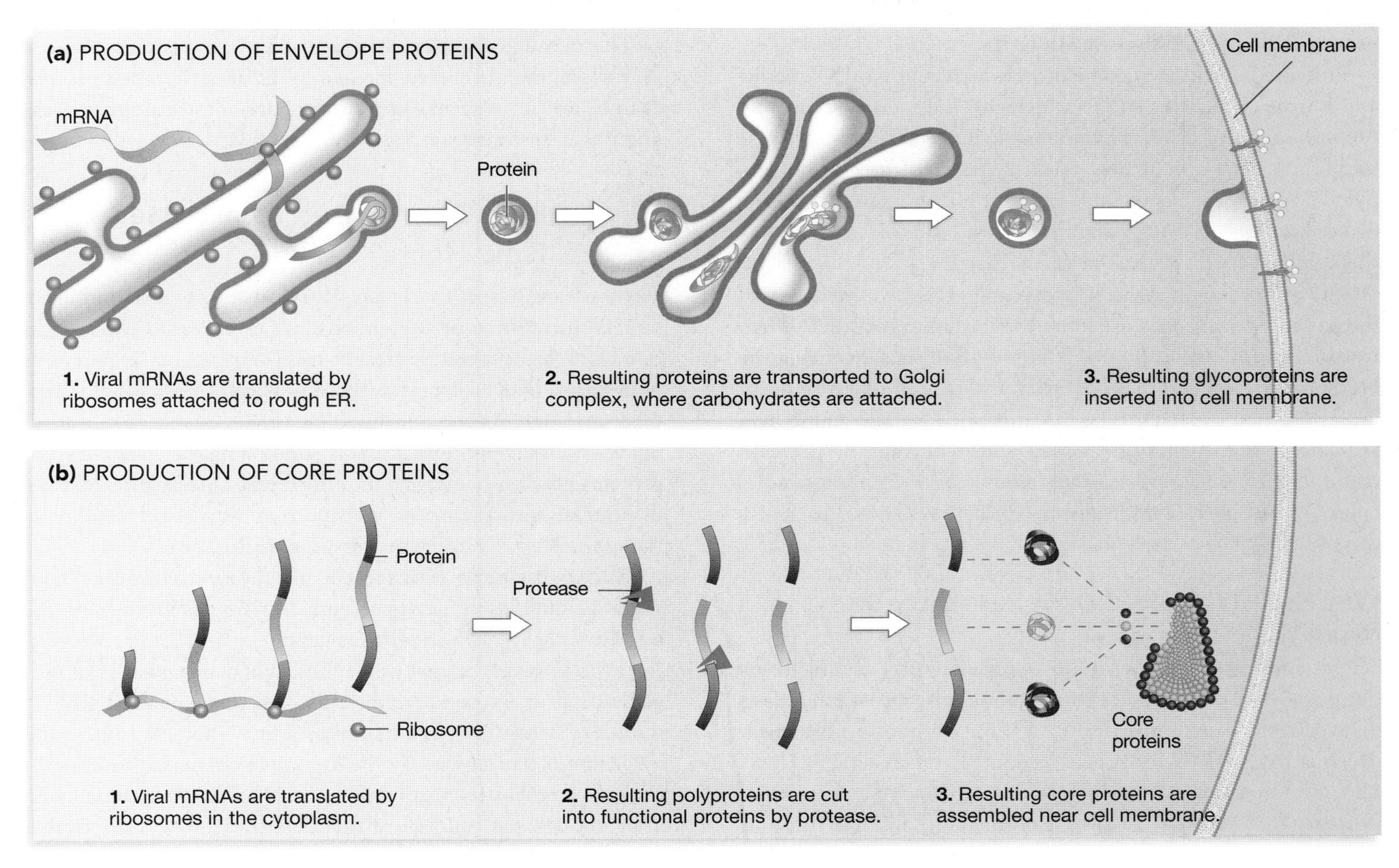

FIGURE 26.10 Production of Viral Proteins
(a) After being synthesized on the rough ER, envelope proteins are inserted into the cell membrane. **(b)** After being synthesized on smooth ER and being processed in the cytoplasm, core proteins assemble near the boundary of the cell.

(a) HIV's protease enzyme

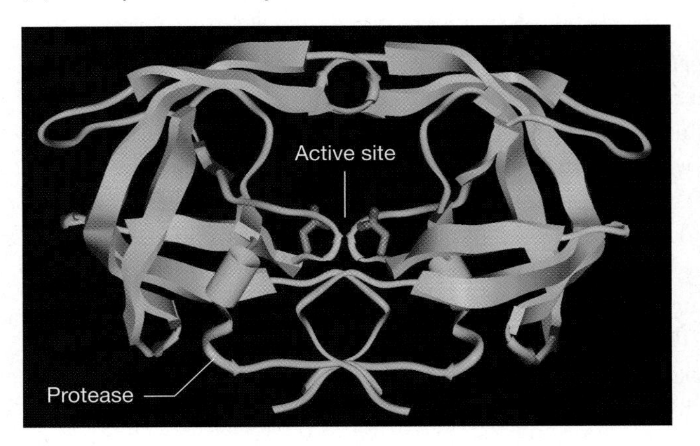

(b) Could a drug block the active site?

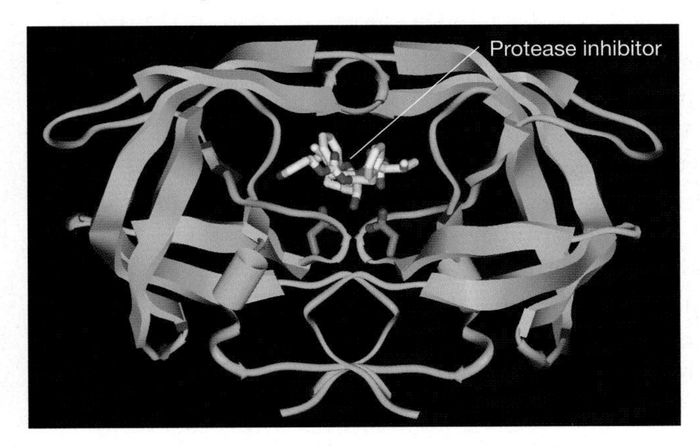

FIGURE 26.11 The Three-Dimensional Structure of Protease
(a) This ribbon diagram describes the three-dimensional shape of HIV's protease enzyme. **(b)** Once protease's structure was solved, researchers began looking for compounds that would fit into the active site and prevent the enzyme from working.

block it. After synthesizing several candidate molecules, researchers tested them by trial and error to determine if any were safe and effective. After six years of work, pharmaceutical companies had produced about a dozen promising molecules. These drugs were extremely expensive to manufacture but proved to be potent agents against HIV.

Protease inhibitors were dispensed widely in North America and Europe, beginning in the mid-1990s. Physicians frequently prescribed the drugs in combination with reverse transcriptase inhibitors, with spectacular results. After therapy, many patients no longer had detectable levels of HIV in their blood. The drugs knocked HIV populations down. But did they knock them out? Were patients completely cured?

The answer is no. Within two years, HIV levels in many of the patients taking protease inhibitors began to rebound. When researchers followed up on this observation by sequencing the HIV protease gene in these patients, they found that a series of mutations had occurred. The mutations led to new amino acid sequences in the enzyme's active site. The mutant protease could function well, even in the presence of the inhibitor molecules. Yet again, HIV had quickly evolved resistance to a new class of drugs.

Why Has HIV Evolved Drug Resistance So Rapidly?

The scientific literature abounds with examples of organisms that have evolved resistance to antibiotics and viruses that have evolved resistance to antiviral drugs. But perhaps no organism or virus has evolved resistance to control agents as quickly as HIV. The question is, why? Does this virus have some special property that helps it to evolve resistance rapidly?

The leading hypothesis to explain the rapid evolution of these populations is that HIV's mutation rate is particularly high. A corollary to this hypothesis is that HIV's reverse transcriptase is particularly inaccurate. This protein is a multifaceted enzyme. It can polymerize DNA from an RNA template, act as a nuclease in digesting the original RNA strand, and then catalyze the formation of double-stranded DNA. But unlike the DNA polymerase analyzed in Chapter 12, it has no 3′ exonuclease activity. That is, the protein cannot back up to revisit a base-pairing mistake, remove the mismatched nucleotide, and insert the correct one. Furthermore, HIV's genome codes for just three enzymes: reverse transcriptase, integrase, and protease. There are no repair enzymes in the virus's tool kit.

When researchers assay transcripts produced by HIV's reverse transcriptase, they find that on average, the wrong base is inserted once every 8000 nucleotides. (In contrast, *E. coli*'s DNA polymerase incorporates the wrong nucleotide about once in every *billion* bases.) This means that on average, a new mutant is generated every time HIV replicates its genome. Genetically, no two HIV particles are alike.

Why is HIV's high mutation rate important? In infected individuals, approximately 10 billion new particles are produced daily. If each contains a random error, it is likely that among the 10 billion there are particles with a mutation in the active sites of reverse transcriptase or protease. As a result, HIV populations are almost certain to contain variants that are at least partially resistant to drugs that cripple most particles in the population.

The message to researchers and physicians is clear. Due to HIV's high mutation rate, the search for drug therapies promises to be an "arms race"—a constant battle between novel drugs and novel, resistant strains of the virus.

26.6 How Are Viruses Transmitted to New Hosts?

Viruses leave a host cell in one of two ways: by budding from the cell membrane or by bursting out of the cell. Viruses that bud from the host-cell membrane take some host-cell membrane with them, and incorporate these host-cell phospholipids into their envelope. As a result, budding viruses include a sample of the host cell's membrane proteins, along with the virus envelope proteins previously inserted into the membrane (**Figure 26.12a**). In contrast, viruses that burst erupt from the plasma membrane, lysing the host cell in the process (**Figure 26.12b**).

Once particles are released, the infection cycle is over. Thousands or millions of newly assembled particles are now in extracellular space. What happens next?

If the host cell is part of a multicellular organism, the new generation of particles begins traveling through the body via the bloodstream or lymph system. There, they may be bound by antibodies produced by the immune system. In vertebrates, this marks them for destruction. But if a particle contacts an appropriate host cell before it is destroyed, it will infect that cell. This starts the replication cycle anew.

What if the virus has infected a unicellular organism, or if the virus leaves a multicellular host entirely? For example, when people cough, sneeze, spit, or wipe a runny nose, they help rid their body of viruses and bacteria. But they also project the pathogens into the environment, sometimes directly onto an uninfected host. From the virus's point of view, this new host represents an unexploited habitat brimming with resources in the form of target cells. The situation is analogous to a multicellular animal that disperses to a new habitat and colonizes it. Viruses that successfully colonize a new host replicate and increase in number. The alleles carried by these successful colonists increase in frequency in the total population. In this way, natural selection favors alleles that allow viruses to do two things: replicate within a host and be transmitted to new hosts.

The task now is to understand this phase of HIV's infection cycle. How is HIV transmitted to new hosts? What can people do to stop transmission and slow the spread of the epidemic?

Modes of Transmission

In every case analyzed by biomedical researchers, transmission of HIV has taken place through exchange of blood or reproductive tract secretions between an uninfected person and an infected person. Like most viruses, HIV is fragile: Particles are quickly destroyed if they are heated or dried. As a result, viruses usually do not survive long outside of host tissues.

At the start of the AIDS epidemic in North America and Europe, public health officials were able to infer that virtually all patients had been infected with HIV by one of three routes: blood transfusions that contained HIV, sharing of bloody needles with infected drug users, or anal intercourse with infected homosexual men. But by the early 1990s, transmission between heterosexual partners through vaginal intercourse began to increase in frequency in the industrialized nations. In Africa and Asia, this has been the dominant mode of transmission throughout the epidemic's history. AIDS is primarily a sexually transmitted disease.

The challenge for public health officials and researchers is to develop a strategy to stop transmission. Historically, the most effective way to end virus epidemics has been through vaccination, an approach explored in detail in Chapter 46. Why haven't researchers been able to design an effective vaccine against HIV?

Prospects for an AIDS Vaccine

Vaccines create immunity to disease-causing agents because they contain protein fragments from those agents. These protein fragments are recognized by the immune system's T cells and are bound by antibodies and destroyed. In addition, the immune system responds to the protein fragments in a vaccine by building a reservoir of T cells and B cells called memory cells, that remain in the body for years—or even for life. If an authentic in-

(a) Budding of enveloped viruses

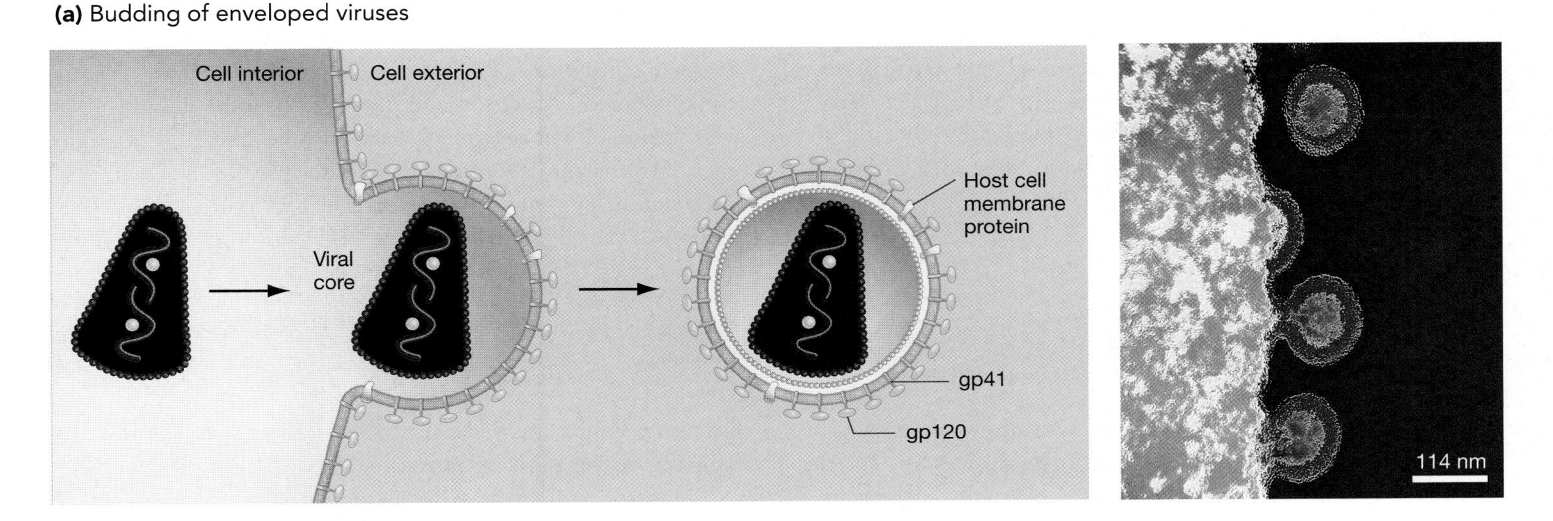

(b) Bursting of nonenveloped viruses

FIGURE 26.12 Completing the Replication Cycle
(a) Enveloped viruses bud from a host cell, taking with them a lipid bilayer containing their envelope proteins. **(b)** Nonenveloped viruses burst from a host cell. This event ruptures the cell membrane (and cell wall, if there is one). QUESTION Budding viruses don't always kill their host cell; bursting viruses always do. Why?

fection occurs after vaccination, these cells spring into action. The B cells produce antibodies and kill infected cells. In many cases, their response is fast enough to wipe out the colonizing population of pathogens. If so, then a state of immunity has been achieved.

What parts of HIV particles might stimulate the immune system? Vaccine researchers have focused their efforts on the glycoproteins found on HIV's surface (called gp41 and gp120) because they are the proteins that T cells encounter most frequently. Designing an effective vaccine would seem straightforward. Researchers could produce gp41 and gp120 proteins in quantity, package them in a pill or liquid, and administer them to individuals at risk of infection with HIV.

To date, this strategy has not worked. HIV's mutation rate is so high that few particles contain identical envelope proteins. A vaccine that primes the immune system to recognize hundreds or even thousands of different protein fragments might still leave vaccinated individuals susceptible to infection by many other strains. In this respect, HIV is very much like the viruses that cause colds and flu. These viruses also have particularly high mutation rates, and the vaccination programs that have been mounted against them to date have met with only moderate success. But unlike the cold and flu viruses, HIV destroys the very cells that are capable of eliminating an infection once it is under way.

Education and Abstinence

Faced with decades of disappointing results in drug and vaccine development, public health officials are aggressively promoting preventive medicine. Needle exchange programs stop HIV from being transmitted through contaminated needles. Condom use reduces sexual transmission of HIV. Aggressive treatment of venereal diseases may also help; the lesions caused by chlamydia, genital warts, and gonorrhea encourage the transmission of HIV-contaminated blood during sexual intercourse. Certainly the most effective forms of preventive medicine are sexual abstinence or monogamy.

These measures hearken back to the pre-antibiotic era when epidemics caused by bacteria or viruses were slowed by quarantining infected people, disposing of bed linens or other items used by infected people, and simple measures like hand-washing.

The effectiveness of preventive medicine underscores one of this chapter's fundamental messages. Viruses are a fact of life. Every organism is victimized by viruses; every organism has defenses against them. But the tree of life will never be free of these parasites. Mutation and natural selection guarantee that viral genomes will continually adapt to the defenses offered by their hosts, regardless of whether those defenses are devised by the immune system or by biomedical researchers. Combating viruses demands eternal vigilance.

Essay Emerging Viruses

With alarming regularity, the front pages of newspapers carry accounts of deadly viruses that are infecting humans for the first time. In 1993, a hantavirus that normally infects mice suddenly afflicted dozens of people in the American Southwest. Nearly half of these infections proved to be fatal. Still higher fatality rates were recorded in 1996 when the Ebola virus, a variant of a monkey virus, caused a wave of infections in the Democratic Republic of Congo. By the time the outbreak subsided, over 200 cases had been reported; 75 percent were fatal.

Determining how a virus is transmitted takes old-fashioned detective work.

When physicians see a large number of patients with identical and unusual disease symptoms in the same geographic area over a short period of time, they become alarmed. The doctors report the cases to public health officials, who take on two urgent tasks: identifying the agent causing the new illness and determining how it is being transmitted.

Several strategies can be used to identify a pathogen. In the case of the hantavirus outbreak, officials recognized strong similarities between the U.S. cases and symptoms caused by the Hantaan virus native to northeast Asia. The Hantaan virus rarely causes disease in humans; its normal host is rodents. To determine if a Hantaan-like virus was responsible for the U.S. outbreak, researchers began capturing mice in the homes and workplaces of afflicted people. About a third of the captured rodents tested positive for the presence of a Hantaan-like virus. DNA sequencing studies confirmed that the virus was a previously undescribed type of hantavirus. Further, the sequences found in the mice matched those found in infected patients. Based on these results, officials were confident that a rodent-borne hantavirus was causing the wave of infections.

The next step in the research program, identifying how the agent is being transmitted, is equally critical. If a virus that normally parasitizes a different species suddenly begins infecting humans, and if it can be transmitted efficiently from person to

(Continued on next page)

(Essay continued)

person, then the outbreak has the potential to become an epidemic. But if transmission takes place only between the normal host and humans, as with rabies, then the number of cases will probably remain low.

Determining how a virus is transmitted takes old-fashioned detective work. By interviewing patients about their activities and movements, researchers decide if each patient could have acquired the virus independently. Were the individuals infected with hantavirus in contact with mice? If so, were they bitten? Did they handle rodent feces or urine during routine cleaning chores or come into contact with contaminated food? What if the illness begins showing up in health-care workers, implying that it is being transmitted from patients?

In the case of the hantavirus outbreak, public health officials concluded that no human-to-human transmission was taking place. Health-care workers did not become ill; and, because patients had not had extensive contact with one another, it was likely that each had acquired the virus independently. The outbreak also coincided with a short-term, weather-related explosion in the local mouse population. The most likely scenario was that people had acquired the pathogen by inhaling dust or handling food that contained remnants of feces or urine. In short, hantavirus did not have the potential to cause an epidemic. The best medicine was preventive: Homeowners were advised to trap mice out of living areas, wear dust masks while cleaning, and store food in covered jars.

The Ebola virus, in contrast, was clearly being transmitted from person to person. Many doctors and nurses were stricken after tending to patients with the virus. Infections that originate in hospitals usually spread when they are carried from patient to patient on the hands of caregivers. But by carefully observing the procedures that were being followed by hospital staff, researchers concluded that transmission was taking place only through direct contact with bodily fluids (blood, urine, feces, or sputum). The outbreak was brought under control when hospital workers raised their standards of hygiene and insisted on the immediate disposal or disinfection of all contaminated bedding, utensils, and equipment. Had the Ebola virus been transmitted by more casual contact, like touching or inhaling, it could have caused a massive epidemic.

Chapter Review

Summary

Viruses are parasites that can infect virtually every cell type known on the tree of life. Their genomes are small and can consist of either RNA or DNA, but not both. Most viral genomes do not code for ribosomes, nor for the enzymes needed to translate their own proteins or to perform other types of biosynthesis. To make copies of themselves, viruses enter a host cell and use its biosynthetic machinery to copy their genomes and manufacture viral proteins. When a new generation of virus particles assembles and emerges from a host cell, that cell usually dies.

Although the genomes and morphologies of viruses are diverse, the basic features of their replication cycle are similar. An infection begins when the contents of a particle enter a host cell. For example, when HIV binds to a transmembrane protein called CD4 and a co-receptor, the virus's membrane-like envelope fuses with the cell's membrane, and the viral genome and proteins spill into the cytoplasm.

The viral genome is replicated in the second phase of the replication cycle. Some RNA viruses use the cell's RNA polymerase to make copies of their genomes, but other viruses code for their own polymerase. HIV, for example, is an RNA virus that reverse-transcribes its genome—to DNA—using a viral enzyme called reverse transcriptase. The DNA then integrates into a chromosome of the host cell, where it is transcribed by the cell's RNA polymerase.

In the third phase of the replication cycle, viral mRNAs are translated and processed. For example, the mRNAs that code for HIV's envelope proteins are translated by ribosomes on the cell's ER. The proteins are then processed in the Golgi complex and inserted into the cell's membrane. The mRNAs that code for HIV's other components are translated by ribosomes in the cytoplasm. While they are still inside the cell, the viral proteins that result assemble into complete particles.

In the final stage of the infection cycle, the new generation of complete particles buds or bursts from the cell. Viruses that bud from the host cell acquire a lipid bilayer, containing viral proteins, during the process. Viruses that burst from the host cell do not have this membrane-like envelope. Instead, they are protected by a protein coat. These coat proteins are coded for by viral genes and assemble inside the host-cell cytoplasm.

Once they are released from the host cell, particles can infect more cells in the same multicellular organism or be transmitted to an entirely new host.

Viruses cause many serious illnesses in humans. Viral diseases are difficult to treat with drugs because molecules that incapacitate enzymes needed by the virus are likely to damage host cells as well. Vaccination is frequently effective in stopping virus epidemics. Unfortunately, it is difficult or impossible to design a vaccine that can prepare the immune system for viruses that mutate very rapidly, like HIV or the cold and flu viruses.

Questions

Content Review

1. What is the hereditary material in viruses?
   a. DNA
   b. RNA
   c. protein
   d. DNA or RNA
2. How do animal viruses enter cells?
   a. They pass through a wound.
   b. They bind to a membrane protein.
   c. They puncture the cell wall.
   d. They bind to the lipid bilayer.
3. Which experimental result convinced researchers that HIV required a co-receptor to enter cells?
   a. Their close relatives, the SIVs, require a co-receptor.
   b. Mouse cells with CD4 in their membranes could not be infected with HIV.
   c. HIV can infect cells other than $CD4^+$ T cells.
   d. Mouse cells with CD4 in their membranes could be infected with HIV.
4. What does reverse transcriptase do?
   a. It synthesizes proteins from mRNA.
   b. It synthesizes tRNAs from DNA.
   c. It synthesizes DNA from RNA.
   d. It synthesizes RNA from DNA.
5. What do host cells provide for viruses?
   a. nucleotides and amino acids
   b. ribosomes
   c. ATP
   d. all of the above
6. When do virus particles acquire a membrane-like envelope, including a lipid bilayer?
   a. during entry
   b. during budding
   c. as they burst
   d. as they integrate into the host chromosome
7. What reaction does protease catalyze?
   a. polymerization of amino acids into peptides
   b. cutting of long peptide chains into functional proteins
   c. folding of long peptide chains into functional proteins
   d. assembly of particles
8. Why is it difficult to design a vaccine for viruses with high mutation rates, such as HIV and the cold and flu viruses?
   a. The vaccines tend to be unstable and deteriorate over time.
   b. So many protein fragments are represented that the immune system overreacts.
   c. They have no protein fragments that can be recognized by a host cell.
   d. New mutations constantly change viral proteins.

Conceptual Review

1. The outer surface of a virus consists of either a membrane-like envelope or a protein coat called a capsid. Look back at Figure 26.3c and at Figure 26.5a to examine these structures. Which type of outer surface does HIV have? Which type do bacteriophages have?
2. Look at Figure 26.3c and at Figure 26.5a again, and compare the morphological complexity of HIV with that of bacteriophage T4. Which virus would you predict has the larger genome? Why?
3. Explain why viral diseases are difficult to treat compared with diseases caused by bacteria.
4. What type of data convinced researchers that HIV originated when a simian immunodeficiency virus "jumped" to humans? Do you agree with this conclusion? Why or why not?

Applying Ideas

1. It is not known whether viruses parasitize archaea. Suppose you discovered a new species of archaea living on the walls of a cave, and you were able to culture the organism. How would you go about determining whether any viruses parasitize this species?
2. Baculoviruses parasitize insects. In general, each species of baculovirus afflicts just one or a few species of insect. Would you expect that the baculoviruses that parasitize beetles are more closely related to one another than they are to the baculoviruses that parasitize other insect groups, such as fruit flies? Explain your reasoning.
3. Suppose you could isolate a virus that parasitizes the pathogenic bacterium *Staphylococcus aureus*. This bacterium causes acne, boils, and a variety of other afflictions in humans. How could you test whether the virus might serve as a safe and effective antibiotic?
4. If you were in charge of the government's budget devoted to stemming the AIDS epidemic, would you devote most of the resources to drug development, vaccine development, or preventive medicine? Defend your answer.
5. Bacteria fight viral infections with restriction enzymes. These are bacterial enzymes that were introduced in Chapter 15. They cut up, or break, viral DNA at specific sequences. The enzymes do not cut a bacterium's own DNA, because the bases in the bacterial genome are protected from the enzyme by a chemical modification called methylation (this is the addition of a $-CH_3$ group). Generate a hypothesis to explain why members of the Eukarya do not have restriction enzymes.
6. Consider these two contrasting definitions of life:
   a. An entity is alive if it is capable of replicating itself via the directed chemical transformation of its environment.
   b. A living organism is an integrated system for the storage, maintenance, replication, and use of genetic information.

   According to these definitions, are viruses alive? Explain.

CD-ROM and Web Connection

CD Activity 26.1: Herpes Virus Replication *(tutorial)*
(Estimated time for completion = 5 min)
Learn how the herpes virus takes over the machinery of the cell.

At your **Companion Website** (http://www.prenhall.com/freeman/biology), you will find self-grading exams and links to the following research tools, online resources, and activities:

Harvard AIDS Institute
Explore a wealth of information about research projects and worldwide efforts to curb the epidemic of AIDS.

Life Cycle of HIV
Trace the steps of HIV replication in this interactive life cycle.

Special Pathogens Branch of the CDC
This site provides information on recent outbreaks of viral diseases, including the Lassa fever virus, the Ebola virus, and the HIV virus and AIDS.

Additional Reading

Editors. 1998. Defeating AIDS: What will it take? *Scientific American* 279 (July): 81–107. A series of 10 articles covering many aspects of the epidemic.

Keese, P., and A. Gibbs. 1993. Plant viruses: Master explorers of evolutionary space. *Current Opinion in Genetics and Development* 3: 873–877. A superb introduction to the plant viruses.

Laver, W. G., N. Bischofberger, and R. G. Webster. 1999. Disarming flu viruses. *Scientific American* 280 (January): 78–87. The story of drug development aimed at combating the flu virus.

Le Guenno, B., and L. Garrett. 1995. Emerging viruses. *Scientific American* 273: 5663. A review of viruses that recently began afflicting humans, and why they have not yet caused widespread epidemics.

O'Brien, S. J., and M. Dean. 1997. In search of AIDS resistance genes. *Scientific American* 277: 44–51. An introduction to the CCR5 mutants that confer immunity to HIV infections.

Protists

27

This chapter introduces the third domain on the tree of life: the Eukarya. These organisms have cells that contain a membrane-bound compartment where the hereditary material resides. The organisms are called eukaryotes, the cell compartment is called the nucleus, and the genetic material consists of long molecules of DNA.

The eukaryotes include the largest and most morphologically complex organisms on the tree of life—fungi, land plants, and animals. This chapter, however, considers questions about the remaining eukaryotes: the protists. These species are typically small and unicellular. Chapter 25 emphasized the amazing metabolic diversity of the Bacteria and Archaea. Chapter 26 highlighted the remarkable diversity of genetic materials found among viruses. This chapter focuses on the extraordinary structural complexity of protist cells.

The first questions about protists are the obvious ones: What are these organisms? What do they look like and how do they live? These issues are considered in sections 27.1 and 27.2. The following section examines the origin of two critically important organelles found in many protists and other eukaryotes: the mitochondrion and the chloroplast. The chapter concludes by investigating how protists affect the daily life of humans. The most spectacular crop failure in history, the Irish potato famine, was caused by a protist. Another protist causes malaria, which ranks as the world's most chronic public health problem.

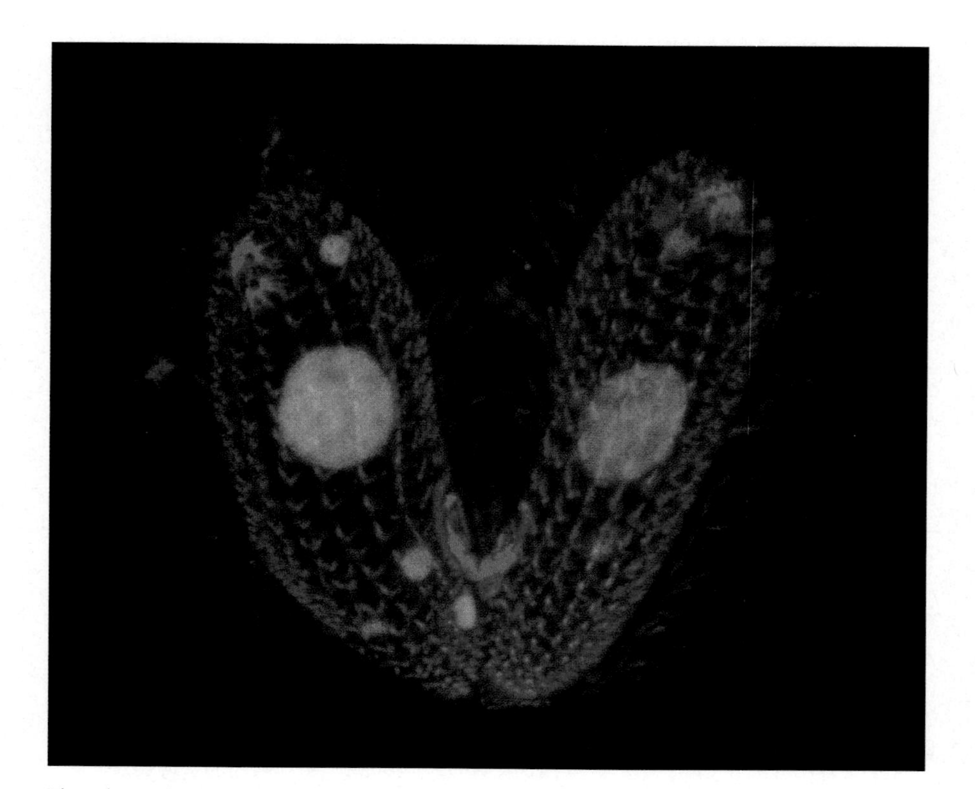

This photo shows a protist called *Tetrahymena thermophila* during sexual reproduction. The tubulin proteins in each cell have been stained red and the nucleic acids have been stained green. Sexual reproduction is one of the signature attributes of protists.

27.1 What Are the Protists?

Protists represent a subset of the eukaryotes (**Figure 27.1**). To understand the nature of this subset, it's necessary to understand the larger group. The domain Eukarya is defined by the

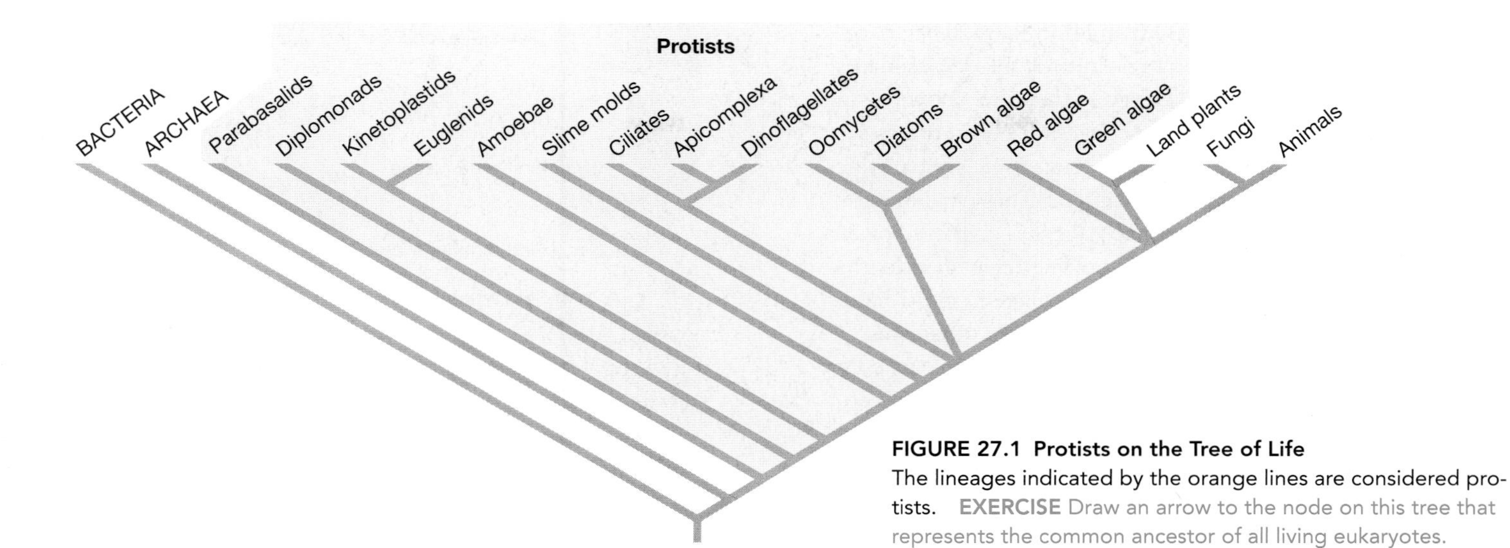

FIGURE 27.1 Protists on the Tree of Life
The lineages indicated by the orange lines are considered protists. **EXERCISE** Draw an arrow to the node on this tree that represents the common ancestor of all living eukaryotes.

presence of a unique, diagnostic feature: the cell nucleus. All members of the domain possess a nucleus, chromosomes where DNA is complexed with proteins called histones, cell division via mitosis, and a system of structural and contractile proteins called the cytoskeleton. As a group, however, protists can be defined only by what they are not. The term **protist** refers to all eukaryotes that are not green plants, animals, or fungi. The array of morphologies and lifestyles found in protists is impressive (**Figure 27.2**).

FIGURE 27.2 Morphologies and Lifestyles Found Among Protists
Protists are abundant in a wide variety of aquatic habitats. In marine environments, they are found in the open ocean as well as in near-shore and intertidal habitats.

In a sense, then, protists represent an informal grab bag of organisms. Because they include many—but not all—descendants of the common ancestor of eukaryotes, they are known as a **paraphyletic** group. Many are microscopic single cells; others are multicellular organisms up to 60 meters long. Some are parasitic while others are predatory or photosynthetic. They can be sessile or motile. Their cells can change shape almost constantly or be encased in rigid, glass-like "tests." Regarding their lifestyle, the protists' common feature is that they tend to live in moist habitats like wet soils, aquatic habitats, or inside other organisms.

Biologists study protists in part because of their intrinsic interest, in part because of their medical, commercial, and ecological importance, and in part because of their evolutionary relationships to today's plants, fungi, and animals. Let's begin with this last issue. When did the earliest protists—the first eukaryotes—appear on Earth? Which of their descendants are alive today?

Fossil Forms

J. William Schopf, an expert on ancient life-forms, says that identifying the earliest fossil eukaryotes is a "vexing problem." This is because the distinguishing features of eukaryotes—the nuclear envelope and the cytoskeleton—do not fossilize. As a result, paleontologists have had to identify early eukaryotes on the basis of distinctive cell forms. Cell size can also be a useful criterion because the cells of most living eukaryotes are larger than most of today's prokaryotes. In some cases, paleontologists have also been able to identify eukaryotic cells by the presence of distinctive molecules not found in bacteria or archaea.

Using these criteria, biologists have pieced together a rough timeline of early eukaryotic evolution, which is shown in **Figure 27.3**. The earliest fossils thought to be eukaryotes are found in marine sediments dated at 2.1 billion years ago. This is not long after the first rocks to be formed in an oxygen-containing ocean and atmosphere were being laid down—about 2.2 billion years ago. This is an important observation because the two events may not be coincidental. The record suggests that eukaryotes may have originated at about the same time that the atmosphere and oceans first became oxygen-rich. Unfortunately these early fossils, shown in Figure 27.3, are so poorly preserved that they are "problematic." This term means that not all paleontologists concur that the fossils actually represent the remains of organisms. The first fossils that scientists agree are conclusively eukaryotes show up much more recently, in sediments dated to about 1.75 billion years ago. As a result, it is not still clear whether the origin of Eukarya coincided with a change in the composition of the atmosphere and ocean.

Where Do Protists Fit on the Tree of Life?

Perhaps the most popular technique for reconstructing the tree of life is to examine small-subunit (SSU) rRNA sequences from living organisms, and then use the similarities and differences to estimate which species are most closely related. (See Box

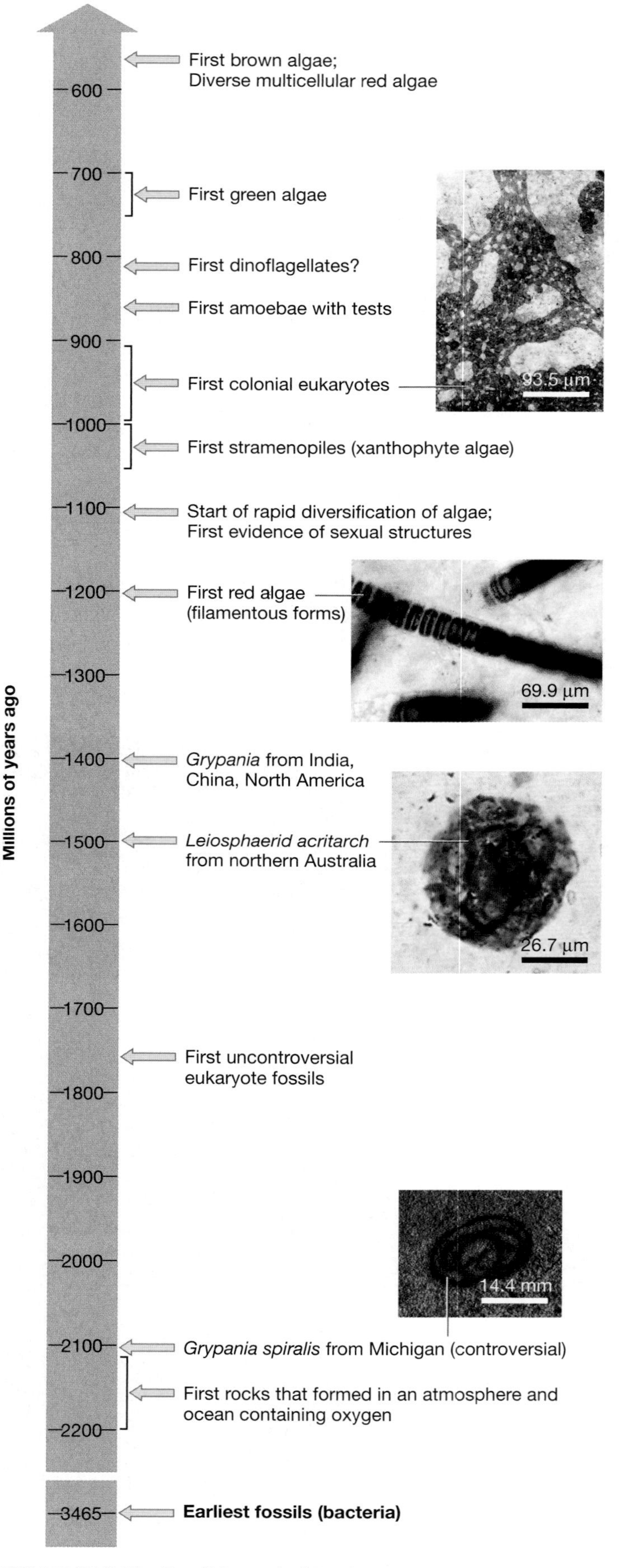

FIGURE 27.3 The Fossil Record of Protists
This timeline indicates milestones in the evolution of eukaryotes.

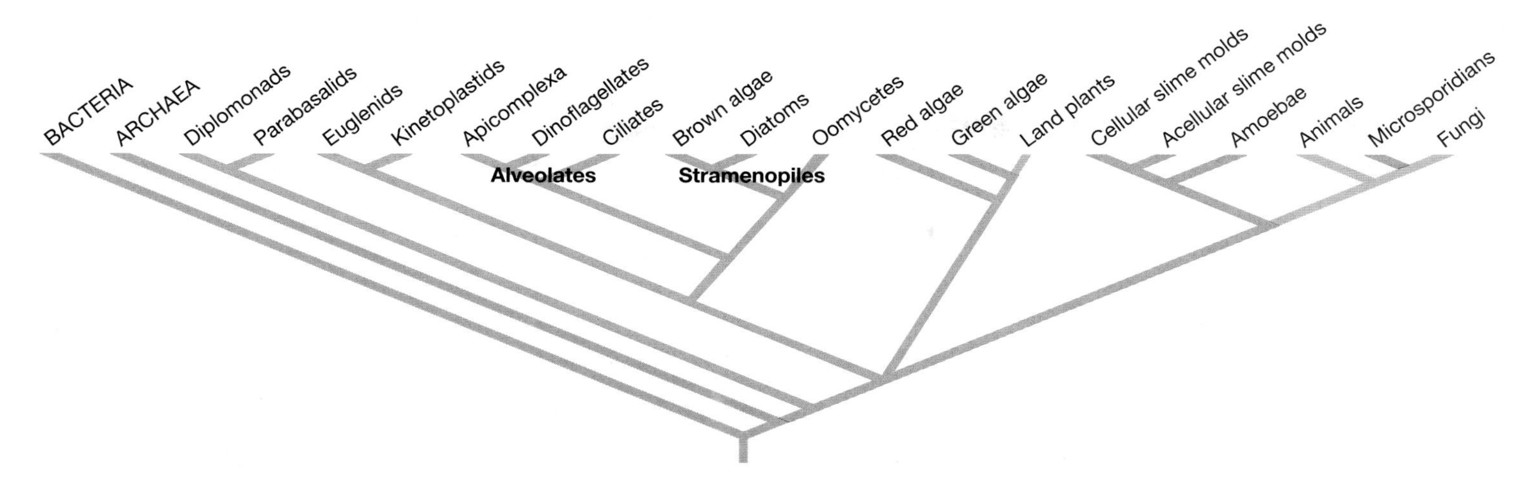

FIGURE 27.4 An Updated Phylogeny of the Protists
This tree includes only the most familiar, species-rich eukaryote lineages. Over 50 major groups of protists were left off the tree in order to make it more interpretable.

25.1 for an introduction to how molecular and morphological data are used to infer evolutionary relationships.) The tree that resulted from a recent analysis of SSU rRNA from eukaryotes is shown in Figure 27.1.

Since that tree was published, however, researchers have examined the relationships implied by a wider variety of genes. In some cases, the results of these analyses conflict with the relationships implied by SSU rRNA data. Most of these conflicts concern the order in which lineages branched off early in the evolution of eukaryotes. As a result, most researchers contend that a tree like that pictured in **Figure 27.4** is much more consistent with the available data than the tree in Figure 27.1.

In many or most instances, the branches on this tree group organisms that have distinctive aspects of cell structure. To drive this point home, consider the characteristics of two of the most species-rich groups.

- *Alveolates.* Members of the alveolates have cells containing sacs under their membranes called alveoli. Although they share this distinctive aspect of cell structure, members of this lineage are very diverse ecologically. The lineage includes unicellular organisms called ciliates that are extremely abundant in many freshwater and marine environments; the photosynthetic, unicellular, and planktonic species called dinoflagellates; and some unusual parasitic forms called apicomplexa, which are also unicellular.
- *Stramenopiles.* At some stage of their life cycle, all of the stramenopiles have flagella with distinctive hair-like projections. The lineage includes a large number of unicellular forms along with three especially important groups: oomycetes (or water molds), diatoms, and brown algae. Many of these organisms are photosynthetic and have chloroplasts containing chlorophylls *a* and *c*. The brown algae are multicellular and include the world's tallest marine organisms, the kelp.

In both cases, distinctive aspects of cell structure are associated with tremendous ecological diversification. Given their rapid and extensive radiation, what aspects of their biology unite protists?

27.2 Themes in the Evolution of Protists

The protists reviewed in section 27.1 are almost bewildering in their morphological and ecological diversity. They do not share unique, defining characteristics that set them apart from other lineages on the tree of life (see **Box 27.1**, page 524). Fortunately, several general themes tie these organisms together. The key to understanding the protists is to recognize that a series of important innovations occurred, often repeatedly, as eukaryotes diversified.

Cell Size and Structure

One of the marked differences between prokaryotic and eukaryotic cells is their size. An average eukaryotic cell is 10 times larger than an average-sized bacterial cell. How is this possible? The question is important, because large size presents significant difficulties. As cells become larger, their volume increases much more rapidly than does their surface area. This is because volume increases as the cube of a sphere's diameter, while surface area increases as the square of a sphere's diameter. (The relationship between surface area and volume is explored in more detail in Chapter 38.) Increases in cell size create a problem because food and waste molecules must diffuse across the surface, while the volume is filled with biosynthetic machinery that requires energy and generates waste. As cell size increases, then, metabolism in the cell interior can outstrip the transport and exchange processes that take place along the surface area.

Eukaryotes solve this dilemma by dividing their cell volume into compartments. As an example of how the compartmentalization of the eukaryotic cell works, consider the protist called

CD ACTIVITY 27.1 Themes in the Evolution of Protists

BOX 27.1 How Should We Name the Tree of Life's Major Branches?

Taxonomy is the branch of biology devoted to describing and naming new species and classifying groups of species. Swedish naturalist and physician Carl Linnaeus, who published the first work in the field in 1735, founded the effort. Linnaeus invented the system of Latin binomials still being used today, in which each organism is given a unique genus and species name. He also invented a hierarchy of more general taxonomic categories that included orders, classes, and kingdoms (see Chapter 1). These units were created in an attempt to organize and describe larger groups of organisms.

Linnaeus worked long before Darwin discovered the principle of evolution by natural selection, however. As a result, Linnaeus viewed the orders, classes, and kingdoms he described as distinct groups that had been created separately and independently from one another. When biologists realized that all organisms are related by common descent, they had to reinterpret the categories that Linnaeus had established. Instead of representing neat "bins" that held increasingly dissimilar species produced by special creation (**Figure 1a**), biologists recognized that the groupings actually designated twigs, branches, and stems on the tree of life (**Figure 1b**).

For almost 100 years after Darwin published, however, few reliable techniques existed for reconstructing evolutionary relationships; that is, for estimating phylogenies. As a result, biologists had to continue grouping organisms by their morphological similarity, much as Linnaeus had done. They were left to hope that these groupings reflected the actual evolutionary relationships on the tree of life.

Did they? Over the past 40 years, increasingly sophisticated techniques have become available for estimating phylogenies. In general, the old groupings were remarkably accurate in reflecting actual lineages. But there are some important exceptions. On the basis of morphology alone, for example, biologists thought that the most fundamental division on the tree of life was between prokaryotes and eukaryotes. This is now known not to be true. According to the data available today, the fundamental splits are between bacteria, archaea, and eukaryotes. The split between bacteria and archaea is as fundamental as the split between either of these groups and the eukaryotes. Among the protists, old groups like "protozoans" and "algae" are now understood to be conglomerations of distantly related, though superficially similar, species. Even the name *protist* will probably be abandoned before long. According to rules currently being adopted, most biologists assign names only to monophyletic groups; that is, to branches on the tree of life that include all the descendants of a common ancestor. In violation of this rule, the group called protists refers to some, but not all, descendants of the first eukaryote (see Figure 27.1).

According to a Chinese proverb, "The first step in wisdom is to call things by their right name." In taxonomy, a massive effort is under way to do just that. The initial task is to produce accurate estimates of where each branch occurs on the tree of life. Then monophyletic groups can be named, with confidence that each represents a distinct stem, branch, or twig on the tree.

FIGURE 1 Contrasting Approaches to Classification
(a) In taxonomic thinking, organisms are grouped according to morphological similarity. **(b)** In phylogenetic thinking, groups are named if they represent distinct lineages on the tree of life.

(a) Taxonomic thinking

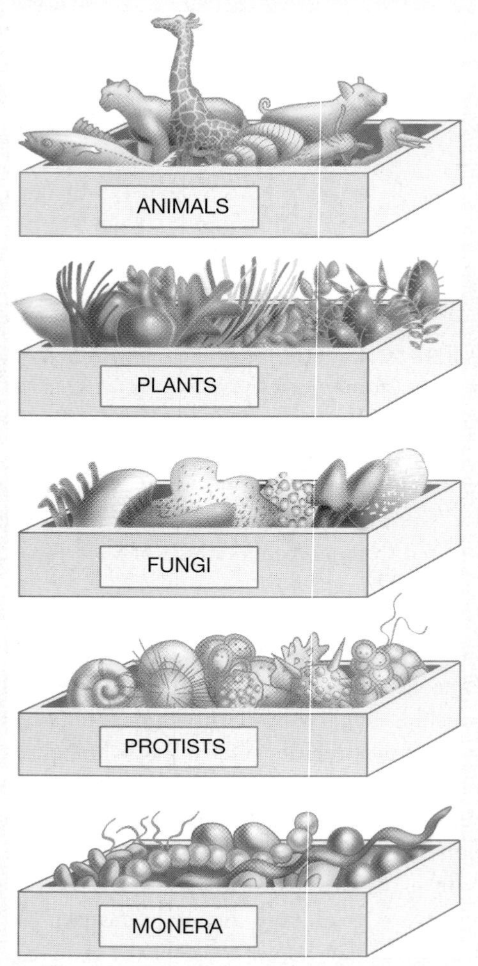

(b) Phylogenetic thinking

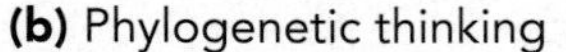

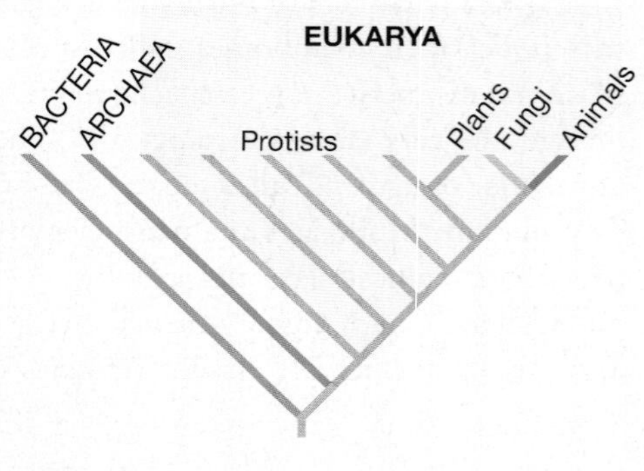

Paramecium (**Figure 27.5**). This organism is a ciliate that makes a living by eating bacteria. After ingesting a bacterium, a *Paramecium* surrounds it with an internal membrane, thus forming a compartment called a food vacuole. Food vacuoles combine with membrane-bound structures called lysosomes, which hold digestive enzymes, and circulate around the cell. When the food has been digested and absorbed, the vacuole merges with the external cell membrane at a special organelle—the cell anus—and expels waste molecules. Due to the large area of internal membrane, food molecules are delivered efficiently throughout the volume of the cell and waste products are expelled rapidly.

Although not all eukaryotic cells contain food vacuoles and lysosomes, they all have numerous other internal compartments. Each of these structures—including the nucleus, peroxisomes, mitochondria, chloroplasts, central vacuole, Golgi, and rough and smooth endoplasmic reticulum (ER) introduced in Chapter 5—has a distinct function. Many are membrane bound and are devoted to the synthesis, transport, and distribution of molecules. Increasing the structural complexity of the cell is a prominent theme in the diversification of protists.

In terms of organizing the cell volume, however, the most important feature of eukaryotes may be the cytoskeleton. This system of filaments and tubules, introduced in Chapter 5, provides scaffolding that supports and organizes the cell. Although some bacteria and archaea have rudimentary cytoskeletal elements, an extensive, highly ordered cytoskeleton is found only in eukaryotes.

In sum, the extensive compartmentalization and differentiation of the eukaryotic cell makes large size possible. Now the question is, what advantage does large size confer that makes it worth the trouble?

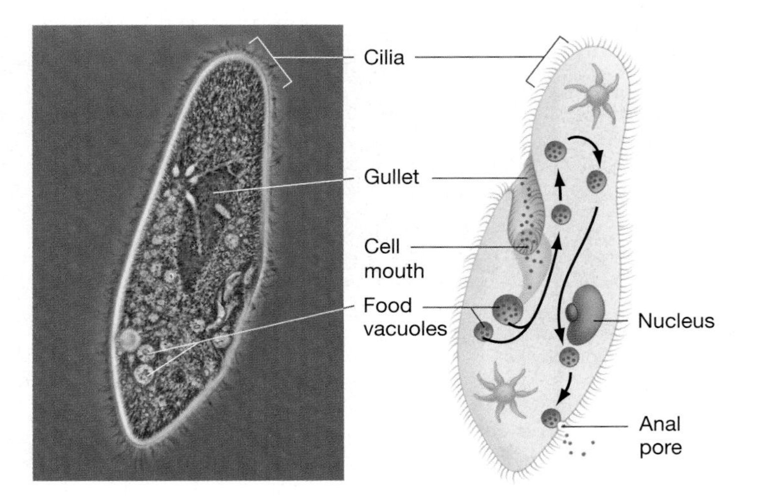

FIGURE 27.5 Protists Have Specialized Intracellular Structures
Paramecium is in the lineage of protists called the ciliates. It feeds by sweeping bacteria into a subcellular structure called its gullet.
EXERCISE Next to the drawing of *Paramecium*, list other organelles and intracellular structures that are found in protist cells.

Ingestive, Absorptive, and Photosynthetic Lifestyles

Bacteria and archaea obtain the energy they need in one of three ways: (1) by absorbing reduced inorganic compounds from the environment; (2) by absorbing reduced organic compounds like sugars, either from a host or from decaying organic matter; or (3) through photosynthesis. Among the protists, many species also absorb reduced organic compounds for nutrition, and many are photosynthetic. But large cell size made a unique feeding strategy possible. Protists can eat bacteria, archaea, and other protists—dead or alive.

Predation and Scavenging Ingestive lifestyles are based on eating live or dead organisms, or on scavenging organic debris. Protists are large enough to engulf bacteria and archaea; many are large enough to surround and ingest other protists or even animals. The engulfing process is possible because protists lack a cell wall, and because their flexible membrane and cytoskeleton make the movements shown in **Figure 27.6a** (page 526) possible. Losing the cell wall and gaining the ability to move their cell membrane around objects was a tremendous innovation during the evolution of protists. Instead of competing with bacteria and archaea for sunlight or food molecules, the protists could simply eat them.

Parasitism Absorptive lifestyles are also common among protists. Some absorptive groups, like the oomycetes (water molds) are decomposers; many others live inside other organisms. If they damage their hosts, the absorptive species are called parasites. The concluding section of this chapter explores the most damaging of all human parasites—the protist that causes malaria—in detail.

Parasitic feeding strategies have evolved in a wide variety of protist groups. Parasitism even occurs in photosynthetic lineages like the red algae. Some parasitic red algae are almost virus-like in their lifestyle. The parasitic cell injects its nucleus and mitochondria into a host cell and "transforms" it, meaning that the host cell's nucleus is inactivated or destroyed and superseded by the parasite's genes. The parasite genome directs the formation of the parasite body and the manufacture of a new generation of infectious parasite cells, using adenosine triphosphate (ATP) and reduced carbon compounds supplied by the host (**Figure 27.6b**).

Photosynthesis Parasitic relationships are actually extremely rare among photosynthetic protists; mutually beneficial, or mutualistic, relationships are much more common. For example, most ciliates that live in marine environments contain algal or dinoflagellate symbionts that perform photosynthesis; many anemones and reef-building animals do as well (**Figure 27.6c**). In this case the host provides protection and access to light; the enclosed algal cell provides oxygen and surplus glucose that the host can use.

The vast majority of photosynthetic protists are free-living, however. The red algae, brown algae, green algae, and other photosynthetic groups are distinguished by the pigments they contain (**Figure 27.6d**). Each of these pigments, in turn, absorbs unique wavelengths of light. As a result, the various groups of photosynthetic protists specialize in harvesting light energy from particular regions of the electromagnetic spectrum. The leading hypothesis to explain this pattern is that photosynthetic species harvest different wavelengths of light to avoid competition. This theme in the diversification of protists echoes the variation in photosynthetic pigments observed in bacteria (see Chapter 25).

Diversity in Lifestyles Before leaving the topic of obtaining food in protists, it is important to recognize that all three lifestyles—ingestive, absorptive, and photosynthetic—occur in many different eukaryote lineages. All three can occur in the same groups of species. To drive this point home, consider the lineage called the alveolates. The three major groups on this branch differ radically in how they make a living. The ciliates include many predators or herbivores. But some ciliates live in the guts of cattle or the gills of fish and absorb nutrients from their hosts. Other ciliate species make a living by holding algae or other types of photosynthetic symbionts inside their cells. In contrast, about half of the dinoflagellates are photosynthetic, while many others are parasitic. Most of the apicomplexa are parasitic.

This diversity in modes of nutrition is reminiscent of the variation in electron donors and acceptors used by bacteria and

(a) Predation and scavenging

Pseudopodia engulf food

Ciliary currents sweep food into gullet

(b) Parasitism

(c) Symbiosis

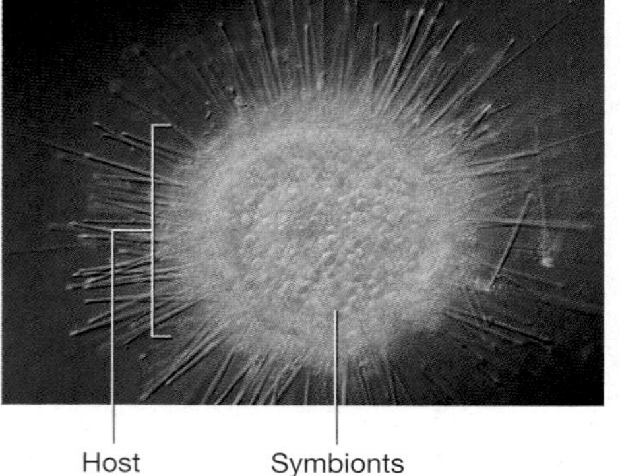

(d) Photosynthetic pigments

FIGURE 27.6 Modes of Feeding in Protists
(a) Methods of prey capture vary widely among ingestive protists. The photos show how predators engulf prey with pseudopodia or sweep them into their gullets with water currents set up by the beating of cilia. **(b)** Parasites get all of their nutrition from their hosts. The nonpigmented cells in this photograph belong to a parasitic species of red algae. Note that the parasite is growing out of the body of the red-colored host. **(c)** Symbioses are common in protists. The greenish cells inside this protist, called a heliozoan, are dinoflagellates that provide their host with food. **(d)** Many photosynthetic groups of algae are distinguished by the accessory pigments they contain, in addition to chlorophyll *a*. Each of these accessory pigments intercepts different wavelengths of light.

archaea. Like the prokaryotes, the diversification of protists was driven by diversification in methods of food-getting.

Locomotion

Many of the ingestive protists are "sit and wait" predators or scavengers. Like the ciliate pictured in Figure 27.6a, they attach themselves to a substrate and feed by sweeping particles of food into their mouths. But many other protists move to find food. How is movement accomplished?

Amoeboid movement is a gliding motion that is observed in some protists. Even after years of study, the mechanics behind this mode of locomotion are not well understood. In the classic amoeboid motion illustrated in **Figure 27.7a**, long, finger-like projections called pseudopodia (false-feet) stream forward over a substrate. The motion involves interactions between proteins called actin and myosin and ATP inside the cytoplasm; the mechanism is closely related to muscle movement in animals (see Chapter 43). But at the molecular level, the precise sequence of events is still unknown.

The other major mode of locomotion in protists involves flagella or cilia (**Figure 27.7b and c**). Flagella and cilia have identical structures, as shown in **Figure 27.7d**. Recall from Chapter 5 that both consist of nine sets of doublet microtubules arranged around two central, single microtubules. Cilia, however, are short and usually occur on a cell in large numbers, while flagella are long and are usually found singly or in pairs. Both structures are markedly different from the flagella found in bacteria and archaea. In the cilia and flagella of protists, the structure is made up of microtubules, and dynein is the major motor protein. An undulating motion occurs as dynein molecules walk down microtubules. The flagella of bacteria and archaea, in contrast, are primarily composed of a protein called flagellin. Instead of undulating, these flagella rotate to produce movement.

Closely related protists can use radically different forms of locomotion. For example, consider again the lineage of protists called the alveolates. The three major groups within this lineage are the ciliates, dinoflagellates, and apicomplexa. The ciliates swim using cilia, dinoflagellates swim by whipping their flagella, and apicomplexa glide and have gametes with flagella.

Structures for Support and Protection

Thanks to their cytoskeleton and organelles, all protists have a complex intracellular structure. But in addition, many protists have a rigid internal skeleton or a hard, external structure. Rigid internal structures, like the silicon-containing skeletons of heliozoans shown in Figure 27.6c, provide support for the cell. Hard external structures, often called a test or a shell, provide support or protection.

Tests and shells are found in groups that are scattered about the eukaryotic tree. The structures themselves are equally diverse. As **Figure 27.8** (page 528) shows, tests vary from the intricate chambers of calcium carbonate ($CaCO_3$) found in foraminifera to the glass-like, silicon-containing structures enclosing diatoms, to the cellulose plates that surround some dinoflagellates. Some amoebae even cover themselves with cobbles.

In sum, protists evolved many different ways to support their large cells and protect themselves from predation. Like the adaptations that allow protists to ingest, absorb, or photosynthesize

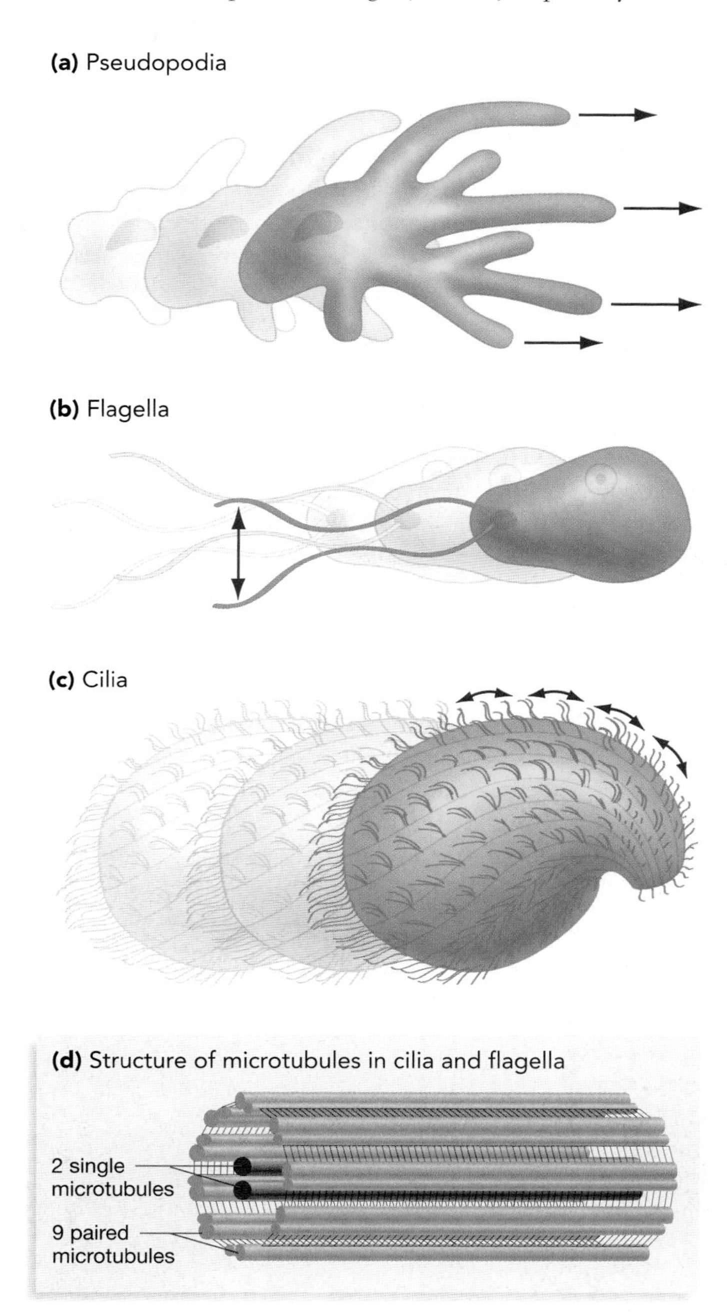

FIGURE 27.7 Modes of Locomotion in Protists
(a) In amoeboid movement, long pseudopodia (false-feet) stream out from the cell. The rest of the cytoplasm, organelles, and external membrane follow. Amoeboid movement has been the object of intense study, in part because certain cells in the human immune system use it. **(b)** Flagella are long, few in number, and power swimming movements. **(c)** Cilia are short and numerous. In many cases they are used in swimming. **(d)** In protists, flagella and cilia are built from the same "9 + 2" arrangement of microtubules.

their food, tests and shells have evolved many times, independently. The same theme holds true for the next trait considered: multicellularity.

Multicellularity

Multicellularity is a difficult trait to define. To understand why, consider the branch of protists called the green algae, and in particular a twig called the Volvocales. As **Figure 27.9a** shows, the organisms in this lineage range from single cells, such as *Chlamydomonas*, to small clumps of cells, such as *Gonium*, to small spheres of cells, such as *Pandorina*, to large colonies like *Volvox*. In this lineage there appears to be a smooth continuum between single-celled and multicelled forms. Where do biologists draw the line between the loose aggregations of cells called colonies and the more highly structured arrangements of cells that create a body and typify multicellular organisms? For example, is *Volvox* multicellular?

Most biologists would answer the latter question yes, for two reasons: Individual *Volvox* cells cannot survive on their own, and each cell has a distinct function. Some *Volvox* cells are strictly vegetative, meaning that they perform photosynthesis and supply the organism with ATP and sugar. Other cells are sexual, meaning that they undergo meiosis and form reproductive cells equivalent to sperm and eggs.

Why is differentiation of cell types such a crucial criterion? The answer is that when differentiation occurs, only a well-defined subset of each cell's genome is transcribed. Different types of cells express different genes. This feature is normally used to distinguish colonial growth—in which cells aggregate but perform the same function—from multicellular growth. Other green algae, like the *Coleochaete orbicularis* shown in **Figure 27.9b**, have the masses of interacting cells, distinct tissue types, and larger size that non-biologists usually associate with multicellularity.

Multicellularity evolved several times, independently, in protists. In many of the multicellular lineages, the first step in the evolution of this trait was the appearance of cells devoted to producing gametes. Gametes, in turn, are a product of a crucial eukaryotic invention: meiosis.

Sex

Sexual reproduction is much more extensive in protists than it is in bacteria and archaea. Before exploring the nature of reproduction in this group, however, it's necessary to address a basic question: What *is* sex, precisely?

Sexual reproduction can best be understood in contrast to asexual reproduction. When an individual reproduces asexually, by mitosis in diploid organisms or through simple cell division in prokaryotes, the resulting offspring are genetically identical to the parent. But when sexual reproduction occurs, offspring are genetically different from their parents.

Chapters 9 and 14 introduced the mechanisms that bacteria use to transfer genes and undergo genetic recombination. These cells exchange short stretches of DNA between individuals, usually via plasmids. In protists, however, the entire genome is involved during sexual reproduction instead of just one or a few genes. This is possible because of meiosis. Meiosis is the reduction division explored in Chapter 9. It was invented by the protists and results in the production of haploid daughter cells from a diploid parent cell. In species where meiosis is followed by a union of gametes from different individuals, individuals exchange half of their genomes. But as **Figure 27.10a** shows, some offspring are genetically different from their parents even when individuals routinely self-fertilize. Genetic differences between parents and offspring are increased when crossing over occurs during meiosis.

Why was the advent of meiosis important? The answer has two parts. First, meiosis may allow individuals to produce better-quality offspring. To understand why, suppose that a debilitating mutation knocks out one of the two alleles in a diploid individual. As **Figure 27.10b** shows, this individual cannot produce offspring that are free of the defective allele unless sexual

(a) Foraminifera

Calcium carbonate test, with chambers

(b) Diatom

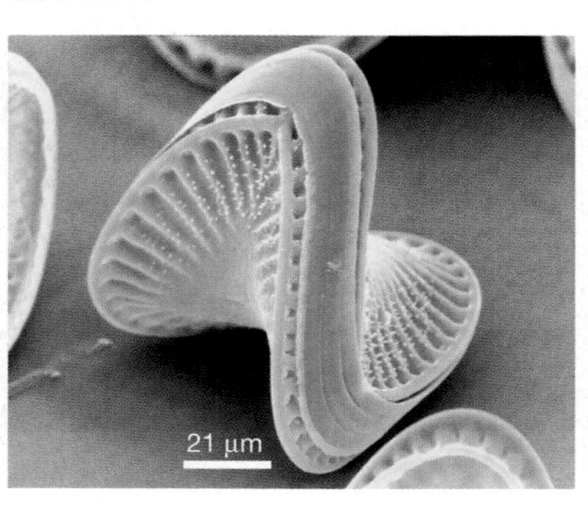

Test made of silicon oxides

(c) Dinoflagellate

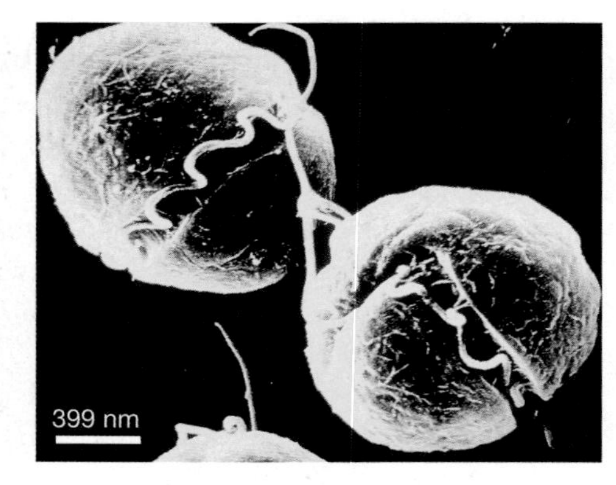

Plates made of cellulose

FIGURE 27.8 Hard Outer Coverings in Protists Vary in Composition

(a) Volvocales species range from unicellular to colonial to multicellular.

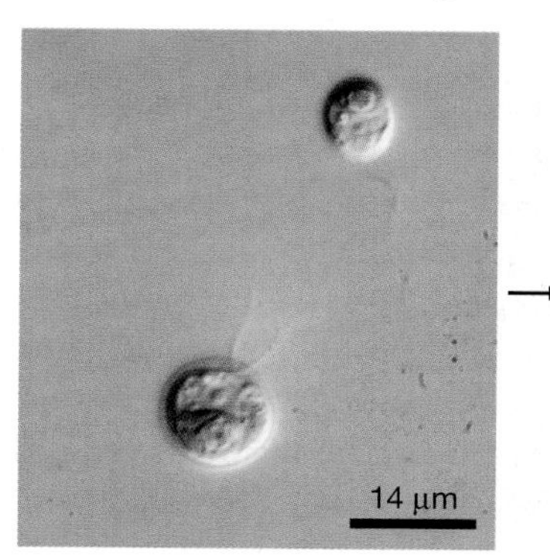

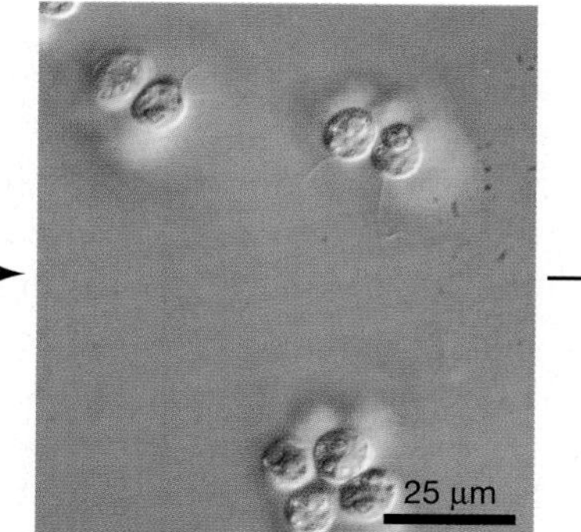

Gonium

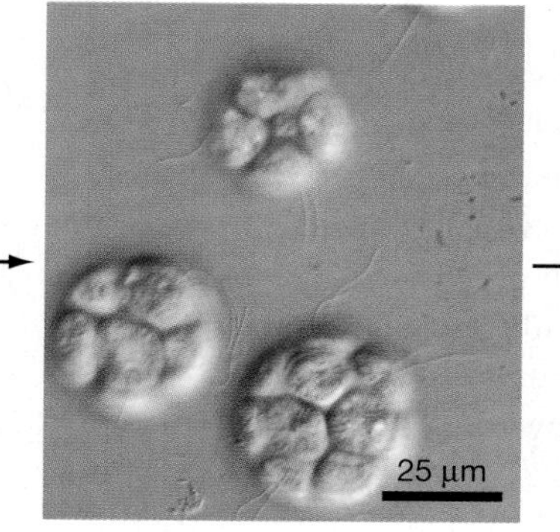

Pandorina

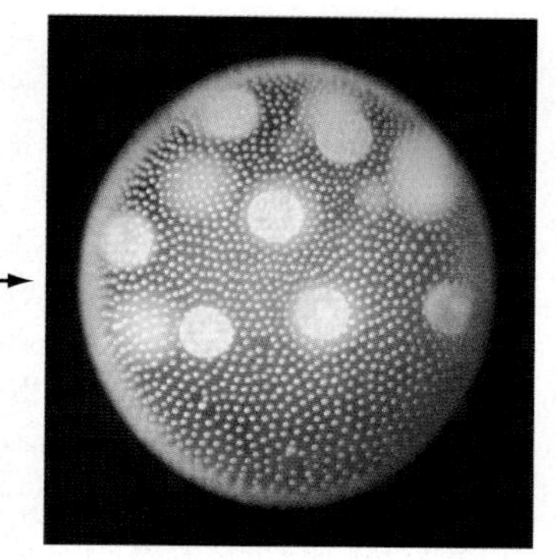

Volvox

(b) Some protists are clearly multicellular.

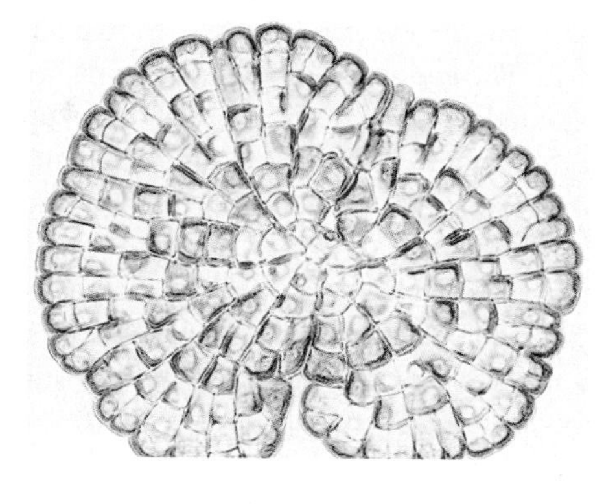

FIGURE 27.9 Multicellularity in Protists
(a) Members of the green algal family called Volvocales are graded in morphology, from unicellular to colonial to multicellular. The large spheres within the *Volvox* are daughter colonies that are being produced asexually. **(b)** This green alga, called *Coleochaete orbicularis*, is clearly multicellular. Its cells are packed tightly together and are differentiated in form and function.

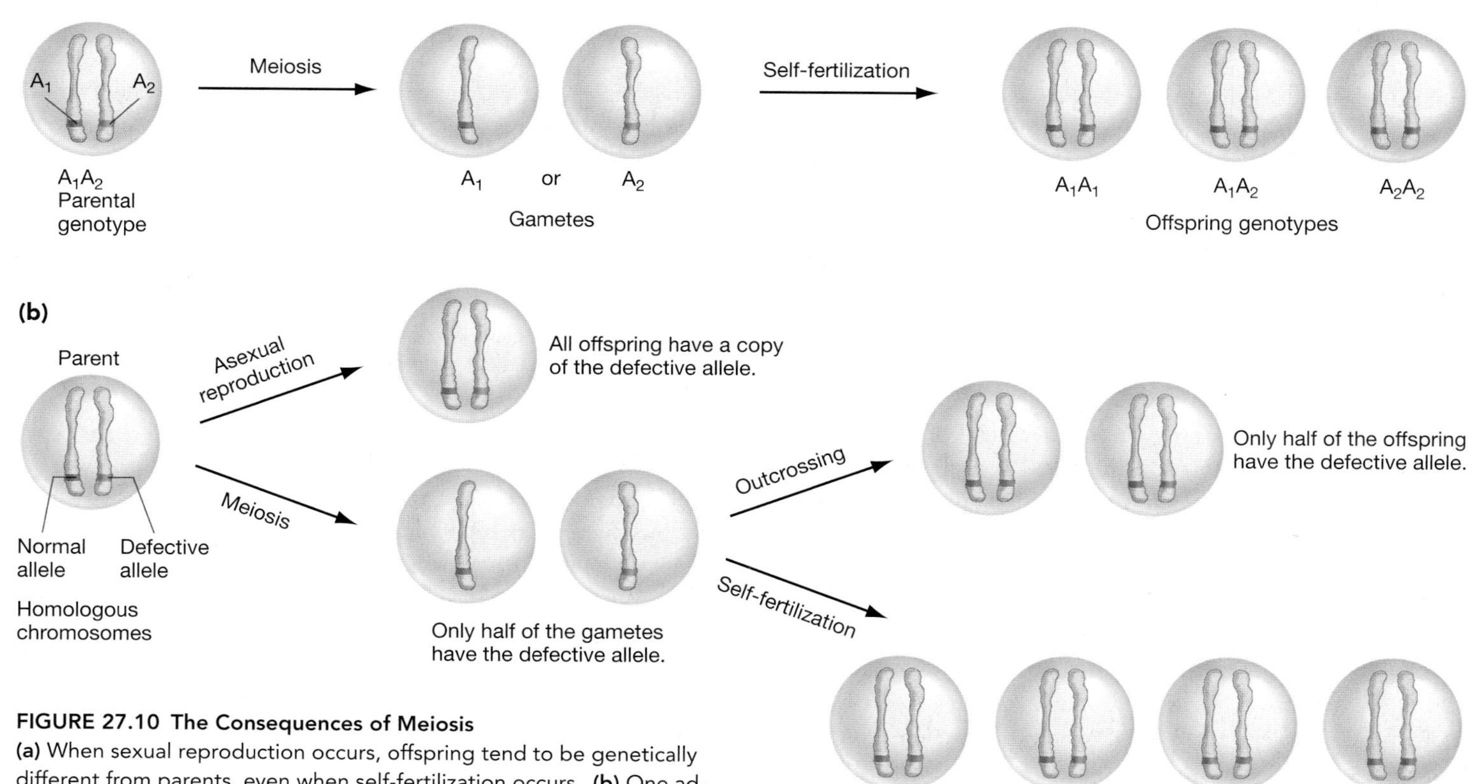

FIGURE 27.10 The Consequences of Meiosis
(a) When sexual reproduction occurs, offspring tend to be genetically different from parents, even when self-fertilization occurs. **(b)** One advantage of sexual reproduction is that it allows parents to produce offspring that are free of defective alleles. **EXERCISE** Circle the offspring genotypes in part (a) that are different from the parental genotype.

reproduction based on meiosis occurs. If it reproduces purely by mitosis, all of its offspring will carry the defective gene. Second, genetically variable offspring have a better chance of surviving if the environment changes from one generation to the next. In particular, offspring with genotypes different than their parents' may be better able to withstand attacks by viruses and other parasites that successfully attacked their parents. Sex was a major innovation during the diversification of protists.

Life Cycles and the Alternation of Generations

A life cycle describes the sequence of developmental events that occurs as individuals grow, mature, and reproduce. The advent of meiosis not only introduced a new event in the life cycle of protists, it also created a distinction between haploid and diploid phases in the life of an individual. These innovations created yet another layer of complexity and variability in the biology of protists. It is not an overstatement to say that every conceivable aspect of a life cycle is variable among protists: whether meiosis occurs; whether asexual reproduction occurs, and if so, how it is carried out; and whether the haploid or the diploid phase of the life cycle is longer and more prominent.

It is not uncommon, for example, for protists to be haploid most of their lives. **Figure 27.11a** illustrates one such haploid-dominated life cycle—in the green alga *Chlamydomonas*.

To further complicate matters, some multicellular protists are multicellular in both the haploid and diploid phases of the life cycle, producing a phenomenon known as **alternation of generations**. **Figure 27.11 (b and c)** diagrams this type of life cycle in a green alga and a brown alga. These illustrations show that the multicellular haploid and diploid forms of the same individual can be identical in morphology or radically different. Why does one type of life cycle prevail over others in certain groups? The answer is not known. Variation in life cycles is a major theme in the diversification of protists, but explaining why that variation exists remains a topic for future research.

27.3 The Origin of Mitochondria and Chloroplasts

The preceding sections provided an overview of the protists, with an emphasis on their morphological complexity and variability. Now let's address a more fundamental question: How did these lineages originate?

A glance back at the tree of life in Figure 27.1 indicates that the eukaryotes share a common ancestor with the archaea. Because the archaea and the earliest-branching protists are unicellular, biologists infer that the first eukaryote was also a single-celled organism. Further, because all eukaryotes alive today have a nucleus and a cytoskeleton, researchers conclude that their common ancestor also had these structures. Finally, because early-branching eukaryote lineages lack cell walls, biologists suggest that their common ancestor also lacked this feature. In sum, the early eukaryotes were probably single-celled organisms with a cytoskeleton and nucleus but no cell wall. What happened next?

The answer is that one of these ancestral eukaryotes acquired a mitochondrion. Recall from Chapter 6 that mitochondria are membrane-bound organelles that generate ATP using pyruvate as an electron donor and oxygen as the ultimate electron acceptor. Because all living eukaryotes have mitochondria, or contain genetic evidence that their ancestors had them, biologists contend that the common ancestor of today's eukaryotes also had mitochondria. Where did these organelles come from?

The Endosymbiotic Theory

Over 30 years ago, Lynn Margulis expanded upon a radical hypothesis—first proposed in the nineteenth century—to explain the origin of mitochondria. The endosymbiotic theory proposes that mitochondria originated when a bacterial cell took up residence inside a eukaryote about 2 billion years ago. Its name is inspired by the Greek word roots *endo* (inside) and *symbio* (living together).

In its current form, the endosymbiotic theory proposes that mitochondria evolved through a series of steps. As **Figure 27.12** (page 532) shows, the process began when large eukaryotic cells began using their cytoskeletal elements to surround and engulf small bacterial or archaeal prey. Next, instead of being digested, an engulfed bacterium began to live symbiotically within its eukaryotic host. Specifically, the theory maintains that the engulfed cell survived by absorbing reduced carbon molecules from its host and metabolizing them, using oxygen as an electron acceptor. The host cell, in contrast, was a predator capable only of anaerobic fermentation. As a result, the relationship was presumed to be stable because a mutual advantage existed: The host supplied the bacterial symbiont with protection and reduced carbon compounds from its other prey, while the symbiont supplied the host with ATP.

The endosymbiotic theory contends that chloroplasts originated in an analogous way. In this case, however, the theory maintains that a photosynthetic, endosymbiotic cyanobacterium provided its eukaryotic host with oxygen and glucose in exchange for protection.

Do the Data Support the Theory?

When biologists assess a theory, they often begin by evaluating its plausibility. For example, is it reasonable to suppose that the steps in the endosymbiotic theory actually happened? Given observations that biologists have made about living organisms today, each step in the endosymbiotic theory passes this test. For example, a member of the α-proteobacteria group, called *Paracoccus*, absorbs reduced carbon compounds from the environment and performs aerobic respiration—much like the

FIGURE 27.11 Life Cycles Vary Widely Among Protists
Note that many protists can reproduce by both asexual and sexual reproduction.

(a) Chlamydomonas

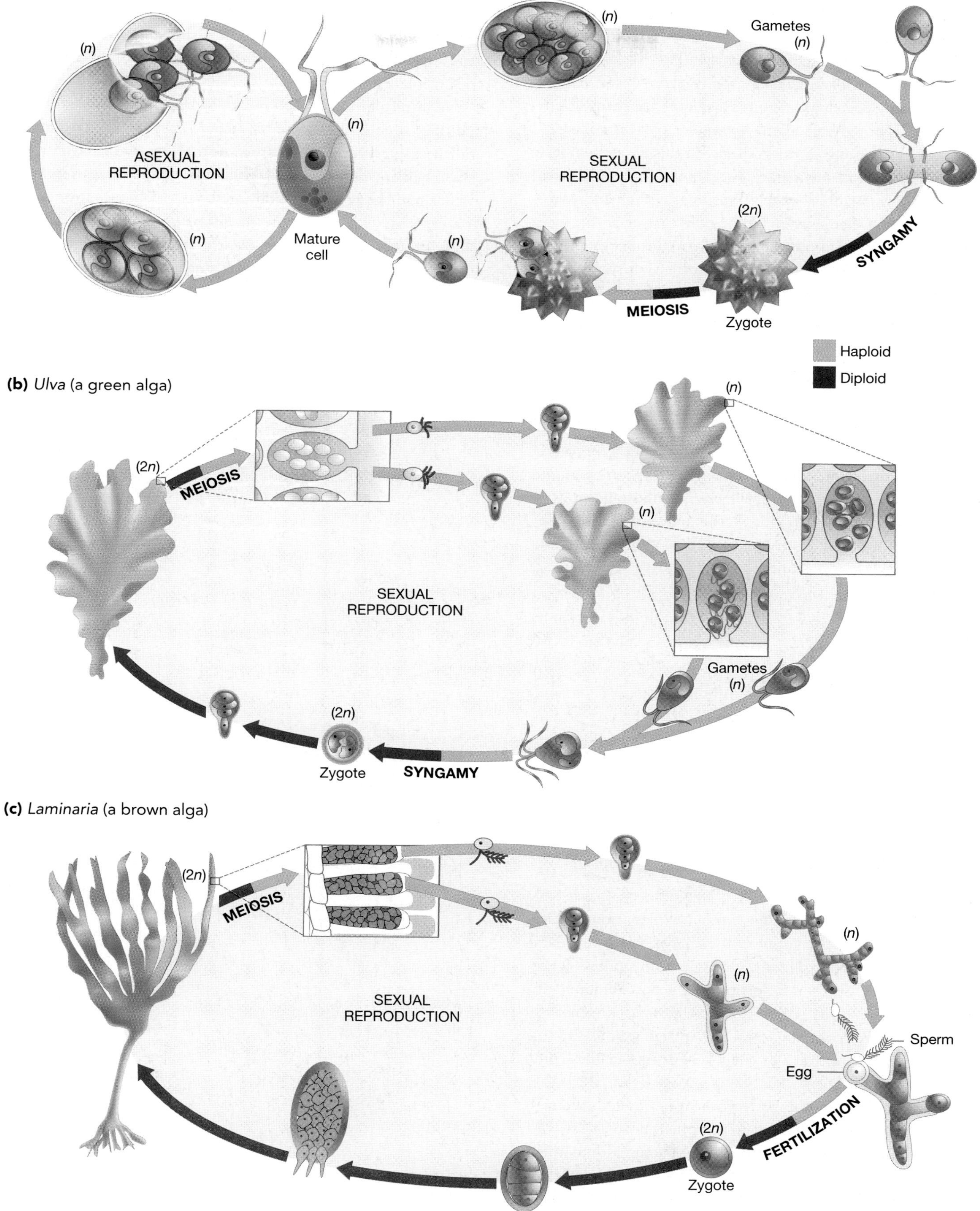
(n)
(n)
Gametes
(n)
(n)
ASEXUAL
REPRODUCTION
SEXUAL
REPRODUCTION
(2n)
(n)
Mature
cell
(n)
SYNGAMY
MEIOSIS
Zygote
Haploid
Diploid
(b) Ulva (a green alga)
(n)
(2n)
MEIOSIS
(n)
SEXUAL
REPRODUCTION
Gametes
(n)
(2n)
Zygote
SYNGAMY
(c) Laminaria (a brown alga)
(2n)
MEIOSIS
(n)
(n)
SEXUAL
REPRODUCTION
Sperm
Egg
(2n)
FERTILIZATION
Zygote

presumed mitochondrial ancestor. Dozens of contemporary species of protists obtain nutrients by ingesting bacteria and are anaerobic—much like the presumed ancestral host. More convincingly, many examples of endosymbiotic relationships between protists and bacteria exist today. Among the α-proteobacteria alone, three major groups are found *only* inside eukaryotic cells. There are also some examples of contemporary endosymbioses based on photosynthesis. In this case, members of the cyanobacteria reside inside protist hosts. In short, there are contemporary examples of relationships that the endosymbiotic theory claims also existed billions of years ago.

Several observations about the structure and function of mitochondria and chloroplasts themselves are also consistent with the theory:

- Mitochondria and chloroplasts are about the size of an average bacterium.
- Both organelles replicate by fission. The duplication of mitochondria and chloroplasts is independent of division by the host cell.
- Mitochondria and chloroplasts have their own ribosomes and manufacture their own proteins. The ribosomes found in the organelles closely resemble bacterial ribosomes in size and composition. Mitochondrial ribosomes are also poisoned by antibiotics like streptomycin that inhibit bacterial, but not eukaryotic, ribosomes.
- The photosynthetic organelle of one algal-like protist, called *Cyanophora*, has a cell wall containing the same constituent (peptidoglycan) found in the cell walls of cyanobacteria.
- Some cyanobacteria even look like chloroplasts. *Prochloron*, for example, contains chlorophyll *a* and *b* and has a system of internal membranes called thylakoids, which are found in chloroplasts (see Chapter 7).
- Mitochondria and chloroplasts have their own genomes, which are organized as circular molecules—much like a bacterial chromosome. The organelles also produce the enzymes needed to replicate and transcribe their own genomes.

Although these data are impressive, they are only consistent with the theory; they do not exclude other explanations. Years after Margulis began to champion the theory, however, data emerged that confirmed it. These data came from studies on the phylogenetic relationships of mitochondrial genes. Beginning in the early 1970s, a series of researchers compared gene sequences isolated from the nuclear DNA of eukaryotes, mitochondrial DNA from the same eukaryote species, and DNA from several species of bacteria.

Exactly as the endosymbiosis hypothesis predicted, the mitochondrial sequences turned out to be much more closely related to the sequences from the α-proteobacteria than they were to sequences from the nuclear DNA of their own cell. The result was overwhelming evidence that the mitochondrial genome came from an α-proteobacterium rather than from a eukaryote. An analogous analysis of SSU rRNA sequences by Stephen Giovannoni and colleagues showed that chloroplast genes are most closely related to sequences from cyanobacteria, not plants.

These results were a stunning vindication of a theory that had once been intensely controversial. They also carried a general message: Protists and other eukaryotes are cut-and-paste jobs. They are chimeras, like the monster from Greek mythology that had the head of a lion, the body of a goat, and the tail of a serpent. This innovative group of organisms was created when an ancient eukaryote—closely related to the archaea—combined with a bacterium. The lineage called the Eukarya was almost undoubtedly created when two cells fused.

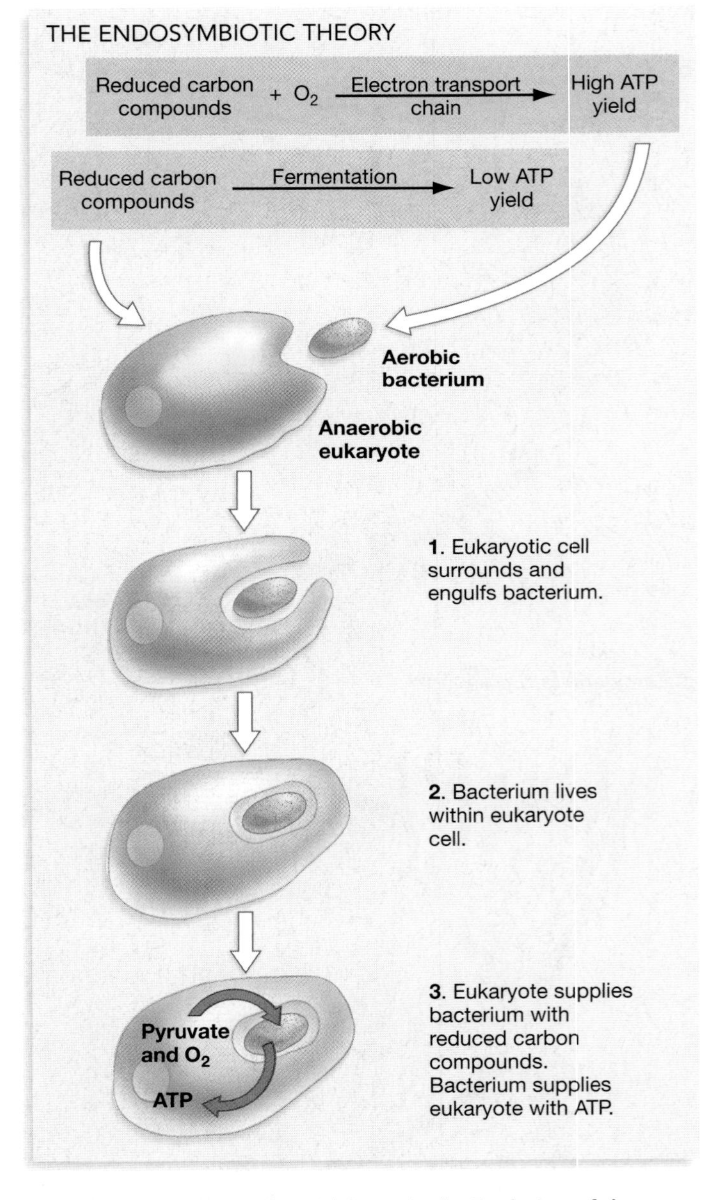

FIGURE 27.12 Proposed Initial Steps in the Evolution of the Mitochondrion by Endosymbiosis

Lateral Transfer of Mitochondrial Genes

When eukaryotes gained the symbionts that became mitochondria and chloroplasts, they acquired new, auxiliary genomes. The symbiotic α-proteobacterium that took up residence in an early eukaryote would have had a genome that coded for at least 500 proteins and rRNAs. But only 13 protein-coding genes are present in the mitochondrial DNA (mtDNA) of animals alive today. Yeast mtDNA has just eight protein-coding sequences, while the mtDNA in the protist that causes malaria has only three. At 62, a protist called *Reclinomonas* holds the current record for the most protein-coding sequences in mtDNA. The question is, what happened to all of the other bacterial genes?

There are two answers. Some were apparently lost over the course of evolution, while others moved from the mitochondrial genome to the nuclear genome (**Figure 27.13a**). How could this lateral transfer happen? Jacqueline Nugent and Jeffrey Palmer answered this question by virtually catching a gene in the act. These researchers studied a locus called *coxII* in 20 different flowering plants. This gene codes for a key component of the electron transport chains, introduced in Chapter 6, that are located in mitochondrial membranes. To determine whether *coxII* is located in the mtDNA of their study organisms, Nugent and Palmer isolated the mitochondrial genome from each species, cut the DNA into fragments using a restriction enzyme, separated the fragments by gel electrophoresis, and then probed the gel for *coxII* fragments using a radioactively tagged piece of the *coxII* DNA from soybeans. (This technique, called Southern blotting, was introduced in Chapter 15.) As Figure 27.13a shows, the probe DNA hybridized to fragments in the mtDNA of every species examined, except for one—mung bean. This means that the coxII gene is found in the mtDNA of every species tested except the mung bean.

Where did the mitochondrial coxII sequences of mung bean go? Nugent and Palmer looked in the nucleus. They repeated their "cut and probe" protocol with the nuclear DNA of mung bean and nuclear DNA from several closely related species. The resulting gel, shown in **Figure 27.13b**, shows that the *coxII* gene sequences are indeed located in the nucleus of mung bean. But *coxII* also showed up in the nuclear DNA of several species that contain the *coxII* gene in their mtDNA. This means that some plants have copies of the same gene in both genomes. The cowpea, in contrast, has just one copy located in its mtDNA. Patrick Covello and Michael Gray followed up on this result by showing that in soybeans, the mtDNA copy of *coxII* is inactive; it is not transcribed. Only the copy in nuclear DNA actually produces an active product.

How can we interpret these observations and results? The diagram in **Figure 27.13c** shows how Nugent and others have attempted to pull all of these data together. Their hypothesis is that a series of steps occurred during the evolution of pea-family plants. These steps resulted in the *coxII* gene moving from the mitochondrial genome to the nuclear genome.

Given the frequent loss of mitochondrial genes since endosymbiosis began, it appears that this sequence of events has occurred hundreds of times in the past and in many different lineages. In some protists, in fact, *all* of the mitochondrial genes that have not been lost have been transferred. *Giardia*, for example, is a human parasite that causes intestinal distress. It does not have mitochondria and makes ATP through anaerobic fermentation. But work by Andrew Rogers and others confirms

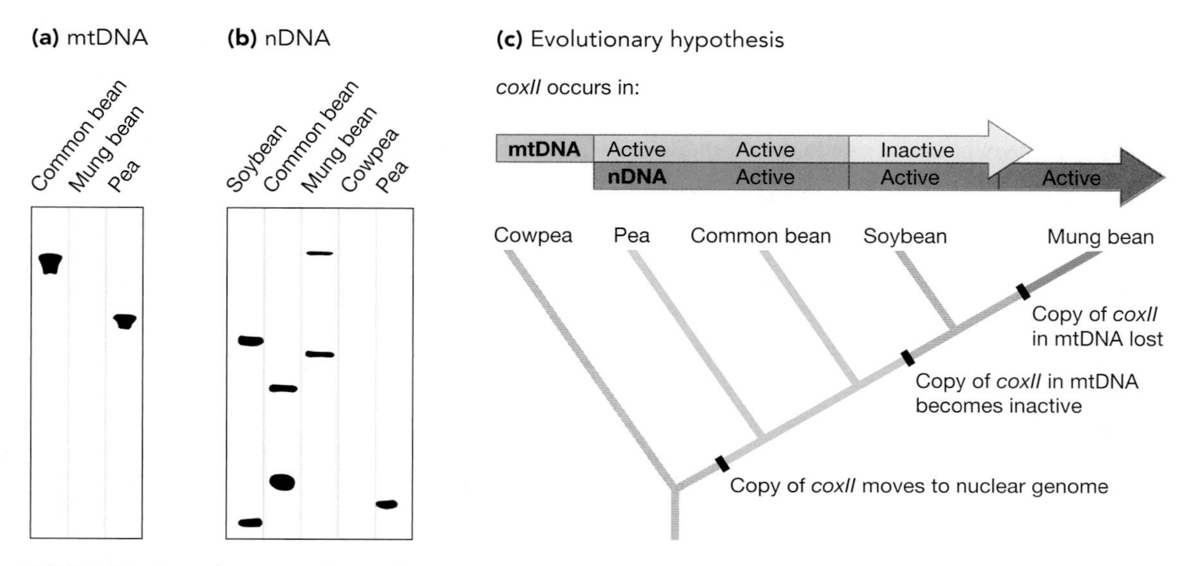

FIGURE 27.13 In Flowering Plants, the Mitochondrial Gene *coxII* Has Recently Been Transferred to the Nucleus
(a) This autoradiograph shows fragments of mtDNA from several plant species that were probed with the *coxII* gene. Note that the lane with mung-bean mtDNA is empty. **(b)** This autoradiograph shows fragments of nuclear DNA that were probed with the *coxII* gene. Note that *coxII* is found in mung-bean nuclear DNA. **(c)** This scenario explains the data in parts (a) and (b).

that its nuclear genome contains several genes that are clearly related to bacterial and mitochondrial sequences. Based on this observation, the researchers conclude that *Giardia's* ancestors had mitochondria.

27.4 How Do Protists Affect Human Health and Welfare?

Researchers who study the numerous innovations observed among protists—traits like sexual reproduction and events like the origin of intracellular organelles—are usually motivated by simple curiosity. When biologists see unusual patterns in the natural world, they want to explain them. Occasionally, however, basic research like this leads to unforeseen applications for human health and welfare.

As an example, consider the recent discovery of an unusual organelle in the group called Apicomplexa. You may recall that the Apicomplexa is a lineage of unusual parasitic protists. The group includes an organism called *Plasmodium falciparum*, which causes malaria.

Malaria

In India alone, over 30 million people each year suffer from debilitating fevers caused by malaria. About 100 million people worldwide are thought to be infected; over one million die from the disease annually.

The causative agent of malaria, *Plasmodium*, is transmitted to humans by mosquitoes. Because the parasite spends part of its life cycle inside these insects, most anti-malaria campaigns have focused on mosquito control. This strategy has become less and less effective over time, however, because natural selection has favored mosquito strains that are resistant to insecticides. Further, few drugs are available to combat the parasite directly, despite intensive efforts by researchers around the globe. This situation may soon change, however, due to the recent discovery of an unusual biosynthetic pathway in *Plasmodium*.

The story revolves around a molecule called chorismate and a strange organelle recently named the apicoplast. Chorismate is required for the synthesis of folate, which in turn is required for the synthesis of almost every carbon-ring-containing compound used by organisms, including the aromatic amino acids introduced in Chapter 3. The apicoplast is an organelle with four membranes whose function was unknown for many years.

Sabine Köhler and colleagues have recently provided insight into the apicoplast's origins by confirming that it has its own circular loop of DNA. These researchers also established that the genes in the apicoplast are very closely related to sequences from the chloroplast DNA of green algae. To interpret these findings, they propose that the ancestor of the Apicomplexa acquired a green algal chloroplast through secondary endosymbiosis. What is a secondary endosymbiosis? Sally Gibbs and others have argued that some groups of protists acquired chloroplast-like organelles not by inheriting them from a parental cell, but by engulfing photosynthetic protists.

As **Figure 27.14** shows, the secondary symbiosis hypothesis neatly explains why the apicoplast has four membranes. According to Köhler and coworkers, the apicoplast gradually lost the genes for photosynthetic enzymes as it switched to a endosymbiotic mode of existence.

What does all this have to do with chorismate and efforts to cure malaria? Fiona Roberts and colleagues provided the key link when they discovered the current function of the apicoplast: It is a center for biosynthesis. Specifically, the apicoplast genome houses the genes that code for enzymes in the shikimate pathway, which is the sequence of reactions that leads to the formation of chorismate. This biosynthetic pathway is found in all land plants, bacteria, and fungi, but is absent in animals. The researchers quickly seized on their discovery to test a radical proposal: *Plasmodium* might be controlled by using an herbicide called glyphosate as a drug. Glyphosate kills plants because it disrupts one of the enzymes in the shikimate pathway found in chloroplasts. In preliminary tests con-

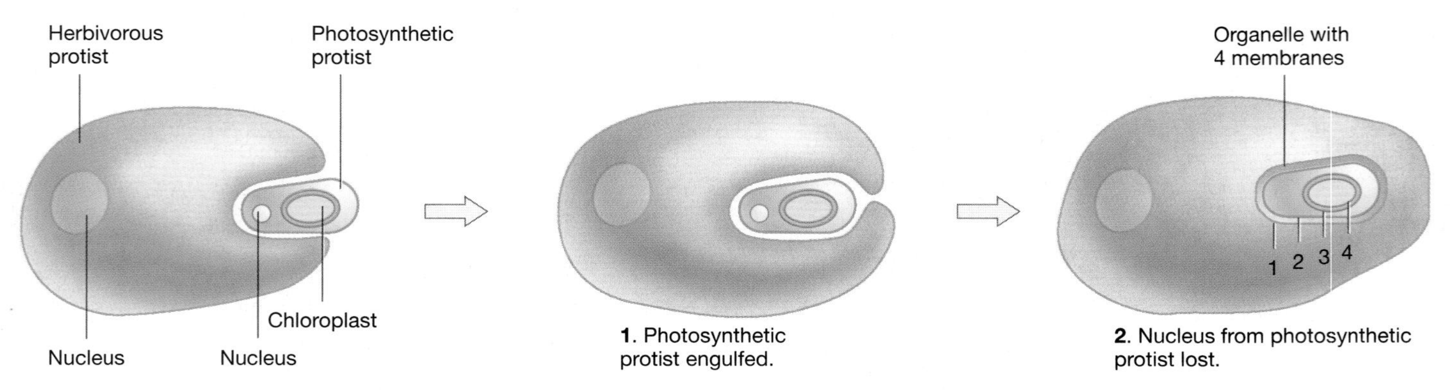

FIGURE 27.14 Secondary Endosymbiosis
A secondary endosymbiosis event leads to an organelle with four membranes, like the apicoplast found in *Plasmodium*.

ducted by Roberts and associates, glyphosate improved the health of mice infected with apicomplexan parasites. This test has not been done in a natural population of hosts or in infected mosquitoes, however. Will an herbicide become an effective anti-malarial drug, due to a secondary endosymbiotic event that occurred millions of years ago? If so, it would be one of the great breakthroughs in medical history.

Unfortunately, malaria is not the only important disease caused by protists. **Table 27.1** lists protists from many different lineages that have been the cause of human misery and economic loss. As the following paragraphs illustrate, however, protists have also been a source of well-being.

Ecological Importance of Protists

As a whole, the protists represent just 10 percent of the total number of named eukaryote species. Although the species diversity of protists may be relatively low, their abundance is extraordinarily high. The numbers of individual protists found in appropriate habitats is truly astonishing. A single teaspoon of pondwater can contain well over 1000 flagellated protists. Under certain conditions, dinoflagellates can reach concentrations of 60 million cells per liter of water.

The abundance of protists is important, for a simple reason. It makes protists, particularly the photosynthetic species, major players in the global carbon cycle. Diatoms, for example, are among the leading primary carbon producers in the oceans. Primary productivity by the world's oceans, in turn, is responsible for almost half of the total carbon that is fixed on the planet.

The dynamics of the oceanic carbon cycle are vastly different from the dynamics in terrestrial environments. On land, it is not unusual for carbon atoms to stay in the same form—say, as a component of cellulose in a tree—for tens if not hundreds of years. The turnover of carbon from one form to another can be extremely slow. In the oceans, however, carbon turns over rapidly. The protists that dominate marine plankton have short life spans. They die or are eaten in a matter of days or weeks. As the diagram in **Figure 27.15** shows, carbon atoms shuttle rapidly between life-forms.

There is one extremely stable repository for carbon in the oceans, however—limestone. The carbon-rich tests of foraminifera and the carbon-containing shells that surround

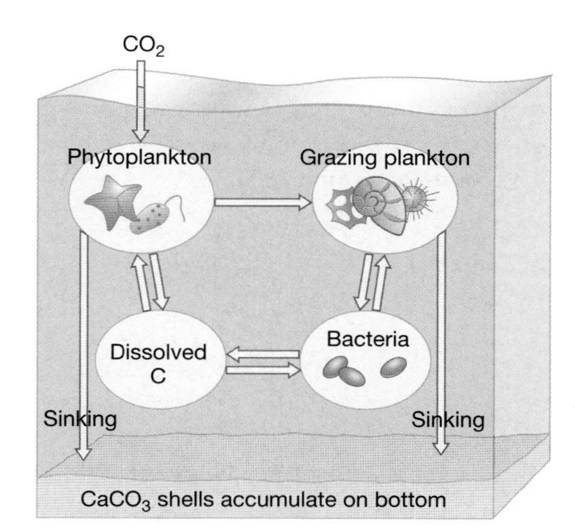

FIGURE 27.15 Protists and the Global Carbon Budget
This diagram of the carbon budget in the oceans shows that carbon atoms shuttle quickly among life-forms.

TABLE 27.1 Human Health Problems Caused by Protists

Lineage	Species	Disease
Apicomplexa	*Plasmodium falciparum*	malaria
Apicomplexa	*Toxoplasma*	infections in AIDS patients
Dinoflagellates	Many different species involved	Toxins released during "red tides" accumulate in clams and mussels and cause paralytic shellfish poisoning in people.
Diplomonads	*Giardia*	diarrhea
Trichomonads	*Trichomonas*	reproductive tract infections
Kinetoplastids	*Leishmania*	leishmaniases
	Trypanosoma gambiense and *T. rhodesiense*	sleeping sickness
Kinetoplastids	*Trypanosoma cruzi*	Chagas disease
Entamoebae	*Entamoeba histolytica*	amoebic dysentery
Oomycetes	*Phytophthora infestans*	An outbreak of this protist wiped out potato crops in Ireland in 1845–1847, causing famine that killed a million people and forced 2 million more to emigrate to the United States.
Microsporidians	*Encephalitozoon cuniculi*	diseases of the nervous system, respiratory tract, and digestive tract—especially in AIDS patients

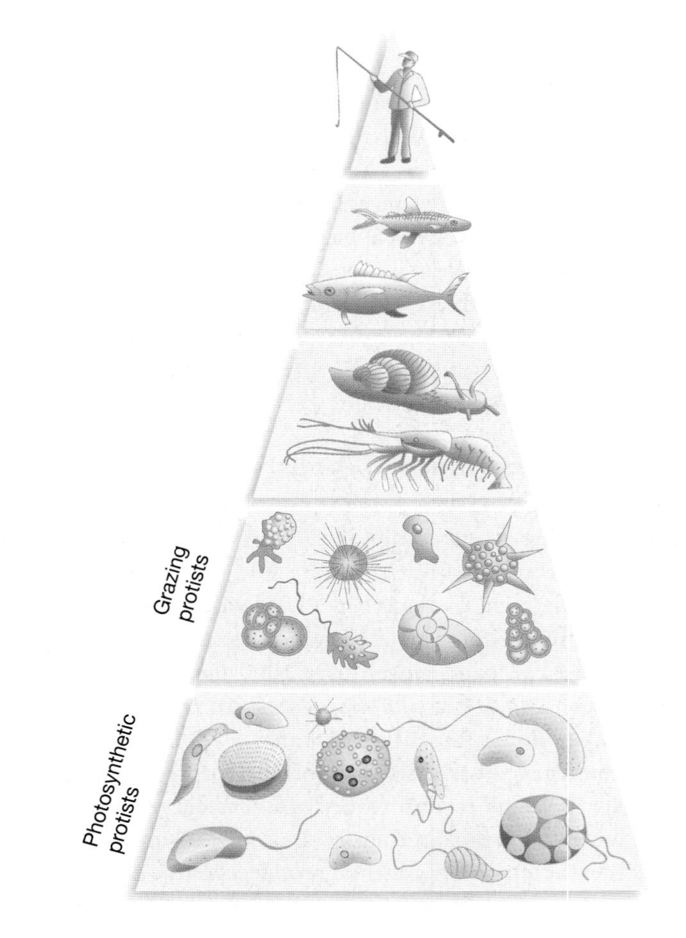

FIGURE 27.16 Protists and Marine Ecosystems
In aquatic ecosystems, photosynthetic protists form the base of the food chain.

some unicellular algae settle out of the ocean after these organisms die. The calcium carbonate ($CaCO_3$) deposits that result eventually turn into rock and lock up carbon atoms for tens of millions of years. Researchers who study the global carbon cycle refer to these deposits of marine shells as a carbon "sink." If techniques can be found to increase the amount of carbon raining down into this sink, it would be an effective way to reduce carbon dioxide levels in the atmosphere and thus reduce global warming.

The productivity that protists provide is important for another reason. The sugars they produce are the basis of food chains in both freshwater and marine environments (**Figure 27.16**). It is not an exaggeration to say that without protists, most aquatic ecosystems would collapse.

Essay Revolutions in Science

In 1962 a philosopher and science historian named Thomas Kuhn published an influential book called *The Structure of Scientific Revolutions*. In it, Kuhn presents a thesis for how the scientific enterprise works. Most scientific work, he contends,

Revolutionary science introduces fundamentally new theories and styles of thinking . . .

represents "normal science." In normal science the questions that scientists ask are inspired by the currently accepted theories and styles of thinking, or what Kuhn calls the prevailing paradigm. In Kuhn's view, most scientific results are also "normal" in the sense that they support the prevailing paradigm.

In contrast, Kuhn maintained that entirely new paradigms emerge during "scientific revolutions." Revolutionary science introduces fundamentally new theories and styles of thinking, which eventually displace the older paradigms.

To Kuhn, the Sun-centered theory of the solar system developed by Nicolaus Copernicus represents the classic example of revolutionary science. This theory, published in the early 1500s, maintains that the planets revolve around the Sun rather than the Earth. In addition to being correct, Kuhn's theory defined an entirely new research program in astronomy and physics. It also inspired new ideas in other fields, including religion and philosophy. Copernicus's paradigm inspired new questions, new techniques, and a new style of thinking in science.

This chapter on protists introduced two ideas that could be considered revolutionary. The first is the endosymbiotic theory for the origin of mitochondria and chloroplasts. Prior to its publication, the prevailing paradigm was that the lateral transfer of genes—transfer of genes from one lineage to another—occurred rarely, if ever. The endosymbiotic theory not only inspired a vigorous series of tests, it also established an entirely new way of thinking about the origin of the eukaryotes and the importance of lateral gene transfer in the history of life.

The second revolutionary idea is the change from taxonomic to phylogenetic thinking, highlighted in Box 27.1. Prior to this revolution, the prevailing paradigm was that prokaryotes and eukaryotes represented the two fundamental types of organisms. Traditional taxonomic schemes aimed to organize and categorize the diversity of life around a handful of kingdoms, often defined as the Monera (prokaryotes), Protista, Plantae, Fungi, and Animalia. Biologists analyzed diversity by thinking about morphological similarities among species rather than thinking about their history and evolutionary relationships.

A change in phylogenetic thinking is currently under way, and is inspiring the development of a new naming system for the diversity of life. The old five-kingdom system is being replaced by a system of three domains with many kingdoms.

Undoubtedly, 20 years from now you will be able to look back at this book and pinpoint paradigms that have been supplanted by new scientific revolutions. Which of the ideas in this book are doomed? What types of data will emerge, showing them to be inadequate? Scientific revolutions are bound to occur. In this sense, they too are "normal."

Chapter Review

Summary

Protists are a paraphyletic grouping that includes all eukaryotes except the land plants, fungi, and animals. The protists include many unicellular organisms as well as multicellular slime molds, red algae, brown algae, and green algae. The first fossil eukaryotes appear in rocks dated to about 1.75 billion years ago, at a time when the first sediments from an oxygen-rich ocean were being deposited.

Several features are common to all eukaryotes: large cell size, a nucleus, a cytoskeleton, cell division by mitosis, and extensive intracellular structuring that includes organelles. Many eukaryotes also undergo meiosis at some phase in their life cycle. In addition to these common features, many aspects of morphology and lifestyle are extremely variable among the protists. For example, multicellularity, swimming via the beating of cilia or flagella, amoeboid movement, hard outer coverings, and predatory, parasitic, or photosynthetic lifestyles have evolved in many different protist groups independently.

Eukaryotes also contain mitochondria or have genes indicating that their ancestors once contained mitochondria. Several types of data support the hypothesis that mitochondria originated as endosymbiotic bacteria, possibly engulfed by an early protist closely related to the archaea. The symbiosis is thought to have been successful because the endosymbiont provided its host with ATP while the host provided the bacterium with reduced carbon compounds and protection. Since the original endosymbiotic event, most genes in mitochondrial DNA have been transferred to the nucleus or lost from the genome entirely. The chloroplast's size, DNA structure, ribosomes, double membrane, and evolutionary relationships are also consistent with the hypothesis that this organelle originated as an endosymbiotic bacterium.

Parasitic protists cause several important diseases in humans, including malaria. Recent research has shown that the malaria parasite contains an organelle derived from the chloroplast, probably obtained through secondary endosymbiosis. As a result, it may be possible to treat malaria by using herbicides as drugs. Concerning their impact on humans, however, the most important aspects of protists result from their tremendous abundance. They provide food for many organisms in aquatic ecosystems and fix so much carbon that they have a large impact on the global carbon budget.

Questions

Content Review

1. When biologists refer to protists, they mean all eukaryotes except which groups?
   a. extinct forms
   b. land plants, fungi, and animals
   c. multicellular forms
   d. those with cells containing a cytoskeleton and nucleus
2. What materials do protists use to manufacture hard, outer coverings?
   a. stiff organic compounds like cellulose
   b. mineral-like molecules compounds such as calcium carbonate ($CaCO_3$)
   c. glass-like compounds that contain silicon (Si)
   d. all of the above
3. The fossil record and molecular phylogenies agree that green algae appeared relatively recently. Which of the following groups shares a recent common ancestor with green algae?
   a. brown algae
   b. red algae
   c. land plants
   d. cryptomonads
4. What does amoeboid movement result from?
   a. interactions between actin, myosin, and ATP
   b. coordinated beats of cilia
   c. the whip-like action of flagella
   d. action by the mitotic spindle
5. According to the endosymbiotic theory, what type of organism is the ancestor of the chloroplast?
   a. a photosynthetic archaea
   b. a photosynthetic bacterium
   c. a primitive photosynthetic eukaryote
   d. a modified mitochondrion
6. Multicellularity is defined in part by the presence of distinctive cell types. At the cellular level, what does this criterion imply?
   a. Individual cells must be extremely large.
   b. The organism must be able to reproduce sexually.
   c. Cells must be able to move.
   d. Different cell types express different genes.
7. Why are protists an important part of the global carbon cycle and marine food chains?
   a. they have high species diversity
   b. they are numerically abundant
   c. they have the ability to parasitize humans
   d. they have the ability to undergo meiosis

Conceptual Review

1. Why is an advanced cytoskeleton and the lack of a cell wall required for ingestive modes of feeding? What did this mode of feeding have to do with increased cell size in protists versus bacteria and archaea? What did it have to do with the acquisition of mitochondria and chloroplasts by endosymbiosis?
2. Compare and contrast sexual reproduction in protists versus bacteria and archaea.
3. What is the relationship between meiosis and the phenomenon known as alternation of generations?
4. Outline the steps in the endosymbiotic theory for the origin of the mitochondrion. What did each partner provide the other, and what did each receive in return? Answer the same questions for the chloroplast.
5. On page 533 the text states that "The symbiotic α-proteobacterium that first took up residence in an early eukaryote would have had a genome that coded for at least 500 proteins and rRNAs." No evidence is provided to back up this assertion. Do you agree with it? Where do you think the estimate of 500 genes came from?
6. The text refers to eukaryotes as "chimeras," and draws an analogy with the monster from Greek mythology that was a combination of parts from a lion, goat, and snake. Why is this analogy appropriate?

Applying Ideas

1. Consider the following:

   All living eukaryotes have mitochondria, or have evidence in their genomes that they once had these organelles. Thus it appears that eukaryotes acquired mitochondria very early in their history.

   The first eukaryotic cells in the fossil record correlate with the first appearance of rocks formed in an oxygen-rich ocean and atmosphere.

   Are these two observations connected? If so, how?
2. Biologists are beginning to draw a distinction between "species trees" and "gene trees." A *species tree* is a phylogeny that describes the actual evolutionary history of a lineage. A *gene tree*, in contrast, describes the evolutionary history of one particular gene, like the gene for chlorophyll *a*. In some cases, species trees and gene trees don't agree with each other. For example, the species tree for green algae says that their closest relatives are protists and plants. But the gene tree for chlorophyll *a* from green algae says that this gene's closest relative is a bacterium, not a protist. What's going on? Why do these types of conflicts exist?
3. Suppose a friend says that we don't need to worry about global warming. Her claim is that increased temperatures will make planktonic algae grow faster, and that carbon dioxide (CO_2) will be taken out of the atmosphere faster as a result. According to her, this carbon will be buried at the bottom of the ocean in calcium carbonate tests. As a result, the amount of carbon dioxide in the atmosphere will decrease and global warming will decline. Comment.

CD-ROM and Web Connection

CD Activity 27.1: Themes in the Evolution of Protists *(animation)*
(Estimated time for completion = 5 min)
Learn about the diversity in this fascinating group of organisms.

At your **Companion Website** (http://www.prenhall.com/freeman/biology), you will find self-grading exams and links to the following research tools, online resources, and activities:

Protists
Explore this image gallery of many different kinds of protists. Follow the links to learn about individual species.

Organelle Genomics
This site describes research into the genomes of mitochondria and chloroplasts.

Toxoplasmosis
Learn about recent treatments developed to fight toxoplasmosis, a serious complication of those infected with the HIV virus.

The Ecological Importance of Protists
Discover why protists are an important part of the trophic structure of marine communities.

Additional Reading

Anderson, D. M. 1994. Red tides. *Scientific American* 274 (August): 62–68. A review of the planktonic protists that spill toxins into the environment and produce paralytic shellfish poisoning.

de Duve, C. 1996. The birth of complex cells. *Scientific American* 274 (April): 50–57. An exploration of how endosymbiosis and the oxygen atmosphere influenced the evolution of the first eukaryotes.

Morell, V. 1997. How the malaria parasite manipulates its host. *Science* 278: 223. A commentary on recent research suggesting that *Plasmodium* infection causes mosquitoes to bite more aggressively.

Land Plants

28

When plants evolved, a major group of organisms began growing on dry land for the first time. Fungi and animals also accomplished the feat of moving from aquatic to terrestrial habitats, but plants were the pioneers—the first truly terrestrial organisms. It is no exaggeration to say that, by colonizing the continents, plants transformed the nature of life on Earth. The biologist Karl Niklas calls it "one of the greatest adaptive events in the history of life."

How and when did this transition occur? Section 28.1 addresses this question by examining the group's fossil record and phylogeny. Section 28.2 follows up by analyzing the adaptive breakthroughs that made the diversification of plants possible. These innovations include **vascular tissue**, the seed, and the flower. But as section 28.3 points out, the fundamental story of plants revolves around photosynthesis. Growing on land gave plants access to massive supplies of the raw material and energy source used for photosynthesis—carbon dioxide from the atmosphere and light from the Sun. In no small part, the group's diversification occurred because numerous strategies exist for capturing photons.

The chapter ends with a look at how plants affect the health and welfare of humans. People depend on plants for the necessities of life—food, fiber, and fuel. Tens of thousands of biologists are employed in research designed to increase the productivity of plants and to create new ways of using them. But humans also depend on plants for important intangible values, such as healthy ecosystems and beauty. Consider that the sale of cut flowers generates over $1 billion each year in the United States alone. How did plants evolve from an alga on a muddy shore that existed 460 million years ago to organisms that enrich the soil, produce the oxygen you breathe and the food you eat, and serve as symbols of health, love, and beauty?

Mosses are common in moist habitats and are among the most ancient of all plant groups. According to data reviewed in this chapter, the earliest land plants evolved from green algae that inhabited ponds, streams, and other freshwater habitats.

FIGURE 28.1 Timeline of Plant Fossil History
This diagram illustrates five major intervals that can be distinguished in the fossil record of plants.

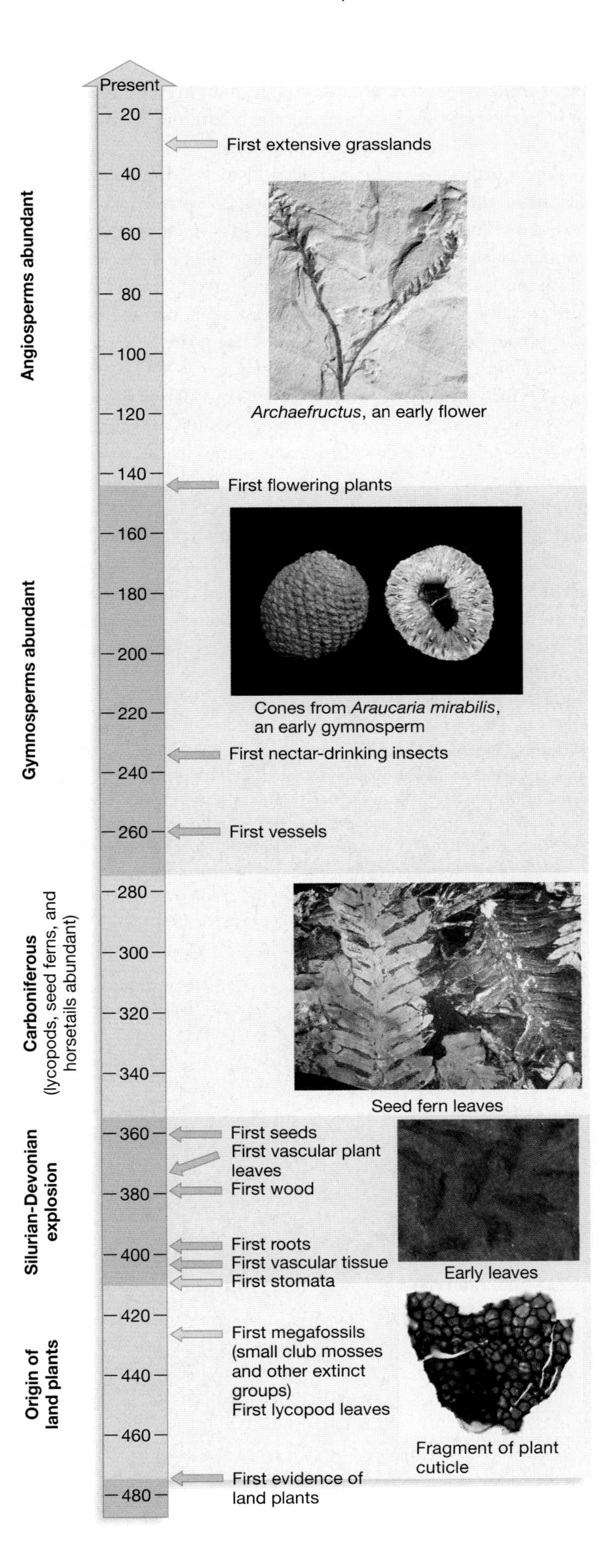

28.1 Phylogenies and the Fossil Record: Origins and Diversification

The land plants living today belong to 11 major lineages, or phyla. As **Table 28.1** (page 542) shows, these groups vary enormously in their species richness. The ancient phylum Ginkgophyta is represented by a single living species, while the most recent phylum to appear, the flowering plants, contains well over 230,000 species. The table also includes some important informal names that botanists use to identify major plant groups. The mosses, liverworts, and hornworts, for example, are often referred to as bryophytes (mossy-plants); the conifers, cycads, ginkgo, and gnetophytes are collectively called gymnosperms.

How and when did these major lineages diversify? To answer this question, biologists turn to the fossil record and phylogenies.

A Fossil Timeline

The fossil record of plants is massive. In an attempt to organize and synthesize the database, **Figure 28.1** breaks it into five parts. Each of these segments of time identifies a major event in the diversification of land plants.

The oldest interval begins 476 million years ago, spans almost 60 million years, and documents the origin of land plants. Three types of microscopic fossils are present during this period: (1) spores that are surrounded by a tough membrane, (2) sheets of a waxy material called cuticle, and (3) small tubes. When paleobotanist Jane Gray found the oldest of these fossil spores in rocks that formed 476 million years ago, she contended that they must be the remains of land plants because the coating surrounding the structures was intact. She reasoned that because the covering had resisted decay, it may have been made of the same tough molecule (sporopollenin) that surrounds the spores and pollen of today's land plants. Subsequently, Wilson Taylor tested this prediction by examining the fossilized material under the electron microscope. He confirmed that it had the sheet-like, laminate structure typical of spores from some modern land plants. Because sporopollenin membranes help pollen and spores resist damage from drying out, the discovery suggests that the spores were adapted to conditions on land.

The second major event in the record is sometimes called the "Silurian-Devonian Explosion." In rocks dated between 410 and 360 million years ago, biologists find macroscopic fossils from most of the major plant lineages. Virtually all of the adaptations that allow plants to occupy dry, terrestrial habitats are present, including water-conducting cells, roots, and wood.

The third segment in plant fossil history is aptly named the Carboniferous period. In sediments dated from about 350 to 290 million years ago, biologists find extensive deposits of coal. Coal, in turn, is a carbon-rich rock packed with fossil spores, branches, leaves, and tree trunks. These fossils are frequently derived from plants called lycopods. Because coal formation is thought to start only in the presence of water, the Carboniferous fossils indicate the presence of extensive, forested swamps.

The fourth segment in land plant history is dominated by fossils from the plant lineages that are collectively called gymnosperms (naked-seeds). These lineages include the cycads, conifers such as pines and spruces, and ginkgos. Because gymnosperms grow readily in dry habitats, biologists infer that both wet and dry environments on the continents became blanketed with green plants for the first time during this interval, which lasted from 250 to 120 million years ago.

The fifth section in the history of land plants is still under way. This is the age of flowering plants, or angiosperms (encased-seeds). The first flowering plants in the fossil record appear about 125 million years ago. The woody plants that produced the first flowers are the ancestors of today's grasses, orchids, daisies, oaks, maples, and roses.

There is an important general trend in this sequence of events. Organisms that appear in the fossil record more recently are much less dependent on moist habitats than are groups that appear earlier. For example, the sperm of mosses and ferns swim to accomplish fertilization, while gymnosperms and angiosperms encase their sperm in pollen grains that are transported via the wind or insect pollinators.

To test the validity of these observations, biologists analyze data sets that are independent of the fossil record. Does the phylogeny of land plants confirm or contradict the patterns in the fossil record?

TABLE 28.1 The Land Plant Phyla

Traditionally, the 11 phyla of land plants with living representatives are grouped as indicated in this table. Vascular plants are distinguished from "nonvascular" plants, or bryophytes, even though some bryophytes have water-conducting cells. Among vascular plants, the gymnosperms and the angiosperms are referred to as the seed plants because they produce seeds instead of spores. **EXERCISE** Circle the phyla that are included in the group called seed plants. Rank the phyla from 1 to 11 according to their species richness.

	Estimated Number of Living Species
Bryophytes (nonvascular plants)	
Mosses	12,000
Liverworts	6,500
Hornworts	100
Vascular plants	
Lycopods (club mosses)	1,000
Horsetails	15
Ferns	12,000
Gymnosperms	
Conifers	550
Cycads	100
Ginkgo	1
Gnetophyta	70
Angiosperms	
Flowering plants	230,000

A Phylogeny of Plants

Understanding the phylogeny of plants is a challenge. Most of the major lineages appear very early in the history of land plants (during the Silurian-Devonian explosion). Much later, the ferns, gymnosperms, and angiosperms underwent adaptive radiations. These later radiations produced tens of thousands of diverse branches and twigs on the tree of life. To understand how plants diversified, then, botanists must determine the relationships among the major groups, which diverged from one another 410–360 million years ago. Only then can they unravel the relationships of species within these groups, which diverged much more recently.

Chapter 23 introduced how biologists use morphological and molecular traits to estimate the evolutionary relationships among species. But whether researchers use morphological characteristics or DNA sequences to understand the history of the major plant groups, they face a common problem. They must find traits that change slowly enough to be informative about what happened during the Silurian-Devonian explosion some 400 million years ago. If the traits they analyze change rapidly, unique features that were once capable of identifying particular lineages will have changed too much to be recognizable.

Fortunately, morphological characters like the presence or absence of vascular tissue, leaves, seeds, and other traits provide reliable data for reconstructing relationships among the deepest branches on the plant phylogeny. The loci that code for small-subunit (SSU) rRNA and the large subunit of the rubisco enzyme introduced in Chapter 7 also change slowly enough to be informative about relationships among ancient lineages.

Paul Kenrick and Peter Crane recently put these morphological and molecular data together and produced the tree in **Figure 28.2**. There are several important points to note about the relationships:

- The oldest branches on the tree lead to groups that are collectively called the green algae. The green algae and all of the later lineages on the tree define a lineage called the green plants. All green plants have important distinguishing characteristics like a cellulose cell wall, chlorophyll *a* and *b* as photosynthetic pigments, and use of starch as a storage product in the chloroplast.
- The ancestor of the land plants descended from a green-alga-like lineage called the Charophyta. The lineages that descended from the Charophyta are collectively called the Land Plants.
- The liverworts, hornworts, and mosses are the earliest-branching, or most "primitive," groups among land plants. Mosses have limited numbers of water-conducting cells; liverworts and hornworts have none at all.

- The next group on the tree, the Rhyniopsida, is known only from fossils. They are leafless; some species possess a simple type of water-conducting cell called a tracheid.
- The lycopods, horsetails, and the ferns have vascular tissue and leaves, but reproduce by making spores.
- The more recently evolved groups such as gymnosperms and angiosperms also have vascular tissue. They produce seeds and have complex leaves.

When combined with recent analyses of the fossil record, this phylogeny supports some important conclusions. In all of the trees generated so far, land plants trace back to a single common ancestor, a member of the green algal group called Charophyta. This result supports the hypothesis that there was only one water-to-land transition in plants. Further, because nearly all Charophyta alive today inhabit lakes and ponds, it is likely that the transition to land occurred from freshwater, and not saltwater, habitats.

FIGURE 28.2 A Phylogeny of the Land Plants
The Charophyta are a lineage of green algae that live in freshwater habitats. A subgroup of the Charophyta called the Charales represents the closest living relative to the land plants.

The phylogeny also supports the fossil record's most striking trend. The most ancient groups of plants are dependent on wet habitats, while more recently evolved groups are tolerant of dry—or even desert—conditions.

Let's turn now to this trend. What adaptations made the water-to-land transition possible? Which innovations led to the evolution of plants that could tolerate drought?

28.2 The Transition to Land: Key Innovations and Trends

For aquatic life-forms, terrestrial environments are deadly. Compared to a habitat where the entire organism was bathed in fluid, terrestrial environments present a situation where only a portion, if any, of their tissues were bathed in fluid.

If this problem could be solved, however, growth on land offered a bonanza of resources. Take light, for example. The water in ponds, lakes, and oceans absorbs light, making it unavailable to drive photosynthesis in plants. In addition, the most important molecule to photosynthetic organisms, carbon dioxide, is much more readily available on land than it is in water. Not only is it more abundant in the atmosphere, it also diffuses more readily.

Solving the water problem was a breakthrough that evolution by natural selection achieved in three steps: (1) preventing water loss from cells and thereby preventing cells from drying out, (2) transporting water from tissues with access to water to tissues without access, and (3) transporting gametes without water. Let's examine each of these steps in turn.

Preventing Water Loss: Cuticle and Stomata

Section 28.1 pointed out that sheets of the waxy substance called cuticle begin showing up early in the fossil record of plants, along with encased spores and tube-like fragments. This is significant because the presence of cuticle in fossils is a diagnostic indicator of land plants. Cuticle is a waxy, watertight sealant that gives plants the ability to survive in dry environments. If biologists had to point to one innovation that made the transition to land possible, it would be the production of cuticle.

Covering surfaces with wax creates a problem, however, in terms of gas exchange. Plants need to take in carbon dioxide from the atmosphere. But cuticle is almost as impervious to this gas as it is to water. Most modern plants solve this problem with a structure called a **stoma** (plural: stomata), consisting of an opening surrounded by specialized **guard cells.** The opening, called a pore, opens or closes as the guard cells change shape. When guard cells become soft, they close stomata down, limiting water loss from the plant. When guard cells become taut, they open the pore, maximizing carbon dioxide intake. (The mechanism behind guard-cell movement is explored in Chapter 35.)

The first stomata in the fossil record appear in a member of the Rhyniopsida that lived about 385 million years ago. As **Figure 28.3** shows, the surface of this plant contained pores framed by cells shaped exactly like modern-day guard cells. Stomata are also present in living members of several of the earliest branching groups on the land plant phylogeny—the hornworts and mosses illustrated in Figure 28.2. The other early-branching lineage—the liverworts—have pores but no guard cells. The presence of guard cells in these groups implies that many of the early land plants had the ability to regulate gas exchange—and control water loss—by opening and closing their pores.

Transporting Water: Vascular Tissue and Wood

Once cuticle and stomata evolved, plants could keep photosynthesizing while exposed to air. Cuticle and stomata allowed plants to grow in the saturated soils of tidelands or pond edges. The next challenge? Defying gravity.

Water is 1000 times denser than air. Aquatic algae can grow erect because they float. They float because their density is similar to water's density. In contrast, biologists hypothesize that the first land plants probably had a low, sprawling growth habit. The logic behind this hypothesis has two elements. To obtain water, the first plants had to grow in a way that kept their tissues in direct contact with moist soil. Also, their algal-like tissues lacked the rigidity required to withstand gravity and remain erect outside of water. If this hypothesis is correct, then competition for space and light would have become intense soon after the first plants began growing on land. To escape competition, plants would have to grow erect.

For plants to adopt erect growth habits on land, two problems had to be overcome. The first is transporting water from tissues that are in contact with wet soil to tissues that are in contact with dry air, against the force of gravity. The second is becoming rigid enough to avoid falling over in response to gravity and wind. As it turns out, vascular tissue solved both problems.

Biologists Paul Kenrick and Peter Crane explored the origin of water-conducting cells and erect growth by examining the extraordinary fossils found in a rock formation from Scotland

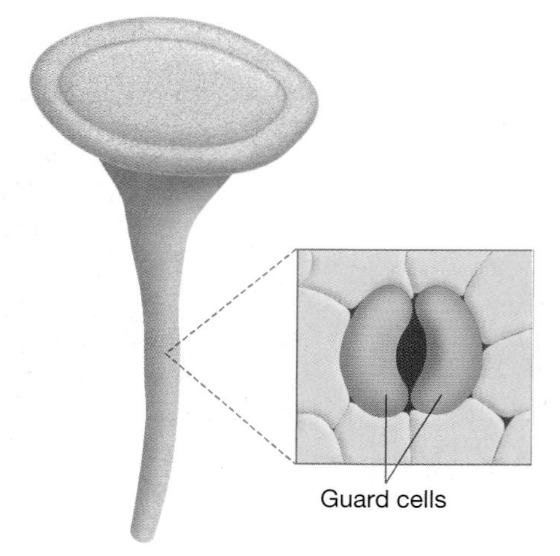

FIGURE 28.3 The First Stomata
Guard cells surround pores found on the stems of plants preserved in the fossil-rich formation called the Rhynie Chert of Scotland.

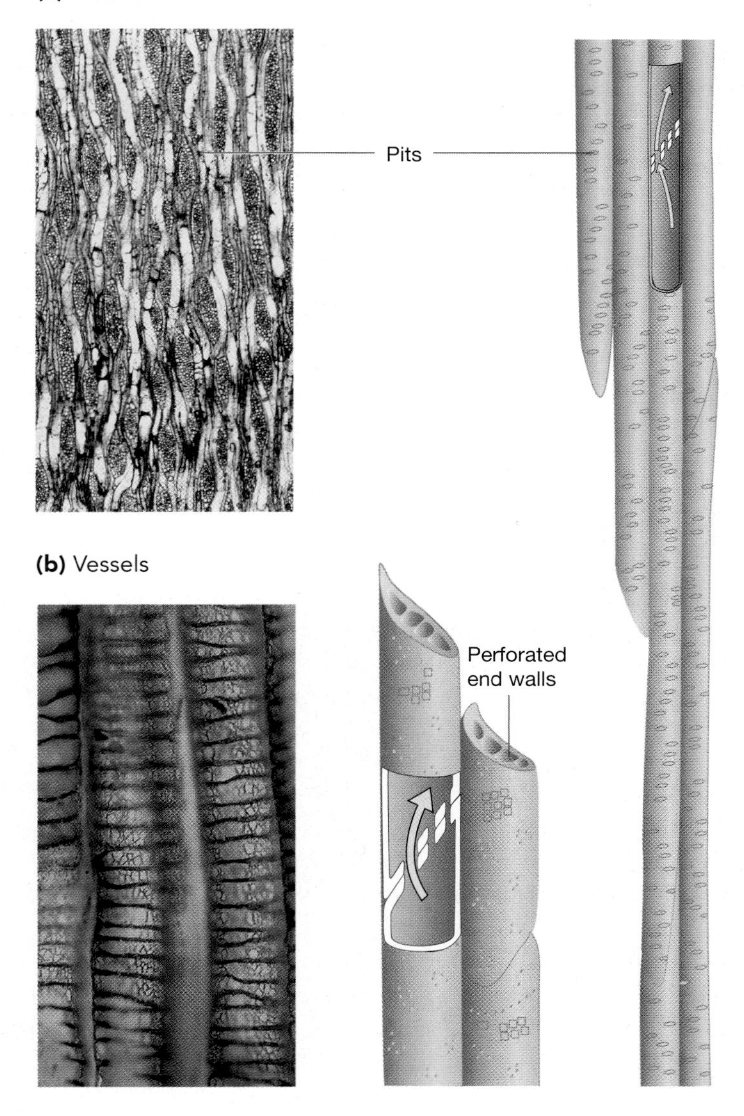

FIGURE 28.4 The Water-Conducting Cells Found in Vascular Tissue **(a)** The photograph shows some of the first tracheids in the fossil record; the diagram shows tracheids from a modern plant. Water flows through pits in the side and end walls. The pits have no thickened secondary wall—only a primary cell wall. **(b)** The photograph shows 50-million-year-old vessels; the diagram shows modern vessels. The cells that make up vessels are shorter and fatter than tracheids and are stacked end to end. Water flows up vessels directly, as well as laterally (from vessel to vessel), through perforations in the primary and secondary cell walls.

called the Rhynie Chert. These rocks formed about 400 million years ago and contain some of the first visible plant specimens in the fossil record (as opposed to the microscopic spores, cuticle, or tube fragments found in older rocks). The Rhynie Chert also contains numerous plants that fossilized in an upright position. This indicates that many or most of the Rhynie plants grew erect. How did they stay vertical?

By examining fossils under the electron microscope, Kenrick and Crane established that four distinct types of water-conducting cells existed in plants from the Rhynie Chert. Apparently these cells were efficient enough to fill cells up with water and make the early land plants rigid due to turgor pressure (see Chapter 32). In addition, some of the water-conducting cells preserved in the Rhynie Chert have thickened rings under their surfaces. These rings contain a molecule called lignin. Lignin is a polymer built from six-carbon rings and is extraordinarily strong for its weight. It is particularly effective in resisting compression. Kenrick and Crane hypothesize that the lignin rings gave stem tissues the strength to remain erect in the face of wind and gravity.

By about 380 million years ago, the fossil record contains the advanced water-conducting cells called tracheids (from the Latin *trachea*, or windpipe). **Tracheids** are the elongated cells shown in **Figure 28.4a** and are found in all modern phyla of vascular plants. Tracheids die after maturing, so they do not contain cytoplasm. They are also packed together to form vascular tissue. These features made the efficient transport of water to aboveground tissues possible. Tracheids that appear later in the fossil record have thickened secondary cell walls with extensive deposits of lignin, in addition to the primary cell wall made of cellulose. In several different lineages the secondary meristems described in Chapter 31 arose, producing tracheids with thickened secondary walls and the material called wood. The combination of vascular tissue and wood made tall, tree-like growth forms possible. The first trees appear in the fossil record 370 million years ago.

In fossils dated to 250–270 million years ago, Hongqi Li and colleagues have found true vessel elements. **Vessels** are the most advanced type of water-conducting tissue known in plants, and are found only in gnetophytes and angiosperms. Water transport is extremely efficient through vessels because their end walls have gaps in the primary and secondary cell walls (**Figure 28.4b**).

Can all of these observations be pulled together into a coherent picture of how plants adapted to dry habitats over time? One way to answer this question is to identify where each innovation occurred as plants diversified. The first step in an analysis like this is to determine whether each trait is present or absent in the groups that form the tips of the land plant phylogeny. Then biologists ask, based on these data, where is it most likely that each of the major innovations occurred? The answers are given by the bars across the branches shown in **Figure 28.5** (page 546). Fundamentally important adaptations to dry conditions, such as cuticle, stomata, and vascular tissues, evolved just once. In contrast, more specialized structures involved in terrestrial growth, like wood and leaves, evolved several times independently.

But for plants to become free of any dependence on wet habitats, one final challenge remained. The bryophytes and the most ancient vascular plant lineages, like ferns, cycads, and ginkgos,

FIGURE 28.5 Where Major Innovations Occurred as Land Plants Diversified
The bars on this phylogeny indicate where some major innovations occurred during the evolution of land plants.

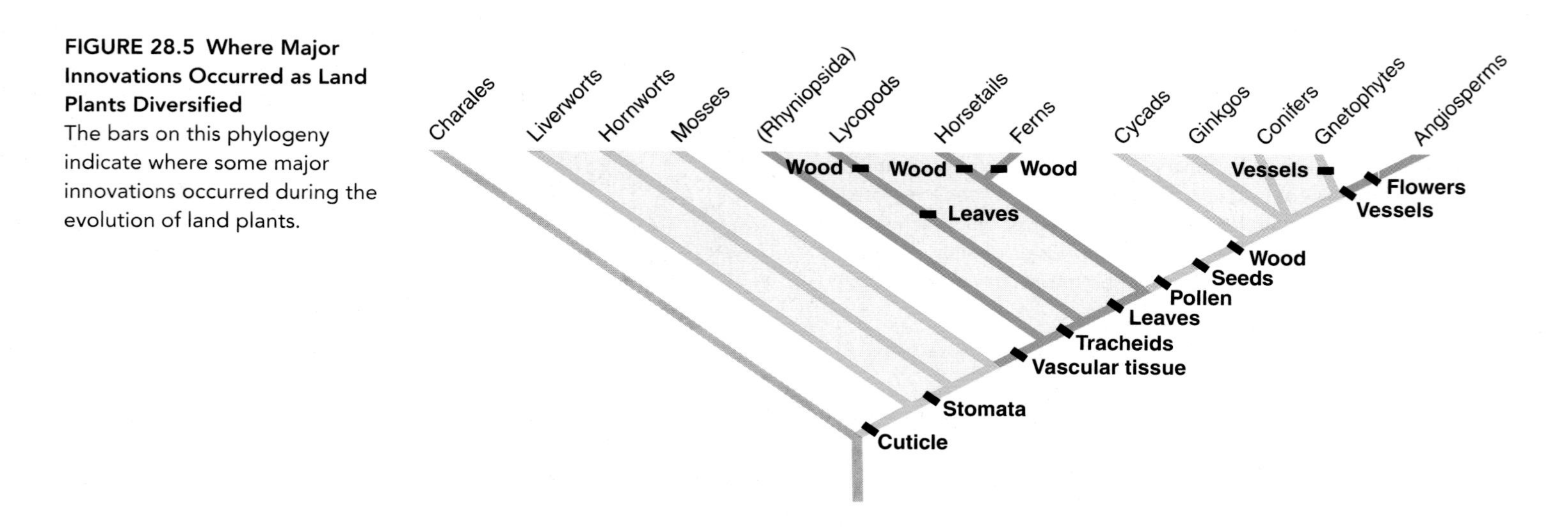

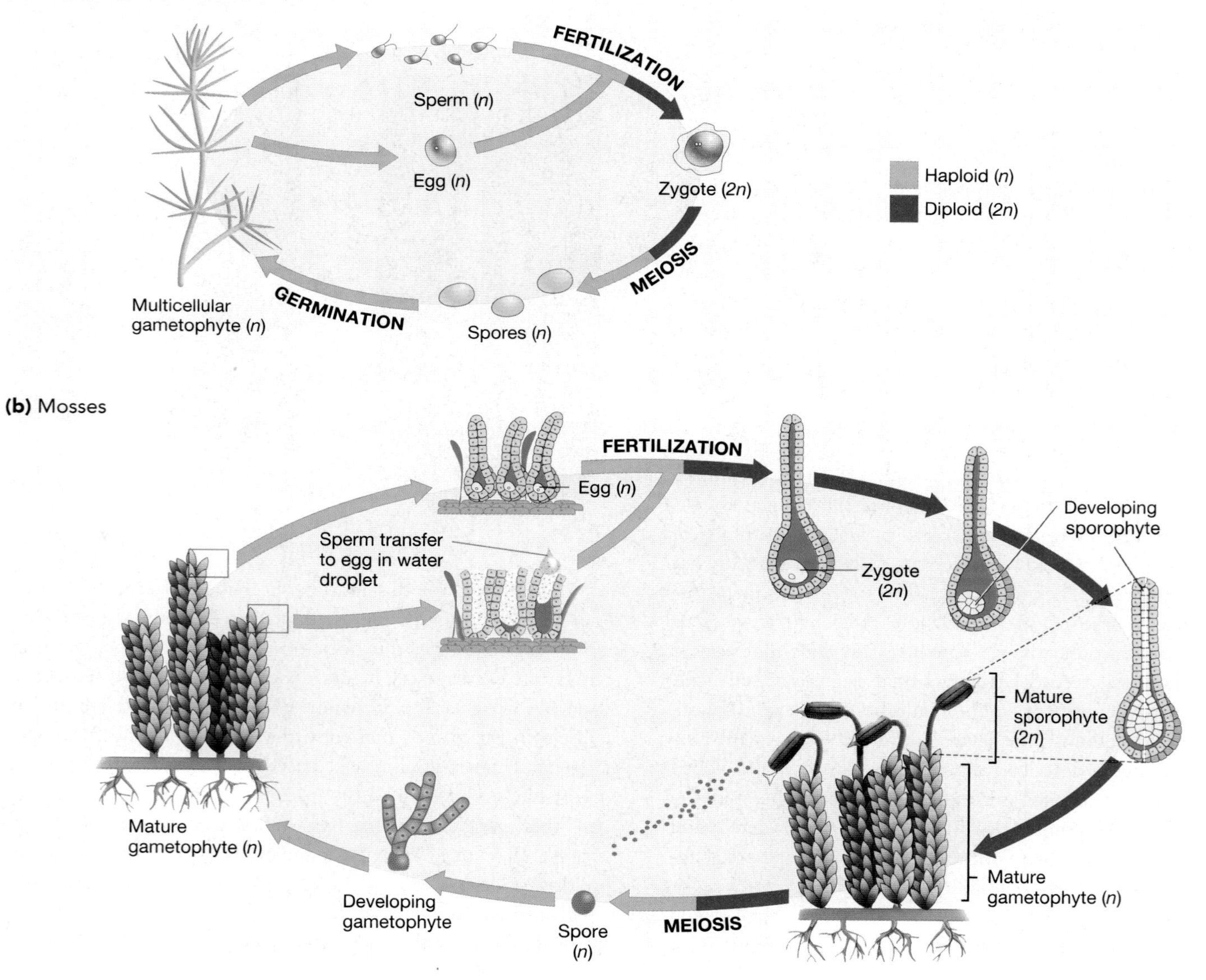

FIGURE 28.6 Life Cycles of Charophyta and Land Plants
These life cycles are presented in order, from earlier- to later-branching lineages.

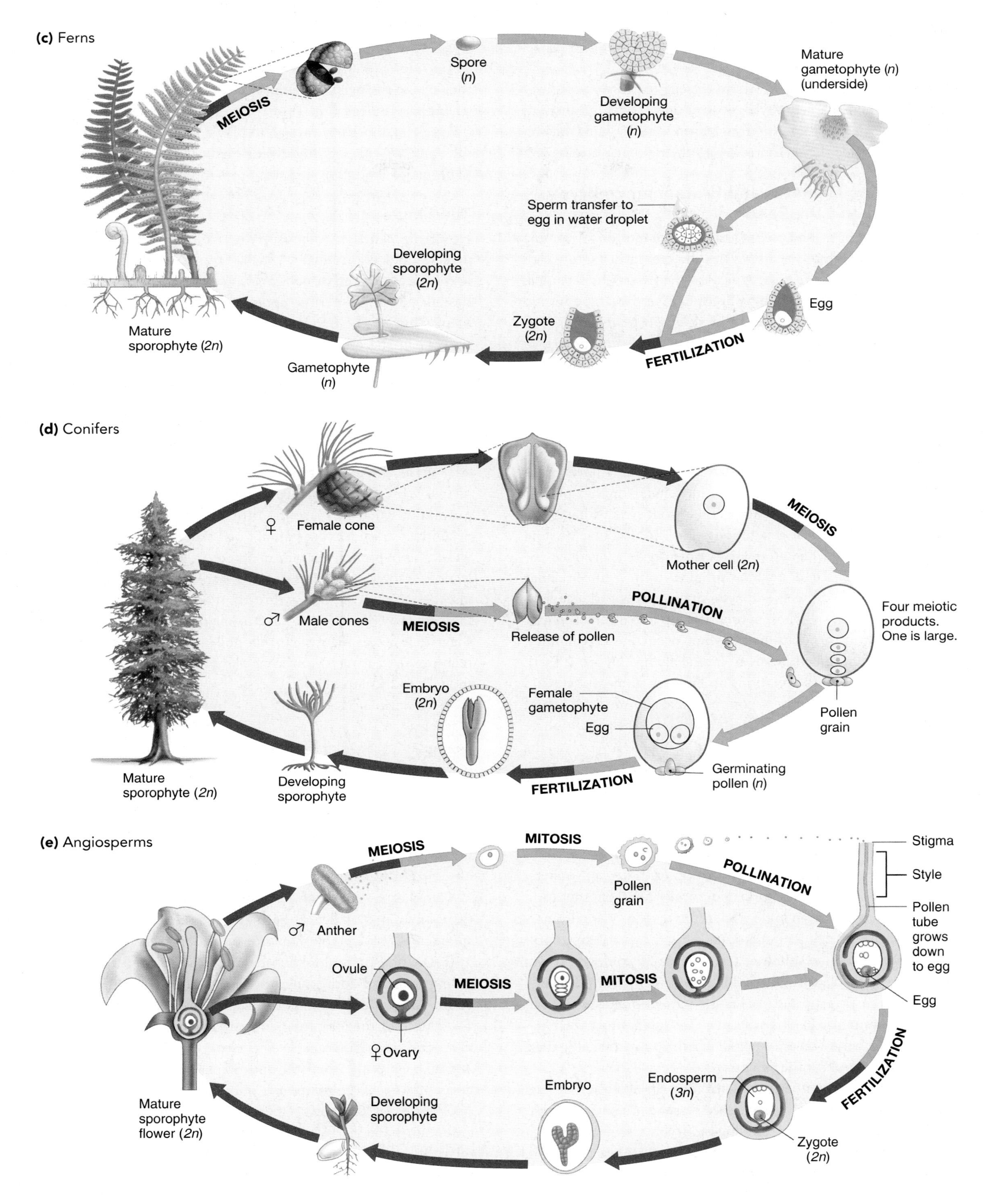

(c) Ferns
MEIOSIS
Spore
(n)
Developing
gametophyte
(n)
Mature
gametophyte (n)
(underside)
Sperm transfer to
egg in water droplet
Egg
FERTILIZATION
Zygote
(2n)
Developing
sporophyte
(2n)
Gametophyte
(n)
Mature
sporophyte (2n)
(d) Conifers
♀ Female cone
Mother cell (2n)
MEIOSIS
♂ Male cones
MEIOSIS
Release of pollen
POLLINATION
Four meiotic
products.
One is large.
Pollen
grain
Embryo
(2n)
Female
gametophyte
Egg
Germinating
pollen (n)
FERTILIZATION
Mature
sporophyte (2n)
Developing
sporophyte
(e) Angiosperms
MEIOSIS
MITOSIS
Pollen
grain
POLLINATION
Stigma
Style
Pollen
tube
grows
down
to egg
♂ Anther
Ovule
MEIOSIS
MITOSIS
Egg
♀ Ovary
FERTILIZATION
Endosperm
(3n)
Embryo
Zygote
(2n)
Mature
sporophyte
flower (2n)
Developing
sporophyte

all have male gametes that swim to the egg to perform fertilization. When male gametophytes evolved into pollen grains that produce sperm and that could be transported by wind or animals, plants made the final break with their aquatic origins.

Transporting Gametes and Protecting Embryos: Pollination and the Seed

To understand how land plants lost their reliance on swimming gametes, let's analyze the life cycles illustrated in **Figure 28.6** (pages 546–547). A life cycle typical of the Charophyta is illustrated in part (a), followed by a bryophyte in part (b), a fern in part (c), a gymnosperm in (d), and an angiosperm in (e). The diagrams are ordered in the sequence that the groups appear in the fossil record and in phylogenies estimated from molecular and morphological data.

Do any patterns exist in these data? If so, what do they suggest about how plants solved the problem of reproducing in dry environments?

As you study the figure, note that all land plants have multicellular haploid phases as well as multicellular diploid phases. That is, they undergo the phenomenon introduced in Chapter 27 known as alternation of generations. The multicellular haploid phase is called the **gametophyte**. Gametophytes produce gametes by mitosis. These gametes unite to form a multicellular diploid phase called the **sporophyte**. Sporophytes produce haploid spores by meiosis. These spores develop into a gametophyte.

Alternation of generations does not occur in the Charales, where only the gametophyte is multicellular (Figure 28.6a). Among land plants, the relationship between gametophyte and sporophyte is variable. In bryophytes like the moss in Figure 28.6b, the sporophyte is small and depends on the gametophyte for nutrition. In ferns the sporophyte is much larger than the gametophyte; each generation produces its own food through photosynthesis (Figure 28.6c). In gymnosperms and angiosperms (Figure 28.6d and e), the sporophyte is dominant. The male gametophyte is reduced to the pollen grain; the female is reduced to a small structure containing the egg(s). Both gametophytes obtain all of their nutrition from the sporophyte.

Why are these observations important? When mutation and selection resulted in male gametophytes that were dramatically reduced in size and surrounded by a coat of sporopollenin, pollen grains resulted. These structures could persist in dry environments without dying from dehydration. They could also be transported to female gametophytes by wind, gravity, or animals. In this way, the seed plants lost their dependence on water to accomplish fertilization. Instead, their gametophytes took to the skies.

The second innovation that occurred in the evolution of seed plants was, logically enough, the seed. A **seed** is a structure that encloses and protects a developing embryo. A seed provides a case for the embryo and a store of nutritive tissue. In addition, seeds are often attached to a structure that aids in dispersal by wind, water, or animals. As **Figure 28.7** shows, these structures range from the wings on a milkweed seed to the sticky barbs on cheatgrass. By the time seeds had evolved, land animals were diverse and abundant enough to play a lead role in their dispersal. Animals also triggered the diversification of the most recent of all major land plant innovations—the flower.

The Flower

Flowering plants, or angiosperms, are far and away the most diverse land plants living today. Over 230,000 species have been described, and more are discovered each day. As a group, though, the angiosperms represent a break with the dominant themes during the first 300 million years of land plant evolution. Their success in terms of geographical distribution, numbers of individuals, and number of species has relatively little to do with increasingly advanced adaptations to dry conditions. Instead, their story revolves around a reproductive organ: the flower.

As **Figure 28.8a** shows, flowers are fabulously diverse in size, shape, and coloration. They also produce a wide range of scents. Why? Biologists have long hypothesized that flowers are adaptations that increase the probability that pollination will occur. Instead of leaving pollination to an undirected force, like wind, the hypothesis is that in some circumstances natural selection favored structures that reward an animal—usually an insect—for carrying pollen directly from one plant to the next. Flowers provide a food reward to the pollinator in the form of sugar-rich nectar or protein-rich pollen grains. As a result, the relationship between flowering plants and their pollinators is thought to be mutually beneficial.

What evidence supports this hypothesis? The first angiosperms in the fossil record appear in rocks that are about 145 million years old; by 100 million years ago, the record contains the first evidence of correlations between the size and shape of flowers and the size and shape of the mouthparts in their insect pollinators (**Figure 28.8b**). Presumably, this correlation makes pollination more efficient. Correlations between flower size and shape and pollinator morphology are routinely observed in contemporary plants as well.

Correlation does not demonstrate causation, however. Is there any direct evidence that natural selection favors a correspondence between the shapes of flowers and their pollinators?

S. D. Johnson and K. E. Steiner recently completed a series of experiments addressing this question. They studied two South African populations of an orchid called *Disa draconis*. The length of the long tube, or spur, located at the back of the flower is very different in the two populations (see **Figure 28.9a**, page 550). Johnson and Steiner began their study by noting that each population is pollinated by a different insect. As Figure 28.9a shows, short-spurred plants that grow in mountain habitats are pollinated by horseflies, which have a relatively short proboscis. (A *proboscis* is a specialized mouthpart found in some insects. When extended, it functions like a straw in sucking.) Long-spurred plants that grow on low-lying sandplain habitats are pollinated by tanglewing flies, which have a particularly long proboscis.

Why might spur length be important? To answer this question, study the diagram in **Figure 28.9b**. Note that pollinators probe the spur to reach the nectar-producing organ at its end. As they do, they brush the flower's stamens and pick up pollen grains. When they visit subsequent flowers to feed, pollen on their body is often deposited against the pistil. The pollen grains germinate, grow down the length of the pistil, and eventually reach the egg-bearing ovules.

(a) Dispersal by wind

(b) Dispersal by water

(c) Dispersal by animals

FIGURE 28.7 Seeds and Seed Dispersal
The structures that surround seeds can facilitate dispersal by wind, water, or animals. **EXERCISE** Next to each example, write the name of other plant species that use that mode of seed dispersal.

FIGURE 28.8 Flowers and Their Pollinators
Flowers with different shapes, colors, and fragrances attract different pollinators. The carrion plant (left) has a foul odor that attracts carrion flies. Flowers that are pollinated by hummingbirds (middle) usually have a long spur with a nectar-producing structure in the back. Large, complex flowers like sunflowers (right) often attract a wide variety of pollinators.

To test the hypothesis that spur length actually affects the fitness of individuals in these populations, Johnson and Steiner experimentally shortened the spurs of individuals in the long-spurred population. They did this by tying the spurs of randomly selected individuals closed with a piece of yarn. They left nearby flowers alone, but tied a piece of yarn near the spur as a control (**Figure 28.9c**).

The data they collected are shown in Figure 28.9c. Individuals with short spurs received much less pollen and set much less seed than the controls. This result strongly supports the hypothesis that spur length is an adaptation that increases the frequency of pollination.

In a parallel experiment, the researchers compared how much fruit was produced by individuals that had been hand-pollinated versus individuals that were pollinated by flies. The results show that hand-pollinated individuals set much more fruit than individuals left to be pollinated by flies. This is a key observation, because it shows that the fitness of individuals is limited by their ability to attract pollinators. Results like this support the hypothesis that plants gain enormous benefits by successfully attracting pollinators. An even more general implication is that the spectacular diversity of angiosperms resulted, at least in part, from natural selection exerted by the equally spectacular diversity of insect, mammal, and bird pollinators.

(a) Orchid spur length and fly proboscis length are correlated.

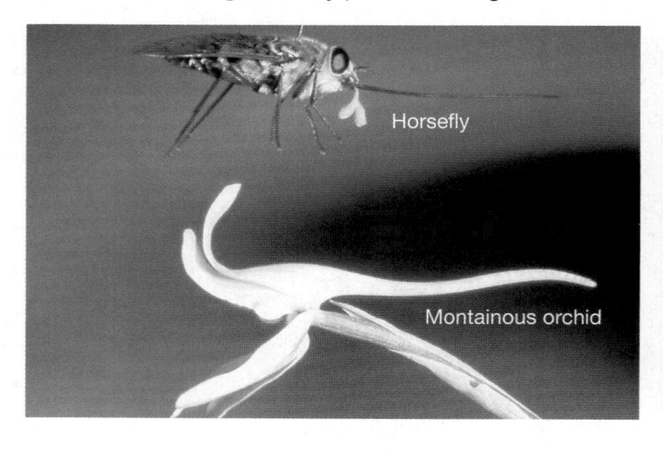

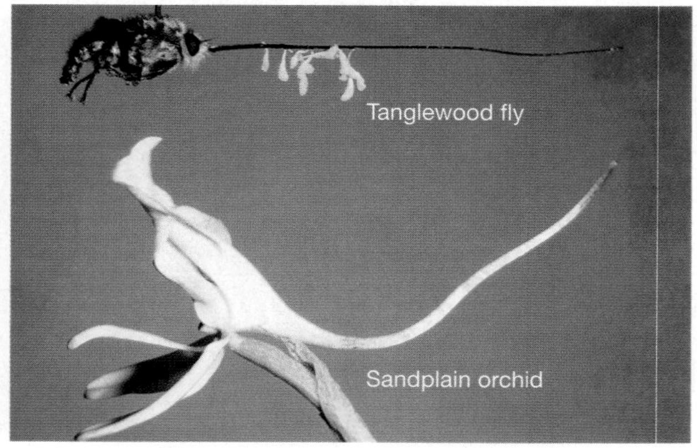

(b) This correlation is important for pollination.

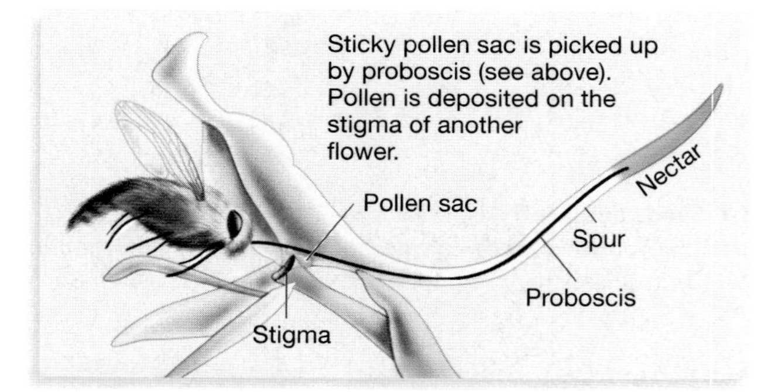

(c) Artificial shortening of spurs lowers pollination success.

	Normal spurs (49 mm)	Shortened spurs (35 mm)
Flowers that received pollen (n = 59)	46%	17%
Flowers that set fruit (n = 56)	41%	18%

FIGURE 28.9 The Adaptive Significance of Flower Shape: An Experimental Test
(a) These photos show an orchid flower from the mountainous habitat with its horsefly pollinator (left) and on orchid flower from the sandplain habitat with its tanglewing fly pollinator (right). Note the correspondence between the length of each flower's spur and the length of each pollinator's proboscis. **(b)** This cutaway diagram of an orchid shows how a visiting fly picks up sacs containing pollen from the anthers and deposits them on the stamen. **(c)** Experimental flowers had their spurs tied off with a piece of yarn, 35 mm from the opening; control flowers had a piece of yarn tied around them, too, but had normal spurs averaging 49 mm. The data show that flowers with shortened spurs received much less pollen and set less fruits. Other experiments showed that, in both orchid populations, hand-pollinated flowers produce much more fruit than naturally pollinated flowers. In cases like this, biologists say that the ability of plants to reproduce is "pollinator limited." **QUESTION** If spur length has no effect on reproductive success, what would the data in this table look like?

28.3 Strategies for Photon Capture

The first two sections of this chapter have emphasized how adaptations for coping with dry conditions and for attracting pollinators triggered the diversification of plants. Now let's get back to basics and consider the diversity of ways that land plants accomplish their most fundamental task, capturing photons and making sugar. Chapter 7 introduced the molecules involved in these processes and explored how they function. Here the goal is to consider how strategies for capturing photons affected the diversification of plants.

The Calvin-Benson Cycle: Variations on the Theme

The light reactions of photosynthesis occur in much the same way in most land plants. Chlorophyll *a* and chlorophyll *b* function as the antenna pigments, just as they do in green algae. These molecules, along with the enzymes and electron transport chains required for photosynthesis, are embedded in the internal membranes of chloroplasts. In contrast to the "standardized" photosynthetic light reactions, however, two major variations have evolved in the sugar-making machinery.

C_4 Photosynthesis The first of these variations in sugar-making mechanisms was discovered, independently, by Hugo Kortschack and colleagues and Y. S. Karpilov and associates. Both groups of scientists were investigating the mechanisms of carbon fixation in plants, and both used the same experimental strategy to follow the fate of CO_2 molecules. They briefly exposed leaves to radioactive carbon dioxide ($^{14}CO_2$) and sunlight, and then characterized the products. They expected to find the first of the radioactive carbon atoms in the 3-carbon sugar called 3-phosphoglycerate (3PG). This is the sugar that feeds the Calvin-Benson cycle analyzed in Chapter 8, and it is the normal product of carbon fixation by the enzyme called rubisco. To the researchers' amazement, however, they found that in some plants the radioactive carbon atom ended up in 4-carbon sugars, such as malate and aspartate. They had inadvertently discovered a novel twist on the usual, or "C_3," photosynthesis. This variation came to be called C_4 photosynthesis, because it involves 4-carbon sugars (**Figure 28.10a**).

How does this variation work, and why is it important? Hal Hatch and Roger Slack followed up on the initial reports by showing that, in the plants studied by Kortschack's and Karpilov's groups, carbon dioxide can be fixed by an enzyme called PEP carboxylase as well as by the rubisco enzyme. They also showed that the two enzymes are found in distinct cell types within the same leaf. PEP carboxylase is common in the mesophyll cells of a leaf (**Figure 28.10b**), while rubisco is found in bundle-sheath cells that surround the vascular tissue.

Based on these observations, Hatch and Slack proposed a model explaining how the carbon fixed to malate or aspartate is involved in the Calvin-Benson cycle. As Figure 28.10c shows, the model consists of three steps, starting at upper left: (1) PEP carboxylase fixes CO_2 in cells near the surface of the leaf; (2) the 4-carbon organic acids that result travel to interior cells;

(a) C_4 plants use PEP carboxylase rather than rubisco for the initial fixation of carbon.

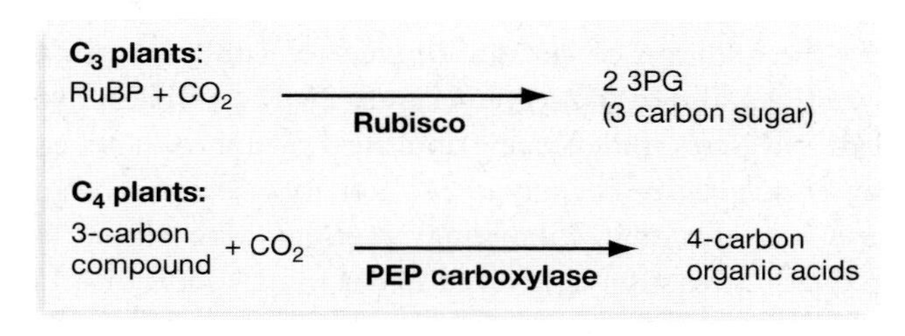

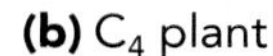

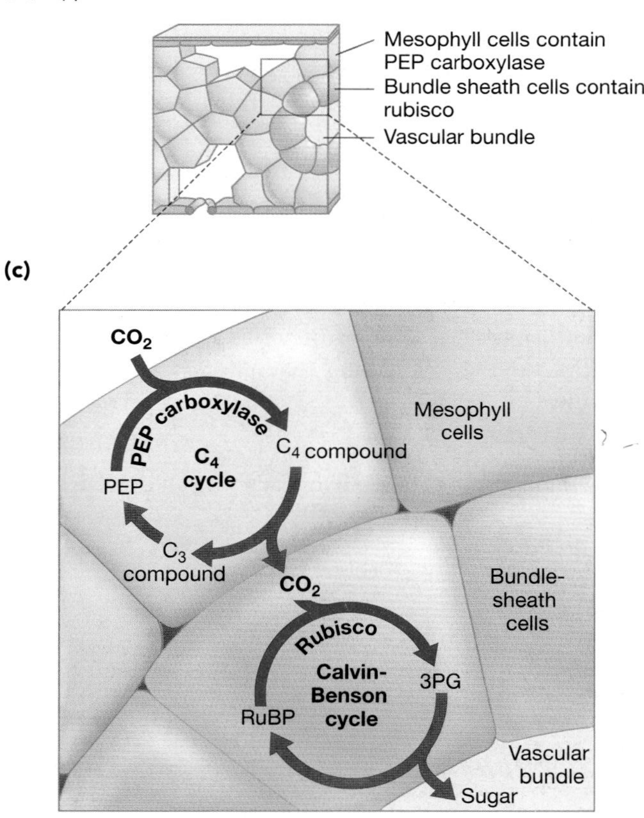

FIGURE 28.10 C_4 Photosynthesis
(a) The carbon fixation step in C_4 plants is different from that in C_3 plants. **(b)** The carbon-fixing enzyme PEP carboxylase is located in mesophyll cells, while rubisco is located in the bundle-sheath cells. **(c)** In the Hatch and Slack model of C_4 photosynthesis, CO_2 is fixed to a 3-carbon compound called PEP by PEP carboxylase. This forms a 4-carbon organic acid. This 4-carbon then releases a CO_2 molecule to rubisco, which fixes it and starts the Calvin-Benson cycle.
QUESTION What predictions does the Hatch and Slack model make? For example, if you gave a leaf a brief "pulse" of radioactively labeled CO_2, where would you expect to find the labeled carbon atom at different time intervals after the pulse?

CD ACTIVITY 28.1 Strategies for Carbon Fixation

and (3) the 4-carbon organic acids release a CO_2 molecule in a reaction, catalyzed by rubisco, that forms 3-phosphoglycerate. This final step initiates the normal Calvin-Benson cycle.

Later experiments supported the Hatch and Slack model in almost every detail. Now the question is, why do some plants have this complicated metabolic pathway at all?

The leading hypothesis for the function of the C_4 pathway focuses on an unusual property of rubisco. In addition to catalyzing the addition of carbon dioxide to RuBP, it can also catalyze the addition of oxygen (**Figure 28.11a**). This activity inspired rubisco's full name: ribulose-1,5-biphosphate carboxylase/*oxygenase*. The oxygenase activity is also important, because it means that competing reactions take place. The carboxylation reaction is favored under most conditions, but the oxygenation reaction dominates when CO_2 levels are low and the temperature is high. These last two factors are mutually reinforcing (**Figure 28.11b**). When the temperature is high, leaves lose water and tend to shut down their stomata. When stomata are closed, gas exchange decreases. Inside photosynthesizing cells, oxygen concentrations increase and carbon dioxide levels drop.

Why is this important? When rubisco catalyzes the addition of oxygen to RuBP, the molecule that results is eventually broken down into carbon dioxide. This series of reactions is called **photorespiration** because it depends on light ("photo") and uses oxygen ("respiration"). Photorespiration reverses carbon fixation and consumes some of the cell's ATP. In effect, photorespiration undoes everything that photosynthesis accomplishes. Why?

No one knows why photorespiration occurs. The leading hypothesis is that it is an unavoidable side effect of rubisco's structure. According to this point of view, rubisco remains a flawed enzyme even after billions of years of evolution. If the hypothesis is valid, it reinforces a crucial point. Evolution by natural selection continuously improves the function of enzymes and morphological structures, but it cannot make them "perfect."

According to Hatch and Slack's model, then, the C_4 pathway is an adaptation to limit the damaging effects of photorespiration. The pathway accomplishes this by building higher concentrations of CO_2 in cells containing rubisco (to see how this happens, look back at Figure 28.10c). Because photorespiration is most prevalent under hot, dry conditions, Hatch and Slack hypothesized that natural selection favors individuals with the C_4 pathway if they inhabit hot, arid environments.

The hypothesis is supported by correlational data. The C_4 pathway is found almost exclusively in plants that thrive in hot, dry habitats. Sugarcane, maize (corn), and crabgrass are all C_4 plants. The pathway is found in several thousand species in 19 different lineages of flowering plants. These observations suggest that it has evolved independently several different times.

To test the "favored when hot and dry" hypothesis more rigorously, however, F. A. Bazazz and R. W. Carlson set up an experiment. They grew seedlings of a C_4 species and a C_3 species in growth chambers. These chambers allowed the researchers to control CO_2 concentration and temperature. Seedlings were placed in six different treatments: wet or dry; and low, medium, or high CO_2 (**Figure 28.12a**). After 35 days of growth, Bazazz and Carlson harvested the plants and measured their dry

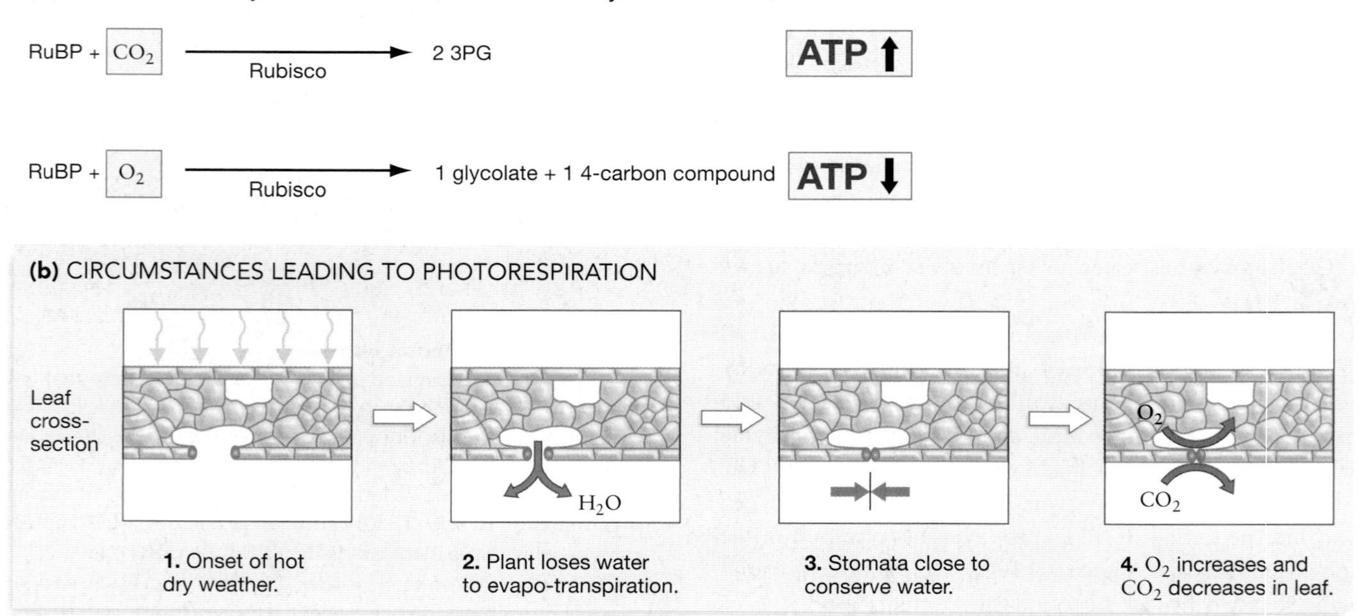

FIGURE 28.11 Photorespiration
(a) Rubisco can fix carbon or lead to a loss of fixed carbon. (b) The oxygenation reaction is favored when O_2 is plentiful and CO_2 is rare in the cell. This flow diagram shows why these conditions occur in hot, dry weather.

weight. **Figure 28.12b** shows some of their results. The C_4 species performed better at all CO_2 concentrations in dry soil, and performed better in wet soil when CO_2 levels were low. But in wet soils with medium or high concentrations of CO_2, seedlings of the C_3 species were able to fix more carbon and grow more quickly. The experiment strongly supports the prediction that the C_4 pathway is favored in dry environments because it limits the loss of carbon and ATP to photorespiration.

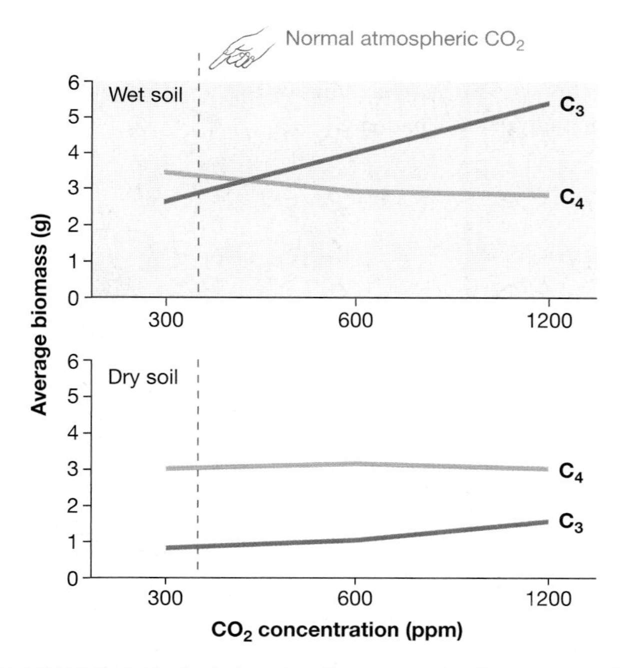

FIGURE 28.12 Under What Conditions Are C_4 Plants Favored Over C_3 Plants?
(a) This diagram shows how seedlings of a C_3 plant and a C_4 plant were grown under six different conditions. **(b)** These data show the results of the experiment. The vertical line indicates the normal atmospheric concentration of CO_2. **QUESTION** As a result of deforestation and the burning of fossil fuels, CO_2 levels in the atmosphere have increased dramatically. Because CO_2 is a "greenhouse gas," temperatures are increasing as well. Does this development favor C_3 plants or C_4 plants?

CAM Plants Some years after the discovery of C_4 photosynthesis, researchers studying a group of flowering plants called the Crassulaceae came across a second mechanism for limiting the effects of photorespiration: **crassulacean acid metabolism**, or **CAM**. Like the C_4 pathway, CAM represents an additional, preparatory step to the Calvin-Benson cycle. It also has the same effect; it increases the concentration of CO_2 inside photosynthesizing cells. But unlike the C_4 pathway, CAM occurs at a different time than the Calvin-Benson cycle—not in a different place.

CAM occurs only in species, such as cactus, that occupy environments that are so hot and dry that individuals routinely keep their stomata closed all day. In C_3 plants, keeping stomata closed is like writing a prescription for photorespiration. But each night CAM plants store up huge quantities of CO_2 in the form of sugars and sequester them in their central vacuoles. During the day, these sugars are metabolized to release CO_2 and feed the Calvin-Benson cycle.

For plant diversification, the consequences of the C_4 pathway and CAM are almost identical. They are adaptations that allow angiosperms to colonize extremely dry habitats. But how do plants cope with conditions where water is not the limiting resource? In many environments the critical resource is not water, but energy—in the form of photons.

Growth Habits

The photosynthetic pigments, as arranged in a plant leaf, are extremely efficient at harvesting energy from photons. For example, virtually no photosynthetically active wavelengths reach the underside of a dandelion plant; the plant absorbs them all. To appreciate why this is important, do the following thought experiment. Imagine that a group of early land plants is growing in the wet soils of a marshy habitat. If the soil contains adequate quantities of nitrogen, phosphorus, and potassium, individuals will grow rapidly and the habitat will quickly fill with a tangle of stems and leaves. Tissues that are overgrown by competing plants will receive few usable wavelengths of light. These cells will not be able to make enough ATP to sustain their basic metabolic processes, and they will die.

Now suppose that in one individual among the tangle, a mutation occurs that results in an increased deposition of cellulose—or perhaps the synthesis of a small amount of lignin—in the walls of its water-conducting cells. Due to this mutation, the individual's stems can stand up under the force of gravity and grow vertically. The photosynthetic cells in its stem and leaves are now above competing individuals. This gives the mutant individual a huge advantage. It can harvest much more energy and make many more spores or seeds than its competitors. Its descendants will spread throughout the habitat. If later mutations result in the formation of a secondary meristem and the deposition of enough lignin to produce wood, a shrubby or tree-like growth habit can result.

The fossil record and the phylogeny of living plants indicate that this event occurred repeatedly over the course of land plant history. Erect growth habits and woody tissues have appeared in many different lineages. Not all trees, for example, descended from the same common ancestor. To use the vocabulary introduced in Chapter 27, trees are not monophyletic. Instead, woody growth and the tree growth habit have originated and been lost many times in the course of evolution (**Figure 28.13**).

Among angiosperms alone, growth habits vary from sprawling types that form mats, to herbs that grow in a variety of profiles, to shrubs, to trees that range in height from 10 feet to almost 300 feet. Biologists interpret these growth forms as adaptations that allow individuals to capture photons efficiently. Other growth forms, like vines and epiphytes (upon-plants), appear to be adaptations for capturing photons high above the ground without expending the ATP required to synthesize wood. Instead, vines and epiphytes take advantage of the structural support provided by trees.

In short, plants have diversified structurally. Biologists hypothesize that some of this diversification was driven by competition for energy in the form of photons.

28.4 Food, Fuel, and Fiber: Human Use of Plants

Humans have always depended on plants for food, fuel, and the fibers used for making clothing, paper, ropes, twine, and baskets. It is not surprising, then, that forestry, agriculture, and horticulture are among the most important endeavors supported by biological science.

In this section you'll review how humans have used plants in the past, and explore how research is currently helping people use plants in innovative and efficient ways.

Domestication and Selective Breeding

Most anthropologists would agree that the single greatest event in the evolution of human culture was the domestication of crops and the invention of agriculture. But this "single greatest event" was not actually a single event. There is compelling evidence that plants were domesticated independently by different people living in widely scattered locations around the world.

As **Figure 28.14a** shows, the grains that form the basis of our current food supply were derived from wild species between 10,000 and 2000 years ago. In each case, archaeologists have observed a distinct set of changes as wild forms were brought under cultivation. Seeds became larger; squash rinds became thicker.

What is the mechanism behind these changes? The leading hypothesis is that humans were actively selecting seeds from individuals that exhibited these traits, and that farmers used only the selected seeds to plant the next generation of crops. By repeating these steps year after year, humans could gradually change the characteristics of the species, resulting in the characters associated with domestication. This process is called artificial selection. It is exactly analogous to the process of natural selection analyzed in Chapter 21.

FIGURE 28.13 Wood
Wood is made from tracheids and/or vessels that have walls reinforced with lignin. Trees with woody stems evolved independently in lycopods, horsetails, and seed plants. The structure of the tracheids and woody stems is different in each group. (A six-foot-tall human is shown for scale.)

LYCOPODS
During the Carboniferous period, *Lepidodendron* trees reached heights of 120 feet.

HORSETAILS
This extinct tree, called *Calamites*, reached heights of 80 feet.

SEED PLANTS
Some conifers, such as this California redwood, can reach heights of 300 feet.

In more recent times, artificial selection was responsible for the diversification of the wild species *Brassica oleracea* into the domesticated varieties of cabbage-family plants (see Chapter 1), the dramatic increases in the oil content of corn kernels illustrated in **Figure 28.14b**, and an explosive diversification in domesticated forms. As an example of this diversification, researchers have identified 178 distinct varieties of potato being cultivated in a single farming community in the Andes of South America.

This diversity is increasingly threatened, however, as the essay at the end of this chapter points out. Agricultural scientists are also concerned about the ability of artificial selection programs to increase yields quickly enough to keep up with the caloric demands of growing human populations. As a result, researchers are increasingly turning to genetic engineering to supplement traditional breeding programs (see Chapter 17). For example, several labs are searching for a rubisco allele that has no oxygenase activity. If such an allele could be found and inserted into crop plants, photorespiration would stop reducing yields by siphoning off sugar and ATP. In soybeans alone, investigators project that productivity could jump as much as 20 percent. Potential benefits like these are measured in the billions of dollars.

How do traditional selection programs and genetic engineering differ? The essence of the issue is this: To improve crops, artificial selection relies on naturally occurring variation produced by random mutations. In genetic engineering, researchers isolate genes with desired properties and insert them into selected individuals that serve as breeding stock. The mutations used in genetic engineering programs are not random mutations at all, but new alleles that have been chosen or created by design. Also, recombinant DNA techniques allow researchers to rapidly move alleles from one species to another—something that may be difficult or impossible to achieve in traditional breeding programs, which rely on fertilization to move alleles from one individual to another.

Will genetic engineering programs augment traditional artificial selection techniques and lead to increasingly diverse and

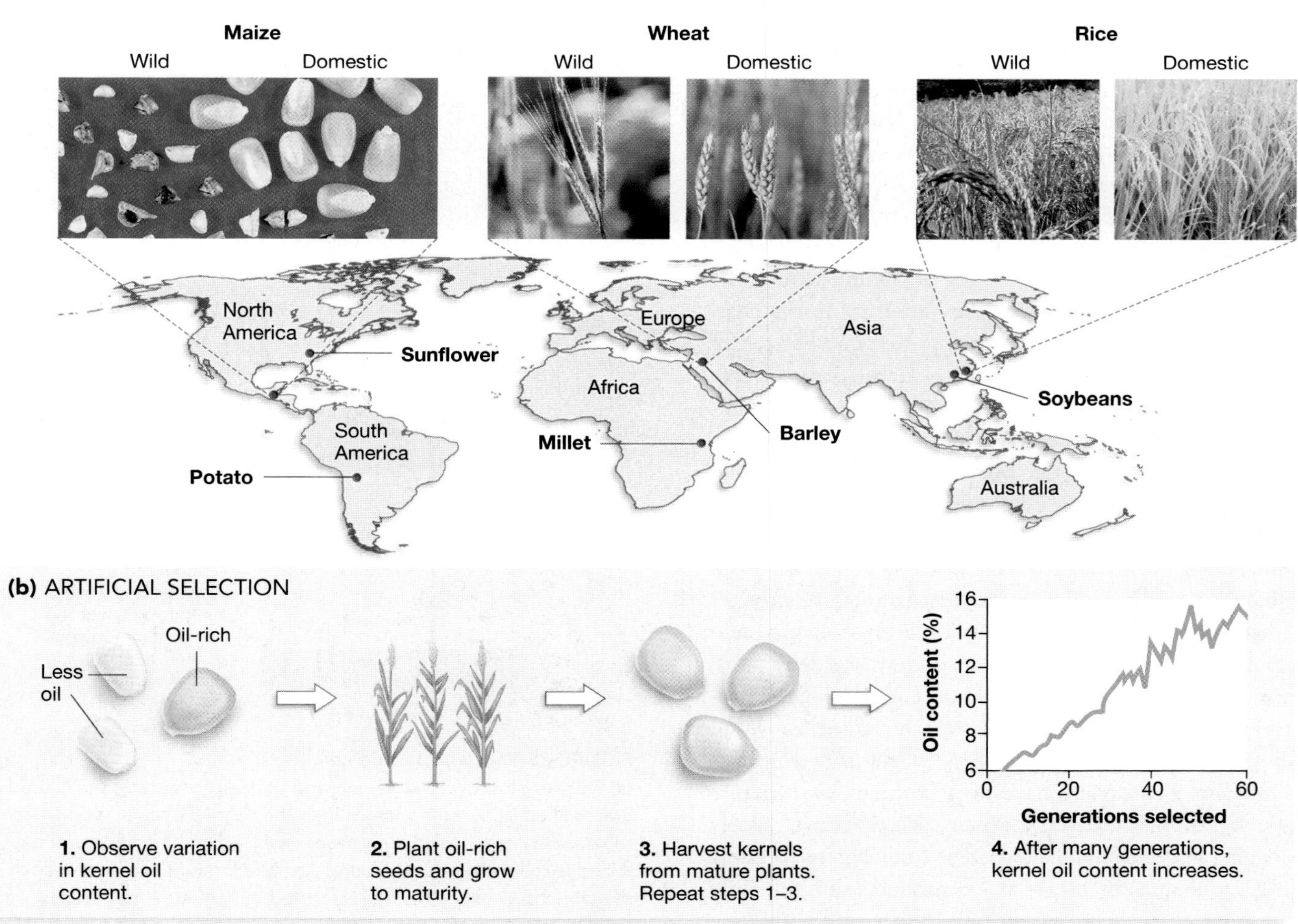

FIGURE 28.14 Domestication and Artificial Selection of Crop Plants
(a) Crop plants were derived from wild relatives in many different parts of the world. **(b)** Artificial selection proceeds in steps, as shown by the example of selecting for increased seed (kernel) oil content.

productive domestic crops? If research and development programs currently under way are fruitful, this question will be answered in the next few years.

Wood and Coal

Was agriculture the innovation with the greatest impact on human culture? To challenge that hypothesis, consider an alternative view: Harnessing fire was the most significant advance in human history. For perhaps 100,000 years, wood burning was the primary source of energy used by humans.

As **Figure 28.15a** shows, wood has recently been replaced by other sources of energy. In economies across the globe, the first fuel to replace wood has usually been coal. Starting in the mid-1800s, for example, extensive coal deposits in England, Germany, and the United States fueled blast furnaces that smelted vast quantities of iron ore into steel and powered the steam engines that sent trains streaking across Europe and North America. It is no exaggeration to claim that the sugars synthesized during the Carboniferous period laid the groundwork for the industrial revolution (**Figure 28.15b**). Coal still supplies about 20 percent of the energy used in Japan and Western Europe and 80 percent of the energy used in China.

Will there be yet another generation of plant-based fuels? The answer is probably no. Outside of efforts to convert corn to liquid fuels like ethanol or methanol, few biologists are working to develop new plant-based fuels. The currently dominant combustible material, petroleum, is derived from the partially decayed remains of protists. Oil is also the source material for nylon and polyester fibers that have supplanted many plant fibers used in traditional clothing and implements. The energy sources that are predicted to be important in the post-petroleum economy are solar and wind power, combustion of hydrogen, and perhaps nuclear fusion. None of these sources are based on the reduced carbon found in plants.

Today, the primary commercial interest in woody plants is for building materials and the fibers used in papermaking. Wood excels in both capacities. Relative to its density, wood is a stiffer and stronger building material than concrete, cast iron, aluminum alloys, or steel. The cellulose fibers refined from trees or bamboo and used in paper manufacturing are stronger under tension (pulling) than nylon, silk, chitin, collagen, tendon, or bone—even though they are 25 percent less dense.

Current research in forestry focuses on maintaining the productivity and diversity of forests that are managed for wood and pulp production. The vast majority of old-growth forests in the Northern Hemisphere have now been cut, and mature forests in the Southern Hemisphere are disappearing rapidly due to logging and burning. The current challenge to foresters is twofold: reducing demand for wood products through recycling and more efficient use of existing timber, and sustaining the productivity of second- and third-growth forests.

Increasingly, the productivity and diversity of regrowing forests is thought to hinge on the productivity and diversity of fungi. As the next chapter shows, the histories of the two lineages have long been intertwined.

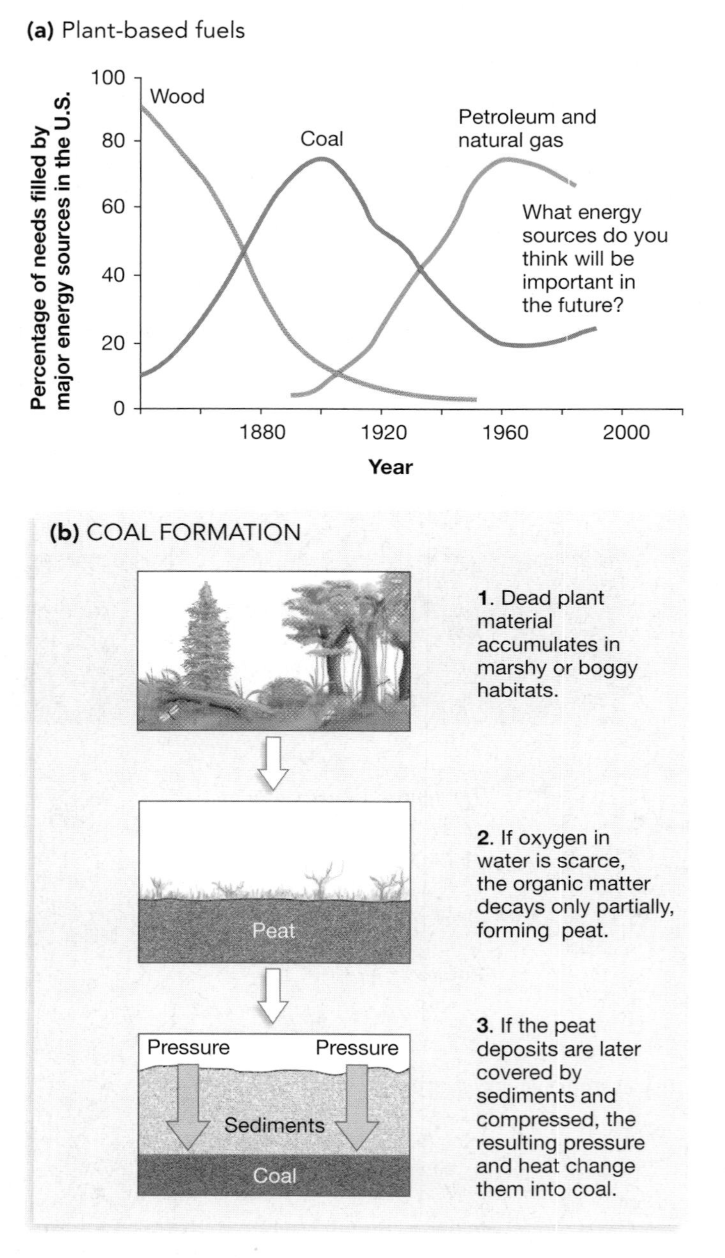

FIGURE 28.15 Plant-Based Fuels
(a) This graph shows the percentage of energy needs in the United States filled by various sources over the past 150 years. Wood was replaced by coal, which in turn was replaced by oil and natural gas. Wood is still the primary cooking and heating fuel in many areas of the world, however. **(b)** Coal is formed by terrestrial deposits of partially decayed organisms; petroleum is formed by *marine* deposits of partially decayed organisms.

Essay Genetic Diversity in Crop Plants

The spread of mechanized agriculture throughout the globe has produced some spectacular success stories. As a result of increased use of fertilizers and high-yielding strains of wheat and rice, India has become self-sufficient in grain production after decades of importing food. This so-called green revolution has a downside, however. Because planting programs in numerous countries rely on just a tiny number of source strains, the genetic diversity of crop systems is extremely low. In the United States, for example, virtually all soybeans that are grown trace their ancestry back to just 12 strains imported from northeast China; most winter wheat is descended from two stocks that originated in Poland and Russia.

To justify their concern about the lack of genetic diversity in crop plants, agricultural scientists point to several recent events:

- In 1970, a corn blight wiped out 15 percent of the U.S. crop, sending prices skyrocketing and numerous farmers out of business.
- In 1980, over a million tons of sugar cane were destroyed by a single fungal disease in Cuba.
- Banana production in Central America was halved during the 1980s by a fungal disease called black sigatoka.
- Potato production in the United States is threatened by the recent evolution of potato blight strains that do not respond to existing treatments.

In response to the scientists' concern, governments from around the world have established seed banks. These are repositories where seeds from dozens or hundreds of different crop plant strains are stored. Worldwide, there are now over 700 seed collections with some 2.5 million entries. The United States alone spends $20 million each year to maintain and expand its seed banks.

At present, however, the banks represent a largely untapped resource. Researchers have only begun to actually use the banks in breeding programs designed to improve the characteristics of commercial strains. The wild relatives of species like tomato and rice have been particularly important in recent breeding programs. Using genetic maps developed in Stephen Tanksley's laboratory, researchers have designed crosses between commercial strains of tomato and wild species that resulted in dramatic improvements in fruit characteristics. With traditional breeding techniques, annual improvement in tomato traits like fruit yield and percentage of soluble solids (important in the production of tomato paste) averages less than 1 percent. After carefully designed crosses with a wild relative, the same traits improved 28 percent and 22 percent, respectively.

Planting programs often rely on just a tiny number of source strains . . .

With accelerating human populations and loss of agricultural lands to urbanization, this research becomes increasingly urgent.

Chapter Review

Summary

The land plants evolved from green algae and were the first multicellular organisms to grow erect in terrestrial habitats. The fossil record of the group can be broken into five intervals: an early period starting about 476 million years ago and marked by the appearance of encased spores, cuticle fragments, and tubes; the Silurian-Devonian explosion, in which most major lineages appeared; the Carboniferous period, made famous by extensive lycopod and horsetail forests whose remains fossilized as coal; the rise to dominance of the seed plant lineages that are collectively called gymnosperms; and finally the appearance and diversification of angiosperms, or flowering plants.

This sequence of events is verified by phylogenies estimated from relationships of DNA sequence data. The three groups that are collectively called "bryophytes" (the liverworts, hornworts, and mosses) branch off the land plant phylogeny first, followed by horsetails and ferns. This latter clade represents the descendants of the first vascular plants. The seed plant lineage, which includes the gymnosperms and angiosperms, is the most recent major group to appear.

Both the fossil record and phylogenetic analyses indicate that land plants became less and less dependent on aquatic habitats and more able to tolerate dry conditions through time. This trend began with the initial transition to land, which in the fossil record correlates with the appearance of waxy cuticle that protected tissues from drying. After the evolution of stomata, plants began to diversify. The next major innovation, which occurred during the Silurian-Devonian explosion, was the evolution of vascular tissue. Due to the deposition of lignin in the secondary cell walls of water-conducting cells, vascular plants were sturdy enough to grow upright. Wood and the tree growth habit appeared independently in several lineages.

The second major trend in land plant diversification is a reduction of the gametophyte generation. By the time gymnosperms and angiosperms appeared, male gametophytes were reduced to pollen grains. Female gametophytes were reduced to a small structure bearing the egg. Because pollen grains are surrounded by sporopollenin, they can resist drying and be transported to female gametophytes by wind or animals instead of having to swim through water. Similarly, seeds can resist drying and be dispersed by wind or by animals because they are encased in a tough coat. Both of these innovations reinforced the trend toward a decreasing dependence on water.

Plants have also diversified in ways that affect their ability to capture photons and make sugar. Biologists interpret the diversity of growth habits found in many plant lineages—sprawling, herbaceous, shrubby, and tree forms—as strategies to facilitate photon capture by leaves. The C_4 pathway and CAM both augment the Calvin-Benson cycle and limit the loss of fixed carbon and ATP to photorespiration in hot, dry habitats. C_4 plants fix carbon in mesophyll cells and store it as organic acids that feed the Calvin-Benson cycle; CAM plants fix carbon at night and also store it in the form of organic acids.

Humans have always been dependent on plants for food, fiber, and fuel. This dependence became extreme after the domestication of plant species beginning about 10,000 years ago, and with the development of agricultural economies beginning about 8000 years ago. Artificial selection techniques have produced huge increases in the yields of these domesticated plants. Plant biologists now use recombinant DNA technologies to supplement the natural variation that forms the basis of artificial selection. Gene transfer techniques allow researchers to move selected alleles into targeted species.

Questions

Content Review

1. During the "Silurian-Devonian explosion," what occurred?
   a. Millions of fragmentary tissues—primarily spores, cuticle, and tubes—dominate the fossil record.
   b. The gymnosperms diversified rapidly and occupied dry habitats.
   c. Most major lineages of land plants appear in the fossil record.
   d. The continents were covered with coal-forming forests, dominated by tree-sized lycopods and horsetails.

2. What is the difference between tracheids and vessels?
   a. The end walls of vessels are highly reduced or even absent, while the end walls of tracheids retain their primary cell wall.
   b. The end walls of tracheids are highly reduced or even absent, while the end walls of vessels retain their primary cell wall.
   c. Only tracheids have a thick, secondary cell wall containing lignin.
   d. Only vessels have a thick, secondary cell wall containing lignin.

3. Circle any of the following statements that are true.
   a. Phylogenies estimated from both morphological traits and DNA sequences indicate that the ancestor of the land plants was a green alga in the lineage called Charophyta.
   b. "Bryophytes" is a name given to the land plant lineages that do not have vascular tissue.
   c. The horsetails and the ferns form a distinct clade, or lineage. They have vascular tissue but reproduce via spores, not seeds.
   d. The angiosperms evolved before the gymnosperms. Angiosperms have a unique reproductive structure called the flower, but lack vascular tissue.

4. The appearance of cuticle and stomata correlated with what event in the evolution of land plants?
   a. The first erect growth forms
   b. The first woody tissues
   c. Growth on land
   d. A drastic reduction in photorespiration

5. What do seeds contain?
   a. male gametophyte and nutritive tissue
   b. female gametophyte and nutritive tissue
   c. embryo and nutritive tissue
   d. mature sporophyte and nutritive tissue

6. What do pollen grains contain?
   a. male gametophyte
   b. female gametophyte
   c. male sporophyte
   d. sperm

7. In what sense are C_4 photosynthesis and CAM alike?
   a. They represent a step that precedes the Calvin-Benson cycle.
   b. They are adaptations to reduce photorespiration.
   c. They lead to the production of organic acids.
   d. They fix carbon dioxide.
   e. All of the above

Conceptual Review

1. Why are biologists convinced that the spores, cuticle, and tube fragments that appear in the fossil record beginning about 476 million years ago represent the remains of the first land plants?

2. The introduction to this chapter claims that the land plants were the first truly terrestrial organisms. But most biologists contend that saturated soils on the continents were undoubtedly teeming with bacteria, archaea, and protists long before land plants evolved. In light of this, what does the phrase "truly terrestrial" mean? To answer this question, list the physiological problems presented by growing on land. Beside each entry in this list, note the adaptation or adaptations that allowed plants to overcome the problem.

3. In the experiments with orchid pollination reviewed in this chapter, why did the researchers put yarn around the flowers of control individuals?

4. The chapter emphasizes that the C_4 pathway represents an addition, or "add-on," to normal C_3 photosynthesis. In the fossil record, the first leaves containing the cell types involved in C_4 photosynthesis appear just 6 million years ago, even though angiosperms originated 125 million years ago. Do these two observations make sense in light of each other? Explain.

5. Why does photorespiration occur? Why is it considered "maladaptive"?

6. In the experiment reviewed in this chapter that compared growth in C_4 versus C_3 plants, why didn't the C_4 plants always do better?

Applying Ideas

1. In a recent textbook that introduces the fossil record of plants, the authors heralded the discovery that some members of the Charales synthesize sporopollenin and lignin. Why are these observations significant?
2. In many vascular plants, the end walls of tracheids form an extreme diagonal, as shown in Figure 28.4. Regarding the cell's ability to transport water, why does this make sense? (Hint: Consider how the cell's ratio of surface area to volume affects transport functions.)
3. Some angiosperms, like grasses, birches, and oaks, are wind pollinated. Their ancestors were probably pollinated by insects, however. As an adaptive advantage, why might a species "revert" to wind pollination?
4. You have been hired as a field assistant for a researcher interested in the evolution of flower shape in orchids. The five species she is working on are each pollinated by a different insect. Design an experiment to determine which parts of the five flower types are most important in attracting the pollinators. Assume that you can change the flower's color with a dye; and that you can remove petals or nectar stores, add nectar by injection, or switch parts among species by cutting and gluing.
5. To stimulate creative thinking, biologists sometimes ask "why not" questions. Try this one: Why has evolution not produced a rubisco that lacks oxygenase activity? The activity is extremely costly to the plant, but natural selection has not eliminated it. In thinking about this question, consider the following:

   Some bacteria that live in environments lacking oxygen contain rubisco. When researchers have examined the activity of rubisco from these organisms, they find that it has an oxygenase activity.

   Rubisco has only one active site for catalyzing the addition of O_2 or CO_2 to the substrate. (In other words, both reactions take place in the same location.)
6. Review the terms *paraphyletic* and *monophyletic* introduced in Chapter 27. Are bryophytes and gymnosperms paraphyletic or monophyletic?

CD-ROM and Web Connection

CD Activity 28.1: Strategies for Carbon Fixation *(animation)*
(Estimated time for completion = 10 min)
Explore some strategies that plants have evolved to make photosynthesis more efficient in hot, dry climates.

At your **Companion Website** (http://www.prenhall.com/freeman/biology), you will find self-grading exams and links to the following research tools, online resources, and activities:

Phylogeny of Plants
Explore the evolutionary history of plants.

Land Conquest
This interactive site focuses on the transition of aquatic plants to land plants.

Photosynthesis
This site provides a great description of the process of photosynthesis.

Genetically Engineered Plants
Read this article that describes the arguments for and against the development of genetically modified agricultural plant resources.

Additional Reading

Edwards, R. 1996. Tomorrow's bitter harvest. *New Scientist* (August): 14–15. A commentary on the loss of genetic diversity in crop plants.

Moore, P. D. 1999. Mixed metabolism in plant pools. *Nature* 399: 109–111. A news report about research on CAM, C_3, and C_4 plants that coexist in ponds.

Niklas, K. 1999. What's so special about flowers? *Natural History* 108 (May): 42–45. An introduction to the structure and origins of the flower.

Pringle, H. 1998. The slow birth of agriculture. *Science* 282: 1446–1450. A look at recent archaeological research on the domestication of plants.

Fungi

29

Fungi are the master decomposers on the tree of life. They are specialists in *reversing* biosynthesis. They seek out gigantic molecules like cellulose, lignin, proteins, and nucleic acids and break them down into hundreds or thousands of smaller compounds. Although bacteria are also important decomposers, fungi are the only organisms that can digest wood completely. Given enough time, fungi can turn even the hardest, most massive trees into a soft bedding medium for seedlings (**Figure 29.1**).

Like the green plants and the animals, fungi form a major lineage in the domain Eukarya. The species in these three groups use radically different strategies when it comes to the basic tasks of living, however. Most animals obtain the energy and carbon they need by eating other organisms. Most plants make their own reduced carbon compounds through photosynthesis. Fungi, in contrast, absorb the nutrients they need from other organisms.

Many fungi absorb their food from dead plant tissues that they digest. But other species absorb nutrients from living plants. Recent experiments have shown that the relationship between fungi and living plants can be mutually beneficial. In exchange for the sugars and other reduced carbon compounds that they absorb, the fungi provide key nutrients like nitrogen or phosphorus. Without these nutrients, the plants grow much more slowly or starve.

Based on these observations, biologists maintain that fungi have a profound influence on the productivity and biodiversity of terrestrial ecosystems. While some fungi recycle the carbon stored in dead and decaying tissues, others provide plants with important nutrients.

Before examining the evidence for these claims, however, it will be helpful to explore the diversity of fungi. Section 29.2 follows up

The scarlet hood (*Hygrocybe coccinea*) is an edible fungus native to northern Europe and North America. The mushrooms pictured here are reproductive structures—most of the organism is underground.

FIGURE 29.1 Fungi Recycle Nutrients
The fungi that are decomposing this section of tree trunk are breaking up its proteins, nucleic acids, lignin, and cellulose. In doing so, the fungi release nitrogen, phosphorus, and other nutrients that can be used by other organisms.

by analyzing the adaptations that make fungi so effective at dismantling the macromolecules found in dead plants. The chapter concludes by investigating how fungi interact with other living organisms, first through the mutually beneficial associations called mutualisms (section 29.3) and then through the exploitative relationships called parasitism (section 29.4).

29.1 What Are the Fungi?

To date, about 70,000 species of fungi have been described and named; and about 1000 more are discovered each year. But the fungi are so poorly studied that the known species are widely regarded as a tiny fraction of the total. To predict the actual number of species alive today, David Hawksworth looked at the ratio of vascular plant species to fungal species in the British Isles—the area where the two groups are the most thoroughly studied. According to Hawksworth's analysis, there are six species of fungus for every species of vascular plant on these islands. If this ratio holds worldwide, then the total of 275,000 vascular plant species implies that there are 1.65 million species of fungi.

Where did the fungi come from? A glance at the tree of life in **Figure 29.2** supports two conclusions about the ancestor of the fungi. It was a protist, and it is closely related to the common ancestor of the animals. The evolutionary relationships shown in the figure were estimated by comparing the DNA sequences that code for ribosomal RNA, but several morphological traits also link the animals and the fungi. For example, most animals and all fungi synthesize the tough structural material called chitin. In fungi that have flagella, the organelle is very similar to those observed in animals: it is single, located at the back of the body, and moves in a whiplash manner. Further, both animals and fungi store carbon by synthesizing the polysaccharide called glycogen. (Plants, in contrast, synthesize starch as their storage product.)

Because all fungi alive today pursue an absorptive lifestyle, meaning that they absorb nutrients from the environment, biologists infer that their common ancestor did as well. Now we can ask: What do the descendants of this ancestral fungus look like? How do biologists organize and analyze the tremendous diversity of species alive today?

Traditional Analyses Identified Four Major Groups

Traditional approaches to classifying the diversity of life are based on grouping species that are similar morphologically. This strategy is difficult to implement with fungi, however, because their bodies are so simple. As **Figure 29.3** (page 562) shows, fungi have only two growth forms: the filamentous structures called **mycelia** and the single-celled forms called **yeast**. The filaments that make up a mycelium are called **hyphae**. Most hyphae are haploid and are similar in structure, so taxonomists have few traits to use in grouping or distinguishing species.

Fortunately, many species of fungi produce distinctive structures when they reproduce sexually. By analyzing these structures, biologists have succeeded in identifying four major groups of fungi.

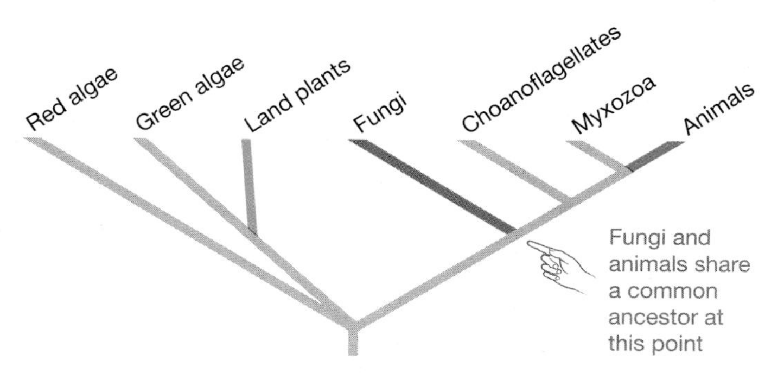

FIGURE 29.2 Where Fungi Are Found on the Tree of Life
This phylogeny shows the evolutionary relationships among the green plants, animals, fungi, and various groups of protists. Choanoflagellates are solitary or colonial protists found in fresh water and are introduced in Chapter 30.

Chytridiomycota The chytrids are the only group of fungi that live in water. They have recently been the focus of intense research because an epidemic of chytrid infections may be responsible for catastrophic declines that have occurred in frog populations all over the world. (For additional details on the amphibian die-off, see the essay at the end of this chapter.)

Like most other fungi, chytrids reproduce both asexually and sexually. Chytrids are unique in an important respect, however. They are the only fungi that have motile cells (**Figure 29.4a**). Gametes that form during sexual reproduction have flagella, as do spores produced during asexual reproduction. These swimming cells are the reproductive structures that distinguish chytrids from other fungi.

Zygomycota Most species of Zygomycota live in soil, and many are key partners in the mutualistic interactions with

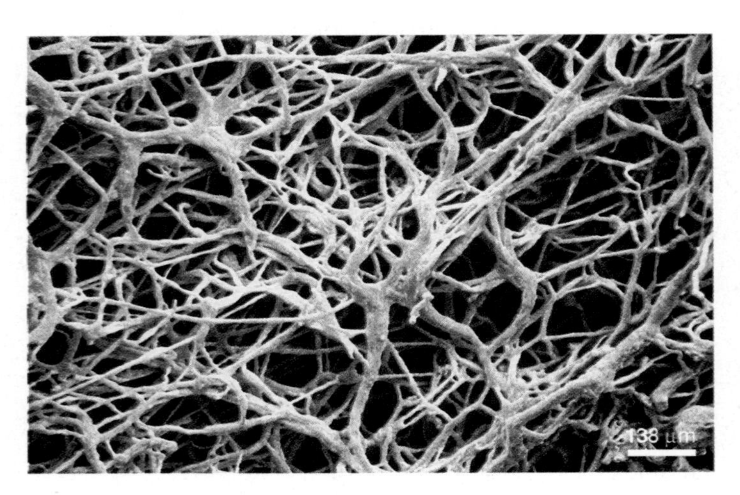

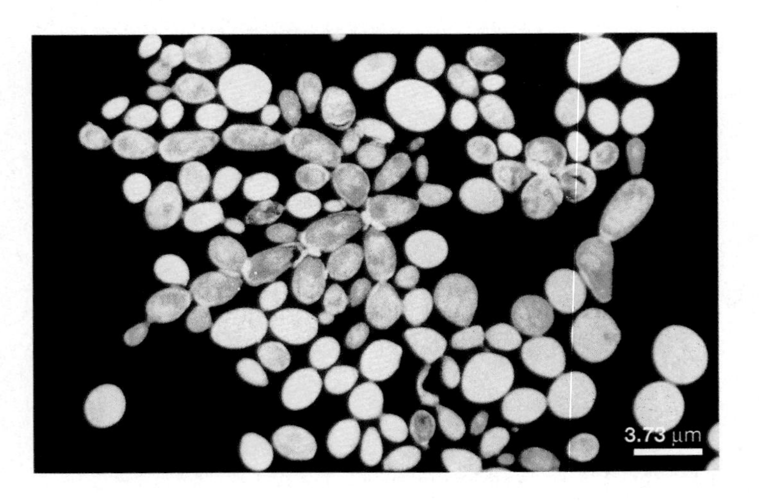

FIGURE 29.3 Growth Habits of Fungi
Fungi grow in one of two ways: as multicellular mycelia made up of long, thin filaments called hyphae (left), or as single-celled yeasts (right). *Yeast* can be a confusing term because it refers to two things, the single-celled growth habit *and* a specific lineage of ascomycetes, all of whose members are unicellular. The yeast pictured here is baker's yeast, *Saccharomyces cerevisiae*. The mycelium is from the rind of a piece of Melbury cheese.

(a) Chytridiomycota

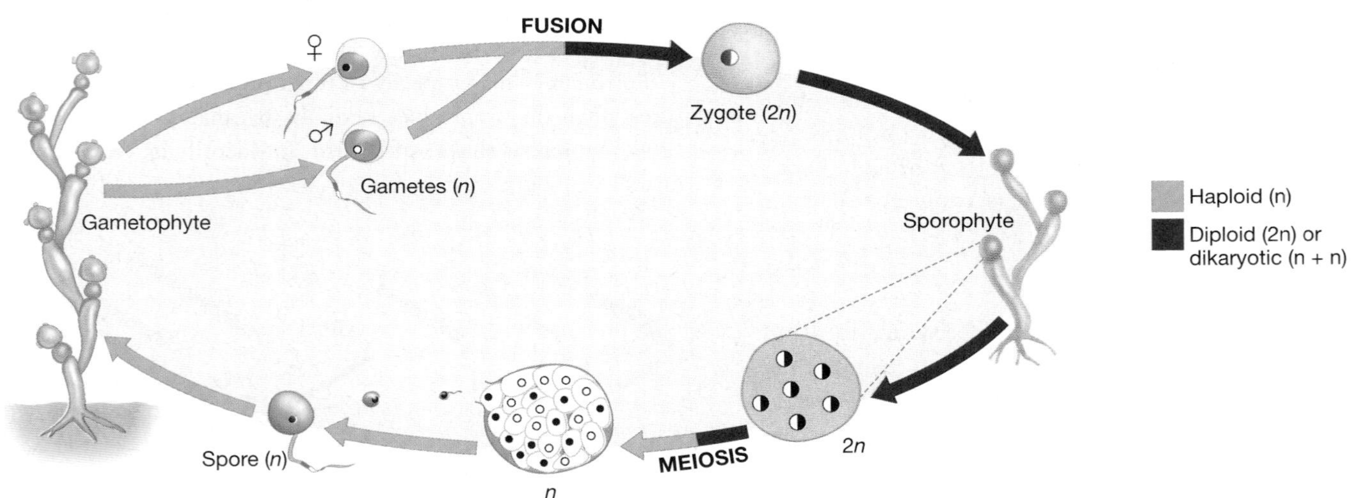

FIGURE 29.4 The Four Major Groups of Fungi
These diagrams illustrate the sexual part of the life cycle in the four major groups of fungi. **(a)** A distinctive feature of chytrids is that they have motile gametes and spores. **(b)** The common bread mold *Rhizopus stolonifer*, a member of the Zygomycota. **(c)** When basidiomycetes mate, hyphae of different mating types fuse. The heterokaryotic hyphae that result eventually differentiate into the spore-producing body called a mushroom, bracket, or puffball. **(d)** Sexual reproduction in ascomycetes is similar to the sequence of events in basidiomycetes. A key difference is the mechanism of spore formation: Ascomycetes form their spores in a sac, while basiodiomycetes push their spores out of structures called basidia.

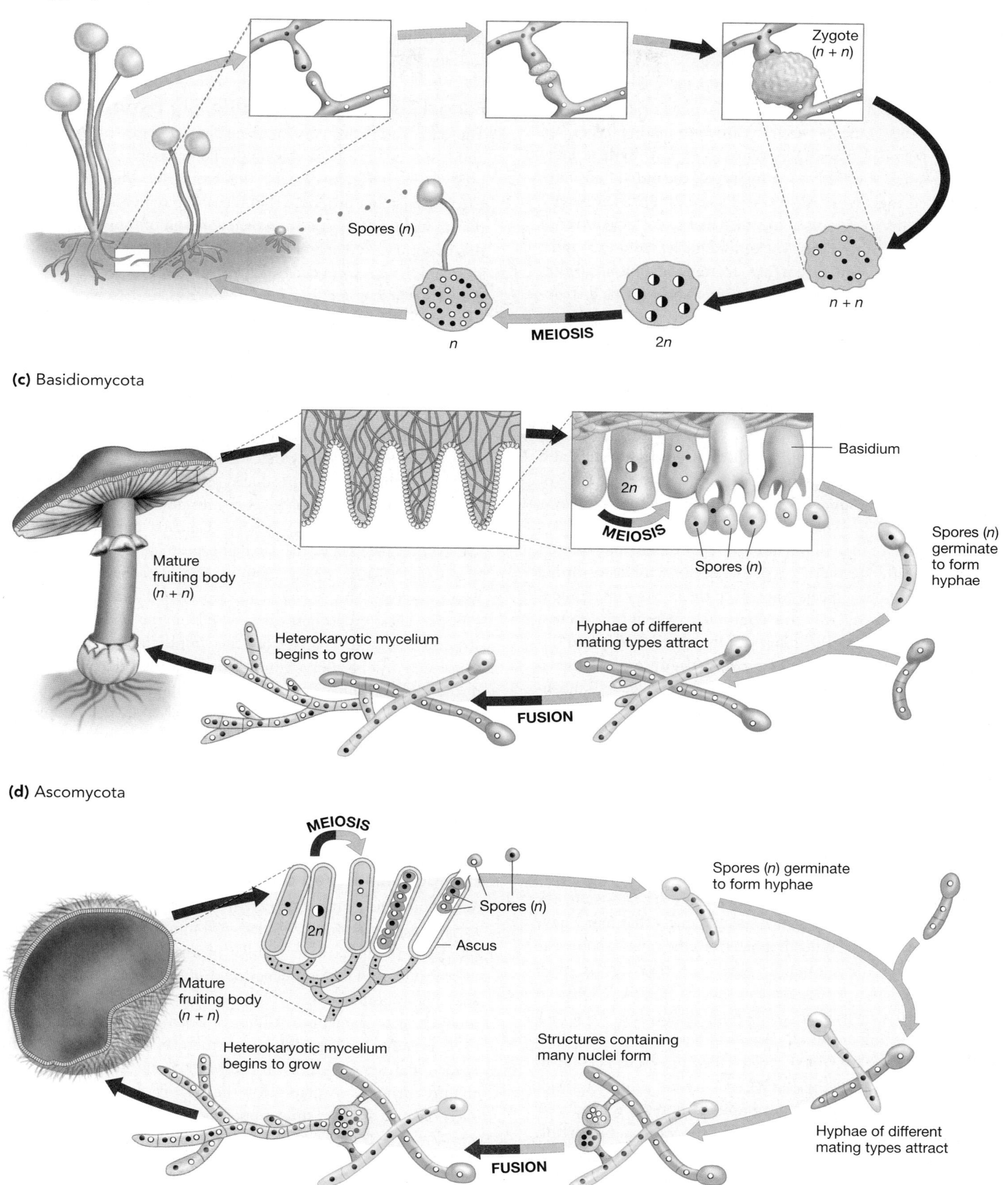
(b) Zygomycota
Zygote
(n + n)
Spores (n)
n + n
2n
MEIOSIS
n
(c) Basidiomycota
Basidium
2n
MEIOSIS
Spores (n)
Spores (n)
germinate
to form
hyphae
Mature
fruiting body
(n + n)
Hyphae of different
mating types attract
Heterokaryotic mycelium
begins to grow
FUSION
(d) Ascomycota
MEIOSIS
2n
Spores (n)
Ascus
Spores (n) germinate
to form hyphae
Mature
fruiting body
(n + n)
Heterokaryotic mycelium
begins to grow
Structures containing
many nuclei form
Hyphae of different
mating types attract
FUSION

plants explored in section 29.2. The zygomycetes that you are most familiar with, however, is neither soil dwelling nor mutualistic. The black bread mold *Rhizopus stolonifer*, pictured in Figure 29.4b, is a common household pest.

The name of this group was inspired by a structure that forms during sexual reproduction. The hyphae of Zygomycota are haploid and come in several different mating types, or "sexes." If chemical messengers released by two hyphae indicate that they are of different mating types, the individuals may become yoked together as shown in Figure 29.4b. (The Greek root zygos means to be yoked together like oxen. *Zygomycota* means "yoked-together fungi.") Haploid nuclei from each individual fuse at the point of contact, forming a diploid zygote. The zygote develops a tough, resistant coat and forms a structure that can persist if conditions become too cold or dry to support growth. When temperature and moisture conditions are again favorable, meiosis occurs. The meiotic products grow into a structure that produces haploid spores. When spores are released and germinate, they grow into new mycelia.

Basidiomycota (Club-Fungi) Mushrooms, bracket fungi, and puffballs are reproductive structures produced by members of the Basidiomycota (**Figure 29.4c**). Their size, shape, and color vary enormously from species to species, but they all originate from the hyphae of mated individuals. After the hyphae of individuals from different mating types fuse, the cytoplasms of the two cells mix. In most species, however, the nuclei remain independent—even as cell division continues. When this occurs, each cell produced after the initial fusion contains a haploid nucleus from each parent. These **heterokaryotic** (different-nucleus) hyphae eventually grow into the fruiting bodies we call mushrooms, brackets, or puffballs. When they do, specialized cells called basidia (little-pedestals) form at the ends of the hyphae (see Figure 29.4c). Inside these club-like cells, the two nuclei finally fuse. The diploid nucleus that results undergoes meiosis; haploid spores mature from the meiotic products. Sexual reproduction concludes when the spores are expelled from the end of the basidium and dispersed by the wind.

Ascomycota (Sac-Fungi) The ascomycetes are similar to the basidiomycetes in an important respect, namely that many have a heterokaryotic stage. When hyphae from the same ascomycete species but from different mating types fuse, they form a structure containing many independent nuclei (**Figure 29.4d**). A short heterokaryotic hyphae, containing one nucleus from each parent, emerges and eventually grows into a complex reproductive structure with a distinctive cell called the ascus (sac) at its tip. After the two nuclei in the heterokaryotic cells fuse inside the sac, meiosis takes place and haploid spores are produced. When the ascus matures, it splits. The spores may be forcibly ejected and are often picked up by the wind and dispersed.

The key difference between ascomycetes and basidiomycetes, then, is where they go through meiosis. Ascomycetes perform meiosis within an ascus, while basidiomycetes go through meiosis in a basidium. Ascomycetes produce their spores inside a sac; basidiomycetes produce their spores at the end of a little pedestal.

Fungal Diversity: An Evolutionary Hypothesis

The data you've just reviewed inspired a series of hypotheses about the evolution of the fungi. For example, based on the presence of flagellated cells, the biologists who defined the Chytridiomycota proposed that the group represents the lineage that branched off first during the evolution of fungi. To support this hypothesis researchers pointed to other traits that indicate a close link between chytrids and protists and that support the hypothesis that zygomycetes, basidiomycetes, and ascomycetes are more derived—meaning that they appeared later in evolution. For example, the protists that are most closely related to fungi and the chrytrids both live in water, and both have centrioles that associate with the nuclear membrane during cell division. In contrast, all of the other fungi lack centrioles. The zygomycetes, basidiomycetes, and ascomycetes have an unusual structure called a spindle pole body that forms during cell division.

This early research produced two additional hypotheses based on the reproductive structures found in zygomycetes, basidiomycetes, and ascomycetes. The first hypothesis is that the zygomycetes were the next lineage to branch off as fungi diversified (**Figure 29.4a**). The second hypothesis is that the Basidiomycota and Ascomycota represent later, more derived groups. These hypotheses are based on the assumption that structures tend to evolve from the simple to the complex. Because sacs and basidia are more complex than the zygotes produced by zygomycetes, biologists hypothesized that the ascomycetes and basidiomycetes evolved later. The simple-to-complex assumption does not always hold, however, and needed to be evaluated rigorously.

To test this assumption and the other hypotheses about fungal evolution, let's turn to an alternative source of information about evolutionary history: phylogenies estimated from molecular data.

Phylogenies Based on Molecular Data

When biologists attempt to reconstruct the evolutionary history of organisms that are relatively simple morphologically—like bacteria, archaea, unicellular protists, or fungi—they often rely heavily on DNA sequence data. Recently a coalition of researchers compared the results of ribosomal DNA sequence analyses with data from studies of cell-wall composition, life-cycle patterns, and reproductive structures. The biologists concluded that the phylogeny in **Figure 29.5** represents the best estimate currently available for the history of the group. The relationships support three important conclusions:

- It is unclear whether a chytrid lineage called Blastocladiales or a zygomycete lineage called Mucorales was the first group to branch off as the fungi diversified. This lack of res-

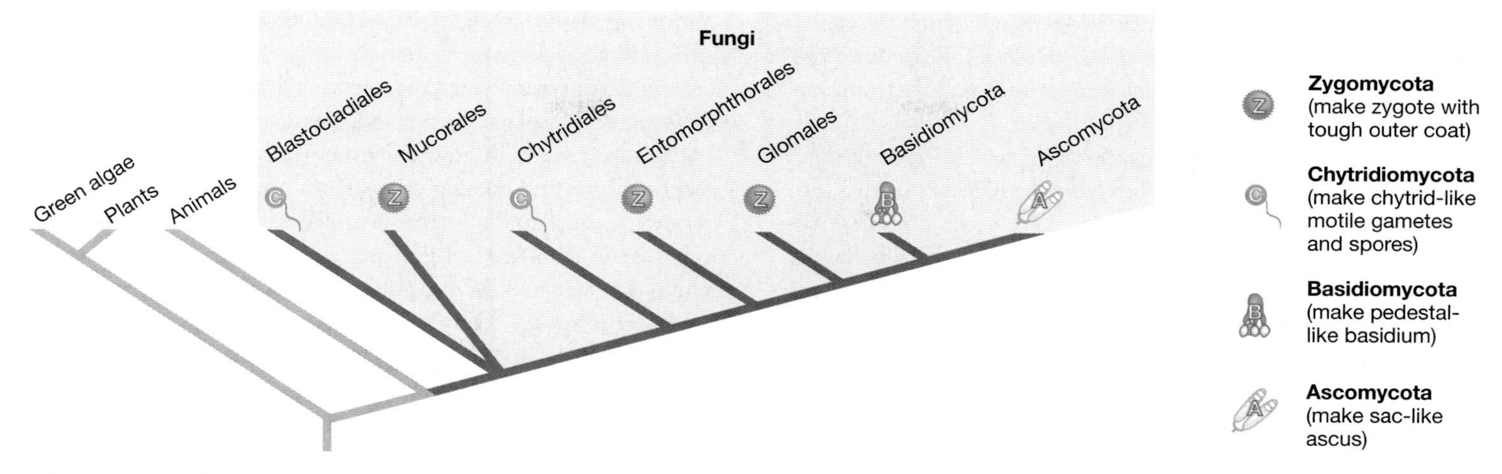

FIGURE 29.5 A Phylogeny of the Fungi
This phylogeny is based on comparisons of morphological traits and analyses of DNA sequence data. The key indicates the types of sexual reproductive structures found in each major lineage of fungi.

olution occurs because very few sequence differences are observed in the ribosomal RNAs of these two groups. To explain this observation, researchers suggest that the two groups branched off very close to one another in time, before a measurable amount of sequence divergence had occurred between them.

- The groupings called Chytridiomycota and Zygomycota do not include all descendants of the same common ancestor. To use the term introduced in Chapter 27, they are not monophyletic. Instead, fungal groups with swimming gametes and yoked hyphae are interspersed throughout the tree. If the relationships implied by the molecular data are correct, these traits are not unique to certain lineages. Instead, they have either evolved more than once or were present in a common ancestor and were retained in several lineages.
- The Basidiomycota and Ascomycota appeared late during the diversification of fungi, just as the traditional hypothesis contended.

In sum, the molecular data support some classical ideas about the relationships among fungi but challenge others. (**Box 29.1**, page 566, provides additional examples of traditional ideas about fungi that were overturned by molecular studies.) To gain an improved understanding of fungal evolution, researchers are currently studying additional DNA sequences and reanalyzing morphological traits. The goal of this research is to clarify the sequence of events that occurred during the early evolution of fungi.

29.2 Growth, Digestion, And Absorption

One basic attribute unites the diversity of fungi: their absorptive lifestyle. Fungi obtain the energy and carbon they need by absorbing it from their environment. But large molecules like starch, cellulose, proteins, and RNA cannot diffuse across the cell membranes of hyphae. Only small molecules like sugars, amino acids, and nucleic acids can enter the cytoplasm. As a result, fungi have to digest most food molecules before they can absorb them.

This section considers a series of questions about the way that fungi make a living. How do these organisms find energy and carbon sources in the first place? What enzymes enable them to break huge macromolecules into monomers that can be transported across the cell wall? Finally, what happens to the enormous quantities of reduced carbon that fungi absorb? To begin answering these questions, let's examine how the body of a fungus develops and grows.

Morphology of Hyphae

Fungi produce spores in such prodigious quantities that it is not unusual for them to outnumber pollen grains in air samples. If a spore falls on a food source and is able to germinate, a mycelium begins to form. As the fungus expands, hyphae grow in the direction in which food is most abundant. If food sources begin to run out, however, hyphae respond by making spores. Why? The leading hypothesis is that spore production is favored by natural selection when hyphae are under nutritional stress, because it allows starving mycelia to disperse offspring to new habitats where more food might be available.

If food sources are plentiful, however, mycelia can be large and long-lived. Researchers in the northwest corner of the United States recently discovered an individual mycelium growing across 1290 acres (6.5 km^2). This is an area substantially larger than most college campuses. They estimated the individual's weight at hundreds of tons and its age at thousands of years, making it one of the largest and most long-lived organisms known.

The hyphae that make up the body of a fungus are long, narrow filaments. In most fungi, these filaments are broken into cell-like compartments by cross-walls called septa. As

Figure 29.6a shows, septa do not close off segments of hyphae completely. Instead, gaps exist. These gaps enable a wide variety of materials, even organelles and nuclei, to flow from one compartment to the next.

Perhaps the most important aspect of hyphae, however, is their shape. Because they are tubes, they have the highest possible surface-area-to-volume ratio. To drive this point home, consider that the hyphae found in any fist-sized ball of rich soil typically have a surface area equivalent to half a page of this book. This surface area is important because it makes absorption extremely efficient.

Based on these observations, biologists interpret the unusual morphology of fungi as an adaptation to their absorptive lifestyle. It makes sense, then, that the only thick, fleshy structures produced by these organisms are reproductive organs. Structures like mushrooms and puffballs do not absorb food, and they are often exposed to the air, where drying is a problem. As **Figure 29.6b** shows, the mass of filaments on the inside of these reproductive structures is protected from drying by the densely packed hyphae that form the surface.

Extracellular Digestion

In the introduction to this chapter, fungi are presented as experts at reversing biosynthesis. It's claimed that they can efficiently cleave macromolecules as diverse as lignin, cellulose, and proteins. But instead of digesting food inside of a stomach or food vacuole, as most animals and some protists do, fungi secrete digestive enzymes outside their hyphae into their food. Digestion takes place outside the organism, or extracellularly.

As a case study in how this process occurs, consider the enzymes responsible for digesting lignin and cellulose. These are the two most abundant organic molecules on Earth. Recall that cellulose is a polymer of glucose, and that "lignin" refers to a family of extremely strong polymers built from monomers that are 6-carbon rings. Only a handful of organisms on Earth can digest cellulose, and members of the basidiomycetes are the only organisms that can degrade lignin completely—to CO_2.

Lignin Degradation How do fungi digest lignin? In addition to its biological significance, this question has enormous practical importance. To make soft, absorbent paper products, manufacturers must find efficient ways to degrade lignin without using caustic, dangerous chemicals. The same problem faces environmental scientists charged with cleaning up waste from old sawmills and paper mills.

To find out how fungi do it, biochemists began analyzing the proteins secreted into the extracellular space by lignin-digesting fungi. After purifying these molecules, they tested each protein for the ability to degrade lignin. Using this approach, investiga-

BOX 29.1 The Problem of Convergence

Before the advent of DNA sequencing, attempts to reconstruct the evolutionary relationships of organisms relied almost exclusively on analyzing morphological traits. This approach was responsible for identifying the four major groups of fungi based on similarities in their sexual structures. It also led to the naming of a fifth group called the Deuteromycetes (second-fungi). This group was analogous to the miscellaneous bin of a filing system; it consisted of species that reproduced only asexually. Because they lacked sexual structures, the Deuteromycetes could not be placed in any of the four main categories.

Grouping fungi by morphological similarity led to other problems. Species in an important group called the water molds look and act a lot like fungi. Water molds obtain their reduced carbon from dead organic matter and are important decomposers in freshwater habitats. They also have feeding tissues that are long and filamentous, much like hyphae (**Figure 1**). Based on these similarities, researchers proposed that water molds were closely related to what were then called the Eumycota (true-fungi).

In contrast, other biologists contended that water molds were actually not a type of fungi. These researchers argued that water molds were actually related to photosynthetic protists like diatoms and brown algae because certain aspects of their cell walls, reproductive structures, and flagella were similar.

FIGURE 1
Water molds look and function like fungi but are not closely related to them.

When DNA sequences became available, the data confirmed a close relationship between water molds and diatoms and brown algae. The molecular data indicated that the similarities among fungi and water molds are due to a common way of life, not common ancestry. The similarities arose independently, and they represent adaptations to the same lifestyle.

The fungi and water molds are examples of what biologists call convergent evolution, or simply convergence. Convergent evolution occurs when species independently evolve similar traits in similar environments. Once biologists were able to use molecular and morphological data in combination, they obtained a much clearer picture of the relationship between fungi and water molds.

tors from two labs independently discovered an enzyme called lignin peroxidase.

What does lignin peroxidase actually do? The researchers who followed up on its discovery found that the enzyme catalyzes the removal of a single electron from an atom in the aromatic rings of lignin. This oxidation step creates a free radical, which is an atom with an unpaired electron (see Chapter 2). This extremely unstable electron configuration leads to a series of uncontrolled and unpredictable reactions that end up splitting the polymer into smaller units.

Kent Kirk and Roberta Farrell refer to this mechanism as "enzymatic combustion." The phrase is apt because the uncontrolled oxidation reactions triggered by lignin peroxidase are analogous to the uncontrolled oxidation reactions that occur when wood burns in a fireplace. In contrast, virtually all of the other reactions catalyzed by enzymes are extremely specific.

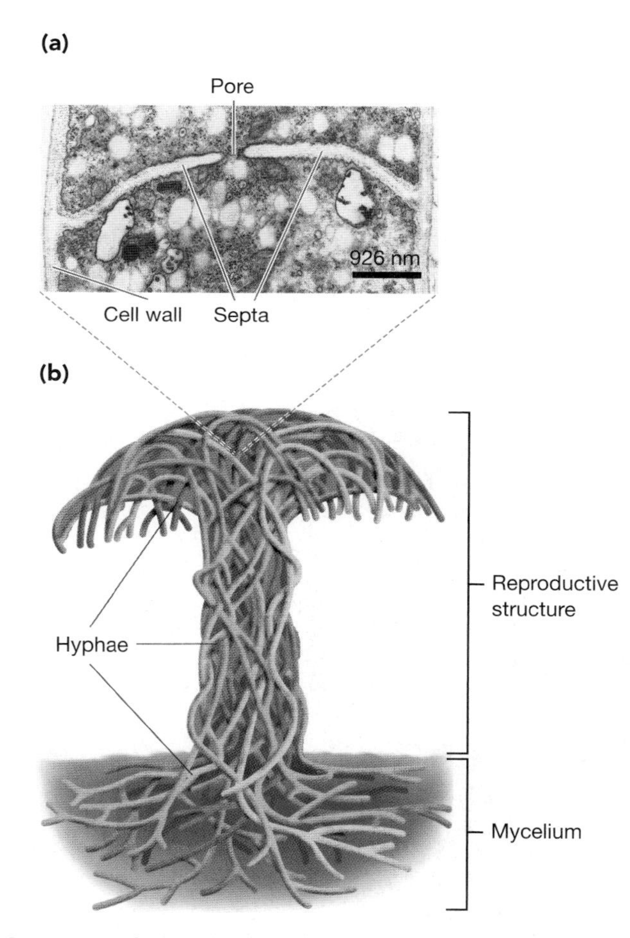

FIGURE 29.6 The Morphology of Hyphae
(a) Hyphae are often broken into cell-like compartments by partitions called septa. These electron micrographs show that septa are broken by pores. As a result, the cytoplasm of different compartments is continuous. **(b)** The hyphae from an individual make up a feeding body called a mycelium. During sexual reproduction, hyphae form fruiting bodies. Frequently, these fruiting bodies erupt from the ground or substrate, so spores can be dispersed by the wind. QUESTION What sorts of environmental changes or cues might cause a feeding body to make a fruiting body?

The lack of specificity in the reactions actually makes sense, however, given the nature of the compound. Unlike proteins, nucleic acids, and most other polymers, lignin is extremely heterogeneous. Over 10 different types of linkages are routinely found between the monomers that make up lignin. Once lignin peroxidase has created a free radical in the aromatic ring, however, any of these linkages can be broken.

The uncontrolled nature of the reactions has an important impact. Instead of being stepped down an orderly electron transport chain, the electrons involved in the reactions lose their potential energy in large, unpredictable jumps. As a result, the oxidation of lignin cannot be harnessed to drive the production of ATP. This prediction accords with the observation made by Kirk and colleagues in 1976 that fungi cannot grow with lignin as their sole carbon source.

If fungi don't use lignin as a carbon or energy source, why do they bother to digest it? The answer is simple. In wood, lignin forms a dense matrix around long strands of cellulose. Degrading the lignin matrix gives hyphae access to huge supplies of energy-rich cellulose. Fungi are like miners. They use lignin peroxidase to blast away waste molecules, exposing rich veins of polysaccharides that can fuel growth and reproduction.

Digesting Cellulose Once lignin peroxidase has softened wood by stripping away its lignin matrix, the long strands of cellulose that remain can be attacked by enzymes called cellulases. Like lignin peroxidase, cellulases are secreted into the extracellular environment by fungi. But unlike lignin peroxidase, cellulases are extremely specific in their action. Biochemists have now purified seven different cellulases from a fungus called *Trichoderma reesei*. Two of these enzymes catalyze a critical early step in digestion—they cleave long strands of cellulose into the disaccharide called cellobiose. The other cellulases are equally specific and also catalyze hydrolysis reactions. In combination, the suite of seven enzymes in *T. reesei* transforms long strands of cellulose into a simple monomer (glucose) that can be used by fungi as a source of reduced carbon.

But do the fungi that degrade cellulose produce all seven of the enzymes all the time? As biologist Fred Singer points out, a fungus that produced cellulases constantly would be like a store that stocked Mother's Day cards year round. It would seem logical to predict that fungi synthesize cellulases only when they are needed. Do they?

To answer this question, Marja Ilmén and colleagues put *T. reesei* into media where either cellulose or glucose was the only carbon and energy source available. After letting the organisms grow for 16 hours, they removed a portion of the cultures and extracted the mRNAs present. By running the mRNAs out on a gel and probing the sequences with single-stranded copies of five different cellulase genes, they could tell whether the loci were being transcribed. (This procedure, using DNA probes to identify mRNAs, is the Northern blot technique introduced in Chapter 15.)

Their results? When glucose was available, none of the cellulase mRNAs were present, meaning that none of the genes were being transcribed. But when cellulose was present, all five cellulase mRNAs appeared, indicating that all cellulase loci were actively being transcribed.

In addition, when the researchers added glucose to cultures growing on cellulose, transcription of the cellulase genes stopped. According to Ilmén and associates, the data support three conclusions about these loci: Genes that produce cellulases are expressed together, cellulose acts as an inducer, and glucose acts as a repressor.

Workers in several labs are testing these ideas by finding and characterizing the promoter that triggers the transcription of cellulase genes. Researchers are also keenly interested in determining how glucose acts to regulate transcription. Is glucose itself the repressor, or are additional molecules involved? Finally, once the genetic system responsible for cellulose degradation is understood, researchers would like to pursue a practical question: Can cellulases be used to make paper production and environmental cleanup efforts more efficient?

The Ecological Impact of Wood-Rotting Fungi

The data just reviewed make it clear that fungi are superbly adapted decomposers, capable of digesting a wide variety of molecules. Fungi synthesize a battery of sophisticated enzymes in a highly regulated manner, and deploy them in a way that is suited to environmental conditions.

It is difficult to overstate the ecological importance of this activity. Consider the role that lignin- and cellulose-degrading fungi play in the carbon cycle illustrated in **Figure 29.7**. Carbon started moving into land plants in huge quantities about 375 million years ago, when species with vascular tissue began synthesizing cell walls containing both lignin and cellulose. If fungi had not evolved the ability to digest these molecules soon after, carbon atoms would have been sequestered in wood for millennia instead of being recycled as CO_2 and glucose molecules. Terrestrial environments would be radically different than they are today, and probably much less productive.

To underscore this point, let's consider two events documented in the fossil record. The first is a remarkable lack of fungi that degrade dead plant material among the fossils pres-

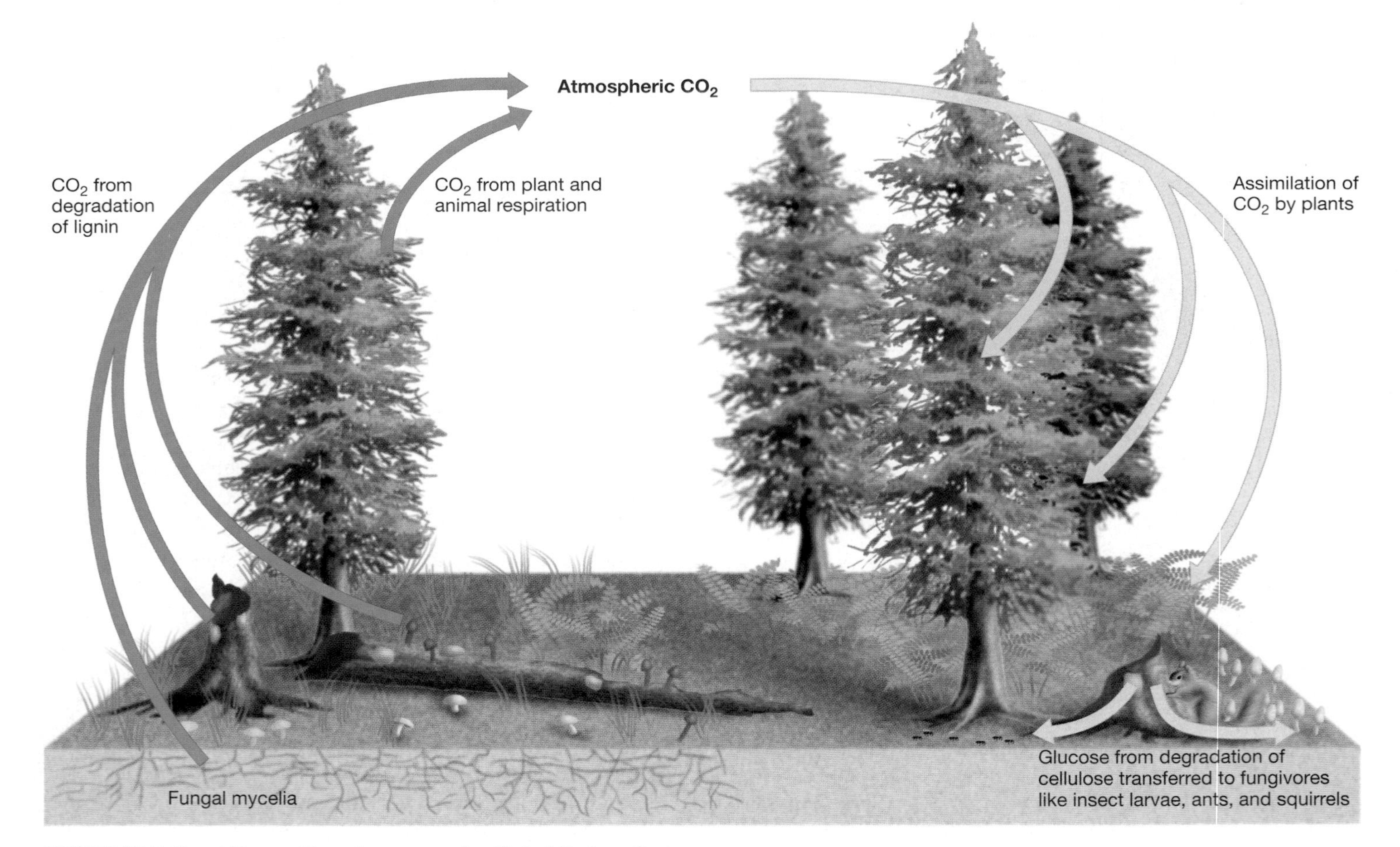

FIGURE 29.7 Fungi Have a Huge Impact on the Global Carbon Cycle
Carbon atoms are cycled through terrestrial ecosystems. If fungi were not capable of degrading lignin to CO_2 and cellulose to glucose, most carbon would eventually be tied up in indigestible woody tissues. The cycle would slow dramatically as a result. **EXERCISE** Draw an X through arrows that would not exist if fungi could not digest lignin and cellulose. Then, on a separate sheet, sketch what this habitat would look like after 1000 years.

ent in coal formed during the Carboniferous period. When Thomas Taylor and Jeffrey Osborn noted this "blip" in the fossil record, they concluded that an absence of saprophytic (rotten-plant) fungi was the driving force behind coal formation. Their logic ran as follows: If saprophytic fungi were rare, an enormous buildup of undigested and partially digested lignin and cellulose would result. Deposits of partially decayed plant material are called peat. When peat is buried under other sediments and subjected to heat and pressure, coal forms. Based on the chemistry of bogs and other peat-forming environments present today, Taylor and Osborn hypothesize that fungi did not grow in the coal-forming swamps because the water was too acidic—even though fungi tend to grow under somewhat acidic conditions. In short, coal-forming swamps formed coal because fungi were absent.

The second episode involving saprophytes is a "fungal spike" in the fossil record that was discovered by Henk Visscher and colleagues. As **Figure 29.8** shows, the spike is a huge but short-lived increase in the number of fungal spores and fossilized hyphae during the great end-Permian extinction about 250 million years ago. The discovery of this sudden increase was important because biologists had thought that the end-Permian extinction primarily affected marine organisms. But according to Visscher and colleagues, the explosion of fungal fossils in terrestrial sediments shows that land plants were also devastated. They suggest that a massive die-off in trees and shrubs produced gigantic quantities of rotting wood, and led to the explosion of fungi documented in the spike.

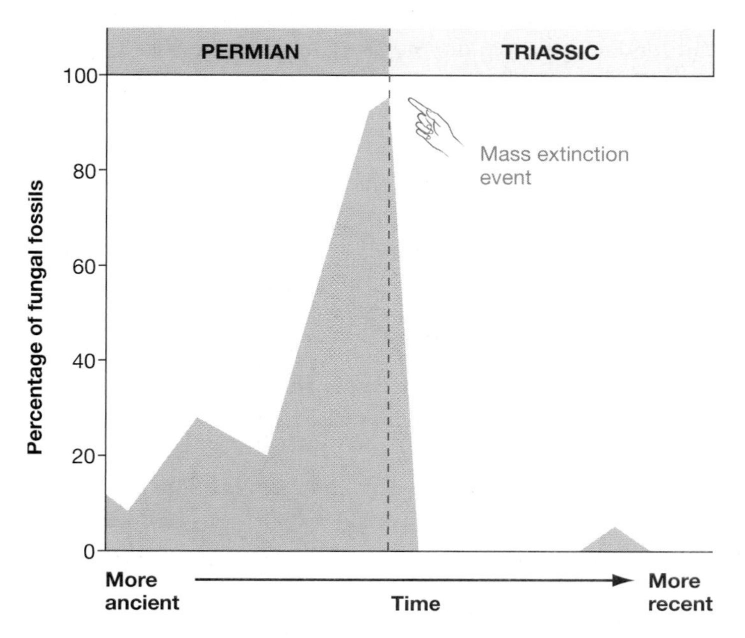

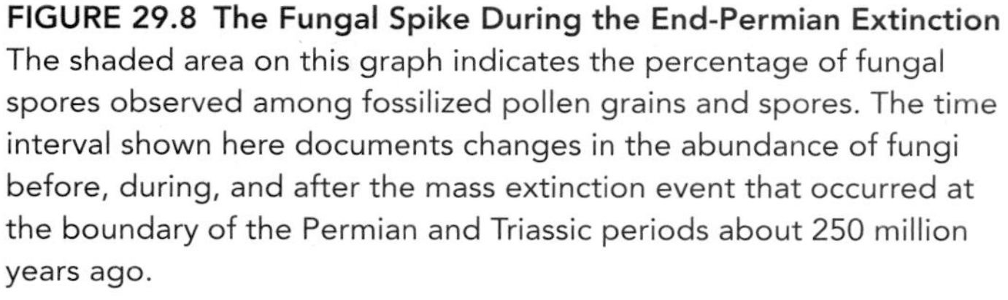

FIGURE 29.8 The Fungal Spike During the End-Permian Extinction The shaded area on this graph indicates the percentage of fungal spores observed among fossilized pollen grains and spores. The time interval shown here documents changes in the abundance of fungi before, during, and after the mass extinction event that occurred at the boundary of the Permian and Triassic periods about 250 million years ago.

To summarize, the saprophytic fungi have a huge impact on the functioning of terrestrial ecosystems. Some very recent research, however, suggests that certain *non*-saprophytic fungi play an even larger ecological role than saprophytes. These fungi interact with living plants, not dead ones.

29.3 Mutualism

Most of the research reviewed in the chapter thus far has focused on fungi that absorb nutrients from dead tissues. But fungi can also absorb food molecules from living cells. When they do, a wide variety of outcomes are possible. At one extreme the fungus can damage or even kill the other organism; at the other, it can provide benefits to the host in return for the food it absorbs. The first relationship is called **parasitism** and the second **mutualism.** This section explores mutualistic relationships that involve fungi.

Mycorrhizae

Mycorrhizal (fungal-root) associations between fungi and the roots of land plants are extremely common. Several researchers now claim that in terms of the productivity and species-richness of terrestrial ecosystems, these plant-fungal associations are the most important cooperative relationships on Earth. Why?

To answer this question, it's necessary to explore the two major types of interactions that have been characterized between fungi and plant roots. Each type of association has a distinctive morphology, geographic distribution, and function.

Ectomycorrhizal Fungi Ectomycorrhizal fungi (EMF) form a dense network of hyphae around roots (see **Figure 29.9a**, page 570). EMF are found on virtually all of the tree species that grow in temperate and boreal forests; at least half of the mushroom-forming basidiomycetes form this type of association with plants.

How and why do these trees and fungi interact? In the cold, northern habitats where EMF are abundant, the growing season is so short that the decomposition of needles, leaves, twigs, and trunks is often extremely sluggish. As a result, nitrogen tends to remain tied up in dead tissues. EMF, however, release peptidases that cleave the peptide bonds between amino acids. The nitrogen released by this cleavage is absorbed by the hyphae and transported close to the tree roots, where it can be absorbed by the plant. In return, the fungi receive sugars and other reduced carbon compounds from the plant.

R. A. Abuzinidah and David Read demonstrated the importance of this fungal-plant connection experimentally. These biologists grew birch seedlings with and without their normal EMF in pots filled with a forest soil. They found that only seedlings with EMF were able to acquire nitrogen. Inspired by data like these, K. W. Cullings and co-workers have referred to EMF as the "dominant nutrient-gathering organs in most temperate forest ecosystems."

Arbuscular Mycorrhizal Fungi (AMF) In contrast to EMF, the arbuscular mycorrhizal fungi (AMF) grow *into* the cells of root tissue. The name *arbuscular* (little-tree) was inspired by the highly branched hyphae, shown in **Figure 29.9b**, that form inside root cells. Thomas Taylor and colleagues have found these same structures in fossils that are 400 million years old (see Figure 29.9b). Taylor's discovery confirms that mycorrhizal associations existed in the most ancient of all land plants.

AMF are members of the Zygomycota. They are found in a whopping 80 percent of all land plant species, and are especially common in grasslands and in the forests of warm or tropical habitats.

Soon after AMF were discovered, biologists hypothesized that AMF had a mutualistic relationship with plants. Further, they suggested that unlike EMF, which supplies nitrogen, AMF supplied phosphorus.

This hypothesis made sense in light of an important ecological contrast. In the grasslands and tropical forests where AMF dominate, the growing season is long and warm and plant tissues decompose quickly. As a result, nitrogen is often readily available. In these habitats phosphorus, not nitrogen, is usually the growth-limiting nutrient. Experiments with radioactive atoms confirmed the hypothesis by showing that host plants supply fungi with reduced carbon, while fungi supply host plants with phosphorus.

(a) Ectomycorrhizal fungi (EMF)

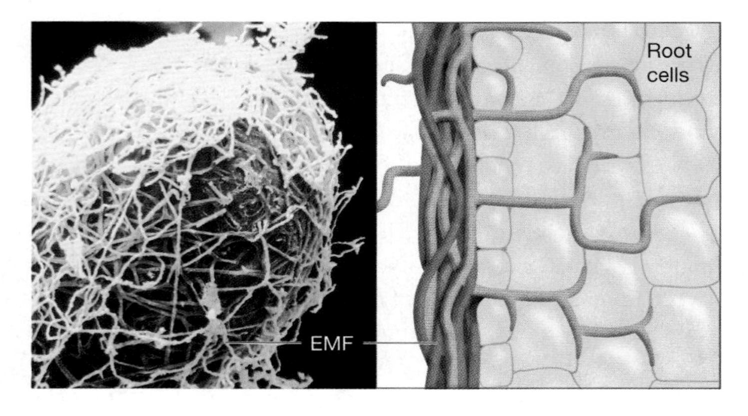

(b) Arbuscular mycorrhizal fungi (AMF)

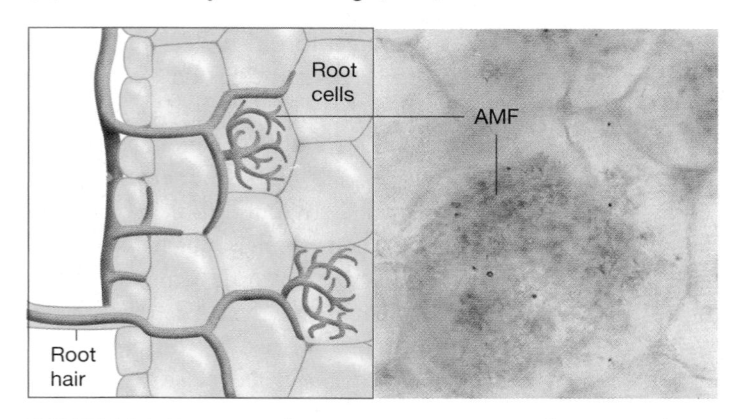

FIGURE 29.9 How Mutualistic Fungi Interact with the Roots of Plants
(a) Ectomycorrhizal fungi (EMF) form a dense network around the roots of plants. The combination of root and fungus is called a mycorrhiza. The drawing shows how the interaction works in EMF. Note that their hyphae penetrate the intercellular spaces of the root, but do not enter the cells themselves. **(b)** Interactions of arbuscular mycorrhizal fungi (AMF) with roots. The fungal hyphae penetrate the root cell walls and contact the plasma membrane, where they branch into bushy structures called arbuscules. The photograph shows a fossilized arbuscule, from an AMF, inside a plant cell. The fossil is over 390 million years old.

The Importance of Mycorrhizal Diversity In recent surveys of the EMF and AMF species that colonize plants, a strong pattern has emerged. Most plants are colonized by multiple individuals and by multiple species of fungi. Two recent experiments, conducted by a team led by Marcel van der Heijden and John Klironomos, show why this diversity of mycorrhizal fungi is important.

In the first experiment, which was conducted in a greenhouse, the biologists set up 48 identical trays filled with sterilized soil. In each tray they planted the same array of 70 seedlings, which represented 11 species of grasses and herbs common in European grasslands. Then they divided the 48 trays into six different groups. In four of these six groups they inoculated the soil with one of four different species of AMF (designated A, B, C, and D). The soil in the fifth group was inoculated with all four species of AMF; the soil in the sixth group received a control inoculation containing heat-killed fungal spores.

After growing the seedlings under identical light, temperature, and moisture conditions for 16 months, they harvested the plants and weighed them. The results were startling: Different plant species showed radically different responses to the presence or absence of particular AMF. The angiosperm *Hieracium piloselia* grew best in the tray inoculated with the AMF designated A, while seedlings of *Centaurium erythrea* were largest when B was the only AMF present. In contrast, *Lotus corniculatus* grew best when all four AMF were present; and the sedge *Carex flacca* was most luxuriant in the tray that lacked AMF.

The general implication is that the AMF present in a habitat have a profound effect on the composition of the plant community. If certain AMF species are missing, it is probable that certain plant species will do poorly.

The second experiment was even more revealing. To investigate the effect that increased AMF diversity has on the diversity and productivity of plant communities, the biologists set up 70 plots in an experimental garden. They sterilized the soil in each plot, and then planted each plot with the same mixture of seeds from 15 grasses and herbaceous plants found in North American old-field habitats. Finally, they inoculated each plot with a specific number of AMF, randomly selected from a pool of 23 different species. To test the possible effects of different numbers and combinations of fungal species, the biologists gave each plot between 0 and 14 different AMF species. After three months, they harvested the plants and analyzed the chemical composition of the soils in each plot. Several patterns jump out of the data. **Figures 29.10a** and **29.10b** show that the diversity of plant species and overall plant productivity increased dramatically with an increased diversity of AMF. The data graphed in **Figure 29.10c** and **29.10d** suggest a reason for these results: The amount of phosphorus left in the soil decreased with increasing AMF diversity, while the amount of phosphorus in the plant material increased.

What is the general message of this work? By taking up phosphorus from the soil and transferring it to plants, AMF hold a key to ecosystem function. They provide a growth-limiting nutrient in a species-specific manner. As a result, a diversity of mycorrhizal fungal species leads to increased plant species diversity and productivity in North American old-field habitats.

Leaf and Stem Endophytes

Fungi also live in and on the leaves and stems of vascular plants. Species with this growth habit are called endophytes (inside-plants). Endophytic fungi are extremely common but are relatively new to science. When biologists in Brazil examine tree leaves for the presence of fungi, they regularly find several previously undiscovered species of endophytes.

Endophytes do not cause disease symptoms in the tissues they invade. Based on this observation, researchers have hypothesized that their relationship with the host plant is a mutualistic one. This hypothesis has been supported in studies involving grasses, where endophytes have been shown to produce compounds that deter or even kill insect herbivores. The relationship is much less conclusive for endophytes found in trees, however. For example, when Stanley Faeth and Kyle Hammon sprayed spores from endophytic fungi onto the leaves of Emory oak trees, they confirmed that this treatment dramatically increased the abundance of endophytes. However, they observed no impact on the survival or growth rate of insect larvae that mine the leaves of these trees. From these data, Faeth and Hammond concluded that an increased number of endophytes did not deter insect herbivores in oaks.

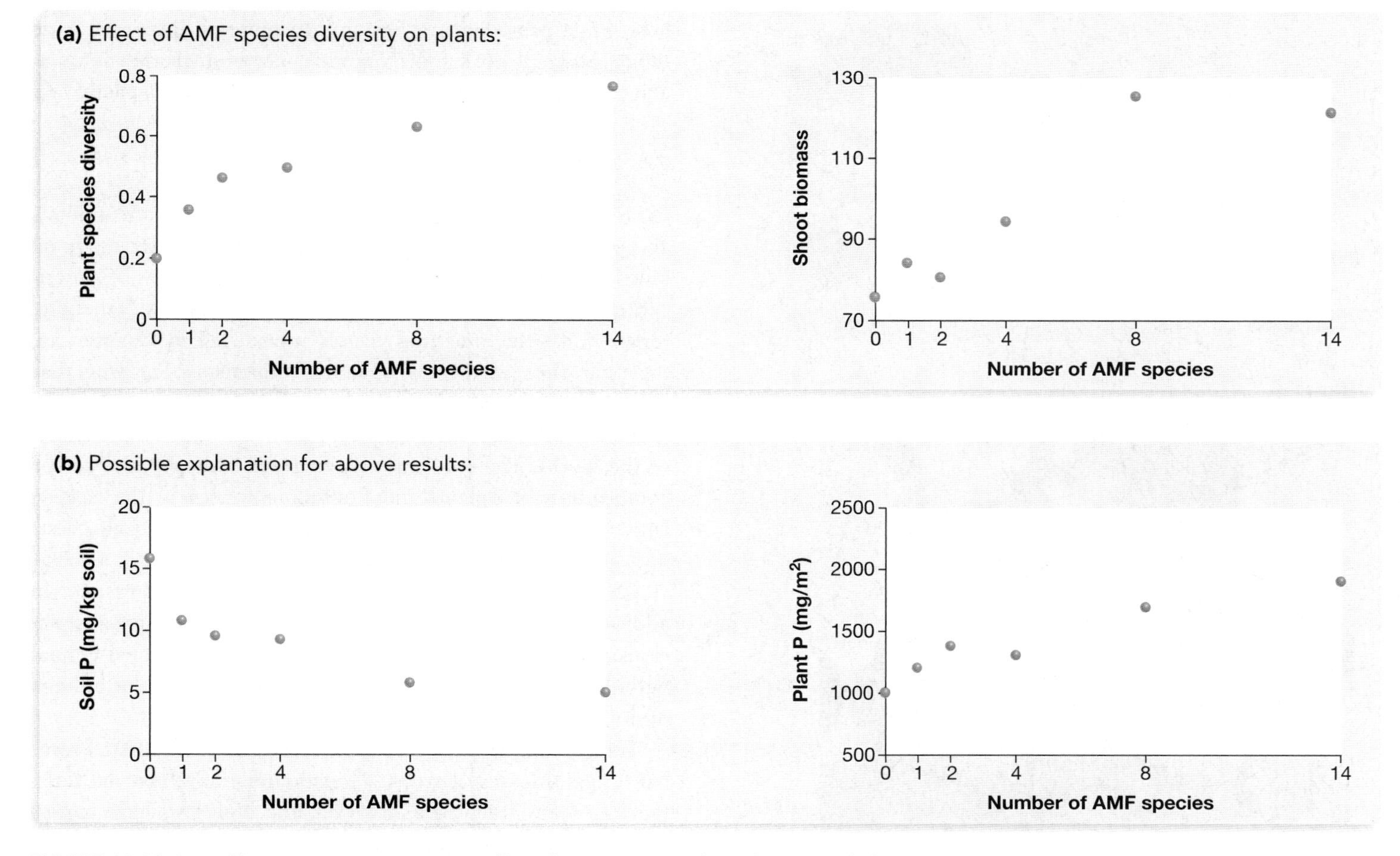

FIGURE 29.10 How Changes in AMF Diversity Affect the Diversity and Productivity of Plants
In all of these scatterplots, the x-axis represents the number of different arbuscular mycorrhizal fungi (AMF) species. The y-axes plot: **(a)** overall diversity of plant species; **(b)** total productivity of plant species (measured as grams of shoot tissue); **(c)** concentration of phosphate ions in the soil; and (d) concentration of phosphate ions in plant material. **EXERCISE** Draw a line between the points in each graph to make the trend in the data clear. Next to each plot, write a brief summary of what the data mean.

Are all endophytes mutualists? From the data collected to date, the answer appears to be no. Are some parasites? No one knows. Biologists are still working to understand the relationship between endophytic fungi and their tree hosts. The current consensus is that many may be commensals—meaning that the two species simply coexist with no large deleterious or beneficial effect.

Lichens

Of all the mutualisms involving fungi, the most visible are their associations with cyanobacteria and with single-celled members of the green algae. When fungi grow in association with these photosynthetic organisms, the resulting structure is called a **lichen** (**Figure 29.11a**). Tens of thousands of different lichens have been described to date; in most, the fungus involved is an ascomycete (although a few basidiomycetes participate as well).

(a)

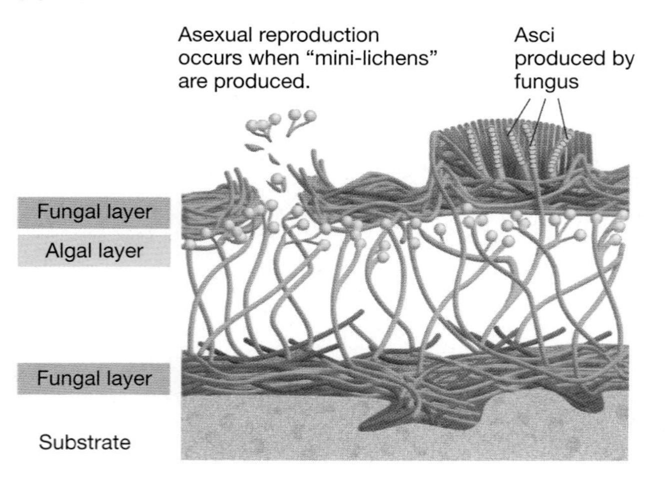

(b)

FIGURE 29.11 Lichens Are Associations between a Fungus and a Green Alga or Cyanobacterium
(a) In a lichen, green algae or cyanobacteria are enmeshed in a dense network of fungal hyphae. In this example, the fungal participant is an ascomycete. **(b)** Lichens often colonize surfaces like tree bark or bare rock, where other organisms are rare.

Lichens are important ecologically. In terms of their abundance and diversity, they dominate the Arctic and Antarctic tundras. They also are the most prevalent colonizers of bare rock surfaces throughout the world (**Figure 29.11b**). These rock-dwelling lichens are especially significant because they break off mineral particles from the rock surface as they grow. By doing so, they launch the first step in soil formation.

Biologists who study lichens are unsure whether the relationship can accurately be described as mutualistic, however. On the one hand, there is abundant evidence that both partners benefit from the relationship. In habitats where lichens are common, for example, neither partner can exist as a free-living organism. The fungi in lichens appear to protect the photosynthetic cells, perhaps because the chitin in the walls of their cells minimizes water loss. In return, the algae or cyanobacteria provide carbohydrate that the fungus uses as a source of carbon and energy. On the other hand, the hyphae of some lichen-forming fungi invade algal cells and kill them. This observation suggests a partially parasitic relationship.

In groups other than lichens, purely parasitic strategies are common. In these relationships, the fungus actively harms the host while providing no benefits. In humans alone, parasitic fungi cause athlete's foot, vaginitis, diaper rash, pneumonia, and thrush, among other miseries (but see **Box 29.2**, page 574).

29.4 Parasitism

Parasitic fungi cause death and disease among both animals and plants. During the last century, epidemics caused by fungi killed four billion chestnut trees and tens of millions of American elm trees. These outbreaks were triggered by "emerging fungi." Like the emerging viruses introduced in Chapter 26, these parasites recently switched host organisms. The fungi that cause chestnut blight and Dutch elm disease were accidentally imported on species of chestnut and elm native to other regions of the world. The epidemics they caused radically altered the composition of upland and floodplain forests in the eastern United States. Before these fungal epidemics occurred, chestnuts and elms dominated these habitats.

Clearly, fungi can be devastatingly effective parasites. In addition to causing illness in humans and trees, fungi cause serious diseases in corn, wheat, barley, and other crops. Fungal pathogens cause annual crop losses computed in the billions of dollars.

Traditionally, researchers have turned to chemicals to combat fungal diseases of crops. Parasitic fungi have evolved resistance to many fungicides, however, and biologists have had to search for alternative approaches to prevent fungal epidemics. One such approach was suggested in a recent study performed by Stanley Freeman and Rusty Rodriguez.

Freeman and Rodriguez analyzed a fungal disease of watermelons and other crops in the cucumber family that is caused by an ascomycete called *Colletotrichum magna*. When the spores of this fungus germinate on pumpkin, squash, cucumber, cantaloupe, or watermelon tissue, they produce hyphae that begin growing throughout the plant body, killing cells as they go (**Figure 29.12a**). Although infections are often lethal, the biologists were able to isolate a mutant form of the fungus that did not cause disease. In the watermelons studied by Freeman and Rodriguez, the mutants grew slightly more slowly and did not form spores. Mutant and wild-type strains affected the same tissues, however. Under the microscope, the two forms appeared identical.

When Freeman and Rodriguez crossed pathogenic, wild-type strains of the fungus with the benign, mutant form, they found that about half of the progeny were pathogenic and half were benign. Because spores are haploid and because the trait segregated in this 1:1 ratio, they inferred that the difference in virulence was caused by a mutation at one locus (**Figure 29.12b**). In short, the data support the hypothesis that a single genetic change converted a serious pathogen into an apparently harmless endophyte.

The biologists completed their initial study of the mutants by performing a series of experiments. In these trials, they exposed watermelon seedlings to the pathogenic form, to the mutant form, or first to the mutant variety and then to the wild-type strain. The results, reported in **Figure 29.12c**, show that seedlings are protected from disease if they are infected by the mutant form before being exposed to the pathogenic strain. In effect, the benign, mutant strain acts as a live vaccine. But how? Does the endophytic form stimulate the host plant's defenses? Or does the mutant fungus itself produce molecules that inhibit the growth of the pathogenic strain? These questions remain unanswered. Research continues on what could become a promising alternative strategy for fighting pathogenic fungi.

(a) A fungus was observed killing crop seedlings.

(b) Experimental crosses suggested that the virulence of the wild-type fungus was caused by a mutation at one locus.

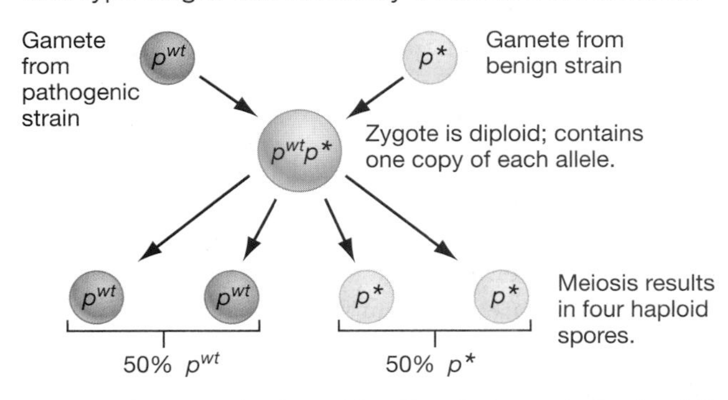

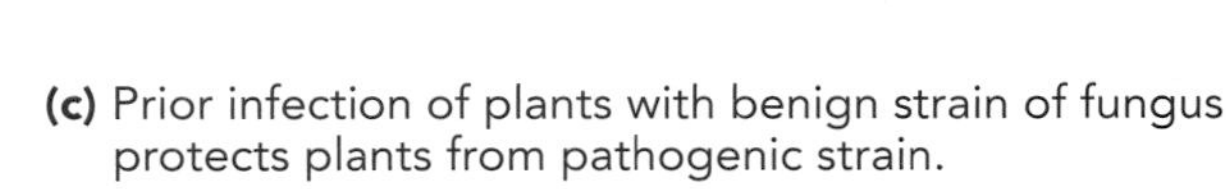
(c) Prior infection of plants with benign strain of fungus protects plants from pathogenic strain.

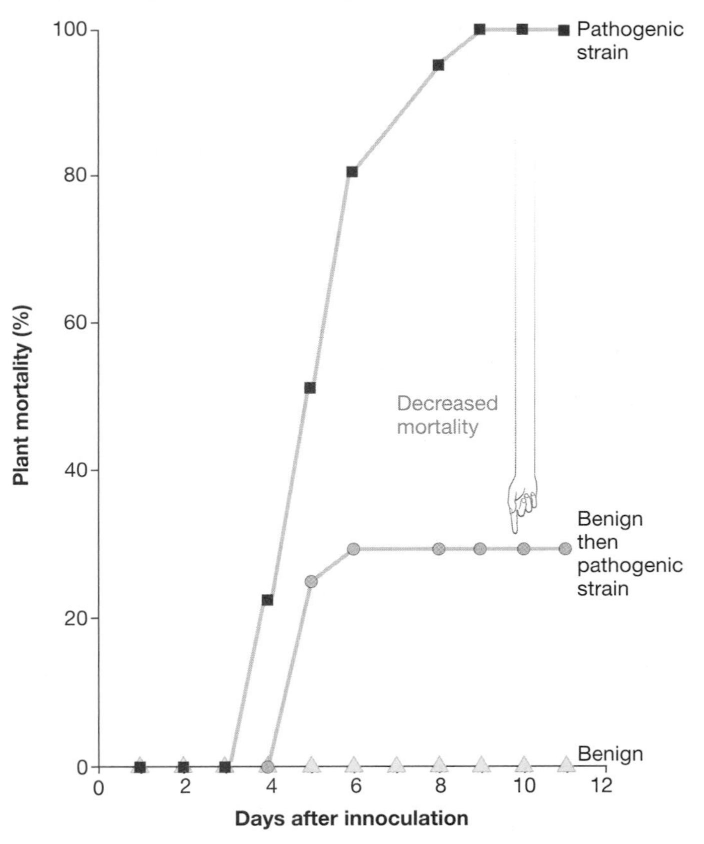

FIGURE 29.12 Prior Infection with a Non-Pathogenic Strain of Fungus Can Protect Plants from Disease
(a) A watermelon plant infected with the wild-type, pathogenic strain of the fungus *Colletotrichum magna*. **(b)** Both pathogenic and non-pathogenic strains of *C. magna* are haploid. If the difference in pathogenicity is the product of a change in just one gene (labeled *p* here), then half of the offspring in this cross should carry the pathogenic allele (p^{wt}) and half should carry the non-pathogenic allele (p^*). Experimental crosses produced just such a 1:1 ratio of pathogenic and non-pathogenic offspring. **(c)** The graph shows the percentage of seedlings that died after three different treatments: inoculation with the pathogenic strain, with the benign strain and then 24 hours later with the pathogenic strain, or with the benign strain only.

QUESTION Speculate on the function of the "pathogenicity locus" in this fungus. What might the gene product do that causes disease in the host?

BOX 29.2 Fungi at Work

Although fungi are responsible for an impressive list of maladies in humans, in reality only about 29 species of fungi cause illness in humans. Further, the incidence of fungal infections in human populations is low when compared to the frequency of diseases caused by bacteria, viruses, and protists. The major destructive impact that fungi have on people is through food spoilage and through rusts, smuts, mildews, wilts, and blights on crops.

It would be terribly misleading to characterize interactions between fungi and humans in wholly negative terms, however. Fungi are vitally important to several industries, including food and beverage production. The yeast called *Saccharomyces cerevisiae*, a member of the Ascomycota, ferments sugars when it grows anaerobically; the CO_2 that forms as a by-product makes bread rise. Several different *Saccharomyces* species are used in brewing, winemaking, and distilling; the ethanol that they produce as a waste product during fermentation acts as the active ingredient in beer, wine, and spirits. Edible mushrooms, from species in the Basidiomycota, are important in many cuisines.

The yeasts *Schizosaccharomyces pombe* and *Saccharomyces cerevisiae* are also workhorses for researchers interested in the basic cell biology and molecular genetics of eukaryotes. Because they are unicellular and easy to culture and manipulate in the lab, these species have become the organisms of choice for research on the cell cycle in eukaryotes. Research has now confirmed that several genes that control cell division and DNA repair in these yeasts have homologs in humans that, when mutated, lead to cancer. Strains of yeast that carry these mutations are now being used to test drugs that might be effective against cancer.

Essay Why Are Frogs Dying?

Over the past two decades conservation biologists have become increasingly alarmed about rapid, and in many cases catastrophic, declines in the numbers of frogs and toads around the world. The trend is particularly worrisome because populations in undisturbed habitats such as national parks are being affected. Biologist David Wake has followed the declines closely and suspects that several species extinctions have already occurred.

This fungus could be the first infectious agent known to cause a species extinction.

Despite the magnitude of the problem, biologists have had frustratingly few clues as to what was killing the organisms. Researchers would re-visit sites where frogs and toads had once been abundant and diverse only to discover that these animals had disappeared. Without dead or dying animals to study, investigators could only speculate about what was happening—until recently.

In the late 1990s a team of Australian, English, and American researchers witnessed simultaneous and massive die-offs of amphibians in the rainforests of Australia and Panama. During the events they collected 120 dead frogs and toads representing 19 different species. Although general autopsies showed no obvious tissue abnormalities or lesions, light and electron microscopy of skin samples were revealing. In every case, the epidermis of the dead frogs was riddled with a parasitic fungus. An analysis of ribosomal RNA sequences isolated from the fungal cells confirmed what researchers had suspected from the morphology of the spores: The parasite was an unusual and undescribed member of the Chytridiomycota. To confirm that the fungus had caused the deaths, the researchers collected spores from the skin of dead frogs, suspended them in water, and added the suspension to aquaria containing six healthy frogs. Within 18 days, all of the experimental frogs had become infected with the fungus and were dead or dying.

Researchers who are working on the frog die-off contend that this chytrid could be the first infectious agent known to cause a species extinction. Biologists caution, however, that the fungus is probably not the only agent involved in the worldwide declines. In California, for example, no evidence of fungal infections has been found to date. Instead, pesticides and herbicides are thought to play an important role. It is also entirely possible that well-documented increases in ultraviolet radiation—or other factors—have compromised the immune systems of these organisms, rendering them susceptible to the disease.

Chapter Review

Summary

Fungi are a diverse and species-rich lineage of organisms that make their living by absorbing reduced carbon from other organisms—dead or alive. Some estimates suggest that over 1.6 million species exist. Phylogenies based on DNA sequence data indicate that fungi evolved from protists, and confirm that some traditional groupings, based on similarities in reproductive structures, represent distinct lineages of fungi. The four traditional groupings are the aquatic fungi with motile gametes called the Chytridiomycota; the soil-dwelling fungi with tough spores called the Zygomycota; the species with club-like, spore-forming structures called the Basidiomycota; and the species with sac-like, spore-forming structures called the Ascomycota.

Several adaptations make fungi exceptionally effective at absorbing nutrients from the environment. The two growth habits found in the lineage—long, filamentous hyphae or single-celled "yeasts"—give fungal cells extremely high surface-area-to-volume ratios. Extracellular digestion, in which enzymes are secreted into food sources, is another hallmark of fungi. Perhaps the most important reactions catalyzed by these enzymes are the degradation of lignin and cellulose found in wood. Lignin decomposes through a series of uncontrolled oxidation reactions triggered by a protein called lignin peroxidase. This "enzymatic combustion" contrasts sharply with the breakdown of cellulose, which occurs in a carefully regulated series of steps, each catalyzed by a specific cellulase. Experimental evidence suggests that transcription of cellulase genes is induced by the presence of cellulose and repressed by the presence of glucose.

Acquiring phosphate is critical to an important mutualistic interaction between certain basidiomycetes and the roots of grasses and trees that grow in warm or tropical habitats. Similar mycorrhizal associations, based on an exchange of nitrogen for reduced carbon, are found between zygomycetes and the roots of trees in cold, northern habitats. Experimental evidence supports the hypothesis that an increased diversity of mycorrhizal fungi leads to increased diversity and productivity in plant communities.

Although mutualistic associations between fungi and living plants are common, so are parasitic relationships. Parasitic fungi are responsible for devastating blights in crops and other plants.

Questions

Content Review

1. Only one group of fungi has any type of motile cells. These are the flagellated gametes found in what group?
   a. Chytridiomycota
   b. Zygomycota
   c. Basidiomycota
   d. Ascomycota

2. The hyphal growth habit leads to a body with a high surface-area-to-volume ratio. Why is this important?
   a. Hyphae are specialized for an absorptive way of life.
   b. Hyphae are long, thin tubes.
   c. Hyphae are broken up into compartments by walls called septa.
   d. Hyphae can infiltrate living or dead tissues.

3. Glucose represses the transcription of cellulase genes. What does this statement mean?
   a. Glucose binds to the cellulase promoter.
   b. Cellulase genes are transcribed when glucose is present.
   c. Cellulase genes are not transcribed when glucose is present.
   d. Lignin peroxidase must be transcribed first.

4. What do researchers hypothesize that the "fungal spike" at the end-Permian boundary results from?
   a. unusual soil conditions, which made it more likely that fungal spores and hyphae could fossilize
   b. an "adaptive radiation" of fungi
   c. a switch to saprophytic lifestyles
   d. a mass extinction event that created a huge quantity of dead plant material

5. The Greek root ecto means "out." Why are ectomycorrizal fungi, or EMF, aptly named?
   a. Their hyphae form tree-like branching structures inside plant cells.
   b. They are mutualistic.
   c. Their hyphae form dense mats that envelope roots, but do not penetrate them.
   d. They release nitrogen outside their walls, where it is acquired by their plant hosts.

6. Which statement best characterizes a parasitic relationship?
   a. Parasites gain benefits, like nutrients, and provide a benefit to their host in return.
   b. Parasites gain benefits, like nutrients, and harm their hosts in the process.
   c. Parasites gain benefits, like nutrients, without an effect on their host.
   d. The participants are not considered symbionts.

Conceptual Review

1. Explain why fungi that degrade dead plant materials are important to the global carbon cycle. Do you accept the text's statement that without these fungi, "Terrestrial environments would be radically different than they are today, and probably much less productive"? Why or why not?

2. Lignin and cellulose provide rigidity to the cell walls of plants. But in fungi, chitin performs this role. Why is it logical that fungi don't have lignin and cellulose in their cell walls?

3. AMF are the mycorrhizal fungi whose hyphae actually penetrate the interior of plant root cells. Experiments show that AMF contribute phosphate ions to their host plants. Why is this important? As part of your answer, make a list of macromolecules found in plants that contain phosphorus.

4. Review the greenhouse experiment that tested the effect of AMF on plant growth. Why was it important that the researchers used sterilized soil?

5. Experiments indicate that cellulase genes are transcribed and translated in a coordinate fashion. If cells are selected to be extremely efficient at digesting cellulose, is this result logical? Would you predict that the gene lignin peroxidase is transcribed along with the cellulase loci? How would you test your prediction?

Applying Ideas

1. After reading the essay about the amphibian die-off, review the material on Koch's postulates from Chapter 25. These postulates present criteria for implicating an organism as a disease-causing agent. Did the studies on the chytrid found in dead frogs fulfill Koch's postulates? Are you convinced that the fungus is the causative agent?
2. Some biologists contend that the ratio of plant species to fungus species worldwide is on the order of 1:6. Does this claim make sense? In formulating your answer, consider the analyses of endophytic, parasitic, lichen-forming, mycorrhizal, and saprophytic strategies presented in the chapter. Also consider the diversity of tissues available in plants.
3. To date there has been little follow-up to the suggestion made in the study of watermelon disease that non-pathogenic forms of fungi could serve as "live vaccines." One reason is that if non-pathogenic strains differ from disease-causing forms by just one or a few mutations, as current data imply, a "back mutation" could cause the non-pathogenic strain to revert to pathogenesis. Comment.
4. In discussing the experiment on AMF and plant species diversity, the text implies that AMF are always mutualistic. But could at least some of the results be explained by an alternative hypothesis, which proposes that at least some of the AMF used in the experiment are parasitic? Comment.
5. Many mushrooms are extremely colorful. Why? Fungi do not see, so colorful mushrooms are obviously not communicating with one another. One hypothesis is that the colors serve as a warning to animals that eat mushrooms, much like the bright yellow-and-black stripes on wasps. Design an experiment capable of testing this hypothesis.

CD-ROM and Web Connection

CD Activity 29.1: Feeding Strategies in Fungi *(Animation)*
(Estimated time for completion = 5 min)
Fungi are not photosynthetic, so how do they feed?

At your **Companion Website** (http://www.prenhall.com/freeman/biology), you will find self-grading exams and links to the following research tools, online resources, and activities:

Cornell Center for Fungal Biology
This site describes the variety of fungal research studies currently being conducted.

Mutualism and Bioremediation
Discover how mutualistic relationships between fungi and heavy metal-eating plants can be used to help clean contaminated soils.

Infection Biology of Fungi
This site is a resource for the study of fungal parasites and the diseases associated with these organisms.

Additional Reading

Hudler, G. W. 1998. *Magical mushrooms, mischievous molds* (Princeton, NJ: Princeton University Press). A well-written introduction to pathogenic fungi.

Johnson, C. N. 1996. Interactions between mammals and ectomycorrhizal fungi. *Trends in Ecology and Evolution* 11: 293–297. Some EMF species produce large reproductive structures underground (the truffles prized by gourmet cooks are an example). This article explores how spores from these fruiting bodies are dispersed by mammals.

Mueller, U. G., S. A. Rehner, and T. R. Schultz. 1998. The evolution of agriculture in ants. *Science* 281: 2034–2038. Some ant species farm fungi for food. This phylogenetic study explores how the relationship evolved.

Schultz, T. R. 1999. Ants, plants, and antibiotics. *Nature* 398: 747–748. A commentary on how ants use antibiotics to control pests in their fungal gardens.

Talbot, N. J. 1999. Coming up for air and sporulation. *Nature* 398: 295–296. Water-repellent proteins produced by fungal hyphae allow reproductive structures to rise aboveground.

Animals

30

As a group, the animals are distinguished by one trait: They eat to live. Many protists also make their living by ingesting other organisms or detritus, but they are limited to eating microscopic prey because they are unicellular. Animals, in contrast, are multicellular. They are the dominant predators, herbivores, and detritivores in virtually every ecosystem known—from the deep ocean to alpine ice fields, and from tropical forests to arctic tundras. They find food by tunneling, swimming, crawling, creeping, walking, running, and flying, and they eat virtually every organism on the tree of life.

Over 1.2 million species of animals have been described and given scientific names; biologists predict that tens of millions more have yet to be discovered.

To analyze this almost overwhelming number and diversity of organisms, let's consider three broad questions. First, what are the animals? Section 30.1 identifies the major groups and analyzes their origins and early diversification. Second, how do they make a living? Section 30.2 analyzes how animals capture food. Third, what traits have made certain lineages particularly successful? Sections 30.3 and 30.4 examine key innovations found in the arthropods and vertebrates. The arthropods include the spiders, insects, and crustaceans; the vertebrates include the fishes along with the amphibians, turtles, snakes, lizards, birds, and mammals.

The chapter concludes by asking the same three questions about an animal that has recently colonized and altered many of the habitats on Earth—our own species, *Homo sapiens*.

The animals in this photo are a small sample of the spiders, mites, and insects that were collected from a single tree in the Amazonian rainforest. The number of different animal species that exists today is unknown, but may number in the tens of millions.

30.1 Origins and Early Diversification

Animals exploded onto the scene over the course of 65 million years, starting about 580 million years ago. Their fossil record begins with the Doushantuo microfossils, Ediacaran faunas, and Burgess Shale deposits introduced in Chapter 24. These fossil assemblages document the origins of most major lineages of animals, from simple forms like the sponges to complex organisms like arthropods.

Sponges are the first animals to appear in the fossil record of animals. Based on similarities in cell size, shape, and function, biologists propose that the closest living relatives to sponges are a group of protists called the choanoflagellates. As **Figure 30.1** shows, the most complex choanoflagellates and the simplest sponges share other characteristics. They are sessile, meaning that adults live permanently attached to a substrate. They also feed in the same way. The figure illustrates how the beating of flagella creates water currents. These currents bring organic debris toward their feeding cells, where the particles are trapped and ingested.

Based on these observations, biologists infer that animals diversified from an ancestor that looked something like a simple contemporary sponge. Biologists currently recognize about 30 different **phyla**, or major groups, of animals. The animal phyla range from species-rich lineages like the arthropods and the molluscs (clams, snails, squid, and octopuses) to lineages represented by less than 10 living species. Sponges, for example, are placed in a phylum of their own, called the Porifera (meaning "pore bearers"); humans are placed in the phylum Chordata. The chordates include the vertebrates, or animals with backbones.

The major groups of animals are defined not by a particular type of reproductive structure, as in fungi, but by a particular body plan. A body plan is an animal's architecture, the major features of its structural and functional design. The architecture of sponges is built around a system of tubes and pores that create channels for water currents. Chordates have a body plan that is organized by a rod-like structure, called a notochord, that appears along the backs of their embryos. Most animal bodies are variations on the simple design called a tube, and many animal phyla have bodies that are worm-like in overall appearance.

In essence, then, each animal phylum represents a different way of putting a multicellular animal together—a different way of designing a feeding machine. To understand the origin and early diversification of animals, it's necessary to understand the changes that created this diversity of body plans. What characteristics of the body plan vary among the major groups of animals? How do these traits affect the type of feeding machine that results?

The Architecture of Animals

Most variation in the body plans of animals can be understood as differences in just a handful of basic design and construction features. This section reviews four of these features: the number of tissue types found in embryos, the type of body symmetry, whether the body includes a fluid-filled cavity, and how the earliest developmental events—specifically cleavage and gastrulation—proceed.

Embryonic Tissues One of the fundamental traits that biologists use to analyze differences in body plans is the number of tissue layers that exist in an embryo. A tissue is a highly organized and functionally integrated group of cells. Animals whose embryos have two types of tissues are called **diploblasts** (two-

(a) Choanoflagellate (a protist)

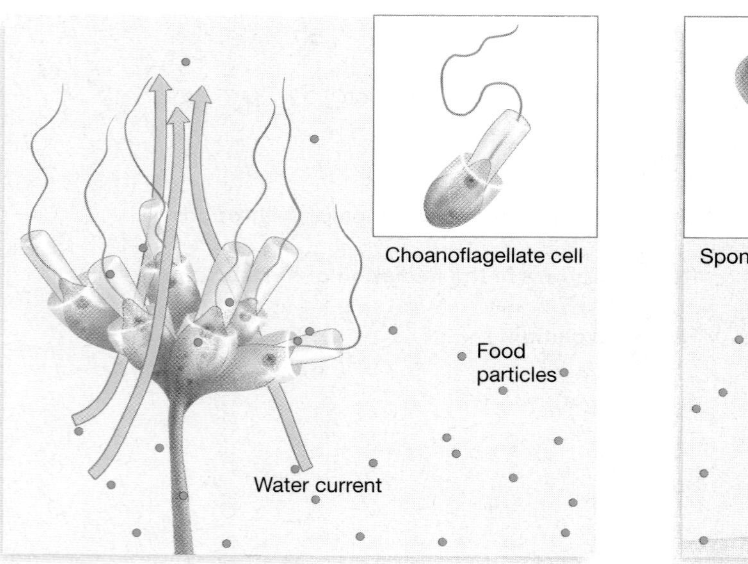

(b) Sponge (an animal)

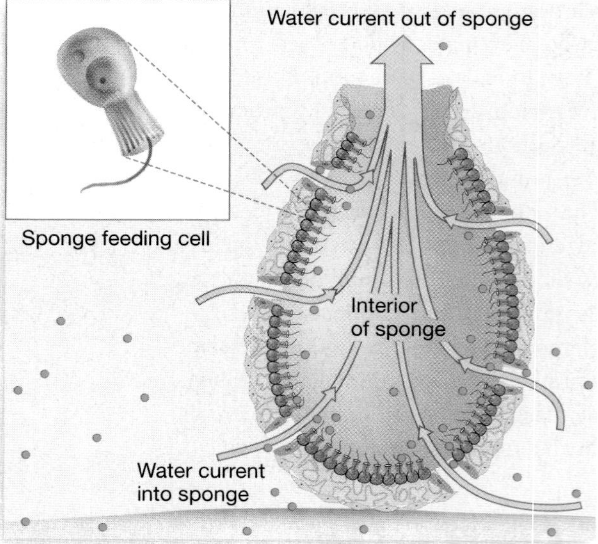

FIGURE 30.1 The First Animals
(a) Choanoflagellates are protists whose cells form colonies. **(b)** The feeding cells of some sponges are similar in form and function to choanoflagellate cells. The cross-sectional view of the sponge shows how the beating of flagella produces a water current. This current brings food into the body of the feeding cells where it can be ingested by feeding cells.

sprouts); animals whose embryos have three are called **triploblasts** (three-sprouts). As Chapter 18 explained, these embryonic tissues are organized in layers. In diploblasts the layers are called ectoderm and endoderm; the third layer in triploblasts is found between these two and is called mesoderm. The Greek roots *ecto*, *meso*, and *endo* refer to outer, middle, and inner, respectively; the root *derm* means skin.

The embryonic tissues found in animals develop into distinct adult tissues. In triploblasts, for example, ectoderm gives rise to skin and the nervous system. Endoderm gives rise to the digestive tract, or gut. The circulatory system, muscle, and internal structures like bone are derived from mesoderm. In general, then, ectoderm produces the covering of the animal and endoderm generates the digestive tract. Mesoderm gives rise to the tissues responsible for locomotion.

Only two groups of diploblastic animals are alive today: the cnidarians and ctenophorans illustrated in **Figure 30.2**. Sponges are the only group of animals that lack tissues. Although sponges have several different cell types, the cells are not organized into the tightly integrated structural and functional units called tissues. All remaining animals, from leeches to humans, are triploblastic.

Radial and Bilateral Body Symmetry A basic feature of a multicellular body is whether it has a plane of symmetry. An object like an animal's body is symmetrical if it can be divided by a plane such that the resulting pieces are similar. Animal bodies can have 0, 1, 2, or more planes of symmetry. Some sponges, like the one illustrated in **Figure 30.3a**, are asymmetrical. They cannot be sectioned in a way that produces similar sides.

(a) Cnidaria

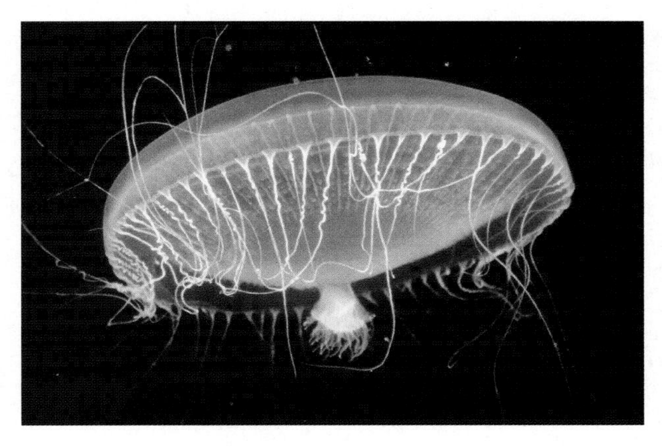

(b) Ctenophora

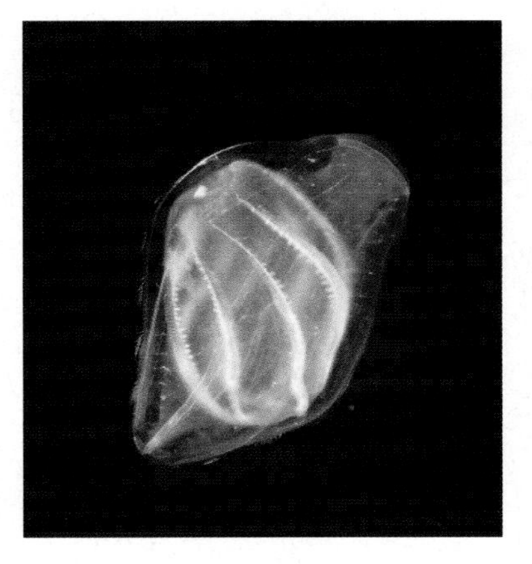

FIGURE 30.2 Diploblastic Animals
(a) The phylum Cnidaria includes the jellyfish, sea anemones, corals, and sea pens. *Cnida* is the Greek root for "nettle"; it refers to the stinging cells, or cnidocytes, that these animals use to capture prey. **(b)** The phylum Ctenophora is composed of comb jellies. Comb jellies are a major component of planktonic communities in the open ocean. The dark blue individual here has just swallowed a whitish comb jelly.

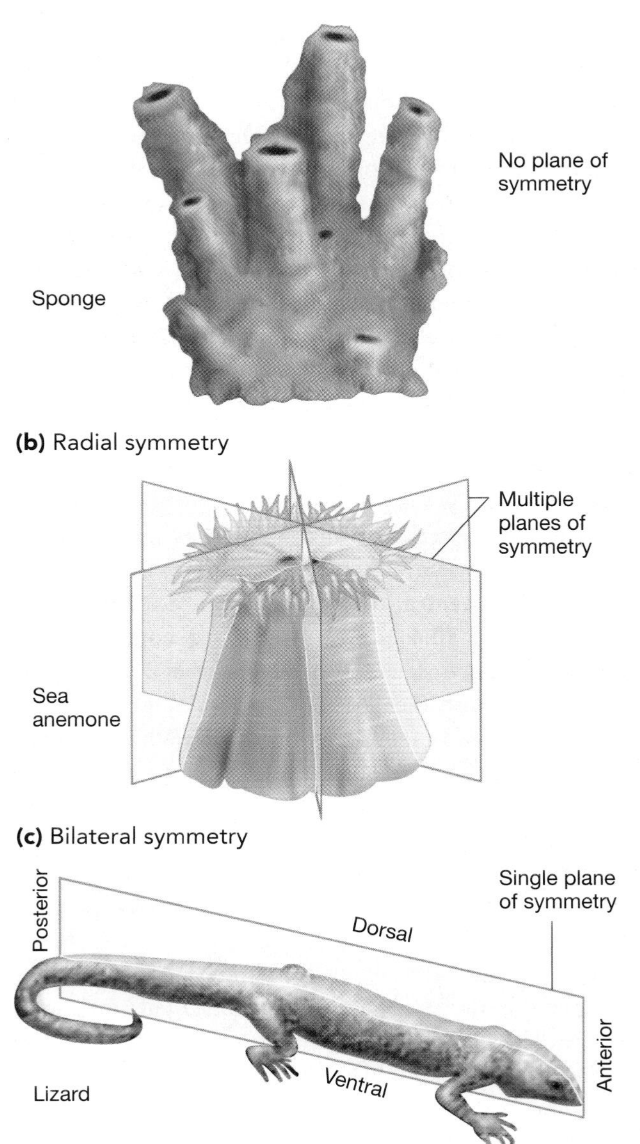

FIGURE 30.3 Types of Body Symmetry in Animals
QUESTION Are the animals in Figure 30.2 asymmetric, radially symmetric, or bilaterally symmetric?

All other animals exhibit radial (spoke) or bilateral (two-sides) symmetry. Radially symmetric animals have at least two planes of symmetry. Most of the radially symmetric organisms living today either float in water or live attached to a substrate. As **Figure 30.3b** shows, their bodies are often cylinder-like. As a result, they can capture prey or react to predators that approach from any direction.

Bilaterally symmetric organisms, in contrast, face their environment in one direction. Because they have one plane of symmetry, they tend to have a long, narrow body with a distinct head and tail region (**Figure 30.3c**). This body plan is important because it makes bilaterally symmetric animals capable of unidirectional movement. Feeding and sensory structures of bilaterally symmetric organisms are concentrated in the head, which faces the environment, while posterior regions are specialized for powering locomotion. With the exception of adult forms of species in the phylum Echinodermata (sea stars, sea urchins, feather stars, brittle stars), all triploblastic animals are bilaterally symmetric.

To explain the pervasiveness of bilateral symmetry, biologists point out that unidirectional movement directed by a distinctive head and tail region is an exceptionally efficient way to locate and capture food. The advent of mesoderm made the evolution of extensive musculature and nerve systems possible; the advent of bilateral symmetry made directed movement and hunting possible. A triploblastic, bilaterally symmetric body had the potential to develop into a formidable eating machine.

The Body Cavity A third architectural element that distinguishes animal phyla is the presence of a fluid-filled cavity in the body. This feature creates a medium for circulation, along with space for internal organs. Even more important, fluid-filled chambers are central to the operation of a **hydrostatic skeleton**. As **Figure 30.4** shows, fluid-filled compartments change shape when muscles push against them. The shape change occurs because the fluid is enclosed (Figure 30.4a), and because water cannot be compressed. When muscles push against a hydrostatic skeleton in a coordinated way, efficient and directed movement results (**Figure 30.4b**). The evolution of the hydroskeleton was a seminal event in the diversification of animals because it gave bilaterally symmetric organisms the ability to move efficiently in search of food and mates.

Diploblasts do not have a fluid-filled body cavity other than a central canal that functions in digestion and circulation, however, and neither does the phylum of triploblasts called flatworms. The remaining triploblasts do have an additional body cavity. In a few of these groups, like the roundworms and rotifers, the cavity forms between the endoderm and mesoderm layers in the embryo. This design is called a pseudocoelom, or "false-hollow." The term is unfortunate, though, because there is nothing false about this cavity—it exists and is the basis for an effective hydroskeleton.

In all other triploblasts, a body cavity forms from within the mesoderm itself and is lined with cells from the mesoderm. As a result, muscle and blood vessels can form on either side of the cavity, or coelom. In this respect, the coelom represents a more advanced design than the pseudocoelom and the gastrovascular cavity produced by the endoderm of diploblasts.

Early Developmental Events in Coelomates With one exception (the echinoderms), all true coelomates are bilaterally symmetric and have three embryonic tissue layers. Despite these similarities, however, this huge group of organisms can be split into two distinct subgroups called the protostomes and the deuterostomes. The vast majority of animals, including the arthropods, molluscs, and segmented worms, are protostomes. Vertebrates and echinoderms are deuterostomes. The division between these subgroups is based on several striking differences in how embryos develop.

Recall from Chapter 18 that an animal embryo's growth begins with cleavage. This is the series of mitotic divisions that

(a) Hydrostatic skeleton of a nematode

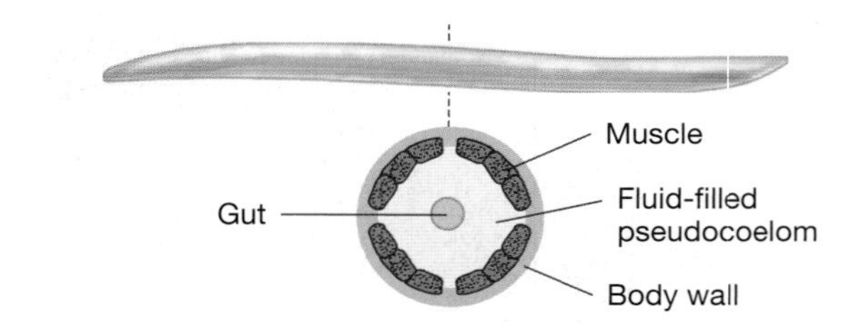

(b) Coordinated muscle contractions result in locomotion.

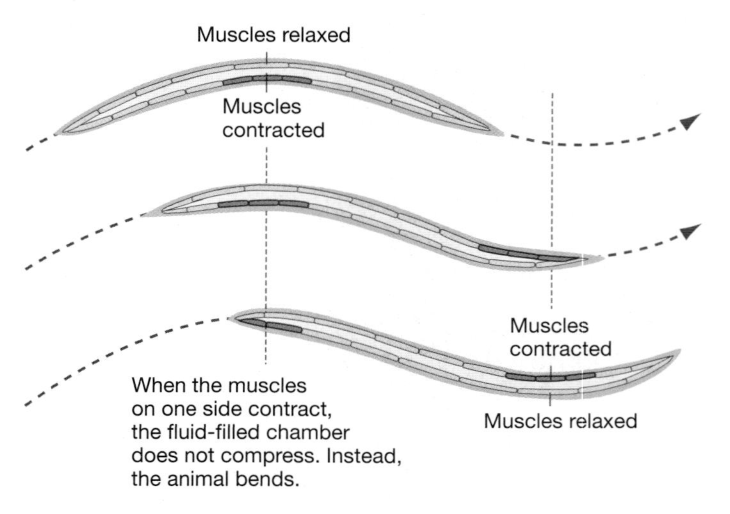

FIGURE 30.4 Hydrostatic Skeletons and Locomotion
The roundworm shown here moves with the aid of its hydrostatic skeleton. This is an enclosed, fluid-filled chamber.

divides the egg into a hollow ball of cells. In protostomes, these cell divisions take place in a pattern known as spiral cleavage. In deuterostomes they take place in a completely different pattern called radial cleavage (**Figure 30.5a**).

After cleavage has created a ball of cells, gastrulation occurs. Gastrulation is a series of cell movements that results in the formation of the three embryonic tissue layers. In both protostomes and deuterostomes, gastrulation begins when cells move into the center of the ball of cells. This invagination of cells creates a pore that opens to the outside (**Figure 30.5b**). In **protostomes**, this pore becomes the mouth. The other end of the gut, the anus, forms later. In **deuterostomes**, however, this initial pore becomes the anus and the mouth forms later. Translated literally, protostome means "first-mouth" and deuterostome means "second-mouth."

The final difference between the groups arises as gastrulation proceeds and the body cavity begins to form. As **Figure 30.5c** indicates, the coelom of protostomes forms within a solid block of mesoderm. In deuterostomes, however, layers of mesodermal cells pinch off from the gut to form the coelom. In terms of cleavage, gastrulation, and coelom formation, protostomes and deuterostomes evolved different ways of achieving the same ends.

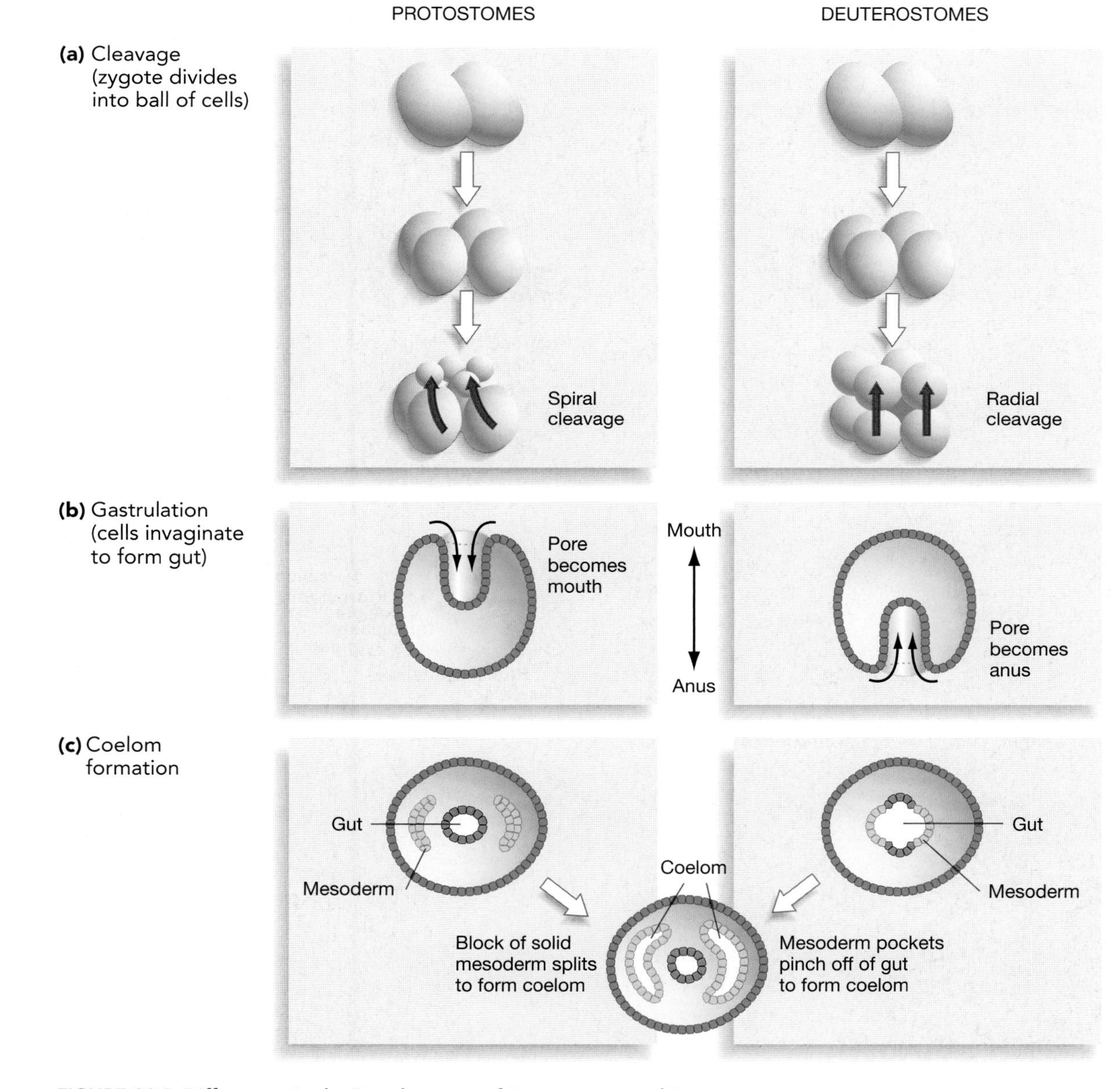

FIGURE 30.5 Differences in the Development of Protostomes and Deuterostomes
This chart illustrates some of the differences between early developmental events in protostomes and deuterostomes. The differences show that there is more than one way to produce a bilaterally symmetric, coelomate body plan.

An Evolutionary Hypothesis The variation in embryonic tissues, body symmetry, types of body cavity, and early development observed in animals can seem overwhelming. To make sense of this diversity, the diagram in **Figure 30.6a** groups phyla that share some or all of the four fundamental characteristics. The illustration also places these groups in an evolutionary framework based on the assumption that complex body plans are derived from simpler forms.

The evolutionary tree that results represents a series of hypotheses about the relationships among animal phyla. For example, sponges are shown as the earliest-branching lineage because they lack tissues, and because some are asymmetrical. **Radially symmetric** phyla are placed next on the tree because their tube-like body plans are relatively simple. Among the bilaterally symmetric phyla, the tree predicts that groups evolved in the following order: acoelomates, then pseudocoelomates, and finally coelomates.

What happened *after* the coelomates split into the protostomes and deuterostomes? Figure 30.6a suggests that two major events occurred. First, radial symmetry evolved in some echinoderms. Second, a type of body architecture called segmentation evolved independently in both protostomes and deuterostomes. When a body is divided into a series of repeated structures, like an earthworm's segments or a fish's vertebral column and ribs, it is said to be segmented. Segmentation is found in protostome lineages like earthworms and insects as well as in a deuterostome lineage, the vertebrates.

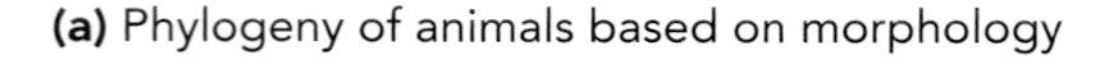

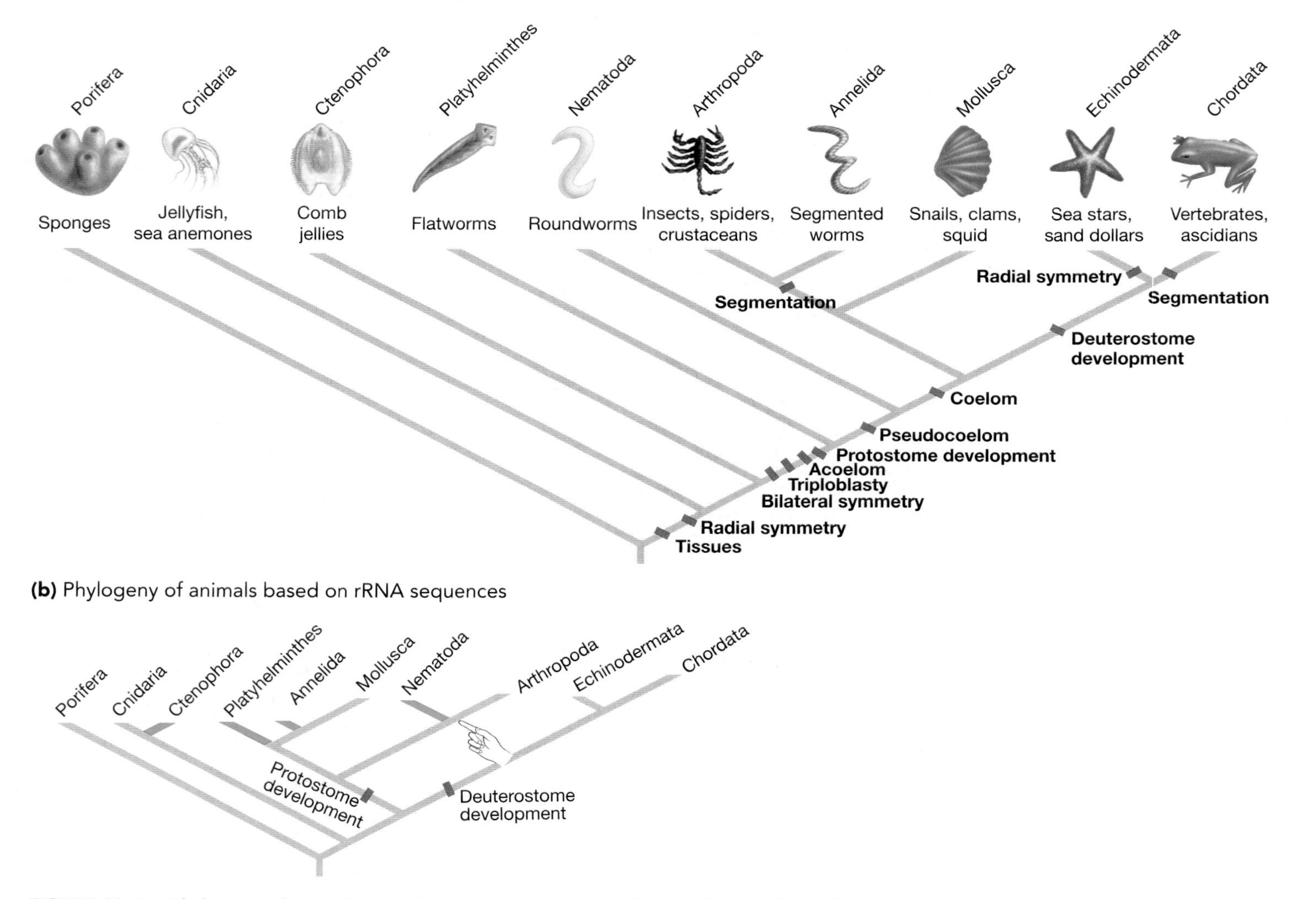

FIGURE 30.6 A Phylogeny of Animal Phyla Based on a Comparison of Body Plans and Developmental Sequences
(a) This phylogeny is based on similarities and differences in the body plans and developmental sequences of various animal phyla. The bars along the branches indicate when certain traits originated. **(b)** This phylogeny is based on similarities and differences in the SSU rRNA sequences of various animal phyla.

How can these hypotheses be tested? One approach is to construct a second tree using a different set of characteristics. For example, when biologists analyze similarities and differences in DNA sequences for the small subunit ribosomal RNA (SSU rRNA) in animals, what does the resulting tree look like? If the relationships implied by the two independent sources of data agree, it would be strong evidence that the hypotheses just examined are correct.

Molecular Phylogenies

The molecular phylogeny of animals presented in **Figure 30.6b** is a work in progress. Many phyla are left off this tree because their locations, based on SSU rRNA data, are still uncertain. One major finding that emerged from comparisons of gene sequences has become widely accepted, however: animals with protostome development form one lineage which split into two major subgroups. The subgroup that contains molluscs and annelids is called Lophotrochozoa, while the subgroup that contains the arthropods and nematodes is called the Ecdysozoa. Work on the phylogeny of major animal groups continues at a furious pace, and major new results are published almost annually.

To summarize the data in this section and the analyses of the Doushantuo, Ediacara, and Burgess Shale fossils presented in Chapter 24, the earliest events in animal evolution featured changes in body architectures. Due to variation in body symmetry, tissue types, body cavities, and developmental patterns, evolution produced feeding machines based on a variety of designs.

30.2 Feeding

Within each animal phylum, the basic features of the body plan do not vary from species to species. All molluscs, for example, have a coelom; all are triploblastic, bilaterally symmetric protostomes. What triggered the diversification of species within the molluscs and other animal phyla?

In most cases, the answer to this question is diversification in habitat use. Diversification in habitat use, in turn, usually occurs as a result of pronounced variation in methods for feeding and moving. Recall that animals obtain nutrients and reduced carbon compounds by ingesting other organisms. Animals are diverse because there are thousands of ways to find and eat the millions of different organisms that exist. The mechanisms of animal movement are explored in some detail in Chapter 43; this section focuses on the diversity of feeding strategies observed in animals.

The feeding tactics observed in animals can be broken into five general types. All five are found among animals fossilized in the Burgess Shale (see Chapter 24) and among species in the most familiar groups living today—the molluscs, arthropods, and vertebrates. Let's examine the five general feeding methods in turn.

Suspension Feeding

A wide variety of animals feed by filtering out food particles suspended in water. The organisms pictured in **Figure 30.7** illustrate a few of the many variations on this theme.

Figure 30.7a shows how clams use a large muscle, called a foot, to burrow into sediments. As they burrow, they extend long tubes called siphons to maintain contact with the surface. Inside their body cavity, cilia on their gills pump water in one siphon and out the other. The incoming water contains food particles that are trapped on the gills and swept toward the mouth by cilia.

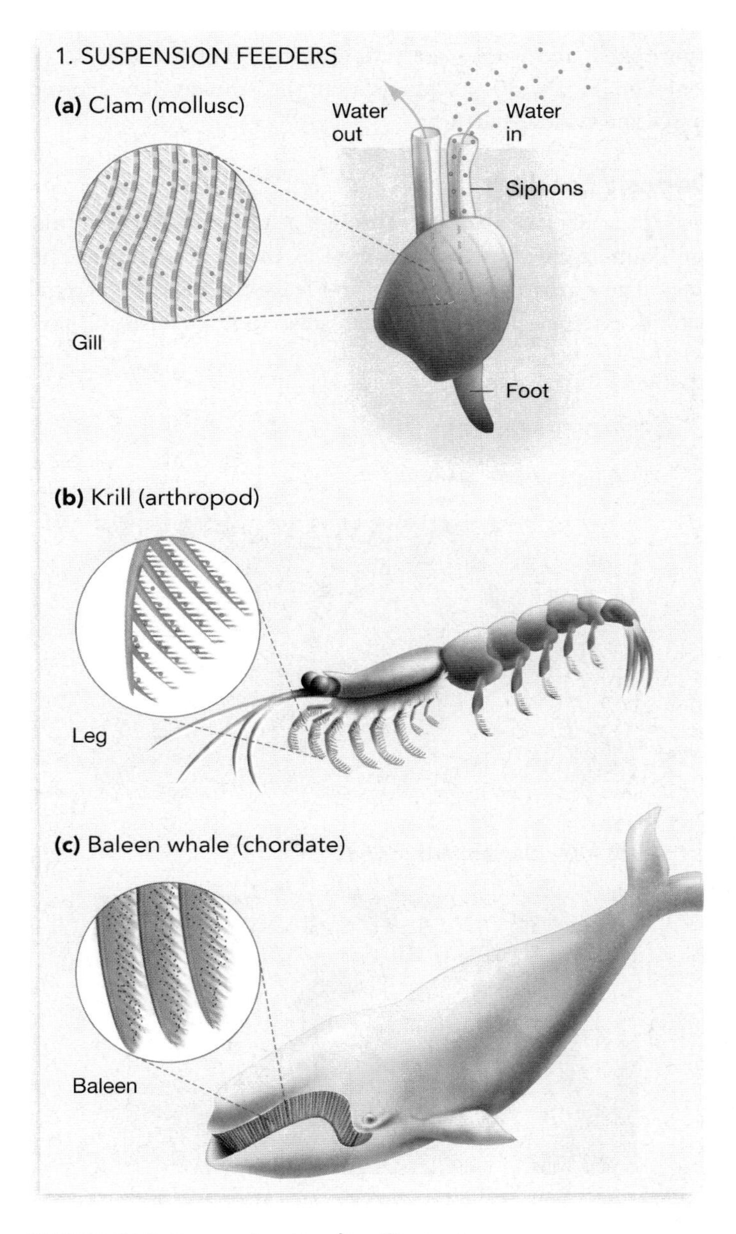

FIGURE 30.7 Suspension-Feeding Strategies
Clams **(a)**, krill **(b)**, and baleen whales **(c)** are all suspension feeders. They filter food particles from water using the trapping structures shown in the close-ups.

Part (b) in the figure shows the small animals called krill, which suspension-feed as they swim. As individuals move forward, their legs wave in and out. Projections on their legs trap food particles as they flow past. The food particles are then moved up the body to the mouth, where they are ingested.

Figure 30.7c shows how krill are eaten by a group of suspension feeders called the baleen whales. These whales have a series of long plates hanging from their jaws, which are made from a horny material called baleen. They feed by gulping water containing krill, squeezing the water out between their baleen plates, and trapping the krill inside their mouths.

It is important to note that suspension feeding is found in a wide variety of animal groups. Clams are molluscs, krill are arthropods, and whales are vertebrates. A glance at the phylogenies in Figure 30.6 suggests that the strategy has evolved many times, independently.

Deposit Feeding

Deposit feeders eat their way through a substrate. Earthworms, for example, swallow soil material as they tunnel through it. They digest any organic matter and leave the mineral material behind as feces. For these organisms, food consists of soil-dwelling bacteria, protists, fungi, and archaea, along with detritus—the dead and often partially decomposed remains of organisms.

Insect larvae that burrow through plant leaves and stems, bore through piles of feces, or mine the carcasses of dead animals or plants can also be considered deposit feeders because they eat their way through a substrate.

Unlike suspension feeders, which are diverse in size and shape and employ a wide variety of trapping or filtering systems, deposit feeders are similar in appearance. They usually have simple mouthparts and a worm-like body shape (**Figure 30.8**). Like suspension feeding, however, deposit feeding is found in a wide variety of taxonomic groups, including roundworms, segmented worms, molluscs, peanut worms, and chordates like hagfish.

2. DEPOSIT FEEDERS

(a) Earthworm (annelid)

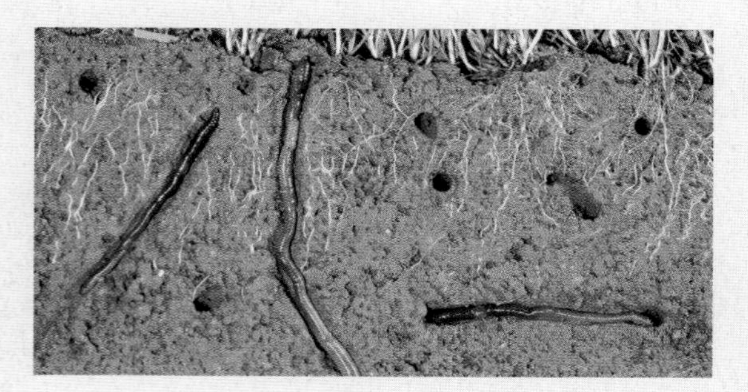

(b) Insect larvae (arthropods)

FIGURE 30.8 Deposit Feeders
Deposit feeders mine their way through a substrate, eating as they go. Like the earthworm and maggots shown here, they generally have long, thin bodies. QUESTION Why does it make sense that deposit feeders tend to have tube-like bodies?

Herbivory

Animals from a diversity of phyla harvest and digest algae or plant tissues. In sharp contrast to filter feeders and deposit feeders, herbivores have complex mouths with structures that make biting and chewing or sucking possible. Let's consider, for example, the structure called a radula, which is found in snails and other molluscs. As **Figure 30.9a** shows, this organ is located next to the mouth and functions like a rasp or a file. In herbivorous species, the sharp plates on the radula move back and forth to scrape away leaf material, which can then be ingested.

The photographs and diagrams in **Figures 30.9b** and **30.9c** illustrate the diversity of mouthparts found in herbivorous insects and vertebrates. Note that there is a general correlation between the structure of the mouthpart and the type of tissue harvested. The proboscises of insects and the long beaks of hummingbirds are used to harvest nectar, while the chewing mouthparts of grasshoppers and grinding molars of horses process leafy tissues. As Chapter 40 explains, mouthparts are a classical example of how the structures found in organisms match their function.

Predation

Animals use a fascinating variety of strategies and structures to capture and eat other animals. One way to categorize these hunting strategies is to consider whether the predator waits for quarry or actively stalks its prey.

Web-spinning spiders are classic sit-and-wait predators (**Figure 30.10a**). But there are many variations on this theme. Sea anemones and corals, for example, rarely move from the substrate, if they move at all. They capture prey that swim by using tentacles lined with specialized cells called cnidocytes (nettle-cells). Each cnidocyte contains a capsule with a coiled, hollow thread that is filled with a reservoir of toxins (**Figure 30.10b**). When prey brush up against these cells, the threads project to the exterior explosively, pierce the prey, and inject the toxins. The dead or paralyzed organism is then pulled to the mouth and eaten.

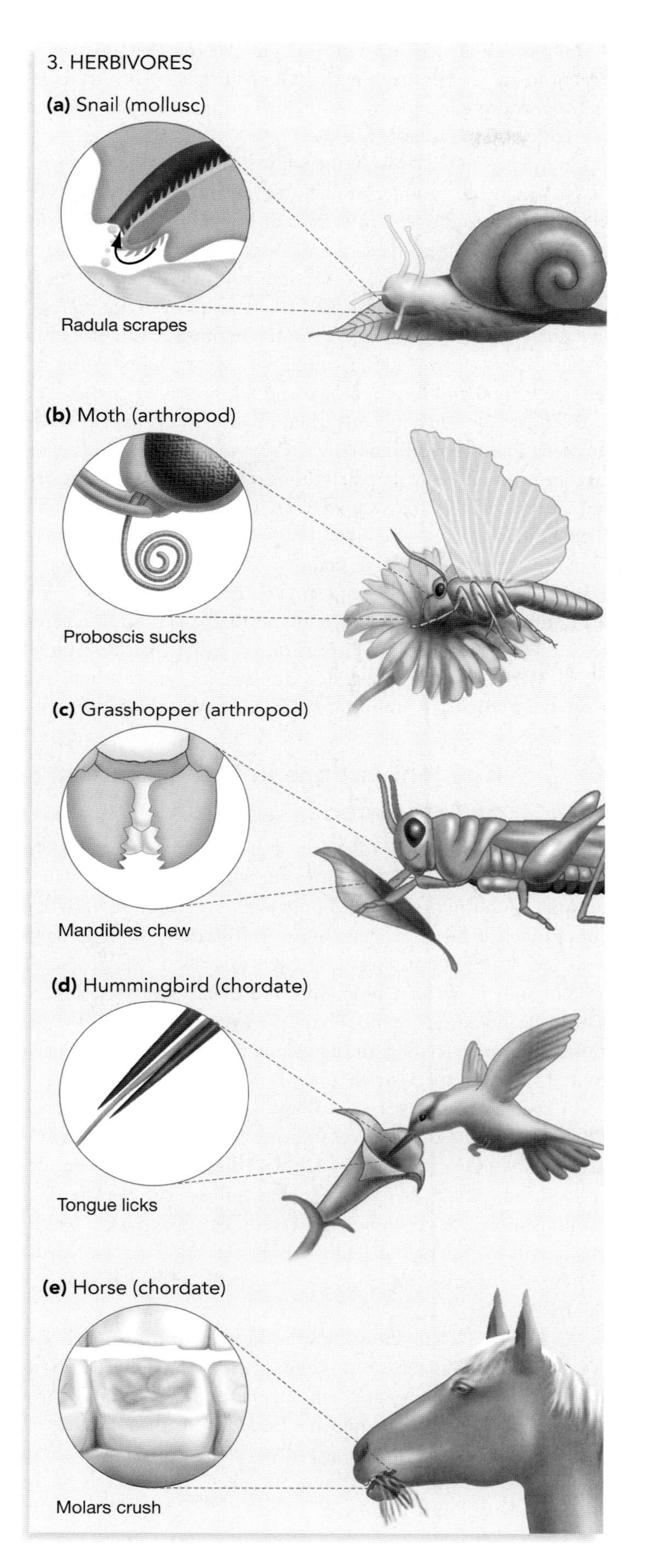

FIGURE 30.9 Herbivory
These examples illustrate a tiny subset of the methods animals use to harvest plant tissues. In these and many other cases, the structure of the mouthparts correlates with their function in harvesting a particular type of plant tissue. In addition to taking food into the body, mouthparts may start the digestive process by breaking food items into small pieces.

4. PREDATORS

(a) Spider (arthropod)

(b) Portugese man-of-war (Cnidaria)

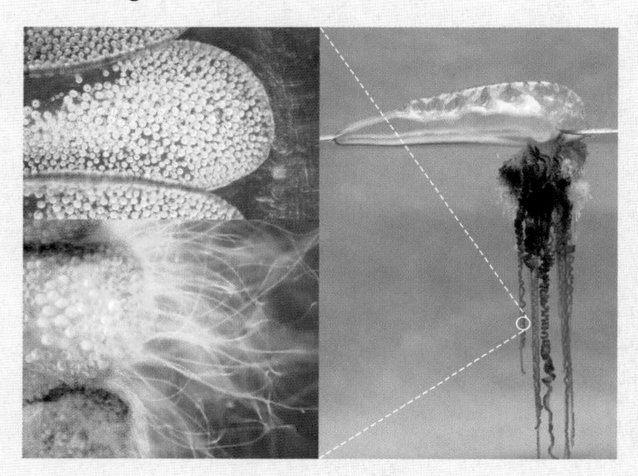

(c) Wolf (chordate)

FIGURE 30.10 Predation
(a) Most spiders are sit-and-wait predators. **(b)** Jellyfish, sea anemones, and hydrozoans like this Portuguese man-of-war have specialized cells called cnidocysts attached to their tentacles. The detailed photos show clusters of cnidocytes before and after firing into a prey animal. **(c)** Wolf families hunt together by chasing prey.

Stalkers also vary in their hunting strategies. Consider wolves and mountain lions. Both organisms are native to North America and prey primarily on members of the deer family. Like other species of dogs, wolves hunt by locating a prey organism and then running it down during an extended, long-distance chase (**Figure 30.10c**). They also live and hunt in family groups called packs. Mountain lions, in contrast, are solitary animals. They hunt by slowly stalking their prey, then pouncing on it or running it down in a short sprint.

Parasitism

It is often difficult to draw a sharp distinction between predation and parasitism. In general, parasites are much smaller than their victims and often harvest nutrients without causing death. Predators, in contrast, are typically larger than their prey or about the same size. Predation almost always leads to the death of the victim.

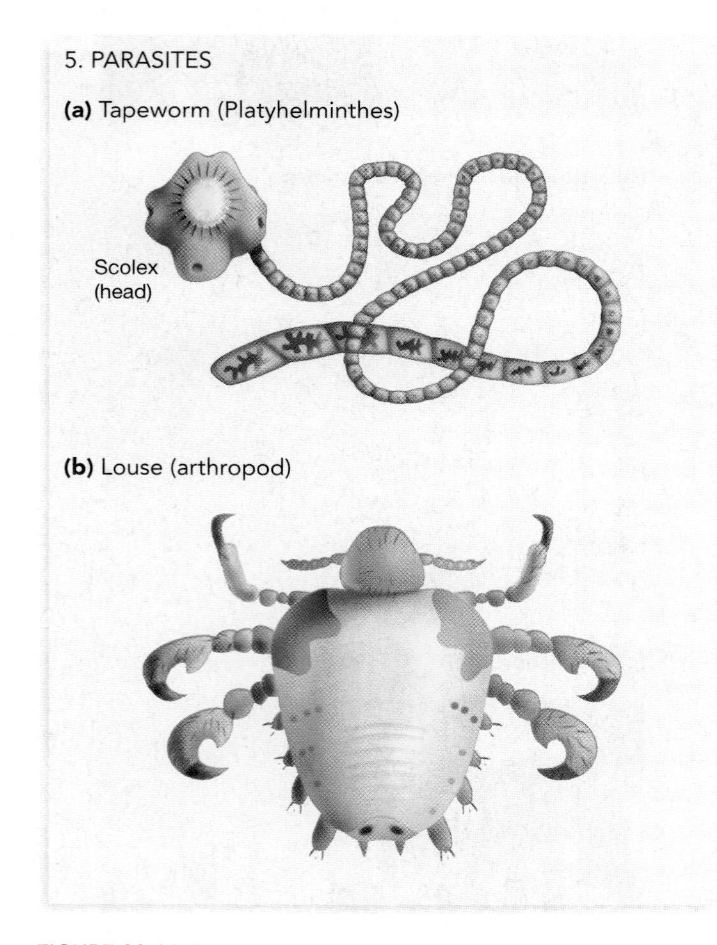

FIGURE 30.11 Parasitism
(a) Tapeworms are common intestinal parasites of humans and other vertebrates. They attach to the wall of the digestive tract using the barbed hooks on their scolex, or head, and absorb nutrition directly across their body wall. **(b)** Lice are insects that parasitize vertebrates and birds. This drawing shows *Phthirus pubis*, a louse that attaches to the pubic region of humans using its claw-like legs. The animal pierces the skin with its mouthparts and feeds by sucking body fluids.

Like the other feeding methods surveyed here, parasitism is practiced by species from a variety of lineages. The strategies employed also vary widely, but can be grouped into two broad categories: endoparasitism and ectoparasitism.

Endoparasites live inside their hosts. They are often worm-like in shape and can be extremely simple morphologically. The tapeworms found in the intestines of humans and other vertebrates, for example, have no digestive system. Instead of a mouth, they have hooks or other structures that attach to their host's intestinal wall (**Figure 30.11a**). Instead of digesting food in a gut, they absorb reduced carbon compounds and other nutrients directly from their surroundings. Most endoparasites ingest their food and have a digestive tract, however.

Ectoparasites live outside their hosts. They usually have grasping mouthparts that allow them to pierce their host's exterior and suck the nutrient-rich fluids inside. The louse pictured in **Figure 30.11b** is an example of an insect ectoparasite that afflicts humans.

A quick glance back at Figures 30.7–30.11 is an effective way to reinforce the message that animals employ a diverse array of methods to capture food and that variation in feeding techniques keyed the diversification of species within animal phyla. With this general overview in place, let's look at how and why two particularly important groups of animals diversified.

30.3 Key Innovations in the Radiation of Arthropods

The number of species found in animal phyla varies widely. There are just 15 species each in the worm-like groups called priapulids and phoronids and only eight species in the phylum containing the beard worms. But over 100,000 species of mollusc and 47,000 chordates have been described. Clearly, some animal body plans have been much more successful than others. This section and section 30.4 have the same goal: to analyze the traits that made feeding and movement particularly efficient in two of the most species-rich animal phyla.

In species numbers and ecological diversity, which lineage of animals is the most successful of all? The answer is clearly arthropods. Over a million species have already been described, and researchers predict that millions or even tens of millions more are yet to be discovered. The group includes the chelicerates (spiders and relatives), an extinct group called the trilobites, insects, and crustaceans (crabs and relatives—see **Figure 30.12a**).

As **Figure 30.12b** shows, the body plan of arthropods has several distinctive features. Like many other phyla, arthropods have **bilateral symmetry** and triploblastic embryos with the protostome pattern of development. In addition, their bodies are segmented. These segments are often grouped into specialized regions like a head, thorax, and abdomen; some or all of the segments have jointed limbs.

What traits have made arthropods so successful? To answer this question, this section analyzes three features of arthropods that relate directly to success in feeding. The evolution of an exoskeleton provided an attachment site for muscles, which in combination with jointed limbs made coordinated, rapid movement in search of food possible. The striking feature of life history called metamorphosis allowed juvenile and adult forms of the same species to exploit different habitats and food sources. As a result, biologists hypothesize that metamorphosis made feeding more efficient by reducing competition from members of the same species. Let's take a closer look at each of these evolutionary innovations.

Exoskeleton

The exoskeleton secreted by ectodermal tissues is a particularly important aspect of arthropod architecture. The arthropod exoskeleton is a hard outer covering that has a water-proof wax coating. The exoskeleton itself consists of layers of tough proteins interwoven by a stiff molecule called chitin. In crustaceans and trilobites, the exoskeleton is further hardened by deposition of calcium carbonate ($CaCO_3$), forming a shell around the animal.

The arthropod exoskeleton serves as a protective device and as an attachment point for muscles. Having a rigid surface for muscle insertion makes an entirely new type of locomotion possible. In most worm-like phyla, movement results when sheets of muscles push against a hydrostatic skeleton. But in arthropods, muscles are organized in short bands and are inserted at precise points on the exoskeleton. As Figure 30.12b shows, the limbs of arthropods move in response to contractions by pairs of these muscles. The arrangement turns joints into fulcrums and limbs into lever arms. Their action results in rapid movement combined with precise control.

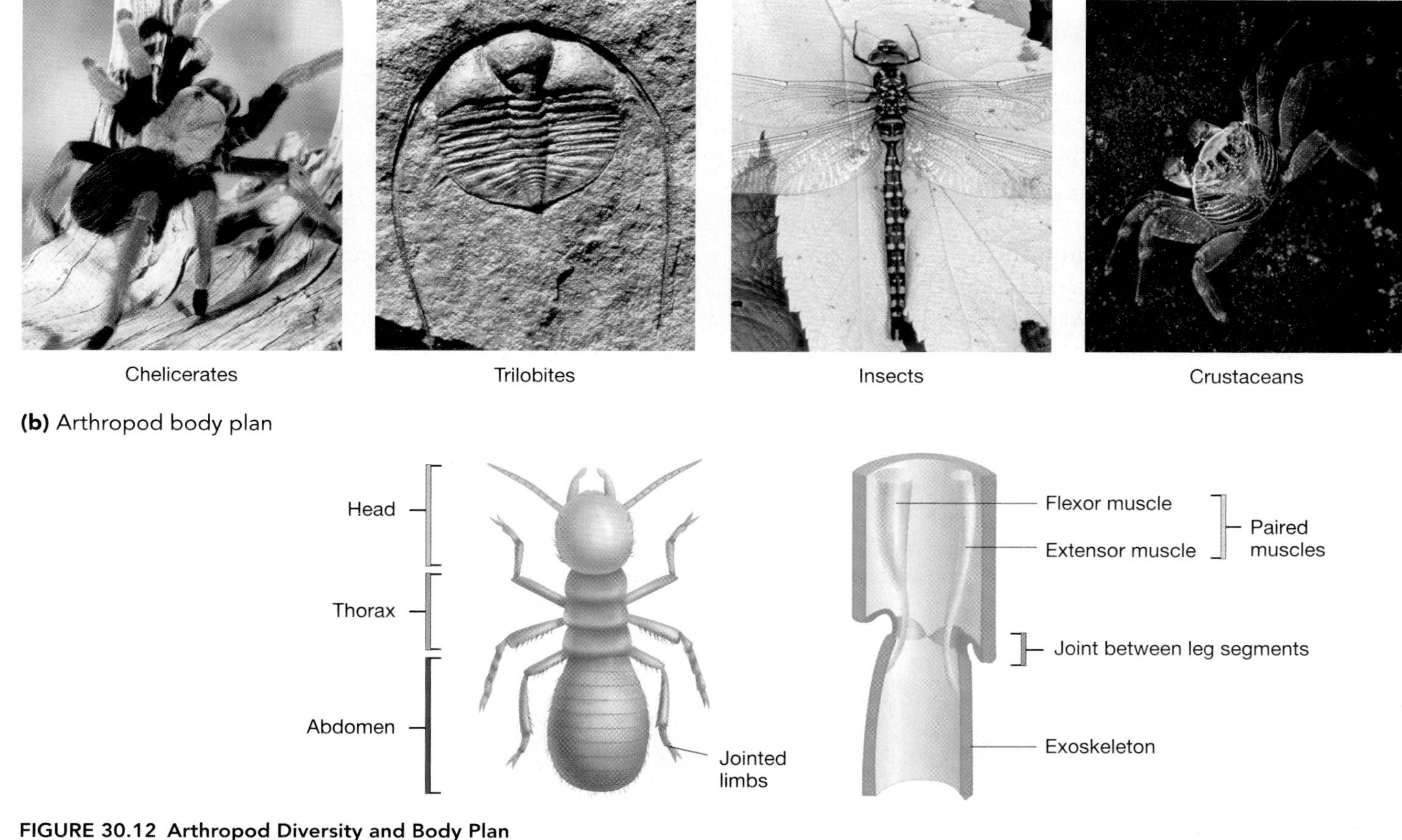

FIGURE 30.12 Arthropod Diversity and Body Plan
(a) From left to right, these photos show representative species from the four major lineages of arthropods: chelicerates (spiders, mites, scorpions, and horseshoe crabs), trilobites, insects, and crustaceans (crayfish, lobsters, shrimp, and crabs). **(b)** This sketch of a termite shows some of the general features of the arthropod body plan: segmentation, organization of the body into distinct regions (like the head, thorax, and abdomen shown here), and jointed legs that move with the aid of paired muscles. **EXERCISE** On the drawing in part (b), add an arrow showing the direction that the lower leg moves when the extensor muscle relaxes (lengthens) and the flexor muscle contracts (shortens).

The key point here is that the exoskeleton provides a stiff surface for muscles to pull against, while jointed limbs translate muscle action into quick, directed movement. Given their importance, it's worthwhile to examine arthropod limbs in more detail.

Limbs

Arthropods use their limbs to walk, run, burrow, jump, swim, court, and fight. But despite this functional diversity, all arthropod limbs represent variations on a few basic designs. Insect limbs are a series of cylinders connected by joints (see Figure 30.12b). Crustacean limbs can be grouped into two general types. They either consist of an insect-like jointed cylinder with a small secondary element attached, or a broad, fleshy pad called a phyllopod (leafy-foot).

Current research on the evolution of arthropod limbs has focused on the question of whether the three limb designs are homologous. Recall from Chapter 21 that homologous structures are similar because they trace their origin back to a common ancestor. Analogous traits, in contrast, are similar in structure but are not derived from a common ancestor; instead the similarities in analogous traits result from convergent evolution.

Are the different types of limbs observed in arthropods homologous, or did they evolve independently? And if they are homologous, what relationship do arthropod limbs have to the simpler appendages found in polychaete worms, onycophoryans, water-bears, and many other animal phyla? Is it possible that all of the different types of animal limbs are homologous—meaning that they are all derived from a simple appendage that existed in a common ancestor?

To answer these questions, Grace Panganiban and colleagues examined the expression of a gene called *Distal-less* in animals with simple appendages as well as in arthropods. *Distal-less*, or *Dll*, is aptly named. In fruit flies that lack the gene's normal product, only the most rudimentary limb buds are formed. The mutant limbs are "distal-less." (Recall that distal means "away from the body.") Because the *Dll* sequence contains the homeodomain introduced in Chapter 19, researchers infer that it is a DNA-binding protein with regulatory functions. Based on the morphology of *Dll* mutants, the protein seems to carry the simple message, "Grow appendage out this way."

Panganiban and co-workers used an antibody to the *Dll* protein to locate the tissues where the gene is expressed in animal embryos. Remarkably, *Dll* is active in tissues that form the simple appendages of phyla with worm-like bodies as well as all leg types found in arthropods (see **Figure 30.13**). Based on this observation, Panganiban and co-workers argue that the various protostome appendages are indeed homologous. The hypothesis here is that *Dll* arose via mutation some 530 million years ago and led to the development of simple limb-like structures in early protostomes. Subsequently, natural selection and the other evolutionary processes introduced in Chapter 22 led to the diversity of limbs observed today.

As T. A. Williams and Lisa Nagy pointed out recently, the challenge now is to understand the genes responsible for the unique features of arthropod limbs—segments, joints, and the muscle groups that power movement. When progress is made on this front, biologists can begin to understand how changes in gene structure or regulation have affected the extensive diversification of arthropod limbs. As the limbs of arthropods diversified over time, so did their methods for moving and feeding.

Insect Metamorphosis

In species numbers and ecological diversification, insects are the dominant lineage on the tree of life. More than 300,000 species of beetles alone have already been described; estimates of the total number of insect species alive today range from 3 to 30 million. Insects have a number of characteristics that appear related to their success. These include a watertight, chitin-rich exoskeleton that allows them to thrive in terrestrial environments; a body with three distinct parts and three pairs of legs; and often, the ability to fly. Here, however, the focus is on just one aspect of their life history, the change from juvenile to adult body type known as **metamorphosis** (change-form).

All insects undergo metamorphosis. (Species from many other animal groups do as well.) This transformation can be subtle or it can be spectacular. For example, a series of molts

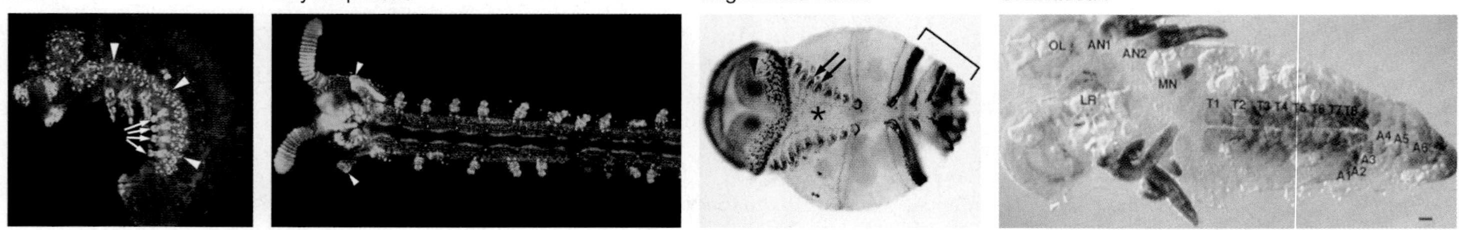

FIGURE 30.13 ***Distal-less*** **Expression in Protostomes**
These embryos have been stained with an antibody to the protein produced by *Dll*. In each case, the *Dll* gene product is localized in cells that will become part of an outgrowth from the body, such as a limb.
QUESTION If the protein produced by *Dll* was not involved in limb development in these species, what would be the predicted result of this experiment?

gradually changes the grasshopper from a wingless, sexually immature juvenile form to a sexually mature adult that is capable of flight (**Figure 30.14a**). Throughout this process, grasshoppers feed on the same food source in the same way—they chew leaves. This type of metamorphosis is called hemimetabolous. Literally translated, the name means half-change; it refers to the limited morphological differences between juveniles and adults. Hemimetabolous development is a one-step process of sexual maturation—from juvenile to adult.

A fruit fly, in contrast, changes from a worm-like larva that burrows through rotting fruit to a flying adult that feeds by lapping up yeast from the surface of fruit (**Figure 30.14b**). This more drastic type of metamorphosis is referred to as holometabolous (*holo* means "whole"). Holometabolous development is a two-step maturation process, from larva to pupa to adult, involving dramatic changes in morphology and habitat use.

The holometabolous transformation is found in 10 times as many species as the hemimetabolous sequence. The leading hypothesis to explain this difference is ecological. Because juveniles and adults of holometabolous species feed on different materials in different ways, they do not compete with each other. In many cases, juveniles and adults even occupy distinct environments. For example, aquatic larvae occur in many species that transform into terrestrial, flying adults; mosquitoes are a familiar example.

It is no exaggeration to say that holometabolous individuals tear one body down and build a completely different one, almost from scratch. How do insects control this drastic change?

In 1922 a biologist named Stefan Kopec showed that metamorphosis is under hormonal control. When Kopec cut the nerve cords of gypsy moth larvae near their brains, the insects metamorphosed normally. But if he removed the brain itself, they could not metamorphose. The result suggested that the insect brain produces a molecule that circulates through the body and triggers transformation.

As a result of follow-up work in a variety of labs around the world, the hormones involved in controlling metamorphosis have now been characterized and their functions worked out. The brain hormone responsible for Kopec's results initiates the production of a steroid hormone called ecdysone. But ecdysone can elicit two very different responses in insects, depending on the concentration of a different molecule called juvenile hormone (JH). If levels of JH are high, the individual responds to

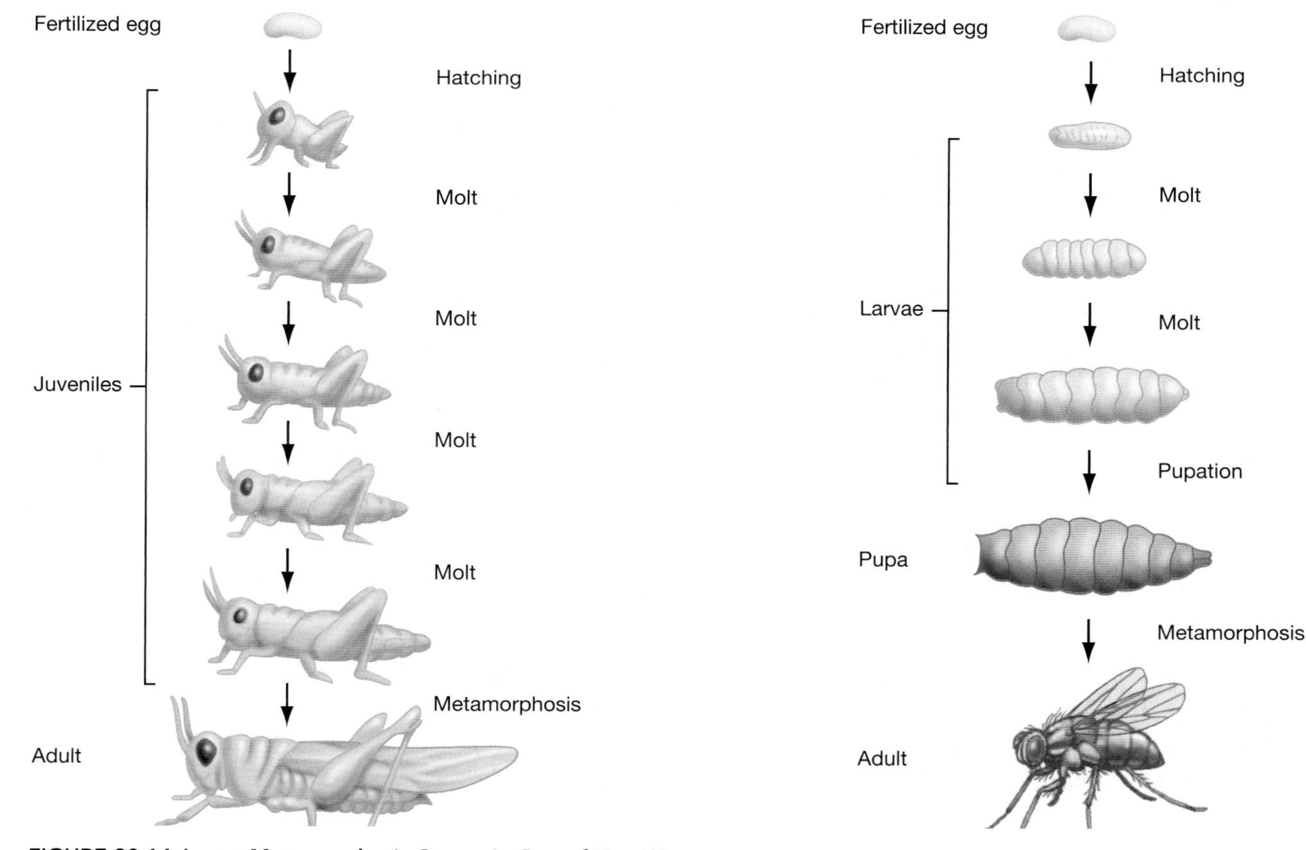

FIGURE 30.14 Insect Metamorphosis Occurs in One of Two Ways
(a) Hemimetabolous development is a one-step process—from juvenile to adult. **(b)** Holometabolous development is a two-step process—from larva to pupa to adult.

FIGURE 30.15 Changes in Gene Expression During Insect Metamorphosis Caused by Ecdysone
The experiment diagrammed here explores how ecdysone regulates the genes that make metamorphosis possible. The graph at the top of step 5 shows how levels of ecdysone change through time from the last larval stage through the formation of the pupa, when metamorphosis is well under way. The histograms show when some of the genes studied were expressed during this interval. The black bars show the level of expression for each gene during each two-hour period, expressed as a percentage of the total transcription observed over the entire time interval studied.

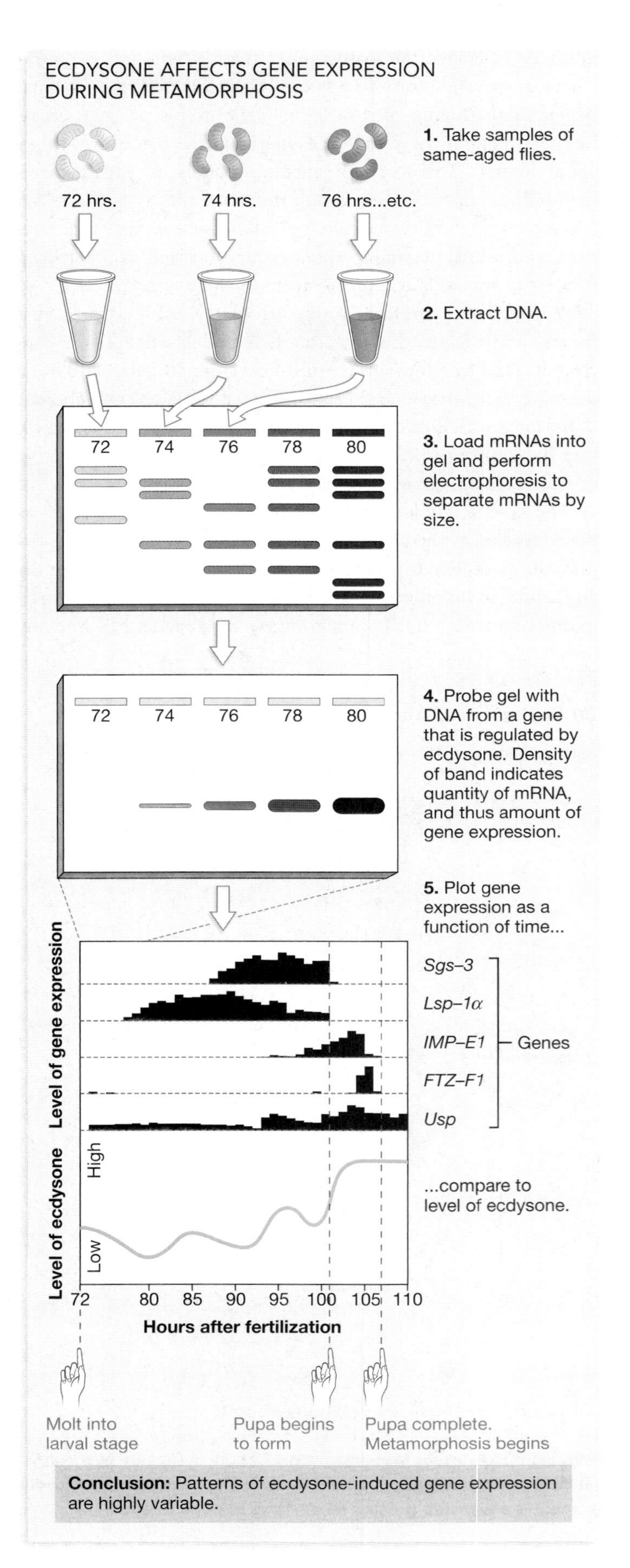

the pulse of ecdysone by molting; if levels of JH are low, the individual responds by initiating metamorphosis.

Like other steroid hormones, including those introduced in Chapter 44, ecdysone affects gene regulation directly. The molecule binds to a receptor inside cells and then to the promoter regions of certain genes. Research on fruit flies has identified over 30 genes whose transcription is affected by ecdysone.

To explore how ecdysone affects gene expression as an insect matures, Andrew Andres and colleagues studied a large cohort of flies that had hatched within an hour of one another. As **Figure 30.15** indicates, the researchers collected a sample of individuals every two hours starting when the larvae were 72 hours old. They purified the mRNAs found in the flies, ran them out on a gel, and then probed the gels with each of the 30 DNA sequences affected by ecdysone. By repeating the sample collections until metamorphosis was well under way and by repeating the complete probing protocol, they learned how ecdysone affected the transcription of each of the 30 genes.

Some of their data are reproduced in Figure 30.15. Note that some genes stop being transcribed when the larvae cease moving and begin secreting the case that encloses the body during metamorphosis. Other loci only begin to be expressed at this stage. Still others are expressed at about the same rate throughout the process. The message of these data? Patterns of ecdysone-induced gene expression are highly variable.

An obvious question is, why do some genes respond differently to the same molecule? Perhaps some of the genes transcribed early in development act as regulators that affect transcription of other genes, along with ecdysone. Perhaps the amount of ecdysone is important. Researchers would also like to know what turns the genes off once ecdysone has turned them on.

Although a great deal of work remains to be done, the data in Figure 30.15 show that ecdysone is not a simple switch that turns larval genes off and adult genes on. Instead, the hormone initiates and manages a complex cascade of gene regulation events. As Chapter 44 will show, this is a common mode of action in animal hormones. Understanding how the products of ecdysone-regulated genes act and interact is currently the focus of intense interest. Stopping ecdysone's effects and halting metamorphosis might be a way to control certain insect pests. But it would also be interesting to understand an adaptation that is central to the success of insects at the level of gene regulation. The advent of metamorphosis, and subsequent changes to its timing and duration, were important in opening up new habitats and food sources for insects to exploit.

30.4 Key Innovations in the Radiation of Vertebrates

The presence of an exoskeleton, jointed limbs, and metamorphosis has allowed arthropods to exploit an extraordinarily wide variety of food sources and habitats. What features have allowed another particularly diverse group of animals—the vertebrates—to flourish?

Let's begin by establishing who the vertebrates are. The Vertebrata is a sub-phylum of the phylum Chordata that includes the fishes, amphibians, lizards, birds, and mammals. These organisms have a unique internal skeleton that is used for attaching muscles and supporting and protecting other tissues. The vertebrate endoskeleton (inside-skeleton) can be constructed of cartilage or bone, and functions much like the exoskeleton of arthropods.

To delve more deeply into the story of vertebrates, this section analyzes five of the key innovations responsible for the origin and diversification of the group: a long, flexible rod running down the back called a **notochord**; the jaw; limbs; a membrane-bound egg; and **endothermy** (inside-heat), the ability to use metabolic energy to maintain a constant body temperature. The notochord gave early chordates the initial component of an internal skeleton; over time, it led to the evolution of the vertebral column. Jaws are the principal tool that vertebrates use to catch and process food; the evolution of limbs allowed members of the group to colonize terrestrial habitats. The evolution of a membrane-bound egg permitted vertebrates to breed away from aquatic habitats, and endothermy made high rates of activity and sustained movement possible.

To appreciate the importance of these traits, consider the phylogeny shown in **Figure 30.16**. This tree was estimated from comparisons of both morphological characters and DNA sequences. It shows the major lineages of vertebrates that have living representatives, along with several prominent groups that are extinct. The points where the five innovations originated are mapped onto the branches.

This analysis of vertebrate diversification begins at the bottom of this tree and works up. Let's take a look at the basic features that define the group.

Understanding the Origins of the Chordates

Species in the phylum Chordata have four distinguishing features: (1) openings into the throat called pharyngeal gill slits; (2) a notochord; (3) a central nervous system characterized by a dorsal hollow nerve cord; and (4) a tail that extends past the anus. To understand why these traits are important, biologists study their function in a variety of chordate groups. To understand how they originated evolutionarily, biologists can compare how these features develop in the most primitive chordates, or identify the genes responsible for them. Let's take a closer look.

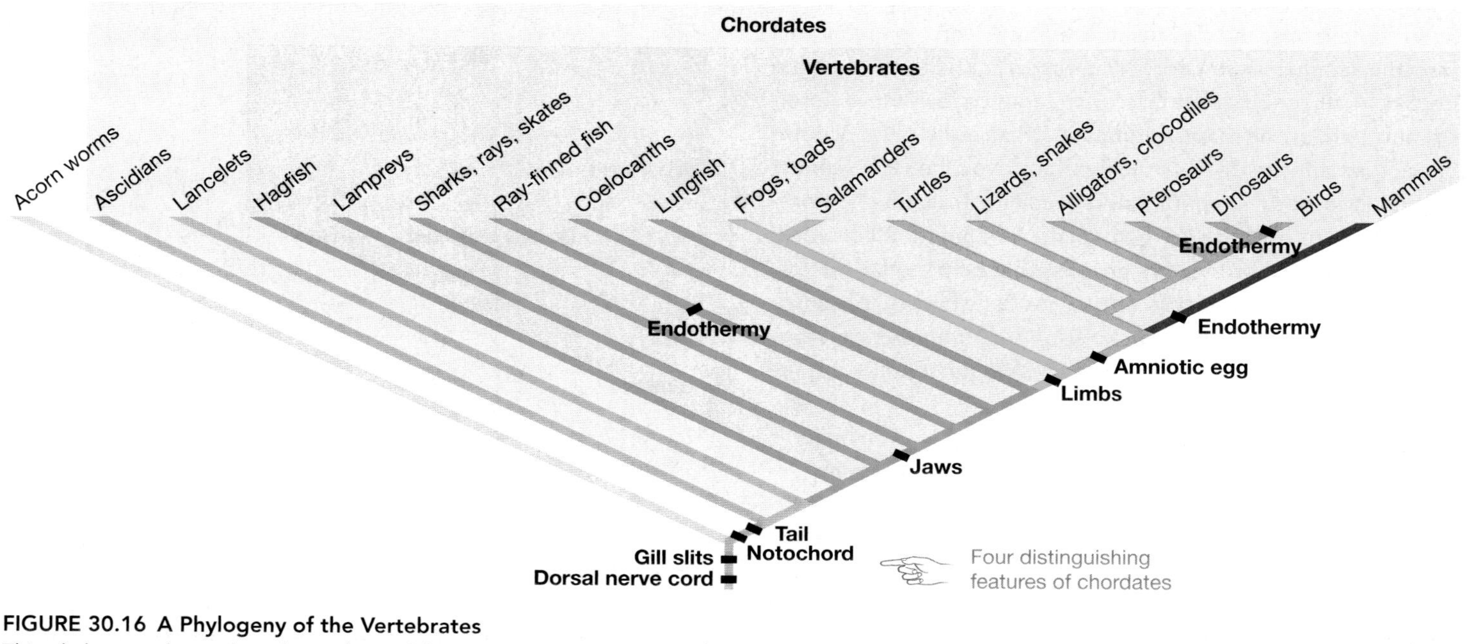

FIGURE 30.16 A Phylogeny of the Vertebrates
This phylogeny shows the relationships of the major groups of vertebrates and their close ancestors: the acorn worms, ascidians, and lancelets. All of the animals included here are chordates, except for the acorn worms. The short bars indicate where major innovations occurred as the lineage diversified. EXERCISE Circle the taxa on the tree that qualify as tetrapods, or limbed vertebrates. Name the groups that are color-coded green, orange, and red.

Comparative Approaches To analyze how a structure originates and changes through time, biologists compare what it looks like in different species. **Figure 30.17** illustrates the pharyngeal gill slits, notochords, dorsal hollow nerve cords, and tails found in vertebrates and the related lineages called acorn worms, ascidians, and lancelets. Note that the four groups are ordered from top to bottom to reflect the evolutionary sequence indicated by the phylogeny of chordates.

What conclusions can be drawn from comparing these organisms? First, acorn worms do not have a notochord, so they are not chordates. They do have a dorsal nerve cord, however, as well as pharyngeal gill slits that function in feeding and respiration. Water enters the mouths of these animals, flows through structures where oxygen and food particles are extracted, and exits through the gill slits.

Like acorn worms, adult ascidians are sessile filter feeders that live in the ocean. Pharyngeal gill slits are present in both the larvae and the adults; the figure shows how water moves through the slits as adults harvest oxygen and food. A notochord, dorsal hollow nerve cord, and tail are present only in larvae. Because the notochord stiffens the tail, muscular contractions on either side wag it back and forth and result in forward movement. Larvae are specialized for floating in the upper water layers of the ocean; there they drift to new habitats where food might be more abundant.

Adult lancelets filter feed with the aid of their pharyngeal gill slits, much like ascidian larvae. They also have a notochord that stiffens their bodies, so that muscle contractions on either side result in fish-like movement. The fossil record indicates that these features evolved over 505 million years ago.

In vertebrates, the dorsal hollow nerve cord is elaborated into the familiar spinal cord. Pharyngeal gill slits or pouches appear in all vertebrate embryos; in aquatic species they develop into part of the main respiratory organ—the gills. A notochord also appears in all vertebrate embryos, but it no longer functions in body support and movement. Instead, the notochord helps to organize the body plan. As Chapter 20 detailed, cells in the notochord secrete proteins that help induce the formation of segmented blocks of tissue called somites. Although the notochord itself disappears, cells in the somites later differentiate into the vertebrae, ribs, and skeletal muscles of the back, body wall, and limbs.

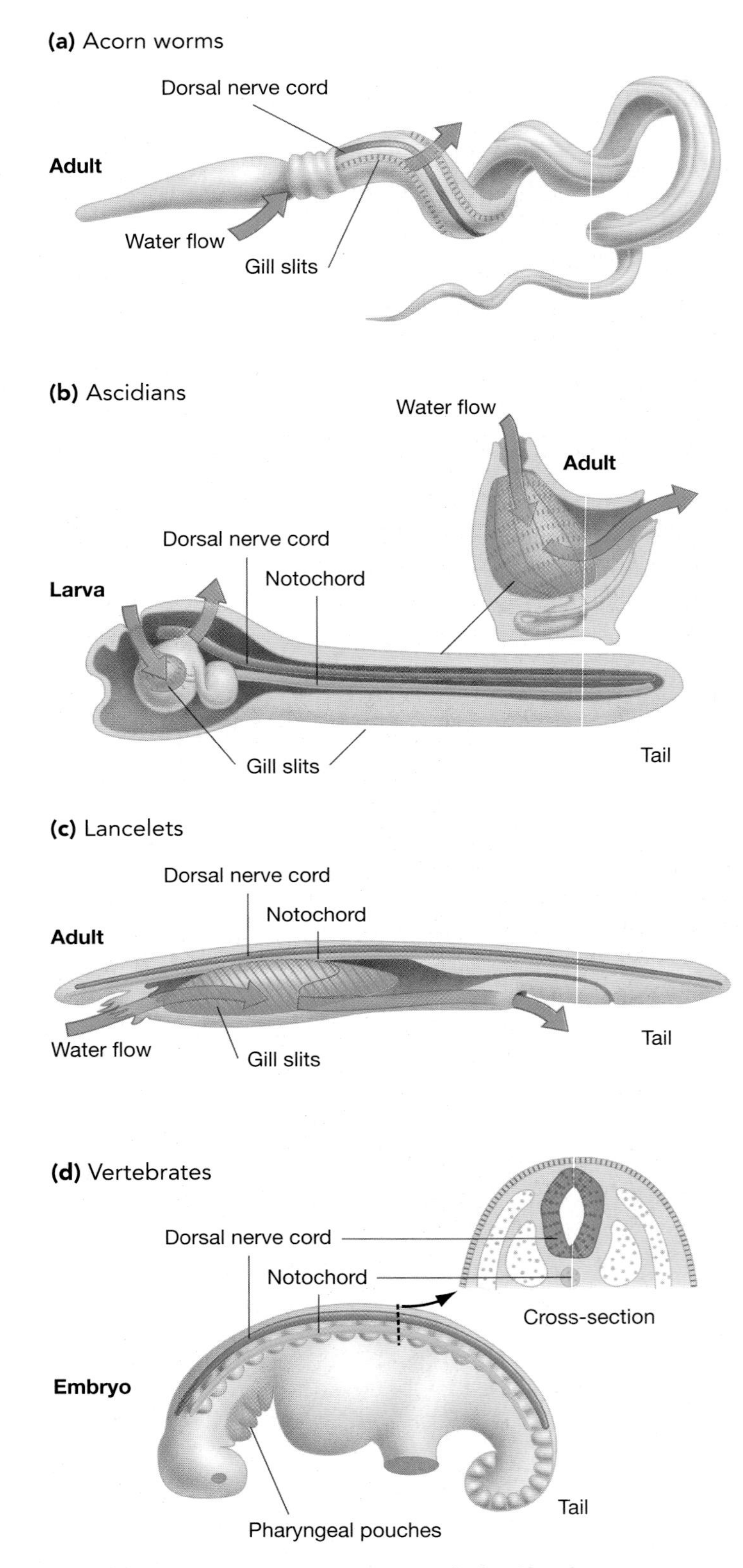

FIGURE 30.17 Four Features Distinguish the Chordates
These drawings illustrate the four distinguishing features of chordates in the three major chordate lineages (the ascidians, lancelets, and vertebrates). Two of the features, a dorsal nerve cord and gill slits, also appear in the acorn worms, which are not chordates.

Data from Molecular Genetics The comparative data just reviewed suggest that the defining features of vertebrates originated as structures that helped worm-like animals feed and move. The function of these traits at the time they first appeared seems clear. Can experiments illuminate how these structures originated in terms of the genetic mechanisms involved?

Recent work by Billie Swalla, William Jeffery, and colleagues suggests that the answer is yes. These researchers have explored how the dorsal hollow nerve cord, notochord, and tail form in ascidian larvae. Their work was inspired by the observation that the larvae of some ascidian species *lack* these structures. Larvae that have tails can swim; larvae that lack them simply float in ocean currents.

Swalla and co-workers hunted for genes that might be responsible for tail formation. To search for these loci, the investigators isolated mRNAs from the embryos of tailed and tailless species. Using a protocol introduced in Chapter 17, they used these mRNAs to make a complementary DNA (cDNA) library. When they compared these libraries, they found several sequences unique to the tailed species. Their attention quickly focused on a gene they christened *Manx*, after a breed of tailless cat. When they performed in situ hybridizations with *Manx* mRNA, they found that the transcripts localized to the nerve cord, notochord, and tail muscles.

To confirm that *Manx* is involved in the formation of tail structures, Swalla and Jeffery disabled *Manx* mRNA. They did this by treating embryos of the tailed species with single-stranded DNA that had a sequence complementary to *Manx* mRNA. Their logic was that this "anti-sense DNA" would bind to *Manx* mRNA and prevent it from functioning. As predicted, the resulting embryos were tailless. This result establishes that *Manx* is involved in the formation of the ascidian tail, and suggests that the origin of *Manx* might correlate with the origin of a signature chordate trait. The research team is now trying to ascertain how *Manx* functions, and whether it is also found in acorn worms and lancelets.

Jaws

The first fish in the fossil record date from the middle part of the Cambrian period, about 530 million years ago. Although the interpretation of these fossils is controversial, the animals appear to have had a skull and gill skeleton made of cartilage. The next good-quality fossil fish come from the early part of the Ordovician period, about 480 million years ago. The small, marine chordates in these deposits have a novel feature—the calcium-phosphate-rich material called bone. Rather than building an internal skeleton, however, bone was deposited in scale-like plates to form an exoskeleton. Based on morphology, biologists infer that these animals swam with the aid of a notochord and both breathed and fed by gulping water and filtering it through their pharyngeal gill slits. Presumably, the bony plates helped provide protection from predators.

Another group of fish that lack vertebrae is still living. These are the burrow-dwelling marine animals called hagfish (**Figure 30.18a**). Hagfish feed by gulping and swallowing material from the carcasses of dead animals. A group that is closely related to hagfish, called lampreys, do have vertebrae as well as a persistent notochord. Lamprey larvae burrow into soft mud and filter feed, while adult lampreys are ectoparasites. As **Figure 30.18b** shows, they latch onto fish, whales, or porpoises and feed by sucking their blood and tissues.

The thread that binds these animals together is their mode of feeding. The Ordovician fish, hagfish, and lampreys all obtain their nutrients by gulping, sucking, or rasping flesh away with a barbed tongue. They are incapable of an enormously important innovation in the radiation of vertebrates: biting.

Capturing prey by biting became possible when jaws evolved. Where did jaws come from? The leading hypothesis is that they resulted from modifications in structures called gill arches. The jawless vertebrates have bars of cartilage that stiffen the tissue between their gills. The gill-support hypothesis proposes that mutation and natural selection increased the size of the first arch and modified its orientation slightly, producing the first working jaw (**Figure 30.19**, page 594).

Three lines of evidence, drawn from comparative anatomy and embryology, suggest that the gill-support hypothesis is correct. Both gill supports and jaws consist of flattened bars of bony or cartilaginous tissue that hinge and bend forward. Unlike most other parts of the vertebrate skeleton, jaws and gill supports are derived from specialized embryonic cells called neural crest cells. The muscles that move the two structures also derive from an identical population of embryonic cells.

(a) Hagfish

(b) Lamprey feeding on fish

FIGURE 30.18 Jawless Vertebrates
(a) Hagfish burrow in sediments and feed by gulping material from carcasses. **(b)** Adult lampreys have hook-like barbs around their mouths. They attach themselves to other vertebrates, like this carp, and suck nutrients.

Taken together, these data support the hypothesis that gill supports and jaws are homologous—meaning that they are derived from the same ancestral structure.

The fossil record shows that teeth appeared soon after the advent of the jaw. Thanks to these structures, vertebrates became armed and dangerous. The fossil record shows that a spectacular radiation of jawed fish followed, filling marine and freshwater habitats. What was the next great event in the diversification of vertebrates? The answer is moving to land.

Limbs

For vertebrates to succeed on land, they had to evolve the ability to breathe, move, and avoid drying out. To understand how this was accomplished, consider the morphology and behavior of their closest living relatives, the lungfish. Most living species of lungfish inhabit shallow, oxygen-poor water. To supplement the oxygen taken in by their gills, they have lungs and breathe air. Some also have fleshy fins supported by bones and are capable of walking along mudflats or the bottoms of ponds. Some species can tolerate intermittent dry periods, as well.

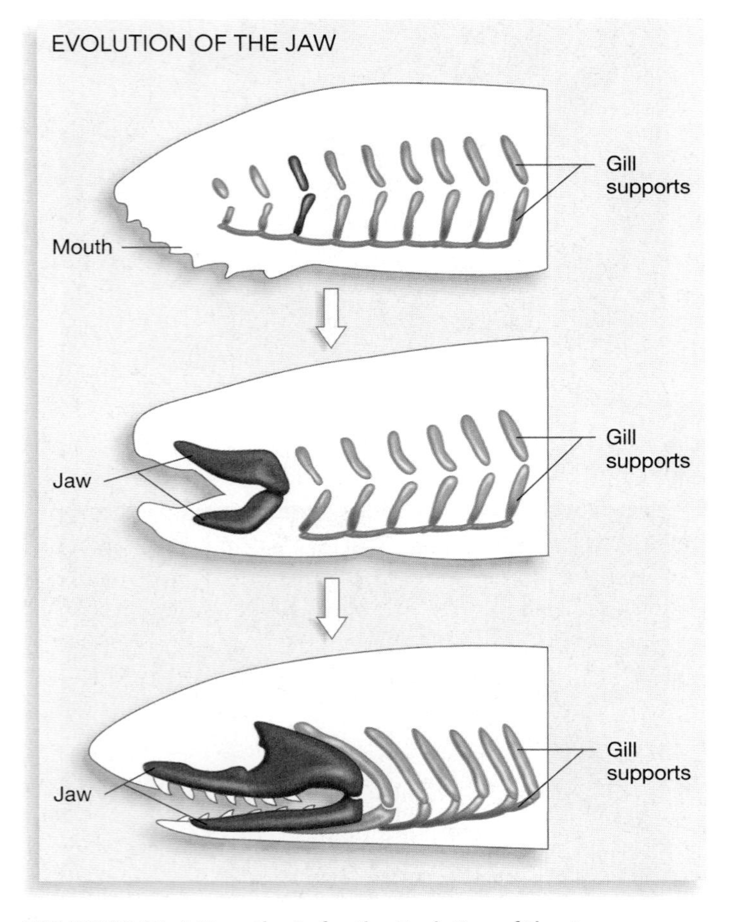

FIGURE 30.19 A Hypothesis for the Evolution of the Jaw
Gill supports are found in jawless vertebrates **(top)**. In the fossil record, jawbones appeared first in fossil sharks **(bottom)**. The leading hypothesis for the evolution of the jaw suggests that the structure evolved from gill supports by intermediate stages **(middle)**.

The ancestors of these fish left fossils that provide strong links to the earliest land-dwelling vertebrates. **Figure 30.20** shows a representative species of fish from the Devonian period about 375 million years ago along with one of the oldest tetrapods, or limbed vertebrates, found to date. This figure compares the arrangement of bones in the fossil fish limb with a generalized version of the ancient tetrapod (four-foot) limb. Because the number and arrangement of bones are so similar, the evidence for homology is strong. Based on the lifestyle of living lungfish, biologists suggest that mutation and natural selection gradually transformed fins into limbs as the first tetrapods became more and more dependent on terrestrial habitats.

Is there any molecular genetic evidence to support the hypothesis of a fin-to-limb transition? To answer this question, recall the experiments on fish fins and mouse limbs reviewed in Chapter 24. These studies showed that several regulatory proteins involved in pattern formation of zebra-fish fins and the upper parts of mouse limbs are homologous. Specifically, the proteins produced by *Hox* genes and the homeotic locus called *Sonic hedgehog* (*Shh*) are found at the same times and in the same locations in fins and limbs. The result suggests that the appendages are patterned by the same genes, and supports the hypothesis that tetrapod limbs evolved from fins.

The Amniotic Egg and Endothermy

After the tetrapods emerged, two other great innovations occurred. The first was the **amniotic egg**. This membrane-bound egg functions well in terrestrial environments (**Figure 30.21**). The first tetrapods, like today's amphibians, lacked amniotic eggs and had to return to aquatic habitats to breed. This requirement limited the range of habitats that these animals could exploit. But because amniotic eggs contain a membrane-bound supply of water and are encased in a leathery or hard shell, they are resistant to drying. Turtles, snakes, lizards, crocodiles, birds, and the egg-laying mammals all produce amniotic eggs.

The other great innovation that occurred during the radiation of vertebrates is endothermy (inside-heat). Endothermy is the ability to maintain a high body temperature using heat supplied by the oxidation of food. Endothermy allows individuals to maintain high levels of activity in cold habitats. Because it requires rapid oxygen intake and use, it also facilitates the high levels of aerobic metabolism needed for sustained running, flying, or rapid swimming. Chapter 38 explores how endotherms go about manufacturing heat and regulating their body temperature.

These features are in sharp contrast to ectothermic (outside-heat) animals, which regulate their body temperature using energy from the environment—for example, by basking in sunlight. Ectotherms are usually inactive in cold weather, and are less capable than endotherms of sustaining rapid movement for long periods.

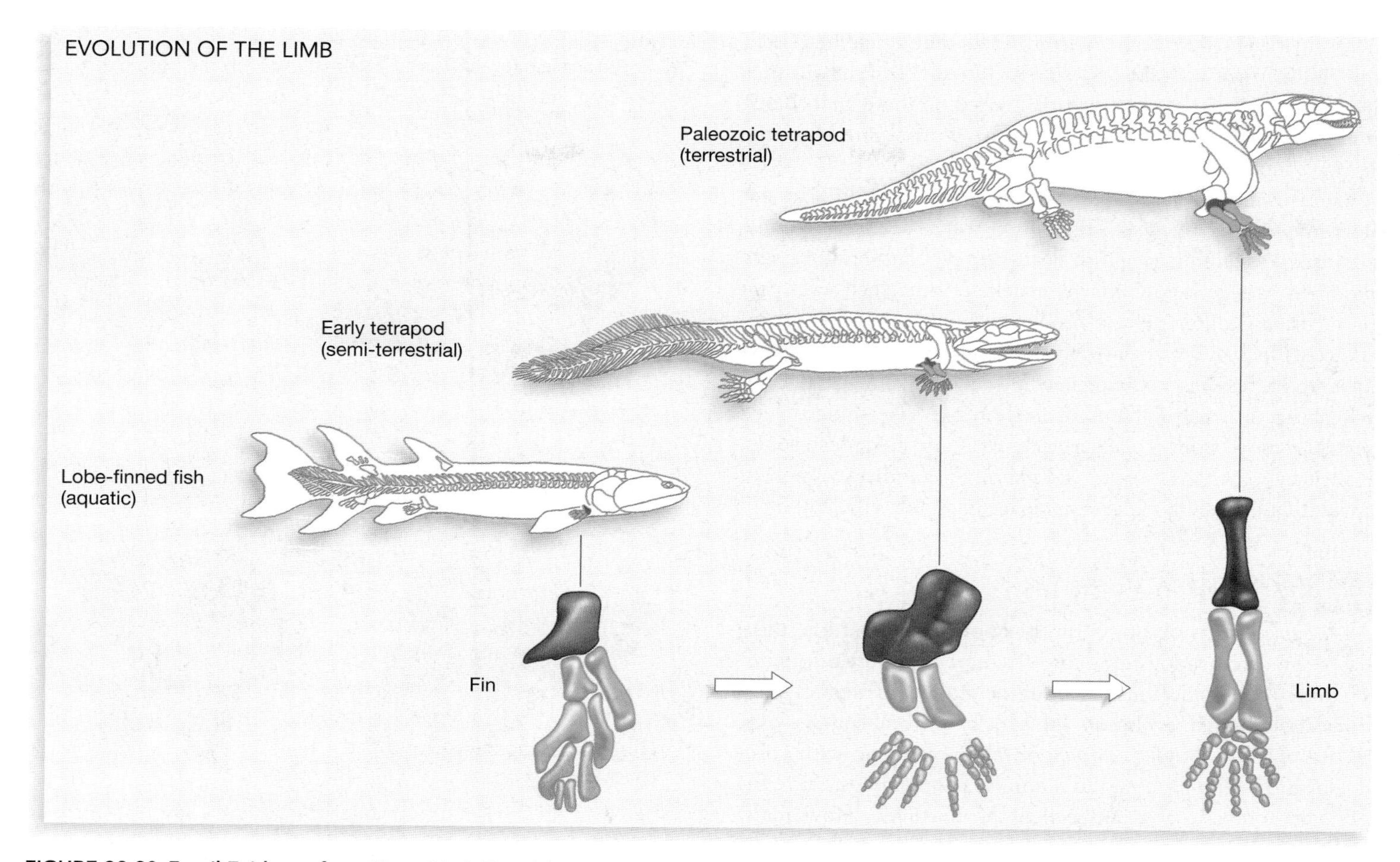

FIGURE 30.20 Fossil Evidence for a Fin-to-Limb Transition
The fossil fish and the early tetrapod are both from the Devonian period, about 375 million years ago. The number and arrangement of bones in the fins and limbs of these two fossil organisms agree with the general form of the modern tetrapod limb. The color coding indicates homologous elements.

The cost of endothermy is equally clear, however. An enormous amount of food has to be oxidized to maintain the high body temperatures observed in endotherms. Based on this observation, it is not surprising that a measure of oxygen consumption called the resting metabolic rate is 7 to 10 times higher in birds and mammals than it is in reptiles. To fuel this activity, endotherms must obtain huge amounts of food. Like the tortoise and hare of Aesop's fable, ectotherms and endotherms represent contrasting life strategies. Energetically, endotherms live in the fast lane. If you've ever owned a pet gerbil or mouse at the same time as an equivalent-sized goldfish or lizard, you'll appreciate the difference in their caloric demands.

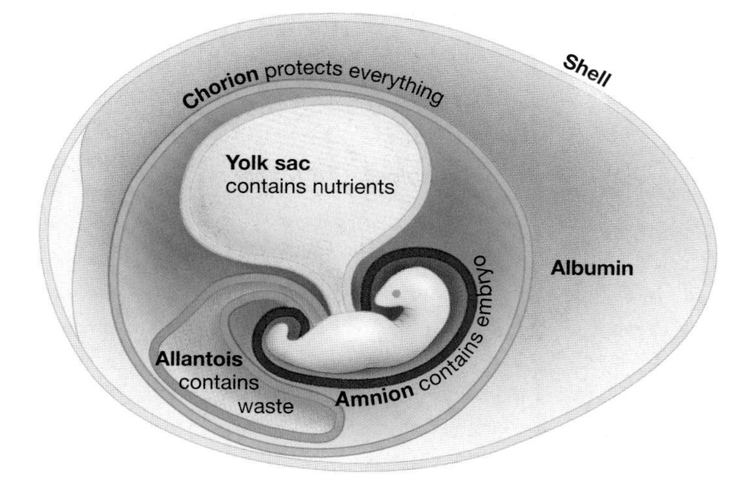

FIGURE 30.21 An Amniotic Egg
Amniotic eggs have membrane-bound sacs that hold nutrients (yolk), water (albumin), waste, and fluid that bathes the embryo.

Barbara Block and colleagues recently asked two fundamental questions about endothermy. How many times has the trait evolved, and why? To find answers, the researchers considered the distribution of endothermy in fish. Among the ray-finned fish, or teleosts, the trait is found in tuna, mackerel, and billfish. Tuna heat muscles in the core of their body that are used for sustained swimming. Billfishes and mackerel, in contrast, heat only their eyes and brain.

When did these types of endothermy originate? To answer this question, the researchers sequenced a gene found in the mitochondrial DNA of endothermic and ectothermic fish species and used the data to estimate their evolutionary relationships.

The phylogeny that resulted confirmed that tuna, mackerel, and billfish are not closely related. Instead, each group has close relatives that are ectothermic. Based on these data, Block and colleagues suggest that endothermy evolved in each of the lineages independently.

The result correlates nicely with the observation that endothermy occurs in three distinct ways: Tunas heat certain muscles used in swimming, billfish heat their eyes and brain using a modified muscle, and butterfly mackerel heat their eyes and brain using a different heat-producing muscle. In all three instances, Block and associates hypothesize that endothermy was an adaptive response to living in cold water. Tuna, mackerel, and billfish live in frigid habitats or feed on prey like squid, which regularly move from warm surface waters into cold, deep zones. If their hypothesis is correct, endothermy is another example of a trait that evolved because it made feeding more efficient.

30.5 Human Evolution

Although humans occupy a tiny twig on the tree of life, there has been a tremendous amount of research on human origins. This section considers just a few of the most fundamental questions about human evolution. It begins by introducing the fossil record of *Homo sapiens* and close relatives. Along with gorillas, pygmy chimpanzees, and common chimpanzees, these organisms comprise a group called the **hominids.** How many other human-like hominids were there, and when and where did they live? When did our species, *Homo sapiens*, appear?

The Hominid Radiation

The fossil record of human-like hominids, though not nearly as complete as investigators would like, is rapidly improving. Almost yearly, scientists discover new fossils that inform the debate about human ancestry. Although naming the extinct species and interpreting their characteristics continues to be intensely controversial, most researchers agree that they can be organized into the four general groups listed here and illustrated in **Figure 30.22.**

1. Three species of small apes called gracile australopithecines have been identified. Radiometric dating analyses like those introduced in Chapter 2 indicate that these species lived from 4.1 to about 2.4 million years ago. Because the hole in the back of their skulls where the spinal cord connects to the brain is oriented downward, just as it is in *Homo sapiens*, researchers infer that these species were bipedal (two-footed). In chimps, gorillas, and other vertebrates that walk on four feet, this hole is oriented backward.

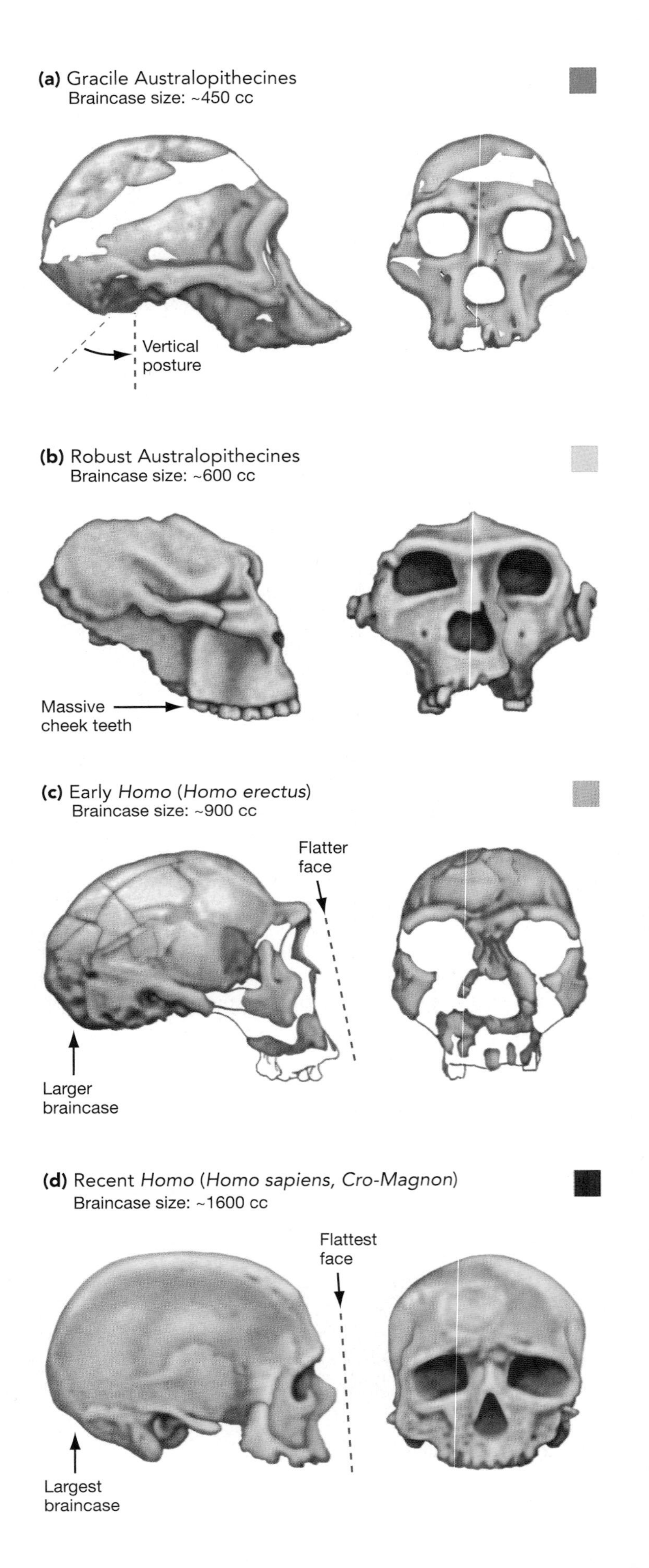

FIGURE 30.22 African Hominids Comprise Four Major Groups
These skulls are from species that represent the four major groups of hominids. **QUESTION** The skulls are arranged as they appear in the fossil record, from most ancient to most recent (top to bottom). How did the forehead and brow ridge of human-like hominids change through time?

2. Most investigators agree that three distinct species of robust australopithecines existed, and that they were also bipedal. These fossils date from 2.7 to 1.0 million years ago. The skull in Figure 30.22b illustrates the key difference between the gracile (slender) and robust species in the genus *Australopithecus*. The robust forms had much more massive cheek teeth and jaws, which may have been used in chewing hard seeds. These traits earned one of the robust species the nickname "nutcracker man."
3. The earliest species in the genus *Homo* date from 2.4 to 1.5 million years ago. The skull in Figure 30.22c highlights several important distinctions between *Australopithecus* (southern-ape) and *Homo* (human). *Homo* species have flatter, narrower faces and smaller jaws, and have braincases that are as much as three times larger than those of the australopithecines.
4. More recent species of *Homo* date from 1.2 million years ago to the present. The skull illustrated in Figure 30.22d is from a population of *Homo sapiens* called the Cro-Magnons. The fossil is dated to about 30,000 years before present and has a braincase that is substantially larger than those found in most humans today. Fossil skulls from another species of recent human, *H. neanderthalensis*, also have braincases that are larger, on average, than those of modern humans. Although Neanderthal skulls are easily distinguished from modern humans by a heavy ridge of bone along the eyebrow, Cro-Magnon skulls are indistinguishable from those of contemporary humans.

How are these hominids related? **Figure 30.23a** shows the distribution through time of some important species from each of the four groups; **Figure 30.23b** shows an estimate of their evolutionary relationships. The phylogeny was constructed by David Straight and colleagues and is based on comparisons of skull and tooth characteristics.

Although the chronology and evolutionary relationships presented here are almost certain to change as the fossil record continues to improve, one conclusion about hominid evolution is clear. Humans did not evolve through a simple, steady progression from a chimpanzee-like ancestor. Instead, a complex radiation of bipedal hominids occurred in Africa during the past 4–5 million years. At one time there may have been as many as five hominid species living in eastern and southern Africa. Most of these lineages went extinct without leaving descendant species. In short, the fossil record shows that *Homo sapiens* is the sole survivor of a remarkable adaptive radiation. How did our species originate?

The Out-of-Africa Hypothesis

The first fossils of *Homo sapiens* appear in African rocks that date to about 130,000 years ago. For perhaps 80,000 years thereafter, our species occupied Africa while other recent members of the genus resided elsewhere—*Homo erectus* in Asia and *H. neanderthalensis* in Europe and the Middle East. Then, in rocks dated between 50,000 and 30,000 years ago, *H. sapiens* fossils are found outside of Africa. Soon thereafter, *H. sapiens* are found throughout the Old World and Australia. But both *H. neanderthalensis* and *H. erectus* have disappeared by this time (see **Figure 30.24a**, page 598).

Phylogenies of *Homo sapiens* that have been estimated from DNA sequence data tend to agree with the pattern in the fossil record, because the trees imply that our species originated in

(a) Ages of fossil hominids

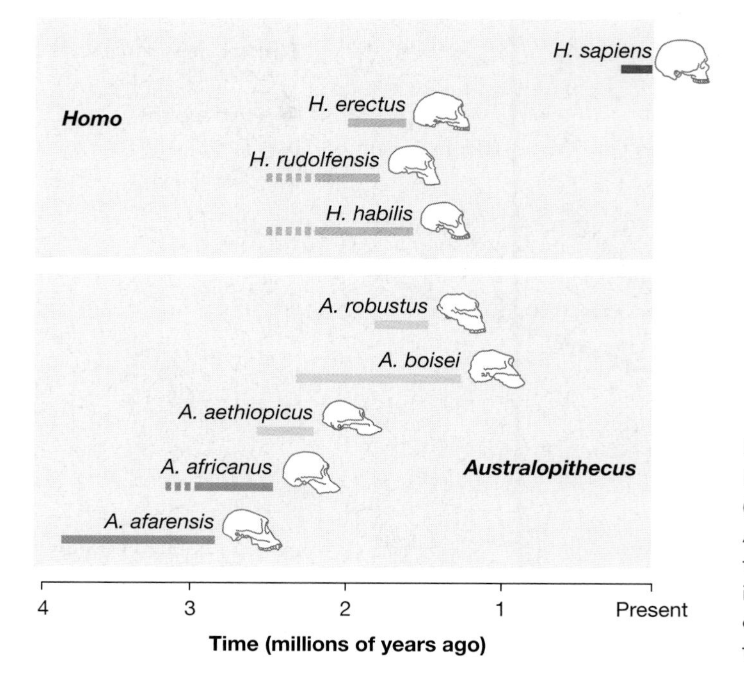

(b) Phylogeny of hominids

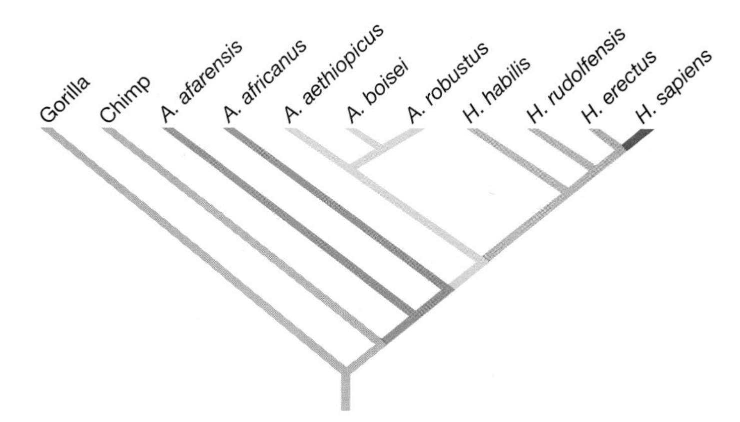

FIGURE 30.23 What Are the Relationships of *Homo sapiens* and the Extinct Hominids?
(a) This chart plots the ages of fossils from human-like hominids from Africa. The solid lines give the range of ages in the fossil record currently; the dashed lines indicate dates when investigators suspect the species existed. **(b)** This phylogeny suggests that the robust australopithecines went extinct without leaving any descendant species. If this tree is correct, then the robust australopithecines represent an evolutionary "dead end."

Africa and then spread throughout the world. Consider, for example, the phylogeny in **Figure 30.24b**. A team of researchers led by Luca Cavalli-Sforza estimated these relationships based on variation in 30 microsatellite loci (see Chapter 16). The study included people representing 14 geographically distinct populations. The pointer at the base of the tree indicates a population that contains the common ancestors of all humans living today.

Because the first lineage to branch off leads to descendant populations that live in Africa today, it is logical to infer that the ancestral population also lived in Africa. The tree shows that lineages subsequently branched off leading to populations occupying Europe, then Asia, the Americas, and the South Pacific.

What happened to the Neanderthals and to *Homo erectus* as *H. sapiens* expanded its range? This simple question has provoked years of heated controversy. Currently, the debate boils down to a dichotomy. Either *H. sapiens* interbred with the other two hominids as it moved into Europe and Asia, or it did not. The first possibility is called the assimilation hypothesis. It implies that the genetic composition and morphological features of *H. sapiens* are an amalgam of ancient traits from Neanderthals and *H. erectus* and recent traits from *H. sapiens*. The second possibility is called the out-of-Africa hypothesis. It contends that *H. sapiens* evolved independently of the European and Asian species of *Homo*—that there was no interbreeding between *H. sapiens* and Neanderthals or *H. erectus*.

As Chapter 12 explained, a team of researchers led by Svante Pääbo recently tested the two hypotheses by extracting DNA from the fossilized bone of a Neanderthal. The bone fragment they used was taken from the first *H. neanderthalensis* fossil ever discovered. The researchers took painstaking measures to make sure that the sample was not contaminated by their own gene sequences as they ground the bone, extracted DNA, and used the polymerase chain reaction to copy a 379-base-pair section of the mitochondrial genome. When they sequenced the Neanderthal DNA and compared it to sequences from 986 humans living today, the results were striking. The Neanderthal DNA was extremely different. When they compared sequences from any two randomly chosen *H. sapiens* to each other, they found an average of 8 bases that were unalike. But when they compared randomly chosen *H. sapiens* sequences to the *H. neanderthalensis* DNA, they found an average of over 25 differences. Further, differences observed

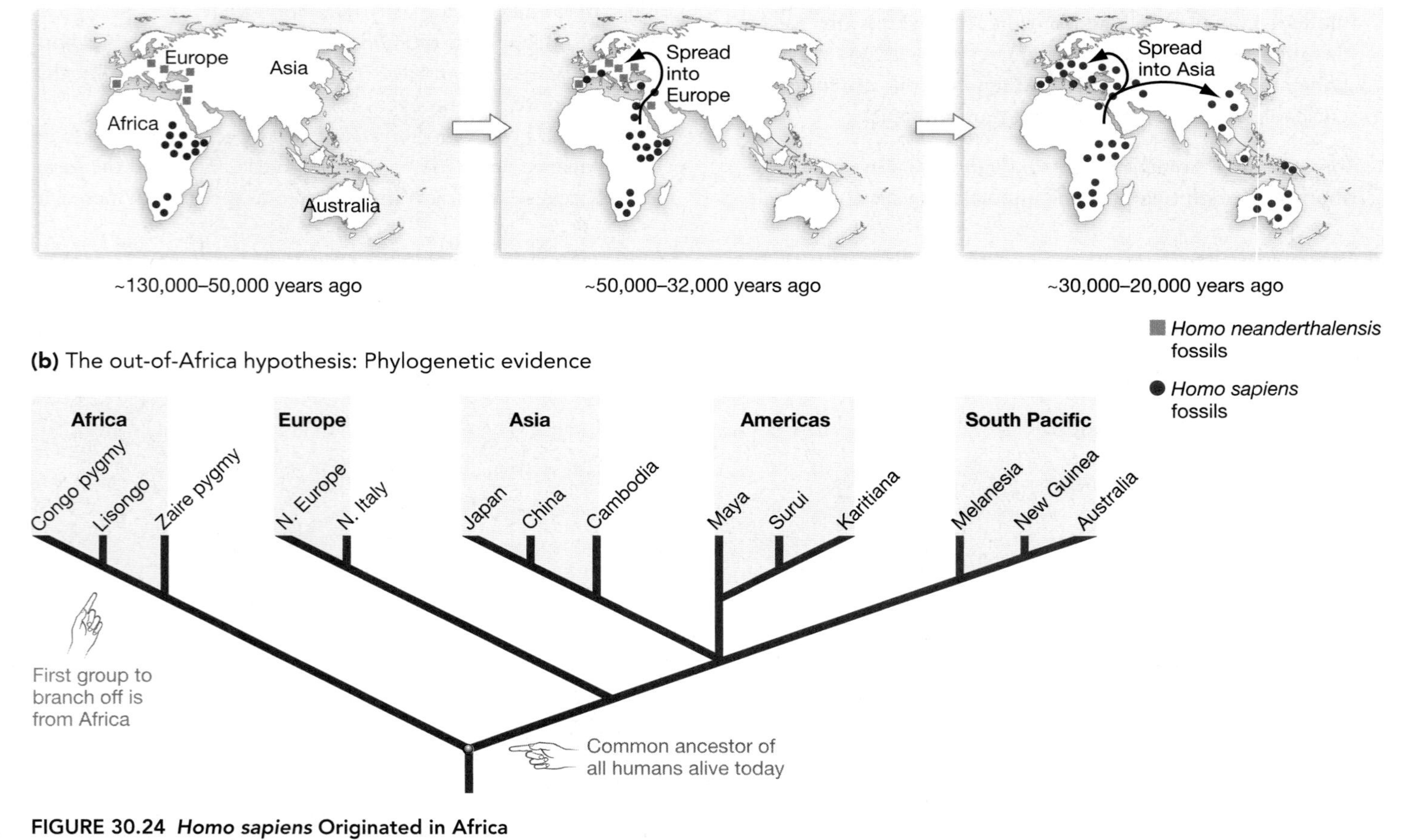

FIGURE 30.24 ***Homo sapiens*** **Originated in Africa**
The phylogeny of living humans is based on comparisons of DNA sequences.

in the Neanderthal DNA were unique; none of the *H. sapiens* sequences contained the nucleotide substitutions found in *H. neanderthalensis*. This result supports the hypothesis that *H. sapiens* and *H. neanderthalensis* did not interbreed.

Unfortunately, it has been impossible so far to extract DNA from *H. erectus* fossils and perform the same test. Although the weight of evidence currently tips the scales in favor of the out-of-Africa hypothesis, research continues.

Essay So Human an Animal

From a biological point of view, what is a human being? Historically, scientists and philosophers have answered this question by arguing that humans have one or more unique, defining characteristics that set us apart from other organisms. For decades this defining characteristic was thought to be tool use. But then Jane Goodall, who pioneered the study of common chimpanzees in the field, observed that chimps collect and modify twigs to "fish" for termites or ants and eat them (**Figure 1**). Later, Goodall and other biologists reported additional examples of tool use by common chimps, pygmy chimps, baboons, sea otters, and a wide variety of birds.

In response, some observers began to argue that the defining characteristic of humans was not toolmaking, but the use of language based on the abstract symbols we call words and letters. But researchers have now been able to teach chimps rudimentary aspects of the American Sign Language. This result has an important message: The ability to learn and to use abstract symbols and grammatical rules is not unique to humans.

Which 1 to 2 percent of the human genome is responsible for the differences between chimps and humans?

In short, there may be no single characteristic that "defines" humans. Instead, most contemporary biologists would characterize our species by listing a suite of traits that are not necessarily unique. Humans are an intensely social, bipedal animal with a brain that is absolutely huge given our body size. A biologist would also point out that humans have a particularly long period of sexual immaturity, or childhood, accompanied by extraordinarily high levels of parental care. Finally, most biologists would emphasize that the changes in human morphology and behavior that have taken place over the past 40,000 years have probably been dominated by cultural evolution—meaning changes fashioned by teaching and learning—rather than evolution by natural selection.

Recently, research in molecular genetics has furnished a new perspective on the question, who are humans? When Svante Pääbo and co-workers sequenced a 10,156-base-pair segment from the X chromosomes of humans and common chimpanzees, they found that the sequences were identical at almost 99 percent of the 10,156 sites.

These findings prompt a question. Which 1% to 2% of the human genome is responsible for the morphological and behavioral differences between chimps and humans? Will it be possible to identify the alleles responsible for our bipedal posture, large brains, slow rate of sexual maturation, and huge capacity for learning? Researchers have only begun to tackle these questions. But, with a variety of genome analysis projects now under way, answers may soon be forthcoming.

FIGURE 1
The adult female chimp in the crotch of the tree is using a stick as a tool to "fish" for ants. Her son is sitting on her back and is also using a stick. Tool-use is a culturally transmitted behavior in chimpanzees, meaning that it is learned. In this case, a mother is teaching her offspring.

Chapter Review

Summary

The animals consist of over 30 major lineages, or phyla. Each animal phylum represents a group of species with a distinctive body plan. The wide diversity of animal body plans results from variation in body symmetry, the number of embryonic tissue layers, the sequence of events during early development, and the type of body cavity.

By assuming that simpler body plans evolved earlier than more complex forms, biologists have been able to estimate the phylogeny of the animals. According to this tree, sponges and radially symmetric groups like ctenophores and coelenterates were the first groups to evolve, and segmented body plans evolved independently in some protostome and deuterostome lineages. These conclusions are supported by phylogenies estimated from molecular data.

A wide variety of feeding strategies occur among animals. Suspension feeders filter organic material or small organisms from water; deposit feeders swallow soils and digest the food particles they contain; herbivores use complex mouthparts to bite, suck, or rasp away plant tissues; predators kill prey using sit-and-wait or stalking strategies; parasites can live inside or outside of their hosts.

The arthropods and the vertebrates are particularly important animal lineages because of their species numbers and ecological and morphological diversity. Their adaptive radiations are associated with a series of important innovations:

- *Arthropod limbs are used for foraging, fighting, and moving.* These appendages develop in response to some of the same regulatory genes, like *Distal-less*, that control the growth of the simple appendages found in phyla with worm-like bodies.
- *Insect metamorphosis allows juveniles and adults of the same species to exploit different food sources.* The genes responsible for this transformation are expressed in complicated patterns in response to the steroid hormone ecdysone.
- *A notochord is a long, flexible rod that aids the swimming movements observed in close relatives of the vertebrates.* The development of the notochord is controlled in part by a regulatory gene called *Manx*.
- *The evolution of jaws allowed early fish to capture food by biting.* Morphological and developmental evidence suggests that jaws originated as modified gill arches.
- *Vertebrate limbs are based on a common number and arrangement of bone elements.* Data from comparative anatomy and genetics suggest that tetrapod limbs evolved from the fins of fish.
- *Endothermy is the ability to maintain a high body temperature from the heat generated by aerobic metabolism.* Endothermy arose independently in a variety of vertebrate lineages and is associated with exploitation of cold habitas or sustained, rapid movement.

The fossil record contains a variety of hominids. Several of these species lived in Africa at the same time, and some lineages went extinct without leaving descendant populations. Thus, *Homo sapiens* is the sole surviving representative of an extensive radiation. The phylogeny of living humans, based on comparisons of DNA sequences, agrees with evidence in the fossil record that *H. sapiens* originated in Africa and later spread throughout Europe, Asia, and the New World. DNA sequences recovered from the fossilized bones of *H. neanderthalensis* suggest that *H. sapiens* replaced this species in Europe without interbreeding.

Questions

Content Review

1. Which of the following is true of *all* bilaterally symmetric animals?
   a. They are triploblastic, meaning that they have three embryonic tissue layers.
   b. They have coeloms.
   c. They exhibit the protostome pattern of development.
   d. They exhibit the deuterostome pattern of development.

2. According to phylogenies based on comparisons of morphology, development, and gene sequences, animals evolved in which of the following patterns?
   a. from suspension feeders to deposit feeders, then herbivores, predators, and parasites
   b. from complex body plans to simple, efficient ones
   c. from simple body plans to more complex
   d. first on land, then in aquatic habitats

3. Why do some researchers maintain that the limbs of all animals are homologous?
   a. Homologous genes, like *Dll*, are involved in their development.
   b. The number and arrangement of elements in all animal limbs is the same.
   c. They all function in the same way.
   d. They all develop at the same rate.

4. Endothermy is the ability to use the heat generated by the oxidation of sugars to maintain a high body temperature. What does it allow individuals to do?
   a. sustain high levels of activity like flying or rapid swimming
   b. function in cold habitats
   c. hunt at night
   d. all of the above

5. What is the main difference between gracile and robust australopithecines?
   a. Gracile forms occupied Africa, while robust species lived in Europe and Asia.
   b. Robust species were much taller.
   c. Robust species had much heavier jaws and larger cheek teeth.
   d. Only species in the gracile group were bipedal.

6. Researchers agree that modern *Homo sapiens* originated in Africa, and then spread throughout Europe, Asia, and eventually the New World. What do they disagree about?
   a. whether the original African population of *H. sapiens* left any descendants
   b. whether members of *H. sapiens* interbred with members of *H. erectus*
   c. whether the Neanderthals represent a distinct species from *H. sapiens*
   d. whether *H. neanderthalensis* and *H. erectus* died out at the same time

Conceptual Review

1. The text maintains that the animal body plans vary in four fundamental aspects of their architecture: the number of embryonic tissues, type of body symmetry, presence and type of body cavity, and protostome or deuterostome pattern of development. How does each aspect of body architecture vary among animal lineages? How does this variation relate to the view that the animal phyla represent different ways of designing a feeding machine?
2. Section 30.3 lists the key innovations of arthropods as exoskeletons, limbs, and insect metamorphosis; section 30.4 lists the key innovations of vertebrates as internal support structures, jaws, limbs, and endothermy. How are the adaptations observed in these two lineages similar and different? How do they relate to the ability of arthropods and vertebrates to feed efficiently?
3. Summarize the types of mouthparts required for suspension feeding, deposit feeding, herbivory, predation, and parasitism.
4. Why is it logical to claim that endothermy evolved independently in birds, in mammals, and in several different species of fish?
5. The text claims that "*Homo sapiens* is the sole survivor of a remarkable adaptive radiation." Do you agree with this statement? Why or why not?

Applying Ideas

1. When juvenile hormone (JH) is present in insect larvae, surges of ecdysone result in molt. When JH is absent, a surge of ecdysone results in metamorphosis to the adult form. Suppose you injected fruit-fly larvae with JH, so that extra molts occurred and metamorphosis was delayed. When metamorphosis finally occurred, predict how the experimental adults would differ from normal adults.
2. Suppose you made a strand of DNA whose sequence was complementary to mRNA transcripts produced by the gene called *Distal-less* in the worm-like onychophorans, polychaete worms, insects, and tetrapods. If you treated embryos from each group with this molecule, what would happen? Why?
3. Suppose you were walking along an ocean beach at low tide and found an animal that was unlike any you had ever seen before. How you would go about determining how the animal feeds? How would you go about determining the major features of its body plan? What features would allow you to determine if it is a vertebrate or arthropod?
4. Figure 30.23a shows that there is a large gap in the fossil record of *Homo* in Africa between about 1.4 million years ago and 0.6 million years ago. Ernesto Abbate and colleagues have recently announced that they found a hominid skull in Ethiopia dated to 1 million years ago. If the fossil came from a species that is part of the lineage leading to *H. sapiens*, predict what the skull looks like.

CD-ROM and Web Connection

CD Activity 30.1: The Architecture of Animals *(Animation)*
(Estimated time for completion = 5 min)
Are morphological adaptations needed for survival related to the length of time available for a population to change?

At your **Companion Website** (http://www.prenhall.com/freeman/biology), you will find self-grading exams and links to the following research tools, online resources, and activities:

Animal Diversification and Comparative Biology and Geology
Take a look at how a combination of techniques can be used to study the evolution of animal diversity.

Evolution of Feeding
An article from the *Journal of Evolutionary Biology* discusses the evolution of feeding behavior in terrestrial animals.

Journal of Human Evolution
Explore journal articles presenting original research on the evolution of humans.

Additional Reading

Knoll, A. H., and S. B. Carroll. 1999. Early animal evolution: Emerging views from comparative biology and genetics. *Science* 284: 2129–2137. A synthesis of data from the fossil record and molecular genetics.

Leakey, M., and A. Walker. 1997. Early hominid fossils in Africa. *Scientific American* 276 (June): 74–79. An examination of *Australopithecus* fossils and what they imply for the origin of bipedalism in humans.

Sereno, P. C. 1999. The evolution of dinosaurs. *Science* 284: 2137–2147. A comprehensive look at a lineage that dominated terrestrial environments for 150 million years.

Tattersall, I. 1997. Out of Africa again . . . again? *Scientific American* 276 (April): 60–67. A look at the evidence for the out-of-Africa hypothesis for the origin of modern humans.

How Plants Work

Unit 7

"Revolutionary" is the best word to describe recent changes in the study of plants. Sophisticated new techniques for probing the structure and function of plant genes and proteins have led to a long list of exciting advances. Progress has been particularly rapid because researchers using molecular techniques are building on a strong foundation of classical studies in plant anatomy and physiology. Over a century's worth of careful descriptive work has documented the diversity of plant anatomical structures in great detail; decades of experimental work using traditional techniques have probed their function.

The goal of this unit is to blend insights from classical and molecular plant biology while exploring fundamental questions about how plants grow and reproduce. **Chapter 31: Plant Form and Function** focuses on how the bodies of plants are put together and how they grow. **Chapter 32: Water and Sugar Transport in Plants** and **Chapter 33: Plant Nutrition** ask how plants obtain the molecules they need to develop and reproduce. **Chapter 34: Sensory Systems in Plants** and **Chapter 35: Communication: Chemical Signals** delve into how plants sense light, gravity, and other aspects of the environment and how they grow or move in response to the information they receive. **Chapter 36: Plant Reproduction** surveys classical and recent work on how flowering plants breed and produce offspring. **Chapter 37: Plant Defense Systems** spotlights cutting-edge research on the systems that plants use to ward off disease-causing viruses, bacteria, and fungi and to defend themselves against insect and mammalian herbivores.

MacFarlane's Four-O'Clock lives only in a few river canyons in northwestern North America. Its beauty belies its toughness—it grows on dry rocky soils and can withstand searing heat. ©1994 Susan Middleton and David Liittschwager

Plant Form and Function

31

This chapter addresses two fundamental questions: How are plants put together, and how do they grow? In size and shape, land plants range from 90 m-tall (300 ft.) redwood trees to sprawling mosses that are just 2 mm (1/16 in.) high. Between these extremes is a bewildering variety of growth forms—from small trees and shrubs to vines, grasses, mat-formers, and herbs. Given this diversity, are there any common internal or external features? How do researchers explore the function of bark, leaves, wood, pollen, thorns, and other structures?

Section 31.1 delves into these questions by examining the diversity of root systems found in plants native to the prairies of North America and the diversity of stems and shoots observed among the Hawaiian silverswords. The silverswords are a group of 30 species that exhibits many of the forms and growth habits found among plants as a whole. This overview is followed by a more detailed investigation of the cells, tissues, and systems found in the flowering plants, and then by a look at the tissues responsible for plant growth. The chapter concludes by exploring how biologists study the function of diverse plant structures.

The data reviewed here are essential not only for understanding subsequent chapters in this unit, which explore the transport, sensory, reproductive, and defense systems found in plants; they are also important for a basic understanding of the world. If you go to a window and look outside, you're almost certain to see an impressive diversity of plants. What are the various parts of these organisms called? What does each structure do? Why are the size and shape of structures like leaves, twigs, and roots so variable among species? Learning about plant form and function is a basic aspect of understanding the world.

This 250-million-year-old fossil shows a cross section from the trunk of a tree related to today's Norfolk Island pine. In this chapter we explore features of plant anatomy, including how wood forms and why growth rings appear in tree trunks.

31.1 The Diversity of Plant Form

The most fundamental attribute of plants is that they make their own food. Chapter 7 detailed how plants—along with algae, cyanobacteria, and a variety of protists—obtain the energy and carbon they need to grow and reproduce. These organisms use the light reactions of photosynthesis to harness the kinetic energy in sunlight and manufacture ATP and NADPH (**Figure 31.1a**). The enzymes of the Calvin cycle then use the chemical energy in ATP and NADPH to convert (or "fix") carbon from an oxidized, inorganic state in the form of carbon dioxide (CO_2) to a reduced, organic state in the form of a sugar called glyceraldehyde 3-phosphate (G3P). The G3P produced by the Calvin cycle feeds biochemical pathways that result in the formation of energy-rich molecules such as sucrose and starch.

To grow and reproduce, then, plants need to obtain light and carbon dioxide. Light and carbon dioxide are just the beginning, however. Plants must also have water as a source of electrons to run photosynthesis. (In Chapter 32 you'll see that plants also need water to keep their cells in proper working order.) Further, plants must obtain nitrogen (N), phosphorus (P), potassium (K), magnesium (Mg), and a host of other nutrients to synthesize nucleic acids, enzymes, phospholipids, and the other macromolecules needed to build and run cells. (Chapter 33 explores how plants acquire these key elements.)

Stripped to its essence, then, a plant's body has two tasks: Obtaining light and nutrients and using these components to grow and reproduce. **Figure 31.1b** illustrates the basic systems that most plants use to accomplish these tasks. In most plants a belowground section called the **root system** takes in water and nutrients from soil, while an aboveground portion called the **shoot system** harvests light and carbon dioxide. Roots also help the stem stay upright in the face of lateral forces such as wind, while the stem holds reproductive structures and leaves and keeps the entire body upright against the pull of gravity and the force of rain or snow. In many plants, **vascular tissues** run continuously through the root and shoot systems and transport materials from one to the other (**Figure 31.1c**).

Section 31.2 is devoted to exploring the details of internal and external anatomy in roots and shoots; here we focus on understanding why so much diversity exists in their overall form. Let's take a closer look.

The Diversity of Roots: North American Prairie Plants

Over a span of four decades from about 1910 to 1950, J. E. Weaver conducted a series of studies on the prairie plants of North America. Prairies are grassland ecosystems found in areas of the world like the Serengeti Plain of East Africa, the Pampas region of Argentina, and the steppes of central Asia. Rain is abundant enough in these areas to support a lush growth of grasses and herbaceous plants but scarce enough to exclude trees and most shrubs (**Figure 31.2a**). Fires that are triggered by lightning or set

(a) Plants need resources for photosynthesis.

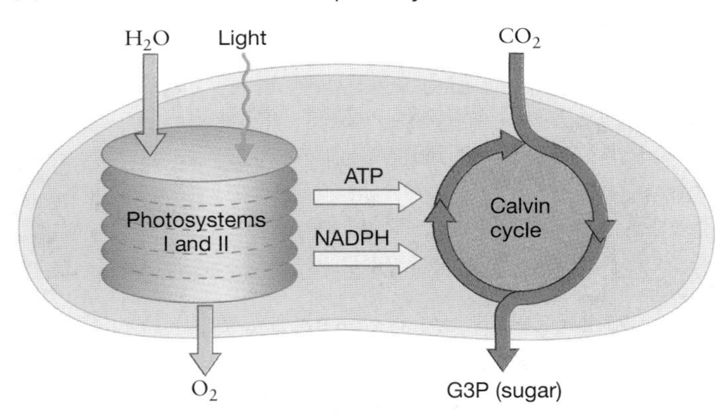

(b) These resources, plus key nutrients, are harvested by the shoot and root systems.

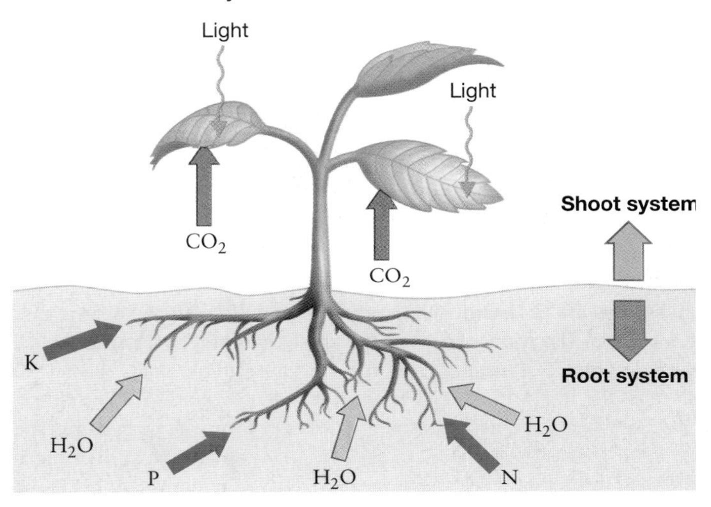

(c) Materials are transported internally by the vascular tissue.

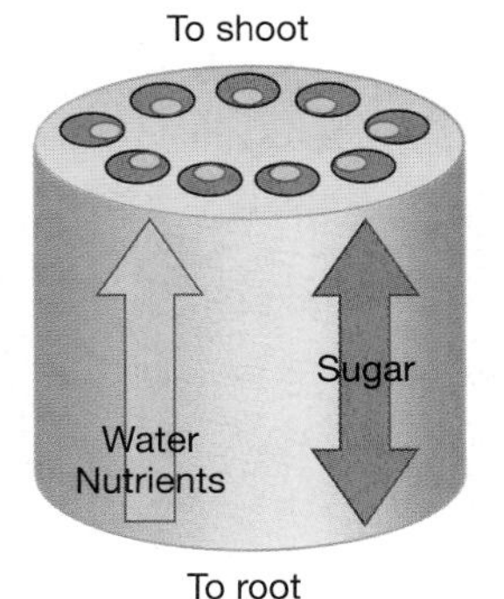

FIGURE 31.1 Plants as Resource Harvesting, Processing, and Transport Structures
(a) Photosystems I and II require light and water and produce ATP and NADPH. The Calvin cycle requires CO_2 and produces sugar. **(b)** Shoot systems are specialized for harvesting light and CO_2; root systems absorb water and key nutrients like nitrogen (N), phosphorus (P), and potassium (K). **(c)** In many plants, vascular tissues transport materials between the root and shoot systems.

by humans sweep through these ecosystems regularly, also tending to eliminate woody species.

To understand the dynamics of prairie ecosystems, Weaver focused much of his time and energy on describing the morphology of roots. He had two reasons for doing so. Deep roots help plants find water in time of drought and, because roots function as starch-storage organs as well as resource-harvesting systems, they also provide the energy required for plants to regenerate themselves after a fire.

Weaver's approach to studying root anatomy was simple. He and co-workers dug deep trenches and examined the exposed root systems, or excavated a deep section of soil around a particular plant, removed the section, and carefully washed the soil away to reveal the roots. Some of the photographs that resulted from this effort are shown in **Figure 31.2b**. The figure shows that the root systems of prairie plants that live side by side can be very different from one another. Compare the dense, fibrous system in the grass called big bluestem with the 6.4 m-long (21-foot) taproot found in the prairie rose. Other species, like the compass plant, have thick roots that contain particularly large reservoirs of food stored in the form of starch.

Why is this diversity of form important? If all root systems were similar in size and shape, then every plant in the ecosystem would be in direct competition for water and nutrients. If root systems are diverse, however, then different species can avoid competing for the same sources of water and nutrients. Due to the diversity of root systems illustrated in Figure 31.2b, a small patch of prairie should be able to sustain a relatively large number of species.

Weaver actually observed this theory in action. During the 1930s a decade-long drought seared his study area, producing the changes in vegetation recorded in **Figure 31.3**. This drawing diagrams a cross section of a trench that Weaver and colleagues dug when the drought was at its height. Note that the soil between 2 feet and 5.5 feet below the surface had become virtual-

(a)

(b) Big bluestem grass Compassplant Prairie rose

FIGURE 31.2 The Root Systems of Prairie Plants
(a) Prairies are grassland ecosystems. The prairie pictured here is in Kansas. **(b)** The roots of prairie plants that live side by side can be very different. For example, the roots of these three plants reach depths of 6 ft., 14 ft., and 21 ft., respectively. EXERCISE Roots function as storage organs as well as resource-harvesting systems and anchoring devices. In part (b) of the figure, draw arrows to thickened areas of roots that store starch. How might you confirm that they have high starch concentrations?

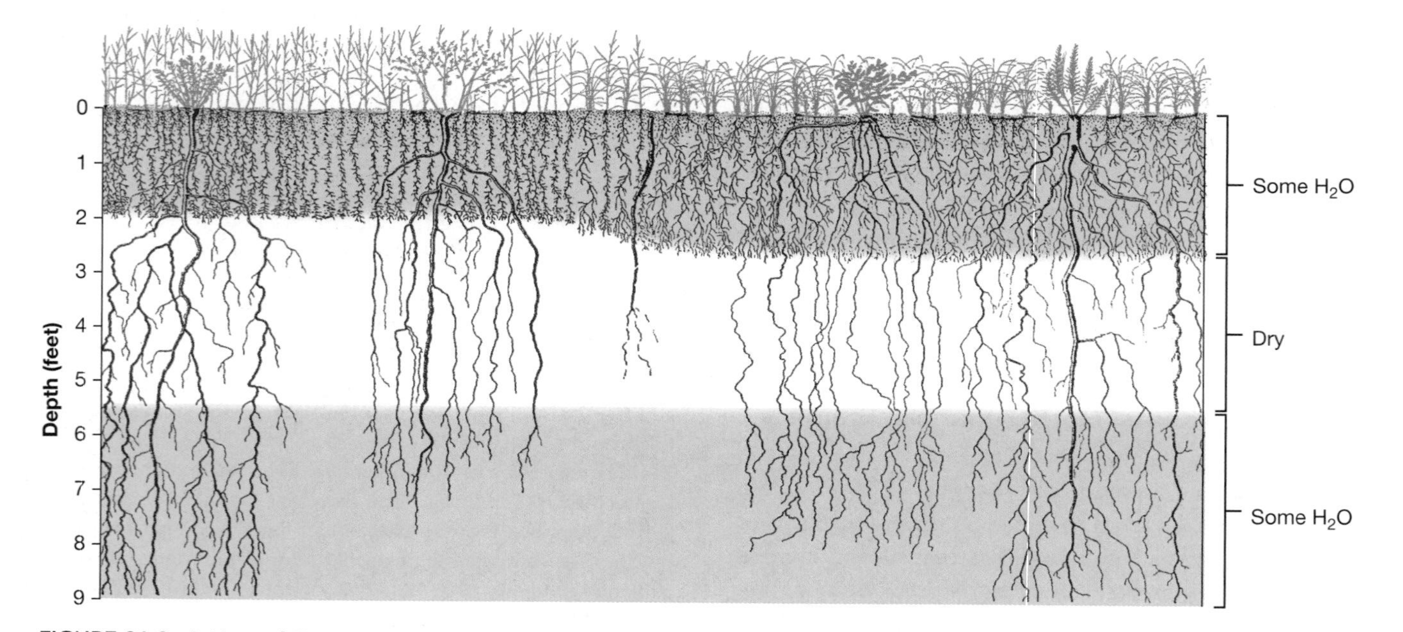

FIGURE 31.3 A Natural Experiment on the Importance of Diverse Root Systems
At the end of a severe drought, prairie plant roots occupied shallow- and deep-soil layers that still held a small amount of water. EXERCISE Near the middle of the drawing is a root system that stops about 5 ft. deep in the soil layer that is depleted of water. Note that the plant is dead, and circle its root system.

ly waterless. Although small amounts of water remained near the surface and in the deep subsurface, this intermediate zone had been pumped dry by the plants. As a result, plants that depended on this soil layer for water and nutrients died. Further, the roots of big bluestem and other grasses, which normally penetrate to depths of 4–6 feet, died back to a mere 2–2.5 feet in depth. Taprooted species were also able to survive the drought, because their roots extended past the waterless zone to layers in the deep subsurface that still contained moisture.

What is the overall message of this natural experiment on the size and shape of root systems? Even though most root systems have similar functions—in support, storage, and nutrient absorption—they are highly diverse in structure. In prairie ecosystems, Weaver's data suggest that natural selection has favored a diverse array of root systems to minimize competition for water and nutrients. Consequently an array of prairie plants can coexist, and most species can survive intense water stress.

(a) Mat-forming silversword

(b) Tree-sized silversword

FIGURE 31.4 Hawaiian Silverswords
(a) *Dubautia scabra* is a low-growing silversword that colonizes lava flows and forms extensive mats. **(b)** Several of the silverswords grow as small trees, including the *Dubautia reticulata* shown here.

The Diversity of Shoots: Hawaiian Silverswords

The Hawaiian Islands are home to a group of closely related plant species called the silverswords, which are found nowhere else in the world. As **Figure 31.4** shows, these species are wonderfully diverse in size, shape, and growth habit. Some form sprawling cushions while as others grow as the compact forms called rosettes. Other silverswords have thick, woody stems and grow as small shrubs, large shrubs, or trees. The silverswords are ecologically diverse as well. Depending on the species, they can be found from sea level to alpine altitudes and from rain forests to the desert-like conditions of exposed lava flows.

To fully appreciate this diversity, consider recent work by Bruce Baldwin and colleagues on the evolutionary history, or phylogeny, of silverswords. These studies were inspired by the observation that the organization of vascular tissue in silverswords is strikingly similar to the vascular tissue of plants called tarweeds, which are native to California. Are the tarweeds of California and the silverswords of Hawaii similar because they are related? To answer this question, Baldwin and co-workers sequenced the same stretch of DNA in a large sample of silverswords, tarweeds, and other plants. By comparing similarities and differences in the sequences, the researchers were able to infer which species are most closely related and which are more distantly related. (Chapter 23 provides more detail on how researchers use DNA sequences and other sources of data to reconstruct the phylogeny of a group of species.) **Figure 31.5** presents their results in the diagrammatic form of a phylogenetic tree.

To interpret Figure 31.5, recall from Chapter 1 that the branches on a phylogenetic tree represent populations through

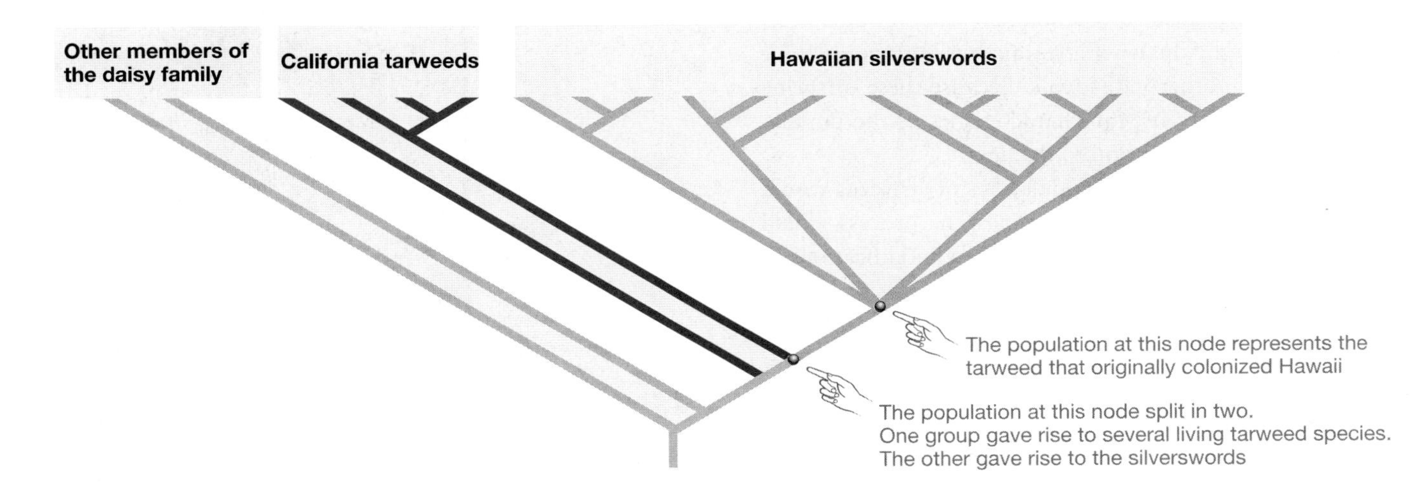

FIGURE 31.5 Tarweeds Are the Closest Living Relatives of Silverswords
The relationships described by this evolutionary tree are based on similarities and differences observed in the DNA sequences of tarweeds, silverswords, and other species that are members of the daisy family of plants.

time. In Figure 31.5, the population of organisms indicated by the upper pointer in the figure represents the common ancestor of all silverswords. The next closest branch to this population (indicated by the lower pointer in the figure) leads to the tarweeds of California and represents the next closest relative of the silverswords. This pattern implies that tarweeds and silverswords recently shared a common ancestor.

How could tarweeds and silverswords be so closely related? According to Baldwin and his group, the answer hinges on the observation that some tarweeds have sticky seeds and can form viable offspring through self-fertilization. Based on these data, the researchers suggest that long ago a tarweed seed adhered to the leg of a seabird that made the trip from California to Hawaii. From the population founded by this seed, the dramatic diversity of silverswords evolved. The question is, how?

Making Sense of Diversity: Natural Selection and Adaptation

To understand how the shoot systems of silverswords and the root systems of prairie plants have come to be so diverse, it is critical to review the principle of natural selection that was introduced in Chapters 1 and 21. Natural selection can be summarized as a two-step process. The first step occurs when mutations result in a population of individuals that are variable in a particular trait. The initial tarweed population in Hawaii, for example, would inevitably accumulate mutations that affect the stem length of individuals. As **Figure 31.6a** shows, some individuals in the population would have longer stems than others.

The second step in natural selection occurs when these heritable variations allow certain individuals to survive or reproduce better in certain environments. For example, short-stemmed individuals in this population might be able to survive dry, exposed environments on lava flows, where tall plants tend to dry out or be damaged by high winds (**Figure 31.6b**). Individuals with long stems, in contrast, might flouish in wetter, more protected sites because their height allows their leaves to avoid being shaded by lush surrounding growth. Natural selection occurs when individuals with certain heritable traits leave more offspring than other individuals. Because these offspring also have the favored traits, the characteristics of the population change from one generation to the next.

Through this process the two populations of tarweeds would tend to change and diverge through time. The process would continue because new mutations produce new modifications in stem length every generation, and because shorter- or longer-stemmed individuals would still be favored in the two habitats. Over time, then, the short-stemmed and long-stemmed populations would become distinct enough to be considered separate species. The evolution of the silverswords would be under way.

It's also important to note that this example is not hypothetical. As **Figure 31.6c** shows, short- and long-stemmed silverswords actually did descend from the original population of tarweeds that colonized the Hawaiian Islands.

One final point is worth noting. The contrasting stem lengths in these silverswords are considered adaptations to their respective environments. Recall from Chapter 21 that an **adaptation** is a trait that allows an individual to produce more offspring in a certain environment than individuals without the trait. The diverse shoot systems of silverswords are adaptations that allow individuals to withstand wind and capture light. The diverse root sys-

(a) Stem height varies among individuals.

(b) Natural selection culls unfit individuals.

(c) Both long- and short-stemmed silverswords descended from the original population of tarweeds.

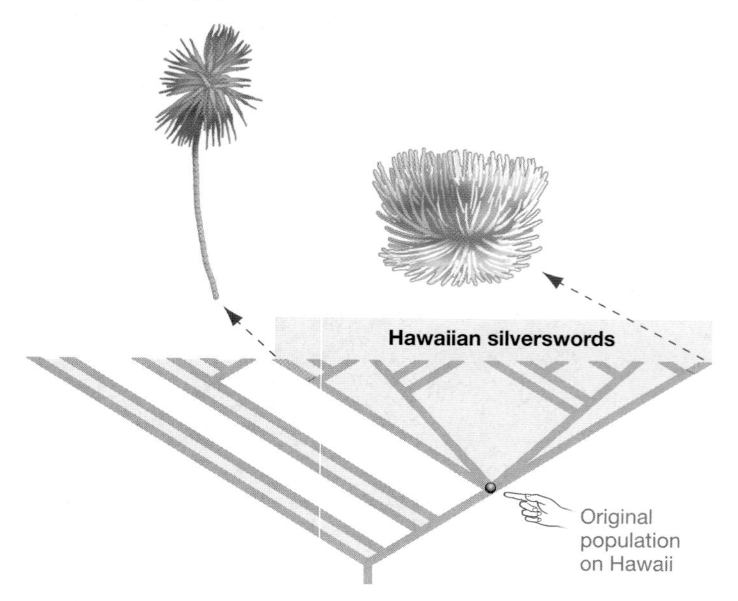

FIGURE 31.6 How Could Natural Selection Produce a Diversity of Shoot Systems in Silverswords?
(a) Due to mutation, the founding population of silverswords on Hawaii would have included both short- and long-stemmed individuals. **(b)** Individuals with different stem heights thrive in different habitats. **(c)** This phylogeny was estimated from DNA sequence data.

tems of prairie plants are adaptations that allow individuals to harvest water and maximize their ability to grow and reproduce.

At their core, the seven chapters in this unit are about the adaptations that allow plants to thrive in a remarkable diversity of habitats. As you study the traits that allow plants to grow upright and efficiently harvest nutrients, water, and light, keep in mind that these traits are the products of natural selection.

31.2 Cells, Tissues, Organs, and Systems

This section explores the internal and external anatomy of plants, starting with individual cells and working up—first to the collections of cells called tissues, then to the collections of tissues called organs, and finally to the collections of organs and tissues called systems. After viewing shoots and roots as a whole, the focus shifts to taking these systems apart and seeing how they are put together.

Instead of surveying the entire diversity of cells, tissues, and systems found among plants, however, our emphasis is on the anatomy of flowering plants, or **angiosperms.** The angiosperms are the most recent major group of plants to appear in the fossil record, meaning that they evolved a relatively short time ago. They are also the most species-rich lineage among the plants and are of huge economic and medical importance to humans. Most of the food and many of the drugs we use are derived from angiosperms. What do their cells, tissues, and systems look like, and how do they work?

Plant Cells

Chapter 5 introduced a generalized version of the plant cell. The sketch in **Figure 31.7a** reviews its most important organelles and structures, including the cellulose-rich wall surrounding the cell membrane. As with any generalization, however, it is important not to take this view of the cell too literally. In reality, the cells that make up an individual plant are enormously diverse in size, shape, composition, and function. Some cells are actually programmed to die at maturity and are only fully functional after death.

Figure 31.7 also offers a sampling of the cell types found in angiosperms, grouped by their function. The **parenchyma cells** found in leaves, for example, are jammed with chloroplasts and are the primary site of photosynthesis; the parenchyma cells of roots serve as storage cells for starch deposits (**Figure 31.7b**). When you eat a salad, potato, or apple, you are ingesting primarily parenchyma cells.

The cells illustrated in **Figure 31.7c** are specialized for support. **Collenchyma cells** have thickened primary walls and serve to stiffen leaves and stems; if you've ever peeled a "string" from a stalk of celery, you have dissected a column of collenchyma cells. **Sclerenchyma cells** also stiffen stems and other structures,

(a) Generalized plant cell

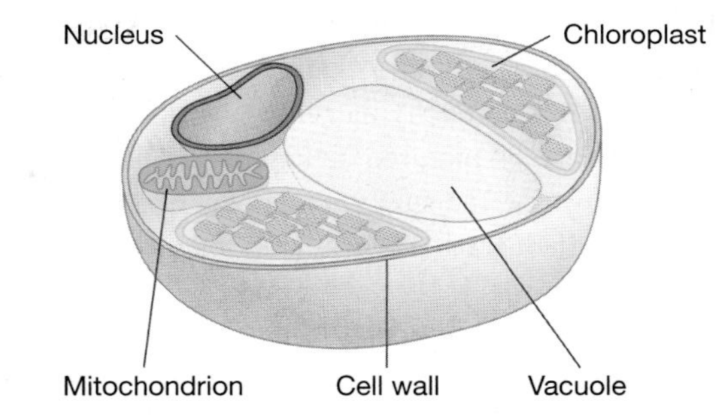

(b) Workhorse cells—parenchyma

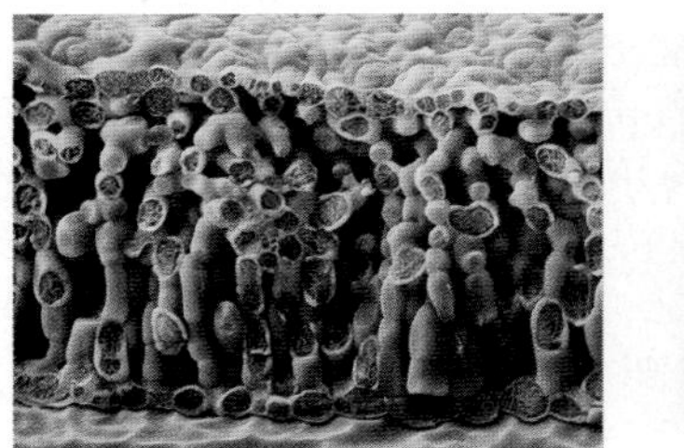

From leaf tissue

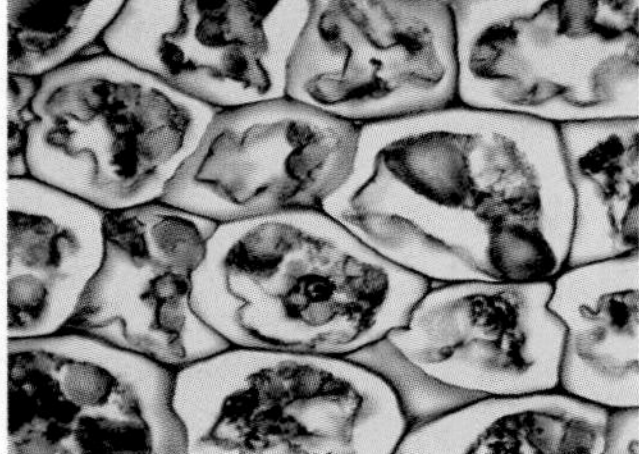

From root tissue

(c) Structural cells

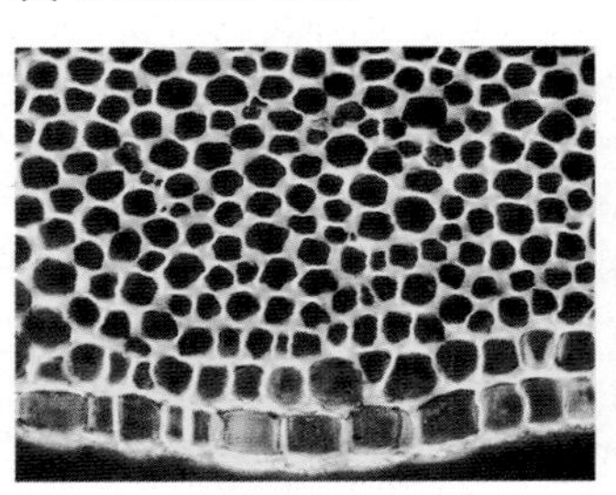

Collenchyma

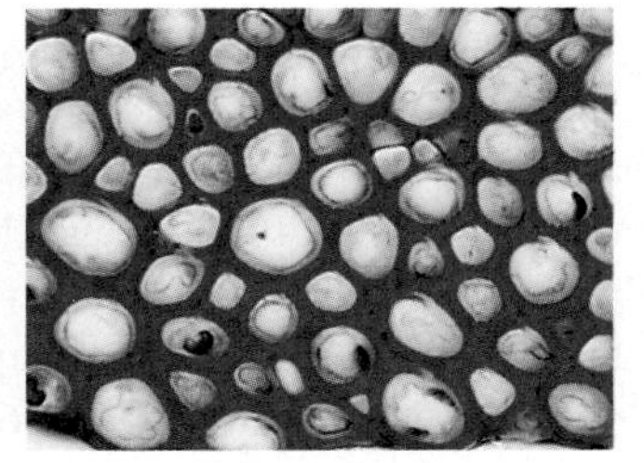

Sclerenchyma

(d) Water- and sugar-conducting cells

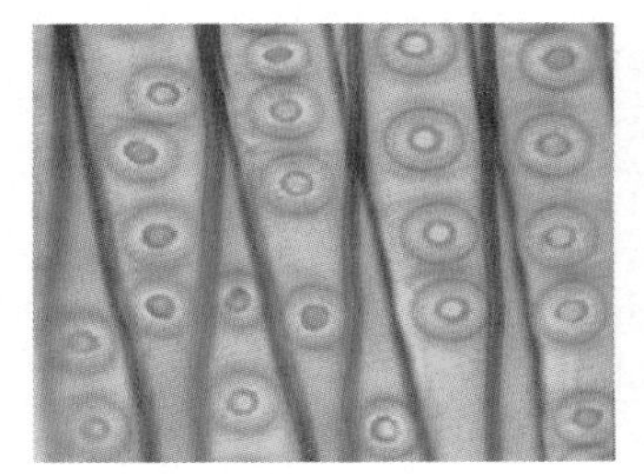

Tracheids

Vessel elements

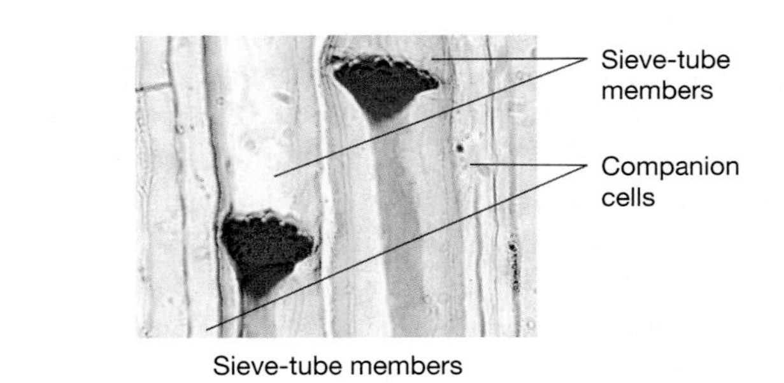

Sieve-tube members

FIGURE 31.7 Plant Cells Are Diverse in Size, Shape, and Function
The sketch in part (a) shows some of the organelles and structures found in plant cells. Not all of these features are found in all plant cells.

but are distinguished by thickened secondary cell walls that are strengthened with tough lignin molecules. The fibers used in cotton and linen fabrics and in hemp and jute ropes are composed primarily of sclerenchyma cells.

The long, thin cells pictured in **Figure 31.7d** are involved in the transport of water and sugar in angiosperms. The water-conducting cells called **tracheids** and **vessel elements** are dead at maturity, while the food-conducting cells called **sieve-tube members** remain alive. Vascular cells like these usually have some sort of opening between cells to facilitate the flow of material up or down the plant body. In essence, then, tracheids and vessel elements are empty channels made of cell walls. Sieve-tube members have cytoplasm but no nuclei and a minimal number of organelles. How is water pulled up tracheids and vessel cells? How is the movement of sugar up and down sieve-tube elements controlled? These questions are the focus of Chapter 32.

Plant Tissues

A **tissue** is defined as a group of cells that functions as a unit. Plants have three fundamental types of tissues. The epidermal tissue, or **epidermis** (over-skin) consists of a single sheet of cells that covers the entire plant body. A layer of wax called the **cuticle** covers the outside of epidermal cells in the shoot and drastically reduces the amount of water that is lost to the surrounding air.

The **ground tissue** is made up of cells below the epidermis and surrounding the vascular tissues. Ground tissue is usually made up of parenchyma stiffened by collenchyma and sclerenchyma cells. The **vascular tissue** is made up of the conducting cells introduced earlier. There are two types of vascular tissue. **Xylem** tissue conducts water; **phloem** tissue conducts sugar. (As **Box 31.1** points out, several major groups of plants do not have vascular tissues.) To fully understand the roles of these tissues, however, we need to take a closer look at how they are organized into functioning systems.

Plant Systems: The Root and Shoot Revisited

As our analysis of prairie plants and the Hawaiian silverswords showed, the size and shape of root and shoot systems are diverse. Here we examine the general morphology of roots and shoots in more detail, and introduce the anatomy of leaves—a particularly important component of the shoot system.

Anatomy of the Root **Figure 31.8** shows a cutaway view of a root and highlights how the three primary tissues—epidermal, ground, and vascular—are organized. The roots of angiosperms have three outstanding features. The vascular tissues are organized into a central region called the **stele, lateral roots** erupt from the outer portion of the stele and project horizontally through the cortex and out into the surrounding soil, and some epidermal cells contain long, thin projections called **root hairs.**

These anatomical elements are all compatible with a root's role as a water- and nutrient-gathering device. Root hairs increase the amount of cell surface area that is in contact with soil. As a result, they greatly increase the system's ability to absorb moisture and nutrients via membrane proteins—a process that will be explored in detail in Chapter 33. Lateral roots allow the system to penetrate new spaces in the environment and exploit untapped sources of water and nutrients. Once materials enter the system, they are transferred to the central plumbing system—the stele—and transported to tissues where they are needed.

Anatomy of the Shoot **Figure 31.9a** shows a twig from an oak tree in winter and summer. Twigs represent the tips of the stem in woody species. Compared to other parts of the shoot, twigs have a unique feature. They contain a set of cells at their apex whose growth lengthens the system. These cells are called the apical meristem and are analyzed in detail in section 31.3. In other respects, however, the arrangement of epidermal, ground, and vascular tissues in a twig is similar to that observed in branches, stems, trunks, and other components of the shoot (**Figure 31.9b**).

As you trace the twig in Figure 31.9a down from its apex, note the structures called **nodes**, where leaves attach. (The portion of twig or stem between nodes is called an **internode.**) At the base of each node is a bud. These **lateral buds** are important because they contain cells capable of producing side branches.

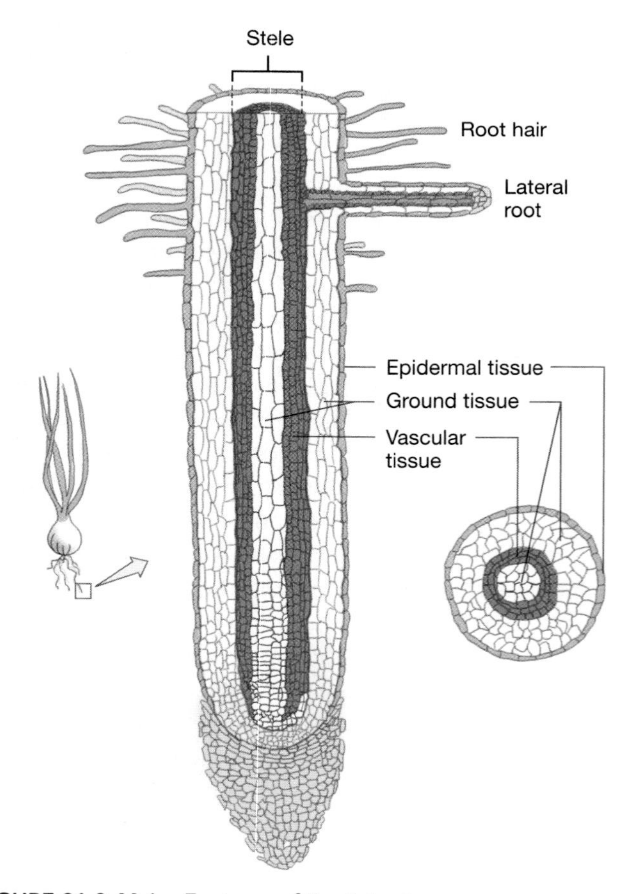

FIGURE 31.8 Major Features of Root Anatomy
The three tissues in roots are organized into concentric circles. The vascular tissues of roots are found in the central section, called the stele.

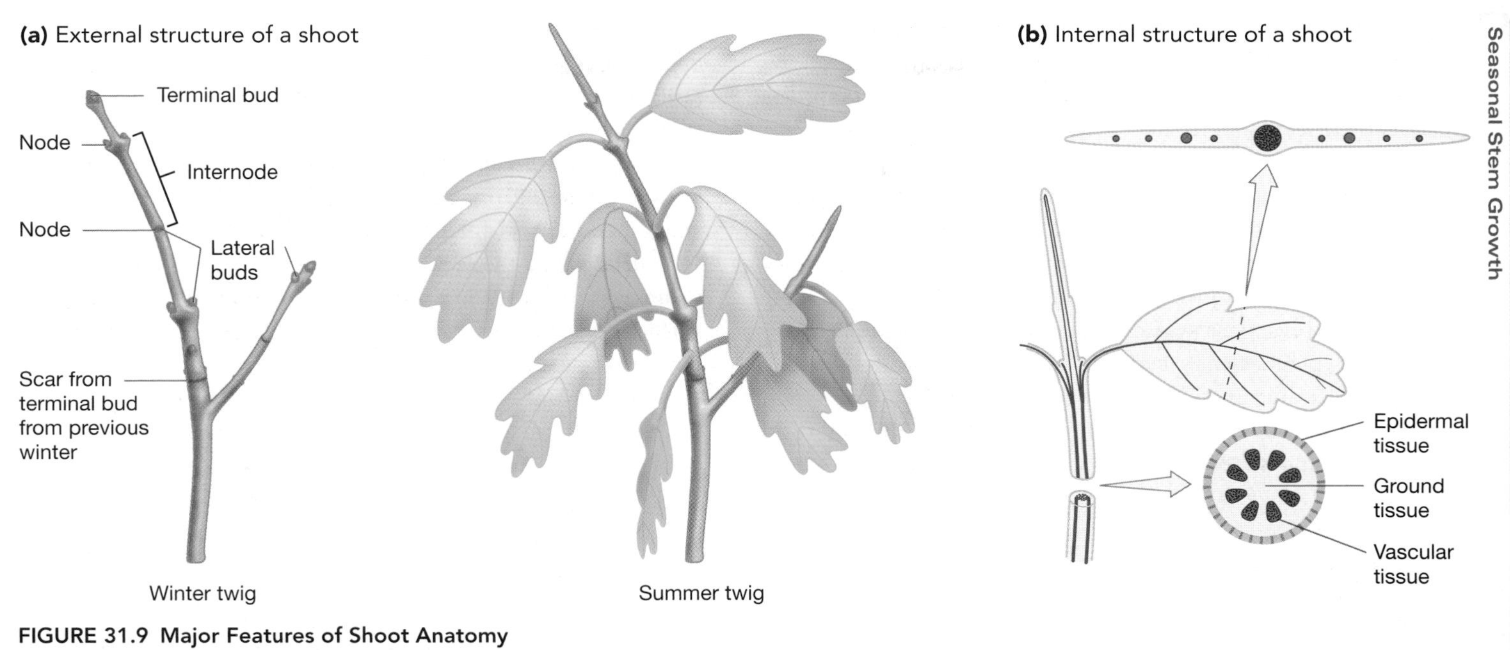

FIGURE 31.9 Major Features of Shoot Anatomy
(a) An oak-tree twig in winter (left) and summer (right). Note that lateral buds are found in the axils of leaves, at nodes. **(b)** In many flowering plants the epidermal, ground, and vascular tissues are arranged as shown here. The cross section of the stem shows that vascular tissues frequently occur in bundles that are arranged in a circle. EXERCISE In part (a), label each part on the summer twig that is shown on the winter twig.

CD ACTIVITY 31.1 Seasonal Stem Growth

BOX 31.1 Nonvascular Plants

Among the land plants, three major groups of species do not have vascular tissues. These lineages are the mosses, liverworts, and hornworts, which are collectively known as the bryophytes or nonvascular plants. Although a few species in these groups have water-conducting cells that are organized into tissue-like strands, the cells do not have the thickened secondary walls—stiffened by lignin—that characterize the tracheids and vessels found in true vascular tissues. Instead of having roots specialized for water and nutrient absorption, bryophytes have rhizoids that anchor the plant in place.

Because the mosses, liverworts, and hornworts are all short in stature, they do not have to transport water or nutrients long distances. Instead, the cells in these plants are capable of absorbing the water and nutrients they need directly from the environment.

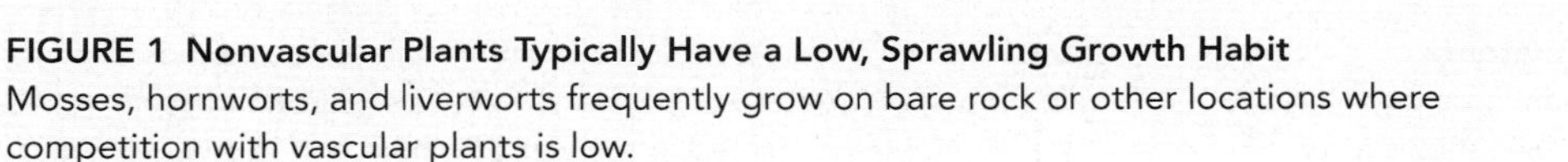
FIGURE 1 Nonvascular Plants Typically Have a Low, Sprawling Growth Habit
Mosses, hornworts, and liverworts frequently grow on bare rock or other locations where competition with vascular plants is low.

To pull these observations together, it's important to realize that the size and shape of twigs and stems have important functional consequences for a plant. Twigs and stems are structures that arrange leaves in space. (They also hold reproductive structures, but we'll save this topic for Chapter 36.) The arrangement of leaves on a plant is significant because it dictates the amount of light received and thus determines the rate of photosynthesis and growth and reproduction. It is not surprising, then, that a shoot's growth is controlled by sophisticated sensory and response systems. Chapter 34, for example, explores how plants sense certain wavelengths of light and respond by bending and growing in that direction.

How about the leaf? Although leaves are enormously diverse in size and shape, they are normally composed of just two major structures: an expanded portion called the blade and a stalk called the **petiole** (**Figure 31.10**). The most important feature of their surface anatomy is a structure called the **stoma** (plural is *stomata*), which consists of a pore surrounded by paired **guard cells** (**Figure 31.11**). Guard cells can change shape by gaining or losing water. With these shape changes, guard cells either open the pore and facilitate gas exchange with cells inside the leaf, or they close the pore and minimize water loss. Stomata are necessary because the rest of the leaf surface is covered with wax that inhibits gas exchange. To see how cells near stomata are in contact with the atmosphere, study Figure 31.11. This figure also shows the general arrangement of epidermal, ground, and vascular tissues inside a leaf.

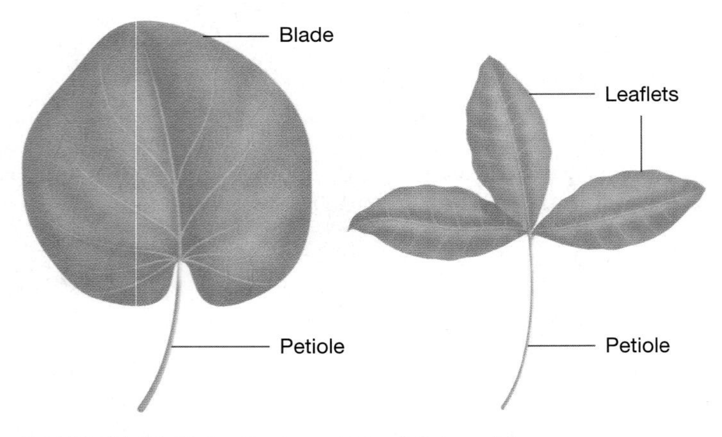

FIGURE 31.10 Major Features of Leaf External Anatomy
The size and shape of leaves varies enormously among species. Leaves contain a petiole (a stalk) and a blade, which is sometimes divided into leaflets.

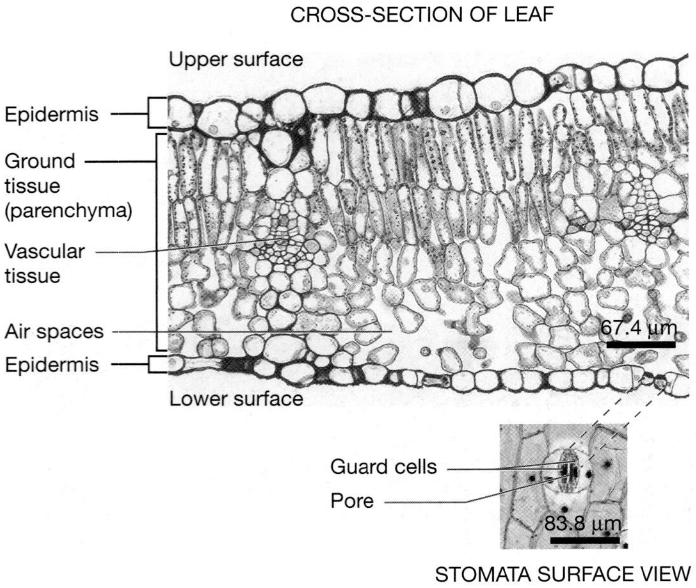

CROSS-SECTION OF LEAF
Upper surface
Epidermis
Ground tissue (parenchyma)
Vascular tissue
Air spaces
Epidermis
Lower surface
67.4 µm
Guard cells
Pore
83.8 µm
STOMATA SURFACE VIEW

FIGURE 31.11 Major Features of Leaf Internal Anatomy
This section through a leaf shows that stomata open into space inside the leaf where gas exchange can take place. The inset shows a surface view of a stoma.

31.3 The Anatomy of Plant Growth

How do plants grow? In many species, two distinct types of tissue are responsible for growth. At the tips of each shoot and root, a group of dividing cells called the **apical meristem** extends the structure (**Figure 31.12**). All plants have apical

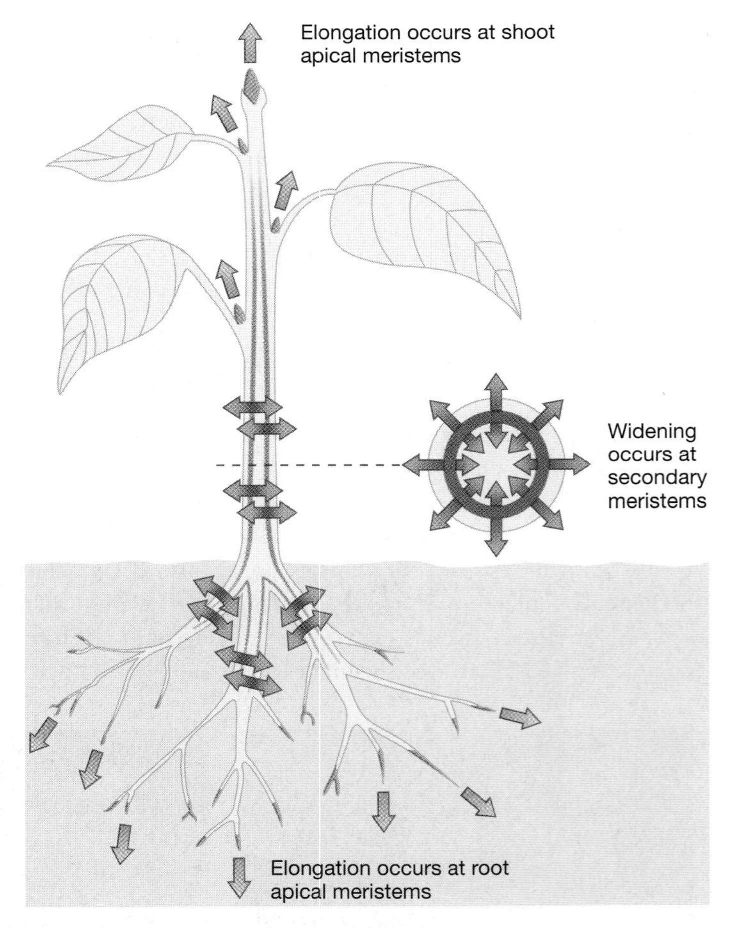

FIGURE 31.12 In Many Plants, Two Types of Growing Tissues Produce Two Types of Growth
The orange arrows indicate the direction of primary growth. Primary growth, which occurs at apical meristems located in the tips of roots and shoots, elongates each system. Secondary growth occurs at secondary meristems located in the walls of roots and shoots. As the brown arrows indicate, secondary growth widens each system.

meristems that produce growth in length—what biologists call **primary growth**. Growth in length also continues to occur throughout the life span of plants. Unlike the determinate growth of birds and mammals, which grow to a certain size and then stop, many plants have indeterminate growth. They add to their root and shoot systems throughout most of their lives. Many species are also able to produce entirely new root and shoot systems in the process of cloning themselves—a topic that is taken up in Chapter 36.

Inside the stems and roots of certain angiosperms, a layer of growing cells also adds girth over time. These layers are secondary meristems that produce **secondary growth** and result in stems and roots that grow wider as they mature. In some groups of plants, however—including grasses, lilies, and orchids—secondary growth does not occur (see **Box 31.2**, page 618).

Let's take a detailed look at the tissues responsible for primary and secondary growth. The key point here is to understand the relationship between these growing points and the three basic tissue types found in mature roots and shoots.

Primary Growth and Apical Meristems

If you were to dissect the terminal bud of an oak twig with the help of a microscope, you would find the structures indicated in **Figure 31.13a**. After removing small, protective structures called bud scales and the nearby precursors of leaves, you would find the rounded dome of cells called the apical meristem. This group of small, similar-looking cells is responsible for producing all of the cells in the stem. As these meristematic cells divide and their daughter cells mature and enlarge, the shoot grows in length.

The cells produced by the apical meristem differentiate into three basic layers. These cell populations conform to the three adult tissues. As the cells of the outside layer mature, they become the epidermis. A second group of cells differentiates into what are termed the primary xylem and phloem tissues. The third population of cells produces the ground tissue of the stem. In this way, the apical meristem produces three tissues that mature into the three tissues found in adult structures.

Apical meristems are also found in roots. Instead of being protected by the structures called bud scales, however, each of the delicate apical meristems in the root system is covered by a group of cells called the **root cap** (Figure 31.13b). As we'll see in Chapter 34, the cells of the root cap sense gravity and determine the direction of root growth. The root apical meristem produces all of the cells for primary growth as well as the cells of the root cap. As in the apical meristems of the shoot, the cells produced by the root apical meristem differentiate into three tissues, which later mature into adult epidermal, vascular, and ground tissue.

Secondary Growth and Cambium Tissues

In some plants, the older parts of the root and shoot undergo secondary growth. Secondary growth occurs when certain cell populations in the root and shoot never fully differentiate into specialized cell types. Instead, these cells continue to grow and produce new cells that, in turn, differentiate and cause the stem

(a) Shoot apical meristem

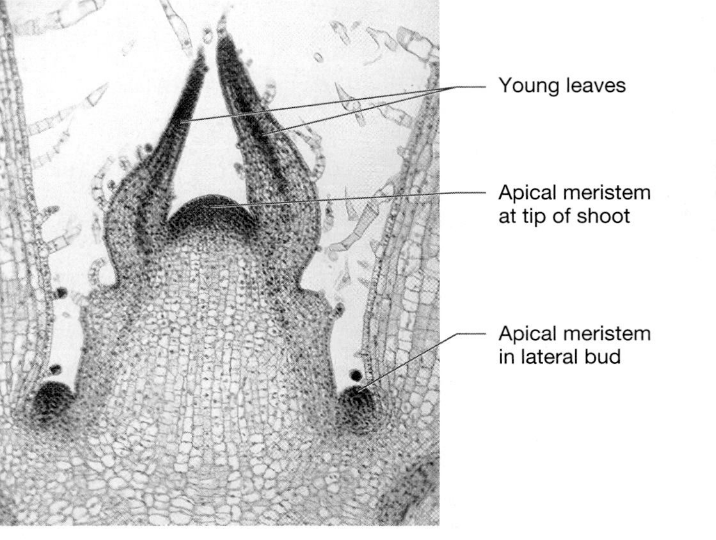

(b) Root apical meristem

FIGURE 31.13 Apical Meristems
Apical meristems consist of small, similar-looking cells that divide rapidly when water and nutrients are plentiful. Growth occurs as cell number and cell size increase. **EXERCISE** Turn back to Figure 31.8 and label the root cap and apical meristem in the drawing.

[ri]se in diameter. The undifferentiated population of cells is called a secondary meristem, or **cambium**. For example, much of the secondary growth that occurs in trees results from mitotic divisions that occur in a layer of meristematic cells between the primary xylem and primary phloem. This secondary meristem is called the vascular cambium.

Figure 31.14a highlights the position of the vascular cambium, and **Figure 31.14b** details how this meristematic tissue works. The first thing to note in Figure 31.14b is that when a cell in the cambium divides, the two daughter cells have different fates. Instead of differentiating, one daughter remains as a meristematic cell and continues to divide. The other daughter differentiates into a vascular cell. A second key observation is that the differentiated cells produced by the vascular cambium occur on both sides of the meristem. The cells to the outside form a tissue called secondary phloem; the cells to the inside form a tissue called secondary xylem. (Recall that primary xylem and primary phloem are produced by the apical meristem during primary growth, early in the tree's life.)

As the wedge-shaped drawing in Figure 31.14a shows, wood is made up of primary and secondary xylem, while bark is made up of cork and primary and secondary phloem. (Recall that xylem conducts water while phloem conducts sugar.) The wedge also indicates that additional cambium layers, called cork cambiums, are located to the outside of the vascular cambium. Cork cambiums produce cells that replace epidermal cells originally produced by the apical meristem. The cells produced by a cork cambium die and form the protective outer layer of bark. As the vascular and cork cambiums grow, the trunk of a tree or shrub widens.

Why do sections of tree trunks have rings? Trees are **perennial plants**, meaning that they live and grow for many years. (Species that die after a single year of growth are called **annual plants.**) In many environments, the vascular cambium of trees ceases growth each year. This period of no growth, or **dormancy**, occurs during the winter in cold climates and during the dry season in tropical habitats. The vascular cambium then resumes growth in the spring or at the start of the rainy season. As growth restarts, water is generally abundant and the first cells to be produced are usually large and thin-walled. But as the growth period approaches the next dormant phase, less water is available and the cells produced become smaller and thicker-walled. This alternation of large, thin-walled cells and small, thick-walled cells gives each year's growth a dark-light appearance. When these bands are juxtaposed with growth from other years and viewed in cross section, the growth rings illustrated in **Figure 31.15a** result. Further, as a tree grows in width, the inner xylem layers begin to look different from the newer layers to the outside. The inner xylem is called the heartwood and the outer xylem is called the sapwood; water movement up the stem is confined to the sapwood (**Figure 31.15b**).

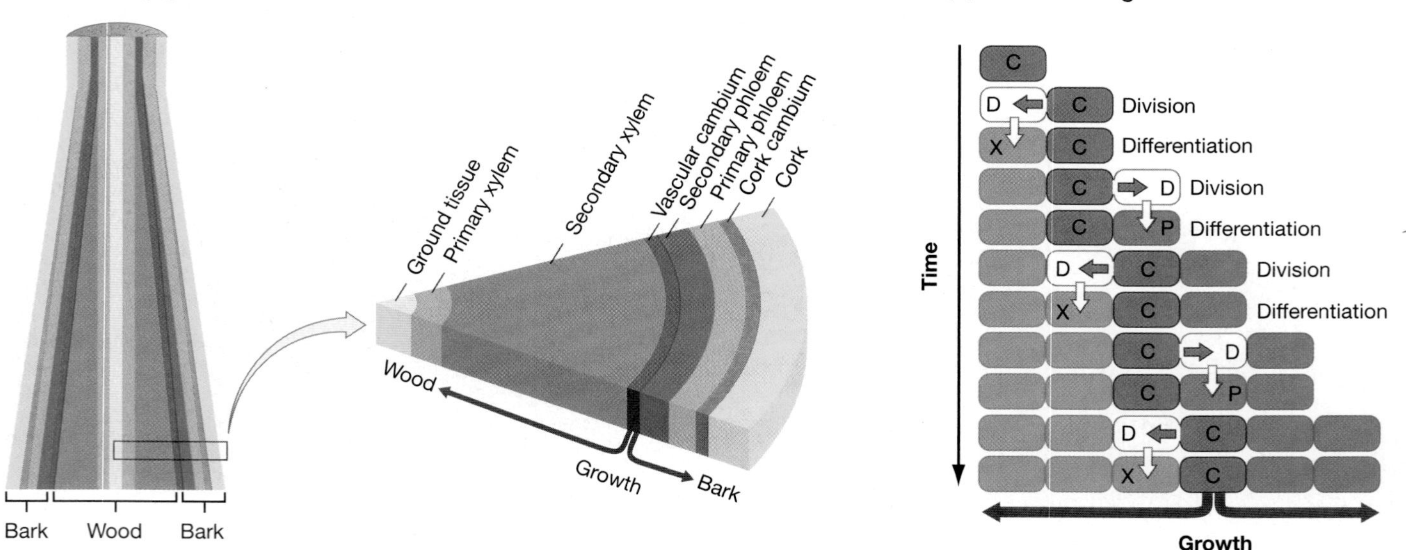

FIGURE 31.14 Secondary Growth in a Tree Trunk
(a) The vascular cambium is located between wood and bark. Vascular cambium is a secondary meristem that produces secondary xylem to the inside and secondary phloem to the outside. Cork cambiums produce part of the protective cork tissue on the outside of the trunk. **(b)** A woody stem grows in width because cells in the vascular cambium undergo a series of mitotic divisions. Each division produces a cell that continues to divide (marked C in the figure) and a derivative cell (D) that later differentiates into a secondary xylem (X) cell or secondary phloem (P) cell.

Analyzing patterns in tree rings is an important field of study in biology. Because trees grow faster when moisture and nutrients are plentiful, wide tree rings are reliable indicators of wet years. Narrow rings, in contrast, signal drought years. By studying the tree rings found in fossil trees and extremely old living trees, biologists can gain a better understanding of climate changes that occurred in the past (see **Box 31.3**, page 621). With continued research, researchers also hope to predict how forests might respond to the global warming that is currently under way.

31.4 Studying Adaptation

The previous two sections explored the anatomy and growth of flowering plants in detail. In many cases, the form of a particular cell or tissue appears to affect its function and result in clear benefits for the individual. Epidermal cells have a waxy cuticle, to prevent water loss. Guard cells open and close, to balance the requirement for carbon dioxide uptake with the need to minimize water loss. Branching allows roots and shoots to fully occupy the space around the individual. A continuous vascular system allows plants to transport water and food efficiently between roots and shoots.

In short, many of the anatomical features we've reviewed appear to be adaptations that help plants function in their environment. But how can researchers support the claim that certain features of plants are adaptive? It is not enough to supply a plausible argument; some sort of experimental or observational data are required to explicitly test the hypothesis that a trait is adaptive.

This section of the chapter is devoted to case studies aimed at understanding how the size and shape of plants affects their ability to grow and produce offspring. How do biologists test the hypothesis that certain body types or growth patterns are adaptive—meaning that they lead to greater survival and reproduction in certain environments? Let's begin with a classical experiment on why the size and shape of individuals varies among populations of the same species.

How Does Genetic and Environmental Variation Affect the Form and Function of Plants?

As any farmer or gardener can tell you, individuals from the same plant species can vary enormously in size and shape depending on the amount of light, nutrients, and water they receive. For example, the leaves in **Figure 31.16a** (page 616) are from the same tree. The differences between them do not arise because of differences in their genetic makeup, but because one developed in shaded portions of the lower body of the tree while another developed in full sun near the top of the tree. **Figure 31.16b** shows that individuals from the same species can have radically different overall shapes. In this case, the robust individual grew in an open habitat while the tall, spindly individual grew up in a crowded forest where competition for light was intense.

The tendency for genetically identical organisms to differ in response to different environmental stimuli, or for individuals to vary over time in response to changing environmental conditions, is called **developmental plasticity**. Because of their indeterminate growth, plants tend to have much more developmental plasticity than do organisms such as bacteria or birds or mammals. This raises an important question. How much of the variation in size and shape that we observe among plants is due to the variation in the environments that individuals experience? How much is due to genetic divergence that is caused by natural selection and that adapts individuals to specific environmental conditions?

Jens Clausen and colleagues explored this issue in a series of influential experiments on a perennial herbaceous plant called

(a) Growth rings result from variation in cell size.

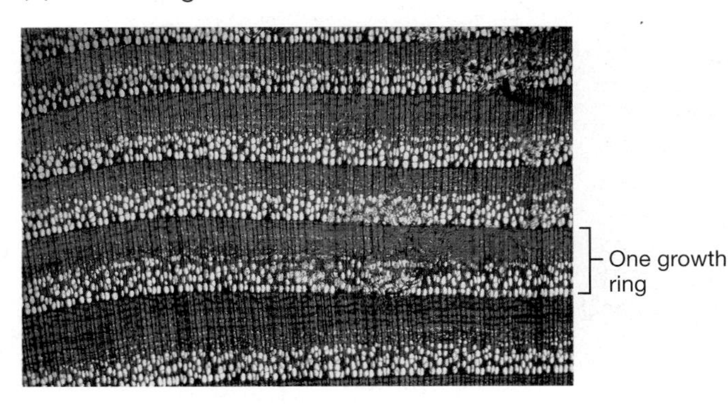

(b) Inner and outer xylem have different functions.

FIGURE 31.15 Anatomy of a Tree Trunk
(a) This section is stained to show individual cells. Growth rings appear because the size of secondary xylem cells that are produced changes during the course of a growing season. Cells produced at the start of the season are large; those produced near the end of the growing season are small. **(b)** This section is not stained; it shows the marked color difference between heartwood and sapwood.

glandulosa. In California, populations of *P. glandulosa* are found in a wide variety of habitats. Clausen and co-workers focused on understanding the cause of morphological variation in the populations at three study sites: along the coast (elevation 30m or 100 ft.), in the foothills of the Sierra Nevada mountains (elevation 1400m or 2500 ft.), and at an alpine site above timberline in the same mountain range (elevation 3050m or 10,500 ft.).

To begin their study, the researchers documented measurable differences in a total of 16 morphological and physiological traits among the three populations. (A few of these variable traits are summarized in **Figure 31.17a**). The question they asked was: Are these differences determined entirely by the environments that prevail at the three locations, or do they represent adaptations resulting from the action of natural selection on genetic variation?

Clausen and colleagues used two experimental strategies to answer this question. One approach was based on making a series of crosses between parents from the three populations. For example, they mated alpine individuals with coastal individuals. To explore the genetic basis of the variable traits they had measured, they reared the F_1 progeny in a common garden. This step minimized the amount of variation among individuals that was due to environmental effects. When these offspring were self-fertilized and the F_2 generation reared to maturity in the same garden, the researchers were able to confirm that each of the 16 characters they scored had segregated independently. You might recall from Chapter 10 that this protocol and result are similar to Mendel's experiments with garden peas. The crosses among *P. glandulosa* populations confirmed that the variation observed in each of the traits is at least partly due to variation in the alleles present at different loci. Different alleles appeared to be more prevalent in different populations.

The second approach pursued by the researchers involved transplanting individuals into gardens established at each of the three sites. The biologists took cuttings from the stems of individuals growing along the coast, in the foothills, and above timberline and then transplanted cuttings from the same plants at each of the three sites. This experimental design is diagrammed in **Figure 31.17b**. The biologists' goal was to compare the size and shape of genetically identical individuals from each population at the three sites. If the morphological differences among populations were due to developmental plasticity only, then individuals from each of the different sites should look alike in each of the three experimental gardens. If the morphological differences among populations were due solely to genetic differences, then plants from the alpine habitat should still have the alpine growth form in the foothills and coastal habitats.

As **Figure 31.17c** shows, the results support the hypothesis that genetic differences explain a great deal of the morphological variation in these populations. Individuals retained their normal growth form in each experimental garden. Further, at each location the individuals that were native to that site grew best. In the alpine garden, for example, plants native to that habitat grew well while individuals native to the foothills habitat died. Combined with the results of the crossing experiments, the results strongly support the hypothesis that many, if not most, of the morphological and physiological differences among *P. glandulosa* populations result from differences in their genetic make-up.

In other species, however, developmental plasticity is just as important as genetic variation in explaining morphological variation among populations. As a result, it would be misleading to make a broad generalization about the relative importance of developmental plasticity versus genetic variation to

(a) Leaves of the same plant

Grown in shade Grown in sun

(b) Individuals of the same species

Grown in the open Grown in crowded conditions

FIGURE 31.16 Developmental Plasticity
(a) Leaves that grow in sunlight versus shade may have very different shapes and sizes, even though they are part of the same individual. **(b)** Individuals from the same species may have very different shapes and sizes, depending on environmental conditions. The tree at the left grew in an open habitat with full sun; the tree at the right grew in a crowded forest where competition for light was intense.

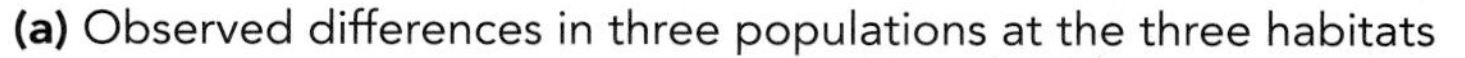

(a) Observed differences in three populations at the three habitats

	Coastal habitat (mild summers, rain and fog in winter)	**Foothills habitat** (warm, dry slopes, cold in winter)	**Alpine habitat** (above timberline, short, dry summers, long winters)
Overall size and shape	Large and robust	Tall with slender highly-branched stems	Dwarf
Leaf size	Large	Large	Small
Flower size	Small	Small	Large
Growing season	Year around; resistant to frost	Dormant in winter; susceptible to frost	Dormant for 9 months; rapid flowering in summer

(b) Transplant experiment

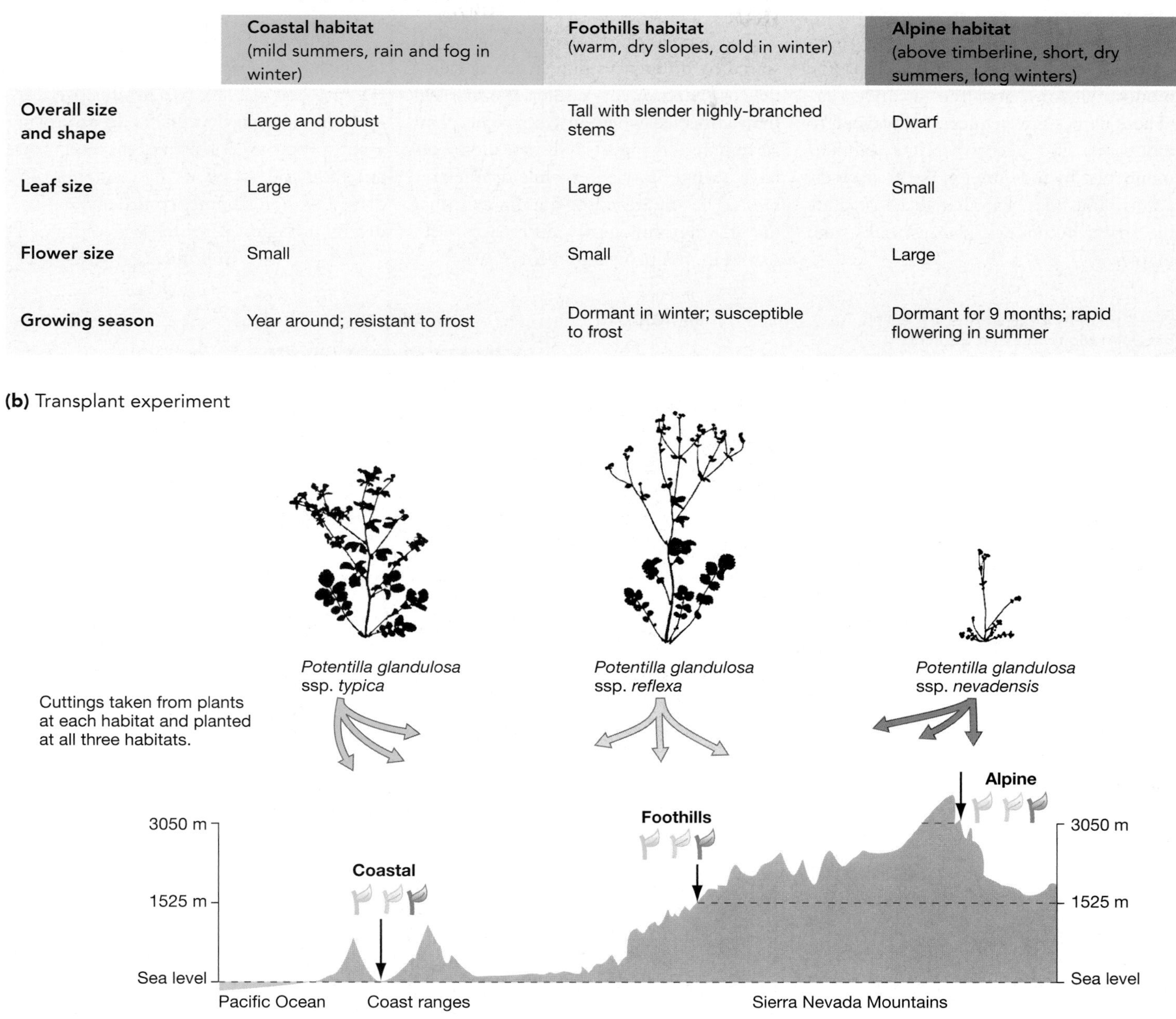

(c) Sample of plants resulting from transplant experiment

FIGURE 31.17 Do Different Populations Vary Because of Developmental Plasticity or Genetic Differences?
(a) Researchers documented significant variation in a total of 16 morphological and physiological traits among populations of *Potentilla glandulosa*. A few of the differences are summarized here. **(b)** To answer the question posed in the title of this figure, biologists took cuttings from individuals in each population and grew them in gardens established at the three habitats indicated. **(c)** Genetically identical individuals have the same general growth form if they grow up in different habitats. But in each habitat, individuals native to that area thrived best. **QUESTION** What is the best answer to the question in the figure title—developmental plasticity, genetic differences, or both?

BOX 31.2 Monocots and Dicots

The angiosperms have traditionally been grouped into two major lineages: the monocotyledons and the dicotyledons. (These names are frequently shortened to monocots and dicots.) Some familiar monocots include the grasses, orchids, palms, and lilies. Familiar dicots include the roses, buttercups. daisies, oaks, and maples.

The names of these groups were inspired by differences in a structure called the cotyledon. A cotyledon is the first leaf that is formed when an embryonic plant germinates; as **Figure 1** shows, monocots have a single cotyledon while dicots have two. The figure also highlights other major morphological differences observed in monocots and dicots.

It would be misleading, however, to think that all species of flowering plants fall into one of these two groups. Recent work has shown that dicots do not form a natural group consisting of a common ancestor and all of its descendants. To drive this point home, consider the phylogeny in **Figure 2**, which was estimated

(Continued on next page)

FIGURE 1 Morphological Differences Between Monocots and Eudicots
These photographs summarize morphological differences between monocots and species in the lineage called eudicots.

MONOCOTS | DICOTS

One cotyledon | Two cotyledons

Cotyledons
Monocots have one cotyledon, or "seed-leaf"; dicots have two.

Parallel veins | Branching veins

Veins in leaves
In monocot leaves, the vascular bundles called veins run side by side; in dicots they fan out.

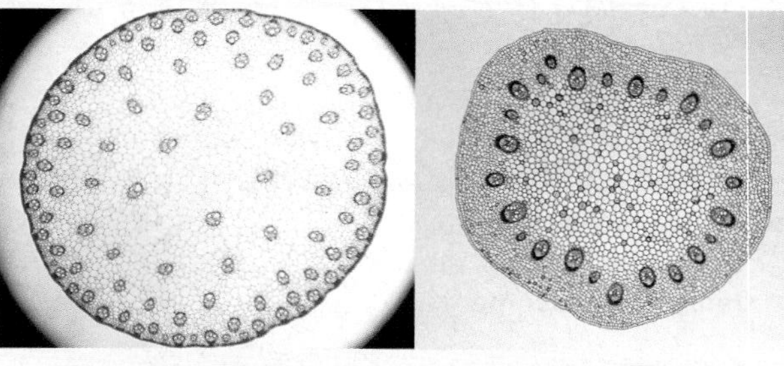

Scattered throughout | Circular arrangement

Vascular tissue in stems
In monocots, bundles of vascular tissue are scattered throughout the ground tissue in stems; in dicots they are found in a circular arrangement.

Multiples of 3 | Multiples of 4 or 5

Flower form
Flower parts are usually found in multiples of three in monocots but multiples of four or five in dicots.

explain why variation in plant anatomy occurs. The real value of the experiments on *P. glandulosa* lies in illustrating the types of approaches that biologists can use to rigorously test the hypothesis that at least some of the variation observed in plant anatomy results from genetic divergence.

Patterns in Leaf Traits

The work on *P. glandulosa* is an excellent example of how well-designed experiments can help biologists understand the relationship between developmental plasticity and genetic differences among individuals. As an example of how researchers address questions about the morphological variation that exists among a broad range of species, instead of within a single species, consider a recent study by Peter Reich and colleagues.

Both within and among habitats, the leaves found in different plant species vary widely in size, life span, nitrogen content, and rates of photosynthesis and respiration. (Leaf life span is defined as the average length of time a leaf exists before it dies and falls from the plant.) In an attempt to make sense of this variation, Reich and co-workers collected data on the leaf characteristics of 111 species from six dramatically different habitats, ranging from tropical rain forest to deserts, prairies, and northern forests. When the biologists analyzed the relationships among these leaf characteristics, several striking patterns emerged. As an example of their results, consider the graph in **Figure 31.18a** (page 620). The x-axis plots leaf life span in months, while the y-axis plots the concentration of nitrogen in leaves in units of milligrams of nitrogen/gram of leaf. Nitrogen concentration provides an index of protein and nucleic acid content, or leaf "richness." Leaf richness reflects the amount of resources invested in each leaf. It also indicates the concentration of enzymes and the productivity of each leaf. A statistical analysis confirms the visual impression in the graph—that the two traits are negatively correlated. In other words, short-lived leaves tend to have a high nitrogen concentration while long-lived leaves tend to have a low nitrogen concentration. **Figure 31.18b**, in contrast, shows that there is a positive correlation between leaf thinness and nitrogen content. Thick leaves tend to have low nitrogen concentration. **Figure 31.18c** summarizes other relationships that Reich and co-workers discovered.

Why do these patterns exist? The biologists considered two hypotheses. The first is that leaf traits depend on habitat. For example, desert species might tend to have short-lived leaves with high nitrogen concentrations, while species from northern forests might tend to have long-lived leaves with low nitrogen concentrations. According to the analyses done by Reich and colleagues, however, this hypothesis is not correct. Short-lived leaves tend to have high nitrogen concentrations whether they are found in desert flowers or

(Box 31.2 continued)

by comparing the sequences of several genes shared by all angiosperms. Note that species with dicot-like characters are actually scattered around the angiosperm tree. To use the vocabulary introduced in Chapter 27, dicots are not a monophyletic group. They are paraphyletic. Based on this observation, many biologists no longer use the term *dicot*. Instead, they refer to the lineage that includes roses, daisies, and maples as the eudicots (true dicots). Magnolias, laurels, and other non-monocot lineages now have names that lack the word *dicot*.

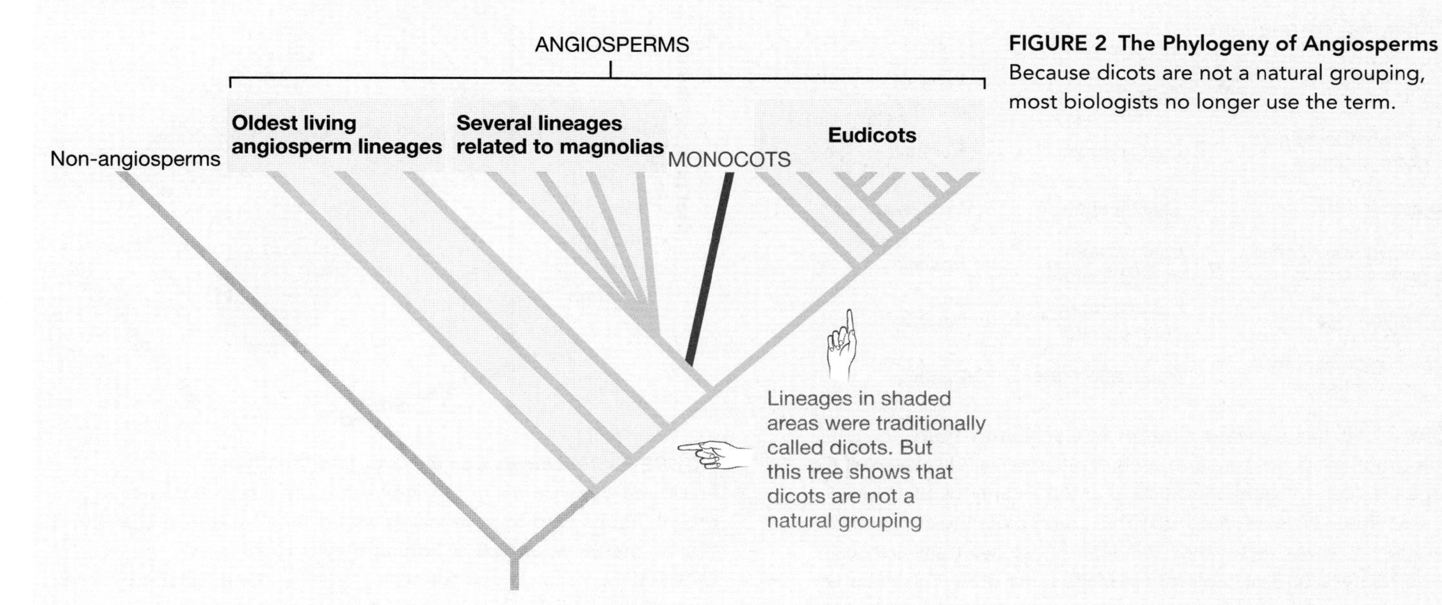

FIGURE 2 The Phylogeny of Angiosperms
Because dicots are not a natural grouping, most biologists no longer use the term.

tropical rain forest trees. An alternative hypothesis is that leaf traits correlate with specific growth habits and lifestyles. For example, certain plant species are considered pioneers because they are the first to colonize an area after the existing vegetation has been disturbed or destroyed by fire, windstorms, or animal activity. These "weedy" plants are usually adapted to the conditions that prevail after disturbance: bright sunlight, low competition, and poor or erratic nutrient availability.

To test the hypothesis that pioneering plants share certain leaf characteristics regardless of the habitat they occupy, Reich and associates placed the species in their study into five groups, defined by distinctive lifestyles, and then plotted leaf traits for the species in each group. **Figure 31.19** shows the three-dimensional plot that resulted. The graph indicates that species with similar growth habits tend to have leaves with similar characteristics—regardless of whether the plants are native to prairies or arctic forests.

The fundamental message of these data is that leaf characteristics appear to be an adaptation to particular growth habits and lifestyles—not temperature or moisture regimes. An even more general message here is that researchers have an array of techniques available for making sense of the enormous variation that exists in plant anatomy. In many cases, plants look the way they do because they inherited certain characteristics from their ancestors. All species of maple, for example, share a distinctive leaf form and seed shape. But in the case of the leaf traits studied by Reich and associates, variation also appears to be an adaptive response to certain growth habits and lifestyles.

(a) Leaf nitrogen is negatively correlated with leaf lifespan.

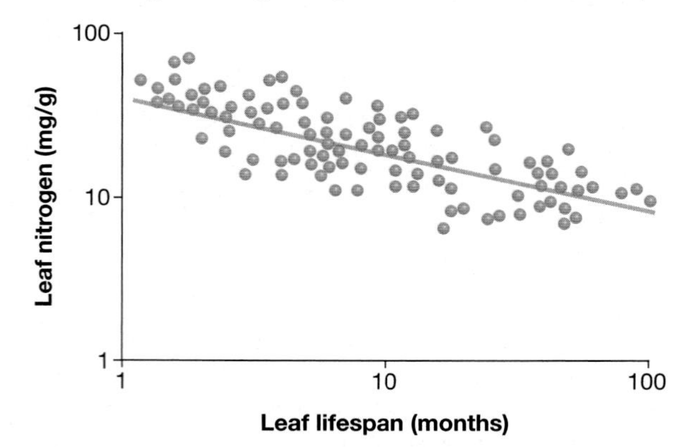

(b) Leaf nitrogen is positively correlated with leaf thinness.

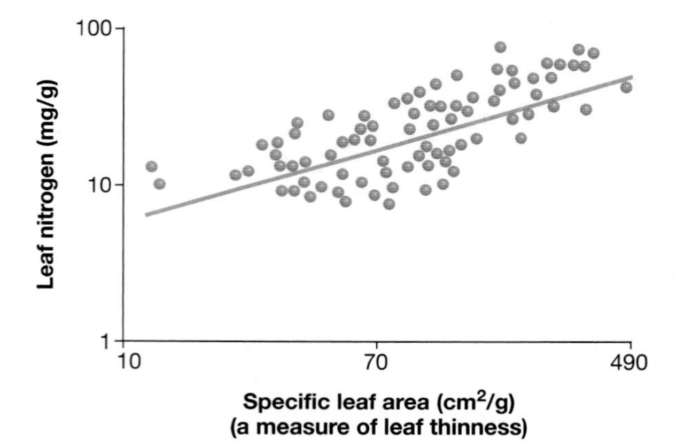

(c) Other leaf correlations

Trait 1	Trait 2	Type of correlation
Rate of photosynthesis per gram of tissue	Leaf lifespan	Negative
Respiration rate	Leaf lifespan	Negative
Rate of photosynthesis per gram of tissue	Leaf nitrogen concentration	Positive
Respiration rate	Leaf nitrogen concentration	Positive
Rate of photosynthesis per gram of tissue	Respiration rate	Positive

FIGURE 31.18 Leaves Have a Series of Correlated Traits
(a) This graph plots the concentration of nitrogen in leaves versus the average life span of a leaf. A statistical analysis confirms that these traits are negatively correlated. **(b)** This graph plots the concentration of nitrogen in leaves versus their thinness. These two traits are positively correlated. **(c)** This table summarizes some other correlations observed between leaf traits.

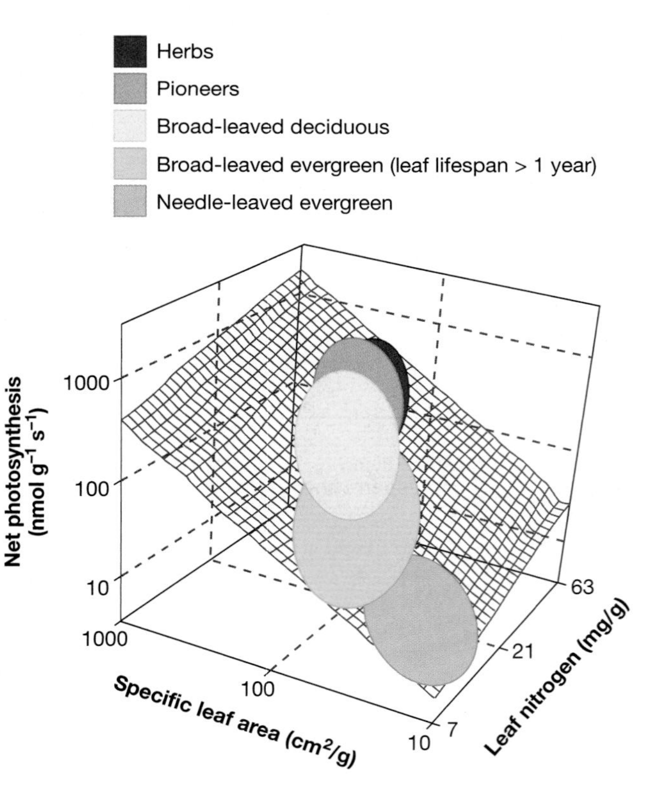

FIGURE 31.19 Leaves Can Be Grouped Into Types
This three-dimensional graph shows that species with particular growth habits tend to have leaves with distinctive sets of characteristics, no matter what type of habitat they occupy.
QUESTION According to the data presented here, what are some prominent characteristics of the leaves found in pioneering species?

BOX 31.3 Tree-Ring Studies

In the Sierra Nevada Mountains of California, dead tree trunks decay so slowly that growth rings can still be discerned in plant material that is 3500 years old. To gain a better understanding of how climate and vegetation have changed at this location over this period, Andrea Lloyd and Lisa Graumlich extracted samples from a large number of these ancient trunks using an instrument called an increment borer. By correlating patterns in the fossil growth rings with patterns present in the living trunks of extremely old trees nearby (**Figure 1**), Lloyd and Graumlich could assign exact or nearly exact ages to each tree in the sample.

The resulting data set represented a nearly continuous record of forest density at these study sites, dating back to the period when the Minoan civilization of Crete dominated the Mediterranean region, Egypt was at the height of its power, and the Shang dynasty ruled China. Although all of the study sites are above the zone where trees are currently able to grow, the tree-ring data showed that the sites were heavily forested during most of the past 3500 years. Will forests once again begin to grow at high altitudes in the Sierra Nevada, now that a period of global warming is under way? According to Lloyd and Graumlich's data, warm periods in the past coincided with heavy forest growth *if* the climate became wetter as well. Although it is too early to make a strong prediction, tree seedlings have recently become established at their study sites—hinting that in this mountain range, forests may soon begin climbing toward the peaks once again.

FIGURE 1 Studying Fossil Growth Rings Using an instrument called an increment borer, researchers can obtain samples of growth rings present in a woody stem.

Essay Wood as a Structural Material

Secondary xylem tissue—the substance called wood—helps trees remain upright in the face of hurricane-force winds and keeps them from snapping under heavy loads of snow and ice. Secondary xylem has also provided humans with building material for thousands of years. Why is wood such an effective structural component in construction? The answer begins at the molecular level. The cellulose fibers that are the primary component of plant cell walls surpass nylon, silk, tendon, and bone in their ability to withstand pulling force, or tension. Further, the strands of cellulose found in the thickened secondary walls of xylem tissue are enmeshed in a network of tough lignin molecules. The addition of lignin greatly strengthens the cell walls.

Why is wood such an effective structural component in construction?

To understand how these tough cell walls help produce wood's most striking structural characteristic, recall that xylem cells are dead at maturity. In effect, then, wood is a honeycombed lattice made up of pockets of empty space surrounded by tough cell walls. This combination of air-filled chambers and strong walls makes wood lumber exceptionally strong for its weight. Relative to its density, wood is better able to resist bending and twisting than are concrete, cast iron, aluminum alloy, or steel.

Combining high strength and low weight is important because every component that is used to stiffen or support a structure also adds weight, increasing the total load that the structure must bear. As a structure grows in height, then, it is important that light materials be used to keep the overall load to a minimum. The combination of strength and lightness is the key to wood's success as a structural material.

An additional feature qualifies tree trunks as particularly well engineered structures. As a tree grows in height, its vascular cambium adds extra width at the base of the trunk. **Figure 1** shows how this flare at the bottom gives large trees a shape that is similar to the architectural design used for tall, thin structures such as the Eiffel tower. A broadened base helps tree trunks withstand the lateral forces exerted by wind. The resulting structure is so stable that it may stand for decades after the tree's death.

From an engineering viewpoint, trees are sophisticated structures. It is not surprising that these tall, central-stemmed plants are Earth's tallest organisms and are among the longest-lived.

FIGURE 1 How Can Tall, Thin Structures Be Made Stronger? The flare at the base of large trees and other tall, thin structures helps them withstand the lateral force of wind.

Chapter Review

Summary

The root and shoot systems of plants are specialized for harvesting the light, water, and nutrients required for performing photosynthesis and synthesizing sugar. Roots extract water and nutrients like nitrogen, potassium, and phosphorus from the soil; shoots capture light and carbon dioxide from the atmosphere.

The overall morphology of root and shoot systems varies widely among plant species. In the prairie plants of North America, for example, root systems range from long, linear taproots to shallow, dense mats. Among the silverswords of Hawaii, shoot systems vary from low rosettes to woody, highly branched tree trunks. Experimental and phylogenetic data support the hypothesis that much of this variation is adaptive, meaning that different sizes and shapes of root and shoot systems help individuals survive and reproduce better in certain habitats. In addition, closely related species of plants tend to be similar in anatomy due to characteristics they inherited from a common ancestor.

Although many different types of cells are found in plants, they can be grouped into three general categories based on their function. Parenchyma cells carry out photosynthesis or store food, structural cells provide stiffness to the body, and vascular cells transport materials between the root and shoot systems.

The groups of cells called tissues can also be placed into three functional categories. The arrangement of these epidermal, ground, and vascular tissues varies among the parts of the same plant—in leaves versus stems versus roots, for example. The relationship among these tissues also varies when a single body part—a stem for example—is compared in different species. Vascular tissues are further divided into xylem that is specialized for carrying water and nutrients from roots to shoots and phloem that is specialized for transporting sugar.

The detailed anatomy of roots and shoots is consistent with their role as resource-harvesting systems. Root hairs increase the membrane surface area available for water and nutrient absorption, and lateral roots increase the amount of soil penetrated by the system. The organization of vascular tissues into a central stele allows efficient transport of water and nutrients. In the shoot, the number of lateral branches and the length of internodes dictate the arrangement of leaves on the plant and the amount of incoming light received by the plant.

All plants grow from apical meristems that are located at the tips of roots and shoots. Apical meristems are responsible for primary growth that results in the elongation of both systems. In some plants, other meristems produce secondary growth that widens shoots and roots. For example, in many trees a vascular cambium produces secondary layers of xylem and phloem (in addition to the primary xylem and phloem produced by the apical meristem), while cork cambiums produce bark.

The anatomy of plants provides a structural framework for solving the problems posed by living on land. As we'll see, these challenges range from absorbing and transporting water and nutrients to sensing and responding to light, gravity, and enemies.

Questions

Content Review

1. Which statement best characterizes secondary growth?
   a. It results from divisions of the vascular cambium cells.
   b. It increases the length of the plant stem.
   c. It results from divisions in the apical meristem cells.
   d. It often produces phloem cells to the inside and xylem cells to the outside of the vascular cambium.

2. Which statement best characterizes primary growth?
   a. It does not occur in roots, only in shoots.
   b. It leads to the development of cork.
   c. It produces epidermal, ground, and vascular tissues.
   d. It produces rings of xylem tissue.

3. What do parenchyma cells do?
   a. They are specialized to transport food.
   b. They are specialized to transport water.
   c. They have thickened cell walls that provide support for roots and shoots.
   d. They contain chloroplasts or store substances such as starch.

4. What features of tracheids makes them effective water-conducting cells? (Choose all that apply.)
   a. They are arranged end to end and form continuous columns.
   b. They have gaps in their primary cell walls that connect to adjacent tracheids.
   c. They are dead at maturity.
   d. They are alive at maturity.

5. What is a sieve-tube element?
   a. the food-conducting cell found in phloem
   b. the advanced water-conducting cell found only in angiosperms
   c. the nutrient- and water-absorbing cell found in root hairs
   d. the stiffened structural cell found in leaves

6. What three basic tissue types are found in plants?
   a. xylem, phloem, and sieve-tube
   b. ground, vascular, and epidermal
   c. stele, cambium, and lateral
   d. meristematic, wood, and parenchyma

Conceptual Review

1. What is indeterminate growth? Why does indeterminate growth make plant growth and development especially responsive—or "plastic"—to environmental conditions?

2. Describe the general function of the shoot system and the root system. What tissues are continuous throughout these two systems? Why are the shoot and root systems of various species so variable in size and shape?

3. Describe how the vascular cambium produces secondary xylem and phloem. Why do growth rings occur when tree trunks are viewed in cross section?

4. What do stomata do? What does cuticle do? Predict how these two anatomical features differ in plants from wet habitats versus those from dry habitats. How would you test your predictions?

Applying Ideas

1. The total number of stomata on a leaf is usually a function of the blade's size. Broad leaves tend to have many stomata. This correlation sets up a trade-off: Large leaves can harvest more light and manufacture more sugar, but they lose more water than small or narrow leaves. Based on this trade-off, would you expect broad-leaved plants to be found in wet or dry habitats? Which environments should be dominated by narrow-leaved plants?

2. Chapter 1 explains how the shoot systems of domesticated broccoli family plants (broccoli, cauliflower, cabbage, kohlrabi, and others) changed in response to artificial selection. Make a sketch of three varieties; then predict which cells, tissues, and systems changed as the ancestral population pictured in Figure 1.3 underwent artificial selection. For example, would you predict that leafy populations like kale and savoy have more or fewer support cells in their stems and leaves than the wild population? How would you test your predictions?

3. Go to the nearest window, look out, and choose a plant that is in view. Sketch it and label the prominent features of its external anatomy. Make a cutaway view that predicts the major features of its internal anatomy. If possible, obtain a twig, stem, or section of root. Examine and label its parts.

4. Choose a favorite plant species. Design an experiment that will reveal how individuals of this species grow in response to environmental conditions such as poor nutrients, low water or light availability, and constant wind. If different populations of this species vary in size or shape, how would you test the hypothesis that some of the variation represents genetically based adaptations to certain environmental conditions?

CD-ROM and Web Connection

CD Activity 31.1: Seasonal Stem Growth *(animation)*
(Estimated time for completion = 5 min)
Learn how to interpret the growth history evident in the scars left on the tree's stem.

CD Activity 31.2: Primary and Secondary Growth *(animation)*
(Estimated time for completion = 5 min)
Follow a plant through time and observe primary growth and secondary growth.

At your **Companion Website** (http://www.prenhall.com/freeman/biology), you will find self-grading exams and links to the following research tools, online resources, and activities:

Hawaiian Silverswords
This site describes the morphological and genetic diversity of the Hawaiian silverswords.

Dendrochronology
This site shows how dendrochronology is used to study the ecology and life histories of the bristlecone pines of the White Mountains of California.

Additional Reading

Niklas, K. 1996. How to build a tree. *Natural History* 105 (February): 48–49.

Norris, S. 2000. Reading between the lines. *BioScience* 50: 389–394. A look at how biologists use tree rings to analyze climate change and fire history.

Water and Sugar Transport in Plants

32

On a hot summer day, a single deciduous tree can lose enough water to fill three 55-gallon drums. Why? Recall from Chapter 31 that the surfaces of leaves are dotted with structures called stomata. Stomata open during the day so that gas exchange can occur between the atmosphere and the cells inside the leaf. This exchange is crucial because, for photosynthesis and food production to continue, leaf cells must acquire carbon dioxide (CO_2) and release a waste product—oxygen. There's a catch, however. While stomata are open so gas exchange can take place, the moist interior of the leaf is exposed to the dry atmosphere. As a result, huge quantities of water evaporate.

In essence, then, water loss is an inevitable consequence of a plant's need to obtain carbon and release oxygen. Water loss is a side effect of conducting photosynthesis. The question is, how do plants make up for this loss? In the case of a redwood tree, the leaves that are losing water may be located 100m from the root hairs that are absorbing water. How do plants transport water a distance equivalent to the length of a football field against the force of gravity? Similarly, how do plants move the sugar they produce from manufacturing sites in leaves to storage sites in roots?

These questions are the heart and soul of this chapter. Section 32.1 begins to answer them by examining how water moves from cell to cell in a plant. Section 32.2 returns to the topic of water loss from leaves, analyzes how it occurs, and considers recent tests of an old theory for how water moves from root to shoot through xylem. The chapter closes by reviewing experiments that explore how the products of photosynthesis move throughout the plant body via phloem tissues.

This giant sequoia is 82.9 meters (272 ft) tall and weighs just under 1300 metric tons—about the same as 10 diesel locomotives. The closely related redwoods, which also grow along the Pacific Coast of North America, are the tallest organisms that have ever lived. They can be over 100 m in height. This chapter explores how trees and other plants move water from their roots to their leaves and how they transport sugars to all of their tissues.

32.1 Water Potential and Cell-to-Cell Movement

To understand how water moves 100 meters up the trunk of a redwood tree, let's begin by analyzing how it moves just a few micrometers—into or out of a single cell. Consider the cell in the beaker on the left of **Figure 32.1a**. Note that it is sitting in a solution that has dissolved substances, or solutes, and that the solute concentrations in the cell and the solution are in equilibrium. Now suppose that this cell is transferred to the beaker of pure water on the right of Figure 32.1a, which has no solutes. You know from reading Chapter 4 that water will begin moving into the cell via osmosis. When solutions are separated by a selectively permeable membrane, water moves from regions of low solute concentration to regions of high solute concentration. Water's tendency to move via osmosis is a function of the difference in solute concentrations between two cells. This difference is called the **solute potential**, or osmotic potential.

As **Figure 32.1b** shows, however, water will not move into the experimental cell until the cell bursts. As the cell swells in response to the incoming water, its cell membrane begins to contact the cell wall and push against it. This force is called **turgor pressure.** The cell wall exerts an equal and opposite force called **wall pressure.** Cells that are firm and that experience wall pressure are said to be **turgid.**

Wall pressure is important because it counteracts the movement of water due to osmosis. In the example illustrated in Figure 32.1, the solute potential favors water moving into the cell but wall pressure favors water moving out of the cell. Cells can experience other forms of pressure as well. As we'll see in section 32.2, the water inside plant cells is often under tension—a force that tends to pull water out. The sum of all the types of pressure on water is called its **pressure potential.**

The question is, how do biologists determine the direction of water movement when osmosis and pressure interact? What is the combined effect of solute potential and pressure potential?

Defining Water Potential

When the solute potential and pressure potential of a cell are added together, the resulting quantity is called the cell's water potential. **Water potential** can be thought of as the tendency of water to move from one location to another. It is actually a form of potential energy, not unlike chemical energy. Recall from Chapter 2 that the potential energy of an electron can be thought of as its tendency to move to a new position, and that it is defined by its location relative to the atomic nucleus.

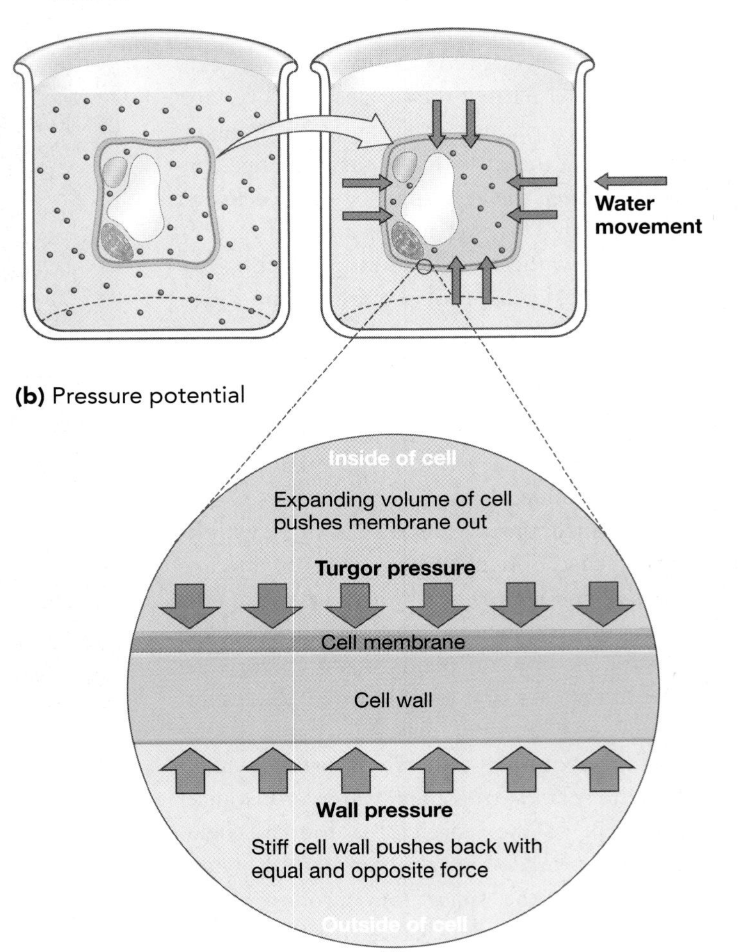

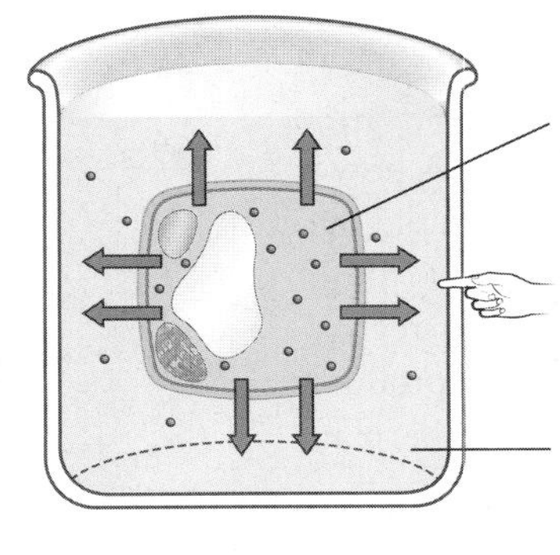

FIGURE 32.1 Solute Potential and Pressure Potential
(a) The solute potential of a solution is defined by its solute concentration relative to pure water. In the drawing at the left, the concentration of solutes in the solutions on either side of the selectively permeable membrane is identical. As a result, the difference in the solute potentials of the two solutions is 0. In the drawing at the right, water moves from an area of high solute potential (the solution surrounding the cell) to an area of low solute potential (the cell), via osmosis.
(b) Wall pressure and other types of pressure define a cell's pressure potential. **(c)** If wall pressure is very high, it can counteract the tendency of water to move via osmosis as in the example given here.

Similarly, the water potential of a cell can be thought of as water's tendency to move to a new position and is defined by the pressure and solute potential at its location relative to other locations.

Water potential is symbolized by the Greek letter ψ (psi, pronounced "sigh"). Its two components—solute potential and pressure potential—are abbreviated as ψ_s and as ψ_p, respectively. Algebraically, then, water potential is defined as

$$\psi = \psi_p + \psi_s$$

In words, water potential is the sum of the pressure potential and the solute potential that it experiences. This tendency is measured with a unit of pressure called the megapascal (MPa).

The water potentials measured in plant cells are usually negative in sign. Solute potentials (ψ_s) are always negative because they are measured relative to the solute potential of pure water. Because it contains no dissolved substances, the solute potential of pure water is defined as 0. The solutes in a cell make the water inside less likely than pure water to move via osmosis. As a result, the water inside a cell has a lower potential energy than the potential energy in a bath of pure water. Increasing the concentration of solutes in a cell lowers its water potential even more. Because its solute potential makes water less likely to move out of the cell, it is logical that its sign is negative. In contrast, the pressure potential (ψ_p) from wall pressure in a turgid cell is positive, because it increases the potential energy of the water inside—it pushes on the water and makes it more likely to move out of the cell.

To help pull these ideas together, suppose that the solute potential of the cell on the left of Figure 32.1a is −1 MPa. When the cell is transferred to the beaker filled with pure water and water begins to enter, the solute potential of the cell becomes less negative and wall pressure begins to increase. When the positive wall pressure plus the negative solute potential equals 0 MPa—the water potential of pure water—there is no additional net movement of water. In this case, wall pressure is high enough to offset the tendency for water to enter the cell via osmosis.

What happens if the cell is then transferred into a beaker with a solute potential of −0.5 MPa, as in **Figure 31.1c**? The key to answering this question is to realize that water moves from regions of high water potential to regions of low water potential. The water potential of the cell is 0 MPa; the surrounding solution has a water potential of −0.5 MPa. As a result, water moves from the cell to the surrounding solution.

What happens when a cell is exposed to a solution that has an extremely low water potential? The photographs in **Figure 32.2** provide the answer. The cells at the left are from normal plant tissue and are turgid; the photograph on the right shows the same cells after being flooded with a highly concentrated sugar solution. The water potential of the cells is higher than the water potential of the surrounding solution, so water flows out of the cells. In this case, the cells have lost so much water that their membranes have shrunk back from the cell wall. This state is called **plasmolysis**. If the condition is not corrected, the cells will die.

The Effects of Water Movement

Although we have been focusing on the water potential of individual cells, it's important to realize that ψ can also be measured for tissues, root and shoot systems, and entire plants. The water contained within a leaf or root system or plant has a pressure potential and a solute potential, just as the water inside a cell does. Likewise, the air around the shoots of plants has a water potential. So does the soil that surrounds the root system. For example, the water in soil has a solute potential because it contains dissolved substances; it may also be under pressure.

It's also important to note that the water potentials of plant tissues, air, and soil are dynamic. They routinely change from day to day or even hour to hour in response to heat, cold, rain, and other conditions.

These observations are important because water moves from regions of high water potential to regions of low water potential. (More specifically, water moves from locations with less-negative water potentials to locations with more-negative water potentials.) The water potential of soil is usually high relative to plants, while the water potential of air is extremely low. These contrasts set up a series of water potential differences, called a **water potential gradient**, between the soil and plants and the atmosphere. Plants tend to gain water from the soil and lose it to the atmosphere.

When the cells inside a leaf or stem lose water, what are the consequences? If the water potential in the space that surrounds a cell drops, water moves out of the cell in response. Water loss makes cell walls contract, so cells begin shrinking like grapes drying into raisins. If the cells do not regain turgor

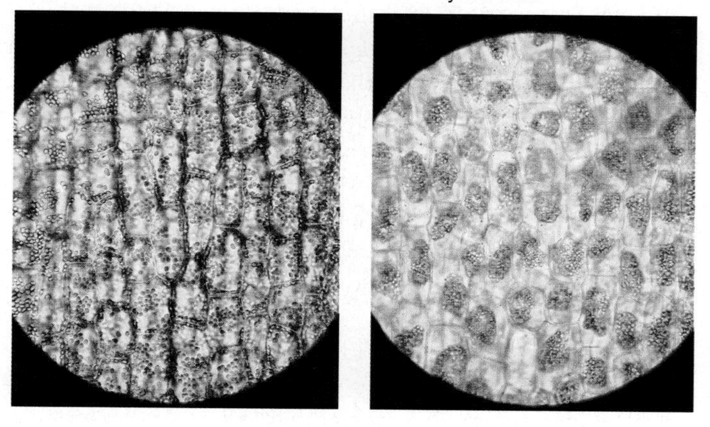

FIGURE 32.2 Turgor and Plasmolysis
The cells on the left are turgid. The cells on the right are plasmolyzed because they have been treated with a solution containing a high concentration of solutes. **QUESTION** Would you predict that photosynthesis, DNA repair and replication, and other normal cell activities occur in plasmolyzed cells? Why or why not?

in a reasonable length of time, they will be threatened with dehydration and death. Further, when an entire tissue loses turgor, it wilts. Unless corrected, wilting leads to the death of the tissue and, eventually, the plant.

In a meaningful sense, then, the water potential of plant tissues is a measure of dehydration and drought stress. If plants are constantly losing water, how do they replace it?

32.2 Transpiration and Water Movement from Roots to Leaves

As you know from the introduction to this chapter, leaves lose water through their stomatal pores as an inevitable by-product of gas exchange. Water loss from the aerial parts of a plant is called **transpiration**. As many as 500 molecules of water are lost from stomatal pores for every molecule of CO_2 that is absorbed from the atmosphere and fixed by the enzyme rubisco. To avoid dehydration and turgor loss, plants must absorb huge quantities of water from soil and transport it to their leaves. How does this happen? Understanding the answer depends on grasping how the water molecules inside leaves behave.

The Cohesion-Tension Theory

Where does transpiration take place? The photograph in **Figure 32.3a** shows that the leaf area just below a stoma is not packed with cells. Instead there is space filled with moist air. When the pore of a stoma opens, this humid air is exposed to the atmosphere. The water potential of the atmosphere is extremely low compared with the water potential of the space inside the leaf, meaning that there is a steep water potential gradient between the leaf interior and the air. Water exits through the stomatal pore as a result (**Figure 32.3b**).

Something interesting begins to happen as water exits the leaf and enters the atmosphere. As the humidity of the gas-filled space inside the leaf drops, water begins to evaporate from the cell walls of leaf cells. The water inside the cell walls, at the interface with the air, forms a concave boundary layer called a meniscus (plural, *menisci*). If the air is very hot and dry and the water potential gradient is especially steep, then water molecules leave the surface rapidly and the menisci in the cell walls become more concave. In 1894, Henry Dixon and John Joly hypothesized that the formation of these menisci produced a force capable of pulling water up from the roots, dozens of meters into the air. At first glance, Dixon and Joly's idea seems fantastic. Is it really possible?

CD ACTIVITY 32.1 Solute Transport in Plants

The Role of Surface Tension in Water Transport A meniscus actually forms at all air-water interfaces. Menisci form because of a property of water that was introduced in Chapter 2. Water molecules are polar and interact with one another through hydrogen bonding. Under an air-water interface, in the body of a solution, all of the water molecules present are surrounded by other water molecules and form hydrogen bonds in all directions. The water molecules on the surface, however, can form hydrogen bonds in only one direction—with the water molecules below them (**Figure 32.4**). As a result, the topmost layer of water molecules is pulled inward and the surface forms a meniscus. If water molecules leave the surface rapidly due to a steep water-potential gradient with the air, fewer molecules are available at the surface and the inward pull on the remaining molecules becomes stronger. This deepens the meniscus.

The pull that occurs on water molecules at the air-water interface is called **surface tension.** (*Tension* refers to a pulling force.) Dixon and Joly proposed that the force generated at the air-water interface is transmitted through the water present in leaf cells to the water present in xylem tissue, on to the water in the vascular tissue of roots, and finally on to the water in the soil. This continuous transmission of pulling force is possible because water is present throughout the plant, and because all of the water molecules present bond to one another in a continuous fashion. In this way, tension at the air-water interface inside the leaves pulls up water molecules from the soil.

(a) Inside a leaf, the area not occupied by cells is filled with moist air.

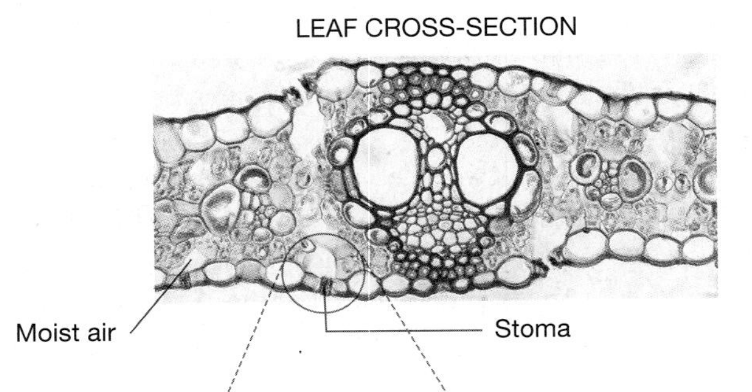

(b) Water moves from the inside of the leaf to the atmosphere, down a water potential gradient.

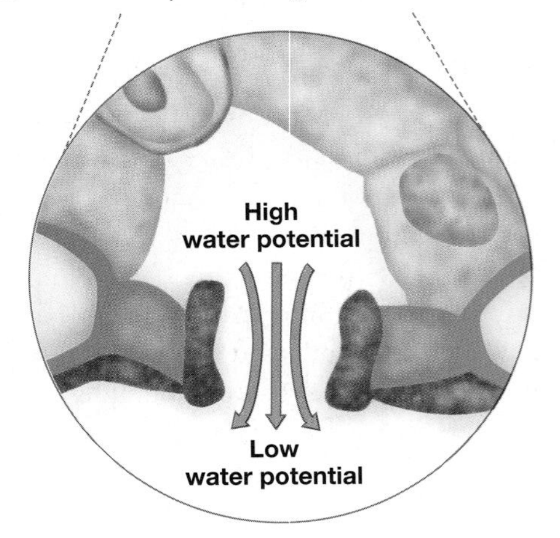

FIGURE 32.3 Where Does Transpiration Occur, and Why?
QUESTION How does the situation diagrammed in part (b) change during a rainstorm or fog? In this case, the air surrounding a leaf is just as humid as the air inside the leaf.

Bonding between similar molecules is called **cohesion**. It is logical, then, that the Dixon and Joly proposal is called the **cohesion-tension theory** of water movement. In effect, the theory claims that because of the hydrogen bonding that occurs between water molecules, the water inside a plants acts as a continuous chain that transmits the pulling force generated by surface tension in leaf cells.

If this hypothesis is true, then the water present in xylem should experience a strong pulling force. Do any data support this prediction?

What Evidence Do Biologists Have That Xylem Sap Is Under Tension? If you take a leaf and cut its petiole, the watery fluid in the xylem, or **xylem sap**, withdraws from the edge. This simple observation confirms that xylem sap is under tension—otherwise the fluid would not have moved. It is exactly the result predicted by the mechanism of transpiration pull just outlined.

In addition to supporting the cohesion-tension theory, the observation that xylem sap is under tension inspired P. F. Scholander, H. T. Hammel, and others to develop an instrument for measuring the water potential of plant tissues. This tool, called a pressure bomb, is diagrammed in **Figure 32.5**. Its design is based on an important insight: The tension experienced by xylem sap should correlate directly with the water potential of the tissue surrounding it. If the tissue is drought stressed and has a low water potential, then the pull on the xylem sap should be greater than it is when the tissue is fully hydrated and has a high water potential.

This insight is translated into data by placing a leaf or branch in an airtight container, applying a steadily increasing external pressure, and recording the pressure required to push the xylem sap back to the cut surface (Figure 32.5). This pressure is equal to the tension experienced by the xylem sap.

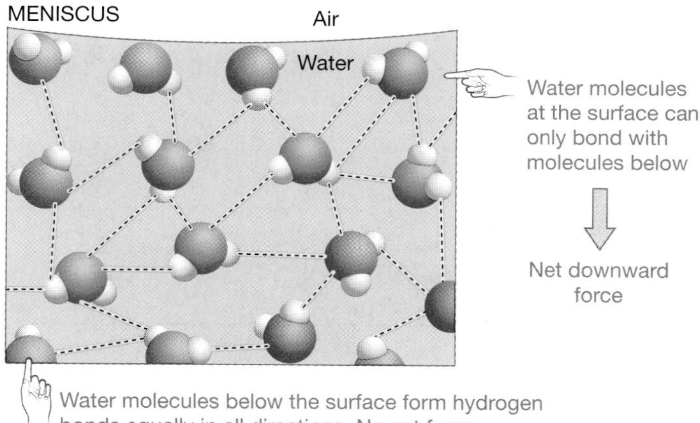

FIGURE 32.4 Why Does Surface Tension Exist?
Water molecules at an air-water interface can form hydrogen bonds only with molecules below them. This asymmetric bonding pulls the molecules at the surface down. Water molecules below the surface, in contrast, hydrogen bond equally in all directions and experience no net force.

To see how a pressure bomb can be used to study water transport, consider data that J. Hellkvist and colleagues collected on the water potential of Sitka spruce leaves. These biologists collected branches from the tops of 10m-tall spruce trees several times a day for several days, used a pressure bomb to measure the water potential of the tissues, and produced the data shown in **Figure 32.6**. The fundamental observation is that the water potential of these tissues dropped during the day and rose at night. This result makes sense given that stomata are open during the day when light is available and photosynthesis is under way. At night, stomata close and transpiration virtually ceases.

Before moving on, it is important to note that the force exerted on xylem sap can be extremely high. The water pressure

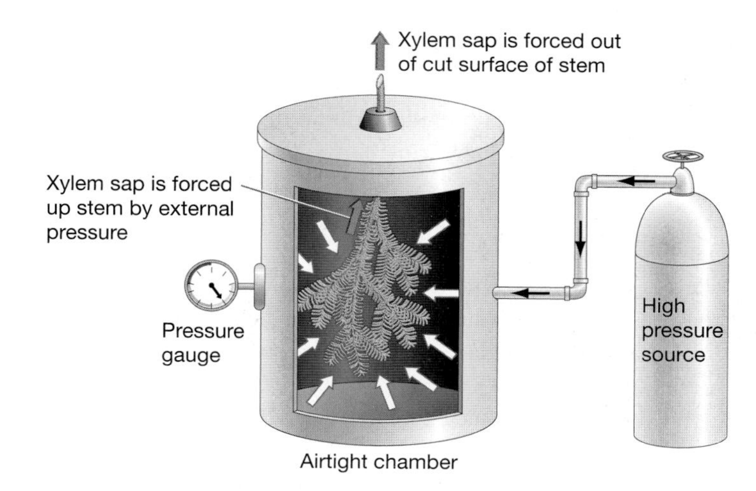

FIGURE 32.5 A Pressure Bomb
When a leaf, stem, root, or other tissue is placed in an airtight chamber and an external pressure is applied, xylem sap is forced up and out. The amount of external pressure required to force the sap out correlates with the water potential of the tissue.

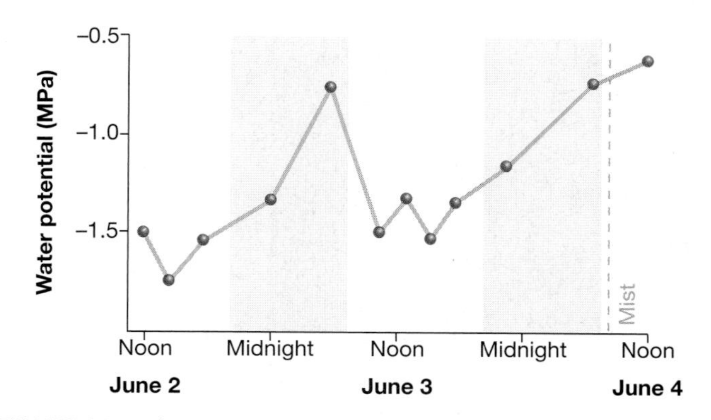

FIGURE 32.6 The Water Potential of Spruce Tree Leaves Varies over the Course of a Day
These data are for branches from the tops of spruce trees, over a two-day period in early summer. QUESTION The note "mist" indicates that a light rain fell on the morning of June 4. Why didn't the water potential of the tree leaves decrease that morning as it had the previous two days when it didn't rain?

in the water pipes where you live is probably about 0.25 MPa; the xylem sap of trees can experience tensions greater than −0.50 MPa. Work by Harold Fritts confirmed that this tension is actually great enough to make tree trunks shrink. Fritts used the instrument illustrated in **Figure 32.7a** to document changes in the diameter of beech trees growing outside of Columbus, Ohio. **Figure 32.7b** shows data from one tree collected over the course of a week. Note that the tree expanded at night and shrank during the day. On the hottest day of that week, July 14, transpiration rates were particularly high and the total change in trunk diameter was over a thousandth of an inch.

Direct Tests of the Cohesion-Tension Theory The observation that xylem sap is under tension has been confirmed hundreds of times. Whether this tension is large enough to account for all water movement in plants—especially movement to the tops of all trees—has been controversial, however. Further, some studies have failed to document rapid changes in xylem pressure when temperature and humidity conditions are changed experimentally. Changes in light availability, temperature, or humidity should change either the degree to which stomata open or the transpiration rate or both, and thus should lead to rapid changes in xylem pressure. Failure to observe these changes undermines biologists' confidence in the cohesion-tension theory, because the model predicts that the tension on xylem sap should change almost immediately when transpiration slows down or speeds up.

It is important to recognize, however, that these experiments represent a negative result. A **negative result** fails to confirm a predicted outcome. Although positive results can be definitive, negative results are by nature problematic. In this case, biologists who support the cohesion-tension theory could claim that the negative result was the product of a flawed experimental design. For example, a critic could argue that the equipment used was not sensitive enough to record rapid changes in xylem pressure, or that the species being tested had peculiar properties that made it inappropriate as an experimental subject. In this case, was the negative result produced by inadequate tools or by some other problem with the experimental design? Or was it due to a fundamental flaw in the cohesion-tension theory?

Chunfang Wei, Melvin Tyree, and Ernst Steudle set out to answer these questions using instruments called a root bomb and a xylem pressure probe. Their experimental apparatus is diagrammed in **Figure 32.8a**. The xylem pressure probe includes an oil-filled glass tube that Wei and co-workers inserted into the xylem tissue of a corn leaf. Using a xylem pressure probe allowed the investigators to record changes in xylem pressure instantly and directly.

To test the cohesion-tension theory, Wei and co-workers altered xylem pressure in two ways. First they added pressure to the root systems of their experimental plants using the root bomb illustrated in Figure 32.8a. As predicted by the cohesion-tension theory, pulses of pressure applied through the root bomb produced sharp decreases in the tension experienced by xylem sap (**Figure 32.8b**). Then Wei and co-workers released pressure on the root system and began to alter light levels. This second experiment was designed to alter transpiration rates in the leaves. The experimenters tested this assumption by weighing the entire plant (including soil and bomb) and confirming that it became lighter after experiencing high light intensities. As **Figure 32.8c** shows, the tension experienced by xylem sap increased each time that light intensity was increased. (While reading this graph, remember that tension is a pull, or a negative pressure.)

Although this result provides strong support for the cohesion-tension theory, opponents could argue that the experiment was

(a) A dendrograph

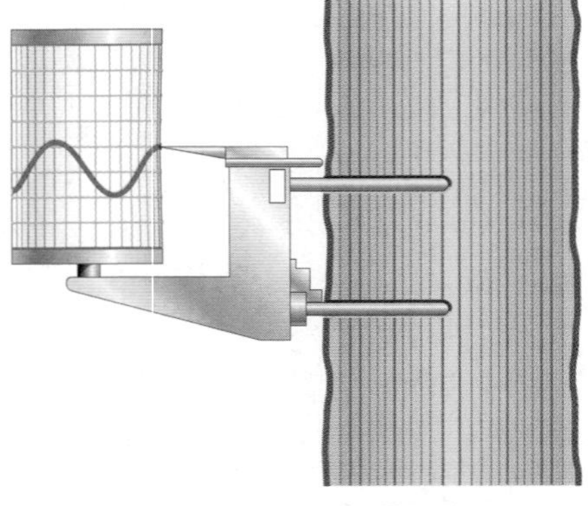

(b) A dendrograph trace shows changes in tree diameter over time.

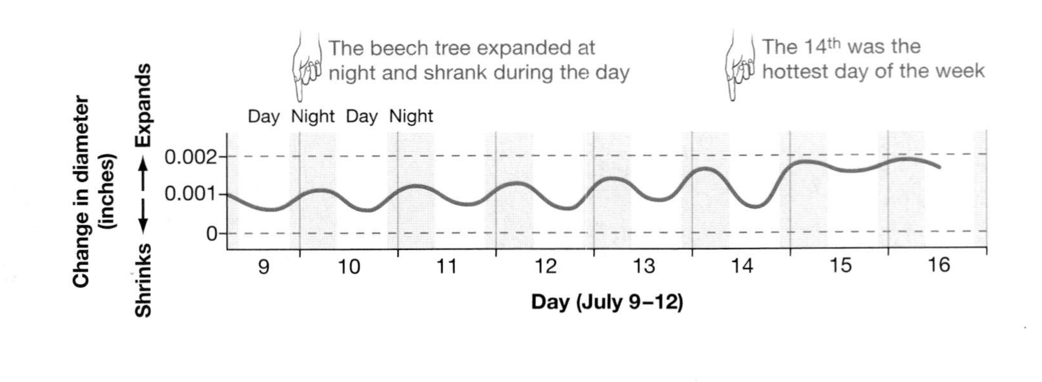

FIGURE 32.7 Trees Shrink When They Transpire
(a) A dendrograph records changes in the diameter of trees. **(b)** These data are a trace from a dendrograph recorded from a beech tree in midsummer. QUESTION The pin on the dendrograph touches the exterior of the tree, while the bolts that hold the instrument are anchored to the interior of the trunk. Why don't the exterior and interior of the tree trunk move together?

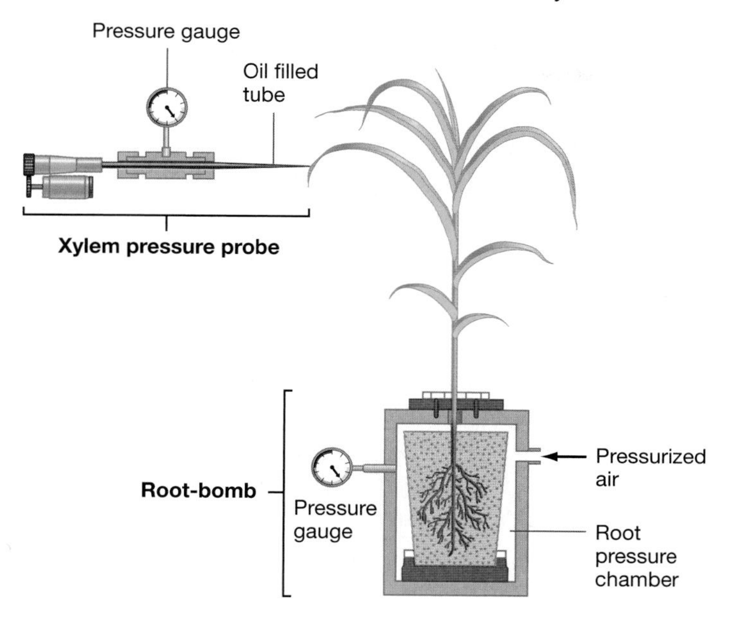

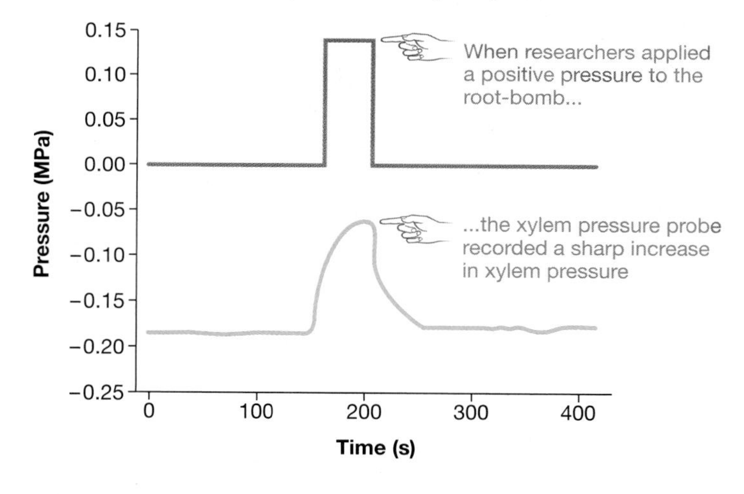

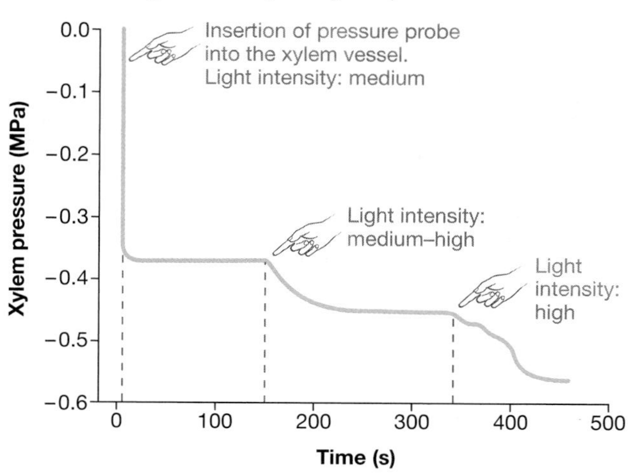

FIGURE 32.8 Testing the Cohesion-Tension Theory
QUESTION Why did the researchers do the experiment in part (b)?

not done on a tall tree, where the theory's adequacy has been challenged most vigorously. The logistics involved in repeating the xylem pressure probe experiment on a redwood tree are daunting. But given the importance of this theory to understanding how plants work, it is likely that someone will attempt it.

Water Absorption and Water Loss

One of the most important features of the cohesion-tension theory is that it does not require plants to expend energy. The Sun furnishes the energy that breaks hydrogen bonds between water molecules at the air-water interface inside leaves and causes transpiration. Rapid transpiration, creates deep menisci in the walls of leaf cells; hydrogen bonding between water molecules at the menisci creates tension that pulls water from root to shoot. Water moves along a water potential gradient by **bulk flow**—a movement in response to pressure (**Figure 32.9**).

It should make sense, then, that a plant does not expend energy to take up water from soil. Rather, water moves from the

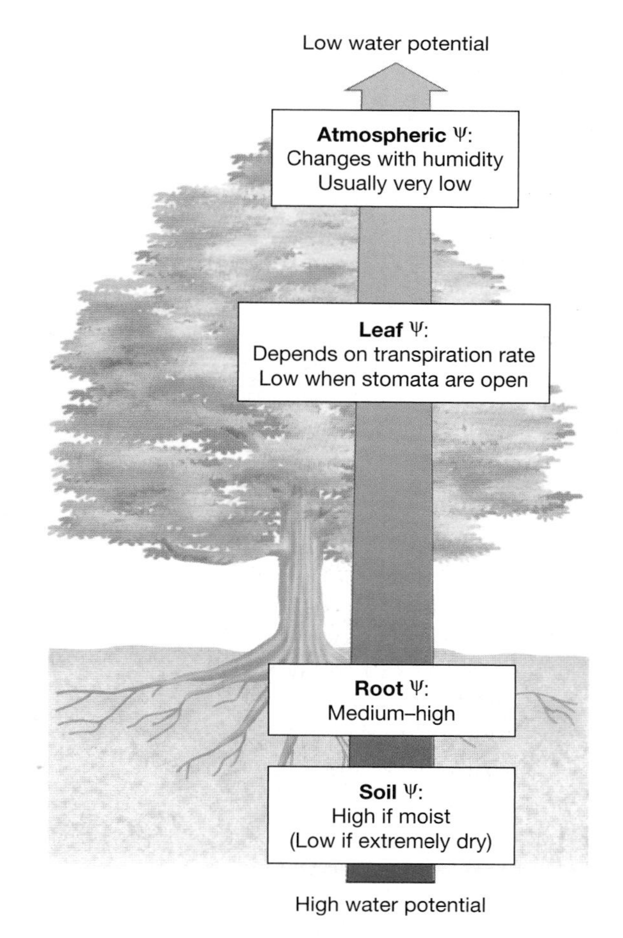

FIGURE 32.9 When Plants Transpire, There Is a Water Potential Gradient Between Soil, Plant, and Air
Water moves from regions of high water potential to low water potential. Xylem conducts water along this gradient.

soil into the root and up the shoot along a water potential gradient. Xylem is merely a set of pipes that allows water to move from a region of high potential to a region of low potential—from soil to leaves.

Water continues to flow from roots to shoots as long as the water potential of the plant is lower than that of soil. In wet soils, plants do not have to be at a very low water potential to extract water. But in dry soils, plants have to be at quite low water potentials to extract water. In effect, they have to be drier than the soil. How do plants cope with environments where soils have extremely low water potentials?

Limiting Water Loss Plants that thrive in dry sites have several ways of coping with soil that has a low water potential. Among the most important features of drought-resistant plants are (1) morphological traits that limit the amount of water lost to transpiration and (2) physiological traits that allow cells to function when their water content is low.

Plants that thrive in dry sites have several morphological features that appear to be adaptations for limiting water loss during transpiration. As an example, consider the oleander plant illustrated in **Figure 32.10a**. This shrub is native to the dry shrub-grassland habitats of southern Eurasia. As the photograph in **Figure 32.10b** shows, the stomata of oleanders are located on the undersides of their leaves, inside deep pits in the epidermis. These pits, in turn, are shielded from the atmosphere by hair-like extensions of nearby epidermal cells. The leading hypothesis to explain these traits is that they serve to slow the loss of water vapor from stomata to the dry air surrounding the leaf. The upper surface of the leaf in part (b) suggests another adaptation to slow water loss. Compared to the cuticles observed in species from wetter habitats, the waxy covering on oleander leaves is extraordinarily thick.

Competing for the Little Water That Is Available Species that thrive in dry environments also have a series of physiological traits that help them cope with chronic water stress. For example, Chapter 28 introduced two novel biochemical pathways that are found in species native to deserts and other hot, dry habitats. These pathways are called crassulacean acid metabolism (CAM) and C_4 photosynthesis. CAM plants are able to continue photosynthesizing even though their stomata close during the heat of the day. C_4 plants use CO_2 so efficiently that they are able to keep their stomata closed more than competing plants. In an important sense, CAM and C_4 photosynthesis are similar to the stomata of oleanders: They are examples of traits that help plants conserve water by limiting transpiration.

Other types of traits, in addition to those that limit transpiration, help plants thrive in dry conditions. As an example, consider data that Richard Cline and Gaylon Campbell collected on trees and shrubs growing in the Rocky Mountains of western North America. Water availability varies enormously in these mountains. Southern exposures are much drier than northern exposures; steep slopes are much drier than gentle slopes. To explore whether physiological differences occur among species that occupy dry sites versus wet sites, Cline and Campbell measured the water potential and solute potential of leaves from several different species over the course of the growing season. They estimated water potential with a pressure bomb and measured solute potential by sampling liquids from tissues and documenting the concentration of dissolved solutes.

Some of their data are shown in **Figure 32.11**. The graph in part (a) documents how the solute potential of leaf tissue in ninebark shrubs varied over the course of a growing season. This species is found primarily on dry sites. Note that the solute potential of the leaf tissue is relatively high in June, at the start of the growing season. In the Rockies, the month of June tends to be rainy and cool. July and August are progressively hotter

(a) Oleander

(b) Cross-section of oleander leaf

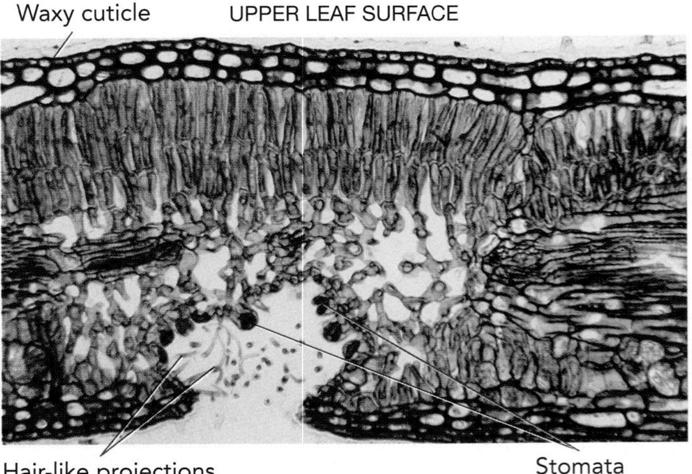

FIGURE 32.10 Adaptations to Dry Habitats
(a) Oleanders grow in desert-like conditions in south-central Asia.
(b) Unlike species from wet habitats, oleander stomata are located in pits instead of on the flat leaf surface. Also, the waxy cuticle on the upper surface of the leaf is especially thick.

and drier, however. As the summer progressed, the data indicate that ninebark shrubs were able to keep growing even though they experienced a huge drop in solute potential. Why is this observation significant?

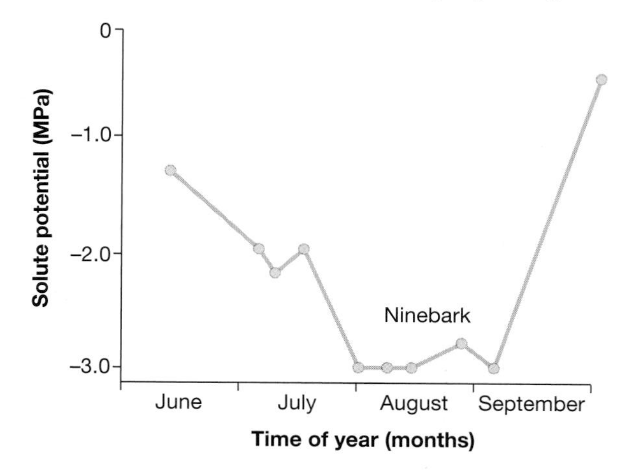

(b) Changes in water potential during a day

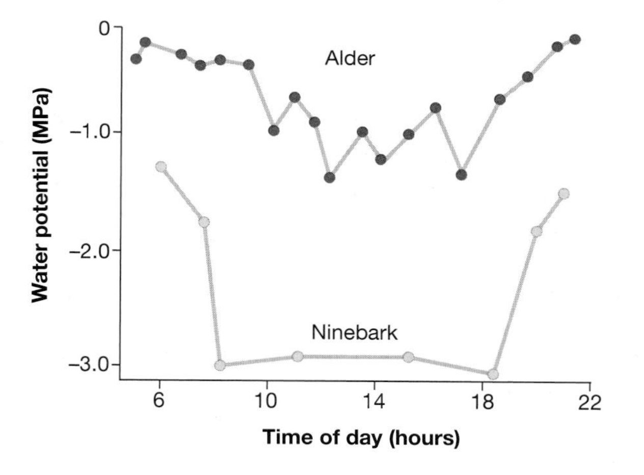

(c) Competition between Sitka alder and ninebark

Sitka alder

Ninebark

Ψ = −1.5

Ψ = −3.0

?

?

Soil Water

The graph in **Figure 32.11b** plots changes in the water potential of ninebarks and Sitka alder shrubs, which grow on wet north-facing slopes, over the course of a single day during August. Note that both plants have high water potentials at night when stomata are closed and low water potentials during the day. By midday, however, the water potential of ninebark is almost 1.5 MPa lower than the water potential of Sitka alder.

Why is this difference important? To answer this question, study the illustration in **Figure 32.11c** and ask yourself which direction soil water would tend to move if a Sitka alder and ninebark were growing side by side. Would water move toward the Sitka alder or toward the ninebark? Because water always moves toward regions of lower water potential, the ninebark would be able to compete for soil water much more efficiently than the Sitka alder. Its cells are able to tolerate low solute potentials, so the ninebark plant can generate extremely low water potentials without dying. Because ninebarks can extract water from dry soils, they can grow in conditions that might kill a Sitka alder.

What is it about the enzymes, membrane proteins, and phospholipids of dry-adapted plants that allows them to tolerate such low solute potentials? Is there a cost to ninebark and similar species, in terms of how fast they can grow when water is readily available? These questions remain unanswered. Like most good research, the data collected by Cline and Campbell generate as many questions as they answer.

32.3 Translocation

Translocation is the movement of sugars through a plant. The guiding principle in translocation research is that sugars move from sources to sinks. In vascular plants a **source** is defined as a tissue where sugar enters the phloem; a **sink** is a tissue where sugar exits the phloem. During the growing season, leaves and stems that are actively photosynthesizing and producing sugar in excess of their own needs act as sources. Sugar moves from these tissues to a variety of sinks: apical meristems, cambium layers (secondary meristems), growing leaves, developing seeds and fruits, and storage cells in roots (**Figure 32.12**, page 634). Early in the growing season, though, the situation is reversed. Storage cells in roots act as sources and developing leaves act as sinks.

FIGURE 32.11 Comparing the Solute and Water Potentials of Plants from Wet and Dry Sites
(a) Over the course of a growing season, the solute potential of ninebark leaves drops dramatically. **(b)** On a hot day in midsummer, the water potential of ninebark leaves drops much more dramatically than the water potential of Sitka alder leaves. **(c)** If a Sitka alder and ninebark competed for water on a hot day in midsummer, which plant would gain water more effectively?

To explore the relationship between sources and sinks in more detail, K. W. Joy exposed sugar-beet plants to carbon dioxide molecules that contained the radioactive isotope ^{14}C. Joy's goal was to track where these carbon atoms moved after they were incorporated into sugars. He determined the location of the ^{14}C atoms by using an instrument to record the number of radioactive emissions emanating from different tissues or by laying plant parts on x-ray film and allowing the radioactivity to expose and blacken the film.

The result? Joy's data support the prediction that fully expanded leaves act as sources of sugar, while actively growing leaves and roots act as sinks. For example, when he enclosed single leaves in a bag and introduced a fixed amount of radioactive CO_2 for a fixed amount of time, he found that mature leaves retained just over 9 percent of the labeled carbon, while growing leaves retained 67 percent. The results of one of these single-leaf experiments are shown visually in **Figure 32.13a.** The leaf that was exposed to the labeled carbon dioxide is shown in green; the leaves that received the labeled sugar are shown in purple. The message in this figure is that young, actively growing tissues acted as sinks. Roots acted as sinks as well. When Joy let all of the leaves on a plant receive labeled carbon, he found that over 16 percent of the total was translocated to root tissue within 3 hours.

Similar experiments have made two important generalizations possible. First, sugars are translocated very rapidly. Second, there is a strong correspondence between the physical location of certain sources and certain sinks. For example, source leaves send sugar to tissues on the same side of the plant. Experiments with tall herbaceous plants show that leaves on the upper part of the stem send sugar to apical meristems, while leaves on the lower part of the plant send sugar to the roots (**Figure 32.13b**). Why would leaves send sugar to tissues on the same side or part of the body? The answer hinges on understanding the anatomy of phloem tissue.

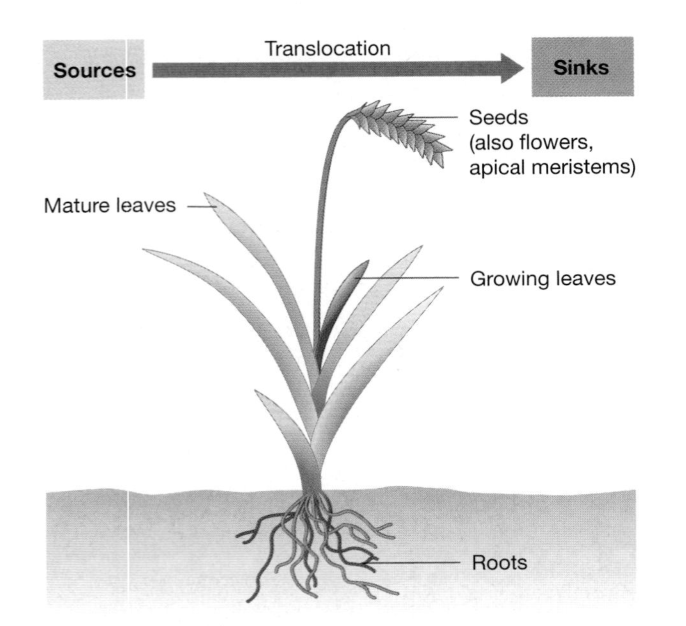

FIGURE 32.12 Sugars Move from Sources to Sinks
Sources are tissues that release sucrose to the phloem; sinks are tissues that take up sucrose from the phloem. QUESTION In species that have secondary meristems (cambium tissues), would these tissues serve as sources or sinks?

The Anatomy of Phloem Tissue

In 1928, T. G. Mason and E. J. Maskell established that sugars move through phloem tissues. They arrived at this conclusion by removing a ring of bark from a small section of a tree trunk and noting that water movement continued while sugar transport stopped. They also documented that the phloem tissues present in bark have a high concentration of sucrose. Later workers painstakingly traced the anatomical connections among phloem tissues in leaves, stems, and roots.

Based on these results, the physical relationships observed between sources and sinks are logical. For example, the phloem in the leaves on one side of a plant connects directly with the phloem of branches, stems, and roots on the same side of the individual (**Figure 32.13c**).

Early studies established other important observations about phloem anatomy. Microscopy confirmed that the phloem of flowering plants includes two particular prominent types of cells: sieve-tube elements and companion cells. Unlike tracheid and vessel elements that make up xylem, both sieve-tube elements and companion cells are alive at maturity. In most plants, sieve-tube elements lack nuclei and many major organelles; they are connected to one another, end to end, by open pores (**Figure 32.14**). These sieve-like pores create a direct connection between the cytoplasm of adjacent cells. Companion cells, in contrast, have nuclei and a large number of ribosomes, mitochondria, and chloroplasts. These morphological differences suggested that sieve-tube elements might be the actual conduit for phloem sap, while companion cells serve some sort of support role.

By 1930 most major components of the sugar transport system in angiosperms had been described in detail, and chemical analyses had shown that phloem sap is dominated by the disaccharide called sucrose (table sugar). The primary question then became, how do all of these elements work? How is sucrose transported from sources to sinks?

The Pressure-Flow Hypothesis

Ernst Münch proposed a mechanism for translocation in 1930. His idea, called the **pressure-flow hypothesis**, is diagrammed in **Figure 32.15** (page 636). The fundamental proposition is that events at source tissues and at sink tissues create a steep pressure potential gradient in phloem; the water in phloem sap moves down this gradient, and sugar molecules follow along by bulk flow. Unlike the cohesion-tension theory for water transport, however, the driving force for movement is not provided by transpiration and hydrogen bonding between water molecules. Instead, the force responsible for movement is generated

(a) Source leaves send sugar to the same side of the plant.

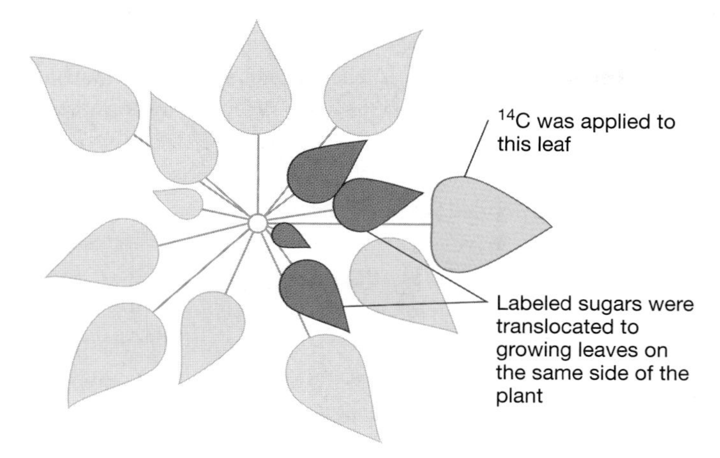

(b) Source leaves send sugar to tissues on the same end of the plant.

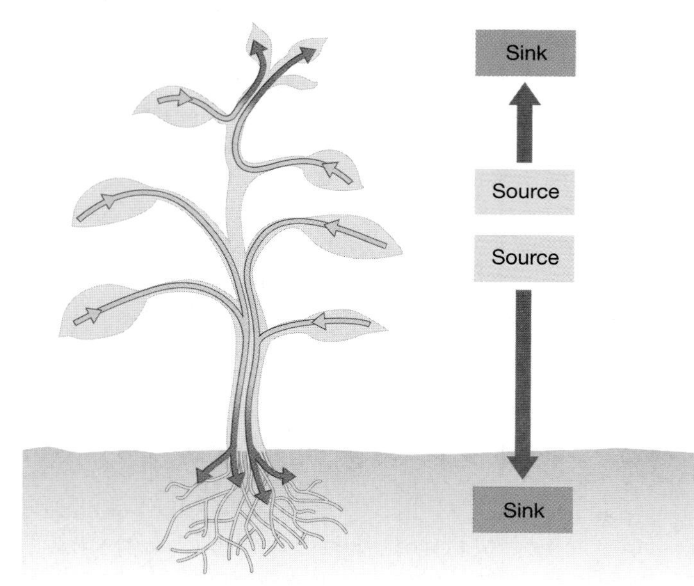

(c) These patterns occur because phloem is arranged in discrete bundles.

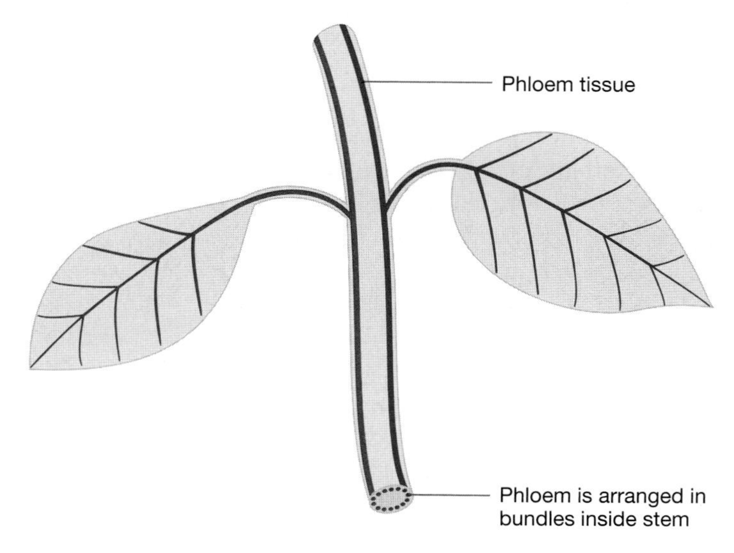

FIGURE 32.13 What Is the Physical Relationship Between Sources and Sinks?
The patterns described in parts (a) and (b) occur because phloem is arranged in discrete bundles, as shown in part (c). Each bundle forms a distinct set of connections inside the plant body.

by large differences in the turgor pressure of phloem sap between source and sink tissues.

To understand how Münch's model works, take a closer look at Figure 32.15 starting with the source cell diagrammed on the upper right. The small red arrows reflect Münch's proposal that sucrose moves from source cells into companion cells and from there into sieve-tube elements. Consequently the phloem sap near the source has a high concentration of sucrose, meaning that it has a very low water potential compared to the adjacent xylem cells. As the blue arrow in the diagram shows, water moves from xylem across the semipermeable membrane of sieve-tube cells along a water potential gradient. In response, turgor pressure begins to build in the sieve-tube elements near the source region.

What is happening at the sink? Münch proposed that cells in the sink remove sucrose from the phloem sap. As a result, the water potential in sieve-tube cells increases until it is higher than the water potential in adjacent xylem cells. As the blue arrow at the bottom of the figure shows, water flows across the semipermeable membrane of sieve-tube cells into xylem along a water potential gradient. Turgor pressure in the sieve-tube elements near the sink drops in response.

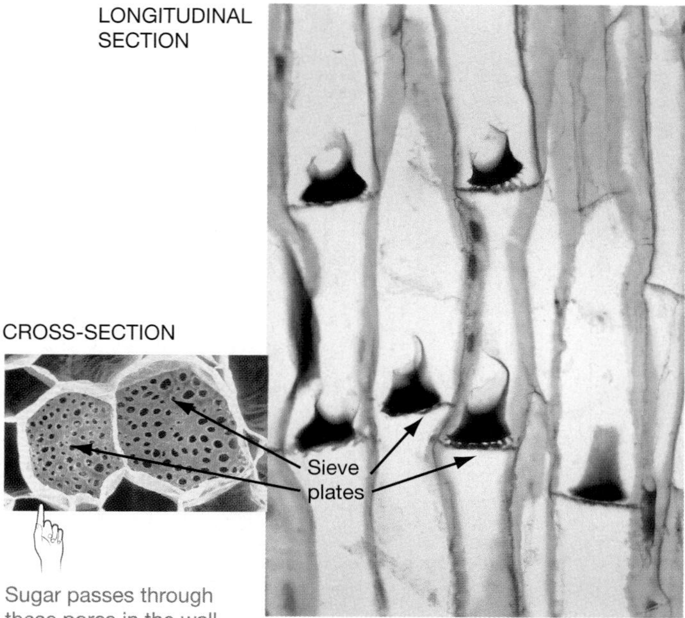

FIGURE 32.14 Sieve-Tube Elements Are Connected by Pores
These micrographs show the connections between sieve-tube elements in cross-section and in longitudinal section.

The net result of these events is high turgor pressure near the source and low turgor pressure near the sink. Because the pores of the sieve plate have no membranes, the difference in pressure potential drives phloem sap from source to sink. The pressure contrast is responsible for a one-way flow of sucrose molecules.

The grand result of all these events is a continuous loop of water flow. The cycle illustrated in Figure 32.15 is driven by water potential gradients between xylem and phloem and pressure potential gradients within phloem.

The pressure-flow hypothesis is logical given the anatomy of vascular tissue and the basic principles that govern water movement. But has any experimental work been able to confirm the theory? To illustrate the types of data that support the model, consider work by Robert Turgeon and Peter Hepler on the phloem of leaves from squash plants. These researchers wanted to test the prediction that sucrose is present at extremely high concentrations in phloem cells near a source. To do this, they set out to document the solute potentials of phloem cells versus nearby cells. When they exposed samples of leaf tissue to solutions with a variety of solute concentrations and then examined the cells under the microscope, the results were striking. The parenchyma cells adjacent to phloem lost water and turgor when exposed to solutions with very low solute concentrations. Phloem cells, in contrast, did not lose turgor even when exposed to extremely concentrated solutions. The result implies that sugars are loaded to very high concentrations in the phloem tissues of leaves, just as the pressure-flow hypothesis predicts. Now the question is, how does this loading occur?

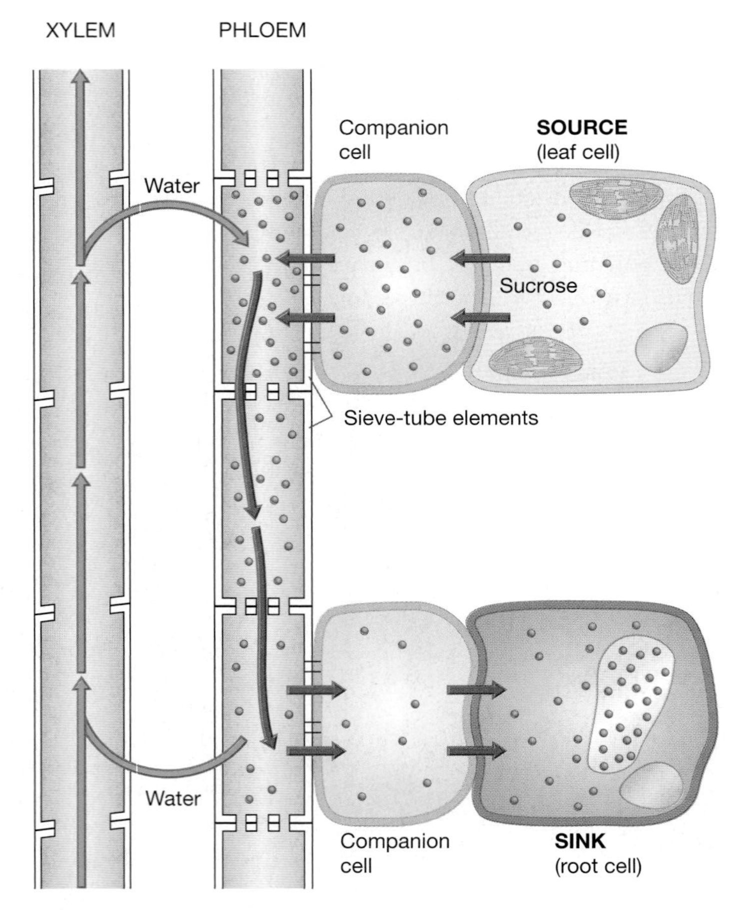

FIGURE 32.15 The Pressure-Flow Hypothesis
The pressure-flow hypothesis predicts that water cycles through xylem and phloem, and that water movement in phloem is a response to a gradient in pressure potential. **EXERCISE** Label parts of the cycle where water moves in response to a water potential gradient. Label parts of the cycle where phloem sap moves in response to a pressure potential gradient.

Phloem Loading

To understand Münch's model, it is critical to recognize that sugar transport depends on an expenditure of energy. At source cells, sugar has to be loaded into sieve-tube elements against a concentration gradient. As a result, loading requires an expenditure of ATP (adenosine triphosphate) and some sort of membrane transport system. Conversely, sugar must be unloaded against its concentration gradient at sinks. Unloading requires another expenditure of ATP and a second membrane transport mechanism. Given that phloem loading and unloading requires an expenditure of ATP, how does each event occur? What specific membrane proteins are involved?

To answer these questions, let's look first at events at source tissues, where the active transport of sucrose into sieve-tube elements results in high water potentials. The chapter concludes by considering how the same molecules are off-loaded to maintain low water potentials at sinks.

How Are Sugars Concentrated in Sieve-Tube Elements at Sources? Molecules and ions can be transported across a concentration gradient in one of two ways. A membrane protein can use the energy released by ATP hydrolysis to shuttle the molecule or ion directly across the membrane, or transport can be indirect. Indirect transport is usually based on two types of membrane proteins. One of these hydrolyzes ATP and uses the energy released to transport protons (H^+) across the membrane to the exterior of the cell. Proteins like these are called proton pumps, or more formally H^+-ATPases (**Figure 32.16, left**). Their activity establishes a large difference in the charge and hydrogen ion concentration on either side of the membrane. This electrochemical gradient drives protons into the cell. A second protein, called a cotransporter, acts as a conduit for these protons and uses the force generated by their passage to bring a second molecule into the cell (**Figure 32.16, right**). In the case of phloem tissue, the second molecule is sucrose.

Which type of transport occurs in source tissues? The observation of strong pH differences between the interior and exterior of phloem cells led researchers to suspect that phloem loading depends on an H^+-ATPase. This enzyme was later purified from phloem cell membranes, providing evidence that transport was indirect. Then the question became, exactly where are these transporters located? How does their physical location lead to the concentration of sucrose in sieve-tube elements?

Where Are the H^+-ATPases Located? Proton pumps are found in the cell membranes of a wide variety of organisms, including bacteria, fungi, and animals. To analyze the proton pumps found in plants, researchers from several different laboratories began focusing on a small member of the mustard family called *Arabidopsis thaliana*. This species was introduced in Unit 3 because it serves as a model organism for studies in plant molecular genetics and plant genomics. It is a superb subject for experimental work because it is small, grows and reproduces quickly, and is inexpensive to maintain.

When researchers determined the amino acid sequence of a proton pump purified from *Arabidopsis* cell membranes and used these data to infer the DNA sequence of the corresponding gene, they found that the *Arabidopsis* genome actually codes for 10 different pump proteins. One of these genes, called *AHA3*, appeared to be expressed primarily in vascular tissues.

These observations led Natalie DeWitt and Michael Sussman to hypothesize that *AHA3* encoded the proton pump responsible for phloem loading. To test this hypothesis they raised antibodies to the *AHA3* protein using the types of techniques introduced in Chapter 46 (see **Figure 32.17a**, page 638). As that chapter points out, an antibody is a polypeptide that binds to a specific protein. DeWitt and Sussman's goal was to treat *Arabidopsis* leaves with the *AHA3* antibody, examine treated leaves under the electron microscope, and determine exactly where the pump proteins are located. To visualize the antibody, they attached gold particles to the polypeptide (a gold particle looks like a black dot when viewed with the electron microscope).

Figure 32.17b shows the data that resulted from this protocol. The proton pumps responsible for phloem loading are found almost exclusively in the membranes of companion cells. This result inspired the model for phloem loading illustrated in **Figure 32.17c**. The important points to note about this model are that (1) sucrose is transported into companion cells from the surrounding cell walls and intercellular spaces, and (2) once inside companion cells, sucrose travels into sieve-tube elements via a direct cytoplasmic connection. How sucrose gets into these extracellular spaces is still a mystery, however. Work on the mechanism of phloem loading in *Arabidopsis* and other species continues.

Phloem Unloading

Research on the mechanism of phloem unloading at sinks has shown that the membrane proteins involved in sugar transport and the mechanism of movement vary among different types of sinks within the same plant, as well as among different species.

Figure 32.18 (page 638) provides an example of this diversity. In sugar beets, sucrose crosses the membranes of cells in young, growing leaves along a concentration gradient. This passive transport occurs because sucrose is rapidly used up inside the cells to provide carbon for the synthesis of proteins, nucleic acids, and phospholipids needed by the growing cells (Figure 32.18a). In the roots of the same plant, however, an entirely different mechanism is responsible for off-loading sucrose. Root cells in this species have a large organelle that stores sucrose. (It's very likely that the table sugar you use was purified from these cells.) The membrane surrounding this organelle contains a protein that hydrolyzes ATP and uses the energy released to transport sucrose into the organelle, against its concentration gradient (Figure 32.18b). This movement keeps the water potential of the nearby phloem sap low, as required by the Münch pressure-flow model.

In short, more than seven decades of research has provided convincing evidence that the pressure-flow hypothesis is fundamentally correct. Understanding how translocation occurs at the molecular level, however, remains an active area of research.

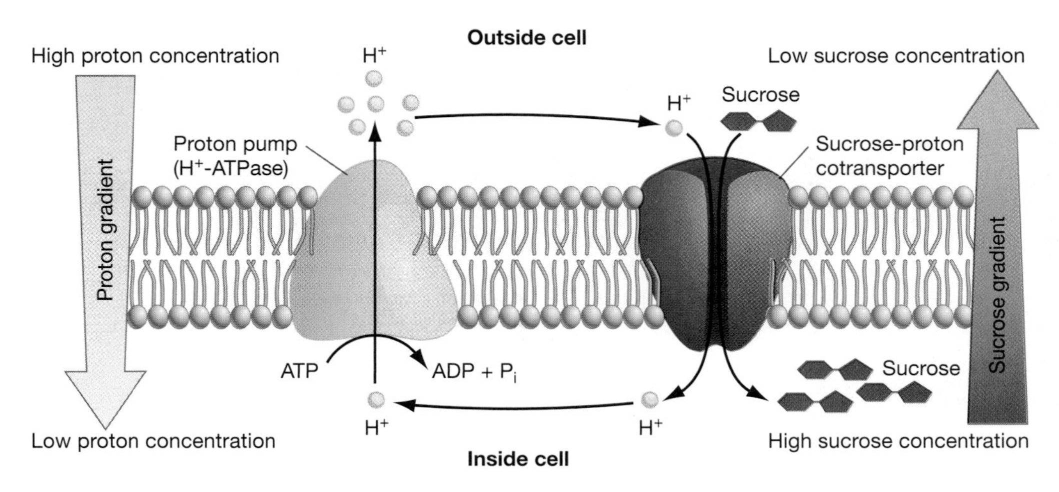

FIGURE 32.16 A Model for Cotransport of Protons and Sucrose
According to the model of cotransport, a proton pump hydrolyzes ATP to move hydrogen ions to the exterior of the cell. The high concentration of H^+ establishes an electrochemical gradient that drives sucrose into the cell against its concentration gradient.

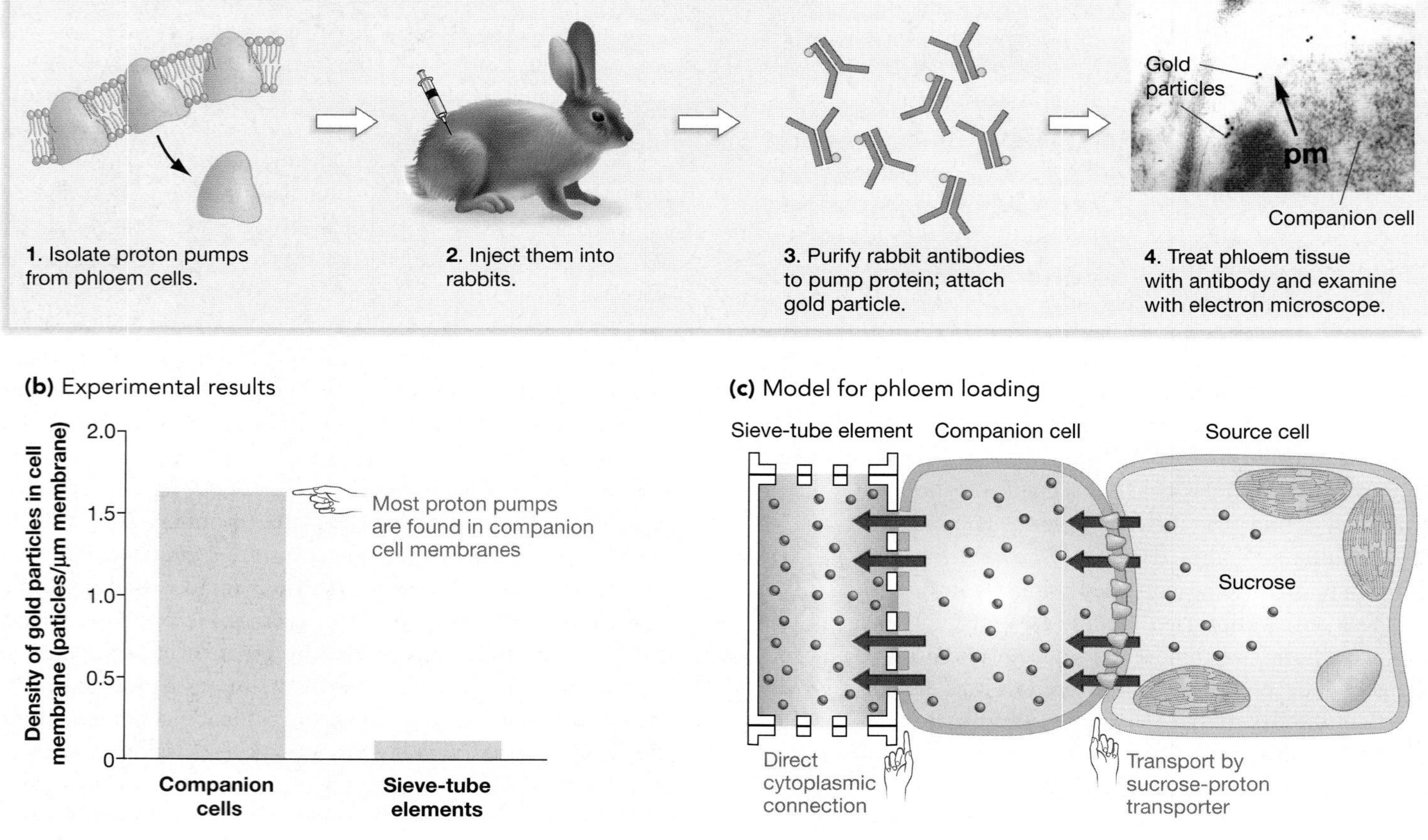

FIGURE 32.17 Where Is the H⁺-ATPase Located?
(a) Researchers raised antibodies to the proton pump found in *Arabidopsis* phloem tissues using the protocol diagrammed here. In the photo at the right, the label "pm" points to the plasma membrane of a companion cell. **(b)** The histogram compares the average density of proton pumps in the membranes of companion cells versus sieve-tube elements. **(c)** The data in part (b) suggest this model for how active transport of sucrose leads to phloem loading.

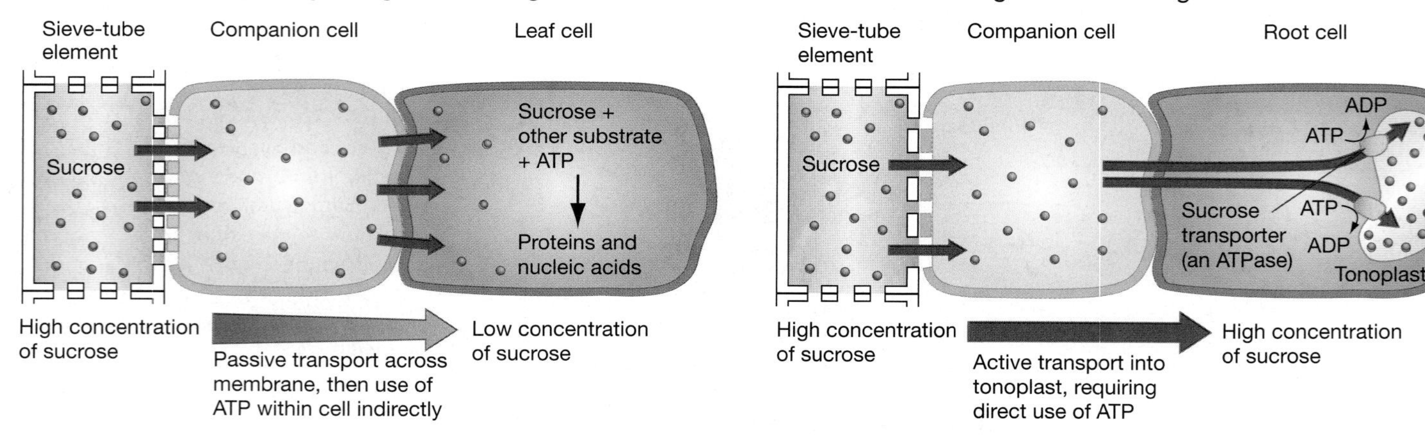

FIGURE 32.18 Phloem Unloading
The mechanism of phloem unloading may vary from sink to sink within the same plant.

Essay Irrigated Agriculture

Most of the crop plant varieties in use today have undergone extensive artificial selection, aimed at increasing their yields. The productivity of plants is largely a function of their photosynthetic rate. The rate of photosynthesis, in turn, hinges on a plant's ability to acquire carbon dioxide through its stomata. This is a long-winded way of making a simple point: Crop plants transpire a lot.

How much do crop plants transpire? Through transpiration and evaporation, one hectare of corn growing in midwestern North America uses about 6 million liters of water during a growing season. A hectare is about the size of two football fields; 6 million liters of water is enough to cover this area about two feet deep. High-yielding varieties of rice require much more water—up to 18 million liters per hectare.

How much do crop plants transpire?

The correlation between water availability and productivity is strong. Just 16 percent of the world's farmland is irrigated, but irrigated crops account for 33 percent of the world's food. Rice grown in flooded fields produces 50 percent more food than rice grown under sprinkler irrigation.

It's not surprising, then, that irrigation is frequently promoted as a way to use farmland efficiently, boost productivity, and raise standards of living in food-poor countries. Irrigation is not without costs, however.

The first issue is simply the availability of water. Agriculture already accounts for 87 percent of the water consumed for human use in the United States, and some sources are drying up. The groundwater reservoir underlying south-central North America is tapped extensively to water corn; it is now half the size it was before irrigation began a few decades ago. The Colorado River is used so intensively that it is barely a trickle by the time it reaches the Pacific Ocean. In the Middle East, water shortages have threatened to trigger armed conflicts among nations.

Irrigation can also cause significant problems with soil quality, especially in arid regions. When ion-rich water is used to irrigate crops, evaporation leaves large quantities of salt behind. These compounds dissolve in the remaining soil water and lower its solute potential. In extreme cases, the water potential of irrigated soil becomes so low that root cells lose water to the soil instead of gaining it. When this occurs, the land is infertile and has to be abandoned. At the end of the third Punic War with Carthage in 146 B.C., the victorious Romans razed the city and salted the soil to make it unproductive.

In response to increasing problems with salt buildups, plant breeders are working to develop varieties of plants that require less water or that can tolerate more salinity. In one approach, breeders have inserted a gene into crop plants that encodes the enzyme required to make a molecule called mannitol. Mannitol is sequestered in plant cells and helps them maintain a low water potential. In salty soils, these plants grow substantially better than normal varieties. Only time will tell, however, whether breeders can successfully outrace the trend toward increasing salt buildups in irrigated soils worldwide.

Chapter Review

Summary

Plants lose water as an inevitable by-product of exchanging gases with the atmosphere. According to the cohesion-tension theory, this water is replaced without an expenditure of energy by the plant. Instead, it is pulled from the soil to roots to shoots, against the force of gravity, by surface tension inside leaves. The surface tension occurs at menisci that form as water evaporates from the walls of leaf cells.

The flow of water from soil to air via plant tissues follows a water potential gradient. Water potential (φ) can be thought of as the tendency of water to move from one location to another. In plants, water potential has two components. The concentration of solutes in a cell or tissue creates a solute potential; the cell wall provides a pressure potential. The water potential of a cell, tissue, or plant is the sum of its solute potential and pressure potential. When semipermeable membranes are present, water moves by osmosis from areas of high potential to low potential; when no membranes are present, water moves by bulk flow from areas of high pressure to low pressure (independently of differences in solute potential).

Several types of data support predictions made by the cohesion-tension theory. For example, on hot, sunny days when transpiration is rapid, xylem sap is under enough tension to make trees shrink slightly in diameter. The measured water potential of leaf tissue drops during the day, when stomata are open; and it rises at night, when stomata are closed. Finally, direct measurements show that the tension on xylem sap changes rapidly in response to changes in light intensity and thus transpiration rate.

Plants that occupy dry habitats have several ways of limiting the amount of water they lose to transpiration. In some species, stomata are located on the undersides of their leaves, in pits that are protected from the dry atmosphere. There is also evidence that the tissues of dry-habitat plants can tolerate higher solute potentials than tissues of wet-habitat plants. As a result, dry-habitat plants have lower water potentials and can compete more successfully for scarce soil water. In Chapter 35 we'll examine how plants can respond to drought stress by closing their stomata and shutting down photosynthesis.

Translocation is the movement of sucrose and other products through the plant and has been studied by labeling sugars with ^{14}C. According to the Münch pressure-flow model, sugars move from sources to sinks via bulk flow along a pressure gradient. The gradient is generated by the transport of sugars into sieve-tube elements in source tissues, coupled with the transport of sucrose out of sieve-tube elements at sink tissues. Water moves from xylem into sieve-tube elements near sources and cycles back to xylem near sinks. The membrane proteins responsible for phloem loading in *Arabidopsis* have now been isolated and studied in detail. They include a proton pump and a proton-sucrose cotransporter.

In the following chapter we explore how plants take up the soil nutrients needed for growth. Water uptake is critical to the mining operations carried out by plant roots.

Questions

Content Review

1. Which of the following observations must be accounted for in any hypothesis to explain how water moves upward in plant stems? (Choose all that apply.)
   a. Xylem is under tension.
   b. Sieve-tube elements are under pressure.
   c. Some trees are 100m tall.
   d. Tracheids and vessel elements are dead at maturity.
2. When does the rate of transpiration increase? (Choose all that apply.)
   a. when stomata open during the day
   b. when stomata close at night
   c. when the weather changes and air becomes wetter
   d. when the weather changes and air becomes drier
3. When researchers treat plants with drugs that poison respiratory chain proteins in mitochondria and that affect ATP production, what would be the effect on translocation?
   a. There should be no effect.
   b. It should speed up.
   c. It should slow down or stop.
   d. There is not enough information provided to answer the question.
4. Plants that live in dry habitats have traits that appear to reduce water losses due to transpiration. Which of the following are true? (Choose all that apply.)
   a. They don't open their stomata.
   b. They don't have stomata.
   c. Many grow close to the ground, where they are less exposed to drying winds.
   d. Their stomata are frequently located on the undersides of leaves.
5. What is a proton pump?
   a. a membrane protein that transports sucrose across a concentration gradient
   b. a membrane protein that transports protons across an electrochemical gradient
   c. a membrane protein that transports protons with an electrochemical gradient and sucrose against a concentration gradient
   d. any membrane protein that acts as a channel—meaning that it does not consume ATP
6. Why is it interesting that species from dry habitats appear to tolerate particularly low solute potentials in their tissues?
   a. It enables them to lose less water to transpiration.
   b. It enables them to translocate sugars with a lower expenditure of ATP.
   c. It is not consistent with the principle of natural selection.
   d. It should help them compete more effectively for the little soil water available.

Conceptual Review

1. Water flows from regions of high water potential to regions of low water potential. Why does this happen?
2. Water transport in plants is a passive process that does not require an expenditure of energy, but translocation is an energy-demanding process. Explain why this important difference exists.
3. Why are "cohesion-tension" and "pressure-flow" sensible names for the hypotheses analyzed in this chapter?
4. Aphids prey on plants by inserting an organ called a stylet into phloem tissue and harvesting sap. Suppose two aphids named Alice and Bernadette are sitting on the same plant. Alice inserts her stylet into the phloem of a large, mature leaf, while Bernadette prefers to probe the phloem of the young growing tissue near the shoot's apical meristem. Which aphid is attacking a source, and which is preying on a sink? Is Alice or Bernadette getting a higher concentration of sugar?
5. How does cotransport result in phloem loading? Where does this process occur? What data support your answers to these questions?
6. How does a pressure bomb work, and what does it measure? Why does a pressure bomb work?

Applying Ideas

1. The text claims that water loss is an inevitable by-product of gas exchange. What data or observations support or challenge this claim? Would the same statement be true in bacteria? In animals?

2. Suppose that you discovered a mutant pea plant that is unable to make the major wax found in cuticle. In terms of water balance, what are the physiological consequences of this mutation?

3. When trees grow on the edge of a forest and a field, one side of their body is partially shaded while the other is exposed to full sun. The growth rings in trunks like these are often asymmetric—they tend to be fatter on sunny side of the trunk and skinnier on shaded side. Why is this observation logical, given what you've learned about how sugars are translocated from sources to sinks?

4. Suppose you discover a mutant sugar beet that has extraordinarily high concentrations of sucrose in its root cells. Generate a hypothesis predicting the nature of the mutation responsible for the observation. How would you test your hypothesis?

5. Sucrose can be labeled with ^{14}C. When researchers treat leaves with solutions containing a low concentration of labeled sucrose, they find that most or all of the label is rapidly incorporated into phloem tissues. Does this observation support or undermine the hypothesis of active uptake at source tissues? Explain.

CD-ROM and Web Connection

CD Activity 32.1: Solute Transport in Plants *(animation)*
(Estimated time for completion = 5 min)
Observe the mechanisms by which plants assure that their tissues receive necessary water and nutrients.

At your **Companion Website** (http://www.prenhall.com/freeman/biology), you will find self-grading exams and links to the following research tools, online resources, and activities:

Water and Solute Potential
This site helps clarify the concept of water potential.

Water and Mineral Transport
This site describes how the different cell types in the plant's water pathway are adapted for water transport.

Phloem
The page describes the cellular components, including the protein structure and biochemistry, that make phloem.

Additional Reading

Ryan, M. G., and B. J. Yoder. 1997. Hydraulic limits to tree height and tree growth. *BioScience* 22: 235–242.

Mohlenbrock, R. H. 1998. This land: Salt of the earth. *Natural History* 107 (January): 57–59.

Plant Nutrition

33

Obtaining carbon-containing molecules and chemical energy is the most urgent task facing all organisms. Plants fulfill both requirements by making sugar through the process of photosynthesis. Yet plants cannot live on sugar alone. To make the molecules required to run their cells, plants must harvest a wide variety of elements and ions. To understand why, think back to the structures of the nucleic acids, amino acids, enzymes, chlorophylls, and cofactors introduced in earlier chapters. In addition to the C, H, and O found in sugar, these macromolecules contain nitrogen, phosphorus, sulfur, magnesium, and other elements.

This chapter focuses on a simple question: How do plants obtain these nutrients? The answer to this question is fundamental to understanding how plants work and how farmers and foresters manage their land.

We begin by analyzing the basic nutritional needs of plants—the equivalent of their minimum daily requirements. Our next step is to examine the composition of soil and the features of roots that allow plants to mine essential nutrients. In section 33.4 we consider how plants obtain nitrogen. Nitrogen is required for every amino acid and nucleotide in the body and is a major component of fertilizers. Our discussion of nitrogen uptake includes an introduction to what may be the most intensively studied cooperative interaction between species. Inside the cells of certain plant species, bacterial cells process nitrogen gas from the atmosphere into a form that is usable by their hosts. This process is called nitrogen fixation and is a key to the global nitrogen cycle introduced in Chapter 25. In the final section of the chapter, we look at some particularly unusual aspects of nutrition in plants, including parasitism and insect eating.

Plants obtain most of the nutrients they need from soil, through their root systems.

33.1 Nutritional Requirements

In addition to carbon dioxide and water, what do plants need to live? W. L. Latshaw and E. C. Miller offered an initial answer to this question by analyzing the chemical composition of the shoot system in corn. They dried the leaves, stem, cob, and kernels of a corn plant, determined which elements were present and in what amounts, and reported the data shown in the middle column of **Table 33.1**.

Latshaw and Miller analyzed dry weight, so the composition of corn shoots reported in the table does not include the hydrogen and oxygen atoms found in water. Even so, hydrogen and oxygen account for about 50 percent of the shoot; adding carbon brings the total accounted for to over 94 percent.

Are these data typical? The short answer is yes and no. When follow-up studies were done that included roots as well as shoots from a variety of vascular plants, the data in the right-hand column of Table 33.1 resulted. The major differences between the two data sets are that silicon and aluminum do not appear to be important components of most plants. Silicon is abundant in grasses like corn and may make their tissues harder for herbivores to chew (silicon is the major ingredient in sand and glass); it is not clear why Latshaw and Miller found so much aluminum in corn tissues. In many other respects, the two data sets are similar.

TABLE 33.1 The Elements Found in Plants

	Element	Corn Shoot (% of Dry Weight)	Vascular Plants; Typical or Average (% of Dry Weight)
Macronutrients	Oxygen	44.40	45
	Carbon	43.60	45
	Hydrogen	6.20	6
	Nitrogen	1.50	1.5
	Potassium	0.92	1.0
	Calcium	0.23	0.5
	Magnesium	0.18	0.2
	Phosphorus	0.20	0.2
	Sulfur	0.17	0.1
Micronutrients	Chlorine	0.14	0.01
	Iron	0.08	0.01
	Manganese	0.04	0.005
	Zinc	not detectable	0.002
	Boron	not detectable	0.002
	Copper	not detectable	0.0006
	Nickel	not detectable	not tested
	Molybdenum	not detectable	0.00001
	Silicon	1.20	not detectable
	Aluminum	0.89	not detectable

What do these data mean in terms of how farmers should evaluate their soils and their fertilizing regimes? How might competition for these elements influence the diversity and productivity of plants in natural habitats? Answering these questions hinges on understanding the concept of essential nutrients.

Essential Nutrients

After surveying early studies on the composition of plants, D. I. Arnon noted that most elements in the periodic table could be found in one species or another. The important thing about plant nutrition, Arnon realized, is to understand which elements are essential for growth and reproduction and in what relative quantities. To follow up on this insight, Arnon proposed that an **essential nutrient** should fulfill the following three criteria:

- It is required for growth or reproduction, meaning that the plant cannot grow to maturity and produce offspring without it.
- No other element can substitute for it. Symptoms that are observed when the element is withheld are corrected only by supplying that element.
- It is required for a specific structure or metabolic function, not because it aids in the uptake of a different essential element.

Based on studies that have accumulated over the past 75 years, the elements listed in the "Elements" column of Table 33.1 are recognized as essential for most vascular plants. It is important to recognize, however, that some elements are essential in certain species but not others. Silicon, for example, is an essential nutrient for rice and corn but not for most vascular plants.

Although different classification schemes for the essential elements have been proposed, perhaps the most common is to group them as micronutrients and macronutrients. **Micronutrients** are required in relatively small quantities and usually function as a cofactor for specific enzymes. Even though several are required in amounts so tiny that measurement is difficult, their importance should not be underestimated. For example, a typical plant contains just 1 molybdenum atom for every 60 million hydrogen atoms in its body (not including water). Yet plants die if they have no molybdenum because it is a cofactor for several enzymes involved in nitrogen processing.

Macronutrients, in contrast, are present in relatively large quantities. They are the building blocks of nucleic acids, proteins, phospholipids, and other key molecules. Plants obtain the carbon, oxygen, and hydrogen they need from the atmosphere and from water, but the other macronutrients have to be extracted from soil. It is no coincidence that the leading ingredients in virtually every commercial fertilizer are nitrogen, phosphorus, and potassium. Now the question is, what happens to plants when one of the essential nutrients is missing?

Nutrient Deficiencies

Using a growth system called hydroponics, Arnon and co-worker P. R. Stout undertook a series of pioneering studies on how nutrient deficiencies affect plants. As **Figure 33.1a** shows, hydroponic growth takes place in liquid culture. This system allowed Arnon, Stout, and subsequent workers to precisely control the availability of nutrients. For example, to explore the effect of copper deficiency on tomatoes, Arnon and Stout grew seedlings in two types of treatments. One treatment consisted of flasks containing water and all the essential nutrients listed in Table 33.1 in the relative concentrations that are optimal for tomato growth. The second treatment was identical, except that the nutrient solution lacked copper (**Figure 33.1b**).

As **Figure 33.1c** shows, copper-deprived individuals have stunted shoots, unnaturally light foliage, curled leaves, and no flowers. Given copper's role as a cofactor or component of several enzymes involved in redox reactions (see Chapter 6), it is understandable that the symptoms were very general in nature as well as life threatening. All tissues in the plant were affected, and all were affected severely. Likewise, it is reasonable to expect that a relatively small amount of copper would cure the deficiency. Arnon and Stout found that the symptoms were prevented if the plants were cultured in a solution containing just 0.002 mg of this element. Analogous studies have now been done on the other essential nutrients.

For farmers, gardeners, and plant ecologists, understanding which nutrients are essential, and why, is basic to understanding why certain plants thrive and others fail. Understanding where nutrients come from, and how plants obtain them, is our next task.

33.2 Soil

In the early 1600s, Jean-Baptiste van Helmont performed a classic experiment on plant nutrition using a willow tree as a study organism. He wanted to know where the mass of a growing plant comes from. He began the study by weighing out 200 pounds of soil and placing it in a pot with a 5-pound willow sapling (**Figure 33.2**). He allowed the plant to grow for five years, adding only water. At the end of the experiment he weighed the willow and the soil. The willow weighed 169 lbs, 3 ounces; the soil weighed 199 lbs, 14 ounces.

Where had the additional 164 pounds, 3 ounces of tree come from? Because he was not aware that gases have mass, van Helmont concluded that the new plant material had to have come from water. He also ignored the loss of two ounces in the soil, chalking it up to error. In the light of the experiments we've just reviewed on essential nutrients, though, those two ounces of soil become critically important. They are the source of all macro- and micronutrients save for C, H, and O.

Soil Composition

The process of soil building begins with solid rock. As **Figure 33.3** shows, water, wind, and organisms continually break tiny pieces off large rocks. Depending on their size and composition, the particles that result are called clay, silt, sand, or gravel.

(a) Hydroponic growth takes place in liquid culture.

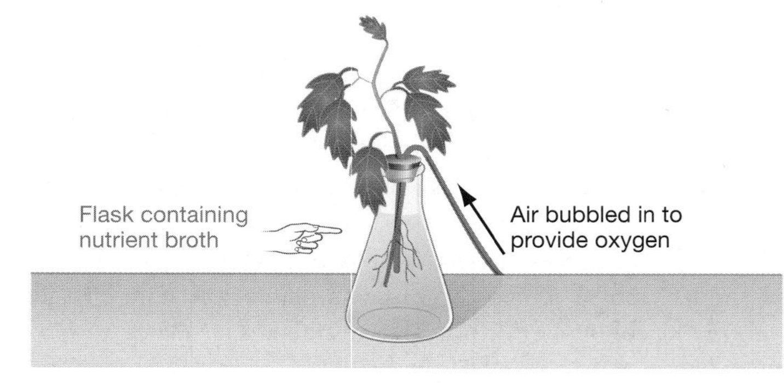

(b) What happens if a nutrient is absent?

(c) Closeup of copper-deficient tomato plant

FIGURE 33.1 Hydroponics as a Method for Studying Nutrient Deficiencies
(a) Culturing plants in a solution instead of soil allows researchers to precisely control which nutrients are available and in what concentrations. **(b)** To determine the symptoms of copper deficiency, biologists set up the experiment described here. The treatment containing all of the essential nutrients served as the control. **(c)** A lack of copper affects most tissues and produces an array of symptoms.

These particles are the first ingredient in soil. As organisms occupy the substrate, they add their waste products and carcasses. This organic matter is called humus. With time, soil eventually becomes an extremely complex and dynamic mixture of inorganic particles, organic particles, and organisms. It is commonplace to find thousands of species living in the top few centimeters of a single square meter of soil. In addition to plants, soil-dwelling organisms include a variety of fungi and animals along with a vast number of bacteria, archaea, microscopic protists, and roundworms called nematodes.

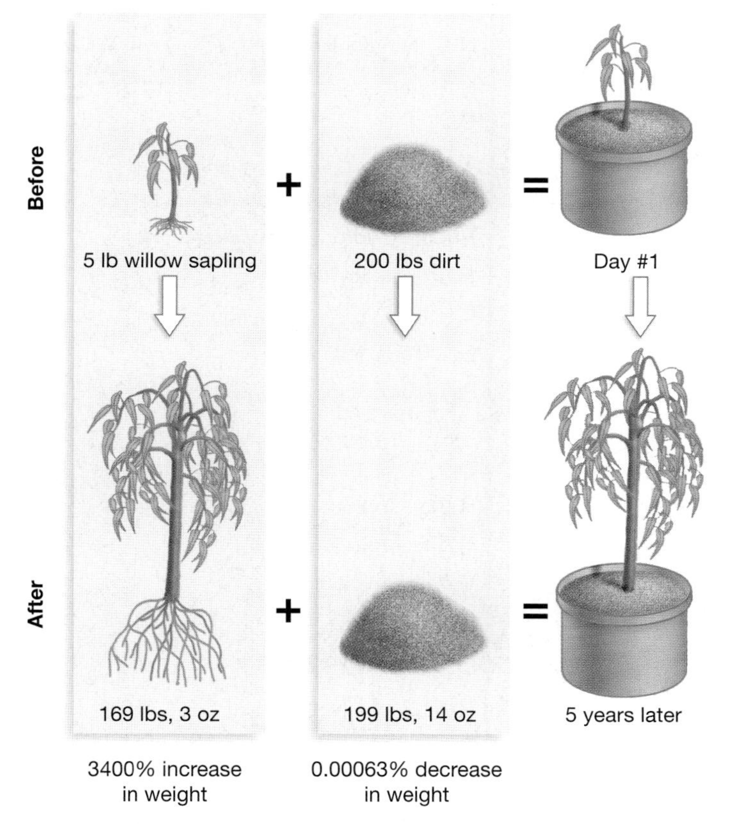

FIGURE 33.2 An Early Experiment in Plant Nutrition
A scientist in the 1600s carefully weighed a tree and the soil it grew in before and after 5 years of growth. To interpret the data, he proposed that most of the mass of plants is derived from water and not soil.
QUESTION Was the researcher correct? Explain why or why not.

Soil provides plants with water, support, and nutrients. It also contains air pockets that supply the oxygen required to fuel cellular respiration. Because the parent rock and organisms that occupy soils vary from one site to another, the texture and chemical composition of the soil vary as well. The properties of a soil are important in determining the types of plant species that will grow there. Why? The texture of a soil affects the availability of oxygen and the ability of roots to penetrate more deeply; its chemical composition dictates which nutrients are available.

Nutrient Availability

The elements required for plant growth are not found in the soil as atoms. Instead, they exist as ions. **Table 33.2** (page 646) lists some of the ions found in soil that contain important plant nutrients.

The ions present in soil tend to behave in one of two ways, depending on their charge (see **Figure 33.4**, page 646). Ions with negative charges usually dissolve in soil water because they interact with water molecules via hydrogen bonding. (Phosphate is an exception to this rule. Phosphate ions tend to bind to soil particles along with positively charged ions.) As solutes, negatively charged ions are available to plants for absorption; however, they are also easily washed out of the soil by rain. The loss of nutrients via washing is called leaching.

Ions with positive charges, in contrast, often interact with the negative charges found in organic matter and on the surfaces of the tiny, sheet-like particles called clay. As a result, the presence of organic matter and clay in a soil slows leaching. Organic soils that contain clay tend to retain nutrients. Clay also makes these ions more difficult for plants to extract and use, however, because they are tightly bound. Before roots can absorb these ions, they must be released from the clay particles and enter the soil solution. How does this occur? In general, what mechanisms do plants have for absorbing the nutrients they need?

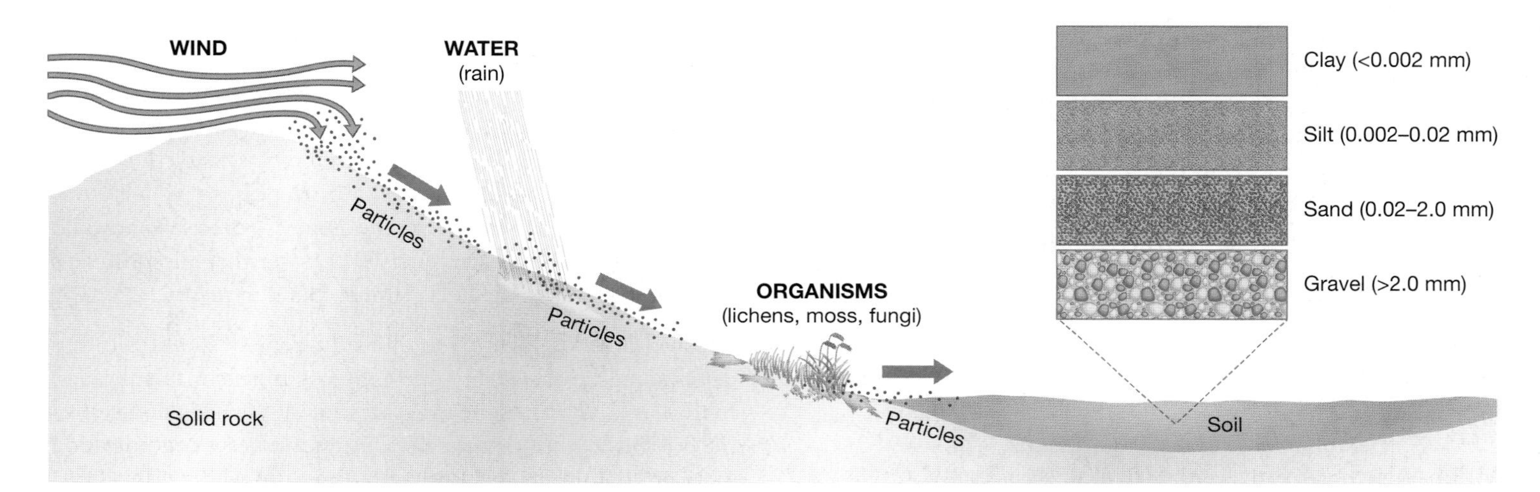

FIGURE 33.3 The Start of Soil Formation
Soil formation begins when wind, rain, and organisms break small particles off solid rock.

CD ACTIVITY 33.1 Soil Formation and Nutrient Uptake

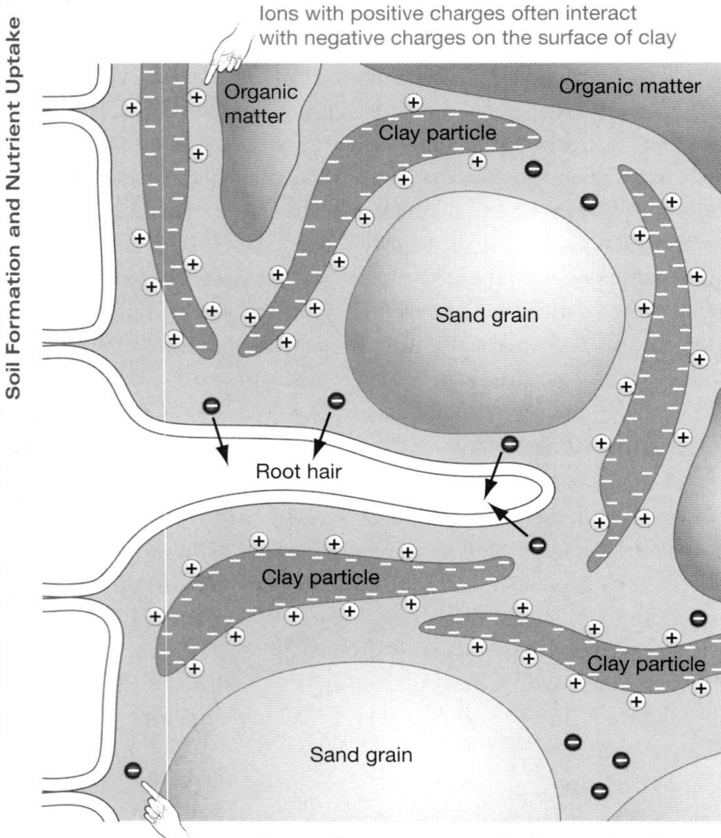

FIGURE 33.4 Ion Interaction with Soil Particles and Soil Water Positively charged ions tend to bind to organic matter in soil as well as clay particles. Negatively charged phosphate ions also tend to bind to soil organic matter. As a result, soils that are poor in organic matter and clay tend to lose nutrients rapidly. EXERCISE Add symbols indicating that positive ions bind to organic matter.

TABLE 33.2 Nutrients in the Soil

Element	Form Available to Plants
Oxygen	CO_2, O_2, H_2O
Carbon	CO_2
Hydrogen	H_2O
Nitrogen	NO_3^- (nitrate)
	NH_4^+ (ammonium ion)
Potassium	K^+
Calcium	Ca^{2+}
Magnesium	Mg^{2+}
Phosphorus	$H_2PO_4^-$ (dihydrogen phosphate ion)
	HPO_4^{2-} (hydrogen phosphate ion)
Sulfur	SO_4^{2-} (sulfate ion)
Chlorine	Cl^-
Iron	Fe^{3+} (ferric ion)
	Fe^{2+} (ferrous ion)
Manganese	Mn^{2+}
Zinc	Zn^{2+}
Boron	$H_2BO_3^-$ (borate ion)
Copper	Cu^+ (cuprous ion)
	Cu^{2+} (cupric ion)
Nickel	Ni^{2+}
Molybdenum	MoO_4^{2-} (molybdate ion)

33.3 Nutrient Uptake

In the vast majority of plants, the root system is the site of nutrient uptake. (Section 33.5 introduces some of the exceptions to this rule.) The general features of root anatomy were introduced in Chapter 31; how roots function in water uptake was analyzed in Chapter 32. Here we focus on features of roots that are directly involved in the transfer of nutrient-containing ions from the soil to the inside of the plant.

All three of the features reviewed here involve the epidermal cells of roots. As **Figure 33.5** shows, these cells are in direct physical contact with soil. Recall from Chapter 31 that the amount of interaction between these cells and soil is greatly increased by the presence of thin projections called root hairs. The root hairs of epidermal cells dramatically increase the surface area available for nutrient and water absorption. Why is this important?

Mechanisms of Nutrient Uptake

The plasma membrane is a fluid, sheet-like structure consisting of a phospholipid bilayer studded with proteins (Figure 33.5). Recall from Chapter 4 that the interior of the phospholipid bilayer is uncharged. As a result, it resists the passage of ions. Some of the proteins found in certain cells span the bilayer, however, and act as channels that allow the transit of specific ions. For example, the essay in Chapter 4 introduced the chloride channel found in some human cells. This channel allows chloride ions—and only chloride ions—to pass through the plasma membrane.

The large surface area of root hairs is important for a very simple reason. It holds large numbers of membrane proteins, which contact the soil and selectively facilitate the passage of ions that contain nutrients into the cell. Based on the earlier analysis of the chloride channel, it would be sensible to predict that each of the ions listed in Table 33.2 passes through a different, specific channel. Now the issue is, precisely how do these membrane proteins work? To answer this question, let's focus on the potassium (K^+) channel found in plants.

Passive Uptake **Figure 33.6** illustrates one of the keys to understanding how nutrients move across root-hair membranes. In the left-hand panel, the concentration of K^+ is high outside of the cell but low inside the cell. This establishes a concentration gradient that results in a flow of K^+ into the cell by diffusion.

In the figure's center panel, this concentration gradient is augmented by an electrical difference between the outside and the inside of the cell. In this case, a large number of hydrogen ions (H^+) have accumulated in the soil. As a result, the outside of the membrane has become positively charged relative to the inside. Note that K^+ flows into the cell along this charge gradient in response to the electrical attraction exerted by the negative charges inside. The presence of a potassium channel ensures that only K^+, and not other ions, follows this gradient. In this way, the presence of H^+ helps move potassium ions into the cell.

The combined effect of concentration and electrical charge on an ion is called its **electrochemical gradient**. When an electrochemical gradient favors the movement of an ion, no energy expenditure is required and the movement is described as passive. For example, if soil water contains a large number of potassium ions, then an electrochemical gradient should favor passage of K^+ into root hairs. By adding a high concentration of radioactive potassium ions to the solution outside of a root cell and following their movement, Emanuel Epstein and co-workers confirmed that K^+ does indeed flow into cells along an electrochemical gradient.

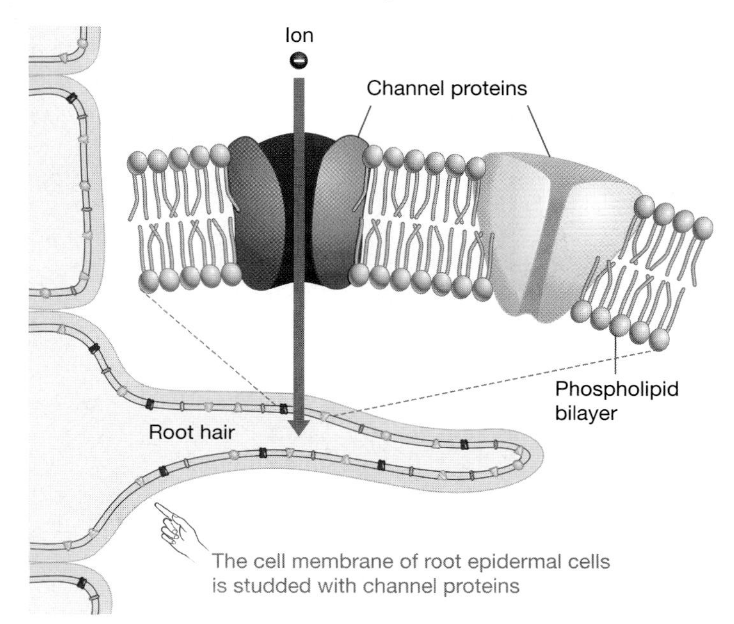

FIGURE 33.5 Root-Cell Hairs and Their Membranes
Plasma membranes are composed of lipid bilayers and an array of proteins. Some of these proteins act as channels that permit the passage of specific nutrients into the root cell. EXERCISE Make a rough estimate of the amount of membrane area exposed to the soil in root cells that have a hair-like projection versus adjacent cells that don't.

Following this work, the membrane protein responsible for passive uptake of K^+ was purified and sequenced. The gene that codes for this potassium channel has also been isolated and sequenced. Similar types of studies have been done for the membrane proteins that allow nitrate (NO_3^-) and other nutrients to move through membranes passively.

The take-home message from these studies is that root-hair membranes are packed with nutrient-specific channels that allow the rapid uptake of important ions whenever a favorable electrochemical gradient exists. But what if an appropriate gradient does not exist for a particular nutrient? Would a plant begin to starve?

Active Uptake For a cell to import an ion or molecule against an electrochemical gradient, it must expend energy in the form of ATP. This is called active uptake. The experiments that Epstein and his associates did on potassium movement suggested that K^+ uptake could be active as well as passive in root cells. This conclusion was reasonable, because ion movement continued into experimental cells even when the researchers made the K^+ concentration outside the cell much lower than the concentration inside. The rate of movement during uptake at low concentrations was strikingly different than the rate at high concentrations, however. This observation suggested that different membrane proteins are responsible for active and passive transport.

Daniel Schachtman and Julian Schroeder set out to test the hypothesis that active transport of K^+ occurs via a distinctive membrane protein. It had already been established that the

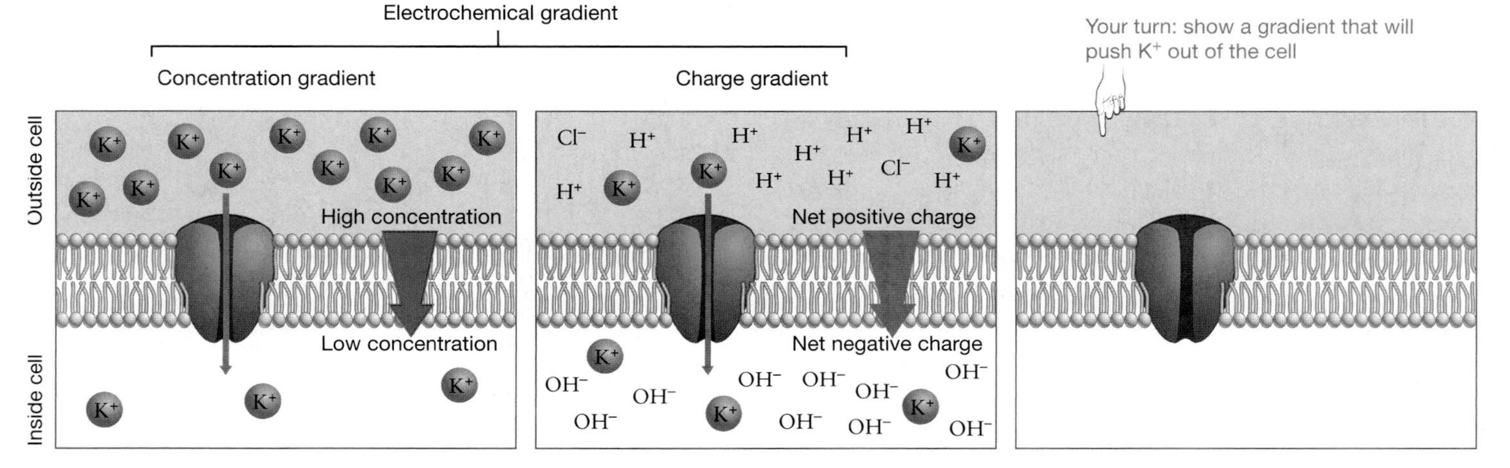

FIGURE 33.6 Ions Move in Response to Concentration and Charge Gradients
These sketches show that ions diffuse through protein channels in response to concentration gradients (left) and charge gradients (middle). The combined effect of concentration and charge is called the electrochemical gradient. EXERCISE In the right-hand panel, make a sketch showing a concentration or charge gradient that tends to push potassium ions out of a cell.

FIGURE 33.7 How Does Potassium Enter Cells Against a Concentration Gradient?
(a) This is the protocol used by researchers to identify the potassium channel protein that is involved in active uptake. **(b)** Because K^+ uptake increased when the extracellular solution was made more acidic, the biologists proposed that active uptake depends on a proton pump, and that the protein identified in part (a) is an H^+-K^+ cotransporter.

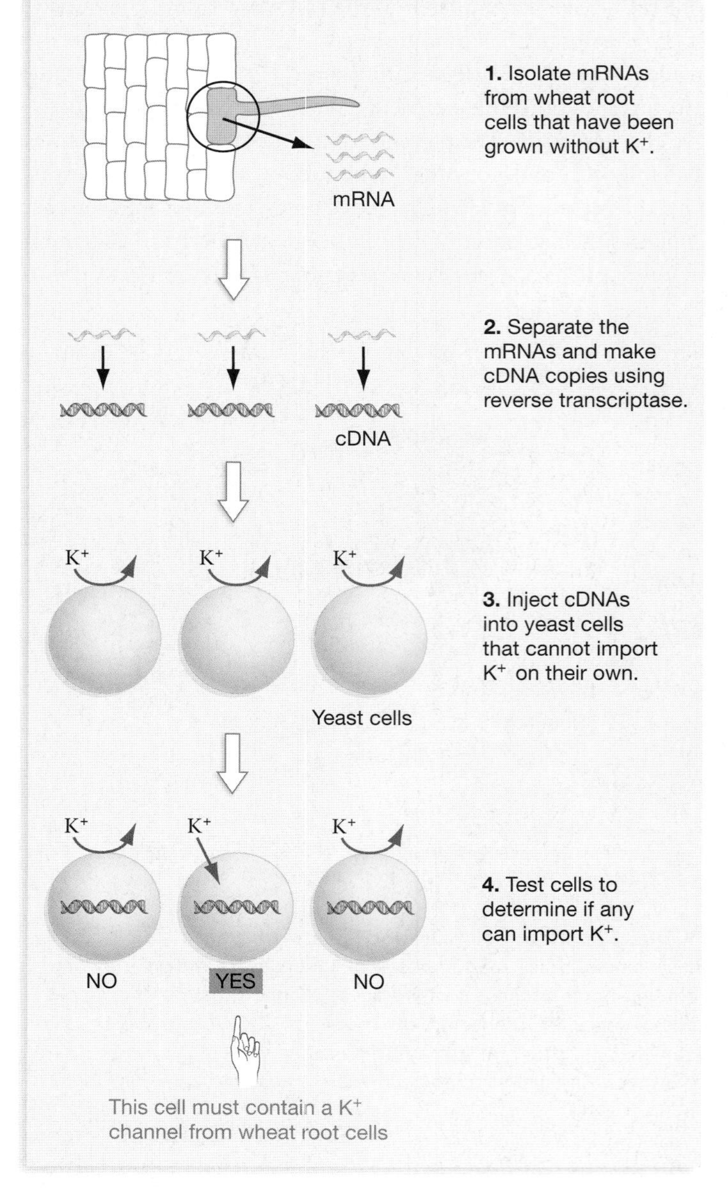

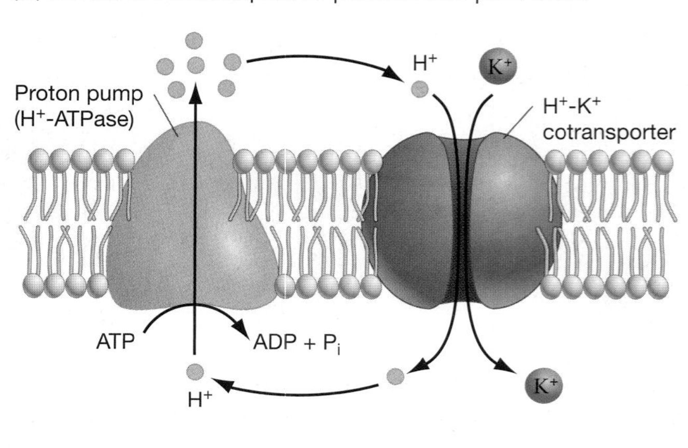

protein responsible for active uptake appears in root cells when potassium is missing from the growth medium. As **Figure 33.7a** shows, Schachtman and Schroeder began their effort to find this protein by isolating mRNAs from wheat roots that had been deprived of potassium. Then they used reverse transcriptase to produce the complementary DNAs, or cDNAs, to these mRNAs (this technique is described in detail in Chapter 17). When they injected these cDNAs into yeast cells that lacked the ability to import potassium, they found one that permitted the experimental cells to successfully absorb K^+.

How does this protein work? Schachtman and Schroeder found that if they made the solution around their experimental yeast cells more acidic, K^+ transport occurred more rapidly. This observation led them to propose that active transport of potassium depends on the action of a membrane protein that pumps protons from the inside to the outside of the cell membrane, using ATP (**Figure 33.7b**). In effect, they proposed that active transport of K^+ is based on the same mechanism responsible for loading sucrose into sieve-tube cells at source tissues (see Chapter 32). In the case of K^+ uptake, a proton pump creates an excess of protons on the exterior of the root-hair membrane. Then an H^+/K^+ cotransporter uses the resulting electrochemical gradient for H^+ to drive K^+ into the cell against its concentration gradient. Current evidence suggests that the H^+-ATPase found in root epidermal cells is different from the H^+-ATPase present in the membranes of companion cells.

Along with the array of channels involved in the passive uptake of nutrients, then, the root-hair membrane also includes proton pumps and cotransporters. The latter are capable of bringing scarce ions into the cell against strong electrochemical gradients.

Nutrient Transfer via Soil-Dwelling Fungi Plants need to extract large quantities of nitrogen and phosphorus from the soil. In many habitats, however, plants don't accomplish all of the N and P uptake on their own. Instead, nitrogen and phosphorus acquisition is a cooperative venture. For example, many plants living in northern forests receive large quantities of nitrogen from fungi that wrap themselves around the epidermal cells of roots and radiate out into the surrounding soil (**Figure 33.8a**). Plants that live in grasslands and in tropical forests receive much of the phosphorus they need from species of fungi whose bodies actually penetrate into the interior of plant root cells.

Fungi that live in close association with roots are called **mycorrhizae** (fungus-root). Chapter 29 introduced the two major types of mycorrhizae and presented evidence that these fungi transfer nitrogen and phosphorus, respectively, from soil to plant roots. That discussion also emphasized the ecological importance of mycorrhizae and their impact on the species diversity and productivity of ecosystems. Here, however, we focus on testing the claim that the association between plants and mycorrhizae is mutually beneficial. Does some sort of equitable partnership actually exist, or are plants really stealing nutrients from these fungi?

The research that answered this question used radioactive isotopes of nitrogen, phosphorus, and carbon. Carbon was included in these experiments to test the hypothesis that the association between mycorrhizae and plants is mutually beneficial. Specifically, the prediction was that radioactively labeled carbon dioxide, fed to plants, would be fixed by the light-independent reactions of photosynthesis and transferred to their mycorrhizal fungi. In return, radioactive phosphorus or nitrogen fed to fungi would show up in plant tissues.

Studies with radioactive C, N, and P confirmed that mycorrhizal fungi receive sugar from plants in exchange for providing nitrogen or phosphorus (**Figure 33.8b**). This is an important example of a mutually beneficial relationship, or **mutualism**—a topic that is explored more thoroughly in Chapter 49.

Mechanisms of Ion Exclusion

Not all nutrient uptake is beneficial. Many of the metal ions found in soils are poisonous to plants; even essential nutrients can become toxic when they are present in high quantity. Sodium is also detrimental at high concentrations. An excess of Na^+ inside cells can directly disrupt enzyme functioning; a surplus of sodium in extracellular spaces creates a solute potential that pulls water out of cells and results in a loss of turgor.

How do plants exclude certain ions? This question is currently the focus of intense research because of its practical applications. Conservationists are attempting to propagate plants that are capable of growing on abandoned mine sites with high concentrations of toxic metals; farmers need to cultivate plants that can grow on fields where poorly managed irrigation programs have resulted in salt buildups.

Passive Exclusion Some crop species are much more tolerant of salty soils than others. Rice is notoriously sensitive to salt buildups, for example, while barley is relatively tolerant. Even within the same species, the degree of salt tolerance can vary widely. Why are high concentrations of sodium deleterious to some populations of plants but not others?

Sven Schubert and André Läuchli were interested in exploring the molecular mechanism responsible for variation in salt tolerance among corn varieties. More specifically, they wanted to test the hypothesis that plants might have passive mechanisms for excluding sodium ions. Corn was an interesting study organism in this case because of its agricultural importance. Could certain populations of plants simply be better at preventing certain ions from entering their roots without expending energy?

Schubert and Läuchli answered this question by growing seedlings from salt-tolerant and salt-intolerant populations in hydroponic cultures containing a high concentration of NaCl. The concentration they used was high enough to ensure that a strong concentration gradient would tend to drive sodium ions into the root hairs. When they harvested the plants and compared their sodium contents, however, they found that the salt-intolerant individuals had taken up almost twice as much Na^+ as salt-tolerant individuals.

To interpret this result, the researchers suggest that individuals from salt-tolerant populations have fewer sodium channels

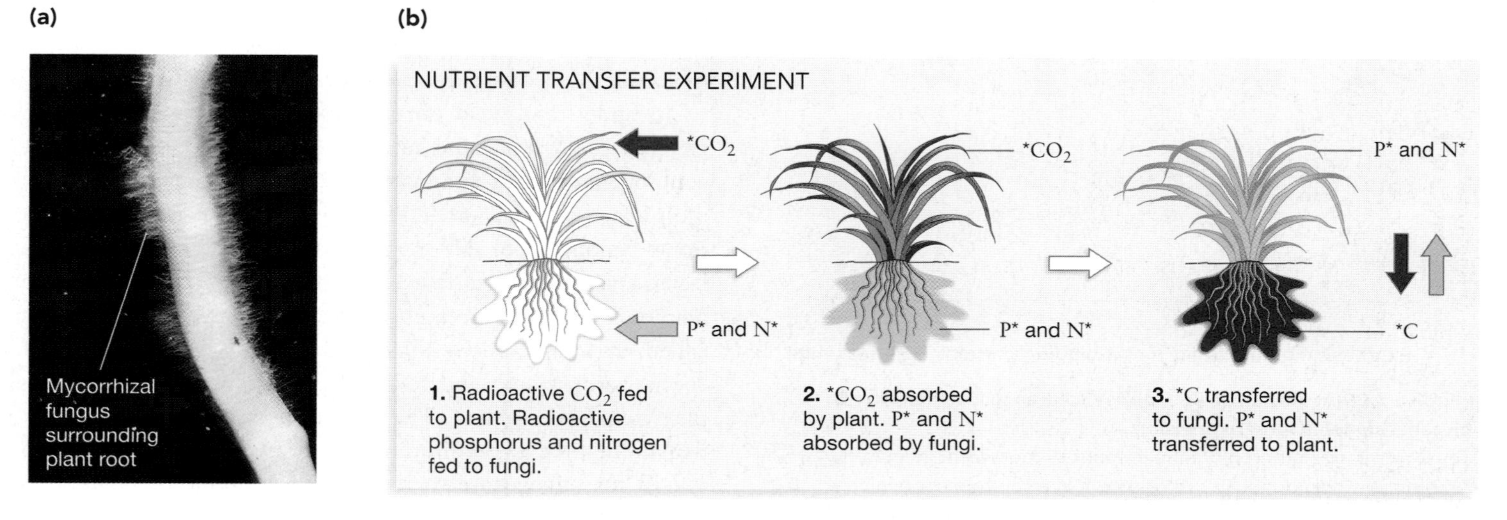

FIGURE 33.8 Mycorrhizae and Plants Have a Mutually Beneficial Relationship
(a) In this grass, a mycorrhizal fungus radiates from the plant's roots into the surrounding soil. **(b)** When radioactively labeled CO_2 ($^*CO_2$) was fed to plants and radioactively labeled P or N (P* or N*) was fed to mycorrhizae, the labeled compounds were transferred from one organism to the other. QUESTION Fungi are made up of long, thin filaments called hyphae. How would these structures affect the amount of membrane surface area available for absorbing nutrients needed by a plant?

in their root hairs than do salt-intolerant individuals. If this prediction is confirmed, it implies that salt tolerance is a genetically based adaptation, and that it is at least partly based on variation in the abundance of membrane proteins.

Active Exclusion Although passive mechanisms of nutrient exclusion have been confirmed for a variety of ions in addition to sodium, it would seem logical to predict that plants would also have a way of coping with toxins once they are inside the body. Consider copper, for example. Plants that grow on soils near copper-mining operations experience large concentration gradients that favor an influx of this nutrient. If copper channels are necessary to admit the small amounts required for normal cell function, it would seem inevitable that a surplus will eventually build up inside the body. How do plants neutralize excess nutrients before they begin to poison key enzymes?

To explore this issue, Angus Murphy and Lincoln Taiz turned to the leading model organism among angiosperms: *Arabidopsis thaliana*. They were able to find 10 different natural populations of *A. thaliana* that showed varying degrees of copper tolerance. To explain this variation, they hypothesized that the individuals involved might vary in the expression of one or more metallothionein genes. Metallothioneins are small proteins that bind to metal ions. Once an excess copper, zinc, or other metal ion binds to a metallothionein, the metal cannot act as a poison. Genes for these proteins have been found in a wide variety of organisms—bacteria, fungi, and animals as well as plants.

Murphy and Taiz grew individuals from the 10 populations in solutions containing a high concentration of copper ions. Using cameras, they monitored the rate of root growth in each plant and compared it to the rate of root growth in plants from the same population that were not exposed to high copper concentrations. After allowing growth to proceed, they harvested a tissue sample of each individual and documented the concentration of metallothionein mRNA it contained.

The data that resulted are shown in **Figure 33.9**. Note that an extremely strong positive correlation exists between the expression level of the metallothionein gene called *MT2* and the growth rate of *Arabidopsis* individuals exposed to copper. In *Arabidopsis*, copper-tolerant individuals produce more *MT2* mRNA than copper-intolerant individuals. This observation suggests that metal tolerance in some plants may be a function of how metallothionein genes are regulated.

Are these differences due to features of the promoter or enhancer elements in metal tolerant plants? If so, could these regulatory sequences be added to other species to produce an array of individuals capable of growing on abandoned mines and other contaminated sites? These questions remain unanswered. Research on MT2 and other metal-binding proteins is pressing forward.

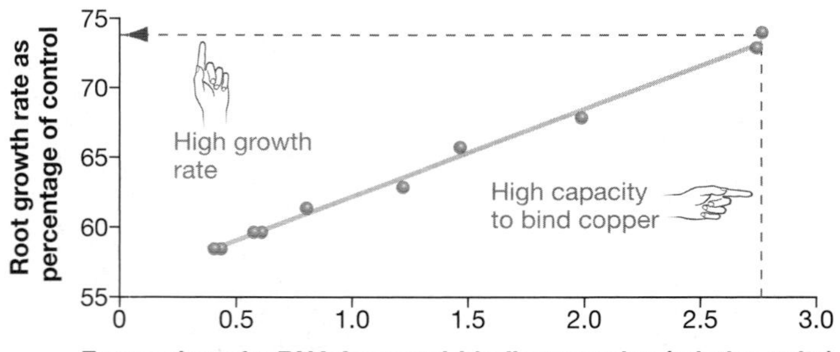

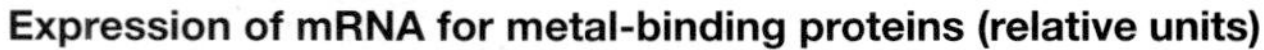

FIGURE 33.9 In *Arabidopsis*, Copper Tolerance Correlates with Metallothionein Gene Expression
The x-axis in this graph plots the amount of metallothionein mRNA in *Arabidopsis* individuals that were grown in a solution containing a high concentration of copper. (The amount of metallothionein mRNA present is expressed relative to the amount of mRNA present from a common cell component called β-tubulin in the same individuals.) The y-axis plots how much the roots of these individuals grew relative to individuals from the same populations that were exposed to normal amounts of copper. Each dot on the graph represents the average values from individuals in one of the 10 populations in the study.

33.4 Nitrogen Fixation

In corn-growing regions of North America, it is not unusual in springtime to see farmers spraying plowed fields from huge bottles labeled liquid ammonia. The chemical formula of ammonia is NH_3. Because nitrogen is required for every amino acid and nucleic acid in the body, plants require it in large quantities. N is also a limiting nutrient in many soils, meaning that it is in short supply. Most plants grow much faster when they receive a nitrogen-containing fertilizer.

The use of nitrogen-based fertilizers has drawbacks, however. In many parts of the world, ammonia-containing fertilizers are too expensive for farmers to use; in more affluent regions, they are used so extensively that they are causing serious pollution problems (Chapter 25 explained why). For these and other reasons, there has been intense interest in understanding the molecular basis of a phenomenon called nitrogen fixation.

To understand what nitrogen fixation is, it is important to recognize that molecular nitrogen, N_2, is the most abundant component of Earth's atmosphere. N_2 is an exceptionally stable molecule, however, and rarely participates in chemical reactions. Among all of the organisms on the tree of life, only selected species of bacteria are able to take up N_2, convert it to ammonia, and use it to fuel growth. This conversion process is called **nitrogen fixation.** Nitrogen fixation requires a series of specialized enzymes and cofactors, including an enzyme complex called nitrogenase. The process is also extremely energy demanding. An expenditure of up to 24 ATP molecules is required for nitrogenase to reduce one molecule of N_2 to two molecules of NH_3.

Here we investigate nitrogen-fixing bacteria in the genus *Rhizobium* and their relationship with plants in the pea family. (Pea-family plants are often called legumes; members of the genus *Rhizobium* are often called rhizobia.) Like mycorrhizal fungi and their host plants, legumes and rhizobia have

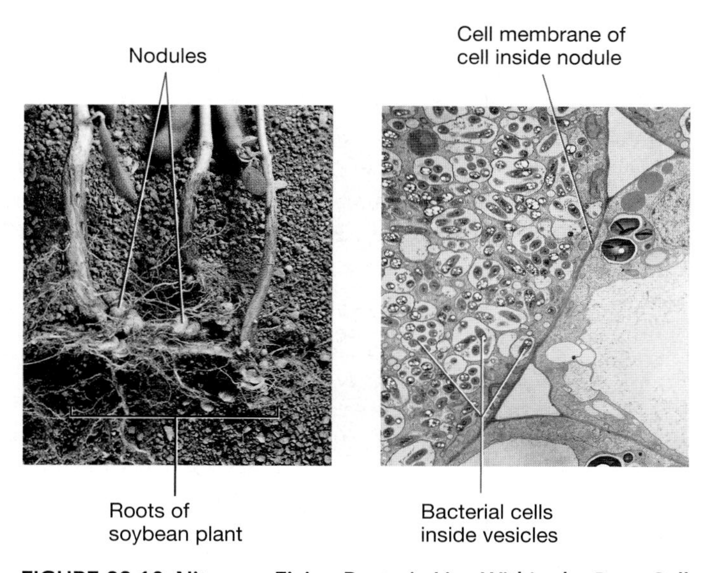

FIGURE 33.10 Nitrogen-Fixing Bacteria Live Within the Root Cells of Pea-Family Plants
(a) A soybean root with nodules containing nitrogen-fixing bacteria. (Roots that are not colonized by rhizobia do not have nodules.)
(b) An electron micrograph showing a cross section through a root cell from a nodule. Note that the bacterial cells are abundant and are enclosed within vesicles.

a mutualistic relationship. The bacteria provide the plant with ammonia, while the legume provides the bacteria with sugar and protection.

As **Figure 33.10a** shows, nitrogen-fixing bacteria and their host cells form distinctive structures in the roots of legumes. These structures are called nodules. Because the bacteria actually occupy the interior of root cells inside nodules, they are considered a **symbiotic** (together-living) organism (**Figure 33.10b**). A symbiosis occurs when members of different species live in close physical contact and affect each other's ability to survive and reproduce.

What genes and proteins are involved as these organisms interact? If the key genes could be identified, can they be introduced into species like wheat or rice using the techniques introduced in Chapter 17, so that their roots can also be occupied by nitrogen-fixing bacteria? If so, perhaps the use of nitrogen-based fertilizers could be reduced.

How Do Nitrogen-Fixing Bacteria Colonize Plant Roots?

When a pea seed germinates, its roots do not contain a population of rhizobia. Instead, the root must make contact with bacterial cells existing in the soil, and the rhizobia must colonize the plant. Colonization is a complex process involving a series of specific interactions between the rhizobia and the legume. What are these steps, and how do they occur?

As **Figure 33.11a** shows, the first event in colonization is recognition. In pea family plants, the surfaces of root hairs contain compounds that are closely related to the pigments found

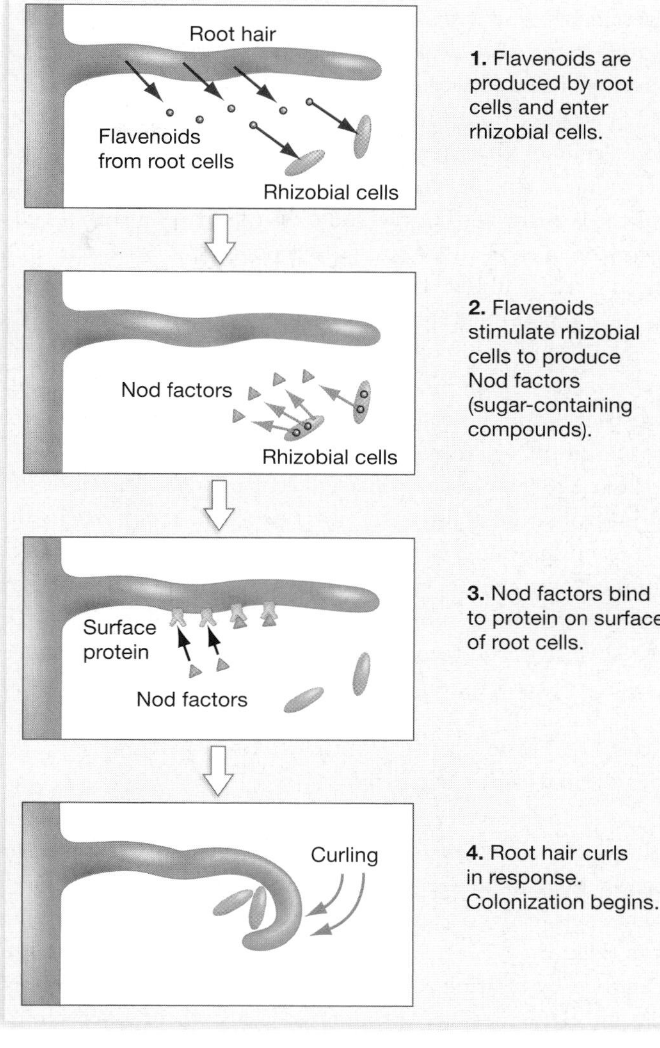

(b) Roots exposed to rhizobial cells

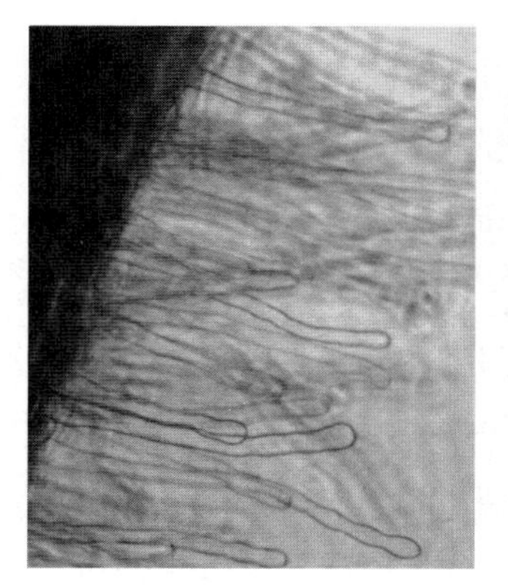

Protein that binds Nod factors disabled by an antibody—no curling occurs.

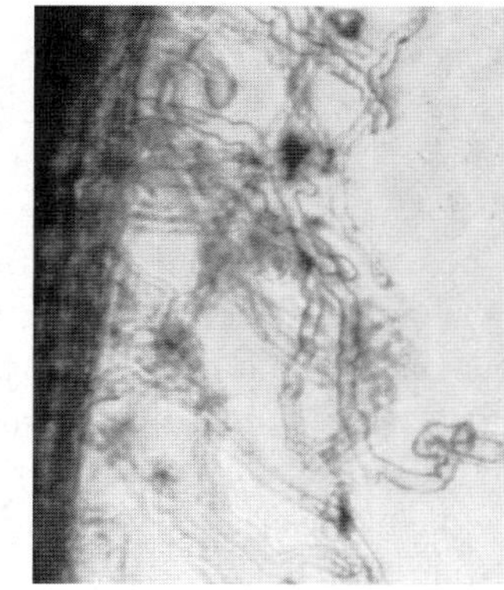

No antibody treatment. (Extensive curling of root hairs)

FIGURE 33.11 Colonization by Nitrogen-Fixing Bacteria Begins with a Recognition Step
(a) Recognition between legumes and nitrogen-fixing bacteria is reciprocal and species-specific. **(b)** When the Nod-factor binding protein is disabled, recognition is blocked.

in flower petals. When rhizobia contact these molecules they respond by producing sugar-containing molecules called Nod factors (for *nod*ule-formation). Nod factors, in turn, bind to proteins on the surface of root hairs.

For years, no one was able to isolate a membrane protein from legumes that binds to Nod factors and completes this recognition step. But recently Marilynn Etzler and co-workers isolated a protein from the roots of a legume called *Dolichos biflorus* and showed that it binds strongly to a Nod factor produced by its symbiotic bacterium. Further, when the researchers disabled this protein by treating root hairs with an antibody to it, they found that colonization by the rhizobium could not proceed (see **Figure 33.11b**). These results support their claim that they have succeeded in identifying a membrane protein that binds to Nod factors.

Before moving on to consider what happens once a legume and rhizobium make contact, it is important to appreciate that this recognition step is species-specific. Each legume produces a different recognition signal, and each rhizobium species responds with one or more unique Nod factors. When investigators have switched recognition signals or Nod factors between species, the recognition step fails.

How Do Host Plants Respond to Contact from a Symbiotic Bacterium?

The species-specific recognition step outlined in Figure 33.11 is followed by a response on the part of the host plant. Once Nod factors bind to the root-hair surface, they set off a chain of events that leads to the transcription of a suite of genes in host cells (**Figure 33.12a**). Some of these response genes have been identified by treating root cells with Nod factors and isolating the mRNAs that are produced. Using this approach, researchers have been able to identify a series of loci called *ENOD* (for *e*arly *nod*ulin) genes that respond to the recognition event.

To find out exactly where the *ENOD* loci are expressed in the root, Yiwen Fang and Ann Hirsch attached the promoter sequence from one of these genes to a "reporter gene." A **reporter gene** is a sequence that produces a detectable compound. Reporter genes are frequently fused to specific genetic regulatory elements, like promoters, so researchers can study where the promoter is activated. In this case, the product of the reporter gene was an enzyme that makes a certain dye turn blue. As a result, the presence of blue color in roots treated with the dye should indicate exactly where *ENOD* genes are being expressed.

Fang and Hirsch tested root cells for the presence of the reporter protein at each phase of the colonization process and found that the blue color was most intense as the nodule itself formed (**Figure 33.12b**). The *ENOD* gene they worked with, however, was also expressed at low levels in roots that were not contacted by rhizobia. This observation suggests that at least some of the plant genes that respond to rhizobia are not unique to the nodule formation process. Instead, they code for normal cell proteins that also have a role in nodule formation.

The overall message of these experiments is that the recognition and response steps leading to colonization and nitrogen fixation are intricate processes. Each phase of the process in-

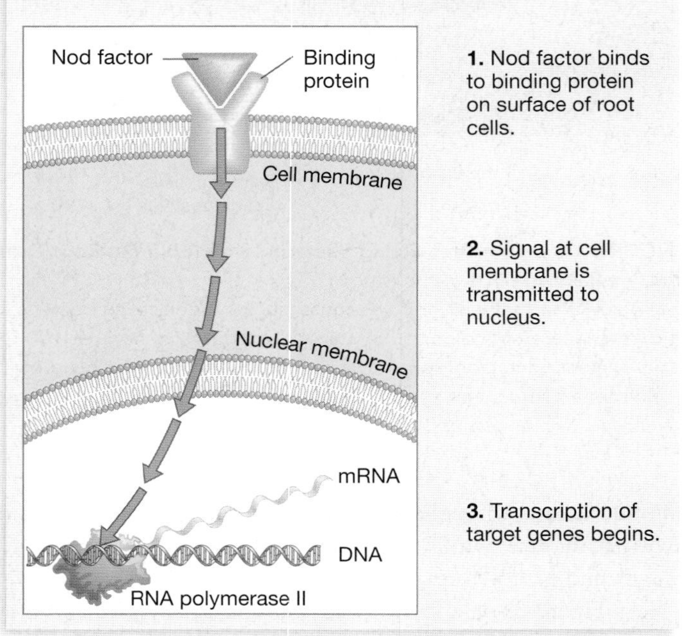

(b) The blue color indicates where in the root the response genes are being transcribed.

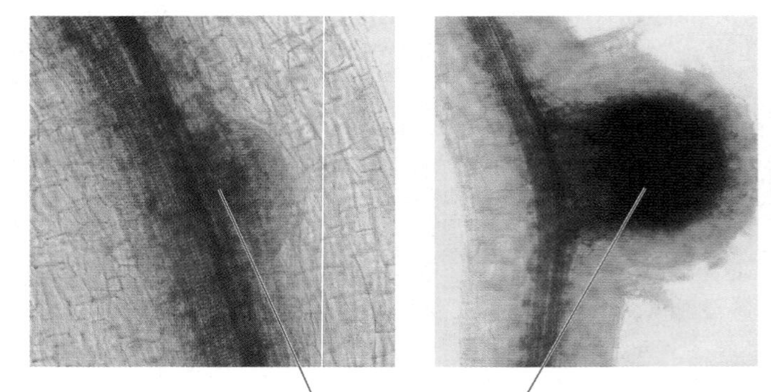

FIGURE 33.12 In Legumes, the Response to Nod Factors Involves Gene Transcription
(a) When Nod factors bind to receptor proteins on the surfaces of root cells, a series of events results in the production of mRNAs from target genes. **(b)** The blue color in these photographs is from the product of a reporter gene. In these plants, the reporter gene was attached to the promoter for an early nodulin gene called *ENOD40*. As a result, the blue color indicates where *ENOD40* is being expressed and in what amount.

volves numerous loci in both the host and symbiont. Identifying all of the critical structural and regulatory genes and introducing them into wheat, rice, or corn will be no easy task. Although developing a strain of nitrogen-fixing rice or corn would be considered a crowning achievement in plant biotechnology research, investigators have realized that it will be extremely difficult to replicate the recognition and response steps in these species.

33.5 Nutritional Adaptations of Plants

Up to this point we have considered how an average plant obtains the nutrients it needs. Most of these species make their own sugar through the process of photosynthesis. Most obtain N, P, K, and other essential nutrients through active or passive transport or with assistance from mutualistic fungi or bacteria.

A substantial number of species don't follow these rules, however. In the tropics and subtropics, for example, species from a diverse array of lineages do not absorb nutrients from soil. In fact, these species never even make contact with soil. As **Figure 33.13** shows, they often grow in the leaves or branches of trees. For this reason they are called **epiphytes** (upon-plants). Familiar examples of epiphytes include the staghorn fern, certain mosses, Spanish moss (which is actually not a moss at all, but a relative of the pineapple), and many other species of bromeliads and orchids. Epiphytes absorb most of the nutrients they need from rainwater that collects in their tissues or in the crevices in bark.

In terms of acquiring nutrients, though, epiphytes are not nearly as remarkable as the plants reviewed next. Some species are parasites that acquire all of their nutrition, including C, H, and O, from other individuals. A few even kill their host in the process. Yet another group of plants is **carnivorous**, meaning that they eat meat. Carnivorous plants make sugar via photosynthesis. But instead of obtaining the nitrogen they need from the soil, mycorrhizal fungi, or symbiotic bacteria, they catch and digest small animals.

FIGURE 33.13 Epiphytes
The Bromeliad family includes epiphytic plants that are common in tropical rain forests. In some species, the leaves curl together at the base to form a tube or basin that catches and holds water. These miniature ponds attract insects and algae that provide nutrients.

Parasitic Plants

Based on the data currently available, biologists estimate that about 3000 species of plants are parasitic. This number represents less than 1 percent of the total number of plant species that have been studied and named to date. Some of these parasites are nonphotosynthetic and obtain all of the nutrition they need from their hosts, but most make their own sugars through photosynthesis. In these cases, parasitism is limited to the root systems of their hosts, which are tapped for water and essential nutrients.

How detrimental are these root parasites? For example, is parasitism more damaging to the host than competition with an individual of the same species?

To answer this question, Dierthart Matthies put two alfalfa seeds into each of a large number of pots. Some of these pots also received a seed from a root parasite called *Odontites rubra*, others got a seed from a root parasite called *Rhinanthus serotinus*, some received an additional alfalfa seed, and some pots were left alone. When the plants had grown to maturity, he dried and weighed them. The resulting data showed that the total amount of plant matter produced, or **biomass**, was much smaller in the pots with parasites than in the pots with only host plants (see **Figure 33.14**, page 654). The result supports two conclusions: Parasitism is much more damaging than competition in this host-parasite system, and parasitism lowers the total productivity of the individuals involved. The second result implies that parasites do not grow as efficiently as their host plants.

Why do parasites grow so poorly? Matthies suggests that the benefits that parasites experience in nutrient acquisition are offset by a cost in competing with the host plant for light. If this point of view is correct, then it is not surprising that parasitism is relatively rare in plants, and that parasites are most common in habitats where nutrients are particularly scarce.

Carnivorous Plants

The Venus flytrap is an angiosperm native to bog habitats in the southeastern United States. It is a green plant that makes carbohydrates via photosynthesis. Bog habitats are notoriously poor in nutrients, however—particularly in nitrogen.

To cope with these conditions, natural selection has led to the development of a modified leaf called a trap. This trap includes hair-like structures that protrude from the epidermis. When an insect lands on the trap and bumps these hairs, the leaf responds by snapping shut. After the insect is trapped, glands near the trap release enzymes that slowly digest the prey. The plant then absorbs the nutrients—primarily iron and nitrogen—released from the digested prey.

How do the hair-like structures on a flytrap sense the presence of an insect? How is the signal from the hair-like structures transmitted to the rest of the leaf, and how is this information translated into a movement that closes the trap? Answering questions like these is our next task. Chapter 34 focuses on how plants sense touch, light, and other stimuli, and how they move in response.

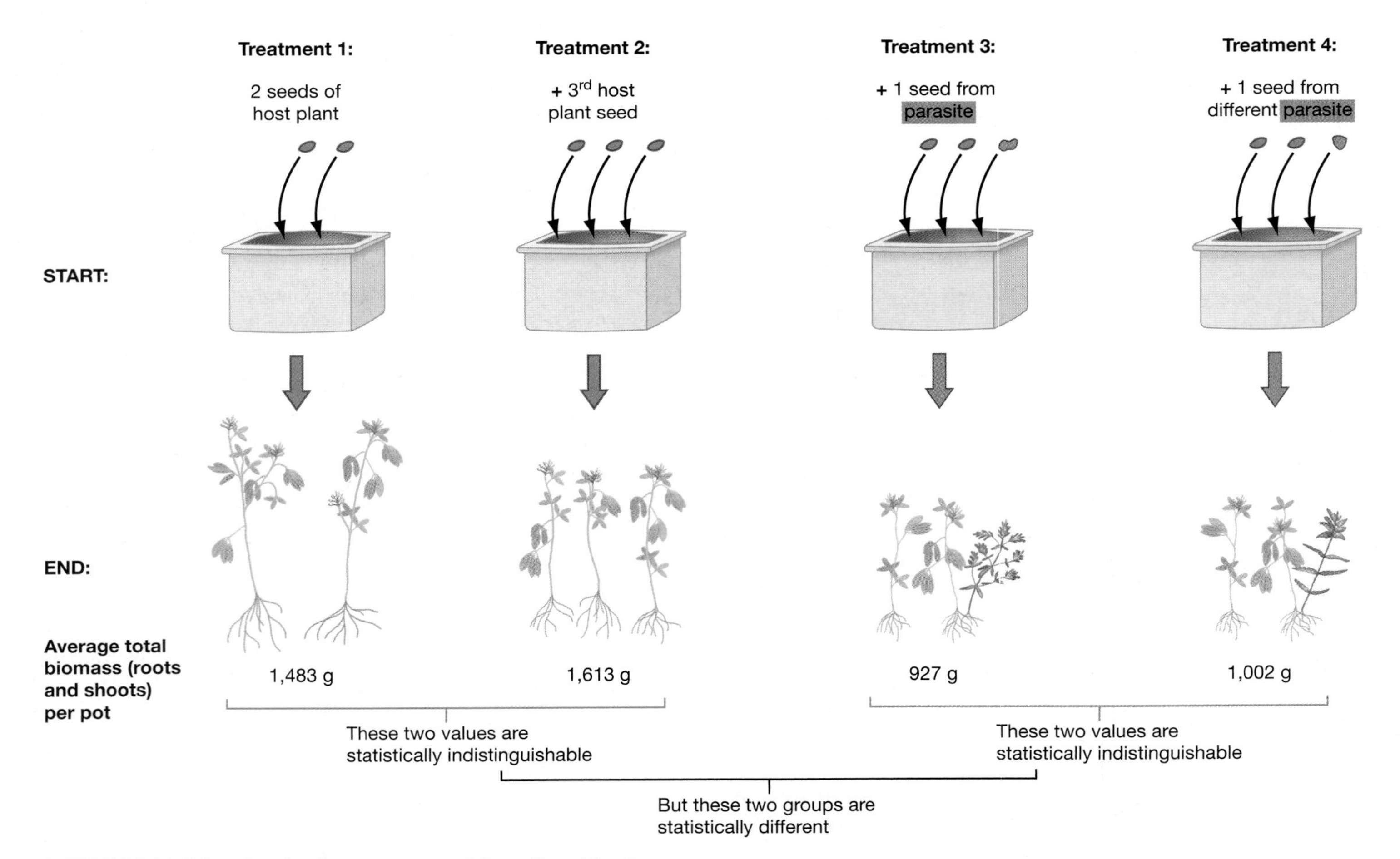

FIGURE 33.14 What Are the Consequences of Root Parasitism?
Each of the four treatment groups shown had a different combination of host plants and parasites, and each was replicated 10 times (giving a total of 40 pots). When the total dry weight of the plants in each pot was determined, statistical tests showed the two treatments without parasites were indistinguishable from one another, meaning that adding another host plant did not suppress overall productivity. The two treatments with parasites were also statistically indistinguishable, meaning that the two parasite species had the same effect on the host. **QUESTION** Why was it important to have so many replicates of each treatment for the experiment to be meaningful?

Essay Tropical Soils

The soils that underlie the Amazon basin and much of equatorial Africa are home to the world's most diverse and productive ecosystems: tropical rain forests. Ironically, these soils are also the most nutrient-poor in the world.

Why? The same features that encourage a luxuriant growth of plants—large amounts of rain and high temperatures—also promote rapid leaching of nutrients. Negatively charged ions that dissolve in soil water tend to wash out of soils that are subjected to large amounts of precipitation. Further, a chain of events triggered by tropical climates causes positively charged ions to leach out of tropical soils. This chain begins when hot, wet weather encourages the growth of soil bacteria. As a by-product of respiration, many of these cells produce CO_2, which reacts with soil water to form carboxylic acid. Acidity also accumulates in tropical soils from phosphate-containing molecules and phenolic compounds that are released from decaying vegetation. These chemical groups react with soil water to form phosphoric acid and an array of organic acids. The accumulation of protons is important because it frees nutrient-containing cations from their binding sites on clay particles and allows them to be leached away.

If nutrients in tropical soils are in short supply, how can plant growth be so lush?

When intensive leaching continues for hundreds of thousands or even millions of years, what is left behind? Instead of presenting a rich mix of components, tropical soils typically become dominated by minerals that do not dissolve readily. In fact, soil scientists classify many tropical soils as oxisols (oxygen-soils), because they contain so many tough iron oxides and aluminum oxides. The iron oxides present in oxisols turn them a bright rust-red color.

The events that led to weathering of tropical soils make sense, but an important question remains. If nutrients in tropical soils are in such short supply, how can plant growth be so lush? The answer is that virtually all of the nutrients in these ecosystems are tied up in living vegetation. When a plant dies in a tropical rain forest, its component molecules are rapidly recycled by the surrounding bacteria, archaea, fungi, and plants.

Unfortunately, one of two things happens when a tropical forest is cleared and burned for agriculture. Most of the nutrients either go up in smoke or sit on the soil surface, in ash, until they are washed away in subsequent rains. Farmers who work tropical soils can get good production from their crops for a year or two after the forest is cleared, but after that the soils are depleted. Trying to raise crops on oxisols is often a losing proposition.

To cope with this situation and yet encourage the development of local economies, land managers are attempting to implement logging practices that protect these soils while delivering a sustainable yield of forest products. It is not at all clear, however, that sustainable forestry practices can produce enough revenue to counter the incessant pressure from farm families for more cropland.

Chapter Review

Summary

Typically about 96 percent of the dry weight of a plant consists of carbon, hydrogen, and oxygen. Plants obtain these elements by absorbing carbon dioxide from the atmosphere and water from the soil. The other 4 percent of the plant body consists of a complex suite of elements. The most important of these are the essential nutrients, which are absolutely required for normal growth and reproduction. Essential elements must be absorbed from the soil in the form of ions like nitrate and phosphate.

Soil is a complex and dynamic mixture of inorganic particles such as clay and sand, organic particles, and living organisms. If nutrients are present at high concentration in a soil, they can be transported into the plant body passively. This is because ions move across plasma membranes in response to electrochemical gradients if an appropriate membrane channel is present. But if a certain nutrient is scarcer in the soil than it is in the plant, or if passage is opposed by a buildup of the opposite charge, active uptake is required. In many cases, plants import nutrients against an electrochemical gradient by pumping protons outside of the root. A strong proton gradient results from this ATP-consuming mechanism and allows nutrients to enter the root via H^+-nutrient cotransporters.

In addition to acquiring nutrients through these active and passive mechanisms, many plants obtain nitrogen or phosphorus from mycorrhizal fungi. Studies with radioactively labeled isotopes have confirmed that the relationship between plants and these fungi is mutually beneficial. The host plant provides the fungi with reduced carbon, while the fungi harvest N or P from the surrounding soil and transport it to the root system.

Passive and active systems also exist for excluding certain ions. This is important because all nutrients are toxic if they are present at high concentration. Passive exclusion is thought to result from a low number of ion channels in the membranes of root hairs. Active exclusion often results from the production of proteins that bind a particular ion, such as the metallothionein proteins found in *Arabidopsis thaliana*.

Certain plants, including those in the pea family, are capable of forming intimate associations with nitrogen-fixing bacteria. This association begins when bacterial cells contact certain proteins found on the surface of root hairs in legumes. When contact is made, the bacterial cell produces sugar-containing molecules called Nod factors, which bind to a different cell-surface protein on the host plant. A suite of host-plant genes is transcribed in response to this binding event, and the proteins that are produced result in the formation of a nodule structure. The bacteria inside the nodule eventually take up residence inside root cells, where they receive protection and sugar in exchange for producing ammonia.

Not all plants take up nutrients through membrane proteins or bacterial or mycorrhizal associations. Some species are parasitic and some are carnivorous. Most parasitic plants produce their own carbohydrates through photosynthesis, but steal water and nutrients by infecting the root systems of host plants. Carnivorous plants have evolved mechanisms for trapping and digesting insects.

Questions

Content Review

1. Which of the following characteristics defines an element as essential for a particular species?
   a. For high seed production, it has to be added as fertilizer.
   b. If it is missing, a plant cannot grow or reproduce normally.
   c. If it is present in high concentration, plant growth increases.
   d. If it is absent, other nutrients may be substituted for it.
2. When do nutrients enter cells through passive transport?
   a. when an electrochemical gradient favors entry
   b. when an electrochemical gradient discourages entry
   c. when a proton pump establishes a proton-gradient
   d. when they are bound by small proteins like metallothionein
3. If an experiment shows that the calcium concentration inside a root cell is thousands of times higher than the calcium concentration of the surrounding soil, what does it suggest?
   a. Passive exclusion of calcium is occurring.
   b. Active exclusion of calcium is occurring.
   c. Passive uptake of calcium is occurring.
   d. Active uptake of calcium is occurring.
4. What is an epiphyte?
   a. a plant that grows in boggy habitats
   b. a plant that is not rooted in soil, but grows in tree branches
   c. a plant that has a particularly deep taproot
   d. a plant that traps and digests insects to obtain nitrogen
5. Why is the presence of clay particles important to soil structure?
   a. They provide macronutrients like nitrogen, phosphorus, and potassium.
   b. They bind metal ions, which would be toxic if absorbed by plants.
   c. The positive charges on clay bind to negatively charged ions and prevent them from leaching.
   d. The negative charges on clay bind to positively charged ions and prevent them from leaching.

Conceptual Review

1. A farmer is concerned that her corn crop is suffering from iron deficiency. Design a greenhouse experiment that uses hydroponic cultures to (1) identify the symptoms of iron deficiency in corn, and (2) identify the minimum amount of iron required to support normal growth rates.
2. Are carnivorous and parasitic plants more common in nutrient-poor or nutrient-rich habitats? Why does your answer make sense?
3. Sandy soils contain very little clay. They are often nutrient-poor because of extensive leaching. Why does this occur? Would you expect plant productivity to be higher on sandy soils or on soils containing both sand and clay?
4. Why is it important for plants to exclude certain ions? Summarize the difference between active and passive exclusion mechanisms.
5. How were radioactive isotopes used to test the hypothesis that a mutualistic relationship exists between mycorrhizal fungi and their host plants?

Applying Ideas

1. There is a conflict between van Helmont's data on willow-tree growth and the data on essential nutrients listed in Table 33.1. According to the table, nutrients other than C, H, and O should make up about 4 percent of a willow tree's weight. Most or all of these nutrients should come from soil. But van Helmont claimed that the soil in his experiment lost just 2 ounces, while the tree gained 2627 ounces. If so, then soil contributed just 0.08 percent of the added weight instead of 4 percent. List three hypotheses to explain the conflict. How would you test these hypotheses?
2. Overwatering is a common cause of death in houseplants. When soils become waterlogged, the air pockets normally found in soil are filled with water. Why would this be detrimental to a plant? Can plants drown?
3. Researchers have observed that in forests, the filamentous bodies of mycorrhizal fungi frequently form a physical connection between individuals from different plant species. Design an experiment to test the hypothesis that sugars can be translocated from one plant

species to another through shared mycorrhizal fungi. Compare your experimental design to the work by Suzanne Simard's group by going to the library and reading their paper. You'll find it in the journal called *Nature*, vol. 388 (1997), pp. 579–582.

4. Design a series of experiments to test the hypothesis that phosphate ions can enter root hairs passively and actively. Next, design an experiment using the drug vanadate, which poisons proton pumps, to test the hypothesis that phosphate ions enter cells via an H^+-PO_4^{2-} cotransporter.

5. Fill in the table at the right based on information you have covered in other chapters.

Nutrient	How Used in Cell	Example of a Molecule That Contains It
Nitrogen		
Phosphorus		
Carbon		
Oxygen		
Hydrogen		
Iron		
Magnesium		

CD-ROM and Web Connection

CD Activity 33.1: Soil Formation and Nutrient Uptake *(animation)*
(Estimated time for completion = 10 min)
Learn about the relationship between plants and soil.

At your **Companion Website** (http://www.prenhall.com/freeman/biology), you will find self-grading exams and links to the following research tools, online resources, and activities:

Soil Science of America
An excellent resource for understanding the study of soil and the role of soils in crop production, environmental quality, ecosystem sustainability, bioremediation, waste management and recycling, and land use.

Modeling of Root Growth and Nutrient Uptake
This site introduces computer modeling of nutrient uptake.

Nitrogen Fixation
A comprehensive overview of nitrogen fixation and its importance to plants.

Additional Reading

Ausubel, J. H. 1996. Can technology spare the Earth? *American Scientist* 84: 166–178.

Meeks, J. C. 1998. Symbiosis between nitrogen-fixing cyanobacteria and plants. *BioScience* 48: 266–276. Considers how the association develops between nitrogen-fixing bacteria and plant hosts.

Sensory Systems in Plants

34

Once a seed germinates, that individual plant's location is largely fixed for life. Like clams, barnacles, corals, and many other animals, plants do not move around to find their food. To feed itself, a plant must grow up, capture photons and nutrients, and make sugar. If conditions at the germination site happen to be harsh or unsuitable, a plant has just two options: cope or die. Unlike animals that can run or swim or fly, plants cannot move away from trouble.

It would be a grave mistake, however, to think that plants are unable to sense conditions in their environment and respond in a way that helps them grow and reproduce effectively. Plants may not have eyes or ears, but they do have sophisticated systems for sensing light, gravity, and pressure. They may not jump or swim or run, but their shoot systems do grow toward light and become shorter and stockier in response to wind. In response to gravity, shoots grow up and roots grow down. In response to touch, the modified leaves of a Venus flytrap snap shut fast enough to catch insects.

How does a plant sense what conditions are like in its physical environment? How is information about surrounding conditions translated into action that helps the individual grow and produce offspring? This chapter considers how plants receive information about light, gravity, and touch, and begins to explore how they respond to this information.

Because many of these responses involve changes in growth patterns, Chapter 35 follows up by introducing the chemical signals called hormones and investigating how they regulate growth and other events. Chapter 37 explores how plants respond to changes in their biological environment, like infections and predation. In that chapter you'll examine data indicating that plants not only know

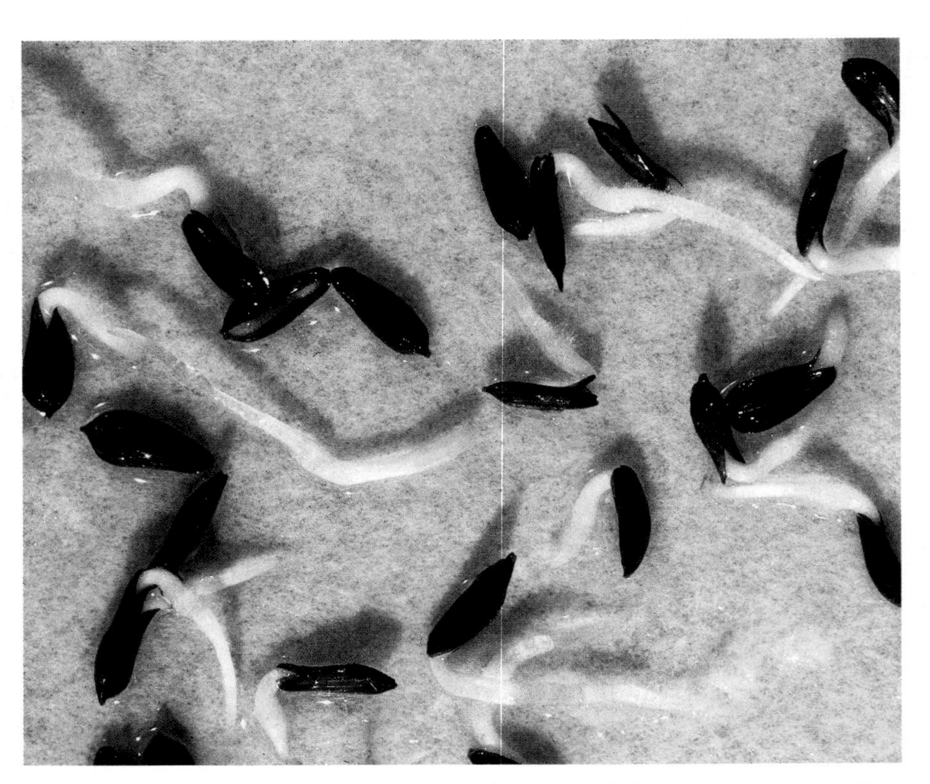

Plants have sophisticated sensory systems. The lettuce seeds shown here germinated because they were exposed to wavelengths of light typical of sunny locations—the preferred habitat of this species. As the young roots and shoots grow, each individual also senses gravity.

when they are being attacked by an enemy, they can also sense when individuals nearby are being eaten and can mount defenses in response to the threat.

34.1 Sensing Light

Plants are not blind. The first evidence supporting this claim came from experiments published by Charles Darwin and his son Francis in 1881. Their initial experiment involved just a candle and the emerging shoots, or coleoptiles, of a plant called reed canary grass (**Figure 34.1a**). Darwin and Darwin germinated reed canary grass seeds in the dark, placed the young, straight shoots next to a candle, and noted that the shoots bent toward the light. This type of directed movement in response to light is called **phototropism** (light-turn). Plant shoots are positively phototropic, meaning that they grow toward light. When the Darwins placed a solution of potassium dichromate between the candle and the coleoptile, however, no bending occurred. Because a potassium dichromate solution filters blue light out of the visible spectrum, the Darwins concluded that phototropism is a response to blue light (**Figure 34.1b**).

What other responses do plants have to light? If you happened to grow bean seeds as a class project in grade school, you might recall that the shoots tend to get long and thin if the plants are given insufficient light. The plants react as though they are being shaded and are attempting to grow high enough to reach open sunlight. To explore this phenomenon more thoroughly, D. C. Morgan and H. Smith set up growth chambers with an array of different light environments. In each chamber, the blue and red wavelengths used in photosynthesis were constant, but the amount of far-red light varied from low to high (**Figure 34.2a**, page 660). Morgan and Smith varied far-red wavelengths because shady habitats are enriched in these wavelengths compared to open, unshaded habitats. (Recall from Chapter 7 that land plants absorb many of the wavelengths in the red and blue portions of the spectrum, but do not absorb wavelengths in the far-red.) Further, the researchers tested species that naturally grow in sunny versus shady locations.

CD ACTIVITY 34.1 Sensing Light

Some of the data collected are shown in **Figure 34.2b**. The message in these graphs is that species from sunny habitats elongate their stems much more strongly in response to shade light (far-red wavelengths) than do species from forest-floor habitats. This result prompts several conclusions. First, it is clear that at least some plants can sense far-red wavelengths. Second, they respond to this light by elongating their stems if they are adapted to open, sunny habitats.

In short, plants monitor the amount of blue and far-red light they receive and moderate their behavior accordingly. To understand how light sensing works, researchers set out to isolate and characterize the molecules that perceive light.

Which Molecules Act as the Light Receptors?

Plants contain a wide variety of pigments that absorb light. These molecules include the chlorophylls and carotenoids involved in photosynthesis (see Chapter 7). As soon as phototropism and other light-dependent responses were observed in plants, however, researchers hypothesized that the receptor molecules involved were different from the pigments responsible for photosynthesis.

This hypothesis was confirmed in a series of studies on how certain varieties of lettuce seeds germinate. In lettuce, seed sprouting is affected by light. By plotting the frequency of germination against the specific wavelengths that seeds were exposed to, researchers discovered that germination rates peak when seeds receive light in the red part of the spectrum—specifically, at wavelengths around 660 nm. This effect disappeared,

(a) Coleoptile bends toward full-spectrum light

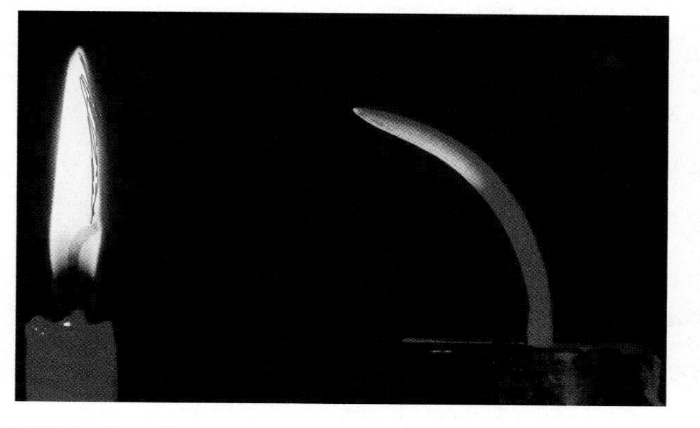

(b) Coleoptile does not bend when blue light removed

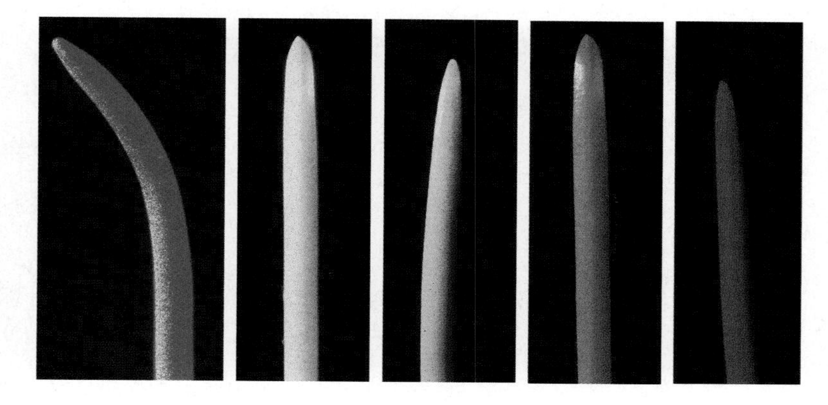

FIGURE 34.1 Early Experiments with Light Responses
(a) The shoot, or coleoptile, on the left is bending toward light. **(b)** When a solution that absorbs blue light is placed between the candle and the coleoptile, the bending response did not occur. Phototropism occurs specifically in response to blue light.

however, if the seeds were later exposed to far-red light. The inhibitory effect was maximized by wavelengths near 735 nm.

Why do lettuce seeds respond so differently to red and far-red light? As the data in **Table 34.1** show, red and far-red light act like an on-off switch for germination. Red light promotes germination and far-red inhibits germination. The last wavelength sensed by the plant determines whether germination occurs at a high frequency. The inhibitory effect of far-red light is logical because lettuce plants need full sun to thrive. A lettuce seed that germinated in shade, with primarily far-red light available, would have poor prospects.

The next challenge for biologists was to understand the molecular mechanism responsible for the on-off nature of the response. During the 1940s and early 1950s, Sterling Hendricks, Vivian Toole, and others addressed this issue by proposing that lettuce seeds contain a light-sensitive molecule distinct from chlorophylls and carotenoids. This pigment came to be called **phytochrome** (plant-color). To explain the switching behavior, they proposed that phytochrome is a protein that can assume two distinct conformations. In one conformation, called P_r (for phytochrome-red), the molecule absorbs red light. In the other, dubbed P_{fr} (for phytochrome-far-red), it absorbs far-red light. According to the model developed by these biologists, when either form absorbs its preferred wavelength, it is converted to the other form. The researchers called this model of pigment behavior photoreversibility (**Figure 34.3**). The hypothesis stimulated a great deal of interest because it suggested a molecular mechanism for an adaptive response to environmental stimuli.

The researchers extended their hypothesis by predicting that the P_r form was biologically inactive. The P_{fr} form, in contrast, was proposed to actively stimulate germination, stem elongation, and other responses to the far-red wavelengths that dominate shady habitats. Then the question became, is the photoreversible hypothesis correct? More specifically, were researchers able to isolate a protein that absorbs red and far-red light reversibly, as the model predicts?

The Red and Far-Red Light Receptors: Phytochromes

The hunt for phytochrome concluded in 1959, when Warren

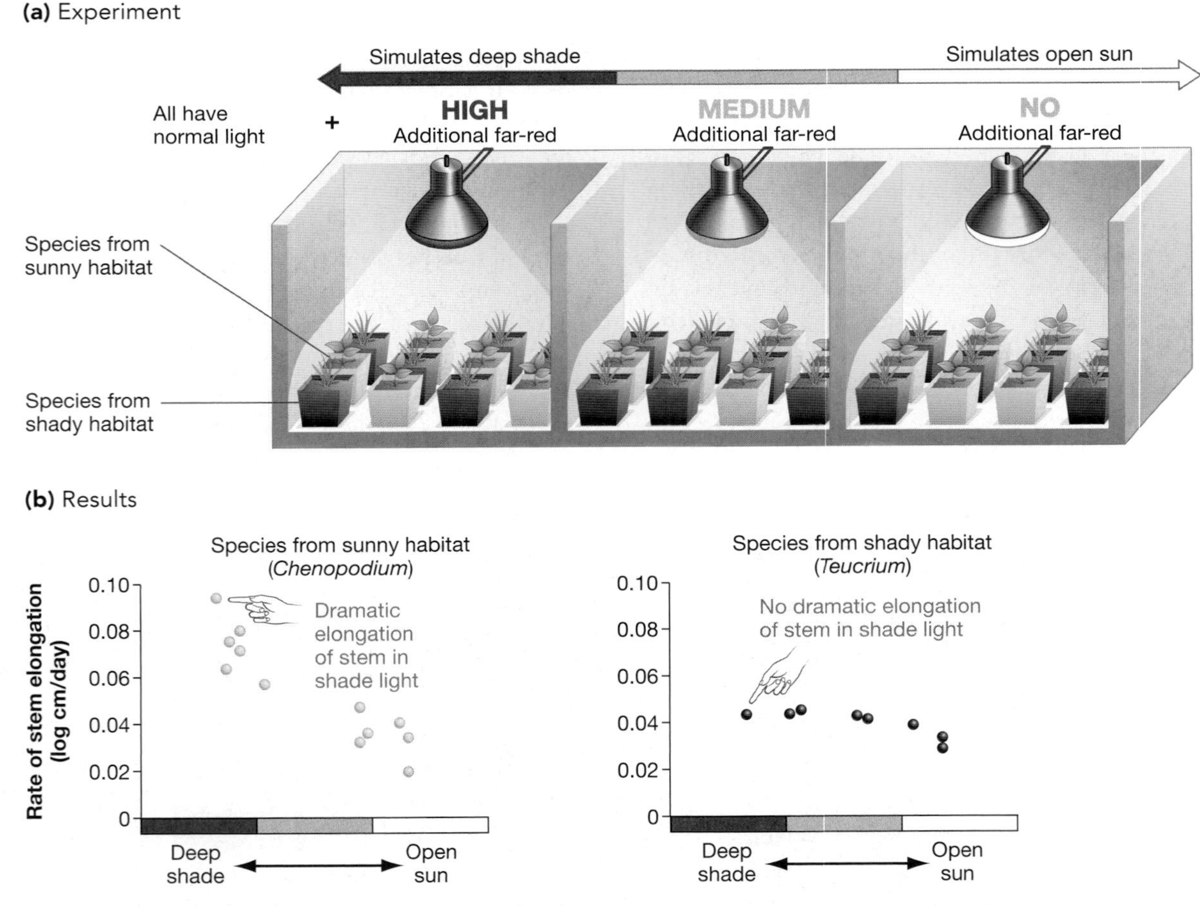

FIGURE 34.2 Only Species from Sunny Habitats Elongate Their Stems in Response to Shade Light
(a) In this experiment, each plant received the same quantity of the light wavelengths used in photosynthesis. The only thing that varied among treatments was the relative amount of far-red light. Shady habitats are dominated by far-red light. **(b)** These graphs plot the relative amount of far-red light on the x-axis versus the rate of stem elongation on the y-axis. QUESTION In part (a), the diagram implies that the seedlings were placed in random locations within each growth chamber. Why is this an important part of the experimental design?

Butler and colleagues published a study confirming the existence of the pigment. Using the types of protein-isolation techniques introduced in Units 1 and 2, these investigators purified a molecule from young corn shoots that was photoreversible. (Like the bean seedlings discussed earlier, young corn plants lengthen their stems in response to light deprivation or when exposed to an excess of far-red light. This behavior suggested that they have a far-red light receptor.) Specifically, when the protein that Butler and co-workers isolated was placed in solution and exposed to alternating red or far-red light, its color switched from blue to blue-green and back. The color switches suggested that the protein absorbs red as well as far-red light and that it is photoreversible. Based on these data, Butler and co-workers claimed that they had succeeded in isolating the phytochrome protein.

Subsequently, the locus that codes for phytochrome has been isolated and sequenced. The genome of *Arabidopsis thaliana*—the small, weedy mustard plant that serves as a model organism in molecular genetic studies of plants—actually has five distinct loci that encode phytochrome proteins. These genes were designated as phytochromes A–E and are symbolized *PHYA*, *PHYB*, and so on. Each of these proteins holds a small pigment molecule that absorbs light in the red and far-red parts of the spectrum. All of the phytochromes appear to be photoreversible. Further, recent data suggest that different phytochromes within the same individual trigger different responses to far-red light. For example, *Arabidopsis* individuals with mutant forms of *PHYB* have abnormal stem elongation responses. Are other phytochromes specifically involved in germination or other types of responses to far-red light? How is the information present in each of the P_r-P_{fr} switches translated into action that affects germination, stem elongation, and other responses? These questions remain unanswered. Interest in the phytochromes is intense, however, and research continues at a rapid pace.

TABLE 34.1 How Do Red and Far-Red Light Affect the Germination of Lettuce Seeds?

Biologists collected these data by exposing lettuce seeds to flashes of light containing one of two wavelengths: red or far-red ("FR"). After exposure to light, the seeds were moistened and held in the dark for several days. QUESTION According to these data, what is the average germination rate of lettuce seeds that were last exposed to red light? To far-red light? How do these values compare to the germination rate of seeds that are buried underground and receive no light at all?

Light Exposure	Germination (%)
None	9
Red	98
Red + FR	54
Red + FR + Red	100
Red + FR + Red + FR	43
Red + FR + Red + FR + Red	99
Red + FR + Red + FR + Red + FR	54
Red + FR + Red + FR + Red + FR + Red	98

A Blue-Light Receptor: NPH1 Phytochrome proteins appear to be specialized for monitoring the amount of shade light that plants receive. But far-red is not the only type of light that is important to plants. For example, the chlorophylls and carotenoids involved in photosynthesis absorb strongly in the blue part of the visible spectrum. It is logical, then, to expect that plants might have a blue-light receptor that triggers phototropism. Plants benefit when they move toward light that contains the wavelengths required for photosynthesis.

What is this "sunlight receptor?" Experiments by Charles Darwin and Francis Darwin in the 1870s established that the blue-light receptors responsible for phototropism in reed canary grass are located in the tips of coleoptiles. (These experiments

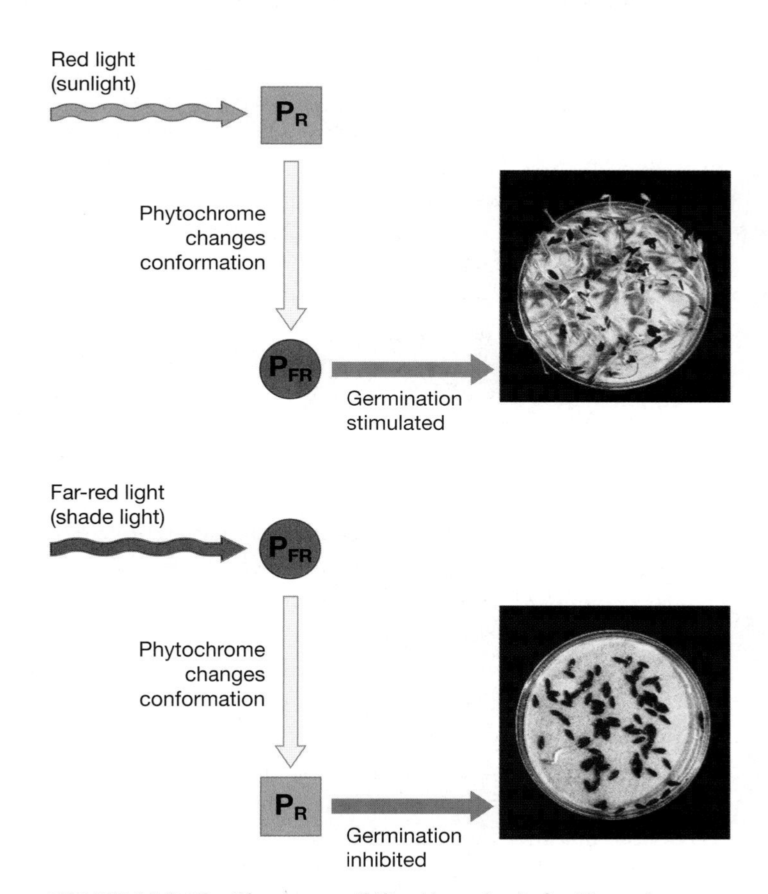

FIGURE 34.3 The Photoreversibility Hypothesis for Phytochrome Behavior

The photoreversible model for phytochrome function maintains that (1) the phytochrome protein has two distinct conformations; (2) the P_r conformation absorbs red light while the P_{fr} conformation absorbs far-red light; (3) when either form absorbs light, it converts to the other form; and (4) the P_{fr} form stimulates germination while the P_r conform inhibits it.

FIGURE 34.4 Evidence That NPH1 Protein Catalyzes Its Own Phosphorylation Reaction
(a) Insect cells were transformed with *nph1* genes from *Arabidopsis*, grown in a medium containing radioactive phosphorus, and exposed to blue light. The photographs at the bottom of the panel show that the transformed cells produced the NPH1 protein and that many copies of the NPH1 protein acquired a radioactive phosphate group if the cells were exposed to blue light. **(b)** This diagram shows the model called autophosphorylation. The idea is that NPH1 phosphorylates itself when it absorbs blue light.
QUESTION In part (a), why did the researchers bother to analyze cells that were not transformed with the *nph1* gene, and that were not exposed to blue light?

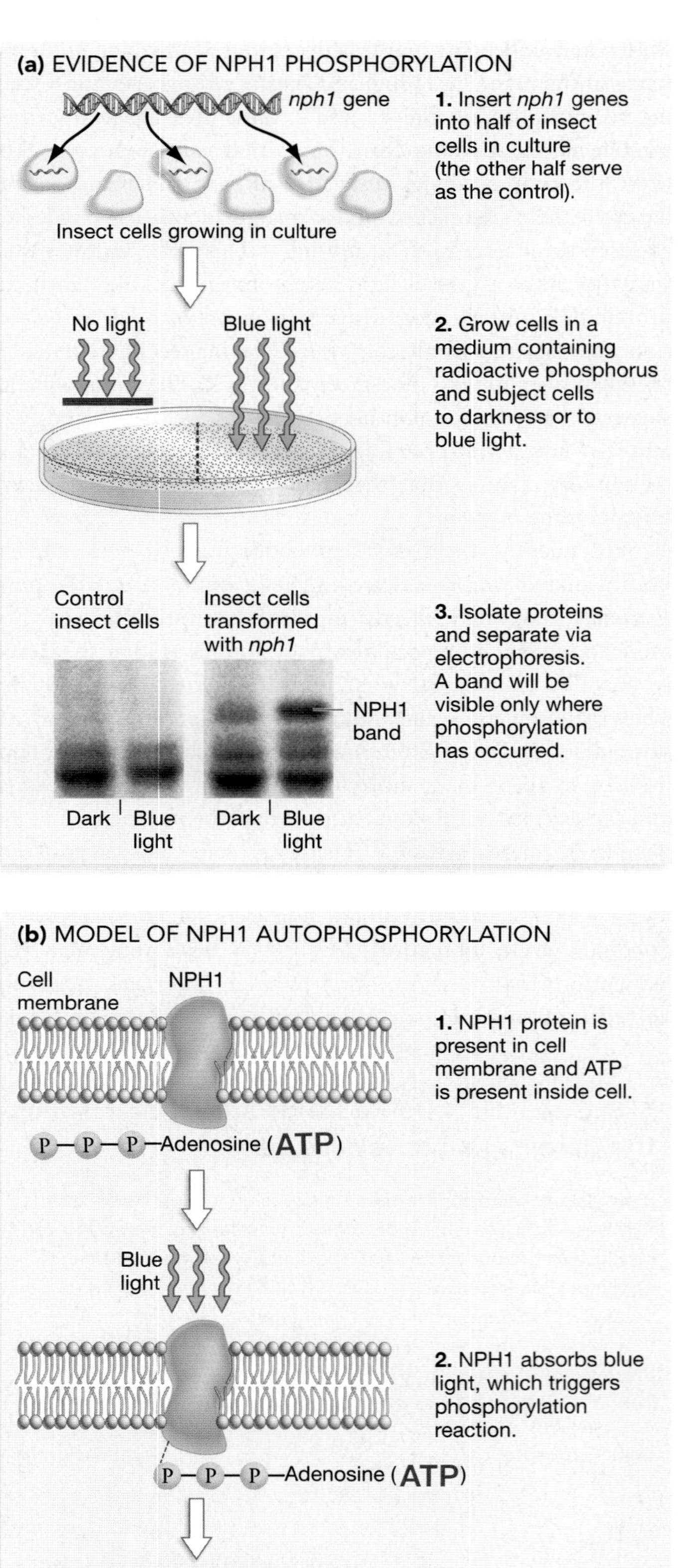

are explained in detail in Chapter 35.) Based on this observation, researchers in Winslow Briggs's laboratory began searching for the sunlight receptor in the tips of young shoots. Their search was also guided by the hypothesis that the blue-light receptor—like the light receptors isolated from animals—might be a membrane protein.

An important breakthrough in the search for the sunlight detector came when Briggs and co-workers found a membrane protein that is abundant in the tips of emerging shoots and that becomes phosphorylated in response to blue light. To understand why this result was so significant, recall from Chapter 15 that many proteins switch from inactive to active states, or vice versa, when a phosphate group (PO_4^{2-}) from ATP is added to them. Based on this observation, the biologists hypothesized that the protein they had found was directly involved in the blue-light response. The logic here is that the membrane protein becomes activated via phosphorylation in response to blue light, and that the activated protein triggers the phototropic response.

But was this membrane protein the receptor itself? This seemed unlikely, because the receptors in many sensory systems function by activating nearby proteins called protein kinases. Protein kinases hydrolyze ATP and catalyze the addition of the phosphate group to another protein, which then activates another protein. In this way, the receptor molecule and its adjacent protein kinase set off a chain of events that eventually leads to a response by the cell.

Was the protein that Briggs and co-workers found the blue-light receptor itself, or was it a protein that is phosphorylated after the blue-light receptor absorbs a photon? The tools to answer this question turned up when researchers in Briggs's laboratory began analyzing *A. thaliana* mutants that do not show a phototropic response to blue light. These plants are called *nph* mutants, for "non-phototrophic hypocotyl." (A hypocotyl is part of the stem in a seedling.) By mating these individuals and analyzing the nature of the phototropic response in their offspring, Emmanuel Liscum and Briggs documented that there were four distinct types of responses to blue light. Based on this observation, the biologists inferred that there are four distinct types of mutants and thus four distinct *nph* loci in *Arabidopsis*. Mutants at one of these loci, called *nph1*, turned out to lack the membrane protein that becomes phosphorylated in stems in response to blue light. In this way, analyzing mutants allowed the biologists to find the gene for a protein they had isolated years earlier.

The next task was to determine whether this gene coded for the blue-light receptor itself. When John Christie and colleagues inserted copies of the *nph1* gene into insect cells that were growing in culture, they found that the NPH1 protein became phosphorylated in response to blue light (**Figure 34.4a**). This was an important result because the insect cells contained no other proteins that might be associated with the blue light response in *Arabidopsis*. Consequently, Christie and co-workers could defend their claim that the NPH1 protein is the long-sought photoreceptor, and that it acts as its own protein kinase (**Figure 34.4b**). Proteins that can catalyze their own phosphorylation are said to autophosphorylate. Based on this result and on follow-up experiments, the current consensus is that NPH1 is indeed the sunlight detector in plants, and that a plant's response to blue light is triggered by autophosphorylation of this protein.

How is this phosphorylation step translated into the phototropic response? On this question, there is no consensus. As the following discussion will show, the steps between light perception and response are not well understood in most cases.

From Perception to Response: Signal Transduction

One or more steps must occur between the sensing of a signal and an organism's response to that signal. In effect, the information that an individual receives about what is going on outside must be translated into some sort of response. The series of events that are responsible for this translation is called **signal transduction**.

Transduction is an appropriate word for this process, because it often involves an energy transformation. (The verb *transduce* means to convert energy from one form to another.) In plants, for example, the energy in blue light is transformed into chemical energy in the form of a phosphorylated NPH1 protein. In this case, the phosphorylation event is the first step in the signal transduction process. Now the question becomes, what happens next?

The diagram in **Figure 34.5** illustrates some possible outcomes of signal transduction. It turns out that in many cases, the response to a signal involves a phosphorylation event and a subsequent change in the activity of a response protein. The phosphorylated protein might increase or decrease the transcription of certain genes or alter the translation of particular mRNAs. In this way, information from outside the cell can lead to changes in gene expression.

Is the phototropic response in *Arabidopsis* and other plants triggered by a change in gene expression? The answer is still unclear. What *is* known is that in addition to *nph1*, the products of genes called *cry1* and *cry2* are also involved in the response to blue light. (The *cry* is short for cryptochrome, or "hidden-color.") Plants with mutant forms of both *cry1* and *cry2* do not show the phototropic response. This observation suggests that their products are involved somewhere in the signal transduction process. Recently, researchers in several laboratories independently showed that the CRY1 and CRY2 proteins localize to the nucleus. This observation suggests that they may be transcription activators or repressors that act in response to blue light.

Results such as these are like pieces of a puzzle. It is clear that the *cry1* and *cry2* products are involved in the phototropic response, but biologists still do not understand exactly where they fit in the sequence of events that are triggered by blue light. As this chapter goes to press, the puzzle pieces concerning phototropism are still scattered about and difficult to interpret. Stated another way, biologists have a good understanding of how plants perceive blue light, but they do not yet have a coherent model for how plants transduce that information into a response.

Frequently in science, experimental results like those concerning *cry1* and *cry2* remain difficult to understand until entirely new and unforeseen pieces of the puzzle come to light. Then scattered pieces quickly come together into a lucid picture.

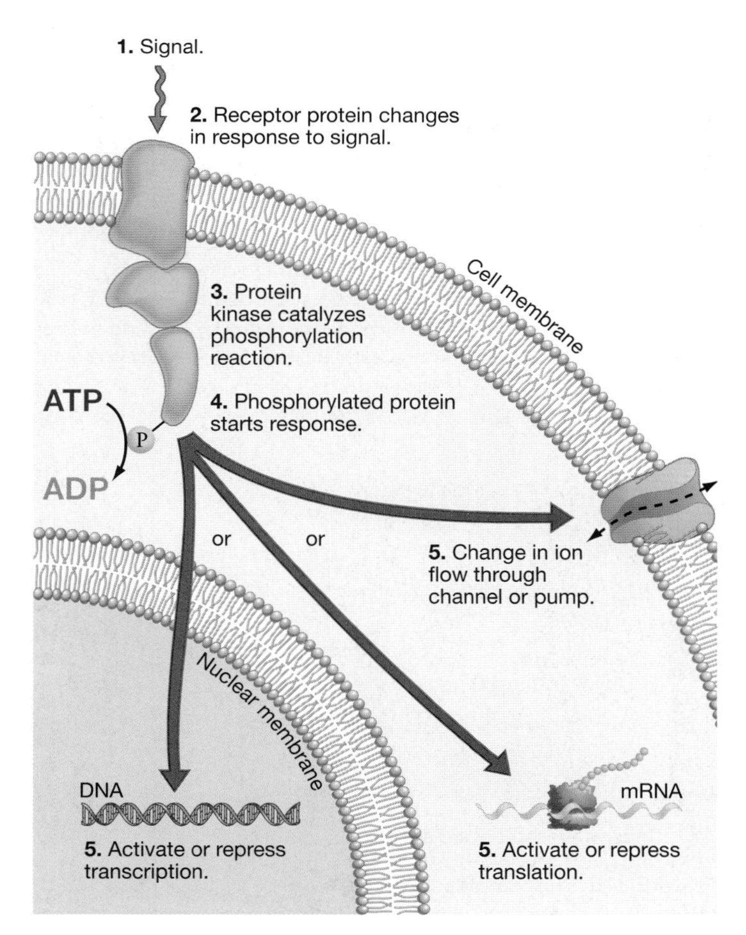

FIGURE 34.5 What Are the Possible Outcomes of Signal Transduction?
Many different responses occur after signal transduction. They range from changes in transcription and translation to alterations in channels, pumps, and other membrane proteins.

Until those key experiments are done, though, researchers keep struggling to make sense of conflicting results and hypotheses.

34.2 How Do Plants Perceive Gravity?

Shoots usually respond to gravity by growing up or horizontally; roots usually respond by growing down (**Figure 34.6a**). Explaining the mechanism responsible for these responses has been a huge challenge to biologists. In 1881, Charles and Francis Darwin published one of the first experimental results about the ability of plants to move in response to gravity, or **gravitropism** (gravity-turn). The Darwins found that roots stop responding to gravity if their caps are removed. This observation suggested that gravity sensing occurs somewhere in the root cap. Recently, Elison Blancaflor and colleagues demonstrated precisely which cells are involved in gravity sensing in *Arabidopsis* roots. By killing tiny blocks of cells with laser beams, Blancaflor and his co-workers were able to demonstrate that the cells illustrated in **Figure 34.6b**, directly under the cells at the tip of the root, are the most important for regulating the gravitropic response.

Given that a small group of root-cap cells is involved in gravity sensing, what is the actual mechanism? Over the past decade, research on gravitropism has focused on two competing hypotheses. Both hypotheses agree that gravity is a unique stimulus because it is constant and unidirectional. Both agree that gravity is sensed by unidentified pressure or stretch recep-

(a) Roots grow down, shoots grow up.

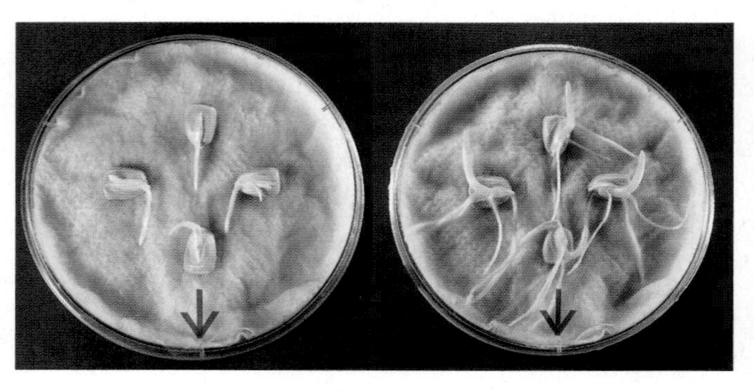

(b) Root tip cells sense gravity.

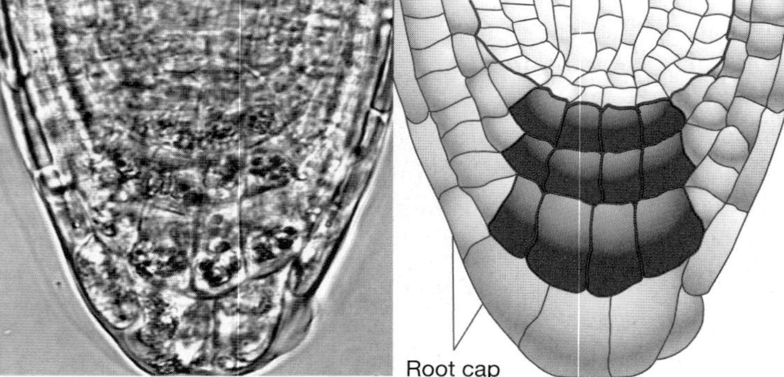

FIGURE 34.6 Gravitropism
(a) The blue arrows on these photos indicate the direction of gravity as these corn seeds germinated. In species with upright growth habits, shoots respond to gravity by growing away from it. In species that form mats or spread laterally, shoots respond to gravity by growing horizontally. **(b)** When the cells marked in red are killed experimentally, the gravitational response in roots is dramatically reduced.

(a) Statolith hypothesis

(b) Support for the statolith hypothesis

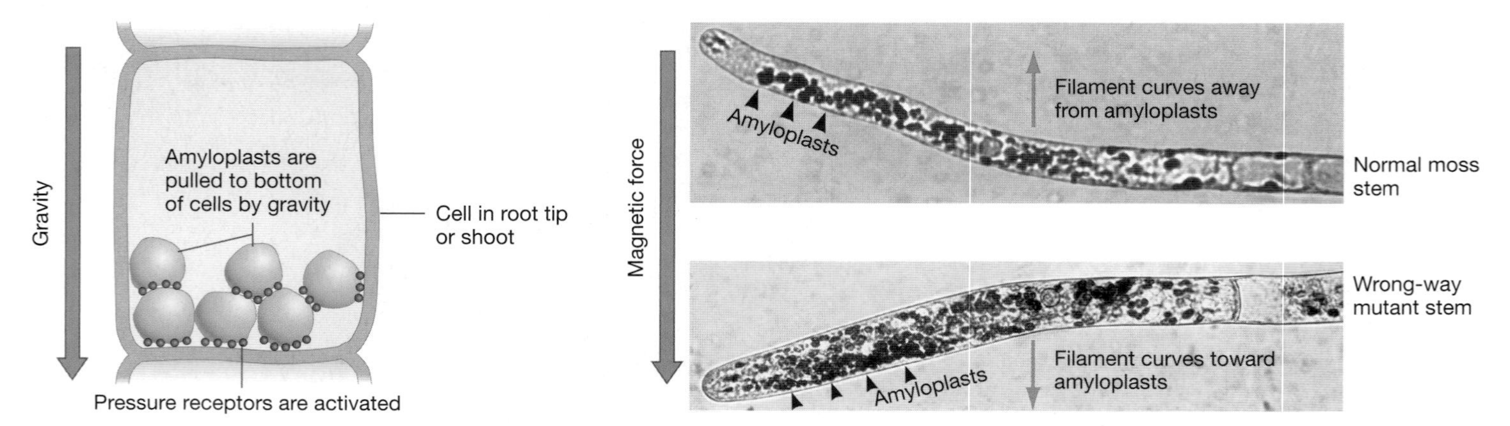

FIGURE 34.7 The Statolith Hypothesis
(a) Amyloplasts are filled with starch. They are dense, so they sediment in response to gravity. The statolith hypothesis predicts that pressure receptors in the amyloplast membrane become activated as a result.
(b) These photographs show the germinating stems of a moss. The stems are from a normal individual (top) and an individual with the *wrong-way response* mutation (bottom). The stems grew in a magnetic field that pulled amyloplasts in the direction indicated, opposite the direction of gravity.

tors located somewhere in the cell. The disagreement is about which part of the cell actually activates the receptor in response to gravity.

The Statolith Hypothesis

Many plant cells, including those in the root cap, contain starch-storing organelles called amyloplasts. The statolith (place-stone) hypothesis, illustrated in **Figure 34.7a**, contends that these dense organelles act as the primary gravity sensor in plants. The idea is that gravity pulls the heavy amyloplasts to the bottom of cells, where the force of the amyloplast on the cytoskeleton or plasma membrane—or the distension of the amyloplast membrane itself—activates pressure or stretch receptors that initiate the gravitropic response.

The statolith hypothesis was inspired by the way that many animals sense gravity. Lobsters, for example, take up grains of sand that become positioned in specialized gravity-sensing organs in their antennae. The grains of sand are called statoliths. When the animal tilts or flips over, the statolith moves in response to gravity. Inside the organ, the sand grain ends up pushing against a new receptor cell, which sends an electrical signal to the brain. Because the location of the responding cell tells the brain how the body is now positioned with respect to gravity, the animal can respond by righting itself.

To test the statolith hypothesis in plants, Oleg Kuznetsov and colleagues devised a way to move amyloplasts to new locations in a cell independent of gravity's pull. They did this by subjecting the thin, germinating filaments of a moss to a directional magnetic field. The amyloplasts in the cells responded by migrating toward the magnetic field. As **Figure 34.7b** shows, the filaments subjected to this treatment bent away from the amyloplasts and toward the pull of gravity—just as the statolith hypothesis predicts. Further, the researchers did the same test using mutant mosses of the same species, which have a gravitropic response opposite to that of normal individuals. When subjected to a directional magnetic field, these *wwr* (wrong-way response) mutants curved toward their amyloplasts, just as they do under gravity.

Although these results strongly support the statolith hypothesis, one outstanding problem remains. In *Arabidopsis*, mutant individuals exist that are incapable of manufacturing starch. As a result, their amyloplasts are small and not very dense. Although these starchless mutants have a dramatically reduced sensitivity to gravity, they still show about 25 percent of the normal response. Why they show any response at all is a question that the statolith hypothesis cannot answer. The gravitational pressure hypothesis, however, offers a possible solution.

The Gravitational Pressure Hypothesis

A key component of the statolith hypothesis is that the amyloplast itself somehow serves as the primary gravity sensor. The gravitational pressure hypothesis, in contrast, contends that the weight of the entire cell triggers a response to gravity via receptor proteins located between the cell membrane and the extracellular matrix (**Figure 34.8a**). According to this point of view, amyloplasts merely serve as ballast, or extra weight, in gravity-sensing cells. It is thus not surprising that starchless mutants have reduced, but nonzero, gravity responses.

To test the gravitational pressure model, Mark Staves and associates experimented with the roots of rice. Because this tissue grows in water, the biologists were able to change the buoyancy of the surrounding solution. (They did this by adding solutes that would not affect turgor or other aspects of the tissues involved.) When they increased the buoyancy of the solution and placed the tips of the roots horizontally, the gravitational response of the root decreased. The amyloplasts in the

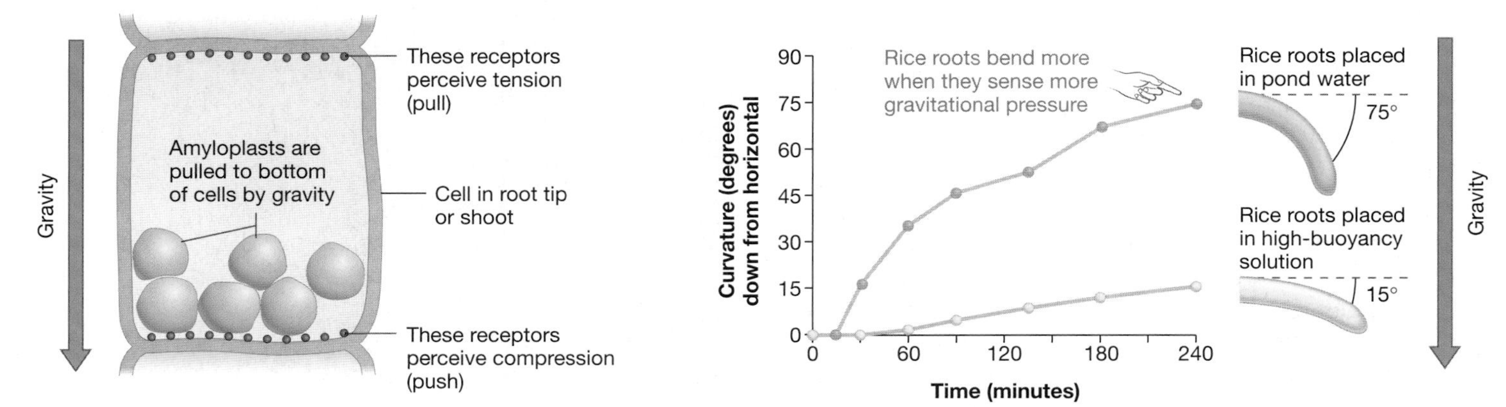

FIGURE 34.8 The Gravitational Pressure Hypothesis
(a) The gravitational pressure hypothesis predicts that in response to gravity pulling on the entire mass of the cell, pressure receptors at the top of the cell sense tension, while pressure receptors at the bottom sense compression. **(b)** This graph shows the differing gravitational response of rice roots placed horizontally in a solution of pond water with normal buoyancy and those placed in a solution with extremely high buoyancy.

root caps still sedimented to the bottom of the root-cap cells, however. According to these researchers, this result is consistent with the gravitational pressure model because the high buoyancy of the experimental solution would lessen the effect of gravity and decrease the pressure that the cell exerted on the extracellular matrix (**Figure 34.8b**).

The debate between proponents of the two hypotheses continues, however, and it is possible that the conflict will not be resolved until the receptor molecule itself is identified and its location and mode of action determined. To understand gravity sensing, it will be essential to discover exactly how the signal is received and translated into a response.

Is the Gravity Sensor a Transmembrane Protein?

Recent research on gravity receptors and signal transduction has focused on membrane proteins called integrins. In animals, integrins form a line of communication between the extracellular environment and the interior of the cell. As **Figure 34.9a** shows, integrins span the plasma membrane. Their extracellular end binds to components of the extracellular matrix such as the fibrous proteins collagen and fibrinogen. Their intracellular ends bind to components of the cytoskeleton, such as the microfilaments introduced in Chapter 5.

Randy Wayne and Mark Staves have proposed that in plants, integrin proteins sense the difference in pressure on the tops and bottoms of root cells, as predicted by the gravitational pressure model. These researchers hypothesize that integrins function as the gravity sensor in roots and shoots, and that changes in integrin binding trigger a signal transduction sequence that results in gravitropism. But when Timothy Lynch and his co-workers made antibodies to the integrin proteins found in chickens and applied them to plant roots, they found that the membranes of amyloplasts also contain integrins (**Figure 34.9b**). (The concept of using antibodies to determine the location of membrane proteins was introduced in Chapter 32.) If experiments confirm that integrins on the cell membrane and amyloplast membranes act as gravity receptors, it could turn out that plants sense gravity in both places. If so, then the gravitational pressure and statolith hypotheses may both be correct.

Signal Transduction

It is important to recognize that integrins are only a candidate gravity receptor. There is no firm evidence to date that they actually serve this function. Given that the receptor molecule has yet to be identified, it is not surprising that researchers describe the subsequent events in signal transduction as an enigma.

Two important things *are* known about the gravitropic response, however. First, the curvature that occurs in reaction to gravity is due to differences in cell elongation on the opposite sides of a root or shoot (**Figure 34.10a**). Further, a signaling molecule called auxin promotes cell elongation in shoots and inhibits it in roots. The current model for the gravitropic response puts these two observations together by proposing that the gravity signal results in an asymmetric distribution of auxin, as shown in **Figure 34.10b**. The idea here is that the gravity sensor triggers a signal transduction pathway that results in a change in auxin transport.

How is auxin transported, and how does it affect the elongation of cell walls? Chapter 35 explores these questions by discussing experiments on auxin and other hormones that control

(a) In animals, integrins connect the extracellular matrix and the cytoskeleton.

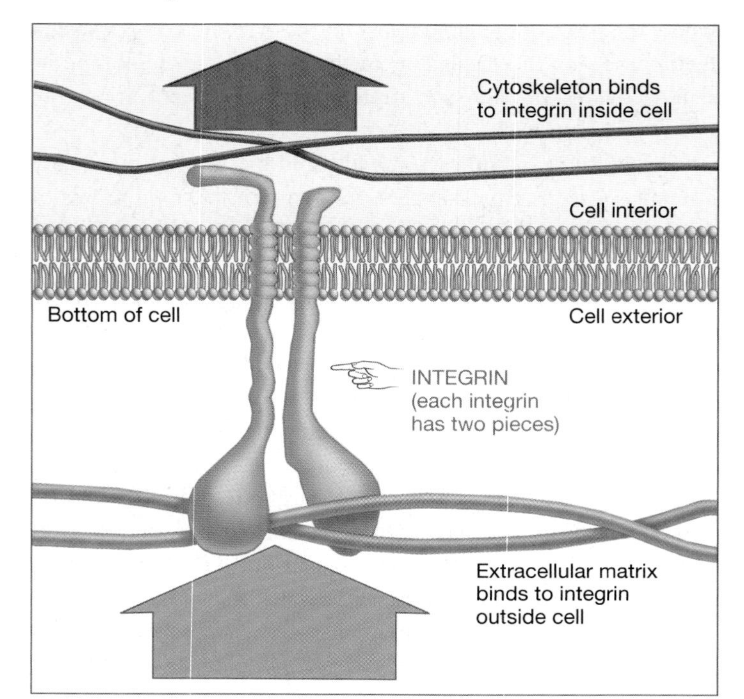

(b) Inside amyloplasts, the membranes surrounding starch grains contain integrins.

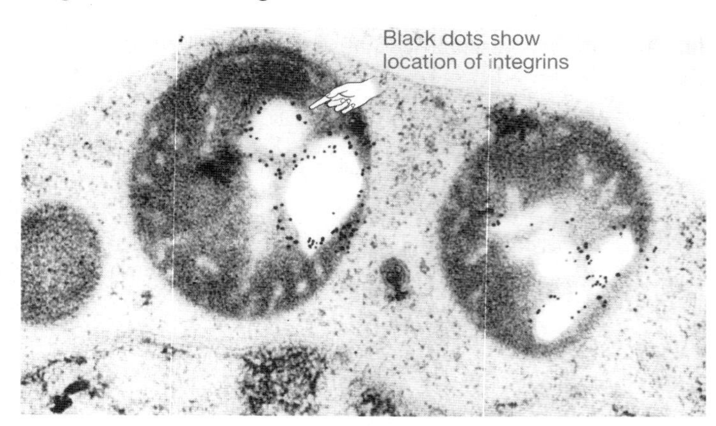

FIGURE 34.9 Are Integrin Proteins Gravity Sensors?
(a) Integrin proteins span the cell membrane and project far into the extracellular space. They bind to components of the extracellular matrix and the cytoskeleton inside the cell. If these binding sites are altered by pressure, the change could indicate the direction of gravity. **(b)** The black dots on this electron micrograph are gold particles that are attached to an antibody. The antibody, in turn, is attached to integrin proteins in the membrane of amyloplasts.

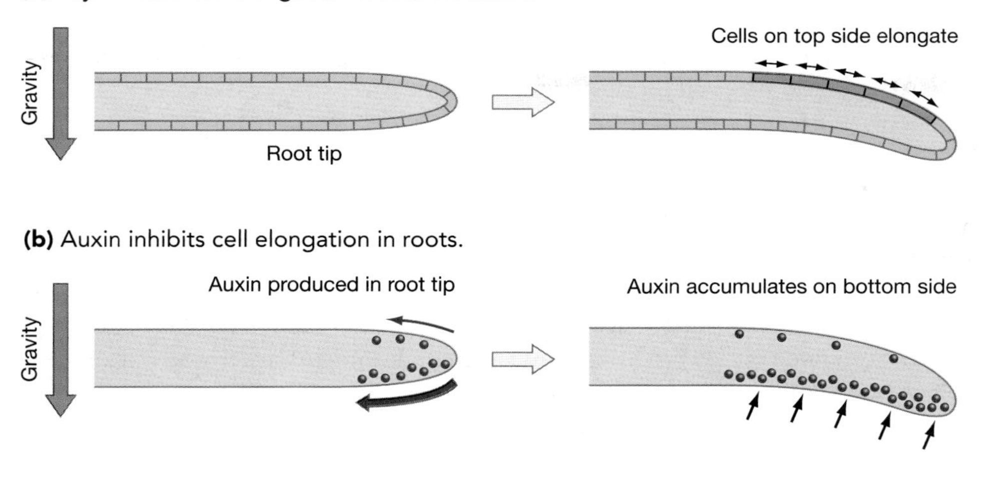

FIGURE 34.10 What Causes Shoots and Roots to Curve in Response to Gravity?
(a) Roots and shoots curve during gravitropism because cells on one side of the structures elongate. **(b)** In response to gravity, a signaling molecule called auxin is distributed asymmetrically in root tips. It is not yet known how gravity sensing results in this asymmetric distribution of auxin.

plant growth. In the meantime, let's consider an entirely different type of sensory and response system in plants. In addition to communicating through chemical signals such as auxin, plants are capable of producing electrical signals that are similar to the electrical impulses streaming through your nervous system right now.

34.3 How Do Plants Respond to Touch?

When plants perceive blue light, red light, far-red light, or gravity, the sensor that is involved responds in part by triggering some sort of chemical change in the cell. Often this change is a phosphorylation event that activates a key protein. In this way, plants transduce the kinetic energy in light or the mechanical energy of gravity into chemical energy.

When plants are buffeted by wind, the mechanical force is also transduced by a chemical event. Although the exact receptor and transduction pathway are not known, studies in *Arabidopsis thaliana* have shown that a large suite of genes are transcribed in response to touch. The protein products of some of these loci act to stiffen cell walls, resulting in plants that are shorter and stockier relative to plants that do not experience repeated touching or high winds (**Figure 34.11**).

In some cases, though, plant sensory cells may also transduce the signal they receive into an electrical change. Here we consider a deceptively simple question: How does this happen?

An Introduction to Electrical Signaling

The key to understanding electrical signaling in plants is to recognize that the interior of most plant cells has a negative charge relative to the exterior. This charge difference occurs because the proton pumps introduced in Chapter 32 are active in many cells. When these H^+-ATPases move protons to the exterior of a plant cell, they create a charge separation across the membrane (**Figure 34.12a**, page 668). Biologists frequently refer to this charge separation as a membrane polarization. (The word *polar* refers to opposing characteristics.)

Any charge separation, including the one created by the proton pump in plants, creates a voltage. Most plant cells, then, have a **membrane voltage**. Voltage is a form of potential energy. Potential energy, in turn, can be thought of as the tendency for something to move. For example, Chapter 2 explained that the chemical potential of an electron describes its tendency to move closer to a nucleus. Chapter 32 pointed out that the water potential of a tissue or substance (like air or soil) expresses the tendency of water to move toward it or away from it. In the same way, an electrical potential is the tendency of charged particles (electrons or ions) to move toward an area of opposite charge.

FIGURE 34.11 Plants Get Shorter and Stockier in Response to Wind or Touch
The tomato plants shown here received 0, 10, or 20 brushings per day each day for 10 consecutive days.

Because a voltage represents a form of potential energy, biologists often refer to the voltage across a cell membrane as a **membrane potential.** The size of this potential is a function of the amount of charge separation. If the interior of the cell is very negative relative to the exterior, then the membrane potential is large.

The size of a membrane potential can be measured with a set of electrodes placed on the inside and outside of a cell, as shown in **Figure 34.12b.** Because membrane potentials are small, they are usually expressed in units called millivolts (mV, or 1/1000 of a volt). By convention, membrane potentials are expressed as the state of a cell's interior relative to the exterior. As a result, the **resting potential** of a plant cell—meaning its normal state—is usually negative.

What does all this have to do with sensory perception and responses? As an example of electrical signaling in plants, consider the Venus flytrap introduced at the end of Chapter 33. The modified leaves of this bog-dwelling species have sensory hairs on their surface. When an insect brushes against at least two of these hairs, the trap responds by slamming shut. The response to this hair trigger takes just half a second. In 1873, J. Burdon-Sanderson showed that the rapid movement is mediated by an electrical signal that spreads across the trap. Let's take a closer look.

(a) Proton pumps send H^+ outside cells.

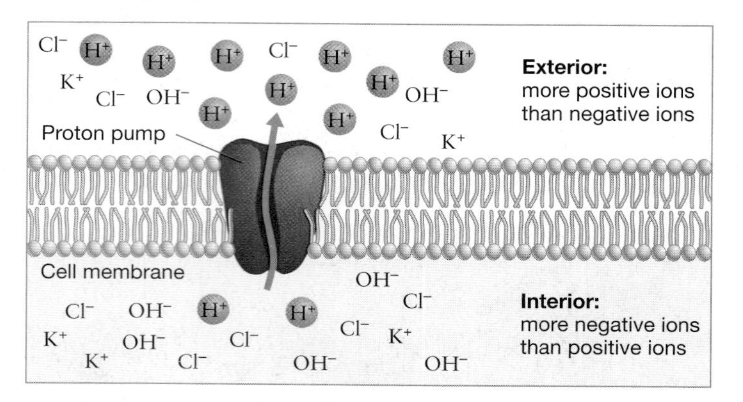

(b) The charge separation can be measured.

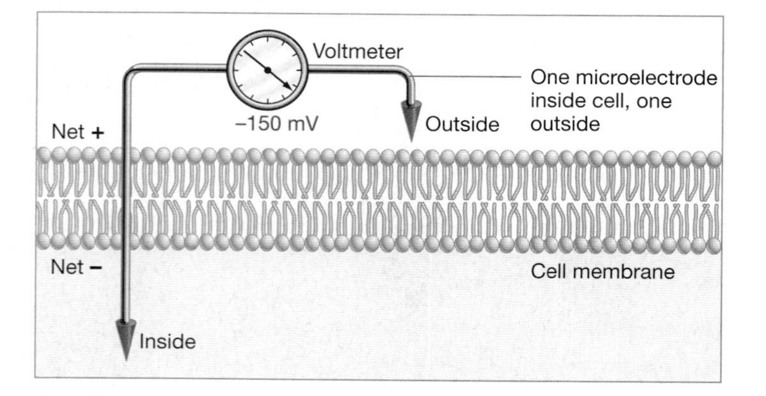

FIGURE 34.12 Proton Pumps Create a Charge Separation Across Plant-Cell Membranes
(a) The insides of plant cells normally have a negative charge relative to the outside of the cell. **(b)** Any charge separation creates a voltage. The voltage across a cell membrane can be read by inserting electrodes into the interior and exterior of a cell.

Action Potentials

In animals, nerve impulses travel down the length of specialized cells called neurons. The impulses consist of a flow of charge in the form of ions across the cell membrane. Because of the charge flow, the voltage across the cell membrane changes drastically. This voltage change is what Burdon-Sanderson recorded when he inserted an electrode into a Venus flytrap cell and triggered the sensory hairs.

The voltage change that occurs in the Venus flytrap has a characteristic pattern, called an **action potential. Figure 34.13a** shows what an electrode records as an action potential passes through a cell; **Figure 34.13b** diagrams the corresponding events at the cell membrane. To interpret these figures, begin at the leftmost point of part (a). Note that the resting potential of the cell where the electrode is placed is −70 mV. At the time marked 0 on the x-axis, the voltage across the membrane starts to become more positive. This type of change is called a **depolarization** because the charges on either side of the membrane become more alike. As the corresponding panel in part (b) shows, depolarization occurs because positively charged ions begin flowing into the cell.

The most prominent part of Figure 34.13a, though, is the action potential itself. An action potential is an extremely rapid change in membrane potential from negative to positive, then back to negative. As part (b) shows, the 2-millisecond-long "spike" in membrane potential occurs because positive charges rush into the cell and then out, causing a dramatic swing in voltage. In Chapter 42 we will examine the molecular mechanisms responsible for the action potential in much more detail. We'll consider which ions flow across the membrane during each part of the response, which membrane channels mediate these ion flows, and how the signal propagates down the length of a cell.

In understanding plant sensory systems, though, it's important to know that at least some plants use electrical signals that are directly analogous to the nerve impulses observed in animals. The great advantage of electrical signaling, compared to chemical messages like phosphorylation events and changes in gene regulation, is speed. Although only a few plants other than the Venus flytrap exhibit rapid movements, in every case the motion is triggered by action potentials.

How Does the Venus Flytrap Close?

For a Venus flytrap to snare an insect, a sequence of three events must occur. First, the mechanical energy that moves the receptor hair on the trap surface needs to be transduced to an electrical signal. Next, this electrical signal must be propagated across the leaf in the form of one or more action potentials. Fi-

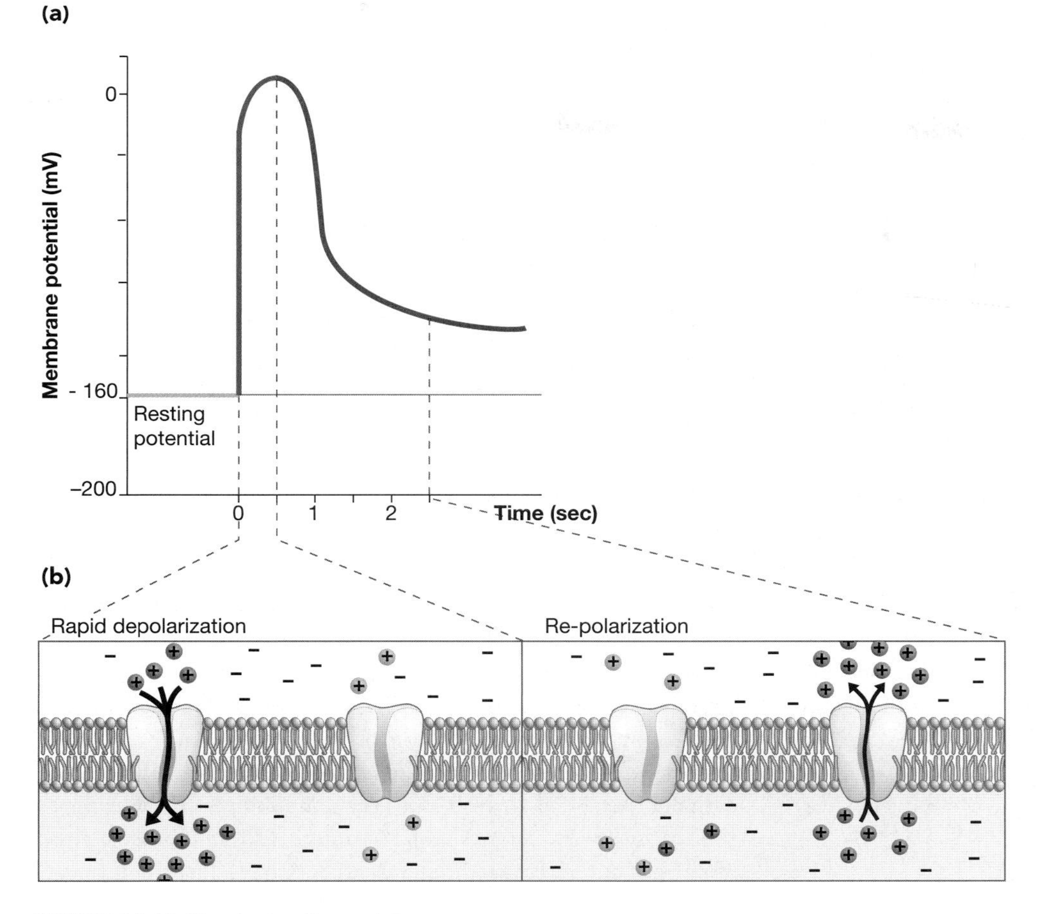

FIGURE 34.13 The Action Potential
(a) This diagram shows the changes in membrane potential that occur during an action potential. **(b)** Each phase of an action potential is caused by a flow of ions across the cell membrane through channels.

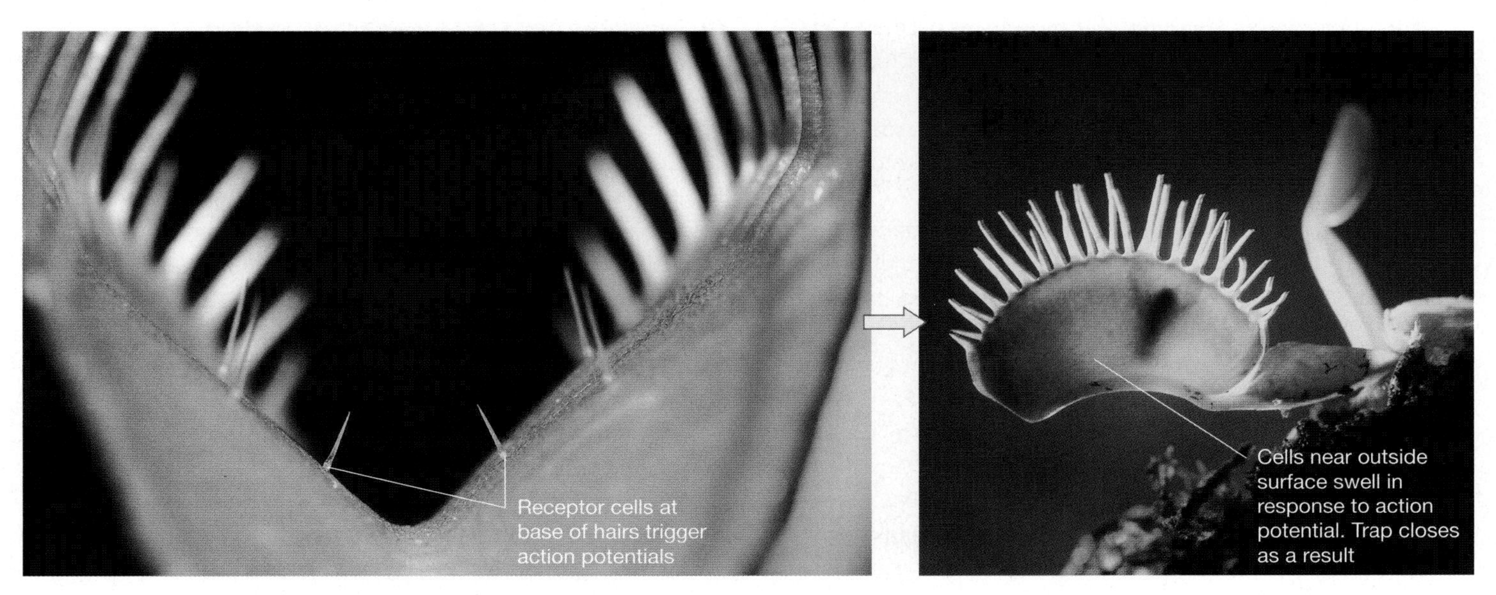

FIGURE 34.14 How Does a Venus Flytrap Close?
Action potentials spread across the Venus flytrap from receptor cells at the base of trigger hairs. When these action potentials reach cells on the outside surface of the trap, the cells swell in response and close the trap.
EXERCISE On the left-hand photo, label the cells that swell in response to the action potential. In the right-hand photo, indicate the insect that has been trapped.

nally, effector cells must respond to the action potential by undergoing a rapid change in turgor that moves the leaf and closes the trap (**Figure 34.14**).

The first and third steps in this process are reasonably well understood. For example, R. M. Benolken and S. L. Jacobson studied dissected receptor hairs from the trap surface and showed that their membranes depolarize in response to touch. For this depolarization to occur, stretch receptors in these membranes must allow an influx of positive ions in response to being pulled or pushed. In this way, the mechanical signal we call touch is transduced to an electrical signal in a sensory cell. The depolarization at the receptor cell then triggers action potentials in other cells. These signals race across the leaf at a rate of about 10 cm/sec. When the action potentials reach effector cells on the outer surface of the trap, the cells swell and push the trap shut.

How this size change occurs is still being studied, however. Stephen Williams and Alan Bennett have shown that the amount of ATP in the effector cells declines dramatically as the trap closes, and that the intracellular pH drops to about 4.5. These observations suggest that an H^+-ATPase is involved in the response. Whether the cells undergo a size increase in response to inrushing water or via a rapid change in the length of their cell walls is still controversial, however. It would be interesting to know exactly how the movement occurs. This group of cells is the closest thing to a muscle found in plants.

Essay Can Plants Tell Time?

An experiment conducted by Jean Jacques d'Ortous de Mairan in 1729 suggested that plants have some sort of internal clock. De Mairan observed that mimosa plants fold their leaves at night and open them during the day. When he put individuals into complete darkness, though, he found that the leaf opening

Do bean leaves have some kind of internal clock?

and closing movements continued on schedule. This result indicated that plants tell time in some way that is independent of direct information from sunlight.

The same type of rhythmic leaf movement occurs in many pea-family plants. Bean leaves, for example, begin to close before nightfall and start to open again before daybreak. The observation that beans anticipate changes in light reinforces de Mairan's result. It implies that plants have some kind of internal clock or timekeeper.

Leaf movements and other events that occur on a 24-hour cycle are said to exhibit a circadian (about-day) rhythm. To establish and maintain this rhythm, a biological clock must have two components. There has to be a system for setting and adjusting the clock, and there has to be a mechanism for keeping time. How do these components work?

The leading hypothesis for the clock mechanism maintains that there are two timekeeping molecules which oscillate in abundance over the course of 24 hours. Specifically, Steve Kay and others have proposed that a "day protein" begins to rise in abundance early in the morning. At high concentrations late in the day, though, this protein begins to inhibit its own production while triggering the manufacture of a second protein. This "night protein" then begins to rise in abundance. At high concentrations late at night, however, it begins to inhibit its own production and trigger synthesis of the day protein.

Is there any evidence that these types of oscillator molecules actually exist? The answer is a very tentative yes. By screening thousands of *Arabidopsis thaliana* mutants for individuals with abnormal circadian rhythms, researchers have identified three different candidate timekeeping genes. A series of studies are now under way to confirm that the products of these genes are essential for timekeeping. Very recent data also suggest that the protein products of these genes interact with each other and change in abundance over the course of the day, as the oscillator hypothesis predicts.

What about the setting mechanism? In habitats far from the equator, day length changes dramatically over the course of a growing season. To adjust the timekeeping component of the clock, plants must have a molecule that (1) monitors day length, and (2) alters the concentration of timekeeping molecules as day length changes. To drive this point home, think about jet lag. If you fly to a location where the night-day period is dramatically different from the one where you live, your internal clock gets out of whack, disrupting your sleep patterns and other rhythms. Over the course of several days, however, your internal clock becomes reset.

In *Arabidopsis*, some evidence indicates this resetting may be done by a combination of phytochromes and cryptochromes. If these light-absorbing proteins are disrupted by mutations, circadian behavior is abnormal. Beyond this hint, however, the sensory system responsible for clock setting is still a mystery.

Chapter Review

Summary

Plants perceive a wide variety of stimuli and respond in ways that help them grow and reproduce. The sensory systems making this behavior possible have three major components: (1) a receptor that changes in response to the stimulus, (2) a signal transduction pathway that transforms the stimulus into a chemical or electrical signal that triggers a response, and (3) movement, transcription of target loci, or other action that helps the organism take advantage of the information in the stimulus. In other words, plants receive information from their environment and translate it into a signal that triggers appropriate action.

This chapter explored the receptors, signal transduction pathways, and response mechanisms involved in sensing blue light (sunlight), far-red light (shade), gravity, and touch or wind. **Table 34.2** summarizes some of the main conclusions from experiments on these sensory systems. In most cases, biologists are still working to understand exactly how signals are received and transduced, and how responses are carried out. Gaining a better understanding of phototropism, gravitropism, and other aspects of plant behavior is an exciting frontier in biology. From the work done to date, however, it is abundantly clear that plants don't just sit there. They actively monitor many aspects of the environment, responding quickly and appropriately.

TABLE 34.2 A Summary of Sensory Systems in Plants

Stimulus	Receptor	Signal Transduction	Response	Adaptive Value
Blue light	NPH1 in stems and leaves	NPH1 autophosphorylates; remainder of signal transduction system is unknown but involves CRY1 and CRY2 proteins.	Phototropism occurs.	Stems grow toward light with wavelengths needed for photosynthesis.
Red light	Phytochrome in seeds and elsewhere	Phytochrome changes to P_{fr} form and activates response.	Seed germinates.	Sunlight triggers sprouting in species that require full sun.
Far-red light	Phytochrome in stem	Phytochrome changes to P_r form and induces response; mechanism is unknown.	Stems lengthen.	Species that require full sun attempt to escape shade.
Gravity	Integrin proteins located in cell membrane or in amyloplast membrane?	Details unknown, but auxin is involved.	Cells on opposite sides of root or shoot elongate; tissue curves.	Roots grow down; shoots grow up.
Touch or wind	Stretch receptors; location unknown	Details unknown, but result is transcription activation in target genes.	Stems grow shorter and thicker.	Individual is more resistant to damage.
Touch	Receptor hair cell in Venus flytrap	Depolarization of receptor-cell membrane triggers action potential.	Effector cells swell; trap shuts.	Plant can capture prey.

Questions

Content Review

1. Which of the following observations has been used to challenge the statolith hypothesis for gravity sensing?
   a. In *Arabidopsis*, starchless mutants still show a reduced response to gravity.
   b. Gravity sensing takes place in a small population of root-cap cells.
   c. Roots and shoots respond to gravity in opposite ways.
   d. The signaling molecule called auxin is involved in the response.
2. Which of the following statements about phytochrome is true? Choose all that apply.
   a. It is photoreversible.
   b. Its function was understood long before the protein itself was isolated.
   c. The P_{fr} form activates the responses to light.
   d. It can undergo autophosphorylation.
3. Which of the following statements about gravitropism is true? Choose all that apply.
   a. Gravity sensing occurs primarily in root hairs.
   b. Plants that grow in water do not need to respond to gravity.
   c. A response occurs because cells on one side of a root or shoot elongate.
   d. If plants are flipped upside down, they are not able to respond to gravity.
4. What three major steps are involved in a sensory system?
   a. light absorption, protein phosphorylation, and one or more action potentials
   b. gravitropism, mechanical stimulation, and light
   c. perception, signal transduction, and response
   d. There are actually only two: stimulus and response.
5. What is an action potential?
   a. a lightning-quick response to touch
   b. a rapid change in membrane potential, from negative to positive and back to negative
   c. anything that depolarizes a membrane
   d. a type of potential energy

Conceptual Review

1. The text refers to phytochrome as a "shade detector" and NPH1 as a "sun detector." Explain.
2. In the experiment that confirmed the autophosphorylation hypothesis for the blue-light receptor, researchers inserted the *nph1* gene into insect cells. In this experiment, why was it important for the gene to be expressed in an organism other than a plant?
3. What does "transduce" mean? Give an example of a transduction event in plant sensory systems.
4. Explain how the following responses to sensory input help plants survive and reproduce more efficiently: elongating stems in response to far-red light; bending toward blue light; bending roots toward gravity and shoots away from gravity.
5. Summarize the statolith and gravitational pressure hypotheses for gravity sensing. What is the critical difference between the two ideas?
6. How do the proton pumps in plant-cell membranes create a membrane potential? What is a voltage, and why is it considered a form of potential energy?

Applying Ideas

1. Suggest a hypothesis for how gravitational pressure on integrins could be transduced into an electrical or chemical signal. How would you test this hypothesis?
2. When an insect lands on a sensitive plant, the leaves collapse in a fraction of a second (see photo). Design an experiment to test the hypothesis that this response is mediated by an action potential.

3. Thanks to a long series of experiments, biologists have now worked out conditions to grow plants in the weightless environment aboard the orbiting space shuttle. Think of a question about gravity sensing in plants; and to answer the question, design an experiment that will be conducted on the space shuttle.
4. The resting potential of a plant-cell membrane can be thought of as a signal that means, "conditions are normal." A change in the resting potential indicates that some event has occurred at the cell membrane. Describe how changes in the resting potential of a plant cell might function in signal transduction.
5. In general, small seeds that have few food reserves must be exposed to light before they will germinate. (Lettuce is an example.) In contrast, large seeds that have substantial food reserves typically do not depend on light as a stimulus to trigger germination. Why is this contrast logical?
6. Depending on the species, bamboo plants flower every 3, 6, 12, 60, or even 120 years. Generate a hypothesis to explain how bamboo plants keep time. (The actual mechanism is unknown.) How would you test this hypothesis?

CD-ROM and Web Connection

CD Activity 34.1: Sensing Light *(animation)*
(Estimated time for completion = 10 min)
How do plants detect different wavelengths of light?

At your **Companion Website** (http://www.prenhall.com/freeman/biology), you will find self-grading exams and links to the following research tools, online resources, and activities:

Light Sensing Protein
The article reviews research on the perception of light by plants and demonstrates how scientists investigate plant structure and function.

Glowing Plants
Read about genetically modified plants that glow to the touch.

Senses and Sensitivity
Explore materials on plant sensing and movement.

Additional Reading

Wassersug, R. J. 1999. Life without gravity. *Nature* 401: 758. An update on recent experiments in the gravity-free environment aboard the space shuttle.

Communication: Chemical Signals

35

Chapter 34 focused on one important concept: Plants constantly monitor their environment and respond to light, gravity, touch, and other stimuli in ways that increase their ability to survive and reproduce. Stems sense whether they are in shade or light and redirect their growth accordingly. Both stems and roots perceive gravity and respond by growing up and down, respectively, or horizontally.

The presentation left out a key observation, however. The tissues that sense environmental changes are not necessarily the tissues that respond to those changes.

To drive this point home, let's revisit a topic introduced in Chapter 34—the experiments done by Charles Darwin and Francis Darwin on the emerging shoots, or coleoptiles, of reed canary grass seedlings. The results showed that the coleoptiles are phototropic, meaning they bend toward light, and that the bending response is specifically triggered by wavelengths in the blue part of the spectrum. The Darwins didn't stop there, however. To find out which part of the coleoptile is responsible for light perception, the Darwins removed the coleoptile tips. As **Figure 35.1a** shows, the decapitated coleoptiles stop bending toward the light. To extend this result, the researchers either covered the tips of coleoptiles with opaque covers or put opaque collars below the tip, in the area where bending occurs. Figure 35.1a illustrates the result. Coleoptiles with caps did not bend, but stems with collars did.

To make sense of these results, the Darwins proposed that phototropism depends on "... some manner in the upper part which is acted on by light, and which transmits its effects to the lower part." Their hypothesis was that some sort of signal or substance is produced at the tip of the coleoptile and then trans-

Dwarfing was one of the inherited traits that Mendel studied in pea plants. Long after Mendel's work was published, investigators discovered the molecular basis of the trait. Dwarfed individuals like those on the left do not synthesize normal quantities of a chemical signal that triggers stem elongation. Normal individuals of the same age are shown at the right.

ported to the area of bending. The analysis of rapid movement in the Venus flytrap, in the last section of Chapter 34, showed that these tissue-to-tissue signals can be electrical in nature. But in this chapter we focus on the molecules that carry messages from one tissue to another. These chemical signals are called hormones. A **hormone** is a small molecule that controls growth or other physiological processes at extremely low concentrations. In many cases, hormones are produced in a specific tissue but exert their effect at another location. Julius von Sachs, who worked before the Darwins, was the first to propose that hormones are involved in regulating plant growth and behavior.

Our exploration of chemical signaling in plants begins with understanding the molecular basis of phototropism. What is the name of the molecule that the Darwins inferred must exist? How is it transported, and how does it act? After examining research on these questions, we turn to a phenomenon called apical dominance. If the tip of a shoot is intact, lateral branches near the tip do not grow. But if the apical meristem is removed, vigorous branching results. What molecular signal causes this dramatic change in shape? The chapter closes with a look at the hormones involved in regulating overall plant growth—from germination to cell division to stem elongation. The general message of this chapter is straightforward: Hormones are small, relatively simple molecules that have huge impacts on the size, shape, and behavior of plants.

35.1 Phototropism

Although the Darwins published their hypothesis about chemical signals and phototropism in 1881, their idea was not confirmed until 1913. In that year, Peter Boysen-Jensen presented the experiments summarized in **Figure 35.2a** (page 676). Boysen-Jensen cut the tips off young oat shoots and placed either a porous block of gelatin or a nonporous block, such as a flake of the mineral mica, between the tip and the shoot. Stems treated with the porous block showed normal phototropism, so Boysen-Jensen concluded that the phototropic signal was indeed a chemical, and that it could diffuse. Further, because he used a water-based gelatin, it was clear that the molecule was water soluble.

Twelve years later, Frits Went extended these experiments. He collected the phototropic hormone in gelatinous blocks made of agar. As **Figure 35.2b** shows, he did this by putting the blocks under the decapitated tips of oat coleoptiles. Then he placed treated or untreated blocks of agar off-center on the decapitated coleoptiles of other individuals. As Figure 35.2b shows, the stems responded by bending *away* from the source of the hormone—even though the shoots were kept in the dark during the entire experiment. Went had succeeded in producing the phototropic response without the stimulus of light.

The physical basis of the phototropic response is cell elongation (**Figure 35.2c**). Cells on the side of the coleoptile

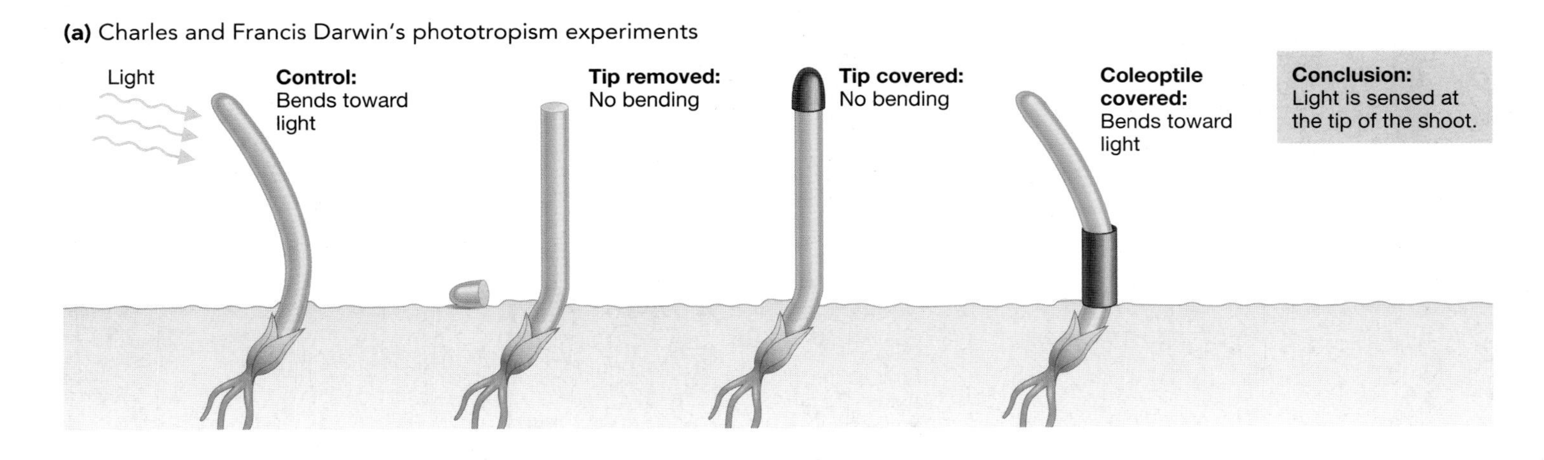

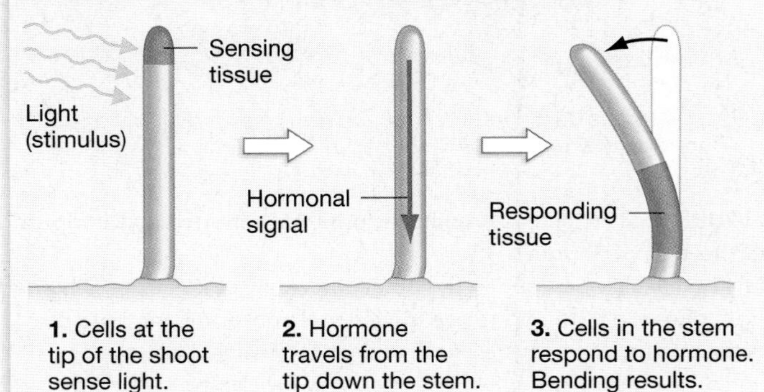

FIGURE 35.1 The Hormone Hypothesis for Phototropism
(a) Decapitated coleoptiles and coleoptiles with their tips covered do not show a phototropic response to light. **(b)** To explain these results, Charles and Francis Darwin proposed that a hormone travels from cells in the tip to cells in the shoot. In this way, hormones function as chemical messengers between tissues.

opposite to the source of light elongate in response to the phototropic hormone. Because it promotes cell elongation in the shoot, Went named the hormone **auxin** (from the Greek *auxein*, to increase).

The Cholodny-Went Hypothesis

Went's experiments were a breakthrough in research on plant hormones. He had accomplished the first step in isolating the hormone—it was somewhere in the agar blocks.

Went's experiments also inspired an important hypothesis for how the hormone produces phototropism. Working independently, N. O. Cholodny and Went both proposed that the bending response results from an asymmetric distribution of the hormone. The Cholodny-Went hypothesis, illustrated in **Figure 35.3a**, contends that auxin produced in the tips of coleoptiles is shunted from one side of the tip to the other in response to light. The auxin is then transported down the shoot. Because auxin concentration is higher on one side than the other, cells on one side of the shoot elongate more than cells on the other side. Bending results.

Other researchers proposed a simpler hypothesis, however. They suggested that auxin is broken down or otherwise inactivated by blue light. If so, then asymmetric distribution would result not from a transport process, but from destruction on one side of the tip (**Figure 35.3b**).

To test these alternatives, Winslow Briggs set up two experiments. He grew corn seedlings in the dark, cut off their tips, and placed the tips on agar blocks. As **Figure 35.4a** shows, he either kept the tips and agar blocks in the dark or exposed them to light from one side. Later he put the agar blocks from each treatment on one side of decapitated coleoptiles (Figure 35.4a). In response, the coleoptiles from each treatment bent the same amount—about 26° from vertical. This result is inconsistent with the auxin destruction hypothesis. If light destroys auxin,

(a) The phototropic signal is a chemical.

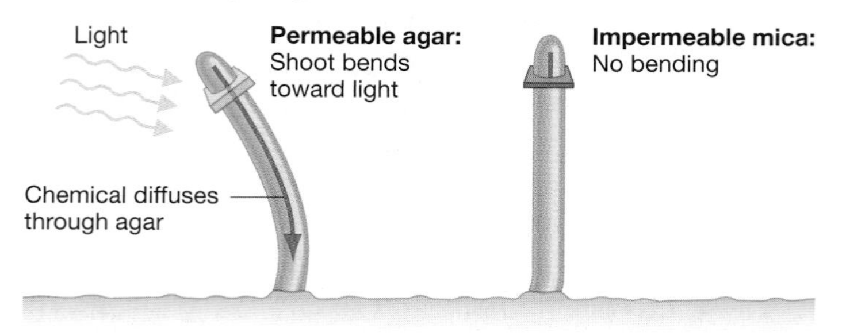

(b) The hormone can cause bending in darkness.

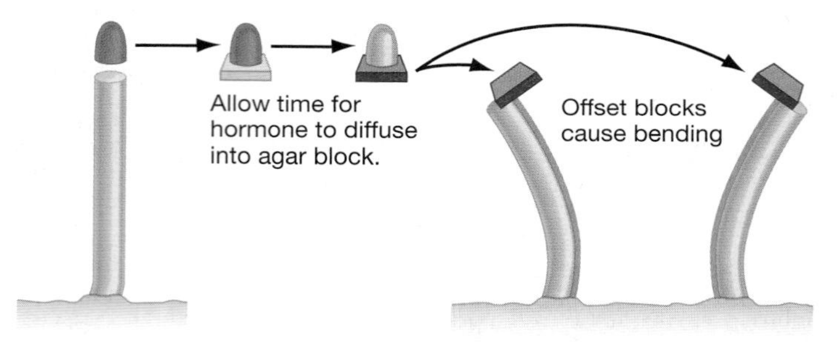

(c) The hormone causes bending by elongating cells.

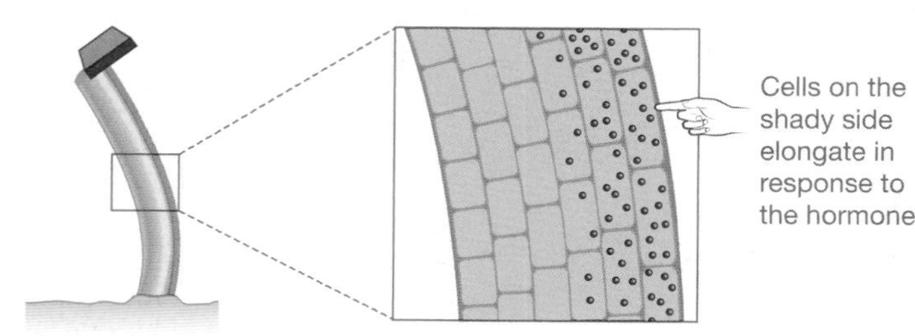

FIGURE 35.2 Confirming the Hormone Hypothesis for Phototropism
(a) If a permeable agar block is placed between a tip and the rest of the coleoptile, the coleoptile still bends in response to light. If an impermeable mica block is used instead, the phototropic response disappears. **(b)** If the tip is cut off a coleoptile, put onto an agar block long enough for hormones to diffuse into the block, and then placed off-center on a decapitated shoot, the coleoptile bends in response. The response takes place in darkness. **(c)** During the phototropic response, bending occurs because cells on the shaded side of the shoot elongate.

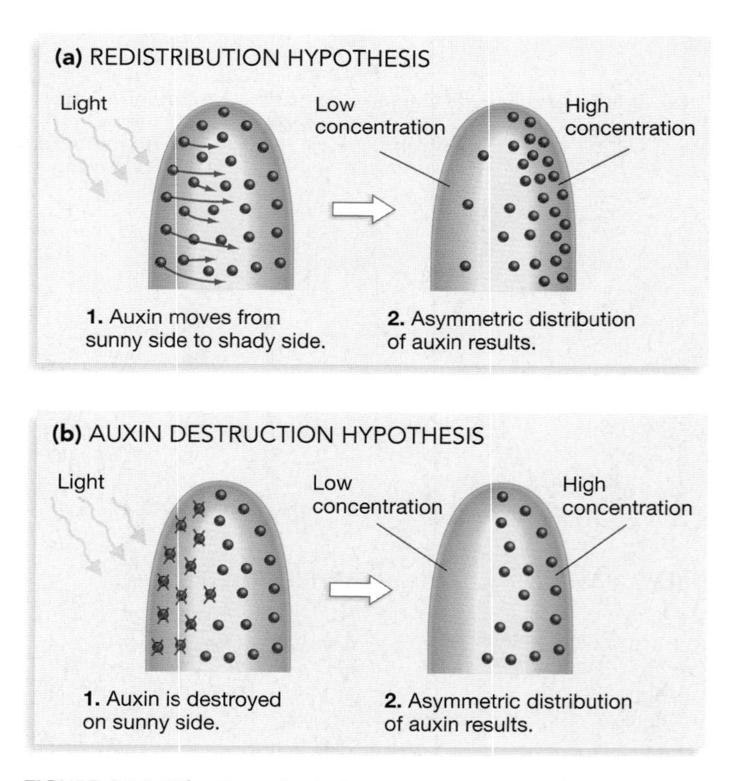

FIGURE 35.3 Why Does Auxin Become Asymmetrically Distributed in the Tips of Shoots?
(a) The auxin redistribution hypothesis maintains that auxin is shunted from one side of the tip to the other, in response to light. **(b)** The auxin destruction hypothesis maintains that auxin is destroyed by light. As a result, auxin concentrations are low on the sunny side of the tip.

then the block exposed to light should be much less effective in inducing bending.

To confirm this result and test the Cholodny-Went model, Briggs carried out the experiment diagrammed in **Figure 35.4b.** Instead of exposing tips to different light conditions, he either divided the tips completely in half or partially in half with a thin piece of mica. (Remember that mica is impermeable to dissolved molecules.) The idea here was that movement of auxin would be stopped in the completely divided tips and agar blocks, but it would not be stopped in the partially divided tips and agar blocks. To test this prediction, Briggs placed the resulting blocks on one side of decapitated shoots and recorded the bending response. The result from the completely divided tip confirmed the earlier result. There was no difference in bending induced by the sunlit or shaded side of the tip. But as the Cholodny-Went hypothesis predicted, the bending responses differed in the partially divided tip. The side away from light induced much more bending, indicating that auxin had been transported from one side of the tip to the other. This result helped to confirm the Cholodny-Went hypothesis.

Although the asymmetric distribution model has been successful in explaining how auxin produces phototropism, a basic question remains. What *is* the hormone, exactly?

Isolating and Characterizing Auxin

Went's work gave researchers a place to look for the phototropic hormone—it was somewhere in the agar blocks used in his experiments—along with an assay to identify it. If a molecule could be found that produced the bending response in decapitated shoots, it must be the auxin hormone.

After years of effort, researchers in laboratories headed by Fritz Kögl and Kenneth Thimann independently succeeded in isolating and characterizing auxin. The hormone turned out to be indole acetic acid, or IAA, which has the chemical structure shown in **Figure 35.5** (page 678). In a typical plant, about 50 nanograms of IAA are present for every 50 grams of fresh tissue.

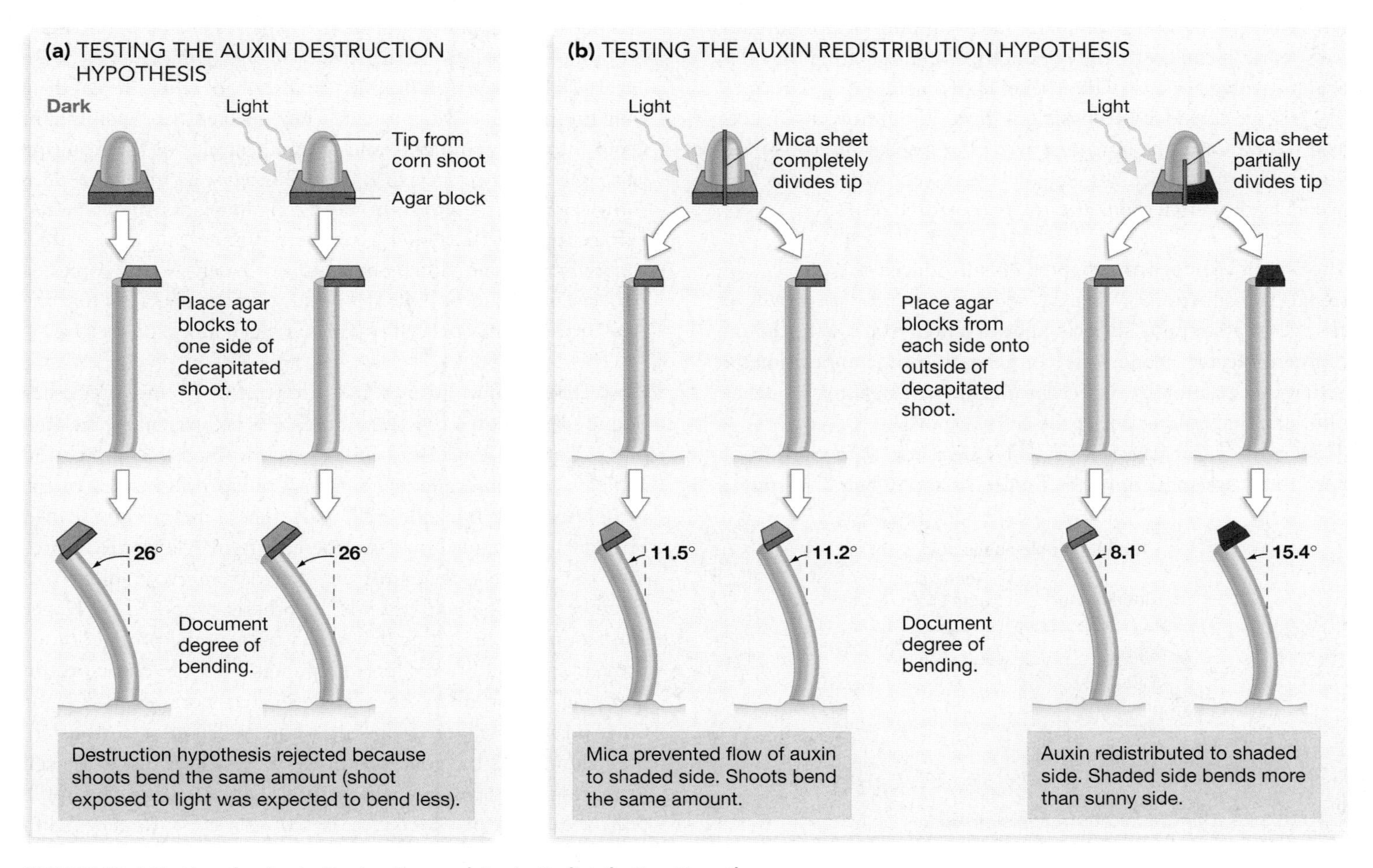

FIGURE 35.4 Testing the Auxin Destruction and Auxin Redistribution Hypotheses
(a) If the auxin destruction hypothesis is correct, much of the auxin in the light-exposed tip should have been destroyed. As a result, the bending response induced by the light-exposed agar block should have been very low. **(b)** The auxin redistribution hypothesis predicts that a mica sheet placed all the way through a tip will prevent a flow of auxin to the shaded side. A mica sheet placed partially through the tip should allow normal flows of auxin to the shaded side.

This is a concentration of about 280 nanomolar (nM). (Recall from Chapter 1 that the prefix *nano-* refers to billionths.)

Why are these observations about auxin's structure and concentration interesting? Like the other plant hormones introduced in this chapter, auxin is a small molecule with a relatively simple structure. It is present in quantities so small that its concentration is difficult to measure. Yet its impact is huge. Auxin can bend stems, leading to tree trunks that are permanently bowed.

How Does Auxin Produce the Phototropic Response?

Although biologists understood where auxin is produced, what its molecular structure is, that it causes cells to elongate, and that its asymmetric distribution produces bending, they also wanted to know how the hormone acts at a molecular level. This is a general goal in biological science. In addition to studying whole-organism responses such as the importance of bending toward blue light, biologists want to know the exact molecular mechanisms involved. Research at the molecular level can not only lead to more complete understanding of how organisms work, it can create the possibility of manipulating the molecules involved in a way that benefits people.

Here we consider two basic questions about how auxin acts as a signal. Once the hormone arrives at a cell, how is its message delivered? Once the signal is received, what molecular events lead to elongation and the phototropic response?

The Auxin Receptor Chapter 34 introduced the concept of signal transduction. When a sensory receptor perceives a change in a stimulus such as light or gravity or touch, the receptor molecule changes its conformation or composition or activity. The action by the receptor is the first step in a sequence of events that culminates in the cell's response.

In many cases, the sensory cell's response is to produce a hormone—a signal for other tissues to act. When a hormone arrives at a target cell, then, its message must be received and transduced to produce the appropriate response. For this reason, understanding receptors is a basic component of understanding hormone action.

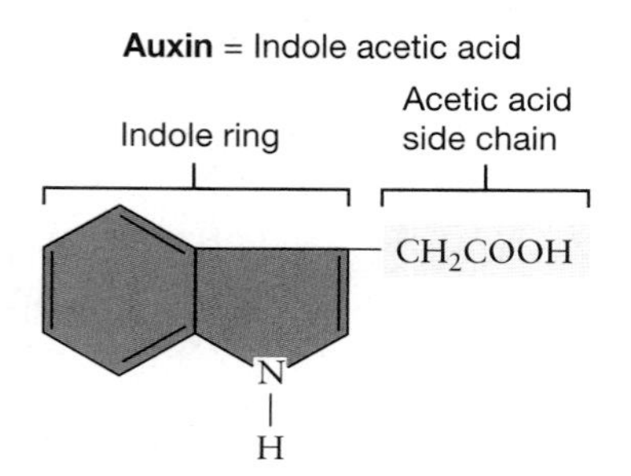

FIGURE 35.5 The Chemical Structure of Auxin
Auxin is indole acetic acid. **EXERCISE** The acetic acid side chain in auxin participates in the following reaction:

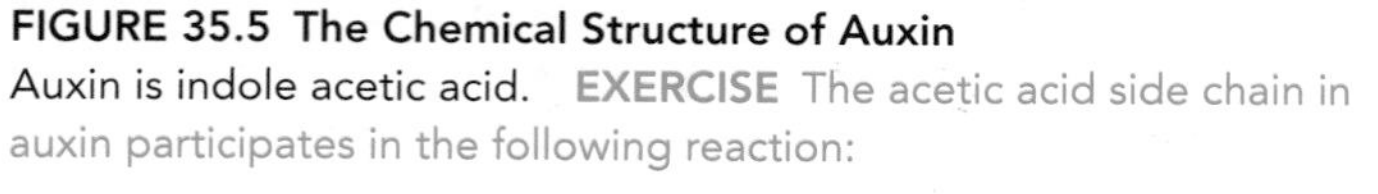

$$CH_2COOH \rightarrow CH_2COO^- + H^+$$

Which side of the equilibrium is favored when auxin is in an acidic environment where hydrogen ions are abundant? In acidic environments, will most of the auxin be in the uncharged form or the negatively charged form?

In the case of auxin, the search for a receptor followed a straightforward strategy. Researchers set out to attach a radioactive label to the hormone, treat cells with the labeled molecule, and purify the protein or proteins to which it binds. Although the purification step turned out to be extraordinarily difficult, M. Löbler and D. Klämbt succeeded in isolating auxin-binding protein 1, or ABP1, from corn plants in 1985. Researchers from several different laboratories followed up by producing antibodies to ABP1 and by determining the protein's amino acid sequence. These studies gave researchers the tools they needed to find the gene that codes for the protein. (Gene-hunting strategies similar to those used in finding the *ABP1* locus were introduced in Chapter 17.)

To confirm that ABP1 is the auxin receptor responsible for phototropism, Alan Jones and co-workers implemented the experimental protocol illustrated in **Figure 35.6**. They began by making many copies of the *ABP1* gene. Then they attached a promoter sequence to each gene copy. The promoter they chose ensured that *ABP1* would be transcribed whenever a drug called tetracycline was present. They injected the recombinant genes into tobacco plants and treated samples of the resulting leaf tissue with tetracycline. The photographs in Figure 35.6 show some of their results. Leaf cells in plants that received extra copies of the *ABP1* gene were dramatically larger than leaf cells from individuals that did not receive extra copies of *ABP1*. These data provide strong support for the claim that the ABP1 protein induces cell expansion when auxin binds to it.

How Does Auxin Induce Cell Elongation? Once auxin has bound to ABP1 in a cell membrane, how is the signal translated into action? A. Hager and colleagues proposed that elongation depends on membrane proteins that pump protons out of the cytoplasm and into the cell wall. These pumps are called H^+-ATPases because they use the energy in ATP to drive protons out of the cell against an electrochemical gradient (see Chapter 32). The acid-growth hypothesis proposed by Hager and associates maintains that auxin triggers the production or activation of additional proton pumps. Why is extra proton pumping important? The answer is that potassium (K^+) or other positively charged ions often enter a cell after protons are pumped out. As the concentrations of these ions inside the cell increases, water follows via osmosis. In the case of coleoptile cells, the influx of water increases turgor pressure and makes cell expansion possible.

To test the acid-growth hypothesis, Hager and co-workers used an antibody tagged with a fluorescent molecule to quantify the number of proton pumps in the cell membranes of corn shoots. (This method for visualizing membrane proteins was introduced in Chapter 32.) Other researchers had shown that treating cells with auxin can lower the pH of the cell wall by as much as a full unit—from a normal value of 5.5 to 4.5. When Hager and colleagues treated shoots with additional auxin, they found that the number of proton pumps increased by 80 percent relative to untreated controls.

Bringing water into the cell is only part of the story, however. For elongation to occur, the cell wall also must expand. As predicted by the acid-growth model, cell-wall expansion occurs in an acidic environment. For example, Daniel Cosgrove has isolated two classes of cell-wall proteins that actively increase cell length when the pH of the wall falls under 4.5. These proteins, called expansins, have now been found in a wide variety of species and tissues. The genes that code for them have been identified and are very similar when compared from species to species. What expansins actually do has yet to be determined, however. One possibility is that they make cell walls extensible by breaking bonds between cellulose strands, pectin fibers, or other wall components.

To summarize, auxin acts as the primary signal in phototropism. It produces cell elongation by triggering the installation of additional proton pumps and the activation of expansin proteins.

In addition to investigating the events that occur after auxin reaches a cell, researchers are looking into another key issue: How is auxin transported from cell to cell so it can act as a signal? We'll look into this question by exploring a pervasive phenomenon called apical dominance.

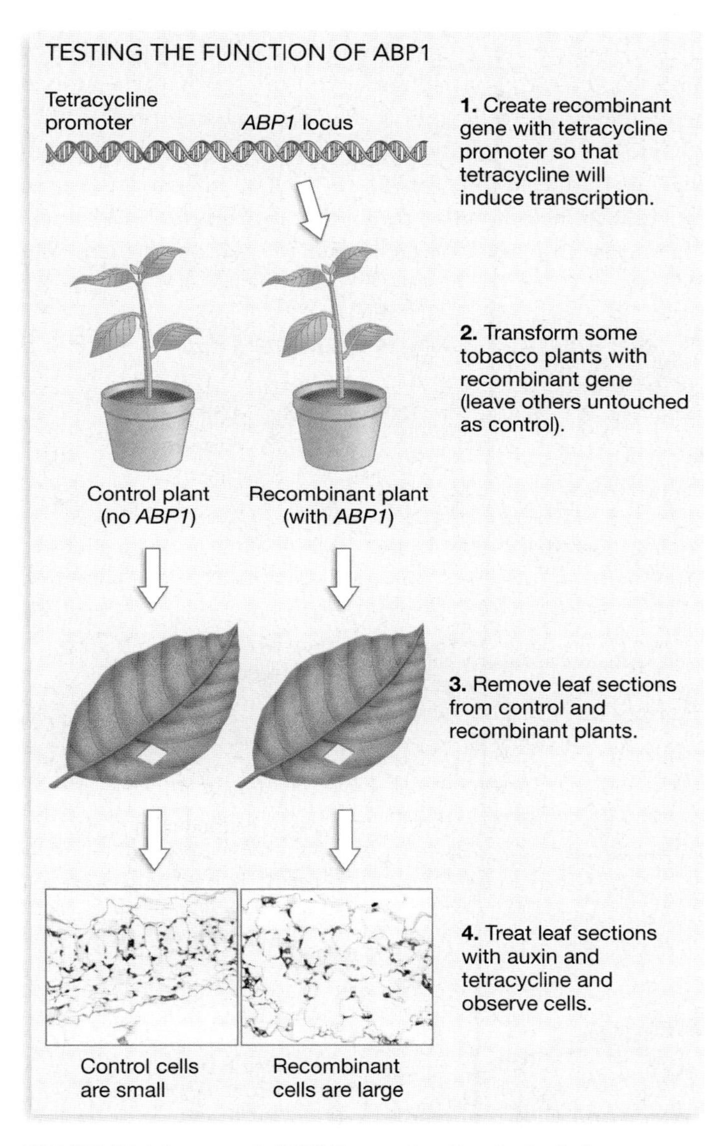

FIGURE 35.6 Increased *ABP1* Expression Results in Cell Enlargement
Cells from the middle of mature tobacco leaves do not normally respond when they are exposed to auxin. But when these cells are given copies of the recombinant *ABP1* gene shown here and are then exposed to auxin, they respond by getting larger. QUESTION Why did the researchers put the tetracycline promoter onto the recombinant gene? Why didn't they just add copies of the *ABP1* locus with the promoter normally found near the gene?

35.2 Apical Dominance

Apical dominance is a growth habit in which the majority of stem elongation occurs at the apical meristem of the shoot. Further, the presence of a topmost meristem inhibits growth by apical meristems that are present lower down on the plant, in nodes. The photographs in **Figure 35.7** show the phenomenon in action. The photo on the left shows an intact bean plant, with one main shoot extending vertically. The photo on the right shows the same plant several weeks after the main shoot was cut. Lateral branches have begun to grow vigorously and vertically.

What caused the change? Because auxin is produced in shoot tips, researchers suspected that it might have a role in apical dominance as well as phototropism. This hypothesis was confirmed when it was shown that apical dominance could be sustained by adding auxin to the ends of shoots whose tips had been cut.

Based on this observation, the mechanism of apical dominance appears clear. Apical dominance occurs because a continuous flow of auxin from the tips of growing shoots to the tissues below signals the direction of growth. If the signal stops, it means that growth has been interrupted. In response, lateral branches sprout and begin to take over for the main shoot. Now the question is, how does this signal move?

Apical meristem intact

Apical meristem cut off

FIGURE 35.7 Apical Dominance
If the apical meristem of a bean plant is cut off, lateral buds respond by growing rapidly. The lateral shoots orient themselves vertically.

Polar Transport of Auxin

Auxin transport is polar, or unidirectional. If radioactively labeled auxin is added to the top of a cut stem, it is transported toward the base. But if labeled auxin is added to the base of a cut stem, it is not transported toward the apex. Further, studies with labeled auxin have shown that the hormone is transported all the way down to the central portion of the root. After it arrives in the root tip, auxin moves out toward the epidermal cells and up—in what M. H. M. Goldsmith called the "fountain model" (**Figure 35.8a**).

This redistribution in the root is important because Cholodny and Went predicted that an asymmetrical distribution of auxin in roots is responsible for gravitropism (see Chapter 34), just as an asymmetrical concentration of auxin in shoots is responsible for phototropism. To follow up, other researchers suggested that gravity-sensing cells in the root tip alter the direction of auxin transport. As **Figure 35.8b** shows, asymmetrical distribution of auxin in roots should lead to asymmetrical elongation and the bending response of gravitropism.

In short, the polar nature of auxin transport is critical to understanding phototropism, gravitropism, and apical dominance. What is the molecular mechanism involved?

The Chemiosmotic Model To explain how polar transport is accomplished, a series of researchers developed the chemiosmotic hypothesis of auxin movement. This hypothesis makes two main claims: (1) auxin travels from cell to cell along an electrochemical gradient established by the proton pumps found in plant cell membranes, and (2) transport is unidirectional because there are two different carrier proteins for auxin, and these carriers are located in different parts of each cell.

The specific steps in the model are illustrated in **Figure 35.9**. Near the top of the figure, you can see that when auxin is in the acidic cell wall, it tends to gain a proton. The chemiosmotic hypothesis proposes that auxin passes through the membrane at the top of the cell in this form. Auxin moves into the cell via a protein that transports the hormone together with protons, which enter the cell along their elec-

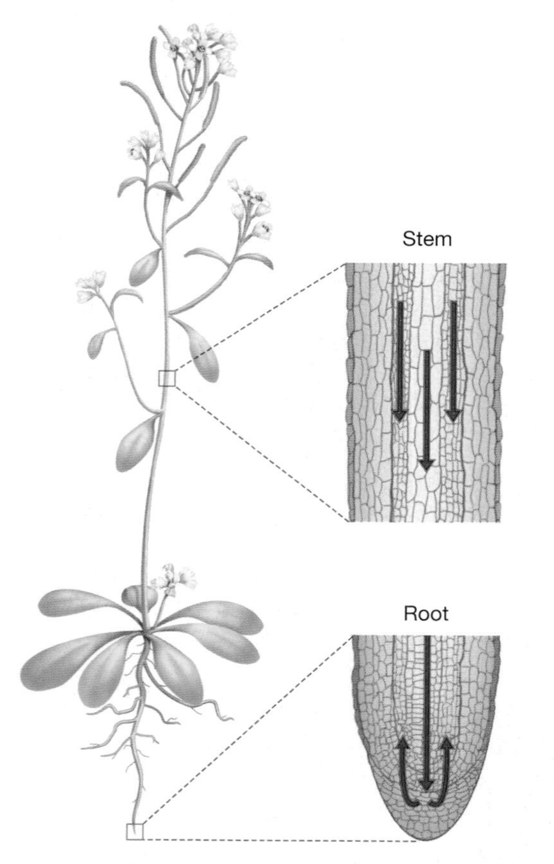

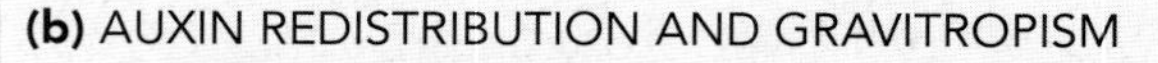

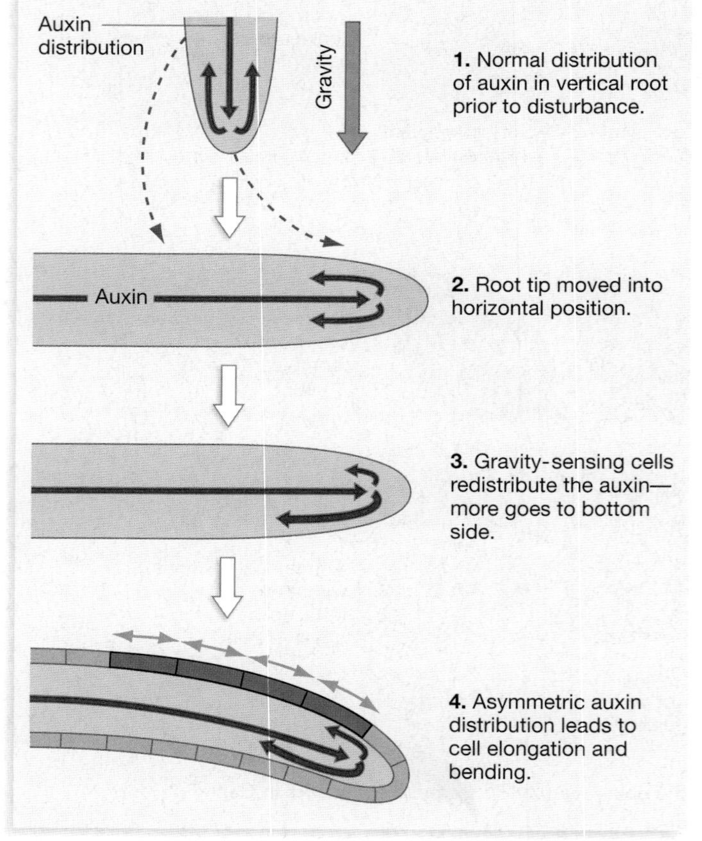

FIGURE 35.8 Auxin Is Transported from the Apical Meristem to Root Tips
(a) This sketch shows how auxin moves through an *Arabidopsis* plant. Note that the hormone is redistributed after reaching a root tip, so that it is transported up and outward. **(b)** The Cholodny-Went hypothesis maintains that auxin redistribution is responsible for gravitropism. The sequence of events shown here might begin when a growing root tip hits a rock and is displaced horizontally or when a plant is tipped in a windstorm and partially uprooted.

trochemical gradient. (Membrane proteins that function in this way are called cotransporters.)

Inside the cell, where the pH is neutral, auxin tends to lose a proton and take its anionic form. The hypothesis contends that carrier proteins specific for negatively charged auxin are located only at the *base* of the cell. Auxin leaves the cell through these carriers along a concentration and electrical gradient. In the acidic cell wall, the molecule again gains a proton.

As this sequence of events repeats, auxin moves steadily from apex to base. In root cells, the hypothesis predicts that the location of the two carrier proteins should be reversed. The "influx carrier" should be located at the base of root cells, while the "efflux carrier" is located at the top. As a result, auxin is transported away from root tips and up the outer surface of roots.

Identifying the Auxin Carriers How can the chemiosmotic model for auxin transport be tested? Perhaps the most direct way would be to find the carrier proteins responsible for auxin movement and to show that they behave as the model predicts. This objective has just been achieved, through a series of studies on *Arabidopsis thaliana* mutants that are unable to transport auxin normally. For example, Leo Gälweiler and colleagues studied individuals like the one shown in **Figure 35.10a**, which have a mutant phenotype called pin-formed. The research team was able to identify a gene named *PIN1* that was disabled in these individuals. When they made an antibody to the PIN1 protein and treated cells of normal individuals with the antibody, they found that the protein is located in the basal part of cells in the stems (**Figure 35.10b**). Based on these results, Gälweiler and co-workers propose that *PIN1* encodes the auxin efflux carrier—the protein that negatively charged auxin passes through as it moves down and out of a cell.

Similar studies have now identified membrane proteins in the root cells of *Arabidopsis* that may be the influx carrier and efflux carrier responsible for the "inverted umbrella" pattern of auxin transport. As predicted by the chemiosmotic hypothesis, the efflux carrier is located at the top of root cells, while the influx carrier is located at the bottom (to see why this makes sense, look back at Figure 35.9).

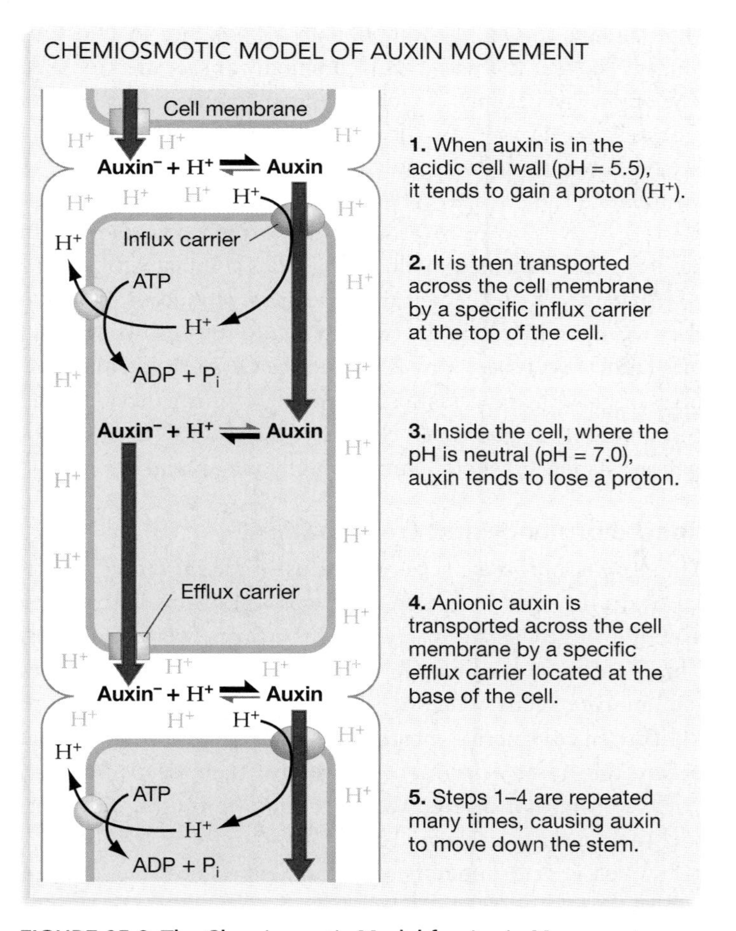

FIGURE 35.9 The Chemiosmotic Model for Auxin Movement
According to the chemiosmotic hypothesis, the influx carrier transports uncharged auxin molecules into the cell along with protons. The movement is driven by the proton gradient set up by H^+-ATPases in the cell membrane. The efflux carrier transports negatively charged auxin molecules out of the cell along an electrochemical gradient.
EXERCISE Make a drawing like this one; in it, show how auxin could be transported up and away from root tips (as in Figure 35.8).

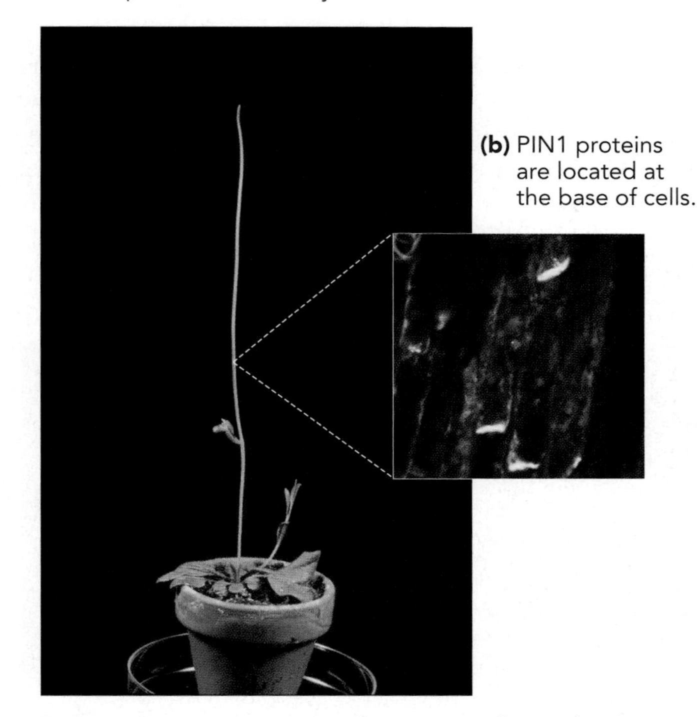

FIGURE 35.10 Pin-Formed Mutants Have a Defect in the Auxin Efflux Carrier
(a) The *Arabidopsis* individual shown here is called pin-formed. It does not have a functioning copy of the *PIN1* gene and cannot transport auxin normally. **(b)** These cells have been treated with an antibody to the PIN1 protein. Because the antibody has a fluorescing subunit attached to it, it vividly marks the location of PIN1 proteins.

Although research is continuing at a furious pace, these recent results have largely confirmed the chemiosmotic model for auxin transport. How auxin is transported through cells, and how its message is received, are increasingly well understood at the molecular level.

An Overview of Auxin Action

Until now, this chapter has emphasized auxin's role in phototropism, gravitropism, and apical dominance. But it is critical to recognize that this chemical messenger has other effects. For example, the presence of auxin in growing roots and shoots is essential for the proper differentiation of xylem and phloem cells and their organization into vascular tissue. In addition, if shoots are cut off and placed in a solution containing auxin, the hormone induces the formation of roots (**Figure 35.11**).

In fact, auxin has so many different effects on plants that it has been difficult for biologists to understand its overall role. Grasping the general role of a chemical messenger is a critical issue in hormone research. For example, consider the diverse effects of the human hormone called cortisol. This molecule stimulates the breakdown of fatty acids and proteins to provide the body with energy, but it also inhibits the immune system so that little swelling occurs in response to short-term insults. These effects seemed completely unrelated until researchers realized that cortisol helps individuals cope with prolonged trauma such as starvation, anxiety, or chronic illness by conserving energy. Cortisol is a stress hormone.

What is auxin's general role? Recently, several investigators have proposed that auxin's overall function is to signal where cells are in space. More specifically, the idea is that auxin concentration identifies where a cell is located relative to the long axis of the plant body. In this way, auxin plays a role in determining the overall shape of a plant. If conditions relating to the long axis change—because of changes in light availability, because a windstorm tips the plant and shifts the position of the roots, or because a deer eats the shoot apex—changes in auxin concentration efficiently signal how the individual's tissues should respond. Phototropism, gravitropism, and apical dominance are all ways of coping with changes in the long axis of the plant.

FIGURE 35.11 Auxins Promote Root Formation
The *Lonicera* leaves in this photo were cut from a plant and placed in pure water (left) or a solution containing an auxin at medium (middle) or high concentration (right).

It is important to recognize, though, that auxin does not act alone. Instead, its effects are modified by the action of other plant hormones. For example, the hormones called cytokinins can override apical dominance. As **Box 35.1** (page 684) explains, cytokinins appear to be a general signal to begin or continue cell division. If a researcher adds cytokinins to a lateral bud, it will begin to grow. The message here is that hormonal signals interrelate in complex ways. Interaction among hormones is a major theme of the next section.

35.3 Growth and Dormancy

When should a plant grow, and when should it stop growing? Plant tissues answer this question by responding to two hormones. The first is abscisic acid, commonly abbreviated to ABA. The second is a family of closely related compounds called the gibberellic acids, or GAs. Although over 100 distinct GAs have been isolated from plants, only a few have been shown to act as hormones. (The rest are probably either intermediates in the synthesis of active GAs or by-products of inactivated hormones). In general, ABA inhibits growth. Gibberellins stimulate it.

Where are these chemical messengers produced? How do they work, and how do the two interact to turn growth on and off? Here we consider how ABA and the GAs regulate three aspects of growth and cessation of growth, or dormancy: (1) seed germination, (2) guard-cell closure (which shuts down photosynthesis), and (3) shoot elongation in young plants.

Seed Dormancy and Germination

Many plants produce seeds that must undergo either a period of drying or a period of cold, wet conditions before they are able to germinate in response to warm, wet conditions. This pattern is sensible. A requirement for drying ensures that mature seeds will not sprout on the parent plant; an obligatory cold period protects seeds from germinating just before the onset of winter. In essence, then, seeds have an "off" setting that discourages germination and an "on" setting that initiates growth. The appropriate state for the on/off switch is determined by environmental cues such as temperature and moisture.

By applying hormones to seeds, researchers learned that in many plants, ABA is the signal that inhibits seed germination and gibberellins are the signal that triggers embryonic development. To understand how these messengers work and how they interact, researchers have concentrated on studying a specific event: the production of an enzyme called α-amylase (alpha-amylase) in germinating oat or barley seeds.

The production of α-amylase is significant because it acts as a digestive enzyme that breaks the bonds between the sugar subunits of starch. (Your saliva contains an amylase; if you hold a piece of bread in your mouth long enough for the enzyme to work, the bread will begin to taste sweet.) **Figure 35.12a** shows that during the germination of a barley seedling, α-amylase is released from a tissue called the aleurone layer. The enzyme diffuses into the carbohydrate-rich endosperm tissue and releases sugars that can be transported and used by the growing embryo. Adding GA to the aleurone layer increases production of α-amylase; adding ABA decreases α-amylase levels. In this way, α-amylase production is a symptom of dormancy or germination and a model system for understanding ABA and GA action.

How Does GA Activate the Production of α-Amylase? Frank Gubler and associates recently explored the molecular basis of GA and ABA action in barley seeds. These researchers noticed that the promoter sequence near the α-amylase gene resembles the DNA sequences targeted by a class of transcription factors called the Myb proteins. Mybs are DNA-binding proteins that turn gene transcription on or off. (To use the vocabulary introduced in Chapter 15, Mybs function as transcription activators or transcription repressors.)

Based on these observations, Gubler's team proposed that a Myb might turn α-amylase on in response to a signal from GA or off in response to a signal from ABA. More specifically, they proposed that the events diagrammed in **Figure 35.12b** occur. It had already been established that, as germination begins, GA produced by the embryo reaches the membranes of cells in the aleurone layer. Gubler and co-workers proposed that a GA receptor on the membrane receives this signal, and that the receptor responds by activating the production of a Myb. The Myb, in turn, travels to the nucleus, binds to the α-amylase promoter, and triggers transcription.

To test their idea, the researchers began a search for a Myb that is activated by GA. Their first step was to isolate all of the mRNAs produced in the aleurone layer of a germinating barley seed. Then they used the enzyme called reverse transcriptase to make a DNA copy of each of these mRNAs. At that point, they had a copy of all the genes being transcribed in activated aleurone tissue. Were any of these sequences related to the Myb family? To answer this question, Gubler's team tested each

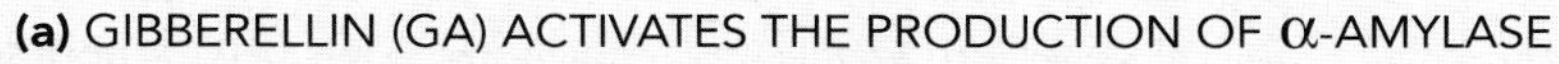

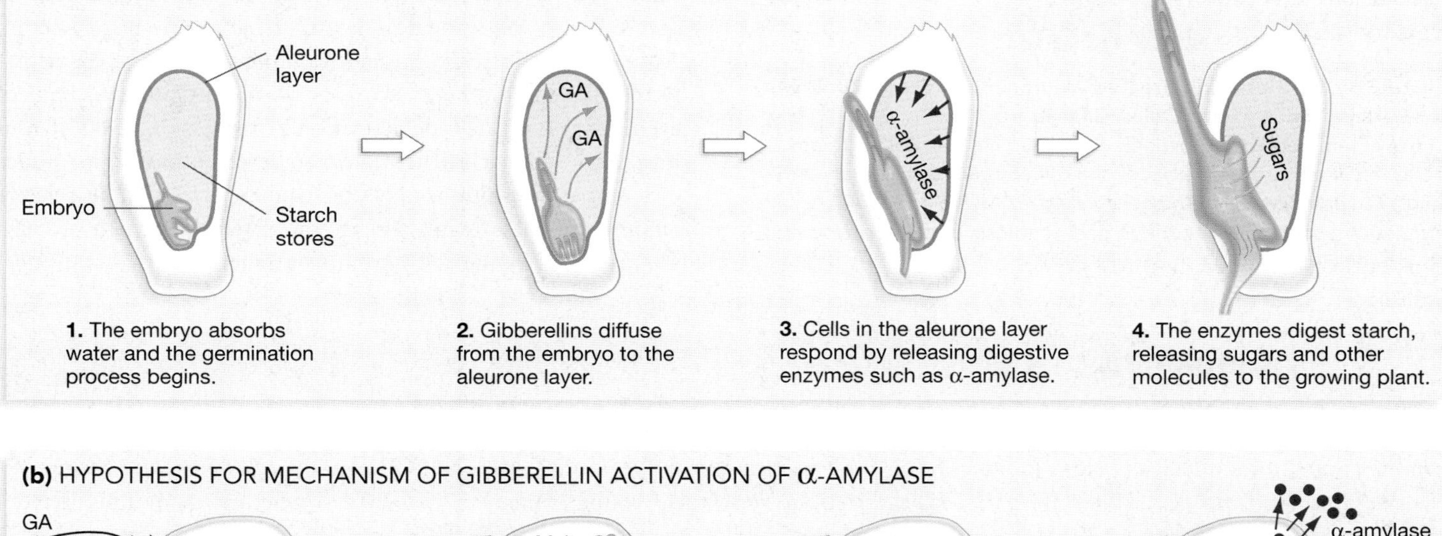

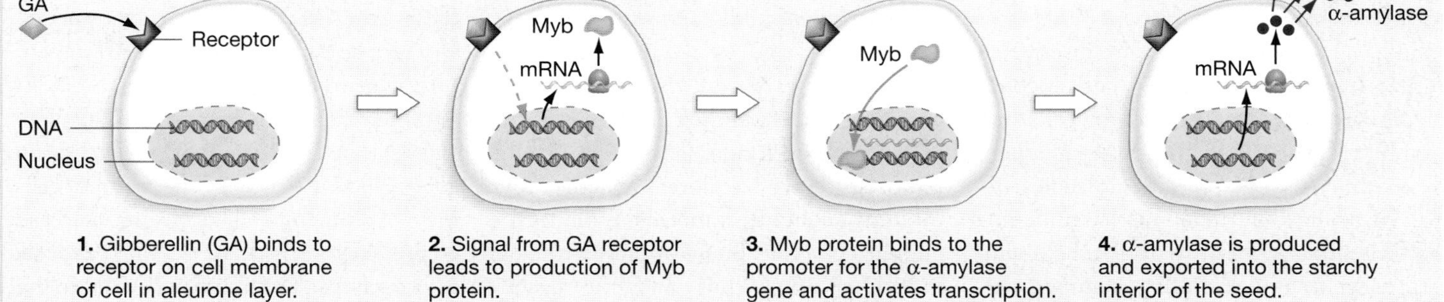

FIGURE 35.12 The Molecular Mechanism of Gibberellin Action
(a) In seeds, release of gibberellin is an important part of the germination process. **(b)** To understand how gibberellin leads to the production of α-amylase, researchers hypothesized that it is responsible for activating a transcription factor called Myb.

DNA sequence to see if it would react with the section of Myb DNA that encodes the DNA-binding region. One DNA sequence from barley did. The DNA sequence had to be from a gene that encodes a Myb. The result confirmed that a Myb protein exists in activated aleurone tissue.

Is this transcription factor produced specifically in response to GA, as the researchers' model contends? To answer this question, they isolated mRNAs from barley aleurone layers that were either exposed or not exposed to GA. They ran these mRNAs out on a gel and then probed the sequences with single-stranded DNA from the *Myb* gene they had discovered in aleurone tissue. (Chapter 15 introduced this experimental technique, which is called Northern blotting.) As the gels in **Figure 35.13** show, the mRNA for the Myb protein appeared only in cells that were exposed to GA. Follow-up experiments confirmed that this protein, which they named GAMyb, binds to the α-amylase promoter. Just as their hypothesis predicted, GAMyb acts as a transcription activator that stimulates α-amylase production.

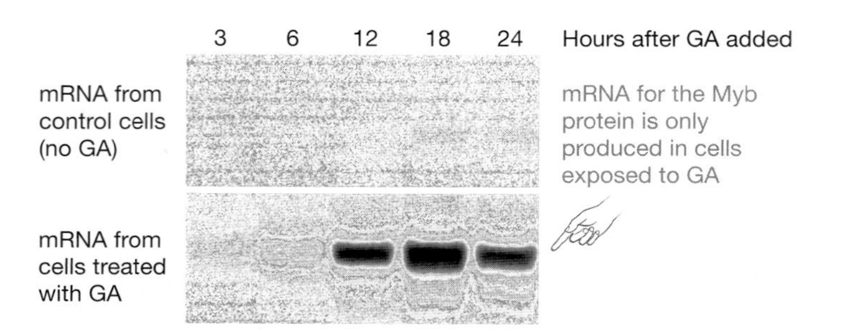

FIGURE 35.13 Gibberellin-Treated Aleurone Cells Produce mRNAs for a Transcription Factor Called GAMyb
These autoradiographs show gels containing mRNAs isolated from barley aleurone layers. The bands show where radioactively labeled, single-stranded copies of the gene called *GAMyb* bind to mRNA. The results show that the *GAMyb* gene is transcribed only after cells have been stimulated by gibberellin (GA).

How Do GA and ABA Interact? Research on GA tells only part of the story about dormancy and seed germination. Studies similar to the one we just reviewed indicate that ABA also induces the production of Myb proteins. Preliminary data suggest that these ABA-dependent transcription factors bind to the α-amylase promoter and shut down transcription. If these experiments are confirmed, it suggests a molecular basis for how hormones with opposite effects interact. The hypothesis is that ABA enforces dormancy by activating transcription repressors that shut down the production of key enzymes, while GA breaks dormancy and supports germination through transcription activators that trigger production of the same enzymes. The key observation is that the transcription activators and repressors compete for the same binding sites near genes. If ABA is present in high concentration, repression dominates and dormancy occurs. If GA is present in high concentration, activators prevail and germination proceeds.

The experiments we've reviewed on α-amylase production in seeds carry several important messages about hormone action. (1) A cell's response to a hormone often occurs because specific genes are turned on or off. (2) Hormones rarely act on DNA directly. Instead, a receptor on the surface of a cell usually receives the message and responds by initiating a chain of events that leads to gene activation or repression. (3) Different hormones interact at the molecular level because they induce different transcription activators and repressors. The relative amounts of these regulatory factors determine which proteins are produced by the cell, and in what quantities.

BOX 35.1 Cytokinins

Cytokinins are a group of plant hormones that promote cell division. (*Cyto* is the Greek root for *cell*; *kinin* refers to *kinesis*, meaning "movement" or "division.") These molecules were discovered after researchers found that coconut milk promoted the growth of cells and embryos growing in culture. Subsequent experiments showed that molecules derived from the nitrogenous base adenine also stimulate the growth of cells in culture; eventually, naturally occurring adenine derivatives that stimulate growth were discovered in corn and apple. These hormones, which biologists found to be similar in structure to the growth-promoting ingredient in coconut milk, were named cytokinins. Cytokinins are synthesized in root tips, young fruits, seeds, growing buds, and other developing organs.

Recent research on cytokinins has focused on understanding how they function at the molecular level. Specifically, biologists have explored whether cytokinins might affect molecules that regulate the cell cycle. Some of these proteins, including the cyclins and the cyclin-dependent kinases (CDKs), were introduced in Chapter 8. Recall that these proteins must be activated for cells to progress through stopping points in the cell cycle. In this way, activated cyclins and CDKs allow cell division to continue.

To assess whether cytokinins affect these genes, C. Riou-Khamlichi and coworkers grew *Arabidopsis* cells in culture, starved them of cytokinins for a day, and then added the hormone. When they assessed the level of mRNA from a cyclin gene called *CycD3* in the treated cells, they documented significant increases compared to cells that were not exposed to cytokinins. This is strong evidence that cytokinins are growth hormones, and that their specific function is to activate genes that keep the cell cycle going.

Closing Guard Cells

Chapter 32 introduced one of the major problems faced by land plants: replacing water that is lost to the atmosphere when guard cells open. Because guard cells open in response to blue light, they allow gas exchange to occur while the plant is receiving the wavelengths of light used in photosynthesis. Chapter 32 also mentioned that if plant roots are unable to obtain enough water to replace the fluid being lost at the leaves, guard cells close. Closing guard cells is a logical response when roots begin to dry, because continued loss of water from the leaves would lead to wilting and potential tissue damage.

The question is, how might roots communicate with leaves about water conditions? Early work on stomatal closing suggested that ABA is involved. For example, applying ABA to the exterior of guard cells causes them to close.

To explore the hypothesis that roots communicate with leaves via ABA, researchers in W. J. Davies's laboratory performed a series of experiments based on the system diagrammed in **Figure 35.14**. The fundamental idea was to grow plants whose roots had been divided into two containers. Experimental plants were watered in one pot, while control plants were watered in both containers. During this treatment, the water potential of the leaves remained the same in both control and experimental plants. The stomata of experimental plants began to close, however. This result suggested that roots from the dry side of the pot were signaling drought stress, even though the leaves were not actually experiencing a water shortage.

FIGURE 35.14 Can Roots Communicate with Shoots?
To assess whether roots communicate information about water availability to shoots, researchers developed the system shown here. Roots are trained into two sides of the same pot, so they can be watered separately. By withholding water from one side, roots can be induced to send a signal indicating dry soil—even though all of the leaves are obtaining sufficient water from the roots on the other side.

In follow-up experiments, the researchers were able to confirm two important predictions: ABA concentrations in roots on the dry side of the pot were extraordinarily high, and ABA concentrations in the leaves of experimental plants were much higher than in the controls. These results suggested that ABA from roots is transported to leaves, and that it actually does serve as an early warning system of drought stress. In this way, ABA fulfills a general role in plants as a dormancy or "no-growth" signal.

Before leaving the topic of guard-cell regulation, though, it is important to realize that ABA does not act in a vacuum. Just as ABA and GAs compete to provide accurate information about the timing and rate of seed germination, ABA has to override information in blue light—the signal that guard cells should be open. To survive and reproduce successfully, plants have to integrate information from a variety of sources.

Shoot Elongation

Over 100 years ago, Japanese farmers noticed that some of their rice seedlings grew exceptionally quickly but fell over before they could be harvested. Biologists who followed up on the observation found that the diseased plants were infected with the fungus *Gibberella fujikuroi*. Treating rice seedlings with an extract from the fungus could induce abnormally long shoots. The active component in the extract was eventually isolated and named gibberellin.

Rice plants produce their own gibberellin, but respond to applications of additional hormone by elongating their stems. In effect, the infected rice seedlings were suffering from a gibberellin overdose.

(As an aside, it's interesting to note that Japanese researchers succeeded in purifying and characterizing the first gibberellins in the 1930s—just when biologists in Europe and the United States were discovering auxin. The first two hormones that affect plant growth were discovered at virtually the same time. Because of language barriers and World War II, however, auxin and gibberellin researchers were unaware of each other's work until the 1950s.)

Analyzing Stem-Length Mutants The strong association between shoot elongation and gibberellin dosage gave researchers an important tool for dissecting how GAs work. By analyzing mutant plants with abnormal stem length, biologists could begin to characterize the genes that are responsible for producing gibberellins and that encode the proteins involved in the response to GA. The fundamental idea here is to understand what specific genes do by observing what happens when they are defective or absent. This research strategy, called forward genetics, was the basis for many of the results analyzed in Unit 3. Forward genetics begins with a mutant phenotype and attempts to characterize the loci responsible for the defect. Recall from section 35.2 that analyzing mutants was instrumental in uncovering the genetic basis of auxin transport.

In the case of stem elongation, how has forward genetics helped biologists understand the gibberellins? As an example, let's consider the dwarfed pea plants introduced in Chapter 10. Gregor Mendel analyzed a series of hereditary traits with strong effects on the phenotypes of peas. One of the traits he investigated was stem length. Mendel analyzed the transmission of two alleles at a single locus. One allele coded for tall stems while the other coded for dwarfed growth. The tall allele was dominant to the dwarf allele.

The locus responsible for the stem-length differences came to be known as *Le* (for *le*ngth). Early work on *Le* mutants showed that they attain normal height if they were treated with the gibberellin called GA_1. This observation suggested that dwarf peas can respond to gibberellins normally, but are unable to manufacture normal quantities of their own GA_1. Follow-up experiments focused on treating dwarf peas with a radioactively labeled precursor to GA_1. Because these plants did not succeed in producing radioactively labeled GA_1, researchers became convinced that the *Le* locus encodes an enzyme involved in GA synthesis.

Recently, Diane Lester and co-workers confirmed this hypothesis by finding a locus in the pea genome that encodes an enzyme called 3β-hydroxylase. The 3β-hydroxylase enzyme adds a hydroxyl group (–OH) to a gibberellin called GA_{20} to produce the biologically active molecule called GA_1. When Lester and associates compared the DNA sequences of this gene from normal and dwarf pea plants, they found an important difference. In a part of the enzyme near the active site, the mutant DNA sequence codes for the amino acid threonine instead of alanine. Follow-up tests showed that the mutant enzymes are unable to convert GA_{20} to GA_1. This is strong evidence that the *Le* gene encodes 3β-hydroxylase, and that a single amino acid change renders the enzyme largely ineffective and causes dwarfing. Some 130 years after Mendel's experiments, researchers finally understand the dwarfing trait at the molecular level.

The Role of Gibberellins in Plant Growth In stems, gibberellins appear to promote both cell elongation and rates of cell division. In seeds, we've reviewed data showing that GAs activate the transcription of digestive enzymes that support germination and growth. Based on these observations, it is clear that different tissues respond to GAs in different ways. The general message, however, is that gibberellins provide a signal to start or continue growth.

How do gibberellins interact with other hormones? In seeds, gibberellins act in direct opposition to ABA, which appears to be a general signal that stops or slows growth due to drought stress. In stems, the situation is different. In these tissues, GA action must be coordinated with the effects of cytokinins and auxin. As Box 35.1 showed, cytokinins regulate rates of cell division. Section 35.1 reviewed data showing that auxin triggers cell elongation; recent data indicate that auxin affects stem elongation by activating the *Le* gene, which leads to increased GA production. Research continues on how GA, cytokinins, and auxin interact on the molecular level to control plant growth and development.

The quest to understand how different hormones interact to change an individual's size and shape brings us to the frontier of research on plant hormones. No one knows how the messages from ABA and GAs are received at the cell membrane. Little is understood about how these signals are transduced into changes in gene expression or protein activation. To date, only a handful of loci that are directly regulated by ABA, GAs, auxin, and the cytokinins have been identified. In short, a great deal remains to be learned about the molecular basis of plant hormone action. Table 35.1 (page 687) summarizes the signaling molecules analyzed in this chapter and introduces other hormones currently being researched by biologists. The effort to understand chemical signaling in plants is an exceptionally dynamic field of inquiry.

Essay Herbicides

The herbicides or "weed-killers" used by farmers and gardeners act in a variety of ways. Some poison specific enzymes. The widely used compound glyphosate, for example, blocks an enzyme in the biosynthetic pathway for a molecule required in the manufacture of some amino acids and a wide variety of other compounds. (Glyphosate is sometimes marketed under the trade name "Roundup.") Paraquat belongs to a family of herbicides that disrupt electron carriers or other components of the photosynthetic apparatus. Trifluralin and oryzalin are examples of herbicides that act by disrupting the polymerization of tubulin during cell division.

Synthetic auxins selectively kill broad-leaved plants but do not harm grasses.

Because their mechanism of action involves interfering with such fundamental cellular processes, most of the herbicides just mentioned are referred to as "broad spectrum"—meaning that they kill all plants. A particularly popular group of herbicides, however, is more selective. These herbicides are synthetic auxins. Their chemical structure is closely related to indole acetic acid—the auxin that functions naturally in plants. When used at the

(Continued on next page)

TABLE 35.1 A Summary of Plant Hormones

Hormone	Function	Notes
Auxin	Helps define long axis of body; involved in phototropism, gravitropism, stem elongation, apical dominance.	First plant hormone ever characterized and isolated.
Gibberellins (GA)	Support growth; involved in germination, stem elongation, flowering.	Fungi that produce gibberellins infect rice plants and induce hyper-elongated stems. Analysis of these fungi led to discovery of gibberellins.
Abscisic acid (ABA)	Inhibits growth; involved in dormancy, closure of stomata in response to water stress.	Acts as a stress hormone analogous to cortisol in humans.
Cytokinins	Promote cell division.	New data indicate that they may act on cell cycle regulators.
Brassinosteroids	Support growth—mutants that lack these hormones or their receptors are dwarfed.	The first steroid hormones discovered in plants. Structurally related to steroid hormones in animals.
Systemin	Triggers certain plant defense responses; discussed in detail in Chapter 37.	The first peptide hormone ever discovered in plants.
Ethylene	Regulates aspects of aging; involved in fruit ripening, aging of leaves and flowers.	A gas; discovered because fruit growers used to ripen fruit in sheds heated with kerosene stoves. Clean-burning stoves did not work; researchers discovered that old stoves produced ethylene as a by-product.

(Essay continued)

recommended dosage, synthetic auxins selectively kill dicots but do not harm monocots such as grasses. As a result, these herbicides have been popular for treating lawns and grass-family crops such as corn, wheat, and rice.

It is not known why the synthetic auxins are selective, however. They appear to be absorbed more readily by dicots compared to grasses, are effective at lower doses, and are transported throughout the body more rapidly. Based on these observations, researchers hypothesize that the synthetic auxins lead to hormone overdoses in broad-leaved species. Why excess auxin levels poison cells is also unknown, however. Apparently, high auxin concentrations overstimulate cells in some way.

Synthetic auxins have achieved notoriety, unfortunately, because of their use by the United States military during the Vietnam War. In an effort to clear vegetation used for cover by soldiers of the North Vietnamese army, U.S. forces sprayed extensive areas of forest with a mix of synthetic auxins known as Agent Orange. Unfortunately, the compounds in Agent Orange turned out to be harmful to humans as well as to plants. Specifically, a common contaminant of the mixture caused severe skin lesions in people who were exposed to the spray. The same compound is strongly suspected to contribute to cancer formation.

The synthetic auxin that is associated with the contaminant is called 2,4,5-T. Although several synthetic auxins are still widely and safely used as herbicides, the manufacture and use of 2,4,5-T is now banned in many countries.

Chapter Review

Summary

Certain plant cells can sense changes in water availability, light availability, gravity, and other aspects of the environment. To trigger a response to the environmental change, receptor cells may produce hormones that are transported to target tissues.

In many cases, plant hormones activate specific proteins or genes that help an individual cope with the environmental change. For example, experiments have shown that cells near the tips of shoots sense changes in blue light and respond by altering the distribution of a hormone called auxin. An asymmetric concentration of auxin causes cells on one side of the shoot to elongate more than cells on the other side. In this way, plants bend toward light.

Auxin initiates cell elongation by binding to a receptor on the cell membrane. The receptor then triggers a series of events that results in an increase in proton pumping by H^+-ATPases. In response, water flows into the cell by osmosis, and the acidification of the cell wall activates proteins called expansins. The combination of water influx and wall extension results in elongation.

Auxin is also involved in the phenomenon known as apical dominance. Auxin transport is polar, because cells in the stem have influx carriers for the hormone on their top end and efflux carriers at their basal end. Auxin transport does not require a direct expenditure of energy, but is possible because of the proton, charge, and pH gradients established by the membrane H^+-ATPase, which does require an expenditure of energy. The general function of auxin is to establish and maintain the long axis of the plant body.

Abscisic acid (ABA) and the gibberellins (GAs) are involved in the regulation of dormancy and growth. In seeds, for example, ABA activates transcription factors that repress the genes required for germination to proceed, such as α-amylase. GAs, in contrast, activate regulatory proteins that increase the transcription of enzymes required for growth. ABA also slows growth by shutting down guard cells, and thus photosynthesis, when soil begins to dry. GAs promote stem elongation and other aspects of overall growth.

Plant growth and behavior are regulated by an extensive array of hormones, which interact in complex ways. Hormones are also prominent features of plant reproductive and defense systems—the topics we take up in the next two chapters.

Questions

Content Review

1. Which of the following statements about hormones are true? Choose all that apply.
   a. They tend to be small molecules.
   b. They exert their effects on the same cells that produce them.
   c. They can exert strong effects even when they are present in low concentrations.
   d. They only function early in development.

2. According to the chemiosmotic hypothesis for auxin transport, what happens when the molecule enters a cell?
   a. It stimulates an increase in proton pumping.
   b. It activates expansin proteins.
   c. It tends to gain a proton and become uncharged, because the cell interior has a low pH.
   d. It tends to lose a proton and become negatively charged, because the cell interior has a high pH.

3. In order for auxin to stimulate cell elongation, two things have to happen. What are these?
   a. The auxin influx and efflux carriers must both be activated.
   b. Myb proteins must be activated and α-amylase transcription increased.
   c. Water must flow into the cell, and the cell wall must become more extensible.
   d. Water must flow out of the cell, and the cell wall must become more extensible.

4. According to the auxin redistribution model, how do light and gravity affect auxin?
   a. Light and gravity receptors cause auxin to move laterally in the tips of shoots and roots.
   b. Light destroys auxin.
   c. Light and gravity receptors increase auxin concentrations throughout the plant.
   d. Light and gravity increase polar transport of auxin.

5. Why do ABA and GA have opposite effects on the gene for α-amylase?
   a. ABA induces a transcription activator, while GA induces a transcription repressor.
   b. ABA induces a transcription repressor, while GA induces a transcription activator.
   c. Both hormones can bind directly to the gene's promoter.
   d. ABA decreases translation of the α-amylase mRNA; GA increases it.

6. What evidence suggests that ABA from roots can signal guard cells to close? (Choose all that apply.)
   a. If roots are given sufficient water, guard cells close anyway.
   b. If roots are dry, guard cells begin to close—even though leaves are not experiencing water stress.
   c. If roots are dry, ABA concentrations in leaves increase.
   d. If roots are dry, ABA concentrations drop drastically.

Conceptual Review

1. What does the chemiosmotic model for auxin movement predict about the location of carriers that transport auxin into and out of cells? Is the location of these carriers the same or different in shoot and root cells?
2. How does polar transport of auxin relate to apical dominance, gravitropism, and phototropism?
3. A plant's response to a given hormone depends on the tissue involved, the plant's developmental stage or age, the concentration of the hormone, and the concentration of other plant hormones that are present. For auxin, ABA, or GA, provide examples that support this statement.
4. Summarize the auxin redistribution hypothesis for phototropic and gravitropic responses. Describe the experiments that confirmed and extended the hypothesis. Suppose you had a supply of radioactively labeled IAA. Design an experiment that would use labeled hormone to test the redistribution hypothesis.
5. Discuss the general role that ABA, GAs, or auxin serves in plants. Provide evidence that supports your claim.
6. What evidence suggests that Mendel's dwarf peas had a defect in gibberellin synthesis?

Applying Ideas

1. To explore how hormones function, researchers have begun to transform plants with particular genes. In one experiment, a gene involved in cytokinin synthesis was introduced into tobacco plants. When the recombinant individuals matured, they produced more than the usual number of lateral branches. How does this experiment inform our understanding of how cytokinins function? Plants can also be transformed with antisense genes. Antisense genes code for products that disable specific mRNAs by binding to them. Suggest an experiment that uses an antisense gene to explore hormone function. Predict the result.
2. Researchers have recently begun to identify genes that are activated in response to auxin. Based on your understanding of auxin's effects, what might these loci be?
3. In many species native to tropical rain forests, seeds do not undergo a period of dormancy. Instead, they germinate immediately. Make a prediction about the role of ABA in these seeds. How would you test your predictions?
4. Researchers have shown that stomata do not close if ABA is injected into guard cells. Stomata do close, however, if ABA is applied to the surface of guard cells. Based on these results, researchers claim that the ABA receptor must be on the surface of guard cells, and not in the interior. Do you agree? Why or why not?
5. Even after years of effort, no one has been able to isolate the gibberellin or ABA receptor. One reason is that individuals with mutant GA or ABA receptors may die prior to germination, making them extremely difficult to study. Why might mutations in these genes be fatal? Suggest an alternative strategy for finding the GA and ABA receptor.
6. Review the experiments diagrammed in Figure 35.2. Suppose a critic argued that the phototropic signal was actually electrical in nature. Do the experiments in part (a) rule out electrical signaling? Do the experiments in part (b)? Explain.

CD-ROM and Web Connection

CD Activity 35.1: Plant Hormones *(animation)*
(Estimated time for completion = 5 min)
Do hormones serve the same role in plants as they do in animals?

At your **Companion Website** (http://www.prenhall.com/freeman/biology), you will find self-grading exams and links to the following research tools, online resources, and activities:

Phototropism
This site discusses the molecular basis of phototropism in *Arabidopsis*.

Gravitropism
Explore the basics of gravitropism and find links to current research about this topic.

Abscisic Research
An overview of abscisic acid, including enzymatic breakdown, function, and specific plant growth-stage activities.

Plant Growth Regulators
This site includes many links to commercial, research, educational, and technical sites related to plant growth and hormones.

Additional Reading

Chang, C. and R. Stadler. 2001. Ethylene hormone receptor action in *Arabidopsis*. *BioEssays* 23: 619–627. Reviews recent results on how plants perceive and respond to ethylene gas.

Jones, A. M. 1998. Auxin transport: down and out and up again. *Science* 282: 2201–2202. Introduces current models of how auxin is transported throughout the plant body.

Palme, K. and L. Gälweiler. 1999. PIN-pointing the molecular basis of auxin transport. *Current Opinion in Plant Biology* 2: 375–381. Explains how analyzing mutant individuals has informed work on the mechanism of auxin transport.

Plant Reproduction

36

For humans, flowers symbolize beauty, love, hope, fertility, and abundance. But for plants, flowers do not have symbolic value. They are reproductive structures—a collection of organs designed to produce gametes, attract gametes from other individuals, and develop seeds. Exploring how these reproductive structures work is a major focus of this chapter.

Plants reproduce in a wide variety of ways. Among the many major groups of plants, only one—the **angiosperms**, or flowering plants—produces flowers. Mosses, liverworts, hornworts, ferns, conifers, and other plant groups do not make flowers. Even among the angiosperms, most species are able to reproduce without making flowers and setting seed.

Our exploration of plant reproduction begins with a look at sexuality in plants, an introduction to the phenomenon called alternation of generations, and a discussion of asexual reproduction. In section 36.2 you'll find a detailed analysis of the structure and function of flowers. The chapter continues with an investigation of how pollination and fertilization are accomplished, and ends by examining how seeds are constructed, how they are dispersed to new habitats, and how they germinate. By the end of this chapter, you'll be able to appreciate the business aspects of bouquets as well as their poetic features.

This chapter focuses on the structure and function of plant reproductive structures, such as this poppy flower painted by Georgia O'Keeffe. *(Georgia O'Keeffe (American, 1887–1986), Poppy, 1927. Oil on canvas, 30 × 36 in. Museum of Fine Arts, St. Petersburg, Florida. Gift of Charles C. and Margaret Stevenson Henderson in memory of Jeanne Crawford Henderson. 71.32.)*

36.1 An Introduction to Plant Reproduction

The roots, shoots, vascular tissues, sensory systems, and hormones introduced in this unit all evolved to achieve the same ultimate goal: enabling individuals to grow, survive, and acquire enough resources to make offspring. The reproductive process is highly variable among plant species, however. Consider just one aspect of reproductive structures—their size. As **Figure 36.1** shows, the reproductive structures of plants vary in size from microscopic to the size of a small child; seeds range in size from dust-like particles to coconuts.

Our first task is to introduce the basic principles that unify this diversity of reproductive systems. Let's begin by defining sex.

Sexual Reproduction

Most plants reproduce sexually. Sexual reproduction is based on the reduction division known as meiosis and on fertilization, the fusion of haploid cells called gametes. Male gametes are called sperm. **Sperm** are small cells that contribute genetic information in the form of DNA but few or no nutrients to the offspring. Female gametes are called **eggs**. In addition to DNA, either the egg itself or cells associated with the egg contribute a store of nutrients to the offspring. The important point is that, while sperm and eggs both contribute DNA, they contribute vastly different amounts of other resources to the offspring.

It is also important to recognize that in the majority of plant species, both male and female reproductive organs are produced on the same individual. In angiosperms, flowers containing both male and female parts are referred to as **perfect**. Flowers can also be **imperfect**, meaning they contain either male or female structures. In some cases, separate male and female flowers occur on the same individual. The corn plant illustrated in **Figure 36.2a** (page 692) is an example—in this case the tassel is the male flower and the ear is the female flower. In other species, each individual produces just one type of reproductive structure, and male and female flowers occur on different individuals. See the hemp plants in **Figure 36.2b**, for example.

Many plants outcross after they form gametes. **Outcrossing** occurs when male and female gametes are exchanged between individuals of the same species. We'll look at how gametes move from one plant to another in section 36.3. In many cases, however, plants self-fertilize, or "self." When **self-fertilization** occurs, a sperm and an egg from the same individual unite to form a progeny. Self-fertilization is the most extreme form of inbreeding. (Inbreeding is defined as mating among relatives.) Chapter 22 explored the consequences of inbreeding in detail.

FIGURE 36.1 Variation in Reproductive Structures
Reproductive structures in plants range in size from the tiny flowers of *Wolfia* (top left) to the gigantic flowers of *Rafflesia arnoldii* (top right). All seeds contain an embryo and a food supply, though they vary in size from orchid seeds (bottom left) to coconuts (bottom right).

Plant Life Cycles

When compared to other multicellular organisms, most aspects of plant sexuality are not unusual. At least some protists, fungi, and animals have male and female forms; in many of these species each individual produces both male and female reproductive structures. Some protists, fungi, and animals even self-fertilize on a regular basis. An important contrast exists in the life cycle, however. The multicellular form of animals is diploid; the multicellular form of many protists is haploid. But plants have both a multicellular form that is diploid *and* a multicellular form that is haploid. (Recall from Chapters 8 and 9 that diploid cells have two copies of each chromosome, while haploid cells have one. The diploid condition is symbolized by $2n$; haploidy is indicated by n.) This type of life cycle is called **alternation of generations.** In multicellular organisms other than plants and certain red, brown, and green algae, there is only one multicellular form, and it is either diploid or haploid.

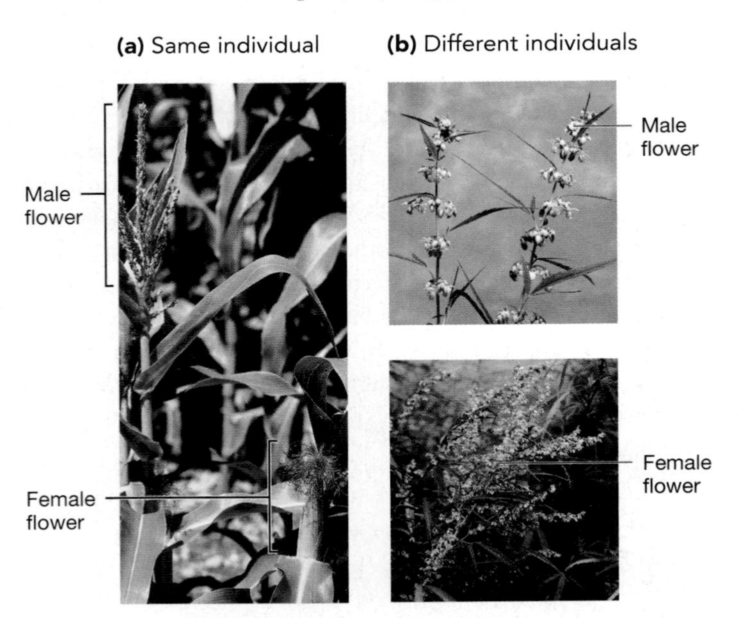

FIGURE 36.2 Male and Female Flowers Can Occur on the Same or Different Individuals
(a) The tassels of corn (*Zea mays*) are male flowers; ears are female flowers. **(b)** In hemp (*Cannabis sativa*), male and female flowers are found on different individuals.

To understand how plants reproduce, then, it is essential to (1) identify which form of a species is diploid and which form is haploid, and (2) grasp how various species make the transition between haploid and diploid generations. As **Figure 36.3** shows, an individual in the diploid phase of a plant life cycle is called the **sporophyte**, while an individual in the haploid phase is referred to as the **gametophyte.** Cells in the reproductive structure of the sporophyte undergo meiosis and produce haploid spores. A **spore** is a reproductive cell that grows into a new individual directly. Thus, a spore divides mitotically to form a gametophyte. Haploid gametophytes produce gametes; thus, plant gametes are produced by mitosis. In contrast to a spore, a **gamete** is a reproductive cell that must fuse with another gamete before growing into a new individual.

It is important to recognize, though, that the relationship between the sporophyte and gametophyte generations is highly variable among plant groups. In mosses and liverworts, for example, the sporophyte generation is tiny, short-lived, and de-

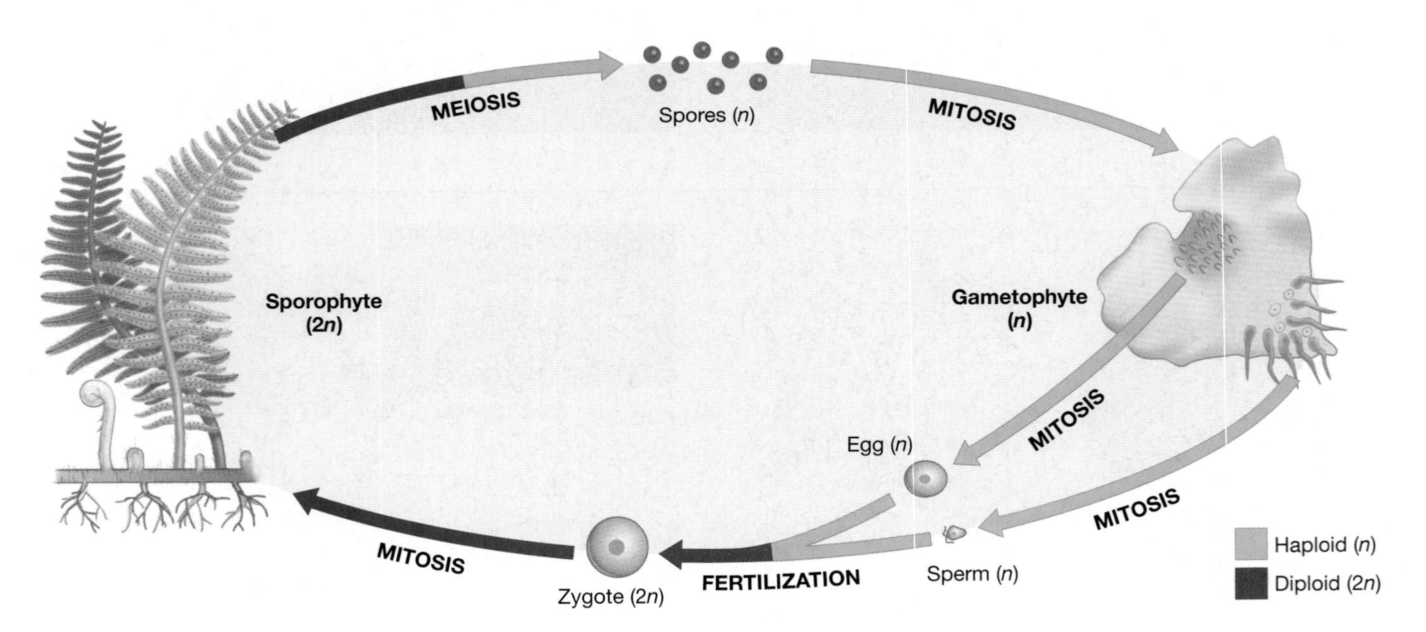

FIGURE 36.3 The Life Cycle of a Plant
Plant life cycles are variations on the elements illustrated here. The sporophyte is a multicellular, diploid stage that produces haploid spores via meiosis. Spores germinate and produce a multicellular, haploid gametophyte via mitosis. The gametophyte produces male and female gametes via mitosis. Gametes fuse to form a diploid zygote, which develops into a sporophyte.

(a) Rhizomes

(b) Stolons

(c) Plantlets

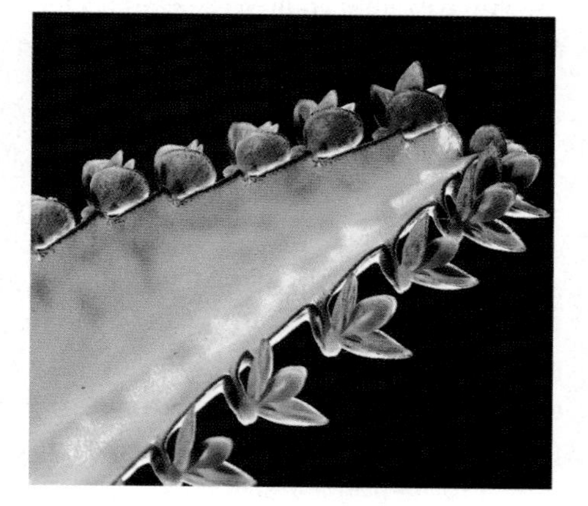

FIGURE 36.4 Mechanisms of Asexual Reproduction Are Diverse
(a) Many grasses generate horizontal underground stems that produce new shoots and roots. **(b)** Strawberry plants send out horizontal stems, called stolons, that produce offspring. **(c)** *Kalanchoe* produce small plantlets at the margins of their leaves.

pendent on the gametophyte for nutrition. In flowering plants, the gametophyte generation is tiny, short-lived, and dependent on the sporophyte for nutrition. Chapter 28 explored the diversity of life cycles found in plants in more depth.

Asexual Reproduction

Figure 36.4a shows how underground stems called **rhizomes** connect many of the individuals in a grass sod. In examining this photo, note that shoots and roots have sprouted from nodes on the rhizomes. Offspring arising from rhizomes are the products of asexual reproduction. **Asexual reproduction** does not involve meiosis or fertilization. By sending out rhizomes, an individual grass plant can produce dozens of copies of itself asexually.

In a similar way, some plants propagate themselves via specialized stems called stolons (or runners) that grow along the surface of the soil (**Figure 36.4b**). Mechanisms of asexual reproduction are diverse, however. Consider the *Kalanchoe* pictured in **Figure 36.4c**. It is producing plantlets from meristematic tissue located along the margins of its leaves. When the plantlets mature, they drop off the parent plant, take root in the soil below, and grow into independent individuals.

The most important point about asexual reproduction is that it leads to offspring that are genetically identical to the parent plant. No crossing over occurs during asexual reproduction, so the same alleles are found in the same combinations along each chromosome in parents and offspring. No independent assortment occurs during meiosis I, so progeny have the same pairs of alleles at each locus as the parent. No fertilization takes place, so no new alleles arrive from other individuals.

The salient characteristic of asexual reproduction is efficiency. If a herbivore or a disease wipes out the plants growing near a big bluestem individual, it can quickly send out rhizomes. Its asexually produced offspring are likely to fill the unoccupied space before seeds from competitors can germinate and grow. The parent plant can also nourish these progeny as they become established.

Although asexual reproduction is extremely common in plants, it does have a downside. If a fungus or other disease-causing agent infects a big bluestem individual, it will probably succeed in infecting the plant's asexual offspring as well—whether or not they are still physically connected. As Chapter 37 will show, plants fight disease with a wide variety of molecules. For sexually reproduced offspring, the combinations of these disease-fighting molecules are often different from those of their parents. Consequently, the offspring may be able to resist infections that devastate their parents. As you saw in Chapter 9, coping with changes in disease-causing agents and other aspects of the environment is a major advantage of reproducing sexually.

Given that sexual reproduction is an important way to produce offspring, how do plants accomplish it?

36.2 Reproductive Structures

Each major group of plants, from mosses to angiosperms, has a characteristic life cycle and characteristic male and female reproductive structures. Our focus here, though, is on the flower. There are two reasons for this. First, most of the food we eat is produced by flowering plants. The staples of human diets—grains such as wheat, corn, and rice—are the fruits of angiosperms. Second, with over 230,000 species named thus far, angiosperms are far and away the most species-rich group of plants. Just two flowering plant families, the orchids and sunflowers, contain over 50 percent more species than all of the non-angiosperm groups combined. (The orchid and sunflower families include over 45,000 species, versus about 28,500 species of all non-flowering plants.)

CD ACTIVITY 36.1
Reproduction in Flowering Plants

In both a practical and a biological sense, then, flowers are enormously important structures. Let's begin our analysis of the flower by asking when plants produce them.

When Does Flowering Occur?

Anatomically, a flower is a stem that develops highly modified leaves. In essence, then, flower formation begins when an apical meristem stops making energy-harvesting stems and leaves and begins to produce the modified leaves that form flowers. Instead of making more food, a plant commits to investing energy into sexual reproduction. Here we ask two questions about this switch: What environmental or hormonal signals trigger it? Which genes respond to the signals and initiate flower formation?

Day Length, Hormones, and Other Cues Early experiments on the environmental signals that promote flowering focused on the number of hours of light and dark during a day. For example, researchers found that exposing some species to artificially lengthened daylight periods triggered flowering. For species occupying northern habitats, this was a sensible result. It is logical for these species to flower during the long days of midsummer, so their seeds have time to ripen and drop before the onset of cold weather. In contrast, other species showed no response to daylight-length changes. Still other plants were actually stimulated to flower by exposure to short periods of daylight.

Why do different species respond to different cues? For the intensively studied mustard plant *Arabidopsis thaliana*, which occupies high latitudes, day length is an external cue containing important information about the time of year. In contrast, for plants growing at the equator, day length provides no information at all because it does not vary over the course of the year. In desert and tropical habitats, then, it is reasonable to expect that external cues other than day length—such as the onset of seasonal rains—might trigger flowering. Based on observations like these, it is not surprising that different species respond to different external cues.

The situation is complex, though, because some species initiate flowering in response to several independent cues. Why would individuals of the same species vary in flowering time? In *Arabidopsis*, flowering can be triggered by many different events, including the long day lengths typical of summer. Day length is an external cue—a signal from the environment. But if the same individuals are exposed to short days, flowering can also be initiated in response to surges of the hormone gibberellin. As you may recall from Chapter 35, gibberellin is a hormone that encourages growth. Gibberellin is an internal cue—a signal about the individual's condition.

The advantage of flowering in response to an external cue like the long days of summer seems clear. But what sorts of internal conditions might trigger a commitment to flowering? One possibility is the individual's nutritional status. In the case of *Arabidopsis*, flowering could be advantageous for the plant when light, water, and nutrient conditions are favorable. An individual that is growing in extremely good conditions might be able to begin flowering long before day length signals the arrival of midsummer. If nutritional status or other internal conditions affect the likelihood of flowering, then gibberellins or other hormones could serve as messengers that trigger the production of reproductive organs. In this way, nutritional status could explain why flowering occurs in the absence of normal external cues.

Which Genes Respond to the Signal to Flower? In *Arabidopsis*, researchers have confirmed that the ability to flower in response to long day lengths and to surges of gibberellin depends on different genes. Some mutant individuals do not flower in response to long days. These same plants respond normally to gibberellin during short days, however. Conversely, individuals that have dysfunctional forms of the enzymes required for gibberellin synthesis do not flower during short days. Yet these individuals still initiate flowering when day lengths are extended.

What genes do these signals act on? To answer this question, consider recent work by Miguel Blázquez and Detlef Weigel on a locus called *LEAFY* (*LFY*). Individuals with mutant copies of this gene are unable to produce normal flowers. Because the locus is involved in flower formation, it must be controlled in some way by the signals that initiate flowering. Blázquez and Weigel hypothesized that this control is direct. As **Figure 36.5a** indicates, they proposed that a transcription factor that is activated by gibberellins and a transcription factor that is activated by day-length changes each bind to the LFY promoter independently, and that each can initiate transcription. Specifically, they hypothesized that the *LFY* promoter has two different binding sites—one for a transcription factor activated by gibberellin and one for a transcription factor activated by a day-length receptor. In this way, one gene could respond to two different signals and trigger flowering.

To test this hypothesis, Blázquez and Weigel followed the protocol outlined in **Figure 36.5b**. They first made many copies of the *LFY* promoter. In most of these copies, they deleted specific and different sections of the promoter DNA sequence. The goal of this step was to delete the binding site for either the gibberellin-induced transcription factor or the day-length-induced transcription factor. The researchers then fused complete or partial promoters to a gene called *GUS*. When *GUS* is transcribed, it produces a product that turns blue when treated with a stain. In this way, *GUS* acts as a **reporter gene**. It identifies where in the plant body a particular promoter has been activated.

Next, the researchers introduced the *LFY* promoter–*GUS* constructs into *Arabidopsis* plants. They exposed the experimental individuals to one of two conditions: long days, or short days combined with applications of gibberellin. The results, shown at the bottom of Figure 36.5b, were striking. Blázquez and Weigel found that if a certain segment of the *LFY* promot-

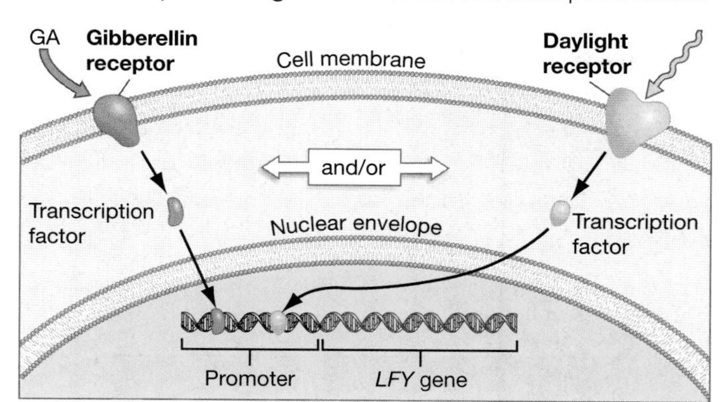

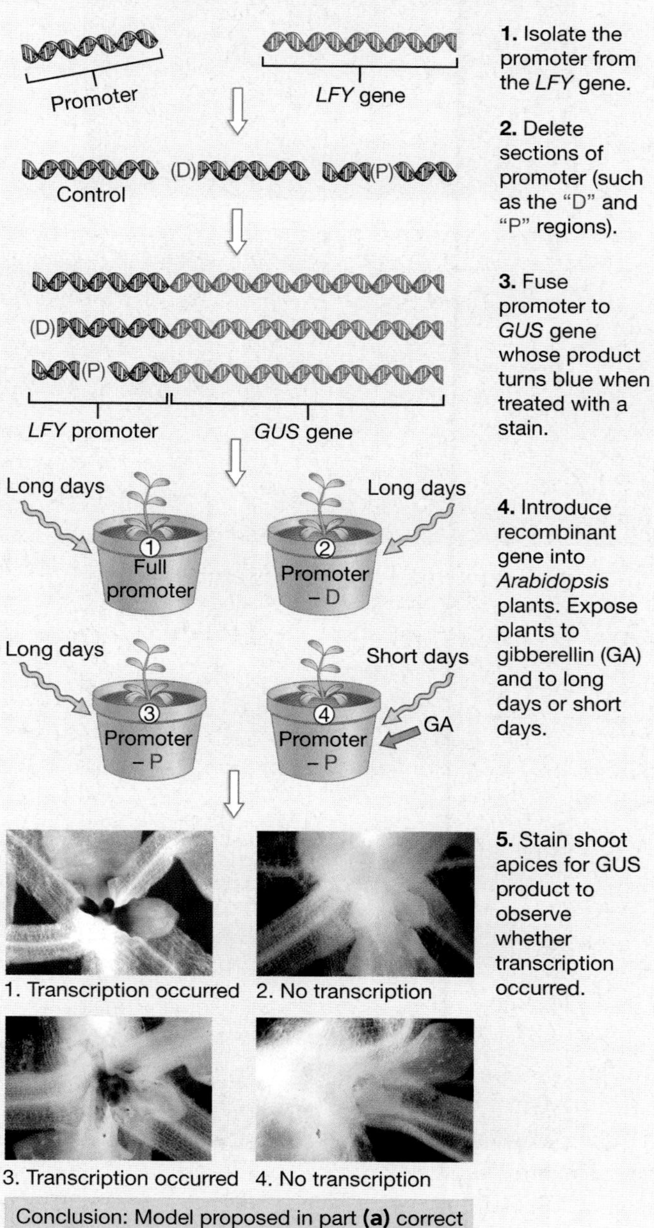

FIGURE 36.5 How Can the Genes Responsible for Flowering Respond to Different Environmental Cues?
(a) The hypothesis illustrated here is an attempt to answer the question posed in the title. The basic idea is that the genes controlling flower development have promoters with binding sites for more than one type of transcription activator. **(b)** To test this hypothesis, researchers followed the protocol diagrammed here. When a portion of the *LFY* promoter nicknamed "D" was deleted, exposure to long day lengths failed to stimulate transcription of the reporter gene. When a portion of the promoter called "P" was deleted, exposure to short days and GA failed to stimulate expression of the reporter gene.

er was deleted, no transcription occurred when plants were stimulated by long days. But if a different section of the *LFY* promoter was deleted, transcription failed when plants were exposed to short days and gibberellin.

These results provide strong support for the researchers' hypothesis that *LFY* and other genes involved in flower formation have two places at which they can be turned on. Transcription activators triggered by long days bind to one of the sites; regulatory proteins activated by gibberellins bind to the other site.

Blázquez and Weigel's work is exciting because it suggests a mechanistic basis for the observation, made decades ago, that flowering can be stimulated by either internal cues, external cues, or both. Now the question is, what does the resulting structure look like? How are flowers put together?

The General Structure of the Flower

Structurally, all flowers are variations on a theme. They are made up of the four basic parts shown in **Figure 36.6a** (page 696). As **Figure 36.6b** shows, however, the coloration, size, and shape of these four elements are fabulously diverse. Let's take a closer look at each part.

Sepals are leaf-like structures that comprise the outermost part of a flower. They are arranged in a whorl around the stem, are usually green, and are relatively thick compared to other parts of the structure. Because they enclose the flower bud as it develops and grows, biologists hypothesize that sepals protect young flower buds from damage by insects or disease-causing agents.

Petals are also arranged around the stem in a whorl. They are often brightly colored and advertise the flower to visually oriented animals such as bees, wasps, and hummingbirds. In some cases, the color of the petals correlates with the visual abilities of particular animals. Bees, for example, respond strongly to wavelengths in the blue and purple parts of the light spectrum. Flowers that attract bees, in turn, often have blue or purple petals with ultraviolet patches. The ultraviolet sections of these petals, which are invisible to humans but visible to bees, frequently highlight the center of the flower (Figure 36.6b). Why? In these flowers, the base of the petals contains a gland called a **nectary**. A nectary produces the sugar-rich fluid called **nectar**, which is harvested by many of the animals that

FIGURE 36.6 Flowers Represent Variations on Four Basic Elements **(a)** Flowers are composed of the four elements illustrated here. **(b)** The characteristics of the four elements in a flower vary enormously from species to species. Color is even more variable than you might think. The photograph at the bottom right, for example, was taken with a camera that records ultraviolet wavelengths that are visible to bees but invisible to humans. **EXERCISE** In the photographs, label as many of the flower parts as you can.

(a) Four basic parts of a flower

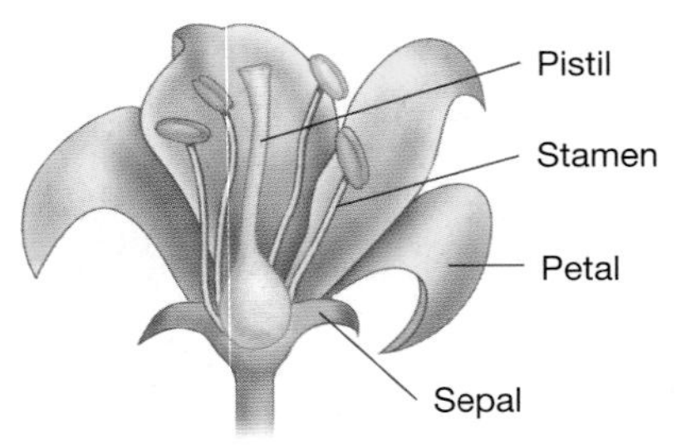

(b) Examples of flower diversity

visit flowers. In short, the function of petals is to advertise the flower's presence to animals. Wind-pollinated angiosperms like oaks, maples, and grasses have flowers with small petals or no petals at all.

The entire group of petals in a flower is called the **corolla.** In some species the petals within the corolla vary in size, shape, and function. For example, flattened petals may provide a landing pad for flying insects, while elongated, tube-like petals frequently have a nectary at their base. Other petals protect the male and female reproductive organs located inside the corolla. The male reproductive structure of angiosperms is called a **stamen,** while the female reproductive structure is called a **pistil.** These structures are crucial. Meiosis and gametophyte formation take place in stamens and pistils. Stamens produce male gametophytes; pistils produce female gametophytes. Let's take a closer look.

Producing the Female Gametophyte

As **Figure 36.7** shows, the pistil consists of three parts: a stigma, style, and ovary. The function of the stigma and style will become clear in section 36.3; for now, let's concentrate on what happens inside the ovary. Figure 36.7 also provides a cross-sectional view of the inside of a typical angiosperm ovary. Note that it contains one or more structures called ovules. The **ovule** is where meiosis takes place and where the female gametophyte is produced. The detailed series of sketches in Figure 36.7 illustrate the steps involved in production of the female gametophyte. There are two important things to note: (1) Four nuclei result from meiosis, but three of these nuclei degenerate. (No one is sure how or why this happens.) The haploid nucleus that remains is called the megaspore. (2) In the species illustrated here, the megaspore divides by mitosis to produce eight nuclei. (The number of nuclei produced by mitosis varies among species.) These eight nuclei segregate to different positions in the structure and form seven cells. These seven cells, which constitute the multicellular female gametophyte, are called the **embryo sac.**

The most important elements of the embryo sac are the egg and the polar nuclei. The egg is located at the base of the structure, near an opening called the **micropyle** (little-gate). The polar nuclei are in the middle of the structure. The cell they occupy is far and away the largest in the female gametophyte.

When development of the embryo sac is complete, the female gametophyte is mature. A gamete (egg cell) has formed and is ready to be fertilized.

Producing the Male Gametophyte

Figure 36.8 gives you a detailed look at the stamen. Note that it consists of two major parts, an anther and a filament. Cells inside the **anther** undergo meiosis. Each of the haploid cells that results is called a microspore. Each microspore becomes a **pollen grain**, which is the male gametophyte. When mature, a pollen grain consists of a small generative cell enclosed within a large vegetative cell. The vegetative cell develops a hard coat. The coat protects the cell contents and the generative cell when pollen grains are shed from the parent plant and enter the environment.

Once pollen grains have matured, they are ready to be carried to female reproductive organs by an animal, the wind, or a water current. It is critical, however, to note that pollen grains

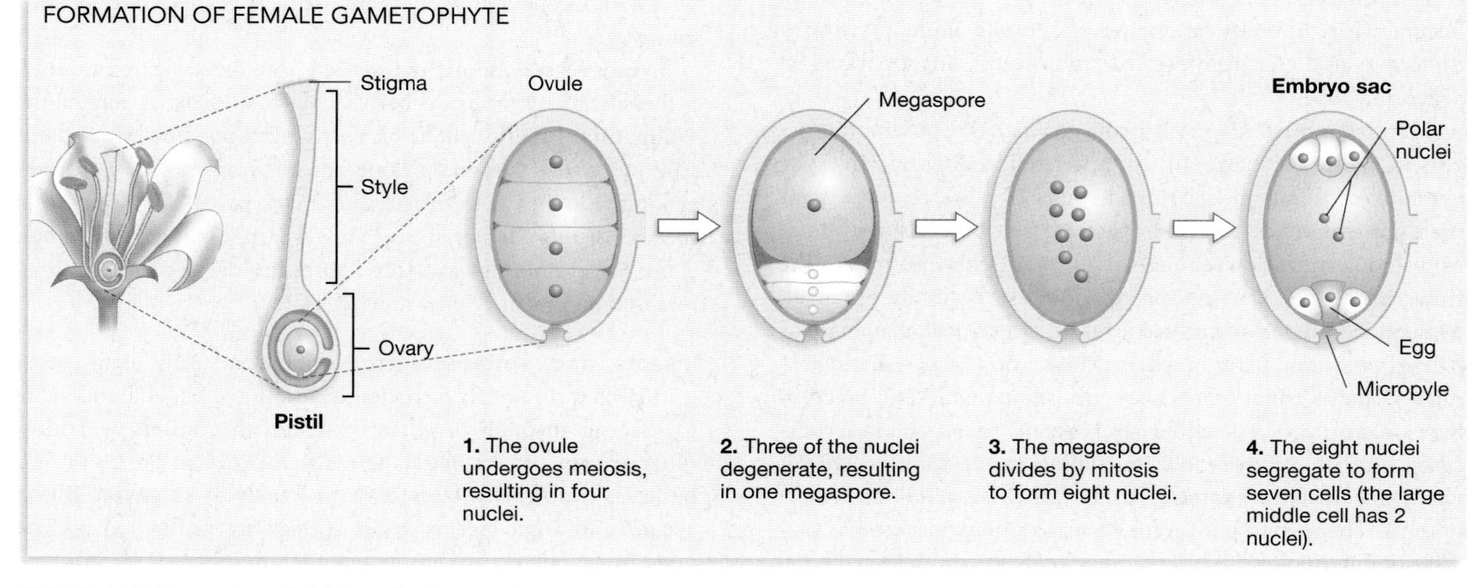

FIGURE 36.7 Formation of the Female Gametophyte
Pistils have three parts: a platform at the top called the stigma, a long tube called a style, and the ovary. The ovule resides in the ovary. A progenitor cell inside the ovule divides by meiosis, but only one of the four products survives. This cell, called a megaspore, divides by mitosis to form the egg and a variety of other cells within the embryo sac.

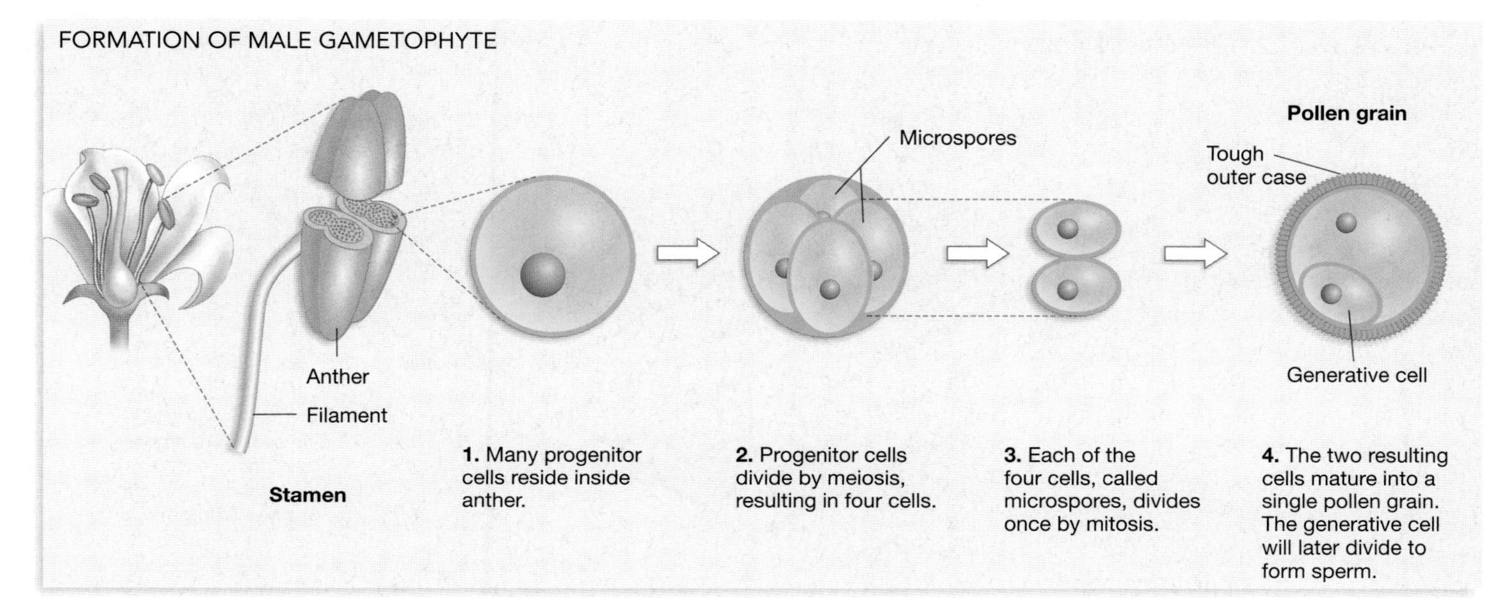

FIGURE 36.8 Formation of the Male Gametophyte
Stamens have two parts. Anthers are perched on top of long filaments. Progenitor cells inside the anther divide by meiosis. The resulting cells, called microspores, each divide once by mitosis then mature into pollen grains.

do not contain sperm. Instead, they are tiny gametophytes. The haploid generative cell will produce sperm cells via mitosis. Now the question is, what happens when a pollen grain arrives at a pistil?

36.3 Pollination and Fertilization

Pollination and fertilization are two different things. **Pollination** is the transfer of pollen from an anther to a stigma. **Fertilization** occurs when a sperm and an egg actually unite to form a diploid zygote. As we'll see, the two events are separated in space and time.

Fertilization occurs in all plant groups, and pollination occurs in many. Pollination is not restricted to angiosperms. The gymnosperms introduced in Chapter 28 also package their male gametophytes into pollen grains. Our analysis of pollination and fertilization, though, will again concentrate on the flowering plants. One reason for doing so is purely practical. Managing pollination and fertilization is a critical challenge for fruit growers and plant breeders. Most crop plants will not produce seeds or fruits unless they are pollinated. Also, breeders develop crop varieties with desirable traits such as disease resistance and productivity by controlling fertilization so that it takes place between gametes from particular male and female parents. The second reason for focusing on angiosperms is that their pollination and fertilization systems are thought to be the key to their evolutionary success.

What is so remarkable about pollination and fertilization in angiosperms? Why have these innovations allowed angiosperms to become so dominant in terms of numbers of species?

Pollination

When pollination occurs, the male gametophyte and female gametophyte are brought together. In many angiosperms, pollen is carried from the anther of one individual to the stigma of a different individual on the body of an insect, bird, or bat that moves from flower to flower. Animals visit flowers to eat pollen grains, harvest nectar, or both. As they are feeding, pollen grains fall onto their bodies incidentally. When they visit a subsequent flower, some of these grains are deposited on a stigma.

In most cases, animal pollination is an example of a mutually beneficial relationship between two species, or **mutualism**. Pollinators usually benefit by receiving food; flowering plants gain by having their male gametophytes transferred to a different individual so that outcrossing takes place. As the essay at the end of this chapter shows, however, pollination by animals is not always mutualistic. Certain species deceive their pollinators and do not provide a reward.

Insects and Angiosperms In mosses, ferns, and other groups that do not form pollen, sperm have flagella and swim to the egg through droplets of water. In conifers and most other groups of non-angiosperms that do form pollen grains, pollen is transmitted from male to female by the wind. But in cycads and angiosperms, most species are pollinated by animals—particularly insects. A smaller number of cycads and angiosperms are wind or water pollinated.

When these observations are placed in an evolutionary context, two important patterns emerge. An evolutionary analysis of pollination is possible because researchers have compared

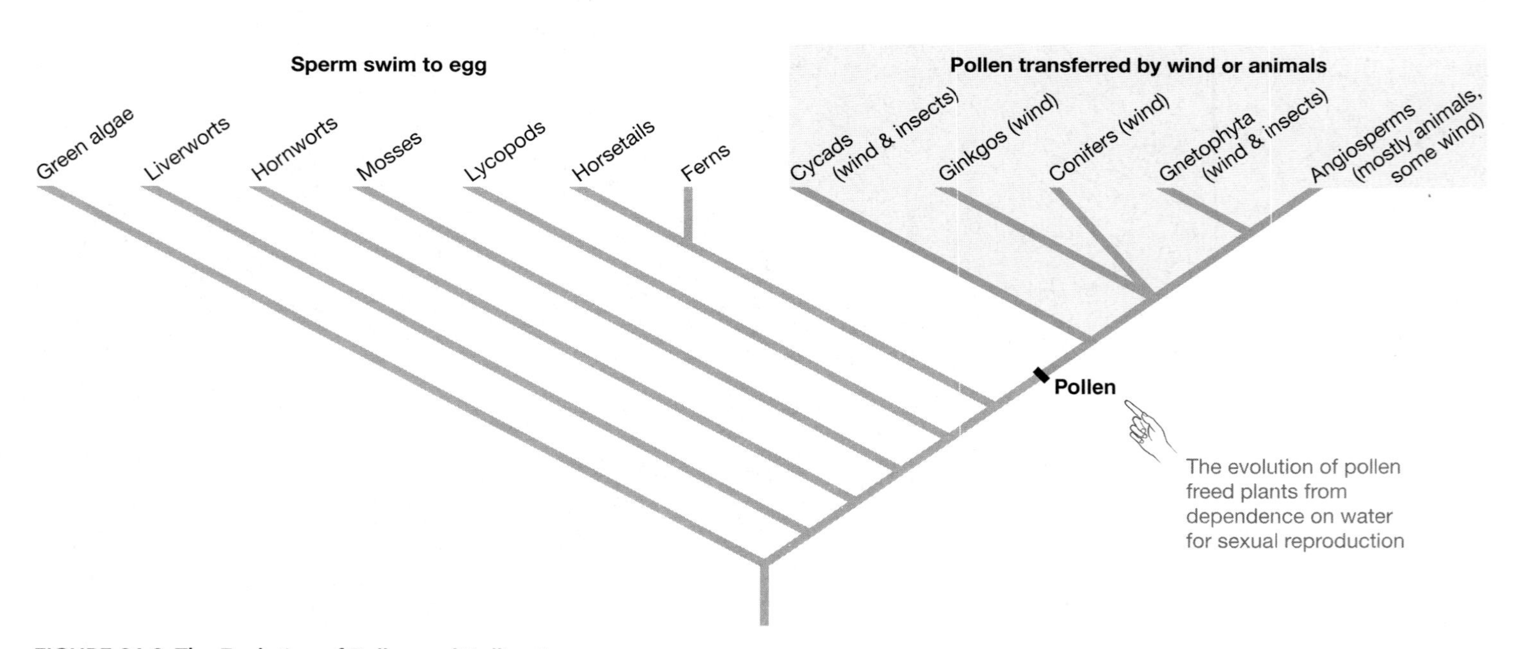

FIGURE 36.9 The Evolution of Pollen and Pollination
This phylogenetic tree shows the evolutionary relationships among the major groups of plants. Descendants of the earliest plant groups to evolve do not produce pollen, and their sperm swim to the egg. The later-evolving groups produce pollen that is transferred to female reproductive structures by wind or animals.

DNA sequences from several genes found in all plants, and then used the observed similarities and differences to reconstruct which plant groups are most closely related and which are most distantly related. **Figure 36.9** shows a typical phylogenetic tree that results from this type of analysis. Note that mosses and other groups that do not form pollen were the first lineages of plants to evolve. This conclusion is supported by the order in which species appear in the fossil record. Conifers and other groups that are strictly or primarily wind pollinated evolved later, but before the angiosperms.

The first important pattern here is that the more recently evolved plant groups do not need water in order for sexual reproduction to occur. As a result, the evolution of pollen made these species much less dependent on wet habitats and paved the way for the colonization of drier environments. The second important pattern is that pollination became a much more precise process when animals began to act as pollinators. Wind-borne pollen grains blow about randomly. Consequently, the probability is small that a grain will land successfully on a flower stigma. Insect-borne pollen, in contrast, is much more likely to be successfully transferred. Insect pollination is an important adaptation because it makes sexual reproduction much more efficient.

Finally, insect pollination appears to be closely associated with the formation of new flower and insect species. As an example of why this association exists, consider the situation shown in **Figure 36.10**. Candace Galen has documented that populations of a mountain-dwelling species called the alpine skypilot have flowers with distinctive characteristics. Individuals that grow in the tundra habitats above timberline have large flowers with long stalks and a sweet smell. Alpine skypilots that grow at or below timberline have small flowers with short stalks and a smell that people describe as "skunky." These differences are interesting, because different insects pollinate the two populations. Large bumblebees pollinate the tundra flowers, while small flies pollinate the timberline individuals. Chapter 21 introduced experimental data showing that bumblebees prefer to pollinate large flowers. Because flies and bumblebees prefer to visit different types of flowers, the two skypilot populations are evolving distinctive characteristics. They are on their way to becoming distinct species.

The message here is that evolutionary changes in the size or eating habits of a pollinator affect the angiosperm populations they pollinate. In return, changes in flower size and shape affect the insects pollinating that population. Because mutation continuously introduces variation in these traits, insect and angiosperm populations frequently change, diverge, and form new species. Based on observations like these, it is no surprise that insects and angiosperms are exceptionally species-rich groups.

It is clear that pollination was a crucial innovation during plant evolution. Now let's get down to mechanics. What happens once a pollen grain is deposited on a stigma?

Self-Incompatibility Even though animals carry pollen between individuals efficiently, in perfect flowers it is likely that at least some pollen from an individual's anthers will fall on the stigmas of its own pistil. In some species this event leads to self-fertilization and inbreeding. But in others, individuals are not able to self-fertilize. Instead, pollen grains are rejected if they land on their own stigma. Species that exhibit this behavior are called **self-incompatible.**

Why does self-incompatibility exist? There are two ways to answer this question. At the evolutionary or "ultimate" level, it is advantageous for some species to avoid self-fertilization because their inbred offspring have poor fitness. (Chapter 22 presented data backing up this claim.) If inbred offspring do not survive or reproduce well, then natural selection should favor individuals with mechanisms to avoid self-fertilization.

FIGURE 36.10 Interactions Between Angiosperms and Animal Pollinators May Lead to Speciation
The alpine skypilot populations in timberline and tundra habitats are very different. Data summarized in Chapter 21 show that the differences between the skypilot populations are due to natural selection exerted by different pollinators. As a result of their interactions with pollinators, the two populations may eventually become different enough to be considered different species.

If it is advantageous to avoid inbreeding, how is it enforced at the molecular or "proximate" level? Recent research has shown that self-incompatibility is enforced by several mechanisms. One of the best-studied examples involves the products of a locus called *S* (for "self"). Research on the *S* locus has attracted a great deal of attention because it has important practical implications. If the genes for self-incompatibility could be introduced into crop species that are normally self-compatible, then plant breeders would no longer have to painstakingly remove the anthers of corn and other self-compatible crop plants to keep them from inbreeding.

June Nasrallah and co-workers have been able to characterize the *S* locus in wild cabbage plants and their domesticated relatives. The locus actually codes for three distinct proteins, called SRK, SLG, and SCR. The three genes are arrayed on the chromosome as diagrammed in **Figure 36.11a**. By sequencing the *S* locus in many different individuals, Nasrallah and colleagues have established that the locus is highly polymorphic. This means that many of the individuals in a given population have different *S* alleles. It is not uncommon, in fact, for a single population to have as many as 50 different alleles at the *S* locus. The observation was remarkable—the *S* locus is the most highly polymorphic known in plants. Is there some sort of connection between polymorphism and self-incompatibility?

To answer this question, Nasrallah and co-workers set out to determine where the *S* gene products are found. As **Figure 36.11b** shows, the SRK protein is located in the membranes of cells covering the surface of the stigma. The SLG protein is found in the walls of these cells, but *SCR* is transcribed only in anthers. Nasrallah and associates suggest that the SCR protein may be secreted by mature pollen grains.

These observations inspired a model for how the *S* system works. To interpret the hypothesis illustrated in **Figure 36.11c**, remember that a stigma cell has two copies of the *S* locus while each pollen grain has one (because pollen are haploid). Now suppose the SRK and SLG proteins on a stigma come from *S* alleles numbered 1 and 2 (out of the 50 in the population). If the SCR protein from a pollen grain is from *S* allele number 3, it must be from a different individual. In this case, the pollen would be allowed to interact with the stigma and fertilization could occur. But if the SCR protein were from *S* allele number 1 or 2, it would very likely represent pollen from the same flower. In this case, Nasrallah and co-workers propose that the SRK and SLG proteins block interactions between the stigma and pollen grain. As a result, self-fertilization and inbreeding are avoided.

Researchers are now focusing on understanding how the S proteins on stigma and pollen interact. Although progress in understanding self-incompatibility has been dramatic, much remains to be learned before the knowledge can be applied to crop systems for which self-compatibility is a problem.

Our next task, though, is to understand a successful interaction between pollen grains and the pistil. What happens when pollen is accepted at the stigma?

Pollen-Tube Growth If pollen grains are not rejected by the self-incompatibility system, they germinate on the stigma. As **Figure 36.12** shows, the male gametophyte produces a long

(a) The self-incompatibility locus ("S")

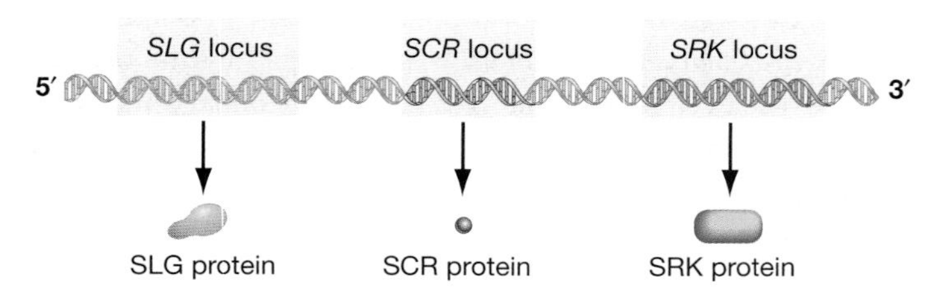

(b) Specific locations of S-proteins

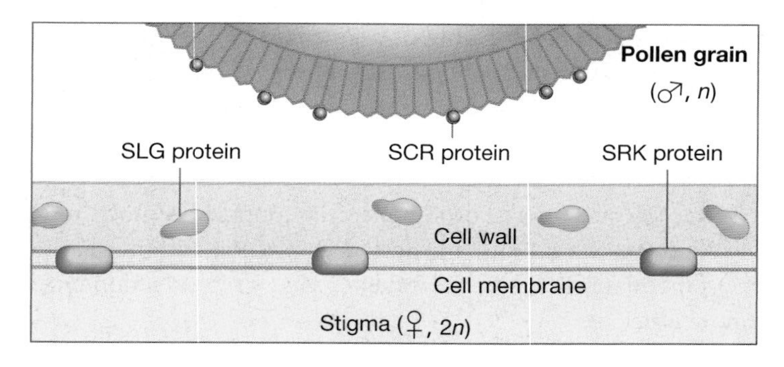

(c) How S-locus polymorphism prevents self-fertilization

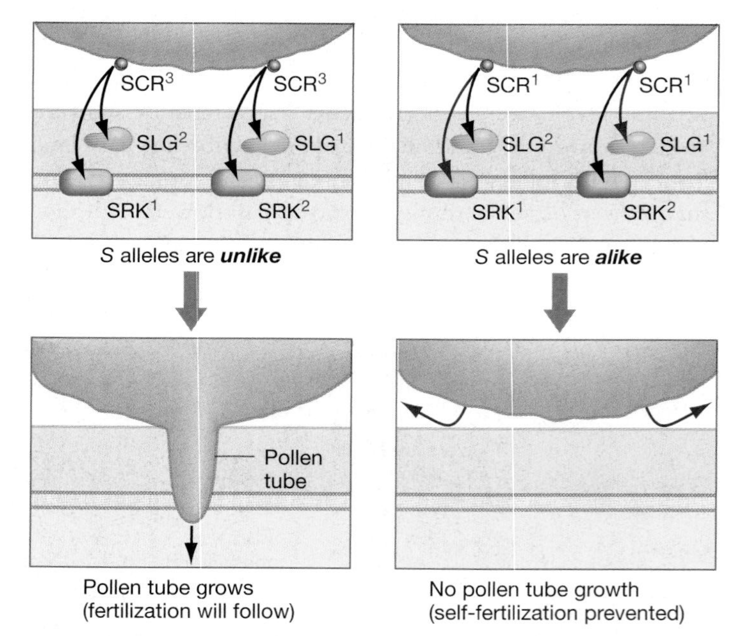

FIGURE 36.11 The Products of the *S* Locus Help Cabbage-Family Plants Avoid Inbreeding
(a) The "*S* locus" is actually three loci. **(b)** The protein products of the *S* locus are found in distinct locations. **(c)** If a pollen grain carries an S protein that is encoded by an allele *different* from the *S* loci found in the female, then the pollen grain is allowed to germinate. If the pollen carries an S protein that matches the *S* alleles found in the female, then a reaction takes place that prevents the pollen grain from germinating. As a result, self-fertilization does not occur. Researchers are now focusing on how the S proteins and stigma interact.

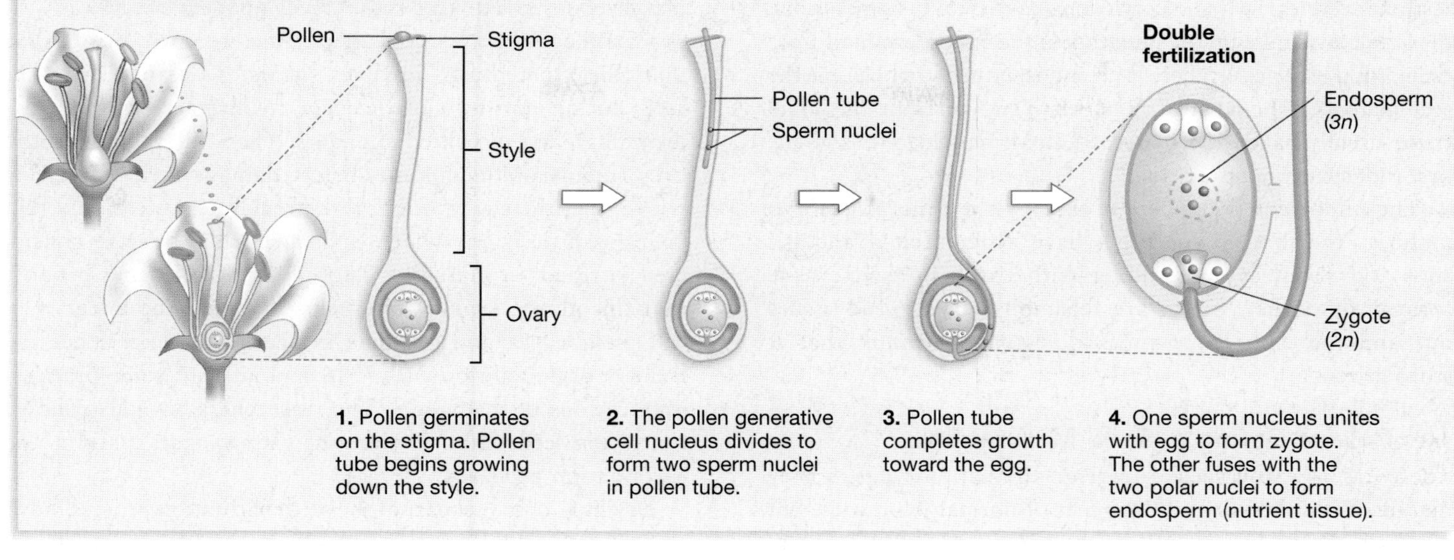

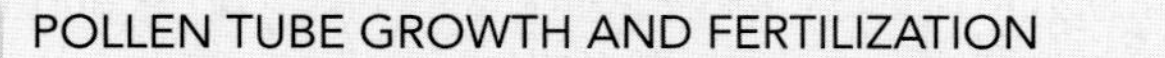

FIGURE 36.12 Pollen-Tube Growth and Fertilization
The male gametophyte germinates on the stigma and grows toward the egg. During this process one of the pollen nuclei divides and forms two sperm, making double fertilization possible. QUESTION Predict what will happen when several pollen grains are deposited on the stigma at the same time. How could you test this prediction?

projection called a pollen tube that grows down the length of the style. The generative cell travels down from the tip of the tube and divides to form two sperm nuclei, or male gametes. When the pollen tube reaches the micropyle, the sperm are discharged into the ovule. This event sets the stage for fertilization.

Fertilization

Fertilization is the fusion of sperm and egg to form a diploid zygote. In most plant groups, fertilization is straightforward—sperm and egg simply combine and a diploid nucleus is formed. In angiosperms, however, an unusual event called **double fertilization** takes place. As the final drawing in Figure 36.12 shows, one sperm nucleus unites with the egg nucleus to form the zygote. The other sperm nucleus moves through the embryo sac and fuses with the two polar nuclei to form a large triploid ($3n$) cell.

The triploid cell resulting from this second fertilization begins a series of mitotic divisions that form a tissue called **endosperm** (inside-seed). The function of endosperm is to store nutrients. Endosperm cells are loaded with starch or oils (lipids) along with proteins and other nutrients that will be needed by the growing plant after it germinates. Long before germination can take place, however, the ovule and ovary must develop into a seed and a fruit, respectively.

36.4 The Seed

Fertilization starts the development of a young sporophyte. In angiosperms and related groups, the first stage in the sporophyte's life is the maturation of the seed. As a seed matures, the embryo and endosperm develop inside the ovule and become surrounded by a tough seed coat. At the same time, the ovary around the ovule develops into a fruit that protects the seed (or seeds, if many ovules are contained within a single ovary) and aids in its dispersal away from the parent plant.

The mature seed can be thought of as an embryo with a food supply. Let's begin our analysis of how seeds work with a closer look at how the embryo develops.

CD ACTIVITY 36.2 Fruit and Seed Structure and Development

Embryogenesis

When a zygote divides for the first time, it produces two daughter cells. The bottom one, called the basal cell, divides to form a row of single cells. This structure provides a route for nutrient transfer from the parent to the developing embryo. The topmost or terminal cell is the parent of all the cells in the embryo. As the terminal cell and its progeny divide, the mass of cells illustrated in **Figure 36.13** (page 702) forms. These cells sort themselves into three groups, each of which conforms to one of the three adult tissue types introduced in Chapter 31. The exterior layer of embryonic cells is the progenitor of the adult epidermis. The cells just within the exterior give rise to the ground tissue, while a group of cells in the core of the embryo becomes the vascular tissue.

As the embryo continues to develop, the long axis of the plant begins to emerge. Figure 36.13 also shows some of the prominent structures that appear in the mature embryos of many plants. Particularly important are the **cotyledons**, or seed leaves. You know from reading Chapter 31 that one prominent group of angiosperms—called the monocotyledons or monocots—has just one seed leaf, while dicotyledons have two. In some dicots, the cotyledons take up the nutrients in the endosperm and store them. (In these species, there is no endosperm left by the time the seed matures.) The other two important structures are the initial stem, usually called the **hypocotyl** (under-cotyledon), and the first root structure or **radicle.**

The important point here is that by the time an embryo matures, the three tissue types have differentiated and the root and shoot systems, along with the first leaves, have formed. Once these events are accomplished, the seed tissues dry and the embryo becomes quiescent—meaning that it stops growing.

The Role of Drying in Seed Maturation

The seeds of many species undergo dramatic drying as they mature. Water makes up 80 percent of normal plant cells, but dried seeds contain just 5–20 percent water. In the short term, loss of water is interpreted as an adaptation that prevents seeds from germinating on the parent plant. Once they are dispersed from the parent plant, the dry condition of seeds ensures that they do not germinate until water is available in the environment and is taken up by the seed. This is logical in the context of adaptation, because water availability is crucial to the survival of germinated seedlings.

The dry state of seeds raises an important question, however. How do the cell membranes and proteins in the embryo and endosperm survive? When researchers reduce the amount of water surrounding isolated membranes and proteins to the levels observed in extremely dry seeds, some of the membranes and proteins disintegrate. Clearly, something is happening at the molecular level in seeds to keep these cell components intact.

Carl Leopold and colleagues have recently established that this "something" involves sugars. As water leaves the seed during drying, sugars begin to interact with the hydrophilic parts of cell components. These interactions stabilize proteins and membranes before damage occurs. If drying is extreme, the sugars form an extremely viscous liquid that contains little if any water. Substances like this are considered vitrified, or glass-like. Leopold and colleagues propose that this glassy, sugary state helps maintain the integrity of cell membranes and proteins in seeds that experience extremely dry conditions. (**Box 36.1** points out some practical implications of this work.) The researchers have also shown that when seeds imbibe water, the glassy sugars dissolve and germination proceeds.

Drying is only one part of the seed maturation process, however. Equally important is the development of tissues surrounding the seed itself. In many cases, these tissues are required for the seed to be dispersed from the parent plant.

Fruit Development and Seed Dispersal

After fertilization occurs in angiosperms, the cells that make up the ovary begin developing into a structure called the **pericarp.** This enclosure protects the seeds. The mature structure, with ripened seeds inside, is called a **fruit.** In some cases, such as pears and apples, much of what we call the fruit actually originates from structures at the base of the flower. Strictly speaking, then, only the structures in the core of a pear or apple are derived from the pistil and would be considered the fruit.

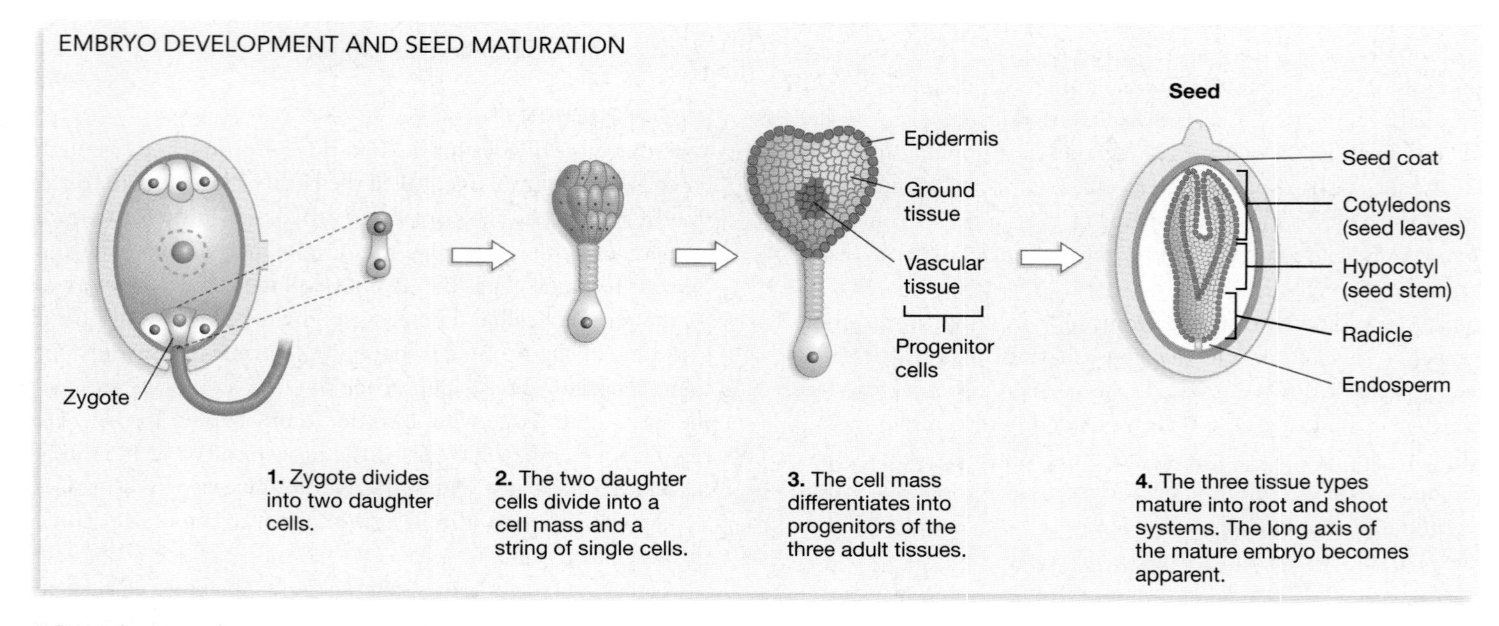

FIGURE 36.13 Embryo Development and Seed Maturation
These sketches summarize the development of an *Arabidopsis* embryo and the structure of a seed. The embryo inside a seed has the beginnings of a root and shoot system and its first leaves, or cotyledons. Inside the embryo, the epidermal, ground, and vascular tissues are organized in distinct layers.

Fruits can be dry, as in nuts, or fleshy, as in cherries and peaches. In addition to providing protection from mechanical damage and seed predators, fruits frequently have structures that aid dispersal. Dispersal is important to the fitness of the young sporophyte. This is especially true in long-lived species where the parent plant would compete with its offspring for light, water, and nutrients.

Most dry fruits are either dispersed by wind or simply fall to the ground. Seeds that are dispersed by wind are usually tiny; the fruits often have external structures to catch the breeze and extend the distance they travel. (The fruits of dandelions and maple trees are familiar examples.) Other dry fruits have hooks or barbs that adhere to passing animals; still others are actually dispersed by the plant via propulsion. The sandbox tree, for example, produces a seed pod that shrinks as it dries. Eventually the pod splits violently, spraying seeds in all directions (see **Figure 36.14**). The pod bursts with so much force that the plant is sometimes called the dynamite tree. The sound of a seed pod bursting resembles a pistol shot, and seeds can be scattered as much as 40 m away from the parent plant. The dwarf mistletoe fruit, in contrast, fills with sugars as it matures. Enough water follows via osmosis to make the fruit explode and shoot seeds as far as 5 m.

FIGURE 36.14 Seed Dispersal via Propulsion
These photos show a ripe seed pod from a sandbox tree, before and after it has burst open and ejected its seeds.

Animals are the most common dispersal agent for fleshy fruits. **Figure 36.15** (page 704) illustrates the steps in the development of some common categories of fleshy fruits. In these species, seed dispersal is another example of a mutualism. The plant provides a fruit rich in sugars and other nutrients; in return, the animal carries the fruit to a new location and excretes the seeds along with a supply of fertilizer.

Even after being dispersed from the parent plant, and even though water and oxygen are available, the seeds may not germinate. This condition is known as **seed dormancy.** Why does dormancy exist? What keeps seeds from germinating, even though conditions appear to be favorable?

Seed Dormancy

In 1995, J. Shen-Miller and co-workers reported a remarkable finding about the seeds of the sacred lotus plant. The seeds they studied were excavated from an old, dried-up lakebed in China. After germinating several of the seeds successfully, the researchers used radiometric-dating techniques (see Chapter 2) to estimate their age. According to their data, the oldest seed that germinated was about 1300 years old. Its parent grew during the height of the Tang dynasty, not long after the founding of Islam and just before Charlemagne ruled France.

Although this example represents an extreme, the important observation is that the seeds of some species may remain dormant for a long time. Dormancy is usually a feature of seeds from species that inhabit seasonal environments, where conditions

BOX 36.1 Pure and Applied Science

People sometimes draw a sharp distinction between pure and applied research. One characterization is that pure science is inspired solely by inquisitiveness about how the natural world works, without regard to commercial payback. Applied science is described as research that is focused on solving a specific problem in business, technology, agriculture, or medicine.

The discovery that seeds enter a glassy state shows just how blurry the line between pure and applied research can be, however. The initial financial support for research on the state of dry seeds was inspired by a practical argument. If biologists could discover why certain seeds have such a long life span, the knowledge could have practical benefits for commercial growers who want to preserve certain seeds for long periods. Yet the actual experiments that were done on dry seeds were "pure" in nature. Investigators simply wanted to know how molecules that normally exist in solution maintain their structural integrity when water is removed. Independently of the plant researchers, Felix Franks discovered that the formation of glassy sugars also helps certain insects tolerate extremely cold temperatures.

The twist to the story is that the discovery of glass-like sugars has potentially important practical applications that are completely different from those originally envisioned. If proteins can be preserved in an extremely dry state by encasing them in sugars, it may be possible to preserve certain drugs, vaccines, and dietary supplements without refrigeration. One of the first applications of this knowledge is now being tested in the United States. The idea is to sugar-coat the insulin manufactured for use by people with diabetes. Instead of injecting the polypeptide in solution, diabetics may be able to take their medicine as an easy-to-use inhalant.

What are the messages of this story? Few research programs turn out to be completely pure or completely applied, and it is next to impossible to predict the practical implications of research.

may be inhospitable for seedlings for extended periods of time. Based on this observation, dormancy is usually interpreted as an adaptation that allows seeds to remain viable until conditions improve. In contrast, dormancy is rare or non-existent in seeds that are produced by plants that inhabit wet tropical forests or other areas where conditions are suitable for germination year round.

Here we address two questions about dormancy. What molecular mechanisms are responsible for the condition? How does dormancy cease so that germination can begin?

Does Abscisic Acid Induce Dormancy? Chapter 35 introduced the hormone abscisic acid (ABA) and described its role in preventing germination. In some species, seeds that enter dormancy have a high concentration of this hormone. The seeds of desert plants, for example, have high concentrations of ABA in their seed coats. When these seeds are exposed to large amounts of water during rare or seasonal rains, the hormone literally washes out of the seed's outer tissues. Once the ABA concentration is reduced in this way, germination proceeds.

In peas and many other species, however, seeds routinely contain high levels of ABA and yet are not dormant. In *Arabidopsis*, ABA concentrations rise as seeds mature and appear to impose dormancy. ABA levels eventually fall, however, so that dormant, mature seeds contain only trace amounts of the hormone.

Faced with observations like these, researchers have concluded that there is no single, universal mechanism for initiat-

FIGURE 36.15 Types of Fruits
Fruits come in many types. The structure of a fruit depends on the number of ovules found in each pistil and whether ovaries fuse during fruit maturation.

ing and maintaining seed dormancy. In some cases, changing ABA levels control the initiation or maintenance of the dormant state. In other cases, changes in sensitivity to ABA, rather than the sheer amount of the hormone present, appear to be important. It is also likely that novel mechanisms for initiating and maintaining dormancy are still to be discovered.

How Is Dormancy Broken? The coats of some seeds are thick enough to physically prevent water and oxygen from reaching the embryo. For germination to occur, these seed coats must be disrupted, or scarified. Crop seeds that require scarification are placed in large, revolving drums along with pieces of sandpaper. The abrasion from the sandpaper scarifies the seeds. In nature, seed coats can be disrupted by a fire, by the passage of the seed through an animal's digestive tract, or by abrasion against wind- or water-driven soil particles. The basic principle is that the seed coat has to be broken for water to enter the seed.

Other types of seeds must experience particular environmental conditions in addition to exposure to water. As Chapter 35 indicated, species native to northern or alpine habitats often produce seeds that have to undergo a period of cool, wet conditions before they will germinate. No one knows how these seeds perceive cold, or what molecular mechanisms are involved in breaking dormancy when the cool, wet period ends.

In many species that produce small seeds, successful germination occurs only at or near the soil surface. This observation is logical because small seeds have few nutrient reserves in their cotyledons or endosperm. As a result, small seeded species need to germinate near the soil surface where individuals are exposed to light. As Chapter 34 showed, lettuce seeds and other small seeds must be exposed to red light before they will break dormancy and germinate. Red light is an important environmental cue because wavelengths in the red portion of the light spectrum support photosynthesis.

Finally, many of the seeds produced by species native to habitats where wildfires are frequent, such as the California chaparral and South African fynbos, have an unusual chemical requirement to break dormancy. These seeds must be exposed to fire or smoke before they will germinate. The active chemical agent that triggers the response is unknown, however. (In fact, the commercial food product "liquid smoke" induces germination in these seeds as well as actual smoke.) Again, it is logical for seeds in this habitat to germinate after fire has cleared away old vegetation and thinned shrubs.

The message here is that dormancy can be broken in response to a wide variety of environmental cues. In general, the cue that triggers germination is a reliable signal that conditions for seedling growth are favorable for a particular species in a particular environment.

Seed Germination

Even if specific environmental signals are required to break dormancy, seeds do not germinate without water. Water uptake is the first event in germination. Once the seed coat allows water penetration, water enters by a steep water-potential gradient because the seed is so dry.

Figure 36.16 shows how water uptake occurs in a typical angiosperm seed. The important point to note is that the curve describing water uptake has three distinct phases. Germination begins with a rapid influx of water. This influx is followed by an extended period in which no further water uptake occurs. The last section of the graph is marked by water intake starting up again and rising steadily.

During the first phase of water uptake, oxygen consumption and protein synthesis in the seed increase dramatically, but no new messenger RNAs are transcribed. Based on these observations, biologists have concluded that the earliest events in germination are driven by mRNAs that are stored in the seed. During the second phase, when water uptake stops, newly transcribed mRNAs appear and are translated into protein products. Mitochondria also begin to multiply. In effect, seeds take up enough water to hydrate their existing proteins and membranes, and then begin to manufacture the proteins and mitochondria needed to support growth. Water uptake resumes as growth begins. This second bout of water uptake enables cells to enlarge and the embryo to burst from the seed coat.

When the radicle emerges from the seed and begins to push into the surrounding soil, germination is complete. The next major event in the seedling's life occurs when the cotyledons commence photosynthesis. The seedling is said to be established when the young plant no longer relies on food reserves in its endosperm or cotyledons; instead, it receives all of its nourishment from its own photosynthetic products. With this, a new generation is under way.

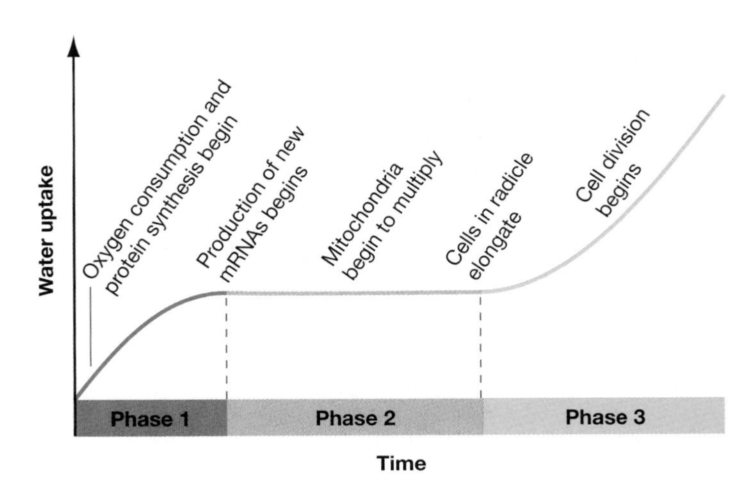

FIGURE 36.16 Water Uptake During Germination
This graph plots the rate of water uptake as a typical seed germinates. (The graph is conceptual, meaning that it represents a general pattern observed in data from many species, so the axes have no units.)

Essay Why Do Wasps Try to Copulate with Hammer Orchids?

Two-thirds of all flowering plant species rely on insects for pollination. In most cases, the relationship between plant and pollinator is mutualistic. But some flowering plants do not reward the animals that visit with nectar or pollen. Instead, they rely on trickery to accomplish pollination.

Deceit pollination is particularly common in orchids. Some biologists estimate that as many as 10,000 orchid species accomplish pollination by swindle. In most cases, the deceit involves food. The flowers of food-deceptive orchids usually have a long spur with a bulbous knob at the end. A similar-looking knob, located at the end of the same type of long spur, is found in many other flowering plant species, including numerous orchids. Typically, the bulb is filled with nectar or oil manufactured by a nearby gland. But in deceit pollinators, there is no nectary. The trick is that a visiting insect picks up the individual's pollen while probing for nectar that isn't there.

In some cases, plants rely on trickery to accomplish pollination.

An even more spectacular form of deceit pollination involves sex. Consider the hammer orchid (*Drakaea*) illustrated in **Figure 1**. This species, which is native to Australia, is pollinated by wasps in the thynnid family. A female wasp from this group is pictured in the figure as well. Female thynnid wasps have no wings and spend the early part of their life feeding on soil-dwelling animals. When they mature, they emerge from the ground, climb up the stem of a plant, and release a chemical signal called a pheromone. Male wasps that detect the pheromone follow the scent to its source, land on the female's back, and copulate.

Male wasps emerge from below ground before females do, however. At that time of year, hammer orchids are in full bloom. The color, size, and shape of a hammer orchid flower bear a striking resemblance to a female thynnid wasp. The orchid flowers also produce a chemical that is extremely similar to the female wasp's attractant. Male wasps are attracted by the signal and may land on the flower. Just above the landing site, the flower presents a curving column that contains its anthers and stigma. As the male attempts to copulate, the flower bends until the male is tipped upside down and his back contacts the reproductive structures. When the wasp flies off, he often does so carrying hammer orchid pollen. If another orchid succeeds in enticing a visit from the same male, it is very likely that the pollen will be deposited on a receptive stigma.

Although hammer orchids save huge amounts of energy by not producing nectar, deceit pollination has a downside. Most insects are extremely good at associative learning. If they are fooled once, they are unlikely to be fooled again. But for deceit pollination to work, the same individual insect must be duped twice. In the case of the hammer orchid, the success of sexual deceit relies on the ability of the flower to closely mimic extremely specific chemical and visual cues from female wasps. In the case of food-deceptive orchids, individuals frequently grow in areas where they are surrounded by high concentrations of similar-looking flowers from species that *do* reward their pollinators with nectar.

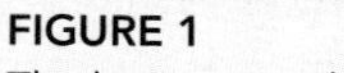

FIGURE 1
The hammer orchid on the left looks and smells like a female thynnid wasp (right).

Chapter Review

Summary

Most plants reproduce asexually as well as sexually. Asexual reproduction is based on mitosis and results in offspring whose genetic makeup is identical to the parent. Sexual reproduction is based on meiosis and results in offspring that are genetically unlike their parent. Perhaps the most important feature of reproduction in plants, though, is that they have a multicellular, haploid stage called a gametophyte and a multicellular, diploid form called a sporophyte. This phenomenon is termed alternation of generations. The gametophyte produces gametes via mitosis, while the sporophyte produces spores via meiosis. An important difference among plant groups is the degree to which the gametophyte and sporophyte depend on each other.

The angiosperms, or flowering plants, initiate flowering and sexual reproduction in response to external cues from the environment as well as internal cues based on the individual's condition. In *Arabidopsis*, a gene that controls flowering has a promoter with two binding sites: one for a transcription factor that is triggered by an external cue (day length), and one for a transcription activator that is activated by an internal cue (the hormone GA). Frequently, the particular cues that a species responds to allow individuals to flower when environmental conditions are favorable.

Flowers are made up of sepals, petals, stamens, and one or more pistils. The pistil contains the ovule. The ovule is where meiosis takes place and the female gametophyte, or embryo sac, develops. The stamens contain anthers. The anthers are where meiosis takes place and the male gametophytes, or pollen grains, develop. Pollination occurs when the male gametophyte is carried to the female gametophyte.

In many angiosperms, proteins on the surfaces of the pollen grain and stigma interact in a way that reduces the chances of self-fertilization. If a pollen grain is allowed to germinate on the stigma, it sends a long pollen tube down the style. Two sperm nuclei are produced by mitosis, and a double fertilization takes place. One sperm fuses with the egg to form a zygote, while the other fuses with two polar nuclei in the embryo sac to produce a triploid endosperm.

The formation of epidermal, ground, and vascular tissue layers is one of the earliest events in the formation of an angiosperm embryo. This event is followed by the formation of the earliest root and shoot structures. As the embryo develops, endosperm cells divide to form a nutrient-rich tissue, cells along the outside of the embryo sac form a tough seed coat, and the ovary develops into a fruit. In many cases, the mature fruit contains structures that aid dispersal of the mature seed via wind, water, propulsion, or animals.

Many seeds do not germinate right away, but experience a period of dormancy. In some species, dormancy is initiated or maintained by the hormone ABA. A wide variety of conditions, ranging from scarification to exposure to red light, may break seed dormancy. In many cases, the event that triggers germination ensures that the seed germinates when environmental conditions are favorable. Germination begins when the seed takes up water, activates existing enzymes, and translates mRNAs already present in the seed. It ends when the radicle breaks the seed coat and begins to penetrate the soil.

Questions

Content Review

1. What does a pollen grain contain when it matures on an anther?
   a. a sperm
   b. two sperm
   c. two sperm and an additional cell
   d. a vegetative cell and a generative cell

2. What happens when double fertilization occurs?
   a. Two zygotes are formed, but only one survives.
   b. Two sperm fertilize the egg, forming a triploid zygote.
   c. One sperm fertilizes the egg, while another sperm fuses with the two polar nuclei.
   d. One sperm fertilizes the egg, while two other sperm fuse with a polar nucleus.

3. In angiosperms, what is the fundamental difference between a male gamete and a female gamete?
   a. A male gamete is small and contributes few nutrients to the offspring. A female gamete is relatively large and contributes many nutrients to the offspring.
   b. A male gamete is relatively large and contributes many nutrients to the offspring. A female gamete is small and contributes few nutrients to the offspring.
   c. Male gametes swim; female gametes do not.
   d. Male gametes are produced by the sporophyte; female gametes are produced by the gametophyte.

4. What happens during embryogenesis? (Choose all that apply.)
   a. The root and shoot systems form.
   b. The long axis of the plant is established.
   c. The first leaves, or cotyledons, form.
   d. Distinct groups of cells form that will become epidermal, ground, and vascular tissue.

5. What is the difference between an ovary and an ovule? (Choose all that apply.)
   a. An ovary encloses one or more ovules.
   b. An ovule encloses one or more ovaries.
   c. An ovule gives rise to the embryo sac; an ovary gives rise to the fruit.
   d. An ovule gives rise to the fruit; an ovary gives rise to the embryo sac.

6. What happens when outcrossing occurs?
   a. Inbred offspring are produced.
   b. The proteins produced by *S* loci block pollen grains from interacting with the stigma.
   c. Gametes from different individuals fuse to form a zygote.
   d. Gametes from the same individual fuse.

Conceptual Review

1. Draw a general version of an angiosperm life cycle. Label the sporophyte and gametophyte stages. Identify ploidy (n vs. $2n$) of each structure in the drawing, and indicate which cells and structures are produced by mitosis versus meiosis. How does the relationship between gametophyte and sporophyte phases vary among plant groups?
2. What is a flower? Draw a generalized flower. Indicate the function of each part during reproduction. Provide detailed diagrams of the male and female reproductive structures. Then diagram flowers that are pollinated by wind, by hummingbirds, and by a wide variety of insects. How is each flower part modified from the generalized flower you first drew, and why?
3. Why is it significant that *S* loci are highly polymorphic? In species that are self-incompatible, why is self-fertilization detrimental?
4. Why is it significant that the promoter for the *LFY* locus in *Arabidopsis* contains sequences where two different transcription activators bind?

Applying Ideas

1. Suppose you discovered an angiosperm that was new to science. The population grows on an island near the equator. The island is dry 10 months of the year but experiences a two-month period of frequent rains. Predict what cues trigger flowering and germination in the newly discovered species. Design experiments to test your predictions.
2. Find a flower and dissect it. Identify as many of the components labeled in Figures 36.6, 36.7, and 36.8 as you can. Based on the flower's size, shape, and color, predict how it is pollinated. How would you test your prediction?
3. The essay at the end of this chapter points out that some flowering plants "cheat" their pollinators. Likewise, certain pollinators cheat plants by removing nectar from flowers without picking up pollen. (In some cases, they do so by chewing through the petals that hold the store of nectar.) Speculate on the types of mutations that might modify insect behavior and/or plant structure in a way that limits cheating and enforces mutualism.
4. Pollinators frequently deposit pollen from more than one plant on a stigma. When they do, pollen grains from different males compete to fertilize the egg. Think of a question about this phenomenon, which is called pollen competition. Design an experiment that would help provide an answer to this question.
5. Find a seed and dissect it. Identify as many of the structures labeled in Figure 36.13 as possible. Based on the structure of the fruit, predict how this seed is dispersed. How would you test this prediction? Design a study that would estimate the average distance that this type of seed is dispersed from the parent plant.

CD-ROM and Web Connection

CD Activity 36.1: Reproduction in Flowering Plants *(animation)*
(Estimated time for completion = 15 min)
Why are flowers such a prominent and important structure in angiosperm plants?

CD Activity 36.2: Fruit and Seed Structure and Development *(tutorial)*
(Estimated time for completion = 10 min)
How are anatomy and development of flowers, fruits, and seeds intertwined?

At your **Companion Website** (http://www.prenhall.com/freeman/biology), you will find self-grading exams and links to the following research tools, online resources, and activities:

Crop Plant Pollination
This virtual book written by the US Department of Agriculture details a number of critical steps in the pollination process.

Seed Germination by Wild Fire Smoke
This article discusses how plants have adapted seeds that are not only fire resistant, but these seeds are actually stimulated to germinate after a fire.

Origin of Flowering Plants
This article discusses ideas on the evolutionary origin of flowering plants.

Additional Reading

Attenborough, D. 1995. *The Private Life of Plants* (Princeton: Princeton University Press), Chapter 3.

Crepet, W. 1999. Early bloomers. *Natural History* 108 (May): 40-41. Examines how certain species of flowers are able to bloom very early in spring, when temperatures are cool.

Small, M. 1999. Floral arrangements. *Natural History* 108 (May): 46-47. Notes on the shape of flowers.

Temeles, E. J., and P. W. Ewald. 1999. Fitting the bill? *Natural History* 108 (May): 52–55. A look at correlations between the size and shape of flowers and hummingbird beaks.

Plant Defense Systems

37

To protect crops from diseases and herbivores, the United Nations Food and Agricultural Organization estimates that the world's farmers apply over 1.5 million metric tons of the active ingredients found in herbicides, fungicides, and insecticides to their fields each year. When these agents fail, the results can be catastrophic. In 1979 and 1980, a bacterial disease wiped out 60 percent of the rice crop in numerous regions of India; in 1980, a fungus killed the entire wheat crop in many parts of Kazakhstan.

Diseases and herbivores constantly threaten crop plants with annihilation. Yet the world is green. This trivial observation is actually interesting. Most plant tissues are *not* eaten by herbivores or destroyed by bacteria, fungi, or parasitic nematodes (roundworms). Why? Plants cannot run away from these nemeses; instead, they must stand and fight. Wild plant species do this fighting on their own—they get no assistance from the chemical agents applied by humans.

Understanding how plants defend against disease and herbivory is among the most active research areas in all of plant biology. The experiments we'll review are motivated not only by the excitement of understanding the fundamental biological questions involved, but also by concern about the future of agriculture. Many crops lack desirable disease-fighting traits because such traits have been bred out in an effort to boost productivity. In many cases, massive applications of pesticides are required to make up for the loss of these natural defense systems.

How do the molecules involved in plant defense work? Could alleles that are important for defense be safely introduced into crop varieties that are particularly productive and nutritious? To begin answering these questions, let's

Diseases and insects cause billions of dollars of crop damage each year. What keeps parasites and herbivores from eating *everything*?

take a look at defense systems found on the exteriors of plants, and then look at the defenses found inside plants.

37.1 Barriers to Entry

The epidermis of a plant is analogous to the skin of an animal. Because it provides a barrier to entry by disease-causing organisms, or **pathogens**, it serves as an individual's first line of defense. Bacteria, viruses, and fungi must break through the epidermal barrier before they can infect cells, reproduce, and cause disease. Thorns or other structures on the epidermis of a plant are important for thwarting animals that eat plant tissues, or **herbivores** (plant-eaters).

Here we consider two questions about the role of the epidermis in preventing disease and damage. How does the plant epidermis resist entry by pathogens and herbivores? How do pathogens and herbivores overcome these defenses?

Cuticle and Cutinase

In stems and leaves, the outside surface of epidermal cells is covered with a substance called cuticle (**Figure 37.1**). **Cuticle** is made up of a matrix of cross-linked lipid molecules impregnated with the extremely long-chained lipids called waxes. Both elements are hydrophobic and function in limiting water loss from the aerial parts of the plant. The waxy sheet of cuticle also functions in defense, forming a physical barrier that resists penetration by virus particles, bacterial cells, and the spores or growing filaments of fungi. The epidermal cells of roots lack a waxy coating, but have a tough lipid matrix similar to that found in cuticle.

Pathogens have several ways of circumventing this barrier. Wounds from herbivores and mechanical damage expose cells to airborne fungal spores, virus particles, and bacterial cells. Bacteria and fungi can also enter stomata when they are open; viruses can be injected directly into plant tissues by aphids, leafhoppers, and other insects that pierce the plant epidermis with their mouthparts to suck phloem sap.

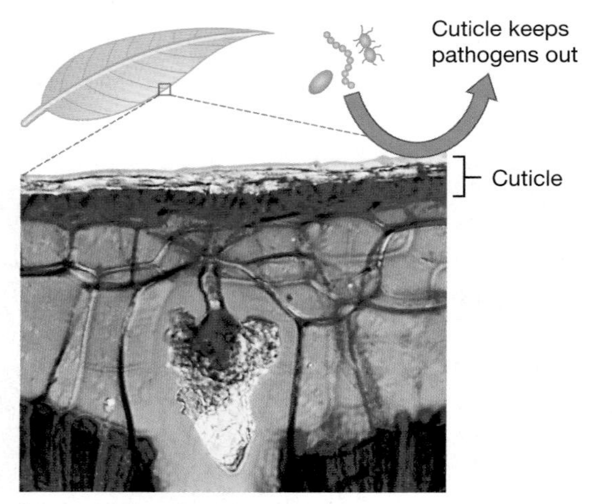

FIGURE 37.1 Cuticle Coats Epidermal Cells
Cuticle forms a barrier that keeps water in and pathogens out.

Work in Pappachan Kolattukudy's laboratory established that certain fungi have an equally direct way of entering plant tissues: They use enzymes to disrupt the cuticle. Specifically, many pathogenic fungi produce an enzyme called cutinase. Cutinase cleaves the lipid cutin, which forms the cross-linked matrix in cuticle. In this way, fungi hack out a hole in cuticle to expose the cells within.

To appreciate the importance of this enzyme, consider work carried out by Linda Rogers and her colleagues. They were able to isolate individuals of the fungus *Fusarium solani* that contain many copies of the cutinase gene, just one copy, or no functional copies. When Rogers and co-workers infected pea plants with spores from each strain, they obtained the data in **Figure 37.2**. The graphs demonstrate a strong correlation between the number of copies of the cutinase gene found in a particular fungal strain and its **virulence**, or its ability to cause disease. In this case, the ability to break through a plant's first line of defense determines the difference between virulent and benign strains of a pathogen. Strains with more copies of the gene for cutinase are more virulent.

Weapons

The spines, thorns, and prickles produced by some plants can impede insect and mammal herbivores, just as the cuticle layer helps to thwart viruses, bacteria, and other microscopic pathogens. In some cases, these structures have other functions as well. As Chapter 32 pointed out, the hair-like struc-

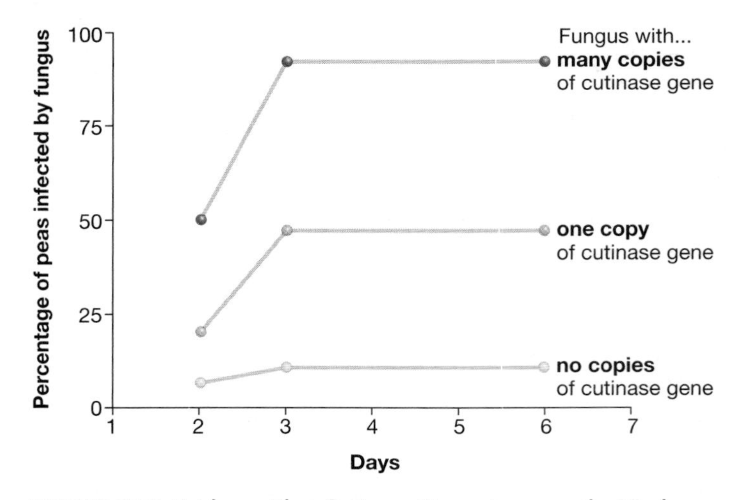

FIGURE 37.2 Evidence That Cutinase Genes Increase the Virulence of Fungi
Cutinase is an enzyme that digests plant cuticle. These data indicate a strong correlation between the number of cutinase genes in *Fusarium* strains and their ability to infect peas and cause disease.
EXERCISE These data show what happened when the researchers applied a high dosage of spores. They repeated the experiment with medium and low doses of spores from each fungal strain. Add lines to the graph predicting the results from medium and low doses.

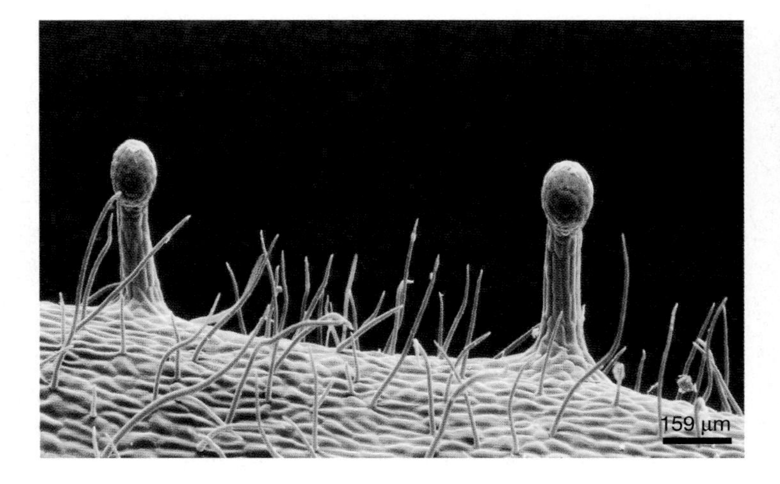

FIGURE 37.3 Structures That Protect Plants
Left: A close-up of a sepal that protects a rose bud. The large bulbous structures projecting from the surface contain chemicals that deter insects and other small herbivores from attacking the young flower. Dozens of simple hairs are visible as well. Right: In adult roses, stems are protected from large herbivores by prickles. QUESTION Why are the leaves of some plants "fuzzy?"

tures called trichomes help desert-dwelling plants limit water loss from stomata.

As the photographs in **Figure 37.3** show, hairs and prickles are weapons analogous to the teeth, pincers, or stingers that animals use to defend themselves from predators. In one case, an army of mercenaries augments these plant weapons. Some *Acacia* trees native to East Africa have large, bulbous spines that house ant colonies (**Figure 37.4a**). When a browsing mammal begins to eat leaves or twigs from the tree, the shaking motion stimulates the ants to attack the animal and deliver painful bites. The relationship between the acacias and ants appears to be mutually beneficial, with the trees providing a safe nesting site and the ants providing defense.

The ferocity of the ants made P. G. Willmer and G. N. Stone wonder how acacias are able to attract pollinators to their flowers. How could a bee spend enough time on an acacia flower to pick up and deliver pollen if it is quickly attacked by biting ants? To answer this question, the researchers proposed that acacia flowers might produce some sort of chemical deterrent that affects the ants but not bees. After documenting that ants spend much more time at old flowers than at newly opened flowers on the same tree, Willmer and Stone were able to formulate a more specific hypothesis. They proposed that a component of newly opened flowers, such as pollen, contains a chemical that deters ants but not bees. To test this idea, they wiped old flowers with newly opened flowers and documented how long ants stayed during each visit to the treated flowers.

The results of their experiment are shown in **Figure 37.4b**. As predicted, old flowers that had been wiped with newly opened flowers were just as unattractive to ants as newly opened flowers. Ant-guarded acacia flowers appear to produce a substance that keeps ants away during the interval when pollination occurs. The study suggests that plants are able to make powerful deterrent compounds, and that in at least some cases these insecticides are species specific.

(a) *Acacia* trees are protected from herbivores by biting ants, which live in large bulbs at the base of the thorns.

(b) Does a substance in young flowers keep the ants away from pollinators?

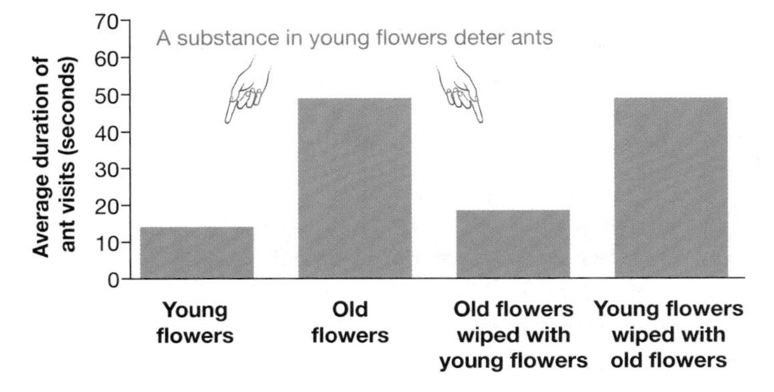

FIGURE 37.4 Ant-Guarded Acacias
(a) In certain species of *Acacia*, ants in the genus *Crematogaster* live in large bulbs at the base of spines. **(b)** These data show the average amount of time that individual ants spend on acacia flowers of various ages and experimental treatments. QUESTION What would the data in part (b) look like if young flowers did *not* produce an ant-deterrent compound? (Your answer is the null hypothesis in this experiment.)

37.2 Plant Poisons

Many plants avoid being eaten by lacing their tissues with poisons. Some of these chemicals are familiar. The flavorful oils in peppermint, lemon, basil, and sage have insect repellent properties. The pitch that oozes from pines and firs contains a molecule called pinene, which is toxic to bark beetles. The pyrethroids produced by *Chrysanthemum* plants are a common ingredient in commercial insecticides. Molecules called tannins accumulate in the cell vacuoles of many different species. Because they bind to proteins, tannins inactivate digestive enzymes in the herbivores that ingest them. As a result, the herbivores get sick. In humans, small doses of tannins cause a sharp, astringent sensation in the mouth that is prized in apples, blackberries, teas, and red wine. (Tannins also cause structural changes that make proteins less likely to disintegrate; tannins are used to tan animal skins and produce leather.)

These plant poisons are similar to cuticle and thorns in an important way: They are always present in the plant. Stated another way, they are produced constitutively—which means they do not need a specific stimulus to initiate their production. This is in sharp contrast to the defense systems we'll explore later (in sections 37.4 and 37.5), which are activated only after an infection or attack has begun.

Plants that produce poisons constitutively are similar to skunks, stinkbugs, monarch butterflies, and other animals that avoid being eaten by tasting bad or spraying their enemies with noxious chemicals. (Plants that sequester poisons also mount a rapid response to pathogens and herbivores using systems introduced later in the chapter.) Here we ask: Where do plant poisons come from, and how do these molecules act?

The Role of Secondary Metabolites

In almost every case, plant defense compounds have structures and compositions that are closely related to the molecules required for basic cell activities, such as amino acids. To explain this resemblance, biologists hypothesize that during the evolution of some plant groups, mutations occurred in the genes for enzymes involved in fundamental biosynthetic pathways. The hypothesis claims that these mutations resulted in the production of altered enzymes and the synthesis of new or "secondary" compounds. If these compounds happened to be toxic to herbivores or pathogens, they would help the mutant individual survive better and reproduce more. Over time, then, the alleles responsible for encoding the required enzymes would increase in frequency in the population. Based on this logic and their chemical similarity to important compounds, plant poisons are referred to as **secondary metabolites.**

To understand how secondary metabolites relate to amino acids and other "primary" compounds, consider recent work by Monika Frey and colleagues. This research group was interested in a molecule called DIMBOA (2,4-dihydroxy-7-methoxy-1,4-benzoxazin-3-one) that is found in corn and wheat. DIMBOA is similar in structure to the amino acid tryptophan, but is extremely effective at deterring several of the animal pests that afflict these crops.

The steps involved in tryptophan synthesis are well known. Is the synthesis of DIMBOA related to these steps? If so, how? Experiments by other researchers had provided an intriguing hint. When corn plants are fed radioactively labeled molecules that are precursors in the pathway for tryptophan synthesis, the label shows up in DIMBOA as well as in tryptophan. But if the same plants are fed radioactively labeled tryptophan, the label does not show up in DIMBOA. These observations suggest that the synthesis of this secondary metabolite is an "offshoot" of tryptophan synthesis, as illustrated in **Figure 37.5a**.

To explore this idea, Frey and co-workers analyzed corn plants that have a mutation called *Bx1*. These individuals cannot synthesize DIMBOA. After an extensive search, the researchers were able to locate the *Bx1* gene. Nearby on the chromosome, they found four related loci that they called *Bx2*, *Bx3*, *Bx4*, and *Bx5*. To determine whether the protein products of these five genes are involved in DIMBOA synthesis, they added the five BX enzymes one by one to cells that were growing in culture and analyzed the products that resulted. Each BX enzyme catalyzed a different reaction, and led to the production of a different intermediate compound in the pathway for DIMBOA synthesis. The experimental results are consistent with the pathway illustrated in **Figure 37.5b**.

These experiments provide strong support for the hypothesis that secondary metabolites evolved as offshoots of basic synthetic pathways. Now that we know where secondary metabolites come from, let's take a look at how they poison their targets.

How Do Caffeine, Nicotine, and Other Alkaloids Act as Poisons?

Morphine, cocaine, nicotine, and caffeine are plant secondary metabolites. They are part of a family of chemicals called alkaloids. Members of the alkaloid family are found in about 20 percent of all plant species, and over 12,000 different molecules have been identified in various plant species to date. Although some researchers propose that alkaloids are merely waste products of normal biosynthetic activities, most biologists hypothesize that they are actively synthesized as defense compounds.

Michael Wink and associates set up a large-scale study to determine how alkaloids work. They assessed the action of 70 different alkaloids and determined whether each of the molecules has a detrimental effect on DNA structure, DNA synthesis, protein synthesis, membrane permeability, bacterial growth, the survival of insects and worms, or the receptors located in the nerve cells and brains of mammals. The researchers' goals were to test whether alkaloids act as poisons—and if so, to determine how alkaloids affect herbivores and pathogenic bacteria, viruses, and fungi.

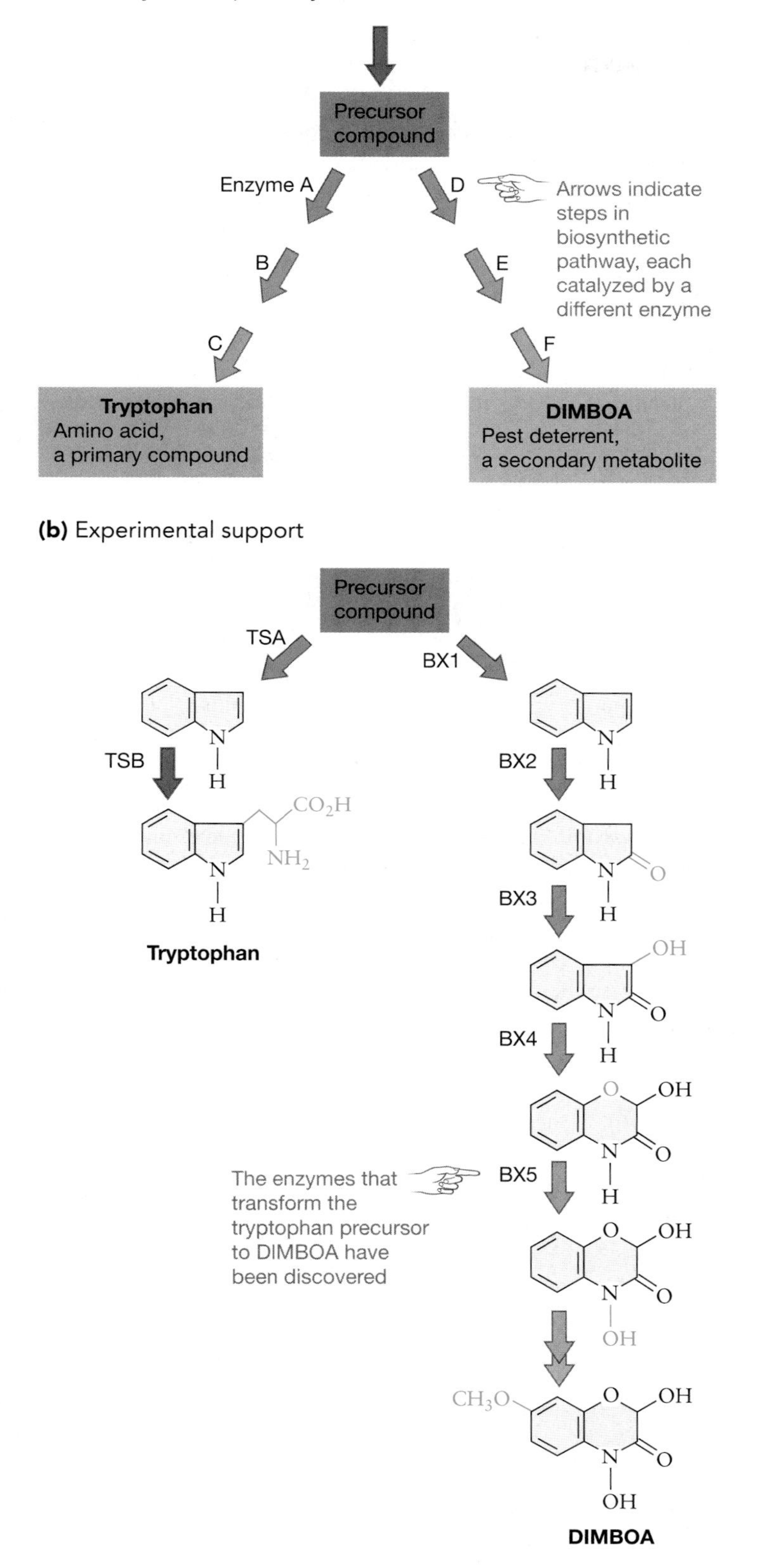

FIGURE 37.5 What Is "Secondary" About Secondary Metabolites? **(a)** A secondary metabolite is a molecule that is synthesized as an "offshoot" of a synthetic pathway for producing a fundamental, or primary, compound. Experimental data suggested that the defense compound DIMBOA is manufactured as an offshoot of tryptophan synthesis. **(b)** In corn, each of the *Bx* gene products catalyzes a different reaction involved in the synthesis of DIMBOA. **QUESTION** Corn seedlings contain over 10 times as much DIMBOA as tryptophan. In this light, are the terms *primary* and *secondary* used appropriately?

The study results showed that most of the alkaloids tested exhibited strong toxic effects, and that most alkaloids affected more than one aspect of cell biology. An alkaloid that poisons a membrane transport protein might also disrupt DNA structure. Quinine, for example, poisons the enzyme called reverse transcriptase that is found in pathogenic plant viruses; but it also has a strong inhibitory effect on protein synthesis. In interpreting these data, the researchers suggest that natural selection favors the synthesis of alkaloids that poison several target enzymes or structures in the enemy at the same time. The key idea here is that pathogens and herbivores are unlikely to evolve resistance to poisons that affect a large number of important enzymes at the same time. Recall that in Chapter 21, we explored the evolution of resistance in detail. Here, it becomes clear that plant secondary compounds are extremely potent defensive weapons.

37.3 The Cost of Defense

If secondary compounds are so effective, why don't all plants produce a lot of them all the time? To address this question, researchers rely on a fundamental observation: Every organism has a finite amount of energy. To understand why this simple fact is important, consider a thought experiment on the weedy mustard plant *Arabidopsis thaliana*. To begin, suppose that every individual in a population manufactures an average of 100 million ATP molecules in the course of its two-month lifetime. Further, suppose that the individuals in the population vary in how they expend this ATP, and that this variation is based on genetic variation among individuals. For example, some individuals might spend a large percentage of their energy from ATP making defensive compounds and a relatively small percentage making seeds. Others might make few secondary metabolites and invest most of their energy from ATP in reproduction instead. The question is, which type of individual will survive and reproduce best?

The short answer to this question is "it depends." If pathogens and herbivores are abundant, then most of the poorly defended plants will die. In this case, well-defended plants produce more offspring than lightly defended plants (**Figure 37.6a**, page 714). As a result, alleles that lead to increased production of secondary metabolites will increase in frequency in the population. But if pathogens and herbivores are rare, then poorly defended plants will produce the most offspring (**Figure 37.6b**). In this case, alleles that lead to high production of defense compounds will decrease in frequency.

To summarize these ideas, biologists say that trade-offs occur in the way that individuals allocate their resources. The concept of trade-offs suggests an answer to the question of why all individuals don't produce abundant compounds for their own defense all the time: Manufacturing poisons is energetically expensive. Producing these molecules means that less energy is available for growth and reproduction.

Even though these theoretical arguments are logical, it is essential that they be tested rigorously. As an example of experimental studies on the costs of defense, consider work performed by Rodney Mauricio and Mark Rauscher on *Arabidopsis thaliana*. This species has two major traits that help in its defense against herbivores. Leaves are covered with sharp, hair-like structures called trichomes (look back at Figure 37.3). In addition, both leaves and seeds contain unpalatable molecules called glucosinolates. To explore how these traits affect the ability of *A. thaliana* to survive and reproduce, Mauricio and Rauscher collected seeds from a large number of individuals growing in a wild population, planted the seeds out in garden plots, and divided each of the plots into two treatments. The researchers regularly sprayed half of the plots with an array of insecticides and fungicides; they did not spray the other half. When the plants were mature, Mauricio and Rauscher collected all of the seeds that had been produced and evaluated the density of trichomes and the concentration of glucosinolates in each individual.

What did their data show?

- Seeds collected from the same parent produced mature individuals with similar levels of trichomes and glucosinolates. Stated another way, related individuals had similar defense traits. This result supports the hypothesis that variation among individuals in these defense characters is at least partly based on variation in their genetic makeup.
- In the unsprayed plot where herbivores were present, individuals with high concentrations of glucosinolates produced the most offspring. This result is consistent with the prediction that well-defended individuals do better when pest pressure is high.
- In the sprayed plot with no herbivores, individuals that produced very few hairs and defense compounds produced the most seeds. This result supports the hypothesis that defense is energetically expensive. Stated another way, the ATP and nutrients devoted to synthesis of defensive compounds and structures reduces the number of resources available for seed production.

To summarize, the experiment confirmed an important series of predictions about the nature of trade-offs between defense and reproduction. The data also produced a puzzle, however. In both the sprayed and unsprayed treatments, the plants with the fewest trichomes produced the most seeds. Even though trichomes have been shown to be effective against many types of herbivores in many plant species, they appeared to have no effect in the *Arabidopsis* population that Mauricio and Rauscher studied. If this pattern continued, a population of hairless *Arabidopsis* would evolve. Alternatively, it is possible that trichomes protect this species against herbivores that did not happen to be present in the year of the study. Further work is needed to distinguish between these two hypotheses.

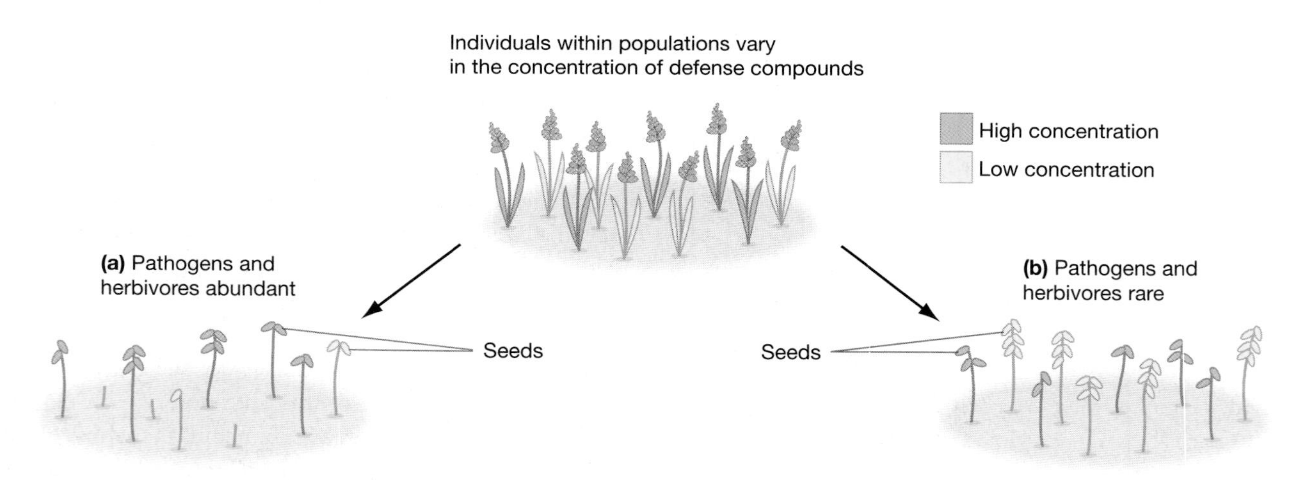

FIGURE 37.6 When Do Heavily Defended Individuals Thrive Best?
Under what conditions will individuals thrive if they invest a large number of resources in defensive compounds? The diagram illustrates a thought experiment to answer this question. Part (a) shows the results in environments where plants are attacked frequently; part (b) shows the results in environments where plants are attacked infrequently. **EXERCISE** Make a two-row by two-column table. Label the rows "high defense" and "low defense"; label the columns "high pest" and "low pest." Fill in each box in the table by counting the number of seeds produced by the individuals in this thought experiment. Then answer the question in the figure title.

37.4 Responding to Pathogens

Manufacturing defensive compounds and structures requires large expenditures of ATP and limits the ability of individuals to grow and reproduce. In light of these findings, it is not surprising that most plants have systems for responding to pathogens and herbivores only *after* an infection or attack has begun.

Response systems are currently the focus of intensive research because of their importance for agriculture. The production of constitutive poisons has been bred out of most crop plants, because toxins taste bad and lower productivity. To fight off diseases and herbivores, then, some crop plants depend solely on their response systems and the application of pesticides. Researchers hope that by gaining a detailed understanding of how response systems work, they will be able to breed or genetically engineer crops that defend themselves more efficiently and are less dependent on pesticides.

Section 37.5 explores how plants respond to attacks by insect herbivores; here, we investigate how plants respond to infections by disease-causing viruses, bacteria, fungi, and nematodes. The response systems triggered by these parasites are analogous to the mammalian immune system introduced in Chapter 46. The responses come in two waves: a rapid sequence of events triggered at the point of infection, and a general or systemic reaction occurring throughout the body.

The Hypersensitive Response

When a virus, bacterium, fungus, or nematode gets inside a plant and begins to grow, the infected cells respond by dying. The rapid and localized death of one or a few infected cells is called the **hypersensitive response (HR)**. If the HR is successful, the pathogen is starved as host cells commit suicide.

In several respects, the hypersensitive response in plants is similar to the cell-mediated immune response in mammals, which leads to the death of infected cells. The HR is also extremely effective. Plants that mount a hypersensitive response rarely succumb to disease. How does this response get started, and how is it sustained?

What Triggers the Hypersensitive Response? The Gene-for-Gene Hypothesis In the early decades of the twentieth century, crop breeders established that plants have disease resistance genes that are inherited according to Mendel's rules. These loci came to be known as resistance (*R*) genes; many are responsible for triggering the HR. The same researchers also established that the fungi found to cause disease in wheat, flax, barley, and other crops have alleles that cause them to be virulent or avirulent (not virulent) on certain strains or varieties of these crops. The loci associated with virulence or avirulence came to be known as avirulence (*avr*) genes.

In 1956, H. H. Flor published data demonstrating a one-to-one correspondence between the resistance alleles found in host plants and the avirulence alleles found in pathogens. Stated another way, Flor showed that certain *R* alleles and certain *avr* alleles match. If an *R* allele in the host matches the *avr* allele in the pathogen, it means that their protein products must also match in some way. In response to the matching event, an HR occurs. If the *R* allele in a host and the *avr* alleles in a pathogen do not match, then no HR occurs and the plant succumbs to disease.

The idea that *R* and *avr* gene products interact in a specific way came to be called the gene-for-gene hypothesis. What molecular mechanism could be responsible for this pattern? Researchers following up on Flor's work suggested that the HR begins when proteins produced by host plants bind to proteins or other molecules produced by the pathogen. The hypothesis was that *R* genes produce receptors and *avr* genes produce ligands—meaning molecules that bind to receptors. **Figure 37.7** provides a general overview of this model.

The first breakthrough in confirming the gene-for-gene hypothesis occurred when researchers from a variety of laboratories around the world were able to clone and sequence a series of *R* genes from crop plants and *avr* loci from bacterial and

CD ACTIVITY 37.1 Plant Defenses

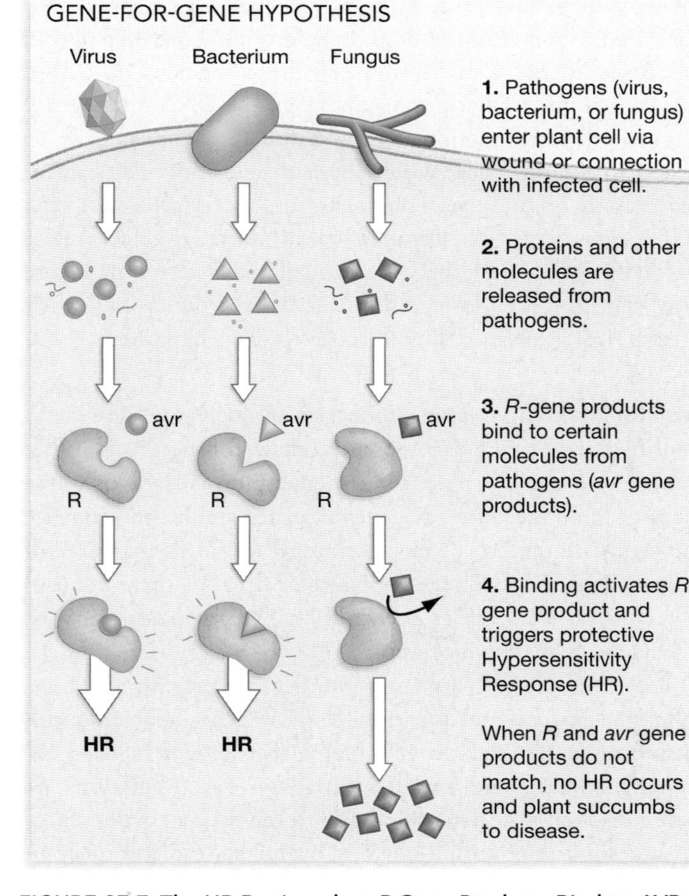

FIGURE 37.7 The HR Begins when *R* Gene Products Bind to *AVR* Gene Products

The gene-for-gene hypothesis predicts that there is a physical interaction between specific *R* and *avr* gene products, and that this interaction initiates the hypersensitive response, which protects plants from pathogens.

FIGURE 37.8 Evidence That *R* and *AVR* Gene Products Bind to One Another

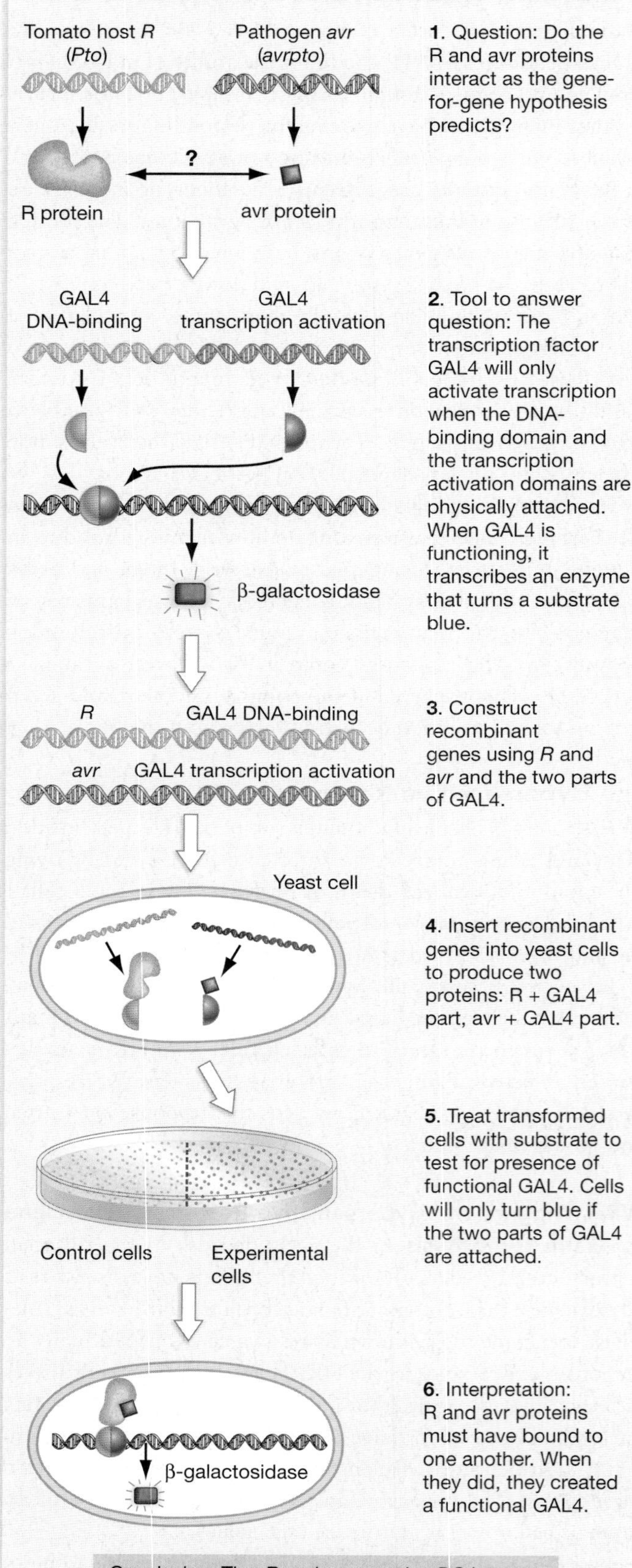

fungal pathogens. These results confirmed that *R* and *avr* genes exist and code for products that could interact at the start of an infection.

A second major advance, published in 1996, confirmed the gene-for-gene hypothesis by showing that the *R* and *avr* products actually do bind to one another. Steven Scofield and co-workers demonstrated this interaction with an *R* gene in tomato called *Pto* and an *avr* gene from *Pseudomonas* bacteria called *avrPto*. **Figure 37.8** diagrams the researchers' experimental strategy. Note that the approach relies on a transcription factor from yeast called GAL4, which was introduced in Chapter 15. In yeast cells, GAL4 activates the transcription of the gene for the β-galactosidase enzyme. This enzyme, in turn, can react with a substrate molecule to produce a bright blue product. If the substrate is added to yeast cells in which GAL4 is active, the cells turn bright blue.

To confirm that *Pto* and *avrPto* actually do bind to one another, the researchers fused the two genes to two different segments of the GAL4 gene. As step 3 in Figure 37.8 shows, *Pto* was fused to the DNA-binding domain of GAL4, while *avrPto* was fused to the transcription activation segment of GAL4. GAL4 activates transcription only when the DNA-binding domain and the transcription activation domains are physically attached to one another. When Scofield and co-workers introduced both of the gene constructs into yeast cells that lack GAL4, a complete GAL4 protein was produced, β-galactosidase was synthesized, and the cells turned blue (step 6 in Figure 37.8). This result confirmed that *R* and *avr* products physically interact, just as predicted by the gene-for-gene hypothesis.

Resistance Loci A large number of *R* loci have now been identified in *Arabidopsis*, tomato, flax, tobacco, and other plants. Two general patterns are emerging as data on these genes accumulate. First, *R* genes that are similar in sequence and structure tend to be clustered together on the same chromosome. Second, within a population of plants, there are usually many different alleles at each *R* locus. (Stated another way, *R* loci are highly polymorphic.)

These observations are important because they provide hints about the history and function of these genes. For example, clusters of similar loci, or what biologists call **gene families**, are thought to originate through errors in recombination. As **Figure 37.9** shows, **gene duplication** events occur when chromosomes misalign during crossing over. The result of this mutation is a chromosome with an extra copy of a gene. Because the organism with this chromosome already has a functioning copy of the original gene, mutations occurring in the extra copy of the gene do not damage the individual. Instead, the new copy may acquire mutations that make a new function possible. In the case of *R* loci, the hypothesis is that mutations in the new copies gave individuals the ability to recognize and respond to novel *avr* products and thus new types of pathogens.

Why is it significant that many different alleles exist at each *R* locus? The hypothesis here is that different alleles allow plants to recognize different proteins from the same pathogen. Plants are diploid, so they have two copies of each *R* gene. If many different alleles exist in a population, then each individual is likely to have two different alleles at each locus. The different alleles allow the host to recognize different *avr* products. This is important because new *avr* products constantly arise in pathogen populations via mutation.

In combination, then, plants that have different alleles at each of many *R* loci should be able to recognize and respond to a wide variety of disease-causing agents. The variability observed in *R* genes supports the hypothesis that plants rely on gene-for-gene interactions to recognize and thwart a diverse variety of invaders. Stated another way, if gene-for-gene interactions underlie some important resistance responses, and if an individual plant has many different *R* genes, then it can recognize and respond to many different pathogen genes and thus many different pathogens.

Before moving on to consider other aspects of the HR, it is interesting to note that mammals have two different sets of genes that are responsible for recognizing pathogens and triggering a

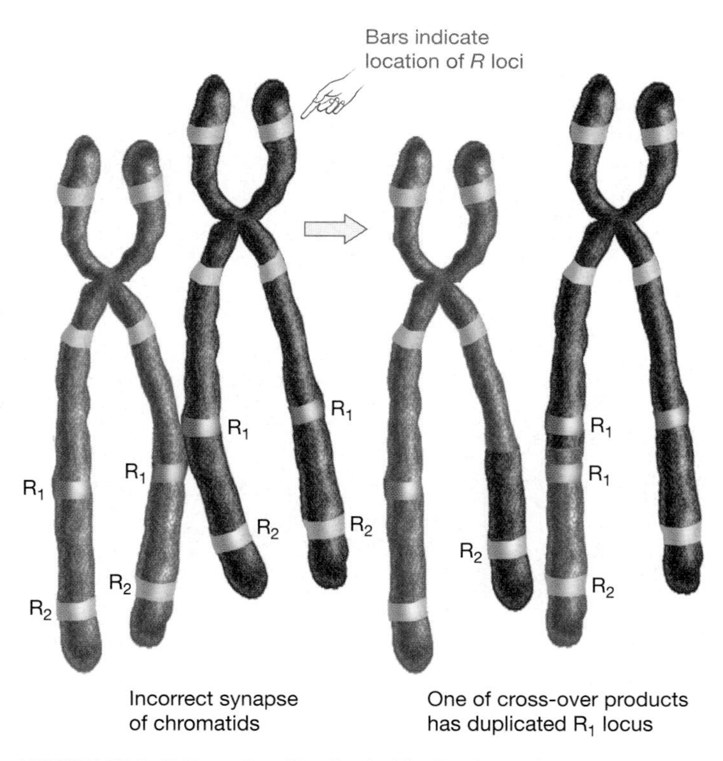

FIGURE 37.9 *R* Gene Families Probably Originated in Gene Duplication Events
When biologists find a series of closely related genes clustered together on the same chromosome, as in *R* loci, they infer that the cluster originated via a series of gene duplication events like the one illustrated here.

response. As we'll see in Chapter 46, both types of recognition loci occur in gene families. These loci are found in clusters on the same chromosome, and they are present in high copy number. The genes, which are called the immunoglobulin and *MHC* loci, are also highly polymorphic. Within any population, many different alleles of each gene exist. The parallels between plant and mammalian disease-response systems are striking.

Reactive Oxygen Intermediates (ROI) The interaction between recognition proteins in plants and avirulence molecules from pathogens triggers a series of events. These responses include the production of hydrogen peroxide (H_2O_2), superoxide (O_2^-), and related molecules that are collectively called reactive oxygen intermediates (ROI). These compounds trigger reactions that help reinforce cell walls. In addition, ROI are extremely unstable molecules. As a result, they may also trigger reactions that are responsible for the death of infected cells or that kill the pathogen directly.

What steps occur between the *R-avr* interaction and the end product of the HR—cell death? When Massimo Delladonne and co-workers treated plant cells with reactants that lead to the production of ROI, they found that some cell death occurred, but nothing like that induced by the actual *R-avr* interaction. This observation suggested that the *R-avr* interaction does not simply lead to the production of compounds required to manufacture ROI. In interpreting their results, the biologists suggested the *R-avr* interaction must lead to the production of some other molecule that increases or augments the ROI response.

Delladonne and associates proposed that the missing molecule might be nitric oxide, or NO. (Nitric oxide is the active ingredient in the anesthetic called laughing gas.) Their hypothesis was inspired by the role that NO plays in the human immune system. The immune-system cells in our bodies frequently use a lethal combination of NO and ROI to kill bacteria and diseased host cells.

To test whether the same killing mechanism occurs in plants, the researchers treated *Arabidopsis* plants with a drug that inhibits the enzyme responsible for producing NO. Then they challenged the plants by spraying the leaves with a bacterial pathogen. **Figure 37.10a** (page 718) documents the result. Plants that are able to produce normal amounts of NO show the localized cell death typical of the HR. Plants with abnormal NO production are not able to stop the bacterial infection and have large diseased areas in their leaves.

The experiment provides strong support for the hypothesis illustrated in **Figure 37.10b**. Both ROI and NO production appear to be triggered by the *R-avr* interaction, and both types of molecules seem to be required for the HR. The involvement of NO in the disease response furnishes another parallel between disease-fighting systems in plants and animals.

Phytoalexin Production A hallmark of the HR is the production of antibiotic compounds called phytoalexins at the site of infection. A **phytoalexin** is defined as a small (low molecular weight) plant product that is induced by infection and that poisons the disease-causing agent. Because phytoalexins are defined by their function and not their structure, it is not surprising that different plant species produce a diverse array of these molecules. The general idea here is that plant cells not only commit suicide when infected by a pathogen; they produce toxic molecules to poison the pathogen.

To get a better appreciation of how phytoalexins work, consider recent research by P. C. Stevenson and colleagues. These researchers were interested in antifungal compounds produced by chickpeas in response to fungi that infect their roots. Chickpeas (also known as garbanzo beans) are an important food in the Middle East and south Asia. However, they are often victimized by a fungal disease called fusarium wilt. Earlier researchers had shown that chickpea roots contain antifungal compounds called maackiain and medicarpin. What Stevenson and co-workers wanted to know is, are these antibiotics effective against fusarium wilt? If so, are they phytoalexins that chickpeas produce in response to infection?

To answer the first question, the researchers collected spores from the *Fusarium* fungus and allowed them to germinate in the presence of various concentrations of maackiain and medicarpin. As the graph in **Figure 37.11** shows, both compounds are effective antifungal agents—especially at high concentration.

Are these phytoalexins produced in response to infection? When the researchers monitored the concentrations of these antibiotics in chickpea roots over time, they found that strong increases occurred over the span of a week if they inoculated the soil with *Fusarium* spores. No such increase occurred if the soil was kept free of fungi. This is strong evidence that maackiain and medicarpin are phytoalexins and are part of the response to infection in chickpeas.

Even more important, Stevenson and co-workers were able to demonstrate a strong correlation between the ability of chickpeas to ward off fusarium wilt and their ability to produce phytoalexins. **Figure 37.12** shows the results of the experiment. Note that the investigators worked with three varieties of chickpeas and two strains of *Fusarium*, and that each of the three crop varieties shows a different susceptibility to the *Fusarium* strains. Some chickpea populations are more resistant to certain fungal strains than others. When the experimenters inoculated individuals from different chickpea varieties with different strains of the fungus and monitored the concentrations of phytoalexins produced, they found that the

(a) Plants must produce NO to stimulate HR

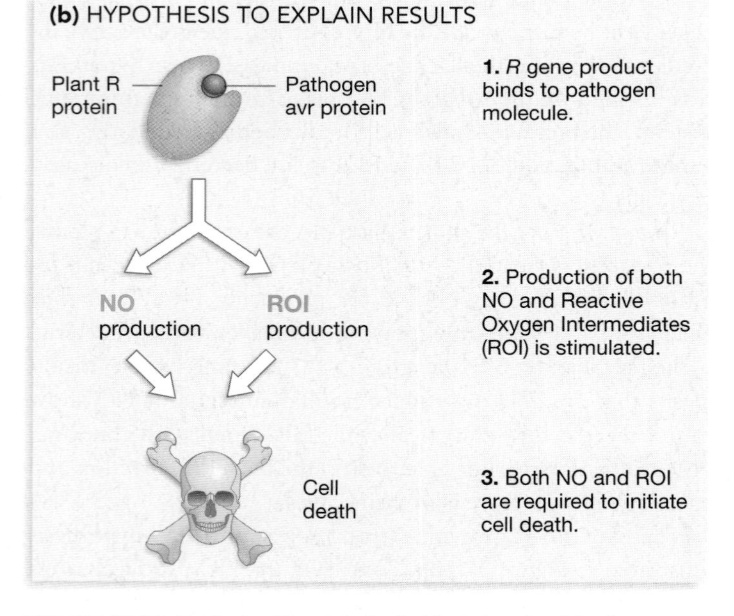

FIGURE 37.10 Evidence That Nitric Oxide Is Involved in the Hypersensitive Response (HR)
(a) If *Arabidopsis* are challenged with a pathogenic bacterium, the HR occurs in normal plants and stops the infection. But if plants are treated with a drug that inhibits nitric oxide (NO) production, the bacterial infection spreads. **(b)** Researchers suggest that both NO and ROI are required for the cell suicide component of the hypersensitive response.

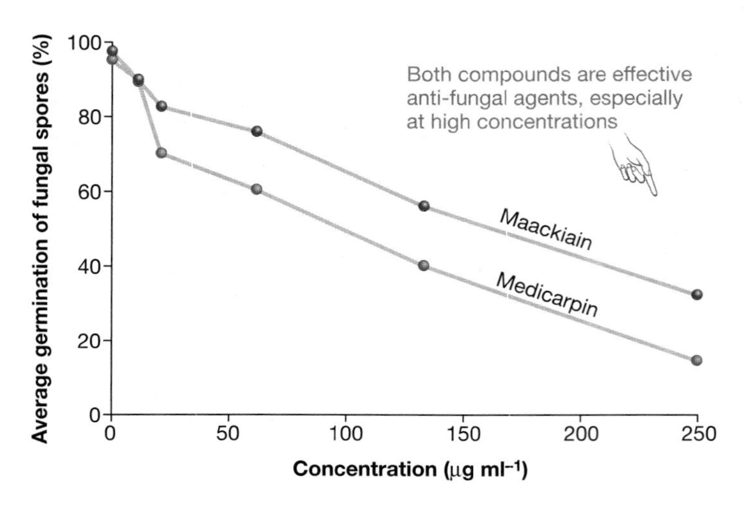

FIGURE 37.11 Phytoalexins Are Effective Pesticides
This graph shows the percentage of fungal spores that germinated when exposed to various concentrations of two different phytoalexins (a type of poison) produced by chickpeas.

ability to produce phytoalexins correlated strongly with the observed level of resistance. Apparently, only certain varieties of chickpeas are able to mount an effective response to certain strains of *Fusarium*.

The data on phytoalexin production in chickpeas are exciting because they suggest a molecular mechanism for the genetic differences in disease resistance among chickpea strains. Time will tell whether this knowledge will lead to the development of new crop varieties with increased disease resistance.

Systemic Acquired Resistance (SAR)

The HR is fast and leads to localized cell death. Phytoalexin production also occurs at the point of infection. These responses are followed by a slower and more widespread set of events called **systemic acquired resistance (SAR)**. Over the course of several days, SAR primes cells throughout the root or shoot system for assault by a pathogen—even in cells that have not been directly exposed to the disease-causing agent.

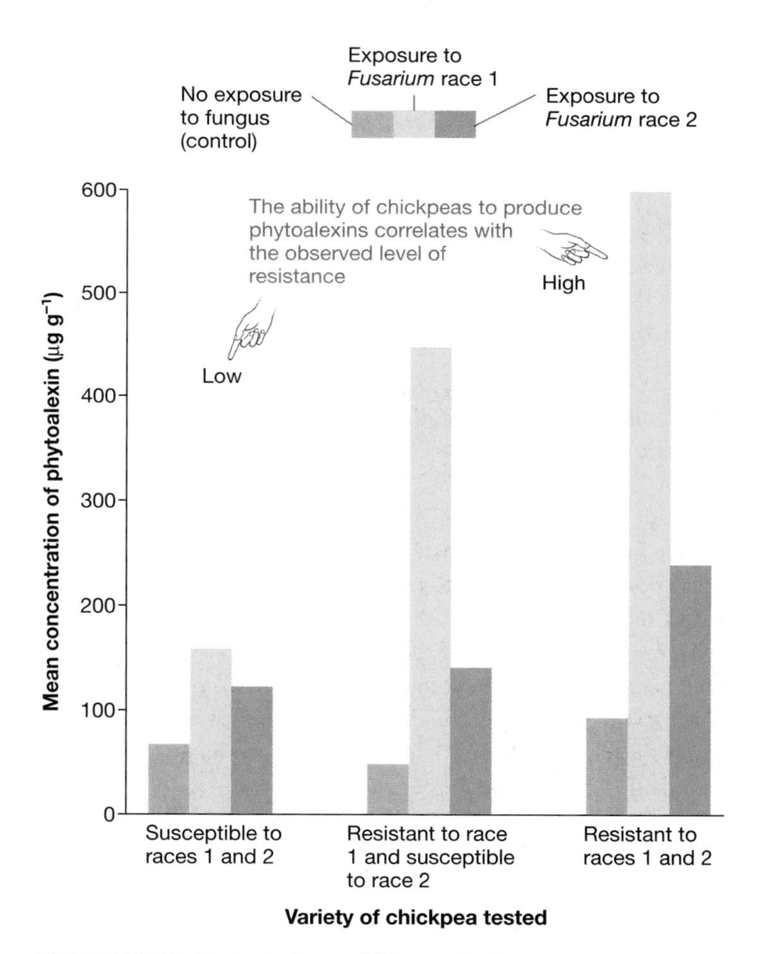

FIGURE 37.12 Phytoalexins and Disease Resistance
To produce the data in these histograms, researchers analyzed production of the phytoalexin medicarpin by three chickpea varieties. One chickpea variety is susceptible to both strains of *Fusarium* tested, one variety is resistant to one *Fusarium* strain but susceptible to the other, and one variety is resistant to both *Fusarium* strains. As the results show, chickpeas that are resistant to certain fungal strains produce phytoalexins only when they are infected with those specific strains.

How Do the SAR and HR Interact? Figure 37.13 illustrates how the HR and SAR are thought to interact. A key point here is that an interaction between *R* and *avr* products leads to the production of a signal that initiates SAR. This signal acts globally as well as locally (at the point of infection), and results in the expression of a large suite of genes called the *PR* (pathogenesis related) loci. The SAR signal qualifies as a hormone because it carries information from one location to another. What is this signal, and what effects does it have?

What Molecule Triggers the SAR? When biologists set out to locate the signaling molecule that triggers the SAR, their attention quickly turned to salicylic acid (SA). Salicylic acid is found in a wide variety of plants* and was found to increase dramatically in concentration after tissues are infected with a pathogen. The SA hypothesis became more convincing when researchers in

*As an aside, it is interesting to note that SA is very closely related to the active ingredient in aspirin. SA is particularly abundant in willow and aspen trees. Long before medical scientists discovered the pain-relieving qualities of aspirin, Native American people used teas made with willow or aspen bark as a remedy for pain.

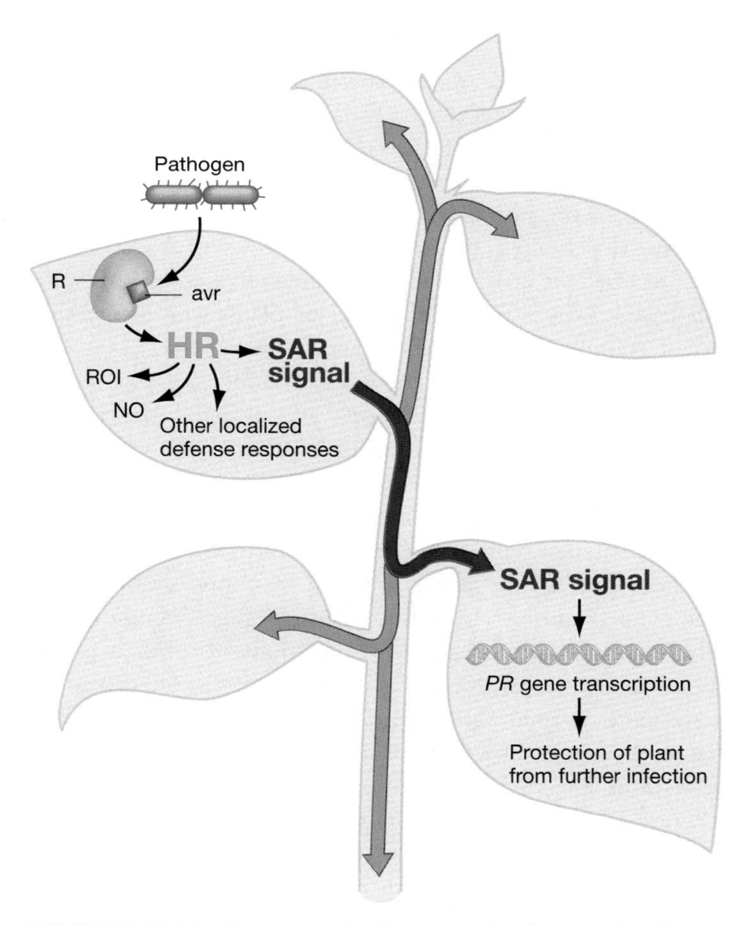

FIGURE 37.13 The Hypersensitive Response Produces a Signal That Induces Systemic Acquired Resistance
This diagram summarizes the current consensus on how the HR and SAR interact.

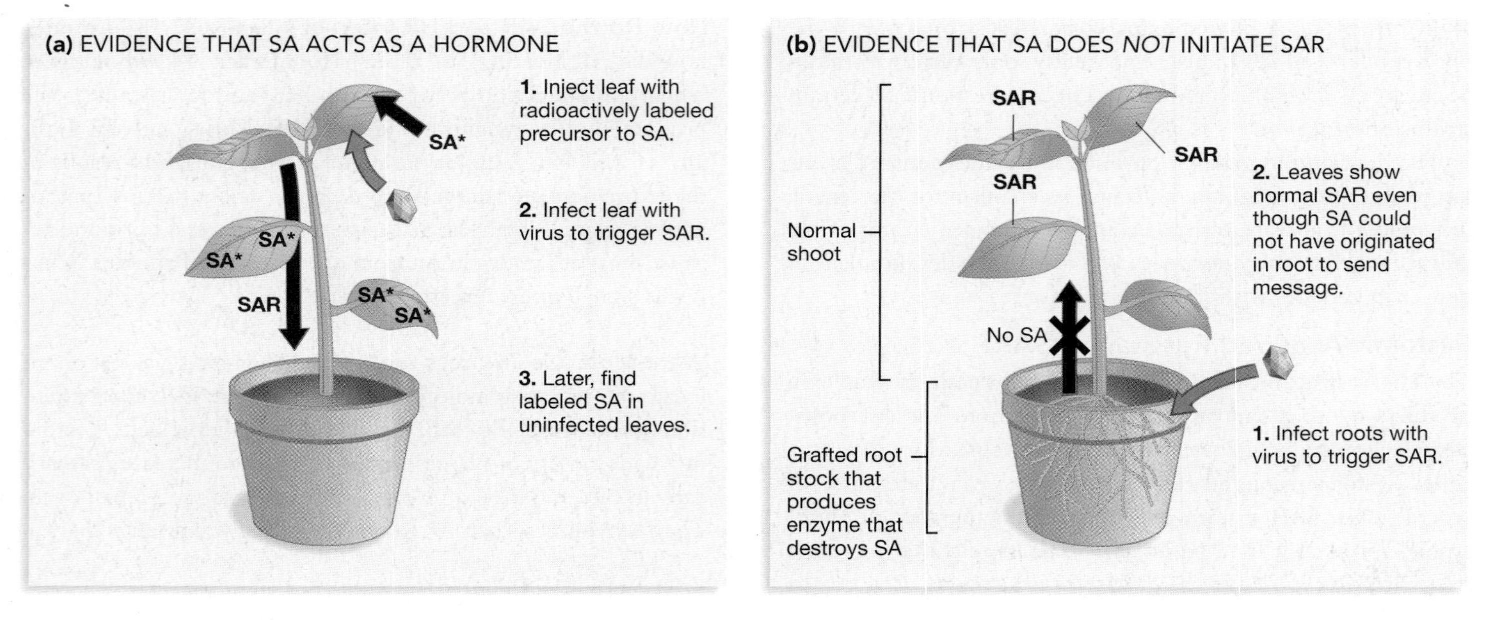

FIGURE 37.14 Is Salicylic Acid the Hormone That Triggers Systemic Acquired Resistance?
(a) Radioactively labeled SA is transported from the site of an infection throughout the plant. This observation suggests SA is the hormone that initiates SAR. **(b)** When roots that do not produce SA are grafted onto normal shoot systems and then infected with a virus, the shoots respond with SAR. This observation suggests SA is not the hormone that initiates SAR.

several different laboratories showed that applying SA directly to tissues triggers the SAR in a variety of plant species.

To confirm SA's role, Thomas Gaffney and co-workers introduced a gene called salicylate hydroxylase into tobacco plants. This gene is not normally found in plants, and it codes for an enzyme that leads to the breakdown of SA. In tobacco plants that received the gene, both SA accumulation and the SAR were abolished. As a result, the transformed plants were susceptible to infection by a wide variety of pathogens.

It is still controversial, however, whether SA acts as the SAR hormone or whether it is only a local signal that triggers the expression of genes involved in the SAR. Biologists from several research groups have labeled SA with a radioactive atom and confirmed that it is transported from infected to uninfected tissues. This observation is consistent with the hypothesis that SA actually is a signal that travels throughout the plant (**Figure 37.14a**). But when investigators in a different laboratory grafted the lower portions of tobacco plants that had been transformed with the salicylate hydroxylase gene onto normal shoots and challenged the roots with a virus, the leaves showed a normal SAR. This observation suggests that SA is not transported throughout the plant body (**Figure 37.14b**). As this book goes to press, the issue remains unresolved. SA is clearly involved in triggering the SAR. It is still not clear, however, whether SA is a local signal, a global signal, or both. Research on the mechanisms of the hypersensitive response and the systemic acquired response continues. It is clear, however, that these systems are only activated in response to a direct attack by bacteria or viruses and thus minimize the cost of defense. Are analogous systems activated in response to attacks by herbivores?

37.5 Responding to Herbivores

Over a million species of insects have already been discovered and named. Most of these species make their living by eating vegetation, seeds, roots, or pollen. How do plants withstand this onslaught of herbivores? In addition to fending off would-be predators with thorns and spines and sequestering secondary metabolites that taste bad, plants respond to herbivore attacks once they have begun. When a grasshopper bites a grass leaf, it sets off a series of carefully orchestrated events.

Here we explore two of the herbivore response systems that have been researched in some detail. The first response results in the synthesis of insecticides at the point of attack and in nearby leaves; the second leads to the production of compounds that attract enemies of the herbivores.

Proteinase Inhibitors

In the course of studies to determine why some foods are more palatable and digestible than others, biochemists discovered that many seeds and some storage organs, such as potato tubers, contain proteins called proteinase inhibitors. Proteinase inhibitors block the enzymes—found in the mouths and stomachs of animals—that are responsible for digesting proteins. When an insect or mammalian herbivore ingests a large dose of a proteinase inhibitor, the herbivore gets sick. As a result, herbivores learn to detect proteinase inhibitors by taste, and avoid plant tissues containing high concentrations of these molecules.

Researchers in Clarence Ryan's laboratory documented that proteinase inhibitors also occur in the leaves of tomatoes and potatoes. During preliminary studies on these compounds, the investigators noticed that the concentration of compounds varied dramatically from plant to plant—sometimes by a factor of 10. To explain the variability, the researchers hypothesized that individuals might produce proteinase inhibitors in response to attack by herbivores. They confirmed this idea by documenting that levels of proteinase inhibitors are relatively low in undamaged tomato leaves, but much higher in leaves that were undergoing damage and in leaves at a distance from where damage was occurring.

T. R. Green and Ryan followed up on this experiment by allowing herbivorous beetles to attack one leaf on each of several potato plants. In leaves on the same plant that were not attacked, proteinase inhibitor concentrations averaged 336 μg per mL of liquid from leaves. In leaves of control plants, where no insect damage had occurred, proteinase inhibitor levels averaged just 103 μg per mL of leaf juice. This result supports the idea that a hormone produced at the site of damage travels to undamaged tissues and induces the production of proteinase inhibitors.

What is this wound-response hormone? After years of effort, biologists in Ryan's laboratory succeeded in isolating the molecule by purifying the compounds found in tomato leaves and testing them for the ability to induce proteinase inhibitor production. The hormone turned out to be a polypeptide, just 18 amino acids long, called systemin. It was the first peptide hormone ever described in plants. When Gregory Pearce and colleagues labeled copies of systemin with a radioactive carbon atom, injected the hormone into plants, and then monitored its location, they confirmed that systemin moves from damaged sites to undamaged tissues.

Currently, work on systemin and proteinase inhibitor production is focused on determining each step in the signal transduction pathway that alerts undamaged cells to danger. As **Figure 37.15** shows, the data indicate that systemin binds to a receptor on the membrane of an undamaged cell. The activated receptor triggers a long series of chemical reactions that eventually result in the synthesis of a molecule called jasmonic acid. Jasmonic acid, in turn, activates the production of at least 15 new gene products, including proteinase inhibitors. In this way, plants build potent concentrations of insecticides in tissues that are in imminent danger of an attack.

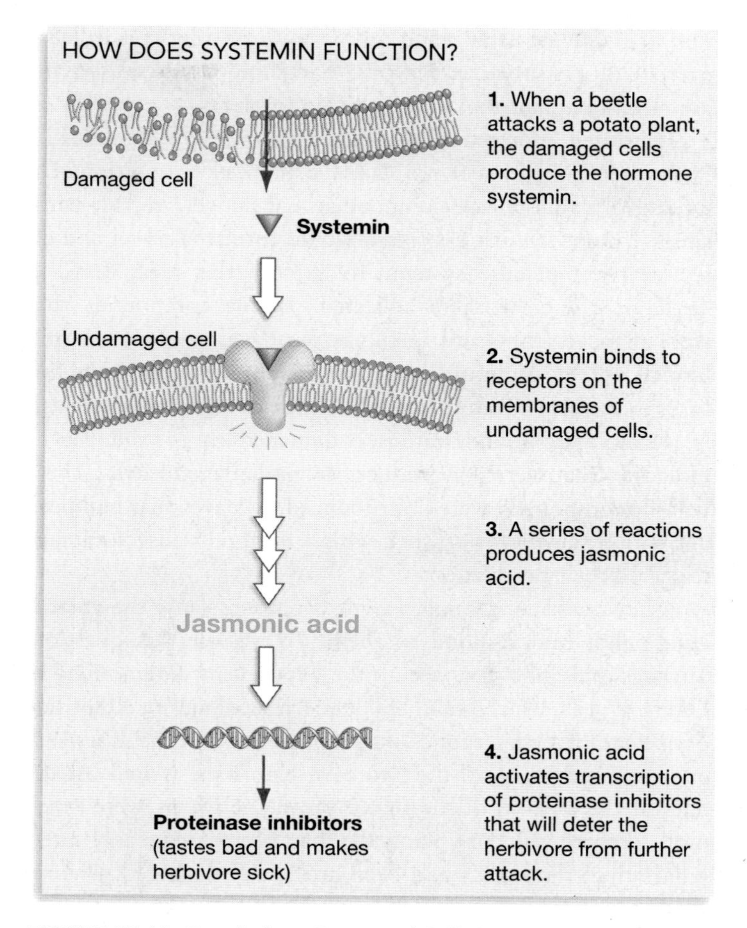

FIGURE 37.15 Signals from Damaged Cells Prepare Other Cells for Attack
Systemin is a hormone produced by herbivore-damaged cells that initiates a protective response in undamaged cells.

Recruiting Parasitoids

In addition to coping with spines, proteinase inhibitors, and other plant defenses, caterpillars and other herbivorous insects have enemies of their own. In many cases, these enemies are wasps that lay their eggs in the bodies of the herbivores. When a wasp egg hatches inside a caterpillar, the wasp larva begins eating its host from the inside out (**Figure 37.16**). An organism

FIGURE 37.16 Parasitoids Kill Herbivores
Parasitoids lay their eggs in caterpillars and other types of herbivores. As the parasitoid larvae grow, they devour the host. This photo shows wasp larvae emerging from a parasitized caterpillar.

that is free-living as an adult but parasitic as a larva is called a **parasitoid**. For obvious reasons, parasitoid attacks limit the amount of damage that herbivores do to plants.

The observation that parasitoids seem to be common during herbivore outbreaks in croplands prompted a question: Do wounded plants release compounds that actively recruit parasitoids? If so, the distress signal would furnish a novel and effective plant defense system. To explore this idea, T. C. J. Turlings and co-workers collected volatile compounds that were released from corn seedlings during attacks by caterpillars. (A volatile molecule is one that evaporates rapidly and diffuses in the air.) When they analyzed the compounds chemically, the biologists found that insect-damaged leaves produced 11 molecules that were not produced by undamaged leaves. These volatile compounds were not produced by leaves that had been cut with a scissors or crushed with a tool; only insect damage triggered their production.

A key question, though, is whether wasps sense the volatile compounds and respond to them. To answer this question, Turlings and colleagues set up the experiment diagrammed in **Figure 37.17a**. The researchers put corn seedlings on either side of one end of a wind tunnel, put a female wasp at the other end, and recorded which of the two corn plants she visited. As the data at the bottom of the figure show, the wasps were much more likely to visit corn plants that had been damaged by caterpillars than those that had been artificially damaged. The researchers could make artificially damaged plants just as attractive to the wasps, however, by adding caterpillar saliva.

H. T. Alborn and colleagues followed up on this result by purifying a large series of substances from the saliva of beet armyworm caterpillars. The researchers were able to show that one of these molecules, called volicitin, induced damaged leaves to emit the volatile compounds attractive to wasps. The message of this work is that plant cells are able to sense a specific molecule that is present in caterpillar saliva. As a result, they are able to recognize when they are being attacked by an herbivore and sound an alarm that recruits help (**Figure 37.17b**).

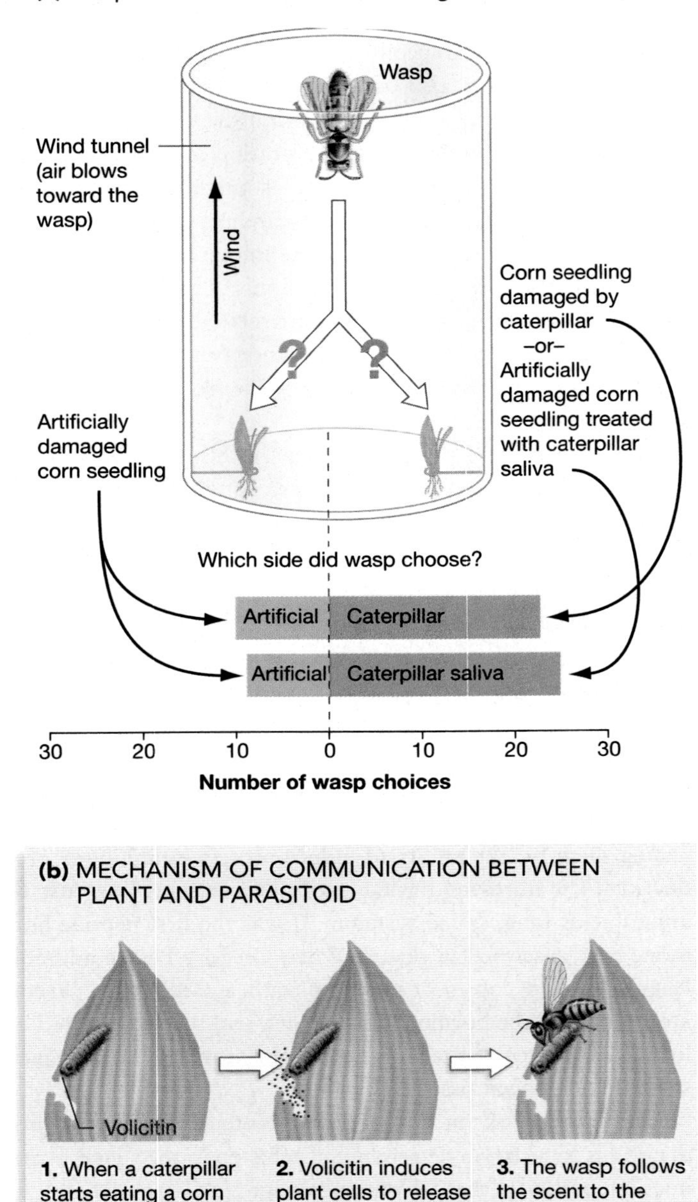

FIGURE 37.17 Wasps Hone In on Volatile Compounds That Plants Release in Response to Volicitin

(a) In choice tests, wasps are attracted to corn seedlings that have been damaged by caterpillars, or to artificially damaged leaves that have been treated with caterpillar saliva. **(b)** When plant cells sense volicitin, they release the volatile compounds sensed by parasitoid wasps. QUESTION In part (a), why did the experimenters use artificially damaged leaves as a comparison treatment? Why not use undamaged leaves?

Essay Chemical Prospecting

The effort to find naturally occurring compounds that can be used as drugs, fragrances, insecticides, herbicides, or fungicides has been called chemical prospecting. Traditionally, plant secondary metabolites—and particularly defense compounds—have been a rich source of these molecules. For example, the prominent anticancer drug vincristin/vinblastine was first synthesized from a plant native to Madagascar called the rosy periwinkle; the anticancer drug taxol was discovered in extracts from the Pacific yew tree, which lives only in old-growth forests of the Pacific Northwest region of North America. These drugs are effective because they kill certain types of rapidly growing cells, but their normal function in periwinkles and Pacific yew may be to harm herbivores. Similarly, the psychoactive properties of alkaloids such as morphine, cocaine, nicotine, and caffeine are almost certainly related to their role in plant defense. The natural function of tetrahydrocannabinol, the active ingredient in marijuana, is probably to injure the plant's enemies.

The psychoactive properties of alkaloids such as morphine, cocaine, nicotine, and caffeine are almost certainly related to their role in plant defense.

Here we consider two recent twists on traditional efforts to purify plant defense compounds and screen them for medical or agricultural use. In one effort, Ilya Raskin and co-workers have been able to stimulate plants that are growing hydroponically—that is, in liquid culture instead of in soil—to churn out large quantities of defense chemicals. The researchers do this by exposing root cells to a fragment of a bacterial cell wall or a toxin from a fungus, and then collecting the defense compounds that are produced in response. To date, Raskin's group has been able to collect 5000 samples of materials exuded by the roots of 700 different plant species. Some of the extracts have been found to kill various types of cancer cells growing in culture. It is still unknown, however, whether this novel strategy for producing and purifying defense compounds will result in the discovery of important new drugs.

The second innovative research program is exemplified by recent work on a group of fungi called mycoparasites (fungiparasites). Mycoparasites infect and parasitize other fungi. Matteo Lorito and associates were able to clone a gene from a mycoparasite that encodes an enzyme called endochitinase. This enzyme breaks down chitin, which is a major component of cell walls in many species of pathogenic fungi. The mycoparasite uses this enzyme to open the cell walls of its hosts so that it can enter and begin to feed, just as pathogenic fungi use cutinase to open the cuticle of plants. When the researchers inserted the gene for endochitinase into tobacco and potato plants, the transgenic individuals manufactured the enzyme and were highly resistant to fungal infection.

The work by Lorito and colleagues illustrates a general theme in current research on plant diseases. Chemical prospecting has a major new thrust: gene prospecting. If important genes—such as the *R* alleles found in disease-resistant species, genes for particularly effective defense compounds, or sequences that encode antibiotics such as endochitinase—can be introduced into crop plants safely, rapid improvements in pest resistance might be possible. If so, then farmers might be able to reduce their dependence on insecticides, fungicides, and other compounds that are expensive and that can be dangerous to humans and wildlife when improperly handled.

Chapter Review

Summary

By providing a barrier to entry, the epidermis and cuticle of plants act as the first line of defense against pathogens. Some fungi pierce this barrier by secreting an enzyme called cutinase that cleaves some of the molecules found in cuticle. Bacteria and some fungi circumvent the barrier by entering tissues through open stomata or wound sites. In many plants, the epidermis contains defensive hairs or spines and serves as the first line of defense against herbivores as well as pathogens.

Many plants discourage herbivores by lacing their tissues with poisons. Analyses of the structures of these poisons have supported the hypothesis that many or most are secondary metabolites. In both structural and evolutionary terms, many secondary metabolites are closely related to fundamentally important compounds such as amino acids. When researchers explored the mode of action of the common secondary metabolites known as alkaloids, they discovered that many of these molecules work by poisoning several different biochemical pathways or processes at the same time.

Many plants do not produce defense compounds all of the time, however. Instead, they produce antibiotics and pesticides only when an infection or herbivore attack has started. This pattern is sensible because defense compounds are energetically expensive to produce, and because plants face trade-offs in allocating resources to defense or to growth and reproduction. The prediction that defense is costly and that trade-offs occur has been confirmed experimentally. For example, when *Arabidopsis thaliana* plants grow in an environment that lacks pathogens and herbivores, lightly defended individuals produce the most seeds. But in environments where pathogens and herbivores are common, heavily defended individuals produce the most offspring.

When an attack by a pathogen begins, resistant plants respond with a rapid series of events called the hypersensitive response (HR). The HR is believed to be triggered when the product of a resistance gene (*R*) located in the host binds to the product of an avirulence (*avr*) gene secreted by the pathogen. Resistance loci are found in large gene families and are highly polymorphic, meaning that many different alleles exist at each locus. Individuals that produce many different *R* gene products have the ability to recognize many different pathogen products. As a result, they are immune to a wide array of pathogens.

During the HR, cells at the point of infection commit suicide. Cell death results from the combined action of nitrous oxide (NO) and reactive oxygen intermediates (ROI). Cell walls around the point of infection are also reinforced through the action of ROI. In some cases, antibiotic compounds called phytoalexins are produced as part of the HR as well.

The HR is commonly followed by a slower response that leads to systemic acquired resistance (SAR). SAR occurs when a hormone, which may be salicylic acid, travels from the infection site to nearby tissues and triggers the expression of a specific set of genes.

When an attack by an herbivore begins, plants respond by producing proteinase inhibitors. The synthesis of these poisons occurs throughout the plant and is triggered by a polypeptide hormone called systemin. Once systemin binds to a receptor on the membrane of undamaged cells, a series of chemical reactions leads to the manufacture of proteinase inhibitors and heightened plant defense. Some plants are also able to sense specific compounds in the saliva of herbivores; in response, they secrete volatile compounds that attract parasitoids.

Questions

Content Review

1. Why is it significant that fungi produce the enzyme called cutinase?
   a. It cleaves molecules in plant cuticle and allows the fungus to enter.
   b. It neutralizes *R* gene products, so the HR does not occur.
   c. It prevents the release of salicylic acid, so SAR does not occur.
   d. It prevents the synthesis of secondary metabolites.
2. What does the gene-for-gene hypothesis claim?
   a. Plant defense systems are heritable, or genetically based.
   b. *R* loci occur in families because of gene duplication events.
   c. Secondary metabolites are the most efficient form of plant defense.
   d. Plant defense systems activate when an *R* gene product binds to an *avr* gene product.
3. What happens if *none* of the molecules produced by a pathogen bind to an *R* gene product?
   a. The HR is triggered, and the infection is stopped in its tracks.
   b. The HR is not triggered, and the infection spreads.
   c. The SAR is triggered, and the infection does not spread to surrounding tissue.
   d. NO and ROI combine to kill cells at the site of infection.
4. What is a phytoalexin?
   a. any plant secondary metabolite
   b. any plant secondary metabolite that is used in defense
   c. any plant secondary metabolite that is used in defense and produced in response to a pathogen attack
   d. any plant secondary metabolite that is used in defense and produced constitutively
5. What happens after volicitin binds to a receptor in a plant cell?
   a. SAR is initiated.
   b. Proteinase inhibitor production is initiated.
   c. Volatiles are released that attract parasitoids.
   d. Phytoalexin production is initiated.

Conceptual Review

1. Summarize the events that occur during the HR and SAR. In what way are these responses different? How are they complementary?
2. Why is it significant that *R* loci are usually clustered on the same chromosome? Why is it important that *R* loci are usually highly polymorphic?
3. What evidence suggested that a hormone travels from the site of insect damage to undamaged tissues? How did investigators discover systemin and confirm that it is the wound-response hormone?
4. Many of the fragrant molecules given off by plants to attract pollinators are secondary metabolites. Suggest a scenario for the evolution of these attractant compounds, analogous to the presentation in the text regarding the evolution of defense compounds.
5. Describe some of the similarities and differences in plant defenses against pathogens and herbivores.
6. The HR has been called "an efficient suicide program." Why?

Applying Ideas

1. Suppose you are walking along a garden path with a friend and notice a plant leaf with several small, white, dead-looking blotches. Nearby, an individual from the same species has leaves that are completely yellow and wilted. Explain to your friend what you think is going on. How could you test your hypothesis?
2. Suppose that you were studying two closely related species of plants and were able to document that one has much higher concentrations of tannins than the other. Which species would you expect to grow more slowly and produce fewer seeds each year? Why?

3. Researchers have recently begun studying a secondary metabolite called myrosinase in the plant *Brassica rapa*. (*Brassica rapa* is an important source for vegetable oils used in cooking.) Myrosinase is an effective deterrent for herbivorous beetles. In one study, biologists mated individuals with high concentrations of myrosinase with one another for several generations and individuals with low concentrations of myrosinase with one another for several generations. In this way, they created populations with high versus low concentrations of myrosinase. Then they documented how much each population was visited by pollinators. Predict which population was more successful in attracting long visits from pollinators. To see if you are correct, check the paper published by Sharon Strauss and co-workers in the journal called *Evolution*, volume 53, pages 1105–1113. In what way does this experiment show a novel cost of defense?
4. Researchers have been able to document that plants undergoing insect attack release jasmonic acid as a volatile. Other plants that are growing nearby synthesize increased amounts of proteinase inhibitors in response. Do these observations surprise you? Why or why not?
5. Review the text section that analyzes the experiments on phytoalexins produced by chickpeas. Note the close correspondence between the amount and type of phytoalexins produced by different chickpea strains in response to different strains of fungi. Suggest a hypothesis that explains this observation in terms of differences among the *R* alleles of chickpea and the *avr* alleles of fungus strains. How would you test your hypothesis?
6. The experimental approach diagrammed in Figure 37.8 is widely used and is called the yeast two-hybrid system. Note that a hybrid is produced when two unlike things combine. Explain why the name is appropriate and why the strategy is an effective way to study protein-protein interactions in cells.

CD-ROM and Web Connection

CD Activity 37.1: Plant Defenses *(animation)*
(Estimated time for completion = 10 min)
How have plants adapted to survive in a world filled with herbivores?

At your **Companion Website** (http://www.prenhall.com/freeman/biology), you will find self-grading exams and links to the following research tools, online resources, and activities:

Gene Silencing
This article explains how a plant virus can alter the normal expression of genes in the plant.

Plant Defense and Reproductive Energy Budgets
This article describes how the energy cost of reproduction must be balanced by the benefits gained by shunting energy into the development of plant defenses.

Plant-Produced Pesticides
This article describes some of the toxic chemicals that plants can actively produce to fight off herbivory.

Additional Reading

Baker, B., P. Zambryski, B. Staskawicz, and S.P. Dinesh-Kumar. 1997. Signaling in plant-microbe interactions. *Science* 276: 726–733. Reviews many of the plant defense systems introduced in this chapter.

Barlow, C. and M. Rothman. 2001. Ghost stories from the ice age. *Natural History* 110 (September): 62–67.

Marchand, P. 2001. In the field: riding the witches broom. *Natural History* 110 (May): 40–41.

Stanton, M., and T. Young. (1999). Thorny relationships. *Natural History* 108 (September): 28–30.

How Animals Work

Unit 8

Physiology is the study of how organisms manage the chemical and physical conditions inside their bodies. Understanding animal anatomy and physiology is fundamental to a wide array of applied fields in biological science—from veterinary medicine and animal husbandry to biomedical research and clinical practice. **Chapter 38: Animal Form and Function** introduces some of the major themes in animal physiology by reviewing ideas and patterns that tie the diversity of physiological systems in animals together. The subsequent eight chapters follow up with a more detailed survey of each major type of animal physiological system.

As a group, **Chapter 39: Water Balance in Animals, Chapter 40: Animal Nutrition**, and **Chapter 41: Gas Exchange and Circulation** focus on the basic chemical conditions required to run the animal body. The following three chapters—**Chapter 42: Electrical Signals in Animals, Chapter 43: Animal Sensory Systems and Movement**, and **Chapter 44: Chemical Signals in Animals**—consider how animals receive, process, and act on information. The unit concludes with **Chapter 45: Animal Reproduction** and **Chapter 46: The Immune System in Animals**. Understanding animal reproductive and defense systems is basic to many applied fields in biology. Reviewing recent research on these systems also provides a convenient way to integrate an array of concepts presented throughout the unit.

This adult Schaus swallowtail butterfly has just emerged from a case where metamorphosis took place. As a caterpillar, this individual crawled, ate leaves, and exuded a noxious fluid to repel predators. ©1993 Susan Middleton and David Liittschwager

Animal Form and Function

38

In many parts of the Old World and North America, it is not unusual to see a bird of prey called an osprey circling over a lake or stream. If you watch one of these birds long enough, you might see it suddenly fold its wings and plummet to the water's surface with talons extended. A few seconds after impact, the osprey will begin to beat its wings again and slowly lift itself from the water. If the osprey is lucky, it will fly away with a trout or another fish locked in those talons.

Many students decide to study biology because they have witnessed events like this and want to know more about what was happening. How does an osprey track a fish moving 20 or 30 meters below it and dive accurately enough to make a strike? How does a trout sense the presence of a hunting bird, and what does it do to avoid being caught?

In essence, these types of questions address the phenomenon called adaptation. As Chapter 21 explained, an adaptation is a trait that allows individuals to survive and reproduce in a certain environment better than individuals that lack that trait. An osprey's keen eyes, hooked beak, and sharp talons are adaptations for hunting prey. A trout's streamlined body, powerful swimming muscles, and flattened tail are adaptations for hunting its own prey and for avoiding predation by ospreys.

The chapters in this unit explore an array of adaptations found in animals—the traits that allow them to hear, see, move, digest, breathe, and fight infections. This chapter serves as an introduction by focusing on the nature of adaptation itself. Before exploring specific adaptations, it's essential to ask how adaptations arise and to identify themes that unify the diversity of adaptive traits observed in animals.

Section 38.1 delves into these issues by reviewing some key points about natural selec-

Osprey are fish-eating birds native to North America and the Old World. Their curved beak, sharp talons, and streamlined body are interpreted as adaptations that help them find prey and produce offspring. This chapter explores the nature of adaptation.

tion—the process that produces adaptations. Section 38.2 follows up by examining factors that constrain the evolution of adaptive traits. Section 38.3 looks at how the structures of tissues and organs correlate with their functions, and section 38.4 introduces some of the ways that overall body size affects animals. The concluding section explores an important phenomenon known as homeostasis (same-state). **Homeostasis** is the maintenance of relatively constant physical and chemical conditions within an organism. When the temperature is 0°C and the air is dry, the tissues within an osprey are still 37°C and fully hydrated. How does homeostasis occur? As a case study, section 38.5 explores how animals achieve homeostasis with respect to body temperature. At the completion of this chapter, you'll see osprey, trout, and other animals in a new light—as efficient machines for gathering resources and producing offspring.

38.1 The Nature of Natural Selection

The process of natural selection was introduced in Chapter 1 and analyzed in detail in Chapter 21. Recall that natural selection occurs whenever individuals that carry certain alleles leave more offspring than individuals with different alleles of the same gene. Because of this difference in reproductive success, the frequency of the selected alleles increases from one generation to the next. In this way, natural selection produces evolution.

As Chapter 22 explained, natural selection is not the only process that leads to changes in allele frequencies over time. Evolution also occurs through the random process called genetic drift, the movement of alleles into and out of populations by migration, and the constant introduction of new alleles by mutation. But natural selection is the only mechanism that produces adaptive evolution. Stated another way, natural selection is the only process that, over time, increases the ability of organisms to survive and reproduce.

Natural Selection in Action: A Case Study

Although natural selection is a deceptively simple process, it is a bit difficult to understand thoroughly. To help clarify how natural selection works, let's consider some of the more common misconceptions about the process in light of a specific example: changes in beak size that have occurred in finches native to the Galápagos Islands off the coast of Ecuador, South America.

The medium ground finch pictured in **Figure 38.1a** makes its living by eating seeds. Individuals grasp seeds at the base of their beak and crack the hulls by applying force. For the past three decades, the population of medium ground finches on Isle Daphne Major has been studied by a team of biologists led by Peter and Rosemary Grant. Because Daphne Major is small—about the size of 80 football fields—the researchers have been able to catch, weigh, and measure all individuals and mark each with a unique combination of colored leg bands.

Early studies on the finch population established that beak size and shape vary among individuals, and that beak morphology is heritable. Stated another way, parents with particularly wide or deep beaks tend to have offspring with wide or deep beaks. (Chapter 21 introduced some of the techniques used by biologists to establish that traits have a genetic basis).

Not long after the team had established these results, a dramatic selection event occurred. In 1977, Daphne Major received just 24 mm of rain during the annual wet season instead

(a) Medium ground finch

(b) Natural selection during a drought

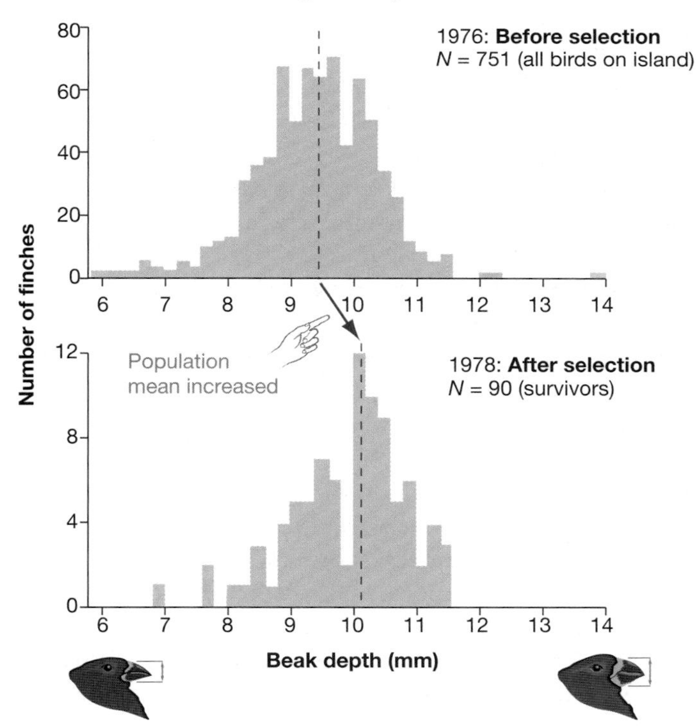

FIGURE 38.1 Natural Selection in Action
(a) Medium ground finches live only on the Galápagos Islands.
(b) These histograms show the distribution of beak depth in medium ground finches before and after natural selection occurred, due to changes in food availability, during the drought of 1977. *N* is the sample size. QUESTION Why was sample size so much smaller in 1978?

of the 130 mm that normally falls. During the drought, 84 percent of the medium ground finch population disappeared.

Two observations support the hypothesis that most or all of these individuals died of starvation. The researchers found a total of 38 dead birds, and all were emaciated. Further, none of the missing individuals reappeared once the drought ended and food supplies returned to normal.

The research team realized that the die-off presented an opportunity to study natural selection. Were the survivors different from nonsurvivors? When the biologists analyzed the characteristics of each group, they found that survivors tended to have much deeper beaks than did the birds that died. This was an important finding, because the type of seeds that were available to the finches had changed dramatically as the drought continued. At the drought's peak, the tough fruits of a plant called *Tribulus cistoides* served as the finches' primary food source. These fruits are so difficult to crack that they are ignored in years when food supplies are normal. Grant's group hypothesized that individuals with particularly large and deep beaks were more likely to crack these fruits efficiently enough to survive.

As the data in **Figure 38.1b** show, natural selection led to an increase in average beak depth in the population. When breeding resumed in 1978, the offspring that were produced had beaks that were half a millimeter deeper, on average, than the population that existed before the drought. In only one generation, natural selection had led to measurable evolution—a change in the characteristics of the population over time. Alleles that led to the development of deep beaks must have increased in frequency. Large, deep beaks were an adaptation for cracking large fruits and seeds.

Now the question is, how can these data help correct common misconceptions about natural selection?

Selection Acts on Individuals, but Evolutionary Change Occurs in Populations

Perhaps the most important point to clarify about natural selection is that individuals do not change during the process—only the population does. During the drought, the beaks of individual finches did not become deeper. Rather, the average beak depth in the population increased over time, because deep-beaked individuals were more likely to survive and produce offspring with deep beaks. Natural selection acted on individuals, but the evolutionary change occurred in the characteristics of the population.

To formalize this point, biologists differentiate between adaptation and a phenomenon called acclimation. Adaptation occurs when a population changes in response to natural selection. **Acclimation** occurs when an individual changes in response to a change in environmental conditions. For example, wood frogs native to northern North America are exposed to extremely cold temperatures as they overwinter. When ice begins to form in their skin, their bodies begin producing molecules that protect their tissues from being damaged by ice crystals.* These individuals are acclimating to cold temperatures.

In frogs, the ability to produce antifreeze molecules is an adaptation. Actually producing the molecules is acclimation. Similarly, you may have observed changes in your own body as you acclimated to high altitude or to particularly hot or cold environments. You did not adapt to these environments in the evolutionary sense, however.

Evolution Is Not Progressive

It is often tempting to think that evolution by natural selection is progressive—that organisms have gotten "better" over time. (In this context, *better* usually means more complex.) It is true that groups that appear late in the fossil record are often more morphologically complex or "advanced" than related groups that appeared earlier. Similarly, groups that branch off early on phylogenetic trees sometimes have simpler characteristics than groups that branched off later. But traits can be lost as well as gained. There are, in fact, thousands of instances in which a group became simpler in form and function than its ancestors as a result of evolution by natural selection. Populations that take up a parasitic lifestyle are particularly prone to this trend. Tapeworms, for example, evolved from species with a sophisticated digestive tract, but lack any sort of gut. They absorb nutrients directly from their environment instead.

Based on this discussion, it should make sense that there is no such thing as a higher or lower organism. As the evolutionary tree in **Figure 38.2** shows, species can belong to lineages that are more ancient or more recent and thus be considered more basal or more derived, but they cannot be higher or lower. All populations have undergone natural selection based on their ability to gather resources and produce offspring. All organisms are adapted to their environments. No organism is any "higher" than another organism. This is an important point to keep in mind as you study the subsequent chapters in this unit. Even though the discussion often focuses on human adaptations—simply because so much is known about our own species—it would be an error to consider human physiology as better or more advanced than fruit-fly or clam or earthworm physiology. Humans are adapted to a particular environment; fruit flies and clams and earthworms are also adapted to their particular environments.

To illustrate that evolution is not progressive, consider the population of finches studied by Grant and colleagues. A series of drought years in the late 1970s and early 1980s produced additional increases in average beak depth of medium ground finches on Daphne Major. But then in 1983, torrential rains fell

*In some species of frogs, so much extracellular fluid freezes during cold snaps that individuals appear to be frozen solid. Their hearts also stop beating. When temperatures warm in the spring, their hearts resume beating, their tissues thaw, and they resume normal activities.

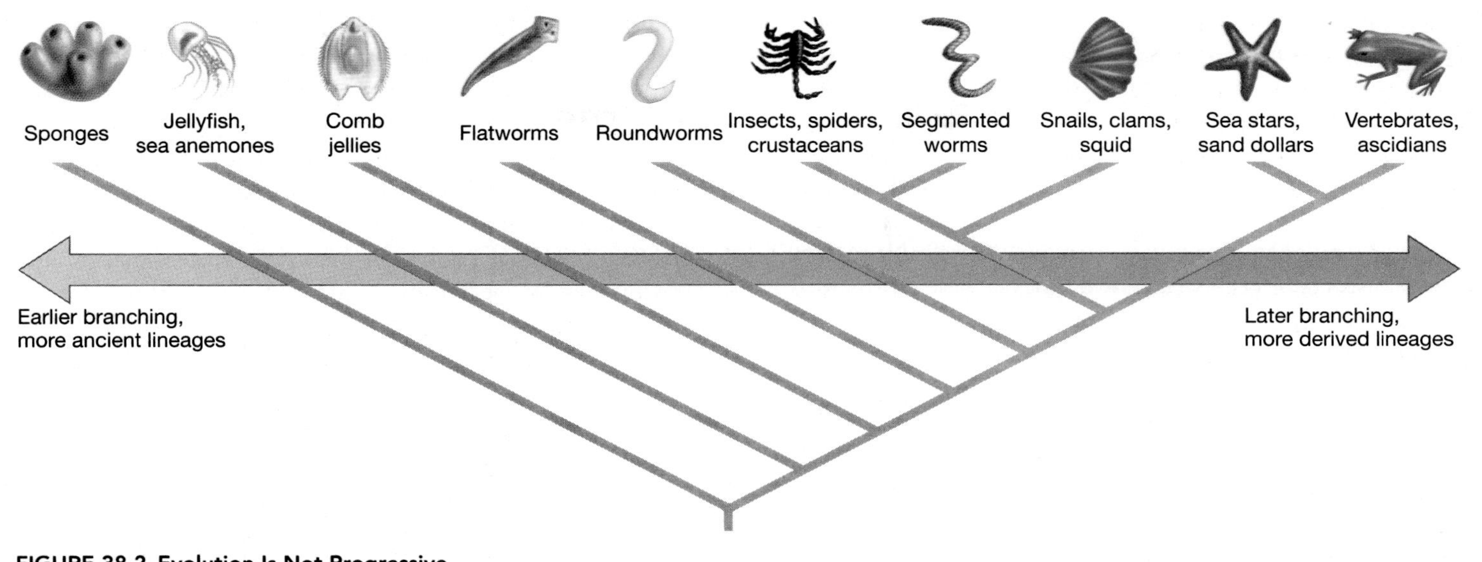

FIGURE 38.2 Evolution Is Not Progressive
This evolutionary tree shows the relationships between the major groups of animals. Although these lineages evolved at different times, none is any higher or lower than any other.

in the area and produced a profusion of small, soft seeds on Daphne Major. In this environment, medium ground finches with small beaks had extraordinarily high reproductive success. As a result, the average beak size of the population *declined*. Instead of producing a progression of ever-larger or more-complex beaks in Galápagos finches, natural selection simply occurred in response to whatever changes happened to occur in the environment.

38.2 The Nature of Adaptation

Natural selection on medium ground finches resulted in changes in beak shape that allowed finches to find food more efficiently and thus survive better and reproduce more in a dry environment. As a result of natural selection, the finch population adapted to its environment.

In essence, each chapter in this unit explores a specific suite of adaptations found in animals. Chapter by chapter, we'll analyze the structures and processes that allow animals to maintain adequate solute concentrations in their cells and tissues, find and digest food, respond effectively to changes in their environment, reproduce, and eliminate parasites. The general impression will be that animals are extremely good at solving the problems posed by the environments they live in.

It would be a serious error, however, to assume that every structure and process found in animals and other organisms is adaptive. Before we plunge into a detailed analysis of animal adaptations, a cautionary note is in order. Adaptation is far from perfect. To drive this point home, let's examine how evolution by natural selection is constrained.

Not All Traits Are Adaptive

Animals possess a variety of structures that were present in ancestral populations but are not currently adaptive. Chapter 21 analyzed the human coccyx (tailbone), goose bumps, and appendix in this context. Vestigial traits like these are widespread. Whales still have pelvic bones—not because their hips are used in swimming, but because their ancestors were related to hippopotamus-like mammals that walked on their legs. Even though they live in complete darkness, many cave-dwelling animals still have rudimentary eyes.

Vestigial traits are not the only types of structures with no function that are found in animals. Some adult traits exist as holdovers from structures that appear early in development. For example, human males have rudimentary mammary glands. They exist only because nipples form in the embryo before sex hormones begin influencing the development of organs.

The general point here is that not all traits are adaptive. In addition, structures that do have functions are constrained in a variety of important ways. Let's take a closer look.

Genetic Constraints

When the Grants' team analyzed data on the characteristics of finches that survived the 1977 drought, they made an interesting observation. Although individuals with deep beaks survived better than individuals with shallow beaks, it was also true that birds with particularly narrow beaks survived better than individuals with wider beaks. This observation made sense because finches crack *Tribulus* fruits by twisting them, and because narrow beaks should concentrate the twisting force more efficiently. But narrower beaks did not evolve in the population. To explain why, the biologists noted that parents with deep beaks tend to have offspring with beaks that are both deep and wide. This is a common pattern in animals. Many alleles that affect body size have an effect on all aspects of size—not just one structure. As a result, selection for increased beak depth overrode selection for narrow beaks, even though a deep and narrow beak would have been more advantageous.

The general point here is that selection was not able to optimize all aspects of beak shape. Wider beaks were not the best possible beak shape for finches living in an arid habitat. They evolved anyway, due to a type of constraint called a genetic correlation. In this case, selection on alleles for one trait (increased beak depth) caused a correlated and nonoptimal increase in another trait (beak width).

Genetic correlations are not the only genetic constraint on adaptation, however. Lack of genetic variation is also important. Consider that salamanders have the ability to regrow severed limbs, and that some eels can sense electric fields. Humans cannot. Even though the traits might confer increased reproductive success, they do not exist in humans because the requisite genes are lacking. Similarly, during the 1960s and 1970s, poisoning from the insecticide DDT devastated osprey populations in many parts of the world. An enzyme that detoxified DDT would have been adaptive, but it did not exist. Consequently, many osprey populations were driven to extinction.

Historical Constraints

In addition to being constrained by genetic correlations and lack of genetic variation, adaptations are also constrained by history. The reason is simple: All traits have evolved from previously existing traits. For example, the tiny hammer, anvil, and stirrup bones found in your middle ear evolved from bones that were part of the jaw and braincase in the ancestors of mammals. These bones now function in the transmission and amplification of sound from your outer ear to your inner ear. Biologists routinely interpret these bones as adaptations that make hearing airborne sounds more efficient. But are these bones a "perfect" solution to the problem of transmitting sound from one part of the ear to the other? The answer is no. They are the best solution possible, given an important historical constraint. Natural selection was acting on structures that originally had a very different function.

The importance of historical constraint will surface again in the discussion of the vertebrate eye in Chapter 43. Although the vertebrate eye is sometimes presented as an example of a "perfect" adaptation, optometrists can attest that the organ is often defective. In addition, vertebrate eyes have a blind spot that results from a historical constraint. As Box 43.2 will detail, the constraint originated in the arrangement of nerve cells and light-receptor cells in the ancestors of vertebrates. It is far from optimal or perfect for an eye to have a blind spot; the trait exists simply because the vertebrate eye evolved from a structure that had cells arranged in a peculiar configuration.

Trade-Offs

Perhaps the most important of all constraints on adaptation involves the phenomenon known as trade-offs. In many cases the trade-offs that animals make involve expenditures of time or energy. For example, every female animal has a defined amount of time and energy to devote to egg production. Given this constraint, there should be a trade-off between the number of eggs produced and their quality. In species that do not care for their young once eggs are laid, egg quality is determined by size—specifically by the amount of nutrient-rich yolk that the mother provides. In species that do provide parental care, quality is determined by the amount of food and other resources that the mother delivers to each offspring. If the number of offspring is large, a mother can provide less food to each individual.

Does the predicted trade-off in egg size and egg number actually exist? To answer this question, Barry Sinervo and coworkers developed techniques for manipulating egg size and egg number in side-blotched lizards, which live in the deserts of western North America (**Figure 38.3a**). These biologists were able to induce the production small eggs by catching females and surgically removing yolk from their eggs early in development. To create clutches with small numbers of eggs, they destroyed all but two or three of the developing eggs in selected females. The biologists also performed the surgery on a large number of control females, whose eggs were left unmanipulated.

The graph in **Figure 38.3b** shows the results of this experiment. Note that the average mass of eggs that were laid by experimental and control females is plotted against clutch size, or the number of eggs laid. The data show a strong increase in egg size as egg number is reduced. The pattern confirms that there is a trade-off between egg size and number in this species. As common sense would predict, it is not possible for a female lizard to produce a large number of large eggs.

What about the assumption that large offspring are higher quality, meaning that they survive better? To test this idea, the team marked 1668 hatchling lizards in the experimental population, released them, and caught them again a month later. The graph in **Figure 38.3c** confirms the prediction that larger offspring survive much better than smaller offspring.

Given this trade-off between offspring number and offspring quality, does natural selection favor mothers that produce large numbers of small offspring or small numbers of large offspring? The data in **Figure 38.3d** show the combined effect of egg number and offspring survival. In this population, mothers that produced an intermediate number of offspring of intermediate size generated the highest total number of surviving offspring.

Trade-offs involving these types of investments in time or energy are pervasive in nature. In a similar vein, it is not possible for an organism to be perfectly adapted to all aspects of its environment at all times. Desert animals that sweat to cool off are threatened with dehydration. An osprey's beak is superbly adapted for tearing meat, but is not useful for weaving nesting materials together.

The general message of this section is simple. All adaptations are compromises, and all adaptations are constrained by both historical and genetic factors.

(a) Side-blotched lizard

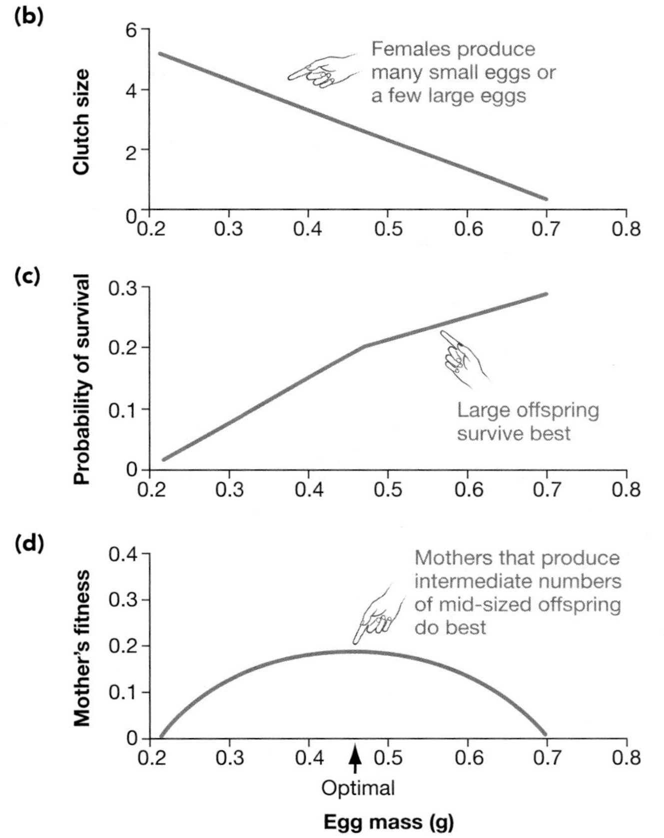

FIGURE 38.3 Female Lizards Face a Quality/Quantity Trade-Off in Offspring Production
(a) Side-blotched lizards live in the deserts of North America. By surgically manipulating clutch size or egg size in this species, researchers were able to augment the natural variation that exists in each trait. **(b)** This graph plots data from surgically manipulated and control clutches. **(c)** Large offspring were much more likely to survive and be recaptured a month after being marked and released into a natural habitat. **(d)** A mother's fitness is calculated as the number of young she produces multiplied by the probability that they survive.

38.3 Tissues, Organs, and Systems: How Does Structure Correlate with Function?

Given that natural selection produces adaptation, and that adaptations are constrained in a variety of ways, how do biologists recognize adaptations? In studies of animal anatomy, biologists often hypothesize that a structure is adaptive if there is a correlation between its function and its size, shape, or composition. For example, **Figure 38.4** highlights the close correspondence existing between the size and shape of the beaks of different species of Galápagos finch and the diet of each species. Birds with large beaks eat large seeds; birds with small beaks eat small seeds. Birds with long, tweezer-like beaks eat insects. Based on observations like these, biologists hypothesize that beak shape has changed over time in response to natural selection.

Correlating structure with function is a pervasive theme in research on how animals work, and it is one of the most basic ways that biologists have to study adaptation. Observations about form and function occur repeatedly throughout this unit. As an introduction to this important theme in biological science, let's take a closer look at the tissues, organs, and systems found in animals.

Levels of Organization in Animal Anatomy

Correlations between form and function occur at many levels in organisms. Earlier chapters emphasized that the shape of proteins often relates to their role as enzymes or structural components of the cell. Similarly, Chapter 5 pointed out that strong correlations exist between the structure and function of the rough ER, Golgi apparatus, mitochondrion, chloroplast, and other organelles. But structure-function correlations also exist at the levels of tissues, organs, and systems.

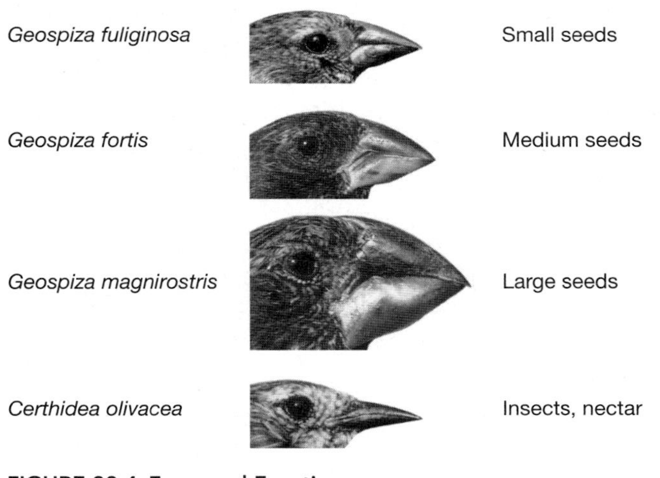

FIGURE 38.4 Form and Function
In Galápagos finch species, there is a strong correlation between the size and shape of the beak and the preferred food items.

Tissues

A **tissue** is a group of cells with the same structure and function. In this unit, we'll encounter the four basic tissue types illustrated in **Figure 38.5**. The **epithelial tissues** cover the outside of the body and line the surfaces of organs. Consistent with its role as a barrier and protective layer, epithelium consists of layers of tightly packed cells. The epithelial cells that line surfaces inside the body are often studded with membrane proteins that regulate the exchange of molecules between organs and tissues.

In contrast to epithelium, **connective tissue** is made up of cells that are loosely arranged in a liquid, jellylike, or solid extracellular matrix. Each type of connective tissue illustrated in Figure 38.5 has a structure that correlates with its function. Loose connective tissue, for example, contains an array of fibrous proteins in a soft matrix. This tissue serves as a packing material between organs. Other connective tissues, such as cartilage and bone, have a firmer extracellular matrix and function in supporting the body or in binding tissues or organs together. Blood is a connective tissue that has a liquid extracellular matrix; it functions in the transport of materials throughout the body.

Muscle tissue functions in movement. As Figure 38.5 shows, muscle is made up of long cells called muscle fibers. Muscle cells are packed with specialized proteins that move when they hydrolyze ATP. ATP hydrolysis, in turn, occurs when electrical signals arrive from nerve cells. Nerve cells or neurons make up **nervous tissue**. Although they vary widely in shape, all neurons have projections that contact other cells and deliver signals in the form of electrical impulses.

Organs and Systems

An **organ** is a structure that serves a specialized function and consists of several tissues. Familiar organs in the human body include the heart, lungs, brain, and liver. A **system** consists of tissues and organs that work together to perform one or more functions. Using the digestive system as an example, **Figure 38.6** illustrates how the structure of organs correlates with their function, and how the components of a system work together in an integrated fashion.

In effect, each chapter in this unit focuses on a different system found in animals, beginning with the excretory system and

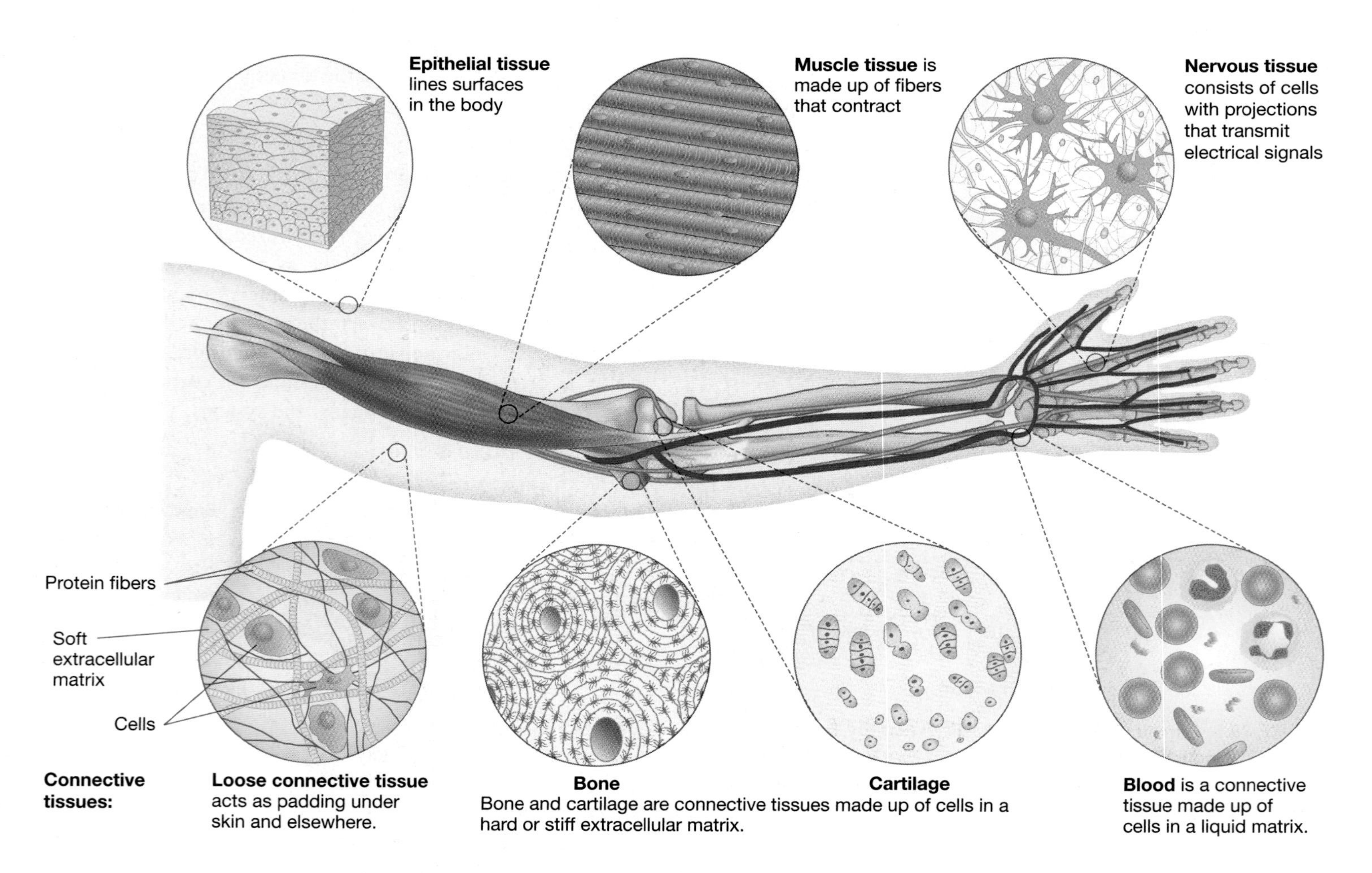

FIGURE 38.5 The Structure of Tissues Correlates with Their Function

QUESTION How does the type of extracellular matrix found in a connective tissue correlate with that tissue's function?

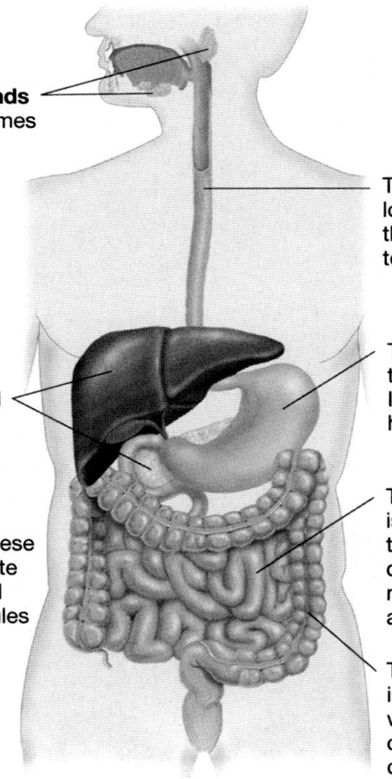

Salivary glands secrete enzymes that begin to digest food

The **esophagus** is a long, muscular tube that transports food to the stomach

The **liver** and **pancreas** contain cells that are packed with endoplasmic reticulum. These organs secrete enzymes and other molecules that aid digestion

The **stomach** is a thick, muscular sac. Its contractions help break up food

The **small intestine** is a long, coiled tube where enzymes digest food and nutrients are absorbed

The **large intestine** is a large tube where wastes are compacted and dried

FIGURE 38.6 The Structure of Organs Correlates with Their Function
The human digestive system shown here is essentially one long tube. The tube is divided into distinct chambers where food is processed and nutrients are absorbed. The salivary glands, liver, and pancreas are organs that secrete specific enzymes or compounds into the tube.

ending with the immune system. The presentations introduce the structures that are found in each system and explore how biologists investigate their function. Each of these systems can be interpreted as a suite of adaptations; each accomplishes a specific task required for survival and reproduction.

Before delving into a detailed look at each system, however, it's essential to examine two general phenomena that affect all of the systems found in animals. Let's look first at how overall body size affects how animals work, and then consider how homeostasis unifies the study of animal systems.

38.4 Body Size and Scaling

Body size has pervasive effects on how animals function. Large animals need more food than small animals. They also produce more waste, take longer to mature, reproduce more slowly, and live longer. Conversely, small animals are more susceptible to damage from cold and dehydration than large animals because they lose heat and water faster. Within the same species, infants and adults face different challenges simply because their body size is different.

Why is body size such an important factor in how animals work? How do biologists study the consequences of size? Let's consider each question in turn.

Surface Area/Volume Relationships

Recall from Chapter 4 that diffusion takes place across the surface of the cell membrane. Nutrients such as glucose must diffuse into the cell, and waste products such as urea and carbon dioxide must diffuse out. As Chapter 41 explains, the rate at which these and other molecules and ions diffuse depends in part on the amount of surface area available for diffusion. In contrast, the rate at which nutrients are used and waste products are produced depends on the volume of the cell.

The contrast between processes that depend on surface area versus volume is important for a simple reason. As a cell gets larger, its volume increases much faster than its surface area. Reviewing a little basic geometry will convince you why this is so. As **Figure 38.7a** (page 736) shows, the surface area of a cube increases as a function of the square of its linear dimension. The volume of the same structure, however, increases as a function of the *cube* of its linear dimension. **Figure 38.7b** graphs the consequences of these relationships. As a cube gets bigger, its volume increases much faster than does its surface area. The same general relationship holds for cells, tissues, organs, and systems.

The relationship between surface area and volume is fundamental. How do changes in this relationship affect the form and function of animals?

CD ACTIVITY 38.1 Surface Area/Volume Relationships

Comparing Mice and Elephants As an example of how surface area/volume relationships affect animals, consider the metabolic rate of mammals. **Metabolic rate** is defined as the overall rate of energy consumption by an individual. Because the consumption and production of energy in mammals depends largely on aerobic respiration, metabolic rate is often measured in terms of oxygen consumption. If so, then metabolic rate is reported in units of milliliters of O_2 consumed per hour.

An elephant consumes a great deal more oxygen per hour than a mouse. But what is going on at the level of cells and tissues? To furnish a more useful comparison of metabolic rates in different species, biologists divide metabolic rate by overall mass and report a mass-specific metabolic rate in units of ml O_2/gram/hour. The mass-specific metabolic rate gives the rate per gram of tissue. Because an individual's metabolic rate varies dramatically with its activity, the accepted convention is to report the basal metabolic rate (BMR)—the rate at which oxygen is consumed at rest, with an empty stomach, in the absence of temperature or water stress.

As **Figure 38.8** (page 736) shows, elephants may be 100,000 times larger than mice, but their mass-specific basal metabolic rate is only one-tenth as large. The leading hypothesis to explain this pattern is based on surface area/volume ratios. Many aspects of metabolism—including oxygen consumption, the digestion of food, the delivery of nutrients to tissues, and the removal of wastes—depend on the exchange of materials across surfaces. As the size of an organism increases, then, its mass-specific metabolic rate must decrease, or the surface area available for exchange of materials would fail to keep up with the metabolic demands generated by the enzymes in the volume.

(a) What is the total surface area and volume of each cube if l (length) = 5 cm?

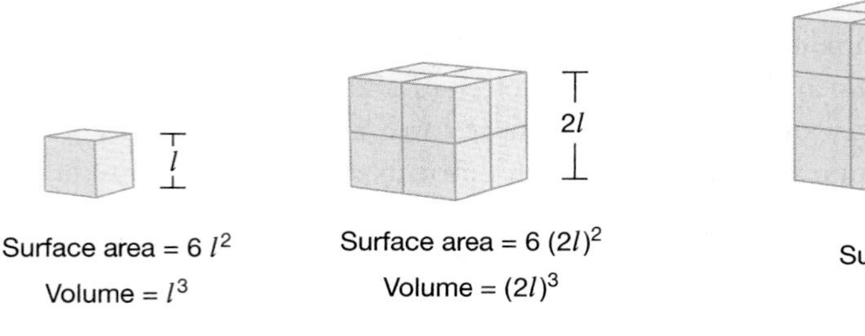

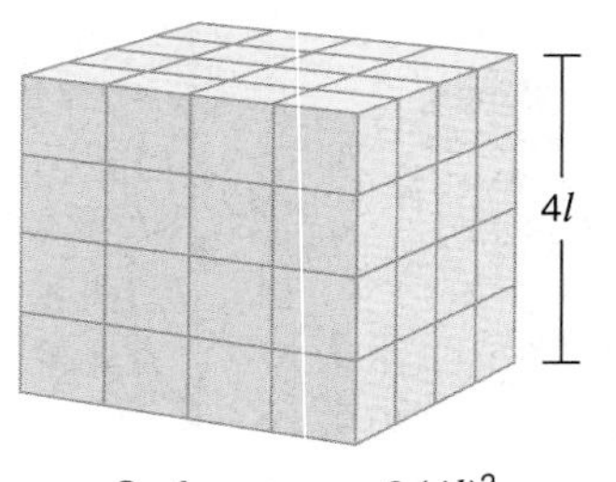

(b) Results are plotted on this graph

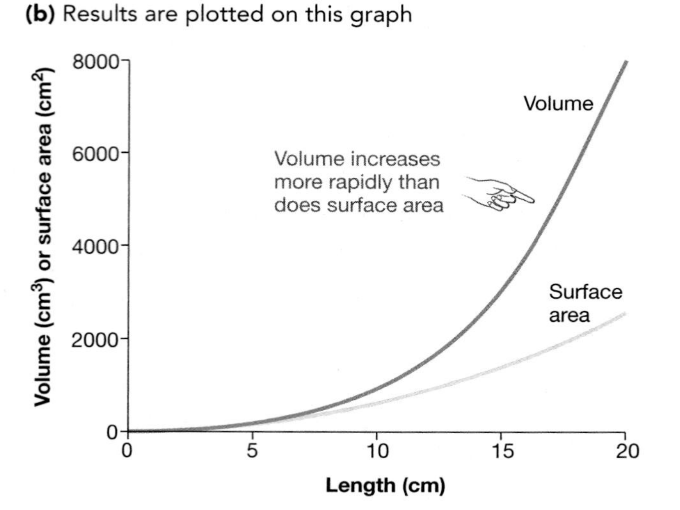

FIGURE 38.7 How Do Surface Area and Volume Change as an Object Gets Larger?
(a) The total surface area of an object increases as the square of the linear dimension (l). The volume, in contrast, increases as the cube of the linear dimension. **(b)** As a result, volume increases much more rapidly than total surface area as linear dimensions increase. **EXERCISE** In part (a), calculate the total surface area and volume of each cube if l (length) is 5 cm.

The same issues influence the development of an individual over the course of its lifetime. A king salmon, for example, weighs a few milligrams or less at hatching but grows into an adult weighing 50 kg or more. This represents a millionfold increase in body mass. To explore the consequences of this change, Patrick Wells and Alan Pinder studied how gas exchange occurs in newly hatched Atlantic salmon. Like most fish species, young salmon have rudimentary gills but also exchange gases across their skin and yolk sac (**Figure 38.9a**). To document the amount of gas exchange that occurs in the gills versus the general body surface, Wells and Pinder inserted individuals' heads through a pin hole in a soft rubber membrane (**Figure 38.9b**). After waiting for the fish to adjust to the apparatus, they recorded the rate of oxygen uptake on either side of the membrane.

Figure 38.9c shows data from experiments using this preparation. The graph shows that newly hatched larvae take up most of the oxygen they need by diffusion across the body surface. As an individual grows, however, its skin surface area decreases in relation to its volume. To avoid suffocation, the fish must develop gills to take over the bulk of the gas exchange activity. The individual breathes through its gills for the rest of its life. This raises an interesting question. What makes gills so effective as the site of oxygen uptake?

Adaptations That Increase Surface Area Figure 38.10a shows a close-up of a gill. Although Chapter 41 explores how this structure functions in detail, the important points to note

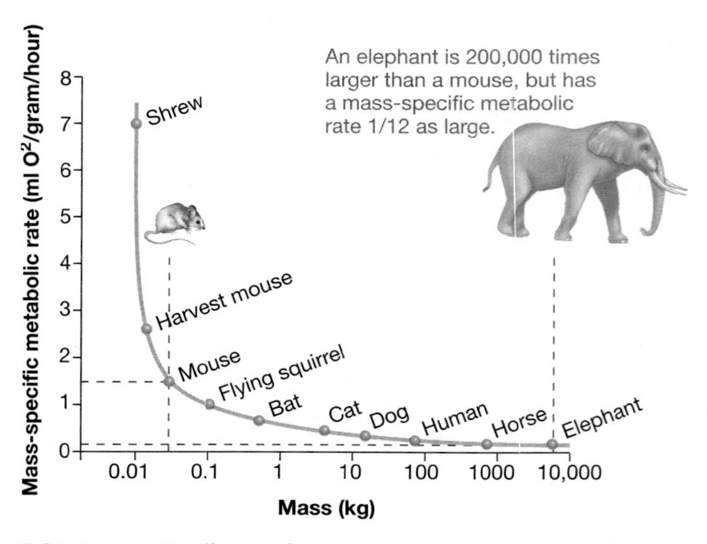

FIGURE 38.8 Small Animals Have Higher Metabolic Rates Than Large Animals
This graph plots overall body mass, on a logarithmic scale, versus metabolic rate per gram of tissue.

(a) Atlantic salmon larvae

(b) Measurement of oxygen uptake

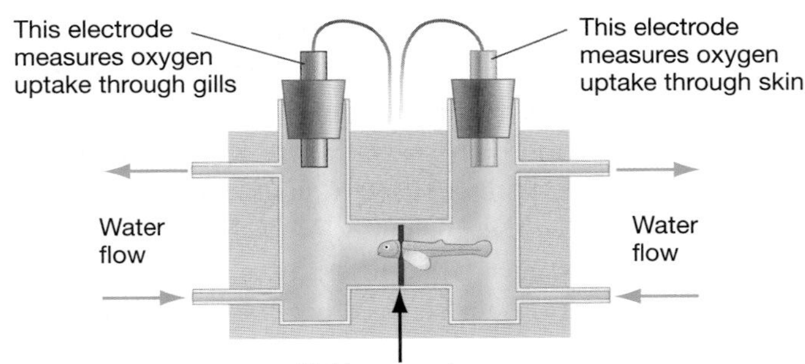

(c) Breathing changes from skin to gills as larvae grow.

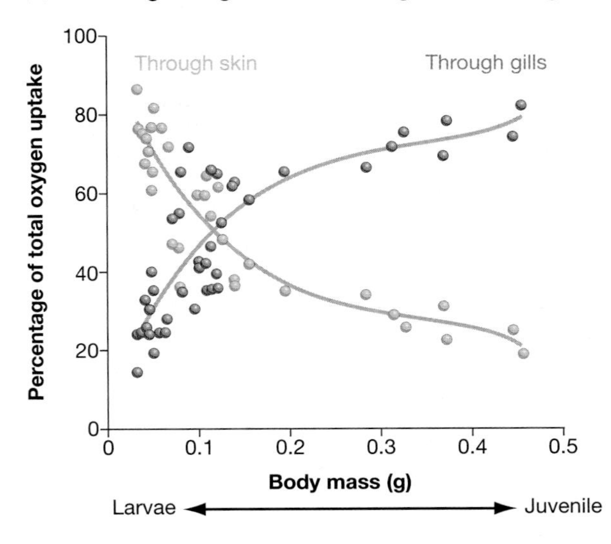

FIGURE 38.9 Do Young Salmon Breathe Through Their Gills or Their Skin?
(a) The bright yellow structure in these Atlantic salmon is a yolk sac. **(b)** This apparatus allowed researchers to measure oxygen uptake by gills versus skin. **(c)** As young salmon increase in size, the percentage of oxygen taken up by gills increases.

here are that the organ consists of sheet-like structures called lamellae, and that the cells on the surface of the sheets are flattened. These features are significant because they produce an organ with an extremely high surface area relative to its volume. Because of these traits, diffusion of gases from water to blood can take place rapidly enough to keep up with the growth in the volume of the developing fish.

In general, if the function of a cell or tissue depends on diffusion, its structure has a shape that increases its surface area relative to its volume. Figure 38.10 also shows some examples from the human body. In addition to flattening, folding and branching are effective ways for structures to have a high surface area to volume ratio. Specifically,

- **Figure 38.10b** illustrates the extensive folding observed in portions of the digestive tract where nutrients diffuse into the body. Because of these folds and the narrow, tube-like projections called villi, the surface area available for diffusion is extremely high. Folded surfaces are common in diffusion-dependent organs.
- The circulatory system shown in **Figure 38.10c** is a highly branched network. The smallest elements in this network are blood vessels called capillaries. Capillaries have a high surface area available for gases, nutrients, and waste products to diffuse. In general, highly branched structures increase the surface area available for diffusion.

(a) Flattened structures

Lamellae of fish gills

(b) Folded surfaces with projections

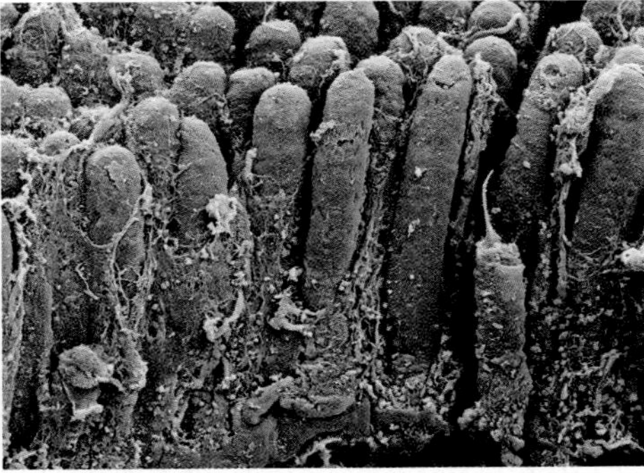

Lining of small intestine

(c) Highly branched structures

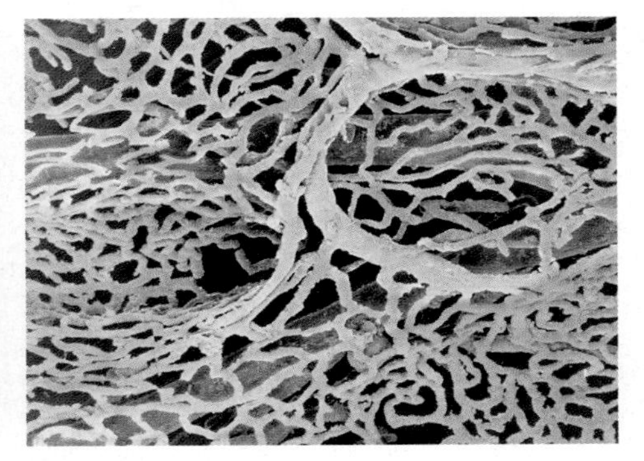

Capillaries

FIGURE 38.10 What Structural Features Produce High Surface Area/Volume Ratios?
All of the cells and structures pictured here have extremely high surface areas relative to their volume.
EXERCISE Identify the diffusion-based process that takes place in each of the structures pictured.

Surface area/volume relationships have a pervasive influence on the structure and function of animals. How do they affect other aspects of animal physiology besides diffusion?

Do All Aspects of an Animal's Body Increase in Size Proportionately?

Biologists have long been fascinated by the fundamental differences that exist between small and large animals. In 1637, for example, Galileo Galilei observed that the bones of large animals are not simply thicker than those of small animals, but disproportionately thicker. **Figure 38.11** shows the relationship between mammals' overall body mass and their skeletal mass. The relationship is considered allometric (different-measure) because the two quantities do not change at the same rate. If they did change at the same rate, their relationship would be described by the dashed line shown in Figure 38.11. Quantities that change at the same rate are considered isometric (same-measure). **Allometry** occurs when changes in body size are accompanied by disproportionate changes in anatomical structures or physiological processes. Galileo was the first to recognize the existence of allometry.

Allometry as a Response to Surface Area/Volume Relationships Why is the relationship between skeletal size and body size in mammals allometric? Answering this question involves another analysis of area and volume relationships. In this case, the volume involved is the mass that must be supported by the skeleton. The area involved is the cross-sectional area of the bones supporting this mass. The strength of these bones, and thus their ability to support weight, is a function of their cross-sectional area. As mass increases, the amount of bone required for support has to increase disproportionately.

Given these arguments, it is not surprising that elephants can't jump. Given its enormous mass and relatively fragile skeleton, an elephant that landed from a height of less than a meter would probably suffer multiple leg fractures.

Allometry as Adaptation Structures that display allometry can sometimes be interpreted as adaptations to a particular lifestyle or environment. Consider, for example, the plot of heart size versus overall body size in **Figure 38.12**. Each point on the graph represents the average value for a species in either the dog family or the cat family. Note that at any given body size, dogs have much larger hearts than do cats. In these two groups, the relationship between heart size and body size is allometric.

To explain this pattern, Ted Garland and Raymond Huey point out that although dogs and cats are both predators, they have very different hunting styles. Dogs run their prey down during long-distance chases, but cats stalk prey and then make a short sprint. Based on this observation, Garland and Huey hypothesize that the large hearts of dogs are an adaptation that increases blood flow to muscles, making long-distance chases possible. In this light, the allometric relationship between heart size and body size in these mammals is logical.

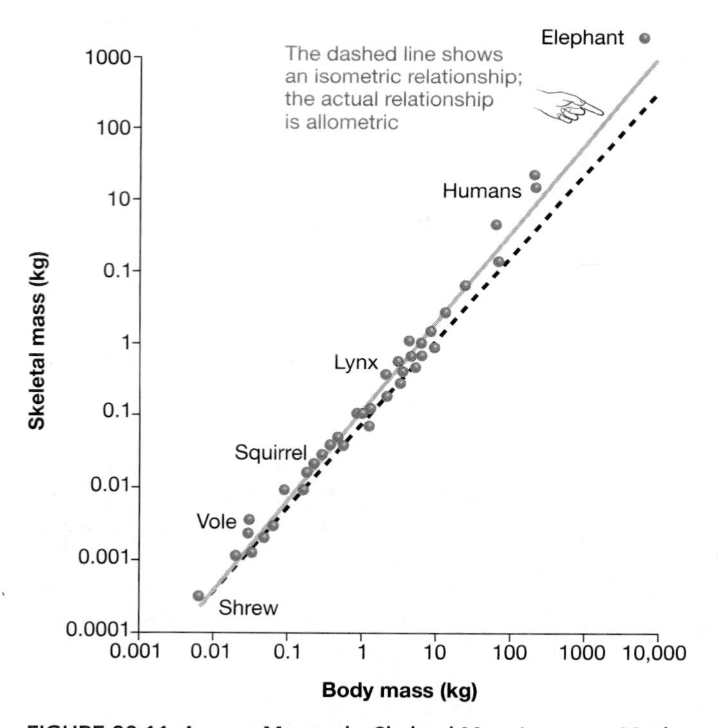

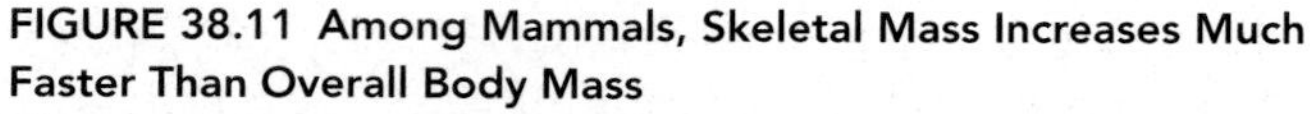

FIGURE 38.11 Among Mammals, Skeletal Mass Increases Much Faster Than Overall Body Mass
This graph plots overall body mass versus total skeletal mass of terrestrial mammals. Note that both axes are logarithmic, and that large land-dwelling mammals have much heavier skeletons than expected for their size. QUESTION Would you expect the same relationship in aquatic mammals, whose bodies are supported by water?

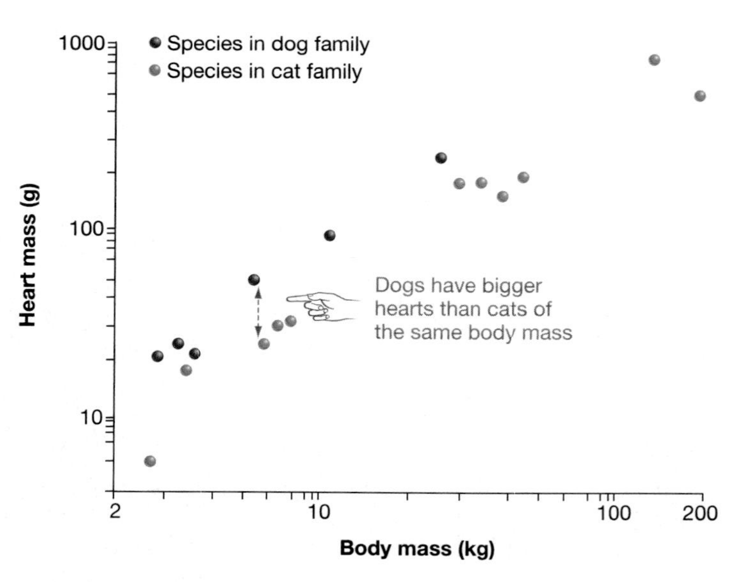

FIGURE 38.12 Compared to Cats, Dogs Have Large Hearts
This graph plots heart size as a function of overall body size. Each data point represents the average value for a species. Note that both axes are logarithmic.

38.5 Homeostasis

In analyzing animal form and function, biologists focus on understanding adaptations as a product of natural selection, exploring the constraints on adaptation, and interpreting the influence of general principles like surface area/volume ratios. In addition, researchers analyze animal physiology in the context of a general principle called homeostasis.

As indicated in the introduction to this chapter, homeostasis is the maintenance of relatively constant chemical and physical conditions in an animal's cells, tissues, and organs. Many of the structures and processes observed in animals can be interpreted as mechanisms for achieving homeostasis with respect to some quantity, ranging from the concentration of nutrients to the pH of blood.

The Role of Epithelium

Epithelium plays a vital role in creating an internal environment that is dramatically different from the external environment, and in maintaining physical and chemical conditions inside an animal that are relatively constant. No matter where it is found in the body, epithelial tissue has a distinctive polarity. As **Figure 38.13** shows, the apical side of epithelium faces away from other tissues and toward the environment. The basolateral side of epithelium, in contrast, always faces the interior of the animal and is connected to other tissues.

Because epithelium exists at the interface between the internal and external environment, the tissue plays a key role in achieving homeostasis. Its most basic function is to control the exchange of materials across surfaces. As we'll see in subsequent chapters, epithelial cells are frequently studded with membrane proteins that regulate the transport of ions, water, nutrients, and wastes. No molecule can enter or leave the body without crossing an epithelium. Achieving homeostasis is possible because epithelia control this exchange.

Why is achieving homeostasis important? A large part of the answer springs from a consideration of enzyme function. Enzymes are proteins that catalyze chemical reactions within cells. Because temperature, pH, and other physical and chemical conditions have a dramatic effect on protein structure, an enzyme functions most efficiently under a fairly narrow range of conditions. Temperature changes also affect membrane permeability and how quickly solutes diffuse. Further, if tissues are allowed to drop much below 0°C, the expansion of water as it freezes can rip cells apart. Conversely, extremely high temperatures can cause proteins to lose their tertiary structure. When proteins denature in this way, they stop functioning.

Subsequent chapters in this unit explore how animals achieve homeostasis with respect to the solute concentrations of their cells and tissues, their oxygen supply, and nutrient availability. To introduce the concept, let's focus on how animals achieve homeostasis with respect to temperature.

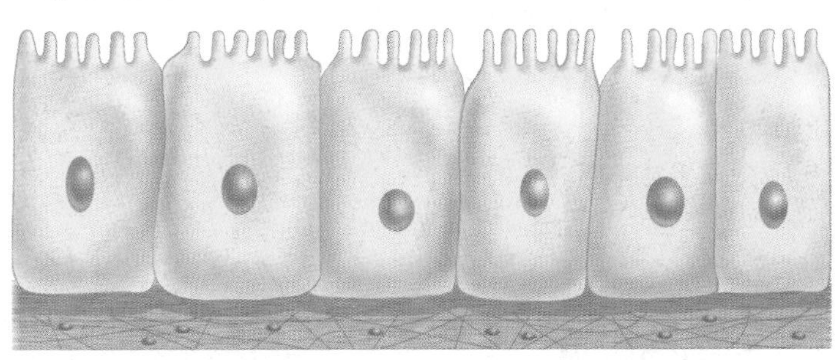

FIGURE 38.13 Epithelium Has a Distinctive Orientation
QUESTION The inner surface of the lungs, stomach, and small intestine are lined with epithelium. What is the nature of the environment faced by the apical surface of these epithelia?

CD ACTIVITY 38.2 Homeostasis

Gaining and Losing Heat

To understand how animals achieve homeostasis with respect to body temperature, it's critical to realize that animals fall into two general categories based on how they gain heat. An **endotherm** (inner-heat) produces heat in its own tissues while an **ectotherm** (outer-heat) relies on heat gained from the environment. Birds and mammals are endotherms; most other animals are ectotherms. Several species lie between these two extremes, however. Some ectothermic pythons can generate body heat when they need to warm their eggs; tuna, mackerel, and certain other species of ectothermic fish generate heat to warm certain sections of their bodies, such as their eyes or specific muscles.

Ectotherms can elevate their body temperatures far above the surrounding air temperature by basking in the sun or by resting on warm surfaces. In addition, ectotherms generate a small amount of heat as a by-product of metabolism. Compared to endotherms, however, ectotherms have low metabolic rates. Endotherms warm themselves because their basal metabolic rates are high, and because they have elaborate insulating structures like feathers or fur.

The contrast in metabolic rate and heat-producing ability is reflected in the structure of tissues in the two groups. When researchers compare the cells of endotherms and ectotherms, they routinely find that mitochondrial density and mitochondrial enzyme activity are 3 to 4 times higher in endotherms than in ectotherms of similar size. The correlation is meaningful because mitochondria are the site of chemical reactions that generate ATP and heat.

Specialized Heat-Generating Tissues In many ectotherms as well as in most endotherms, heat is produced by muscle activity during movement—and sometimes by the involuntary muscle contractions known as shivering—as well as by normal cell metabolism. Specialized heat-generating tissues are also observed, however. For example, many mammals have specialized heat-generating cells called brown adipose tissue. As **Figure 38.14** shows, brown adipose tissue has a high density of mitochondria and stored fats. When fats are oxidized by the mitochondria in brown adipose tissue, no ATP is produced. Instead, all of the stored energy is released as heat. As a result, brown adipose tissue produces almost 10 times as much heat as do other tissues.

Brown adipose tissue is particularly common in small animals and in the infants of large species. In humans, for example, patches of brown adipose tissue are found in the neck and chest regions of newborns. Because infants are small, they have a high surface area/volume ratio and lose heat quickly. Based on observations like this, brown adipose tissue is interpreted as an adaptation for small endotherms to achieve homeostasis with respect to temperature.

Exchanging Heat with the Environment All animals constantly exchange heat with their environment. Heat "flows downhill" from regions of higher temperature to regions of lower temperature. If an individual is warmer than its surroundings, it will lose heat; if it is cooler than its environment, it will gain heat.

As **Figure 38.15** shows, animals exchange heat with the environment in four ways: conduction, convection, radiation, and evaporation. **Conduction** is the direct transfer of heat between two physical bodies that are in contact with each other. The rate at which conduction occurs depends on the surface area of transfer, the steepness of the temperature difference between the two bodies, and how well each body conducts heat. Water, for example, conducts heat much better than air. As a result, a person immersed in 15°C water loses heat much faster than a person exposed to air at the same temperature.

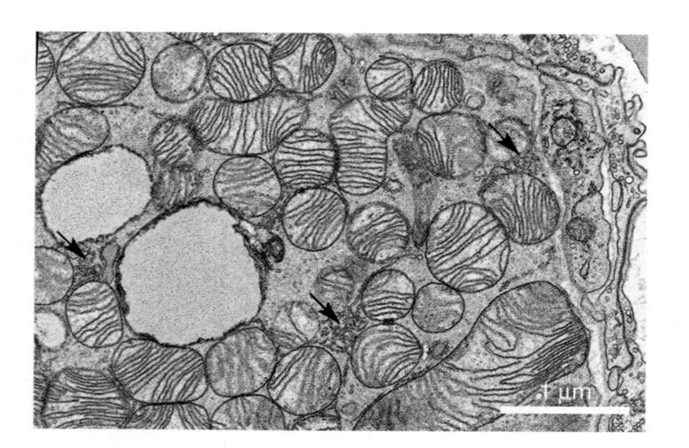

FIGURE 38.14 The Structure of Cells in Brown Adipose Tissue Correlates with Their Function
The adipose (fat) cells in brown adipose tissue are packed with mitochondria. Instead of producing ATP, these mitochondria primarily produce heat.

Convection occurs when air or water moves over the body surface. In effect, convection increases the rate of heat exchange by conduction. The increase occurs because the air or water in direct contact with the body surface is constantly replaced. As a result, convection maintains a large temperature gradient for conduction. As the speed of the air or water flow increases, so does the rate of heat transfer.

Radiation is the transfer of heat between two bodies that are not in direct physical contact. All objects, including animals, radiate energy as a function of their temperature. The major source of radiant energy, however, is the Sun.

Evaporation is the phase change that occurs when liquid water becomes a gas. Unlike conduction, convection, and radiation, evaporation leads to heat loss only—not loss or gain. Because of the extensive hydrogen bonding in liquid water, it takes a large amount of energy to heat water up and produce evaporation. Consequently, water is an efficient coolant on a hot day. Conversely, getting wet can be deadly on a cold day.

Conserving Heat In terrestrial environments, heat is readily lost through evaporation, convection, and radiation. Air conducts heat poorly, however; it is a good insulator as a result. In this light, it is no coincidence that endothermic animals have elaborate external structures that trap air, slow the rate of heat exchange through conduction, and conserve body heat. Birds have down feathers; mammals have underfur that forms a fine mesh, traps air, and prevents heat lost by conduction.

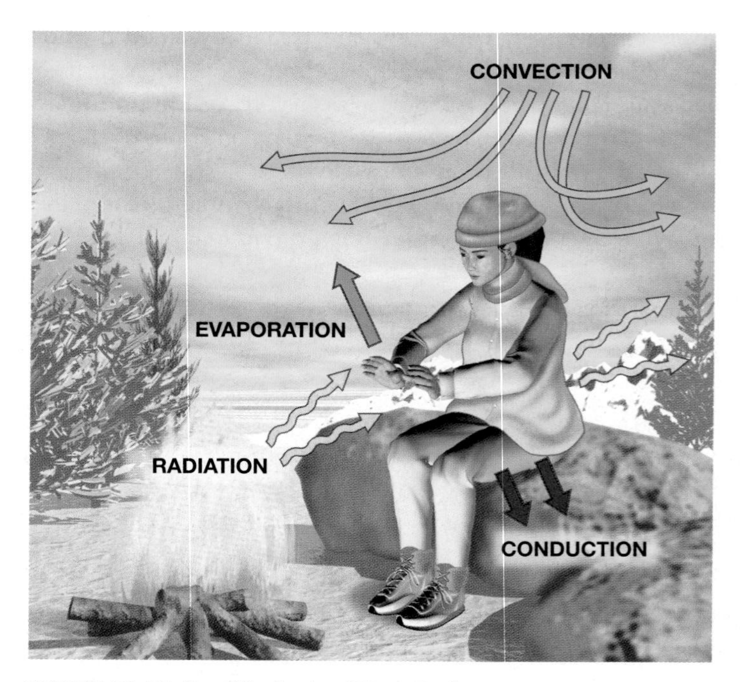

FIGURE 38.15 Four Methods of Heat Exchange
The arrows indicate the direction of heat exchange from the warmer element to the cooler element. **EXERCISE** Sketch a similar diagram that features a butterfly in summer.

In the aquatic environment, there is no heat loss due to evaporation. Similarly, little heat is lost or gained due to radiation because water transmits radiation poorly. But water is such an effective conductor of heat that metabolic heat produced by aquatic organisms is lost very rapidly—particularly through structures like the gills, which have a large surface area and a high rate of blood flow. As a result, the body temperature of most aquatic invertebrates and fish is the same as that of the water they inhabit.

How do endotherms that live in aquatic habitats cope with heat loss due to conduction? If you've ever gone swimming in cold water, you can appreciate the problem faced by seals, otters, and whales. To conserve heat, otters have dense, water-repellent fur that maintains a layer of trapped air next to the skin. Seals and whales are insulated by thick layers of fatty blubber. In addition, at least some aquatic mammals have a feature that minimizes heat loss from their tongue, which is exposed to cold water during feeding. John Heyning and James Mead have shown that the arteries and veins in the tongue of gray whales are juxtaposed, as shown in **Figure 38.16a**. The arteries carry warm, oxygenated blood to the tongue tissues while the veins carry cooled, deoxygenated blood back to the body. Because the two types of blood vessel are arranged in an anti-parallel fashion, heat is transferred efficiently between them. This arrangement is called a **countercurrent heat exchanger** and is explored in more detail in Chapter 39. The key point is that heat is transferred from the arteries to the veins and returned to the body instead of being lost to the surrounding water. As **Figure 38.16b** shows, similar heat-conserving arrangements of arteries and veins are found in the flippers of whales and dolphins and in the legs of arctic fox.

Regulating Body Temperature

To achieve homeostasis, animals have regulatory systems that constantly monitor internal conditions like temperature, blood pressure, and blood pH. If one of these parameters changes, a homeostatic system acts quickly to modify it. Like the thermostat in a home heating system, each of these systems has a **set point**—a normal or target value for the controlled variable. In mammals, the set point for body temperature is about 37°C. How does an individual maintain tissues at this set point in the face of changes in activity and the environment?

The key to answering this question is to recognize that homeostatic systems are based on the three general components illustrated in **Figure 38.17** (page 742). A **sensor** is a structure that senses some aspect of the external or internal environment. An **integrator** is a component of the nervous system that evaluates the incoming sensory information and "decides" if a response is necessary to achieve homeostasis. An **effector** is any structure that helps to restore the desired internal condition. Without these three elements, homeostatic control systems are unable to maintain a desired set point and homeostasis is impossible.

Figure 38.17 notes how these three components work in the case of body temperature. Receptors located in the brain constantly monitor body temperature. This information is integrated by nearby neurons. If an endotherm starts becoming too cold, the integrator initiates signals that induce shivering to generate

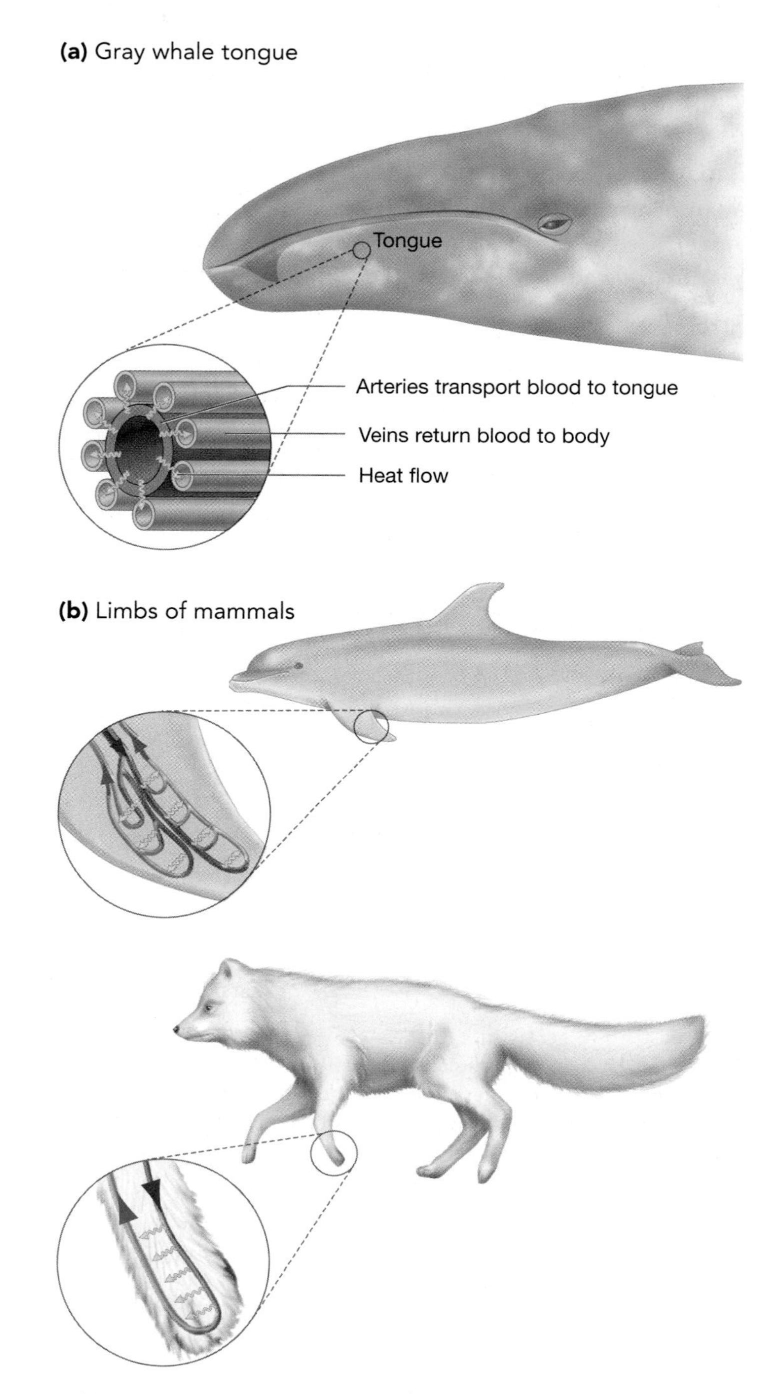

FIGURE 38.16 Countercurrent Exchangers Conserve Heat
(a) In the tongue of a gray whale, blood runs in opposite directions through juxtaposed veins and arteries. As a result, heat flows from arteries to veins and is returned to the body instead of being lost to the surrounding water. **(b)** Similar countercurrent arrangements are common in the limbs of mammals and birds that live in cold habitats.

more heat and fluffing up fur or feathers to improve insulation and retain heat. Many ectotherms position their bodies at right angles to the sun in response to cold, to maximize the surface area exposed to solar radiation. Ectotherms may also sprawl on or under sun-baked rocks to gain heat by conduction. If an individual becomes too hot, in contrast, the integrator triggers signals that initiate evaporative cooling by sweating or panting along with seeking shade or a cool burrow. In both cases, behavior (like shade seeking) and physiological responses (like sweating) move the variable in question back toward the set point.

Ectothermy Versus Endothermy

Analyzing animal thermoregulation provides a convenient way to review some of the major themes in this chapter. Biologists interpret homeostatic systems as adaptations that have arisen

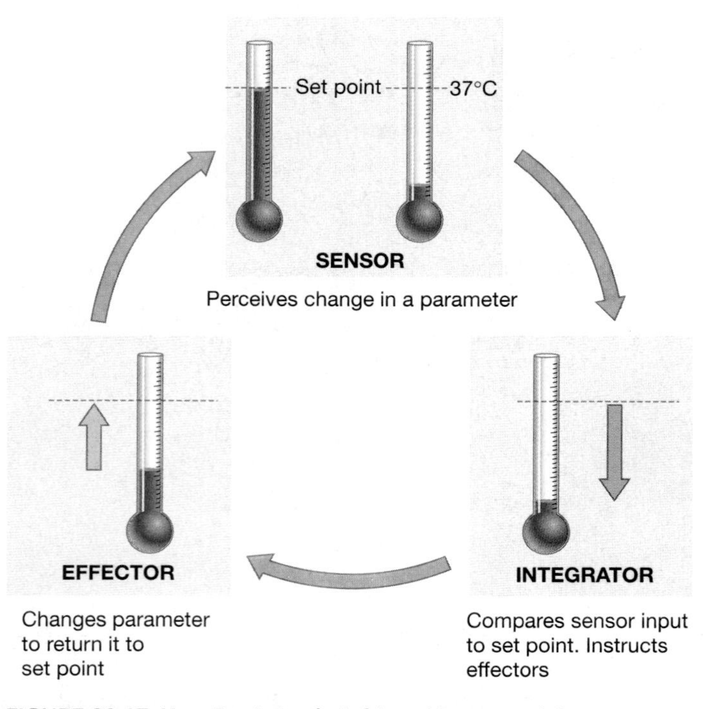

FIGURE 38.17 How Do Animals Achieve Homeostasis?
This diagram shows the interactions that occur between the three components of a generalized system for attaining homeostasis. In the case of maintaining a preferred body temperature, many animals use homeostatic systems similar to this one.

through natural selection. The structure of cells and tissues involved in homeostasis, such as mitochondria-rich brown adipose tissue and countercurrent heat exchangers, correlates with their function. Because heat is gained and lost across surfaces, surface area/volume relationships are important. Small objects lose heat much faster than large objects, so it is difficult for small endotherms to stay warm without expending prohibitive amounts of energy. Surface area and volume relationships explain why there are no endotherms smaller than a shrew, even though there are millions of centimeter- or millimeter-sized ectotherms.

Contrasting endothermy and ectothermy also provides a superb example of trade-offs in adaptation. Because endotherms maintain enzymes at optimal temperatures at all times, mammals and birds are able to remain active in winter and at night. Due to their high metabolic rates and insulation, endotherms are also able to sustain very high levels of aerobic activities like running or flying. These abilities come at a cost, however. To fuel their high metabolic rates, endotherms have to obtain large quantities of energy-rich food. The energy used to produce heat is then unavailable for other energy-demanding processes such as reproduction and growth.

Ectotherms, in contrast, are able to thrive with much lower intakes of food. They can also use a greater proportion of their total energy intake to support reproduction. But because chemical reactions are temperature dependent, muscular activity and digestion slow down dramatically as the body temperature of an ectotherm drops. As a result, ectotherms are more vulnerable to predation in cold weather, and in general are not as successful at inhabiting cold environments or maintaining activity in cool nighttime temperatures.

In short, each suite of adaptations has advantages and disadvantages. Like all adaptations, endothermy and ectothermy involve trade-offs. Analyzing these trade-offs has also inspired a spirited debate on why endothermy evolved in the first place. Birds evolved from ectothermic dinosaurs and mammals from a lineage of ectothermic reptiles that is now extinct. What was the selective advantage of endothermy in each case? Was it to make sustained activity like flying possible? Or did endothermy allow birds and mammals to colonize cold habitats or remain active at night? These questions are unresolved. Research on animal form and function and the nature of adaptation continues.

Essay Is Fever Adaptive?

If you've ever experienced a sustained elevation in body temperature called fever, you've probably taken aspirin, acetaminophen, or ibuprofen to reduce that fever. But was taking the medicine a good idea? Does fever help or hinder your ability to fight off an infection?

High fevers are clearly not adaptive. In adult humans, tissue damage begins to take place if body temperature reaches 40°C (104°F); serious brain injury can occur if temperatures elevate to 41°C (106°F). But it is possible that lower fevers, in the range of 38–39°C, could inhibit the growth of infectious agents.

Does fever help fight an infection?

To test this hypothesis, J. J. Schall and colleagues performed an experiment on western fence lizards. In natural populations of this species, it is not uncommon to observe malaria infections in 15 percent or more of the individuals. Malaria is caused by a eukaryotic parasite in the genus *Plasmodium*. Because *Plasmodium* is transmitted to new hosts by mosquitoes, and because mosquitoes seek hosts in part by sensing heat, it is reasonable to suggest that the parasite induces fever. The logic here is that a fever would increase the parasite's chances of being taken up by a mosquito and transmitted to a new host.

The experiment that Schall and co-workers set up was simple in concept. They brought a large number of lizards into the laboratory and infected them with malaria. Then they divided the infected individuals into two groups. One group was maintained in constant conditions at the preferred body temperature for the species. Individuals in the second group were allowed to change their body temperature by seeking colder or warmer areas in their enclosure. The researchers documented that individuals in the second group actively sought out warm environments, and that this behavior induced fever.

Schall and colleagues claimed that fever is adaptive in this species because individuals that experienced fever recovered from infection better than individuals that were maintained at a normal body temperature. Experiments on other ectotherms have confirmed this result, and have led to a strong consensus that fever is an adaptive response in ectotherms. Unfortunately, experiments with humans have produced much less conclusive results. As this book goes to press, the role of fever in the human disease-fighting response is still not clear. The wisdom of using fever-suppressing medications remains controversial.

Chapter Review

Summary

An adaptation is a trait that allows individuals to produce more offspring in a particular environment than individuals without the trait. Adaptations are genetic changes that result from natural selection exerted by the environment. Although the genetic characteristics of populations change through time in response to natural selection, the genetic characteristics of individuals do not. If the phenotype of an individual changes in response to environmental change, the change in the individual is due to acclimation—not adaptation.

There is often a strong correlation between the structures found in animals and their function. Not all structures have functions, however, and not all traits are adaptive. Some traits exist because they are remnants of structures that were present in ancestors or that appeared in the individual as an embryo. Further, all adaptations are constrained by genetic correlations with other traits, lack of genetic variation, historical constraints, and trade-offs. In reproduction, for example, females must make a trade-off between the number and quality (size) of eggs that they produce. Because trade-offs are inevitable, all adaptations are compromises.

Body size has pervasive effects on how animals work. Many of these effects result from the relationship between the surface area of a structure and its volume. For example, large animals have low metabolic rates because they have a relatively small surface area for exchanging the oxygen and nutrients required to support metabolism. The relatively high surface area of small animals, in contrast, means that they lose heat extremely rapidly. As a result, there are no tiny endothermic animals.

When structures are compared across species, most are found to change allometrically—that is, in a nonlinear fashion—with body mass. Among mammals, for example, skeletal mass increases much faster than overall body mass. Basal metabolic rate, in contrast, increases much more slowly than overall body mass.

Homeostasis refers to the maintenance of relatively constant physical and chemical conditions inside the body. Animals have a set point or preferred value for blood pH, tissue oxygen concentration, nutrient availability, and other parameters. For example, mammals have a set point for body temperature of about 37°C. If an individual starts to overheat, it will pant or sweat and seek a cool environment in response. If its body temperature begins to drop, it will respond by shivering, basking in the sun, or fluffing its fur.

Questions

Content Review

1. Which of the following statements about natural selection and adaptation is not correct?
   a. Natural selection acts on individuals, but only populations change in response.
   b. Populations evolve in response to environmental change; individuals acclimate.
   c. Natural selection inevitably leads to more complex structures and physiological systems over time.
   d. Evolution does not produce "higher" and "lower" organisms.
2. Which of the following observations supports the claim that not all traits are adaptive?
   a. Vestigial traits are functionless holdovers from ancestral traits.
   b. Some traits are holdovers from traits that appeared in embryos.
   c. During the drought on Daphne Major, average beak width in the finch population increased even though narrower beaks were favored by natural selection.
   d. All of the above support the claim.
3. How do biologists measure an animal's metabolic rate?
   a. by taking its temperature
   b. by measuring how rapidly it uses oxygen
   c. by measuring how rapidly it uses glucose
   d. by measuring how rapidly it produces wastes
4. How is the structure of a connective tissue correlated with its function?
   a. The density of cells in the tissue correlates with its function.
   b. The type of cell in the tissue correlates with its function.
   c. The type of protein fiber in the tissue correlates with its function.
   d. A jellylike, solid, or liquid extracellular matrix correlates with the tissue's function in padding, support, or transport.
5. As an animal gets larger, which of the following occurs?
   a. Its surface area grows more rapidly than its volume.
   b. Its volume grows more rapidly than its surface area.
   c. Its volume and surface area increase in perfect proportion.
   d. Its volume increases, but its total surface area actually decreases.
6. Which of the following describes the set point in a homeostatic system?
   a. the cells that collect and transmit information about the state of the system
   b. the cells that receive information about the state of the system and direct changes
   c. the elements that produce appropriate changes in the system
   d. the target or "normal" value of the variable in question

Conceptual Review

1. A pie can be cut into many small pieces or a few large pieces. But the same pie cannot be cut into many large pieces. (a) Explain how this example illustrates the concept of trade-offs. Relate the pie example to trade-offs that occur as a female is producing eggs. (b) For a female dragonfly, what environmental factors might determine whether she would produce more surviving offspring by making many small eggs versus a few large eggs?
2. The metabolic rate of a frog in summer (at 35°C) is about eight times higher than in winter (at 5°C). Compare and contrast the individual's ability to move, exchange gases, and digest food at the two temperatures. During which season will the frog require more food energy, and why?
3. How does the function of the following cells, tissues, and organs relate to their structural characteristics?

   Nerve cells — Absorptive sections of digestive tract
   Brown adipose tissue — Capillaries
   Lung tissue in humans — Galápagos finch beaks
   Gill tissue in fish
4. When researchers graph overall body mass versus skeletal mass for mammals ranging from mice to elephants, skeletal mass increases much more rapidly than body mass. Explain why this relationship occurs, and why it is considered allometric.
5. Consider a day in which daytime temperatures reach 30°C and nighttime temperatures drop to 18°C. Analyze how an ant might gain and lose heat by conduction, convection, radiation, and evaporation to avoid overheating during the day and escape cold-induced lethargy in the early morning and evening.
6. Why is epithelium a particularly important tissue in achieving homeostasis?
7. Explain how a countercurrent heat exchanger works.

Applying Ideas

1. When people move from sea level to high altitude, their bodies undergo acclimation. During this period, heart and lung activity and blood chemistry changes to compensate for the low availability of oxygen. Design a study to test the hypothesis that human populations that have lived at high altitudes for many centuries are not acclimated to oxygen scarcity, but are adapted. Also test the idea that they are both acclimated and adapted.

2. The fossil record documents the existence of tortoises that were the size of small cars, dragonflies that had wingspans of two feet, and predatory sharks that were larger than some of today's whales. Using these examples, discuss the advantages and disadvantages of extremely large size in terms of natural selection.

3. An engineer friend of yours has to design a system for dissipating heat from a new type of car engine that runs particularly hot. Recall that heat is gained and lost as a function of surface area. Suggest ideas for her to consider that are inspired by biological structures with exceptionally high surface area/volume ratios.

4. Biologists often refer to lower and higher plants and animals as a shorthand for differentiating more ancient lineages from more recently derived ones. According to the text, however, the adjectives *higher* and *lower* are inaccurate and misleading. Suggest better terms.

5. Suppose that you are the veterinarian at a zoo and have just been put in charge of caring for six individuals of a new mammal species. According to the scientific literature, almost nothing is known about the physiology of this animal. Describe how you would go about determining the set points for body temperature, blood pH, blood glucose, blood pressure, and heart rate in this species. Is it reasonable to assume that the set points for adult males, adult females, and juveniles are identical? Why or why not?

CD-ROM and Web Connection

CD Activity 38.1: Surface Area/Volume Relationships *(animation)*
(Estimated time for completion = 10 min)
Explore the relationship between surface area and volume, which determines many aspects of anatomy and physiology in animals.

CD Activity 38.2: Homeostasis *(animation)*
(Estimated time for completion = 10 min)
How do animals achieve homeostasis despite changing external conditions?

At your **Companion Website** (http://www.prenhall.com/freeman/biology), you will find self-grading exams and links to the following research tools, online resources, and activities:

Evolving Backwards
Jared Diamond, author of the book *Guns, Germs, and Steel*, writes about the fate of functionless (vestigial) organs.

Growing New Organs
This article describes medical research intended to artificially prompt cells to differentiate into new tissue types in an effort to regenerate organs.

Of Mice and Mammoths
This editorial discusses recent thought and research on the relationship of body surface to volume.

Are Vomiting and Pain Adaptive?
This article describes the adaptive significance of neurological responses and physiological behaviors that help the body maintain homeostasis.

Additional Reading

Harvey, P. H. and C. J. Godfray. 2001. A horn for an eye. *Science* 291: 1505–1506. Comments on recent research about constraints on sexual selection—the type of natural selection introduced in Chapter 22.

Körtner, G., R. M. Brigham, and F. Geiser. 2000. Winter torpor in a large bird. *Nature* 407: 318. Data suggest that Australian frogmouths lower their body temperature at night to conserve resources.

Mills, C. 1998. Blood feud. *The Sciences* 38: 34–38. Introduces adaptations and counter-adaptations in ticks and their human hosts.

Storey, K. B. and J. M. Storey. 1999. Lifestyles of the cold and frozen. *The Sciences* 39: 32–37. Reviews research on how frogs survive cold weather by freezing and thawing.

Water and Electrolyte Balance in Animals

39

The chemical reactions that make life possible occur in an aqueous solution. If the balance of water and solutes in that aqueous solution is disturbed, those chemical reactions—and life itself—may stop. Although most animals can tolerate at least some short-term deviation from normal water and solute conditions in their bodies, longer-term problems have serious consequences. A human may be able to stay alive for weeks without eating, but will probably survive less than 2 or 3 days without drinking water. Hurricanes occasionally introduce so much freshwater to the ocean shore that normal salt concentrations are disrupted and marine animals die. Maintaining water balance is a matter of life or death.

An animal has achieved water balance when its intake of water equals its loss of water. Maintaining water balance is an important element in homeostasis—the ability to keep cells and tissues in constant and favorable conditions—and is intimately associated with sustaining a balanced concentration of solutes throughout the body. In many animals, the most abundant solutes are the ions sodium (Na^+), chloride (Cl^-), potassium (K^+), and calcium (Ca^{2+}). As a group, these substances are called electrolytes. An **electrolyte** is a compound that dissociates into ions when dissolved in water. Because cells require precise amounts of Na^+, Cl^-, K^+, and Ca^{2+} to function normally, maintaining electrolyte balance is crucial to survival. In humans, severe electrolyte imbalances can lead to muscle spasms, confusion, irregular heart rhythms, fatigue, or even paralysis.

How do animals maintain water and electrolyte balance? The answer to this question depends on the taxonomic group and environment being considered. In freshwater fish, for example, the central problem in maintaining water balance is to rid the body of excess water

Terrestrial animals lose water every time they breathe. For many animals, drinking is an important way to gain water and achieve homeostasis. This chapter explores how terrestrial and aquatic animals maintain water balance.

and prevent cells from rupturing. But in marine fish and in terrestrial animals, the central problem is to conserve water and prevent cells from dehydrating.

Our investigation into water and electrolyte balance begins with a look at how water moves in and out of cells. We'll consider why freshwater, saltwater, and terrestrial environments affect animals so differently, and then we'll delve into research on the cellular and molecular mechanisms responsible for maintaining water and electrolyte balance in each of these environments. Although aquatic and land animals face very different challenges, in several cases these challenges are met by common molecular mechanisms.

39.1 Osmotic Stress and Osmoregulation

Chapter 4 introduced the processes called diffusion and osmosis. Diffusion describes the movement of substances from regions of higher concentration to regions of lower concentration. **Figure 39.1a** illustrates diffusion of a dissolved substance, or solute. Water, like dissolved substances, can also move down its concentration gradient. When dissolved substances are separated by a semipermeable membrane and the solutes cannot cross the membrane, water will move from areas where it is more concentrated—meaning that solute concentrations are lower—to areas where it is less concentrated, meaning that solute concentrations are higher. The movement of water from areas of higher water concentration to areas of lower water concentration is called osmosis.

The concentration of dissolved substances in a solution, measured in moles per liter, is referred to as its **osmolarity**. Water moves across semipermeable membranes from a region of lower osmolarity to a region of higher osmolarity. How does osmosis affect water balance in animals?

Water Balance in Freshwater, Marine, and Terrestrial Environments

Consider the freshwater fish illustrated in **Figure 39.2a** (page 748). The epithelial cells on the surfaces of its gills are exposed to the surrounding water. (Chapter 41 explores how gills, like lungs,

(a) DIFFUSION

Solutes move from areas of higher concentration to areas of lower concentration.

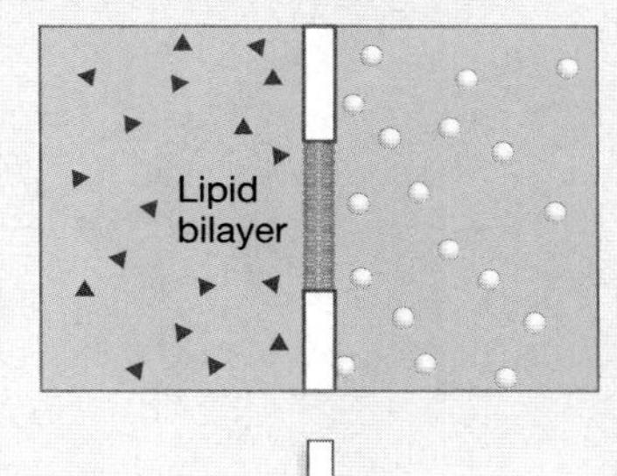

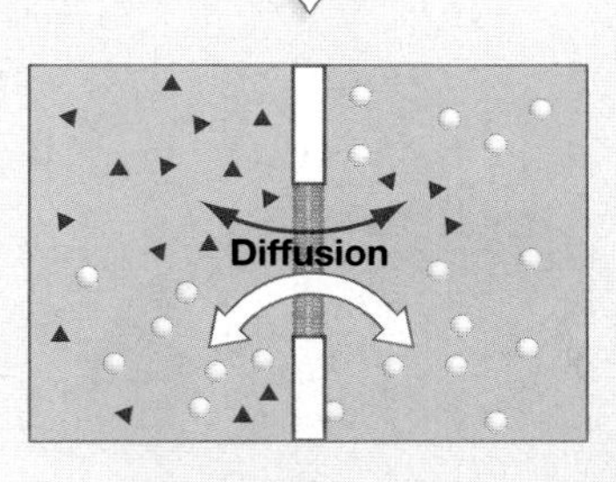

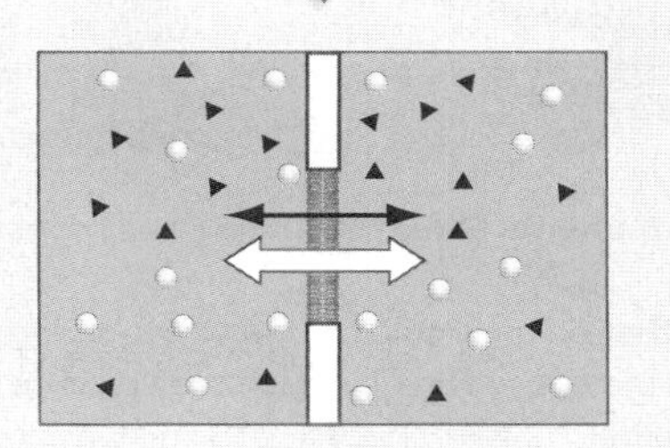

(b) OSMOSIS

Water moves from areas of higher concentration to areas of lower concentration.

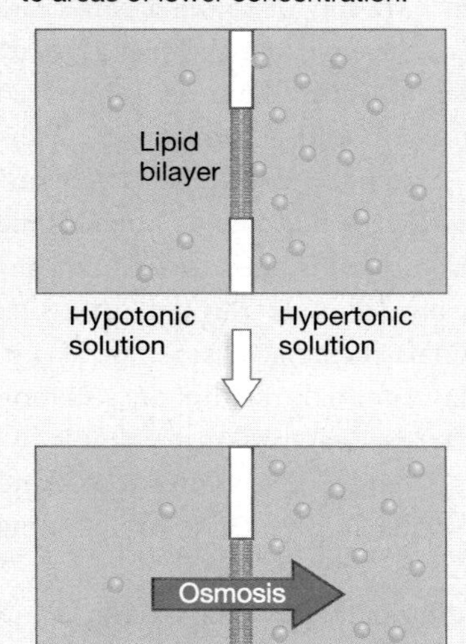

FIGURE 39.1 Diffusion and Osmosis
(a) This diagram illustrates the diffusion of two solutes across a membrane that is permeable to both substances. Note that the solutes are initially at different concentrations on either side of the membrane. Because of diffusion, they reach equilibrium, meaning that the total solute concentration is the same on either side of the membrane and no further net movement occurs. **(b)** Osmosis is a special case of diffusion. It involves the movement of water across a semipermeable membrane.

function in gas exchange.) The exposed cell membranes are semipermeable, and there is a large difference in the concentration of solutes between the inside of the cell and the fresh water outside. To use the vocabulary introduced in Chapter 4, the gill epithelium is hypertonic relative to the surrounding water. As a result, epithelial cells gain water through osmosis. This water will then move from the epithelium into the adjacent tissues.

Cells and tissues that are gaining water are under osmotic stress. **Osmotic stress** means that the concentration of dissolved substances in a cell or tissue is abnormal. If a freshwater fish does not get rid of the incoming water and maintain homeostasis, its cells will eventually burst and the individual will die. The ability to achieve homeostasis with respect to water and electrolyte balance in the face of osmotic stress is called **osmoregulation.** To survive, freshwater fish must osmoregulate. They do so both by excreting large amounts of water in their urine and by not drinking.

Now consider the situation in marine environments. As **Figure 39.2b** shows, fish are hypotonic relative to saltwater. As a result, water tends to flow out of the gill epithelium. If this water is not replaced, the cells will shrivel and die. Marine fish also must osmoregulate. They do so by drinking large quantities of seawater and by secreting salt.

What about land animals? From the standpoint of water balance, terrestrial environments are similar to the ocean. Land animals constantly lose water to the environment, just as many marine animals do. In this case, however, the process involved is not osmosis but evaporation. Evaporation threatens terrestrial animals because they must exchange gases across a wet surface (**Figure 39.2c**). The epithelial cells that line a human's lung and a fruit fly's tracheal system are moistened to facilitate the exchange of oxygen and carbon dioxide through diffusion (Chapter 41 explores how animals exchange gases in detail.) In this respect, water balance in land animals mirrors the situation in land plants. Land plants also lose water as an inevitable byproduct of gas exchange. As Chapter 32 pointed out, land plants lose huge quantities of water in the course of obtaining CO_2 through their stomata.

To complicate matters of water balance, some terrestrial animals lose water when they sweat or pant to lower their body temperature. And all marine and terrestrial animals lose water in the process of excreting soluble wastes. Urine formation has a significant impact on water balance in these groups, just as it does in freshwater fish.

Electrolyte Balance in Freshwater, Marine, and Terrestrial Environments

The discussion thus far has focused on the movement of water by osmosis. The movement of ions by diffusion is equally important. To drive this point home, consider **Figure 39.3**. Because fresh water is hypotonic to the gill epithelium of freshwater fish, ions and other solutes tend to diffuse out of gill cells (Figure 39.3a). Freshwater animals must replace electrolytes that are lost by obtaining them in food or by actively transporting them in from the surrounding water.

As Figure 39.3b shows, the opposite situation holds true for many marine animals. Because seawater is hypertonic to the tissues of ocean-dwelling fish, they tend to constantly gain ions and other solutes via diffusion. Much of this gain occurs across the gills. Marine animals also ingest excess electrolytes in the water they drink and may acquire excess electrolytes in their food. Their challenge is to rid the body of these surplus solutes.

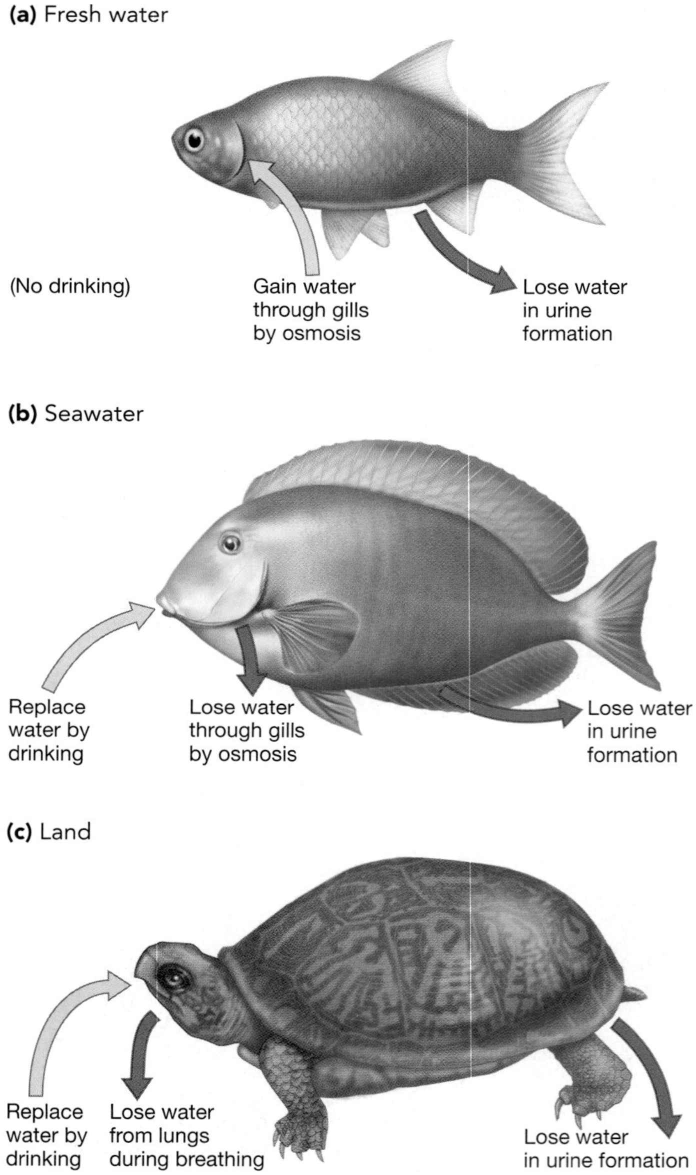

(a) Fresh water

(No drinking)
Gain water through gills by osmosis
Lose water in urine formation

(b) Seawater

Replace water by drinking
Lose water through gills by osmosis
Lose water in urine formation

(c) Land

Replace water by drinking
Lose water from lungs during breathing
Lose water in urine formation

FIGURE 39.2 Different Environments Offer Different Challenges to Water Balance
These diagrams indicate whether animals gain or lose water during respiration, and summarize some of the ways in which they replace lost water or excrete excess water. **EXERCISE** In parts (a) and (b), label whether each environment is hypotonic or hypertonic to the organism illustrated.

Terrestrial animals also ingest electrolytes in food; land animals that are capable of sweating lose electrolytes when they do so. Depending on conditions, then, terrestrial animals may either need to conserve or to excrete electrolytes to maintain homeostasis (Figure 39.3c).

How do animals cope with these diverse challenges? Section 39.2 focuses on how fish from freshwater or marine environments either excrete or absorb electrolytes through their gills. The remaining two sections in the chapter focus on water and electrolyte balance in terrestrial organisms. Specifically, the question we'll explore is how terrestrial insects and mammals regulate the amounts of electrolytes and water that are excreted in urine.

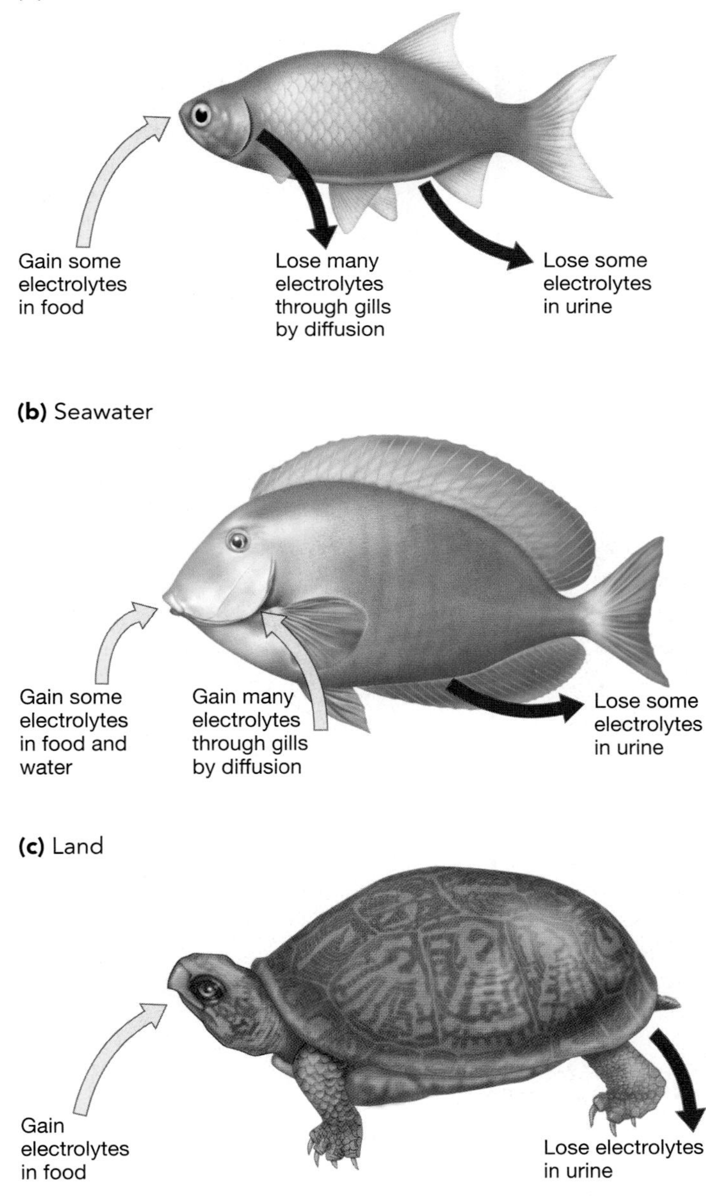

FIGURE 39.3 Different Environments Offer Different Challenges to Electrolyte Balance
These diagrams indicate whether animals gain or lose electrolytes as they breathe, drink, eat, and urinate.

39.2 Water and Electrolyte Balance in Aquatic Environments

Early research on how aquatic animals maintain water and electrolyte balance focused on an unusual organ in sharks called the rectal gland (**Figure 39.4a**, page 750). The rectal gland secretes salt. Research on how this secretion occurs turned out to be extraordinarily important. After researchers had worked out the molecular mechanisms of salt transport in the shark rectal gland, they found that the same processes occur in a wide array of marine animals, including birds and reptiles. As we'll see, understanding the shark rectal gland also turned out to have an unforeseen benefit for human biomedical research.

Why must sharks secrete salt? Shark tissues are actually isotonic to seawater. In this respect, they are similar to corals, sea urchins, sea anemones, sponges, clams, and many other marine invertebrates. (Marine fish, in contrast, have tissues that are strongly hypotonic to saltwater.) The cells and extracellular fluids of sharks are isotonic to seawater because they contain large quantities of soluble compounds called urea and trimethylamine oxide (TMAO), in addition to moderate concentrations of sodium ions, potassium ions, and chloride ions. Because the concentration of TMAO, urea, and ions is high enough to match the osmolarity of seawater and prevent the loss of water by osmosis, biologists interpret the synthesis of urea and TMAO as an adaptation that prevents osmotic stress. Sharks still need to excrete salt (NaCl), however, because sodium and chloride ions diffuse into their gill cells from seawater along their concentration gradients. How do they do this?

The Shark Rectal Gland

In the course of studying shark anatomy, J. W. Burger discovered that an organ called the rectal gland secretes a concentrated salt solution. To study how this gland works, Patricio Silva and co-workers studied them in vitro—outside of the shark's body. As **Figure 39.4b** shows, these researchers connected a tube to the blood vessel that carries blood into the rectal organ and attached a collecting tube to the end of the gland that empties to the environment. This experimental setup allowed the researchers to introduce solutions of defined composition and evaluate the fluid that the rectal gland produced in response.

To establish that the rectal gland functioned normally in this preparation, Silva and co-workers perfused the organ with a solution containing Na^+, K^+, and Cl^- in concentrations that might be observed in an intact animal: 280 mM, 5 mM, and 270 mM, respectively. The fluid that left the gland contained the same ions in concentrations of 449 mM, 12 mM, and 446 mM. Because the product solution was more concentrated than the initial solution, the result gave the researchers confidence that the organ was functioning normally in vitro. The data were also consistent with earlier evidence that secreting salt is an energy-demanding activity. Ions can be concentrated in this way only if they are actively transported against a concentration gradient.

The Role of Na^+/K^+-ATPase An energy-demanding mechanism for salt excretion implies that a membrane protein is involved in pumping Na^+, Cl^-, or both ions out of the epithelial cells that line the surface of the gland. The best-characterized membrane protein pump involving these ions is the sodium-potassium pump, also known as Na^+/K^+-ATPase. Chapter 42 discusses how this enzyme was discovered and explores its function in the transmission of electrical signals in animals. That discussion also introduces a plant defense compound called ouabain (pronounced "waa-bane"). Ouabain is toxic to animals because it binds to Na^+/K^+-ATPase, prevents it from functioning, and poisons the transmission of nerve impulses.

Prior to the experiments with the preparation illustrated in Figure 39.4b, Silva and associates had determined that epithe-

(a) Spiny dogfish shark

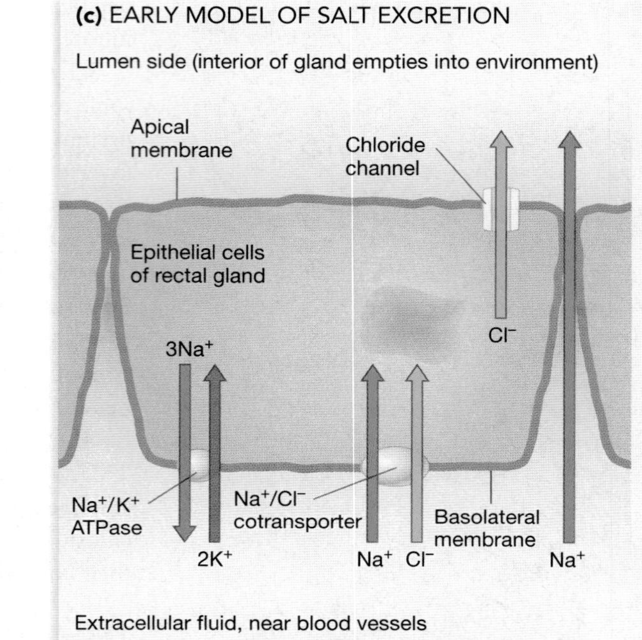

(b) Rectal gland experiment

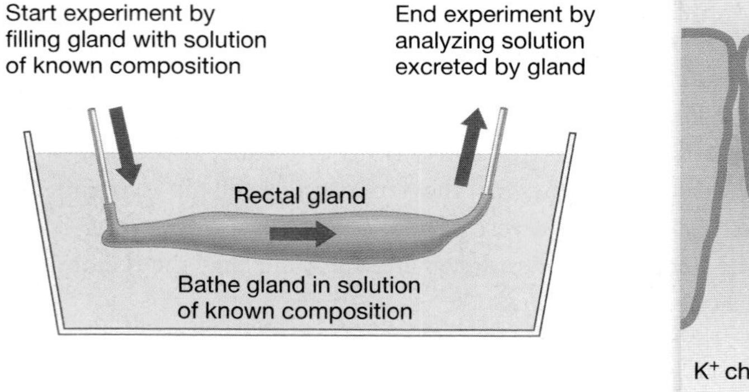

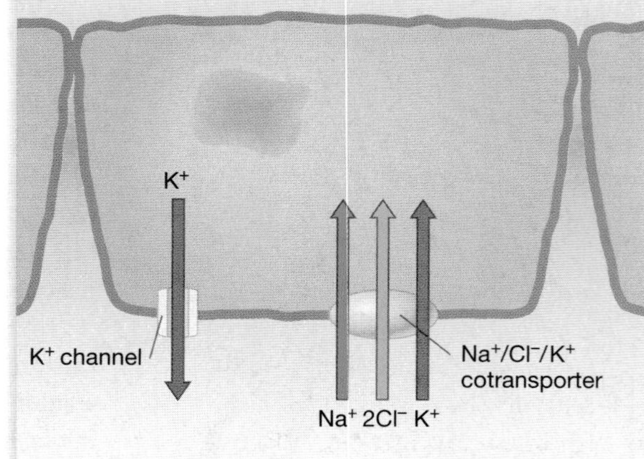

FIGURE 39.4 The Shark Rectal Gland
(a) Sharks, such as this spiny dogfish, belong to a lineage of animals called the Elasmobranchii. Sharks are closely related to rays and ratfishes, which also have skeletons constructed from cartilage. (Fish have bony skeletons and belong to different lineages.) **(b)** Researchers used the preparation diagrammed here to explore how the shark rectal gland excretes salt. **(c)** Based on early experimental results, researchers offered this model to explain how Na^+ and Cl^- move from the body fluids to the lumen of the rectal gland. **(d)** More recent studies have required researchers to modify the model for salt excretion slightly as shown here.

lial cells along the inner space, or **lumen**, of the rectal gland contain Na^+/K^+-ATPase. (The cavity inside any tube- or sac-shaped organ is called the lumen.) They discovered this by attaching a radioactive atom to ouabain, treating cells from the rectal gland with the radioactively labeled poison, and then examining the distribution of the radioactive particles with the electron microscope. These studies showed that the ouabain bound to Na^+/K^+-ATPase localized along the basolateral (bottom-side) membrane of the cells, which faces the inside of the body and thus the blood supply (**Figure 39.4c**). No pumps appeared in the apical (top) membrane, which faces the lumen and the secreted solution.

The location of the pumps was paradoxical because the Na^+/K^+-ATPase pumps sodium ions *out* of the cell and potassium ions in. (Three Na^+ go out for every two K^+ that come in.) The paradox arose because the rectal gland cells pump sodium *opposite* to the direction in which it is secreted. Silva and colleagues confirmed that the Na^+/K^+-ATPase is critical to salt secretion, however, by treating the gland with ouabain in vitro. When rectal glands were treated with the drug, they stopped producing a concentrated salt secretion. What was happening?

A Molecular Model for Salt Secretion The researchers were able to suggest a resolution to the paradox when they documented that chloride ion concentrations inside rectal gland cells are much higher than sodium ion concentrations. Based on this observation, they developed the model for salt secretion shown in Figure 39.4c. Their hypothesis was published in 1977, and hinges on the coordinated action of three membrane proteins. To understand their proposal, follow the steps diagrammed in part (c).

1. Na^+/K^+-ATPase pumps sodium ions out of epithelial cells across the basolateral surface and into the surrounding extracellular fluid. The pump creates a large electrochemical gradient favoring the diffusion of Na^+ into the cell.
2. A Na^+/Cl^- cotransporter, powered by the gradient favoring Na^+ diffusion, brings these two ions from the extracellular fluid into epithelial cells across their basolateral surfaces. **Cotransporters** are membrane proteins that transport more than one type of ion or molecule at a time.
3. Although sodium ions are pumped back out by Na^+/K^+-ATPase, chloride ion concentrations build up inside the cell as a result of the co-transport process. A chloride channel located in the apical membrane of the epithelial cells allows Cl^- to diffuse out along its concentration gradient.
4. Sodium ions also diffuse into the lumen of the gland, following their charge and concentration gradient. But instead of passing through the epithelial cells as Cl^- does, Na^+ diffuses out through spaces between the cells.

Subsequent research has shown that this model was essentially correct—with two important modifications. As **Figure 39.4d** shows, a K^+ channel located on the basolateral membrane allows potassium ions to diffuse back out after they have been pumped in, and the cotransporter in the basolateral membrane actually brings a Na^+, $2Cl^-$, and a K^+ into the cell together. In addition, subsequent work has shown that ionic gradients such as those found in the shark rectal gland drive membrane transport processes in a wide array of animal cells, tissues, organs, and systems.

A Common Molecular Mechanism Underlies Many Instances of Salt Secretion The mechanism of salt transport illustrated in Figure 39.4d is widespread in animals.

- Marine birds and reptiles drink salt water and must excrete NaCl. They have salt-excreting glands in their nostrils that function much like the shark rectal gland.
- Because marine fish with bony skeletons are hypotonic to seawater, salt constantly diffuses in through their gills. Their gills contain specialized cells that are configured precisely like the cells lining the rectal gland. These chloride cells are responsible for excreting excess salt and maintaining electrolyte balance.
- Cells with the same configuration of pumps, cotransporters, and channels are responsible for transporting salt in the kidneys of mammals.

Research on the shark rectal gland also turned out to have an unforeseen benefit for biomedical research. Several years after the shark chloride channel was isolated and characterized, investigators succeeded in identifying a human protein called cystic fibrosis transmembrane regulator (CFTR). Although it was known that the disease cystic fibrosis is associated with defects in this protein, its function was undetermined. Cystic fibrosis is the most common genetic disease in populations of northern European extraction. When investigators realized that the amino acid sequence of CFTR is 80 percent identical to the shark chloride channel, it was their first hint that CFTR is involved in chloride ion transport. Subsequent studies confirmed this hypothesis and clarified the molecular basis of the disease (see Chapter 4 essay). In this way, studies on water and electrolyte balance in sharks shed light on an important human disease.

How Do Salmon Osmoregulate?

Because of the success of research on the shark rectal gland and the salt-secreting cells found in the gills of marine fish, the molecular mechanisms of salt balance in marine animals are now well known. How freshwater fish achieve homeostasis with respect to electrolytes is still unclear, however. Recall from Figure 39.2 that freshwater fish do not drink, and that they secrete large volumes of watery urine to rid themselves of water that enters their gills via osmosis. But no one knows exactly how freshwater fish obtain enough ions to make up for the electrolytes they lose to diffusion from their gill cells.

Recently, an interesting hypothesis to explain how freshwater fish maintain electrolyte balance has emerged from research on salmon. Salmon are a family of fish that include familiar species like rainbow trout; the species called chum salmon is illustrated in **Figure 39.5a.** As **Figure 39.5b** shows, many of the species in the family are anadromous (up-running)—meaning that they hatch in freshwater, migrate to the ocean where they spend several years feeding and growing, then return to freshwater to breed (and die). The question is, how do individuals make the transition from a hypotonic to a hypertonic medium and back again?

In an effort to answer this question, Katsuhisa Uchida and colleagues examined the gill epithelia of young chum salmon in fresh water. Researchers in other laboratories had already established that in many salmon species, there is a dramatic increase in the number of chloride cells found in the gills of young fish as they prepare to migrate to salt water. There is also a significant increase in Na^+/K^+-ATPase activity. These results are logical, given that the Na^+/K^+-ATPase in chloride cells is an adaptation to a hypertonic environment. Did the same pattern occur in chum salmon?

When Uchida and co-workers analyzed the gills of young salmon that were maintained in fresh water versus salt water, they made an interesting observation. As **Figure 39.6** shows, the two populations of fish appeared to have two distinct populations of chloride cells. Individuals that were living in salt water had a large number of chloride cells at the base of their gill filaments. In contrast, most of the chloride cells observed in individuals from fresh water were located in the sheet-like lamellae that extend from the base of the gill filaments.

To follow up on this study, the researchers examined adult salmon that were migrating from the ocean back to fresh water. These fish showed the same pattern observed in young individuals. In adults that had just returned from the open ocean, most chloride cells were located at the base of the gill filaments. In adults that were moving into fresh water, most chloride cells were located in the gill lamellae. These observations inspired the hypothesis that salmon have two distinct types of cells involved in maintaining electrolyte balance. More specifically, the proposal is that chloride cells at the base of gills secrete salt, while chloride cells in the gill lamellae import electrolytes.

The chloride-cell-switch hypothesis implies that salmon sense changes in the osmolarity of their environment and respond by producing or destroying the appropriate populations of chloride cells. The molecular mechanisms involved in the sensory and switching response are a mystery, however. Fur-

(a) Chum salmon

(b) Migration to the ocean and back

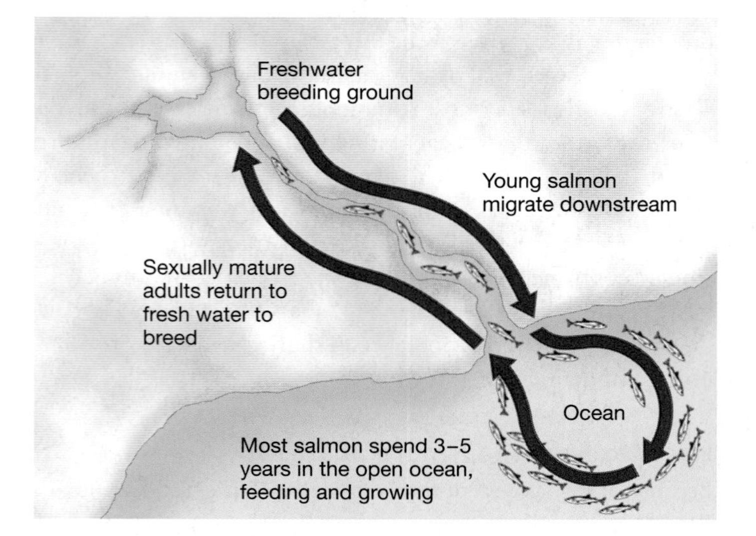

FIGURE 39.5 Salmon Need to Osmoregulate in Both Fresh and Salt Water
(a) Chum salmon live in the North Pacific. Breeding populations can be found from Japan and Siberia to Alaska, and from Alaska to the western coast of Canada and the United States. **(b)** During their life cycle, salmon move from fresh water to salt water and back to fresh water.

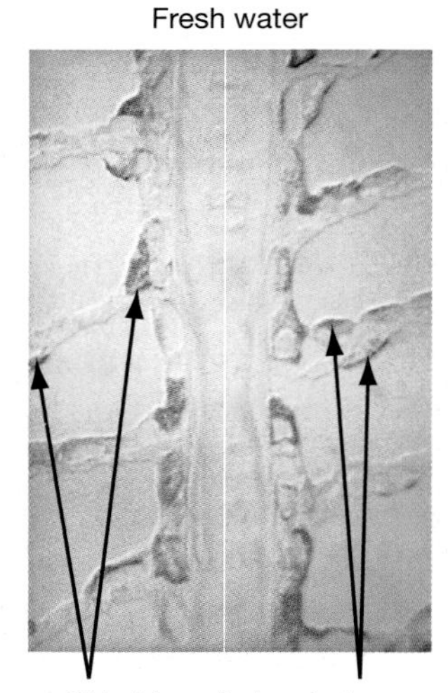

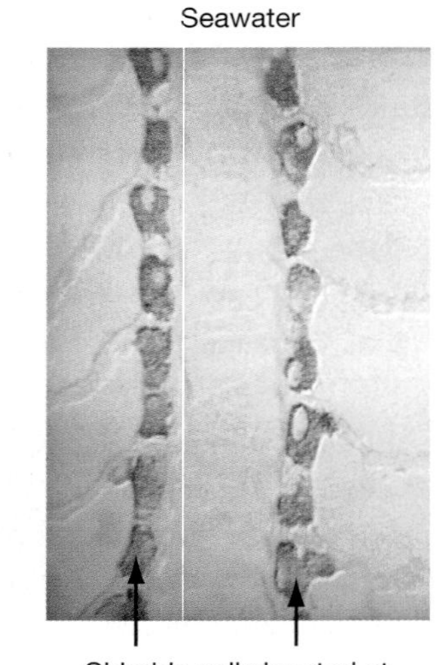

FIGURE 39.6 Chloride Cells of Salmon
These photographs show gills from young salmon that were maintained in fresh water versus salt water. The gills have been treated with a molecule that binds to Na^+/K^+-ATPase and leaves a dark stain, labeling the chloride cells.

ther, it is still unknown exactly how lamellar chloride cells take up electrolytes from fresh water.

Researchers from a variety of laboratories are currently working on these questions, in part because similar changes have been observed in the chloride cells of other fish species that switch between freshwater and saltwater habitats. Identifying the mechanisms of electrolyte uptake in freshwater fish is an important challenge for researchers who want to know how aquatic organisms maintain water and electrolyte balance.

39.3 Water and Electrolyte Balance in Terrestrial Invertebrates

To understand how terrestrial invertebrates maintain water and electrolyte balance, researchers have focused on studying species that inhabit particularly dry environments. The success of this work has validated a general research strategy in biological science. By studying extreme situations or unusual organisms, biologists can often gain insight into how organisms cope with more moderate environments.

In studies on the molecular mechanisms of water and electrolyte balance in terrestrial insects, the most fruitful model organisms have been the desert locust shown in **Figure 39.7a** and the common household pest called the flour beetle. (You may be familiar with the larvae of flour beetles, called mealworms.) Desert locusts and flour beetles rarely, if ever, drink—simply because little or no water is available in the habitats they occupy.

These insects live in environments where osmotic stress is severe. How do they maintain water and electrolyte balance? The answer actually has two parts. They minimize water loss from their body surface, and they carefully regulate the amount of water and electrolytes that they excrete in their urine and feces. Let's look at each issue in turn.

How Do Insects Minimize Water Loss from the Body Surface?

As Chapter 41 will show, terrestrial animals must expose an extremely thin surface to the atmosphere in order to take up oxygen and rid themselves of carbon dioxide efficiently. Because water constantly leaks across the thin respiratory surface, desert locusts, flour beetles, and other terrestrial invertebrates lose large amounts of water to evaporation. Stated another way, water loss is an inevitable by-product of respiration. Evaporation from the body surface itself is another threat, and is particularly challenging to insects because they are small. As Chapter 38 emphasized, small organisms have a high surface-area-to-volume ratio. Insects have a relatively large surface area with which to lose water, and a small volume in which to retain it.

How do insects minimize water loss during respiration? To answer this question, study **Figure 39.7b** and note that gas exchange occurs across the membranes of an extensive system of tubes called tracheae ("tray-kee-ee"). The tracheal system connects with the atmosphere at openings called spiracles. Muscles just inside each spiracle close or open the pore, much as guard cells close or open the stomata of plants. When investigators manipulated bugs called *Rhodnia* so that their spiracles stayed open, and then placed them in a dry environment, the individuals died within three days. These data support the hypothesis that the ability to close spiracles is an important adaptation for minimizing water loss. If an insect is under osmotic stress, it has

(a) Desert locust

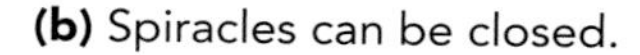

(b) Spiracles can be closed.

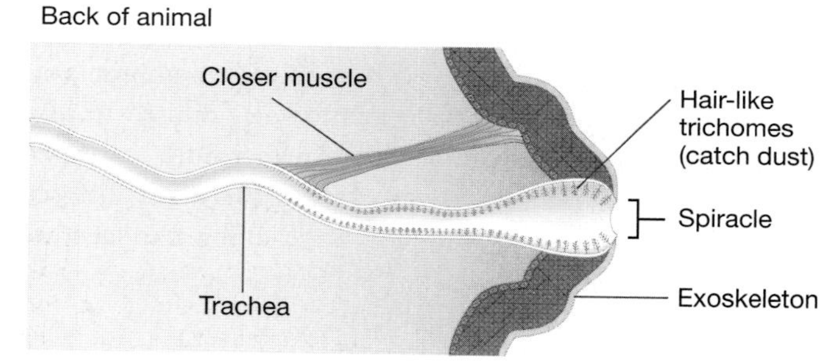

(c) The insect body is covered with wax.

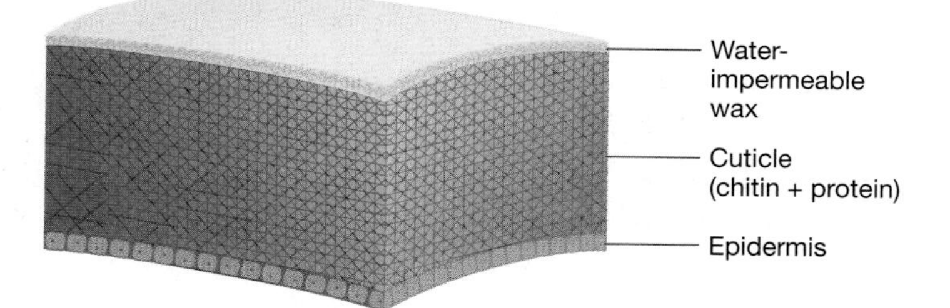

FIGURE 39.7 Adaptations That Limit Water Loss in Insects
(a) The desert locust has been a model organism for studying how insects limit water loss. **(b)** Spiracles are openings in the insect exoskeleton that lead to the tube-like trachea where gas exchange takes place. When muscles that are connected to the trachea contract, the spiracle closes. **(c)** This cross section through the insect exoskeleton shows that wax, cuticle, and epidermal cells form layers.
QUESTION In desert grasshoppers, spiracles are located in a row along the underside of the abdomen. Why would this location help minimize water loss?

the option of closing its spiracles and waiting until conditions improve before resuming activity.

To understand how desert locusts, flour beetles, and other insects minimize evaporation from their body surface, examine **Figure 39.7c**. This diagram is a cross-sectional view of the insect exoskeleton, which consists of a tough, nitrogen-containing polysaccharide called chitin and layers of protein. As the figure shows, the surface of the exoskeleton is covered with a layer of wax. Waxes are a type of lipid and are highly impermeable to water. When researchers have removed the wax from insect exoskeletons, they've confirmed that the rate of water loss from the body surface increases sharply. Based on this observation, the wax layer is interpreted as an adaptation that minimizes evaporative water loss.

Types of Nitrogenous Wastes: Impact on Water Balance

Animal cells contain a pool of amino acids and nucleic acids that are used to synthesize proteins, RNA, and DNA. If the monomers are present in excess of a cell's needs, they are broken down. These reactions produce ammonia (NH_3). Ammonia is a strong base and readily gains a proton to form an ammonium ion (NH_4^+). As a result, it is highly toxic. High concentrations of ammonia raise the pH of intracellular and extracellular fluids enough to poison enzymes.

Different species of animals rid themselves of ammonia in different ways. In freshwater fish, ammonia is diluted to low concentration and excreted in a watery urine. Both saltwater and freshwater fish rid themselves of ammonia as it diffuses across their gills into the surrounding water along a concentration gradient. In humans, enzyme-catalyzed reactions convert ammonia to a nontoxic compound called urea, which is excreted in urine. In birds, reptiles, and terrestrial arthropods, reactions convert ammonia to uric acid. Compared with urea and ammonia, uric acid is extremely insoluble in water. As a result, it can be excreted along with a minimum amount of water. Many birds and some insects, in fact, do not produce any urine at all. (You may have seen uric acid without knowing it. It is the white, paste-like substance in bird feces. Its lack of solubility in water explains why washing bird droppings off car windshields is so difficult.)

Why do different types of animals have different ways of detoxifying ammonia and excreting nitrogenous wastes? As **Table 39.1** shows, the type of nitrogenous waste produced by an animal correlates with the amount of osmotic stress it endures. For example, most small animals and animals that live in dry habitats minimize the amount of water they must excrete by synthesizing uric acid as their principal waste product. The uric acid produced by locusts, flour beetles, and other insects is interpreted as another important adaptation to minimize water loss.

Maintaining Homeostasis: The Excretory System

For desert locusts, flour beetles, and other insects, minimizing water loss is only half the battle in avoiding osmotic stress. To maintain homeostasis, insects must carefully regulate the composition of a blood-like fluid called **hemolymph**. Unlike the situation in vertebrates, hemolymph does not flow through blood vessels in a closed circulatory system. Instead, the heart pumps hemolymph into an open body cavity. In this way, hemolymph bathes tissues directly. Nutrients pass from the hemolymph into cells; waste products such as ammonia diffuse out of cells and

TABLE 39.1 Attributes of Nitrogenous Wastes Produced by Animals

Attribute	Ammonia	Urea	Uric acid
Solubility in water	high	medium	very low
Water loss (amount required for excretion of waste)	high	medium	very low
Energetic cost (amount of ATP required to synthesize)	low	high	high
Toxicity	high	low	medium
Groups where it is the primary waste	fish aquatic invertebrates	mammals* sharks	birds† reptiles most terrestrial insects and spiders
How synthesized?	breakdown of amino acids and nucleic acids	synthesized in liver, starting with amino groups from amino acids	synthesis starts with amino acids and nucleic acids
How excreted?	diffuses across gills	in urine (mammals); diffuses across gills (sharks)	in feces

*Mammals also excrete a small amount of uric acid, synthesized from excess nucleic acids.

†Birds also excrete a small amount of ammonia.

into the surrounding fluid. Hemolymph also contains a wide variety of electrolytes.

How do insects regulate the composition of the hemolymph? This is an important question, because the nitrogenous wastes present in hemolymph must be removed before they build up to toxic concentrations. Similarly, if excess electrolytes enter the hemolymph after a meal, the ions must be excreted before they lead to osmotic stress. Water balance also must be regulated constantly.

Pre-Urine Formation in the Malpighian Tubules To maintain water and electrolyte balance, insects rely on an organ called the Malpighian tubules and their hindgut (**Figure 39.8**). As Figure 39.8a shows, Malpighian tubules have a large surface area and are in direct contact with the hemolymph. When J. A. Ramsay and colleagues collected fluid from inside the Malpighian tubules of mealworms and compared its composition to hemolymph from the same individuals, they found that the two solutions were roughly isotonic. The concentration of sodium ions inside the tubule was low relative to the hemolymph, however, while K^+ concentrations were extraordinarily high. To follow up on this observation, the researchers dissected the tubules, rinsed them, and bathed their interior and exterior in a solution containing a known concentration of potassium ions. When they measured the concentration of electrolytes on either side of the membrane over time, they found that K^+ accumulated inside the tubules against a concentration gradient.

To explain their data, Ramsay and colleagues hypothesized that cells in the membranes of Malpighian tubules contain a pump that transports potassium ions into the organ. Subsequent work confirmed this hypothesis. A high concentration of potassium ions brings water into the tubules by osmosis. Other electrolytes and nitrogenous wastes then diffuse into the structure along their concentration gradients.

The Hindgut: Selective Reabsorption of Electrolytes and Water The "pre-urine" that accumulates inside the Malpighian tubules flows into the hindgut, where it joins material emerging from the digestive tract. Because the urine that is finally expelled from the anus is often strongly hypertonic to the pre-urine, a large amount of water must be reabsorbed between the start and end of the hindgut. In desert locusts, flour beetles, and other species that live in extremely dry environments, 80 to 95 percent of the water in the pre-urine is recovered and kept inside the body. The ability to recover this water allows these insects to live in dry habitats like deserts and flour bins. How does reabsorption happen?

Experiments by John Phillips and co-workers helped establish the molecular mechanisms responsible for water and electrolyte recovery. These researchers succeeded in devising a system for mounting the rectal epithelium from desert locusts as a sheet dividing two solutions. This preparation allowed them to manipulate electrolyte concentrations on either side of the rectal wall and measure changes in the solutions over time. For example, when they removed K^+ and Na^+ from the solution on the lumen side of the organ, they observed that water reabsorption stopped. These data established that the hindgut's ability to recover water from urine is dependent on ion movement. Stated another way, the epithelial cells transport ions out of the hindgut and water follows by osmosis.

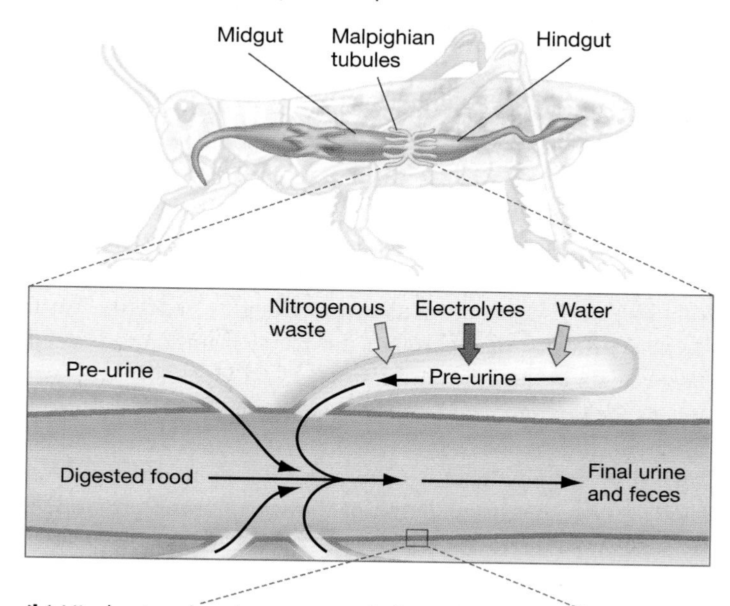

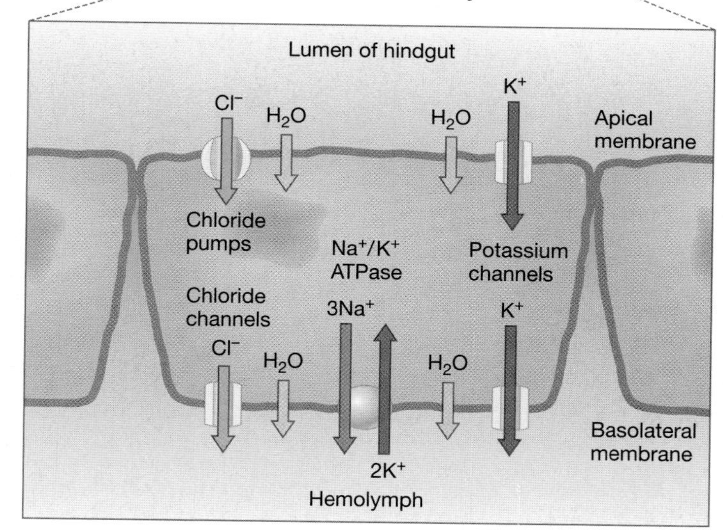

FIGURE 39.8 Urine Formation in Insects
(a) The Malpighian tubules present a large surface area to the hemolymph. The isotonic pre-urine that forms in the Malpighian tubules empties into the hindgut. **(b)** In the hindgut, the primary driving forces for reabsorption of electrolytes and water are chloride pumps in the apical membrane and Na^+/K^+-ATPases in the basolateral membrane. There is strong experimental evidence for the existence of several important channels, pumps, and cotransporters that are not shown here. These proteins transport protons, ammonia, amino acids, and other molecules across the epithelium of the hindgut.

How do the ions move? By poisoning the membrane with ouabain, they confirmed that Na^+/K^+-ATPase is involved in moving ions out of the lumen and into the hemolymph. Ouabain-treated membranes continued to transport chloride ions, however. When later experiments confirmed that the rectal epithelium transports Cl^- against electrical and concentration gradients, they hypothesized that the cells contain a chloride pump in addition to the Na^+/K^+-ATPase. Years of experiments on the locust hindgut resulted in the model diagrammed in Figure 39.8b. Ion and water movement in the insect hindgut depends on a complex suite of pumps and channels.

Regulating Water and Electrolyte Balance: An Overview

Fortunately, the overall pattern in regulation of water and electrolyte balance in insects is much simpler than the details of the mechanisms involved. One key point is that formation of the pre-urine is not particularly selective. Most of the molecules present in the hemolymph are also present in the Malpighian tubules. Reabsorption, in contrast, is highly selective. Waste products do not pass through the rectal membrane. Instead, they remain in the urine and feces.

An equally important point is that water and electrolyte recovery in the hindgut is precisely controlled to maintain homeostasis. If an individual is dehydrated and is under severe osmotic stress, then virtually all of the water in the pre-urine is reabsorbed. But if the same individual has plenty to drink, reabsorption does not occur. As a result, a well-hydrated insect produces watery urine that is hypotonic to hemolymph. Are the signaling molecules called hormones responsible for regulating the pumps and channels that line the hindgut? If so, what hormones are involved? Where are they produced and how do they act? Answering these questions is an important focus of current research in insect physiology.

39.4 Water and Electrolyte Balance in Terrestrial Vertebrates

On land, vertebrates face the same hazards as terrestrial insects with respect to water loss. Crocodiles, turtles, lizards, frogs, birds, and mammals lose water from their body surfaces and from the surface of their lungs every time they breathe. Like all organisms, they must carefully regulate the osmolarity of their tissues. How do land-dwelling vertebrates maintain water and electrolyte balance?

Due to intensive research on the mammalian kidney, an answer to this question began to emerge in the mid-1970s. The kidney is the organ responsible for regulating water and electrolyte balance in terrestrial vertebrates. In addition, the kidney is where nitrogenous wastes—usually in the form of urea—are removed from the blood and eliminated in urine.

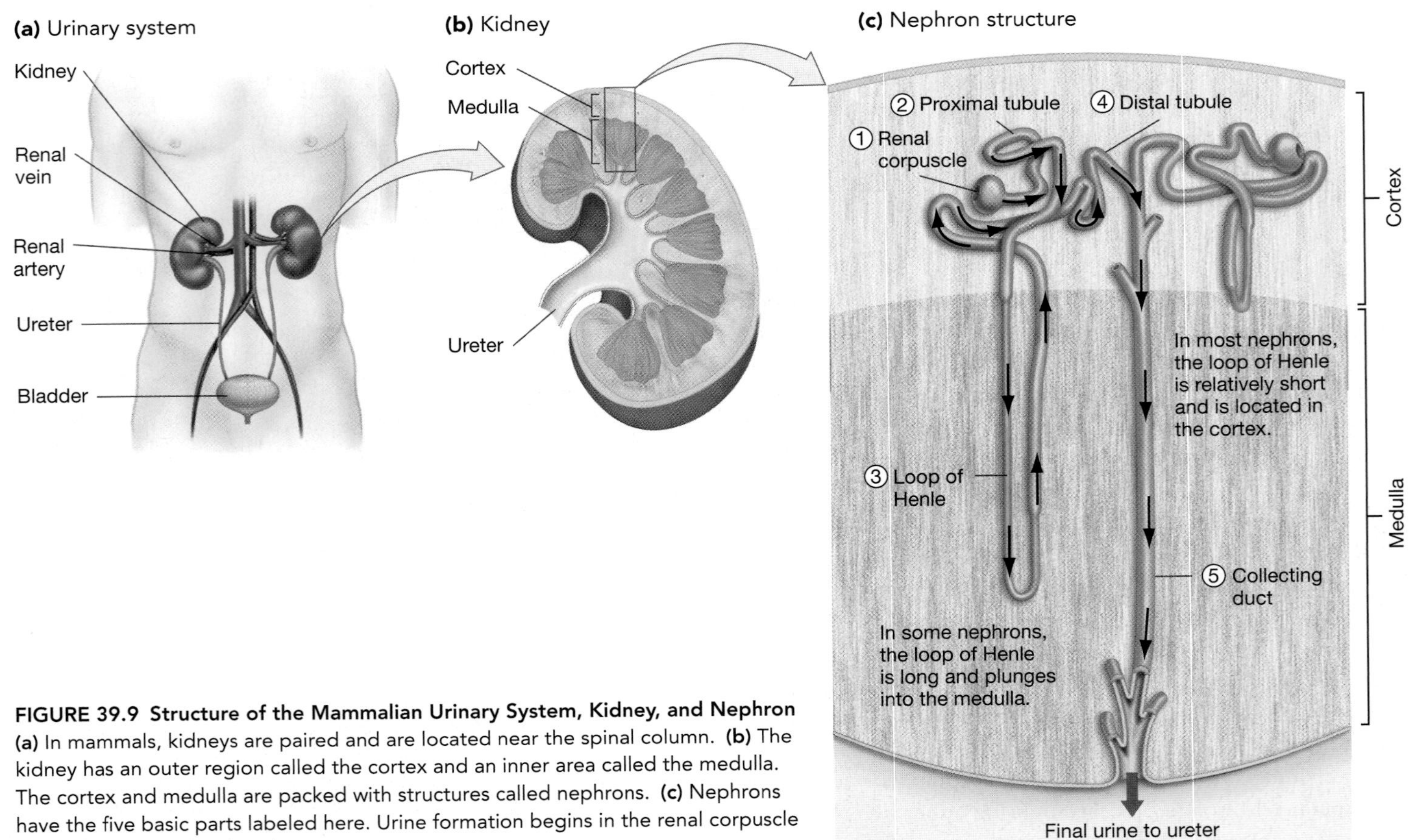

FIGURE 39.9 Structure of the Mammalian Urinary System, Kidney, and Nephron **(a)** In mammals, kidneys are paired and are located near the spinal column. **(b)** The kidney has an outer region called the cortex and an inner area called the medulla. The cortex and medulla are packed with structures called nephrons. **(c)** Nephrons have the five basic parts labeled here. Urine formation begins in the renal corpuscle and ends in the collecting duct.

The general anatomy of the mammalian kidney was described in the late nineteenth century. Kidneys occur in pairs, tend to be bean-shaped, and are usually located near the dorsal (back) side of the body (**Figure 39.9a**). Large blood vessels called the renal artery and vein carry blood into and out of the organ; a long tube called the ureter transports the urine that is formed in the kidney to the bladder for storage prior to excretion.

As the diagrams in **Figure 39.9b** and **Figure 39.9c** show, most of the kidney's mass is made up of small structures called nephrons. The **nephron** is the basic functional unit of the kidney—nephrons perform the work involved in maintaining water and electrolyte balance. Most of the approximately one million nephrons in a human kidney are located in the outer region of the organ, or cortex. As Figure 39.9c shows, however, some nephrons extend from the cortex into the kidney's inner region, or medulla.

To understand how the kidney maintains water and electrolyte balance, it is essential to understand how the nephron functions. Let's follow the flow of material through a nephron, starting with the blood that arrives at the structure and ending with the urine that exits it. The changes that occur are not unlike those observed as insect hemolymph passes into the Malpighian tubule and through the hindgut.

Filtration: The Renal Corpuscle

In vertebrates, urine formation begins in a structure called the renal corpuscle. To understand how this part of the nephron works, examine Figure 39.9c again and note that the nephron is actually a tube that is closed at one end and open at the other. As the detailed diagram in **Figure 39.10a** shows, the closed end forms a capsule around a cluster of tiny blood vessels, or capillaries, that bring blood to the nephron. The cluster of capillaries is called the glomerulus (*glomerulus* is Latin for "ball of yarn"). The region of the nephron that surrounds the glomerulus is named Bowman's capsule. The entire structure is called the renal corpuscle (little-body).

By carefully sectioning and examining the renal corpuscle, biologists established that the glomerular capillaries have large pores. These blood vessels are also surrounded by unusual cells whose membranes fold into a series of slits and ridges (**Figure 39.10b**).

Based on these observations, researchers working in the late 1800s proposed that the renal corpuscle serves as a filtration device. The hypothesis was that water and small solutes are pushed out of the capillary's pores, through the slits in the surrounding cells, and into the fluid-filled space inside Bowman's capsule. Because proteins, cells, and other large components of blood would not fit through the pores, they would not enter the nephron. They would stay in the blood instead. Stated another way, urine formation was thought to start with a size-selective filtration step. According to this hypothesis, the force required to perform filtration is supplied by blood pressure, which is created by the heart and the closed circulatory systems of vertebrates.

In mammals, the filtration hypothesis was confirmed almost 100 years after it was initially proposed. In 1971 Barry Brenner and co-workers were able to make direct measurements of the pressure inside the glomerular capillaries and in the tubule leading away from Bowman's capsule in rats. They found that pressure was much higher inside the capillaries than it was in the surrounding capsule. As the filtration hypothesis predicts, the pressure differential forces water and solutes out of the blood and into the capsule space.

Reabsorption: The Proximal Tubule

What happens to the filtrate produced in the renal corpuscle? As Figure 39.9c shows, fluid leaves Bowman's capsule and enters a convoluted structure called the proximal tubule. When

CD ACTIVITY 39.1 The Mammalian Kidney

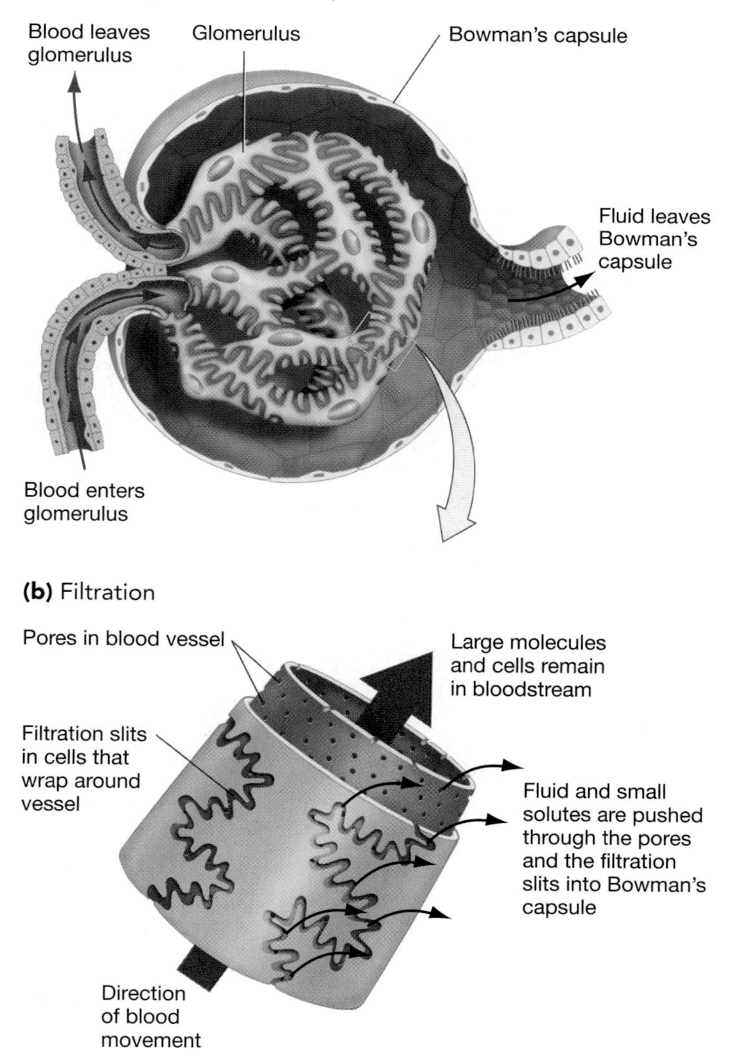

FIGURE 39.10 The Renal Corpuscle
(a) The renal corpuscle consists of Bowman's capsule and the glomerulus. **(b)** The capillaries in the glomerulus have pores and are surrounded by cells that have filtration slits. Blood pressure forces water and small molecules out of the capillaries, through the slits, and into Bowman's capsule. The filtrate then enters the proximal tubule.

researchers analyzed samples of the fluid inside this tubule, they found that it contained the components predicted by the filtration hypothesis: water and small solutes like urea, glucose, amino acids, vitamins, and electrolytes. Some of these molecules are waste products; others are valuable nutrients.

The first clue to the proximal tubule's function came from anatomical studies. As **Figure 39.11a** shows, the epithelial cells of the tubule have a prominent series of small projections called **microvilli** (literally, "little shaggy hairs") facing the lumen. The presence of microvilli greatly expands the surface area of an epithelium. Greater surface area provides more space for pumps, channels, and cotransporters; logically enough, cells with microvilli are often associated with transport processes. The observation that these cells are also packed with mitochondria suggested that ATP-demanding active transport is occurring. Based on these observations, anatomists hypothesized that the proximal tubule functions in the active transport of molecules into and out of the filtrate.

When researchers were able to isolate the proximal tubules of rabbits and rats in vitro and inject solutions with known compositions, they confirmed that electrolytes and nutrients are actively reabsorbed from the filtrate. As solutes moved from the filtrate into the epithelial cells, water followed along the osmotic gradient.

Using techniques similar to those employed during research on the shark rectal gland (see section 39.2), investigators confirmed that the basal membrane of cells in the proximal tubule contains Na^+/K^+-ATPase, and that the apical membrane contains a variety of cotransporters. **Figure 39.11b** summarizes the current model for how these cells function. The key to the model is that the Na^+/K^+-ATPase in the basolateral membranes removes intracellular Na^+ and creates a gradient for Na^+ entry. In the apical membrane, Na^+-dependent cotransporters simultaneously bind Na^+ and another solute such as glucose, an amino acid, or Cl^-. As with other examples of secondary active transport, the movement of Na^+ into the cell, *down* its concentration gradient, provides the means for moving these other solutes *against* a concentration gradient. Water follows the movement of these solutes by osmosis. It leaves the proximal tubule through membrane proteins called **aquaporins** (water-pores).

(a) Lumen has large surface area

173 nm

(b) Model of water and solute reabsorption

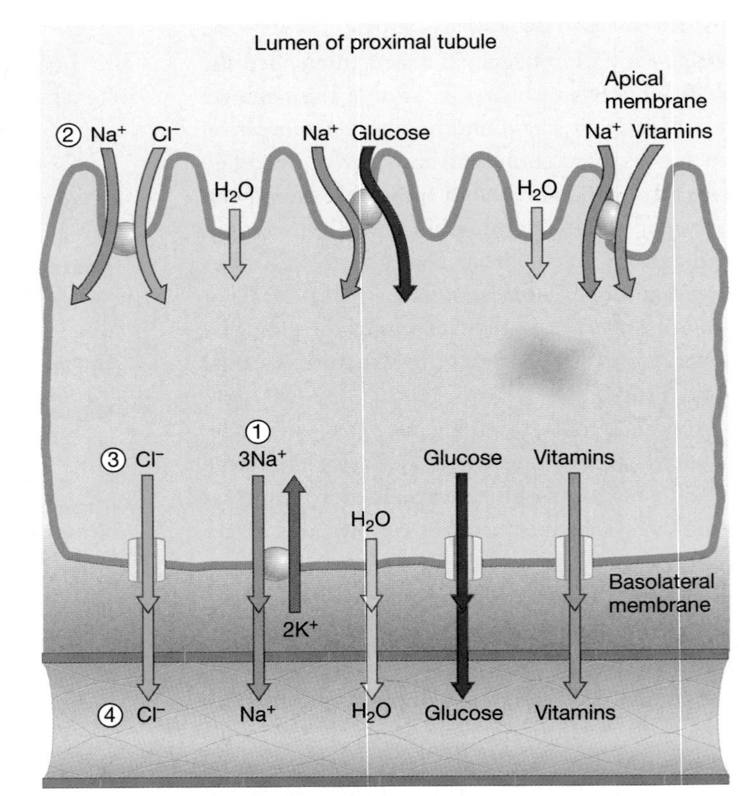

FIGURE 39.11 The Proximal Tubule
(a) Microvilli from epithelial cells extend into the lumen of the proximal tubule. **(b)** Reabsorption of water and solutes occurs because (1) Na^+/K^+-ATPase in the basolateral membrane of epithelial cells sets up a strong concentration gradient to bring Na^+ into the cell. (2) Na^+ cotransporters in the apical membrane result in reabsorption of nutrients and electrolytes, with water following via osmosis. (3) The solutes exit the cell through channels, with water following, and (4) enter nearby blood vessels.

Careful measurements have confirmed that two-thirds of the NaCl and water that is filtered by the renal corpuscle is reabsorbed in the proximal tubule. The osmolarity of the tubular fluid is unchanged despite this huge change in volume, however, because water reabsorption is proportional to solute reabsorption. The electrolytes, nutrients, and water that are reabsorbed eventually enter blood vessels on the basolateral side of the tubule and are returned to the body.

To summarize, the cells that line the proximal tubule act as a recycling center. In effect, the pumps and cotransporters in their membranes allow the proximal tubule to recover water and valuable nutrients and electrolytes from the raw filtrate. Compared to the fluid that enters, the pre-urine that leaves is isotonic but greatly reduced in volume. As it flows into the next part of the nephron, the pre-urine has a relatively high concentration of waste molecules and a relatively low concentration of nutrients.

Creating an Osmotic Gradient: The Loop of Henle

In mammals, the fluid that emerges from the proximal tubule enters a long loop. In some nephrons, this loop plunges from the cortex of the kidney deep into the medulla (look back at Figure 39.9c). Although Jacob Henle described this part of the mammalian nephron in the early 1860s, it was not until 1942 that Werner Kuhn offered a hypothesis to explain what the structure does.

Kuhn's proposal was inspired by the use of systems called heat exchangers in chemistry and physics. A heat exchanger is diagrammed in **Figure 39.12a**. The design principle at work in the system is that two adjacent currents flow through pipes in opposite directions. As the illustration shows, the water or air in one of the pipes is initially cold, while the water or air in the countercurrent is initially hot. As the currents pass, heat is transferred from one to the other.

One of the key observations about a countercurrent heat exchanger is that the gradient between the two currents stays high even as one current cools and the other heats. If the two solutions ran the same way, the gradient would disappear as the source current cooled and the recipient current heated. Even though the temperature differential at any point along the pipes is small, there is a large temperature differential from the start of each pipe to its end. In effect, small differences in heat are multiplied along the length of the exchanger to create a large overall temperature gradient. The longer the pipes, the greater the overall differential. Logically enough, systems like this are known as **countercurrent multipliers**.

Kuhn proposed that the loop of Henle functions as a countercurrent multiplier. But instead of setting up a heat gradient, Kuhn hypothesized that the loop sets up an osmotic gradient. As **Figure 39.12b** shows, he proposed that the osmolarity of the fluid inside the loop of Henle is low in the cortex and high in the medulla. Further, Kuhn maintained that the osmolarity in tissues surrounding the loop mirrors the gradient inside the loop. This is a key point. Kuhn's hypothesis proposed that an exchange of water and solutes occurs between each of the segments in the loop and the cells outside. This is in contrast to a conventional countercurrent system, in which exchange occurs between the two segments themselves. Was Kuhn correct?

CD ACTIVITY 39.2 The Loop of Henle: A Countercurrent Multiplier

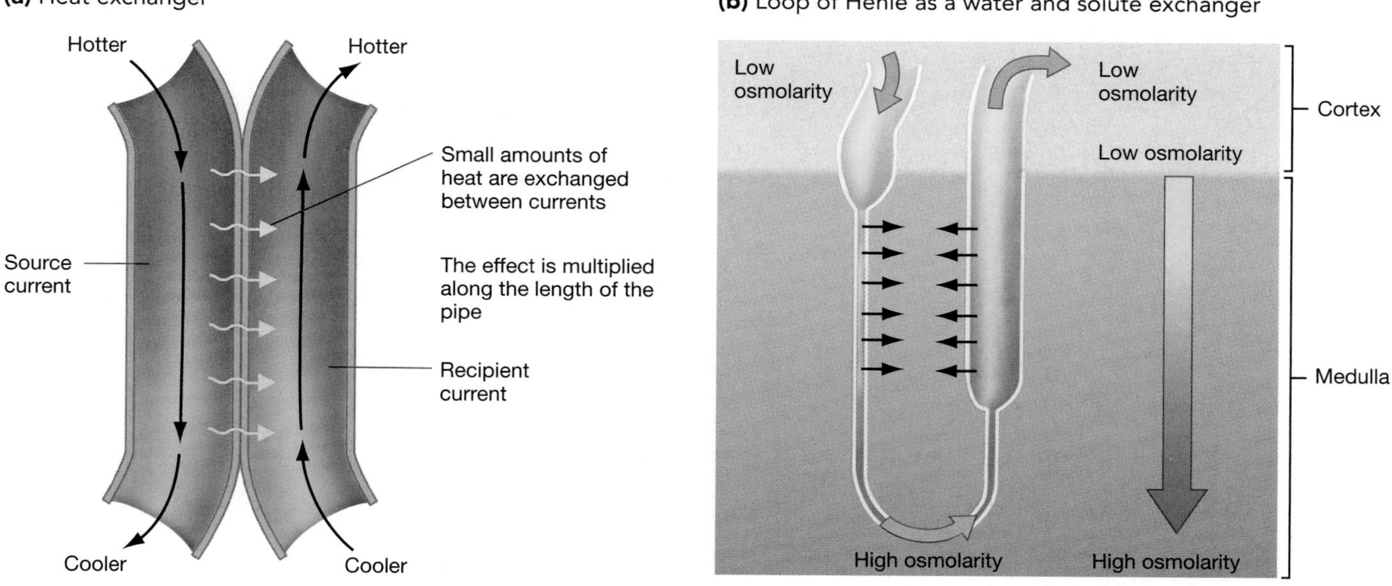

FIGURE 39.12 Countercurrent Multipliers
(a) This diagram shows how a heat exchanger works. **(b)** The loop of Henle was proposed to act as a countercurrent multiplier. The hypothesis was that the looping tubule exchanged water and solutes with the surrounding tissue in a way that set up an osmotic gradient.

Testing Kuhn's Hypothesis A series of papers published during the 1950s by Karl Ullrich, H. Wirz, and others supplied important experimental support for the countercurrent multiplier model. Two particularly important data sets, obtained by analyzing the osmolarity of tissue slices cut in sections perpendicular to the loop of Henle, are reproduced in **Figure 39.13**. Part (a) shows data on the osmolarity of fluid inside the tubule. As predicted by Kuhn's model, a strong gradient in osmolarity exists from the cortex to the medulla. The data in Figure 39.13b show that outside the tubule, the concentrations of Na^+, Cl^-, and urea also increase sharply from the cortex to the medulla. This was an important observation. It not only confirmed the prediction that an osmotic gradient exists in the tissue surrounding the loop of Henle, but also suggested that the solutes responsible for the gradient are Na^+, Cl^-, and urea. As we'll see, the concentration change in urea turned out to be particularly important.

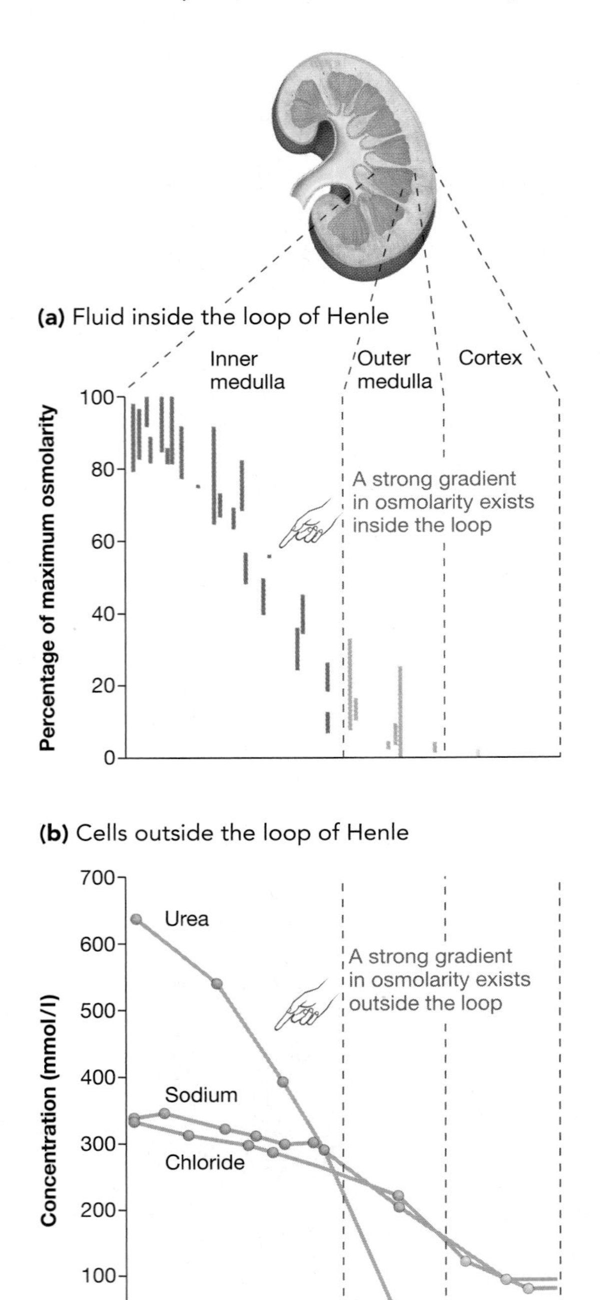

FIGURE 39.13 Data Confirm the Existence of a Strong Osmotic Gradient Along the Loop of Henle
(a) This graph plots the relative osmolarity of the fluid inside the rat nephron as a function of its location in the kidney. **(b)** This graph plots the concentration of urea, Na^+, and Cl^- in the tissue surrounding the loop of Henle in dogs versus the tissue's location in the kidney.

Before we consider how the steep osmotic gradient created by the loop of Henle affects urine formation, let's take a look at research on the molecular mechanisms involved in setting up the countercurrent multiplier. How do water and solutes move in and out of the loop in a way that establishes the dramatic osmotic gradient that researchers observed?

How Is the Osmotic Gradient Established? **Figure 39.14a** provides a detailed look at the morphology of the loop of Henle. Note that there are three distinct regions in the structure: the descending limb of Henle, the thin ascending limb, and the thick ascending limb. The thin and thick ascending regions differ in the thickness of their walls. Do the three regions also differ in their permeability to water and solutes?

It took over 15 years of experiments performed in laboratories around the world to formulate a definitive answer to this question. One important result began to emerge after C. W. Gottschalk and M. Mylle were able to puncture Henle's loop with a micropipette, analyze the composition of the fluid inside, and compare it to the nephron's final product—the urine inside the collecting duct. Although the two solutions were isotonic at the same cross-sectional level, the solutes involved were very different. In the ascending limb of Henle, Na^+ and Cl^- constituted at least 60 percent of the solutes and urea constituted about 10 percent. In contrast, both ions were rare in the collecting duct; urea was the major solute instead. Because Na^+ and Cl^- are also present at high concentration in the surrounding tissue, the data suggested that sodium might be actively pumped out of the ascending limb (**Figure 39.14b**).

Follow-up experiments that utilized ouabain and other poisons confirmed that sodium ions are actively transported out of the thick ascending limb. The epithelial cells responsible for sodium excretion are configured like the epithelium of the shark rectal gland. The combination of Na^+/K^+-ATPase, Cl^- channels, and $Na^+/2Cl^-/K^+$ cotransporters results in the active transport of salt out of the loop and into the surrounding tissue.

An important series of experiments by Juha Kokko and others established how water and solutes move across the two remaining segments of the loop. By injecting solutions of known concentration into the nephrons of rabbits, they documented that the descending limb is highly permeable to water but almost completely impermeable to solutes. The thin ascending limb, in contrast, is highly permeable to Na^+ and Cl^-, moderately permeable to urea, and almost completely impermeable to water (Figure 39.14b).

All of these observations about the loop of Henle were pulled together in 1972 when two papers—published simultaneously and independently—proposed the same model for how the loop of Henle works. The best way to understand the hypothesis is to follow the events that occur as fluid leaves the proximal tubule and moves through the loop, using Figure 39.14b. Note that as urine flows down the descending limb, it loses water to the tissue surrounding the nephron. This movement of water is passive (in the sense of not requiring an expenditure of ATP) because it follows an osmotic gradient. Once the hypertonic fluid at the bottom of the loop begins to flow up the ascending limb, however, it stops losing water because the membrane is now impermeable. Instead, the fluid begins to lose Na^+ and Cl^-. These solutes also move out of the loop passively, following their concentration gradient. In the thick ascending limb, additional Na^+ and Cl^- are actively transported out of the nephron.

The overall result of this complex system is the creation and maintenance of a strong osmotic gradient. Due to the countercurrent multiplier effect, the gradient is maintained with a minimal expenditure of energy. The water and salt that move out of the loop diffuse into a blood vessel that runs along the structure and are returned to the body (**Figure 39.14c**).

Regulating Water and Electrolyte Balance: The Distal Tubule and the Collecting Duct

The first three steps in urine formation—filtration, reabsorption, and establishment of an osmotic gradient—result in the production of a fluid that is slightly hypotonic to blood. The major solutes in this fluid are urea and other wastes. To this point, however, the events that have occurred in the nephron have been largely unregulated. Although the amount of blood filtered at the renal corpuscle can change dramatically depending on conditions, the proximal tubule and loop of Henle function in about the same way at all times. As a result, the fluid that enters the distal tubule is relatively constant in composition over time. In contrast, the urine that leaves the collecting duct is highly variable in osmolarity and in Na^+ and Cl^- concentration. Why?

The activity of the distal tubule and the collecting duct can change dramatically over time. Although electrolytes and water are always reabsorbed in the distal tubule, the amount of Na^+, Cl^-, and water that is reabsorbed in this structure and in the collecting duct varies with the animal's condition and is under hormonal control. If Na^+ levels in the blood are low, the hormone aldosterone is released and leads to the reabsorption of Na^+ and Cl^- in the distal tubule. Water follows these ions via osmosis. If an individual is dehydrated, a molecule called antidiuretic hormone (ADH) is released from the brain. When ADH interacts with the cells lining the collecting duct, aquaporin channels are inserted into the membrane. As a result the cells become highly permeable to water, and large amounts of water are reabsorbed. As **Figure 39.15a** (page 762) shows, water leaves along the concentration gradient established by the loop of Henle and results in the formation of urine that is strongly hypertonic to blood. When ADH is absent, however, the collecting duct is relatively impermeable to water. In this case, a hypotonic urine is produced (**Figure 39.15b**). People with defective forms of ADH or aquaporins

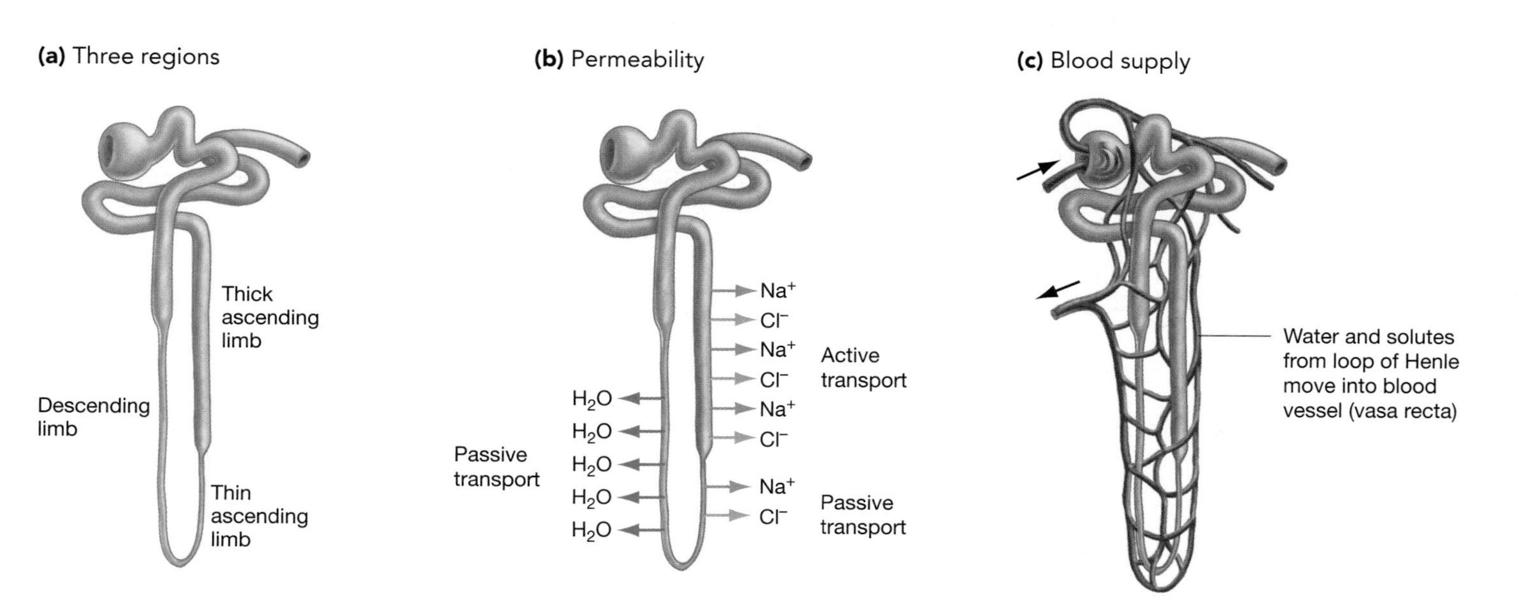

FIGURE 39.14 How Does the Loop of Henle Establish an Osmotic Gradient?
(a) The loop of Henle has three distinct regions. **(b)** The permeability of the descending limb, thin ascending limb, and thick asending limb to water and solutes varies dramatically, as shown here. The only active transport occurs in the thick ascending limb. **(c)** A blood vessel called the vasa recta runs along the loop of Henle in a countercurrent fashion.

produce copious amounts of urine—up to 30 liters per day—and suffer from a condition known as diabetes insipidus. (*Diabetes* means "to run through"; *insipidus* means "tasteless.")

To summarize, the first three segments of the nephron accomplish two objectives: They effectively concentrate nitrogenous wastes, and they create the possibility for Na^+, Cl^-, and water to be either excreted or reabsorbed by the final two segments in the structure. As a unit, the nephron is a remarkably effective mechanism for regulating water and electrolyte balance and achieving homeostasis.

(a) ADH present: Collecting duct is highly permeable to water

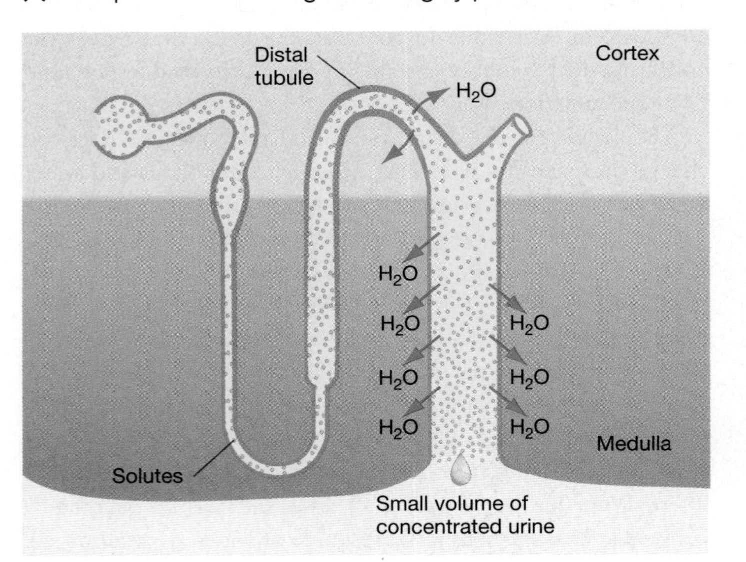

(b) No ADH present: Collecting duct is not permeable to water

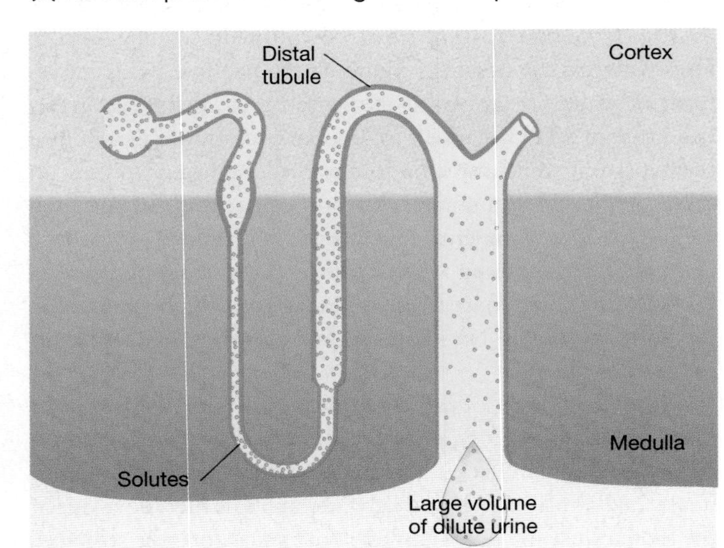

FIGURE 39.15 Water Reabsorption by the Collecting Duct Is Regulated by ADH
(a) The term *diuresis* refers to increased urine production. Antidiuretic hormone (ADH) leads to the production of small volumes of concentrated urine, because it makes the collecting ducts permeable to water. The water that leaves the collecting ducts passes into the vasa recta (see Figure 39.14c). **(b)** In the absence of ADH, urine volumes increase and contain a low concentration of solutes.

Essay Life in the Desert

The combination of extreme heat and severe dryness presents extraordinary challenges for water and electrolyte balance in desert-dwelling animals. To understand how these organisms cope, researchers have focused on studying the desert kangaroo rat (**Figure 1**). These rodents do not drink because no water is available. How do they obtain and conserve water?

Kangaroo rats do not drink water.

Unlike leaves, roots, and other plant tissues that contain a large amount of water, the seeds that kangaroo rats feed on are extremely dry. Almost by process of elimination, then, the major source of water for these animals is what biologists call metabolic water. As Chapter 6 showed, the oxidation of sugar during cellular respiration results in the release of water. More specifically, the oxidation of 1 g of carbohydrate yields 0.56 g of metabolic water.

Kangaroo rats have two major ways of conserving this precious resource. The first mechanism involves the production of an extremely hypertonic urine. Kangaroo rats have exceptionally long loops of Henle. These loops produce steep osmotic gradients in the medulla and make it possible for their kidneys to produce urine that is 30 times more concentrated than blood. This is the most hypertonic urine known among animals. As a result, kangaroo rats eliminate nitrogenous wastes with a minimal loss of water.

The second major water conservation mechanism involves respiration. During the heat of the day, kangaroo rats stay in their underground burrows where the air is cooler than their body temperature. As this cool air is drawn in through the rats'

(Continued on next page)

(Essay continued)

nasal passages, it is warmed and picks up water vapor (**Figure 2a**). The loss of water and heat from the nasal passages cools the tissue. Conversely, when warm, moist air returns from the lungs, it cools as it flows through the extensive nasal passages. Moisture condenses during this process, because cool air can hold much less water vapor than warm air (**Figure 2b**). As a result, most of the water that is lost from the nasal passages during inhalation is returned during exhalation.

In effect, the nasal passages of a kangaroo rat act as an unusual sort of countercurrent exchanger. Instead of exchanging heat or molecules across a space, as heat exchangers and the loop of Henle do, the nasal passages of kangaroo rats exchange heat and water over time.

FIGURE 1 The Kangaroo Rat
Kangaroo rats are native to the deserts of southwestern North America.

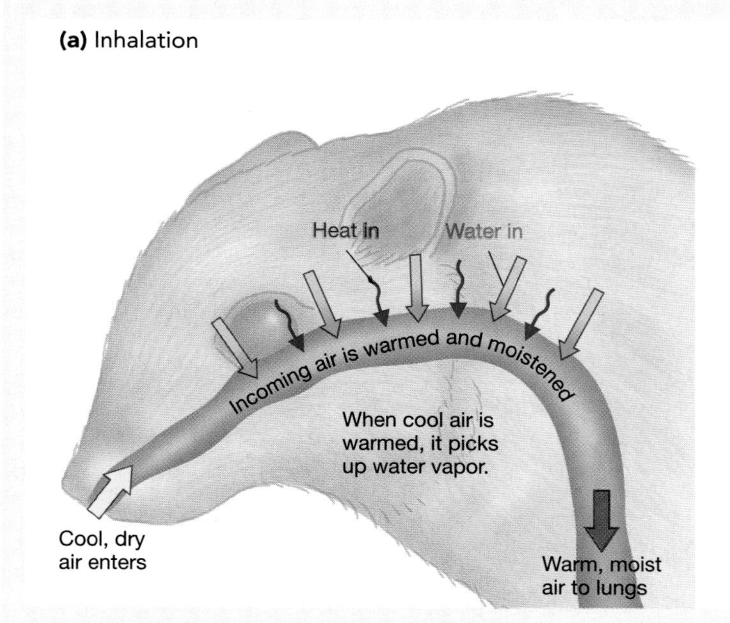

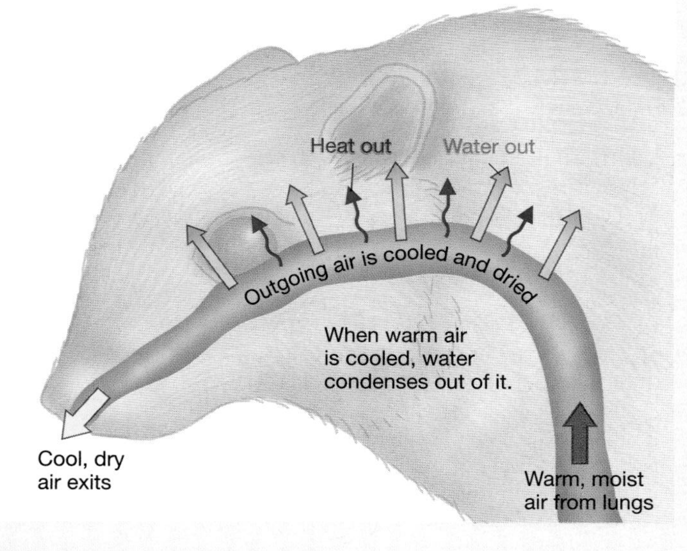

FIGURE 2 Countercurrent Exchange Limits Water Loss During Respiration
The nasal passages of the kangaroo rat are particularly long and convoluted. There is a strong temperature and moisture gradient from the tip of the nose to the end of the nasal passages.

Chapter Review

Summary

Homeostasis with respect to water and electrolyte balance is critical to the health of animals. The mechanisms involved in regulating water and electrolyte balance vary widely among animal groups, in part because different habitats present different challenges. Freshwater fish, for example, are strongly hypertonic to their environment and tend to gain water and lose electrolytes. Marine fish, in contrast, are strongly hypotonic to seawater and tend to lose water and gain electrolytes. Terrestrial animals lose water every time they breathe.

Research on the shark rectal gland established how marine animals excrete excess salt. Sodium-potassium pumps are located on the basolateral side of epithelial cells in this gland. These pumps establish an electrochemical gradient that carries Na^+ and Cl^- from the blood into the lumen of the gland, where it is excreted. Later research revealed that a similar arrangement of pumps, channels, and cotransporters occurs in a wide variety of animal cells that transport salt. These cells occur in the gills of saltwater fish, the salt glands of marine birds and reptiles, and the kidneys of mammals.

Despite extensive studies, it is still not clear how freshwater fish take up electrolytes from their environment to replace the ions that they lose during respiration. Research on salmon has shown that the gills of freshwater fish contain cells with sodium-potassium pumps. Some researchers suggest that these cells may be involved in the active transport of ions into the gills.

Terrestrial insects limit water loss in several ways. Their exoskeleton is covered with wax to limit evaporation, and the openings to their respiratory organs close when water stress is severe. They are also capable of forming a hypertonic urine that minimizes water loss during excretion of nitrogenous wastes. A pre-urine that is isotonic with the hemolymph forms in the Malpighian tubules of insects. If hormones trigger activity by K^+ and Cl^- pumps in the epithelium of their hindgut, then electrolytes and water are reabsorbed from the pre-urine and returned to the hemolymph. A hypertonic final urine results.

Terrestrial mammals have an analogous system for regulating water and electrolyte balance during urine formation. Nephrons in the mammalian kidney form a raw filtrate and then reabsorb nutrients, electrolytes, and water. The resulting solution flows through the loop of Henle, where changes in water and salt permeability create a steep osmotic gradient. If ADH triggers changes in the water permeability of the final sections of the nephron, water is reabsorbed along the osmotic gradient and a hypertonic urine is produced.

In animals, maintaining water and electrolyte balance is an active, energy-demanding process that is based on the action of membrane proteins. Water and electrolyte excretion or reabsorption is carefully controlled to achieve homeostasis. As a result, animals are able to occupy habitats that present a diversity of osmotic challenges.

Questions

Content Review

1. Which of the following statements is true of freshwater fish? Which is true of marine fish?
   a. Their tissues are hypotonic to their environment.
   b. They lose water to their environment primarily through the gills.
   c. They must prevent a potential increase in the volume of their cells.
   d. They actively pump Na^+ and Cl^- from blood into external water.
2. Na^+/K^+-ATPase plays an important role in ion transport in which of the following structures? (Circle all that apply.)
   a. the shark rectal gland
   b. the proximal tubule in mammal kidneys
   c. gills of marine fish
   d. gills of freshwater fish
   e. the loop of Henle
   f. Malpighian tubules
   g. the salt gland in birds
   h. the salt gland of marine reptiles
3. Which of the following places the regions of the nephron in their correct sequence with respect to flow of tubular fluid?
   a. distal tubule → ascending limb of Henle → descending limb of Henle → proximal tubule → collecting duct
   b. ascending limb of Henle → descending limb of Henle → proximal tubule → distal tubule → collecting duct
   c. proximal tubule → ascending limb of Henle → descending limb of Henle → distal tubule → collecting duct
   d. proximal tubule → descending limb of Henle → ascending limb of Henle → distal tubule → collecting duct
4. Which of the following statements about kidney function is correct? (Circle all that apply.)
   a. The loop of Henle acts as countercurrent multiplier system.
   b. The thin ascending limb of Henle is highly permeable to water.
   c. The movement of water and solutes is entirely passive—no active transport mechanisms are involved.
   d. The descending limb of Henle is highly permeable to salt.
   e. Reabsorption of water and solutes takes place primarily in the distal tubule.
5. What effect does antidiuretic hormone (ADH) have on the nephron?
   a. It increases water permeability of the descending limb of the loop of Henle.
   b. It decreases water permeability of the descending limb of the loop of Henle.
   c. It increases water permeability of the collecting duct.
   d. It decreases water permeability of the collecting duct.

Conceptual Review

1. Study the diagram of a salmon's life cycle in Figure 39.5b. Explain the changes that need to occur in water and electrolyte balance as an individual moves from freshwater to salt water and back. Specifically, state when the animal should drink or not drink, when cells in the gill epithelium should excrete or import electrolytes, and why each change occurs.
2. The chloride cells of fish gills are sometimes called mitochondria-rich cells due to their high density of mitochondria. How does high

mitochondrial density relate to the functional role of chloride cells? Would you expect other epithelial cells involved in ion transport to contain high numbers of mitochondria? Explain.

3. Why is it significant that cells involved in transport processes often have microvilli?

4. This chapter introduced a number of features that help terrestrial animals reduce water loss. These traits include the layer of wax found on insect exoskeletons, the ability of insects to close the openings to their respiratory passages, secretion of nitrogenous wastes as insoluble uric acid, exceptionally long loops of Henle in the mammalian kidney, and retention of water in the nasal passages of desert mammals. Predict how each of these traits differs in animals that live in very humid versus very dry habitats. How would you test your predictions?

5. In insects, active transport of electrolytes into the Malpighian tubules leads to the formation of a pre-urine that is isotonic with the hemolymph. In mammals, urine formation begins with blood pressure in the renal corpuscle that leads to the formation of a filtered pre-urine that is isotonic with the blood. In insects, the pre-urine is processed in the hindgut; in mammals the pre-urine is processed in the remainder of the nephron. How are the processing steps in insects and mammals similar? How are they different?

Applying Ideas

1. Examine Figure 39.14b again, and note that the blood vessel that runs along the mammalian nephron is arranged so that blood flows in a countercurrent arrangement relative to the flow of fluid in the loop of Henle. Recall that the water and electrolytes that leave the loop of Henle and collecting duct diffuse into this blood vessel and are returned to the body. Why is the countercurrent arrangement for blood flow important?

2. You have isolated a segment of a rat nephron and introduced a solution of known composition. When you compare the fluid collected at the end of the segment with the test solution introduced at the beginning, you find that the volume has decreased by 30 percent, the Na^+ concentration has decreased by 30 percent, but the urea concentration has increased by 50 percent. What processes in this segment might account for these changes?

3. According to the data reviewed in Chapter 24, the first animals that appear in the fossil record lived in marine habitats. What changes occurred in the regulation of water and electrolyte balance in animal populations that first began to occupy freshwater and terrestrial habitats? Why would natural selection favor these changes?

4. In some areas of the world, highway maintenance crews make extensive use of salt to keep roads free of ice in winter. Biologists are concerned about the effect of this salt on terrestrial and freshwater organisms. Why?

5. Ethanol is the active ingredient in alcoholic beverages. One of its many toxic effects is to inhibit the release of antidiuretic hormone (ADH) by the brain. How does the ingestion of ethanol affect urine formation? Why do people who ingest large amounts of ethanol suffer from headache, body ache, and other symptoms of dehydration?

CD-ROM and Web Connection

CD Activity 39.1: The Mammalian Kidney *(tutorial)*
(Estimated time for completion = 15 min)
How do your kidneys work?

CD Activity 39.2: The Loop of Henle: A Countercurrent Multiplier *(animation)*
(Estimated time for completion = 5 min)
What is the role of the loop of Henle in producing concentrated urine?

At your **Companion Website** (http://www.prenhall.com/freeman/biology), you will find self-grading exams and links to the following research tools, online resources, and activities:

Osmotic Stress
Learn about current research in the field of osmotic stress and homeostasis.

Freshwater vs. Saltwater Fish
Explore the differences in osmotic regulation between freshwater and saltwater fish.

Gravity and Electrolyte Balance
This article describes how gravity—or lack of gravity—affects the body's ability to maintain electrolyte balance.

Additional Reading

Schmidt-Nielsen, K. 1959. Salt glands. *Scientific American* 200 (January): 109–116. Recounts the discovery of salt-excretng glands in marine birds.

Walsberg, G. E. 2000. Small mammals in hot deserts: some generalizations revisited. *BioScience* 109: 50: 109–120. Explores how desert rodents maintain water balance and normal body temperature in extremely arid and hot environments.

Animal Nutrition

40

If you're like most people, you've probably given a fair amount of thought to the types of food that you choose to eat, but no thought at all to what happens to that food in your digestive tract. As a meal moves through the digestive system, its chemical composition and physical characteristics change dramatically. Structurally and chemically, complex raw materials are broken down into simple, usable components. In this respect, the digestive system is analogous to an oil refinery or a lumber mill. How is food processed? Which substances are used as nutrients? How do humans and other animals maintain appropriate levels of key nutrients in their bodies?

One of the keys to answering these questions is to recognize that most animal bodies are based on the simple design of a tube. For example, the earthworm body in the chapter-opening photograph is essentially a tube within a tube. The interior tube starts at the mouth and ends at the anus. Food enters one end of this internal tube; waste products exit the other end.

To explore what happens to food on its way from mouth to anus, the first section in this chapter introduces the nutrients that animals need to stay alive. The next two sections investigate how animals extract nutrients from food, starting with questions about how teeth, beaks, and other mouthparts work and ending with research on how feces forms. The chapter closes by introducing research on a particularly important aspect of achieving homeostasis with respect to nutrition: regulating the concentration of glucose in the blood. Throughout the discussion, important practical issues will arise. For example, can certain nutrients be used to boost athletic performance? Why are nutrition-related diseases like type II diabetes mellitus and heart disease increasing in frequency in some human populations? Just as

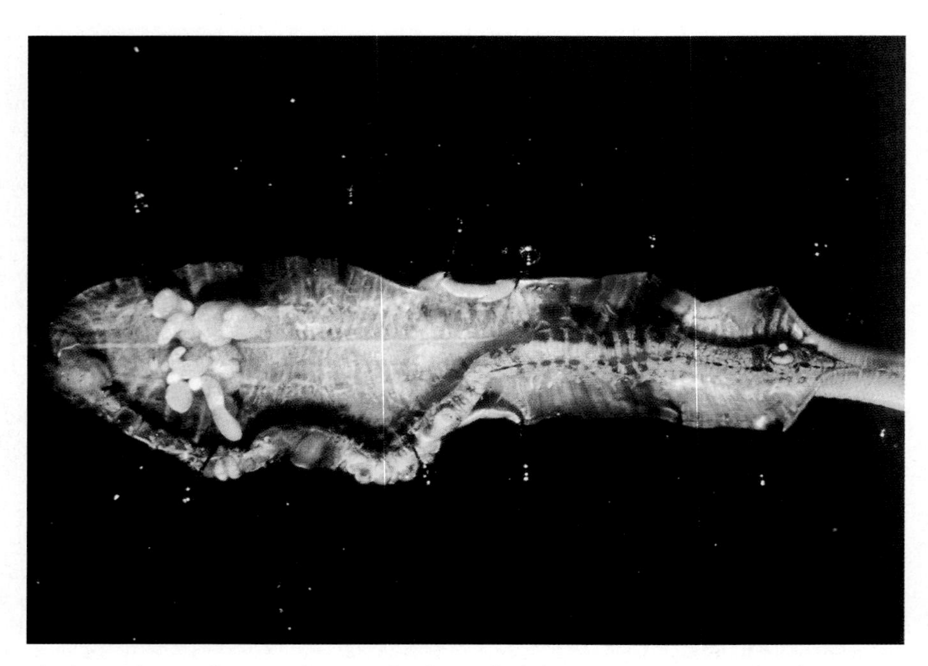

The long tube inside an earthworm's body is called the alimentary canal. This chapter explores what happens to food during its journey down the length of an animal's alimentary canal.

plant nutrition is fundamental to the productivity of agricultural and natural ecosystems, nutrition is a basic component of human health and welfare.

40.1 Nutritional Requirements

A **nutrient** is a substance that an organism needs to remain alive. Food is any material that contains nutrients. When an animal takes in enough nutrients to perform day-to-day tasks and stay healthy, it is said to be in nutritional balance. To maintain this state, an animal must acquire two general types of nutrients: Reduced carbon compounds that can be oxidized to produce ATP, and the elements and molecules that are needed to synthesize body components and sustain cells.

Understanding which nutrients an individual needs, and in what amounts, are basic issues in research on animal nutrition. Determining nutritional requirements is seldom simple, however. To illustrate how biologists go about the task, let's first consider what humans need to maintain good health, and then what they need in the special case of enhancing athletic performance.

Meeting Basic Needs

The United States government instituted a program to establish basic nutritional requirements as part of its war effort during World War II. In 1943, the Food and Nutrition Board of the National Academy of Sciences* published the first recommended daily allowances (RDAs). The goal of the RDAs was to specify the amount of each essential nutrient that must be ingested to meet the needs of practically all healthy people. The recommendations focused on several types of nutrients:

- Proteins provide amino acids and are sometimes oxidized to provide energy. Of the 20 amino acids required to manufacture proteins, humans can synthesize 12. The other eight must be obtained from food and are called **essential amino acids.** The word *essential* refers to nutrients that cannot be synthesized and must be obtained in the diet.
- **Vitamins** are carbon-containing molecules that have a variety of roles; several function as coenzymes in critical reactions. Vitamins are defined as molecules that are vital for health in minute amounts. **Table 40.1a** (page 768) introduces a few of the vitamins for which RDAs have been established.
- **Essential elements** serve a wide variety of functions (see **Table 40.1b**). Some, such as calcium and phosphorus, are needed in relatively large quantities. Others, like iron and magnesium, are required in small or trace amounts.
- **Electrolytes** form ions in solution. As ions, they influence osmotic balance and are required for normal membrane function. Sodium (Na^+), potassium (K^+), and chloride (Cl^-) are the major ions in the body.

*The National Academy of Sciences is a group of scientists and engineers that advises the U.S. Congress on scientific and technical matters. Its Food and Nutrition Board is made up of biologists who specialize in the area of animal nutrition.

RDAs for these nutrients are familiar to people who have either visited or lived in the United States because food packaging is required to state how a serving of the contents meets the RDA (**Figure 40.1**). In many cases, labels must also furnish information on the amount of chemical energy available in the food. In nutrition research, the calorie is used as a unit of energy. If a single helping of food contains 150 calories, it means that 150 kilocalories of energy are released when the food is oxidized by cells. Most energy in food is in the form of carbohydrates and fats, which are reduced carbon compounds. Because fats and other lipids are much more highly reduced than carbohydrates such as starch and sugar, fats and other lipids provide over twice the amount of energy on a per-gram basis.

In reading and interpreting information about the RDAs, it is important to recognize that the recommendations frequently change as research advances. In many cases, members of the Food and Nutrition Board are cautious about making recommendations due to a lack of reliable data. Studying nutrient requirements in humans is difficult. The types of experiments that would be most enlightening would assign different, carefully controlled diets to several groups of subjects. Because some of these diets might lead to illness, however, they are unethical. As a result, researchers rely on observational studies that document the health of people whose intake levels for specific nutrients are known or that monitor the intake of a specific nutrient versus the level maintained in the body over time.

The scientists who publish the RDAs constantly remind the public that a balanced diet fulfills all nutrient requirements. In the vast majority of cases, nutritional supplements are not needed; if abused, supplements can lead to health problems. They note one exception: Some women may need to take iron

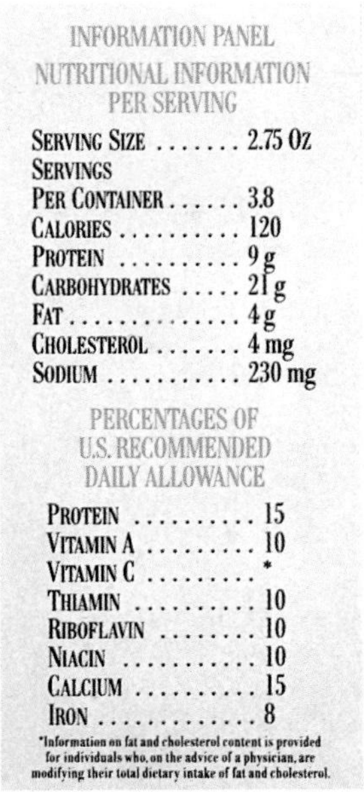

INFORMATION PANEL
NUTRITIONAL INFORMATION
PER SERVING

SERVING SIZE	2.75 OZ
SERVINGS PER CONTAINER	3.8
CALORIES	120
PROTEIN	9 g
CARBOHYDRATES	21 g
FAT	4 g
CHOLESTEROL	4 mg
SODIUM	230 mg

PERCENTAGES OF
U.S. RECOMMENDED
DAILY ALLOWANCE

PROTEIN	15
VITAMIN A	10
VITAMIN C	*
THIAMIN	10
RIBOFLAVIN	10
NIACIN	10
CALCIUM	15
IRON	8

*Information on fat and cholesterol content is provided for individuals who, on the advice of a physician, are modifying their total dietary intake of fat and cholesterol.

FIGURE 40.1 Nutrition Labeling on Food Packaging
Food packaging is often required to indicate the percentage of recommended daily allowances furnished by a standard-sized serving.

supplements as recommended by their physician. These are beneficial in some cases because hemoglobin—the protein that carries oxygen in the blood—contains iron. Consequently, significant amounts of iron are lost during menstruation.

Nutrition and Athletic Performance

RDAs identify the nutrients that healthy people need to stay healthy and the amounts that are required to maintain nutrient balance. These recommendations target people with normal needs and activities. How do researchers determine nutrient requirements for extraordinary circumstances? To answer this, let's consider research on athletes. These individuals do extraordinary amounts of physical work. What nutrients do they require to perform at an optimal level?

For a researcher interested in questions about the nutrient requirements of athletes, at least some types of experiments are possible. As long as experimental subjects are provided with a diet that fulfills RDAs and as long as safe dosages of experimental compounds are used, studies should be able to meet accepted ethical guidelines.

In 1967, Jonas Bergström and colleagues published one of the classic experiments in nutrition research. They were interested in the performance of endurance athletes such as cross-country skiers, bicyclists, and distance runners. At the time, most researchers supported the hypothesis that fatty acids provide the fuel required for extended physical labor or athletic activity. This hypothesis was logical because fatty acids provide more than twice as much energy as carbohydrates on a per-gram basis.

As Bergström and co-workers were aware, however, glycogen is an important energy storage molecule in animals. Recall from Chapter 6 that glycogen is a polysaccharide made up of glucose molecules, and that glucose is the preferred starting compound for the production of ATP through cellular respiration. Most of the body's glycogen is stored in skeletal muscles

TABLE 40.1 Essential Nutrients

(a) Some vitamins required by humans

Name	Source in Diet	Function	Symptoms if Deficient
Vitamin B_1 (thiamine)	legumes, whole grains, potatoes, peanuts	formation of coenzyme in Krebs cycle	Beriberi (fatigue, nerve disorders, anemia)
Vitamin B_{12}	red meat, eggs, dairy products; also synthesized by bacteria in intestine	coenzyme in synthesis of proteins and nucleic acids; formation of red blood cells	Anemia
Niacin	meat, whole grains	component of coenzymes NAD^+ and $NADP^+$	Pellagra (digestive problems, skin lesions, nerve disorders)
Folate	green vegetables, oranges, nuts, legumes, whole grains; also synthesized by bacteria in intestine	coenzyme in nucleic acid and amino acid metabolism	Anemia
Vitamin C (ascorbic acid)	citrus fruits, tomatoes, broccoli, cabbage, green peppers	used in collagen synthesis, prevents oxidation of cell components, improves iron absorption	Scurvy (degeneration of teeth and gums)
Vitamin D	fortified milk, egg yolk; also synthesized in skin exposed to sunlight	aids absorption of calcium and phosphorus in small intestine	Rickets (bone deformities) in children; bone softening in adults

(b) Some elements required by humans

Name	Source in Diet	Function	Symptoms if Deficient
Calcium (Ca)	dairy products, green vegetables, legumes	bone and tooth formation, nerve signaling, muscle response	loss of bone mass, slow growth
Phosphorus (P)	dairy products, meat, grains	bone and tooth formation, synthesis of nucleotides and ATP	weakness, loss of bone
Sulfur (S)	any source of protein	amino acid synthesis	swollen tissues, degeneration of liver, mental retardation
Magnesium (Mg)	whole grains, green leafy vegetables	enzyme cofactor	nerve disorders
Iron (Fe)	meat, eggs, whole grains, green leafy vegetables, legumes	enzyme cofactor; synthesis of hemoglobin and electron carriers	anemia, weakness
Fluorine (F)	fluoridated water, seafood	maintenance of tooth structure	higher frequency of tooth decay

and the liver. During exercise, glycogen is catabolized by both types of cells. Muscle cells use the glucose that is released to manufacture ATP; glucose from the liver enters the bloodstream and is delivered to cells that are doing work.

Based on these observations, Bergström and associates hypothesized that an individual's ability to run, ski, or bicycle fast and for a long time might depend primarily on the amount of glycogen stored in the body rather than on the amount of fatty acids. They also predicted that a diet rich in carbohydrates would maximize the amount of stored glycogen.

To test these ideas, Bergström and co-workers set up the experiment outlined in **Figure 40.2a**. Nine student volunteers began by eating a mixed diet of carbohydrate, fat, and protein for a day. After the researchers had removed a tiny sample of

(a) Protocol for diet experiment

		6 volunteers	3 volunteers
First test period	Day 1	**Diet:** MIxed carbohydrate, protein, fat	**Diet:** MIxed carbohydrate, protein, fat
	Day 2	**Morning test:** ride bike to exhaustion	**Morning test:** ride bike to exhaustion
Second test period		**Diet:** 1500 kcal protein 1300 kcal fat	**Diet:** 2300 kcal carbohydrate 500 kcal protein
	Day 3		
	Day 4		
	Day 5	**Morning test:** ride bike to exhaustion	**Morning test:** ride bike to exhaustion
Third test period		**Diet:** 2300 kcal carbohydrate 500 kcal protein	**Diet:** 1500 kcal protein 1300 kcal fat
	Day 6		
	Day 7		
	Day 8	**Morning test:** ride bike to exhaustion	**Morning test:** ride bike to exhaustion

(b) Results: A diet high in carbohydrates supports optimal performance

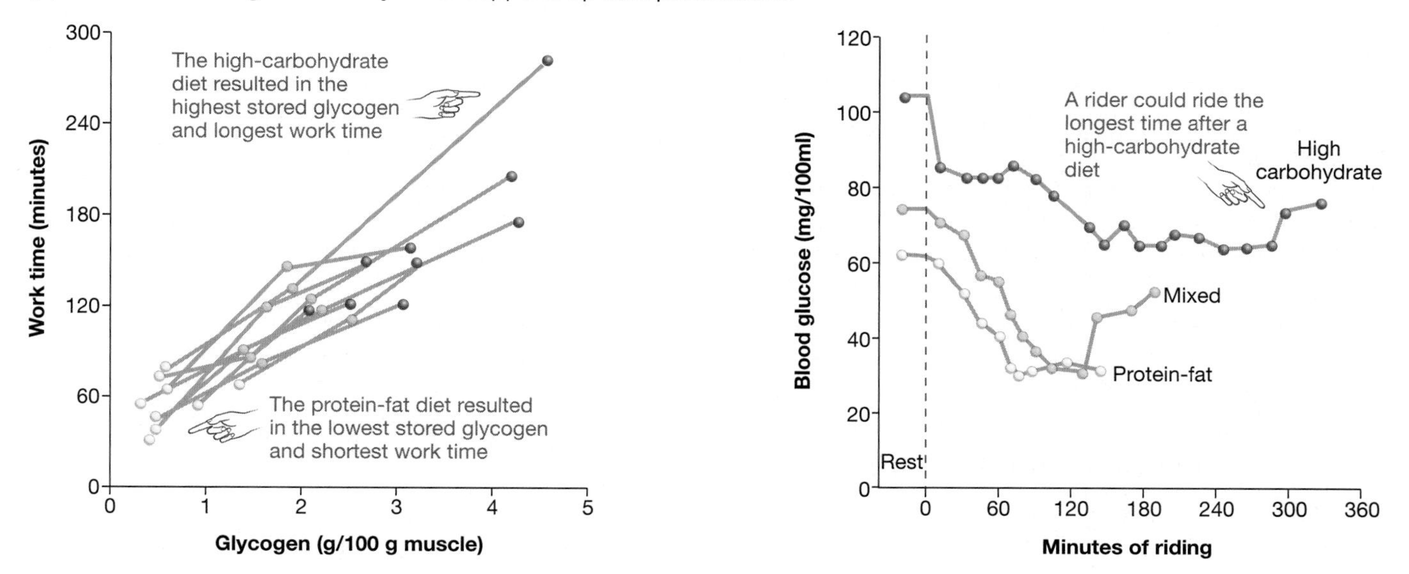

FIGURE 40.2 Does Carbohydrate Loading Increase Glycogen Stores and Endurance?
(a) The protocol for an experiment on diet and athletic performance. During each test, volunteers rode exercise bikes at 75 percent of their maximum work rate until feeling exhausted. **(b)** The graph on the left shows the relationship between the amount of glycogen stored in muscle and riding time after consuming each of the three diets in the study. The data on the right indicate how blood glucose changed over the course of the three riding tests experienced by one volunteer. QUESTION Why were the researchers careful to provide the same amount of energy, in kcal, in each diet?

leg muscle from each volunteer to measure the amount of stored glycogen, the volunteers rode exercise bikes at the same relative work rate until they were exhausted. During the exercise period, the biologists removed blood samples at regular intervals to check the concentration of glucose.

Six individuals then switched to a diet high in fat and protein for three days. The remaining three individuals switched to a high-carbohydrate diet. After the researchers checked their glycogen stores, each volunteer again rode a stationary bike to exhaustion while their blood glucose levels were recorded. Finally, each volunteer again switched diets for a three-day period. The individuals who had been eating fat and protein switched to carbohydrates; the three students who had been eating carbohydrates switched to fat and protein. To conclude the experiment, each volunteer had their glycogen reserves documented, then rode the stationary bike to exhaustion as their blood glucose concentrations were monitored.

Figure 40.2b shows the results. The graph on the left plots the amount of glycogen present in leg muscle versus the amount of time each volunteer was able to work on each ride. Note that in every case, both the amount of stored glycogen and the work time were much higher after individuals had eaten a high-carbohydrate diet. Now examine the graph on the right, which plots the changes in blood glucose levels that occurred in one of the subjects during each of the three rides. These data also support the hypothesis that a diet high in carbohydrate supports optimal performance.

Subsequent experiments, performed in a variety of labs around the world, confirmed that performance in endurance sports is enhanced by a high-carbohydrate diet. Does ingesting glucose or other carbohydrates just before a race also improve performance? A thorough set of studies has rejected this hypothesis. The current consensus among nutrition researchers is that ingesting sugars right before a contest actually reduces performance.

The experiments with student volunteers revolutionized how athletes all over the world train for endurance events. The dietary regime known as carbohydrate loading is now a routine part of race preparations. For endurance athletes, a high-carbohydrate meal is the breakfast of champions.

40.2 Obtaining Food: The Structure and Function of Beaks, Teeth, and Mouthparts

Whether an animal is eating to meet basic needs or a specialized nutritional requirement, the food that has been found or captured must be ingested. The transition between acquiring food and taking it into the body occurs in the mouth. Given the diversity of food sources that animals exploit, it is not surprising that they have a wide variety of structures in and around the mouth to accomplish this.

Some animals, such as the snake and maggot illustrated in **Figure 40.3a**, lack extensive mouthparts. They simply open their mouths wide and ingest food items whole. In contrast, the insects in **Figure 40.3b** have a diversity of mouthpart structures, ranging from the needle-like proboscis of mosquitoes to the pinching mandibles of leaf-cutting ants. The function of these mouthparts varies. A mosquito's proboscis allows it to pierce the hide of an animal and suck up blood; the mandibles of leaf-cutting ants enable them to cut large leaves into small pieces. In other insect species, mouthparts are specialized for grasping small particles, filtering small particles, or sponging, siphoning, or lapping up liquids.

The correlation between the structure and function of insect mouthparts is akin to the correspondence between the beaks and food sources of Galápagos finches highlighted in Chapter 38. Similarly, **Figure 40.3c** shows the mouthparts of two mammals that are adapted to different diets. Humans eat both plants and animals and have a variety of tooth sizes and shapes. Human teeth include sharp canines for tearing meat and flattened molars for crushing seeds, roots, and other sources of carbohydrate. Wolves, in contrast, are primarily meat-eaters. They lack molars and have an array of sharp teeth that tear flesh efficiently.

Across animal groups, there is a general and strong correlation between the size and shape of mouthparts and the size and shape of food sources. Let's pursue this correlation further by analyzing the structure and function of jaws and teeth in what may be the most diverse lineage of all vertebrate lineages: the cichlid fishes of Africa (**Figure 40.4**).

The cichlids that inhabit the Rift Lakes of east Africa are a spectacular example of adaptive radiation. Recall from Chapter 24 that adaptive radiation refers to the diversification of a single ancestral population into many species, each of which lives in a different habitat or utilizes a distinct feeding method. Lake Tanganyika, for example, is home to 300 endemic cichlids. (*Endemic* means that the species lives nowhere else.) Most are specialist feeders; as a group, they utilize almost every conceivable food source in the lake: plankton, algae, eggs, fish scales, fish fins, whole fish, plants, insects, and snails. Because geologic evidence indicates that the lake dried up completely during the most recent ice age, biologists contend that the 300 species descended from a single colonizing population in less than 10,000 years.

How could so much evolution take place so fast? Karl Liem contends that cichlids were uniquely situated to exploit a diversity of foods because they have an extra set of jaws. As **Figure 40.5a** (page 772) shows, these pharyngeal (throat) jaws are located well behind the normal, or oral jaws. Embryologically, pharyngeal jaws are derived from the stiff gill arches that are found in all fish. Pharyngeal jaws are actually found in many types of fish besides cichlids, but in non-cichlids the extra jaws are simply used to transport food from the mouth to the back of the throat. In cichlids, however, pharyngeal jaws are capable of biting and are used to process food by crushing, tearing, or compacting it. In effect, then, cichlids are able to feed with particular efficiency because they have two structures for ingesting and processing food instead of just one.

How did the pharyngeal jaws of cichlids originate? To answer this question, Liem compared the structure of the pharyngeal jaws in cichlids to the same structure in a closely related non-cichlid. He found that cichlid pharyngeal jaws have several unique features:

- One end of the pharyngeal jaws connects with the skull. This is important because the connection forms a hinge and allows the structure to act as a lever.

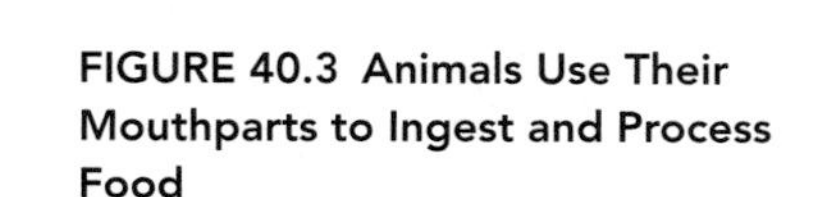
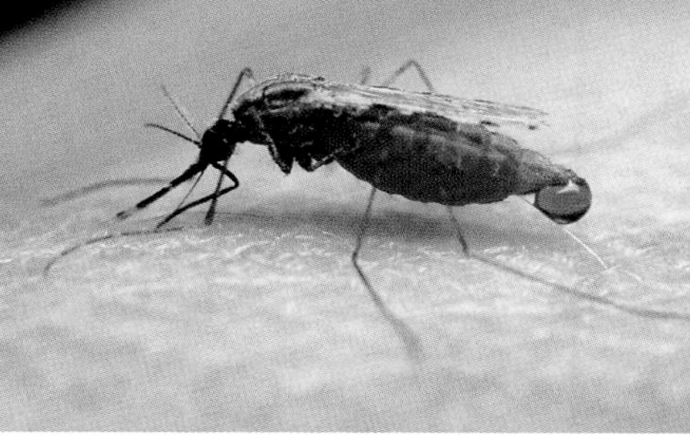
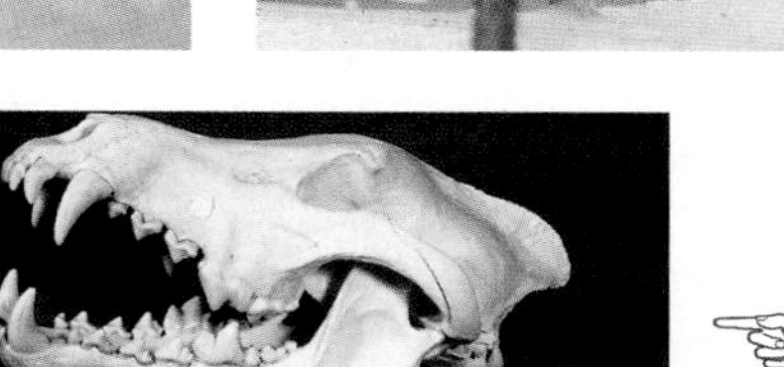

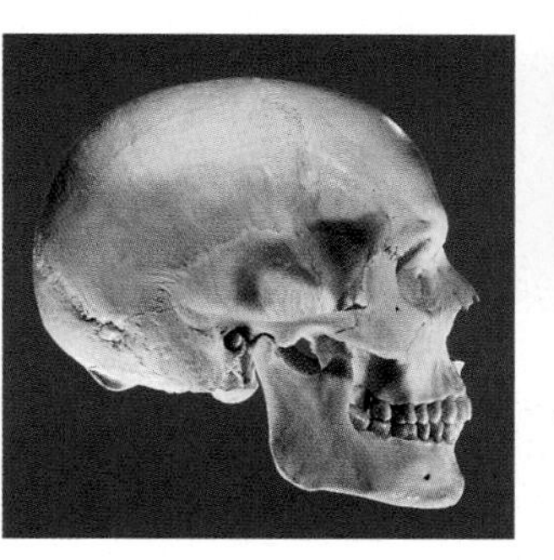
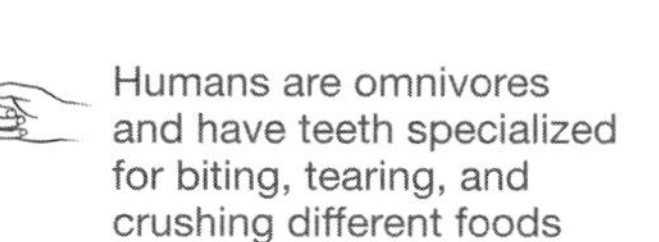

FIGURE 40.3 Animals Use Their Mouthparts to Ingest and Process Food
(a) Snake jaws are held together by elastic ligaments that allow the mouth to open wide enough to swallow large prey. Poisonous snakes have highly modifed teeth called fangs that hold and deliver venom. Maggots (right) lack complex mouthparts and also swallow their food whole. **(b)** As the mosquito (left) and leaf-cutting ant (right) show, there is a strong correlation between the structure of insect mouthparts and their function in obtaining food. **(c)** Humans (left) and wolves (right) have very different teeth. **QUESTION** What tooth shapes would you expect to find in species (elephants, cows) that eat only carbohydrate-rich foods, such as grass stems and leaves?

FIGURE 40.4 Rift Lake Cichlids
These species live in Lake Malawi. Unfortunately, dozens if not hundreds of cichlids have recently gone extinct because of water pollution and the impact of introduced predators.

(a) Pharyngeal (throat) jaws are located behind the oral jaws.

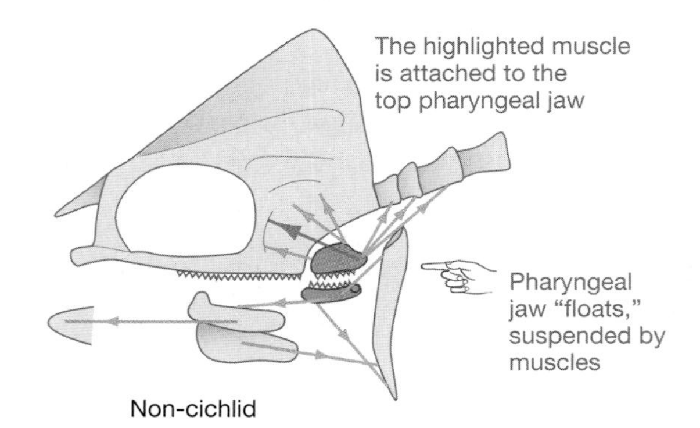

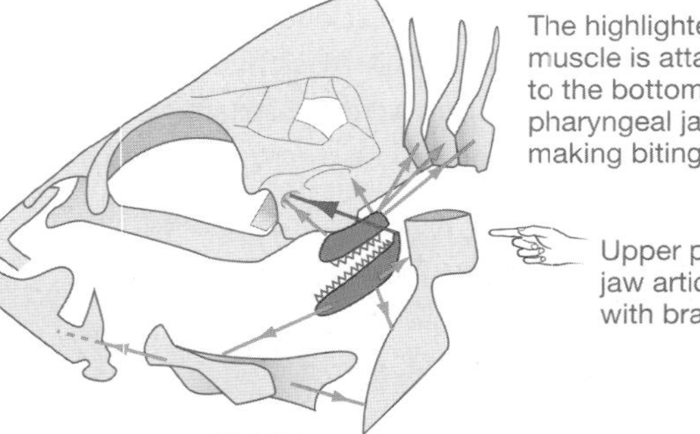

(b) Teeth on pharyngeal jaws vary among species.

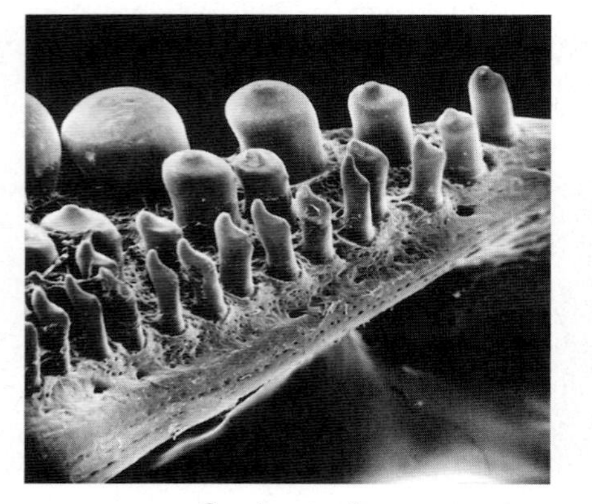

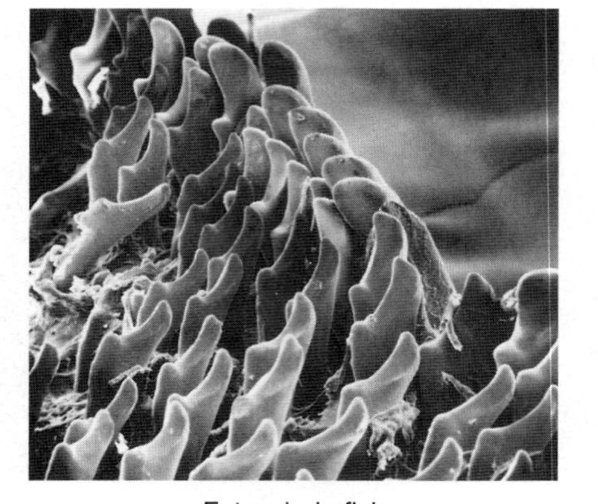

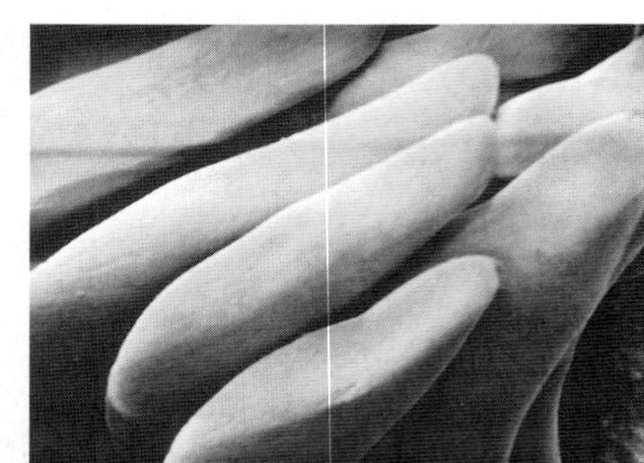

FIGURE 40.5 The Pharyngeal Jaw
(a) The anatomy of the pharyngeal jaw in non-cichlids (left) and cichlids (right). Arrows indicate where muscles are inserted in and around the pharyngeal jaws. **(b)** Scanning electron micrographs illustrate how the morphology of pharyngeal jaw teeth correlates with the food utilized. **EXERCISE** In the drawing in part (a) right, indicate where the teeth illustrated in part (b) are located. Also label the oral jaw in the skull in part (a) right.

- Several muscles that bypass the pharyngeal jaws in other groups attach to the pharyngeal jaws of cichlids. This is a key trait because the added musculature makes the pharyngeal jaw in cichlids capable of biting.
- The lower jaw is sutured into one solid structure instead of existing as two independent pieces. This feature makes the jaw more effective as a biting device.

Finally, Liem and others have documented that tooth-like protuberances on the pharyngeal jaws of cichlids vary in size and shape. In addition to having an extra pair of jaws, cichlids have an extra set of "teeth." As the scanning electron micrographs in **Figure 40.5b** show, the structure of these "teeth" correlates with their function in compacting algae, tearing meat, or crushing snail shells. In combination with analyses of pharyngeal jaw structure, these observations provide strong support for the hypothesis that the diversity of animal mouthparts is an adaptive response to a diversity of food sources.

40.3 Digestion

The first two sections of this chapter introduced the nutrients that animals require and examined how those nutrients are ingested. Unlike plants and unicellular organisms, animals do not acquire nutrients as individual molecules. Instead, they ingest large packets of food that are digested internally. In many cases, large chunks of material are taken in and must be broken down into small pieces (**Figure 40.6**). Nutrients must be extracted from the meal and waste materials must be eliminated. How and where does this processing occur?

Digestion takes place in a structure called the **alimentary** (nourishment) **canal**—also known as the digestive tract or gastrointestinal (GI) tract. As **Figure 40.7** shows, the alimentary canal starts at the mouth and ends at the anus. The interior of this tube communicates directly with the external environment. Embryologically, it derives from the hollow tube that forms as cells invaginate during gastrulation (see Chapter 19).

The alimentary canal is only part of the entire digestive system, though. Several vital organs and glands are connected to the digestive tract. These structures contribute digestive enzymes and other products to specific portions of the canal. They include the salivary glands, liver, gallbladder, and pancreas (Figure 40.7).

Before looking at the function of each component in detail, let's consider the general changes that happen to food on its way through the alimentary canal. Throughout the discussion humans will serve as a model species—simply because so much is known about human digestion.

Digestion begins with the tearing and smashing activity of teeth. Chewing reduces the size of food particles and softens them. In humans, the mechanical breakdown of food is augmented by the use of knives and cooking. (The invention of tools and cooking is the leading hypothesis to explain why average tooth size has declined steadily over the past several million years of human evolution.) The chemical breakdown of carbohydrates also begins in the mouth, through enzymes in saliva. Chemical processing continues in the stomach and is completed in the small intestine. Note that three major types of macromolecules must be broken down during digestion: carbohydrates, proteins, and lipids. Absorption of breakdown products from digestion of these macromolecules, along with water,

FIGURE 40.6 Most Animals Take Food in Large Packets
This tiger rattlesnake is ingesting a mouse. Although this is an extreme example, most animals take food in large bites that must be broken down into small pieces.

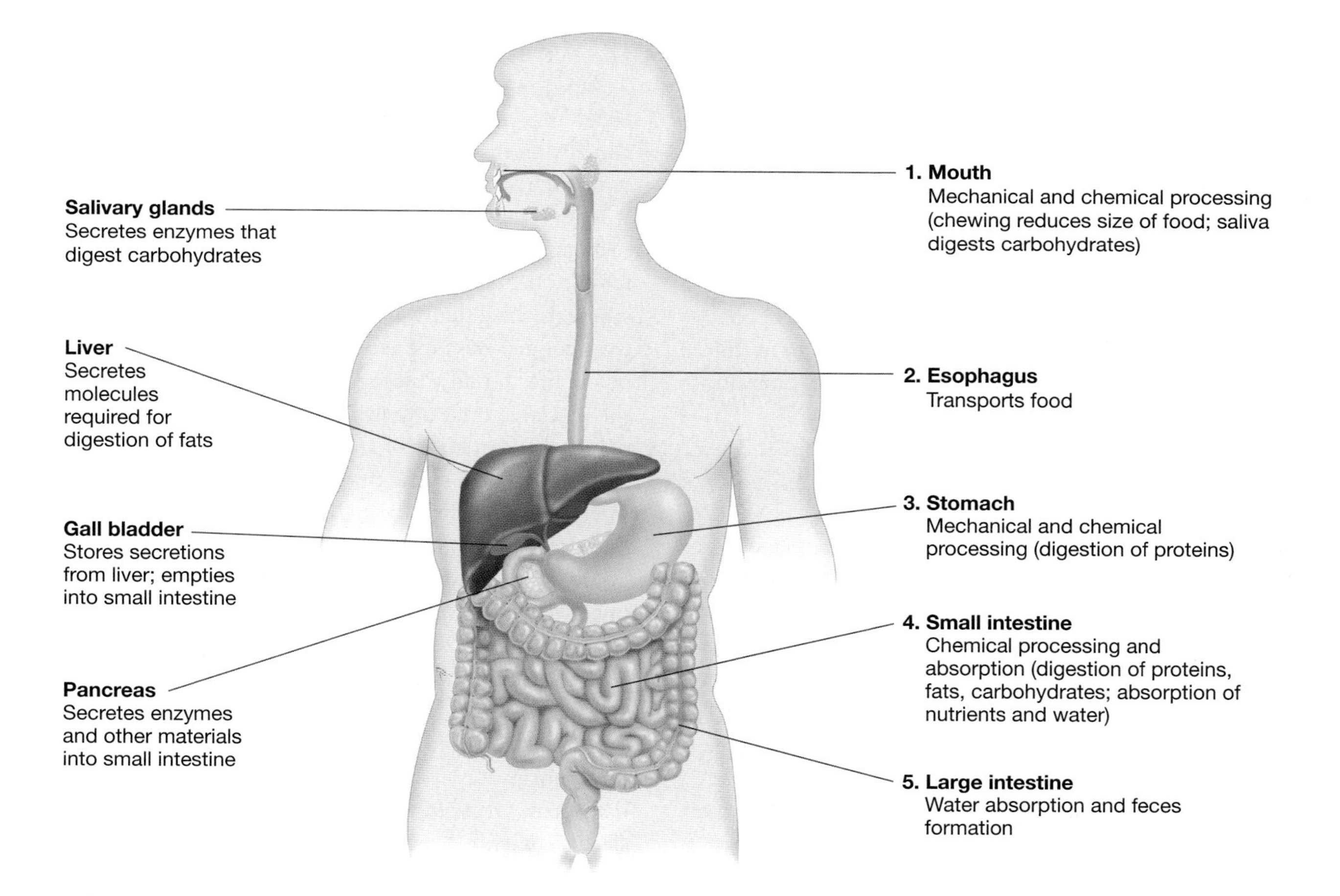

FIGURE 40.7 The Alimentary Canal
The alimentary canal is a tube that runs from the mouth to the anus. The salivary glands, liver, gallbladder, and pancreas are not part of the canal itself. Instead, they secrete material into the tube at specific points.

CD ACTIVITY 40.1
The Digestion and Absorption of Food

vitamins, and ions, occurs in the small intestine. In the large intestine, or colon, more water is absorbed. The feces that result are held in the rectum until they can be excreted.

For an animal to stay healthy, each step in this process must be completed correctly. Problems in the alimentary canal can lead to heartburn, ulcers, nausea, constipation, and other maladies. By far the most serious of these is diarrhea, which kills more than four million children each year. Because digestion is so important to how animals work, let's analyze each step in more detail.

The Mouth and Esophagus

If you hold a cracker in your mouth long enough, it will start to taste sweet. The sensation occurs because starch molecules in the cracker are being hydrolyzed to glucose by an enzyme in saliva called α-amylase. Starch breakdown was actually the first enzyme-catalyzed reaction ever discovered. In the early 1800s several researchers found that a component of certain plant extracts digested starch; in 1831, the same activity was discovered in human saliva. The enzyme responsible is called amylase. Amylase ranks as one of the best studied of all enzymes; in animals it is the most important catalyst in the breakdown of carbohydrates. Amylase cleaves the bonds that link glucose monomers in starch, glycogen, and other glucose polymers.

Amylase is not the only important ingredient in saliva, however. Salivary glands also release water and glycoproteins called mucins. When mucins contact water, they form the slimy substance called mucus. The combination of water and mucus makes food soft and slippery enough to be swallowed.

Once food is swallowed, it enters the esophagus. About six seconds later, the swallowed food reaches the bottom of the esophagus, having been propelled by a wave of muscular contractions called **peristalsis**. Peristalsis in the esophagus is the reason that you can bend down to drink from a water fountain. How does it occur?

Anatomically, the esophagus is a long, muscular tube. The upper third consists of skeletal muscle; the lower third is composed of smooth muscle; the middle third contains a mix of both muscle types. (Chapter 43 explains the differences between the structure and function of these two types of muscle.) By cutting selected nerves in experimental animals, researchers established that peristalsis in the upper third of the esophagus is a response to electrical signals that originate at the base of the brain. A series of nerve cells fire during peristalsis; the skeletal muscle that is innervated by each nerve contracts in response. In this way, a wave of contractions propagates down the tube, propelling the food mass. The action of these nerves is unconscious—the system is a reflex that is stimulated by the act of swallowing.

Even after decades of experiments, however, the mechanism of peristalsis in the lower third of the esophagus is still not well understood. Researchers can induce normal peristalsis in this section of the esophagus in the absence of stimulation by nerves from the brain or spinal cord. Based on these results, most researchers agree that peristalsis in the lower region of the esophagus is coordinated at least in part by cells in the wall of the tube itself. A clear understanding of the mechanism has yet to emerge, however.

The Stomach

Regarding digestion, little if anything happens in the esophagus. When food reaches the stomach, though, the situation changes dramatically. Early anatomical studies showed that the stomach is a tough, muscular pouch in the alimentary canal, bracketed on either end by valves called sphincters (**Figure 40.8a**). When the stomach is filled by a meal, muscular contractions result in churning that mixes the contents. As a result, a certain amount of mechanical breakdown of food occurs. By far the most important function of the stomach, though, is the digestion of proteins.

Compared with the mouth or esophagus (or virtually any other tissue for that matter), the stomach is highly acidic. Researchers in the late 1700s and early 1800s documented this fact by analyzing vomit or the contents of sponges that were tied to strings, swallowed, and pulled back up. In the early 1800s, chemists were able to document that the predominant acid in the stomach is hydrochloric acid (HCl).

Not long after, a physician named Samuel Beaumont established that digestion takes place in the stomach. He reached this conclusion through an extraordinary series of experiments on a young man named Alexis St. Martin. In 1822, when St. Martin was 19 years old, a shotgun accidentally discharged into his belly area and created a series of wounds. Despite repeated attempts, Beaumont was unable to close a hole in the patient's stomach. Eventually Beaumont inserted a small tube through the opening; the tube remained in St. Martin's body for the rest of his life. (Today, biologists insert tubes into various parts of the digestive tract of cows or sheep to study how these animals digest different types of feed.) With the tube in place, Beaumont was able to tie a string onto small pieces of meat or vegetables, insert the food directly into St. Martin's stomach, and draw it out after various intervals. Beaumont was also able to remove liquid from inside the stomach and observe how this gastric juice* acted on food in vitro. His experiments showed that gastric juice digests food—particularly meat.

Theodor Schwann later purified the enzyme that is responsible for digesting proteins in the stomach and named it **pepsin**. Because it destroys proteins, biologists hypothesized that pepsin must be synthesized and stored in cells in an inactive form. Otherwise it would kill the cells where it is synthesized. Through careful microscopy, Rudolf Heidenhain established in 1870 that granules occur in specialized stomach cells

*The Greek root *gastro* refers to the stomach. The root is used in the terms *gastric juice* and *gastrointestinal tract*.

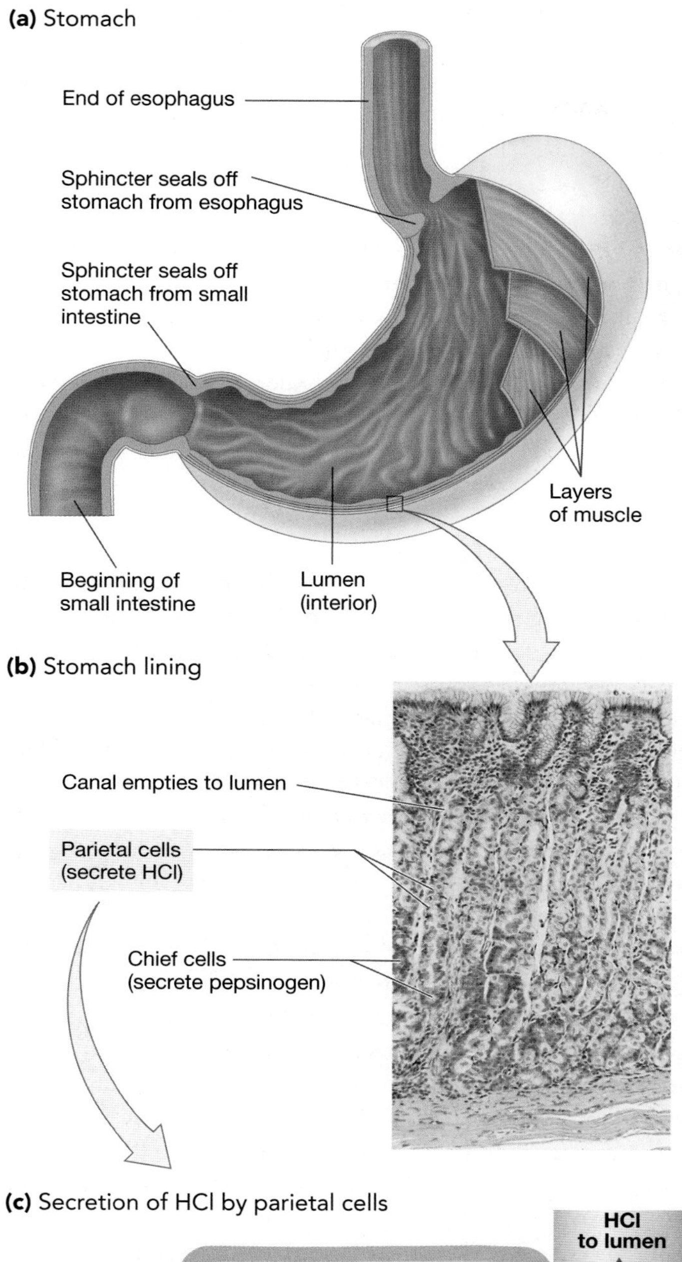

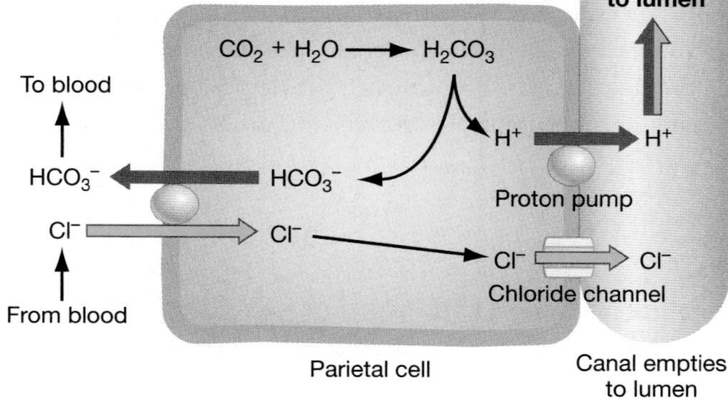

FIGURE 40.8 How Does the Stomach Work?
(a) The stomach is a muscular outpocketing of the alimentary canal. Muscle contractions mix food and break it into smaller pieces.
(b) This transmission electron micrograph shows a section through the stomach wall. **(c)** According to available data, this is how parietal cells secrete hydrochloric acid. **EXERCISE** In part (b), draw the route of molecules and ions secreted by parietal and chief cells. In part (c), label the reaction catalyzed by carbonic anhydrase.

called chief cells. These granules were hypothesized to be a pepsin precursor. Follow-up work showed that this precursor compound, which came to be called **pepsinogen**, is converted to active pepsin by contact with the acidic environment of the stomach.

Which Cells Produce Stomach Acid? In the stomach wall, Heidenhain also found distinctive cells which came to be called parietal cells. He noted that they are found along canals that communicate with the lumen of the stomach (**Figure 40.8b**). Further, the shape and activity of these cells appeared to vary as the digestion of a meal proceeded. Based on these observations, Heidenhain inferred that parietal cells are the source of the HCl in gastric juice, whose pH can be as low as 1.5. Earlier microscopists had shown that a third type of cell, called a goblet cell, secretes the mucus found in gastric juice. Mucus lines the gastric epithelium and protects the stomach from damage by HCl. These anatomical studies showed that the epithelium of the stomach contains several types of secretory cells, each of which is specialized for a particular function.

To test whether parietal cells secrete HCl directly or in an inactive form, Mabel FitzGerald injected a dye into the stomach of dogs. This dye turns blue in the presence of acid. When she examined the stained stomach tissue with the microscope, she found blue strands running through the canals in the gastric epithelium and connecting to parietal cells. Based on these results, FitzGerald claimed that parietal cells secrete protons and chloride ions directly. But because other researchers found inconclusive results with dye experiments, the hypothesis of direct secretion remained controversial for decades. It was resolved only when researchers began to work out the molecular mechanism of HCl production.

How Do Parietal Cells Secrete HCl? The first clues about how parietal cells manufacture hydrochloric acid emerged in the late 1930s, when Horace Davenport found a high concentration of an enzyme called carbonic anhydrase in parietal cells. This result was interesting because carbonic anhydrase catalyzes the formation of carbonic acid (H_2CO_3) from carbon dioxide and water. In solution, the carbonic acid that is formed immediately dissociates to form the bicarbonate ion (HCO_3^-) and a proton.

$$CO_2 + H_2O \rightleftharpoons H_2CO_3 \rightarrow H^+ + HCO_3^-$$

A second clue to the formation of HCl came in the 1950s, when the development of the transmission electron microscope allowed researchers to analyze parietal cells at high magnification; they observed that parietal cells are packed with mitochondria. Based on this observation, researchers suspected that HCl secretion involves active transport. Later work confirmed this hypothesis by showing that the protons formed by the dissociation of carbonic acid are actively pumped into the lumen of the stomach. Subsequent studies showed that chloride ions from the blood enter parietal cells in exchange for bicarbonate ions, and then move into the lumen through a chloride channel. The current model for HCl production is diagrammed in **Figure 40.8c**.

The Small Intestine

After proteins have been partially digested in the stomach, peristalsis in the stomach wall moves small amounts of material through the valve at the base of the stomach and into the small intestine. There, the partially digested material mixes with secretions from the pancreas and the liver (Figure 40.7) and begins a journey of 6 m (20 feet). At the end of the small intestine, digestion is complete and most nutrients—along with large quantities of water—have been absorbed. To understand how these changes in the composition of food occur, let's explore what happens to proteins, lipids, and carbohydrates as they move through this section of the digestive tract.

Protein Processing by Pancreatic Enzymes The acidic environment of the stomach denatures proteins—meaning that their secondary and tertiary structures are destroyed. In addition, pepsin cleaves enough peptide bonds to reduce the macromolecules to relatively small polypeptides. How is protein digestion completed so that individual amino acids can enter the bloodstream and be transported to cells throughout the body?

By the end of the nineteenth century, biologists had isolated a series of digestive enzymes from the pancreas and established that they are synthesized in an inactive form. Like the pepsinogen produced by chief cells in the stomach, the production of digestive enzymes in an inactive conformation prevents pancreatic cells from digesting themselves.

In 1900, N. P. Schepowalnikov showed that pancreatic enzymes are activated by contact with juice from the upper part of the small intestine. Because activation did not occur when he heated the intestinal juice, and because heat (like acid) denatures proteins, Schepowalnikov hypothesized that the agent responsible for activating the pancreatic enzymes was also an enzyme. He called the activating enzyme **enterokinase**. Decades later, M. Kunitz succeeded in purifying a pancreatic enzyme called trypsinogen and demonstrated that enterokinase activates it in vitro. Follow-up work showed that enterokinase activates trypsinogen by phosphorylating it. The active enzyme that results from this reaction is called **trypsin** (**Figure 40.9**). Trypsin triggers the activation of other protein-digesting enzymes secreted by the pancreas (Figure 40.9).

In short, enterokinase triggers the activation of a suite of digestive enzymes. The pancreas secretes digestive enzymes in an inactive form. When these enzymes arrive in the upper reaches of the small intestine, they are activated by enterokinase and begin digesting proteins and carbohydrates.

What regulates the secretion of these proteins from the pancreas? Because these proteins are needed only when food reaches the small intestine, investigators inferred that the release of digestive enzymes is carefully controlled. The leading hypothesis was that pancreatic secretion is controlled by the nervous system—possibly in response to the sight or smell of food. This hypothesis was logical. An extensive series of experiments based on cutting specific nerves in experimental animals had established that digestive processes like the release of saliva, swallowing, and the secretion of gastric juice are at least partially controlled by signals from the nervous system.

The hypothesis of nervous control was refuted, however, in a classic experiment conducted by William Bayliss and Ernest Starling. To study the mechanism responsible for stimulating secretion from the pancreas, Bayliss and Starling cut the nerves that innervate the pancreas and small intestine of a dog. When they introduced a weak HCl solution into the upper reaches of the animal's small intestine to simulate the arrival of material from the stomach, the pancreas secreted in response. This observation was startling. The small intestine had successfully signaled the pancreas, even though the nerves connecting the two organs had been cut. To explain the observation, Starling hy-

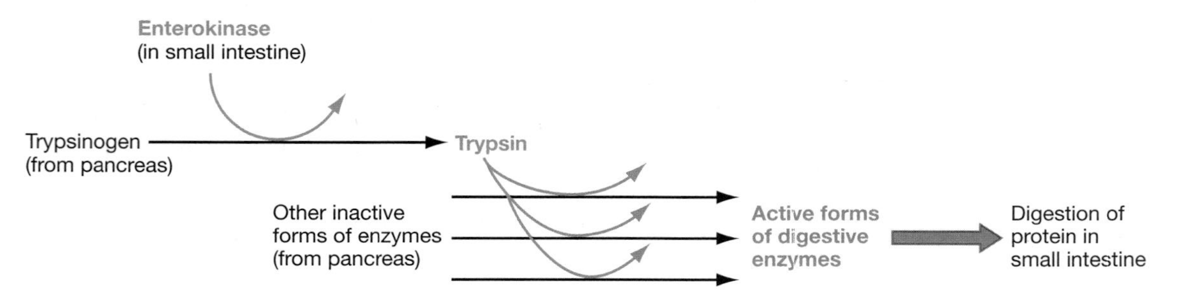

FIGURE 40.9 Enterokinase Triggers an Enzyme-Activation Cascade
Enterokinase activates trypsin; in turn, trypsin activates other digestive enzymes secreted by the pancreas.

pothesized that a chemical messenger must be involved. To test this idea, he cut off a small piece of the small intestine, ground it up, and injected the resulting solution into a vein in the animal's neck. Minutes later, the pancreas secreted.

Bayliss and Starling had discovered a hormone—a chemical messenger that influences physiological processes at very low concentrations. The molecule they detected, which they called secretin, is produced by the small intestine in response to the arrival of food from the stomach. The discovery of secretin was important because it confirmed that digestion is under both neural and hormonal control.

Follow-up work showed that secretin induces a flow of bicarbonate ions (HCO_3^-) from the pancreas to the small intestine, which neutralizes the acid arriving from the stomach. Researchers also discovered a second hormone produced in the small intestine, called cholecystokinin, that induces secretion from the liver as well as the pancreas. Cholecystokinin stimulates the secretion of digestive enzymes from the pancreas and the secretion of molecules from the liver that are involved in the digestion of lipids. Hormones are involved in stomach function as well. For example, certain stomach cells produce the hormone gastrin after being stimulated by nerves or the arrival of food. In response, parietal cells begin secreting HCl.

Digesting Lipids: Bile and Transport The pancreatic secretions include digestive enzymes that act on fats and carbohydrates as well as proteins. Prominent among these molecules is an enzyme called **lipase.** Lipase breaks certain bonds present in complex fats. The fatty acids and other small lipids that result can be taken up by the bloodstream.

Before lipase can act, however, the fats that emerge from the stomach must be modified. Recall from Chapter 4 that fats are insoluble in water. As a result, they tend to congregate in large globules as they are churned in the stomach. For lipase to act, these large droplets must be broken up, or emulsified. In the small intestine, emulsification results from the action of small lipids called bile salts. As **Figure 40.10** shows, bile salts function like the detergents introduced in Chapter 4 that researchers use to break up cell membranes. Bile salts are synthesized in the liver and secreted in a complex solution called **bile,** which is stored in the gallbladder (see Figure 40.7). When bile enters the small intestine, it raises the pH and emulsifies fats so they can be digested.

What happens to the fatty acids that are released by lipase activity? An answer to this question began to emerge when Robert Ockner and Joan Manning injected radioactively labeled fatty acids into the small intestines of laboratory rats. They analyzed the epithelial cells that line the small intestine and found that most of the radioactive label had entered the cells and attached to a protein they called fatty-acid binding protein. Later, researchers in other laboratories established that a fatty-acid binding protein also occurs in the membranes of these cells. Based on these and other results, fatty-acid binding proteins are thought to bring lipids into the cell. There they are processed into protein-coated complexes and released for transport to fat-storage cells and other tissues.

Digestion and Absorption of Carbohydrates In addition to manufacturing lipase and protein-digesting enzymes, the pancreas produces nucleases that digest the RNA and DNA in food and an amylase that is similar to the salivary enzyme introduced earlier. Pancreatic amylase completes the digestion of carbohydrates that began in the mouth.

The sugars that are released by amylase digestion are absorbed by cells that line the small intestine. The surface area available for nutrient absorption in the small intestine is nothing short of remarkable. As **Figure 40.11** (page 778) shows, the epithelium of this organ is folded and covered with finger-like projections called villi. In turn, the cells that line the surface of villi have projections on their apical surfaces called microvilli.

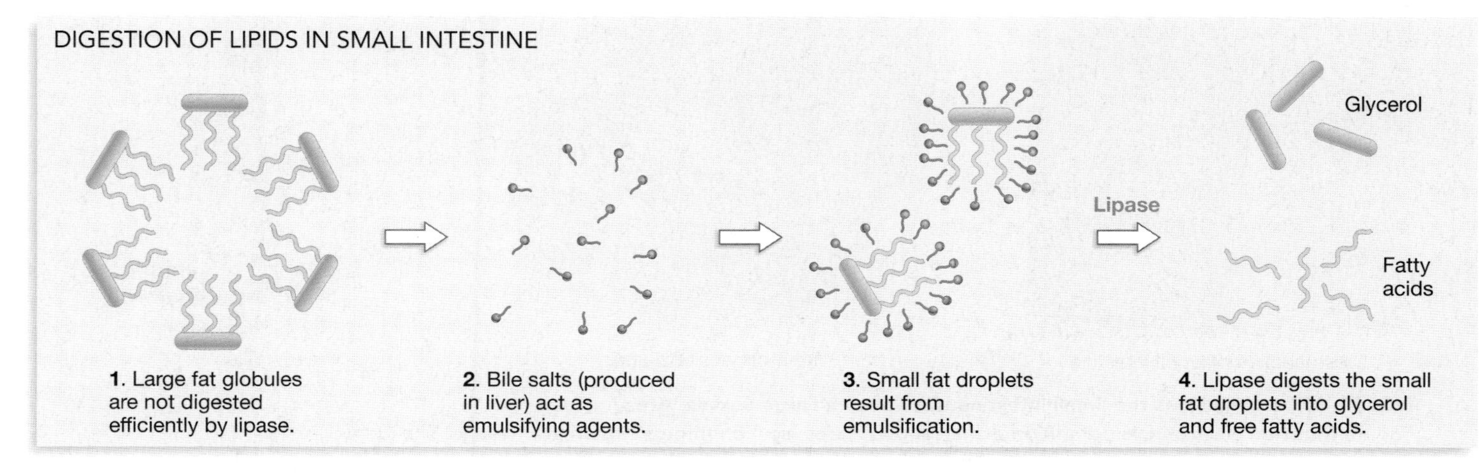

FIGURE 40.10 How Are Lipids Digested in the Small Intestine?
Bile salts are small lipids that act as emulsifying agents. Once they break up large fat globules, lipase can digest fats efficiently.

If the small intestine lacked folds, villi, and microvilli, it would have a surface area of about 3300 cm^2 (3.6 ft^2). Instead, the epithelium covers about 2 million cm^2 (over 2200 ft^2)—an area larger than a singles court in tennis. This enormous area increases the efficiency of nutrient absorption.

How does nutrient absorption actually occur? Specifically, how are glucose and other nutrients transported from the lumen of the small intestine into the microvilli of epithelial cells? A series of experiments by Ernest Wright and co-workers during the 1980s established that nutrient absorption depends on the presence of an electrochemical gradient favoring an influx of sodium ions into the epithelium. Based on this observation, Wright and others hypothesized that the apical membrane of these cells must contain a series of cotransporters. To confirm that a sodium-glucose cotransporter exists, the researchers set out to find the gene that codes for the hypothesized membrane protein.

The strategy that Wright and co-workers used is outlined in **Figure 40.12a**. They began by purifying mRNAs from rabbit intestinal cells, which presumably were transcribing the cotransporter genes. Then they injected one of each type of message into a series of frog eggs—a cell that does not normally transport glucose. The frog cells translated the rabbit mRNAs. When tested, one of the experimental eggs was able to import Na^+ and glucose in tandem. Wright's group inferred that this egg had received the mRNA for the rabbit Na^+-glucose cotransporter. They used the enzyme reverse transcriptase to make a DNA copy of the mRNA (see Chapter 16), and sequenced the gene using techniques introduced in Chapter 12. Based on this sequence of experiments, the researchers were able to infer the amino acid sequence of the protein.

The discovery of the Na^+-glucose cotransporter inspired a flurry of activity. Kuniaki Takata and colleagues raised an antibody to the cotransporter protein and confirmed that it is abundant in the villi of the small intestine (**Figure 40.12b**). Wright's group proposed the model for glucose absorption summarized in **Figure 40.12c**. Researchers in other labs discovered that the same cotransporter occurs in the proximal tubule of the kidney, where it is responsible for the reabsorption of sodium and glucose from urine, and that similar cotransporters were responsible for the absorption of other nutrients.

Understanding how glucose, sodium, and other nutrients are absorbed by the epithelium of the small intestine also revealed the major mechanism of water reabsorption in the small intestine. Water follows solutes into the epithelium passively, by osmosis (Figure 40.12c). This movement of water is the mechanistic basis of an extremely important medical strategy called oral rehydration therapy. If a patient is suffering from diarrhea, clinicians frequently prescribe dilute solutions of glucose to be taken orally. When the glucose in the drink is absorbed in the small intestine through the sodium-glucose cotransporter, enough water follows in most cases to prevent life-threatening dehydration. This simple medication saves many lives every year.

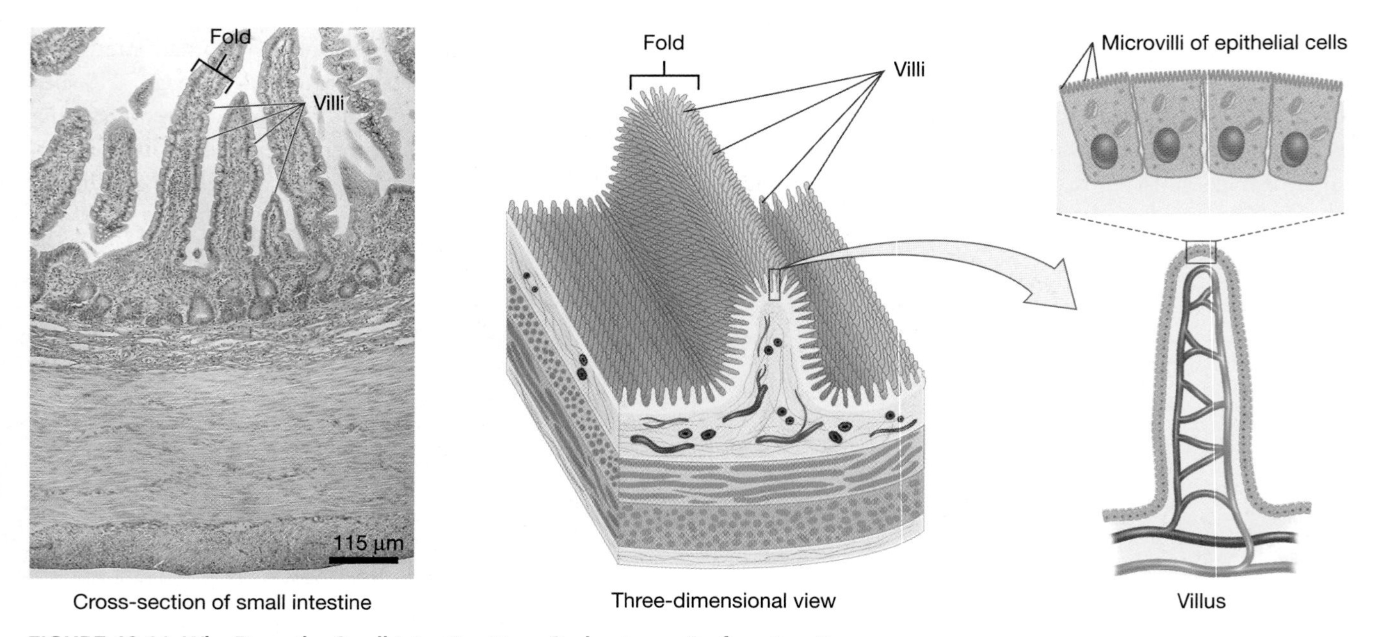

FIGURE 40.11 Why Does the Small Intestine Have Such a Large Surface Area?
The transmission electron micrograph on the left shows a cross section through the small intestine. The drawing in the middle shows a three-dimensional view of one of the folds in the lining of the small intestine. The drawings on the right illustrate a cross-section of a villus, showing its blood supply and a detail of epithelial cells. **EXERCISE** Draw a cross section and a three-dimensional view showing what the small intestine would look like if folds and villi did not exist.

The Large Intestine

By the time digested material reaches the large intestine, a large amount of water and virtually all of the available nutrients have been absorbed. The primary function of the large intestine is to compact the wastes that remain and absorb enough water to form feces. Unlike the situation in the small intestine, however, the mechanism of water movement in the large intestine is not well understood.

In an attempt to identify the mechanism of water absorption in the large intestine, researchers have recently focused on the membrane proteins called aquaporins, which were introduced in Chapter 39. Recall that aquaporins are water channels. The best studied of these proteins, called AQP1 for aquaporin 1, is common in the descending limb of Henle of the kidney. It is the protein responsible for water reabsorption from urine along the osmotic gradient described in Chapter 39.

To date, four distinct aquaporins have been found in the large intestine of rats, mice, or humans. The aquaporins called AQP3 and AQP4, for example, are located in the basolateral membrane of cells in the epithelium of the large intestine in mice. (Recall that the basolateral membrane of epithelial cells faces away from the lumen and toward the blood.) To test the

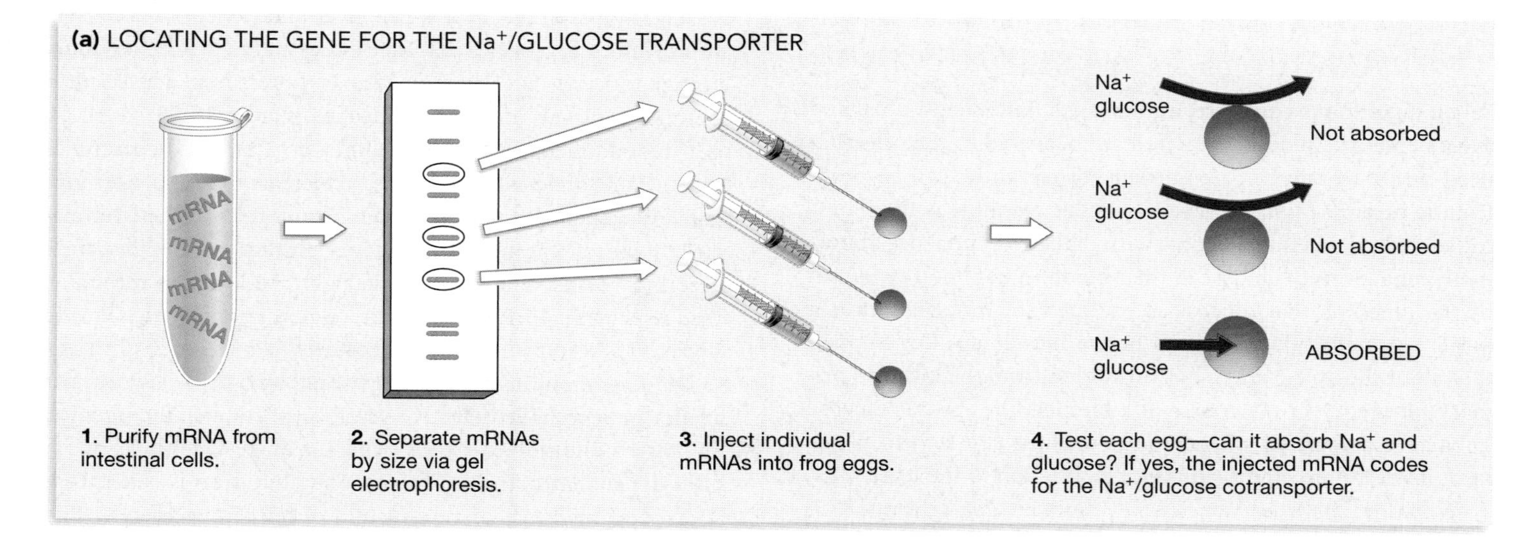

(b) Cotransporter is located in apical membrane

Experimental treatment: Red antibody binds to Na$^+$/glucose cotransporter; blue antibody binds to DNA

Control: Green antibody binds to actin (cytoskeletal protein); blue antibody binds to DNA

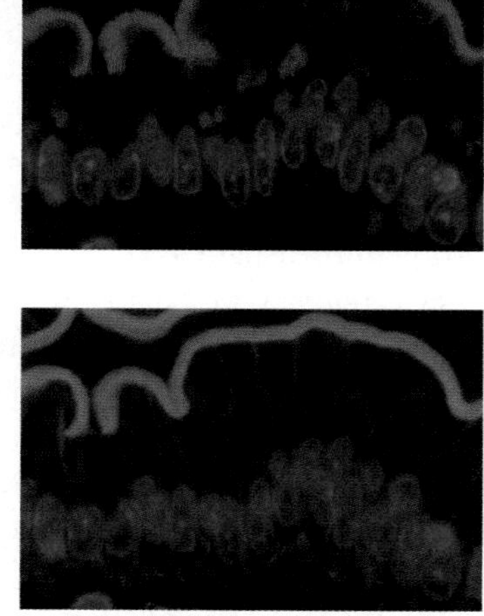

(c) Model of glucose absorption

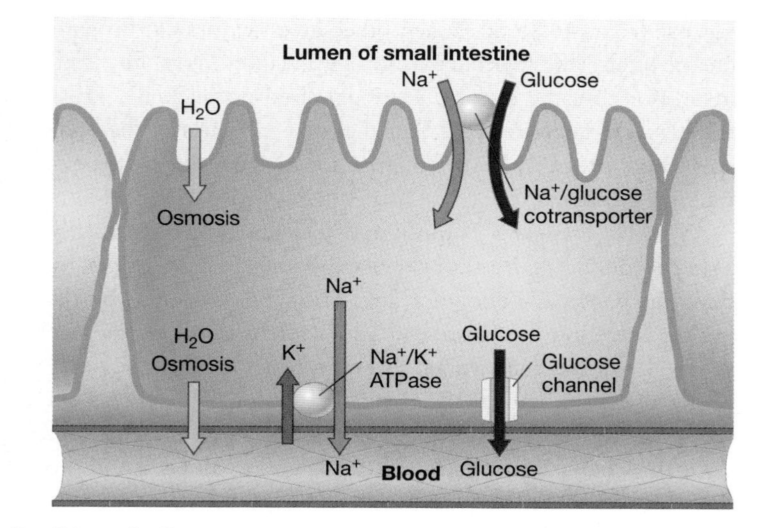

FIGURE 40.12 How Are Glucose and Other Nutrients Absorbed in the Small Intestine?
(a) Researchers followed this experimental protocol to find the gene for the Na$^+$-glucose cotransporter. **(b)** These micrographs show the same cross section through intestinal epithelium treated with two different fluorescing antibodies. The red in the top photo shows the location of antibodies that bind to the Na$^+$-glucose cotransporter. The green in the bottom photo shows the location of antibodies that bind to a cytoskeletal protein called actin. Note that the cotransporter is found only in the apical membranes of epithelial cells. **(c)** This diagram summarizes the current understanding of how glucose is absorbed in the small intestine.

hypothesis that these proteins are essential for water absorption and normal feces formation, investigators in A. S. Verkman's lab generated mice that have two abnormal copies of the gene for AQP4. To the researchers' surprise, however, the experimental mice were able to form feces that were only slightly wetter than normal. As this book goes to press, the researchers are working to create mice with abnormal copies of other aquaporins. If aquaporins are essential for water reabsorption, the mutant mice should have a clear defect in feces formation. In the meantime, the challenge of understanding water absorption in the large intestine remains.

40.4 Nutritional Homeostasis—Glucose as a Case Study

When digestion is complete, amino acids, fatty acids, ions, and sugars enter the bloodstream and are delivered to the cells that need them. Too much of a nutrient, or too little, can be problematic or even fatal, however. A classic example is the illness called diabetes mellitus. People with diabetes experience abnormally high levels of glucose in their blood. A common symptom of the illness is that glucose is excreted in urine just after a meal. Normally, glucose levels in the blood stay low enough that all of the nutrient can be reabsorbed from the kidney's primary filtrate. Over the course of a lifetime, the chronic glucose imbalance associated with diabetes mellitus can lead to blindness, heart failure, and a failure of circulation in the legs.

Why does diabetes occur? A key hint came to light in 1879, when researchers removed the pancreas from a dog and observed that diabetes developed. This experiment suggested that the pancreas secretes a compound involved in the removal of glucose from the blood. When other investigators cut up pancreatic tissues and injected extracts into diabetic dogs, however, it did not cure the disease. They observed no response. Frustrated, researchers realized that digestive enzymes in the pancreas were probably destroying the active agent during the extraction process.

In 1921, Frederick Banting and Charles Best were finally able to conduct a breakthrough experiment. They began by tying the pancreatic duct of a dog. Their logic was that blocking the secretion of digestive enzymes might kill the cells that synthesize them. They waited several weeks for the tissue to atrophy and then removed the gland. They froze the pancreas tissue, ground it up, and injected an extract into a diabetic dog. To their delight, the dog's blood-sugar levels stabilized and the dog became more active and healthy looking. After they repeated the experiment and observed the same result, Banting and Best grew increasingly confident that they had found the molecule responsible for controlling diabetes. The molecule came to be called insulin.

Insulin's Role in Homeostasis

Within a few years after Banting and Best's discovery, insulin had been isolated in pure form. Drug companies began purifying it from the pancreas of calves and other domestic animals in large enough quantities to administer to people suffering from diabetes mellitus. Later, insulin's mode of action was revealed. **Insulin** is a hormone that is produced in the pancreas when blood-glucose levels are high. It travels through the bloodstream and binds to receptors on cells throughout the body. In response, the cells increase their rate of glucose uptake and processing. For example, cells in the liver and skeletal muscle synthesize more glycogen; cells that store lipids synthesize more storage forms of fat, using glucose as a precursor. The result is that glucose levels in the blood decline (**Figure 40.13**).

If blood-glucose levels fall, as they do after hard exercise or with a lack of food, cells in the pancreas secrete a hormone called **glucagon**. In response to glucagon, cells in the liver and skeletal muscle catabolize glycogen; cells that store lipids catabolize fatty acids. The result is that glucose levels in the blood rise. As Figure 40.13 shows, the combination of insulin and glucagon is responsible for homeostasis.

Diabetes mellitus develops in people who do not synthesize insulin or who have defective versions of the insulin receptor. The first condition is called type I diabetes mellitus, or insulin-dependent diabetes. The second condition is called type II diabetes mellitus, or non-insulin-dependent diabetes. Both types of diabetes mellitus lead to the production of large volumes of urine as the body attempts to clear excess glucose. Recall from Chapter 39 that *diabetes* means "to run through." That chapter also introduced the illness diabetes insipidus, which results from lack of water reabsorption in the collecting ducts of the kidney. *Insipidus* means tasteless, while *mellitus* means honeyed (sweet). Before they had chemical methods of analyzing urine, physicians would distinguish between diabetes insipidus and diabetes mellitus by tasting the urine of their patients.

Currently, type I diabetes mellitus is treated with insulin injections and careful attention to diet; type II diabetes is managed primarily through prescribed diets and monitoring blood-glucose levels. For physicians and patients, the challenge is to achieve homeostasis with respect to blood glucose in the absence of the body's normal regulatory mechanisms.

The Type II Diabetes Mellitus Epidemic

The discovery of insulin qualifies as one of the great medical advances of the twentieth century; it led to dramatic improvements in the quality of life for people who suffer from type I diabetes mellitus. Unfortunately, an epidemic of type II diabetes mellitus is currently under way in certain human populations. In the United States, about 6.6 percent of people aged 20 to 74 have type II diabetes. The frequency of this illness is much higher in African American, Hispanic, and Native American people, however, and the frequency of type II diabetes mellitus in teenagers of all ethnic backgrounds is increasing rapidly. To date, the Pima Indians of the southwestern United States have the highest prevalence ever recorded. Among the Pima, almost 60 percent of adults over 35 years old have type II diabetes mellitus. As **Figure 40.14a** (page 782) shows, prevalence rates in this population have increased rapidly over the past several decades.

Because there is a strong association between the prevalence of type II diabetes mellitus in parents and children, researchers have concluded that some individuals have a genetic predisposition for developing the disease. There is also strong evidence, however, that environmental conditions have an important impact as well. For example, in the Pima and many other populations, type II diabetes mellitus is more prevalent in people who are obese (**Figure 40.14b**). The incidence of obesity has been rising in many populations because the fat content of diets is increasing, and because motorized transportation and labor-saving devices are encouraging individuals to become less physically active. Because of these environmental changes, nutrition-related diseases like obesity and diabetes are rapidly becoming a major public health concern—especially in young people.

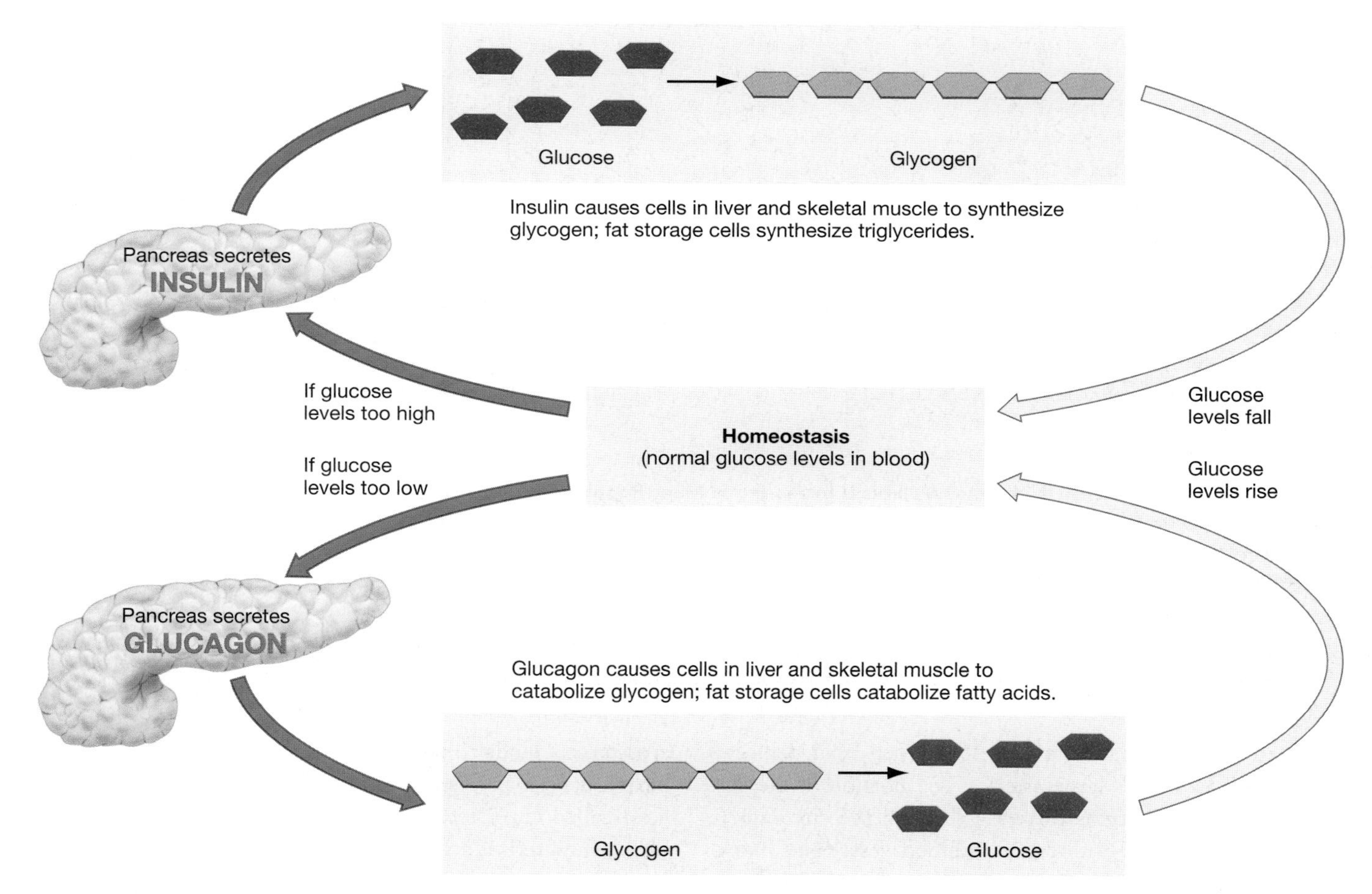

FIGURE 40.13 How Do Insulin and Glycogen Act to Maintain Homeostasis?
Insulin and glucagon are both secreted by cells in the pancreas, but have opposite effects on blood glucose concentrations. **EXERCISE** Draw a large X through the arrow that is disrupted in individuals with diabetes mellitus. Next to the X, jot notes indicating the nature of the defect in individuals with type I diabetes mellitus versus type II diabetes mellitus.

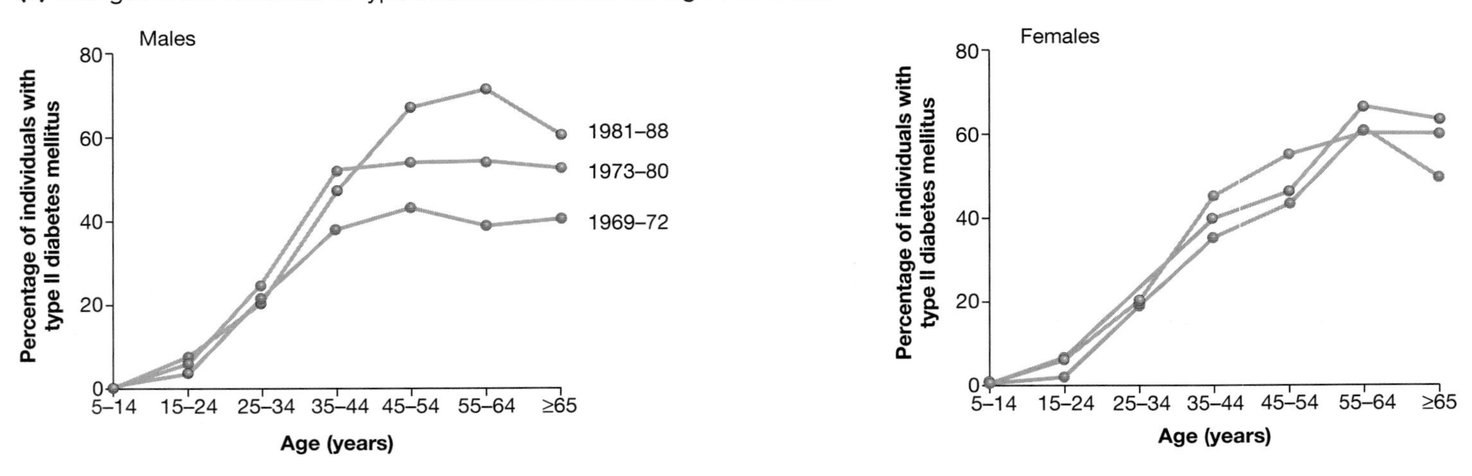

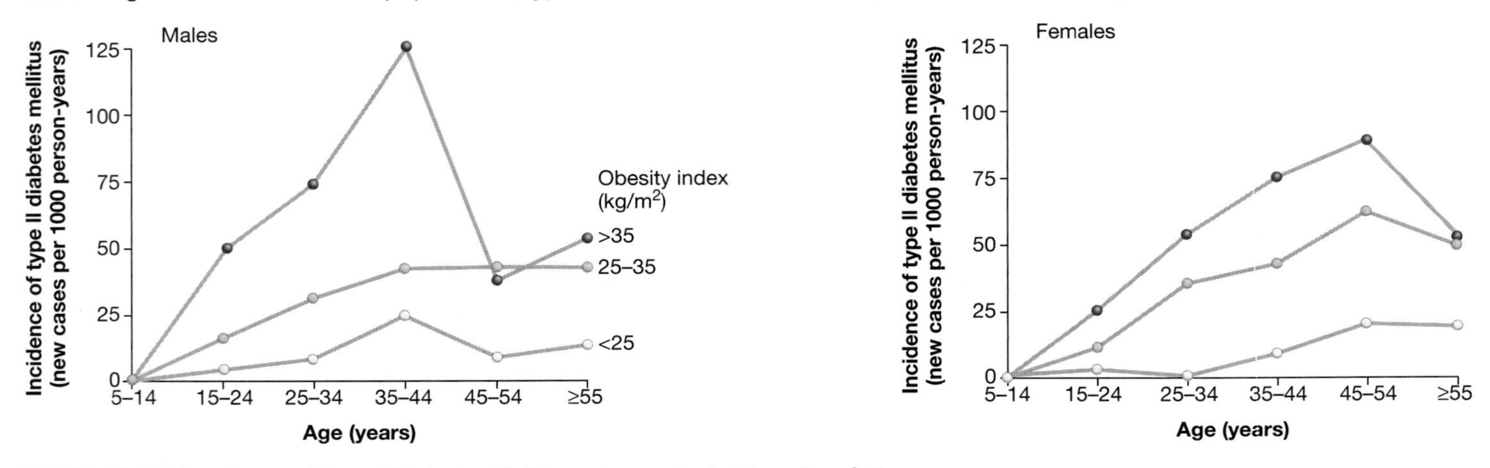

FIGURE 40.14 Prevalence of Type II Diabetes Mellitus Is Increasing in Many Populations
(a) These graphs show the percentage of Pima Indian males and females who suffer from type II diabetes mellitus. **(b)** These data document a strong association between the incidence of diabetes mellitus and obesity. Obesity in this case was evaluated as weight divided by height-squared, in units of kg/m^2.

Essay Cholesterol, Heart Disease, and Diet

Atherosclerosis is a disease that develops when lipid deposits accumulate in the body's large arteries. If a portion of the deposit breaks free and begins traveling through the circulatory system, the clot is likely to lodge in a blood vessel and block blood flow. Heart attacks occur when the blockage affects an artery that serves the heart. Strokes result from blockages in an artery serving the brain. In Europe and North America, atherosclerosis is responsible for about half of all deaths.

What causes atherosclerosis?

What causes the condition? By the late 1800s, researchers had found two important clues. First, the observation that atherosclerosis tends to run in families implied that some people have a genetic predisposition to the disease. Second, the small lipid called cholesterol is a major component of the deposits that characterize atherosclerosis. Cholesterol is a common component of animal cell membranes and is also required as a substrate for the synthesis of certain molecules. In 1971, William Kannel and co-workers showed that there is a close linkage between having high levels of cholesterol in blood and having a high risk of heart disease.

Kannel and co-workers made the connection between cholesterol and heart disease by analyzing data that had been collected since 1949 on men and women from Framingham, Massachusetts. Volunteers entered the Framingham study by

(Continued on next page)

(Essay continued)

undergoing a physical examination. If they were found to be free of heart disease, they joined the study and were examined every two years thereafter. When Kannel and associates analyzed data on over 5000 people who had participated in the study for at least 14 years, they found a strong correlation between the concentration of blood cholesterol when individuals entered the study and their risk of subsequently developing coronary heart disease. The correlation remained significant when the researchers controlled for other factors associated with heart disease, like smoking, high blood pressure, and diabetes mellitus.

The results inspired a massive effort to monitor and reduce cholesterol levels in the general population and an explosion of research on cholesterol metabolism. Joseph Goldstein and Michael Brown, for example, were instrumental in establishing that cholesterol is transported in the blood in two different forms, called low-density lipoprotein (LDL) and high-density lipoprotein (HDL). Cholesterol is synthesized in the liver and transported to cells inside LDL particles. In contrast, HDL particles take up excess cholesterol and return it to the liver. In the liver, cholesterol is used to manufacture the bile salts that emulsify fats in the small intestine.

One of the more important findings since the Framingham study is that high LDL levels are associated with a high rate of coronary heart disease, while high HDL levels are associated with a low rate of coronary heart disease. It is still not clear, however, exactly how HDL takes up excess cholesterol or how it exerts its protective effects.

A general message of research on cholesterol is that atherosclerosis, like diabetes mellitus, has a strong environmental as well as genetic component. Exercise raises HDL while obesity lowers it, and people who eat a high-fat, high-cholesterol diet risk developing heart disease. Like diabetes mellitus, heart disease results from a lack of homeostasis with respect to a key nutrient.

Chapter Review

Summary

Animals need a variety of nutrients to meet their basic needs. These nutrients include fats and carbohydrates that provide energy, proteins that furnish amino acids, vitamins that serve as co-factors for enzymes and perform other functions, ions required for water balance and for nerve and muscle function, and selected elements that are incorporated into certain molecules synthesized by cells. To determine the levels of nutrients that are needed to sustain normal activities, researchers monitor the relationship between nutrient intake, the levels of nutrients maintained in the body, and health. To determine how nutrition can affect athletic performance, researchers alter the intake of specific nutrients and assess how athletes respond. For example, investigators have shown that athletes who ingest a high-carbohydrate diet for three days outperform individuals who ingest a high-fat, high-protein diet.

Once food is ingested, it is processed in an alimentary canal that begins at the mouth and ends at the anus. Food processing frequently begins with tearing or chewing action by a beak, teeth, or other mouthparts. Among animal species, there is a strong correlation between the size and shape of mouthparts and their function in capturing and processing food.

In many species, chemical digestion of food also begins in the mouth. A salivary enzyme called α-amylase hydrolyzes the bonds linking glucose monomers in starch, glycogen, and other carbohydrates. Once food is swallowed, it passes down the esophagus via peristalsis. Digestion continues in the stomach, where a highly acidic environment denatures proteins and the enzyme pepsin begins the cleavage of peptide bonds linking amino acids.

From the stomach, food passes into the small intestine, where it is mixed with secretions from the pancreas and liver. Carbohydrate digestion is completed by pancreatic amylase; fats are emulsified by bile salts and digested by lipase; protein digestion is completed by a suite of pancreatic enzymes that are activated by enterokinase. Secretion from the liver and pancreas is triggered by the hormones cholecystokinin and secretin, which are produced in the small intestine.

Cells that line the small intestine absorb the nutrients released by digestion through transporters or cotransporters in their membranes. In many cases, uptake is driven by an electrochemical gradient favoring a flow of Na^+ into the cell. As solutes leave the lumen of the intestine and enter cells, water follows by osmosis. Water reabsorption is completed in the large intestine, where feces forms. Although researchers suspect that water channels called aquaporins are involved in water uptake by epithelial cells in the large intestine, the detailed mechanism of water transport in this part of the alimentary canal is still unknown.

Maintaining homeostasis with respect to nutrients is critical to health. For example, diabetes mellitus develops when concentrations of glucose in the blood are chronically too high. The illness is caused by a defect in the production of insulin—a hormone secreted by the pancreas that promotes uptake of glucose from the blood—or by a defect in the insulin receptor on the surface of cells. Development of type II diabetes is correlated with obesity and is reaching epidemic proportions in some populations.

Questions

Content Review

1. What does secretin stimulate?
   a. secretion of bile salts from the liver and HCO_3^- from the pancreas
   b. secretion of HCl from the stomach epithelium
   c. secretion of digestive enzymes from the pancreas
   d. uptake of glucose from the bloodstream by cells throughout the body
2. Why does carbohydrate loading increase the performance of distance athletes?
   a. It increases glycogen stores in muscle and liver.
   b. It increases fat storage.
   c. It increases the blood's capacity to deliver oxygen to tissues.
   d. It increases fatty-acid levels in the blood.
3. Why are the pharyngeal jaws of cichlid fish unique?
   a. No other group of fish has pharyngeal jaws.
   b. They are made of bone instead of cartilage, so they are harder and more durable.
   c. They are extremely specialized, and function only in transporting food from the front of the mouth to the throat.
   d. They articulate with the braincase and have extra muscles attached, so they are capable of biting.
4. How are carbohydrates digested?
   a. by lipases in the small intestine
   b. by pepsin and HCl in the stomach
   c. by aquaporins in the large intestine
   d. by amylases in the mouth and small intestine
5. What role do bile salts play in the digestion of complex fats?
   a. They catalyze the cleavage of bonds leading to the release of fatty acids and other small lipids.
   b. They emulsify lipids, meaning that large masses of fat molecules are broken into smaller masses.
   c. They include fatty acid binding proteins, which are involved in fat absorption.
   d. They activate the enzymes that are responsible for digesting fats.
6. How is water absorbed in the small intestine?
   a. through water channels called aquaporins
   b. exact mechanism is not known
   c. by sodium cotransporters
   d. following solutes through osmosis

Conceptual Review

1. Why is it logical that digestive enzymes are produced in an inactive form and then activated in the lumen of the digestive tract?
2. The text claims that cichlid pharyngeal jaws are adaptations that increase feeding efficiency. Do you accept this conclusion? Why or why not?
3. How was it established that gastric juice is acidic and that the stomach's primary function is the digestion of proteins?
4. What features are responsible for the large surface area of the small intestine? Why is it important that this organ have such a large surface area?
5. Diabetes mellitus is a disease that results from disruption of mechanisms for maintaining homeostasis in blood-glucose concentrations. Explain this statement.

Applying Ideas

1. Predict how the nutritional requirements of female mammals change during pregnancy and breastfeeding. How would you test your predictions? Design a study that uses humans as a study subject and one that uses laboratory mice.
2. Predict the physical symptoms that would result from defects in each of the following molecules: pancreatic amylase, pepsin, Na^+-glucose cotransporter, fatty-acid binding protein, aquaporin.
3. Give an example of a correlation between the structure of an animal mouthpart and its function in obtaining or processing food. Design a study to test the hypothesis that the structure is an adaptation that increases feeding efficiency.
4. According to Figure 40.9, enterokinase catalyzes the production of trypsin from trypsinogen, and trypsin then catalyzes the activation of other digestive enzymes secreted by the pancreas. The text did not provide evidence to back up these claims, however. Design an in-vitro study to test whether pancreatic enzymes other than trypsinogen are activated by enterokinase.
5. Scientists who backed the hypothesis that secretion from the pancreas is under nervous control strenuously objected to the experiment that led to the discovery of hormonal control. They claimed that the experiment was inconclusive because it was very likely that not all of the nerves innervating the small intestine had been cut. The biologists who did the experiment replied that even if it were true that not all nerves had been cut, their result was still valid. In your opinion, who is correct? Why?
6. Among vertebrates, the large intestine occurs only in lineages that are primarily terrestrial (amphibians, reptiles, birds, and mammals). Propose a hypothesis to explain this observation.
7. Insulin is thought to be an adaptation that allows animals to eat large meals intermittently, rather than having to eat small meals constantly. Why might it be advantageous for an animal to feed less often? Why would insulin action make intermittent feeding possible?

CD-ROM and Web Connection

CD Activity 40.1: The Digestion and Absorption of Food *(tutorial)*
(Estimated time for completion = 20 min)
What happens to food after we eat it?

CD Activity 40.2: Understanding Diabetes Mellitus *(tutorial)*
(Estimated time for completion = 15 min)
How is blood glucose regulated, and how is that regulation affected in individuals with diabetes mellitus?

At your **Companion Website** (http://www.prenhall.com/freeman/biology), you will find self-grading exams and links to the following research tools, online resources, and activities:

Garlic and Eggs
Read a summary of several recent advances in understanding nutrition and its role in health.

Nutrition and Good Health
This article discusses the state of nutrition and health in the United States.

Gastrointestinal and Liver Physiology
This journal focuses on medical research in gastrointestinal and liver physiology. Abstracts are available for review; access to the complete articles requires a subscription.

American Diabetes Association
This American Diabetes Association site provides a wide variety of links about diabetes.

Additional Reading

Lasley, E.N. 1999. Having their toxins and eating them too. *BioScience* 49: 945–950. Explores how some animals protect themselves by sequestering toxins obtained in the diet.

Sherman, P.W. and S.M. Flaxman. 2001. Protecting ourselves from food. *American Scientist* 89: 142–151. Investigates how morning sickness during pregnancy and the use of spices in cooking may help protect embryos and adults from bacteria.

Gas Exchange and Circulation

41

In the process of producing ATP, the mitochondria inside animal cells consume oxygen (O_2) and produce carbon dioxide (CO_2). To support continued ATP production, it is essential to obtain additional oxygen and expel excess carbon dioxide. If an animal is deprived of oxygen long enough, death by suffocation or drowning may result. If carbon dioxide is allowed to build up in tissues, debilitating disease symptoms appear.

How does gas exchange between an animal's environment and its mitochondria occur? In most cases, gas exchange involves the sequence diagrammed in **Figure 41.1**: (1) moving air or water through a specialized gas-exchange organ such as gills or lungs; (2) exchanging O_2 and CO_2 between air or water and the blood; (3) transporting the dissolved gases throughout the body via the circulatory system, and (4) exchanging gases between the blood and mitochondria.

The goal of this chapter is to examine each of these steps in depth. The first section investigates aspects of gas behavior that affect respiration. Section 41.2 follows up by exploring how gills, lungs, and other respiratory organs work. The focus here is on how air or water move through these organs and how they facilitate gas exchange. The next two sections address the structure and function of blood and circulatory systems, which transport blood between respiratory organs and tissues. The concluding section outlines how homeostasis is achieved with respect to O_2 and CO_2.

Along the way, it is important to recognize the role of gas exchange and circulation in human health. In most human populations, tuberculosis, heart disease, lung cancer, malaria, and other diseases of the respiratory tract and circulatory system are leading causes of illness and death.

During intense exercise, animal circulatory systems must deliver large amounts of oxygen to tissues and remove large amounts of carbon dioxide. This chapter explores how gas exchange occurs in animals that live in both aquatic and terrestrial environments.

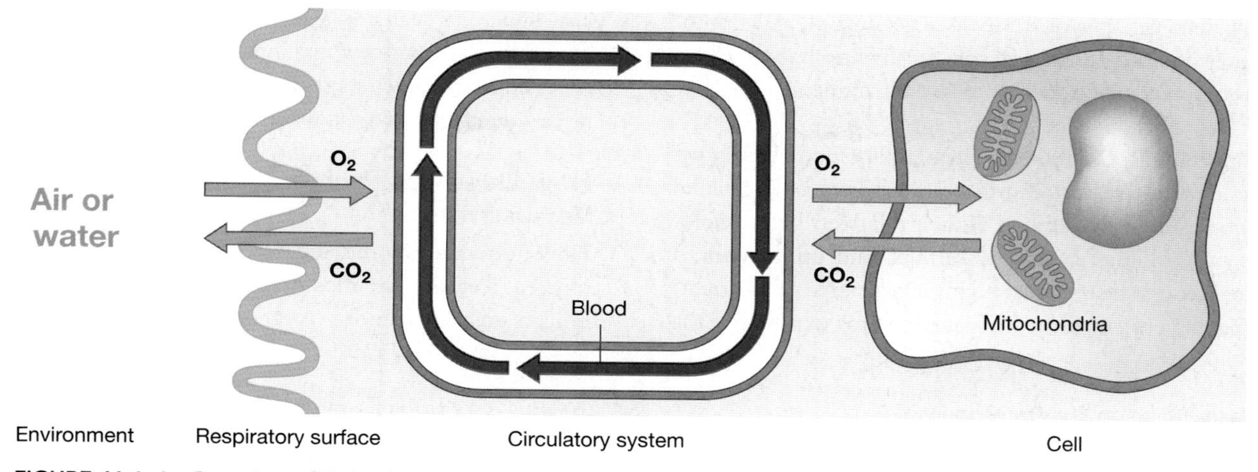

FIGURE 41.1 An Overview of Animal Gas Exchange
In most animals, oxygen and carbon dioxide are exchanged across the surface of a lung, gill, trachea, or other organ. Once the gases are dissolved in blood or hemolymph, they are transported to cells—where gas exchange again takes place.

41.1 Air and Water as Respiratory Media

Gas exchange between the environment and mitochondria depends on diffusion. Under normal conditions, oxygen concentrations are relatively high in the environment and low in tissues, while carbon dioxide levels are relatively high in cells and low in the environment.

To get a more precise understanding of how gas exchange occurs by diffusion, it's important to consider some fundamental questions about how diffusion occurs in terrestrial and aquatic environments. How much oxygen and carbon dioxide are present in the atmosphere versus the ocean? What factors influence how quickly these gases move by diffusion?

How Do Oxygen and Carbon Dioxide Behave in Air?

As **Figure 41.2a** (page 788) shows, the atmosphere is composed primarily of nitrogen (N_2) and oxygen, with trace amounts of CO_2. Nitrogen and the other atmospheric gases are not important to animals and are routinely ignored in studies of gas exchange.

The data in Figure 41.2a are slightly misleading, however. To understand why, consider that the percentage of O_2 in the atmosphere does not vary with altitude. The atmosphere at the top of Mt. Everest is composed of 21 percent O_2 just as it is at sea level. The key difference is that there are many fewer molecules of oxygen and other atmospheric gases present at high altitude than at sea level. Air at the top of Mt. Everest is much less dense than air at sea level, so much less oxygen is present.

To understand how gases move by diffusion, then, it is important that their presence is expressed in terms of a partial pressure instead of a percentage. Pressure is a type of force. A **partial pressure** is the pressure of a particular gas in a mixture. To calculate the partial pressure of a particular gas, you multiply the percent composition of that gas by the total pressure exerted by the entire mixture.* For example, **Figure 41.2b** shows that the total atmospheric pressure at sea level is 760 mm Hg (millimeters of mercury). If you multiply this value by 0.21, which is the fraction of air that is O_2, you obtain a partial pressure of oxygen (P_{O_2}) at sea level of 160 mm Hg. Because the atmospheric pressure is only about 250 mm Hg at the top of Mt. Everest, the P_{O_2} is $0.21 \times 250 = 53$ mm Hg.

Oxygen and carbon dioxide diffuse between the environment and cells along a partial pressure gradient. In both air and water, O_2 and CO_2 move from regions of high partial pressure to regions of low partial pressure.

How Do Oxygen and Carbon Dioxide Behave in Water?

How much oxygen is available in different types of aquatic habitats? Gases are soluble in liquids, but the amount of gas that dissolves depends on several factors.

- The solubility of the gas in that liquid. For example, CO_2 is almost 30 times more soluble in water than is O_2. As a result, fish rid themselves of carbon dioxide much more easily than they can obtain oxygen.
- The temperature of the liquid. As the temperature of water increases, the amount of gas that dissolves in it decreases. Other things being equal, cold-water habitats have much more oxygen available than warm-water habitats.
- The presence of other solutes. Because seawater has a much higher concentration of solutes than does freshwater, it can hold much less dissolved gas. At 10°C, up to 8.02 ml of O_2

*This calculation is valid because the total pressure in a mixture of gases is the sum of the partial pressures of all the individual gases. This is called Dalton's law.

can be present per liter of freshwater versus only 6.35 ml of O_2 per liter of seawater. As a result, freshwater habitats tend to be more oxygen-rich than marine environments.

Another important aspect of oxygen availability involves mixing. Shallow ponds and streams tend to be much better oxygenated than deep bodies of water because of their relatively high ratio of surface area to volume. Rapids, waterfalls, and other types of "whitewater" tend to be the most highly oxygenated of all aquatic environments because of the high surface area exposed to the atmosphere as water splashes over rocks and logs, and because air bubbles are incorporated into the water (**Figure 41.3**). In contrast, bogs and other stagnant-water habitats often become completely deoxygenated, or anaerobic.

How Does Gas Exchange Compare in Air Versus Water?

How do terrestrial and aquatic habitats compare as a source of oxygen? Perhaps the most important difference is that air contains a great deal more oxygen than does water. At 15°C, for

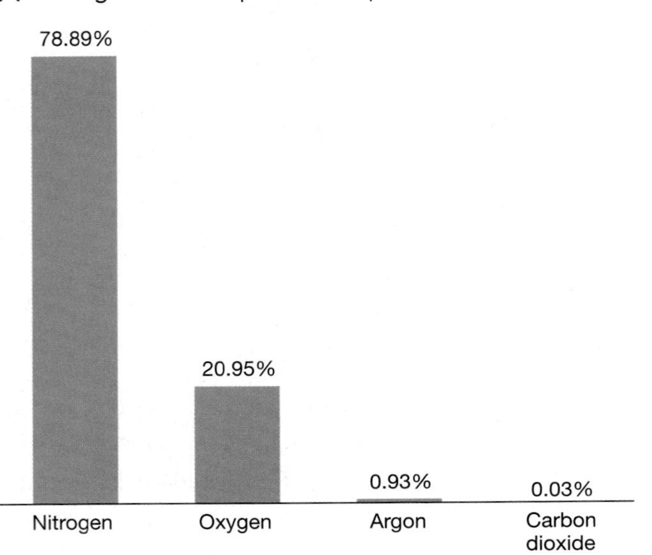

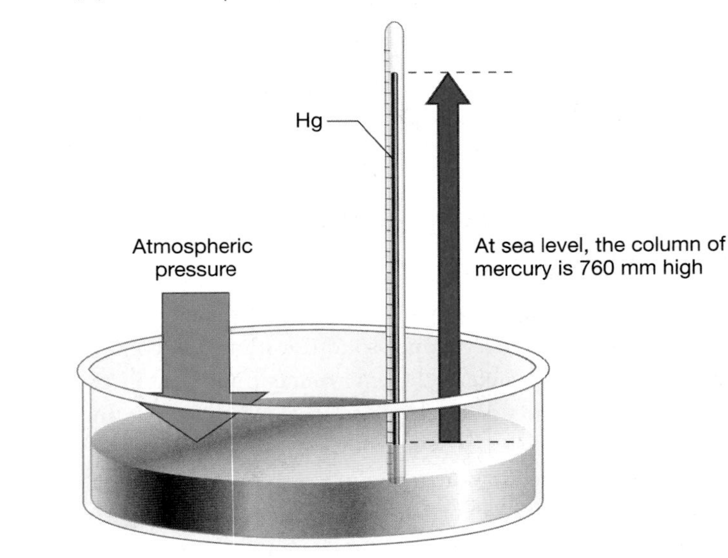

FIGURE 41.2 Some Characteristics of Air
(a) Nitrogen and oxygen are the primary components of Earth's atmosphere. **(b)** This diagram shows how a mercury barometer works. Atmospheric pressure is reported in a unit called millimeters of mercury, which is abbreviated mm Hg. (Hg is the chemical symbol for mercury.)

Natural aeration

Artificial aeration

FIGURE 41.3 The Amount of O_2 in Water Varies Widely
The amount of oxygen that diffuses into a lake, stream, or pond increases with the surface area of the water. In the rapids shown on the left, water breaks up into droplets that have a high surface area. In the stagnant pond on the right, engineers have added an aerator to increase the surface area of the water.
QUESTION How are oceans aerated?

example, a liter of air can contain up to 209 ml of O_2 while a liter of water may contain up to 7 ml of O_2. To extract the same amount of oxygen from its respiratory medium, an animal must process 30 times more water than air. This is important because water is also about a thousandfold denser than air. As a result, water breathers expend considerably more energy to ventilate their respiratory surfaces than do air breathers.

How do air and water compare as a medium for taking up carbon dioxide? D. J. Randall and co-workers compared the fraction of CO_2 that was removed from human blood versus rainbow trout blood during normal respiration. They found that over the same amount of time, humans are able to remove just 9 percent of the CO_2 in their blood, while trout remove 50–60 percent. To explain the difference, the researchers point to carbon dioxide's extremely high solubility in water and blood. Air breathers have a difficult time ridding themselves of CO_2, and they carry much more dissolved CO_2 in their blood than water breathers do. The implications of high CO_2 concentrations in blood are addressed in detail in section 41.4. Now let's consider how respiratory organs are designed. How do gills and lungs cope with the differences between water and air as respiratory media?

41.2 Organs of Gas Exchange

Many unicellular and small multicellular animals do not have specialized organs of gas exchange such as gills or lungs. Instead, these animals obtain O_2 and eliminate CO_2 by diffusion directly across the body surface. Most of these animals are aquatic or, like the earthworm, live in terrestrial habitats that are moist. For these animals, diffusion across the body surface is rapid enough to fulfill their requirements for taking in O_2 and expelling CO_2. Living in wet environments allows them to exchange gases across their outer surface while avoiding catastrophic amounts of water loss.

In contrast, animals that are large or that live in dry habitats have some sort of specialized respiratory organ. To understand why the presence of gills or lungs is associated with large size, recall from Chapter 38 that the volume of an animal's body increases much faster than does its surface area. Respiratory organs provide a vastly increased surface area for gas exchange—enough to meet the demands of a large volume filled with cells.

Biologists have long marveled at the efficiency of gills and lungs. To appreciate why, let's first examine the physical factors that control diffusion rates and then look at the structure and function of these respiratory organs.

Design Parameters: The Law of Diffusion

In 1855, Adolf Fick derived the law of diffusion based on the results of experiments he had performed on the rate at which gases diffuse. As **Figure 41.4** shows, Fick's law states that the rate of diffusion depends on five parameters: the solubility of the gas being considered, the temperature, the surface area available for diffusion, the difference in partial pressures of the gas across the respiratory surface, and the thickness of the barrier to diffusion.

Fick's law provides a compact way of identifying traits that allow animals to maximize the rate at which oxygen and carbon dioxide diffuse across the surfaces of their gills or lungs. Specifically, the law states that O_2 and CO_2 diffuse most rapidly when a large, extremely thin surface area is available for gas exchange.

Based on the insights provided by Fick's law, it is not surprising that the respiratory surface in the human lungs would cover about 100 m^2—or almost a quarter of a basketball court—if the epithelium were spread flat. In addition, the blood vessels and lung epithelium are so thin that it is relatively easy to damage them. For example, when John West and Odile Matheiu-Costello asked bicyclists to climb a steep hill for seven minutes, they observed a marked increase in blood on the respiratory surface of the athletes' lungs. To explain the appearance of blood inside the lung, the researchers proposed that the cyclists' high heart rates (up to 177 beats/minute during their sprint up the hill) increased blood pressure to the point where thin-walled vessels in the lungs ruptured and leaked blood into the lung cavity. The implication here is that natural selection has favored the evolution of respiratory surfaces that are exceptionally thin and thus optimized for rapid gas exchange.

What other aspects of gill and lung design conform to the dictates of Fick's law? What other factors affect the structure of these respiratory organs? To answer these questions, let's delve into the anatomy of gills and lungs.

The Gill

Gills are outgrowths of the body surface that are used for gas exchange in aquatic animals. Among invertebrates, the structure of gills is extremely diverse. In some species, such as the nudibranch

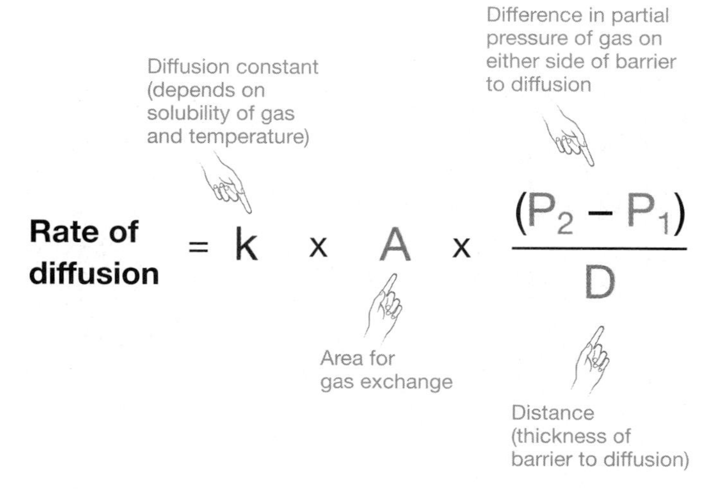

FIGURE 41.4 The Law of Diffusion

QUESTION Suppose you were going to design a system for rapid diffusion of oxygen across a phospholipid bilayer. According to Fick's law, which aspects of the system should be maximized in order to maximize the diffusion rate? Which should be minimized?

pictured on the left of **Figure 41.5**, gills project from the body surface and contact the surrounding water directly. In other species, such as the crayfish illustrated on the right of **Figure 41.5**, gills are located inside the exoskeleton or body wall. If gills are located internally, water must be driven over them by specialized limbs or cilia.

In contrast to the diversity of gills found in invertebrates, the gills of bony fishes are similar in structure. As **Figure 41.6** shows, fish gills are located on either side of the head and consist of four arches. To move water through these structures, most fish open and close their mouths and the stiff flap of tissue that covers the gills. The pumping action of these two structures creates a pressure gradient that moves water over the gills. In tuna and other fish that are particularly fast swimmers, however, this pumping action does not provide enough water to meet the demand for oxygen. Instead, these fish force water through their gills by swimming with their mouths open. This process is called ram ventilation.

Regardless of how they are ventilated, the flow of water over gills is unidirectional. This is a crucial point. To see why, examine the detailed sketch in Figure 41.6 and note that each of the gill arches is made up of a series of ridged filaments. Each of the ridges contains a bed of small blood vessels called capillaries. The flow of blood through the capillary bed is in the opposite direction to the flow of water over the gill surface. As a result, a countercurrent exchange system exists.

Why is countercurrent exchange in the fish gill important? Countercurrent flow ensures that the difference in the amount of oxygen and carbon dioxide in water versus blood is large over the *entire* respiratory surface. Stated another way, the effect of countercurrent exchange is to maximize the $P_2 - P_1$ term in Fick's law of diffusion over the entire gill surface (see Figure 41.4). Based on this observation, biologists cite countercurrent exchange as another example of how gills are optimized for rapid gas exchange.

Gas Transfer in Air

Air and water are dramatically different respiratory media because their density and their ability to hold oxygen and carbon dioxide differ. In addition, the consequences of exposing the respiratory surface to air versus water differ. In aquatic habitats, breathing disrupts water and electrolyte balance. As Chapter 39 explained, many marine animals lose water across their respiratory surface through osmosis; animals that live in freshwater lose ions and other solutes by diffusion. In terrestrial environments, breathing leads to a loss of water by evaporation.

How do terrestrial animals minimize water loss while maximizing the efficiency of gas exchange? Part of the answer is that the respiratory organs of terrestrial animals are located well within the body. In addition, the openings to these organs tend to be small. Biologists interpret these features as adaptations that help minimize the loss of water by evaporation. Let's take a close look at the respiratory systems found in terrestrial insects and vertebrates.

The Insect Tracheal System Chapter 39 introduced the trachea of insects. Recall that this respiratory system consists of a series of tubes that branch throughout the body. The key point about the tracheal system is that it transports air close enough to cells for gas exchange to take place directly. Consequently, insects do not require a circulatory system to transport gases from a respiratory structure to the tissues. They also lack an oxygen-carrying protein like the hemoglobin found in vertebrates.

Trachea connect to the exterior through an opening called a spiracle that can be closed. As Chapter 39 explained, biologists have found that when spiracles are kept open experimentally, insects are likely to die of dehydration. Based on results like these, spiracles are interpreted as adaptations to minimize water loss.

Recent research on the tracheal system has focused on the question of how air moves in and out of the tubes. Is simple dif-

External gills

Internal gills

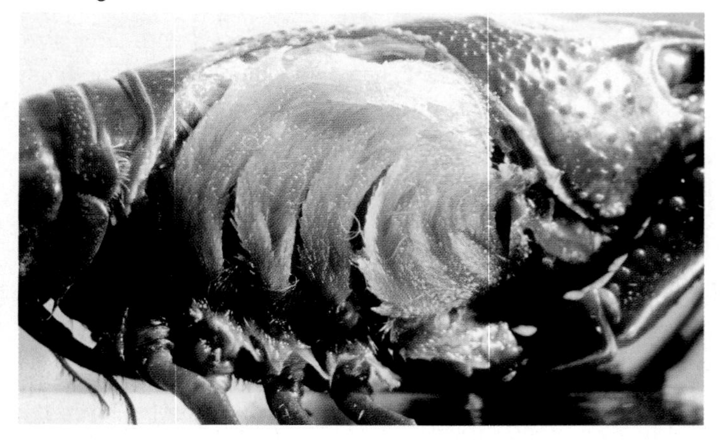

FIGURE 41.5 Gills Can Be External or Internal
Nudibranchs (left) are a group of marine snails. Many nudibranch species have gills that project outside of the main body wall. Crayfish, in contrast, have gills that are located inside the main body wall (right).
QUESTION In terms of fitness, what are the advantages and disadvantages of external versus internal gills?

fusion efficient enough to ventilate the system, or is some type of breathing mechanism involved?

Yutaka Komai set out to answer this question for an insect called the sweet potato hawkmoth (**Figure 41.7a**). More specifically, Komai wanted to investigate how the amount of O_2 delivered to muscles changes during flight. To document partial pressure of oxygen (P_{O_2}) in flight muscles, he inserted a needle-like electrode into the wing muscles of a hawkmoth. The electrode was attached to an instrument that measured the partial pressure of oxygen, and the hawkmoth was tethered to a stand.

When the hawkmoth had sufficiently recovered from the procedure to beat its wings normally, Komai began recording P_{O_2} as the insect rested. Then he stimulated the moth to fly by exposing it to wind. **Figure 41.7b** shows how P_{O_2} changed during one such experiment. (He repeated the procedure on several different individuals and observed the same pattern.) Because flight is an energetically demanding activity, it is not surprising that P_{O_2} levels in the flight muscles dropped initially. As flying continued, however, P_{O_2} levels recovered until they were nearly as high as they were at rest.

To explain these results, Komai invoked a hypothesis that T. Weis-Fogh had proposed to explain how desert locusts ventilate their trachea during flight. As **Figure 41.7c** shows, the trachea of flying insects are sandwiched between muscles. When these muscles contract and relax, trachea constrict and expand in response. As a result, wing beats lead to efficient ventilation of the respiratory surface.

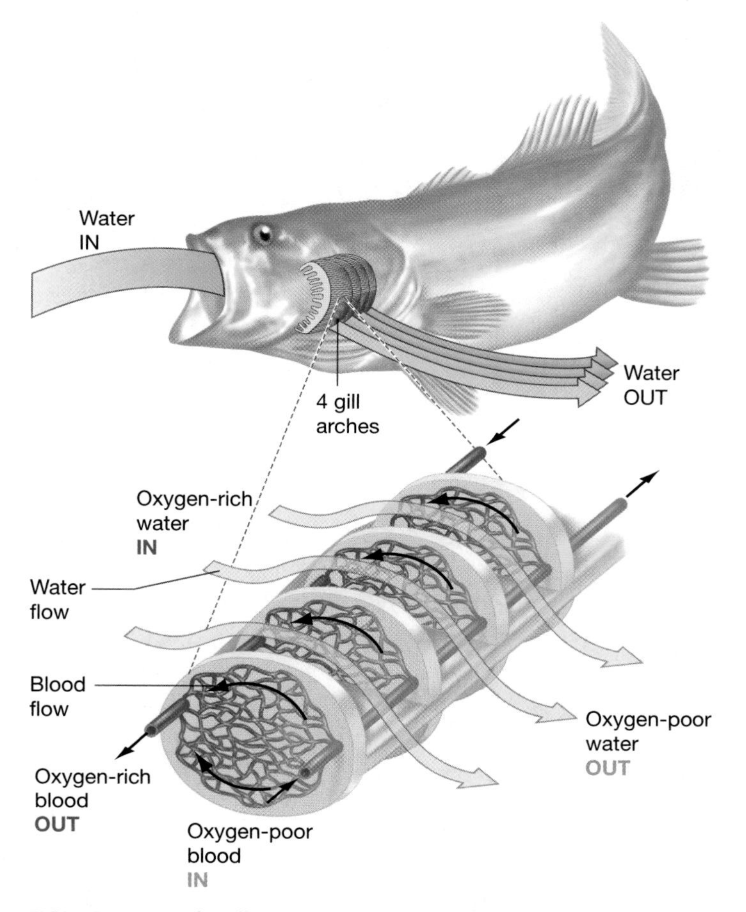

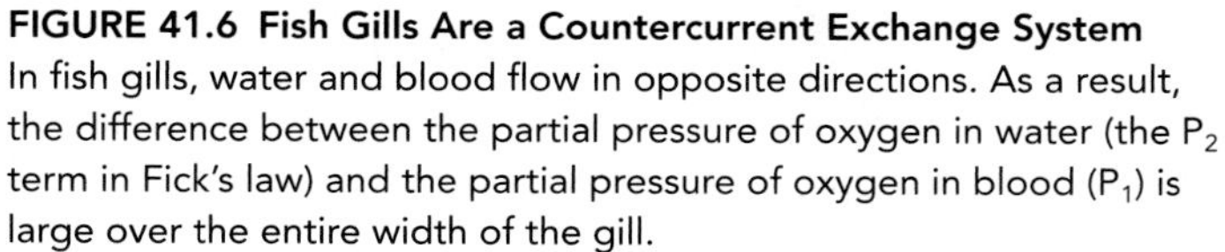
FIGURE 41.6 Fish Gills Are a Countercurrent Exchange System
In fish gills, water and blood flow in opposite directions. As a result, the difference between the partial pressure of oxygen in water (the P_2 term in Fick's law) and the partial pressure of oxygen in blood (P_1) is large over the entire width of the gill.

(a) Potato hawkmoth

(b) Measurement of oxygen in flight muscle

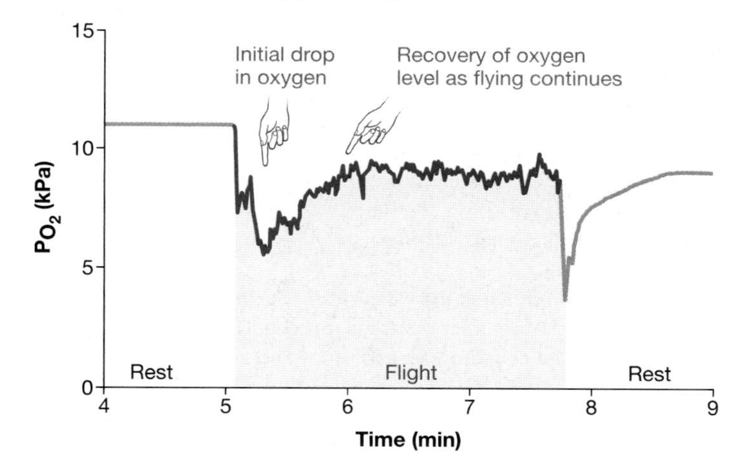

(c) Model of oxygen delivery in desert locusts

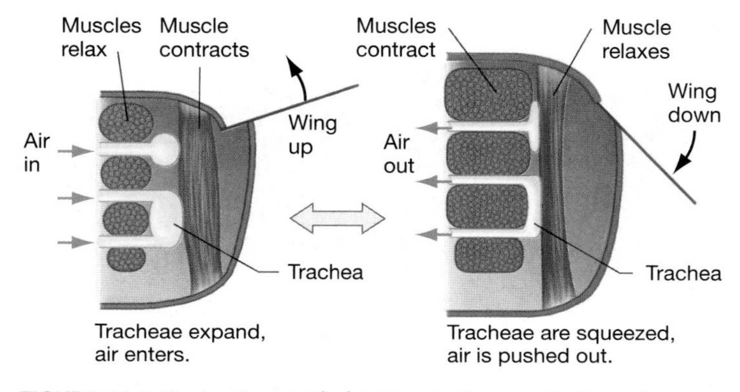

FIGURE 41.7 During Insect Flight, How Is Oxygen Delivered to Muscles?
(a) Potato hawkmoths live throughout the Old World and in the Hawaiian islands. **(b)** This graph plots the partial pressure of oxygen in flight muscles (reported here in a unit called the kilopascal—see Chapter 32) before, during, and after flight. **(c)** These diagrams show how the trachea of desert locusts change shape as their wings beat.

The Lung Lungs are infoldings of the body surface that are used for gas exchange. They occur in amphibians, lizards, birds, mammals, and certain fish and invertebrates. The amount of surface area available for gas exchange varies a great deal among species with lungs, however. In frogs and other amphibians, the lung is a simple sac lined with blood vessels. The lungs of mammals, in contrast, are finely divided into tiny sacs called **alveoli** (**Figure 41.8**). Each human lung contains 150 million of these structures. Due to the presence of alveoli, mammal lungs have about 40 times more surface area for gas exchange than an equivalent volume of frog lung tissue.

As the right-hand side of Figure 41.8 shows, an alveolus provides an interface between air and blood that consists of a layer of epithelial cells, some extracellular material, and the wall of a capillary. In the human lung, this barrier to diffusion is only 0.2 μm thick—about 1/200th of the thickness of this page.

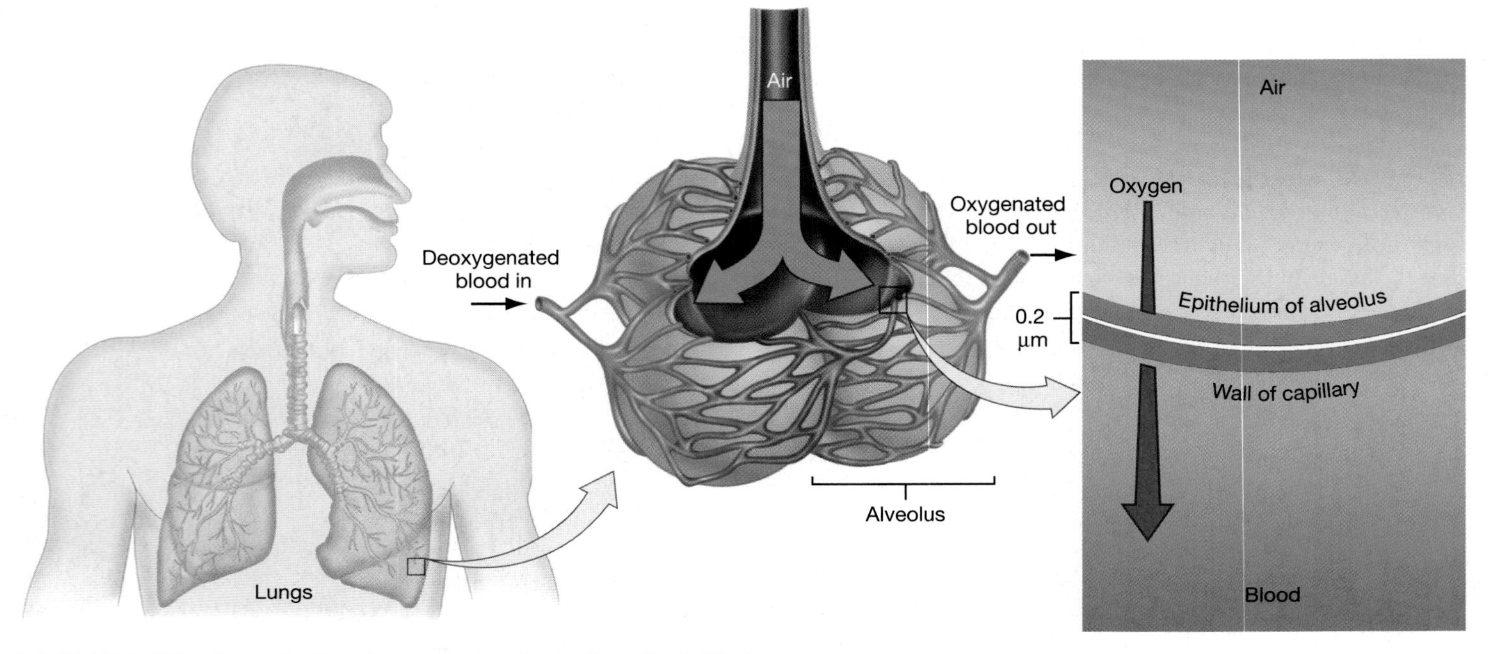

FIGURE 41.8 What Does the Respiratory Surface in the Lung Look Like?
The human respiratory tract branches repeatedly. Eventually, each branch ends in a cluster of tiny sacs called alveoli. Alveoli are covered with capillaries and are the site of gas exchange.

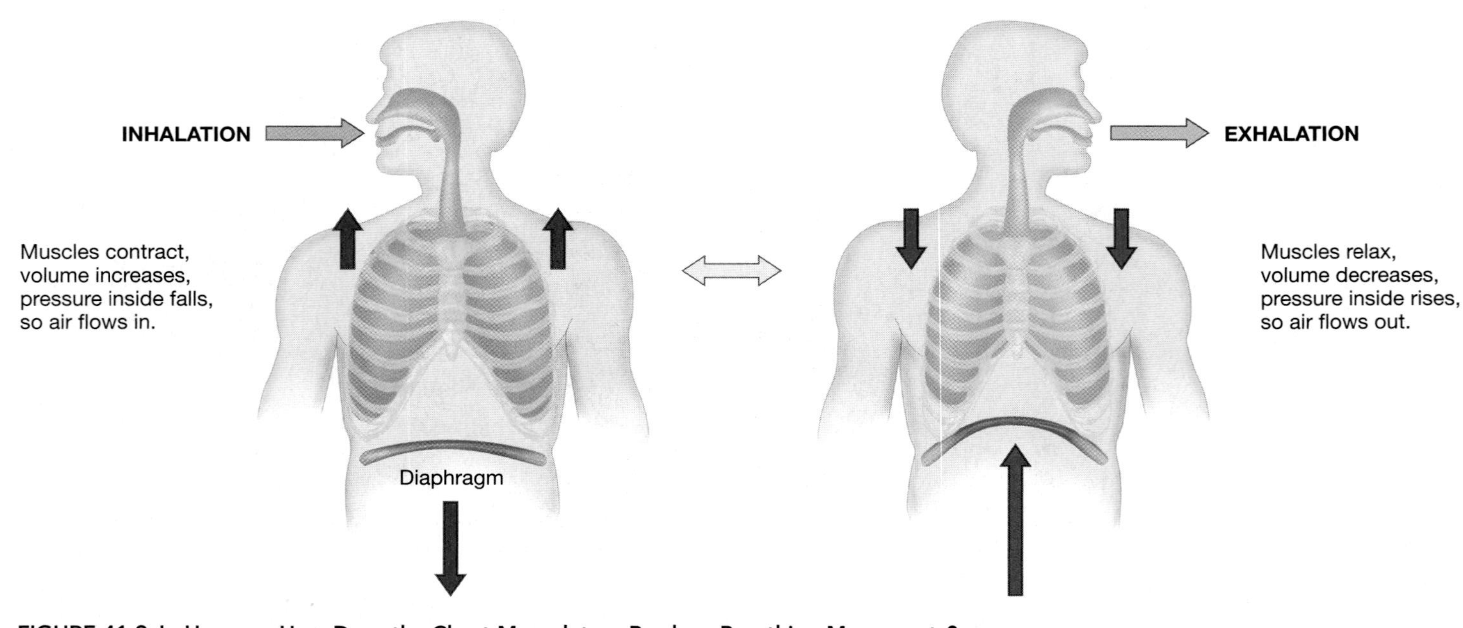

FIGURE 41.9 In Humans, How Does the Chest Musculature Produce Breathing Movements?
The arrows in this diagram show the movements that result from contraction (left) and relaxation (right) of the diaphragm and muscles in the ribs. QUESTION What happens when a person sighs or takes a deep breath?

In addition to total surface area, the other major feature of lungs that varies among species is ventilation. In the lungs of snails and spiders, air movement takes place by diffusion only. Vertebrates, in contrast, actively ventilate their lungs by pumping air via muscular contractions. In humans, for example, air enters the lungs when muscles in the ribs and a sheet of muscle under the lungs, called the diaphragm, contract. These contractions lift the rib cage and breastbone and lower the diaphragm. As **Figure 41.9** shows, the volume of the chest cavity and the lungs increases as a result.

As the lungs expand, the air pressure inside them drops below atmospheric pressure. In response to this pressure gradient, external air enters the lungs. Exhalation occurs when the diaphragm and rib muscles relax. Relaxing reduces the volume of the chest cavity and lungs. As a result, air pressure inside the lungs increases and produces a flow of air to the atmosphere.

When air enters an alveolus, it is brought close to blood that has a relatively high partial pressure of carbon dioxide and relatively low P_{O_2}. What is blood, and how does it transport O_2 and CO_2 between the respiratory surface and body tissues?

41.3 Blood

Blood is a tissue—a collection of cells that functions as an integrated unit. More specifically, blood is a connective tissue that consists of cells in a fluid, extracellular matrix. In an average human, about half of the blood volume is composed of an extracellular matrix called **plasma**. The other half of the volume comprises a variety of cells and cell fragments that are collectively called the formed elements.

Given the wide variety of functions it serves, it is not surprising that blood is complex. In addition to carrying oxygen and carbon dioxide between mitochondria and lungs, blood transports nutrients from the digestive tract to other tissues in the body, moves waste products to the kidney and liver for processing, conveys hormones from glands to target tissues, delivers immune system cells to sites of infection, and distributes heat from deeper organs to the surface.

The formed elements in blood include cell fragments called platelets, several types of white blood cells, and red blood cells. White blood cells are part of the immune system and fight infections; platelets minimize blood loss from ruptured blood vessels by releasing material that forms clots. In terms of volume, though, the formed elements are dominated by red blood cells. In human blood, red blood cells make up 99.9 percent of the formed elements.

Red blood cells transport oxygen from the lungs to mitochondria and return to the lungs with carbon dioxide. In humans, new red blood cells are synthesized at the rate of 2.5 million per second to replace old red blood cells, which are dying at the same rate. In humans, red blood cells live about 120 days. They develop along with white blood cells and platelets from stem cells located in the tissue inside bone (bone marrow). As new red blood cells mature, they lose their nuclei, mitochondria, and most other organelles but become filled with about 280 million molecules of the oxygen-carrying molecule hemoglobin.

Hemoglobin consists of four polypeptide chains, each of which binds to a non-protein group called heme. Each heme molecule, in turn, contains an iron ion (Fe^{2+}) that can bind to an oxygen molecule. As a result, each hemoglobin molecule can bind up to four oxygen molecules.

Because of its low solubility, only 1.5 percent of the O_2 carried in blood is dissolved in plasma; 98.5 percent is bound to hemoglobin. What happens when red blood cells carrying oxygen-rich hemoglobin reach cells that contain little oxygen?

Capillary Exchange: The Bohr Effect

Blood leaving the lungs has a P_{O_2} of 100 mm Hg, while muscles and other tissues have P_{O_2} levels of about 40 mm Hg at rest. This partial pressure difference drives the unloading of O_2 from the hemoglobin to the tissues through diffusion.

When researchers studied the dynamics of O_2 unloading in tissues, they found the pattern illustrated in **Figure 41.10a** (page 794). This graph is called the oxygen-hemoglobin dissociation curve; it plots the percent saturation of hemoglobin in red blood cells versus the P_{O_2} levels in tissues. The remarkable feature of the relationship is that it is sigmoidal, or S-shaped. This pattern occurs because the binding of an oxygen molecule to one subunit of hemoglobin causes a conformational change in the protein that makes the other three subunits much more likely to bind to oxygen. This phenomenon is known as cooperative binding. If cooperative binding did not occur, all four subunits of hemoglobin would simply load or unload oxygen all at once. The hyperbolic relationship given by the gray line in Figure 41.10a would result.

Why is cooperative binding important? Two points are involved in answering this question. The first is that if cooperative binding did not occur, the gray curve in Figure 41.10a shows that tissues would have to be almost devoid of oxygen before oxygen would begin to be released from hemoglobin. The second point springs from the realization that a sigmoidal curve has the three general regions illustrated in **Figure 41.10b.** In the sections to the lower left and upper right of an S-curve, the quantity plotted on the y-axis changes relatively little for a given change in the quantity plotted on the x-axis. The opposite is true in the middle section of the curve. With these points in mind, examine the hemoglobin dissociation curve in part (a) again. Note that when tissue P_{O_2} ranges between about 25 and 50 mm Hg, which are the values routinely observed in tissues, a relatively small change in tissue P_{O_2} results in a large change in hemoglobin saturation.

To pull these observations together, consider that, at a resting tissue P_{O_2} level of about 40 mm Hg, hemoglobin unloads only 25 percent of the O_2 it carries to the tissues. At rest, then, an average hemoglobin molecule is still 75 percent saturated as it returns to the lungs. But if you start to exercise and tissue

P_{O_2} levels drop to 30 mm Hg, so much more oxygen is released that the average saturation of a hemoglobin molecule drops to less than 40 percent. The message here is that, because of the sigmoidal relationship, hemoglobin responds quickly and effectively to small changes in oxygen demand.

Cooperative binding is only part of the story behind oxygen delivery. Hemoglobin—like other proteins—is sensitive to changes in pH and temperature. During exercise, the temperature and the partial pressure of CO_2 in tissues rises. The CO_2 produced by exercising tissues reacts with the water in blood to form carbonic acid (H_2CO_3). Carbonic acid, in turn, dissociates and releases a hydrogen ion (H^+) and bicarbonate ion (HCO_3^-).

$$CO_2 + H_2O \rightleftharpoons H_2CO_3 \rightleftharpoons H^+ + HCO_3^-$$

As a result, increases in P_{CO_2} lead to an increase in hydrogen ion concentration and a drop in pH.

Decreases in pH and increases in temperature alter hemoglobin's conformation. As the graph in **Figure 41.10c** shows, these shape changes make hemoglobin more likely to release O_2 at all values of tissue P_{O_2}. This phenomenon is known as the Bohr shift. The name honors Christian Bohr, who published on the phenomenon in 1904, and highlights the rightward shift of the hemoglobin dissociation curve with decreasing pH. The Bohr shift is important because it makes hemoglobin more likely to release oxygen during exercise or other conditions where P_{CO_2} is high and tissues are under oxygen stress.

To see the effect of cooperative binding and the Bohr shift in action, consider an experiment on rainbow trout conducted by Joe Kiceniuk and David Jones. These biologists wanted to determine how the oxygen transport system in trout responds to sustained exercise. First, the fish swam continuously against a current in a water tunnel. As Kiceniuk and Jones increased the speed of the current, they periodically sampled the O_2 content of both the arterial and venous blood of experimental individuals. Not surprisingly, they found that arterial O_2 levels remained fairly constant as swimming speed increased. In contrast, the O_2 content of venous blood—which had undergone gas exchange with the tissues—dropped steadily as swimming speed increased. When the fish had reached their maximum sustainable speed, virtually all of the oxygen available in hemoglobin had been extracted. The data show that in hard-working tissues, the combination of increased temperature, lower pH, and lower P_{O_2} caused hemoglobin to become almost completely deoxygenated.

We've yet to consider the other side of gas exchange, however. What happens as CO_2 is released from tissues to the blood? How does the body cope with the decline in pH that results? After all, if hemoglobin's conformation and activity are affected so dramatically by pH changes, it is likely that many other proteins will be altered as well. If these proteins function less well at low pH, then carbon dioxide buildups should result in illness.

(a) How does the amount of O_2 bound to hemoglobin change as a function of P_{O_2} in tissues?

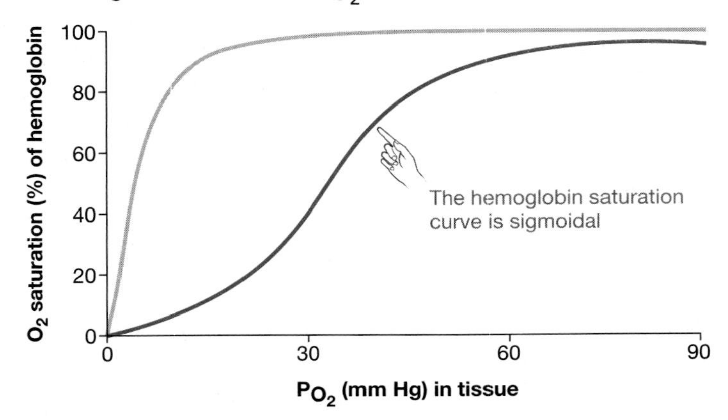

(b) What are the salient features of a sigmoidal curve?

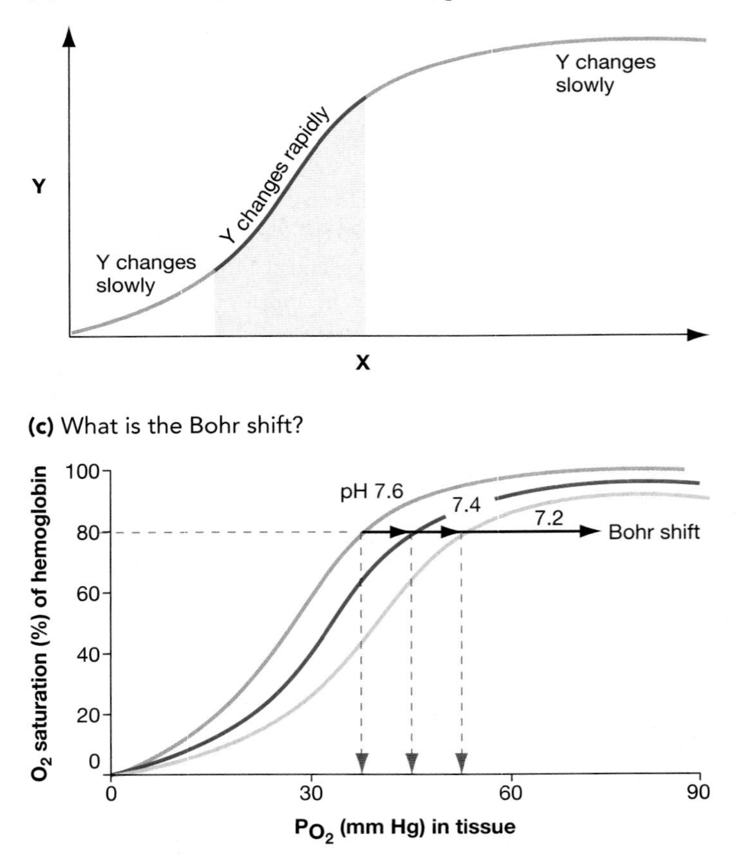

FIGURE 41.10 Hemoglobin Dissociation Curves
(a) In this graph, the y-axis plots the percentage of binding sites on hemoglobin that hold oxygen molecules. The x-axis plots the partial pressure of oxygen in tissues. The gray line gives the relationship when cooperative binding is absent. **(b)** Sigmoidal curves have three distinct regions. **(c)** As pH drops, oxygen becomes less likely to stay bound to hemoglobin at all values of tissue P_{O_2}. EXERCISE Suppose that three different tissues in a body have a P_{O_2} of 40 mm Hg, but that their pHs are 7.2, 7.4, and 7.6. Using the graph in part (c), determine the percentage of hemoglobin binding sites that hold oxygen in the capillaries serving each tissue.

CO_2 Exchange and the Regulation of Blood pH

How is blood P_{CO_2} and pH regulated? The answer to this question became apparent when researchers discovered large amounts of the enzyme carbonic anhydrase in red blood cells. Recall, from Chapter 40, that carbonic anhydrase catalyzes the formation of carbonic acid from carbon dioxide in water. Consequently, CO_2 that diffuses into red blood cells is quickly converted to bicarbonate ions and protons (and induces the Bohr shift as a result). The same reaction occurs much more slowly in the plasma surrounding red blood cells.

Why is the carbonic anhydrase in red blood cells so important? The key insight was that its activity maintains a strong partial pressure gradient favoring the entry of CO_2 into red blood cells (**Figure 41.11**). Once bicarbonate ions form, they diffuse into the blood plasma along a concentration gradient. In this way, most CO_2 is transported in plasma in the form of $H_{CO_3^-}$. Further, because the catalyzed reaction in red blood cells occurs so much faster than the uncatalyzed reaction in plasma, relatively small amounts of H^+ build up in solution. Instead, large amounts of H^+ build up inside red blood cells.

What happens to the protons inside red blood cells? The answer to this question emerged when researchers studied the composition of deoxygenated hemoglobin. When hemoglobin is not carrying oxygen molecules, it has a high affinity for protons. As a result, much of the H^+ that is produced by the dissociation of carbonic acid is taken up by hemoglobin. In this way, hemoglobin acts as a **buffer**—a compound that minimizes changes in pH. In addition, analyses of hemoglobin in deoxygenated blood revealed that the CO_2 that is not transported as HCO_3^- binds to amino groups on the protein portion of the molecule. The overall effect of these events is efficient transport of CO_2 with a minimal effect on blood pH.

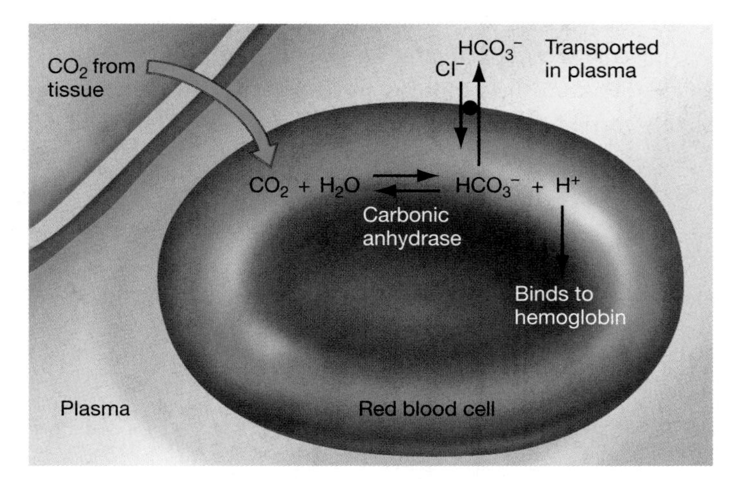

FIGURE 41.11 How Is Carbon Dioxide Transported in Blood? When CO_2 diffuses into red blood cells, carbonic anhydrase quickly converts it to bicarbonate ion (HCO_3^-) and a proton (H^+). As a result, a strong partial pressure gradient is maintained favoring entry of CO_2 into red blood cells. The protons produced by the reaction bind to deoxygenated hemoglobin. Most CO_2 in blood is transported to the lungs in the form of the bicarbonate ion (HCO_3^-). QUESTION This diagram shows the sequence of events in tissues. Explain what happens when the red blood cell in the diagram reaches the lungs.

When deoxygenated blood reaches the alveoli, however, the environment changes dramatically. In the lungs, a strong concentration gradient favors diffusion of CO_2 from plasma and red blood cells to the atmosphere. As CO_2 diffuses from the blood into the alveoli, P_{CO_2} in blood declines. The drop promotes the reversal of the chemical reactions that occurred in tissues. Bicarbonate is converted back to CO_2, which then diffuses into the alveoli and is exhaled from the lungs. Hemoglobin picks up O_2 during inhalation, and the cycle begins anew. Now the question becomes, how is blood transported between the lungs and the tissues?

41.4 The Circulatory System

Animal circulatory systems move blood throughout the body. Most have the same two components: one or more muscular pumps (usually called a heart) that provide the force to move blood, and vessels that convey nutrients and O_2 to the tissues and carry away waste products, including CO_2.

Virtually all animals that are bigger than a few millimeters in size have some sort of circulatory system. If a large animal lacked a circulatory system, oxygen, nutrients, and wastes could not diffuse fast enough to service the tissues.

Animal circulatory systems are of two general types: open or closed. After examining the nature of these systems, let's consider some recent research on a central component of the human circulatory system—the heart.

Open Circulatory Systems

Many invertebrates have an open circulatory system in which a blood-like fluid, called hemolymph, is not confined to blood vessels but, instead, comes into direct contact with the tissues. **Figure 41.12** (page 796) illustrates the anatomy of the open circulatory system found in grasshoppers and in clams. Note that in both cases, one or more hearts pump hemolymph into an artery that empties into an open, fluid-filled space. Hemolymph is returned to the heart when it relaxes and creates suction. General body movements also serve to move hemolymph to and from the heart(s).

The salient characteristic of an open circulatory system is that hemolymph is under low pressure. To understand why, examine the grasshopper circulatory system in Figure 41.12 more closely. When the muscles of the tubular hearts contract, the hemolymph inside is put under pressure. Consequently, the fluid is pushed out of the hearts and into the adjacent vessels, where pressure is lower. These vessels empty into the open body cavity, where pressure is lower still. When the heart muscles relax, the volume inside the organs expands. The expansion phase lowers pressure inside the hearts below the pressure found in the body cavity. As a result, hemolymph follows a pressure gradient from the cavity through the vessels and back toward the heart.

The key point is that hemolymph is not tightly constrained inside vessels during this entire cycle; thus, the overall pressure in the system is low. This is important because low pressure means that the amount and rate of blood movement are also relatively low. In addition, open circulatory systems do not precisely direct hemolymph toward tissues that have a high oxygen demand and CO_2 buildup. (Recall that hemolymph does not contain an oxygen-carrying protein like hemoglobin.) These features are in stark contrast to closed circulatory systems, where pressure tends to be high throughout the system, and where blood delivery to specific tissues can be precisely controlled.

Closed Circulatory Systems

In animals with closed circulatory systems, blood is completely contained within blood vessels and flows in a continuous circuit through the body under pressure generated by the heart. The major types of blood vessels in closed circulatory systems are arteries, capillaries, and veins. Arteries conduct blood away from the heart to tissues; veins return blood to the heart. Molecules and ions are exchanged between the blood and tissues in capillaries.

Figure 41.13 shows the anatomy of a typical vein, artery, and capillary. Because blood pressure in veins is low, these vessels are thin-walled and contain valves that prevent backflow of blood. The relatively thick walls of arteries, in contrast, allow them to withstand the high blood pressures that exist near the heart. Capillaries are extremely thin walled. Because distance is critical to the efficiency of diffusion, the thinness of these vessels correlates with their function in materials exchange.

Interstitial Fluid and the Lymphatic System In addition to the blood found inside arteries, capillaries, and veins, animals with closed circulatory systems have interstitial fluid that directly surrounds their cells. Interstitial fluid is constantly augmented by water as well as by electrolytes from blood, which leak out between the cells that form the walls of capillaries. The excess interstitial fluid that results from this leakage collects in the vessels of the lymphatic system. The **lymphatic system** is a collection of vessels that branches throughout the body. The fluid inside lymph vessels, called lymph, is eventually returned to the blood.

In this way, the lymphatic system acts as a type of circulatory system. Because its primary function is in defense against disease-causing agents, however, the lymphatic system is analyzed in more detail in Chapter 46.

Partitioning Blood Flow Compared to open circulatory systems, closed circulatory systems have high overall pressure and thus deliver more blood to tissues faster. In addition, blood can be directed to certain tissues and away from others. As an example of how blood can be shunted from one part of the body to another in a closed circulatory system, consider research that Vera Cherepanova and colleagues have conducted on Baikal seals. Lake Baikal is the world's deepest lake, and the seals are capable of extended dives in search of food. In some species of seals, dives can last 20–60 minutes and take individuals to depths of 500m. How do seals deliver enough oxygen to the brain to maintain consciousness during extremely long dives?

An answer to this question began to emerge when biologists found that the muscle and blood of seals store huge amounts of oxygen. This observation led to the hypothesis that these stores are carefully rationed during long dives, with priority for oxygen delivery given to the heart and brain. The logic here is that if oxygen levels in the blood become low during particularly long dives, only the most vital organs should receive it. Restricting blood flow to the digestive system might lead to indigestion, but restricting blood flow to the heart or brain results in unconsciousness and death.

To test this idea in Baikal seals, Cherepanova and co-workers injected a radioactive atom into several individuals. Because the atom they chose is physiologically inactive, it was simply carried around in the blood as a marker of flow and did

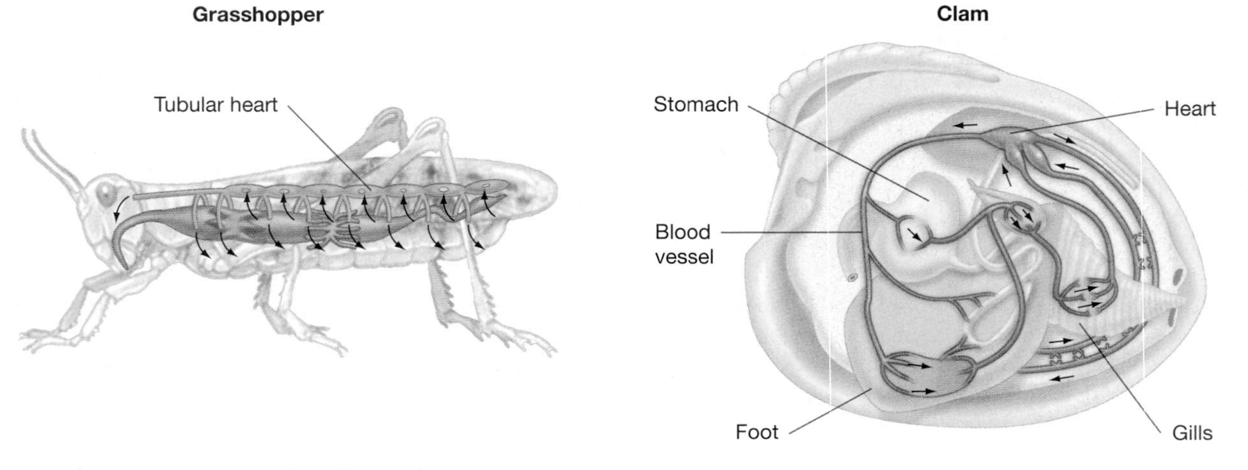

FIGURE 41.12 Open Circulatory Systems
The arrows in these diagrams indicate the direction of blood flow. **QUESTION** What is "open" about an open circulatory system?

Vein and artery

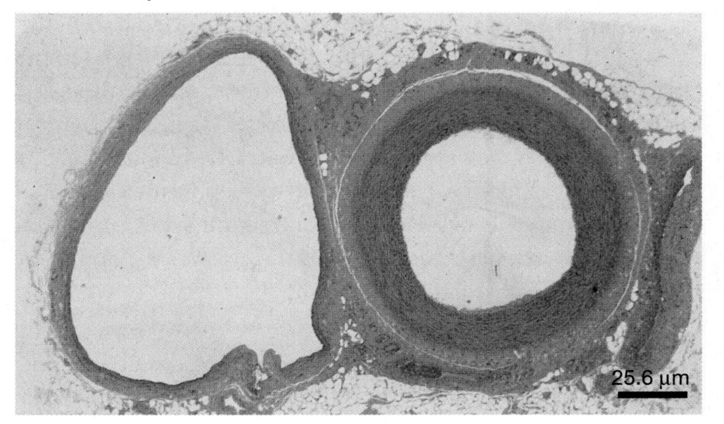

Capillary

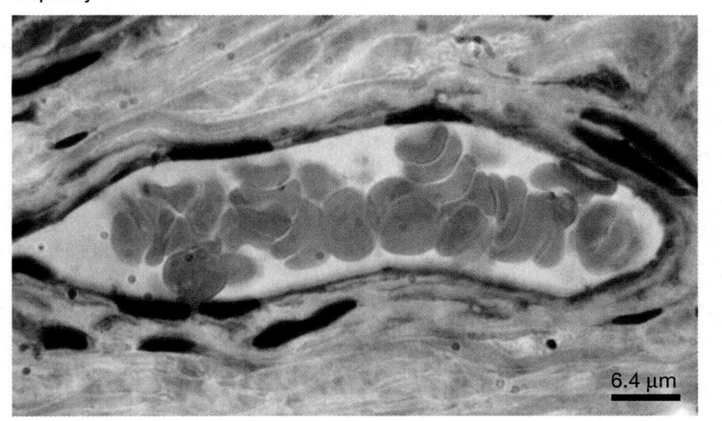

FIGURE 41.13 Veins, Arteries, and Capillaries
These micrographs show cross sections through a vein, an artery, and a capillary. Note the contrasts in the relative wall thickness and overall size of each type of vessel. The photo of the capillary is magnified four times as much as the photo of the vein and artery.

not harm the animals. The researchers then extracted blood samples from different tissues at intervals and measured the amount of radioactivity present.

As **Figure 41.14** shows, the researchers' data showed a strong contrast in circulation during resting and diving. When the seals were inactive, the amount of radioactive marker did not vary among tissue groups. This observation indicated that blood was flowing evenly throughout the body. When the biologists sampled the same tissues during dives, however, they found that blood flow to neck muscles was normal but declined dramatically—in several locations to zero—everywhere else they sampled. This result suggests that virtually all of the blood vessels in the animal's trunk and limb constricted during the dive, and that blood flow was highest in the circuit between the heart and brain.

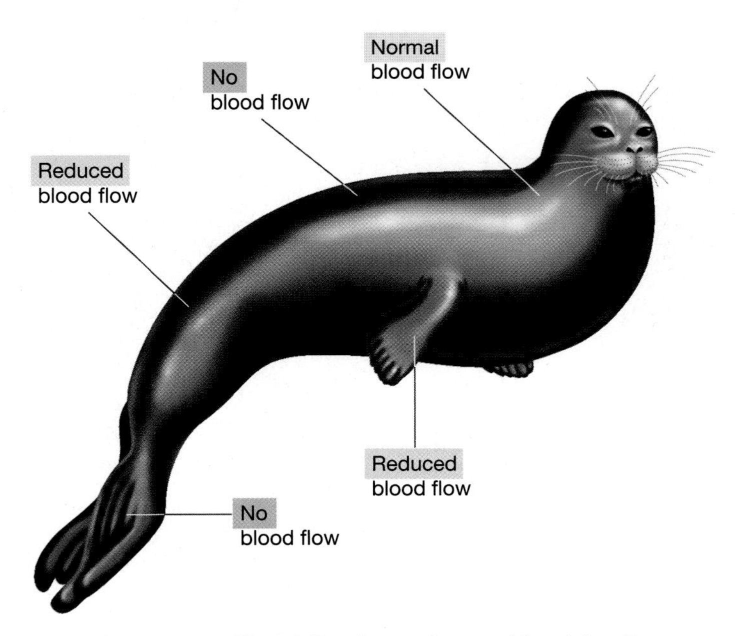

FIGURE 41.14 In a Closed Circulatory System, Blood Can Be Shunted to Certain Locations
The labels indicate how blood flow to different parts of a Baikal seal's body during a dive compares to blood flow in the same locations at rest. QUESTION In humans, blood is shunted between three major compartments: the brain, digestive system, and muscles. How would you predict that blood flow changes after a large meal?

A similar redirection in the blood supply occurs in humans exposed to situations that induce fear. As part of the fight-or-flight response analyzed in Chapter 44, blood is directed away from the skin and digestive system and toward the heart, brain, and muscles. In this way, the body delivers oxygen to the brain and muscles in preparation for rapid decisions and movement. The resulting lack of blood flow to the skin produces the feeling of being in a "cold sweat." Lack of blood flow to the digestive tract induces the sick feeling associated with fear.

The Heart CD ACTIVITY 41.2

The Heart

In animals with closed circulatory systems, the heart contains at least two chambers. There is at least one thin-walled **atrium,** which receives blood returning from circulation, and at least one thick-walled chamber called a **ventricle**, which generates the force to propel blood through the system.

When studies of vertebrate hearts and circulatory systems had advanced enough to make comparisons among groups possible, biologists noticed several important patterns. To understand these, study the diagram in **Figure 41.15** (page 798). This phylogenetic tree shows the evolutionary relationships among some of the major vertebrate groups; the sketches above each group offer a simplified version of their heart and circulatory system. Two points are particularly important to note.

- The number of distinct chambers in the heart increased as vertebrates diversified. Fish hearts have two chambers; amphibians have three; crocodiles, birds, and mammals have four.

- In fish, the circulatory system forms a single loop. In other groups, there are separate circuits to the lungs and body.

To understand how the chambers within the heart function together and how the two circulatory loops work, let's look more closely at how the human heart interacts with the circulatory system.

The Human Heart Your heart is roughly the size of your fist and is located in the chest cavity between your lungs. As **Figure 41.16** shows, the circulatory system returns blood from the body to the right atrium of the heart. This blood is low in oxygen. When the muscles that line this chamber contract, deoxygenated blood is sent to the right ventricle. The right ventricle, in turn, contracts and sends blood out to the lungs, via the pulmonary arteries. In this way, the right ventricle powers the movement of blood through the lungs and back to the heart. This loop is called the **pulmonary circulation.**

The flow of blood from atrium to ventricle to artery goes only one way, because flaps of tissue, called **valves**, separate the heart's chambers from each other and from the adjacent blood vessels. As you can see in Figure 41.16, the flaps are oriented in a way that ensures a one-way flow of blood with little or no backflow.

After blood has circulated through the capillary beds in the lung's alveoli and become oxygenated, it returns to the heart through the pulmonary veins. The oxygenated blood enters the left atrium. When this chamber contracts, blood is pushed into the left ventricle. The walls of the left ventricle are so thick with muscle cells that their contraction sends oxygenated blood into arteries and capillaries throughout the body—or into the **systemic circulation**.

The Cardiac Cycle Although the preceding discussion suggests that each chamber in the heart acts independently, this is actually not the case. Instead, the four chambers alternate between contraction and relaxation in a concerted fashion. Contraction generates the pressure required to move blood into the systemic and pulmonary circuits; relaxation allows each chamber to be refilled before the next contraction. A cardiac cycle consists of one complete contraction phase, called **systole**, and one complete relaxation phase, called **diastole.**

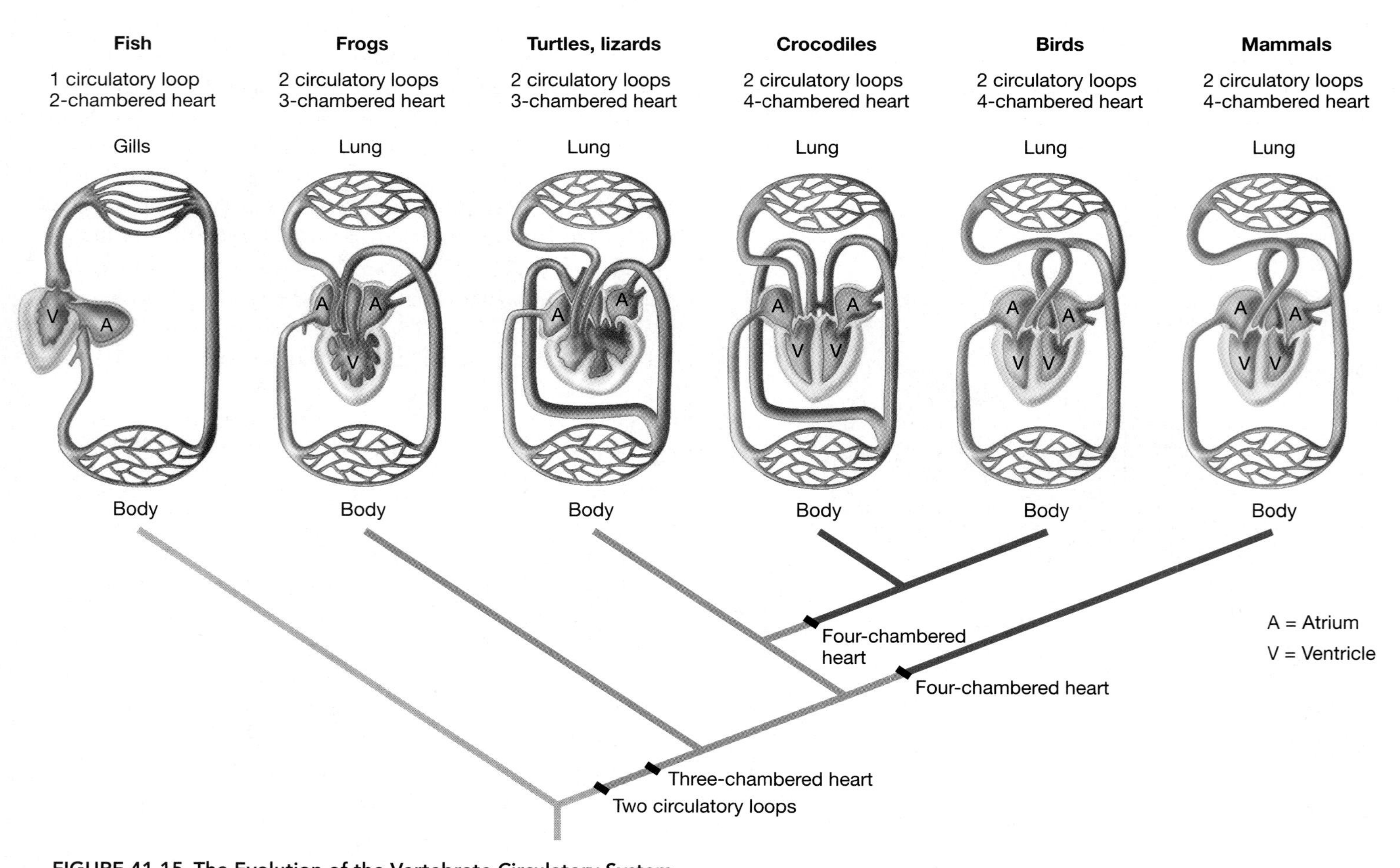

FIGURE 41.15 The Evolution of the Vertebrate Circulatory System
In fish, the heart and circulatory system are simpler than in vertebrate groups that evolved later. For example, later groups have separate circulatory loops to serve the lungs and body instead of a single loop.
EXERCISE Turtles (and relatives) and crocodiles have blood vessels that shunt blood between their two circulatory loops, as needed. Label these shunts on the diagram.

How is the activity of each chamber coordinated? Cardiac muscle cells, like other muscle cells, contract in response to electrical signals. In invertebrates, the electrical impulses that trigger contraction come from specific cells in the nervous system. In vertebrates, however, a group of cells in the heart itself is responsible for generating the initial signal. A vertebrate heart will beat even if all nerves supplying it are severed.

The "pacemaker cells" that initiate contraction are located in a region of the right atrium called the **SA (sinoatrial) node**. The electrical impulse generated by the SA node is rapidly conducted throughout the right and left atria. The signal spreads quickly from cell to cell because of a striking property of cardiac muscle cells: They branch to make contact with several other cells, are joined end to end with these neighboring cells, and are physically connected by specialized structures called intercalated discs. Because these discs contain numerous gaps, electrical signals pass directly from one cell to the next. (Chapter 42 explores how electrical signals in animals are initiated and propagated.) As a result, the atria contract simultaneously and fill the ventricles.

Once the electrical signal has swept over the atria, however, it is conducted to an area called the AV (atrioventricular) node and delayed slightly before passing to the ventricles. The delay allows the ventricles to fill completely before they contract. After the delay, the electrical impulse is rapidly conducted through the walls of the ventricles. In response, the ventricles contract as the atria relax.

Heart Rate How fast does a heart beat? Heart rate is determined by the pace at which the SA node initiates electrical signals. This pace, in turn, is regulated by electrical signals from the brain. In this way, heart rate is controlled by the nervous system, even though the beat itself is generated in the heart muscle. As section 41.5 will show, signals from the brain change heart rate in response to changes in blood pressure or blood chemistry.

As you are probably aware from personal experience, heart rate can also change during certain emotional states. Heart rate may be altered by electrical signals from emotional centers in the brain or by signals from the chemical messengers called epinephrine and norepinephrine. These hormones are released from the adrenal glands in response to danger and initiate the fight-or-flight reaction analyzed in Chapter 44. (Note that epinephrine is also called adrenaline.)

Heart rate can also vary with exposure to drugs. For example, physicians have long noted that smoking and the use of smokeless tobacco are associated with a variety of abnormal heart rhythms. Early work established that nicotine—a prominent component of tobacco—stimulates the release of epinephrine and norepinephrine from the adrenal glands. But some studies indicated that nicotine has a direct effect on the heart as well—independent of hormone action. To explain these results, Huizhen Wang and colleagues hypothesized that nicotine might inhibit a membrane protein that is involved in electrical signaling by cardiac cells. Specifically, they proposed that nicotine might block a potassium ion channel called Kir2.

Why would a drug that disrupts a K^+ channel affect heart rate? As explained in Chapter 34, differences in ion concentrations across a cell membrane create a charge separation, or voltage. When membrane proteins allow ions to flow in response to this electrical potential, an electric current results. Wang and coworkers proposed that nicotine blocks a K+ channel that helps establish the normal charge separation in cardiac cells. If so,

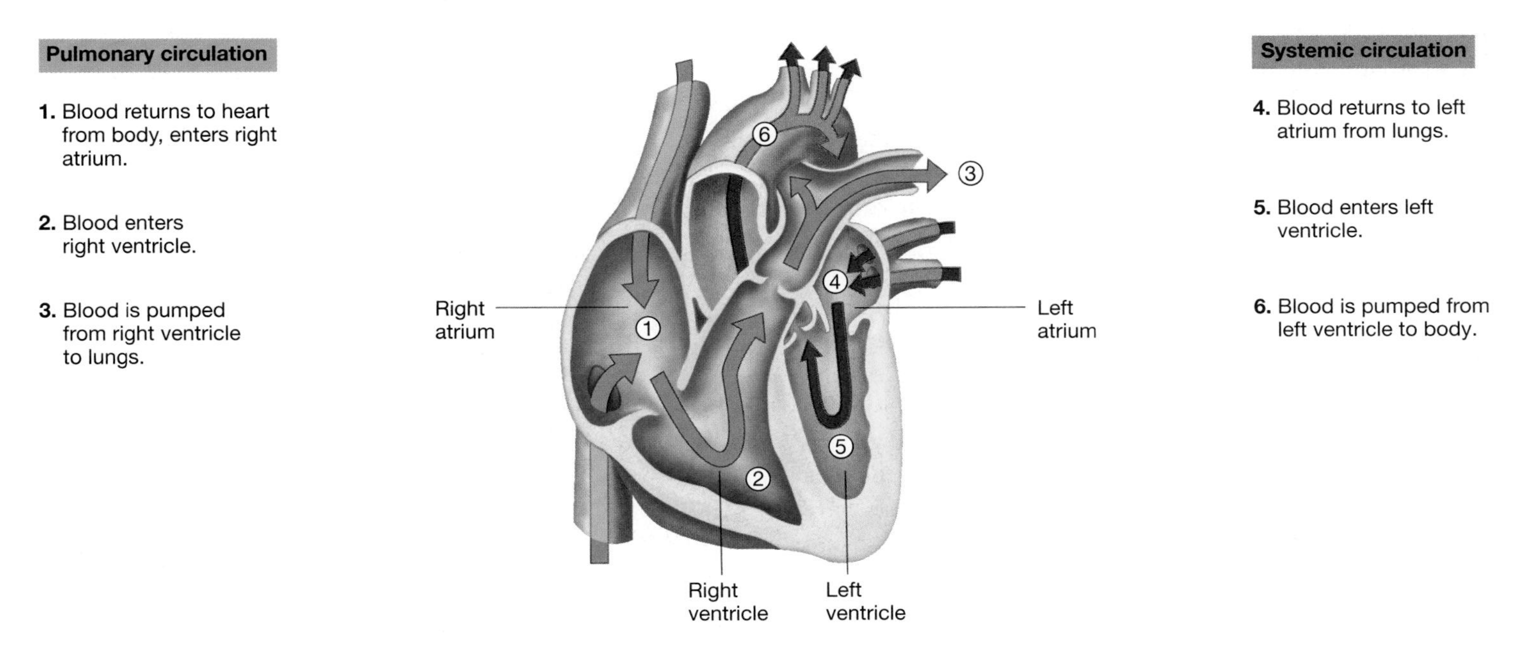

FIGURE 41.16 The Human Heart
Blood flows through a four-chambered heart in the sequence shown here. QUESTION As this sketch shows, the left ventricle is by far the largest and most muscular chamber in the heart. Why?

then cardiac cells would be less likely to transmit electrical signals properly. In this way, nicotine could throw off the cardiac cycle and result in abnormal heart rhythms.

To test this hypothesis, the biologists isolated copies of the Kir2 channel from humans and inserted them into frog eggs (**Figure 41.17a**). In one set of experiments, they compared the amount of electrical current that passed through the experimental cells when nicotine was present at different concentrations, compared to the amount of current that passed through the same cells in the absence of nicotine. As the data in **Figure 41.17b** show, increases in nicotine had a dramatic effect on the behavior of the Kir2 channel. Because nicotine blocks the channel, it is likely to disrupt normal electrical potentials in cardiac muscle. Based on results like these, it is not surprising that long-term exposure to nicotine can lead to an array of heart abnormalities.

Blood Pressure **Blood pressure** is the force that blood exerts on the walls of arteries, capillaries, and veins. As the top drawing in **Figure 41.18** shows, blood pressure changes dramatically as blood flows through the circulatory system. To understand why, note that arteries branch, re-branch, and eventually form the networks of capillaries diagrammed in the middle drawing of Figure 41.18. As a result, the total surface area of blood vessels in the circulatory system increases as shown in the bottom drawing of Figure 41.18, even though the size of each blood vessel decreases as the blood moves away from the heart. Because the same amount of fluid fills a larger and larger area, the pressure exerted by the fluid drops. In addition, the pressure exerted by blood is resisted by the walls of arteries and capillaries and leads to continuous losses in pressure due to friction.

Overall, blood pressure is highest in the artery that leads away from the left ventricle and lowest in the veins that return blood from the body to the right atrium. The pressure difference is typically about a factor of 10, from 100 mm Hg in the artery leading away from the heart to 10 mm Hg in the vein that returns blood from the body.

In arteries, blood pressure also varies in response to the cardiac cycle. Recall that when the left ventricle contracts, blood is ejected into the arterial system. The pulse of blood causes the arterial walls to stretch. When the ventricle relaxes, the artery returns to its original diameter. As a result, a pulse of high and low blood pressure propagates through the arterial system. Peak pressure is due to contraction of the ventricle and is called systolic pressure; the lowest pressure occurs at the end of ventricular relaxation and is called diastolic pressure. By the time the blood reaches the capillaries, though, pressure losses have dampened the systolic-diastolic difference. Consequently, blood pressure in capillaries and veins is constant instead of pulsating.

Arterial blood pressure is a particularly valuable diagnostic tool for physicians and is always reported as two values (see **Box 41.1**). The first number indicates the systolic pressure in mm Hg; the second figure is the diastolic pressure in mm Hg. For a healthy person, a typical blood pressure might be 120/80. People with blood pressures of 150/90 and greater are considered to be suffering from high blood pressure, or hypertension.

Hypertension is serious because it can lead to a variety of defects in the heart and circulatory system. Because of the stress it puts on arteries, abnormally high pressure increases the risk of heart attack, stroke, and burst or dilated blood vessels. As a result, a blood pressure reading—especially when combined with

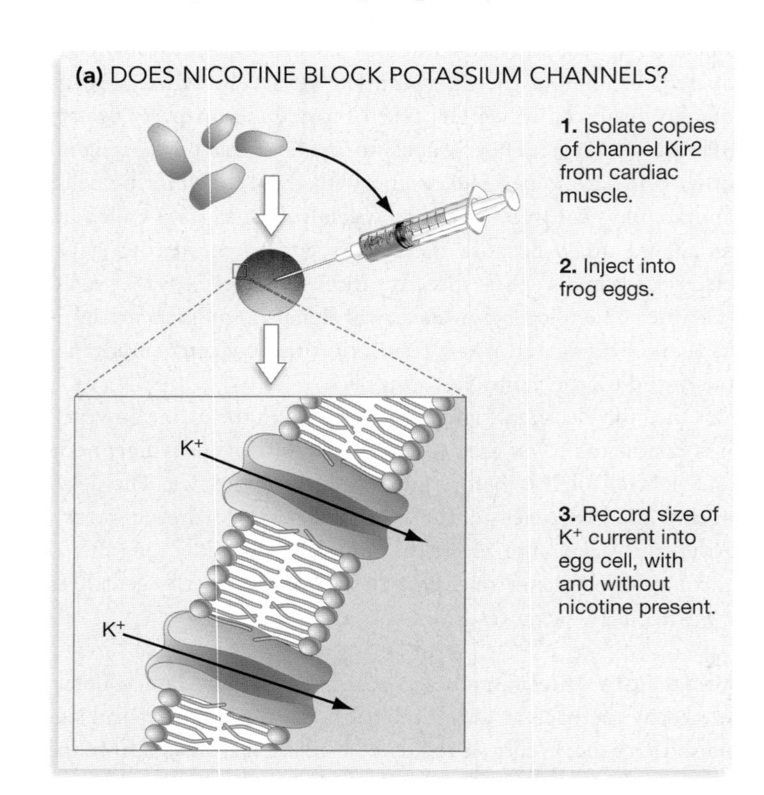

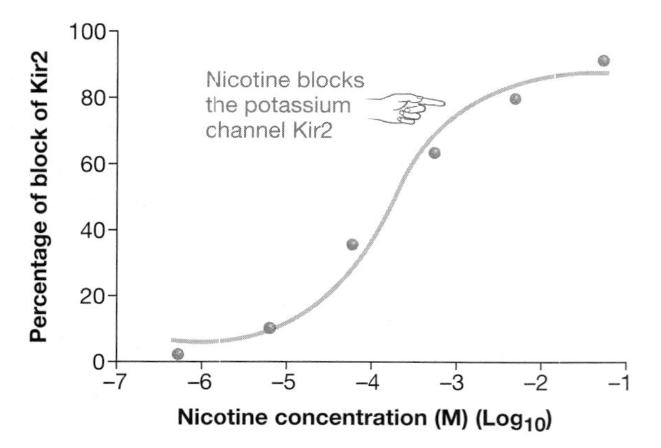

FIGURE 41.17 Does Nicotine Affect Electrical Signals in the Heart by Blocking Potassium Channels?
(a) Researchers used this protocol to test the hypothesis that nicotine disrupts normal heart rhythms by blocking certain potassium channels. **(b)** The data points from the experiment are plotted in orange; the gray line is a function that fits the data well. QUESTION Why did the researchers use frog eggs to test their hypothesis? Why didn't they use cardiac muscle cells?

a measurement of resting pulse rate and the cholesterol counts introduced in Chapter 40—gives clinicians relatively quick, accurate, and inexpensive data on the state of a patient's heart and circulatory systems.

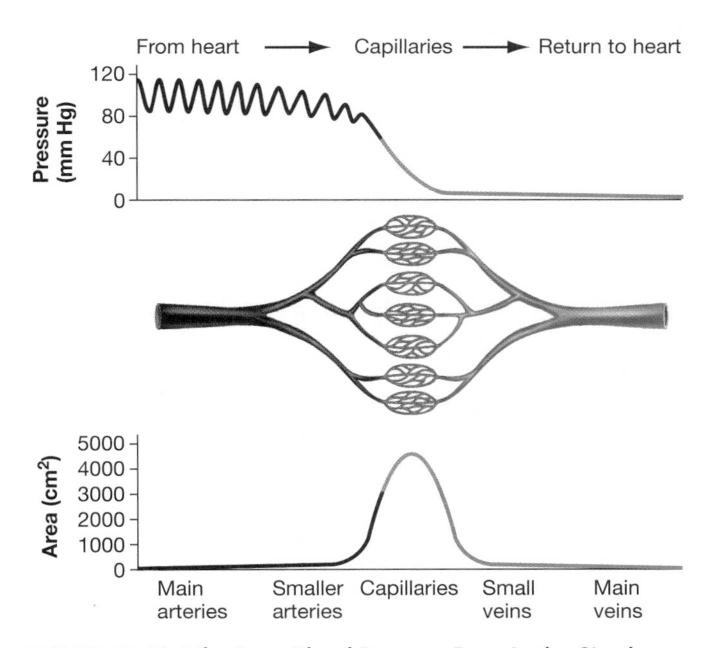

FIGURE 41.18 Why Does Blood Pressure Drop in the Circulatory System?
The graph at the top shows how blood pressure changes as blood leaves the heart, travels through a succession of branching arteries to capillaries, and moves from capillaries into a succession of veins that eventually return blood to the heart. Note that in arteries near the heart, each heartbeat makes blood pressure go up and down. These data were taken from blood vessels that show the branching pattern diagrammed in the middle drawing. The graph at the bottom plots the total area of blood vessels shown in the diagram.
EXERCISE After reading Box 41.1, label the systolic and diastolic pressures in the top graph.

41.5 Homeostasis in Blood Pressure and Blood Chemistry

An animal is in trouble if homeostasis with respect to blood pressure or blood chemistry fails. Chronic high blood pressure can cause a variety of problems; low blood pressure can lead to oxygen deficits in the brain and other tissues. Changes in blood chemistry—especially P_{O_2} and P_{CO_2}—are important because of their impact on ATP production and tissue pH.

Figure 41.19a (page 802) summarizes how the human body regulates homeostasis with respect to blood pressure. A key component of this system consists of sensory cells that are located next to an artery in the neck, an artery near the heart, and the wall of the heart. These cells act as baroreceptors, or pressure receptors. When they detect a change in blood pressure, they trigger electrical signals that change the heart rate, the stroke volume or amount of blood ejected with each contraction, and the diameter of blood vessels. As a result, homeostasis is restored.

Figure 41.19b shows the major components of the system that regulates homeostasis with respect to P_{O_2}, P_{CO_2}, and pH in humans. Changes in blood P_{O_2}, P_{CO_2}, and pH are detected by specialized nerve cells in the neck, near the heart, and in the brain region just above the spinal cord. When these cells detect a change in P_{O_2}, P_{CO_2}, or pH, they initiate electrical signals to the lungs and heart. Subsequent changes in the heart's output and the respiratory rate restore homeostasis.

How are the electrical signals responsible for achieving homeostasis transmitted? Chapter 42 is devoted to answering this question and to exploring how the brain and other components of the vertebrate nervous system are organized.

BOX 41.1 Measuring Blood Pressure

To measure a person's blood pressure, a clinician wraps an inflatable cuff around the upper arm. When the cuff is fully inflated, the pressure it exerts closes down the main artery in this part of the body. The clinician then listens for the sounds of bloodflow downstream from the cuff using a stethoscope.

At first, the clinician hears no pulse sounds through the stethoscope because no blood is flowing downstream of the cuff. As the clinician slowly releases air from the cuff, however, the pressure in the cuff begins to fall until it is below the systolic pressure in the artery. At this point, blood begins to spurt through with each systolic peak. The flow is turbulent, so the clinician can detect a thumping pulse in the stethoscope. The cuff pressure when sound is first detected is equal to systolic pressure. As more air is released from the cuff, its pressure falls below the diastolic pressure in the artery. At this point, blood flow in the artery is smooth and the sound of turbulent flow has faded away. The cuff pressure when the sound becomes muffled or disappears represents diastolic pressure.

(a) Homeostasis with respect to blood pressure

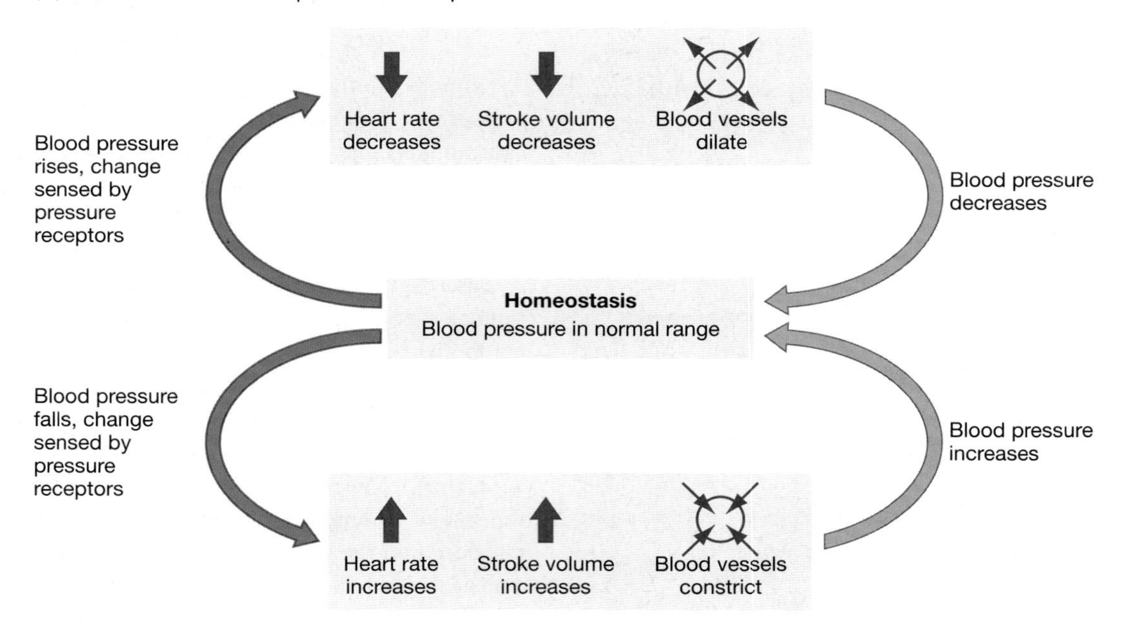

(b) Homeostasis with respect to P_{CO_2}, P_{O_2}, pH

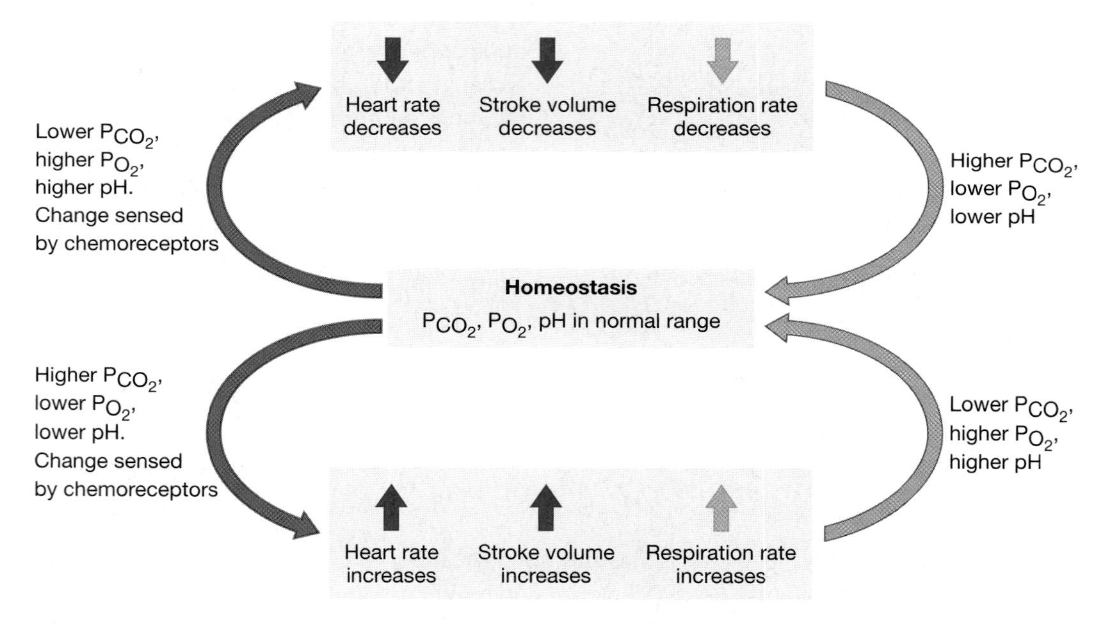

FIGURE 41.19 Homeostatic Systems Regulate Blood Pressure and Blood Chemistry

QUESTION Each of the red or orange arrows in this diagram indicates that an alteration in heart rate and stroke volume, blood vessel diameter, or respiration rate produces a change in blood pressure or blood chemistry. In each case, why do these changes occur?

Essay Smoking and Lung Function

People use tobacco primarily because they are addicted to a molecule called nicotine. Addiction is characterized by compulsive drug seeking, even when the user recognizes adverse health consequences. Recent research has shown that within 10 seconds of inhaling cigarette smoke, blood transports nicotine from the lungs to the brain. Once in the brain, nicotine activates cells associated with feelings of reward or pleasure. Because these effects are short-lived, cigarette smokers dose themselves repeatedly throughout the day. A typical addict takes 10 puffs from 30 cigarettes a day, meaning that he or she acquires an average of 300 "hits" of nicotine.

At least 43 ingredients in cigarette smoke are carcinogenic.

How do the other 4000 ingredients in cigarette smoke affect the body? Carbon monoxide, ammonia, nitrogen oxides, and formaldehyde are among the many toxic ingredients of tobacco smoke. Using test systems that quantify the tendency of particular compounds to cause mutations, researchers have determined that at least 43 ingredients in cigarette smoke are carcinogenic.

Tobacco smoke also has a dramatic effect on the respiratory surface. Soon after a person begins to smoke regularly, the lung tissue begins to swell. As smoking continues and the swelling worsens, airflow into the alveoli is reduced and gas exchange rates decrease. Mucus-producing cells that line the airway become hyperactive, producing excessive amounts of mucus. Toxins in smoke also kill or paralyze cilia that normally sweep debris and mucus out of the respiratory passages. As a result, mucus and debris build up in the lungs, leading to smoker's cough—a chronic and futile effort to clear the airway. Because accumulations of mucus and debris promote the growth of bacteria, smokers suffer from an increased incidence of respiratory infections.

Lifelong smokers are at high risk for developing emphysema, which develops when alveoli break down and lungs lose their elasticity. As emphysema advances, breathing becomes so impaired that physical activity becomes difficult. Eventually, the patient slowly suffocates to death. Male smokers are also 22 times more likely to develop lung cancer than nonsmokers; females are 12 times more likely. In the United States, lung cancer is the most deadly of all cancers.

When all the effects of nicotine addiction are taken into account, public health officials in the United States attribute over 430,000 deaths each year to tobacco use. This figure is larger than deaths attributed to alcohol, cocaine, heroin, homicide, suicide, car accidents, fire, and AIDS combined. Nicotine addiction also continues to grow as a public health problem. Even though it is not legal for minors in the United States to obtain tobacco products, surveys indicate that the prevalence of cigarette smoking among high school students in that country increased from 27.5 to 36.4 percent between 1991 and 1997. Currently, it is estimated that nearly 3000 minors start smoking each day in the United States.

Chapter Review

Summary

To support the production of ATP in mitochondria, animals have to obtain oxygen and rid themselves of carbon dioxide. As media for exchanging these gases, air and water are dramatically different. Compared to water, air contains much more oxygen and is much less dense. As a result, terrestrial animals have to process a much smaller volume to extract the same amount of O_2 and don't have to work as hard to do so. But because CO_2 is much more soluble in water than in air, aquatic animals have a much easier time ridding themselves of this waste product.

Both aquatic and terrestrial animals pay a price for exchanging gases. Land-dwellers lose water to evaporation during respiration; freshwater animals lose ions through diffusion; many marine animals lose water by osmosis.

Natural selection has resulted in the evolution of gills, trachea, lungs, and other gas-exchange organs that minimize the cost of respiration while maximizing the rate at which O_2 and CO_2 diffuse. Consistent with predictions made by Fick's law of diffusion, respiratory epithelia tend to be extremely thin and are highly folded to increase surface area. In fish gills, countercurrent exchange ensures that the partial pressure difference between O_2 and CO_2 in water and blood are high over the entire length of the respiratory surface.

Blood is a complex tissue that performs many functions. About half of the blood volume in a typical human consists of red blood cells that are specialized for gas exchange. Red blood cells lack most organelles but contain hundreds of millions of hemoglobin molecules. Each hemoglobin molecule can carry up to four oxygen molecules. The tendency of hemoglobin to give up these oxygen molecules varies as a function of the P_{O_2} in surrounding tissue in a sigmoidal fashion. As a result, a relatively small change in tissue P_{O_2} causes a large change in the amount of oxygen released from hemoglobin. In addition, oxygen binds less tightly to hemoglobin when pH is low. Because CO_2 tends to react with water to form carbonic acid, the existence of high CO_2 partial pressures in exercising tissues lowers their pH and makes oxygen less likely to stay bound to hemoglobin. The CO_2 that diffuses into red blood cells from tissues is rapidly converted to carbonic acid by the enzyme carbonic anhydrase. The protons that are released as carbonic acid dissociates bind to deoxygenated hemoglobin. In this way, hemoglobin acts as a buffer that takes protons out of solution and prevents large swings in blood pH.

In many animals, blood or hemolymph moves through the body via a circulatory system consisting of a pump (heart) and vessels. In open circulatory systems, overall pressure is low and tissues are bathed directly in hemolymph. In closed circulatory systems, blood is contained in vessels that form a continuous circuit and overall pressure is high. In humans, a four-chambered heart pumps blood into two circuits, which separately serve the lungs and the rest of the body. Although heart rate is controlled by electrical signals that originate in the heart itself, these signals can be modified by hormones or electrical impulses triggered by exercise, sensory information, or an emotional state. In addition, signals from receptor cells modify both heart rate and respiratory rate in a way that restores homeostasis with respect to blood pressure and blood chemistry.

Questions

Content Review

1. O_2 will diffuse from blood to tissue faster in response to which of the following conditions?
   a. an increase in the P_{O_2} of the tissue
   b. a decrease in the P_{O_2} of the tissue
   c. an increase in the thickness of the capillary wall
   d. a decrease in the surface area of the capillary

2. Which of the following does blood *not* do?
   a. transport O_2 and CO_2
   b. distribute body heat
   c. produce red blood cells (red blood cells) and other formed elements
   d. buffer against pH changes

3. How is most carbon dioxide transported in the blood?
   a. It is dissolved as a gas in the plasma.
   b. It is bound to amino groups on hemoglobin.
   c. It is bound to the heme group of hemoglobin.
   d. It is in the form of bicarbonate ion (HCO_3^-).

4. Which of the following are advantages of breathing air over breathing water?
   a. Air is less dense than water, so it takes less energy to move during respiration.
   b. Oxygen diffuses faster through air than it does through water.
   c. The oxygen content of air is greater than that of an equal volume of water.
   d. All of the above are advantages.

5. Which of the following promotes the release of oxygen from hemoglobin?
   a. a decrease in temperature
   b. a decrease in CO_2 levels
   c. a decrease in pH
   d. a decrease in carbonic anhydrase

6. An open circulatory system is less efficient than a closed circulatory system in what respect?
   a. It is harder to deliver O_2 to specific tissues based on need.
   b. Hemolymph does not contain respiratory pigments like hemoglobin.
   c. There is no heart to pump the blood.
   d. In closed systems, body movements cannot help circulate the blood.

Conceptual Review

1. Compare and contrast the respiratory and circulatory system of an insect and a human. What are some advantages and disadvantages of each system?

2. Why is the sigmoidal shape of the oxygen dissociation curve significant? What is the Bohr shift? Describe the changes in oxygen delivery that occur as a person changes from a resting state to intense exercise.

3. Explain how carbon dioxide is transported in the blood. In humans, why doesn't intense exercise and rapid production of CO_2 lead to a rapid reduction in blood pH?

4. When researchers compared the total amount of respiratory surface area in the gills of various species of fish, they found that species that do a lot of high-speed swimming have a larger total surface area than slow-moving species of the same size. Interpret this pattern in light of the theory of evolution by natural selection.

5. Review Fick's law of diffusion. Explain how each parameter in the equation is reflected in the structure of respiratory organs.

Applying Ideas

1. If a person at sea level is given pure oxygen to breathe (100 percent rather than 21 percent O_2), is transport of oxygen by the blood greatly increased? Explain why or why not.
2. At high altitude, it is not uncommon for people to suffer from pulmonary edema. This is an increase in the amount of fluid at the respiratory surface of the lungs. How might pulmonary edema affect gas exchange between the air and blood?
3. At high altitude, certain cells in humans increase production of 2,3-diphosphoglycerate (DPG). In blood, DPG shifts the oxygen-hemoglobin dissociation curve to the right. How does this shift affect O_2 unloading and loading?
4. Carp are fish that thrive in stagnant-water habitats with low oxygen partial pressures. Compared to many other fish species, carp hemoglobin has an extremely high affinity for O_2. Is this trait adaptive? If so, why?
5. Review the molecular structure of O_2 and CO_2 given in Chapter 2, and recall that solubility depends on electrical attraction between polar molecules. Why is CO_2 so much more soluble in water than O_2? Why are both gases much more soluble in cold water than warm water?
6. The carbon monoxide (CO) found in furnace and engine exhaust and in cigarette smoke binds to the heme groups in hemoglobin 210 times more tightly than does O_2. Explain why exposure to large doses of CO can lead to suffocation.

CD-ROM and Web Connection

CD Activity 41.1: Gas Exchange in the Lung and Tissues *(animation)*
(Estimated time for completion = 10 min)
What are the mechanisms that allow oxygen and carbon dioxide to move from the air to the deep tissues of our body?

CD Activity 41.2: The Heart (tutorial)
(Estimated time for completion = 15 min)
How does the heart circulate blood?

At your **Companion Website** (http://www.prenhall.com/freeman/biology), you will find self-grading exams and links to the following research tools, online resources, and activities:

Lung Disease Resources
This National Institutes of Health site provides descriptions of diseases that affect the lungs as well as information on other organizations and research sites.

Highlanders
This article features research into the biological differences of peoples living in high altitudes in an attempt to explain their adaptability to these extreme conditions.

The Heart
This site provides an online exploration of the anatomy and function of the heart.

Bug Blood
This article focuses on the circulatory system of insects, showing how different organisms with different physiological needs have evolved different anatomies and physiologies based on the niche they occupy.

Additional Reading

Lusis, A. J. 2000. Atherosclerosis. *Nature* 407: 233–241. Examines the genetic defects and environmental risk factors that can lead to heart disease and stroke.

Rubin, E. M., and A. Tall. 2000. Perspectives for vascular genomics. *Nature* 407: 265–269. Explores how genome-sequencing projects are helping researchers identify the genes responsible for vascular diseases.

Shadwick, R. E. 1998. Elasticity in arteries. *American Scientist* 86: 535–541. A look at how the anatomy of arteries allows them to respond to changes in blood pressure.

Electrical Signals in Animals

42

If you are unfortunate enough to touch a hot stove burner, your hand will jerk away before your brain perceives pain. When a trout realizes that it is about to be attacked by an osprey, its tail thrashes so rapidly that the movement is a blur. These and other animal movements are triggered by electrical impulses. Sensory cells in your finger and in the trout's eye send information to the brain and spinal cord, which then relay electrical signals to muscles in your arm or the trout's tail. Electrical signals are fast and precise. They are conducted from point to point by cells called **neurons** at speeds of up to 100 m/sec (225 mph). The result is movement that saves your finger from extensive tissue damage—or saves the trout's life.

The goal of this chapter is to explore how electrical signaling occurs. Section 42.1 introduces the general anatomy of a nerve cell, discusses why its membrane creates a voltage, and investigates how electrical currents are produced in cells. The next three sections focus on how neurons generate and propagate electrical signals, how information is transferred from cell to cell, and how neurons work in groups to integrate information from different sources. As we'll see, processes such as seeing, moving, and thinking are based on flows of ions across cell membranes.

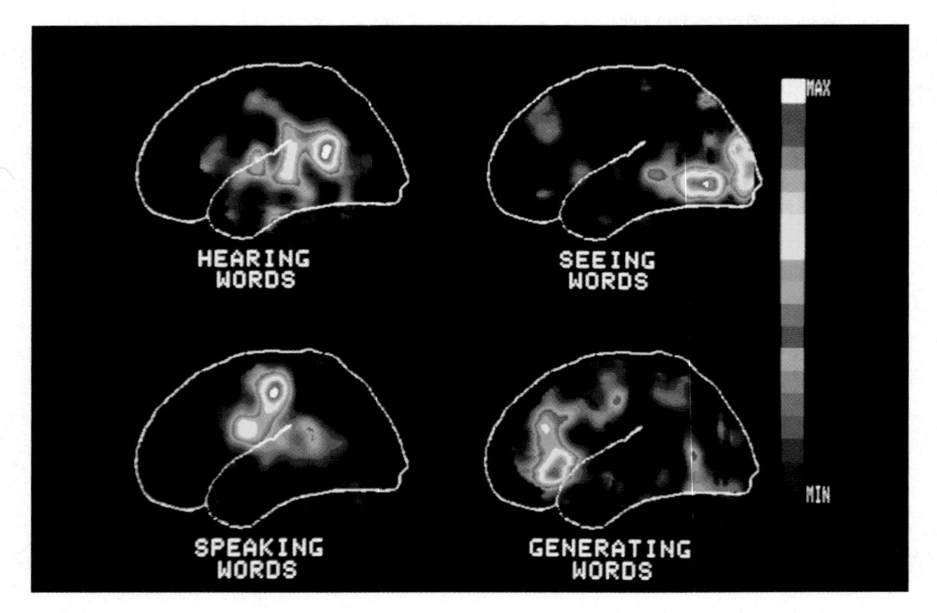

These positron-emission tomography (PET) scans show changes in the activity of brain cells during different tasks such as remembering words and speaking them aloud. Different brain areas are specialized for performing certain tasks. This chapter explores electrical signaling at a variety of levels, from individual ion channels to whole cells to regions of the brain.

42.1 Principles of Electrical Signaling

Nerve cells, or neurons, transmit information. In many cases the information they carry originates in sensory cells that are located on the periphery of the body. Receptors inside the skin, eyes, ears, and nose transmit streams of data about the external environment in the form of electrical signals. Similarly, sensory cells inside the body monitor conditions that are important in homeostasis, such as blood pH and oxygen levels. In this way, sensory cells monitor conditions inside and outside the body.

As **Figure 42.1a** shows, a receptor cell transmits the information it receives by means of a **sensory neuron**. (In many cases, the receptor cell also acts as a sensory neuron.) In vertebrates, the sensory neuron sends the information to neurons in the brain or spinal cord. Together, the brain and spinal cord form the **central nervous system**, or **CNS**. The function of the CNS is to integrate information from many sensory neurons and then, using **motor neurons**, send signals to effector cells in glands or muscles. All of the components of the nervous system that are outside the central nervous system are considered part of the **peripheral nervous system**, or **PNS**. Section 42.4 describes the structure and function of the PNS in detail.

In some cases, information from a sensory neuron bypasses the brain and travels directly to an effector. This is what happens when you touch a hot burner and jerk your hand back (**Figure 42.1b**). The **reflex** is fast and automatic. Later, when your brain integrates information from pain receptors and from looking at the burn, other motor neurons might carry out a response such as plunging your hand into cold water. What do the cells that carry and process all this information look like?

The Anatomy of a Neuron

Compared to studying cells in the kidney, digestive system, or lungs, documenting the anatomy of neurons was an extraordinarily difficult task. Neurons are small, transparent, and morphologically complex. So when Camillo Golgi discovered that some neurons become visible when samples of preserved tissue were treated with a solution containing silver nitrate, his finding was a major advance (**Figure 42.2a**, page 808).

Golgi published his initial observations in 1898; through the early decades of the twentieth century, his work and that of Santiago Ramón y Cajal revealed several important points about the anatomy of neurons. As **Figure 42.2b** shows, most neurons have the same three parts: a cell body, a highly branched group of short projections called dendrites, and one or more long projections called axons. Dendrites are rarely more than 2 mm long, but axons can be over a meter in length. The number of dendrites and their arrangement vary greatly from cell to cell, however. Further, many brain cells have only dendrites and lack axons.

What do axons and dendrites do? Long before the anatomists began their work, it had been established that neurons carry electrical signals. Based on their observations of stained cells, researchers suggested that a **dendrite** receives electrical signals from the axons and dendrites of adjacent cells, and that a neuron's **axon** then sends the signal to the dendrites of other neurons. Dendrites collect electrical signals; axons pass them on.

Golgi and Cajal disagreed about how these signals pass from neuron to neuron, however. Golgi thought that neurons were directly connected. His hypothesis was that neurons form a continuous network. In contrast, Cajal maintained that the cell membrane of each neuron is distinct and that the membranes of axons and dendrites meet at junctions called synapses. The dispute wasn't settled until the 1950s, when the advent

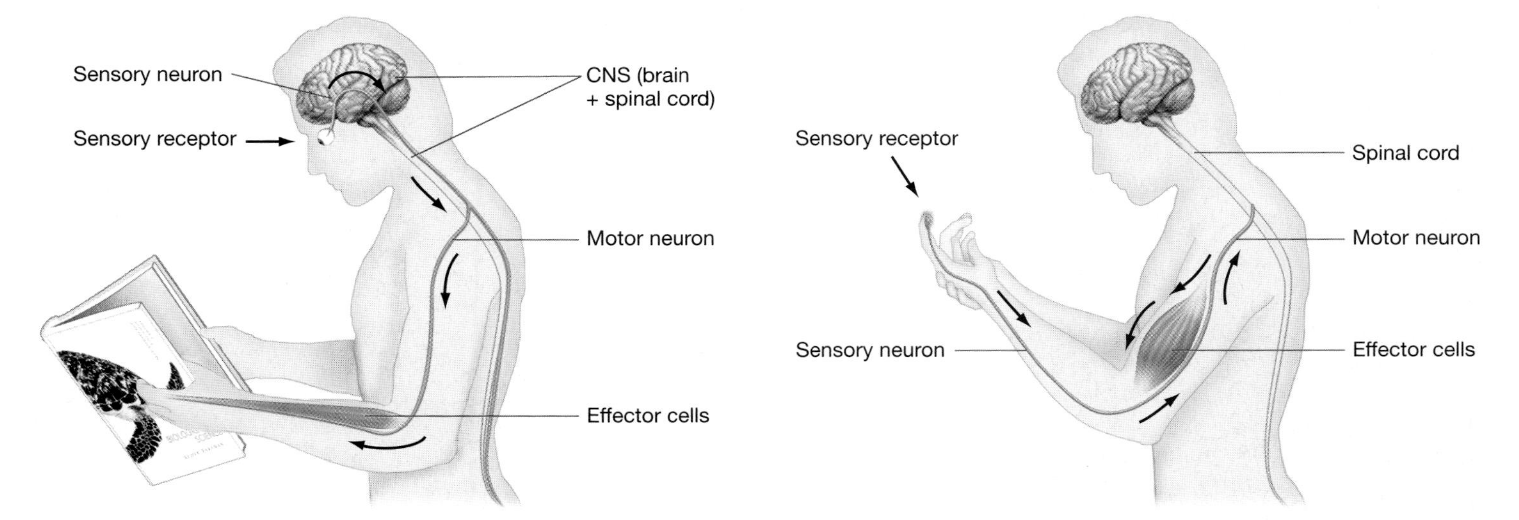

FIGURE 42.1 How Does Information Flow Through the Nervous System?
(a) In most cases, sensory neurons send information to the brain. There, the electrical signals are integrated with information from other sources. Once integration is complete, a response is sent to effector cells through motor neurons. **(b)** In a spinal reflex, sensory information triggers a response by effector cells directly—without integration by the brain. As a result, spinal reflexes are fast and involuntary.

of the electron microscope confirmed that only a small subset of neurons are directly connected; most interactions occur where the cell membranes of two neurons meet.

The basic message of early anatomical studies was that neurons are highly branched, and that each neuron makes many connections with other neurons. With these results in hand, the focus of research shifted to a new question. If these cells transmit electrical signals, how do they create these signals?

An Introduction to Membrane Potentials

Ions carry an electric charge. Because cytoplasm and extracellular fluids contain ions, cells are inherently electrical in nature. For example, if the positive and negative charges on ions that exist on either side of a cell membrane do not balance each other out, the membrane will have an electrical potential or voltage. A separation of charge across a cell membrane is called a **membrane potential.** If the difference between charges on either side of the membrane is large, so is the membrane voltage.

When charges are separated by a membrane, the ions on either side have potential energy. To convince yourself that this is true, consider that if the membrane were removed, ions would spontaneously move from the area of like charge to the area of unlike charge. A flow of charge is called an electric current. This movement would occur because like charges repel and unlike charges attract.

As Chapter 4 emphasized, however, ions move across membranes in response to concentration gradients as well as charge gradients. The combination of an electric gradient and a concentration gradient is called an electrochemical gradient. What do all of these facts have to do with neurons?

The Resting Potential

When a neuron is not transmitting an electrical signal but is merely sitting in extracellular fluid at rest, its membrane has a voltage called the **resting potential.** Why? To answer this question, consider that most of the large molecules inside cells are acids that tend to release a proton. Proteins, for example, are made up of amino acids. When acids release a proton, they acquire a negative charge. The major positively charged ion inside neurons is the potassium ion (K^+). In the extracellular fluid, however, the sodium ion (Na^+) and chloride ion (Cl^-) predominate.

Figure 42.3a shows the relative concentrations of organic anions, K^+, Na^+, and Cl^- inside and outside a resting neuron. Note that if each type of ion diffused across the membrane in accordance with its concentration gradient, organic anions and K^+ would leave the cell, while Na^+ and Cl^- would enter. But as earlier chapters indicated, ions cross membranes readily in only three ways: (1) by flowing along their electrochemical gradient through an ion channel; (2) by being carried along, via a membrane cotransporter, with an ion that experiences a strong electrochemical gradient; or (3) by being pumped against an electrochemical gradient by a membrane protein that hydrolyzes ATP. In resting neurons, only K^+ can cross the membrane easily along its concentration gradient. It does so through potassium channels.

As K^+ moves from the interior of the cell to the exterior through potassium channels, the inside of the cell becomes more and more negatively charged relative to the outside. Eventually, the buildup of negative charge inside the cell begins to attract K^+ and counteract the concentration gradient that favors movement of K^+ out. As a result, the membrane reaches a

(a) Neurons

(b) Information flow through neurons

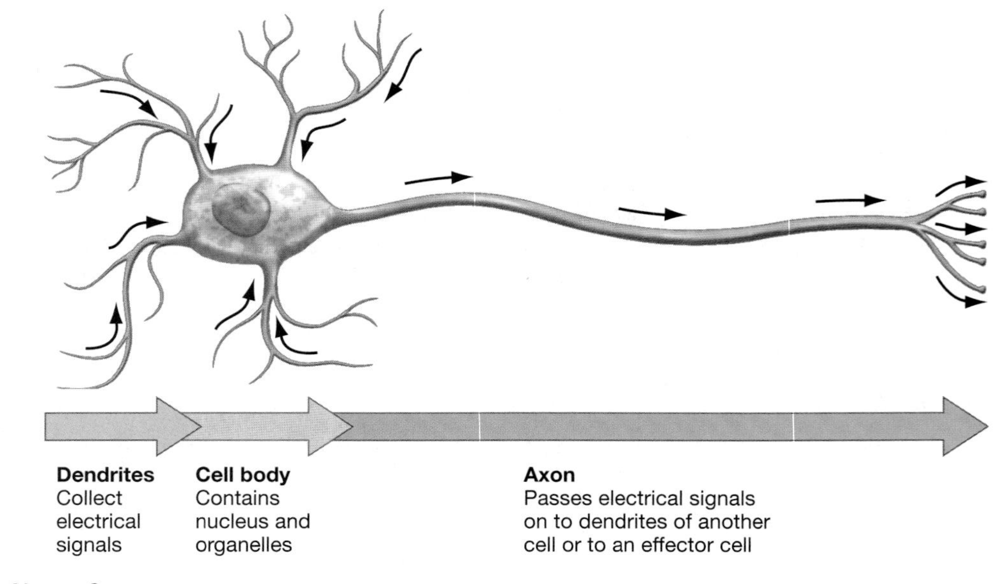

FIGURE 42.2 How Does Information Flow in a Neuron?
(a) Silver nitrate staining allowed researchers to visualize neurons and their connections. This slide shows neurons from the brain of a cat. **(b)** The structure of a generalized neuron. **EXERCISE** Choose a neuron in part (a). Label the cell body and axon. Circle points where the cell makes contact with other neurons.

voltage where there is an equilibrium between the concentration gradient that moves K^+ out and the electrical gradient that moves K^+ in. At this voltage, there is no longer a net movement of K^+. This voltage is called the **equilibrium potential** for K^+. **Box 42.1** (page 810) explains how the equilibrium potential for an ion is calculated.

Although Cl^- and Na^+ cross the cell membrane much less readily than K^+, some movement of these ions also occurs. The large anions inside the cell, in contrast, are not able to cross the membrane at all. In combination, Cl^-, Na^+, and K^+ each cross the membrane in response to their concentration and charge gradients. When the concentrations of the three ions are at equilibrium, the resting potential results. How do biologists measure it?

Using Microelectrodes to Measure Membrane Potentials

During the 1930s and 1940s, A. L. Hodgkin and Andrew Huxley helped pioneer the study of electrical signaling in animals. Their work focused on what has become a classic model system in the study of electrical signaling: the giant axons of squid. Squid are ocean-dwellers that are preyed upon by fish and whales. When a squid is threatened, electrical signals travel down the giant axon to muscle cells. When these muscles contract, water is expelled from a cavity in the squid's body. As a result, the squid lurches away from danger by jet propulsion.

Hodgkin and Huxley decided to study the squid's giant axon simply because it is so large. Many of the axons found in humans are a mere 2 μm in diameter, but the squid giant axon is about 500 μm in diameter. The squid axon was large enough that the researchers could record membrane potentials by inserting a wire down its length. Later, Gilbert Ling and Ralph Gerard developed glass microelectrodes that were small enough to be inserted into the axon and record membrane voltage.

To record membrane voltages in the squid's giant axon, Hodgkin and Huxley would dissect a section of the axon, bathe it in seawater or a solution of known composition, and insert a microelectrode as shown in **Figure 42.3b**. Their earliest recordings suggested that the squid's giant axon has a resting potential of about −45 mV (millivolts), but later work documented that most neurons have a resting potential of around −70 mV. (Membrane potentials are always reported as inside-cell relative to outside-cell.) In their earliest experiments, Hodgkin and Huxley also confirmed that the resting potential can be disrupted by an event called the action potential.

What Is an Action Potential?

Figure 42.4 (page 810) shows the form of the action potential that Hodgkin and Huxley recorded from the squid's giant axon. As the drawing to the right of the trace indicates, the action potential has three distinct phases. The initial event is a rapid **depolarization** of the membrane. Current flow causes the inside of the membrane to become less negative and then positive with respect to the outside. (A membrane is said to be polarized if the charges on either side are different. Depolarization means that the membrane becomes less polarized.) When the membrane potential reaches about +40 mV, an abrupt change occurs. During the second phase of the action potential,

(a) Separation of charge creates the resting potential.

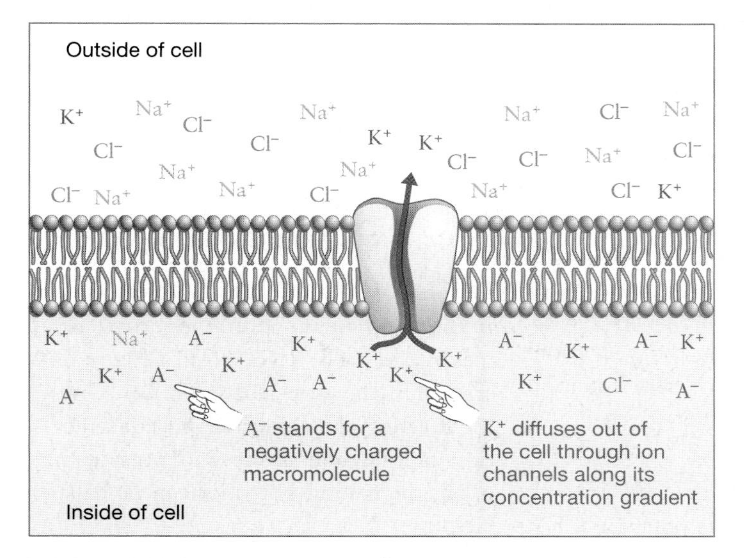

(b) Measuring membrane potentials

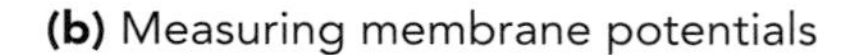

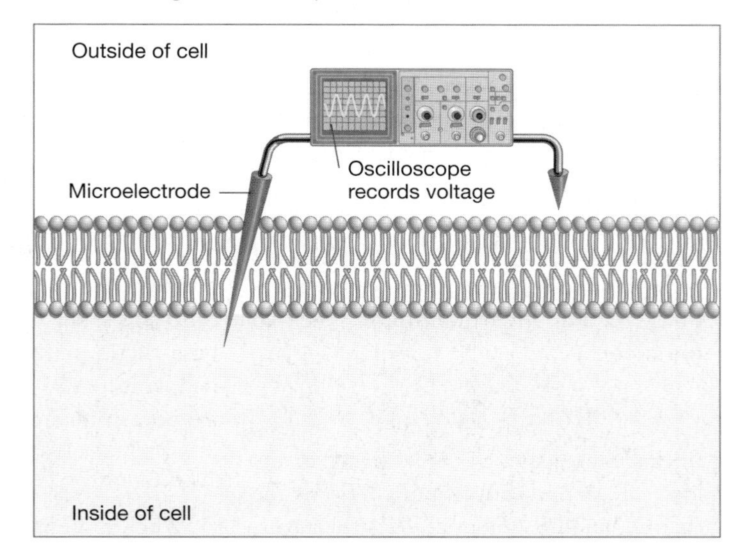

FIGURE 42.3 The Resting Potential
(a) In resting neurons, the inside of the cell is negative relative to the outside because the membrane is selectively permeable to K^+. As K^+ leaves the cell along its concentration gradient, the inside becomes negatively charged relative to the outside. **(b)** To measure a neuron's membrane potential, researchers insert a microelectrode into the cell and compare the reading with the voltage outside the cell.

the membrane experiences a rapid repolarization. The repolarization event actually results in the membrane becoming more negative than the resting potential. As a result, biologists call the final part of the action potential the undershoot. The entire event takes less than a millisecond.

Hodgkin and Huxley made other important observations about the action potential. In addition to being fast, it is an all-or-none event. There is no such thing as a partial action potential, and all action potentials for a given neuron are identical in size and shape. The biologists also found that action potentials could be triggered by artificially depolarizing the membrane with an injection of electrical current through a microelectrode. Finally, action potentials are propagated down the length of the axon. When an impulse was recorded at a particular point on a squid axon, an action potential that was identical in shape and size would be observed farther down the same axon soon afterward.

Taken together, these observations suggested a mechanistic basis for electrical signaling. In the nervous system, information is coded in the form of action potentials that travel down neurons. The frequency of action potentials—not their size—is the meaningful signal. In the squid's giant axon, action potentials signal muscles to contract. As a result, the animal escapes from danger.

42.2 Dissecting the Action Potential

Once intracellular recording was possible, an entirely new field of research opened up. The most urgent question was to understand which ions are involved in the currents that form the action potential. Are Na^+, Cl^-, or K^+ responsible for the depolarization

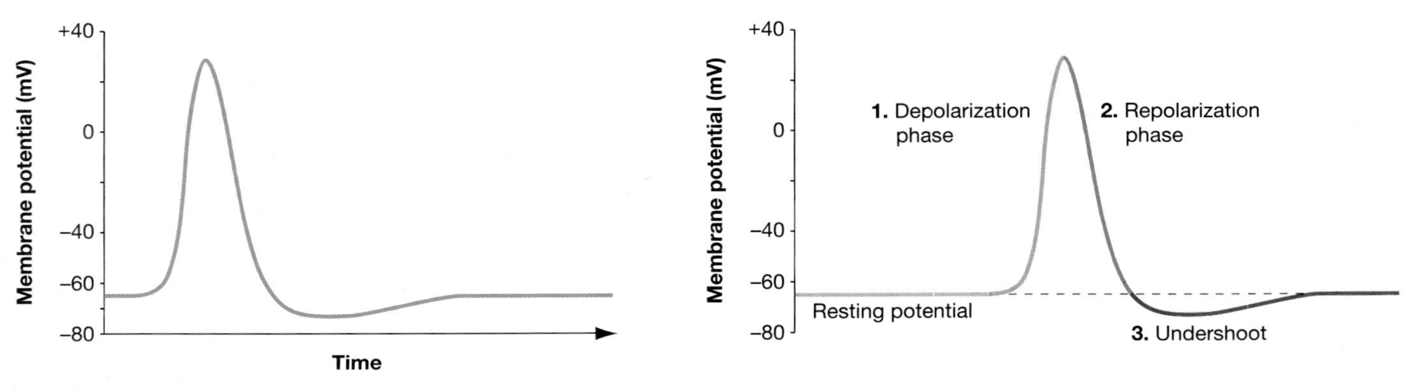

FIGURE 42.4 An Action Potential
The trace at the left shows a typical action potential. All action potentials have the same general shape shown here.

BOX 42.1 The Nernst Equation and the Goldman Equation

The equilibrium potential is the voltage that is required to counteract the tendency for an ion to cross a membrane in response to its concentration gradient. At the equilibrium potential for a given ion, there is no net movement of that ion across the membrane. The Nernst equation specifies the equilibrium potential for an ion as:

$$E_{ion} = \left(\frac{RT}{zF}\right) \ln \frac{[ion]_o}{[ion]_i}$$

In this expression, z is the valence of the ion (e.g., +1 for potassium), RT/F is the thermodynamic potential, *ln* is the natural logarithm, and $[ion]_o$ and $[ion]_i$ are the concentration of the ion outside and inside the cell, respectively. The three terms in the thermodynamic potential are the gas constant (R), which acts as a constant of proportionality, the absolute temperature (T), and the faraday (F), which specifies the amount of charge carried by a mole of a univalent ion. The thermodynamic potential specifies the voltage required to balance an e-fold concentration ratio across the membrane, where e is the base of the natural logarithm. At room temperature (25°C), the Nernst equation for potassium simplifies to:

$$E_{ion} = 58 \log \left(\frac{[K^+]_o}{[K^+]_i}\right)$$

What is the membrane potential when K^+, Na^+, and Cl^- are each at their equilibrium potential? David Goldman realized that the answer depended on the permeability of the membrane to the three ions. If these permeabilities are symbolized as p_K, p_{Na}, and p_{Cl}, then the membrane potential is given by the following expression:

$$V_m = 58 \log \left(\frac{[K^+]_o + (p_{Na}/p_K)[Na^+]_o + (p_{Cl}/p_K)[Cl^-]_i}{[K^+]_i + (p_{Na}/p_K)[Na^+]_i + (p_{Cl}/p_K)[Cl^-]_o}\right)$$

In the squid's giant axon, the relative permeabilities for K^+, Na^+, and Cl^- are described by the ratio 1.0 : 0.03 : 0.1. Given these ratios and the observed concentration of each ion inside and outside the axon, the resting potential can be calculated as:

$$V_m = 58 \log \frac{10 + (0.03)460 + (0.1)40}{400 + (0.03)50 + (0.1)540} = -70 \text{ mV}$$

This is very close to the observed value.

and repolarization phases of the event? One early hypothesis was that the action potential resulted from a temporary breakdown in the membrane's selective permeability. The idea was that channels opened and allowed a free flow of Na^+ and Cl^- as well as K^+. But based on observed calculations for the concentration of each ion on either side of the squid's giant axon, the all-channels-open hypothesis predicted that the peak of the action potential should occur at about 0 mV, not the +50 mV observed.

In contrast, Hodgkin noted that +50 mV corresponded to the equilibrium potential for Na^+. This insight led to the hypothesis that the depolarization phase of the action potential resulted from an influx of sodium ions. How could this hypothesis be tested?

Distinct Ion Currents Are Responsible for Depolarization and Repolarization

To dissect the currents responsible for the action potential, Hodgkin and Huxley recorded electrical activity in a squid axon bathed in seawater. Then they replaced the seawater with an isotonic solution containing an uncharged and metabolically inert sugar called dextrose. **Figure 42.5** shows the result of this experiment. Washing Na^+ out of the solution surrounding the axon abolished action potentials. But when the biologists replaced the dextrose solution with seawater, action potentials again began traveling down the neuron. Further, when they did the same experiment using solutions with various concentrations of Na^+ instead of seawater, they found that the peak of the action potential tracked the equilibrium concentration of Na^+ based on its concentration outside the cell. Taken together, these experiments furnished strong evidence that the action potential begins when Na^+ flows into the neuron.

What ion flow is responsible for the repolarization phase? Using radioactive K^+, Hodgkin and Huxley showed that there was a strong flow of potassium ions out of the cell during the repolarization phase. Thus, the action potential consists of a strong inward flow of Na^+ followed by a strong outward flow of K^+. The questions then became, which ion channels are responsible for these currents, and how do they work?

Voltage-Gated Channels

When ion channels were introduced in Chapter 4, they were presented as simple, static proteins that facilitate a constant stream of ions. But research on the action potential suggests that certain ion channels open and close in response to changes in membrane voltage. To capture this point, biologists refer to them as **voltage-gated channels**. The idea is that the conformation of these proteins changes in response to the charges present at the membrane surface, and that conformational changes lead to the channel opening or closing.

To study the relationship between membrane voltage and channel activity, Kenneth Cole and co-workers came up with a way to hold the membrane potential of a neuron constant, even while ions were flowing through channels in the membrane. Their apparatus employed an electrical feedback system, which injected an artificial current that was equal in size and opposite in direction to the current that flowed naturally across the membrane. Using this technique, called **voltage clamping**, Cole and associates could hold an axon at any voltage they desired and record the electrical currents generated by flows of Na^+ and K^+ across the membrane.

CD ACTIVITY 42.2 Action Potential

Hodgkin and Huxley quickly adopted the new equipment and produced data such as those shown in **Figure 42.6** (page 812). The top record is from an experiment in which they voltage-clamped a squid axon at −9 mV. The next record shows the current that resulted from the depolarization. As you study this trace, note that there is a rapid inward-directed current followed by a slower outward-directed current. The two records at the bottom dissect this overall current into the components due to an inward flow of Na^+ and outward flow of K^+, respectively.

This experiment, and others like it, confirmed that Na^+ and K^+ channels open in response to changes in voltage. The data also answered an important question. If Na^+ flows in and K^+ flows out, why don't the two currents cancel each other out? The key observation is that the potassium current is delayed. Because potassium channels open more slowly than sodium channels, the outward flow of K^+ occurs after sodium ions have entered the cell and eliminated the electrochemical gradient favoring Na^+ entry.

To summarize, early work on electrical signaling showed that action potentials result from the staggered activity of

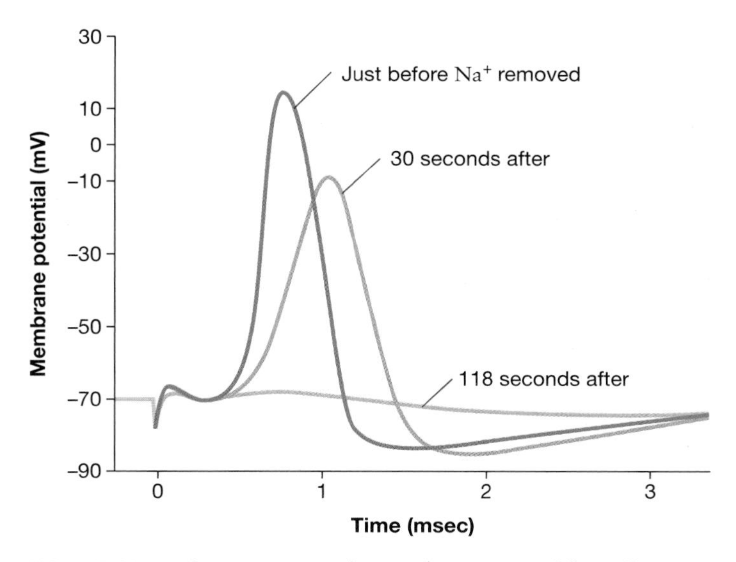

FIGURE 42.5 What Happens When Na^+ Is Removed from the Solution Surrounding a Neuron?
These action potentials were recorded at three different intervals during removal of Na^+ from the solution surrounding a squid's giant axon. The word *seconds* in the labels for the middle and lower traces refers to the number of seconds after the solution containing Na^+ was flushed away with a solution containing isotonic dextrose.
QUESTION Why was it important that the researchers used an isotonic solution of dextrose?

voltage-gated Na^+ and K^+ channels. Once these results were established, attention turned to the channels themselves. How do these proteins work?

Patch Clamping and Studies of Single Channels Studying individual ion channels became possible when Erwin Neher and Bert Sakmann perfected a technique known as patch clamping. As **Figure 42.7a** shows, the researchers patch clamped a membrane by touching it with a fine-tipped microelectrode and then applying suction. The suction seals the membrane against the glass so that no current leaks out. Using this technique, they documented the currents that flowed through individual channels. The recordings shown in **Figure 42.7b** are typical, and they make several important points about the nature of voltage-gated channels.

- The shape of the current traces supports the hypothesis that voltage-gated channels are either open or closed. There is no gradation in channel behavior.
- Sodium channels open quickly after depolarization. They stay open for about a millisecond and then close. Once Na^+ reaches its equilibrium potential, sodium channels are inactivated. **Figure 42.7c** shows a simplified model for how voltage-gated sodium channels open.
- Potassium channels open slowly after depolarization. They continue to blink open and closed until the membrane repolarizes. Once the membrane returns to the resting potential, they remain closed.

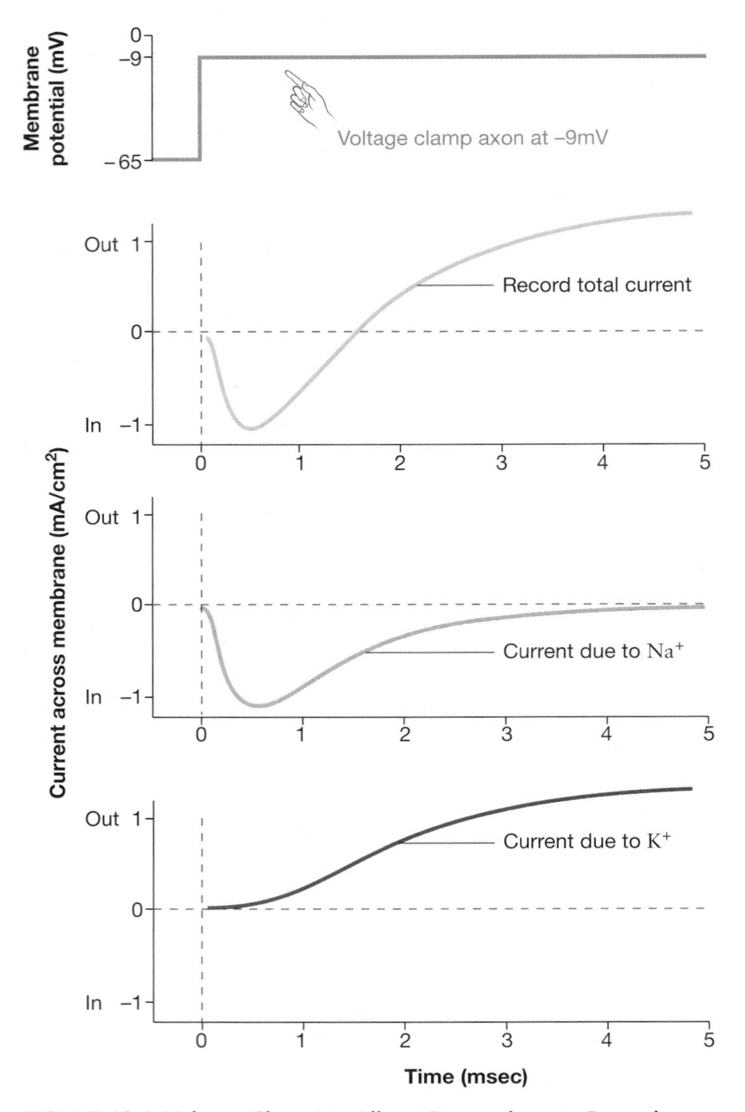

FIGURE 42.6 Voltage Clamping Allows Researchers to Record Current Flows Across Membranes
When Hodgkin and Huxley voltage clamped a squid's giant axon at −9 mV, they recorded the flow of current shown in the second trace. Current initially flowed into the axon and then back out. As the bottom two traces show, this overall current is due to a rapid flow of Na^+ into the cell and a slower flow of K^+ out of the cell.

Observations on the behavior of sodium channels also helped to explain why the action potential is an all-or-none event. The key observation is that sodium channels are more likely to open as a membrane depolarizes. An initial depolarization thus leads to the opening of more sodium channels, which depolarizes the membrane further and leads to the opening of additional Na^+ channels. In this way, the opening of sodium channels exhibits positive feedback. **Positive feedback** takes place when the occurrence of an event makes the same event more likely to recur. When a fuse is lit, for example, the heat generated by the oxidation reaction accelerates the reaction, which generates still more heat and leads to additional reactions.

Patch clamping was not the only technical advance that helped researchers explore the dynamics of voltage-gated channels. The discovery of neurotoxins from sources as diverse as venomous snakes and foxglove plants also served as important experimental tools.

Using Neurotoxins to Identify Channels and Dissect Currents Many of the toxins produced by poisonous animals and plants cause convulsions, paralysis, or unconsciousness when they are ingested. Based on these symptoms, physicians and researchers suspected that at least some poisons affect neuron function. Toshio Narahashi and colleagues helped confirm this hypothesis by treating giant axons from lobsters with the tetrodotoxin found in puffer fish.* Although the researchers' experiments showed that the resting potential was normal, action potentials were abolished. More specifically, they documented that the outward-directed K^+ current was normal, but that the inward-directed Na^+ flow was wiped out. Based on this result, they concluded that puffer-fish toxin specifically blocks the voltage-gated Na^+ channel.

*Puffer fish are considered a delicacy in Japan. The toxin that these fish synthesize and sequester in their tissues is so dangerous that professional chefs are trained and certified in safe preparation techniques. In one recent 10-year study period, however, there were 646 reported cases of puffer-fish poisoning in Japan, with 179 fatalities. Most of the poisoning events involved home chefs who were attempting to prepare puffer fish for dinner.

(a) Patch clamping isolates ion channels.

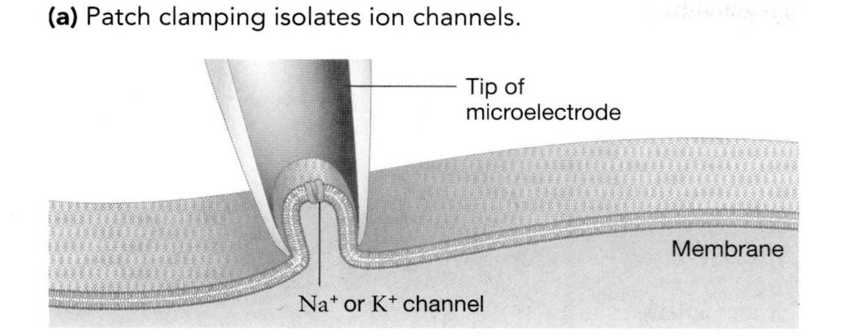

(b) The current of isolated channels can be measured.

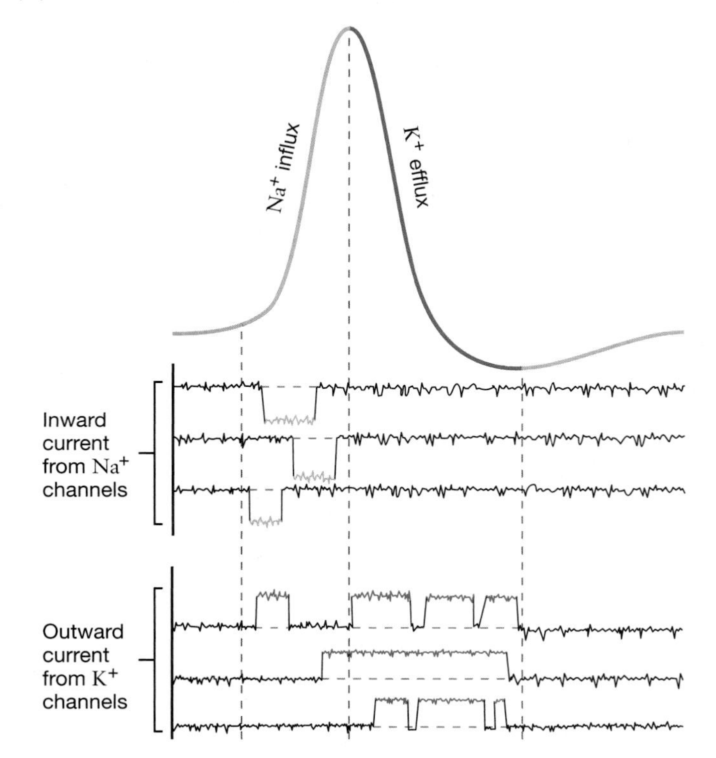

(c) At the resting potential, voltage-gated Na+ channels are closed.

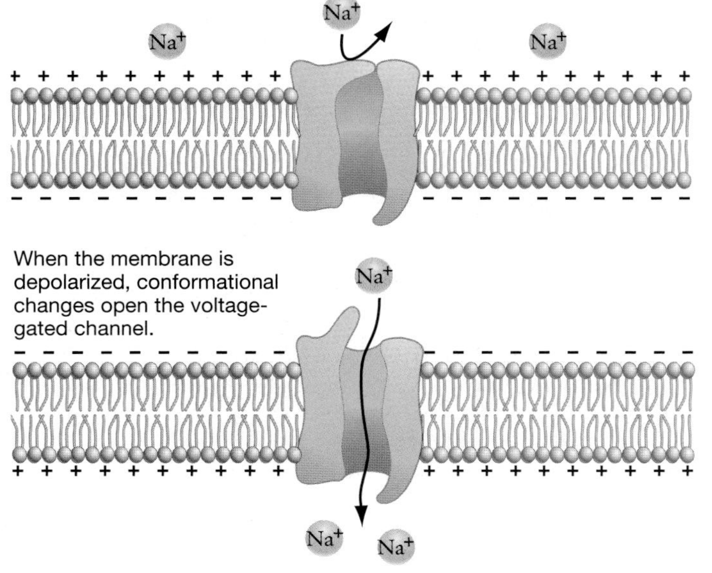

FIGURE 42.7 Patch Clamping Allows Researchers to Record from Individual Channels
(a) Patch clamping depends on the use of extremely fine-tipped microelectrodes. The goal is to isolate one channel and record from it. **(b)** The current traces shown here are from three voltage-gated sodium channels and three voltage-gated potassium channels. Note that no current flows through either type of channel at the resting potential. **(c)** Changes in the conformation of voltage-gated channels are responsible for changes in a neuron's permeability to Na^+ and K^+.

In contrast, when researchers treated neurons with the venom of black mamba snakes (**Figure 42.8a**, page 814), preliminary data hinted that the K^+ channel was blocked. To confirm this result, Evelyne Benoit and Jean-Mark Dubois voltage-clamped neurons that had been dissected from frogs, bathed the neurons in a solution containing black-mamba poison, varied the membrane potential, and studied the K^+ currents that resulted. As predicted, the poison had a dramatic effect on permeability to K^+. But as the data in **Figure 42.8b** show, the venom did not stop the current completely. Instead, the graph suggests that a specific K^+ channel was poisoned, but that other potassium channels still allowed an outflow of current. The existence of multiple K^+ channels in axons was confirmed by patch-clamping studies that documented K^+ flowing through several distinct types of potassium channels.

The Role of the Sodium-Potassium Pump

Experiments with neurotoxins helped establish that most neurons contain just one type of sodium channel but several different potassium channels. But of all the toxins that poison ion channels, perhaps the most important is a molecule from foxglove plants. This compound is called ouabain and was introduced in Chapter 39. Ouabain wipes out the resting potential of neurons. This was an extremely important observation. To understand why, recall that sodium ions enter the neuron during an action potential and that potassium ions leave. Although the number of ions that flow across the membrane during an action potential is tiny, the concentration gradients that favor an inflow of Na^+ and an outflow of K^+ would eventually decline if enough action potentials occurred. In addition, K^+ leaks out of neurons when they are at rest. To maintain the concentration gradients in these ions, then, a protein that hydrolyzes ATP must pump Na^+ out and K^+ in. This is an energy-demanding process because the ions are moved against their electrochemical gradients.

Does ouabain poison this hypothetical pump? Jens Skou was able to answer this question when he succeeded in isolating a protein from crab neurons that began hydrolyzing ATP if both Na^+ and K^+ were present (**Figure 42.9a**, page 814). When Skou added ouabain to this reaction mix, the enzyme's activity stopped. Soon after, Robert Post and co-workers used radioactive Na^+ and K^+ to establish that Skou's enzyme, which is called **Na^+/K^+-ATPase**, transports three sodium ions out of cells for every two potassium ions transported inward (**Figure 42.9b**).

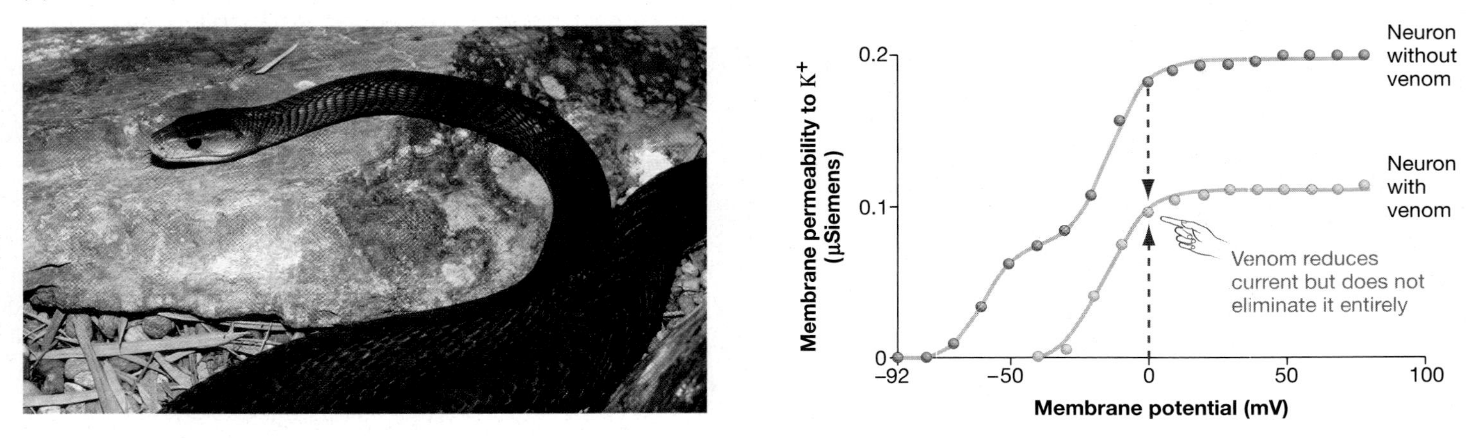

FIGURE 42.8 Using Poisons to Dissect Ion Currents
(a) Black mambas are among the most dangerous snakes in the world. Untreated bites are almost always lethal—often within 30 minutes. **(b)** The graphs plot the permeability of neurons to K^+ as a function of membrane voltage. Each data point was generated by measuring membrane permeability while the membrane was voltage-clamped at a particular potential.

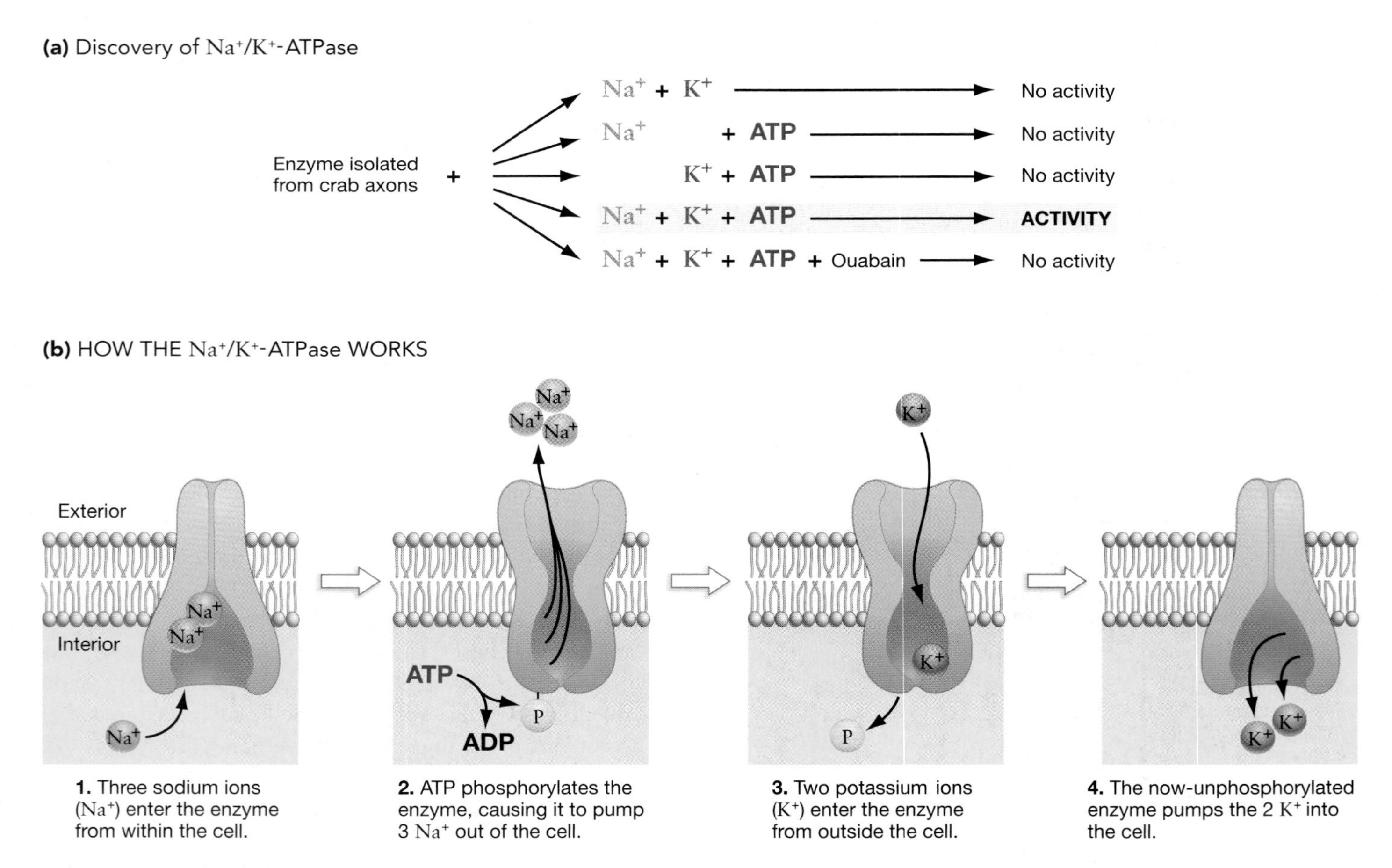

FIGURE 42.9 Na^+/K^+-ATPase
(a) The reactions listed here tested whether an enzyme isolated from crab neurons was a sodium-potassium pump. **(b)** By following radioactive Na^+ and K^+, biologists found that the pump transports 3 Na^+ out of the cell for every 2 K^+ brought in. This diagram is a model of how the pump works.

With the discovery and characterization of Na^+/K^+-ATPase, researchers added an important piece to the puzzle of how action potentials are generated. The sodium-potassium pump restores normal concentrations of Na^+ and K^+ after action potentials have brought sodium ions into the cell and sent potassium ions out. The pump is responsible for maintaining the concentration gradients required for the resting potential.

The experiments reviewed thus far haven't considered a fundamental question about electrical signaling, however. How do action potentials move down the axon?

How Is the Action Potential Propagated?

To explain how action potentials travel down an axon, Hodgkin and Huxley hypothesized that the influx of Na^+ causes charge to spread away from sodium channels. As **Figure 42.10a** shows, positive charges inside the cell are repulsed by the influx of Na^+ and negative charges are attracted. As a result, sections of the membrane close to the site of an action potential are depolarized. Nearby voltage-gated Na^+ channels pop open in response, positive feedback occurs, and a full-fledged action potential results. In this way, an action potential is continuously regenerated as it moves down the axon.

Why don't action potentials propagate back up the axon as well? The answer is that Na^+ channels are refractory. Once they have opened and closed, they are less likely to open again for a short period of time. The presence of the undershoot phase, in which the membrane is briefly hyperpolarized relative to the resting potential, also keeps the charge that spreads "upstream" from triggering an action potential in that direction.

Understanding how the action potential propagates helped researchers explain why the squid's giant axons are so large. Because charge spreads farther in a membrane with a large surface area, the squid's giant axon and other large neurons transmit action potentials much faster than small axons.

Analyzing charge spread also helped biologists explain the phenomenon called myelination. Vertebrates do not have giant

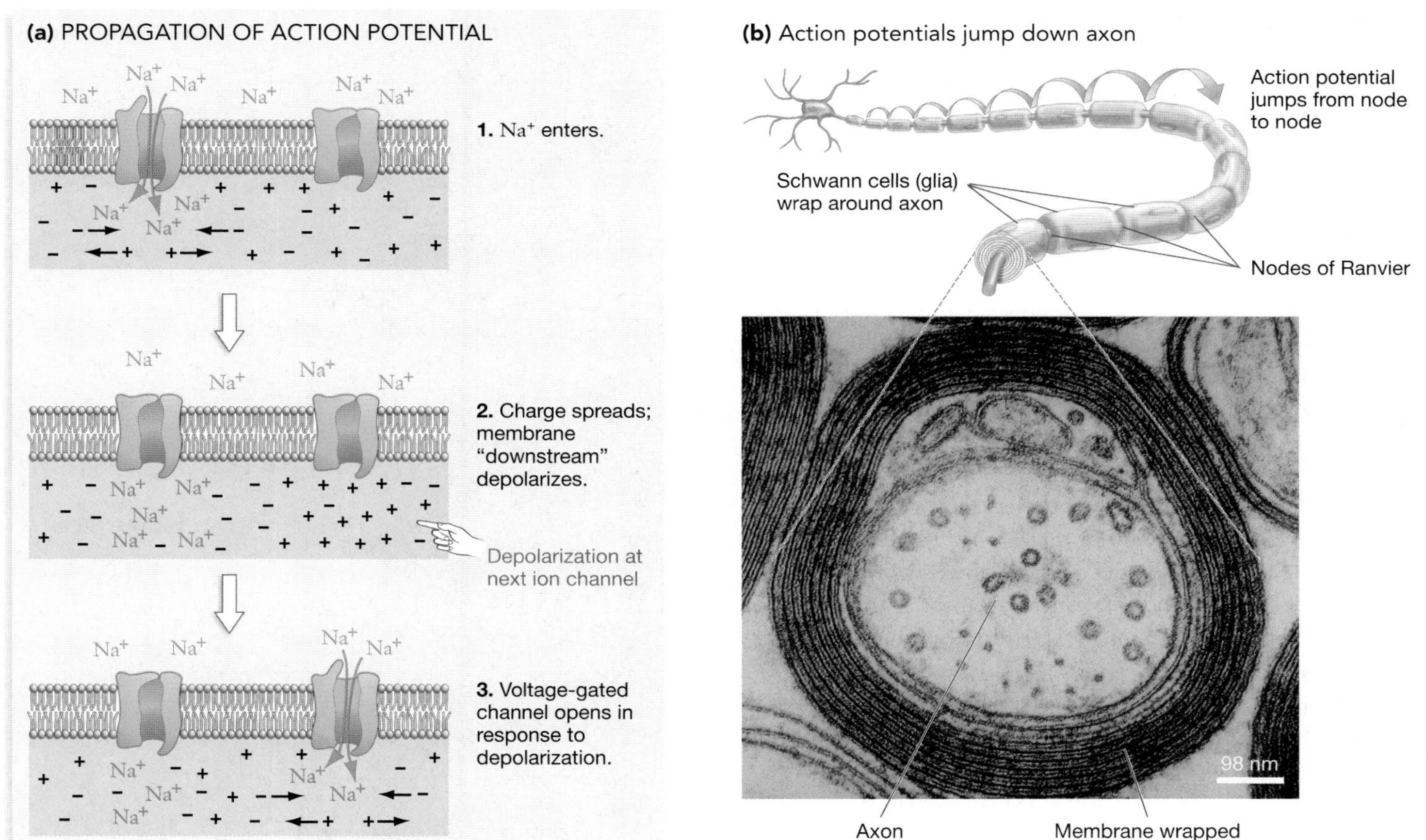

FIGURE 42.10 How Is the Action Potential Propagated?
(a) An action potential starts with an inflow of Na^+. The influx of positive charge attracts negative charges inside the cell and repulses positive charges. As a result, positive charge spreads away from the channel where the Na^+ enters and depolarizes nearby regions of the neuron. Voltage-gated Na^+ channels open in response. **(b)** Schwann cells wrap their membranes around axons as shown here. In myelinated axons, no charge leaks across the membrane as it spreads down the axon. Consequently, action potentials jump down the axon rapidly.

axons. Instead, the membranes of specialized accessory cells, called Schwann cells or glia, wrap around the axons of their neurons (**Figure 42.10b**). **Myelination** acts as insulation. As charge spreads down an axon, myelination prevents charge from leaking out. Consequently, the influx of charge that results from an action potential is able to spread unimpeded until it hits an unmyelinated section of the axon, called the node of Ranvier. The node has a dense concentration of voltage-gated channels and supports action potentials. In this way, electrical signals jump down the axon much faster than they can move down an unmyelinated cell. To drive this point home, consider what happens when myelination is decreased. If myelin degenerates, the transmission of electrical signals slows considerably. The disease multiple sclerosis (MS) develops as muscles weaken and coordination lessens, in response to impaired electrical signaling.

What happens once an action potential has traveled the length of the axon? Cajal proposed, and electron microscopy confirmed, that in most neurons the membrane at the end of the axon presses up against the membrane of another neuron's dendrite. The surfaces of the two membranes are separated by a tiny gap. What happens when an action potential arrives at this interface between cells?

42.3 The Synapse

Most neurons are not directly connected. Based on this observation, there must be some indirect mechanism that transmits electrical signals from cell to cell. In the 1920s, Otto Loewi showed that the indirect mechanism involves molecules called **neurotransmitters.** Loewi was interested in how the nervous system affects heart rate, and knew that signals from the vagus nerve slow the heart. To test his hypothesis that the signal from nerve to muscle is delivered by a chemical, Loewi performed the experiment diagrammed in **Figure 42.11**. First, he isolated the vagus nerve and heart of a frog. As predicted, the heartbeat slowed when he stimulated the nerve electrically. But when he took the solution that bathed the first heart and applied it to another, isolated heart, it slowed as well. This result provided strong evidence for the chemical transmission of electrical signals. The vagus nerve had released a neurotransmitter into the bath.

How are neurotransmitters delivered? When transmission electron microscopy (TEM) became available in the 1950s, biologists finally understood the physical nature of the interface between neurons. This interface is called a **synapse.** As **Figure 42.12a** shows, the membranes of axons and dendrites juxtapose closely. Just inside the synapse, the axon contains numerous sac-like structures called **synaptic vesicles.** These were hypothesized to be storage sites for neurotransmitters.

Anatomical observations like these, combined with chemical studies of the synapse, led to the model of synaptic transmission illustrated in **Figure 42.12b**. The sequence begins when the arrival of action potentials opens calcium channels and causes an inflow of calcium ions. In response to the increase in calcium concentration, synaptic vesicles fuse with the presynaptic membrane and release neurotransmitters into the synaptic cleft.

What happens next? Biologists hypothesized that neurotransmitters bind to receptors in the membrane of the next cell in the circuit. As step 3 in Figure 42.12b shows, the claim was that interactions between neurotransmitters and receptors on

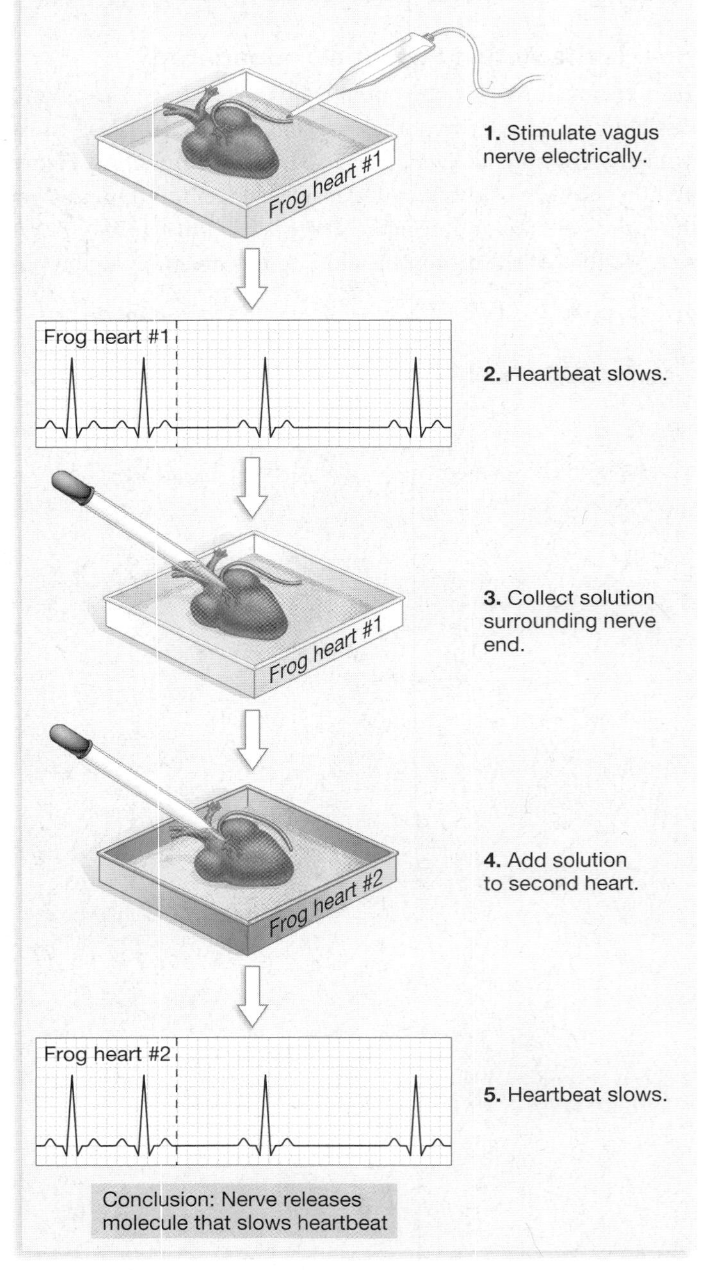

FIGURE 42.11 The Discovery of Neurotransmitters

QUESTION Prior to this experiment, some researchers contended that most neurons communicated with each other electrically, not chemically. How does this experiment refute the hypothesis of electrical transmission? (Note: It turns out that a few neurons do make electrical connections with one another.)

the postsynaptic cell would cause a change in the membrane potential of the postsynaptic cell and possibly trigger an action potential. Is this part of the model correct?

What Do Neurotransmitters Do?

Once Loewi had established that neurotransmitters exist, investigators were able to isolate and purify them from a variety of neurons. They accomplished this by stimulating a neuron, collecting the neurotransmitter that was released, and analyzing it chemically. To find the receptor for a particular neurotransmitter, researchers could add a radioactive label to a neurotransmitter and add it to neurons. Once the labeled transmitter had bound to its receptor, the protein could be isolated and analyzed. Using techniques like these, biologists have identified and characterized a wide array of neurotransmitters.

By patch-clamping receptors, biologists confirmed that many function as **ligand-gated ion channels.** (Any molecule that binds to a specific site on a receptor molecule is referred to as a **ligand.**) When a neurotransmitter binds to a ligand-gated ion channel in the postsynaptic membrane, the channel opens. In this way, the neurotransmitter's chemical signal is transduced to a change in the membrane potential of the postsynaptic cell.

Not all receptors are directly linked to ion channels, however. Instead, some receptors activate enzymes that lead to the

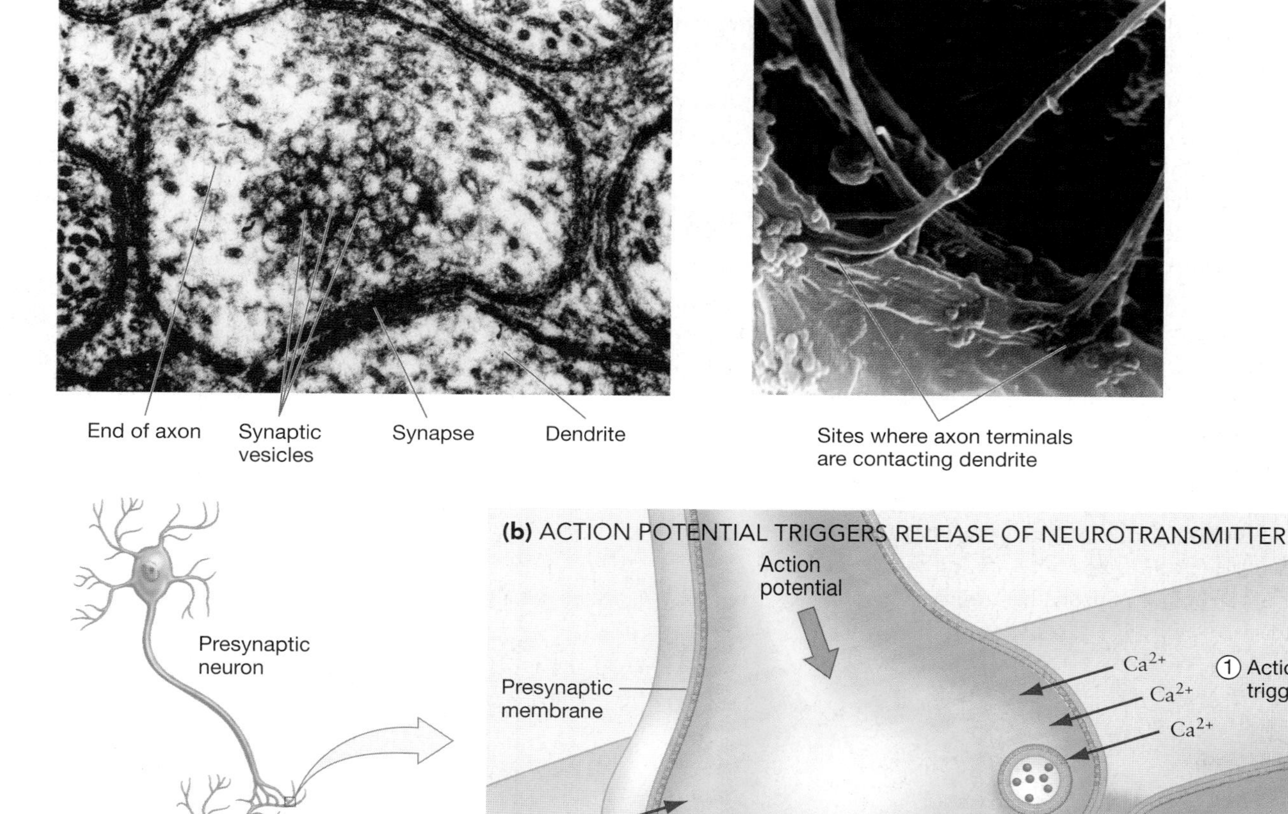

FIGURE 42.12 The Synapse
(a) This transmission electron micrograph is a cross section of the site where an axon meets a dendrite.
(b) These diagrams illustrate the sequence of events that occurs when an action potential arrives at a synapse.
(c) This scanning electron micrograph shows synaptic vesicles fusing with the presynaptic membrane.

production of a "second messenger" molecule in the postsynaptic cell. The second messenger may induce changes in enzyme activity, gene transcription, or membrane potential. The role of second messengers in cells is explored in detail in Chapter 44.

Postsynaptic Potentials, Summation, and Integration

What happens when a neurotransmitter binds to a receptor and opens an ion channel? The scanning electron micrograph in **Figure 42.12c** indicates that numerous synapses can be found in the same region of a dendrite or cell body. And as **Figure 42.13** illustrates, these synapses can be one of two general types. If the receptors at the synapse admit an influx of sodium ions in response to the arrival of the neurotransmitter, then the postsynaptic membrane depolarizes. In most cases, depolarization makes an action potential in the postsynaptic cell more likely. Changes in the postsynaptic cell that make action potentials more likely are called **excitatory postsynaptic potentials (EPSPs).** Other receptors, in contrast, lead to an efflux of K^+ or an influx of Cl^- in the postsynaptic cell. These events hyperpolarize the membrane and make action potentials less likely to occur in the postsynaptic cell. Changes in the postsynaptic cell that make action potentials less likely are called **inhibitory postsynaptic potentials (IPSPs).**

It is critical to realize that EPSPs and IPSPs are not all-or-none events. Instead, they are graded in size and short-lived. The size of an EPSP or IPSP depends on the amount of neurotransmitter that is released at the synapse; both types of signal are short-lived because neurotransmitters are quickly inactivated or taken up by the presynaptic cell and recycled. If either the amount or life span of neurotransmitters is altered, the normal functioning of neurons is altered. The street drugs cocaine and amphetamine, for example, exert their effects by inhibiting the uptake and recycling of particular neurotransmitters.

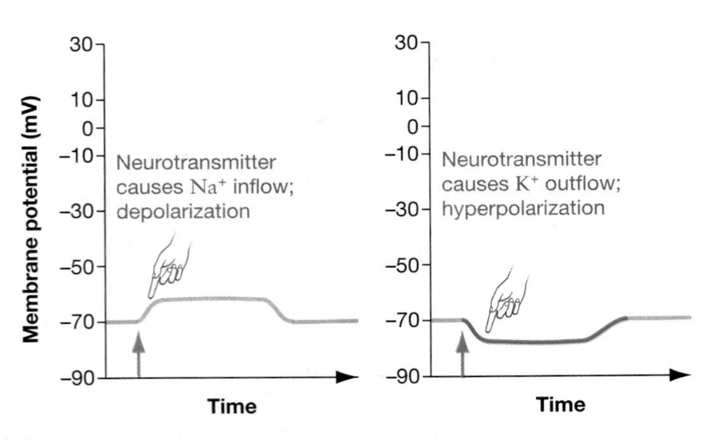

FIGURE 42.13 Postsynaptic Potentials
As these records show, the arrival of an action potential causes a short-lived depolarization or hyperpolarization of the postsynaptic cell membrane. The change in postsynaptic potential can be excitatory or inhibitory, depending on whether it makes action potentials more or less likely.

What effect do EPSPs and IPSPs have on the postsynaptic cell? As **Figure 42.14a** shows, the dendrites and cell body of a neuron typically make hundreds or thousands of synapses with other cells. At any instant, the EPSPs and IPSPs that are occurring at each of these synapses lead to short-lived surges of charge in the dendrites and cell body of the postsynaptic cell. These changes in membrane potential are additive, as shown in **Figure 42.14b**. If an IPSP and EPSP occur close together in space or time, the changes in membrane potential tend to cancel each other out. But if a series of EPSPs occur close together in space or time, they sum and make the neuron likely to fire an action potential. The additive nature of postsynaptic potentials is termed **summation.**

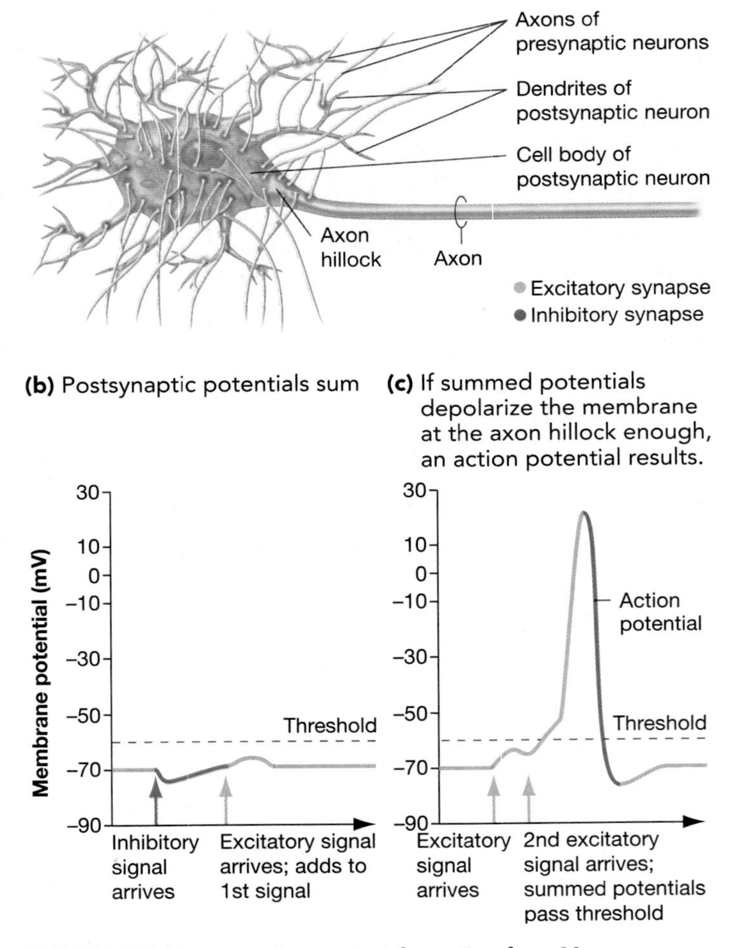

FIGURE 42.14 Neurons Integrate Information from Many Synapses
(a) As this sketch shows, the dendrites and cell body of a neuron typically receive signals from hundreds or thousands of other neurons. **(b)** When action potentials arrive from the same axon close together in time or from different axons that have synapses close to one another, the postsynaptic potentials are additive. **(c)** If excitatory postsynaptic potentials depolarize the axon hillock past a point called threshold, enough Na^+ channels open to trigger an action potential.

The sodium channels that trigger action potentials in the postsynaptic cell are located near the start of the axon at a site called the **axon hillock** (Figure 42.14a). As IPSPs and EPSPs interact throughout the dendrites and cell body, charge spreads to the axon hillock. If the membrane at the axon hillock depolarizes past a point termed the threshold, enough sodium channels open to trigger positive feedback and an action potential (**Figure 42.14c**).

The general message is that the postsynaptic cell integrates information from hundreds or thousands of other neurons. The information arrives in the form of action potentials that produce EPSPs and IPSPs. If the combination of postsynaptic potentials depolarizes the membrane at the axon hillock sufficiently, the postsynaptic cell fires an action potential in response. But if the events lead to a depolarization that is below threshold, the information is not passed on to other neurons.

What Happens When Ligand-Gated Channels Are Defective?

Recent research on synaptic transmission has focused on whether certain neurotransmitters and receptors are involved in drug addiction or mental illness. As an example of these studies, let's focus on the hypothesis that defects in a particular ligand-gated channel contribute to the development of schizophrenia.

Schizophrenia is a mental illness that is highly variable in nature. In most cases the disease includes what physicians call positive and negative symptoms. A positive symptom is something that a patient does; a negative symptom is something that a patient cannot do. The positive symptoms of schizophrenia are often episodic in nature and may include delusions, hallucinations, and paranoia. The negative symptoms tend to be chronic and may include social withdrawal, cognitive deficits, impaired attention, and a restricted range of emotions or a lack of emotional response. In men, the first symptoms usually appear between the ages of 18 and 25; in women, disease onset usually occurs between 26 and 45 years of age.

In the 1950s researchers noticed that people who were intoxicated with the street drug phencyclidine (PCP, also called angel dust) exhibited some of the symptoms of schizophrenia. Based on this observation, investigators began using PCP to induce the symptoms of schizophrenia in lab animals and document the neurological changes that occurred in response. Several of these studies indicated that PCP induced significant changes in the number or function of a ligand-gated channel called the *N*-methyl-D-aspartic acid (NMDA)-sensitive glutamate receptor. Glutamate is an amino acid that functions as a neurotransmitter in certain neurons. The NMDA-sensitive receptor is one of three distinct types of ligand-gated channel that respond to glutamate.

Are defects in the NMDA receptor at least partially responsible for the onset of schizophrenia? Data from several recent studies argue that the answer is yes. Amy Mohn and coworkers, for example, succeeded in generating mice that produce just 5 percent of the normal amount of a key subunit of the NMDA receptor. (The technical details involved in creating these mutant mice are not important here.) The behavior of these mice was similar to those treated with PCP. They perform a large number of repetitive, stereotyped movements and are socially withdrawn compared to normal mice (**Figure 42.15a**, page 820).

In related work, Xue-Min Gao and colleagues observed differences between the NMDA receptors present in brain slices preserved from the bodies of normal and schizophrenic humans. Even though the number of NMDA receptors per neuron did not differ in the two groups, brain cells from schizophrenics had abnormal levels of mRNA for two of the subunits that make up the receptor. The same subunit that Mohn and colleagues had studied in mice was present in lower than normal concentrations, while a second subunit was present in higher than normal concentrations. To make sense of these data, Gao and associates suggest that normal amounts of NMDA receptor are present, but that many do not work properly because they lack the key subunit (**Figure 42.15b**).

Although these data are suggestive, they are far from conclusive. It remains to be shown decisively that schizophrenia is associated with defects in a subunit of the NMDA receptor. Research on the molecular basis of schizophrenia is continuing in labs around the world.

42.4 The Vertebrate Nervous System

The first three sections of this chapter examined electrical signaling at the level of molecules, membranes, and individual cells. The goal of this section is to step back and consider electrical signaling at the level of tissues, organs, and systems.

To begin, let's consider the overall anatomy of the vertebrate nervous system. Then we can ask how researchers go about exploring the function of the most complex organ known—the human brain. The chapter concludes by returning to the molecular level and introducing recent work on learning and memory.

A Closer Look at the Peripheral Nervous System

Biologists routinely distinguish the central nervous system, or CNS, from the peripheral nervous system, or PNS. Recall that the CNS is made up of the brain and spinal cord and is primarily concerned with integrating information; the PNS is made up of neurons outside the CNS.

What specific functions do the cells of the PNS control? Anatomical and functional studies indicate that the PNS consists of two distinct systems: a **somatic system** that controls the skeletal muscles and an **autonomic system** that controls internal processes like digestion and heart rate. In effect, the somatic system responds to external stimuli and results in movement, while the autonomic system responds to internal stimuli and controls the activity of internal organs. Many organs and

(a) Schizophrenia-like symptoms in mice

Normal mice | Mice with faulty NMDA subunit

(b) Hypothesis to explain link between NMDA receptors and schizophrenia

Observation 1:
NMDA receptor is made up of several subunits.

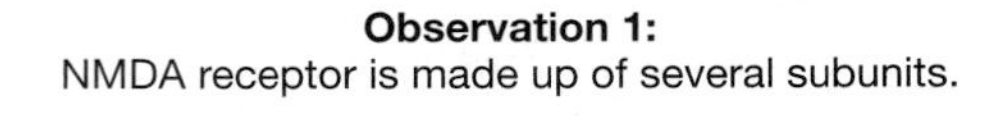

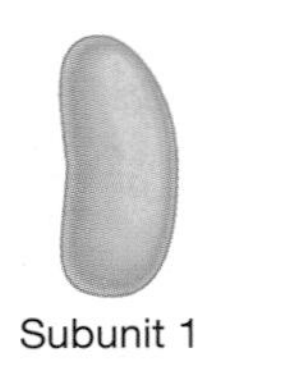

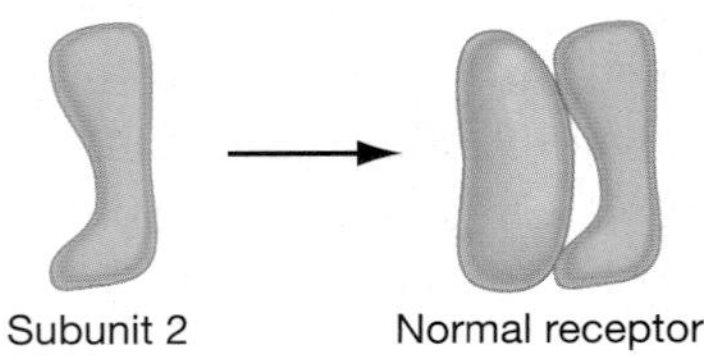

Subunit 1 + Subunit 2 → Normal receptor

Observation 2:
In schizophrenics, subunit 1 concentrations are abnormally low; subunit 2 concentrations are abnormally high.

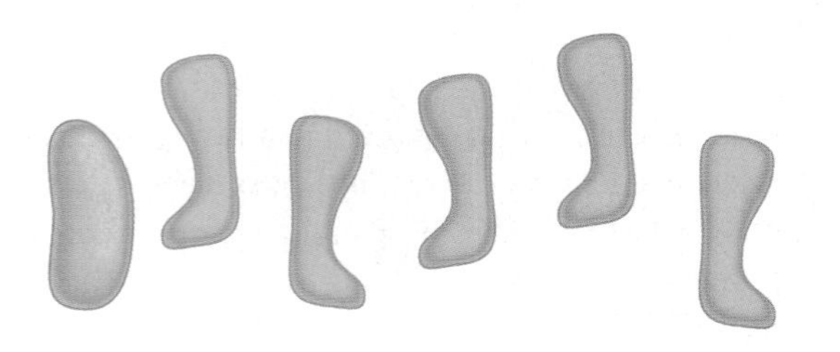

Subunits available in cell, after translation of NMDA mRNAs

Hypothesis:
In schizophrenics, NMDA receptors are faulty because subunit composition is incorrect.

A few normal receptors... ...many faulty receptors

FIGURE 42.15 Is There a Link Between the NMDA Receptor and Schizophrenia?
(a) Mice that have a faulty NMDA subunit show symptoms reminiscent of schizophrenia. These photos, taken a few minutes apart, show typical behavior of normal and experimental mice. **(b)** When researchers compared brain tissue of normal and schizophrenic people, they observed that mRNAs for subunit 1 of the NMDA receptor were rare while mRNAs for subunit 2 were abnormally common.

glands are served by two distinct types of autonomic nerves, however. In some cases, a **parasympathetic nerve** signal promotes relaxation and digestion. For example, the parasympathetic nerves that project to the heart slow it down, while those that serve the digestive tract stimulate its activity. Some **sympathetic nerves,** in contrast, prepare organs for stressful situations. Sympathetic nerves speed up heart rate, stimulate the release of glucose from the liver, and inhibit action by digestive organs. **Figure 42.16** summarizes the effects of signals from parasympathetic and sympathetic nerves.

Functional Anatomy of the CNS

Parasympathetic nerves originate at the base of the brain or the base of the spinal cord; most sympathetic nerves also originate in the spinal cord, but from the central part of the structure. Similarly, most sensory and motor neurons in the somatic nervous system project to or from the spinal cord. In effect, then, the spinal cord serves as an information conduit. It collects and transmits information from throughout the body. With the exception of spinal reflexes such as the one illustrated in Figure 42.1, virtually all of the information that travels to or from the spinal cord is sent to the brain for processing.

The brain is far and away the most complex organ found in animals. In the human brain, researchers estimate there are 100 billion neurons, each of them making an average of 1000 synaptic connections with other neurons. How do researchers even begin to study how this structure functions? They begin with general anatomy. Nineteenth-century anatomists established that the brain is made up of the regions labeled in **Figure 42.17a** (page 822). The largest of these areas, the cerebrum, is divided into left and right hemispheres (**Figure 42.17b**). Each cerebral hemisphere has four major areas or lobes; the hemispheres themselves are connected by a thick band of neurons called the corpus callosum. What do each of these structures do?

Mapping Functional Areas: Lesion Studies Early work on brain function focused on studying people with specific mental deficits. Paul Broca, for example, studied an individual who could understand language but could not speak. After the individual's death in 1861, Broca examined the person's brain and discovered a damaged area in the left front lobe of the cerebrum. Based on this observation, Broca hypothesized that this region is responsible for speech. More generally, he claimed that specific regions in the brain are specialized for coordinating particular functions.

Broca's claim that functions are localized to specific brain areas has been verified through extensive efforts to map the cerebrum. In some cases, advances were made by studying people who had to have portions of their brains removed. In 1953, for example, surgeons treated a 27-year-old man for life-threatening seizures by removing a small portion of his temporal lobe. The man recovered (he is still alive, in fact), has normal intelligence, and can vividly remember his childhood. He has no short-term memory, however. Brenda Milner, a researcher who has worked with this individual for over 40 years, still has to introduce herself to him every time they meet; he cannot recognize a recent picture of himself. Based on case histories like this and a great deal of supporting evidence from studies of memory in laboratory animals, a consensus has emerged that several aspects of memory localize to interior sections of the temporal lobe.

Electrical Stimulation of Conscious Patients Studying the mental abilities of people who have suffered accidental or

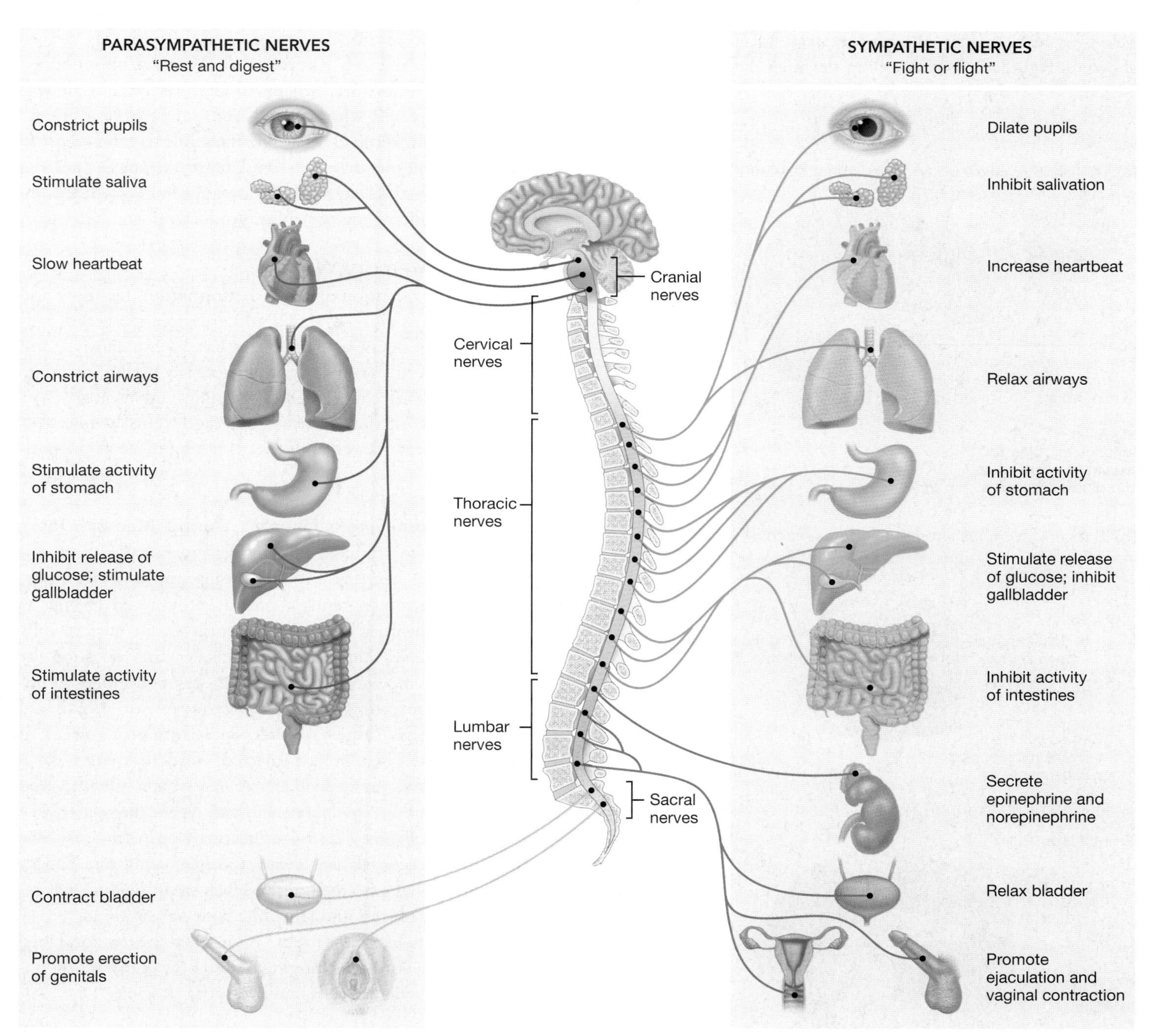

FIGURE 42.16 The Autonomic Nervous System

QUESTION Many parasympathetic nerves promote digestion and lack of movement; many sympathetic nerves support energy utilization and muscle activity. How do each of the responses listed here support these functions?

surgical brain damage has been extraordinarily fruitful. An entirely different approach to studying brain function, however, was pioneered by Wilder Penfield, who worked with severe epileptics scheduled to have seizure-prone areas of their brains surgically removed. While the patients were awake and under local anesthetic, Penfield electrically stimulated portions of their cerebrum. His goal was to map areas that are particularly important and should be spared from removal if possible. When he stimulated specific areas, patients reported sensations or movement in particular regions of the body. In this way, Penfield was able to map the sensory regions of the cerebrum shown in **Figure 42.18b**, as well as the adjacent motor regions.

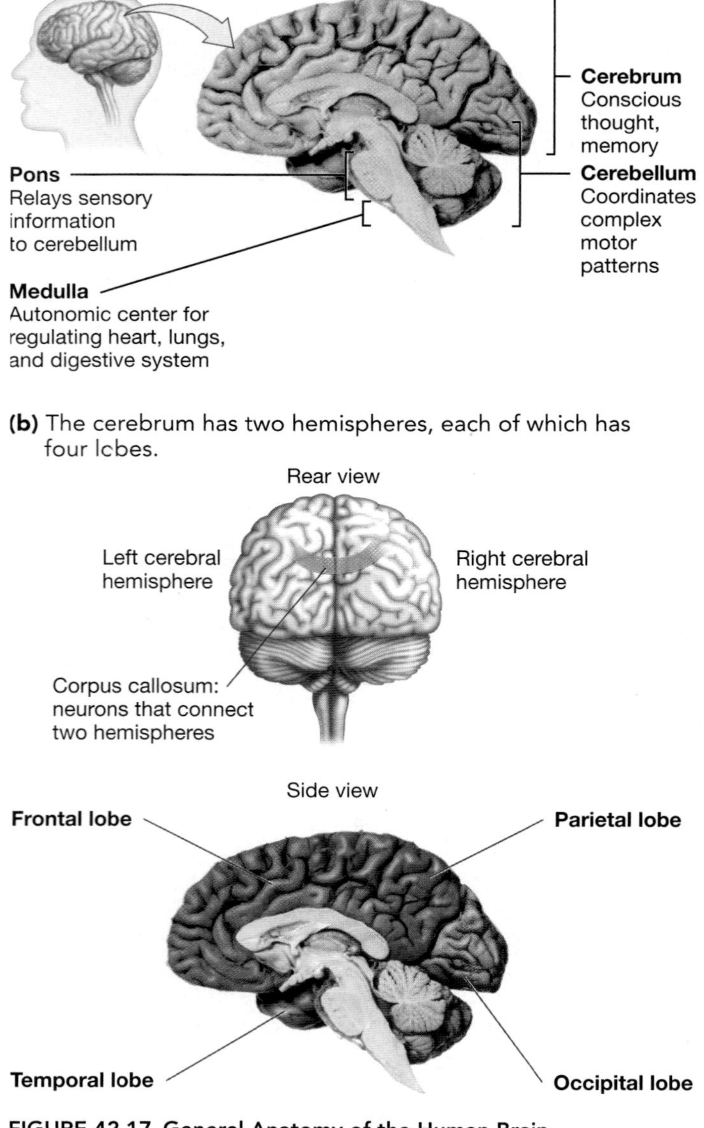

FIGURE 42.17 General Anatomy of the Human Brain
EXERCISE On your own skull, point to each of the areas labeled here.

The technique is still used by brain surgeons to map critical areas near tumors and seizure-prone areas.

But perhaps the most striking of Penfield's findings was that, on occasion, patients would respond to stimulation of their temporal lobe by having what appeared to be flashbacks. After one region was stimulated a woman said, "I hear voices. It is late at night around the carnival somewhere—some sort of traveling circus. . . . I just saw lots of big wagons that they used to haul animals in."

Was this a memory, stored in a small set of neurons? The hypothesis that memories are stored in these types of "grandmother cells" is intensely controversial. As critics have pointed out, Penfield's results are difficult to interpret because he was working with people who suffered from severe brain dysfunction. In addition, Penfield was sometimes able to get patients to replay the same memory by stimulating other cells after the original area had been surgically removed. Have other approaches to studying memory been productive?

How Does Memory Work?

Memory is the retention of learned information. Learning and memory are thus closely related and are often studied in tandem. Research on these phenomena has been extensive, and progress has been particularly rapid recently. As an introduction to how researchers explore learning and memory, let's first explore work that is focused at the level of whole neurons and then review research that is taking place at the molecular level.

Recording from Single Neurons During Memory Tasks

One approach to studying learning and memory is to record from individual neurons during learning and memory tasks. How do the action potentials generated by a cell change as learning and memory take place? George Ojemann and Julie Schoenfield-McNeill attempted to answer this question by recording from individual neurons in the temporal lobes of humans. Before operating on patients who were still awake and about to undergo surgery to remove seizure-prone areas of their brain, the researchers projected words or names on a screen and asked the individuals to read them silently, read them aloud, and/or remember them and repeat them later. The data shown in **Figure 42.19** are informative. In this case, the neuron being recorded was relatively quiet while the patient named objects but extremely active when the individual was remembering the objects and repeating their names aloud.

What do data like these mean? Neurons in the temporal lobe are more active during memory tasks. So how could action potentials from particular cells make memory possible?

Documenting Changes in Synapses Research on the molecular basis of memory is based on two fundamental ideas. First, learning and memory must involve some type of short- or long-term change in the neurons responsible for these processes.

(a) Top view of cerebrum

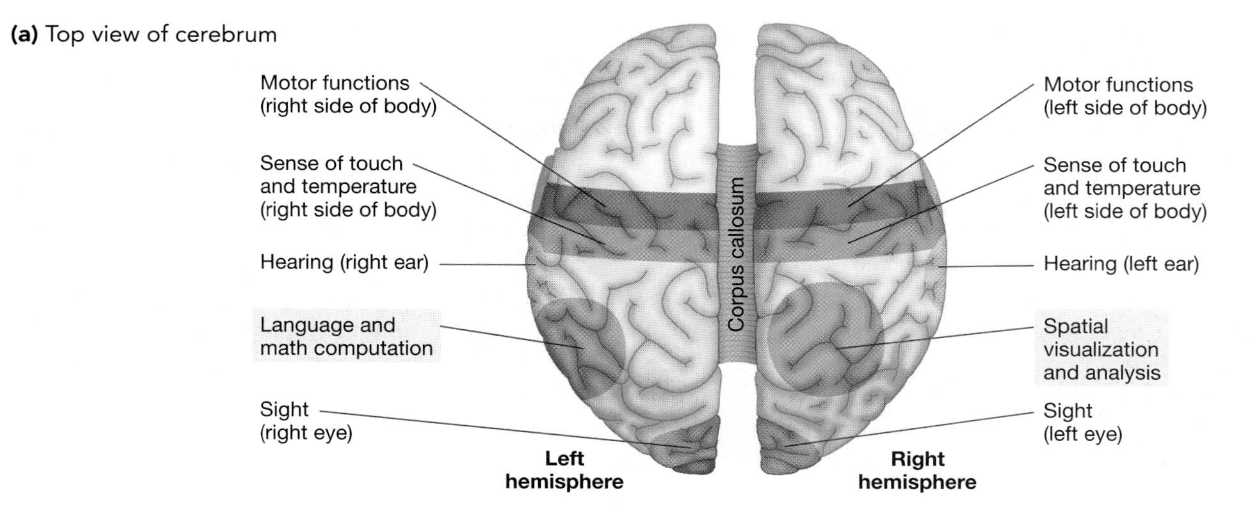

(b) Cross-section through area responsible for sense of touch and temperature

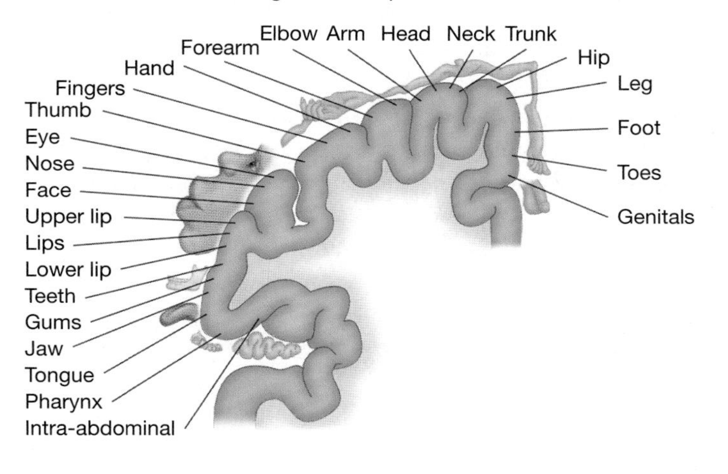

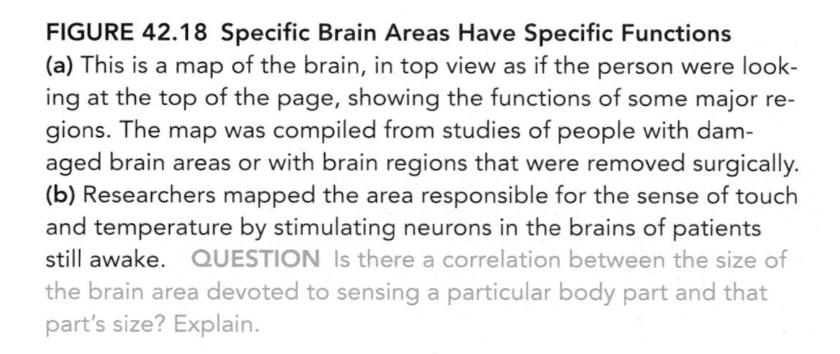

FIGURE 42.18 Specific Brain Areas Have Specific Functions
(a) This is a map of the brain, in top view as if the person were looking at the top of the page, showing the functions of some major regions. The map was compiled from studies of people with damaged brain areas or with brain regions that were removed surgically. **(b)** Researchers mapped the area responsible for the sense of touch and temperature by stimulating neurons in the brains of patients still awake. **QUESTION** Is there a correlation between the size of the brain area devoted to sensing a particular body part and that part's size? Explain.

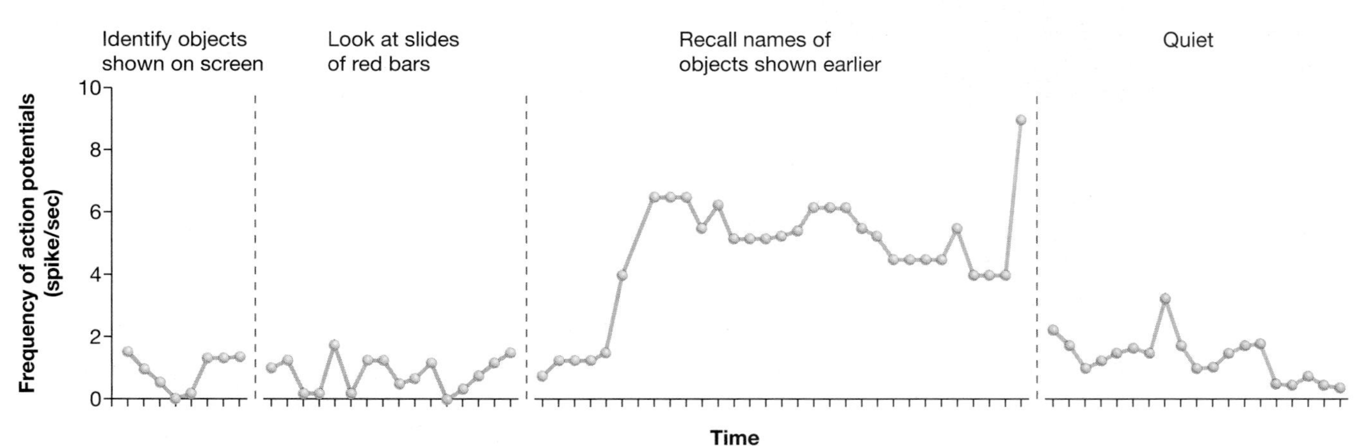

FIGURE 42.19 The Frequency of Action Potentials from Brain Neurons Varies with Activity
These data show the frequency of action potentials from a single brain neuron from the temporal lobe as a patient identified a series of objects shown on slides, looked at a series of red bars shown on slides, recalled the names of the objects shown previously, and looked at a blank screen while doing nothing.

This change could be structural or chemical in nature. Structural changes might include modifications in the number of synapses that a particular neuron makes; molecular changes might involve alterations in the amount of neurotransmitter released at certain synapses. Second, it will be much easier to understand what these changes are if an extremely simple system of neurons can be studied.

To explore the molecular basis of learning and memory, Eric Kandel's group has focused on the sea slug *Aplysia californica* (**Figure 42.20a**). Much of their work has explored the reflex diagrammed in **Figure 42.20b**. When an *Aplysia*'s siphon is touched, it responds by withdrawing its gill. The reflex involves a sensory neuron that is activated by touch and a motor neuron that projects to a gill muscle. Retracting the gill protects it from predators.

Early work established that this simple reflex is modified by learning. For example, *Aplysia* also withdraw their siphons when their tails are given an electrical shock. If shocks to the tail are given in conjunction with a very light touch to the siphon—a touch that is too light to get a response on its own—an *Aplysia* will learn to withdraw its gills in response to a light siphon touch alone.

Follow-up studies on this reflex showed that the neurons involved in learning release the neurotransmitter serotonin, and that serotonin causes an EPSP in the motor neuron to the gill. (Because serotonin is 5-hydroxytryptamine, it is also called 5-HT.) Later experiments documented that repeated application of serotonin, which mimics what happens at the synapse during learning, leads to increased sensitivity on the part of the serotonin-secreting cell (**Figure 42.21a**). Specifically, more serotonin is released after learning has taken place. As a result, EPSPs are higher and the motor neuron is more likely to generate action potentials. These results implied that in *Aplysia*, changes in the nature of the synapse form the molecular basis of learning and memory.

Kelsey Martin and others in Kandel's laboratory were recently able to replicate these results with sensory and motor neurons growing in culture. **Figure 42.21b** shows two *Aplysia* motor neurons on a culture plate. Note that each motor neuron receives synapses from a sensory neuron, and that a researcher is artificially applying serotonin to the synapse. To mimic the learning process, Martin and co-workers applied serotonin to the synapse five times over a short period. When they stimulated the sensory neuron a day later, they found a huge increase in

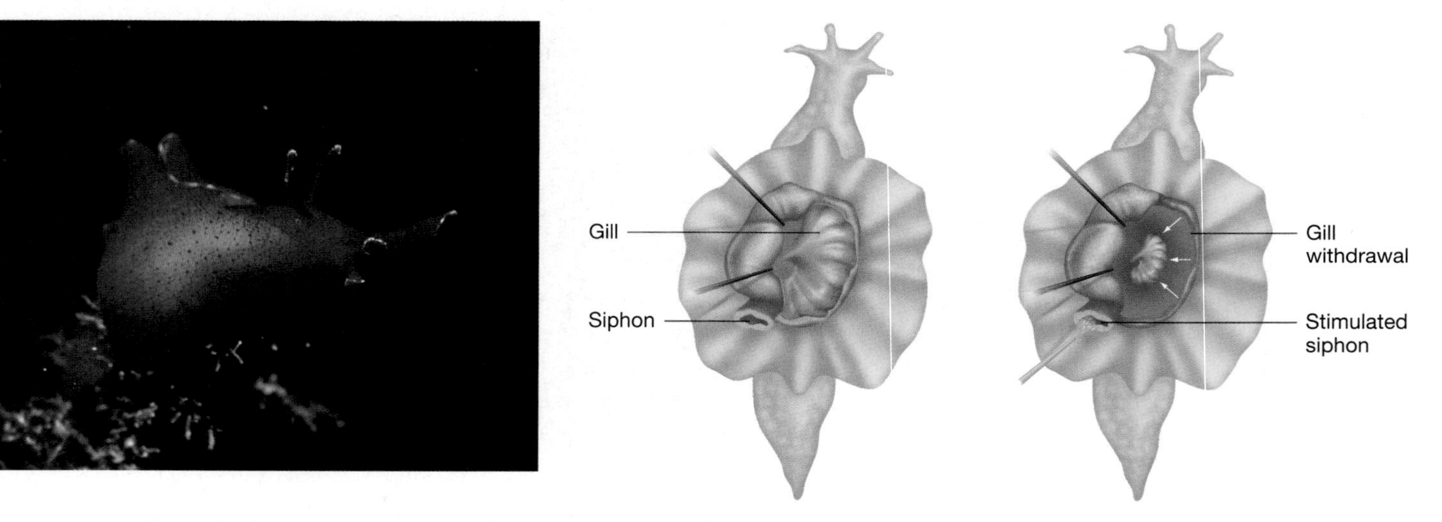

FIGURE 42.20 Studying the Gill-Withdrawal Reflex in *Aplysia*
(a) The sea slug *Aplysia californica.* **(b)** The gill-withdrawal reflex shown here is an important model system in the study of learning and memory.

FIGURE 42.21 Learning and Memory Involve Changes in Synapses **(a)** This experiment showed that repeated application of serotonin changes the behavior of post-synaptic neurons. **(b)** Interactions between *Aplysia* motor neurons and sensory neurons can be studied in culture, as shown here. **(c)** These histograms document the percentage increase in postsynaptic potentials that occurs after repeated application of serotonin, as in part (a).

(a) Postsynaptic potentials change as a result of learning.

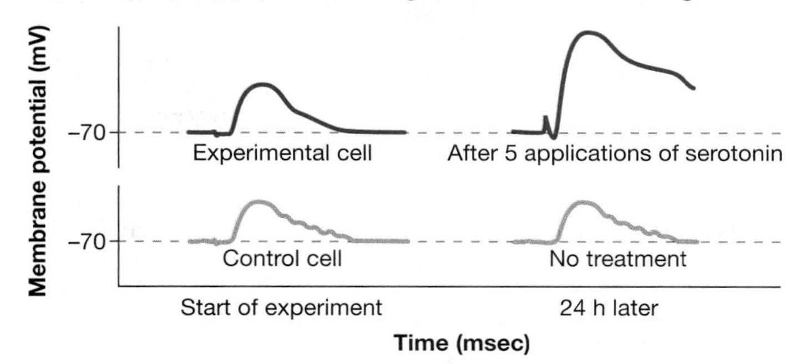

(b) Growing neurons in culture

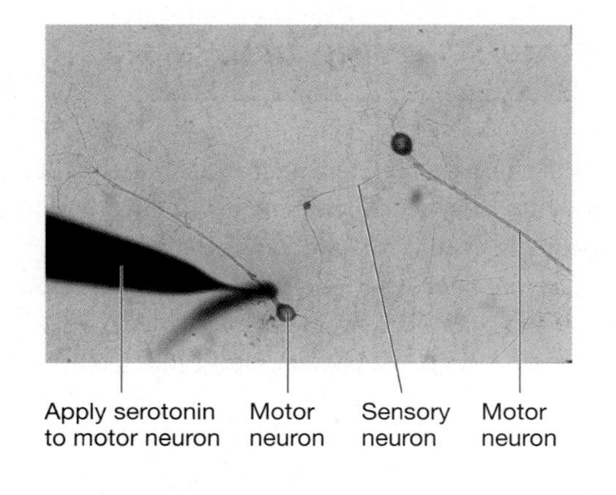

(c) Repeating *Aplysia* experiment in culture

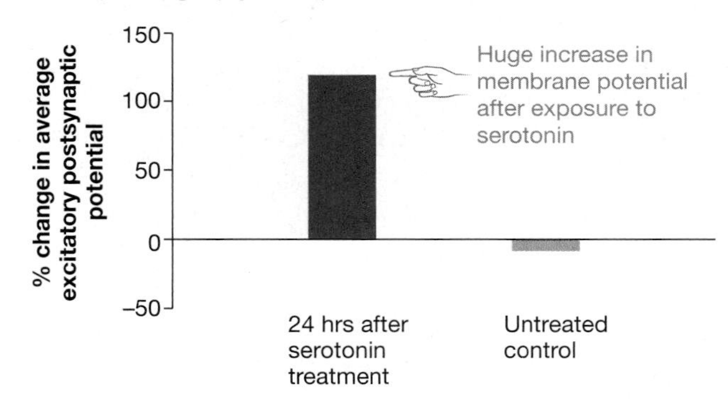

EPSPs (**Figure 42.21c**). The cells had also established additional synapses. Neurons that had not received the repeated stimulation with serotonin had normal postsynaptic responses and numbers of synapses.

Such results reinforce a growing consensus that learning and memory involve molecular and structural changes in the nature of synapses. Further, most researchers now agree that at least some aspects of long-term memory involve changes in gene expression. Chapter 44 explores how the chemical messengers called hormones cause changes in gene expression in target cells. But before investigating how hormones work, let's focus on the electrical signals involved in vision, hearing, taste, and movement—the subject of Chapter 43.

Essay Can Brain Tissue Transplants Help People with Parkinson's Disease?

Parkinson's disease (PD) is a disorder that causes progressive deterioration of motor function. The cellular basis of the disease is the inactivation or destruction of neurons in a region called the substantia nigra, which is located at the base of the brain. The neurons in this area secrete a neurotransmitter called dopamine, and are part of a neural circuit that inhibits action potentials in motor neurons that travel down the spinal cord.

One approach involves the transplantation of dopamine-secreting neurons.

As dopamine-secreting neurons in the substantia nigra begin to inactivate, motor neurons become more excitable. As a result, muscles throughout the body become more likely to contract. Chapter 43 explains that muscles work in pairs. For a movement to occur, one muscle must contract and another must relax. People suffering from PD have a difficult time relaxing their muscles. They experience tremors as paired muscles alternately twitch. They also have a great deal of difficulty making smooth, coordinated movements and have a mask-like expression because they are unable to move facial muscles normally. Millions of people have become familiar with these symptoms by following the careers of former heavyweight champion boxer Muhammad Ali and actor Michael J. Fox, both of whom suffer from PD.

In some cases, physicians have been able to alleviate these symptoms by administering dopamine-like molecules orally. Researchers continue to look for alternative therapeutic approaches, however, because many of the existing drug protocols stop working after several years of use. One promising effort involves the direct transplantation of dopamine-secreting neurons into the brains of PD patients.

Initially, the cells used in these transplants were harvested from aborted human fetuses. Although a significant number of PD patients have shown marked improvement after receiving transplants, the use of tissue from aborted human embryos has been intensely controversial. To alleviate this concern, researchers have recently focused on the possibility of using dopamine-secreting neurons from fetal pigs as a source of cells. Initial experiments involved transplanting fetal pig neurons into rats and other experimental animals that had been induced to develop PD-like symptoms. Because the procedure appeared to be both safe and helpful in alleviating PD-like symptoms, Stephen Fink and co-workers were given approval to transplant these cells into 10 humans suffering from PD. One year after the operation, most of the patients were exhibiting small but important improvements in their condition, and no measurable side effects.

The medical use of tissues from non-human animals, which is called xenografting or xenotransplantation (*xeno* is a Greek root meaning "stranger" or "guest"), is still in its infancy. Time will tell whether the techniques improve enough to make the approach a mainline therapy for a serious degenerative disease.

Chapter Review

Summary

The animal nervous system is made up of a central nervous system (CNS) and peripheral nervous system (PNS). The CNS consists of the brain and spinal cord (in vertebrates); the PNS consists of all nervous system components outside the CNS. Although the neurons in the CNS and PNS are highly variable in size and shape, all have a cell body and multiple, short dendrites that receive electrical signals from other cells; most have axons that transmit electrical signals to other neurons or to effector cells in glands or muscles.

Studies on the squid's giant axon established that neurons have a resting potential due to differences in the concentrations of ions on either side of the membrane and the selective permeability of the membrane to potassium ions. The difference in concentration of ions is created by the sodium-potassium pump. When Na^+/K^+-ATPase hydrolyzes ATP, it transports 3 Na^+ out of the cell and 2 K^+ in.

Studies on the squid's giant axon also established that the action potential is a rapid and stereotyped change in membrane potential. An action potential begins with an influx of Na^+ that depolarizes the membrane. An outflow of K^+ follows and repolarizes the membrane. As charge spreads from the site of an action potential, the nearby membrane is depolarized enough to trigger additional Na^+ inflows and propagate the signal. Propagation takes place most rapidly in large axons or myelinated axons.

The development of voltage clamping allowed researchers to hold membranes at specific potentials and record the current flows that resulted. Voltage clamping confirmed that both Na^+ and K^+ flow through voltage-gated channels. Patch clamping allowed researchers to record the currents that flow through individual Na^+ and K^+ channels. Patch-clamp studies showed that the opening of K^+ channels is delayed relative to Na^+ channels. Experiments with neurotoxins allowed researchers to block these channels and observe the effect on the resting potential and action potential.

When action potentials arrive at a synapse, synaptic vesicles fuse with the axon membrane and deliver neurotransmitters that bind to receptors on the membrane of a postsynaptic cell. One class of receptors functions as ligand-gated channels. In response to binding by a neurotransmitter, the channels open and admit ions that depolarize or hyperpolarize the dendrite membrane. Postsynaptic potentials from nearby synapses add together; if the membrane at the axon hillock depolarizes to a threshold value, then an action potential is triggered. Defects in specific ligand-gated channels have been linked to the development of schizophrenia and other illnesses.

In vertebrates, the PNS contains somatic and autonomic components. The somatic PNS is responsible for sensing external stimuli and effecting movement; the autonomic system monitors internal conditions and effects changes in the activity of organs.

Although the CNS is enormously complex, researchers have succeeded in mapping the function of brain structures. Early mapping studies depended on individuals with lesions in certain areas or on the direct stimulation of certain regions of the cerebral cortex. Efforts to understand higher brain functions like learning and memory form the current focus of research on the CNS. To date, research has established that learning and memory are based on modifications in synapses. After learning takes place, certain neurons release more or less neurotransmitter in response to stimulation. In the case of long-term memory, these changes depend on changes in gene expression.

Questions

Content Review

1. Why does the resting potential exist?
   a. because cells are inherently electrical, due the presence of ions
   b. because ion concentrations differ on either side of the membrane, and because the membrane is selectively permeable to K^+
   c. because ion concentrations differ on either side of the membrane, and because the membrane is selectively permeable to Na^+
   d. because ion concentrations differ on either side of the membrane, and because the membrane is selectively permeable to Cl^-

2. Why did the squid's giant axon become a model system for studying electrical signaling in animals?
   a. Its action potentials are particularly large.
   b. It is the tissue from which researchers initially isolated Na^+/K^+-ATPase.
   c. They are abundant and easy to obtain.
   d. It was large enough to support intracellular recording by the first microelectrodes.

3. How does myelination affect the propagation of an action potential?
   a. It speeds propagation by increasing the density of voltage-gated channels.
   b. It speeds propagation by increasing electrochemical gradients favoring Na^+ entry.
   c. It speeds propagation because charge does not leak out of the membrane as it spreads down the axon.
   d. It slows down propagation because Na^+ channels exist only at unmyelinated nodes.

4. In a neuron, what creates the electrochemical gradients favoring entry of Na^+ and outflow of K^+?
   a. Na^+/K^+-ATPase
   b. voltage-gated K^+ channels
   c. voltage-gated Na^+ channels
   d. ligand-gated Na^+/K^+ channels

5. What do biologists mean when they say that an action potential is "all or none?"
   a. Partial action potentials are rare.
   b. Action potentials can start and then stop before they go to completion.
   c. The size of action potentials varies. Larger action potentials trigger the release of large amounts of neurotransmitter at the synapse.
   d. All action potentials are alike.

6. Why is memory thought to involve changes in particular synapses?
   a. In some systems, an increased release of neurotransmitters occurs after learning takes place.
   b. In some systems, the type of neurotransmitter released at the synapse changes after learning takes place.
   c. When researchers stimulated certain neurons electrically, individuals replayed memories.
   d. People who lack short-term memory have specific deficits in synapses within the brain regions responsible for memory.

Conceptual Review

1. Explain why the resting potential exists. Be sure to differentiate between the role of K^+ channels and Na^+/K^+-ATPase in the membrane's selective permeability and ion concentration gradients.
2. Draw a graph of an action potential and label the axes. Label the parts of the graph and explain which ion flow(s) is responsible for each part.
3. Draw a diagram of a synapse. Label the parts. Then make a series of diagrams showing what happens when an action potential arrives at a synapse. Be sure to explain the events that occur at both the presynaptic and postsynaptic cell.
4. Why do summation and integration occur in postsynaptic cells?
5. Compare and contrast the somatic and autonomic components of the PNS.
6. Compare and contrast the sympathetic and parasympathetic components of the autonomic nervous system.

Applying Ideas

1. Compare and contrast the circuitry in the nervous system diagrammed in Figure 42.1 to the generalized homeostatic systems diagrammed in Figure 38.17. How can nervous system function be interpreted as a mechanism for achieving homeostasis?
2. Discuss the pros and cons of lesion studies in determining the function of particular brain structures.
3. During an epileptic fit, muscles convulse spasmodically and the person is unable to think, see, or hear even though they are awake. The cause of epilepsy is unknown. Speculate on what is happening at the level of the synapse in seizure-prone areas of the brain. How would you test your hypothesis?
4. Suppose that research establishes a causative relationship between defects in an NMDA-sensitive glutamate receptor and the development of schizophrenia. What new therapeutic strategies would this finding suggest for treatment of this disease?

CD-ROM and Web Connection

CD Activity 42.1: Membrane Potential *(tutorial)*
(Estimated time for completion = 10 min)
Learn how electrochemical gradients are formed across the plasma membranes of neurons.

CD Activity 42.2: Action Potential *(animation)*
(Estimated time for completion = 5 min)
How do neurons transmit information via action potentials?

At your **Companion Website** (http://www.prenhall.com/freeman/biology), you will find self-grading exams and links to the following research tools, online resources, and activities:

Tetrodotoxin
This site describes the puffer fish toxin—tetrodotoxin—and offers information about its effect on the nervous system.

Stem Cells
This site at CNN.com discusses the political, scientific, and moral issues surrounding the stem cell controversy.

Basic Neuropharmacology of Anti-Depressants
Explore the actions of the neurotransmitters that are commonly used as therapies for mood disorders.

Additional Reading

Azari, N. P. and R. J. Seitz. 2000. Brain plasticity and recovery from stroke. *American Scientist* 88: 426–431. Explores how the brain shifts functions to new areas in response to damage.

Gazzaniga, M. S. 1998. The split brain revisited. *Scientific American* 279 (July): 50–55. Reviews classic and recent work on interactions between the brain hemispheres.

Kempermann, G., and F. H. Gage. 1999. New nerve cells for the adult brain. *Scientific American* 280(May): 48–53. Considers recent work on regeneration of neurons in the human brain.

Powledge, T. M. 1999. Addiction and the brain. *BioScience* 49: 513–519. Examines the role of dopamine receptors and dopamine transporters in addiction—especially to cocaine.

Animal Sensory Systems and Movement

43

Adult moths are active at night when it is difficult or impossible to see. It is logical, then, that sexually mature females and males do not go looking for each other. Instead, females release a chemical attractant called a pheromone into the air. Males of the same species that are flying in the area can detect even a single molecule of the pheromone due to the receptor cells located on their large, feathery antennae. In response to an airborne gradient of pheromone molecules, males fly toward a female.

As male moths are patrolling in search of these airborne pheromones, however, they are hunted by bats. Like moths, bats are active almost exclusively at night. Instead of hunting by sight, like an osprey or a cheetah, they hunt with the aid of sonar. Bats emit a train of high-pitched sounds as they fly, and then listen for echoes that indicate the direction and shape of objects in their path. If the object is a moth, the bat flies toward it, catches the moth in its mouth, and eats it (**Figure 43.1**, page 830). Moths, however, can hear bat calls. When moths detect sounds from an onrushing bat, they tumble out of the sky in chaotic escape flights.

If you were out at night as these dramas unfolded, at best you might be dimly aware that bats and moths were flying about. Humans cannot smell moth pheromones or hear the sounds that bats emit when flying. It took decades of careful experimentation for biologists to understand how moths and bats sense the world around them and how they move in response to the information they receive.

Movement is fundamental to how animals work. But without accurate sensory information, animals cannot move effectively. To explore the mechanisms of sensation and movement, section 43.1 introduces how animals receive information from the environment and

In many species of moths, males have much larger antennae than females. Receptor cells on the antennae detect airborne chemical signals that are produced by sexually mature females. As a result, males can locate females in complete darkness.

FIGURE 43.1 Bats Hunt Moths
The moth near the center of this photograph is about to be captured by an onrushing bat.

respond to it. The next three sections delve into the molecular processes involved in hearing, vision, and taste and smell. The chapter concludes by investigating the cellular and molecular mechanisms responsible for movement.

43.1 How Do Sensory Organs Convey Information to the Brain?

As a moth flies through the night, its brain receives streams of signals from an array of sensory organs. Antennae provide information about the concentration of pheromones; ears send data on the presence of high-pitched sounds; balance and gravity detectors transmit signals about the body's orientation in space. Each type of sensory information is detected by a sensory neuron or by a specialized receptor cell that makes a synapse with a sensory neuron. As **Figure 43.2** shows, the moth's brain integrates the information from sensory neurons and responds with electrical impulses to specific muscle groups.

How does this information processing and response system get started? Specifically, how do sensory cells receive information from the environment, and how do they report it to the brain?

Sensory Transduction

Chapter 42 introduced how electrical signals are generated and propagated in nerve cells. **Figure 43.3a** reviews one of the most basic principles from that presentation—the nature of the membrane voltage. Recall that an electrical potential is created whenever charges are separated across a cell membrane in the form of different concentrations of ions. In the resting state, the inside of a sensory cell membrane is more negative than its exterior.

As **Figure 43.3b** shows, researchers can record changes in a cell's membrane potential with the aid of microelectrodes. Recall from Chapter 42 that if ion flows cause the interior to become more positive (less negative), the membrane is said to be depolarized. If changes in ion channels cause the cell interior to become more negative than the resting potential, the membrane is said to be hyperpolarized.

What do these principles have to do with sensory systems? To answer this question, examine the voltage recording from a human sound-receptor cell in **Figure 43.3c.** Note that when the experimenter played a sound, the sound-receptor cell depolarized in response.

Other sensory cells work in a similar way. Although animals have sensory receptors that detect a remarkable variety of stimuli (see **Box 43.1**), they all transduce sensory input—including light, sounds, touch, and odors—to a change in membrane potential. In this way, different types of information are transduced to a common type of signal.

If a sensory stimulus induces a large change in a sensory receptor's membrane potential, action potentials are sent to the

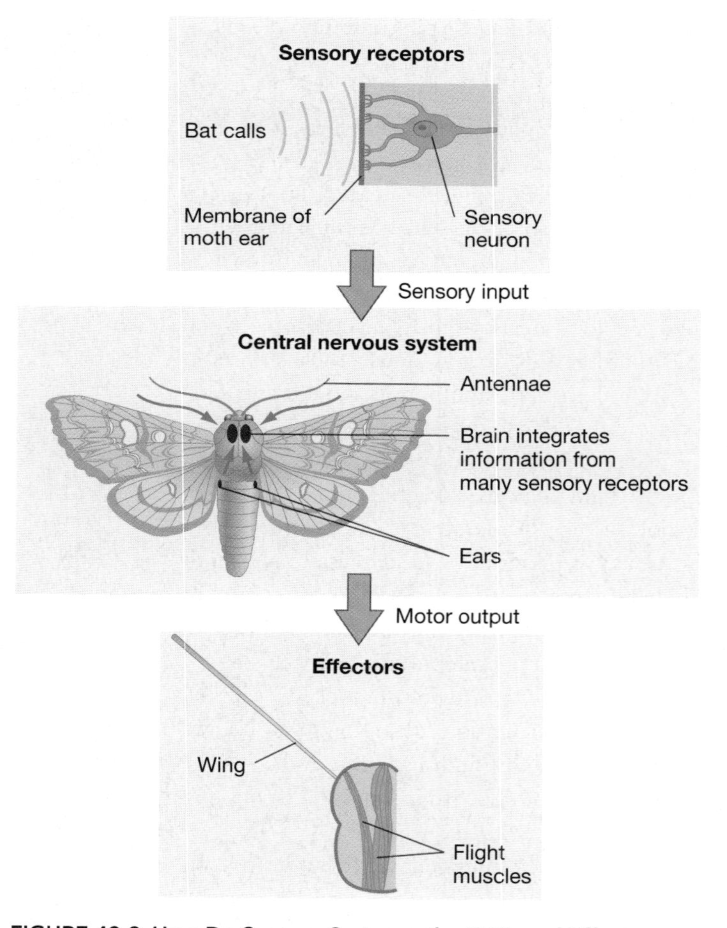

FIGURE 43.2 How Do Sensory Systems, the CNS, and Effectors Like Muscles Interact?
Sensory neurons relay information about conditions inside and outside an animal to the central nervous system. After integrating information from many sensory neurons, the CNS sends signals to muscles through motor neurons.

brain. For example, the amount of depolarization that occurs in a sound-receptor cell is proportional to the loudness of the sound. If the depolarization passes threshold, enough voltage-gated sodium channels open to trigger action potentials that are sent to the brain. Recall from Chapter 42 that all action potentials are identical in size and shape. As Figure 43.3d shows, however, louder sounds induce a higher frequency of action potentials than do softer sounds. In this way, receptor cells provide information on the intensity of a stimulus.

The universal nature of sensory transduction raises an important question. If all types of external stimuli are converted to electrical signals in the form of action potentials, how is it possible for the brain to interpret the information properly?

Transmitting Information to the Brain

There are two keys to understanding how the brain interprets sensory information. First, receptor cells tend to be highly specific. For example, each receptor cell in a human ear responds best to

BOX 43.1 Senses That Humans Don't Have

This chapter introduces animals that can sense wavelengths of light, frequencies of sound, and odors that humans cannot perceive. But in addition to having eyes, ears, noses, and taste buds that are more acute than ours, many species have completely different sensory capacities. Some aquatic predators are so sensitive to electric fields that they can detect electrical activity in the muscles of passing prey. Sharks are so sensitive to electric fields that they become badly disoriented in captivity by electric water pumps near their tanks. Similarly, many birds, sea turtles, and other animals can detect magnetic fields and use Earth's magnetic field as an aid in navigation. Homing pigeons, for example, tend to become disoriented if small magnets are glued to their heads. Biologists who work on species like these are faced with the challenge of studying stimuli that they themselves cannot sense. In addition, it is possible that animals have senses that are yet to be discovered.

(a) Electrical signals originate in sensory cells.

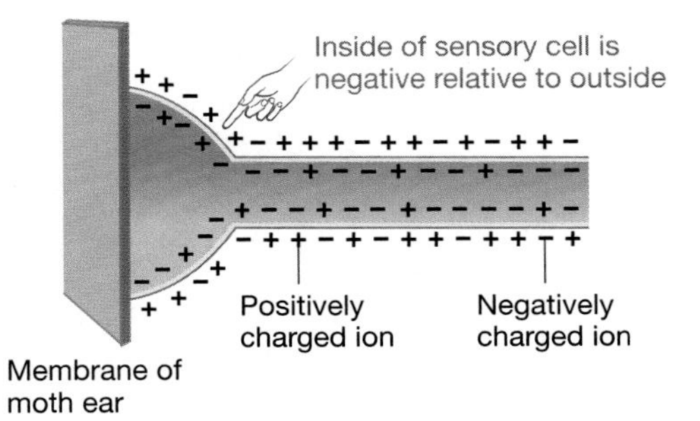

(b) Microelectrodes measure voltage across membrane.

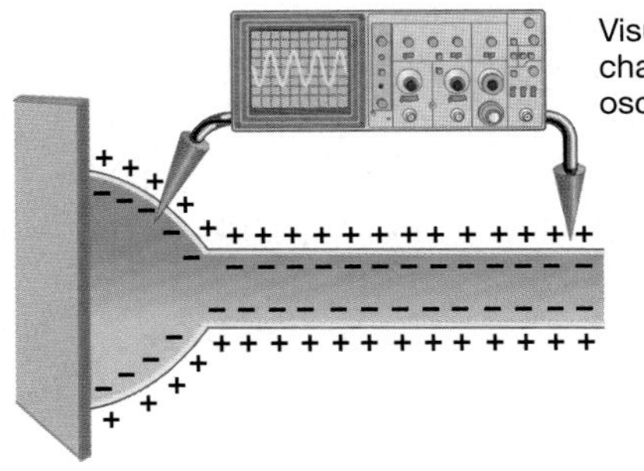

(c) Sound receptor cells depolarize in response to sound.

(d) Sound receptor cells respond more strongly to loud sounds.

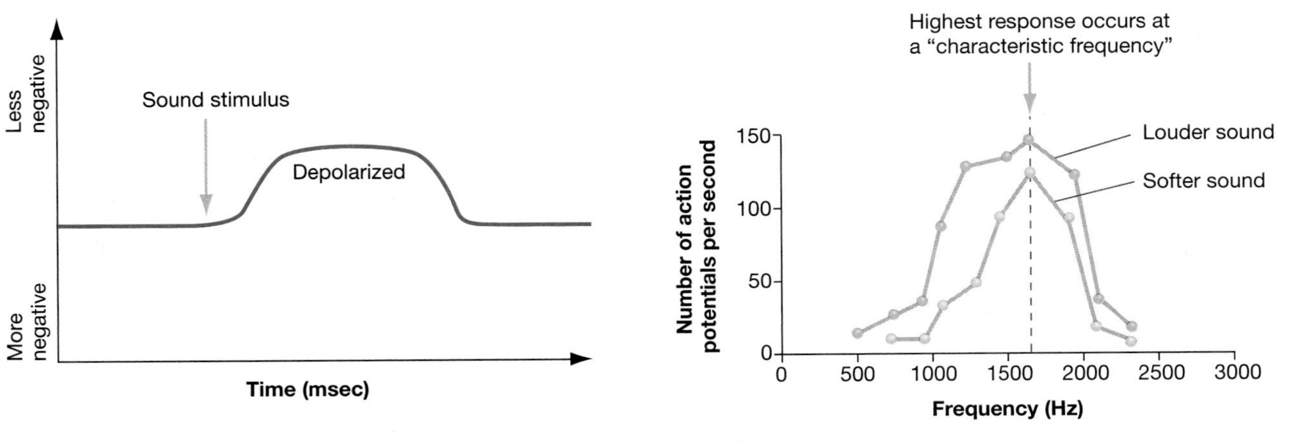

FIGURE 43.3 Sensory Inputs Change the Membrane Potential of Receptor Cells
In response to sensory stimuli, ions flow across the membranes of receptor cells and change the membrane potential. Depending on the type of receptor involved, the change may be a depolarization or a hyperpolarization.

certain frequencies of sound. Some receptors are more sensitive to low-pitched sounds at 1000 Hz (hertz, or cycles per second) while others respond best to high-pitched sounds at 8000 Hz. The receptor depicted in Figure 43.3d responds most strongly to sounds at about 1600 Hz. In this way, the train of action potentials from a cell contains information about the frequency of sound that is being received, its intensity, and how long the stimulus lasts.

Second, each type of sensory neuron sends its signal to a specific portion of the brain. Axons from sensory neurons in the human ear project to a particular area in the side of the brain, but axons from sensory receptors in the eye deliver action potentials to a specific area on the back of the brain. Different regions of the brain are specialized for interpreting different types of stimuli.

With this introduction to the basic principles of sensory reception and transduction, let's delve into the details of four sensory systems that are particularly well understood: hearing, vision, taste, and smell.

43.2 Hearing

Hearing is the ability to sense the changes in pressure called sound. A sound consists of waves of air or water pressure. The number of pressure waves that occur in one second is called the **frequency** of the sound. We perceive different sound frequencies as different pitches.

Animals actually have a wide variety of systems for sensing changes in pressure in addition to sensing air- or waterborne sound waves. One particularly important system is responsible for monitoring changes in pressure caused by gravity. Crabs, for example, have a fluid-filled organ that is lined with pressure-receptor cells and contains a small stone that rests on the bottom of the organ. If the crab is tipped or flipped over, the stone presses against receptors that are not on the bottom of the organ. When the brain receives action potentials from these cells, it responds by activating muscles that restore normal posture. In crabs and other animals, pressure-receptor cells are found in a diverse array of tissues and organs. Some are responsible for detecting direct physical pressure on skin; others monitor how far muscles or blood vessels are stretched.

Virtually all of the pressure-sensing systems found in animals are based on the same mechanism, however. Before investigating the specific structures involved in hearing, let's examine the general nature of a pressure-receptor cell.

How Do Sensory Cells Respond to Sound Waves and Other Forms of Pressure?

Pressure receptors are fairly simple in concept. In every case, direct physical pressure on a cell membrane or distortion by bending causes ion channels in the membrane to open or close. In response to a change in ion flows, the membrane depolarizes or hyperpolarizes. The result is a new pattern of action potentials from a sensory neuron.

In ears and several other pressure-sensing organs, the receptors are similar to the hair cell illustrated in **Figure 43.4**.

Hair cells are named for a set of stiff outgrowths, called **stereocilia**, that occur at one end of the cell. Note that stereocilia are arranged in order of increasing height, and that they extend into a fluid-filled chamber. If stereocilia bend in one direction in response to pressure, potassium ion (K^+) channels open and depolarize the cell.* If stereocilia bend in response to pressure from the other direction, the K^+ channels close and the cell hyperpolarizes.

How can the bending of stereocilia affect ion channels? Electron micrographs show that tiny threads connect the tips of stereocilia to each other. One hypothesis contends that when the stereocilia are bent, the threads somehow pull open ion channels in the wall of the next tallest stereocilium—like tiny trapdoors. This hypothesis remains to be confirmed, however. Researchers still do not fully understand how the ion channels involved in pressure reception work.

The Mammalian Ear

How does hearing occur? To answer this question, let's focus on the human ear as a case study. **Figure 43.5** shows that the structure has three general sections—the outer ear, middle ear, and inner ear. Each of the three regions is separated from the others by a membrane. The hair cells that respond to sounds are located in the inner ear, in a fluid-filled chamber called the **cochlea.**

What do the other parts of the ear do? When sound waves reach a human's head, they pass through a canal and strike the

*Recall from Chapter 42 that the opening of K^+ channels hyperpolarizes neurons. Hair-cell membranes respond differently because they are bathed by extracellular fluid with an extraordinarily high K^+ concentration. As a result, the equilibrium potential for K^+ in hair cells is 0 mV instead of the −80 mV in a typical neuron. The resting potential of the hair-cell membrane is −70 mV, so K^+ rushes in and causes a depolarization.

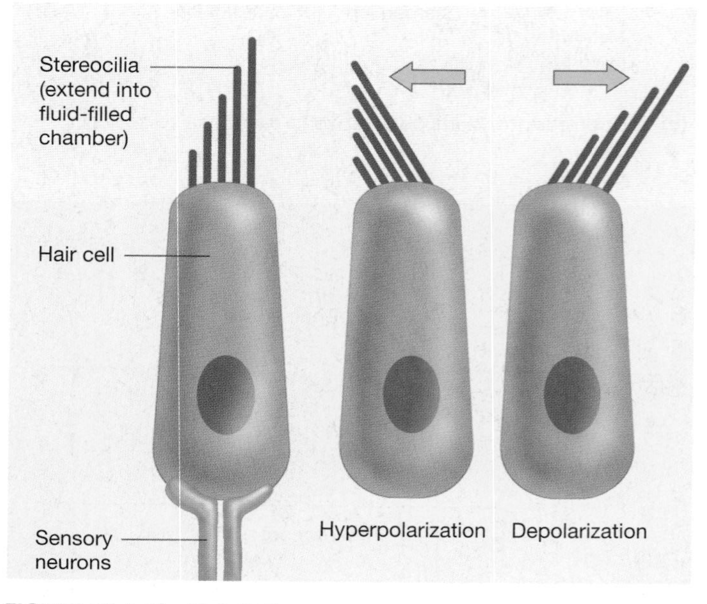

FIGURE 43.4 The Hair Cell
Hair cells respond to deflection of their stereocilia with either depolarization or hyperpolarization, depending on the direction of deflection.

tympanic membrane or eardrum, which separates the outer ear from the middle ear. The repeated cycles of air compression cause this membrane to vibrate back and forth with the same frequency as the sound wave. The vibrations are passed to three tiny bones called the ear ossicles. In response, these bones vibrate against one another. The last ossicle, called the **stapes** (pronounced STAY-peez), vibrates against a membrane called the oval window that separates the middle ear and inner ear. The oval window oscillates in response. In doing so, it generates waves in the fluid inside the cochlea. These pressure waves are sensed by hair cells in the cochlea.

In effect, the ear translates airborne waves to waterborne waves. The system seems extraordinarily complex, though, for such a simple result. Why doesn't the outer ear canal lead directly to the oval window? Why have a middle ear at all?

The Middle Ear Amplifies Sounds Biologists began to understand the function of the middle ear when they recognized two key aspects of its structure. First, the size difference between the tympanic membrane and oval window is important. As the detailed drawing in Figure 43.5 shows, the ossicles in the middle ear receive vibrations from a relatively large surface and transfer them to a relatively small surface. As a result, the amount of vibration induced by sound waves is increased by a factor of 17. In addition, the three ossicles act as levers that further amplify the vibrations at the tympanic membrane. If there were just one ear ossicle instead of three, this levering action would not be possible. The overall effect in the mammalian middle ear is to amplify sound by a factor of 22. Biologists interpret the middle ear as an adaptation for increasing sensitivity to sound.

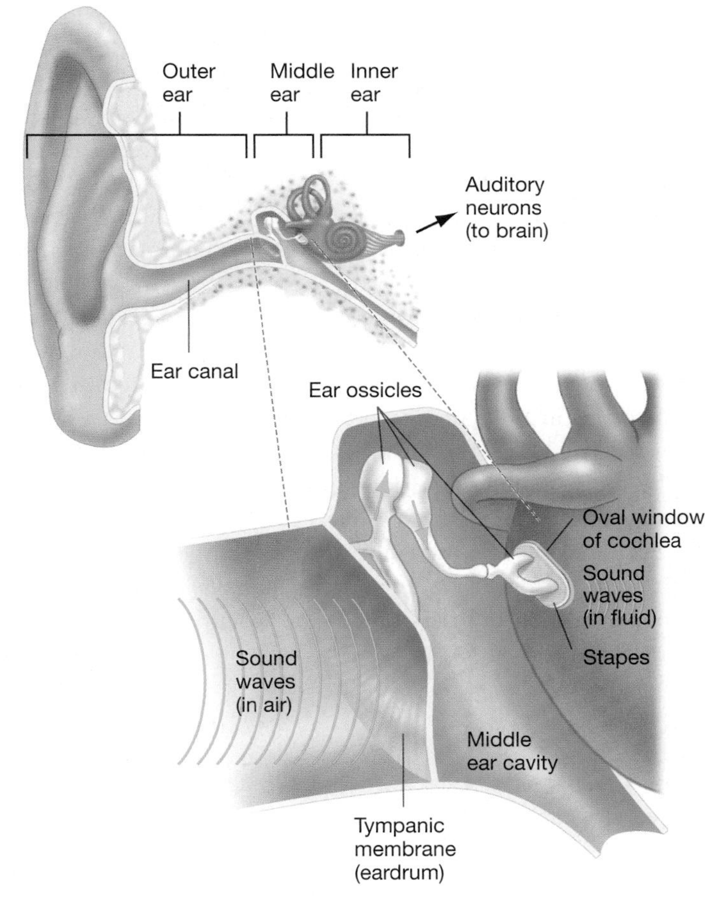

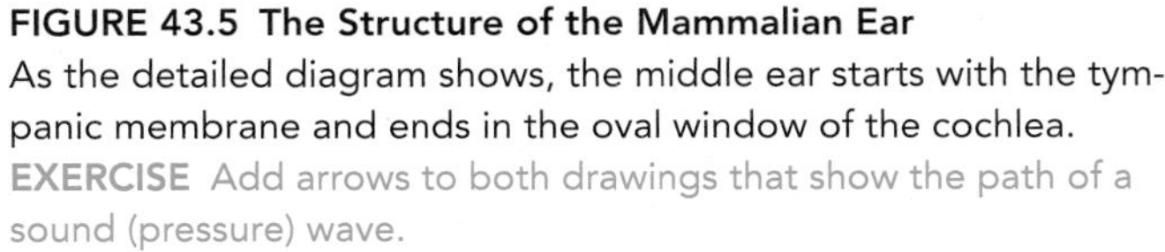

FIGURE 43.5 The Structure of the Mammalian Ear
As the detailed diagram shows, the middle ear starts with the tympanic membrane and ends in the oval window of the cochlea.
EXERCISE Add arrows to both drawings that show the path of a sound (pressure) wave.

To summarize, the outer ear transmits sound waves from the environment to the inside of a mammal; the middle ear amplifies these waves enough to stimulate the hair cells that line the cochlea. Now the critical question is, how can hair cells distinguish different frequencies of sound? If all hair cells responded equally to all frequencies of sound, we would be able to perceive only one pitch. Everyone's voice—indeed, every sound—would sound the same.

The Cochlea Detects the Frequency of Sounds Careful anatomical studies and experimental work have revealed why different hair cells respond to specific frequencies of sound. As the cross section in **Figure 43.6a** (page 834) shows, the cochlea has a set of internal membranes that divide it into three chambers. Hair cells form rows in the middle chamber. As **Figure 43.6b** indicates, the bottom of each hair cell connects to a structure called the basilar membrane; the hair cells' stereocilia touch yet another, smaller surface called the tectorial membrane. In effect, hair cells are sandwiched.

Researchers struggled for decades to understand how these membranes affect hair-cell function. Unfortunately, it is virtually impossible to study cochleas in living organisms because the cochleas are tiny, complex, coiled, and buried deep inside the skull. During the 1920s and 1930s, however, Georg von Békésy was able to perform experiments on cochleas that he dissected from human cadavers. Once a cochlea was isolated, von Békésy was able to vibrate the oval window and record how the cochlea's internal membranes moved in response. He found that when a pressure wave traveled down the fluid in the upper and lower chambers, the basilar membrane vibrated in response. His key finding, though, was that sounds of different frequencies caused the basilar membrane to vibrate in specific spots along its length. When the basilar membrane vibrated in a particular location, the stereocilia of the hair cells there were bent one way and then the other by the tectorial membrane.

An additional observation allowed von Békésy to understand cochlea function in detail. He noted that the basilar membrane varies in stiffness. Specifically, it is very stiff near the oval window and very flexible at its end. Thus each segment of the membrane vibrates in response to a different frequency of sound. Just as a stiff drumhead produces a high-pitched sound and a loose drumhead yields a low-pitched sound, high-frequency sounds cause the stiff part of the basilar membrane to vibrate, and low-frequency sounds cause the more flexible part to vibrate.

To summarize, certain portions of the basilar membrane vibrate in response to specific pitches and result in the bending of hair-cell stereocilia. In this way, hair cells in a particular place on the membrane respond to sounds of a certain frequency. The brain, in turn, receives action potentials from the neurons associated with each hair cell and interprets them as a particular frequency of sound. Complex sounds contain a wide variety of frequencies and cause particular combinations of hair cells to fire in response. Through experience, the brain learns which combinations of frequencies represent music, a siren, or a parent's voice.

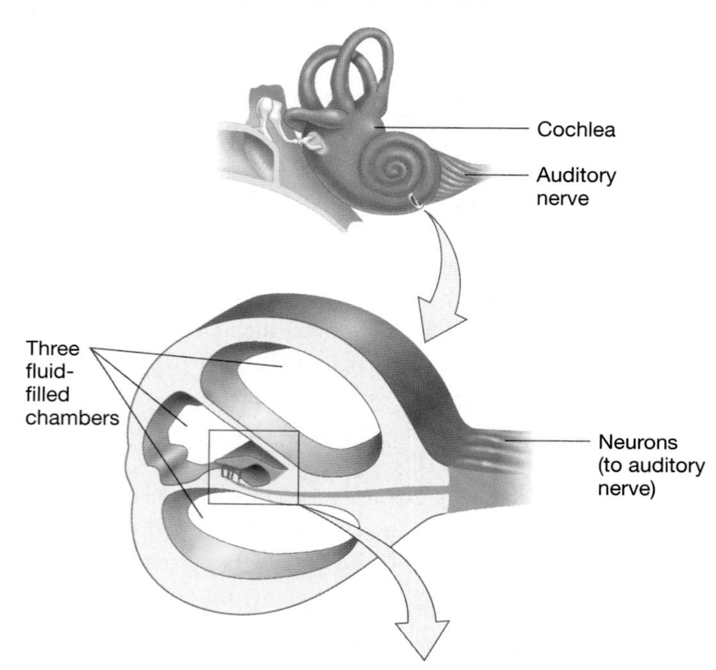

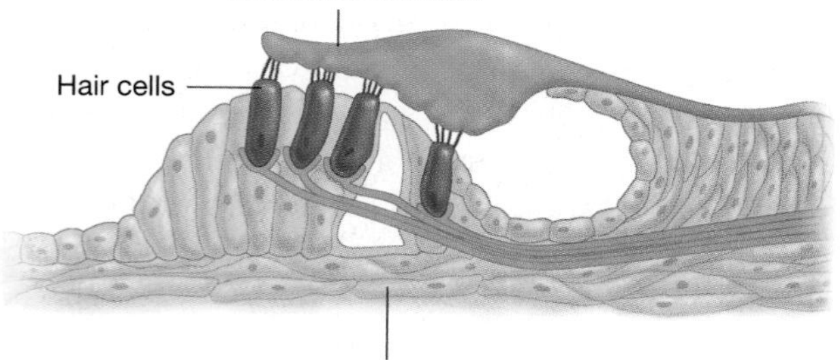

FIGURE 43.6 Where Are Hair Cells Located in the Human Cochlea?
(a) The cochlea contains fluid-filled chambers separated by membranes. **(b)** Hair cells are located in the middle chamber and are sandwiched between the basilar membrane and tectorial membrane.

Sensory Worlds: What Do Other Animals Hear?

Compared to many mammals, human hearing is not particularly acute. For example, as Katherine Payne was observing elephants at the Washington Park Zoo in Portland, Oregon, she noticed a throbbing sensation in the air. Payne knew that **infrasound**, or sound frequencies that are too low for humans to hear, can produce such sensations. To test the hypothesis that the elephants were producing infrasonic vocalizations, she returned to the zoo with microphones that were capable of picking up extremely low-frequency sounds. At normal speed, the tape Payne made was silent. But when she raised the pitch of the sounds by speeding the tape up, she heard a chorus of cow-like noises. The elephants were calling to each other using extremely low frequency sounds.

According to recent follow-up research, elephants have the best infrasonic hearing of any land mammal. Why? Infrasound can travel exceptionally long distances. Based on this observation, biologists hypothesize that infrasonic calls allow wild elephants to coordinate their movements when they are miles apart.

The bats introduced at the start of the chapter furnish another example of an animal perceiving sounds that humans can't. But in this case, the sounds in question have frequencies *above* the range of human hearing. Ultrasonic hearing in bats was discovered in the late 1930s, when Donald Griffin borrowed the only ultrasonic apparatus then in existence from a fellow graduate student named Robert Galambos. Griffin used the machine to demonstrate that flying bats constantly emit ultrasounds. In follow-up experiments, he documented that a bat with cotton in its ears, or with its mouth taped shut, frequently crashes into walls when released in a room. Blindfolded bats, in contrast, never crashed.

Based on these data, Griffin and Galambos concluded that bats use sound echoes to navigate. This was an outlandish idea at the time. When Galambos described the use of sonar by bats at a meeting in 1940, another scientist shook him by the shoulders and said, "You can't really mean that!" Decades later, it is well established that bats, dolphins, shrews, and certain other animals use sonar. In fact, it is likely that at least some of these species perceive shapes with their ears better than they do with their eyes. But for the vast majority of animals, vision is critical to understanding the size, shape, and location of the objects around them.

43.3 Vision

Most animals have some way to sense light. The organs involved range from simple light-sensitive eyespots found in flatworms to the sophisticated, image-forming eyes of vertebrates, cephalopod molluscs, and arthropods. Insects, for example, have a **compound eye** that is composed of hundreds or thousands of light-sensing columns called **ommatidia** (**Figure 43.7**). Each ommatidium contributes information about one small piece of the visual field—not unlike a single pixel on a computer monitor. Thus, the more ommatidia in a compound eye, the better the resolution. Vertebrates and cephalopods, in contrast, have a **camera eye**. This is a structure that focuses incoming light onto a layer of receptor cells. Let's examine this structure more closely.

The Vertebrate Eye

Figure 43.8 shows a typical vertebrate eye. Note that the outermost layer of the structure is a tough rind of white tissue called the sclera. This is the "white of the eye." The front of the sclera is transparent and forms the **cornea**. Just inside the cornea is a colored, round muscle called the **iris**. The iris can contract or expand to control the amount of light entering the eye. The hole in the center of the iris is the **pupil**. Light enters the eye through the cornea, passes through the pupil, and strikes a curved, clear lens. Together, the cornea and lens focus incoming light onto the retina in the back of the eye. The **retina** contains a layer of photoreceptors and several layers of neurons.

CD ACTIVITY 43.1 The Vertebrate Eye

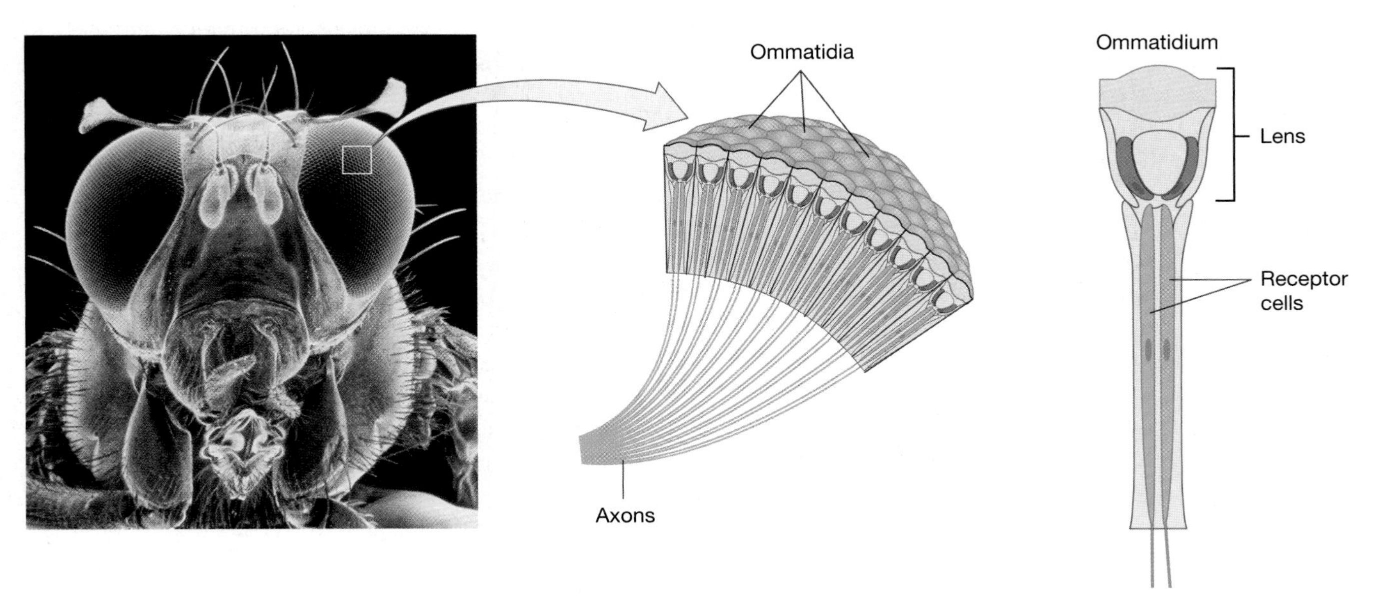

FIGURE 43.7 The Compound Eye of an Insect
Insect eyes are made up of units called ommatidia. Receptor cells inside each ommatidium send axons to the brain.

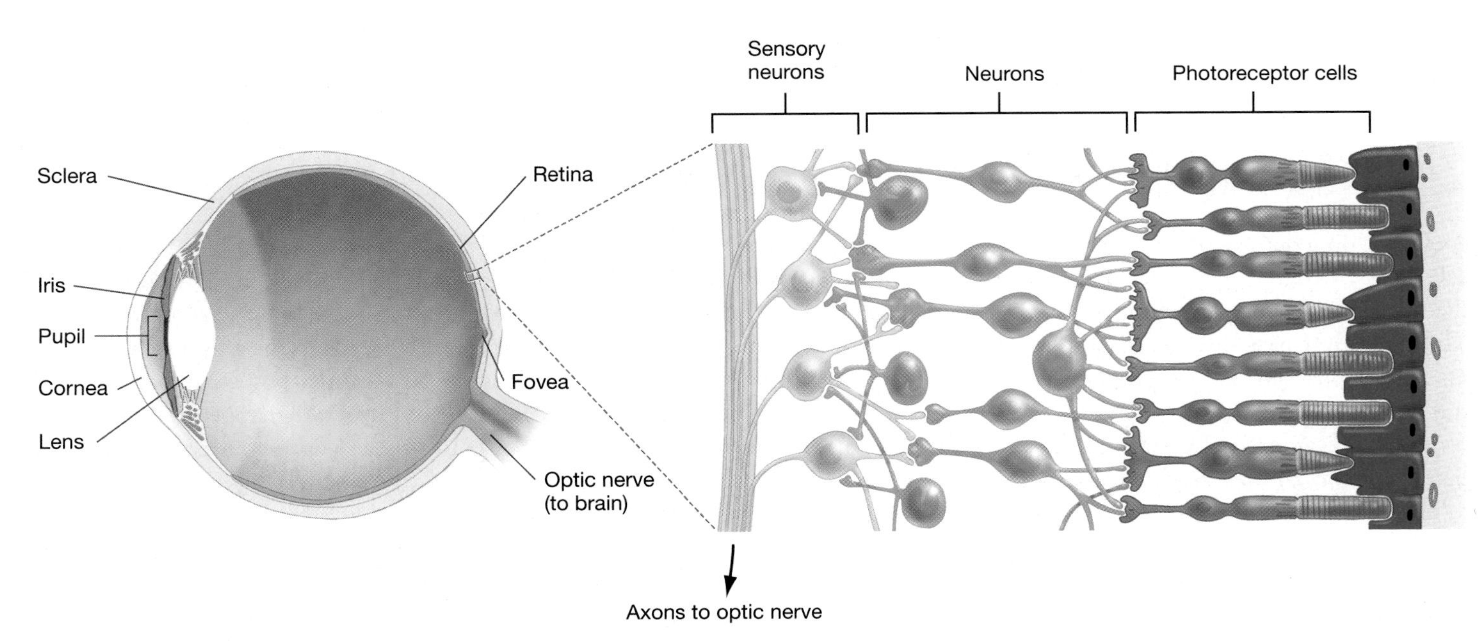

FIGURE 43.8 The Structure of the Vertebrate Eye
Light passes through the pupil of the eye and is focused onto the retina by the cornea and lens. The photoreceptor cells that respond to light are in the outermost layer of the retina. QUESTION Are there any photoreceptor cells located in the spot where the optic nerve leaves the retina?

The detailed diagram in Figure 43.8 provides a closer look at the retina. The sensory cells that respond to light are called **photoreceptors** and form a layer at the back of the retina. Photoreceptors synapse with an intermediate layer of neurons. Cells in the intermediate layer connect with one another and with the neurons of the innermost layer, which project to the brain via the optic nerve. This is an unusual arrangement of neurons, with consequences that are explored in **Box 43.2**. Here, though, let's explore how photoreceptors work.

What Do Photoreceptor Cells Do? Early anatomists established that the photoreceptors in vertebrate eyes are small rod- or cone-shaped cells called **rods** and **cones**. When technical advances allowed researchers to record changes in the membrane potentials of these cells, it became clear that rods are very sensitive to dim light but not to color. Cones, in contrast, are much less sensitive to faint light but are stimulated by different colors. These discoveries explained why night vision is largely black and white.

Early descriptions of the human retina also established that rods dominate most of the structure. There is one small spot in the center of the retina, however, that has only cones. This is the **fovea**. Researchers noticed that when people focus on an object, their eyes move so that the image falls on the fovea of each eye. Based on these observations, biologists concluded that the high density of cones in the fovea maximizes the resolution of the image.

How Do Rods and Cones Detect Light? How do rods and cones actually work? Answering this question required a combination of careful anatomical and biochemical studies. As **Figure 43.9a** shows, rods and cones have segments that are packed with membrane-rich disks. The membranes of these disks contain large quantities of a transmembrane protein called **opsin**. Each opsin molecule is associated with a much smaller molecule called **retinal**. In rod cells, the two-molecule complex is called **rhodopsin** (**Figure 43.9b**).

Is rhodopsin the light receptor? If so, how does it work? Experiments with isolated retinal, opsin, and rhodopsin molecules confirmed that retinal changes shape when it absorbs a photon of light and leads to a change in opsin's conformation (**Figure 43.9c**). Studies of whole rod and cone cells confirmed that the shape change in rhodopsin triggers a series of events that culminates in a change in the cell's membrane potential. If enough photons strike enough rhodopsins, the change in the cell's membrane potential is large enough to alter the amount of neurotransmitter released at the synapse with the neighboring neuron. As a result, a new pattern of action potentials is sent to the brain.

Color Vision: The Puzzle of Dalton's Eye Studies of rhodopsin revealed the molecular mechanism responsible for light perception. But it remained to be determined how humans and other animals can respond to specific wavelengths. How do we perceive color?

To answer this question, consider the research program initiated by eighteenth-century chemist John Dalton. At the age of 26, Dalton realized that he and his brother saw colors differently than did other people. To them, red sealing wax and green laurel leaves appeared to be the same color, and a rainbow exhibited only two hues. Dalton and his brother could not differentiate the colors red and green. This condition is called red-green color blindness.

In a lecture delivered in 1794, Dalton explained his perceptions by hypothesizing that red wavelengths failed to reach his retinas. Further, he hypothesized that because a normal eyeball is filled with clear fluid, and because blue fluids absorb red light, his defective vision resulted from the presence of bluish

BOX 43.2 Vertebrate Versus Cephalopod Eyes

As Figure 43.8 shows, the photoreceptor cells in the vertebrate eye are located in the outermost layer of the retina. As a result, light must pass through several layers of neurons before reaching the photoreceptors. What are the consequences of this arrangement? The major issue is that the axons of the ganglion cells have to disrupt the layer of photoreceptor cells in order to exit the eye at the optic nerve. This creates a blind spot—a portion of the retina where there are no photoreceptors.

To convince yourself that this blind spot exists, point a finger up and hold it out at arm's length. Close your left eye, and focus your right eye on a location just beyond the tip of your finger. Now slowly move your finger to the right. Can you find a place where the tip of your finger vanishes? Light from this region is falling on the blind spot of your right eye. If an arrow or other projectile approached from this direction, you would never detect the danger.

Squid, octopus, and other cephalopods have camera eyes analogous to those found in vertebrates, but their photoreceptors are located in the innermost layer of the retina. As a result, light strikes the photoreceptors directly and they have no blind spot. Why do vertebrates have what appears to be an inferior eye design? As Chapter 38 noted, adaptations are not perfect. To explain the existence of blind spots, biologists hypothesize that the ancestor of vertebrates happened to have an eye-like organ with photoreceptors located in the rearmost cell layer. As a camera eye evolved in the descendants of this species, the ancestral arrangement was retained. If this hypothesis is correct, our blind spots furnish another example of historical constraints on adaptation.

fluid in his eyes. To test this hypothesis, Dalton left explicit instructions that his eyes should be removed after his death and examined to see if the fluid inside was blue. When he died 50 years later, an assistant dutifully removed the eyes from his corpse and examined them. The fluid inside the eye was not blue at all, however, but slightly yellow. This is the normal color for an older person. Further, when the back was cut off of one eye and colored objects viewed through the lens, the objects looked perfectly normal. Dalton's hypothesis was incorrect.

What caused Dalton's color blindness? The key to answering this question was the hypothesis, developed by Thomas Young and others, that the human retina contains just three types of color-sensitive photoreceptors: red, green, and blue cones, named for the colors that they best perceive. When researchers analyzed the activity of these three different cone cells, they found that each absorbed short, medium, or long wavelengths of light (**Figure 43.10**, page 838). To follow up on this result, biologists analyzed opsin molecules from the three cell types and found that each had a distinctive amino acid sequence. The three proteins are called the blue, green, and red (or S, M, and L, for short-, medium-, and long-wavelength) opsins.

Based on these results, biologists hypothesized that the brain distinguishes colors by combining signals initiated by the three classes of opsins. For example, note in Figure 43.10 that a wavelength of 560 nm stimulates L cones strongly, M cones to an intermediate degree, and S cones hardly at all. In response to the corresponding signals from these cells, the brain would perceive the color yellow.

Does this hypothesis explain Dalton's color blindness? According to the data in Figure 43.10, wavelengths from green to red do not stimulate S opsin at all. It is thus unlikely that S opsin is involved in red-green color blindness. Did Dalton fail to distinguish red and green because his M or his L cones were defective? Research has shown that color-blind people lack either functional M or L cones, or both. Was the same true of Dalton? Answering this question became possible in the 1990s, when the genes for the M and L opsins were identified and sequenced. A team led by David Hunt realized that if they could get a sample of Dalton's DNA,

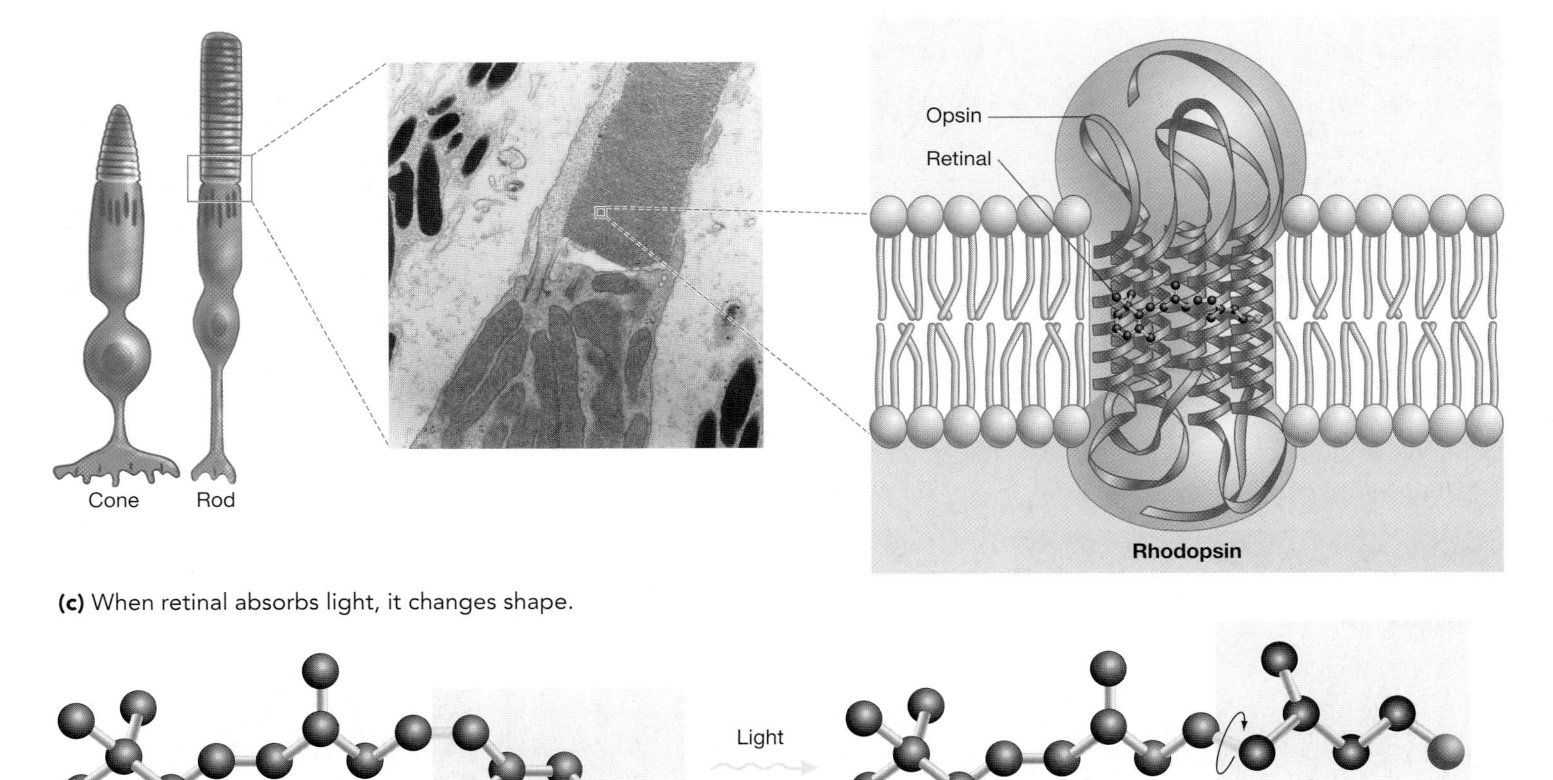

FIGURE 43.9 How Do Rods and Cones Respond to Light?
Rods and cones contain membranous disks containing thousands of opsin molecules. Each opsin holds one retinal molecule. Retinal changes shape when it absorbs a photon of light. In response, opsin also changes conformation.

they could analyze his M opsin genes and L opsin genes directly. Remarkably, Dalton's eyes had been preserved. Hunt's team managed to extract DNA from the 150-year-old tissue and found that Dalton had a perfectly normal L allele but lacked a functional M allele. As a result, he did not have green-sensitive cones. The puzzle of Dalton's color vision was finally solved.

Sensory Worlds: Do Other Animals See Color?

What about other animals—do they see color the way we do? The short answer to this question is probably not. Recent research has shown that different species have opsins with different peak light sensitivities. A marine fish called the coelacanth, which lives in water 200 meters deep, has two opsins that respond to the blue region of the spectrum (wavelengths of 478 nm and 485 nm). Based on this observation, it is likely that coelocanths perceive several distinctive hues of blue that we would perceive as a single color. The existence of these opsins is logical because wavelengths in the yellow and red part of the spectrum do not penetrate well into deep water. Only blue light exists in their habitat, so biologists interpret coelocanth opsins as an adaptation to life in a deepwater environment.

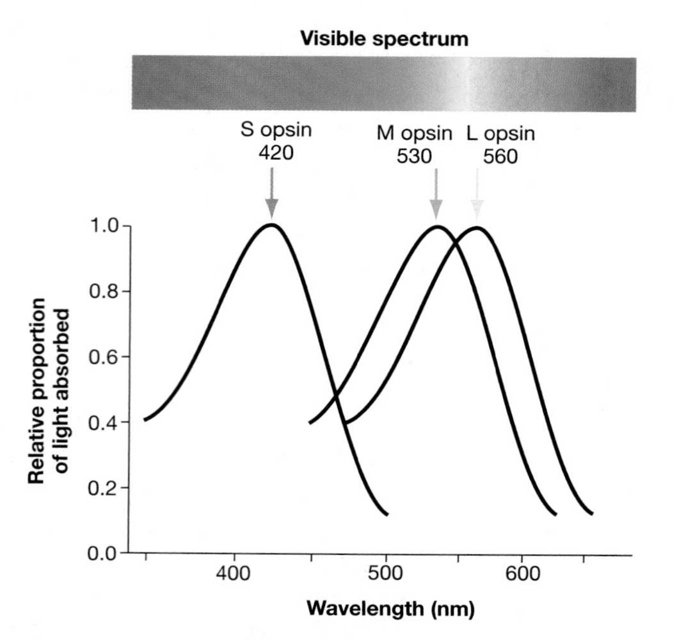

FIGURE 43.10 Different Opsins Have Different Absorption Spectra
Human cone cells contain three different opsins. Each of the opsins has a different absorption spectrum.

Correlations between opsin structure and visual function are extensive. For example, humans and other primates that eat fruit have opsins that are sensitive to wavelengths around the 550 nm range of the spectrum. The presence of these opsins allows individuals to make fine distinctions between the greens, yellows, and reds of unripe and ripe fruits.

In addition, numerous vertebrate and invertebrate species have four or more different types of opsins and probably perceive a world of colors that is much richer than ours. Birds and insects can also see **ultraviolet light**, which has shorter wavelengths than humans can see. This ability is adaptive. As Chapter 36 explained, certain flowers have ultraviolet patterns that serve as signals for pollinating insects. In addition, many birds have strong ultraviolet patterns in their plumage that are invisible to us. Recent experiments have confirmed that in some species, females use these ultraviolet patches as a criterion for selecting mates. At the other end of the visible spectrum, rattlesnakes and other pit vipers can sense **infrared light** using specialized infrared-sensing organs on their snouts (**Figure 43.11a**). Infrared light has longer wavelengths than humans can see. Because endothermic animals radiate heat in the form of infrared radiation, pit vipers can detect and strike at prey in complete darkness (**Figure 43.11b**).

(a) Pit vipers can detect infrared radiation.

(b) Warm animals give off infrared radiation.

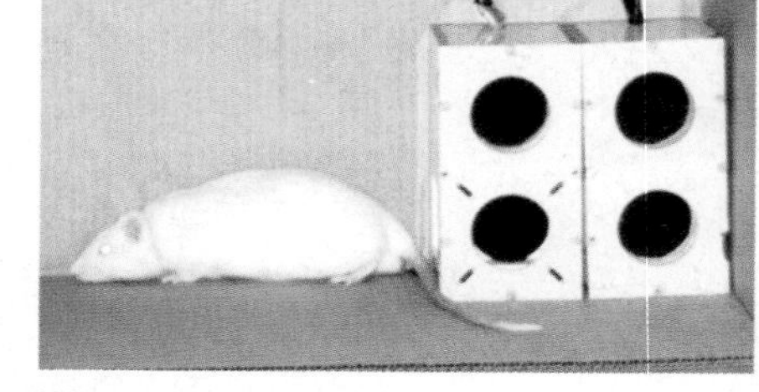

What we see in the light

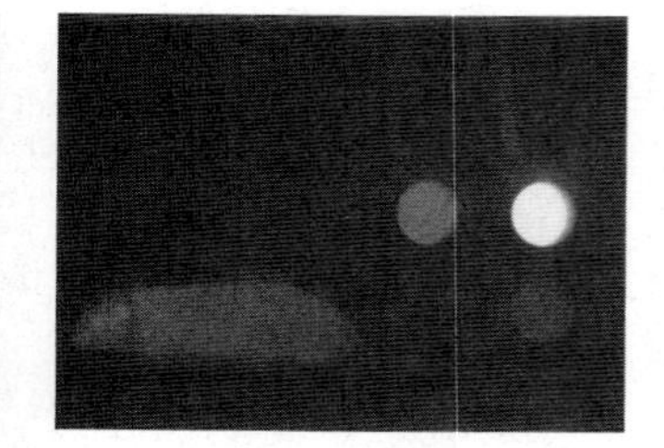

What pit vipers "see" in the dark

FIGURE 43.11 Pit Vipers Detect Infrared Radiation
(a) Pit vipers are a group of snakes named for the "pits" on their heads. These pits contain receptor cells that detect infrared radiation. **(b)** Pit vipers are able to "see" a fuzzy image of warm objects. As a result, a rattlesnake can strike accurately at a mouse even in darkness. (They will even strike at warm, covered light bulbs.) The snake loses this ability if the pits are covered experimentally.

43.4 Taste and Smell

The senses of taste and smell (olfaction) originate in chemoreceptors. These cells detect the presence of particular molecules. Until recently, taste and smell were poorly understood compared to vision and hearing. Eyes and ears respond to relatively simple stimuli—light and sound waves—that can be reproduced in the lab. The tongue and nose, in contrast, respond to thousands of different chemicals. Consequently, taste and smell were difficult to study until techniques were available for identifying how particular molecules bind to certain receptors. But in the last 10 years, research on the chemosenses has exploded.

Understanding taste and smell is important because animals use these senses to find food, assess mates, and avoid danger. For researchers in the food and cosmetic industries, understanding particular aspects of chemoreception, such as the sensation of sweetness and the activity of perfumes, is critical to developing new products.

Taste: Detecting Molecules in the Mouth

Although humans have taste buds scattered in the mouth and throat, most taste buds are located on the tongue. As **Figure 43.12** shows, a taste bud contains about 100 spindle-shaped **taste cells** that synapse to taste neurons. How do these receptors work on a molecular level, and how do they produce the sensation of taste?

Early taste research focused almost exclusively on the hypothesis that four "basic tastes" existed—salty, sour, sweet, and bitter. When researchers analyzed the membrane proteins found in taste cells, they found strong evidence that salt and sour sensations result from the activity of ion channels. Specifically, the sensation of saltiness is primarily due to sodium ions (Na^+) dissolved in food. These ions flow into certain taste cells through open Na^+ channels and depolarize their membranes. Similarly, sourness is due to the presence of protons (H^+), which flow directly into certain taste cells through H^+ channels. The sour taste of grapefruit and other citrus fruits, for example, is due to protons released by citric acid. In general, the lower the pH of a food, the more it depolarizes a taste-cell membrane and the more sour the food tastes.

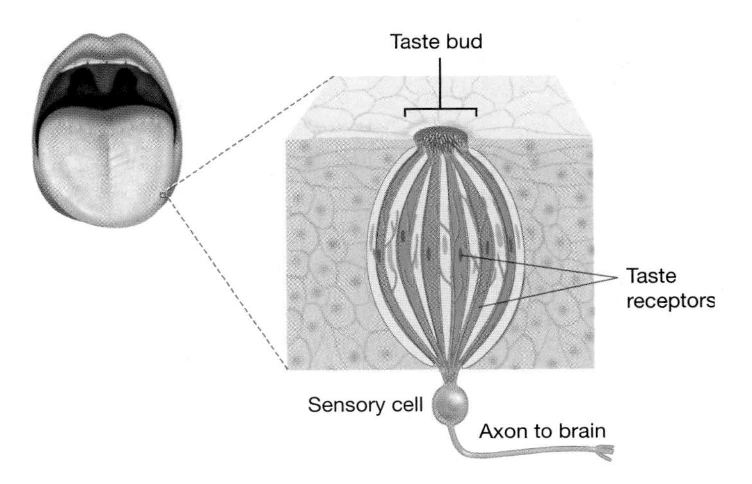

FIGURE 43.12 The Human Tongue and Its Taste Buds
The human tongue has a high density of taste cells located in structures called buds.

The molecular mechanisms responsible for the sensations of bitterness and sweetness have been much more difficult to identify. Researchers have only recently been able to document that certain food molecules actually bind to specific receptors on taste cells to cause bitter and sweet tastes.

Why Do Many Different Foods Taste Bitter? Bitterness has been a particularly difficult taste to understand. The problem is that molecules with very different structures are perceived as bitter. How is this possible? An answer began to emerge after researchers confirmed that some humans genetically lack the ability to taste certain bitter substances. This phenomenon came to light in 1931, when Arthur Fox was synthesizing some phenylthiocarbamide (PTC) and accidentally blew some into the air. A nearby colleague complained of a bitter taste in his mouth. Fox could not taste anything, however. Follow-up research confirmed that the ability to taste PTC is inherited and polymorphic; about 25 percent of United States citizens cannot sense the molecule. To find the gene responsible for the trait, biologists compared the distribution of genetic markers observed in "tasters" and "nontasters." The effort to map the gene narrowed its location down to several candidate chromosomal segments. (Chapter 17 introduced how this type of gene hunt is done.)

At about the same time, researchers who were working on taste reception in mice found that some individuals are better than others at perceiving bitter substances and avoiding them. Genetic mapping, analogous to the effort in humans, narrowed the location of the bitter receptor gene down to a specific region of a mouse chromosome.

These independent lines of research converged when different research teams searched a database of known genes to look for sequences similar to those from the candidate regions of the human and mouse chromosomes. Independently, the two teams identified a family of 40–80 genes that encode transmembrane receptor proteins. Follow-up work documented that each protein in the family binds to just one particular type of bitter molecule. A taste cell, however, can have many different receptor proteins from this family. As a result, many different kinds of molecules can depolarize the same cell and cause the taste of bitterness. These results explained why so many different molecules give rise to the same sensation.

Why do animals have such extensive molecular machinery to detect bitterness? Many of the molecules that bind to bitter receptors are found in toxic plants. Also, most animals react to bitter foods by spitting them out and avoiding them in the future. Based on this, biologists hypothesize that bitter receptors evolved in response to selection for avoiding toxic molecules. In essence, bitterness means "this food might be dangerous; don't swallow it." The proliferation of genes responsible for detecting bitterness hints that the trait is extremely important.

What Is the Molecular Basis of Sweetness and Other Tastes? Inspired by the rapid recent progress on bitter receptors, research teams are using the same techniques to search for sweetness receptors. As this book goes to press, however, membrane receptors that respond to sugars have yet to be identified. Membrane receptors for amino acids have recently been discovered, however. One of these proteins is responsible for the sensation called umami, which is the meaty taste of monosodium glutamate (MSG).

In addition to searching for other types of taste receptors, several teams are studying the question of how the different taste sensations are conveyed to the brain and interpreted. Although taste is beginning to reveal its secrets, the complete story will probably not be known for many more years.

Olfaction: Detecting Molecules in the Air

Taste allows animals to assess the quality of their food before swallowing it. Smell, in contrast, allows animals to monitor airborne molecules that convey information. Wolves and domestic dogs, for example, can distinguish millions of different odors at vanishingly small concentrations. The molecules that cause odor contain information about the movements and activities of prey and other members of their own species.

How does the sense of smell work? When odor molecules, or odorants, reach the nose, they diffuse into a mucus layer in the roof of the structure. There, they activate olfactory receptor neurons via membrane-bound receptor proteins (**Figure 43.13**). Axons from these neurons project up to the olfactory bulb of the brain.

Understanding the anatomy of the odor recognition system was a relatively simple task. Understanding how receptor neurons distinguish one molecule from another turned out to be much more difficult. Initially, investigators hypothesized that receptors respond to a small set of "basic odors" such as musky, floral, minty, and so on. The idea was that each basic odor would be detected by its own type of receptor.

In 1991, however, Linda Buck and Richard Axel demolished the basic odor hypothesis with an astonishing result. Their studies of odor recognition in mice culminated in the discovery of a gene family—containing *hundreds* of distinct coding regions—that encodes receptor proteins on the surface of olfactory receptor neurons. Follow-up experiments confirmed that each receptor protein binds to a small set of molecules. Further work established that most, if not all, vertebrates have this family of genes. Mammals typically have between 500 and 1000 of these genes, meaning that they can produce 500–1000 different odor-receptor proteins. In humans, however, about half of these genes have mutations that probably render them nonfunctional. This observation may explain why our sense of smell is so poor compared to that of other mammals.

Buck and Axel's announcement inspired a series of questions. How many different receptors occur in the membrane of each neuron involved in odor reception? How does the brain make sense of the input from so many different receptors? Recently, Buck determined that each olfactory neuron has only one type of receptor, and that neurons with the same type of receptor project to distinct regions in the olfactory region of the

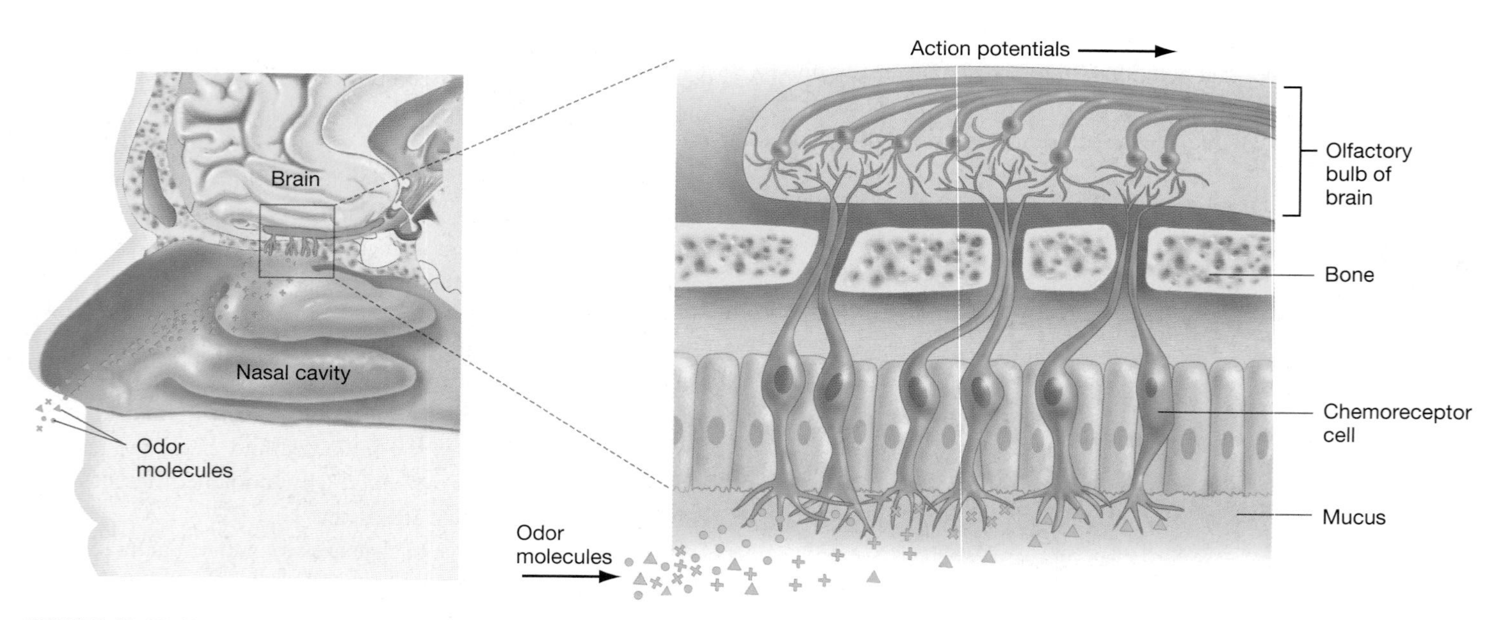

FIGURE 43.13 How Do Mammals Sense Odors?
Odor molecules are detected in the nose by chemoreceptor cells. Each of these sensory neurons carries one type of odorant-receptor protein on its dendrites. Further, sensory neurons with the same receptor project to a particular section, called a glomerulus, within the olfactory bulb of the brain.

brain. These regions are called glomeruli (meaning "little balls"). Further, recent results from Axel's lab indicate that particular smells in mice are associated with the activation of a certain subset of the 2000 glomeruli in the brain. For example, activation of clumps 130, 256, and 1502 might be perceived as the smell "cinnamon." In essence, then, the sense of smell is similar to the visual system's use of three cones to perceive many colors, but on a much larger scale. Research on this complex and impressive sense continues at a furious pace.

43.5 Movement

The first four sections of this chapter focused on how animals sense different aspects of their environment. Acquiring information is only a first step, however. Animals also must respond to this information—usually by moving. Moths and bats not only hear ultrasonic frequencies, they alter their flight paths and activities in response. Flight and other types of motion are based on muscle contractions.

To explore how movement occurs, let's begin by examining the diversity of muscle cells found in vertebrates (**Figure 43.14**). After exploring the molecular mechanisms of contraction, we can return to the question of how muscles work together to produce movement.

How Do Muscles Contract?

Early microscopists established that the muscle tissue found in vertebrate limbs and heart is composed of slender fibers. A **muscle fiber** is a long, thin muscle cell. Within each cell there are many small strands called **myofibrils**. When workers examined the sheets or bands of muscle that surround blood vessels and most organs, however, they found that fibers and myofibrils were absent.

Further anatomical studies confirmed that vertebrates have three distinctive types of muscle tissue. The tissues are called skeletal, cardiac, and smooth muscle and are compared and contrasted in Figure 43.14. Although smooth muscle is essential to the function of the lungs, blood vessels, digestive system, urinary bladder, and reproductive system, most research on muscle function has focused on how skeletal and cardiac muscle work. How do these muscle cells contract to produce movement?

The Sliding-Filament Theory As **Figure 43.15** (page 842) shows, the myofibrils inside skeletal muscle cells look striped. The pattern is caused by the repeating light-dark units called **sarcomeres**. Comparative studies showed that the sarcomeres

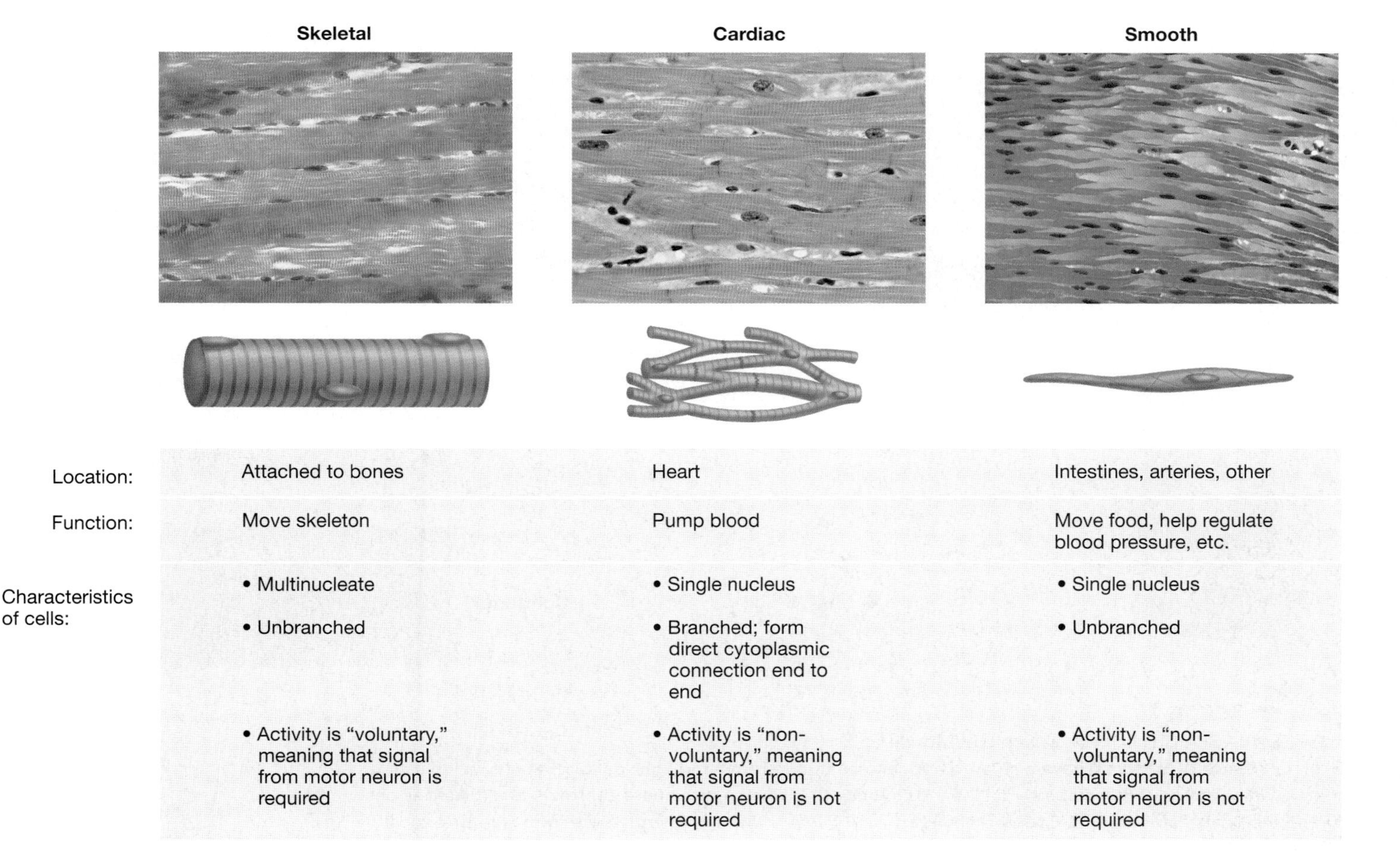

FIGURE 43.14 The Three Types of Vertebrate Muscle

lengthen when a muscle is stretched and shorten when the muscle is contracted (Figure 43.15).

Based on these observations, the question of how muscles contract simplifies to the question of how sarcomeres contract. Andrew Huxley and Jean Hanson proposed an answer to this question in 1954. These biologists had noticed that the relationship between different parts of the sarcomere—specifically, the points labeled A–D in the photographs in Figure 43.15—change during contraction. Points A and B move closer to each other during contraction; points C and D also move closer together. But the distance from point A to point C does not change, nor does the distance from point B to point D.

To explain these observations, Huxley and Hanson hypothesized that the banding patterns in the sarcomere are actually caused by two types of long filaments—thick and thin—and that the filaments slide past one another during contraction. **Figure 43.16** illustrates the model. Note that thin filaments extend from A to C and thick filaments extend from B to D. According to Huxley and Hanson, the thick and thin filaments slide past one another during a contraction. This explanation became known as the **sliding-filament theory.**

Follow-up research has shown that the Huxley-Hanson model is correct in almost every detail. Thin filaments are composed of two coiled chains of a globular protein called actin (Figure 43.16). One end of each actin chain is bound to the wall of the sarcomere; the other end interacts with a thick filament. Thick filaments are composed of multiple strands of a long protein called **myosin** (Figure 43.16). The thick strands are anchored to the middle of the sarcomere. When researchers isolated actin and myosin and mixed them together on a glass slide in the presence of ATP, they observed myosin crawling along actin. As predicted, thick and thin filaments slide past one another.

How Do Actin and Myosin Interact? Once the sliding-filament theory was confirmed, researchers focused on the question of how the crawling action occurs on the molecular level. Early work on the three-dimensional structure of myosin revealed that each molecule has a head region that projects from the main body of the thick filament (Figure 43.16). Follow-up experiments established that the myosin head can bind to actin and catalyze the hydrolysis of ATP into ADP and a phosphate ion. In addition, electron microscopy revealed that myosin and actin are locked together shortly after an animal

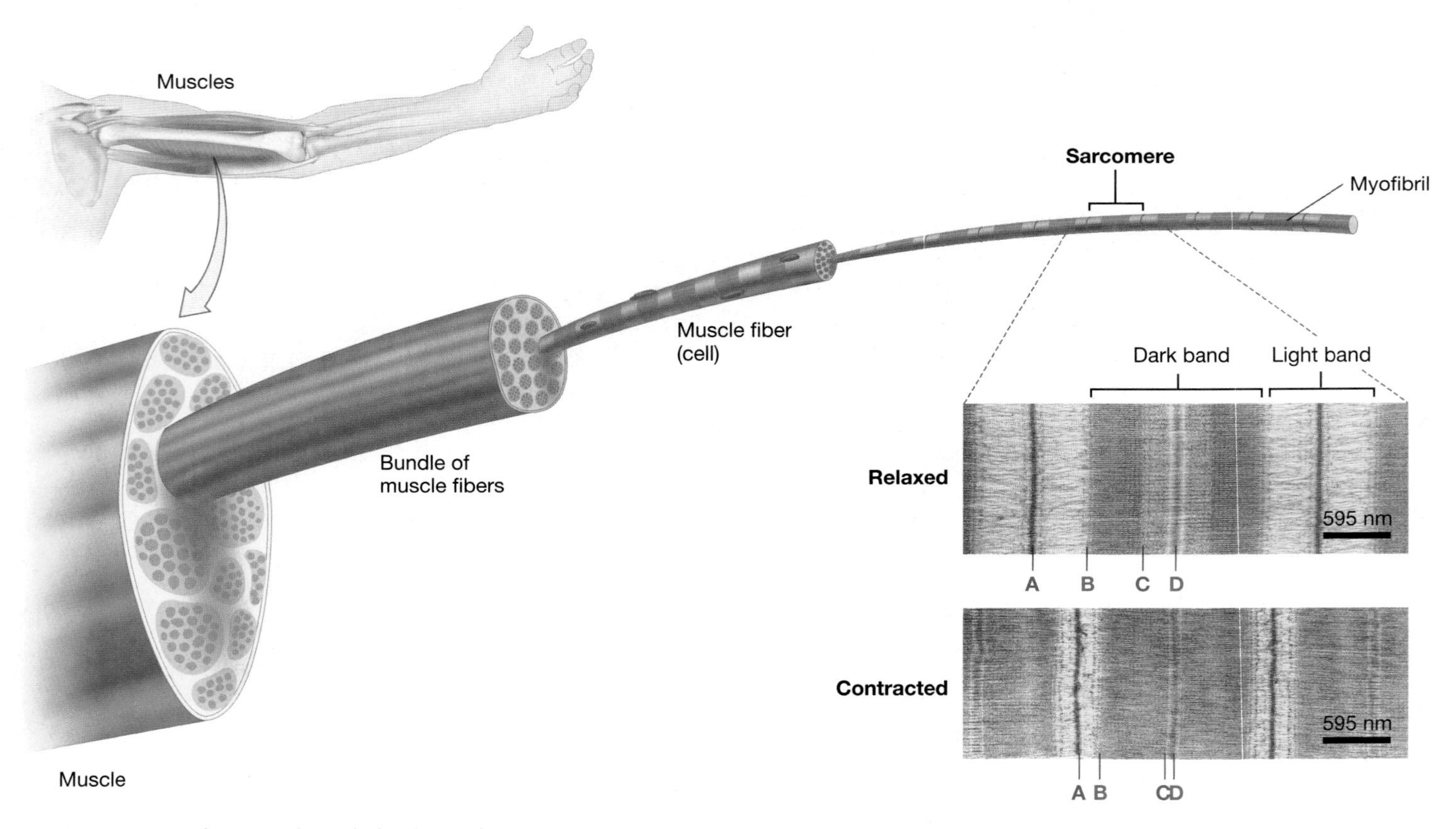

FIGURE 43.15 A Closer Look at Skeletal Muscle
Skeletal muscle cells (fibers) have a striated appearance due to repeating light-dark-light units called sarcomeres, which are bounded by heavy dark bands. Notice that when a sarcomere contracts, the distances between the points labeled A, B, C, and D change. QUESTION Measure the distance from A to B, A to C, A to D, B to C, B to D, and C to D. Which points get closer together? Which points stay the same distance from each other?

dies and its muscles enter the stiff state known as rigor mortis. This observation suggested that ATP must be present for myosin to release from actin.

The next major advance in understanding how actin and myosin interact to produce movement was achieved when Ivan Rayment and colleagues solved the three-dimensional structure of the myosin head using the x-ray crystallographic techniques introduced in Chapter 13. **Figure 43.17a** (page 844) shows the structure of the head, with the actin binding site and ATP binding site labeled. Rayment's group also examined how the protein's structure changed when ATP or ADP was bound to it, and they noted significant changes in the protein's conformation.

Based on their data on the structure of the myosin head, Rayment and co-workers proposed the model for actin-myosin interaction illustrated in **Figure 43.17b**. Step 1 shows that when a molecule of ATP binds to the myosin head, the myosin releases from actin. When ATP is subsequently hydrolyzed to ADP and inorganic phosphate in step 2, the conformation of the protein changes. Specifically, the neck of the myosin straightens out, the head pivots, and the myosin head then binds to actin in a new location. The third step in the sequence occurs when inorganic phosphate is released and the protein's conformation changes again. Specifically, the neck bends back into its original position. This bending is called the power stroke because it moves the entire thin filament. In step 4, ADP is released. A new ATP molecule then binds to the myosin, causing myosin to release from actin, and the cycle starts anew.

As ATP binding, hydrolysis, and release continues, the two ends of the sarcomere are pulled closer together. The same basic mechanism is observed in cardiac and smooth muscle cells. Most animal movement, in fact, is powered by similar interactions between actin and myosin. The two proteins are also responsible for the amoeboid movement observed in amoebae and slime molds (see Chapter 27) and the streaming of cytoplasm observed in algae and land plants. The message of these observations is that actin and myosin are evolutionarily ancient proteins. Because they make movement possible in the absence of cilia and flagella, they have played a critical role in the diversification of eukaryotes.

How Does Relaxation Occur? The results summarized in Figures 43.16 and 43.17 were obtained by studying myosin and actin in isolation. Although elegant and powerful, experiments on simplified muscle raised an important question. In vitro, thick filaments from muscle continue to crawl along thin filaments as long as ATP and a thin filament are available. ATP is almost always available in living muscles. Given these observations, how do our muscles ever stop contracting and relax?

Researchers were able to answer this question when they confirmed that sarcomeres contain proteins called tropomyosin and troponin in addition to actin and myosin. Tropomyosin and troponin work together to block the myosin binding sites on actin. As a result, thick and thin filaments cannot slide past each other. Now the question is, how are tropomyosin and troponin moved out of the way so that contraction can begin?

An Overview of Events at the Neuromuscular Junction **Figure 43.18** (page 845) summarizes what happens when an action potential from a motor neuron arrives at a muscle cell and initiates contraction.

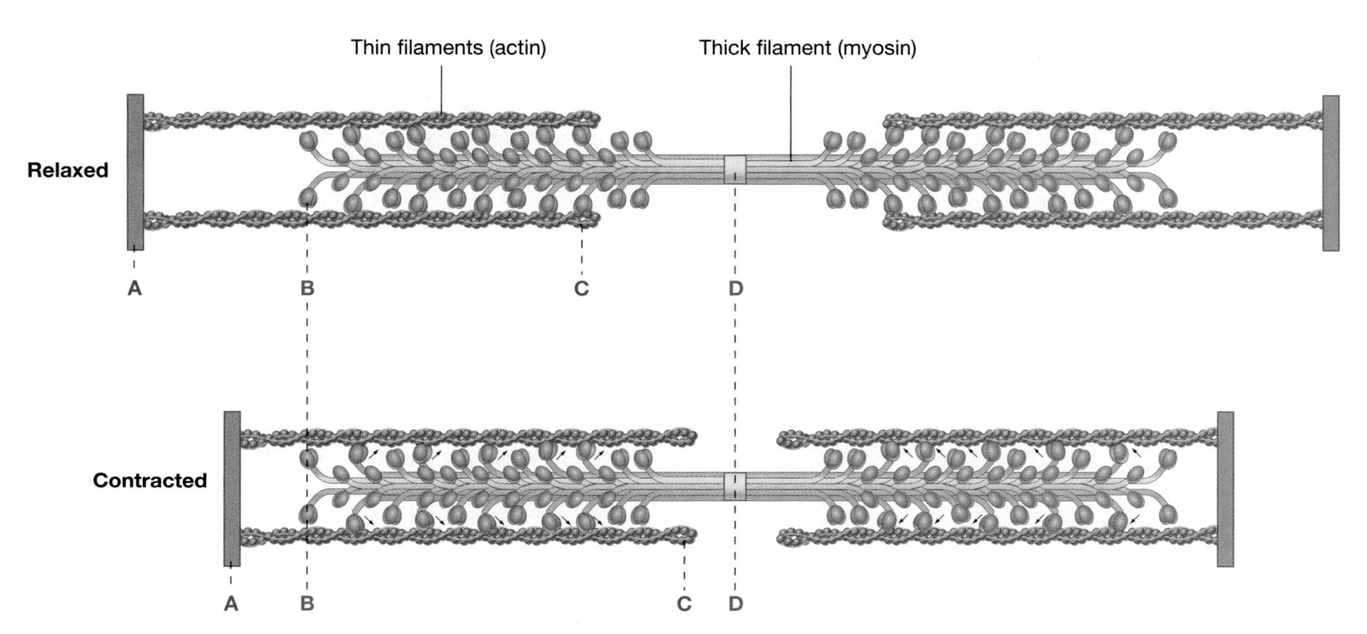

FIGURE 43.16 The Sliding-Filament Model of Sarcomere Contraction
The points labeled A, B, C, and D on this drawing are the same as those illustrated in the photographs of a sarcomere in Figure 43.15. **QUESTION** Does this model explain the changes in distances that you observed in Figure 43.15? Explain.

1. Action potentials trigger the release of acetylcholine (ACh) from the motor neuron onto the muscle cell. Chapter 42 introduced the experiments that documented this event in cardiac muscle; follow-up research confirmed that ACh is also released by motor neurons in skeletal muscle.
2. By recording voltage changes in muscle cells, Stephen Kuffler showed that a membrane depolarization occurs in response to ACh release. If enough ACh is applied to a muscle cell, depolarization triggers a series of action potentials in the fiber. The action potentials propagate throughout muscle cells via axon-like structures called the T tubules. (The *T* stands for *transverse*, meaning "extending across.")
3. T tubules intersect extensive sheets of smooth endoplasmic reticulum called the sarcoplasmic reticulum. As an action potential passes down a T tubule, a protein in the T tubule membrane changes conformation and opens a calcium channel in the sarcoplasmic reticulum.
4. Calcium ions (Ca^{2+}) are released from the sarcoplasmic reticulum. Ca^{2+} causes a conformational change in troponin, which then moves tropomyosin away from the myosin binding sites on actin.

The grand result of these events is that myosin binds to actin and contraction begins. Muscle contractions are impor-

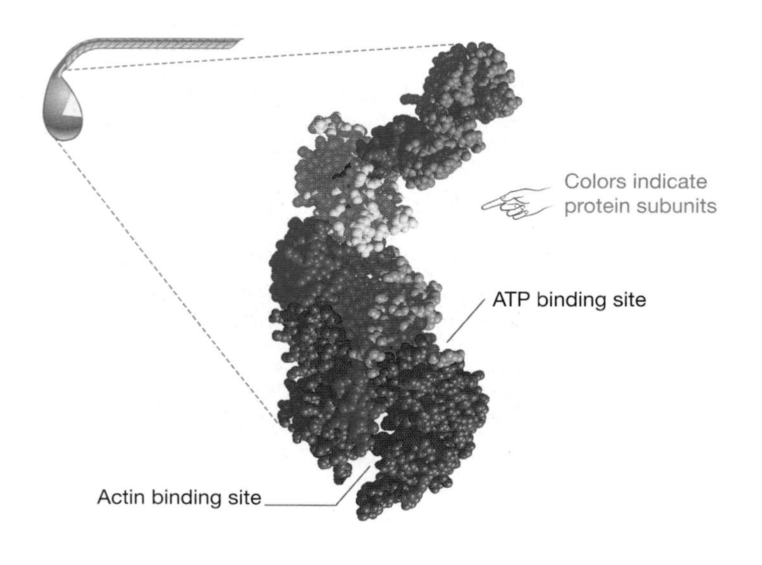

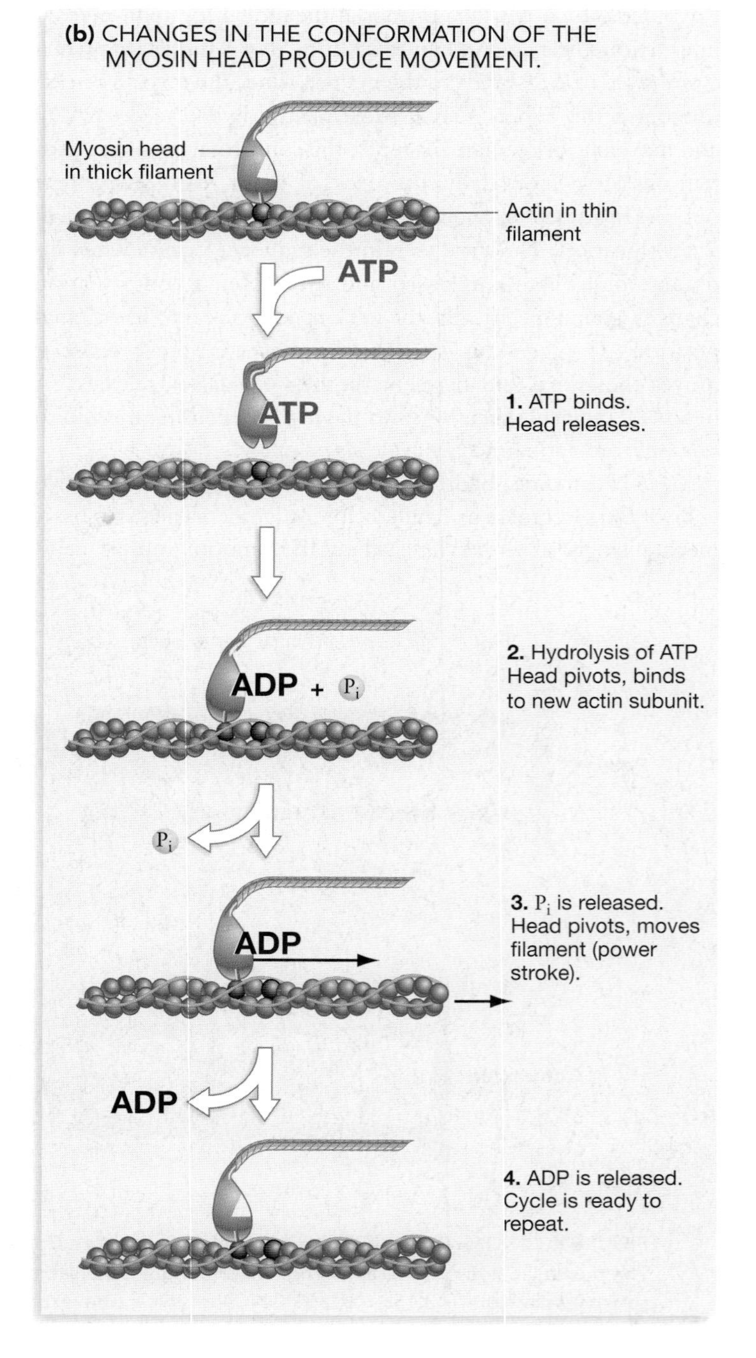

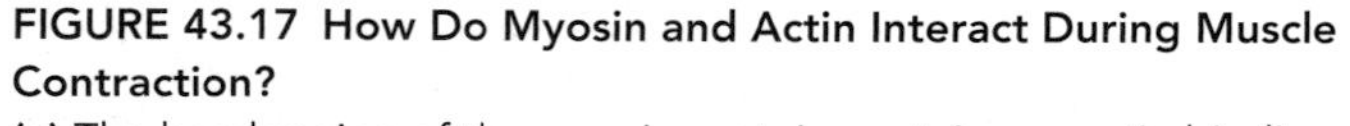

FIGURE 43.17 How Do Myosin and Actin Interact During Muscle Contraction?
(a) The head region of the myosin protein contains an actin binding site and an ATP binding site. **(b)** This diagram summarizes the current model for how myosin and actin interact as a sarcomere contracts. The four steps repeat rapidly.

tant because they result in movement. Movement is important because it allows animals to escape predators and acquire the resources needed to reproduce. How do muscle contractions result in the beating of a moth's wing or the opening of a bat's jaw? To answer this question, we need to explore the nature of animal skeletons.

Skeletons

Skeletons provide attachment sites for muscles and a support system for the body's soft tissues. As Chapter 30 indicated, three types of skeletons are found in animals. Exoskeletons are hard, hollow structures that envelop the body; hydrostatic skeletons use the pressure of internal body fluids to support the body; endoskeletons are hard structures inside the body. Because Chapter 30 introduced the design of exoskeletons and hydrostatic skeletons in some detail, our focus here is on the structure and function of endoskeletons.

As **Figure 43.19a** (page 846) shows, endoskeletons are composed of connective tissues called cartilage and bone. **Cartilage** is made up of cells scattered in a gelatinous matrix of polysaccharides and protein fibers. In the skeletal system, cartilage provides padding between bones. **Bone**, in turn, is made up of cells in a hard extracellular matrix of calcium phosphate ($CaPO_4$) with small amounts of calcium carbonate ($CaCO_3$) and protein fibers. When bones articulate at joints, they do so in ways that allow limbs to swivel, hinge, or pivot (**Figure 43.19b**).

How do skeletal muscles attach to and move elements of an endoskeleton? In vertebrates, the ends of a skeletal muscle are often attached to two different bones by tendons. **Tendons** are bands of tough, fibrous connective tissue. Because muscles can exert force only by contracting, pairs of muscles must work together to move a bone back and forth. In the case of a limb, one muscle pulls the limb in one direction; the other muscle pulls it in the opposite direction (**Figure 43.20a**, page 847). This pairing of muscles is called an **antagonistic muscle group**. The muscle that swings long bones in an arc toward each other is called a **flexor**; the muscle that straightens them out is called an **extensor**. For example, the hamstring in the back of your thigh flexes your lower leg; the quadriceps in the front of your thigh extends it.

As **Figure 43.20b** and **c** show, animals with exoskeletons have a similar flexor-extensor arrangement inside of their hollow leg joints; animals with hydrostatic skeletons have circular and longitudinal muscles that work in concert. The movements of these paired muscles are coordinated by motor neurons that originate in the brain. These motor neurons, in turn, project from processing centers that receive input from sensory systems. The activity of these motor neurons changes in response to information about balance, smells, sights, and sounds. In this way, an animal's movements are directly tied to the functioning of its sensory systems. The interplay of sensory input and motor output results in the coordinated behaviors we call running, eating, and flying.

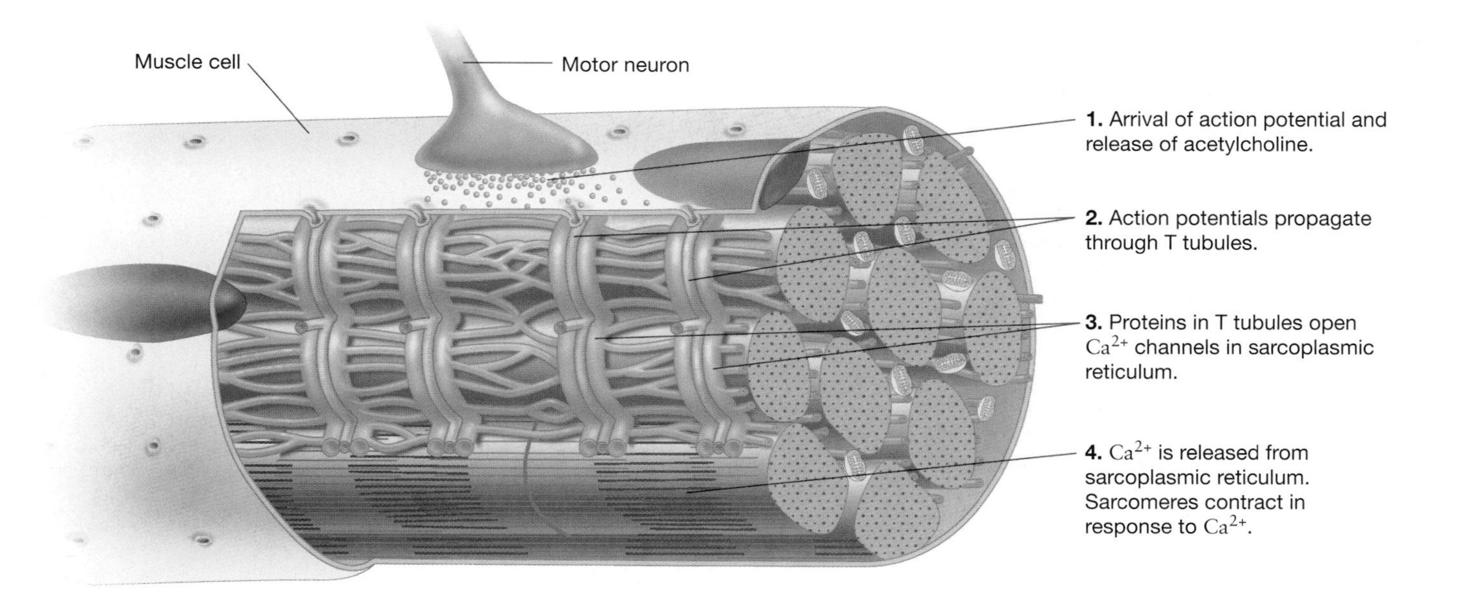

FIGURE 43.18 How Do Action Potentials Trigger Muscle Contraction?
As this diagram shows, the arrival of action potentials from a motor neuron triggers a series of events in a muscle cell. The sequence ends with changes in troponin-tropomyosin proteins that allow contraction to begin.

(a) Endoskeletons contain two major tissue types.

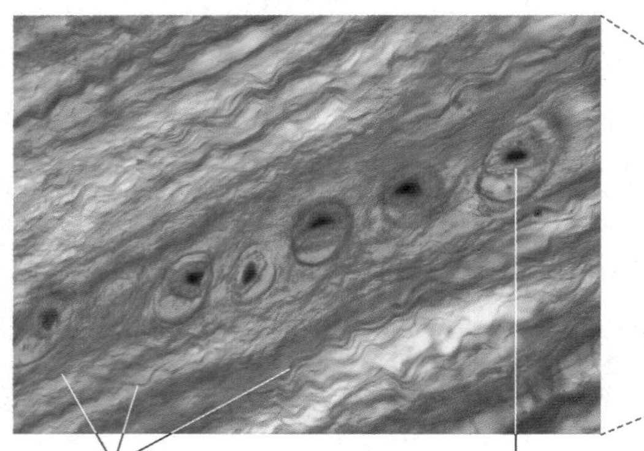

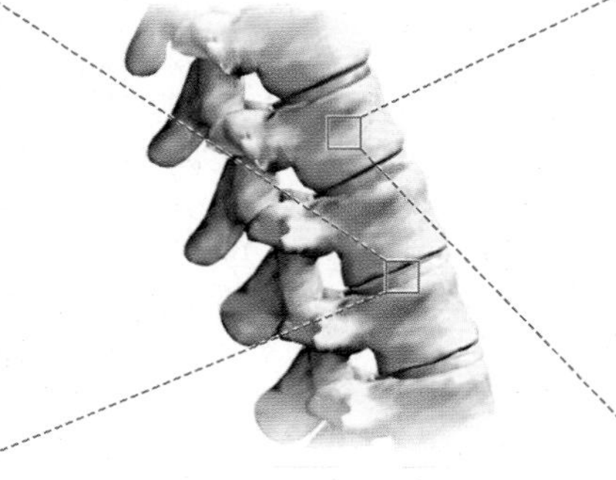

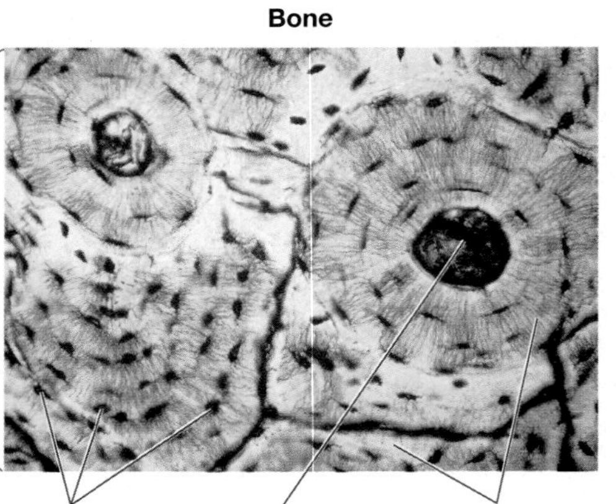

(b) Joints allow limbs to move.

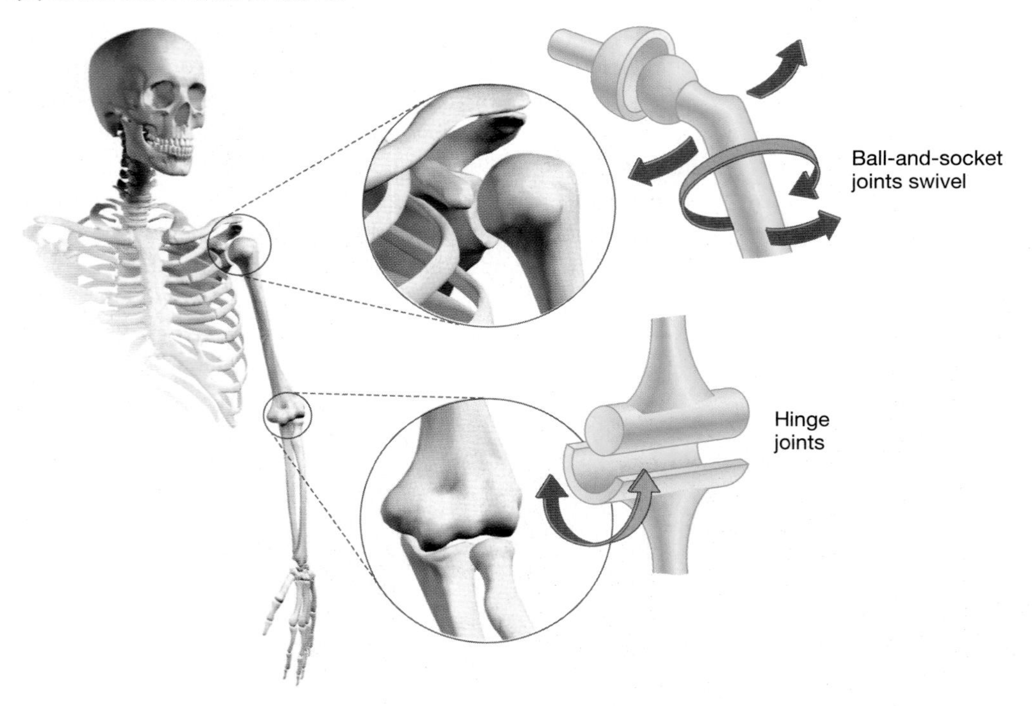

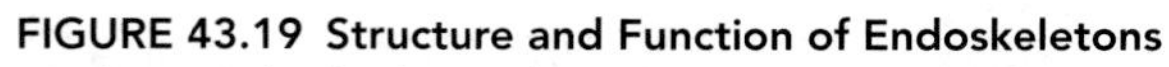

FIGURE 43.19 Structure and Function of Endoskeletons
(a) The vertebral column shown here is made up of bones separated by disks of cartilage. **(b)** Bones articulate in ways that make specific types of movement possible.

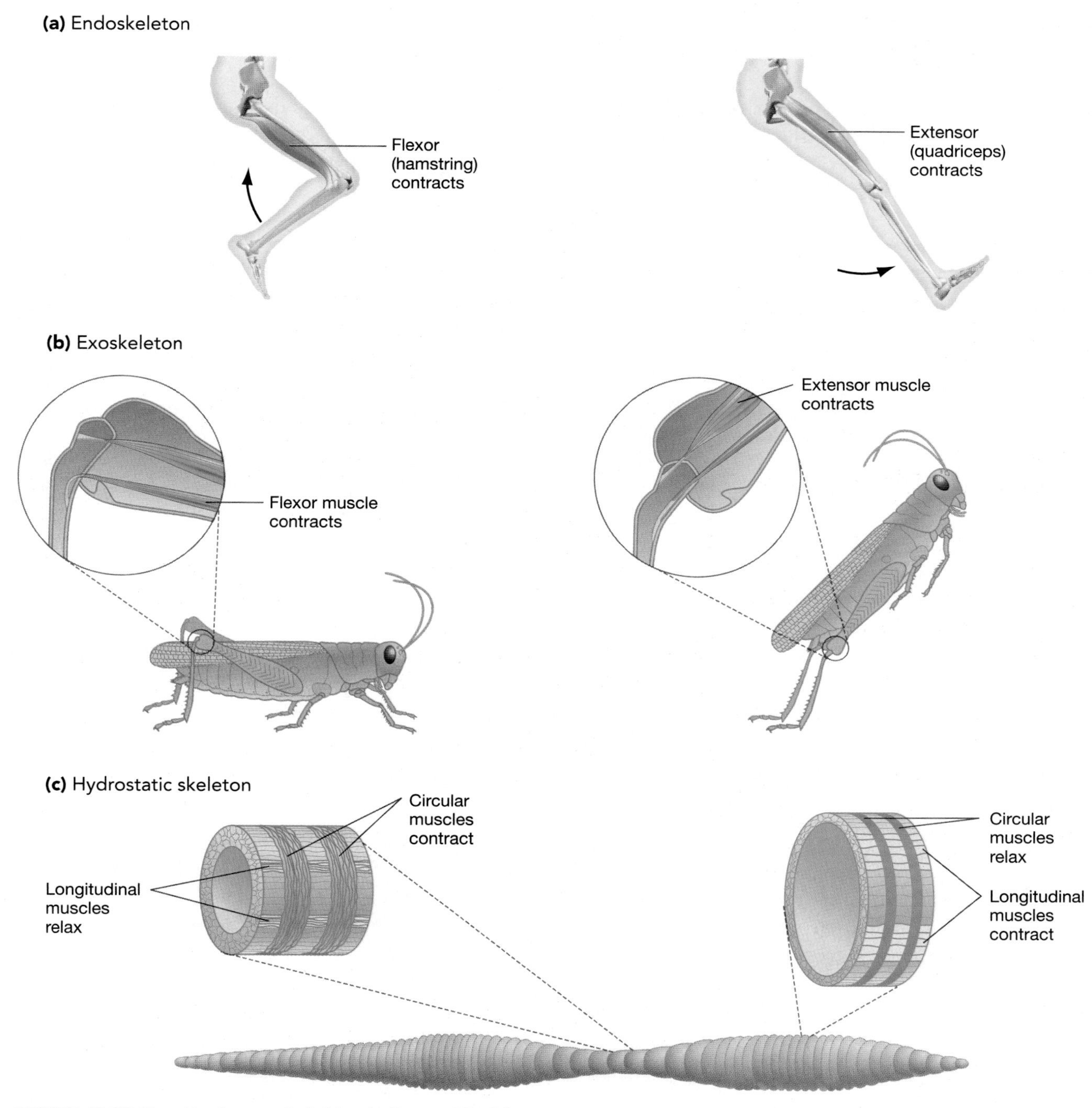

FIGURE 43.20 How Do Antagonistic Muscle Groups Work?
(a) In vertebrates, muscles are arranged in pairs of flexors and extensors. Together, a flexor and extensor can control a particular joint. **(b)** Animals with exoskeletons also have flexors and extensors inside their hollow joints. **(c)** In animals with hydrostatic skeletons, the coordinated movements of longitudinal and circular muscles shorten and lengthen segments of the body.

Essay Sprinters and Marathoners—Born or Made?

Vertebrate muscle fibers come in two basic varieties: slow-twitch and fast-twitch. As **Table 1** shows, slow-twitch fibers excel at sustained activity such as distance running. Slow-twitch fibers generate most of their ATP by aerobic respiration and have large stores of a protein called myoglobin, which holds oxygen. Because myoglobin is red, slow-twitch muscle is reddish. In contrast, fast-twitch muscle is paler. Fast-twitch fibers contract rapidly and excel at quick bursts of motion like sprinting. Because they generate most of their ATP by anaerobic metabolism, fast-twitch fibers fatigue rapidly due to lactic acid buildups.

Training can increase the number of myofibrils per cell.

Why does contraction speed vary so dramatically between these cell types? The answer is that each type of fiber contains a different type of myosin. Slow-twitch fibers make type I myosin; fast-twitch fibers make either type IIa or IIx, with IIx being faster. Type II myosin supports faster contraction because it hydrolyzes ATP more rapidly than does type I myosin.

When researchers compare the muscle fibers of sprinters and endurance athletes, they find strong contrasts in myosin types. The leg muscles of elite sprinters are dominated by type IIx myosin, while endurance athletes have mostly type I myosin. Is this difference due to training, or are people born with a certain myosin type?

To answer this question, biologists have surveyed myosin types in people who do not train their muscles. Untrained individuals have substantial differences in their percentages of I, IIa, and IIx fibers. This result suggests that at least some of the observed phenotypic variation is due to genetic variation. But because everyone has all three myosin genes in all of their muscle fibers, it is clear that genetic variation occurs at the level of gene regulation—meaning, variation in which genes are actually expressed as muscles develop.

It would be incorrect, however, to conclude that all differences in fiber types are genetic. Studies of sprinters and endurance athletes have shown that the proportion of muscle cross-sectional area devoted to each myosin type changes with training. For example, sprint training causes a significant increase in the percentage of fast-twitch fibers after just two months. Long-distance running takes more time to have an effect, but after 2–6 months the proportion of slow-twitch fibers increases. How do these changes occur? Although new muscle cells are never created, training can increase the number of myofibrils per cell. The change in proportion of muscle cross-sectional area devoted to each myosin type occurs as cell size changes.

Some recent data suggest that muscle fibers may also change their myosin type with training. Preliminary results indicate training with weights can lead to a conversion of type IIx fibers to IIa. In addition, some data imply that type I fibers can be induced to switch to type II with a vigorous sprinting regimen. But according to the 2- to 3-month-long studies that have been conducted to date, type II fibers do not convert to type I.

Given these data, how do elite endurance athletes end up with up to 95 percent of their muscle fibers being type I? The answer is not known. Marathoners could have a strong genetic predisposition for expressing the type I myosin gene. Alternatively, it is possible that type II fibers gradually shift to type I over many years of training. Research on the nature and extent of myosin type conversion continues.

TABLE 1 Muscle-Fiber Types in Vertebrates

Type	Contraction Speed	Fatigue Rate	ATP Generation	Myosin type
Fast-twitch	rapid	fast—within minutes	primarily anaerobic	type II (IIx faster than IIa)
Slow-twitch	relatively slow	slow—active for hours	primarily aerobic	type I

Chapter Review

Summary

Sensory stimuli cause changes in the membrane potential of specific receptor cells. If the membrane potential of a sensory cell is altered substantially enough, the pattern of action potentials that it sends to the brain is altered. In this way, sensory stimuli as different as sound and light are transduced to electrical signals. The brain is able to distinguish different types of stimuli because axons from different types of sensory neurons project to different regions of the brain.

Pressure receptors detect direct physical stimulation. A hair cell, for example, undergoes a change in membrane potential in response to bending of its stereocilia. Hair cells are the major sensory detector in the vertebrate ear. Experiments with isolated cochleas showed that sound waves of a certain frequency cause a certain part of the membrane underlying hair cells to vibrate. Because of this specificity in hair cell response, mammals can discriminate different pitches.

Photoreceptors detect light through conformational changes in rhodopsin molecules. In rod cells, rhodopsin consists of retinal paired with an opsin protein. Color vision is possible because different types of opsin respond to different wavelengths of light absorbed by retinal. Humans distinguish different colors based on the pattern of stimulation of three types of opsins found in cone cells. People who lack one of these cone opsins cannot distinguish as many colors.

Chemoreceptors detect the presence of certain molecules. Taste buds, for example, contain taste-receptor cells with membrane proteins that process toxins, salt, acid, and other types of molecules in food. Sodium ions and protons enter taste cells via channels and depolarize the membrane directly. Toxic compounds bind to membrane receptors and trigger action potentials that are interpreted by the brain as bitterness. Smell, or olfaction, is used to scan molecules from the outside environment. Airborne chemicals are detected by hundreds of different odorant-receptor proteins located in the membranes of receptor cells in the nose.

In many cases, animals respond to sensory stimuli by moving. All animal muscles use the same basic mechanism for contraction. Muscles shorten when thick filaments slide past thin filaments in a series of binding events mediated by hydrolysis of ATP. Calcium ions play an essential role in making the actin in thin filaments available for binding. In animals with exoskeletons or endoskeletons, muscles are usually arranged in opposing pairs of flexors and extensors. Animals with hydrostatic skeletons have muscles arranged in opposing pairs of circular and longitudinal bands. In all cases, muscles work in concert with skeletons.

Animals can also respond to sensory information by secreting chemical signals. As the next chapter will show, chemical signals allow animals to coordinate responses to sensory input from many different systems.

Questions

Content Review

1. What is the major function of the middle ear?
   a. It amplifies the sound energy transmitted to the cochlea.
   b. It transmits sound vibrations to the tympanic membrane.
   c. It circulates fluid through the cochlea.
   d. It contains hair cells that detect specific frequencies of sound.

2. In the human ear, why do different hair cells respond to different frequencies of sound?
   a. Waves of pressure move through the fluid in the cochlea.
   b. Hair cells are "sandwiched" between membranes.
   c. Receptors in the stereocilia of each hair cell are different; each receptor protein responds to a certain range of frequencies.
   d. Because the basilar membrane varies in stiffness, it vibrates in certain places in response to certain frequencies.

3. Which of the following comparisons of rods and cones is *false*?
   a. Most humans have one type of rod and three types of cones.
   b. Rods are more sensitive to dim light than cones are.
   c. There are more rods than cones in the fovea.
   d. Rods and cones both use retinal and opsins to detect light.

4. Which of the following statements about taste is true?
   a. Sweetness is a measure of the concentration of H^+ ions in food.
   b. Na^+ from foods can directly depolarize certain taste cells.
   c. All bitter-tasting compounds have a similar chemical structure.
   d. Membrane receptors are involved in detecting acids.

5. In muscle cells, myosin molecules continue moving along actin molecules as long as:
   a. ATP is present and troponin is not bound to Ca^{2+}.
   b. ADP is present and tropomyosin is released from intracellular stores.
   c. ADP is present and intracellular ACh is high.
   d. ATP is present and intracellular Ca^{2+} is high.

6. Which of the following is critical to the function of exoskeletons, endoskeletons, and hydrostatic skeletons?
   a. Muscles interact with the skeleton in antagonistic groups.
   b. Muscles attach to the skeleton via tendons.
   c. Muscles extend joints through forceful pushing.
   d. Segments of the body or limbs are extended when paired muscles relax in unison.

Conceptual Review

1. Describe how a sound, a light, and an odor are transduced into a change in the pattern of action potentials. How does the brain know which sense is which when the action potentials finally reach the brain?
2. Give three examples of how the sensory abilities of an animal correlate with its habitat or method of finding food and mates.
3. How did the discovery of odorant-receptor genes affect our understanding of how the sense of smell works?
4. Compare and contrast the structure and function of slow-twitch and fast-twitch muscle fibers. In making the comparison, use the race between the hare and the tortoise of Aesop's fable as a metaphor.
5. Scientists generally think that a "good hypothesis" is one that is reasonable, is testable, and inspires further research into the field. Using these criteria, was Dalton's hypothesis about color vision a good hypothesis?

Applying Ideas

1. Myasthenia gravis is a disease that develops in humans when the immune system produces proteins that bind to the acetylcholine (ACh) receptors in muscles. The primary symptom of myasthenia gravis is muscle weakness. Why?
2. Houseflies have about 800 ommatidia in each of their compound eyes. Dragonflies, in contrast, have up to 10,000 ommatidia per eye. Houseflies feed by lapping up watery material from piles of excrement or rotting carcasses, which they locate by scent. Dragonflies are aerial predators and hunt by sight. Suggest a hypothesis to explain the difference in the structure of their eyes. How would you test your hypothesis?
3. Muscles that are dominated by slow-twitch fibers have a reddish color because they have a high concentration of myoglobin, while fast-twitch muscles are light in color. Chickens, turkeys, and other ground-dwelling birds can run long distances but escape from predators by flying in short, explosive bursts. Based on these observations, explain the distribution of "dark meat" and "light meat" in these birds. Most other birds do not have "white meat." Why?
4. When looking at faint stars through a telescope, astronomers will focus their vision just to the side of the object. Instead of landing on the fovea, then, the star's image falls next to the fovea. Using this technique, faint objects pop into view. They vanish if looked at directly, however. Explain what is going on.
5. Recall that fast-twitch fibers come in two types—IIa and IIx, and that IIx fibers contract more rapidly. Recent data have shown that sedentary people tend to have high amounts of IIx fibers, and that paralyzed people have even more IIx fibers than world-class sprinters. Based on these observations, some researchers are proposing that making type IIx fiber is the "default" setting during muscle cell development. How could you test this hypothesis? If it is correct, how might it affect training programs for sprinters and marathoners?

CD-ROM and Web Connection

CD Activity 43.1: The Vertebrate Eye *(animation)*
(Estimated time for completion = 5x min)
Learn how eyes sense light and pass signals to the brain.

CD Activity 43.2: Muscle Contraction *(animation)*
(Estimated time for completion = 10 min)
Explore how muscles are specialized for generating force by contracting.

At your **Companion Website** (http://www.prenhall.com/freeman/biology), you will find self-grading exams and links to the following research tools, online resources, and activities:

Transplanted Cell Technology
This article describes modern methods that may enable doctors to repair spinal cord damage.

Circadian Rhythm Photoreceptor
This site details research identifying a photoreceptor involved in establishing and maintaining the body's circadian rhythm.

Fly Genes
This article describes the elucidation of a set of genes responsible for taste and smell in *Drosophila*.

Additional Reading

Andersen, J. T., P. Schjerling, and B. Saltin. 2000. Muscles, genes and athletic performance. *Scientific American* (September): 48–55. A look at recent research on how fast-twitch and slow-twitch fibers respond to training.

Ben-Ari, E. T. 1999. A throbbing in the air. *BioScience* 49: 353–358. Recounts how the ability to detect infrasound was discovered in elephants.

Gadsby, P. 2000. Tourist in a taste lab. *Discover* (July): 70–75. An introduction to recent research on the sense of taste.

Geisler, C. D. 1998. *From Sound to Synapse: Physiology of the Mammalian Ear.* New York: Oxford University Press. An overview of the ear's structure and function.

Chemical Signals in Animals

In response to sights, sounds, and other sensory stimuli, an animal's nervous system sends rapid messages to precise locations in the body. Frequently these messages result in muscle contractions and movement.

As conditions inside and outside an animal change, however, cells in the central nervous system (CNS) or cells in a specialized suite of organs called the endocrine system may also trigger the release of molecules that produce longer-term responses in a broad range of tissues and organs. A chemical signal that circulates through the blood or other body fluids and affects target cells is called a **hormone.** As you read this, a large suite of hormones is coursing through your circulatory system. They are regulating the production of sperm or eggs by your reproductive system, changing the osmolarity of the urine forming in your kidneys, and controlling the release of digestive enzymes in your gastrointestinal tract. Earlier in your life, changes in hormone concentrations led to the dramatic suite of changes associated with puberty.

Animal hormones are present in tiny concentrations but have huge effects on their target cells. Compared to action potentials, the messages they carry are received by many different cells and have a relatively long-lasting effect. In combination, electrical and chemical signals allow animals to coordinate the activities of hundreds of billions of cells within the body.

The goal of this chapter is to explore how chemical signals work in animals. Section 44.1 reviews the types of experiments that led to the discovery of specific hormones and introduces some general principles about how hormones act. The following section explores how hormones direct the development of embryos and juveniles, function in homeostasis, and trigger responses to environmental stimuli

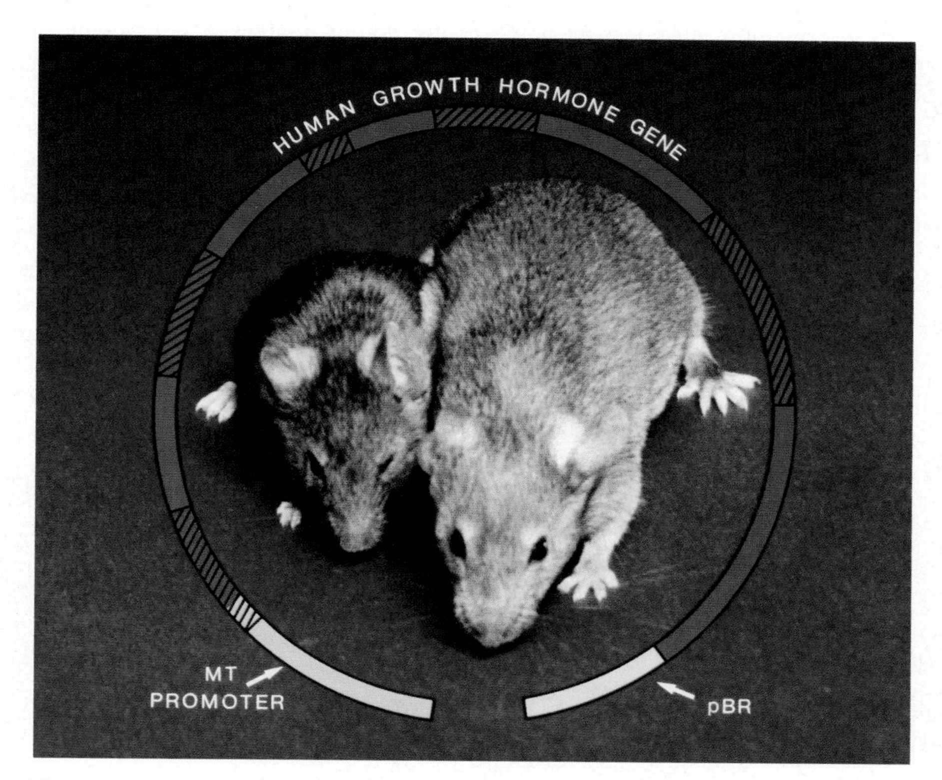

These two mice are siblings from the same litter. As an embryo, the individual on the right was given a copy of the gene for human growth hormone. The structure of this recombinant gene is diagrammed in the red loop around the mice. Overexpression of this gene leads to abnormally high concentrations of a chemical signal that supports growth, and gigantism results. This chapter explores the nature and function of growth hormone and other chemical signals in animals.

such as dangerous situations. Section 44.3 investigates the close relationship between the nervous system and the **endocrine system** that is responsible for the production and secretion of hormones.

The chapter closes by investigating research on how hormones affect target cells at the molecular level. Starting with classical work from the 1950s and 1960s, we'll delve into the question of how the signal carried by a hormone is translated into action by a target cell. Section 44.4 will bring us to the forefront of research on chemical signaling, and pave the way for exploring how hormones regulate reproduction by humans and other animals in Chapter 45.

44.1 Cataloging Hormone Structure and Function

Research on animal hormones began in earnest in the early 1900s with the discovery of a molecule called secretin. Recall from Chapter 40 that researchers who were investigating how secretion from the pancreas is regulated had cut all of the nerves connected to the upper part of the small intestine and pancreas of a dog. When they added a small amount of dilute hydrochloric acid (HCl) to the small intestine, they found that the pancreas secreted compounds that neutralize the acid in the small intestine (**Figure 44.1a**).

How could stimulating the small intestine lead to a response by the pancreas? The biologists proposed that a chemical had traveled from the small intestine to the pancreas to signal the arrival of acid. To test this hypothesis, they did the experiment diagrammed in **Figure 44.1b**. The key step was to inject an extract from the small intestine into blood vessels in a dog's neck. A short time later, they observed that the pancreas secreted an alkaline solution. This was strong evidence that the extract from the small intestine contained a hormone. The molecule was later purified and named secretin.

Over the ensuing century, researchers have isolated dozens of other animal hormones, located their sources in glands or other organs, and documented the responses that they induce. Let's take a closer look at how these studies were done and the catalog of animal hormones that resulted.

How Do Researchers Identify a Hormone?

Frequently, the first clue that a tissue produces a hormone comes from studies of laboratory animals in which the tissue has been

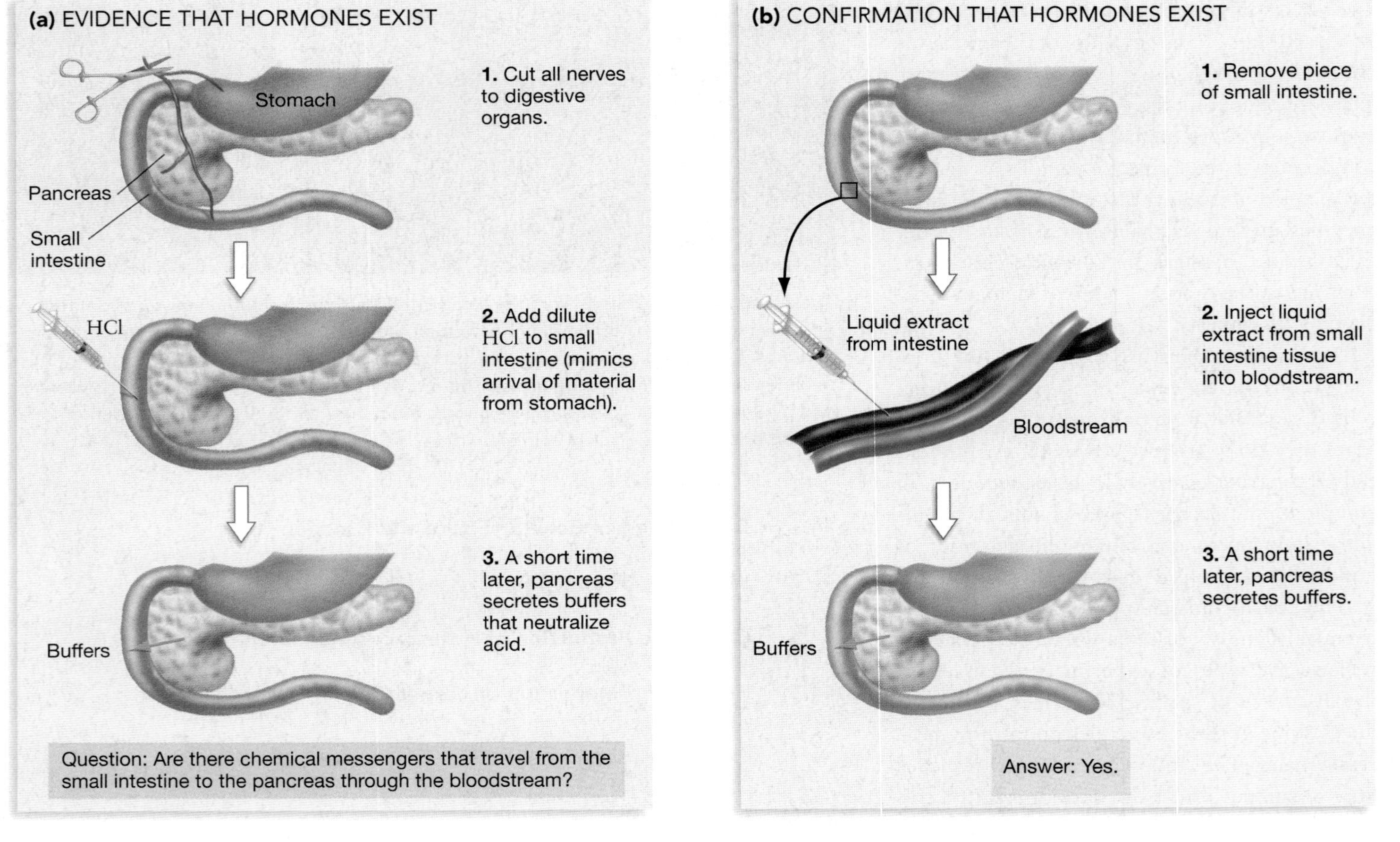

FIGURE 44.1 The Discovery of Secretin
These diagrams illustrate how the first animal hormone was discovered. The molecule secretin was eventually isolated from the soluble intestinal extract shown in part (b).

removed. For example, when researchers removed the adrenal glands of dogs they noted that the animals had a greatly reduced ability to accumulate glycogen in their livers. (Recall that glycogen is the major storage form of glucose in animals.) This observation suggested that the adrenals might produce a hormone that triggers glycogen formation in the liver (**Figure 44.2a**).

Once investigators suspect that a tissue produces a hormone, a typical next step is to extract a solution from that tissue from intact animals, inject the extract into animals that lack the tissue, and observe whether their symptoms are alleviated. In the case of the adrenal gland, K. Paschkis and co-workers extracted compounds from the blood leaving the adrenal glands of intact dogs. When they injected the solution into dogs that lacked adrenal glands, increased glycogen deposition resulted (**Figure 44.2b**). As with the injection experiment that led to the discovery of secretin, Paschkis's result provided strong evidence that the adrenals produce a hormone involved in regulating glycogen formation.

The final step to identify a hormone is to purify the active ingredient in the raw tissue extract, inject the hormone into animals that cannot produce the molecule, and determine if their symptoms are cured. In this case, Don Nelson and associates purified a molecule called corticosterone from the adrenal blood vessels of dogs; later work identified a closely related compound called cortisol that is also secreted from the adrenal glands. When researchers injected these molecules into experimental animals whose adrenal glands had been removed, they observed increased blood-glucose levels and glycogen accumulation in the liver (**Figure 44.2c**).

These experiments identified cortisol and corticosterone as adrenal hormones that increase glucose concentrations in the blood and increase glycogen formation in the liver. Follow-up studies documented that these compounds, which together are referred to as glucocorticoids, are produced in the outer section of the adrenal glands. This tissue is called the adrenal cortex.

Analogous studies showed that the central part of the gland, called the adrenal medulla, produces hormones called epinephrine and norepinephrine. (Epinephrine is also known as adrenaline. The Greek word roots *epi* and *nephron* mean "top-kidney"; the Latin word roots *ad* and *renal* also mean "top-kidney.") In contrast to cortisol, epinephrine promotes the breakdown of glycogen in the liver. In addition to having different functions, the two molecules are very different structurally. On a molecular level, what do hormones look like?

Chemical Characteristics of Hormones

Secretin, glucocorticoids, and epinephrine represent the three general types of molecules that act as chemical messengers in

(a) Evidence that adrenal glands produce a hormone

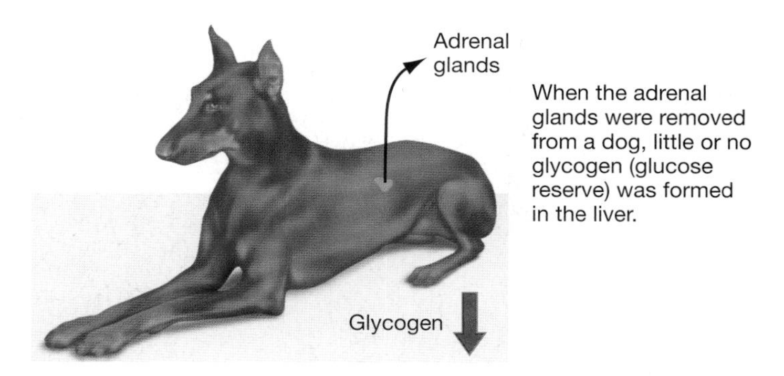

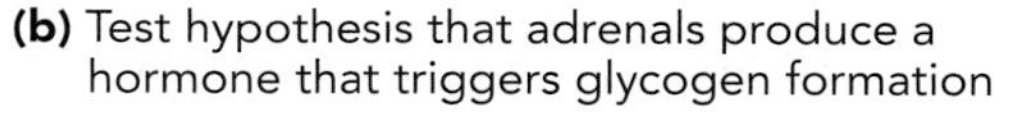

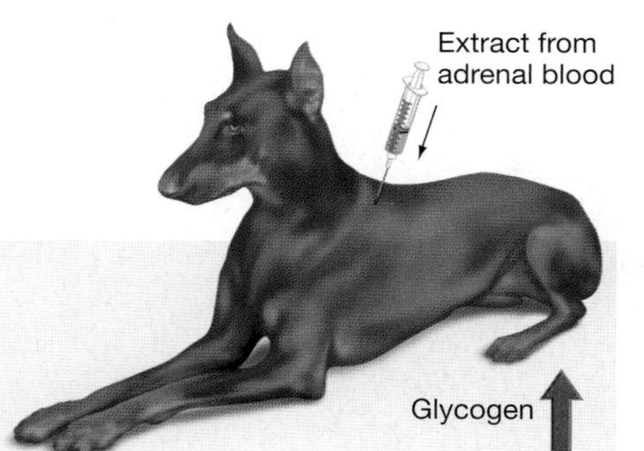

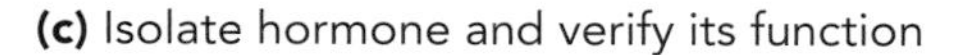

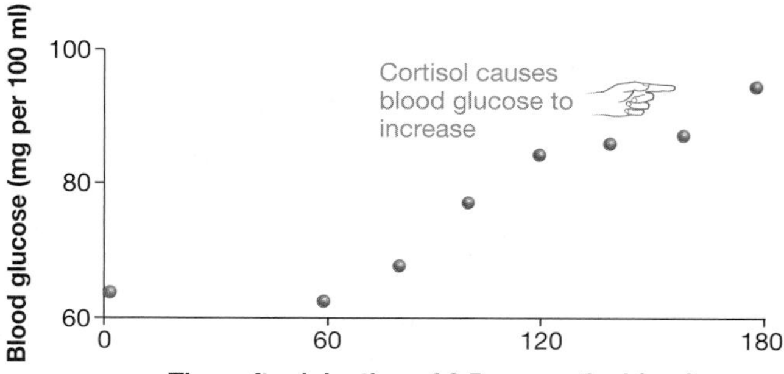

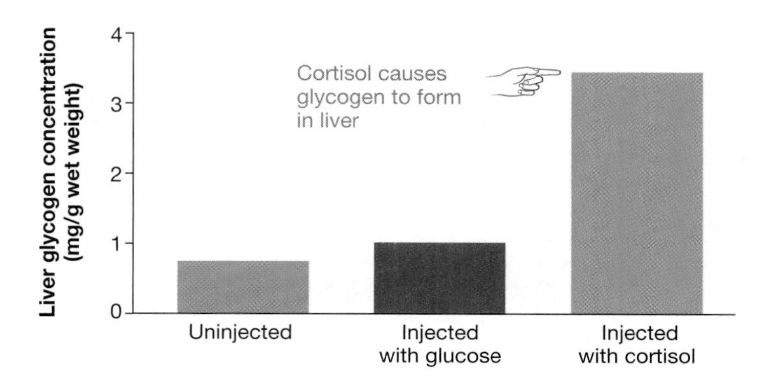

FIGURE 44.2 The Discovery of Glucocorticoids
A series of experiments suggested that the adrenal glands produce a hormone involved in increasing both blood glucose and glycogen formation in the liver and confirmed that cortisol is the hormone.

animals. As **Figure 44.3** shows, secretin is a polypeptide, epinephrine is derived from the amino acid tyrosine, and cortisol is synthesized from cholesterol and belongs to the group of lipids called steroids.

These three families of hormones have important similarities as well as important differences. All are organic compounds, all are secreted from the cells where they are synthesized, and all act on target cells remote from their point of origin. In addition, all are present at extremely small concentrations, yet have large effects. To drive this last point home, consider work that Choh Hao Li and co-workers did on growth hormone (GH). Several researchers had noted that rats and other laboratory animals stopped growing when their pituitary glands were removed. Based on this observation, it was widely suspected that the pituitary produces a chemical signal that promotes cell division and other aspects of growth. Li's group was able to purify a polypeptide from cow pituitary glands, inject it into lab rats, and document rapidly accelerated growth. When the researchers injected 0.01 mg of the molecule a day for nine days into rats that lacked pituitary glands, the width of the growth plates in the leg bones increased by 50 percent, and the individuals gained an average of 10 grams compared with rats that lacked a pituitary and did not receive the hormone treatment. Stated another way, the addition of a total of 0.09 mg of hormone led to a weight gain of 10 grams. Further, 1 kilogram of cow pituitary tissue yielded a grand total of 0.04 gram of growth hormone. By weight, the hormone makes up four one-thousandths of 1 percent of the cow pituitary.

Given that small amounts of polypeptide, amino acid derivative, and steroid hormones have large effects, how do the three types of hormones differ? The most important contrast is that steroids are lipid soluble while polypeptides and amino acid derivatives are not. As a result, steroids cross cell membranes much more readily than do other types of hormones. To affect a target cell, all polypeptides and most amino acid derivatives bind to a receptor on the cell surface. Steroid hormones, in contrast, diffuse through the membrane and bind to receptors inside the cell. We'll explore this contrast in much more detail in section 44.4. Now let's move ahead by surveying the diversity of steroids, polypeptides, and amino acid derivatives that act as hormones in humans.

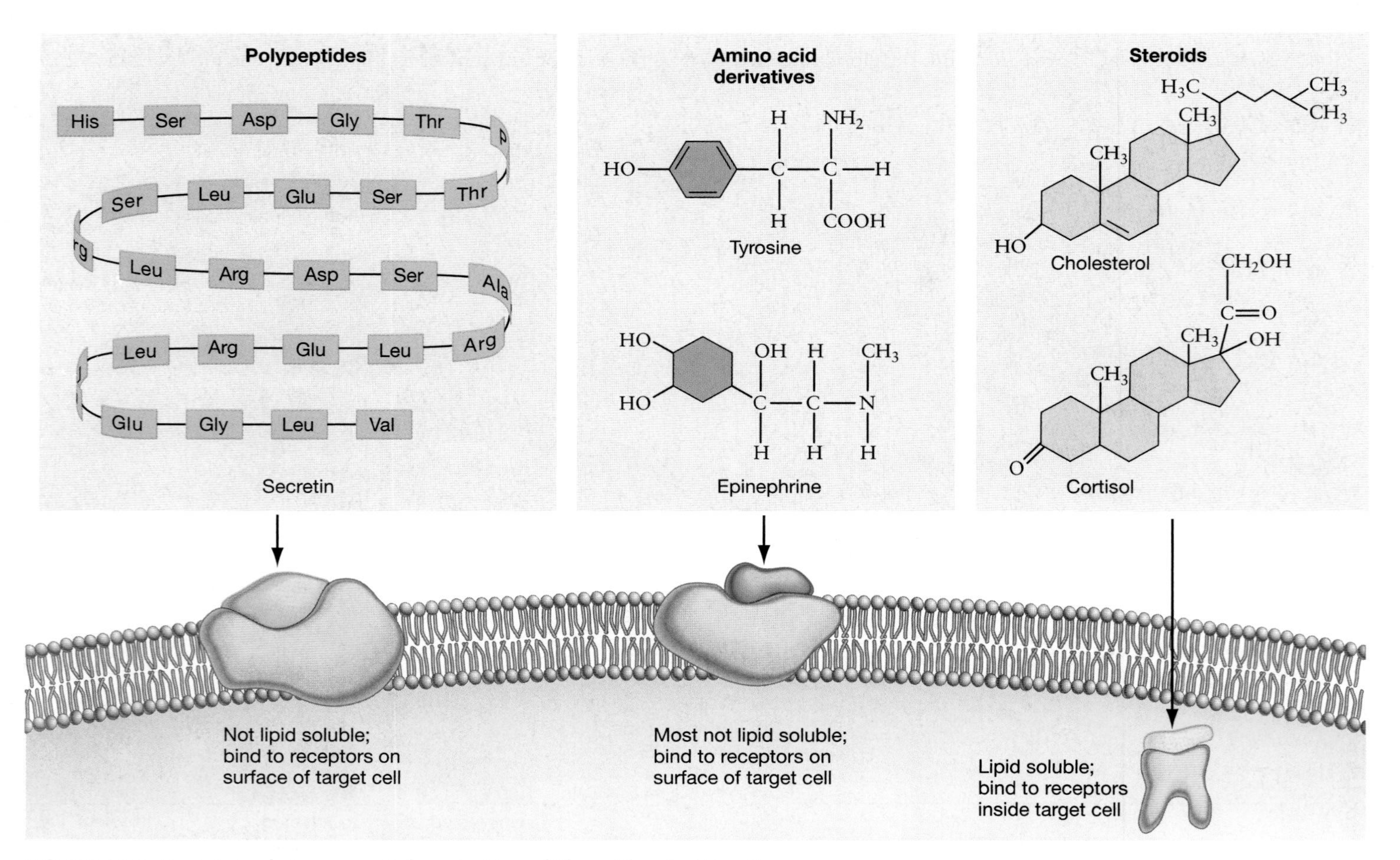

FIGURE 44.3 Most Animal Hormones Belong to One of Three Chemical Families

QUESTION Secretin and epinephrine are not lipid soluble and do not cross cell membranes readily. Cortisol, in contrast, is lipid soluble and diffuses through cell membranes efficiently. How do the size and charge of these molecules explain the difference in lipid solubility?

The Human Endocrine System—An Overview

Figure 44.4 illustrates some of the components of the human endocrine system. The endocrine system is the collection of cells, tissues, and organs responsible for hormone production and secretion. Hormone-secreting organs are called **endocrine glands.** The basic message of the illustration is that endocrine glands are diverse in their size, shape, and location.

The figure presents selected hormones produced by each component of the endocrine system, along with their target cells or effect. At first glance the diversity of hormones, glands, and effects can seem almost overwhelming, especially considering that the figure represents just a partial catalog for a single species. Several unifying themes emerge from the diversity of molecules and functions, however.

As you may recall, most animal hormones belong to one of just three major structural families (polypeptides, amino acid derivatives, and steroids). Further, most hormones share a basic functional characteristic. They are released into the bloodstream and act on target cells that are distant from the source gland. Finally, all hormones fulfill the same general task. Their role is to coordinate the activities of diverse groups of cells in response to changes in the internal or external environment. More specifically, the

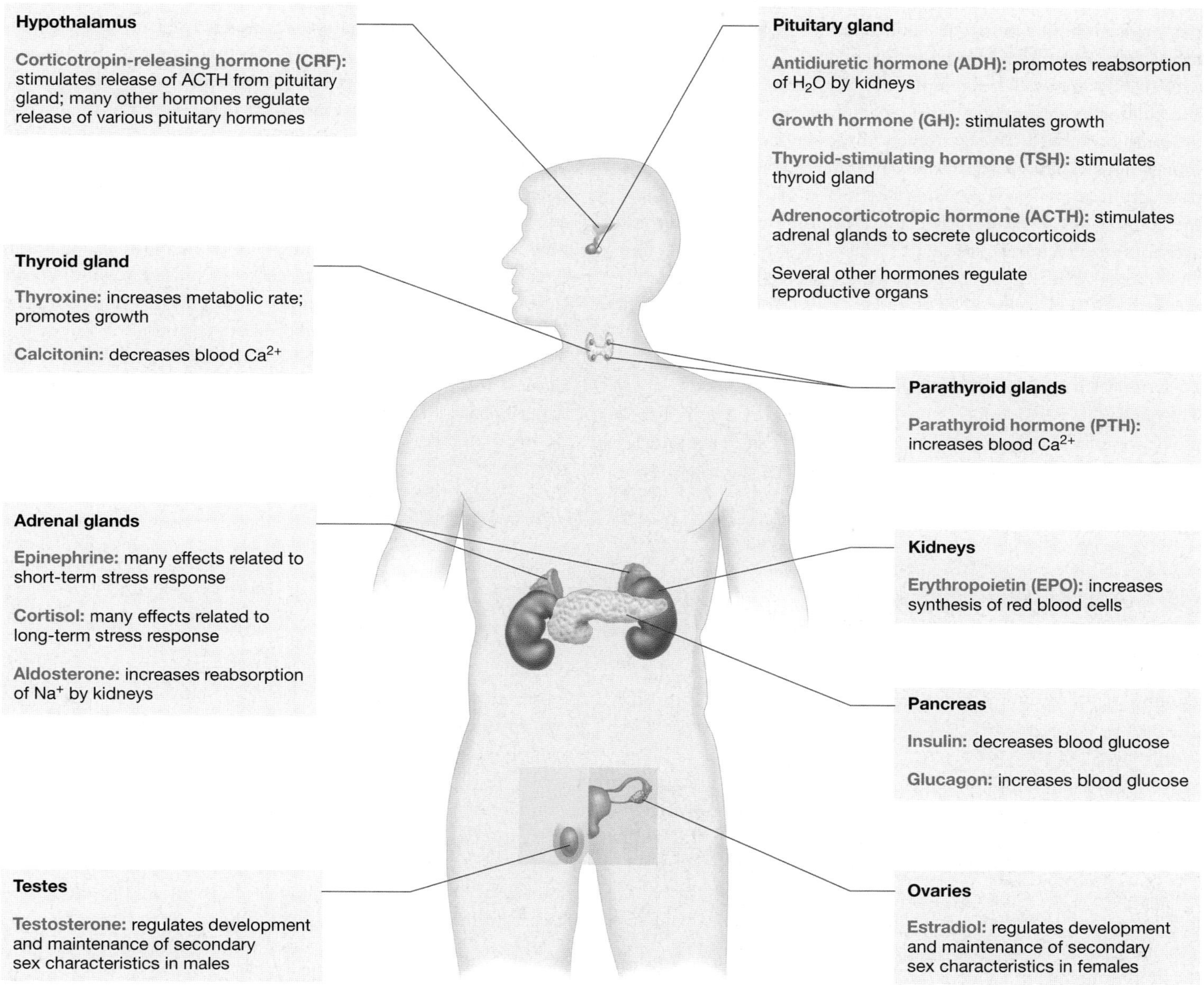

CD ACTIVITY 44.1 Endocrine System Anatomy

Hypothalamus

Corticotropin-releasing hormone (CRF): stimulates release of ACTH from pituitary gland; many other hormones regulate release of various pituitary hormones

Pituitary gland

Antidiuretic hormone (ADH): promotes reabsorption of H_2O by kidneys

Growth hormone (GH): stimulates growth

Thyroid-stimulating hormone (TSH): stimulates thyroid gland

Adrenocorticotropic hormone (ACTH): stimulates adrenal glands to secrete glucocorticoids

Several other hormones regulate reproductive organs

Thyroid gland

Thyroxine: increases metabolic rate; promotes growth

Calcitonin: decreases blood Ca^{2+}

Parathyroid glands

Parathyroid hormone (PTH): increases blood Ca^{2+}

Adrenal glands

Epinephrine: many effects related to short-term stress response

Cortisol: many effects related to long-term stress response

Aldosterone: increases reabsorption of Na^+ by kidneys

Kidneys

Erythropoietin (EPO): increases synthesis of red blood cells

Pancreas

Insulin: decreases blood glucose

Glucagon: increases blood glucose

Testes

Testosterone: regulates development and maintenance of secondary sex characteristics in males

Ovaries

Estradiol: regulates development and maintenance of secondary sex characteristics in females

FIGURE 44.4 An Overview of the Human Endocrine System
This is a partial list of the endocrine glands and hormones found in humans. The heart and gastrointestinal tract could be added, as both produce hormones. EXERCISE Circle the hormones responsible for decreasing or increasing the levels of glucose in the blood and thus achieving homeostasis. Put a box around the hormones involved in modifying the concentration of calcium ions (Ca^{2+}) in blood.

functions of animal hormones fall into a handful of broad categories. What are these? In general, how does the action of hormones help an individual to survive and reproduce?

44.2 What Do Hormones Do?

At the beginning of this chapter, you read that hormones are chemical messengers. If so, what do hormones "say?" It is clear from the data in Figure 44.4 that even a single type of hormone can exert a diversity of effects. Consider thyroxine, for example. In humans, thyroxine stimulates metabolism and thus oxygen consumption throughout the body, increases heart rate, and promotes red blood cell production. (**Box 44.1** explores the medical implications of abnormalities in thyroxine.) In other cases, several different hormones may affect the same aspect of physiology. For example, blood-glucose concentration is regulated by insulin, glucagon, epinephrine, and cortisol.

Some hormones have extremely diverse effects; other hormones have functions that appear to overlap. These observations begin to make sense when hormone action is viewed in the context of the whole organism. Hormones coordinate the activities of cells in response to three general situations: (1) environmental challenges; (2) growth, development, and reproduction; or (3) homeostasis. Let's take a closer look.

Hormones Coordinate Responses to Environmental Challenges

The challenges or stimuli that hormones respond to can be simple or complex. Digestive hormones are a good example of how hormones function in simple stimulus and response circuits. When acidic food material passes from the stomach to the upper reaches of the small intestine, it triggers the release of secretin and cholecystokinin. Secretin induces the pancreas to secrete an alkaline solution and the gallbladder to release bile salts that neutralize acid and emulsify fats. Cholecystokinin activates secretion of digestive enzymes from the pancreas into the small intestine. In this way, digestive hormones signal the arrival of food and regulate the release of molecules that aid digestion. But what about more complex environmental stimuli?

Short-Term Responses to Stress When a person is thrust into a dangerous or unpredictable situation, hormones are involved in both the short-term and long-term response. The short-term reaction, called the fight-or-flight response, occurs in conjunction with activation of the sympathetic nervous system. If you were being chased by a grizzly bear, action potentials from sympathetic nerves would stimulate the adrenal medulla and lead to the release of epinephrine.

To understand the responses that occur, consider the experiment reported in **Figure 44.5**. These data were collected from human volunteers who received injections of epinephrine or other molecules. The graphs indicate dramatic increases in the concentration of free fatty acids and glucose in the blood following injection of epinephrine or norepinephrine; the data in the table confirm significant increases in pulse rate, blood pressure, and oxygen consumption by the brain. In addition, the volunteers in the blood circulation study reported strong subjective feelings of anxiety and excitement. Other experiments showed that epinephrine leads to dramatic changes in the distribution of blood. Specifically, epinephrine leads to a redirection of blood toward the heart, brain, and muscles and away from the skin and digestive system.

Taken together, the responses to epinephrine lead to a state of heightened alertness and rapid energy utilization that pre-

BOX 44.1 A Closer Look at Thyroxine and the Thyroid Gland

The thyroid hormone thyroxine is an amino acid-derived hormone that is synthesized from tyrosine. Thyroxine is an unusual molecule, however, because it contains four iodine atoms. This feature inspired its alternative name, T_4. A closely related thyroid hormone called triiodothyronine, or T_3, contains three iodine atoms.

In mammals, T_3 is the more active of the two hormones. Although the thyroid releases much more T_4 than T_3, cells in the liver and elsewhere frequently convert T_4 to T_3.

T_3's primary effect on target cells is to trigger increased cellular metabolism. People who produce inadequate amounts of thyroid hormones are lethargic and unable to tolerate cold. In contrast, people who produce excessive amounts of T_3 and T_4 are restless, excitable, and prone to mood swings, but have limited energy reserves and tend to tire easily. Former President George Bush and First Lady Barbara Bush both suffer from a syndrome called Graves' disease, which is caused by an overactive thyroid.

Two other medical conditions have been traced to problems with the iodine atoms incorporated into T_3 and T_4. When radioactive iodine atoms are released into the environment, as during the accident at the Chernobyl nuclear power plant in the Ukraine, people living in the area become highly susceptible to thyroid cancers. The pattern occurs because radioactive iodine molecules are sequestered in the thyroid and because radiation may damage DNA in a way that makes cells cancerous.

If iodine is deficient in the diet, the thyroid gland may swell to huge proportions in response and produce the condition called goiter. Goiter is now rare in many parts of the world because iodized salt is available for consumption.

pares the body for rapid, intense action like fighting or fleeing. By coordinating the activities of cells in many organs and systems throughout the body, epinephrine prepares an individual to cope with a life-threatening situation.

Long-Term Responses to Stress If you have ever been in an acutely stressful situation and experienced the fight-or-flight response, you may recall that the state is short-lived. Once an epinephrine "rush" has worn off, most people feel exhausted and want to rest and eat.

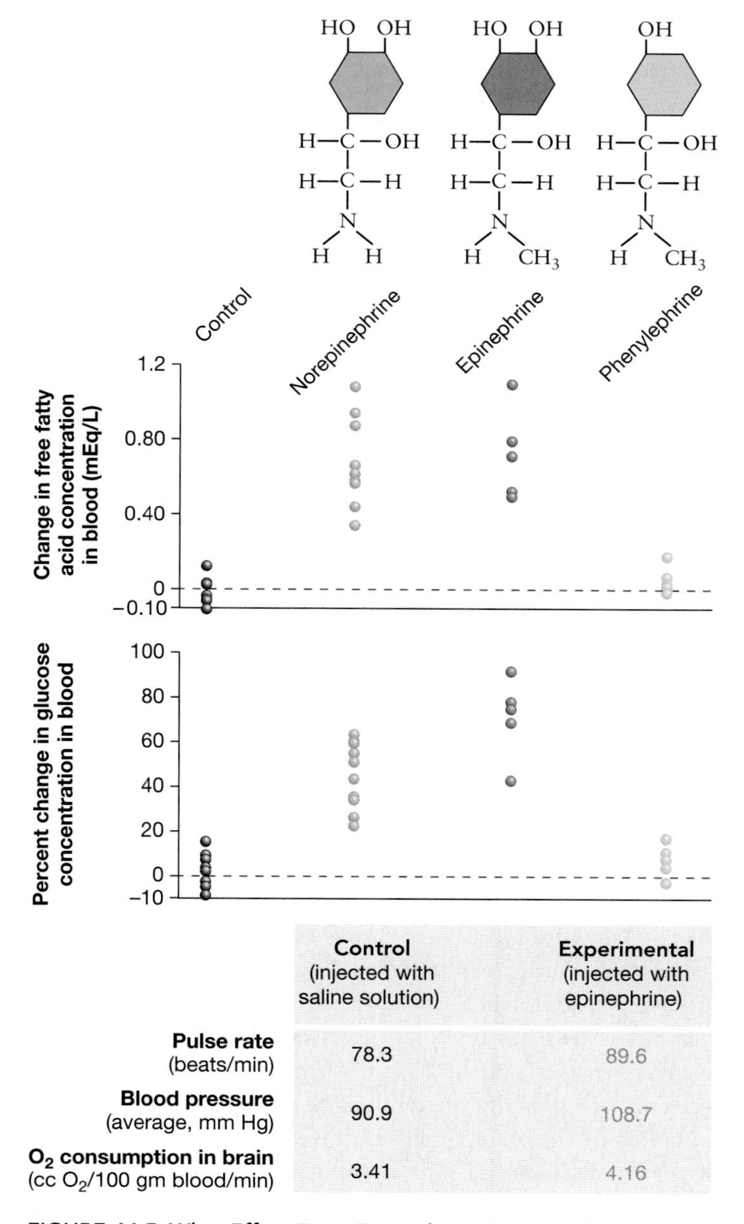

	Control (injected with saline solution)	Experimental (injected with epinephrine)
Pulse rate (beats/min)	78.3	89.6
Blood pressure (average, mm Hg)	90.9	108.7
O_2 consumption in brain (cc O_2/100 gm blood/min)	3.41	4.16

FIGURE 44.5 What Effect Does Epinephrine Have on the Body? When researchers inject human volunteers with epinephrine or related compounds, they observe an array of responses. Each data point on the graphs represents a human volunteer; the data in the table are average values from 7 volunteers. QUESTION Examine the chemical structures of norepinephrine, epinephrine, and phenylephrine. How do they differ? How do the effects of these molecules differ?

What happens if the trauma continues and turns into a long-term condition? In the course of a lifetime, it is not unusual for a person to experience periods of starvation, prolonged emotional distress, or chronic illness. How do hormones help humans and other animals cope with extended stress?

Early studies of long-term stress in human subjects suggested a role for the glucocorticoids produced in the adrenal cortex. Increased levels of cortisol or corticosterone were found in airplane pilots and crew members during long flights, athletes who were training for intense contests, the parents of children undergoing treatment for cancer, and college students who were preparing for final exams.

What do these glucocorticoids do? As Figure 44.2c shows, glucocorticoid injection in laboratory animals leads to increased glycogen formation in the liver and increased blood-glucose concentrations. Other studies documented that glucocorticoids induce the release of amino acids and lipids from muscle and fat cells, respectively, which promotes their use in ATP production instead of glucose. Taken together, these observations suggest that the glucocorticoids prepare an individual for long-term stress by conserving glucose. The general hypothesis here is that during long-term stress, glucocorticoids reserve glucose for use by the brain at the expense of other tissues and organs. In addition, glucocorticoids conserve glucose by actively suppressing wound healing and other aspects of immune-system function. As in the short-term response to stress, important trade-offs are involved. During the long-term response, wound healing and other valuable activities are repressed.

In combination, then, the hormones produced by the adrenal glands coordinate an individual's response to short- and long-term trauma. As with digestive hormones, these molecules alert cells and tissues to a particular environmental stimulus or challenge. Other hormones, however, have dramatically different effects.

Hormones Direct Developmental Processes

Growth hormones play a crucial role in promoting cell division and increasing overall body size as an individual matures. In addition, certain hormones direct the development of particular cells and tissues at critical junctures in an individual's life. Which hormones are involved in regulating development, and what specific effects do they have?

- *Early embryonic development* Events early in development dictate whether the sex organs, or gonads, of an embryo become male or female. This process is called primary sex determination and is not dependent upon hormone action. Once testes or ovaries develop, though, they begin producing male- or female-specific hormones. In humans, the early testes produce two hormones. A steroid hormone called testosterone induces early development of the male reproductive tract and a protein hormone called MIS inhibits development of the female reproductive tract. Early development of female embryos occurs without the involvement of hormones.

- *Juvenile-to-adult transition* When humans reach early adolescence, surges of sex hormones lead to the physical and emotional changes associated with puberty. These developmental changes create the adult phenotype and the ability to produce offspring. In boys, surges of sex hormones lead to changes that include enlargement of the penis and testes and growth of facial and body hair. In girls, increased concentrations of estradiol—a steroid hormone in the family of molecules called estrogens—lead to the enlargement of breasts, the onset of menstruation, and other changes. Hormonal control over sexual maturation is not limited to vertebrates, however. As Chapter 30 showed, insect metamorphosis is controlled by hormones. If the steroid called juvenile hormone (JH) is present at a high concentration, surges of the steroid hormone ecdysone induce growth of a juvenile insect via molting. But if JH levels are low, ecdysone triggers metamorphosis and the transition to adulthood and sexual maturity.
- *Seasonal or cyclical sexual activity* Most long-lived animals reproduce seasonally. In many species, environmental cues like increasing day length, warmth, or the onset of seasonal rains trigger the release of sex hormones. Chapter 47 details how this flush of testosterone or estrogen induces the development of seasonal traits like singing in male birds and sexual receptivity in female lizards. Even though humans do not breed seasonally, sex hormones are instrumental in regulating sperm production and the menstrual cycle. Chapter 45 explores these processes in detail.

To summarize, hormones play key roles in growth and development. Growth hormones support overall size increases in juveniles. Sex hormones coordinate the activities of diverse groups of cells and tissues at three critical life stages: development of sexual structures in embryos, maturation of secondary sexual structures at the juvenile-to-adult transition, and regulation of sexual activity in adults.

How Are Hormones Involved in Homeostasis?

Chapter 38 introduced the concept of homeostasis, or the maintenance of relatively constant physical and chemical conditions inside the body. Recall that homeostatic systems depend on (1) a sensory receptor that monitors conditions relative to a preferred value or set point, (2) an integrator that processes information from the sensor, and (3) effector cells that return conditions to the set point. (To review these ideas, glance back at Figure 38.17.) What Chapter 38 did not mention is that messages often travel from integrators to effectors in the form of hormones.

Several hormones that act as messengers in homeostatic systems have already been introduced. Recall from Chapter 39 that when an individual is dehydrated, antidiuretic hormone (ADH) is released from the pituitary gland. ADH increases the permeability of the kidney's distal tubules and collecting ducts to water. As a result, water is reabsorbed from the urine and saved. In this way, ADH is instrumental in achieving homeostasis with respect to water balance. If ADH's action is inhibited, homeostasis fails and illness may occur. For example, the ethanol in alcoholic beverages inhibits the release of ADH from the pituitary. People who imbibe large quantities of these beverages produce large quantities of dilute urine. The resulting water loss can lead to dehydration and nausea—symptoms associated with an alcoholic hangover.

Chapter 39 also mentioned that aldosterone is released from the adrenal cortex when ion concentrations in body fluids are low. Aldosterone increases the reabsorption of Na^+ in the distal tubules of the kidney. Because aldosterone results in an increase in Na^+ and blood and interstitial fluids, it plays a key role in homeostasis with respect to electrolyte concentrations and the osmolarity of body fluids.

Several of the hormones introduced in Figure 44.4 also are involved in homeostatic systems. For example, erythropoietin (EPO) is released from the kidneys and other tissues when blood oxygen levels are low. EPO acts to restore oxygen homeostasis by stimulating the production of red blood cells. Larger numbers of red blood cells increase the oxygen-carrying capacity of blood. In this way, EPO is a crucial element in the homeostatic system for blood oxygen levels.

Figure 44.6 diagrams how the homeostatic system for blood calcium concentration works. As you can see, the hormones called calcitonin and parathyroid hormone (PTH) work in tandem to keep Ca^{2+} levels in the blood close to the set point, much as insulin and glucagon interact to maintain blood-glucose concentrations at preferred levels (see Chapter 40).

To summarize, hormones stimulate actions by effector cells that restore homeostasis. Hormones also direct key developmental processes and coordinate responses to specific stimuli. But in addition, certain hormones control the release of other hormones. Let's explore in more detail how hormone secretion is turned on and off.

44.3 How Is the Production of Hormones Regulated?

As you know from reading section 44.2, most hormones are released in response to either an environmental cue or a message from an integrator in a homeostatic system. In both cases, the nervous system may be closely involved. For example, environmental cues that signal the onset of the breeding season or the presence of a predator are received by sensory receptors and interpreted by the brain. Similarly, the integrators in most homeostatic systems consist of groups of neurons in the CNS.

Based on these observations, the short answer to the question posed in the title of this section is simple. In many cases, hormone production is directly or indirectly controlled by the nervous system. How, then, does the nervous system direct the endocrine system? To explore this question, let's focus on how cortisol and epinephrine are regulated by the hypothalamus and pituitary.

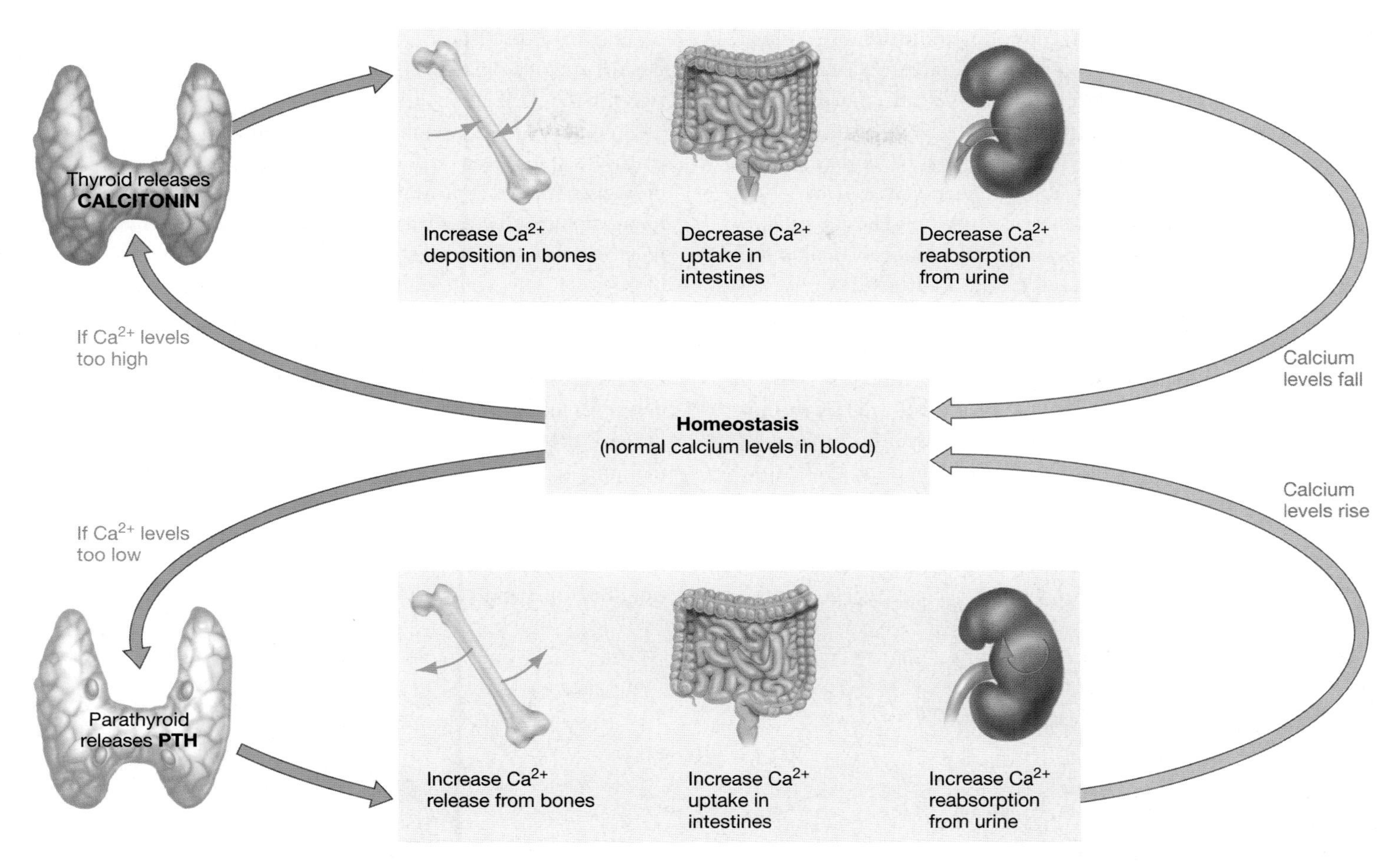

FIGURE 44.6 Hormones Act to Maintain Homeostasis
The system for maintaining homeostasis with respect to calcium ion concentration does not involve the CNS. Instead, sensor cells in the thyroid and parathyroid glands respond directly to changes in Ca^{2+} levels.
EXERCISE Draw a similar type of diagram to illustrate the action of EPO, ADH, aldosterone, or glucagon and insulin. (Hint: Look back at Figure 44.4.)

The Hypothalamus and Pituitary

Researchers who were interested in how hormone production is controlled quickly focused on the pituitary gland. As **Figure 44.7** shows, the pituitary is located at the base of the brain and is directly connected to a brain region called the hypothalamus. In 1930, Philip Smith showed that laboratory rats suffer from a variety of debilitating symptoms when their pituitary is removed. The animals stop growing and cannot maintain a normal body temperature. In addition, their genitals, thyroid glands, and adrenal cortexes atrophy (shrink). Not surprisingly, their life span is also dramatically shortened.

Smith's work suggested that in addition to growth hormone, the pituitary secretes substances that affect the gonads, thyroid, and adrenals. Based on this observation, the pituitary took on the nickname of the "master gland." To explore the pituitary's role in regulating hormone release, let's take a closer look at the pituitary hormone that acts on the adrenal glands.

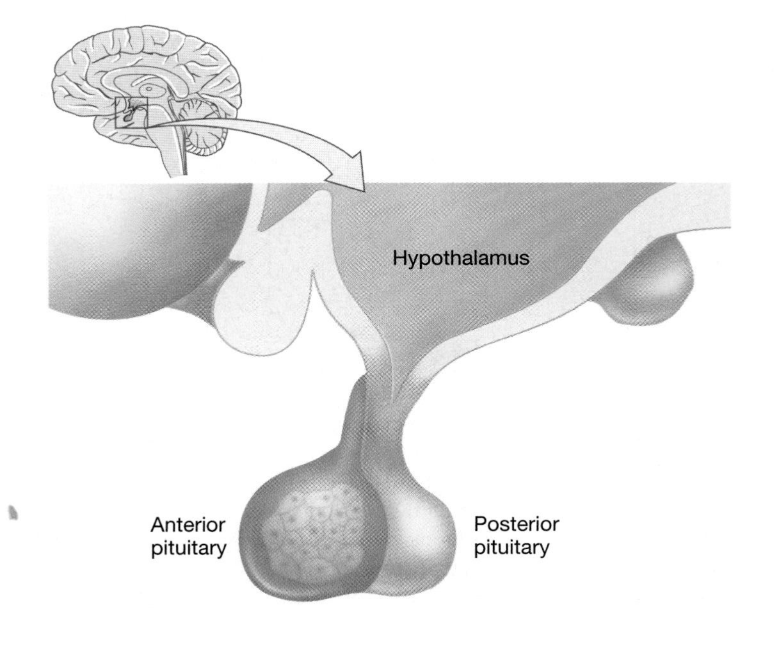

FIGURE 44.7 The Hypothalamus and the Pituitary Gland
The pituitary gland is directly connected to the brain region called the hypothalamus. The pituitary has two distinct regions designated the anterior and posterior pituitary; each secretes different hormones in response to different signals from the hypothalamus.

Controlling the Release of Glucocorticoids Smith's work proposed the existence of an unidentified molecule that affects the adrenal gland. This molecule came to be called adrenocorticotropic hormone, or ACTH. (*Adreno* refers to the adrenal glands; *cortico* refers to the outer portion or cortex of the gland; and *tropic* means "affecting the activity of.") In 1943, ACTH was purified and characterized by Choh Hao Li and co-workers. These biologists confirmed that glucocorticoids are secreted from the adrenal cortex in response to ACTH released from the pituitary. Thus, ACTH is a regulatory hormone.

Not long after ACTH was isolated, biologists from two different laboratories simultaneously and independently showed that it is released in response to a molecule produced by the hypothalamus. Researchers called the hypothetical hormone produced by the brain corticotropin releasing factor, or CRF. After years of effort, Wylie Vale and associates finally succeeded in purifying CRF. It is a peptide—just 41 amino acids long. When CRF is released from the hypothalamus, cells in the anterior portion of the pituitary are stimulated to secrete ACTH into the bloodstream. In response to ACTH, the adrenal glands release glucocorticoids. **Figure 44.8a** diagrams the relationship between CRF, ACTH, and glucocorticoids.

Work by G. V. Upton and colleagues confirmed that this regulatory sequence has additional components, however. These biologists studied four human patients with a rare illness called lipoatrophic diabetes. A hallmark of the disease is that individuals secrete CRF constantly rather than intermittently. To explore how CRF production is regulated, the researchers injected ACTH into the patients and a normal individual. As expected, cortisol levels in the normal individual shot up. The lipotrophic diabetes patients showed no change in their cortisol levels, however. Presumably, they showed no response to additional ACTH because they were secreting CRF constantly and were already maintaining high levels of ACTH. Upton and co-workers made a key observation, though, as documented by the data in **Figure 44.8b.** Three of the four patients experienced a dramatic drop in the levels of CRF in their blood in response to ACTH injection; the fourth had a very slight drop. To make sense of this re-

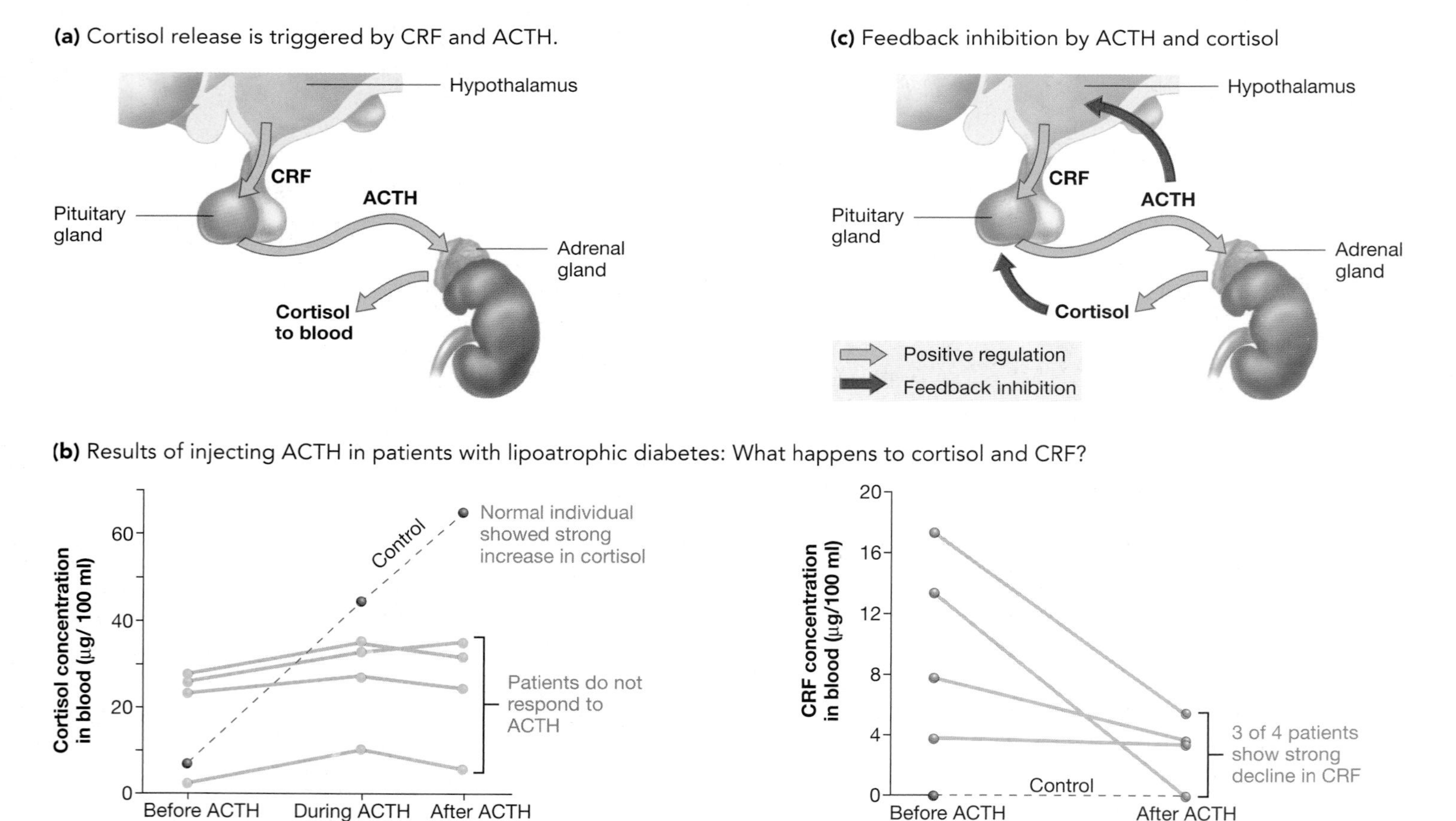

FIGURE 44.8 How Is the Secretion of Cortisol Regulated?
(a) The release of CRF from the hypothalamus triggers the release of ACTH from the pituitary, which in turn induces the release of cortisol from the adrenal glands. **(b)** These graphs show the concentration of cortisol (left) and CRF (right) in blood after five humans were injected with ACTH. The data are from a normal individual and four patients with chronic overproduction of CRF. **(c)** Combined with results from other experiments, the data in part (b) suggest that cortisol production is under both positive and negative control. Feedback inhibition shuts down continued secretion of the hormone.

sult, the investigators proposed that the injected ACTH inhibited the production of CRF. Follow-up experiments in other laboratories confirmed this result and showed that the presence of cortisol inhibits the release of ACTH.

When the presence of a molecule inhibits its production in this way, **feedback inhibition** is said to occur. **Figure 44.8c** illustrates this concept using the example of glucocorticoid regulation. The key to understanding the system is to realize that all of the hormones involved act as regulators, and that all are involved in feedback inhibition. CRF triggers ACTH production and ACTH triggers glucocorticoid release, but ACTH also inhibits CRF secretion through a feedback loop, and the glucocorticoids provide feedback by inhibiting ACTH release. As a result, glucocorticoids serve a regulatory role in addition to acting as the physiologically active hormones.

The existence of multiple regulatory elements is vital, because it is these elements that make precise control possible. In this system, control is reinforced because cortisol is extremely short-lived. Based on studies of molecules with radioactive atoms attached, biologists estimate that the half-life of a cortisol molecule in the bloodstream is only about 12 minutes. Because feedback inhibition occurs and because the life span of glucocorticoids is so short, the system is constantly shutting itself off. For the long-term stress response to continue, the brain must continually integrate information and support further release of CRF.

The Hypothalamic-Pituitary Axis—An Overview Six decades of research have revealed that the CRF-ACTH-glucocorticoid relationship is part of an array of hormone systems based on interactions between the hypothalamus, pituitary, and target glands or cells. As **Figures 44.9a** and **b** illustrate, the hypothalamus and pituitary actually form two anatomically distinct systems. The posterior and anterior sections of the pituitary gland are each influenced by different populations of neurons in the hypothalamus. These populations of hypothalamic neurons are called neurosecretory cells because they synthesize and release hormones. The release of hormones by cells in the hypothalamus is under the control of brain regions responsible for integrating information about the

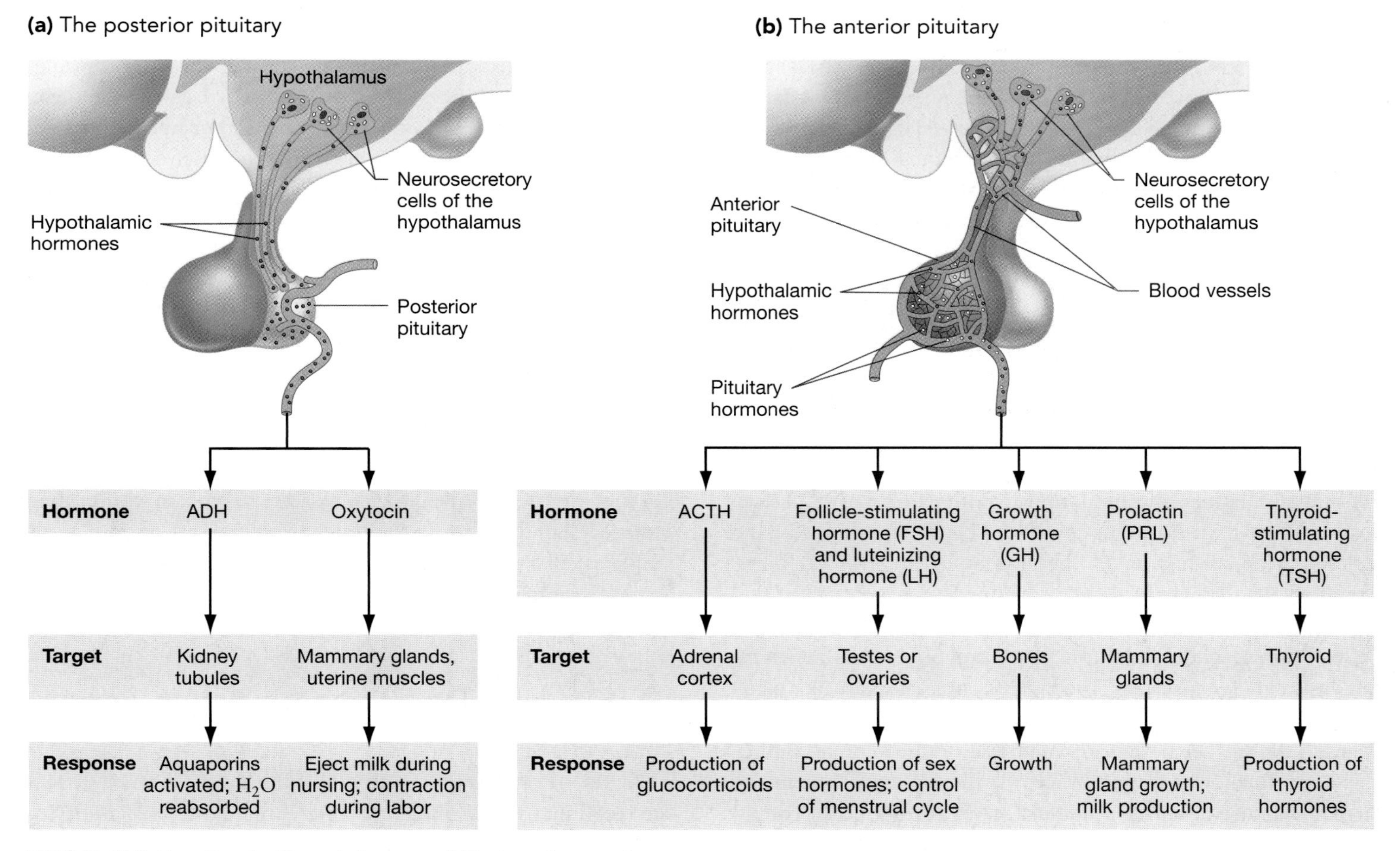

FIGURE 44.9 How Do the Hypothalamus and Pituitary Interact?
(a) Developmentally and anatomically, the posterior pituitary is an extension of the hypothalamus. Neurosecretory cells in the hypothalamus extend directly into the posterior pituitary and secrete ADH and oxytocin.
(b) The hypothalamus and the anterior pituitary communicate indirectly, via blood vessels. Hormones produced by neurosecretory cells in the hypothalamus travel to the anterior pituitary in blood.

BOX 44.2 Oxytocin and ADH

The amino acid sequences of the posterior pituitary hormones oxytocin and ADH are extremely similar. Each molecule is a mere nine amino acids long, and seven are identical when the two sequences are compared. Based on this similarity, biologists infer the two hormones are derived from the same ancestral molecule.

Functionally, however, the two hormones appear very different. ADH triggers water conservation in the kidney, while oxytocin triggers contractions in the uterus during labor. Some biologists suspect that these functional differences are actually related. According to this hypothesis, production of oxytocin was originally an adaptation that allowed desert mammals to abort embryos in times of drought. This hypothesis is supported by the phenomenon of Braxton Hicks contractions, which pregnant women may experience if they become dehydrated. In at least some cases, labor triggered by oxytocin could play a role in water conservation.

external or internal environment. For example, information about the approach of an important exam or athletic contest might trigger action potentials that lead to the release of CRF from neurosecretory cells in the hypothalamus.

It's important to recognize that the anterior and posterior pituitary function in distinct ways. As Figure 44.9a indicates, the posterior portion of the pituitary is actually an extension of the hypothalamus itself. Neurosecretory cells that project from the hypothalamus produce the hormones ADH and oxytocin, which are then stored in the posterior pituitary. From there, ADH and oxytocin are released into the bloodstream. (**Box 44.2** introduces some interesting similarities between ADH and oxytocin.)

Neurosecretory cells in the hypothalamus do not project into the anterior portion of the pituitary, however. As Figure 44.9b illustrates, the hypothalamus and anterior pituitary are connected indirectly, by blood vessels. What happens when neurosecretory cells from the hypothalamus secrete hormones into these blood vessels? In response to the arrival of releasing hormones from the hypothalamus, the anterior pituitary secretes hormones that enter the bloodstream and act on target tissues or glands. Hypothalamic inhibitory hormones, in contrast, stop the secretion of certain pituitary hormones. In several cases, the production of a pituitary hormone is controlled by both a stimulatory and inhibitory hypothalamic hormone. Because many of the hormones produced by the anterior pituitary stimulate the production of other hormones, the structure's designation as "the master gland" is justified.

To summarize, interactions between the hypothalamus and pituitary furnish a key mechanism for control of hormone action by the central nervous system. The hypothalamic-pituitary axis actually consists of two anatomically and functionally distinct systems, however. The posterior pituitary serves as a storage and release site for hormones produced by neurosecretory cells in the hypothalamus; the anterior pituitary serves as a hormone production and release site that is controlled by hypothalamic hormones.

Control of Epinephrine by Sympathetic Nerves

When biologists analyze how the nervous system and endocrine system interact to control the release of epinephrine, the distinction between the two systems begins to blur. Section 44.2 introduced how epinephrine acts as a hormone. During the fight-or-flight response, sympathetic nerves trigger the release of epinephrine and norepinephrine from the adrenal medulla into the bloodstream. But in addition, some sympathetic nerves release epinephrine directly onto target cells. In effect, the nervous system delivers a chemical messenger to particular cells while the endocrine system broadcasts the same messenger by secreting it into the bloodstream.

Epinephrine and norepinephrine function as neurotransmitters as well as hormones. The close similarity between the chemical messengers called hormones and neurotransmitters does not end there, however. In some cases, the mode of action of hormones and neurotransmitters is similar. To drive this point home, recall from Chapter 42 that neurotransmitters have two major types of effects on target cells. All neurotransmitters trigger a postsynaptic potential, which can make the postsynaptic neuron more or less likely to deliver an action potential. In addition, certain neurotransmitters may initiate changes in gene expression in neurons. Chapter 42 illustrated the effect of neurotransmitters on transcription by reviewing experiments on the sea slug *Aplysia*. Repeated application of serotonin led to gene activation and changes in the behavior of the synapse. By altering synapses in this way, neurotransmitters play a central role in learning and memory. Similarly, many hormones exert their effects by activating particular genes in target cells. Understanding how gene activation occurs in response to hormones is currently the subject of intense research at laboratories around the world.

44.4 How Do Hormones Act on Target Cells?

The key to understanding how hormones act on target cells is to recall that some animal hormones are lipid soluble and cross cell membranes readily, while others are not. More specifically, steroid hormones are small lipids that enter cells without difficulty; the peptide hormones and most amino acid derivatives do not cross cell membranes easily because of their large size and electrical charge.

Differences in lipid solubility are important because they influence where a target cell receives the chemical message. Most amino acid derivatives and all peptides act at the cell surface; steroids act inside the cell. To explore these two distinct paths of hormone action, let's consider how estrogens and epinephrine affect target cells. As a steroid and non-steroid, they serve as model systems for hormone action.

Steroid Hormones and Intracellular Receptors

Estrogens are steroids that direct the development of female secondary sex characteristics in many animal species. In humans and other mammals, the most important estrogen is a molecule called estradiol (formally, 17β-estradiol). Because of its importance in reproduction by humans and domesticated animals, estradiol's mode of action has been the topic of intense investigation for over 50 years.

Identifying the Estrogen Receptor A key advance in research on hormone action occurred in 1964, when David Toft and Jack Gorski succeeded in isolating the estradiol receptor in laboratory rats. Their research strategy is outlined in **Figure 44.10**. They began by labeling purified estradiol with a radioactive atom and injecting a small amount into five adult female rats. Twenty minutes later, they removed the uterus from each female and homogenized the tissue. After removing the particulate matter, they added the soluble fraction to a centrifuge tube that contained a sucrose density gradient. As Chapter 5 explained, sucrose density centrifugation allows researchers to separate molecules by density. When centrifugation was complete, the biologists found that the radioactivity was concentrated in a narrow band in the tube (see the graph in Figure 44.10). The band contained radioactive estradiol bound to the estradiol receptor. When Toft and Gorski purified the receptor molecule, they found that it could readily be destroyed by treatment with proteinase enzymes. As a result, they inferred that the estradiol receptor was a protein.

Follow-up experiments confirmed that the estradiol receptor is located in the nucleus, but is not associated with the nuclear envelope. Further, it is found only in estradiol target tissues including the uterus, hypothalamus, and mammary glands. The latter finding was particularly important because it clarified how hormones act in a tissue-specific way. Hormones are broadcast throughout the body via the bloodstream, but they act only on cells that express the appropriate receptor.

Later, biologists led by Bert O'Malley found that the gene for the estradiol receptor has a sequence and structure that are similar to the receptors for the glucocorticoids, testosterone, and other steroids. This result suggested that all of the steroid receptors are related by descent from an ancestral receptor molecule, and that the binding of any steroid hormone to its receptor might affect the target cell in a similar way. Once the hormone-receptor complex forms inside the nucleus of a target cell, what happens?

Documenting Changes in Gene Expression During the 1970s and 80s, work in several different laboratories suggested that estradiol and other steroid hormones affect gene transcription after they bind to their receptor. When researchers injected laboratory animals with estradiol, they documented changes in the mRNAs and proteins produced in target cells. As the data reviewed in Chapter 30 showed, however, steroid hormones do not

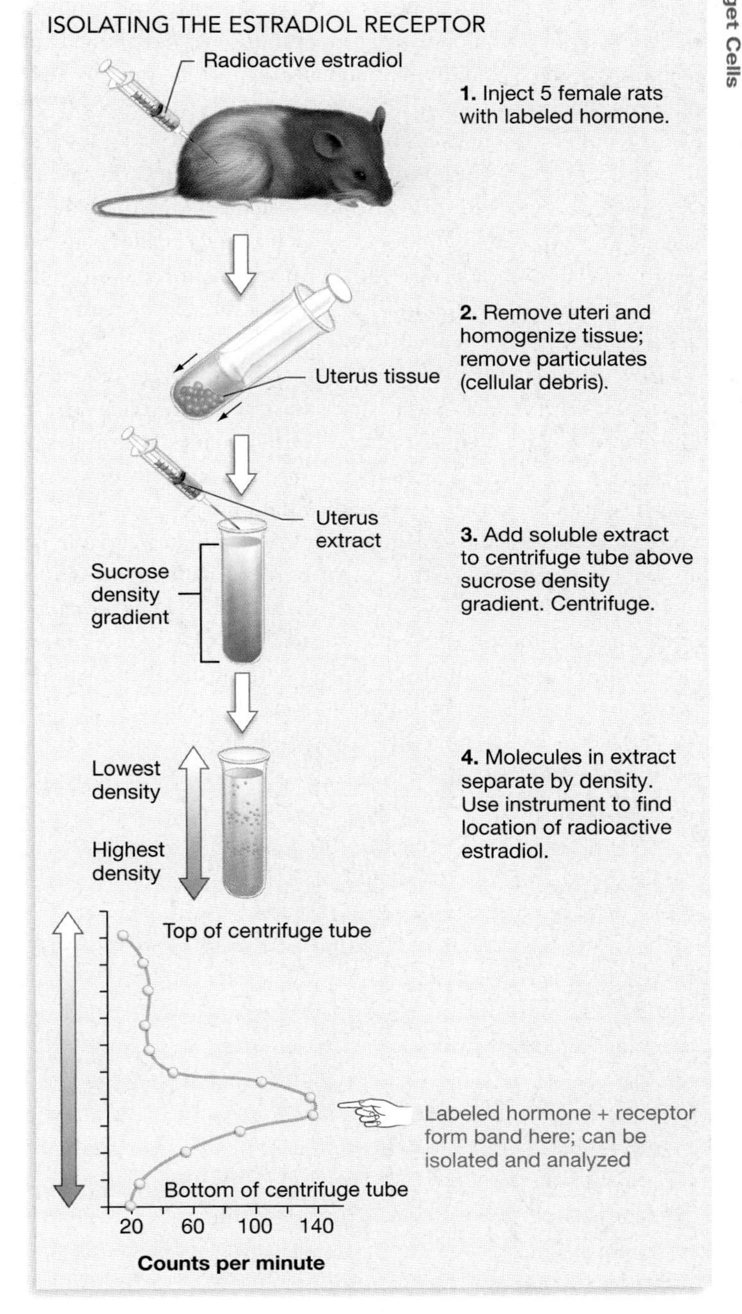

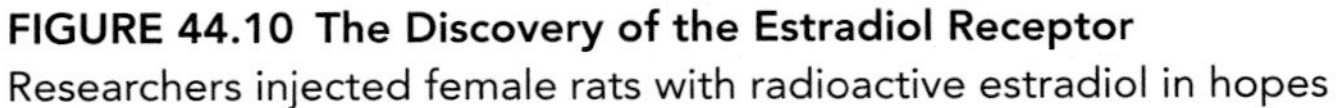

FIGURE 44.10 The Discovery of the Estradiol Receptor
Researchers injected female rats with radioactive estradiol in hopes that the labeled molecule would bind to its receptor in the uterus. After molecules from uterine cells were separated by centrifugation, investigators documented the amount of radioactivity at each level in the centrifuge tube.

CD ACTIVITY 44.2 Hormone Actions on Target Cells

act as simple all-on/all-off switches for target genes. Recall that when investigators analyzed the mRNAs produced in response to the insect hormone ecdysone, they found that the timing and amount of expression varied dramatically from gene to gene.

An important insight into the molecular mechanism of hormone action came when researchers in Bert O'Malley's laboratory found a distinctive DNA-binding region in the steroid hormone receptor. The region codes for the DNA-binding domain called a zinc finger. Zinc fingers are found in all of the proteins in the steroid hormone receptor family. As explained in Chapter 14, DNA-binding domains are protein regions that make physical contact with DNA. Investigators in O'Malley's group confirmed that steroid hormone-receptor complexes bind to specific sites in DNA. They called these sites hormone-response elements. Hormone-response elements, in turn, function as the type of regulatory DNA sequence called an enhancer. As Chapter 15 showed, transcription of a gene begins when a regulatory protein like a steroid-receptor complex binds to an enhancer element for that locus.

How do all these mechanisms fit together? The model in **Figure 44.11** summarizes the current state of thinking about how steroid hormones affect target cells. When estradiol or another steroid hormone enters a target cell, it binds to its receptor. The binding event causes a conformational change in the receptor that makes the hormone-receptor complex capable of binding to DNA. Once the activated regulatory protein has bound to DNA, transcription of specific genes begins. Because each hormone-receptor complex leads to the production of many copies of the gene product, the signal from the hormone is amplified. In this way, a small number of hormone molecules produces a large change in the activity of target cells and tissues.

Currently, research on steroid hormone action is focused on the details of how the hormone-receptor protein interacts with the transcription initiation complex, and how interactions with other regulatory proteins alter the timing and amount of expression of target genes.

Hormones That Bind to Cell-Surface Receptors

Epinephrine, norepinephrine, and the peptide hormones are not lipid soluble. For these molecules to affect a cell, they have to bind to receptors on the cell surface. The messenger never enters the target cell, so its message must be transduced—meaning, changed into a form that is active inside the cell. This phenomenon is known as **signal transduction.**

To explore how signal transduction occurs, we first need to examine the nature of the hormone receptors residing in the cell membrane. Then we can explore research on the molecules that are responsible for processing the message inside the cell. In both cases, we'll focus on epinephrine as a model system.

Identifying the Epinephrine Receptor In 1948, Raymond Ahlquist published the results of an exhaustive set of studies on how epinephrine and five related molecules affect dogs, cats, rats, and rabbits. The responses he observed fell into two distinct categories, depending on the tissue being considered. To explain the data, Ahlquist proposed that the epinephrine-like molecules were acting as **agonists**, meaning that they bind to the same receptor as the hormone itself. Further, he suggested that epinephrine and its agonists produce two distinct patterns of response because two types of receptor exist. He called these hypothetical proteins the alpha receptor and the beta receptor.

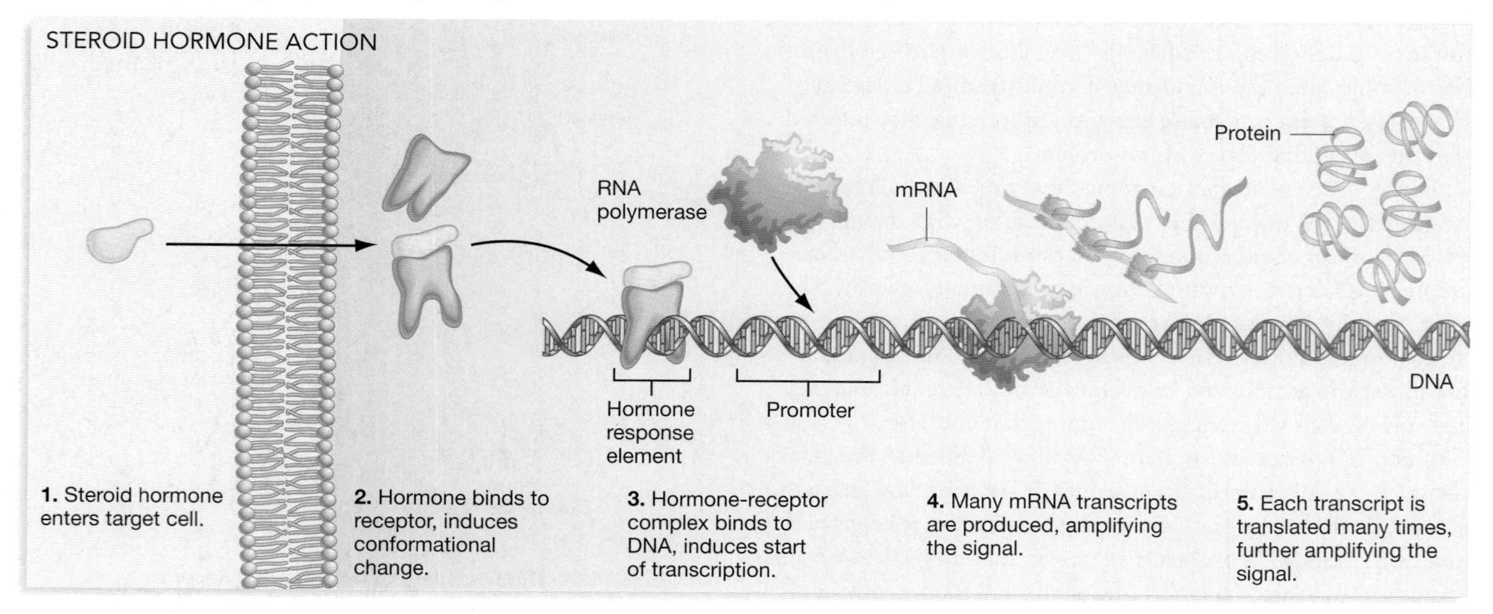

FIGURE 44.11 How Do Steroid Hormones Act?
The current model holds that steroid hormones bind to receptor proteins inside a cell, and that hormone-receptor complexes trigger the transcription of certain genes by binding to regulatory DNA segments called hormone-response elements.

Follow-up work with agonists and with molecules that block epinephrine receptors confirmed Ahlquist's hypothesis. In fact, these experiments expanded his suggestion by documenting that there are two types of alpha receptor and two types of beta receptor. In short, there are four distinct epinephrine receptors. Each is found on a distinct tissue type, and each induces a different response from the cell.

The discovery of four epinephrine receptors reinforces the concept of tissue specificity observed in experiments on the estradiol receptor. Hormones are transmitted throughout the body, not unlike a radio or television signal that is broadcast through the atmosphere. Their message is received only by cells with the appropriate receptor, however—just as a TV or radio signal is received only by equipment with an appropriate antenna. In the case of epinephrine, there are four distinct receptors instead of just one. As a result, the same hormone can trigger different effects in different cells. Now let's analyze what happens once epinephrine binds to one of these receptors.

Signal Transduction and the Role of Second Messengers

Signal transduction occurs when a chemical message at the cell surface elicits a response inside the cell. Signal transduction was introduced in Chapter 35 and is currently an exceptionally active area of research. Research on epinephrine has produced fundamental insights into how the process works at a molecular level.

Recall from section 44.2 that one of epinephrine's major effects is to increase glucose levels in the blood. To understand how epinephrine triggers the release of glucose from target cells, biologists focused on an enzyme called phosphorylase. Phosphorylase was discovered in the mid-1940s by Gerty Cori and Carl Cori, who were studying how cells cleave glucose molecules off glycogen. As **Figure 44.12a** shows, phosphorylase catalyzes this reaction. The Coris found that phosphorylase exists in active and inactive forms, and that the enzyme switches between these states when it is phosphorylated or dephosphorylated by another enzyme. Phosphorylase was the first of many enzymes shown to be activated by the addition of a phosphate group.

Researchers in Earl Sutherland's laboratory followed up on these results by exploring whether phosphorylase is activated in response to epinephrine. During the fight-or-flight response, the primary source for the glucose that enters the bloodstream is glycogen that is stored in the liver. Does epinephrine stimulate glucose production by activating phosphorylase in liver cells where glycogen is stored?

To answer this question, Sutherland's group worked out techniques for homogenizing liver cells and studying the production of glucose from glycogen in vitro. As **Figure 44.12b** shows, they found that when they added epinephrine to their cell-free system, large amounts of phosphorylase were activated. The effect disappeared, however, when they removed the particulate fraction from the liver extract. This observation suggested that there was something in the homogenate that activated phosphorylase in response to epinephrine. By purifying the components of the homogenate and testing them in the cell-free system, they eventually found the ingredient that activated phosphorylase. It was a molecule called cyclic adenosine monophosphate, or cAMP.

cAMP's role in signal transduction was confirmed when researchers in Sutherland's lab studied the effects of epinephrine on the activity of the rat heart. During the fight-or-flight response, both heart rate and the contractile force of the heart increase dramatically. As the graphs in **Figure 44.13** (page 866) show, Sutherland's group found that there is a striking increase in cAMP levels inside heart cells soon after treating them with epinephrine. In addition to observing an increase in phosphorylase activity in these cells, the investigators also confirmed that the contractile force of the heart-muscle cells increased.

To capture the importance of cAMP in triggering these effects, the biologists referred to it as a second messenger. A **second messenger** is a signaling molecule that increases in concentration inside a cell in response to a molecule that binds at the surface. What does cAMP do? Researchers in Edwin Krebs' lab found that cAMP binds to a protein called cAMP-dependent protein kinase A. This enzyme in turn responds by phosphorylating an enzyme, which then phosphorylates phosphorylase. This extensive chain of events is initially triggered by the synthesis of cAMP. cAMP is produced from ATP in a reaction that is catalyzed by the enzyme adenylyl cyclase.

To pull all of these details and reactions together, researchers proposed the model for epinephrine action diagrammed in

(a) Phosphorylase catalyzes the production of glucose from glycogen.

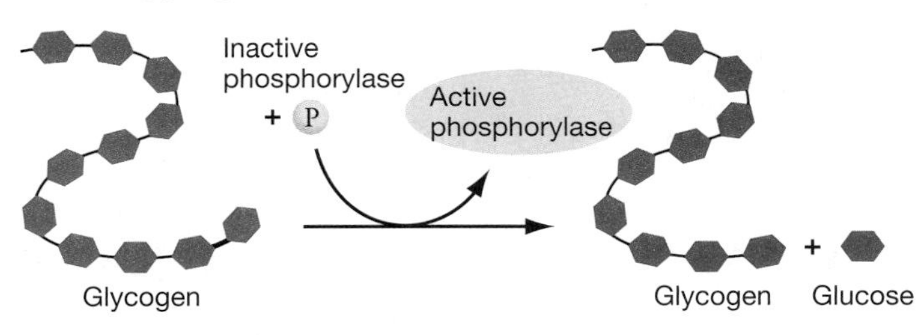

(b) Phosphorylase must be activated by epinephrine.

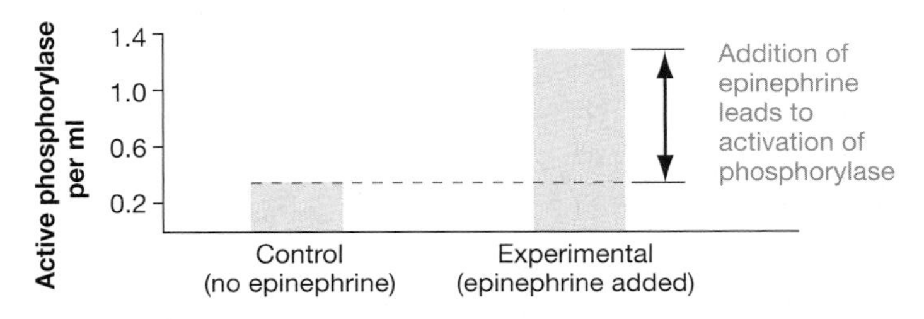

FIGURE 44.12 Epinephrine Activates the Enzyme That Catalyzes Formation of Glucose from Glycogen
(a) Phosphorylase is activated when an enzyme adds a phosphate group to it. The activated protein catalyzes the breakdown of glycogen shown here. **(b)** When epinephrine is added to cell-free extracts from liver tissue, the amount of activated phosphorylase increases dramatically.

Figure 44.14. In studying this model, it is crucial to recognize two points. First, the second messenger cAMP transmits the signal from the cell surface to the signaling cascade. Second, the other events in the sequence amplify that signal. To drive this point home, consider that in response to stimulation by the hormone-receptor complex, adenylyl cyclase is thought to catalyze the formation of at least 100 molecules of cAMP. In turn, each of these cAMP molecules activates many molecules of cAMP-dependent protein kinase. Subsequently, each protein kinase molecule activates many molecules of phosphorylase kinase, and so on. In this way, the binding of a single epinephrine molecule may trigger the release of millions or billions of glucose molecules. This is an extremely important point. Amplification through a signal transduction cascade explains why tiny amounts of hormones can have such huge effects on an individual.

Since the model in Figure 44.14 was published, researchers have uncovered a rapidly growing catalog of signal transduction systems. Sutherland and Krebs and co-workers happened to be studying an epinephrine receptor called the beta-1 receptor. But other groups showed that when epinephrine binds to an alpha-1 receptor, a completely different signal transduction event occurs. In this and many other receptor systems, calcium ions (Ca^{2+}) serve as the second messenger in conjunction with a second messenger molecule called IP3. Diacylglycerol (DAG) and 3′, 5′ cyclic GMP (cGMP) are also common second messengers in hormone response systems. In some cases, target cells that have the same receptor protein have different second messengers or enzyme systems. In this way, the same hormone and receptor can give rise to different responses in different target cells.

Recent work on signal transduction has delivered one fundamental message. Different target cells may contain different receptors, second messengers, protein kinases, and amplification steps. The feature that unites these systems is their endpoint: the activation of a specific protein or set of proteins, usually by phosphorylation.

How do polypeptide, amino-acid-derivative, and steroid hormones differ in their mode of action? All three classes of hormones are present in tiny concentrations, yet exert large ef-

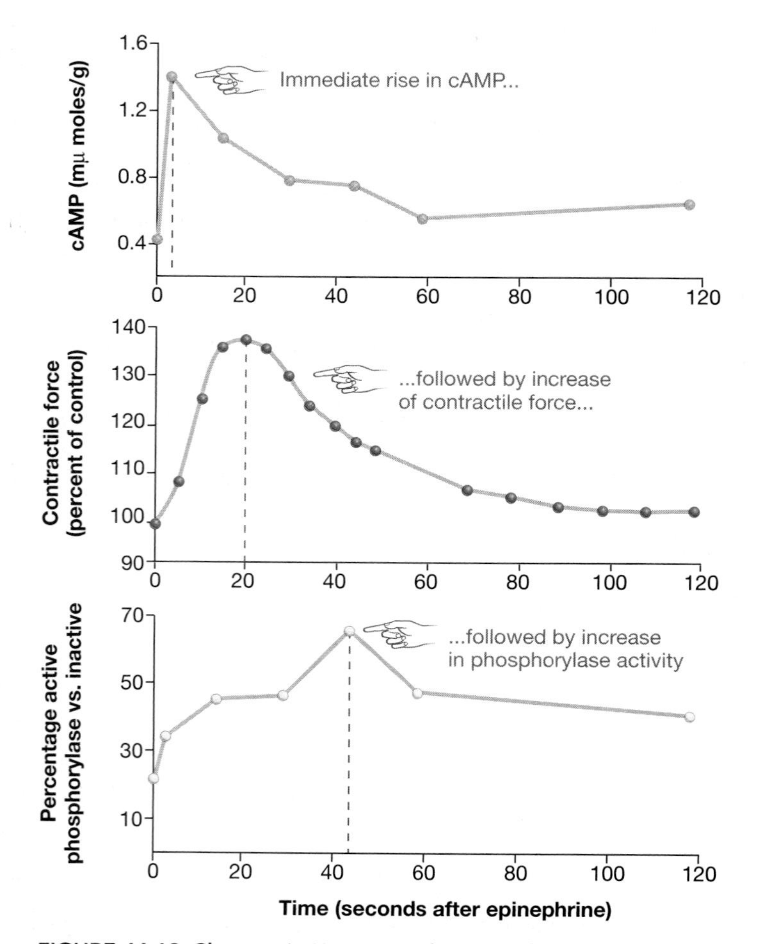

FIGURE 44.13 Changes in Heart-Muscle Cells After Addition of Epinephrine
When epinephrine is added to heart-muscle cells, there is a rapid and dramatic rise in intracellular cAMP concentration. The contractile force of muscle fibers shows a sharp increase soon thereafter; slightly later, phosphorylase activity increases significantly.

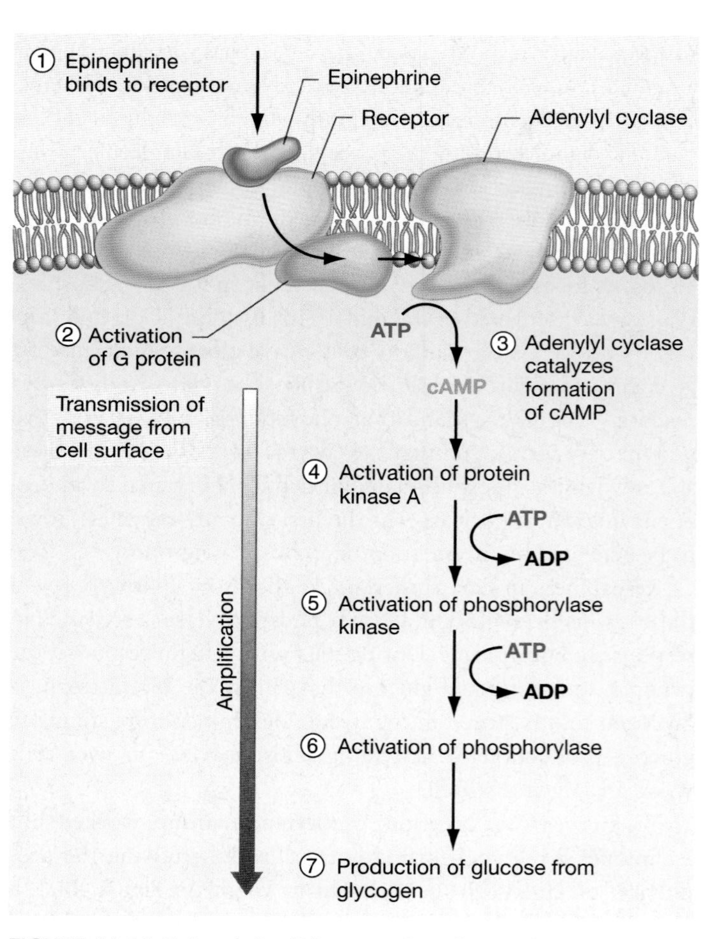

FIGURE 44.14 Epinephrine Triggers a Signal Transduction Cascade
This diagram illustrates the sequence of events that occurs after epinephrine binds to a receptor on the cell membrane. Several steps involve phosphorylation events that activate proteins. Because each molecule in the cascade acts on many molecules downstream, the effect of the initial binding event is greatly amplified.

fects through amplification. The key contrast is that steroid hormones tend to exert their effect through changes in gene expression, while non-steroid hormones tend to change cells by activating existing proteins. Steroid hormones activate transcription factors; non-steroid hormones activate signal transduction cascades.

Essay Do Humans Produce Pheromones?

A pheromone is a chemical messenger synthesized by an individual and released into the environment, which subsequently elicits a response from members of the same species. Pheromones are odorants that act as chemical messengers by changing the behavior or physiology of another individual.

Pheromones have been studied most intensively in insects and mice. To date, over a thousand different insect pheromones have been isolated and characterized. Most of these molecules are produced by females and serve as airborne, species-specific sex attractants. In mice, pheromones have been isolated from the urine of males and females and from glands that secrete to the skin. These molecules have a variety of effects, ranging from the synchronization of ovulation and sexual receptivity in females that are living together to delaying or accelerating the onset of puberty in females.

Is a pheromone responsible for menstrual synchrony?

Recently, Kathleen Stern and Martha McClintock collected data suggesting that pheromones also exist in humans. In addition to studying the role of pheromones in triggering ovulation in rodents, these researchers had documented a phenomenon in humans known as menstrual synchrony. When human females live in close proximity—as in college dormitories or other communal situations—their menstrual cycles tend to synchronize over time. (Chapter 45 explains the phases of the menstrual cycle and details how hormones control the timing and duration of each stage.) What Stern and McClintock wanted to know is whether a pheromone might be responsible for changing these phases in a way that leads to their synchronization among individuals.

To test the hypothesis that a pheromone is responsible for menstrual synchrony, Stern and McClintock recruited a group of female volunteers at a university. For several months they had nine donor individuals wear cotton pads under their armpits. The researchers collected these pads daily and wiped them or fresh cotton pads under the noses of recipient individuals. Then they monitored whether exposure to armpit secretions led to a change in the length of menstrual phases in recipients. Their data showed that recipients experienced significant changes in their menstrual cycle. Because these changes corresponded to the menstrual stage being experienced by donors, Stern and McClintock concluded that recipients had responded to a pheromone.

Is there any anatomical or molecular evidence that supports this result? In mice, the organ, neurons, and receptor molecules responsible for sensing pheromones are well characterized. Near the main olfactory epithelium, mice have a specialized structure called the vomeronasal organ, or VNO. This organ contains sensory neurons that project to a portion of the hypothalamus that is independent of the brain area responsible for the sense of smell. Although most adult humans lack a VNO, Ivan Rodriguez and co-workers recently found a gene in humans that is extremely similar to the genes for receptor proteins in the VNO of mice. Based on this result, they claim that humans have at least one functioning VNO receptor.

Does this receptor respond to the pheromone that induces menstrual synchrony? Although research continues at a rapid pace, this question remains unanswered.

Chapter Review

Summary

Hormones are chemical messengers that are released from cells in the endocrine system and that trigger a response in target cells containing an appropriate receptor. Most hormones were discovered when researchers removed a gland from a laboratory animal and were able to correct the resulting symptoms by injecting the individual with a solution extracted from that gland. By purifying molecules in these extracts and testing their activity, biologists have been able to isolate hormones with a variety of chemical structures. Most animal hormones are steroids, amino acid derivatives, or polypeptides. Although they are produced in tiny concentrations, hormones have large effects because they trigger gene expression or because their message is amplified through a signal transduction cascade.

In conjunction with the nervous system, hormones coordinate the activities of diverse cells and tissues. Many hormones function in one of three situations.

- Responding to environmental challenges or stimuli. In humans and other mammals, for example, epinephrine and cortisol activate the short-term and long-term response to stress.
- Directing growth or development. Estradiol, for example, triggers the formation of female sex characteristics in human embryos and the maturation of these tissues in adolescence.
- Achieving homeostasis. A suite of hormones is involved in directing cells that modify the concentrations of glucose, Na^+, Ca^{2+}, O_2, water, and other molecules in the blood and interstitial fluid.

In many cases, the release of a hormone is regulated by chemical messengers from the anterior pituitary gland. The pituitary is located near the base of the brain in humans and other mammals. Hormone-secreting cells in the anterior pituitary are regulated by hormones released by the hypothalamus. For example, the brain responds to long-term stress by triggering the release of the hypothalamic hormone CRF. CRF activates the release of ACTH by the pituitary gland, which stimulates the production of cortisol by cells in the adrenal cortex. This chain of events is dampened by feedback inhibition. Cortisol inhibits the production of ACTH, and ACTH inhibits the continued release of CRF. In general, the synthesis and release of hormones are highly regulated.

Animal hormones have two basic modes of action. Steroid hormones are lipid soluble, they cross cell membranes readily, and they bind to receptors inside cells. Amino acid derivatives and polypeptides are not lipid soluble and bind to receptors located in the membranes of target cells. In both cases, the response to a hormone is tissue specific because only certain cells express certain receptors.

Work on estradiol and ecdysone indicates that most steroid hormones act by inducing a change in gene expression. Steroid hormone receptors are closely related structurally and have a distinctive DNA-binding domain. The hormone-receptor complex binds to enhancer-like hormone-response elements in DNA. In response, the transcription of specific genes is activated or repressed.

Research on epinephrine has revealed that amino acid derivatives and polypeptide hormones trigger a complex sequence of events when they bind to a receptor on the cell membrane. These signal transduction cascades induce a change in the concentration of a second messenger, like cAMP, inside the target cell. In many cases the endpoint of the process is activation of a target protein by phosphorylation and a dramatic change in the cell's activity.

Questions

Content Review

1. Epinephrine and cortisol are both involved in the body's response to stress. How do the two molecules differ?
   a. Cortisol is an amino acid derivative; epinephrine is a steroid.
   b. Cortisol mediates the short-term response; epinephrine mediates the long-term response.
   c. Epinephrine mediates the short-term response; cortisol mediates the long-term response.
   d. Epinephrine is produced in the adrenal cortex; cortisol is produced in the adrenal medulla.
2. Why are the structural differences between amino acid derivatives, polypeptides, and steroid hormones significant?
   a. Steroids are lipid soluble and cross cell membranes readily.
   b. Amino acid derivatives and polypeptide hormones are lipid soluble and cross cell membranes readily.
   c. Polypeptide hormones are structurally complex, and thus more difficult to synthesize.
   d. Only polypeptides and steroids bind to receptors in the cell membrane.
3. What is signal transduction?
   a. the binding of a steroid hormone-receptor complex to DNA
   b. release of a hormone from the anterior pituitary, in response to a hypothalamic hormone
   c. release of a hormone from the posterior pituitary, in response to action potentials from the hypothalamus
   d. production of a second chemical messenger inside a cell, in response to hormone binding at the cell surface
4. Which of the following developmental processes are controlled by hormones? (Circle all that apply.)
   a. initial development of male and female gonads, soon after fertilization
   b. maturation of the male and female reproductive tract and other secondary sex characteristics at puberty
   c. molting in insects and other invertebrate animals
   d. metamorphosis in insects and other invertebrate animals
   e. overall growth
5. What is a hormone response element?
   a. a receptor for a steroid hormone
   b. a receptor for a polypeptide hormone
   c. a segment of DNA where a hormone-receptor complex binds
   d. a protein that is phosphorylated in response to hormone binding
6. In hormone systems, what is feedback inhibition?
   a. when the presence of a hormone inhibits its synthesis
   b. when the presence of a hormone stimulates its synthesis
   c. when a second messenger triggers phosphorylation of a repressor protein
   d. when a hormone from the hypothalamus inhibits release of a hormone from the anterior pituitary

Conceptual Review

1. Compare and contrast neurotransmitters and hormones. What are these molecules, and what do they do? Use epinephrine as an example to highlight the similarities and differences between the two functional categories of chemical messengers.
2. Describe the steps involved in the discovery of a hormone. Which step would you predict is the most difficult, and why?
3. The pituitary is often referred to as "the master gland." Why?
4. Compare and contrast the mode of action of steroid and nonsteroid hormones.
5. The text claims that hormones are present in tiny concentrations, yet have large effects on target cells and on the individual as a whole. How is this possible?

6. Compare and contrast the short-term and long-term responses to stress in humans. Which hormones are involved? Why are their effects logical, in terms of maximizing an individual's ability to survive a potential life-threatening situation?

Applying Ideas

1. Suppose that during a detailed anatomical study of a marine invertebrate, you found a small, previously undescribed structure. How would you test the hypothesis that the structure is a gland that releases one or more hormones?

2. Cortisone is a glucocorticoid that suppresses inflammation and other aspects of wound healing. Cortisone was once widely used to treat athletes with joint injuries and people who suffered from arthritis. Short-term, it is extremely effective in reducing swelling and pain. Over time, however, physicians found that repeated large doses of cortisone had damaging side effects. Predict what these side effects are. Explain your logic.

3. You are a physician supervising a patient's recovery from surgical removal of the posterior pituitary. Name a hormone that you will have to administer to this patient artificially. Which symptom(s) will you monitor to assess whether the dosage and timing of your injections is having the desired effect?

4. Suppose that a researcher announces the discovery of a hormone that affects the metabolism of fats in lab rats. Preliminary data indicate that the hormone is a polypeptide, about 50 amino acids long. How would you go about isolating the receptor for this hormone? Using fat cells growing in vitro, how could you test the hypothesis that hormone binding causes a change in a second messenger?

5. Discuss the trade-offs involved in the short-term response to stress in humans. What are the advantages and disadvantages of activating certain systems in the body while shutting others down?

6. Study Figure 44.4 and note that the components of the endocrine system are all located along the midline of the body. Suggest a hypothesis for why this pattern exists.

7. Turn back to Chapter 14 and briefly review the role of cAMP in the regulation of the *lac* operon in *Escherichia coli*. How does cAMP's role in glucose utilization by bacteria compare to the role of cAMP in the production of glucose by human cells in response to epinephrine?

CD-ROM and Web Connection

CD Activity 44.1: Endocrine System Anatomy *(tutorial)*
(Estimated time for completion = 15 min)
What are the components of the endocrine system, and how are hormones regulated?

CD Activity 44.2: Hormone Actions on Target Cells *(animation)*
(Estimated time for completion = 5 min)
Different hormones have different chemical characteristics. Do all hormones act on their target cells in a similar fashion?

At your **Companion Website** (http://www.prenhall.com/freeman/biology), you will find self-grading exams and links to the following research tools, online resources, and activities:

Health Pills, Patches, and Shots: Can Hormones Prevent Aging?
Explore how hormones work and learn about some possible implications of taking hormone supplements to slow the aging process.

Treating Eating Disorders
This article discusses the hormones leptin and neuropeptide and how they might be used in the treatment of eating disorders.

The Role of Estrogen in Sexual Differentiation
Learn how estrogen plays a role in the sexual development of women.

Additional Reading

Ben-Ari, E. T. 1998. Pheromones: What's in a name? *BioScience* 48: 505–511. What is the evidence that humans produce pheromones?

Jordan, V. C. 1998. Designer estrogens. *Scientific American* 279 (October): 60–67. Introduces molecules that bind to estrogen receptors and may be useful as drugs.

Scott, J. D. and T. Pawson. 2000. Cell communication: the inside story. *Scientific American* 282 (June): 72–79. Explores how hormones, receptors, and signal transduction systems work.

Animal Reproduction

45

All of the cells, tissues, organs, and systems introduced in this unit exist for one reason. They allow animals to survive long enough and gather enough resources to reproduce. The shark rectal gland, the fish gill, the squid giant axon, the infrared sensors of pit vipers, and the human adrenal glands are all adaptations that help individuals survive in particular environments, acquire energy, and produce offspring. Stated another way, producing offspring is the reason that adaptations exist; reproduction is the unconscious goal of virtually everything that an animal does.

Evolution by natural selection explains why reproduction occurs; the goal of this chapter is to explore *how* animals reproduce. Section 45.1 begins by introducing research on animals that cycle between asexual and sexual reproduction. In addition, this introductory section reviews how both asexual and sexual reproduction occur at the cellular level. Section 45.2 surveys the diversity of ways in which animals accomplish fertilization once gametes have formed. The final three sections focus on mammalian reproduction, using humans as the primary model organism. The discussion begins by analyzing male and female reproductive organs, continues by exploring how hormones control changes in those organs, and concludes with a description of pregnancy and birth.

In the course of investigating research on the molecular and cellular mechanisms of reproduction, we'll also consider topics of urgent practical interest. These include the dangers of steroid hormone use by athletes and bodybuilders, recent and dramatic declines in the death rate of mothers during childbirth, and methods of contraception. Understanding and manipulating reproductive systems is an important issue in the working life of physicians,

The swollen, red rump of this female baboon indicates that she is about to produce an egg and will accept courtship from males and possibly copulation. She will probably mate with several males before the egg is fertilized. This chapter discusses an array of questions about animal reproduction, including how hormones control the female reproductive cycle and the consequences of multiple mating.

veterinarians, farmers, zoo keepers, and many others in biology-related professions.

45.1 Asexual and Sexual Reproduction

Several other chapters have explored how asexual and sexual reproduction differ. Recall that when asexual reproduction occurs, offspring are genetically identical to their parent. But when sexual reproduction occurs, offspring are genetically different from their parents. Before reviewing the details of how these two modes of reproduction occur in animals, let's consider a species that does both. Why would an organism switch between reproducing sexually and asexually?

Switching Reproductive Modes: A Case History

Daphnia are crustaceans that live in freshwater habitats throughout the world. In a typical year, *Daphnia* reproduce asexually and produce only diploid female offspring throughout the spring and summer. The production of offspring from unfertilized eggs is called **parthenogenesis** (virgin-birth). As **Figure 45.1a** shows, the eggs produced by parthenogenesis develop in a brood pouch; they are released when the female molts her exoskeleton.

In late summer or early fall, however, many females begin producing male offspring parthenogenetically. Sperm from these males fertilize haploid eggs that females produce via meiosis. The fertilized eggs are released into a durable case that falls to the bottom of the pond or lake for the winter. In spring, the sexually produced offspring hatch and begin reproducing asexually.

What cues trigger the change between modes of reproduction? For decades, most researchers contended that day length cued the asexual-sexual switch. The idea was that the shortening days of late summer or fall induced the production of males and haploid eggs in anticipation of the need for an overwintering stage. Then in 1965, Raymond Stross and Jeanne Hill showed that high-population densities are also a factor in causing the switch. When Stross and Hill brought *Daphnia* populations into the lab and kept day length constant, they found a strong correlation between population density and how much sexual reproduction occurred (**Figure 45.1b**).

In 1992, Ole Kleiven and colleagues extended this result by pinpointing the specific aspects of crowding that were affecting the animals. To do this work, they brought a different population of *Daphnia* into the laboratory and altered day length, the amount of food available to individuals, and the quality of the water they occupied. (To vary water quality, they used clean water or water taken from tanks where *Daphnia* were being maintained at high density.) As **Figure 45.1c** shows, individuals in their study population switched to sexual reproduction only if they were exposed to short day lengths, low food availability, and water from crowded populations. In short, *Daphnia* need three different cues from the environment to switch to sexual reproduction.

These experiments don't address the question of why *Daphnia* make the switch, however. Why is it adaptive to reproduce sexually under the adverse conditions dictated by shortened

(a) *Daphnia* produce diploid eggs without sex.

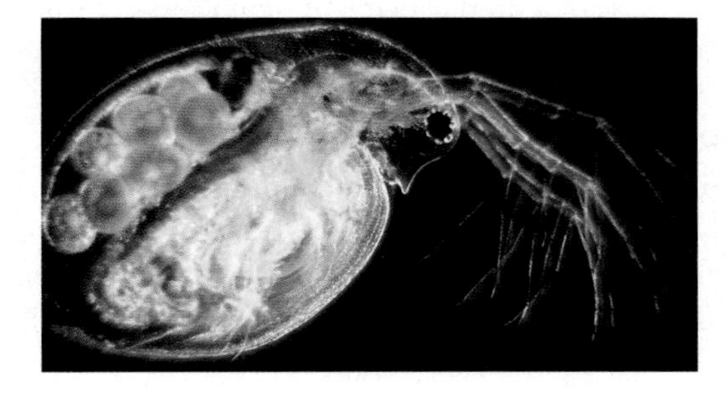

(b) But crowded *Daphnia* often switch to sexual reproduction.

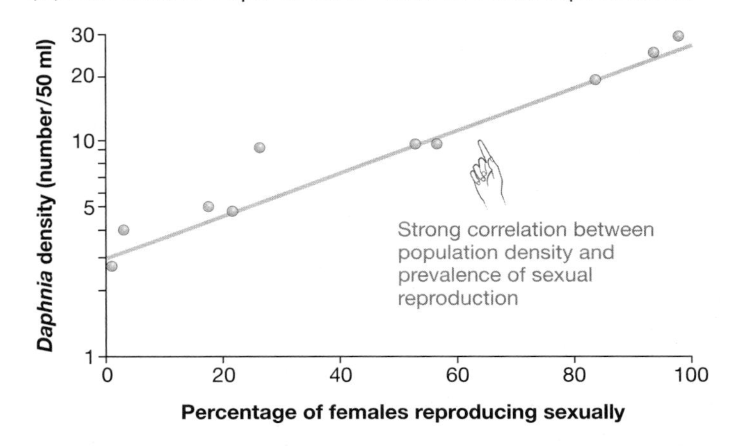

(c) Which environmental cues trigger the switch to sexual reproduction?

Water quality	Food concentration	Day length	% Sexual broods
Clean	Low	Short	0
Crowded	Low	Short	0.44
Clean	Low	Long	0
Crowded	Low	Long	0
Clean	High	Short	0
Crowded	High	Long	0
Clean	High	Short	0
Crowded	High	Long	0

FIGURE 45.1 Switching Between Asexual and Sexual Reproduction
(a) The brood pouch of this *Daphnia* female is filled with asexually produced embryos. **(b)** This graph plots the percentage of *Daphnia pulex* females that reproduce sexually versus *Daphnia* density. Density is reported on a logarithmic scale. **(c)** In a series of experiments, environmental conditions experienced by *Daphnia magna* individuals were varied. "Crowded" water was taken from tanks containing dense populations. QUESTION How would you go about determining which molecule or molecules in crowded water serve as a signal that triggers sexual reproduction?

days, nutrient stress, and high population density? The leading hypothesis to answer this question is that because of their genetic diversity, sexually produced offspring are more capable of overwintering and thriving the following spring than asexually produced offspring. To date, however, this hypothesis has yet to be tested rigorously. Research on the adaptive significance of sex in *Daphnia* and other organisms continues.

Mechanisms of Asexual Reproduction

Which animals produce offspring asexually, and how is it accomplished? **Figure 45.2** addresses both questions. The hydra in Figure 45.2a, for example, is generating offspring by budding. When budding occurs, an offspring begins to form within or on a parent. Budding is completed when the offspring—a miniature version of the parent—breaks free and begins to grow on its own. Figure 45.2b illustrates a flatworm undergoing fission, in which an individual simply splits into two or more descendants. In some taxonomic groups, offspring can develop from unfertilized eggs. The dark structures along the back of the *Daphnia* in Figure 45.2c are unfertilized eggs that will be released from the mother and hatch into offspring.

The general point here is that asexual reproduction occurs in a variety of ways and in a wide variety of animal groups. To reinforce this point, consider the following observations about parthenogenesis—the type of asexual reproduction that takes place in *Daphnia*.

- Some groups, like the bdelloid rotifers, reproduce exclusively by parthenogenesis, never sexually. The same is true for selected species of fishes and lizards.
- In some parthenogenetic species of fish, eggs do not develop unless they are fertilized by a sperm. None of the father's genes are incorporated into the offspring, however.
- In ants, honeybees, and related insects, males are produced by parthenogenesis; but females develop from fertilized eggs produced through sexual reproduction.

With the exception of the bdelloid rotifers and a few fish and lizard species, though, animals that reproduce asexually also reproduce sexually on occasion. To introduce how sexual reproduction occurs, let's take a brief look at the processes responsible for producing male and female gametes. We'll then discuss the intricacies of fertilization and care of eggs.

(a) Budding

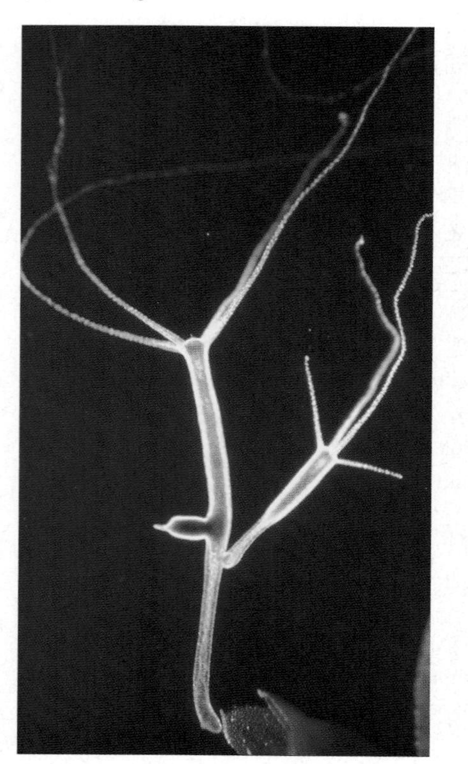

(b) Fission

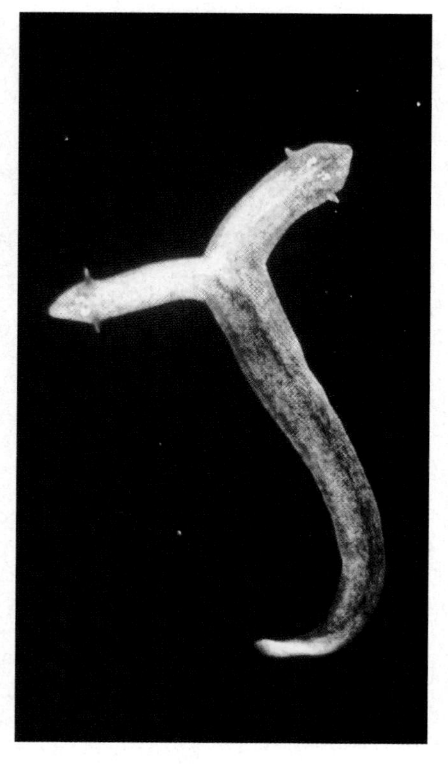

(c) Parthenogenesis

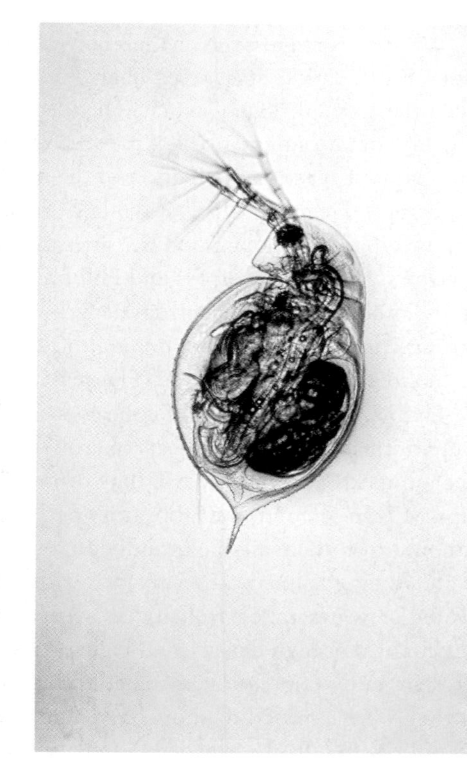

FIGURE 45.2 There Are Many Ways to Reproduce Asexually

Mechanisms of Sexual Reproduction: Gametogenesis

The mitotic cell divisions, meiotic cell divisions, and developmental events that result in the production of male and female gametes, or sperm and eggs, are collectively called gametogenesis. In the vast majority of animals, gametogenesis occurs in a sex organ, or **gonad**. Male gonads are called **testes**; female gonads are called **ovaries**.

Figure 45.3 summarizes the events that take place during gametogenesis in humans. It's important to note that in the male and female gonad, diploid cells called spermatogonia and oogonia divide by mitosis to generate the cells that undergo meiosis. In males, the primary spermatocytes that are produced when a spermatogonium divides go on to produce four haploid cells by meiosis. In turn, each of these haploid products goes on to mature into a sperm. The production of spermatogonia and primary spermatocytes occurs continuously throughout adult life.

Gametogenesis is markedly different in human females. For example, the oocyte that is produced when an oogonium divides goes on to produce just one haploid cell that matures into an egg. The other cells produced by meiosis in females have a tiny amount of cytoplasm and do not mature into eggs. Because the distribution of cytoplasm is so unequal during each of these meiotic divisions in females, the smaller cells are called polar bodies. (Recall that *polar* refers to inequality or opposites.) Further, the production of oogonia stops early in development in many species; in humans it stops before birth. And in humans and many other mammals, primary oocytes begin the first meiotic division but then stop for a period of months or years.

Once mature sperm and eggs are produced, how do these gametes accomplish fertilization? What does the mother do with the egg that results?

CD ACTIVITY 45.1 Human Gametogenesis

45.2 Fertilization and Egg Development

Fertilization is the joining of a sperm and egg to form a diploid zygote. Chapter 18 introduced the process by describing how a sperm makes contact with an egg and penetrates its membrane. The discussion in that chapter focused on the molecular mechanisms of fertilization using the sea urchin as a model

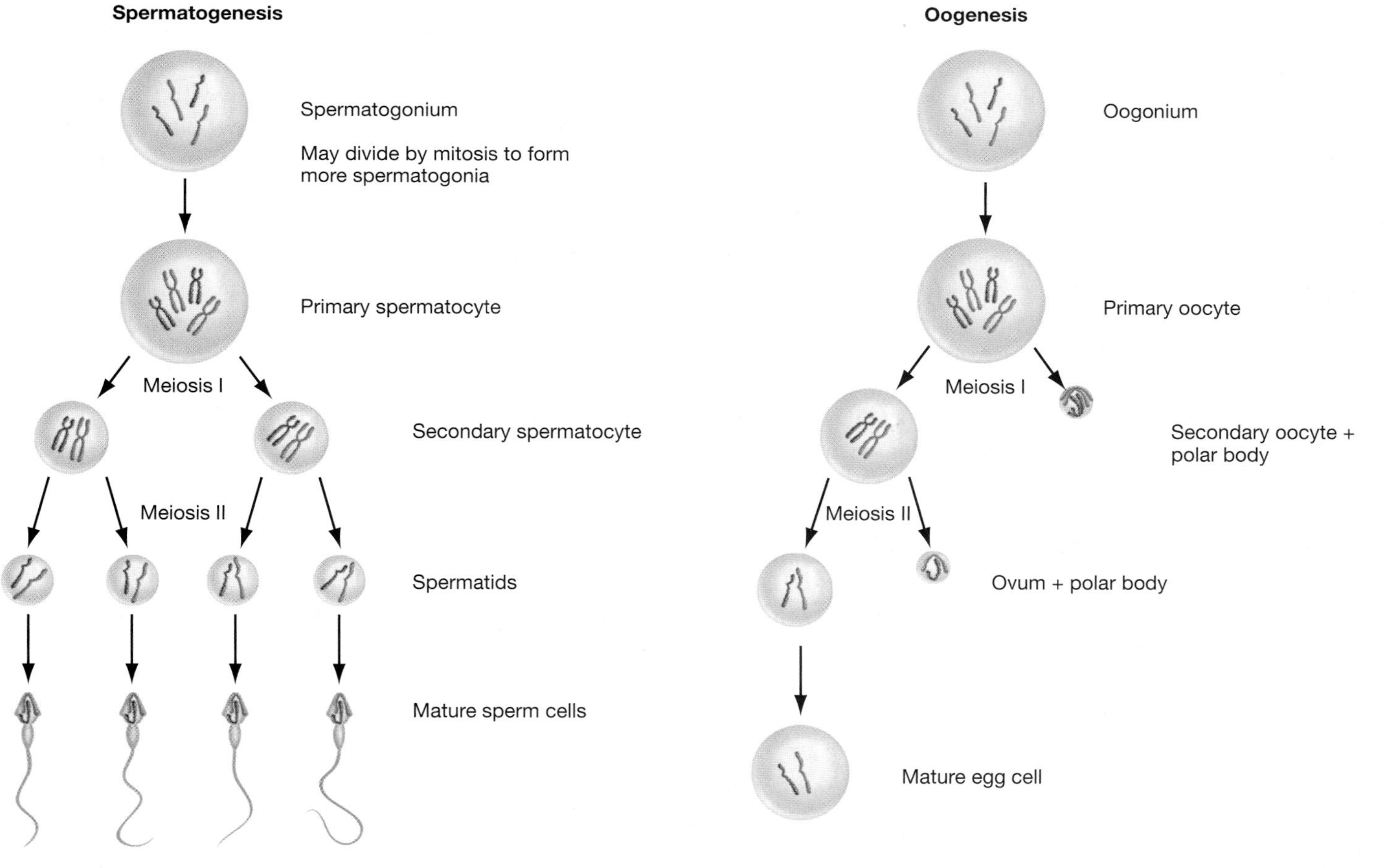

FIGURE 45.3 Gametogenesis in Humans
Early in development, spermatogonia and oogonia arise by mitosis from germ cells in the testes and ovaries, respectively. EXERCISE In each diagram, label the haploid cells and the diploid cells.

organism. To complement that material, the view here is broader. The objectives of this section are to introduce the variety of fertilization mechanisms and systems of offspring care observed among animals.

The most basic aspect of diversity in fertilization is where the union of sperm and egg takes place. In many species, individuals release their gametes into their environment and external fertilization occurs. In other animals, males deposit sperm in the reproductive tracts of females and internal fertilization occurs. Let's take a closer look.

External Fertilization

Most animals that rely on external fertilization live in aquatic environments. The correlation between external fertilization and aquatic environments is logical because gametes and embryos must be protected from drying. If external fertilization occurred in a terrestrial environment, either the gametes or the resulting zygote would likely die of desiccation.

Another basic observation is that species with external fertilization tend to produce exceptionally large numbers of gametes. For example, a female sea star *Asterias amurensis* typically releases 100,000,000 eggs into the surrounding seawater during spawning. Males release many times that number of sperm. The leading hypothesis to explain this pattern is that the probability of a sperm and egg meeting in an ocean or lake is extremely small unless huge numbers of gametes are present.

If sperm and eggs from different individuals must be released into the environment synchronously for external fertilization to work, how is gamete release coordinated? The answer to this question has two parts. Most research to date indicates that gametogenesis occurs in response to environmental cues like lengthening days and warmer water temperatures. These results are logical because the cues indicate a favorable season for breeding. But what triggers the actual release of gametes? In fishes and other aquatic animals with well-developed eyes and external fertilization, spawning is often the culmination of an elaborate courtship ritual between a male and female. In contrast, courtship behavior appears to be much less important—or even absent—in species such as clams, sea urchins, and sea cucumbers. How do animals that cannot see their mates time the release of their gametes?

Researchers have long hypothesized that the chemical messengers called pheromones might be involved in synchronizing gamete release. The idea is that a sea star will release a pheromone when it is ready to mate, and that the pheromone then triggers gamete release in other individuals. Data to back this claim have begun to emerge only recently, however. As an example, consider data collected by Jean-François Hamel and Annie Mercier. To test the pheromone hypothesis for spawning synchronization, these biologists maintained two groups of sea cucumbers under natural conditions of light and temperature for 15 months. In one treatment, individuals were kept in isolated tanks; in another they were kept in tanks with other sea cucumbers. Hamel and Mercier found that all of the individuals maintained in groups released gametes during the normal spawning period. In contrast, only about 10 percent of the individuals maintained in isolation released gametes.

Although these data suggest that pheromones might be involved in coordinating external fertilization in sea stars, they are not definitive. Most biologists will not be convinced that the pheromone hypothesis is valid until a chemical messenger is purified that induces spawning in isolated individuals.

Internal Fertilization and Sperm Competition

Internal fertilization occurs in the vast majority of terrestrial animals as well as in a significant number of aquatic animals. To accomplish internal fertilization, males deposit sperm directly into the female reproductive tract with the aid of a copulatory organ, usually called a **penis**. Alternatively, males may package their sperm into a structure called a **spermatophore**, which is then placed into the female's reproductive tract by the male or female.

Although internal fertilization appears to be straightforward, research on the topic has produced some of the most striking or even bizarre observations in all of biology (see **Box 45.1** for some examples). Perhaps the most important insight about internal fertilization, though, originated with Geoff Parker. Parker realized that in many animal species, females mate with more than one male before fertilization occurs. As a result, sperm from different males should compete to fertilize the eggs.

In 1970, Parker published the results of experiments on dung flies that confirmed the existence of sperm competition. The experiments consisted of a series of carefully controlled matings between one female and two males. In each experiment, the two males were selected in such a way that Parker could distinguish their offspring. As predicted by the sperm competition hypothesis, the proportion of offspring fathered by each male was not 50:50. Instead, whichever male was last to copulate fathered an average of 85 percent of the offspring produced.

Follow-up research has confirmed that this "second-male advantage" is widespread, although not quite universal, in insects and other animal groups. How does it occur? To answer this question, Catherine S. C. Price and co-workers studied sperm competition in the fruit fly *Drosophila melanogaster*. This research group was able to introduce a gene into male flies that resulted in the production of sperm with green tails (**Figure 45.4a**). When a mating by a green-spermed male was followed with a mating by a male having normal-colored sperm, most of the green-tailed sperm disappeared from the female's sperm-storage area (**Figure 45.4b**). To interpret this result, Price and associates suggest that the second male's sperm physically dislodged the first male's gametes and inserted themselves in their place. The researchers also showed that the fluid that accompanies sperm during fertilization is able to poison stored sperm from competing males. These two mechanisms resulted in the second male's sperm fertilizing most of the eggs laid in the next clutch.

In addition to documenting the mechanisms responsible for second-male advantage, another major result has recently emerged from research on sperm competition. In species where multiple mating is common, males have extraordinarily large testes for their size and produce proportionately larger numbers of sperm. Although this pattern was first documented in primates, it has now been observed in a wide variety of animal groups. According to data presented in Chapter 38, for example, males from species of fruit bats that live in large social groups produce larger ejaculates, on average, than male fruit bats from species that live in small groups. The idea here is that large ejaculates are an adaptive response to sperm competition.

Before moving on, it is important to note that most work on sperm competition has been guided by the view that fertilization is similar to a lottery in which each sperm represents a ticket. The hypothesis is that the more tickets a male enters in

BOX 45.1 Unusual Aspects of Fertilization

Studies on animal sexual reproduction have documented some of the most remarkable behavior and structures observed in nature. The list that follows is by no means exhaustive. It is a sample, offered simply because the diversity of mating arrangements in animals is fascinating to biologists and laypeople alike.

- When Australian redback spiders mate, the male does a somersault after inserting his penis-like intromittent organ. The somersault behavior places his dorsal surface in front of the female's mouthparts. The female then eats the male. Experiments by Maydianne C. B. Andrade show that suicidal males copulate longer and fertilize more eggs than males that survive copulation, probably because sperm transfer continues until the meal is over.
- In the fruit fly *Drosophila bifurca*, males average 1.5 mm in total body length. Their sperm, however, are each 6 cm long. No one knows what advantage, if any, these extraordinarily long sperm confer.
- When researchers use genetic markers to assess paternity in birds that appear to be monogamous, they find that up to 60 percent of nests contain at least one offspring fathered by a male that is not mated to the resident female. In most cases, cuckoldry occurs because females actively solicit copulations from males on nearby territories.
- Many snails and slugs are hermaphroditic, meaning that an individual has both male and female gonads. During mating, two individuals simultaneously receive and deposit sperm to fertilize each other's eggs. In some species, individuals also fire "love darts" of unknown function from their genitalia into the mating partner. In other species, the penis frequently becomes stuck in the female reproductive tract and is bitten off by one of the individuals.
- Some bedbugs have hypodermic penises. Males force the organ through the female's abdominal wall and deposit sperm directly into her body cavity.

Did the process of sexual selection, introduced in Chapter 22, lead to the evolution of these structures and behaviors? If so, how and why? In many or even most cases, the answers to these questions are not known.

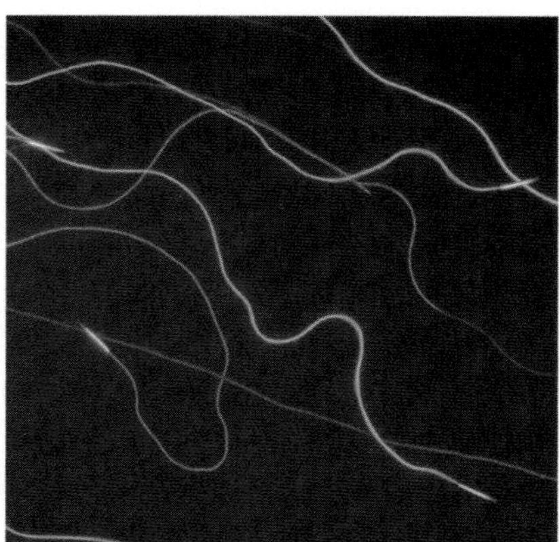

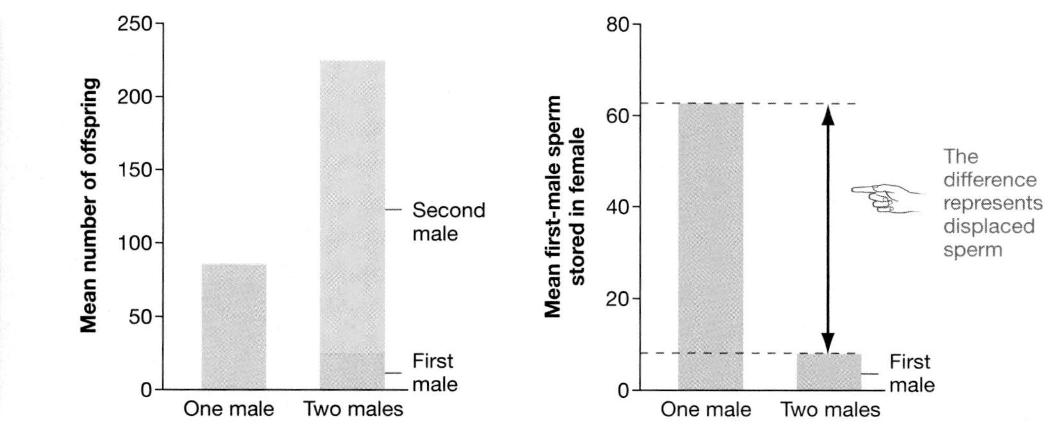

FIGURE 45.4 What Is the Physical Basis of Second-Male Advantage?
(a) Researchers created fruit-fly males with green-tailed sperm to explore why the second male to mate fathers more offspring than the first male in many species. **(b)** These graphs compare results when males with green sperm were the only male to mate or the second male to mate. When sperm competition occurred, the first male to mate fathered a tiny percentage of offspring (left), probably because most of their sperm disappeared from the sperm-storage organ in females (right).

the lottery, the higher his chance of "winning" fertilizations and passing his alleles on to the next generation. As a result, it is logical to observe exceptionally large testes size in species where sperm competition occurs.

The lottery model has recently been challenged, however, by evidence that at least in some species, females exert control over which sperm are successful in fertilization. In other words, females do not always accept the results of sperm competition passively. For example, females of some species actively choose which male performs the last copulation before fertilization takes place. In other species, females physically eject ejaculates from undesirable males. This phenomenon has been dubbed cryptic female choice. The name is appropriate because the selection of sperm by females is hidden from males.

Oviparity and Viviparity

Once fertilization is accomplished, the embryo is either laid as an egg outside the mother's body or retained so it can develop inside. In **oviparous** (egg-bearing) animals, the embryo develops in the external environment. In oviparous species such as sea stars, sea urchins, and most insects, neither the male nor the female provides any further care; the eggs and embryos are left to fend for themselves. But a substantial number of oviparous species continue to care for their young after eggs have emerged from the mother's body. For example, birds incubate their eggs and feed the young after hatching; fish may guard their eggs from predators and fan the clutches to oxygenate them.

In **viviparous** (live-bearing) species, development takes place within the mother's body. Viviparity is possible because the embryo attaches to the female's reproductive tract and receives nutrition directly from the mother's circulatory system. In section 45.5 we will explore how viviparity occurs, using humans as a model organism.

Why does oviparity exist in some groups and viviparity in others? Miriam Benabib and colleagues tackled this question by studying six species from the lizard genus *Sceloporus* (**Figure 45.5a**). Some of the populations they studied are oviparous while others are viviparous; all live in the highlands of central Mexico. To better understand how oviparity and viviparity evolved, the biologists sequenced extensive regions of DNA from many individuals. By analyzing similarities and differences in the sequences, they were able to estimate the phylogeny, or evolutionary history, of each population.

Figure 45.5b shows the phylogenetic tree that resulted. One of the researcher's first conclusions was that because most *Sceleporus* populations are oviparous, egg laying probably represents

(a) Some *Sceloporus* lizards lay eggs, others bear live young.

(b) Phylogeny of *Sceloporus* from central Mexico

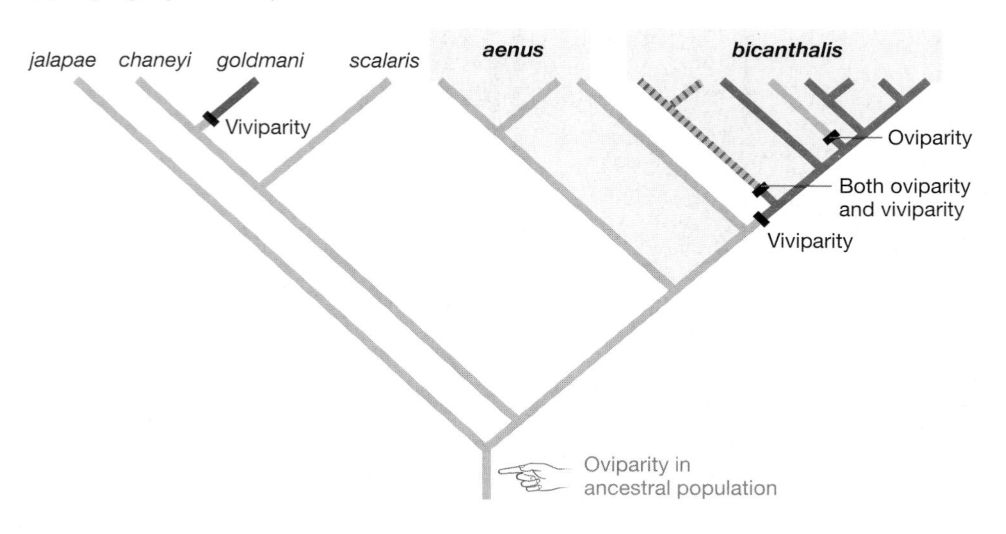

FIGURE 45.5 Evolution of Oviparity and Viviparity
(a) *Sceloporus* is a genus of lizards from desert areas in North America. **(b)** Each twig on this tree represents a *Sceloporus* species or population from central Mexico. The bars on the tree indicate the most logical explanation for when viviparity and oviparity evolved as the populations diversified into new habitats.

the original or ancestral condition. As the red branches on the tree show, however, viviparity evolved independently in two groups. And as a careful reading of the tree indicates, some populations where individuals apparently gave birth to live young in the past have reverted to egg laying. Using a similar research strategy, Michael Hart and colleagues found similar patterns in sea stars. According to Hart and co-workers, viviparous populations of sea stars have evolved from oviparous populations on several different occasions.

Why did natural selection favor these changes between egg laying and live birth over time? Researchers have long hypothesized that viviparity is favored in cold or high-altitude habitats if it leads to higher survival of young. As yet, however, the proposition has not been tested rigorously. As this book goes to press, researchers are collecting data on the habitats used by the *Sceloporus* species studied by Benabib and co-workers. If these efforts are successful, it should be possible to test the proposed correlation between cold weather and the evolution of viviparity.

45.3 Reproductive Structures and Their Functions

Now that we've considered the contrast between asexual and sexual reproduction and had an overview of fertilization and egg care, let's explore the mechanics of sexual reproduction in more detail. Our first task is to understand the anatomy of the male and female reproductive systems. Then, the next two sections follow up by exploring the action of male and female sex hormones and by reviewing the events that occur during pregnancy and birth in viviparous species.

The Male Reproductive System

In humans, the external anatomy of the reproductive system is simple. It consists of a sac-like scrotum and the penis. The scrotum holds the testes; the penis functions as the organ of intromission prior to fertilization.

When biologists have compared these structures among animal species, though, they have been struck by their variability and complexity. For example, a scrotum occurs only in certain mammal species. Whales, elephants, hedgehogs, moles, and many other mammal groups lack the structure entirely and have testes that are located well within the abdominal cavity. Among species that do have a scrotum, the structure's size and shape vary. In numerous species of primates, the scrotum is brightly colored and appears to function in sexual display.

Variation in External Anatomy Is Important in Sperm Competition Diversity in scrotal morphology pales in comparison with variation in the structure of the penis or other types of male genitalia, however. (*Genitalia* is the appropriate term when the organ of intromission has several distinct parts.) In many genera of insects and spiders, the only way that biologists can distinguish different species is by examining their genitalia. **Figure 45.6a** illustrates the type of elaborate structure typically observed in spiders. Even among closely related species, the size and shape of these organs tends to be strikingly different. Why are insect genitalia so diverse?

Gören Arnqvist and Ingela Danielsson recently offered data that may help answer this question. They hypothesized that certain genital shapes are advantageous when sperm competition occurs. To test this idea, they studied how much second-male advantage varied among males with different shapes of genitalia when water-strider females mated with two males. When sperm competition occurred, Arnqvist and Danielsson found that some males consistently fathered a higher percentage of offspring than others. As the data in **Figure 45.6b** show, there was a strong correlation between fertilization success and the size and shape of two parts of the male genitalia. In this and

(a) Male genitalia in the spider *Zosis geniculatus*

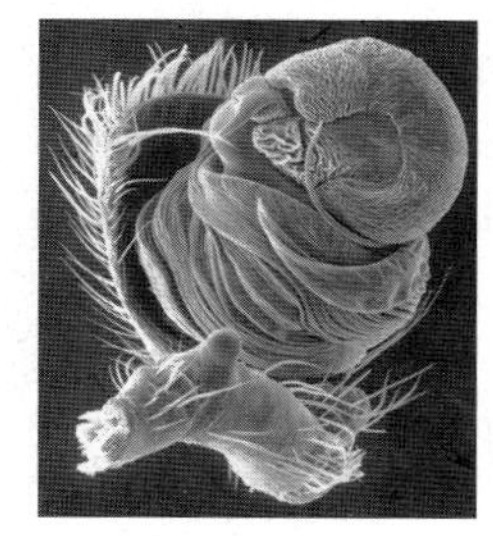

(b) Fertilization success depends on shape of male genitalia

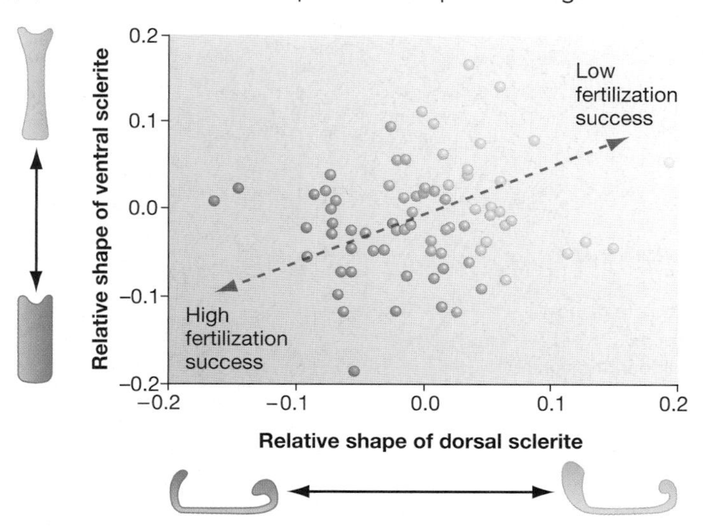

FIGURE 45.6 Variation in Insect Genitalia
(a) This elaborate male reproductive structure is found in a species of spider and is used to transfer sperm to the female. In insects and spiders, species that look identical often have distinctive genitalia. **(b)** This three-dimensional plot shows the relationship between the size and shape of two parts of male genitalia in water striders and success during sperm competition. Second-male advantage was higher for males with genitalia of a certain size and shape.

perhaps many other species, variation in the size and shape of male genitalia has a strong impact on male reproductive success.

These data suggest that in insects and perhaps other animal groups, sexual selection has been responsible for the rapid diversification of male genitalia among species. Does the same type of variation exist in the *internal* anatomy of the male reproductive system?

Internal Anatomy Figure 45.7 furnishes front and side views of the internal reproductive structures in a human male. To make sense of the numerous structures involved, note that the system has just three basic functional components.

- *Spermatogenesis and sperm storage* Sperm are produced in the testes and stored in a nearby structure called the epididymis.
- *Production of additional fluids* Complex solutions that form in the seminal vesicles, prostate gland, and bulbourethral gland are added to sperm prior to ejaculation, or expulsion from the body. The table in Figure 45.7 lists some components of these **accessory fluids.** The combination of sperm and accessory fluids is referred to as **semen.**
- *Transport and delivery* These functions are the domain of the vas deferens, urethra, and penis.

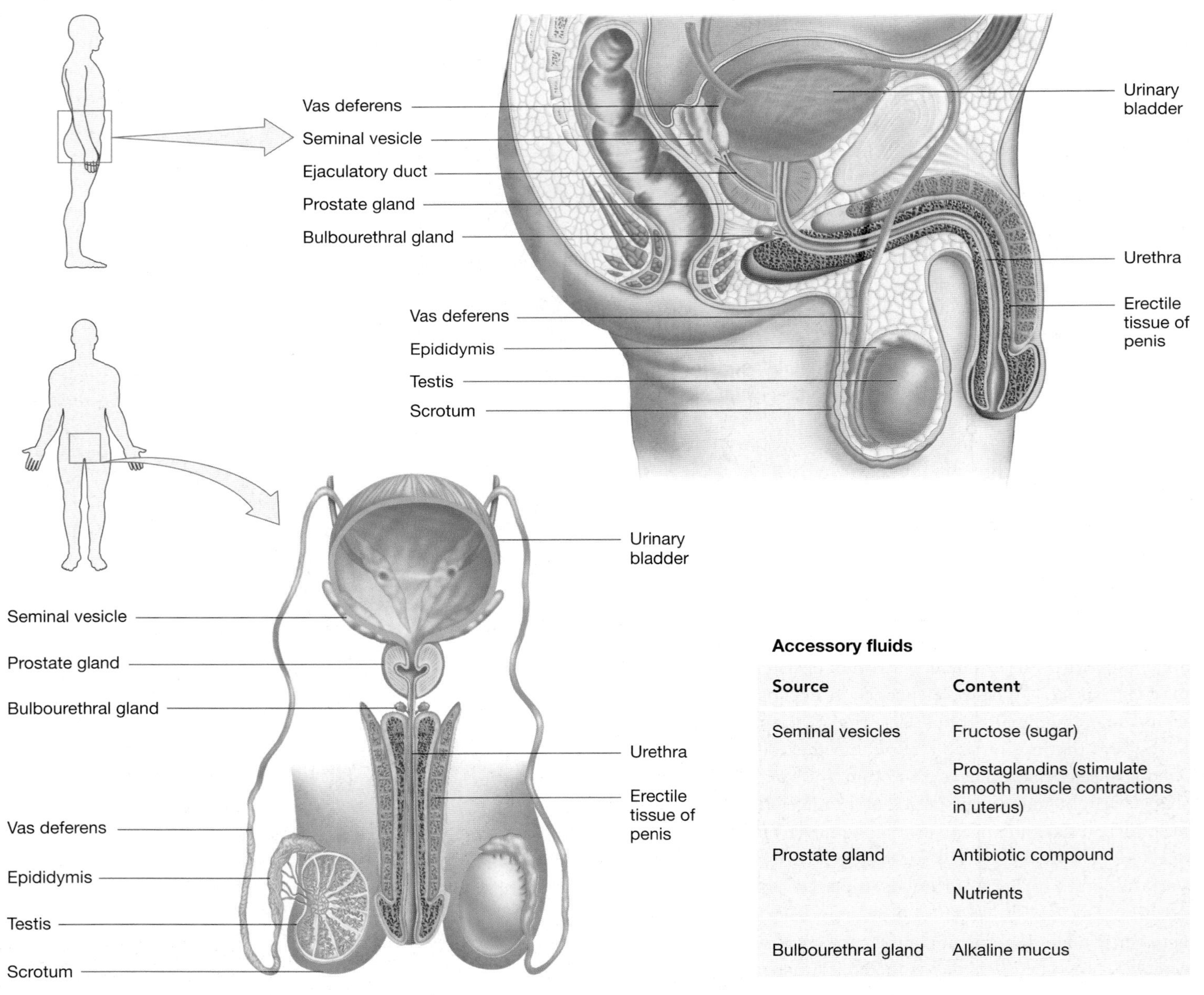

Accessory fluids

Source	Content
Seminal vesicles	Fructose (sugar)
	Prostaglandins (stimulate smooth muscle contractions in uterus)
Prostate gland	Antibiotic compound
	Nutrients
Bulbourethral gland	Alkaline mucus

FIGURE 45.7 Reproductive Tract of the Human Male

EXERCISE Add arrows to the diagrams indicating the direction of flow of sperm, accessory fluids, and semen during ejaculation.

How do these structures vary among animals? Variation in relative testes size among mammal species has already been discussed. In addition, the composition of the accessory fluids varies widely. In many insects, spiders, and vertebrates, molecules in the accessory fluids congeal after they arrive in the female reproductive tract and plug it. Experiments have shown that these copulatory plugs can serve as an effective deterrent to future matings; in some species, though, females actively remove them.

Another diverse aspect of male internal anatomy is a bone inside the penis called the baculum. Some mammal species, including humans, lack the feature entirely. But among rodents, the shape of the baculum is so variable that it can be used as a characteristic that distinguishes species. And in seals, baculum size appears to correlate with mating system. In species where females routinely mate with several males before becoming pregnant, males not only have large testes for their size but a large baculum as well. In all species where it appears, the baculum helps stiffen the penis during copulation.

The Female Reproductive System

Figure 45.8 illustrates the anatomy of the human female reproductive system, in front and side view. The external anatomy features two folds of skin that cover the clitoris, the opening of the urethra, and the opening of the vagina. The **clitoris** develops from the same population of embryonic cells that gives rise to

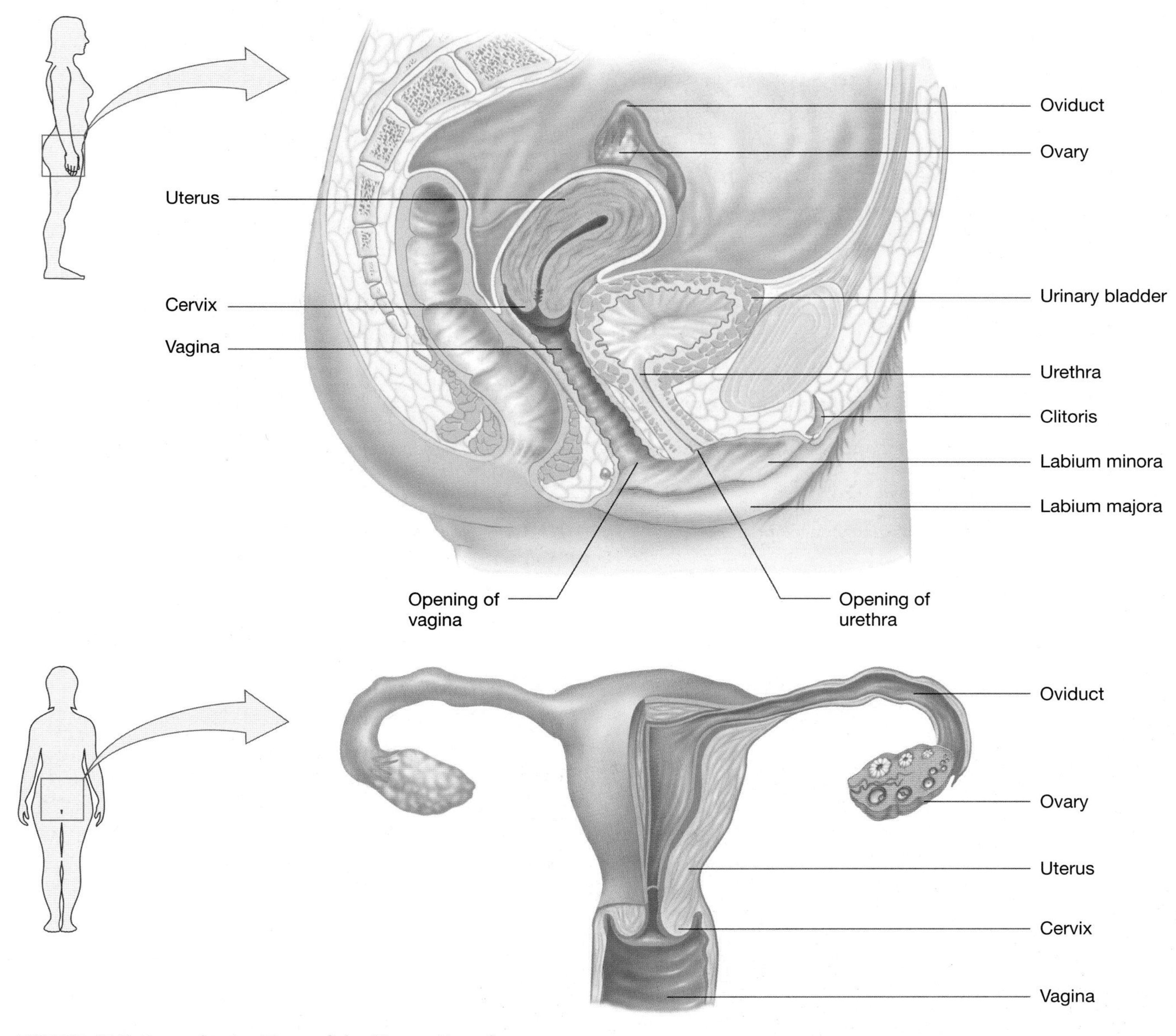

FIGURE 45.8 Reproductive Tract of the Human Female

QUESTION Generate a hypothesis to explain why male and female gonads are paired in mammals. How would you test your hypothesis?

the penis in males, and likewise becomes erect during sexual stimulation. Unlike the situation in males, however, the urethral opening where urine is expelled is separate from the reproductive structures. The **vagina**, or birth canal, is where semen is deposited during sexual intercourse and where the embryo is delivered during birth.

The internal anatomy of the female reproductive system is dominated by structures with two functions: the production and transport of eggs, and the development of offspring. Eggs are produced in the paired ovaries. When **ovulation** occurs a developing egg, or oocyte, is expelled from the ovary and enters the oviduct, where fertilization may take place. As section 45.5 will detail, fertilized eggs are transported to the muscular sac called the **uterus**, where embryonic development takes place.

To appreciate how the ovaries and uterus work, it's essential to explore the monthly menstrual cycle and pregnancy in much more detail. It is not possible to understand either event, though, without investigating the hormones that regulate them. Let's turn now to research on the sex hormones of mammals.

45.4 The Role of Sex Hormones in Mammalian Reproduction

Chapter 44 introduced the female sex hormone, estradiol, and the male sex hormone, testosterone. Recall that both are steroids, that both bind to receptors inside the nucleus of target cells, and that the resulting hormone-receptor complexes bind to DNA and trigger changes in gene expression. Testosterone is synthesized in specialized cells inside the testes; estradiol and other estrogens are synthesized in the ovaries. More specifically, the female sex hormones are produced by cells that surround each developing egg and form a structure called a **follicle**.

According to Chapter 44, the human sex hormones play a key role in three events: the development of the male reproductive tract in embryos, the maturation of the reproductive tract during the transition from childhood to adulthood, and the regulation of spermatogenesis and oogenesis in adults. To begin exploring sex-hormone action in more detail, let's take a closer look at their role in the juvenile-adult transition. How do hormones influence the changes that take place during puberty?

Puberty

Table 45.1 lists some of the changes that occur in boys and girls as they undergo puberty. Physicians and researchers have long known that these changes are triggered by increased levels of testosterone and estradiol in boys and girls, respectively. Miles Yu and colleagues recently offered dramatic evidence for testosterone's role. These physicians were presented with a two-year old boy who showed signs of entering puberty. His symptoms included increased penis size, development of pubic hair, and facial acne. Through careful interviewing, the doctors determined the cause for his condition. The child's father was a bodybuilder who was smearing a testosterone cream on his shoulders and arms in attempt to build muscle mass. The child

TABLE 45.1 Changes That Take Place During Puberty

Pediatricians use the five stages of puberty listed here to diagnose the developmental stage of their patients.

	Boys	Girls
Stage 1 (Prepubertal)	No sexual development	No sexual development
Stage 2	Testes enlarge Body odor	Breast budding First pubic hair Body odor Height spurt
Stage 3	Penis enlarges First pubic hair Ejaculation (wet dreams)	Breasts enlarge Pubic hair darkens, becomes curlier Vaginal discharge
Stage 4	Continued enlargement of testes and penis Penis and scrotum deepen in color Pubic hair curlier and coarser Height spurt Male breast development (temporary)	Onset of menstruation Nipple is distinct from surrounding areola Pelvis begins to widen
Stage 5 (Fully mature adult)	Pubic hair extends to inner thighs Increases in height slow, then stop Increased muscle mass	Pubic hair extends to inner thighs Increases in height slow, then stop Deposition of fat in hips, buttocks, thighs, and breasts

was apparently absorbing enough testosterone from being carried around by his father to trigger puberty-like symptoms. Fortunately for the two-year old, all of his symptoms except for penile enlargement subsided once his father discontinued use of testosterone creams. **Box 45.2** introduces other unfortunate side effects of steroid use by athletes and bodybuilders.

Although the role of testosterone and estradiol in puberty is well documented, understanding how the process is initiated was a stubborn problem. Some researchers hypothesized that the sex hormones are regulated by the hypothalamic-pituitary axis, introduced in Chapter 44. Recall that chemical signals from the brain region called the hypothalamus lead to the release of regulatory hormones from the pituitary gland, which then cause the release of hormones from other glands. Are the hypothalamus and pituitary involved in regulating the release of sex hormones?

Two advances made it possible to answer this question definitively. First, researchers isolated a hormone from the hypothalamus called gonadotropin-releasing hormone (GnRH). Second, investigators noted that boys and girls who were entering puberty experienced distinctive pulses in the concentration of a pituitary hormone called luteinizing hormone (LH) and follicle-stimulating hormone (FSH). Based on these observations, researchers hypothesized that pulses of LH and FSH occurred in response to release of GnRH, and were responsible for increases in testosterone and estradiol. To test the hypothesis of hypothalamus and pituitary control diagrammed in **Figure 45.9**, R. Stanhope and co-workers administered pulses of GnRH to boys and girls who had deficits in their hypothalamus and were experiencing delays in the onset of puberty. As predicted, the GnRH treatment induced puberty.

What triggers GnRH increases at the appropriate age? This is the most pressing question facing researchers and physicians interested in hormonal control of sexual development. Although the question remains unanswered, there is some evidence that nutritional state is involved. For example, girls with large fat stores tend to enter puberty earlier than girls who are thin. In addition, the current average age for onset of puberty in females in the United States is slightly over 12 years. This is much earlier than the average of 17 years of age observed in the

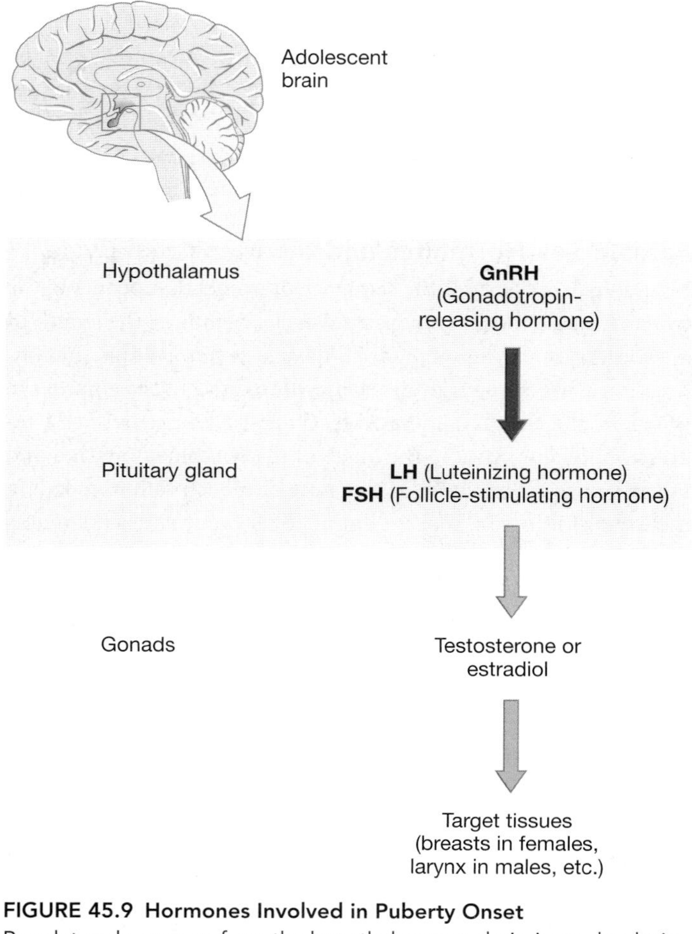

FIGURE 45.9 Hormones Involved in Puberty Onset
Regulatory hormones from the hypothalamus and pituitary gland trigger puberty. QUESTION How does control of testosterone and estradiol production compare to control of cortisol release by the adrenal gland? (See Figure 44.8a.)

BOX 45.2 Abuse of Synthetic Steroids

Among bodybuilders and athletes, the use of synthetic testosterone-like molecules, commonly called androgens, has recently exploded in popularity. Both men and women use these drugs because they support the development of increased muscle mass during weight training, just as testosterone supports increased muscle mass during puberty.

Research on bodybuilders has confirmed a wide variety of damaging side effects, however. Women who use synthetic androgens routinely experience voice deepening and growth of facial hair. Both male and female users may develop acne, the emotional changes dubbed 'roid rage (for steroid-rage), fluid retention, and liver ailments. There is some evidence that males may also experience fertility problems. The leading hypothesis to explain infertility is that the synthetic androgens mimic the negative feedback that testosterone exerts on LH production from the pituitary. Because LH is required to support spermatogenesis, high doses of androgens may reduce LH concentrations to a level that precludes sperm production.

Unfortunately, opinion polls have shown that most bodybuilders and athletes are willing to tolerate these side effects in the hope of increasing their performance in competitions. In response, some regulatory agencies actively test for synthetic androgens in the urine of competitors and ban individuals who are using them.

United States during the eighteenth and nineteenth centuries, when the general nutritional state of the population was poorer.

If you recall the discussion in Chapter 44 of how the adrenal hormone cortisol is controlled, however, you might suspect that the model of sex-hormone regulation diagrammed in Figure 45.9 is probably simplified. Chapter 44 emphasized that many hormones participate in negative feedback, meaning that the presence of a hormone inhibits the factor that triggers its release. Do sex hormones participate in negative feedback? The short answer to this question is yes. To appreciate the details, let's investigate hormonal control of the human menstrual cycle.

Female Sex Hormones and the Menstrual Cycle

Figure 45.10 illustrates the sequence of events that occurs in the ovary during a monthly menstrual cycle. Although the length of a cycle varies among women, 28 days is about average. In conjunction with changes in the ovary illustrated in the figure, the lining of the uterus undergoes a dramatic thickening and regression. By convention, the onset of the combined ovarian and uterine cycles is taken to be the start of **menstruation**, meaning the expulsion of the uterine lining. This event is designated as day 0 in the menstrual cycle.

The cycle itself has two distinct phases. A follicle matures during the follicular phase, which lasts an average of 14 days. Ovulation occurs when the follicle is mature and releases its oocyte into the oviduct. Although the timing of ovulation is highly variable, the subsequent luteal phase also averages 14 days in length. Its name was inspired by the formation and subsequent degeneration of a structure called the corpus luteum (yellowish-body) from the ruptured follicle.

How Do Pituitary and Gonadal Hormones Change During a Cycle? Signals from hormones are responsible for the changes that occur in the uterus and ovary during a menstrual cycle. When biologists began to probe how hypothalamic, pituitary, and ovarian hormones interact to regulate the menstrual cycle, their first task was to document how each hormone involved changed over time. By monitoring hormone concentrations in the blood or urine of a large number of women, researchers were able to produce the graphs in **Figure 45.11**. Note that in addition to estradiol, several other hormones exhibit dramatic changes in concentration during the cycle. Luteinizing hormone (LH) and follicle-stimulating hormone (FSH) are produced in the pituitary in response to GnRH; progesterone is produced along with estradiol in the ovary.

Several observations jump out of these data. LH levels are fairly constant except for a spike just prior to ovulation, suggesting that LH might be the trigger for this event. FSH concentrations, in contrast, are relatively high during the follicular phase and low during the luteal phase, though they also make a small spike prior to ovulation. Progesterone is present at very low levels during the follicular phase but high levels during the luteal phase. This correlation suggested that progesterone might support the maturation of the thickened uterine lining. Estradiol concentrations, though, change in a much more complex way. What's the effect of all these changing hormone concentrations?

How Do the Pituitary and Gonadal Hormones Interact? Experiments by Charles March and colleagues helped establish that changes in the concentration of estradiol and progesterone

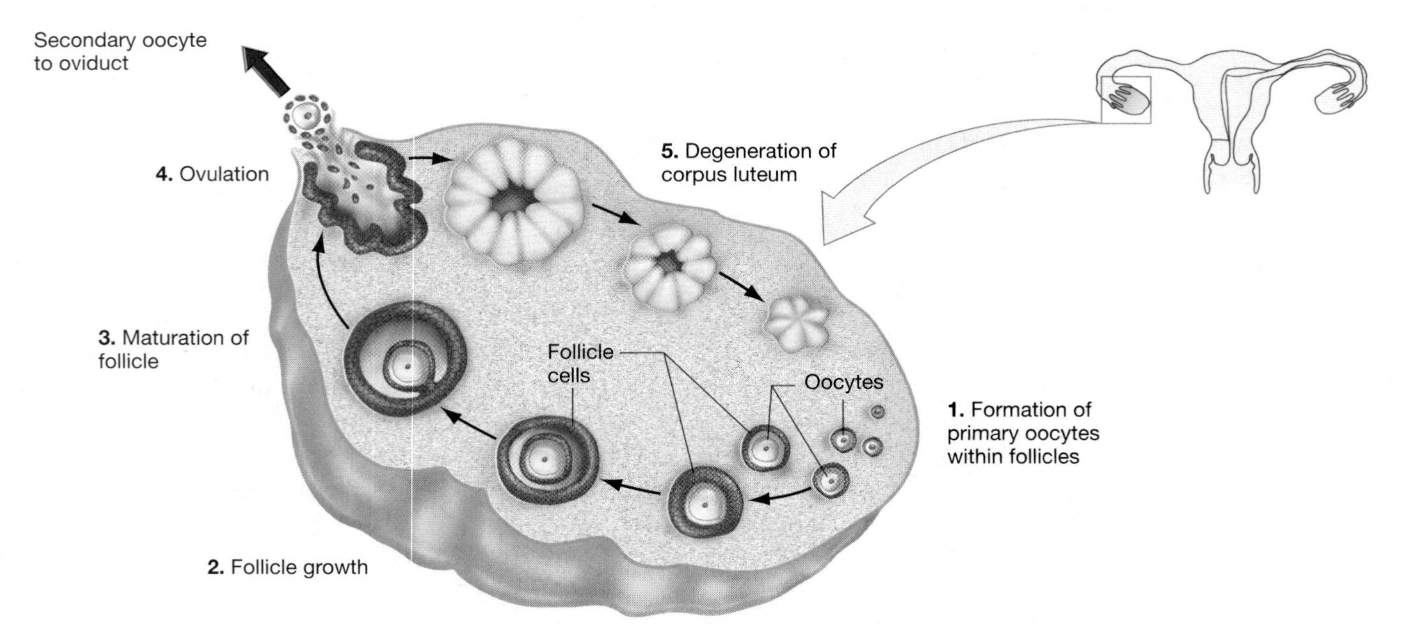

FIGURE 45.10 Cyclical Changes in the Ovary
As a follicle matures, primary oocytes complete meiosis I. The secondary oocyte that results is expelled from the ovary and then travels down the oviduct. Note that follicles don't actually move in a circle during the menstrual cycle; they are shown that way here for illustrative purposes only. **EXERCISE** Label the events that occur during the follicular phase of the menstrual cycle. Then label the events that occur during the luteal phase.

affect the release of the pituitary hormones LH and FSH. These researchers worked with three volunteers whose ovaries had been removed because of cancerous growths or other problems. The women were receiving low, maintenance-level doses of estradiol, which appeared to exert negative feedback on LH and FSH. But when March and co-workers injected them with larger doses of estradiol or with progesterone, the situation changed dramatically. For example, a large increase in estradiol stimulated a dramatic spike in LH levels. This suggested that positive feedback was occurring—meaning that high levels of estradiol actually increased release of its regulatory hormone, even though low doses of estradiol suppressed it. Injections of progesterone, in contrast, appeared to inhibit both FSH and LH. Progesterone exerted only negative feedback on the pituitary hormones.

Thanks to dozens of experiments like these, conducted in a wide variety of labs over the past several decades, researchers now have a fairly good grasp of how the various hormones interact. To summarize the interplay between LH, FSH, estradiol, and progesterone, let's start at day 0 and follow key events as the cycle progresses.

As a cycle begins, the uterus is shedding much of its lining during menstruation. In the ovary, a follicle is being stimulated to develop under the influence of FSH. As the follicle grows, its production of estradiol gradually begins to increase; it also produces a small amount of progesterone. While estradiol levels are still relatively low, the hormone suppresses LH secretion through negative feedback inhibition. Once the follicle has grown enough to produce large quantities of estradiol, however, the molecule begins to exert the positive feedback observed by March and colleagues. As the graphs in Figure 45.11 show, the positive feedback loop between LH and estradiol produces a spike in both hormones. The LH surge triggers ovulation and ends the follicular phase.

As the corpus luteum develops from the remains of the ruptured follicle, it begins to secrete large amounts of progesterone and small quantities of estradiol in response to LH. This rise in progesterone converts the thickened uterine lining to an actively secreting tissue with a well-developed blood supply. In this way, progesterone creates an environment that supports embryonic development if fertilization occurs.

If fertilization does not occur, however, the corpus luteum degenerates approximately 12 days after ovulation. Progesterone levels fall as a result, which causes the thickened lining of the uterus to degenerate. Once progesterone levels have declined, the hypothalamus and pituitary are released from inhibitory control. LH and FSH levels rise, and a new cycle begins.

If fertilization occurs, however, the corpus luteum does not degenerate; and progesterone and estradiol levels stay high. How does the corpus luteum "know" that fertilization has taken place? What other hormonal and development changes occur during pregnancy? To answer these questions, let's investigate the major events during pregnancy and birth in humans.

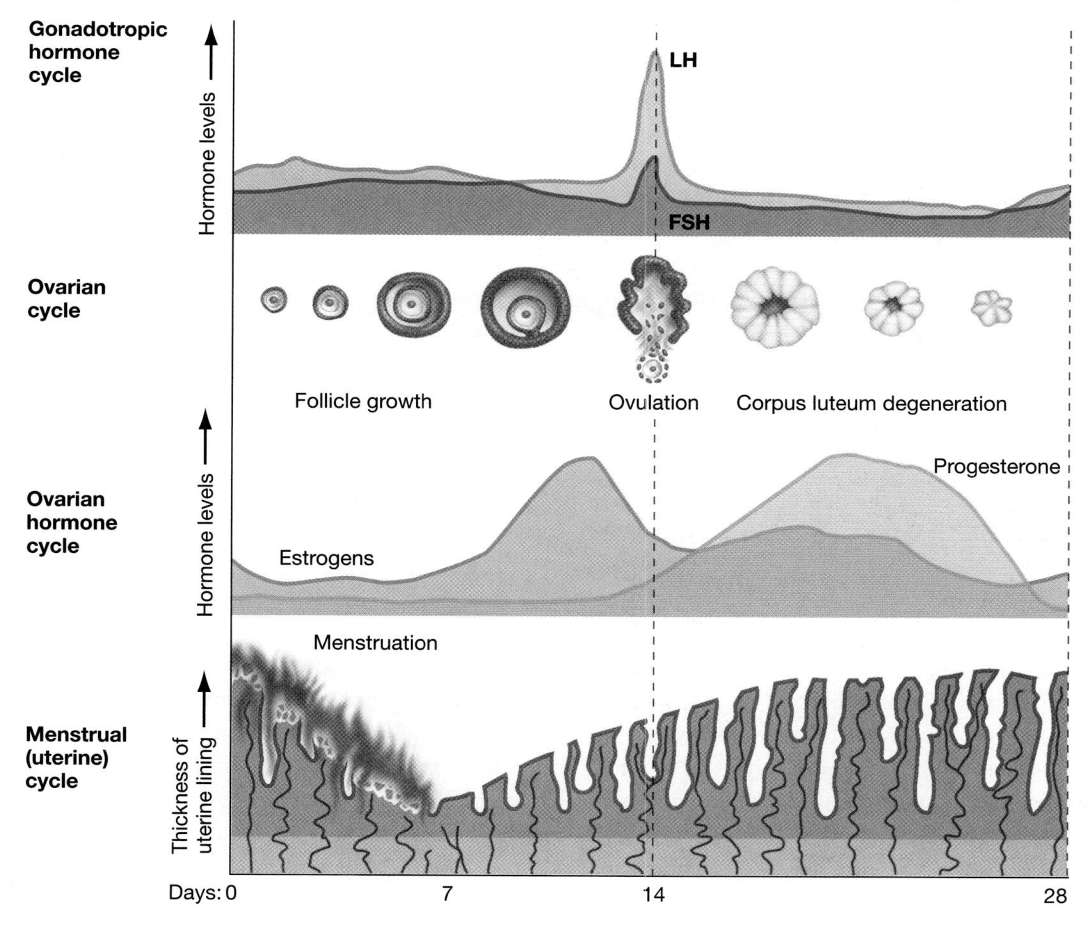

FIGURE 45.11 Reproductive Cycle of the Human Female
Shown here are the cyclical changes occurring in the pituitary hormones LH and FSH, the ovary, the ovarian hormones estradiol and progesterone, and the lining of the uterus. During the menstrual flow phase of the cycle, the thickened lining of the uterus sloughs off and is expelled through the vagina.

45.5 Human Pregnancy and Birth

Once an egg is released from the human ovary, it is viable for less than 24 hours. Human sperm, in contrast, remain capable of fertilizing an egg for up to five days. Therefore, sexual intercourse has to occur less than five days prior to ovulation for pregnancy to result.

Although an ejaculate may contain hundreds of millions of sperm, most die as they travel through the uterus. Only 100 to 300 actually reach the oviduct, where fertilization takes place.

Chapter 18 detailed how sperm and egg interact when they meet. Recall that contact between sperm and egg triggers the acrosomal reaction, in which enzymes released from the head of the sperm chew a path through the material surrounding the egg membrane. Once the membranes of the egg and sperm have fused, the egg nucleus completes meiosis II. The two nuclei then unite to form a diploid zygote.

Major Events During Pregnancy

As smooth muscle contractions in the oviduct gradually move the zygote toward the uterus, the cell begins to divide by mitosis. By the time it reaches the lining of the uterus, the embryo consists of a hollow ball of cells. It then becomes embedded in the thickened, vascularized wall of the uterus, where it will stay for the full period of development, or **gestation**, of approximately 270 days (9 months).

Once the embryo has become implanted in the uterine lining, its cells begin synthesizing and secreting a hormone called **human chorionic gonadotropin (hCG)**. hCG is a chemical messenger that prevents the corpus luteum from degenerating. When hCG is present, the ovary continues secreting progesterone and the menstrual cycle is arrested. (Enough hCG is produced by the embryo and excreted in the mother's urine to be used as an indicator in pregnancy tests.)

The First Trimester In humans, gestation is divided into 3-month stages called trimesters. The first trimester is particularly eventful. Not long after implantation is complete, mass movements of cells result in the formation of the embryonic tissues called ectoderm, endoderm, and mesoderm (see Chapter 18). By 8 weeks of age, these tissues have differentiated into the various organs and systems of the body. By this time, the heart has begun pumping blood through a circulatory system. The embryo is now called a **fetus**.

The three embryonic tissues also contribute to several important membranes. One of these completely surrounds the embryo and is called the amnion. The amnion eventually fills with amniotic fluid, which provides the embryo with a protective cushion.

The other key event in the first trimester is the formation of the **placenta**. This structure starts to form on the uterine wall a few weeks after implantation. The placenta is comprised of tissues from both the mother and embryo and is the primary source of nutrition for the growing fetus. Arteries transport blood from the circulatory system of the fetus to an extensive capillary bed in the placenta. In this way, the placenta provides a large surface area for the exchange of gases, nutrients, and wastes between maternal and fetal blood, even though maternal and fetal blood do not co-mingle. The placenta also secretes a variety of hormones. For example, by the end of the first trimester placental cells synthesize and secrete enough progesterone and estrogens to maintain the pregnancy, even though the corpus luteum has degenerated by this time.

The Second and Third Trimesters Figure 45.12 contains photographs of a typical first-, second-, and third-trimester fetus. After the organs and placenta form during the first trimester, the remainder of development focuses on growth. During the last weeks of pregnancy, the brain and lungs undergo particularly dramatic growth and development. If a baby is born prematurely, the primary challenge to caregivers is to maintain an adequate oxygen supply. In many cases, dramatic intervention is required to keep the baby alive until the lungs can complete their development.

The machinery and level of hospital care required by premature infants emphasizes just how superbly adapted mothers are to nourishing a growing fetus in the uterus. Let's take a closer look at this critical aspect of pregnancy.

How Does the Mother Nourish the Fetus?

In viviparous species, the developing embryo is completely dependent on the mother for oxygen, chemical energy in the form of sugars, amino acids and other raw materials for growth, and waste removal. What physiological changes occur in human mothers to accommodate these demands?

To answer this question, researchers have focused on the body's basic transport and delivery systems—the heart, lungs, and blood vessels. For example, a woman's total blood volume expands by as much as 50 percent during pregnancy. To accommodate this increase, maternal blood vessels dilate (widen) and blood pressure drops. In addition, dramatic increases occur in the pumping capacity or stroke volume of the heart as well as in the heart rate. The mother's heart enlarges and begins to beat faster. The result is that a mother's total cardiac output increases by almost 50 percent over the course of a pregnancy.

The mother's breathing rate and breathing volume also increase to accommodate the fetus' demand for oxygen and its production of carbon dioxide. Important adaptations heighten the efficiency of gas exchange between the mother and the embryo as well. As **Figure 45.13a** shows, maternal and fetal arteries in the placenta are arranged in countercurrent fashion. As explained in Chapters 38 and 39, countercurrent flows maintain a concentration gradient that makes diffusion efficient. The data for partial pressures of oxygen given in Figure 45.13a are

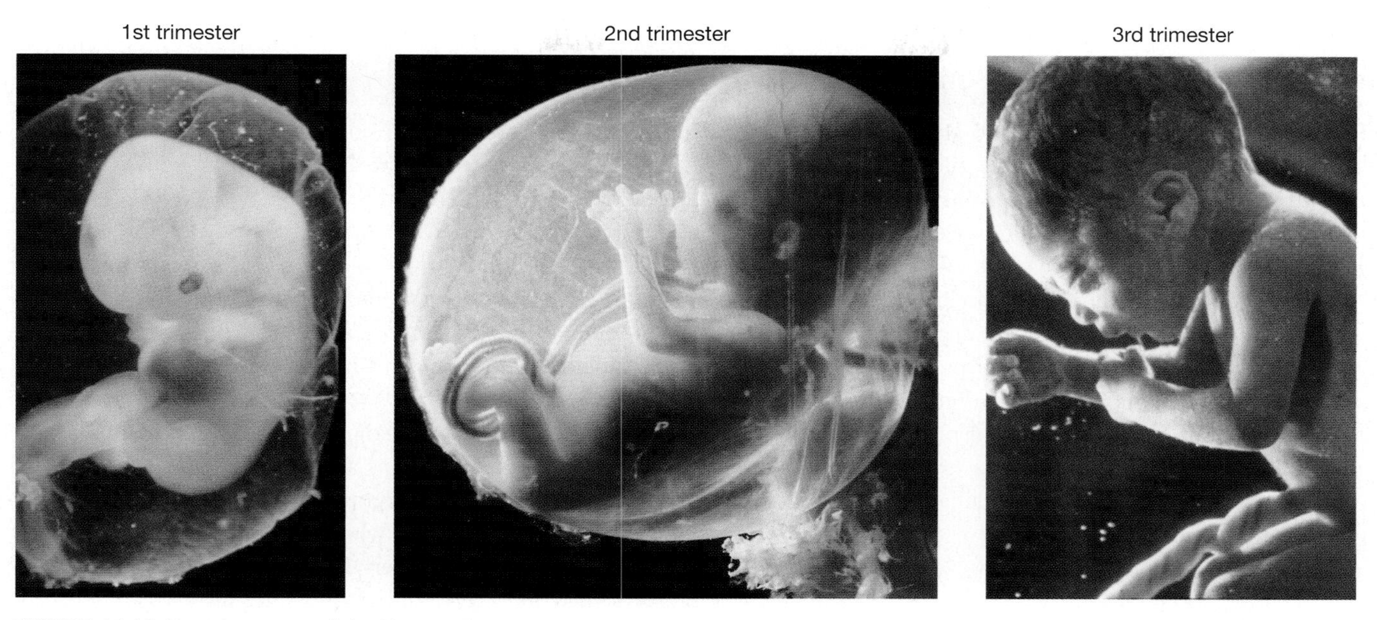

FIGURE 45.12 Development of the Human Fetus
EXERCISE On each photograph, label the hands, feet, eye, and umbilical cord. Estimate the ratio of head length to total body length in each photograph.

from experiments on sheep, but similar differences between the maternal and fetal circulation have been observed in humans.

Figure 45.13b illustrates a second adaptation that increases the rate of oxygen delivery to the fetus. Note that the axes on this graph give the partial pressure of oxygen in blood or tissue versus the percentage of hemoglobin that holds oxygen at that pressure. (You might recall, from Chapter 41, that this type of graph is called an oxygen-hemoglobin dissociation curve.) The key point here is that the data for fetal hemoglobin are shifted to the left of the graph for adult hemoglobin. As a result, the fetus's blood always has a higher affinity for oxygen than does the mother's blood. Biologists interpret this pattern as an adaptation. The high oxygen affinity of fetal hemoglobin ensures that the fetus is always able to acquire oxygen from the mother, no matter what the concentration of oxygen in maternal blood.

Before leaving the topic of pregnancy, it is important to acknowledge that mothers and embryos exchange more than nutrients and wastes. For example, think back to the thalidomide tragedy introduced in Chapter 3. Recall that during the 1950s, hundreds of children were affected by their mothers' consumption of the tranquilizer thalidomide, which diffused into the fetal bloodstream and caused birth defects. Although thalidomide is now banned for use by pregnant women, alcohol use continues to take a similar toll. Children of mothers who use alcohol are at high risk of experiencing hyperactivity, severe learning disabilities, and depression. Collectively, these symptoms are termed fetal alcohol syndrome (FAS).

(a) Countercurrent blood flow in the placenta

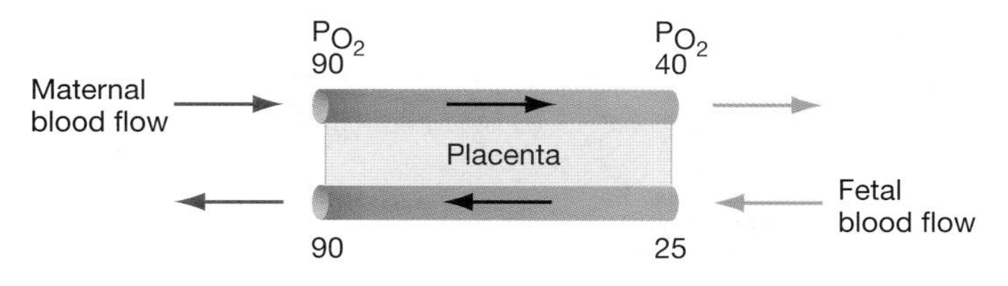

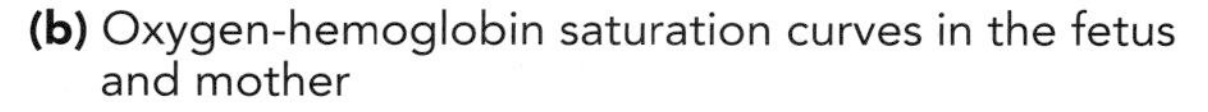

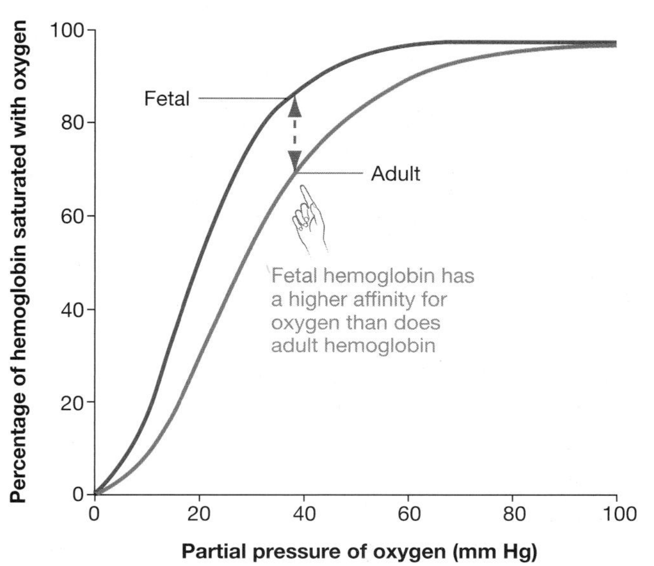

FIGURE 45.13 Adaptations That Increase Delivery of Oxygen to the Fetus
(a) Countercurrent blood flow maximizes the efficiency of oxygen diffusion from mother to fetus. **(b)** As these graphs show, fetal hemoglobin has a much higher affinity for oxygen than does adult hemoglobin.

To understand why FAS exists, recall that the brain undergoes especially rapid growth in the third trimester, and then consider the data in **Figure 45.14** collected by Chrysanthy Ikonomidou and co-workers. The photos show two slices from the brains of newborn rats. A day before the photos were taken, one individual was given two injections of a harmless salt solution at a dosage of 2.5 g per kg body weight; the other was given the same injections except that the salt solution contained 20 percent ethanol. Note the contrast in brain size and in the number of black specks, which indicate degenerating neurons. The histograms in Figure 45.14 confirm the visual impression in the photos. The data indicate marked differences in the average weights of the brain in groups of rats subjected to each treatment. The messages of these data are that ethanol decimates growing neurons, and that a single large dose of alcohol can have a devastating effect.

Birth

The rapid growth of the human brain during the last trimester makes birth challenging. Although the mechanisms responsible for triggering the birthing process are not completely understood, the pituitary hormone **oxytocin** is clearly important in stimulating smooth-muscle cells in the uterine wall to begin contractions.

During the first stage of birth, the uterus contracts at relatively low frequency. As labor progresses, the cervix at the base of the uterus begins to open, or dilate (**Figure 45.15a**). Once the cervix is fully dilated, uterine contractions become more forceful, longer, and more frequent. Eventually, the fetus is expelled through the cervix and into the vagina.

After the baby is delivered, the placenta remains attached to the uterine wall. At this point, caregivers clamp and then cut the umbilical cord that connects the child and the placenta. By gently tugging on the cord, the caregiver helps the mother deliver the placenta and accompanying membranes, and birth is complete.

Although this description sounds straightforward, in reality a large number of complications are possible. To drive this point home, consider the data in **Figure 45.15b**. The graph shows the number of Swedish mothers who died, per 100,000 live births, in five-year intervals between 1750 and 1980. In 1750, 1.1 mothers died for every 100 infants successfully delivered. In most cases, the cause of death was blood loss or infection following delivery. As a result of sterile techniques, antibiotics, and blood transfusion technology, Sweden's mortality rate has now declined to less than 0.007 percent. Although it is generally unheralded, this huge reduction in the rate of death in childbirth qualifies as one of the great triumphs of modern medicine.

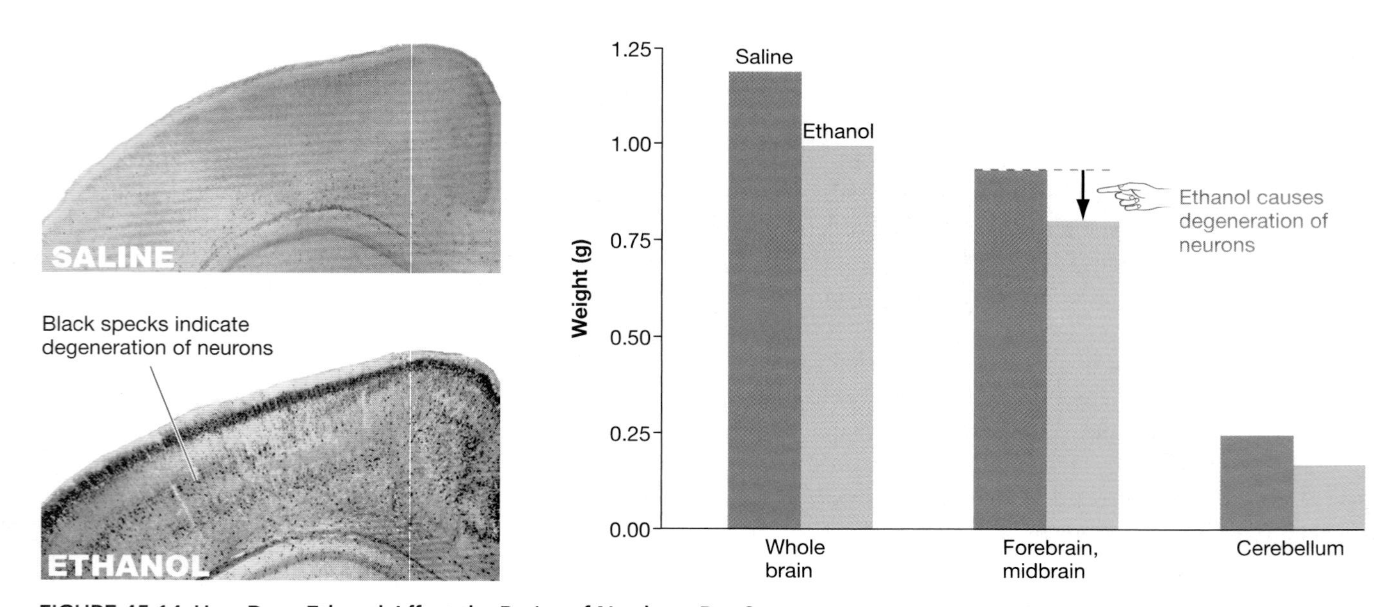

FIGURE 45.14 How Does Ethanol Affect the Brains of Newborn Rats?
These photographs and histograms document degeneration of neurons in the brains of newborn rats that were injected with ethanol. QUESTION Why did the researchers measure the two different brain areas shown in the histograms instead of just assessing weight changes in the whole brain?

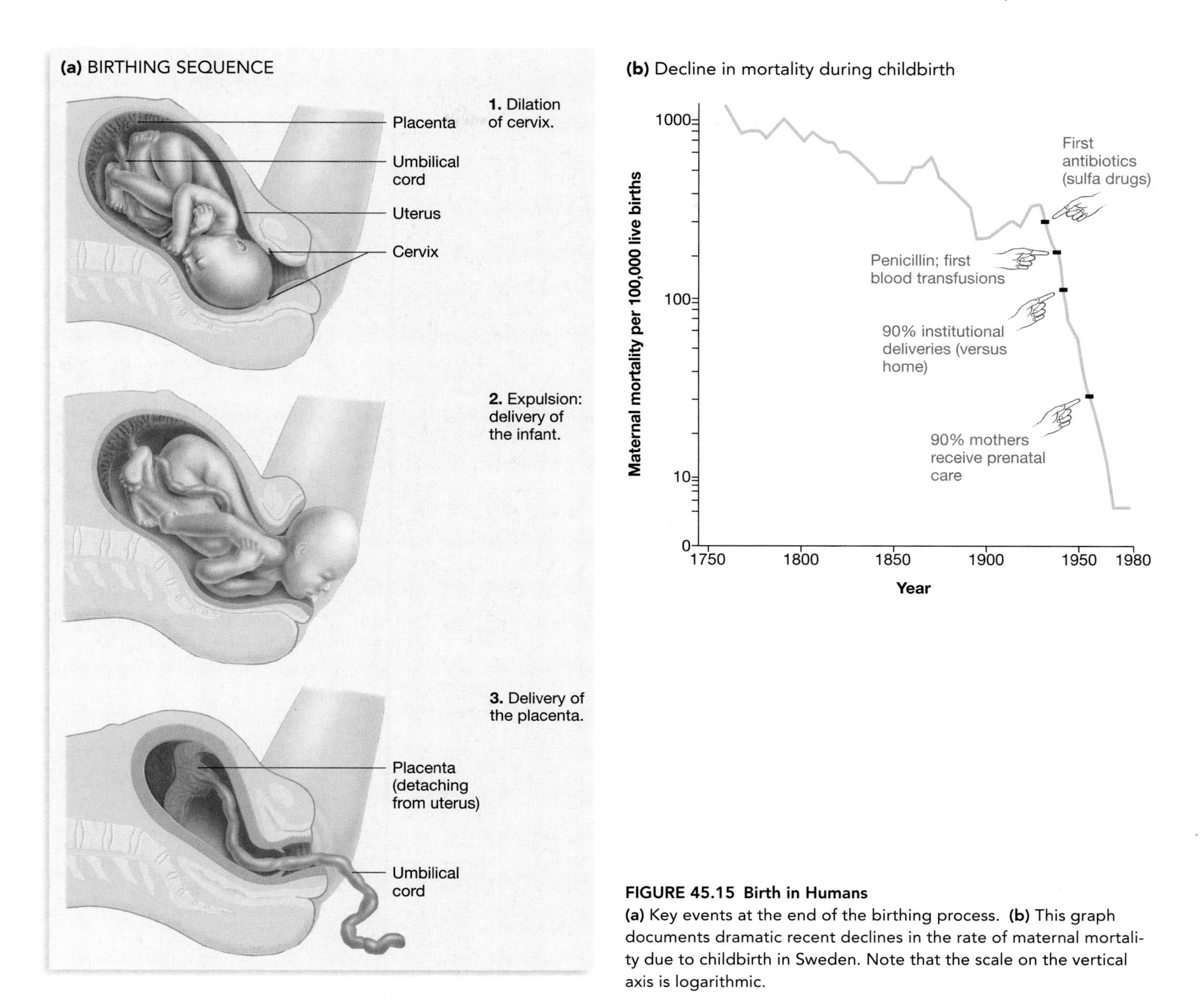

FIGURE 45.15 Birth in Humans
(a) Key events at the end of the birthing process. **(b)** This graph documents dramatic recent declines in the rate of maternal mortality due to childbirth in Sweden. Note that the scale on the vertical axis is logarithmic.

Essay Contraception

This chapter detailed the remarkable adaptations that make fertilization, pregnancy, and birth possible. The purpose of contraception is to circumvent these events and prevent pregnancy. In general, birth control methods work by (1) inhibiting ovulation, (2) preventing sperm from contacting the egg, or (3) interfering with implantation of the embryo.

Birth control pills inhibit ovulation because they contain synthetic versions of estradiol and progesterone. These hormones suppress the release of GnRH and FSH through negative feedback inhibition and thus prevent the maturation of a follicle. A woman uses these hormones in a cyclic fashion, taking the pills for three weeks and then stopping for one week to allow menstruation to occur. In the United States, birth control pills are currently the most widely used contraceptive method.

Is there a completely effective method of birth control?

A variety of barrier methods prevent sperm from reaching the egg. A condom is a sheath of latex or rubber. The male version of the condom covers the penis and retains the ejaculate; the female condom covers the cervix, vagina, and the labia. In contrast, a diaphragm or cervical cap covers just the cervix of the uterus. Surgical sterilization methods can also be thought of as barrier methods for birth control. When a vasectomy is performed, the vas deferens is cut to prevent sperm from moving into the urethra along with the rest of the ejaculate. In a tubal ligation, both oviducts are cut to prevent the passage of eggs to the uterus.

Two methods of preventing the implantation of the embryo are currently in widespread use. Intrauterine devices (IUDs) are inserted through the cervix and into the uterus, where they can remain for 3 to 4 years. Although it is not completely understood how IUDs work, the presence of one in the uterus effectively prevents the implantation of the embryo. The pill RU-486, in contrast, acts by blocking the action of progesterone. As a result, menstruation occurs despite the presence of a zygote.

Although many of these methods are effective at preventing pregnancy, only abstinence offers a 100-percent guarantee. Further, only abstinence is certain to prevent the transmission of HIV, genital warts, genital herpes, and other sexually transmitted diseases.

Finally, each of the methods reviewed here entails trade-offs in convenience, expense, and side effects. In the case of the hormone-based treatments, the extent and severity of side effects may vary among individuals. As a result, there is no completely effective method of birth control that is right for everyone. Each approach has advantages and disadvantages. The choice of a method, in consultation with a health-care professional, is an important decision for couples to make.

Chapter Review

Summary

Animals reproduce asexually or sexually. Only a few animal species reproduce exclusively by asexual means, but many alternate between asexual and sexual reproduction. Asexual reproduction results in offspring that are genetically identical to their parent. In contrast, every sexually produced offspring is genetically unique because of recombination during meiosis and the fusion of haploid gametes from different parents during fertilization.

In humans, the production of sperm and eggs differs in several respects. Spermatogenesis is continuous throughout the adult life of males, but in females all of the cells that divide mitotically to begin oogenesis are formed very early in development. Meiosis in females is also arrested for long periods of time, whereas it occurs as a continuous process in males. Finally, cell division during meiosis is so unequal in females that just one egg (instead of four) is produced from each oocyte.

Fertilization is external in many aquatic animals, but internal in almost all terrestrial species. Release of gametes during external fertilization may be synchronized by courtship displays or possibly by pheromones. Research on internal fertilization has documented that multiple mating and sperm competition are widespread. Two important generalizations have emerged from studies of sperm competition. First, the last male to mate usually fathers a disproportionately large number of offspring. Second, males of species that engage in sperm competition have relatively large testes for their overall size.

Once eggs are fertilized, females may lay the eggs or retain them and give birth to live offspring. By estimating the evolutionary relationships among closely related species with viviparity and oviparity, researchers have determined that populations have changed between the two modes relatively frequently, and that viviparity may be an adaptation that increases survival of young in cold habitats.

In humans, the external anatomy of the male reproductive tract consists of the penis and scrotum. Biologists are unsure why the scrotum evolved in mammals, but have documented that sexual selection has produced striking variation in the genital morphology of male insects and spiders. The internal anatomy of the human male reproductive tract includes structures devoted to sperm production and storage, synthesis of accessory fluids, and transport and delivery of semen. The external anatomy of the human female reproductive tract consists of the vaginal opening and clitoris. The internal reproductive structures are devoted to egg production or to the nourishment and delivery of offspring.

The development and activity of male and female reproductive structures are under hormonal control. In mammals, experiments confirmed that GnRH from the hypothalamus triggers the release of FSH and LH from the pituitary. The pituitary hormones, in turn, regulate the production of testosterone and estradiol in the testes and ovaries, respectively. Research on the human menstrual cycle has confirmed that estradiol and progesterone exert positive and negative feedback on production of FSH and LH. Interactions between the pituitary and ovarian hormones are responsible for regulating cyclical changes in the ovary and uterus.

If fertilization occurs, the developing embryo secretes a hormone called hCG that arrests the menstrual cycle and allows pregnancy to continue. During the first trimester, the embryo becomes implanted in the thickened uterine wall, organs develop, and the nutritive tissue called the placenta forms. To make rapid growth possible during the second and third trimesters, the mother's heart rate and pumping volume increase. Nutrients and gases are exchanged efficiently via countercurrent flow in the placenta's capillary beds.

Questions

Content Review

1. What term describes the mode of asexual reproduction in which offspring develop from unfertilized eggs?
   a. parthenogenesis
   b. budding
   c. regeneration
   d. fission
2. In sperm competition, what is "second-male advantage"?
   a. the observation that when females mate with two males, each male fertilizes the same number of eggs
   b. the observation that when females mate with two males, the second male fertilizes most of the eggs
   c. the observation that females routinely mate with at least two males before laying eggs or becoming pregnant
   d. the observation that accessory fluids form copulatory plugs and prevent matings by second males
3. How are the human penis and clitoris similar? (Circle all that apply.)
   a. Both develop from the same population of embryonic cells.
   b. Both become erect in response to sexual stimulation.
   c. Both contain the urethra.
   d. Both produce accessory fluids.
4. In terms of hormone production, how do the follicle and corpus luteum compare?
   a. Both primarily produce estradiol.
   b. Both primarily produce progesterone.
   c. The follicle produces more estradiol than progesterone; the corpus luteum produces more progesterone than estradiol.
   d. The follicle produces mostly progesterone; the corpus luteum produces estradiol and progesterone.
5. What pituitary hormones are involved in regulating the human menstrual cycle?
   a. GnRH and LH
   b. estradiol and progesterone
   c. oxytocin and LH
   d. FSH and LH
6. The corpus luteum is retained upon fertilization due to the presence of what hormone?
   a. LH
   b. estradiol
   c. progesterone
   d. human chorionic gonadotropin (hCG)

Conceptual Review

1. Summarize the experimental evidence that *Daphnia* require three cues to trigger sexual reproduction. Discuss what these cues indicate about the environment. Generate a hypothesis for why sexual reproduction is adaptive for these animals.
2. Compare and contrast spermatogenesis with oogenesis. How do these two processes differ with respect to numbers of daughter cells produced, gamete size, and timing of the second meiotic division?
3. The drug RU-486 blocks progesterone receptors in the uterus. Why does this drug end pregnancy in humans, even if implantation has occurred?
4. Explain the difference between negative and positive feedback in hormonal control of the human menstrual cycle. How is it possible that estrogen inhibits LH release at one point of the reproductive cycle but stimulates its release at another point?
5. Why are pregnant mothers advised to refrain from smoking and from drinking alcoholic or caffeinated beverages?

Applying Ideas

1. Researchers have recently developed methods for cloning mammals. In effect, biologists now have the capacity to induce asexual reproduction in species that do not normally reproduce asexually. Suppose this practice becomes so widespread in the future that most of the sheep in the world become genetically identical. Discuss some possible consequences of this development.
2. The text claims that species with external fertilization produce extraordinarily large numbers of gametes. How would you test this hypothesis rigorously? In answering, assume that you have data on the average number of gametes produced by different species, along with data on their average body size.
3. Suppose that you've been given a chance to study the viviparous and oviparous *Sceloporus* populations in central Mexico. Which populations would you expect to produce especially large eggs, relative to their overall body size? How would you go about testing this prediction?
4. HIV, hepatitis B, genital herpes, and other sexually transmitted diseases (STDs) are transferred from one person to another through semen, blood, or other body fluids that come into contact during sexual intercourse. Only one of the contraceptive methods described in the essay is effective in preventing the transmission of STDs. Which method is it, and why?

5. You are studying an endangered species of clam and notice that individuals release their gametes synchronously, even though they live in burrows. Design a study to determine what cues the animals are using to synchronize gamete release. Be sure to test the possibility that a pheromone is involved.

6. Sperm competition has never been studied rigorously in humans. Design a research program to test whether the phenomenon occurs. Assume that you can obtain DNA from many adult males and adult females and offspring, and that differences in DNA among individuals allow you to determine which male is the father of each offspring.

CD-ROM and Web Connection

CD Activity 45.1: Human Gametogenesis *(animation)*
(Estimated time for completion = 10 min)
How are the processes of spermatogenesis and oogenesis similar, and how do they differ?

CD Activity 45.2: Human Reproduction *(tutorial)*
(Estimated time for completion = 20 min)
How do the male and female reproductive systems work and how is the menstrual cycle regulated?

At your **Companion Website** (http://www.prenhall.com/freeman/biology), you will find self-grading exams and links to the following research tools, online resources, and activities:

Flying Apart: Mating Behavior and Speciation
Explore the relationship among sexual reproduction, mating behavior, and speciation.

Medline Health Information—Pregnancy and Reproduction
This site contains information on reproduction and pregnancy from the National Institute of Health.

The Center for Human Reproduction Home Page
This site discusses human reproductive biology and current technologies associated with it.

Additional Reading

Dunbar, R. I. M. 2001. What's in a baboon's behind? *Nature* 410: 150. Comments on the function of sexual swellings in female baboons. The individual featured in this chapter's opening photograph has such a sexual swelling.

Smith, R. 1999. The timing of birth. *Scientific American* 280 (March): 68–75. Reviews recent research on the hormonal control of the timing of delivery and the causes of premature birth in humans.

The Immune System in Animals

46

Disease threatens every animal. Humans alone are victimized by hundreds of different disease-causing bacteria and viruses and a wide array of parasitic worms, fungi, and protists. Given the ability of these pathogens to cause illness and death, it is remarkable that so many animals stay healthy for most of their lives.

To understand how humans and other animals are protected against disease, biologists have focused on explaining three observations. First, wounds usually heal even if they become infected. Second, most people who contract a bacterial or viral illness eventually recover even without help from medications. Third, people who acquire bacterial or viral infections and recover are frequently immune (safe) from contracting the same disease in the future.

This last observation is particularly intriguing. Why do people who get measles or chicken pox fail to get these diseases again later, during a second exposure? The Greek historian Thucydides commented on this pattern. In 430 B.C. he wrote that when plague struck Athens, only people who had recovered from the illness could nurse the sick, because they would not become ill a second time. In the middle ages, Chinese and Turkish practitioners protected people from smallpox by intentionally exposing individuals to the dried crusts of smallpox pustules taken from infected people.

In the late 1700s, the English physician Edward Jenner had a key insight that allowed him to refine this immunization technique. In Jenner's day, milkmaids were considered pretty because their faces were not pockmarked with scars from smallpox infections. Jenner knew that cows suffered from a smallpox-like disease called cowpox, and reasoned that milkmaids were immune from smallpox because they had been exposed to cowpox while milking cows.

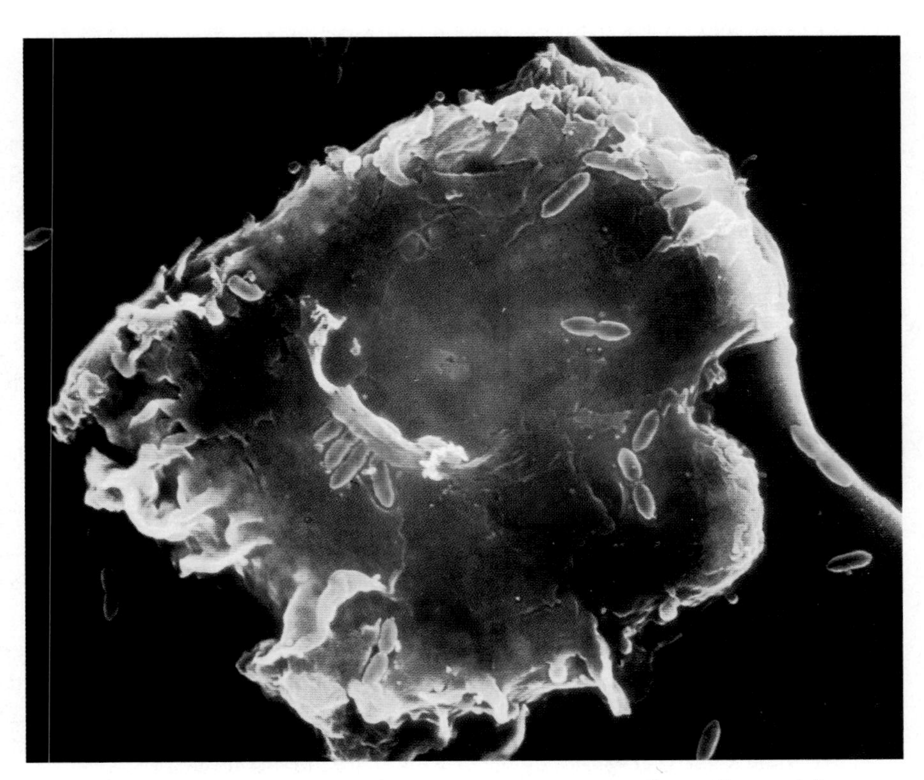

This scanning electron micrograph shows an immune system cell engulfing and destroying bacteria (the small green cells). This chapter explores how immune system cells are able to recognize bacteria and viruses as foreign and how they respond once disease-causing agents are identified.

To test this hypothesis, he inoculated a boy with fluid taken from a cowpox pustule, then later inoculated the same child with fluid from a smallpox pustule. As predicted, the boy did not contract smallpox. Jenner's technique was quickly adopted throughout Europe and was later named vaccination. (The Latin root *vacca* means "cow.")

Why did Jenner's technique work? More precisely, what is the molecular basis of a person's ability to fight off a bacterial or viral infection and achieve immunity? This chapter answers that question by exploring how the human immune system recognizes and eliminates pathogens. We'll consider the body's response to two events: a bacterial infection and an assault by a virus. Both infections could begin on any given day. For example, suppose that you found yourself on a crowded bus next to someone with a persistent cough and runny nose, and that 30 minutes later you stumbled while sprinting across campus and skinned your elbow. Several days later, you realize that the wound on your arm has gotten sore and red. Worse yet, you are developing a sore throat, runny nose, and fever. Your elbow has become infected with bacteria, and your upper respiratory tract is supporting a growing population of influenza or "flu" virus.

Section 46.1 introduces your body's response to these invaders by considering the events that occur at the site of a wound. The concluding three sections explore what happens when bacteria in a wound begin multiplying beyond the initial site, and how immune cells respond when a virus in your trachea and lungs begins exploding in numbers. More specifically, Section 46.2 introduces classical experiments that established how immune cells are able to recognize foreign invaders. Sections 46.3 and 46.4 follow up by considering how immune-system cells are activated once they recognize a pathogen as foreign, and what specific actions they take to eliminate the threat. After reading this chapter, you're likely to have a newfound appreciation for the sophisticated disease-fighting systems in your body, as well as an increased interest in one of the most rapidly developing and medically important fields in all of biology.

46.1 Innate Immunity

When biologists first began analyzing the immune system, they realized that certain immune-system cells are ready to respond to foreign invaders at all times, while other components must be activated first. Cells of the first type confer **innate immunity;** cells of the second type contribute **acquired immunity.** The key to understanding the two types of immunity is to recognize that the cells involved provide different responses to foreign substances, or antigens. An **antigen** is any foreign molecule. Most antigens are proteins or glycoproteins from bacteria or viruses or other invaders, but foreign carbohydrates and lipids can also function as antigens. As we'll see later, the immune system actually recognizes and responds to specific sites on antigens, not the entire foreign molecule.

The cells involved in innate immunity are nonspecific in their response to antigens. Stated another way, the innate immune system responds in the same way to all antigens. The innate immune response to a wound in your elbow, for example, is the same no matter what species of bacteria or fungi enter the tissue. In contrast, cells involved in acquired immunity respond in an extremely specific way to the particular strains of bacteria or viruses or fungi involved.

To launch our investigation into immune-system function, let's focus on how the body prevents entry by foreign invaders. Then we can consider how the innate response functions once they get in.

Barriers to Entry

The most effective way to avoid getting sick is to avoid contact with pathogens. In humans and other animals, the most important barrier to pathogen entry is the skin. In addition to providing a physical barrier, human skin offers a chemical deterrent. Skin cells secrete lactic acid and fatty acids that lower the pH of the surface and prevent bacterial growth.

The body has gaps in this barrier, however, where the digestive tract, reproductive tract, respiratory surface, and sensory organs make contact with the environment. As **Figure 46.1** shows, these surfaces have a protective physical barrier in the form of mucus or other features that discourage pathogen entry.

With regularity, though, preventive measures fail and pathogens gain entry to tissues beneath the skin. Flu viruses, for example, have an enzyme on their surface that disrupts the mucous lining of the respiratory tract. When the outer surface of the virus makes contact with a host cell beneath the mucous layer, it is able to enter and begin an infection. When a fall or other physical trauma breaks your skin, bacteria and other pathogens enter and gain direct access to the tissues inside. What happens then?

The Innate Response

When bacteria enter your body at the site of a wound, the cells pictured in **Figure 46.2a** (page 894) implement the innate response. As a group, these cells are called **leukocytes** (white-cells); all reside in the blood. Like the red blood cells that carry oxygen in the bloodstream, leukocytes are produced in the bone marrow.

The cells involved in the innate response are alerted to the presence of foreign invaders by the presence of certain molecules. All bacteria, for example, have proteins that begin with the amino acid N-formylmethionine instead of methionine. Bacteria in the gram-negative group also express a distinctive type of lipopolysaccharide on their surface, and certain cell-wall components in pathogens terminate with the sugar mannose. Cells in the innate system have proteins on their cell membranes that bind to these distinctive compounds. When these **pattern-recognition receptors** detect specific shapes or patterns present in foreign molecules, the leukocytes respond.

Figure 46.2b summarizes the major steps in the innate response. When skin breaks, blood components called platelets release proteins that form clots and lessen bleeding; other clotting proteins in the blood form cross-linked structures that help wall off the wound and reduce blood loss. In addition, the leukocytes called mast cells release chemical messengers that cause blood vessels near the wound to decrease blood flow by constricting. Other molecules released by mast cells induce blood vessels slightly further away from the wound to dilate and increase blood delivery to the general region. Vascular constriction occurs at the wound; vasodilation occurs nearby. Cells in nearby tissues then release chemical signals called chemokines (chemical-movers). Chemokines create a chemotactic gradient that neutrophils and other leukocytes use in migrating toward the site.

Neutrophils are phagocytic, meaning that they engulf foreign particles and digest them. (Translated literally, *phagocytic* means "eat-cell.") Neutrophils also secrete the enzyme lysozyme, which degrades bacterial cell walls, along with nitric oxide (NO), hydrogen peroxide (H_2O_2), and other reactive oxygen intermediates (ROI) that attack bacteria and fungi. In this way, neutrophils begin to kill and degrade the foreign material present in the wound.

The leukocytes called macrophages arrive after neutrophils and perform many of the same functions. In addition, macrophages secrete chemical messengers called **cytokines** (cell-movers). Cytokines have an array of effects. Cytokines from macrophages stimulate the bone marrow to make and release additional neutrophils and macrophages. They also attract other immune-system cells to the site, activate cells involved in tissue repair and wound healing, and induce fever (elevated body temperature).

In total, the events diagrammed in Figure 46.2b are referred to as the **inflammatory** (in-flames) **response.** The site of inflammation often becomes swollen due to increased numbers of cells and fluids, red and warm due to increased blood flow, and painful due to signals from pain receptors. The inflammatory response continues until all of the foreign material is eliminated and the wound is repaired.

If the pathogens that entered the wound are able to replicate quickly, however, the innate immune response may not be sufficient to contain and eliminate them. What happens then? And what happened to the virus that blasted through the mucous membrane of the respiratory tract and entered the cells below? As Chapter 26 showed, viruses actually enter cells to infect them. As a result, they are largely hidden from the innate response. To keep infections at wound sites from spreading throughout the body and to limit the impact of viral invasions, the acquired immune response has to come into play.

46.2 The Acquired Immune Response: Recognition

Section 46.1 claimed that the acquired immune response is based on interactions between specific immune-system cells and specific pieces of antigens. If so, the observation raises an important question. Given the array of different pathogens that exist, every individual animal is almost certain to be exposed to an enormous variety of antigens in the course of its lifetime. How many different antigens can the system respond to? Are there limits to the number of antigens that the acquired immune system can recognize?

Eyes
Blinking wipes tears across the eye. Tears contain the anti-bacterial enzyme lysozyme.

Nose
The nasal passages are lined with mucus secretions and hairs that trap pathogens.

Digestive tract
Pathogens are trapped in saliva and mucus, then swallowed. Most are destroyed by the low pH of the stomach.

Ears
Hairs and ear wax trap pathogens in the passageway of the external ear.

Respiratory tract

The trachea is lined with ciliated cells and mucus-secreting cells that keep pathogens out of the lungs.

FIGURE 46.1 How Does the Body Keep Pathogens Out?
QUESTION Why is a combination of hair or cilia and sticky secretions such as mucus effective at trapping pathogens?

Research conducted in the early 1920s provided an initial answer to these questions. The basic protocol was to synthesize organic compounds that do not exist in nature, inject the novel molecules into rabbits, and observe whether the acquired immune system in these animals was activated in response. To the amazement of the researchers involved, the rabbits were able to mount an immune response to the antigens. More specifically, the animals produced antibodies against each of the antigens. An **antibody** is a protein that binds to a specific part of a specific antigen. As we'll see, antibodies are produced by certain cells in the acquired immune system. The binding of an antibody to an antigen leads to the destruction of the foreign cell or molecule. The take-home message of these early injection experiments was that the immune system can produce an almost limitless array of antibodies.

How can the immune system produce antibodies to antigens that had never before existed? It took over 60 years of experimentation and debate to answer that question. To begin exploring this work, let's review the cells and organs involved in the acquired immune system.

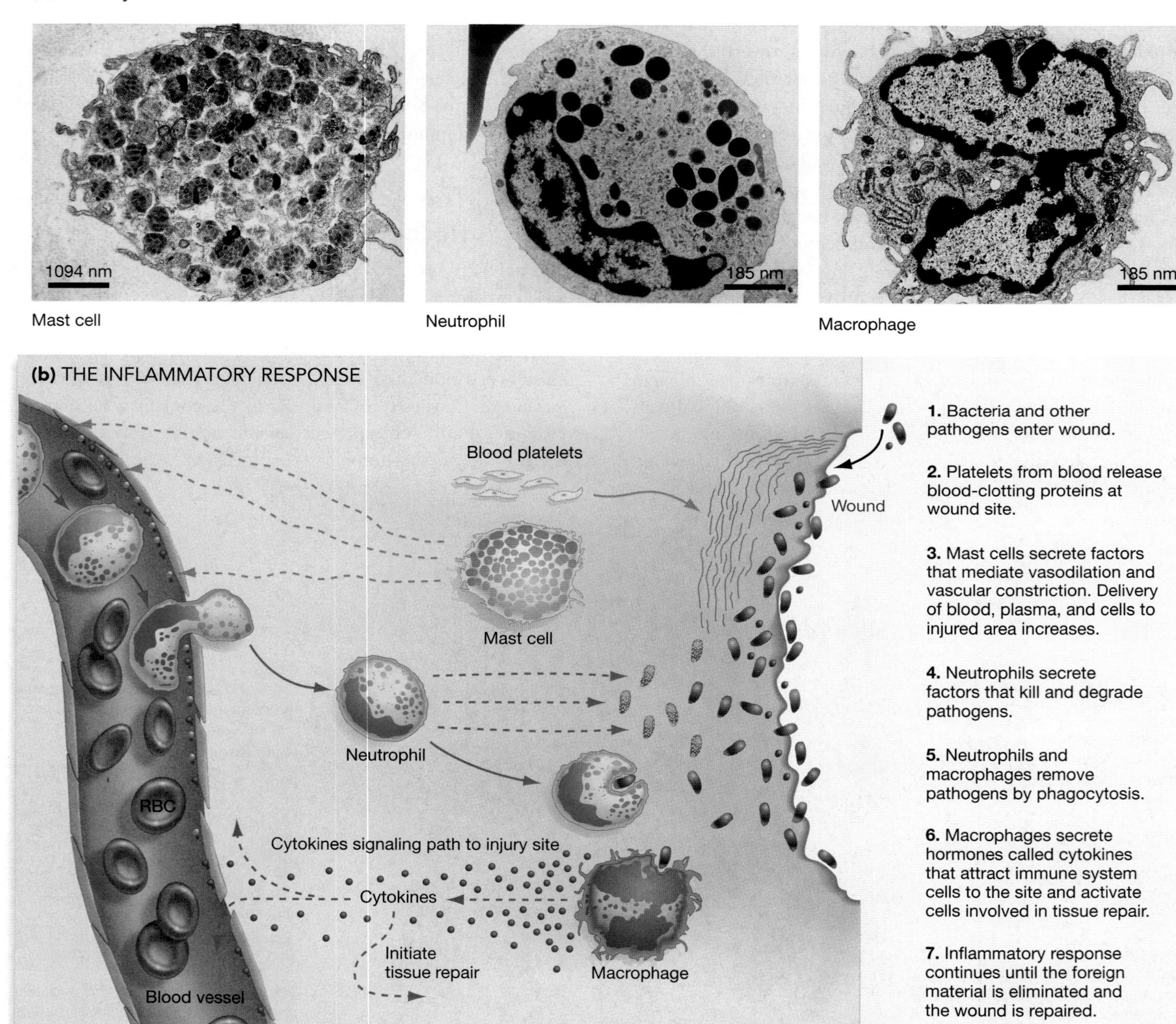

FIGURE 46.2 The Innate Immune Response
(a) These transmission electron micrographs illustrate just a few of the types of leukocytes in the innate immune system. **(b)** The inflammatory response occurs at sites of tissue injury.

An Introduction to Lymphocytes and the Immune System

The cells involved in the acquired immune response are called **lymphocytes.** Lymphocytes form and mature in the primary organs of the immune system: the bone marrow and thymus (**Figure 46.3a**). Cells in the innate system circulate between the blood and tissues, while lymphocytes circulate through the blood and the lymph nodes, spleen, and lymphatic ducts—structures that are termed the secondary organs of the immune system. The lymph nodes and spleen are important sites where lymphocytes encounter antigens. The mixture of fluid and lymphocytes present in the lymph nodes and ducts is called **lymph.**

Circulating lymphocytes are in a resting or inactivate state. As **Figure 46.3b** shows, circulating lymphocytes have a large nucleus, little cytoplasm, few mitochondria, and a ruffled membrane. Over the course of a day, a resting lymphocyte might migrate through the spleen, enter the blood, cross over into the

(a) Components of the immune system

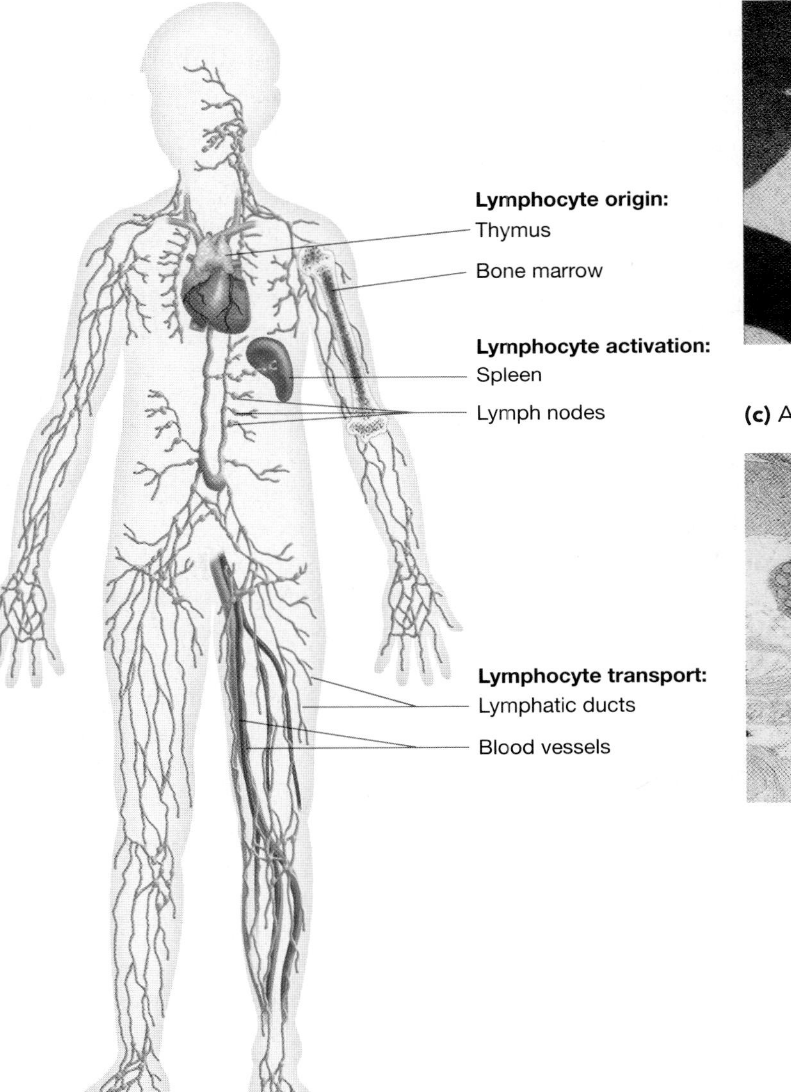

(b) Inactive lymphocyte

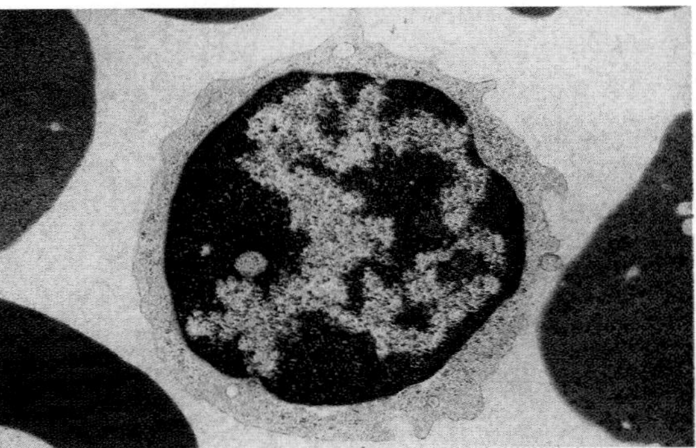

(c) Activated lymphocyte

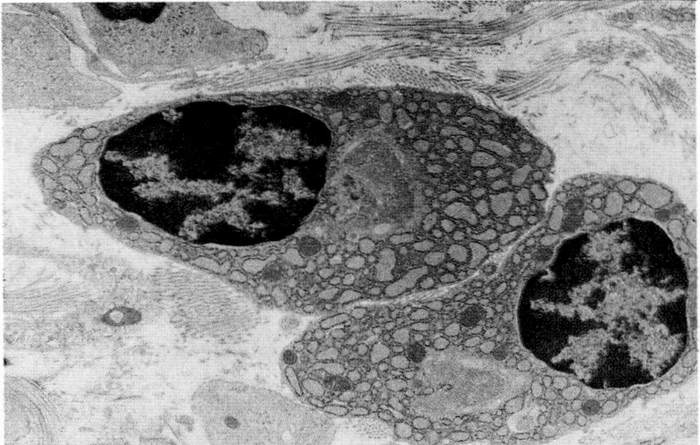

FIGURE 46.3 The Immune System
(a) The immune system has three functional components. The thymus and bone marrow are primary organs where lymphocytes mature. Lymphocytes are presented with antigen and become activated in the lymph nodes and spleen. The lymphatic ducts and blood vessels transport lymphocytes throughout the body.
(b) Inactive lymphocytes have a small amount of cytoplasm and few organelles. **(c)** Active lymphocytes have extensive rough endoplasmic reticulum. This observation suggests that a great deal of protein synthesis is taking place.

lymphatic vessels, migrate through a lymph node, return to the blood, and so on.

If a resting lymphocyte does not encounter an antigen that it is programmed to respond to, it eventually dies. But if a lymphocyte encounters an appropriate antigen, it becomes activated. As **Figure 46.3c** shows, an activated lymphocyte has a large population of mitochondria and a massive amount of rough ER. You might recall, from Chapter 5, that many of the proteins synthesized in the rough ER are inserted into the cell membrane or secreted from the cell. The increased rough ER in activated lymphocytes suggests that these cells are manufacturing and possibly secreting proteins.

The contrast in morphology between inactive and active lymphocytes is remarkable. How are lymphocytes programmed to recognize particular antigens, and what does an activated lymphocyte do?

The Discovery of B Cells and T Cells

In 1956, Bruce Glick and colleagues provided an important insight into the acquired immune system, quite by accident. These biologists were investigating the immune system's response to the bacterium *Salmonella typhimurium*, which is a common cause of food poisoning in humans. To do this work, they needed to produce and isolate antibodies to a particular antigen from *S. typhimurium* that is toxic. Their plan was to inject a large number of chickens with the antigen and collect the antibodies that the treated individuals produced in response.

In addition to injecting many normal chickens, though, Glick and co-workers happened to include some chickens that had earlier undergone experimental removal of an organ called the bursa. To the researchers' surprise, 6 of the 9 chickens that lacked a bursa and that were injected with the antigen died. The other 3 birds without a bursa failed to produce antibodies to the antigen. In contrast, all of the intact chickens used in the experiment produced large quantities of antibodies and survived. To make sense of these results, the researchers proposed that the bursa is critical for antibody production, and that antibodies are important in neutralizing antigens.

Not long after the observation was published, three groups of scientists independently conducted a related experiment. To explore the function of the thymus, these groups removed the organ from newborn mice. Mice lacking a thymus developed pronounced defects in their immune systems. For example, when pieces of skin from other mice were grafted onto the experimental individuals, their immune systems did not recognize the tissue as foreign. In contrast, individuals with an intact thymus quickly mounted an immune response to the foreign skin cells and killed them.

The results of these and follow-up experiments showed that lymphocytes from the bursa and thymus have different functions. The two types of lymphocytes became known as bursa-dependent and thymus-dependent lymphocytes, or B cells and T cells. **B cells** produce antibodies. **T cells** are involved in an array of functions, including recognizing and killing infected host cells. Later work showed that in humans and other species that lack a bursa, B cells mature in bone marrow.

More recent research has shown that important subtypes of B cells and T cells exist. For example, T cells come in two major types that are distinguished by distinctive proteins on their cell membranes. These membrane proteins are called CD4 and CD8. T cells with the CD4 protein on their surface are known as $CD4^+$ cells. As we'll see in section 46.3, $CD4^+$ cells have different functions than $CD8^+$ T cells. Right now, however, we need to return to the question of how B cells and T cells are able to recognize so many different antigens.

Antigen Recognition and Clonal Selection

By the 1950s a series of important general observations had been made about the immune system. It was clear that antibodies to a seemingly limitless number of antigens could be produced, and that each antibody is specific to an antigen. By studying the time course of antibody production in rabbits and other experimental animals, researchers established that the immune response increased through time after an infection had begun. In addition, the response is remembered—meaning that individuals do not get sick at all or recover extremely quickly if they are exposed to the same pathogen again in the future.

To explain these patterns, researchers developed the **clonal-selection theory** of immune-system function. This theory made three central claims about how the acquired immune system works. First, each lymphocyte formed in the bone marrow or thymus was hypothesized to have a unique receptor on its surface that recognizes one antigen. Stated another way, there is a one-to-one correspondence between the receptors found on B cells and T cells and the antigens they respond to. Second, when the receptor on a lymphocyte binds to an antigen, the lymphocyte is activated. An activated lymphocyte divides and makes many identical copies of itself. In this way, specific cells are selected and cloned in response to an infection. Third, some of these cloned cells persist after the pathogen is eliminated. As a result, they are able to respond quickly and effectively if the infection recurs in the future.

The clonal selection theory provided a coherent explanation for most major attributes of the acquired immune system. Just as important, it made a number of specific predictions that could be tested. Which aspects of the theory have been confirmed, and which rejected?

The Discovery of B-Cell Receptors and T-Cell Receptors

A fundamental prediction of the clonal-selection theory was confirmed when researchers used radioactively labeled antigens to find and isolate the protein on the surface of B cells that antigens bind to. When this **B-cell receptor (BCR)** was analyzed chemically, investigators found that the protein has the same structure as the antibodies produced by B cells.

Both the BCR and antibodies belong to a family of proteins called the gamma globulins. The molecules are referred to as **immunoglobulins.**

As **Figure 46.4a** shows, the BCR has three distinct components. The first is a protein with a molecular weight of about 25,000 daltons called the **light chain.** The second component is roughly twice the size of the light chain and is called the **heavy chain.** Each BCR has two copies of the light chain and two copies of the heavy chain. The final section is a transmembrane domain that anchors the protein in the B-cell membrane. Antibodies lack this transmembrane domain and are secreted from the cell. As **Table 46.1** (page 899) explains, there are actually five distinct classes of immunoglobulin proteins that act as receptors or antibodies. Each class is distinguished by unique amino acid sequences in the heavy-chain region, and each has a distinct function in the immune response.

It took much longer for researchers to isolate and characterize the **T-cell receptor (TCR).** It turns out that the TCR only binds to antigens after they have been processed by other immune-system cells and presented on their cell membranes. B cells can bind to antigens directly; T cells only bind to antigens that are displayed by other cells. When investigators finally succeeded in isolating the TCR, they found that it is composed of a single **alpha** (α) chain and a single **beta** (β) chain. As **Figure 46.4b** shows, the overall shape of the TCR is very similar to the "arm" of an antibody or BCR molecule.

Antibodies and lymphocyte receptors do not bind to the entire antigen, but to a selected region called an **epitope.** To understand the relationship between an antigen and an epitope, recall that every bacterium, virus, fungus, and protist is made up of a large number of different molecules. Each of these molecules is an antigen, because each is foreign. An antigen may have many different epitopes, however, where binding by antibodies and lymphocyte receptors actually takes place. To drive this point home, **Figure 46.5** (page 898) illustrates a protein from flu virus that has six epitopes. It is not unusual for an antigen to have between 10 and 100 different epitopes. Each epitope is recognized by a particular antibody, B-cell receptor, or T-cell receptor.

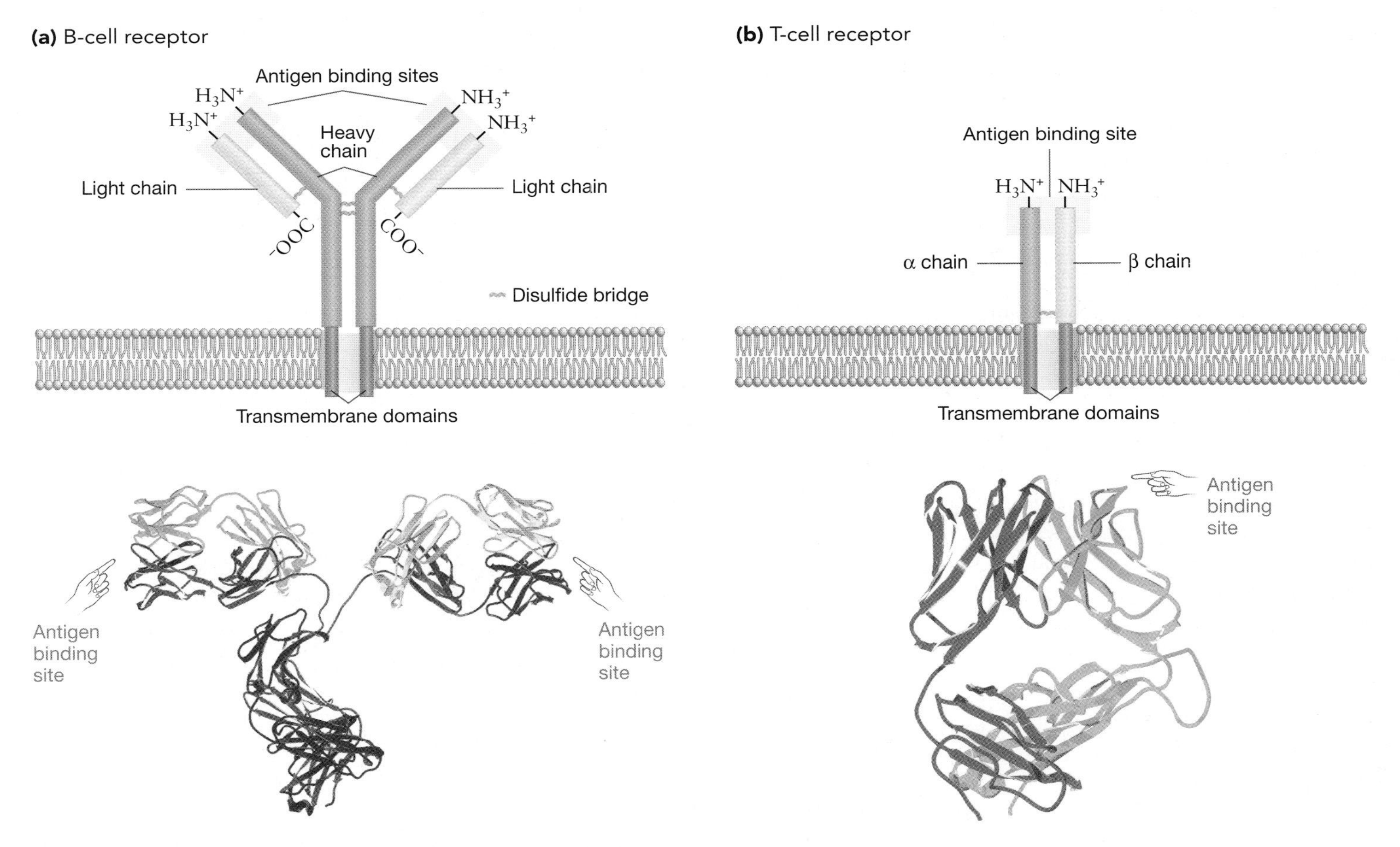

FIGURE 46.4 What Do the Receptors on Lymphocytes Look Like?
(a) The schematic and ribbon models on the left show that the B-cell receptor is shaped like a Y. The antibodies and receptors produced by each B cell are identical, except that antibodies lack the transmembrane domain and are secreted. **(b)** The shape of the T-cell receptor resembles one "arm" of the Y-shaped B-cell receptor.

How are the immunoglobulins able to recognize specific epitopes? The answer to this question came through detailed studies of the BCR's heavy and light chains.

What Is the Molecular Basis of Antigen Specificity? In the 1950s, biologists developed an important model system for studying the BCR and antibody production. The cells involved were B-cell tumors, or **myelomas**, that could be grown in laboratory culture. Like other cancers, myelomas grow in an uncontrolled fashion. Further, each type of myeloma produces a single type of antibody. (**Box 46.1**, page 900, explains how myelomas are used to produce antibodies used in research.) By isolating and comparing the light chains produced by different myelomas, researchers were able to document that each light chain is different, and infer why certain antibodies bound to certain antigens.

The initial studies that explored the composition of BCRs and antibodies used enzymes to chop the light chains from different B cells into pieces. When researchers separated these pieces by electrophoresis, they noticed that light chains from different B cells appeared to have one segment in common. In addition, there seemed to be light-chain regions that were unique to each B cell. These light-chain segments came to be known as the constant (C) and variable (V) regions, respectively.

When techniques to determine the amino acid sequence of proteins became available, the data that resulted confirmed that the light chain and the heavy chain consist of a variable region at the amino terminus and a conserved region at the carboxy terminus. Consider, for example, the data in **Figure 46.6**, compiled by Tai Te Wu and Elvin Kabat. These biologists compared the amino acid sequences of 77 different light chains. To identify exactly which points in the chain varied the most, Wu and Kabat counted how many of the 77 proteins had a different amino acid at each position in the chain, and then divided this number by the frequency of the most common amino acid at that position. As the resulting histogram shows, certain amino acids in the light chain are extremely variable among B cells.

Why is this observation important? First, the presence of unique amino acids at certain positions in the light chain and heavy chain is responsible for the ability of the BCR and antibodies to bind to unique epitopes. Proteins with different amino acid sequences bind to different epitopes. Second, the binding occurs at the tips of the antibody and BCR molecule's arms (Figure 46.6b). Analogous studies showed that the TCR also has a variable region near its amino terminus of both the α and β chains (Figure 46.6b).

Analyses of the variable region raised a paradox, however. The acquired immune system responds to an almost limitless number of antigens because there is a virtually limitless number of different immunoglobulins and TCRs. But the genome of animals is not limitless. How does the genome code for so many different immune-system proteins?

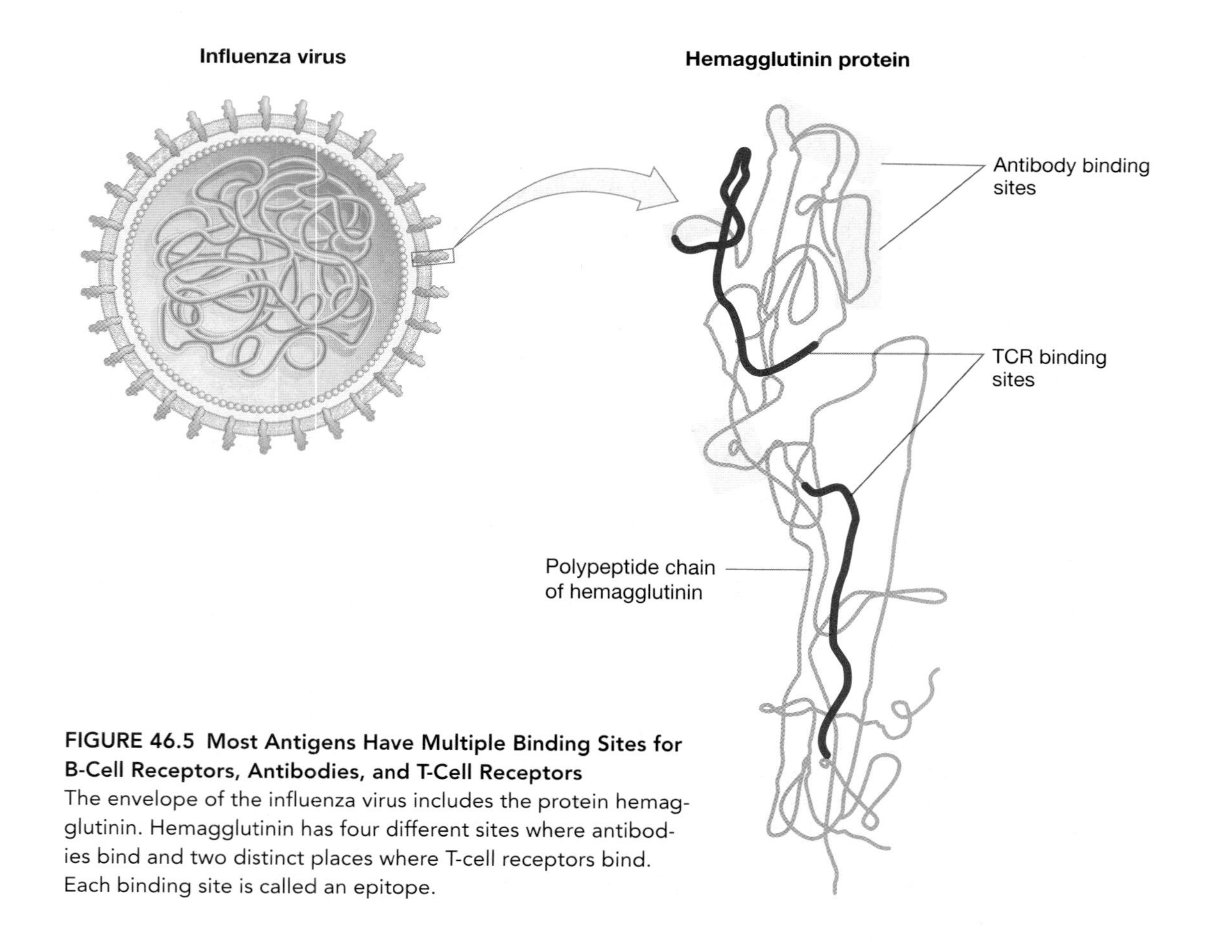

FIGURE 46.5 Most Antigens Have Multiple Binding Sites for B-Cell Receptors, Antibodies, and T-Cell Receptors
The envelope of the influenza virus includes the protein hemagglutinin. Hemagglutinin has four different sites where antibodies bind and two distinct places where T-cell receptors bind. Each binding site is called an epitope.

TABLE 46.2 Immunoglobulins Come in Five Classes

Name	Structure	Function
IgD	Monomer	Present on membranes of mature B cells; probably involved in activation of B cells.
IgM	Pentamer	First type of secreted antibody to appear during an infection. Binds many antigens at once; effective at clumping viruses and bacteria so they can be killed.
IgG	Monomer	The most abundant type of secreted antibody. Circulates in blood and interstitial fluid. Protects against bacteria, viruses, and toxins.
IgA	Dimer	Most common antibody in breast milk, tears, saliva, and mucus that lines the respiratory and digestive tracts. Prevents bacteria and viruses from attaching to mucus membranes; helps immunize breastfed newborns.
IgE	Monomer	Rarest type of antibody. Involved in hypersensitive reaction that produces allergies. (See essay at end of chapter.)

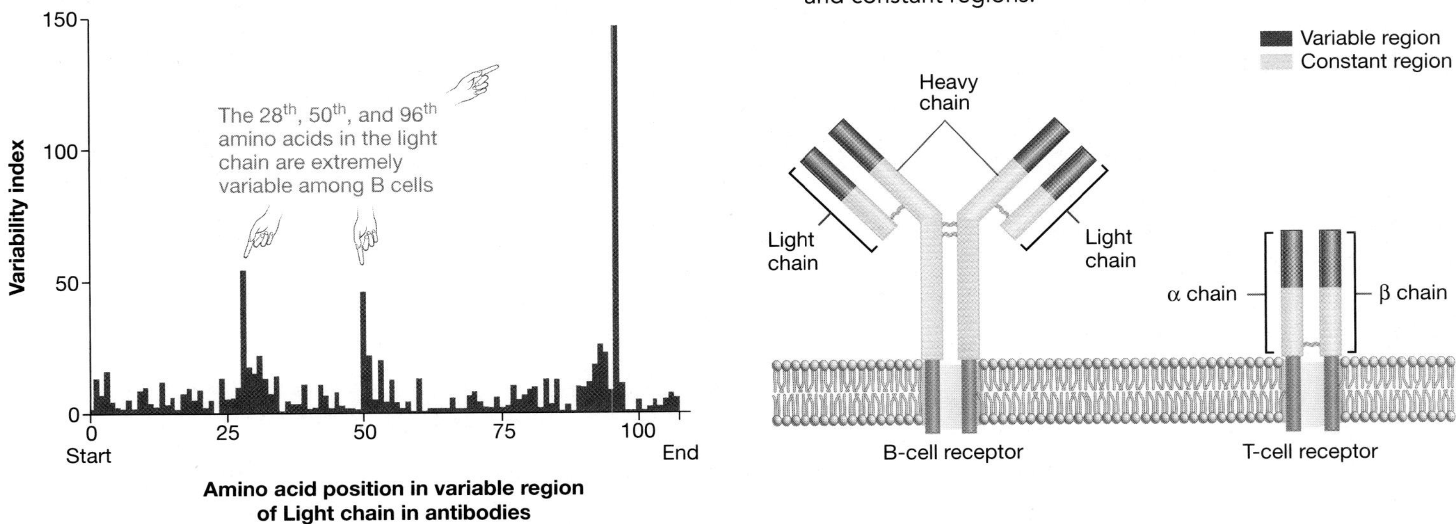

FIGURE 46.6 How Does the Amino Acid Composition of Antibodies Vary?
(a) These data indicate which amino acids in the immunoglobulin L-chain are variable when many different antibody molecules are compared. The text explains how the "Variability index" was computed. When researchers performed the same analysis for the constant region from the same group of molecules, they observed almost no variation. Similar patterns are found in the variable and constant regions of the heavy chain. **(b)** The variable regions of a B-cell receptor and T-cell receptor face away from the cell membrane.

The Discovery of Gene Recombination In 1965, W. J. Dryer and J. Claude Bennett proposed a controversial explanation for how immunoglobulin genes code for so many different variable regions and thus so many different proteins. They suggested that a single gene codes for the C region of the light chain and that a separate gene codes for the V region. In addition, they hypothesized that early in the life of a lymphocyte, sequences from the V gene are combined with sequences from the the C gene. In effect, Dryer and Bennett hypothesized that the two immunoglobulin genes are recombined in a way that results in the production of a sequence unique to that particular cell. This type of DNA shuffling had never been observed, however. (Introns and alternative splicing had not been discovered yet.) As a result, most researchers considered the hypothesis wildly implausible. The only evidence that Dryer and Bennett could offer in support of their gene recombination idea was that viral genes had been shown to insert themselves into the genomes of host cells (see Chapter 26).

BOX 46.1 Producing Monoclonal Antibodies

When an animal is injected with an antigen, many different B cells respond. Some of these B cells produce antibodies to different epitopes on the antigen, and some produce slightly different antibodies that bind to the same epitope. As a result, biologists say that the immune response is polyclonal (many-clones).

In many situations, though, it is extremely valuable for investigators to have antibodies from a single B cell that bind to the same epitope of a single antigen in the same way. For example, having an antibody to the insulin receptor would allow you to analyze cells from different parts of the body or at different stages in development and understand where and when the insulin receptor protein is expressed.

How can such monoclonal (one-clone) antibodies be produced? In 1975, Georges Köhler and Cesar Milstein announced that they had developed an approach for producing monoclonal antibodies. The approach relied on the use of myelomas, which can grow continuously in culture and produce a monoclonal antibody. **Figure 1** outlines the protocol that Köhler and Milstein used to create myelomas that produced large quantities of a specific, desired antibody.

The process begins by injecting a mouse with the antigen of interest. In the example used here, the antigen is a protein called hemagglutinin that is found on the surface of influenza virus. In response, B cells in the animal's spleen activate and produce antibodies to hemagglutinin. Spleen cells are then removed from the animal and added to a test tube containing myeloma cells. A compound that encourages cells to stick together is added to the mixture. The resulting solution is cultured under conditions that allow growth only by cells that have fused together. The resulting combination of a myeloma cell and B cell is called a hybridoma.

To complete the process, each hybridoma is isolated, grown separately, and tested for its ability to produce antibodies to the antigen of interest. Once the most effective hybridomas are identified, researchers have a virtually limitless supply of a monoclonal antibody.

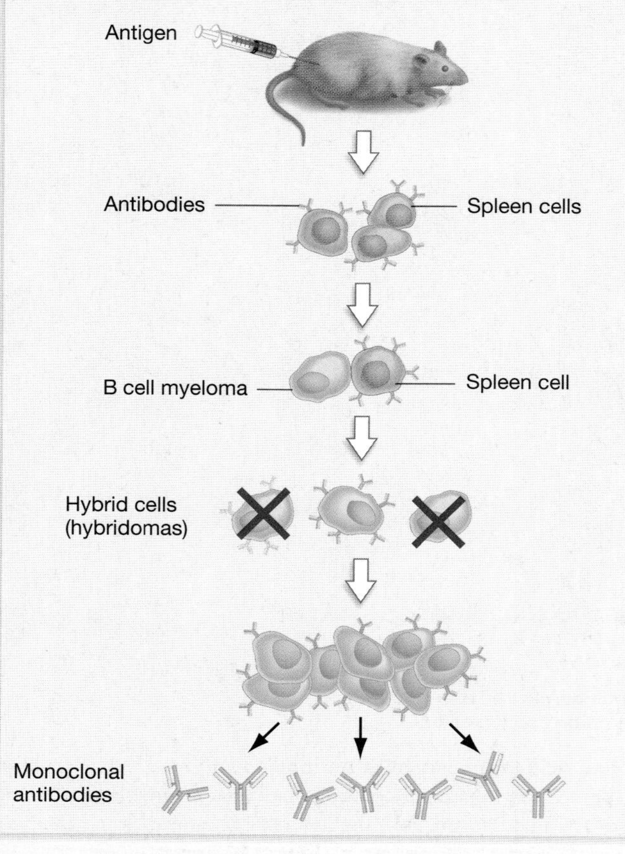

FIGURE 1 Steps to Produce Monoclonal Antibodies

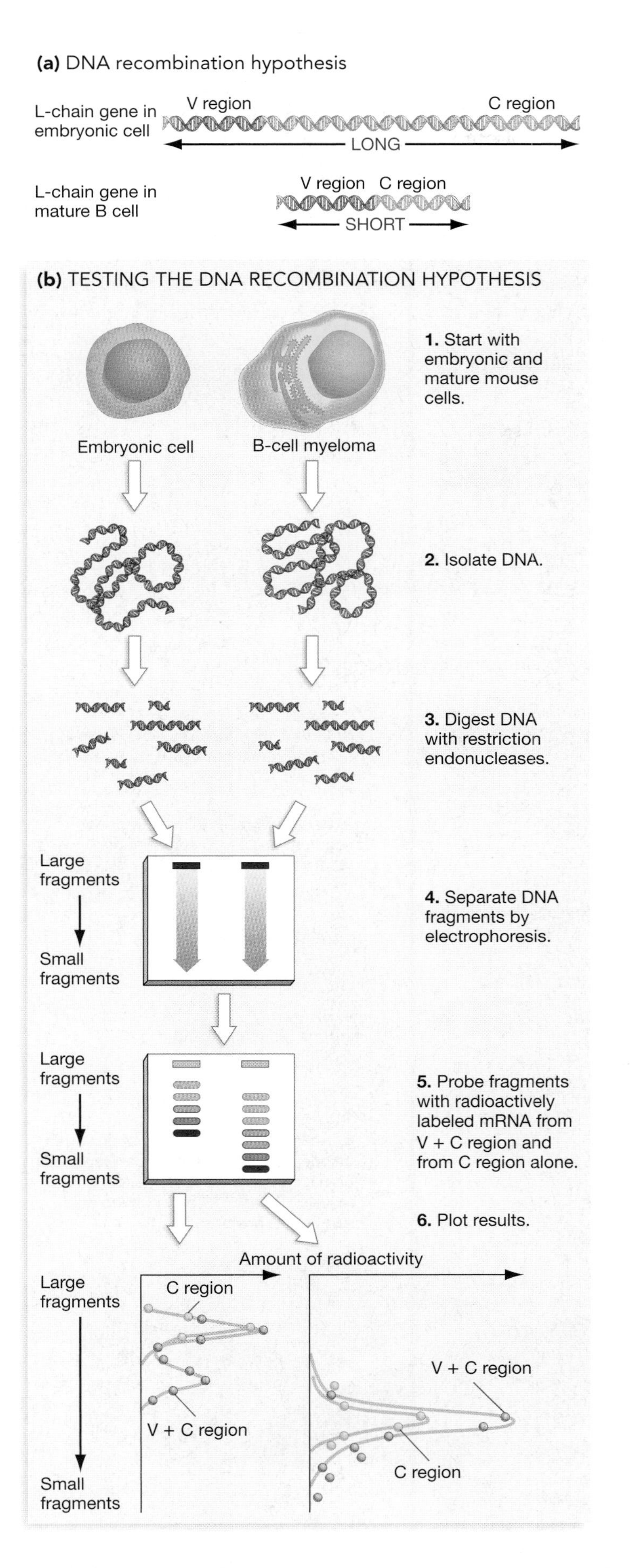

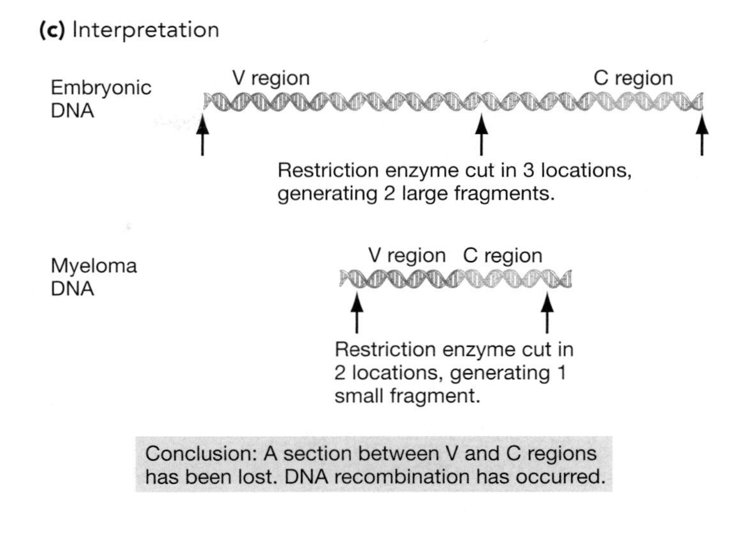

FIGURE 46.7 Experimental Evidence for DNA Recombination in B Cells

QUESTION If DNA recombination had not occurred, what would the data reported in part **(b)** look like?

In 1976, however, Nobumichi Hozumi and Susumu Tonegawa published data supporting Dryer and Bennett's radical idea. Hozumi and Tonegawa reasoned that if light-chain diversity was produced by DNA recombination, then the variable and constant regions of mature B cells—meaning, B cells that were producing BCR and antibodies—should be shorter than the same region in immature cells that are not producing antibodies (**Figure 46.7a**). Stated another way, the V + C region should be smaller in adult cells than it is in embryonic cells.

To test this prediction, the researchers implemented the protocol diagrammed in **Figure 46.7b**. As the illustration shows, they began by isolating DNA fragments from a mouse B-cell myeloma and from embryonic mouse tissue. Then they probed the DNA fragments from each type of cell with radioactive mRNA from either the entire light chain or from just the constant region.

The data at the bottom of Figure 46.7b show their results. Note that two large DNA fragments from the immature cell hybridized with the entire light chain, and that the larger of these two fragments also hybridized with the constant-region mRNA. To interpret this result, the biologists proposed that the smaller fragment contained the variable region and that the larger fragment included the constant region (Figure 46.7c). In contrast, only one small fragment from the mature B-cell DNA hybridized with the entire light chain. This fragment also hydridized with the constant-region mRNA.

The researchers viewed this result as strong evidence that DNA recombination brings V and C genes closer together as B cells mature. Subsequent studies confirmed their result, and showed that the heavy chain of the immunoglobulins and the α and β chain of the TCR are also produced by DNA recombination.

Hozumi and Tonegawa's paper inspired a flurry of studies on the mechanism of DNA recombination. **Figure 46.8** summarizes these results by presenting the current model for how the immunoglobulin light-chain locus changes as a B cell matures. One central finding is that the light-chain gene contains distinctive regions called gene segments. In humans, the locus includes about 70 distinct variable (V) segments, 9 joining (J) segments, and a single constant (C) segment. Three additional gene sequences code for the variable region of heavy chain. The heavy-chain locus includes about 51 V segments, 27 diversity (D) segments, and 6 J segments. A similar situation occurs in the genes that encode the α and β chains of the TCR.

Early in lymphocyte development, the gene segments are brought closer together by DNA rearrangement to create a functional gene. For example, in B cells any one of the 70 V light-chain segments can recombine with any one of the 9 J segments. Thus, $70 \times 9 = 630$ different light chains can be produced by recombination. In the heavy chain, any one of the 51 V segments, 27 D segments, and 6 J segments can recombine, giving a total of $51 \times 27 \times 9 = 8{,}262$ possible H chains. Be-

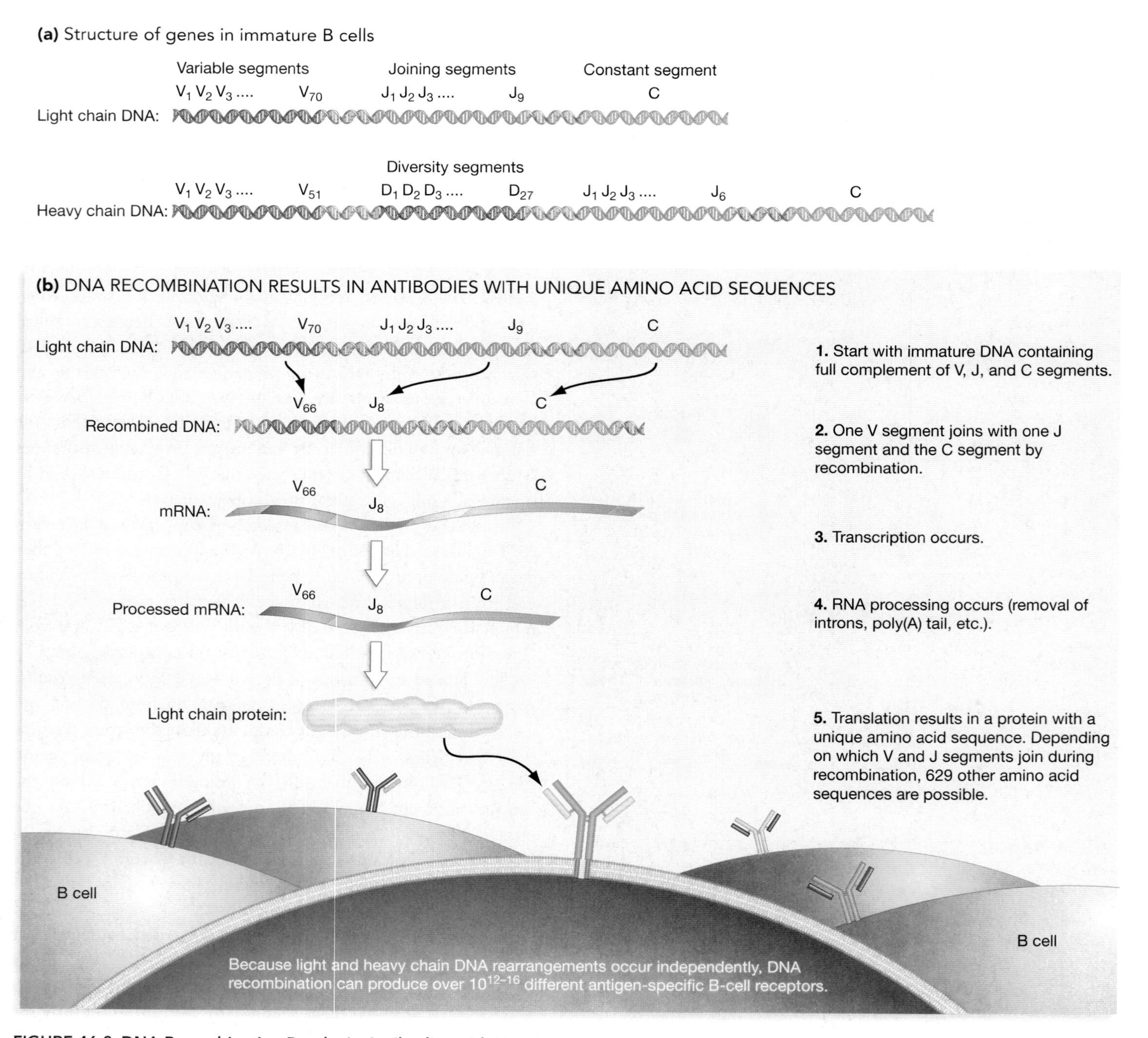

FIGURE 46.8 DNA Recombination Results in Antibodies with Unique Amino Acid Sequences
This diagram summarizes current data on how light-chain proteins are produced as a B cell matures.

cause L and H rearrangement occurs independently, DNA recombination can produce $630 \times 8{,}262 = 5{,}205{,}060$ or 5.2×10^6 different antigen-specific BCRs.

The story does not end here, however. Gene segments are not always ligated precisely during DNA recombination. Some variation occurs where the V and D segments join and where the D and J segments join. This **junctional diversity** results in many additional novel sequences. Thus, by recombining only 163 small DNA segments, an estimated 10^{12-16} different BCRs can be created. A similar number pertains for the TCRs.

To summarize, each B cell and T cell in the body has a receptor on its surface. Each receptor has a unique amino acid sequence, and thus the ability to bind a unique epitope of a unique antigen. Before maturation is complete, the finished receptors are checked to make sure that they do not recognize and bind to normal cellular proteins found in the host. (See **Box 46.2** for more detail on this process.)

If you are nursing an infected elbow and a blossoming case of the flu, then, your blood and lymph are being flooded with antigens from invading bacteria and viruses. Presumably, B cells and T cells capable of recognizing these antigens exist somewhere in your body. How do the lymphocytes that are circulating through the immune system encounter their specific antigen? And what happens after a receptor successfully binds to an antigen?

46.3 The Acquired Immune Response: Activation

When the cell-surface receptor of a B cell or T cell binds to an antigen, the event is like pushing the button that launches a "smart bomb"—a missile that is programmed to destroy a specific target. The lethal power of activated B cells and T cells is nothing short of awesome.

Perhaps because lymphocytes are so destructive, their interaction with antigens is a carefully controlled, step-wise process that eventually leads to activation. The mechanism is reminiscent of the precautions that nations with powerful missiles take to avoid accidental deployment. For the most dangerous weapons, a signal to launch must come from the chief executive, and it is checked and crosschecked using a series of codes and signals.

For lymphocytes, the sequence that leads to activation begins when antigens are taken up by a specific type of leukocyte, cut into pieces, packaged with specific cell proteins, and then transferred to the cell surface. Once antigens are presented in this way, T cells can bind to them via the TCR and begin their transformation from an inactive to active state. The antigen-presentation step is crucial to the activation of the acquired response.

The goal of this section is to explore a series of questions about this interaction between antigens and lymphocyte receptors. How does the presentation step work? How are B cells and T cells activated? To begin, let's return to the wound in your elbow, where bacterial growth has outpaced the innate response.

Antigen Presentation by MHC Proteins: Activating T Cells

If bacteria begin to multiply rapidly at a wound site, leukocytes called dendritic (tree-like) cells are recruited to the area. Dendritic cells take up some of the antigens present and migrate to the nearest lymph node.

Figure 46.9 (page 904) details how antigens and lymphocytes interact in the lymph node. As Figure 46.9a shows, the

BOX 46.2 How Does the Immune System Distinguish Self from Non-Self?

As B cells and T cells mature and begin producing receptors for specific antigens, an important problem arises. How does the immune system ensure that these cells don't recognize molecules that are part of normal host cells? If a B-cell receptor or T-cell receptor responds to a self-molecule, it would trigger a lethal response. Stated another way, immune-system cells could turn on the host and begin destroying parts of the body. An anti-self reaction like this is known as autoimmunity.

When researchers introduce B cells and T cells with anti-self receptors into mice, they find that the experimental lymphocytes are eliminated. Follow-up work showed that if B cells and T cells that are maturing in the bone marrow and thymus have anti-self receptors, the cells are destroyed before they leave these organs. In the case of T cells, it appears that the mechanism involved in this negative selection is virtually identical to the process of T-cell activation introduced in section 46.3. But if a T cell begins to activate in response to a self-molecule presented in the thymus, it is destroyed. Based on the work done to date, it appears that about 90% of all immature B cells and T cells that are produced in the bone marrow and thymus are destroyed before they mature, because they are anti-self.

Occasionally the process of negative selection breaks down, however, and lymphocytes are produced that react to self-proteins. For example, the disease multiple sclerosis (MS) results from the production of anti-self T cells. The T cells involved attack the myelin sheath of nerve fibers. Because insults to myelin reduce the efficiency of nerve signaling, a variety of muscular and coordination problems result. MS tends to develop in young adults aged 20–30 years and can be debilitating in certain cases. Other examples of autoimmune diseases include rheumatoid arthritis, type-1 diabetes mellitus, myasthenia gravis, and lupus (systemic erythematosus).

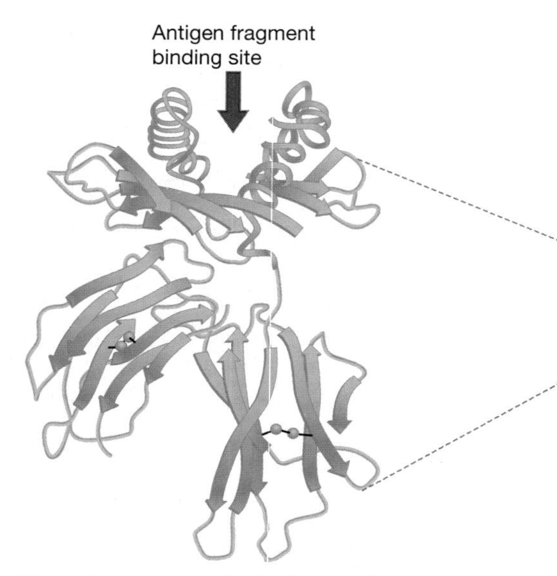

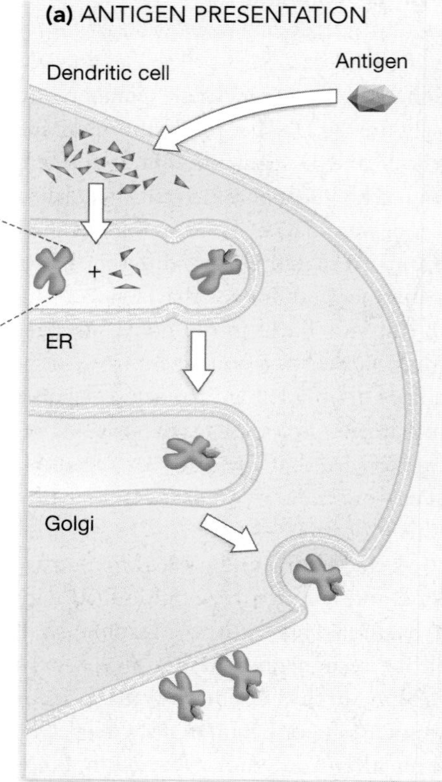

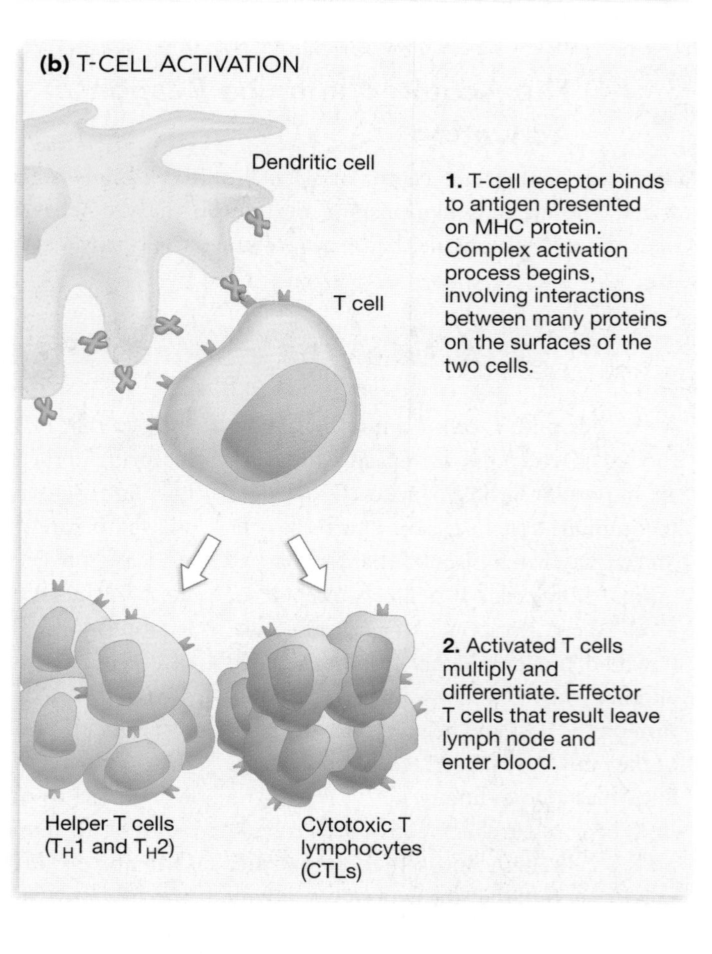

FIGURE 46.9 T Cells Are Activated After Interacting with Antigens Presented by Dendritic Cells
(a) Dendritic cells take in a wide variety of antigens, break them into pieces, and then present the fragments in the cleft of an MHC protein. **(b)** If a T cell has a receptor that is complementary to the antigen presented on the surface of a dendritic cell, the T cell will begin the activation process.

process begins when dendritic cells process the antigens they have ingested. Enzymes inside the cell break the antigens into small pieces. These pieces then become bound to a major histocompatibility (MHC) protein. As the figure shows, MHC proteins have a groove where small antigen fragments, typically from 8 to 20 amino acids in length, bind. MHC proteins come in two types, called the Class I and Class II MHC proteins. In dendritic cells, antigen fragments are attached to both Class I and Class II proteins. Once an MHC-antigen complex forms inside a dendritic cell, the molecules are transported to the cell surface. Antigen presentation is now complete.

Figure 46.9b illustrates what happens next. As $CD4^+$ T cells move through the lymph node in search of antigen, those with complementary receptors interact with their corresponding epitope on dendritic cells. To a $CD4^+$ T cell, the combination of an MHC protein and bound antigen carries the message "self cell has found antigen—response required."

The initial binding event between a TCR and an MHC-bound epitope is critically important because it signals that self-cells are presenting an antigen that is recognized by the $CD4^+$ T cell. The binding event is also the start of a long and involved activation process. But before activation occurs, additional receptors on the same T cell have to bind to additional MHC-antigen complexes on the same dendritic cell. These receptor-antigen complexes are taken into the T cell and supply what biologists call signal 1. This preliminary signal is followed by a second, co-stimulatory signal 2. Signal 2 is caused by an interaction between other receptors on the T cell and other proteins on the dendritic cell. Signal 1 and signal 2 furnish a check and cross-check on the activation process.

Once a T cell receives signals 1 and 2, it begins to divide and produce a series of daughter cells. This event is called **clonal expansion.** It is a crucial step in the acquired immune response because it leads to a large population of lymphocytes capable of responding specifically to the antigen that has entered the body.

If the original T cell is $CD4^+$, the daughters differentiate into one of two types of **helper T cells.** (As we'll see, the term *helper* is appropriate because activated $CD4^+$ cells assist with the activation of other lymphocytes.) The two types of helper T cells are designated T_H1 and T_H2 and have distinctive functions.

If the T cell is $CD8^+$, however, its daughter cells develop into **cytotoxic T lymphocytes (CTLs).** One additional detail is important to note: $CD4^+$ T cells interact only with antigens presented on Class II MHC protein molecules; $CD8^+$ T cells interact only with antigens presented on Class I MHC proteins.

Helper T cells and CTLs are often referred to as **effector T cells.** Following clonal expansion and maturation, effector T cells leave the lymph node through the lymphatic ducts. Later they enter the blood and migrate to the site of infection. A key part of the acquired immune response is now under way.

B-Cell Activation and Antibody Secretion

If T cells are activated by interactions with dendritic cells that present antigens, how are B cells activated? As **Figure 46.10** shows, the answer has two parts. Receptors on B cells interact directly with the bacterial or viral antigens that are floating free in lymph or blood. Once a free antigen is bound, B cells internalize the molecule and process it via the same Class II pathway used by dendritic cells. As a result, a B cell that encounters its antigen displays the epitope on its surface, cradled in the groove of a Class II MHC protein. The second part of the activation process occurs when an activated $CD4^+$ T_H2 cell with a complementary receptor arrives. As the diagrams show, this helper cell binds to the receptor-MHC complex on the B cell. The interaction between a B cell and helper T cell supplies an initial activation signal, which is followed by a co-stimulatory signal 2 analogous to the one that occurs during T-cell activation.

After receiving signals 1 and 2, a B cell begins to divide and form daughter cells. Some of the daughters differentiate into activated B lymphocytes called **plasma cells.** Plasma cells produce large quantities of antibodies. Recall that antibodies are identical to the BCR on the cell's surface, except that antibodies lack a transmembrane domain and are secreted into the blood and lymph.

Once B cells are activated, then, the acquired immune response gains an important dimension. Antibodies specific to the invading bacterium or virus begin to circulate in the blood. When they encounter antigen, an antibody binds tightly. Antigens that are coated with antibody are marked for destruction.

Antigen Presentation by Infected Cells: A Signal for Action by $CD8^+$ T Cells

Once T cells and B cells have been activated, most of the major elements of the acquired immune response are in place. Another fundamental event in the acquired response is occurring as the receptors on $CD8^+$ T cells interact with antigens on the surface of infected host cells, however. To understand this part of the immune response, recall from Chapters 11 and 26 that viruses enter cells when they infect them.

Infected cells respond to the arrival of a virus by processing antigens from the invader. Viral antigens are attached to MHC

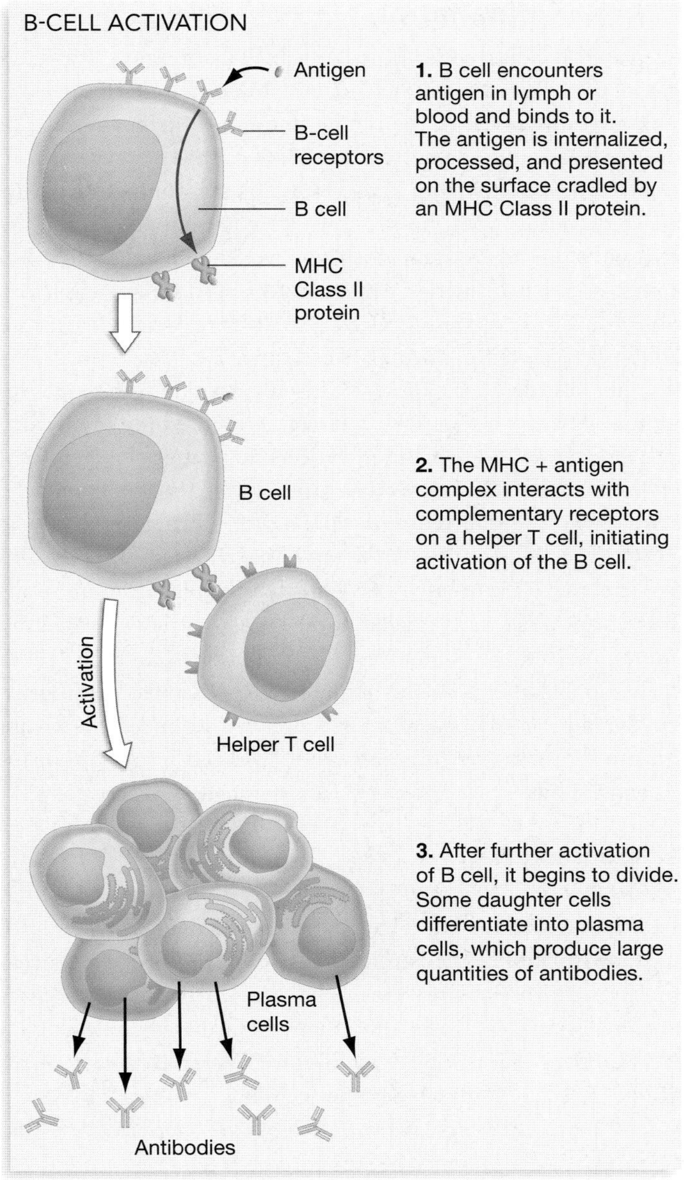

FIGURE 46.10 B-Cell Activation
A series of steps are involved in activating B cells. In addition to interactions between many MHC-antigen complexes on a B cell and complementary receptors on a helper T cell, other proteins on the surface of a B cell and T cell must interact.

Class I proteins and presented on the surface of the infected cell. Every nucleated cell in the body expresses MHC Class I proteins and has the ability to signal that it is infected. Cells that display viral antigens bound to Class I MHC molecules are effectively waving a flag that says, "I'm infected. Kill me."

How are afflicted cells disposed of before the viruses inside replicate and spread the infection? What happens to viruses and bacteria that become coated with antibodies? And finally, what events follow a successful immune response and provide protection against future infections by the same pathogen?

46.4 The Acquired Immune Response: Culmination

Activated B cells, helper T cells, and CTLs present a formidable response to invading pathogens. In combination with the leukocytes involved in the innate immune system, the cells of the acquired immune system are almost always successful in eliminating threats from bacteria, parasites, fungi, and viruses. To drive this point home, consider what happens when the acquired immune system does *not* work. For example, children who are born with a genetic defect in the enzyme responsible for DNA recombination in maturing lymphocytes are unable to generate normal T-cell and B-cell receptors. As Chapter 17 detailed, these children's acquired immune response is badly impaired as a result. Afflicted individuals suffer debilitating illness from infections that other children fight off easily, and die before they are two years old. Similarly, people who are infected with HIV suffer a progressive loss of $CD4^+$ T cells, because HIV parasitizes $CD4^+$ T cells and macrophages. Eventually, HIV-infected people succumb to illnesses that physicians almost never see in people with healthy immune systems.

To understand how the acquired response kills pathogens, let's again return to a bacterial infection in a wounded elbow and an upper respiratory tract infection sustained by the flu virus. How do activated B cells and T cells eliminate these invaders?

Killing Bacteria

During the innate response to bacteria that enter a wound, macrophages at the site phagocytose some of the invaders. In addition to killing the foreign cells, these leukocytes process and present antigens via the MHC Class II pathway. As a result, macrophages at the site of infection display epitopes on their surfaces that can be recognized by helper T cells. If an activated T_H1 cell binds to these antigen-laden macrophages, two things happen. First, the phagocytic activity of the macrophages is enhanced. Second, the T_H1 cells secrete cytokines that kill bacteria and viruses, recruit additional phagocytic cells to the site, and increase the inflammatory response.

Bacteria at the infection site also begin to be coated with antibodies from plasma cells. Bacteria that are tagged with antibodies are readily destroyed by macrophages, either by phagocytosis or by a stream of reactive oxygen intermediates released onto their cell walls. Antibodies that are bound to antigens also stimulate a group of proteins called the complement system. Complement proteins circulate in the bloodstream and assemble at antigen-antibody complexes. When the molecules activate, they punch lethal holes in the cell membranes of bacteria. Within a few days, this combination of killing mechanisms usually eliminates all invading bacteria.

Destroying Viruses

The immune system has two major ways to eliminate viruses, just as it has an array of mechanisms to dispatch bacteria. One major route involves CTLs (activated $CD8^+$ cells) and is called the cell-mediated response; the other involves antibodies and is called the humoral response. (The Latin root *humor* means "fluid." The humoral response takes place in blood and lymph; the cell-mediated response takes place at the surface of cells.) Let's take a closer look at each.

To understand how CTLs act, recall that influenza-infected cells present viral antigens on their cell surface in conjunction with MHC Class I proteins. As $CD8^+$ cells migrate into the area, they recognize and bind to these epitopes. Once binding occurs, a circle of molecules forms that holds the CTL tightly to the infected cell. A hole in the infected cell's membrane forms inside this seal, and signaling molecules from the CTL enter the cell and activate a self-destruct response (see **Figure 46.11a**). The CTL then releases and seeks out another infected cell to kill. Over time, all virus-infected cells are eliminated. In this way, the cell-mediated response limits the spread of the infection by preventing new generations of virus particles from maturing.

Even as CTLs swing into action, plasma cells are producing antibodies to viral proteins. In most cases, the most effective antibodies are those that bind to epitopes on the surface of the virus. As **Figure 46.11b** shows, two things happen when antibodies bind to the outside of a virus. First, the virus is blocked from making contact with cell membranes and entering new host cells. Second, macrophages recognize the antibody-coated particles and phagocytose them. Because of this humoral response, the number of virus particles floating free in lymph and blood is eventually reduced to zero.

Responding to Future Infections: Immunological Memory

When B cells and T cells are activated and undergo clonal expansion, their daughter cells differentiate into plasma cells, helper T cells, and cytotoxic T lymphocytes. In addition, specialized daughter cells called memory cells are produced. **Memory cells** do not participate in the initial, or primary, immune response. Instead, they provide a surveillance service after the original infection has been cleared. Memory cells circulate through blood and tissue for years or decades, ready to provide an extremely rapid response should an infection with the same antigen recur.

CD ACTIVITY 46.2 The Immune Response

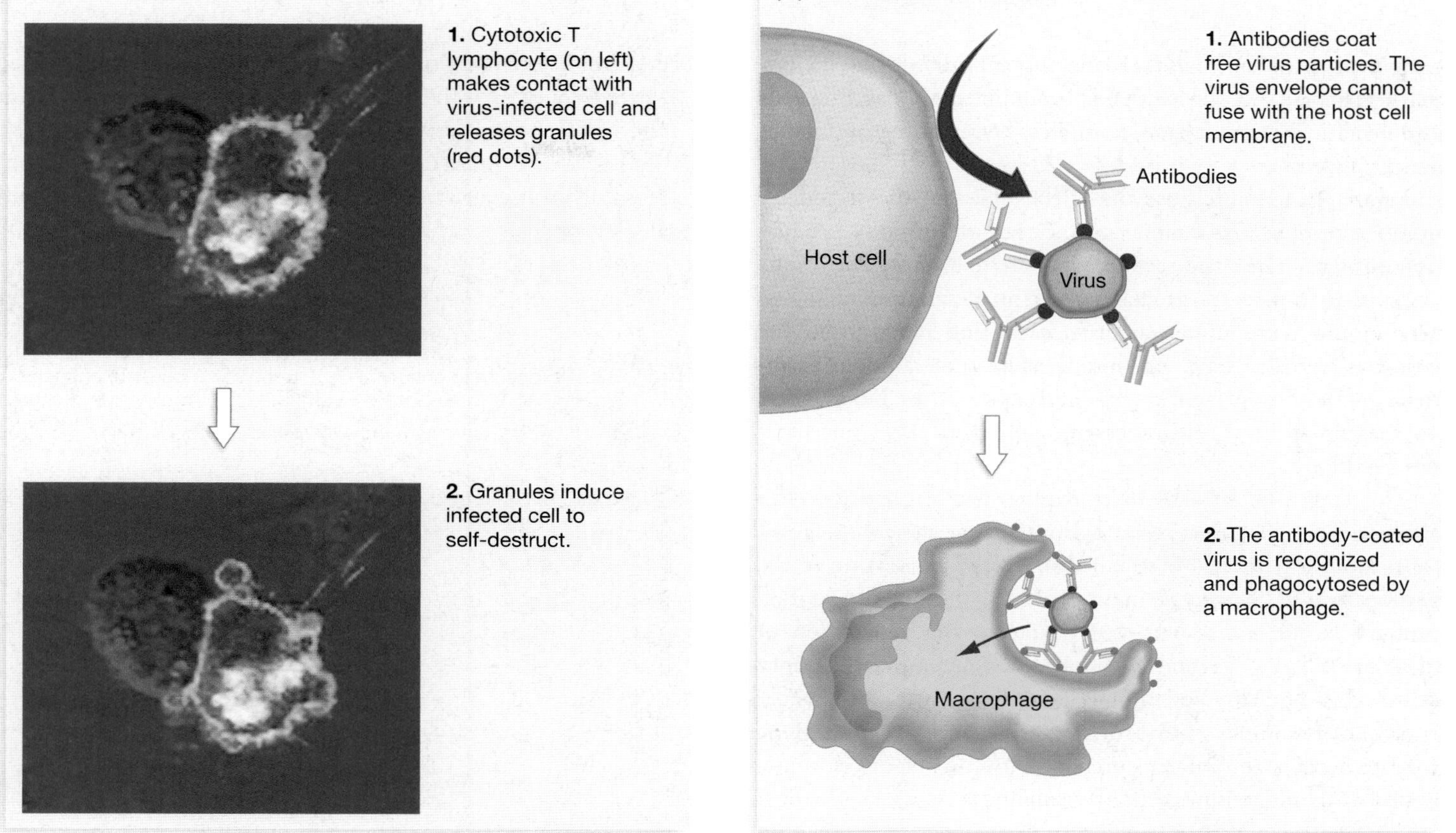

FIGURE 46.11 How Do Lymphocytes Kill Viruses?
Lymphocytes kill viruses via the cell-mediated response and the humoral response. (a) These photographs show CTLs killing a virus-infected cell. The red-stained granules are loaded with molecules that activate a self-destruct program in the infected cell. (b) After antibodies bind to intact virus particles, the particles are unable to attach to cells and begin an infection. Antibody-coated virions are readily phagocytosed by macrophages.

If the same antigen enters the body a second time, memory cells recognize the antigen and trigger a secondary acquired immune response. The secondary response is faster and more efficient than the primary response. It is faster because circulating memory T and B cells increase the likelihood that lymphocytes with the correct antigen-specific receptors will find the antigen and activate quickly. It is more efficient because some of the memory B cells that respond migrate to a specialized area in the lymph node called the germinal center. There the DNA sequences that code for the variable region of the immunoglobulin gene begin to undergo rapid mutation. As **Figure 46.12** shows, the mutations in the DNA of the variable region modify the receptors and antibodies produced by the memory cell. Memory B cells with receptors that end up binding better to the antigen live and produce daughter cells; those that bind to antigens less well die.

In effect, this process of **somatic hypermutation** leads to a fine-tuning of the immune response. The B cells that result from somatic hypermutation produce antibodies that bind to the

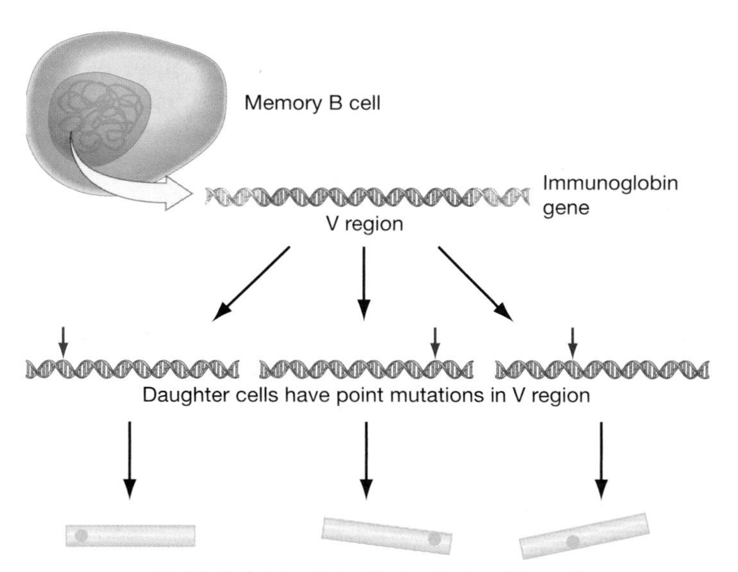

FIGURE 46.12 Somatic Hypermutation
Somatic hypermutation occurs as memory B cells divide in response to an influx of antigen. Because the variable regions of L-chain and H-chain DNA tend to contain point mutations, the antibodies produced by each daughter cell have a slightly different amino acid sequences. Some of these novel antibodies will bind more tightly to the antigen than the original antibody did.

antigen more tightly than their ancestor cells did during the primary response. As the secondary immune response proceeds and somatic hypermutation continues, better-fitting antibodies are produced.

Figure 46.13 underlines the effectiveness of the secondary immune response by comparing the rate of antibody production during the first and second exposure to a virus. The results are similar to data researchers collect by inoculating a laboratory mouse with influenza virus, collecting blood from the mouse every other day, and then measuring the amount of antiviral antibodies present in the fluid phase of the blood. (**Box 46.3** explains how researchers measure antibody concentrations in blood.)

The data in Figure 46.13 also explain why vaccination is an effective defense against certain viruses. A **vaccine** contains epitopes from a pathogen or a killed or weakened version of the pathogen itself. After vaccination occurs, the body mounts a primary immune response that results in the production of memory cells. If an actual infection occurs later, these memory cells respond quickly and eliminate the threat before illness occurs. For example, Edward Jenner's vaccination strategy worked because the antigens presented by cowpox are extremely similar to antigens presented by smallpox. As a result, exposing people to cowpox virus elicited the production of memory cells that effectively thwarted future infections by smallpox virus.

Unfortunately, viruses such as influenza, HIV, and the rhinoviruses that cause colds mutate so rapidly that they present the immune system with a constantly changing array of epitopes. Memory cells that were effective during a previous infection by these viruses are unlikely to respond to a later infection. As a result, it is extremely difficult for biologists to design an effective vaccine for these agents. Currently, the only "cure" for HIV is prevention; the most effective treatment for the common cold is rest and nourishment. To keep up with rapidly mutating flu populations, flu vaccines must be redesigned and administered every year.

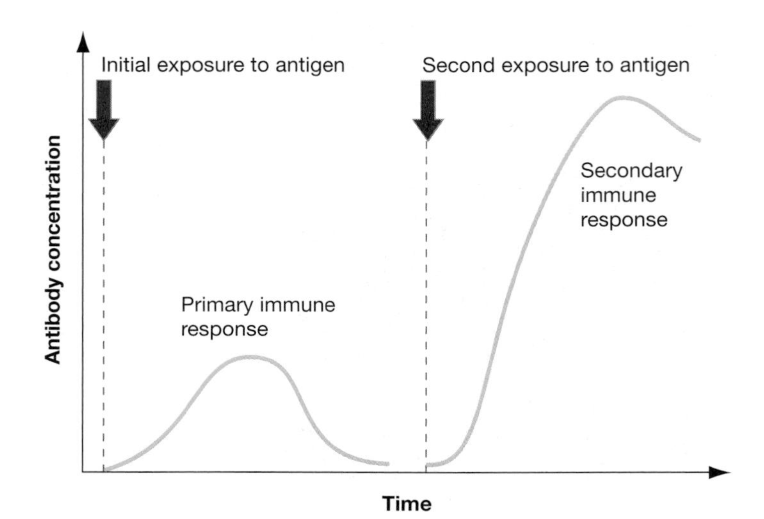

FIGURE 46.13 The Secondary Immune Response Is Faster and Stronger Than the Primary Response
These data show the types of results that biologists get when they inject a mouse with an antigen, document changes in antibody concentration over time, then inject the same individual with the same antigen later. QUESTION In terms of the data presented here, why is vaccination effective?

BOX 46.3 The ELISA Test

One of the common assays used to measure antibody concentrations in blood is called an indirect or sandwich ELISA (enzyme-linked immunosorbent assay). The standard screening test for HIV infection, for example, is an ELISA for antibodies that the immune system produces in response to this virus. The diagram in **Figure 1** illustrates how you might use an ELISA to determine the amount of antibody present in a mouse that has been infected with influenza.

To begin an ELISA, each well in a plate containing many wells is coated with antigen. In our example, the antigen is purified hemagglutinin protein from the influenza virus. In the next step, a defined amount of plasma from an infected mouse is added. (Recall, from Chapter 41, that plasma is the liquid or non-cell fraction of the blood.) If antibodies are present, they bind to the antigen. To continue the assay, a secondary antibody is added to the mixture of antigen and plasma. A secondary antibody is one that attaches to an antibody. In ELISA tests, the secondary antibody that is used contains an enzyme such as alkaline phosphatase. Alkaline phosphatase is useful because it catalyzes a reaction with a substrate, and because the product of the reaction changes the color of the solution. As a result, the presence of color in a well indicates a positive reaction—meaning that antibody to hemagglutinin is present.

To determine the concentration of antibody, researchers make a series of dilutions of the plasma. The most diluted solution that gives a positive response is then recorded as the antibody titer, meaning the relative amount of antibody produced.

(Continued on next page)

(Box 46.3 continued)

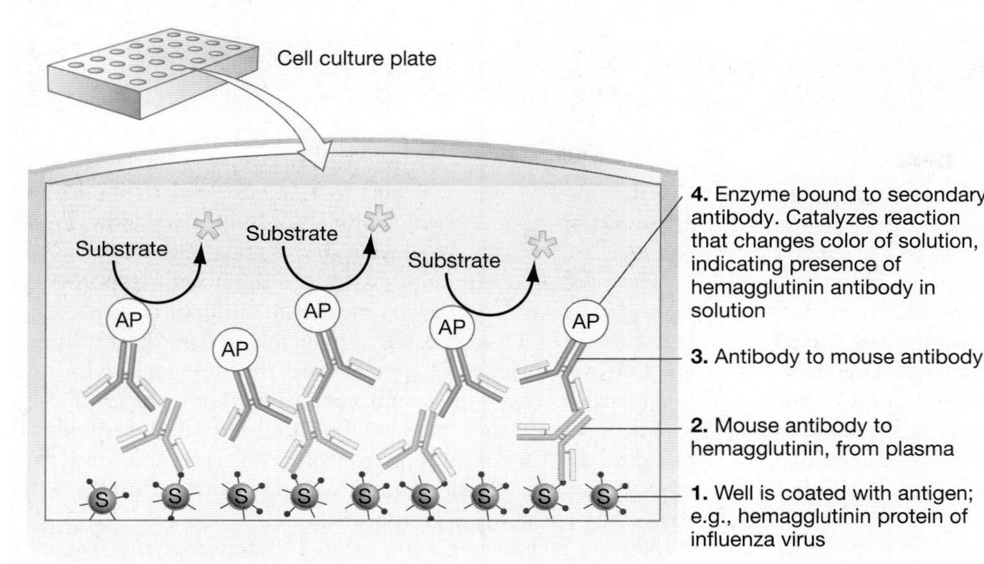

FIGURE 1 How Does the ELISA Test Work?
This diagram shows the relationship between the antigen, antibody, and secondary antibody-test enzyme complex in an ELISA test. The three components are added sequentially. The sample is washed after addition of the mouse antibody to remove any protein that is not bound to the antigen. After the secondary antibody and test enzyme are added to a well, the sample is washed again to remove any free-floating secondary antibody. Then the substrate is added and the resulting color observed. As a result, the amount of color produced is an index of the amount of mouse antibody present.

Essay Allergies

For some people, walking through a field of ragweed or petting a cat is a prescription for developing a runny nose, itchy eyes, and labored breathing. For other people, being stung by a bee or eating a peanut is a life-threatening experience. Why are some people allergic to certain molecules?

An allergic reaction is an abnormal response to an antigen. For reasons that are still not clear, certain people produce the IgE class of antibodies in response to specific molecules found in cat dander, nuts, plant pollen, or other products. Molecules that trigger this response are called allergens instead of antigens. In people who are not allergic, IgE antibodies are produced only in response to infections by worms.

What happens during an allergic reaction?

The presence of IgE antibodies is important because it triggers a series of events known as the hypersensitive reaction. When a person is first exposed to an allergen, receptors on leukocytes called mast cells and basophils bind to the IgE antibodies that are produced. Once this binding event occurs, the cells (and the person) are said to be sensitized. If the person is later exposed to the allergen, the sensitized cells rapidly produce large quantities of histamine, cytokines, chemokines, and other compounds. In response to these molecules, blood vessels dilate and become more permeable, smooth-muscle cells contract, and mucus-producing cells secrete.

If the immune response is relatively mild, symptoms like runny nose, watery eyes, and mild wheezing develop. The syndrome known as hay fever, for example, is a common allergic response to the allergens in grass or ragweed or birch-tree pollen. In other cases, tissues may respond to an allergen by reddening, swelling, and itching. (These symptoms are called hives.) If the smooth-muscle constrictions triggered by the hypersensitive response are localized to the respiratory passages and lead to constriction of the airway, the condition known as asthma develops.

If the immune response is more severe, however, blood vessels can dilate to the point where blood pressure plummets and the person loses consciousness. Severe smooth-muscle contractions can induce vomiting, diarrhea, and complete constriction of the respiratory passages. This combination of events is lethal, and is known as anaphylactic shock. The most effective treatment for anaphylactic shock is to inject the affected individual with epinephrine. Because epinephrine is a hormone that relaxes smooth muscle, it opens up blood vessels and counteracts the effects of histamine and other agents of the hypersensitive response. People who are hypersensitive to bee stings or other allergens routinely carry small syringes, called epi-pens, filled with epinephrine.

What can be done about allergies? The number one approach is to avoid the allergen. If exposure does occur, drugs called antihistamines can sometimes reduce histamine release from mast cells effectively; in some cases corticosteroids applied to skin can reduce swelling. Longer-term solutions are still in the experimental stage, however. Developing effective treatments for allergies is an important research frontier in biomedicine.

Chapter Review

Summary

The immune system employs two major responses to invading pathogens. The innate immune system responds in the same way to all antigens, but the acquired immune system mounts a highly targeted response to specific antigens.

During the innate immune system's inflammatory response, mast cells release chemical messengers that result in increased blood flow to wounds or areas of tissue damage. After neutrophils migrate to the site, they respond to bacteria that stimulate their pattern-recognition receptors by releasing toxic molecules and phagocytosing the invading cells. Macrophages also phagocytose pathogens and release chemical messengers that raise body temperature.

The acquired immune response begins when T-cell receptors and B-cell receptors recognize an epitope on an antigen. The receptor protein on the surface of a T cell or B cell is generated through a rearrangement of gene segments called DNA recombination. Because every receptor that results from this process is slightly different, the immune system is able to recognize and respond to an enormous array of antigens. **Table 46.2** summarizes how T cells and B cells are activated and what the resulting cells do.

The acquired immune system's humoral response to pathogens is based on tagging them with antibodies. Viruses that become coated with antibodies are unable to enter host cells and are destroyed by macrophages. Bacteria that are tagged with antibodies are destroyed by complement proteins or by macrophages.

The acquired immune system's cell-mediated response is based on destruction of infected cells. Host cells that are infected with a virus display antigen on their surface and are destroyed by cytotoxic T lymphocytes.

Because memory cells are created as activated B cells and T cells undergo clonal expansion, the immune system is able to respond rapidly and effectively to future infections by the same pathogen. Vaccines trigger the production of memory cells because they contain epitopes from pathogens.

TABLE 46.2 A Summary of the Acquired Immune System

Type of Lymphocyte	How Activated?	What Cells Result from Activation and Clonal Expansion?	What Do These Cells Do?
B cell	Receptor binds to free antigen, then interacts with T_H2 cell.	Plasma cells Memory cells	Secrete antibodies Participate in secondary response
$CD4^+$ T cell	Receptor binds to antigen/MHC Class II protein complex on dendritic cell.	T_H1 (helper T cell)	Activate CTLs, regulate inflammatory response
		T_H2 (helper T cell) Memory cells	Activate B cells Participate in secondary response
$CD8^+$ T cell	Receptor binds to antigen/MHC Class I protein complex on dendritic cell. May also interact with T_H1 cell.	Cytotoxic T lymphocytes (CTLs) Memory cells	Kill infected host cells Participate in secondary response

Questions

Content Review

1. *Innate* means "inborn or untrained." Why is the innate immune response appropriately named?
   a. It occurs throughout life.
   b. It depends primarily on barriers to pathogen entry, such as skin and mucous linings in tissues.
   c. It does not require dramatic changes in immune-system cells, as the acquired response does.
   d. It does not improve in response to medicine or other external treatments.

2. Which of the following events are involved in the inflammatory response? (Select all that apply.)
   a. CTLs kill infected host cells.
   b. Neutrophils phagocytose pathogens that stimulate their pattern-specific receptors.
   c. Mast cells secrete chemical messengers that lead to increased blood flow.
   d. B cells are stimulated to divide and undergo clonal expansion.
   e. Macrophages release chemical messengers that lead to increased body temperature.
   f. Plasma cells begin to secrete antibodies.

3. What is the difference between an epitope and an antigen?
   a. An epitope is any foreign substance; an antigen is a foreign protein.
   b. An epitope is a part of an antigen where an antibody or lymphocyte receptor binds.
   c. An antigen is a part of an epitope where an antibody or lymphocyte receptor binds.
   d. Antigens are recognized by B cells and antibodies; epitopes are recognized by T cells.

4. How do B-cell receptors and antibodies differ?
   a. B-cell receptors are made up of heavy chains; antibodies are made up of light chains.
   b. B-cell receptors are made up of light chains; antibodies are made up of heavy chains.
   c. Only antibodies include a variable region.
   d. Antibodies lack a transmembrane domain and are secreted.

5. How do memory cells become activated?
   a. They undergo somatic hypermutation.
   b. They encounter antigen.
   c. They undergo clonal expansion.
   d. They are stimulated by cytokines and chemokines.

6. In terms of their function, T cells come in two major types. How are they distinguished?
   a. They have a CD4 or CD8 protein on their surface.
   b. They have extensive ER and many mitochondria.
   c. Only one type has pattern-recognition receptors.
   d. One type functions in the primary response; the other in the secondary response.

Conceptual Review

1. To a physician, the classical signs of the inflammatory response are ruber (reddening), calor (heat), dolor (pain), and tumor (swelling). Explain why each symptom occurs. Be sure to mention at least two of the leukocytes and two of the chemical messengers that are involved in the inflammatory response.
2. Sketch the general structure of a B-cell receptor. Explain how this protein interacts with an antigen.
3. Why does vaccination work? Why don't we have vaccines for HIV or cold viruses? Why do people need to get a new flu shot every year?
4. Summarize the clonal-selection theory of acquired immune-system function. How was the theory tested and verified?
5. Explain why helper T cells and cytotoxic T cells are aptly named.
6. Explain how DNA recombination leads to the production of almost limitless numbers of different B-cell receptors, T-cell receptors, and antibodies.

Applying Ideas

1. If you were being treated for a badly skinned knee from a bicycle accident, health care workers would irrigate the wound with large volumes of sterile water and scrub it as thoroughly as possible with warm, soapy water. Explain why these measures are effective, in the context of how the innate and acquired immune responses operate.
2. The signals that are required to activate B cells and T cells are exceedingly complex. Some researchers maintain that the requirement for multiple signals precludes lymphocytes from activating in response to antigens on the surface of self-cells. For example, suppose that a B cell and a T cell in your body each have a receptor that recognizes a protein found in your heart, and that these cells are mistakenly released from the bone marrow and thymus. Summarize the signals that are needed to activate the B cell and T cell. Explain why the lymphocytes are likely or unlikely to actually attack your heart.
3. According to the text, it is astonishing that the immune system can produce antibodies to compounds that were recently synthesized for the first time in the lab. But because viruses and other pathogens are constantly undergoing mutation and producing new antigens over time, this observation should not be surprising at all. Explain.
4. During World War II, physicians discovered that if they removed undamaged skin from a patient and grafted it onto the site of a burn on the same patient, the tissue healed well. But if the donated tissue came from a different individual, the graft was rejected. Explain why non-self tissues are killed by the immune system.

CD-ROM and Web Connection

CD Activity 46.1: The Inflammatory Response *(animation)*
(Estimated time for completion = 5 min)
How does the inflammatory response repair a wound and slow the spread of pathogens?

CD Activity 46.2: The Immune Response *(animation)*
(Estimated time for completion = 10 min)
How does the immune response fight infections by specific pathogens?

At your **Companion Website** (http://www.prenhall.com/freeman/biology), you will find self-grading exams and links to the following research tools, online resources, and activities:

Ancient "Jumping DNA" and the Human Immune System
Explore research on how the transposable genetic elements introduced in Chapter 16 might have been responsible for the origin of genetic recombination in the antibody genes.

Immunizations
This site describes this process of immunization, and tackles some basic misconceptions about the process.

Lymphocyte Tracking
Read an article that describes how lymphocytes move in a directed way.

Cytokines
This site describes the structure and molecular biology of many cytokines found in the immune response.

Additional Reading

Beck, G., and G. S. Habicht. 1996. Immunity and the invertebrates. *Scientific American* 275 (November): 60–66. An introduction to "comparative immunology," or how invertebrate animals fight disease.

Gulbins, E., and F. Lang. 2001. Pathogens, host-cell invasion, and disease. *American Scientist* 89: 406–413. Explores recent research on how pathogens and host cells interact.

Litman, G. W. 1996. Sharks and the origins of vertebrate immunity. *Scientific American* 275 (November): 67–71. A review of the humoral immune system in sharks.

Ecology

Ecology is the study of how organisms interact with their environment. Because humans are drastically altering most environments, ecological research has an increasingly important applied component.

Ecological research can be organized into levels in a hierarchy, with each category representing a different way that organisms interact with their environment. For example, **Chapter 47: Behavior** considers how animals interact with members of their own species during one-on-one interactions. **Chapter 48: Population Dynamics** focuses on how the groups of individuals called populations change through time. **Chapter 49: Species Interactions** investigates how predation, competition, and other types of relationships between species affect the organisms involved.

Above the species level, ecologists routinely distinguish two categories of research. **Chapter 50: Community Ecology** asks how events like an environmental disturbance affect groups of co-existing species, and how factors such as species diversity influence their future. **Chapter 51: Ecosystem Ecology** probes how the species in biological communities interact with their physical or abiotic environment, with a special emphasis on energy flow and nutrient cycling.

In this unit, then, the levels of ecological analysis are arranged from small to big—from individual to population, species, community, and ecosystem. Taken together, these first five chapters lay the groundwork for delving into applied ecology. **Chapter 52: Biodiversity and Conservation** serves as the capstone of this unit and text.

The black-footed ferret is an endangered species. It preys only on prairie dogs, which have been exterminated from most of their former range. ©1996 Susan Middleton and David Liittschwager

Behavior

47

At its most fundamental level, an organism's behavior is based on receiving information about conditions in the environment and acting in response to that information. Viewed in this context, most organisms behave. Plants sense light and gravity and move or grow in response; many can also sense predators and release defense compounds in response. Numerous species of bacteria perceive changes in light, magnetic fields, or chemical concentrations and react by swimming to or from the stimulus. There are even predatory fungi that sense the presence of roundworms and then capture and eat them. In this chapter, however, we limit ourselves to examining the behavior from a single part of the tree of life: the animals.

Biologists analyze how animals behave at two levels. At the proximate, or mechanistic level, researchers ask how actions occur in terms of the neurological, hormonal, and skeletal-muscular mechanisms that cause the behavior. At the ultimate, or evolutionary level, investigators ask why actions occur in terms of their effect on reproductive success. Is a particular behavior adaptive—meaning that it increases an individual's ability to survive and reproduce in a particular environment? Analyses at the ultimate level also seek to explain the evolutionary origins of particular behavior patterns and analyze how they have changed through time.

As an example of proximate and ultimate causation, let's consider singing by male birds. As section 47.2 shows, researchers study birdsong at the proximate level by asking how hormones, neurons, and muscles interact to produce a vocalization. An experiment at the proximate level might involve manipulating the concentration of the sex hormones testosterone and estrogen and then analyzing whether the areas of the brain responsible for

This mated pair of red-crowned cranes is performing a courtship display named the unison call. In addition to exploring the genetic and cellular-level underpinnings of behavior, this chapter investigates how the unison call and other types of behavior evolve.

song change in response. At this level, an explanation for birdsong runs as follows. Sex hormones are released in response to environmental cues that indicate the onset of the breeding season. These hormones cause certain areas of the brain to grow; when neurons in these song control regions fire, they stimulate muscles around an organ called the syrinx. When these muscles respond, the syrinx vibrates. Song results.

At the ultimate level, research takes a different tack. As section 47.4 details, the function of a bird's song is to defend a territory for breeding and to attract a mate. A breeding territory contains food that birds need to raise offspring. Research at the ultimate level might look for correlations between the size of a male's song repertoire and his success in attracting mates, or involve adding food to a territory and documenting whether the resident male's singing rate increases in response. At this level, an explanation for birdsong runs as follows. Defending a reliable food supply and attracting a mate, through the use of the signal called song, is critical for successful reproduction. Birds that cannot or do not sing produce few offspring.

The important point to grasp about proximate and ultimate levels of analysis is that they are complementary. Studies at the proximate level explain how the behavior occurs; research at the ultimate level explains why. The goal of this chapter is to introduce how biologists study behavior at both levels.

To begin, let's explore some research on how genes affect behavior. Genes provide the link between proximate mechanisms and ultimate function. They code for the molecules involved in proximate mechanisms, and the frequency of certain alleles changes over time in response to mutation, natural selection, and other evolutionary processes.

47.1 The Role of Genes

Until the 1970s and 1980s, the genes that influence animal behavior represented a "black box." Biologists assumed that at least some behavioral traits were influenced by particular genes and alleles in addition to being affected by environmental conditions, but they knew almost nothing about the extent of this influence. Experimental and analytical tools for testing the assumption were simply lacking.

The situation is changing dramatically, however. Powerful techniques are now available to investigate behavioral genetics. This section begins by introducing two approaches for analyzing the genetic basis of behavior, and closes by investigating how genotypes and environmental conditions interact to produce behavior.

Are Behavioral Traits Influenced by Genes?

Does some of the variation in behavior observed among individuals result from variation in their genotypes? To introduce how biologists answer this question, let's consider alcohol abuse in humans. Alcoholism is a behavioral disorder that tends to run in families. As a result, physicians have long suspected that it has a significant genetic component. But a plausible alternative hypothesis exists. Alcoholism could run in families not because family members share a genetic predisposition to the behavior, but simply because they share an environment that promotes alcohol abuse. How can researchers distinguish the two?

In an attempt to tease apart the effects of heredity and environment, Robert Cloninger and colleagues studied the children of alcoholics and of nonalcoholics. The researchers focused on a large group of Swedish men who had been adopted by nonrelatives at a very early age. The genetic parents of these men included both alcoholics and nonalcoholics. The boys had been assigned to their foster homes randomly, meaning without respect to either their biological or foster parents' behavior. Also, the vast majority of the boys had been placed as newborns or infants. Presumably, this was before their parental environment could exert a strong influence on their behavior.

In effect, these adoptive placements created a natural experiment. An experiment like this—where young are raised by nonrelatives in randomly assigned environments—is called a cross-fostering experiment. Cross-fostering experiments are routinely performed on nonhuman animals. The logic is simple. If the cross-fostered offspring show the same type of behavior as their genetic parents, then the trait has a strong genetic basis; if offspring behave more like their foster parents, then the trait has a strong environmental basis. Here is the question that Cloninger and co-workers asked: Were the biological children of alcoholic parents more likely to become alcohol abusers than the biological children of nonalcoholic parents?

The data that answer this question are summarized in **Figure 47.1a** (page 916). Boys who had alcoholic biological fathers were almost nine times as likely to abuse alcohol as were boys whose biological fathers were nonalcoholics. This is strong evidence that certain alleles create a predisposition to alcoholism. Cloninger and co-workers realized, however, that this effect occurred in one type of alcoholism only. They termed this behavior teen-onset alcohol abuse.

A strikingly different pattern occurred in boys who did not develop alcoholism until they became adults. As the data in **Figure 47.1b** show, boys experienced an increased frequency of this syndrome only when alcohol abuse was also favored by their foster environment; that is, if their adoptive parents abused alcohol, too.

In short, the study revealed that there are two distinct types of alcoholism, and that the interaction between genetic background and environment is different in these two types. In teen-onset alcoholism, the expression of the trait is largely independent of whether alcohol abuse occurs in the home environment. In contrast, development of adult-onset alcoholism is heavily influenced by alcohol abuse at home.

Can We Find Specific Genes or Alleles Responsible for Certain Behavior Patterns?

Cross-fostering experiments can establish that a particular behavior has a genetic basis. Once this has been confirmed, researchers turn their attention to the next level of analysis, which is to locate and analyze the specific genes and alleles involved. When these have been identified and their mode of action has been explained, the genetic basis of behavior is largely understood. The black box is open.

In classical genetics, researchers begin a study by identifying a trait of interest. Then they search for the genes responsible for the trait by finding a mutant that does not show the trait. By carrying out the types of breeding experiments analyzed in Chapter 17, researchers can map the physical location of the mutant locus. Once the locus is pinpointed, investigators attempt to sequence the gene, produce the protein, and perform experiments focused on explaining how the gene product influences the phenotype. This approach is called **forward genetics** (**Figure 47.2**). The researcher starts with a normal phenotype and a mutant phenotype and attempts to determine the underlying genotype. (**Box 47.1**, page 918, explains how a newer approach, called **reverse genetics**, is used in behavior genetics. In reverse genetics, investigators start with a mutant gene and attempt to determine how it affects the phenotype.)

Normally, it is extremely difficult to do forward genetics with behavioral traits. This is because in most cases, dozens of different genes influence the expression of a behavior. For example, there is no single gene for behavioral disorders such as alcoholism. Instead, many genes are involved in the illness. The products of these genes might affect the development of the brain, the behavior of individual neurons, the output of sensory organs, and so on. When many genes are involved, and when each has a small effect on the trait, it is difficult to map the loci involved.

Marla Sokolowski, however, was fortunate enough to discover an important behavioral trait controlled by a single gene. Consequently, she was able to use a forward genetics approach to uncover what may be the best example to date of a gene that influences behavior.

As an undergraduate research assistant, Sokolowski noticed that some of the fruit-fly larvae she was studying tended to move several centimeters after feeding in a particular location, while others tended to remain in place. She reared the two types of larvae separately and confirmed that as adults, the two forms also acted as either "rovers" or "sitters." Further, the same behavior was found in the wild. Sokolowski studied a population of fruit flies living in an apple orchard and found that rovers made up 70 percent of the population while sitters comprised 30 percent.

By breeding rovers and sitters that possessed other distinct genetic markers, Sokolowski and colleagues were able to map the gene responsible for the behavior. (Genetic mapping was explained in detail in Chapter 17.) The researchers named the locus *foraging* (*for*). Rovers have the for^R allele, while sitters have the for^S allele; heterozygotes act like rovers.

Exactly which gene is at this locus? Fortunately, the *Drosophila* genome has been sequenced. The locus called *for* happened to map at or near a previously identified gene called *dg2*. The protein product of *dg2* is a protein kinase that is acti-

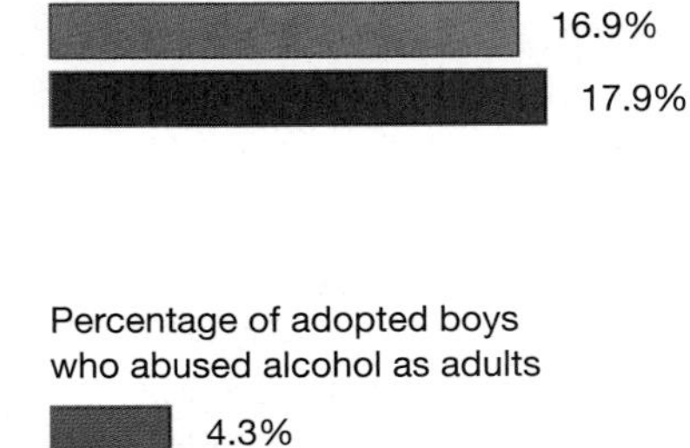
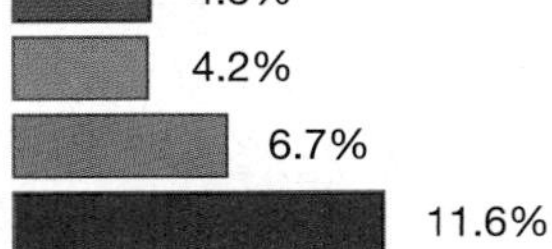

FIGURE 47.1 Analyzing Genetic Predispositions to Alcoholism
These data are for adopted boys who were placed in foster homes shortly after birth. **(a)** Teen-onset alcoholism was most likely to develop in boys whose biological parents were alcoholics, regardless of whether their foster parents abused alcohol. **(b)** Adult-onset alcoholism was more likely to develop only if boys had biological parents who abused alcohol and if their foster parents also abused alcohol. (A statistical test indicated that the observed differences between the first three categories probably arose by chance.)

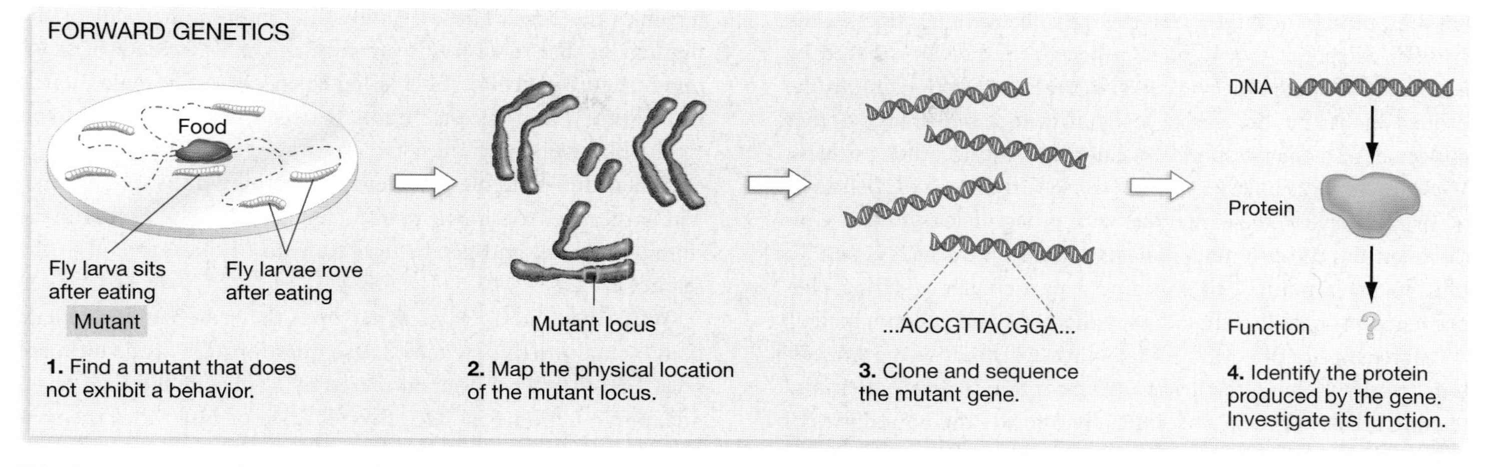

FIGURE 47.2 Forward Genetics
Forward genetics is based on identifying, analyzing, and mapping mutants. In effect, a researcher works from phenotype to genotype.

vated by cyclic guanosine monophosphate (cGMP). As several earlier chapters pointed out, protein kinases activate or deactivate certain proteins by phosphorylating them. The change in the protein often leads to the transcription or repression of certain genes and a dramatic change in the cell's activity.

Are *for* and *dg2* actually identical? To answer this question, Sokolowski and co-workers inserted extra copies of the *dg2* gene into eggs that were homozygous for the for^S allele (**Figure 47.3**). The control group in the experiment, which had the same sitter genotype, did not receive the extra copies of *dg2*. These eggs developed into sitters. But the transformed eggs hatched into larvae and then metamorphosed into adults that acted like rovers. This is convincing evidence that *for* and *dg2* are in fact the same gene. Follow-up experiments confirmed that the for^R allele produces more of the protein kinase than the for^S allele.

In section 47.3 we'll consider how roving and sitting affect the ability of flies to survive and reproduce in different habitats. In the meantime, let's take a closer look at how the genes involved in behavioral traits interact with various aspects of the environment.

Interactions Between Genes and the Environment

A common misconception about the genetic basis of behavior is to assume that traits can be "genetically determined." This is almost never the case. In the rover/sitter example just examined, the associations between certain alleles and certain types of behavior are statistical in nature—not deterministic. Fruit flies with the for^R allele tend to move more after feeding, on average, than fruit flies with the for^S allele. In this and most other cases, behavioral responses are not an either-or proposition.

The message here is that certain alleles create a disposition for a certain behavior, but they do not dictate or determine it. There are several reasons for this. To begin with, genes and alleles must be expressed. Their expression depends on regulatory molecules—often hormones or other developmental signals. The

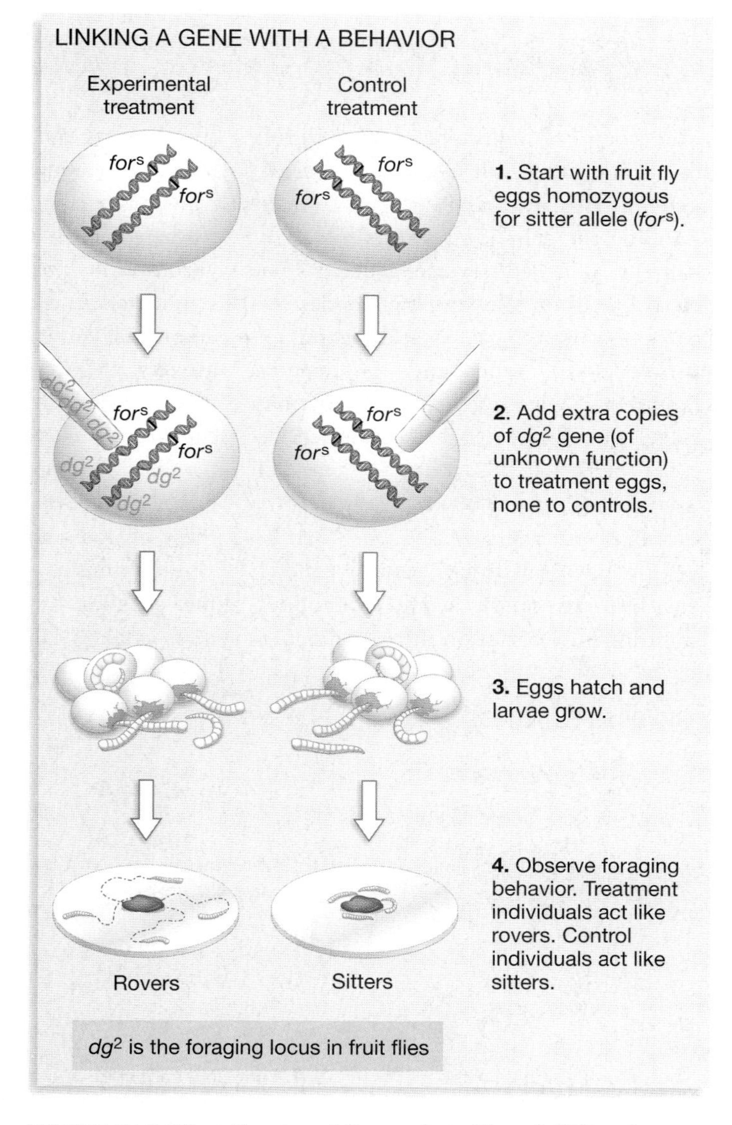

FIGURE 47.3 What Convinced Researchers That *dg2* Was the *Foraging* Locus in Fruit Flies?
Researchers were able to produce the rover phenotype by injecting fly eggs with extra copies of the *dg2* gene. QUESTION In the second step, why did the researchers inject the eggs in the control treatment?

amount, timing, or duration of the signal has a huge effect on the amount of protein produced. Further, the proteins produced by genes interact with other proteins. In the rover/sitter example, the protein kinase has to receive a signal from a hormone or other molecule, and then respond by phosphorylating other proteins. Variation in these interactions will cause variation in the behavior.

In short, every allele interacts with other alleles within the individual. But the external environment also influences genes. To illustrate this point, let's examine a particularly startling phenomenon: female fish that change their sex and become male.

Many species of coral-reef fish have a distinctive mating system. Males defend territories that contain nesting sites and feeding areas. A group of females lives inside the boundaries of this territory. When these females spawn (lay eggs), the male fertilizes the eggs. Thus, a single male monopolizes all of the matings in that territory. Invariably, this male is the largest individual in the group. This is logical because the male guards the territory, and because fights between fish are usually won by the biggest contestant.

When this male dies, however, something unusual happens. The largest female in the group changes sex. She becomes the dominant male, and begins fertilizing all of the eggs laid in the territory.

Why does she do this? What does the sex-changing female gain by making this switch? These questions are at the ultimate level, and they have been answered by Michael Ghiselin's "size-advantage hypothesis" (see **Box 47.2**). But how does the sex-switching female switch from an egg producer to a sperm pro-

BOX 47.1 Using Reverse Genetics to Study Behavior

The growth of molecular genetics during the 1970s and 1980s inspired an increasingly popular research strategy for understanding the genetic basis of traits. This strategy is called reverse genetics. The tactic hinges on knowing the identity of a particular protein and then designing experiments to understand its function (**Figure 1**). Reverse genetics is sometimes called the candidate gene approach because the first step is to find the gene that is hypothesized to be responsible for a certain phenotype.

In behavioral studies, reverse genetics has been especially useful for understanding how variation in certain neurotransmitters and receptors affects behavior. For example, consider the neurotransmitter called serotonin that was introduced in Chapter 42. In humans, variation in serotonin function has been implicated in alcohol abuse and in the emotional states called depression and anxiety. (Many common antidepressant medications act on serotonin or its receptors.) At least 14 different receptor proteins bind serotonin and mediate its effects, however. How do variations in these proteins affect behavior?

Lora Heisler and co-workers answered this question for a serotonin receptor in mice called 5-HT$_{1A}$. The researchers disrupted the DNA sequences for this receptor. Mice that were homozygous for this mutation had no functioning copies of the 5-HT$_{1A}$ protein. When these individuals were tested in mazes, open-field environments, or cages with novel objects, they were much more likely to cower, hide, or show other anxiety-like behavior than normal mice. The result supports the hypothesis that this receptor is involved in modulating specific types of behavior. The experiment also illustrates the value of reverse genetics as a research strategy in behavioral genetics.

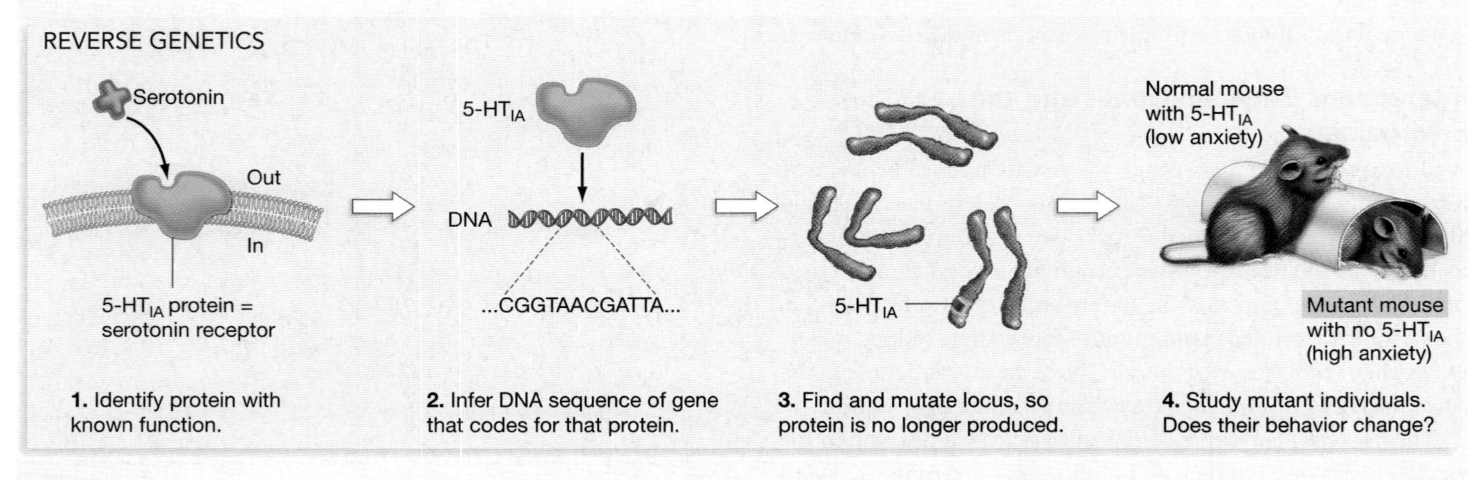

FIGURE 1
Reverse genetics is also called the candidate gene approach. In effect, a researcher works from genotype to phenotype.

ducer? How does her behavior change from submissive to aggressive? These questions are at the proximate level, and they are still unanswered.

Sex change in fish illustrates an important point. Organisms frequently have the genetic "tools" to act in several different ways, but their physical or social environment dictates which behavior they exhibit. Sex-switching fish do not change their genotype, just their physiology and behavior. This is a classic case of condition-dependent behavior. Now let's consider how the genes that influence behavior act, via nerve cells and hormonal signals.

47.2 How Animals Act: Neural and Hormonal Control

If genes create the potential for a certain behavior, how is this potential realized? Answering this question is the essence of the proximate approach to behavior. Because the nervous and endocrine systems initiate and modify behavior, answering proximate-level questions involves analyzing neuronal connections and hormone signaling pathways. Here we examine interactions among neurons, hormones, and behavior.

Birdsong

In 1981 Francisco Nottebohm shocked the scientific world by suggesting that each spring, adult male canaries grow new neurons in the brain centers responsible for producing song. Canaries sing in the spring during the breeding season, but they do not sing in the fall or winter. Nottebohm wanted to know whether these changes in song behavior might be correlated with changes in the size or shape of the brain's song centers. By examining and comparing the song centers in the brains of male canaries in spring and fall, Nottebohm demonstrated that dramatic changes occur. The song centers are much larger in spring than in fall. He proposed that neurons in these areas die each fall, and are replaced by new neurons generated each spring.

Nottebohm's hypothesis caused a sensation, for two reasons. First, if his hypothesis proved to be correct, similar neuron-generating areas might be found in humans and offer an avenue for treating patients with stroke damage or spinal cord injuries. Second, the idea contradicted accepted dogma, which was that the brains of adult vertebrates make few, if any, neurons.

Within three years, however, the hypothesis was confirmed in spectacular fashion. To establish that new neurons do indeed form, Steven Goldman and Nottebohm exploited a finding made by Mark Gurney and Masakazu Konishi. Gurney and Konishi had shown that when female birds are given injections of testosterone, the song centers in their brains enlarge and they begin to sing. Goldman and Nottebohm also injected females with testosterone, but they added radioactively labeled molecules of the deoxyribonucleic acid, thymidine. Because thymidine is incorporated into DNA when it is synthesized, these radioactive bases would specifically label the DNA in new nerve

CD ACTIVITY 47.1 Proximate Causes of Behavior

BOX 47.2 Conditional Strategies and the Nature/Nurture Debate

The size-advantage hypothesis, developed by Michael Ghiselin, states that fish living in a harem breeding system, dominated by a single male, should switch from female to male when they become very large. (Note that fish have indeterminate growth, which means that they continue growing throughout their lives.) To understand why this switch should occur, suppose that a small female can lay 10 eggs a year, while a large female can lay 20 eggs a year. If six small females and two large females live in a harem, the male that owns the territory thus fertilizes 100 eggs each year. If the male dies, the largest female can increase the number of offspring she produces each year from 20 to 80 by changing sex and taking over the role of dominant male. Alleles that allow females to do this will increase rapidly in the population. It does not pay for smaller females to switch sex, though, because they would be defeated in fights and have 0 offspring per year instead of 10.

Biologists refer to this type of behavior as a conditional strategy. The term is apt because the individuals involved are capable of behaving in several different ways. The behavior that they adopt depends on conditions—in this case, on the size and sex of other individuals in their social group. The critical point here is that individuals usually adopt the behavior that allows them to produce the most surviving offspring. This is an issue we return to in section 47.3.

Conditional strategies are also noteworthy because they shed light on an old controversy called the nature-nurture debate. This conflict, which raged for decades, was based on a stark dichotomy. Some researchers claimed that differences in the behavior of individuals were due primarily to differences in their genetic makeup (or "nature"). Other investigators argued that behavioral differences among individuals were due to differences in the environments they experienced (or "nurture").

As the data analyzed in this chapter show, both points of view are incorrect. Do genes or the environment determine behavior? The short answer is both. The longer, more accurate answer is that in most cases genetic backgrounds create the potential for various types of behavior; the behavior exhibited depends on environmental conditions.

Instead of arguing about the primacy of nature or nurture, then, biologists have become more interested in collecting data that show how genetic predispositions and environmental conditions interact, and how the behaviors adopted by individuals allow them to survive and reproduce.

cells. As predicted, Goldman and Nottebohm found many radioactively labeled neurons in and around the growing song centers. This observation supported the hypothesis that new neurons are being formed, that the song center enlarges due to the synthesis of new cells, and that the expansion of song centers is contingent upon stimulation by testosterone.

To verify that these new neurons were actually involved in generating song, John Paton and Nottebohm did the experiment diagrammed in **Figure 47.4**. They began by injecting radioactively labeled thymidine into adult canaries to mark new neurons. Thirty days later, they recorded electrical impulses generated by individual neurons in the song centers. After making a recording and confirming that a particular cell functioned as a normal neuron, Paton and Nottebohm injected the cell with an enzyme called horseradish peroxidase. The chemical reaction catalyzed by this enzyme results in a product that acts as a dark stain. In this way, working neurons in the song center became permanently marked. When Paton and Nottebohm analyzed the resulting cells under the microscope, they found that some cells contained both radioactive thymidine and black stain (see Figure 47.4, right). This result corroborated that new neurons, which form in the song centers in response to stimulation by testosterone, actually function.

Figure 47.5 summarizes these experiments on the proximate control of birdsong. The figure contains a crucial step that we've yet to explore, however. What aspect of the social or physical environment signals the arrival of spring and triggers the release of testosterone? David Crews performed a series of ingenious experiments to investigate how different types of environmental cues affect sexual behavior in the green anole lizard.

Courtship, Copulation, and Egg-Laying in *Anolis* Lizards

Anolis carolinensis (**Figure 47.6a**) live in the woodlands of the southeastern United States. After spending the winter under a log or rock, males emerge in January and establish breeding territories. Females become active a month later; by May, they are laying an egg every 10–14 days. By the time the breeding season is complete three months later, females will have produced an amount of eggs equaling twice their body mass. **Figure 47.6b** maps this series of events, along with the changes occurring in the male and female reproductive systems.

What causes these dramatic, seasonal changes in behavior? The proximate answer is sex hormones—testosterone in males and estradiol in females. Testosterone is produced in the testes and estradiol in the ovaries. The evidence for these statements is direct. Testosterone injections induce sexual behavior in castrated males, while estradiol injections induce sexual activity in females whose ovaries have been removed.

Now the question becomes, what environmental cues trigger the production of sex hormones? To answer this question for females, Crews brought a large group of inactive adult lizards into the laboratory and divided them into five treatment groups. The environment was exactly the same in all treatments. Each individ-

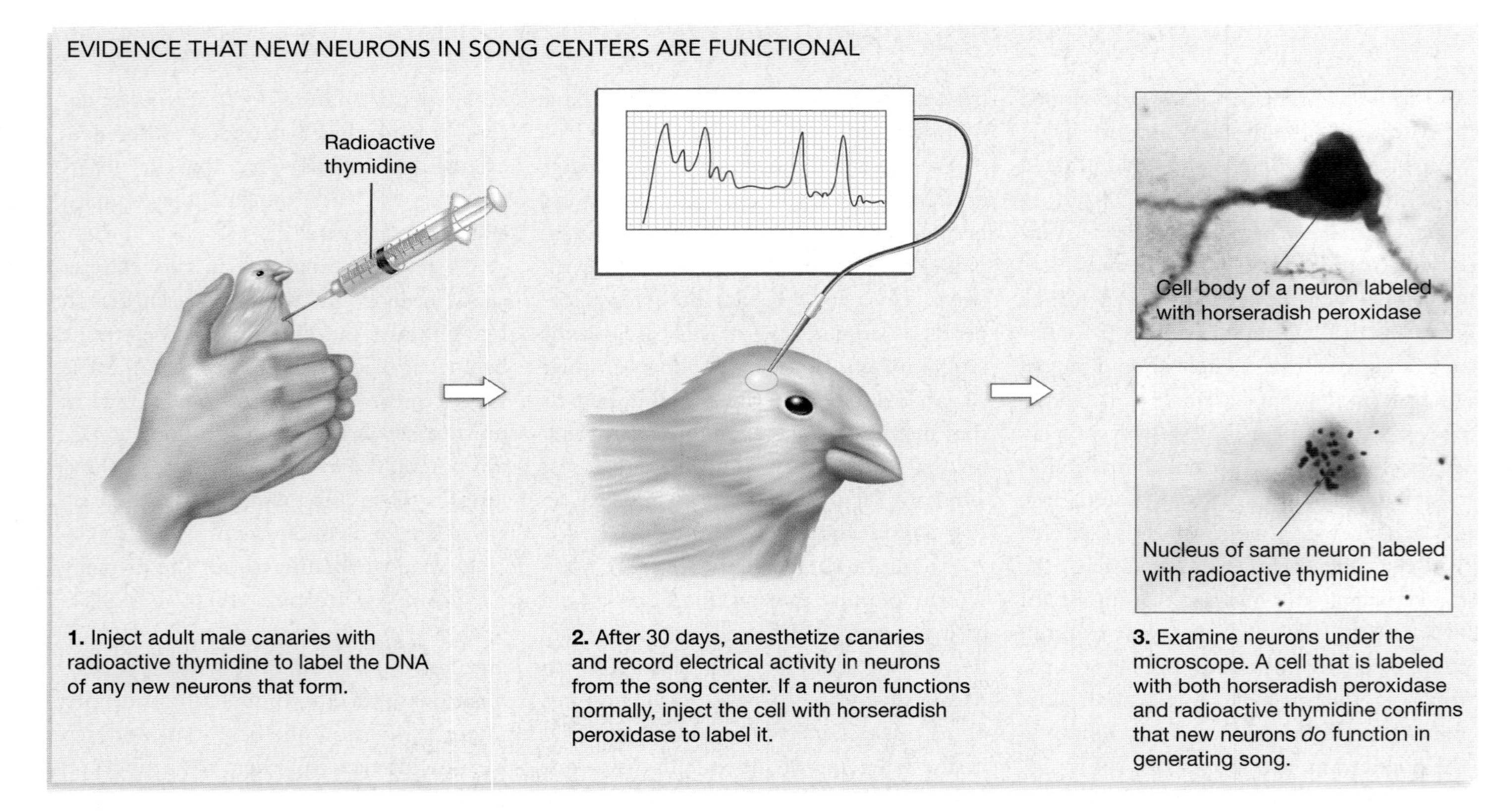

FIGURE 47.4 Evidence That New Neurons in Song Centers Are Functional

ual received identical food, and in all treatments artificial lighting simulated the long days and short nights of spring. Further, high "daytime" temperatures were followed by slightly lower "nighttime" settings. Only the social setting varied among treatment groups: (1) single isolated females, (2) groups of females only, (3) single females each with a single male, (4) single females each with a group of castrated (non-breeding) males, or (5) single females each with a group of uncastrated (breeding) males.

Each week, Crews examined the ovaries of females in each group. He also monitored the ovaries of females in nearby natural habitats, since those females were not exposed to spring-like conditions. As **Figure 47.7** (page 922) shows, the differences in their reproductive systems were dramatic. Females exposed to breeding males began producing eggs much earlier than the females placed in the other treatment groups. The experiment showed that two types of stimulation are necessary to produce the hormonal changes that lead to sexual behavior. Females need to experience spring-like light and temperatures *and* exposure to breeding males.

What aspect of male behavior causes the difference in female egg production? Crews suspected that visual stimulation from the male's courtship display was important. Specifically, he hypothesized that a flap of skin, called the dewlap, played a role in stimulating females to produce eggs. To court females, male anoles bob up and down and extend their dewlaps (see Figure 47.6a). To test this idea, Crews repeated the previous experiment, but

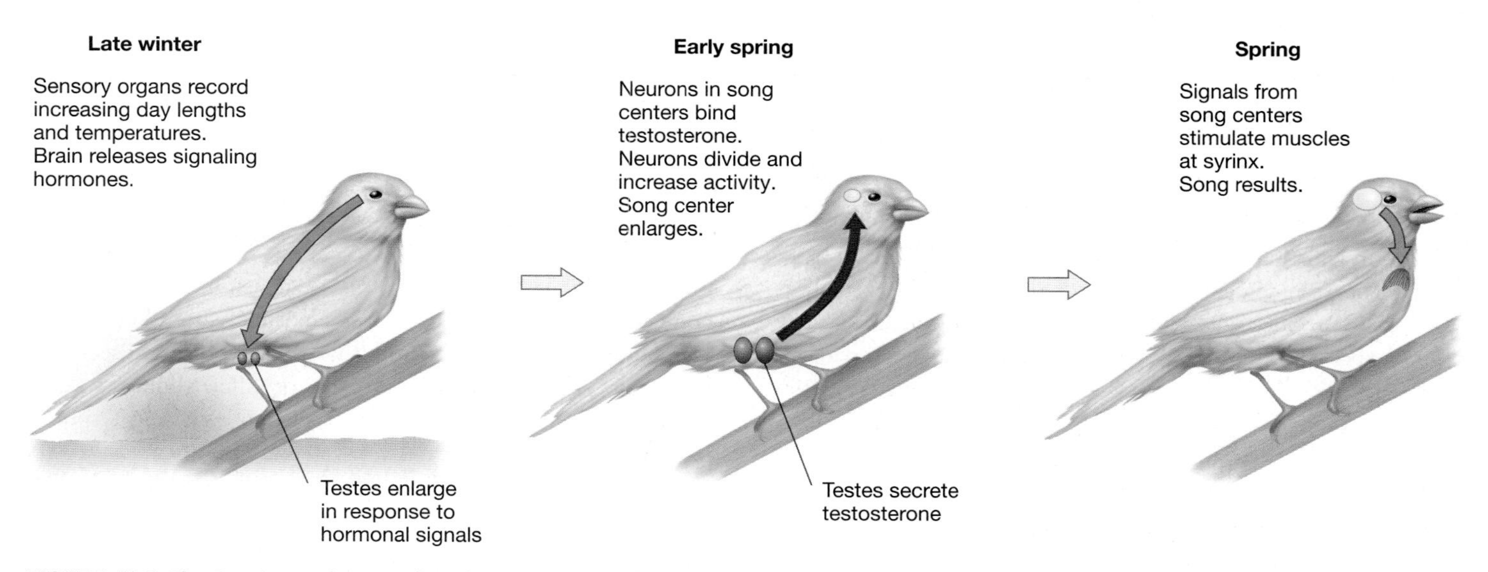

FIGURE 47.5 The Proximate Causes of Birdsong—A Summary Diagram
In male canaries, the capacity to sing results from interactions between environmental cues, hormonal signals, and the nervous system.

(a) Courting of *Anolis* lizards

(b) Changes in sexual organs through the year

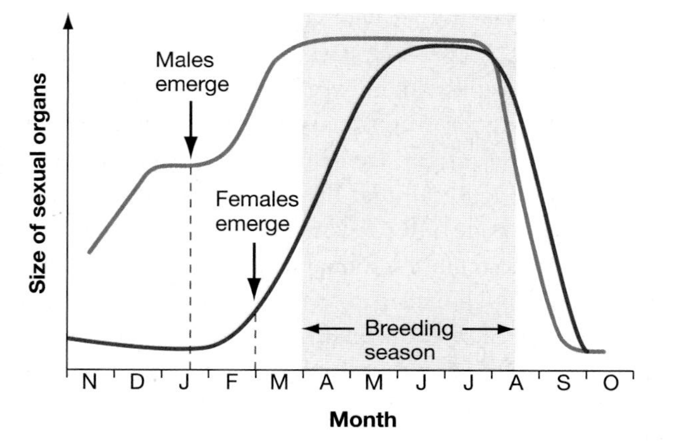

FIGURE 47.6 Sexual Behavior in *Anolis* Lizards
(a) A male *Anolis* lizard courting a nearby female. While bobbing up and down, the male is extending a flap of skin called a dewlap. **(b)** These graphs show changes in the sexual organs of *Anolis* lizards through the year. The red line indicates the relative size of the follicles inside a female's ovary. The blue line indicates the relative size of a male's testicles.

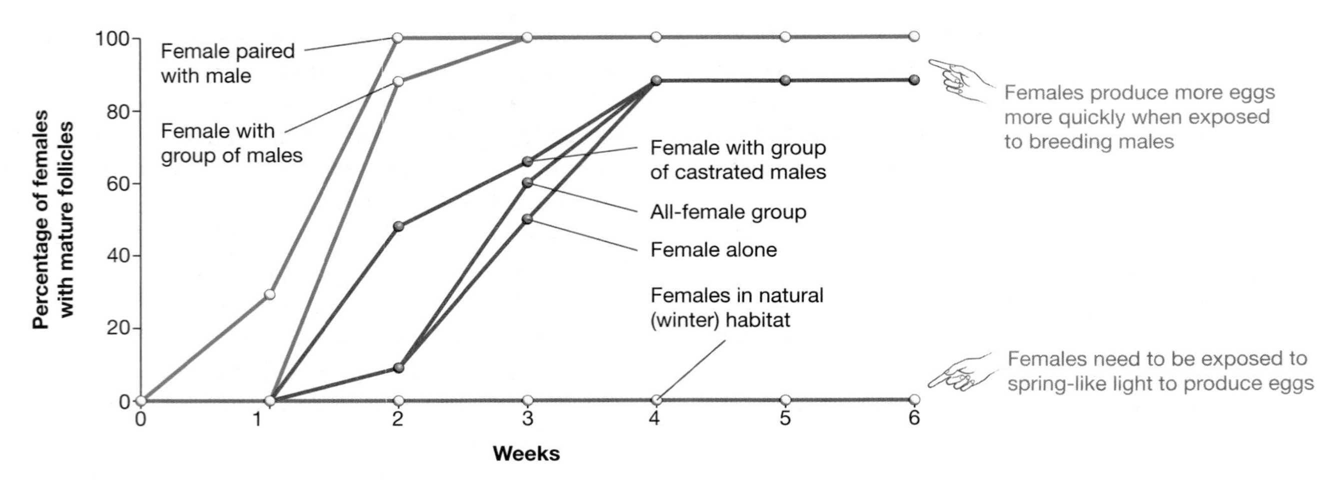

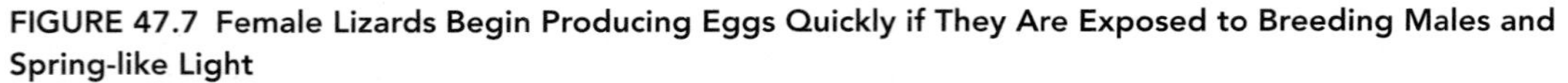

FIGURE 47.7 Female Lizards Begin Producing Eggs Quickly if They Are Exposed to Breeding Males and Spring-like Light
The percentage of female lizards with mature follicles is given on the y-axis, with data plotted for each of five treatment groups exposed to spring-like light and temperature and females left in natural (winter) habitat. Each data point represents the average value for a group of 6–10 females.

added a twist. He placed females with males that had intact dewlaps or with males whose dewlaps had been surgically removed. The result? Females grouped with dewlap-less males were slow to produce eggs—just as slow, in fact, as the females in the first experiment that had been grouped with castrated males. These females had not been courted at all. The result suggests that the dewlap is a key visual signal. The experiments succeeded in identifying the environmental cues that trigger hormone production and the onset of sexual behavior.

47.3 The Adaptive Consequences of Behavior

In addition to understanding how the nervous and endocrine systems control behavior at the proximate level, biologists would like to know why behavior exists at all. What is its function? How did different types of behavior come to be? At this level of analysis—the ultimate level—the questions are no longer about *how* animals do what they do, but why they do what they do.

To begin exploring the evolutionary significance of behavior, let's return to the research on fruit fly foraging that was introduced in section 47.1. Recall that flies with the *for*^{R} allele tend to rove, or move about, after feeding; while individuals with the *for*^{S} allele tend to remain stationary. Also recall that natural selection occurs if individuals with certain alleles survive better or reproduce more in certain environments than individuals with other alleles. Does natural selection favor *for*^{R} in some environments and *for*^{S} in others?

Natural Selection and Behavior

Marla Sokolowski and co-workers hypothesized that roving and sitting are feeding strategies that work best at different population densities, meaning in crowded or uncrowded conditions. The logic here was that sitting should be favored when population density is low because sitters conserve energy and grow faster than rovers. Thus, in low-density situations where competition for food is virtually nonexistent, sitters should survive better and reproduce more than rovers do. Because sitters have more offspring than rovers, the *for*^{S} allele should increase in frequency under these conditions. But in high-density populations, the opposite is true. Food is extremely scarce. Rovers should be favored because they are more likely to find uneaten food as they move. In high-density situations, the *for*^{R} allele should increase in frequency.

To test these predictions, Sokolowski and colleagues did the experiment diagrammed in **Figure 47.8**. They began by placing mixtures of 50 rovers and sitters in jars with a prescribed amount of food and mixtures of about 1000 rovers and sitters in jars with the same amount of food. They let the flies in each treatment group breed. Then they randomly selected offspring—50 from the jar with 50 adults and 1000 from the jar with 1000 adults—to start the next generation. They repeated this sequence for a total of 74 generations.

What happened? Over time, a strong pattern emerged. The average distance traveled by larvae after feeding decreased in the low-density treatment groups but increased in the high-density groups. These results suggest that as predicted, sitting is favored when population density is low, while roving is advantageous when population density is high.

In general, if some variations in behavior allow individuals to survive better or to reproduce more than other individuals, then alleles associated with the advantageous behavior will increase in frequency in the population. As a result, the behavior of natural populations will change through time, or evolve, just as it did in the laboratory experiment by Sokolowski and colleagues.

If natural selection favors a certain behavioral trait in a certain environment, it is called an **adaptation.** An adaptation is a characteristic that increases the ability of individuals to survive and reproduce compared to individuals without the trait. In fruit flies, roving and sitting are adaptive in certain environments.

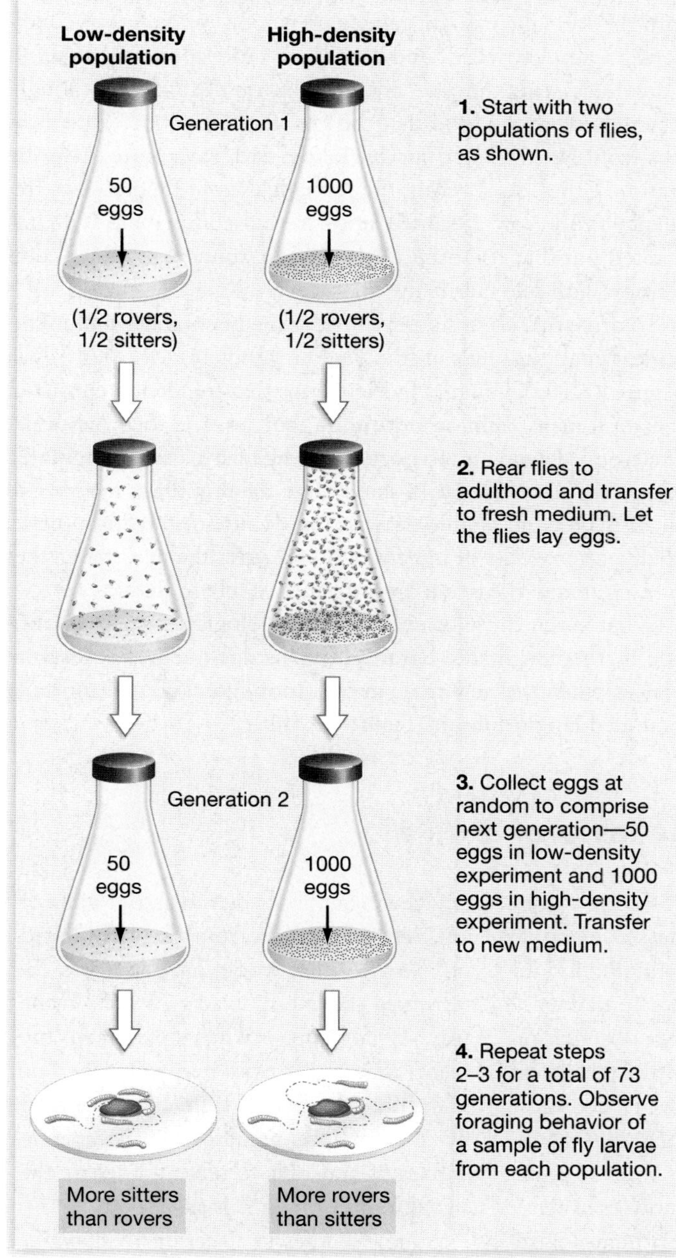

FIGURE 47.8 Do Fruit Fly Rovers or Sitters Have an Advantage at Different Population Densities?
This type of protocol is called an artificial selection experiment. In this case, researchers selected flies that were able to thrive in low- or high-density populations. QUESTION In addition to changes in foraging behavior, what other traits might change in response to selection for the ability to survive and reproduce in low- and high-density populations? How would you test your hypotheses?

In many species, however, behavior is much more flexible than it is in fruit flies. Instead of tending to either rove or sit, an individual might be able to rove and sit. If individuals are capable of acting in several different ways, how do they choose what to do? To answer this question, let's consider territorial defense behavior in hummingbirds.

Flexible Behavior: Territorial Defense in Hummingbirds

Many types of animal behavior involve "decision-making." The phrase is in quotes because the decisions that animals make are probably not conscious in the human sense. Yet they are decisions nonetheless. In coral-reef fish that have the capacity to be either male or female, individuals "decide" which sex to become based on their social situation.

The question is, are these decisions adaptive? Let's explore this question by considering territorial behavior in hummingbirds. These tiny birds defend territories that contain the nectar and insects they eat. They use several different types of behavior in territory defense, including displaying the iridescent feathers found on their head or neck (**Figure 47.9**), calling, chasing, and fighting. Chasing away a competitor is considered an expensive behavior because it requires a great deal of energy, entails a risk of injury, and leaves the territory undefended. Calling or displaying, in contrast, is relatively inexpensive.

Which type of territory defense behavior should a hummingbird use when an intruder approaches? Paul Ewald and Gordon Orians proposed that hummingbirds should make choices based on the value of their territory at the time. This prediction was based on the observation that hummingbirds empty the nectar stores when they feed on flowers. Just after feeding, then, flowers offer few benefits to hummingbirds. If the choice of territorial defense behavior is adaptive, then hummingbirds should call or display at intruders soon after feeding, but chase

Hummingbirds have feeding territories.

FIGURE 47.9 Hummingbird Territorial Behavior Depends on the Value of the Territory
Hummingbirds defend feeding territories by chasing intruders or by calling at them.

them if enough time had passed for flowers to refill with nectar and for the territory owner to be hungry again.

To test these ideas, the biologists put feeders filled with a sugary solution on hummingbird territories. The feeders were designed so that only a small amount of food would be delivered at a time, however. Once the feeders were in place, the investigators recorded what happened on a minute-by-minute basis. (**Box 47.3** highlights the importance of observation in behavioral studies.) When they compared how individuals behaved during different time intervals after feeding, they found interesting patterns. In the interval just after feeding, territory owners used displays more frequently when intruders approached, chased a lower percentage of intruders, and reduced the duration of chases. These results support the prediction that hummingbirds choose among behavioral options according to current conditions, and make choices in an adaptive manner.

Why Is Behavior Selfish?

When biologists analyze why an animal behaves in a certain way in a certain situation, they focus on the benefit to the individual. But when nonbiologists consider why animals behave in certain ways, they frequently suggest that individuals will do things "for the good of the species." Territorial behavior in birds, for example, was once thought to exist for the good of the species. The idea was that by spacing themselves out in a habitat, members of a species avoided depleting their food supplies and causing a population crash. Individuals that did not own a territory were thought to honor the existing territorial boundaries for the good of the species.

According to this view, individuals sacrifice themselves to benefit others. Over 35 years ago, however, George Williams pointed out that this type of behavior does not exist for a simple reason. Natural selection eliminates alleles that allow organisms to behave this way.

To understand Williams's logic, let's work through the following thought experiment. Suppose that the for^R allele in fruit flies acted for the good of the species. For example, individuals with the for^R allele might reduce their food intake when conditions became crowded, and thus leave more food for others. If individuals with the for^S allele ate normal amounts, though, they would eat the food that the for^R flies left alone. Thus, for^S flies would grow faster, survive better, and leave more offspring than for^R flies. As a result, the for^S allele would increase in frequency while for^R declined until it was eliminated. Williams pointed out that "selfish" alleles are guaranteed to win a competition with self-sacrificing alleles every time.

To drive this point home, consider the phenomenon of infanticide (infant-killing) in the langur monkeys of south Asia (**Figure 47.10a**). Langurs live in troops that frequently consist of one adult male, eight or more adult females, and their offspring. The troop defends a territory, often against bands of unmated males that roam about. In some cases the dominant member of an all-male band is able to evict the dominant territorial male, usually after weeks of intense fighting. After the new male takes over, infants in the troop are extremely likely to disappear.

What is the cause of death? Most biologists suspect infanticide by the new males. Sarah Hrdy has documented several instances when the new male in the troop attacked and mortally wounded langur infants (**Figure 47.10b**).

BOX 47.3 The Importance of Observation in Behavioral Studies

Virtually every study reviewed in this chapter is based on observation. Biologists watch what animals do and then ask how or why the behavior occurs. In many cases, finding an answer involves performing relatively simple experiments and carefully observing the outcome. For example, to test the hypothesis that hummingbirds use different types of territorial displays under different conditions, a biologist and an undergraduate assistant set feeders out on territories, recorded when territory owners fed, and observed how the owner responded to intruders.

Although the ability to make insightful observations is basic to all of the biological sciences, behavior may be the most observation-intensive field of study. Jane Goodall's pioneering work on chimpanzee behavior is a classic example. Goodall had little formal training in biology and was able to make fundamental contributions solely through observation. Her discoveries included tool use, nest-making, the existence of coalitions among females and among males, and hunting. Her observations were so astute and accurate that she is now considered one of the world's leading primatologists.

One of the attractive aspects of observational studies of behavior is that they are relatively inexpensive and accessible. At least some behavioral biology can even be done at home or on a college campus. Ants, wasps, spiders, beetles, birds, and bees can be found and observed almost anywhere. Suppose you ran across a line of ants retrieving pieces from a scrap of bread on a lawn. What would happen if you interrupted the line by placing a rock in the way? What happens when you remove any scents that might exist by wiping the line with rubbing alcohol? If the line is marked by a scent, are predators or competing species of ants able to sense it and use the signal to prey on ants or steal food?

Simple studies like these provide valuable practice in making observations, forming hypotheses, making predictions, and carrying out experimental tests. More than a few professional biologists got their start as undergraduate assistants on behavioral studies that emphasized observation.

CD ACTIVITY 47.2 Observing Behavior: Homing in Digger Wasps

Why does infanticide occur when new males take over? A group-selection argument would contend that the death of the infants benefits the species, by reducing population size and conserving resources. Hrdy's hypothesis, in contrast, is that the behavior benefits the individual perpetrators. She suggests that incoming males receive a reproductive benefit because the mothers of the dead infants respond by coming into estrus. (Nursing mothers do not ovulate. If their babies die and they stop nursing, they respond by ovulating.) This means that they are again ready to be fertilized and bear young. Instead of spending months or years helping to rear unrelated young, the new, dominant male kills them and quickly fathers his own offspring.

Data from lions collected by Craig Packer and Anne Pusey strongly support Hrdy's hypothesis. Lions live in social groups called prides. Prides are sometimes taken over by groups of incoming males. Packer and Pusey have shown that mothers whose offspring are victims of infanticide come into estrus eight months earlier, on average, than mothers with offspring who are not killed by males. The upshot is that males that commit infanticide do better, in terms of producing offspring, than males that do not commit infanticide.

Behavior like infanticide does not exist for the good of the species. It exists because alleles that encourage selfish behavior tend to increase in frequency.

Kin Selection and the Evolution of Cooperation: Alarm Calling in Prairie Dogs

The preceding discussion emphasizes that much of animal behavior is selfish—at least regarding its effects on the animal exhibiting the behavior and the recipient of this behavior. But this is by no means the entire story of behavior. Hundreds of examples of altruistic behavior exist in nature. In biology, **altruism** is defined as an act that has a cost to the actor, in terms of his or her ability to survive and reproduce, and a benefit to the recipient. It is the formal term for self-sacrificing behavior.

How can altruism exist, if theory proposes that selfish alleles are always favored? William Hamilton created a mathematical model that provided the answer. Hamilton asked: How could an allele that contributes to altruistic behavior increase in frequency in a population? To model the fate of these alleles, he represented the cost of the altruistic act to the actor as C and the benefit to the recipient as B. Both C and B are measured in units of offspring produced. His mathematical proof showed that the allele could spread if

$$Br - C > 0$$

where r is the **coefficient of relatedness.** The coefficient of relatedness is a measure of how closely the actor and beneficiary are related. Specifically, r measures the fraction of alleles in the actor and beneficiary that are identical by descent—that is, inherited from the same ancestor (see **Box 47.4**, page 926).

This result is called Hamilton's rule. It is important because it confirms that individuals can pass their genes on to the next generation not only by having their own offspring but also by helping close relatives produce more offspring. According to Hamilton's rule, if the benefits of altruistic behavior are high, if the benefits are dispersed to close relatives, and if the costs are low, then alleles associated with altruistic behavior will be favored by natural selection and will spread throughout the population. Biologists use the term **kin selection** to refer to natural selection that acts through benefits to relatives.

Does Hamilton's rule work? Do animals really favor relatives when they act altruistically? John Hoogland tested the

FIGURE 47.10 Infanticide Occurs in Nature When It Benefits the Perpetrators
(a) Two female langur monkeys intervene as a male, who has recently taken over the troop, attacks an infant (the infant's tail is visible above the male's head). **(b)** The infant is protected by its mother on the following morning. Note the wound in the infant's abdomen. QUESTION One of the females intervening on the infant's behalf is presumably its mother. Why would the other female help her in such a dangerous situation? What data would you gather to test your hypothesis?

theory by studying a behavior called alarm calling in Gunnison's prairie dogs. These burrowing mammals live in large communities, called towns, in the Four Corners region of the American Southwest (**Figure 47.11a**). When a badger, coyote, hawk, or other predator approaches a town, some prairie dogs give alarm calls that alert others to run to mounds and scan for the threat. Giving these calls is risky. In several species of ground squirrels and prairie dogs, researchers have shown that alarm-callers draw attention to themselves by calling and are in greater danger of being attacked.

To assess whether this selfless act is consistent with Hamilton's rule, Hoogland recorded the identity of the callers and listeners during 125 experiments. In these studies, a student assistant dragged a stuffed badger through the colony on a sled. Did alarm-callers preferentially help relatives, did non-relatives usually receive the benefit, or was everyone alerted equally with no difference in benefits? The data in **Figure 47.11b** illustrate that Gunnison's prairie dogs are much more likely to call if relatives are nearby. This same pattern—of preferentially dispensing help to close relatives—has been observed in many other species of social mammals and birds. Most cases of altruism that have been analyzed to date are consistent with Hamilton's rule.

47.4 The Evolution of Behavior

As noted in the introduction to this chapter, biologists have two goals when they analyze behavior at the ultimate level. They want to understand why the behavior helps an individual produce more offspring, and they want to know the evolutionary

BOX 47.4 Calculating the Coefficient of Relatedness

The coefficient of relatedness, *r*, varies between 0.0 and 1.0. If individuals have no alleles that are identical because they were inherited from the same ancestor, then their *r* is 0.0. Because every gene in pairs of identical twins is identical by descent, their coefficient of relatedness is 1.0.

What about other relationships? **Figure 1a** shows the general scheme for calculating coefficients of relatedness between individuals whose place in a genealogy is known, using first cousins as an example. The thick arrows in **Figure 1b** mark the path of common descent, from one cousin (marked A) to its parent (marked B), to the parent's sibling (marked C), to the other cousin (marked D). In each case, the individuals on either side of an arrow share half of their alleles by descent. To understand why, recall how the process of meiosis distributes alleles to gametes. Meiosis distributes alleles from the diploid genome to the haploid gamete randomly. As a result, if you pick any two gametes at random, the probability of them having the same allele is 50 percent.

To find *r*, then, we have to calculate the probability that the same allele is shared by A and B, *and* B and C, *and* C and D. The calculation is $1/2 \times 1/2 \times 1/2$, or 1/8. Using the same logic and rules for combining probabilities, you can calculate *r*'s for any two individuals. (Biologists who study altruism in nature have to do this routinely.)

(a) Structure of a pedigree

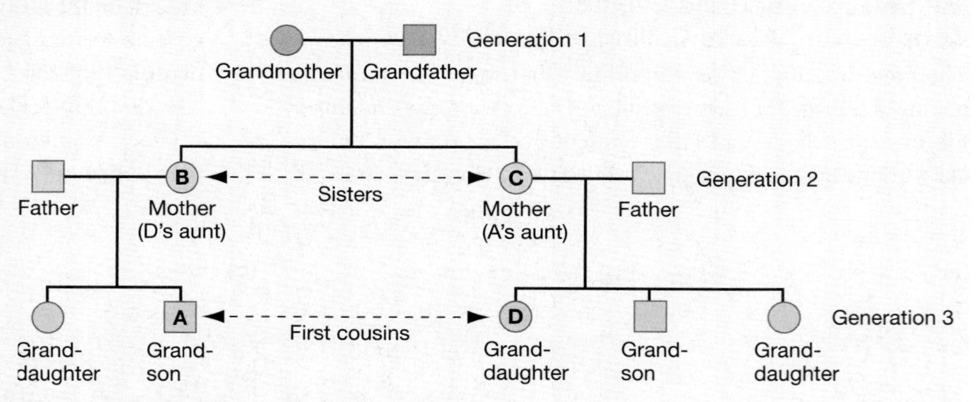

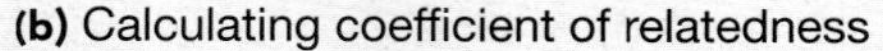

(b) Calculating coefficient of relatedness

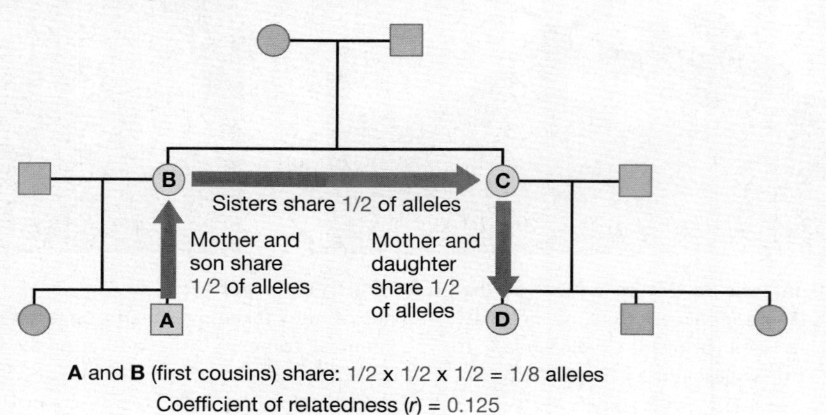

FIGURE 1
(a) This figure is the standard way of depicting human pedigrees, or family trees. **(b)** The thick arrows represent the genetic connections, or paths of common descent, between first cousins.

(a) Prairie dogs give alarm calls. **(b)** When do prairie dogs give an alarm call?

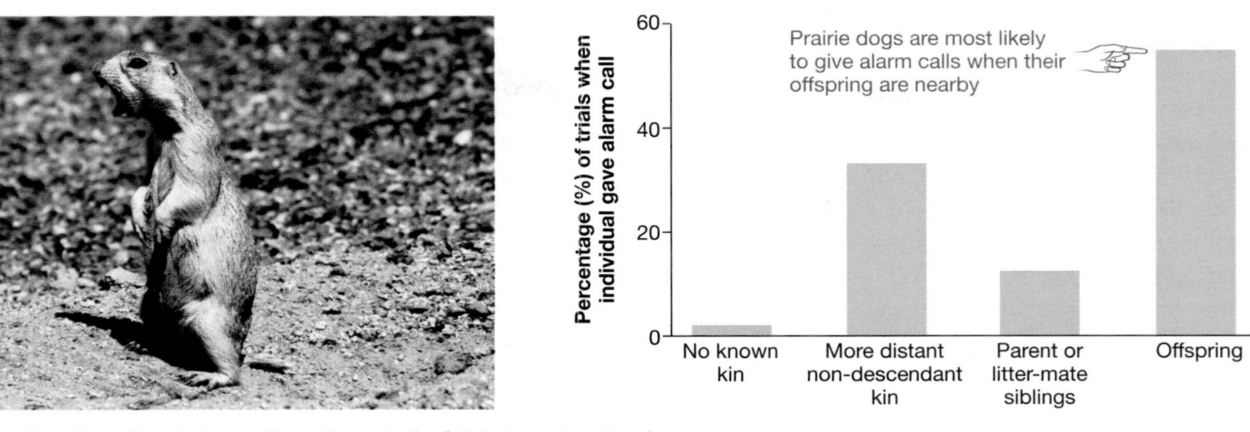

FIGURE 47.11 Prairie Dogs Give Alarm Calls if Relatives Are Nearby
(a) Prairie dogs are social animals that live in the grasslands of western North America. **(b)** This graph shows how likely prairie dogs are to give alarm calls, after detecting a predator, in four different situations: when no relatives are nearby, when cousins or more distant kin are around, when parents or siblings are within earshot, and when offspring are close. QUESTION The researcher who gathered these data is not sure why prairie dogs are so unlikely to give alarm calls when parents or siblings are close by. Generate a hypothesis to explain the observation. How would you test your hypothesis?

origins of the behavior. Stated another way, how did the behavior originate, and how has it changed through time?

To appreciate why biologists want to study the evolutionary origins of behavior, consider facial expressions in humans. Distinct muscle movements produce changes in the position of our lips, eyebrows, and other features. We use these poses to communicate our emotional state. Across all human cultures, people use the same facial expressions to indicate greeting, happiness, surprise, anger, and fear. (In contrast, hand gestures and other aspects of body language are learned. They vary from culture to culture.) A few of these expressions are strikingly similar to behavior used by our close relatives, the chimpanzees, in the same contexts (**Figure 47.12**). Is this similarity coincidental, or did humans and chimps inherit some of their behavior from a common ancestor, who used the same types of facial expressions to communicate?

As a case study of how biologists answer questions like this, let's consider the evolution of a spectacular territorial bird display carried out by cranes.

Cranes are large, marsh-dwelling birds found everywhere but Antarctica and South America. Individuals can live over 80 years in captivity, and they mate for life. Mated pairs sing a set of alternating songs, called the unison call, as a duet (**Figure 47.13a**, page 928). The display is used in the context of courtship and defending a breeding territory.

All 15 crane species perform a unison call, and in all cranes the display is an **innate behavior**. It is not modified by learning. If a crane chick is reared by humans with no contact from other cranes, or if a chick is reared by a different crane species that acts as a foster parent, the individual will still perform a fully functional unison call—of its own species—as an adult. Among the 15 species, however, the display varies widely in complexity.

How did the behavior evolve, and change through time? George Archibald answered this question by breaking the unison calls of the 15 species into a series of components. These components consist of motions and vocal inflections, some of which are unique to certain species and some of which are shared. Then Archibald proposed a hypothesis for how the display evolved. He assumed that complex displays are derived from simpler displays. For example, he proposed that the unison originated from a display called the alarm call. The alarm call is a short, loud blast given by a male or female crane when a predator approaches. It is closely related to the guard call, which is given by

FIGURE 47.12 Chimpanzee Facial Expressions
Some of the facial expressions used by chimpanzees are strikingly similar to those of humans.

either males or females when another crane enters their territory. The alarm calls and guard calls of all crane species are similar. Further, Archibald proposed that they are ancestral—meaning that these displays predated the advent of the unison call.

In two species of cranes that are native to Africa, the unison call is actually very similar to the guard-call and alarm-call. In these species the unison call consists of guard calls given alternately by a male and female that are paired. The vocalizations are accompanied by a simple visual display, consisting of head movements. In other species, however, the display is much more complex. Depending on the species, pairs may stand side by side or walk together as they call. In many species females begin the display, and male and female calls and postures are different.

Archibald grouped species whose motions and vocal inflections were similar and then ordered the display components from simple to complex. As a result of this analysis, he proposed that the unison call evolved in the order proposed in **Figure 47.13b**. The relationships establish a hypothesis for how complex behavior might evolve, in a step-by-step fashion, from simpler behavior.

Carey Krajewski and colleagues tested this hypothesis by reconstructing the evolutionary history of the 15 crane species. These researchers compared the composition of DNA sequences in both nuclear and mitochondrial genomes, and used these data to group species whose genes are similar. This analysis resulted in the relationships shown in **Figure 47.13c**. Compare parts (b) and (c), and note that the evolutionary relationships estimated from the genetic data are remarkably similar to the relationships of the unison-call display. Stated another way, the assumption that complex unison calls are derived from simpler behavior leads to a match between the history of the species and the history of the display. This result supports the validity of the assumption about how the behavior actually evolved. By reconstructing history in this way, researchers can understand how particular behavior patterns originated and changed through time.

(a) Crane courtship display

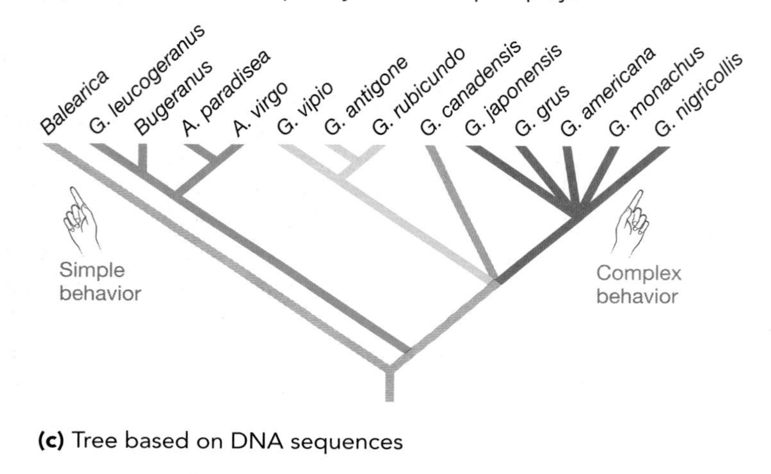

(c) Tree based on DNA sequences

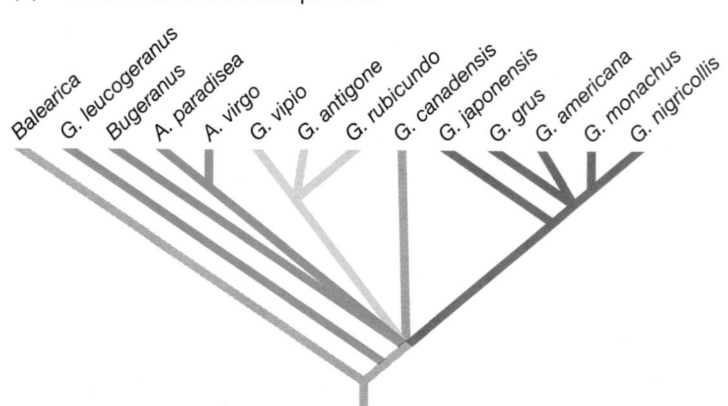

FIGURE 47.13 The Evolution of Crane Courtship Displays
(a) Cranes are long-legged, long-necked birds that live in marshes. This photo shows a mated pair of red-crowned cranes giving a unison call. **(b)** This tree illustrates a hypothesis for how the unison call evolved, based on the assumption that simpler displays originated earlier and more complex displays arose later. **(c)** This tree shows the evolutionary relationships of cranes. **EXERCISE** The text claims that the two trees are "remarkably similar." Do you agree?

Essay Children at Risk

Some of the concepts discussed in this chapter—including altruism, Hamilton's rule, kin selection, and natural selection—may seem a little abstract at first. Yet Martin Daly and Margo Wilson have shown that they can be enormously important tools for understanding certain aspects of human behavior.

Daly and Wilson study the "dark side" of human behavior—homicide, domestic violence, and adultery—from a biological perspective. They try to determine whether some of these events are consistent with the types of actions favored by natural selection.

Can we study the "dark side" of human behavior?

For example, let's consider data that Daly and Wilson have collected on child abuse. They hypothesized that natural selection should favor parents who invest resources in biological children, but not in stepchildren—with whom they have no genetic relationship. Based on this hypothesis, they predicted that child abuse should be much more common in households containing a stepparent than it is in homes containing only biological parents. Further, infants should be most at risk if stepparents choose to abuse their stepchildren because very young children demand the most time and resources and are least able to defend themselves.

Are the data consistent with the prediction? Daly and Wilson have analyzed the most extreme form of child abuse, which is the killing of children by parents. Using a database on all homicides reported in Canada between 1974 and 1983, they found 341 cases in which a child was killed by a biological parent and 67 in which the perpetrator was a stepparent. Because households containing only biological parents were much more common, however, Daly and Wilson realized that they needed to compare the rate of violence in the two types of households, rather than the absolute number of incidents.

When they expressed the data as a rate—specifically, as the number of children killed per million years by their parents—the histograms in **Figure 1** resulted. The pattern is striking. Children who live with a stepparent are at much higher risk of abuse than children who live with biological parents. Kids who are less than 2 years old are, in fact, 70 times more likely to be killed. To appreciate the magnitude of this relative risk, consider that smokers are 11 times more likely to develop lung cancer than are nonsmokers.

The study points up the value of applying "selection thinking" to human behavior. It also has an urgent practical message. Society should be especially alert to indications of violence in households where very young stepchildren are present. Several points about the data are worth noting, however. It is *extremely* important to recognize that violence occurs among biological kin as well as non-kin, and that the vast majority of stepparents are solicitous and generous with their stepchildren. Further, the behavior is undoubtedly pathological rather than adaptive. Killing "stepcubs" may improve the reproductive success of male lions, but human perpetrators are jailed and treated as abhorrent. It is extremely unlikely, then, that the act of killing a stepchild improves the reproductive success of the murderer.

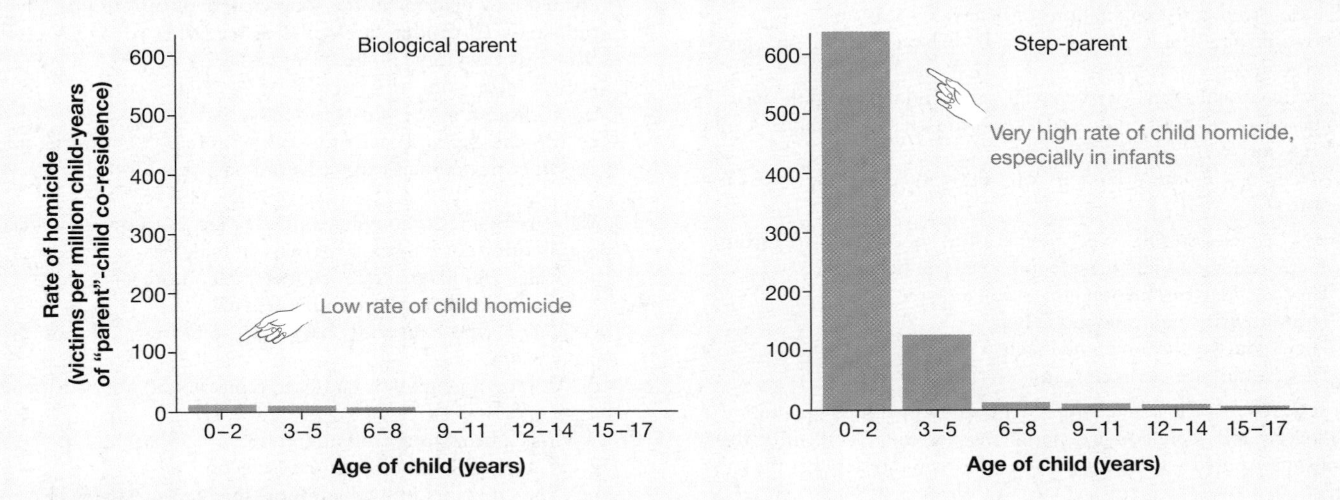

FIGURE 1 Rate of Child Homicide
Here, rates are expressed as victims per million child-years. A child-year is a year that a child of a particular age lived in a particular family situation. For example, the histogram to the left indicates that if a million children aged 0–2 years lived with their biological parents for one year, on average about 10 would be murdered.

Chapter Review

Summary

Biologists study behavior at the proximate and ultimate levels. At the proximate level, experiments and observations focus on understanding how neurotransmission and hormonal signals cause behavior. At the ultimate level, researchers seek to understand the adaptive significance of behavior, or how it enables individuals to survive and reproduce. Studying the genes and alleles responsible for behavior provides an important link between proximate and ultimate levels of explanation, because different proximate mechanisms for behavior result from the presence of certain genes or alleles, and because alleles that affect behavior change in frequency over time in response to natural selection.

In the field of study called behavior genetics, cross-fostering experiments are an important technique for establishing that certain types of behavior have a genetic basis. Investigators can also examine the specific genes involved in behavior by identifying a variable behavioral phenotype, such as roving and sitting in fruit flies, and doing experiments designed to map and characterize the locus where the mutation occurred.

In virtually all cases, however, the presence of certain alleles only creates a predisposition for certain types of behavior rather than others. The actual expression of a behavior is contingent upon environmental conditions. For example, biologists study condition-dependent behavior like sex change in fish to explore how an individual's social environment affects its actions.

Proximate-level studies usually focus on the neuronal and hormonal mechanisms responsible for certain behavior. Studies of song centers in bird brains have revealed that some changes in behavior are associated with the generation of new neurons and new synaptic connections. In this case, the changes in neurons are instigated by the sex hormone testosterone, which in songbirds is produced in response to the long day lengths and warm temperatures that signal the onset of spring. Hormone release triggered by environmental change is also responsible for increased sexual activity in *Anolis* lizards. But in this case environmental cues are not sufficient for full expression of the behavior. Female lizards must also be stimulated by males, which exhibit their dewlaps in courtship display.

At the ultimate level of analysis, one fundamental question is whether animals choose among behavioral options in a way that maximizes their ability to survive and reproduce. In hummingbirds, for example, experiments have shown that individuals defend their territories with low-cost calls or visual displays when food resources on the territory are depleted, but with high-cost chases when food resources are plentiful.

Studies of phenomena like infanticide have helped convince biologists that individuals do not do things for the good of the species. But when close relatives benefit from a behavior, like alarm calling in prairie dogs, natural selection can favor individuals who act altruistically. To understand how certain types of behavior originated and changed through time, biologists examine phylogenies and analyze the sequence in which the behavior evolved.

By combining proximate and ultimate viewpoints and studying the genetic basis of behavior, biologists can seek a comprehensive understanding of how and why animals do what they do.

Questions

Content Review

1. What do proximate explanations of behavior focus on?
   a. how displays and other types of behavior have changed through time, or evolved
   b. the functional aspect of a behavior, or its "adaptive significance"
   c. neurological, hormonal, and skeletomuscular mechanisms of behavior
   d. psychological interpretations of behavior—especially motivation.

2. Why do cross-fostering experiments allow researchers to distinguish environmental and genetic influences on behavior?
   a. They keep environmental effects constant.
   b. They intensify environmental effects.
   c. They reduce environmental effects.
   d. They randomize environmental effects.

3. If a researcher starts a behavior genetic study by noticing a mutant phenotype and then doing experiments designed to identify the gene(s) associated with the behavior, what is the researcher pursuing?
   a. forward genetics
   b. reverse genetics
   c. a candidate gene approach
   d. a molecular genetic study

4. When researchers wanted to determine if cell division events had created new neurons in the song centers of bird brains, they used radioactive thymidine molecules as a marker for DNA synthesis. Why was thymidine an excellent choice?
   a. It is more stable than uracil.
   b. It is found only in DNA, not RNA.
   c. It is cheaper and easier to work with than other nucleotides.
   d. Other nucleotides cannot be radioactively labeled.

5. Why are biologists convinced that the sex hormone testosterone is required for normal sexual activity in male *Anolis* lizards?
   a. Males with larger testes court females more vigorously.
   b. The molecule is not found in females.
   c. Males whose gonads had been removed did not develop dewlaps.
   d. Males whose gonads had been removed did not court females.

6. When is a behavioral trait adaptive?
   a. When it is favored by natural selection.
   b. When it makes individuals more likely to survive, relative to individuals without the trait.
   c. When it helps individuals produce more offspring than individuals without the trait.
   d. All of the above.

7. What does Hamilton's rule specify?
   a. Why "selfish" alleles spread at the expense of alleles that function "for the good of the species."
   b. Why infanticide is adaptive.
   c. When alleles that favor altruism increase in frequency.
   d. More complex behaviors evolve from simpler behaviors.

Conceptual Review

1. This chapter outlined four reasons why behavior is not determined by genes in the sense of "if you have this allele, you will act this way":
   - The transcription of genes is regulated, so expression of an allele is a function of action by hormones, neurotransmitters, and other signals.
   - Gene products function only in concert with many other gene products, so their effect varies with the amount and identity of these other molecules.
   - Individuals frequently have a large repertoire of different types of behavior; which behavior is used depends on the social environment and other types of conditions.
   - Genetic differences among individuals invariably produce statistical differences in behavior, not either-or responses.

   Using the data reviewed in this chapter, or other examples you are aware of, provide a case that illustrates each of these four points.
2. Compare and contrast proximate and ultimate explanations for behavior.
3. To study how displays or other types of behavior have changed through time, why must researchers know the phylogeny, or evolutionary history, of the species involved?
4. What data convinced researchers that *for* and *dg2* were the same gene? How might changes in the activity or amount of a protein kinase affect foraging behavior in fruit flies?

Applying Ideas

1. The data reviewed in section 47.1 suggest that the development of teen-onset alcoholism in boys does not depend on whether alcohol abuse occurs at home. Yet less than 17 percent of the boys in the study who were genetically predisposed to teen-onset alcoholism actually developed the illness. What other factors might be involved in the development of this behavior? How would you test your hypotheses?
2. In many species of songbirds native to the tropics, both females and males sing (often in duets). How would you expect the song centers in the brains of these female birds to change as the breeding season approaches? What molecule would you predict is responsible for causing these changes? Design experiments that can test your predictions.
3. Most tropical habitats are highly seasonal. But instead of alternating warm and cold seasons, there are alternating wet and dry seasons. Most animal species breed during the wet months. If you were studying a species of *Anolis* native to the tropics, what environmental cue would you simulate in the lab to bring them into breeding condition? How would you simulate this cue? Further, think about the sensory organs that lizards use to receive this cue. Are they the same or different than the receptors that *Anolis carolinensis* uses to recognize that spring has arrived in the southeastern United States?
4. Biologists use the following terms to describe the four types of interactions that are possible between two individuals:

	Benefits Recipient	Costly to Recipient
Benefits actor	cooperative	selfish
Costly to actor	altruistic	spiteful

   This chapter discussed only altruism and selfishness, but cooperation and spite are also interesting. Can these types of behavior evolve? If so, would you expect them to be more or less common than selfishness and altruism? Why or why not?
5. A friend of yours argues that the unison call of cranes is genetically determined because it is stereotyped and completely uninfluenced by learning. Another friend argues it is a condition-dependent behavior, because only mated pairs do it and because the tendency to give unison calls varies among pairs (some pairs are more aggressive than others, and give unison calls more frequently in territorial contexts). Who's right?

CD-ROM and Web Connection

CD Activity 47.1: Proximate Causes of Behavior *(animation)*
(Estimated time for completion = 5 min)
Observe how the canary's seasonal breeding song is influenced by neurological and hormonal changes.

CD Activity 47.2: Observing Behavior: Homing in Digger Wasps *(tutorial)*
(Estimated time for completion = 10 min)
How does a wasp find its nest after returning from a foray for food?

At your **Companion Website** (http://www.prenhall.com/freeman/biology), you will find self-grading exams and links to the following research tools, online resources, and activities:

Behavioral Genetics
This article from *Science* Magazine discusses research on the genetics of behavior.

Behavior and the Judicial System
This provocative article addresses the ethical and sociological ramifications of scientific advances in behavioral genetics.

Seasonal Neuroplasticity
This article from *Nature* magazine discusses the ways in which hormonal development affects bird songs.

The Egotist and the Altruist
This article from *Nature* magazine reviews an attempt by Jacob Koella to understand behavior and its effect on evolution using computerized models.

Duplicitous Ducks
This article from *Scientific American* describes a relationship called "conspecific brood parasitism" within some duck species.

Additional Reading

Diamond, J. 1992. *The Third Chimpanzee* (New York: HarperCollins). An insightful look at adaptive aspects of human behavior.

Heinrich, B. 1989. *Ravens in Winter* (New York: Summit Books). Introduces how behavioral biologists do field observations and experiments.

Population Ecology

48

In ecology and evolutionary biology, the basic unit of analysis is the population. A **population** is a group of individuals from the same species that live in the same area at the same time. Evolution is the study of how the characteristics of populations change through time; ecology is the study of how populations interact with their environment. Chapter 47 explored how organisms respond to environmental stimuli and how they interact with members of their own population. Chapters 49 through 51 investigate how different populations interact with one another and with their physical environment. But here our focus is on the population itself—specifically its size.

How do the sizes of different populations change through time, and why? With the explosion of human populations across the globe, the massive destruction of natural habitats, and the resulting threats to species from throughout the tree of life, this question has become vitally important. To answer it we need to review the basic mathematical models and concepts that biologists use to study population growth. Section 48.2 follows up with four examples of how biologists study changes in the overall size of populations over time. Section 48.3 introduces how the age structure and geographic distribution of a population affects its fate. The chapter concludes by asking how all of these elements fit together in efforts to save endangered species. Sophisticated models of population growth are helping biologists predict changes in population size and design management strategies accordingly.

Human populations have been growing rapidly for the past 500 years. This chapter explores how and why growth rates in populations of humans and other organisms change through time.

48.1 Population Growth

The fate of a population depends on four factors: its birth rate, death rate, immigration rate, and emigration rate. Births and immigration add individuals to the population; deaths and emigration remove them. In this section we consider only the impact of births and deaths on population growth; for the moment we'll ignore the movement of individuals into and out of populations.

To understand how biologists model changes in population size, let's consider data on whooping cranes. Whooping cranes are large, wetland-breeding birds native to North America (**Figure 48.1a**). Although the total population was reduced to about 20 individuals in the mid-1940s, intensive conservation efforts have resulted in a current total of about 410. That number includes a population of 187 individuals that breeds in Wood Buffalo National Park in the Northwest Territories of Canada.

Figure 48.1b shows changes in the size of the Wood Buffalo population over the past 60 years. How fast is this population growing? Whooping cranes breed once per year, so the simplest way to express the population's growth rate is to compare the number of individuals at the start of one breeding season to the number at the start of the following year's breeding season. The current total is 187. Suppose that biologists counted 198 cranes on the breeding grounds next year. The population growth rate could be calculated as $198/187 = 1.06$. Stated another way, the population grew at 6 percent per year.

(a) Whooping cranes

(b) Changes in Wood Buffalo Park population

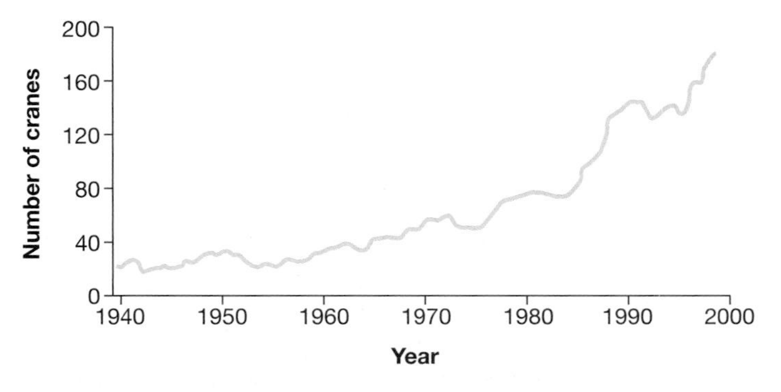

FIGURE 48.1 Growth in a Whooping Crane Population
(a) Whooping cranes once bred in marshes throughout northern North America. **(b)** This graph shows how the whooping crane population that breeds in Wood Buffalo National Park, Northwest Territories, Canada, has increased since the 1940s.

Basic Models of Population Growth

When populations breed during discrete seasons, their growth rate can be calculated as for whooping cranes. To create a general expression for how these populations grow, biologists use N to symbolize population size. N_0 is the population size at time zero (the starting point) and N_1 the population size one breeding interval later. In equation form,

Equation 48.1 $$N_1/N_0 = \lambda$$

The parameter λ (lambda) is called the **finite rate of increase.** (In mathematics, a *finite rate* refers to an observed rate over a given period of time. A *parameter* is a variable or constant term that affects the shape of a function, but not its general nature.) In the whooping crane example, λ was 1.06. Rearranging the expression in Equation 48.1 gives

Equation 48.2 $$N_1 = N_0\lambda$$

Stated more generally, the size of the population at the end of time t will be given by

Equation 48.3 $$N_t = N_0\lambda^t$$

This equation summarizes how populations grow when breeding takes place seasonally. In species with discrete reproduction, λ works like the interest rate at a bank. For species that breed once per year, the "interest" on the population is compounded annually. A savings account with a 6 percent annual interest rate increases by a factor of 1.06 per year.

Frequently, though, biologists want to express growth rates on a per-capita basis—meaning, per individual—instead of in terms of the overall population. A population's **per-capita rate of increase** is symbolized by r and is defined as the per-capita birth rate minus the per-capita death rate.

The parameters λ and r have a simple mathematical relationship. The best way to understand the relationship is to recognize that λ expresses a population's growth rate over a discrete interval of time. In contrast, r gives the population's per-capita growth rate at any particular instant. As a result, r is also called the instantaneous rate of increase. The relationship between the two parameters is given by

Equation 48.4 $$\lambda = e^r$$

where e is the base of the natural logarithm, or about 2.72. (In general, the relationship between any finite rate and any instantaneous rate is given by finite rate $= e^{\text{instantaneous rate}}$.) Substituting Equation 48.4 into Equation 48.3 gives

Equation 48.5 $$N_t = N_0 e^{rt}$$

This equation summarizes how populations grow when they breed continuously, as do humans and bacteria, instead of at defined intervals. For species that breed continuously, the "interest" on the population is compounded continuously. When the growth rates λ and r are equivalent, however, the differences between discrete and continuous growth are negligible. As **Figure 48.2a** shows, graphs for discrete and continuous growth can be superimposed.

Because r represents the growth rate at any given time, and because r and λ are so closely related, biologists routinely calculate r for species that breed seasonally. For seasonal breeders, r is the same as the growth rate expressed as the percent increase (or decrease) in the population per year. In the whooping crane example, $\lambda = 1.06 = e^r$. To solve for r, take the natural logarithm of both sides. In this case, $r = 0.058$.

Applying the Models

To get a better feel for r and Equation 48.5, consider the following series of questions about whooping cranes. The key to answering these questions is to realize that Equation 48.5 has just four parameters. Given three, you can calculate the fourth.

1. If 20 individuals were alive in 1941 and 410 existed in 2001, what is r? Here $N_t = 410$, $N_0 = 20$, and $t = 60$ years. Substitute these values into Equation 48.5 and solve for r. Then check your answer below.[1]
2. In the most recent report issued by the biologists working on the recovery program, it is estimated that the Wood Buffalo flock should be able to sustain an r of 0.046 for the foreseeable future. If the flock currently contains 187 individuals, how long will it take to double? Here $N_0 = 187$ and $N_t = 2 \times 187 = 374$. In this case, you solve Equation 48.5 for t. Then check your answer.[2]
3. Suppose that conservationists agree to take whooping cranes off the endangered species list when the total number of individuals reaches 1,000. If populations are able to sustain an r of 0.05, how long will it take to reach this goal? Recalling that $N_0 = 410$, solve for t again.[3]

Exponential Growth

The graphs in **Figure 48.2b** plot changes in population size for various values of r. The per-capita growth rate tends to be very high in species like fruit flies, which breed at a young age and produce many offspring; but it is low in species like elephants (or whooping cranes), which take years to mature and produce few offspring.

The sweeping curves in Figure 48.2b illustrate exponential growth. **Exponential growth** occurs when r does not change over time. The key point about exponential growth is that the per-capita rate of increase does not depend on the number of individuals in the population. Growth continues indefinitely because increases in the size of the population do not affect the per capita rate of increase. As a result, biologists say that growth is density independent.

It's also important to note that exponential growth adds an increased number of individuals as N gets larger. As an extreme example, an r of 0.02 per year in a population of 1 billion adds over 20 million individuals. The same growth rate in a population of 100 adds just over 2 individuals per year.

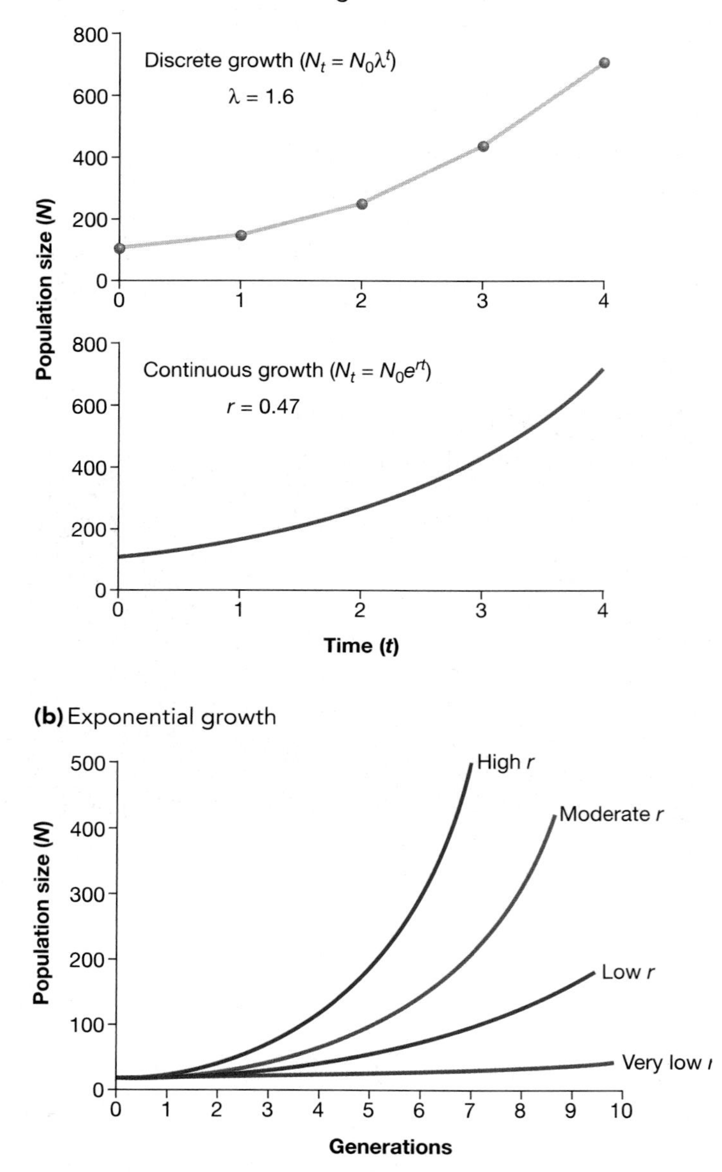

FIGURE 48.2 Density-Independent Growth
(a) These curves illustrate discrete and continuous growth when λ and r are equivalent. In both of these cases, the growth rate has been held constant over time. **(b)** When the per-capita growth rate r does not change over time, exponential growth occurs. Population size may increase slowly or rapidly, depending on the size of r. Curves like those illustrated in parts (a) and (b) occur when growth is not affected by the population's density.

[1] $r = 0.05$
[2] The doubling time is 15 years.
[3] $t = 17.8$ years.

In reality, however, it is not possible for population growth to continue indefinitely. If whooping crane populations continued to increase exponentially, they would eventually fill all available breeding habitat or begin to exhaust food supplies on the wintering grounds. If population density gets very high, we would expect the population's per-capita birthrate to decrease and the per-capita death rate to increase, causing r to decline. Stated another way, growth should become density dependent.

Figure 48.3 illustrates density-dependent growth graphically. Note that the graph has three sections. Initially, r is constant, meaning that growth is exponential. With time, however, the population's size increases to the point where competition for resources begins to affect birth rates and death rates. As a result, r begins to decline. Eventually, it reaches 0. When a population stabilizes at the maximum number that can be supported by the resources available in a habitat over a sustained period of time, biologists say that the population has reached the habitat's **carrying capacity**. Resources that define carrying capacity include the availability of food and the number of suitable nesting sites.

If the per-capita death rate actually begins to exceed the per-capita birthrate, then r becomes negative and the population declines. (When a population is declining, λ is less than 1.0. At zero population growth, $\lambda = 1.0$). The general point here is that even though exponential growth may occur over short intervals, it cannot be sustained. Within populations, r varies through time and can be positive, negative, or 0. To understand why density dependence and other types of population dynamics occur, let's explore four case studies of how populations change through time.

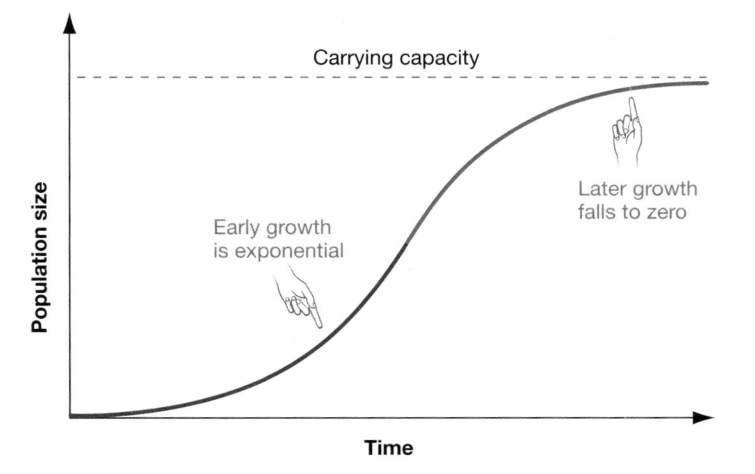

FIGURE 48.3 Density-Dependent Growth
This hypothetical curve illustrates density-dependent growth. The pattern often occurs when a small number of individuals colonizes an unoccupied habitat because competition for resources is initially low to nonexistent. Carrying capacity depends on the quality of the habitat. 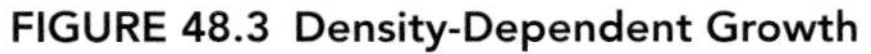 EXERCISE Label the portion of the graph where growth rates begin to slow. This is the inflection point of the curve, which signals the end of exponential growth.

48.2 How Do the Sizes of Populations Change over Time?

The theory introduced in section 48.1 provides a foundation for exploring how biologists study changes in populations through time. The goal of this section is to review case studies that illustrate some key aspects of density-independent and density-dependent growth. The examples explore how populations respond to the colonization of a new habitat, whether humans are beginning to experience density dependence, why some populations undergo regular cycles, and how populations recover (or don't recover) from a catastrophic drop in numbers. Let's take a closer look.

Density-Dependent Growth in a Coral-Reef Fish

In the Caribbean Sea, the abundance of certain fish species varies dramatically from one coral reef to another. Some researchers propose that the variation reflects density dependence. The hypothesis here is that if the resources available to a particular fish species vary among reefs, then the density of fish populations will also vary.

An important alternative to the density-dependence hypothesis exists, however; it is based on variation in immigration rates. Recall that population sizes depend on rates of births, deaths, immigration, and emigration. Although we ignored immigration and emigration in section 48.1, it is plausible to suspect that immigration plays a large role in this case. Juvenile coral-reef fish usually disperse from their place of origin by floating in the plankton and then colonizing a different reef. Thus fish density might vary among reefs purely because of the reef's location with respect to source populations and prevailing ocean currents.

To test the viability of the density-dependent and immigration hypotheses, Graham Forrester performed an experiment on a common reef resident called the bridled goby (**Figure 48.4a**, page 936). He began by creating eight artificial reefs with coral rubble and identifying eight nearby natural reefs with varying densities of gobies. Then Forrester stocked each of the artificial reefs with different densities of adult gobies, ranging from an average of 0.7 gobies per m^2 to 10.7 per m^2. The fish were marked to allow later identification. After 2.5 months, he captured all of the gobies on each reef and computed the growth rate of individuals, the survival rate of marked adults, and the number of juveniles that immigrated successfully.

The results of the experiment provide a clear example of density dependence in population growth. As **Figure 48.4b** shows, both survival and recruitment decreased as initial density increased. Due to these relationships, the final population density of gobies on each of the artificial reefs was remarkably similar, despite dramatic differences in initial density. Presumably, the artificial reefs were so alike that their carrying capacity was nearly identical.

Was lack of food responsible for halting population growth in the artificial reefs? Forrester claims that the answer is no, because he found no relationship between initial population density and the growth rates of individual fish. To explain his results, he suggests that higher rates of predation or disease, and not food limitation, must occur in higher-density populations of gobies. If Forrester's hypothesis is correct, then researchers should be able to document marked differences in predation rates or disease prevalence among natural reefs where the goby density varies.

To date, however, this prediction has not been tested. Meanwhile, the experiment stands as a good example of how researchers can document density-dependent effects in natural populations.

(a) Bridled gobie

(b) Survival and recruitment at different population densities

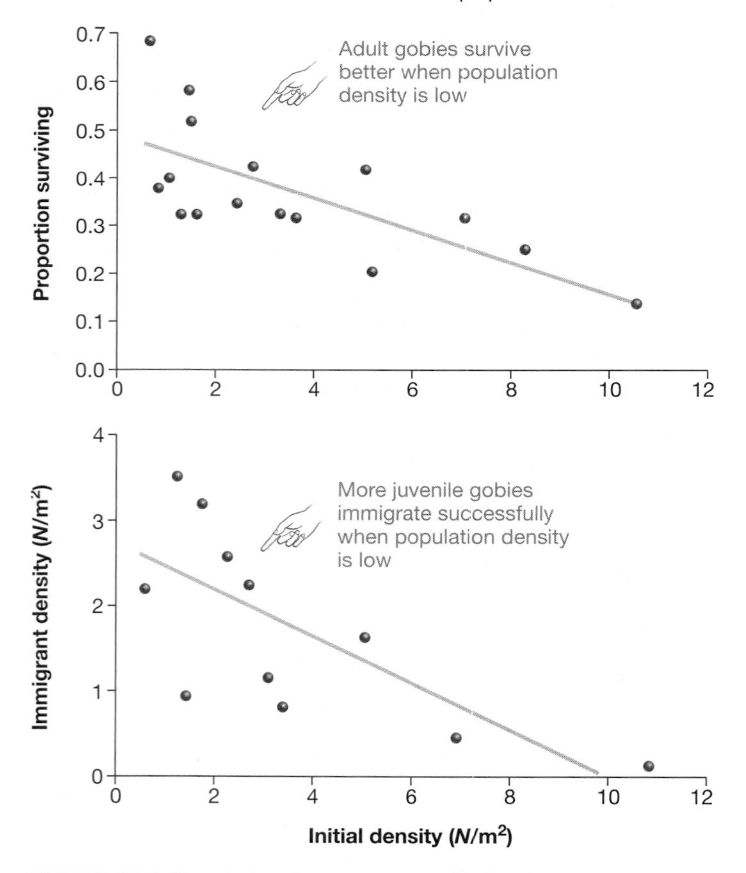

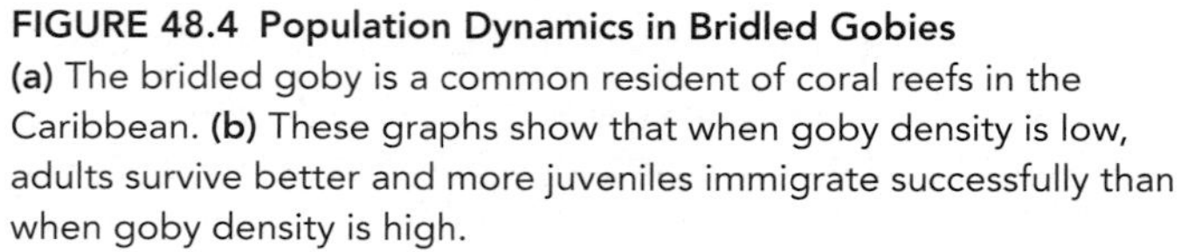

FIGURE 48.4 Population Dynamics in Bridled Gobies
(a) The bridled goby is a common resident of coral reefs in the Caribbean. **(b)** These graphs show that when goby density is low, adults survive better and more juveniles immigrate successfully than when goby density is high.

Analyzing Change in the Growth Rate of Human Populations

How have human populations grown over the past 12,000 years? Although census-based estimates of human population size are available for only the past few decades, archeological and anthropological data have been used to estimate the size of the human population in the past. **Figure 48.5a** shows the estimated size of the human population from the end of the last ice age, about 12,000 years ago, to the present. You can see that the shape of the curve is similar to the examples of exponential growth in Figure 48.5. In reality, though, the growth rate for humans has increased over time since about 1400, leading to a very steeply rising curve over the past few centuries. The highest values occurred between 1965 and 1970, when population growth averaged 2.04 percent per year.

Since 1970, however, this growth rate has been dropping. Between 1990 and 1995 the growth rate in human populations averaged 1.46 percent per year; currently the rate is 1.33 percent per year. The rate of growth in human populations is slowing, for what may be the first time in history. What will human population size be at its peak?

To answer this question, the UN Population Division makes regular projections based on three different scenarios. These scenarios, in turn, hinge on different values for the average number of children that each woman has during her lifetime. Currently, the worldwide average for fertility is 2.7. (This is a huge reduction from fertility rates during the 1950s, which averaged 5.0 children per woman.) **Figure 48.5b** shows how total population size is expected to change between now and 2050 if average fertility drops to 2.5, 2.1, or 1.7 children per woman. The middle number, 2.1, is particularly important because it represents the replacement rate—the average fertility rate required for each woman and man to produce exactly one offspring that survives long enough to breed. (At this fertility rate, $r = 0$.)

Even a glance at the graphs should convince you that the three scenarios are starkly different. The high projection predicts that world population will reach 10.7 billion in 2050 and show no signs of peaking. That's many more than today's total, which is slightly above six billion. The low projection, in contrast, predicts that the total human population in 2050 will be at 7.3 billion and will have already peaked.

The table at the right in Figure 48.5b makes another important point. The UN had to alter its population projections dramatically between 1992 and 1998, primarily to account for the impact of AIDS. Growth rates are dropping dramatically in areas of the world that are being hard hit by the epidemic.

To summarize, it appears clear that humans are ending a period of rapid growth that lasted well over 500 years. How quickly growth rates decline and how large the maximum population eventually becomes will be decided by changes in fertility rates and the course of the AIDS epidemic.

Are density-dependent processes like disease and resource shortages responsible for these changes? In other words, are humans approaching Earth's carrying capacity? This question is controversial, and is explored in more depth in the essay at the end of the chapter. For now, let's turn to questions about a different pattern of population change.

Population Cycles: The Case of the Red Grouse

Why do some animal populations show regular fluctuations in size? For example, over three-fourths of red grouse populations (**Figure 48.6a**, page 938) in Britain rise and fall in regular cycles (**Figure 48.6b**). These cycles average between four and eight years in length. Explaining why cycles like these occur has been a particularly stubborn problem in population biology. Most hypotheses to explain them hinge on some sort of density-dependent factor. The idea is that predation, disease, or food shortages intensify at high population density and cause population numbers to crash. It has been extremely difficult, however, to document a causal relationship between population cycles and factors such as disease outbreaks, predation intensity, and food availability.

What factor causes red grouse populations to cycle? Peter Hudson and colleagues suspected that a parasitic roundworm called *Trichostrongylus tenuis* is responsible. They contended that infection rates skyrocket when grouse densities are high, causing populations to drop rapidly. They also hypothesized that infection rates lessen when grouse densities are low, allowing populations to recover. The idea here is that the parasite is transmitted readily at high population densities, and that individuals in dense populations are less well-fed and thus less able to fend off an infection.

To test these propositions experimentally, the researchers caught from 1000 to 3000 red grouse in each of several populations, treated the individuals with a drug that kills roundworms, and released them. Then they monitored the number of birds shot by hunters in the ensuing years in each of the treatment populations. Finally, they compared these data to hunting success in two nearby populations where hunting intensity was comparable, and where birds were not treated with the anti-roundworm drug.

Some results of the experiment are shown in Figure 48.6b. As you can see, the control populations showed a typical dramatic four-year cycle in numbers, but the treated populations maintained high densities. This is strong evidence that red grouse cycles are driven by density-dependent changes in disease rates. The study also illustrates the power of a well-designed experiment to solve long-standing controversies in ecology.

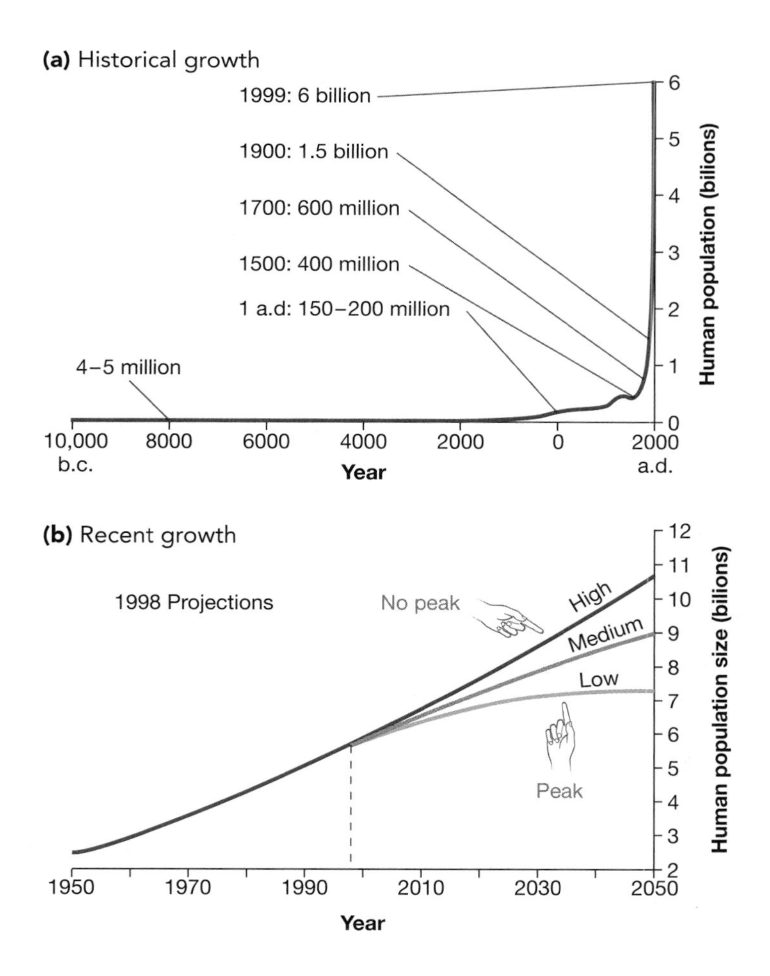

FIGURE 48.5 Human Population Growth
(a) This graph is based on estimates from archaeological, historical, and census data. **(b)** Each time the UN Population Division updates its database on population growth, it makes a 50-year projection based on high, medium, and low average-fertility rates over the interval. The projections shown at lower left were released in 1998 and were based on an average of 2.5, 2.1, or 1.7 children being born per woman. The table on the right summarizes projections made in 1992.

1992 Projections

Fertility rate	Projected population in 2050
High	12.5 billion
Medium	10.15 billion
Low	7.8 billion

The 1992 projections for 2050 are higher than those from 1998 primarily because the earlier projections did not account for the impact of AIDS.

Recovering from Trauma: The *Exxon Valdez* Oil Spill

Natural and human-caused disasters provide biologists with opportunities to study population growth in the context of recovery. If a habitat is suddenly degraded or if large numbers of individuals are killed, what is the impact on subsequent birth, death, immigration, and emigration rates? Do some species bounce back better than others? If so, why?

The *Exxon Valdez* oil spill may be the most intensively studied of all environmental disasters. On March 24, 1989, the oil tanker *Exxon Valdez* ran aground on Bligh Reef in Alaska's Prince William Sound. About 41 million liters (over 10 million gallons) of oil leaked out, making the event the largest marine oil spill in U.S. history.

In the wake of the spill, biologists set out to determine its impact on marine life. Unfortunately, data on population sizes before the oil spill were unavailable for most types of organisms. Some pre-spill data had been collected for seabirds, however. Contact with oil had an immediate impact on these species. About 35,000 carcasses of marine birds were collected from oil-affected areas (**Figure 48.7**). The current consensus estimate of total seabird mortality—including bodies that were never recovered—is that 250,000 individuals perished.

Based on these data, researchers hypothesized that most or all seabird populations would decline rapidly in oil-affected areas relative to pre-spill levels. To test these predictions, Stephen Murphy and co-workers set out to estimate seabird populations by cruising along precise lines, or transects, that corresponded to the areas surveyed in a count conducted in 1984–1985. As the researchers moved along each transect, they counted the number of seabirds that were resting or flying within a 200 m-wide region extending from the high-tide line on shore out onto the water. Using this protocol, they estimated population numbers of seabirds in 10 bays within Prince William Sound. They also repeated the censuses so they could document any changes in population sizes that occurred in 1989, 1990, and 1991.

According to their data, 7 of the 11 species in the study showed no significant response to oiling, indicating that their numbers in Prince William Sound did not change as a result of

(a) Red grouse

(b) Population cycle experiment

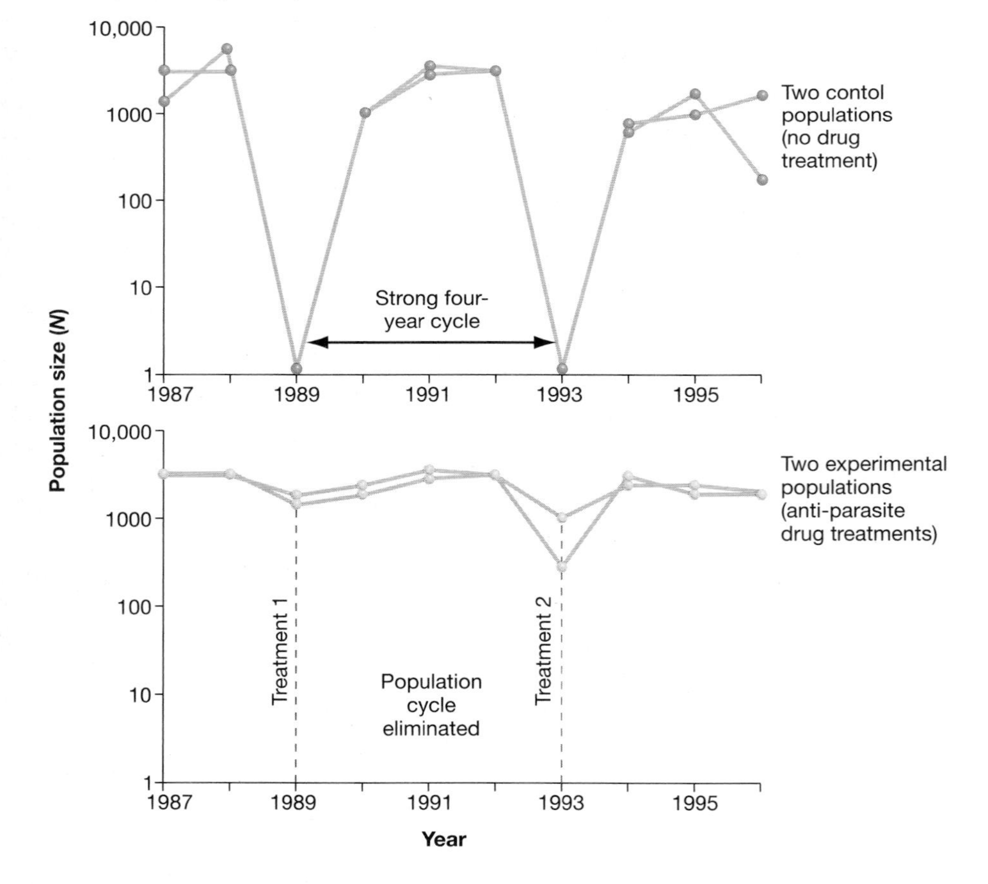

FIGURE 48.6 Population Cycles in Red Grouse
(a) Red grouse are native to heather moorlands in Britain. **(b)** The top graph shows the number of red grouse shot in two control populations. The bottom graph shows the number of birds shot each year in two populations treated with a drug that kills roundworms. Large numbers of individuals were caught and treated in 1989 and 1993.

the oil spill. The leading hypothesis to explain these data is that enough individuals immigrated from nearby populations to make up for the losses in oiled areas. One species, the black-legged kittiwake, even exhibited a *positive* response to the event, probably because of being attracted to food garbage left by cleanup crews. (**Box 48.1**, page 940, summarizes research on aspects of the cleanup effort that worked or did not work.) Three of the seabird species in the study had a negative initial response to oiling, but had started to show some signs of recovery by the third year after the spill.

Other types of organisms were not so fortunate, however. Detailed studies of bird, mammal, and fish species have supported two general conclusions about the spill's impact.

- Recovery has been very slow for species that depend on habitats where oil has persisted. For example, oil tended to collect and persist under beds of mussels like those shown in **Figure 48.8a**. Because mussels are filter feeders, they have continued to ingest these oil molecules as they feed. Species that feed heavily on mussels, like the Barrow's goldeneyes and sea otters pictured in **Figure 48.8b**, have yet to recover to their pre-spill numbers.
- The availability of immigrants from nearby populations affected the recovery of many species. Enough seabirds were present in nearby populations to furnish immigrants and dampen some effects of the spill. Pink salmon, in contrast, did not benefit from immigration. These fish breed in freshwater streams or in protected beach areas. Young pink salmon move to the open ocean to feed for two years but then return to their natal habitat to breed. In oiled areas, many salmon embryos failed to develop, and growth rates of juveniles were low. Pink salmon populations were particularly slow to recover from the spill because other salmon populations continued to return to their natal areas to breed, and thus did not augment populations in Prince William Sound by immigration.

FIGURE 48.7 Seabird Mortality After an Oil Spill
Tens of thousands of seabirds died after being coated with oil from the *Exxon Valdez* spill in Alaska.

(a) Persistent oil in mussel beds...

(b) ...led to slow recovery in other species.

FIGURE 48.8 Long-Term Effects of the *Exxon Valdez* Oil Spill
(a) This oiled mussel bed was photographed five years after the spill. **(b)** Populations of Barrow's goldeneye (above), sea otters (below), and other species that feed heavily on mussels had not recovered within a decade after the spill.

BOX 48.1 What's the Best Way to Clean Up an Oil Spill?

The tenth anniversary of the *Exxon Valdez* oil spill provided an opportune moment for researchers to summarize the lessons of the disaster. One important topic was to assess which cleanup methods worked well, and which were less effective. If and when a major oil spill recurs in this or similar environments, biologists hope that the cleanup effort will be faster and more cost-effective. Data on the efficacy of cleanup techniques supported a variety of important conclusions.

- Rinsing oiled beaches with high-pressure streams of hot water (**Figure 1**) probably did more harm than good. Pressure-washing slowed recovery because it removed virtually all of the multicellular organisms from the beaches. It also removed most of the sand and other fine sediments that provided a substrate for clams and other burrowing organisms, leaving just gravel and large cobbles. As a result, this cleanup technique radically changed the nature of these environments.
- Because so many seabirds and mammals eat mussels, most mussel beds were left uncleaned. The logic here was that predators would starve if mussel densities were reduced during the cleaning process. Unfortunately, for years mussels continued to ingest the oil that collected at their bases. As a result, mussels and the animals that eat them were poisoned.
- Bacteria in beach sediments were extremely effective at degrading the oil. Efforts to encourage the growth of oil-eating bacteria were probably the most effective aspect of the cleanup operation. Storms that dispersed oil from beaches and the ocean surface were another particularly valuable cleanup agent.

FIGURE 1 A Failed Technique?
Members of this cleanup crew used high-pressure hoses to spray hot water and clean oiled beaches after the *Exxon Valdez* spill.

The general message here is that populations are able to recover from catastrophic losses if the habitat is allowed to recover. The recovery process is rapid if nearby populations serve as a source of immigrants.

48.3 Population Structure

To predict the fate of endangered species and other populations, biologists not only have to understand how populations grow and change through time, but how populations are structured. Understanding the history of a population and predicting its future entails more than simply tallying the overall number of individuals. It is also critical to know how many reproductively active individuals there are versus juveniles that are too young to reproduce. Is a large proportion of the population old and likely to die soon? Are groups of individuals arranged in space in a way that makes emigration and immigration important?

Age Structure in Developed Versus Developing Nations

To understand why documenting the age of the individuals in a population is informative, consider the human populations of developed and developing nations. In nations where industrial and technological development is advanced and average incomes are relatively high, the age distribution of the population tends to be even. As the horizontal bars in **Figure 48.9a** show, developed nations have similar numbers of people in most age classes. This type of age structure results from long periods of no or very slow population growth.

In addition to shedding light on a population's history, though, age distributions can help predict its future. For example, the populations of developed nations are not expected to grow especially quickly, because only modest numbers of individuals will be reaching reproductive age in the near future. The green lines in Figure 48.9a show the projected age structure of developed nations in 2050. They highlight a major public policy concern in developed countries—how to care for an increasingly aged population.

In contrast, the age distribution is bottom-heavy in the less-developed nations. As **Figure 48.9b** shows, these populations are dominated by the very young. This type of age distribution occurs when populations have undergone a period of rapid growth. Because most people in the younger age classes are expected to survive to reproductive age, extremely rapid population growth is expected to continue. The projected age distribution in 2050, shown in green, illustrates major public policy concerns in less-developed countries—providing education and jobs for an enormous group of young people who will want to be starting families. Understanding age structure is a key component of analyzing population dynamics.

Geographic Structure

If you browse through an identification guide for trees or birds or butterflies, you'll likely find range maps indicating that the species occupies a broad range. In reality, however, the habitat preferences of many species are restricted, and individuals occupy only isolated patches within the broad area. To capture this point, biologists say that many species exist as a metapopulation, or a population of populations. As **Figure 48.10a** (page 942) shows, a **metapopulation** is made up of many small populations isolated in fragments of habitat.

Figure 48.10b shows why this geographic structure is important. Over time, each population within the larger metapopulation is likely to be wiped out. The cause could be catastrophic, like an oil spill or violent storm; it could also be a disease outbreak or a sudden influx of predators. Migration from nearby populations can reestablish populations in these empty habitat fragments, however. In this way, the overall population is maintained at a stable number of individuals, even though subpopulations blink on and off over time.

The metapopulation concept originated as a purely theoretical construct—an abstraction with interesting population dynamics. A group of researchers led by Ilkka Hanski has shown that metapopulations actually exist in nature, however. These biologists work on the Glanville fritillary butterfly (**Figure 48.11**, page 943), which now resides only on the Åland islands off the coast of Finland. Hanski and colleagues began their research on the fritillary with a survey of meadow habitats on the islands. Because the butterfly feeds on just two host plants, *Plantago lanceolata* and *Veronica spicata*, they were able to pinpoint potential butterfly habitats. They estimated the population size within each patch by counting the number of larval webs in each. Of the 1502 meadows that contained the host plants, 536 contained Glanville fritillaries. Most had only a single larval group; the largest population they found contained 3450 larvae.

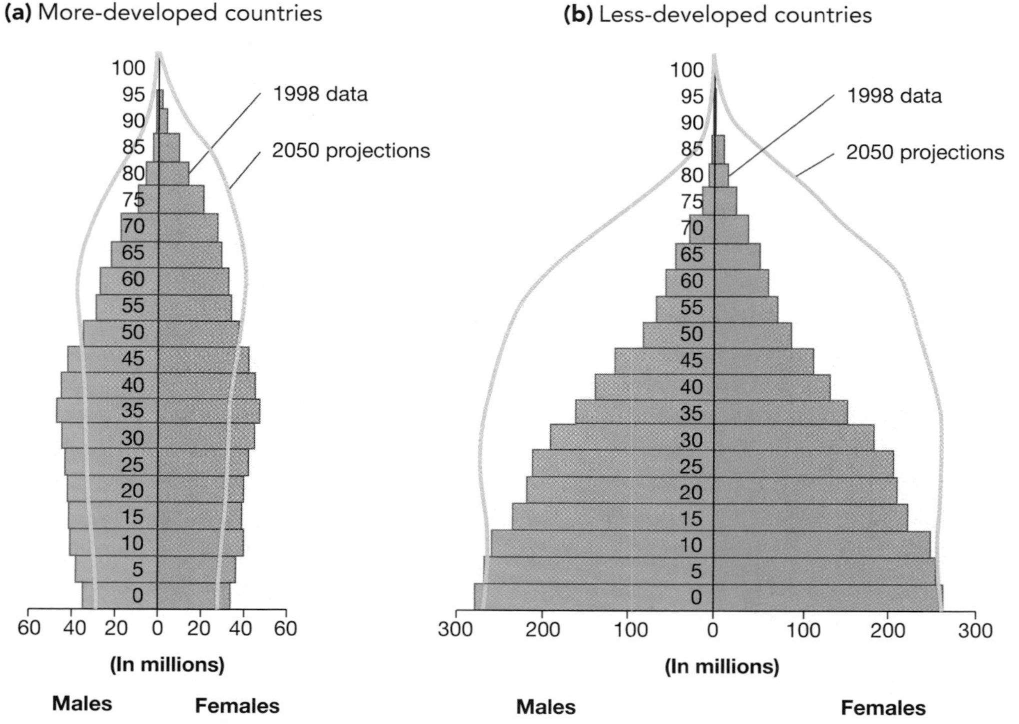

FIGURE 48.9 The Age Structure of Human Populations Varies Dramatically These age distributions show the number of males and females in 5-year age increments from 0 to 100. The horizontal bars represent data from 1998; the green lines show projections for 2050.

FIGURE 48.10 Metapopulation Dynamics
The overall population size of a metapopulation stays relatively stable even if subpopulations go extinct. These populations may be restored by migration, or unoccupied habitats might be colonized.

(a) A metapopulation is made up of small, isolated populations.

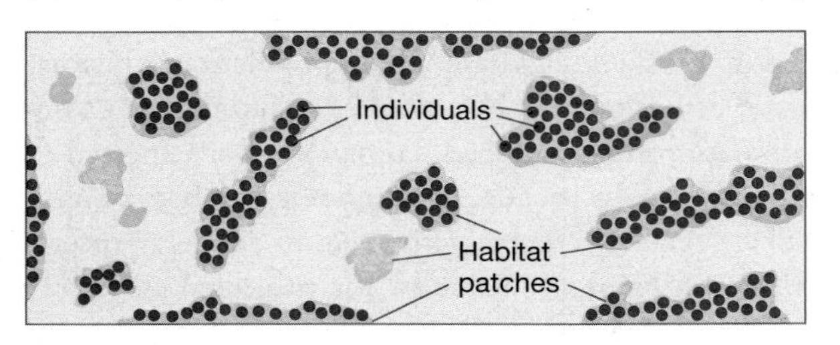

(b) Although some subpopulations go extinct over time...

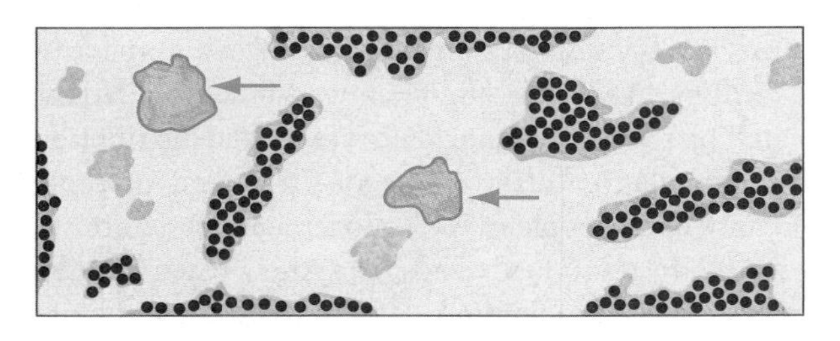

...migration can restore or establish subpopulations.

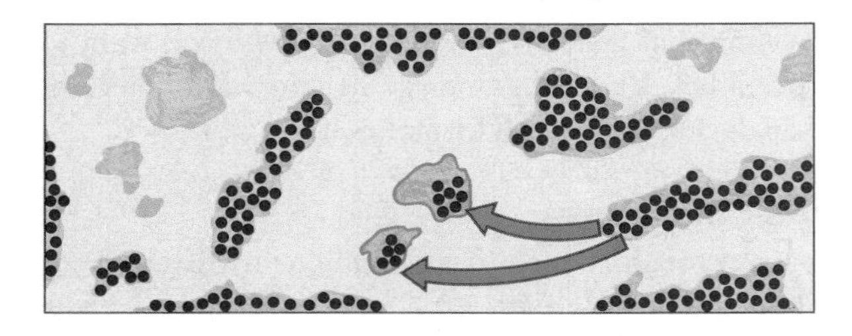

To determine whether migration among these patches occurs, the investigators conducted a mark-recapture study. (**Box 48.2**, provides details on how a mark-recapture study is done.) Of the 1731 butterflies that they marked and released, 741 were recaptured over the course of the summer. Of the recaptured individuals, 9 percent were found in a new patch. This migration rate is high enough to suggest that patches where a population has gone extinct will eventually be recolonized.

Hanski and co-workers repeated the survey two years after their initial census. Just as the metapopulation model predicts, some populations had gone extinct and others had been created. In addition to confirming that the metapopulation model is valid, the work has important implications for endangered species like the Glanville fritillary. A small, isolated population—even one within a nature reserve—is unlikely to survive over the long term.

Based on this realization, conservation biologists are attempting to design reserves for threatened species that are sizeable enough to maintain large populations. When this is not possible, an alternative is to establish systems of small reserves that are connected by corridors of habitat, so that migration between patches is possible. Hanski also emphasizes that it is crucial to preserve at

BOX 48.2 Mark-Recapture Studies

To estimate the population size of sedentary organisms, researchers sample appropriate habitats by counting the individuals that occur along transects or inside rectangular plots—called quadrats—set up at randomized locations in the habitat. These counts can be extrapolated to the entire habitat area to estimate the total population size. In addition, they can be compared to later censuses to document trends in population size.

In contrast, estimating the total population size of organisms that move, like the Glanville fritillary, is much more of a challenge. In these species, the population inside sample quadrants or along transects changes constantly as individuals move in and out. Further, it can be difficult to track whether a particular individual has already been counted.

If individuals can be captured and tagged in some way, then the total population size of a mobile species can be estimated using the mark-recapture technique. To begin a mark-recapture study, researchers catch individuals in live traps and mark them with leg bands, ear tags, or some other identification system. After the marked individuals are released, they are allowed to mix with the unmarked animals in the population for a period of time. Then a second trapping effort is conducted, and the percentage of marked individuals captured is recorded.

To estimate the total population from these data, researchers make the assumption that the percentage of marked and recaptured individuals is equal to the percentage of marked individuals in the entire population. This assumption should be upheld when no bias exists regarding which individuals are caught in each sample attempt. It is important that individuals do not avoid traps after being caught once, or that they do not emigrate after being trapped.

The relationship between marked and unmarked individuals can be concisely expressed algebraically.

$$\frac{m_2}{n_2} = \frac{n_1}{N}$$

In this equation, m_2 is the number of marked animals in the second sample, n_2 is the total number of animals (marked and unmarked) in the second sample, n_1 is the number of individuals caught in the first sampling attempt, and N is the total population size. Having measured m_2, n_2, and n_1, the researcher can estimate N.

FIGURE 48.11 The Glanville Fritillary
This butterfly is extinct in much of its former range, and is now found only on the Åland islands.

least some patches of currently unoccupied habitat as well, to provide future homes for immigrants in the metapopulation.

To summarize, understanding a population's age structure and geographic structure is critical to predicting its future. But to make detailed predictions and to design management plans for threatened species, even more information is required. Let's take a closer look.

48.4 Demography and Conservation

Demography is the study of factors that determine the size and structure of populations through time. To make detailed predictions about the future of a population, biologists need to know more than the general age structure and geographic structure of a population. They also need to know how likely it is that individuals in each age group will survive to the following year. They need to know how many offspring are produced by females of different ages, and how many individuals of different ages immigrate and emigrate each generation.

Demographic data like these provide an important tool for biologists charged with designing management programs for endangered species. To understand the nature of these data and how they are used in conservation programs, let's examine a classical tool for describing the demography of a population.

Life Tables

Formal demographic analyses of populations are based on a type of data set called a life table. Life tables were invented almost 2000 years ago; in ancient Rome they were used to predict food needs. In modern times, life tables have been the domain of life insurance actuaries, who need to predict the likelihood of a person dying at a given age. More recently, biologists have employed life tables to study the demographics of endangered species and other nonhuman populations. A life table summarizes the probability that an individual will survive and reproduce in any given year over the course of its lifetime.

To understand how researchers use a life table to predict the future of a population, consider data that Henk Strijbosch and R. C. M. Creemers collected on the lizard *Lacerta vivipara* (**Figure 48.12**). *L. vivipara* is a common resident of open, grassy habitats in Western Europe. Strijbosch and Creemers' goal was to estimate the life table of a low-elevation population in the Netherlands, and compare their results to data that other researchers had collected from *L. vivipara* populations in the mountains of Austria and France and from lowland populations in Britain and Belgium. In this sense, their research was "pure." Their work was not motivated by a desire to solve a specific, applied problem like saving an endangered species, but simply to find interesting patterns in the comparative demography of populations.

To complete their study, the researchers visited their study site daily during the seven months that lizards are active during the year. Each day, Strijbosch and Creemers captured and marked as many individuals as possible. Because they continued this program of daily monitoring for seven years, they were able to document the number of young produced by each female in each year of its life. If a marked individual was not recaptured in a subsequent year, they assumed that it had died sometime during the previous year. These data allowed them to calculate the number of individuals that survived each year in each particular age group. Let's take a closer look.

FIGURE 48.12 ***Lacerta vivipara***
Translated literally, *vivipara* means "live birth." Females in some populations lay eggs; in other populations, females bear live young.

Survivorship and Age-Specific Mortality Survivorship is a key component of a life table. **Survivorship** is defined as the proportion of offspring produced that survive, on average, to a particular age. For example, if 1000 *L. vivipara* are born in a particular year, how many would survive to age 1, age 2, age 3, and so on? Survivorship is symbolized as l_x, where x represents the age class being considered. Survivorship for an age class is calculated by dividing the number of individuals in that age class by the number of individuals in the first age class.

Equation 48.6
$$l_x = \frac{N_x}{N_0}$$

As **Table 48.1** shows, Strijbosch and Creemers calculated survivorship from birth to age 1 as 0.424. If 1000 individuals were born in a particular year, on average 424 would still be alive one year later.

When the number of survivors is plotted versus age, a survivorship curve results. Studies on a wide variety of species indicate that three general types of survivorship curves exist. As **Figure 48.13a** shows, humans have what biologists call a type I survival curve. In this pattern, survivorship throughout life is high and most individuals approach the species' maximum life span. Songbirds, in contrast, experience relatively constant mortality throughout their lives, resulting in a type II survivorship curve. Many plants have type III curves due to extremely high death rates for seeds and seedlings. **Figure 48.13b** provides a graph for you to plot survivorship for *Lacerta vivipara*.

Fecundity The number of female offspring produced by each female in the population is termed **fecundity.** (In most cases, biologists keep track of only females when calculating life-table data because only females produce offspring.) Because Strijbosch and Creemers documented the reproductive output of the same females year after year, they were able to calculate a quantity called age-specific fecundity (m_x). Age-specific fecundity is the number of female offspring produced by an individual in age class x.

Documenting age-specific fecundity allows researchers to calculate the net reproductive rate, R_0, of the population.

Equation 48.7
$$R_0 = \sum_{i=0}^{x} l_x m_x$$

The logic behind this calculation is that the growth rate of a population per generation is identical to the average number of female offspring that each female produces over the course of her lifetime. Lifetime reproduction is a function of fecundity at each age (m_x) and survivorship to each age class (l_x).

If R_0 is greater than 1, the population is increasing in size. If R_0 is less than 1, then the population is declining. The cell in the

TABLE 48.1 Life Table for *Lacerta vivipara*

x	N_x	l_x	m_x	$l_x m_x$
0	1000	1.000	0.00	0.00
1	424	0.424	0.08	0.03
2	308	0.308	2.94	0.91
3	158	0.158	4.13	0.65
4	57	0.057	4.88	0.28
5	10	0.010	6.50	0.07
6	7	0.007	6.50	0.05
7	2	0.002	6.50	0.01
				$R_0 = 2.00$

Data are from Strijbosch and Creemers, 1988.

(a) Three general types of survivorship curves

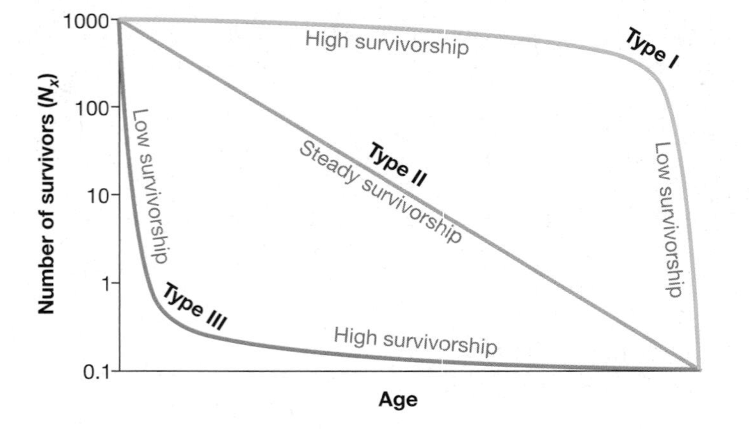

(b) Exercise: Survivorship curve for *Lacerta vivipara*

FIGURE 48.13 Three General Types of Survivorship Curves
These generalized graphs show the number of survivors, on a logarithmic scale, plotted against age.
EXERCISE Fill in the graph in part **(b)** with data on survivorship of *Lacerta vivipara* given in Table 48.1. Compare the shape of the curve to the generalized graphs in part (a).

bottom right of Table 48.1 shows that an average *L. vivipara* female produces 2.00 female offspring over the course of her lifetime. This population is growing rapidly.

Using Life-Table Data to Make Population Projections The data in Table 48.1 are interesting because they contrast with results from other populations of *Lacerta vivipara*. For example, note that almost no 1-year-old females breed in the Netherlands. But in Brittany, France, 50 percent of 1-year-old females breed. In contrast, most females in the mountains of Austria do not begin breeding until they are 4, and most live much longer than individuals in lowland populations. If these differences are due at least in part to genetic differences among populations, it suggests that life-history characteristics like age at first reproduction and life span are adaptations to particular environments.

Why is an understanding of age-specific survivorship and fecundity important for applied problems like saving endangered species? To answer this question, suppose that you were in charge in reintroducing a population of lizards to a nature reserve. Suppose further that research conducted when the species occupied the site previously documented the survivorship and fecundity data in **Figure 48.14a.** Your initial plan is to take 1000 1-year-old females from a captive breeding center and release them into the habitat just before the breeding season begins. What will happen to this population?

To answer this question, you need to calculate (1) how many adults will survive to each age class each year and (2) how many juveniles will be produced by each adult age class each year over the course of several years. **Figure 48.14b** starts these calculations for you. Note that the fate of the original 1000 females is indicated in red. In the second year, just 200 of these individuals have survived; 40 are left as 3-year-olds.

How many offspring did this generation produce? As the first purple number in the table indicates, the 1000 females

(a) Life table

Age (x)	Survivorship (l_x)	Fecundity (m_x)
0 (birth)	----	0.0
1	0.33	3.0
2	0.2	4.0
3	0.04	5.0

(b) Fate of first-generation females

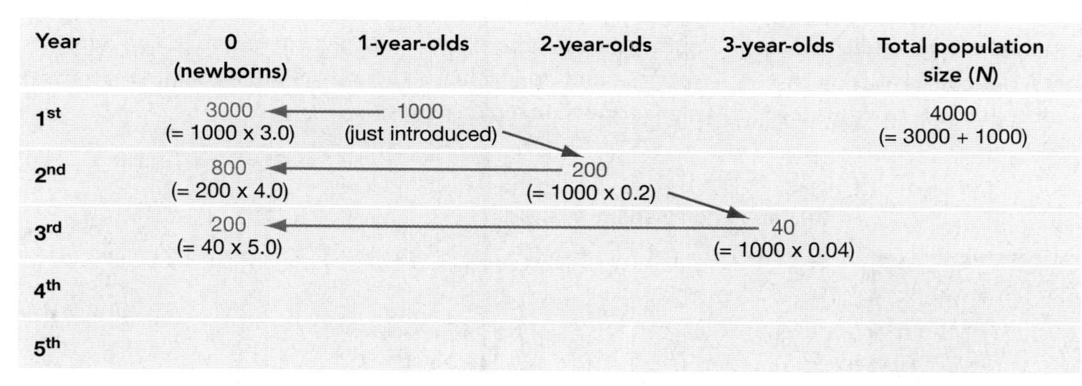

Year	0 (newborns)	1-year-olds	2-year-olds	3-year-olds	Total population size (N)
1st	3000 (= 1000 x 3.0)	1000 (just introduced)			4000 (= 3000 + 1000)
2nd	800 (= 200 x 4.0)		200 (= 1000 x 0.2)		
3rd	200 (= 40 x 5.0)			40 (= 1000 x 0.04)	
4th					
5th					

(c) Fate of first- and second-generation females

Year	0 (newborns)	1-year-olds	2-year-olds	3-year-olds	Total population size (sum across all rows)
1st	3000	1000			4000 (= 3000 + 1000)
2nd	800 + 3000 (3000 = 1000 x 3.0)	1000 (= 3000 x 0.33)	200		5000 (= 3800 + 1000 + 200)
3rd	200 + 3200 (3200 = 800 x 4.0)		800 (= 3000 x 0.2)	40	
4th	600 (600 = 120 x 5.0)			120 (= 3000 x 0.04)	
5th					

FIGURE 48.14 Doing a Population Projection
(a) This life table provides age-specific survivorship and fecundity for a hypothetical population.
(b) This table predicts the fate of 1000 1-year-old females introduced to a habitat just before the breeding season. The number of individuals in this cohort is shown in red; the offspring they produce each year is indicated in purple. **(c)** This table extends the data in part (b) by indicating how many of the offspring produced by the original females in their first year survived in subsequent years, shown in purple numbers, and how many offspring they produced in each of the subsequent years, shown in green numbers. EXERCISE Assume that all 4-year-old females die after producing 3 young. Fill in the 4th and 5th years in the table.

have an average of 3 female offspring apiece, so they contribute 3000 new juveniles to the population. In the second year, the 200 females that are left have an average of 4 female offspring each and contribute 800 juveniles. In their third year, the 40 surviving females average 5 offspring and contribute 200 juveniles.

Figure 48.14c extends the calculations by showing what happens as the offspring of the original females begin to breed. Their offspring are shown in green. By adding subsequent generations and continuing the analysis, you could predict whether the population will reach a steady state, decline, or increase over time. In this way, life-table data can be used to predict the fate of populations.

Using Life-Table Data to Guide Conservation Programs

Part of the value of a population projection like the one begun in Figure 48.14 is that it allows biologists to alter values for survivorship and fecundity at particular ages and assess the consequences. For example, suppose that a predatory snake began preying on juvenile lizards. According to the model in Figure 48.14, what would the impact of a change in juvenile mortality rate be? Analyses like this allow biologists to determine which aspects of survivorship and fecundity are particularly sensitive for particular species. The studies done to date support some general conclusions.

- Whooping cranes, sea turtles, spotted owls, and many other endangered species have high juvenile mortality, low adult mortality, and low fecundity. In these species, the fate of a population is extremely sensitive to increases in adult mortality. Based on this insight, conservationists have recently begun an intensive campaign to reduce the loss of adult female sea turtles in fishing nets. Previously, most conservation action had focused on protecting eggs and nesting sites.
- In humans and other species with high survivorship in most age classes, rates of population growth are extremely sensitive to changes in age-specific fecundity. Based on this result, programs to control human population growth focus on two issues: lowering fertility rates through the use of birth control, and delaying the age of first reproduction by improving women's access to education.

In some or even most cases, however, the population projections made from life-table data may be too simple to be useful. For example, conservationists may need to expand the basic demographic models to account for occasional disturbances like fires or storms or even oil spills. If the population exists in the fragmented habitats typical of a metapopulation, planners also need to assess the impact of emigration and immigration from nearby populations.

Population Viability Analysis

A population viability analysis, or PVA, is a model that estimates the likelihood that a population will avoid extinction for a given time period. In most cases, PVAs attempt to combine basic demographic models for the species in question with data on geographic structure and the rate and severity of habitat disturbance. Typically, a population is considered viable if the analysis predicts that it has a 95 percent probability of surviving for at least 100 years. Natural resource managers are currently using PVA to assess the effects of logging, development, and other land management practices on sensitive species, and to evaluate the merits of alternative recovery plans for endangered species.

(a) Leadbeater's possum

(b) Population viability analysis

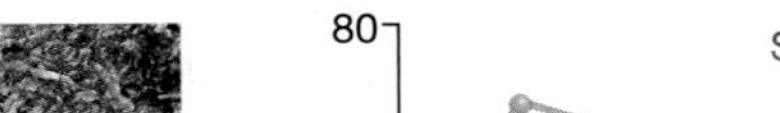

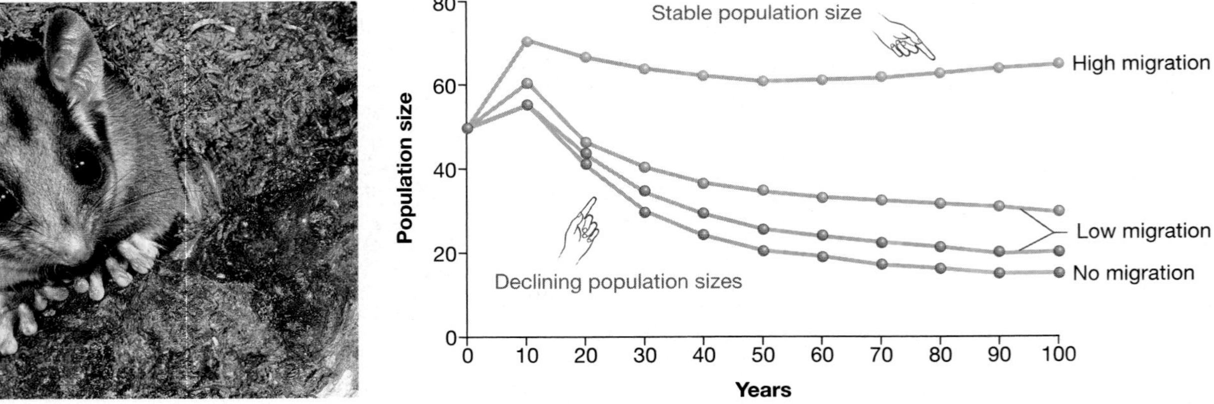

FIGURE 48.15 How Will Loss and Fragmentation of Habitat Affect Leadbeater's Possum?
(a) Leadbeater's possum is a marsupial that breeds in the old-growth forests of southeast Australia. **(b)** This graph plots the fate of populations under four different scenarios in a PVA. The scenarios allow different rates of migration among habitat patches, ranging from a case where patches are completely isolated and no migration occurs to a case where 5 percent of the individuals migrate each year. The no-migration scenario simulates what might happen if extensive deforestation occurred and created isolated patches of old-growth forest.

A recent study—conducted by David Lindenmayer and Robert Lacy on an endangered marsupial called Leadbeater's possum—illustrates the value of a carefully constructed PVA. Leadbeater's possum (**Figure 48.15a**) inhabits old-growth forests in southeastern Australia and relies on dead trees for nest sites. The goal of the PVA was to assess the effects that logging these habitats might have on the viability of the species.

Life-table data are difficult to obtain for Leadbeater's possum because the species is rare, lives in trees, and is active at night. Enough data were available from field studies, however, to provide an estimate for age-specific survival and the average fecundity per female per year. Lindenmayer and Lacy used these data to make population projections while varying the spatial configuration of habitat patches. The analysis allowed them to assess the impact of migration and to simulate the effects of logging, fires, storms, and other types of disturbances.

Figure 48.15b illustrates their results. The four lines on the graph describe the changes in population size predicted to occur over time, based on different assumptions about the migration rates between groups of possums in fragments of old-growth forest. When migration is high, as in the top line, the overall population size is predicted to stabilize at approximately 70 individuals. In the scenario where no migration among fragments occurs, the final population size is predicted to be fewer than 20 individuals. Based on these results, Lindenmayer and Lacy concluded that extensive timber harvesting would pose a serious threat to the species. The model showed that reducing the size and number of remaining old-growth fragments would reduce migration and lead to rapid population declines.

Like all analyses based on simulations, though, a PVA makes many assumptions about future events and is only as accurate as the data entered into it. In particular, the usefulness of PVAs has been challenged because basic demographic information is lacking or poorly documented in many endangered species. Barry Brooks and colleagues recently defended the approach, however, by analyzing 21 long-term studies that have been completed on threatened populations of animals. To do this analysis, the researchers broke each of the 21 data sets in two. Data from the first half of each study was used to run a PVA on each of the 21 species. When the researchers compared the predictions of the PVAs to the actual data from the second half of each study, they found that the correspondence was extremely close. This result strengthens confidence that a PVA has good predictive value for land managers.

Essay What Limits Human Population Growth?

On October 12, 1999, the United Nations Population Division estimated that the human population exceeded six billion (6,000,000,000) for the first time. Why has the growth rate of the human population increased over time? Is our species at or near the planet's carrying capacity?

Has our species surpassed Earth's carrying capacity?

Population biologists point to several key events in human history that have contributed to the rising growth rate. The first event occurred about 10,000 years ago, when groups of humans began to develop agriculture. The transition from a nomadic existence to a settled, agricultural one was associated with dramatic increases in fecundity. More recently, improvements in medicine and hygiene have led to dramatic declines in death rates and especially in infant mortality. These technological advances are responsible for the dramatic rise in human populations over the past 100 years.

Until recently, most people considered population growth as a good thing for religious, economic, and political reasons. As the twentieth century progressed and human population size continued to explode, however, a few people began to express concern about the ecological consequences of population growth. Paul Ehrlich, who began his career studying butterflies, has been one of the most prominent voices arguing that human population growth must cease. His concern is that humans will exceed Earth's carrying capacity and cause an ecological collapse that might lead to our own downfall. At the root of Ehrlich's arguments is the concern that exponential growth has shot us past Earth's carrying capacity for humans, and is causing us to do irreparable harm to the environment. But what is Earth's carrying capacity, and how do we know that we have surpassed it?

Population demographers who have attempted to calculate Earth's carrying capacity for humans have produced estimates that range from 2 billion people to over 1 trillion people. The low number is much less than the current population; the high number assumes that people would live in floating cities above the planet's surface. These estimates are variable because carrying capacity depends on the standard of living that people are willing to accept, and on technological advancements. Could each of us live with only 100 m^2 of living space and 1500 calories per day? Could we build floating cities in time?

Joel Cohen argues that the classical concept of carrying capacity has little meaning for humans. Although he acknowledges that density-dependent factors like competition and disease are already affecting human populations, Cohen em-

(Continued on page 948)

(Essay continued)

phasizes that our future hinges much more directly on how many children the 6 billion of us decide to have, and how many resources each of us demands. His point is that human choice is a key element in defining carrying capacity.

From this point of view, Earth's carrying capacity is what we decide it will be. For example, the human population is already too large for all 6 billion of us to live with the same standard of living as Americans and Western Europeans. Increased demand for this level of resource use, or fertility rates that exceed the replacement rate, would almost certainly bring about the type of ecological collapse envisioned by Ehrlich. Over the next few decades, the decisions made by 6 billion people will determine the future of life on Earth.

Chapter Review

Summary

A population is a group of individuals of the same species that occupy the same area at the same time. One of the basic characteristics of a population is its growth rate. Exponential growth occurs when the per-capita growth rate, r, does not change over time. Eventually, however, growing populations approach the carrying capacity of their environment. As population density increases, competition for resources, disease, or other factors succeed in slowing and then stopping growth.

Field studies of natural populations have confirmed the existence of exponential and density-dependent growth in species ranging from coral-reef fish to humans. Density dependence is also a key feature of the regular population cycles observed in certain species. In the case of red grouse, roundworm infections spread rapidly when population density is high and cause the population to crash. Environmental catastrophes like the *Exxon Valdez* oil spill have allowed biologists to study how populations recover from catastrophic losses in numbers or habitat degradation. In the case of the *Exxon Valdez*, the rate of recovery depended on the rate of habitat recovery and the availability of immigrants from nearby populations.

Researchers are also interested in the structure of populations—particularly the number of individuals in various age classes. A population with few juveniles and many adults past reproductive age, like the human populations of the developed world, may be declining or stable in size. In contrast, a population with a large proportion of juveniles is likely to increase rapidly in size. Human populations in the developing nations currently have this type of age distribution.

The realization that human activities are isolating populations in small, fragmented habitats has inspired interest in the dynamics of metapopulations. The history of a metapopulation is driven by the birth and death of populations, just as the dynamics of a single population are driven by the birth and death of individuals. Because migration among habitat patches is essential for the stability of a metapopulation, conservationists are trying to establish corridors that link small fragments of preserved habitat.

Demography is the study of patterns in births and deaths in populations. Demographic data are the basis of population projections that are fundamental to population viability analysis, or PVA. Most PVAs attempt to model the effects that different management strategies might have on populations of endangered species. A PVA estimates the probability that a population will persist for a certain number of years under a prescribed set of demographic and habitat conditions.

Although this chapter has considered the dynamics of populations as if they existed in a vacuum, in reality every population shares its environment with other species. Chapter 49 investigates the kinds of interactions that occur between species and explores how those interactions affect the fate of populations.

Questions

Content Review

1. What is the defining feature of exponential growth?
   a. It lasts indefinitely.
   b. The growth rate is constant.
   c. The growth rate increases rapidly over time.
   d. The growth rate is very high.

2. What four factors define population growth?
   a. age-specific birthrates and death rates, age and metapopulation structure
   b. survivorship, age-specific mortality, fecundity, death rate
   c. mark-recapture, census, quadrat sampling, transects
   d. births, deaths, immigration, emigration

3. When does exponential growth tend to occur?
   a. in populations that colonize new habitats
   b. in populations that experience intense competition
   c. in populations that experience high rates of predation
   d. in declining populations

4. Why does population growth decline as population size approaches carrying capacity? (Circle all correct statements.)
   a. Climate becomes unfavorable.
   b. Competition for resources increases.
   c. Predation rates increase.
   d. Disease rates increase.

5. If most individuals in a population are young, why is the population likely to grow rapidly in the future?
   a. Many individuals will begin to reproduce soon.
   b. The population has a skewed age distribution.
   c. Immigration and emigration can be ignored.
   d. Death rates will be low.

6. Why have population biologists become particularly interested in the dynamics of metapopulations?
   a. Because they are an interesting theoretical construct.
   b. Because whooping cranes exist as a metapopulation.
   c. Because they explain why populations occupying large, contiguous areas are vulnerable to extinction.
   d. Because many populations are becoming restricted to small islands of habitat.

Conceptual Review

1. Explain equations 48.3 and 48.5 in words.
2. Draw type I, II, and III survivorship curves on a graph with labeled axes. Explain why the growth rate of species with type I survivorship curves depends primarily on fertility rates. Explain why the growth rate of species with type III survivorship curves is extremely sensitive to changes in adult survivorship.
3. Offer a hypothesis for why humans have undergone near-exponential growth for over 500 years. Why can't exponential growth continue indefinitely? Give two examples of density-dependent factors that influence population growth.
4. Compare and contrast the dynamics of a population that resides in a large contiguous habitat to those of a metapopulation. Assume that the total amount of area occupied is the same in the metapopulation as in the contiguous population.
5. Make a rough sketch of the age distribution in developing versus developed countries, and explain the significance of the differences. How is AIDS, which is a sexually transmitted disease, affecting the age distribution in countries hard-hit by the epidemic?

Applying Ideas

1. The chapter described the effects of the *Exxon Valdez* oil spill on birds and other vertebrates. Design an experiment or observational study to determine the effects of the spill on algae that live in the intertidal zone (the area between high and low tide). What effect would you expect an oil spill to have on these organisms, which do not move? How would your study test this hypothesis?
2. When wild plant and animal populations are logged, fished, or hunted, only the oldest or largest individuals tend to be taken. What impact does harvesting have on a population's age structure? How might this affect the population's life table and growth rate?
3. Design a system of nature preserves for an endangered species of beetle whose larvae feed only on one species of sunflower. The sunflowers tend to be found in small patches that are scattered throughout dry grassland habitats. Explain the rationale behind your proposal.
4. Snowshoe hares are preyed upon by lynx. Both species show pronounced population cycles. The cycles are roughly synchronized, although the rise and fall of lynx populations slightly lags the rise and fall of hare populations. Design an experiment to test the hypothesis that the cycles are driven by predation of hares by lynx.
5. In most species the sex ratio is at or near 1.00, meaning that there is an approximately equal number of males and females. In contemporary China, however, there is a strong preference for male children; the sex ratio of infants is 1.15, meaning that there are 115 boys for every 100 girls. Do you expect this skewed sex ratio to affect the population growth rate in China in the future? Explain your answer.

CD-ROM and Web Connection

CD Activity 48.1: Modeling Population Growth *(animation)*
(Estimated time for completion = 5 min)
How does a population grow with unlimtied resources compared to growth with limited environmental resources?

CD Activity 48.2: Human Population Growth and Regulation *(tutorial)*
(Estimated time for completion = 10 min)
Examine the effects of birth and death rates on the growth of the human population.

At your **Companion Website** (http://www.prenhall.com/freeman/biology), you will find self-grading exams and links to the following research tools, online resources, and activities:

Population Growth and Balance
Explore current models of population growth rates and issues of population balance.

Cyclic Populations
This site describes research on voles and their predator populations in the United Kingdom.

Metapopulations
Review the concept of the metapopulation and its relationship to issues in population biology, and learn how to apply the Geographical Information System (GIS) to questions on metapopulations.

Society for Conservation Biology
This site describes the research, advocacy, and education programs supported by the Society for Conservation Biology.

Additional Reading

Cohen, J. E. 1995. *How Many People Can the Earth Support?* (New York: W.W. Norton). A demographer evaluates Earth's carrying capacity for humans.

Grant, R. B., and P. R. Grant. 1989. *Evolutionary Dynamics of a Natural Population* (Chicago: University of Chicago Press). Summarizes long-term data on the demography of the large cactus finches native to the Galápagos islands.

Wiens, J. A. 1996. Oil, seabirds, and science. *BioScience*. 46: 587–597. A look at seabirds that did and did not recover from the *Exxon Valdez* oil spill.

Species Interactions

49

Chapter 48 explored the dynamics of populations—how and why they grow or decline, and how they change over time and space. Although that chapter considered populations in isolation, the reality is that individuals of different species constantly interact. As a result, the fate of a population may be tightly linked to the other species that share its habitat. Members of different species eat one another, pollinate each other, exchange nutrients, compete for resources, and provide habitats for each other.

Biologists organize the array of possible interactions between species by considering their effects on the individuals involved. Does the relationship benefit the members of one species in terms of their ability to survive and reproduce (a "+" interaction) but hurt members of the other species (a "−" interaction)? Or does the association have no effect? This chapter investigates three broad categories of interaction: the +/− interactions called parasitism, predation, and herbivory, the −/− relationship known as competition, and the +/+ association termed mutualism.

In addition to introducing the array of tools that biologists use to study these interactions, a major theme in the chapter is that the outcome of interactions between species is dynamic and conditional. Consider the seemingly straightforward +/− relationship between wolves and moose. How does predation by wolves affect moose populations? The answer depends on the relative abundance of each species, the presence of other predators and prey, and how diseases and parasites are affecting each of the partners in the interaction. The answer is also likely to change over time, as conditions change and each species evolves. Let's begin with a particularly dynamic and conditional type of relationship: parasitism.

This aerial photograph shows members of a wolf pack battling a moose. This chapter introduces a variety of species interactions, including predation.

49.1 Parasitism

As **Figure 49.1** shows, parasites are an extraordinarily diverse group. They range in size from subcellular viruses to unicellular organisms like *Mycobacterium*, which causes tuberculosis, to larger, multicellular forms like lice and tapeworms. Parasites may be generalists, meaning that they use many host species, or more commonly specialists that afflict just one or a few types of hosts. If they attack the outside of the body they are **ectoparasites**; if they live inside the body they are termed **endoparasites.** Some live their entire lives inside the same host species; others have complex life cycles that involve several host species. Some even spend parts of their lives as free-living organisms.

The feature that unites parasites is their ability to acquire resources from a host. This transfer of resources is costly to the host, so parasitism is a +/− interaction. Parasitism differs from predation because parasites are usually small relative to their host and because parasitism is not necessarily fatal. The degree of harm caused to the host depends on the type of parasite. Bloodsucking ectoparasites like ticks or lice have a relatively minor impact; *Plasmodium* infections can cause severe illness or death from malaria.

How do parasites grow, and how do hosts defend themselves? Let's consider these questions by focusing on the most serious of all human parasites.

Adaptations and Arms Races

Over 50 years ago, J. B. S. Haldane suggested that the interaction between a parasite and its host species should resemble an arms race. In humans, an arms race is said to occur when one nation develops a new weapon, which prompts a rival country to develop a defensive weapon, which pushes the original country to manufacture a more powerful weapon, and so on.

According to Haldane, a **coevolutionary arms race** between parasites and hosts begins when a parasitic species develops a trait that allows it to survive and reproduce in a host. In response, natural selection favors host individuals that are able to defend themselves against the parasite. As the frequency of resistant host individuals increases in the population, natural selection favors parasitic individuals with novel traits that allow them to escape host defenses. Recent data on *Plasmodium*—the parasite that causes malaria—suggest that Haldane's arms race metaphor is an accurate description of how parasites and hosts interact over time.

The four species of *Plasmodium* that afflict humans cause at least 100 million cases of malaria each year. The annual death toll from the parasite is more than 1 million people, most of them under the age of five. Along with AIDS and tuberculosis, malaria ranks as the most devastating of all infectious diseases. How does *Plasmodium* parasitize humans, and how is it transmitted from one individual to another? Does the arms race analogy help us understand the evolution of the interactions between this parasite and its host?

CD ACTIVITY 49.1 Life Cycle of the Malaria Parasite

Life Cycle of the Malaria Parasite As **Figure 49.2** (page 952) shows, *Plasmodium* has a complex life cycle. It is a unicellular organism, but has several distinctive cell types that correspond to the host and tissue being infected. To trace the parasite's life cycle, begin with the haploid cells labeled "sporozoites"—step 1 in the diagram. Note that these cells live in the salivary glands of mosquitoes, and that they are transferred to a human host when the infected mosquito bites a human. The sporozoites travel to the human liver, where they reproduce asexually and form a large population of cells called merozoites. Merozoites are released into the bloodstream. These *Plasmodium* cells invade the host's red blood cells and reproduce asexually until they cause the blood cells to rupture. Loss of red blood cells causes anemia in the human host; the immune system's response to the release of new merozoites induces debilitating fevers.

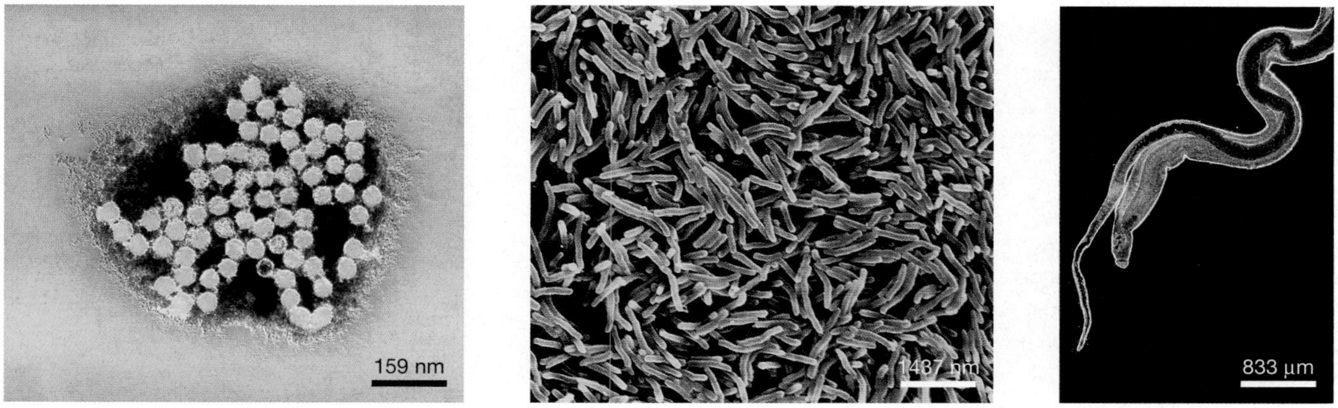

159 nm

1487 nm

833 μm

FIGURE 49.1 An Introduction to Parasite Diversity
Parasites are very diverse in their taxonomy, life cycles, size, and habitat requirements. The cold virus (left), mycobacteria (center), and blood fluke (right) shown here parasitize humans.

As you can see in step 5 of Figure 49.2, *Plasmodium* cells that survive the immune response develop into male and female gametocytes. When another mosquito bites the infected person, haploid gametocytes enter the mosquito as part of the blood meal. Male and female gametocytes then fuse in the mosquito's gut to form a diploid cell. After meiosis occurs, the resulting haploid cells develop into sporozoites that migrate to the mosquito's salivary glands. At this point, the cycle of infection is ready to be repeated. How do hosts respond to infection?

Adaptation and Counter-Adaptation The cells of the human immune system do not sit passively by while the events diagrammed in Figure 49.2 are taking place. As Chapter 46 showed, the immune system wields potent weapons against pathogens. In West Africa, for example, Adrian Hill and co-workers have found that there is a strong association between a certain allele in the major histocompatibility complex (MHC), called *HLA-B53*, and protection against malaria. In liver cells that are infected by *Plasmodium*, HLA-B53 proteins bind to a particular protein found in sporozoites (**Figure 49.3a**). The protein-protein complex is displayed on the surface of the infected cells, marking them for destruction by cytotoxic T lymphocytes before they can produce merozoites. In this way, people who have at least one copy of the *HLA-B53* allele are better able to beat back malaria infections before the infection progresses.

Follow-up research by Sarah Gilbert and colleagues showed that the situation is complex, however. *Plasmodium* populations in West Africa now have a variety of alleles for the protein recognized by HLA-B53. Some of these variants bind to HLA-B53 and trigger an immune response, while others escape detection. Furthermore, many people in West Africa are infected with several different strains of *Plasmodium*. In some cases, the recognition step by HLA-B53 breaks down when certain strains are found together (**Figure 49.3b**). To make sense of these observations, Gilbert and co-workers suggest that natural selection has favored the evolution of *Plasmodium* strains with weapons to counter HLA-B53.

To summarize, the arms race envisioned by Haldane occurs because of evolutionary interactions between *Plasmodium* and immune-system cells. Certain human proteins act as antimalarial weapons, but *Plasmodium* has evolved effective responses, as predicted.

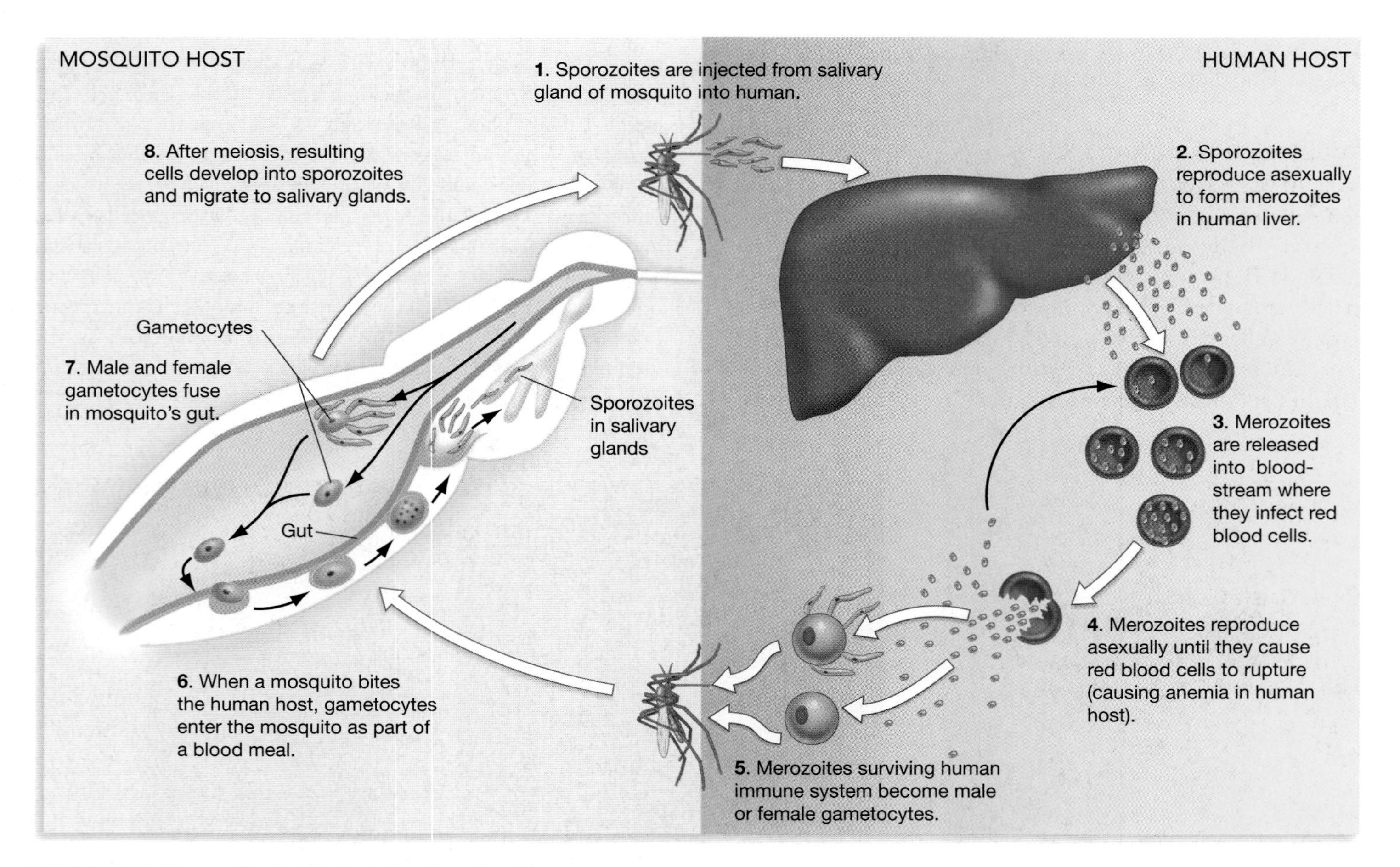

FIGURE 49.2 *Plasmodium*'s Life Cycle Involves Two Hosts
Plasmodium are eukaryotes that belong to a lineage of organisms called the apicomplexans (see Chapter 27). This diagram illustrates the life cycle of the *Plasmodium* species that infect humans. Mosquitoes transmit the parasite from one human to another.

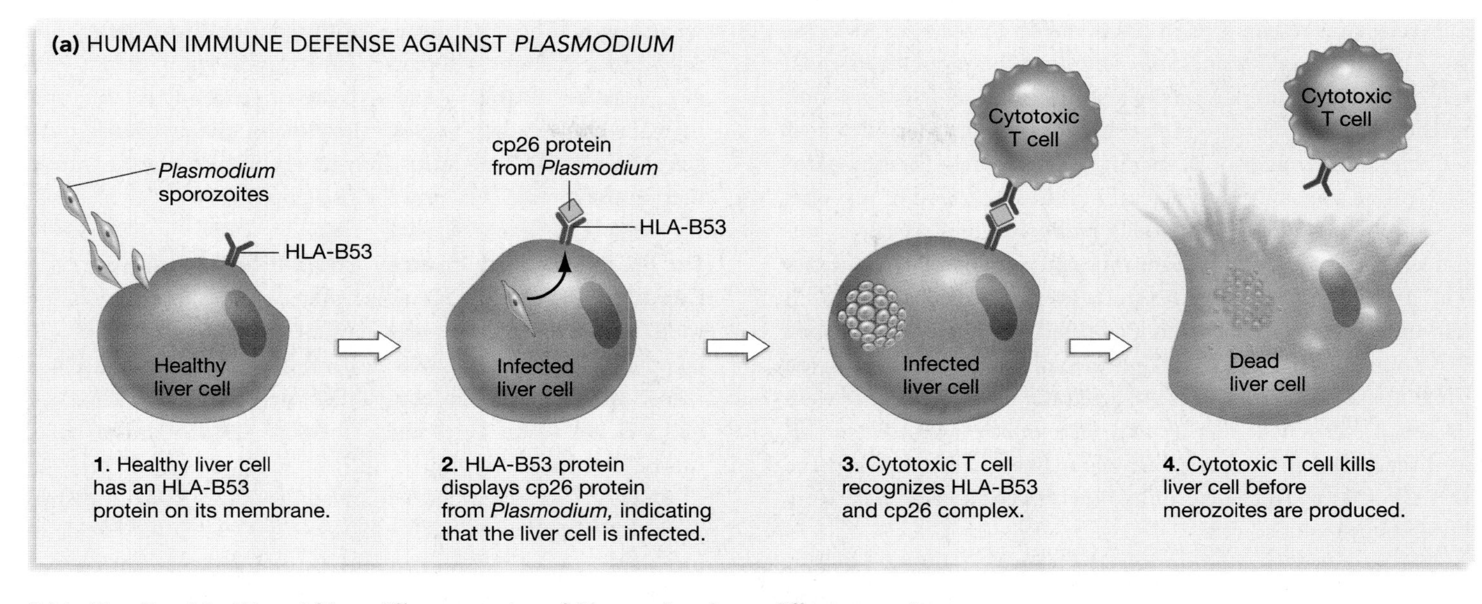

(b) In The Gambia, West Africa, different strains of *Plasmodium* have different versions of the cp protein. How successful are these different strains at infecting people?

Plasmodium strain	Infection rate	Interpretation
cp26	Low	HLA-B53 binds to these proteins. Immune response is effective.
cp29	Low	
cp26 and cp29 strains together	High	Immune response fails when these strains infect the same person.
cp27	High	HLA-B53 does not bind to these proteins. Immune response is not as effective.
cp28	Average	

FIGURE 49.3 Interactions Between the Human Immune System and *Plasmodium*
(a) If HLA-B53 binds to a particular *Plasmodium* protein, then infected cells are recognized and destroyed. **(b)** Some strains of *Plasmodium* appear to avoid detection by the immune system better than others. Also, certain strains are able to defeat the immune response by infecting the same person at the same time.

Parasites do not just have to invade tissues and grow while evading defensive responses by their host, however. They also have to be transmitted to new hosts. To a parasite, an uninfected host represents uncolonized habitat, teeming with resources. What have biologists learned about how parasites are transmitted to new hosts?

Transmission Dynamics

Several species of land snails are parasitized by flatworms called flukes. Researchers who studied the association discovered something unusual. When the flukes have matured and are ready to be transmitted to their next host—a bird—they burrow into the snail's tentacles and wriggle. Infected snails also become attracted to light, even though uninfected snails avoid sunlit areas and prefer dark, shady environments (**Figure 49.4**). To interpret these observations, biologists suggest that the worms manipulate the behavior of the snail, and that the change in snail behavior makes the parasite more likely to be transmitted to a new host.

FIGURE 49.4 A Parasite That Manipulates Host Behavior
The behavior of snails that are infected with flukes in the genus *Leucochloridium* is dramatically different from the behavior of uninfected snails. **EXERCISE** Design an experiment to test the hypothesis that infected snails are more likely to be eaten by birds.

Recently, a research team led by Jacob Koella set out to investigate whether *Plasmodium* is able to manipulate its mosquito hosts in a similar way. They proposed that *Plasmodium* might somehow increase the probability that its mosquito host will bite numerous people, making it more likely that the parasite will be transmitted to several new hosts instead of just one.

To test this hypothesis, Koella and colleagues recruited volunteers from a village in Tanzania where malaria is common. The 19 men included in the study were chosen because each had unique genetic characteristics, which gave the researchers a way to identify their blood. After the volunteers had slept in the same house, the researchers collected the mosquitoes that were present, determined whether they were infected with *Plasmodium*, measured the size of their blood meal, and analyzed the DNA in the blood to determine how many different men were bitten by each mosquito.

The data that resulted support the biologists' hypothesis that infected mosquitoes bite more human hosts than do uninfected mosquitoes. Only 10 percent of uninfected mosquitoes bit more than one of the volunteers, but 22 percent of the infected mosquitoes bit multiple volunteers. It is yet to be determined, however, how this change in mosquito behavior occurs, and whether it actually does lead to increased transmission rates for the parasite.

Studies on how parasites are transmitted to new hosts reinforce the general message that host-parasite interactions may be extremely specific. Predation, in contrast, is usually a much less specific interaction than parasitism. Unlike parasites, predators often exploit many different species. How do predators affect prey populations, and how do prey respond?

49.2 Predation

When predation occurs, the predator usually kills and consumes the prey individual. Although some **herbivores** (plant-eaters) completely consume the plants they eat, the term *predation* generally refers only to animals eating other animals. At first glance, predation is a simple +/− interaction: It is costly for the prey species and beneficial for the predator. This first impression, though, leads to several questions about predator-prey dynamics and the conditions affecting the interaction. Are all predators equally detrimental for a given prey species? Why don't predators eat their prey to extinction (or do they)? If predation is a negative interaction from the prey's perspective, what kinds of strategies have prey species evolved to avoid it?

Effects of Predators on Prey Populations

At the level of the individual, the answer to the question, "what effect do predators have on prey?" is simple. Predators eat prey, and therefore the effect is that individuals die. At a population level, however, the answer is more complex. One question that is central to the study of predator effects on prey populations is whether predators reduce the size of the prey population below what it would be in the absence of predation. If predators prey only on the sick and old members of a population, then the answer may be no. If they select young or reproductive-age prey, then the answer may be yes. More likely, the answer is not a simple yes or no—predation may have complex effects on population dynamics. Let's take a closer look.

Do Predators Reduce Prey Populations Below Carrying Capacity? The population size that can be supported by the food available is called the habitat's carrying capacity (see Chapter 48). To determine whether moose populations are held below carrying capacity by timber wolves, François Messier and Michel Crête determined the age and condition of 62 moose consumed by wolves during the winters of 1981–1984 in Québec. The researchers found that wolves killed a disproportionate number of older moose. Older individuals made up just 9 percent of the total population but 34 percent of wolf prey. Nonetheless, almost two-thirds of the total kills were either calves or moose of reproductive age. This observation suggests that predation could have a strong effect on the moose population size. It supports the hypothesis that predation holds populations below carrying capacity.

In some cases, data from predator control programs also support the same conclusion. For example, a wolf control program in Alaska during the 1970s decreased predator abundance to 55–80 percent below pre-control density. Concurrently, the population of moose tripled. This observation suggests that predators reduced this moose population far below the maximum number that the habitat can support.

In general, most predators probably do have some effect on the size of prey populations. The magnitude of that effect depends on the degree to which predators selectively remove old and sick individuals that contribute little to population growth. Predators that selectively remove such individuals will have a much lower impact on prey population size than predators that prey on younger age classes.

Do Predators Regulate Prey Populations? A second question that is central to the study of predation is whether predators regulate prey populations. Populations are considered regulated if predation or some other process maintains their size within a limited range, below the habitat's carrying capacity, as shown in **Figure 49.5a**. For predation to act as a regulatory force, two events have to occur. The per capita predation rate (number killed/density) must increase when prey populations increase, to prevent prey from exceeding carrying capacity. Predation rate also has to decrease when the prey population is low, to prevent the prey population from going extinct (**Figure 49.5b**).

Messier and Crête set out to determine whether wolves regulate the moose populations they studied in Québec. After documenting that moose populations fluctuated over time, they considered three alternative explanations: (1) Predation has little effect on moose populations because moose numbers fluctuate with availability of their food; (2) Predation's primary effect is to further reduce population size when food availability is low and moose populations decline; and (3) Predators regulate moose populations below carrying capacity.

To evaluate these hypotheses, the researchers compared three moose populations. The first, which was not affected by human hunting, had a density of 0.37 moose/km^2. The second and third had been intensively hunted, and moose density had decreased to 0.22 and 0.17 moose/km^2, respectively. If food scarcity were responsible for limiting moose populations, then the biologists predicted that individuals from the densest population should show strong signs of physiological stress from food shortage. They found no such evidence when they examined these individuals, so they concluded that the moose populations do not vary with food availability.

To study the effects of predation, Messier and Crête returned to the same populations and conducted an intensive study of predator behaviors and predation rate. If predation is regulatory, then the highest predation rate should occur in the densest population and the lowest predation rate should occur in the least-dense population. As the data in **Figure 49.6** show, this is exactly what they found. Predation rate was 19.3 percent in the high-density moose population and 6.1 percent in the low-density population. As a result, moose survival was highest in the lowest-density population. These data have two important implications: (1) When moose are scarce, wolves appear to switch to other prey items; and (2) higher moose survival should lead to increased population size in the lowest-density population. These conclusions will be strengthened if the same trend continues when additional studies add to the sample size of populations analyzed.

In contrast to the situation with wolves and moose in Québec, Bruce Dale and co-workers found that Alaskan wolves that prey on caribou, dall sheep, moose, snowshoe hares, and beaver show exactly the opposite pattern. As the data in **Table 49.1** indicate, wolf packs in this area appear to specialize on caribou as they become scarcer. The contrast carries an important message. A particular predator may respond differently to fluctuations in the size of different prey species. The same predator may also respond differently to the same prey species in different areas and at different times.

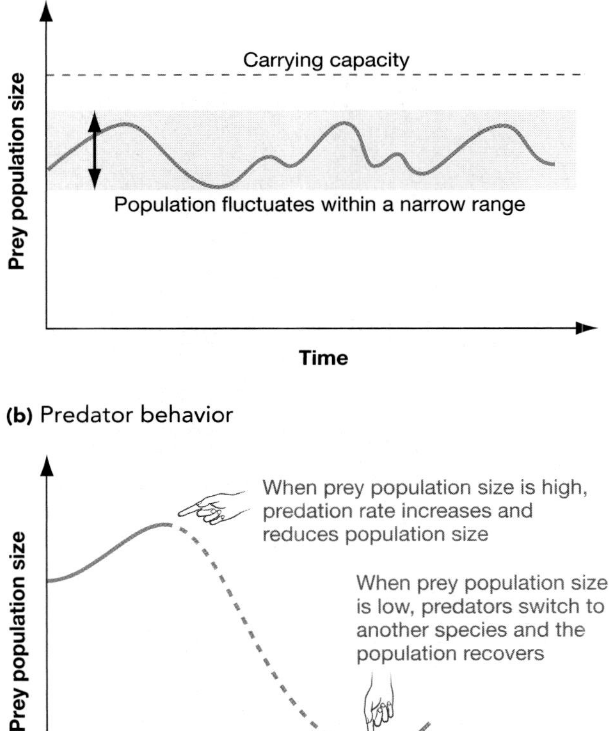

FIGURE 49.5 Can Predators Regulate Prey Populations?
(a) Prey populations are regulated if they fluctuate within a narrow range over time and remain below carrying capacity. **(b)** For predation to result in regulation, predators must behave in the ways indicated here. **EXERCISE** In part (b), add a line to the graph showing what will happen if predators do *not* switch to different prey species when the prey population is low.

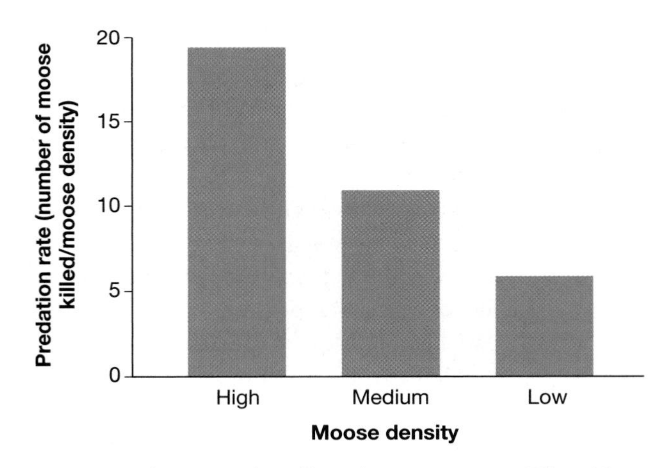

FIGURE 49.6 The Rate of Wolf Predation Increases When Moose Population Density Is High
In moose populations from Québec, the predation rate by wolves was highest in the highest-density moose population and lowest in the population with lowest density. The differences in predation rate resulted in higher moose survival in the lowest-density population.

TABLE 49.1 Caribou Density and Prey Selection for Wolves in Gates of the Arctic National Park, Alaska

Pack	Caribou (per km^2)	Caribou in Kill (%)
Walker Lake	2.34	85
Iniakuk	0.31	86
Unakserak	0.08	93
Sixty Mile	0.07	100

Messier and Crête concluded that their third hypothesis was correct—that the wolf packs they studied may regulate moose populations. The Alaska study, in contrast, supports their second hypothesis—that predation's primary effect may be to lower the size of low-density populations.

These contrasts have important public policy consequences. As the essay at the end of the chapter notes, efforts to kill or sterilize predators are usually justified by the claim that instead of regulating prey populations, predators threaten them with extinction. In some cases, at least, this claim may not be correct.

Responses of Prey to Predators

Prey individuals do not passively give up their lives to increase the fitness of their predators, any more than the hosts victimized by parasites acquiesce to parasitism. Instead, prey hide or run away when they sense the presence of a predator, sequester or spray toxins, or bristle with spines or other weapons (**Figure 49.7**). Moose, for example, are capable of killing or maiming wolves by kicking them. Traits like these are sometimes called standing defenses, because they are always present.

Recent research on prey responses has focused on the role of defenses that are triggered by the presence of predators. Chapter 37 introduced the array of inducible defenses that plants produce in response to herbivores. A major theme in that discussion was that inducible defenses may be energetically cheaper than standing defenses because they are produced only when needed. Do inducible defenses also occur in animals, in response to predators?

George Leonard and co-workers hypothesized that inducible defenses might be important for blue mussels living in an estuary along the coast of Maine (**Figure 49.8a**). In a previous study these researchers documented that predation pressure varies between two areas within the estuary. Predation by crabs was high in an area with relatively slow tidal currents (the "low-flow" area) and low in an area with relatively rapid tidal currents (the "high-flow" area). The researchers hypothesized that if blue mussels possess inducible defenses, they should occur in the low-flow area where predation pressure is higher.

To evaluate this hypothesis, the biologists measured mussel shell characteristics in the two areas. They found that mussels in the high-predation area had thicker shells and were more strongly attached to their substrate than mussels in the low-predation area (**Figure 49.8b**). These traits make the mussels more difficult to remove from the substrate and harder to crush, and so function as effective antipredator defenses.

Are these traits actually induced by predation? To answer this question, Leonard and co-workers did the experiment diagrammed in **Figure 49.8c**. The tanks on the left allowed them to measure shell growth in mussels that were "downstream" from crabs that were fed fish. As predicted by the induced-defenses hypothesis, the mussels exposed to a crab in this way developed significantly tougher shells than individuals that were not exposed to a crab. The other tanks, on the right of the diagram, allowed the researchers to measure shell growth in mussels that were exposed to broken mussel shells—simulating the debris left by predators. As predicted, mussels growing in the presence of broken conspecifics had tougher shells than the control individuals.

These results show that molecules that are shed by crabs or by killed conspecifics induce blue mussel defenses. What are these chemical cues, and how are they sensed? These questions remain unanswered. It is clear, however, that inducible defenses do occur in animals as well as plants.

Camouflage

Mimicry

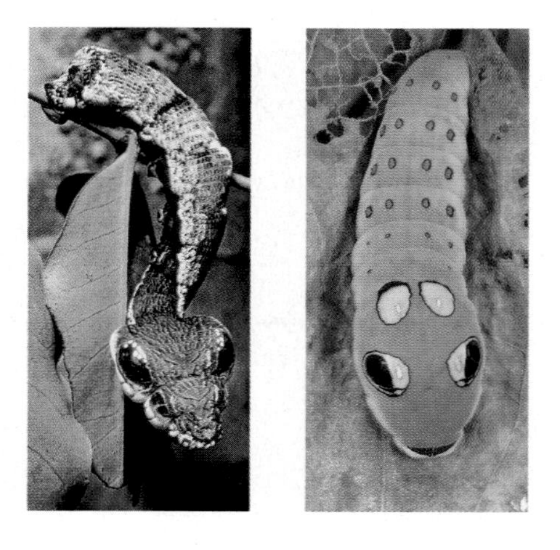

Weapons

FIGURE 49.7 Prey Defend Themselves from Predators
Prey avoid predation in several different ways. The flounder in the top photo is camouflaged. The caterpillars in the middle photos may startle potential predators because they resemble snakes. In the bottom photo, the porcupine on the right is brandishing its quills at the fisher attacking from the left.

(a) Prey and predator

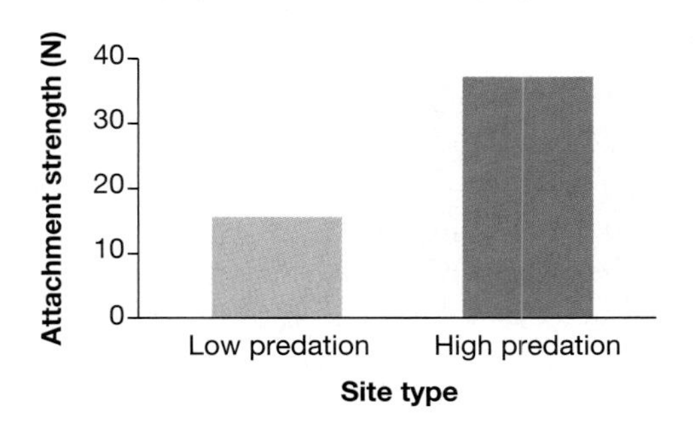

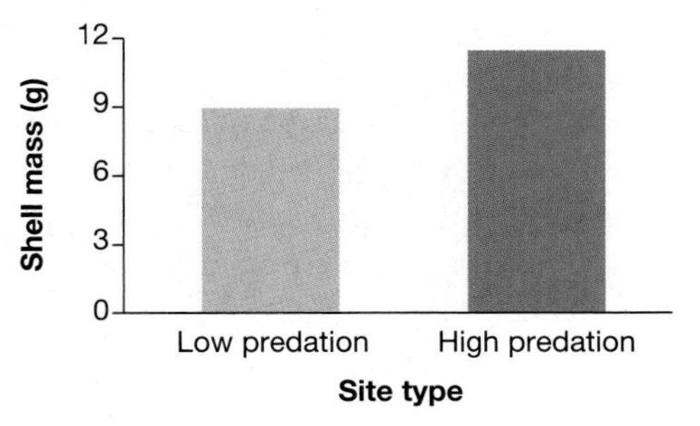

(b) Correlation between predation rate and prey defense

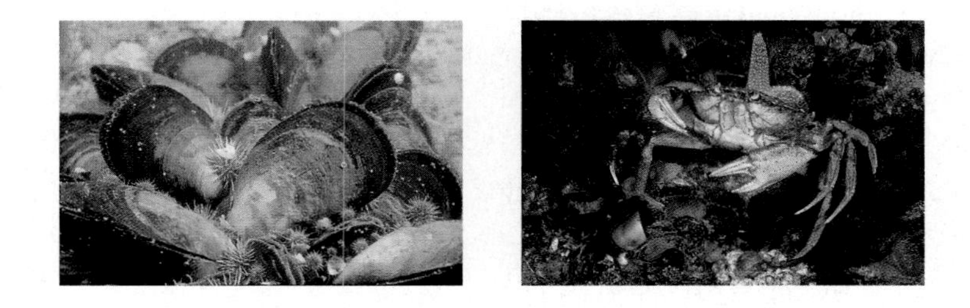

(c) Is prey defense induced by presence of predator?

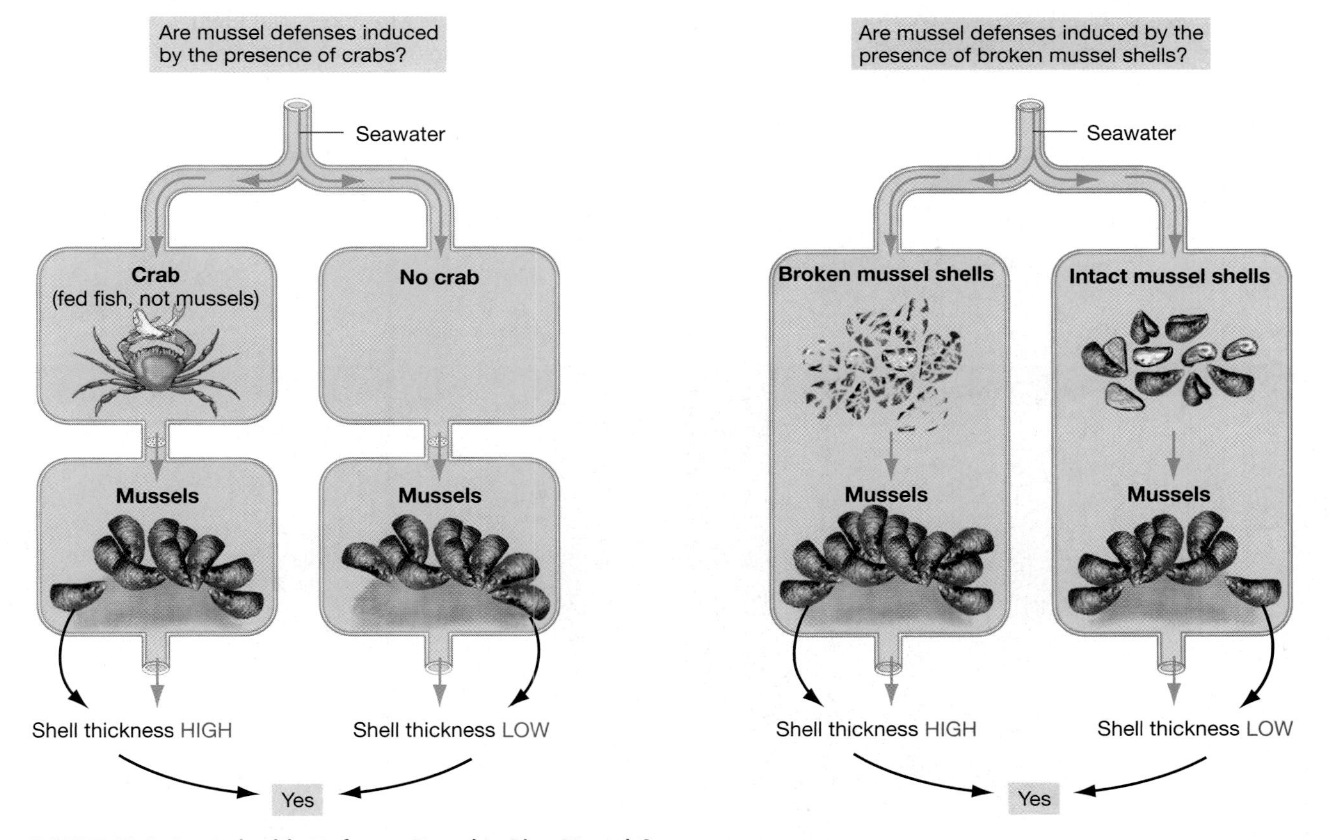

FIGURE 49.8 Are Inducible Defenses Found in Blue Mussels?
(a) In shallow-water environments in Maine, blue mussels (left) are preyed upon by crabs (right). **(b)** To measure how strongly mussels are attached to the substrate, researchers drilled a hole in their shells and measured the force, in newtons (N), required to pull them off. They also measured how large mussel shells were relative to an individual's size by dividing shell weight by soft tissue weight. As these histograms show, attachment strength and relative shell size are higher in high-predation areas. **(c)** This diagram shows the design of an experiment intended to determine whether blue mussels can sense the presence of a crab and/or killed mussels.

Keystone Predators

Up to this point, we have considered how predators affect the size of prey populations and how prey individuals respond to the presence of predators. Beyond these direct interactions, how do predators affect the broader community of species in which they live?

Robert Paine did a series of classical experiments on this question. Paine studied intertidal habitats in the Pacific Northwest of North America, where the sea star *Pisaster ochraceous* is an important predator (**Figure 49.9**). When Paine removed *Pisaster* from experimental areas, what had been diverse communities of algae and invertebrates became overgrown with solid stands of the mussel *Mytilus californianus*. Although *M. californianus* is a dominant competitor, its populations had been held in check by sea-star predation. When this predator was gone, the species diversity and structural complexity of the habitat changed radically. Predation was affecting a second type of species interaction—competition.

To capture the effect that a predator like *Pisaster* can have on a community, Paine coined the term **keystone species.** A keystone species is one that has an exceptionally great impact on the surrounding species relative to its abundance. Sea otters furnish another example of a keystone predator. Although sea otters are not particularly common members of nearshore, marine habitats, they feed heavily on large herbivorous invertebrates such as sea urchins. When otters keep sea urchin populations low, the huge algae called kelp grow more readily and form forests that are home to a diversity of fish and invertebrates (**Figure 49.10**). Recently, killer whales in the Aleutian islands of the North Pacific began feeding heavily on sea otters. In response, sea urchin populations have increased and kelp have declined. The general message here is that the relationship between predators and prey is rarely, if ever, isolated from impacts on a wide variety of other species.

49.3 Herbivory

Herbivory is similar to predation. Both are +/− interactions where one organism eats another. So why separate the two? Probably the most important distinction has to do with the fate of the organism that gets eaten. Predators kill their prey. Herbivores, in contrast, remove tissue from their prey but rarely kill them outright.

Chapter 37 introduced the question that inspires most recent research on herbivory: Why don't herbivores eat more plants than they do? Plants cannot run away from danger, and they cannot hide. As a result, they would seem to make perfect prey items. But when Hélène Cyr and Michael Pace tallied the results of more than 100 studies on herbivory, they found that the median percentage of biomass removed by herbivores was 79 percent for aquatic algae but just 30 percent for aquatic plants and only 18 percent for terrestrial plants. Why, with the exception of ecosystems dominated by aquatic algae, do herbivores eat so little of the available food?

Biologists routinely consider three possible explanations for this observation. Herbivores could be kept in check by predation and disease; plant tissues could offer poor or incomplete nutrition; or plants could defend themselves effectively against attack.

Keystone predator present

Keystone predator absent

FIGURE 49.9 Keystone Predation in a Rocky Intertidal Habitat
When the sea star *Pisaster ochraceous* is present, rocky intertidal habitats have a large diversity of species with varied forms (left). When *Pisaster* is excluded, the same habitats are dominated by beds of California mussels (right).

FIGURE 49.10 Keystone Predation in a Nearshore Habitat
When sea otters are present, kelp forests are common in the nearshore habitats of western North America (left). When sea otters are not present, herbivory by sea urchins causes drastic reductions in the size and frequency of kelp forests (right).

The Top-Down Hypothesis

Top-down control over herbivores occurs when herbivore populations are limited by predation or disease. The term *top-down* is inspired by the concept of a food chain, which is explained in detail in Chapter 51. As an example, consider European grassland habitats like the one illustrated in **Figure 49.11**. In this environment, the food chain begins with grasses and other plants that harvest sunlight and make carbohydrates. Rabbits and other herbivores eat these plants, forming a second link in the food chain. Rabbits, in turn, are hunted by owls, foxes, and other predators. These predators are above rabbits in the food chain.

To support the hypothesis of top-down control, biologists point to a series of natural experiments that have taken place throughout the world. For example, after a dozen pairs of European rabbits were introduced to Australia in 1859, the population exploded. Biologists estimate that the population was over 250,000 individuals by 1865. As a result, vast areas of the continent were virtually stripped of vegetation. Similar consequences have occurred when goats and pigs have escaped from captivity and established large, wild populations in areas that lack their natural predators and diseases.

Critics of the top-down hypothesis point out that these uncontrolled experiments are unnatural situations and difficult to interpret. Critics thus advise caution in drawing broad conclusions from these data sets. When these observations are combined with data on the control of moose, sea urchins, and other herbivores by predators, however, it becomes reasonable to conclude that in some situations herbivory may be limited in a top-down fashion.

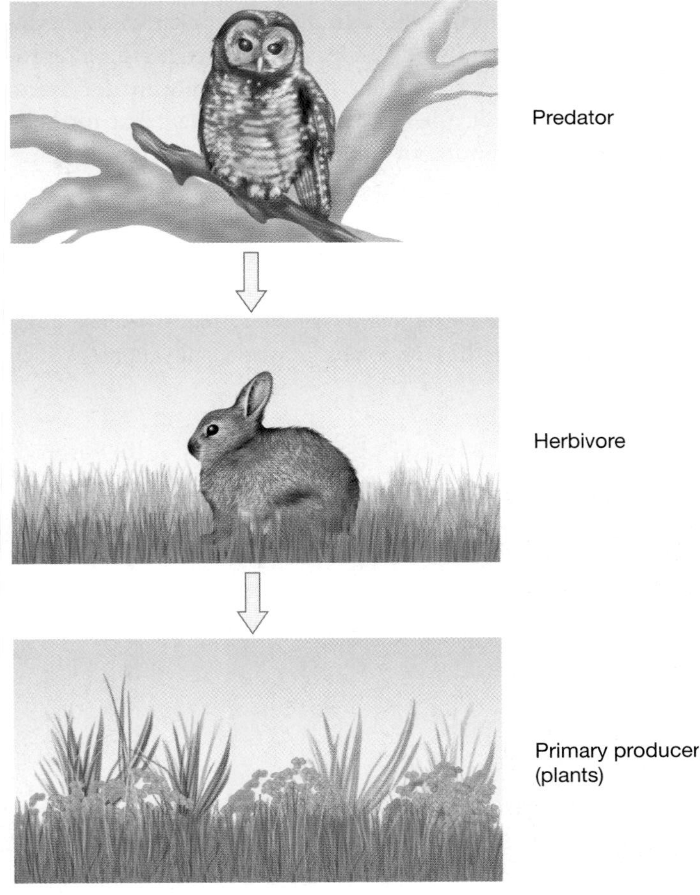

FIGURE 49.11 Do Predators Exert "Top-Down" Control of Herbivores?
This simple diagram shows three components in a food chain found in European grasslands. Predators eat herbivores, and herbivores eat a variety of grasses and herbaceous plants. The hypothesis of top-down control is named for these relationships.

The Poor Nutrition Hypothesis

A second hypothesis to explain low levels of herbivory is that plants are a poor food source. According to this viewpoint, the quality of the resources available at the base of the food chain limits herbivore density.

Why would plants be considered low-quality food? In addition to the presence of the toxins introduced in Chapter 37 and reviewed later in this section, plants contain very low concentrations of nitrogen. This element is found in all proteins and nucleic acids and is particularly critical for animals. By weight, many plant tissues have less than 10 percent of the amount of nitrogen found in animal tissues. This observation led to the hypothesis that the growth and reproduction of herbivores is limited by the availability of nitrogen. Consequently, herbivore populations are kept low enough to prevent the consumption of more than a fraction of the available food.

According to the nitrogen-limitation hypothesis, increasing a plant's tissue nitrogen concentration, by fertilizing it, should allow its herbivores to grow faster and reproduce more. To test this prediction, G. L. Waring and N. S. Cobb performed a type of inquiry called a meta-analysis. A **meta-analysis** is a study of studies—an analysis of a large number of data sets on a particular question. In this case, Waring and Cobb examined 185 studies of how insect herbivores had responded to plant fertilization. In over half of the cases, herbivores showed a significant increase in growth rate or reproduction when the plants that they feed on were fertilized. Based on this result, the researchers concluded that nitrogen limitation is an important factor in a large fraction of plant-herbivore interactions.

The Plant Defense Hypothesis

In addition to being low in nitrogen, plants produce toxins to repel herbivores. As Chapter 37 showed, plant defense compounds range from the component of wood called lignin, which reduces the digestibility of tissues, to compounds such as nicotine, which are potent poisons.

Manufacturing defensive compounds seems like the perfect solution to the problem of herbivory. In practice, however, plants face a complex challenge in defending themselves. To drive this point home, consider a recent study conducted by G. D. Martinsen and colleagues. These researchers wanted to know how plant defenses function when individuals are threatened by two very different herbivores. Will one set of defenses work against different attackers?

To answer this question, Martinsen and co-workers studied interactions between cottonwood trees, beavers, and a leaf beetle called *Chrysomela confluens* in northern Utah. From the perspective of the cottonwood, beavers are exceptionally dangerous herbivores. They cut trees down and remove virtually all of the aboveground biomass. In response, cottonwoods may resprout from the surviving rootstock (**Figure 49.12a**). Leaf beetles are dangerous in a different way. They skeletonize cottonwood leaves, meaning that they eat the soft tissue and leave the veins. *C. confluens* is capable of completely defoliating small cottonwoods.

To begin their study, the biologists compared the chemical composition of two types of cottonwood trees: individuals that had resprouted in response to browsing by beavers and individuals of similar size and age that had never been browsed. As the data in **Figure 49.12b** show, they found that the browsed cottonwoods contained a significantly higher concentration of several defensive compounds. Because previous studies had documented that these specific compounds protect plants against other mammalian herbivores, the researchers concluded that beaver browsing induced the synthesis of chemicals to prevent repeated browsing.

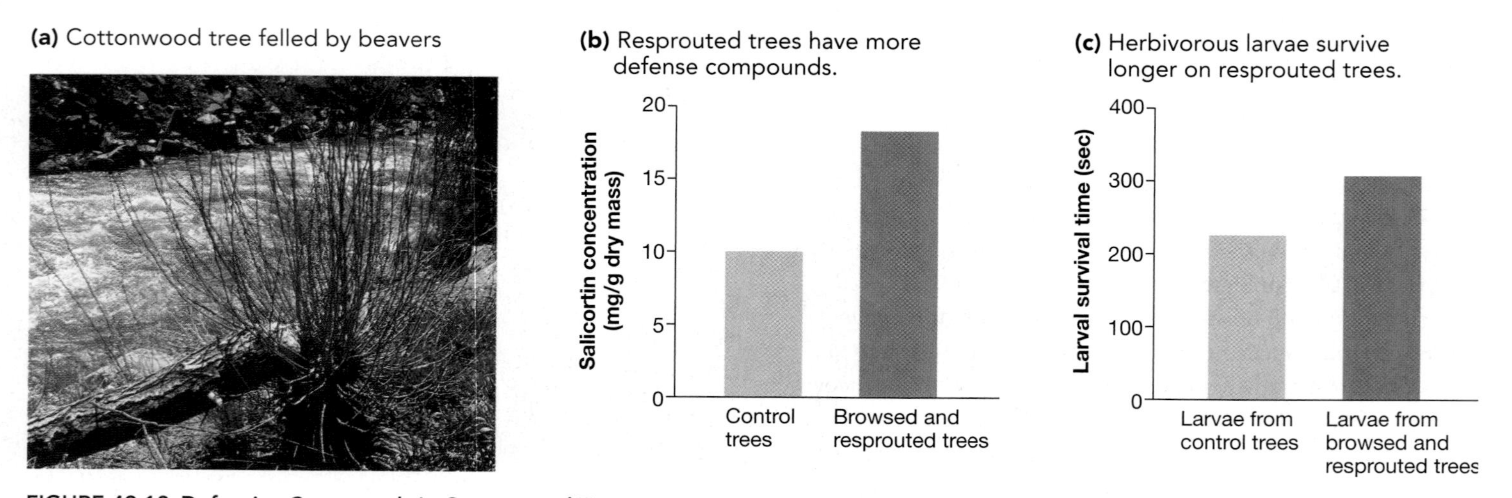

FIGURE 49.12 Defensive Compounds in Cottonwood Trees
(a) If they are felled by a beaver, most cottonwood trees respond by sending up new shoots from their roots. **(b)** Cottonwoods that have resprouted following attack by beavers have significantly higher concentrations of defensive compounds than cottonwoods that were never attacked by beavers. **(c)** Leaf beetle larvae placed on the mounds of ants (their primary predator) survived significantly longer if they had been reared on cottonwood leaves with high levels of defensive compounds.

The biologists also discovered, however, that leaf beetles were several times more abundant on the resprouted cottonwood trees than on control cottonwoods—even though the resprouted individuals had higher concentrations of defense compounds. Why? Martinson and co-workers were aware that some insect larvae tolerate certain plant defensive compounds and actually use them to protect themselves against predators.

Were leaf beetles sequestering cottonwood defense compounds to thwart their major predator—ants? To test this hypothesis, the biologists grew larvae on resprouted cottonwoods and on control cottonwoods, then placed the young beetles on an ant mound. As the data in **Figure 49.12c** show, larvae that had eaten resprouted cottonwood tissue survived longer than larvae that had grown up on the less defended, control cottonwood. These data support the hypothesis that leaf beetle larvae sequester the anti-beaver compounds and use them as a defense against ants. This is an example of an indirect effect in species interactions. The response by cottonwoods to herbivory by beavers benefits another herbivore—the leaf beetle.

What is the message of this research? First, the interactions that were documented support the conclusion that herbivory, like predation, affects more than two species at a time. Second, herbivores have evolved mechanisms to cope with the array of defensive compounds produced by plants. The net result is that there is no perfect, one-size-fits-all defensive strategy. Natural selection should favor plants that evolve an ever-changing suite of compounds to deter the ever-changing array of herbivores they face.

Finally, the data reviewed in this section suggest that there is no single answer to the question of why herbivores don't eat a greater fraction of the available plant food. All three of the hypotheses we have examined are correct. Top-down control, nitrogen limitation, and effective defense are all important factors in limiting the impact of herbivory, although the particular mix of factors will vary from plant species to plant species.

49.4 Competition

Ever since ecological studies began, researchers have focused on competition as an important interaction within and between species. The attention is justified in part by the central place that competition holds in Darwin's theory of evolution by natural selection. Darwin pointed out that individuals within a population compete for the resources that are required to grow and reproduce. Further, some individuals are more successful in this competition and leave more offspring than others. If the traits that lead to success are heritable, then the frequency of alleles in the population changes and evolution by natural selection occurs.

How does competition affect members of different species? To answer this question, it is important to recognize that every species has a unique **niche**, or set of habitat requirements. G. Evelyn Hutchinson proposed that a species' niche could be envisioned by plotting these habitat requirements along a series of axes. **Figure 49.13a**, for example, shows one niche axis for a hypothetical species. In this case, the habitat requirement plotted is a food item utilized by this organism; the size of seeds used by members of this population might be a function of mouth or tooth size. Other niche axes could represent other types of foods

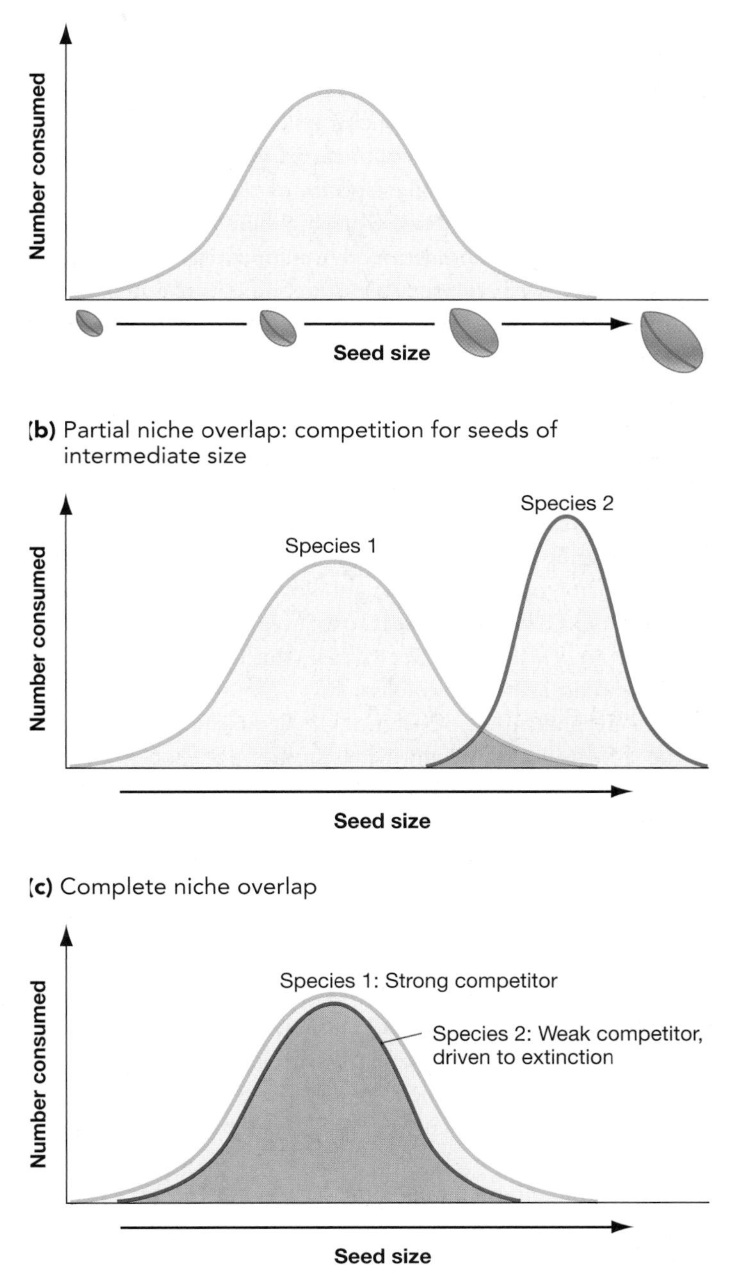

FIGURE 49.13 Niche Overlap Leads to Competition
(a) This graph describes one aspect of a species' fundamental niche, meaning the range of resources that it is able to use. **(b)** Competition occurs when the niches of different species overlap. In this case, both species use seeds of intermediate size. If species 2 is a weaker competitor than species 1, then species 2 will have a realized niche of only large seeds. **(c)** If two species have completely overlapping fundamental niches, there is no refuge for the weaker competitor and it may be driven to extinction.

used or the temperatures, humidity, and other environmental conditions tolerated by the species.

Competition occurs when the niches of two species overlap. According to **Figure 49.13b**, for example, the two species plotted compete for seeds of intermediate size. Competitive interactions reduce the amount of resources available to each species. This is why competition is considered a −/− interaction.

The Consequences of Competition—Theory

According to G. F. Gause, it is not possible for species with the same niche to coexist. This conclusion is called the principle of competitive exclusion, and was inspired by a series of competition experiments with similar species of the unicellular pond-dweller *Paramecium*. Gause's work highlights the central question in the study of competition: If competition is a common interaction, why haven't the superior competitors outcompeted all other species? Is coexistence possible despite competition?

Before considering experimental work on this question, let's consider the predictions made by theory. For example, imagine a hypothetical scenario in which two species are competing and one is a much stronger competitor than the other. If the two species have completely overlapping niches, as diagrammed in **Figure 49.13c**, then the stronger species is likely to drive the weaker species to extinction, just as Gause predicted. If the niches do not overlap completely, however, then the weaker species should be able to retreat to an area of non-overlap, as illustrated in Figure 49.13b. When this occurs, competing species should be able to coexist.

This mechanism of coexistence is called **niche differentiation**. Niche differentiation occurs when species partition the use of resources to reduce overlap in habitat use and thus minimize competition. When niche differentiation occurs, an important distinction arises between a species' **fundamental niche**, which is the combination of conditions that it will occupy in the absence of competitors, and its **realized niche**, which is the portion of resources or areas used when competition occurs. However, the actual mechanism of competition depends on the species and resource involved, as shown in **Figure 49.14.** Only occasionally does competition involve direct physical interaction.

These theoretical predictions are clear, logical, and compelling, but is there evidence that niche differentiation actually occurs in nature? What types of experiments can reveal the existence of competition and identify its consequences?

The Consequences of Competition—Experimental Studies

Joseph Connell initiated a classic study of competition after observing an interesting pattern in an intertidal rocky shore in Scotland. He noticed that there were two species of barnacles with intriguing distributions. Barnacle larvae are mobile, but adults are sessile—they live attached to rocks. The adults of one species, *Chthamalus stellatus*, occurred in the upper intertidal zone while the adults of the other, *Balanus balanoides*, were restricted to the lower intertidal zone (**Figure 49.15a**, page 964). The upper intertidal is a more severe environment for barnacles because it is exposed to the air for longer periods each day. The young of both species were found together in the lower intertidal zone, however.

To explain these observations, Connell hypothesized that adult *Chthamalus stellatus* were competitively excluded from the lower intertidal zone. His view was that *Chthamalus* larvae occupy the species' fundamental niche, but that adults are constrained by competition for attachment space. The alternative hypothesis is that adult *Chthamalus* are absent from the lower intertidal zone because they do not thrive in the physical conditions there.

To test these hypotheses, Connell performed the experiment diagrammed in **Figure 49.15b.** He began by removing a number of rocks from the upper intertidal zone that had been colonized by *Chthamalus* and transplanting them into the lower intertidal zone. He screwed the rocks into place and allowed *Balanus* larvae to colonize them. Once the spring colonization period was over, Connell divided each rock into two treatments. In one half, he removed all *Balanus* that were in contact with or next to a *Chthamalus*.

As the diagram shows, the experimental design allowed Connell to compare *Chthamalus* survival in the absence of competition with *Balanus* with survival during competition. This is a common experimental strategy in competition studies: One of the competitors is removed, and the response by the remaining species is observed.

Connell's results support the hypothesis of competitive exclusion. In the unmodified areas, *Balanus* killed many of the young *Chthamalus* by growing against them and lifting them off the substrate. As **Figure 49.15c** shows, *Chthamalus* mortality was much lower when all of the *Balanus* were removed.

Connell's experiment was one of the first examples of how competition can modify the distribution of a species. In this case, the poorer competitor (*Chthamalus*) is restricted to the upper intertidal zone, which is a poor habitat because it is prone to drying. *Chthamalus* is able to grow there, however, because it is more resistant to drying than *Balanus*. In these species, the overall outcome of competition is coexistence by habitat partitioning. If both species had identical fundamental niches, *Chthamalus* would probably be eliminated entirely from the area.

To summarize, competition is a common interaction that has a wide range of ecological outcomes. In extreme cases, competition can lead to the complete exclusion of one species. When the two competitors have slightly different fundamental niches, however, the poorer competitor can take refuge in areas that are beyond the better competitor's tolerance, thus achieving coexistence. Finally, as you saw in the experiments on keystone predators reviewed in section 49.3, competitors can coexist if some other factor (predation, bad weather, disturbance) reduces the population of the better competitor.

Consumptive competition occurs when organisms compete for the same resources. These trees are competing for nitrogen and other nutrients.

Preemptive competition occurs when individuals occupy space and prevent access to resources by other individuals. The space preempted by these barnacles is unavailable to competitors.

Overgrowth competition occurs when one organism grows over another, blocking its access to resources. This large fern has overgrown other individuals and is shading them.

Chemical competition occurs when one species produces toxins that negatively affect another. Note how few plants are growing under these *Salvia* shrubs.

Territorial competition occurs when mobile organisms protect a feeding or breeding territory. These red-winged blackbirds are displaying to each other at a territorial boundary.

Encounter competition occurs when organisms interfere directly with each other's access to specific resources. Here, spotted hyenas and vultures fight over a kill.

FIGURE 49.14 Mechanisms of Competition

The mechanisms of competition vary widely. These photographs illustrate a few of the major types of competition.

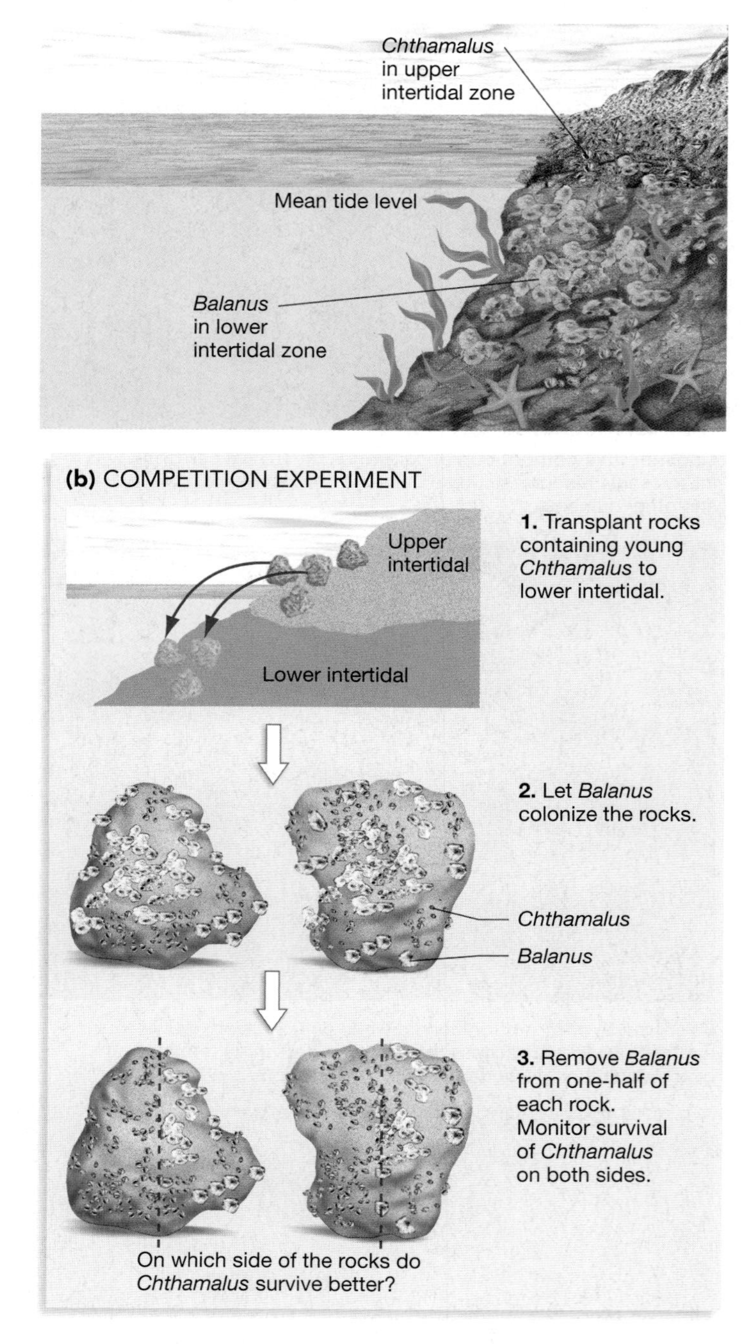

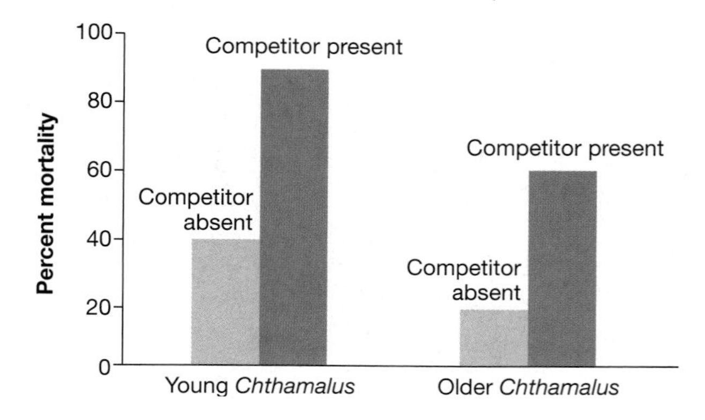

FIGURE 49.15 An Experiment on Competition
(a) In natural habitats, adult *Chthalamus* and *Balanus* barnacles do not coexist. **(b)** This experiment tested the hypothesis that *Chthamalus* is competitively excluded from the lower intertidal zone by *Balanus*. **(c)** The results of the experiment indicate that *Balanus* may exclude *Chthamalus* from certain habitats. QUESTION In part (b), why was it important to carry out the experimental and control treatments on the same rock? Why not use separate rocks?

49.5 Mutualism

Mutualisms are +/+ interactions that involve a wide variety of organisms and rewards. Many species of bees, for example, visit flowers to harvest nectar and pollen. Bees benefit because nectar is used as a food source for adult bees while the pollen is fed to larvae. Flowering plants also benefit because in the process of visiting flowers, foraging bees carry pollen from one plant to another and accomplish pollination.

Figure 49.16 illustrates other interesting mutualisms. The photograph in part (a) shows the interior of an ant nest in the New World tropics. In this part of the world, many species of ants cut up leaves, carry them to their underground nests, and add them to gardens where particular species of fungi are cultivated. The ants provide the fungus with food and protection; the fungi furnish the primary food source for the colony. Figure 49.16b illustrates cleaner fish in action. These fish pick external parasites from other fish species. In this mutualism, one species receives dinner while the other obtains medical attention.

As these examples show, the rewards from mutualistic interactions range from transportation of gametes to food, housing, medical help, and protection. It is important to note, however, that even though mutualisms benefit both species, the interaction does not involve individuals from different species being altruistic or "nice" to each other. Judith Bronstein describes mutualisms as "a kind of reciprocal parasitism; that is, each partner is out to do the best it can by obtaining what it needs from its mutualist at the lowest possible cost to itself." Her point is that the benefits received in a mutualism are a byproduct of each individual pursuing its own self-interest, in terms of maximizing its ability to survive and reproduce.

In this light, it is not surprising that some species that are closely related to mutualists "cheat" on the system. For example, Chapter 36 introduced deceit pollination, in which certain species of plants produce a showy flower but no nectar reward. Pollinators have to be deceived to make a visit and carry out pollination. Evolutionary studies show that deceit pollinators evolved from ancestral species that did provide a reward. These species changed a +/+ interaction into a +/− one.

The Costs and Benefits of Mutualism

From Bronstein's perspective, it is logical to predict that mutualisms vary widely in the magnitude of the benefit received by each participant, and that the costs and benefits might vary through time. For each individual involved, the net benefit of a mutualism can be measured by subtracting the costs of the interaction against the benefits, in terms of producing offspring.

What are the costs of participating in a mutualism? Consider insect pollinators, which expend energy extracting nectar from plants. The arrangement of petals, anthers, and nectaries in a flower may force pollinators to take a circuitous path through reproductive structures to reach the rewards they seek. If the costs of entering and leaving the flower are high enough relative to the amount of energy obtained from the visit, it may be more efficient for the insect to approach the nectary from the outside and chew through its housing. Plants, in turn, expend energy producing the nectar that attracts their pollinators. During its flowering period, the milkweed *Asclepias syriaca* spends up to 37 percent of the energy it produces through photosynthesis to manufacture nectar for its pollinators. If the cost of producing nectar is high enough, it may be more efficient for plants to reduce or eliminate the reward and rely on deception.

In addition, the magnitude of costs and benefits experienced by the participants may change over time as environmental conditions change. For example, the energy spent by the milkweed on nectar might be more costly in a year in which conditions are poor and total growth is low, causing the mutualism to be less beneficial. The costs and benefits of mutualisms are not necessarily equal for each partner; nor are the costs and benefits static through time or constant from one geographic area to another.

Studying Mutualisms Experimentally

What is the range of outcomes in a mutualism? Hall Cushman and Thomas Whitham set out to answer this question for a mutualism that occurs between ants and a species of treehopper called *Publilia modesta*. Treehoppers are small, herbivorous insects that feed by sucking sugar out of the phloem of plants. As **Figure 49.17a** (page 966) shows, treehoppers excrete a sugary solution known as honeydew from their posteriors. The honeydew, in turn, is harvested for food by ants.

It is clear that ants benefit from this association. But do the treehoppers? Cushman and Whitman hypothesized that the ants might protect the treehoppers from their major predator, salticid spiders. These spiders feed heavily on juvenile treehoppers.

To test the hypothesis that ants protect treehoppers, the researchers studied ant-treehopper interactions over a three-year period. As **Figure 49.17b** shows, they began by marking out a 1000-m^2 study plot. Each year they randomly assigned the treehopper host plants inside the plot to one of two groups. They removed the ants from one group, but left the others alone to serve as a control. Then they compared the growth and survival of treehoppers on plants with and without ants.

In both the first and third years of the study, the number of treehopper young on host plants increased in the treatment with ants but declined slightly in the treatment with no ants (**Figure 49.17c**). This result strongly supports the hypothesis that treehoppers benefit from the interaction with ants because they are protected from predation by salticid spiders.

In the second year of the study, however, the researchers found a very different pattern. There was no difference in nymph survival, adult survival, or overall population size between treehopper populations with ants and those without ants. Why? The researchers were able to answer this question because they also measured the abundance of spiders in each of

(a) Mutualism between ants and fungus

(b) Mutualism between fish

FIGURE 49.16 Mutualisms Take Many Forms
(a) Leaf-cutter ants farm certain species of fungi and harvest them for food. The fungi are the fuzzy, white structures in this photo. **(b)** Notice the small fish with a black stripe inside the mouth of this brindlebass. It is a cleaner wrasse that is picking parasites off the brindlebass's mouth and gills.

FIGURE 49.17 Is the Treehopper-Ant Interaction Mutualistic?
(a) The treehopper *Publilia modesta* is a small herbivorous insect that sucks sugar out of the phloem of plants. The droplet emerging from this individual is honeydew. **(b)** This diagram shows the design of an experiment intended to test whether the presence of ants is beneficial for treehoppers. **(c)** In some years, the number of young treehoppers increased over time when ants were present and declined over time when ants were absent.

(a) Treehopper excreting honeydew, which is harvested by ants

(b) Are ants beneficial to treehoppers?

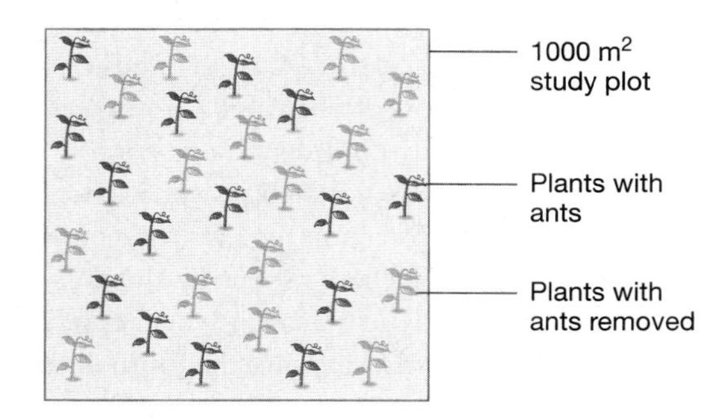

(c) Which treatment contained more treehoppers?

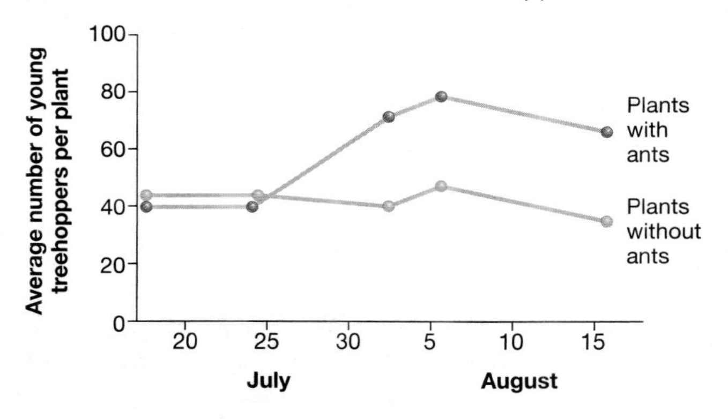

the three years. Their census data showed that in the second year of the study, spider populations were very low.

Based on these results, Cushman and Whitman concluded that the benefits of the ant-treehopper interaction depend entirely on predator abundance. Treehoppers benefit from their interaction with ants in years when predators are abundant, but are unaffected in years when predators are scarce. If producing honeydew is costly to treehoppers, then the +/+ mutualism changes to a +/− interaction when spiders are rare.

To summarize, mutualism is like parasitism, competition, and other types of species interactions in an important respect—the outcome of the interaction depends on current conditions. Because the costs and benefits of species interactions are fluid, an interaction between the same two species may vary from parasitism to mutualism to competition.

Essay Predator Control

In September 1999, Alaskan state legislators reauthorized the aerial shooting of wolves as a means of predator control. In promoting the bill, legislators argued passionately that humans should prevent "the carnage wrought by wolves" on prey species. The law is the latest episode in a long struggle over the necessity of predator control and the acceptable means for accomplishing it. At the heart of this struggle are important biological questions. Are predators responsible for the decline of some prey species? If so, can predator control effectively reduce predation and allow prey populations to rebound?

Can caribou populations rebound if wolf predation is controlled?

The controversy over predator control began when a number of the state's caribou herds began declining in the 1960s and 1970s. The Nelchina herd in south-central Alaska, for example, dropped from a high of 71,000 individuals in 1961 to just 7500 individuals in 1972. Biologists concluded that the fall-offs were caused by a series of severe winters in the mid-1960s, wolf predation, and overhunting by humans. Because caribou are a major source of food for people in some areas, pressure developed for action to reverse the declines.

In response, wildlife managers initiated programs to shoot or relocate wolves. Coincident with the onset of predator control, many of the herds rebounded from their decline. The Nelchina herd, for example, increased from 2700 individuals at the start of wolf control in 1976 to 4962 individuals in 1981. Predator control programs were initiated again in the early 1990s in order to further increase herd sizes.

Controversy continued over the necessity of these programs, however. The problem was that no one knew whether wolf removal was directly responsible for the increases in herd size or whether it was coincidental. The Nelchina herd, for example, continued to increase *after* the cessation of predator control. Finally, Alaska Governor Tony Knowles commissioned a National Academy of Sciences (NAS) study to determine whether predator control programs work.

The NAS panel evaluated 11 predator control programs in Alaska and Canada. Of these, only five actually measured whether prey density had increased following predator control, and only two analyzed whether the objective of the control programs—improved hunting success—had been met. In reporting on the evaluation, the NAS panel stated that the existing evidence was simply not adequate to conclude that predator control is effective. To remedy the situation, panel members proposed that the following guidelines be used to design predator control programs:

- Identify the reason for wanting to increase prey populations.
- Determine how much of an increase in prey population is needed to meet the stated objective.
- Analyze the costs and benefits of meeting the goal.
- If benefits outweigh costs, conduct research to determine the likelihood that predator control will work.
- Explore a variety of options for reaching the goal, including experimenting with nonlethal methods of control like sterilization.
- Conduct long-term monitoring to assess the impacts of the action.

Although the biological questions at the heart of predator control programs remain unanswered, the guidelines proposed by the National Academy of Sciences panel may make it possible to obtain a definitive answer in the future.

Chapter Review

Summary

Organisms interact with members of other species in many different ways. They eat and are eaten; parasitize and are parasitized; and pollinate, provide protection, or create habitat for each other. To categorize the different types of interactions that occur among species, biologists consider whether each participant experiences a net cost or benefit from the interaction, in terms of their ability to survive and reproduce. The fundamental messages of studies on species interactions is that their costs and benefits depend on the conditions that prevail at a particular time and place, and that the costs and benefits may change through time.

Parasitism is a +/− interaction that is characterized by an exceptionally close association between participants. A parasite generally spends all or part of its life cycle in or on its host (or hosts), and usually has traits that allow it to escape host defenses. Because hosts have evolved counter-adaptations that help them fight off parasites, biologists liken the interaction between parasites and hosts to an evolutionary arms race.

Predators can have a variety of effects on their prey. If predation rates increase when the density of a prey species increases and decrease when the density of the prey species decreases, predation may regulate the prey population below the habitat's carrying capacity. To counter predation, prey employ standing defenses or induced defenses. Keystone predators have a disproportionately large effect on their community, usually by suppressing the population of a dominant competitor.

Herbivory differs from predation because herbivores rarely kill their prey outright. Levels of herbivory are very low in terrestrial ecosystems because predation and disease limit herbivore populations, because plants provide little nitrogen, and because many plants contain toxic compounds or other types of defenses.

Competition occurs when the niches of two species overlap—meaning that they use the same resources. Competition may result in the complete exclusion of one species. It may also result in the type of niche differentiation exemplified by certain barnacle species, in which the better competitor occupied the preferred habitat while the poorer competitor occupied less desirable sites where the better competitor was unable to survive.

Mutualism is a +/+ interaction that provides participating individuals with food, shelter, transport of gametes, or defense against predators. For each species involved, the costs and benefits of a mutualism may vary over time, and from place to place. In the mutualism between ants and treehoppers, for example, the benefit of the interaction to treehoppers depends on the abundance of its predators.

Within a particular habitat, suites of interacting species make up a community. Chapter 50 explores the structure and dynamics of these multi-species assemblages. What are the dominant lifeforms in the biological communities found around the world? How do communities of organisms respond to disturbances like fires, floods, and disease outbreaks?

Questions

Content Review

1. How do parasites differ from predators? (Circle all that apply.)
   a. Parasites do not affect the density of their host species, but predators affect the density of their prey.
   b. Parasites tend to be very small relative to their host, and eat only certain tissues.
   c. Predators kill their prey more quickly and consume all of it.
   d. Parasites affect how host populations evolve; predators do not affect how prey species evolve.
2. How do predators differ from herbivores? (Circle all that apply.)
   a. Herbivores do not affect the density of their host species, but predators affect the density of their prey.
   b. Herbivores tend to eat only certain tissues of their host species.
   c. Predators kill their prey more quickly and consume all of it.
   d. Herbivores affect how their host populations evolve; predators do not affect how prey species evolve.
3. What distinguishes a keystone predator?
   a. It is extremely abundant.
   b. It regulates its prey below the carrying capacity of the habitat.
   c. It is a specialist, meaning that it preys on only one species.
   d. It has a large impact on the community, even though it is not particularly abundant.
4. What is an inducible defense?
   a. any defensive response that is produced in response to the presence of a predator
   b. any defensive response that is always present
   c. a chemical toxin that is particularly effective
   d. a weapon that an individual uses only against predators—never against members of its own species
5. What is competitive exclusion?
   a. interactions that cause a species to occupy a realized niche that is different from its fundamental niche
   b. interactions that allow species to occupy their fundamental niche
   c. the degree to which the niches of two species overlap
   d. the claim that species with the same niche cannot coexist
6. What is niche differentiation?
   a. interactions that cause a species to occupy a realized niche that is different from its fundamental niche
   b. interactions that allow species to occupy their fundamental niche
   c. the degree to which the niches of two species overlap
   d. the claim that species with the same niche cannot coexist

Conceptual Review

1. What is an arms race? In what sense do evolutionary arms races occur between parasites and their hosts? Did the use of this analogy help you understand host-parasite interactions?
2. Review the experiment on inducible defenses diagrammed in Figure 49.8c. Do you agree with the researchers' claim that this experimental design allowed them to test the effect of the presence of crabs and the presence of broken mussels independently? Explain.
3. What are the three hypotheses to explain the low level of herbivory in terrestrial plant communities? Are these hypotheses mutually exclusive? (In other words, can more than one be true?)
4. What is the difference between a fundamental niche and a realized niche? How could you show, experimentally, that the fundamental and realized niches for a particular species are different?
5. The text claims that species interactions are conditional and dynamic. Do you agree with this statement? Why or why not? Cite specific examples to support your answer.

Applying Ideas

1. Sneezing might be a way for humans to expel viruses that infect the respiratory tract, or the behavior could be induced by viruses as a way to increase the probability that they will be transmitted to new hosts. What data would you collect to test these alternative hypotheses?
2. You are the manager of a wildlife preserve that contains a small herd of bison. To make the preserve a more complete ecosystem, you have been asked to reintroduce mountain lions or wolves, which prey on old bison and on calves. A large series of studies from other localities have shown how the predation rate of lions and wolves varies as a function of bison density. What characteristics of lion versus wolf predation would make one or the other predator the better choice for your reintroduction project?
3. Two graduate students are engaged in a heated discussion. One says that honeydew production is an adaptation that supports the mutualism between treehoppers and ants. The other says that honeydew is simply a type of feces, and that its role in the mutualism is completely incidental. Design a study that would distinguish between these two hypotheses.
4. Some insects harvest nectar by chewing through the wall of the nectary. As a result, they obtain a nectar reward but no pollination occurs. Suppose that you observed a certain bee species obtaining nectar in this way from a particular orchid species. Over time, how would you expect the characteristics of the orchid population to change in response to this bee behavior?

CD-ROM and Web Connection

CD Activity 49.1: Life Cycle of the Malaria Parasite *(animation)*
(Estimated time for completion = 5 min)
Explore how a single parasite requires two host species to complete its life cycle.

At your **Companion Website** (http://www.prenhall.com/freeman/biology), you will find self-grading exams and links to the following research tools, online resources, and activities:

Herbivory and Mutualism
This essay discusses herbivory and its relationship to mutualism.

Competition Among Pasture Species
This article discusses how prairie plants compete for limited resources above and below the ground.

Endosymbiosis
This site outlines a dissertation presenting new methods to approach the topic of mutualism at the molecular level.

Additional Reading

Currie, C. R., J. A. Scott, R. C. Summerbell, and D. Malloch. 1999. Fungus-growing ants use antibiotic-producing bacteria to control garden parasites. *Nature* 498: 701–704. New insights into the mutualism between ants and the fungi they farm.

Knapp, A. K., J. M. Blair, J. M. Briggs, and S. L. Collins. 1999. The keystone role of bison in North American tallgrass prairie. *Bioscience* 49: 49–50. A look at how an herbivore affects the distribution and abundance of plants.

McClintock, J. B., and B. J. Baker. 1998. Chemical ecology in Antarctic seas. *American Scientist* 86: 254–263. An analysis of the chemical cues that fish, sponges, crustaceans, and other marine organisms use in competition and predation.

Ostfeld, R. 1997. The ecology of Lyme-disease risk. *American Scientist* 85: 338–346. An overview of research on the life cycle of the parasite that causes Lyme disease, with insight into strategies for limiting its spread.

Wikel, S. K. 1999. Modulation of the host immune system by ectoparasitic arthropods. *Bioscience* 49: 311–320. Explores how ticks, mosquitoes, and other parasites manipulate host defenses.

Community Ecology

50

Chapters 47 and 49 explored how individuals interact with members of their own species and with members of other species. Here we move beyond these pairwise interactions to consider how large numbers of species interact with each other in the assemblages called communities. A biological **community** consists of interacting species, usually living within a defined area. What major types of plant and animal assemblages, from deserts to rain forests, have been identified around the world? How do communities change through time? Why is the number of species higher in some areas than others, and why is species diversity important?

To explore these questions, the chapter opens with a brief survey of some major community types found throughout the world. The chapter's middle section explores two of the central questions in community ecology: Do communities represent tightly integrated assemblages of species, and do communities respond to disturbance in a predictable way? The chapter closes with a look at why species diversity varies among communities, and how it affects their productivity and stability.

Biological communities exist at many different spatial scales. In addition to this frog, the inside of this bromeliad teems with an array of microscopic species that interact with one another.

50.1 Climate and the Distribution of Ecological Communities

If you were to put on a pair of hiking boots and start walking from equatorial South America to the North Pole, you would notice startling ecological changes. Lush tropical forests would give way to seasonally dry forests and then to deserts. The deserts would yield to the vast grasslands of central North America, which end at the boreal forests of the subarctic. If you pressed on, you would reach the end of the trees and the beginning of the most northerly community—the arctic tundra.

A study by R. H. Whittaker documented that similar types of changes occur with increased altitude rather than latitude. In the Siskiyou Mountains of Oregon, Whittaker found that Douglas fir and other drought-tolerant tree species dominate the forests at low elevations, where the climate is warm and dry. At higher elevations, where the climate is cooler and moister, a few white fir trees join the Douglas fir. Douglas fir drops out on higher and cooler slopes, leaving just white fir. The highest-elevation forests, where the climate is cold and wet, are dominated by noble fir and mountain hemlock. Above timberline, the landscape is covered with alpine tundra.

These two journeys have a simple message. Areas with different climatic characteristics contain different biomes. A **biome** is a general term for a broad type of community, such as a grassland or desert. Within each biome there are many distinct communities, defined by the populations that interact in a particular area.

Let's start our exploration of community ecology by characterizing some of the world's major climatic regimes and biomes. Then we can turn to questions at a finer spatial scale by investigating how the communities within biomes are structured.

How do biologists identify climatic regimes and characterize biomes? The most common approach, called the Koeppen classification system, categorizes climate types according to two major features: (1) average annual temperature and precipitation, and (2) the variations occurring in temperature and precipitation throughout the year. Classification schemes based on temperature and precipitation data have identified approximately 24 distinct climate regions around the world. Here we survey six of the most abundant, ranging from the wet tropics to the arctic (see **Figure 50.1**).

Tropical Wet Forests

Tropical wet forests, or rain forests, are found in equatorial regions where temperatures and rainfall are high and variation is low. As an example of this climatic regime, consider the data shown in **Figure 50.2a** (page 972), from Belém, Brazil. In the driest month of the year, this region receives over 5 cm of rainfall. This is considerably more than the *annual* rainfall of many desert areas. Further, mean monthly temperatures never drop below 25°C and never exceed 30°C.

Thanks to this relatively small amount of variation, plants grow all year long. Although individuals shed older leaves throughout the year, there is no complete, seasonal loss of leaves as in temperate climates and many other regions. In the wet tropics, plants are broad-leaved evergreens.

As **Figure 50.2b** shows, the favorable year-round growing conditions in wet tropical forests produce riotous growth, leading to extremely high productivity and aboveground biomass. **Productivity** is defined as the total amount of photosynthesis per unit area per year; **aboveground biomass** is defined as the total mass of living plants, excluding roots.

Equatorial forests are also renowned for their species diversity. Researchers routinely find over 200 tree species in a single 10 m × 100 m plot in wet tropical sites. Based on counts of the insects and spiders collected from single trees, some biologists contend that the world's tropical forests may hold up to 30 million species of arthropods alone.

Further, the diversity of plant sizes and growth forms in these communities produces extraordinary structural diversity.

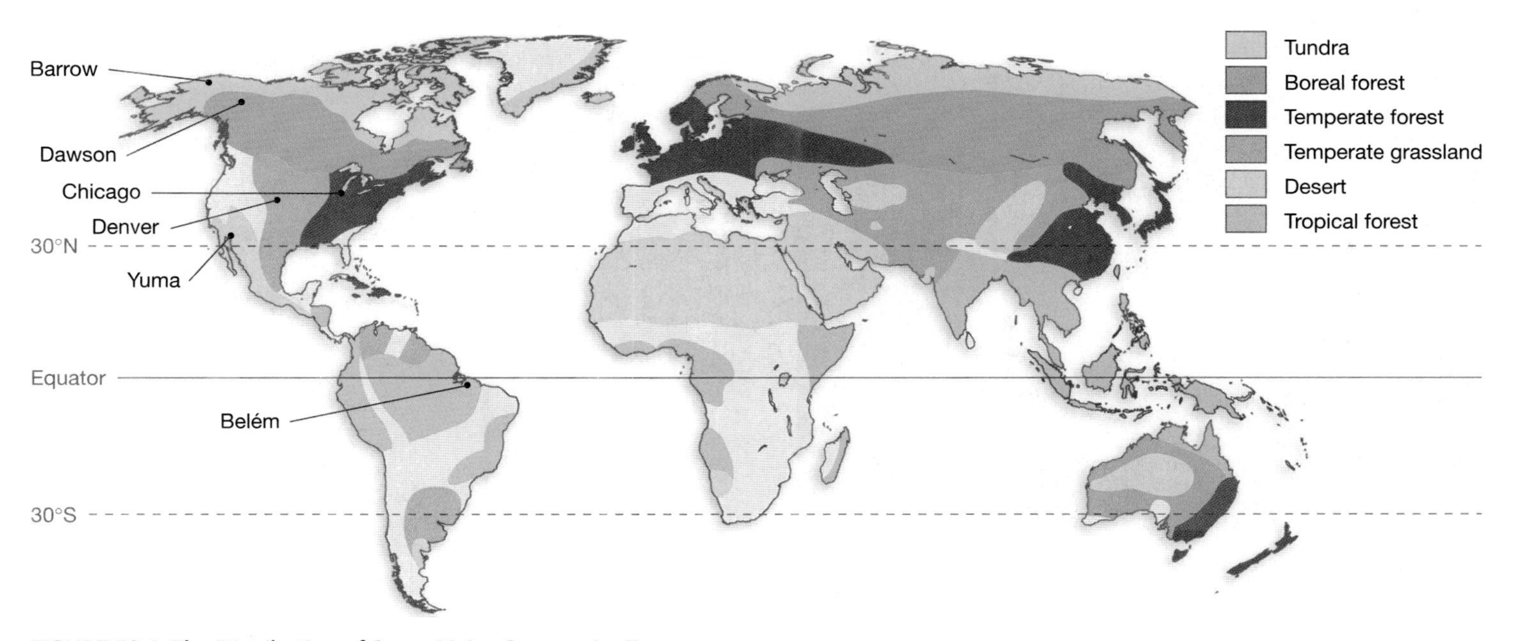

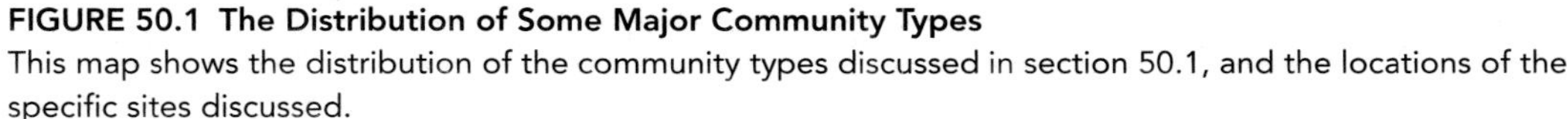
FIGURE 50.1 The Distribution of Some Major Community Types
This map shows the distribution of the community types discussed in section 50.1, and the locations of the specific sites discussed.

In tropical wet forests, a multilayered canopy of trees is intermingled with vines, **epiphytes** (plants that grow entirely on other plants), shrubs, and herbs. The diversity of growth forms presents a diverse array of habitat types for animals, which in turn pursue an impressive array of feeding strategies. In tropical wet forests, frugivores, insectivores, herbivores, carnivores, and detritivores all coexist.

Subtropical Deserts

As Figure 50.1 shows, tropical forests are found along or near the equator. Subtropical deserts, in contrast, exist in bands about 30 degrees of latitude north and south of the equator.

(a) Climate characteristics

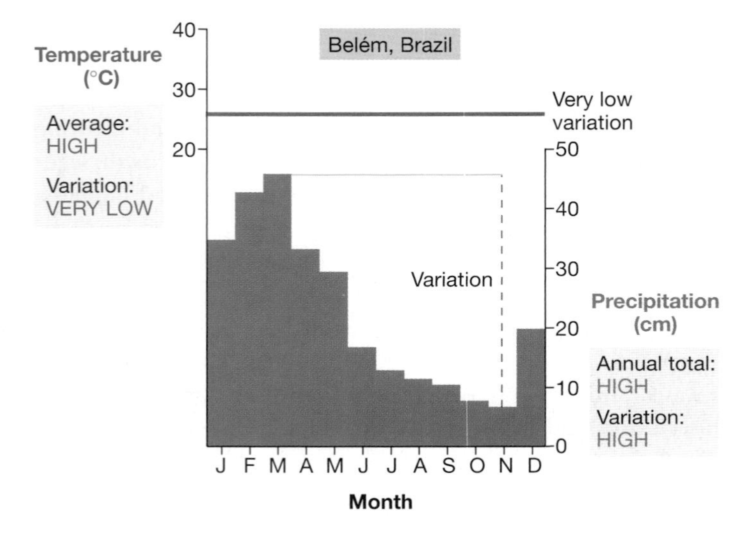

(b) Appearance

FIGURE 50.2 Tropical Wet Forest
(a) This climograph illustrates the climate of coastal Brazil, typical of a wet tropical climate. Climographs have months on the x-axis and precipitation and temperature on the y-axes. The graph shows the average monthly temperature and precipitation for an area. **(b)** Tropical wet forests are extremely species rich.

CD ACTIVITY 50.1 Tropical Atmospheric Circulation

Why? A major pattern in global air circulation, called a Hadley cell, is responsible. As **Figure 50.3** indicates, air that is heated along the equator expands and rises. As the air rises, it radiates heat to space and begins to cool. As it cools, its ability to hold water declines. The result? High levels of precipitation occur along the equator.

As more air is heated along the equator, the cooler, "older" air is pushed poleward. At about 30° latitude (north, in Figure 50.3), this air mass has cooled so much that its density increases. It begins to sink (Figure 50.3). As it does, it warms. The warming air also gains water-holding capacity, because water molecules tend to stay in vapor form instead of condensing into droplets. As a result, little rain occurs and the area is bathed in warm air. This pattern of air movement is responsible for the world's great deserts, including the Sahara, Gobi, Kalahari, and Australian Outback.

To appreciate the climatic regime in subtropical deserts, look at **Figure 50.4a**, which provides temperature and precipitation data for Yuma, Arizona, in the Sonoran Desert of southwestern North America. Note that mean monthly temperatures vary more than in tropical wet climates, but never fall below freezing. The most striking feature of the climate of these regions, though, is the low precipitation. The average annual precipitation in Yuma is just 7.5 cm.

The scarcity of water in deserts has profound implications. Because conditions are rarely good enough to support photosynthesis, the productivity of desert communities is a tiny fraction of average values for tropical forest communities. Further, as **Figure 50.4b** shows, individual plants are

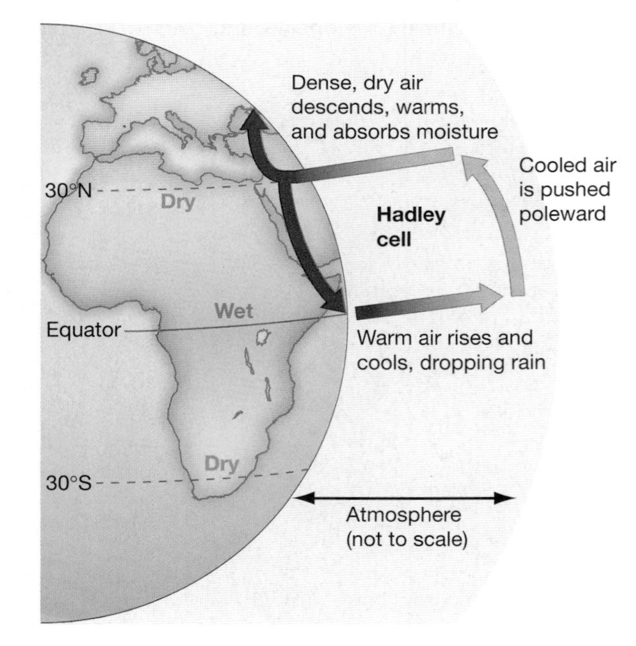

FIGURE 50.3 A Hadley Cell
A Hadley cell exists on both sides of the equator. Hadley cells are named after George Hadley, who conceived the idea of enormous air circulation patterns in 1735. **EXERCISE** Draw in the Hadley cell on the southern side of the equator.

widely spaced—a pattern that may reflect intense competition for water.

Desert species are adapted to the extreme temperatures and aridity in one of two ways. Cacti are examples of plants that have features like small leaves, waxy leaves, and CAM photosynthesis that allow them to tolerate the conditions and grow at an extremely low rate throughout the year. Other desert species have traits that allow them to escape drought. Some spend much of the year in a dormant state, resuming growth only after rain falls. The shrub called ocotillo, for example, sprouts leaves within days of a rainfall, then drops them a week or two later when soils dry out again. Each individual produces and sheds leaves in this way several times a year. Similarly, desert amphibians like the spadefoot toad spend much of the year dormant in subterranean burrows. They respond to rain by emerging for a brief time to feed and mate.

In short, deserts provide a stark contrast to tropical forests. Although both communities experience high temperatures that rarely or never drop below freezing, the absence of water makes species diversity and productivity dramatically lower in deserts.

(a) Climate characteristics

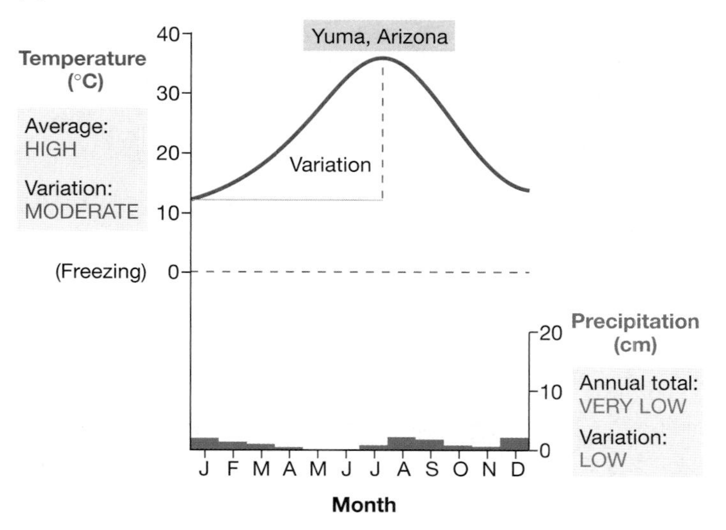

(b) Appearance

FIGURE 50.4 Subtropical Desert
(a) The climate of Yuma, Arizona, is typical of a subtropical desert. **(b)** Saguaro cacti are a prominent feature of the Sonoran Desert in the southwestern United States and northern Mexico.

Temperate Grasslands

Figure 50.5a presents climate data for a typical grassland community—in this case, Denver, Colorado. As you can see, precipitation in Denver is over four times greater than in Yuma, Arizona. Nonetheless, conditions are still quite dry; no month has more than 5 cm of precipitation. In comparison with regions to

(a) Climate characteristics

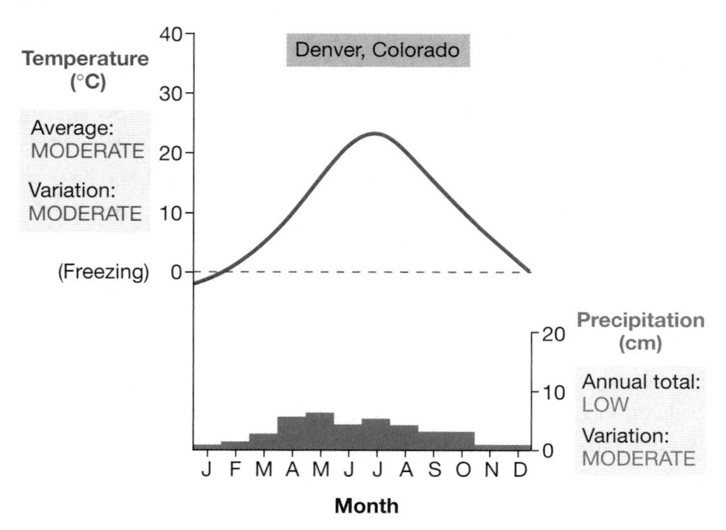

(b) Appearance

FIGURE 50.5 Temperate Grassland
(a) Denver, Colorado, has a climate typical of a mid-latitude steppe. **(b)** Grasses are the dominant life-form in the mid-latitude steppe.

the south and north, temperatures in the area are moderate, or temperate. Temperate grassland communities are found throughout central North America and the heartland of Eurasia (see Figure 50.1).

Temperature profiles in these regions are also highly seasonal. In Denver, for example, the mean monthly temperature exceeds 20°C in the summer but drops below freezing in the winter. Temperature variation is important because it dictates a well-defined growing season. (In many areas of the tropics, variation in precipitation produces a well-defined growing season.) In the temperate zone, plant growth is possible only in months when moisture and warmth are adequate.

Temperate grasslands, also called prairies or steppes, usually exist because conditions are too dry to support tree growth but too cold, wet, and seasonal for drought-adapted desert species. In temperate areas where rainfall is high enough to support forests, grasslands may develop if recurring fires burn out encroaching trees (**Figure 50.5b**). Prairie fires, which begin naturally as a result of lightning strikes, were also set intentionally by native people. The plants that dominate steppe and prairie communities tolerate fire because unlike many plants, they have meristematic tissue, which produces new growth, at the base of their stems. As a result, they resprout quickly after burning.

Although the productivity of temperate grasslands is generally lower than that of forest communities, grassland soils are often highly fertile. The subsurface is packed with roots, which add organic material to the soil as they die and decay. Further, grassland soils retain nutrients because rainfall is low enough to keep key ions from dissolving and leaching out of the soil. It is no accident, then, that the grasslands of North America and Eurasia are the breadbaskets of these continents. The conditions that give rise to natural grasslands are ideal for growing wheat, corn, and other cultivated grasses.

Temperate Forests

In temperate areas with high precipitation, grasslands give way to forests. A typical climate graph for an area near a grassland-forest border is shown in **Figure 50.6a**. Most of the area around Chicago, Illinois, was forested prior to the arrival of European settlers. Temperate forests are found in eastern North America, most of Europe, Chile, and New Zealand (see Figure 50.1).

Temperate forests experience a period in which mean monthly temperatures fall below freezing and plant growth stops. Unlike grassland climates, however, precipitation is high and relatively constant throughout the year. Chicago, for example, has an annual precipitation of 85 cm; most months record more than 5 cm.

The abundance of moisture allows trees to dominate the landscape (**Figure 50.6b**). Unlike many regions in the tropics, however, plants in temperate forests experience a dormant period. In North America and Europe, temperate forests are dominated by deciduous species that drop their leaves in the autumn and grow new leaves in the spring. Evergreens, which also enter dormancy each fall, are common. In the temperate forests of New Zealand and Chile, however, broad-leaved evergreens predominate.

Although some southern temperate forests rival or exceed the productivity of tropical forests, most have productivity levels that are higher than deserts and grasslands but lower than tropical forests. The level of diversity is also moderate. A temperate forest in southeastern North America may have more than 20 tree species, while those to the north may have fewer than 10.

Boreal Forests

The boreal forest, or taiga, stretches across most of Canada, Alaska, Russia, and northern Europe. Because these regions

(a) Climate characteristics

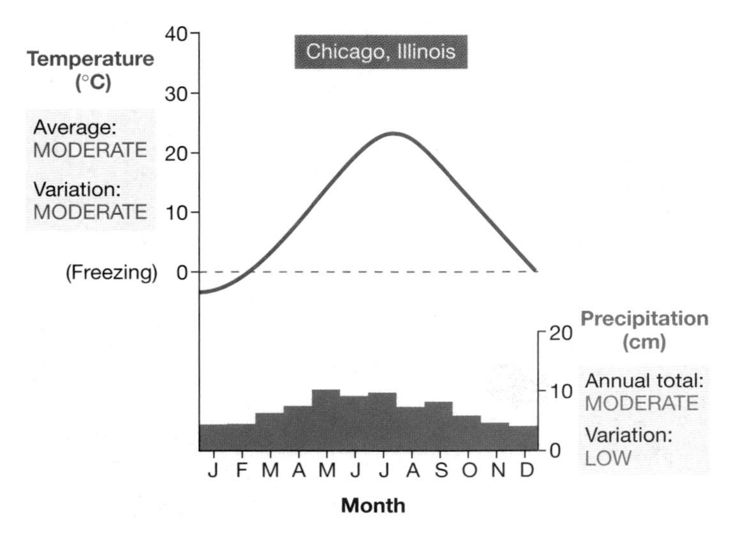

(b) Appearance

FIGURE 50.6 Temperate Forest
(a) Climograph for Chicago, Illinois, typical of a mid-latitude cold and humid climate. **(b)** Temperate forests are dominated by broad-leaved deciduous trees.

are located just under the Arctic Circle, they are referred to as the subarctic. Their climate is characterized by very cold winters, cool, short summers, and extraordinarily high annual variation in temperature. In the course of a year, subarctic areas may experience daily temperatures differing by more than 70°C.

Figure 50.7a shows a subarctic climate diagram from Dawson, in the Yukon Territory of Canada. Note that annual precipitation, at 36 cm, is virtually identical to the temperate grassland climate of western North America. The subarctic is so cold, however, that evaporation is minimal; moisture is usually abundant enough to support tree growth as a result. The climate diagram also documents the extreme temperature swings typical of the subarctic. Summer temperatures hover around 10°C, but hot spells in which temperatures exceed 20°C are not uncommon. Winters are cold, with mean monthly temperatures dropping well below freezing for nearly half the year.

As **Figure 50.7b** shows, the subarctic landscape is dominated by highly cold-tolerant conifers, including spruce, pine, fir, and larch trees. With the exception of larch, these species are evergreen. Two hypotheses have been offered to explain why evergreens predominate in cold environments, even though they do not photosynthesize in winter. The first is that they can begin photosynthesizing early in the spring, even before the snow melts, when sunshine warms their needles. The second hypothesis is based on the observation that subarctic soils tend to be acidic and contain little available nitrogen. Because leaves are nitrogen-rich, species that have to produce an entirely new set each year might be at a disadvantage. To date, however, these hypotheses have not been tested rigorously.

Based on these observations, it is not surprising that the productivity of boreal forests is low. Aboveground biomass is high, however, because slow-growing tree species may be long-lived and gradually accumulate large standing biomass. Boreal forests also have exceptionally low species diversity. The boreal forests of Alaska, for example, typically contain seven or fewer tree species.

(a) Climate characteristics

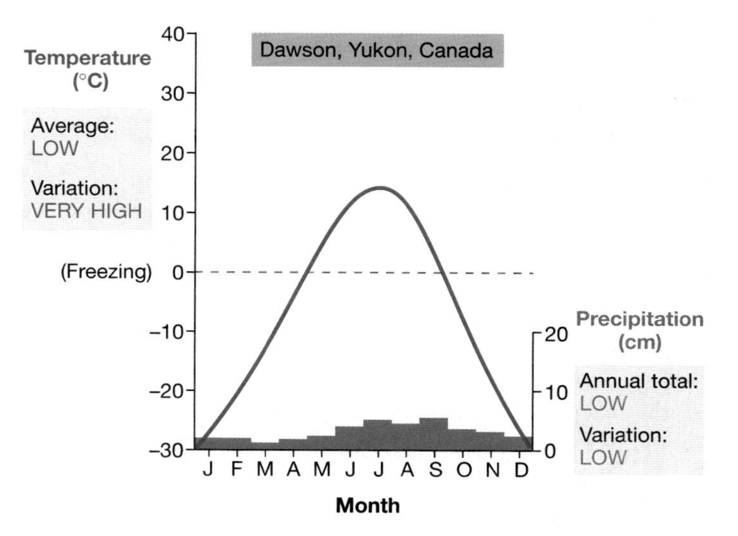

(b) Appearance

FIGURE 50.7 Boreal Forest
(a) Climograph for Dawson, Yukon Territory, Canada, in the heart of the boreal forest. **(b)** The boreal forest is dominated by needle-leaved evergreens like spruce and fir.

Tundra

Lying poleward from the subarctic is the arctic. This region has climatic conditions similar to those shown in **Figure 50.8a** (page 976) for Barrow on the northern coast of Alaska. The growing season is 6–8 weeks long at most; otherwise temperatures are below freezing. Precipitation is also low. The annual precipitation in Barrow is actually less than that in the Sonoran Desert of southwestern North America. Because of the extremely low evaporation rates, however, arctic soils are saturated year round.

As **Figure 50.8b** shows, the arctic is treeless. The leading hypothesis to explain the lack of trees is that the growing season is too short and cool to support the production of large amounts of non-photosynthetic tissue. Also, tall plants that poke above the snow in winter may experience substantial damage from blowing snow and ice crystals. Woody shrubs, such as willows, birch, and blueberries, are common but rarely exceed the height of a child. Most tundra species hug the ground.

Tundra has low species diversity, productivity, and aboveground biomass. Most soils are in the perennially frozen state known as permafrost, and the cold temperatures inhibit both the release of nutrients from decaying organic matter and the uptake of nutrients into live roots. Unlike the desert, however, plants in many tundra communities may completely cover the ground. Animal diversity also tends to be low in the tundra, although insect abundance—particularly of biting flies—can be staggeringly high.

Observations About the Distribution of Community Types

The preceding paragraphs highlighted just six of the world's many ecological communities, and focused exclusively on terrestrial systems and plant species. From that small sample, however, a few key observations emerge about the distribution of community types.

1. Trees dominate every area where it is not too cold or too dry.
2. Evergreen trees dominate climatic regions at two extremes: tropical forests and subarctic forests. Evergreens are broad-leaved in the tropics and needle-leaved at subarctic latitudes.

(a) Climate characteristics

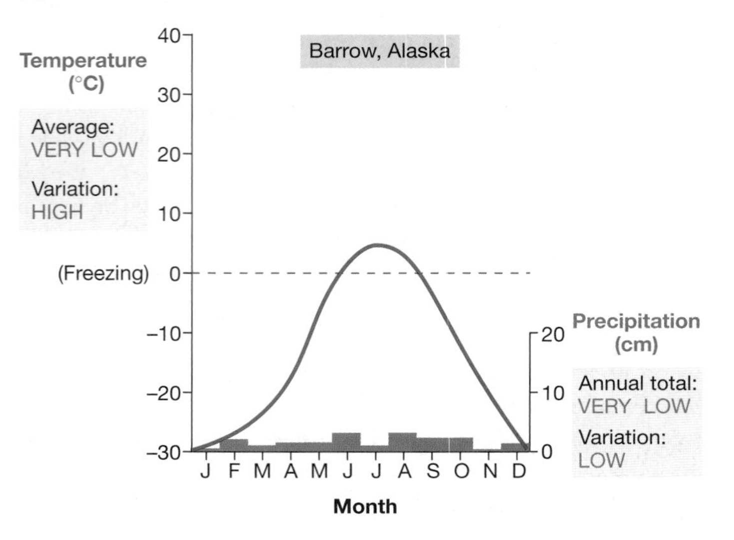

(b) Appearance

FIGURE 50.8 Arctic Tundra
(a) Climograph for Barrow, Alaska, on the north slope of Alaska.
(b) Arctic tundra is treeless. Very cold-tolerant shrubs and herbaceous plants dominate plant communities.

3. Productivity is positively correlated with temperature and precipitation. Communities in warm, wet regions are more productive than those in cold or dry regions.
4. Species diversity generally declines from the tropics to the poles. Section 50.3 explores this pattern in more detail.

One point left out of the preceding discussion is that most communities have a characteristic pattern or type of disturbance. In tropical and temperate forests, for example, windstorms blow down large trees and create openings or gaps in the canopy. Tropical, temperate, and boreal forests may burn at intervals. Grasslands are subjected to frequent burning and to heavy grazing by herbivores. Tundras are also subject to heavy grazing by geese, caribou, and other species. How do plant and animal communities recover from these types of disturbances?

50.2 How Predictable Are Community Assemblages?

To understand how communities respond to disturbance, we need to explore a question that occupied researchers for much of the twentieth century. Do biological communities have a tightly prescribed organization and composition, or are they merely loose assemblages of species? If communities are highly structured entities, then their makeup is predictable. Specifically, the diversity and abundance of species at a particular site should be identical before a disturbance and after recovery. But if communities are merely artificial constructs that give biologists a convenient way to categorize the vast diversity of species in nature, then community composition is unpredictable. The diversity and abundance of species found in a region may vary substantially before and after disturbance.

Clements Versus Gleason: Two Views of Community Dynamics

Beginning with a paper published in 1936, Frederick Clements promoted the view that biological communities are stable, integrated, and orderly entities with a highly predictable composition. To drive this point home, he likened the development of a plant community to the development of an individual organism. He argued that communities develop by passing through a series of predictable stages, and that this development culminates in a stable **climax community**. According to Clements, the nature of the climax community is determined by the area's climate and does not change over time. Further, he held that if a fire or other disturbance destroys the climax community, it will reconstitute itself by repeating the developmental stages.

Henry Gleason, in contrast, contended that the community found in a particular area is neither stable nor predictable. He claimed that plant and animal communities are ephemeral associations of species that just happen to share similar climatic re-

FIGURE 50.9 Do Identical Communities Develop in Identical Habitats?
If biological communities are static entities whose composition is determined by climate, then identical habitats should develop identical communities. But if the development of biological communities is contingent upon historical or chance events, then identical habitats should often develop distinctive communities.

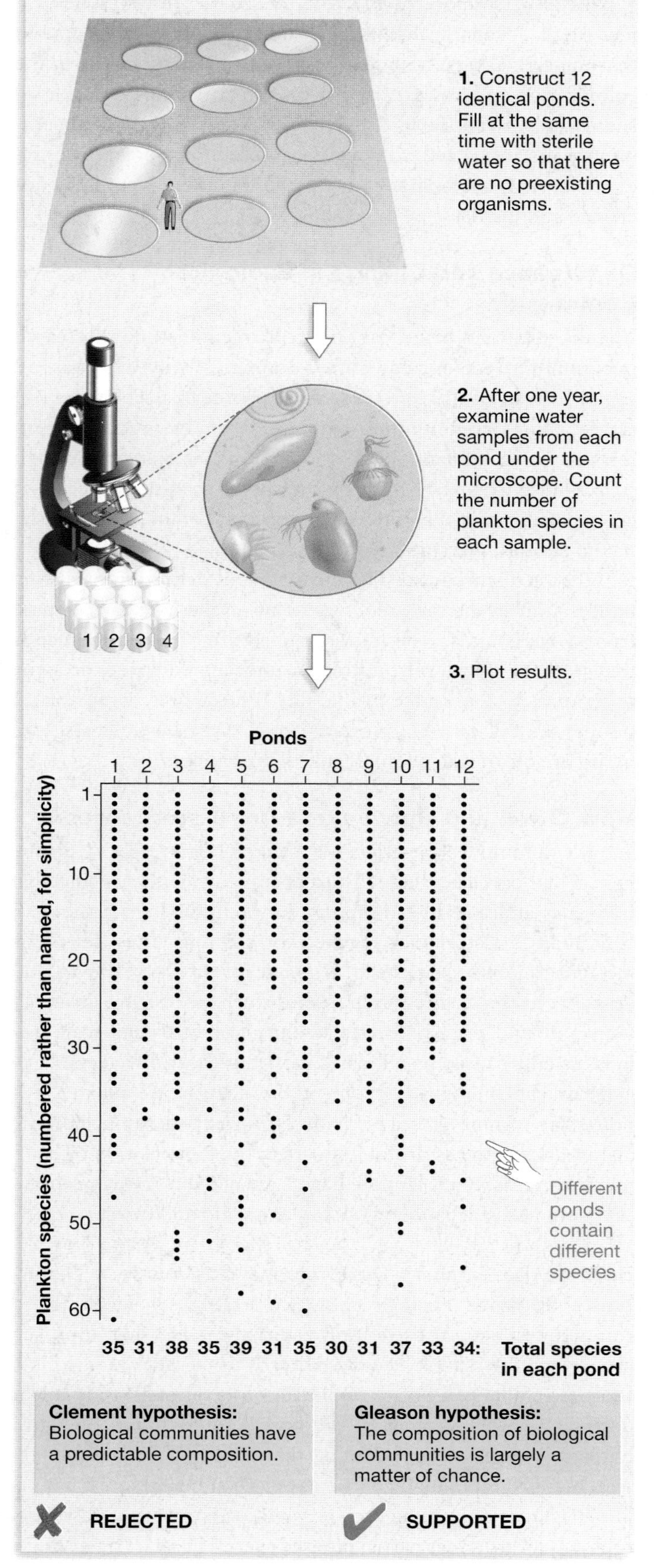

quirements. According to Gleason, it is largely a matter of chance whether a similar community develops in the same area after a disturbance occurs.

Which viewpoint is more accurate?

Historical Data on Community Structure Data on the historical composition of plant communities began to accumulate during the 1970s and 1980s. Studies of plant fossils, for example, confirmed that the distribution of species and communities in North America has changed radically since the end of the last ice age about 15,000 years ago. An important pattern emerged from these data: Groups of species in the same communities did not change their distributions in close association, as Clements would predict. Instead, individual species tended to change their ranges independently of one another.

The general message of these historical studies is that the composition of plant communities has been dynamic and unpredictable rather than static. Do experimental tests, which document community change over a much shorter time scale, agree?

Experimental Tests on Community Structure As an example of experimental work on how communities are structured, consider a recent study by David Jenkins and Arthur Buikema. These researchers asked whether similar communities of animals develop in the plankton of similar ponds. To answer this question, Jenkins and Buikema constructed 12 identical ponds (**Figure 50.9**). They filled the ponds at the same time with water that contained enough chlorine to kill any preexisting organisms. At the outset, then, the ponds were sterile. If Clements' view of community structure is correct, each pond should develop the same community of species once the chlorine vaporized and made the water habitable.

To test this prediction, the researchers sampled water from the ponds repeatedly for one year. They measured temperature, chemical makeup, and other physical characteristics of the water, recording the diversity and abundance of animals by examining the samples under the microscope. They found a total of 61 species in all of the ponds during the study, but discovered that individual ponds had only 31 to 39 species (see Figure 50.9). This observation is important. Each pond contained just half to two-thirds of the total number of species that lived in the experimental area and that were available for colonization. Why would each pond develop a different species assemblage?

To explain their results, Jenkins and Buikema contended that some species are particularly good at dispersing and are likely to colonize all or most of the available habitats. Other species disperse more slowly, and tend to reach only one or a few of the available habitats. Further, the authors proposed that the arrival of certain competitors or predators early in the colonization process greatly affects which species are able to successfully invade later. As a result, the specific details of community assembly and composition are contingent and unpredictable. As Gleason maintained, communities are a product of chance and history.

Disturbance and Change in Ecological Communities

Thanks to careful historical and experimental studies, research in community ecology has shifted dramatically over the past 50 years. Instead of trying to describe the orderly development of stable, climax communities, biologists are now much more interested in documenting how communities change through time, and particularly how they respond to disturbance. The change was inspired by the realization that communities are dynamic entities, not static ones.

What is disturbance? The most general definition of disturbance is an event that removes some individuals or biomass from a community. The important feature of a disturbance is that it alters light levels, nutrients, unoccupied space, or some other aspect of resource availability. Forest fires, windstorms, floods, the fall of a large canopy tree, disease epidemics, and herbivore outbreaks all qualify as disturbances.

How Often and How Severe Are Disturbances?

Most communities experience a characteristic type of disturbance. In most cases, these disturbances occur with a predictable frequency and severity. To capture this point, biologists refer to a community's disturbance regime. For example, fires kill all or most of the existing trees in a boreal forest every 100 to 300 years on average. In contrast, the disturbance regime in many temperate and tropical forests is that numerous small-scale tree falls, usually caused by windstorms, occur every few years.

How can biologists document the disturbance regime in a particular community? This is an important question, because disturbance patterns profoundly affect the composition of communities. As a result, land managers are actively trying to duplicate natural disturbance regimes in grassland and forest reserves.

How do Researchers Determine a Community's Disturbance Regime? Ecologists use two strategies to determine a community's natural pattern of disturbance. The first approach is based on inferring long-term patterns from data obtained in a short-term analysis. For example, an observational study might document that 1 percent of the boreal forest found on Earth burns in a given year. Assuming that fires occur randomly, researchers project that any particular piece of boreal forest has a 1 in 100 chance of burning each year. According to this reasoning, fires will recur in that particular area every 100 years on average.

This extrapolation approach is straightforward to implement, but it has important drawbacks. In boreal forests, for example, fires do not occur randomly in either space or time. They are more likely in some areas than others, and they tend to occur in particularly dry years. Unless sampling is extensive, it is difficult to avoid errors caused by extrapolating from particularly disturbance-prone or disturbance-free years or areas.

The second approach for determining disturbance regimes is based on reconstructing the history of a particular site. Flooding frequency, for example, can be estimated by analyzing sediments, because floods deposit distinctive waterborne groups of particles. Researchers estimate the frequency and impact of storms by finding wind-killed trees and determining their date of death. (They do so by comparing patterns in the growth rings of the dead trees with living individuals nearby; see Chapter 31.) The disturbances that have been most extensively studied using historical techniques, however, are forest fires.

Forest fires often leave a layer of burned organic matter and charcoal on the surface of the ground. As a result, researchers can dig a soil pit, find charcoal layers, and use radiocarbon dating to establish when the fires occurred. It is also possible to date the death of trees killed by fire by comparing their growth rings to living trees. Further, trees that are not killed by fire are often scarred. When a fire burns close enough to kill a patch of cambium tissue, a scar forms. These fire scars occur most often at the tree's base, where dead leaves and twigs accumulate and furnish fuel. Fire scars can be dated by analyzing growth rings.

Why Is It Important to Understand Disturbance Regimes?

To appreciate why biologists are so interested in understanding disturbance regimes, consider a recent study by Tom Swetnam on the fire history of giant sequoia groves in California (see **Figure 50.10a**. Giant sequoias grow in small, isolated groves on the west side of the Sierra Nevada range. Individuals live more than a thousand years, and many have been scarred repeatedly by fires. Swetnam obtained samples of cross sections through the bases of 90 trees in five different groves. As Figure **50.10b** shows, the cross sections contained numerous rings that had been scarred by fire. To determine the date of each disturbance, Swetnam counted rings back from the present. He found that in most of the groves, 10 to 50 fires had occurred each century for the past 1,500 to 2,000 years (**Figure 50.10c**). The data indicated that each tree had been burned an average of 64 times.

Swetnam's study established that fires are extremely frequent and of low severity in this community. Partly because of his work, the biologists responsible for managing sequoia groves now set controlled fires or let low-intensity natural fires burn instead of suppressing them immediately. (For more information about let-it-burn policies in natural areas, see the essay at the end of this chapter.) Similarly, studies of disturbance regimes along the Colorado River in southwestern North America inspired land managers to release a huge pulse of water from the reservoirs behind dams on the waterway recently. The flood that resulted was designed to mimic a natural disturbance event, and it appears to have beneficially affected the plant and animal communities downstream.

Succession: The Development of Communities After Disturbance

Severe disturbances remove all or most of the organisms from an area. The recovery that follows is called **succession. Primary succession** occurs when a disturbance removes the soil and its organisms as well as organisms that live above the surface. Glaciers, floods, volcanic eruptions, and landslides often initiate primary succession. **Secondary succession** occurs when a disturbance removes some or all of the organisms from an area but leaves the soil intact. Fire and logging are examples of disturbances that initiate secondary succession.

As **Figure 50.11** (page 980) shows, a sequence of communities develops as succession proceeds. Early successional communities are dominated by species that disperse their seeds long distances and are short-lived and small. Late successional communities are dominated by species that tend to be long-lived, large, and good competitors for resources such as light and nutrients. The specific sequence of species that appears over time is called a successional pathway. What role do chance and history play as a successional pathway develops? More specifically, what determines the pattern and rate of species replacement during succession at a particular time and place?

Theoretical Considerations Biologists focus on three factors to predict the outcome of succession in a community: (1) the particular traits of the species involved, (2) how species interact, and (3) environmental circumstances such as the size of the area involved and short-term weather conditions. Before going on to consider a detailed case history, we need to explore these factors in more detail.

Species traits, such as dispersal capability and the ability to withstand extreme dryness, are particularly important early in succession. As common sense would predict, recently disturbed sites tend to be colonized by plants and animals with good dispersal ability. When these organisms arrive, however, they often have to endure harsh environmental conditions. As a result, pioneering species tend to have "weedy" characteristics such as

(a) Giant sequoias after a fire

(b) Fire scars in the growth rings

(c) Reconstructing history from fire scars

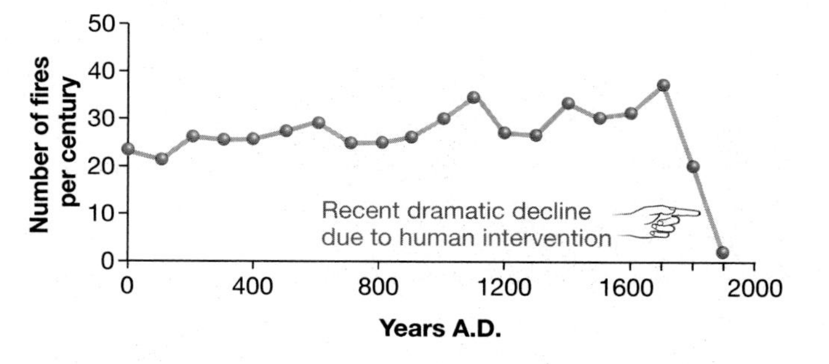

FIGURE 50.10 Reconstructing the History of Disturbance
(a) A giant sequoia grove. **(b)** The thick black arcs in this trunk cross-section are fire scars. **(c)** Because trees form one ring (light band/dark band) every year, researchers can count the rings and determine when fires have occurred during the last 2,000 years.

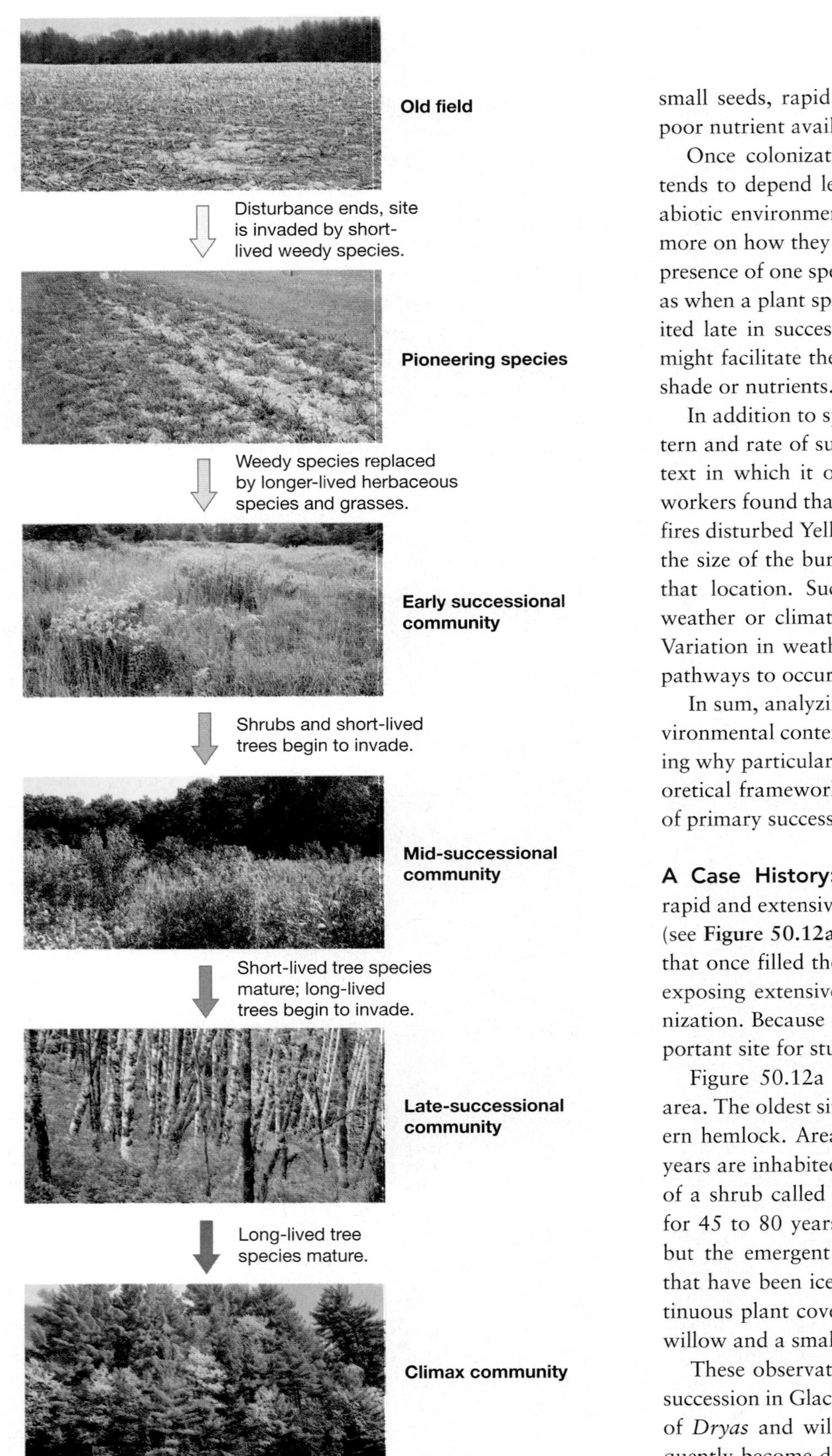

FIGURE 50.11 Succession in Mid-Latitude Temperate Forests This sequence of photos shows how succession leads to the development of a temperate forest from a disturbed state (in this case, an agricultural field).

small seeds, rapid growth, and tolerance of high light levels, poor nutrient availability, and drying.

Once colonization is under way, the course of succession tends to depend less on how species cope with aspects of the abiotic environment like dryness and nutrient availability and more on how they interact with other species. For example, the presence of one species can inhibit the establishment of another, as when a plant species that requires light to germinate is inhibited late in succession. Alternatively, an early-arriving species might facilitate the arrival of certain later species by providing shade or nutrients.

In addition to species traits and species interactions, the pattern and rate of succession depends on the environmental context in which it occurs. For example, M. G. Turner and coworkers found that the communities that developed after forest fires disturbed Yellowstone National Park in 1988 depended on the size of the burned patch and how hot the fire had been at that location. Succession is also affected by the particular weather or climate conditions that occur during the process. Variation in weather and climate causes different successional pathways to occur in the same place at different times.

In sum, analyzing species traits, species interactions, and environmental context provides a useful structure for understanding why particular successional pathways occur. To see this theoretical framework in action, let's examine data on the course of primary succession in Glacier Bay, Alaska.

A Case History: Glacier Bay, Alaska An extraordinarily rapid and extensive glacial recession is occurring at Glacier Bay (see **Figure 50.12a**, pages 982–983). In just 200 years, glaciers that once filled the bay have retreated approximately 100 km, exposing extensive tracts of barren glacial sediments to colonization. Because of this event, Glacier Bay has become an important site for studying succession.

Figure 50.12a shows the plant communities found in the area. The oldest sites are dense forests of Sitka spruce and western hemlock. Areas that have been deglaciated for about 100 years are inhabited by scattered spruce trees and dense thickets of a shrub called Sitka alder. Sites that have been deglaciated for 45 to 80 years are also covered with dense alder thickets, but the emergent trees are primarily cottonwood. Locations that have been ice-free for 20 years or less do not have a continuous plant cover. Instead, they host scattered individuals of willow and a small shrub called *Dryas*.

These observations inspired a hypothesis for the pattern of succession in Glacier Bay: With time, the youngest communities of *Dryas* and willow succeed to alder thickets, which subsequently become dense spruce-hemlock forests (Figure 50.12a). A recent study by Chris Fastie refuted this hypothesis, however. When Fastie reconstructed the history of each community by studying tree rings, he found that three distinct successional pathways have occurred (**Figure 50.12b**). In the lower part of the bay, soon after the ice retreated, Sitka spruce began growing and quickly formed dense forests. Western hemlock arrived after spruce, and is now common in the understory. At middle-aged sites in the upper part of the bay, alder thickets were dominant for several decades, and spruce are just beginning to become common. These forests will probably never be as dense as the ones in the lower bay, however, and there is no sign that western hemlock has begun to establish itself. In contrast, the youngest sites in the uppermost part of the bay may be following a third pathway altogether. Alder thickets became dominant fairly early, but spruce trees are scarce. Instead, cottonwood trees are abundant.

These data present a challenge. How do species traits, species interactions, and dispersal patterns interact to generate the three observed pathways?

Species traits may be especially important in explaining certain details about the successional pathways. Western hemlock, for example, is abundant on older sites but largely absent from young ones. This is logical because its seeds germinate and grow only in soils containing a substantial amount of organic matter. Western hemlock's intolerance of early-successional conditions explains why none of the three pathways began with colonization by this species.

Species interactions have been important in all three pathways. For example, Terry Chapin and colleagues confirmed that alder facilitates the growth of Sitka spruce. The facilitating effect occurs because symbiotic bacteria that live inside nodules on the roots of alder convert atmospheric nitrogen (N_2) to nitrogen-containing molecules that alder can use to build proteins and nucleic acids. Although spruce trees are capable of invading and growing without the presence of alder, they grow faster when alder stands have added nitrogen to the soil.

Competition is another important species interaction. For example, shading by alder reduces the growth of spruce until the trees are tall enough to protrude above the alder thicket. Once the trees breach the alder canopy, however, alder dies out because it is unable to compete with spruce trees for light.

Environmental context also clearly influences succession at Glacier Bay. For example, geologists have found evidence that the ice was more than 1,100 m thick in the upper part of Glacier Bay during the mid-1700s. Because forests grow to an elevation of only 700 m or 800 m in this part of Alaska, the glacier eliminated all of the existing forests. In the lower part of the bay, however, the ice was substantially thinner. As a result, some forests remained on the mountain slopes above the ice. As the glacier retreated from these areas, the forests on the slopes provided a source of spruce and hemlock seeds and set a dramatically different successional pathway in motion. In this way, environmental context—in this case distance to existing forests—helped to determine how the community developed.

To summarize, successional pathways are determined by an array of factors. These factors include the adaptations that certain species have to the abiotic environment, interactions among species, and the particular environmental context. As a result, the outcome of succession is not entirely predictable. To a certain extent, it is dependent on history and chance. Successional pathways may differ from time to time and from place to place.

50.3 Species Diversity in Ecological Communities

Species diversity is a key feature of biological communities. It can be quantified as a simple count of how many species are present, or it can involve more complex calculations that incorporate a species' abundance as well as its presence or absence (see **Box 50.1**, page 984). Research on diversity has focused on two questions. First, why are some communities more diverse than others? Second, why is diversity important? As human populations increase and species losses mount, answering these questions becomes increasingly urgent.

Global Patterns in Species Diversity

When European naturalists began cataloguing the flora and fauna of the tropics in the mid-1800s, they immediately recognized that these communities contained many more species than temperate or subarctic environments. Data compiled in the intervening years have confirmed the existence of a latitudinal gradient in species diversity for many taxa. In birds, mammals, fish, reptiles, many aquatic and terrestrial invertebrates, orchids, and trees, species diversity declines as latitude increases (**Figure 50.13**, page 984). Although this pattern is not universal—a number of marine groups, including shorebirds, show a *positive* relationship between latitude and diversity—it is widespread. Why does it occur?

One hypothesis to explain the gradient is that high productivity in the tropics promotes high diversity. The idea is that increased biomass production supports more herbivores and thus more predators and parasites and scavengers. Although this hypothesis is supported by the global-scale correlation between productivity and diversity, experimental studies refute it. For example, when researchers add fertilizer to aquatic or terrestrial communities, they routinely observe significant increases in productivity but decreases in diversity. Productivity alone is probably not a sufficient explanation for the higher diversity in the tropics.

A second hypothesis to explain the latitudinal gradient in diversity is inspired by the impressive physical complexity of tropical forests. The idea here is that habitats with complex physical structures have more niches, or places and ways to make a living, than habitats with simple physical structures. To drive this point home, **Figure 50.14** (page 985) compares the structural diversity of tropical forests and boreal forests. Boreal forests have a layer of herbaceous plants and a canopy, with tree stems between the two. Tropical forests are much more complex because their canopies are uneven in height, and their understory is laced with epiphytes and vines. Although this hypothesis may explain why insects, spiders, and some other groups are more diverse, it does not explain why there are more tree species in tropical forests.

A third hypothesis is that tropical regions have had more time for speciation to occur. Temperate and arctic latitudes were repeatedly scoured by ice sheets over the last two million years, but tropical regions were not. Recent data suggest, however, that tropical forests were dramatically reduced in size by widespread drying trends during the ice ages. Existing forests may be much younger than originally thought. If so, then the contrast in the age of northern and southern habitats may not be enough to explain the dramatic difference in species diversity.

A final hypothesis is that tropical areas experience fewer disturbances than temperate or subarctic regions. The idea here is that species accumulate slowly following a disturbance, and that frequent disturbances hold species numbers down. This hypothesis has also been challenged. Joseph Connell, for example, proposed that species diversity should be highest at intermediate levels of disturbance, because communities would contain pioneering species as well as species better adapted to late-successional conditions. Second, recent studies have confirmed that treefalls and canopy gaps occur as regularly in tropical forests as they do in temperate forests, and that occasional fires have been important sources of disturbance. According to these data, it is naïve to view tropical forests as undisturbed. In fact, Connell has argued that tropical forests and coral reefs are diverse precisely because they undergo disturbance of intermediate frequency and severity.

To summarize, there is no simple answer to the question of why some communities are more diverse than others. Each of the factors discussed here may influence diversity, but no single one offers a convincing explanation for the global diversity gradient.

FIGURE 50.12 Succession Pathways at Glacier Bay
(a) Several distinctive plant communities are found in Glacier Bay, Alaska. The first hypothesis about succession at Glacier Bay proposed that these communities represented stages in the same successional pathway.

CD ACTIVITY 50.2 Primary Succession

The Role of Diversity in Ecological Communities

Research on species diversity has recently taken a practical turn. Does species diversity affect the stability and productivity of communities? If so, then the rapid loss of diversity that is occurring now (see Chapter 52) will have serious consequences.

Researchers have been particularly interested in measuring the effect of diversity on two characteristics of communities. **Resistance** is a measure of how much a community is affected by a perturbation such as a drought. **Resilience** is a measure of how quickly a community recovers following a disturbance.

How does species diversity affect resistance, resilience, and productivity? Several biologists have proposed that diversity is positively correlated with stability, meaning that highly diverse communities are more resistant and more resilient. The logic behind this hypothesis is that when a large number of species are present, some are likely to be redundant in terms of their role in the community. For example, there are likely to be many different types of herbivores, carnivores, pollinators, and so on. Because of this redundancy, wiping out one or a few species is unlikely to profoundly affect the community as a whole. In addition, biologists have hypothesized that high diversity leads to high productivity. The reasoning here is that if species differ, even slightly, in the way they use resources, then the total resource base available in a particular habitat will be used more completely and productivity will be higher.

To test whether high species diversity actually does lead to greater resistance, resilience, and productivity, David Tilman and co-workers have conducted a series of experiments using plants native to the prairies of North America. **Figure 50.15** (page 985) illustrates an experiment designed to test whether higher diversity leads to higher productivity. The researchers grew seedlings from 24 species and established 147 study plots.

(b) Hypothesis 2: Three distinct successional pathways occur in Glacier Bay.

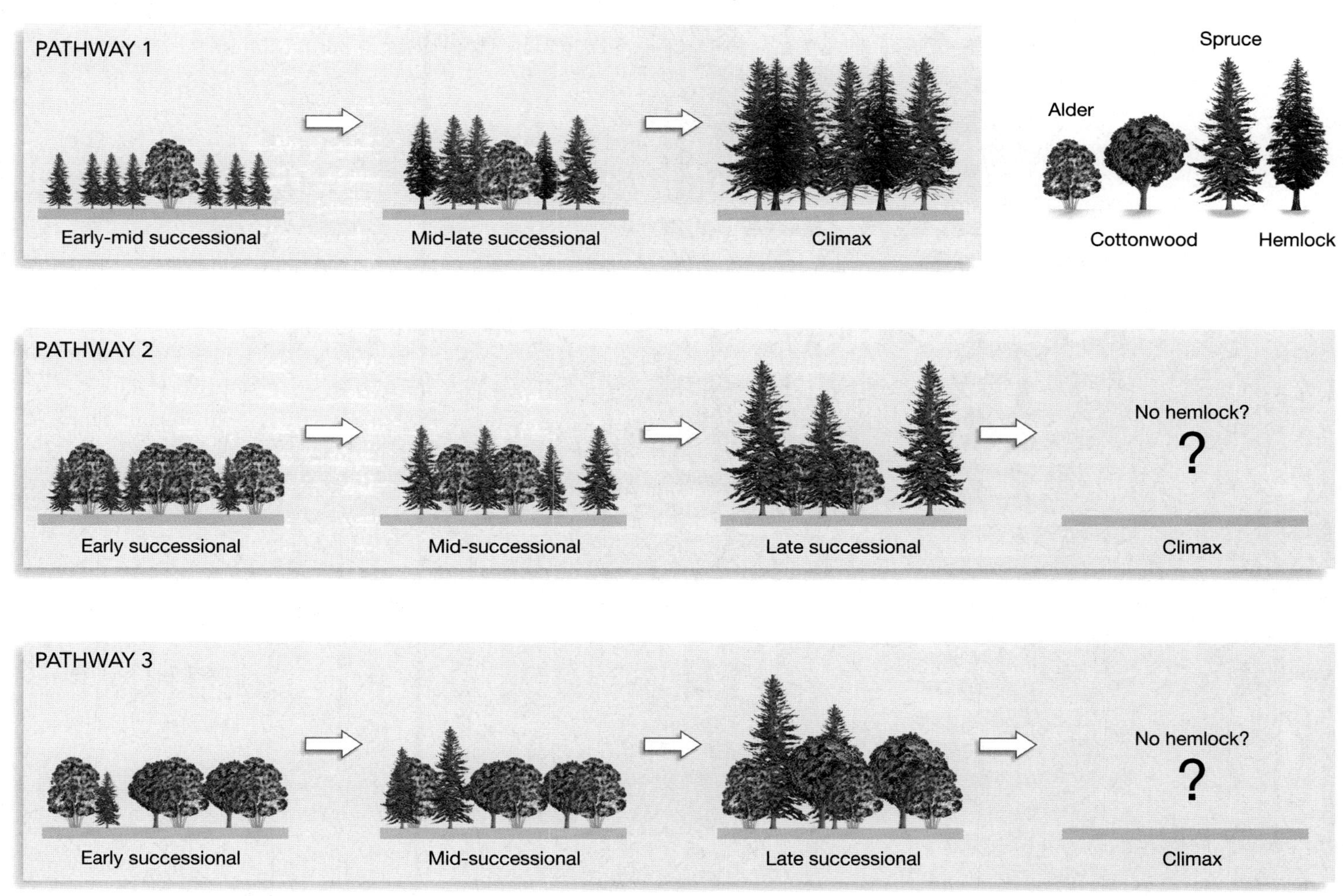

(b) Recent data indicate that three distinct successional pathways exist. In Pathway 1, spruce arrives early and in high densities. Alder never forms continuous groves, and western hemlock arrives after spruce. In Pathway 2, alder arrives early and forms continuous thickets. Spruce arrives early as well, but develops low-density stands under the alder thicket. Western hemlock is absent. In Pathway 3, alder arrives early and forms continuous thickets. Cottonwood and spruce form a low-density forest in the alder thickets. Western hemlock is absent.

Each plot received a similar number of individuals but a different number of species: 1, 2, 4, 6, 8, 12, or 24. The research team allowed the plants to grow for a year and weeded the plots to remove any invading species. In the second year, they measured productivity by quantifying the area of the plot covered by plants. As the data in Figure 50.15 show, plant cover was higher in the more species-rich plots. This result supports the hypothesis that more diverse communities are more productive.

To test the hypothesis that diversity leads to resistance and resilience, Tilman and John Downing followed up on a natural experiment. In 1987–1988, a severe drought hit sites in Minnesota where the researchers had been measuring species richness and other characteristics in a series of permanent study plots. When the drought ended, they were able to ask whether resistance and resilience correlated with the number of species that existed before the disturbance. Some of their results are shown in **Figure 50.16a**. This graph documents the change in total biomass that occurred from the year prior to the drought to the height of the drought. A completely resistant community would show no change in biomass. As predicted, drought resistance appeared to be higher in more diverse communities. Tilman and Downing also measured resiliency by comparing biomass four years after the drought to biomass prior to the drought. A completely resilient community would recover quickly from the disturbance and have the same biomass at both times. As the data plotted in **Figure 50.16b** show, communities with more species seemed to exhibit higher resiliency.

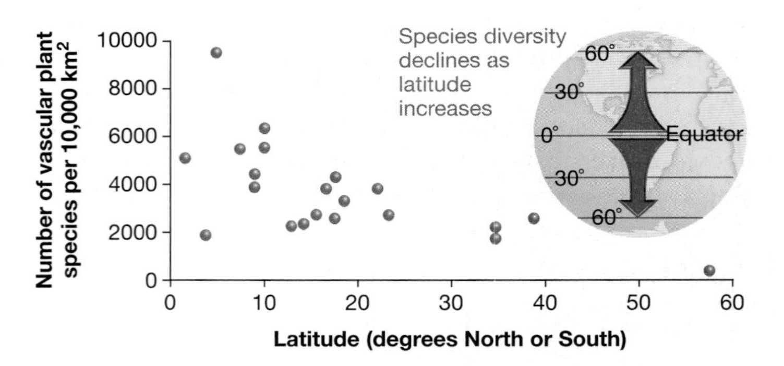

FIGURE 50.13 Latitudinal Gradients in Species Diversity
The number of plant species in a 10,000-km² area declines from several thousand in equatorial regions to less than 1,000 at very high latitudes. This is one example of a latitudinal gradient in species diversity.

BOX 50.1 Measuring Species Diversity

To measure species diversity, a biologist could simply count the number of species present in a community. The problem is that simple counts provide an incomplete picture of diversity. The relative abundance of species is also an important component of diversity. To grasp this point, consider the composition of the three hypothetical communities shown in **Figure 1**. These communities are nearly identical in species composition, but differ greatly in the relative abundance of each species.

We can use these data to compare two measures of species diversity. Species richness is simply the number of species found in a community. In this case, communities 1 and 2 have equal species richness and community 3 is lower in richness by one species. It is important to note, however, that communities 2 and 3 have similar relative abundances of each species, or what biologists call high evenness. Community 1, in contrast, is highly uneven. Fifty-five percent of the individuals in community 1 belong to species A, and other species are relatively rare. An uneven community has lower effective diversity than its species richness would indicate.

To take evenness into account, other diversity indices have been developed. A simple example is the Shannon-Wiener index, given by Equation 50.1:

$$H' = -\sum p_i \log(p_i)$$

In this equation, p_i is the proportion of individuals in the community that belong to species *i*. Note that the index is summed over the *i* species in the study. The Shannon-Weaver index for the three hypothetical communities is shown in the figure. Notice that while communities 1 and 2 have the same species richness, community 2 has higher diversity because of its greater evenness. Community 3 has lower species richness than community 1, but higher diversity.

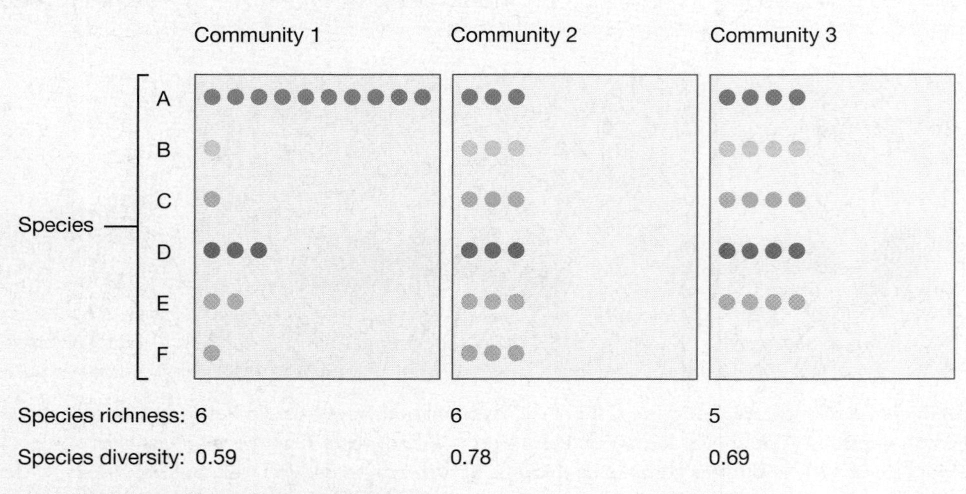

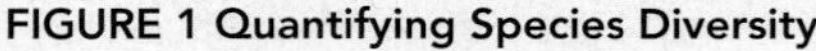
FIGURE 1 Quantifying Species Diversity

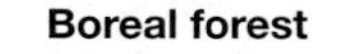

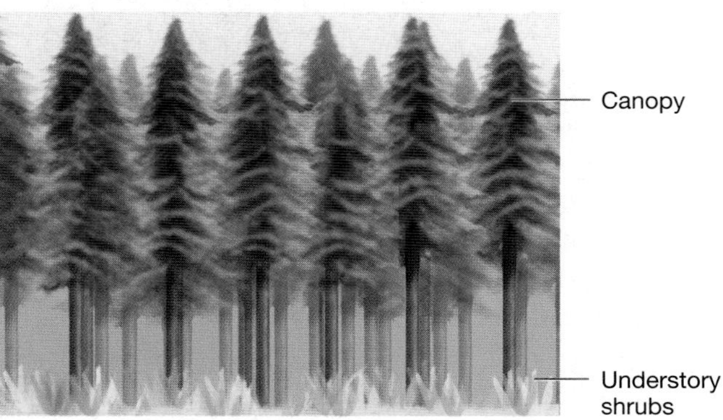

FIGURE 50.14 Structurally, Tropical Forests Are More Complex Than Boreal Forests
Tropical forests support a range of understory trees, shrubs, vines, and epiphytes. Boreal forests, in contrast, typically have just a canopy and understory. This difference in structural diversity may partly explain the gradient in species diversity.

These experiments support the hypothesis that species richness affects how communities function. In North American prairies, at least, communities that are more diverse appear to be more productive, more resistant, and more resilient than communities that are less diverse. Why this pattern occurs is not clear, however. More experiments are needed to confirm these results and determine the mechanisms responsible for these patterns.

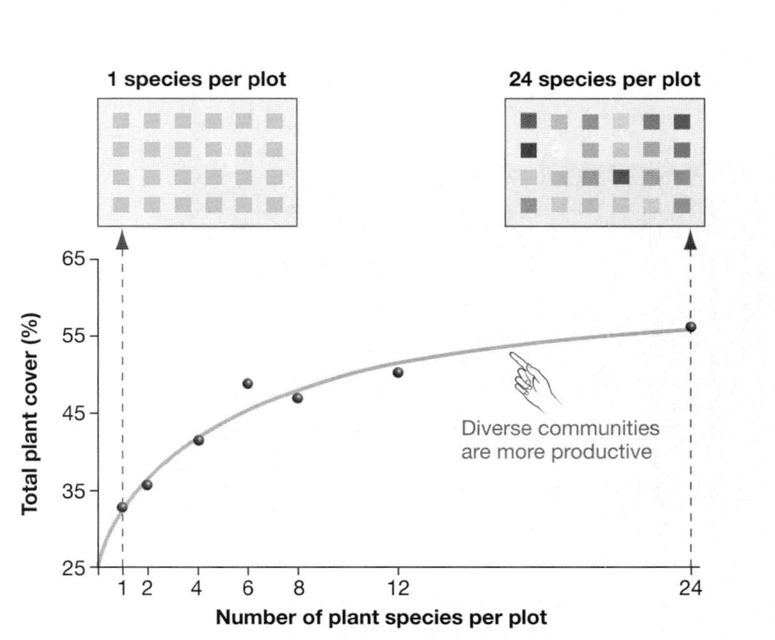

FIGURE 50.15 Species Diversity and Community Productivity
Total plant cover is defined as the percentage of a study plot's area that is covered by plants. It is interpreted as an index of productivity. **QUESTION** Extrapolating from these data, what would total plant cover be in a plot that contained 50 plant species?

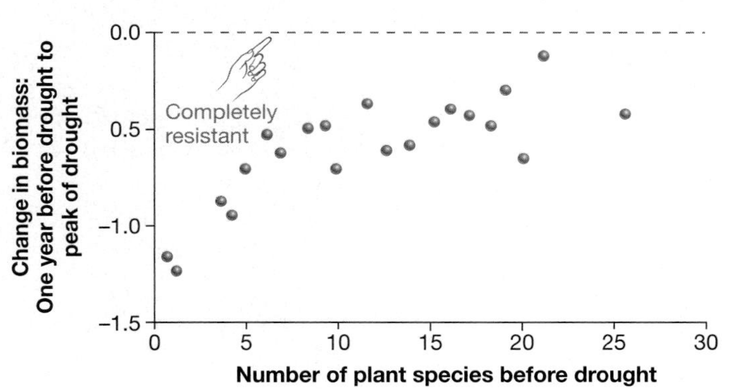

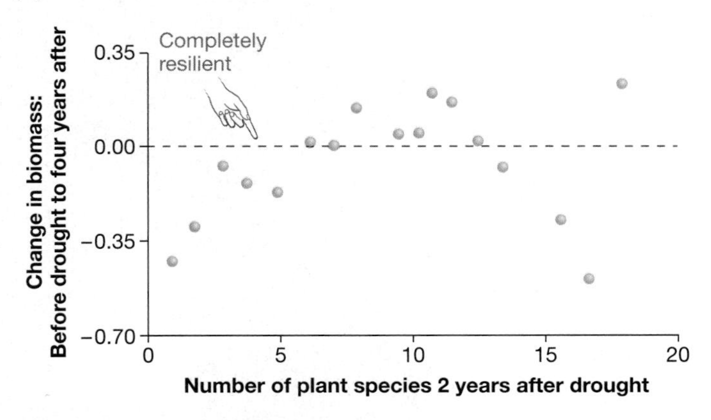

FIGURE 50.16 Species Diversity and Community Resilience
(a) The data graphed here support the hypothesis that diverse communities are more stable, meaning that they change less during a disturbance. **(b)** The data graphed here support the hypothesis that diverse communities recover from a disturbance more quickly than less-diverse communities. **QUESTION** The 0.0 values on the y-axes of these graphs are labeled "completely resistant" or "completely resilient?" Why?

Essay Let-It-Burn Policies

The summer of 1988 was extremely dry in the western United States. In the Greater Yellowstone Area (GYA), which includes Yellowstone National Park, almost no rain fell from mid-July through August. Thunderstorms rolled through with ample lightning but no rain. As a result of the severe conditions, fires ignited. Park managers allowed the fires to burn, because research had shown that the GYA experienced frequent fires before the arrival of European settlers, and because many of the region's communities depend on fires to thin dense stands of trees and allow regeneration. By late July, however, the blazes had grown so large that the National Park Service decided to suppress them. The conflagration continued to grow, however. By September 1, the fire perimeter enclosed approximately 350,800 ha—an area substantially larger than the state of Rhode Island (**Figure 1**).

As a result of the severe conditions, fires ignited.

The onset of winter, combined with a $120-million firefighting effort, eventually extinguished the fires. Winter did not extinguish a firestorm of controversy, however. Critics of the let-it-burn policy decried the destruction of the National Park Service's crown jewel.

But was the park destroyed? The species traits of trees that are particularly abundant in the GYA suggests that the answer should be no. For example, both aspen and lodgepole pine require fire in order to regenerate. Aspen cannot tolerate shade and does not sprout beneath its own canopy. Lodgepole pine is also shade-intolerant, and has cones that are glued shut with resin. The cones open and shed their seeds only in the heat of a fire.

Since 1988, the regeneration promised by ecologists has started to become visible. Tourists have continued to visit the park, and the controversy over the fires has quieted somewhat. In other areas where fires have historically been frequent, most land managers continue their policies of letting natural fires burn. In some areas where fires have been suppressed for decades, such as the sequoia preserves introduced earlier in the chapter, biologists are attempting to reintroduce fire through controlled burning.

Controversy will undoubtedly continue around the globe, however, because land managers often have two conflicting mandates. They are expected to provide recreational opportunities *and* maintain plant and animal communities in a more-or-less natural condition. Although fire and other disturbances are a normal and necessary component of many communities, these disturbances clash with the need to maintain a safe and aesthetically pleasing landscape for recreation. Should aesthetic demands take precedence over ecological needs? Or is it possible that our aesthetic values could change, so that we begin to see fires as a sign of renewal instead of disaster? The controversy remains unresolved.

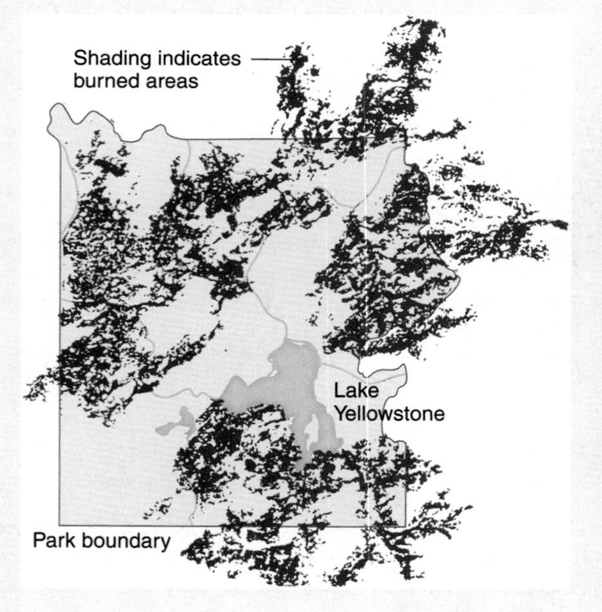

FIGURE 1
This map of the Greater Yellowstone Area shows the area burned in the 1988 fires; the photo shows one of the blazes in progress.

Chapter Review

Summary

A community is an assemblage of interacting species. Ecologists have debated whether communities are fixed, predictable entities or simply places where the distributions of various species overlap. Historical and experimental evidence support the view that communities are dynamic rather than static and that their composition is neither entirely predictable nor stable.

Nonetheless, it is clear that some of the very general features of plant communities are predictable, even though the details of species composition vary among regions and through time. Specifically, areas with similar climates have communities with the same dominant life-form. Tropical wet climates, for example, predictably support tropical wet forests. Similarly, mid-latitude steppe climates predictably support grasslands, although the dominant species of grass varies from place to place.

In addition to climate, the composition of a community is dictated by disturbance. Disturbances remove individuals and biomass and change aspects of resource availability. In extreme cases, disturbance may remove all organisms and all soil from a large area. Each community has a characteristic disturbance regime—meaning a type, severity, and frequency of disturbance that it experiences.

The process by which communities develop after a disturbance is called succession. Three types of factors influence the pattern of succession. First, a species' physiological traits influence the kinds of abiotic environmental conditions it can tolerate, and dictate when it can successfully invade a community. Second, interactions between species influence when and if a species appears during succession. Finally, the environmental context in which succession occurs can be an important determinant of the outcome.

One of the most widely studied patterns in community ecology is the latitudinal gradient in species diversity. In many different taxonomic groups, species diversity declines from the equator to the poles. Various hypotheses have been proposed to account for this pattern. These explanations suggest that the tropics have high species diversity because of their greater productivity, greater structural heterogeneity, greater evolutionary age, and low level of disturbance.

Biologists are increasingly interested in the role that species diversity plays in the fate of ecological communities. Two major hypotheses have been proposed to describe how diversity influences plant communities. Diverse communities are predicted to be more resilient and resistant in the face of disturbance, because greater diversity is likely to be associated with greater redundancy in terms of how species function in the community. Diverse communities are also predicted to be more productive, because differences among species in resource use allow the nutrients and space in a community to be used more completely. Recently, both of these hypotheses have received some experimental support.

In the next chapter, we consider how communities exchange energy and nutrients with the atmosphere, oceans, and land. Together, biological communities and the environment interact to form ecosystems.

Questions

Content Review

1. Temperate forests are found in which of the following climate types?
   a. wet tropical
   b. subtropical dry
   c. mid-latitude steppe
   d. mid-latitude cold and humid
2. Which of the following disturbances is frequently responsible for preventing trees from establishing in grasslands?
   a. volcanoes
   b. windstorms
   c. fire
   d. glaciation
3. Why are trees absent from arctic latitudes?
   a. Sunlight is too scarce in winter.
   b. Snow is too deep.
   c. Temperatures are too low.
   d. Herbivory is too intense.
4. Which of the following is not correlated with species diversity?
   a. latitude
   b. productivity
   c. longitude
   d. resilience
5. In an ecological sense, what is productivity?
   a. an individual's lifetime reproductive success
   b. an individual's average annual reproductive success
   c. the total amount of photosynthesis that occurs in an area of a given size, per year
   d. the amount of energy that is stored in standing bioma

Conceptual Review

1. State Clements' and Gleason's ideas about plant communities as hypotheses that make testable predictions. (Write sentences that read something like this: "If Clements' view is correct, then X should happen in biological communities in response to Y.") Which hypothesis appears to be more accurate? Why?
2. Why are deserts found at 30° N and 30° S latitude?
3. Why do evergreen trees dominate communities in the tropics and in the subarctic?
4. What is a disturbance? List five types of disturbance. Compare and contrast their effects.
5. Describe the latitudinal gradient in species diversity that exists for most taxonomic groups. Discuss the pros and cons of one hypothesis to explain this pattern.

Applying Ideas

1. Describe the kind of ecological community you would expect to find in an area with climate shown in the figure. Explain why.

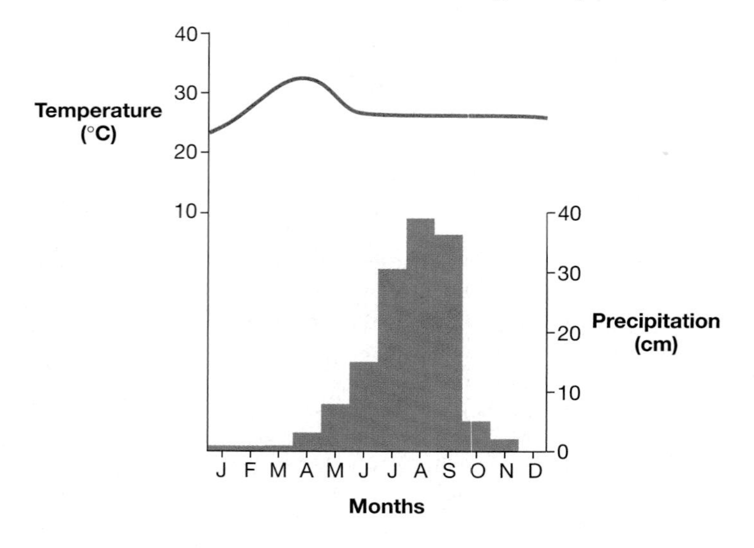

2. Using the information in the text on fire regimes in giant sequoia groves, propose a management plan for Sequoia National Park.

3. Think of a natural or semi-natural area near your home. Explain how you would document the characteristics of the plant and animal communities there. Be sure to include aspects such as species diversity, productivity, disturbance regime, and response to disturbance.

4. An historical approach has been effective for determining the frequency of disturbance events in many different communities. The same approach has been less useful for determining their spatial extent. Why?

5. Suppose that a two-acre lawn on your college's campus is allowed to undergo succession. Describe how species traits, species interactions, and environmental circumstances might affect the community that develops.

CD-ROM and Web Connection

CD Activity 50.1: Tropical Atmospheric Circulation *(animation)*
(Estimated time for completion = 5 min)
Examine the influence of atmospheric phenomena on global ecological patterns.

CD Activity 50.2: Primary Succession *(tutorial)*
(Estimated time for completion = 10 min)
Learn how succession varies by exploring succession at Glacier Bay, Alaska.

CD Activity 50.3: Community Stability *(tutorial)*
(Estimated time for completion = 10 min)
Explore how resilience and resistance can be used to measure stability.

At your **Companion Website** (http://www.prenhall.com/freeman/biology), you will find self-grading exams and links to the following research tools, online resources, and activities:

Mission: Biomes
Hosted by NASA, this website provides information about biomes, including average temperatures, precipitation, and key vegetation.

Ecological Policy
Learn about attempts to understand and remedy biodiversity loss in Australia.

The Yellowstone Fires
This website, at NASA's Classroom of the Future, has a great overview of research and history related to the Yellowstone fires.

Additional Reading

Baskin, Y. 1999. Yellowstone fires: A decade later. *Bioscience* 49: 93. An update on the fires discussed in the chapter's essay.

Blaney, C. 1995. Rooting for causes of succession. *BioScience* 45: 741.

Collins, S. L. et al. 1998. Modulation of diversity by grazing and mowing in native tallgrass prairie. *Science* 280: 745–747. Discusses how disturbance affects species diversity in a grassland community.

Turner, M. G. et al. 1997. Fires, hurricanes, and volcanoes: Comparing large disturbances. *BioScience* 47: 758–768. Explores "catastrophic" disturbances.

Yarie, J. et al. 1998. Flooding and ecosystem dynamics along the Tanana River. *BioScience* 48: 690–695. Investigates how disturbance by flooding affects streamside communities.

Ecosystems

51

Over the past 40 years, a new type of environmental problem has emerged. Global warming, acid rain, the development of a hole in the ozone layer over Antarctica, the formation of anaerobic "dead zones" in the oceans due to nitrate pollution, and the development of algal blooms and eutrophication in lakes due to phosphorus pollution all have a common thread. Severe problems can develop when humans change the chemical and physical characteristics of the environment.

How are organisms and the physical and chemical characteristics of the environment linked? To answer this question we need to explore how organisms interact with the abiotic, or inorganic, environment to produce an ecosystem. An **ecosystem** consists of the organisms that live in an area along with certain nonbiological components. The relevant abiotic components include nutrients and energy. The focus of ecosystem studies is to understand how molecules cycle between the abiotic and biotic world, and how the energy in sunlight dissipates as it flows through an ecosystem. As humans disrupt nutrient cycles and change the chemistry of lakes, oceans, and the atmosphere, ecosystem ecology has taken on an increasingly applied aspect.

How does an ecosystem differ from a community? In most cases, ecosystems are composed of multiple communities along with their chemical and physical environment. For example, ecologists who study lakes recognize a number of communities within a lake. There is a distinct community of interacting species along the lake bottom, near the surface of the water, and in other locations. Because energy and matter flow among those different communities, they are studied as a unit called the lake ecosystem.

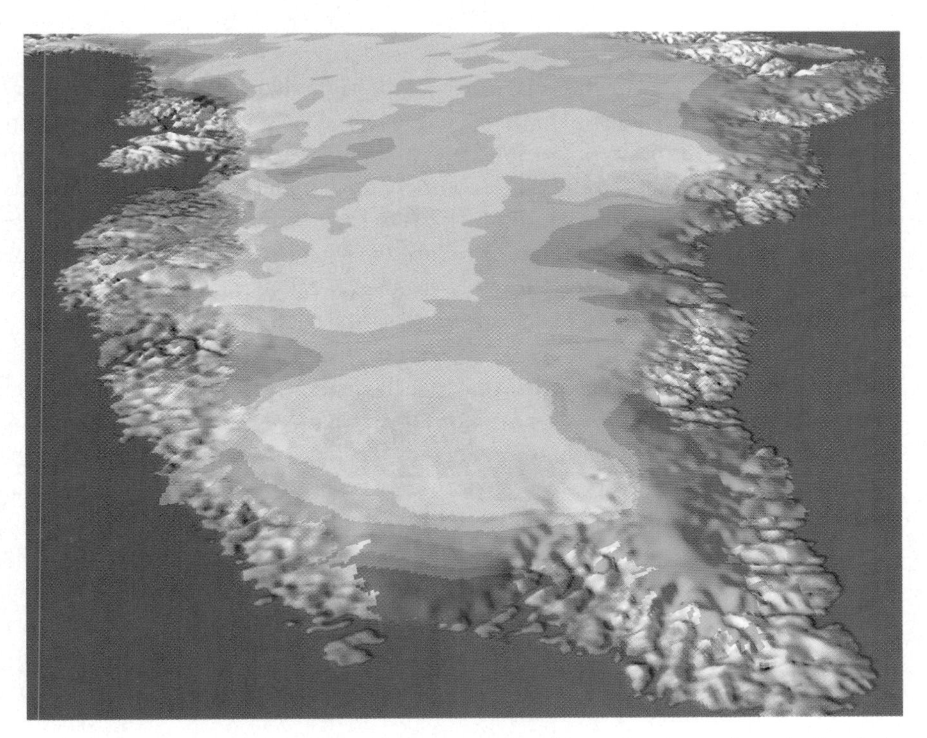

This image summarizes the results of a recent study of Greenland's ice sheet. In the darkest blue areas, ice is being lost at a rate of about a meter of thickness per year. Overall, the entire ice sheet is losing about 51 cubic kilometers per year. This chapter examines how energy and nutrients flow through Earth's oceans, atmosphere, soils, and organisms, and closes with a look at the causes and possible consequences of global warming.

This chapter introduces ecosystem studies with a look at how energy flows among the components of an ecosystem; it ends by exploring how key elements like carbon and nitrogen cycle through organisms, sediments, the oceans, and atmosphere. As we'll see, farming, logging, urbanization, and other human activities are having a massive impact on these energy flows and nutrient cycles. Understanding ecosystem ecology is fundamental to managing the future of our planet.

51.1 Energy Flow and Trophic Structure

If an ecosystem is an economy, then energy is its currency. As **Figure 51.1** shows, ecosystems have four components that are linked by a flow of energy. The first component consists of **primary producers.** Producers use solar energy or the chemical energy contained in reduced inorganic compounds to manufacture their own food. Producers form the basis of ecosystems by transforming the energy in sunlight or reduced inorganic compounds into the chemical energy stored in sugars. As Chapter 6 pointed out, producers are called autotrophs (self-feeders).

The second component of an ecosystem consists of consumers. Consumers eat other organisms. Herbivores are consumers that eat plants; carnivores are consumers that eat animals. **Decomposers**, which obtain energy by feeding on the dead remains of other organisms or waste products, form the third component of ecosystems. The fourth and final element is the abiotic environment, which includes the soil, climate, atmosphere, and the particulates and solutes in water.

Energy enters ecosystems in the form of sunlight and reduced inorganic compounds such as hydrogen (H_2), methane (CH_4), and hydrogen sulfide (H_2S). How much of this energy goes into the production of new plant tissue? How much goes into the tissues of consumers and decomposers?

Energy Flow: A Case History

James Gosz and co-workers have assembled one of the best data sets available on how energy flows through ecosystems. These researchers performed an exhaustive series of measurements on how energy flows through a temperate forest ecosystem. Their study site was in the Northeast region of the United States—the Hubbard Brook Experimental Forest in the state of New Hampshire.

As **Figure 51.2** shows, energy flow in this ecosystem begins when plants capture the energy in solar radiation via photosynthesis. (In some ecosystems, energy flow begins not with photosynthesis but with chemosynthesis. For example, bacteria that live near the deep-sea vents introduced in Chapter 2 derive energy from hydrogen sulfide—H_2S—rather than sunlight.) At Hubbard Brook, the amount of energy entering the ecosystem from sunlight varies throughout the year and, to a lesser extent, from year to year. From 1 June 1969 to 31 May 1970, which Gosz and colleagues describe as a typical year, 1,254,000 kilocalories (kcal) of solar radiation per square meter (m^2) reached the forest. If this amount of energy were available in the form of electricity, it would easily power two 75-watt lightbulbs that burned continuously for one year.

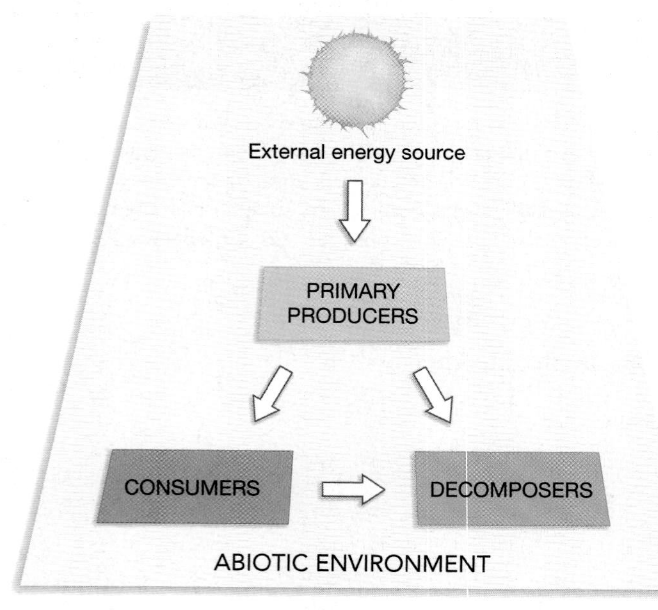

FIGURE 51.1 The Four Components of an Ecosystem Interact
Primary producers harness an external energy source to manufacture ATP and reduced carbon compounds, which are then available to consumers. When primary producers and consumers die, their remains are digested by decomposers. Primary producers, consumers, and decomposers all exchange energy and matter with the soil, atmosphere, water, and other aspects of the abiotic environment.

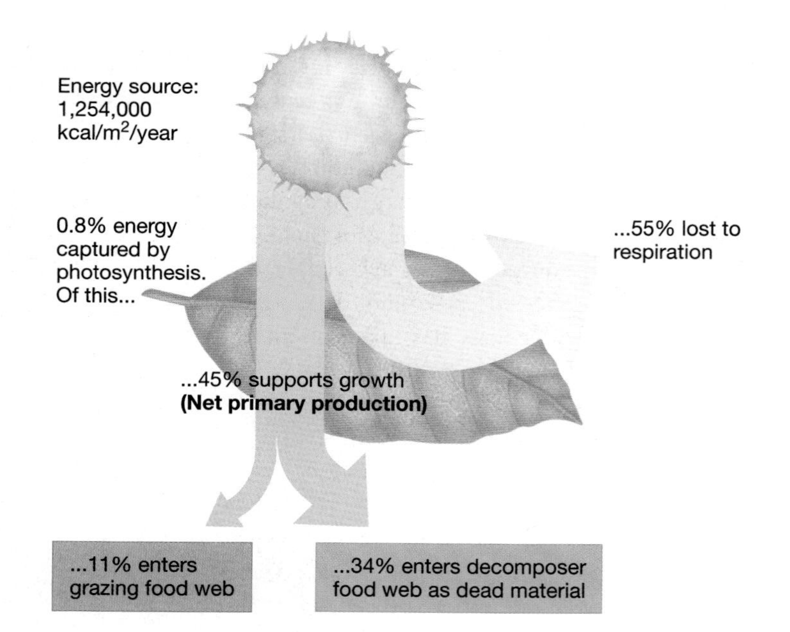

FIGURE 51.2 Energy Flow Through the Hubbard Brook Forest Ecosystem
In a temperate forest ecosystem, energy from sunlight is transformed to chemical energy by photosynthesis. The products of photosynthesis go, in part, to fuel new plant growth. Plant tissue is either consumed by herbivores in the grazing food web or falls into the decomposer food web when the plant dies.

By documenting rates of photosynthesis in a large sample of forest plants, Gosz and colleagues calculated that the plants used 10,400 kcal/m^2 of energy in photosynthesis. This figure represents **gross photosynthesis,** which is defined as the total amount of photosynthesis in a given area and time period. They also calculated **gross photosynthetic efficiency,** or the efficiency with which plants use the total amount of energy available to them, as the ratio of gross photosynthesis to solar radiation in kcal/m^2. At Hubbard Brook, efficiency was $10{,}400 \div 1{,}254{,}000 = 0.8\%$. This value is typical of other ecosystems as well. As Chapter 7's discussion of photosynthesis showed, plants are able to use only a tiny fraction of the total radiation received.

As Figure 51.2 shows, primary producers used the energy captured by photosynthesis in two ways: About 45 percent supported the synthesis of new tissue, while the remaining 55 percent was used for maintenance or respiratory costs. The energy invested in new tissue is called **net primary production.** Net primary production represents the amount of energy that is available to herbivores—the organisms that eat primary producers.

How does net primary production vary among ecosystems? As **Figure 51.3a** indicates, the terrestrial ecosystems with highest productivity are located in the wet tropics. Tropical rain forests and tropical seasonal forests cover just 16 percent of Earth's land surface but together account for 43 percent of

(a) Terrestrial productivity

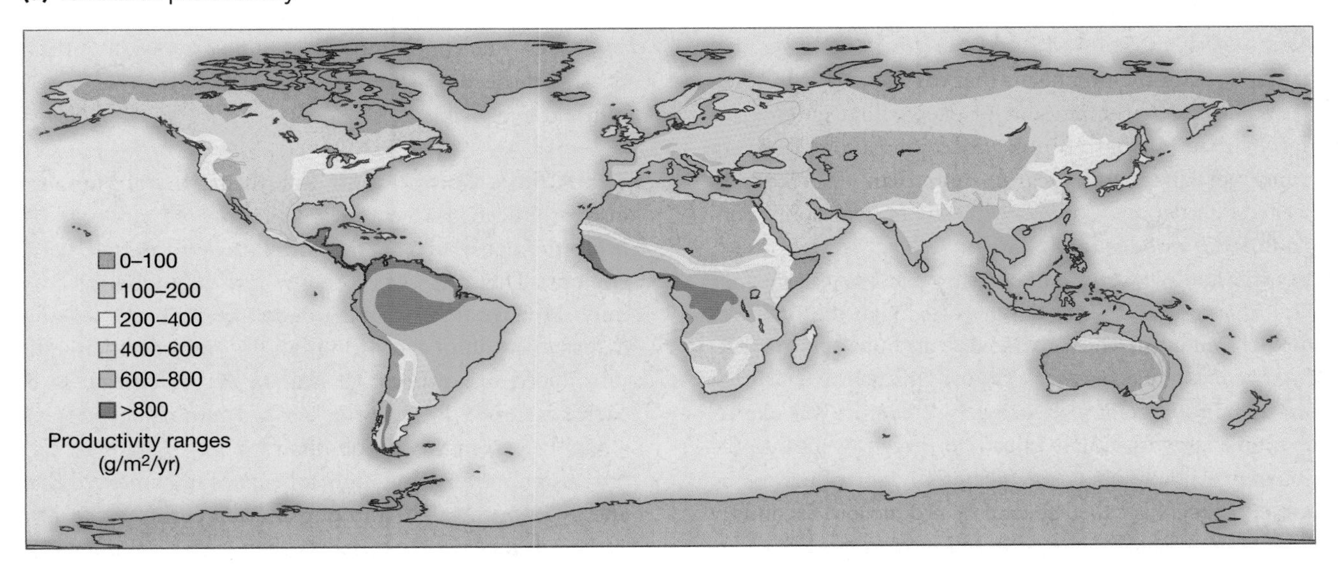

(b) Marine productivity

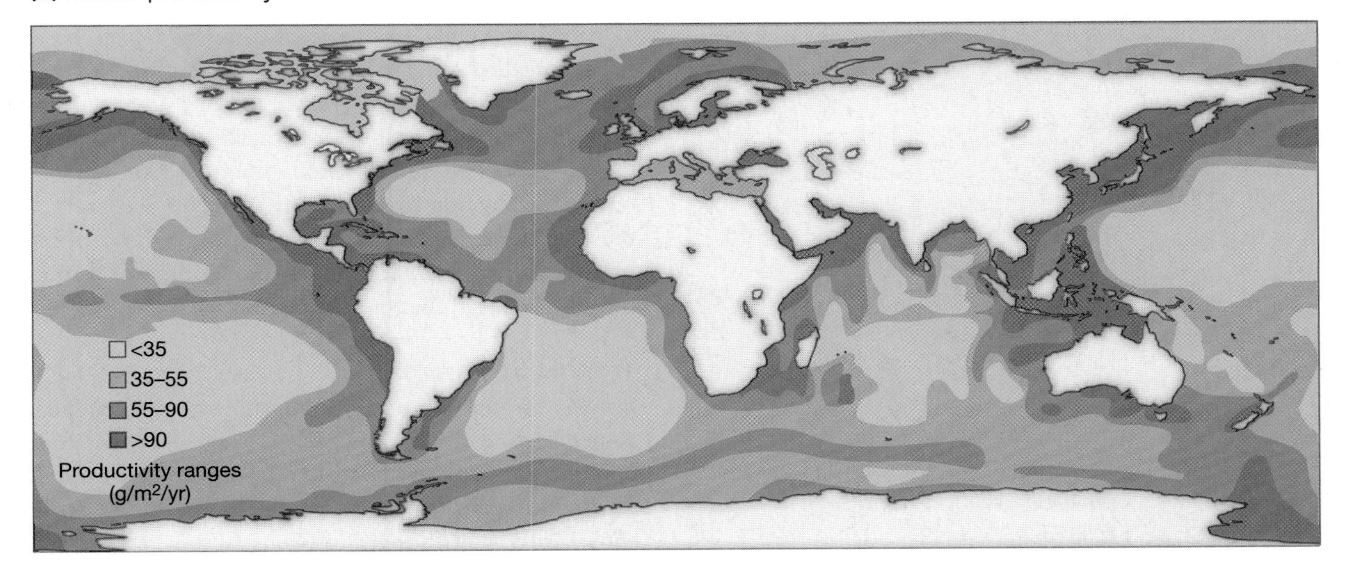

FIGURE 51.3 Primary Productivity Varies Among Regions
(a) The terrestrial ecosystems with the highest primary productivity are found in the tropics, where warm temperatures and high moisture encourage high photosynthetic rates. Tundras and deserts have the lowest productivity. **(b)** The highest productivity in the oceans occurs in nutrient-rich coastal areas.

terrestrial net primary productivity. Deserts and arctic regions have the lowest productivity. This pattern suggests that the productivity of terrestrial ecosystems is limited by a combination of temperature and water availability. Productivity patterns in marine ecosystems are somewhat different, however, as **Figure 51.3b** shows. Marine productivity is highest along coasts and in areas where water wells up from the ocean bottom to the surface. This pattern occurs because nutrient levels are high at these locations. Coastal areas receive inputs of nutrients from terrestrial ecosystems via rivers, and upwelling brings nutrients that have rained down into cold, deep water back to the surface.

Primary Consumers The energy contained in plant tissue moves into the **grazing food web** if it is eaten by an herbivore. A grazing food web is made up of the network of organisms that eat plants, along with the organisms that eat the herbivores. Between 1969 and 1971, the amount of energy that entered the grazing food web at Hubbard Brook varied from about 1 percent of net primary production per year to more than 40 percent. In general, leaves, seeds, and fruits are the most commonly consumed plant parts; wood is rarely eaten.

What is the fate of energy obtained by primary consumers? **Figure 51.4** illustrates how energy flowed through the small rodents called chipmunks, which eat seeds and nuts. On average, these primary consumers harvested 31 kcal/m^2 each year. Of that total, 82.3 percent was assimilated and 17.7 percent was excreted; of the total energy assimilated, just 1.6 percent went into the production of new chipmunk tissue. The production of new tissue by primary consumers is called **secondary production.** Secondary production was much higher in ectothermic consumers like caterpillars, which transformed about 5.4 percent of the energy they ingested into new tissue. Because ectotherms do not oxidize sugars to keep warm, they devote much less energy to cellular respiration. Even so, it is clear that only a tiny fraction of the available solar radiation is involved in secondary production.

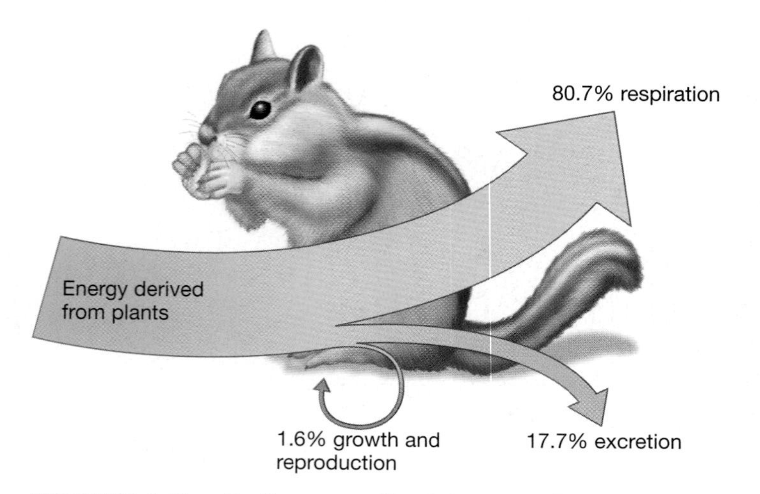

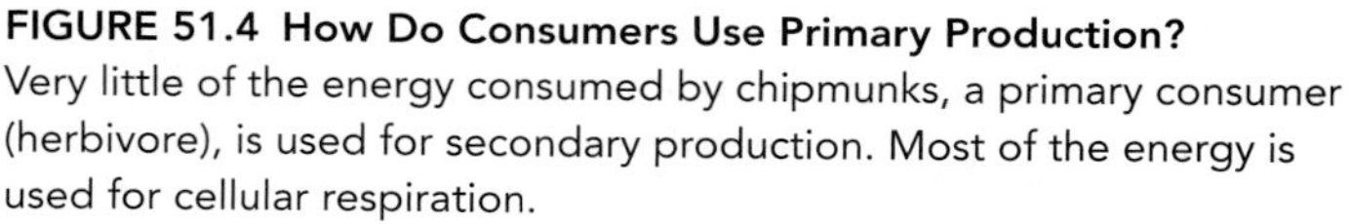
FIGURE 51.4 How Do Consumers Use Primary Production?
Very little of the energy consumed by chipmunks, a primary consumer (herbivore), is used for secondary production. Most of the energy is used for cellular respiration.

Decomposers Not all plant tissue is consumed by herbivores. Tissues that are not consumed eventually die. When they do, they enter a **decomposer food web** composed of species that eat the dead remains of organisms. At Hubbard Brook, about 75 percent of net primary production enters the decomposer food web. Plant litter and dead animals, collectively known as **detritus**, are consumed by a variety of primary decomposers: bacteria, archaea, fungi, protozoa, and millipedes (**Figure 51.5**). These primary decomposers are in turn consumed by carnivores such as centipedes, spiders, salamanders, and shrews.

Decomposers rarely use all of the detritus that is available. Instead, some detritus accumulates as organic matter in the soil or leaves the community in sediments. We'll return to the process of decomposition later in the chapter, when we discuss nutrient cycling.

The Abiotic Component The researchers at Hubbard Brook also produced data confirming that large amounts of energy leave the ecosystem in the form of detritus that washes into streams. This transfer of energy into the aquatic ecosystem is important because it is a major source of energy for aquatic organisms. At Hubbard Brook, photosynthesis by aquatic plants introduced only about 10 kcal/m^2/yr of energy. In contrast, each year 6039 kilocalories washed into each square meter of streambed from the surrounding forest.

To summarize, Gosz and colleagues documented the flow of energy into, through, and out of a temperate-forest ecosystem. Although the specific numbers they found are unique to a particular site and time, the general patterns have turned out to be typical. Energy enters ecosystems through the actions of primary producers. Only a fraction of the energy fixed by these organisms, called net primary production, actually becomes available to consumers. Most net primary production that is consumed enters the decomposer food web, but some enters the grazing food web. From there, a small fraction is used for secondary production by herbivores and their predators. Overall, most of the energy fixed during photosynthesis is ultimately used not for the synthesis of new tissues, but for respiration by plants, animals, fungi, bacteria, and archaea.

Trophic Structure

Primary producers, primary consumers, and secondary consumers define the feeding levels, or trophic structure, of an ecosystem. Organisms that obtain their energy from the same type of source occupy the same **trophic level**. As **Figure 51.6a** shows, all plant-eating organisms occupy the same trophic level. Other trophic levels in the herbivore food chain are defined by carnivores that eat herbivores and by the secondary carnivores that eat them. Organisms at the top trophic level are not killed for food by any other organisms.

FIGURE 51.5 The Decomposer Food Web
Dead leaves, sticks, bodies of dead animals, and other types of detritus are fed upon by an enormous variety of organisms. These decomposers, in turn, are fed upon by salamanders, shrews, spiders, centipedes, and other predators.

(a) Trophic levels

Trophic level	Feeding strategy	Grazing food chain	Decomposer food chain
4	Secondary carnivore	Cooper's hawk	Owl
3	Carnivore	Robin	Shrew
2	Herbivore	Cricket	Earthworm
1	Autotroph	Maple tree leaves	Dead maple leaves

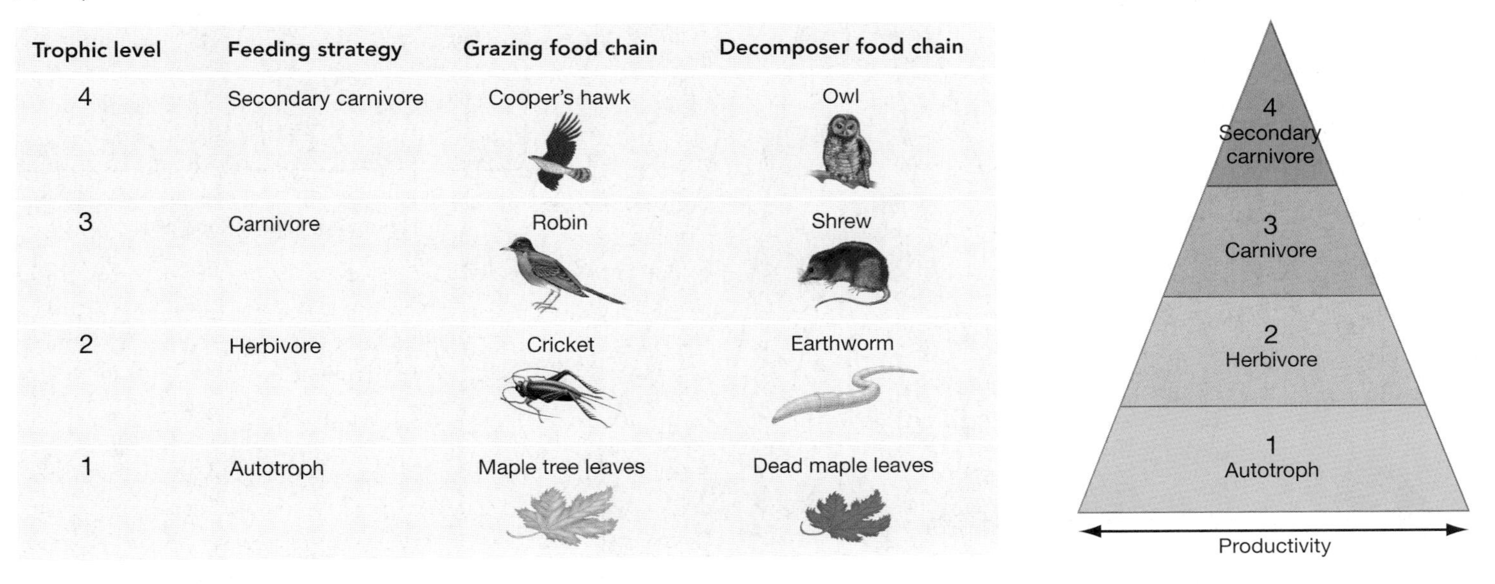

FIGURE 51.6 Organisms That Share a Trophic Level Obtain Their Energy from the Same Source
(a) Each trophic level in an ecosystem is defined by a distinct feeding strategy. The organisms illustrated in this table furnish an example for each trophic level in the grazing and decomposer food chains of a temperate-forest ecosystem. Many other species exist at each trophic level in this ecosystem. **(b)** In all ecosystems, productivity is highest at the first trophic level and declines at higher levels. This pattern is called the pyramid of productivity.

As **Figure 51.6b** shows, ecosystems share a characteristic pattern of productivity at each trophic level. Productivity declines from one trophic level to the next. To understand why the pattern occurs, consider the interface between primary producers and herbivores. Much of the net primary productivity that exists is unavailable to herbivores because it resides in indigestible substances like lignin (wood) or because it is protected by noxious defensive compounds. Even if the material is ingested, some of the energy stored in the chemical bonds of reduced carbon compounds is lost as heat as it is metabolized. At the next trophic level, many herbivores are never consumed by carnivores because they hide effectively or sequester toxins. Of the energy that is successfully consumed by carnivores, some is again lost as heat or used up in capturing prey.

The general point here is simple. Productivity at the second trophic level (herbivores) must be less than productivity at the first trophic level (plants); productivity at the third trophic level (carnivores) must be less than that at the second. This pattern holds true for the entire food chain and produces a pyramid of productivity. Productivity is highest at the lowest trophic level.

Food Chains and Food Webs

A **food chain** describes the species occupying each trophic level in a particular ecosystem. As an example, **Figure 51.7a** illustrates part of the food-chain relationships discovered by Robert Paine for intertidal organisms in the Gulf of California. As **Figure 51.7b** shows, food chains are often embedded in more complex **food webs**, because multiple species are present at several trophic levels. Food chains and webs are among the most basic ways to describe the structure of an ecosystem.

It is interesting to note that none of the food chains, food webs, or trophic structures presented in this chapter thus far have more than four trophic levels. When Thomas Schoener reviewed the characteristics of food webs that had been documented by researchers working in a wide variety of ecosystems, he found that the maximum number of links ranged from 1 to 6 (**Figure 51.7c**). Each link joins two trophic levels. In Schoener's study, terrestrial and lake ecosystems had 3.7 links per food web on average, while streams had an average of 3.2 links. Why don't ecosystems have 8 or 9 or 10 trophic levels? Why does the overall average number of levels seem to be about 3.5? Several competing hypotheses have been offered to explain this observation. Let's take a closer look.

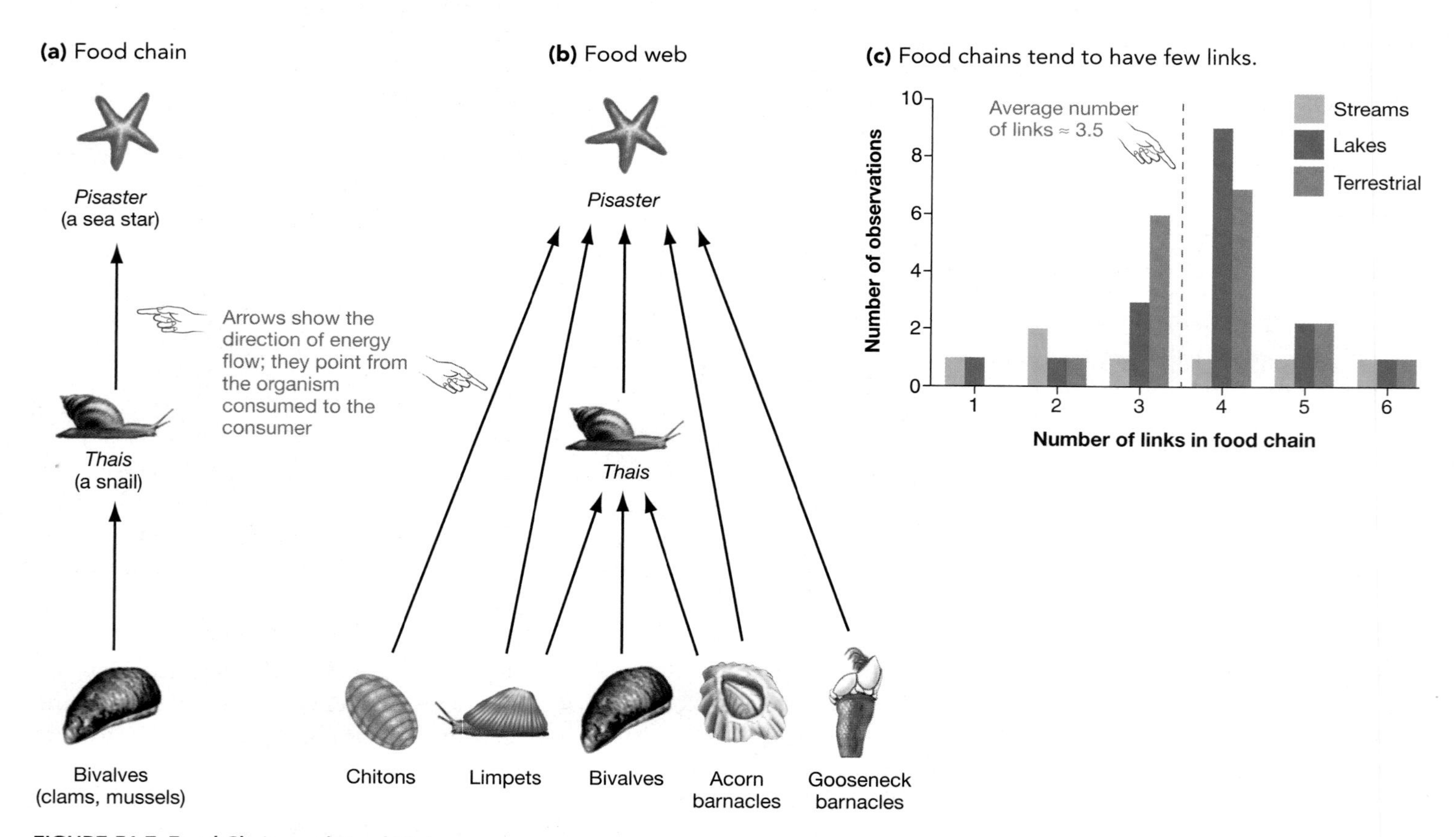

FIGURE 51.7 Food Chains and Food Webs
(a) An example of a food chain in an intertidal zone. **(b)** Food chains are embedded in food webs that include more species and different feeding relationships. **(c)** The y-axis on this graph plots the number of research studies that described food chains with from 1 to 6 links in stream, lake, and terrestrial habitats.

Does Energy Transfer Limit Food-Chain Length? As energy is transferred up the food chain, a large fraction is lost. By the time energy reaches the top trophic level, there may not be sufficient amounts left to support an additional suite of consumers. To better understand this explanation, suppose that the efficiency of energy transfer between trophic levels is 10 percent. If the initial trophic level in a hypothetical ecosystem produces 10,000 kcal per day, then the second, third, and fourth levels will produce 1000, 100, and 10 kcal per day, respectively. Can any organism obtain enough energy to survive at a fifth trophic level?

The hypothesis that food-chain length is limited by productivity leads to a strong prediction: There should be more trophic levels in ecosystems with higher productivity or higher energy-transfer efficiency. When Stuart Pimm analyzed data on the productivity of four aquatic and four terrestrial ecosystems, however, he found that low-productivity ecosystems were as likely to contain four trophic levels as high-productivity ones. To date, research has not supported the prediction that more productive ecosystems have longer food chains.

Are Long Food Chains Fragile? Stuart Pimm proposed an alternative hypothesis to the energy-limitation explanation for why food chains are short. His idea was that long food chains are easily disrupted by environmental perturbations and thus tend to be eliminated. Using mathematical models, Pimm demonstrated that long food chains took longer to return to their previous state following a disturbance than did short food chains. He then proposed that long food chains are unlikely to persist in a variable environment.

Pimm's hypothesis also predicts that food chains should be longer in more stable environments. R. L. Kitching tested this prediction by comparing the animal communities that develop inside tree holes in Australia and Great Britain. After water accumulates in depressions or in the holes found in tree branches or roots, leaf litter from the trees falls in and forms the basis of a food web. Kitching found that tree-hole communities in Australia and Great Britain were similar in many respects, except that annual leaf fall was much more variable in the British communities than in the Australian ones. As predicted by the hypothesis, the British ecosystems had only two trophic levels while Australian habitats supported three.

Other researchers have challenged the assumptions behind Pimm's theoretical analysis, however, and support for the stability hypothesis remains very tentative. Experimental and observational tests of the hypothesis are continuing.

Does Food-Chain Length Depend on Environmental Complexity? Frederic Briand and Joel Cohen hypothesized that food-chain length is a function of an ecosystem's physical structure. Specifically, Briand and Cohen proposed that tundra, grasslands, lake and sea bottoms, streambeds, and intertidal zones offer a largely flat, or two-dimensional, surface to the organisms living there. In contrast, forests and open-water environments in lakes and rivers offer three-dimensional volumes. Briand and Cohen predicted that three-dimensional ecosystems should have longer food chains than two-dimensional sites.

To test this hypothesis, they examined 113 publications that described food webs in a wide variety of aquatic and terrestrial environments. These data supported the prediction that food webs are significantly longer in three-dimensional ecosystems. As a result, the researchers concluded that dimensionality does influence food-web structure. The mechanism for the pattern remains to be determined, however. What is it about three-dimensional ecosystems that allow longer food webs to develop?

There is unlikely to be a single, simple answer to the question of what limits food-chain length. Inefficiencies of energy transfer, environmental variability, and environmental complexity may all influence the number of trophic levels that can be supported in a given ecosystem.

51.2 Biogeochemical Cycles

Energy is not the only quantity that is transferred when one organism eats another. The organisms that are eaten also contain carbon (C), nitrogen (N), phosphorus (P), calcium (Ca), and other elements that act as nutrients. Energy flows through ecosystems; but atoms cycle through trophic levels and air, water, and soil. The path that an element takes as it moves from abiotic systems through organisms and back again is referred to as its **biogeochemical cycle.**

Interest in biogeochemistry has increased recently because humans are now disturbing biogeochemical cycles on a global scale. To drive this point home, consider the three basic aspects of biogeochemical cycling that researchers study:

1. The nature and size of the pools or reservoirs where elements are stored for a period of time. In the case of carbon, the biomass of living organisms is an important pool, along with sediments and soils. Another significant carbon reservoir is buried in the form of coal and oil.
2. The rate of movement between pools and the factors that influence these rates. The global photosynthetic rate, for example, measures the rate of carbon flow from the atmosphere into living biomass. Massive quantities of buried carbon have recently moved into the atmosphere in the form of carbon dioxide (CO_2) as a result of fossil-fuel burning.
3. How different biogeochemical cycles interact. How do changes in the biogeochemical cycle of nitrogen affect the fate of carbon? How do changes in the sulfur cycle affect that of nitrogen?

Biogeochemical Cycles in Ecosystems

Figure 51.8 (page 996) shows a simplified version of a general terrestrial nutrient cycle. Nutrients are taken up from the soil by plants and incorporated into plant tissue. If the plant is eaten, the nutrients pass to the animal members of the ecosystem; if the plant dies, the nutrients and the plant biomass within them become litter, or dead biomass. Once ingested by an animal, nutrients are excreted in fecal matter or urine, taken up by a parasite or predator, or enter the dead biomass pool when the individual dies.

The nutrients in plant litter, animal excretions, and dead animal bodies are broken down by bacteria, archaea, roundworms, fungi, and other organisms. The combination of breakdown products and microscopic decomposers forms the soil organic matter. Soil organic matter is a complex mixture of partially decomposed detritus, rich in a family of carbon-containing molecules called humic acids (soil organic matter is also called humus). Eventually, the nutrients in soil organic matter are converted to an inorganic form. For example, decomposition converts the nitrogen present in amino acids in detritus to ammonium (NH_4^+) or nitrate (NO_3^-) ions. Once this step is accomplished, the nutrients are available for uptake by plants.

A key feature of this process is that nutrients are reused. Reuse is not total, however. Nutrients leave the ecosystem whenever plant or animal biomass is removed. For example, plant biomass is removed when an herbivore enters the ecosystem, eats a plant, leaves the ecosystem, and travels elsewhere before excreting nutrients or dying; quantities of nutrients also leave ecosystems when flowing water or wind removes particles or inorganic ions. Soil erosion has a huge impact on nutrient cycles.

What Factors Control the Rate of Nutrient Cycling in Ecosystems? Of the many links in a nutrient cycle, decomposition of detritus most often limits the overall rate at which nutrients move through an ecosystem. The decomposition rate, in turn, is influenced by two types of factors—abiotic conditions such as temperature and precipitation, and the quality of the detritus as a nutrient source for the fungi, bacteria, and archaea that accomplish decomposition.

To appreciate the importance of abiotic conditions on decomposition rate, consider the difference in detritus accumulation between boreal forests and tropical rain forests. Chapter 50 indicated that boreal forests occur in areas where temperature is low. As a result, soils in these ecosystems are cold and wet. Tropical rain forests, in contrast, occur in areas where temperatures and rainfall are high. Soils tend to remain moist and warm all year long.

Figure 51.9 illustrates typical soils from boreal forests and tropical rain forests. Note that the uppermost part of the soil in a boreal forest consists of partially decomposed detritus and organic matter. There is no such layer at the top of the soil in a tropical forest. The contrast occurs because the cold

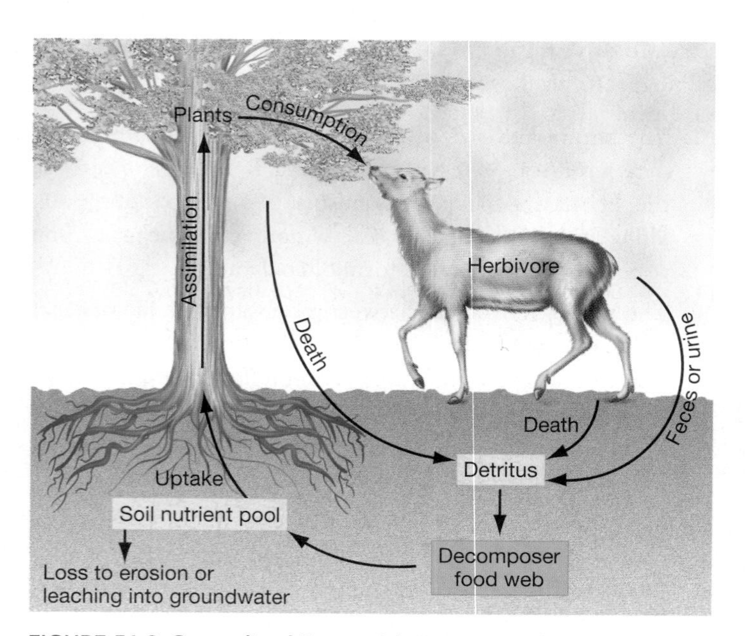

FIGURE 51.8 Generalized Terrestrial Nutrient Cycle
Nutrients cycle from organism to organism in an ecosystem as a result of assimilation by primary producers, consumption, and decomposition. Nutrients are exported from ecosystems through the migration of organisms out of the area or, more commonly, in flowing water or groundwater.

Boreal forest

Tropical rainforest

FIGURE 51.9 Temperature and Moisture Affect Decomposition Rates
In boreal forests, decomposition rates are limited by cold soil temperatures. The input of detritus into the soil thus exceeds the decomposition rate, and organic matter builds up. In tropical rain forests, warm temperatures allow decomposition to proceed rapidly so that organic matter does not build up.

FIGURE 51.10 Land-Use Changes Affect the Rate of Nutrient Loss from Ecosystems
(a) An experiment to test the effects of vegetation removal on nutrient cycling. **(b)** Nutrient export increased dramatically in the devegetated watershed. QUESTION In effect, this experiment removed one of the arrows in Figure 51.8. Which one?

(a) Devegetation experiment

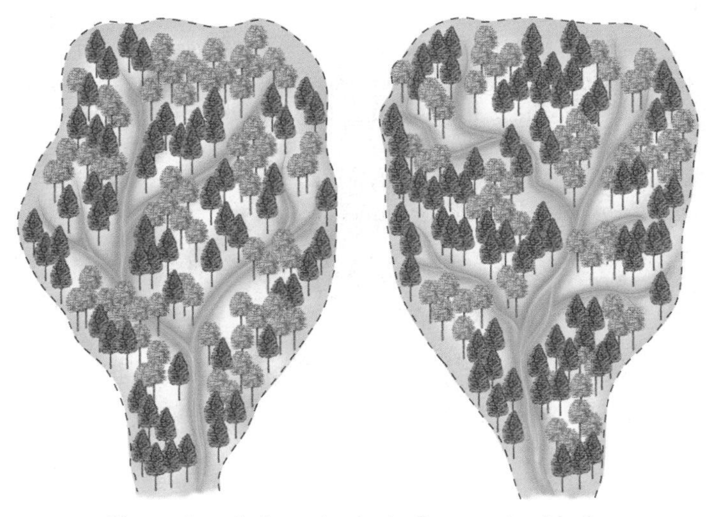

Choose two similar watersheds. Document nutrient levels in soil organic matter, plants, and streams.

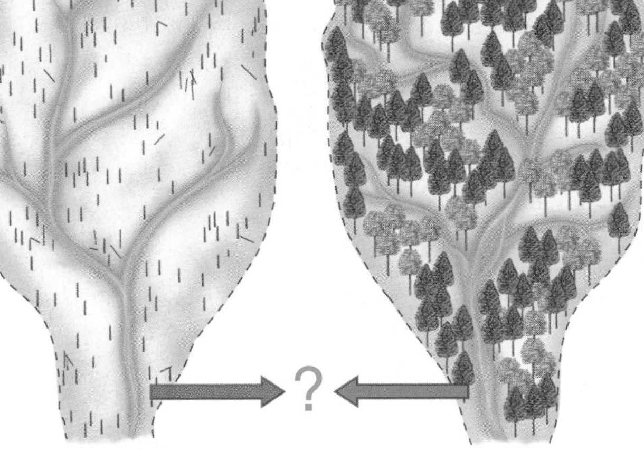

Devegetate one watershed and leave the other intact. Monitor the amount of dissolved substances in streams.

(b) Nutrient runoff results

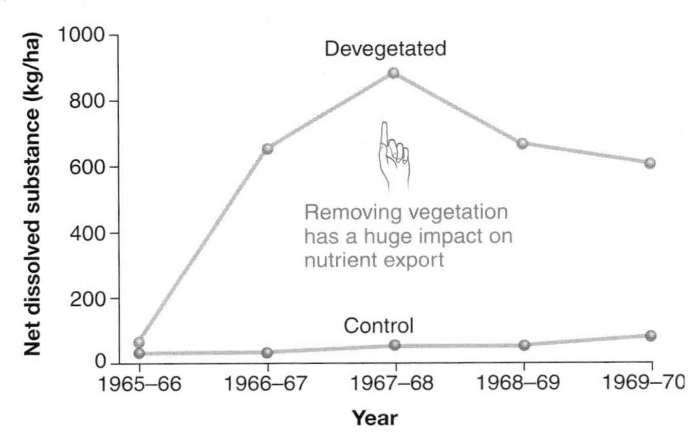

and wet conditions in boreal forests limit the metabolic rates of decomposers. As a result, decomposition fails to keep up with the input of detritus. In the tropics, conditions for fungi, bacteria, and archaea are so favorable that decomposition keeps pace with detrital inputs.

The quality of detritus also exerts a powerful influence on decomposition rate and thus on nutrient availability. For example, decomposers can be limited by the presence of large compounds in detritus that are difficult to digest, such as lignin. Lignin is one of the primary constituents of wood, and is one reason why wood takes much longer to decompose than leaves. The growth of decomposers is also inhibited if detritus is low in nitrogen.

What Factors Influence the Rate at Which Nutrients Are Exported from Ecosystems? Nutrient availability has a profound effect on productivity. The rate of nutrient loss is thus an important characteristic of an ecosystem. Several of the major impacts that humans have on ecosystems—such as farming, logging, burning, and soil erosion—accelerate nutrient loss.

To test the effect of vegetation removal on nutrient cycling, Gene Likens and Herbert Bormann initiated a large-scale experiment at the Hubbard Brook Experimental Forest. They chose two watersheds—areas drained by streams—for study (**Figure 51.10a**). They then removed all vegetation, including the trees, from the forests in one of the two watersheds. In the following two years, the devegetated (clear-cut) area was treated with an herbicide to prevent vegetation from regrowing. The untreated watershed served as a control.

Before removing the vegetation, the researchers had documented that 90 percent of the nutrients in the ecosystems were in soil organic matter with an additional 9.5 percent in plant biomass. After the vegetation was removed, they monitored the concentrations of nutrients in the streams exiting the two watersheds. **Figure 51.10b** documents the amount of dissolved substances that subsequently washed out of the stream in each watershed over the course of four years. Losses were typically 10 times as high in the devegetated site versus the control site.

Based on these data, the researchers concluded that devegetation has a huge impact on nutrient export. Nutrients in the soil may dissolve in water or be attached to small particles. Without plants to recycle them, nutrients are quickly washed out of the soil and lost to the ecosystem. Consequently, the amount of productivity that an area can support may decline over time if it is kept in a devegetated state.

Global Biogeochemical Cycles

When nutrients leave one ecosystem, they enter another. In this way, the movement of nutrients among ecosystems links local biogeochemical cycles into one massive global system. As an introduction to how these global biogeochemical cycles work, let's consider the global carbon and nitrogen cycles. These cycles recently have been heavily modified by human activities—with serious ecological consequences.

The Carbon Cycle As **Figure 51.11** shows, the **global carbon cycle** involves the movement of carbon between terrestrial ecosystems, the oceans, and the atmosphere. Of these three pools, the ocean is by far the largest. Despite its small size, the atmospheric pool is also important because carbon moves in and out of it rapidly.

The arrows in Figure 51.11 emphasize that carbon frequently moves in and out of the atmospheric pool through organisms. In both terrestrial and aquatic ecosystems, for example, photosynthesis is responsible for taking carbon from the atmosphere and incorporating it into tissue. Respiration, in contrast, releases carbon that has been incorporated into living organisms to the atmosphere.

How have humans changed the carbon cycle? The fossil fuels found on Earth, which are derived from carbon-rich sediments, are estimated to contain a total of 5000–10,000 gigatons (Gt) of carbon. (A gigaton is a billion tons). This is one-eighth to one-fourth the size of the oceanic pool and 2.5 to 5 times the size of the terrestrial pool. In effect, burning fossil fuels moves carbon from an inactive geological reservoir to an active pool—the atmosphere.

Land-use changes have also altered the global carbon cycle. Deforestation, for example, reduces the area's net primary productivity. It also releases CO_2 when fire is used for clearing or when dead limbs, twigs, and stumps are left to decompose. When R. A. Houghton and colleagues reconstructed fluxes to the atmosphere from data on the amount of land cleared for agriculture and forestry, they found that at a global scale, a net movement of carbon to the atmosphere has been occurring from terrestrial ecosystems for at least the past 100 years.

Figure 51.12a shows the amount of carbon released from fossil-fuel burning and land-use changes over the past century; **Figure 51.12b** shows the consequences. In just 40 years, CO_2 in the atmosphere at Mauna Loa Observatory on the island of Hawaii has increased from about 315 ppm (parts per million—meaning that there were 315 milligrams of CO_2 per kilogram of air) to over 360 ppm. The same trend has been observed at sites around the globe.

Why are these changes in the global carbon cycle important? Carbon dioxide functions as a **greenhouse gas** because it traps heat that has been radiated from Earth and keeps it from being lost to space. More specifically, it is one of several gases in the atmosphere that absorb and reflect the infrared wavelengths radiating from Earth's surface. As the essay at the end of this chapter details, increases in greenhouse gases are warming Earth's climate by increasing the atmosphere's heat-trapping potential.

The Global Nitrogen Cycle **Figure 51.13a** illustrates the **nitrogen cycle.** A key aspect of this biogeochemical cycle is that plants are able to use nitrogen only in the form of ammonium or

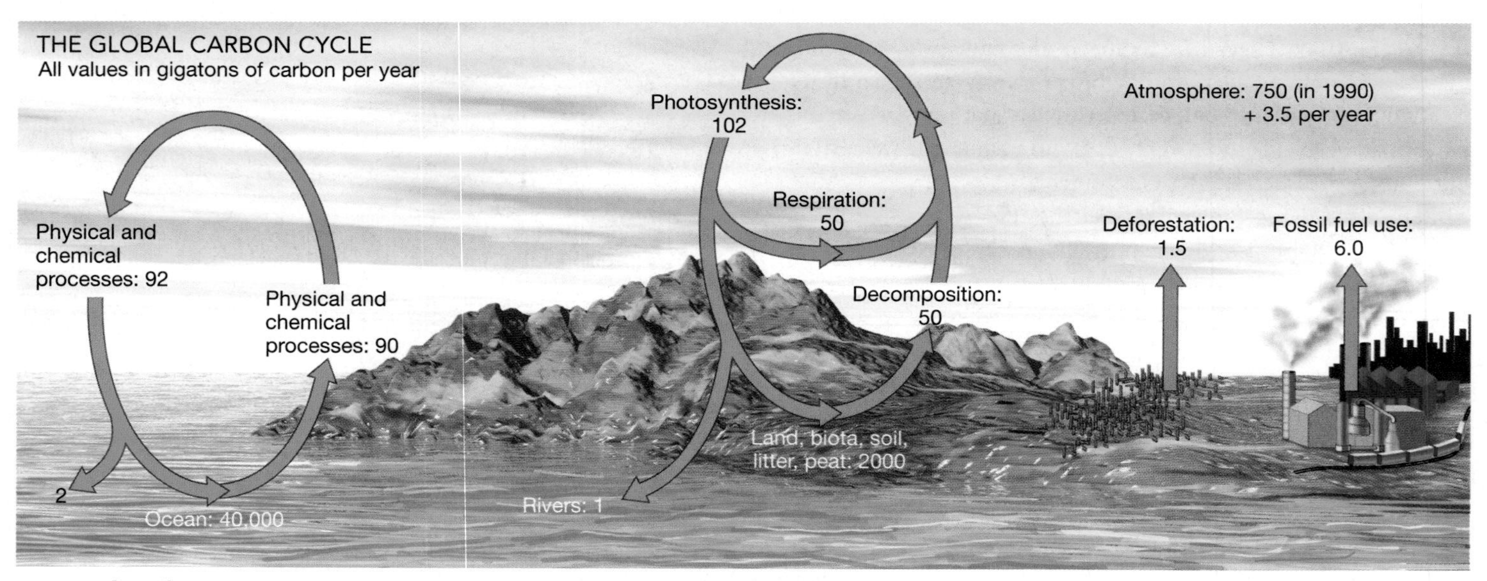

FIGURE 51.11 The Global Carbon Cycle
The arrows in this diagram indicate how carbon moves into and out of ecosystems. Note that deforestation and fossil fuel use are adding 7.5 gigatons of carbon to the atmosphere each year. Of this 7.5 gigatons, two are fixed by photosynthesis in terrestrial ecosystems and two are fixed by physical and chemical processes in the oceans.

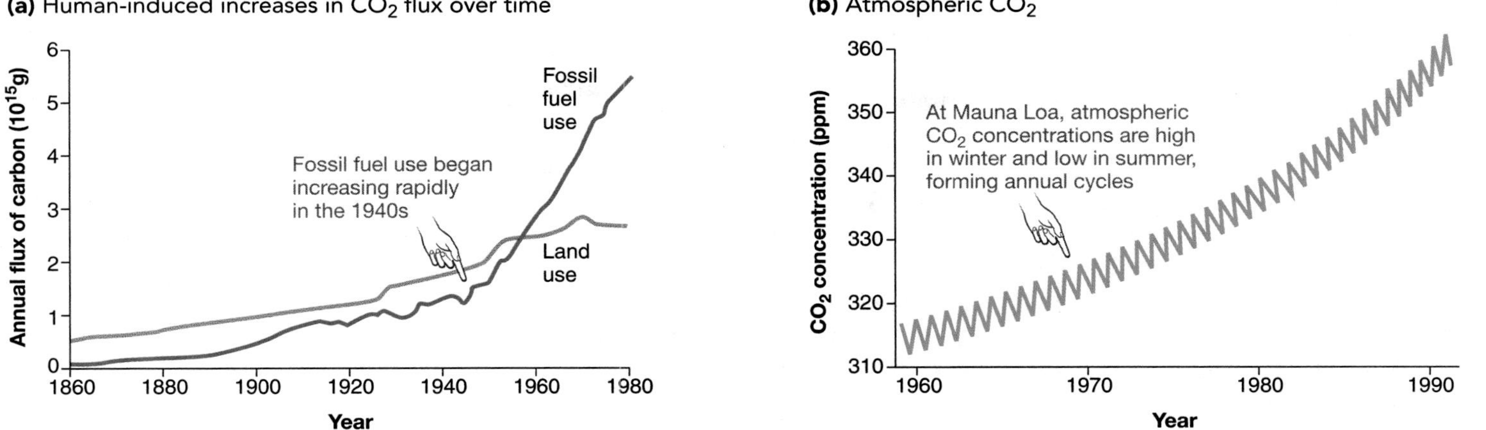

FIGURE 51.12 Humans Are Causing Increases in Atmospheric Carbon
(a) Carbon fluxes from fossil-fuel burning and land-use changes have been increasing since 1860. **(b)** Scientists have been collecting air samples and measuring CO_2 concentrations at the Mauna Loa Observatory in Hawaii since 1958. Because the site is far from large-scale human influence, it should accurately represent the average condition of the atmosphere in the northern hemisphere. QUESTION Why are atmospheric CO_2 concentrations low in the northern hemisphere during the summer and high in winter? What pattern would you expect in the southern hemisphere?

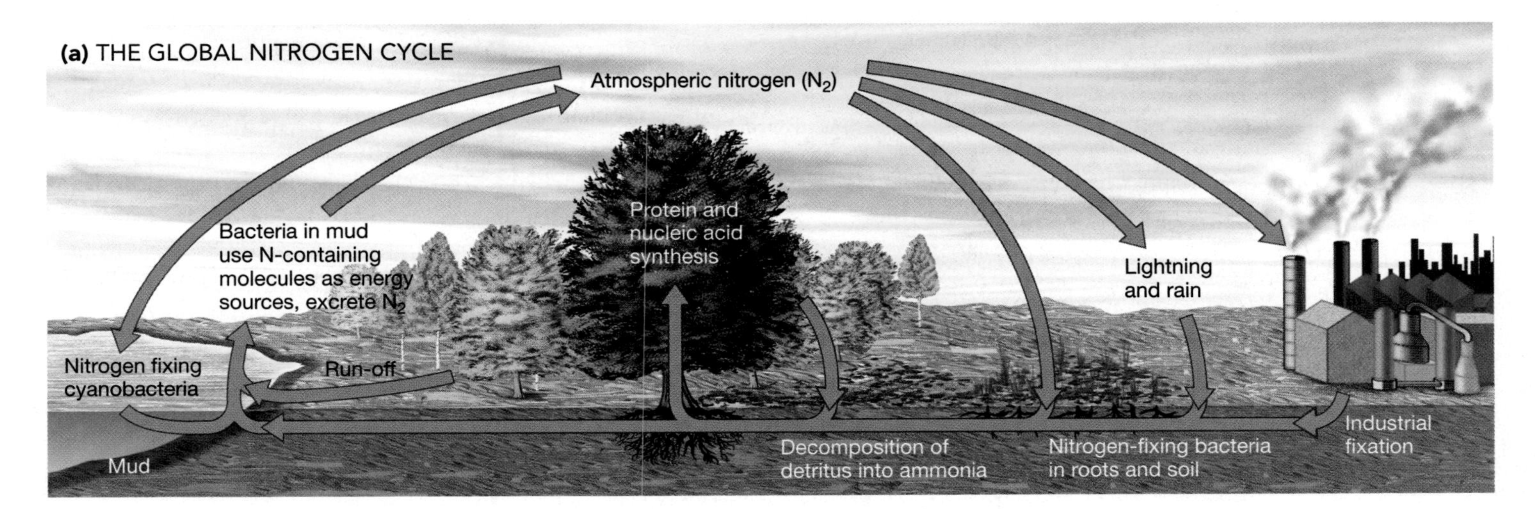

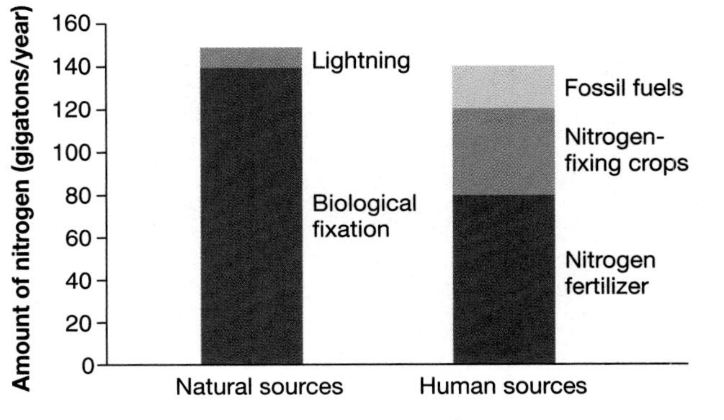

FIGURE 51.13 The Global Nitrogen Cycle
(a) Nitrogen enters ecosystems as ammonia or nitrate via fixation from atmospheric nitrogen. Within ecosystems, nitrogen cycles through producers and consumers. Eventually it reaches the decomposer food web. Nitrogen is exported from ecosystems in runoff and as nitrogen gas given off by bacteria that use nitrogen-containing compounds as an electron acceptor. **(b)** Human activities now fix almost as much nitrogen each year as natural sources. Thus, human activities have almost doubled the total amount of nitrogen available to organisms.

nitrate ions (NH_4^+ or NO_3^-). As a result, the vast pool of molecular nitrogen (N_2) in the air blanketing Earth—N_2 makes up 78 percent of the atmosphere—is unavailable to plants. Nitrogen is added to ecosystems in a usable form only when it is "fixed," meaning when it is converted from N_2 to NH_3. Nitrogen fixation results from lightning-driven reactions in the atmosphere or enzyme-catalyzed reactions in bacteria that live in the soil and oceans. (Chapter 33 explained how bacteria fix nitrogen.)

The nitrogen cycle has been profoundly altered by human activities. As **Figure 51.13b** shows, nitrogen fixation from human sources is now approximately equal to the amount of nitrogen fixation from natural sources. There are three major sources of this human-fixed nitrogen: industrially produced fertilizers, release of nitric oxide during the combustion of fossil fuels, and the cultivation of crops such as soybeans and peas that harbor nitrogen-fixing bacteria.

Adding nitrogen to terrestrial ecosystems usually increases productivity. In some cases, then, a massive increase in nitrogen availability is beneficial. But in other cases, it is not. Chapter 25, for example, detailed how excessive applications of nitrogen-containing fertilizers on farmlands have produced nitrogen-laced runoff, which has had disastrous impacts on numerous aquatic ecosystems. David Tilman has also documented that nitrogen inputs can lead to a significant loss of biodiversity in terrestrial ecosystems. Tilman added nitrogen to plots of native grassland in midwestern North America and found that a few competitively dominant species tended to take over. As they grew rapidly, they displaced other species that did not respond to nitrogen inputs as strongly. In this ecosystem, increased nitrogen boosted productivity but decreased species diversity, by altering the balance of competitive interactions.

To summarize, several of the most pressing environmental problems facing our species result from recent and massive alterations in biogeochemical cycles. Even local changes in biogeochemical cycles tend to have large-scale consequences, because nutrients are transported among ecosystems.

Essay Global Warming

Carbon dioxide concentration in the atmosphere has been increasing throughout the twentieth century. Most analyses indicate that the increase is due to fossil-fuel burning and clearing of land, particularly forests, for agriculture. Whether the increase in CO_2 is producing global warming has been intensely controversial, however.

In 1988, an international group of scientists called the Intergovernmental Panel on Climate Change was formed to evaluate the consequences of rising CO_2. The group has since produced a series of reports summarizing the state of scientific knowledge on the issue. The most recent of these reports concluded that current evidence suggests a "discernible human influence on climate." The panel has now taken the position that rising concentrations of greenhouse gases are at least partially responsible for global warming.

How much will average temperatures rise in our lifetimes?

How much will average temperatures rise in our lifetimes? Predicting the future state of a system as complex and variable as Earth's climate is extremely difficult. Global climate models are the primary tools that scientists use to make these projections. A global climate model is based on a large series of equations that describe how the concentrations of various gases in the atmosphere, solar radiation, evapotranspiration rates, and other parameters interact to affect climate. The models currently being used suggest that average global temperature will increase between 1.5 and 3.5°C over the next 50 years.

How will ecosystems respond to this increase? Answering this question is difficult, because ecosystems respond to warming in ways that increase or decrease CO_2 concentrations and thus exacerbate or mitigate warming. Positive feedbacks, for example, occur when warmer and drier climate conditions lead to more fires, which in turn release more CO_2 and lead to more warming. Walter Oechel and associates recently documented that a form of positive feedback is already occurring in arctic tundras. Traditionally, tundras sequester carbon in the form of soil organic matter because decomposition rates are extremely low. During a series of warm summers in the 1980s, however, the researchers documented that decomposition rates increased sufficiently to release carbon from stored soil organic matter and produce a net flow of carbon to the atmosphere.

Negative feedbacks, in contrast, arise when warmer conditions lead to increased photosynthetic rates and hence an increase in the uptake of CO_2. In addition, several tree species and some agricultural crops have been shown to increase their growth rates in direct response to increasing atmospheric CO_2.

Will ecosystem responses reduce or increase global warming? Currently, it is not clear whether positive or negative feedbacks will predominate. Researchers are working to answer this question using computer models, experiments, and analyses of how ecosystems responded to past climate changes.

Chapter Review

Summary

An ecosystem consists of one or more communities of interacting species and their abiotic environment. Energy and elements flow through ecosystems and are exchanged between biotic and abiotic components. Ecosystems share a common structure, including the following components: primary producers, consumers, decomposers, and an abiotic environment. Energy flows into ecosystems as a result of photosynthesis or conversion of reduced inorganic molecules by bacteria and archaea. It then enters the grazing food web or the decomposer food web.

Organisms that acquire energy from the same type of source are said to occupy the same trophic level. Most ecosystems have at least three trophic levels: primary producers, herbivores or primary decomposers, and carnivores. Because energy transfer from one trophic level to the next is inefficient, ecosystems have a pyramid of productivity. Productivity is highest at the lowest trophic level and lower at higher trophic levels.

The feeding relationships between species in a particular ecosystem are described by a food chain or food web. Food chains rarely exceed five or six trophic levels. Various hypotheses have been proposed to explain why food webs and food chains are not longer. The energy-flow hypothesis contends that there simply is not enough energy left at the top of the food chain to support an additional trophic level. The dynamic stability hypothesis states that long food webs are uncommon because they are unstable in variable environments. The dimensionality hypothesis proposes that structurally complex ecosystems have longer food webs.

Nutrients move through ecosystems in biogeochemical cycles. The rate of nutrient cycling is strongly affected by the rate of decomposition of detritus. Decomposition rate, in turn, is affected by abiotic environmental conditions like temperature and by the quality of the detritus. Nutrients are also lost from ecosystems. Work in the Hubbard Brook Experimental Forest demonstrated that loss of vegetation greatly increased the rate of nutrient loss.

Humans have major impacts on biogeochemical cycles. Land-use changes and fossil-fuel burning have increased the flow of carbon in the form of CO_2 into the atmosphere, with dramatic effects on global climate. Fertilizer production and planting of nitrogen-fixing crops has almost doubled total nitrogen fixation on Earth. These increases have led to increased productivity, but also to pollution and loss of biodiversity.

Questions

Content Review

1. Which of the following is *not* one of the four components of ecosystems?
   a. primary producers
   b. the abiotic environment
   c. decomposers
   d. secondary carnivores

2. Which of the following ecosystems would you expect to have the highest primary productivity?
   a. subtropical desert
   b. temperate grassland
   c. boreal forest
   d. tropical wet forest

3. Most of the net primary production that is consumed in an ecosystem is used for what?
   a. respiration by herbivores
   b. respiration by carnivores
   c. growth by herbivores
   d. growth by consumers (i.e., secondary production)

4. According to the dynamic stability hypothesis for food-chain length, food chains will be shorter in which kind of environment?
   a. cold
   b. constant
   c. variable
   d. low nutrient availability

5. Which of the following is a pool for carbon?
   a. plant biomass
   b. petroleum deposits
   c. oceans
   d. all of the above

6. Scientists at Hubbard Brook demonstrated that clear-cutting had what effect on ecosystem dynamics?
   a. It increased aboveground biomass.
   b. It increased secondary production.
   c. It increased nutrient export.
   d. It increased the pool of soil organic matter.

Conceptual Review

1. What is the difference between a community and an ecosystem?
2. Why is primary production always greater than secondary production in an ecosystem?
3. How does the decomposer food web regulate nutrient availability in an ecosystem?
4. Compare and contrast the energy-flow, dynamic stability, and dimensionality (environmental complexity) hypotheses for food-chain length.
5. Draw a diagram of a generalized terrestrial nutrient cycle, indicating major pools and fluxes.

Applying Ideas

1. Suppose you had a small set of experimental ponds at your disposal, and an array of pond-dwelling algae, plants, and animals. How could you use radioactive isotopes of carbon or phosphorus to study energy flows or nutrient cycling in these experimental ecosystems?
2. Figure 51.3b describes the pattern of marine primary productivity. The hypothesis to explain the pattern is that the productivity of marine ecosystems is limited by nutrient availability. Design an experiment to test this hypothesis.
3. What effect do herbivores have on the rate of nutrient cycling in terrestrial ecosystems? How would you go about testing your hypothesis? (For example, predict what would happen in a particular ecosystem if herbivores were removed experimentally.)
4. Explain why human-caused changes to the global carbon cycle are affecting Earth's climate. State why you think these changes are beneficial or detrimental, and then list something that you, local institutions, and state and national governments can do to either augment or mitigate these changes.

CD-ROM and Web Connection

CD Activity 51.1: The Global Carbon Cycle *(animation)*
(Estimated time for completion = 5 min)
Animals generate carbon dioxide as a natural byproduct of respiration. Why is carbon dioxide production such a concern to so many people?

At your **Companion Website** (http://www.prenhall.com/freeman/biology), you will find self-grading exams and links to the following research tools, online resources, and activities:

Global Warming
This article discusses how global warming affects ecosystems and potentially human health.

Wetland Biogeochemical Database
This database collects and compiles biogeochemical information on wetlands in the United States.

Global Nitrogen Cycles
This article discusses how human activity can lead to increased nitrogen inputs that impact ecosystems.

Additional Reading

Covich, A. P., M. A. Palmer, and T. A. Crowl. 1999. The role of benthic invertebrate species in freshwater ecosystems. *Bioscience* 49: 119.

Houghton, J. 1997. *Global Warming, the Complete Briefing,* 2nd ed. Cambridge, U.K.: Cambridge University Press.

McNaughton, S. J., R. W. Ruess, and S. W. Seagle. 1988. Large mammals and process dynamics in African ecosystems. *Bioscience* 38: 794–800.

Post, W. M., T. Peng, W. R. Emanuel, A. W. King, V. H. Dale, and D. L. DeAngelis. 1990. The global carbon cycle. *American Scientist* 78: 310–326.

Biodiversity and Conservation

52

Most biologists choose their profession for two reasons: They love organisms and they love answering questions about organisms. But even as the extent of our knowledge about life explodes and the science of biology expands, the number of species decreases. For people who have devoted their lives to the study of life, the irony is cruel. In response, many students cite a third reason for becoming a biologist: Learning how to preserve biodiversity. This chapter focuses on exactly this goal.

Let's first get a better understanding of the problem—its scope and its causes. Chapter 48 examined data indicating that human populations and resource use began to increase extremely rapidly around 1600. Chapter 24 analyzed evidence that in response to human impacts, extinction rates are now accelerating to between 10 to 100 times the normal, or "background" extinction rates, indicating that a mass extinction event may currently be under way. In the first section of this chapter, you'll delve more deeply into data on the extent and causes of extinction.

Section 52.2 focuses on why biodiversity matters. As biologists, what reasons can we provide for preserving species? Are these reasons compelling enough to influence the farmers, fishermen, loggers, miners, and others whose lives are directly affected by conservation efforts?

Most biologists accept that human populations and technological capabilities have grown to the point that widespread extinctions are inevitable. Confronted with this reality, how can biologists influence the agonizing decisions that must be made about which species and habitats will be lost and which will be saved? Establishing conservation priorities is the subject of section 52.3.

In discussing the biological aspects of conservation work, a central point is that biologists must do more than collect and analyze

This photograph shows lowland rainforest along the Segama River in Borneo, Indonesia. Most of the world's biodiversity is found in tropical rainforests.

data. They must also be effective in the economic and public policy arenas. A conservation biologist is similar to a physician managing the recovery of an acutely ill patient. Success demands more than the clinical skills of making a correct diagnosis and prescribing medication; it requires an empathy with the person affected, the ability to manage the cost and implementation of care, and a devotion to collaborating with other professionals to achieve a cure. Biologists have become the doctors of biodiversity.

52.1 How Many Species Are Being Lost, and Why?

No one knows exactly how many species exist on Earth. About 1.7 million organisms have been studied and formally named so far, but this is widely acknowledged as only a tiny fraction of the total number of organisms. Why? Let's look at a few examples.

First, biologists know the most about groups like the vertebrates and flowering plants, which are large and relatively easy to study. Species-rich lineages like bacteria, archaea, mites, and insects, in contrast, are relatively poorly studied. Consider that a mere 5000 species of bacteria and archaea have been named to date. But when microbiologists sample small areas of habitat intensively, using the direct sequencing techniques introduced in Chapter 25, they routinely discover dozens or even hundreds of new species (**Figure 52.1**). Based on results like these, researchers suspect that millions of bacteria and archaea exist worldwide.

Further, most fieldwork in biology has been concentrated on habitats in the northern, temperate regions of the world. But biodiversity is concentrated in the tropics. Tropical rain forests cover just 7 percent of Earth's total land area, but they are estimated to contain 80 percent of its species. To drive this point home, biologist E. O. Wilson recounts that he once collected 43 species of ants belonging to 26 different genera from a single tree in a Peruvian rain forest. The ant fauna on this tree was equivalent in diversity to the ant fauna of the entire British Isles.

FIGURE 52.1 Finding New Species of Bacteria and Archaea Researchers take samples from a hot spring called Obsidian Pool, in Yellowstone National Park, Wyoming, USA. During one recent survey of bacteria and archaea present in the pool, the researchers found 54 previously undescribed species.

Finally, some habitats, like the deep ocean, are virtually unstudied to date, and life-forms have only recently been discovered in exotic habitats such as hot springs, rocks beneath Earth's surface, and frozen Antarctic lakes.

How Many Species Are There?

To estimate the actual number of species on Earth, biologists analyze species diversity in a well-studied taxon or habitat and extrapolate the results to less well-studied groups and areas. For example, one classic study focused on insects that live in the top, or canopy, of tropical trees. Terry Erwin and J. C. Scott used an insecticidal fog to knock down species from the top of a *Luehea seemannii* tree. They identified over 900 species of beetles among the individuals that fell. Most of these were new to science. To use these data as an indicator of total arthropod species diversity, Erwin used the following train of logic. Based on his earlier work with insects on this tree, he estimated that 160 of the 900 beetle species live only on *L. seemannii*. Worldwide, beetles represent about 40 percent of the known arthropods. Thus, it was reasonable to suggest that 400 species of arthropods live only in the canopy of *L. seemannii*. By adding an estimate of arthropods specializing on the trunk and roots of this tree, Erwin projected that it is host to 600 specialist arthropods. If each of the 50,000 species of tropical tree harbors the same number of arthropod specialists, then the world total of arthropod species is in excess of 30 million species. Based on studies like this, biologists estimate that at least 10 million, and possibly as many as 100 million, species of all types exist today.

To obtain a more direct estimate of total species numbers, the first effort to find and catalog *all* forms of life present at a single site is now under way. The location is the Great Smoky Mountains National Park in the southeastern United States (**Figure 52.2**). A consortium of biologists and research organizations initiated this all-taxon survey in 1999. When it is complete, in 2015, biologists will have a much better database to use in estimating the extent of biodiversity.

Extinction Rates: Data and Projections

Biologists are working on the question of how quickly species are going extinct, even as they work on the question of how many species exist. There are two components to this effort. The first is to understand how many species are currently threatened with extinction. The second is to estimate how many will go extinct in the near future.

Conservationists use several devices to track current trends in biodiversity. One of the most prominent is a "red list" compiled by the International Union for the Conservation of Nature. This is a book-length catalog of plant and animal species that are extremely rare or whose populations are in steep decline. The latest red lists for animals and plants estimate that of known species, 11 percent of birds, 18 percent of mammals, 5 percent of fish, and 10 percent of plant species are already threatened. (To a conservationist, "threatened" means that there is a 10 percent to 99 percent probability of extinction within 100 years.)

These percentages are large, and the number of species being added to the red lists is growing rapidly. What does this increase mean in terms of how many species will become threatened over the next 60 years? Can biologists predict how many species will go extinct during your lifetime?

Species-Area Curves One tactic for answering these questions is to analyze **species-area relationships.** This approach relates measured rates of habitat destruction to projected rates of species loss. For example, to quantify the extent of habitat destruction in the world's largest continuous rain forest and most species-rich region—the Amazon River basin—David Skole and Compton Tucker compared satellite images of the region made in 1978 and in 1988. Their analysis shows that an average of 15,000 km^2, an area the size of the state of Connecticut, was deforested in the Amazon each year during this decade (**Figure 52.3**). Based on studies like these, conservationists estimate that about 10 to 20 percent of the original area covered by tropical forests worldwide is being lost each decade. It is not unreasonable, then, to project that 90 percent of the world's tropical forest habitat will be lost over the next 60 years.

To predict how many species will be lost as a result, biologists use species-area curves like the one shown in **Figure 52.4** (page 1006). This graph was generated by Jared Diamond, who analyzed the number of bird species found on islands in the Bismarck Archipelago near New Guinea, in the South Pacific. The graph shows the number of species on islands of various sizes. Note that both axes on the graph are logarithmic. The solid line drawn through the points is described by the function $S = (18.9)\ A^{0.15}$, where S is the number of species and A is the area. In this island group, the function indicates that each tenfold increase in island area increases the number of bird species present by about 40 percent; a tenfold decline in area reduces bird species by about 30 percent.

Diamond's data turn out to be typical for other habitats and taxonomic groups as well. When biologists have plotted species-area relationships for plants, butterflies, mammals, or birds from islands or continental habitats around the globe, the relationship is always described by a function of the form $S = c\ A^z$. The c term is a constant that scales the data. It is high in species-rich areas like coral reefs or tropical rain forests and low in species-poor areas like arctic tundra. The exponent z represents the slope of the line on a log-log plot. As a result, z describes how rapidly species numbers change with area. In analyses of species-area relationships in island groups, z is typically 0.25–0.30. For habitats on continents, z usually ranges from 0.10 to 0.20.

To understand how biologists use these analyses to project extinction rates, study Figure 52.4 again. Ask yourself, if 90 percent of the habitats were destroyed—for example, if A were reduced from 10,000 km^2 to 1000 km^2—what percentage of species would disappear? According to the graph, the number of bird species in the Bismarck Archipelago would drop from about 75 to about 53. Thus, the answer is roughly 30 percent. This is an important conclusion. If z is about 0.15 for taxa that inhabit tropical rain forests and other threatened habitats, the prediction is that 30 percent of all species will be wiped out during your lifetime.

Testing the Theory Are the predictions made by species-area curves valid? Do they give a realistic estimate of extinction rates based on measured rates of habitat destruction?

To answer these questions, Thomas Brooks and co-workers studied bird species found only on the islands of the Philippines and Indonesia. Their first step was to use species-area curves to predict the number of endemic bird species that will soon go extinct given current rates of habitat destruction. When they compared this prediction to recent data on the actual number of bird species threatened with extinction on these islands, they found a strong correlation. The prediction and the actual data

FIGURE 52.2 The Location of the First All-Taxon Survey
The first attempt to document every species living in a prescribed area is under way at Great Smoky Mountains National Park. The park is located along the border of Tennessee and North Carolina, in the southeastern United States.

FIGURE 52.3 Deforestation in Tropical Rain Forests
Deforestation is occurring in the tropics due to logging, conversion of forest to pasture or cropland, and wildfires set by humans.

corresponded almost exactly. This is strong evidence that species-area relationships are a legitimate tool to use in predicting extinction rates, at least on islands.

The "Human Asteroid"

Recent analyses of extinction rates have a simple message: Over the next several centuries, human impacts on the planet could rival the gigantic asteroid that smashed into Earth 65 million years ago. The reasons for human-induced extinctions range from direct exploitation, such as hunting, to indirect or even unintentional causes, such as the introduction of exotic organisms that outcompete, eat, or parasitize native species. **Figure 52.5** illustrates a few examples from the long list of human-induced threats.

Two broad trends emerge from data on the recent extinctions caused by humans. Until about 1950, overhunting and introduced species were the predominant cause of extinction, and the extinctions occurred primarily on islands. Although overhunting and introduced species are still causing problems, most newly threatened species live on continents and are at risk from deforestation and other forms of habitat destruction. Let's take a closer look at this shift in human impacts.

Island Extinctions Most human-induced extinctions that have occurred over the past 400 years took place on islands. As a case study of this phenomenon, let's consider data collected by David Steadman, who studied bird fossils on islands in the Pacific Ocean (**Figure 52.6a**, page 1008). His data show that, on the seven best-studied islands in central Polynesia, at least 20 species or populations of birds were wiped out soon after the arrival of human colonists. The presence of humans was indicated by fire pits, rat bones, and other archaeological evidence (**Figure 52.6b**). There are 800 major islands in the Pacific. Based on this sample of seven, Steadman claims that humans may have been responsible for eliminating at least 2000 bird species in the region between about 500 and 1600 A.D. Because about 9000 bird species exist today, Steadman's estimate means that almost 20 percent of the world's bird species went extinct during the period.

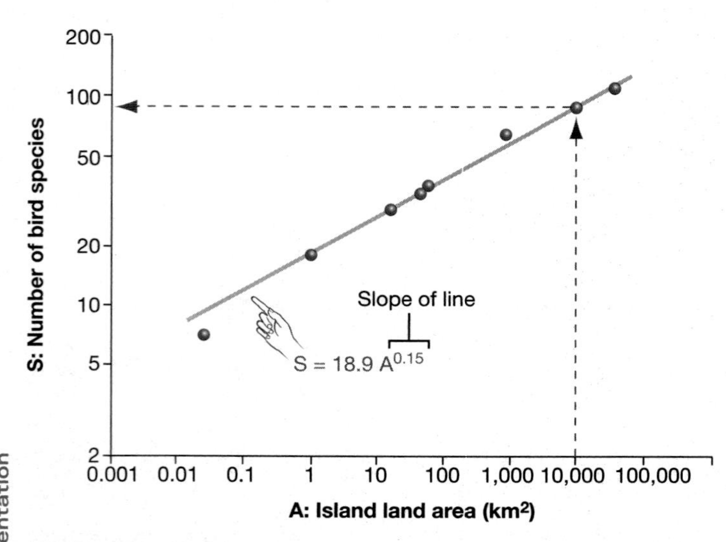

FIGURE 52.4 Species-Area Plots
The abscissa (x-axis) on this graph plots the sizes of islands in the south Pacific; the ordinate (y-axis) plots the number of different bird species present on each island. Note the log-log scale. The text explains the solid line through the points; the dotted lines show the number of bird species that are expected to live on an island of 10,000 km^2.

CD ACTIVITY 52.2 Habitat Fragmentation

To support this claim, Steadman points to data on the family of birds called rails (**Figure 52.6c**). After rails colonize islands, they frequently evolve flightlessness. Presumably this is because most islands lack ground-dwelling predators like rats and cats, which makes energetically costly flights to escape danger unnecessary. To quantify the extent of extinction among flightless rails, Steadman considered the 19 Pacific islands that have been studied thoroughly enough to yield at least 50 bones from fossilized land birds. On each of these islands, researchers have found one to four endemic species of flightless rail. (Endemic species are found in just one location.) But today, only four species of flightless rail remain in the entire region. If this sample is representative of the other major islands in the Pacific, then some 1500 rail species alone were extinguished by human activities prior to the arrival of Europeans. These activities included hunting and the introduction of exotic predators such as dogs and rats.

Habitat Destruction—The New Threat Habitat destruction dominates today's conservation concerns. You have already examined Skole and Tucker's data on deforestation in the Amazon basin. Another important message of that study, however, was that in addition to the huge area that was cleared, twice as much acreage was affected by **habitat fragmentation.** This is the transformation of large, continuous blocks of habitat into isolated parcels surrounded by pasture, cropland, or urban development.

When habitats become fragmented, biodiversity receives a "double whammy." Not only is the total area of forest reduced, but the habitat that remains is adversely affected by what biologists call edge effects. Compared to forest interiors, the forest edge habitats are drier and more exposed to high winds. They are also inundated by weed seeds.

A long-term experiment initiated by Thomas Lovejoy and colleagues in 1980 quantified these effects in the Brazilian rain forest. In an area near Manaus, Brazil, that was slated for clear-cutting, the group set up 66 square, 1-hectare experimental plots that remained uncut. (A hectare is 100 meters by 100 meters, or about the size of two football fields. The abbreviation for *hectare* is ha.) Thirty-nine of these study plots were located in fragments designed to contain either 1-, 10-, or 100-hectares of intact forest. Twenty-seven of the plots were set up nearby, in continuous rain forest. As **Figure 52.7a** (page 1009) shows, the distribution of the study plots allowed the research team to monitor changes inside forest fragments of different sizes, and compare these changes to conditions in unfragmented forest.

When the research group surveyed the plots 10–17 years after the initial cut, they recorded two predominant effects: a rapid loss of species diversity, especially from the smaller fragments, and a startling drop in **biomass** in the study plots located near the edges

(1) Overexploitation

Overhunting led to the extinction of the passenger pigeon.

(2) Introduced species

When species that are not native are introduced to an area, a number of different problems can occur.

Competition: In North American marshes, purple loosestrife is crowding out native organisms.

Disease: An introduced fungus has virtually wiped out the American chestnut.

Predation: The brown tree snake has extinguished dozens of bird species on the island of Guam.

(3) Habitat destruction

Logging in the Pacific Northwest of North America has removed so many large nesting trees that the northern spotted owl is now endangered.

(4) Habitat fragmentation

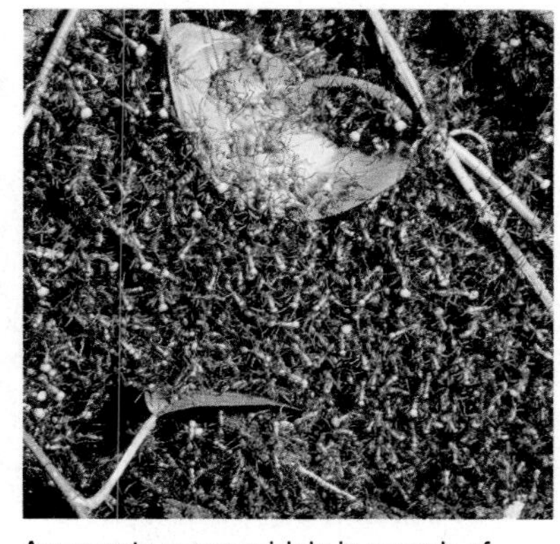

Army ants range widely in search of suitable food. Small forest plots are not large enough to support them.

(5) Domino effects

Antbirds feed on the insects stirred up by army ants.When army ants disappear from forest fragments, so do antbirds.

(6) Pollution

In the late 1940s, DDT was widely used to control mosquitos in marshy areas. By the mid-1950s, fish were accumulating large quantities of the molecule in their bodies. Fish-eaters like bald eagles accumulated even larger quantities of DDT, which turned out to be toxic.

(7) Global warming

Due to increased levels of CO_2 in the atmosphere, average temperatures are expected to increase by 5.5°C over the next 100 years. Species that disperse slowly and take a long time to grow to maturity, like beech trees, may not be able to shift their ranges quickly enough to survive.

FIGURE 52.5 Seven Deadly Sins That Affect Biodiversity
These photos illustrate seven reasons why humans have caused species or populations to become endangered.
(Photo (2) middle © Gary Braasch; photo (5) © Doug Wechsler/VIREO.)

of logged parcels (**Figure 52.7b**). *Biomass* is the total amount of fixed carbon in an area; in tropical rain forests, most of the biomass is concentrated in large trees. The edges of the experimental plots contained many downed and dying trees. Based on this observation, the researchers inferred that the decrease in biomass occurred because large trees near the edges of fragments died from exposure to high winds and dry conditions. This is important because it eliminates treetop, or canopy, habitats. Because many organisms live in the forest canopy, the quality as well as the quantity of habitat in the study area declined drastically (**Figure 52.7c**). This is fragmentation's double whammy.

The Genetics and Demography of Small Populations

Habitat loss and fragmentation interact in another important way. Habitat loss reduces the total size of a population, while fragmentation changes how the overall population is structured. Instead of existing as a large, continuous group, the two events combine to produce small populations that occupy isolated patches.

This type of population structure is known as a metapopulation, or a population of populations, and was introduced in Chapter 48. Metapopulations are actually common in nature. Butterflies, for example, frequently lay their eggs on a specific host plant, which their larvae eat. As a result, butterfly populations occupy distinct patches of habitat where their host plant happens to grow (**Figure 52.8a**, page 1010). These patches are often scattered around a particular area.

As an example of how metapopulation dynamics can affect the fate of endangered species, consider work that a research team led by Ilkka Hanski has done on a butterfly called the Glanville fritillary. In the Åland islands of southwest Finland, this research team found a total of 1600 meadows that contain host plants where females of this species lay eggs (**Figure 52.8b**). These patches range from 6 m^2 to 3 ha in size and can contain

(a) Over the last 400 years, most human-induced extinctions have occurred on islands, such as in the South Pacific.

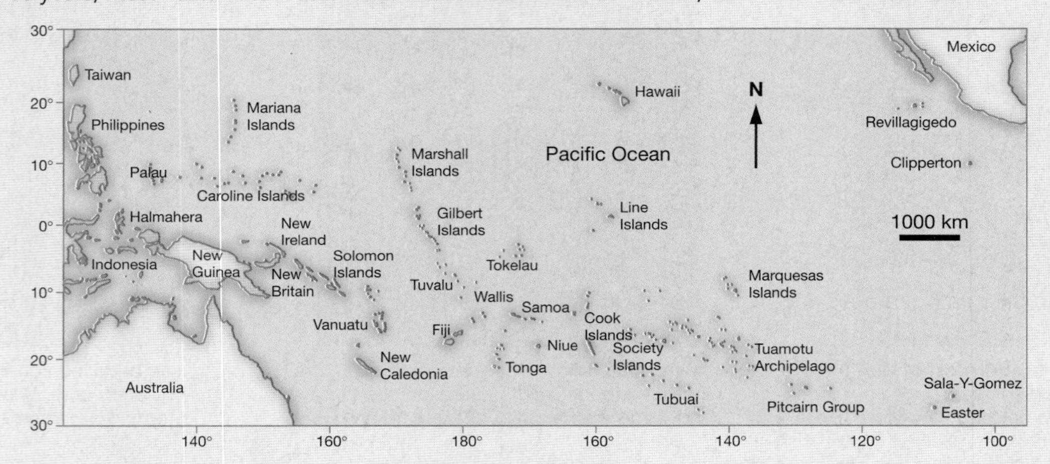

(b) Archaeological evidence suggests that the arrival of humans coincides with the disappearance of birds and other organisms.

(c) On islands, endemic species are vulnerable to hunting by humans and predation by introduced species.

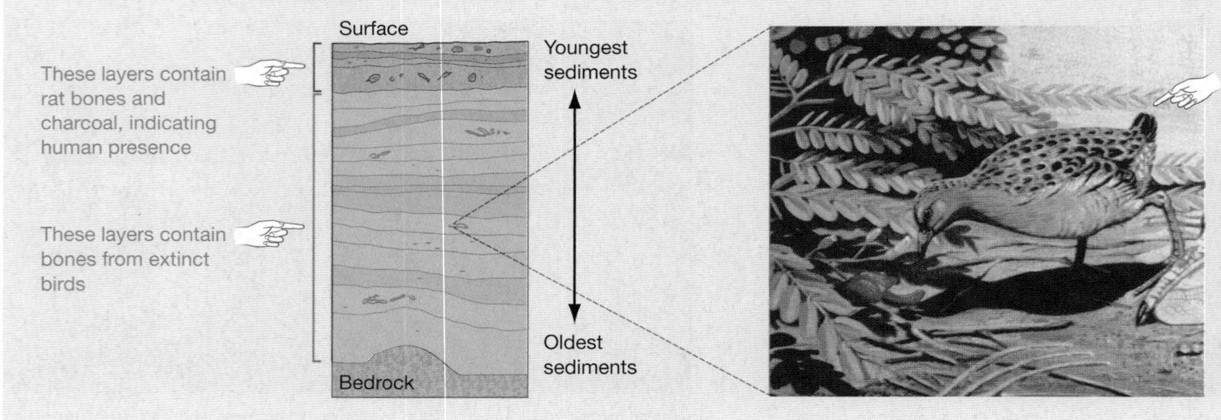

FIGURE 52.6 Evidence of Extensive Extinctions in Polynesia
(a) The tropical Pacific Ocean includes over 800 islands. Major island groups are shown here. **(b)** The diagram shows how sediments and fossils build up at these sites. The labels indicate that evidence of human occupation is routinely found in younger layers of sediments, which are located above layers that contain the remains of extinct birds. **(c)** An artist's interpretation of what an extinct, flightless rail—from the island of Hawaii—may have looked like.

(a) Forest fragmentation experiment in Brazil

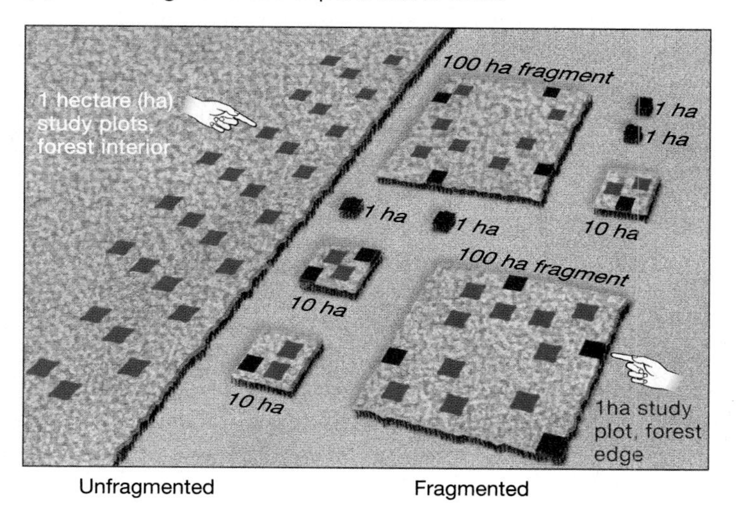

(b) Resulting decline in biomass along forest edges

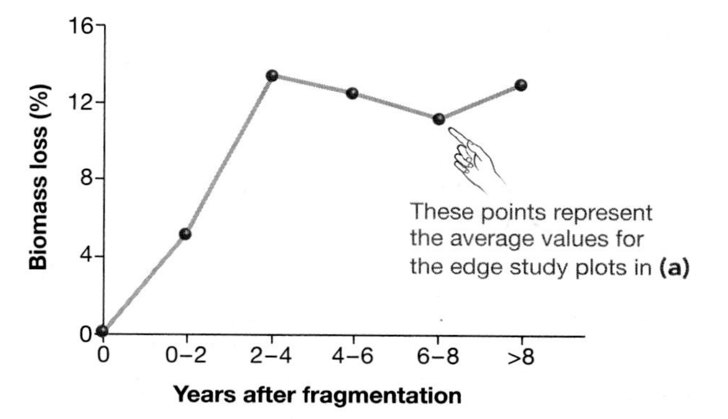

(c) Hypothesis: Decline in biomass is due to death of large trees near forest edges

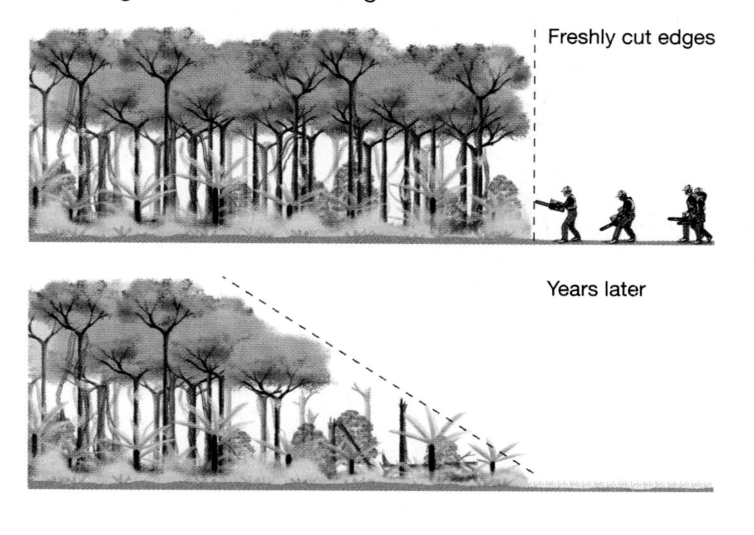

FIGURE 52.7 Edge Effects in Fragmented Forests
(a) Biologists track the results of the Manaus, Brazil, forest fragmentation experiment by following the fate of 66 study plots. Many of these plots are in forest fragments created for the experiment. There are four 1-hectare fragments, three 10-hectare fragments, and two 100-hectare fragments. **(b)** This graph shows the changes in biomass that have occurred in 16 study plots located near forest edges. **(c)** The loss of large trees reduces the number of habitat layers in a forest. Beetles, orchids, and birds that live in the top, or canopy, layer disappear.

from one to hundreds of pairs of breeding adults. The researchers monitored each meadow for four years and found that the number of patches containing butterfly populations changed dramatically each year. On average, butterflies were lost from 200 patches each year, while 114 unoccupied patches were colonized and newly occupied each year.

Why do so many of these fragmented populations go extinct each year? One hypothesis is that demographic factors, or changes in birth and death rates, play a large role. For example, it is well established that small populations are frequently destroyed by random events like storms, unusual weather conditions, or disease outbreaks. If demographic forces like these are at work, then larger populations should be less likely to go extinct simply because a few individuals might survive by chance. In accordance with this prediction, Saccheri and Hanski's group found that populations of Glanville fritillaries that were large, that occupied larger areas, and that were closer to neighboring populations (and hence more likely to be colonized) were less likely go extinct.

Did genetic factors also play a role in the fate of the butterfly populations? Population genetics theory suggests that they should. You may recall from Chapter 22 that small populations tend to become inbred, and that inbreeding is usually detrimental to the fitness of organisms. This is because inbred individuals are more likely than outbred individuals to be homozygous for deleterious recessive alleles (for more detail, see **Box 52.1**, page 1012). To determine whether inbreeding had occurred in their metapopulation, Saccheri and Hanski's team analyzed the amount of heterozygosity present in individuals from 42 of the study populations. As expected, individuals from the smaller populations were much less heterozygous, on average, than individuals from the larger populations. Further, their data showed that the inbred populations were much more likely to go extinct than the outbred populations, even when demographic factors like their size were taken into account.

Finland's fritillary butterflies may seem far removed from the crisis in the tropics, but these organisms have delivered a poignant message: If habitat destruction brings a species to the brink of extinction, the demographic and genetic consequences of habitat fragmentation can push it over that brink.

52.2 Why Should We Care?

No species lasts forever. Extinction, climate change, competition from newly arrived species, and habitat alteration are all natural processes. They have been happening since life began. Extinction, like death, is a fact of life. Even mass extinctions, where 50 percent or more of all species are wiped out in less than a million years, have occurred repeatedly during Earth's history.

Why, then, are conservationists ringing alarm bells and calling for immediate, and often expensive, action on behalf of endangered species and habitats? The answer is that biodiversity offers benefits to humans and to natural ecosystems.

Benefits to Humans

Earlier chapters explored how the diversity of bacteria, archaea, protists, plants, and fungi benefit people. But there are many other examples of how people profit from biodiversity. Wild fish and shellfish provide 19 percent of the animal protein consumed by humans. "Ecotourism," which focuses on viewing wildlife and wild places, is a major industry in East Africa, Costa Rica, and other regions of the world. The anticancer drugs vincristine/vinblastine and taxol were discovered in extracts from plants native to Madagascar and the old-growth forests in the Pacific Northwest, respectively; about a quarter of the prescription drugs dispensed in the United States contain active ingredients derived from plants. The widely used family of insecticides called the Bt-toxins were first isolated from the bacterium *Bacillus thuringiensis*. Domestic plant breeders regularly turn to wild relatives of corn, tomato, and other plants to introduce the alleles needed to resist new diseases or pests.

Preserving biodiversity has moral and aesthetic aspects as well as direct economic impacts. Although nothing could have prevented the asteroid that smashed into Earth 65 million years ago, humans have the capacity to anticipate the results of their own actions and act accordingly. Many people are troubled by the possibility that humans could be responsible for a mass extinction event; most prefer to live in an environment that is clean and healthy due to intact ecosystems and feel that the presence of wildlife and wild places adds values like beauty, tranquility, and adventure to their lives. Regions that have been traumatized by severe deforestation or pollution are often seen as dreary or ugly and of little value.

In short, there are several reasons to argue that biodiversity is good for human health and welfare. But is biodiversity valuable in a biological sense? Is species diversity inherently important?

The Biological Value of Biodiversity

For decades ecologists have debated whether species diversity increases the productivity and stability of ecosystems. If biodiversity makes ecosystems more productive and stable,

(a) Meadow habitat occupied by Glanville fritillaries

(b) A metapopulation is a population of populations.

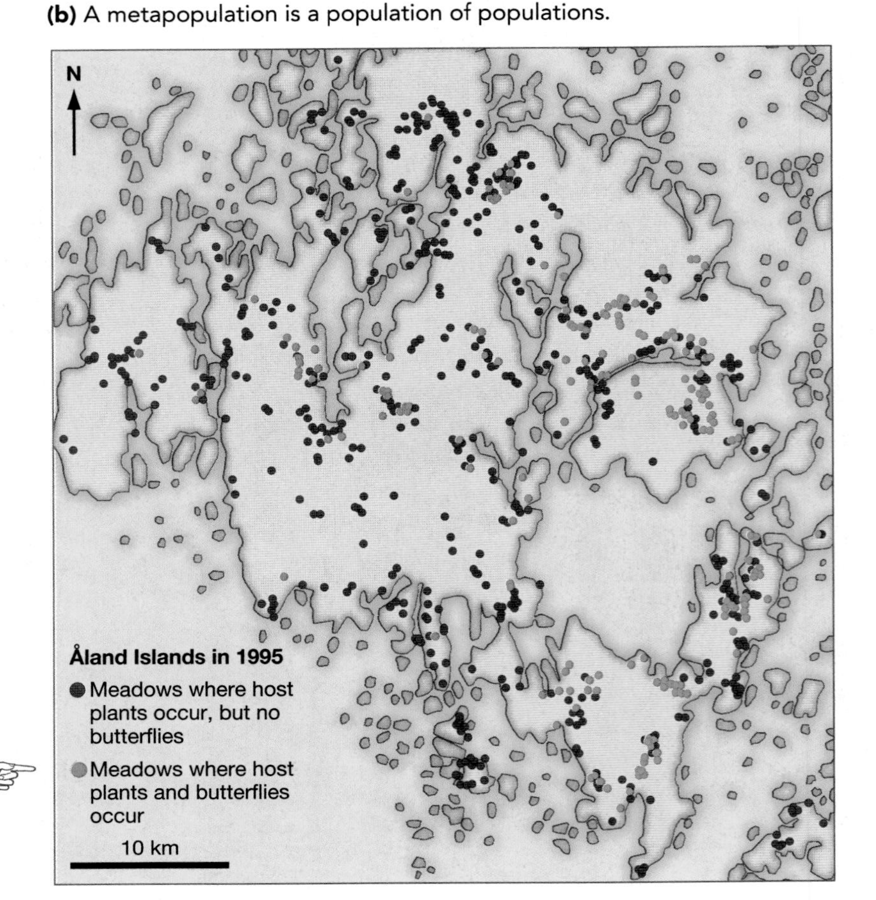

FIGURE 52.8 Studying a Metapopulation of Butterflies
This map shows the study sites monitored by Hanski and colleagues.

then increased species numbers have a quantifiable biological value.

Recently the debate about biodiversity and ecosystem function has focused on a specific question: Does the productivity of ecosystems depend primarily on the number of species present or on the types of species present?

To address this question, David Tilman and colleagues classified 32 grassland plant species into the five types, or functional categories, described in **Figure 52.9a**. Note that some of the types differ by the timing of their growing season, while others are defined by whether they allocate most of their resources to manufacturing woody stems or seeds. To test the effect that these ecological differences have on productivity, the researchers planted 3m × 3m plots with a mixture of between 0 and 32 randomly chosen species, representing from 0 to 5 different functional categories (**Figure 52.9b**). After 2 years of growth, they harvested and weighed all the aboveground tissues from all the plots.

When the researchers compared the productivity of plots with different mixes of functional groups and species numbers, they found that both the number and type of species present had important effects (**Figure 52.9c**). Further, certain functional groups had a greater effect on productivity than others. Overall, plots with a wider diversity of functional groups and with more species were more productive. The leading hypothe-

(a) Five functional groups of plants:

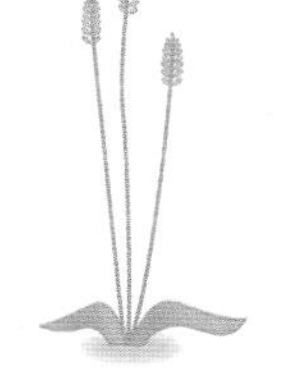

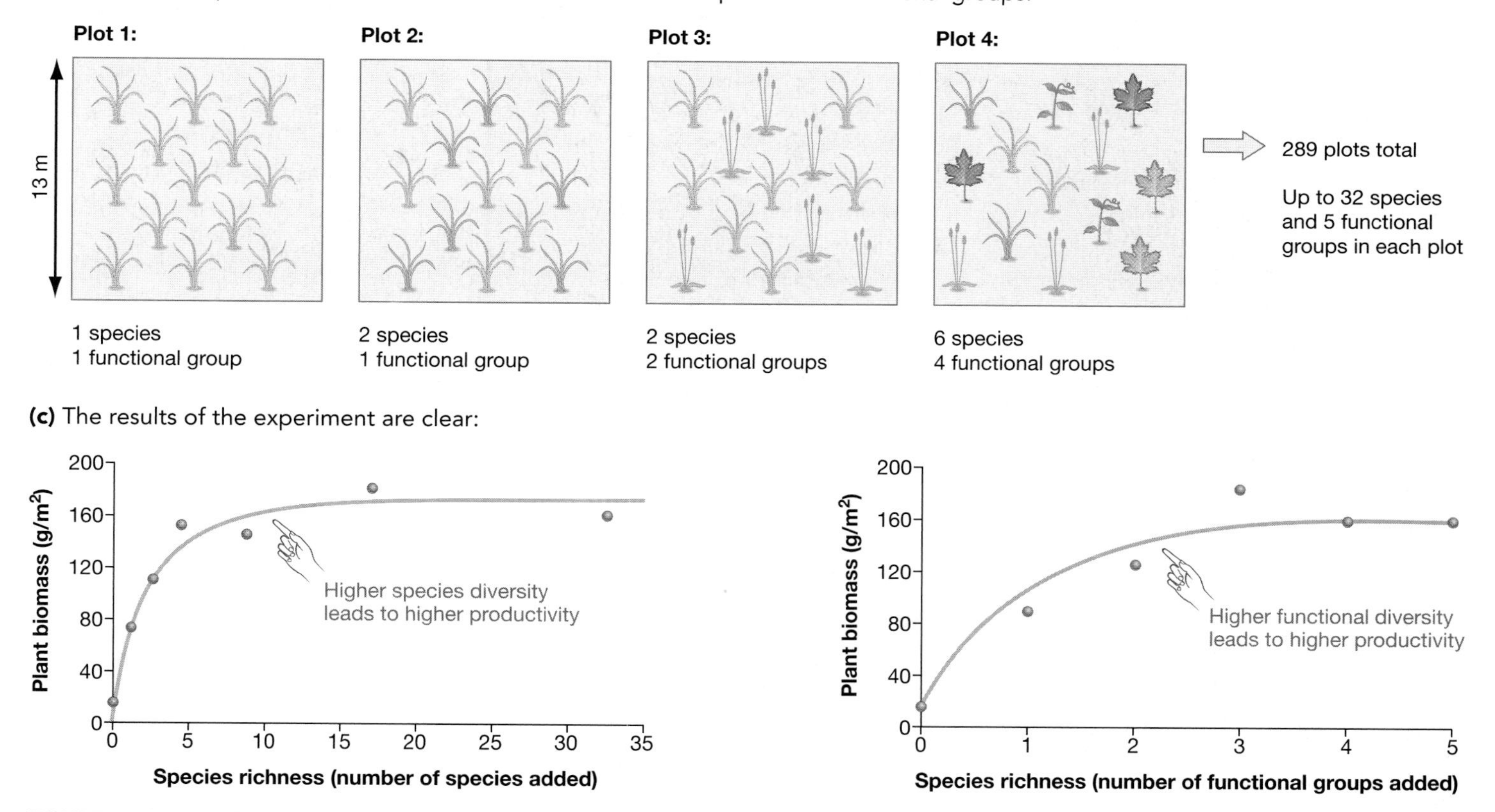

FIGURE 52.9 Does the Productivity of Ecosystems Depend on the Number or the Types of Species Present?
(a) To answer the question posed in the title, a group of researchers worked with five distinct types of plants. Each category represents a particular type of growth pattern. **(b)** The researchers set up experimental plots that were seeded with different numbers of species and different numbers of functional groups. **(c)** After two years of growth, plots with a larger number of species and a larger number of functional groups had higher productivity.

ses to explain this pattern are that a diverse assemblage of plant species makes more efficient use of the resources available, and that certain species or functional groups provide nutrients or other benefits to other species.

The message of this research is that species diversity and functional diversity may influence ecosystems in important ways. By implication, biologists can infer that if ecosystems are simplified by extinctions, their productivity may decrease.

52.3 Setting Conservation Priorities

In assessing the future of biodiversity, biologists tend to be pessimistic about the short-term outlook for preserving species and optimistic about the long-term outlook. The pessimism is caused by data on current rates of habitat loss and newly threatened species. The optimism is grounded in a view of human beings as adaptable and rational, data on recent successes in cleaning up pollution and saving endangered species, and studies indicating that growth rates in human populations have begun to stabilize or even decrease.

If this assessment is accurate, then the agenda for biologists is to intervene in the biodiversity crisis, assist as many species as possible through the trauma that is projected to occur over the next 50 to 100 years, and hope that conditions improve in the future.

BOX 52.1 Why Do Small Populations Become Inbred, and Why Is Inbreeding Harmful?

Chapter 22 provided a detailed explanation of why small populations tend to become inbred, and why this trend worries biologists concerned about the fate of endangered species. Here is a brief review of how genetic drift and inbreeding interact to put small populations in danger.

The frequencies of the alleles at any particular locus change constantly due to chance. This process is especially pronounced in small populations. Because allele frequencies can change randomly until they reach fixation or loss, the process of genetic drift tends to reduce overall allelic diversity. In very small populations, drift can quickly eliminate most alleles, leaving individuals extremely similar genetically.

Further, all of the individuals in small populations eventually become related to one another because so few mates are available. When this happens, some degree of inbreeding becomes inevitable. As the self-pollination example in Chapter 22 demonstrates, inbreeding increases the frequency of homozygotes.

When drift reduces allelic diversity and inbreeding increases homozygosity, problems can arise. The data in the following table illustrate this point. They show the percentage of young children that died—in two different human populations during two different time intervals—as a function of whether their parents were related to each other. Offspring of first-cousin marriages are two to three times more likely to die in early childhood than are the offspring of unrelated spouses. Because socioeconomic status and other variables were controlled in these studies, the results suggest that inbred offspring are much less fit than outbred offspring.

Deaths	Period	Children of First Cousins	Children of Unrelated Parents
Children under 10 (U.S.)	1920–1956	8.1%	2.4%
Children aged 1–8 (Japan)	1948–1954	4.6%	1.5%

Why? Recall that drift is just as likely to fix deleterious, loss-of-function mutations as it is to fix beneficial alleles. When inbreeding makes these mutations homozygous, individuals suffer from genetic diseases and a general loss of vigor.

When conservation biologists suspect that inbreeding and drift are lowering the fitness of an endangered population, they take steps to encourage outbreeding and increase population size. For example, a team led by Ronald Westemeier recently documented a 35-year decline in the population size of greater prairie chickens in southern Illinois. The researchers suspected that inbreeding and drift were reducing the fitness of this fragmented population, because a smaller and smaller percentage of eggs were hatching over time (**Figure 1**). In an attempt to counteract the trend, the team brought in individuals from thriving populations in nearby states and released them. As the graph shows, the influx of genetically unrelated birds was correlated with a restoration of hatching success. Based on this observation, the researchers claimed that the transplanting program served as an effective antidote to inbreeding and drift.

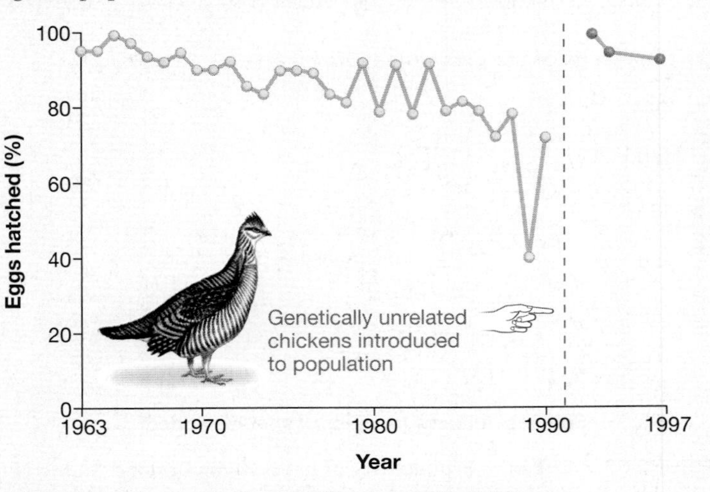

FIGURE 1 Fitness Declines in Illinois Prairie Chickens
When genetically unrelated individuals were introduced into this population, hatching success increased.

Given that the short-term outlook is bleak, however, it is critical that priorities be established for spending scarce resources and for preserving the most important habitats left. Land and money are scarce. How can biologists design systems of nature reserves that will maximize the number of species protected?

Designing Nature Reserves

To understand how researchers approach the problem of designing efficient nature reserves, let's consider the task faced by Jared Diamond. In the late 1970s, officials from the government of Indonesia asked for Diamond's assistance in setting up a series of forest reserves in the province of Irian Jaya. The province is located on the island of New Guinea in the South Pacific. This was an urgent task because the government was being pressed to grant extensive logging concessions in old-growth rain forest. As a result, Diamond had no time to recruit other biologists, catalog biodiversity throughout the nation, and recommend that reserves be established in the habitats that maximized protection for all groups on the tree of life.

Instead, Diamond urged the government to protect habitats that would maximize protection for birds. Designing reserves around this goal was feasible, because far more was known about the birds of Indonesian New Guinea than about any other taxonomic group. (This is often true, in fact. Because humans perceive flowering plants, butterflies, large mammals, and songbirds as beautiful and interesting, they are almost always the best-studied species in any given area of the world.) Diamond's hypothesis was that maximizing bird diversity would efficiently preserve a diversity of other organisms as well.

Diamond's proposal is illustrated in **Figure 52.10**. The reserves shown on the map were designed to protect habitats ranging from mangrove swamps at sea level to glaciers at altitudes of 16,500 feet. The reserves included habitat types ranging from dry forest to wet forest as well as lakes and marshes. Further,

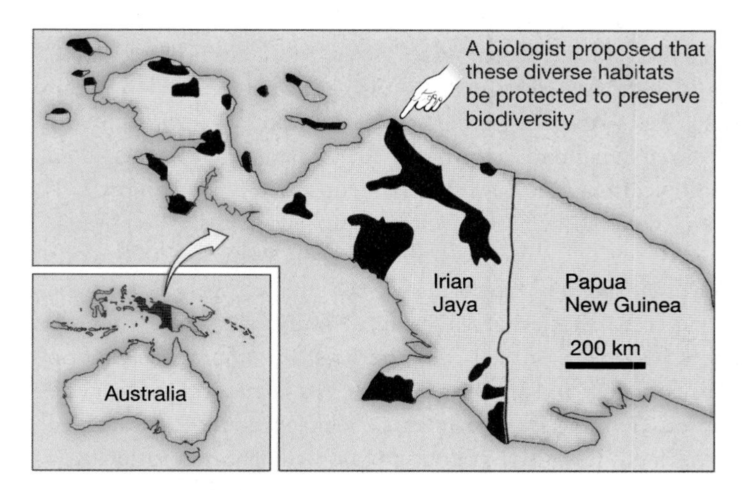

FIGURE 52.10 Forest Reserves in Irian Jaya
The government of Indonesia adopted this proposed system of reserves in 1983.

they incorporated all prominent "centers of endemism," meaning the areas where large numbers of endemic birds are found.

Are the reserves effective? The jury is still out. Follow-up studies still need to be conducted as development on Irian Jaya proceeds.

Complementary Sets The general approach that Diamond pioneered was later extended, formalized, and evaluated by A. T. Lombard and colleagues. These biologists set about to design a system of reserves capable of protecting the rich and highly endemic plant diversity of the Agulhas Plain in South Africa.

As a first step, Lombard and co-workers divided maps of the region into a grid of cells 3 km × 3 km. The researchers identified the flowering plant species found in each cell, and then used a computer to search for the smallest set of cells that would succeed in protecting the majority of the endemic populations. The resulting group of cells is called a complementary set. The name is apt, because the selected cells complement one another. Individually, none of them contains a majority of the endemic flowering plants. But as a group, they contain all of the species in question. In effect, complementary sets are a way to formalize and quantify the conservation strategy that Diamond employed in a qualitative way.

Evaluating Reserve Designs Lombard and colleagues are now working with government officials and local landowners to protect the selected areas on the Agulhas Plain. Their study raised an important question, however: Do complementary sets for one group of species, like flowering plants, succeed in protecting other groups as well? This hypothesis has now been tested rigorously in several other areas of the world.

In Uganda, complementary sets work well. Peter Howard and colleagues came to this conclusion after cataloging the woody plants, large moths, butterflies, birds, and small mammals in government-owned forest reserves throughout the country. The government wanted to designate 20 percent of the forests as strict nature reserves, which would be off limits to logging as well as to other forms of commercial use.

The question faced by Howard and co-workers was, which 20 percent would protect the maximum number of species? To answer this question, they broke the forest area into plots of equal area and recorded how many woody plants, large moths, butterflies, birds, and small mammals were found in each plot. Then they ranked the plots from highest to lowest, according to how many butterfly species they contained. Next they determined what percentage of *all* species would be saved if they saved plots one by one, starting with the plot that was highest in butterfly diversity and ending with the plot that was lowest. This allowed the researchers to predict how many moths, woody plants, birds, and small mammals would also be saved if reserves were designed to maximize the diversity of butterflies.

Finally, Howard and colleagues repeated the ranking and analysis procedure four times, using moths, woody plants, birds, or small mammals as the focal group. The goal of this effort was to determine whether one taxonomic group was particularly valuable as an indicator of diversity in other lineages.

When butterflies were used as a focal group, the graph in **Figure 52.11** resulted. Analyses for other groups were similar, and support two important conclusions. First, designing nature preserves around one group, such as butterflies, would be effective at protecting other groups as well, although only slightly better than choosing plots randomly. Second, saving just 20 percent of the total forest reserve area would protect over 75 percent of the total species studied. In this case, complementary sets based on a single taxon worked efficiently.

In contrast, complementary sets based on a single taxon work poorly in the Transvaal region of South Africa. Albert van Jaarsveld and colleagues arrived at this conclusion after studying detailed data on the distributions of over 9000 species of plants and animals. On average, the overlap in complementary sets of cells between any two taxonomic groups, like vascular plants and birds, was less than 10 percent. Equally discouraging reports have been filed when researchers analyzed data from England, the United States, and several locations in Africa outside of Uganda. Why the differences?

In reviewing studies where complementary sets both failed and succeeded, Stuart Pimm and John Lawton concluded that in general, distinct reserves will have to be set up for different taxonomic groups. Indicator species seem to work well in Ugandan forests, because groups of plants, birds, mammals, insects, and other species occur together in highly distinct habitats. In other parts of the world, habitats are not as distinct in character and the pattern does not hold. To conserve as many branches on the tree of life as possible, then, conservationists have a tremendous amount of work to do. It will not be enough to save habitats for one group such as birds or trees and expect that this will preserve beetles, spiders, flowers, and frogs as well.

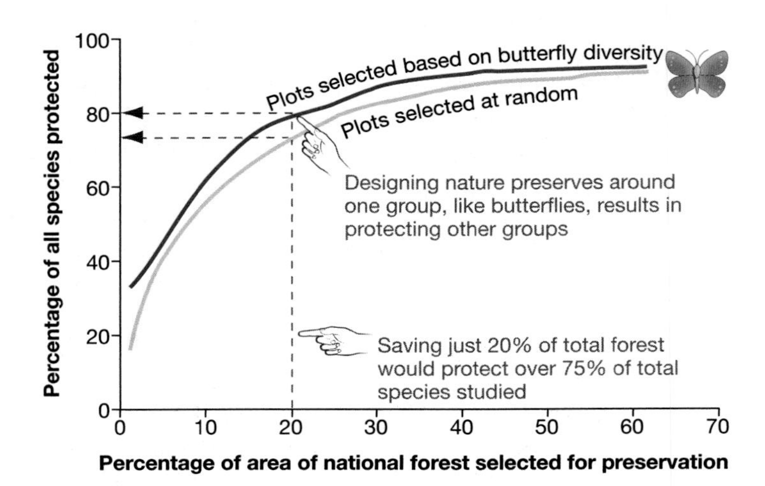

FIGURE 52.11 Designing Forest Reserves in Uganda
The red line plots the percentage of species from several groups (woody plants, moths, birds, and small mammals) that would be saved if plots within Ugandan forests were selected for preservation on the basis of containing the highest number of butterflies. The light blue line indicates the number of species that would be preserved in cells that were chosen randomly.

Essay Metaphors for the Future: Easter Island and Guanacaste

Emotionally, conservation biology can be a tough business. Predictions for the future are not optimistic and lessons from the past can be discouraging.

Consider the story of the most isolated, inhabited place on Earth: Easter Island (**Figure 1**). When European explorers "discovered" Easter Island in 1722, about 1000 people lived there. The island was treeless and dotted with the gigantic stone statues shown in the figure. When John Flenley and Sarah King analyzed pollen cores taken from swamps on the island, however, they discovered that it had once been covered with lush forest dominated by palm trees. A steady decline in tree pollen coincided with the arrival of the first human settlers around 400 A.D.

What is "restoration ecology?"

Fossil digs by David Steadman and colleagues confirmed that the fauna of the island underwent drastic changes at the same time. In the oldest human garbage piles excavated by the biologists, bones from dolphins, sea birds, and land birds are abundant. But these species dropped out of the fossil record by about 1200 A.D.—about the time that deforestation was complete.

Jared Diamond interpreted these data as evidence of an ecological disaster. In his view, people arrived to find a lush tropical island brimming with natural resources. The population flourished, possibly numbering as many as 7000 at its peak. During the flowering of the culture, people had the leisure time and resources to carve the gigantic statues and roll them into place on beds of palm logs. But after deforestation was complete and local extinctions had begun, the system collapsed. Without palm trunks to make canoes, Easter Island natives did less dolphin hunting and fishing. Soil erosion may

(Continued on next page)

(Essay continued)

have cut into the productivity of banana and sweet potato plantations. The population crashed and the great statues fell.

To keep history from repeating itself, conservationists have launched dozens of large- and small-scale restoration and reforestation projects around the globe. In the tropics, one of the most successful programs is the restoration of dry forest at the Area de Conservación Guanacaste in northwestern Costa Rica. The primary task facing conservation area staff there was to stop human-caused fires. In just 15 years, their efforts have succeeded in transforming a vast swath of marginal ranching land into an increasingly popular tourist destination and water source for neighboring farms and ranches (**Figure 2**).

Conservationists contend that, to preserve biodiversity over the long term, it will not be enough to save patches of undisturbed habitat in parks and preserves; work to restore damaged ecosystems is also crucial. From small-scale restorations of woodland and prairie habitats in North America to efforts the size of Guanacaste, the field of "restoration ecology" is burgeoning.

FIGURE 2 The Guanacaste Reserve
These before-and-after photos were taken of the same view, 17 years apart, in the Area de Conservación Guanacaste, Costa Rica. To appreciate how quickly the trees grew, notice the person standing near the middle of the photo on the bottom.

(a) Start of restoration project

(b) 17 years later

FIGURE 1 Easter Island
Easter Island is located in a remote area of the southeast Pacific. Many of the gigantic statues found on the island have fallen.

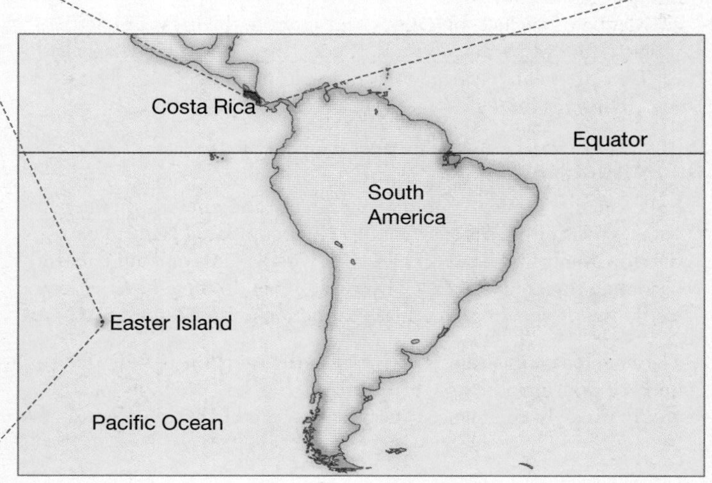

Chapter Review

Summary

To estimate how many species will go extinct over the next several decades, biologists combine data on current rates of habitat loss—often estimated from satellite images taken over time—with data on the average number of species found in habitats of a given size. These species-area analyses suggest that if 90 percent of habitats are destroyed, then over 30 percent of all species will become extinct.

Historically, the majority of human-caused extinctions have occurred on islands because of direct exploitation or the introduction of exotic herbivores and predators. For example, data from fossil digs indicate that hunting and introduced predators have devastated the endemic birds of Pacific islands over the past 1000 years.

Habitat loss is projected to be the leading cause of extinctions over the next 100-plus years, however, especially in the tropics. Experiments in the Brazilian Amazon have shown that habitat loss leads not only to a rapid decline in biodiversity but also to a decline in the quality of the remaining habitats as well, due to fragmentation. Populations in small, isolated fragments of habitat are of special concern because genetic drift and inbreeding are amplified and can lead to widespread declines in fitness.

Biologists give several reasons for trying to stem the loss of biodiversity. In addition to the aesthetic value of natural areas, humans gain direct economic benefits from fishing, forestry, agriculture, tourism, and other activities that depend on biodiversity. Species diversity is also important to maintaining the productivity of natural ecosystems.

Faced with increasingly ominous threats to biodiversity, however, biologists have debated how to best utilize scarce resources. An early strategy preserved areas for well-studied species or groups, with the assumption that other lineages would be protected in those areas as well. More recent analyses have shown, however, that this approach does not work in many, or perhaps even most, cases. In designing conservation programs, then, each taxonomic group presents a unique challenge. Preserving biodiversity may well be the most important—and difficult—enterprise our species has ever undertaken.

Questions

Content Review

1. What does a species-area plot show?
   a. The overall distribution, or area, occupied by a species.
   b. The relationship between the body size of a species and the amount of territory or home range it requires.
   c. The number of species found, on average, in tropical versus northern areas.
   d. The number of different species found, on average, in a habitat of a given size.

2. When forested habitats are fragmented, why does their quality decline?
   a. Trees on the edge of the fragments are frequently blown down.
   b. The edges of the forest are exposed to more sunlight, which increases temperatures and decreases humidity.
   c. Weedy species from the deforested areas invade the edge.
   d. All of the above.

3. Why are island species particularly vulnerable to extinction after human colonization?
   a. Islands often lack predators and large herbivores, but settlers introduce rats, cats, dogs, pigs, goats, and other exotic species.
   b. They are relatively easy to hunt to extinction because their distribution is limited.
   c. They are relatively easy to hunt to extinction, because their total population size is often small.
   d. All of the above.

4. What did research find out about inbred populations of butterflies?
   a. They are more likely to be found in large expanses of habitat where butterflies are common.
   b. They are more likely to go extinct.
   c. They are never found.
   d. They are identified by obvious abnormalities, such as albinism.

5. To design efficient nature preserves, biologists often divide a region into a grid of cells and map the species found in each cell. What does a complementary set of these cells do?
   a. It indicates where species in other, more poorly studied, lineages—like mosses or spiders—are also likely to be found.
   b. It consists of the 10 cells with the most species.
   c. It identifies the group of cells that includes one of each species in the lineage being analyzed—for example, birds.
   d. It consists of the cells containing the rarest species.

Conceptual Review

1. In percentage terms, the number of plants and animals threatened with extinction is relatively small when compared with mass extinction events in the past. The *rate* at which plants and animals are becoming threatened is large, however. How do these observations justify the projection that a mass extinction event is under way?

2. The chapter claims that the rate of extinctions on islands due to hunting and introduction of exotic species has probably peaked, and that newly endangered species are more likely to be found on continents and be threatened by habitat destruction. Why is this trend occurring?

3. Biologists claim that the all-taxa survey, now under way at the Great Smoky Mountains National Park in the United States, will improve their ability to estimate the total number of species living today. Why?

4. How are species-area curves used to relate rates of habitat destruction to projected extinction rates?

Applying Ideas

1. The introduction to this chapter stated that its goal was to introduce the information needed to help preserve biodiversity. To test whether this goal was achieved, think of a species that is endangered in the region surrounding your college or university. Then answer the following questions: What are the primary factors responsible for the species' decline? Are these factors similar to or different from the factors responsible for concern about diversity in tropical rain forests? If you were a conservation manager, what steps would you take to ensure that the remaining population did not become inbred? How would you go about designing a regional conservation plan for this species and for other organisms that occupy the same habitat?
2. In your opinion, of the several reasons for preserving biodiversity listed in this chapter, which is the most persuasive? Explain your reasoning.
3. Projections for ecological disaster are contingent on the continuation of both the present trends in human population growth and in habitat destruction. Do you believe these trends will continue? In general, are you optimistic or pessimistic about the future of biodiversity?
4. Suppose you were assigned to assist a team of biologists commissioned to design a series of reserves in a tropical country. List the series of steps you would recommend for gathering data and creating a plan that would protect a large number of species in a small amount of land.
5. The maps shown here chronicle the loss of old-growth (> 200 years old) forest that occurred in Warwickshire, England, and in the United States during recent intervals. In your opinion, under what conditions is it ethical for conservationists who live in these countries to lobby government officials in Brazil, Indonesia, and other tropical countries to slow the rate of loss of old-growth forest?

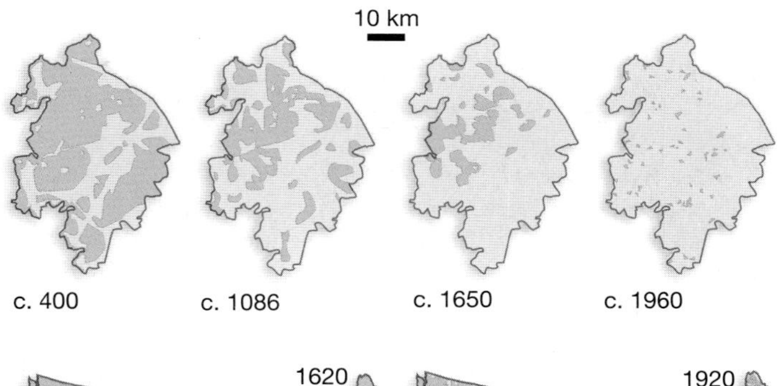

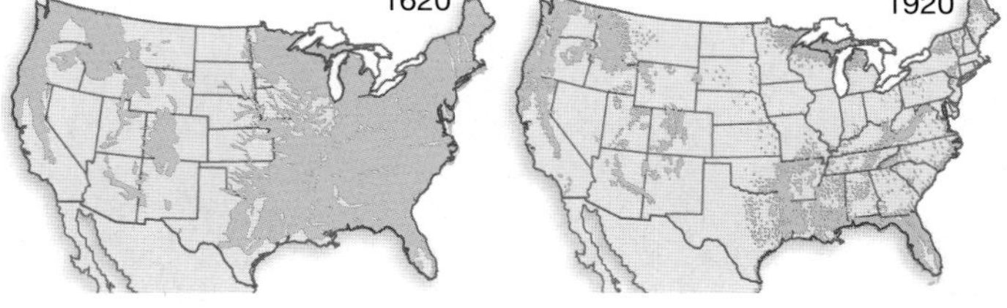

6. Make a list of characteristics that would render a species particularly vulnerable to extinction by humans. Make a list of characteristics that would render a species particularly resistant to pressure from humans. Try to think of an example of each.

CD-ROM and Web Connection

CD Activity 52.1: Species-Area Curves *(tutorial)*
(Estimated time for completion = 5 min)
What is the relationship between species diversity and island size? Examine this question using a graph of the diversity of bird species on islands in Southeast Asia.

CD Activity 52.2: Habitat Fragmentation *(animation)*
(Estimated time for completion = 5 min)
What are the effects of a reduction in habitat size—called habitat fragmentation—on biomass in a tropical forest?

At your **Companion Website** (http://www.prenhall.com/freeman/biology), you will find self-grading exams and links to the following research tools, online resources, and activities:

The Extinction Files
This is an excellent resource sponsored by the BBC to educate the public and scientists about extinction.

Environmental Protection Agency
The EPA website offers a tremendous amount of information that directly relates to concerns about environmental degradation and the destruction of species.

Conservation Genetics
This site discusses how the field of conservation genetics combines research in molecular biology and conservation biology to preserve endangered species.

Additional Reading

Diamond, J. 1995. Easter's end. *Discover* (August): 62–69. An account of deforestation and its consequences on Easter Island.

Laurance, W. F. 1998. A crisis in the making: responses of Amazonian forests to land use and climate change. *Trends in Ecology and Evolution* 13: 411–415. A superb overview of recent data on deforestation and protection efforts in the Brazilian Amazon.

Ward, Peter. 1994. *The End of Evolution* (New York: Bantam). A paleontologist's perspective on a human-induced mass extinction event.

Appendix

TABLE 1.1 The Metric System

Measurement	Unit of Measurement and Abbreviation	Metric System Equivalent	Converting Metric Units to English Units
Length	kilometer (km)	1 km = 1000 m = 10^3 m	1 km = 0.62 miles
	meter (m)	1 m = 100 cm	1 m = 1.09 yards = 3.28 feet = 39.37 inches
	centimeter (cm)	1 cm = 0.01 m = 10^{-2} m	1 cm = 0.3937 inch
	millimeter (mm)	1 mm = 1000 μm = 10^{-3} m	1 mm = 0.039 inches
	micrometer (μm)	1 μm = 1000 nm = 10^{-6} m	
	nanometer (nm)	1 nm = 10^{-9} m	
	angstrom (Å)	1 Å = 0.1 nm = 10^{-10} m	
Area	hectare (ha)	1 ha = 10,000 m^2	1 ha = 2.47 acres
	square meter (m^2)	1 m^2 = 10,000 cm^2	1 m^2 = 1.196 square yards
	square centimeter (cm^2)	1 cm^2 = 100 mm^2 = 10^{-4} m^2	1 cm^2 = 0.155 square inches
Mass	kilogram (kg)	1 kg = 1000 g	1 kg = 2.20 pounds
	gram (g)	1 g = 1000 mg	1 g = 0.035 ounces
	milligram (mg)	1 mg = 1000 μg = 10^{-3} g	
	microgram (μg)	1μg = 10^{-6} g	
Volume	liter (L)	1 L = 1000 mL	1 L = 1.06 quarts
	milliliter (mL)	1 mL = 1000 μL = 10^{-3} L	1 mL = 0.034 fluid ounces
	microliter (μL)	1 μL = 10^{-6} L	
Temperature	†Kelvin (K)		K = °C + 273.15
	Degrees Celsius (°C)		°F = $\frac{9}{5}$°C + 32
	Degrees Fahrenheit (°F)		°C = $\frac{5}{9}$(°F − 32)

†Absolute zero is −273.15°C = 0 K

TABLE 1.2 Prefixes Used in the Metric System

Prefix	Abbreviation	Definition
micro-	μ	0.000001 = 10^{-6}
milli-	m	0.001 = 10^{-3}
centi-	c	0.01 = 10^{-2}
deci-	d	0.1 = 10^{-1}
		1 = 10^0
kilo-	k	1000 = 10^3

QUESTIONS

(1) Some friends of yours just competed in a 5-kilometer run. How many miles did they run?
(2) An American football field is 100 yards long, while rugby fields are 140 meters long. In yards, how much longer is a rugby field than a football field?
(3) What is your normal body temperature in degrees Celsius? (Normal body temperature is 98.6°F.)
(4) What is your current weight in kilograms?
(5) A friend asks you to buy a gallon of milk. How many liters would you buy to get approximately the same volume?

Content Review Answers

Chapter 1
1. d, 2. a, 3. a, 4. c, 5. b, 6. a.

Chapter 2
1. b, 2. d, 3. b and d, 4. c, 5. d, 6. c, 7. c, 8. c, 9. b.

Chapter 3
1. c, 2. d, 3. a, 4. b, 5. c, 6. c, 7. a, 8. d.

Chapter 4
1. b, 2. a, 3. b, 4. d, 5. c, 6. d.

Chapter 5
1. b, 2. d, 3. a, 4. a, 5. c, 6. b, 7. a, 8. d.

Chapter 6
1. d, 2. b, 3. b, 4. d, 5. d, 6. c, 7. a.

Chapter 7
1. d, 2. c, 3. a, 4. c, 5. c, 6. d.

Chapter 8
1. b, 2. c, 3. b, 4. d, 5. d, 6. a.

Chapter 9
1. b, 2. a, 3. d, 4. b, 5. a, 6. b.

Chapter 10
Answers to Chapter 10 genetics problems can be found on p. A-4.

Chapter 11
1. a, 2. d, 3. b, 4. a, 5. d, 6. d.

Chapter 12
1. b, 2. d, 3. a, 4. c, 5. d, 6. b.

Chapter 13
1. c, 2. b, 3. c, 4. c, d, 5. c, 6. d.

Chapter 14
1. b, 2. d, 3. c, 4. d, 5. a, 6. c.

Chapter 15
1. c, 2. a, 3. d, 4. a, 5. b, 6. a.

Chapter 16
1. c, 2. a, 3. a, 4. b, 5. d, 7. d.

Chapter 17
1. c, 2. b, 3. a, 4. b, 5. d, 6. d.

Chapter 18
1. c, 2. d, 3. a, 4. d, 5. b, 6. a.

Chapter 19
1. b, 2. d, 3. c, 4. a, 5. c, 6. c.

Chapter 20
1. d, 2. b, 3. c, 4. a, 5. b, 6. b.

Chapter 21
1. b, 2. d, 3. a, 4. c, 5. a, 6. a.

Chapter 22
1. b, 2. d, 3. a, 4. b, 5. c, 6. d, 7. d.

Chapter 23
1. a, 2. d, 3. c, 4. d, 5. b, 6. d.

Chapter 24
1. c, 2. d, 3. d, 4. a, 5. b, 6. a.

Chapter 25
1. c, 2. a, 3. b, 4. b, 5. d, 6. a, 7. b, 8. c.

Chapter 26
1. d, 2. b, 3. b, 4. c, 5. d, 6. b, 7. b, 8. d.

Chapter 27
1. b, 2. d, 3. c, 4. a, 5. b, 6. d, 7. b.

Chapter 28
1. c, 2. a, 3. a and b, 4. c, 5. c, 6. a, 7. e.

Chapter 29
1. a, 2. a, 3. c, 4. d, 5. c, 6. b.

Chapter 30
1. a, 2. c, 3. a, 4. d, 5. c, 6. b.

Chapter 31
1. a, 2. c, 3. d, 4. b and c, 5. a, 6. b.

Chapter 32
1. a, c, and d, 2. a, c, and d, 3. c, 4. c and d, 5. b, 6. d.

Chapter 33
1. b, 2. a, 3. d, 4. b, 5. d.

Chapter 34
1. a, 2. a, b, and c, 3. c, 4. c, 5. b.

Chapter 35
1. a and c, 2. d, 3. c, 4. a, 5. b, 6. b and c.

Chapter 36
1. d, 2. c, 3. a, 4. b, c, d, 5. a, c, 6. c, d.

Chapter 37
1. a, 2. d, 3. b, 4. c, 5. c.

Chapter 38
1. c, 2. d, 3. b, 4. d, 5. b, 6. d.

Chapter 39
1. For freshwater fish: "c" is true. For marine fish: "a" and "b" are true. 2. a, b, c, e, g, and h, 3. d, 4. a and e, 5. c.

Chapter 40
1. a, 2. a, 3. d, 4. d, 5. b, 6. d.

Chapter 41
1. b, 2. c, 3. d, 4. d, 5. c, 6. a.

Chapter 42
1. b, 2. d, 3. c, 4. a, 5. d, 6. a.

Chapter 43
1. a, 2. c, 3. c, 4. b, 5. a, 6. a.

Chapter 44
1. d, 2. a, 3. d, 4. b, c, d, e, 5. c, 6. a.

Chapter 45
1. a, 2. b, 3. a, b, d, 4. c, 5. d, 6. d.

Chapter 46
1. c, 2. c, 3. b, 4. d, 5. b, 6 a.

Chapter 47
1. c, 2. d, 3. a, 4. b, 5. d, 6. d, 7. c.

Chapter 48
1. a, b, and d, 2. d, 3. a, 4. b, c, and d, 5. a, 6. d.

Chapter 49
1. b and c, 2. b and c, 3. d, 4. a, 5. d, 6. a.

Chapter 50
1. d, 2. c, 3. c, 4. c, 5. c.

Chapter 51
1. d, 2. d, 3. a, 4. c, 5. d, 6. c.

Chapter 52
1. d, 2. d, 3. d, 4. b, 5. c.

Answers to Genetics Problems

Chapter 10

1. Some attributes of a model organism can include that it is easy to study, has a relatively short reproductive cycle, and serves as a model for processes and patterns that occur in other species as well.

2. The allele is considered recessive because the trait associated with the allele is not expressed in heterozygotes.

3. Haploid organisms have only a single allele at each locus and it is expressed.

4. Indicating that individuals are homozygotes gives us all the information we need. It is more easily understood and clearer than "pure line."

5. If Mendel had selected some trait pairs that showed linkage and some that did not, he might have been unable to reach the conclusions that led him to formulate the principle of independent assortment. The loci that were linked would not have yielded the 9:3:3:1 ratios that he obtained with the unlinked loci.

6. This statement is valid because of the following: The traits he analyzed were discrete and depended on the alleles at a single gene; just two alleles were present at each gene he analyzed, and the alleles were completely dominant and recessive with respect to one another; and, there were no epistatic effects—meaning that the traits were not affected by interactions with alleles at other loci.

7. Individuals with PKU cannot metabolize phenylalanine to tyrosine. Drinking beverages sweetened with NutraSweet™ will increase their blood levels of phenylalanine. The makers of the sweetener need to warn consumers with PKU about this possibility.

   Organisms interact with aspects of the environment in countless ways every day. We consume food, breathe air, sit in the sun, take medications, etc. In many cases our genes determine our response to that environment. Sun exposure initiates the production of melanin by activating genes that catalyze the reactions in the pathway. Among Caucasians in general, the amount of melanin produced varies widely as we have all observed when one person tans quickly and evenly while another gets a few more freckles. And albinos are unable to produce melanin at all because they lack the necessary enzymes. Eating carbohydrates triggers the production of the enzymes (gene products) that are part of the pathway of glycolysis and the Krebs cycle. Bacteria that can metabolize lactose produce the necessary enzymes when lactose is present in the environment and do not when it is absent.

8. The four types of gametes should be observed at equal frequencies.

9. Half of their offspring should have the genotype iI^A and have the type A blood phenotype. Half of their offspring should have the genotype iI^B and have the type B blood phenotype.

10. This is a dihybrid cross that yields progeny phenotypes in a 9:3:3:1 ratio. Let *O* stand for the allele for orange petals and *o* the allele for yellow petals; let *S* stand for the allele for spotted petals and *s* the allele for unspotted petals. Start with the hypothesis that *O* is dominant to *o*, that *S* is dominant to *s*, that the two genes are found on different chromosomes so that they assort independently, and that the parent individual's genotype is *OoSs*. If you do a Punnett square for the *OoSs* × *OoSs* mating, you'll find that progeny phenotypes should be in the observed 9:3:3:1 proportions.

11. Let *D* stand for the normal allele and *d* stand for the allele responsible for Duchenne-type muscular dystrophy. The woman's family has no history of the disease so her genotype is almost certainly *DD*. The man is not afflicted, so he must be *DY*. (The trait is X-linked, so he has only one allele; the '*Y*' stands for the Y chromosome.) Their children are not at risk. The man's sister could be a carrier, however—meaning that she has the genotype *Dd*. If so, then half of this couple's male children are likely to be affected.

12. If the F_2 individuals have one dominant and one recessive trait, it means that their seeds are round and green or yellow and wrinkled. Individuals with round, green seeds have one of two genotypes: 1/3 of them are *RRyy* and 2/3 of them are *Rryy*. Similarly, individuals with yellow, wrinkled seeds have one of two genotypes: 1/3 of them are *rrYY* and 2/3 of them are *rrYy*. If you do the Punnett squares for each of these genotypes when they self-fertilize, you should be able to convince yourself that individuals with *RRyy* or *rrYY* genotypes are true-breeding; individuals with *Rryy* or *rrYy* genotypes are not. Thus, the answer is 1/3.

    If the F_2 individuals have two dominant traits, it means that their seeds are round and yellow. Individuals that produce round, yellow seeds have one of four genotypes: 1/9 of them are *RRYY*; 2/9 are *RRYy*, 2/9 of them are *RrYY*; and 4/9 of them are *RrYy*. If you do the Punnett squares for each of these genotypes when they self-fertilize, you should be able to convince yourself that only individuals with the *RRYY* genotype will yield offspring that only produce round, yellow seeds. Thus, the answer is 1/9.

13. Your diagram should be similar to Figure 10.11(b) with *A* and *a* replacing *R* and *r* and *B* and *b* replacing *Y* and *y*. The gametes would be *AB*, *Ab*, *aB*, and *ab* and would be equally frequent. The two long chromosomes are homologous to each other and the two shorter chromosomes are homologous to each other. Each chromosome should consist of two sister chromatids. At the locus for *A*, the two alleles are *A* and *a*. At the locus for *B*, the two alleles are *B* and *b*. Segregation occurs at anaphase of meiosis I when the chromosome containing *A* separates from the one containing *a* and the chromosome containing *B* separates from the one containing *b*. Independent assortment occurs because non-homologous chromosomes can arrange themselves on the metaphase plate differently in different gametes as they undergo meiosis.

14. Presumably, Vulcans and earthlings are homozygous for both traits. Spock, then, should be heterozygous at both loci. Because he has the Vulcan phenotype, though, we can infer that pointed ears and a right-sided heart are dominant to rounded ears and a left-sided heart. Let *P* be the allele for pointed ears and *p* be the allele for rounded ears; let *R* be the allele for a right-sided heart and *r* the allele for a left-sided heart. Spock is *PpRr*; his earthling wife is *pprr*. Mendel would predict that their offspring would have the genotypes *PpRr*, *Pprr*, *ppRr*, and *pprr* in equal proportions. Thus a quarter of their children should have pointed ears and left-sided hearts or rounded ears and right-sided hearts. The only logical explanation for the actual results is that the two genes are closely linked. If so, then Spock's gametes are either *PR* or *pr* and his wife's gametes are all *pr*.

15. Four phenotypes should be observed in the ratio 9:3:3:1, as follows: curly and large: curly and smooth: silky and large: silky and smooth. One-quarter of the progeny should be heterozygous for both traits. One-eighth should be homozygous for both traits.

16. According to Mendel's model, palomino individuals should be heterozygous at the locus for coat color. If you mated palomino individuals, you would expect to see a combination of chestnut, palomino, and cremello offspring. If blending inheritance occurred, however, all of the offspring should be palomino.

17. Let *H* stand for the allele for normal blood clotting and *h* for the allele for hemophilia A. The genotypes of couple #1 must be *H*- and *HY* because neither is afflicted; their son is *hY*. If the mother is *HH* there is no chance that they would have an *hY* son, but if she is *Hh* then half her sons should be *hY*. The genotypes of couple #2 are *H*- and *hY*; their son is *HY*. The father did not transmit an X chromosome to his son so his genotype is irrelevant. If the mother is *HH* then all of her sons will be *HY*, but if she is *Hh* then half her sons should be *hY*. You should advise the jury that either the hospital or the parents could be correct. The answer depends on the genotypes of the mothers.

18. Curved wing is autosomal recessive; lozenge eye is X-linked recessive. Let w^+ be the allele for long wings and w be the allele for curved wings; let r^+ be the allele for red eyes and r the allele for lozenge eyes. The female parent is w^+wr^+r; the male parent is w^+wr^+Y.

19. Albino is the absence of pigment, so let *b* stand for an allele that gives absence of blue and *y* for an allele that gives absence of yellow pigment. If blue and yellow pigment blend to give green, then the green parents are both *BbYy*. The phenotypes of the offspring should be in the ratio 9:3:3:1 as green:blue:yellow:albino. The green phenotype is found in *BBYY*, *BBYy*, *BbYY*, and *BbYy* offspring. The blue phenotype is found in *BByy* or *Bbyy* offspring. The yellow phenotype is observed in *bbYY* or *bbYy* offspring. Albino offspring are *bbyy*.

Two types of crosses yield *BbYy* F_1 offspring: *BByy* × *bbYY* (blue × yellow), or *BBYY* × *bbyy* (green × albino).

Glossary

–10 box The six-pair base sequence (usually TATAAT) in most prokaryotic promoters (DNA binding sites for RNA polymerase), located 10 bases upstream from the start of messenger RNA transcription. Also called the Pribnow box.

A site The site in a ribosome at which an aminoacyl transfer RNA pairs with a messenger RNA codon, in preparation for adding its amino acid to the growing peptide chain. (See also E site, P site.)

aboveground biomass The total mass of living plants in an area, excluding roots.

accessory fluids The fluid portion of semen, which contains nutrients, enzymes, and other substances. It is added to the spermatozoa by male reproductive glands.

acclimation A gradual adjustment to new environmental conditions.

acid Any compound that gives up protons or accepts electrons during a chemical reaction, or that releases hydrogen ions when dissolved in water.

acid-base reaction A chemical reaction that involves a transfer of protons.

acquired immunity The acquisition of antibodies that confer immunity to a certain pathogen. The antibodies may be produced by previous exposure to the pathogen or to a vaccine, or acquired passively through the placenta or breast milk.

acrosomal reaction A set of events occurring in a sperm cell upon encountering an egg cell. These events include release of acrosomal enzymes and formation of microfilaments that bind the sperm to the egg.

acrosome A packet of enzymes found on the head of a sperm cell.

action potential A rapid, temporary change in electrical potential across a membrane.

activation energy The amount of energy required to cause a chemical reaction; specifically, the energy required to reach the transition state.

activator Any regulatory protein that enhances transcription of certain genes, typically by binding to a certain enhancer.

active transport The movement of ions or molecules across a cell membrane against a concentration gradient (from a region of lower concentration to a region of higher concentration) or electrochemical gradient. Such movement requires energy such as ATP.

adaptation Any heritable trait that increases the fitness of an individual with that trait compared to individuals without that trait.

adaptive radiation Rapid evolutionary diversification within a single lineage, producing numerous descendant species with a wide range of adaptive forms.

adenosine triphosphate (ATP) A molecule consisting of adenine, a sugar, and three phosphate groups that can be hydrolyzed to release free energy. Universally used by cells to store and transfer energy.

aerobic respiration Cellular respiration using oxygen as the electron receptor, typically using the Krebs cycle and the electron transport chain of mitochondria. (See also anaerobic respiration.)

agonist A compound that can bind to and activate a hormone receptor.

alimentary canal The part of the digestive system that forms a muscular tube starting at the mouth and ending at the anus. Also called digestive tract or gastrointestinal tract.

allele A particular version of a gene. Example: the allele for black hair pigment and the allele for red hair pigment are two versions of the mammalian gene for hair pigment color.

allometry A disproportionate relationship of body size to another feature, such as limb size or heart rate. Also may refer to the general study of the effects of body size on biological processes.

allopatric speciation The divergence of populations into different species by physical isolation of populations in different geographic areas.

allosteric regulation Regulation of gene activity by a change in shape of a large regulatory protein, usually induced by a small molecule.

alpha chain (α chain) One of the two polypeptide chains that makes up a T-cell receptor protein, enabling the T cells of the immune system to bind to antigens on other cells. (The other polypeptide is the β chain.)

alternation of generations A life cycle that involves alternating a multicellular haploid stage (the gametophyte) with a multicellular diploid stage (the sporophyte). Occurs in most plants and some protists.

altruism Any act or behavior by an individual that has a cost to that individual, in terms of survival and reproduction, and a benefit to the recipient; self-sacrifice. (See also kin selection.)

alveolus (plural = alveoli) One of the tiny air-filled sacs of a mammalian lung.

aminoacyl tRNA A transfer RNA molecule that is covalently bound to an amino acid.

amniotic egg An egg containing a membrane-bound supply of water (the amnion), a membrane-bound supply of food (yolk sac), and a waste sac (allantois), all encased in a leathery or hard shell.

amphipathic Containing hydrophilic and hydrophobic elements. Amphipathic molecules such as phospholipids are often used as components of cell membranes.

amplify Make numerous copies of a particular piece of DNA, typically via the polymerase chain reaction (PCR) or some other laboratory technique.

anabolic pathway Any set of chemical reactions that synthesizes larger molecules from smaller ones.

anaerobic respiration Cellular respiration using an electron acceptor other than oxygen, such as nitrate or sulfate.

analogous trait Similarity between different species that is not due to inheritance from a common ancestor, but instead is due to adaptation to a similar way of life. Often the result of convergent evolution.

anaphase A stage in cell division (mitosis or meiosis) during which chromosomes are moved to opposite ends of the cell.

angiosperm A flowering plant; a member of the lineage of plants that produces seeds within mature ovaries (fruits).

animal A multicellular, complex eukaryote that obtains nutrients by eating other organisms and is typically mobile.

animal hemisphere The upper half of an amphibian egg cell, containing little of the yolk. Gives rise to most of the animal's body.

anion A negatively charged ion.

annual plant A plant whose life cycle normally lasts only one growing season; i.e., less than one year.

antagonistic muscle group A set of muscles whose actions oppose each other and whose control must be coordinated. Often a pair of muscles, such as a flexor and an extensor, which move a body part back and forth.

anther The pollen-producing structure at the end of a flower stamen.

antibiotic A substance that can kill or inhibit the growth of another microorganism.

antibody A Y-shaped protein that can bind to a specific part of a certain antigen (foreign molecule), tagging it for attack by the immune system; a type of immunoglobulin that is secreted from B cells.

anticodon Three bases of a transfer RNA molecule that bind to a messenger RNA codon with a complementary sequence.

antigen Any foreign molecule that can stimulate a specific response by the immune system.

apical dominance Inhibition of lateral bud growth by the apical meristem at the tip of the plant branch.

apical meristem A group of undifferentiated cells found at the tip of a stem or root of a vascular plant. Responsible for growth in length.

apoptosis Programmed cell death. Occurs frequently during embryonic development and later may occur in response to infections or cell damage.

aquaporin A membrane protein that facilitates movement of water across a cell membrane.

artificial selection Deliberate manipulation of the hereditary traits of a population by allowing reproduction of only certain individuals. Examples include animal and plant breeding.

asexual reproduction A reproductive process in which an individual inherits all of its chromosomes from a single parent, making it genetically identical to that parent. Examples include spore formation and budding. (See also sexual reproduction.)

assay Any procedure that measures the amount or purity of a component, such as the concentration of a hormone, or measures an event, such as the rate of a reaction.

atomic mass unit A unit of weight equal to 1/12 the mass of one carbon-12 atom; approximately the weight of 1 proton or 1 neutron.

atomic number The number of protons in the nucleus of an atom, giving the atom its identity as a certain chemical element.

ATP (adenosine triphosphate) A molecule consisting of adenine, a sugar, and three phosphate groups that can be hydrolyzed to release free energy. Universally used by cells to store and transfer energy.

ATP synthase A membrane-bound, ATP-synthesizing enzyme found in the membranes of chloroplasts and mitochondria. Consists of a proton channel and a knob that synthesizes ATP when protons flow through the channel.

atrium A thin-walled chamber of the heart that receives blood from veins and pumps it to a neighboring chamber (the ventricle).

autonomic nervous system The part of the peripheral nervous system (outside the brain and spinal cord) that controls internal organs and processes, such as stomach contraction, hormone release, and heart rate. Not under voluntary control. Includes parasympathetic and sympathetic nerves. (See also somatic nervous system.)

autosome One of any pair of chromosomes that do not carry the gene(s) that determine gender. (See also sex chromosomes.)

autotroph Any organism that can synthesize its own food—complex organic compounds—by using simple inorganic sources such as CO_2 or CH_4. Most plants and some bacteria and archaea are autotrophs. In ecology, may also be called a primary producer.

auxin A plant hormone—indoleacetic acid—that stimulates cell elongation and some other responses.

axon A long projection from a neuron that propagates an action potential.

axon hillock The location at which an axon joins the cell body of a neuron; the site at which action potentials are first triggered.

axoneme An arrangement of two central microtubules surrounded by nine doublet microtubules, found in eukaryotic cilia and flagella and responsible for their motion.

BAC (bacterial artificial chromosome) A laboratory-created loop of DNA that can be inserted into a living bacterial cell, which can be grown in a cell culture to create many copies of the BAC. BACs are commonly used to amplify and study a particular DNA sequence.

background extinction The average rate of extinctions that occurs normally rather than during mass extinction events. (See also mass extinction.)

basal transcription factor A protein that binds to eukaryotic promoters and helps to initiate transcription. Part of the core transcription complex.

base Any compound that acquires protons or gives up electrons during a chemical reaction, or that accepts hydrogen ions when dissolved in water.

B cell A type of lymphocyte that produces antibodies. Produced by the bursa of birds, or the bone marrow of mammals. If activated, may give rise to plasma cells and memory cells. Also called B lymphocyte. (See also T cell.)

B-cell receptor (BCR) A Y-shaped immunoglobulin protein found on the surfaces of B cells, and to which antigens bind. (See also antibody.)

benign Relatively harmless. A benign tumor is a growth that is not cancerous; a benign microorganism does not cause severe disease. (See also malignant and virulent.)

benign tumor A mass of abnormal tissue that is growing slowly or not at all, does not disrupt surrounding tissues, and does not metastasize to other organs. Benign tumors are not cancers. (See also malignant tumor.)

beta chain (β chain) One of the two polypeptide chains that make up a T-cell receptor protein, enabling the T cells of the immune system to bind to antigens on other cells. (The other polypeptide is the α chain).

bilateral symmetry An animal body pattern in which there is one plane of symmetry dividing the body into a left side and a right side. Typically the body is long and narrow with a distinct head end and tail end.

bile A greenish fluid produced by the liver, stored in the gall bladder, and secreted into the intestine, where it emulsifies fats during digestion. Contains bile salts, bile pigments, and other components.

biogeochemical cycle The pattern of circulation of a chemical element between living organisms and the environment.

biological species concept The concept that species are groups reproductively isolated from each other; different species cannot crossbreed in nature to produce viable and fertile hybrid offspring.

biomass The total mass of all organisms in a given population or geographical area; usually expressed as total dry weight.

blastopore A small pore in the surface of an early vertebrate embryo, through which cells move during gastrulation.

blastula An early stage of embryonic development in vertebrate animals, consisting of a ball of cells enclosing a fluid-filled space.

blood pressure The pressure exerted by blood against the walls of the blood vessels. In medical usage, refers specifically to pressure in arteries at the level of the heart, measured while the heart is contracting (systolic pressure) and again while it is relaxing (diastolic pressure).

bone An animal tissue consisting of living cells and blood vessels within a hard extracellular matrix of calcium phosphate ($CaPO_4$) with small amounts of calcium carbonate ($CaCO_3$) and protein fibers.

buffer A substance that, in solution, acts to minimize changes in the pH of that solution.

bulk flow The directional movement of a substantial volume of fluid. Examples: the movement of water through plant phloem; the movement of blood in animals.

Burgess shale fauna A characteristic set of soft-bodied animal fossils found in several early Cambrian rock formations (525–515 million years old), particularly the Burgess Shale in Canada and the Chengjiang deposits in China.

C_4 photosynthesis A variant of photosynthesis in which atmospheric carbon is first fixed into four-carbon sugars, rather than the three-carbon sugars of classic C_3 photosynthesis. Enhances photosynthetic efficiency in hot, dry environments, by reducing loss of oxygen due to photorespiration.

cadherins A class of cell-surface proteins involved in cell adhesion and important for coordinated movements of cells during embryological development.

Calvin cycle The light-independent component of photosynthesis. A set of reactions that use the NADPH and ATP formed earlier (in the light-dependent reactions of photosynthesis) to drive the reduction of atmospheric CO_2, ultimately producing sugars.

CAM (crassulacean acid metabolism) A variant of photosynthesis in which CO_2 is stored in sugars at night when stomata are open, and then released to enter the Calvin cycle during the day when stomata are closed. This helps reduce water loss and oxygen loss by photorespiration in hot, dry environments.

cambium (plural = cambia) A layer of undifferentiated plant cells found in older stems and roots that is responsible for secondary growth (increase in width).

Cambrian explosion The rapid diversification of animal body types that began about 543 million years ago and spanned approximately 40 million years at the beginning of the Cambrian Period.

camera eye An eye found in vertebrates and cephalopods that consists of a hollow chamber with a hole at one end through which light enters and a sheet of light-sensitive cells against the opposite wall. (See also compound eye.)

capsid A shell of protein enclosing the genome of a virus particle.

carbon cycle The circulation of carbon between living organisms and the environment (e.g., oceans, atmosphere).

carcinogen Any cancer-causing agent.

carnivore An animal, such as dogs, cats, weasels, and bears, that eats other animals.

carotene A class of carotenoid pigments found in many orange, yellow, and red vegetables, such as carrots and tomatoes.

carotenoid A class of plant pigments that absorbs wavelengths of light not absorbed by chlorophyll.

carrier A heterozygous individual carrying a normal allele and a recessive allele for an inherited condition; a carrier does not display the phenotype of the condition but can pass the recessive gene on to offspring.

carrying capacity The maximum population size of a certain species that a habitat can support.

cartilage A type of connective tissue in the skeletons of vertebrates, which consists of relatively few cells scattered in a fibrous or elastic matrix of polysaccharides and protein fibers.

catabolic pathway Any set of chemical reactions that breaks down larger, complex molecules into smaller ones, releasing energy in the process.

catabolism All of the metabolic reactions that result in the breakdown of large molecules into smaller subunits, producing energy.

catabolite activity protein (CAP) A prokaryotic protein that binds near the promoter of an operon, facilitating the binding of RNA polymerase and stimulating gene expression.

catabolite repression A type of inhibition of gene transcription, in which a gene codes for an enzyme in a catabolic pathway, and the end product of that pathway inhibits further transcription of that gene.

catalyst Any substance that lowers the activation energy of a particular reaction (by lowering the potential energy of the transition state), thus increasing the rate of the reaction. The catalyst itself does not take part in the reaction and its chemical composition is not changed during the reaction.

cation A positively charged ion.

cdk (cyclin-dependent kinase) A protein kinase that is active only when bound to a cyclin. Involved in control of the cell cycle.

cDNA (complementary DNA) DNA created in vitro from an RNA transcript using the enzyme reverse transcriptase. A cDNA corresponds to a gene, but lacks introns.

cDNA library A set of DNA sequences created by reverse transcription of a set of RNAs from a particular organism or cell type. Often used to identify and study the genes for all the proteins made by that particular organism or cell type.

cell culture A collection of cells that grows and divides in a lab, typically in liquid suspension or in a petri dish on a solid food medium.

cell cycle The sequence of stages that a dividing eukaryotic cell goes through from the time it is created (by division of a parent cell) to the time it undergoes mitosis. Includes M, G_1, S, and G_2 phases.

cell plate A double layer of new cell membrane that appears in the middle of a dividing plant cell, which ultimately divides the cytoplasm into two separate cells.

cell theory The theory that all organisms are made of cells, and that all cells come from pre-existing cells.

cell-cycle checkpoint A point in the cell cycle that is regulated, and at which the cell cycle can be stopped.

cellular respiration A common and efficient pathway to produce ATP, which involves a transfer of electrons from a reduced compound to an electron transport chain, and ultimately to an electron acceptor.

central nervous system (CNS) The brain and spinal cord of vertebrate animals. (See also peripheral nervous system.)

central vacuole A large, water-filled organelle in plant cells that stores toxins, degrades enzymes, disposes of wastes, and is involved in cell growth.

centriole One of two small, cylindrical structures found together near the nucleus of a eukaryotic cell. Together they are called the centrosome and serve as a microtubule hub for the cell's cytoskeleton. (See also centrosome.)

centromere The region joining two sister chromatids during meiosis.

centrosome A structure consisting of two small, cylindrical structures (centrioles) found together near the nucleus of a eukaryotic cell. Serves as a microtubule organizing center for the cell's cytoskeleton. During cell division, it replicates into two daughter centrosomes, each of which organizes one end of the mitotic spindle.

chemical energy The potential energy stored in covalent bonds between atoms.

chemical equilibrium A dynamic but stable state of a reversible chemical reaction in which the forward reaction and the reverse reaction proceed at the same rate; this means that the concentrations of reactants and products remain constant.

chemical evolution The hypothesis that simple chemical compounds in the ancient atmosphere and ocean combined by natural and spontaneous chemical reactions to form larger, more complex substances that eventually led to the origin of life and the start of biological evolution.

chemical reaction An event in which one compound or element is combined with others, or is broken down.

chemiosmotic hypothesis The hypothesis that ATP synthesis in mitochondria and chloroplasts occurs indirectly via proton movement across a membrane.

chiasma (plural = chiasmata) The X-shaped structure formed during meiosis by crossing over between adjacent chromatids of a pair of homologous chromosomes.

chiral Describing a molecule that does not have a plane of symmetry, such that the molecule can occur in two mirror-image forms that cannot be superimposed on each other.

chlorophyll A green pigment molecule found in plant cells and in photosynthetic protists that absorbs light energy to power photosynthesis.

chloroplast A chlorophyll-containing organelle in plant cells in which photosynthesis occurs. Also the location of amino acid, fatty acid, purine, and pyrimidine synthesis.

chromatid One of the daughter strands of a chromosome that has recently been copied (during mitosis or meiosis), and which is still connected to the other daughter strand. Upon separation from each other, the chromatids become chromosomes.

chromatin The entire complex of protein and DNA that makes up a eukaryotic chromosome.

chromosome A single long molecule of DNA and any associated proteins (e.g., histone proteins in eukaryotes).

chromosome theory of inheritance The theory that Mendel's rules of genetics can be explained by the independent segregation of homologous chromosomes during meiosis.

cilia (singular = cilium) Short, numerous filamentous projections of some eukaryotic cells used to move the cell and/or to move fluid or particles across a stationary cell.

cisterna (plural = cisternae) One of the flattened, membrane-bound compartments of the Golgi apparatus.

cleavage Rapid cell division, without production of new cytoplasm, which is seen only in early embryonic development in animals. Cleavage transforms a zygote into a similar-sized ball of tiny cells (a blastula).

cleavage furrow A pinching-in of the cell membrane that occurs as the cytoplasm of an animal cell begins to divide; one of the final events of cell division. (See also cell plate.)

climax community The final stage of an ecological community that develops from ecological succession.

clitoris A small rod of erectile tissue in the external genitalia of female mammals. Forms from the same embryonic tissue as the male penis and has a similar function in sexual arousal.

clonal expansion Rapid cell division by a particular T cell or B cell of the immune system in response to a particular antigen. This process produces a large population of descendant cells to combat the infection.

clonal-selection theory The dominant theory for the development of acquired immunity in vertebrates, it proposes that the immune system retains a vast pool of inactive lymphocytes, each of which has a unique receptor for a unique antigen. Lymphocytes that encounter their particular antigen are stimulated to divide (e.g., they are selected and cloned), producing daughter cells that combat infection and confer immunity.

CNS (central nervous system) The brain and spinal cord of vertebrate animals. (See also peripheral nervous system.)

cochlea A coiled, fluid-filled tube in the inner ear of mammals, birds, and crocodilians. Contains nerve cells that detect sounds of different pitches.

codominance An inheritance pattern in which heterozygotes have a phenotype that is a combination of the two different alleles. Example: pink four o'clock flowers (red + white alleles); AB blood type (A + B alleles).

codon A sequence of three nucleotides of DNA or RNA that codes for a certain amino acid, or initiates or terminates protein synthesis.

coefficient of relatedness A measurement of how closely two individuals are related. Calculated as the probability that an allele in two individuals is inherited from the same ancestor. Often used to estimate the effects of kin selection.

coenzyme A non-protein molecule or ion that is a required co-factor for an enzyme-catalyzed reaction. Often transfers or receives electrons or functional groups.

coevolutionary arms race A pattern of evolution seen in parasites and their hosts, in which the host evolves a defense to the parasite, the parasite then evolves a counter-defense, and so on.

cohesion A phenomenon seen in some liquids, such as water, in which attractive forces between molecules cause the liquid molecules to cling together and resist disruption—often through hydrogen bonds.

cohesion-tension theory The theory that water movement upward through plant vascular tissues is due to transpiration (loss of water from leaves), which pulls a cohesive column of water upward.

collenchyma cell An elongated type of plant cell with cell walls thickened at the corners that provides support to growing plant parts; usually found in strands along leaf veins and stalks.

complementary base pairing The specific pairing that occurs between nitrogenous bases of nucleic acid, such that each base will form hydrogen bonds only with one other base. Adenine pairs only with thymine (in DNA) or uracil (in RNA) and guanine pairs only with cytosine. This allows accurate replication of DNA and RNA sequences.

complementary DNA (cDNA) DNA created in the lab from an RNA transcript, using reverse transcriptase. Corresponds to a certain gene, but lacks introns.

compound eye A type of arthropod eye formed of many independent light-sensing columns (ommatidia). (See also camera eye, ommatidium.)

condensation reaction A class of chemical reactions involving bonding of two molecules by removal of an –OH from one subunit and an -H from another. (The –OH and –H combine to form a free water molecule.) Also called a dehydration reaction. The reverse reaction is hydrolysis.

conduction Direct transfer of heat between two objects that are in physical contact. (See also convection, radiation, evaporation.)

cone A photoreceptor cell with a cone-shaped outer portion that is particularly sensitive to bright light of a certain color. Found in eyes of vertebrates and some other animals. (See also rod.)

connective tissue A class of animal tissues consisting of scattered cells in a liquid, jellylike, or solid extracellular matrix. Includes bone, cartilage, tendons, ligaments, and blood. (See also epithelial tissue, muscle tissue, nervous tissue.)

constitutive mutant A mutant in which certain genetic loci are constantly transcribed due to flaws in gene regulation.

continuous strand In DNA replication, the strand of new DNA that is synthesized in one continuous piece, with nucleotides added to the 3′ end of the growing molecule. Also called leading strand. (See also lagging strand.)

convection Transfer of heat by movement of large volumes of a gas or a liquid. (See also conduction, radiation, evaporation.)

convergent evolution Evolution of similar traits in distantly related organisms due to adaptation to similar environments and a similar way of life. Often produces analogous traits.

core transcription complex A set of eukaryotic proteins that act together to remodel chromatin and initiate transcription. Includes RNA polymerase II and proteins bound to it, as well as basal transcription factors (proteins that bind to the promoter).

cornea The transparent sheet of connective tissue at the very front of the eye in vertebrates and some other animals. Protects the eye and focuses light.

corolla All of the petals of a flower.

cortical granules Small vesicles full of enzymes that are found in the outer region (cortex) of an egg cell. Involved in formation of the fertilization envelope after fertilization.

cotyledon A small leaf of a plant embryo. Used for storing and digesting nutrients and/or for early photosynthesis. Embryos of flowering plants have either one or two cotyledons. (See also hypocotyl, radicle.)

countercurrent heat exchanger A specialized network of blood vessels that recirculates body heat within a certain part of the body. A type of countercurrent multiplier.

countercurrent multiplier An arrangement of parallel tubes carrying fluids in opposite directions, maximizing transfer of heat or a soluble substance between the tubes. Examples: countercurrent heat exchanger of a dolphin flipper; osmotic multiplier of a vertebrate kidney.

covalent bond A type of molecular bond in which two atoms share one pair of electrons.

crassulacean acid metabolism (CAM) A variant of photosynthesis in which CO_2 is stored in sugars at night when stomata are open, and then released to feed the Calvin cycle during the day when stomata are closed. Helps reduce water loss and oxygen loss by photorespiration in hot, dry environments.

crista (plural = cristae) An invagination of the inner membrane of a mitochondrion. Location of the electron transport chain and ATP synthase.

crossing over The exchange of segments of non-sister chromatids between a pair of homologous chromosomes that occurs during meiosis.

CTL (cytotoxic T lymphocyte) A T cell that destroys infected cells and cancer cells. Cytotoxic T lymphocytes are descendants of an activated $CD8^+$ T cell. Also called killer T cell. (See also helper T cell, B cell.)

cuticle A protective coating secreted by the outermost layer of cells in an animal or plant. Examples: the waxy coating of plants; the thin fibrous coating of earthworms; the hard outer layer of insects.

cyclin-dependent kinase (cdk) A protein kinase that is active only when bound to a cyclin. Involved in control of the cell cycle.

cytokine Generally, any substance that stimulates cell division. Many cytokines are secreted by macrophages and helper T cells during an immune response; these have a variety of effects, including stimulation of leukocyte production, tissue repair, and fever.

cytokinesis Division of the cytoplasm to form two daughter cells. Typically occurs immediately after division of the nucleus by mitosis or meiosis.

cytoplasm All of the contents of a cell, excluding the nucleus of eukaryotic cells.

cytoplasmic determinant A regulatory molecule that is distributed unevenly in the cytoplasm of the egg cells of many animals and that directs the differentiation of embryonic cells.

cytoskeleton A network of protein fibers embedded in the cytoplasm of eukaryotic cells that are involved in cellular shape, support, locomotion, and transport of materials within the cell. The cytoskeleton includes microtubules, intermediate filaments, and microfilaments.

cytotoxic T lymphocyte (CTL) A T cell that destroys infected cells and cancer cells. Cytotoxic T lymphocytes are descendants of an activated $CD8^+$ T cell. Also called killer T cell. (See also helper T cell, B cell.)

Darwinian fitness The ability of an organism to produce a larger number of surviving fertile offspring than other individuals in the same population.

decomposer In an ecosystem, a species that feeds on the dead bodies of other organisms. Decomposers include various bacteria and fungi. (See also consumer, primary producer.)

decomposer food web An ecological network of detritus (dead tissues), decomposers that eat detritus, and predators and parasites of the decomposers. (See also grazing food web.)

demography The study of factors that determine the size and structure of populations through time.

dendrite A short extension from a neuron's cell body that receives neurotransmitters from other neurons.

deoxyribonucleic acid (DNA) A double-stranded, double-helix molecule of deoxyribonucleotides that contains the genetic information of a cell; each molecule contains the sugar deoxyribose, a phosphate group, and one of four possible nitrogenous bases: adenine, guanine, cytosine, or thymine.

deoxyribonucleotide A nucleotide consisting of the five-carbon sugar deoxyribose, a phosphate group, and one of four nitrogen-containing bases (adenine, cytosine, guanine, or thymine). Can be polymerized to form deoxyribonucleic acid (DNA).

depolarization A change in membrane potential from its resting negative state to a less negative, or even positive, state.

depolarization A reduction or reversal of resting potential.

detergent Any amphipathic molecule that forms micelles in water and which can cleanse by suspending hydrophobic molecules (such as oily dirts) in water. Most detergents have a long hydrocarbon tail attached to a hydrophilic sulfate head.

determinant An instruction for cell differentiation, inherited only by certain cells in embryos. Once thought to be genes, but now known to be molecules distributed unevenly in the cytoplasm of the egg cell. (See cytoplasmic determinant.)

determination The irreversible commitment during development of a cell to becoming a particular cell type (e.g., liver cell, brain cell).

detritus A loose mixture of plant litter, dead animals, and soil particles. Consumed by decomposers.

deuterostomes A lineage of animals that share a pattern of embryological development including radial cleavage, formation of the anus before the mouth, and formation of the coelom by pinching off of layers of mesoderm from the gut. Includes echinoderms and chordates.

developmental homology Similarities in embryonic form, or in the fate of embryonic tissues, that are due to inheritance from a common ancestor.

developmental plasticity Variation in development or growth in response to environmental conditions.

diastole The portion of the heartbeat cycle during which the heart is relaxed.

dideoxy sequencing A laboratory technique of determining the exact nucleotide sequence of a sequence of DNA or RNA. Relies on the use of dideoxynucleotide triphosphates (ddNTPs), which lack a 3′ hydroxyl group and thus terminate transcription of DNA or RNA.

differential gene expression Expression of different genes in different cell types.

differentiation The process by which a cell becomes a particular cell type (e.g., liver cell, brain cell) by differential gene expression.

diffusion Spontaneous movement of molecules and ions along a concentration gradient, from a region of high concentration to a region of low concentration.

dimer An association of two molecules. Example: Some proteins form a dimer of two similar subunits.

diploblast An animal whose body develops from two basic embryonic cell layers, the ectoderm and the endoderm. Cnidarians and ctenophores are diploblasts. (See also triploblast.)

diploid Having two sets of chromosomes. Most plants and animals are diploid.

direct sequencing A laboratory technique for discovery and study of unknown bacteria and archaea that will not grow easily in the lab. Relies on detecting and amplifying copies of their DNA from samples of soil, air, or water.

directional selection A pattern of natural selection in which individuals with one particular extreme phenotype have higher fitness than individuals with average phenotypes or with the other extreme of the phenotype.

disaccharide A carbohydrate consisting of two monosaccharides (sugar subunits) linked together. Most disaccharides are soluble, sweet crystals. Examples: fructose, sucrose.

discontinuous strand In DNA replication, the strand of new DNA that is synthesized discontinuously in a series of short pieces that are later joined together. Each segment is synthesized in the 5′ to 3′ direction. Also called lagging strand. (See also leading strand.)

discrete trait A phenotypic trait that exhibits distinctly different forms, such as green peas vs. yellow peas, rather than the smooth variation seen in quantitative traits.

disruptive selection A pattern of natural selection in which individuals with extreme phenotypes (at either end of the range of phenotypic variation) have higher fitness than individuals with an average phenotype.

DNA (deoxyribonucleic acid) A double-stranded, double-helix molecule of deoxyribonucleotides that contains the genetic information of a cell; each molecule contains the sugar deoxyribose, a phosphate group, and one of four possible nitrogenous bases: adenine, guanine, cytosine, or thymine.

DNA polymerase Any of several enzymes that catalyze synthesis of DNA from deoxyribonucleotides.

domain (1) A section of a protein that has a distinctive tertiary structure. (2) A fundamental taxonomic group of organisms sharing similarities in basic cellular biochemistry, such as bacteria, archaea, or eukaryotes.

dominant Refers to an allele that determines the phenotype of a heterozygous individual. Example: The *R* allele is said to be dominant to the *r* allele if a diploid with an *Rr* genotype has the same phenotype as an individual with an *RR* genotype and this is distinct from the phenotype of an *rr* individual.

dormancy A temporary state of greatly reduced, or no, metabolic activity. Example: Many plants and insects experience a period of dormancy during the winter.

double fertilization An unusual form of reproduction seen in flowering plants, in which one sperm nucleus fuses with an egg to form a zygote, and the other sperm nucleus fuses with two polar nuclei to form the triploid endosperm.

double helix The characteristic three-dimensional shape of DNA, consisting of two antiparallel DNA strands wound around each other.

Doushantuo fossils A characteristic assemblage of microscopic fossils found in pre-Cambrian rocks of the Doushantuo geologic formation in China from around 580 million years old, including evidence of sponges, different cell types, and multicellular embryos.

dynein A motor protein that produces movement of cilia and flagella. Dynein bridges use the chemical energy of ATP to walk along the adjacent microtubule doublets.

E site The site in a ribosome into which an empty transfer RNA is shunted after addition of its amino acid to the growing peptide chain. (See also A site, P site.)

ecosystem All of the organisms that live in a geographic area, together with the non-biological components that affect or exchange materials with the organisms (e.g., climate, energy, soil, nutrients).

ectoderm One of the three basic embryonic cell layers of a triploblast animal. Forms the outer covering and nervous system. (See also endoderm, mesoderm.)

ectoparasite A parasite that lives on the outer surface of the body. Examples: lice, ticks.

ectotherm An animal that does not use internally generated heat to regulate its body temperature.

Ediacaran fauna A characteristic set of animal fossils found in various pre-Cambrian rocks around 565–544 million years old, containing sponges, jellyfish, comb jellies, and other filter-feeding, shallow-water marine animals.

effector Any structure, cell, or organ with which an animal can respond to external or internal stimuli. Usually under the control of the nervous system. Examples: muscles, hormone glands, cilia.

effector T cell T cells (lymphocytes from the thymus) that are actively involved in combating an infection; e.g., are descendants of activated T cells. Includes helper T cells and cytotoxic T lymphocytes. (See also B cell.)

egg A mature female gamete and any associated external layers (such as a shell). Larger and less mobile than the male gamete. In animals, the egg cell is also called an ovum.

electrochemical gradient The combined effects of a concentration gradient and an electrical gradient, which affects the movement of charged ions across cell membranes.

electrolyte Any compound that dissociates into ions when dissolved in water.

electron shell A set of atomic orbitals of similar energies. Arranged in roughly concentric layers around the nucleus of an atom, with electrons in outer shells having more energy than those in inner shells. Atoms with partially filled outer shells tend to form covalent bonds.

electronegativity The tendency of an atom to attract electrons toward itself.

elongation (1) The process by which messenger RNA strands lengthen during transcription. (2) The process by which polypeptide chains lengthen during translation.

embryo sac The multicellular female gametophyte of a flowering plant. It consists of a small seven- or eight-celled structure inside an ovule of a flower ovary.

embryogenesis The process by which a single-celled zygote becomes a multicellular embryo.

emerging virus Any of several pathogenic viruses that have recently become dramatically more common, often due to changes in host species or human population movements.

enantiomer One form of a chiral molecule; e.g., the left-handed form or the right-handed form. (See also chiral.)

endocrine gland An organ that secretes hormones directly into the bloodstream.

endocrine system All of the glands that produce and secrete hormones into the bloodstream.

endoderm One of the three basic embryonic cell layers of a triploblast animal. Forms the digestive tract and organs that connect to it (liver, lungs, etc.). (See also mesoderm, ectoderm.)

endoparasite Any parasite that lives inside the body. Examples: tapeworms, the malaria parasite (*Plasmodium*).

endoplasmic reticulum (ER) A network of interconnected membranous sacs and tubules found inside eukaryotic cells. This system includes rough ER, with ribosomes attached, and smooth ER.

endosperm A triploid tissue in the seed of a flowering plant that contains food for the plant embryo.

endosymbiotic theory The theory that mitochondria and chloroplasts evolved from prokaryotes that were engulfed by host cells and took up a symbiotic existence within those cells.

endotherm An animal that uses internally generated heat from metabolic reactions to maintain a body temperature higher than that of the environment.

endothermic (1) Refers to a chemical reaction that absorbs heat. This type of reaction occurs if the products have a higher potential energy than the reactants. (2) Having the ability to maintain a high body temperature using internally generated heat.

endothermy The ability to maintain a high body temperature using internally generated heat from metabolic reactions, primarily oxidation of sugars.

energy The capacity to do work or to supply heat.

enhancer A regulatory DNA sequence in eukaryotes to which certain proteins can bind, which enhances the transcription of certain genes. The affected genes may be located at a distance from the enhancer.

enterokinase An intestinal enzyme that converts trypsinogen (from the pancreas) to the active enzyme trypsin.

entropy The amount of disorder in any system, such as in a group of molecules, commonly symbolized by S.

envelope A membrane-like covering that encloses a virus particle. The envelope shields some viruses from immune attack directed against the proteins of their capsid coats.

enzyme Any molecule, usually a protein, that can catalyze a chemical reaction.

epidermis The outermost layer of cells of any multicellular organism.

epigenesis The embryological development that produces a complex body from a simple, formless embryo. (See also preformation.)

epiphyte A plant that grows on trees or other solid objects, and is not rooted in soil; it is not parasitic.

epistasis An interaction of independently inherited genes, such that alleles at one locus alter the phenotypic effect of alleles at another locus.

epithelial tissue A class of animal tissues consisting of layers of tightly packed cells that line an organ, duct, or body surface.

epitope The unique region of a particular antigen (foreign molecule) to which antibodies or lymphocytes bind.

EPSP (excitatory postsynaptic potential) A change in membrane potential at a neuron dendrite that makes an action potential more likely. Usually a depolarization. (See also IPSP.)

equilibrium potential The membrane potential at which there is no net movement of a particular ion in or out of a cell. Generally, this is the voltage at which the electrical gradient equals and opposes the concentration gradient of that ion.

ER (endoplasmic reticulum) A network of interconnected membranous sacs and tubules found inside eukaryotic cells. This system includes rough ER, with ribosomes attached, and smooth ER.

essential amino acid An amino acid that an animal cannot synthesize and must obtain in the diet. May also refer specifically to the eight essential amino acids of adult humans: isoleucine, leucine, lysine, methionine, phenylalanine, threonine, tryptophan, and valine. A category of macronutrient.

essential element A chemical element that an organism must obtain from its environment. Examples: carbon (a macronutrient), zinc (a micronutrient). (See also essential amino acid, macronutrient, micronutrient.)

essential nutrient Any chemical element or compound required for normal growth, reproduction, and maintenance of a living organism. Includes micronutrients and macronutrients.

eukaryote A member of the domain Eucarya; an organism with complex cells characterized by a variety of distinctive internal cell features, including the presence of membrane-bound organelles such as the nucleus, and the presence of introns in genes. May be unicellular (e.g., yeast) or multicellular (e.g., humans).

evaporation The energy-absorbing phase change from a liquid state to a gaseous state. Many organisms use evaporation of water as a mechanism of heat loss. (See also conduction, convection, radiation.)

evolution (1) The theory that all organisms on Earth are related by common ancestry, and that they have changed over time predominantly in response to natural selection. (2) Any change in the genetic characteristics of a population over time; especially, a change in allele frequencies.

excitatory postsynaptic potential (EPSP) A change in membrane potential at a neuron dendrite that makes an action potential more likely. Usually a depolarization. (See also IPSP.)

exon A region of a eukaryotic gene that is translated into a peptide or protein. (See also intron.)

exonuclease Any enzyme that can remove nucleotides from the end of a strand of DNA or RNA.

exoskeleton A system of hard, jointed layers secreted on the outside of the body, used for body support, protection, and muscle attachment.

exothermic Refers to a chemical reaction that releases heat. Occurs if the products have lower potential energy than the reactants.

experimental control A group of organisms in an experiment that does not receive the experimental treatment, but is identical in all other ways to the group that is receiving the experimental treatment. Also called the control group.

exponential growth A constantly accelerating increase in population size that occurs when the size of the population does not affect the growth rate. In nature, occurs only when populations are well below carrying capacity.

extensor A muscle that pulls two bones further apart from each other, such as when extending a limb or the spine. (See also flexor, antagonistic muscle group.)

F_1 generation (first filial generation) The first generation of offspring produced from a mating; i.e., the offspring of the parental generation.

F_2 generation (second filial generation) The second generation of offspring produced from a mating; the offspring of members of the F_1 generation.

Fallopian tube A narrow tube connecting the uterus to the ovary in humans, and through which the egg travels after ovulation. Fertilization and cleavage occur in the fallopian tube. In nonhuman animals, this tube is called the oviduct.

fate map A hypothetical map drawn on an embryo, showing the eventual fate of every cell in that embryo—the ultimate location and the tissues each cell will give rise to.

fatty acid A type of lipid consisting of a hydrocarbon chain bonded to a carboxyl functional group (–COOH) at one end. Used by many organisms as a store of chemical energy; a major component of animal and plant fats.

fecundity The average number of female offspring produced by a single female per unit time.

feedback inhibition Inhibition of an enzyme in a metabolic pathway by high concentrations of the product of that metabolic pathway.

fermentation Any of several metabolic pathways that allow continued production of ATP in the absence of oxygen by transferring electrons from a reduced compound (such as glucose) to an electron acceptor other than oxygen, with production of ATP by substrate-level phosphorylation.

fertilization The fusion of the nuclei of two haploid gametes to form a zygote with a diploid nucleus.

fertilization envelope A physical barrier that forms around a fertilized egg in amphibians and some other animals. Formed by an influx of water under the vitelline membrane.

fetus A later developmental stage of an embryo, usually developed sufficiently to be recognizable as a certain species. In humans, an embryo that is at least eight weeks old.

finite rate of increase The rate of increase of a population over a given period of time. Calculated as the ending population size divided by the starting population size, and symbolized by lambda (λ).

fitness The ability of an organism to produce a larger number of surviving fertile offspring than other individuals in the same population. Also called Darwinian fitness.

flagellum (plural = flagella) A long, cellular projection that undulates (in eukaryotes) or rotates (in prokaryotes) to move the cell through an aqueous environment.

flexor A muscle that pulls two bones closer together, such as when flexing a limb or the spine. (See also extensor, antagonistic muscle group.)

flower The reproductive structure of angiosperm plants. Includes a calyx, corolla, and one or more stamens or carpels.

fluorescence The spontaneous emission of light from an excited electron falling back to its normal (ground) state.

follicle An egg cell and its surrounding ring of supportive cells in a mammalian ovary.

food chain A simple pathway of energy through a few species in an ecosystem; e.g., a primary producer, a primary consumer, a secondary consumer, and a decomposer.

food web Any complex pathway along which energy moves among many species at different trophic levels of an ecosystem.

forward genetics An experimental approach for identifying genes that affect a certain phenotypic trait. Involves starting with known variations of a trait and using classical breeding experiments and genetic mapping techniques to pinpoint the location of the gene(s) affecting the trait. (See also reverse genetics.)

fossil record All of the fossils (traces of organisms that lived in the past) that have been found anywhere on Earth, and that have been formally described in the scientific literature.

fovea The small region of the vertebrate retina in which the photoreceptors are very tightly packed, producing the most acute vision.

frequency The number of waves per second traveling past a stationary point. For sound, frequency is the number of pressure waves in air per second, and is perceived as pitch. For light, frequency is the number of electromagnetic waves per second, and is perceived as color.

fruit A mature, ripened plant ovary (or group of ovaries), along with the seeds it contains and any adjacent fused parts.

functional group A small group of atoms bonded together with a characteristic shape and reactivity, and which can be attached to an organic molecule. Often contains hydrogen, nitrogen, or oxygen bonded to carbon in a specific way. Examples: hydroxyl group (–OH), amine group ($-NH_2$).

fundamental niche The ecological space that a species occupies in its habitat in the absence of competitors.

G_1 phase The phase of a cell cycle that constitutes the first part of interphase before DNA synthesis.

G_2 phase The phase of a cell cycle between synthesis of DNA (S phase) and mitosis (M phase); the last part of interphase.

galactoside permease A membrane transport protein responsible for importing lactose into bacterial cells; coded for by the gene *lacY*.

gamete A haploid reproductive cell that can fuse with another haploid cell to form a zygote. Most multicellular eukaryotes have two distinct forms of gametes: egg cells (ova) and sperm cells.

gametogenesis The production of gametes (eggs or sperm).

gametophyte The multicellular haploid phase of an organism that exhibits alternation of generations. Arises from a single spore and produces gametes. (See also sporophyte, alternation of generations.)

gap genes A class of fruit fly segmentation genes that organize embryonic cells into major groups of segments. Active in broad regions of the embryo. (See also pair-rule genes, segment polarity genes.)

gastrulation The process by which some cells on the outside of a young embryo move to the interior of the embryo, resulting in the three distinct cell layers of endoderm, mesoderm, and ectoderm.

gel electrophoresis A laboratory technique for separating molecules on the basis of size and electric charge, based on their differing rates of movement through a gelatinous substance in an electric field.

gene A section of DNA (or RNA, for some viruses) that encodes information for building a protein or a functional molecule of RNA (such as ribosomal RNA or transfer RNA). Sometimes used loosely for any section of DNA under study.

gene duplication The creation of an additional copy of a gene, typically by misalignment of chromosomes during crossing over.

gene family A set of genetic loci whose DNA sequences are extremely similar to each other, thought to have arisen by duplication of a single ancestral gene.

gene pool All of the alleles of all the genes in a population.

gene therapy The treatment of an inherited disease by replacing the defective gene(s) with normal gene(s).

genetic bottleneck A brief, extreme reduction in population size. Often reduces allelic diversity via genetic drift, which has pronounced effects in small populations.

genetic diversity The diversity of alleles in a population, species, or group of species.

genetic drift Any change in allele frequencies due to random events. Causes allele frequences to drift up and down randomly over time, and eventually can lead to the fixation or loss of alleles.

genetic homology Similarities in DNA sequences or amino acid sequences of proteins among certain organisms that are due to inheritance from a common ancestor.

genetic map A map of the relative locations of specific genes on a certain chromosome.

genetic marker Any genetic locus that can be identified and traced in populations, either by laboratory techniques or by a distinctive phenotype. Includes reporter genes.

genome All of the hereditary genetic information in an organism.

genomics The science of sequencing, interpreting, and comparing whole genomes.

genotype The entire genetic make-up of an individual; i.e., all of the alleles on every chromosome. May be used to specifically refer to the alleles of a particular set of genes under study. (See also phenotype.)

genus A taxonomic category of closely related species.

germ line cell A reproductive cell that passes its genes on to the next generation. In a multicellular body, germ line cells include the sperm and egg cells and their parent cells, spermatogonia and oogonia. (See also somatic cells.)

germ plasm theory The now-disproved theory that embryonic cells differentiate into different cell types because they inherit different genes.

germ theory of disease The theory that infectious diseases are caused by bacteria, viruses, and other microorganisms.

gestation The duration of embryonic development from fertilization to birth, in those species that have live birth.

Gibbs free-energy change A measure of the change in potential energy and entropy that occur in a given chemical reaction at a given temperature. Determines whether or not the reaction will be spontaneous. Depicted as ΔG.

gill An organ in aquatic animals that exchanges gases and other dissolved substances between the blood and the surrounding water. Typically a filamentous outgrowth of a body surface.

glucagon A protein hormone produced by the pancreas in response to low blood glucose that raises blood glucose by triggering breakdown of glycogen. (See also insulin.)

glycogen A polysaccharide that is the major form of stored carbohydrate in animals. Consists of α-glucose monomers joined end-to-end in highly branched chains.

glycolysis A series of chemical reactions that oxidize glucose to produce pyruvate and ATP. Used by all organisms as part of fermentation or cellular respiration.

glycoprotein A protein with one or more covalently bonded carbohydrate groups.

glycosylation Addition of a carbohydrate group to a molecule.

Golgi apparatus A stack of flattened membranous sacs in eukaryotic cells that processes the proteins and lipids that will be secreted or directed to other organelles.

gonad An organ that produces reproductive cells; e.g., a testis or an ovary.

granum (plural = grana) A stack of the flattened, membrane-bound thylakoid disks inside plant chloroplasts.

gravitropism The growth or movement of a plant in a particular direction in response to gravity.

gray crescent A region of an amphibian zygote that becomes visible shortly after fertilization opposite to the point of sperm entry. Eventually gives rise to the blastopore and the organizer, which determines major body axes.

grazing food web The ecological network of herbivores and the predators and parasites that consume them. (See also decomposer food web.)

green fluorescent protein (GFP) A jellyfish protein that spontaneously emits green light after stimulation. Widely used in laboratory research to mark the location of certain molecules or cells.

greenhouse gas An atmospheric gas that absorbs and reflects infrared radiation, so that heat radiated from Earth is retained in the atmosphere instead of being lost to space.

gross photosynthesis The total amount of energy captured by photosynthesis by all the plants in a given area during a given time.

gross photosynthetic efficiency A measure of the efficiency with which all the plants in a given area use the light energy available to them to produce sugars.

ground tissue A plant tissue consisting of all cells beneath the outer protective layers of epidermis and cork except for vascular tissue. May compose most of the plant's body. Contains parenchyma, collenchyma, and sclerenchyma cells.

growth factor Any of several compounds that are secreted by certain cells and which stimulate other cells to divide or to differentiate.

guard cell A specialized cell forming the border of a plant stoma. Guard cells can change shape to open or close the stoma. (See also stoma.)

habitat fragmentation The breaking up of a large region of a habitat into many smaller regions, each separated from the others by a different habitat.

hairpin A stable loop formed in an RNA molecule by hydrogen bonding between purine and pyrimidine bases on the same strand.

half-life The characteristic time taken for one-half of any amount of a particular radioactive isotope to decay.

haploid Having one set of chromosomes. Bacteria, archaea, animal gametes, plant gametophytes, and many algae are haploid.

Hardy-Weinberg principle A principle of population genetics that states that if mutation, migration, genetic drift, random mating, and selection do not occur, then allele frequencies will not change from generation to generation; in other words, evolution will not occur.

heavy chain One of two identical polypeptides that forms the base of immunoglobulin proteins (antibodies, B-cell receptors, etc.) Differences in the heavy chain determine the different classes of immunoglobulins.

helper T cell A T cell that assists with the activation of other lymphocytes. Helper T cells are descendants of an activated $CD4^+$ T cell. (See also cytotoxic T lymphocyte, B cell.)

hemolymph The circulatory fluid of animals with open circulatory systems in which the fluid is not confined to blood vessels, such as insects.

herbivore An animal that primarily eats plants, and rarely or never eats meat. (See also carnivore, omnivore.)

heredity The transmission of traits from parents to offspring via genetic information.

heritability The proportion of total variation in a trait that is due to the genetic variation in a certain population that was studied in a certain environment. A measured heritability value cannot be generalized to other populations or to other environments.

heritable Refers to traits that are influenced by hereditary genetic material (DNA, or RNA for some viruses).

heterokaryon A hybrid cell that contains two genetically distinct nuclei. Occurs naturally in many fungi; also used as a research tool.

heterokaryotic Containing nuclei that are genetically distinct. Occurs naturally in many fungi.

heterotroph Any organism that cannot synthesize reduced organic compounds from inorganic sources, and instead obtains them by eating other organisms. Some bacteria, some archaea, and virtually all fungi and animals are heterotrophs.

heterozygote advantage A pattern of natural selection in which heterozygotes have greater fitness than either parental homozygote. Also called heterozygote superiority.

heterozygous Having two different alleles of a certain gene.

homeosis The phenomenon that certain mutations can result in extra, fully formed segments or appendages normally found elsewhere in the body, replacing the structures usually formed at that location.

homeostasis The maintenance of a relatively constant physical and chemical environment within an organism.

homeotic complex A set of eight fruit fly genes, closely linked on the same chromosome, that are active in different regions of the fly embryo, and that specify the identity of particular body segments.

hominid Any member of the Family Hominidae of bipedal, tailless great apes. Includes human beings and extinct related forms, chimpanzees, gorillas, and orangutans. This taxon was once restricted to humans and extinct related forms, but has recently been broadened.

homologous chromosomes Chromosomes of the same type, with the same genes in the same locations. Also called homologs.

homologous trait Any trait showing marked similarity between different species due to inheritance from a common ancestor. (See also analogous trait.)

homology Similarity between organisms that is due to genetic inheritance from a common ancestor.

homozygous Having two identical alleles of a certain gene.

hormone Any molecule used as a chemical signal between cells of a multicellular organism. Can trigger pronounced responses in certain target cells at very low concentrations.

human chorionic gonadotropin A glycoprotein hormone produced by the human placenta from about week 3 to week 14 of pregnancy. Maintains the corpus luteum (the ruptured follicle in the ovary), which produces hormones that maintain the uterine lining.

hybrid The offspring of parents from two different strains, populations, or species.

hybrid zone A geographic area where interbreeding occurs between two species, sometimes producing fertile hybrid offspring.

hydrocarbon A molecule that contains only hydrogen and carbon, usually with the carbon atoms bonded covalently to form chains or rings.

hydrogen bond A weak electrical attraction between two molecules, due to partially positive hydrogen atoms of one molecule attracting partially negative atoms of the other molecule.

hydrolysis A type of chemical reaction in which a compound reacts with water to break down into smaller molecules. In biology, most hydrolysis reactions involve polymers breaking down into monomers. The reverse reaction is called a condensation, or dehydration, reaction.

hydrophilic Having the property of mixing readily with water. Hydrophilic compounds are typically polar compounds, with charged or electronegative atoms.

hydrophobic Having the property of not mixing readily with water. Hydrophobic compounds are typically nonpolar compounds, without charged or electronegative atoms, and often contain many C–H bonds.

hydrostatic skeleton A system of body support involving fluid-filled compartments that can change in shape, but cannot easily be compressed.

hypersensitive response The rapid death of a plant cell that has been infected by a pathogen. Thought to be a strategic mechanism that protects the rest of the plant.

hypertonic Having a greater solute concentration, and therefore a lower water concentration, relative to another solution.

hypha (plural = hyphae) One of the strands of a fungal mycelium (the mesh-like body of a fungus). Also found in some protists.

hypocotyl The stem of a mature plant embryo; the region between the cotyledon (embryonic leaf) and the radicle (embryonic root).

hypothesis A proposed explanation for a phenomenon or for a set of observations.

hypotonic Having a lower solute concentration, and therefore a higher water concentration, relative to another solution.

immunoglobulin A class of Y-shaped protein capable of binding to specific antigens (foreign molecules) and responsible for acquired immunity. Each immunoglobulin contains two light chains and two heavy chains. Examples: the five types of antibodies (IgA, IgD, IgE, IgG, IgM) and B-cell receptors.

imperfect Refers to flowers that contain either male parts (stamens) or female parts (carpels), but not both. (See also perfect.)

inbreeding Mating between closely related individuals, or within a genetically homozygous strain.

inbreeding depression A loss of fitness that occurs when homozygosity increases in a population.

incomplete dominance An inheritance pattern in which heterozygotes have a phenotype that is a blend or a combination of both homozygote phenotypes. Examples: pink four o'clock flowers (red + white alleles); AB blood type (A + B alleles).

indicator plate A laboratory technique for detecting mutant cells by observing color changes due to cells cleaving (or not cleaving) the bonds in a pigmented substance.

inducer A molecule that triggers transcription of a specific gene.

induction The process by which one embryonic cell, or group of cells, alters the differentiation of neighboring cells.

inflammatory response A general response of the immune system seen in most cases of infection or tissue injury, in which the infected or injured tissue becomes swollen, red, warm, and painful. Part of the innate immune response.

infrared light Light with a wavelength longer than visible red light.

infrasound Sound frequencies that are too low for humans to hear—lower than about 20 Hertz.

inheritance of acquired characters The now-disproved theory that traits acquired during the lifetime of an organism due to non-inherited causes (injuries, muscular training, learning of habits or skills) will be passed genetically to its offspring.

inhibitory postsynaptic potential (IPSP) A change in membrane potential of a neuron dendrite that makes an action potential less likely. Usually a hyperpolarization. (See also EPSP.)

innate behavior Behavior that is inherited genetically and does not have to be learned.

innate immunity A set of nonspecific defenses against pathogens that occurs even without previous exposure to the pathogen and that does not involve antibodies. Includes responses by mast cells, neutrophils, and macrophages, and typically results in an inflammatory response. (See also innate immunity, inflammatory response.)

insulin A protein hormone produced by the pancreas in response to high levels of glucose (or amino acids) in blood. Enables cells to absorb glucose and coordinates synthesis of fats, proteins, and glycogen. (See also glucagon.)

integrator A component of an animal nervous system that evaluates sensory information and triggers appropriate responses.

intermediate filaments Long, thin cellular fibers composed of thin filaments of various protein polymers wound into thicker cables that form part of a cell's cytoskeleton. Help maintain cell shape and anchor the nucleus and other organelles.

internode The section of a plant stem between two nodes (sites where leaves attach).

interphase The part of the cell cycle during which no cell division occurs. Includes the G_1 phase, the S phase, and the G_2 phase.

intron A region of a eukaryotic gene that is transcribed into RNA but is later excised from the messenger RNA transcript before translation into protein.

ion An atom or molecule that has gained or lost electrons and carries an electric charge.

ionic bond An association of atoms in which an electron is completely transferred from one atom to another. The atoms are not bound by covalent bonds, but typically remain associated due to their electric charges.

ionophore Any compound that increases membrane permeability for a certain ion. Many ionophores are membrane proteins that form a channel across the membrane, or bind to the ion and carry it through the membrane.

IPSP (inhibitory postsynaptic potential) A change in membrane potential at a neuron dendrite that makes an action potential less likely. Usually a hyperpolarization. (See also EPSP.)

iris A ring of pigmented muscle just inside the vertebrate eye that contracts or expands to allow different amounts of light into the eye.

isotonic Having the same solute concentration and water concentration relative to another solution.

isotope Any of several forms of an element that have the same number of protons but differ in the number of neutrons.

junctional diversity Genetic diversity created by variations in the joining of gene segments; occurs during gene recombination in lymphocytes of the immune system.

karyotype The distinctive appearance of all of the chromosomes in an individual, including the number of chromosomes and their length and banding patterns (after staining with dyes).

keystone species A species that has an exceptionally great impact on the other species in its ecosystem relative to its abundance.

kin selection A form of natural selection that favors traits that increase survival or reproduction of an individual's kin at the expense of the individual. Can result in evolution of altruistic behaviors.

kinesin A motor protein that uses the chemical energy of ATP to transport vesicles, particles, or chromosomes along microtubules.

kinetic energy The energy of motion.

kinetochore The attachment that forms during cell division between the microtubule of a spindle fiber and a chromosome. It contains motor proteins that move the chromosome along the microtubule.

knock-out mutant An organism homozygous for a mutation that eliminates the function of a specific gene.

Koch's postulates Four criteria used to determine whether a putative infectious agent causes a particular disease.

Krebs cycle A series of chemical reactions in cellular respiration that take place in the mitochondria. These reactions break down acetyl CoA, the product of glycolysis, to CO_2 and produce ATP and reduced compounds to feed into the electron transport chain.

***lac* operon** The operon in *E. coli* that includes genes responsible for metabolism of lactose.

lacY The genetic locus that codes for galactoside permease, a membrane transport protein that imports lactose into bacterial cells.

lagging strand In DNA replication, the strand of new DNA that is synthesized discontinuously in a series of short pieces that are later joined together. Each segment is synthesized in the 5′ to 3′ direction. Also called discontinuous strand. (See also leading strand.)

lateral bud A bud on a plant stem that is capable of producing a new side branch.

lateral gene transfer Transfer of DNA between two different species, especially distantly related species. Commonly occurs among bacteria and archaea via plasmid exchange; also can occur in eukaryotes via viruses and some other mechanisms.

lateral root A plant root extending from another, older root.

leading strand In DNA replication, the strand of new DNA that is synthesized in one continuous piece with nucleotides added to the 3′ end of the growing molecule. Also called continuous strand. (See also lagging strand.)

leukocyte Any of several types of immune system cells that circulate in the blood or lymph. Also called white blood cells. Examples: neutrophils, macrophages, B cells, T cells, etc.

lichen A symbiotic association of a fungus and a photosynthetic alga.

life cycle The sequence of developmental events and phases that occurs as organisms grow, mature, and reproduce.

ligand-gated channel An ion channel that opens or closes in response to the presence of a certain molecule. (See also voltage-gated channel.)

ligation A chemical reaction that bonds two molecules together; ligation often refers to the bonding together of DNA nucleotides.

light chain One of two identical short polypeptides that, together with two identical longer polypeptides called heavy chains, form a Y-shaped antibody molecule that can bind a specific antigen. (See also heavy chain.)

linkage A physical association between two genes because they are on the same chromosome; also, the inheritance patterns resulting from this association.

lipase A fat-digesting enzyme produced by the pancreas and secreted into the intestine.

lipid A subtance that is hydrophobic, and thus does not dissolve in water but dissolves well in nonpolar organic solvents. Lipids include fats, oils, phospholipids, and waxes.

lipid bilayer Two adjacent layers of phospholipid molecules, with the hydrophobic tails oriented toward the inside of the two layers and the hydrophilic heads oriented toward the outside; forms spontaneously when phospholipids are mixed in water. A lipid bilayer is the fundamental component of cell membranes.

liposome A tiny spherical structure composed of a phospholipid bilayer membrane and formed in vitro.

locus (plural = loci) A gene's physical location on a chromosome.

loss-of-function mutant A mutation that eliminates the function of a specific gene.

lumen The interior space of any hollow organ.

lung One of two respiratory organs used for gas exchange between blood and air.

lymph The mixture of fluid and lymphocytes that circulate through the ducts and lymph nodes of the lymphatic system in vertebrates.

lymphatic system In vertebrates, a network of thin-walled tubes that collects excess fluid from body tissues and returns it to the veins of the circulatory system. Includes lymph nodes, which filter the fluid and check it for infection.

lymphocyte A type of leukocyte (white blood cell) that circulates through blood and through the lymphatic system, and that is responsible for the development of acquired immunity. Includes B cells and T cells.

lysogeny A type of viral replication in which the viral DNA is inserted into the host's chromosome, remaining there indefinitely and passively replicating whenever the host cell divides. (See also lytic growth.)

lysosome A small organelle in an animal cell that contains acids and enzymes, which can digest large molecules.

lytic growth A type of viral replication in which new virus particles are made inside a host cell and eventually burst out of the cell, killing it in the process. (See also lysogeny.)

M phase The phase of the cell cycle during which mitosis occurs. Includes mitosis and cytokinesis.

M-phase promoting factor (MPF) A complex of two proteins—kinase cdc2 and cyclin—that causes cells to enter the M phase (mitosis) of the cell cycle. (Also called mitosis promoting factor.)

macromolecule Any very large molecule. Common macromolecules include proteins, nucleic acids, and polysaccharides.

macronutrient An essential nutrient required in large quantities. Usually a major component of many organic molecules. Examples: carbon, oxygen, nitrogen. (See also micronutrient.)

malignant tumor A tumor that is actively growing and is disrupting local tissues, and/or is spreading to other organs. A cancer consists of one or more malignant tumors. (See also benign tumor.)

mass extinction Rapid extinction of a unusually large number of diverse evolutionary groups across a wide geographic area. May occur due to sudden and extraordinary environmental changes. (See also background extinction.)

mass number The total number of protons and neutrons in an atom.

maternal effect inheritance A pattern of inheritance in which an individual's phenotype is determined by its mother's genotype. Common in embryological development, in which egg components made by the mother influence development of the offspring.

meiosis A type of cell division in which one diploid parent cell produces four haploid reproductive cells (gametes). During meiosis, chromosome pairs synapse and can exchange genes via crossing over.

membrane potential A difference in electric charge across a cell membrane; a form of potential energy. (Also called membrane voltage.)

membrane voltage A difference in electric charge in the inside of a cell compared to the outside. A form of potential energy. (Also called membrane potential.)

memory cell A type of lymphocyte responsible for maintenance of immunity for years or decades after an infection. Memory cells are descendants of B or T cells activated during a previous infection.

menstruation The periodic shedding of the uterine lining through the vagina of a female if fertilization of an egg does not occur.

meristem A group of undifferentiated plant cells that can produce cells that differentiate into specific adult tissues.

mesoderm One of the three basic, embryonic cell layers of a triploblast animal. Forms the middle tissues between skin and gut: muscles, internal organs, bones. (See also ectoderm, endoderm.)

messenger RNA (mRNA) An RNA molecule that carries encoded information, transcribed from DNA, for the synthesis of one or more proteins.

meta-analysis An analysis that combines and compares the results of many smaller, previously published studies.

metabolic pathway A series of distinct chemical reactions that build up or break down a particular molecule. Often each reaction is catalyzed by a different enzyme.

metabolic rate The total energy consumption of all the cells of an individual. For aerobic organisms, this rate is often measured as the amount of oxygen consumed per hour.

metabolism All the chemical reactions occurring in a living cell or organism.

metamorphosis A dramatic change from the larval to the adult form of an animal.

metaphase A stage in cell division (mitosis or meiosis) during which chromosomes line up in the middle of the cell.

metaphase plate The imaginary plane along which chromosomes line up during metaphase of cell division (mitosis or meiosis).

metapopulation A population of a single species that is divided into many smaller populations.

metastasis The process by which cancerous cells leave the primary tumor and establish additional tumors elsewhere in the body.

micelle A small droplet of similar molecules clumped together in a solution.

microbe Any microscopic organism. Includes bacteria, archaea, and various tiny eukaryotes.

microbiology The study of microscopic organisms, such as bacteria, archaea, and various tiny eukaryotes.

microfilament A long, thin fiber composed of two intertwined strands of polymerized actin that is involved in several types of cell movement.

micronutrient An essential nutrient required in very small quantities. Usually is a cofactor for an enzyme. Examples: zinc, vitamin B_{12}. (See also macronutrient.)

micropyle The tiny pore in a plant ovule through which the pollen tube reaches the embryo sac.

microtubule A long, tubular polymer of protein subunits called α-tubulin and β-tubulin. Involved in several types of cell movement and transport of materials within the cell.

microvilli (singular = microvillus) Tiny protrusions from the surface of an epithelial cell that increase the surface area for absorption of substances. Common in certain animal organs, such as the vertebrate kidney and intestine.

migration (1) A predictable seasonal movement of organisms from one geographic location or habitat to another. (2) In population genetics, movement of individuals from one population to another, and the resulting contribution of their alleles to the gene pool.

mitochondrion (plural = mitochondria) An organelle of eukaryotic cells that contains the enzymes that catalyze the reactions of aerobic respiration, oxidizing carbohydrates and fatty acids and producing ATP (along with CO_2, water, and heat).

mitochondrial matrix The solution inside the inner membrane of a mitochondrion. Contains the enzymes of the Krebs cycle.

mitosis Nuclear division in eukaryotes producing two daughter nuclei that are genetically identical to the parent.

mitotic spindle An array of microtubules formed during cell division that moves chromosomes to opposite sides of the cell.

model organism Any organism selected for intensive scientific study based on particular features that make it easy to work with (e.g., body size, life span), and with the idea that the findings will also apply to many other species.

molarity The number of moles of a dissolved solute in one liter of solution; a unit of concentration.

mole 6.022×10^{23} molecules of a substance. This number of atoms will weigh the same as the molecular weight of that substance expressed in grams.

molecular chaperone A class of protein that facilitates the folding of newly synthesized proteins.

molecular formula A notation system that indicates the numbers and types of atoms in a molecule. Example: The molecular formula of water is H_2O; methane is CH_4.

molecular phylogeny A phylogeny entirely derived from the comparison of sequences of amino acids or DNA.

molecule Two or more atoms held together by covalent bonds.

monomer A small molecular subunit that can bond to other subunits to form long macromolecules, or polymers. Examples: monosaccharides, nucleotides, amino acids.

monophyletic group An evolutionary unit that includes an ancestral population, all of its descendants, and only its descendants.

monosaccharide A single sugar monomer. Formally, a small carbohydrate of the chemical formula $(CH_2O)_n$ that cannot by hydrolyzed to form any smaller carbohydrates. Examples: glucose, galactose, ribose.

morphogenesis A process of embryologic development during which cells become organized into recognizable tissues, organs, and other structures.

morphospecies concept The concept that species can be defined as groups with measurably different anatomical features.

motif A domain (section of a protein with a distinctive tertiary structure) found in many different proteins. Examples: helix-turn-helix motif, zinc-finger motif.

motor neuron A nerve cell that carries signals from the central nervous system (brain and spinal cord) to an effector, such as a muscle or gland.

motor protein Any protein whose major function is to convert the chemical energy of ATP into motion.

MPF (M-phase promoting factor) A complex of two proteins (protein kinase cdc2 and cyclin) that causes cells to enter the M phase (mitosis) of the cell cycle. Also called mitosis promoting factor.

mRNA (messenger RNA) An RNA molecule that carries the encoded information, transcribed from DNA, for building a protein.

multiple allelism The occurrence of more than two alleles of a gene in a given population.

muscle fiber A single muscle cell.

muscle tissue A class of animal tissue consisting of bundles of long, thin contractile cells (muscle fibers).

mutant An individual that carries a mutation, particularly a new or rare mutation.

mutation Any change in the hereditary material of an organism (DNA for most organisms, RNA for some viruses).

mutualism A type of symbiotic relationship between two species that is beneficial to both species.

mycelium (plural = mycelia) A mass of filaments (hyphae) that form the body of a fungus. Also found in some protists.

mycorrhiza (plural = mycorrhizae) A mutualistic association between certain fungi and most vascular plants, sometimes visible as nodules or nets in or around plant roots.

myelination Electrical insulation of nerve axons by myelin, a lipid. The myelin is contained in the cell membranes of neighboring cells, which wrap around the axons.

myoblast Embryonic cells that are committed to becoming muscle cells.

MyoD A regulatory protein involved in differentiation of muscle cells during embryological development. It enhances transcription of muscle-specific genes.

myofibril A bundle of strands of contractile proteins organized into repeating units called sarcomeres. It is found in vertebrate heart and skeletal muscle.

myosin A eukaryotic protein that polymerizes to form thick filaments that are used in muscle contraction and intracellular movement.

natural selection One of the major mechanisms of evolution, this is the process by which individuals with certain heritable traits tend to produce more surviving offspring than individuals without those traits, resulting in a change in the genetic make-up of the population.

nectar The sugary fluid produced by flowers to attract and reward pollinating animals.

nectary The nectar-producing gland at the base of a flower.

negative control A type of gene regulation in which a repressor protein binds to a control sequence in the DNA and prevents transcription. (See also positive control.)

negative result An experimental result that does not show a predicted difference between two groups or conditions.

nephron One of the tiny tubes within the vertebrate kidney that filters blood and concentrates salts to produce urine.

nervous tissue A class of animal tissue consisting of nerve cells (neurons) and various supportive cells.

net primary productivity The amount of new organic material produced by all the plants (or other autotrophs) in a given area over a given time.

neuron A nerve cell; a cell that is specialized for the transmission of nerve impulses. Typically, it has dendrites, a cell body, and a long axon that forms synapses with other neurons.

neurotransmitter A molecule that conveys information from one neuron to the next, or from a neuron to a muscle or gland. Neurotransmitters are released from the end of an axon and diffuse a very short distance to the next cell, in which they can trigger an action potential. (See also hormone.)

niche The particular set of habitat requirements of a certain species, and the role it plays in its ecosystem.

niche differentiation A phenomenon in which competing species use different ecological niches to reduce competition.

nitrogen cycle The pattern of circulation of nitrogen between living organisms and the environment (e.g., oceans, atmosphere).

nitrogen fixation The incorporation of atmospheric nitrogen (N_2) to make ammonia (NH_3), which can then be incorporated into organic compounds.

node (1) Any small thickening, e.g., a lymph node; (2) The part of a stem where leaves or leaf buds are attached.

nondisjunction An error that can occur during meiosis or mitosis in which both homologous chromosomes of a pair move to the same side of the dividing cell. One daughter cell receives two copies of this chromosome and the other daughter cell receives none.

notochord A defining feature of chordate animals, this is a long, gelatinous rod extending down the back of a chordate embryo and providing support.

nuclear envelope The double-layered membrane enclosing the nucleus of a eukaryotic cell.

nuclear matrix The protein scaffolding inside a eukaryotic nucleus that defines its shape and holds the chromosomes in place.

nuclear pore An opening in the nuclear envelope that connects the inside of the nucleus with the cytoplasm, and through which molecules can pass, such as messenger RNA and some enzymes.

nuclear pore complex A large complex of dozens of proteins lining a nuclear pore, defining its shape and transporting substances through the pore.

nucleolus The structure in a eukaryotic nucleus where ribosomal RNA processing occurs and ribosomal subunits are assembled.

nucleosome A repeating, bead-like structure of eukaryotic chromosomes, consisting of about 200 nucleotides of DNA wrapped twice around eight histone proteins.

nucleotide A monomer that can be polymerized to form the nucleic acids DNA or RNA. One nucleotide consists of a five-carbon sugar (ribose or deoxyribose), a phosphate group, and one of several possible nitrogen-containing bases.

nutrient Any substance needed in an organism's diet for normal growth, maintenance, and reproduction.

Okazaki fragments Short fragments of DNA produced during DNA replication; pieces of the lagging strand.

oligonucleotide A short nucleic acid molecule, containing fewer than about 25 nucleotides.

ommatidium (plural = ommatidia) One of the light-sensing columns of an arthropod's compound eye.

omnivore An animal whose diet regularly includes both meat and plants. (See also carnivore, herbivore.)

one-gene, one-enzyme hypothesis The proposal that each gene is responsible for making one (and only one) certain protein, and that most of these proteins are enzymes that catalyze specific reactions. Many exceptions to this hypothesis are now known.

operon A region of bacterial DNA that codes for a series of functionally related genes.

opsin One of several proteins involved in animal vision. A particular opsin joins with retinal to form a light-detecting pigment (called rhodopsin in rod cells).

organ Several animal tissues organized into a functional and structural unit. Examples: lung, liver, heart, skin.

organelle Any discrete structure in the cytoplasm of a eukaryotic cell. Examples: nucleus, ribosome, mitochondrion, etc.

organic Any compound containing carbon and hydrogen, and usually containing carbon-carbon bonds. Organic compounds are widely used by living organisms.

organizer A region of an amphibian embryo (around the upper side of the blastopore) that can organize the development of an entire embryo, even when transplanted to an unusual location.

organogenesis A stage of embryonic development in vertebrate embryos, during which major organs develop from the three embryonic cell layers. Organogenesis occurs just after gastrulation.

osmoregulation The process by which a living organism controls the concentration of water and salts in its body.

osmosis Diffusion of water across a selectively permeable membrane from regions of high water concentration (or low solute concentration) to regions of low water concentration (or high solute concentration).

osmotic stress A condition in which there are abnormal concentrations of water and salts in an organism's cells or tissues.

outcrossing Reproduction by fusion of the gametes of different individuals, rather than self-fertilization. Typically refers to plants.

ovary The egg-producing organ of a female animal, or the seed-producing structure of the female part of a flower.

oviparous Refers to animals that lay eggs, rather than giving live birth. (See also viviparous.)

ovulation The process by which an egg cell (ovum) is released from the ovary of a female vertebrate.

ovule A structure inside a flower ovary that produces the female gametophyte, and eventually (if fertilized), becomes a seed.

oxidation The loss of electrons during a redox reaction, either by donation of an electron to another atom or by the shared electrons in covalent bonds moving farther away from the atomic nucleus; results in a loss of potential energy.

oxidative phosphorylation Production of ATP molecules from the redox reactions of an electron transport chain, starting with reduced compounds (such as NADH or $FADH_2$) and ending with oxygen as the final electron acceptor. (See also substrate-level phosphorylation and photophosphorylation.)

oxytocin A protein hormone produced by the pituitary gland. Oxytocin causes uterine contractions during labor, induces release of milk from mammary glands, and in both sexes is associated with bonding behavior, parental behavior, and pleasure.

P site The site in a ribosome where peptide bonds are formed between amino acids. (See also A site, E site.)

pair-rule genes A class of fruit fly segmentation genes that organize embryonic cells into particular segments. They are active in alternating segments. (See also gap genes, segment polarity genes.)

paper chromatography A technique for separating molecules on the basis of their size and solubility, by wicking them through a piece of filter paper using a certain solvent.

paraphyletic group A group of animals that is not monophyletic; i.e, a group that includes some but not every descendant of the last common ancestor of the group. Paraphyletic groups are not meaningful evolutionary groups.

parasitism A long-term relationship between two organisms that is beneficial to one organism (the parasite) but detrimental to the other (the host). A type of symbiosis. (See also mutualism.)

parasitoid An organism has a parasitic larval stage and a free-living adult stage. Most parasitoids are insects that lay their eggs in the bodies of other insects.

parasympathetic nervous system The portion of the autonomic nervous system that stimulates activities of relaxation, repair, and rebuilding, such as slowing heart rate and digesting food.

parenchyma cell A thin-walled type of plant cell found in leaves, in the centers of stems and roots, and in fruits. Involved in photosynthesis, starch storage, and new growth. The most common type of cell in plants.

parental generation The adult organisms used in the first experimental cross in a formal breeding experiment. (The offspring will be the F_1 generation.)

parthenogenesis Development of offspring from unfertilized eggs. A form of asexual reproduction.

partial pressure The pressure of one particular gas in a mixture; the contribution of that gas to the overall pressure.

pathogen Any microorganism capable of causing disease.

pattern formation The series of events that determines the spatial organization of an embryo. Example: alignment of the major body axes and orientation of the limbs.

pattern-recognition receptor Leukocyte membrane proteins that bind to molecules found in many bacteria. Part of the innate immune response.

PCR (polymerase chain reaction) A laboratory technique for rapidly generating millions of identical copies of a specific stretch of DNA (or

RNA). PCR involves incubation of the original DNA sequence with primers, free nucleotides, and DNA polymerase.

PDGF (platelet-derived growth factor) A growth factor secreted by platelets, usually at the site of an injury, which promotes wound healing. It is also secreted by other cell types.

pedigree A family tree of parents and offspring, showing inheritance of particular traits of interest.

penis The copulatory organ of male mammals, used to insert sperm into a female.

pepsin A protein-digesting enzyme produced in the stomach.

pepsinogen The precursor to the digestive enzyme pepsin. Converted to pepsin by the acidic environment of the stomach.

peptide Two or more amino acids bonded together. (Peptides of three or more amino acids may also be called polypeptides, and very large polypeptides may also be called proteins.)

peptide bond The carbon-nitrogen bond between two amino acid residues in a peptide or protein.

per-capita rate of increase The growth rate of a population expressed per individual. Calculated as the per-capita birth rate minus the per-capita death rate, and symbolized as r.

perennial plant A plant that normally lives for more than one year.

perfect Refers to flowers that contain male parts (stamens) and female parts (carpels). (See also imperfect.)

pericarp The part of a fruit that surrounds the seeds, formed from the ovary wall. Examples: the flesh of most edible fruits; the hard shell of most nuts.

peripheral nervous system (PNS) All the components of the nervous system that are outside the brain and spinal cord. Includes the somatic nervous system and the autonomic nervous system. (See also central nervous system.)

peristalsis Rhythmic waves of muscular contraction that push food along the alimentary canal.

peroxisome A eukaryotic organelle that degrades fatty acids and amino acids, and also degrades the resulting hydrogen peroxide.

petal One of the modified leaves arranged around the reproductive structures of a flower. Often colored to attract pollinating animals.

petiole The stalk of a leaf.

pH scale A measurement of the concentration of protons in a solution; a measure of acidity or alkalinity. Defined as the negative of the base-10 log of the concentration of protons.

phenotype Any of the observable traits of an individual. Commonly includes physical, physiological, and behavioral traits.

phloem A plant vascular tissue that conducts sugars. Contains sieve cells and sieve tubes. (See also xylem.)

phloem sap The sugary fluid found in phloem tissue of plants.

phosphodiester bond The bond formed in a polymerization reaction between the phosphate bond of one nucleotide and the hydroxyl group on the sugar component of another nucleotide.

phospholipids A class of lipid having a hydrophilic head (a phosphate group) and a hydrophobic tail (one or more fatty acids). The phosphate group and fatty acids are often linked by a glycerol molecule. Phospholipids are major components of cell membranes.

phosphorylation Addition of a phosphate group to a molecule. Commonly refers to phosphorylation of proteins to control protein shape or function.

photon A discrete packet of light energy; a particle of light.

photophosphorylation Production of ATP molecules using the energy of light to excite electrons, which are then passed down an electron transport chain. Occurs during photosynthesis. (See also oxidative phosphorylation and substrate-level phosphorylation.)

photoreceptor A molecule, cell, or organ that is specialized for detection of light.

photorespiration A series of light-driven chemical reactions that consumes oxygen and releases carbon dioxide, basically reversing the work of photosynthesis. Usually occurs in bright, hot, dry environments when stomata must be kept closed.

photosynthesis A series of chemical reactions and electron transfer events that converts the energy of light into the chemical energy stored in glucose. Occurs in plant chloroplasts and in various prokaryotes.

photosystem I A system of molecules and enzymes in chloroplasts that absorbs the energy of red light with wavelength of 700 nm, and uses it to produce NADPH.

photosystem II A system of molecules and enzymes in plant chloroplasts that absorbs the energy of red light with a wavelength of 680 nm, and uses it to produce ATP and to split water into protons and oxygen.

phototropism Growth or movement in a particular direction in response to light.

phylogenetic species concept The concept that species can be defined as the smallest monophyletic group in a tree diagram representing populations.

phylogenetic tree A diagram that depicts the evolutionary history of a group of organisms.

phylogeny The evolutionary history of a group of organisms.

phylum (plural = phyla) A taxonomic category, typically consisting of a major evolutionary lineage with a distinctive body plan and/or cellular biochemistry. In plants, the phylum category may be called a division.

phytoalexin Any small plant compound that is produced to combat an infection (usually a fungal infection).

phytochrome A light-sensitive plant protein involved in detecting light and timing certain physiological processes, such as flowering, germination, or dormancy.

pistil The female reproductive structure of a flower. Consists of one or more fused structures (carpels), each of which consists of an ovary (which produces a fruit and seeds) and a stalk with a sticky end that receives pollen.

placenta An organ formed by a union of maternal and fetal tissue. Exchanges nutrients and wastes between mother and fetus, anchors the fetus to the uterine wall, and produces some hormones. Occurs in most mammals and in a few other vertebrates.

plasma The fluid portion of blood that remains when the red blood cells, white blood cells, and platelets are removed. Contains water, dissolved gases, hormones, food molecules, other soluble substances, and clotting factors.

plasma cell A type of leukocyte that produces large quantities of antibodies to combat an ongoing infection. Plasma cells are descendants of activated B cells.

plasma membrane A membrane that surrounds a cell, separating it from the external environment and selectively regulating passage of molecules and ions into and out of the cell.

plasmolysis The shrinkage of the cytoplasm of a plant cell away from its cell wall due to loss of water.

plastocyanin A small protein that shuttles electrons from photosystem II to photosystem I during photosynthesis.

platelet-derived growth factor (PDGF) A growth factor secreted by platelets, usually at the site of an injury. Promotes wound healing. Also is secreted by other cell types.

PNS (peripheral nervous system) All the components of the nervous system that are outside the brain and spinal cord. Includes the voluntary nervous system and the autonomic nervous system. (See also central nervous system, sympathetic nervous system, parasympathetic nervous system.)

point mutation A mutation that results in a change in a single nucleotide pair of DNA.

pol (DNA polymerase) Any of several enzymes that catalyze the synthesis of DNA.

polar Carrying a partial positive charge on one side of a molecule and a partial negative charge on the other. Electrons in a polar covalent bond spend more of their time near one atom than they do near the other.

pollen grain A male gametophyte of a flowering plant. It consists of two cells in a protective coat.

pollination The process by which pollen reaches the pistil of a flower (in flowering plants) or reaches the ovule directly (in conifers and relatives).

polymer Any long molecule composed of small repeating subunits (monomers) bonded together. Examples: polysaccharide, DNA, protein.

polymerase chain reaction (PCR) A laboratory technique for rapidly generating millions of identical copies of a specific stretch of DNA (or RNA). PCR involves incubation of the original DNA sequence with primers, free nucleotides, and DNA polymerase.

polymerization The process by which a monomer (a small subunit molecule) is bound to other monomers to form a polymer (a long chain molecule).

polymorphism (1) The occurrence of more than one allele at a certain genetic locus in a population. (2) The occurrence of more than two distinct phenotypes of a trait in a population.

polypeptide A peptide of three or more amino acids linked together in a chain.

polyploid Having more than two sets of chromosomes.

polyploidy The state of having more than two sets of chromosomes.

polysaccharide A large carbohydrate polymer consisting of many monosaccharides (sugar subunits) linked together in a chain. Examples: starch, cellulose, glycogen.

polyspermy Fertilization of an egg by multiple sperm. This is usually an abnormal situation.

population A group of individuals of the same species living in the same area at the same time.

pore Any small opening, such as the small opening in the stoma of a plant leaf.

positive control A type of gene regulation in which an activator protein binds to a control sequence in the DNA and promotes transcription. (See also negative control.)

positive feedback Stimulation of a reaction or process by the end result of that process. Tends to accelerate processes rapidly. Examples: the opening of sodium channels during membrane depolarization; the stimulation of LH release and ovulation by estrogen.

post-transcriptional modification Any modification of messenger RNA after it has been transcribed. Examples: splicing, addition of a poly-A tail or a cap.

post-transcriptional regulation Regulation of gene expression via any process that occurs after transcription. Examples: modifications to RNA or the efficiency of translation (translational regulation) or to proteins (post-translational regulation).

post-translational modification Any chemical alteration of a protein after it has been synthesized. Examples: cleavage of side chains, phosphorylation.

post-translational regulation Regulation of gene expression via chemical modification of proteins after translation. A type of post-transcriptional regulation.

potential energy Energy stored in matter through its position or its shape.

prebiotic soup A hypothetical solution of sugars, amino acids, nitrogenous bases, and other building blocks of larger molecules that may have formed in shallow waters or deep-ocean vents of the ancient Earth, and may have led to formation of larger biological molecules.

predator Any organism that kills and consumes other organisms.

preformation A now-disproved hypothesis that embryological development consisted simply of an increase in size of a tiny, fully formed organism. (See also epigenesis.)

pressure potential Potential energy of water caused by pressure differences. Equals the sum of all the types of pressure that affect water, such as atmospheric pressure, wall pressure, and tension. Influences the movement of water. (See also solute potential and water potential.)

pressure-flow hypothesis The hypothesis that sugar movement through plant phloem tissue is due to differences in the turgor pressure of phloem sap.

Pribnow box The six-pair base sequence (usually TATAAT) in most prokaryotic promoters (DNA binding sites for RNA polymerase) located 10 bases upstream from the start of messenger RNA transcription. Also called −10 box.

primary growth Plant growth that results in an increase in length of stems and roots. Produced by apical meristems. (See also secondary growth.)

primary producer An autotroph; a species that creates its own food through photosynthesis or from reduced inorganic compounds, and is a source of food for other species in its ecosystem.

primary structure The sequence of amino acids in a peptide or protein.

primary succession The gradual colonization of a habitat of bare rock or gravel. Example: After an environmental disturbance such as a glacier or a lava flow that removes all soil and previous organisms. (See also secondary succession.)

primer A short, single-stranded sequence of RNA that enables the start of replication of a DNA sequence that is synthesized from the 3 end of the primer.

principle of independent assortment One of Mendel's two principles of genetics, this states that each pair of hereditary elements (alleles of the same gene) behaves independently of other genes during meiosis. This holds true only for genes that are on different chromosomes.

product inhibition Inhibition of a chemical reaction or metabolic process due to high concentrations of the product of that reaction.

productivity The total amount of new organic material produced per unit area per unit time. (See also net primary productivity, secondary productivity.)

promoter A short sequence of DNA that facilitates binding of RNA polymerase to enable transcription of downstream genes.

prophase The first stage of cell division (mitosis or meiosis) during which chromosomes become visible, the mitotic spindle forms, and the nuclear membrane breaks down. In addition, synapsis and crossing over occur during prophase of meiosis I.

proplastid A colorless organelle found in undifferentiated plant cells that matures to become a plastid, such as a chloroplast, a leucoplast, or a chromoplast.

protein A long chain of amino acids linked together; a large polypeptide.

protein kinase Any enzyme that catalyzes the addition of a phosphate group to another protein, typically activating or inactivating the other protein.

proteomics The study of the three-dimensional structure and function of proteins.

protist Any microscopic eukaryote that is not a plant, animal, or fungus.

proton-motive force The combined effect of a proton gradient and an electric potential gradient across a membrane, which can drive protons across the membrane. Used by mitochondria and chloroplasts to power ATP synthesis.

pseudogene A DNA sequence that closely resembles a working gene but is not transcribed. They are thought to have arisen by gene duplication of the working gene, followed by accidental inactivation due to a mutation.

pulmonary circulation The part of the circulatory system that sends oxygen-poor blood to the lungs. Occurs to some degree in most terrestrial vertebrates. In mammals and birds, the pulmonary circulation is separate from the rest of the circulatory system. (See also systemic circulation.)

pupil The hole in the center of the iris through which light enters the vertebrate or cephalopod eye.

quantitative trait A phenotypic trait that exhibits variation along a smooth, continuous scale of measurement (for example, human height), rather than the distinct forms seen in discrete traits.

quaternary structure The overall three-dimensional shape of several polypeptide chains, and sometimes other small functional groups, arranged together to form a large, multi-unit protein.

radial symmetry A pattern of animal body symmetry in which the body has at least two planes of symmetry. Typically the body is in the form of a cylinder or disk, with body parts radiating from a central hub. Examples: Echinoderms, cnidarians, and ctenophores.

radiation Production of electromagnetic energy (light). One of the mechanisms of heat transfer between organisms and the environment. (See also conduction, convection, evaporation.)

radicle The root of a plant embryo. (See also cotyledon, hypocotyl.)

radioactive isotope An isotope (i.e., an element with a characteristic number of neutrons) that spontaneously decays to form a different isotope or element, by emitting radiation or a particle.

radiometric dating A technique for determining the age of a rock by measuring the amount of radioactive decay that has occurred since the rock solidified; this method relies on measuring the amount of new daughter element that has formed. Also called radioactive dating.

rational drug design Development of new drug treatments based on specific and detailed information of the shape, structure, and function of a certain target molecule, rather than by trial and error.

reading frame The division of a sequence of DNA or RNA into a particular series of 3-nucleotide codons. There are three possible reading frames for any sequence. In functional genes, the start codon defines which reading frame is to be used.

realized niche The ecological niche that a species occupies in the presence of competitors.

receptor (1) A molecule, cell, or group of cells specialized for detecting environmental signals (e.g., light, sound). (2) A molecule that binds to a particular chemical (hormone, sperm protein) and triggers a cellular response.

receptor molecule A cell-surface molecule that is necessary for detection of and response to a molecule, such as a neurotransmitter, hormone, or other signaling molecule; also a molecule necessary for a virus particle to gain entry to a cell.

recessive Refers to an allele whose influence on phenotype can be entirely hidden by the presence of another, dominant allele.

reciprocal cross A breeding experiment in which the mother's phenotype and father's phenotype are the reverse of that done in a prior breeding experiment.

recombinant Possessing a new combination of alleles. May refer to a single chromosome or an entire organism.

recombinant DNA Any DNA that is changed by exchange with, or inclusion of, foreign DNA. May occur via meiosis, viruses, laboratory techniques, etc.

recombinant DNA technology A variety of laboratory techniques for isolating certain DNA fragments and introducing them into a different host organism. Also called biotechnology or genetic engineering.

recombination Any change in the combination of genes or alleles found on a given chromosome or in a given individual.

redox reaction A class of chemical reactions that involves the loss and gain of electrons. An abbreviation of reduction-oxidation reaction.

reduction The gain of electrons during a redox reaction, either by accepting an electron from another atom, or by the electrons in covalent bonds moving closer to the atomic nucleus. Results in a gain in potential energy.

reduction-oxidation reaction A class of chemical reactions that involves the loss and gain of electrons. Also called redox reaction.

redundant code Any code in which several different sequences can code for the same information. The genetic code is a redundant code because some amino acids are coded for by two or three different codons.

reflex An involuntary response to environmental stimulation. May involve the brain (e.g., conditioned reflex) or not (e.g., spinal reflex).

regulatory protein Any protein that affects gene transcription by binding to specific enhancers, silencers, or other sites in DNA. Includes transcription activators and repressors.

reinforcement Natural selection for traits that prevent interbreeding between recently diverged species.

replication fork The Y-shaped site at which a double-stranded molecule of DNA is separated into two single strands for replication.

reporter gene A gene whose activity is easy to determine, either by a distinctive phenotype or a simple test. A type of genetic marker.

repressor Any regulatory protein that inhibits transcription of certain genes, typically by binding to a silencer upstream of the promoter.

resilience The ability of a cell, organism, or community to tolerate a perturbation and quickly return to normal. In community ecology, a measure of the speed with which a community returns to its original state following an environmental disturbance.

resistance (1) The ability of an organism to defend itself against drugs, pathogens or parasites. Example: antibiotic resistance in bacteria. (2) The ability of a species community to tolerate environmental perturbations such as droughts, fires, etc.

resting potential The membrane potential of a cell in its resting, or normal, state.

restriction enzyme Any of a number of bacterial endonucleases (enzymes that cut double-stranded DNA) that cuts DNA at a specific basepair sequence. This class of enzyme is used extensively in laboratory experiments involving molecular genetics.

retina A thin layer of light-sensitive cells (rods or cones) and neurons at the back of a camera-type eye, such as in cephalopods or vertebrates.

retinal A carotenoid pigment that, together with opsin, forms rhodopsin, the light-detecting pigment of the rod cells in animal eyes. Also occurs with cone opsins in cone cells. A derivative of vitamin A. (See also opsin.)

retrovirus A virus with an RNA genome that reproduces by transcribing its RNA into a DNA sequence, then inserting it into the host's genome for replication.

reverse genetics An experimental approach for identifying the phenotypic traits affected by a certain gene of unknown function. Recombinant DNA techniques are used to produce mutated versions of a DNA sequence of unknown function, the mutated versions are introduced into organisms, and resulting phenotypes are inspected for any changes. (See also forward genetics.)

reverse transcriptase A enzyme of retroviruses (RNA viruses) that can synthesize DNA using an RNA template.

rhizome A plant stem extending horizontally underground.

rhodopsin A combination of two molecules (retinal and one of various opsins) instrumental in detection of light by vertebrate cones and rods.

ribonucleic acid (RNA) A polymer consisting of a string of ribonucleotides. May have versatile roles such as structural components of ribosomes (rRNA), transporters of amino acids (tRNA), or translaters of the message of the DNA code (mRNA). Usually single-stranded.

ribosomal RNA (rRNA) A class of specialized RNA molecules that forms part of the structure of a ribosome.

ribosome A complex of specialized ribosomal RNA (rRNA) and proteins that mediates protein synthesis from messenger RNA strands, facilitating placement of transfer RNA, and catalyzing formation of the peptide bonds between amino acids.

ribozyme Any RNA molecule that acts as a catalyst for a chemical reaction.

RNA (ribonucleic acid) A nucleic acid consisting of a sequence of ribonucleotides. RNA plays several roles in a cell, including as a structural component of ribosomes (rRNA), a transporter of amino acids (tRNA), or a translater of the message of the DNA code (mRNA).

RNA polymerase An enzyme that catalyzes synthesis of RNA from ribonucleotides, using a DNA template.

rod A photoreceptor cell with a rod-shaped outer portion found in vertebrate retinas. Particularly sensitive to dim light, but not used to distinguish colors. (See also cone.)

root apical meristem A group of undifferentiated plant cells at the tip of a plant root.

root cap A small group of cells that covers and protects the tip of a plant root. Responsible for sensing gravity and determining the direction of root growth.

root hair A long, thin outgrowth of the epidermal cells of plant roots. Root hairs increase surface area and allow the root to obtain more water and nutrients.

root system The part of a plant that is below ground.

rough ER (rough endoplasmic reticulum) A type of endoplasmic reticulum dotted with ribosomes. Involved in processing of membrane proteins and secretory proteins.

rRNA (ribosomal RNA) A class of specialized RNA molecules that forms part of the structure of a ribosome.

rubisco The enzyme that catalyzes the first step of the Calvin cycle during photosynthesis, the addition of one molecule of CO_2 to the five-carbon sugar ribulose biphosphate (RuBP). Rubisco is an abbreviation of the formal name D-ribulose 1,5-bisphosphate carboxylase/oxygenase.

S phase The phase of the cell cycle during which DNA is synthesized and chromosomes are duplicated.

SA node (sinoatrial node) A cluster of heart muscle cells that initiates the heart beat and determines the heart rate. Located in the wall of the right atrium.

SAR (systemic acquired resistance) A set of events through which plant tissues prepare to combat a possible infection.

sarcomere A single contractile unit of a skeletal muscle cell.

saturated Containing the maximum possible amount of some substance. A saturated fat contains the maximum possible number of hydrogen atoms such that there are only single bonds between carbon atoms (no double bonds), resulting in a straight chain (no kinks) and a high melting point.

sclerenchyma cell A thick-walled type of plant cell that provides support and often contains the tough structural polymer lignin.

second messenger A signaling molecule that is produced or activated inside a cell in response to stimulation on the outer surface of the cell. Commonly used to relay the message of a protein hormone throughout the cell.

secondary growth Plant growth produced by cambia that results in an increase in width of stems and roots. (See also primary growth.)

secondary metabolite Any poison produced by a plant that is synthesized by a variation of another biosynthetic pathway. Typically found only in some parts of the plant.

secondary productivity The total amount of new body tissue produced by animals that eat plants. May involve growth and/or reproduction. (See also net primary production.)

secondary structure The three-dimensional structure of a polymer that is due to hydrogen bonds between adjacent components of the polymer strand. Example: The α-helix and β-pleated sheet structures found in proteins.

secondary succession Gradual colonization of a habitat after an environmental disturbance that removes some or all previous organisms but leaves the soil intact (e.g., fire, windstorm, logging). (See also primary succession.)

seed A plant embryo with nutritive tissue (endosperm) to fuel its early growth, and surrounded by an outer protective layer (seed coat). Formed from the fertilized ovule of a flower.

seed dormancy A state of suspended development of a plant seed. Can be terminated by certain environmental cues indicating that favorable environmental conditions have arrived (sunlight, water), or that unfavorable ones have passed (fire, cold).

segment polarity genes A class of fruit fly segmentation genes that establish the anterior-posterior orientation of each embryonic segment. Active in particular regions of each segment. (See also gap genes, pair-rule genes.)

self-fertilization The fusion of two gametes from the same individual to form a diploid offspring. This is common in some plants and is also called selfing. (See also outcrossing.)

self-incompatible Incapable of self-fertilization.

semen The combination of sperm and accessory fluids that is released by male mammals and reptiles at ejaculation.

sensor A cell, organ, or structure that senses some aspect of the external or internal environment.

sensory neuron A nerve cell that carries sensory information to the central nervous system.

sepal One of the protective leaf-like structures enclosing a flower bud and later supporting the blooming flower.

serum The liquid that remains when clotted cells are removed from blood; plasma without the clotting factors. Serum contains water, dissolved gases, hormones, food molecules, and other soluble substances.

set point A normal or target value for a regulated internal factor, such as body heat or blood pH.

sex chromosome One of the pair of chromosomes that carries the gene(s) that determine gender. (See also autosomes.)

sex-linked inheritance The various inheritance patterns that can occur when genes are carried on the sex chromosomes, such that females and males have different numbers of alleles of that gene, and may pass the trait only to one sex of offspring.

sexual dimorphism Any morphological difference between males and females that is not directly related to the reproductive organs. Examples: plumage coloration, body size.

sexual reproduction Reproduction in which genes from two parents are combined, producing offspring that are genetically distinct from both parents. (See also asexual reproduction.)

sexual selection The process by which individuals with certain heritable traits leave more offspring than other individuals due specifically to superiority in competing for mating opportunities.

shoot apical meristem A group of undifferentiated plant cells at the tip of a plant stem.

shoot system The part of a plant that is above ground.

sieve-tube member An elongated food-conducting plant cell with a cytoplasm but no nucleus. Has clusters of pores at both ends, allowing sap to flow to adjacent cells. Sieve-tube members are stacked end-to-end to form sieve tubes in the phloem.

signal hypothesis The hypothesis that proteins destined for secretion are directed to the rough ER by a signal consisting of a certain sequence of amino acids at the very beginning of newly synthesized proteins.

signal transduction The process by which a stimulus (e.g., hormone, neurotransmitter, or sensory information such as light or sound) outside of a cell is translated into a response by the cell.

silencer A regulatory DNA sequence in eukaryotes to which repressor proteins can bind, inhibiting transcription of nearby genes.

silent mutation A mutation that does not detectably affect the phenotype of the organism. This is typically a point mutation in the third position of certain codons, such that the resulting amino acid sequence is unchanged.

simple sequence repeat A stretch of DNA consisting of repeats of a short, simple DNA sequence that does not code for any protein or RNA.

single nucleotide polymorphism (SNP) A genetic locus where a single base pair often varies between individuals. Often used as a genetic marker to help track the inheritance of nearby genes.

sink A location where a substance is consumed or taken out of circulation. (See also source.)

sinoatrial node (SA node) A cluster of cardiac muscle cells located in the wall of the right atrium that initiate the heart beat and determine the heart rate.

sister chromatids The paired strands of a recently replicated chromosome that has not yet divided.

sliding-filament theory The theory that the contraction of muscle cells is caused by filaments of actin and myosin sliding past each other.

smooth ER Endoplasmic reticulum that does not have ribosomes attached to it. Involved in synthesis and secretion of lipids.

SNP (single nucleotide polymorphism) (Pronounced snip) A genetic locus where a single base pair often varies between individuals. Often used as a genetic marker to help track the inheritance of nearby genes.

snRNPs (small nuclear ribonucleoproteins) A complex of proteins and small RNA molecules that catalyzes splicing (removal of introns from messenger RNA). Components of spliceosomes.

solute potential The potential energy of water caused by a difference in solute concentrations at two locations, which influences the movement of water by osmosis. May refer specifically to the high concentration of solutes in plant roots (sugars, ions, etc.) relative to the soil. (See also pressure potential and solute potential.)

solution A liquid containing one or more dissolved solids or gases in a homogeneous mixture.

solvent Any liquid in which some substance will dissolve.

somatic cell Any type of cell that does not pass its genes on to the next generation. In a multicellular organism, all cells except eggs, sperm, and their parent cells (oogonia and spermatogonia) are somatic cells.

somatic hypermutation Rapid DNA mutation that occurs in a somatic cell (e.g., not in reproductive cells). Occurs in certain cells of the immune system, such as memory cells that are fine-tuning antibody performance.

somatic nervous system The part of the peripheral nervous system (outside the brain and spinal cord) that controls skeletal muscles; it is under voluntary control. (See also autonomic nervous system.)

source A location where a substance is produced or enters circulation. (See also sink.)

Southern blot A laboratory technique developed by E. M. Southern for the isolation and analysis of pieces of DNA, by cleavage with restriction enzymes, separation with gel electrophoresis, and visualization with autoradiography.

speciation The evolution of two or more distinct species from a single ancestral species.

species A distinct, identifiable group of individuals that regularly breeds together and is thought to be an evolutionarily independent group. Generally distinct from other species in appearance, behavior, habitat, ecology, genetic characteristics, etc. (See also biological species concept, morphological species concept, phylogenetic species concept.)

species-area relationship The mathematical relationship between the area of a certain habitat and the number of species that it can support.

specific heat capacity The amount of energy required to raise the temperature of 1 gram of a substance by 1 degree C; a measure of the capacity of a substance to absorb energy.

sperm A mature male gamete, smaller and more mobile than the female gamete.

spermatophore A gelatinous package of sperm cells that is produced by males of species that have internal fertilization without copulation. The spermatophore is either given to the female (cephalopods and insects) or placed where she can pick it up (salamanders).

spliceosome A complex of snRNPs (small nuclear ribonucleoproteins) that mediates the breaking of the 5′ ends of introns during splicing.

splicing The process of removing introns from messenger RNA molecules and connecting the remaining exons together.

spontaneous generation The now-disproved hypothesis that living organisms can develop spontaneously and rapidly from non-living, noncellular materials under certain conditions.

spore A single cell produced by mitosis or meiosis (not by cell fusion) that is capable of developing into an adult organism without fusing with another cell.

sporophyte The multicellular diploid phase of an organism that exhibits alternation of generations; arises from two fused gametes and produces spores. (See also gametophyte, alternation of generations.)

stabilizing selection A pattern of natural selection in which individuals with an average phenotype have higher fitness than those with extreme phenotypes.

stamen The male reproductive structure of a flower. Consists of an anther, which produces pollen, and a filament, which supports the anther.

stapes A stirrup-shaped bone in the middle ear. Receives vibrations from the tympanic membrane (via two other bones) and passes them to the cochlea.

star radiation A star-like shape on a phylogenetic tree, indicating a rapid adaptive radiation.

starch Any of several polysaccharides used primarily for food storage in plants. Consists of a certain form of glucose (α-glucose) attached end-to-end to form helices, sometimes branched. May also refer to glycogen, a similar branched polysaccharide found in animals.

start codon The messenger RNA sequence AUG, which induces the beginning of protein synthesis and codes for the amino acid methionine.

stele The central region of a plant root or stem, containing the vascular tissue.

stereocilia A set of stiff outgrowths jutting from the surface of a hair cell that is involved in detection of sound by terrestrial vertebrates, or of water-borne vibrations by fishes.

stoma (plural = stomata) A pore or opening. In plants, refers to a microscopic pore on the surface of a leaf or stem, through which gas exchange occurs for photosynthesis. Also can refer to the entire apparatus of the pore and the adjacent guard cells.

stop codon One of three messenger RNA triplets (UAG, UGA, or UAA) that cause termination of protein synthesis.

stroma The fluid between a chloroplast's outer membrane and its thylakoid disks.

structural formula A graphical representation of the atoms and bonds in a molecule, often with covalent bonds represented by straight lines.

structural homology Similarities in organismal structures (e.g., limbs, shells, flowers, etc.) that are due to inheritance from a common ancestor.

substrate-level phosphorylation Production of ATP molecules via an enzyme-catalyzed transfer of a phosphate group from an intermediate substrate directly to ADP without using an electron transport chain. (See also oxidative phosphorylation and photophosphorylation.)

succession Gradual colonization of a habitat after an environmental disturbance (such as by fire, landslide, flood, etc.), usually by a series of species assemblages. (See also primary succession, secondary succession.)

summation The additive effect of different postsynaptic potentials at a nerve or muscle cell such that several sub-threshold stimulations can cause an action potential.

surface tension The attractive force between liquid molecules that causes the liquid to form a rounded surface at an air-liquid interface.

survivorship The proportion of offspring produced that survive to a particular age, on average.

symbiosis Any close, prolonged relationship between individuals of two different species. Includes mutualism and parasitism.

sympathetic nervous system The portion of the autonomic nervous system that stimulates fight-or-flight responses, such as increased heart rate, increased blood pressure, and decreased digestion. (See also parasympathetic nervous system.)

sympatric Living in the same geographic area.

synapse The connection between two neurons or between a neuron and a muscle cell, consisting of a tiny space into which neurotransmitters are released.

synapsis The physical pairing of two homologous chromosomes during prophase I of meiosis. Crossing over occurs during synapsis.

synaptic vesicle Tiny neurotransmitter-containing vesicles just inside the end of an axon. Can fuse with the axon membrane to release neurotransmitter outside the axon, stimulating the next neuron (or muscle cell).

system A group of organs that work together to perform a function. Examples: digestive system, immune system.

systemic acquired resistance (SAR) A set of events through which plant tissues prepare to combat a possible infection.

systemic circulation The part of the circulatory system that sends oxygen-rich blood from the lungs out to the rest of the body. In mammals and birds, the systemic circulation is separate from the lung (pulmonary) circulation. (See also pulmonary circulation.)

systole The portion of the heartbeat cycle during which the heart muscles are contracting.

Taq polymerase A DNA polymerase commonly used in PCR due to its stability at high temperatures; this enzyme is derived from *Thermus aquaticus*, a bacterium found in hot springs.

taste cell A spindle-shaped cell found in taste buds of the mammalian tongue that responds to chemical stimuli.

TBP (TATA-binding protein) A protein that binds to eukaryotic promoters and helps initiate transcription.

T cell A class of lymphocytes that matures in the thymus (hence T cell), and that is involved in acquired immunity in vertebrates. May be involved in activation of other lymphocytes (helper T cells) or destruction of infected cells (cytotoxic T cells). Also called T lymphocyte. (See also B cell.)

T-cell receptor (TCR) A T cell membrane protein that can bind to antigens displayed on the surfaces of other cells. Contains one α chain and one β chain; similar to one arm of an antibody.

telophase The final stage in cell division (mitosis or meiosis) during which chromosomes finish moving and new nuclear envelopes begin to form around each set of daughter chromosomes. Often occurs concurrently with cytokinesis, division of the cytoplasm.

temperature A measurement of thermal energy present in an object or substance; e.g., how much the object's molecules are moving. Informally, a measurement of heat.

temperature-sensitive mutations Mutations that affect the phenotype of an organisms only at certain temperatures. Often involve enzymes that perform either unusually well or unusually poorly at certain temperatures.

tendon A band of tough, fibrous connective tissue that connects a muscle to a bone.

tertiary structure The overall three-dimensional shape of a single polypeptide chain (or other large molecule), caused by a variety of interactions including hydrogen bonds, ionic bonds and hydrophobic interactions among R-groups and the peptide backbone.

testcross A breeding experiment in which an individual of unknown genotype is bred with an individual that has only recessive alleles for the traits of interest. The purpose is to determine the unknown genotype by examining phenotypic ratios seen in offspring.

testis (plural = testes) The sperm-producing organ of a male animal.

tetrapod Any member of the taxon Tetrapoda, including all descendants of the first four-footed animals to move onto the land. Living tetrapods include amphibians, mammals, turtles, lizards and snakes, crocodilians, and birds. Many modern tetrapods still have four legs, but some do not (e.g., humans, snakes, birds).

thermal energy The kinetic energy of molecular motion. Measured as temperature and perceived as heat.

thylakoid A flattened, membrane-bound disk inside a plant chloroplast. A stack of thylakoids is called a granum.

tissue A group of similar cells that functions as a unit.

top-down control The limitation of herbivore population size by predation or disease rather than by limited or toxic nutritional resources.

tracheid An elongated water-conducting plant cell that has gaps (pits) in its secondary cell wall to allow water movement from one cell to the next, though the water still must cross the primary cell wall.

transcription The process by which messenger RNA is made from a DNA template.

transcriptional regulation Regulation of gene expression via changes in the rate at which genes are transcribed to form messenger RNA.

transfer RNA (tRNA) A class of specialized RNA molecules responsible for matching amino acids to messenger RNA codons during translation. They have a cloverleaf shape, with the amino acid binding site at one end and an anticodon at another site.

transition state A brief intermediate stage in a chemical reaction during which there is a combination of new and old chemical bonds. Typically this is a high-energy state that determines the necessary activation energy for the reaction to proceed.

translation The process by which proteins and peptides are synthesized from messenger RNA molecules.

translational regulation The regulation of gene expression by altering the lifespan of the messenger RNA or the efficiency of translation. A type of post-transcriptional regulation.

translocation (1) The movement of water, food or minerals through a plant by bulk flow. May refer specifically to the movement of food. (2) A type of mutation in which a piece of a chromosome moves to a nonhomologous chromosome.

transmission genetics A field of study concerning the genotypic and phenotypic patterns that occur as genes pass from one generation to the next.

transpiration Water loss from aboveground plant parts that occurs primarily through the stomata.

transporter Any membrane protein that enables specific molecules to cross plasma membranes, sometimes by causing a conformational change in the protein.

transposable elements Any of several kinds of parasitic DNA sequences that are capable of moving themselves, or copies of themselves, to other locations in the genome. Typically contain genes for one or two enzymes needed for the relocation. Include LINEs, SINEs, transposons, etc.

tree of life A diagram depicting the genealogical relationships of all living organisms on Earth, with a single ancestral species at the base.

triploblast An animal whose body develops from three basic embryonic cell layers: ectoderm, mesoderm, and endoderm. All animals except sponges, cnidarians, and ctenophores are triploblasts. (See also diploblast.)

triploid Having three sets of chromosomes.

tRNA (transfer RNA) A class of specialized RNA molecules responsible for matching amino acids to messenger RNA codons during translation. They have a cloverleaf shape, with the amino acid binding site at one end and an anticodon at another site.

trophic level A feeding level within an ecosystem. Examples: primary producers, primary consumers.

trypsin A protein-digesting enzyme produced by the pancreas, secreted into the intestine, and activated by enterokinase. Breaks the bond between lysine and arginine.

tumor A mass of cells formed by uncontrolled cell division.

tumor suppressor Any gene that (when functioning normally) prevents cell division, particularly if the cell has DNA damage. Also refers to the protein produced by such a gene.

turgid Refers to a plant cell that is firm; i.e., contains enough water for the cell cytoplasm to press against the cell wall.

turgor pressure The outward pressure exerted by the fluid contents of a plant cell against its cell wall. (See also wall pressure.)

tympanic membrane The eardrum. A membrane separating the middle ear from the outer ear in vertebrates. Also may refer to analogous structures in insects.

ultraviolet light Light with a wavelength shorter than visible blue light.

unsaturated Containing less than the maximum possible amount of some substance. An unsaturated fat has fewer than the maximum possible number of hydrogen atoms, due to the occurrence of one or more double bonds between carbon atoms, resulting in a kink in the chain and a decreased melting point.

uterus The organ in which developing embryos are housed in vertebrates that give live birth. Common in most mammals, and in some lizards, sharks, and other vertebrates.

vaccine A medical treatment designed to stimulate development of immunity to a pathogen without causing illness. Commonly includes epitopes from the pathogen, and sometimes entire killed or weakened pathogens.

vagina The birth canal of female mammals; a muscular tube that extends from the uterus through the pelvis to the exterior.

valence Relating to the outermost electron shell of an atom. Valence electrons are those in the outermost shell, and valency is a number indicating how many more electrons would fill up the outermost shell (e.g., how many covalent bonds the atom can form).

vascular tissue In plants, a tissue involved in conducting water or solutes from one part of the plant to another, or that gives rise to such tissues. Examples: xylem, phloem, vascular cambium.

vector (1) A plasmid or other vehicle used to transfer recombinant genes to a new host. (2) A biting insect or other organism that transfers pathogens between two other species.

vegetal hemisphere The lower half of an amphibian egg cell, containing most of the yolk. This portion eventually becomes part of the gut.

vein (1) Any blood vessel that carries blood from capillaries to the heart. A vein has thinner walls and lower pressure than an artery. (2) A strip of vascular tissue in a plant leaf. (3) A supporting filament of an insect wing.

ventricle (1) A thick-walled chamber of the heart that receives blood from an atrium and pumps it to the body or to the lungs. (2) One of several small fluid-filled chambers in the vertebrate brain.

vessel An efficient water-conducting apparatus found in the xylem of certain advanced plants (gnetophytes and angiosperms). Consists of many vessel elements (elongated cells with gaps in the cell walls at both ends) arranged end-to-end.

vessel element An elongated water-conducting plant cell found in the xylem of certain advanced plants (gnetophytes and angeiosperms). Has gaps through both the primary and secondary cell walls, allowing unimpeded passage of water from one cell to the next.

vestigial trait Any rudimentary structure with no known function, or minimal function, that is homologous to functioning structures in other species. Vestigial traits are thought to reflect evolutionary history.

vicariance Speciation that occurs by the splitting of a population into smaller, isolated populations by a geographic barrier.

virulence The ability to cause disease and death.

virulent Tending to cause severe disease rather than mild disease. (See also benign.)

virus A tiny infectious parasitic entity consisting of DNA or RNA enclosed in a protective covering (a capsid and sometimes an envelope). The DNA or RNA contain the necessary instructions to make more viruses, but must use the machinery of a host cell to do so.

vitamin Any micronutrient that is a carbon-containing compound rather than a single chemical element. Several vitamins function as coenzymes.

vitelline envelope A fibrous layer surrounding a mature egg cell in many vertebrates.

viviparous Refers to animals that reproduce by live birth rather than by laying eggs. (See also oviparous.)

voltage clamping A laboratory technique for imposing a certain constant membrane potential on a cell. Widely used to investigate ion channels.

voltage-gated channel An ion channel that opens or closes in response to changes in membrane voltage. (See also ligand-gated channel.)

wall pressure The inward pressure exerted by a cell wall against the fluid contents of a plant cell. (See also turgor pressure.)

water potential The total potential energy of water. In living organisms, it is equal to the sum of solute potential and pressure potential. The water potential determines the movement of water into or out of cells. (See also solute potential and pressure potential.)

water potential gradient A difference in water potential in one location compared to another, which determines the movement of water through plant tissues.

wild type The most common phenotype seen in a population, especially the most common phenotype in wild populations as compared to inbred lab strains of the same species.

wobble hypothesis The hypothesis that some transfer RNA molecules can pair with more than one codon of messenger RNA, tolerating some wobble in the third base as long as the first and second base are correctly matched.

X-linked inheritance Any inherited trait that is coded for by a gene located on the X chromosome; females have two X chromosomes while males have one.

xanthophyll A class of carotenoid pigments found in many algae, typically appearing yellow.

xylem A plant vascular tissue that conducts water and minerals; it contains tracheids and vessels. (See also phloem.)

xylem sap The watery fluid found in xylem tissue of plants.

yeast Any fungus growing as a single-celled form. Also may refer to a specific lineage of ascomycetes.

yolk The nutrient-rich cytoplasm inside an egg cell. Example: The yolk of a hen's egg is the cytoplasm of a single huge egg cell.

Z scheme A proposed model for the passage of electrons from photosystem II to photosystem I during photosynthesis. Refers to the Z shape traced by the electrons on a graph of energy level and location.

zygote The diploid cell formed by the union of two haploid gametes. Capable of undergoing embryological development to form an adult organism.

Image Credits

FRONTMATTER

ii © 1994 Susan Middleton & David Liittschwager vii Scott Freeman xvii Ted Clutter/Photo Researchers, Inc. xviii (left) Biophoto Associates/Photo Researchers, Inc. xviii (right Bill Longcore/Photo Researchers, Inc. xix CNRI/Phototake NYC xx (left) Biophoto Associates/Photo Researchers, Inc.
xx (right) S. Nielsen/DRK Photo xxi Dr. Gopal Murti/Science Photo Library/Photo Researchers, Inc. xxii (left) Michael Freeman/Phototake NYC xxii (right) Edward M. Laufer, Columbia University; Chang-Yeol Yeo and Clifford J. Tabin, Harvard Medical School xxiii Gerard Lacz/Peter Arnold, Inc. xxiv (left) Robert & Linda Mitchell Photography xxiv (right) Alfred Pasieka/Science Photo Library/Photo Researchers, Inc. xxv Gopal Murti/Science Photo Library/Photo Researchers, Inc.
xxvi (left) Gerry Ellis/Minden Pictures xxvi (right) Robert & Linda Mitchell Photography xxvii (left) Gary Gray/DRK Photo xxvii (right) Jerome Wexler/Visuals Unlimited xxviii (left) Frans Lanting/Minden Pictures xxviii (right) Mark Moffett/Minden Pictures xxix (top) Reproduced from Kuniaki Takata et al., Immunohistochemical localization of Na^+-dependent glucose transporter in rat jejunum. *Cell & Tissue Research* 267:3-9 (1992), fig. 2c. Copyright 1992 Springer-Verlag GmbH & Co KG. Image courtesy of Kuniaki Takata, Gunma University School of Medicine, Gunma, Japan. xxix (bottom) Reproduced from Kuniaki Takata et al., Immunohistochemical localization of Na^+-dependent glucose transporter in rat jejunum. *Cell & Tissue Research* 267:3-9 (1992), fig. 2d. Copyright 1992 Springer-Verlag GmbH & Co KG. Image courtesy of Kuniaki Takata, Gunma University School of Medicine, Gunma, Japan. xxx (left) Frans Lanting/Minden Pictures xxx (right) Michael Fogden/Animals Animals/Earth Scenes xxxi Ben Osborne/Getty Images Inc. xxxii Mark Moffett/Minden Pictures

CHAPTER 1

Opener John Eskenazi Ltd, 15 Old Bond Street, London W1X 4JL England
1.1a Burndy Library/Omikron/Photo Researchers, Inc. 1.1b M.I. Walker/Photo Researchers, Inc. 1.3a Horticulture Research International, Wellesbourne, Warwickshire, U.K. 1.3c Jane Grushow/Grant Heilman Photography, Inc. 1.5a Samuel F. Conti and Thomas D. Brock 1.5b Kwangshin Kim/Photo Researchers, Inc.
1.6 (top left) Jorg Overmann, University of Oldenburg, Oldenburg, Germany
1.6 (top middle) Dennis Kunkel/Dennis Kunkel Microscopy, Inc. 1.6 (top right) Biophoto Associates/Photo Researchers, Inc. 1.6 (bottom left) Biophoto Associates/Photo Researchers, Inc. 1.6 (bottom middle) Darwin Dale/Photo Researchers, Inc.
1.9b Renee Lynn/Photo Researchers, Inc. 1.10 Erick Greene, University of Montana

UNIT 1

© 1994 Susan Middleton & David Liittschwager

CHAPTER 2

Opener Science Photo Library/Photo Researchers, Inc. 2.1 (right) Mitchell Layton/Duomo Photography Incorporated 2.3b J. William Schopf, University of California at Los Angeles 2.10 (top) Hugh S. Rose/Visuals Unlimited 2.16c Geostock/Getty Images, Inc/PhotoDisc, Inc.

CHAPTER 3

Opener Ralph Wetmore/Getty Images Inc. 3.2b NOAA Vents Program 3.11b Clare Sansom, Birkbeck College, University of London, London, England 3.17b (right) Michael Freeman/Phototake NYC Box 3.2 Figure 1 autoradiograph Reproduced by permission from J.P. Ferris et al., 1996, Synthesis of long prebiotic oligomers on mineral surfaces, *Nature* 381:59.61, Fig. 2. Image courtesy of James P. Ferris, Rensselaer Polytechnic Institute. Box 3.3 Figure 1a T.A. Steitz, Yale University Box 3.4 Figure 1b A. Barrington Brown/Science Source/Photo Researchers, Inc.

CHAPTER 4

Opener Kit Pogliano and Marc Sharp, University of California at San Diego 4.1a Alec D. Bangham, M.D., F.R.S./Alec D. Bangham, M.D., F.R.S. 4.1b Fred Hossler/Visuals Unlimited 4.6a (left) James J. Cheetham, Carleton University 4.14 (step 4) Don W. Fawcett/Photo Researchers, Inc. 4.18(1) Andrew Syred/Getty Images Inc 4.18(2) David Phillips/Science Source/Photo Researchers, Inc. 4.18(3) Joseph F. Hoffman, Yale University School of Medicine Box 4.1 Figure 1 (left) /JEOL USA Inc.

UNIT 2

© 1994 Susan Middleton & David Liittschwager

CHAPTER 5

Opener Andrew Syred/Science Photo Library/Photo Researchers, Inc. 5.1 (bottom) M. Wurtz/Biozentrum, University of Basel/Science Photo Library/Photo Researchers, Inc. 5.4a (top) Don W. Fawcett/Photo Researchers, Inc. 5.4b P. Motta & T. Naguro/Science Photo Library/Photo Researchers, Inc. 5.5 Carl M. Feldherr, University of Florida College of Medicine 5.7a (left) K.G. Murti/Visuals Unlimited 5.7a (right) Don W. Fawcett/Visuals Unlimited 5.7b Don W. Fawcett/Visuals Unlimited 5.8a James D. Jamieson, M.D., Yale University School of Medicine 5.11a Herb Segars/Animals Animals/Earth Scenes 5.11b Reproduced by permission from J.F. Presley et al., ER to Golgi traffic visualized in living cells. Nature 389:81-85 (1997), fig. 1b. Copyright © 1997 Macmillan Magazines Limited. Images courtesy of John F. Presley, McGill University. 5.14a K.G. Murti/Visuals Unlimited 5.14b Peter Dawson/Science Photo Library/Photo Researchers, Inc.
5.15 Reproduced by permission of the American Society for Cell Biology *from Molecular Biology of the Cell* 9(12), December 1998, cover. Copyright © 1998 by the American Society for Cell Biology. Image courtesy of Bruce J. Schnapp, Oregon Health Sciences University. 5.16a (left) Deep-etch electron micrograph kindly provided by Dr. John Heuser of Washington University School of Medicine, St. Louis, Missouri. 5.17a (left) Dennis Kunkel/Phototake NYC 5.17a (right) Dennis Kunkel/Phototake NYC 5.17b Dr. Gopal Murti/Science Photo Library/Photo Researchers, Inc. 5.18 Reproduced from C.J. Brokaw, Microtubule sliding in swimming sperm flagella: direct and indirect measurements on sea urchin and tunicate spermatozoa. *Journal of Cell Biology* 114:1201-1215 (1991), cover. Reproduced by copyright permission of The Rockefeller University Press. Image courtesy of Charles J. Brokaw, California Institute of Technology.

CHAPTER 6

Opener Stephen J. Mojzsis/NGS Image Collection 6.5 Clare Sansom, Birkbeck College, University of London, London, England. 6.8 (right) K.R. Porter/Photo Researchers, Inc. 6.13a and 6.13b Yasuo Kagawa, Jichi Medical School, Tochigi, Japan.

CHAPTER 7

Opener Alfred Pasieka/Science Photo Library/Photo Researchers, Inc.
7.2a (right) John Durham/Science Photo Library/Photo Researchers, Inc.
7.2b (left) E.H. Newcomb & W.P. Wergin/Getty Images Inc 7.2c Wanner/Eye of Science/Photo Researchers, Inc. 7.4b Sinclair Stammers/Science Photo Library/Photo Researchers, Inc. 7.6b David Newman/Visuals Unlimited 7.12b James A. Bassham/James A. Bassham 7.14 I. Andersson, Oxford Molecular Biophysics Laboratory/Science Photo Library/Photo Researchers, Inc. 7.15a Dr. Jeremy Burgess/Science Photo Library/Photo Researchers, Inc. Box 7.1 Figure 1 (left) Dr. Jeremy Burgess/Photo Researchers, Inc. Box 7.1 Figure 1 (right) Richard Green/Photo Researchers, Inc.

CHAPTER 8

Opener Michael Whitaker/Science Photo Library/Photo Researchers, Inc. 8.1 Originally published in Walter Flemming, Zellsubstanz, Kern, und Zelltheilung. Leipzig: Verlag von F.C.W. Vogel, 1882. Image courtesy of Conly Rieder, Wadsworth Center, New York State Department of Health. 8.2 Biodisc/Visuals Unlimited
8.3a (bottom) Mark E. Warchol, Washington University School of Medicine, and Jeffrey T. Corwin, University of Virginia School of Medicine 8.9a David M. Phillips/Visuals Unlimited 8.9b R. Calentine/Visuals Unlimited 8.10 Ed Reschke/Peter Arnold, Inc. 8.13a E.R. Degginger/Color.Pic, Inc. 8.17 Dr. E. Walker/Science Photo Library/Photo Researchers, Inc. Box 8.1 Figure 1 National Institutes of Health/Science Source/Photo Researchers, Inc.

UNIT 3

© 1994 Susan Middleton & David Liittschwager

CHAPTER 9

Opener David Phillips/The Population Council/Photo Researchers, Inc. 9.4a Clare A. Hasenkampf/Biological Photo Service 9.4b-d /Biological Photo Service
9.7 (bottom) Dennis Drenner/Visuals Unlimited 9.8 (bottom) Doug Sokell/Visuals Unlimited 9.10a /G.R. Dick Roberts Photo Library Box 9.1 Figure 1 Michael Speicher and David C. Ward

CHAPTER 10

Opener Dr. Madan K. Bhattacharyya/Dr. Madan K. Bhattacharyya. 10.12 Carolina Biological Supply Company/Phototake NYC 10.13a Reproduced from Nettie M. Stevens (1905), Studies in Spermatogenesis with Especial Reference to the Accessory Chromosome. pp. 1.32, fig. 21. 10.17a Robert Calentine/Visuals Unlimited
10.18a David Cavagnaro/Peter Arnold, Inc. 10.19a Albert Blakeslee, Journal of Heredity, 1914. Reproduced by permission of Oxford University Press.

CHAPTER 11

Opener Oliver Meckes/Max.Planck.Institut.Tubingen/Photo Researchers, Inc.
11.1a Reproduced from O.T. Avery, C.M. MacLeod, and M. McCarty, Studies on the chemical nature of the substance inducing transformation of pneumcoccal types, *The Journal of Experimental Medicine*, 1944, 79:137.157, plate 1, by copyright permission of The Rockefeller University Press. Photo by Joseph B. Haulenbeek, courtesy of Maclyn McCarty, Rockefeller University.

CHAPTER 12

Opener Dr. Gopal Murti/Science Photo Library/Photo Researchers, Inc.
12.2a (step 4) Reproduced by permission of Matthew S. Meselson, Harvard University, from M. Meselson and F.W. Stahl, The replication of DNA in *Escherichia coli*. *Proceedings of the National Academy of Sciences* 44(7):671-682 (July 1958), p. 675, fig. 4. 12.6 autoradiograph Dr. John Cairns, Harvard School of Public Health 12.10 Dr. Gopal Murti/Science Photo Library/Photo Researchers, Inc. 12.12c gel Loida J. Escote-Carlson, Penn State University 12.16b Bill Longcore/Photo Researchers, Inc.

CHAPTER 13

Opener I. Andersson, Oxford Molecular Biophysics Laboratory/Science Photo Library/Photo Researchers, Inc. 13.3b H. Fernandez Moran, M.D./Omikron/Photo Researchers, Inc. 13.7a Bert W. O'Malley, M.D., Baylor College of Medicine 13.12a Oscar Miller/Science Photo Library/Photo Researchers, Inc. 13.20 (left) Reprinted by permission from B.T. Wimberly et al., Structure of the 30*S* ribosomal subunit, *Nature* 407:327-339, Fig. 2c (September 21, 2000). Copyright 2000 Macmillan Magazines Limited. Image courtesy of V. Ramakrishnan. 13.20 (right) Reprinted with permission from T.R. Cech, The ribosome is a ribozyme, *Science* 289:878-879 (August 11, 2000). Copyright 2000 American Association for the Advancement of Science. Image courtesy of Thomas R. Cech, Howard Hughes Medical Institute. Box 13.1 Figure 1 (right) Science Source/Photo Researchers, Inc. Box 13.2 Figure 1 Modified from Mestel, 1996. *Science* 273:184-189. Drawing by F. Cohen.

CHAPTER 14

Opener EM Unit, VLA/Science Photo Library/Photo Researchers, Inc. 14.15 Modified from Kercher et al., 1997. *Current Biology* 7:76-85. Box 14.3 Figure 1 autoradiograph Reproduced by permission from A. Schmitz and D.J. Galas, The interaction of RNA polymerase and *lac* repressor with the lac control region. *Nucleic Acids Research* 6:111-137 (1979), fig. 2b. Copyright © 1979 by Oxford University Press. Image courtesy of David J. Galas, Keck Graduate Institute Applying Ideas Question 14.4 Michael Gabridge/Visuals Unlimited

CHAPTER 15

Opener (top) CNRI/Photo Researchers, Inc. Opener (bottom) Quest/Science Photo Library/Photo Researchers, Inc. 15.1a Juan Jose Sanz-Ezquerro and Cheryll Tickle, University of Dundee, Dundee, Scotland 15.1b Richard Hutchings/Photo Researchers, Inc. 15.3a Ada Olins/Don Fawcett/Photo Researchers, Inc. 15.4b gels Mark T. Groudine, M.D., Fred Hutchinson Cancer Research Center 15.6 autoradiograph Reproduced by permission of Elsevier Science from S.D. Gillies et al., A tissue-specific transcription enhancer element Is located in the major intron of a rearranged immunoglobulin heavy chain gene. *Cell* 33: 717-728 (1983), fig. 3. Copyright © 1983 by Elsevier Science Ltd. Image courtesy of Cell Press.

CHAPTER 16

Opener David Parker/Science Photo Library/Photo Researchers, Inc. 16.10a (bottom) Camilla M. Kao and Patrick O. Brown, Stanford University.

CHAPTER 17

Opener Keith V. Wood/Science VU/Visuals Unlimited 17.4a Science VU/G.R. Sutherland/Visuals Unlimited 17.9a Brad Mogen/Visuals Unlimited 17.10b Peter Beyer, University of Freiburg, Freiburg, Germany

UNIT 4

© 1994 Susan Middleton & David Liittschwager

CHAPTER 18

Opener C. Eldeman/Petit Format/Science Source/Photo Researchers, Inc. 18.1a Holt Studios International/Photo Researchers, Inc. 18.12b Roslin Institute/PA News 18.13a David S. Addison/Visuals Unlimited

CHAPTER 19

Opener Carolina Biological Supply Company/Phototake NYC 19.1a Microworks/Phototake NYC 19.1b Cabisco/Visuals Unlimited 19.3a Michael V. Danilchik, Oregon Health and Science University 19.3b Douglas A. Melton, Harvard University 19.5b Gregory Ochocki/Photo Researchers, Inc. 19.6b Reproduced by permission from K.R. Foltz and W.J. Lennarz, Identification of the sea urchin egg receptor for sperm using an antiserum raised against a fragment of its extracellular domain. *Journal of Cell Biology* 116:647-658 (1992), fig. 6d. Copyright © 1992 by The Rockefeller University Press. Image courtesy of Kathleen R. Foltz, University of California at Santa Barbara. 19.7a Michael Whitaker/Science Photo Library/Photo Researchers, Inc. 19.7b Victor D. Vacquier, Scripps Institution of Oceanography, University of California at San Diego 19.8b Biodisc/Visuals Unlimited 19.9c David M. Phillips/Visuals Unlimited

CHAPTER 20

Opener Gary C. Schoenwolf, University of Utah School of Medicine 20.1a F. Rudolf Turner, Indiana University 20.1b Gary Grumbling, Indiana University 20.2 photos Wolfgang Driever, University of Freiburg, Freiburg, Germany 20.3 (left) Jim Langeland, Stephen Paddock, and Sean Carroll, University of Wisconsin at Madison 20.3 (middle) Stephen J. Small, New York University 20.3 (right) Jim Langeland, Steve Paddock, and Sean Carroll 20.5 (top) Oliver Meckes/Photo Researchers, Inc. 20.5 (bottom) Edward B. Lewis, California Institute of Technology 20.9a John L. Bowman, University of California at Davis 20.11a Kathryn W. Tosney, University of Michigan Box 20.2 Figure 1 Reproduced by permission of Elsevier Science from K. Kuida et al., Reducedapoptosis and cytochrome *c*-mediated caspase activation in mice lacking caspase 9. *Cell* 94:325-337 (1998), figs. 2E and 2F. Copyright © 1998 by Elsevier Science Ltd. Image courtesy of Keisuke Kuida, Vertex Pharmaceuticals.

UNIT 5

© 1994 Susan Middleton & David Liittschwager

CHAPTER 21

Opener Tui De Roy/The Roving Tortoise Nature Photography 21.1a (top left) Mickey Gibson/Animals Animals/Earth Scenes 21.1a (top right) Tui De Roy/Bruce Coleman Inc. 21.1a (bottom left) Marie Read/Animals Animals/Earth Scenes 21.1a (bottom right) George D. Lepp/Photo Researchers, Inc. 21.2b (left) Photo by Michael K. Richardson, reproduced by permission from *Anatomy and Embryology* 305, Fig. 7. Copyright © Springer-Verlag GmbH & Co KG, Heidelberg, Germany. 21.02b (right) Photo by Ronan O'Rahilly, reproduced by permission from *Anatomy and Embryology* 305, Fig.8. Copyright © Springer-Verlag GmbH & Co KG, Heidelberg, Germany. 21.3a Robert Lubeck/Animals Animals/Earth Scenes 21.4 (top left) Vincent Zuber/Custom Medical Stock Photo, Inc. 21.4 (top right) /Custom Medical Stock Photo, Inc. 21.4 (bottom left) CMCD/Getty Images, Inc./PhotoDisc, Inc. 21.4 (bottom right) Mary Beth Angelo/Photo Researchers, Inc. 21.5 Arthur C. Aufderheide, M.D., University of Minnesota School of Medicine, Duluth 21.7 Candace Galen, University of Missouri-Columbia Box 21.1 Figure 1a /CORBIS Box 21.01 Figure 1b /Getty Images Inc.

CHAPTER 22

Opener Gerald and Buff Corsi/Visuals Unlimited 22.5 gel Reprinted with permission from S.J. O'Brien et al., *Science*, March 22, 1985, 227:1428-14__, p. 1429, Fig. 2(A). Copyright 1985 American Association for the Advancement of Science. Photo courtesy of Stephen J. O'Brien. 22.10a (top and bottom) Robert and Linda Mitchell/Robert & Linda Mitchell Photography 22.10b (top) John Muagge/Visuals Unlimited 22.10b (bottom) S.J. Krasemann/Peter Arnold, Inc. 22.10c (top and bottom) Roger K. Doyle/Roger K. Doyle 22.11a Francois Gohier/Photo Researchers, Inc.

CHAPTER 23

Openers Scott P. Carroll, University of California at Davis 23.3a sparrows Diane Pierce/NGS Image Collection 23.4 (top) Joseph T. Collins/Photo Researchers, Inc. 23.4 (bottom) Alvin E. Staffan/National Audubon Society/Photo Researchers, Inc. 23.10a H. Douglas Pratt/NGS Image Collection 23.11 (left) Jason Rick/Loren H. Rieseberg

CHAPTER 24

Opener © 1985 David L. Brill 24.1b(1) Reproduced by permission from P.S. Herendeen, W.L. Crepet, and K.C. Nixon, Chloranthus.like stamens from the Upper Cretaceous of New Jersey, *American Journal of Botany* 80(8):865-871. © Botanical Society of America. Micrograph courtesy of William L. Crepet, Cornell University. 24.1b(2) Martin Land/Science Photo Library/Photo Researchers, Inc. 24.1b(3) Monte Hieb & Harrison Hieb/Geocraft/www.geocraft.com 24.1b(4) John Gerlach/DRK Photo 24.2 (bottom) /American Museum of Natural History 24.5(1-3) Shuhai Xiao, et al., Three-dimensional preservation of algae and animal embryos in a Neoproterozoic phosphorite. *Nature* vol. 391:553-558 (Feb. 5, 1998). 24.5(4) Shuhai Xiao, Tulane University 24.6 (left) Simon Conway Morris, University of Cambridge, Cambridge, United Kingdom 24.6 (right) Ken Lucas/Visuals Unlimited 24.7 (left) A. Flowers and L. Newman/Photo Researchers, Inc. 24.7 (right) Ken Lucas/Visuals Unlimited 24.9(a) and (b) Denis Duboule, University of Geneva, Geneva, Switzerland 24.11a Jonathan B. Losos, Washington University in St. Louis 24.13 Glen A. Izett/U.S. Geological Survey, Denver

UNIT 6

© 1994 Susan Middleton & David Liittschwager

CHAPTER 25

Opener Dr. Tony Brain/Science Photo Library/Photo Researchers, Inc. 25.1a (left) CNRI/Science Photo Library/Photo Researchers, Inc. 25.1a (right) Reprinted with permission from Schulz et al., *Science* 284:493-495, Fig. 1b, August 4, 1999. Copyright 1999 American Association for the Advancement of Science. Image courtesy of Dr. Heide Schulz, Max Planck Institute for Marine Microbiology, Bremen, Germany. 25.1b (left) Cabisco/Visuals Unlimited 25.1b (right) David M. Phillips/Visuals Unlimited 25.1c (left) Lee D. Simon/Photo Researchers, Inc. 25.1c (right) Linda Stannard, University of Cape Town/Science Photo Library/Photo Researchers, Inc. 25.7a Spence Titley/Peter L. Kresan Photography 25.7b Donald E. Canfield, Odense University, Odense, Denmark Box 25.1 Figure 1 Reprinted with permission from S.V. Liu et al., Thermophilic Fe(III)-reducing bacteria from the deep subsurface, *Science* 277:1106-1109, Fig. 2, 1997. Copyright 1997 American Association for the Advancement of Science. Image courtesy of Yul Roh, Oak Ridge National Laboratory.

CHAPTER 26

Opener Hans Gelderblom/Eye of Science/Photo Researchers, Inc. 26.2 Jed Fuhrman, University of Southern California 26.3a (left) Biophoto Associates/Photo Researchers, Inc. 26.3a (middle) Dennis Kunkel/Phototake NYC 26.3a (right) David M. Phillips/Visuals Unlimited 26.3b (top) Omikron/Photo Researchers, Inc. 26.3b (right) Oliver Meckes/Eye of Science/Max Planck Institut Tubingen/Photo Researchers, Inc. 26.3b (bottom left) Biophoto Associates/Photo Researchers, Inc. 26.3b (bottom right) K.G. Murti/Visuals Unlimited 26.5a Oliver Meckes/Hans Gelderblom/Photo Researchers, Inc. 26.11 Photos provided by Abbott Laboratories. 26.12a (right) National Institute for Biological Standards and Control (U.K.)/Science Photo Library/Photo Researchers, Inc. 26.12b (right) Reproduced from R.H. Meints, J.L. Van Etten, D. Kuczmarski, K. Lee, and B. Ang, Viral infection of the symbiotic Chlorella-like alga present in *Hydra viridis*, *Virology* 113:698-703 (1981), Fig. C. © 1981 Academic Press, Inc. Photo courtesy of James L. Van Etten, University of Nebraska.

CHAPTER 27

Opener Reproduced by permission of David J. Asai and Amy Walanski, Purdue University. Image courtesy of Molecular Probes, Inc. 27.2 (left top) Norman T. Nicoll/Norman T. Nicoll 27.2 (left middle) Gregory Ochocki/Photo Researchers, Inc. 27.2 (left bottom) Tom and Therisa Stack/Tom Stack & Associates, Inc. 27.2 (right) John Anderson/Animals Animals/Earth 27.3 (right, 1-3) Andrew H. Knoll, Harvard University Scenes 27.3 (right, 4) Bruce Runnegar, University of California at Los Angeles 27.5 (left) M.I. Walker/Photo Researchers, Inc. 27.6a (left) Biophoto Associates/Photo Researchers, Inc. 27.6a (middle) /Bruce Coleman, Ltd. 27.6b Reproduced by permission from L.J. Goff, J. Ashen, and D.A. Moon, The evolution of parasites from their hosts: A case study in the parasitic red algae. *Evolution* 51(4):1068-1078, Fig. 1 (August 1997). Image courtesy of Lynda J. Goff, University of California at Santa Cruz. 27.6c M.I. Walker/Science Source/Photo Researchers, Inc. 27.6d (left) Astrid and Hanns-Frieder Michler/Science Photo Library/Photo Researchers, Inc. 27.6d (middle) Cabisco/Visuals Unlimited 27.6d (right) Sherman Thomson/Visuals Unlimited 27.8a Andrew Syred/Science Photo Library/Photo Researchers, Inc. 27.8b Andrew Syred/Science Photo Library/Photo Researchers, Inc. 27.8c David M. Phillips/Visuals Unlimited 27.9ai David L. Kirk, Washington University 27.9a David L. Kirk, Washington University 27.9b Linda Graham, University of Wisconsin-Madison

CHAPTER 28

Opener Darrell Gulin/DRK Photo 28.1 (right, 1) David L. Dilcher and Ge Sun 28.1 (right, 2) Thomas A. Wiewandt/DRK Photo 28.1 (right, 3) Robert & Linda Mitchell/Robert & Linda Mitchell Photography 28.1 (right, 4) William L. Crepet, Cornell University 28.1 (right, 5) Paul K. Strother, Boston College 28.2 (top, 1) K.G. Vock/Okapia/Photo Researchers, Inc. 28.2 (top, 2) Alvin E. Staffan/Photo Researchers, Inc. 28.2 (top, 3) Milton Rand/Tom Stack & Associates, Inc. 28.2 (top, 4) Rod Planck/Photo Researchers, Inc. 28.2 (bottom, 1) Walter H. Hodge/Peter Arnold, Inc. 28.2 (bottom, 2) Biophoto Associates/Photo Researchers, Inc. 28.2 (bottom, 3) Art Wolfe/Getty Images Inc. 28.2 (bottom, 4) Stephen J. Krasemann/DRK Photo 28.4a (left) Courtesy of H. Kerp & H. Hess, Palaeobotanical Research Group, University of Munster, Munster, Germany. 28.4b(left) Imogen Poole, University of Utrecht, Utrecht, The Netherlands 28.8 (left) Dr. James L. Castner/Dr. James L. Castner 28.8 (middle) G.C. Kelley/Photo Researchers, Inc. 28.8 (right) G.A. MacLean/Oxford Scientific Films/Animals Animals/Earth Scenes 28.9a Steven D. Johnson, University of Natal, Pietermaritzburg, South Africa 28.14a(1) Runk/Schoenberger/Grant Heilman Photography, Inc. 28.14a(2) D. Cavagnaro/Visuals Unlimited 28.14a(3) Ed Reschke/Peter Arnold, Inc. 28.14a(4) Dell R. Foutz/Visuals Unlimited 28.14a(5) Akira Kaede/Getty Images, Inc./PhotoDisc, Inc.

CHAPTER 29

Opener Biophoto Asociates/Photo Researchers, Inc. 29.1 Bruce M. Herman/Photo Researchers, Inc. 29.3 (left) Tony Brain/Science Photo Library/Photo Researchers, Inc. 29.3 (right) George Musil/Visuals Unlimited 29.6a Biophoto Associates/Photo Researchers, Inc. 29.9a (left) Stan Flegler/Visuals Unlimited 29.9b (right) Courtesy of H. Kerp & H. Hess, Palaeobotanical Research Group, University of unster, Munster, Germany. 29.11b Pat O'Hara/DRK Photo 29.12a Courtesy Stanley Freeman, Agricultural Research Organization, Volcani Center, Bet Dagan, Israel. Box 29.1 Figure 1 James W. Richardson/Visuals Unlimited

CHAPTER 30

Opener Edward S. Ross/Edward S. Ross 30.2a Herb Segars/Animals Animals/Earth Scenes 30.2b David J. Wrobel/Visuals Unlimited 30.8a J.P. Ferrero/Jacana/Photo Researchers, Inc. 30.8b Robert and Linda Mitchell/Robert & Linda Mitchell Photography 30.10a /DRK Photo 30.10b Kelvin Aitken/Peter Arnold, Inc. 30.10b detail shots /Robert & Linda Mitchell Photography 30.10c Jim Brandenburg/Minden Pictures 30.12a(1) /Robert & Linda Mitchell Photography 30.12a(2) Sinclair Stammers/Science Photo Library/Photo Researchers, Inc. 30.12a(3) Art Wolfe/Photo Researchers, Inc. 30.12a(4) Nicole Galeazzi/Omni-Photo Communications, Inc. 30.13a Reproduced from Grace Panganiban et al., The origin and evolution of animal appendages, *Proceedings of the National Academy of Sciences* 94:5162-5166 (May 1997), p. 5163, Fig. 1a. © 1997 The National Academy of Sciences of the U.S.A. 30.13b Reproduced from Grace Panganiban et al., The origin and evolution of animal appendages, *Proceedings of the National Academy of Sciences* 94:5162-5166

(May 1997), p. 5163, Fig. 1c. © 1997 The National Academy of Sciences of the U.S.A. 30.13c Reproduced from Grace Panganiban et al., The origin and evolution of animal appendages, *Proceedings of the National Academy of Sciences* 94:5162-5166 (May 1997), p. 5163, Fig. 1f. © 1997 The National Academy of Sciences of the U.S.A. 30.13d Reprinted by permission from Grace Panganiban et al., The development of crustacean limbs and the evolution of anthropods, *Science* 270:1363 (1995). © American Association for the Advancement of Science. 30.18a Tom McHugh/Photo Researchers, Inc. 30.18b Tom Stack/Tom Stack & Associates, Inc. Essay Figure 1 Gunter Ziesler/Peter Arnold, Inc.

UNIT 7
© 1994 Susan Middleton & David Liittschwager

CHAPTER 31
Opener John D. Cunningham/Visuals Unlimited 31.2a Tim Hauf Photography/Visuals Unlimited 31.2b (left) Reproduced from John E. Weaver, *North American Prairie*, fig. 32. Lincoln, Nebraska: Johnsen Pub. Co. (1954). J.E. Weaver, Papers, Archives and Special Collections, University of Nebraska. Lincoln Libraries. 31.2b (middle) Reproduced from John E. Weaver, *North American Prairie*, fig. 40. Lincoln, Nebraska: Johnsen Pub. Co. (1954). J.E. Weaver, Papers, Archives and Special Collections, University of Nebraska. Lincoln Libraries. 31.2b (right) Reproduced from John E. Weaver, *North American Prairie*, fig. 39. Lincoln, Nebraska: Johnsen Pub. Co. (1954). J.E. Weaver, Papers, Archives and Special Collections, University of Nebraska. Lincoln Libraries. 31.4a and 31.4b Gerald D. Carr/Gerald D. Carr 31.7b (left) Dr. Jeremy Burgess/Photo Researchers, Inc. 31.7b (right) George Wilder/Visuals Unlimited 31.7c (left) George Wilder/Visuals Unlimited 31.07c (right) Biophoto Associates/Science Source/Photo Researchers, Inc. 31.07d (left) Bruce Iverson/Bruce Iverson 31.07d (right) G. Shih and R. Kessel/Visuals Unlimited 31.07d (bottom) Ken Wagner/Phototake NYC 31.11 Ken Wagner/Phototake NYC 31.11 (inset) Ed Reschke/Peter Arnold, Inc. 31.13a Ed Reschke/Peter Arnold, Inc. 31.13b Walker/Photo Researchers, Inc. 31.15a James W. Richardson/Visuals Unlimited 31.15b Stephen J. Krasemann/Photo Researchers, Inc. 31.16a RDF/Visuals Unlimited 31.16b (left) Walter H. Hodge/Peter Arnold, Inc. 31.16b (right) David Cavagnaro/Peter Arnold, Inc. 31.17c Reproduced by permission from J.C. Clausen, D.D. Keck, and W.M. Hiesey, *The American Naturalist* 81:114-133 (1947), fig. 5, published by The University of Chicago Press. Copyright © 1947 by the American Society of Naturalists. Image courtesy of JSTOR, New York, New York. 31.18a and 31.18b Modified from Reich et al. *Proceedings of the National Academy of Sciences* 94:13730-13734. Fig. 1. 31.19 Modified from Reich et al., *Proceedings of the National Academy of Sciences* 94:13730-13734. Fig. 3. Box 31.1 Figure 1 (left) Charles McRae/Visuals Unlimited Box 31.1 Figure 1 (middle) Robert & Linda Mitchell/Robert & Linda Mitchell Photography Box 31.1 Figure 1 (right) Robert & Linda Mitchell/Robert & Linda Mitchell Photography Box 31.2 Figure 1 (top left) Holt Studios International Ltd/Photo Researchers, Inc. Box 31.2 Figure 1 (top right) Robert & Linda Mitchell/Robert & Linda Mitchell Photography Box 31.2 Figure 1 (middle top, left) F. Stuart Westmorland/Photo Researchers, Inc. Box 31.2 Figure 1b (middle top, right) Getty Images, Inc./PhotoDisc, Inc. Box 31.2 Figure 1 (middle bottom, left) Ed Reschke/Peter Arnold, Inc. Box 31.2 Figure 1 (middle bottom, right) Runk/Schoenberger/Grant Heilman Photography, Inc. Box 31.2 Figure 1 (bottom left) Robert & Linda Mitchell/Robert & Linda Mitchell Photography Box 31.2 Figure 1 (bottom right) Kim Heacox Photography/DRK Photo Box 31.3 Figure 1a Andrea H. Lloyd, Middlebury College Box 31.3 Figure 1b Henri D. Grissino-Mayer, Department of Geography, University of Tennessee Essay Figure 1 (left) Gerry Ellis/ENP Images Essay Figure 1 (right) Richard Phelps Frieman/Photo Researchers, Inc.

CHAPTER 32
Opener David Nunuk/Science Photo Library/Photo Researchers, Inc. 32.2 Runk/Schoenberger/Grant Heilman Photography, Inc. 32.3a Ken Wagner/Phototake NYC 32.10a and 32.10b John D. Cunningham/Visuals Unlimited 32.14 (right) Carolina Biological Supply Company/Phototake NYC 32.14 (left) Jean Claude Revy/Phototake NYC 32.17a (right) Reproduced by permission from N.D. DeWitt and M.R. Sussman, Immunocytological localization of an epitope-tagged plasma membrane proton pump (H^+-ATPase) in phloem companion cells. Plant Cell 7:2053.2067 (1995), fig. 6. Copyright © 1995 American Society of Plant Biologists. Image courtesy of Michael R. Sussman, University of Wisconsin.

CHAPTER 33
Opener John Eastcott/Yva Momatiuk/Animals Animals/Earth Scenes 33.1b and 33.1c Emanuel Epstein, University of California at Davis 33.8a E.R. Degginger/Color.Pic, Inc. 33.10 (left) Runk/Schoenberger/Grant Heilman Photography, Inc. 33.10 (right) E.H. Newcomb and S.R.Tandon, *Science* 212:1394 (1981)/Biological Photo Service 33.11b Reproduced by permission from M.E. Etzler et al., A nod factor binding lectin with apyrase activity from legume roots, *Proceedings of the National Academy of Sciences* 96:5856-5861 (May 1999), p. 5860, Figs. 5A and 5B. Copyright 1999 National Academy of Sciences. Images courtesy of Marilynn E. Etzler, University of California at Davis. 33.12b (left) Reproduced by permission of the American Society of Plant Biologists from Y. Fang and A.M. Hirsch, Studying early nodulin gene ENOD40 expression and induction by nodulation factor and cytokinin in transgenic alfalfa. *Plant Physiology* 116:53-68 (1998), fig. 4G. Copyright © 1998 by the American Society of Plant Biologists. 33.12b (right) Reproduced by permission of the American Society of Plant Biologists from Y. Fang and A.M. Hirsch, Studying early nodulin gene ENOD40 expression and induction by nodulation factor and cytokinin in transgenic alfalfa. *Plant Physiology* 116:53-68 (1998), fig. 4I. Copyright © 1998 by the American Society of Plant Biologists. 33.13 Gerry Ellis/Minden Pictures

CHAPTER 34
Opener Malcolm B. Wilkins, University of Glasgow, Glasgow, Scotland, U.K. 34.1a and 34.1b Malcolm B. Wilkins, University of Glasgow, Glasgow, Scotland, U.K. 34.3 (right) Malcolm B. Wilkins, University of Glasgow, Glasgow, Scotland, U.K. 34.4a gels Reprinted with permission from John M. Christie et al., *Arabidopsis* NPH1: A Flavoprotein with the Properties of a Photoreceptor for Phototropism. *Science* 282:1698-1701, Fig. 1c (1998). Copyright 1998 American Association for the Advancement of Science. Photo courtesy of John M. Christie. 34.6a Runk/Schoenberger/Grant Heilman Photography, Inc. 34.6b (left) Reproduced by permission from E.B. Blancaflor et al., Mapping the functional roles of cap cells in the response of *Arabidopsis* primary roots to gravity, *Plant Physiology* 116:213-222 (1998), Fig. 1A. Copyright © 1998 American Society of Plant Biologists. Image courtesy of Simon Gilroy, Penn State University. 34.7b (top and bottom) Reproduced by permission from O.A. Kuznetsov et al., Curvature induced by amyloplast magnetophoresis in protonemata of the moss *Ceratodon purpureus*. *Plant Physiology* 119:645-650 (1999), figs. 3B and 3C. Copyright © 1999 by the American Society of Plant Biologists. Image courtesy of Oleg A. Kuznetsov, University of Louisiana. 34.9b Reproduced by permission from T.M. Lynch, P.M. Lintilhac, and D.S. Domozych, Mechanotransduction molecules in the plant gravisensory response: amyloplast/statolith membranes contain a beta 1 integrin-like protein. *Protoplasma* 201:92-100 (1998), fig. 2e. Copyright © 1998 by Springer-Verlag GmbH & Co KG. Image courtesy of Timothy M. Lynch. 34.11 Thomas Bjorkman, Cornell University, from the experiments published in L.C. Garner and T. Bjorkman, Mechanical conditioning for controlling excessive elongation in tomato transplants: sensitivity to dose, frequency, and timing of brushing. *Journal of the American Society for Horticultural Science* 121:894-900 (1996). 34.14 (left) Carolina Biological Supply Company/Phototake NYC 34.14 (right) Runk/Schoenberger/Grant Heilman Photography, Inc. Applying Ideas Question 2 David Sieren/Visuals Unlimited

CHAPTER 35
Opener Malcolm B. Wilkins, University of Glasgow, Glasgow, Scotland, U.K. 35.6 (bottom) Reprinted with permission from A.M. Jones et al., Auxin-dependent cell expansion mediated by overexpressed auxin-binding protein 1. *Science* 282:1114-1117 (November 6, 1998), fig. 2e. Copyright © 1998 American Association for the Advancement of Science. Image courtesy of Alan M. Jones, University of North Carolina at Chapel Hill. 35.7a and 35.7b Malcolm B. Wilkins, University of Glasgow, Glasgow, Scotland, U.K. 35.10a Reproduced by permission from K. Palme and L. Gälweiler, PIN-pointing the molecular basis of auxin transport. Current Opinion in Plant Biology 2:375-381 (1999), p. 376, fig. 1. Copyright © 1999 by Elsevier Science Ltd. Image courtesy of Leo Gälweiler, BASF-Lynx Bioscience AG, Heidelberg, Germany. 35.10b Reproduced by permission from K. Palme and L. Gälweiler, PIN-pointing the molecular basis of auxin transport. Current Opinion in Plant Biology 2:375-381 (1999), p. 378, fig. 3. Copyright © 1999 by Elsevier Science Ltd. Image courtesy of Leo Gälweiler, BASF-Lynx Bioscience AG, Heidelberg, Germany. 35.11 Joe Eakes/Color Advantage/Visuals Unlimited 35.13 Reproduced by permission from F. Gubler et al., Gibberellin-regulated expression of a *myb* gene in barley aleurone cells: evidence for Myb transactivation of a high-p1 alpha-amylase gene promoter. *Plant Cell* 7:1879-1891, p. 1883, fig. 5A (1995). Copyright © 1995 American Society of Plant Biologists. Image courtesy of Frank Gubler, CSIRO Division of Plant Industry, Canberra, ACT, Australia.

CHAPTER 36
Opener Georgia O'Keeffe (American, 1887.1986), Poppy, 1927. Oil on canvas, 30 x 36 in. Museum of Fine Arts, St. Petersburg, Florida. Gift of Charles C. and Margaret Stevenson Henderson in memory of Jeanne Crawford Henderson. 71.32. 36.1 (top left) Wayne P. Armstrong, Palomar College 36.1 (top right) BIOS/Peter Arnold, Inc. 36.1 (bottom left) Jerome Wexler/Photo Researchers, Inc. 36.1 (bottom right) C.C. Lockwood/Animals Animals/Earth Scenes 36.2a Walter H. Hodge/Peter Arnold, Inc. 36.2b (top) Tom McHugh/Photo Researchers, Inc. 36.2b (bottom) Stephen J. Krasemann/Nature Conservancy/Photo Researchers, Inc. 36.4a Dan Suzio/Photo Researchers, Inc. 36.4b Kenneth W. Fink/Photo Researchers, Inc. 36.4c Jerome Wexler/Photo Researchers, Inc. 36.5b1 Reprinted by permission from M.A. Blazquez and D. Weigel, Integration of floral inductive signals in *Arabidopsis*, *Nature* 404:889-892, p. 890, fig. 2a (April 20, 2000). Copyright 2000 Macmillan Magazines Limited. Image courtesy of Detlef Weigel, Max Planck Institute for Developmental Biology, Tubingen, Germany. 36.5b2 Reprinted by permission from M.A. Blazquez and D. Weigel, Integration of floral inductive signals in *Arabidopsis*, *Nature* 404:889-892, p. 890, fig. 2f (April 20, 2000). Copyright 2000 Macmillan Magazines Limited. Image courtesy of Detlef Weigel, Max Planck Institute for Developmental Biology, Tubingen, Germany. 36.5b3 Reprinted by permission from M.A. Blazquez and D. Weigel, Integration of floral inductive signals in *Arabidopsis*, *Nature* 404:889-892, p. 890, fig. 2c (April 20, 2000). Copyright 2000 Macmillan Magazines Limited. Image courtesy of Detlef Weigel, Max Planck Institute for Developmental Biology, Tubingen, Germany. 36.5b4 Reprinted by permission from M.A. Blazquez and D. Weigel, Integration of floral inductive signals in *Arabidopsis*, *Nature* 404:889-892, p. 890, fig. 2d (April 20, 2000). Copyright 2000 Macmillan Magazines Limited. Image courtesy of Detlef Weigel, Max Planck Institute for Developmental Biology, Tubingen, Germany. 36.6b (top) Tom & Therisa Stack/Tom Stack & Associates, Inc. 36.6b (middle top, left) Rod Planck/Photo Researchers, Inc. 36.6b (middle top, right) John Gerlach/DRK Photo 36.6b (middle bottom) Walter H. Hodge/Peter Arnold, Inc. 36.6b (bottom, left and right) Leonard Lessin/Photo Researchers, Inc. 36.14a and 36.14b Wayne P. Armstrong, Palomar College 36.15 (top left) R.J. Erwin/Photo Researchers, Inc. 36.15 (top center) Mark Stouffer/Animals Animals/Earth Scenes 36.15 (top right) Gerard Lacz/Animals Animals/Earth Scenes 36.15 (middle left) Tom Edwards/Visuals Unlimited 36.15 (middle center) David Stuckel/Visuals Unlimited 36.15 (middle right) Gregory K. Scott/Photo Researchers, Inc. 36.15 (bottom left) Fritz Prenzel/Peter Arnold, Inc. 36.15 (bottom center) David M. Schleser/Nature's Images, Inc./Photo Researchers, Inc. 36.15 (bottom right) Sylvan Wittwer/Visuals Unlimited Essay Figure 1 (left) B. & B. Wells/Oxford Scientific Films/Animals Animals/Earth Scenes Essay Figure 1 (right) G.H. Thompson/Oxford Scientific Films/Animals Animals/Earth Scenes

CHAPTER 37
Opener Dr. Morley Read/Science Photo Library/Photo Researchers, Inc. 37.1 (bottom) Biophoto Associates/Photo Researchers, Inc. 37.3 (left) Andrew Syred/Science Photo Library/Photo Researchers, Inc. 37.3 (right) Richard Shiell/Animals Animals/Earth Scenes 37.4a Mark Moffett/Minden Pictures 37.10a Reproduced by permission from M. Delledonne et al., Nitric oxide functions as a signal in plant disease resistance. *Nature* 394:585-588 (1998), fig. 4b. Copyright © 1998 Macmillan Magazines Limited. Image courtesy of Massimo Delledonne, Università Cattolica, Piacenza, Italy. 37.16 Nigel Cattlin/Holt Studios International/Photo Researchers, Inc.

UNIT 8
© 1993 Susan Middleton & David Liittschwager

CHAPTER 38
Opener Paul J. Fusco/Photo Researchers, Inc. 38.1a Frans Lanting/Minden Pictures 38.3a James E. Gerholdt/Peter Arnold, Inc. 38.4 Reproduced by permission from K. Petren et al., A phylogeny of Darwin's finches based on microsatellite DNA length variation. *Proceedings of the Royal Society of London*, Series B, 266:321-329 (1999), p. 327, fig. 3. Copyright © 1999 Royal Society of London. Image courtesy of

Kenneth Petren, University of Cincinnati. 38.9a Natalie Fobes/Getty Images Inc 38.10a Fred Hossler/Visuals Unlimited 38.10b Oliver Meckes & Nicole Ottawa/Photo Researchers, Inc. 38.10c P.M. Motta, A. Caggiati, G. Macchiarelli/Science Photo Library/Photo Researchers, Inc. 38.14 Reproduced by permission from Saverio Cinti, *The Adipose Organ.* Milan: Editrice Kurtis S.r.l., 1999. Copyright © 1999 by Editrice Kurtis S.r.l., Milan, Italy.

CHAPTER 39

Opener Frans Lanting/Minden Pictures 39.4a A.B. Joyce/Photo Researchers, Inc. 39.5a Kim Heacox/DRK Photo 39.6 Reproduced by permission from K. Uchida et al., Morphometrical analysis of chloride cell activity in the gill filaments and lamellae and changes in Na^+,K^+-ATPase activity during seawater adaptation in chum salmon fry. *Journal of Experimental Zoology* 276:193-200 (1996), p. 196, figs. 4A and 4B. Copyright © 1996 by Wiley-Liss, Inc. Image courtesy of Toyoji Kaneko, University of Tokyo. 39.7a Stephen Dalton/Photo Researchers, Inc. 39.11a Fred Hossler/Visuals Unlimited Essay Figure 1 Richard R. Hansen/Photo Researchers, Inc.

CHAPTER 40

Opener Biophoto Associates/Photo Researchers, Inc. 40.1 Michael Newman/PhotoEdit 40.3a (left) Michael Fogden/DRK Photo 40.3a (right) Nuridsany et Perennou/Photo Researchers, Inc. 40.3b (left) Sinclair Stammers/Science Photo Library/Photo Researchers, Inc. 40.3b (right) Mark W. Moffett/Minden Pictures 40.3c (left) David Bassett/Photo Researchers, Inc. 40.3c (right) Wayne Lynch/DRK Photo 40.4 Mark Smith/Photo Researchers, Inc. 40.5b (left) Reproduced by permission from K.F. Liem, Evolutionary strategies and morphological innovations: cichlid pharyngeal jaws. *Systematic Zoology* 22:425-441 (1974), p. 436, fig. 15B. Copyright © 1974 by the Society of Systematic Biologists. Image courtesy of Karel F. Liem, Harvard University. 40.05 (middle) Reproduced by permission from K.F. Liem, Evolutionary strategies and morphological innovations: cichlid pharyngeal jaws. *Systematic Zoology* 22:425-441 (1974), p. 436, fig. 15A. Copyright © 1974 by the Society of Systematic Biologists. Image courtesy of Karel F. Liem, Harvard University. 40.5b (right) Reproduced by permission from K.F. Liem, Evolutionary strategies and morphological innovations: cichlid pharyngeal jaws. *Systematic Zoology* 22:425-441 (1974), p. 436, fig. 15D. Copyright © 1974 by the Society of Systematic Biologists. Image courtesy of Karel F. Liem, Harvard University. 40.6 Craig K. Lorenz/Photo Researchers, Inc. 40.8b (right) John D. Cunningham/Visuals Unlimited 40.11 (left) Ed Reschke/Peter Arnold, Inc. 40.12b (top) Reproduced from Kuniaki Takata et al., Immunohistochemical localization of Na^+-dependent glucose transporter in rat jejunum. *Cell & Tissue Research* 267:3-9 (1992), fig. 2c. Copyright 1992 Springer-Verlag GmbH & Co KG. Image courtesy of Kuniaki Takata, Gunma University School of Medicine, Gunma, Japan. 40.12b (bottom) Reproduced from Kuniaki Takata et al., Immunohistochemical localization of Na^+-dependent glucose transporter in rat jejunum. *Cell & Tissue Research* 267:3-9 (1992), fig. 2d. Copyright 1992 Springer-Verlag GmbH & Co KG. Image courtesy of Kuniaki Takata, Gunma University School of Medicine, Gunma, Japan.

CHAPTER 41

Opener Tony Freeman/PhotoEdit 41.3 (left) Russell Illig/Getty Images, Inc./PhotoDisc, Inc. 41.3 (right) Inga Spence/Visuals Unlimited 41.5 (left) Walter E. Harvey/Photo Researchers, Inc. 41.5 (right) John D. Cunningham/Visuals Unlimited 41.7a Tony Pittaway/Tony Pittaway 41.13 (left) Carolina Biological Supply/Visuals Unlimited 41.13 (right) Richard Kessel/Visuals Unlimited

CHAPTER 42

Opener Courtesy Marcus E. Raichle, M.D., Washington University School of Medicine, from research based on S.E. Petersen et al., Positron emission tomographic studies of the cortical anatomy of single-word processing. *Nature* 331:585-589 (1988). 42.2a Innerspace Imaging/Science Photo Library/Photo Researchers, Inc. 42.8a Glenn M. Oliver/Visuals Unlimited 42.10b (bottom) C. Raines/Visuals Unlimited 42.12a David M. Phillips/Visuals Unlimited 42.12c Oliver Meckes & Nicole Ottawa/Photo Researchers, Inc. 42.15a Reproduced by permission from A.R. Mohn et al., Mice with reduced NMDA receptor expression display behaviors related to schizophrenia. *Cell* 98:427-436 (August 20, 1999). Copyright © 1999 Cell Press. Image courtesy of Amy R. Mohn, Duke University. 42.17a (center) Ralph T. Hutchings/Ralph T. Hutchings 42.20a Norbert Wu/Peter Arnold, Inc. 42.21b Reproduced by permission of Elsevier Science from K.C. Martin et al., Synapse-specific, long-term facilitation of aplysia sensory to motor synapses: a function for local protein synthesis in memory storage. *Cell* 91:927-938 (1997), fig. 2a. Copyright © 1998 by Elsevier Science Ltd. Image courtesy of Kelsey C. Martin, University of California, Los Angeles.

CHAPTER 43

Opener Stanley Breeden/DRK Photo 43.1 Stephen Dalton/Photo Researchers, Inc. 43.7 (left) David Scharf/Peter Arnold, Inc. 43.9a (right) Steven K. Fisher, University of California at Santa Barbara 43.11a Michael and Patricia Fogden/Fogden Natural History Photographs 43.11b John A. Pearce, University of Texas at Austin. Research support from Air Force Office of Scientific Research. 43.14 (top left) Brian Eyden/Science Photo Library/Photo Researchers, Inc. 43.14 (top middle) Innerspace Imaging/Science Photo Library/Photo Researchers, Inc. 43.14 (top right) Brian Eyden/Science Photo Library/Photo Researchers, Inc. 43.15 (bottom right) James E. Dennis/Phototake NYC 43.17a Reprinted with permission from I. Rayment et al., Three-dimensional structure of myosin subfragment.1: a molecular motor. *Science* 261:50-58 (July 2, 1993). Copyright 1993 American Association for the Advancement of Science. Image courtesy of Ivan Rayment, University of Wisconsin-Madison. 43.19a (left) Ed Reschke/Peter Arnold, Inc. 43.19a (right) Fred Hossler/Visuals Unlimited

CHAPTER 44

Opener Ralph L. Brinster and Robert E. Hammer, School of Veterinary Medicine, University of Pennsylvania

CHAPTER 45

Opener Irwin DeVore/Anthro-Photo File 45.1a Oxford Scientific Films/Animals Animals/Earth Scenes 45.2a Nova Scientific Corp./Oxford Scientific Films Ltd. 45.2b Tom Adams/Visuals Unlimited 45.2c T.E. Adams/Visuals Unlimited 45.4a Reprinted by permission from Catherine S.C. Price et al., Sperm competition between *Drosophila* males involves both development and incapacitation. *Nature* 400:449-452 (1999), figs. 2 and 3. Copyright © 1999 Macmillan Magazines Limited. Image courtesy of Jerry A. Coyne, University of Chicago. 45.5a G. & C. Merker/B. Tomberlin/Visuals Unlimited 45.6a Brent D. Opell, Virginia Polytechnic Institute and State University. 45.12 (left) Claude Edelmann/Photo Researchers, Inc. 45.12 (middle) Photo Lennart Nilsson/Bonnier Alba AB, *A CHILD IS BORN*, Dell Publishing Company. 45.12 (right) Petit Format/Nestle/Science Source/Photo Researchers, Inc. 45.14 (left) Reprinted with permission from C. Ikonomidou et al., Ethanol-induced apoptotic neurodegeneration and fetal alcohol syndrome. *Science* 287:1056-1060 (February 11, 2000), p. 1057, fig. 1. Copyright © 2000 by the American Association for the Advancement of Science. Image courtesy of Chrysanthy Ikonomidou, Humboldt University, Berlin, Germany.

CHAPTER 46

Opener Dennis Kunkel/CNRI/Phototake NYC 46.1 (right) Fred Hossler/Visuals Unlimited 46.2a (left) Rosalind King/Science Photo Library/Photo Researchers, Inc. 46.2a (middle) Chapman-Williams-Blake/Visuals Unlimited 46.2a (right) Don Fawcett/Visuals Unlimited 46.3b and 46.3c David M. Phillips/Visuals Unlimited 46.4a (bottom) and 46.4b (bottom) Barry Fields, University of Sydney, Sydney, Australia 46.11a Giovanna Bossi and Gillian Griffiths, University of Oxford, Oxford, U.K.

UNIT 9

© 1996 Susan Middleton & David Liittschwager

CHAPTER 47

Opener Mitsuaki Iwago/Minden Pictures 47.4 (right) Reprinted with permission from J.A. Paton and F. Nottebohm, Neurons generated in the adult brain are recruited into functional circuits. *Science* 225:1046-1048 (September 7, 1984), p. 1047, figs. 2A and 2B. Copyright 1984 American Association for the Advancement of Science. Image courtesy of Fernando Nottebohm, Rockefeller University. 47.6a Manuel Leal, Union College 47.9 Robert Lee/Photo Researchers, Inc. 47.10a and 47.10b Sarah Blaffer Hrdy, University of California at Davis/Anthro-Photo File 47.11a Richard Forbes, Portland State University 47.12 Manoj Shah/Getty Images Inc 47.13a Steve Kaufman/Peter Arnold, Inc. 47.13b Modifed from Figure 1 in Krajewski, 1989. *Auk* 106: 603-618. 47.13c Modified from Figure 2 in Krajewski and Fetzner, 1994. *Auk* 111: 351-365.

CHAPTER 48

Opener Ken Cavanagh/Photo Researchers, Inc. 48.1a Stephen A. Nesbitt, Florida Fish and Wildlife Conservation Commission 48.4a Fred McConnaughey/Photo Researchers, Inc. 48.6a Alan Williams/Natural History Photographic Agency/Photo Researchers, Inc. 48.7 Jack Smith/AP/Wide World Photos 48.8a Natalie B. Fobes/Natalie B. Fobes Photography 48.8b (left) S. Fried/Photo Researchers, Inc. 48.8 (right) John Warden/Getty Images Inc 48.11 Marko Nieminen/Ilkka Hanski, University of Helsinki, Helsinki, Finland 48.12 Robert Maier/Animals Animals/Earth Scenes 48.15a BIOS (F. Mercay)/Peter Arnold, Inc. 48.15 b Figures 2a and 2c in Hudson et al. 1998. *Science* 282:2256-2258. Box 48.1 Figure 1 Natalie Fobes/NGS Image Collection

CHAPTER 49

Opener Rolf O. Peterson/Rolf O. Peterson 49.1 (left) A.B. Dowsett/Science Photo Library/Photo Researchers, Inc. 49.1 (middle) S. Lowry/University of Ulster/Getty Images Inc 49.1 (right) Oliver Meckes/Eye of Science/Photo Researchers, Inc. 49.7 (top) Dave B. Fleetham/Visuals Unlimited 49.7 (midde left) Stephen J. Krasemann/DRK Photo 49.7b (middle right) Jeff Lepore/Photo Researchers, Inc. 49.7 (bottom) Gerard Fuehrer/DRK Photo 49.8a (left) Andrew J. Martinez/Photo Researchers, Inc. 49.8a (right) Zig Leszczynski/Animals Animals/Earth Scenes 49.9 (left) Nancy Sefton/Photo Researchers, Inc. 49.9 (right) Thomas Kitchin/Tom Stack & Associates, Inc. 49.10 (left) Norbert Wu/Getty Images Inc 49.10 (right) Norbert Wu/DRK Photo 49.12a Thomas G. Whitham, Northern Arizona University 49.14 (top left) G.R. Roberts/G.R. Dick Roberts Photo Library 49.14 (top right) Fred Bavendam/Minden Pictures 49.14 (middle left) Tom Till/DRK Photo 49.14 (middle right) D. Cavagnaro/Visuals Unlimited 49.14 (bottom left) Lior Rubin/Peter Arnold, Inc. 49.14 (bottom right) Hal Beral/Visuals Unlimited 49.16a Mark W. Moffett/Minden Pictures 49.16b Mark Smith/Photo Researchers, Inc. 49.17a Michael Fogden/DRK Photo

CHAPTER 50

Opener Stephen J. Krasemann/DRK Photo 50.2b Hans Reinhard/OKAPIA/Photo Researchers, Inc. 50.4b C.K. Lorenz/Photo Researchers, Inc. 50.5b Tom Bean/DRK Photo 50.6b Tom Edwards/Animals Animals/Earth Scenes 50.7b Francis Lepine/Animals Animals/Earth Scenes 50.8b Tom Bean/DRK Photo 50.10a Larry Ulrich/DRK Photo 50.10b Tony C. Caprio/Tony C. Caprio 50.11(1) Breck P. Kent/Animals Animals/Earth Scenes 50.11(2) G. Carleton Ray/Photo Researchers, Inc. 50.11(3) Michael P. Gadomski/Animals Animals/Earth Scenes 50.11(4) James P. Jackson/Photo Researchers, Inc. 50.11(5) Bruce Heinemann/Getty Images, Inc./PhotoDisc, Inc. 50.11(6) Michael P. Gadomski/Photo Researchers, Inc. 50.12a(1) Glacier Bay National Park Photo. 50.12a(2) Glacier Bay National Park Photo. 50.12a(3) Christopher L. Fastie, Middlebury College 50.12a(4) Glacier Bay National Park Photo. Essay Figure 1 (right) Jeff Henry/Peter Arnold, Inc.

CHAPTER 51

Opener NASA Goddard Space Flight Center Wallops Flight Facility, Wallops Island, Virginia. Image courtesy of William B. Krabill. 51.5 (bottom left) Jeremy Burgess/Science Photo Library/Photo Researchers, Inc. 51.5 (bottom right) R.F. Ashley/Visuals Unlimited 51.9 (top) Stephen J. Krasemann/DRK Photo 51.9(bottom) James P. Rowan/DRK Photo

CHAPTER 52

Opener Frans Lanting/Minden Pictures 52.1 Norman R. Pace, University of Colorado 52.2 Carr Clifton/Minden Pictures 52.3 Jacques Jangoux/Peter Arnold, Inc. 52.5(1) Stephen J. Krasemann/DRK Photo 52.05(2) (left) John Mitchell/Photo Researchers, Inc. 52.5(2) (middle) © Gary Braasch 52.5(2) (right) David Dennis/Animals Animals/Earth Scenes 52.5(3) Gerry Ellis/ENP Images 52.5(4) Norbert Wu/Peter Arnold, Inc. 52.5(5) © Doug Wechsler/VIREO 52.5(6) Tom & Pat Leeson/Photo Researchers, Inc. 52.5(7) N. Benvie/Animals Animals/Earth Scenes 52.6a and 52.6b Modeified from D.W. Steadman, 1995. *Science* 267: 1123-1131. Fig. 1. 52.6c /Julian Pender Hume 52.8a Marko Nieminen/Ilkka Hanski, University of Helsinki, Helsinki, Finland Figure 52.8b Modified from Saccheri et al., 1998. Nature 392: 491-494. Fig. 1 52.10 Modified from Primack, 1998. Essentials of Conservation Biology, second ed. (Sunderland, OR: Sinauer Associates). Fig. 15.12, p. 423. Essay Figure 1 (left) Fred Bruemmer/DRK Photo Essay Figures 2a and 2b Daniel H. Janzen, University of Pennsylvania

Index